s Block elements	d Block elements
p Block elements	f Block elements

10	11	12	13	14	15	16	17	18
								4.002602 2 He
			10.811 3 2.0 5 B ●	12.011 4, 2, −4 2.6 6 C ●	14.00674 5, 4, 3, 2, −3 3.0 7 N ●	15.9994 −2, −1 3.4 8 O ●	18.9984032 −1 4.0 9 F ●	20.1797 10 Ne
			26.981539 3 1.6 13 Al	28.0855 4, −4 1.9 14 Si ●	30.973762 5, 3, −3 2.2 15 P ●	32.066 6, 4, 2, −2 2.6 16 S ●	35.4527 7, 5, 3, 1, −1 3.2 17 Cl ●	39.948 18 Ar
58.6934 3, 2, 0 1.9 28 Ni ●	63.546 2, 1 1.9 29 Cu ●	65.39 2 1.7 30 Zn ●	69.723 3 1.8 31 Ga	72.61 4 2.0 32 Ge	74.92159 5, 3, −3 2.2 33 As ●	78.96 6, 4, −2 2.6 34 Se ●	79.904 7, 5, 3, 1, −1 3.0 35 Br	83.80 2 3.0 36 Kr
106.42 4, 2, 0 2.2 46 Pd	107.8682 2, 1 1.9 47 Ag	112.411 2 1.7 48 Cd	114.818 3 1.8 49 In	118.710 4, 2 2.0 50 Sn	121.757 5, 3, −3 2.1 51 Sb	127.60 6, 4, −2 2.1 52 Te	126.90447 7, 5, 1, −1 2.7 53 I ●	131.29 8, 6, 4, 2 2.6 54 Xe
195.08 4, 2, 0 2.3 78 Pt	196.96654 3, 1 2.5 79 Au	200.59 2, 1 2.0 80 Hg	204.3833 3, 1 2.0 81 Tl	207.2 4, 2 2.3 82 Pb ●	208.98037 5, 3 2.0 83 Bi	208.9824* 6, 4, 2 2.0 84 Po	209.9871* 7, 5, 3, 1, −1 85 At	222.0176* 2 86 Rn

157.25 3 1.2 64 Gd	158.92534 4, 3 1.2 65 Tb	162.50 3 1.2 66 Dy	164.93032 3 1.2 67 Ho	167.26 3 1.2 68 Er	168.93421 3, 2 1.3 69 Tm	173.04 3, 2 70 Yb	174.967 3 1.0 71 Lu
247.0703* 4, 3 96 Cm	247.0703* 4, 3 97 Bk	251.0796* 4, 3 98 Cf	252.0829* 3 99 Es	257.0951* 3 100 Fm	258.0986* 3 101 Md	259.1009* 3, 2 102 No	260.1053* 3 103 Lr

Organic Chemistry

Organic Chemistry

SECOND EDITION

K. Peter C. Vollhardt

University of California, Berkeley

Neil E. Schore

University of California, Davis

W. H. Freeman and Company

NEW YORK

Cover Image by Tomo Narashima

About the Cover: Calicheamicin (*at right*), one of the most potent cancer fighters ever discovered, is shown approaching a strand of DNA, the genetic material of living cells. This anticancer agent has only recently been found in nature. The cover is adapted from a computer-generated image provided by K. C. Nicolaou (The Scripps Research Institute and the University of California, San Diego) and Michael Peak (The Scripps Research Institute).

Calicheamicin acts by undergoing an extraordinary transformation—into a short-lived chemical species called a *radical,* which then attacks the DNA of tumor cells. As you will see in Chapter 8, radicals underlie the course of many organic reactions, including damage to normal cells that promotes aging. Chapter 14 explains in detail the action of calicheamicin and other naturally occurring antibiotics, and Chapter 25 discusses chemical defenses against damage to human cells.

Library of Congress Cataloging-in-Publication Data

Vollhardt, K. Peter C.
Organic chemistry / K. Peter C. Vollhardt, Neil E. Schore.—2nd ed.
p. cm.

Includes index.
ISBN 0-7167-2010-8
1. Chemistry, Organic. I. Schore, Neil Eric, 1948–
II. Title.
QD251.2.V65 1994
547—dc20 93-15648
CIP

Printed in the United States of America

Second printing 1995, RRD

Contents Overview

Contents

Preface

This book was written to stress basic concepts rather than rote memorization. Our classroom experience had confirmed that a clear introduction to organic chemistry should communicate the excitement of scientific discovery. The first edition therefore presented organic chemistry not as an imposing body of knowledge, but as a field that is rapidly changing. It carefully developed the reasoning students need to understand reactions, featured the working methods of organic chemistry and its diversity of current applications, and introduced the use of color to reinforce principles—an aid praised by overwhelming numbers of students.

In this second edition, a coauthor, Neil Schore, has joined Peter Vollhardt, and our collaboration has offered us a chance to better address the great variety of student interests and preparation. One change will be evident simply by picking up the new edition—a shorter text—and that change meant carefully reexamining the selection of reactions, the depth of theory, and the clarity of topic coverage, down to the sentence structure.

Features That Build on the First Edition: A Modern, Systematic Introduction

We believe that organic chemistry is easiest to learn if it is seen as a language: Reactions are the vocabulary and mechanisms the grammar. We introduce the notion of a mechanism as early as Chapter 3, teach reaction mechanisms step by step in Chapters 6 and 7, and throughout the remaining chapters juxtapose chemical reactions to the mechanisms by which they proceed.

Chapters are designed to unfold systematically. A typical chapter begins with the definition of a functional group, along with a look at sources and uses, before turning successively to its nomenclature, bonding and structure, characteristic spectra, reactions and syntheses, and finally optional applications. We believe that this consistent, logical development makes organic chemistry simpler to learn.

Functional Use of Color

One of the most innovative features of this text is consistent, functional color use. This visual aid ensures easy mastery of basic principles—including nomenclature, orbitals, sequence rules in stereochemistry, and the relation of spectral lines to functional groups. In mechanism schemes, *color specifies the reactivity of transforming centers,* helping students remember, for example, the electrostatic basis of polar reactions.

These applications of color are described more fully for the student immediately following this Preface, and they are again explained within the text wherever necessary. Color is suspended in exercises, chapter reviews, and problems, however, since it is important to learn how *not* to rely on it.

In this edition we carefully avoid confusion that might be caused by closely juxtaposing different uses of color. We also add many marginal notes as reminders of color designations in reactions.

Early Presentation of Spectroscopy

Our first edition also broke ground by introducing spectroscopy right after alcohol chemistry. Early coverage, beginning with NMR in Chapter 10, offers opportunities to practice applying its methods to many kinds of compounds. Since the first edition, spectroscopic characterization of new compounds has assumed an ever more prominent role in organic chemistry and related fields. We therefore introduce IR and UV-visible spectroscopy earlier in this second edition, in Chapters 11 and 14 in a context of functional groups. Classes may now cover each of the principal types of spectroscopy in the first half of the text.

Emphasis on Synthetic Strategy

Retrosynthetic analysis, another key aspect of chemical reasoning introduced in the first edition, is again continually reinforced. The goal of producing target molecules is stressed from the very first chapter, and separate sections describe crucial synthetic strategies. Common pitfalls in synthesis and the expanding role of organometallic reagents are outlined in Chapter 8, polymer synthesis in Chapter 12, and benzene synthesis in Chapter 16. Throughout, reactions illustrating organic transformations are those in common use in synthetic laboratories.

Use of Diverse Biological and Industrial Processes

As in the first edition, we have selected reactions that display the relevance of organic chemistry to the life sciences, to our everyday lives, and to the economy. We also take special care to indicate common sources and uses of organic compounds as they are introduced and to provide additional applications in problems.

Other natural and industrial products are treated at greater length in separate sections. For example, Chapter 9 ends with the physiological effects of alcohols and ethers, Chapter 12 with insect pheromones, and Chapter 21 with industrial uses of amines.

Applications are also featured in the many *boxes*. Most draw on the biological sciences, from antibiotics to anticancer agents and the chemistry of vision; others address environmental and industrial issues.

Further Changes for the Second Edition: A More Accessible Organization

This new edition takes great care in developing concepts. With the advice of many organic chemistry instructors and students, we have substantially clarified the content and organization.

New Order of Functional Groups

We follow a sequence that is widely accepted for its compelling logic, from simple alkanes to molecules with complex functional groups. For this second edition, coverage of the main groups is now complete with Chapter 21, to suit the requirements of both semester and quarter systems.

Much of aromatic chemistry has been moved forward; Chapters 15 and 16 also take up Hückel's rule and polycyclic benzenoid hydrocarbons. Carbonyl chemistry, contained in Chapters 17 through 20, is immediately followed by amines in Chapter 21.

More Manageable Coverage of Reactions

In this second edition, coverage of reactions and mechanisms has often been simplified. We consistently begin with experimental observations, use major reactions to illustrate transformations, streamline theoretical presentations, and avoid long digressions. The result is fewer pages and much reorganization within topics.

Thus, the review of bonding and other concepts that students may have forgotten has been clarified in Chapter 1. Although mechanisms are later accompanied by simplified orbital pictures, some free-energy descriptions are replaced by thermodynamic data that may be more familiar from first-year chemistry.

The presentations of substitution and elimination reactions of haloalkanes and alcohols are also more slowly paced. Thus, the introductions to acid-base chemistry and leaving groups now occupy separate sections in Chapter 6, and methods for converting the alcohol group OH into a good leaving group are deferred to Chapter 9.

Chapters 15 and 16 have been rewritten to concentrate on the fundamentals of aromatic chemistry. We focus on how the relative stability of intermediates influences the reactivity of aromatic compounds and the selectivity of reactions, with advanced topics left to Chapters 22 and 25, which can now be covered at virtually any time in the second half of the course.

Careful Development of New Ideas

We have checked and rechecked to ensure that points are presented forcefully. Most sections open with a question, followed quickly by key conclusions for ease of understanding. The text has been revised to avoid undue complexity in sentence structure, and we have broken up many long discussions, so the number of sections within chapters has increased.

For example, we simplified sections on synthesis in Chapter 6, deferring to later chapters routes to alcohols that involve hydride reduction or Grignard addition to carboxylic acids and their derivatives. Again to improve continuity between topics, Chapters 17 through 20 now group mechanistically related processes more consistently, and patterns of reactivity are introduced before individual reactions, to orient the student.

Increased Sense of Scientific Discovery

Sections new to this edition further show organic chemistry as a process of constant discovery. Boxes are consistently chosen so that students can enjoy reading these discussions on their own. Many added footnotes identify the chemists whose names are associated with reactions, again as a reminder that chemistry is not a static body of knowledge.

One new section may be of particular interest: Section 22-9 discusses radicals and their potential for damage to human health—although biological oxidation and reduction are discussed as early as Chapter 8. Other added examples include Box 15-1, on the growing class of molecules called fullerenes, and Box 26-4, on AZT, a drug that has been used with AIDS patients.

Improved Pedagogical Features

The numerous teaching aids of the first edition have also been extended and improved.

Overviews and Summaries

Every chapter now begins with a brief overview, subheadings *always* take the form of complete sentences that introduce the topic under discussion, and most sections end with a summary. Chapters conclude with Important Concepts and, beginning in Chapter 7, a second summary lists New Reactions with typical reagents and solvents. Chapter 7 also contains a full summary of the haloalkane chemistry that has been presented in the preceding chapters.

Within the lists of New Reactions, this edition has added references to the appropriate section within the chapter. Section references also appear in a detailed summary of the chief functional groups that has been added to the inside back cover, while the periodic table of elements that appears inside the front cover has been brought up to date and made easier to read.

Highlighting and Other Reinforcement

New terms are noted in bold type, reactions are labeled so that students can quickly separate general transformations from particular examples, and key properties are highlighted with a blue vertical line for ease of reference. Some sections lend themselves to further highlighting of steps or experiments, such as the review of acids and bases in Section 6-7.

Because topics in organic chemistry are highly interrelated, cross-references are provided throughout, and key ideas are again reinforced and generalized in new contexts. Many legends have been expanded to clarify the relation of the diagram to the text.

Exercises and Problems

This text will give students ample opportunity to practice what they have learned. Of the more than 400 in-chapter exercises, roughly half are new to the second edition, as are many of the 650 end-of-chapter problems. Every in-text exercise is answered at the back of the book.

We have reevaluated the exercises and initial end-of-chapter problems, to ensure that they apply the concepts in a straightforward manner to molecules of low and moderate complexity. Once this drill is mastered, students will again find more substantial challenges in the remaining problems, which often combine several concepts and reactions. The most advanced problems, derived from the research literature, should test the critical-thinking skills of even the best students. With practice and guidance, students will gain deeper insights into organic chemistry and come to appreciate its role among the sciences.

Nomenclature and Molecular Representations

Students face a confusing abundance of common names and systems of nomenclature, and this edition relies on those established in *Chemical Abstracts*. Of course, IUPAC names are cited, and if a common name has become firmly entrenched in the literature, it is always given in parentheses following the systematic name.

The use of Fischer projections has generally been restricted in this edition to Chapter 24, on carbohydrates, when students have the experience to master them. However, they are defined in Chapter 5, which has earned praise for its clear presentation of stereochemistry, and they are only one among the representations that benefit from a consistent, functional use of color.

Classroom Testing and Accuracy

The second edition of *Organic Chemistry* has benefited from an extensive process of review, by both students and instructors, including class testing at three separate institutions. We have ourselves continued to work with the text in large classes—one of us in sections composed exclusively of nonmajors.

Our students also helped us ensure an accurate, reliable text. Almost all spectra were recorded by us and our students on state-of-the-art equipment. Most ^{1}H NMR spectra were measured on a Varian Associates EM-390 device at 90 MHz, which is becoming the standard frequency for routine use. IR spectra were recorded on a Perkin-Elmer 681 spectrometer with a 580B data station.

More generally, all reactions were checked in the literature or in the laboratory by us and our students. Solvents and other conditions are carefully noted in the text.

Supplemental Materials

The *Study Guide and Solutions Manual* is again written by one us (Neil Schore); his coauthor is gratified that he now brings his lively, understandable writing to the textbook as well! The study guide summarizes each chapter from a different perspective than the text. Sample problems are worked out and the solutions to the end-of-chapter problems given. Hints to the student point out pitfalls of faulty logic and help those who find it hard to visualize the solution steps for various exercises. Tables summarize conveniently the spectral features associated with each functional group. For this edition, the study guide also adds a glossary of key terms.

Overhead Transparencies are available from the publisher to aid instructors in lecture presentations. This selection of illustrations from the text follows its functional use of color in reproductions of spectra, orbitals, other diagrams, and mechanisms. This edition greatly improves the legibility of many transparencies.

The *Maruzen Molecular Structure Model* set is also available for student purchase. This essential tool allows the representation of orbitals in double and triple bonding, as well as the location of atoms.

Acknowledgments

The construction of a textbook that is true to its subject and that meets the needs and desires of both teachers and students is a formidable task. Without the criticism and suggestions of many reviewers, it would have been all but impossible.

The comments of the following instructors helped shape the first edition: Harold Bell, Virginia Polytechnic Institute; Peter Bridson, Memphis State University; William Closson, State University of New York at Albany; Fred Clough, formerly of University of Wisconsin, Parkside; Otis Dermer, Oklahoma State University; Thomas Fisher, Mississippi State University; Marye Anne Fox, University of Texas at Austin; Raymond Funk, University of Nebraska, Lincoln; Roy Garvey, North Dakota State University; Edward Grubbs, San Diego State University; Gene Hiegel, California State University at Fullerton; Earl Huyser, University of Kansas; Taylor Jones, The Master's College, Newhall, California; George Kenyon, University of California, San Francisco; Robert Kerber, State University of New York at Stony Brook; Karl Kopecky, University of Alberta; James Moore, University of Delaware; Harry Pearson, Bedales School, England; William A. Pryor, Louisiana State University; William Rosen, University of Rhode Island; Jay Siegel, University of California, San Diego; Richard Sundberg, University of Virginia; Michael S. Tempesta, University of Missouri, Columbia; Jack Timberlake, University of New Orleans; William Tucker, North Carolina State University; Desmond Wheeler, University of Nebraska, Lincoln; Joseph Wolinsky, Purdue University; Steven Zimmerman, University of Illinois, Urbana.

These reviewers' analysis of the first edition guided the revision: Neil T. Allison, University of Arkansas; Raymond E. Chamberlain III, Arizona State University; Toby M. Chapman, University of Pittsburgh; Brian P. Coppola, University of Michigan; Trudy A. Dickneider, University of Scranton; Keith S.

Kyler, University of Miami; Don S. Matteson, Washington State University; M. Mark Midland, University of California, Riverside; Bruce E. Norcross, State University of New York at Binghamton; Morton Raban, Wayne State University; David W. Seybert, Duquesne University; Michael S. Tempesta, University of Missouri, Columbia; Larry S. Trzupek, Furman University; Carl C. Wamser, Portland State University. Paul Depovere's painstaking reading of the first edition, in preparation for his excellent French translation, has saved us from countless mistakes.

Three chemists were kind enough to test major portions of the manuscript in class and to share with us the reactions of their students: Edward Biehl, Southern Methodist University; Dennis H. Burns, Wichita State University; and Frank S. Guziec, Jr., New Mexico State University. We are grateful to them and to the reviewers who carefully read draft manuscript:

Fred J. Ablenas, *Concordia University*

Merle Battiste, *University of Florida*

John L. Belletire, *University of Cincinnati*

Silas Blackstock, *Vanderbilt University*

James M. Bobbitt, *University of Connecticut*

Albert W. Burgstahler, *University of Kansas*

Clair J. Cheer, *University of Rhode Island*

Carl Dirk, *University of Texas, El Paso*

Edwin C. Friedrich, *University of California, Davis*

John R. Grunwell, *Miami University*

Frank S. Guziec, Jr., *New Mexico State University*

Edwin F. Hilinski, *Florida State University*

Mark Hollingsworth, *University of Indiana*

C. H. Issidorides, *University of California, Davis*

Philip W. LeQuesne, *Northeastern University*

M. Mark Midland, *University of California, Riverside*

J. E. Mulvaney, *University of Arizona*

Linda L. Munchausen, *Southeastern Louisiana University*

David Nelson, *University of Wyoming*

Bruce E. Norcross, *State University of New York at Binghamton*

Charles A. Panetta, *University of Mississippi*

William A. Pryor, *Louisiana State University*

York E. Rhodes, *New York University*

David W. Seybert, *Duquesne University*

Maurice Shamma, *Pennsylvania State University*

Ricardo Silva, *California State University at Northridge*

We wish to express a special debt of gratitude to Professor Edwin Friedrich, who graciously reviewed manuscript at several stages for both scientific accuracy and pedagogical sense. His numerous comments were extraordinarily helpful in our achieving the most "user friendly" arrangement of chapters possible.

We are grateful to those graduate students who helped to record spectral data and to ensure the accuracy of the text and problem material—especially Brian A.

Siesel and David S. Brown at the University of California, Berkeley, who also prepared the index. We wish to thank John Haber, our development editor at W. H. Freeman, for his endlessly useful additions, deletions, and suggestions that kept the project going in the right direction. Finally, our thanks go to Jodi Simpson, who copyedited the second edition; to Alison Lew, its designer; to Lisa Douglis, its page make-up artist; and to Georgia Lee Hadler, our project editor, who tirelessly and skillfully shepherded the book through its final stages.

K. Peter C. Vollhardt
Neil E. Schore

To the Student

Students sometimes see organic chemistry as a formidable subject, with an overwhelming number of facts to be memorized and many difficult concepts to be digested. It does have a fairly rigid structure, because each new topic builds on the preceding ones. However, there is nothing inherently difficult about the subject. Having spent much of our lives studying and teaching organic chemistry, we have some advice that may be of help to you.

Using Your Textbook

We have designed many features of this textbook to help you organize your thoughts and to provide you with an easy review. The order of topics is the same in most chapters. You will learn first how to name compounds; then their physical properties and spectral characterizations; subsequently, the methods used to make these compounds and how they react; and finally, other useful applications. The reactions that you will encounter are likewise reported in a consistent fashion: first, an outline of reagents, substrates, and reaction conditions; and second, the mechanistic details.

Other aids to study and review are similarly designed to appear consistently from chapter to chapter.

We turn now to the chemistry of the carbon–oxygen double bond, the **carbonyl group.** In this and the next chapter we focus on two classes of carbonyl compounds: **aldehydes,** in which the carbon atom of the carbonyl group is bound to at least one hydrogen atom, and **ketones,** in which it is bound to two carbons. These compounds occur widely in nature, contributing to the flavors and aromas of many foods and assisting in the biological functions of a number of enzymes. In addition, industry makes considerable use of aldehydes and ketones, both as reagents and as solvents in synthesis. Indeed, the carbonyl function is frequently considered to be the most important in organic chemistry.

You will find that each chapter opens with a short introductory paragraph, as in the example shown here, which is actually taken from a chapter late in the book. Here and throughout the text, **bold type** will help you locate the definitions of important terms, all of which are again defined in the glossary found in your *Study Guide and Solutions Manual*.

Relative Stability of Carbocations

$$CH_3CH_2CH_2\overset{+}{C}H_2 < CH_3CH_2\overset{+}{C}HCH_3 < (CH_3)_3\overset{+}{C}$$

Primary < **Secondary** < **Tertiary**

Reactions are given titles, so that you can more easily put them to use. Chemical properties that you will use most often are also marked with a blue vertical line in the margin.

of ester enolates is the Claisen condensation, in which the enolate attacks the carbonyl carbon of another ester. This process will be discussed in Chapter 23.

EXERCISE 20-14

Give the products of the reaction of ethyl cyclohexanecarboxylate with the following compounds or under the following conditions (and followed by acidic aqueous work-up, if necessary). **(a)** H^+, H_2O; **(b)** HO^-, H_2O; **(c)** CH_3O^-, CH_3OH; **(d)** NH_3, Δ; **(e)** 2 CH_3MgBr; **(f)** $LiAlH_4$; **(g)** 1. LDA, 2. CH_3I.

In summary, with acidic or basic water, esters hydrolyze to the corresponding carboxylic acids or carboxylates; with alcohols, they undergo transesterification; and with amines at elevated temperatures, they furnish amides. Grignard reagents

Every chapter is frequently interrupted by exercises, all of which are answered at the back of the book. They are designed to help you become proficient in the preceding concepts, and you cannot learn organic chemistry as readily without working many of them. Each section then ends with a paragraph that summarizes its main points.

Often sections contain boxes exploring interesting applications of the main topic. We hope you will enjoy reading them even if they are not assigned to you.

BOX 5-1 Chiral Substances in Nature

Many organic compounds exist in nature as only one enantiomer, some as both. For example, natural *alanine* (systematic name: 2-aminopropanoic acid) is an abundant amino acid that is found in only one form. *Lactic acid* (2-hydroxypropanoic acid), however, is present in blood and muscle fluid as one enantiomer but in sour milk and some fruits and plants as a mixture of the two.

2-Methyl-5-(1-methylethenyl)-2-cyclohexenone (Carvone)

Chapters also conclude with an extensive summary of new reactions and important concepts.

NEW REACTIONS

1. BIMOLECULAR SUBSTITUTION—S_N2 (Sections 6-3 through 6-9, 7-5)

Primary and secondary substrates only

$$H_3C(H)(CH_2CH_3)C{-}I \xrightarrow{:Nu^-} Nu{-}C(CH_3)(H)(CH_2CH_3) + I^-$$

Direct backside displacement with 100% inversion of configuration

2. BIMOLECULAR ELIMINATION–E2 (Section 7-7)

$$CH_3CH_2CH_2I \xrightarrow{:B^-} CH_3CH{=}CH_2 + BH + I^-$$

Simultaneous elimination of leaving group and neighboring proton

IMPORTANT CONCEPTS

1. Secondary haloalkanes undergo slow and tertiary haloalkanes fast unimolecular substitution in polar media. When the solvent serves as the nucleophile, the process is called solvolysis.
2. The slowest, or rate-determining step, in unimolecular substitution is dissociation of the C–X

5. Unimolecular elimination to form an alkene accompanies substitution in secondary and tertiary systems. Elimination is favored by the addition of base.
6. High concentrations of strong base may bring about bimolecular elimination. Expulsion of the

Perhaps the feature that most distinctively sets this book apart is the use of color as a teaching tool. You may not fully comprehend all of its uses until you have learned more of the vocabulary of organic chemistry, but you might find it helpful later if we show them to you here. Each use will be introduced within the text, and marginal notes on many pages will remind you how they work.

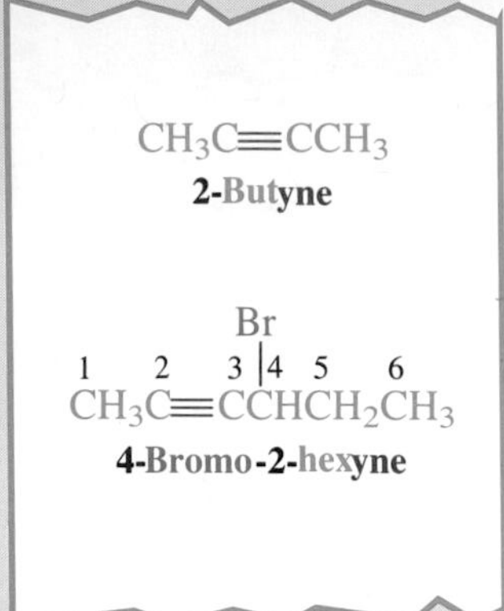

First, color is used to show the relation of the names of organic molecules to their structures. The stem, the functional group that gives the molecule its unique chemical behavior, and other substituents are each labeled by color so as to match the corresponding components of the name itself. In the illustration shown here, which is actually from Chapter 13, you can more easily remember, for example, that the triple bond shown in red gives the name of the molecule its characteristic *-yne* ending.

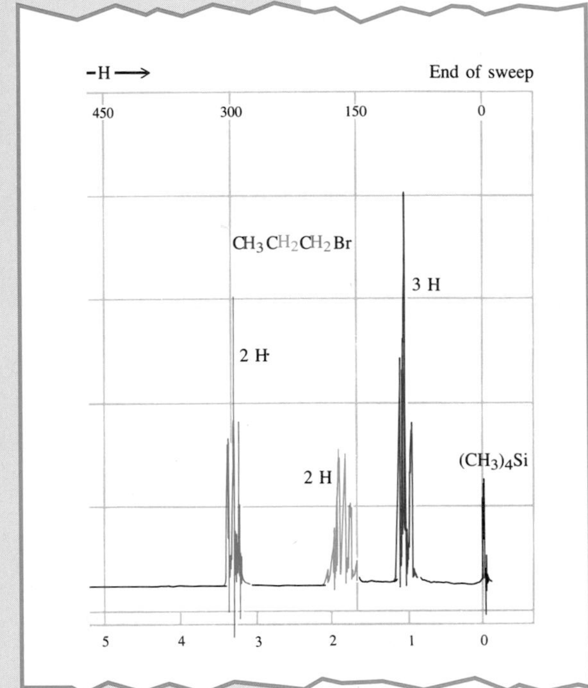

Second, color is used as a ''marker'' to indicate either the fate of atoms in a reaction or, as illustrated here, the association of spectral features with certain molecular units. Here, for example, the three colors show how each pair of hydrogens gives rise to a distinct ''peak''—an observation that will help you identify a molecule once you know its spectrum.

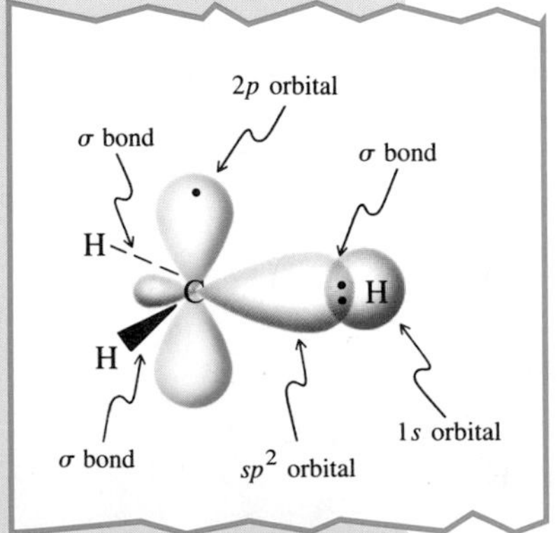

Third, color acts as a ''highlighter'' of a molecule's orbital structure, the basis of the molecule's special shape and properties. Wherever possible, *s* orbitals are shown in red, 2*p* orbitals in blue, sp^n hybrids in purple, 3*p* orbitals in green.

Fourth, color offers clues to a molecule's stereochemistry—the arrangement of its atoms in space. You will see in Chapter 5 that substituents in three dimensions can be assigned a priority according to certain "sequence rules," and this assignment has been indicated, in diminishing order of priority, by red, blue, green, or black.

Remember the use of color to denote group priorities:
Highest—red
Second highest—blue
Third highest—green
Lowest—black

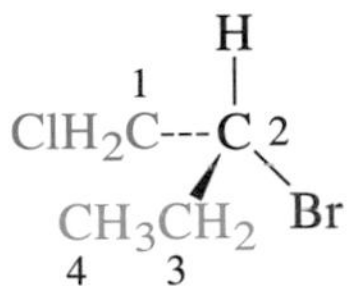

Optically active
2*R*

Most important, color frequently shows how the functional groups transform in the reaction mechanism. Electron-rich, or "nucleophilic," parts are shown in red; electron-deficient, or "electrophilic," fragments are blue; and radicals and leaving groups are green. Red arrows in these transformations indicate the movements of electrons.

STEP 4. Trapping by bromide

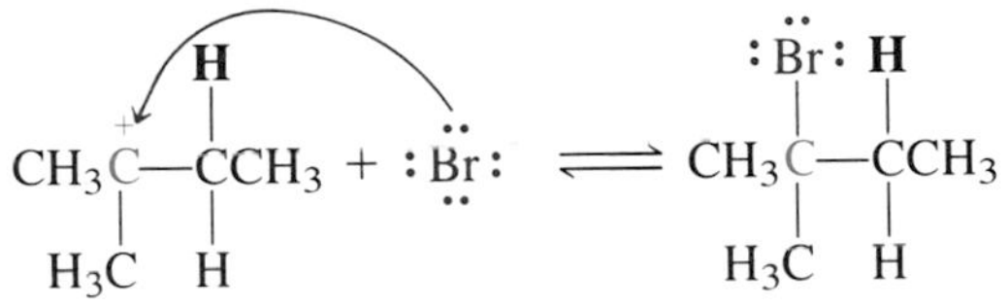

In other words, colors reveal the reactivities of the different functional groups. Because a group's reactivity may change at each step in a reaction mechanism, the color of units may also change as the overall reaction progresses. Do not be confused by these changes: They allow you to visualize in detail the special fate of reactive centers as they transform the starting materials into products.

Although we use color consistently within sections, you will note that its application may change from section to section (and chapter to chapter), because of the concept highlighted. Often color is applied sparingly and dropped when it serves no further purpose. You will quickly grasp the ideas behind the functional use of color and exploit it to your advantage. After you have done so, you will be better able to apply what you have learned *without* the reminder of color. For this reason, the summary sections, as well as almost all exercises and all end-of-chapter problems, are presented without color—just as in a test.

Using Other Resources

Make use of the office hours of your instructor and teaching assistants. They have specifically arranged their schedules to help you deal with difficult material, instruct you on how to solve exercises, and inspire you to *think* organic chemistry.

Avoid falling behind. Make sure that your schedule allows you to set aside a short period every day for reading the book, working the problems, reviewing material from class, and practicing what we like to think of as the *language* of organic chemistry. As you learn the reactions, you will be building a working ''vocabulary'' of organic chemistry; you will also master its ''grammar,'' as you understand the step-by-step history of a chemical reaction known as the reaction mechanism. You will enjoy organic chemistry if you are not under pressure because you have fallen behind, and you will gain a fresh, stimulating view of the chemical world that surrounds you.

We urge you to acquire a molecular model-making kit, such as the *Maruzen Molecular Structure Model* set available from W. H. Freeman. It is an *invaluable* tool for visualizing stereochemistry, the shapes of molecules, their interactions, and molecular mobility. Its utility is not limited to this course; it can be used in your studies whenever understanding the structure of molecules is crucial.

You may also wish to obtain the *Study Guide and Solutions Manual* that accompanies this book, in which all topics are extensively reviewed. The guide also summarizes new reactions and concepts, reminds you of material covered earlier, presents you with alternative explanations, and helps you to solve problems.

Enjoy organic chemistry, and good luck!

K. Peter C. Vollhardt
Neil E. Schore

Organic Chemistry

1

Structure and Bonding in Organic Molecules

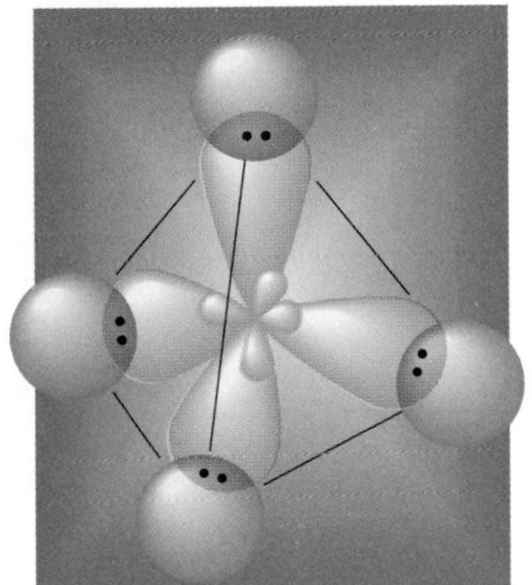

Chemistry is the study of the structure of molecules and the rules that govern their interactions. What, then, is organic chemistry? What distinguishes it from other disciplines, such as physical, inorganic, or nuclear chemistry? A common definition provides one partial answer: *Organic chemistry is the chemistry of carbon and its compounds*. These compounds are called **organic molecules.**

Organic molecules constitute the very essence of life. Fats, sugars, proteins, and the nucleic acids are compounds in which the principal component is carbon. So are countless substances that we take for granted in everyday use. Practically all the clothes we wear are made of organic molecules—some of natural fibers, such as cotton and silk; others artificial, such as polyester. Toothbrushes, toothpaste, soaps, shampoos, deodorants, perfumes—all these contain organic compounds, as do furniture, carpets, the plastic in light fixtures and cooking utensils, paintings, food, and countless other items.

Organic substances such as gasoline, medicines, pesticides, and polymers have improved the quality of our lives. Yet the uncontrolled disposal of organic chemicals has polluted the environment, causing deterioration of animal and plant life, as well as injury and disease to humans. If we are to create useful molecules—*and* learn to control their effects—we need a knowledge of their properties and an understanding of their behavior. We must be able to apply the principles of organic chemistry. This chapter explains how the basic ideas of chemical structure and bonding apply to organic molecules.

1-1 The Scope of Organic Chemistry: An Overview

A goal of organic chemistry is to relate the structure of a molecule to the reactions it can undergo. We can then study the steps by which each type of reaction takes place, and we can learn to create new molecules by applying those processes.

Thus, it makes sense to classify organic molecules according to the subunits and bonds that determine their chemical reactivity: These determinants are groups of atoms called **functional groups.** The study of the various functional groups and their respective reactions provides the structure of this book.

Functional groups determine the reactivity of organic molecules

$CH_3—CH_3$
Ethane

We begin with the alkanes. The alkanes are simple **hydrocarbons,** organic compounds composed of only hydrogen and carbon. As with other classes of molecules, we discuss the systematic rules for naming them, their structures, and their physical properties. An example of an alkane is ethane. Its structural mobility will be the starting point for a review of thermodynamics and kinetics. This review is then followed by a discussion of the strength of alkane bonds, which can be broken by heat, light, or chemical reagents. We shall illustrate these processes with the chlorination of alkanes.

A Chlorination Reaction

$$CH_4 + Cl_2 \xrightarrow{\text{Energy}} CH_3—Cl + HCl$$

(Cyclohexane ring: CH_2, CH_2, CH_2, CH_2, H_2C, H_2C)
Cyclohexane

Next we shall look at cyclic alkanes, which contain carbon atoms in a ring. This arrangement can lead to new properties and changes in reactivity. These molecules exhibit **stereoisomerism,** that is, a variety of ways in which their atoms can be positioned in space.

We shall then study the haloalkanes, our first example of compounds containing a functional group—the carbon–halogen bond. The haloalkanes participate in two types of organic reactions: substitution and elimination. In a **substitution** reaction, atoms in a functional group are replaced by atoms from a reagent; in an **elimination** process, atoms are removed from a molecule and new bonds formed in their place.

A Substitution Reaction

$$CH_3—Cl + K^+I^- \longrightarrow CH_3—I + K^+Cl^-$$

An Elimination Reaction

$$\underset{\substack{|\\ H}}{CH_2}-\underset{\substack{|\\ I}}{CH_2} + K^{+}\,{}^{-}OH \longrightarrow H_2C{=}CH_2 + HOH + K^{+}I^{-}$$

Like the haloalkanes, each of the major classes of organic compounds is characterized by a particular functional group. For example, the carbon–carbon triple bond is the functional group of alkynes; ethyne, a well-known alkyne, is the chemical burned in a welder's torch. A carbon–oxygen double bond fulfills this role for aldehydes and ketones, the starting materials in many industrial processes; and the amines, which include drugs such as nasal decongestants and amphetamines, contain nitrogen in their functional group. We shall study a number of tools for identifying these molecular subunits, including various forms of spectroscopy.

$HC{\equiv}CH$
An alkyne

$H_2C{=}O$
An aldehyde

$$CH_3-\overset{\substack{O\\ \|}}{C}-CH_3$$
A ketone

CH_3-NH_2
An amine

Subsequently we shall study several important classes of organic molecules that are especially crucial in biology and industry. Many of these, such as the carbohydrates and amino acids, contain multiple functional groups. However, in *every* class of organic compounds, the principle remains the same: *The structure of the molecule is related to the reactions it can undergo.*

Synthesis is the making of new molecules

Carbon compounds are called "organic" because it was originally thought that they could be produced only from living organisms. In 1828 Friedrich Wöhler* proved this idea to be false when he converted the inorganic salt lead cyanate into urea, an organic product of protein metabolism in mammals. [The average human excretes 30 g (grams) of urea each day!]

Wöhler's Synthesis of Urea

$$\underset{\textbf{Lead cyanate}}{Pb(OCN)_2} + \underset{\textbf{Water}}{2\,H_2O} + \underset{\textbf{Ammonia}}{2\,NH_3} \longrightarrow \underset{\textbf{Urea}}{2\,H_2N\overset{\substack{O\\ \|}}{C}NH_2} + \underset{\textbf{Lead hydroxide}}{Pb(OH)_2}$$

Synthesis, or the making of molecules, is a very important part of organic chemistry. Since Wöhler's time, nearly 10 million organic substances have been synthesized from simpler materials, both organic and inorganic. These substances include many that also occur in nature, such as the penicillin antibiotics, as well as entirely new compounds. Some, like cubane, which gave chemists the opportunity to study special kinds of bonding and reactivity, are of largely theoretical interest. Others, like the artificial sweetener saccharin, have become a part of everyday life.

Typically, the goal of synthesis is to construct complex organic chemicals from simpler, more readily available ones. To be able to convert one molecule into another, chemists must know organic reactions. They must also know the physical conditions that govern such processes, such as temperature, pressure,

*Professor Friedrich Wöhler, 1800–1882, University of Göttingen, Germany. In this and subsequent biographical notes, only the scientist's last known location of activity will be mentioned, even though much of his or her career may have been spent elsewhere.

BOX 1-1 Saccharin: One of the Oldest Synthetic Organic Compounds in Commercial Use

Saccharin was synthesized in the course of a study of the oxidation of organic chemicals containing sulfur and nitrogen. Its sweetness was discovered in 1879, a time when chemists routinely *tasted* every new compound they made. (You may wonder what effect this practice had on the life expectancy of chemists in those days. In fact, Remsen,* who was 33 when he made this discovery, lived to the ripe old age of 81.) Saccharin is 300 times sweeter than sugar and virtually nontoxic. It has proved to be a lifesaver for countless diabetics and of great value to individuals who need to control their caloric intake. The possibility that saccharin may be *carcinogenic,* that is, capable of causing cancer, was raised in the 1960s. In the 1970s a connection was found between high doses of saccharin and bladder tumors in rats. Experiments completed in 1990 demonstrated that it does not cause cancer directly, but at very high doses it promotes accelerated cell division, which may increase the likelihood of cell mutation and tumor formation. Warning labels are required on saccharin-containing products in the United States. These studies illustrate how society must balance the benefits that synthetic substances bring to our daily lives with the possible risks associated with their use.

*Professor Ira Remsen, 1846–1927, Johns Hopkins University.

solvent, and molecular structure. This knowledge is equally important in analyzing biological transformations.

As we study the chemistry of each functional group, we shall develop the tools both to plan effective syntheses and to predict the processes that take place in nature. But how? The answer will involve looking at reactions step by step.

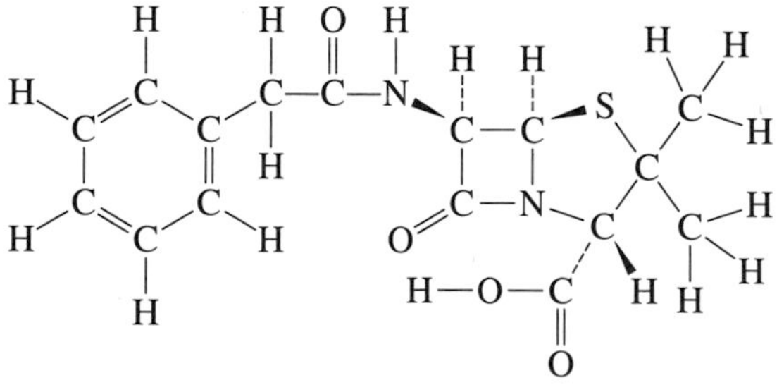

Benzylpenicillin

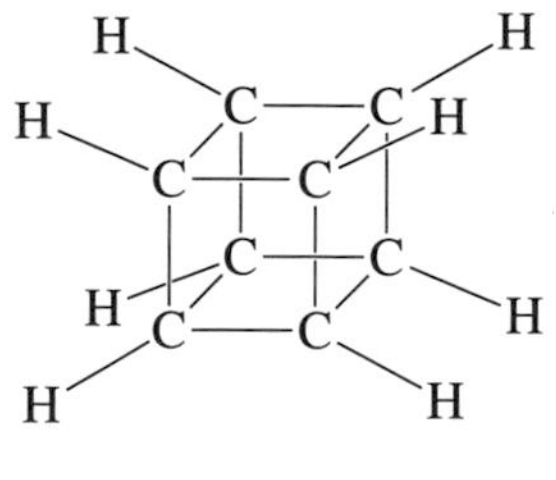

Cubane

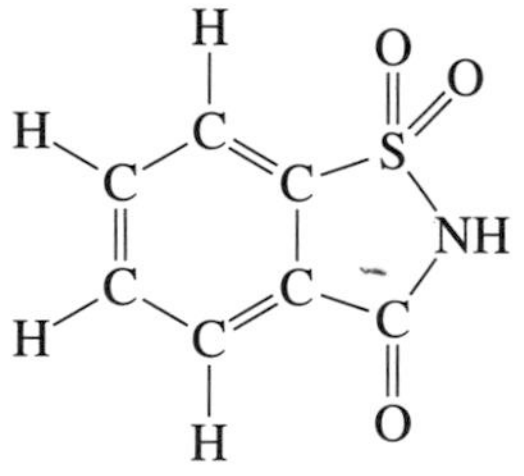

Saccharin

Reactions are the vocabulary and mechanisms are the grammar of organic chemistry

When we introduce a chemical reaction, we will first show just the starting compounds or **reactants** (also called **substrates**) and the products. In the chlorination process that we cited above, the substrates—methane, CH_4, and chlorine, Cl_2—may undergo a reaction to give chloromethane, CH_3Cl, and hydrogen chloride, HCl. The overall transformation was described as $CH_4 + Cl_2 \rightarrow CH_3Cl + HCl$. However, even a simple reaction like this one may actually proceed through a complex sequence of steps. The reactants could have first formed one or more *unobserved* substances—call these X—that rapidly changed into the observed products. These details underlying the reaction constitute the **reaction mechanism.** In our example the mechanism consists of a two-step sequence: $CH_4 + Cl_2 \rightarrow X$ followed by $X \rightarrow CH_3Cl + HCl$. Each step may have a part in determining whether the overall reaction will proceed.

How can we determine reaction mechanisms? Organic molecules are no more than collections of bonded atoms. We shall therefore study how, when, and how fast bonds break and form, in which way they do so in three dimensions, and how changes in substrate structure affect the outcome of reactions.

The substance X cited above is an example of a **reaction intermediate,** a species formed on the pathway between reactants and products. We shall learn the mechanism of this chlorination process and the true nature of the reaction intermediates in Chapter 4.

In a way, the ''learning'' and ''using'' of organic chemistry is much like learning and using a language. You need the vocabulary (i.e., the reactions) to be able to use the right words, but you also need the grammar (i.e., the mechanisms) to be able to converse intelligently. Neither one on its own gives complete knowledge and understanding, but together they form a powerful means of communication, rationalization, and predictive analysis.

Before we begin our study of the principles of organic chemistry, let us review some of the elementary principles of bonding. We shall find these concepts useful in understanding and predicting the chemical reactivity and the physical properties of organic molecules.

1-2 Coulomb Forces: A Simplified View of Bonding

The bonds between atoms hold a molecule together. But why are there bonds? Two atoms form a bond only if their interaction is energetically favorable; that is, energy—heat, for example—is released when the bond is formed. Conversely, breaking that bond requires the input of the same amount of energy.

The two main causes of the energy release associated with bonding are based on fundamental laws of physics:

1. Opposite charges attract each other, and
2. Electrons spread out in space.

Bonds are made by coulombic attraction and electron exchange

Each atom consists of a nucleus, containing electrically neutral particles, or neutrons, and positively charged protons. Surrounding the nucleus are negatively charged electrons, equal in number to the protons so that the net charge is zero. As two atoms approach each other, the positively charged nucleus of the first attracts the electrons of the second; similarly, the nucleus of the second attracts the electrons of the first. This sort of bonding is described by **Coulomb's* law:** Opposite charges attract each other with a force inversely proportional to the square of the distance between the centers of the charges.

*Lieutenant-Colonel Charles Augustin de Coulomb, 1736–1806, *Inspecteur Général* of the University of Paris, France.

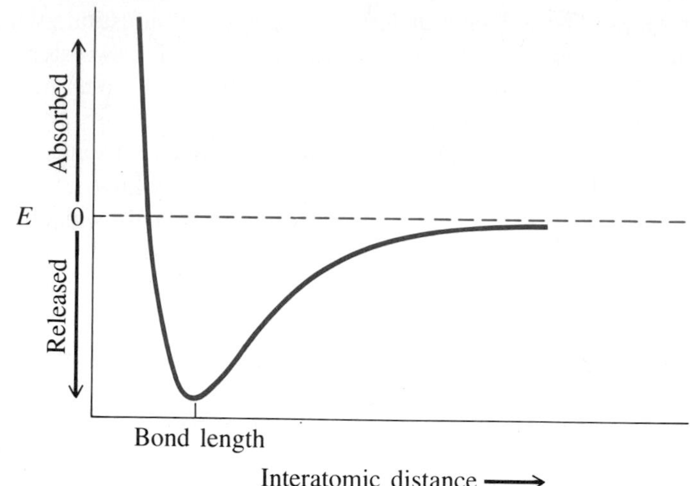

FIGURE 1-1 The changes in energy, E, that result when two atoms are brought into close proximity. At the separation defined as bond length, maximum bonding is achieved.

Coulomb's Law

$$\text{Attracting force} = \text{constant} \times \frac{(+)\text{charge} \times (-)\text{charge}}{\text{distance}^2}$$

This attractive force causes energy to be released as the atoms are brought together.

When the atoms reach a certain closeness, no more energy is released. The distance between the two nuclei at this point is called the **bond length** (Figure 1-1). Bringing the atoms closer together than this distance results in a sharp *increase* in energy. Why? Just as opposite charges attract, like charges repel. If the atoms are too close, the electron–electron and nuclear–nuclear repulsions become stronger than the attractive forces. When the nuclei are the appropriate bond length apart, the electrons are spread out around both nuclei, and attractive and repulsive forces balance for maximum bonding. The energy content of the two-atom system is then at a minimum, the most stable situation (Figure 1-2).

An alternative to this type of bonding results from the complete transfer of an electron from one atom to the other. The result is two charged *ions:* one posi-

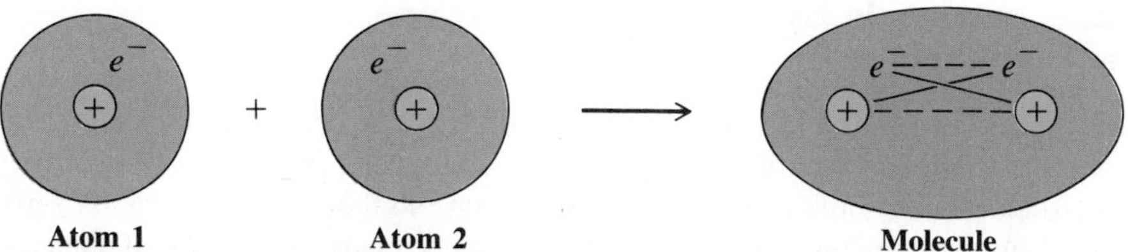

FIGURE 1-2 Covalent bonding. Attractive (solid line) and repulsive (dashed line) forces in the bonding between two atoms. The large circles represent areas in space in which the electrons are found around the nucleus. The small circle around the plus sign stands for the nucleus.

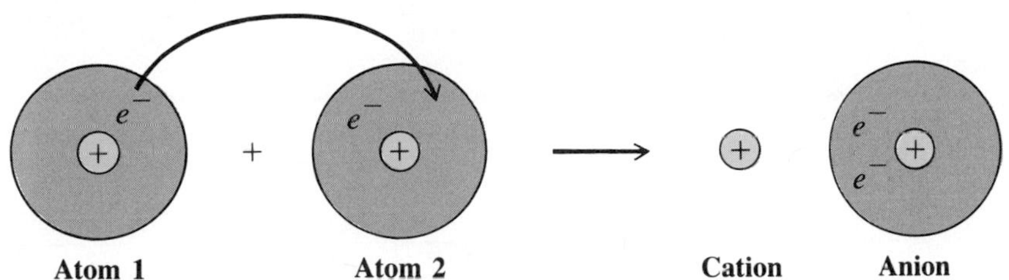

FIGURE 1-3 Ionic bonding. An alternative mode of bonding results from the complete transfer of an electron from atom 1 to atom 2.

tively charged, a *cation,* and one negatively charged, an *anion* (Figure 1-3). Again, the bonding is based on coulombic attraction, this time between two ions.

The coulombic bonding models of attracting and repelling charges shown in Figures 1-2 and 1-3 are highly simplified views of the interactions that take place in the bonding of atoms. Nevertheless, these models explain many of the properties of organic molecules.

We have seen that attraction between negatively and positively charged particles is a basis for bonding. How does this concept work in real molecules?

1-3 Ionic and Covalent Bonds: The Octet Rule

Two extreme types of bonding explain the interactions between atoms in organic molecules:

1. A **covalent bond** is formed by the sharing of electrons (as shown in Figure 1-2).
2. An **ionic bond** is formed by the transfer of one or more electrons from one atom to another (as shown in Figure 1-3), thereby generating ions.

We shall see that many atoms bind to carbon in a way that is intermediate between these extremes: Some ionic bonds have covalent character, and some covalent bonds are partly ionic.

What are the factors that account for the two types of bonds? To answer this question, let us return to the atoms and their compositions. We shall start by looking at the periodic table and at how the electronic makeup of the elements changes as the atomic number increases.

The periodic table underlies the octet rule

The partial periodic table depicted in Table 1-1 includes those elements most widely found in organic molecules: carbon (C), hydrogen (H), oxygen (O), nitrogen (N), sulfur (S), chlorine (Cl), bromine (Br), and iodine (I). Certain reagents, indispensable for synthesis and commonly used, contain elements such as lithium (Li), magnesium (Mg), boron (B), and phosphorus (P). (If you are not familiar with these elements, you should learn Table 1-1.)

TABLE 1-1 Partial Periodic Table

Period							Halogens	Noble gases
First	H^{1}							He^{2}
Second	$Li^{2,1}$	$Be^{2,2}$	$B^{2,3}$	$C^{2,4}$	$N^{2,5}$	$O^{2,6}$	$F^{2,7}$	$Ne^{2,8}$
Third	$Na^{2,8,1}$	$Mg^{2,8,2}$	$Al^{2,8,3}$	$Si^{2,8,4}$	$P^{2,8,5}$	$S^{2,8,6}$	$Cl^{2,8,7}$	$Ar^{2,8,8}$
Fourth	$K^{2,8,8,1}$						$Br^{2,8,18,7}$	$Kr^{2,8,18,8}$
Fifth							$I^{2,8,18,18,7}$	$Xe^{2,8,18,18,8}$

Note: The superscripts indicate the number of electrons in each principal shell of the atom.

EXERCISE 1-1

(a) Redraw Figure 1-1 for a weaker bond than the one depicted. **(b)** Write Table 1-1 from memory.

The elements in the periodic table are listed according to their atomic number, or nuclear charge (number of protons), which also equals their number of electrons. This number increases by one with each element listed. The electrons occupy energy levels, or ''shells,'' each with a fixed capacity. For example, the first shell has room for two electrons; the second, eight; and the third, eighteen. Helium, with two electrons in its shell, and the other noble gases, with eight electrons (called **octets**) in their outermost shells, are especially stable. These elements show very little chemical reactivity. All other elements lack octets in their outermost electron shells. *They will tend to form molecules in such a way as to reach an octet in the outer electron shell and attain a noble-gas configuration.*

In ionic bonds, electron octets are formed by transfer of electrons

Sodium (Na), a reactive metal, interacts with chlorine, a reactive gas, in a violent manner to produce a stable substance: sodium chloride. Similarly, sodium reacts with fluorine (F), bromine, or iodine to give the respective salts. Other alkali metals, such as lithium and potassium (K), undergo the same reactions. These transformations succeed because both reaction partners attain noble-gas character by the *transfer of outer-shell electrons,* called **valence electrons,** from the alkali metals on the left side of the periodic table to the halogens on the right.

Let us see how this works for the ionic bond in sodium chloride. Why is the interaction energetically favorable? First, it takes energy to remove an electron from an atom. This energy is the **ionization potential** (IP) of the atom. For sodium gas, the ionization energy amounts to 119 kcal mol^{-1}.* Conversely,

*This book will cite energy values in the traditional units of kcal mol^{-1}, where a mol is the abbreviation for mole and a kilocalorie (kcal) is the energy required to raise the temperature of 1 kg (kilogram) of water by 1°C. In SI units energy is expressed in joules (kg m s^{-2}, or kilogram-meter per $second^2$). A joule (J) is the energy required to raise a mass of 1 kg every second by 1 m s^{-1}. The conversion factor is 1 kcal = 4184 J = 4.184 kJ (kilojoule).

energy may be released when an electron attaches itself to an atom. For chlorine, this energy, called its **electron affinity** (EA), is −83 kcal mol^{-1}. These two processes result in the transfer of an electron from sodium to chlorine. Together, they require a net energy *input* of 119 − 83 = 36 kcal mol^{-1}.

$$Na^{2,8,1} \xrightarrow{-1e} [Na^{2,8}]^{+} \qquad IP = 119 \text{ kcal mol}^{-1}$$

Sodium cation (Neon configuration) — Energy input required

$$Cl^{2,8,7} \xrightarrow{+1e} [Cl^{2,8,8}]^{-} \qquad EA = -83 \text{ kcal mol}^{-1}$$

Chloride anion (Argon configuration) — Energy released

$$Na + Cl \longrightarrow Na^{+}Cl^{-} \qquad \text{Total} = 36 \text{ kcal mol}^{-1}$$

Why, then, do the atoms readily convert to NaCl? The reason is their electrostatic attraction, which pulls them together in an ionic bond. At the most favorable interatomic distance [about 2.8 Å (angstroms) in the gas phase], this attraction releases about 120 kcal mol^{-1} (see Figure 1-1). This energy release is enough to make the reaction of sodium with chlorine energetically highly favorable (+36 − 120 = −84 kcal mol^{-1}).

More than one electron may be donated (or accepted) to achieve favorable electronic configurations. Magnesium, for example, has two valence electrons. Donation to an appropriate acceptor produces the corresponding doubly charged cation with the electronic structure of neon. In this way, the ionic bonds of typical salts are formed.

Formation of Ionic Bonds by Electron Transfer

$$Na^{2,8,1} + Cl^{2,8,7} \longrightarrow [Na^{2,8}]^{+}\,[Cl^{2,8,8}]^{-}, \text{ or NaCl}$$

A more convenient way of depicting valence electrons is by means of dots around the symbol for the element. In this case, the letters represent the nucleus and all the electrons in the inner shells, together called the **core configuration.**

Valence Electrons as Electron Dots

Li·	Be:	·Ḃ·	·Ċ·	·N̈·	:Ö·	:F̤̈·
Na·	Mg:	·Ȧl·	·Ṡi·	·P̈·	:S̈·	:C̤̈l·

Electron-Dot Picture of Salts

$$Na\cdot + \cdot\ddot{\underset{..}{Cl}}: \xrightarrow{e \text{ transfer}} Na^{+}:\ddot{\underset{..}{Cl}}:^{-}$$

$$Mg: + 2\ \cdot\ddot{\underset{..}{Cl}}: \xrightarrow{2e \text{ transfer}} Mg^{2+}[:\ddot{\underset{..}{Cl}}:]^{-}_{2}$$

The hydrogen atom may either lose an electron to become a bare nucleus, the **proton,** or accept an electron to form the **hydride ion,** [H:]$^{-}$, which possesses the helium configuration. Indeed, the hydrides of lithium, sodium, and potassium (Li^+H^-, Na^+H^-, and K^+H^-) are commonly used reagents.

$$H\cdot \xrightarrow{-1\,e} [H]^+ \quad \text{Bare nucleus} \quad IP = 314 \text{ kcal mol}^{-1}$$

Proton

$$H\cdot \xrightarrow{+1\,e} [H:]^- \quad \text{Helium configuration} \quad EA = -18 \text{ kcal mol}^{-1}$$

Hydride ion

EXERCISE 1-2

Draw electron-dot pictures for ionic LiBr, Na_2O, BeF_2, $AlCl_3$, and MgS.

In covalent bonds, electron octets are formed by sharing electrons

Formation of ionic bonds between two identical elements is difficult because the electron transfer is usually very unfavorable. For example, in H_2, formation of H^+H^- would require an energy input of nearly 300 kcal mol^{-1}. For the same reason, none of the halogens, F_2, Cl_2, Br_2, and I_2, has an ionic bond. The high IP of hydrogen also prevents the bonds in the hydrogen halides from being ionic. For elements nearer the center of the periodic table, the formation of ionic bonds is unfeasible because it becomes more and more difficult to donate or accept enough electrons to attain the noble-gas configuration. Such is the case for carbon. This element would have to shed four electrons to reach the helium electronic structure or add four electrons for a neonlike arrangement. The large amount of charge that would develop makes these processes very energetically unfavorable.

$$C^{4+} \xleftarrow{-4\,e} \cdot\dot{\underset{\cdot}{C}}\cdot \xrightarrow{+4\,e} :\ddot{\underset{\cdot\cdot}{C}}:^{4-}$$

Helium configuration | **Neon configuration**

Instead, **covalent bonding** takes place: The elements *share* electrons so that each attains a noble-gas configuration. Typical products of such sharing are H_2 and HCl. In HCl, the chlorine atom assumes an octet structure by sharing one of its valence electrons with that of hydrogen. Similarly, the chlorine molecule, Cl_2, is diatomic because both component atoms gain octets by sharing two electrons. Such bonds are called **covalent single bonds.**

Electron-Dot Picture of Covalent Single Bonds

$$H\cdot + \cdot H \longrightarrow H:H$$

$$H\cdot + \cdot\ddot{\underset{\cdot\cdot}{Cl}}: \longrightarrow H:\ddot{\underset{\cdot\cdot}{Cl}}:$$

$$:\ddot{\underset{\cdot\cdot}{Cl}}\cdot + \cdot\ddot{\underset{\cdot\cdot}{Cl}}: \longrightarrow :\ddot{\underset{\cdot\cdot}{Cl}}:\ddot{\underset{\cdot\cdot}{Cl}}:$$

Because carbon has four valence electrons, it must acquire a share of four electrons through four single bonds to gain the neon configuration, as in methane. Nitrogen has five valence electrons and needs three to share, as in ammonia; and oxygen, with six valence electrons, requires only two to share, as in water.

```
   H
   ..
H:C:H      H:N:H
   ..         ..        H:O:H
   H          H          ..
Methane    Ammonia      Water
```

It is possible for one atom to supply both of the electrons required for covalent bonding. This occurs upon addition of a proton to ammonia, thereby forming NH_4^+, or to water, thereby forming H_3O^+.

```
  H              [   H   ]+                         [  ..   ]+
  ..             [   ..  ]        ..                [ H:O:H ]
H:N: + H+  --->  [ H:N:H ]      H:O: + H+  --->     [   ..  ]
  ..             [   ..  ]        ..                [   H   ]
  H              [   H   ]        H
                  Ammonium                           Hydronium
                    ion                                 ion
```

Besides two-electron **(single)** bonds, atoms may form four-electron **(double)** and six-electron **(triple)** bonds to gain noble-gas configurations. Atoms that share more than one electron pair are found in ethene and ethyne.

```
 H     H
  .C::C.        H:C:::C:H
 H     H
  Ethene          Ethyne
(Ethylene)*     (Acetylene)*
```

EXERCISE 1-3

Draw electron-dot structures for F_2, CF_4, CH_2Cl_2, PH_3, BrI, OH^-, NH_2^-, and CH_3^-. (Where applicable, the first element is at the center of the molecule.) Make sure that all atoms have noble-gas electron configurations.

In polar covalent bonds, the electrons are not shared equally

The elements on the left of the periodic table are often called **electropositive,** electron donating, or ''electron pushing,'' because their electrons are held by the nucleus less tightly than those associated with elements to the right. The latter are therefore described as being **electronegative,** electron accepting, or ''electron pulling.'' Table 1-2 lists the relative electronegativity of some elements. On this scale, fluorine, the most electronegative of them all, is assigned the value 4. Covalent bonds between atoms of differing electronegativity are said to be **polarized.** Polarization of a bond is the consequence of a shift of the center of electron density in the bond toward the more electronegative atom. The separation of opposite charges is called an electric **dipole,** symbolized by an arrow crossed at its tail and pointing from positive to negative.

*In labels of molecules, systematic names (to be introduced in Section 2-3) will be given first, followed in parentheses by so-called common names that are still in frequent usage.

TABLE 1-2 Electronegativities of Selected Elements

H 2.2						
Li 1.0	Be 1.6	B 2.0	C 2.6	N 3.0	O 3.4	F 4.0
Na 0.9	Mg 1.3	Al 1.6	Si 1.9	P 2.2	S 2.6	Cl 3.2
K 0.8						Br 3.0
						I 2.7

Note: Values established by L. Pauling and updated by A. L. Allred (see *Journal of Inorganic and Nuclear Chemistry*, 1961, *17*, 215).

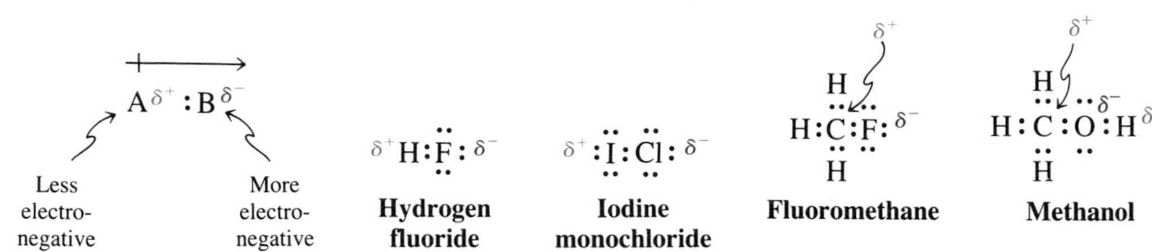

Although the molecule as a whole stays electrically neutral, one end may have a partial positive charge, designated δ^+, the other a partial negative charge, δ^-. Consequently, a polarized bond can impart polarity to a molecule as a whole, as in HF, HCl, and CH_3F. In symmetrical structures, however, the polarizations of the individual bonds may cancel, thus leading to molecules with no net polarization, such as CO_2 and CCl_4. To know whether a molecule is polar, we have to know its shape.

Molecules Can Have Polar Bonds but No Net Polarization

$\delta^-\ \ddot{O}::C::\ddot{O}\ \delta^-$ ($2\delta^+$ on C)

CCl_4: δ^- on each Cl, $4\delta^+$ on C

Electron repulsion controls the shapes of molecules

Molecules adopt shapes in which electron repulsion is minimized. In diatomic species such as H_2 or LiH, there is only one bond and one possible arrangement of the two atoms. However, beryllium fluoride, BeF_2, is a triatomic species. Will it be bent or linear? Electron repulsion is at a minimum in a **linear** structure.* Linearity is also expected for other derivatives of beryllium, as well as of other elements in the same column of the periodic table.

*This is true only in the gas phase. At room temperature, BeF_2 is a solid (it is used in nuclear reactors!) that exists as a complex network of linked Be and F atoms, not as a distinct linear triatomic structure.

BeF_2 Is Linear

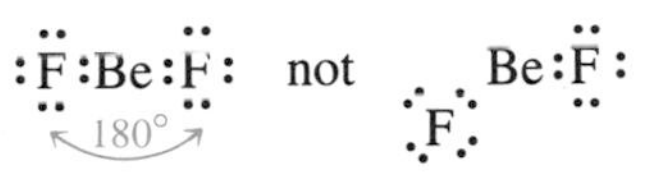

Electrons are farthest apart

Electrons are closer

BCl_3 Is Trigonal

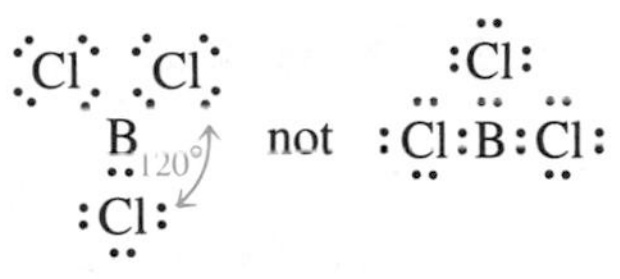

In boron trichloride, the three valence electrons of boron allow it to form covalent bonds with three chlorine atoms. Electron repulsion enforces a regular **trigonal** arrangement—that is, the three halogens are at the corners of an equilateral triangle, the center of which is occupied by boron. Other derivatives of boron, and the analogous compounds with other elements in the same column of the periodic table, are again expected to adopt trigonal structures.

Applying this principle to carbon, we can see that methane, CH_4, has to be **tetrahedral.** Pointing its four valences toward the vertices of a tetrahedron is the best arrangement for minimizing electron repulsion.

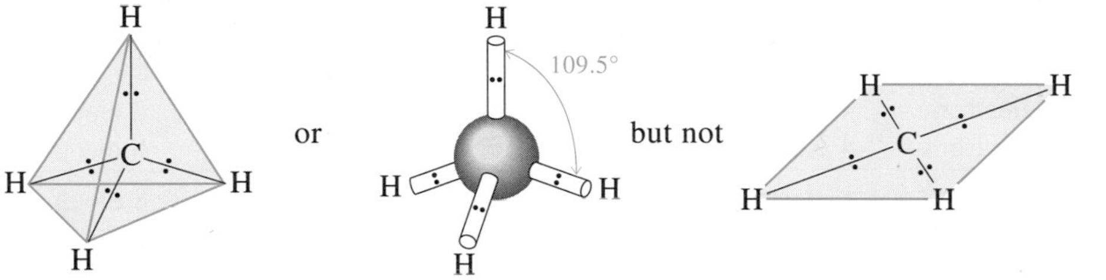

This method for determining molecular shape by minimizing electron repulsion is called the *valence shell electron pair repulsion* (VSEPR) method. Note that we often draw molecules like BCl_3 and CH_4 as if they were flat and had 90° angles. *This depiction is for ease of drawing only*. Do *not* confuse such drawings with the true molecular shapes (trigonal for BCl_3 and tetrahedral for CH_4).

EXERCISE 1-4

Show the bond polarization in H_2O, SCO, SO, IBr, CH_4, $CHCl_3$, CH_2Cl_2, and CH_3Cl by use of dipole arrows to indicate separation of charge. (In the last four examples place the carbon in the center of the molecule.)

EXERCISE 1-5

Ammonia, NH_3, is not trigonal but pyramidal, with bond angles of 107.3°. Water, H_2O, is not linear but bent (104.5°). Why? (See, for example, D.A. McQuarrie and P.A. Rock, *General Chemistry,* 3d ed., W.H. Freeman and Company, New York, 1991, pp. 445–456.)

To summarize, there are two extreme types of bonding, ionic and covalent. Both derive favorable energetics from Coulomb forces and the attainment of noble-gas electronic structures. Most bonds are better described as a combination of the two types: the polar covalent (or covalent ionic) bonds. Polarity in bonds may give rise to polar molecules. The outcome depends on the shape of the molecule, which is determined in a simple manner by arranging its bonds and nonbonding electrons to minimize electron repulsion.

1-4 Electron-Dot Model of Bonding: Lewis Structures

The drawings in the preceding section using pairs of electron dots to represent bonds are also called **Lewis* structures.** In this section, rules are given for writing such structures correctly and for keeping track of valence electrons.

Lewis structures are drawn by following simple rules

H
H C H
H
Correct

H H C H H
Incorrect

The procedure for drawing correct electron-dot structures is straightforward, as long as the following rules are observed.

RULE 1. *Draw the molecular skeleton.* As an example, consider methane. The molecule has four hydrogen atoms bonded to one central carbon atom.

RULE 2. *Count the number of available valence electrons.* Add up all the valence electrons of the component atoms. Special care has to be taken with charged structures (anions or cations), in which case the appropriate number of electrons has to be added or subtracted to account for extra charges.

CH_4	4 H	4×1 electron =	4 electrons
	1 C	1×4 electrons =	4 electrons
		Total	8 electrons
H_3O^+	3 H	3×1 electron =	3 electrons
	1 O	1×6 electrons =	6 electrons
	Charge	+1 =	−1 electron
		Total	8 electrons

H_2O	2 H	2×1 electron =	2 electrons
	1 O	1×6 electrons =	6 electrons
		Total	8 electrons
NH_2^-	2 H	2×1 electron =	2 electrons
	1 N	1×5 electrons =	5 electrons
	Charge	−1 =	+1 electron
		Total	8 electrons

RULE 3. (The octet rule) *Depict all covalent bonds by two shared electrons, giving as many atoms as possible a surrounding electron octet, except for H, which requires a duet.* Make sure that the number of electrons used is *exactly* the number counted according to rule 2. Elements at the right in the periodic table may contain pairs of valence electrons not used for bonding, called **lone electron pairs,** or just **lone pairs.**

Nonbonding or lone electron pairs
H:F: (with lone pairs)
Duet
Octet

Octet
H, H:C:H, H
Duets

Consider, for example, hydrogen fluoride. The shared electron pair supplies the hydrogen atom with a duet, the fluorine with an octet, because the fluorine carries three lone electron pairs. Conversely, in methane, the four C–H bonds satisfy the requirement of the hydrogens, and at the same time furnish the octet for carbon. Examples of correct and incorrect Lewis structures are shown at the top of page 15.

Frequently, the number of valence electrons is not sufficient to satisfy the octet rule with only single bonds. In this event, double (two shared electron pairs) and even triple bonds (three shared pairs) are necessary to obtain octets. An example is the nitrogen molecule, N_2, which has ten valence electrons (shown in the margin on page 15). An N–N single bond would leave both atoms with electron sextets, and a double bond provides only one nitrogen atom with an octet. It is the molecule with a triple bond that satisfies both atoms.

*Professor Gilbert N. Lewis, 1875–1946, University of California, Berkeley.

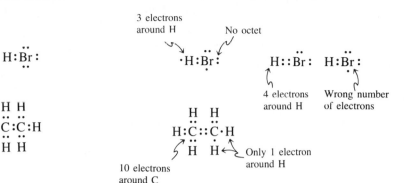

Correct Lewis Structures	Incorrect Lewis Structures
H:Br:	·H:Br: (3 electrons around H; No octet) — H::Br: (4 electrons around H) — H:Br: (Wrong number of electrons)
H:C:C:H (ethane, with H above and below each C)	H:C::C·H (10 electrons around C; Only 1 electron around H)

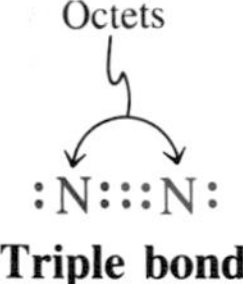

Further examples of such molecules are shown below.

Correct Lewis Structures

Ethene (Ethylene)	**Ethyne (Acetylene)**	**Formaldehyde**

EXERCISE 1-6

Draw Lewis structures for the following molecules: HI, $CH_3CH_2CH_3$, CH_3OH, HSSH, SiO_2 (OSiO), O_2, CS_2 (SCS).

RULE 4. *Assign charges to atoms in the molecule.* Each lone pair contributes two electrons to the valence electron count of an atom in a molecule, and each bonding (shared) pair contributes one. An atom is charged if this total is different from the outer-shell electron count in the free, nonbonded atom. Thus we have the formula

$$\text{Charge} = \begin{pmatrix}\text{number of outer-shell} \\ \text{electrons on the} \\ \text{free, neutral atom}\end{pmatrix} - \begin{pmatrix}\text{number of unshared} \\ \text{electrons on the atom} \\ \text{in the molecule}\end{pmatrix} - \frac{1}{2}\begin{pmatrix}\text{number of bonding} \\ \text{electrons surrounding the} \\ \text{atom in the molecule}\end{pmatrix}$$

As an example, which atom bears the positive charge in the hydronium ion? Each hydrogen has a valence electron count of 1 from the shared pair in its bond to oxygen. Because this value is the same as the electron count in the free atom, the charge on each hydrogen is zero. The electron count on the oxygen in the hydronium ion is 2 (the lone pair) + 3 (half of 6 bonding electrons) = 5. This value is one short of the number of outer-shell electrons in the free atom, thus giving the oxygen a charge of +1. Hence the positive charge is assigned to oxygen.

H:Ö:H (with + on O, H below)

Hydronium ion

Another example is the nitrosyl cation, NO^+. The molecule bears a lone pair on nitrogen, in addition to the triple bond connecting the nitrogen to the oxygen

:N:::Ö: (with + on O)
Nitrosyl cation

atom. This gives nitrogen five valence electrons, a value that matches the count in the free atom; therefore the nitrogen atom has no charge. The same number of valence electrons (5) is found on oxygen. Because the free oxygen atom requires 6 valence electrons to be neutral, the oxygen in NO^+ possesses the $+1$ charge. Other examples are shown below.

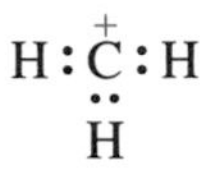

Methyl cation 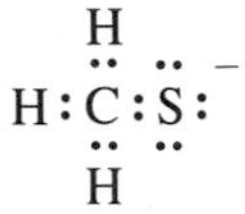 **Methanethiolate ion** 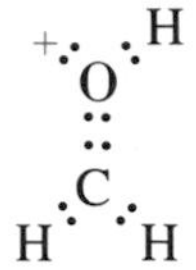 **Protonated formaldehyde**

Atoms in neutral molecules also may be charged

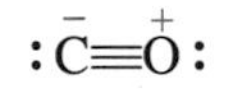

Carbon monoxide

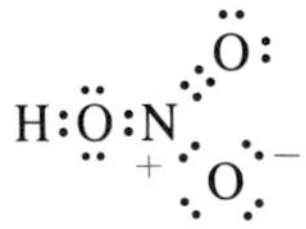

Sometimes the octet rule leads to charges on atoms even in neutral molecules. The Lewis structure is then said to be **charge separated.** Examples include carbon monoxide, CO, and some compounds containing nitrogen–oxygen bonds, such as nitric acid, HNO_3.

Nitric acid

Covalent bonds can be depicted as straight lines

Electron-dot structures can be cumbersome, particularly for larger molecules. It is simpler to represent covalent single bonds by single straight lines; double bonds are represented by two lines and triple bonds by three. Lone electron pairs can either be shown as dots or simply omitted. The use of such notation was first suggested by the German chemist August Kekulé,* long before electrons were discovered; structures of this type are often called **Kekulé structures.**

Straight-Line Notation for the Covalent Bond

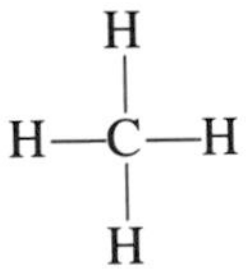

:N≡N:

Methane 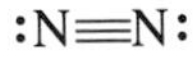 **Diatomic nitrogen** 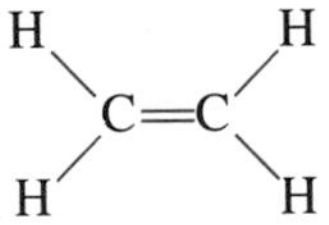 **Ethene** 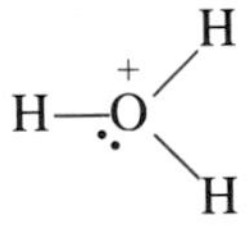 **Hydronium ion** 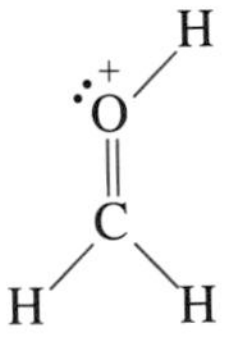 **Protonated formaldehyde**

EXERCISE 1-7

Draw Lewis structures of the following molecules, including the assignment of any charges to atoms (the order in which the atoms are attached is given in parentheses when it may not be obvious from the formula as it is commonly written): SO, F_2O (FOF), $HClO_2$ (HOClO), BF_3NH_3 (F_3BNH_3), $CH_3OH_2^+$ ($H_3COH_2^+$), $Cl_2C{=}O$, CN^-, C_2^{2-}.

*Professor F. August Kekulé von Stradonitz, 1829–1896, University of Bonn, Germany.

In summary, Lewis structures describe bonding by the use of electron dots or straight lines. Whenever possible, they are drawn so as to give hydrogen an electron duet and other atoms an electron octet. Charges are assigned to each atom by evaluating its electron count.

1-5 Resonance Structures

In organic chemistry we also encounter molecules for which there are *several* correct Lewis structures.

The carbonate ion has several correct Lewis structures

Let us consider the carbonate ion, CO_3^{2-}. Following our rules, we can easily draw a Lewis structure (A) in which every atom is surrounded by an octet. The two negative charges are located on the bottom two oxygen atoms; the third oxygen is neutral, connected to the central carbon by a double bond and bearing two lone pairs. But why pick the bottom two oxygen atoms as the charge carriers? There is no reason at all—it is a completely arbitrary choice. We could equally well have drawn structures B or C to describe the carbonate ion. All three Lewis pictures are *equivalent* and are called **resonance structures.** The individual resonance structures are connected by double-headed arrows and all placed within square brackets. They have the characteristic property of being interconvertible by *electron-pair movement only,* the nuclear positions in the molecule remaining *unchanged.* Note that, to turn A into B and then into C, we have to shift two electron pairs in each case. Such movement of electrons can be depicted by curved arrows, a procedure informally called "electron pushing."

The use of curved arrows to depict electron-pair movement is a useful technique to help us avoid making the common mistake of changing the total number of electrons when we draw resonance structures.

Resonance Structures of the Carbonate Ion

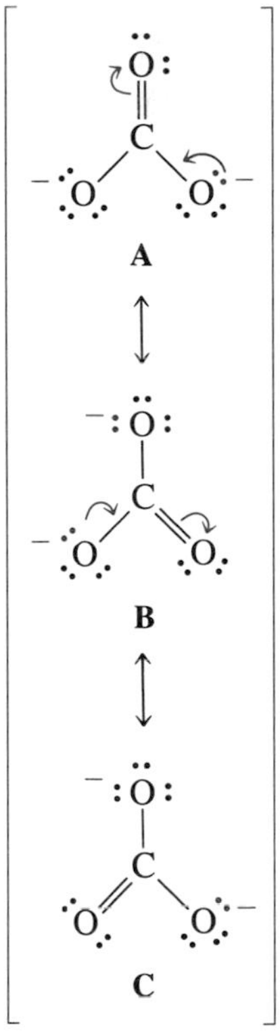

A

B

C

But what is its true structure?

Does the carbonate ion have one uncharged oxygen atom bound to carbon through a double bond and two other oxygen atoms bound through a single bond each, both bearing a negative charge as suggested by the Lewis structures? *The answer is no.* If that were true, the carbon–oxygen bonds would be of different lengths, because double bonds are normally shorter than single bonds. But the carbonate ion is *perfectly symmetrical* and contains a trigonal central carbon, all C–O bonds being of equal length, between that of a double and a single bond. The negative charge is evenly distributed over all three oxygens: It is said to be **delocalized.** In other words, none of the individual Lewis representations of this molecule is correct on its own. Rather, *the true structure is a composite of A, B, and C.* The resulting picture is called a **resonance hybrid.** Because A, B, and C are equivalent, they contribute equally to the true structure of the molecule, but none of them by itself accurately represents it.

The word *resonance* may imply to you that the molecule vibrates or equilibrates from one form to another. This inference is incorrect. The molecule never looks like any of the individual resonance forms; it has only one structure, the resonance hybrid. Unlike substances in ordinary chemical equilibria, resonance forms are *not* real, although each makes a partial contribution to reality.

An alternative convention used to describe resonance hybrids such as carbonate is to represent the bonds as a combination of solid and dotted lines. The $\frac{2}{3}-$ sign here indicates that a partial charge ($\frac{2}{3}$ of a negative charge) resides on each oxygen atom. The equivalence of all three carbon–oxygen bonds and all three oxygens is clearly indicated by this convention. Other examples of resonance hybrids are the acetate anion and the 2-propenyl (allyl) cation.

Dotted-Line Notation of Carbonate as a Resonance Hybrid

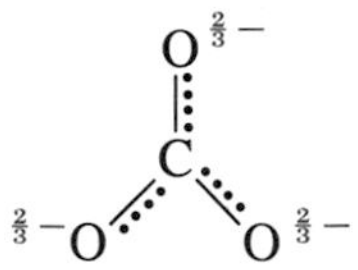

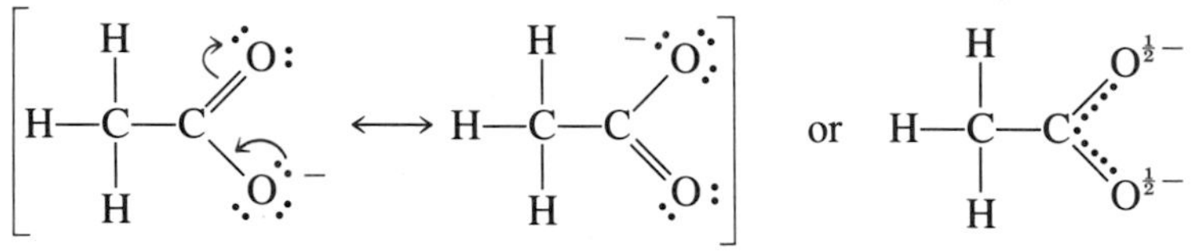

Acetate anion

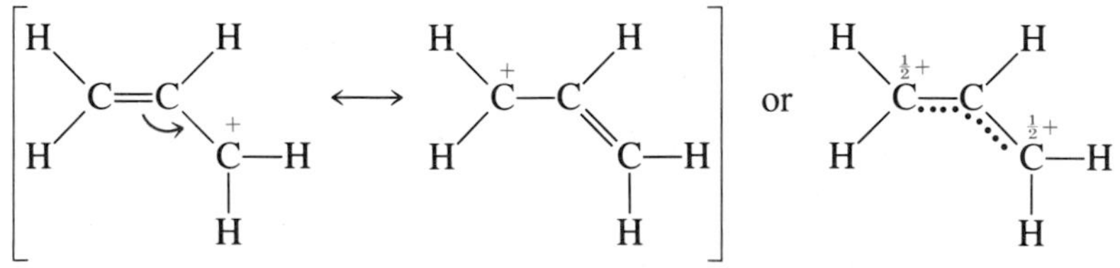

2-Propenyl (allyl) cation

When drawing resonance structures, keep in mind that (1) pushing one electron pair toward one atom and away from another results in a movement of charge; (2) the relative positions of all the atoms stay unchanged; (3) equivalent resonance structures contribute equally to the resonance hybrid; and (4) the arrows connecting resonance structures are double headed (↔).

EXERCISE 1-8

Draw two resonance structures for nitrite ion, NO_2^-. What can you say about the geometry of this molecule?

Not all resonance structures are equivalent

The carbonate and acetate anions and the 2-propenyl cation all have equivalent resonance structures. However, many molecules are described by resonance forms that are not equivalent. An example is the enolate anion. The two resonance structures differ in the locations of both the double bond and the charge.

The Two Nonequivalent Resonance Structures of the Enolate Ion

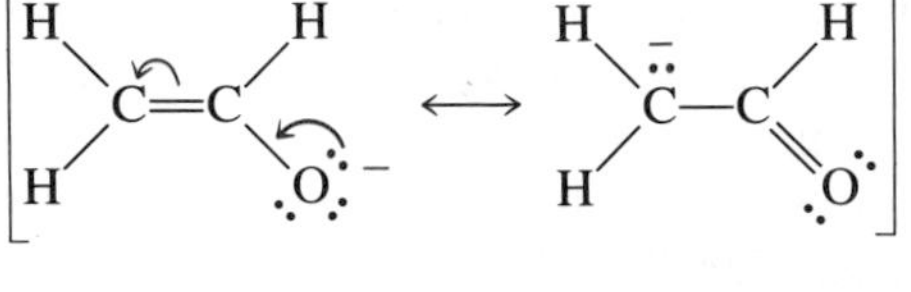

Although both forms are contributors to the true structure of the anion, one is more important than the other. The question is, which one? Several guidelines help us to recognize preferred structures, which are also called **major resonance contributors.**

GUIDELINE 1. *Structures with a maximum of octets are preferred.* In the enolate ion, all component atoms in either structure are surrounded by octets. Consider, however, the nitrosyl cation, NO^+: The better resonance form has a positive charge on oxygen with electron octets around both atoms; the other places the positive charge on nitrogen, thereby resulting in an electron sextet on this atom. Because of the octet rule, the second structure contributes less to the hybrid. Thus, the N–O linkage is closer to being a triple than a double bond and more of the positive charge is on oxygen than on nitrogen.

$$\left[:N{\equiv}\overset{+}{O}: \longleftrightarrow :\overset{+}{N}{=}\ddot{\underset{..}{O}} \right]$$

Major resonance contributor — Minor resonance contributor

Nitrosyl cation

GUIDELINE 2. *Charges should be preferentially located on atoms with compatible electronegativity.* Consider again the enolate ion. Which is the preferred resonance structure? Guideline 2 requires that it is the first, in which the negative charge resides on the more electronegative oxygen atom.

Looking again at NO^+, you might find guideline 2 confusing. The major resonance contributor to NO^+ has the positive charge on the more electronegative oxygen. In cases such as this, *the octet rule overrides the electronegativity criterion;* that is, guideline 1 takes precedence over guideline 2.

GUIDELINE 3. *Structures with a minimum of charge separation are preferred.* This rule is a simple consequence of Coulomb's law: Separating charges requires energy; hence neutral structures are better than dipolar ones.

$$\left[\text{H}{-}\overset{\overset{:O:}{\|}}{C}{-}\ddot{\underset{..}{O}}{-}\text{H} \longleftrightarrow \text{H}{-}\overset{\overset{:\ddot{O}:^-}{|}}{C}{=}\overset{+}{\underset{..}{O}}{-}\text{H} \right]$$

Major — Minor

Formic acid

In some cases, to ensure octet Lewis structures, charge separation is acceptable; that is, guideline 1 takes precedence over guideline 3. An example is carbon monoxide.

$$\left[:C{=}\ddot{O}: \longleftrightarrow :\overset{-}{C}{\equiv}\overset{+}{O}: \right]$$

Minor — Major

Carbon monoxide

When there are several charge-separated resonance structures that comply with the octet rule, the most favorable is the one in which the charge distribution best accommodates the relative electronegativities of the component atoms (guideline 2). In diazomethane, for example, nitrogen is more electronegative than carbon, thus allowing a clear choice between the two resonance contributors.

$$\left[\begin{array}{c} H_2C{=}\overset{+}{N}{=}\overset{-}{\underset{..}{N}}: \\ \text{Major} \\ \updownarrow \\ H_2\overset{-}{\ddot{C}}{-}\overset{+}{N}{\equiv}N: \\ \text{Minor} \end{array} \right]$$

Diazomethane

EXERCISE 1-9

Draw resonance structures for the following two molecules. Indicate the more favorable resonance contributor in each case. **(a)** CNO^-; **(b)** NO^-.

In summary, there are molecules that cannot be described accurately by one Lewis structure but exist as hybrids of several extreme resonance forms. To find the most important resonance contributor, consider the octet rule, make sure that there is a minimum of charge separation, and place on the relatively more electronegative atoms as much negative and as little positive charge as possible.

1-6 Atomic Orbitals: A Quantum Mechanical Description of Electrons Around the Nucleus

We are now ready to delve a little deeper into the nature of atoms and bonds. We will start by learning more about the way in which the electrons are distributed around the nucleus, both spatially and energetically. The simplified treatment presented here has as its basis the theory of quantum mechanics developed independently in the 1920s by Heisenberg, Schrödinger, and Dirac.* In this theory, the movement of an electron around a nucleus is expressed in the form of equations that are very similar to those characteristic of waves. The solutions to these equations, called **atomic orbitals,** allow us to describe the probability of finding the electron in a certain region in space. The shape of these domains depends on the energy of the electron.

The electron is described by wave equations

The classical description of the atom (Bohr† theory) assumed that electrons move on more or less defined trajectories around the nucleus. Their energy was thought to relate to their distance from the nucleus. This view is intuitively appealing because it coincides with our physical understanding of classical mechanics. Yet it is incorrect for several reasons.

First, the classical picture of an electron moving in an orbit requires the emission of electromagnetic radiation. The resulting energy loss from the system would cause the electron to spiral toward the nucleus, a prediction that is completely at odds with reality.

Second, in the classical picture, an electron can have any energy, so it can have any of an infinite number of orbits of differing radii. This, again, is not what is observed. Rather, only certain defined energies called **energy states** are possible for an electron around a nucleus. Thus, classical mechanics does not satisfactorily explain atomic structure and, ultimately, bonding.

A better model is afforded by considering the wave nature of moving particles. Matter of mass m that moves with velocity v has a wavelength λ.

de Broglie‡ Wavelength

$$\lambda = \frac{h}{mv}$$

in which h is Planck's§ constant. As a result, an orbiting electron can be described by equations that are the same as those used in classical mechanics to

*Professor Werner Heisenberg, 1901–1976, University of Munich, Germany, Nobel Prize 1932 (physics); Professor Erwin Schrödinger, 1887–1961, University of Dublin, Ireland, Nobel Prize 1933 (physics); Professor Paul Dirac, 1902–1984, Florida State University, Nobel Prize 1933 (physics).

†Professor Niels Bohr, 1885–1962, University of Copenhagen, Sweden, Nobel Prize 1922 (physics).

‡Prince Louis-Victor de Broglie, 1892–1987, Nobel Prize 1929 (physics).

§Professor Max K. E. L. Planck, 1858–1947, University of Berlin, Germany, Nobel Prize 1918 (physics).

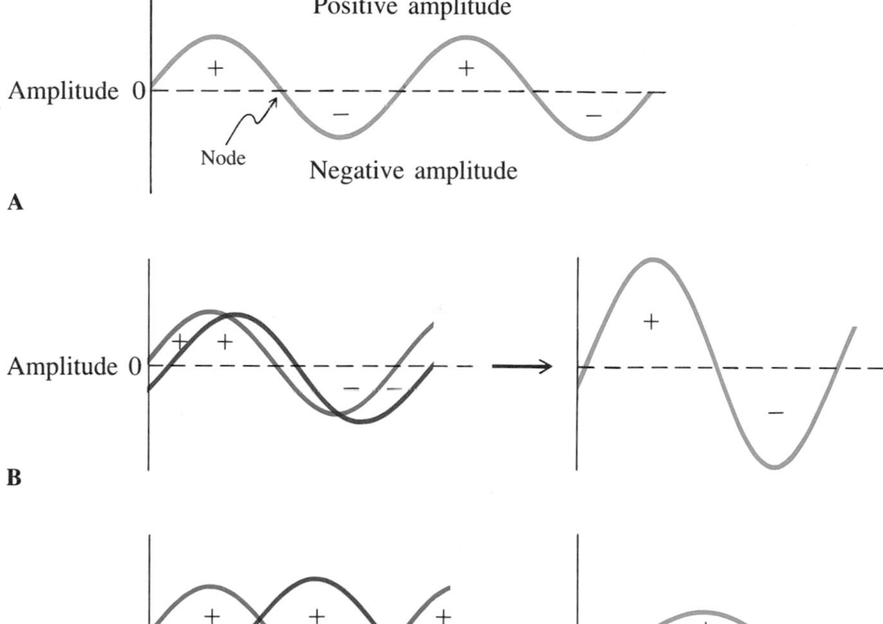

FIGURE 1-4 (A) A wave. The signs of the amplitude are assigned arbitrarily. At points of zero amplitude, called nodes, the wave changes sign. (B) Waves with amplitudes of like sign (in phase) reinforce each other to make a larger wave. (C) Waves out of phase subtract from each other to make a smaller wave.

describe waves (Figure 1-4). The latter have amplitudes with alternating positive and negative signs. Points at which the sign changes are called **nodes.** Waves that interact in phase reinforce each other, as shown in Figure 1-4B. Those out of phase interfere with each other to make smaller waves (and possibly even cancel each other), as shown in Figure 1-4C.

Note: The + and − signs in Figure 1-4 refer to signs of the mathematical functions describing the wave amplitudes and have nothing to do with electrical charges.

This theory of electron motion is called **quantum mechanics.** The equations developed in this theory, the **wave equations,** have a series of solutions called **wave functions,** usually described by the Greek letter psi, ψ. Their values around the nucleus are not directly identifiable with any observable property of the atom. However, *the squares* (ψ^2) *of their values at each point in space describe the probability of finding an electron at that point*. The physical realities of the atom make solutions attainable only for certain *specific energies*. The system is said to be **quantized.**

EXERCISE 1-10

Draw a picture similar to Figure 1-4 of two waves overlapping such that their amplitudes cancel each other.

Atomic orbitals have characteristic shapes

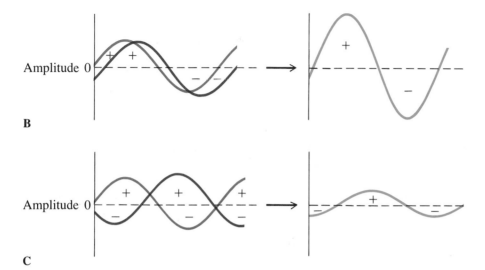

Plots of wave functions in three dimensions typically have the appearance of spheres or dumbbells with flattened or teardrop-shaped lobes. For simplicity, we may regard artistic renditions of atomic orbitals as indicating the regions in space in which the electron is likely to be found. Nodes separate portions of the wave function with opposite mathematical signs. The value of the wave function at a

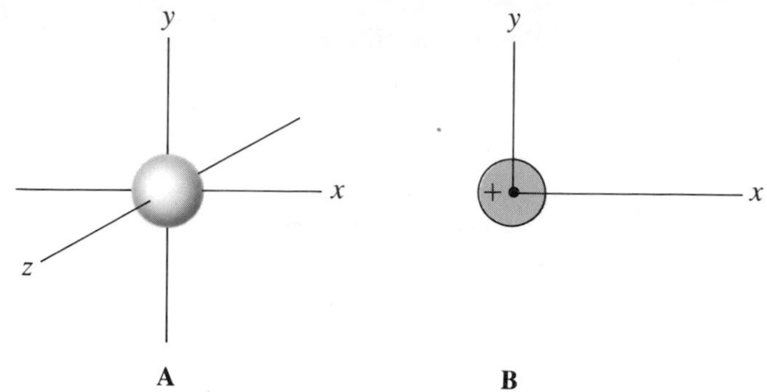

FIGURE 1-5 Representations of a 1s orbital. (A) The orbital is spherically symmetric in three dimensions. (B) A simplified two-dimensional view. The plus sign denotes the mathematical sign of the wave function and is *not a charge*.

node is zero; therefore the probability of finding electron density there is zero. Higher energy wave functions have more nodes than those of low energy.

Let us consider the shapes of the atomic orbitals for the simplest case, that of the hydrogen atom, consisting of a proton surrounded by an electron. The single lowest energy solution of the wave equation is called the 1s orbital, the number one referring to the first (lowest) energy level. An orbital label also denotes the shape and number of nodes of the orbital. The 1s orbital is *spherically symmetric* (Figure 1-5) and has no nodes. This orbital can be represented pictorially as a sphere (Figure 1-5A) or simply as a circle (Figure 1-5B).

The next higher energy wave function, the 2s orbital, is also unique and, again, spherical. The 2s orbital is larger than the 1s orbital; the higher energy 2s electron is on the average farther from the positive nucleus. In addition, the 2s orbital has one node, a spherical surface of zero electron density separating regions of the wave function of opposite sign (Figure 1-6). Like that of classical waves, the sign of the wave function on either side of the node is arbitrary, as long as it changes at the node.

After the 2s orbital, the wave equations for the electron around a hydrogen atom have three energetically equivalent solutions, the $2p_x$, $2p_y$, and $2p_z$ orbitals. Solutions of equal energy of this type are called **degenerate** (*degenus,* Latin, without genus or kind). As shown in Figure 1-7, p orbitals consist of two lobes that resemble a solid figure eight. A p orbital is characterized by its directionality in space. The orbital axis can be aligned with any one of the x, y, and z axes,

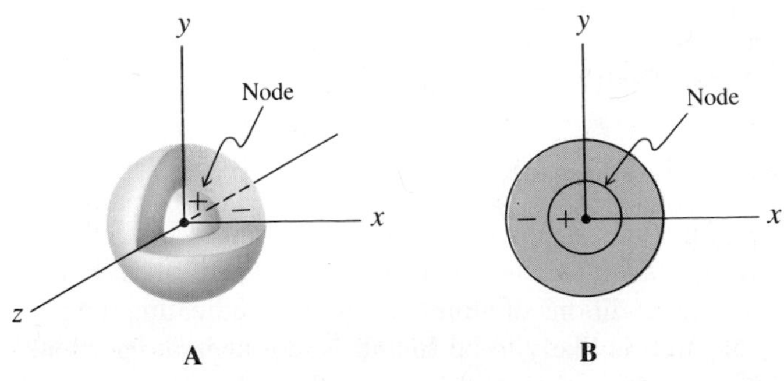

FIGURE 1-6 Representations of a 2s orbital. Notice that it is larger than the 1s orbital and that a node is present. The + and − denote the sign of the wave function. (A) The orbital in three dimensions, with a section removed to allow the visualization of the node. (B) The more conventional two-dimensional representation of the orbital.

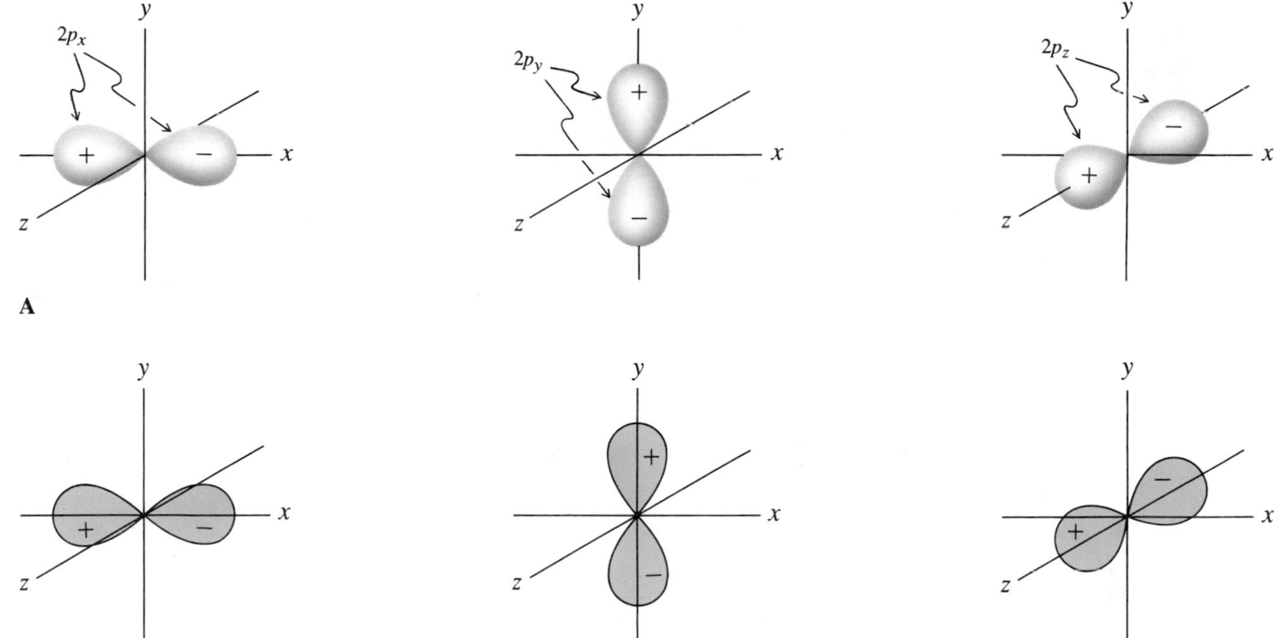

FIGURE 1-7 Representations of $2p$ orbitals (A) in three dimensions and (B) in two dimensions. Remember that the + and − signs refer to the wave functions and *not* to electrical charges. Lobes of opposite sign are separated by a nodal plane that is perpendicular to the axis of the orbital. For example, the p_x orbital is divided by a node in the yz plane.

hence the labels p_x, p_y, p_z. The two lobes of opposite sign of each orbital are separated by a nodal plane through the atom's nucleus and perpendicular to the orbital axis.

The next higher solutions are the $3s$ and $3p$ atomic orbitals. They are similar in shape to but more diffuse than their lower energy counterparts and have two nodes. Still higher energy orbitals are characterized by an increasing number of nodes and a variety of shapes. They are of much less importance in organic chemistry than are the lower orbitals. To a first approximation, the shapes and nodal properties of the atomic orbitals of other elements are very similar to those of hydrogen. Therefore, we may use s and p orbitals in a description of the electronic configurations of helium, lithium, and so forth.

The Aufbau principle assigns electrons to orbitals

Approximate relative energies of the atomic orbitals up to the $5s$ level are shown in Figure 1-8. With its help, we can give an electronic configuration to every atom in the periodic table. To do so, we follow three rules for assigning electrons to atomic orbitals:

1. Lower energy orbitals are filled before those with higher energy.

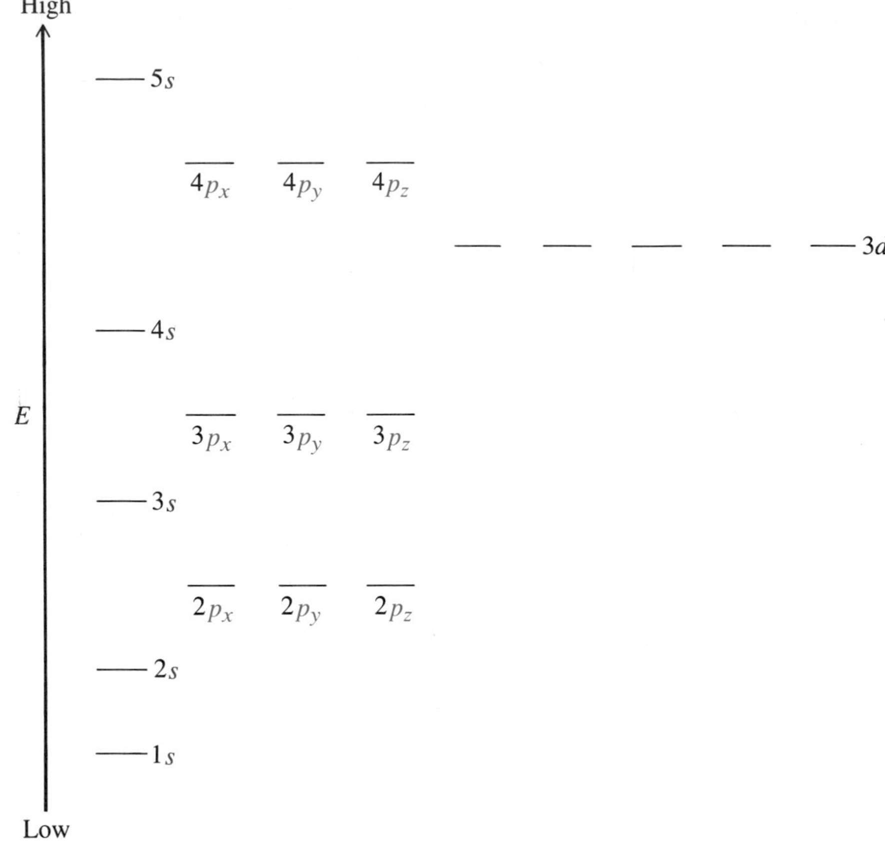

FIGURE 1-8
Approximate relative energies of atomic orbitals, corresponding roughly to the order in which they are filled in atoms. Orbitals of lowest energy are filled first; degenerate orbitals are filled according to Hund's rule.

2. No orbital may be occupied by more than two electrons, according to the **Pauli* exclusion principle.** Furthermore, these two electrons must differ in the orientation of their intrinsic angular momentum, their **spin.** There are two possible directions of the electron spin, usually depicted by vertical arrows pointing in opposite directions. An orbital is filled when it is occupied by two electrons of opposing spin, frequently referred to as **paired electrons.**
3. Degenerate orbitals, such as the *p* orbitals, are first occupied by one electron each, all of these electrons having the same spin. Subsequently, three more, each of opposite spin, are added to the first set. This assignment is based on **Hund's† rule.**

With these rules in hand, the determination of electronic configuration becomes simple. Helium has two electrons in the 1*s* orbital and its electronic structure is abbreviated $(1s)^2$. Lithium $[(1s)^2(2s)^1]$ has one and beryllium $[(1s)^2(2s)^2]$ two additional electrons in the 2*s* orbital. In boron $[(1s)^2(2s)^2(2p)^1]$, we begin the filling of the three degenerate 2*p* orbitals. This pattern continues with carbon and nitrogen, and then the addition of electrons of opposite spin for oxygen,

*Professor Wolfgang Pauli, 1900–1958, ETH Zürich, Switzerland, Nobel Prize 1945 (physics).
†Professor Friedrich Hund, b. 1896, University of Göttingen, Germany.

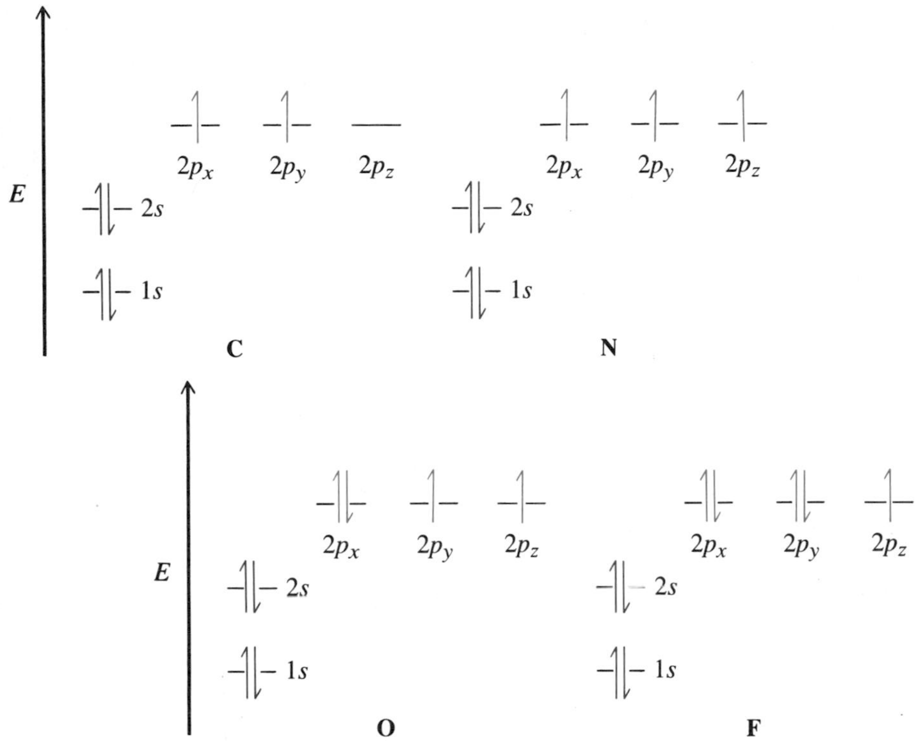

FIGURE 1-9 The most stable electronic configurations of carbon, $(1s)^2(2s)^2(2p)^2$; nitrogen, $(1s)^2(2s)^2(2p)^3$; oxygen, $(1s)^2(2s)^2(2p)^4$; and fluorine $(1s)^2(2s)^2(2p)^5$. Notice that the unpaired electron spins in the p orbitals are in accord with Hund's rule, and the paired electron spins in the filled 1s and 2s orbitals in accord with the Pauli principle and Hund's rule. The order of filling the p orbitals has been arbitrarily chosen as p_x, p_y, and then p_z. Any other order would have been equally good.

fluorine, and neon fills all p levels. The electronic configurations of four of the elements are depicted in Figure 1-9. Atoms with completely filled sets of atomic orbitals are said to have a **closed-shell configuration.** For example, helium and neon have this attribute. Carbon, in contrast, has an **open-shell configuration.**

The process of adding electrons one by one to the orbital sequence shown in Figure 1-8 is called the **Aufbau principle** (*Aufbau,* German, build up). It is easy to see that the Aufbau principle affords a rationale for the stability of the electron octet and duet. These numbers are required for closed-shell configurations. For helium, the closed-shell configuration is a 1s orbital filled with two electrons of opposite spin. In neon, the 2s and 2p orbitals are occupied by an additional eight electrons; in argon, the 3s and 3p levels accommodate eight more (Figure 1-10).

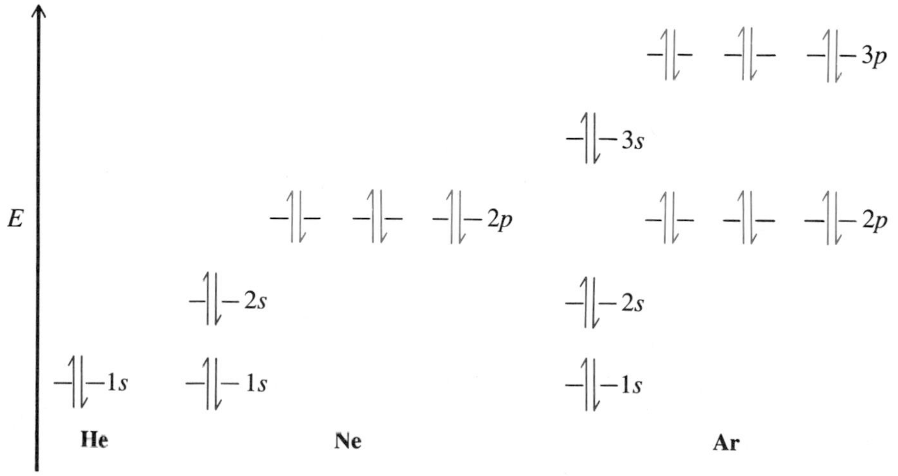

FIGURE 1-10 Closed-shell configurations of the noble gases helium, neon, and argon.

EXERCISE 1-11

Using Figure 1-8, draw the electronic configurations of sulfur and phosphorus.

To summarize, the motion of electrons around the nucleus is described by wave equations. Their solutions, atomic orbitals, can be symbolically represented as regions in space, with each point given a positive, negative, or zero (at the node) numerical value, the square of which represents the probability of finding the electron there. The Aufbau principle allows us to assign electronic configurations to all atoms.

1-7 Molecular Orbitals and Covalent Bonding

We shall now see how covalent bonds are constructed from the overlap of atomic orbitals.

The bond in the hydrogen molecule is formed by the overlap of 1*s* atomic orbitals

Let us begin by looking at the simplest case: the bond between the two hydrogen atoms in H_2. In a Lewis structure of the hydrogen molecule, we would write the bond as an electron pair shared by both atoms to give each a helium configuration. How do we construct H_2 using atomic orbitals? An answer to this question was developed by Pauling*: *Bonds are made by the in-phase overlap of atomic orbitals*. What is meant by that? Recall that atomic orbitals are solutions of wave equations. Like waves, they may interact in a reinforcing way (Figure 1-4B) if the overlap is between areas of the wave function of the same sign, or *in phase*. They may also interact in a destructive way if the overlap is between areas of opposite sign, *out of phase* (Figure 1-4C).

The in-phase overlap of the two 1*s* orbitals results in a new orbital of lower energy called a **bonding molecular orbital** (Figure 1-11). In the bonding combination, the wave function in the space between the nuclei is strongly reinforced.

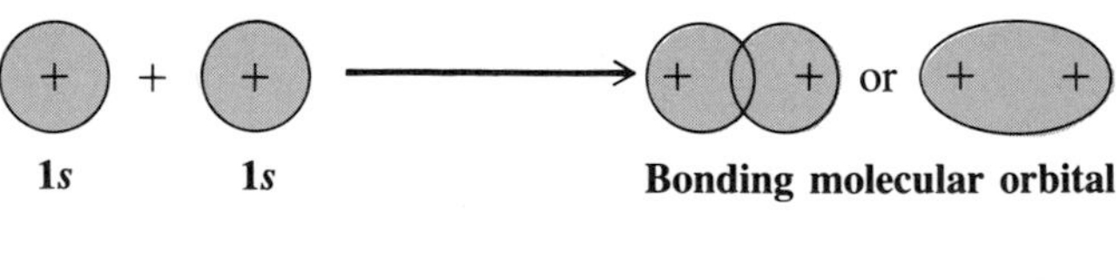

FIGURE 1-11 In-phase (bonding) and out-of-phase (antibonding) combinations of 1*s* atomic orbitals. The + and − signs denote the *sign* of the wave function, not charges. Electrons in bonding molecular orbitals have a high probability of occupying the space *between* the atomic nuclei, as required for good bonding (compare Figure 1-2). The antibonding molecular orbital has a nodal plane, where the probability of finding electrons is zero. Electrons in antibonding molecular orbitals are most likely to be found *outside* the space between the nuclei and therefore do not contribute to bonding.

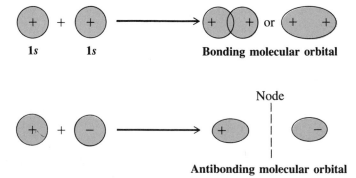

*Professor Linus Pauling, 1901–1994, Stanford University, Nobel Prizes 1954 (chemistry) and 1963 (peace).

Thus the probability of finding the electrons occupying this molecular orbital in that region is very high: a condition for bonding between the two atoms. This picture is strongly reminiscent of that shown in Figure 1-2. The use of two wave functions with *positive* signs for representing the in-phase combination of the two 1*s* orbitals in Figure 1-11 is arbitrary. Overlap between two *negative* orbitals would give identical results. In other words, it is overlap between *like* lobes that makes a bond, regardless of the sign of the wave function.

On the other hand, out-of-phase overlap between the same two atomic orbitals results in a destabilizing interaction and formation of an **antibonding molecular orbital.** In the antibonding molecular orbital, the amplitude of the wave function is canceled in the space between the two atoms, thereby giving rise to a node.

Thus, the net result of the interaction of the two 1*s* atomic orbitals of hydrogen is the generation of two molecular orbitals. One is bonding and lower in energy; the other is antibonding and higher in energy. Because the total number of electrons available to the system is only two, they are placed in the lower energy molecular orbital. The result is a decrease in total energy, thereby making H_2 more stable than two free hydrogen atoms. This difference in energy levels corresponds to the strength of the H–H bond. The interaction can be depicted schematically in an energy diagram (Figure 1-12A).

It is now readily understandable why hydrogen exists as H_2, whereas helium prefers to be monatomic. The overlap of two filled atomic orbitals, as in helium, leads to bonding and antibonding orbitals, *both of which are filled* (Figure 1-12B). Therefore making a He–He bond does not decrease the total energy.

The overlap of atomic orbitals gives rise to sigma and pi bonds

The formation of molecular orbitals by an overlap of atomic orbitals applies not only to the 1*s* orbitals of hydrogen but also to other atomic orbitals. The amount of energy by which the bonding level drops and the antibonding level is raised is called the **energy splitting.** It reflects the strength of the bond being made and depends on a variety of factors. For example, overlap is best between orbitals of similar size and energy. Therefore two 1*s* orbitals will interact with each other more effectively than a 1*s* and a 3*s*.

Geometric factors also influence the degree of overlap. This consideration is important for orbitals with directionality in space such as *p* orbitals. Such orbitals give rise to two types of bonds: one in which the atomic orbitals are aligned along the internuclear axis (parts A, B, C, and D in Figure 1-13) and one in which they

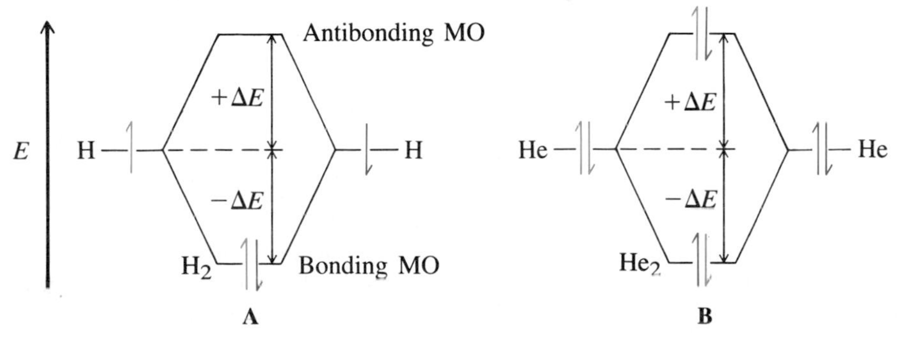

FIGURE 1-12 Schematic representation of the interaction of two (A) singly (as in H_2) and (B) doubly (as in He_2) occupied atomic orbitals to give two molecular orbitals (MO). (Not drawn to scale.) Formation of an H–H bond is favorable because it stabilizes two electrons. Formation of an He–He bond stabilizes two electrons (in the bonding MO) but destabilizes two others (in the antibonding MO). Bonding between He and He thus results in no net stabilization. Therefore, helium is monatomic.

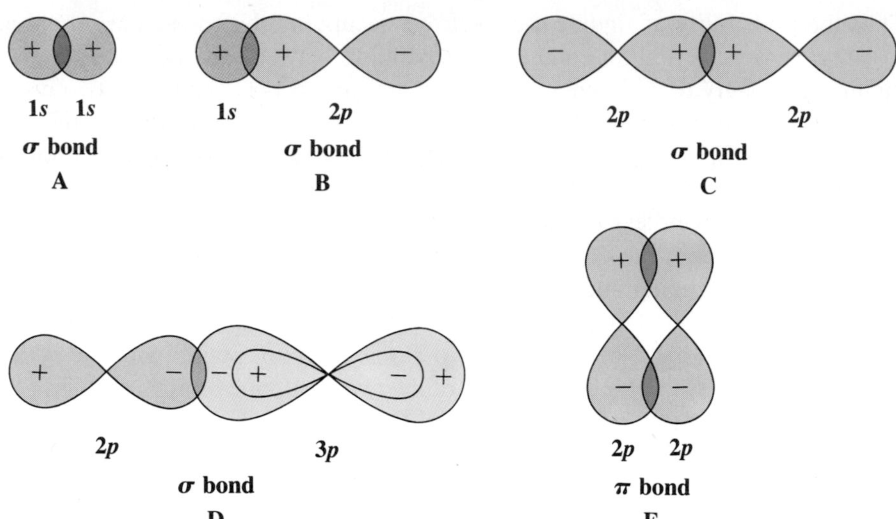

FIGURE 1-13 Bonding between atomic orbitals. (A) 1*s* and 1*s* (e.g., H_2), (B) 1*s* and 2*p* (e.g., HF), (C) 2*p* and 2*p* (e.g., F_2), (D) 2*p* and 3*p* (e.g., FCl) aligned along internuclear axes, σ bonds; (E) 2*p* and 2*p* perpendicular to internuclear axis, a π bond. Note the arbitrary use of + and − signs to indicate in-phase interactions of the wave functions.

are perpendicular to it (part E). The first are called **sigma (σ) bonds,** the second **pi (π) bonds.** All carbon–carbon single bonds are of the σ type; however, we shall find that double and triple bonds also have π components (Section 1-9).

EXERCISE 1-12

Construct a molecular orbital and energy-splitting diagram of the bonding in He_2^+. Is it favorable?

We have come a long way in our description of bonding. First, we thought of bonds in terms of Coulomb forces, then in terms of covalency and shared electron pairs, and now we have a quantum mechanical picture. Bonds are a result of the overlap of atomic orbitals. The two bonding electrons are placed in the bonding molecular orbital. Because it is stabilized relative to the two initial atomic orbitals, energy is given off during bond formation. This decrease in energy represents the bond strength.

1-8 Hybrid Orbitals: Bonding in Complex Molecules

Let us now construct bonding schemes for more complex molecules by using quantum mechanics. How can we use atomic orbitals to build linear (as in BeH_2), trigonal (as in BH_3), and tetrahedral molecules (as in CH_4)?

Mixing orbitals in a single atom gives hybrid orbitals

Consider the molecule beryllium hydride, BeH_2. Beryllium has two electrons in the 1*s* orbital and two electrons in the 2*s* orbital. With no unpaired electrons, this arrangement does not appear to allow for bonding.

However, it takes a relatively small amount of energy to promote one electron from the 2*s* orbital to one of the 2*p* levels (Figure 1-14). In the $1s^22s^12p^1$ configuration, beryllium is ready to enter into bonding, because there are now two

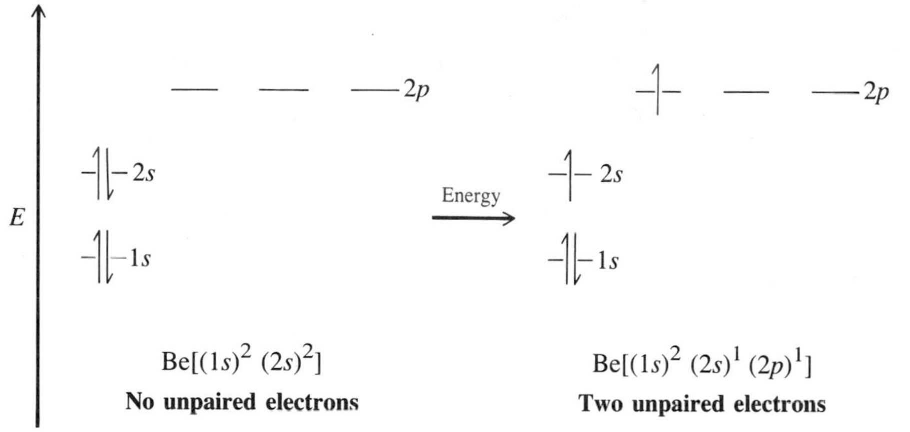

2p 2p

2s 2s

E Energy

1s 1s

Be$[(1s)^2 (2s)^2]$
No unpaired electrons

Be$[(1s)^2 (2s)^1 (2p)^1]$
Two unpaired electrons

FIGURE 1-14 Promotion of an electron in beryllium to allow the use of both valence electrons in bonding.

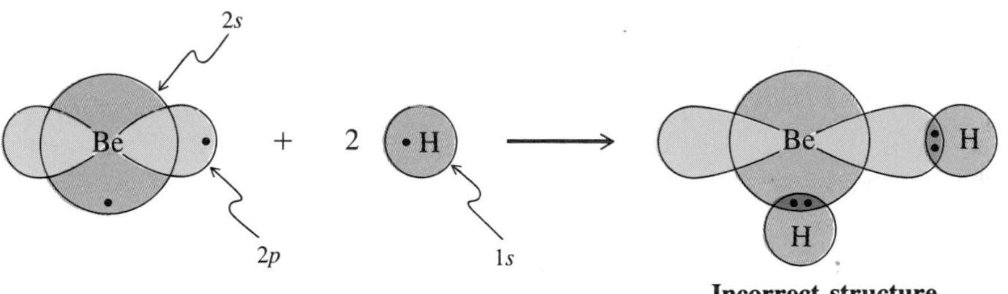

FIGURE 1-15 Possible but incorrect bonding in BeH_2 by separate use of a 2*s* and a 2*p* orbital on beryllium. The node in the former is not shown. Moreover, the other two empty *p* orbitals and the lower energy filled 1*s* orbital are omitted for clarity. The dots indicate the valence electrons.

singly filled atomic orbitals available for overlap. One could propose bond formation by overlap of the Be 2*s* orbital with the 1*s* orbital of one H, on the one hand, and the Be 2*p* orbital with the second H, on the other (Figure 1-15). This scheme predicts two different bonds of unequal length, probably at an angle. However, the theory of electron repulsion predicts that compounds such as BeH_2 should have *linear* structures (Section 1-3). Experiments on related compounds confirm this prediction and also show that the bonds to beryllium are of *equal* length.*

sp Hybrids produce linear structures

How can we explain this geometry in orbital terms? To answer this question, we use a quantum mechanical approach called **orbital hybridization.** Like the mixing of atomic orbitals on different atoms to form molecular orbitals, the mixing of atomic orbitals on the same atom forms new **hybrid orbitals.**

When we mix the 2*s* and one of the 2*p* wave functions on beryllium, we obtain two new hybrids, called *sp* orbitals, made up of 50% *s* and 50% *p* character. This treatment rearranges the orbital lobes in space as shown in Figure 1-16: The major parts of the orbitals, also called front lobes, point away from each other at an angle of 180°. There are two additional minor back lobes (one for each *sp* hybrid) with opposite sign. The remaining two *p* orbitals are left unchanged.

*These predictions cannot be tested for BeH_2 itself, which exists as a complex network of Be and H atoms. However, both BeF_2 and $Be(CH_3)_2$ exist as individual molecules in the gas phase and possess the predicted structures.

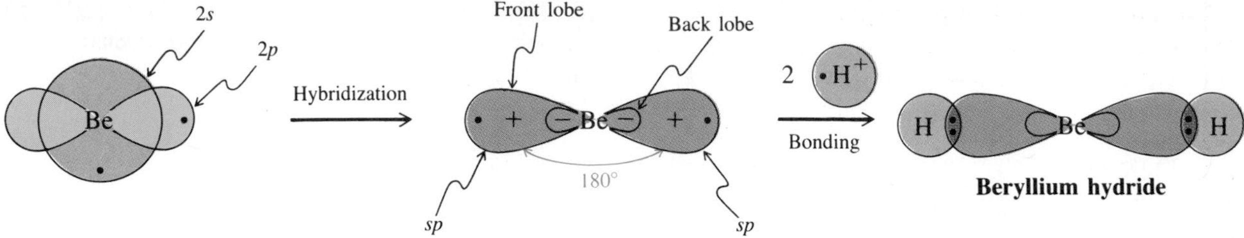

FIGURE 1-16 Hybridization in beryllium to create two *sp* hybrids. The resulting bonding gives BeH_2 a linear structure. Again, both remaining *p* orbitals and the 1*s* orbital have been omitted for clarity. The sign of the wave function for the large *sp* lobes is opposite that for the small lobes.

Overlap of the *sp* front lobes with two hydrogen 1*s* orbitals yields the bonds in BeH_2. The 180° angle that results from this hybridization scheme minimizes electron repulsion. The oversized front lobes of the hybrid orbitals also overlap better than do lobes of unhybridized orbitals; the result is improved bonding.

Note that hybridization does not change the overall number of orbitals available for bonding. Hybridization of the four orbitals in beryllium gives a new set of four: two *sp* hybrids and two essentially unchanged 2*p* orbitals. We will see shortly that carbon uses *sp* hybrids when it forms triple bonds.

sp^2 Hybrids create trigonal structures

Now let us consider the group of elements in the periodic table with three valence electrons. What bonding scheme can be derived for borane, BH_3? Promotion of a 2*s* electron in boron to one of the 2*p* levels gives the three singly filled atomic orbitals (one 2*s*, two 2*p*) needed for forming three bonds. Mixing these atomic orbitals creates *three* new hybrid orbitals, which are designated sp^2 to indicate the component atomic orbitals (Figure 1-17). The third *p* orbital is left unchanged, so the total number of orbitals stays the same, namely, four.

The front lobes of the three sp^2 orbitals of boron overlap the respective 1*s* orbitals of the hydrogen atoms to give trigonal planar BH_3. Again, hybridization minimizes electron repulsion and improves overlap, conditions giving stronger bonds. The remaining unchanged *p* orbital is perpendicular to the plane incorporating the sp^2 hybrids. It is empty and does not enter significantly into bonding.

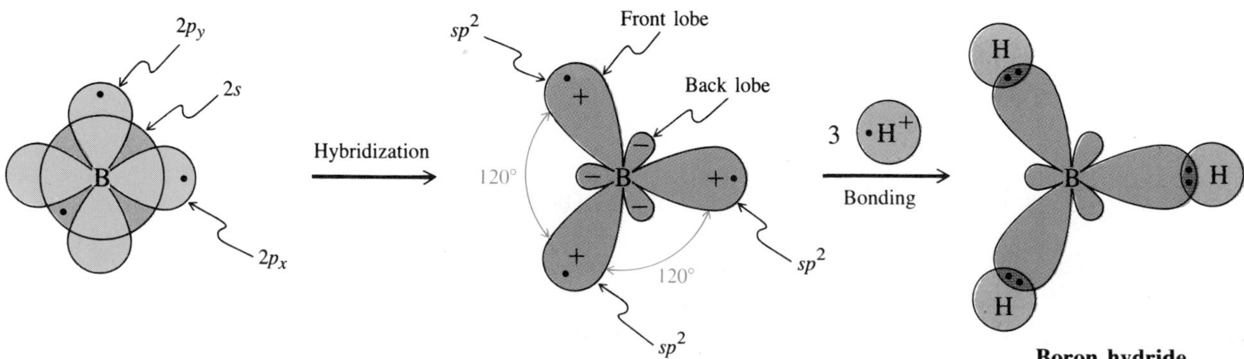

FIGURE 1-17 Hybridization in boron to create three sp^2 hybrids. The resulting bonding gives BH_3 a trigonal planar structure. There are three front lobes of one sign and three back lobes of opposite sign. The remaining *p* orbital (p_z) extends perpendicular to the molecular plane (the plane of the page; one p_z lobe is above, the other below that plane) and has been omitted.

The molecule BH_3 is **isoelectronic** with the methyl cation, CH_3^+; that is, they have the same number of electrons. Bonding in CH_3^+ involves three sp^2 hybrid orbitals, and we shall see shortly that carbon uses sp^2 hybrids in double bond formation.

sp^3 Hybridization explains the shape of tetrahedral carbon compounds

Consider the element whose bonding is of most interest to us: carbon. Its electronic configuration is $(1s)^2(2s)^2(2p)^2$, with two unpaired electrons residing in two $2p$ orbitals. Promotion of one electron from $2s$ to $2p$ results in four singly filled orbitals for bonding. We have learned that the arrangement of the four C–H bonds of methane in space that would minimize electron repulsion is tetrahedral (Section 1-3). To be able to achieve this geometry, the $2s$ orbital on carbon is hybridized with *all three* $2p$ orbitals to make *four* equivalent sp^3 orbitals with tetrahedral symmetry, each occupied by one electron. Overlap with four hydrogen $1s$ orbitals furnishes methane with four equal C–H bonds. The HCH bond angles are typical of a tetrahedron: 109.5° (Figure 1-18).

Any combination of atomic and hybrid orbitals may overlap to form bonds. For example, the four sp^3 orbitals of carbon combine with four chlorine $2p$ orbitals to result in tetrachloromethane, CCl_4. Carbon–carbon bonds are generated by overlap of hybrid orbitals. In ethane, CH_3–CH_3 (Figure 1-19), this bond

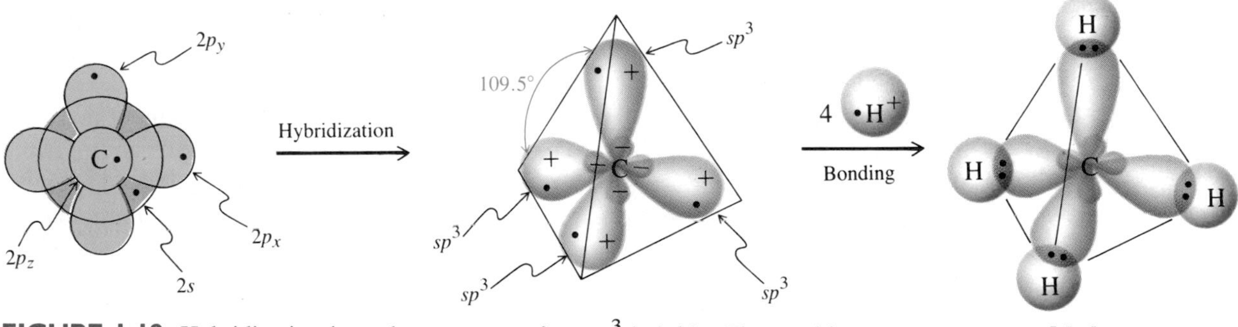

FIGURE 1-18 Hybridization in carbon to create four sp^3 hybrids. The resulting bonding gives CH_4 and other carbon compounds tetrahedral structures. The sp^3 hybrids contain small back lobes of sign opposite that of the front lobes.

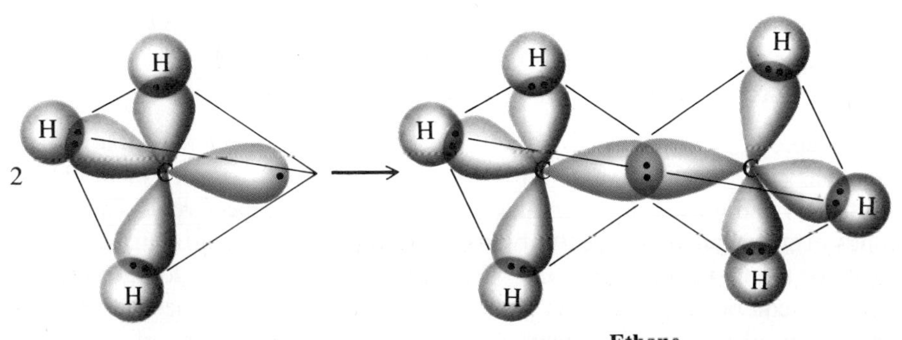

FIGURE 1-19 Overlap of two sp^3 orbitals to form the carbon–carbon bond in ethane.

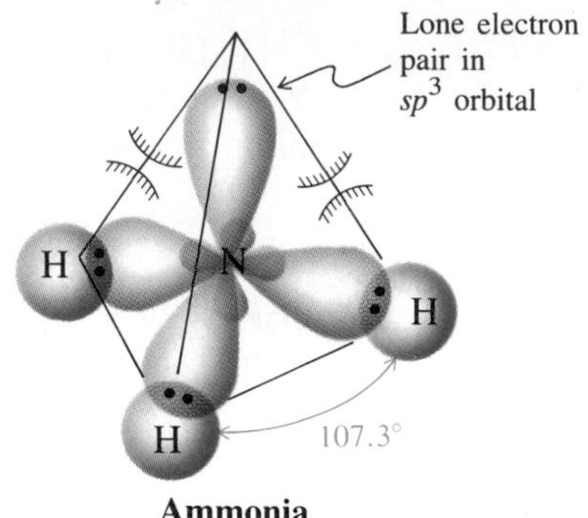

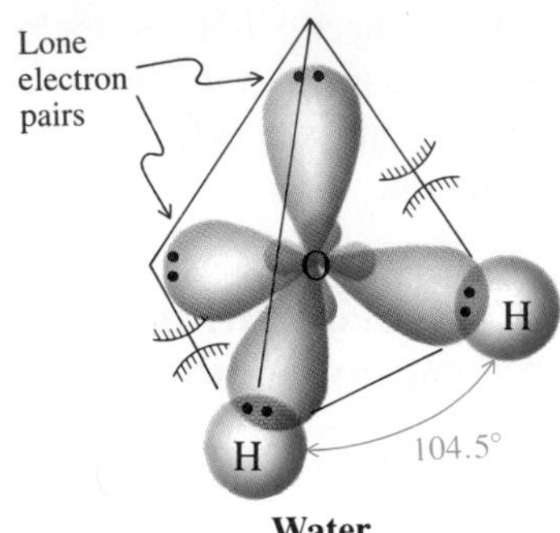

FIGURE 1-20 Bonding and electron repulsion in ammonia and water. The arcs indicate increased electron repulsion by the lone pairs located close to the central nucleus.

involves two sp^3 hybrids, one from each of two CH_3 units. Any hydrogen atom in methane and ethane may be replaced by CH_3 or other groups to give new combinations.

In all of these molecules, and countless more, *carbon is approximately tetrahedral*. It is this ability of carbon to form chains of atoms bearing a variety of additional substituents that gives rise to the extraordinary diversity of organic chemistry.

Hybrid orbitals may contain lone electron pairs: ammonia and water

What sort of orbitals describe the bonding in ammonia and water (see Exercise 1-5)? Let us begin with ammonia. The electronic configuration of nitrogen, $(1s)^2(2s)^2(2p)^3$, explains why nitrogen is trivalent, three covalent bonds being needed for octet formation. We could use p orbitals for overlap, leaving the nonbonding electron pair in the $2s$ level. However, this arrangement does not minimize electron repulsion. The best solution is again sp^3 hybridization. Three of the sp^3 orbitals are used to bond to the hydrogen atoms, while the fourth contains the lone electron pair. The HNH bond angles (107.3°) in ammonia are almost tetrahedral (Figure 1-20).

Similarly, the bonding in water is best described by sp^3 hybridization on oxygen. The HOH bond angle is 104.5°, again close to tetrahedral.

The influence of the lone electron pairs explains why the bond angles in NH_3 and H_2O are reduced below the tetrahedral value of 109.5°. Because they are not shared, the lone pairs are relatively close to the nitrogen or oxygen. As a result, they exert increased repulsion on the electrons in the bonds to hydrogen, thereby leading to the observed bond angle compression.

Pi bonds are present in ethene (ethylene) and ethyne (acetylene)

The double bond in alkenes, such as ethene (ethylene), and the triple bond in alkynes, such as ethyne (acetylene), are the result of the ability of the atomic orbitals of carbon to adopt sp^2 and sp hybridization, respectively. Thus, the σ bonds in ethene are derived entirely from carbon-based sp^2 hybrid orbitals: Csp^2–Csp^2 for the C–C bond, and Csp^2–$H1s$ for holding the four hydrogens (Figure 1-21). In contrast to BH_3, with an empty p orbital on boron, the leftover unhybridized p orbitals on the ethene carbons are occupied by one electron each, overlapping to form a π bond (recall Figure 1-13E). In ethyne, the σ frame is

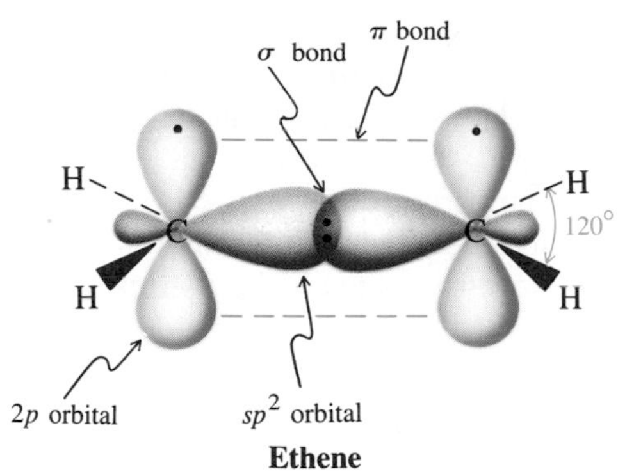

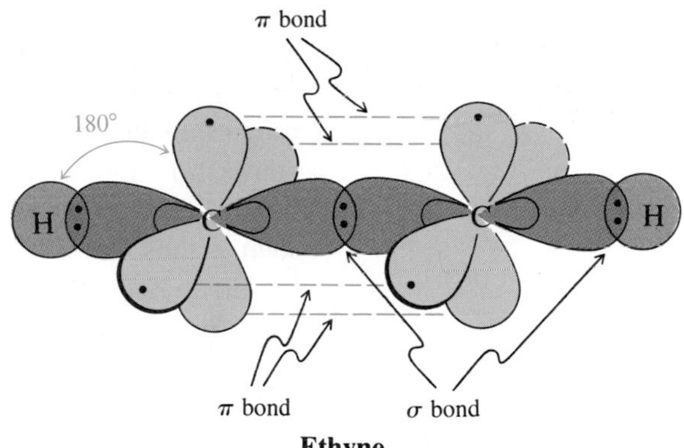

made up of bonds involving C*sp* hybrid orbitals. This arrangement leaves *two* singly occupied *p* orbitals on each carbon and allows the formation of two π bonds (Figure 1-21).

FIGURE 1-21 The double bond in ethene (ethylene) and the triple bond in ethyne (acetylene).

EXERCISE 1-13

Draw a scheme for the hybridization and bonding in methyl cation, CH_3^+, and methyl anion, CH_3^-.

In summary, to minimize electron repulsion and maximize bonding in triatomic and larger molecules, we apply the concept of atomic orbital hybridization to construct orbitals of appropriate shape. Combinations of *s* and *p* atomic orbitals create hybrids. Thus, a 2*s* and a 2*p* orbital mix to furnish two linear *sp* hybrids, the remaining two *p* orbitals being unchanged. Combination of the 2*s* with two *p* orbitals gives three sp^2 hybrids used in trigonal molecules. Finally, mixing the 2*s* with all three *p* levels results in the four sp^3 hybrids that produce the geometry around tetrahedral carbon.

1-9 Structures and Formulas of Organic Molecules

A good understanding of the nature of bonding allows us to learn how chemists determine the identity of organic molecules and depict their structures. Do not underestimate the importance of the latter task. Sloppiness in drawing molecules has been the source of many errors in the literature and, perhaps of more immediate concern, in organic chemistry examinations.

To establish the identity of a molecule, we determine its structure

Organic chemists have many diverse techniques at their disposal with which to determine molecular structure. **Elemental analysis** reveals the **empirical formula,** which summarizes the kinds and ratios of the elements present. However, other procedures are usually needed to determine the molecular formula and to distinguish between structural alternatives. For example, the molecular formula C_2H_6O corresponds to *two* known substances: ethanol and methoxymethane

Ethanol and Dimethyl Ether: Two Isomers

```
     H  H
     |  |
  H—C—C—O—H
     |  |
     H  H
```

Ethanol

(b.p. 78.5°C)

```
     H       H
     |       |
  H—C—O—C—H
     |       |
     H       H
```

Methoxymethane (Dimethyl ether)

(b.p. −23°C)

(dimethyl ether). But we can tell them apart on the basis of their physical properties—for example, their melting points, boiling points (b.p.), refractive indexes, specific gravities, and so forth. Thus ethanol is a liquid (b.p. 78.5°C) commonly used as a laboratory and industrial solvent and present in alcoholic beverages. In contrast, methoxymethane is a gas (b.p. −23°C) used as a refrigerant in place of Freon. Their other physical and chemical properties differ as well. Molecules such as these, which have the same molecular formula but differ in the sequence **(connectivity)** in which the atoms are held together, are called **constitutional** or **structural isomers.**

EXERCISE 1-14

Construct as many isomers with the molecular formula C_4H_{10} as you can.

Two naturally occurring substances illustrate the biological consequences of such structural differences. Prostacyclin I_2 prevents blood inside the circulatory system from clotting. Thromboxane A_2, which is released when bleeding occurs, *induces* platelet aggregation, causing clots to form over wounds. Incredibly, these compounds are constitutional isomers (both have the molecular formula $C_{20}H_{32}O_5$) with only relatively minor connectivity differences. Indeed they are so closely related that they are synthesized in the body from a common starting material (see Section 19-14 for details).

When a compound is isolated in nature or from a reaction, a chemist may attempt to identify it by matching its properties with those of known materials. Suppose, however, that the chemical under investigation is new. In this case, structural elucidation requires the use of other methods, most of which are various forms of spectroscopy. These methods will be dealt with and applied often in later chapters.

The most complete methods for structure determination are X-ray diffraction of single crystals and microwave spectroscopy of gases. These techniques reveal the exact position of every atom, as if viewed under very powerful magnification. The structural details that emerge in this way for the two isomers ethanol and methoxymethane are depicted in the form of ball-and-stick models in Figure 1-22. Note the tetrahedral bonding around the carbon atoms and the bent arrangement of the bonds to oxygen, which is hybridized as in water.

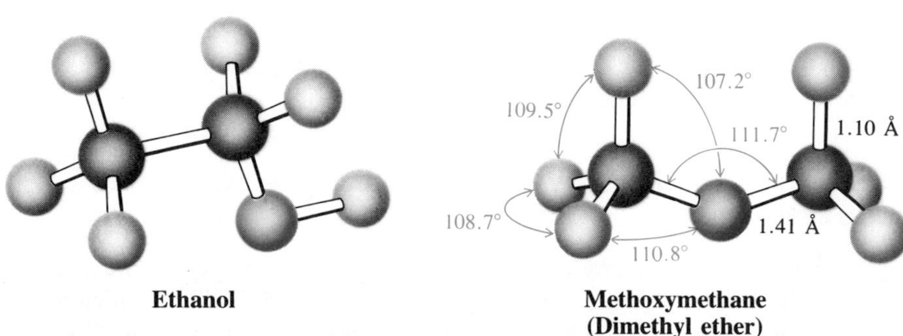

FIGURE 1-22 Three-dimensional ball-and-stick representations of ethanol and dimethyl ether. Bond lengths are given in angstrom units, bond angles in degrees.

The perception of organic molecules in three dimensions is essential for understanding their structures and frequently their reactivities. You may find it difficult to visualize the spatial arrangements of the atoms in even very simple systems. A good aid is a molecular model kit. You should acquire one and practice the assembly of organic structures.

EXERCISE 1-15

Repeat Exercise 1-14, using your molecular model kit to construct as many isomers with the molecular formula C_4H_{10} as you can. Draw each isomer.

Several types of drawings are used to represent molecular structures

The representation of molecular structures is not new to us. It was first addressed in Section 1-4, which outlined rules for drawing Lewis structures. We learned that bonding and nonbonding electrons are depicted as dots. A simplification is the straight-line notation (Kekulé structure), with lone pairs (if present) added again as dots. To simplify even further, chemists use **condensed formulas** in which most single bonds and lone pairs have been omitted. The main carbon chain is written horizontally, the attached hydrogens usually to the right of the associated carbon atom. Other groups (the **substituents** on the main stem) are added through connecting vertical lines.

Condensed Formulas

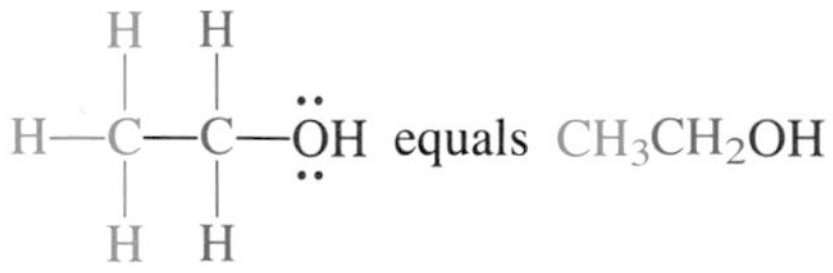

$H_3C{-}CH_2{-}\ddot{O}H$ (Kekulé structure with all H shown) equals CH_3CH_2OH

Ethanol

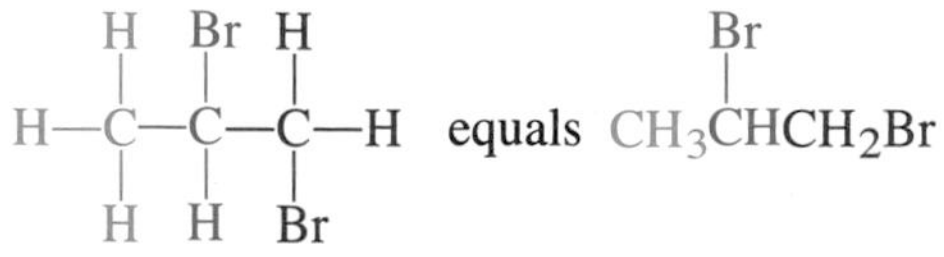

$H{-}CH_2{-}CH(Br){-}CH(H)(Br){-}H$ (Kekulé structure) equals $CH_3\underset{}{\overset{Br}{C}}HCH_2Br$

1,2-Dibromopropane

$CH_3CH_2\overset{CH_2Cl}{\underset{CH_3}{C}}CH_2CH_3$ equals $(CH_3CH_2)_2\underset{CH_3}{C}CH_2Cl$

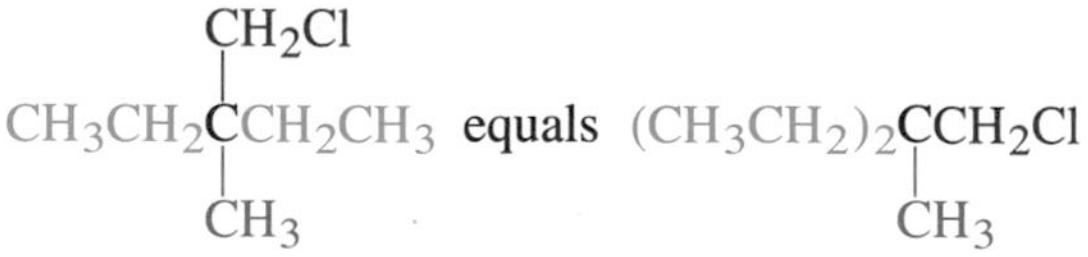

3-(Chloromethyl)-3-methylpentane

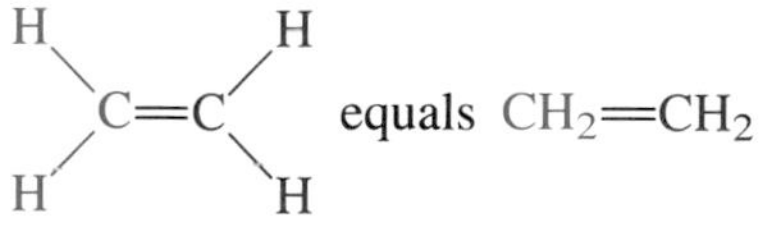

$H_2C{=}CH_2$ (Kekulé structure) equals $CH_2{=}CH_2$

Ethene (Ethylene)

$H{-}C{\equiv}C{-}H$ equals $HC{\equiv}CH$

Ethyne (Acetylene)

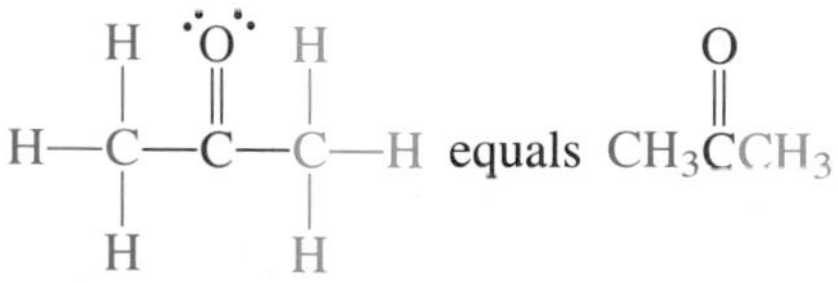

$H{-}CH_2{-}\overset{\ddot{O}}{\overset{\|}{C}}{-}CH_2{-}H$ (Kekulé structure) equals $CH_3\overset{O}{\overset{\|}{C}}CH_3$

2-Propanone (Acetone)

The most economical notation of all is the **bond-line formula.** It portrays the carbon frame by zigzag straight lines, omitting all hydrogen atoms. Each terminus represents a methyl group and each apex a carbon atom.

Bond-Line Formulas

OH O Br Br

Ethanol **2-Propanone (Acetone)** **1,2-Dibromopropane**

EXERCISE 1-16

Draw condensed and bond-line formulas for each C_4H_{10} isomer.

Figure 1-22 calls attention to a problem: How can we draw the three-dimensional structures of organic molecules accurately, efficiently, and in accord with generally accepted conventions? For tetrahedral carbon, this problem is solved by the **dashed-wedged line notation.** It uses a zigzag convention to depict the main carbon chain, now defined to lie *in the plane* of the page. Each apex (carbon atom) is then connected to two additional lines, one dashed and one wedged, both pointing away from the chain. These represent the remaining two bonds to carbon; the dashed line corresponds to the bond that lies *below the plane* of the page and the wedged line to that lying *above that plane* (Figure 1-23). Substituents are placed at the appropriate termini. This convention is applied to molecules of all sizes, even methane (Figure 1-23B–E).

EXERCISE 1-17

Draw dashed-wedged line formulas for each C_4H_{10} isomer.

In summary, determination of organic structures relies on the use of several experimental techniques, including elemental analysis and various forms of spectroscopy. Molecular models are useful aids for the visualization of the spatial arrangements of the atoms in structures. Condensed and bond-line notations are useful shorthand approaches to drawing two-dimensional representations of molecules, whereas dashed-wedged line formulas provide a means of depicting the atoms and bonds in three dimensions.

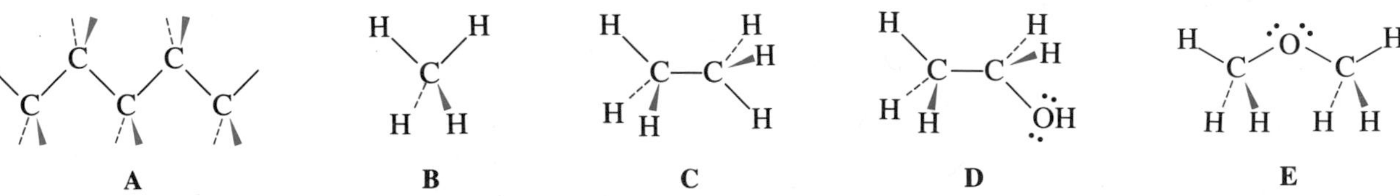

FIGURE 1-23 Dashed (red) and wedged (blue) line notation for (A) a carbon chain; (B) methane; (C) ethane; (D) ethanol; and (E) methoxymethane. Atoms attached by ordinary straight lines lie in the plane of the page. Groups at the ends of dashed lines lie below that plane; groups at the ends of wedges lie above it.

IMPORTANT CONCEPTS

1. Organic chemistry is the chemistry of carbon and its compounds.
2. Coulomb's law relates the attractive force between particles of opposite electrical charge to the distance between them.
3. Ionic bonds result from coulombic attraction of oppositely charged ions. These ions are formed by the complete transfer of electrons from one atom to another, typically to achieve a noble-gas configuration.
4. Covalent bonds result from electron sharing between two atoms. Electrons are shared to allow the atoms to attain noble-gas configurations.
5. Bond length is the average distance between two covalently bonded atoms. Bond formation releases energy; bond breaking requires energy.
6. Polar bonds are formed between atoms of differing electronegativity (a measure of an atom's ability to attract electrons).
7. The shape of molecules is strongly influenced by electron repulsion.
8. Lewis structures describe bonding by the use of valence electron dots. They are drawn so as to give hydrogen an electron duet, the other atoms electron octets (octet rule). Charge separation should be minimized, but may be enforced by the octet rule.
9. When two or more Lewis structures differing only in the positions of the electrons are needed to describe a molecule, they are called resonance forms. None correctly describes the molecule, its true representation being an average (hybrid) of all its Lewis structures. If the resonance forms of a molecule are unequal, those which best satisfy the rules for writing Lewis structures and the electronegativity requirements of the atoms are more important.
10. The de Broglie relationship describes the wavelength of a moving particle in terms of its mass and velocity.
11. The motion of electrons around the nucleus can be described by wave equations. The solutions to these equations are atomic orbitals, which roughly delineate regions in space in which there is a high probability of finding electrons.
12. An *s* orbital is spherical; a *p* orbital looks like two touching teardrops, or a "spherical figure eight." The mathematical sign of the orbital at any point can be positive, negative, or zero (node). With increasing energy, the number of nodes increases. Each orbital can be occupied by a maximum of two electrons of opposite spin (Pauli exclusion principle, Hund's rule).
13. The process of adding electrons one by one to the atomic orbitals, starting with those of lowest energy, is called the Aufbau principle.
14. A molecular orbital is formed when two atomic orbitals overlap to generate a bond. Atomic orbitals of the same sign overlap to give a bonding molecular orbital of lower energy. Atomic orbitals of opposite sign give rise to an antibonding molecular orbital of higher energy and containing a node. The number of molecular orbitals equals the number of atomic orbitals from which they derive.
15. Bonds made by overlap along the internuclear axis are called σ bonds; those made by overlap of *p* orbitals perpendicular to the internuclear axis are called π bonds.
16. The mixing of orbitals on the same atom results in new hybrid orbitals of different shape. One *s* and one *p* orbital mix to give two linear *sp* hybrids, used, for example, in the bonding of BeH_2. One *s* and two *p* orbitals result in three trigonal sp^2 hybrids, used, for example, in BH_3. One *s* and three *p* orbitals furnish four tetrahedral sp^3 hybrids, used, for example, in CH_4. The orbitals that are not hybridized stay unchanged. Hybrid orbitals may overlap with each other. Overlapping sp^3 hybrid orbitals on different carbon atoms forms the carbon–carbon bonds in ethane and other organic molecules. Hybrid orbitals may also be occupied by lone electron pairs, as in NH_3 and H_2O.
17. The composition (i.e., ratios of types of atoms) of organic molecules is revealed by elemental analysis. The molecular formula gives the number of atoms of each kind.
18. Molecules that have the same molecular formula but different connectivity order of their atoms are called constitutional or structural isomers. They have different properties.
19. Condensed and bond-line formulas are abbreviated representations of molecules. Dashed-wedged line drawings illustrate molecular structures in three dimensions.

PROBLEMS

1. Draw a Lewis structure for each of the following molecules and assign charges where appropriate. The order in which the atoms are connected is given in parentheses.
(a) ClF **(b)** BrCN **(c)** $SOCl_2$ ($OSCl_2$)
(d) CH_3NH_2 **(e)** $(CH_3)_2O$ **(f)** N_2H_2 (HNNH)
(g) CH_2CO **(h)** HN_3 (HNNN) **(i)** N_2O (NNO)

2. Draw a Lewis structure for each of the following species. Again, assign charges where appropriate.
(a) H^- **(b)** CH_3^- **(c)** CH_3^+
(d) CH_3 **(e)** $CH_3NH_3^+$ **(f)** CH_3O^-
(g) CH_2 **(h)** HC_2^- (HCC) **(i)** H_2O_2 (HOOH)

3. Several of the compounds in Problem 1 cannot be represented adequately by a single Lewis structure and are best described by resonance hybrids. Identify these molecules and write an additional resonance Lewis structure for each. In each case, indicate the major contributor to the resonance hybrid.

4. Draw two or three resonance structures for each of the following species. Indicate the major contributor or contributors to the hybrid in each case.
(a) OCN^- **(b)** CH_2CHNH^- **(c)** $HCONH_2$ ($HC(=O)NH_2$)
(d) O_3 (OOO) **(e)** $CH_2CHCH_2^-$ **(f)** SO_2 (OSO)
(g) $CH_2NH_2^+$ **(h)** $HOCHNH_2^+$ **(i)** CH_3CNO

5. Use a molecular orbital analysis to predict which species in each of the following pairs has the stronger bonding between atoms. (Hint: Refer to Figure 1-12.)
(a) H_2 or H_2^+ **(b)** He_2 or He_2^+ **(c)** O_2 or O_2^+ **(d)** N_2 or N_2^+

6. Describe the hybridization of each carbon atom in each of the following structures. Base your answer on the geometry about the carbon atom.
(a) CH_3Cl **(b)** CH_3OH **(c)** $CH_3CH_2CH_3$ **(d)** $CH_2{=}CH_2$ (trigonal carbons)
(e) HC≡CH (linear structure) **(f)** $H_3C{-}C(=O){-}H$ **(g)** $[\,^-H_2C{-}C(=O){-}H \longleftrightarrow H_2C{=}C(O^-){-}H\,]$

7. Depict the following condensed formulas in Kekulé (straight-line) notation.
(a) CH_3CN **(b)** $(CH_3)_2CHCH(NH_2)C(=O)OH$ **(c)** $CH_3CH(OH)CH_2CH_3$
(d) $CH_2BrCHBr_2$ **(e)** $CH_3C(=O)CH_2C(=O)OCH_3$ **(f)** $HOCH_2CH_2OCH_2CH_2OH$

8. Convert the following dashed-wedged line formulas into condensed formulas.

(a) 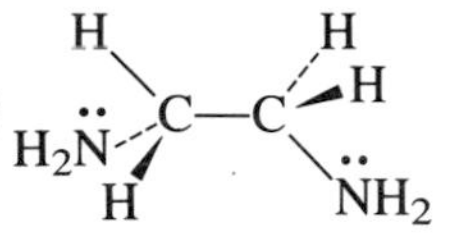 (b) 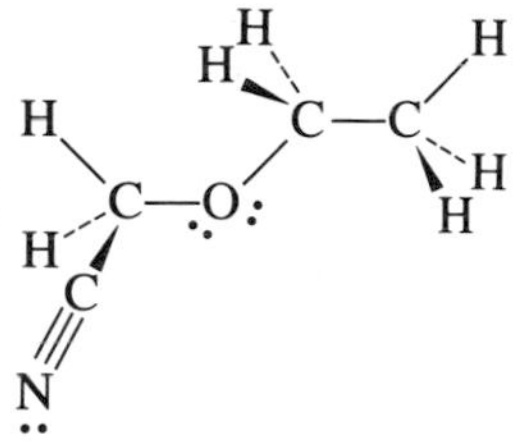 (c)

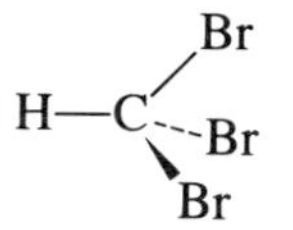

9. Depict the following straight-line formulas in their condensed form.

(a) 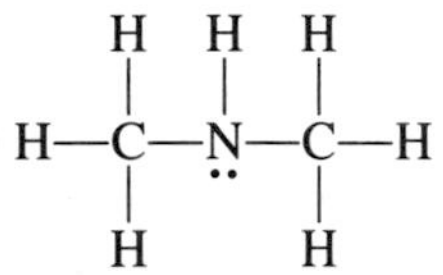 (b) 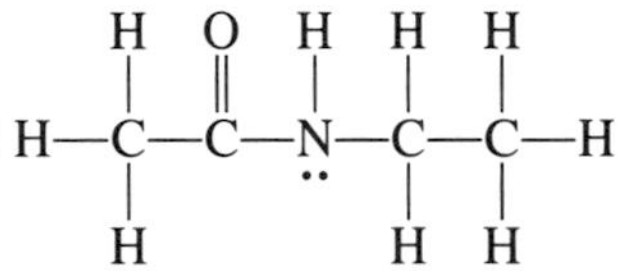 (c)

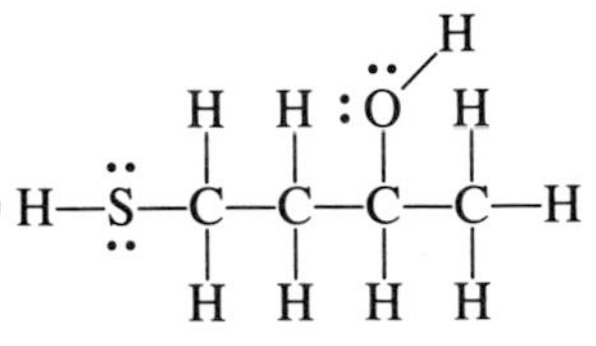

(d) 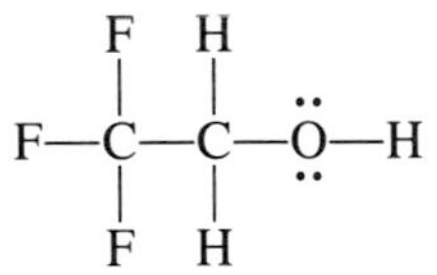 (e) 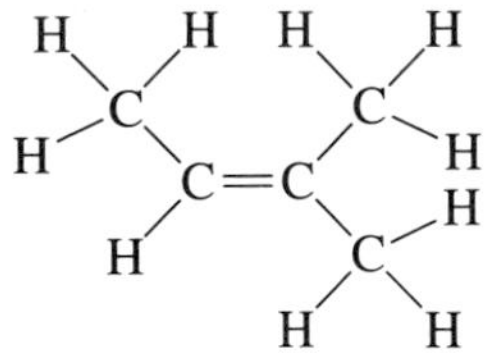 (f)

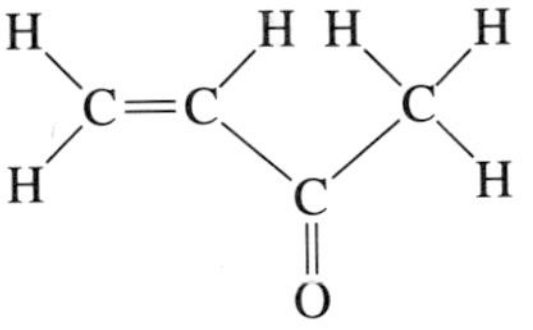

10. Convert the following condensed formulas into dashed-wedged line structures.

(a) $CH_3CH(CN)OCH_3$ **(b)** $CHCl_3$ **(c)** $(CH_3)_2NH$ **(d)** $CH_3CH(SH)CH_2CH_3$

11. Construct as many isomers of each molecular formula as you can for **(a)** C_5H_{12}; **(b)** C_3H_8O. Draw both condensed and bond-line formulas for each isomer.

12. Draw condensed formulas showing the multiple bonds, charges, and lone electron pairs (if any) for each molecule in the pairs of constitutional isomers shown below. (Hint: First make sure you can draw a proper Lewis structure for each molecule.) Do any of these pairs consist of resonance structures?

(a) $HCCCH_3$ and H_2CCCH_2 **(b)** CH_3CN and CH_3NC **(c)** $CH_3\overset{O}{C}H$ and H_2CCHOH

13. Two resonance forms can be written for a bond between trivalent boron and an atom with a lone pair of electrons. **(a)** Formulate them for (i) R_2BNR_2; (ii) R_2BOR; (iii) R_2BF (in each case $R = CH_3$). **(b)** Using the guidelines in Section 1-5, determine which one in each pair of resonance forms is more important. **(c)** How do the electronegativity differences between N, O, and F affect the relative importance of the resonance forms in each case? **(d)** Predict the hybridization of N in (i) and O in (ii) above.

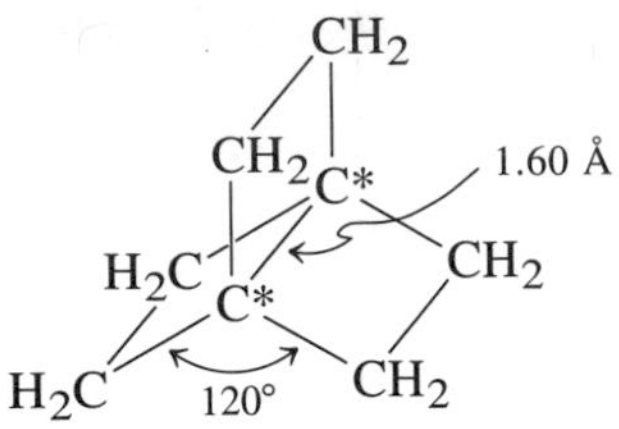

CH_2 / CH_2 C* 1.60 Å / H_2C C* CH_2 / H_2C 120° CH_2

[2.2.2]Propellane

14. The unusual molecule [2.2.2]propellane is pictured in the margin. On the basis of the given structural parameters, what hybridization scheme best describes the carbons marked by asterisks? (Make a model to help you visualize its shape.) What types of orbitals are used in the bond between them? Would you expect this bond to be stronger or weaker than an ordinary carbon–carbon single bond (which is usually 1.54 Å long)?

15. **(a)** Based on the information in Problem 6, give the likely hybridization of the orbital that contains the unshared pair of electrons (responsible for the negative charge) in each of the following species: $CH_3CH_2^-$; $CH_2{=}CH^-$; $HC{\equiv}C^-$. **(b)** Electrons in *sp*, sp^2, and sp^3 orbitals do not have identical energies. Because the 2*s* orbital is lower in energy than a 2*p*, the more *s* character a hybrid orbital has, the lower will be its energy. Therefore the sp^3 ($\frac{1}{4}$ *s* and $\frac{3}{4}$ *p* in character) is highest in energy, and the *sp* ($\frac{1}{2}$ *s*, $\frac{1}{2}$ *p*) lowest. Use this information to determine the relative abilities of the three anions in (a) to accommodate the negative charge. **(c)** The strength of an acid HA is related to the ability of its conjugate base A^- to accommodate negative charge. In other words, the ionization $HA \rightleftharpoons H^+ + A^-$ is favored for a more stable A^-. Although CH_3CH_3, $CH_2{=}CH_2$, and $HC{\equiv}CH$ are all weak acids, they are not equally so. Based on your answer to (b), rank them in order of acid strength.

16. A number of substances containing positively polarized carbon atoms have been labeled as "cancer suspect agents" (i.e., suspected carcinogens or cancer-inducing compounds). It has been suggested that the presence of such carbon atoms is responsible for the carcinogenic properties of these molecules. Assuming that the extent of polarization is proportional to carcinogenic potential, how would you rank the compounds below with regard to cancer-causing potency?

(a) CH_3Cl **(b)** $(CH_3)_4Si$ **(c)** $ClCH_2OCH_2Cl$
(d) CH_3OCH_2Cl **(e)** $(CH_3)_3C^+$

(Note: Polarization is only one of the many factors known to be related to carcinogenicity. Moreover, none of them shows the type of straightforward correlation implied in this question.)

17. The structure of the substance lynestrenol, a component of certain oral contraceptives, is presented below. Locate an example of each of the following types of bonds or atoms: **(a)** a highly polarized covalent bond; **(b)** a nearly unpolarized covalent bond; **(c)** an *sp*-hybridized carbon atom; **(d)** an sp^2-hybridized carbon atom; **(e)** an sp^3-hybridized carbon atom; **(f)** a bond between atoms of different hybridization.

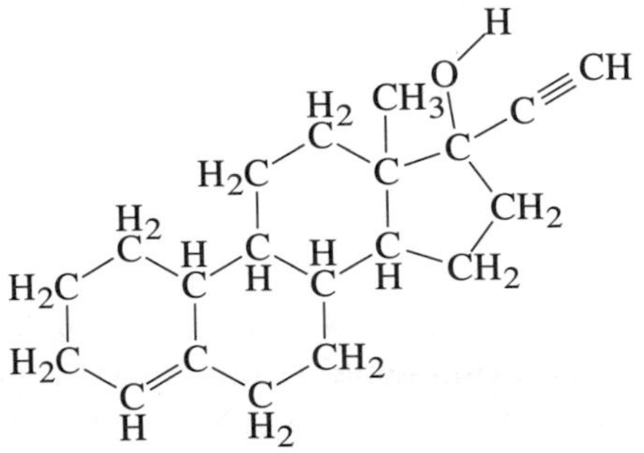

Lynestrenol

2

Alkanes

Molecules Lacking Functional Groups

Organic molecules are classified on the basis of their chemical reactivity. We shall see that molecules with certain groups of atoms called functional groups show characteristic kinds of chemical reactivity. Following a brief description of these groups, we shall consider the simplest class of molecules, the alkanes: their names; physical properties; and their mobility, a phenomenon based on the ability of tetrahedral carbon to rotate in space.

2-1 Functional Groups: Centers of Reactivity

Many organic molecules consist predominantly of a backbone of carbons linked by single bonds, with only hydrogen atoms attached. However, they may also contain doubly or triply bonded carbons, as well as other elements. These atoms or groups of atoms tend to be sites of comparatively high chemical reactivity and are labeled **functional groups.** Such groups have characteristic properties, and *they control the reactivity of the molecule as a whole.*

Hydrocarbons are molecules that contain only hydrogen and carbon

We begin our study with hydrocarbons, which have the general empirical formula C_xH_y. Those containing only single bonds, such as methane, ethane, and propane, are called **alkanes.** Molecules such as cyclohexane, whose carbons form a ring, are called **cycloalkanes.** *Alkanes lack functional groups;* as a result, they are relatively nonpolar and unreactive. The properties and chemistry of the alkanes are described in the next section and in Chapters 3 and 4.

Alkanes

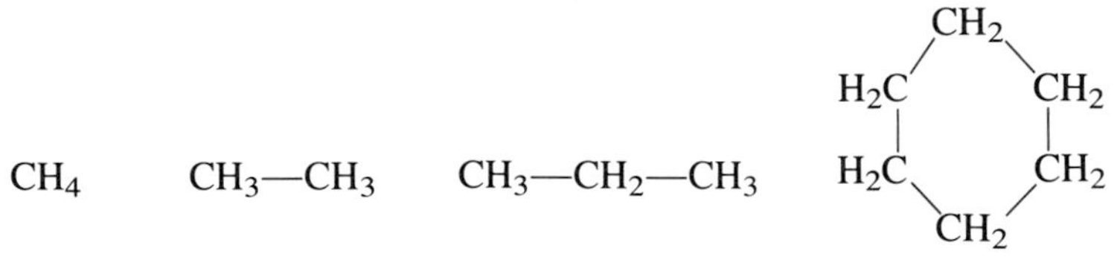

CH_4 — **Methane**

$CH_3—CH_3$ — **Ethane**

$CH_3—CH_2—CH_3$ — **Propane**

CH_2, H_2C, CH_2, H_2C, CH_2, CH_2 (ring) — **Cyclohexane**

Double and triple bonds are the functional groups of **alkenes** and **alkynes,** respectively. Their properties and chemistry are the topics of Chapters 11–13.

Alkenes and Alkynes

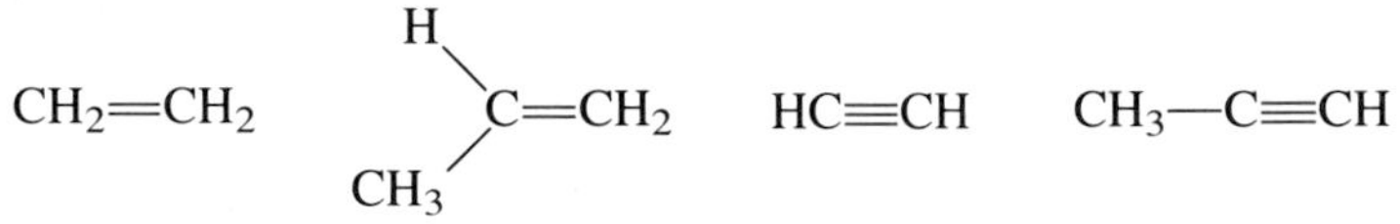

$CH_2{=}CH_2$ — **Ethene (Ethylene)**

$(H)(CH_3)C{=}CH_2$ — **Propene**

$HC{\equiv}CH$ — **Ethyne (Acetylene)**

$CH_3—C{\equiv}CH$ — **Propyne**

A special hydrocarbon is **benzene,** C_6H_6, in which three double bonds are incorporated into a six-membered ring. Benzene and its derivatives are traditionally called **aromatic,** because some substituted benzenes do have a strong fragrance. Aromatic compounds are discussed in Chapters 15, 16, 22, and 25.

Aromatic Compounds

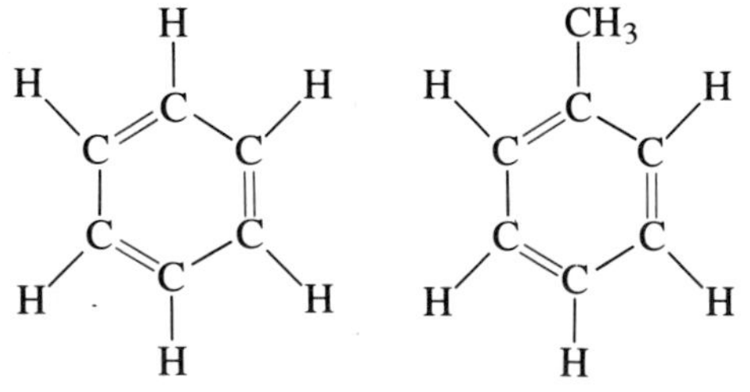

Benzene

Methylbenzene (Toluene)

Many functional groups contain polar bonds

Polar bonds determine the behavior of many classes of molecules. (Recall that polarity is due to a difference in the electronegativity of two atoms bound to each other.) Chapters 6 and 7 will introduce the **haloalkanes,** which contain polar carbon–halogen bonds as their functional groups. Another example is the **hydroxy** group, –O–H, characteristic of **alcohols.** The symbol R (for "radical" or "residue") is commonly used to describe a hydrocarbon-derived molecular fragment. Such fragments are called **alkyl** groups. Therefore a general formula for a haloalkane is R–X, where X stands for any halogen. Alcohols are similarly represented as R–O–H. The **alkoxy** group, –O–R, is the characteristic functional unit of **ethers,** which have the general formula R–O–R′. The functional group in alcohols and those in some ethers can be converted into a large variety of other functionalities and are therefore important in synthetic transformations. This chemistry is the subject of Chapters 8 and 9.

Haloalkanes		**Alcohols**		**Ethers**	
CH_3Cl	CH_3CH_2Cl	CH_3OH	CH_3CH_2OH	CH_3OCH_3	$CH_3CH_2OCH_2CH_3$
Chloromethane (Methyl chloride)	**Chloroethane (Ethyl chloride)**	**Methanol**	**Ethanol**	**Methoxymethane (Dimethyl ether)**	**Ethoxyethane (Diethyl ether)**
(Topical anesthetics)		(Wood alcohol)	(Grain alcohol)	(A refrigerant)	(An inhalation anesthetic)

The **carbonyl** function, C=O, is found in **aldehydes** and **ketones,** and, in conjunction with an attached –OH, in the **carboxylic acids.** Aldehydes and ketones are discussed in Chapters 17 and 18, the carboxylic acids and their derivatives in Chapters 19 and 20.

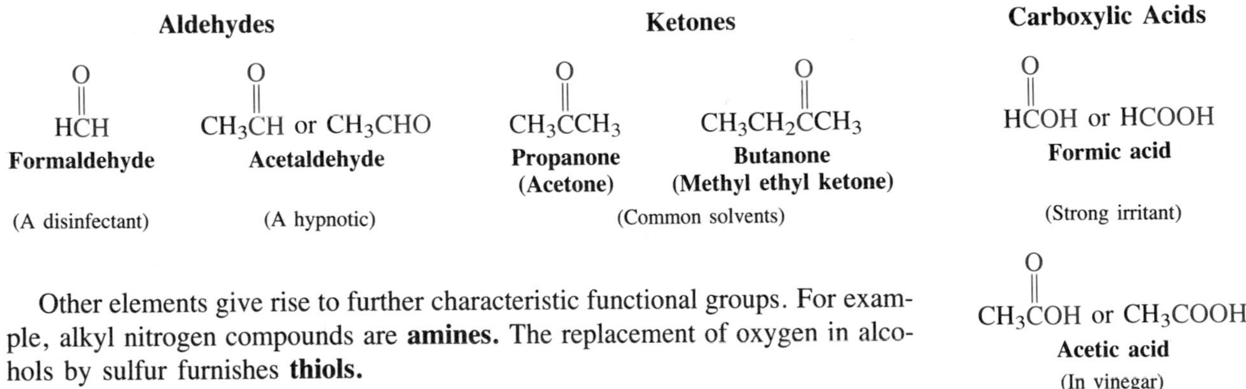

Other elements give rise to further characteristic functional groups. For example, alkyl nitrogen compounds are **amines.** The replacement of oxygen in alcohols by sulfur furnishes **thiols.**

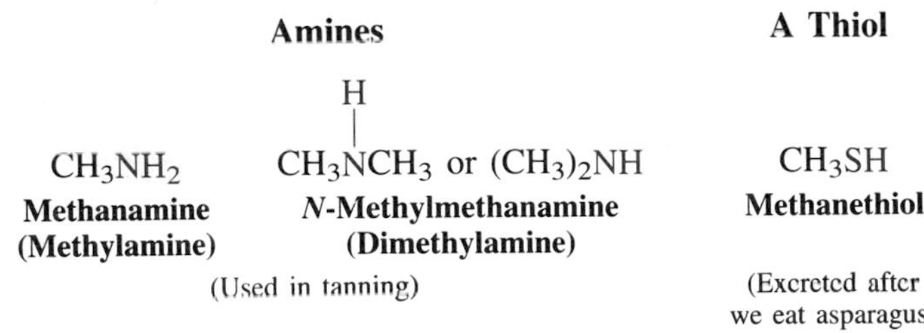

Table 2-1 (on the next two pages) depicts a selection of common functional groups, the class of compounds to which they give rise, a general structure, and an example.

TABLE 2-1 Common Functional Groups

Compound class	General structure[a]	Functional group	Example
Alkanes	R—H	None	$CH_3CH_2CH_2CH_3$ **Butane**
Haloalkanes	R—X (X = F, Cl, Br, I)	—X	CH_3CH_2—Br **Bromoethane**
Alcohols	R—OH	—OH	$(CH_3)_2CH$—OH **2-Propanol (Isopropyl alcohol)**
Ethers	R—O—R′	—O—	CH_3CH_2—O—CH_3 **Methoxyethane (Ethyl methyl ether)**
Thiols	R—SH	—SH	CH_3CH_2—SH **Ethanethiol**
Alkenes	(H)R(H)R C=C R(H)R(H)	>C=C<	$(CH_3)_2C{=}CH_2$ **2-Methylpropene**
Alkynes	(H)R—C≡C—R(H)	—C≡C—	$CH_3C{\equiv}CCH_3$ **2-Butyne**
Aromatic compounds	C_6 ring with six R(H) substituents	C_6 ring	$C_6H_5CH_3$ **Methylbenzene (Toluene)**
Aldehydes	R—C(=O)—H	—C(=O)—H	CH_3CH_2CHO **Propanal**
Ketones	R—C(=O)—R′	—C(=O)—	$CH_3CH_2COCH_2CH_3$ **3-Pentanone**
Carboxylic acids	R—C(=O)—O—H	—C(=O)—OH	CH_3CH_2COOH **Propanoic acid**
Anhydrides	R—C(=O)—O—C(=O)—R′ (H)	—C(=O)—O—C(=O)—	$CH_3CH_2COOCOCH_2CH_3$ **Propanoic anhydride**

a. The letter R denotes an alkyl group. Different alkyl groups can be distinguished by adding primes to the letter R: R′, R″, and so forth.

TABLE 2-1 *(Continued)*

Compound class	General structure[a]	Functional group	Example
Esters	$(H)R-\overset{O}{\overset{\Vert}{C}}-O-R'$	$-\overset{O}{\overset{\Vert}{C}}-O-$	$CH_3CH_2\overset{O}{\overset{\Vert}{C}}OCH_3$ **Methyl propanoate (Methyl propionate)**
Amides	$R-\overset{O}{\overset{\Vert}{C}}-\underset{R''(H)}{\underset{\vert}{N}}-R'\ (H)$	$-\overset{O}{\overset{\Vert}{C}}-\underset{\vert}{N}-$	$CH_3CH_2CH_2\overset{O}{\overset{\Vert}{C}}NH_2$ **Butanamide**
Nitriles	$R-C\equiv N$	$-C\equiv N$	$CH_3C\equiv N$ **Acetonitrile**
Amines	$R-\underset{R''}{\underset{\vert}{N}}-R'$	$-N\langle$	$(CH_3)_3N$ ***N,N*-Dimethylmethanamine (Trimethylamine)**

2-2 Straight-Chain and Branched Alkanes

We begin with the alkanes, hydrocarbons that contain only single bonds. They are classified into several types according to structure: the linear **straight-chain alkanes;** the **branched alkanes,** in which the carbon chain contains one or several branching points; and the cyclic alkanes, or **cycloalkanes,** which will be covered in Chapter 4.

A Straight-Chain Alkane	A Branched Alkane	A Cycloalkane
$CH_3-CH_2-CH_2-CH_3$	$CH_3-\underset{CH_3}{\underset{\vert}{\overset{CH_3}{\overset{\vert}{C}}}}-H$	$\begin{array}{c} CH_2-CH_2 \\ \vert \quad\quad \vert \\ CH_2-CH_2 \end{array}$
Butane, C_4H_{10}	**2-Methylpropane, C_4H_{10} (Isobutane)**	**Cyclobutane, C_4H_8**

The alkanes form a homologous series

In the straight-chain alkanes each carbon is bound to its two neighbors and to two hydrogen atoms. Exceptions are the two terminal carbon nuclei, which are bound to only one carbon atom and three hydrogen atoms. Several general formulas may be written for the series:

$$H-(CH_2)_n-H \qquad CH_3-(CH_2)_{n-1}-H \qquad CH_3-(CH_2)_{n-2}-CH_3$$

Each member of this series differs from the next lower one by the addition of a methylene group, $-CH_2-$. Molecules that are related in this way are **homologs** of each other (*homos,* Greek, same as), and the series is a **homologous series.** Methane ($n = 1$) is the first member of the homologous series of the alkanes, ethane ($n = 2$) the second, and so forth.

Branched alkanes are constitutional isomers of straight-chain alkanes

Branched alkanes are derived from the straight-chain systems by removal of a hydrogen atom from a methylene group and replacement with an alkyl group. Both branched and straight-chain alkanes have the same general formula, C_nH_{2n+2}. The smallest branched alkane is 2-methylpropane, C_4H_{10}, with the same molecular formula as (and therefore isomeric to) butane.

For the higher alkane homologs ($n > 4$), more than two isomers are possible. There are three pentanes, C_5H_{12}, as shown below. There are five hexanes, C_6H_{14}; nine heptanes, C_7H_{16}; and eighteen octanes, C_8H_{18}.

The Isomeric Pentanes

$CH_3—CH_2—CH_2—CH_2—CH_3$

Pentane

$$CH_3—CH_2—\underset{CH_3}{\overset{CH_3}{C}}—H$$

2-Methylbutane

$$CH_3—\underset{CH_3}{\overset{CH_3}{C}}—CH_3$$

2,2-Dimethylpropane (Neopentane)

The number of possibilities in connecting n carbon atoms to each other and to $2n + 2$ surrounding hydrogen atoms increases dramatically with the size of n (Table 2-2).

TABLE 2-2 Number of Possible Isomeric Alkanes C_nH_{2n+2}

n	Isomers
1	1
2	1
3	1
4	2
5	3
6	5
7	9
8	18
9	35
10	75
15	4,347
20	366,319

EXERCISE 2-1

(a) Draw the structures of the five isomeric hexanes. **(b)** Draw the structures for all the possible next higher and lower homologs of 2-methylbutane.

2-3 Naming the Alkanes

The multiplicity of ways of assembling carbon atoms and attaching various substituents accounts for the existence of the very large number of organic molecules. This diversity poses a problem: How can we systematically differentiate by name all these compounds? Is it possible, for example, to name all the C_6H_{14} isomers so that information on any of them (such as boiling points, melting points, chemical reactions) might easily be found in a compound index of an appropriate handbook? And is there a way to name a compound that we have never seen in such a way as to be able to draw the structure of that compound?

This problem of naming organic molecules has been with organic chemistry from its very beginning, but the initial method was far from systematic. Compounds have been named after their discoverers (''Nenitzescu's hydrocarbon''),

TABLE 2-3 Names and Physical Properties of Straight-Chain Alkanes, C_nH_{2n+2}

n	Name	Formula	Boiling point (°C)	Melting point (°C)	Density at 20°C (g ml^{-1})
1	Methane	CH_4	−161.7	−182.5	0.466 (at −164°C)
2	Ethane	CH_3CH_3	−88.6	−183.3	0.572 (at −100°C)
3	Propane	$CH_3CH_2CH_3$	−42.1	−187.7	0.5853 (at −45°C)
4	Butane	$CH_3CH_2CH_2CH_3$	−0.5	−138.3	0.5787
5	Pentane	$CH_3(CH_2)_3CH_3$	36.1	−129.8	0.6262
6	Hexane	$CH_3(CH_2)_4CH_3$	68.7	−95.3	0.6603
7	Heptane	$CH_3(CH_2)_5CH_3$	98.4	−90.6	0.6837
8	Octane	$CH_3(CH_2)_6CH_3$	125.7	−56.8	0.7026
9	Nonane	$CH_3(CH_2)_7CH_3$	150.8	−53.5	0.7177
10	Decane	$CH_3(CH_2)_8CH_3$	174.0	−29.7	0.7299
11	Undecane	$CH_3(CH_2)_9CH_3$	195.8	−25.6	0.7402
12	Dodecane	$CH_3(CH_2)_{10}CH_3$	216.3	−9.6	0.7487
13	Tridecane	$CH_3(CH_2)_{11}CH_3$	235.4	−5.5	0.7564
14	Tetradecane	$CH_3(CH_2)_{12}CH_3$	253.7	5.9	0.7628
15	Pentadecane	$CH_3(CH_2)_{13}CH_3$	270.6	10	0.7685
16	Hexadecane	$CH_3(CH_2)_{14}CH_3$	287	18.2	0.7733
17	Heptadecane	$CH_3(CH_2)_{15}CH_3$	301.8	22	0.7780
18	Octadecane	$CH_3(CH_2)_{16}CH_3$	316.1	28.2	0.7768
19	Nonadecane	$CH_3(CH_2)_{17}CH_3$	329.7	32.1	0.7855
20	Eicosane	$CH_3(CH_2)_{18}CH_3$	343	36.8	0.7886

after localities (''sydnones''), after their shapes (''cubane,'' ''basketane''), and after their natural sources (''vanillin''). Many of these **common** or **trivial names** are still widely used. However, there now exists a precise system for naming the alkanes. **Systematic nomenclature,** in which the name of a compound describes its structure, was first introduced by a chemical congress in Geneva, Switzerland, in 1892. It has continually been revised since then, mostly by the International Union of Pure and Applied Chemistry (IUPAC). Table 2-3 gives the systematic names of the first twenty straight-chain alkanes. Their stems, mainly of Latin or Greek origin, reveal the number of carbon atoms in the chain. For example, the name heptadecane is composed of the Greek word *hepta,* seven, and the Latin word *decem,* ten. The first four alkanes have special names that have been accepted as part of the systematic nomenclature but also all end in **-ane.** It is important to know these names, because they serve as the basis for naming a large fraction of all organic molecules. A few smaller branched alkanes have common names that still have widespread use. They make use of the prefixes **iso-** and **neo-,** as in isobutane, isopentane, and neohexane.

$$CH_3-\underset{H}{\overset{CH_3}{\underset{|}{\overset{|}{C}}}}-(CH_2)_n-CH_3$$

An isoalkane (e.g., *n* = 1, Isopentane)

$$CH_3-\underset{CH_3}{\overset{CH_3}{\underset{|}{\overset{|}{C}}}}-(CH_2)_n-H$$

A neoalkane (e.g., *n* = 2, Neohexane)

EXERCISE 2-2

Draw the structures of isohexane and neopentane.

Primary alkyl groups

CH_3—
Methyl

CH_3—CH_2—
Ethyl

CH_3—CH_2—CH_2—
Propyl

Alkyl groups are formed by the removal of a hydrogen from an alkane. They are named by replacing the ending -ane in the corresponding alkane by **-yl,** as in methyl, ethyl, and propyl. Table 2-4 shows a few branched alkyl groups having common names. Note that some use the prefixes *sec-* (or *s-*), which stands for secondary, and *tert-* (or *t-*), for tertiary. To apply these prefixes, we must first see how to classify carbon atoms in organic molecules. A **primary** carbon is one attached to only one other carbon atom. For example, all carbon atoms at the ends of alkane chains are primary. The hydrogens attached to such carbons are designated primary hydrogens, and an alkyl group created by removing a primary hydrogen is also called primary. A **secondary** carbon is attached to two other carbon atoms, and a **tertiary** carbon to three others. Their hydrogens are labeled similarly. As shown in Table 2-4, removal of a secondary hydrogen results in a secondary alkyl group, and removal of a tertiary hydrogen in a tertiary alkyl group. Finally, a carbon bearing four alkyl groups is called **quaternary.**

TABLE 2-4 Branched Alkyl Groups

Structure	Common name	Systematic name	Derived from	Designation
CH_3—C(H)(CH_3)—	Isopropyl	1-Methylethyl	Propane	Secondary
CH_3—C(H)(CH_3)—CH_2—	Isobutyl	2-Methylpropyl	2-Methylpropane (Isobutane)	Primary
CH_3—C(H)(CH_2CH_3)—	*sec*-Butyl	1-Methylpropyl	Butane	Secondary
CH_3—C(CH_3)(CH_3)—	*tert*-Butyl	1,1-Dimethylethyl	2-Methylpropane (Isobutane)	Tertiary
CH_3—C(CH_3)(CH_3)—CH_2—	Neopentyl	2,2-Dimethylpropyl	2,2-Dimethylpropane (Neopentane)	Primary

Primary, Secondary, and Tertiary Carbons and Hydrogens

Primary C

Secondary C → CH_3 ← Tertiary C

$CH_3CH_2\overset{|}{C}CH_2CH_3$

Primary H → H ← Secondary H

Tertiary H

3-Methylpentane

EXERCISE 2-3

Label the primary, secondary, and tertiary hydrogens in 2-methylpentane (isohexane).

Using the information in Table 2-3 allows us to name the first twenty straight-chain alkanes. How do we go about naming branched systems? A set of IUPAC rules makes this a relatively simple task, as long as they are followed carefully and in sequence.

IUPAC RULE 1. *Find the longest chain in the molecule and name it.* This task is not as easy as it seems. The problem is that, in the condensed formula, complex alkanes may be written in ways that mask the identity of the longest chain. In the following examples, the longest chain, or **stem chain,** is clearly marked; the alkane stem gives the molecule its name. Groups other than hydrogen attached to the stem chain are called **substituents.**

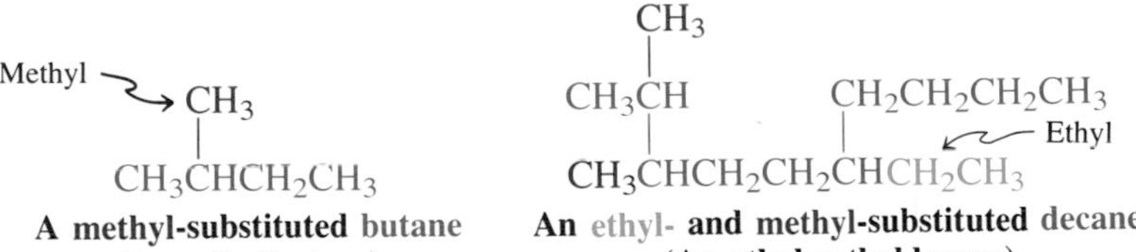

A methyl-substituted butane (A methylbutane)

An ethyl- and methyl-substituted decane (An ethylmethyldecane)

The stem chain is shown in blue in the examples in this section.

If a molecule has two or more chains of equal length, the chain with the largest number of substituents is the base stem chain.

CH_3 CH_3 (on C-2 and C-4)

$CH_3CHCHCHCHCH_2CH_3$

CH_3 $CH_2CH_2CH_3$ (on C-2 and C-3)

4 substituents

A heptane

Correct stem chain

not

CH_3 CH_3

$CH_3CHCHCHCHCH_2CH_3$

CH_3 $CH_2CH_2CH_3$

3 substituents

A heptane

Incorrect stem chain

Two more examples, drawn using bond-line notation, are shown on the next page.

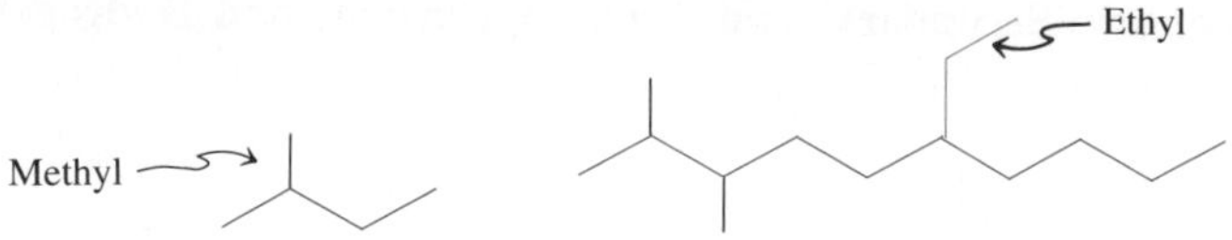

Methyl
A methylbutane

Ethyl
An ethylmethyldecane

IUPAC RULE 2. *Name all groups attached to the longest chain as alkyl substituents.* For straight-chain substituents, Table 2-3 can be used to derive the alkyl name. However, what if the substituent chain is branched? In this case, the same IUPAC rules apply to such complex substituents: First, find the longest chain in the substituent; next, name all its substituents.

IUPAC RULE 3. *Number the carbons of the longest chain beginning with the end that is closest to a substituent.*

$$\begin{array}{l} \quad\;\; CH_3 \\ \quad\;\; | \\ CH_3CHCH_2CH_3 \end{array}$$

1 2 3 4
not 4 3 2 1

not 1 2 3 4 5 6 7 8

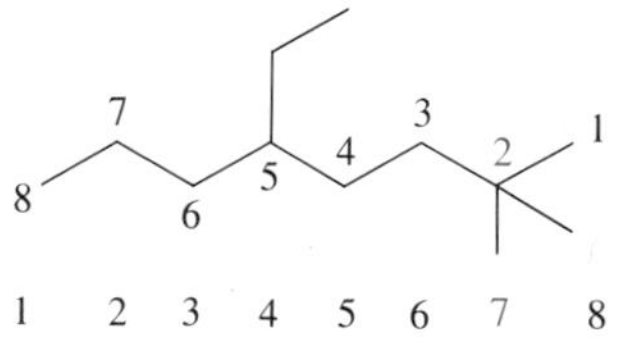

If there are two substituents at *equal* distance from the two ends of the chain, use the alphabet to decide how to number the stem. The substituent to come first in alphabetical order is attached to the carbon with the lower number.

$$\begin{array}{l} \qquad\quad CH_3CH_2 \qquad\quad CH_3 \\ \qquad\qquad\; | \qquad\qquad\quad | \\ CH_3CH_2CHCH_2CH_2CHCH_2CH_3 \end{array}$$

1 2 3 4 5 6 7 8

Ethyl before methyl

16 14 12 10 8 6 4 2
17 15 13 11 9 7 5 3 1

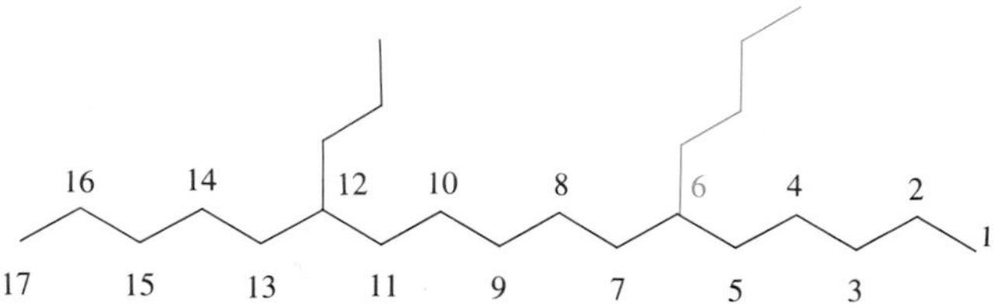

Butyl before propyl

What if there are three or more substituents? Then number the chain in the direction that gives the lower number at the *first difference* between the two possible numbering schemes. This procedure follows the **first point of difference principle.**

$$\begin{array}{l} \qquad\quad CH_3 \qquad\qquad\qquad\quad CH_3 \quad CH_3 \\ \qquad\quad | \qquad\qquad\qquad\qquad\; | \qquad\; | \\ CH_3CH_2CHCH_2CH_2CH_2CH_2CHCH_2CHCH_2CH_3 \end{array}$$

1	2	3	4	5	6	7	8	9	10	11	12	← 3, 8, and 10 (incorrect)
12	11	10	9	8	7	6	5	4	3	2	1	← 3, 5, and 10 (correct; 5 lower than 8)

Numbers for substituted carbons:

3,5,10-Trimethyldodecane

IUPAC RULE 4. *Write the name of the alkane by first arranging all the substituents in alphabetical order (each preceded by the carbon number to which it is attached and a hyphen) and then add the name of the stem.*

Should a molecule contain more than one of a particular substituent, its name is preceded by the prefix di, tri, tetra, penta, and so forth. The positions of attachment to the stem are given collectively before the substituent name and are separated by commas. These prefixes, as well as *sec-* and *tert-*, are not considered in the alphabetical ordering, except when they are part of a complex substituent name.

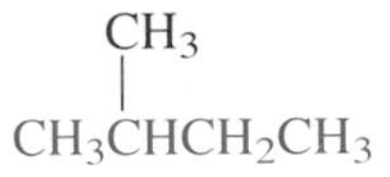

CH_3
$CH_3CHCH_2CH_3$
2-Methylbutane

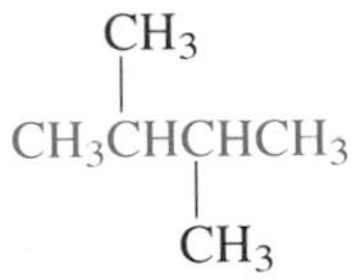

CH_3
$CH_3CHCHCH_3$
CH_3
2,3-Dimethylbutane

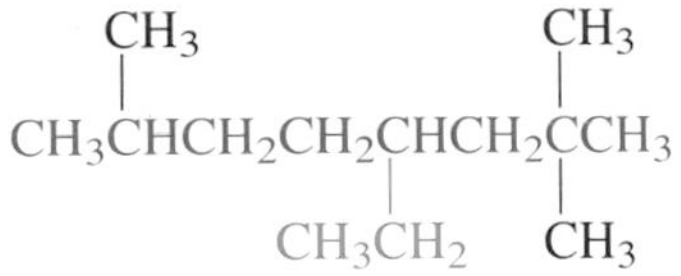

CH_3 CH_3
$CH_3CHCH_2CH_2CHCH_2CCH_3$
CH_3CH_2 CH_3
4-Ethyl-2,2,7-trimethyloctane

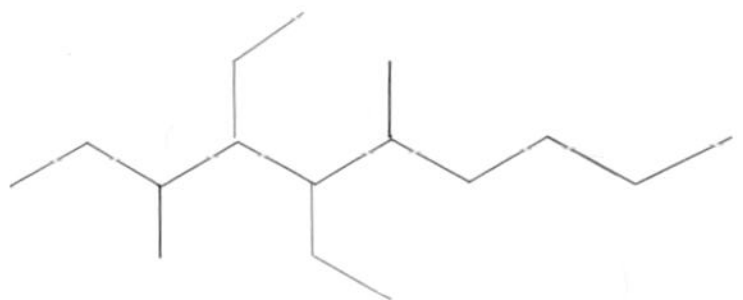

4,5-Diethyl-3,6-dimethyldecane

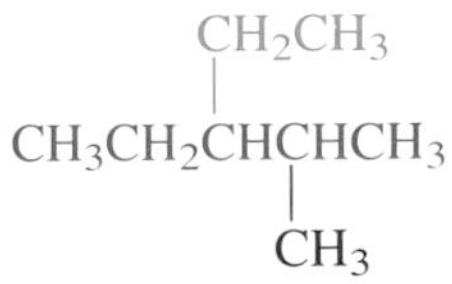

CH_2CH_3
$CH_3CH_2CHCHCH_3$
CH_3
3-Ethyl-2-methylpentane

Although the common group names in Table 2-4 are permitted by IUPAC, it is preferable to use systematic names. Such names are usually enclosed in parentheses, to avoid possible ambiguities. If a particular complex substituent occurs repeatedly, its name is preceded by the prefix bis, tris, tetrakis, pentakis, and so on. In a substituent chain, the carbon numbered one is *always* the carbon atom bound to the principal chain.

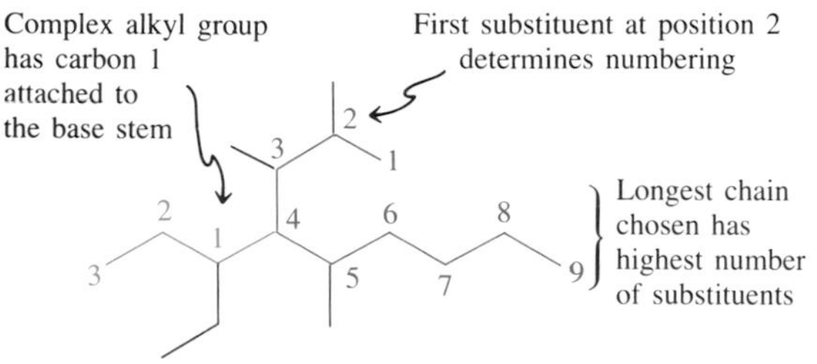

Complex alkyl group has carbon 1 attached to the base stem

First substituent at position 2 determines numbering

Longest chain chosen has highest number of substituents

4-(1-Ethylpropyl)-2,3,5-trimethylnonane

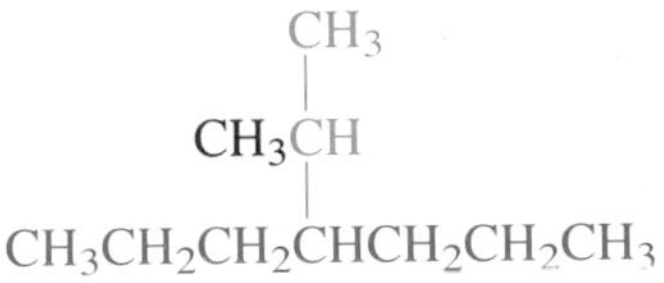

CH_3
CH_3CH
$CH_3CH_2CH_2CHCH_2CH_2CH_3$
4-(1-Methylethyl)heptane
(4-Isopropylheptane)

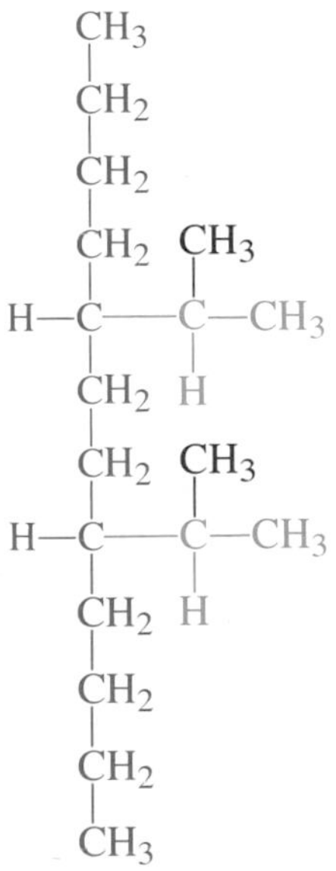

CH_3
CH_2
CH_2
CH_2 CH_3
H—C——C—CH_3
CH_2 H
CH_2 CH_3
H—C——C—CH_3
CH_2 H
CH_2
CH_2
CH_3
5,8-Bis(1-methylethyl)-dodecane

Further instructions on nomenclature will be presented when new classes of compounds, such as the cycloalkanes and haloalkanes, are introduced.

EXERCISE 2-4

Write down the names of the preceding eight branched alkanes, close the book, and reconstruct their structures from those names.

To summarize, four rules should be applied in sequence when naming a branched alkane: (1) find the longest chain; (2) find the names of all the alkyl

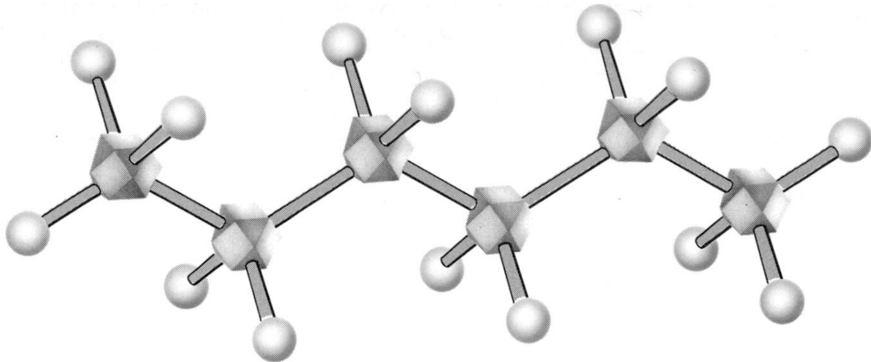

FIGURE 2-1 Molecular model of hexane, showing the zigzag pattern of the carbon chain typical of the alkanes. (Model set courtesy of Maruzen Co., Ltd., Tokyo.)

groups attached to the stem; (3) number the chain; (4) name the alkane, with substituent names in alphabetical order and preceded by numbers to indicate their locations.

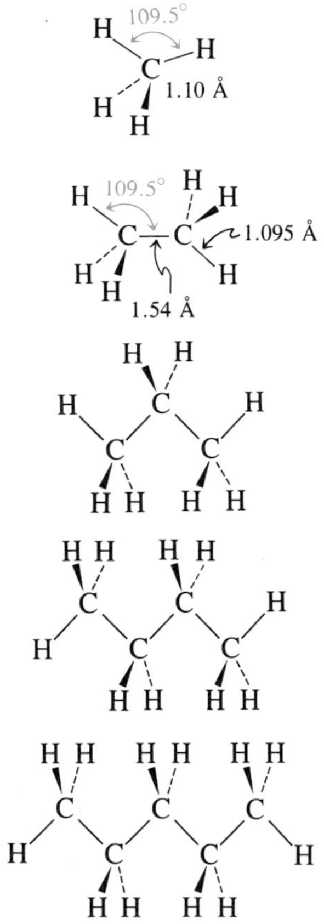

2-4 Structural and Physical Properties of Alkanes

What do the structures of alkanes look like in three dimensions? What are their physical appearances, and what are their physical properties? These questions will be addressed next.

Alkanes exhibit regular molecular structures and properties

The structural features of the alkanes are remarkably regular. The carbon atoms are tetrahedral, with bond angles close to 109° and with regular C–H ($\approx$ 1.10 Å) and C–C ($\approx$ 1.54 Å) bond lengths. Alkane chains often adopt the zigzag patterns used in bond-line notation (Figure 2-1). To depict three-dimensional structures, we shall make use of the dashed-wedged line notation (Figure 1-22). The main chain and a hydrogen at each end are drawn in the plane of the page (Figure 2-2).

FIGURE 2-2 Dashed-wedged line structures of methane through pentane. Note the zigzag arrangement of the principal chain and two terminal hydrogens.

EXERCISE 2-5

Draw zigzag dashed-wedged line structures for 2-methylbutane and 2,3-dimethylbutane.

The regularity in alkane structures suggests that their physical constants would reveal predictable trends. Indeed, inspection of the data presented in Table 2-3 reveals regular incremental increases along the homologous series. For example, at room temperature (25°C), the lower homologs of the alkanes are gases or colorless liquids, the higher homologs waxy solids. From pentane to pentadecane, each additional CH_2 group causes a 20–30°C increase in boiling point (Figure 2-3).

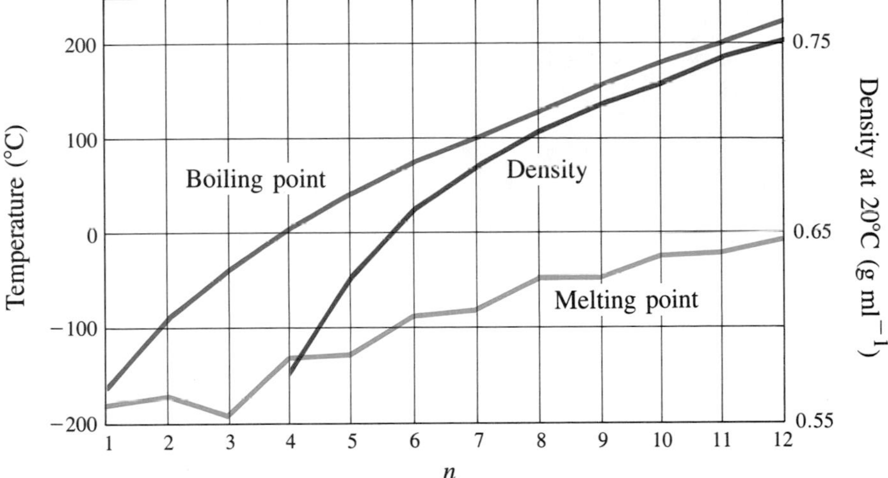

FIGURE 2-3 The physical constants of straight-chain alkanes. Their values increase with increasing size because molecular weights and London forces increase. Also note that "even-numbered" systems have somewhat higher melting points than expected; these are more tightly packed in the solid state (notice their higher densities), thus allowing for stronger attractions between molecules.

Attractive forces between molecules govern the physical properties of alkanes

Why are there such trends? They exist because of **intermolecular** or **van der Waals* forces.** Molecules exert several types of attractive forces on each other, causing them to aggregate into organized arrangements as solids and liquids. Most solid substances exist as highly ordered crystals. *Ionic* compounds, such as salts, are rigidly held in a crystal lattice, mainly by strong Coulomb forces. Nonionic but *polar* molecules, such as chloromethane (CH_3Cl), are attracted by weaker dipole–dipole interactions, also of coulombic origin (Sections 1-2 and 6-2). Finally, the *nonpolar* alkanes attract each other by **London† forces,** which are due to **electron correlation.** A simplified explanation of this effect considers alkanes as electron clouds. When one alkane molecule approaches another, the electrons in the two molecules affect one another and begin to correlate their movements. This correlation results in attraction. Figure 2-4 is a simple picture comparing ionic, dipolar, and London attractions.

London forces are very weak. In contrast with Coulomb forces, which change with the square of the distance between charges, London forces fall off as the sixth power of the distance between molecules. There is also a limit as to how close these forces can bring molecules together. At small distances, nucleus–nucleus and electron–electron repulsions outweigh these attractions.

---

*Professor Johannes D. van der Waals, 1837–1923, University of Amsterdam, Netherlands, Nobel Prize 1910 (physics).

†Professor Fritz London, 1900–1954, Duke University. Note: In older references the term "van der Waals forces" referred exclusively to what we now call "London forces;" "van der Waals forces" now refers collectively to *all* intermolecular attractions.

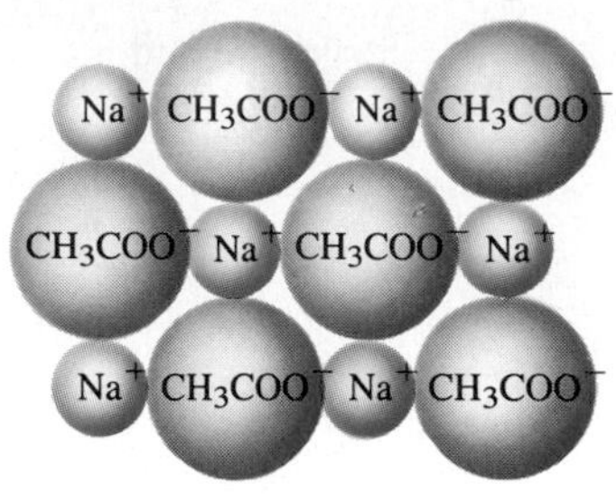

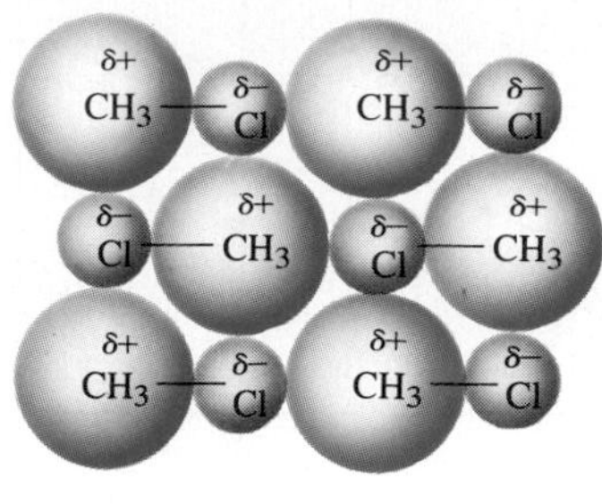

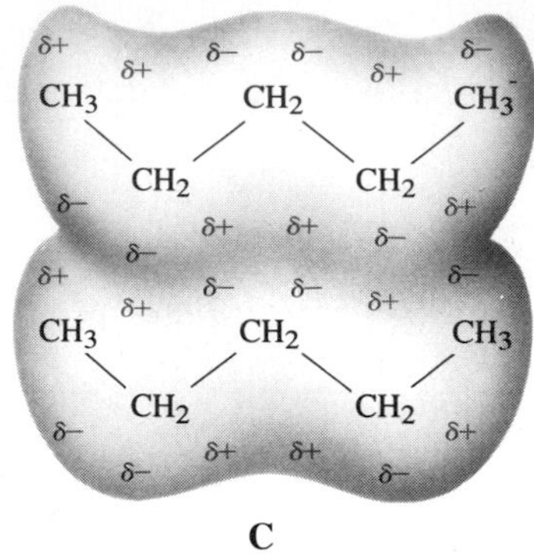

FIGURE 2-4
(A) Coulombic attraction in an ionic compound: crystalline sodium acetate, the sodium salt of acetic acid (in vinegar). (B) Dipole–dipole interactions in solid chloromethane. The polar molecules arrange to allow for favorable coulombic attraction. (C) London forces in crystalline pentane. In this simplified picture, the electron clouds as a whole mutually interact to produce partial charges of opposite sign. The charge distributions in the two molecules change continually as the electrons continue to correlate their movements.

How do these forces account for the physical constants of elements and compounds? The answer is that it takes energy, usually in the form of heat, to melt solids and boil liquids. For example, to cause melting, the attractive forces responsible for the crystalline state must be overcome. In an ionic compound, such as sodium acetate (Figure 2-4A) the strong interionic forces require a rather high temperature (324°C) for the compound to melt. In alkanes, melting points rise with increasing molecular size: Molecules with relatively large surface areas experience greater London attractions. However, these forces are still relatively weak, and even high molecular weight alkanes have rather low melting points. For example, the straight-chain alkanes $C_{29}H_{60}$ and $C_{31}H_{64}$, waxy solids present in the protective coatings of plant leaves, have melting points below 70°C.

For a molecule to escape these same attractive forces in the liquid state and enter the gas phase, more heat has to be applied. When the vapor pressure of a liquid equals atmospheric pressure, boiling occurs. Like melting points, boiling points rise with increasing molecular weight: Heavy compounds require more kinetic energy to leave the liquid state. Boiling points of compounds are also relatively high if the intermolecular forces are relatively large. These effects lead to the smooth increase in boiling points seen in Figure 2-3.

Branched alkanes have smaller surface areas than their straight-chain isomers. As a result, they generally experience smaller London attractions and are unable to pack as well in the crystalline state. The weaker attractions result in lower melting and boiling points. Crystal packing differences also account for the slightly lower than expected melting points of odd-membered straight-chain alkanes relative to those of even-membered systems (Figure 2-3).

In summary, straight-chain alkanes have regular structures. Their melting points, boiling points, and densities increase with molecular weight because of increasing attraction between molecules.

2-5 Rotation about Single Bonds: Conformations

We have considered how intermolecular forces can affect the physical properties of molecules. These forces act *between* molecules. In this section, we shall examine how the forces present *within* molecules (i.e., intramolecular forces) make some arrangements in space energetically more favorable than others.

Rotation interconverts the conformations of ethane

If we build a molecular model of ethane, we can see that the two methyl groups are readily rotated with respect to each other. The energy required to move the

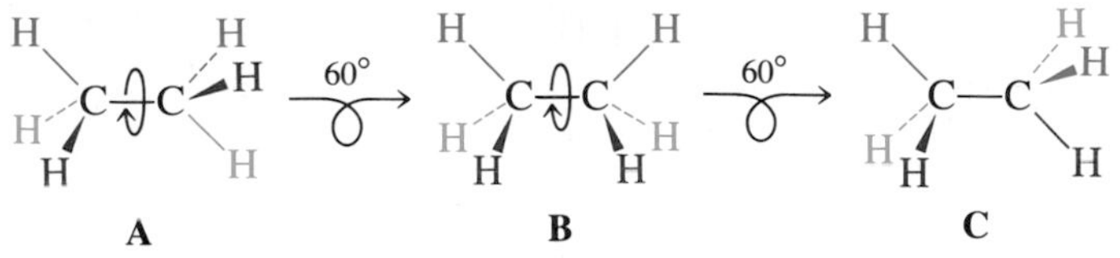

FIGURE 2-5 Rotation in ethane. (A and C) Staggered conformations; (B) eclipsed. There is virtually "free rotation" between conformers.

hydrogen atoms past each other, the *barrier to rotation,* is only 3 kcal mol^{-1}. This value turns out to be so low that chemists speak of "free rotation" of the methyl groups. In general, *there is free rotation around all single bonds.*

Figure 2-5 depicts the rotational movement in ethane by the use of dashed-wedged line structures (Section 1-9). There are two extreme ways of drawing ethane: the staggered conformation and the eclipsed one. If the **staggered conformation** is viewed along the C–C axis, each hydrogen atom on the first carbon is seen to be positioned perfectly between two hydrogen atoms on the second. The second extreme is derived from the first by a 60° turn of one of the methyl groups around the C–C bond. Now, if this **eclipsed conformation** is viewed along the C–C axis, all hydrogen atoms on the first carbon are directly opposite those on the second—that is, those on the first eclipse those on the second. A further 60° turn converts the eclipsed form into a new but equivalent staggered arrangement. Between these two extremes, rotation of the methyl group results in numerous additional positions, referred to collectively as **skew conformations.**

The many forms of ethane (and, as we shall see, substituted analogs) created by such rotations are **conformations** (also called **conformers** or **rotamers**). All of them rapidly interconvert at room temperature. The study of their thermodynamic and kinetic behavior is **conformational analysis.**

Newman projections depict the conformations of ethane

A simple alternative to the dashed-wedged line structures for illustrating the conformers of ethane is the **Newman* projection.** We can arrive at a Newman projection from the dashed-wedged line picture by turning the molecule out of the plane of the page toward us and viewing it along the C–C axis (Figure 2-6A and B). In this notation, the front carbon obscures the back carbon, but the bonds

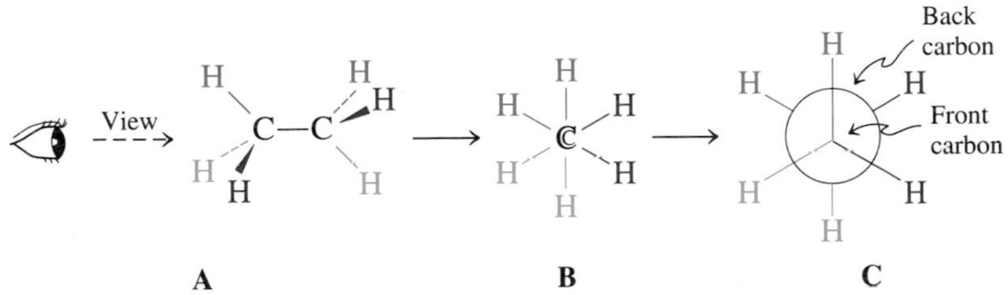

FIGURE 2-6 (A) Dashed-wedged line representation of ethane. (B) End-on view of the ethane molecule, showing the carbon atoms directly in front of each other and the staggered positions of the hydrogens. (C) Newman projection of ethane derived from the view shown in (B). The "front" carbon is represented by the intersection of the bonds to its three attached hydrogens. The bonds from the remaining three hydrogens connect to the large circle, which represents the "back" carbon.

*Professor Melvin S. Newman, 1908–1993, Ohio State University.

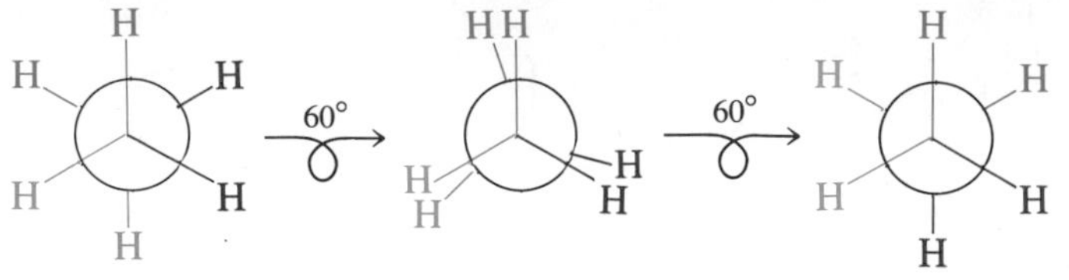

Staggered Eclipsed Staggered

FIGURE 2-7 Newman projections of staggered and eclipsed rotamers of ethane. In these projections, the back carbon is rotated clockwise in increments of 60°.

emerging from both are clearly seen. The front carbon is depicted as the point of juncture of the three bonds attached to it, one of them usually drawn vertically and pointing up. The back carbon is a circle (Figure 2-6C). The bonds to this carbon project from the outer edge of the circle. The extreme conformational shapes of ethane are readily drawn in this way (Figure 2-7). To make the three rear hydrogen atoms more visible in eclipsed conformations, they are drawn somewhat rotated out of the perfectly eclipsing position.

2-6 Potential-Energy Diagrams

As mentioned earlier, about 3 kcal mol^{-1} of heat is required to rotate the methyl groups in ethane. What is the reason for this requirement?

The rotamers of ethane have different potential energies

The various rotamers of ethane do not all have the same potential energies. A simple explanation is based on electron repulsion. As one methyl group turns around the C–C axis, starting from a staggered conformation, the distance between the hydrogen atoms of the respective methyl groups begins to diminish, resulting in increasing repulsion between the bonding pairs of electrons in the C–H bonds. Thus, the potential energy of the system rises steadily as the methyl group rotates from staggered through skew to eclipsed conformations. At the point of eclipsing, the molecule has its highest energy content, because at this stage the two sets of six bonding electrons are closest. This point is 3 kcal mol^{-1} above the lowest energy state of the molecule, the staggered rotamer. The change in energy resulting from bond rotation is called **rotational** or **torsional energy.**

Potential-energy diagrams are a convenient way to depict energy changes

The differences in potential energy between rotamers can be pictured by plotting the energy changes against the degree of rotation (Figure 2-8). Such a plot is called a **potential-energy diagram.** Potential-energy diagrams are useful in the description of other chemical processes as well. Changes in potential energy are plotted against a **reaction coordinate,** which describes the progress of the process or reaction. In the diagram for rotation of ethane, the reaction coordinate is degrees of rotation (Figure 2-8). Ethane is best described in its staggered conformation. In fact, the eclipsed rotamer has only a fleeting lifetime (of the order of

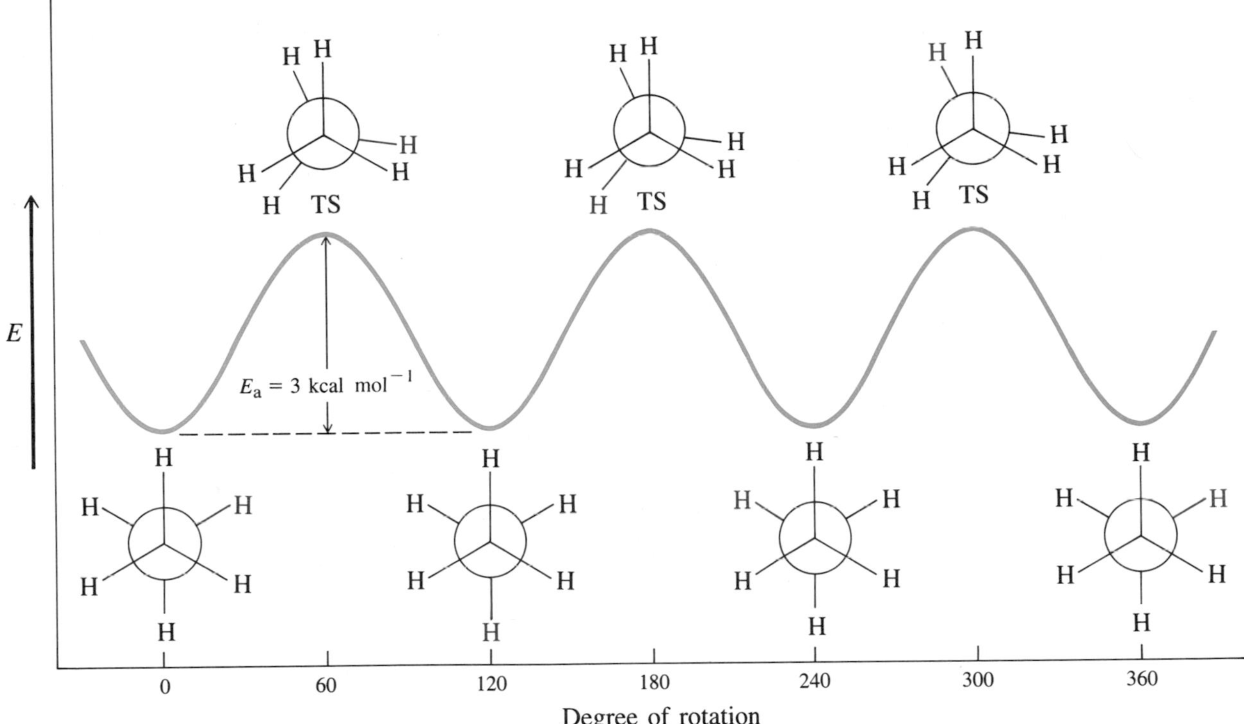

FIGURE 2-8 Potential-energy diagram of the rotational isomerism in ethane. Because the eclipsed conformations have the highest energy, they correspond to peaks in the diagram. These maxima may be viewed as transition states (TS) between the more stable staggered rotamers. The activation energy (E_a) corresponds to the barrier to rotation.

10^{-12} s) as the hydrogens rapidly move past each other, equilibrating one staggered arrangement with another. Because eclipsed conformations have the highest energy, they are maxima in potential-energy diagrams. Such points are called **transition states** (TS), marking the transition from one staggered rotamer to another. The energy of the transition state can be viewed as the barrier to be overcome when the molecule goes from one staggered arrangement to the next. This energy is called the **activation energy,** E_a, for the rotational process. The lower its value, the faster the rotation.

Collisions supply the energy to get past the activation-energy barrier

Where do organic molecules get the energy to overcome the barrier to rotation? Molecules have *kinetic energy* as a result of their motion, but at room temperature the average kinetic energy is only about 0.6 kcal mol^{-1}, far below the activation-energy barrier. To pick up enough energy, molecules must collide with each other or with the walls of the container. Each collision transfers energy from one molecule to another.

A graph called a **Boltzmann* distribution curve** depicts the distribution of kinetic energy. Figure 2-9 shows that, although most molecules have only average speed, at room temperature some molecules have kinetic energies as high as 25 kcal mol^{-1}. In ethane, part of this energy may be used to overcome the activation-energy barrier. Because continual collisions rapidly redistribute the

*Professor Ludwig Boltzmann, 1844–1906, University of Vienna, Austria.

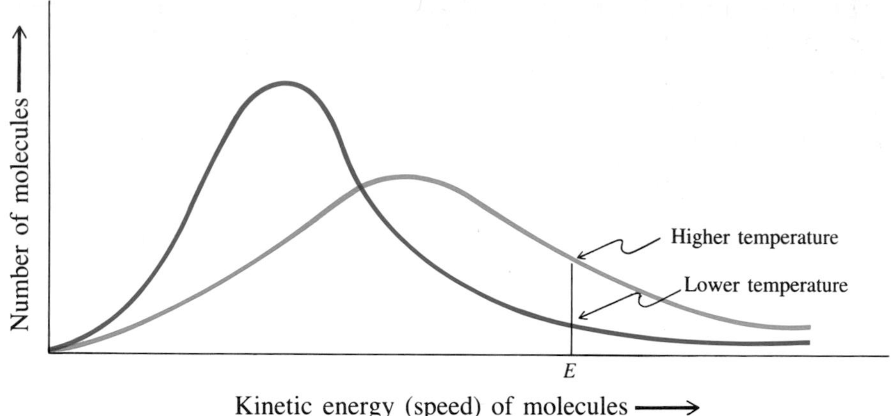

FIGURE 2-9 Boltzmann curves at two temperatures. At the higher temperature (green curve), there are more molecules of kinetic energy E than at the lower temperature (blue curve). Molecules with higher kinetic energy can more easily overcome the activation-energy barrier.

kinetic energy, all molecules eventually get past this barrier. That is why we can speak of "free rotation."

The shape of the Boltzmann curve depends on the temperature. We can see that at higher temperatures, as the average kinetic energy increases, the curve flattens and shifts toward higher energies. Now more molecules have energy higher than is required by the transition state, so the speed of rotation increases. Conversely, at lower temperatures, the rate of rotation decreases.

In summary, intramolecular forces control the arrangement of substituents on neighboring and bonded saturated carbon atoms. In ethane the relatively stable staggered conformations are interconverted through higher energy transition states in which substituents are eclipsed. To reach these transition states, molecules have to absorb the kinetic energy of others through collisions. The energy distribution of a collection of molecules at any given temperature is depicted by a Boltzmann curve. The energetics of rotation around the C–C bond is conveniently pictured in a potential-energy diagram.

2-7 Rotation in Substituted Ethanes

How does the potential-energy diagram change when a substituent is added to ethane? Consider, for example, propane, whose structure is similar to that of ethane, except that a methyl group replaces one of ethane's hydrogen atoms.

Steric hindrance raises the energy barrier to rotation

A potential-energy diagram for the rotation around a C–C bond in propane is shown in Figure 2-10. The Newman projections of propane differ from those of ethane only by the substituted methyl group. Again, the extreme conformations are staggered and eclipsed. However, the activation barrier separating the two is 3.4 kcal mol^{-1}, slightly higher than that for ethane. This difference is due to an unfavorable steric interaction between the methyl substituent and the eclipsing hydrogen in the transition state, a phenomenon called **steric hindrance.** To a first approximation, this effect can be attributed to bulk: Two molecular fragments cannot occupy the same region in space.

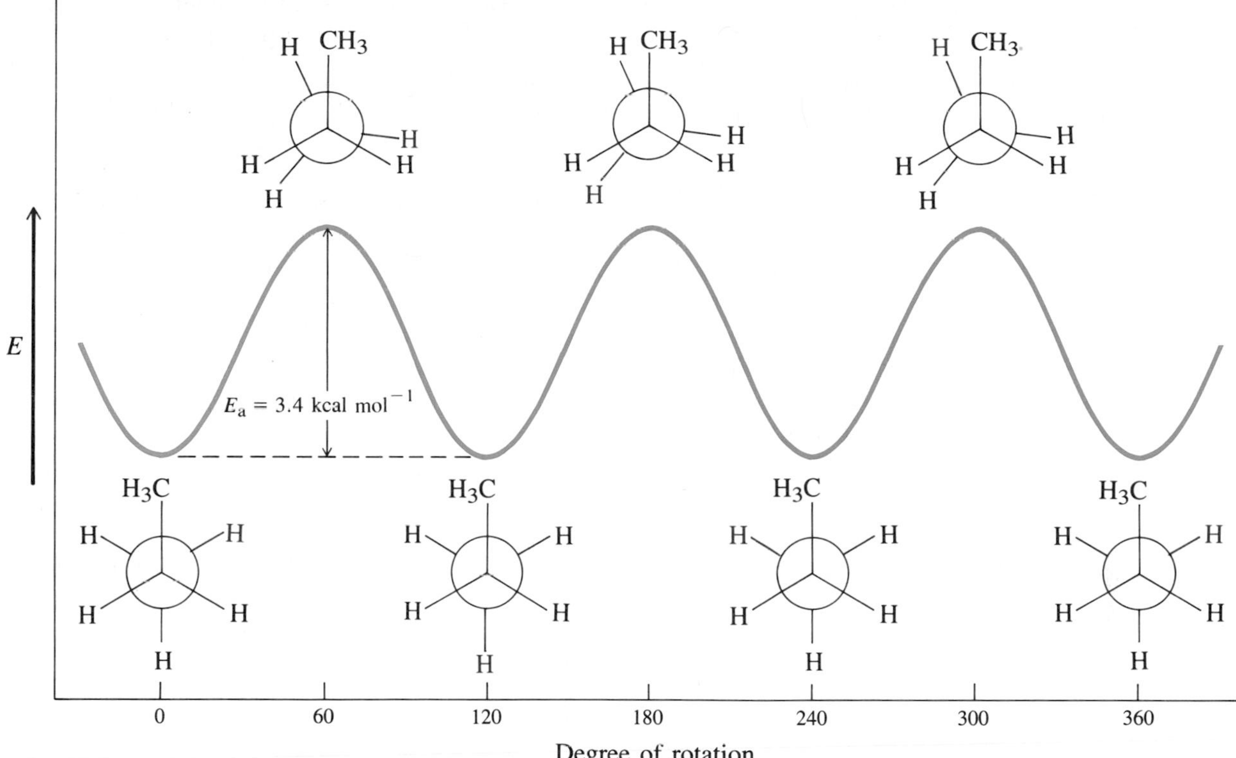

Steric hindrance in propane is actually worse than the E_a value for rotation indicates. Methyl substitution raises the energy, not only of the eclipsed conformation, but also of the staggered (lowest energy, or *ground* state) one, the latter to a lesser extent because of less steric interaction. Because the activation energy reflects the *difference* between ground and transition states, the net result is a small increase in E_a.

FIGURE 2-10 Potential-energy diagram of the rotational isomerism in propane. Steric hindrance increases the relative energy of the eclipsed form.

There can be more than one staggered and one eclipsed conformation: conformational analysis of butane

If we build a model and look at the rotation around the central C–C bond of butane, we find that there are more conformations than one staggered and one eclipsed (Figure 2-11). Consider the staggered conformer in which the two methyl groups are as far away from each other as possible. This arrangement, called ***anti*** (i.e., opposed), is the most stable because steric hindrance is minimized. Rotation of the rear carbon in the Newman projections in either direction (in Figure 2-11, the direction is clockwise) produces an eclipsed conformation with two CH_3–H interactions. This rotamer is 3.8 kcal mol^{-1} higher in energy than the *anti* precursor. Further rotation furnishes a *new* staggered structure in which the two methyl groups are closer than they are in the *anti* conformation. To distinguish this conformer from the others, it is named ***gauche*** (*gauche,* French, in the sense of awkward, clumsy). As a consequence of steric hindrance, the *gauche* conformer is higher in energy than the *anti* conformer by about 0.9 kcal mol^{-1}.

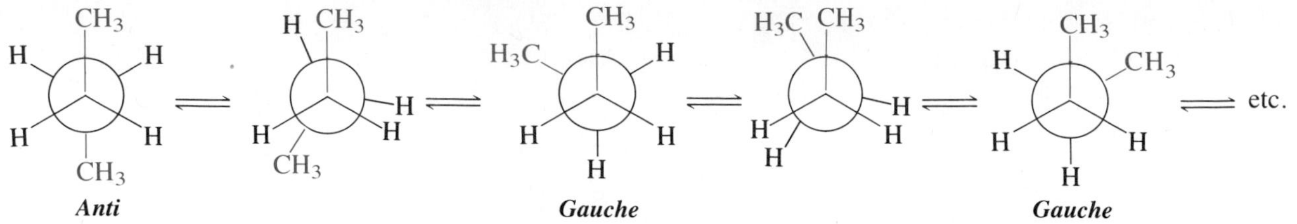

FIGURE 2-11 Clockwise rotation of the rear carbon along the C2–C3 bond in a Newman projection of butane.

Further rotation (Figure 2-11) results in a *new* eclipsed arrangement in which the two methyl groups are superposed. Because the two bulkiest substituents eclipse in this rotamer, it is energetically highest, 4.5 kcal mol^{-1} higher than the most stable *anti* structure. Further rotation produces another *gauche* conformer. The activation energy for *gauche* $\rightleftharpoons$ *gauche* interconversion is 3.6 kcal mol^{-1}. A potential-energy diagram summarizes the energetics of the rotation (Figure 2-12). The most stable *anti* conformer is the most abundant in solution (about 80% at 25°C). Its less stable *gauche* counterpart is present in lower concentration (20%).

We can see from Figure 2-12 that knowing the difference in thermodynamic stability of two conformers (e.g., 0.9 kcal mol^{-1} between the *anti* and *gauche*

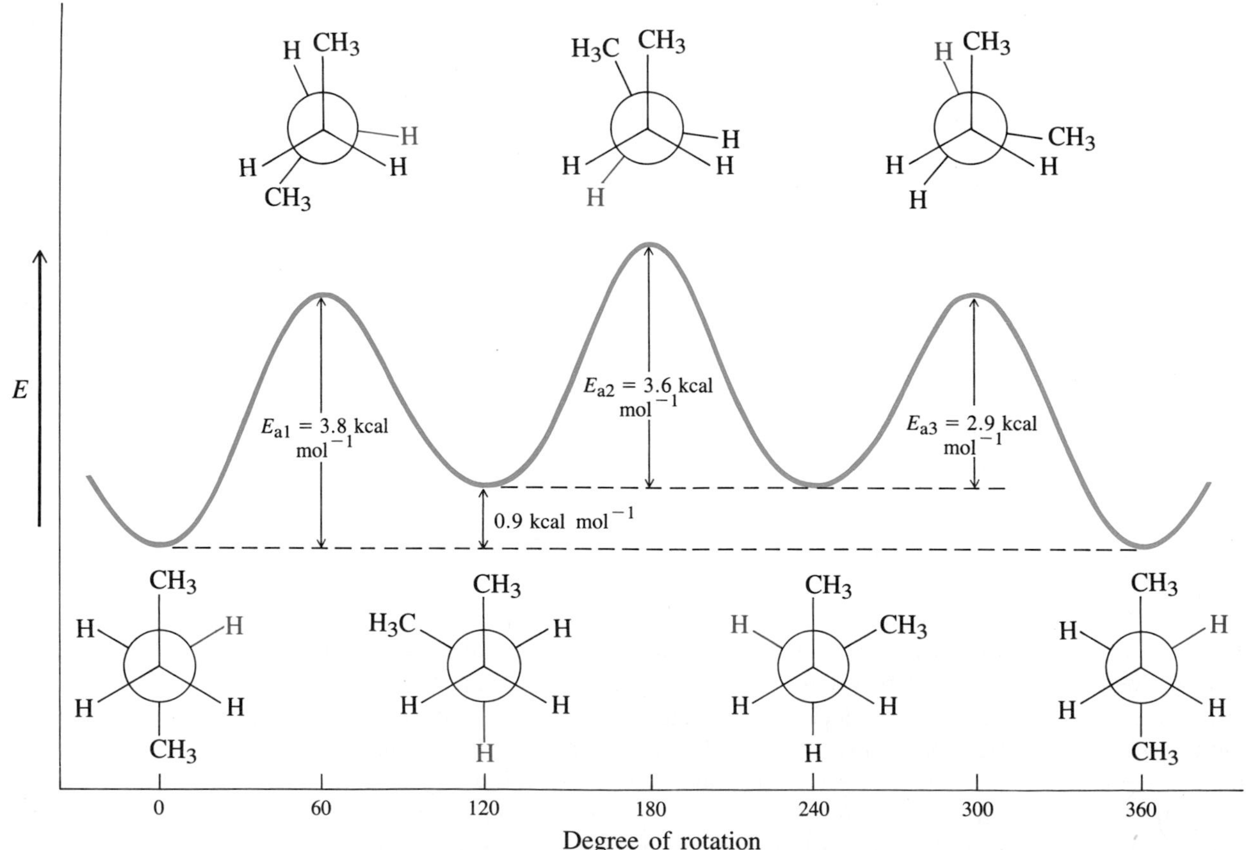

FIGURE 2-12 Potential-energy diagram of the rotation around the C2–C3 bond in butane. There are three processes: *anti* → *gauche* conversion with E_{a1} = 3.8 kcal mol^{-1}; *gauche* → *gauche* rotation with E_{a2} = 3.6 kcal mol^{-1}; and *gauche* → *anti* transformation with E_{a3} = 2.9 kcal mol^{-1}.

isomers) and the activation energy for proceeding from the first to the second (e.g., 3.8 kcal mol^{-1}) allows us to estimate the activation barrier of the reverse reaction. In this case, E_a for the *gauche*-to-*anti* conversion is $3.8 - 0.9 =$ 2.9 kcal mol^{-1}.

EXERCISE 2-6

Draw the expected potential-energy diagram for the rotation around the C2–C3 bond in 2,3-dimethylbutane. Include the Newman projections of each staggered and eclipsed conformation.

In summary, if two substituents (one on each carbon atom) are 180° apart in a staggered Newman projection, they belong to an *anti* conformer. If they are 60° apart, the conformer is *gauche*. *Gauche* conformations are usually less stable than their *anti* counterparts. Conformational analysis is the study of the changes in potential energy that take place on rotation around single bonds.

2-8 Kinetics and Thermodynamics of Conformational Isomerism and of Simple Reactions

The *anti* $\rightleftharpoons$ *gauche* conformational isomerism is a typical example of an equilibrium between two distinct species. Although no bonds are broken or made, as in the usual chemical reaction, the process is controlled by the same physical criteria as are ordinary reactions.

We shall review two basic principles governing chemical reactions:

1. **Chemical thermodynamics** deals with the changes in energy that take place when molecules react, a feature that controls the *extent* to which a reaction will go to completion.
2. **Chemical kinetics** concerns the velocity or rate at which the concentrations of reactants and products change, in other words the *speed* at which a reaction will go to completion.

The two phenomena are frequently related. Reactions that are thermodynamically very favorable often proceed faster than do less favorable ones. Conversely, some reactions are faster than others even though they result in a comparatively less stable product. A transformation that yields the most stable products is said to be under **thermodynamic control.** Its outcome is determined by the net favorable change in energy in going from starting materials to products. A reaction in which the product obtained is the one formed fastest is defined as being under **kinetic control.** Let us put these statements on a more quantitative footing.

Equilibria are governed by the thermodynamics of chemical change

All chemical reactions are reversible, and reactants and products interconvert to various degrees. When the concentrations of reactants and products no longer change, the reaction is in a **state of equilibrium.** In many cases, equilibrium lies extensively (say, more than 99.9%) on the side of the products. When this occurs, the reaction is said to have *gone to completion*. (In such cases, the arrow indicating the reverse reaction is usually omitted.) Equilibria are described by equilibrium constants, K. To find an equilibrium constant, divide the arithmetic product of the concentrations of the components on the right side of the reaction by that of the components on the left, all given in units of moles per liter (mol L^{-1}). A large value for K indicates that a reaction will go to completion; it is said to have a large **driving force.**

Typical Chemical Equilibria

$$\mathrm{A} \underset{}{\overset{K}{\rightleftharpoons}} \mathrm{B} \qquad K = \frac{[\mathrm{B}]}{[\mathrm{A}]}$$

$$\mathrm{A} + \mathrm{B} \underset{}{\overset{K}{\rightleftharpoons}} \mathrm{C} + \mathrm{D} \qquad K = \frac{[\mathrm{C}][\mathrm{D}]}{[\mathrm{A}][\mathrm{B}]}$$

If a reaction has "gone to completion," a certain amount of energy has been released. The equilibrium constant can be related directly to the thermodynamic function of the **Gibbs* standard free energy change, $\Delta G°$,**† at equilibrium:

$$\Delta G° = -RT \ln K = -2.303\, RT \log K \text{ (in kcal mol}^{-1}\text{)}$$

in which R is the gas constant (1.986 cal $\deg^{-1}$ mol^{-1}) and T is the absolute temperature in kelvins‡ (K). A negative value for $\Delta G°$ signifies a release of energy. The equation shows that a large value for K indicates a large favorable free energy change. At room temperature (25°C, 298 K), the preceding equation becomes

$$\Delta G° = -1.36 \log K \text{ (in kcal mol}^{-1}\text{)}$$

This expression tells us that an equilibrium constant of 10 would have a $\Delta G°$ of -1.36 kcal mol^{-1}, and, conversely, a K of 0.1 would have a $\Delta G° = +1.36$ kcal mol^{-1}. Because the relation is logarithmic, changing the $\Delta G°$ value affects the K value exponentially. When $K = 1$, starting materials and products are present in equal concentrations and $\Delta G°$ is zero (Table 2-5).

EXERCISE 2-7

Calculate the equilibrium concentration of *gauche*-butane at 25°C and at 100°C. Use data from Figure 2-12.

*Professor Josiah Willard Gibbs, 1839–1903, Yale University.

†The descriptor $\Delta G°$ refers to the free energy of a reaction with the molecules in their standard states (e.g., ideal molar solutions) after it has reached equilibrium.

‡Temperature intervals in kelvins and degrees Celsius are identical. Temperature units are named after Lord Kelvin, Sir William Thomson, 1824–1907, University of Glasgow, Scotland, and Anders Celsius, 1701–1744, University of Uppsala, Sweden.

TABLE 2-5 Equilibria and Free Energy for A $\rightleftharpoons$ B; K = [B]/[A]

K	Percentage B	Percentage A	$\Delta G°$ (kcal mol^{-1} at 25°C)
0.01	0.99	99.0	+2.73
0.1	9.1	90.9	+1.36
0.33	25	75	+0.65
1	50	50	0
2	67	33	−0.41
3	75	25	−0.65
4	80	20	−0.82
5	83	17	−0.95
10	90.9	9.1	−1.36
100	99.0	0.99	−2.73
1,000	99.9	0.1	−4.09
10,000	99.99	0.01	−5.46

The free energy change is related to changes in bond strengths and the degree of order in the system

The Gibbs standard free energy change is related to two other thermodynamic quantities, the change in **enthalpy,** $\Delta H°$, and change in **entropy,** $\Delta S°$.

Gibbs Standard Free Energy Change

$$\Delta G° = \Delta H° - T\Delta S°$$

In this equation, T is again in kelvins and $\Delta H°$ in kcal mol^{-1}, whereas $\Delta S°$ is in cal deg^{-1} mol^{-1}, also called entropy units (e.u.).

The **enthalpy change,** $\Delta H°$, is defined as the heat of a reaction at constant pressure. Enthalpy changes in an organic chemical reaction relate mainly to changes in bond strengths in the course of the reaction. Thus, the value of $\Delta H°$ can be estimated by subtracting the sum of the strengths of the bonds formed from those broken.

Enthalpy Change in a Reaction

$$\begin{pmatrix}\text{Sum of strengths}\\ \text{of bonds broken}\end{pmatrix} - \begin{pmatrix}\text{sum of strengths}\\ \text{of bonds formed}\end{pmatrix} = \Delta H°$$

If the bonds formed are stronger than those broken, the value of $\Delta H°$ is negative and the reaction is defined as **exothermic** (releasing heat). In contrast, a positive $\Delta H°$ is characteristic of an **endothermic** process. An example of an exothermic reaction is the combustion of methane, the main component of natural gas, to carbon dioxide and liquid water. This process has a $\Delta H°$ value of -213 kcal mol^{-1}.

$$CH_4 + 2\,O_2 \longrightarrow CO_2 + 2\,H_2O_{liq} \qquad \Delta H° = -213 \text{ kcal mol}^{-1}$$

The exothermic nature of this reaction is due to the very strong bonds formed in the products. Many hydrocarbons release a lot of energy on combustion and are therefore valuable fuels.

If the enthalpy of a reaction strongly depends on changes in bond strength, what is the significance of $\Delta S°$? The **entropy change,** $\Delta S°$, is a measure of the changes in the order of a system. The value of $S°$ increases with increasing disorder. Because of the negative sign in front of the $T\Delta S°$ term in the equation for $\Delta G°$, a positive value for $\Delta S°$ makes a negative contribution to the free energy of the system. In other words, going from order to disorder is thermodynamically favorable.

What is meant by disorder in a chemical reaction? Consider a transformation in which the number of reacting molecules differs from the number of product molecules formed. For example, upon strong heating, 1-pentene undergoes cleavage to ethene and propene. This process, in which two molecules are made from one, has a relatively large positive $\Delta S°$. The increased number of particles present after bond cleavage means greater freedom of motion, thus representing an increase in disorder for the system.

$$\underset{\textbf{1-Pentene}}{CH_3CH_2CH_2CH{=}CH_2} \longrightarrow \underset{\substack{\textbf{Ethene}\\\textbf{(Ethylene)}}}{CH_2{=}CH_2} + \underset{\textbf{Propene}}{CH_3CH{=}CH_2} \qquad \begin{array}{l}\Delta H° = +22.4 \text{ kcal mol}^{-1}\\ \Delta S° = +33.3 \text{ e.u.}\end{array}$$

EXERCISE 2-8

Calculate the $\Delta G°$ at 25°C for the preceding reaction. Is it thermodynamically feasible at 25°C? What is the effect of increasing T on $\Delta G°$? What is the temperature at which the reaction becomes favorable?

In contrast, disorder and entropy decrease when the number of product molecules is less than the number of molecules of starting materials. For example, the reaction of ethene (ethylene) with hydrogen chloride to give chloroethane is exothermic by -15.5 kcal mol^{-1}, but the entropy makes an unfavorable contribution to the $\Delta G°$, $\Delta S° = -31.3$ e.u.

$$CH_2{=}CH_2 + HCl \longrightarrow CH_3CH_2Cl \qquad \begin{array}{l}\Delta H° = -15.5 \text{ kcal mol}^{-1}\\ \Delta S° = -31.3 \text{ e.u.}\end{array}$$

EXERCISE 2-9

Calculate the $\Delta G°$ at 25°C for the preceding reaction. In your own words, explain why a reaction that combines two molecules into one should have a large negative entropy change.

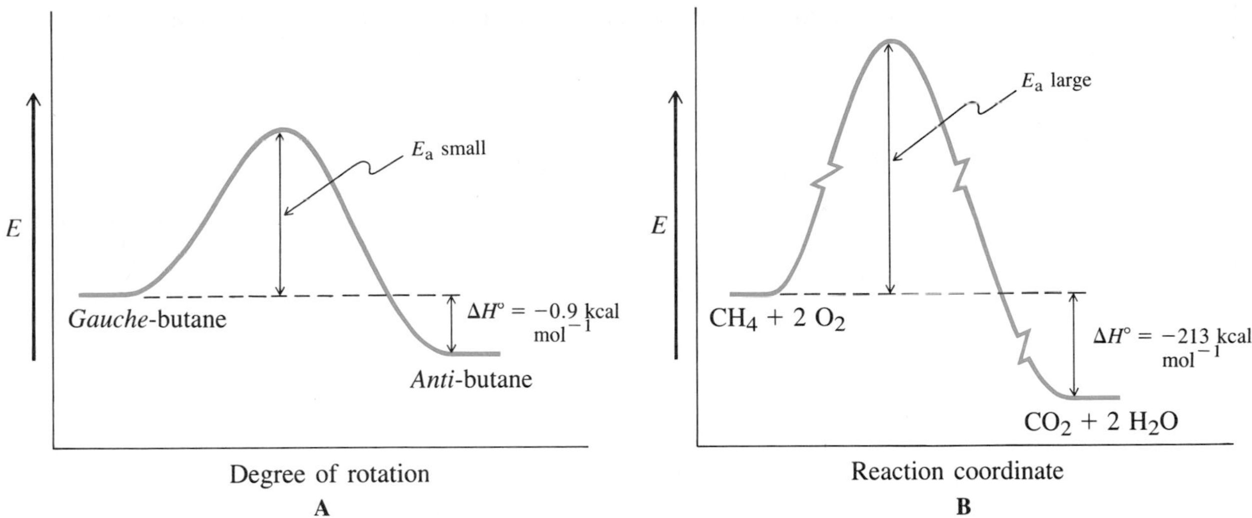

FIGURE 2-13
Comparison of potential-energy diagrams for (A) *gauche–anti* conversion in butane and (B) the combustion of methane. Comparison of the activation energies explains why bond rotation in butane is so much faster, even though the combustion reaction of methane is so much more thermodynamically favorable, as shown by its large negative $\Delta H°$. The diagrams are not drawn to scale.

The rate of a chemical reaction depends on the activation energy

How fast is equilibrium established? The thermodynamic features of chemical reactions do not by themselves tell us anything about their rates. Consider the conversion of *gauche* butane into the *anti* rotamer (Figure 2-12). This process is thermodynamically favorable by only a small amount, and yet equilibrium is established exceedingly rapidly, even at very low temperatures. Now compare that with the combustion of methane considered earlier. This process releases 213 kcal mol^{-1}, a huge amount of energy, but we know that methane does not spontaneously ignite in air at room temperature. Why is the much more favorable combustion process so much slower? The answer is that rates of chemical processes are controlled by activation energies. We have already seen that E_a for bond rotation in butane is very low. This situation reflects a low-energy transition state, through which the molecule may pass very rapidly. Conversely, the transition state for methane combustion is very high in energy, corresponding to a high E_a and a very low rate (Figure 2-13).

How can there be such high activation energies for exothermic reactions? A simple answer is that bond breaking usually precedes bond formation. Thus, before energy is released through bonding, energy must be expended to break bonds. The transition state is the point at which the initial energy input is compensated by a corresponding release of energy.

The concentration of reactants can affect reaction rates

The concentration of reactants can influence the rate of a reaction. Consider the addition of reagent A to reagent B to give C:

$$A + B \longrightarrow C$$

In many transformations of this type increasing the concentration of either reactant increases the rate of the reaction. In such cases, the transition-state structure incorporates both molecules A and B. The rate is expressed by

$$\text{Rate} = k[\text{A}][\text{B}] \text{ in units of mol L}^{-1}\ \text{s}^{-1}$$

in which the proportionality constant, k, is also called the **rate constant** of the reaction. A reaction for which the rate depends on the concentrations of two molecules in this way is said to be of **second order.**

There are processes whose rate depends on the concentration of only one reactant, such as the hypothetical reaction

$$\text{A} \longrightarrow \text{B}$$
$$\text{Rate} = k[\text{A}] \text{ in units of mol L}^{-1}\ \text{s}^{-1}$$

A reaction of this type is said to be of **first order.** Rotation around a carbon–carbon bond follows such a rate law.

EXERCISE 2-10

The dependence of reaction rates on reactant concentrations means that reactions slow down as starting material is used up. For example, for a process following the first-order rate law, rate = k[A], when half of A is consumed (that is, after 50% conversion of the starting material), the reaction rate is reduced to half of its initial value. What would be the reduction in rate of a second-order reaction (in which rate = k[A][B]) after 50% conversion of the starting material?

The Arrhenius equation describes how temperature affects reaction rates

Temperature also greatly affects reaction rates. An increase in temperature leads to faster reactions. The kinetic energy of molecules increases when they are heated, which means that a larger fraction of them have sufficient energy to overcome the activation barrier (Figure 2-9). A useful rule of thumb applies to many reactions: Raising the reaction temperature by 10°C causes the rate to increase by a factor of 2 or 3. The Swedish chemist Arrhenius* noticed the dependence of reaction rate on temperature. He found that his measured data conformed to the equation

Arrhenius Equation

$$k = Ae^{-E_a/RT}$$

The A term reflects the maximum rate constant the reaction would have if every molecule had sufficient collisional energy to overcome the activation barrier: At very high temperature, E_a/RT is small, $e^{-E_a/RT}$ approaches 1, and k nearly

*Professor Svante Arrhenius, 1859–1927, Technical Institute of Stockholm, Sweden, Nobel Prize 1903 (chemistry), director of the Nobel Institute from 1905 until shortly before his death.

equals A. Each reaction has its own characteristic value for A. The Arrhenius equation describes how rates of reactions with different activation energies vary with temperature.

EXERCISE 2-11

(a) Calculate $\Delta G°$ at 25°C for the reaction $CH_3CH_2Cl \rightarrow CH_2{=}CH_2 + HCl$ (the reverse of the reaction in Exercise 2-9. **(b)** Calculate $\Delta G°$ at 500°C for the same reaction. (Hint: Apply $\Delta G° = \Delta H° - T\Delta S°$ and do not forget to first convert °C into kelvins.)

EXERCISE 2-12

For this same reaction, $A = 10^{14}$ and $E_a = 58.4$ kcal mol^{-1}. Using the Arrhenius equation, calculate k at 500°C for this reaction. $R = 1.986$ cal deg^{-1} mol^{-1}.

This completes our review of the thermodynamic and kinetic relations governing many organic transformations. All reactions are described by equilibrating starting materials and products. On which side the equilibrium lies depends on the size of the equilibrium constant, in turn related to the Gibbs free energy changes, $\Delta G°$. An increase in the equilibrium constant by a factor of 10 is associated with a change in $\Delta G°$ of about -1.36 kcal mol^{-1} at room temperature. The free energy change of a reaction is composed of changes in enthalpy, $\Delta H°$, and entropy, $\Delta S°$. Contributions to the former stem mainly from variations in bond strengths, to the latter from the relative disorder in starting materials and products. Whereas these terms define the position of an equilibrium, the rate at which it is established depends on the concentration of starting materials, the activation barrier separating reactants and products, and the temperature. The relation between rate, E_a, and T is expressed by the Arrhenius equation.

IMPORTANT CONCEPTS

1. Organic molecules may be viewed as being composed of a carbon skeleton with attached functional groups.

2. Hydrocarbons are made up of carbon and hydrogen only. Hydrocarbons possessing only single bonds are also called alkanes. They do not contain functional groups. They may exist as a single continuous chain or they may be branched or cyclic. The empirical formula for the straight-chain and branched alkanes is C_nH_{2n+2}.

3. Molecules that differ only by the number of methylene groups, CH_2, in the chain are called homologs and are said to belong to a homologous series.

4. A primary carbon is attached to only one other carbon. A secondary carbon is attached to two, a tertiary to three other carbon atoms. The hydrogen atoms bound to such carbon atoms are likewise designated primary, secondary, or tertiary.

5. The IUPAC rules for naming saturated hydrocarbons are (a) find the longest continuous chain in the molecule and name it; (b) name all groups attached to the longest chain as alkyl substituents; (c) number the carbon atoms of the longest chain; (d) write the name of the alkane, citing all substituents as prefixes arranged in alphabetical order and preceded by numbers designating their positions.

6. Alkanes attract each other through weak London forces, polar molecules through stronger dipole–dipole interactions, and salts mainly through very strong ionic interactions.

7. Rotation around carbon–carbon single bonds is relatively easy and gives rise to conformations (conformers, rotamers). Substituents on adjacent carbon atoms may be staggered or eclipsed. The eclipsed conformation is a transition state between staggered conformers. The energy required to reach the eclipsed state is called the activation energy for rotation. When both carbons bear alkyl or other groups, there may be additional conformers: Those in which the groups are in close proximity (60°) are *gauche;* those in which the groups are directly opposite (180°) each other are *anti*. Molecules tend to adopt conformations in which steric hindrance, as in *gauche* conformations, is minimized.

8. Chemical reactions can be described as equilibria controlled by thermodynamic and kinetic parameters. The change in the Gibbs free energy, $\Delta G°$, is related to the equilibrium constant by $\Delta G° = -RT \ln K = -1.36 \log K$ (at 25°C). The free energy has contributions from changes in enthalpy, $\Delta H°$, and entropy, $\Delta S°$: $\Delta G° = \Delta H° - T\Delta S°$. Changes in enthalpy are mainly due to differences between the strengths of the bonds made compared with those broken. A reaction is exothermic when the former is larger than the latter. It is endothermic when there is a net loss in combined bond strengths. Changes in entropy are controlled by the relative degree of order in starting materials compared with that in products. The greater the increase in disorder, the larger a positive $\Delta S°$.

9. The rate of a chemical reaction depends mainly on the concentrations of starting material(s), the activation energy, and temperature. These correlations are expressed in the Arrhenius equation: rate constant $k = Ae^{-E_a/RT}$.

10. If the rate depends on the concentration of only one starting material, the reaction is said to be of first order. If it depends on the concentrations of two reagents, it is of second order.

PROBLEMS

1. For each example in Table 2-1, identify all polarized covalent bonds and label the appropriate atoms with partial positive or negative charges. (Do not consider carbon–hydrogen bonds.)

2. On the basis of electrostatics (Coulomb attraction), predict which atom in each of the following organic molecules is likely to react with the indicated reagent. Write "no reaction" if none seems likely. (See Table 2-1 for the structures of the organic molecules.) **(a)** Bromoethane, with the oxygen of HO^-; **(b)** propanal, with the nitrogen of NH_3; **(c)** methoxyethane, with H^+; **(d)** 3-pentanone, with the carbon of CH_3^-; **(e)** ethanenitrile, with the carbon of CH_3^+; **(f)** butane, with HO^-.

3. Name the following molecules according to the IUPAC system of nomenclature.

```
    CH3CH2CHCH3                      CH3    CH3CHCH2CH3
          |                          |           |
(a)       CH                 (b) CH3CHCH2CH2CCH2CH2CH2CH3
        /    \                                   |
     CH3      CH3                            CH3CHCH3
```

```
          CH3
          |
          CH2                         H    CH3   CH3
          |                           |    |     |
(c) CH3CH2CCH2CH3            (d) CH3—C————C————C—CH2CH2CH2CH2CH3
          |                           |    |     |
          CH2                         CH2  CH2   CH2
          |                           |    |     |
          CH3                         CH3  CH3   CH—CH3
                                                 |
                                                 CH3
```

(e) $CH_3CH(CH_3)CH(CH_3)CH(CH_3)CH(CH_3)_2$

(f) 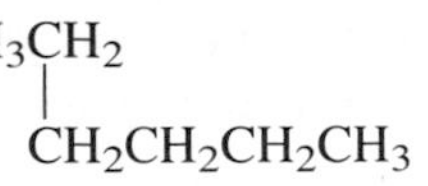 CH_3CH_2 attached to $CH_2CH_2CH_2CH_3$ (vertical bond from first CH_2 of the lower chain)

(g) 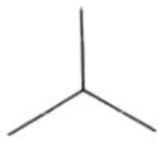 (h) 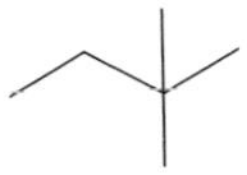 (i) 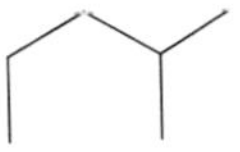 (j) 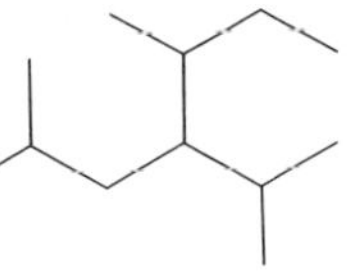 [bond-line structures]

4. Convert the following names into the corresponding molecular formulas. After doing so, check to see if the name of each molecule as given here is in accord with the IUPAC system of nomenclature. If not, name the molecule correctly. **(a)** 2-Methyl-3-propylpentane; **(b)** 5-(1,1-dimethylpropyl)nonane; **(c)** 2,3,4-trimethyl-4-butylheptane; **(d)** 4-*tert*-butyl-5-isopropylhexane; **(e)** 4-(2-ethylbutyl)decane; **(f)** 2,4,4-trimethylpentane; **(g)** 4-*sec*-butylheptane; **(h)** isoheptane; **(i)** neoheptane.

5. Draw and name all possible isomers of C_7H_{16} (isomeric heptanes).

6. Identify the primary, secondary, tertiary carbon atoms and hydrogen atoms in each of the following molecules. **(a)** Ethane; **(b)** pentane; **(c)** 2-methylbutane; **(d)** 3-ethyl-2,2,3,4-tetramethylpentane.

7. Identify each of the following alkyl groups as being primary, secondary, or tertiary, and give it a systematic IUPAC name.

(a) 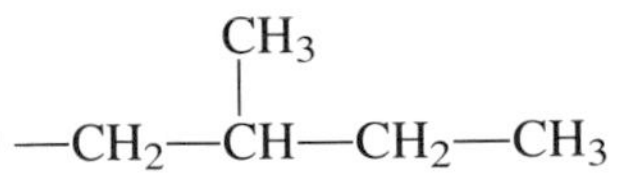 $-CH_2-CH(CH_3)-CH_2-CH_3$

(b) 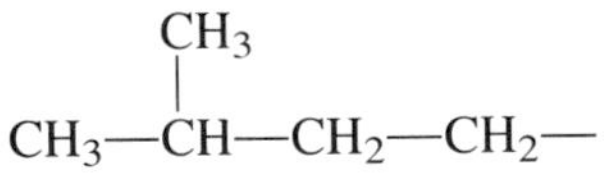 $CH_3-CH(CH_3)-CH_2-CH_2-$

(c) 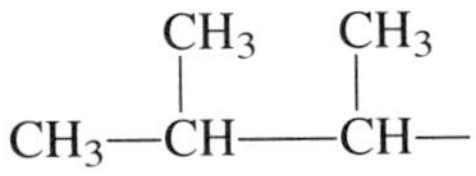 $CH_3-CH(CH_3)-CH(CH_3)-$

(d) 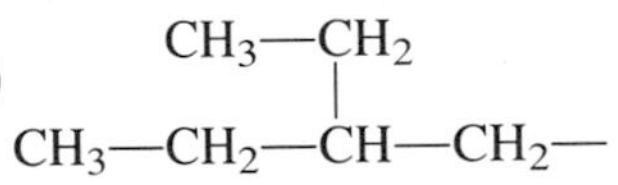 $CH_3-CH_2-CH(CH_2-CH_3)-CH_2-$

(e) 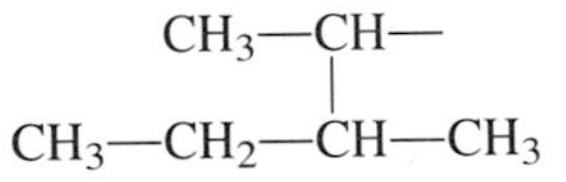 $CH_3-CH_2-CH(CH_3-CH-)-CH_3$

(f) 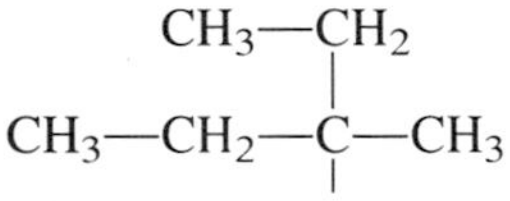 $CH_3-CH_2-C(CH_2-CH_3)(-)-CH_3$

8. Rank the following molecules in order of increasing boiling point (*without* looking up the real values). **(a)** 2-Methylhexane; **(b)** heptane; **(c)** 2,2-dimethylpentane; **(d)** 2,2,3-trimethylbutane.

9. Draw dashed-wedged line structures for the following molecules in the conformations indicated: **(a)** staggered propane; **(b)** eclipsed propane; **(c)** *anti*-butane; **(d)** *gauche*-butane.

10. Using Newman projections, draw each of the following molecules in its most stable conformation with respect to the bond indicated. **(a)** 2-Methylbutane, C2–C3 bond; **(b)** 2,2-dimethylbutane, C2–C3 bond; **(c)** 2,2-dimethylpentane, C3–C4 bond; **(d)** 2,2,4-trimethylpentane, C3–C4 bond.

11. At room temperature, 2-methylbutane exists primarily as two alternating conformations of rotation around the C2–C3 bond. About 90% of the molecules exist in the more favorable conformation and 10% in the less favorable one. **(a)** Calculate the free energy change ($\Delta G°$, more favorable conformation − less favorable conformation) between these conformations. **(b)** Draw a potential-energy diagram for rotation around the C2–C3 bond in 2-methylbutane. To the best of your ability, assign relative energy values to all the conformations on your diagram. **(c)** Draw Newman projections for all staggered and eclipsed rotamers in (b) and indicate the two most favorable ones.

12. For each of the following naturally occurring compounds, identify the compound class(es) to which it belongs, and circle all functional groups.

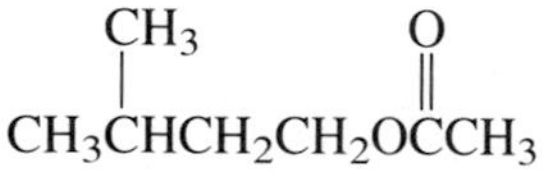

$CH_3CH(CH_3)CH_2CH_2OC(=O)CH_3$
3-Methylbutyl acetate
(In banana oil)

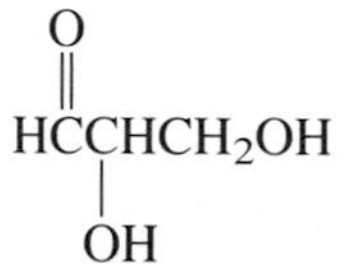

$HC(=O)CH(OH)CH_2OH$
2,3-Dihydroxypropanal
(The simplest sugar)

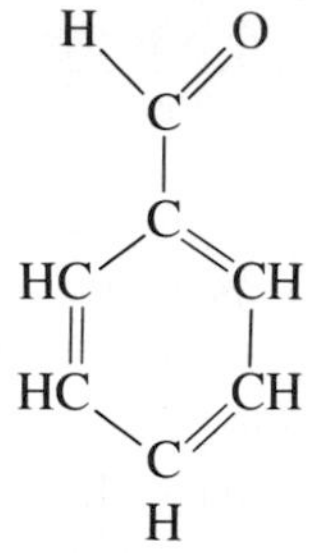

C_6H_5CHO
Benzaldehyde
(In fruit pits)

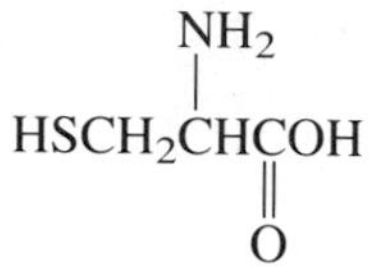

$HSCH_2CH(NH_2)C(=O)OH$
Cysteine
(In proteins)

$CH_3CH{=}CHC{\equiv}CC{\equiv}CCH{=}CHCH_2OH$
Matricarianol
(From chamomile)

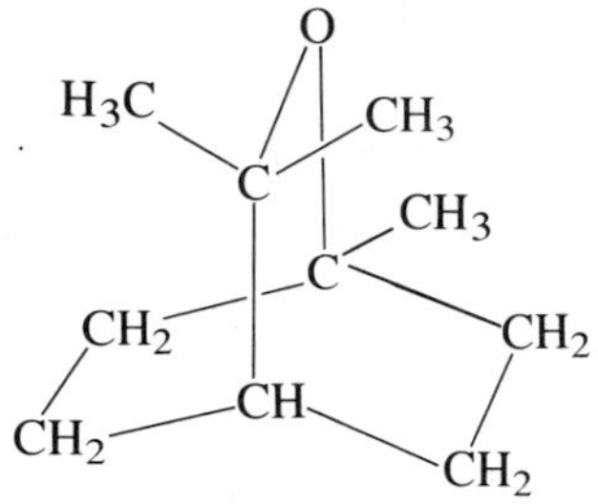

Cineole
(From eucalyptus)

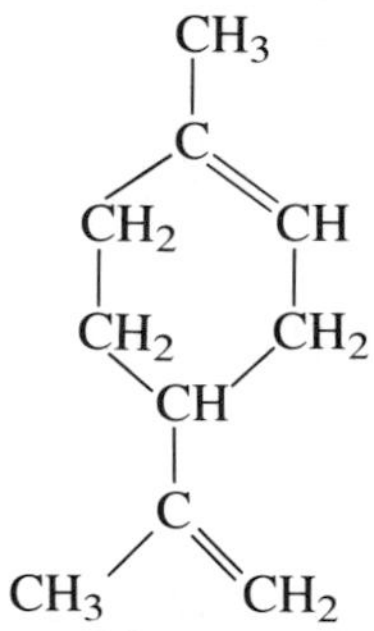

Limonene
(In lemons)

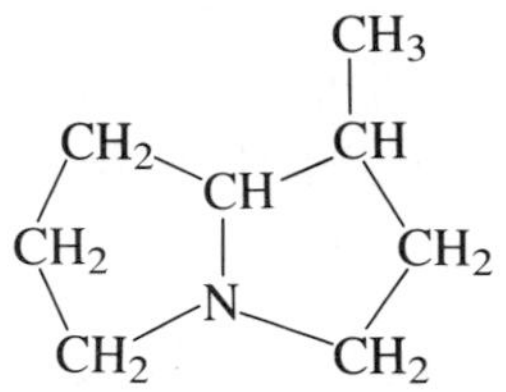

Heliotridane
(An alkaloid)

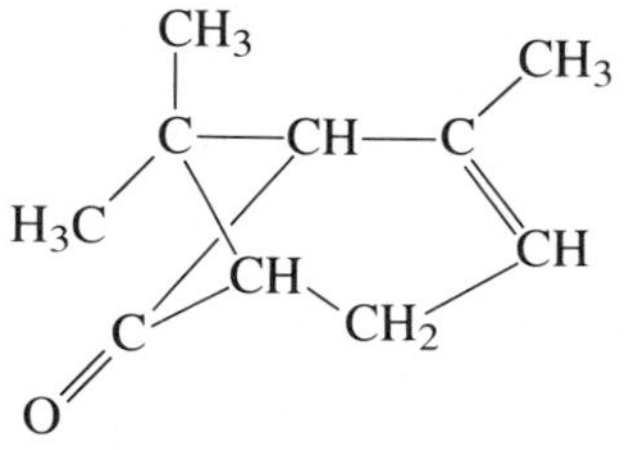

Chrysanthenone
(In chrysanthemums)

13. Give IUPAC names for all alkyl groups marked by dashed lines in each of the following biologically important compounds. Identify each group as a primary, secondary, or tertiary alkyl substituent.

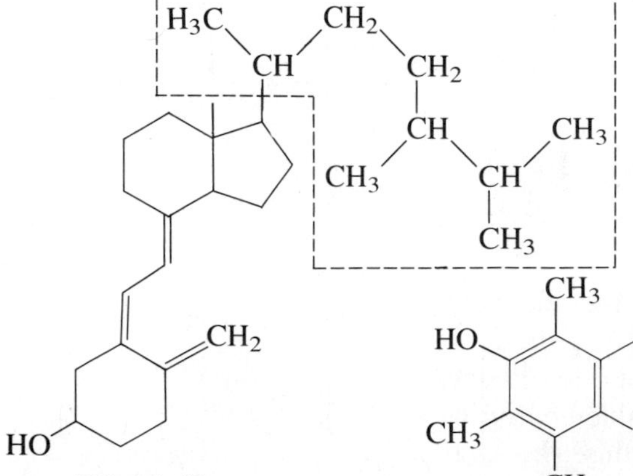

Vitamin D$_4$

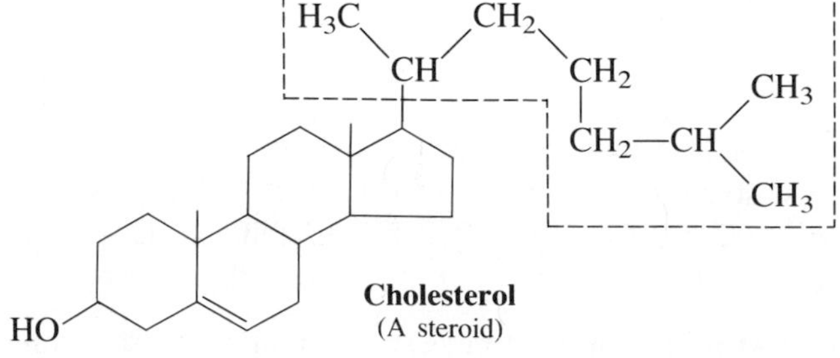

Cholesterol
(A steroid)

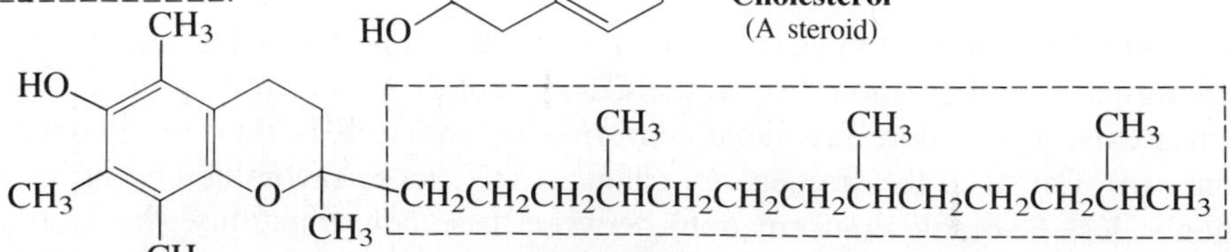

Vitamin E

$$\begin{array}{ccc} \boxed{\begin{matrix} CH_3 & & CH_3 \\ & CH & \end{matrix}} & \boxed{\begin{matrix} CH_3 & & CH_3 \\ & CH & \\ & CH_2 & \end{matrix}} & \boxed{\begin{matrix} & & CH_3 \\ CH_3 & & CH_2 \\ & CH & \end{matrix}} \\ NH_2\text{—}CH\text{—}CO_2H & NH_2\text{—}CH\text{—}CO_2H & NH_2\text{—}CH\text{—}CO_2H \end{array}$$

Valine (An amino acid) | **Leucine** (Another amino acid) | **Isoleucine** (Still another amino acid)

14. The equation relating $\Delta G°$ with K contains a temperature term. Refer to your answer to Problem 11(a) to calculate the answers to the questions that follow. You will need to know that $\Delta S°$ for the formation of the more stable conformer of 2-methylbutane from the next most stable conformer is $+1.4$ cal deg^{-1} mol^{-1}. **(a)** Calculate the enthalpy difference ($\Delta H°$) between these two conformers from the equation $\Delta G° = \Delta H° - T\Delta S°$. How well does this agree with the $\Delta H°$ calculated from the number of *gauche* interactions in each conformer? **(b)** Assuming that $\Delta H°$ and $\Delta S°$ do not change with temperature, calculate $\Delta G°$ between these two conformations at the following three temperatures: -250°C; -100°C; $+500$°C. **(c)** Calculate K for these two conformations at the same three temperatures.

15. The hydrocarbon propene ($CH_3\text{–}CH\text{=}CH_2$) can react in two different ways with bromine (Chapters 12 and 14).

$$\text{(i)}\quad CH_3\text{—}CH\text{=}CH_2 + Br_2 \longrightarrow CH_3\text{—}\underset{|}{\overset{Br}{CH}}\text{—}\overset{Br}{\underset{}{CH_2}}$$

$$\text{(ii)}\quad CH_3\text{—}CH\text{=}CH_2 + Br_2 \longrightarrow \underset{Br}{\underset{|}{CH_2}}\text{—}CH\text{=}CH_2 + HBr$$

(a) Using the bond strengths (kcal mol^{-1}) given in the margin, calculate $\Delta H°$ for each of these reactions. **(b)** $\Delta S° \approx 0$ cal deg^{-1} mol^{-1} for one of these reactions and -35 cal deg^{-1} mol^{-1} for the other. Which reaction has which $\Delta S°$? Briefly explain your answer. **(c)** Calculate $\Delta G°$ for each reaction at 25°C and at 600°C. Are both of these reactions thermodynamically favorable at 25°C? At 600°C?

Bond	Average strength
C—C	83
C=C	146
C—H	99
Br—Br	46
H—Br	87
C—Br	68

16. Using the Arrhenius equation, calculate the effect on k of 10°, 30°, and 50°C increases in temperature for the following activation energies. Use 300 K (approximately room temperature) as your initial T value, and assume A is a constant. **(a)** $E_a = 15$ kcal mol^{-1}; **(b)** $E_a = 30$ kcal mol^{-1}; **(c)** $E_a = 45$ kcal mol^{-1}.

17. The Arrhenius equation can be reformulated in a way that permits the experimental determination of activation energies. For this purpose, we take the natural logarithm of both sides and convert into the base 10 logarithm.

$$\ln k = \ln (Ae^{-E_a/RT}) = \ln A - E_a/RT \quad \text{becomes} \quad \log k = \log A - \frac{E_a}{2.3\ RT}$$

The rate constant k is measured at several temperatures T and a plot of $\log k$ versus $1/T$ prepared, a straight line. What is the slope of this line? What is its intercept (i.e., the value of $\log k$ at $1/T = 0$)? How is E_a calculated?

3

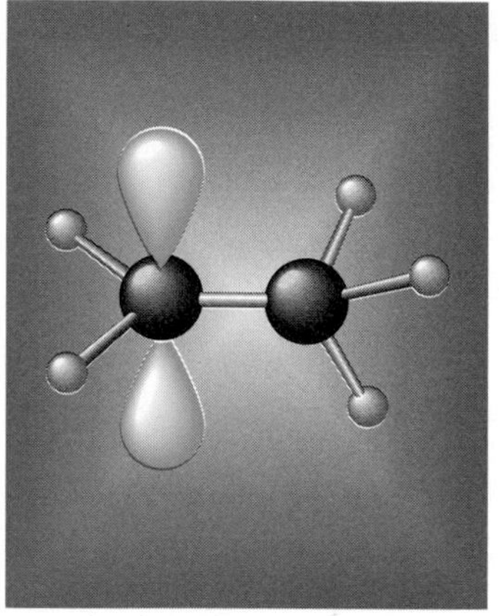

Reactions of Alkanes

Bond-Dissociation Energies, Radical Halogenation, and Relative Reactivity

As stated in Chapter 2, alkanes are organic chemicals that lack functional groups. To turn alkanes into compounds useful for synthesis, we must learn to introduce such groups, a process called **functionalization.**

Many liquid and solid alkanes are obtained cheaply from petroleum (Section 3-3). Thus nature has given us large quantities of hydrocarbons that can be used as ''chemical feedstocks'' or starting materials for the synthesis of other organic molecules. Alkanes are also produced naturally, by the slow decomposition of animal and vegetable matter in the presence of water but the absence of oxygen, a process lasting millions of years. The smaller alkanes—methane, ethane, propane, and butane—are present in natural gas, methane being by far its major component. In the United States, natural gas is a major source of energy, with annual production in the hundreds of millions of tons.

The burning of these fuels and the functionalization of the alkanes rely on the same essential step: In both cases bonds must be broken. This process is called **bond dissociation** and requires energy. The chapter begins by explaining what happens when alkanes are heated to high temperatures. Next, an important functionalization reaction of the alkanes is introduced: **halogenation,** in which a hydrogen atom is replaced by halogen. For each of these processes, we shall use a discussion of the mechanism involved to explain the conditions under which each occurs. Finally, a discussion of alkane combustion leads to a description of the methods used to establish the heat contents and relative stabilities of molecules.

3-1 Strength of Alkane Bonds: Radicals

Chapter 1 explained how bonds are formed and that energy is released on bond formation. For example, bringing two hydrogen atoms into bonding distance produces 104 kcal mol^{-1} of heat (refer to Figures 1-1 and 1-12).

$$H\cdot + H\cdot \longrightarrow H{-}H \qquad \Delta H° = -104 \text{ kcal mol}^{-1}$$

Consequently, breaking such a bond *requires* heat, in fact the same amount of heat that was released when the bond was made. This energy is called **bond-dissociation energy, *DH*°**, or **bond strength.**

$$H{-}H \longrightarrow H\cdot + H\cdot \qquad \Delta H° = DH° = 104 \text{ kcal mol}^{-1}$$

Radicals are formed by homolytic cleavage

In our example the bond breaks in such a way that the two bonding electrons divide equally between the two participating atoms or fragments. This process is called **homolytic cleavage** or **bond homolysis.**

Homolytic Cleavage

$$A{-}B \longrightarrow A\cdot + \cdot B$$

Radicals

The fragments that form have an unpaired electron, for example, $H\cdot$, $Cl\cdot$, $CH_3\cdot$, and $CH_3CH_2\cdot$. When these species are composed of more than one atom, they are called **radicals.** Because of the unpaired electron, radicals and free atoms are very reactive and usually cannot be isolated. However, radicals and free atoms are present in low concentration as unobserved *intermediates* in many reactions, such as the oxidation of fats that leads to the spoilage of perishable foods (Chapter 22).

In an alternative way of breaking a bond, the entire bonding electron pair is donated to one of the atoms. This process is **heterolytic cleavage** and results in the formation of **ions.**

Heterolytic Cleavage

$$A{-}B \longrightarrow A^+ + :B^-$$

Ions

$:\ddot{\underset{..}{Cl}}\cdot$

Chlorine atom

Methyl radical: H at top, H—C—H, with an unpaired electron (·) on carbon

Methyl radical

Ethyl radical: H_3C—C· with H above and H below

Ethyl radical

TABLE 3-1 Bond-Dissociation Energies of Various A–B Bonds ($DH°$ in kcal mol^{-1})

	B in A–B						
A in A–B	**–H**	**–F**	**–Cl**	**–Br**	**–I**	**–OH**	**–NH_2**
H—	104	135	103	87	71	119	107
CH_3—	105	110	85	71	57	93	80
CH_3CH_2—	98	107	80	68	53	92	77
$CH_3CH_2CH_2$—	98	107	81	68	53	91	78
$(CH_3)_2CH$—	94.5	106	81	68	53	92	93
$(CH_3)_3C$—	93	110	81	67	52	93	93

Note: These numbers are being revised continually because of improved methods for their measurement. Some of the values given here may be in (small) error.

Dissociation energies, $DH°$, refer only to homolytic cleavages. They have characteristic values for the various bonds that can be formed between the elements. Table 3-1 lists dissociation energies of some common bonds. Note the relatively strong bonds to hydrogen, as in H–F and H–OH. Even though these bonds have high $DH°$ values, they readily undergo *heterolytic* cleavage in water to H^+ and F^- or HO^-; *do not confuse homolytic with heterolytic processes.*

Bonds are strongest when made by overlapping orbitals that are closely matched in energy and size. For example, the strength of the bonds between hydrogen and the halogens decreases in the order F > Cl > Br > I, because the *p* orbital of the halogen contributing to the bonding becomes larger and more diffuse along the series. Thus the efficiency of its overlap with the relatively small 1*s* orbital on hydrogen diminishes. A similar trend holds for bonding between the halogens and carbon.

EXERCISE 3-1

Compare the bond-dissociation energies of CH_3–F, CH_3–OH, and CH_3–NH_2. Why do the bonds get weaker along this series even though the orbitals participating in overlap become better matched in size and energy? (Hint: Consider Figure 1-2 and Table 1-2 for a simple explanation.)

The stability of radicals determines the C–H bond strengths

How strong are the C–H and C–C bonds in alkanes? The bond-dissociation energies of various alkane bonds are given in Table 3-2. Note that bond energies generally decrease with the progression from methane to primary, secondary, and tertiary carbon. For example, the C–H bond in methane has a high $DH°$ value of 105 kcal mol^{-1}. In ethane, this bond energy is less: $DH° = 98$ kcal mol^{-1}. This latter number is typical for primary C–H bonds, as can be seen for the bond in propane. The secondary C–H bond is even weaker, with a $DH°$ of 94.5 kcal mol^{-1}, and a tertiary carbon atom binds to hydrogen with only 93 kcal mol^{-1}.

TABLE 3-2 Bond-Dissociation Energies for Some Alkanes

Compound	$DH°$ (kcal mol^{-1})	Compound	$DH°$ (kcal mol^{-1})
CH_3–H	105	CH_3–CH_3	90
C_2H_5–H	98	C_2H_5–CH_3	86
C_3H_7–H	98	C_3H_7–CH_3	87
$(CH_3)_2CHCH_2$–H	98	C_2H_5–C_2H_5	82
$(CH_3)_2CH$–H	94.5	$(CH_3)_2CH$–CH_3	86
$(CH_3)_3C$–H	93	$(CH_3)_3C$–CH_3	84
		$(CH_3)_3C$–$C(CH_3)_3$	72

Note: See footnote for Table 3-1.

C–H bond is weaker and easier to break ↓			Radical formed is more stable ↓
	$CH_4 \longrightarrow CH_3\cdot + H\cdot$	$DH° = 105$ kcal mol^{-1}	
	$R{-}H \longrightarrow R\cdot + H\cdot$		
	R primary	$DH° = 98$ kcal mol^{-1}	
	secondary	$DH° = 94.5$ kcal mol^{-1}	
	tertiary	$DH° = 93$ kcal mol^{-1}	

A similar trend is seen for C–C bonds. The extremes are the central linkages in ethane ($DH° = 90$ kcal mol^{-1}) and 2,2,3,3-tetramethylbutane ($DH° =$ 72 kcal mol^{-1}).

Why do all of these dissociations exhibit different $DH°$ values? One explanation is that the radicals formed have different energies. Radical stability *increases* along the series from primary to secondary to tertiary; consequently, the energy required to create them *decreases* (Figure 3-1).

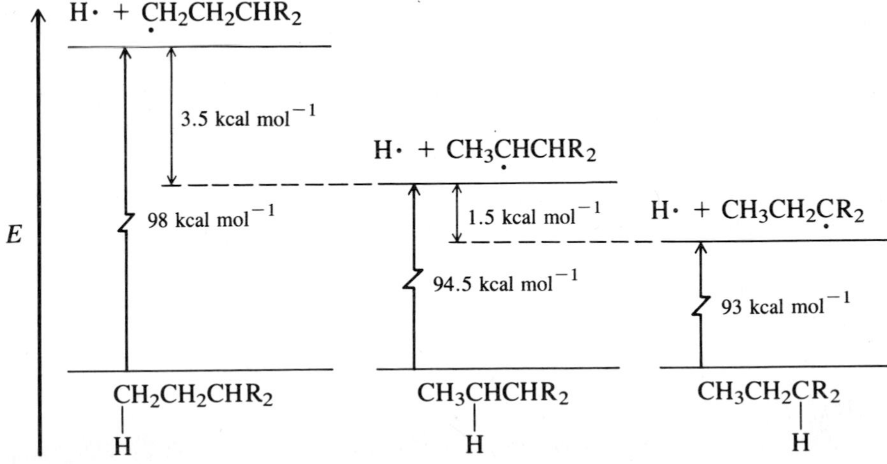

FIGURE 3-1 The different energies needed to form radicals from an alkane $CH_3CH_2CHR_2$. Radical stability increases from primary to secondary to tertiary.

Stability of Alkyl Radicals

$CH_3\cdot$ < primary < secondary < tertiary

EXERCISE 3-2

Which C–C bond would break first, the bond in ethane or that in 2,2-dimethylpropane?

In summary, bond homolysis in alkanes yields radicals and free atoms. The heat required to do so is called the bond-dissociation energy, $DH°$. Its value is characteristic only for the bonds between the two elements involved. Bond breaking that results in tertiary radicals demands less energy than that furnishing secondary radicals; the latter are in turn formed more readily than primary radicals. The methyl radical is the most difficult to obtain in this way.

3-2 Structure of Alkyl Radicals: Hyperconjugation

What is the reason for the ordering in stability of alkyl radicals? To answer this question, we need to inspect their structure more closely. Radicals are stabilized by electron delocalization involving overlap of the *p* orbital at the radical center with a neighboring C–H σ bond.

Consider the structure of the methyl radical, formed by removal of a hydrogen atom from methane. Its bonding could be described as involving an sp^3-hybridized carbon with three sp^3 C–H bonds, the odd electron occupying the fourth sp^3 molecular orbital. Spectral measurements, however, have shown that the methyl radical, and probably other alkyl radicals, adopt a *nearly planar* configuration, more accurately described by sp^2 hybridization (Figure 3-2). The unpaired electron occupies the remaining *p* orbital perpendicular to the molecular plane.

Let us see how the planar structures of alkyl radicals help explain their relative stabilities. As we move from methyl to primary to secondary to tertiary, additional alkyl groups replace hydrogen atoms. For example, in the ethyl radical, a methyl group substitutes for one of the hydrogen atoms in the methyl radical. Figure 3-3A shows that there is a conformer in which a C–H bond in the methyl substituent is aligned with and overlaps one of the lobes of the singly occupied *p* orbital. This arrangement allows the bonding pair of electrons to delocalize into the partly empty *p* lobe, a phenomenon called **hyperconjugation.** The interac-

FIGURE 3-2 The hybridization change upon formation of a methyl radical from methane. The nearly planar arrangement is reminiscent of the hybridization in BH_3 (Figure 1-17).

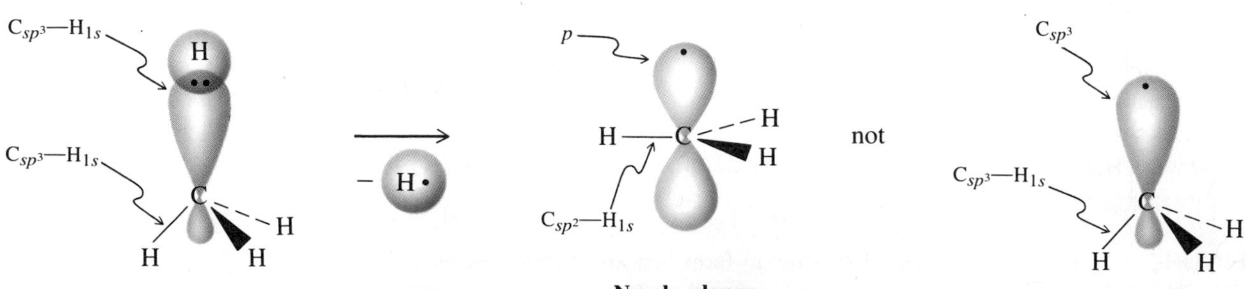

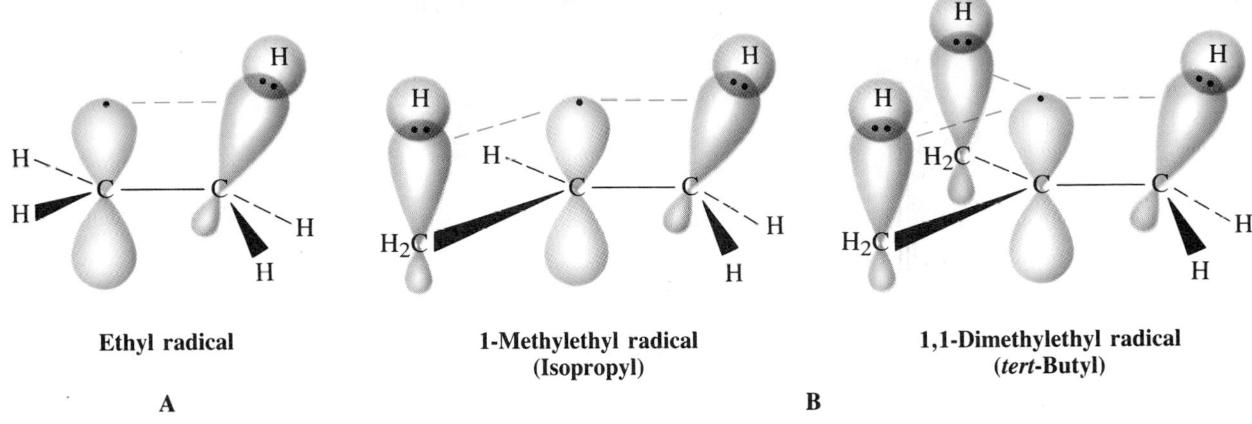

FIGURE 3-3 Hyperconjugation (green dashed lines) resulting from the donation of electrons in filled sp^3 hybrids to the partly filled p orbital in (A) ethyl, (B) 1-methylethyl and 1,1-dimethylethyl radicals. The resulting delocalization of electron density has a net stabilizing effect.

tion between a filled orbital and a singly occupied orbital has a net stabilizing effect (recall Exercise 1-12).

As further hydrogen atoms on the radical carbon are replaced successively by alkyl groups, the number of hyperconjugation interactions increases (Figure 3-3B). The order of stability of the radicals is a consequence of this increased hyperconjugation. Another contribution to the relative stability of secondary and tertiary radicals is the greater relief of steric crowding between the substituent groups as the geometry changes from tetrahedral in the alkane to planar in the radical.

3-3 Conversion of Petroleum: Pyrolysis

A knowledge of bond-dissociation energies helps us understand the high-temperature reactivity of hydrocarbons. Consider, for example, the conversion of crude petroleum into gasoline and other volatile materials. Distillation alone does not meet the demand for these desired lower molecular weight hydrocarbons. Additional heating is required to break up longer carbon chains to smaller fragments. How does this occur? Let us look first at the behavior of simple alkanes under these conditions and then move on to petroleum.

High temperatures cause bond homolysis

When alkanes are heated to a high temperature, both C–H bonds and C–C bonds rupture, a process called **pyrolysis.** In the absence of oxygen, the resulting radicals can combine to form new higher or lower alkanes. They can also remove hydrogen atoms from the carbon atom adjacent to another radical center to give alkenes, a process called *hydrogen abstraction*. Indeed, very complicated mixtures of alkanes and alkenes form in pyrolyses. Under special conditions, however, these transformations can be controlled to obtain a large proportion of hydrocarbons of a defined chain length.

Pyrolysis of Hexane

Examples of cleavage to radicals

$$\underset{\textbf{Hexane}}{\overset{1\ \ 2\ \ 3\ \ 4}{CH_3CH_2CH_2CH_2CH_2CH_3}} \begin{cases} \xrightarrow{\text{C1, C2 cleavage}} CH_3\cdot + \cdot CH_2CH_2CH_2CH_2CH_3 \\ \xrightarrow{\text{C2, C3 cleavage}} CH_3CH_2\cdot + \cdot CH_2CH_2CH_2CH_3 \\ \xrightarrow{\text{C3, C4 cleavage}} CH_3CH_2CH_2\cdot + \cdot CH_2CH_2CH_3 \end{cases}$$

Examples of radical combination reactions

$$CH_3\cdot + \cdot CH_2CH_3 \longrightarrow \underset{\textbf{Propane}}{CH_3CH_2CH_3}$$

$$CH_3CH_2CH_2CH_2CH_2\cdot + \cdot CH_2CH_2CH_3 \longrightarrow \underset{\textbf{Octane}}{CH_3CH_2CH_2CH_2CH_2CH_2CH_2CH_3}$$

Examples of hydrogen abstraction reactions

$$CH_3CH_2\cdot + CH_3\overset{H}{\overset{|}{C}}H{-}CH_2\cdot \longrightarrow \underset{\textbf{Ethane}}{CH_3\overset{H}{\overset{|}{C}}H_2} + \underset{\textbf{Propene}}{CH_3CH{=}CH_2}$$

$$\overset{H}{\overset{|}{C}}H_2CH_2\cdot + CH_3CH_2CH_2\cdot \longrightarrow \underset{\textbf{Ethene}}{CH_2{=}CH_2} + \underset{\textbf{Propane}}{CH_3CH_2\overset{H}{\overset{|}{C}}H_2}$$

BOX 3-1 The Function of a Catalyst

What is the function of the zeolite catalyst? As shown in the illustration, a *catalyst* is a substance that speeds up a reaction; that is, it increases the rate at which equilibrium is established. It does so by allowing reactants and products to be interconverted by a new pathway that has a lower activation energy (E_{cat}) than that of the reaction in the absence (E_a) of the catalyst. Apart from zeolites and other mineral-derived surfaces, many metals act as catalysts. In nature, *enzymes* usually fulfill this function (Chapter 26). The presence of catalysts allows many transformations to take place at lower temperatures and under generally milder conditions.

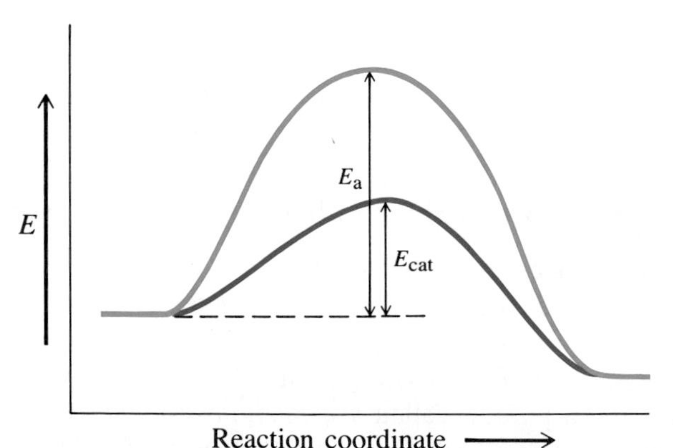

The attainment of such control frequently involves the use of special catalysts, such as crystalline sodium aluminosilicates, also called zeolites. For example, zeolite-catalyzed pyrolysis of dodecane yields a mixture in which hydrocarbons containing three to six carbons predominate.

$$\text{Dodecane} \xrightarrow{\text{Zeolite, 482°C, 2 min}} \underset{17\%}{C_3} + \underset{31\%}{C_4} + \underset{23\%}{C_5} + \underset{18\%}{C_6} + \underset{11\%}{\text{other products}}$$

Petroleum is an important source of alkanes

Breaking an alkane down into smaller fragments is also known as **cracking.** Such processes are important in the oil-refining industry for the production of gasoline and other liquid fuels from petroleum.

As mentioned in the introduction to this chapter, petroleum is believed to be the product of microbial degradation of living organic matter that existed several hundred million years ago. Crude oil, a dark viscous liquid, is primarily a mixture of several hundred different hydrocarbons, particularly straight-chain alkanes, some branched alkanes, and varying quantities of aromatic hydrocarbons. Distillation yields several fractions with a typical product distribution, as shown in Table 3-3. The composition of petroleum varies widely, depending on the origin of the oil.

To increase the quantity of the much-needed gasoline fraction, the higher boiling oils are cracked by pyrolysis. Originally (in the 1920s), this process required high temperatures (800–1000°C), but modern cracking processes use catalysts, such as zeolites, at relatively low temperatures (500°C). Cracking the residual oil from crude petroleum distillation gives approximately 30% gas, 50% gasoline, and 20% higher molecular weight oils and a residue called coke.

Another process converts alkanes into aromatic hydrocarbons with approximately the same number of carbon atoms. The aromatics are highly efficient fuels and are used as feedstocks for the chemical industry. Because the process reforms

TABLE 3-3 Product Distribution in a Typical Distillation of Crude Petroleum

Amount (% of volume)	Boiling point (°C)	Carbon atoms	Products
1–2	<30	C_1–C_4	Natural gas, methane, propane, butane, liquefied petroleum gas (LPG)
15–30	30–200	C_4–C_{12}	Petroleum ether ($C_{5,6}$), ligroin (C_7), naphtha, straight-run gasoline[a]
5–20	200–300	C_{12}–C_{15}	Kerosene, heater oil
10–40	300–400	C_{15}–C_{25}	Gas oil, diesel fuel, lubricating oil, waxes, asphalt
8–69	>400 (Nonvolatiles)	$>C_{25}$	Residual oil, paraffin waxes, asphalt (tar)

a. This refers to gasoline straight from petroleum, without having been treated in any way.

BOX 3-2 Petroleum and Gasoline: Our Main Energy Sources

Oil and natural gas supply most of the U.S. energy requirement. Other industrialized nations have similar dependence on petroleum as a source of energy. U.S. yearly production of natural gas approximates 500 million liters. In 1990, U.S. energy sources apart from gas (24%) and oil (41%) were coal (23%), nuclear power (8%), and hydroelectric power (4%). Domestic annual oil production peaked in 1971 at 4.2 billion barrels. In 1990, it had dropped below 3 billion barrels, with nearly 2 billion barrels imported. (One barrel equals 42 gallons, about 158 liters.)

The dependence of many countries on imported oil has had important economic and political consequences, as demonstrated by the wild swings in oil prices that accompanied the Iraqi invasion of Kuwait in 1990 and the subsequent Persian Gulf War. Renewed efforts are being undertaken to decrease economic dependence on imported oil and to develop new energy sources to satisfy demand when oil and gas reserves are depleted.

$CH_3CH_2CH_2CH_2CH_2CH_2CH_3$
Heptane

$\xrightarrow{\text{Pt-SiO}_2\text{-Al}_2\text{O}_3,\ 500°\text{C},\ 20\text{ atm H}_2}$

CH_3-substituted benzene ring (H, H, H, H, H) $+ 4\ H_2$

Methylbenzene (Toluene)

a new hydrocarbon from an old one, it is referred to as **reforming,** and the product as **reformate.** An example of reforming is the conversion of heptane into methylbenzene (toluene). Hundreds of millions of liters of reformate gasoline are produced in the United States alone.

To summarize, the pyrolysis of alkanes often leads to complex mixtures of hydrocarbons, control being attained in the presence of special catalysts. Are there other reactions of alkanes? If so, can they be carried out with even greater control, and, moreover, introduce functional groups into the chain as well? The sections that follow will answer these questions.

3-4 Chlorination of Methane: The Radical Chain Mechanism

We have seen that alkanes undergo chemical transformations when subjected to pyrolysis. They also undergo other reactions. This section deals with the effect of exposing an alkane, methane, to a halogen, chlorine. A **chlorination reaction** takes place to produce chloromethane and hydrogen chloride. We shall analyze each step in this transformation to establish the *mechanism* of the reaction.

Chlorine converts methane into chloromethane

When methane and chlorine gas are mixed in the dark at room temperature, there is no reaction. The mixture must be heated to temperatures above 300°C *or* irradiated with ultraviolet light before reaction occurs. One of the two initial products is chloromethane, derived from methane in which a hydrogen atom is removed and replaced by chlorine. The other product of this transformation is hydrogen chloride. Further substitution leads to dichloromethane (methylene chloride), CH_2Cl_2; trichloromethane (chloroform), $CHCl_3$; and tetrachloromethane (carbon tetrachloride), CCl_4.

Why should this reaction proceed? A clue may be obtained from a consideration of its $\Delta H°$. Note that a C–H bond in methane ($DH° = 105$ kcal mol^{-1}) and a Cl–Cl bond ($DH° = 58$ kcal mol^{-1}) are broken, whereas the C–Cl bond of chloromethane ($DH° = 85$ kcal mol^{-1}) and an H–Cl linkage ($DH° = 103$ kcal mol^{-1}) are formed. The net result is a release of 25 kcal mol^{-1} in forming stronger bonds: The reaction is substantially *exothermic*.

$$\underset{105}{CH_3\text{—}H} + \underset{58}{:\ddot{C}l\text{—}\ddot{C}l:} \xrightarrow{\Delta \text{ or } h\nu} \underset{85}{CH_3\text{—}\ddot{C}l:} + \underset{103}{H\text{—}\ddot{C}l:}$$

$DH°$ (kcal mol^{-1}) **Chloromethane**

$$\begin{aligned}\Delta H° &= \text{energy input} - \text{energy output}\\ &= \Sigma DH° \text{ (bonds broken)} - \Sigma DH° \text{ (bonds formed)}\\ &= (105 + 58) - (85 + 103)\\ &= -25 \text{ kcal mol}^{-1}\end{aligned}$$

Why then does the thermal chlorination of methane not occur at room temperature? The fact that a reaction is exothermic does not necessarily mean that it should proceed rapidly and spontaneously. Remember (Section 2-8) that the rate of a chemical transformation is dependent on its activation energy, in this case evidently high. Why is this so? What is the function of irradiation when the reaction does proceed at room temperature? Answering these questions requires an investigation of the mechanism of the reaction.

The mechanism explains the experimental conditions required for reaction

A **mechanism** is a detailed, step-by-step description of all the changes in bonding that occur in a chemical reaction (Section 1-1). Even simple reactions may consist of several separate steps. The mechanism shows the sequence in which bonds are broken and formed, as well as the energy changes associated with each step. This information is of great value in both analyzing possible transformations of complex molecules and understanding the experimental conditions required for reactions to occur.

The mechanism for the chlorination of methane consists of three stages: initiation, propagation, and termination. Let us look at these stages in more detail.

The chlorination of methane can be studied step by step

The mechanism of the chlorination of methane includes the formation of free atoms and radicals as intermediates. In the first step of the reaction sequence, the weakest bond in the mixture, the Cl–Cl link, is broken homolytically. This event is caused by heating to about 300°C (Δ) or by the absorption of a photon of light ($h\nu$). In the latter case excitation of a bonding electron to an antibonding level (Section 1-7) leads to bond rupture. Whatever the energy source, the initial chlorine dissociation, also called the **initiation** step, requires a minimum of 58 kcal mol^{-1}.

Note: In this scheme and in those that follow, all radicals and free atoms are in green.

INITIATION.

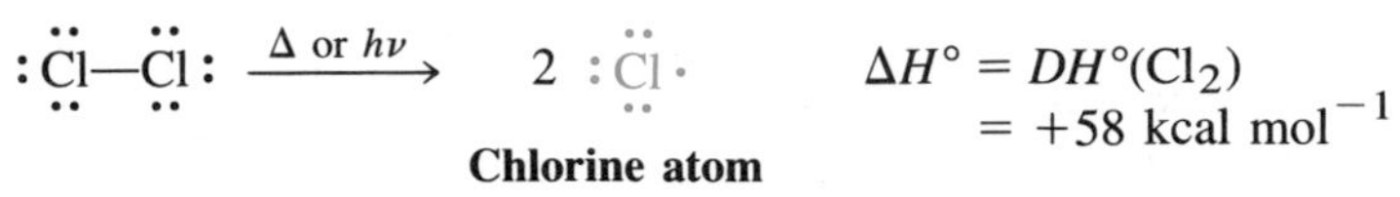

$$:\ddot{\underset{\cdot\cdot}{Cl}}-\ddot{\underset{\cdot\cdot}{Cl}}: \xrightarrow{\Delta \text{ or } h\nu} 2\ :\ddot{\underset{\cdot\cdot}{Cl}}\cdot \qquad \Delta H^\circ = DH^\circ(Cl_2) = +58 \text{ kcal mol}^{-1}$$

Chlorine atom

Next, in the first of two **propagation** steps, the chlorine atom attacks methane by abstracting a hydrogen atom. The resulting products are hydrogen chloride and a methyl radical.

PROPAGATION STEP 1.

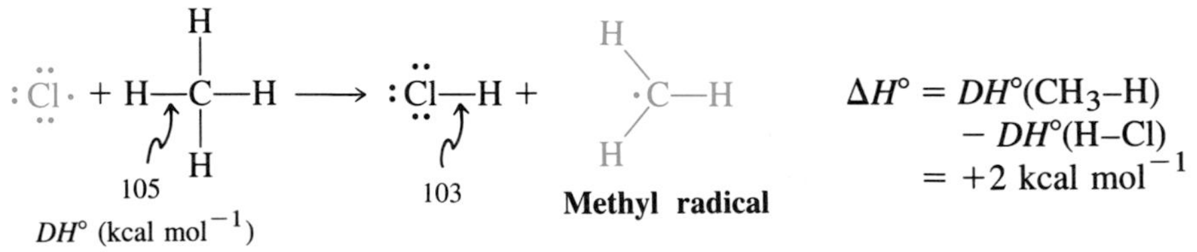

$$:\ddot{\underset{\cdot\cdot}{Cl}}\cdot + H-CH_3 \longrightarrow :\ddot{\underset{\cdot\cdot}{Cl}}-H + \cdot CH_3 \qquad \Delta H^\circ = DH^\circ(CH_3\text{–}H) - DH^\circ(H\text{–}Cl) = +2 \text{ kcal mol}^{-1}$$

105 (C–H), 103 (Cl–H) — DH° (kcal mol^{-1})

Methyl radical

The ΔH° for this transformation is positive; its equilibrium is slightly unfavorable. What is its activation energy, E_a? Is there enough heat to overcome this barrier? In this case the answer is yes. A molecular orbital description of the transition state (Section 2-6) of hydrogen removal from methane (Figure 3-4) reveals the details of the process. The reacting hydrogen is positioned between the carbon and the chlorine, partly bound to both: H–Cl bond formation has occurred to about the same extent as C–H bond breaking. The transition state, which is labeled by the symbol ‡, is located only about 4 kcal mol^{-1} above the starting materials. A potential-energy diagram describing this step is shown in Figure 3-5.

Propagation step 1 gives one of the products of the chlorination reaction: HCl. What about the organic product, CH_3Cl? Chloromethane is formed in the *second*

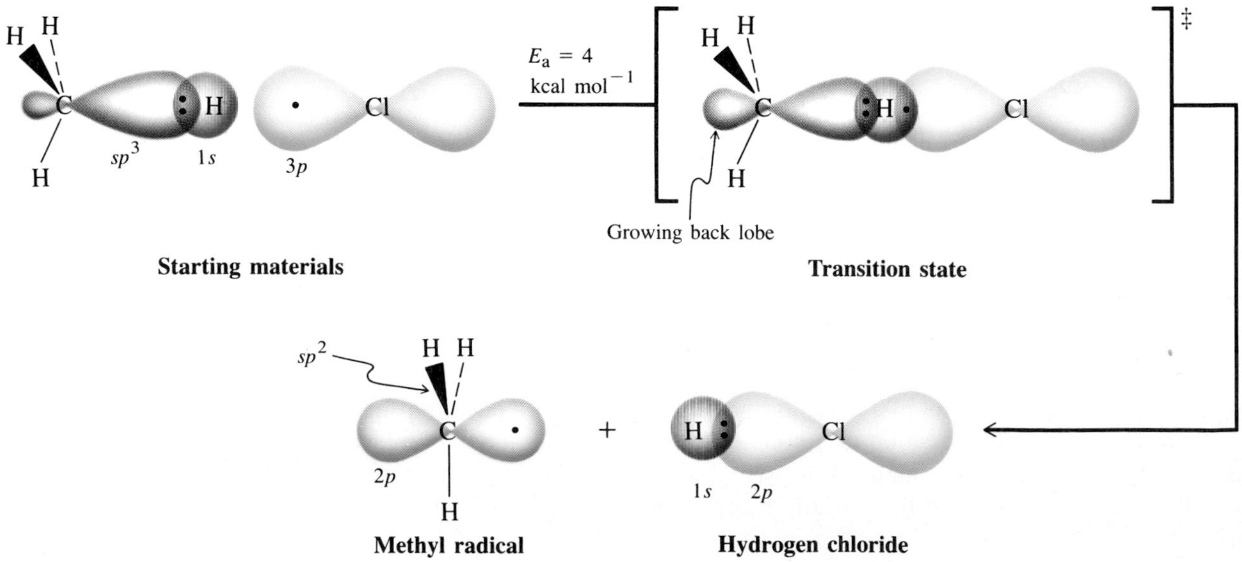

FIGURE 3-4 Approximate molecular-orbital description of the abstraction of a hydrogen atom by a chlorine atom to give a methyl radical and hydrogen chloride. Notice the rehybridization at carbon in the planar methyl radical. The additional three nonbonded electron pairs on chlorine have been omitted. The orbitals are not drawn to scale. The symbol ‡ identifies the transition state.

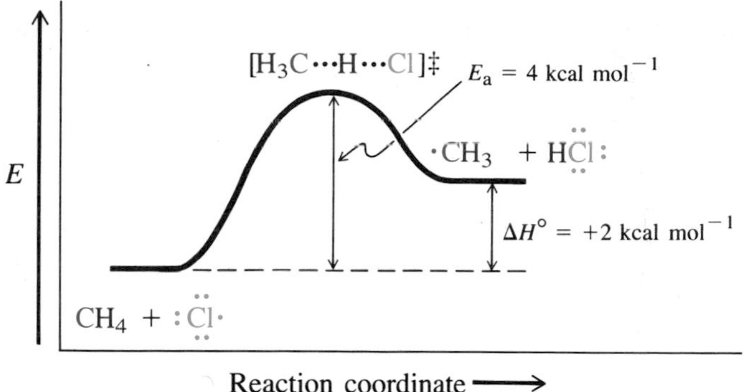

FIGURE 3-5 Potential-energy diagram of the reaction of methane with a chlorine atom. Partial bonds in the transition state are depicted by dotted lines. This process, propagation step 1 in the radical chain chlorination of methane, is slightly endothermic.

propagation step. Here the methyl radical abstracts a chlorine atom from one of the starting Cl_2 molecules, thereby furnishing chloromethane and a *new* chlorine atom. The latter reenters propagation step 1, thus closing a cycle. Note how exothermic this transformation is, -27 kcal mol^{-1}. It supplies the overall driving force for the reaction of methane with chlorine.

PROPAGATION STEP 2.

$$H_3C\cdot + :\ddot{Cl}-\ddot{Cl}: \longrightarrow H_3C-Cl + \cdot\ddot{Cl}: \qquad \Delta H^\circ = DH^\circ(Cl_2) - DH^\circ(CH_3\text{–}Cl) = -27 \text{ kcal mol}^{-1}$$

DH° (kcal mol^{-1}): 58 (Cl–Cl), 85 (C–Cl)

Because propagation step 2 is exothermic, the unfavorable equilibrium in the first propagation step is pushed toward the product side by the rapid depletion of its methyl radical product in the subsequent reaction.

$$CH_4 + Cl\cdot \underset{\text{Slightly unfavorable}}{\rightleftharpoons} CH_3\cdot + HCl \underset{\text{Very favorable; ``drives'' first equilibrium}}{\overset{Cl_2}{\rightleftharpoons}} CH_3Cl + Cl\cdot + HCl$$

The chlorine atom formed in propagation step 2 is now available to reenter step 1, thus closing a **propagation cycle.** Together, these steps result in the formation of products. The potential-energy diagram in Figure 3-6 further illustrates this point by continuing the progress of the reaction begun in Figure 3-5. Propagation step 1 has the higher activation energy and is therefore slower than step 2. The diagram also shows that the overall ΔH° of the reaction is made up of the ΔH° values of the propagation steps: $+2 - 27 = -25$ kcal mol^{-1}. You can see that this should be so by adding the equations for the two.

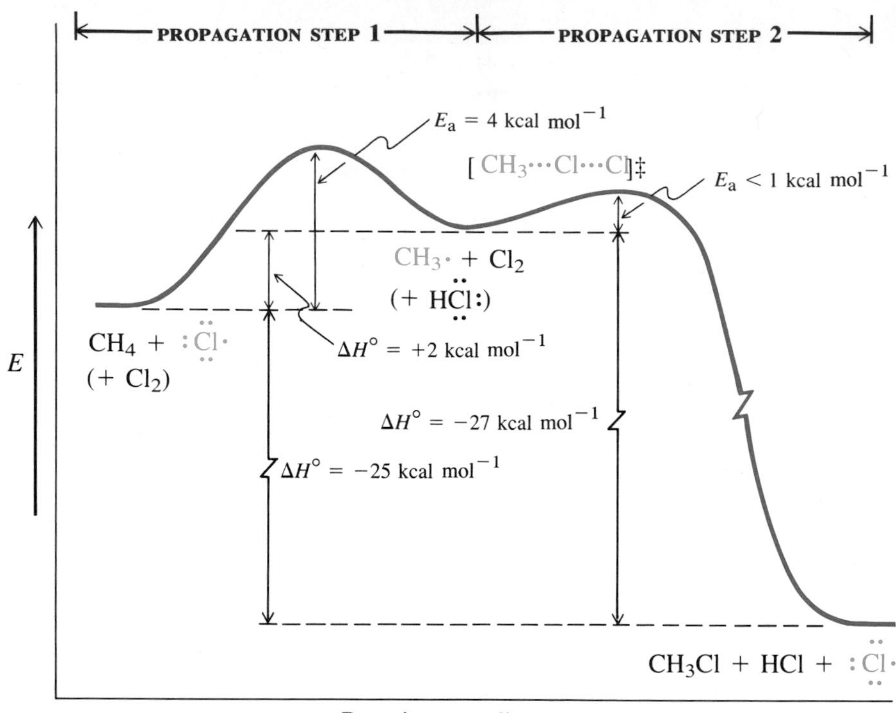

FIGURE 3-6 Complete potential-energy diagram for the formation of CH_3Cl from methane and chlorine. Propagation step 1 has the higher transition state energy and is therefore slower. The $\Delta H°$ of the overall reaction $CH_4 + Cl_2 \rightarrow CH_3Cl + HCl$ amounts to -25 kcal mol^{-1}, the sum of the $\Delta H°$ values of the two propagation steps.

	$\Delta H°$ (kcal mol^{-1})
$:\ddot{Cl}\cdot + CH_4 \longrightarrow CH_3\cdot + H\ddot{Cl}:$	+2
$CH_3\cdot + Cl_2 \longrightarrow CH_3\ddot{Cl}: + :\ddot{Cl}\cdot$	−27
$CH_4 + Cl_2 \longrightarrow CH_3\ddot{Cl}: + H\ddot{Cl}:$	−25

CHAIN TERMINATION.

$$:\ddot{Cl}\cdot + :\ddot{Cl}\cdot \longrightarrow Cl_2$$

$$:\ddot{Cl}\cdot + CH_3\cdot \longrightarrow CH_3\ddot{Cl}:$$

$$CH_3\cdot + CH_3\cdot \longrightarrow CH_3{-}CH_3$$

How is the chain terminated? The answer is, mainly by the covalent bonding of radicals and free atoms with one another. However, the concentration of radicals and free atoms in the reaction mixture is very small, and hence the chance of one radical or free atom finding another is also small. Therefore, *chain termination is relatively infrequent.*

The mechanism for the chlorination of methane is an example of a **radical chain mechanism.**

A Radical Chain Mechanism

Initiation	Propagation steps	Chain termination
$X_2 \longrightarrow 2\ :\ddot{X}\cdot$	$:\ddot{X}\cdot + RH \longrightarrow R\cdot + H\ddot{X}:$	$:\ddot{X}\cdot + :\ddot{X}\cdot \longrightarrow X_2$
	$X_2 + R\cdot \longrightarrow R\ddot{X}: + :\ddot{X}\cdot$	$R\cdot + :\ddot{X}\cdot \longrightarrow RX$
		$R\cdot + R\cdot \longrightarrow R_2$

Only a few halogen atoms are necessary for initiating the reaction, because the propagation steps are self-sufficient in $:\ddot{X}\cdot$. The first propagation step consumes a halogen atom, the second produces one. The newly generated halogen atom then reenters the propagation cycle in the first propagation step. In this way, a *radical chain* is set in motion that can drive the reaction for many thousands of cycles.

EXERCISE 3-3

Chlorination of ethane furnishes chloroethane. Write a mechanism for this transformation and calculate $\Delta H°$ for each step (see Tables 3-1 and 3-2).

One of the practical problems of chlorinating methane is the control of product selectivity. As mentioned earlier, the reaction does not stop at the formation of chloromethane but continues to form di-, tri-, and tetrachloromethane by further substitution. A practical solution to this problem is the use of a large excess of methane in the reaction. Under such conditions, the reactive intermediate chlorine atom is at any given moment surrounded by many more methane molecules than product CH_3Cl. Thus, the chance of $Cl\cdot$ finding CH_3Cl to eventually make CH_2Cl_2 is greatly diminished, and product selectivity is achieved.

In summary, chlorine transforms methane into chloromethane. The reaction proceeds through a mechanism in which heat or light causes a small number of Cl_2 molecules to undergo homolysis to chlorine atoms (initiation). The latter induce and maintain a radical chain sequence consisting of two (propagation) steps: (1) hydrogen abstraction to generate the methyl radical and HCl, and (2) conversion of $CH_3\cdot$ by Cl_2 into CH_3Cl and regenerated $Cl\cdot$. The chain is terminated by various combinations of radicals and free atoms. The heats of the individual steps are calculated by comparing the strengths of the bonds that are being broken with those being formed.

3-5 Other Radical Halogenations of Methane

Fluorine and bromine, but not iodine, also react with methane by similar radical mechanisms to furnish the corresponding halomethanes. The dissociation energies of X_2 (X = F, Br, I) are lower than that of Cl_2, thus ensuring ready initiation of the radical chain.

	F_2	Cl_2	Br_2	I_2
$DH°$ (X_2) in kcal mol^{-1}	37	58	46	36

TABLE 3-4 Enthalpies of the Propagation Steps in the Halogenation of Methane (kcal mol^{-1})

Reaction	F	Cl	Br	I
$:\ddot{\underset{..}{X}}\cdot + CH_4 \longrightarrow \cdot CH_3 + H\ddot{\underset{..}{X}}:$	−30	+2	+18	+34
$\cdot CH_3 + X_2 \longrightarrow CH_3\ddot{\underset{..}{X}}: + :\ddot{\underset{..}{X}}\cdot$	−73	−27	−25	−21
$CH_4 + X_2 \longrightarrow CH_3\ddot{\underset{..}{X}}: + H\ddot{\underset{..}{X}}:$	−103	−25	−7	+13

Relative Reactivities of X· in Hydrogen Abstractions

F· > Cl· > Br· > I·

Fluorine is most reactive, iodine least reactive

Let us compare the enthalpies of the two propagation steps in the different halogenations of methane (Table 3-4). It is apparent that there are quite striking differences in the driving force for hydrogen abstraction. For fluorine, this step is exothermic by −30 kcal mol^{-1}. We have already seen that, for chlorine, the same step is slightly endothermic; for bromine, it is substantially so, and for iodine even more so. This trend has its origin in the decreasing bond strengths of the hydrogen halides in going from fluorine to iodine (see Table 3-1). The strong hydrogen–fluorine bond is the cause of the high reactivity of fluorine atoms in hydrogen abstraction reactions. Fluorine is more reactive than chlorine, chlorine is more reactive than bromine, and the least reactive halogen atom is iodine.

The contrast between fluorine and iodine is illustrated by comparing potential-energy diagrams for their respective hydrogen abstractions from methane (Figure 3-7). The highly exothermic reaction of fluorine has a negligible activation barrier. Moreover, in its transition state the fluorine atom is relatively far from the hydrogen that is being transferred, and the H–CH_3 distance is only slightly greater than that in CH_4 itself. Why should this be so? Remember (Section 2-8) that at the transition state the energy needed for (partial) bond breaking exactly equals that gained by (partial) bond making. In the present case the full H–CH_3

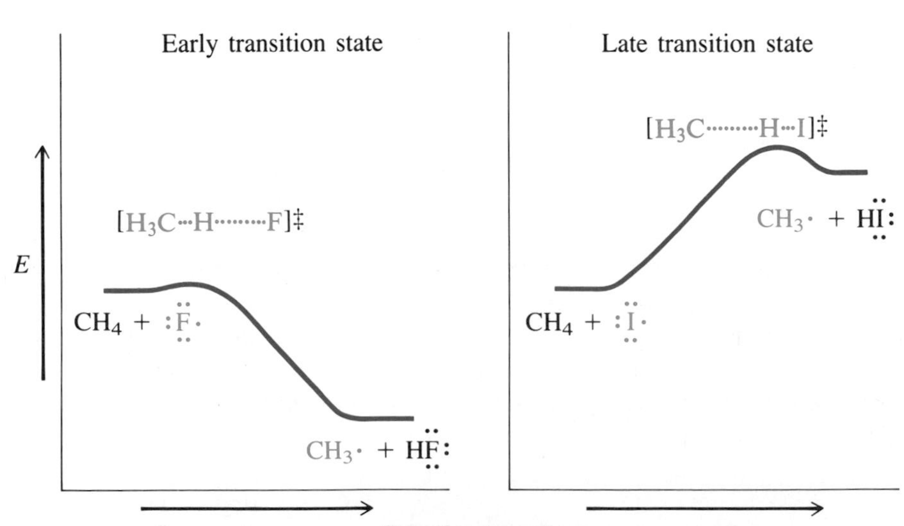

FIGURE 3-7 Potential-energy diagrams: *(left)* the reaction of a fluorine atom with CH_4, an exothermic process with an early transition state; and *(right)* the reaction of an iodine atom with CH_4, an endothermic transformation with a late transition state. Both are thus in accord with the Hammond postulate.

bond is 30 kcal mol^{-1} weaker than that of H–F (Table 3-1). Only a small shift of the H toward the F· is necessary for bonding between the two to overcome that between hydrogen and carbon. If we view the reaction coordinate as a measure of the degree of hydrogen shift from C to F, the transition state is reached *early* and is much closer in appearance to the starting materials than to the products. *Early transition states are frequently characteristic of fast, exothermic processes.*

On the other hand, reaction of I· with CH_4 has a very high E_a (at least as large as its endothermicity, +34 kcal mol^{-1}; Table 3-1). Thus, the transition state is not reached until the H–C bond is nearly completely broken and the H–I bond is almost fully formed. The transition state is said to be *late:* It is substantially further along the reaction coordinate and is much closer in structure to the products of this process, CH_3· and HI. *Late transition states are frequently typical of relatively slow, endothermic transformations.* Together these rules are known as the **Hammond* postulate.**

The second propagation step is exothermic

Let us now consider the second propagation step for each halogenation in Table 3-4. This process is exothermic for all the halogens. Again, the reaction is fastest and most exothermic for fluorine. The combined enthalpies of the two steps for the fluorination of methane result in a $\Delta H°$ of −103 kcal mol^{-1}. Indeed, this value is so large that, at sufficiently high concentrations of methane and fluorine gas, a violent reaction occurs. The formation of chloromethane is less exothermic, and that of bromomethane even less so. In the latter case, the appreciably endothermic nature of the first step ($\Delta H°$ = +18 kcal mol^{-1}) is barely overcome by the enthalpy of the second ($\Delta H°$ = −25 kcal mol^{-1}), resulting in an energy change of only −7 kcal mol^{-1} for the overall substitution. Finally, inspection of the thermodynamics of iodination reveals why iodine does not react with methane to furnish methyl iodide and hydrogen iodide. The first step costs so much energy that the second step, although exothermic, still cannot drive the reaction.

EXERCISE 3-4

Predict the product distribution of the reaction of methane with an equimolar mixture of chlorine and bromine at low conversion.

In summary, fluorine, chlorine, and bromine react with methane to give halomethanes. All three reactions follow the radical chain mechanism described for chlorination. In these processes the first propagation step is always the slower of the two. It becomes more exothermic and its activation energy decreases as we proceed from bromine to chlorine to fluorine. This trend explains the relative reactivity of the halogens, fluorine being the most reactive. Iodination of methane is endothermic and does not occur. Strongly exothermic reaction steps are often characterized by early transition states. Conversely, endothermic or relatively less exothermic steps typically have late transition states.

---

*Professor George S. Hammond, b. 1921, Georgetown University.

3-6 Chlorination of Higher Alkanes: Relative Reactivity and Selectivity

What happens in the radical halogenation of other alkanes? Will the different types of R–H bonds—namely, primary, secondary, and tertiary—react in the same way as those in methane? Let us consider the chlorination of ethane, then propane, and finally 2-methylpropane.

The monochlorination of ethane gives chloroethane as the product.

Chlorination of Ethane

$$CH_3CH_3 + Cl_2 \xrightarrow{\Delta \text{ or } h\nu} \underset{\textbf{Chloroethane}}{CH_3CH_2Cl} + HCl \qquad \Delta H° = -27 \text{ kcal mol}^{-1}$$

This reaction proceeds by a radical chain mechanism analogous to the one observed for methane. The propagation steps include formation of the ethyl radical from ethane and the chlorine atom, followed by generation of chloroethane and the release of another chlorine atom (Exercise 3-3). The greatest difference between the two mechanisms lies in the change in values for $\Delta H°$. Thus the abstraction of a hydrogen from ethane is no longer endothermic, as in methane, but favorable by -5 kcal mol^{-1}. The reason is the weaker C–H bond of ethane ($DH° = 98$ kcal mol^{-1}).

Propagation Steps in the Mechanism of the Chlorination of Ethane

$$CH_3CH_3 + :\ddot{\underset{..}{Cl}}\cdot \longrightarrow CH_3CH_2\cdot + H\ddot{\underset{..}{Cl}}: \qquad \Delta H° = -5 \text{ kcal mol}^{-1}$$

$$CH_3CH_2\cdot + Cl_2 \longrightarrow CH_3CH_2\ddot{\underset{..}{Cl}}: + :\ddot{\underset{..}{Cl}}\cdot \qquad \Delta H° = -22 \text{ kcal mol}^{-1}$$

What can be expected for the next homolog, propane?

Secondary C–H bonds are more reactive than primary ones

In propane, two kinds of hydrogen atoms are available for reaction with chlorine, six primary and two secondary. If the two types of hydrogen atoms were to react at equal rates, we could expect three times as much 1-chloropropane as 2-chloropropane. After all, statistically there are three times as many primary hydrogen atoms (6) as there are secondary ones (2).

Conversely, secondary C–H bonds are weaker than primary ones ($DH° =$ 94.5 versus 98 kcal mol^{-1}). Abstraction of a secondary hydrogen is therefore more exothermic and proceeds with a smaller activation barrier (Figure 3-8). We might thus expect secondary hydrogens to react faster, leading to more 2-chloro- than 1-chloropropane. What is actually observed? At 25°C, the experimental ratio of 1-chloropropane:2-chloropropane is found to be 43:57. This result indicates that both statistical and bond energy factors determine the product formed.

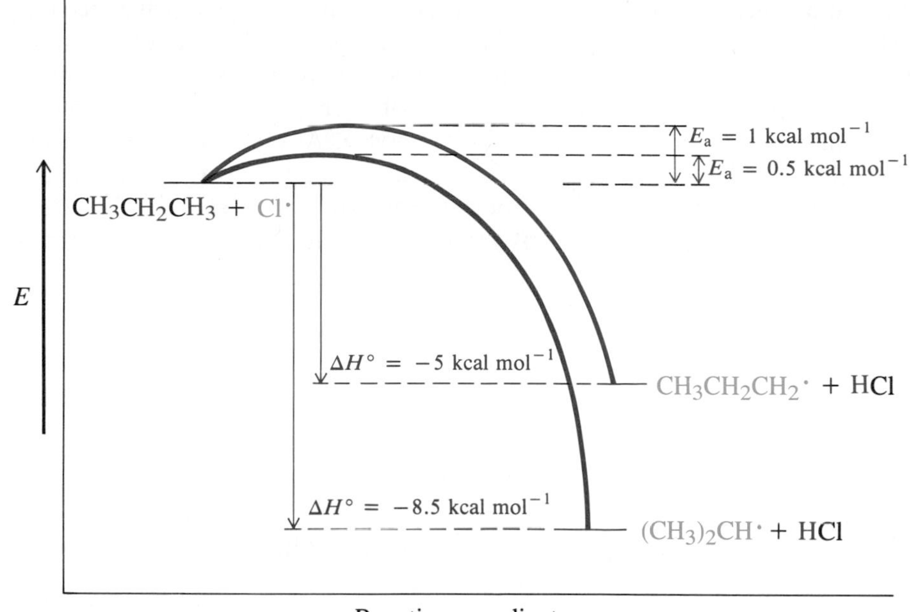

FIGURE 3-8 Hydrogen abstraction by a chlorine atom from the secondary carbon in propane is more exothermic and faster than that from the primary carbon.

Chlorination of Propane

$$Cl_2 + CH_3CH_2CH_3 \xrightarrow{h\nu} \underset{\textbf{1-Chloropropane}}{CH_3CH_2CH_2Cl} + \underset{\textbf{2-Chloropropane}}{CH_3\overset{\overset{Cl}{|}}{C}HCH_3} + HCl$$

Expected statistical ratio	3	:	1
Expected C–H bond reactivity ratio	Less	:	More
Experimental ratio (25°C)	43	:	57
Experimental ratio (600°C)	3	:	1

We can calculate the *relative reactivity of secondary and primary hydrogens* in chlorination by factoring out the statistical contribution to the product ratio.

$$\frac{\text{Relative reactivity of a secondary hydrogen}}{\text{Relative reactivity of a primary hydrogen}} = \frac{\left(\begin{array}{l}\text{yield of product from}\\ \text{secondary hydrogen abstraction}\end{array}\right) \Big/ \left(\begin{array}{l}\text{number of}\\ \text{secondary hydrogens}\end{array}\right)}{\left(\begin{array}{l}\text{yield of product from}\\ \text{primary hydrogen abstraction}\end{array}\right) \Big/ \left(\begin{array}{l}\text{number of}\\ \text{primary hydrogens}\end{array}\right)} = \frac{57/2}{43/6} \approx 4$$

In other words, each secondary hydrogen in the chlorination of propane at 25°C is about four times more reactive than each primary one. We say that chlorine exhibits a **selectivity** of 4:1 in the removal of secondary versus primary hydrogens at 25°C.

We could predict from this analysis that *all* secondary hydrogens are four times more reactive than all primary ones in *all* radical chain reactions. Is this prediction true? Unfortunately, no. Although secondary C–H bonds generally react at a faster rate than do their primary counterparts, their relative reactivity very much depends on the nature of the attacking species, X·, the strength of the resulting H–X bond, and the temperature. For example, at 600°C, the chlorination of propane results in the statistical distribution of products: roughly three times more 1-chloropropane than 2-chloropropane. At such a high temperature, virtually every collision between a chlorine atom and any hydrogen in the propane molecule takes place with sufficient energy to lead to successful reaction. Chlorination is said to be **unselective** at this temperature, and the product ratio is governed by statistical factors.

EXERCISE 3-5

What do you expect the products of monochlorination of butane to be? In what ratio will they be formed at 25°C?

Tertiary C–H bonds are more reactive than secondary ones

Let us now find the relative reactivity of a *tertiary* hydrogen in the chlorination of alkanes. For this purpose, we expose 2-methylpropane, a molecule containing one tertiary and nine primary hydrogens, to chlorination conditions at 25°C. The resulting two products, 2-chloro-2-methylpropane (*tert*-butyl chloride) and 1-chloro-2-methylpropane (isobutyl chloride), are formed in the relative yields of 36 and 64%, respectively.

Chlorination of 2-Methylpropane

$$Cl_2 + CH_3C(CH_3)_2H \xrightarrow{h\nu} ClCH_2C(CH_3)_2H + CH_3C(CH_3)_2Cl + HCl$$

1-Chloro-2-methylpropane (Isobutyl chloride) — **2-Chloro-2-methylpropane (*tert*-Butyl chloride)**

Expected statistical ratio	9	:	1
Expected C–H bond reactivity ratio	Less	:	More
Experimental ratio (25°C)	64	:	36
Experimental ratio (600°C)	80	:	20

Factoring out the statistical contribution of nine primary hydrogens gives a relative reactivity of tertiary:primary = 36/1:64/9 = 5.1:1. This selectivity, again, decreases at higher temperatures. However, we can say that, at 25°C, the relative reactivities of the various C–H bonds in chlorinations are

Tertiary:secondary:primary = 5:4:1

The result agrees well with the relative reactivity expected from consideration of bond strength: The tertiary C–H bond is weaker than the secondary, and the latter in turn is weaker than the primary.

We can verify this ordering by looking at the competition among all three types of hydrogens within a single substrate. 2-Methylbutane, which contains nine primary hydrogens, two secondary hydrogens, and one tertiary hydrogen, is an example. Because this molecule has two types of primary hydrogens, one set of six and one set of three, reaction with chlorine yields a total of four different monochlorination products.

Chlorination of 2-Methylbutane

$$Cl_2 + CH_3-\underset{\underset{\mathbf{H}\,\leftarrow\,D}{|}}{\overset{\overset{A\,\rightarrow\,CH_3}{|}}{C}}-\overset{C}{CH_2}-\overset{B}{CH_3} \xrightarrow[\text{HCl}]{h\nu}$$

$$\underset{\substack{27\% \\ \textbf{1-Chloro-2-methylbutane} \\ \text{(Chlorination at A)}}}{ClCH_2-\underset{\underset{H}{|}}{\overset{\overset{CH_3}{|}}{C}}-CH_2-CH_3} + \underset{\substack{14\% \\ \textbf{1-Chloro-3-methylbutane} \\ \text{(Chlorination at B)}}}{CH_3-\underset{\underset{H}{|}}{\overset{\overset{CH_3}{|}}{C}}-CH_2-CH_2Cl} + \underset{\substack{36\% \\ \textbf{2-Chloro-3-methylbutane} \\ \text{(Chlorination at C)}}}{CH_3-\underset{\underset{H}{|}}{\overset{\overset{CH_3}{|}}{C}}-\underset{\underset{Cl}{|}}{\overset{\overset{H}{|}}{C}}-CH_3} + \underset{\substack{23\% \\ \textbf{2-Chloro-2-methylbutane} \\ \text{(Chlorination at D)}}}{CH_3-\underset{\underset{\mathbf{Cl}}{|}}{\overset{\overset{CH_3}{|}}{C}}-CH_2-CH_3}$$

The combined yield of the two primary halide products is 41% (1-chloro-2-methylbutane plus 1-chloro-3-methylbutane), the secondary halide 2-chloro-3-methylbutane is formed in 36%, and the tertiary halide in 23%. Therefore,

$$\begin{aligned} &\text{Primary:secondary:tertiary halide} && = 41:36:23 \\ &\textit{Relative reactivity}\ \ \text{primary:secondary:tertiary} && = 41/9:36/2:23/1 \\ & && = 1:4:5 \end{aligned}$$

as expected.

EXERCISE 3-6

Give products and the ratio in which they are expected to form for the monochlorination of methylcyclohexane at 25°C.

To summarize, the relative reactivity of primary, secondary, and tertiary hydrogens follows the trend expected on the basis of their relative C–H bond strengths. Relative reactivity ratios can be calculated by factoring out statistical considerations. These ratios are temperature dependent, with greater selectivity at lower temperatures.

3-7 Selectivity in Radical Halogenation with Fluorine and Bromine

How selectively do halogens other than chlorine halogenate the alkanes? Table 3-4 and Figure 3-7 show that fluorine is the most reactive halogen: Hydrogen abstraction is highly exothermic and has negligible activation energy. Conversely, bromine is much less reactive, since the same step has a large positive $\Delta H°$ and a high activation barrier. Does this difference affect their selectivity in halogenation of alkanes?

To answer this question, consider the reactions of fluorine and bromine with 2-methylpropane. Single fluorination at 25°C furnishes two possible products, in a ratio very close to that expected statistically.

2-Fluoro-2-methylpropane	:	1-Fluoro-2-methylpropane
(*tert*-Butyl fluoride)	:	(Isobutyl fluoride)
Observed 14	:	86
Expected 1	:	9

Fluorine thus displays very little selectivity. Why? Because the transition states for the two competing processes are reached very early, their energies and structures are similar to each other, as well as similar to those of the starting material (Figure 3-9).

Fluorination of 2-Methylpropane

$$F_2 + (\ H_3)_3CH \xrightarrow{h\nu} (CH_3)_3CF + FCH_2\text{—}\overset{\displaystyle CH_3}{\underset{\displaystyle CH_3}{C}}\text{—}H + HF$$

14% **2-Fluoro-2-methylpropane (*tert*-Butyl fluoride)**

86% **1-Fluoro-2-methylpropane (Isobutyl fluoride)**

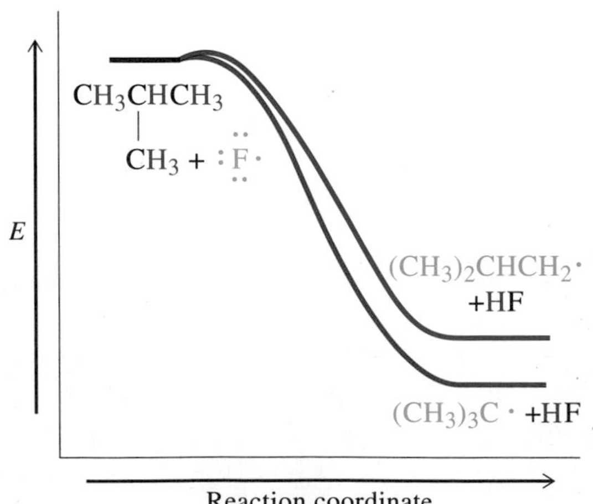

FIGURE 3-9
Potential-energy diagram for the abstraction of a primary or a tertiary hydrogen by a fluorine atom from 2-methylpropane. The energies of the respective early transition states are almost the same and barely higher than that of starting material (i.e., both E_a values are close to zero), resulting in little selectivity.

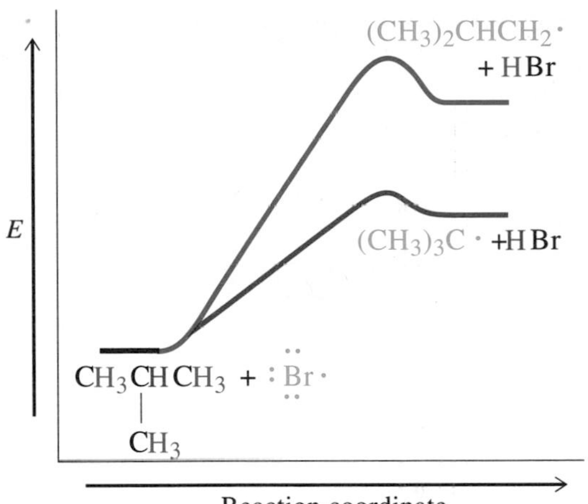

FIGURE 3-10 Potential-energy diagram for the abstraction of a primary or a tertiary hydrogen by a bromine atom from 2-methylpropane. The two late transition states are dissimilar in energy, reflecting the energy difference between the resulting primary and tertiary radicals, respectively, leading to the products with increased selectivity.

Conversely, *bromination of the same compound is highly selective,* giving the tertiary bromide almost exclusively. Hydrogen abstractions by bromine involve *late* transition states in which extensive C–H bond breaking and H–Br bond making have occurred. Thus their respective structures and energies reflect those of the corresponding radical products. As a result, the activation barriers for the reaction of bromine with primary and tertiary hydrogens, respectively, will differ by almost as much as the difference in stability between primary and tertiary radicals (Figure 3-10), a difference leading to the observed high selectivity (over 1700:1!).

Bromination of 2-Methylpropane

$$Br_2 + (CH_3)_3CH \xrightarrow{h\nu} (CH_3)_3CBr + BrCH_2{-}\overset{\displaystyle CH_3}{\underset{\displaystyle CH_3}{C}}{-}H + HBr$$

$(CH_3)_3CBr$: >99% — **2-Bromo-2-methylpropane (*tert*-Butyl bromide)**

$BrCH_2C(CH_3)_2H$: <1% — **1-Bromo-2-methylpropane (Isobutyl bromide)**

TABLE 3-5 Relative Reactivities of the Four Types of Alkane C–H Bonds in Halogenations

C–H bond	F· (25°C, gas)	Cl· (25°C, gas)	Br· (150°C, gas)
CH_3—H	0.5	0.004	0.002
RCH_2—H[a]	1	1	1
R_2CH—H	1.2	4	80
R_3C—H	1.4	5	1700

a. For each halogen, reactivities with four types of alkane C–H bonds are normalized to the reactivity of the primary C–H bond.

In summary, increased reactivity goes hand in hand with reduced selectivity in radical substitution reactions. The more reactive halogens fluorine and chlorine discriminate between the various types of C–H bonds much less than does the less reactive bromine (Table 3-5).

3-8 Synthetic Radical Halogenation

How can we devise a successful and cost-effective alkane halogenation? We must take into account selectivity, convenience, efficiency, and price.

Fluorinations are unattractive, because fluorine is relatively expensive and corrosive; and, even worse, its reactions often are violently uncontrollable. Radical iodinization, at the other extreme, fails because of unfavorable thermodynamics.

Chlorinations are important, particularly in industry, simply because chlorine is cheap. (It is prepared by electrolysis of sodium chloride, ordinary table salt.) The drawback to chlorination is low selectivity, so the process results in mixtures of isomers that are difficult to separate. To circumvent the problem, one can use as a substrate alkanes that contain a single type of hydrogen, thus giving (at least initially) a single product. Cyclopentane is one such alkane.

Chlorination of a Molecule with Only One Type of Hydrogen

$$\text{Cyclopentane} + Cl_2 \xrightarrow{h\nu} \text{Chlorocyclopentane} + HCl$$

92.7%

Cyclopentane (Large excess) **Chlorocyclopentane**

To minimize overhalogenation, chlorine is used as the limiting reagent (Section 3-4). Even then, multiple substitution can complicate the reaction. However, the more chlorinated products have higher boiling points, so they can be separated by distillation.

On an industrial scale, alkanes are chlorinated in large vessels fitted with elaborate controls to ensure smooth, safe operation. In the research laboratory, use of chlorine gas is often avoided, because it is toxic, corrosive, and difficult to measure accurately. Other chlorinating agents have been developed that can be handled more safely and accurately. These are usually liquids or solids, such as sulfuryl chloride (SO_2Cl_2) and *N*-chlorobutanimide (*N*-chlorosuccinimide, NCS).

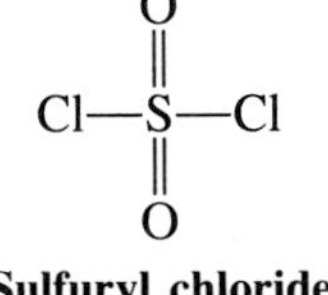

Sulfuryl chloride
(b.p. 69°C)

N-Chlorobutanimide (N-Chlorosuccinimide)
(m.p. 148°C)

EXERCISE 3-7

Which of the following compounds would give a monochlorination product with reasonable selectivity: propane, 2,2-dimethylpropane, cyclohexane, methylcyclohexane?

Because bromination is selective (and bromine is a liquid), it is frequently the method of choice for halogenating an alkane in the research laboratory. Reaction occurs at the more substituted carbon, even in statistically unfavorable situations.

Typical solvents are chlorinated methanes (CCl_4, $CHCl_3$, CH_2Cl_2), which are comparatively unreactive with bromine.

Bromine is obtained from aqueous sodium bromide solutions (found in natural brines) by treatment with chlorine. It is used less in industry because of its relatively greater cost per mole. A popular solid brominating agent is *N*-bromobutanimide (*N*-bromosuccinimide, NBS; see Section 14-2), an analog of NCS.

In summary, even though more expensive, bromine is the reagent of choice for selective radical halogenations. Chlorinations furnish product mixtures, a problem that can be minimized by choosing alkanes with only one type of hydrogen and treating them with a deficiency of chlorine. For the convenience of easier handling, research chemists often prefer solid reagents, such as *N*-bromobutanimide, or liquids, such as sulfuryl chloride.

3-9 Combustion and the Relative Stabilities of Alkanes

Let us review what we have learned in this chapter so far. We started by defining the bond strength as the energy required to effect homolysis into two component radicals or free atoms. Some typical values were then presented in Tables 3-1 and 3-2 and explained through a discussion of relative radical stabilities, a major factor being the varying extent of hyperconjugation. We then used this information to calculate the $\Delta H°$ values of the steps making up the mechanism of radical halogenation, a discussion leading to an understanding of its reactivity and selectivity aspects. It is clear that knowing bond-dissociation energies is a great aid in

BOX 3-3 Chlorination, Chloral, and DDT

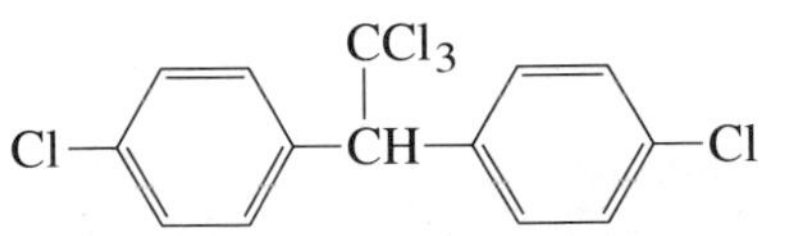

CCl_3
Cl—C₆H₄—CH—C₆H₄—Cl

1,1,1-Trichloro-2,2-bis(4-chlorophenyl)ethane (DDT)

Chlorination of ethanol in the production of trichloroacetaldehyde, CCl_3CHO, was first described in 1832. The hydrated form is commonly called *chloral* and is a powerful hypnotic with the nickname *knockout drops*. Chloral is also a key reagent in the synthesis of the powerful insecticide DDT (an abbreviation derived from the nonsystematic name *d*ichloro*d*iphenyl*t*richloroethane).

The use of DDT in the control of insect-borne diseases has saved many millions of lives over the past half-century, chiefly through the decimation of the *Anopheles* mosquito, the main carrier of the parasite that causes malaria. Although its toxicity toward mammals is low (the fatal human dose is about 500 mg kg^{-1} of body weight), DDT is very resistant to biodegradation. Its accumulation in the food chain makes it a hazard to birds and fish, and consequently it has been banned by the U.S. Environmental Protection Agency since 1972.

the thermochemical analysis of organic transformations, a facet we shall exploit on numerous occasions later on. How are these numbers found experimentally?

Chemists determine bond strengths in two complementary ways: *directly,* by measuring heats of dissociation of the weakest bonds in molecules, and *indirectly,* by recording $\Delta H°$ values of simple reactions. For example, we can readily ascertain the C–Cl bond strength in chloromethane from the heat of halogenation of methane—provided we know $DH°$ values for methane, chlorine, and HCl (Section 3-4). Both approaches have limitations: Pyrolysis often furnishes mixtures, not all radicals are easily accessible, and $\Delta H°$ measurements are meaningful only if the reactions under investigation are selective and give yields close to 100%.

A general solution to these problems is provided by establishing the relative heat contents of entire molecules, or their relative positions along the energy axis in our potential-energy diagrams. The reaction chosen for this purpose is complete oxidation (literally, "burning"), or **combustion,** a process common to almost all organic structures, in which all carbon atoms are converted to CO_2 (gas) and all of the hydrogens to H_2O (liquid).

Both products in the combustion of alkanes have an exceedingly low energy content, and hence their formation is associated with a large negative $\Delta H°$, released as heat.

$$2\ C_nH_{2n+2} + (3n + 1)O_2 \longrightarrow 2nCO_2 + (2n + 2)H_2O + \text{heat}$$

The heat released in the burning of a molecule is called its **heat of combustion,** $\Delta H°_{comb}$, many of which have been measured with high accuracy, thus allowing for comparisons of their relative energy contents (Table 3-6).

TABLE 3-6 Heats of Combustion (kcal mol^{-1} Normalized to 25°C) of Various Organic Compounds

Compound (state)	Name	$\Delta H°_{comb}$
CH_4 (gas)	Methane	−212.8
C_2H_6 (gas)	Ethane	−372.8
$CH_3CH_2CH_3$ (gas)	Propane	−530.6
$CH_3(CH_2)_2CH_3$ (gas)	Butane	−687.4
$(CH_3)_3CH$ (gas)	2-Methylpropane	−685.4
$CH_3(CH_2)_3CH_3$ (gas)	Pentane	−845.2
$CH_3(CH_2)_3CH_3$ (liquid)	Pentane	−838.8
$CH_3(CH_2)_4CH_3$ (liquid)	Hexane	−995.0
⬡ (liquid)	Cyclohexane	−936.9
CH_3CH_2OH (gas)	Ethanol	−336.4
CH_3CH_2OH (liquid)	Ethanol	−326.7
$C_{12}H_{22}O_{11}$ (solid)	Cane sugar (sucrose)	−1348.2

Note: Combustion products are CO_2 (gas) and H_2O (liquid).

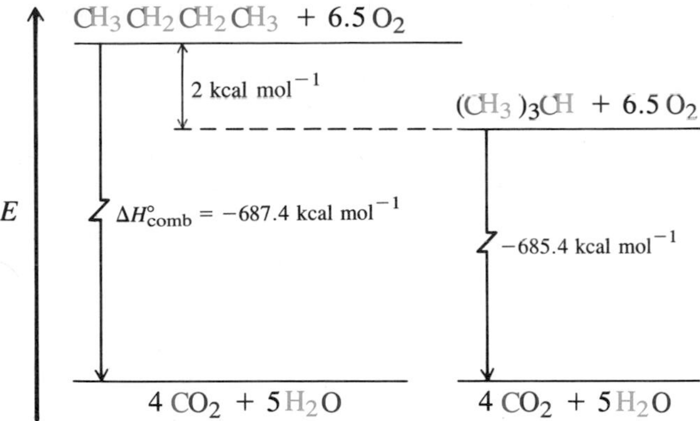

FIGURE 3-11 Butane has a higher energy content than does 2-methylpropane, as measured by the release of energy on combustion. Butane is therefore thermodynamically less stable than its isomer.

Such comparisons have to take into account the physical state of the compound undergoing combustion (gas, liquid, solid). For example, the difference between the heats of combustion of liquid and gaseous ethanol corresponds to its heat of vaporization, $\Delta H^\circ_{vap} = 9.7$ kcal mol^{-1}.

It is not surprising that the ΔH°_{comb} of alkanes increases with chain length, simply because there is more carbon and hydrogen to burn along the homologous series. Conversely, isomeric alkanes contain the same number of carbons and hydrogens, and one might expect that their respective combustions would be equally exothermic, right? Wrong!

A comparison of the heats of combustion of isomeric alkanes reveals that their values are usually *not* the same. Consider butane and 2-methylpropane. The combustion of butane has a ΔH°_{comb} of -687.4 kcal mol^{-1}, whereas its isomer releases $\Delta H^\circ_{comb} = -685.4$ kcal mol^{-1}, 2 kcal mol^{-1} less (Table 3-6). This finding shows that 2-methylpropane has a *smaller* energy content than butane, because combustion yielding the identical kind and number of products produces less energy (Figure 3-11). Butane is said to be *thermodynamically less stable* than its isomer.

EXERCISE 3-8

The hypothetical thermal conversion of butane to 2-methylpropane should have a $\Delta H^\circ = -2.0$ kcal mol^{-1}. What value do you obtain by using the bond-dissociation data in Table 3-2?

To summarize, the heats of combustion of alkanes and other organic molecules afford quantitative estimates of their energy content and, therefore, their relative stabilities.

3-10 Heats of Formation

We were able to determine the relative energy contents of the butanes by comparing their heats of combustion, but only because they have identical molecular formulas, C_4H_{10}. Where would you place propane in Figure 3-11? Its combus-

BOX 3-4 Enzymatic Oxidation of the Alkanes

Combustion is one way to activate the ordinarily quite unreactive alkane. Unfortunately, at high temperatures, oxidation is relatively unselective and destroys much of the starting molecule. A milder approach, called *enzymatic activation,* uses enzymes, the catalysts in living systems (see Chapter 26).

Enzymatic Alkane Activation

$$\underset{\substack{\textbf{Alkanes}\\ (\mathbf{C_1-C_8})}}{\text{R—H}} \xrightarrow{\text{Enzyme, } O_2} \underset{\textbf{Alcohols}}{\text{R—OH}}$$

The monooxygenases, found in mammalian tissue, catalyze the oxidation of drugs, steroids (Chapter 4), and fatty acids (Chapter 19). In microbial systems these same enzymes catalyze the oxidation of alkanes. Thus an enzyme from *Methylococcus capsulatus* inserts oxygen into a number of hydrocarbons, to give alcohols.

Environmental pollutants can promote *non*enzymatic oxidations of molecules of biological importance, such as lipids, the building blocks of fatty tissue (Chapter 20). This chemistry will be discussed in Chapter 22.

tion releases about 156 kcal mol^{-1} less energy than that of its higher homologs. Does that mean that it is more stable by that amount? Clearly not, because both the formula of the alkane and the stoichiometry of the reaction have changed ($C_3H_8 + 5\ O_2 \rightarrow 3\ CO_2 + 4\ H_2O$).

What we seem to require is the relative energy contents of O_2, CO_2, and H_2O. To fill needs of this sort, chemists have established a definition of the heat content of all elements, relative to which all other compounds (and hence the energies of their interconversions) can be placed: the **heat of formation,** ΔH_f°. *The heat of formation of a molecule is the enthalpy of its assembly from the component elements.*

Direct measurement of ΔH_f° is possible only in simple cases, so most data are derived indirectly. To estimate ΔH_f°, each element is taken in its **standard state,** defined as its stable form at room temperature and normal pressure. By definition, the enthalpy of formation of every element in its standard state is zero. Hydrogen, oxygen, and nitrogen have diatomic gaseous standard states, whereas that of carbon is the solid graphite.

Table 3-7 gives the heats of formation of several elements and molecules. We can see that the values for ΔH_f° can be positive or negative. Most organic molecules have negative heats of formation; that is, energy is released when they are formed from their component elements. For example, the reaction of graphite with two molecules of H_2 to give methane is exothermic by 17.9 kcal mol^{-1}.

$$C_{\text{graphite}} + 2\ H_2 \longrightarrow CH_4 \qquad \Delta H^\circ = -17.9\ \text{kcal mol}^{-1}$$

The different heat contents of butane and 2-methylpropane are again revealed by their differing heat of formation: -30.4 and -32.4 kcal mol^{-1}, respectively. Less heat is released in the formation of butane than in the formation of 2-methyl-

TABLE 3-7 Heats of Formation of Selected Elements and Compounds ($kcal\ mol^{-1}$ Normalized to 25°C)

Substance (state)	ΔH_f°	Substance (state)	ΔH_f°
C (graphite)	0	$(CH_3)_3CH$ (gas)	−32.4
CO_2 (gas)	−94.1	$CH_3(CH_2)_3CH_3$ (gas)	−35.1
H_2O (gas)	−57.8	$CH_3(CH_2)_3CH_3$ (liquid)	−41.4
H_2O (liquid)	−68.3	H_2, O_2, N_2 (gases)	0
CH_4 (gas)	−17.9	H (atom)	52.1
CH_3CH_3 (gas)	−20.2	$CH_2{=}CH_2$ (gas)	12.5
$CH_3CH_2CH_3$ (gas)	−24.8	$HC{\equiv}CH$ (gas)	54.2
$CH_3(CH_2)_2CH_3$ (gas)	−30.4		

propane even though the same number of graphite carbon atoms (4 C) and hydrogen molecules (5 H_2) are consumed in both reactions. Again, we can conclude that 2-methylpropane is slightly more stable than butane. The difference in the amount of energy released in both the formation and the combustion of butane and 2-methylpropane, then, is exactly the same: 2 $kcal\ mol^{-1}$. These relations are summarized in Figure 3-12.

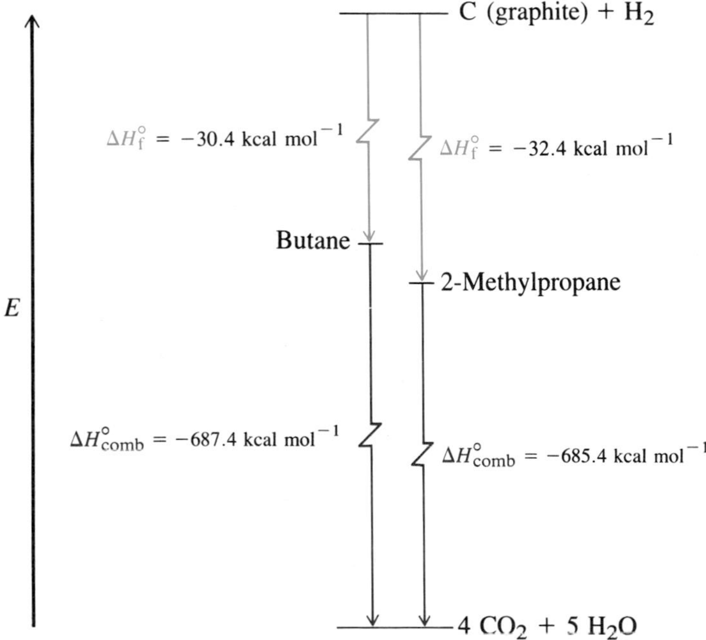

FIGURE 3-12 Formation of butane and 2-methylpropane from the elements shows that the second is more stable than the first. Consequently, combustion of the latter releases less heat.

EXERCISE 3-9

Most heats of formation, including those of alkanes, are obtained indirectly. Show how the ΔH_f° values for butane and 2-methylbutane may be obtained from their experimental heats of combustion and the heats of formation of CO_2 and H_2O (which are simply the heats of combustion of C and H_2, respectively). (Hint: Figure 3-12 shows the relationship between the data you are given and the ΔH_f° values you are to determine.) Heats of formation for reactive species such as atoms and radicals are useful information and may be calculated from bond-dissociation energies and other data. The dissociation of $H_2 \rightarrow 2H\cdot$ has $\Delta H^\circ = +104$ kcal mol^{-1} = 2 (ΔH_f° for $H\cdot$).

EXERCISE 3-10

Enthalpies of chemical reactions may be calculated from heats of formation by using the formula

$$\Delta H^\circ_{\text{reaction}} = \Sigma\ \Delta H_f^\circ(\text{products}) - \Sigma\ \Delta H_f^\circ(\text{reactants})$$

(a) Using ΔH_f° values for $H\cdot$ (Exercise 3-9) and CH_4 (Table 3-7), and DH° for a C–H bond in the latter, calculate ΔH_f° for $CH_3\cdot$. **(b)** Using data from Table 3-1 and Section 3-5, calculate ΔH_f° values for $Br\cdot$ and CH_3Br. Next, calculate ΔH° for the reaction $CH_3\cdot + Br_2 \rightarrow CH_3Br + Br\cdot$, using only the heats of formation for the four species involved. How does your result agree with the value in Table 3-4?

In summary, the heat of formation of a molecule relates its heat content to that of the constituent elements in their standard states, the latter being defined as zero. Most organic molecules have negative heats of formation: They are more stable than their constituent elements.

IMPORTANT CONCEPTS

1. The ΔH° of bond homolysis is defined as the bond-dissociation energy, DH°. Bond homolysis gives radicals or free atoms.
2. The C–H bond strengths in the alkanes decrease in the order

$$\underset{\text{Methyl (Strongest)}}{CH_3{-}H} > \underset{\text{Primary}}{RCH_2{-}H} > \underset{\text{Secondary}}{R{-}\overset{R}{\overset{|}{C}}H{-}H} > \underset{\text{Tertiary (Weakest)}}{R{-}\overset{R}{\overset{|}{\underset{R}{\underset{|}{C}}}}{-}H}$$

because the order of stability of the corresponding alkyl radicals is

$$\underset{\text{Methyl (Least stable)}}{CH_3\cdot} < \underset{\text{Primary}}{RCH_2\cdot} < \underset{\text{Secondary}}{R{-}\overset{R}{\overset{|}{C}}H\cdot} < \underset{\text{Tertiary (Most stable)}}{R{-}\overset{R}{\overset{|}{\underset{R}{\underset{|}{C}}}}\cdot}$$

This is the order of increasing hyperconjugative stabilization.
3. Catalysts speed up the establishment of an equilibrium between starting materials and products.
4. Alkanes react with halogens (except iodine) by a radical chain mechanism to give haloalkanes. The mechanism consists of initiation to create a

halogen atom, two propagation steps, and various termination steps.

5. In the first propagation step, the slower of the two, a hydrogen atom is abstracted from the alkane chain, a reaction resulting in an alkyl radical and HX. Hence, reactivity increases from I_2 to F_2. Selectivity decreases along the same series, as well as with increasing temperature.
6. The Hammond postulate states that fast, exothermic reactions are typically characterized by early transition states, which are similar in structure to the starting materials. On the other hand, slow, endothermic processes usually have late (productlike) transition states.
7. The $\Delta H°$ for a reaction may be calculated from the $DH°$ values of the bonds involved in the process as follows: $\Delta H° = \Sigma\ DH°_{\text{bonds broken}} - \Sigma\ DH°_{\text{bonds formed}}$.
8. The $\Delta H°$ for a radical halogenation process equals the sum of the $\Delta H°$ values for the propagation steps.
9. The relative reactivities of the various types of alkane C–H bonds in halogenations can be estimated by factoring out statistical contributions. They are roughly constant under identical conditions and follow the order

$$CH_4 < \underset{\text{CH}}{\text{primary}} < \underset{\text{CH}}{\text{secondary}} < \underset{\text{CH}}{\text{tertiary}}$$

In radical chlorinations of alkanes at 25°C, the approximate relative reactivities of the tertiary : secondary : primary positions are 5:4:1. In fluorinations, these ratios are about 1.4:1.2:1, whereas in brominations (150°) they are 1700:80:1.
10. Chemists often use halogenating agents other than the halogens. Examples are sulfuryl chloride, SO_2Cl_2, and *N*-chloro- and *N*-bromobutanimide.
11. The $\Delta H°$ of the combustion of an alkane is called the heat of combustion, $\Delta H°_{\text{comb}}$. The heats of combustion of isomeric compounds provide an experimental measure of their relative stabilities.
12. The $\Delta H°$ of the formation of a molecule from its component elements is called its heat of formation, $\Delta H°_f$. The heats of formation of isomeric compounds also allow an assessment of their relative stabilities.

PROBLEMS

1. Label the primary, secondary, and tertiary hydrogens in each of the following compounds.

(a) $CH_3CH_2CH_3$ **(b)** $CH_3CH_2CH_2CH_3$

(c) (methylcyclopentane: cyclopentane ring bearing CH_3) **(d)** $(H_3C)(H_3C)CHCH_2CH_3$

2. Within each of the following sets of alkyl radicals name each radical; identify each as either primary, secondary, or tertiary; rank in order of decreasing stability; and sketch an orbital picture of the most stable radical, showing the hyperconjugative interaction(s).

(a) $CH_3CH_2\dot{C}HCH_3$ and $CH_3CH_2CH_2CH_2\cdot$

(b) $(CH_3CH_2)_2CHCH_2\cdot$ and $(CH_3CH_2)_2\dot{C}CH_3$

(c) $(CH_3)_2CH\dot{C}HCH_3$, $(CH_3)_2\dot{C}CH_2CH_3$, and $(CH_3)_2CHCH_2CH_2\cdot$

3. Write as many products as you can think of that might result from the pyrolytic cracking of propane. Assume that the only initial process is C–C bond cleavage.

4. Answer the question posed in Problem 3 for **(a)** butane and **(b)** 2-methylpropane. Use the data in Table 3-2 to determine the bond most likely to cleave homolytically, and use that bond cleavage as your first step.

5. Calculate $\Delta H°$ values for the following reactions. **(a)** $H_2 + F_2 \rightarrow 2\ HF$; **(b)** $H_2 + Cl_2 \rightarrow 2\ HCl$; **(c)** $H_2 + Br_2 \rightarrow 2\ HBr$; **(d)** $H_2 + I_2 \rightarrow 2\ HI$; **(e)** $(CH_3)_3CH + F_2 \rightarrow (CH_3)_3CF + HF$; **(f)** $(CH_3)_3CH + Cl_2 \rightarrow (CH_3)_3CCl + HCl$; **(g)** $(CH_3)_3CH + Br_2 \rightarrow (CH_3)_3CBr + HBr$; **(h)** $(CH_3)_3CH + I_2 \rightarrow (CH_3)_3CI + HI$.

6. **(a)** Using the information given in Sections 3-6 and 3-7, write the products of the radical monochlorination of (i) pentane and (ii) 3-methylpentane. **(b)** For each, estimate the ratio of the isomeric monochlorination products that would form at 25°C. **(c)** Using the bond-strength data from Table 3-1, determine the $\Delta H°$ values of the propagation steps for the chlorination of 3-methylpentane at C3. What is the overall $\Delta H°$ value for this reaction?

7. Write a mechanism for the radical bromination of the hydrocarbon benzene, C_6H_6 (for structure, see Section 2-1). Use propagation steps similar to those in the halogenation of alkanes, as presented in Sections 3-4 through 3-6. Calculate $\Delta H°$ values for each step and for the reaction as a whole. How does this reaction compare thermodynamically with the bromination of other hydrocarbons? Data: $DH°_{C_6H_5\text{–}H} = 111\ \text{kcal mol}^{-1}$; $DH°_{C_6H_5\text{–}Br} = 81\ \text{kcal mol}^{-1}$.

8. Write the major organic product(s), if any, of each of the following reactions.

(a) $CH_3CH_3 + I_2 \xrightarrow{\Delta}$

(b) $CH_3CH_2CH_3 + F_2 \longrightarrow$

(c) Methylcyclopentane (CH_3 on cyclopentane ring) $+ Br_2 \xrightarrow{\Delta}$

(d) $CH_3CH(CH_3)\text{—}CH_2\text{—}C(CH_3)_2CH_3 + Cl_2 \xrightarrow{h\nu}$

(e) $CH_3CH(CH_3)\text{—}CH_2\text{—}C(CH_3)_2CH_3 + Br_2 \xrightarrow{h\nu}$

9. Calculate product ratios in each of the reactions in Problem 8. Use relative reactivity data for F_2 and Cl_2 at 25°C and for Br_2 at 150°C (Table 3-5).

10. Which, if any, of the reactions in Problem 8 give the major product with reasonable selectivity (i.e., are useful "synthetic methods")?

11. Predict the major product(s) of radical monobromination of each of the following compounds (identified by their common names). Point out any reaction that gives the major product with reasonable selectivity. Except for twistane, all the hydrocarbons shown are derived from molecules representative of the class of naturally occurring compounds called terpenes (see Section 4-7).

(a) 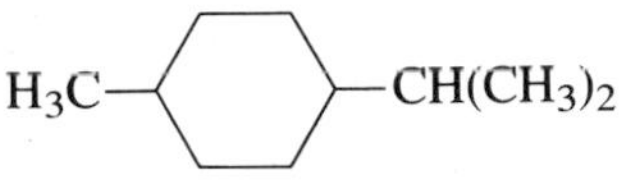 H_3C—(cyclohexane)—$CH(CH_3)_2$

Menthane

(b) 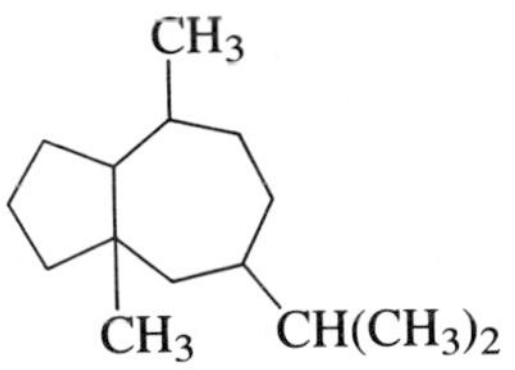 CH_3 ... CH_3 $CH(CH_3)_2$

Pseudoguaiane

(c)

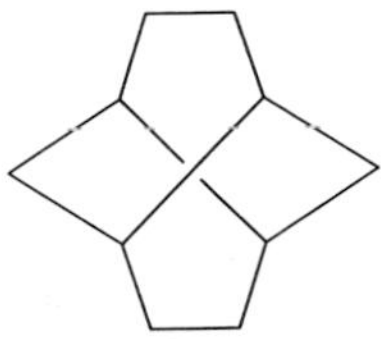

Twistane

(d) 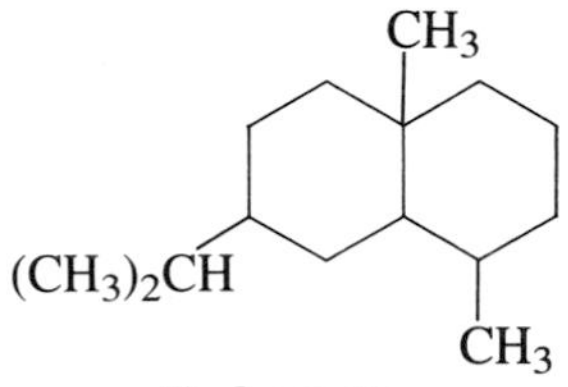 CH_3 ... $(CH_3)_2CH$ ... CH_3

Eudesmane

12. Write balanced equations for the combustion of each of the following substances (molecular formulas may be obtained from Table 3-6). **(a)** Methane; **(b)** propane; **(c)** cyclohexane; **(d)** ethanol; **(e)** sucrose.

13. Propanal ($CH_3CH_2\overset{O}{\overset{\|}{C}}H$) and propanone (acetone; $CH_3\overset{O}{\overset{\|}{C}}CH_3$) are isomers with the formula C_3H_6O. The heat of combustion of propanal is -434.1 kcal mol^{-1}, that of propanone -427.9 kcal mol^{-1}. **(a)** Write a balanced equation for the combustion of either compound. **(b)** What is the energy difference between propanal and propanone? Which has the lower energy content? **(c)** Which substance is more thermodynamically stable, propanal or propanone? (Hint: Draw a diagram similar to that in Figure 3-11.)

14. A hypothetical alternative mechanism for the halogenation of methane has the following propagation steps.

$$\text{(i)}\ X\cdot + CH_4 \longrightarrow CH_3X + H\cdot$$
$$\text{(ii)}\ H\cdot + X_2 \longrightarrow HX + X\cdot$$

(a) Using $DH°$ values from appropriate tables, calculate $\Delta H°$ for these steps for any one of the halogens. **(b)** Compare your $\Delta H°$ values with those for the accepted mechanism (Table 3-4). Do you expect this alternative mechanism to compete successfully with the accepted one? (Hint: Consider activation energies.)

15. The addition of certain materials called radical inhibitors to halogenation reactions causes the reactions to come virtually to a complete stop. An example is the inhibition by I_2 of the chlorination of methane. Explain how this inhibition might come about. (Hint: Calculate $\Delta H°$ values for possible reactions of the various species present in the system with I_2, and evaluate the possible further reactivity of the products of these I_2 reactions.)

16. If a gaseous mixture of CH_3I and HI were to be heated, what would you expect to be the likely result? Suggest a detailed mechanism and calculate $\Delta H°$ for each required step. (Hint: Begin by breaking the weakest bond of all those present in the starting materials, and then examine possible radical chain processes that might follow.)

17. Typical hydrocarbon fuels (e.g., 2,2,4-trimethylpentane, a common component of gasoline) have very similar heats of combustion when calculated in kilocalories *per gram*. **(a)** Calculate heats of combustion per gram for several representative hydrocarbons in Table 3-6. **(b)** Make the same calculation for ethanol (Table 3-6). **(c)** In evaluating the feasibility of "gasohol" (90% gasoline and 10% ethanol) as a motor fuel, it has been estimated that an automobile running on pure ethanol would get approximately 40% fewer miles per gallon than would an identical automobile running on standard gasoline. Is this estimate consistent with the results in (a) and (b)? What can you say in general about the fuel capabilities of oxygen-containing molecules relative to hydrocarbons?

18. Using the $\Delta H_f°$ data in Table 3-7, calculate heats of combustion for **(a)** carbon (as graphite) and **(b)** hydrogen (as H_2 gas). (Hint: First write balanced equations for the combustion processes.) Use the formula

$$\Delta H°_{reaction} = \Sigma\ \Delta H_f°\ (\text{products}) - \Sigma\ \Delta H_f°\ (\text{reactants})$$

19. Two simple organic molecules that are in use as fuel additives are methanol (CH_3OH) and 2-methoxy-2-methylpropane [*tert*-butyl methyl ether, $(CH_3)_3COCH_3$]. The $\Delta H_f°$ values for these compounds in the gas phase are -48.1 kcal mol^{-1} for methanol and -70.6 kcal mol^{-1} for 2-methoxy-2-methylpropane. **(a)** Write balanced equations for the complete combustion of each of these molecules to CO_2 and H_2O. **(b)** Calculate $\Delta H°_{comb}$ for each of these molecules. **(c)** Using Table 3-6, compare the $\Delta H°_{comb}$ values for these compounds with those of alkanes with similar molecular weights.

20. Calculate $\Delta H°$ for the reaction of ethyne (acetylene, HC≡CH) with hydrogen to produce ethane.

21. Ordinary glassblowing torches are fueled by natural gas. However, welders require much hotter temperatures for their work and often use torches fueled by ethyne (acetylene, HC≡CH). **(a)** Write a balanced equation for the combustion of ethyne and calculate its heat of combustion from data in Tables 3-6 and 3-7. **(b)** Compare your result with the heat of combustion of propane, a typical component of natural gas, both per mole and per gram. Does this explain the hotter flame of ethyne?

Species	$\Delta H_f°$ (kcal mol^{-1})
$CH_3CH_2CH_3$	−24.8
$CH_3CH_2CH_2\cdot$	+22.8
$CH_3\dot{C}HCH_3$	+19.2
HBr	−8.7
Br·	+26.7

Approximate E_a for Br· + primary C–H: 13 kcal mol^{-1}. Approximate E_a for Br· + secondary C–H: 10 kcal mol^{-1}.

22. Figure 3-8 compares the reactions of Cl· with the primary and secondary hydrogens of propane. Draw a similar diagram comparing the reactions of Br· with the primary and secondary hydrogens of propane. Use the gas-phase data given in the margin and answer the following questions. **(a)** Calculate $\Delta H°$ for both the primary and the secondary hydrogen abstraction reactions. **(b)** Which among the transition states of these reactions would you call "early," and which "late"? **(c)** Judging from the locations of the transition states of these reactions along the reaction coordinate, should they show greater or lesser radical character than do the corresponding transition states for chlorination (Figure 3-8)? **(d)** Is your answer to (c) consistent with the selectivity differences between Cl· reacting with propane and Br· reacting with propane? Explain.

23. Chlorofluorocarbons such as dichlorodifluoromethane, CCl_2F_2 ("Freon-12"), are widely used refrigerants because of the large quantity of heat they can absorb upon vaporization. When released into the atmosphere, these ordinarily very stable substances diffuse to the stratosphere where intense ultraviolet radiation causes C–Cl bond cleavage. This layer of the atmosphere contains ozone, O_3, which normally absorbs this UV light, preventing it from penetrating to lower levels of the atmosphere. When Cl atoms from Freon decomposition interact with ozone, the latter is destroyed in a radical chain reaction. Substantial and increasing losses of ozone from the atmosphere have been documented repeatedly, and it is feared that the resulting increased penetration of high-energy UV radiation to Earth's surface will lead to significantly higher rates of skin cancer in the coming decades. As one response to this situation, alternative refrigerants that degrade before diffusing into the stratosphere are being developed. Indeed, annual production of Freon-12 in 1992 had dropped by 90% relative to the amounts produced in the late 1980s.

Two of the propagation steps in the $Cl\cdot/O_3$ system consume ozone and oxygen atoms (which are necessary for the production of ozone), respectively.

$$Cl + O_3 \longrightarrow ClO + O_2$$
$$ClO + O \longrightarrow Cl + O_2$$

Using the ΔH_f° data in the margin, calculate ΔH° values for each of these propagation steps. Write the overall reaction described by the combination of these steps and calculate its ΔH°. Comment on the thermodynamic favorability of the process.

Species	ΔH_f° (kcal mol^{-1})
Cl	+29
ClO	+33
O	+60
O_2	0
O_3	+34

4

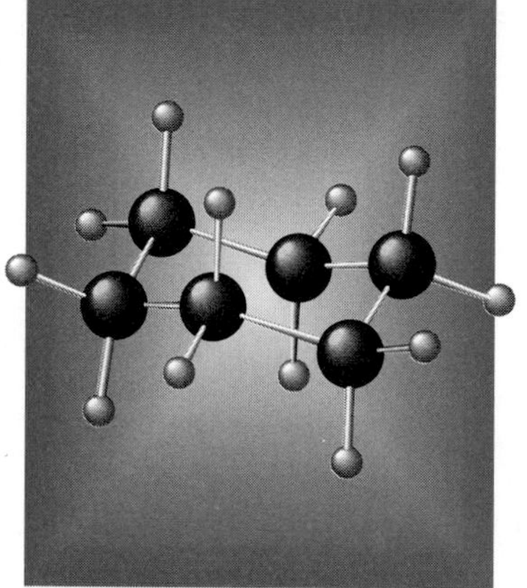

Cyclic Alkanes

Hydrocarbons containing single-bonded carbon atoms arranged in rings are known as **cyclic alkanes, carbocycles,** or **cycloalkanes.** The great majority of organic compounds occurring in nature contain rings. Indeed, so many fundamental biological functions depend on the chemistry of ring-containing compounds that life as we know it could not exist in their absence. This chapter takes up the names, physical properties, structural features, and conformational characteristics of the cycloalkanes. Members of this class of compounds will be used to review and amplify some of the principles presented in Chapter 2 regarding the linear and branched alkanes. We end with the biochemical significance of selected carbocycles and their derivatives, including some common flavoring agents, cholesterol, and other biological regulators.

4-1 Names and Physical Properties of Cycloalkanes

How do the cycloalkanes differ in their names and physical properties from their noncyclic (also called *acyclic*) analogs containing the same number of carbons?

The names of the cycloalkanes follow IUPAC rules

We can construct a molecular model of a cycloalkane by removing two terminal hydrogen atoms from a model of a straight-chain alkane and allowing the terminal carbons to form a bond. The general formula of a cycloalkane is C_nH_{2n}. The system for naming members of this class of compounds is straightforward: Alkane names are preceded by the prefix **cyclo-.** Three members in the homologous series, starting with the smallest, cyclopropane, are shown in the margin, written both in condensed form and in bond-line notation.

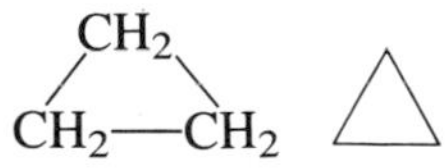

Cyclopropane

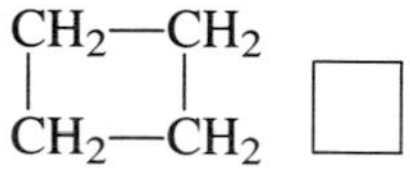

Cyclobutane

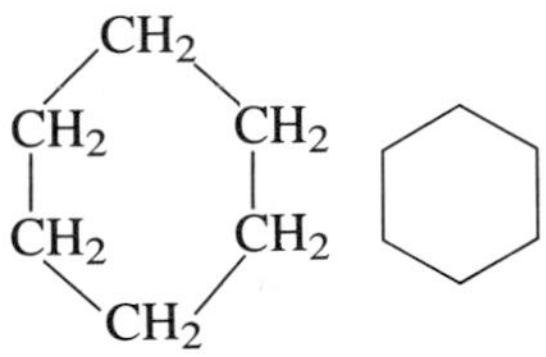

Cyclohexane

EXERCISE 4-1

Make molecular models of cyclopropane through cyclododecane. Compare the relative conformational flexibility of each ring with others within the series and with that of the corresponding straight-chain alkanes.

Naming a substituted cyclic alkane requires a numbering of the individual ring carbons only if more than one substituent is attached to the ring. In monosubstituted systems, the carbon of attachment is defined as carbon 1 of the ring. For polysubstituted compounds, take care to provide the lowest possible numbering sequence. When two such sequences are possible, the alphabetical order of the substituent names takes precedence. Radicals derived from cycloalkanes by abstraction of a hydrogen atom are **cycloalkyl radicals.** Substituted cycloalkanes are therefore sometimes named as cycloalkyl derivatives (large rings take precedence over small rings).

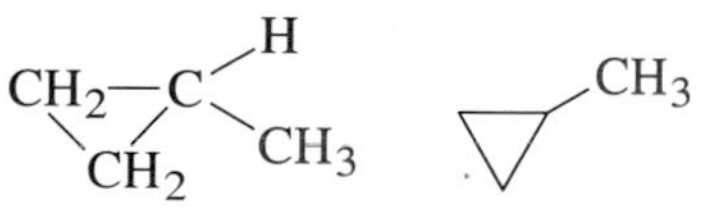

Methylcyclopropane

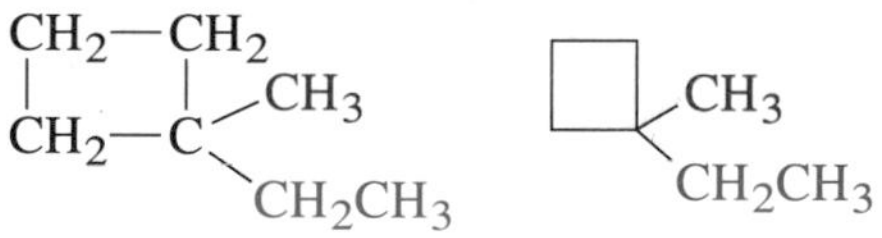

1-Ethyl-1-methylcyclobutane

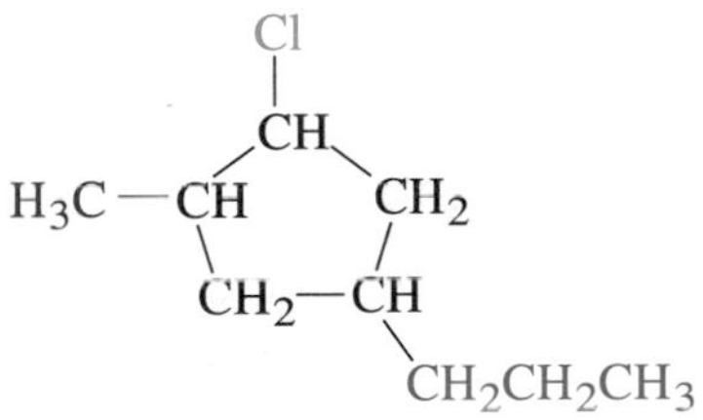

1-Chloro-2-methyl-4-propylcyclopentane
(Not 2-chloro-1-methyl-4-propylcyclopentane)

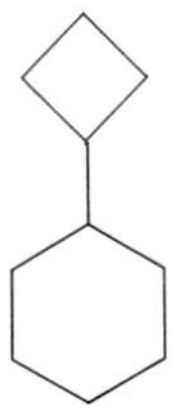

Cyclobutylcyclohexane

Disubstituted cycloalkanes possess isomers

Inspection of molecular models of disubstituted cycloalkanes in which the two substituents are located on different carbons shows that *two isomers are possible* in each case. In one, the two substituents are positioned on the *same* face, or side, of the ring; in the other, on *opposite* faces. Substituents on the same face are called **cis** (*cis,* Latin, on the same side); those on opposite sides, **trans** (*trans,* Latin, across).

H_3C CH_3 or CH_3 CH_3 H H

cis-1,2-Dimethylcyclopropane

CH_3 H_3C or H CH_3 H_3C H

trans-1,2-Dimethylcyclopropane

Br Cl or Cl Br

cis-1-Bromo-2-chlorocyclobutane

Cl Br or Cl Br

trans-1-Bromo-2-chlorocyclobutane

Cis and trans isomers are **stereoisomers**—compounds that have identical connectivities (i.e., their atoms are attached in the same sequence) but differ in the arrangement of their atoms in space. They are distinct from constitutional or structural isomers (Sections 1-9 and 2-2), which are compounds with differing sequences of atoms. Conformations (Sections 2-5 through 2-7) also are stereoisomers by the above definition. However, unlike cis and trans isomers, which can be interconverted only by *breaking* bonds (try it on your models), conformers are readily equilibrated by *rotation* around bonds. The subject of stereoisomerism will be discussed in more detail in Chapter 5.

Dashed-wedged line structures can be used to depict the three-dimensional arrangement of the substituted cycloalkanes. The positions of any remaining hydrogens are not always shown. Structural and cis-trans isomerisms give rise to a variety of structural possibilities in substituted cycloalkanes. For example, there are eight isomeric bromomethylcyclohexanes (three of which are shown below), all with different and distinct physical and chemical properties.

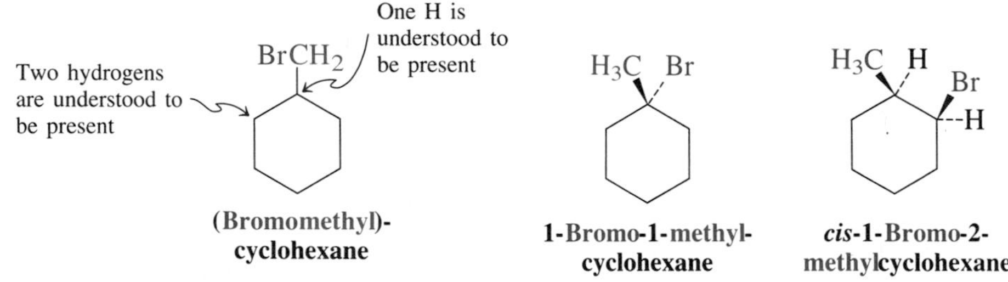

(Bromomethyl)-cyclohexane **1-Bromo-1-methyl-cyclohexane** ***cis*-1-Bromo-2-methylcyclohexane**

EXERCISE 4-2

Give the structures and names of the other five isomeric bromomethylcyclohexanes.

The properties of the cycloalkanes differ from those of their straight-chain analogs

The physical properties of a few cycloalkanes are recorded in Table 4-1. Note that, compared with the corresponding straight-chain alkanes (Table 2-3), the cycloalkanes have higher boiling and melting points, as well as higher densities. These differences are due in large part to increased London interactions of the relatively more rigid and more symmetric cyclic systems. In comparing lower cycloalkanes possessing an odd number of carbons with those having an even

TABLE 4-1 Physical Properties of Various Cycloalkanes

Cycloalkane	Boiling point (°C)	Melting point (°C)	Density at 20°C (g mL^{-1})
Cyclopropane	−32.7	−127.6	
Cyclobutane	−12.5	−50.0	0.720
Cyclopentane	49.3	−93.9	0.7457
Cyclohexane	80.7	6.6	0.7786
Cycloheptane	118.5	−12.0	0.8098
Cyclooctane	148.5	14.3	0.8349
Cyclododecane	160 (100 torr)	64	0.861
Cyclopentadecane	110 (0.1 torr[a])	66	0.860

a. Sublimation point.

number, we find a pronounced alternation in their melting points. This phenomenon has been ascribed to differences in crystal packing forces between the two series.

In summary, names of the cycloalkanes are derived in a straightforward manner from those of the straight-chain alkanes. In addition, the position of a single substituent is defined to be C1. Disubstituted cycloalkanes can give rise to cis and trans isomers, depending on the location of the substituents. Physical properties parallel those of the straight-chain alkanes, except that the individual values for boiling and melting points, and for densities, are higher for the cyclic compounds of equal carbon number.

4-2 Ring Strain and the Structure of Cycloalkanes

The molecular models made for Exercise 4-1 reveal obvious differences between cyclopropane, cyclobutane, cyclopentane, and so forth, and the corresponding straight-chain alkanes. One notable feature of the first two members in the series is how difficult it is to close the ring without breaking the plastic tubes used to represent bonds. This problem is called **ring strain.** The reason for it lies in the tetrahedral carbon model. The C–C–C bond angles in, for example, cyclopropane and cyclobutane differ considerably from the tetrahedral value. As the ring size increases, this difference diminishes. Thus, cyclohexane can be assembled without distortion or strain. A second feature has to do with the eclipsing of the hydrogens: A large degree of eclipsing occurs in the lower cycloalkanes. As the ring size increases, however, the hydrogens can adopt staggered conformations.

Do these observations tell us anything about the relative stability of the cycloalkanes—for example, as measured by their heats of combustion, ΔH°_{comb}? How do the lower cycloalkanes accommodate the distortions at tetrahedral carbon that are necessary for the structures to exist? Are there conformational effects? This section and the following one address these questions.

The heats of combustion of the cycloalkanes reveal the presence of ring strain

Sections 3-9 and 3-10 introduced one measure of the stability of a molecule: its heat content. We also learned that the heat content of an alkane could be estimated by measuring its heat of combustion, ΔH°_{comb} (Table 3-6). To determine the stability of each cycloalkane, we can compare its heat of combustion to the value measured for the analogous straight-chain molecule. The (negative) ΔH°_{comb} values for the straight-chain alkanes increase by about the same amount with each successive member of the series.

ΔH°_{comb} Values for the Series of Straight-Chain Alkanes

$CH_3CH_2CH_3$ (gas)	−530.6		
		−156.8	
$CH_3CH_2CH_2CH_3$ (gas)	−687.4		kcal mol^{-1}
		−157.8	
$CH_3(CH_2)_3CH_3$ (gas)	−845.2		

There appears to be a regular increment of about 157 kcal mol^{-1} for each additional CH_2 group. When averaged over a large number of alkanes, this value approaches 157.4 kcal mol^{-1}.

What does this tell us about cycloalkanes? Because these molecules have the general formula $(CH_2)_n$, we should be able to predict their approximate ΔH°_{comb}: It should just be $-n \times 157.4$ kcal mol^{-1} (Table 4-2, column 2). However, the measured heats of combustion turn out to be *larger in magnitude* (Table 4-2, column 3). For example, cyclopropane should have a ΔH°_{comb} of −472.2 kcal mol^{-1}, but the experimental value is −499.8 kcal mol^{-1}. The difference between expected and observed values is 27.6 kcal mol^{-1} and is attributed to a property of cyclopropane of which we are already aware because of having built a model: *ring strain*. The strain per CH_2 group in this molecule is 9.2 kcal mol^{-1}. A similar calculation for cyclobutane (Table 4-2) reveals a ring strain of 26.3 kcal mol^{-1}, or about 6.6 kcal mol^{-1} per CH_2 group. In cyclopentane, this effect is much smaller, the total strain amounting to only 6.5 kcal mol^{-1}, and cyclohexane is virtually strain free. However, succeeding members of the series again show considerable strain until we reach very large rings. Because of these trends, organic chemists have loosely defined four groups of cycloalkanes.

1. *Small rings* (cyclopropane, cyclobutane)
2. *Common rings* (cyclopentane, cyclohexane, cycloheptane)
3. *Medium rings* (from eight- to twelve-membered)
4. *Large rings* (thirteen-membered and larger)

What kinds of effects contribute to the ring strain in cycloalkanes? We shall answer this question by exploring the detailed structures of several of these compounds.

Strain affects the structures and conformations of the smaller cycloalkanes

As we have just seen, the smallest cycloalkane, *cyclopropane,* is much less stable than expected for three methylene groups. Why should this be? Two factors are involved: torsional strain and bond-angle strain.

TABLE 4-2 Calculated and Experimental Heats of Combustion (kcal mol^{-1}) of Various Cycloalkanes

Ring size (C_n)	ΔH°_{comb} (calculated)	ΔH°_{comb} (experimental)	Total strain	Strain per CH_2 group
3	−472.2	−499.8	27.6	9.2
4	−629.6	−655.9	26.3	6.6
5	−787.0	−793.5	6.5	1.3
6	−944.4	−944.5	0.1	0.0
7	−1101.8	−1108.2	6.4	0.9
8	−1259.2	−1269.2	10.0	1.3
9	−1416.6	−1429.5	12.9	1.4
10	−1574.0	−1586.0	14.0	1.4
11	−1731.4	−1742.4	11.0	1.1
12	−1888.8	−1891.2	2.4	0.2
14	−2203.6	−2203.6	0.0	0.0

Note: The calculated numbers are based on the value of −157.4 kcal mol^{-1} for a CH_2 group.

The structure of the cyclopropane molecule is represented in Figure 4-1. We notice first that all methylene hydrogens are eclipsed, much like the hydrogens in the eclipsed conformation of ethane (Section 2-5). We know that the energy of the eclipsed form of ethane is raised above that of the more stable staggered conformation because of **eclipsing** or **torsional strain.** This effect is also present in cyclopropane. Moreover, the carbon skeleton in cyclopropane is by necessity flat and quite rigid, and bond rotation that might relieve eclipsing strain is very difficult.

Second, we notice that cyclopropane has C–C–C bond angles at 60°, a significant deviation from the "natural" tetrahedral bond angle of 109.5°. How is it possible for three supposedly tetrahedral carbon atoms to maintain a bonding relation at such highly distorted angles? The problem is perhaps best illustrated in Figure 4-2, in which the bonding in the strain-free "open cyclopropane," the trimethylene diradical $\cdot CH_2CH_2CH_2 \cdot$, is compared with that in the closed form. We can see that the two ends of the trimethylene diradical cannot "reach" far enough to close the ring without "bending" the two C–C bonds already present. However, if all three C–C bonds in cyclopropane adopt a bent configuration (interorbital angle 104°, see Figure 4-2B), overlap is sufficient for bond formation. The energy needed to distort the tetrahedral carbons enough to close the ring is called **bond-angle strain.** The ring strain in cyclopropane is derived from a combination of eclipsing and bond-angle contributions.

As a consequence of its structure, cyclopropane has relatively weak C–C bonds (DH° = 65 kcal mol^{-1}). This value is low (recall that C–C strength in ethane is 90 kcal mol^{-1}) because breaking the bond opens the ring and relieves ring strain. Therefore, cyclopropane undergoes several unusual reactions. For example, reaction with hydrogen in the presence of a palladium catalyst opens the ring to give propane.

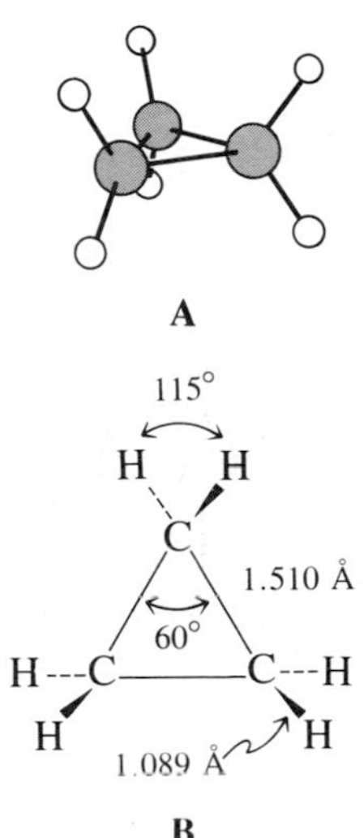

FIGURE 4-1 Cyclopropane. (A) molecular model; (B) bond lengths and angles.

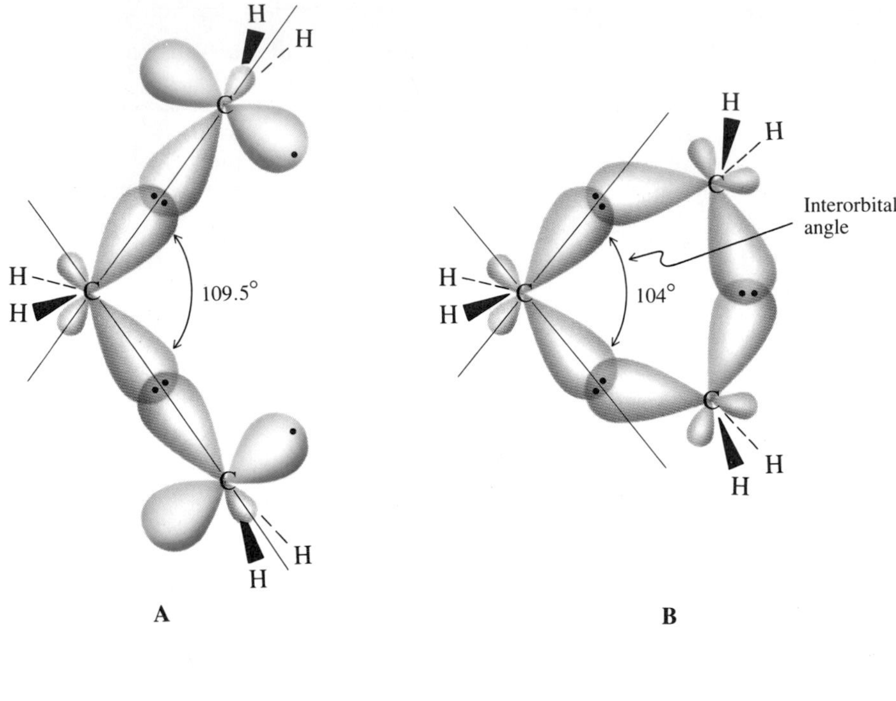

FIGURE 4-2 Molecular-orbital picture of (A) the trimethylene diradical and (B) the bent bonds in cyclopropane. Only the hybrid orbitals forming C–C bonds are shown. Note the interorbital angle of 104° in cyclopropane.

$$\triangle + H_2 \xrightarrow{\text{Pd catalyst}} CH_3CH_2CH_3 \qquad \Delta H^\circ = -37.6\ \text{kcal mol}^{-1}$$

Propane

EXERCISE 4-3

Trans-1,2-dimethylcyclopropane is more stable than *cis*-1,2-dimethylcyclopropane. Why? Draw a picture to illustrate your answer. Which isomer liberates more heat upon combustion?

What about higher cycloalkanes? The structure of *cyclobutane* (Figure 4-3) reveals that this molecule is not planar but puckered, with an approximate bending angle of 26°. The nonplanar structure of the ring, however, is not very rigid. The molecule ''flips'' rapidly from one puckered conformation to the other. Construction of a molecular model shows why distorting the four-membered ring from planarity is favorable: It partly relieves the strain introduced by the eight eclipsing hydrogens. Moreover, bond-angle strain is considerably reduced relative to that in cyclopropane, although maximum overlap is, again, only possible by the use of bent bonds. The C–C bond strength in cyclobutane also is low

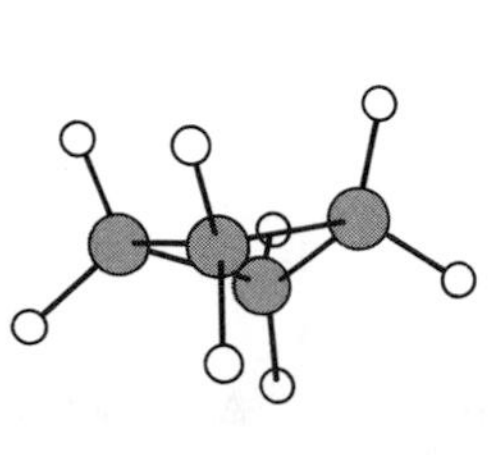

FIGURE 4-3 Cyclobutane. (A) molecular model; (B) bond lengths and angles. The nonplanar molecule ''flips'' rapidly from one conformation to another.

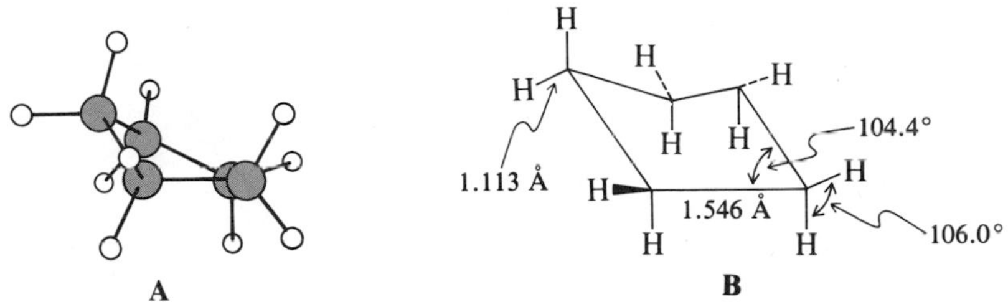

FIGURE 4-4 Cyclopentane. (A) molecular model of the half-chair conformation; (B) bond lengths and angles. The molecule is flexible, with little strain.

(about 63 kcal mol^{-1}) because of the release of ring strain on ring opening and the consequences of relatively poor overlap in bent bonds. Cyclobutane is less reactive than cyclopropane but undergoes similar ring-opening processes.

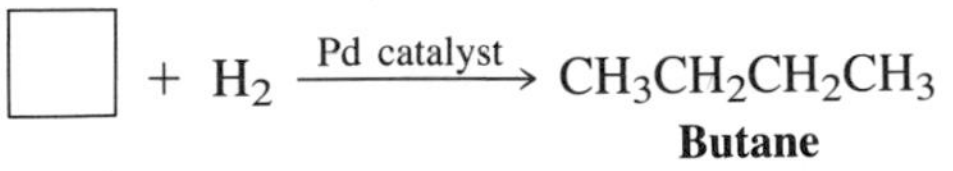

$$\square + H_2 \xrightarrow{\text{Pd catalyst}} CH_3CH_2CH_2CH_3$$

Butane

Cyclopentane might be expected to be planar because the angles in a regular pentagon are 108°, close to tetrahedral. However, such a planar arrangement would have *ten* H–H eclipsing interactions. The puckering of the ring reduces this effect, as indicated in the structure of the molecule (Figure 4-4). Although puckering relieves eclipsing, it also increases bond-angle strain. The conformation of lowest energy is a compromise in which the energy of the system is minimized. There are two puckered conformations possible for cyclopentane: the **envelope** and the **half chair.**

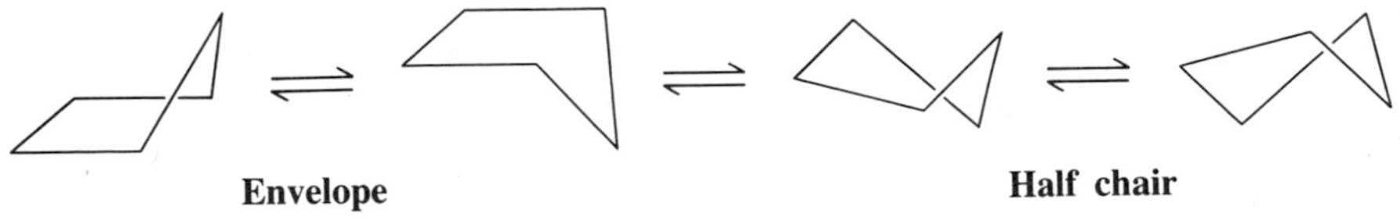

There is little difference in energy between them, and the activation barriers for rapid interconversion are low. Overall, cyclopentane has relatively little ring strain and hence does not show the unusual reactivity of three- or four-membered rings.

4-3 Cyclohexane: A Strain-Free Cycloalkane?

The cyclohexane ring is one of the most abundant and important structural units in organic chemistry. Its substituted derivatives exist in many natural products (see Section 4-7), and an understanding of its conformational mobility is an important aspect of organic chemistry. Table 4-2 reveals that cyclohexane is unusual in that it is almost free of bond-angle or eclipsing strain. Why?

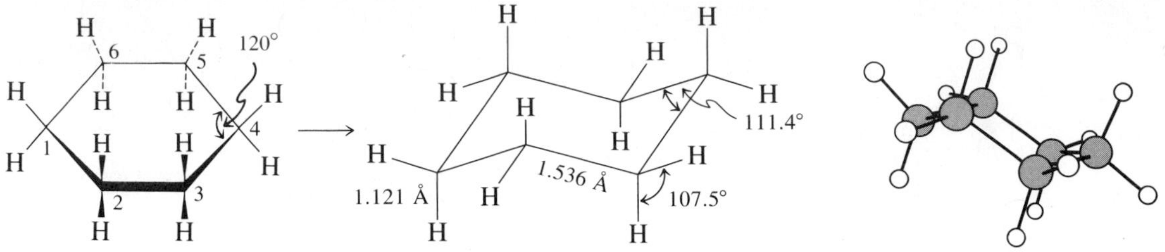

A

Planar cyclohexane

(120° bond angles;
12 eclipsing hydrogens)

B

Chair cyclohexane

(Nearly tetrahedral bond angles;
no eclipsing hydrogens)

C

FIGURE 4-5 Conversion of the (A) hypothetical planar cyclohexane into the chair conformation; (B) bond lengths and angles; (C) molecular model. The chair conformation is nearly strain free.

The chair conformation of cyclohexane is nearly strain free

A hypothetical planar cyclohexane would suffer from twelve H–H eclipsing interactions and sixfold bond-angle strain (a regular hexagon requires 120° bond angles). However, one conformation of cyclohexane, obtained by moving carbons 1 and 4 out of planarity in opposite directions, is in fact almost strain free (Figure 4-5). This structure is called the **chair conformation** of cyclohexane (because it resembles a chair), in which eclipsing is completely prevented and the bond angles are very nearly tetrahedral. As seen in Table 4-2, the calculated ΔH°_{comb} of cyclohexane (-944.4 kcal mol^{-1}) based on a strain-free $(CH_2)_6$ model is very close to the experimentally determined value (-944.5 kcal mol^{-1}).

Looking at the molecular model of cyclohexane enables us to observe the conformational behavior of the molecule. If we view it along (any) one C–C bond, we can see the staggered arrangement of all substituent groups along it. We can visualize this arrangement by drawing a Newman projection of that view (Figure 4-6). Because of its lack of strain, cyclohexane is as inert as a straight-chain alkane.

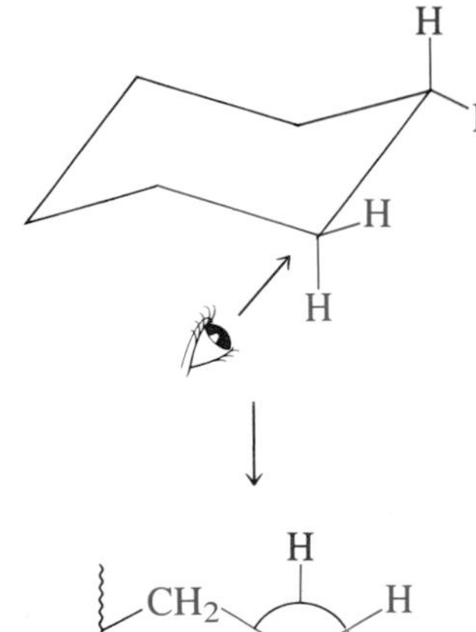

FIGURE 4-6 View along one of the C–C bonds in the chair conformation of cyclohexane. Note the staggered arrangement of all substituents.

EXERCISE 4-4

Draw Newman projections of the carbon–carbon bonds in cyclopropane, cyclobutane, and cyclopentane in their most stable conformations. Use the models you prepared for Exercise 4-1 to assist you and refer to Figure 4-6. What are the approximate torsional angles between the C–H bonds in each?

Cyclohexane also has several less stable conformations

There are other, less stable conformations of cyclohexane, which are nevertheless readily accessible to the molecule. One is the **boat form,** in which carbons 1 and 4 are out of the plane in the *same* direction (Figure 4-7). The boat is less stable than the chair form by 6.5 kcal mol^{-1}. One reason for this difference is the eclipsing of eight hydrogen atoms at the base of the boat. Another is steric

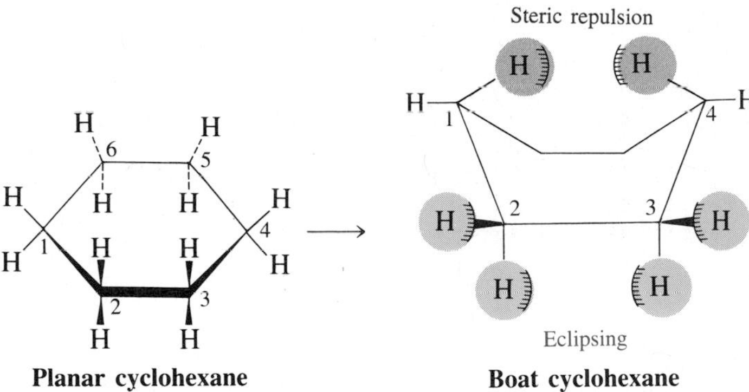

FIGURE 4-7 Conversion of the hypothetical planar cyclohexane into the boat form. In the latter form the hydrogens on carbons 2, 3, 5, and 6 are eclipsed, thereby giving rise to torsional strain. The "inside" hydrogens on carbons 1 and 4 interfere with each other sterically in a transannular interaction.

hindrance (Section 2-7) due to the close proximity of the two inside hydrogens in the boat framework. The distance between these two hydrogens is only 1.83 Å, small enough to create an energy of repulsion of about 3 kcal mol^{-1}. This effect is an example of **transannular strain:** strain resulting from steric crowding of two groups across a ring (*trans,* Latin, across; *annulus,* Latin, ring).

Boat cyclohexane is fairly flexible. If one of the C–C bonds is twisted relative to another, this form can be somewhat stabilized by partial removal of the transannular interaction. The new conformation obtained is called the **twist-boat** (or **skew-boat) conformation** of cyclohexane (Figure 4-8). The stabilization relative to the boat form amounts to about 1.5 kcal mol^{-1}. As shown, there are two possible twist-boat forms. They interconvert rapidly, with the boat conformer acting as a *transition state* (verify this with your model). Thus, the boat cyclohexane is not a normally isolable species, the twist-boat form is present in very small amounts, and the chair form is the major conformer (Figure 4-9). As we shall see, there are also two rapidly interconverting chair conformers of cyclohexane.

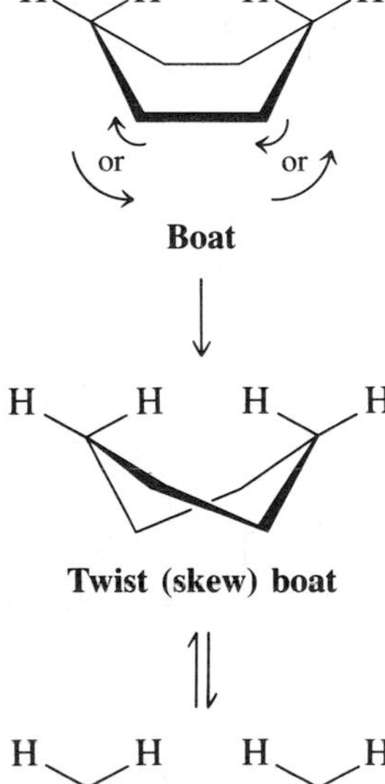

FIGURE 4-8 The conversion of a boat into a twist-boat cyclohexane.

Cyclohexane has axial and equatorial hydrogen atoms

The chair conformation model of cyclohexane reveals that the molecule has two types of hydrogens. Six carbon–hydrogen bonds are parallel to the principal molecular axis (Figure 4-10) and hence are **axial;** the other six are perpendicular to that axis and are therefore **equatorial.***

The ability to draw cyclohexane chair conformations correctly assists in learning the chemistry of six-membered rings. Several rules are useful.

*An equatorial plane is defined as being perpendicular to the axis of rotation of a rotating body and equidistant from its poles, such as the equator of the planet Earth. The equatorial hydrogens in cyclohexane are in its equatorial plane.

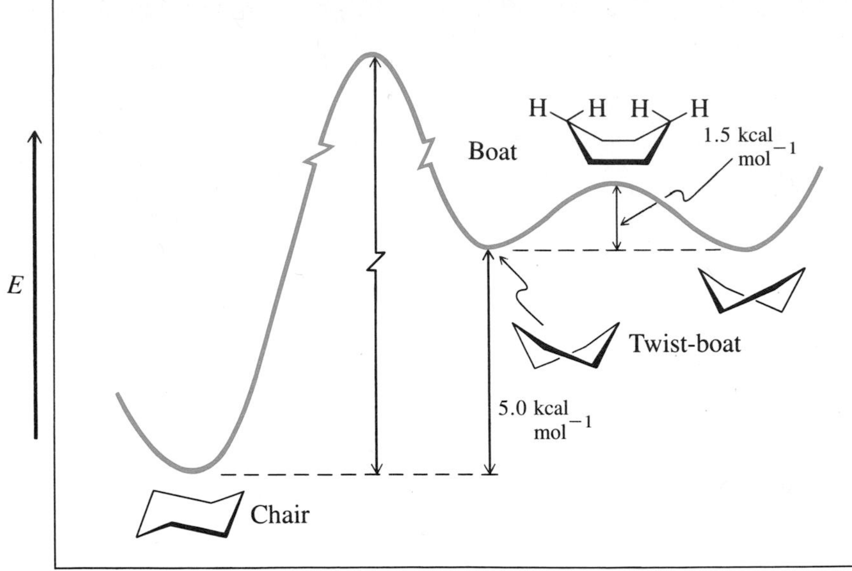

FIGURE 4-9 Potential-energy diagram for the interconversion of the various conformers of cyclohexane. The chair is most stable, but twist-boat conformations, which are 5 kcal mol^{-1} higher in energy, are readily accessible. The twist boats interconvert easily through a boat transition state.

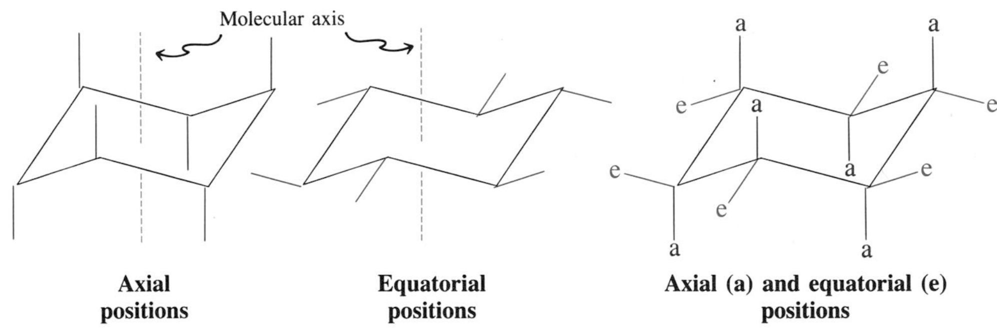

FIGURE 4-10 The axial and equatorial positions of hydrogens in the chair form of cyclohexane.

How to Draw Chair Cyclohexanes

1. Draw the chair so as to place the C2 and C3 atoms to the right of C5 and C6, with apex 1 pointing downward on the left and apex 4 pointing upward on the right.

The bond between C1 and C6 is also parallel to that between C3 and C4.

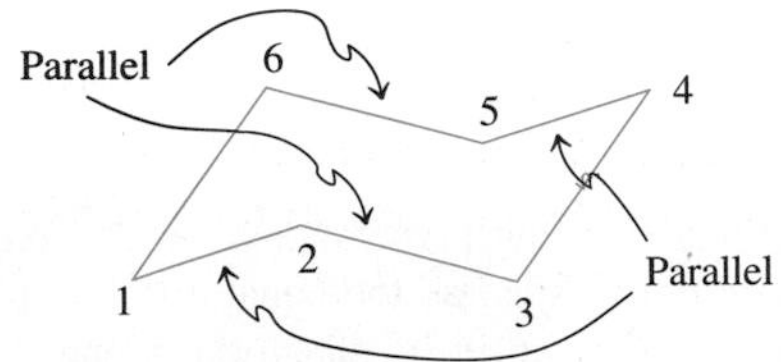

2. Add all the axial bonds as vertical lines, pointing downward at C1, C3, and C5 and upward at C2, C4, and C6.

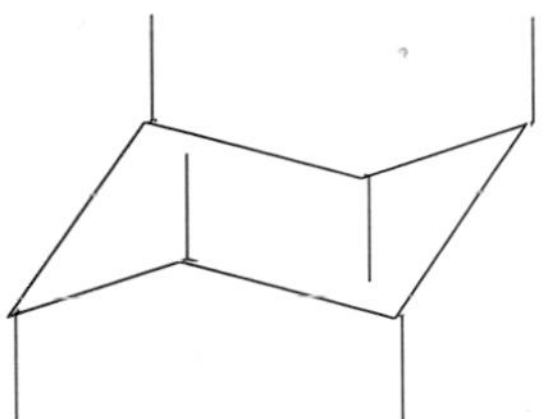

3. Draw the two equatorial bonds at C1 and C4 at a slight angle from horizontal, pointing upward at C1 and downward at C4, parallel to the bond between C2 and C3 (or C5 and C6).

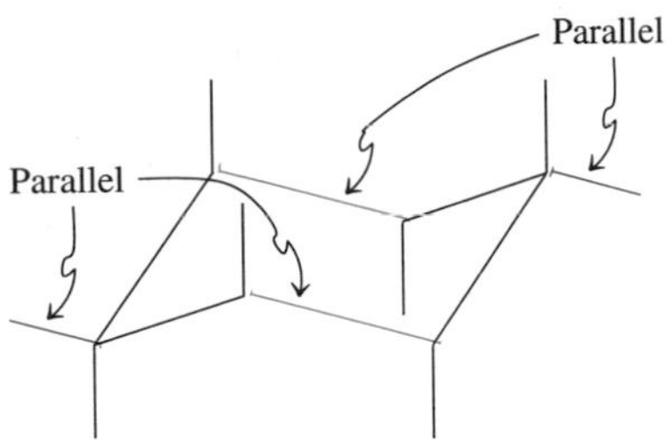

4. This rule is the most difficult to follow: Add the remaining equatorial bonds at C2, C3, C5, and C6 by aligning them *parallel* to the C–C bond "once removed," as shown below.

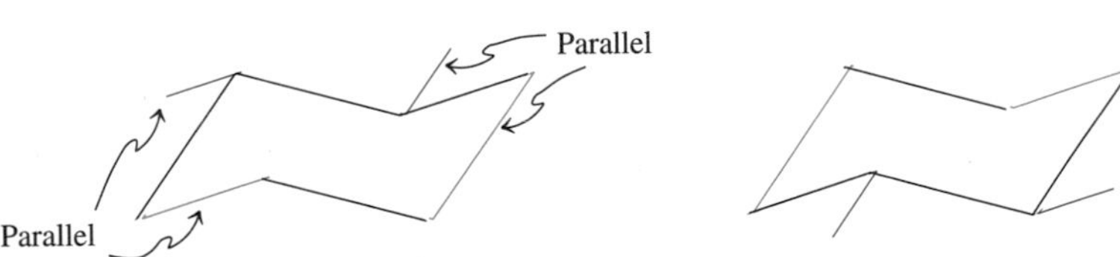

Conformational flipping interconverts axial and equatorial hydrogens

Cyclohexane is not conformationally rigid. It is capable of interconverting one chair conformation with another, thus equilibrating axial with equatorial hydrogens. In this process ("flipping"), all axial hydrogens in one chair become equatorial in the other and vice versa (Figure 4-11). The activation energy for this process is 10.8 kcal mol^{-1}. As suggested in Sections 2-5 through 2-7, this value is so low that at room temperature there is very rapid interconversion of the two equivalent chair forms (approximately 100,000 times per second). Only when solutions of the molecule are cooled to very low temperatures (about $-100°C$) is this equilibration stopped.

To summarize Sections 4-2 and 4-3, the discrepancy between calculated and measured heats of combustion in the cycloalkanes can be largely attributed to three forms of strain: bond-angle (deformation of tetrahedral carbon), eclipsing (torsional), and transannular (across the ring). Because of strain, the small cycloalkanes are chemically reactive, undergoing ring-opening reactions. Cyclohex-

FIGURE 4-11 Chair–chair interconversion ("ring flipping") in cyclohexane. In the process, which is rapid at room temperature, a carbon at one end of the molecule (green) moves up, while its counterpart at the other end (also green) moves down. All groups originally in axial positions (red in the structure at the left) become equatorial, and those that start in equatorial positions (blue) become axial.

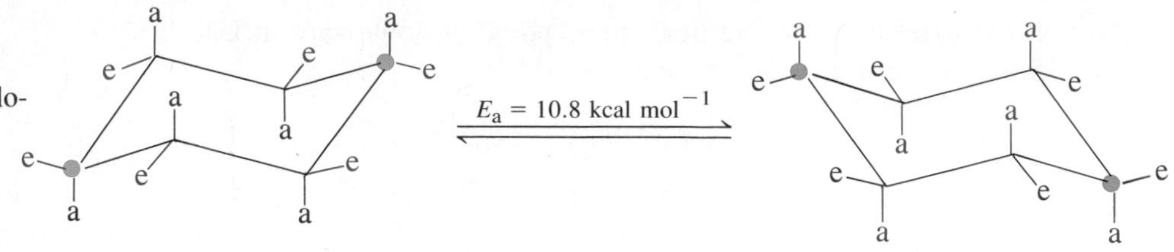

E_a = 10.8 kcal mol^{-1}

ane is almost strain free. It has a lowest energy chair, as well as additional higher energy conformations, particularly the boat and twist-boat structures. Chair–chair interconversion is rapid at room temperature; it is a process in which equatorial and axial hydrogen atoms interchange their positions.

4-4 Substituted Cyclohexanes

We can now apply our knowledge of conformational analysis to substituted cyclohexanes. Let us look at the simplest alkylcyclohexane, methylcyclohexane.

Axial and equatorial methylcyclohexanes are not equivalent in energy

In methylcyclohexane, the methyl group occupies either an equatorial or an axial position.

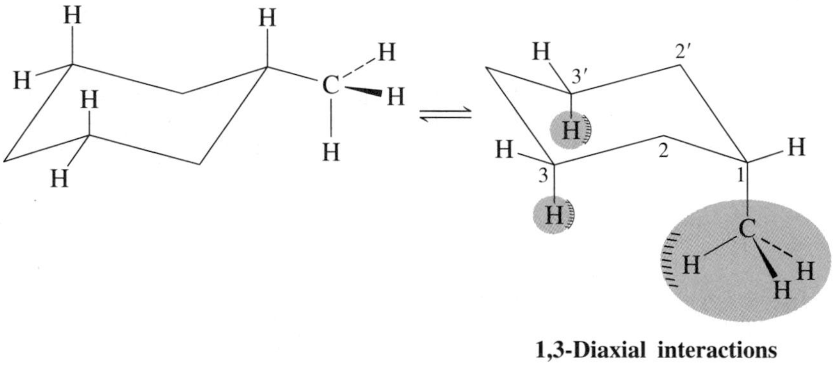

1,3-Diaxial interactions

More stable　　Less stable

Ratio = 95:5

Are the two forms equivalent? Clearly not. In the equatorial conformer, the methyl group extends into space away from the remainder of the molecule. In contrast, in the axial conformer, the methyl substituent is close to the other two axial hydrogens on the same side of the molecule. The distance to these hydrogens is small enough (about 2.7 Å) to result in steric repulsion, another example of transannular strain. Because this effect is due to axial substituents on carbon atoms that have a 1,3-relation (in the drawing, 1,3 and 1,3′), it is called a **1,3-diaxial interaction.**

The two forms of chair methylcyclohexane are in equilibrium. *The equatorial conformer is more stable* by 1.7 kcal mol^{-1} and is favored by a ratio of 95:5 at 25°C (Section 2-8). The activation energy for chair–chair interconversion is simi-

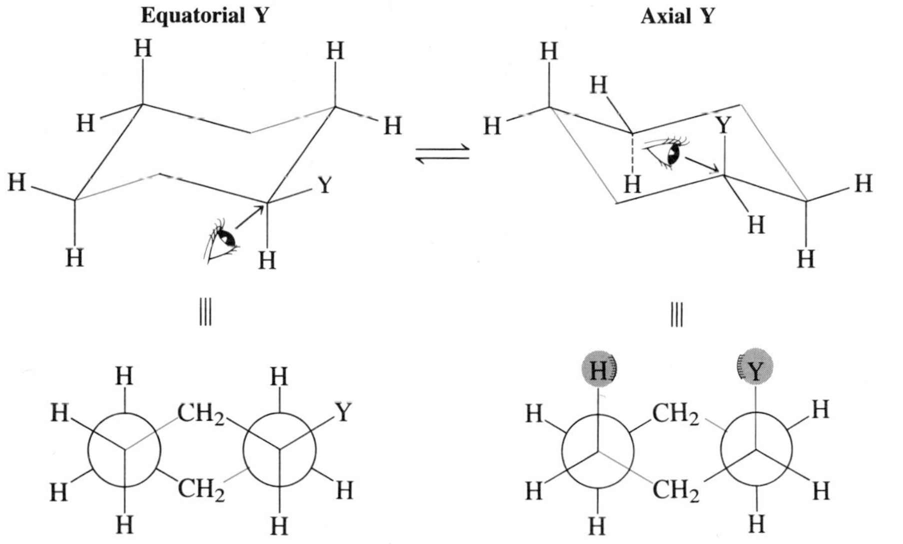

FIGURE 4-12 Newman projections of a substituted cyclohexane. The conformation with axial Y is less stable because of 1,3-diaxial interactions.

lar to that in cyclohexane itself (about 11 kcal mol^{-1}), and equilibrium between the two conformers is established rapidly at room temperature.

The unfavorable 1,3-diaxial interactions to which an axial substituent is exposed are readily seen in Newman projections of the ring C–C bond bearing that substituent. In contrast with that in the axial form, the substituent in the equatorial conformer is away from the axial hydrogens (Figure 4-12).

EXERCISE 4-5

Calculate K for equatorial versus axial methylcyclohexane from the $\Delta G°$ value of 1.7 kcal mol^{-1}. Use the expression $\Delta G°$ (in kcal mol^{-1}) = $-1.36 \log K$. (Hint: If $\log K = x$, then $K = 10^x$.) How well does your result agree with the 95:5 conformer ratio stated in the text?

The energy difference between the axial and the equatorial forms of many monosubstituted cyclohexanes has been measured; several are given in Table 4-3. In many cases (but not all), particularly for alkyl substituents, the energy difference between the two forms increases with the size of the substituent, a direct consequence of increasing unfavorable 1,3-diaxial interactions. This effect is particularly pronounced in (1,1-dimethylethyl)cyclohexane (*tert*-butylcyclohexane). The energy difference here is so large (about 5 kcal mol^{-1}) that very little (about 0.01%) of the axial conformer is present at equilibrium.

The chair with more equatorial groups is usually more stable

In disubstituted cyclohexanes, the conformation with the maximum number of equatorial substituents is generally the most abundant. Let us look at a few simple examples.

There are several isomers of dimethylcyclohexane. In 1,1-dimethylcyclohexane, one methyl group is always equatorial and the other axial. The two chair forms are identical, and hence their energies are equal.

TABLE 4-3 Change in Free Energy upon Flipping from the Cyclohexane Conformer with the Indicated Substituent Equatorial to the Conformer with the Substituent Axial

Substituent	$\Delta G°$ (kcal mol^{-1})	Substituent	$\Delta G°$ (kcal mol^{-1})
H	0	F	0.25
CH_3	1.70	Cl	0.52
CH_3CH_2	1.75	Br	0.55
$(CH_3)_2CH$	2.20	I	0.46
$(CH_3)_3C$	≈5	HO	0.94
HO—C(=O)	1.41	CH_3O	0.75
CH_3O—C(=O)	1.29	H_2N	1.4

Note: In all examples, the equatorial form is more stable.

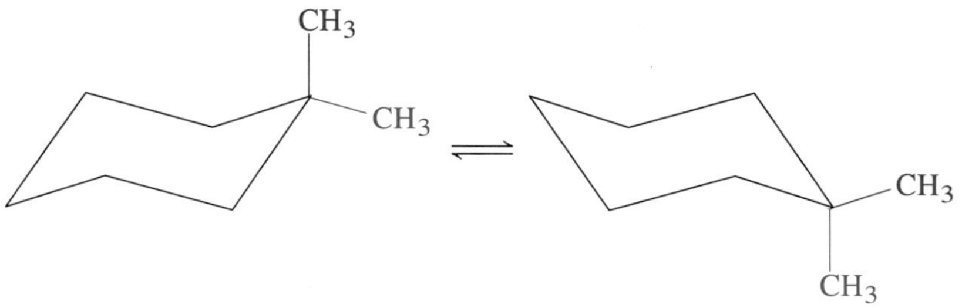

One CH_3 axial
One CH_3 equatorial

One CH_3 axial
One CH_3 equatorial

1,1-Dimethylcyclohexane

(Conformations equal in energy, equally stable)

For 1,2-, 1,3-, and 1,4-dimethylcyclohexane, there are cis and trans isomers of different conformations. For example, in *cis*-1,4-dimethylcyclohexane, both chairs have one axial and one equatorial substituent and are of equal energy.

The bonds to both methyl groups point downward; they are cis.

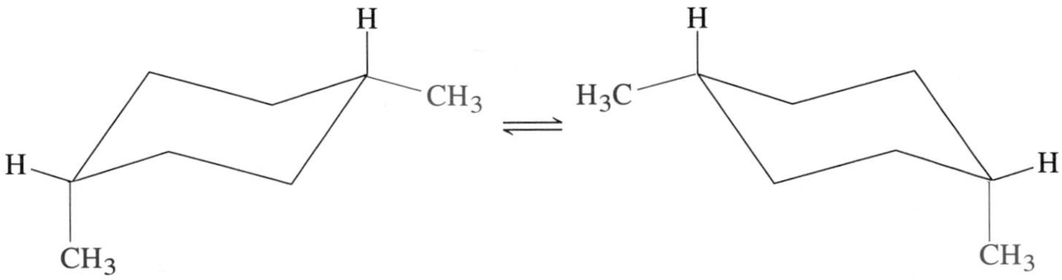

One axial, one equatorial

One axial, one equatorial

***cis*-1,4-Dimethylcyclohexane**

On the other hand, the trans isomer can exist in two different chair conformations: one with two axial methyl groups (diaxial) and the other with two equatorial ones (diequatorial).

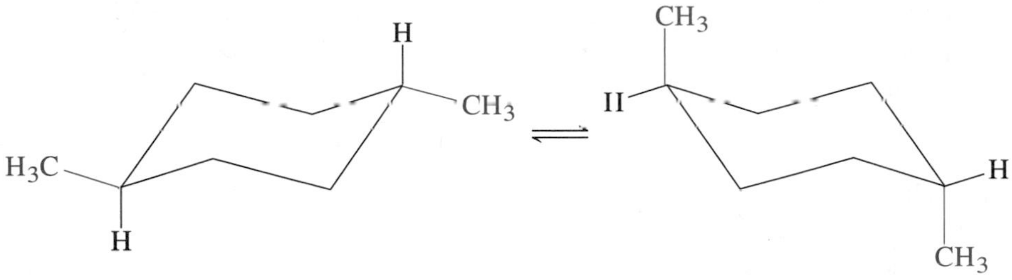

H₃C H H CH₃ CH₃ H CH₃

Diequatorial methyls
More stable

Diaxial methyls
Less stable: $+3.4$ kcal mol^{-1}

***trans*-1,4-Dimethylcyclohexane**

The bond to one methyl group points down; the other up. They are trans, *regardless* of conformation.

In the diaxial arrangement, both methyl groups are subjected to an approximately equal degree of 1,3-diaxial strain. This form is therefore destabilized about twice as much ($+3.4$ kcal mol^{-1}) as the axial monomethylcyclohexane, relative to the diequatorial conformer. This finding indicates that the substituent values in Table 4-3 may be added to calculate approximate $\Delta G°$ values for a number of chair–chair equilibria of substituted cyclohexanes.

When the two chair conformers of a substituted cyclohexane have equal numbers of axial and equatorial substituents, the predominance of one over the other will depend on how likely each substituent is to occupy an equatorial or axial position. For example, in *cis*-1,2-disubstituted cyclohexanes, one substituent is axial, the other equatorial. In *cis*-1-(1,1-dimethylethyl)-2-methylcyclohexane, the size of the 1,1-dimethylethyl (*tert*-butyl) group results in a preference for the conformation in which the methyl group is axial to allow the larger group to be equatorial.

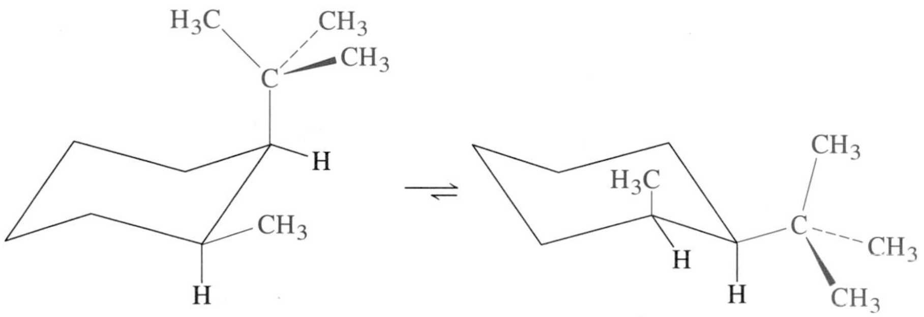

Large group axial
Small group equatorial
Less stable

Small group axial
Large group equatorial
More stable

***cis*-1-(1,1-Dimethylethyl)-2-methylcyclohexane**

EXERCISE 4-6

Calculate $\Delta G°$ for the equilibrium between the two chair conformers of **(a)** 1-ethyl-1-methylcyclohexane; **(b)** *cis*-1-ethyl-4-methylcyclohexane; **(c)** *trans*-1-ethyl-4-methylcyclohexane.

EXERCISE 4-7

Draw both chair conformations for each of the following. **(a)** *cis*-1,2-Dimethylcyclohexane; **(b)** *trans*-1,2-dimethylcyclohexane; **(c)** *cis*-1,3-dimethylcyclohexane;

(d) *trans*-1,3-dimethylcyclohexane. Which of these always have equal numbers of axial and equatorial substituents? Which exist as equilibrium mixtures of diaxial and diequatorial forms?

In summary, the conformational analysis of cyclohexane enables us to predict the relative stability of its various conformers and even to approximate the energy differences between two chair conformations. Bulky substituents, particularly a 1,1-dimethylethyl group, tend to shift the chair–chair equilibrium toward the side in which the large substituent is equatorial.

4-5 Larger Cycloalkanes

Do similar relations hold for the larger cycloalkanes? Table 4-2 shows that cycloalkanes with rings larger than that of cyclohexane also have more strain. This strain is due to a combination of bond-angle distortion, partial eclipsing of hydrogens, and transannular steric repulsions. It is not possible for medium-sized rings to relieve all of these strain-producing interactions in a single conformation. Instead, a compromise solution is found in which the molecule equilibrates among several geometries that are very close in energy.

Essentially strain-free conformations are attainable only for large-sized cycloalkanes, such as cyclotetradecane (Table 4-2). In such rings, the carbon chain adopts a structure very similar to that of the straight-chain alkanes (Section 2-4), having staggered hydrogens and an all-*anti* configuration. However, even in these systems, the attachment of substituents usually introduces various amounts of strain. Most cyclic molecules described in this book are not strain free.

4-6 Polycyclic Alkanes

The cycloalkanes discussed so far contain only one ring and therefore may be referred to as monocyclic alkanes. In more complex structures—the bi-, tri-, tetra-, and higher polycyclic hydrocarbons—two or more rings share carbon atoms. We shall see the structural variety possible in these compounds, many of which, when bearing alkyl and functional groups, occur in nature.

Polycyclic alkanes may contain fused or bridged rings

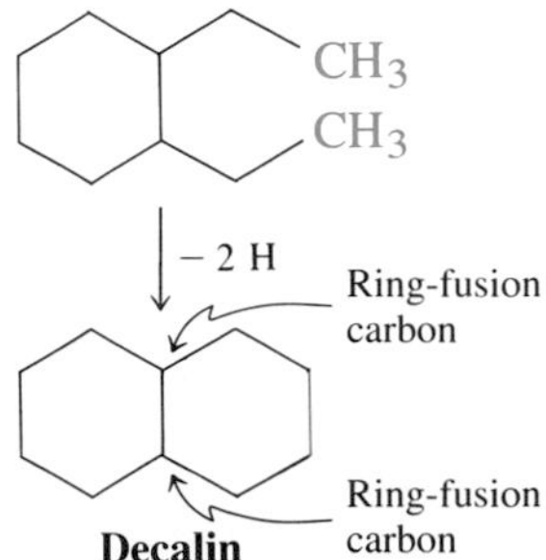

Molecular models of polycyclic alkanes can be readily constructed by linking the carbon atoms of two alkyl substituents in a monocyclic alkane. For example, if two hydrogen atoms are removed from the methyl groups in 1,2-diethylcyclohexane, thereby allowing a new C–C bond to form, the result is a new molecule with the common name decalin. In decalin, two cyclohexanes share two adjacent carbon atoms, and the two rings are said to be **fused.** Compounds constructed in this way are called **fused bicyclic** ring systems, and the shared carbon atoms are called the **ring-fusion** carbons. Groups attached to ring-fusion carbons are called **ring-fusion substituents.**

If we treat a molecular model of *cis*-1,3-dimethylcyclopentane in the same way, we obtain another carbon skeleton, that of norbornane.

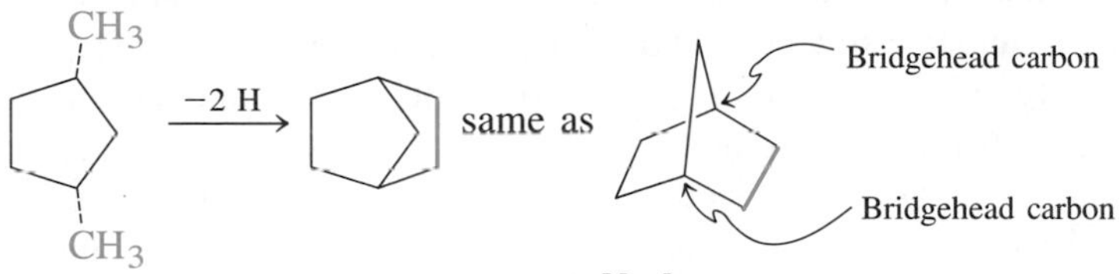

Norbornane

Norbornane is an example of a **bridged bicyclic** ring system. In bridged bicyclic systems two nonadjacent carbon atoms, the **bridgehead** carbons, belong to both rings.

If we think of one of the rings as a substituent on the other, we can identify stereochemical relationships at ring fusions. In particular, bicyclic ring systems can be cis or trans fused. The stereochemistry of the ring fusion is most easily determined by inspecting the ring-fusion substituents. For example, in *trans*-decalin the ring-fusion hydrogens are trans with respect to each other, whereas in *cis*-decalin these hydrogens have a cis relationship (Figure 4-13).

EXERCISE 4-8

Construct molecular models of both *cis*- and *trans*-decalin. What can you say about their conformational mobility?

Is there a strain limit in hydrocarbons?

The search for the limits of strain in hydrocarbon bonds is an area of fascinating research that has resulted in the synthesis of many exotic molecules. What is surprising is how much bond-angle distortion a carbon atom is able to tolerate. A case in point in the bicyclic series is bicyclobutane, whose strain energy is estimated to be 66.5 kcal mol^{-1}. Considering that its heat of formation is +51.9 kcal mol^{-1}, it is remarkable that the molecule exists at all.

Bicyclobutane

A series of compounds attracting the attention of synthetic chemists possess a carbon framework geometrically equivalent to the Platonic solids: the *tetrahedron* (tetrahedrane), the *hexahedron* (cubane), and the pentagonal *dodecahedron*

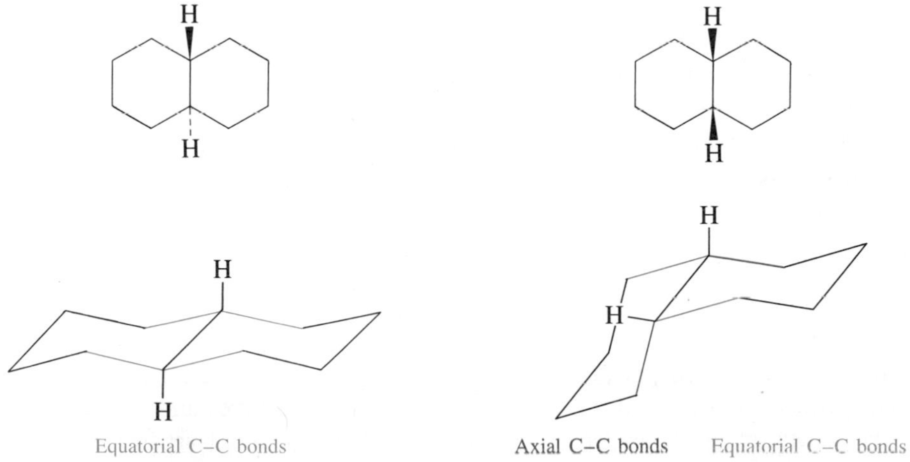

***trans*-Decalin**

***cis*-Decalin**

FIGURE 4-13 Conventional drawings and chair conformations of *trans*- and *cis*-decalin. The trans isomer contains only equatorial carbon–carbon bonds at the ring fusion, whereas the cis possesses two equatorial C–C bonds (green) and two axial C–C linkages, one with respect to each ring (red).

(dodecahedrane). Of them, the hexahedron was synthesized first in 1964, a C_8H_8 hydrocarbon shaped like a cube and accordingly named cubane. The experimental strain energy (157 kcal mol^{-1}) is approximately equal to the total strain of six cyclobutanes. Although tetrahedrane itself is unknown, a tetra(1,1-dimethylethyl) derivative was synthesized in 1978. Despite the estimated strain of 129 to 137 kcal mol^{-1}, the compound is stable and has a melting point of 135°C. The synthesis of dodecahedrane was achieved in 1982. It required 23 synthetic operations, starting from a simple cyclopentane derivative. The last step gave 1.5 mg of pure compound. Although small, this amount was sufficient to permit complete characterization of the molecule. Its melting point at 430°C is extraordinarily high for a C_{20} hydrocarbon and is indicative of the symmetry of the compound. For comparison, eicosane, also with 20 carbons, melts at 36.8°C (Table 2-5).

$C(CH_3)_3$

$(CH_3)_3C$ — — $C(CH_3)_3$

$C(CH_3)_3$

Tetrakis(1,1-dimethylethyl)-tetrahedrane

Tetrahedrane

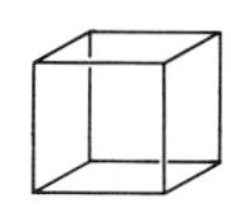

Cubane

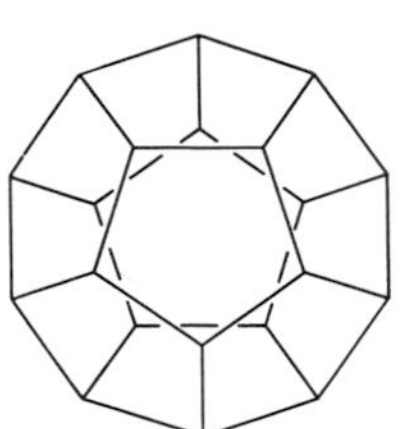

Dodecahedrane

In summary, carbon atoms in bicyclic compounds are shared by rings in either fused or bridged arrangements. A great deal of strain may be tolerated by carbon in its bonds, particularly to other carbon atoms. This capability has allowed the preparation of molecules in which carbon is severely deformed from its tetrahedral shape.

4-7 Carbocyclic Products in Nature

Let us now take a brief look at the variety of cyclic molecules created in nature. **Natural products** are organic compounds produced by living organisms. Some of these compounds are extremely simple, such as methane; others have great structural complexity. Scientists have attempted to classify the multitude of natural products in various ways. Generally, four schemes are followed, in which these products are classified according to (1) chemical structure, (2) physiological activity, (3) organism or plant specificity (taxonomy), and (4) biochemical origin.

There are many reasons why organic chemists are interested in natural products. Many of these compounds are powerful drugs, others function as coloring or flavoring agents, and yet others are important raw materials. A study of animal secretions furnishes information concerning the ways in which animals use chemicals to mark trails, harm their predators, and attract the opposite sex. Investigations of the biochemical pathways by which an organism metabolizes and otherwise transforms a compound are sources of insight into the detailed workings of the organism's bodily functions. Two classes of natural products, terpenes and steroids, have received particularly close attention from organic chemists.

Terpenes are constructed in plants from isoprene units

Most of you have smelled the strong odor emanating from freshly crushed plant leaves or orange peels. This odor is due to the liberation of a mixture of volatile compounds called **terpenes,** usually containing 10, 15, or 20 carbon atoms. Terpenes are used as food flavorings (the extracts from cloves and peppermint), as perfumes (roses, lavender, sandalwood), and as solvents (turpentine).

Terpenes are synthesized in the plant by the linkage of at least two molecular units containing five carbon atoms. The structure of these units is like that of

2-methyl-1,3-butadiene (isoprene), so they are referred to as **isoprene units.** Depending on how many isoprene units are incorporated into the structure, terpenes are classified as mono- (C_{10}), sesqui- (C_{15}), and diterpenes (C_{20}). (The isoprene building units are shown in color in the examples given here.)

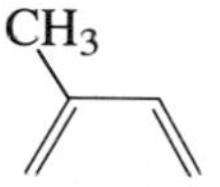

CH_3

2-Methyl-1,3-butadiene (Isoprene)

Isoprene unit in terpenes (Some contain double bonds)

Chrysanthemic acid is a monocyclic terpene containing a three-membered ring. Its esters are found in the flower heads of pyrethrum *(Chrysanthemum cinerariae-folium)* and are naturally occurring insecticides. A cyclobutane is present in grandisol, the sex-attracting chemical used by the male boll weevil *(Anthonomus grandis)*. Menthol (peppermint oil) is an example of a substituted cyclohexane natural product, whereas camphor (from the camphor tree) and β-cadinene (from juniper and cedar) are simple bicyclic terpenes, the first a norbornane system, the second a decalin derivative.

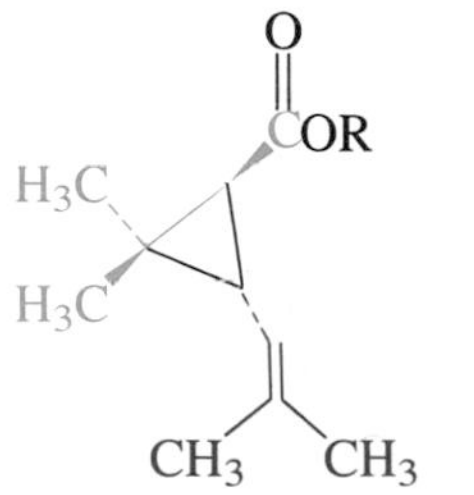

O, COR, H_3C, H_3C, CH_3, CH_3

***trans*-Chrysanthemic acid (R = H)**
***trans*-Chrysanthemic esters (R ≠ H)**

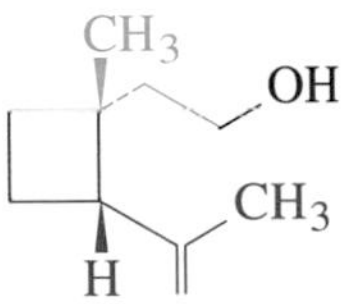

CH_3, OH, CH_3, H

Grandisol

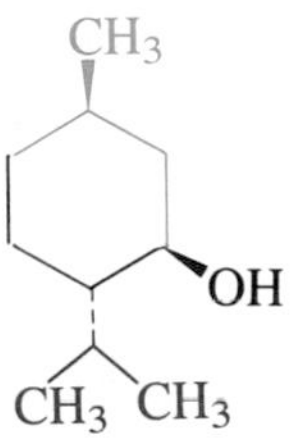

CH_3, OH, CH_3, CH_3

Menthol

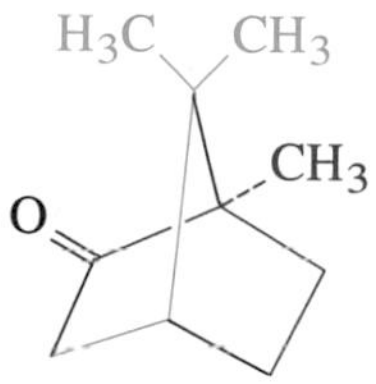

H_3C, CH_3, CH_3, O

Camphor

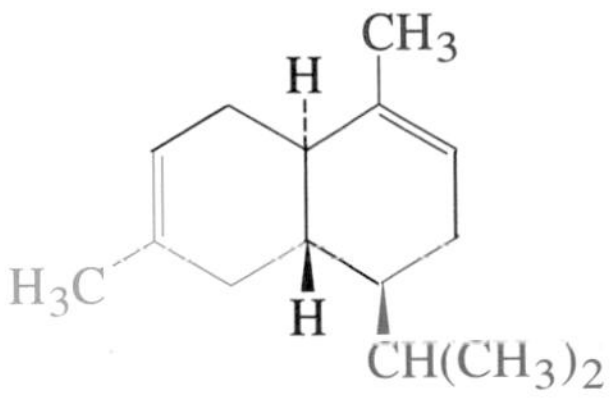

CH_3, H, H_3C, H, $CH(CH_3)_2$

β-Cadinene

EXERCISE 4-9

Draw the preferred chair conformation of menthol.

EXERCISE 4-10

The structures of two terpenes utilized by insects in defense are shown in the margin. Classify them as mono-, sesqui-, or diterpenes. Identify the isoprene units in each.

EXERCISE 4-11

After reviewing Section 2-1, specify the functional groups present in the terpenes shown in Section 4-7.

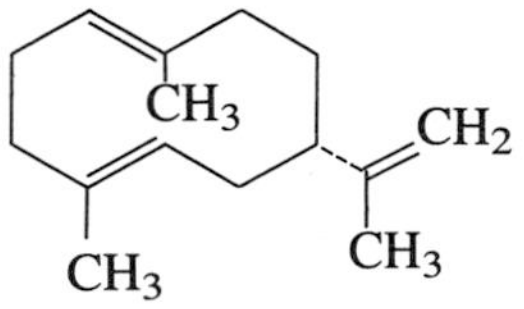

CH_3, CH_2, CH_3, CH_3

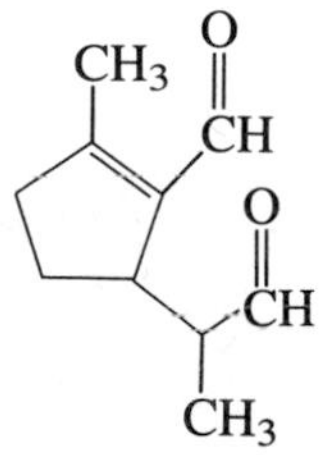

CH_3, O, CH, O, CH, CH_3

Steroids are tetracyclic natural products with powerful physiological activities

Steroids are abundant in nature, and many derivatives have physiological activity. Steroids frequently function as **hormones,** which are regulators of biochemical activity. In the human body, for example, they control sexual development and fertility, in addition to other functions. Because of this feature, many steroids, often the products of laboratory synthesis, are used in medicines in, for example, the treatment of cancer, arthritis, or allergies, and in birth control.

In the steroids, three cyclohexane rings are fused in such a way as to form an angle. The ring junctions are usually trans, as in *trans*-decalin. The fourth ring is a cyclopentane; its addition gives the typical tetracyclic structure. The four rings are labeled A, B, C, D, and the carbons are numbered according to a scheme specific to steroids. Many steroids have methyl groups attached to C10 and C13 and oxygen at C3 and C17. In addition, longer side chains may be found at C17. The trans fusion of the rings allows for a least strained all-chair configuration in which the methyl groups and hydrogen atoms at the ring junctions occupy axial positions.

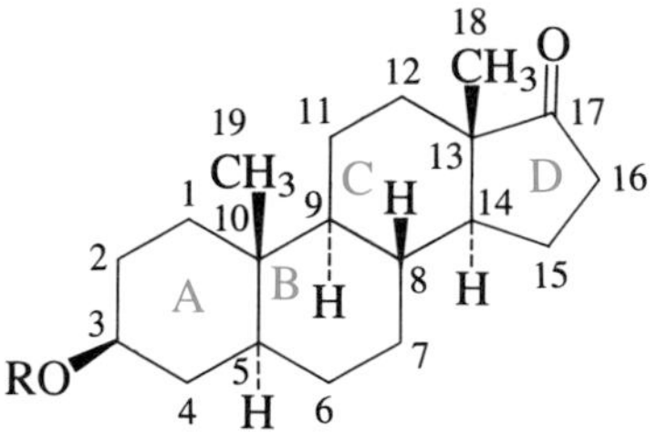

Steroid nucleus
(R = H, Epiandrosterone)

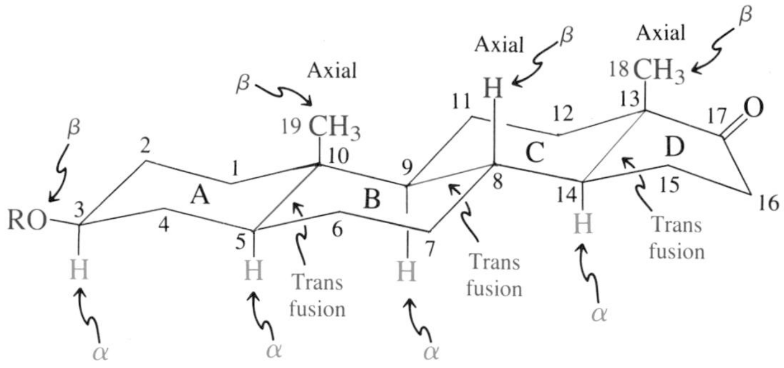

Groups attached above the plane of the steroid molecule as written are β substituents, whereas those below are referred to as α. Thus, the structure shown above has a 3β-OR, 5α-H, 10β-CH_3, and so forth. The axial methyl groups are also called **angular** methyls, because they protrude sharply from the general framework (*angulus,* Latin, at an angle; being at a sharp corner).

Among the most abundant steroids is cholesterol. It is present in almost all human and animal tissue, particularly in the brain and the spinal cord. In fact, it

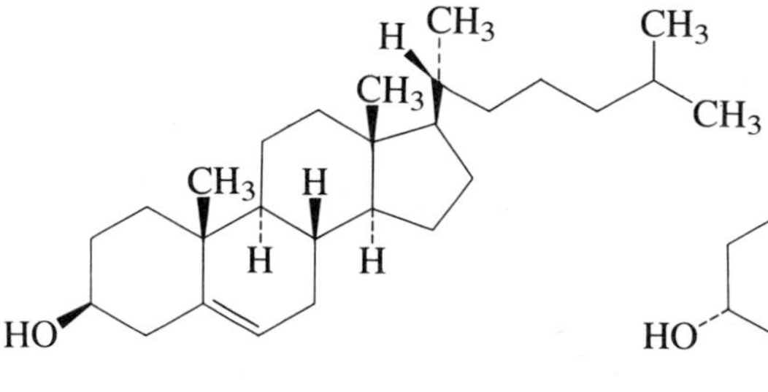

Cholesterol

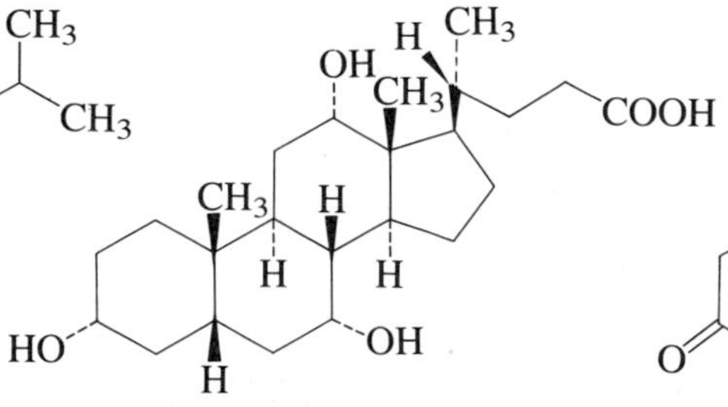

Cholic acid

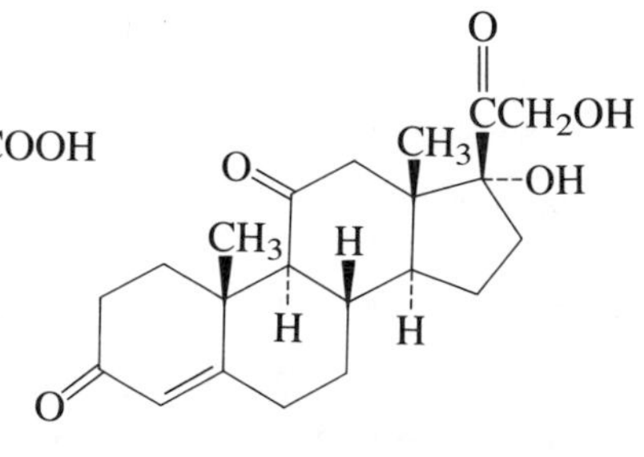

Cortisone

is so concentrated in the spinal cord of cattle that it is isolated from that tissue by simple extraction. The adult human body contains from 200 to 300 g of cholesterol; gallstones may consist entirely of it. This steroid has been implicated in several circulatory diseases because it precipitates in the arteries, thereby causing arteriosclerosis and heart diseases. Although its biological function in the body is not completely understood, it is a precursor of steroid hormones and bile acids. Bile acids are produced in the liver as part of a fluid delivered to the duodenum to aid in the emulsification, digestion, and absorption of fats. An example is cholic acid.

Cortisone, used extensively in the treatment of rheumatoid inflammations, is one of the adrenocortical hormones produced by the outer part (cortex) of the adrenal glands. These hormones participate in regulating the electrolyte and the water balance in the body, as well as protein and carbohydrate metabolism.

The sex hormones are divided into three groups: (1) the male sex hormones, or *androgens;* (2) the female sex hormones, or *estrogens;* and (3) the pregnancy hormones, or *progestins*. Testosterone is the principal male sex hormone. Produced by the testes, it is responsible for male (masculine) characteristics (deep voice, facial hair, general physical constitution). Estradiol is the principal female sex hormone. It was first isolated by extraction of four tons of sow ovaries, yielding only a few milligrams of pure steroid. Estradiol is responsible for the development of the secondary female characteristics and participates in the control of the menstrual cycle. An example of a progestin is progesterone, responsible for preparing the uterus for implantation of the fertilized egg.

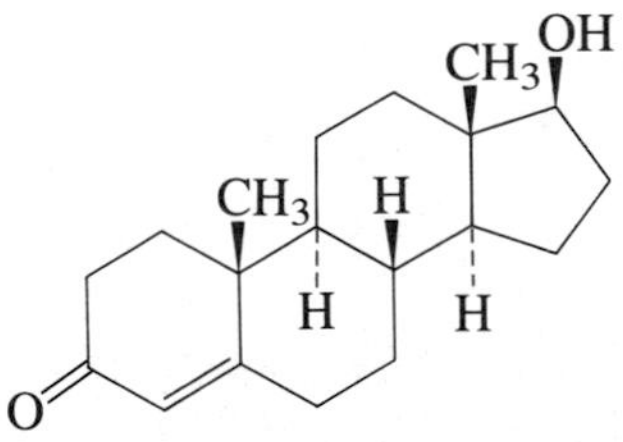

CH₃ OH CH₃ H H H O — **Testosterone**

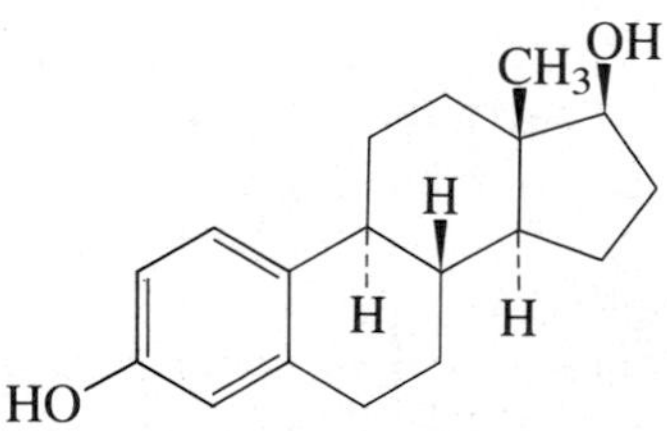

CH₃ OH H H H HO — **Estradiol**

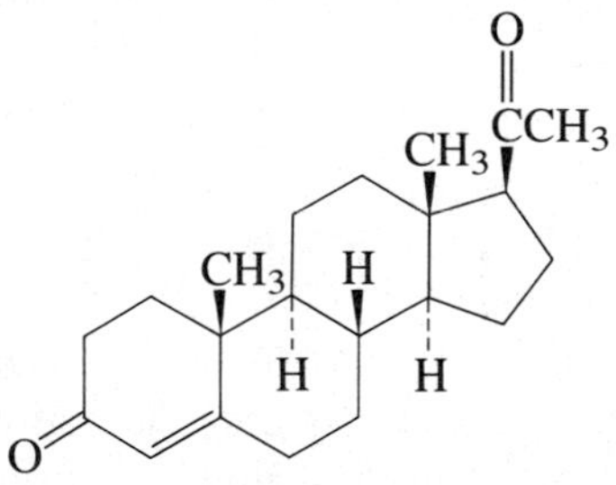

O CCH₃ CH₃ CH₃ H H H O — **Progesterone**

The structural similarity of the steroid hormones is remarkable considering their widely divergent activity. Steroids are the active ingredients of "the pill," functioning as an antifertility agent for the control of the female menstrual cycle and ovulation. At the peak of its use, more than 100 million women throughout the world took "the pill" as the primary form of contraception.

In summary, there is great variety in the structure and function of naturally occurring organic products, as manifested in the terpenes and the steroids. Natural products are frequently introduced in subsequent chapters to illustrate the presence and chemistry of a functional group, to demonstrate synthetic strategy or the use of reagents, to picture three-dimensional relations, and to exemplify medicinal applications. Several classes of natural products will be discussed more extensively: fats (Sections 19 12 and 20 4), carbohydrates (Chapter 24), alkaloids (Section 25-7), and amino and nucleic acids (Chapter 26).

BOX 4-1 Controlling Fertility: From "the Pill" to RU-486

The menstrual cycle is controlled by three protein hormones from the pituitary gland. The follicle-stimulating hormone (FSH) induces the growth of the egg and the luteinizing hormone (LH), its release from the ovaries. The third pituitary hormone (luteotropic hormone) induces formation of an ovarian tissue called the *corpus luteum*.

As the cycle begins and egg growth is initiated, the tissue around the egg secretes increasing quantities of estrogens. Once a certain concentration of estrogen in the bloodstream is reached, the production of FSH is turned off. The egg is released at this stage in response to LH. At the time of ovulation, LH also triggers the formation of the corpus luteum, which in turn begins to secrete increasing amounts of progesterone. The latter hormone suppresses any further ovulation by turning off the production of LH. If the egg is not fertilized, the corpus luteum regresses and the ovum and the *endometrium* (uterine lining) are expelled (menstruation). Pregnancy, on the other hand, leads to increased production of estrogens and progesterone to prevent pituitary hormone secretion and thus renewed ovulation.

The birth control pill consists of a mixture of synthetic potent estrogen and progesterone derivatives (more potent than the natural hormones), which, when taken throughout most of the menstrual cycle, prevent both development of the ovum and ovulation by turning off production of both FSH and LH. The female body is essentially being tricked into believing that it is pregnant. One of the commercial pills contains a combination of norethynodrel (2.5 mg) and

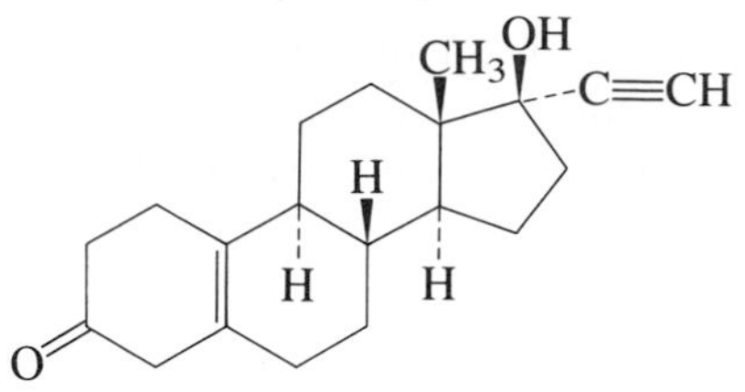

Norethynodrel

Mestranol

mestranol (0.1 mg). Other preparations consist of similar analogs with minor structural variations.

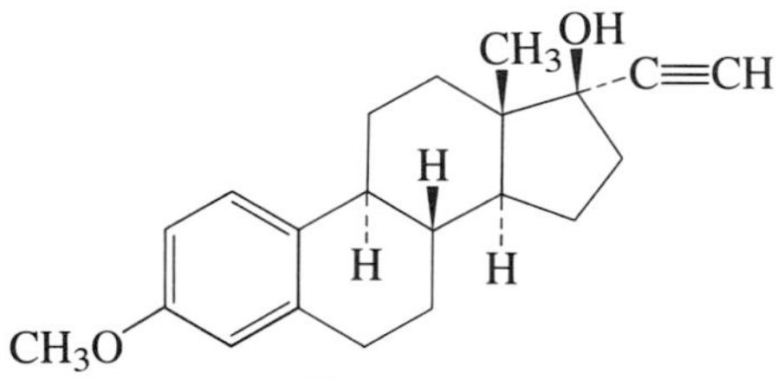

RU-486 is a synthetic steroid that blocks the effects of progesterone. Implantation of the fertilized egg does not occur because the necessary preparation of the endometrium is prevented. RU-486 has been used in France since 1988 as a "morning after" pill.

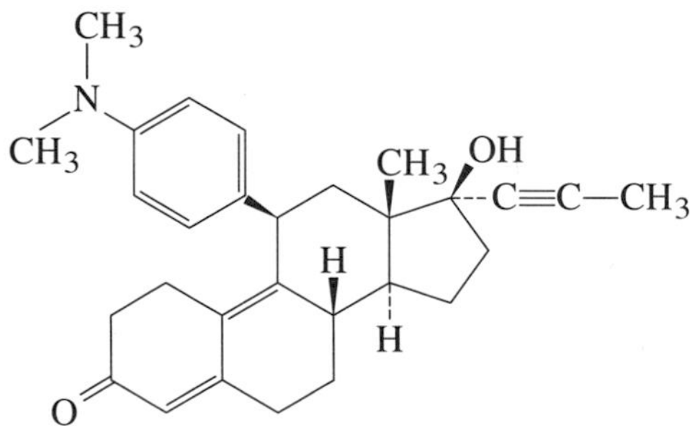

RU-486

IMPORTANT CONCEPTS

1. Cycloalkane nomenclature is derived from that of the straight-chain alkanes.
2. All but the 1,1-disubstituted cycloalkanes exist as two isomers: If both substituents are on the same face of the molecule, they are cis; if they are on opposite faces, they are trans. Cis and trans isomers are stereoisomers—compounds that have identical connectivities but differ in the arrangement of their atoms in space.

3. Some cycloalkanes are strained. Distortion of the bonds about tetrahedral carbon introduces bond-angle strain. Eclipsing or torsional strain results from inability of a structure to adopt staggered conformations about C–C bonds. Steric repulsion between atoms across a ring leads to transannular strain.
4. Bond-angle strain in the small cycloalkanes is largely accommodated by the formation of bent bonds.
5. Bond-angle, eclipsing, and other strain in the cycloalkanes larger than cyclopropane (which is by necessity flat) can be derived by deviations from planarity.
6. Ring strain in the small cycloalkanes gives rise to reactions that result in opening of the ring.
7. Deviations from planarity lead to conformationally mobile structures, such as chair, boat, and twist-boat cyclohexane. Chair cyclohexane is almost strain free.
8. Chair cyclohexane contains two types of hydrogens: axial and equatorial. These interconvert rapidly at room temperature by a conformational chair–chair ("flip") interconversion, with an activation energy of 10.8 kcal mol^{-1}.
9. In monosubstituted cyclohexanes, the $\Delta G°$ of equilibration between the two chair conformations is substituent dependent. Axial substituents are exposed to 1,3-diaxial interactions.
10. In more highly substituted cyclohexanes, substituent effects are often additive, the bulkiest substituents being the most likely to be equatorial.
11. Completely strain-free cycloalkanes are those that can readily adopt an all-*anti* conformation and lack transannular interactions.
12. Bicyclic ring systems may be fused or bridged. Fusion can be cis or trans.
13. Natural products are generally classified according to structure, physiological activity, taxonomy, and biochemical origin. Examples of the last class are the terpenes, of the first the steroids.
14. Terpenes are made up of isoprene units of five carbons.
15. Steroids contain three angularly fused cyclohexanes (A, B, C rings) attached to the cyclopentane D ring. Beta substituents are above the molecular plane, alpha substituents below.
16. An important class of steroids are the sex hormones, which have a number of physiological functions, including the control of fertility.

PROBLEMS

1. Write as many structures as you can that have the formula C_5H_{10} and contain one ring. Name them.

2. Name the following molecules according to the IUPAC nomenclature system.

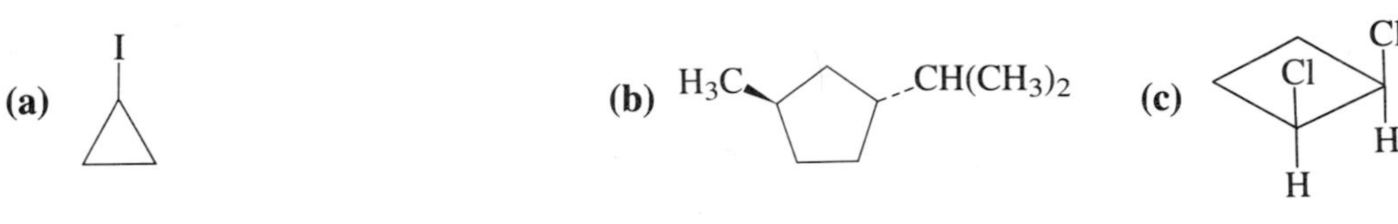

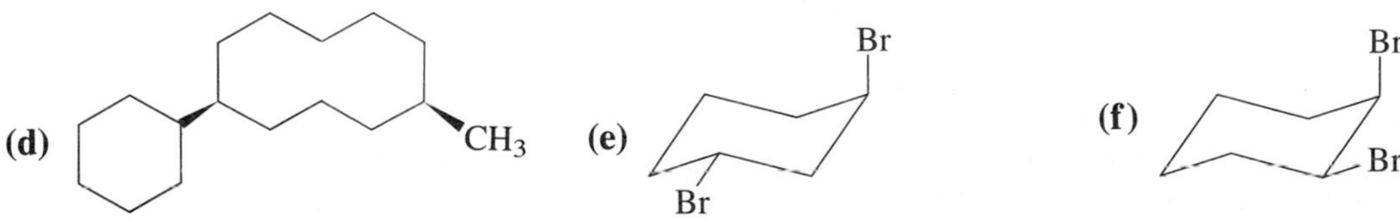

Reactivity per hydrogen toward Cl·

Cycloalkane	Reactivity
Cyclopentane	0.9
Cyclobutane	0.7
Cyclopropane	0.1

Note: Relative to cyclohexane = 1.0; at 68°C, $h\nu$, CCl_4 solvent

3. The kinetic data for the radical chain chlorination of several cycloalkanes (see the table on the left) illustrate that the C–H bonds of cyclopropane and, to a lesser extent, cyclobutane are somewhat abnormal. **(a)** What do these data tell you about the strength of the cyclopropane C–H bond and the stability of the cyclopropyl radical? **(b)** Suggest a reason for the stability characteristics of the cyclopropyl radical. (Hint: Consider bond-angle strain in the radical relative to cyclopropane itself.)

4. Use the data in Tables 3-2 and 4-2 to estimate the $DH°$ value for a C–C bond in **(a)** cyclopropane; **(b)** cyclobutane; **(c)** cyclopentane; and **(d)** cyclohexane.

5. Draw each of the following substituted cyclobutanes in its two interconverting "puckered" conformations (Figure 4-3). When the two conformations differ in energy, identify the more stable shape and indicate the form(s) of strain that raise the relative energy of the less stable one. (Hint: Puckered cyclobutane has axial and equatorial positions similar to those in chair cyclohexane.)

(a) Methylcyclobutane
(b) *cis*-1,2-Dimethylcyclobutane
(c) *trans*-1,2-Dimethylcyclobutane
(d) *cis*-1,3-Dimethylcyclobutane
(e) *trans*-1,3-Dimethylcyclobutane

Which is more stable: *cis*- or *trans*-1,2-dimethylcyclobutane; *cis*- or *trans*-1,3-dimethylcyclobutane?

6. Estimate $\Delta H°$ for the following interconversions. **(a)** Planar cyclopentane $\rightleftharpoons$ envelope cyclopentane; **(b)** planar cyclohexane $\rightleftharpoons$ chair cyclohexane. Consider only differences in eclipsing H–H interactions (ignore bond-angle strain differences).

7. For each of the following cyclohexane derivatives, indicate (i) whether the molecule is a cis or trans isomer and (ii) whether it is in its most stable conformation. If your answer to (ii) is no, "flip" the ring and draw its most stable conformation.

(a)
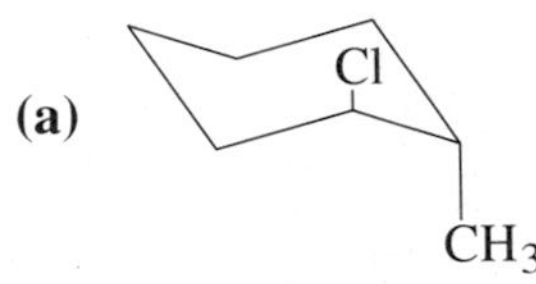

(b) OCH_3, NH_2

(c) HO, $CH(CH_3)_2$

(d) CH_3, C, O, OCH_3

(e) CH_2CH_3, I

(f) Br, NH_2

(g) CH_3, OCH_3

(h) F, HO

(i) O, OH, C, Br

(j) Cl, $CH(CH_3)_2$

8. Draw the most stable conformation for each of the following substituted cyclohexanes; then, in each case, "flip" the ring and redraw the molecule in the higher energy chair conformation. **(a)** Cyclohexanol; **(b)** *trans*-3-methylcyclohexanol; **(c)** *cis*-1-(1-methylethyl)-3-methylcyclohexane; **(d)** *trans*-1-ethyl-3-methoxycyclohexane; **(e)** *trans*-1-chloro-4-(1,1-dimethylethyl)-cyclohexane.

9. For each molecule in Problem 8, estimate the energy difference between the most stable and next best conformation. Calculate the approximate ratio of the two at 300 K.

10. Draw all the possible all-chair conformers of cyclohexylcyclohexane.

11. What is the most stable of the four *boat* conformations of methylcyclohexane, and why?

12. The most stable conformation of *trans*-1,3-bis(1,1-dimethylethyl)-cyclohexane is not a chair. What conformation would you predict for this molecule? Explain.

13. The bicyclic hydrocarbon formed by the fusion of a cyclohexane ring with a cyclopentane ring is known as hexahydroindane. Using the drawings of *trans*- and *cis*-decalin for reference (Figure 4-13), draw the structures of *trans*- and *cis*-hexahydroindane, showing each ring in its most stable conformation.

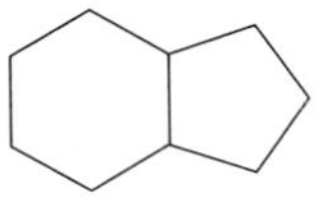

Hexahydroindane

14. On viewing the drawings of *cis*- and *trans*-decalin in Figure 4-13, which do you think is more stable? Estimate the energy difference between the two isomers. Answer the same question for the cis and trans dimethyl-substituted decalins shown below.

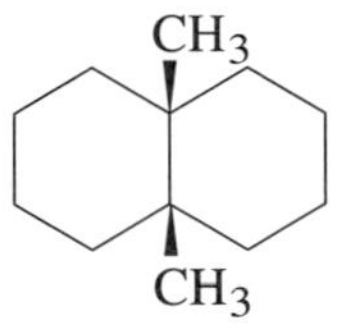

CH_3 CH_3 CH_3 CH_3

***cis*-9,10-Dimethyldecalin** 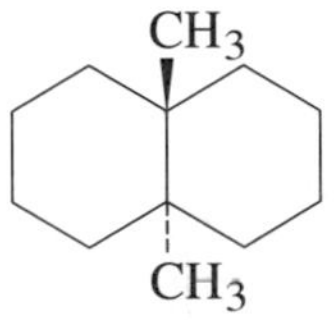 ***trans*-9,10-Dimethyldecalin**

15. Identify each of the molecules below as a monoterpene, a sesquiterpene, or a diterpene (all names are common).

(a) H_3C H_3C CH_3 CH_2OH **Geraniol**

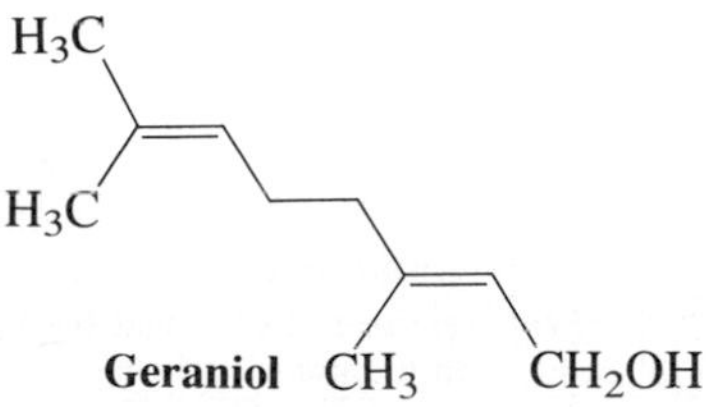

(b) CH_3 CH_2 O $=CH_2$ O **Eremanthin**

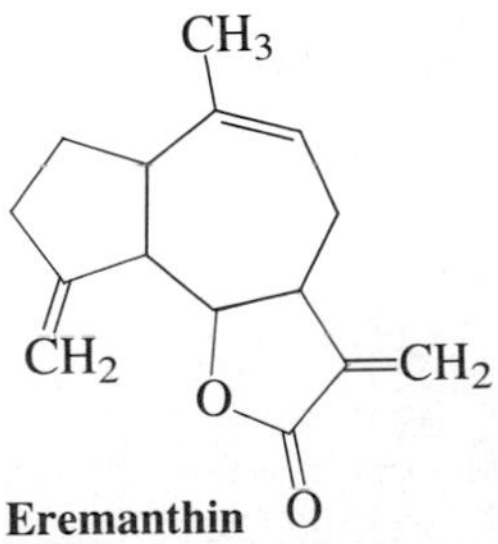

(c) CH_3 CH_3 OH CH_3 CH_2 **Eudesmol**

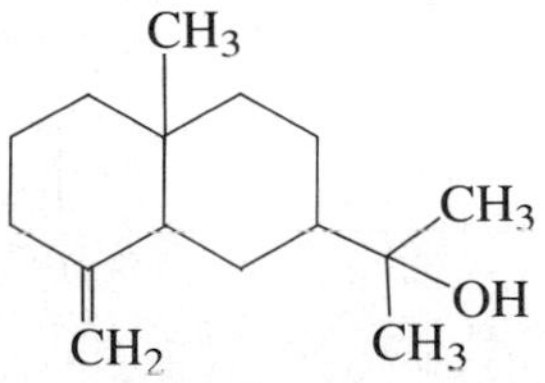

(d)

Ipomeamarone

(e)

Genipin

(f)

Castoramine

(g)

Cantharidin

(h)

Vitamin A

16. Find the 2-methyl-1,3-butadiene (isoprene) units in each of the naturally occurring organic molecules pictured in Problem 15.

17. Circle and identify by name all the functional groups in any three of the steroids illustrated in Section 4-7. Label any polarized bonds with partial positive and negative charges (δ^+ and δ^-).

18. Several additional examples of naturally occurring molecules with strained ring structures are shown below.

1-Aminocyclopropane-carboxylic acid
(Present in plants, this molecule plays a role in the ripening of fruits and the dropping of autumn leaves)

α-Pinene
(Present in cedar-wood oil)

Africanone
(Also a plant-leaf oil)

Thymidine dimer
(A component of DNA that has been exposed to ultraviolet light)

Identify the terpenes (if any) in the preceding group of structures. Find the 2-methyl-1,3-butadiene units in each structure and classify them as mono-, sesqui-, or diterpene.

19. If cyclobutane were flat, it would have exactly 90° C–C–C bond angles and could conceivably use pure *p* orbitals in its C–C bonds. What would be a possible hybridization for the carbon atoms of the molecule that would allow all the C–H bonds to be equivalent? Exactly where would the hydrogens on each carbon be located? What are the real structural features of the cyclobutane molecule that contradict this hypothesis?

20. Compare the structure of cyclodecane in an all-chair conformation (illustrated below) with that of *trans*-decalin. Explain why all-chair cyclodecane is highly strained, and yet *trans*-decalin is nearly strain free. Make models.

All-chair cyclodecane **trans-Decalin**

21. Fusidic acid is a steroidlike microbial product that is an extremely potent antibiotic with a broad spectrum of biological activity. Its molecular shape is most unusual and has supplied important clues to researchers investigating the methods by which steroids are synthesized in nature.

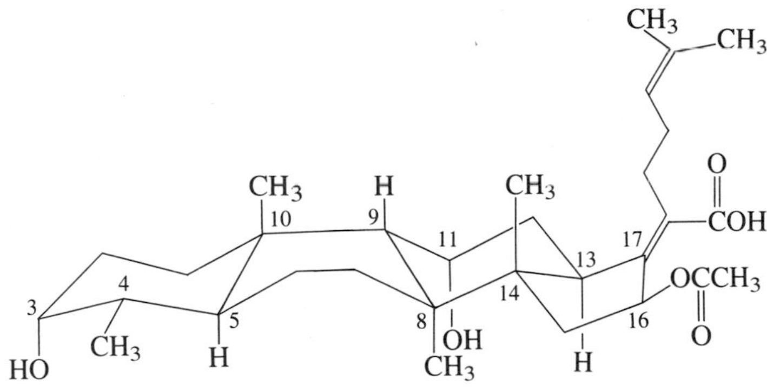

Fusidic acid

(a) Locate all the rings in fusidic acid and describe their conformations. **(b)** Identify all ring fusions in the molecule as having either cis or trans geometry. **(c)** Identify all groups attached to the rings as being either α- or β-substituents. **(d)** Describe in detail how this molecule differs from the typical steroid in structure and stereochemistry. (As an aid to answering these questions, the carbon atoms of the framework of the molecule have been numbered.)

22. The enzymatic oxidation of alkanes to produce alcohols (Section 3-9) is a simplified version of the reactions that produce the adrenocortical steroid hormones. In the biosynthesis of corticosterone from progesterone (Section 4-7), two such oxidations take place successively. It is thought that the monooxygenase enzymes act as complex oxygen-atom donors in these reactions. A suggested mechanism, as applied to cyclohexane, consists of the two steps shown below the biosynthesis on the next page.

Progesterone

Steroid 21-monooxygenase, O_2 →

Steroid 11β-monooxygenase, O_2 →

Corticosterone

$$\text{cyclohexane} + \text{O (atom)} \xrightarrow{\text{Enzyme}} \text{cyclohexyl radical} + \cdot\text{OH} \longrightarrow \text{cyclohexanol}$$

Calculate $\Delta H°$ for each step and for the overall oxidation reaction of cyclohexane. Use the following $\Delta H_f°$ values: cyclohexane, -29.5 kcal mol^{-1}; cyclohexanol, -68.4 kcal mol^{-1}; cyclohexyl radical, $+15$ kcal mol^{-1}; $HO\cdot$, $+9.4$ kcal mol^{-1}; O (atom), $+59.6$ kcal mol^{-1}.

23. Iodobenzene dichloride, formed by the reaction of iodobenzene and chlorine, is, like sulfuryl chloride and NCS (Section 3-8), a reagent for the chlorination of alkane C–H bonds. Chlorinations using iodobenzene dichloride are initiated by light.

$$C_6H_5\text{–I} + Cl_2 \longrightarrow C_6H_5\text{–I}Cl_2$$

Iodobenzene dichloride

(a) Propose a radical chain mechanism for the chlorination of a typical alkane RH by iodobenzene dichloride. To get you started, the overall equation for the reaction is given below, as is the initiation step.

$$RH + C_6H_5\text{–}ICl_2 \longrightarrow RCl + HCl + C_6H_5\text{–I}$$

Initiation: $C_6H_5\text{–}ICl_2 \xrightarrow{h\nu} C_6H_5\text{–}\dot{I}\text{–Cl} + Cl\cdot$

(b) Radical chlorination of typical steroids by iodobenzene dichloride gives, predominantly, three isomeric monochlorination products. On the basis of both reactivity (tertiary, secondary, primary) considerations and steric effects (which might hinder the approach of a reagent toward a C–H bond that might otherwise be reactive), predict the three major sites of chlorination in the steroid molecule. Either make a model or carefully analyze the drawings of the steroid nucleus in Section 4-7.

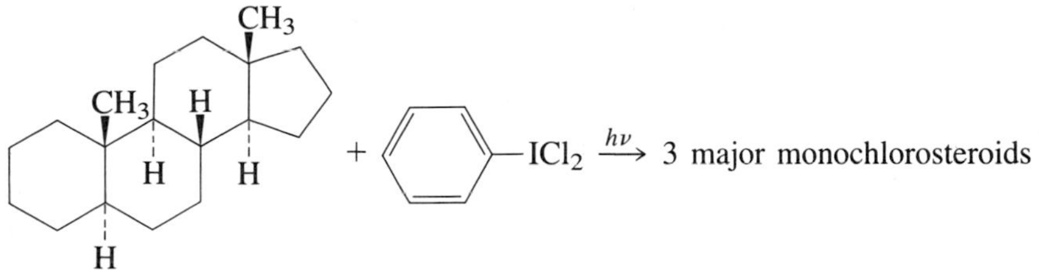

$$\text{steroid} + C_6H_5\text{–}ICl_2 \xrightarrow{h\nu} \text{3 major monochlorosteroids}$$

24. As Problem 22 indicates, the enzymatic reactions that introduce functional groups into the steroid nucleus in nature are highly selective, unlike the laboratory chlorination described in Problem 23. However, by means of a clever adaptation of this reaction, it is possible to partly mimic nature's selectivity in the laboratory. Two such examples are illustrated below.

(a) $\xrightarrow{Cl_2,\ h\nu}$

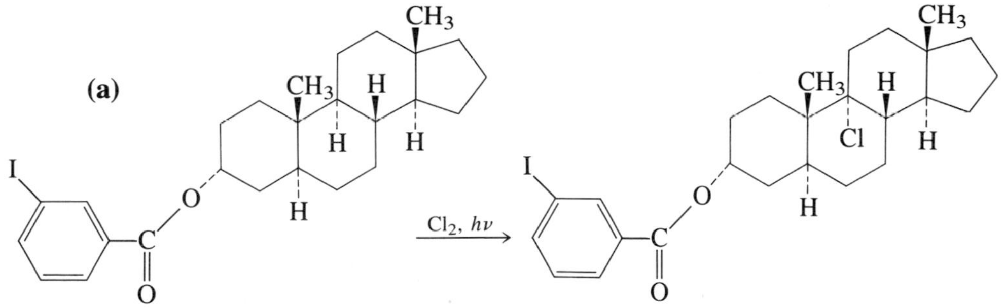

(b) $\xrightarrow{Cl_2,\ h\nu}$

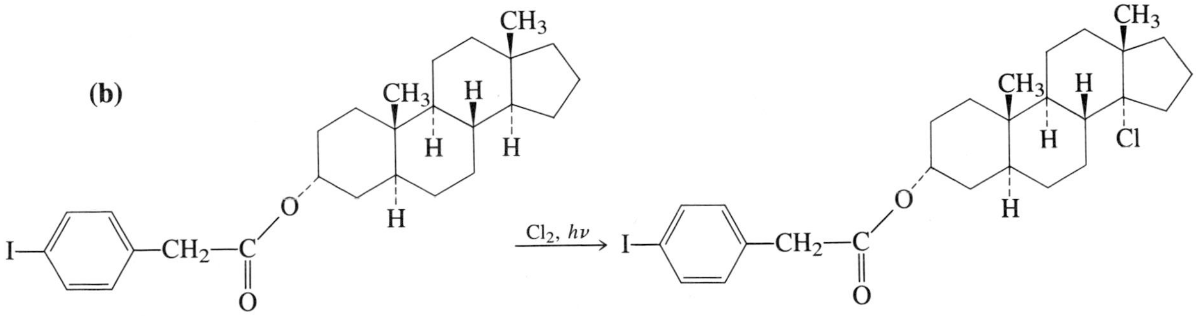

Propose reasonable explanations for the results of these two reactions. Make a model of the product of the addition of Cl_2 to each iodo-compound (compare Problem 23) to help you analyze each system.

5

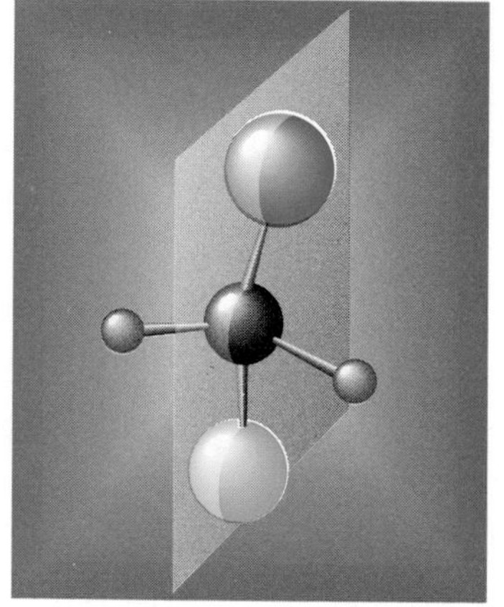

Stereoisomers

The preceding chapters have dealt with two kinds of isomerism: constitutional (also called structural) and stereo. **Constitutional isomerism** describes compounds that have identical molecular formulas but differ in the order in which the individual atoms are connected.

Constitutional Isomers

C_4H_{10}	$CH_3CH_2CH_2CH_3$ **Butane**	$H_3C{-}CH(CH_3){-}CH_3$ **2-Methylpropane**
C_2H_6O	CH_3CH_2OH **Ethanol**	CH_3OCH_3 **Methoxymethane (Dimethyl ether)**

EXERCISE 5-1

Are cyclopropylcyclopentane and cyclobutylcyclobutane isomers?

Stereoisomerism describes isomers whose atoms are connected in the same order but differ in their spatial arrangement. Examples of stereoisomers include the relatively stable cis-trans isomers and the rapidly equilibrating conformational ones (Section 4-1).

Stereoisomers

***cis*-1,3-Dimethylcyclopentane** ***trans*-1,3-Dimethylcyclopentane**

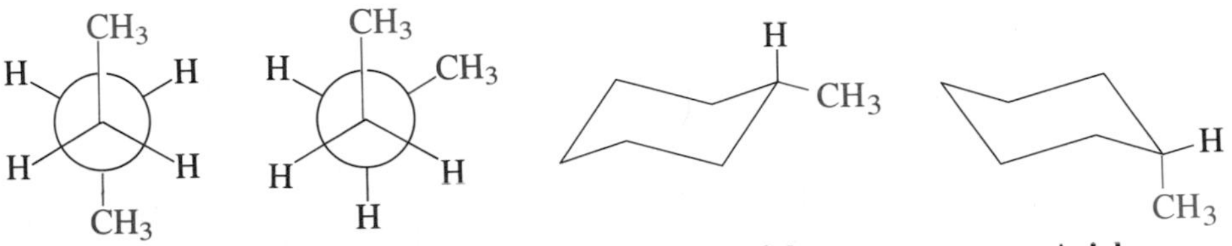

***Anti* rotamer of butane** ***Gauche* rotamer of butane** **Equatorial methylcyclohexane** **Axial methylcyclohexane**

EXERCISE 5-2

Draw additional stereoisomers of methylcyclohexane.

This chapter introduces another type of stereoisomerism, one that has as its basis the "handedness" of certain molecules. We shall see that there are structures that are not superimposable on their mirror images, just as your left hand is not superimposable on your right hand. The two are therefore different objects; they may have different properties and may react differently. Their odors and other biological characteristics also may be entirely different.

5-1 Chiral Molecules

How can a molecule exist as two nonsuperimposable mirror images? Consider the radical bromination of butane. This reaction proceeds mainly at one of the secondary carbons to furnish 2-bromobutane. A molecular model of the starting material *seems* to show that either of the two hydrogens on that carbon may be replaced to give only one form of 2-bromobutane (Figure 5-1). Is this really true, however?

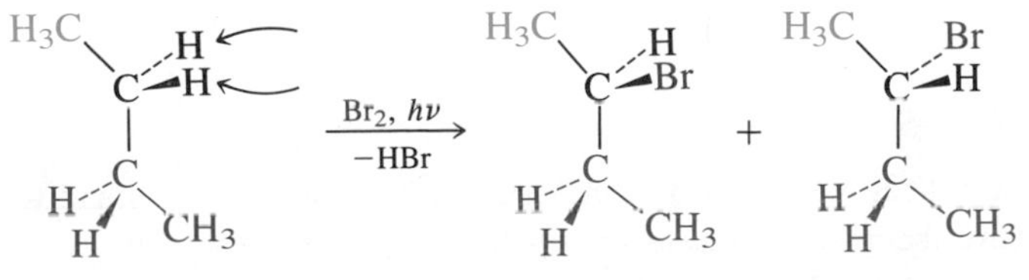

Br_2, $h\nu$ / $-HBr$

FIGURE 5-1 Replacement of one of the secondary hydrogens in butane results in two stereoisomeric forms of 2-bromobutane.

Chiral molecules cannot be superimposed on their mirror images

Look more closely at the 2-bromobutanes obtained by replacing either of the methylene hydrogens with bromine. In fact, the two structures are nonsuperimposable and therefore *not identical*. The two molecules are related as object and mirror image, and to convert one into the other would require the breaking of bonds. Pairs of molecules that exist as nonsuperimposable mirror images of each other are **enantiomers** (*enantios*, Greek, opposite). Such compounds are **chiral.** In our example of the bromination of butane, a 1:1 mixture of enantiomers is formed.

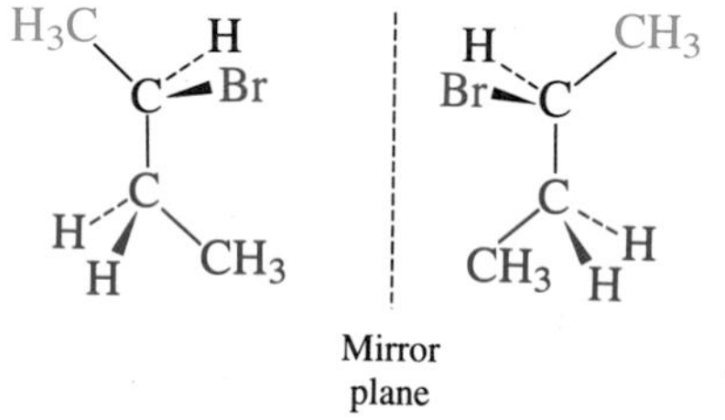

Mirror plane

In contrast with chiral molecules, such as 2-bromobutane, compounds having structures that *are* superimposable on their mirror images are **achiral.** Examples of chiral and achiral molecules are shown below. The first two chiral structures depicted are enantiomers of each other.

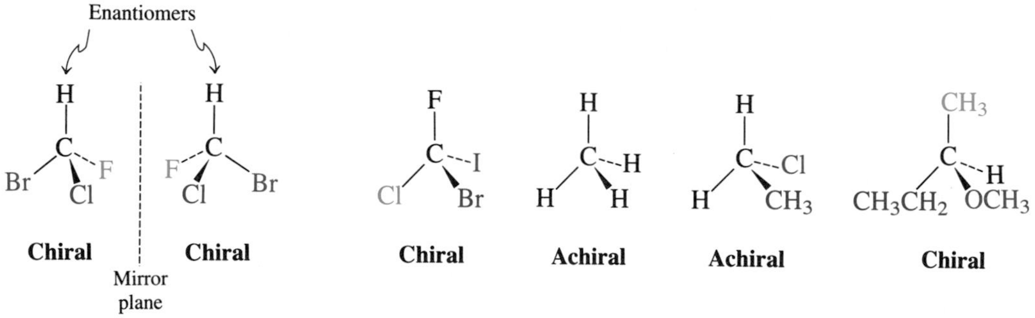

Enantiomers

Chiral | Chiral | Chiral | Achiral | Achiral | Chiral

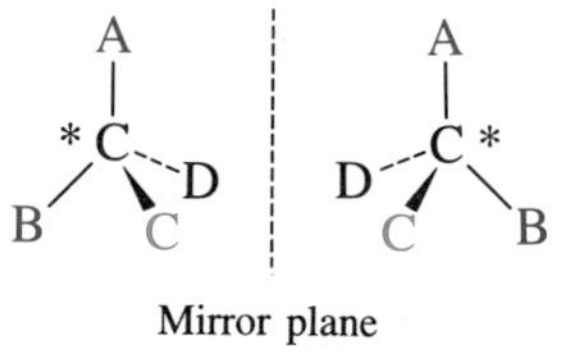

Mirror plane

Mirror plane

(C^* = a stereocenter based on asymmetric carbon)

All the chiral examples contain an atom that is connected to four *different* substituent groups. Such a nucleus is called an **asymmetric atom** (e.g., asymmetric carbon), or a **stereocenter.** Centers of this type are sometimes denoted by an asterisk. *Molecules with one stereocenter are always chiral.* (We shall see in Section 5-5 that structures incorporating more than one such center need *not* be chiral.)

EXERCISE 5-3

Among the natural products shown in Section 4-7, which are chiral and which are achiral? Give the number of stereocenters in each case.

BOX 5-1 Chiral Substances in Nature

Many organic compounds exist in nature as only one enantiomer, somc as both. For example, natural *alanine* (systematic name: 2-aminopropanoic acid) is an abundant amino acid that is found in only one form. *Lactic acid* (2-hydroxypropanoic acid), however, is present in blood and muscle fluid as one enantiomer but in sour milk and some fruits and plants as a mixture of the two.

H
C*
H_2N CH_3 COOH

2-Aminopropanoic acid (Alanine)

H
C*
HO CH_3 COOH

2-Hydroxypropanoic acid (Lactic acid)

Another case is *carvone* [2-methyl-5-(1-methylethenyl)-2-cyclohexenone], which contains a stereocenter in a six-membered ring. This carbon atom may be thought of as bearing four different groups if we consider the ring itself to be two separate and different substituents. They are different because, starting from the stereocenter, the clockwise sequence of atoms differs from the counterclockwise sequence. Carvone is found in nature in both enantiomeric forms. The characteristic odor of caraway and dill seed is due to the enantiomer shown, whereas the flavor of spearmint is due to the other enantiomer.

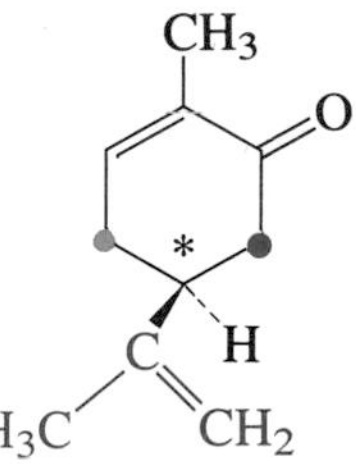

2-Methyl-5-(1-methylethenyl)-2-cyclohexenone (Carvone)

The symmetry in molecules helps to distinguish chiral structures from achiral ones

The word *chiral* is derived from the Greek *cheir,* meaning "hand" or "handedness." Human hands have the mirror-image relation typical of enantiomers (Figure 5-2). Among the many other objects that are chiral are shoes, ears, screws, and spiral staircases. On the other hand, there are many achiral objects, such as balls, ordinary water glasses, hammers, and nails.

Many chiral objects, such as spiral staircases, do not have stereocenters. This statement is true for many chiral molecules. *Remember that the only criterion for*

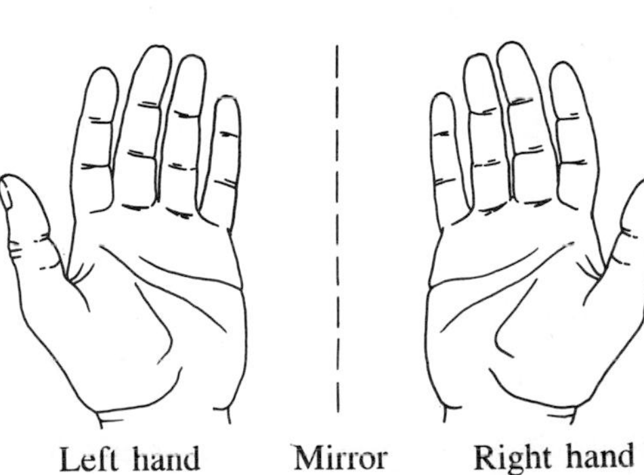

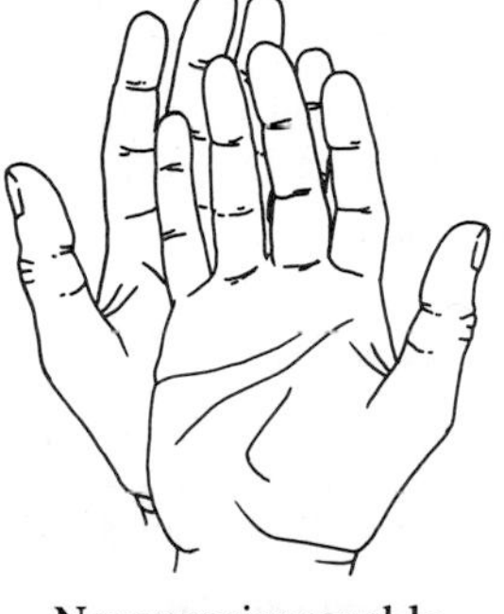

FIGURE 5-2 Left and right hands as models for enantiomeric relations. Like these mirror images, chiral molecules cannot be superimposed on their corresponding enantiomers.

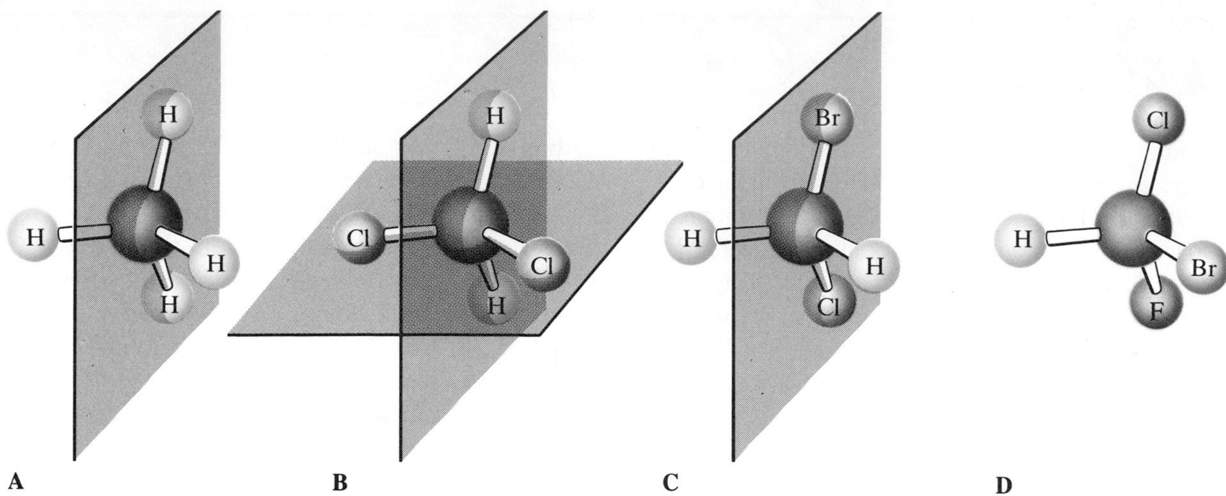

FIGURE 5-3 Examples of planes of symmetry in (A) methane (only one is shown), (B) dichloromethane, and (C) bromochloromethane. Bromochlorofluoromethane (D) has none. Chiral molecules cannot have a plane of symmetry.

chirality is the nonsuperimposable nature of object and mirror image. In this chapter we shall confine our discussion to molecules that are chiral as a result of the presence of stereocenters. But how do we determine whether a molecule is chiral or not? As you have undoubtedly already noticed, it is not always easy to tell. A foolproof way is to construct molecular models of the molecule and its mirror image and look for superimposability. However, this procedure is very time consuming. A simpler method is to look for symmetry in the molecule under investigation.

For most organic molecules, we have to consider only one test for chirality: the presence of a plane of symmetry. A **plane of symmetry** (or **mirror plane**) is one that bisects the molecule so that the part of the structure lying on one side of the plane mirrors the part on the other side. For example, methane has six planes of symmetry, dichloromethane has two, bromochloromethane has one, and bromochlorofluoromethane has none. In Figure 5-3 only one of those mirror planes is shown for methane.

How do we use this idea to distinguish a chiral molecule from an achiral one? *Chiral molecules cannot have a plane of symmetry*. For example, the first three methanes in Figure 5-3 are clearly achiral because of the presence of a mirror plane. You will be able to classify most molecules in this book as chiral or achiral simply by identifying the presence or absence of a plane of symmetry.

EXERCISE 5-4

Draw pictures of the following common achiral objects, indicating the plane of symmetry in each: a ball, an ordinary water glass, a hammer, a chair, a suitcase, a toothbrush.

EXERCISE 5-5

Write the structures of all dimethylcyclobutanes. Specify those that are chiral. Show the mirror planes in those that are not.

To summarize, a chiral molecule exists in either of two stereoisomeric forms called enantiomers. These are related as object and nonsuperimposable mirror image. Most chiral organic molecules contain stereocenters, although chiral structures that lack such centers do exist. A molecule that contains a plane of symmetry is achiral.

5-2 Optical Activity

Considering their close similarity, we may wonder how it is possible to distinguish one enantiomer from another. This task is indeed a very difficult one, because most *physical* properties of enantiomers are identical. A notable exception is the interaction with a special type of light.

Our first example of a chiral molecule was the two enantiomers of 2-bromobutane. If we were to isolate each enantiomer in pure form, we would find that we cannot distinguish between them on the basis of their physical properties, such as boiling points, melting points, and densities. This result should not surprise us: Their bonds are identical and so are their energy contents. However, when plane-polarized light (which will be defined shortly) is passed through a sample of one of the enantiomers, the plane of polarization of the incoming light is *rotated* in one direction (either clockwise or counterclockwise). When the same experiment is repeated with the other enantiomer, the plane of the polarized light is rotated by exactly the same amount, *but in the opposite direction.*

An enantiomer that rotates the plane of light in a clockwise sense as the viewer faces the light source is **dextrorotatory** (*dexter,* Latin, right) and the compound is (arbitrarily) labeled the (+) enantiomer. Consequently, the other enantiomer, which will effect counterclockwise rotation, is **levorotatory** (*laevus,* Latin, left) and labeled the (−) enantiomer. This special interaction with light is called **optical activity,** and enantiomers are frequently called **optical isomers.**

Optical rotation is measured with a polarimeter

What is plane-polarized light, and how is its rotation measured? Light may be thought of as **electromagnetic radiation,** which consists of oscillating electric- and magnetic-field vectors at right angles and perpendicular to the light path (Figure 5-4A). In ordinary light, the fields extend in all directions (Figure 5-4B). The situation changes when ordinary light is passed through a **Nicol prism,** which allows only one of the infinite number of planes of light to pass through.

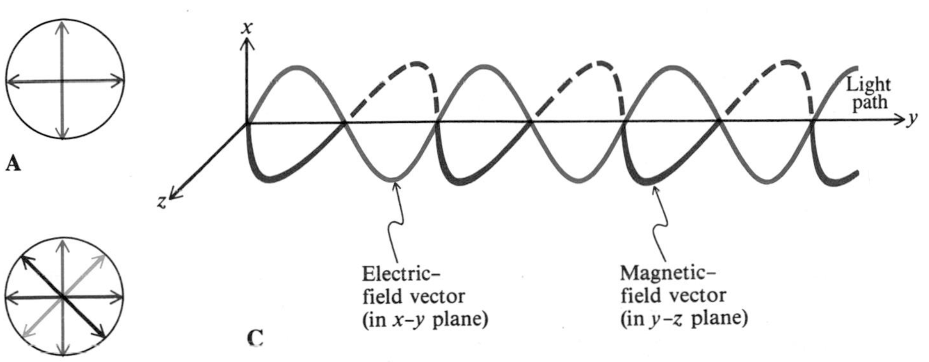

FIGURE 5-4 Representation of electromagnetic radiation. (A) One electric-field vector and the perpendicular magnetic field. In this representation the path of the light is perpendicular to the plane of the page, and the field vectors extend into the plane. (B) Ordinary light has field vectors in all directions. (C) A different picture of (A) in which the light travels from the left to the right side of the page. (A) and (C) are also representative of plane-polarized light. The x–y plane is the plane of polarization.

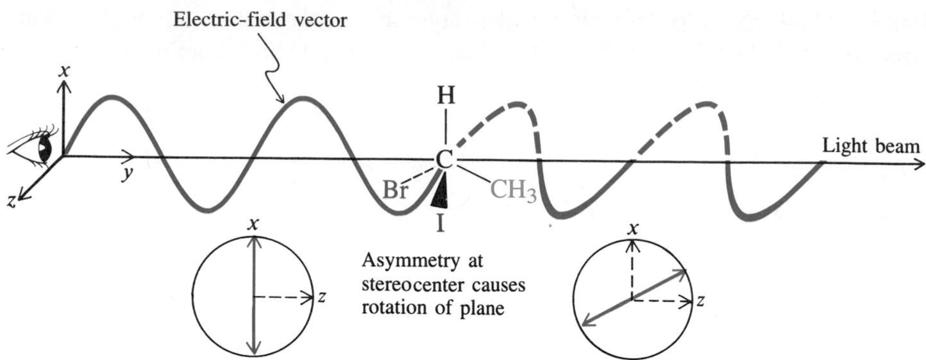

FIGURE 5-5 Unequal interaction of the oscillating electric field of plane-polarized light with a chiral molecule. "Left is not equal to right"; the consequence is a rotation of the plane of polarization, and the molecule is said to be optically active.

The resulting beam is said to be **plane polarized,** meaning that the electric field vectors lie in one plane, called the **plane of polarization** (Figure 5-4C).

When light travels through a molecule, the electrons around the nuclei and in the various bonds interact with the electric field of the light beam. If the light is plane polarized, the electric-field vector may or may not change, depending on the molecule. If the substance is achiral, the vector remains the same, and the molecule is **optically inactive.** Conversely, when a beam of plane-polarized light is passed through a chiral substance, the electric field interacts differently with, say, the "left" and "right" halves of the molecule (Figure 5-5). This interaction results in a rotation of the plane of polarization, called **optical rotation;** the sample giving rise to it is referred to as **optically active.**

Optical rotations are measured by using a **polarimeter** (Figure 5-6). The light source used most frequently is a **monochromatic** (only one wavelength) sodium D lamp (λ = 5890 Å). The light is first plane polarized by a Nicol prism. It subsequently traverses a cell containing the sample. Rotation of the plane is detected by rotating another Nicol prism, the analyzer, to maximize transmittance of the light beam to the eye of the observer. The measured rotation (in degrees) is the **observed optical rotation,** α, of the sample. Its value depends on the concentration and structure of the optically active molecule, the length of the sample cell, the wavelength of the light, the solvent, and the temperature. To

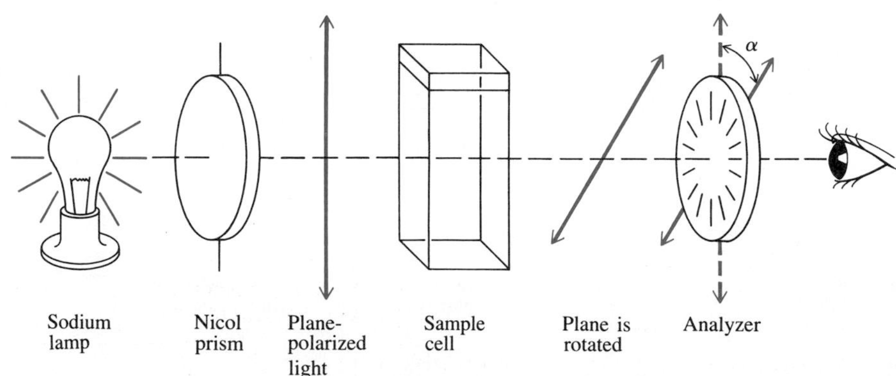

FIGURE 5-6 A polarimeter, which is used to measure optical rotation. When the analyzer is rotated to maximize intensity of the transmitted beam, it is aligned with the plane of polarization of the light emerging from the sample cell.

avoid ambiguities, chemists have agreed on a standard value of α, the **specific rotation,** $[\alpha]$, for each compound. This quantity (which is solvent dependent) is defined as

Specific Rotation

$$[\alpha]_{\lambda}^{t°} = \frac{\alpha}{l \cdot c}$$

where $[\alpha]$ = specific rotation
t = temperature in °C
λ = wavelength of incident light; for the sodium D lamp, indicated simply by ''D''
α = observed optical rotation in degrees
l = length of sample container in decimeters; its value is frequently 1 (i.e., 10 cm)
c = concentration (g mL^{-1} of solution)

EXERCISE 5-6

A solution of 1 g of common table sugar (the naturally occurring form of sucrose) in 10 mL of water in a 10-cm cell exhibits a clockwise optical rotation of 6.65°. Calculate $[\alpha]$. Does this information tell you $[\alpha]$ for the enantiomer of natural sucrose?

The specific rotation of an optically active molecule is a physical constant characteristic of that molecule, just like its melting point, boiling point, and density. Four specific rotations are recorded in Table 5-1.

Optical rotation indicates enantiomeric composition

As mentioned, enantiomers rotate plane-polarized light by equal amounts but in opposite directions. Thus, in 2-bromobutane the (−) enantiomer rotates this plane counterclockwise by 23.1°, its mirror image (+)-2-bromobutane clockwise

TABLE 5-1 Specific Rotations of Various Chiral Compounds $[\alpha]_D^{25°C}$

Compound	$[\alpha]_D^{25°C}$	Compound	$[\alpha]_D^{25°C}$
CH_2CH_3, H, Br, CH_3 on C **(−)-2-Bromobutane**	−23.1°	CH_2CH_3, H, CH_3, Br on C **(+)-2-Bromobutane**	+23.1°
H, H_2N, CH_3, COOH on C **(+)-2-Aminopropanoic acid [(+)-Alanine]**	+8.5°	H, OH, HOOC, CH_3 on C **(−)-2-Hydroxypropanoic acid [(−)-Lactic acid]**	−3.8°

Note: Pure liquid for the haloalkane; in aqueous solution for the acids.

by 23.1°. It follows that a 1:1 mixture of (+) and (−) enantiomers shows no rotation and is therefore optically inactive. Such a mixture is called a **racemate** or a **racemic mixture.** If one enantiomer equilibrates with its mirror image by some process, it is said to undergo **racemization,** or to **racemize.** For example, optically active acids such as (+)-alanine (Table 5-1) have been found to undergo very slow racemization in fossil deposits, a process resulting in reduced optical activity.

Optical activity can be measured in a mixture of enantiomers, but only if the enantiomers are present in unequal amounts. Using the value of the measured rotation, we can calculate the composition of such a mixture. For example, if a solution of (+)-alanine from a fossil exhibits an $[\alpha]$ of only +4.25° (i.e., one-half the value for the pure enantiomer), we can deduce that 50% of the sample is pure (+) isomer and the other 50% is racemic. Because the racemic portion consists of equal amounts of (+) of (−), the actual composition of the sample is 75% (+) and 25% (−).

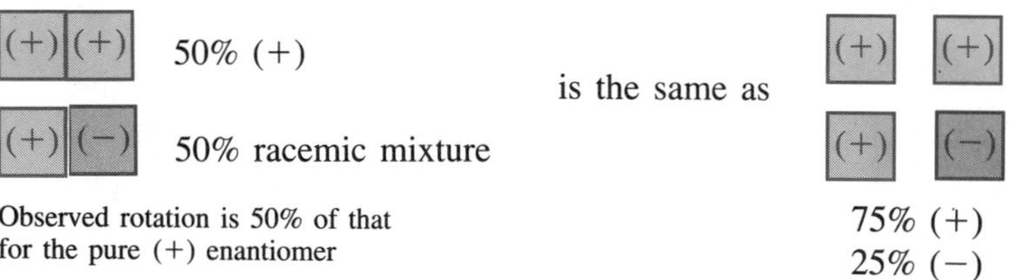

Observed rotation is 50% of that for the pure (+) enantiomer

75% (+)
25% (−)

The 25% (−) enantiomer cancels the rotation of a corresponding amount of (+). The mixture is said to be 50% *optically pure:* The observed optical rotation is one-half that of the pure dextrorotatory enantiomer.

Optical Purity

$$\% \text{ optical purity} = \left(\frac{[\alpha]_{\text{observed}}}{[\alpha]} \cdot 100\right)$$

EXERCISE 5-7

What is the optical rotation of a sample of (+)-2-bromobutane that is 75% optically pure? What percentages of (+) and (−) enantiomers are present in this sample? Answer the same questions for samples of 50 and 25% optical purity.

In summary, two enantiomers can be distinguished by their optical activity, that is, their interaction with plane-polarized light as measured in a polarimeter. One enantiomer always rotates such light clockwise (dextrorotatory), the other counterclockwise (levorotatory) by the same amount. The specific rotation, $[\alpha]$, is a physical constant possible only for chiral molecules. The interconversion of enantiomers leads to racemization and the disappearance of optical activity.

5-3 Absolute Configuration: *R–S* Sequence Rules

How do we establish the structure of one pure enantiomer of a chiral compound? And, once we know the answer, is there a way to name it unambiguously and distinguish it from its mirror image?

X-ray diffraction can establish the absolute configuration

Virtually all the physical characteristics of one enantiomer are identical with those of its mirror image, except for the sign of optical rotation. Is there a correlation between the sign of optical rotation and the actual spatial arrangement of the substituent groups, the **absolute configuration?** Is it possible to determine the structure of an enantiomer by measuring its $[\alpha]$ value? The answer to both questions is, unfortunately, no. *There is no straightforward correlation between the sign of rotation and the structure of the particular enantiomer*. For example, although a specific stereostructure of 2-bromobutane has been assigned a positive $[\alpha]$ (Table 5-1) and its mirror image a negative $[\alpha]$, this assignment was in fact based on additional structural information. Such information can be obtained through a special type of *X-ray diffraction* that directly furnishes the three-dimensional arrangement of the atoms of a molecule. Absolute configuration can also be established by chemical conversion into a structure whose own absolute configuration is known from an X-ray study.

Stereocenters are labeled *R* or *S*

To name enantiomers unambiguously, we need a system that allows us to indicate the handedness in the molecule, a sort of "left hand" versus "right hand" nomenclature. Such a system was developed by three chemists, R. S. Cahn, C. Ingold, and V. Prelog.*

Let us see how the handedness around an asymmetric carbon atom is labeled. The first step is to rank all four substituents in the order of decreasing priority. Priority is established by applying sequence rules. Substituent *a* has the highest priority, *b* the second highest, *c* the third, and *d* the lowest. Next, we position the molecule (mentally, on paper, or by using a molecular model set) so that the lowest priority substituent is placed as far away from us as possible (Figure 5-7). This process results in two (and only two) possible arrangements of the remaining substituents. If the progression from *a* to *b* to *c* is counterclockwise, the configuration at the stereocenter is named *S* (*sinister,* Latin, left). Conversely, if the progression is clockwise, the center is *R* (*rectus,* Latin, right). The symbol *R* or

*Dr. Robert S. Cahn, 1899–1981, Fellow of the Royal Institute of Chemistry, London; Professor Christopher Ingold, 1893–1970, University College, London; Professor Vladimir Prelog, b. 1906, Swiss Federal Institute of Technology, Zürich, Nobel Prize 1975 (chemistry).

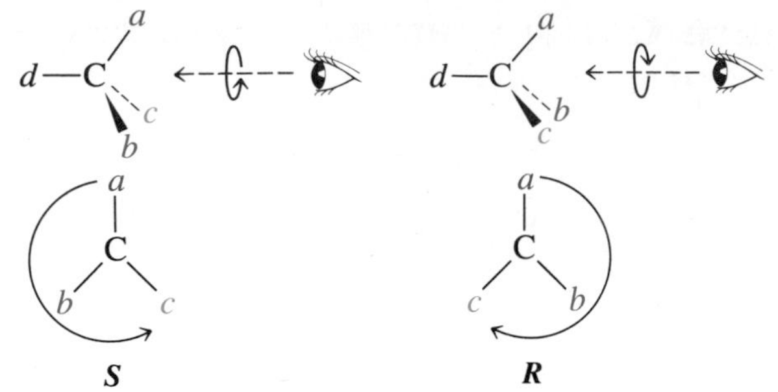

FIGURE 5-7 Assignment of *R* or *S* configuration at a tetrahedral stereocenter. The lowest priority group is placed as far away from the observer as possible. In many of the structural drawings in this chapter, the color scheme shown here is used to indicate the priority of substituents—in decreasing order: red, blue, green, black.

BOX 5-2 Absolute Configuration: A Historical Note

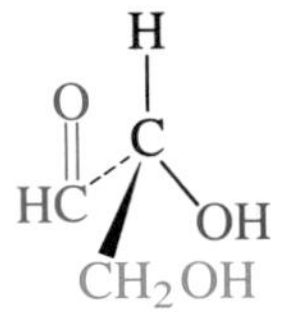

$[\alpha]_D^{25°C} = +8.7°$

D-(+)-2,3-Dihydroxypropanal [D-(+)-Glyceraldehyde]

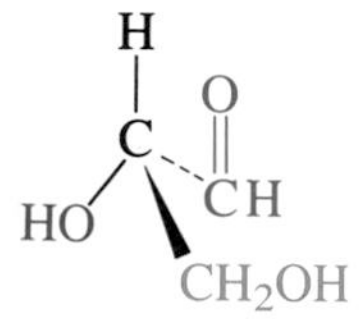

$[\alpha]_D^{25°C} = -8.7°$

L-(−)-2,3-Dihydroxypropanal [L-(−)-Glyceraldehyde]

Before X-ray diffraction, the absolute configurations of chiral molecules were unknown. Amusingly, the first assignment of a three-dimensional structure to a chiral molecule was a *guess* made over a century ago! The naturally occurring dextrorotatory enantiomer of 2,3-dihydroxypropanal (glyceraldehyde) was arbitrarily assigned the structure shown above and labeled "D-glyceraldehyde." The label "D" was not used to refer to the sign of rotation of plane-polarized light but to the relative arrangement of the substituent groups.

The other isomer was called L-glyceraldehyde. All chiral compounds that could be converted into D-(+)-glyceraldehyde by reactions that did not affect the configuration at the stereocenter were assigned the D configuration, and their mirror images the L. In 1951 the absolute configurations of these compounds became known and the original guess found to be correct. D,L nomenclature is still used for sugars (Chapter 24) and amino acids (Chapter 26).

D-Configurations

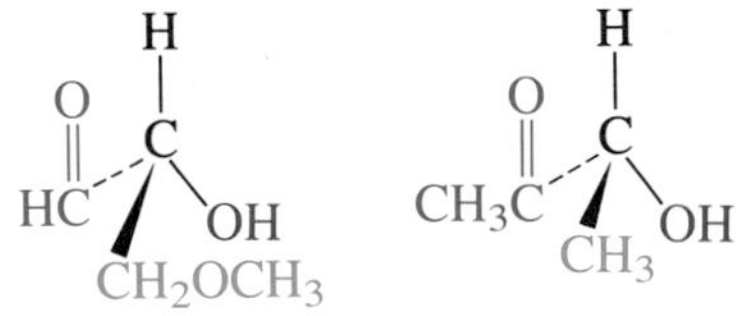

L-Configurations

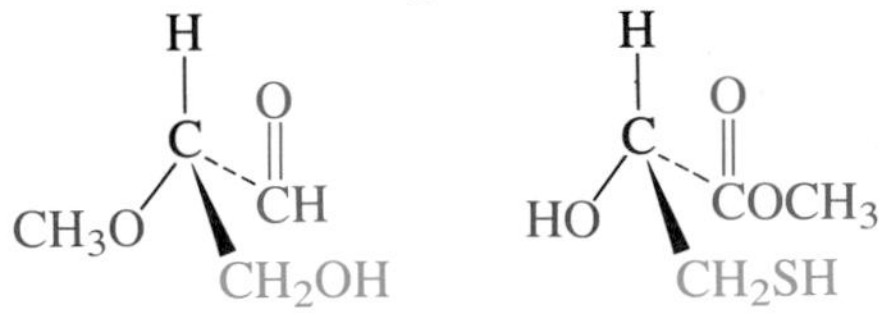

S is added as a prefix in parentheses to the name of the chiral compound, as in (*R*)-2-bromobutane and (*S*)-2,3-dihydroxypropanal. A racemic mixture is designated *R*,*S*, as in (*R*,*S*)-bromochlorofluoromethane. The sign of the rotation of plane-polarized light may be added if it is known, as in (*S*)-(+)-2-bromobutane and (*R*)-(+)-2,3-dihydroxypropanal. It is important to remember, however, that the symbols *R* and *S* are *not* necessarily correlated with either sign of α.

Sequence rules assign priorities to substituents

Before applying the *R*,*S* nomenclature to a stereocenter, we must first assign priorities by using sequence rules.

RULE 1. We look first at the atoms attached directly to the stereocenter. A substituent atom of higher atomic number takes precedence over one of lower atomic number. Consequently, the substituent of lowest priority is hydrogen. In the case of isotopes, the atom of higher atomic mass receives higher priority.

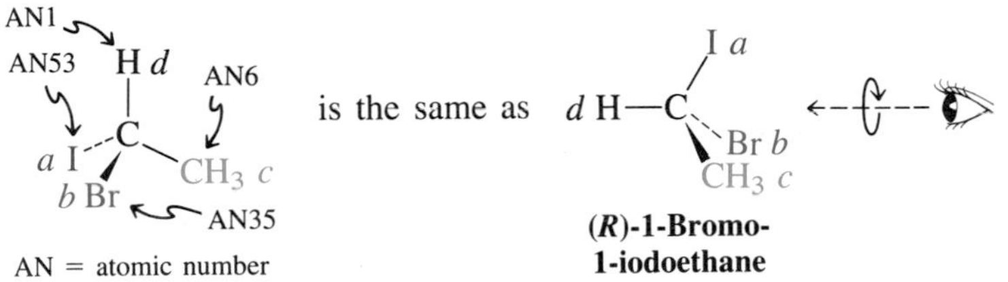

AN1 AN53 H *d* AN6 *a* I C CH_3 *c* *b* Br AN35 is the same as *d* H—C I *a* Br *b* CH_3 *c*

AN = atomic number

(*R*)-1-Bromo-1-iodoethane

RULE 2. What if two substituents have the same rank when we consider the atoms directly attached to the stereocenter? In such a case, we proceed along the two respective substituent chains until we reach a point of difference.

For example, an ethyl substituent takes priority over methyl. Why? At the point of attachment to the stereocenter, each substituent has a carbon nucleus, equal in priority. Farther from that center, however, methyl has only hydrogen atoms, but ethyl has a carbon atom (higher in priority).

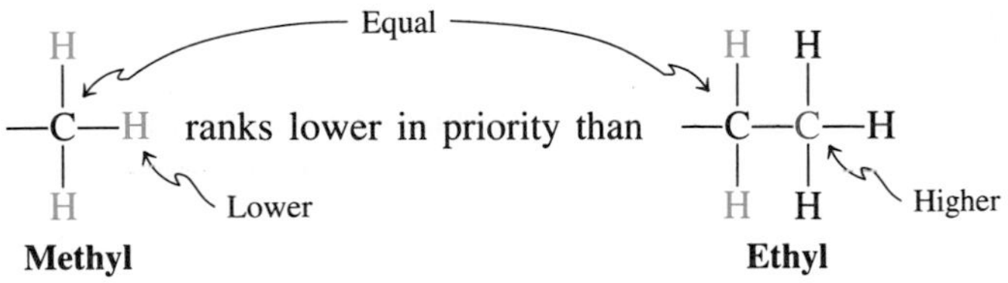

Equal

—C(H)(H)—H ranks lower in priority than —C(H)(H)—C(H)(H)—H

Lower Higher

Methyl **Ethyl**

However, 1-methylethyl takes precedence over ethyl because, at the first carbon, ethyl bears only one other carbon substituent, but 1-methylethyl bears two. Similarly, 2-methylpropyl takes priority over butyl but ranks lower than 1,1-dimethylethyl.

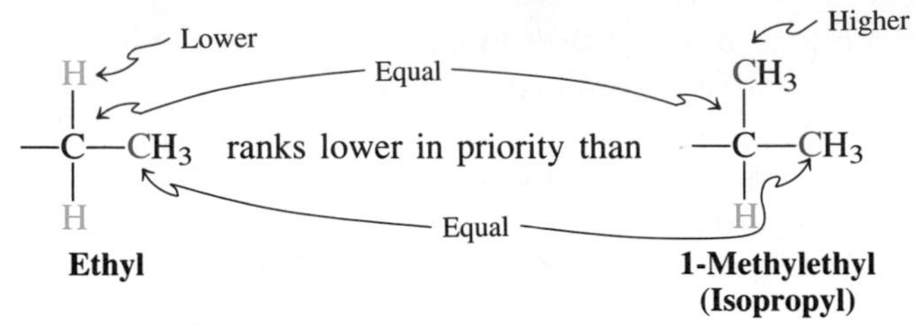

Butyl: $-CH_2-CH_2-CH_2CH_3$ ranks lower in priority than 2-Methylpropyl: $-CH_2-CH(CH_3)-CH_3$

(Lower; Higher; Equal)

2-Methylpropyl: $-CH_2-CH(CH_3)-CH_3$ ranks lower in priority than 1,1-Dimethylethyl (*tert*-Butyl): $-C(CH_3)_2-CH_3$

We must remember that the decision on priority is made at the *first* point of difference along otherwise similar substituent chains. Once that point is reached, the constitution of the remainder of the chain is irrelevant.

$-CH_2-CH_2OH$ ranks lower in priority than $-CH(CH_3)-CH_3$ (First point of difference: H vs. CH_3)

$-CH_2-CH_2CH_2CCl_3$ ranks lower in priority than $-CH(CH_3)-CH_3$ (First point of difference: H vs. CH_3)

When we reach a point along a substituent chain at which it branches, choose the branch that is higher in priority. When two substituents have similar branches, we rank the elements in those branches until we reach a point of difference.

$-CH(CH_2CH_2CH_3)-CH_2-SH$ ranks lower in priority than $-CH(CH_2CH_2CH_3)-CH_2-S-CH_3$ (First point of difference: H vs. CH_3)

Some examples are shown below.

I *a*; C; CH$_2$CH$_3$ *b*; *d* H; CH$_3$ *c*

(*R*)-2-Iodobutane

C(CH$_3$)$_3$ *a*; C; CH$_2$CH$_3$ *c*; *d* H; CH(CH$_3$)$_2$ *b*

(*S*)-3-Ethyl-2,2,4-trimethylpentane

RULE 3. Double and triple bonds are treated as if they were single, and the atoms in them duplicated or triplicated.

HC=CHR (H, H, R) is treated as —C(H)(C)—C(H)(C)—R

—C≡C—R is treated as —C(C)(C)—C(C)(C)—R

—C(=O)—H is treated as —C(O)(H)—O(C)

—C(=O)—OH is treated as —C(O)(OH)—O(C)

The red atoms shown in the groups on the right side of the display are not really there! They are added only for the purpose of assigning a relative priority to each of the corresponding groups to their left.

Examples are shown in the margin.

CH=CH$_2$ *b*; C; CH$_3$ *c*; *d* H; OH *a*

R

OH *a*; C; C(=O)H *b*; *d* H; CH$_2$OH *c*

R

EXERCISE 5-8

Draw the structures of the following substituents and within each group rank them in order of decreasing priority. **(a)** Methyl, bromomethyl, trichloromethyl, ethyl; **(b)** 2-methylpropyl (isobutyl), 1-methylethyl (isopropyl), cyclohexyl; **(c)** butyl, 1-methylpropyl (*sec*-butyl), 2-methylpropyl (isobutyl), 1,1-dimethylethyl (*tert*-butyl); **(d)** ethyl, 1-chloroethyl, 1-bromoethyl, 2-bromoethyl.

EXERCISE 5-9

Assign the absolute configuration of the molecules depicted in Table 5-1.

EXERCISE 5-10

Draw one enantiomer of your choice (specify which) of 2-chlorobutane, 2-chloro-2-fluorobutane, and (HC≡C)(CH$_2$=CH)C(Br)(CH$_3$).

To correctly assign the stereostructure of stereoisomers, we must develop a fair amount of three-dimensional "vision" or "stereoperception." In the structures that have been shown to illustrate the priority rules, the lowest priority substituent has been located at the left of the carbon center and in the plane of the page and the remainder of the substituents at the right, the upper-right group also being positioned in this plane. However, this is not the only way of drawing dashed-wedged line structures; others are equally correct. Consider some of the structural drawings of (*S*)-2-bromobutane. These are simply different views of the same molecule.

Six Ways of Drawing (*S*)-2-Bromobutane

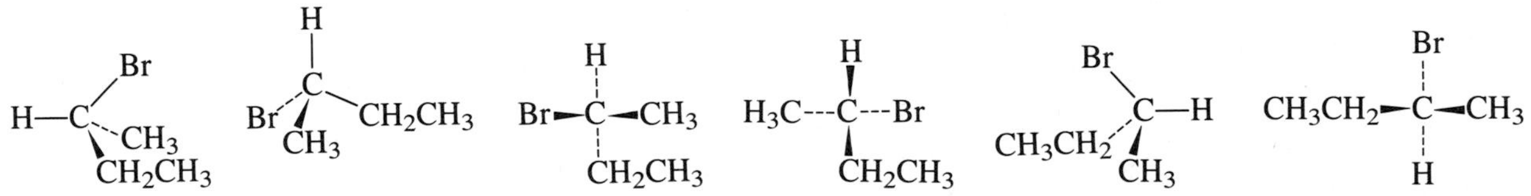

To summarize, the sign of optical rotation cannot be used to establish the absolute configuration of a stereoisomer. Instead, special methods of X-ray diffraction (or chemical correlations) must be used. We can express the absolute configuration of the chiral molecule as *R* or *S* by applying the sequence rules, which allow us to rank all substituents in order of decreasing priority. Turning the structures so as to place the lowest priority group at the back causes the remaining substituents to be arranged in clockwise (*R*) or counterclockwise (*S*) fashion.

5-4 Fischer Projections

A **Fischer* projection** is a standard way of depicting tetrahedral carbon atoms and their substituents in two dimensions. With this method, the molecule is drawn in the form of a cross, the central carbon being at the point of intersection. The horizontal lines signify bonds directed *toward* the viewer; the vertical lines are pointing *away*. Dashed-wedged line structures have to be arranged in this way to facilitate their conversion into Fischer projections.

Conversion of the Dashed-Wedged Line Structures of 2-Bromobutane into Fischer Projections

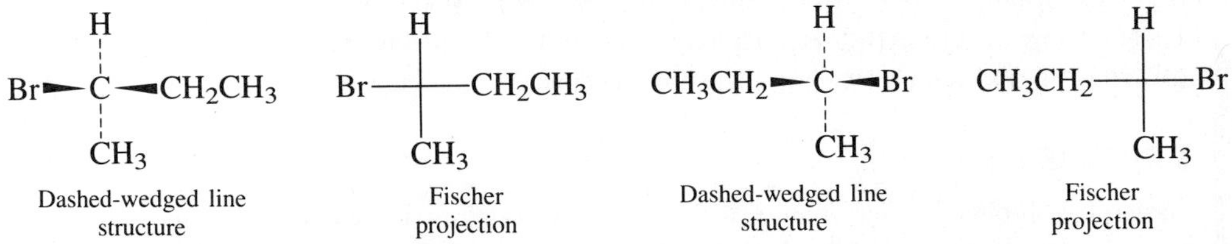

Dashed-wedged line structure

Fischer projection

(*R*)-2-Bromobutane

Dashed-wedged line structure

Fischer projection

(*S*)-2-Bromobutane

*Professor Emil Fischer, 1852–1919, University of Berlin, Nobel Prize 1902 (chemistry).

Rotating a Fischer projection can change the absolute configuration

We must be very careful in manipulating Fischer projections. For example, rotation in the plane by 90° gives the projection of the enantiomer. It follows that rotation by 180° then gives back the original structure. This statement can be verified by drawing conventional dashed-wedged line structures or by looking at molecular models. It is best *not* to rotate Fischer projections in this manner because we may quickly lose the correct absolute configuration.

H; Br—CH$_2$CH$_3$; CH$_3$ **R** —90° turn→ Br; H$_3$C—H; CH$_2$CH$_3$ **S** is the same as Br; H$_3$C—C—H; CH$_2$CH$_3$ **S**

changes into

H; CH$_3$CH$_2$—Br; CH$_3$ **S** —180° turn→ CH$_3$; Br—CH$_2$CH$_3$; H **S** is the same as CH$_3$; Br—C—CH$_2$CH$_3$; H **S**

remains

EXERCISE 5-11

Draw Fischer projections for all the molecules in Exercises 5-9 and 5-10.

Exchanging substituents in a Fischer projection also changes the absolute configuration

As is the case for dashed-wedged line structures, there are several Fischer projections of the same enantiomer, a situation that may lead to confusion. How can we quickly ascertain whether two Fischer projections are depicting the same enantiomer or two mirror images? We have to find a sure way to convert one Fischer projection into another in a manner that either leaves the absolute configuration unchanged or converts it into its opposite. It turns out that this task can be achieved by simply making substituent groups trade places. As we can readily verify by using molecular models, any *single* such exchange turns one enantiomer into its mirror image. Two such exchanges (we may select different substituents every time) produce the original absolute configuration. As the dashed-wedged line structures below show, this operation merely generates a different view of the same molecule, rotated 120° about the C–Cl bond.

Changes of Absolute Configuration on Switching Substituents in Fischer Projections

Cl—C(CH$_3$)(Br)(CH$_2$CH$_3$) = CH$_3$; Cl—Br; CH$_2$CH$_3$ **S** —CH$_3$ ⇄ Br→ Br; Cl—CH$_3$; CH$_2$CH$_3$ **R** —CH$_3$CH$_2$ ⇄ Br→ CH$_2$CH$_3$; Cl—CH$_3$; Br **S** = Cl—C(CH$_2$CH$_3$)(CH$_3$)(Br)

changes into ... changes back to

(The double arrow denotes two groups trading places)

We now have a simple way to establish whether two different Fischer projections depict the same or opposite configurations. If the conversion of one structure into another takes an even number of exchanges, the structures are identical. If it requires an odd number of such moves, the structures are mirror images of each other.

Consider, for example, the two Fischer projections A and B. Do they represent molecules having the same configuration? The answer is found quickly. We convert A into B by two exchanges; so A equals B.

A (top Cl; left H; right CH_3; bottom CH_2CH_3) $\xrightarrow{\text{H} \leftrightarrow \text{Cl}}$ (top H; left Cl; right CH_3; bottom CH_2CH_3) $\xrightarrow{\text{Cl} \leftrightarrow CH_3}$ B (top H; left H_3C; right Cl; bottom CH_2CH_3)

A **B**

EXERCISE 5-12

Draw the dashed-wedged line structures corresponding to Fischer projections A and B, above. Is it possible to transform A into B by means of a rotation about a single bond? If so, identify the bond and the degree of rotation involved. Use models if necessary.

Fischer projections tell us the absolute configuration

When dealing with stereochemical problems, an accurate perception of space is very useful. However, Fischer projections allow us to assign absolute configurations without having to visualize the three-dimensional arrangement of the atoms. For this purpose, we first draw the molecule as a (any) Fischer projection. Next, we rank all the substituents in accord with the sequence rules. Finally, we exchange two groups so that the lowest priority substituent is at the top, and then exchange any other pair (to make sure that the absolute configuration stays unchanged from the original). On completion of these manipulations, we find that the three groups of priority, *a*, *b*, and *c*, turn out to be arranged in either clockwise or counterclockwise fashion, in turn corresponding to either the *R* or the *S* configuration.

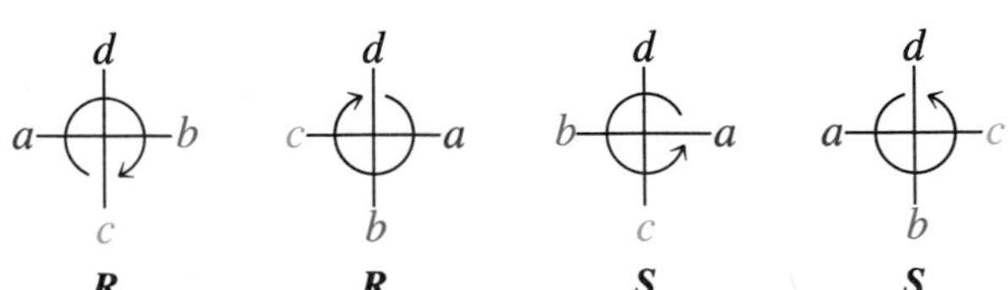

EXERCISE 5-13

What is the absolute configuration of the following molecules?

(top Br; left H; right D; bottom CH_3) (top Cl; left F; right Br; bottom I) (top CH_3; left H_2N; right $\underset{\|}{C}OH$ with =O; bottom H)

H O

EXERCISE 5-14

Convert the Fischer projections in Exercise 5-13 to dashed-wedged line formulas and determine their absolute configurations by using the procedure described in Section 5-3. When the lowest priority group is at the top in a Fischer projection, is it in front of or behind the plane of the page? Does this explain why the procedure outlined above for determination of configuration from Fischer projections succeeds?

In summary, a Fischer projection is a convenient way of drawing chiral molecules. We must not rotate such projections in the plane. Switching substituents reverses absolute configuration, if done an odd number of times, but leaves it intact when the number of such exchanges is even. By placing the lowest priority substituent on top, we can readily assign the absolute configuration.

5-5 Molecules Incorporating Several Stereocenters: Diastereomers

Many molecules contain several stereocenters. Because the configuration around each center can be *R* or *S*, several possible structures emerge, all of which are isomeric.

Two stereocenters can give four stereoisomers: chlorination of 2-bromobutane at C3

Section 5-1 described how a carbon-based stereocenter can be created by the radical halogenation of butane. Let us now consider the chlorination of racemic 2-bromobutane to give (among other products) 2-bromo-3-chlorobutane. The introduction of a chlorine atom at C3 produces a new stereocenter in the molecule. This center may have either the *R* or the *S* configuration, assignable by using the same sequence rules that apply to molecules with only one such center.

How many stereoisomers are possible for 2-bromo-3-chlorobutane? There are four, as can be seen by completing a simple exercise in permutation. Each stereocenter can be either *R* or *S*, and, hence, the possible combinations are *RR*, *RS*, *SR*, and *SS* (Figure 5-8).

Because all horizontal lines in Fischer projections signify bonds directed toward the viewer, the result is a representation of a molecule in an *eclipsed* conformation. Therefore, the first step in converting a staggered Newman or dashed-wedged line representation of a molecule into a Fischer projection is to rotate the molecule to form an eclipsed rotamer. To make stereochemical assignments, one treats each stereocenter *separately*, and the group containing the other stereocenter is regarded as a simple substituent (Figure 5-9).

By looking closely at the structures of the four stereoisomers (Figure 5-8), we see that there are two related pairs of compounds: an *R,R/S,S* pair and an *R,S/S,R* pair. The members of each individual pair are mirror images of each other and therefore enantiomers. Conversely, each member of one pair is not a mirror image of either member of the other pair; hence, they are not enantiomeric with respect to each other. Stereoisomers that are not related as object and mirror image are called **diastereomers** (*dia*, Greek, across).

H
|
$CH_3\overset{*}{C}CH_2CH_3$
|
Br

One stereocenter

$\downarrow$ Cl_2, $h\nu$ / $-HCl$

H Cl
| |
$CH_3\overset{*}{C}-\overset{*}{C}CH_3$
| |
Br H

Two stereocenters

2-Bromo-3-chlorobutane

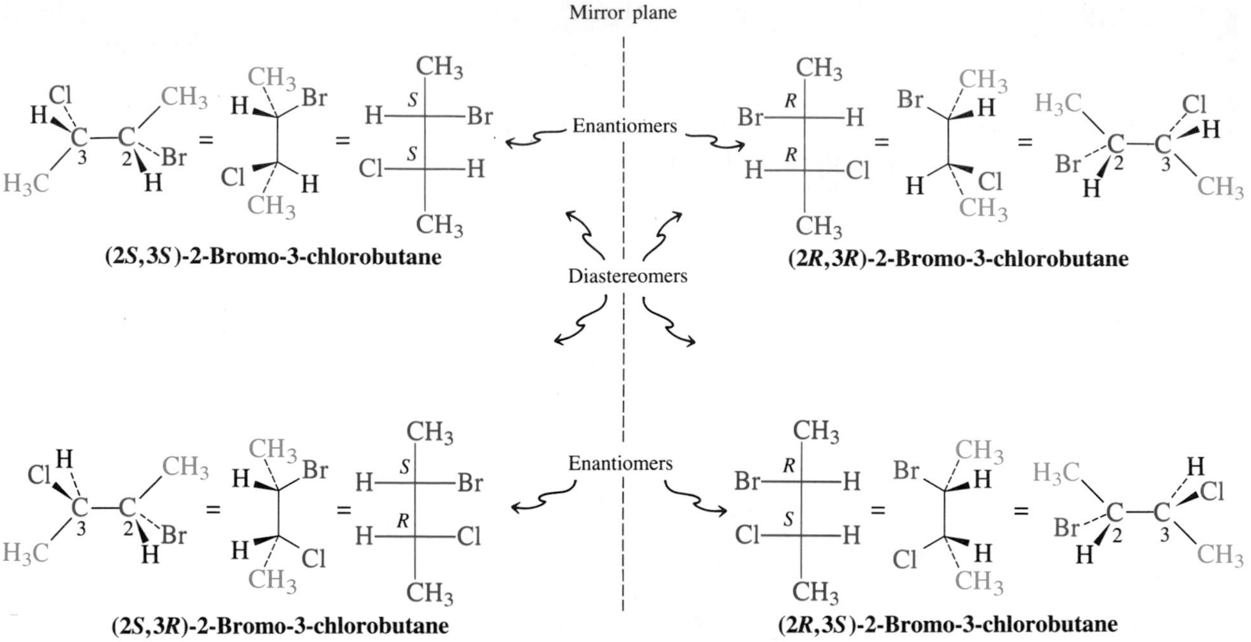

(2S,3S)-2-Bromo-3-chlorobutane

(2R,3R)-2-Bromo-3-chlorobutane

(2S,3R)-2-Bromo-3-chlorobutane

(2R,3S)-2-Bromo-3-chlorobutane

FIGURE 5-8 The four stereoisomers of 2-bromo-3-chlorobutane. Each is the enantiomer of one of the other three (its mirror image) and is at the same time a diastereomer of each of the remaining two. For example, the 2R,3R isomer is the enantiomer of the 2S,3S compound and is diastereomerically related to both the 2S,3R and 2R,3S structures. Notice that two structures are enantiomers only when they possess the opposite configuration at *every* stereocenter.

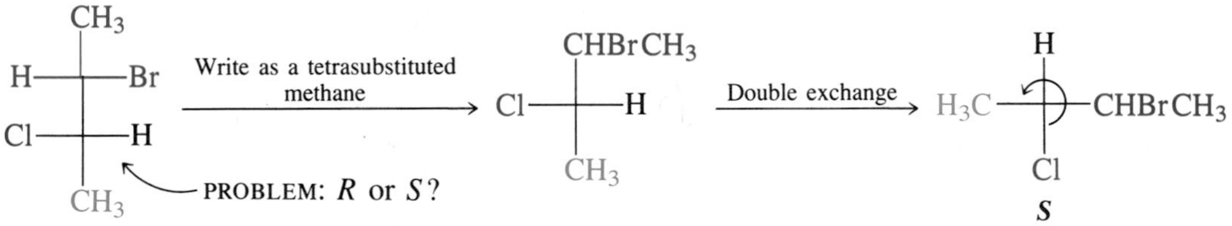

SOLUTION: The center under scrutiny is S.

FIGURE 5-9 Assigning the absolute configuration at C3 in 2-bromo-3-chlorobutane. We consider the group containing the stereocenter C2 merely as one of the four substituents. Priorities (also noted in color) are assigned in the usual way ($Cl > CHBrCH_3 > CH_3 > H$), giving rise to the representation shown in the center. Two exchanges place the lowest priority hydrogen at the top of the Fischer projection to facilitate assignment.

EXERCISE 5-15

The two amino acids isoleucine and alloisoleucine are depicted below in staggered conformations. Convert both to Fischer projections. (Keep in mind that Fischer projections are views of molecules *in eclipsed conformations*.) Are these two compounds enantiomers or diastereomers?

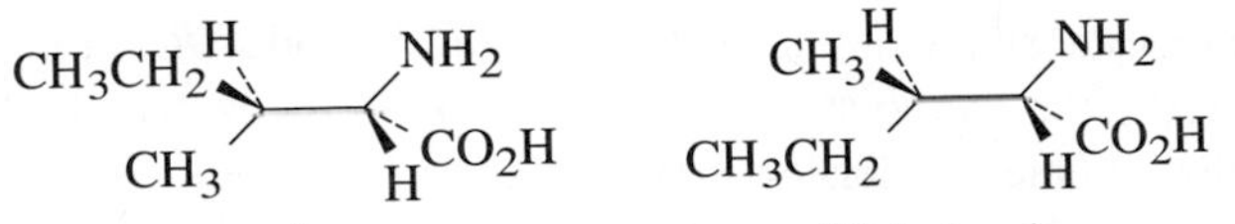

Isoleucine **Alloisoleucine**

In contrast with enantiomers, diastereomers, because they are *not* mirror images of each other, are distinct molecules with *different physical and chemical properties*. Their steric interactions and energies differ. They can be separated by fractional distillation or crystallization or by chromatography. They have different melting and boiling points and different densities, just as constitutional isomers do. In addition, they have different specific rotations.

EXERCISE 5-16

What are the stereochemical relations (identical, enantiomers, diastereomers) of the following four molecules? Assign absolute configurations at each stereocenter.

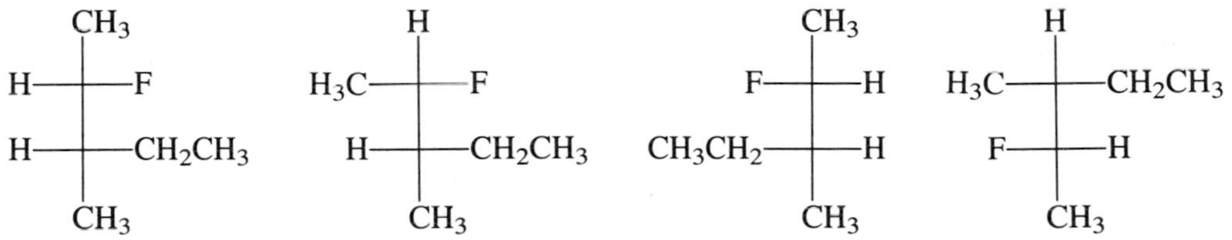

1 **2** **3** **4**

Cis and trans isomers are cyclic diastereomers

It is instructive to compare the stereoisomers of 2-bromo-3-chlorobutane with those of a cyclic analog, 1-bromo-2-chlorocyclobutane (Figure 5-10). In both

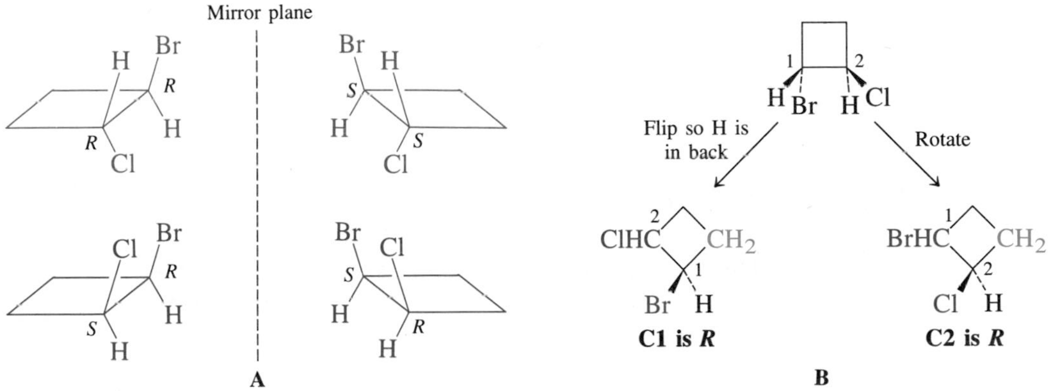

FIGURE 5-10 (A) The diastereomeric relation of *cis*- and *trans*-1-bromo-2-chlorocyclobutane. (B) Stereochemical assignment of the *R*,*R* stereoisomer. Recall that the color scheme indicates the priority order of the groups around each stereocenter: red > blue > green > black.

cases there are four stereoisomers: *R,R*, *S,S*, *R,S*, and *S,R*. In the cyclic compound, however, the stereoisomeric relation of the first pair to the second is easily recognized: One pair has cis stereochemistry, the other trans. Cis and trans isomers (Section 4-1) in cycloalkanes are in fact diastereomers.

More than two stereocenters means still more stereoisomers

What structural variety do we expect for a compound having three stereocenters? We may again approach this problem by permuting the various possibilities. If we label the three centers consecutively as either *R* or *S*, the following sequence emerges:

RRR *RRS* *RSR* *SRR* *RSS* *SRS* *SSR* *SSS*

a total of eight stereoisomers. They can be arranged to reveal a division into four enantiomer pairs of diastereomers.

Image	*RRR*	*RRS*	*RSS*	*SRS*
Mirror image	*SSS*	*SSR*	*SRR*	*RSR*

Generally, *a compound with* n *stereocenters can have a maximum of* 2^n *stereoisomers*. Therefore, a compound having three such centers gives rise to a maximum of eight stereoisomers; one having four produces sixteen; one having five, thirty-two; and so forth. The structural possibilities are quite staggering for larger systems.

EXERCISE 5-17

Draw all the stereoisomers of 2-bromo-3-chloro-4-fluoropentane.

In summary, the presence of more than one stereocenter in a molecule gives rise to diastereomers. These are stereoisomers that are not related to each other as object and mirror image. Whereas enantiomers have opposite configurations at every respective stereocenter, two diastereomers do not. A molecule with n stereocenters may exist in as many as 2^n stereoisomers. In cyclic compounds, cis and trans isomers are diastereomers.

5-6 Meso Compounds

We saw that the molecule 2-bromo-3-chlorobutane contains two distinct stereocenters, each with a *different* halogen substituent. How many stereoisomers are to be expected if both centers are identically substituted?

Two identically substituted stereocenters give rise to only three stereoisomers

Consider, for example, 2,3-dibromobutane, which can be obtained by the radical bromination of 2-bromobutane. As we did for 2-bromo-3-chlorobutane, we have

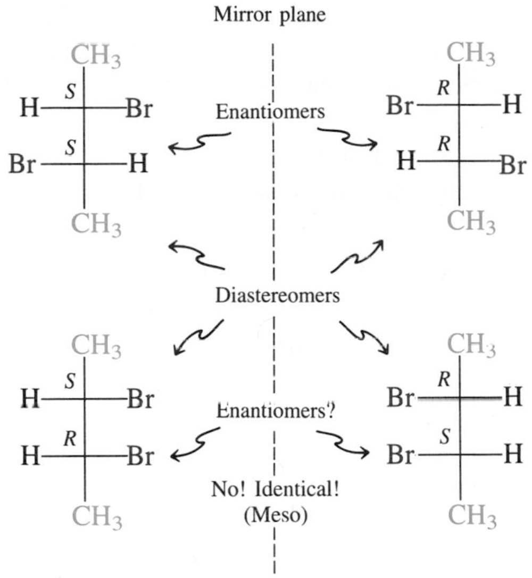

FIGURE 5-11 The stereochemical relationships of the stereoisomers of 2,3-dibromobutane. The bottom pair consists of identical structures. (Make a model!)

to consider four structures, resulting from the various permutations in R and S configurations (Figure 5-11).

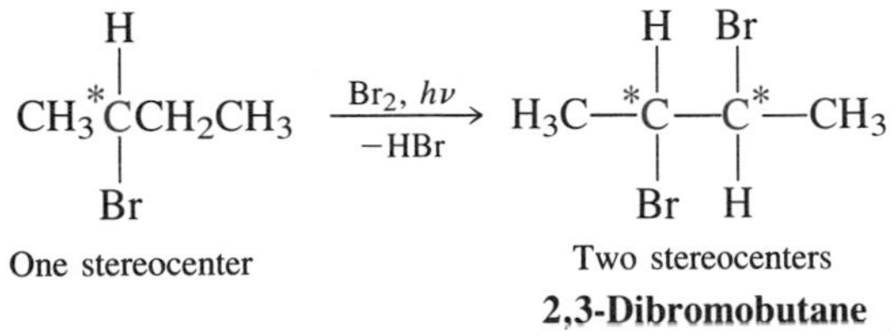

$$\mathrm{CH_3\overset{*}{C}H(Br)CH_2CH_3} \xrightarrow[-\mathrm{HBr}]{\mathrm{Br_2},\ h\nu} \mathrm{H_3C{-}\overset{*}{C}H(Br){-}\overset{*}{C}H(Br){-}CH_3}$$

One stereocenter — Two stereocenters

2,3-Dibromobutane

The first pair of stereoisomers, with R,R and S,S configurations, is clearly recognizable as a pair of enantiomers. However, a close look at the second pair reveals that (S,R) and mirror image (R,S) are superimposable and therefore identical. Thus, the S,R diastereomer of 2,3-dibromobutane is achiral and not optically active, even though it contains two stereocenters. The identity of the two structures can be confirmed readily by using molecular models.

A compound that contains two (or, as we shall see, even more than two) stereocenters but is superimposable with its mirror image is a **meso compound** (*mesos,* Greek, middle). A characteristic feature of a meso compound is the *presence of a mirror plane,* which divides the molecule such that one half is the mirror image of the other half. For example, in 2,3-dibromobutane, the 2R center is the reflection of the 3S center. This arrangement is best seen in an eclipsed dashed-wedged line structure (Figure 5-12). The presence of a mirror plane in

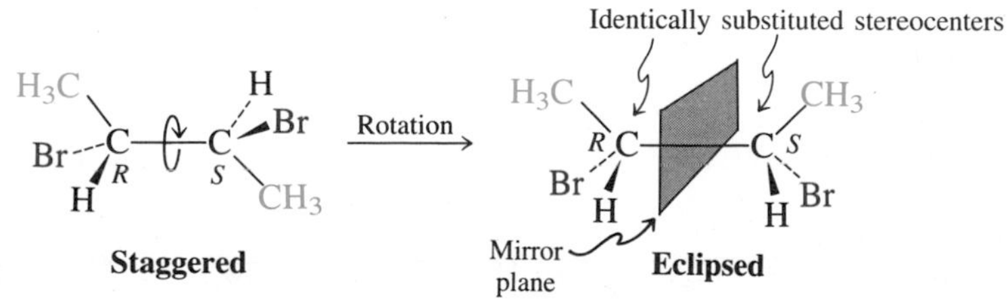

FIGURE 5-12 *meso*-2,3-Dibromobutane contains a mirror plane when rotated into the eclipsed conformation shown. A molecule with more than one stereocenter is meso and achiral as long as it contains a mirror plane in any readily accessible conformation. Meso compounds possess identically substituted stereocenters.

any energetically accessible conformation of a molecule (Sections 2-5 and 2-7) is sufficient to make it achiral. As a consequence, 2,3-dibromobutane exists in the form of three stereoisomers only: a pair of (necessarily chiral) enantiomers and an achiral meso diastereomer.

Meso diastereomers can exist in molecules with more than two stereocenters. Examples are 2,3,4-tribromopentane and 2,3,4,5-tetrabromohexane.

Meso Compounds with Multiple Stereocenters

Mirror plane

EXERCISE 5-18

Draw all the stereoisomers of 2,4-dibromo-3-chloropentane.

Cyclic compounds may also be meso

It is again instructive to compare the stereochemical situation in 2,3-dibromobutane with that in an analogous cyclic molecule: 1,2-dibromocyclobutane. We can see that *trans*-1,2-dibromocyclobutane exists as two enantiomers (*R*,*R* and *S*,*S*) and may therefore be optically active. The cis isomer, however, has a mirror plane and is meso, achiral, and optically inactive (Figure 5-13).

Notice that we have drawn the ring in a planar shape in order to illustrate the mirror symmetry, although we know from Chapter 4 that cycloalkanes with four or more carbons in the ring are not flat. Is this justifiable? Generally yes, because such compounds, like their acyclic analogs, possess a variety of conformations that are readily accessible at room temperature (Sections 4-2 through 4-4). At least one of these will contain the necessary mirror plane to render achiral any cis-disubstituted cycloalkane with identically constituted stereocenters. For simplicity, cyclic compounds usually may be treated *as if they were planar* for the purpose of identifying a mirror plane.

Mirror plane

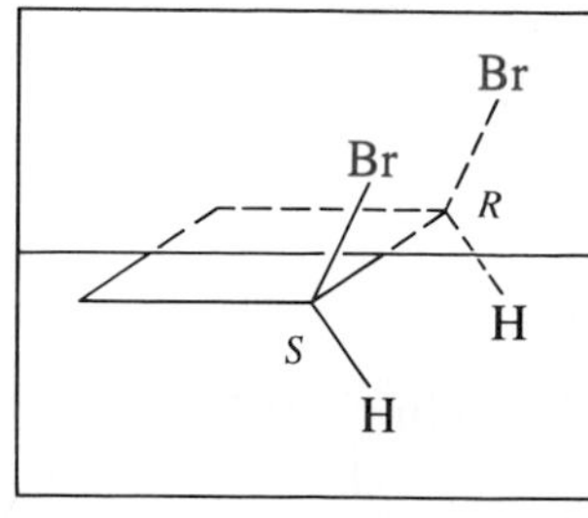

Mirror plane
1*R*,2*S* same as 1*S*,2*R*

FIGURE 5-13 Trans isomer of 1,2-dibromocyclobutane is chiral; cis isomer is a meso compound and optically inactive.

BOX 5-3 Stereoisomers of Tartaric Acid

(+)-Tartaric acid	(−)-Tartaric acid	*meso*-Tartaric acid
COOH; H–C(*R*)–OH; HO–C(*R*)–H; COOH	COOH; HO–C(*S*)–H; H–C(*S*)–OH; COOH	COOH; H–C(*R*)–OH; H–C(*S*)–OH; COOH
$[\alpha]_D^{20°C} = +12.0°$	$[\alpha]_D^{20°C} = -12.0°$	$[\alpha]_D^{20°C} = 0°$
m.p. 168–170°C	m.p. 168–170°C	m.p. 146–148°C
Density (g mL^{-1}) $d = 1.7598$	$d = 1.7598$	$d = 1.666$

Tartaric acid (systematic name: 2,3-dihydroxybutanedioic acid) is a naturally occurring dicarboxylic acid containing two stereocenters with identical substitution patterns. Therefore it exists as a pair of enantiomers (which have identical physical properties but which rotate plane-polarized light in opposite directions) and an achiral meso compound (with different physical and chemical properties from those of the chiral diastereomers).

The dextrorotatory enantiomer of tartaric acid is widely distributed in nature. It is present in many fruits (fruit acid) and its monopotassium salt is found as a deposit during the fermentation of grape juice. Pure levorotatory tartaric acid is rare, as is the meso isomer.

Tartaric acid is of historical significance, because it was the first chiral molecule whose racemate was separated into the two enantiomers. This happened in 1848, long before it was recognized that carbon could be tetrahedral in organic molecules. By 1848, natural tartaric acid had been shown to be dextrorotatory, and the racemate had been isolated from grapes. The words ''racemate'' and ''racemic'' are in fact derived from an old common name for this form of tartaric acid, *racemic acid* (*racemus,* Latin, cluster of grapes). The French chemist Louis Pasteur* obtained a sample of the sodium ammonium salt of this acid and noticed that there were two types of crystals: One set was the mirror image of the second. In other words, the crystals were chiral.

By manually separating the two sets, dissolving them in water and measuring their optical rotation, Pasteur found one of the crystalline forms to be the pure salt of (+)-tartaric acid and the other to be the levorotatory form. Remarkably, the chirality of the individual molecules in this rare case had given rise to the macroscopic property of chirality in the crystal. He concluded from his observation that the molecules themselves must be chiral. These findings and others led in 1874 to the first proposal, by van't Hoff and Le Bel† independently, that saturated carbon has a tetrahedral—and not, for example, a square planar—bonding arrangement. (Why is the idea of a planar carbon incompatible with that of a stereocenter?)

*Professor Louis Pasteur, 1822–1895, Sorbonne, Paris.
†Professor Jacobus H. van't Hoff, 1852–1911, University of Amsterdam, Nobel Prize 1901 (chemistry); Dr. J. A. Le Bel, 1847–1930, Ph.D., Sorbonne, Paris.

EXERCISE 5-19

Draw each of the following compounds, representing the ring as planar. Which ones are chiral? Which are meso? Indicate the location of the mirror plane in each meso compound. **(a)** *cis*-1,2-Dichlorocyclopentane; **(b)** its trans isomer; **(c)** *cis*-1,3-dichlorocyclopentane; **(d)** its trans isomer; **(e)** *cis*-1,2-dichlorocyclohexane; **(f)** its trans isomer; **(g)** *cis*-1,3-dichlorocyclohexane; **(h)** its trans isomer.

EXERCISE 5-20

For each meso compound in Exercise 5-19, draw the conformation that contains the mirror plane. Refer to Sections 4-2 and 4-3 to identify energetically accessible conformations of these ring systems.

In summary, molecules with two or more identically substituted stereocenters may exist as meso stereoisomers. Meso compounds are superimposable on their mirror images and therefore achiral.

5-7 Stereochemistry in Chemical Reactions

We have seen that a chemical reaction can introduce chirality into a molecule. Let us examine more closely the conversion of achiral butane into chiral-2-bromobutane, which gives racemic material. We shall also see that the chiral environment of a stereocenter already present in a molecule exerts control on the stereochemistry of a reaction that introduces a second one. We begin with another look at the radical bromination of butane.

The radical mechanism explains why the bromination of butane results in a racemate

The radical bromination of butane at C2 creates a chiral molecule (Section 5-1). This happens because one of the methylene hydrogens is replaced by a new group, furnishing a stereocenter—a carbon atom with four different substituents.

In the first step of the mechanism for radical halogenation (Sections 3-6 and 3-7), one of these two hydrogens is abstracted by the attacking bromine atom. It does not matter which of the two is removed: This step does not generate a stereocenter. It furnishes a planar, sp^2-hybridized, and therefore achiral radical. The radical center has two equivalent reaction sites—the two lobes of the *p* orbital (Figure 5-14)—equally susceptible to attack by bromine in the second step. We can see that the two transition states resulting in the respective enantiomers of 2-bromobutane are mirror images of each other. They are enantiomeric and therefore energetically equivalent. The rates of formation of *R* and *S* products are hence equal, and a racemate is formed. In general, *the formation of chiral compounds* (e.g., 2-bromobutane) *from achiral reactants* (e.g., butane and bromine) *yields racemates*. Or, *optically inactive starting materials furnish optically inactive products*.*

The presence of a stereocenter affects the outcome of the reaction: chlorination of (*S*)-2-bromobutane

Now we understand why the halogenation of an achiral molecule gives a racemic halide. What products can we expect from the halogenation of a chiral, enantiomerically pure molecule?

---

*We shall see later that it *is* possible to generate optically active products from optically inactive starting materials, if we use an optically active reagent.

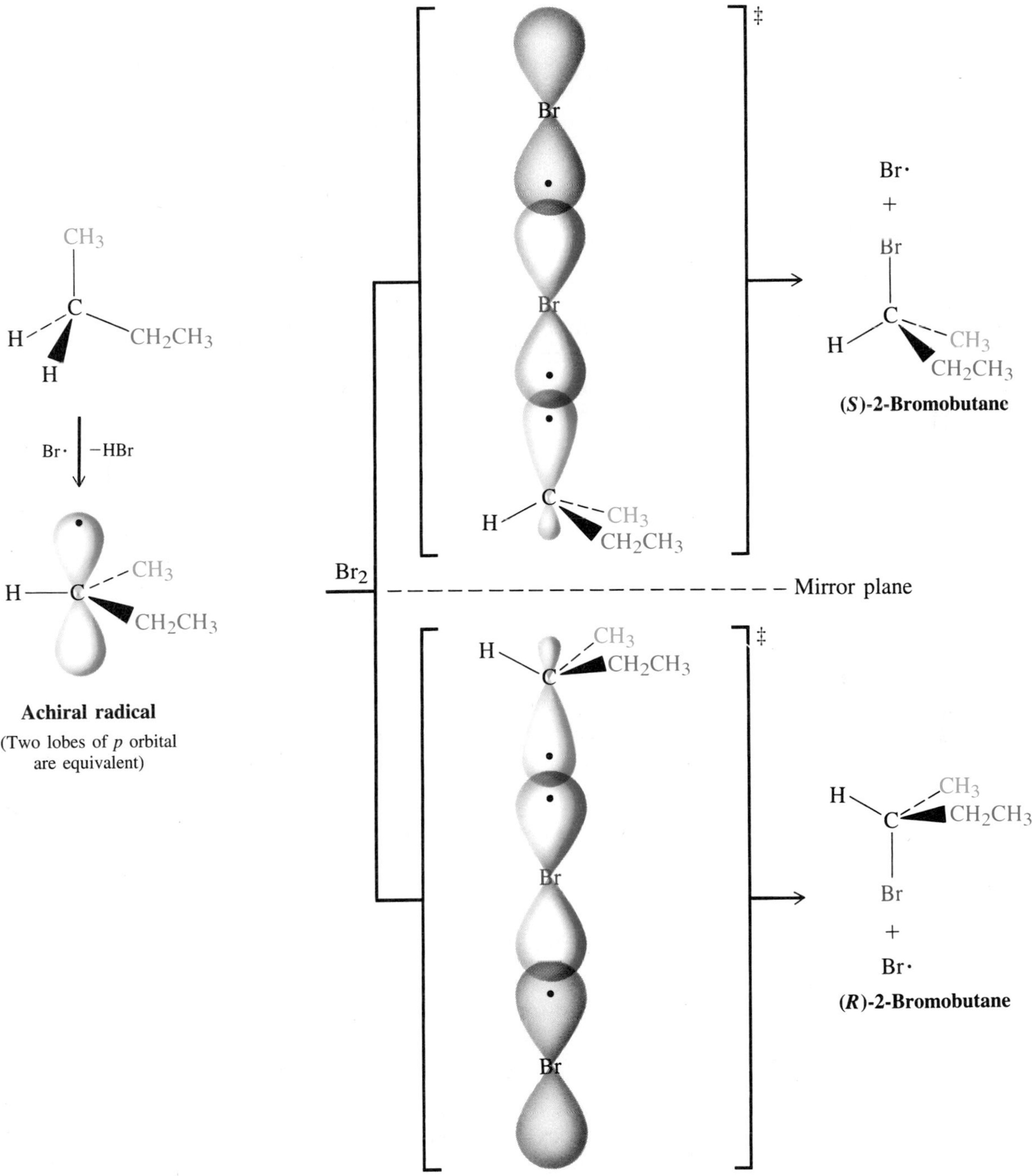

FIGURE 5-14 The creation of racemic 2-bromobutane from butane by radical bromination at C2. Abstraction of either methylene hydrogen by bromine gives an achiral radical. Reaction of Br_2 with this radical is equally likely at either the top or bottom face, a condition leading to a racemic mixture of products.

For example, consider the radical chlorination of the *S* enantiomer of 2-bromobutane. In this case the chlorine atom has several options for attack: the two terminal methyl groups, the single hydrogen at C2, and the two hydrogens on C3. Let us examine each of these reaction paths.

Remember the use of color to denote group priorities:
Highest—red
Second highest—blue
Third highest—green
Lowest—black

Chlorination of (*S*)-2-Bromobutane at Either C1 or C4

ClH_2C–C2(H)(Br)(CH_3CH_2) ← (Cl_2, $h\nu$ at C1; $-HCl$) — H_3C–C2(H)(Br)(CH_3CH_2) — (Cl_2, $h\nu$ at C4; $-HCl$) → H_3C–C3(H)(Br)($ClCH_2CH_2$)

Optically active **2*R*** | Optically active **2*S*** | Optically active **3*S***

Chlorination of either terminal methyl group is straightforward, proceeding at C1 to give 2-bromo-1-chlorobutane, and at C4 to give 3-bromo-1-chlorobutane. In the latter the original C4 has now become C1, to maintain the lowest possible substituent numbering. *Both of these chlorination products are optically active because the original stereocenter is left intact.* Note, however, that conversion of the C1 methyl into a chloromethyl unit changes the sequence of priorities around C2. Thus, although the stereocenter itself does not participate in the reaction, its designated configuration changes from *S* to *R*.

What about halogenation at C2, the stereocenter? The product from chlorination at C2 of (*S*)-2-bromobutane is 2-bromo-2-chlorobutane. Even though the substitution pattern at the stereocenter has changed, the molecule remains chiral. However, an attempt to measure the α value for the product would reveal the absence of optical activity: *Halogenation at the stereocenter leads to a racemic mixture*. How can this be explained? We must look again at the structure of the radical formed in the course of the reaction mechanism for the answer.

A racemate forms in this case because hydrogen abstraction from C2 furnishes a planar, sp^2-hybridized, achiral radical.

H_3C, CH_3CH_2, H, Br–C →($Cl\cdot$; $-HCl$) H_3C, CH_3CH_2–C•–Br →(Cl_2; $-Cl\cdot$) H_3C, CH_3CH_2, Cl, Br–C + H_3C, CH_3CH_2, Br, Cl–C

Optically active **2*S*** | **Achiral** | Optically inactive (A racemate) **50% 2*S*** **50% 2*R***

Chlorination can occur from either side through enantiomeric transition states of equal energy, as in the bromination of butane (Figure 5-14), producing (*S*)- and (*R*)-2-bromo-2-chlorobutane at equal rates and giving a racemic mixture. The reaction is an example of a transformation in which an optically active compound leads to an optically inactive product (a racemate).

EXERCISE 5-21

What other halogenations of (*S*)-2-bromobutane would furnish optically inactive products?

The chlorination of (*S*)-2-bromobutane at C3 does not affect the existing chiral center. However, *the formation of a second stereocenter gives rise to diastereomers*. Specifically, attachment of chlorine to the left side of C3 in the drawing below gives (2*S*,3*S*)-2-bromo-3-chlorobutane, whereas attachment to the right side gives its 2*S*,3*R* diastereomer.

Chlorination of (*S*)-2-Bromobutane at C3

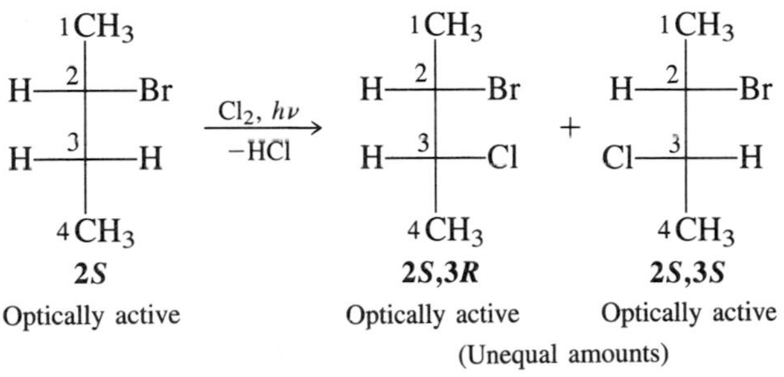

The chlorination at C2 gives a 1:1 mixture of enantiomers. Does the reaction at C3 also give an equimolar mixture of diastereomers? The answer is no. This finding is readily explained on inspection of the two transition states leading to the product (Figure 5-15). Abstraction of either one of the hydrogens results in a radical center at C3. In contrast with the radical formed in the chlorination at C2, however, the two faces of this radical are not mirror images of each other, because the radical retains the asymmetry of the original molecule as a result of the presence of the chiral C2. Thus, the two sides of the *p* orbital are not equivalent.

What are the consequences of this nonequivalency? If the rate of attack at the two faces of the radical differ, as one would predict on steric grounds, then the rates of formation of the two diastereomers should be different, as is indeed found: (2*S*,3*R*)-2-Bromo-3-chlorobutane is preferred over the 2*S*,3*S* isomer by a factor of 3 (Figure 5-15). The two transition states leading to products are not mirror images of each other and are not superimposable: They are diastereomeric. They therefore have different energies and represent different pathways.

EXERCISE 5-22

Write the structures of the products of monobromination of (*S*)-2-bromopentane at each carbon atom. Name the products and specify whether they are chiral or achiral, whether they will be formed in equal or unequal amounts, and which will be in optically active form.

Stereoselectivity is the preference for one stereoisomer

A reaction that leads to the predominant (or exclusive) formation of one of several possible stereoisomeric products is **stereoselective.** For example, the chlorination of (*S*)-2-bromobutane at C3 is stereoselective, as a result of the chirality of the radical intermediate. The corresponding chlorination at C2, however, is not stereoselective: The intermediate is achiral and a racemate is formed.

How much stereoselectivity is possible? The answer depends very much on substrate, reagents, the particular reaction in question, and conditions. Enzymes

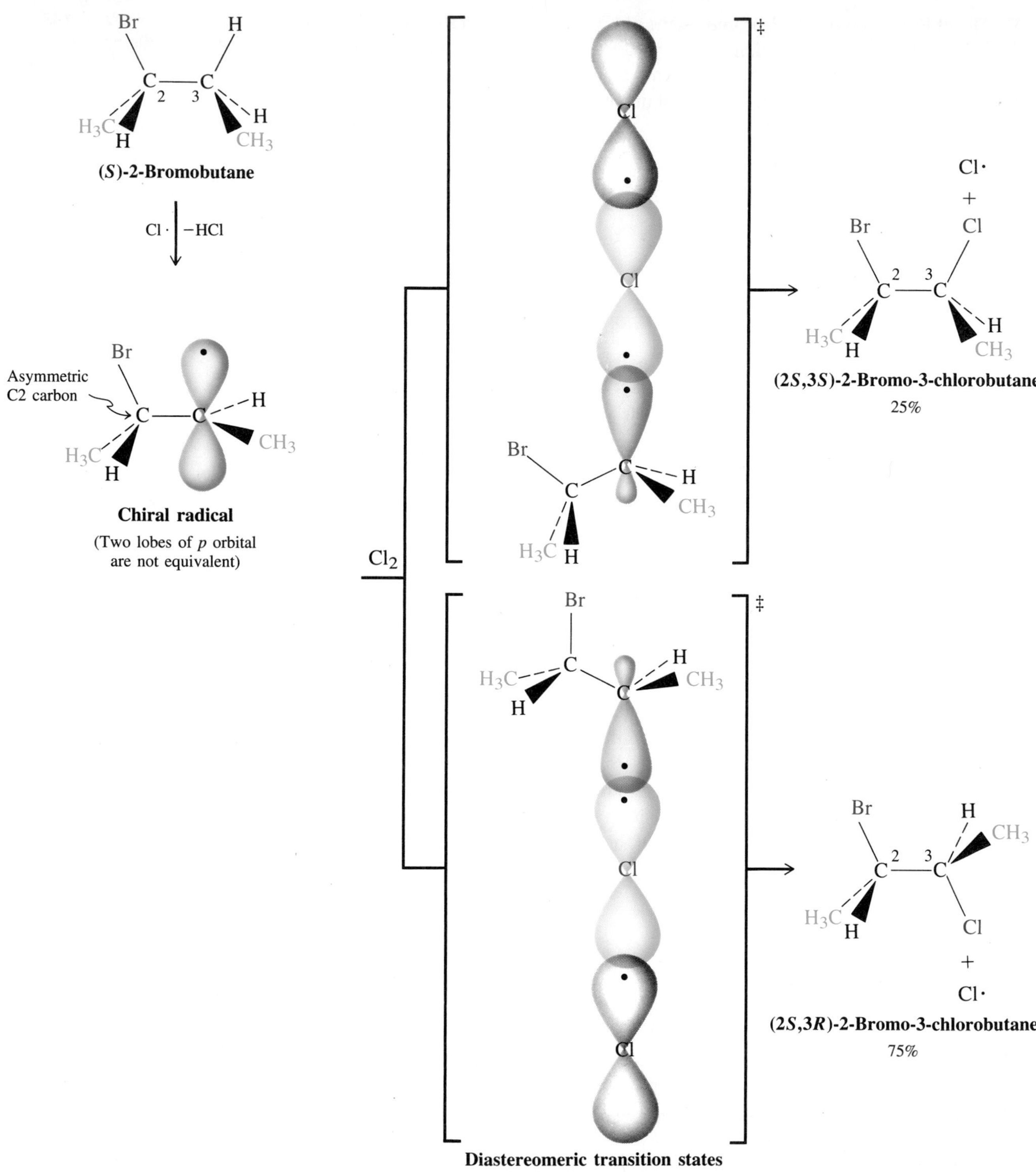

FIGURE 5-15 The chlorination of (*S*)-2-bromobutane at C3 produces the two diastereomers of 2-bromo-3-chlorobutane in unequal amounts as a result of the chirality at C2.

in nature manage to convert achiral compounds into chiral molecules with very high stereoselectivity. They are capable of this task because enzymes themselves have handedness and therefore convert achiral materials into those that are compatible with their own chirality. An example is the enzyme-catalyzed oxidation of dopamine to (−)-norepinephrine discussed in detail in Problem 26 at the end of the chapter. The chiral reaction environment created by the enzyme gives rise to 100% stereoselectivity in favor of the enantiomer shown. The situation is very similar to shaping flexible achiral objects with your hands. For example, clasping a piece of modeling clay with your left hand furnishes a shape that is the mirror image of that made with your right hand.

HO, HO, $CH_2CH_2NH_2$ —Dopamine β-monooxygenase, O_2→ HO, HO, HO, H, C, CH_2NH_2

Dopamine **(−)-Norepinephrine**

In summary, chemical reactions, as exemplified by radical halogenation, can be stereoselective or not. Starting from achiral materials, such as butane, a racemic (nonstereoselective) product is formed by halogenation at C2. The two hydrogens at the methylene carbons of butane are equally susceptible to substitution, the halogenation step in the mechanism of radical bromination proceeding through an achiral intermediate and two enantiomeric transition states of equal energy. Similarly, starting from chiral and enantiomerically pure 2-bromobutane, chlorination of the stereocenter also gives a racemic product. However, stereoselectivity is possible in the formation of a new stereocenter, because the chiral environment retained by the molecule results in two unequal modes of attack on the intermediate radical. The two transition states have a diastereomeric relationship, a condition that leads to the formation of products at unequal rates.

5-8 Resolution: Separation of Enantiomers

As we know, the generation of a chiral structure from an achiral starting material furnishes a racemic mixture. How, then, are pure enantiomers of a chiral compound obtained?

One possible approach is to start with the racemate and separate one enantiomer from the other. This process is called the **resolution** of enantiomers. Some enantiomers, like those of tartaric acid, crystallize into mirror-image shapes, which can be manually separated. However, this process is time consuming, uneconomic for anything but minute-scale separations, and applicable only in rare cases.

A better strategy for resolution is based on the difference in the physical properties of diastereomers. Suppose we can find a reaction that converts a racemate into a mixture of diastereomers. All the *R* forms of the original enantiomer mixture should then be separable from the corresponding *S* forms by fractional

BOX 5-4 Most Synthetic Chiral Pharmaceuticals Are Marketed as Racemic Mixtures

Nearly 90% of synthetic chiral substances in current use in medicine are prepared as racemic mixtures. Why? One reason is that the resolution of such mixtures on a large scale is very expensive and would add substantially to the costs associated with drug development. Fortunately, in many cases both enantiomers have comparable biological activity, or the "wrong" mirror-image isomer merely is biologically inactive. A terrible exception occurred in the case of the sedative *thalidomide,* which was marketed in 1960 as a racemic mixture to pregnant women in Europe and led to serious birth defects in hundreds of babies. In this case the "wrong" enantiomer appears to have been to blame for this tragedy.

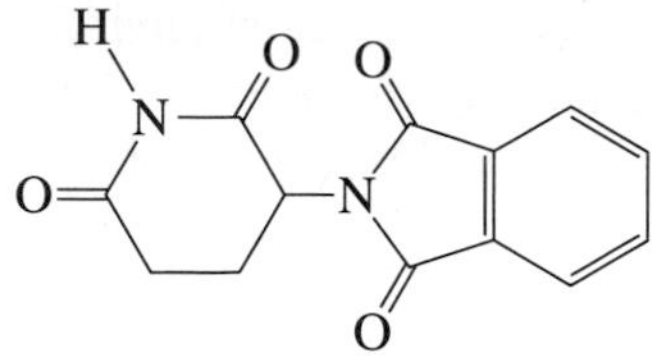

Recent advances in synthetic organic chemistry are changing this situation. By the early 1990s, industrially useful methods had been developed for the conversion of achiral alkenes into single enantiomers of chiral addition products (Chapter 12) by using *optically active catalysts*. Like the enzyme in the oxidation of dopamine (Section 5-7), these catalysts create a chiral environment around the reacting molecules, thereby leading to high stereoselectivity. As shown below, methods such as these have been applied to the syntheses of drugs such as the antiarthritic *naproxen* and the antihypertensive *propanolol* in high enantiomeric purity.

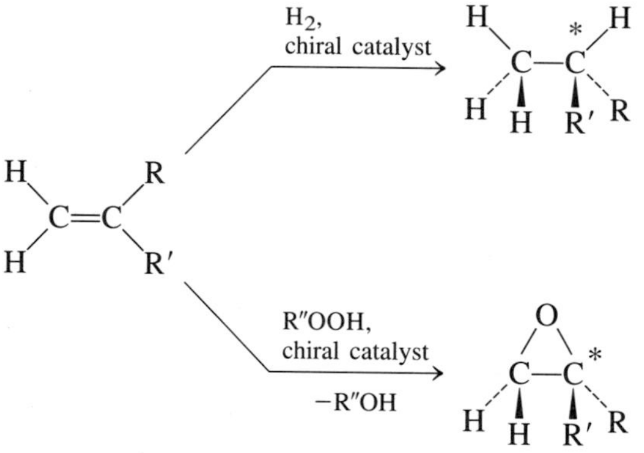

Thalidomide

H_2, chiral catalyst

$R''OOH$, chiral catalyst

$-R''OH$

(C^* = a new stereocenter)

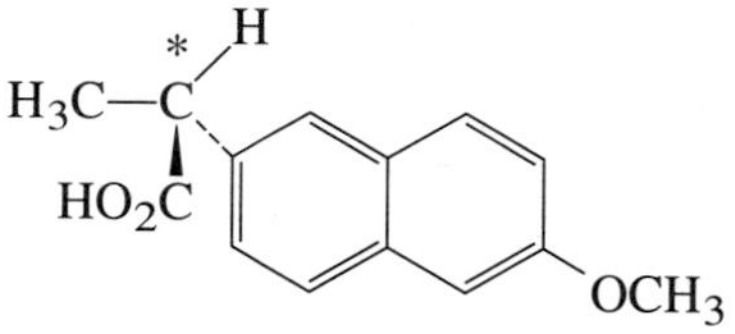

(*R*)-Naproxen

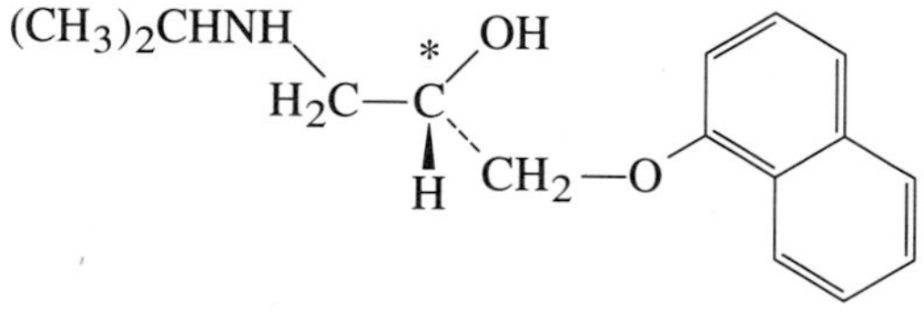

(*S*)-Propanolol

crystallization, distillation, or chromatography of the diastereomers. How can such a process be developed? The trick is to add an enantiomerically pure reagent that will attach itself to the components of the racemic mixture. For example, we can imagine reaction of a racemate, $X_{R,S}$ (in which X_R and X_S are the two enantiomers), with an optically pure compound Y_S (the choice of the *S* configuration is arbitrary; the pure *R* mirror image would work just as well). The reaction produces two optically active diastereomers, X_RY_S and X_SY_S, separable by standard techniques (Figure 5-16). Now the bond between X and Y in each of the

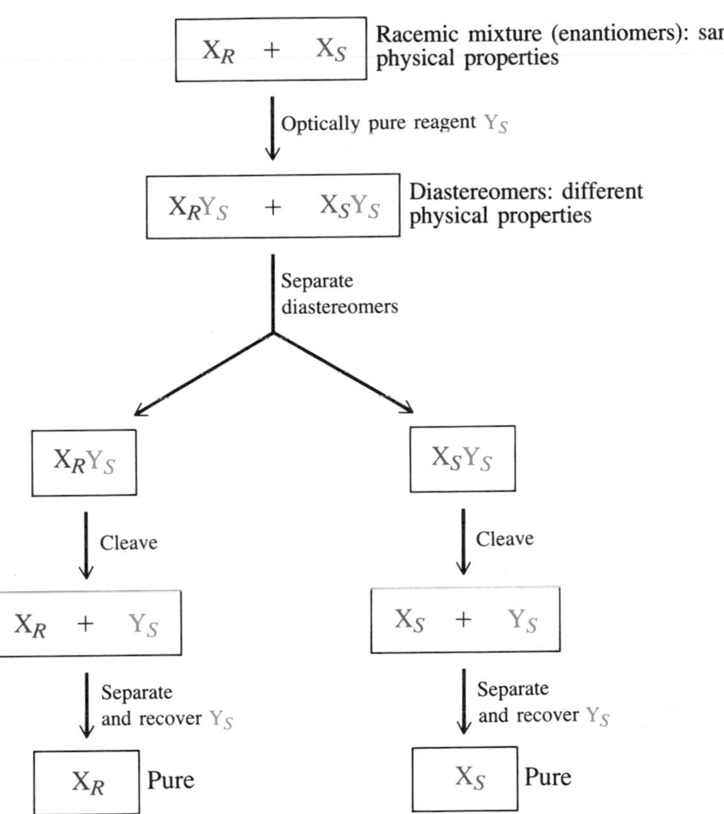

FIGURE 5-16 Flowchart for the separation (resolution) of two enantiomers. The procedure is based on conversion to separable diastereomers by means of reaction with an optically pure reagent.

separated and purified diastereomers is broken, liberating X_R and X_S in their enantiomerically pure state. In addition, the optically active agent Y_S may be recovered and reused in further resolutions.

What we need, then, is a readily available, enantiomerically pure compound, Y, that can be attached to the molecule to be resolved in an easily reversible chemical reaction. In fact, nature has provided us with a large number of pure optically active molecules that can be used. An example is (+)-2,3-dihydroxybutanedioic acid [(+)-(*R*,*R*)-tartaric acid]. A popular reaction employed in the resolution of enantiomers is salt formation between acids and bases. For example, (+)-tartaric acid functions as an effective resolving agent of racemic amines. Figure 5-17 shows how this works for 3-butyn-2-amine. The racemate is first treated with (+)-tartaric acid to form two diastereomeric tartrate salts. That incorporating the *R*-amine crystallizes on standing and can be filtered away from the solution, which contains the more soluble salt of the *S*-amine. Treatment of the (+)-salt with aqueous base liberates the free amine, (+)-(*R*)-3-butyn-2-amine. Similar treatment of the solution gives the (−)-*S* enantiomer (evidently slightly less pure: Note the slightly lower optical rotation). This process is just one of many ways in which the formation of diastereomers can be used in the resolution of racemates.

$$\begin{array}{c} NH_2 \\ | \\ CH_3CHC{\equiv}CH \end{array}$$

Racemic (*R*,*S*)-3-butyn-2-amine

↓ (+)-Tartaric acid, H_2O, several days

(+)-Tartrate salt of *R*-amine (Fischer projections: tartrate COO^-, H—OH, HO—H, COOH; $H_3\overset{+}{N}$—C(H)(*R*)(C≡CH)(CH_3))
+
(−)-Tartrate salt of *S*-amine (Fischer projections: tartrate COO^-, H—OH, HO—H, COOH; $H_3\overset{+}{N}$—C(H)(*S*)(CH_3)(C≡CH))

(+)-Tartrate salt of *R*-amine
$[\alpha]_D^{22°C} = +24.4°$
Crystallizes from solution

↓ K_2CO_3, H_2O

H_2N—C(H)(C≡CH)(CH_3)
47%
(+)-(*R*)-3-Butyn-2-amine
$[\alpha]_D^{22°C} = +53.2°$
b.p. 82–84°C

(−)-Tartrate salt of *S*-amine
$[\alpha]_D^{22°C} = -24.1°$
Stays in solution

↓ K_2CO_3, H_2O

H_2N—C(H)(CH_3)(C≡CH)
51%
(−)-(*S*)-3-Butyn-2-amine
$[\alpha]_D^{20°C} = -52.7°$
b.p. 82–84°C

FIGURE 5-17 Resolution of 3-butyn-2-amine with (+)-2,3-dihydroxybutanedioic [(+)-tartaric] acid. It is purely accidental that the [α] values for the two diastereomeric tartrate salts are similar in magnitude.

IMPORTANT CONCEPTS

1. Isomers have the same molecular formula but are different compounds. Constitutional (structural) isomers differ in the order in which the individual atoms are connected. Stereoisomers have the same connectivity but differ in the three-dimensional arrangement of the atoms.
2. An object that is not superimposable on its mirror image is chiral.
3. A carbon atom bearing four different substituents (asymmetric carbon) is an example of a stereocenter.
4. Two stereoisomers in which one member of the pair is related to the other as an object is to its mirror image are called enantiomers.
5. A compound containing one stereocenter is chiral and exists as a pair of enantiomers. A 1:1 mixture of enantiomers is a racemate (racemic mixture).
6. Chiral molecules cannot have a plane of symmetry (mirror plane). If a molecule has a mirror plane, then it is achiral.
7. Diastereomers are stereoisomers that are not related to each other as object to mirror image. Cis and trans isomers of cyclic compounds are examples of diastereomers.
8. Two stereocenters in a molecule result in up to four stereoisomers—two diastereomerically related pairs of enantiomers. The maximum number of stereoisomers that a compound with n stereocenters can have is 2^n. This number is reduced when equivalently substituted stereocenters give rise to a plane of symmetry. A molecule containing stereocenters and a mirror plane is identical to its mirror image (achiral) and is called a meso compound. The presence of a mirror plane in any energetically accessible

conformation of a molecule is sufficient to make it achiral.

9. Most of the physical properties of enantiomers are the same. A major exception is their interaction with plane-polarized light: One enantiomer will rotate the polarization plane clockwise (dextrorotatory), the other counterclockwise (levorotatory). This phenomenon is called optical activity. The extent of the rotation is measured in degrees and is expressed by the specific rotation, $[\alpha]$. Racemates and meso compounds show zero rotation. The optical purity of an unequal mixture of enantiomers is given by

$$\% \text{ optical purity} = \left(\frac{[\alpha]_{\text{observed}}}{[\alpha]} \cdot 100\right)$$

10. The "handedness" of a stereocenter (its absolute configuration) is revealed by X-ray diffraction and can be assigned as *R* or *S* by using the sequence rules of Cahn, Ingold, and Prelog.

11. Fischer projections are line drawings that simplify the depiction of stereocenters and help in describing their absolute configuration.

12. Chirality can be introduced into an achiral compound by radical halogenation. When the transition states are enantiomeric (related as object and mirror image), the result is a racemate because the faces of the planar radical react at equal rates.

13. Radical halogenation of a chiral molecule containing one stereocenter will give a racemate if the reaction takes place at the stereocenter. When reaction elsewhere leads to two diastereomers, they will be formed in unequal amounts.

14. The preference for the formation of one stereoisomer, when several are possible, is called stereoselectivity.

15. The separation of enantiomers is called resolution. It is best achieved by the reaction of the racemate with the pure enantiomer of a chiral compound to yield separable diastereomers. Cleavage of the chiral reagent frees both enantiomers of the original racemate.

PROBLEMS

1. Classify each of the common objects listed below as being either chiral or achiral. Assume in each case that the object is in its simplest form, without decoration or printed labels. **(a)** A ladder; **(b)** a door; **(c)** an electric fan; **(d)** a refrigerator; **(e)** the earth; **(f)** a baseball; **(g)** a baseball bat; **(h)** a baseball glove; **(i)** a flat sheet of paper; **(j)** a fork; **(k)** a spoon; **(l)** a knife.

2. Each part of this problem lists two objects or sets of objects. As precisely as you can, describe the relation between the two sets, using the terminology of this chapter; that is, specify whether they are identical, enantiomeric, or diastereomeric. **(a)** An American toy car compared with a British toy car (same color and design but steering wheels on opposite sides); **(b)** two left shoes compared with two right shoes (same color, size, and style); **(c)** a pair of skates compared with two left skates (same color, size, and style); **(d)** a right glove on top of a left glove (palm to palm) compared with a left glove on top of a right glove (palm to palm; same color, size, and style).

3. For each pair of molecules on the next page, indicate whether its members are identical, structural isomers, conformers, or stereoisomers. How would you describe the relationship between conformations when they are maintained at a temperature too low to permit them to interconvert?

(a) 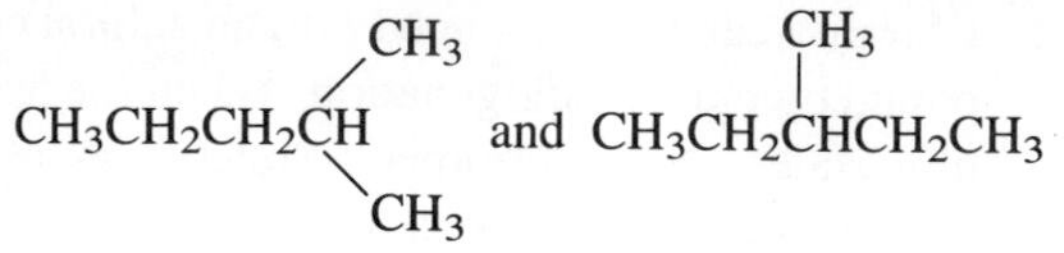 $CH_3CH_2CH_2CH(CH_3)_2$ and $CH_3CH_2CH(CH_3)CH_2CH_3$

(b) 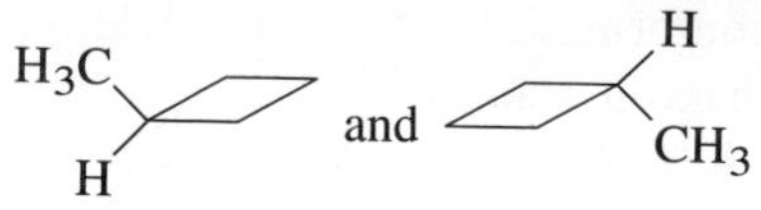 and

(c) 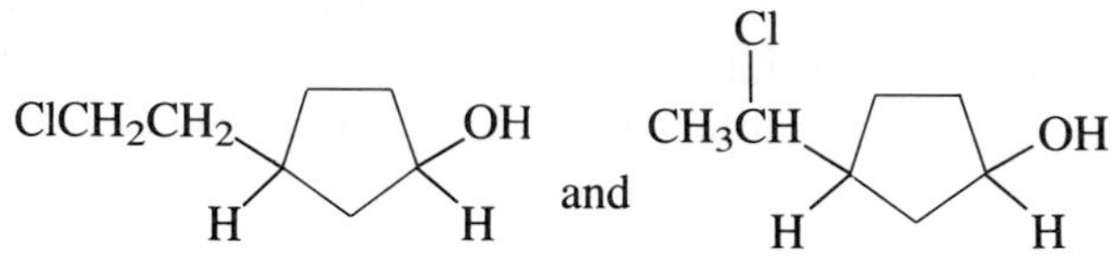 and

(d) 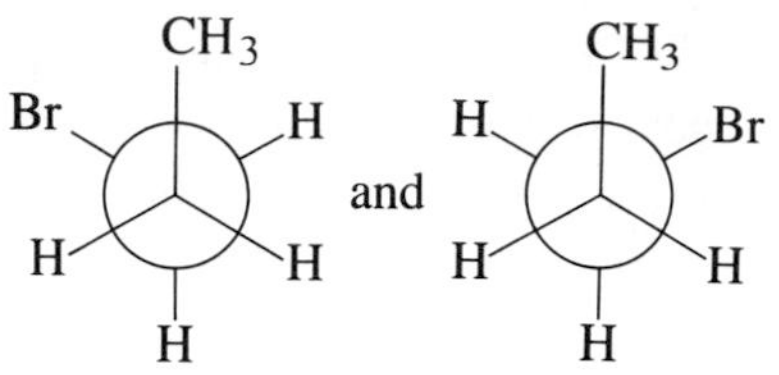 and

(e) 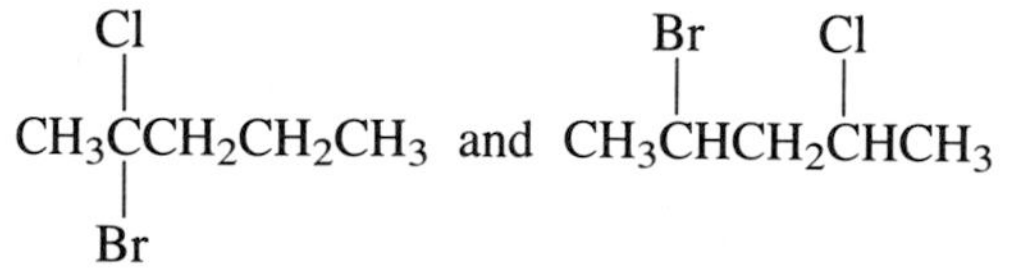 $CH_3CCl(Br)CH_2CH_2CH_3$ and $CH_3CHBrCH_2CHClCH_3$

(f) 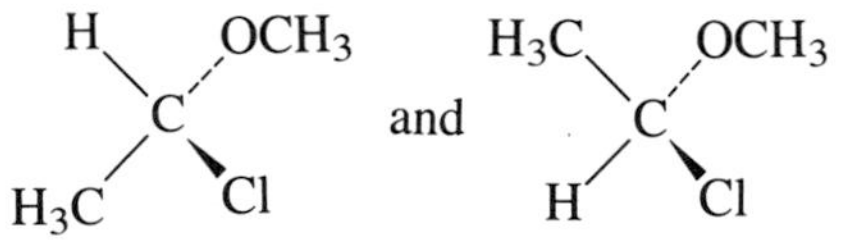 and

(g) 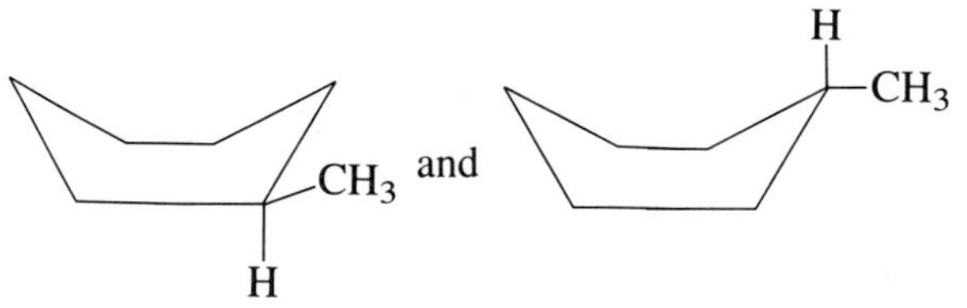 and

(h) 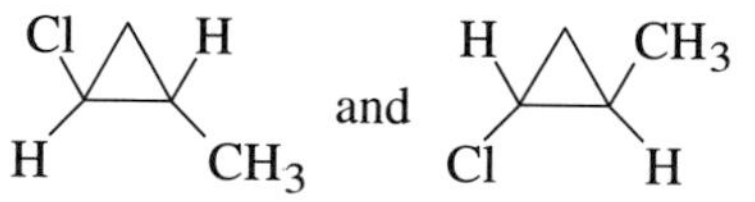 and

4. Which of the following compounds are chiral?

(a) 2-Methylheptane
(b) 3-Methylheptane
(c) 4-Methylheptane
(d) 1,1-Dibromopropane
(e) 1,2-Dibromopropane
(f) 1,3-Dibromopropane
(g) Ethene, $H_2C{=}CH_2$
(h) Ethyne, $HC{\equiv}CH$
(i) Benzene,  (Note: Like ethene, benzene contains all sp^2-hybridized carbons and therefore is planar.)
(j) Epinephrine, 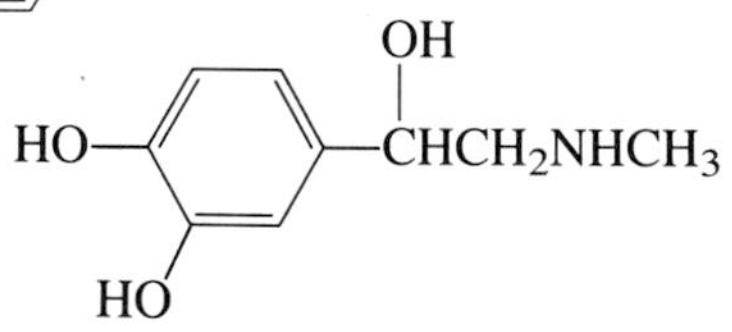
(k) Vanillin, 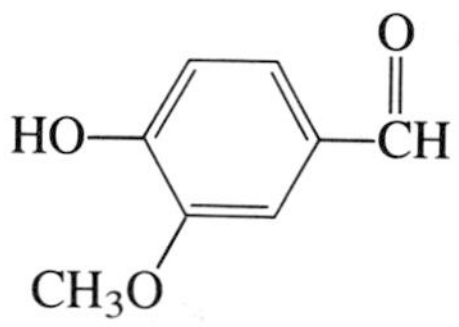
(l) Citric acid, 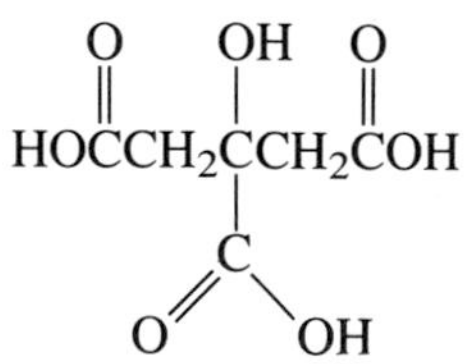
(m) Ascorbic acid, 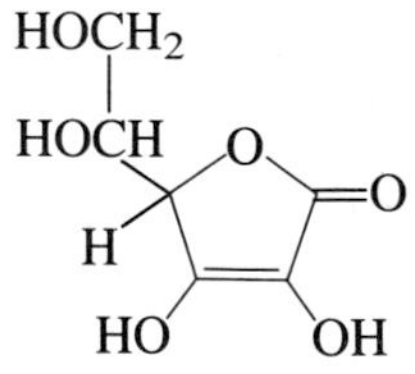
(n) *p*-Menthane-1,8-diol (terpin hydrate), 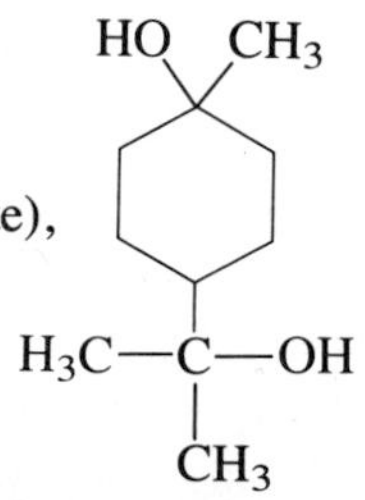
(o) Meperidine (demerol),

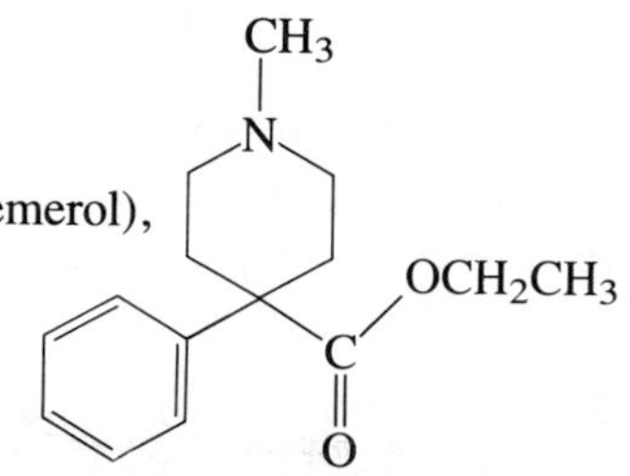

5. Which of the following cyclohexane derivatives are chiral? For the purpose of determining the chirality of a cyclic compound, the ring may generally be treated as if it were planar.

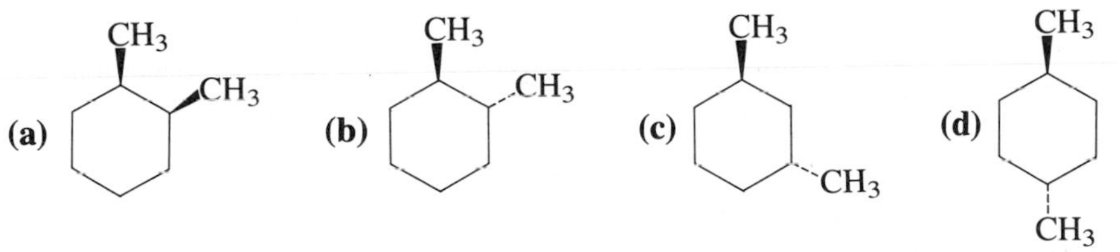

(a) (b) (c) (d)

6. For each formula below, identify every structural isomer containing one or more stereocenters, give the number of stereoisomers for each, and draw and fully name at least one of the stereoisomers in each case.
(a) C_7H_{16} **(b)** C_8H_{18} **(c)** C_5H_{10}, with one ring

7. Mark the stereocenters in each of the chiral molecules in Problem 4. Draw any single stereoisomer of each of these molecules, and assign the appropriate designation (*R* or *S*) to each stereocenter.

8. The two isomers of carvone [systematic name: 2-methyl-5-(1-methylethenyl)-2-cyclohexenone] are drawn below. Which is *R* and which is *S*?

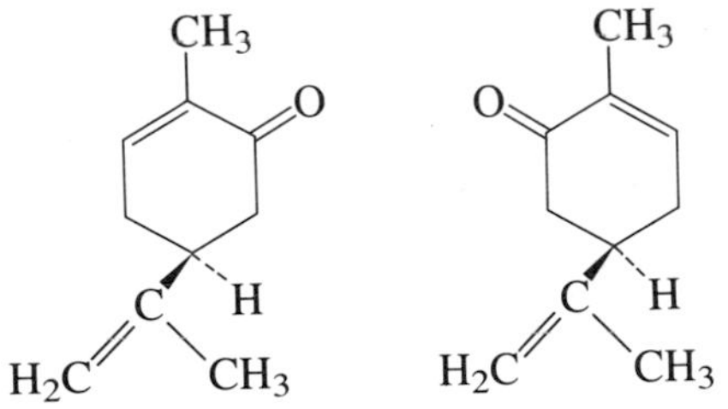

(+)-Carvone
(In caraway seeds)

(−)-Carvone
(In spearmint)

9. Draw structural representations of each of the following molecules. Be sure that your structure clearly shows the configuration at each stereocenter. **(a)** (*R*)-3-Bromo-3-methylhexane; **(b)** (3*R*,5*S*)-3,5-dimethylheptane; **(c)** (2*R*,3*S*)-2-bromo-3-methylpentane; **(d)** (*S*)-1,1,2-trimethylcyclopropane; **(e)** (1*S*,2*S*)-1-chloro-1-trifluoromethyl-2-methylcyclobutane; **(f)** (1*R*,2*R*,3*S*)-1,2-dichloro-3-ethylcyclohexane.

10. For each of the following questions, assume that all measurements are made in 10-cm polarimeter sample containers. **(a)** A solution of 0.4 g of optically active 2-butanol in 10 mL of water displays an optical rotation of −0.56°. What is its specific rotation? **(b)** The specific rotation of sucrose (common sugar) is +66.4°. What would be the observed optical rotation of a solution containing 3 g of sucrose in 10 mL of water? **(c)** A solution of pure (*S*)-2-bromobutane in ethanol is found to have an observed $\alpha = 57.3°$. If $[\alpha]$ for (*S*)-2-bromobutane is 23.1°, what is the concentration of the solution?

11. Natural epinephrine, $[\alpha]_D^{25°C} = -50°$, is used medicinally. Its enantiomer is medically worthless and is, in fact, toxic. You, a pharmacist, are given a solution said to contain 1 g of epinephrine in 20 mL of liquid, but the optical purity is not specified. You place it in a polarimeter (10-cm tube) and get a reading of −2.5°. What is the optical purity of the sample? Is it safe to use medicinally?

12. Sodium hydrogen (*S*)-glutamate [(*S*)-monosodium glutamate], $[\alpha]_D^{25°C} = +24°$, is the active flavor enhancer known as MSG. The condensed formula of MSG is shown in the margin. **(a)** Draw the structure of the *S* enantiomer

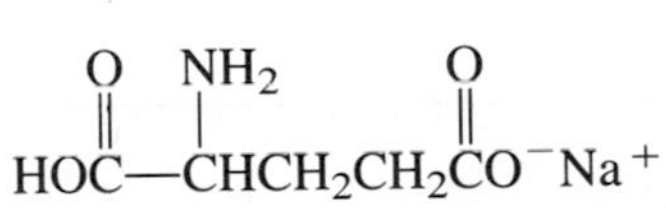

$HO\overset{O}{\overset{\|}{C}}-\overset{NH_2}{\overset{|}{C}}HCH_2CH_2\overset{O}{\overset{\|}{C}}O^-\,Na^+$

of MSG. **(b)** If a commercial sample of MSG were found to have a $[\alpha]_D^{25°C} = +8°$, what would be its optical purity? What would the percentages of the *S* and *R* enantiomers be in the mixture? **(c)** Answer the same questions for a sample with $[\alpha]_D^{25°C} = +16°$.

13. For each compound below, mark each stereocenter, assign an *R* or *S* designation, and draw a clear picture of the molecule's enantiomer.

(a) 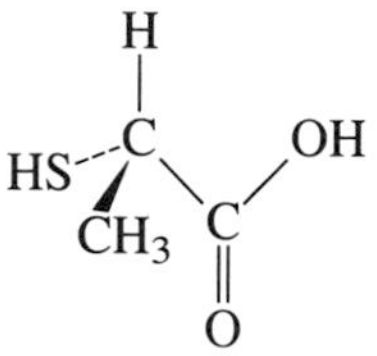 HS, H, CH_3, OH, C=O
(b) CH_3, H, $C(CH_3)_3$, $CH_2CH_2OCH_2CH_3$
(c) CH_3, H_3C, H, Cl, H, CH=CH_2
(d) 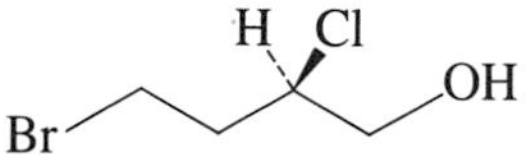 Br, H, Cl, OH
(e) 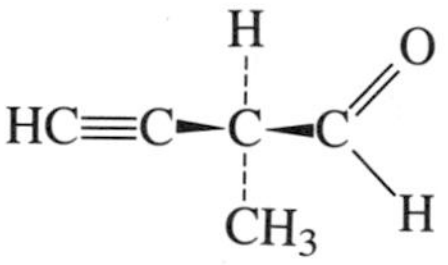 HC≡C–C(H)(CH_3)–CHO
(f) 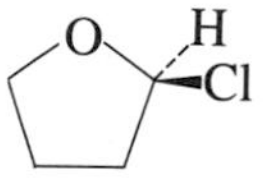 O, H, Cl
(g) 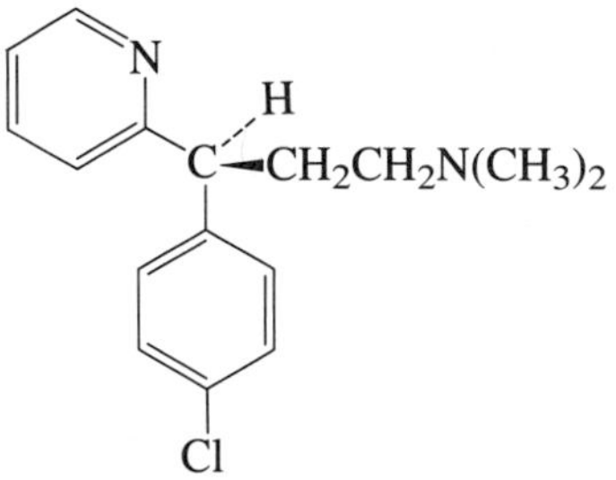 N, H, C–$CH_2CH_2N(CH_3)_2$, Cl (Note: The carbons in benzene or benzenelike rings are treated in the same way as those in alkenes. Use sequence rule 3 from Section 5-3.)

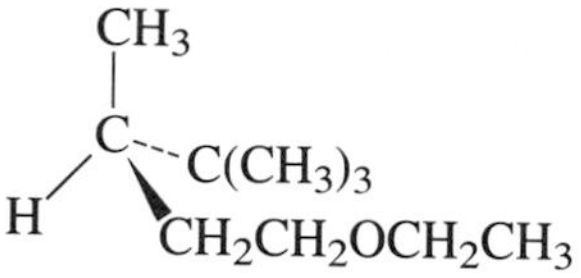

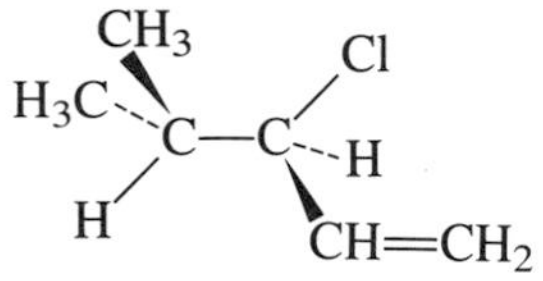

Chlorpheniramine
(As in Coricidin decongestant)

(h) 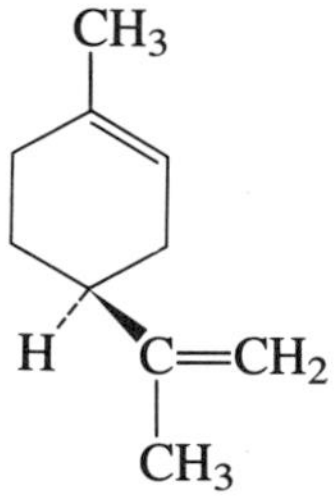 CH_3, H, C=CH_2, CH_3

Limonene
(From trees, fruits, etc.)

14. For each pair of structures below, indicate whether the two compounds are identical or are enantiomers of each other.

(a) 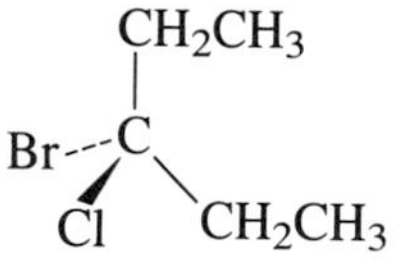 CH_2CH_3, Br, Cl, CH_2CH_3 and CH_2CH_3, Br, CH_3CH_2, Cl
(b) 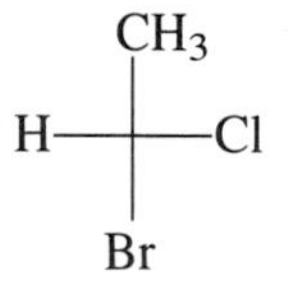 CH_3, H, Cl, Br and CH_3, Cl, H, Br
(c) 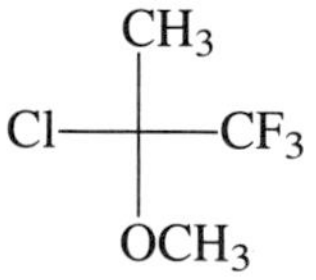 CH_3, Cl, CF_3, OCH_3 and OCH_3, F_3C, CH_3, Cl
(d) 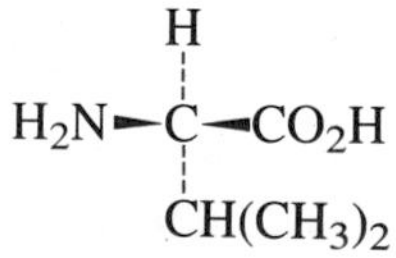 H, H_2N–C–CO_2H, $CH(CH_3)_2$ and NH_2, H, $CH(CH_3)_2$, CO_2H

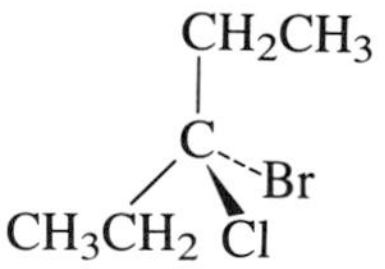

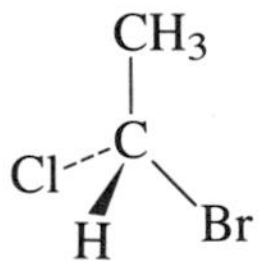

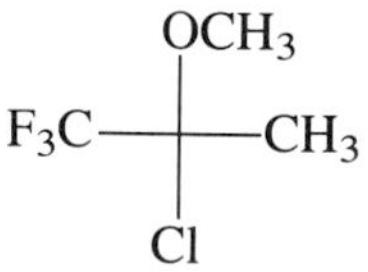

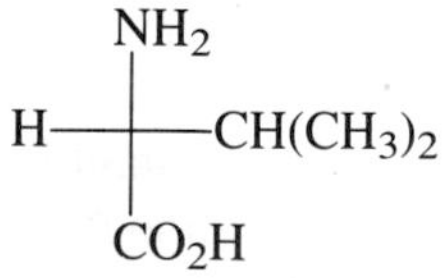

15. Determine the *R* or *S* designation for each stereocenter in the structures in Problem 14.

16. Redraw each of the following molecules as a Fischer projection; then assign *R* or *S* designations to each stereocenter.

(a) 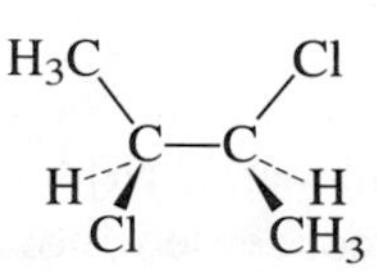 H_3C, Cl, H, Cl, H, CH_3
(b) 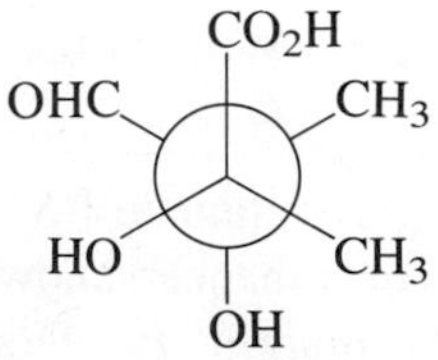 CO_2H, OHC, CH_3, HO, CH_3, OH

(c) 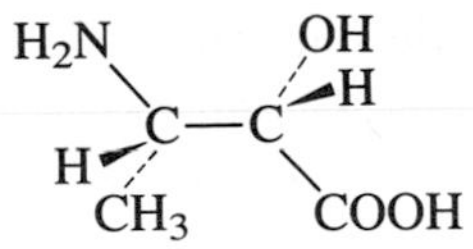 H_2N, OH, H, C—C, H, CH_3, COOH

(d) 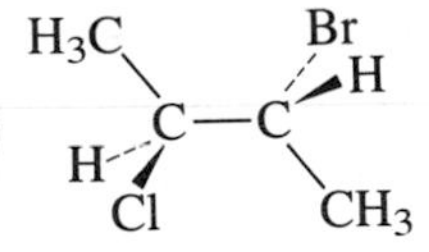 H_3C, Br, H, C—C, H, Cl, CH_3

17. The compound pictured in the margin is a sugar called (−)-arabinose. Its specific rotation is −105°. **(a)** Draw an enantiomer of (−)-arabinose. **(b)** Does (−)-arabinose have any other enantiomers? **(c)** Draw a diastereomer of (−)-arabinose. **(d)** Does (−)-arabinose have any other diastereomers? **(e)** If possible, predict the specific rotation of the structure that you drew for (a). **(f)** If possible, predict the specific rotation of the structure that you drew for (c). **(g)** Does (−)-arabinose have any optically inactive diastereomers? If it does, draw one.

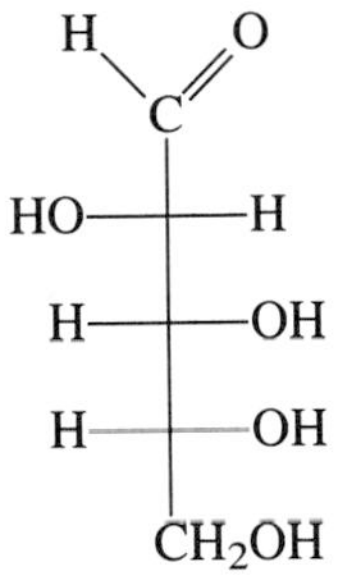

H, O, C; HO—H; H—OH; H—OH; CH_2OH

(−)-Arabinose

18. Write the complete IUPAC name of the following compound (do not forget stereochemical designations).

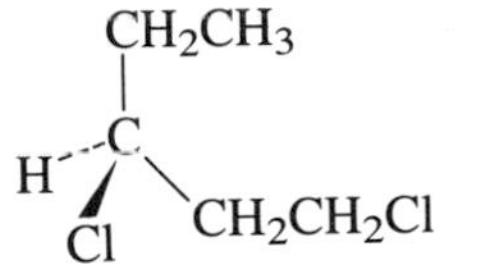

CH_2CH_3, H, C, Cl, CH_2CH_2Cl $C_5H_{10}Cl_2$

Reaction of this compound with 1 mol of Cl_2 in the presence of light produces several isomers of the formula $C_5H_9Cl_3$. For each part of this problem, give the following information: How many stereoisomers are formed? If more than one is formed, are they formed in equal or unequal amounts? Designate every stereocenter in each stereoisomer as *R* or *S*.
(a) Chlorination at C3 **(b)** Chlorination at C4 **(c)** Chlorination at C5

19. Monochlorination of methylcyclopentane can result in several products. Give the same information as that requested in Problem 18 for the monochlorination of methylcyclopentane at C1, C2, and C3.

20. Draw all possible products of the chlorination of (*S*)-2-bromo-1,1-dimethylcyclobutane. Specify whether they are chiral or achiral, whether they are formed in equal or unequal amounts, and which are optically active when formed.

21. Illustrate how to resolve racemic 1-phenylethanamine (shown in the margin), using the method of reversible conversion into diastereomers.

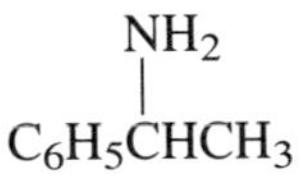

NH_2, $C_6H_5CHCH_3$

22. Draw a flowchart that diagrams a method for the resolution of racemic 2-hydroxypropanoic acid (lactic acid), using (*S*)-1-phenylethanamine (for structure, see Problem 21).

23. How many different stereoisomeric products are formed in the monobromination of **(a)** racemic *trans*-1,2-dimethylcyclohexane and **(b)** pure (*R*,*R*)-1,2-dimethylcyclohexane? **(c)** For your answers to (a) and (b), indicate whether you expect equal or unequal amounts of the various products to be formed. Indicate to what extent products can be separated on the basis of having different physical properties (e.g., solubility, boiling point).

24. Make a model of *cis*-1,2-dimethylcyclohexane in its most stable conformation. If the molecule were rigidly locked into this conformation, would it be chiral? (Test your answer by making a model of the mirror image and checking for superimposability.)

Flip the ring of the model. What is the stereoisomeric relation between the original conformation and the conformation after flipping the ring? How do the results that you have obtained in this problem relate to your answer to Problem 5(a)?

25. Morphinane is the parent substance of the broad class of chiral molecules known as the morphine alkaloids. Interestingly, the (+) and (−) enantiomers of the compounds in this family have rather different physiological properties. The (−) compounds, such as morphine, are "narcotic analgesics" (painkillers), whereas the (+) compounds are "antitussives" (ingredients in cough syrup). Dextromethorphan is one of the simplest and most common of the latter.

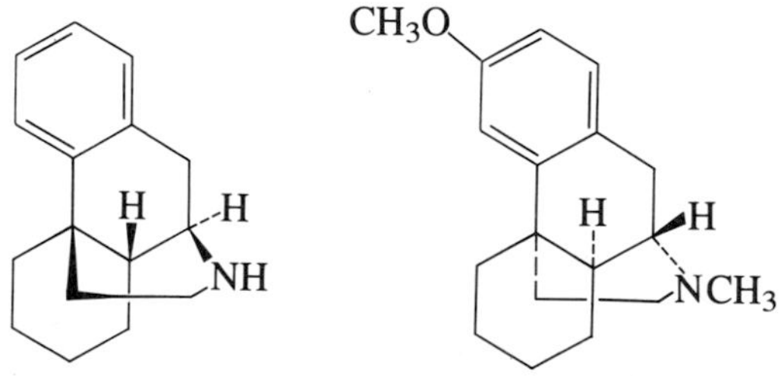

CH_3O H H NH H H NCH_3

Morphinane **Dextromethorphan**

(a) Locate and identify all the stereocenters in dextromethorphan. **(b)** Draw the enantiomer of dextromethorphan. **(c)** As best you can (it is not easy), assign *R* and *S* configurations to all the stereocenters in dextromethorphan.

26. The enzymatic introduction of a functional group into a biologically important molecule is not only specific with regard to the location of the reaction in the molecule (see Chapter 4, Problem 22), but also usually specific in the stereochemistry obtained. The biosynthesis of epinephrine first requires that a hydroxy group be introduced specifically to produce (−)-norepinephrine from the achiral substrate dopamine. (The completion of the synthesis of epinephrine will be presented in Problem 34 of Chapter 9.) Only the (−) enantiomer is functional in the appropriate physiological manner; so the synthesis must be highly stereoselective.

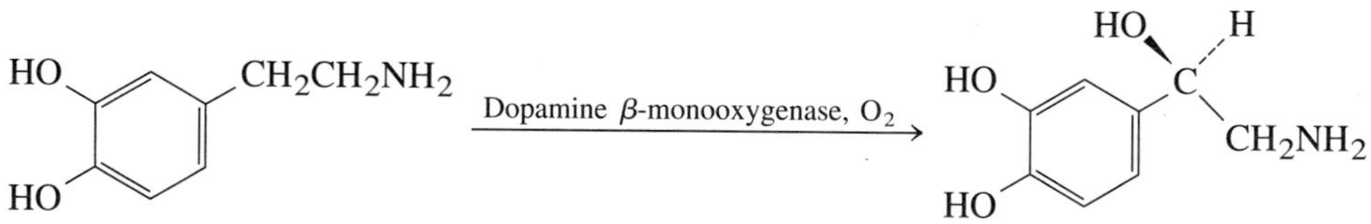

HO HO $CH_2CH_2NH_2$ —Dopamine β-monooxygenase, O_2→ HO HO HO H C CH_2NH_2

Dopamine **(−)-Norepinephrine**

(a) Is the configuration of (−)-norepinephrine *R* or *S*? **(b)** In the absence of an enzyme, would the transition states of a radical oxidation leading to (−)- and (+)-norepinephrine be of equal or unequal energy? What term describes the relation between these transition states? **(c)** In your own words, describe how the enzyme must affect the energy of these transition states to favor production of the (−) enantiomer. Does the enzyme have to be chiral or can it be achiral?

6

Properties and Reactions of Haloalkanes

Bimolecular Nucleophilic Substitution

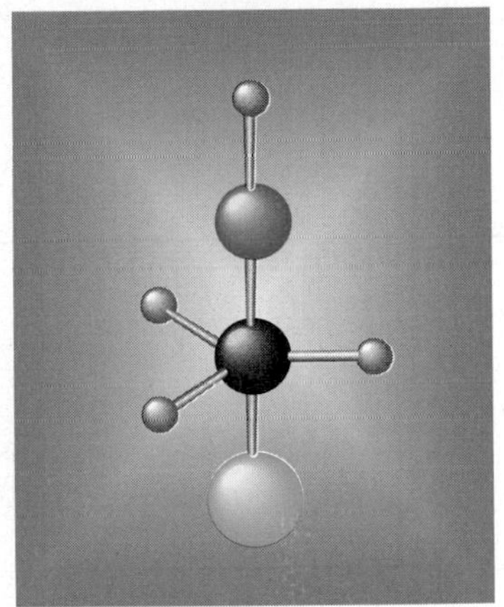

We have looked at the properties, reactions, and structures of the alkanes, which are the building blocks of organic chemistry. In particular, we learned that halogenation leads to haloalkanes, and we examined the stereochemistry of this reaction. We now turn to the chemistry of these substances.

A haloalkane possesses a polarized carbon–halogen bond, which constitutes the center of haloalkane reactivity and dictates its mechanistic behavior. In this chapter we shall introduce a new mechanism to explain the kinetics observed for a common reaction of haloalkanes, and we shall learn how different solvents affect this process. We begin, however, with the rules for naming haloalkanes.

6-1 Naming the Haloalkanes

We saw in Chapter 2 that alkanes are depicted by the general formula R–H, where R denotes an alkyl group. In a similar way, the **haloalkanes** are represented as R–X, in which X corresponds to a halogen atom.

In the systematic (IUPAC) nomenclature the halogen is treated as a substituent to the alkane framework.

CH_3I — **Iodo**methane

Fluorocyclohexane (F on a cyclohexane ring) — **Fluoro**cyclohexane

$(CH_3)_3C{-}Br$ — **2-Bromo-2-methyl**propane

The longest alkane chain is numbered so that the first substituent from either end receives the lowest number. As usual, substituents are ordered according to the alphabet. Complex appendages are named according to the same rules.

$ICH_2CH(CH_3)CH_3$ — **1-Iodo-2-methyl**propane

Cyclooctyl–$CHICH_3$ — **(1-Iodoethyl)cyclooctane**

$CH_3CH_2CH_2CH_2CH_2CH(CH_2CCl(CH_3)C(CH_3)_3)CH_2CH_2CH_2CH_2CH_3$ — **6-(2-Chloro-2,3,3-trimethylbutyl)undecane**

Common names are based on the older term *alkyl halide*. For example, the three structures at the top of this page have the common names methyl iodide, cyclohexyl fluoride, and *tert*-butyl bromide, respectively. Some chlorinated solvents have common names, for example, carbon tetrachloride, CCl_4; chloroform, $CHCl_3$; and methylene chloride, CH_2Cl_2.

EXERCISE 6-1

Draw the structures of (2-iodoethyl)cyclooctane and 5-butyl-3-chloro-2,2,3-trimethyldecane.

In summary, haloalkanes are named in accord with the rules that apply to naming the alkanes (Section 2-3), the halo substituent being treated the same as alkyl groups.

6-2 Physical Properties of Haloalkanes

The physical properties of the haloalkanes are quite distinct from those of the corresponding alkanes. To understand these differences, we must consider the size of the halogen substituent and the polarity of the carbon–halogen bond. Let us see how these factors affect bond strength, bond length, molecular polarity, and boiling point.

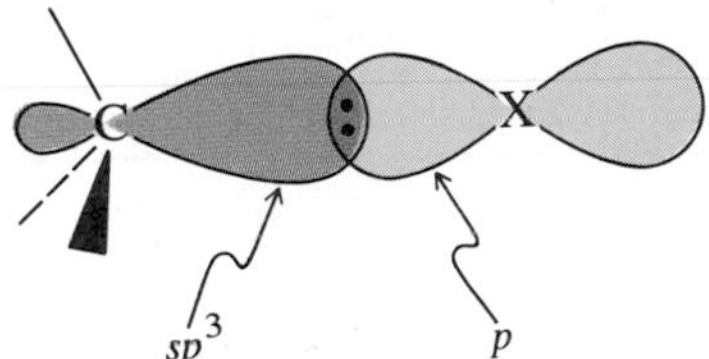

FIGURE 6-1 Bond between an alkyl carbon and a halogen. The size of the *p* orbital is substantially larger than that shown for X = Cl, Br, or I.

The bond strength of C–X decreases as the size of X increases

The bond between carbon and a halogen is made up mainly by the overlapping of an sp^3 hybrid orbital on carbon with a nearly pure *p* orbital on the halogen (Figure 6-1). In the progression from fluorine to iodine in the periodic table, the size of the halogen *p* orbital increases, and the electron cloud around the halogen atom becomes more diffuse. Consequently, its overlap with the carbon orbitals and hence the C–X bond strength diminish. For example, the C–X bond dissociation energies in the halomethanes, CH_3X, decrease along the series; at the same time, the C–X bond lengths increase (Table 6-1).

TABLE 6-1 C–X Bond Lengths and Bond Strengths in CH_3X

Halomethane	Bond length (Å)	Bond strength (kcal mol^{-1})
CH_3F	1.385	110
CH_3Cl	1.784	85
CH_3Br	1.929	71
CH_3I	2.139	57

The C–X bond is polarized

A characteristic of the haloalkanes is their polar C–X bond. Recall from Section 1-3 that the halogens are more electronegative than is carbon. Thus, the electron density along the C–X bond is displaced in the direction of X, thereby giving the halogen a partial negative charge (δ^-) and the carbon a partial positive charge (δ^+). How does this bond polarization govern the chemical behavior of the haloalkanes? We shall see, for example, that anions and other electron-rich species can attack the positively polarized carbon atom. Cations and other electron-deficient species, however, attack the halogen.

The Polar Character of the C–X Bond

$\overset{\delta^+}{C}$—$\overset{\delta^-}{X}$:

Haloalkanes have higher boiling points than the corresponding alkanes

Does the polarity of the C–X bond affect the physical properties of the haloalkanes? Yes, their boiling points are generally higher than those of the corresponding alkanes (Table 6-2). The most important contributor to this effect is coulombic attraction between the δ^+ and δ^- ends of C–X bond dipoles in the liquid state *(dipole–dipole interaction)*.

Dipole-Dipole Attraction

$\overset{\delta^+}{C}$—$\overset{\delta^-}{X}$ ········ $\overset{\delta^+}{C}$—$\overset{\delta^-}{X}$

Boiling points also rise with increasing size of X, the result of increased molecular weight and greater London interactions (Section 2-4). Recall that London forces arise from mutual correlation of electrons among molecules. This

TABLE 6-2 Boiling Points of Haloalkanes (R–X)

R		Boiling point (°C)				
	X =	H	F	Cl	Br	I
CH_3		−161.7	−78.4	−24.2	3.6	42.4
CH_3CH_2		−88.6	−37.7	12.3	38.4	72.3
$CH_3(CH_2)_2$		−42.1	−2.5	46.6	71.0	102.5
$CH_3(CH_2)_3$		−0.5	32.5	78.4	101.6	130.5
$CH_3(CH_2)_4$		36.1	62.8	107.8	129.6	157.0
$CH_3(CH_2)_7$		125.7	142.0	182.0	200.3	225.5

effect is strongest when the outer electrons are not held very tightly around the nucleus, as in the heavier atoms. To measure it, we define the **polarizability** of an atom as the capability of its electrons to respond to a changing electric field. The more polarizable an atom, the more effectively it will enter into London interactions, and the higher will be the boiling point.

To summarize, the halogen orbitals become increasingly diffuse along the series F, Cl, Br, I. Hence, (1) the C–X bond strength decreases; (2) the C–X bond becomes longer; (3) for the same R, the boiling points increase; (4) the polarizability of X becomes greater; and (5) London interactions improve. We shall see next that these interrelated effects also play an important role in the reactions of haloalkanes.

6-3 Nucleophilic Substitution

Reactions of haloalkanes often involve a substance with an unshared electron pair. This reagent could be an anion, such as iodide ($:\ddot{\underset{\cdot\cdot}{I}}:^-$), or a neutral species, such as ammonia ($:NH_3$). Each reagent can attack the haloalkane and replace the halide, a process called **nucleophilic substitution.** A great many species are transformed in this way, particularly in solution. The process occurs widely in nature and can be controlled effectively even on an industrial scale. Let us see how it works.

Nucleophiles attack electrophilic centers

We noted that in the polarization of the carbon–halogen bond, the carbon atom acquires a partial positive charge. As a result, this center exhibits a tendency to react with species possessing unshared pairs of electrons: The carbon is said to be **electrophilic** (literally "electron loving;" *philos,* Greek, loving). In turn, atoms bearing lone pairs are described as **nucleophilic** ("nucleus loving"). Nucleophiles, often denoted by the abbreviation Nu, may be negatively charged or neutral.

The nucleophilic substitution of a haloalkane is described by either of two general equations.

Nucleophilic Substitutions

$$\underset{\textbf{Nucleophile}}{\text{Nu}:^-} + \underset{\text{Electrophile}}{\overset{\delta^+}{\text{R}}}\!-\!\overset{\delta^-}{\ddot{\underset{\cdot\cdot}{\text{X}}}}: \longrightarrow \text{R—Nu} + \underset{\textbf{Leaving group}}{:\ddot{\underset{\cdot\cdot}{\text{X}}}:^-}$$

or

$$\underset{\textbf{Nucleophile}}{\text{Nu}:} + \underset{\text{Electrophile}}{\text{R}}\!-\!\ddot{\underset{\cdot\cdot}{\text{X}}}: \longrightarrow [\text{R—Nu}]^+ + \underset{\textbf{Leaving group}}{:\ddot{\underset{\cdot\cdot}{\text{X}}}:^-}$$

Color code:
Nucleophiles—red
Electrophiles—blue
Leaving groups—green

In the first example a negatively charged nucleophile reacts with a haloalkane to yield a neutral substitution product. In the second example a neutral Nu produces a positively charged product. In both cases the group displaced is the halide ion, $:\ddot{\underset{\cdot\cdot}{\text{X}}}:^-$, which is called the **leaving group.** As will be the case in many equations and mechanisms that follow, nucleophiles, electrophiles, and leaving groups are here shown in red, blue, and green, respectively.

In this section we have described nucleophilic substitution as a process in which the nucleophile is the attacking species and the haloalkane the target, or the **substrate** (*substratus,* Latin, to have been subjected). However, we could just as well view the transformation as an electrophilic attack by the alkyl group; the haloalkane is then said to **alkylate** (attach an alkyl group to) the nucleophile.

Nucleophilic substitution exhibits considerable diversity

Table 6-3 depicts reactions of typical nucleophiles with several haloalkanes. Note that only primary and secondary halides are shown. Tertiary substrates behave differently toward these nucleophiles, and even secondary halides may give other products in addition to those of substitution. These reactions will be addressed in Chapter 7. The "cleanest" nucleophilic substitutions are obtained with primary haloalkanes.

Let us inspect these transformations in greater detail. In reaction 1, a hydroxide ion, typically derived from sodium or potassium hydroxide, displaces chloride from chloromethane to give methanol. This substitution is a general synthetic method for converting a haloalkane into an alcohol.

A variation of this transformation is reaction 2. Methoxide ion reacts with iodoethane to give methoxyethane, an example of the synthesis of an ether (Section 9-6).

In both reactions 1 and 2, the species attacking the haloalkane is an anionic oxygen nucleophile. Reaction 3 shows that a halide ion may function not only as a leaving group but also as a nucleophile. In this case the reverse reaction (replacement of iodide in 2-iodobutane by bromide) is also possible. To obtain the iodobutane, chemists frequently use propanone (acetone) as the solvent. Sodium iodide, the source of the nucleophile, is soluble in propanone, whereas sodium bromide is not. As the reaction proceeds, the latter precipitates from solution,

TABLE 6-3 The Diversity of Nucleophilic Substitution

Reaction number	Substrate	Nucleophile		Product	Leaving group
1.	CH_3Cl **Chloromethane**	$+ HO^-$	⟶	CH_3OH **Methanol**	$+ Cl^-$
2.	CH_3CH_2I **Iodoethane**	$+ CH_3O^-$	⟶	$CH_3CH_2OCH_3$ **Methoxyethane**	$+ I^-$
3.	$CH_3CH(Br)CH_2CH_3$ **2-Bromobutane**	$+ I^-$	⟶	$CH_3CH(I)CH_2CH_3$ **2-Iodobutane**	$+ Br^-$
4.	$CH_3CH(CH_3)CH_2I$ **1-Iodo-2-methylpropane**	$+ N{\equiv}C^-$	⟶	$CH_3CH(CH_3)CH_2C{\equiv}N$ **3-Methylbutanenitrile**	$+ I^-$
5.	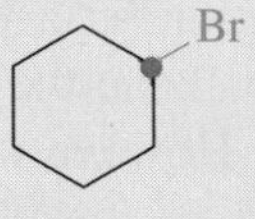 $C_6H_{11}Br$ **Bromocyclohexane**	$+ CH_3S^-$	⟶	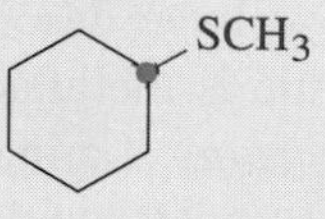 $C_6H_{11}SCH_3$ **Methylthiocyclohexane**	$+ Br^-$
6.	CH_3CH_2I **Iodoethane**	$+ :NH_3$	⟶	$CH_3CH_2\overset{+}{N}H_3$ **Ethylammonium iodide**	$+ I^-$
7.	CH_3Br **Bromomethane**	$+ :P(CH_3)_3$	⟶	$CH_3\overset{+}{P}(CH_3)_3$ **Tetramethylphosphonium bromide**	$+ Br^-$

Note: Remember that nucleophiles are red, electrophiles are blue, and leaving groups are green.

driving the equilibrium to the right. Sodium chloride is also insoluble in propanone, and chloroalkanes may therefore be used in place of bromoalkanes. This method is valuable in the preparation of iodoalkanes, because iodine cannot be introduced directly into alkanes but bromine and chlorine can (Section 3-5). The examples also illustrate how equilibria may be controlled by adjustment of reaction conditions to produce good yields of the desired product.

Reaction 4 depicts a carbon nucleophile, cyanide (often supplied as sodium cyanide, $Na^{+\,-}CN$), and leads to the formation of a new carbon–carbon bond, an important means of lengthening the carbon chain.

Reaction 5 shows the sulfur analog of reaction 2, demonstrating that nucleophiles in the same column of the periodic table react similarly to give analogous products. This conclusion is also borne out by reactions 6 and 7. However, these two reactions involve *neutral* nucleophiles, and the expulsion of the negatively charged leaving group results in a cationic species, an ammonium or phosphonium salt.

EXERCISE 6-2

What are the substitution products of the reaction of 1-bromobutane with **(a)** $:\ddot{\underset{..}{I}}:^-$; **(b)** $CH_3CH_2\ddot{\underset{..}{O}}:^-$; **(c)** $N_3:^-$; **(d)** $:As(CH_3)_3$; **(e)** $(CH_3)_2\ddot{\underset{..}{Se}}$?

EXERCISE 6-3

Suggest starting materials for the preparation of **(a)** $(CH_3)_4N^+I^-$; **(b)** $CH_3SCH_2CH_3$.

In summary, nucleophilic substitution is a fairly general reaction for primary and secondary haloalkanes. The halide functions as the leaving group, and there are several types of nucleophilic atoms that enter into the process.

6-4 A First Look at the Substitution Mechanism: Kinetics

Many questions can be raised at this stage. What are the kinetics of the reaction, and how does this information help us determine the underlying mechanism? What happens with optically active haloalkanes? Can we predict relative rates of substitution? These questions will be addressed one by one in the remainder of this chapter.

When a mixture of chloromethane and sodium hydroxide in water is heated (denoted by the uppercase Greek letter *delta,* Δ, to the right of the arrow in the equation in the margin), a high yield of two compounds—methanol and sodium chloride—is the result. This outcome, however, does not tell us anything about *how* starting materials are converted into products. What experimental methods are available for answering this question?

$$CH_3Cl + NaOH \xrightarrow{H_2O,\ \Delta} CH_3OH + NaCl$$

One of the most powerful techniques employed by chemists is the measurement of the *kinetics* of the reaction (Section 2-8). By comparing the rate of product formation beginning with several different concentrations of the starting materials, we can establish the **rate law** for a chemical process. Let us see what this experiment tells us about the rate law for the reaction of chloromethane with sodium hydroxide.

The reaction of chloromethane with sodium hydroxide is bimolecular

We can monitor rates by following either the disappearance of one of the reactants or the appearance of one of the products. When we apply this method to the reaction of chloromethane with sodium hydroxide, we find that the rate depends on the initial concentrations of *both* of the reagents. For example, doubling the concentration of hydroxide doubles the rate at which the reaction proceeds. Likewise, at a fixed hydroxide concentration, doubling the concentration of chloromethane has the same effect. Doubling the concentrations of both increases the rate by a factor of 4. These results are consistent with a *second-order* process (Section 2-8), which is governed by the following rate equation.

$$\text{Rate} = k[CH_3Cl][HO^-] \text{ mol L}^{-1} \text{ s}^{-1}$$

A reaction following such a rate law is described as **bimolecular.** All the examples given in Table 6-3 exhibit such kinetics: Their rates are directly proportional to the concentration of both substrate and nucleophile.

EXERCISE 6-4

When a solution containing 0.01 M sodium azide ($Na^+N_3^-$) and 0.01 M iodomethane in methanol at 0°C is monitored kinetically, the results reveal that iodide ion is produced at a rate of 3.0×10^{-10} mol L^{-1} s^{-1}. Write the formula of the organic product of this reaction and calculate its rate constant k. What would be the rate of appearance of I^- for each of the following initial concentrations of reactants? **(a)** $[NaN_3]$ = 0.02 M; $[CH_3I]$ = 0.01 M. **(b)** $[NaN_3]$ = 0.02 M; $[CH_3I]$ = 0.02 M. **(c)** $[NaN_3]$ = 0.03 M; $[CH_3I]$ = 0.03 M.

The general term applied to these processes is **bimolecular nucleophilic substitution,** abbreviated as **S_N2** (S stands for substitution, N for nucleophilic, and 2 for bimolecular).

Bimolecular nucleophilic substitution is a concerted, one-step process

What kind of mechanism is consistent with a bimolecular rate law? The simplest is one in which the two reactants interact in a one-step process: The nucleophile attacks the haloalkane, with simultaneous expulsion of the leaving group. Bond-making occurs *at the same time* as bond-breaking. Because the two events take place "in concert," we call this process a **concerted** reaction.

One can envisage two extreme alternatives for such concerted displacements. The nucleophile could approach the substrate from the same side as the leaving group, one group exchanging for the other. This pathway is called **frontside displacement** (Figure 6-2). The second possibility is a **backside displacement,**

The symbol ‡ denotes a transition state, which is very short-lived and cannot be isolated (recall Sections 2-6 and 3-4).

$$H_3C\text{–}Cl + {}^-{:}\ddot{O}H \longrightarrow \left[H_3C \begin{smallmatrix} \cdots \overset{\delta^-}{OH} \\ \cdots \underset{\delta^-}{Cl} \end{smallmatrix} \right]^{\ddagger} \longrightarrow H_3C\text{–}OH + {:}\ddot{Cl}{:}^-$$

FIGURE 6-2 Frontside nucleophilic substitution. The concerted nature of bond-making (to OH) and bond-breaking (from Cl) is indicated by the dotted lines. Note how curved arrows are used in the first structure to indicate *electron flow*. This "electron pushing" technique (Section 1-5) will help us keep track of the electrons in reactions. In the present case, an electron pair on the hydroxide ion is used to form a bond to carbon, and the electrons between carbon and chlorine depart with the leaving group.

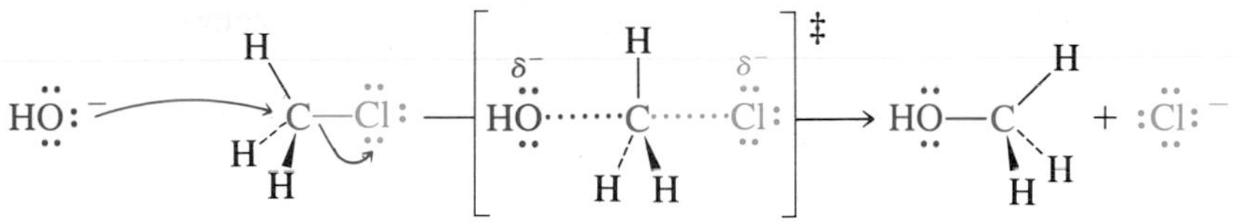

$$H\ddot{O}:^- + CH_3Cl \longrightarrow \left[\overset{\delta^-}{H\ddot{O}} \cdots C(H)(H)(H) \cdots \overset{\delta^-}{\ddot{C}l:} \right]^{\ddagger} \longrightarrow HO{-}CH_3 + :\ddot{C}l:^-$$

FIGURE 6-3 Backside nucleophilic substitution. Attack is from the side *opposite* the leaving group.

in which the nucleophile approaches carbon from the side opposite the leaving group (Figure 6-3). In both equations we use curved arrows to denote the *movement of electron pairs.* An electron pair from the negatively charged hydroxide oxygen moves toward carbon, creating the C–O bond, while that of the C–Cl linkage shifts onto chlorine, thereby expelling the latter as $:\ddot{C}l:^-$. In either of the two respective transition states, the negative charge is distributed over both the oxygen and the chlorine atoms.

EXERCISE 6-5

Draw representations of the hypothetical frontside and backside displacement mechanisms for the S_N2 reaction of sodium iodide with 2-bromobutane (Table 6-3). Use arrows like those used in Figures 6-2 and 6-3 to represent electron-pair movement.

In summary, the reaction of chloromethane with hydroxide to give methanol and chloride, as well as the related transformations of a variety of nucleophiles with haloalkanes, are examples of the bimolecular process known as the S_N2 reaction. Two single-step mechanisms—frontside attack and backside attack—may be envisioned for the reaction. Both are concerted processes, consistent with the second-order kinetics obtained experimentally. Can we distinguish between the two? To answer this question, we return to a topic we have previously discussed in detail: stereochemistry.

6-5 Frontside or Backside Attack? Stereochemistry of the S_N2 Reaction

When we compare the structural drawings in Figures 6-2 and 6-3 with respect to the arrangement of their component atoms in space, we note immediately that in the first conversion the three hydrogens stay put and to the left of the carbon, whereas in the second they have ''moved'' to the right. In fact, the two methanol pictures are related as object and mirror image! Of course, in this example the two are superimposable and therefore indistinguishable—properties of an achiral molecule. The situation is entirely different when we employ a chiral haloalkane in which the electrophilic carbon is a stereocenter.

The S_N2 reaction is stereospecific

Consider the reaction of (S)-2-bromobutane with iodide ion. Frontside displacement should give rise to 2-iodobutane with the *same* configuration as that of the starting material; backside displacement should furnish a product with the *opposite* configuration.

Stereochemistry of a Frontside Displacement Mechanism

(Chiral and optically active)

Frontside displacement

(Chiral and optically active; configuration retained)

Stereochemistry of a Backside Displacement Mechanism

(Chiral and optically active)

Backside displacement

(Chiral and optically active; configuration inverted)

What is actually observed? It is found that (*S*)-2-bromobutane gives (*R*)-2-iodobutane on treatment with iodide: This and all other S_N2 reactions proceed with **inversion of configuration.** A transformation in which a single pure stereoisomer of starting material is converted into a single pure stereoisomer of product is described as **stereospecific.** The S_N2 reaction is therefore a stereospecific process, proceeding by backside displacement to give inversion of configuration at the site of the reaction.

EXERCISE 6-6

Write the products of the following S_N2 reactions. **(a)** (*R*)-3-chloroheptane + $Na^{+-}SH$; **(b)** (*S*)-2-bromooctane + $N(CH_3)_3$; **(c)** (3*R*,4*R*)-4-iodo-3-methyloctane + $K^{+-}SeCH_3$.

EXERCISE 6-7

Write the structures of the products of the S_N2 reactions of cyanide ion with **(a)** *meso*-2,4-dibromopentane (double S_N2 reaction); **(b)** *trans*-1-iodo-4-methylcyclohexane.

The transition state of the S_N2 reaction can be described in an orbital picture

The transition state for the S_N2 reaction can be described in orbital terms, as shown in Figure 6-4. As the nucleophile approaches the back lobe of the sp^3 hybrid orbital used by carbon to bind the halogen atom, the molecule becomes planar at the transition state, by changing the hybridization at carbon to sp^2. The negative charge is no longer located entirely on the nucleophile but also partially on the leaving group. As the reaction proceeds to products, the inversion motion is completed, the carbon returns to the tetrahedral sp^3 configuration, and the leaving group becomes a fully charged anion.

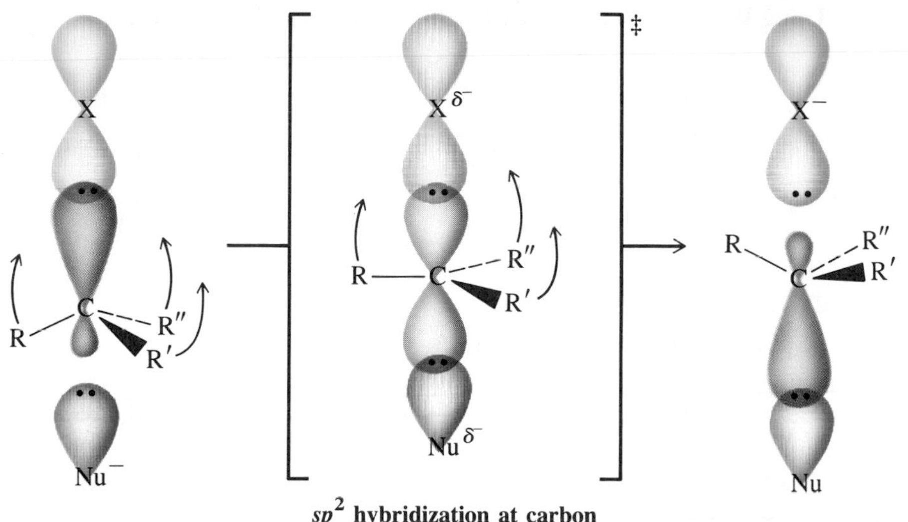

FIGURE 6-4 Molecular-orbital description of the S_N2 reaction. The process is reminiscent of the inversion of an umbrella exposed to gusty winds.

6-6 Consequences of Inversion in S_N2 Reactions

What are the consequences of the inversion of stereochemistry in the S_N2 reaction? Because the reaction is stereospecific, we can design ways to use displacement reactions to synthesize a desired stereoisomer.

We can synthesize a specific enantiomer by using S_N2 reactions

Consider the conversion of 2-bromooctane into 2-octanethiol in its reaction with hydrogen sulfide ion, HS^-. If we were to start with optically pure *R* bromide, we would obtain only *S* thiol and none of its *R* enantiomer.

Inversion of Configuration of an Optically Pure Compound by S_N2 Reaction

$$CH_3(CH_2)_4CH_2\text{—}C(H)(CH_3)\text{—}Br \xrightarrow{HS^-} HS\text{—}C(H)(CH_3)\text{—}CH_2(CH_2)_4CH_3 + Br^-$$

(*R*)-2-Bromooctane ($[\alpha] = -34.6°$) → **(*S*)-2-Octanethiol** ($[\alpha] = +36.4°$)

Color code for priorities (see Section 5-3):
Highest—red
Second highest—blue
Third highest—green
Lowest—black

But what if we wanted to convert (*R*)-2-bromooctane into the *R* thiol? One technique uses a sequence of *two* S_N2 reactions, each resulting in inversion of configuration at the stereocenter. For example, an S_N2 reaction with iodide would first generate (*S*)-2-iodooctane. We would then use this halide with an inverted configuration as the substrate in a second displacement, now with HS^- ion, to furnish the *R* thiol. This double inversion sequence of two S_N2 processes gives us the result we desire, a net **retention of configuration.**

Using Double Inversion to Give Net Retention of Configuration

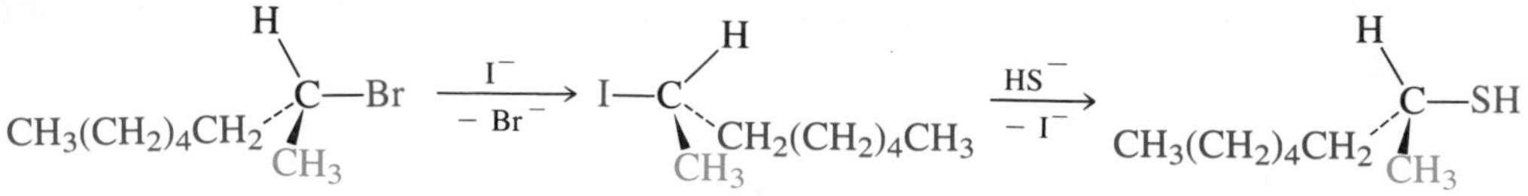

$$CH_3(CH_2)_4CH_2\text{—}C(H)(CH_3)\text{—}Br \xrightarrow[-\ Br^-]{I^-} I\text{—}C(H)(CH_3)\text{—}CH_2(CH_2)_4CH_3 \xrightarrow[-\ I^-]{HS^-} CH_3(CH_2)_4CH_2\text{—}C(H)(CH_3)\text{—}SH$$

(*R*)-2-Bromooctane ($[\alpha] = -34.6°$)

(*S*)-2-Iodooctane ($[\alpha] = +46.3°$)

(*R*)-2-Octanethiol ($[\alpha] = -36.4°$)

EXERCISE 6-8

As we saw in the case of carvone (Chapter 5, Problem 8), enantiomers can sometimes be distinguished by odor and flavor. 3-Octanol and some of its derivatives are examples: The dextrorotatory compounds are found in natural peppermint oil, whereas their (−) counterparts contribute to the essence of lavender. Show how you would synthesize optically pure samples of each enantiomer of 3-octyl acetate (shown below), starting with (*S*)-3-iodooctane.

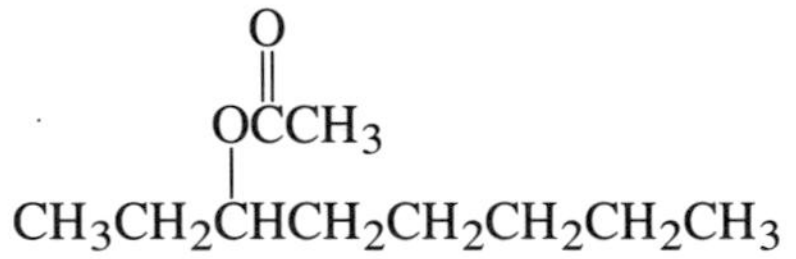

$$CH_3CH_2\underset{|}{\overset{OCCH_3\ (C{=}O)}{C}}HCH_2CH_2CH_2CH_2CH_3$$

EXERCISE 6-9

Treatment of (*S*)-2-iodooctane with NaI causes the optical activity of the starting material to disappear. Explain.

S_N2 reactions of diastereomers may produce diastereomers

In substrates bearing more than one stereocenter, inversion will take place at all the primary and secondary carbons that undergo reaction with the incoming nucleophile. Note that the reaction of (2*S*,4*R*)-2-bromo-4-chloropentane with excess cyanide ion results in a meso product.

S_N2 Reactions of Molecules with Two Stereocenters

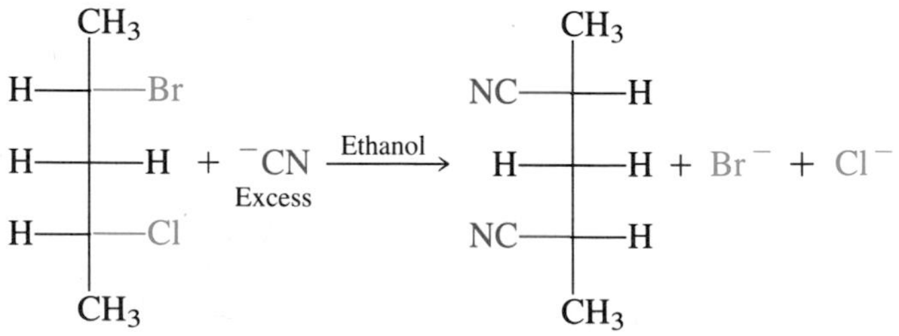

Fischer projection (2*S*,4*R*): CH_3 (top); H—C—Br; H—C—H; H—C—Cl; CH_3 (bottom)

$$+\ {}^-CN\ (\text{Excess}) \xrightarrow{\text{Ethanol}}$$

Fischer projection (2*R*,4*S*: Meso): CH_3 (top); NC—C—H; H—C—H; NC—C—H; CH_3 (bottom) $+\ Br^- + Cl^-$

2*S*,4*R* → **2*R*,4*S*: Meso**

Fischer projection (2*S*,3*R*): $^1CH_3$ (top); H—C_2—Br; CH_3CH_2—C_3—H; CH_3 (bottom)

$$+\ I^- \xrightarrow{\text{Propanone (acetone)}}$$

Fischer projection (2*R*,3*R*): $^1CH_3$ (top); I—C_2—H; CH_3CH_2—C_3—H; CH_3 (bottom) $+\ Br^-$

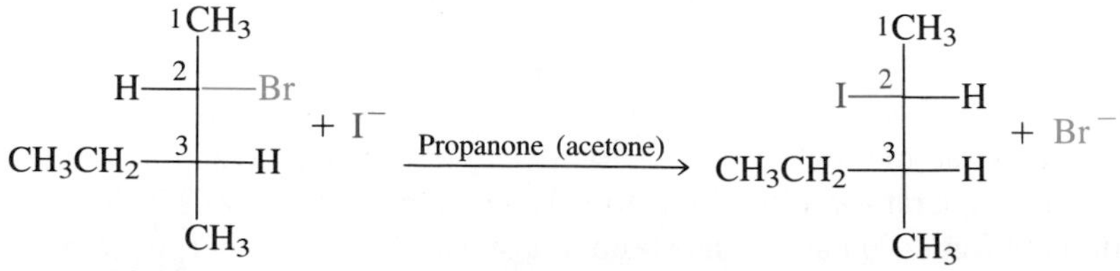

2*S*,3*R* → **2*R*,3*R***

EXERCISE 6-10

As an aid in the prediction of stereochemistry, organic chemists often use the guideline "diastereomers produce diastereomers." Replace the starting compound in each of the two examples above with one of its diastereomers and write the product of S_N2 displacement with the nucleophile shown. Are the resulting structures in accord with this "rule"?

Similarly, nucleophilic substitution of a substituted halocycloalkane may change the stereochemical relationship between the substituents.

$$\text{Cis (H, Br; H, CH}_3\text{)} \xrightarrow{\text{NaI, propanone (acetone)}} \text{Trans (I, H; H, CH}_3\text{)} + \text{NaBr}$$

In summary, inversion of configuration in the S_N2 reaction has distinct stereochemical consequences. Optically active substrates give optically active products, unless the nucleophile and the leaving group are the same or meso compounds are formed. In cyclic systems, cis and trans stereoisomers may be interconverted.

6-7 S_N2 Reactivity and Leaving-Group Ability

The relative facility of S_N2 displacements depends on a variety of factors. These include the nature of the leaving group, the relative reactivity of the nucleophile, and the structure of the alkyl group in the substrate. We can gain further insight into the mechanism of the process by measuring relative reaction rates as we systematically vary each component. We shall begin with leaving groups, first discussing the halides and then generalizing to other groups that can function in this capacity as well. Subsequent sections will address the nucleophile and the alkyl portion of the substrate.

Leaving-group ability is a measure of the ease of its displacement

As a general rule, nucleophilic substitution will occur only when the group being displaced, X, is readily able to depart, taking with it the electron pair of the C–X bond. Are there structural features that might allow us to predict, at least qualitatively, whether a leaving group is "good" or "bad"? Not surprisingly, the relative ease with which it can be displaced, its **leaving-group ability,** can be correlated with its capacity to accommodate a negative charge. Remember that a certain amount of negative charge is transferred to the leaving group in the transition state of the reaction (Figure 6-4).

For the halogens, leaving-group ability increases along the series from fluorine to iodine. Thus, iodine is regarded as a "good" leaving group; fluoride, however, is so "poor" that S_N2 reactions of fluoroalkanes are rarely observed.

Leaving-Group Ability

$I^- > Br^- > Cl^- > F^-$

Best Worst

EXERCISE 6-11

Predict the product of the reaction of 1-chloro-6-iodohexane with one equivalent of sodium methylselenide ($Na^{+}\ {}^{-}SeCH_3$).

Halides are not the only groups that can be displaced by nucleophiles in S_N2 reactions. Other examples of good leaving groups are sulfur derivatives of the type $ROSO_3^-$ and RSO_3^-, such as methyl sulfate ion, $CH_3OSO_3^-$, and various sulfonate ions. Alkyl sulfate and sulfonate leaving groups are used so often that trivial names, such as mesylate, triflate, and tosylate, have found their way into the chemical literature.

Sulfate and Sulfonate Leaving Groups

$CH_3O-S(=O)_2-O^-$

Methyl sulfate ion

$CH_3-S(=O)_2-O^-$

Methanesulfonate ion (Mesylate ion)

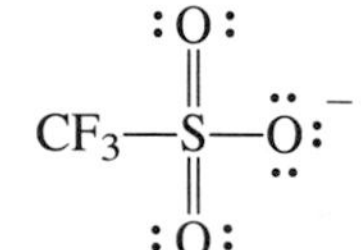

Trifluoromethanesulfonate ion (Triflate ion)

$CH_3-C_6H_4-S(=O)_2-O^-$

4-Methylbenzenesulfonate ion (*p*-Toluenesulfonate ion, tosylate ion)

TABLE 6-4 Base Strengths and Leaving Groups

Leaving group	K_b
Good leaving groups (weaker bases)	
I^-	6.3×10^{-20}
HSO_4^-	1.0×10^{-19}
Br^-	2.0×10^{-19}
Cl^-	6.3×10^{-17}
H_2O	2.0×10^{-16}
$CH_3SO_3^-$	6.3×10^{-16}
Poor leaving groups (stronger bases)	
F^-	1.6×10^{-11}
$CH_3CO_2^-$	5.0×10^{-10}
CN^-	1.6×10^{-5}
CH_3S^-	1.0×10^{-4}
CH_3O^-	32
OH^-	50
NH_2^-	1.0×10^{21}
CH_3^-	$\sim 1.0 \times 10^{36}$

Weak bases are good leaving groups

Is there some characteristic property that distinguishes good leaving groups from poor ones? Yes: *Leaving-group ability is inversely related to base strength.* Weak bases are best able to accommodate negative charge and are the best leaving groups. Among the halides, iodide is the weakest base and therefore the best leaving group in the series. Sulfates and sulfonates are weak bases as well. Table 6-4 lists a variety of species in ascending order of base strength. The latter is quantified by a **basicity constant,** K_b.

Is there a way to recognize weak bases readily? The weaker X^- is as a base, the stronger is its conjugate acid HX. Therefore, *good leaving groups are the conjugate bases of strong acids.* This rule applies to the four halides: HF is the weakest of the conjugate acids, HCl is stronger, and HBr and HI are stronger still. Because a knowledge of acid–base properties is so useful in predicting leaving-group ability, a refresher course in acids and bases is in order.

Acid and base strengths are measured by equilibrium constants

Brønsted and Lowry have given us a simple definition of acids and bases: An **acid** is a proton donor and a **base** is a proton acceptor. Acidity and basicity are commonly measured in water. An acid donates protons to water to give the hydronium ion, whereas a base removes them to give the hydroxide ion. Examples are hydrogen chloride for the former and sodium methoxide for the latter.

$$H-\ddot{\underset{..}{Cl}}: + H\ddot{\underset{..}{O}}H \rightleftharpoons H_3\ddot{O}^{+} + :\ddot{\underset{..}{Cl}}:^{-}$$

Hydronium ion

$$CH_3\ddot{\underset{..}{O}}{:}^-Na^+ + H\ddot{\underset{..}{O}}H \rightleftharpoons CH_3OH + Na^+ + {}^-{:}\ddot{\underset{..}{O}}H$$

Hydroxide ion

Water itself is neutral. It forms an equal number of hydronium and hydroxide ions by self-dissociation. The process is described by the equilibrium constant K_w, the self-ionization constant of water. At 25°C,

$$H_2O + H_2O \overset{K_w}{\rightleftharpoons} H_3O^+ + OH^- \qquad K_w = [H_3O^+][OH^-] = 10^{-14}\ \text{mol}^2\ \text{L}^{-2}$$

From the value for K_w, it follows that the concentration of H_3O^+ in pure water is 10^{-7} mol L^{-1}.

The pH is defined as the negative logarithm of the value for $[H_3O^+]$.

$$\text{pH} = -\log\ [H_3O^+]$$

Thus, for pure water, it is +7. An aqueous solution with a pH lower than 7 is acidic; that with a pH higher than 7 is basic.

The acidity of a general acid, HA, is expressed by the general equation shown below, together with its associated equilibrium constant.

$$HA + H_2O \overset{K}{\rightleftharpoons} H_3O^+ + A^- \qquad K = \frac{[H_3O^+][A^-]}{[HA][H_2O]}$$

Because, in aqueous solution, $[H_2O]$ is constant at 55 mol L^{-1}, that number may be incorporated into a new constant, the **acidity constant,** K_a.

$$K_a = K[H_2O] = \frac{[H_3O^+][A^-]}{[HA]}\ \text{mol L}^{-1}$$

Like the concentration of H_3O^+ and its relationship to pH, this measurement may be put on a logarithmic scale by the corresponding definition of pK_a.

$$pK_a = -\log K_a$$

An acid with a pK_a lower than 1 is defined as strong, one with a pK_a higher than 4 as weak. The acidities of several common acids are compiled in Table 6-5 and compared with those of compounds with higher pK_a values. Sulfuric acid and, with the exception of HF, the hydrogen halides, are very strong acids. Hydrogen cyanide, water, methanol, ammonia, and methane are decreasingly acidic, the last two being exceedingly weak.

Like acid dissociation, the protonation of bases and their basicity can be described by a corresponding set of equations. The basicity of a base, A^-, is governed by the equilibrium constant K'.

$$A^- + H_2O \overset{K'}{\rightleftharpoons} HO^- + HA \qquad K' = \frac{[HO^-][HA]}{[A^-][H_2O]}$$

TABLE 6-5 Relative Strengths of Common Acids (25°C)

Acid	K_a	pK_a
Hydrogen iodide, HI	1.6×10^5	−5.2
Sulfuric acid, H_2SO_4	1.0×10^5	−5.0[a]
Hydrogen bromide, HBr	5.0×10^4	−4.7
Hydrogen chloride, HCl	160	−2.2
Hydronium ion, H_3O^+	50	−1.7
Methanesulfonic acid, CH_3SO_3H	16	−1.2
Hydrogen fluoride, HF	6.3×10^{-4}	3.2
Acetic acid, CH_3COOH	2.0×10^{-5}	4.7
Hydrogen cyanide, HCN	6.3×10^{-10}	9.2
Methanethiol, CH_3SH	1.0×10^{-10}	10.0
Methanol, CH_3OH	3.2×10^{-16}	15.5
Water, H_2O	2.0×10^{-16}	15.7
Ammonia, NH_3	1.0×10^{-35}	35
Methane, CH_4	$\sim 1.0 \times 10^{-50}$	~50

Note: $K_a = [H_3O^+][A^-]/[HA]$ mol L^{-1}.
a. First dissociation equilibrium.

By incorporation of the constant value for $[H_2O]$, this equilibrium constant transforms into the definition for the basicity constant, K_b, and gives rise to a set of pK_b values.

$$K_b = K'[H_2O] = \frac{[HO^-][HA]}{[A^-]} \text{ mol L}^{-1}$$

For an acid HA and its conjugate base A^-, K_a and K_b are related by simple multiplication.

$$K_a \times K_b = \frac{[H_3O^+][A^-]}{[HA]} \times \frac{[HO^-][HA]}{[A^-]} = [H_3O^+][HO^-] = K_w = 10^{-14}$$

We see that the product of the two is equal to the self-ionization constant of water. Hence,

$$\text{p}K_a + \text{p}K_b = 14$$

Therefore, if we know the pK_a of an acid HA, we automatically know the pK_b of A^-. Because of this relation, the species A^- derived from HA is frequently referred to as its **conjugate** base (*conjugatus,* Latin, joined). Conversely, a species HA would be the conjugate acid of base A^-. For example, Cl^- is the conjugate base of HCl, and CH_3OH is the conjugate acid of CH_3O^-. Or, HCl may be viewed as the conjugate acid of Cl^-, and CH_3O^- as the conjugate base of CH_3OH. It follows from this discussion that the conjugate base of a strong acid is weak, as is the conjugate acid of a strong base.

We can predict relative acid and base strengths

Are there structural features that allow us to predict, at least qualitatively, the strength of an acid HA (and hence the weakness of its conjugate base)? Yes, there are several. Prominent among them are two:

1. The increasing *size* of A as we proceed down a column in the periodic table. This trend is seen in the ordering of the acid strengths of the hydrogen halides: HI > HBr > HCl > HF. Consequently I^- is the weakest conjugate base and, as we have seen, the best leaving group, whereas F^-, the strongest base in the series, is difficult to displace in S_N2 processes.

2. The ability of the conjugate base A^- to accommodate the negative charge in either or both of two ways:

(a) The increasing *electronegativity* of A as we proceed from left to right across a row in the periodic table. The more electronegative the atom to which the acidic proton is attached, the more acidic the latter will be. For example, the decreasing order of acidity in the series $HF > H_2O > H_3N > H_4C$ parallels the decreasing electronegativity of A (Table 1-2). In the hydrogen halides, this trend is outweighed by the size of A.

(b) The *resonance* in A^- that allows delocalization of charge over several atoms. For example, acetic acid is more acidic than methanol. In both cases, an O–H bond is broken heterolytically but, unlike methoxide, the acetate ion has two resonance structures to better accommodate the charge (Section 1-5) and is the weaker base.

Acetic Acid Is More Acidic Than Methanol Because of Resonance

$$\underset{\text{Weaker acid}}{CH_3\ddot{O}{-}H} + H_2O \rightleftharpoons \underset{\text{Stronger base}}{CH_3{-}\ddot{O}{:}^-} + H_3O^+$$

$$\underset{\text{Stronger acid}}{CH_3C({=}\ddot{O}{:}){-}\ddot{O}{-}H} + H_2O \rightleftharpoons \underset{\text{Weaker base}}{\left[CH_3C({=}\ddot{O}{:}){-}\ddot{O}{:}^- \longleftrightarrow CH_3C({-}\ddot{O}{:}^-){=}\ddot{O}\right]} + H_3O^+$$

The effect of resonance is even more pronounced in sulfuric acid. The availability of *d* orbitals on sulfur enables us to write "valence shell expanded" Lewis structures containing as many as 12 electrons. Alternatively, charge-separated structures with one or two positive charges on sulfur can be used. Both representations indicate that the pK_a of H_2SO_4 should be low.

$$\left[HO{-}\overset{2+}{S}({-}\ddot{O}{:}^-)_3 \longleftrightarrow HO{-}\overset{+}{S}({=}O)({-}\ddot{O}{:}^-)_2 \longleftrightarrow HO{-}S({=}O)_2{-}\ddot{O}{:}^- \longleftrightarrow HO{-}S({=}O)_2({-}\ddot{O}{:}^-) \longleftrightarrow \text{etc.}\right]$$

Hydrogen sulfate ion

The sulfonic acids are also quite strong, for the same reason. Consequently their conjugate bases, the sulfonates, are weak bases and excellent leaving groups. As a rule, the acidity of HA increases to the right and down in the periodic table. Therefore, the basicity of A^- *decreases* in the same fashion.

EXERCISE 6-12

Predict the relative basicities within each of the following groups. **(a)** ^-OH, ^-SH; **(b)** $^-PH_2$, ^-SH; **(c)** I^-, ^-SeH; **(d)** $CH_3CO_2^-$, $CH_3OCO_2^-$ (draw Lewis structures); **(e)** $HOSO_2^-$, $HOSO_3^-$. Predict the relative acidities of the conjugate acids within each group.

In summary, the leaving-group ability of a substituent is roughly proportional to the strength of its conjugate acid. Both depend on the ability of the leaving group to accommodate negative charge. In addition to the halides Cl^-, Br^-, and I^-, sulfates and sulfonates (such as methane- and 4-methylbenzenesulfonates) are good leaving groups. Good leaving groups are weak bases, the conjugate bases of strong acids. We shall return to uses of sulfates and sulfonates as leaving groups in synthesis in Chapter 9.

6-8 Effect of Nucleophilicity on the S_N2 Reaction

Now that we have looked at the effect of the leaving group, let us turn to a discussion of nucleophiles. How can we predict their relative nucleophilic strength, their **nucleophilicity?** We shall see that nucleophilicity depends on a variety of factors: charge, basicity, solvent, polarizability, and the nature of substituents. To grasp the relative importance of these effects, let us analyze the outcome of a series of comparative experiments.

Increasing negative charge increases nucleophilicity

If the same nucleophilic atom is used, does charge play a role in the reactivity of a given nucleophile as determined by the rate of its S_N2 reaction? The following experiments answer this question.

EXPERIMENT 1.

$$CH_3Cl + HO^- \longrightarrow CH_3OH + Cl^- \qquad \text{Fast}$$
$$CH_3Cl + H_2O \longrightarrow CH_3OH_2^+ + Cl^- \qquad \text{Very slow}$$

EXPERIMENT 2.

$$CH_3Cl + H_2N^- \longrightarrow CH_3NH_2 + Cl^- \qquad \text{Very fast}$$
$$CH_3Cl + H_3N \longrightarrow CH_3NH_3^+ + Cl^- \qquad \text{Slower}$$

Conclusion: Of a pair of nucleophiles containing the same reactive atom, the species with a negative charge is the more powerful nucleophile. Or, of a base and its conjugate acid, the base is always more nucleophilic. This finding is intuitively very reasonable. Because nucleophilic attack is characterized by the formation of a bond with an electrophilic carbon center, the more negative the attacking species the faster the reaction should be.

EXERCISE 6-13

Predict which member in each of the following pairs is a better nucleophile. **(a)** HS^- or H_2S; **(b)** CH_3SH or CH_3S^-; **(c)** CH_3NH^- or CH_3NH_2; **(d)** HSe^- or H_2Se.

Nucleophilicity decreases to the right in the periodic table

Experiments 1 and 2 compared pairs of nucleophiles containing the same nucleophilic element (e.g., oxygen in H_2O versus HO^- and nitrogen in NH_3 versus H_2N^-). What about nucleophiles of similar structure but with different nucleophilic atoms? Let us examine the elements along one row of the periodic table.

EXPERIMENT 3.

$$CH_3CH_2Br + H_3N \longrightarrow CH_3CH_2NH_3^+ + Br^- \qquad \text{Fast}$$
$$CH_3CH_2Br + H_2O \longrightarrow CH_3CH_2OH_2^+ + Br^- \qquad \text{Very slow}$$

EXPERIMENT 4.

$$CH_3CH_2Br + H_2N^- \longrightarrow CH_3CH_2NH_2 + Br^- \qquad \text{Very fast}$$
$$CH_3CH_2Br + HO^- \longrightarrow CH_3CH_2OH + Br^- \qquad \text{Slower}$$

Conclusion: Nucleophilicity again appears to correlate with basicity: The more basic species is seemingly the more reactive nucleophile. Therefore, in proceeding from the left to the right of the periodic table, nucleophilicity decreases. The approximate order of reactivity for nucleophiles in the first row is

$$H_2N^- > HO^- > NH_3 > F^- > H_2O$$

EXERCISE 6-14

In each of the following pairs of molecules, predict which is the more nucleophilic. **(a)** Cl^- or CH_3S^-; **(b)** $P(CH_3)_3$ or $S(CH_3)_2$; **(c)** $CH_3CH_2Se^-$ or Br^-; **(d)** H_2O or HF.

Should basicity and nucleophilicity be correlated?

There appears to be a fairly good correlation between basicity and nucleophilicity, in accord with intuitive expectation. However, these two properties are inherently dissimilar. Basicity (as well as acidity) is a measure of a *thermodynamic* phenomenon, namely, the equilibrium between a base and its conjugate acid in water.

$$A^- + H_2O \overset{K}{\rightleftharpoons} AH + HO^- \qquad K = \text{equilibrium constant}$$

Conversely, nucleophilicity is a measure of a *kinetic* phenomenon, namely, the rate of reaction of a nucleophile with an electrophile.

$$Nu^- + R{-}X \xrightarrow{k} Nu{-}R + X^- \qquad k = \text{rate constant}$$

It is interesting to discover that, despite these inherently different features of basicity and nucleophilicity, such a good correlation exists in the cases examined so far: charged versus neutral nucleophiles, and nucleophiles along a row of the periodic table. Can this relationship be carried one step further?

Solvation impedes nucleophilicity

If it is a general rule that nucleophilicity correlates with basicity, then the elements considered from top to bottom of a column of the periodic table should show decreasing nucleophilic power. Recall (Section 6-7) that basicity decreases in an analogous fashion. To test this prediction, let us consider another series of experiments.

EXPERIMENT 5.

$$CH_3CH_2CH_2O\overset{\overset{O}{\|}}{\underset{\underset{O}{\|}}{S}}CH_3 + Cl^- \xrightarrow{CH_3OH} CH_3CH_2CH_2Cl + {}^-O_3SCH_3 \qquad \text{Slow}$$

$$CH_3CH_2CH_2O\overset{\overset{O}{\|}}{\underset{\underset{O}{\|}}{S}}CH_3 + Br^- \xrightarrow{CH_3OH} CH_3CH_2CH_2Br + {}^-O_3SCH_3 \qquad \text{Faster}$$

$$CH_3CH_2CH_2O\overset{\overset{O}{\|}}{\underset{\underset{O}{\|}}{S}}CH_3 + I^- \xrightarrow{CH_3OH} CH_3CH_2CH_2I + {}^-O_3SCH_3 \qquad \text{Fastest}$$

EXPERIMENT 6.

$$CH_3CH_2CH_2Br + CH_3O^- \xrightarrow{CH_3OH} CH_3CH_2CH_2OCH_3 + Br^- \qquad \text{Not very fast}$$

$$CH_3CH_2CH_2Br + CH_3S^- \xrightarrow{CH_3OH} CH_3CH_2CH_2SCH_3 + Br^- \qquad \text{Very fast}$$

Conclusion: Nucleophilicity *increases* when proceeding down the periodic table, a trend *directly opposing* that expected from the basicity of the nucleophiles tested. Sulfur nucleophiles are more reactive than the analogous oxygen systems but less so than their selenium counterparts. Similarly, among the halides, iodide is the fastest, although it is the weakest base.

How can these trends be explained? What accounts for the increasing nucleophilicity of *negatively charged nucleophiles* from the top to the bottom of a column of the periodic table? An important reason is the interaction of solvent with the different anions.

When a solid dissolves, the intermolecular forces that held it together (Section 2-4) are replaced by molecule–solvent interactions. This phenomenon is called **solvation.** A molecule or ion in solution is surrounded by solvent; it is said to be **solvated** by the solvent molecules. Salts dissolve only in very polar solvents, because only these solvents can interact strongly enough with the ions in the salt to separate the nucleophilic anion from its cationic partner (its **counterion**).

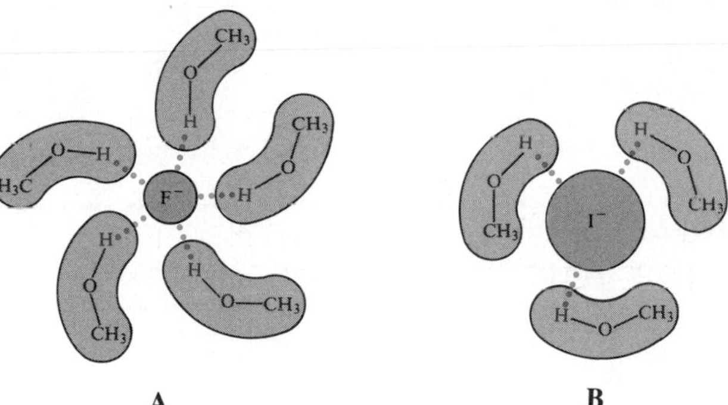

FIGURE 6-5 Approximate representation of the difference between solvation of (A) a small anion (F^-) and (B) a large anion (I^-). The tighter solvent shell around the smaller F^- impedes its ability to participate in nucleophilic substitution reactions.

How does the solvent affect the strength of a nucleophile? Generally it weakens it. Moreover, *smaller anions are more heavily and more tightly solvated than are larger ones* because their charge is more concentrated. Solvation results in a solvent-induced barrier that impedes nucleophilic attack. Figure 6-5 depicts this effect; note that methanol is the solvent. Is this true of all solvents, or are there some that do not exhibit this effect so strongly?

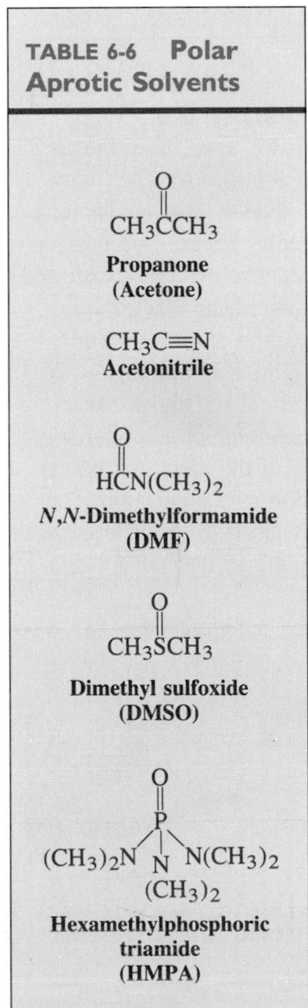
TABLE 6-6 Polar Aprotic Solvents

$CH_3\overset{O}{\overset{\|}{C}}CH_3$
Propanone (Acetone)

$CH_3C{\equiv}N$
Acetonitrile

$H\overset{O}{\overset{\|}{C}}N(CH_3)_2$
***N,N*-Dimethylformamide (DMF)**

$CH_3\overset{O}{\overset{\|}{S}}CH_3$
Dimethyl sulfoxide (DMSO)

$[(CH_3)_2N]_3P{=}O$
Hexamethylphosphoric triamide (HMPA)

Aprotic solvents do not hydrogen-bond to nucleophiles

In some polar solvents, such as methanol, ethanol, and water, a hydrogen is attached to an electronegative atom Y. These solvents contain highly polarized $^{\delta+}H{-}Y^{\delta-}$ bonds, in which the hydrogen has protonlike character and can interact particularly strongly with anionic nucleophiles (Figure 6-5). (We shall study these interactions, called **hydrogen bonds,** more closely in Chapter 8.) Such solvents, called **protic,** are commonly used in nucleophilic substitutions.

Solvents not containing positively polarized hydrogens are called **aprotic.** Polar, aprotic solvents such as propanone (acetone) are also useful in S_N2 reactions. Their polarity keeps anionic nucleophiles well separated from their counterions. However, because hydrogen bonds are not formed, the nucleophiles are considerably less solvated and their reactivity is raised, sometimes dramatically. For example, bromomethane is converted into iodomethane by potassium iodide about 500 times faster in propanone (acetone) than in methanol.

Table 6-6 lists several polar aprotic solvents. They are characterized by the absence of protons capable of hydrogen bonding. Table 6-7 compares the rates of S_N2 reactions of chloride with iodomethane in several solvents. The first three—methanol, formamide, and *N*-methylformamide—are protic. The relatively minor rate differences arise from increasing solvent polarity. The last, *N,N*-dimethylformamide (DMF), is polar and *aprotic*. The reaction rate is more than a million times greater in DMF than it is in methanol.

We saw earlier that small ions often are more encumbered by solvents than are large ones, an observation explaining why the nucleophilicity of the halide ions increases from the top to the bottom of the periodic table. Now we can see that this effect should be most apparent in *protic* solvents, in which the nucleophile is

TABLE 6-7 Relative Rates of S_N2 Reactions of Iodomethane with Chloride Ion in Various Solvents

$$CH_3I + Cl^- \xrightarrow[k_{rel}]{\text{Solvent}} CH_3Cl + I^-$$

Solvent			
Formula	**Name**	**Classification**	**Relative rate, k_{rel}**
CH_3OH	Methanol	Protic	1
$HCONH_2$	Formamide	Protic	12.5
$HCONHCH_3$	*N*-Methylformamide	Protic	45.3
$HCON(CH_3)_2$	*N,N*-Dimethylformamide	Aprotic	1,200,000

heavily solvated. *In aprotic solvents this situation changes.* As solvation is reduced, the two opposing factors of basicity and polarizability (see below) substantially reduce the reactivity differences among the halides.

Increasing polarizability improves nucleophilic power

The solvation effects just described should be very pronounced only for charged nucleophiles. Nevertheless, the degree of nucleophilicity increases down the periodic table, even for *uncharged nucleophiles,* for which solvent effects should be much less strong: for example, $H_2Se > H_2S > H_2O$, and $PH_3 > NH_3$. Therefore, there must be an additional explanation for the observed trend in nucleophilicity.

This explanation lies in the polarizability of the nucleophile. Larger elements have larger, more diffuse, and more polarizable electron clouds. These allow for more effective orbital overlap in the S_N2 transition state (Figure 6-6). The result is a lower transition state energy and faster nucleophilic substitution.

EXERCISE 6-15

Which species is more nucleophilic: **(a)** CH_3SH or CH_3SeH; **(b)** $(CH_3)_2NH$ or $(CH_3)_2PH$?

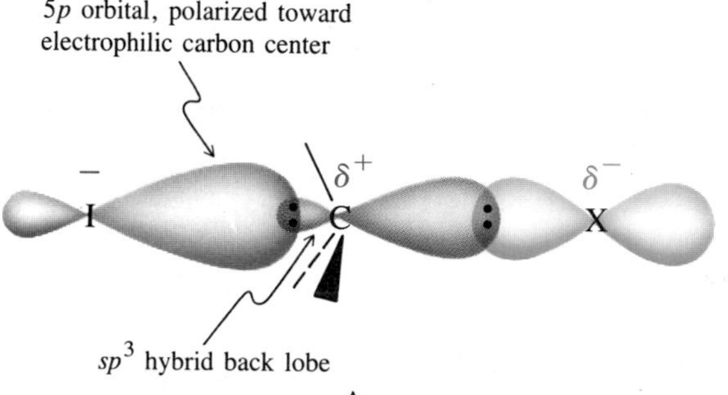

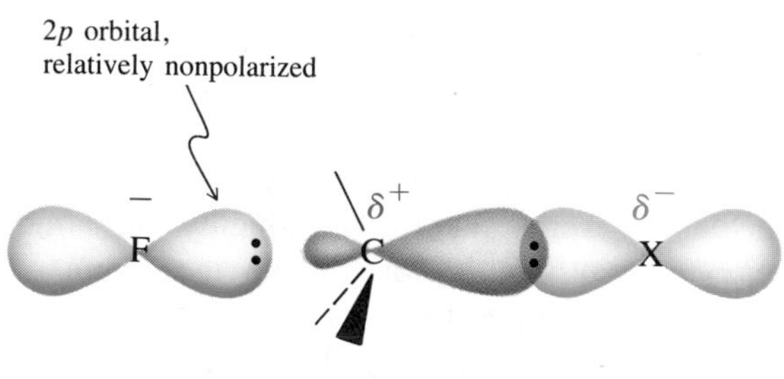

FIGURE 6-6 Comparison of I^- and F^- in the S_N2 reaction. (A) The larger iodide is a better nucleophile, because its polarizable $5p$ orbital is distorted toward the electrophilic carbon atom. (B) The tight, less polarizable $2p$ orbital on fluoride does not interact as effectively with the electrophilic carbon at a point along the reaction coordinate comparable to the one for (A).

Sterically hindered nucleophiles are poorer reagents

We have seen that the bulk of the surrounding solvent may adversely affect the power of a nucleophile, another example of steric hindrance (Section 2-7). Such hindrance may also be built into the nucleophile itself in the form of bulky substituents. The effect on the rate of reaction can be seen in Experiment 7.

EXPERIMENT 7.

$$CH_3I + CH_3O^- \longrightarrow CH_3OCH_3 + I^- \quad \text{Fast}$$

$$CH_3I + CH_3\overset{CH_3}{\underset{CH_3}{\overset{|}{\underset{|}{C}}}}O^- \longrightarrow CH_3O\overset{CH_3}{\underset{CH_3}{\overset{|}{\underset{|}{C}}}}CH_3 + I^- \quad \text{Slower}$$

Conclusion: Sterically bulky nucleophiles react more slowly.

EXERCISE 6-16

Which of the two nucleophiles in the following pairs will react more rapidly with bromomethane?

(a) CH_3S^- or $CH_3\overset{CH_3}{\overset{|}{C}}HS^-$ **(b)** $(CH_3)_2NH$ or $(CH_3\overset{CH_3}{\overset{|}{C}}H)_2NH$

To summarize, nucleophilicity is controlled by a number of factors. Increased negative charge and progression from right to left and down the periodic table generally increase nucleophilic power. Table 6-8 compares the reactivity of a range of nucleophiles relative to that of methanol (arbitrarily set at 1). We can confirm the validity of the conclusions of this section by inspecting the various entries. The use of aprotic solvents improves nucleophilicity, especially of smaller anions, by eliminating hydrogen bonding.

TABLE 6-8 Relative Rates of Reaction of Various Nucleophiles with Iodomethane in Methanol

Nucleophile	Relative rate
CH_3OH	1
NO_3^-	~32
F^-	500
$CH_3\overset{O}{\overset{\|}{C}}O^-$	20,000
Cl^-	23,500
$(CH_3CH_2)_2S$	219,000
NH_3	316,000
CH_3SCH_3	347,000
N_3^-	603,000
Br^-	617,000
CH_3O^-	1,950,000
CH_3SeCH_3	2,090,000
CN^-	5,010,000
$(CH_3CH_2)_3As$	7,940,000
I^-	26,300,000
HS^-	100,000,000

6-9 Effect of the Alkyl Group on the S_N2 Reaction

Finally, does the structure of the substrate, particularly in the vicinity of the center bearing the leaving group, influence the rate of nucleophilic attack? Once again we can get a sense of comparative reactivities by looking at relative rates of reaction. A particularly easy way of obtaining information on relative reactivities is through a **competition experiment.** In this experiment we begin with equal amounts of two substrates whose reactivities are of interest. These are exposed to a small amount (typically 0.05–0.1 molar equivalents) of reagent. Because two starting materials compete for the nucleophile, there are two possible products. The ratio of the products gives an indication of the relative rates of the two competing reactions.

Lengthening the chain by one or two carbons reduces S_N2 reactivity

Let us first try adding methylene groups to a halomethane. Will this affect the rate of its S_N2 reactions? We can set up a competition experiment in which a 1 : 1 mixture of chloromethane and chloroethane is exposed to a small amount of iodide ion.

EXPERIMENT 8.

$$\underset{\text{Ratio 1:1}}{\boxed{CH_3Cl + CH_3CH_2Cl}} + \underset{\text{Small amount}}{I^-} \longrightarrow \underset{\text{Ratio 80:1}}{\boxed{CH_3I + CH_3CH_2I}} + \text{starting chloroalkanes}$$

The outcome of this experiment tells us that chloromethane is approximately 80 times more reactive than chloroethane in its reaction with iodide. Will adding yet another methylene group increase the ratio? To find out, let us study the competition between chloromethane and 1-chloropropane in a similar experiment.

EXPERIMENT 9.

$$\underset{\text{Ratio 1:1}}{\boxed{CH_3Cl + CH_3CH_2CH_2Cl}} + \underset{\text{Small amount}}{I^-} \longrightarrow \underset{\text{Ratio 150:1}}{\boxed{CH_3I + CH_3CH_2CH_2I}} + \text{starting chloroalkanes}$$

We see that the further chain extension reduces the reactivity of the chloroalkane by almost a factor of 2. This result can be verified by an independent control experiment in which chloroethane is found to be roughly two times more reactive than 1-chloropropane.

Will this trend continue? The answer to this question is *no*. All of the higher haloalkanes have approximately the same reactivity toward nucleophiles.

We can find an explanation by comparing the transition states for these three substitutions. Figure 6-7A shows this structure for the reaction of chloromethane with hydroxide ion. The carbon is surrounded by the incoming nucleophile, the outgoing leaving group, and three substituents (all hydrogen in this case). Although the presence of these five groups increases the crowding about the carbon relative to that in the starting halomethane, the hydrogens do not give rise to

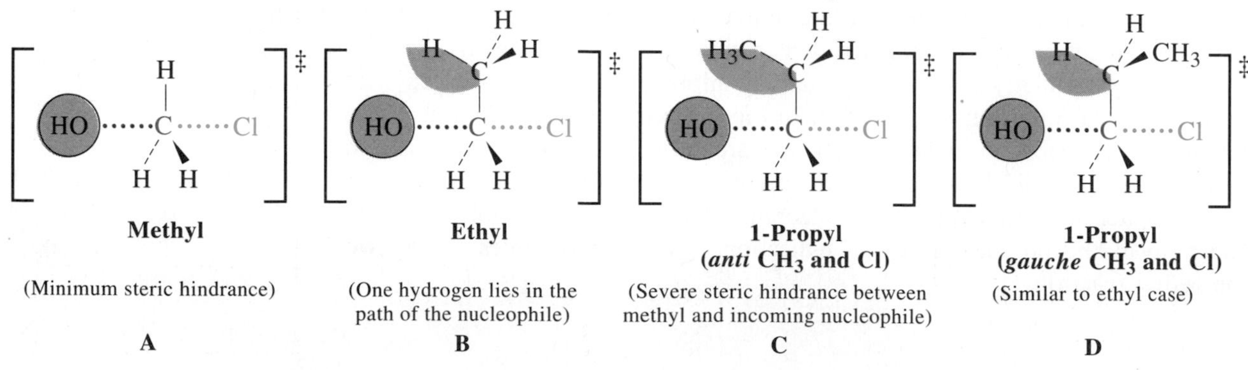

FIGURE 6-7 Transition states for S_N2 reactions of hydroxide ion with (A) chloromethane, (B) chloroethane, and two rotamers of 1-chloropropane: (C) *anti* and (D) *gauche*. The shaded areas highlight steric interference with the nucleophile. Partial charges have been omitted for clarity. (See Figure 6-3.)

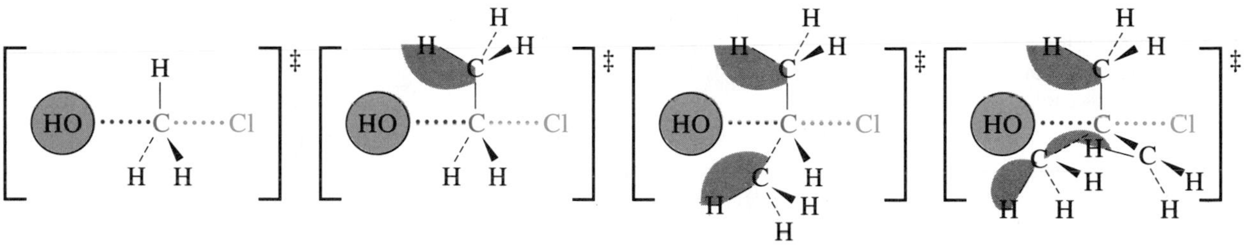

Methyl

A

Primary

B

Secondary

(Slow reaction: two hydrogens interfere with the nucleophile)

C

Tertiary

(No S_N2 reaction; too much steric hindrance)

D

FIGURE 6-8 Transition states for S_N2 reactions of hydroxide ion with (A) chloromethane, (B) chloroethane, (C) 2-chloropropane, and (D) 2-chloro-2-methylpropane. Increasing steric hindrance eventually precludes access to the electrophilic carbon. (See also Figure 6-7.)

serious steric interactions with the nucleophile because of their small size. However, replacement of one hydrogen by a methyl group, as in a haloethane, creates substantial steric repulsion with the incoming nucleophile, thereby raising the transition state energy (Figure 6-7B). This effect significantly retards nucleophilic attack. The 1-halopropanes have an additional methyl group near the reacting carbon center. If reaction occurs from the most stable *anti* rotamer of the substrate, the incoming nucleophile faces severe steric hindrance (Figure 6-7C). However, rotation to a *gauche* conformation before attack gives an S_N2 transition state similar to that derived from a haloethane (Figure 6-7D). The propyl substrate exhibits only a small decrease in reactivity relative to the ethyl, the decrease resulting from the energy input needed to attain a *gauche* conformation. Further chain elongation has no effect, because the added carbon atoms do not increase steric hindrance around the reacting carbon in the transition state.

Branching at the reacting carbon decreases the rate of the S_N2 reaction

As we have seen, the replacement of one hydrogen atom in a halomethane by a methyl group (Figure 6-7B) causes significant steric hindrance in S_N2 reactions. What happens if additional hydrogens are successively replaced by methyl groups? In other words, what are the relative bimolecular nucleophilic reactivities of methyl, primary, secondary, and tertiary halides? Competition experiments show that reactivities rapidly decrease in the order shown in Table 6-9. Thus, the introduction of alkyl groups at the carbon bearing the leaving group produces steric hindrance that impedes bimolecular substitution (Figure 6-8).

TABLE 6-9 Relative Rates of S_N2 Reaction of Branched Bromoalkanes with Iodide

Bromoalkane	Rate
CH_3Br	145
CH_3CH_2Br	1
$(CH_3)_2CHBr$	0.0078
$(CH_3)_3CBr$	Negligible

EXERCISE 6-17

Predict the relative rates of the S_N2 reaction of cyanide with the following pairs of substrates.

(a) 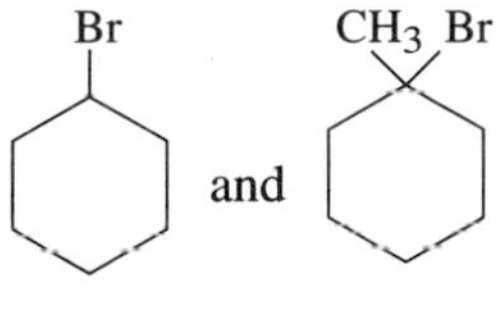 bromocyclohexane (Br) and 1-bromo-1-methylcyclohexane (CH_3 Br)

(b) $CH_3CH_2C(CH_3)_2Br$ and $CH_3CH_2CH_2Br$

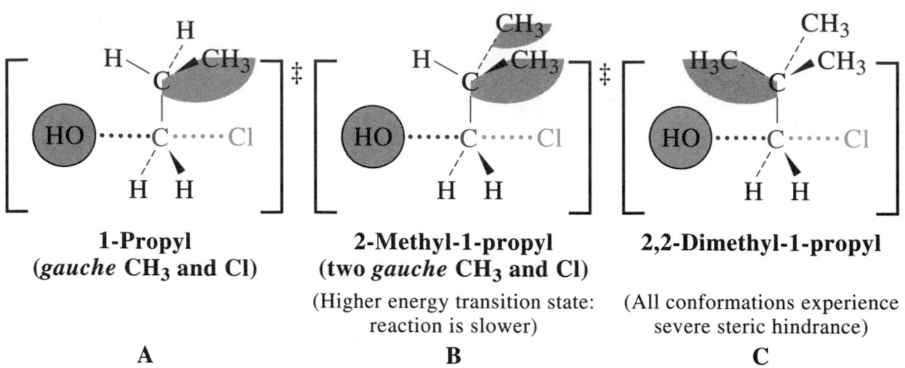

FIGURE 6-9 Transition states for S_N2 reactions of hydroxide ion with (A) 1-chloropropane, (B) 1-chloro-2-methylpropane, and (C) 1-chloro-2,2-dimethylpropane. Increasing steric hindrance from a second *gauche* interaction reduces the rate of reaction in (B). S_N2 reactivity in (C) is eliminated almost entirely because a methyl group prevents backside attack by the nucleophile in all accessible conformations of the substrate. (See also Figures 6-7 and 6-8.)

Branching next to the reacting carbon also retards substitution

TABLE 6-10 Relative Reactivities of Branched Bromoalkanes with Iodide

Bromoalkane	Relative rate
$H{-}CH_2CH_2Br$ (H—C(H)(H)CH$_2$Br)	1
$CH_3CH_2CH_2Br$ ($CH_3C(H)(H)CH_2Br$)	0.8
$CH_3C(CH_3)(H)CH_2Br$	0.003
$CH_3C(CH_3)(CH_3)CH_2Br$	1.3×10^{-5}

What about multiple substitution at the position *next to* the electrophilic carbon? Let us compare the reactivities of bromoethane and its derivatives (Table 6-10). A dramatic decrease in rate is seen on further substitution: 1-Bromo-2-methylpropane is two orders of magnitude less reactive toward iodide than is 1-bromopropane, and 1-bromo-2,2-dimethylpropane is virtually inert. Branching at positions further removed from the site of reaction has a much smaller effect.

Recalling Figure 6-7, we know that rotation into a *gauche* conformation is necessary to permit nucleophilic attack on a 1-halopropane. We can use the same picture to understand the data in Table 6-10. For a 1-halo-2-methylpropane, the only conformation that permits the nucleophile to approach the back side of the reacting carbon experiences *two gauche* methyl–halide interactions, a considerably worse situation (Figure 6-9B). With the addition of a third methyl group, as in a 1-halo-2,2-dimethylpropane, backside attack is blocked almost completely (Figure 6-9C).

EXERCISE 6-18

Predict the order of reactivity in the S_N2 reaction of

(cyclohexyl)CH_2Br versus (1-methylcyclohexyl)CH_2Br

In summary, the structure of the alkyl part of a haloalkane can have a pronounced effect on nucleophilic attack. Competition experiments establish that simple chain elongation beyond three carbons has little influence on the rate of the S_N2 reaction. However, increased branching leads to strong steric hindrance and rate retardation.

IMPORTANT CONCEPTS

1. The haloalkanes, commonly termed alkyl halides, consist of an alkyl group and a halogen.
2. The physical properties of the haloalkanes are strongly affected by the polarization of the C–X bond and the polarizability of X.
3. Reagents bearing lone electron pairs are called nucleophilic when they attack positively polarized centers (other than protons). The latter are called electrophilic. When such a reaction leads to displacement of a substituent, it is a nucleophilic substitution. The group being displaced by the nucleophile is the leaving group.
4. The kinetics of the reaction of nucleophiles with primary (and most secondary) haloalkanes are second order, indicative of a bimolecular mechanism. This process is called bimolecular nucleophilic substitution (S_N2 reaction). It is a concerted reaction, one in which bonds are simultaneously broken and formed. Curved arrows are typically used to depict the flow of electrons as the reaction proceeds.
5. The S_N2 reaction is stereospecific and proceeds by backside displacement, thereby producing inversion of configuration at the reacting center.
6. An orbital description of the S_N2 transition state includes an sp^2-hybridized carbon center, partial bond-making between the nucleophile and the electrophilic carbon, and simultaneous partial bond-breaking between that carbon and the leaving group. Both the nucleophile and the leaving group bear partial charges.
7. Leaving group ability, a measure of the ease of displacement, is roughly proportional to the strength of the conjugate acid. Especially good leaving groups are weak bases such as chloride, bromide, iodide, and the sulfonates.
8. Nucleophilicity increases (a) with negative charge, (b) for elements further to the left and down the periodic table, and (c) in polar aprotic solvents.
9. Polar aprotic solvents accelerate S_N2 reactions because the nucleophiles are well separated from their counterions but are not tightly solvated.
10. Branching at the reacting carbon or at the carbon next to it in the substrate leads to steric hindrance in the S_N2 transition state and decreases the rate of bimolecular substitution.

PROBLEMS

1. Name the following molecules according to the IUPAC system.

(a) CH_3CH_2Cl **(b)** $BrCH_2CH_2Br$ **(c)** $CH_3CH_2\underset{|}{C}HCH_2F$ with CH_2CH_3 on the CH

(d) $(CH_3)_3CCH_2I$ **(e)** (cyclohexane ring)–CCl_3 **(f)** $CHBr_3$

2. Draw structures for each of the following molecules. **(a)** 3-Ethyl-2-iodopentane; **(b)** 3-bromo-1,1-dichlorobutane; **(c)** *cis*-1-(bromomethyl)-2-(2-chloroethyl)cyclobutane; **(d)** (trichloromethyl)cyclopropane; **(e)** 1,2,3-trichloro-2-methylpropane.

3. Draw and name all possible structural isomers having the formula C_3H_6BrCl.

4. Draw and name all structurally isomeric compounds having the formula $C_5H_{11}Br$.

5. For each structural isomer in Problems 3 and 4, identify all stereocenters and give the total number of stereoisomers that can exist for the structure.

6. For each reaction in Table 6-3, identify the nucleophile, its nucleophilic atom (draw its Lewis structure first), the electrophilic atom in the organic substrate, and the leaving group.

7. A second Lewis structure can be drawn for one of the nucleophiles in Problem 6. **(a)** Identify it and draw its alternate structure (which is simply a second resonance form). **(b)** Does this second resonance form predict the presence of another nucleophilic atom in the nucleophile? If so, rewrite the reaction of Problem 6 using the new nucleophilic atom, and write a correct Lewis structure for the product.

8. A solution containing 0.1 M CH_3Cl and 0.1 M KSCN in DMF reacts to give CH_3SCN and KCl with an initial rate of 2×10^{-8} mol L^{-1} s^{-1}. **(a)** What is the rate constant for this reaction? **(b)** Calculate the initial reaction rate for each of the following sets of reactant concentrations: (i) $[CH_3Cl] = 0.2$ M, $[KSCN] = 0.1$ M; (ii) $[CH_3Cl] = 0.2$ M, $[KSCN] = 0.3$ M; (iii) $[CH_3Cl] = 0.4$ M, $[KSCN] = 0.4$ M.

9. Write the product of each of the bimolecular substitutions shown below. The solvent is indicated above the reaction arrow.

(a) $CH_3CH_2CH_2Br + Na^+I^- \xrightarrow{\text{Propanone (acetone)}}$

(b) $(CH_3)_2CHCH_2I + Na^{+-}CN \xrightarrow{\text{DMSO}}$

(c) $CH_3I + Na^{+-}OCH(CH_3)_2 \xrightarrow{(CH_3)_2CHOH}$

(d) $CH_3CH_2Br + Na^{+-}SCH_2CH_3 \xrightarrow{CH_3OH}$

(e) 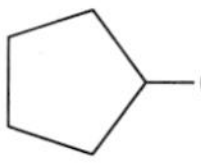 (cyclopentyl)$-CH_2Cl + CH_3CH_2SeCH_2CH_3 \xrightarrow{\text{Propanone (acetone)}}$

(f) $(CH_3)_2CHOSO_2CH_3 + N(CH_3)_3 \xrightarrow{(CH_3CH_2)_2O}$

10. Determine the *R/S* designations for both starting materials and products in the following S_N2 reactions. Which of the products are optically active?

(a) 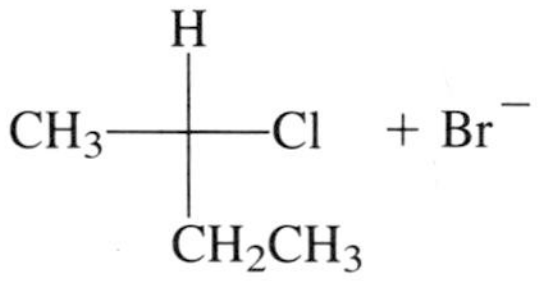 CH_3–C(H)(Cl)(CH_2CH_3) $+ Br^-$ (Fischer projection: H top, CH_2CH_3 bottom, CH_3 left, Cl right)

(b) H_3C–CH(Cl)–CH(H)(Br)–$CH_3 + 2\ I^-$

(c) 3-hydroxycyclohexyl chloride (Cl wedge, HO wedge) $+ {}^-OCCH_3$ (C=O)

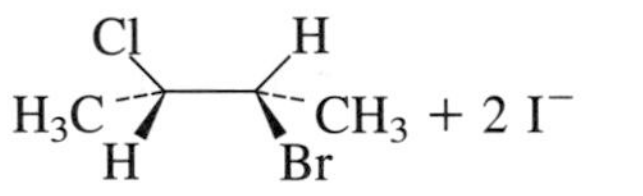

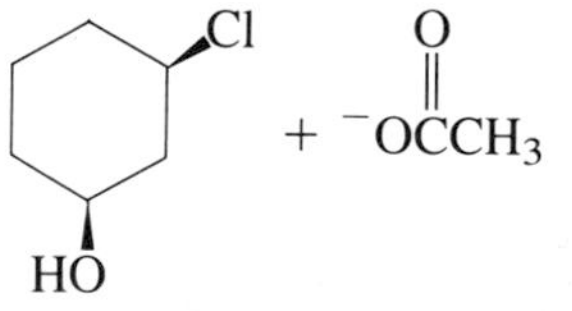

(d) 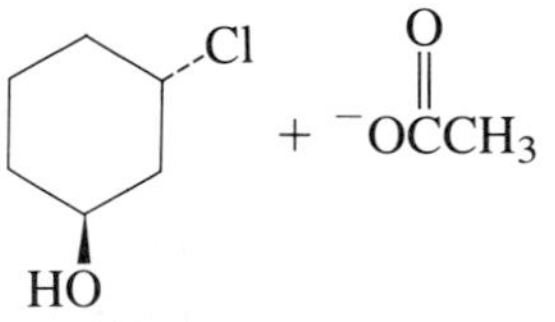 3-hydroxycyclohexyl chloride (Cl hashed, HO wedge) $+ {}^-OCCH_3$ (C=O)

11. List the product(s) of the reaction of 1-bromopropane with each of the following reagents. Write "no reaction" where appropriate. (Hint: Carefully evaluate the nucleophilic potential of each reagent.)

(a) H_2O **(b)** H_2SO_4 **(c)** KOH **(d)** CsI **(e)** NaCN
(f) HCl **(g)** $(CH_3)_2S$ **(h)** NH_3 **(i)** Cl_2 **(j)** KF

12. Formulate the potential product of each of the following reactions. As you did in Problem 11, write "no reaction" where appropriate. (Hint: Identify the expected leaving group in each of the substrates and evaluate its ability to undergo displacement.)

(a) $CH_3CH_2CH_2CH_2Br + K^{+-}OH \xrightarrow{CH_3CH_2OH}$

(b) $CH_3CH_2I + K^+Cl^- \xrightarrow{DMF}$

(c) C_6H_5–$CH_2Cl + Li^{+-}OCH_2CH_3 \xrightarrow{CH_3CH_2OH}$

(d) $(CH_3)_2CHCH_2Br + Cs^+I^- \xrightarrow{CH_3OH}$

(e) $CH_3CH_2CH_2Cl + K^{+-}SCN \xrightarrow{CH_3CH_2OH}$

(f) $CH_3CH_2F + Li^+Cl^- \xrightarrow{CH_3OH}$

(g) $CH_3CH_2CH_2OH + K^+I^- \xrightarrow{DMSO}$

(h) $CH_3I + Na^{+-}SCH_3 \xrightarrow{CH_3OH}$

(i) $CH_3CH_2OCH_2CH_3 + Na^{+-}OH \xrightarrow{H_2O}$

(j) $CH_3CH_2I + K^{+-}OC(=O)CH_3 \xrightarrow{DMSO}$

13. Show how each of the transformations below might be achieved.

(a) (R)-$CH_3CH(OSO_2CH_3)CH_2CH_3 \longrightarrow (S)$-$CH_3CH(N_3)CH_2CH_3$

(b) Fischer projection: CH_3 (top); H–C–Br; CH_3O–C–H; CH_3 (bottom) $\longrightarrow$ CH_3 (top); H–C–CN; CH_3O–C–H; CH_3 (bottom)

(c) Bicyclic (cyclopentane fused to cycloheptane, ring-fusion H, H) with ---Br $\longrightarrow$ same bicyclic with ◄SCH$_3$

(d) N-methylpiperidine (N–CH_3) $\longrightarrow$ N,N-dimethylpiperidinium ($\overset{+}{N}$, CH_3, CH_3)

14. Rank the members of each group of species below in the order of basicity, nucleophilicity, and leaving-group ability. Briefly explain your answers. **(a)** H_2O, HO^-, $CH_3CO_2^-$; **(b)** Br^-, Cl^-, F^-, I^-; **(c)** $^-NH_2$, NH_3, $^-PH_2$; **(d)** ^-OCN, ^-SCN; **(e)** F^-, HO^-, $^-SCH_3$; **(f)** H_2O, H_2S, NH_3.

15. Write the product(s) of each of the following reactions. Provide "no reaction" as your answer if appropriate.

(a) $CH_3CH_2CH_2CH_3 + Na^+Cl^- \xrightarrow{CH_3OH}$

(b) $CH_3CH_2Cl + Na^{+-}OCH_3 \xrightarrow{CH_3OH}$

(c) Newman projection (front: Br, H_3C, H_3C; back: H, H, H) $+ Na^+I^- \xrightarrow{\text{Propanone (acetone)}}$

(d) H–$C(Cl)(CH_2CH_3)(CH_3)$ $+ Na^{+-}SCH_3 \xrightarrow{\text{Propanone (acetone)}}$

(e) $CH_3CH(OH)CH_3 + Na^{+-}CN \longrightarrow$

(f) $CH_3CH(OSO_2CH_3)CH_3 + HCN \xrightarrow{CH_3CH_2OH}$

(g) $CH_3CH(OSO_2CH_3)CH_3 + Na^{+\ -}CN \xrightarrow{CH_3CH_2OH}$

(h) H_3C–C_6H_4–$S(=O)_2OCH_2CH_2CH(CH_3)_2 + K^{+-}SCN \xrightarrow{CH_3OH}$

(i) $CH_3CH_2NH_2 + Na^+Br^- \xrightarrow{DMSO}$ **(j)** $CH_3I + Na^{+-}NH_2 \xrightarrow{NH_3}$

(k) Product of (j) + more $CH_3I \longrightarrow$ **(l)** (iodocycloheptane) $+ Na^{+-}SH \xrightarrow{CH_3OH}$

(m) (iodocycloheptane bearing HO and OCH_3 substituents) $+ Na^{+-}SH \xrightarrow{CH_3OH}$

(n) $CH_3\underset{|}{\overset{CH_3}{C}}HCH_2Br$ + $(C_6H_5)_3P \xrightarrow{CH_3CH_2OH}$

16. Using the information in Chapters 3 and 6, propose the best possible synthesis of each of the following compounds with propane as your organic starting material and any other reagents you need. [Note: On the basis of the information in Section 3-7, you should not expect to find very good answers for (a), (c), and (e). There is one general approach that is best, however.]
(a) 1-Chloropropane **(b)** 2-Chloropropane **(c)** 1-Bromopropane
(d) 2-Bromopropane **(e)** 1-Iodopropane **(f)** 2-Iodopropane

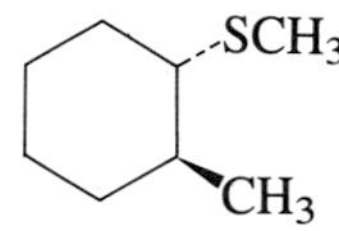

17. Propose two syntheses of *trans*-1-methyl-2-(methylthio)cyclohexane (shown in the margin), beginning with the starting compound **(a)** *cis*-1-chloro-2-methylcyclohexane; **(b)** *trans*-1-chloro-2-methylcyclohexane.

18. Rank each set of molecules below in order of increasing S_N2 reactivity.

(a) CH_3CH_2Br, CH_3Br, $(CH_3)_2CHBr$

(b) $(CH_3)_2CHCH_2CH_2Cl$, $(CH_3)_2CHCH_2Cl$, $(CH_3)_2CHCl$

(c) CH_3CH_2Cl, CH_3CH_2I, (chlorocyclohexane)

(d) $(CH_3CH_2)_2CHCH_2Br$, $CH_3CH_2CH_2\underset{\underset{CH_3}{|}}{C}HBr$, $(CH_3)_2CHCH_2Br$

19. Predict the effect of the changes given below on the rate of the reaction $CH_3Cl + {}^-OCH_3 \xrightarrow{CH_3OH} CH_3OCH_3 + Cl^-$. **(a)** Change substrate from CH_3Cl to CH_3I; **(b)** change nucleophile from CH_3O^- to CH_3S^-; **(c)** change substrate from CH_3Cl to $(CH_3)_2CHCl$; **(d)** change solvent from CH_3OH to $(CH_3)_2SO$.

20. The following table presents rate data for the reactions of CH_3I with three different nucleophiles in two different solvents. What is the significance of these results regarding relative reactivity of nucleophiles under different conditions?

Nucleophile	k_{rel}, CH_3OH	k_{rel}, DMF
Cl^-	1	1.2×10^6
Br^-	20	6×10^5
$NCSe^-$	4000	6×10^5

21. Rings are readily prepared by means of intramolecular S_N2 reactions. An example is shown below, with its mechanism.

$$ClCH_2CH_2CH_2CH_2O{-}H + {}^-OH \xrightleftharpoons{\text{Acid-base reaction}} H_2O + \underset{Cl}{CH_2}CH_2CH_2CH_2O^- \longrightarrow \text{(tetrahydrofuran)} + Cl^-$$

Intramolecular displacement reaction

Explain the outcome of the following transformations mechanistically. (Hint: Notice in the example how an acid-base reaction leads to a stronger nucleophile at one end of the molecule. Use this in your answers.)

(a) $HSCH_2CH_2Br + NaOH \xrightarrow{CH_3CH_2OH}$ (thiirane, three-membered ring with S)

(b) $BrCH_2CH_2CH_2CH_2CH_2Br + NaOH \xrightarrow{CH_3OH}$ (six-membered ring with O)

(c) $BrCH_2CH_2CH_2CH_2CH_2Br + NH_3 \xrightarrow{CH_3CH_2OH}$ (six-membered ring with N–H)

22. S_N2 reactions of halocyclopropane and halocyclobutane substrates are very much slower than those of analogous acyclic secondary haloalkanes. Suggest an explanation for this finding. (Hint: Consider the effect of bond angle strain on the energy of the transition state; see Figure 6-4.)

23. Nucleophilic attack on halocyclohexanes is also somewhat retarded compared to that on acyclic secondary haloalkanes, even though in this case bond angle strain is *not* an important factor. Explain. (Hint: Make a model, and refer to Chapter 4 and Section 6-9.)

7

Further Reactions of Haloalkanes

Unimolecular Substitution and Pathways of Elimination

Haloalkanes can undergo transformations other than S_N2 reactions in the presence of nucleophiles, particularly when the haloalkanes are tertiary or secondary. We shall see in this chapter that bimolecular substitution is only one of *four* possible modes of reaction. The other three are unimolecular substitution and two types of elimination processes, the latter giving rise to double bonds through loss of HX.

7-1 Solvolysis of Tertiary and Secondary Haloalkanes

We have seen that the rate of the S_N2 reaction diminishes drastically when the reacting center changes from primary to secondary to tertiary. For example, although the S_N2 reactivity of bromomethane and bromoethane with iodide ion in propanone (acetone) is high, 2-bromopropane is much less reactive, and 2-bromo-2-methylpropane is essentially inert. However, these observations pertain only to *bimolecular* substitution. Secondary and tertiary halides do undergo substitution, but by another mechanism. This section will show that, in fact, these substrates transform readily, even in the presence of weak nucleophiles, to give the corresponding substitution products.

For example, when 2-bromo-2-methylpropane (*tert*-butyl bromide) is dissolved in aqueous propanone (acetone), it is rapidly converted into 2-methyl-2-propanol (*tert*-butyl alcohol) and hydrogen bromide. The nucleophile here is water, even though it is poor in this capacity. Such a transformation, in which a substrate undergoes substitution by solvent molecules, is called **solvolysis.** When the solvent is water, the term **hydrolysis** is applied.

Reminder:
Nucleophile—red
Electrophile—blue
Leaving group—green

An Example of Solvolysis: Hydrolysis

$$(CH_3)_2\overset{}{C}(CH_3)Br + H{-}OH \xrightleftharpoons[\text{Relatively fast}]{\text{Propanone (acetone)}} (CH_3)_3COH + HBr$$

2-Bromo-2-methylpropane (*tert*-Butyl bromide) → **2-Methyl-2-propanol (*tert*-Butyl alcohol)**

2-Bromopropane is hydrolyzed similarly, albeit much more slowly, whereas 1-bromopropane, bromoethane, and bromomethane are relatively unaffected by these conditions.

Methyl and Primary Haloalkanes: Unreactive in Solvolysis

CH_3Br
CH_3CH_2Br
$CH_3CH_2CH_2Br$

Essentially no reaction with H_2O at room temperature

Hydrolysis of a Secondary Haloalkane

$$(CH_3)_2CHBr + H{-}OH \xrightleftharpoons[\text{Relatively slow}]{\text{Propanone (acetone)}} (CH_3)_2CHOH + HBr$$

2-Bromopropane (Isopropyl bromide) → **2-Propanol (Isopropyl alcohol)**

When the solvent is an alcohol, we use the word **alcoholysis** (more specifically, methanolysis, ethanolysis, and so on).

Methanolysis of 2-Chloro-2-methylpropane

$$(CH_3)_3CCl + CH_3OH \xrightleftharpoons[\text{Solvent}]{} (CH_3)_3COCH_3 + HCl$$

2-Chloro-2-methylpropane → **2-Methoxy-2-methylpropane**

The relative rates of reaction of 2-bromopropane and 2-bromo-2-methylpropane with water to give the corresponding alcohols are shown in Table 7-1 and are compared with the corresponding rates of hydrolysis of their unbranched counterparts. Although the process gives the products expected from an S_N2 reaction, the order of reactivity is *reversed* from that found under typical S_N2 conditions. Thus, primary halides are very slow in their reaction with water, secondary halides are more reactive, and tertiary systems are about *one million times* faster than primary ones.

TABLE 7-1 Relative Reactivities of Various Bromoalkanes with Water

Bromoalkane	Relative rate
CH_3Br	1
CH_3CH_2Br	1
$(CH_3)_2CHBr$	12
$(CH_3)_3CBr$	1.2×10^6

These observations suggest that the mechanism of solvolysis of secondary and, especially, tertiary haloalkanes must be different from that of bimolecular substitution. In an effort to understand the details of this transformation, we can use techniques similar to those employed in the elucidation of the mechanistic features of S_N2 process: kinetics, stereochemistry, and the effect of substrate structure and solvent on reaction rates.

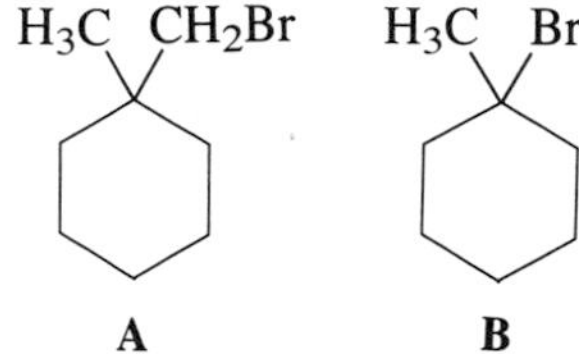

EXERCISE 7-1

Whereas compound A (shown in the margin) is completely stable in ethanol, B is rapidly converted into another compound. Explain.

7-2 Unimolecular Nucleophilic Substitution

In this section we shall learn about a new pathway for nucleophilic substitution. Recall that the S_N2 reaction has second-order kinetics, generates products stereospecifically, and speeds up in the substrate order tertiary-secondary-primary. In contrast, solvolyses follow a *first-order* rate law, are *not* stereospecific, and are characterized by the *opposite* order of reactivity. Let us see how these findings can be accommodated mechanistically.

Solvolysis follows first-order kinetics

In Chapter 6 the reaction's kinetics revealed a bimolecular transition state: The rate of the S_N2 reaction is proportional to the concentration of both the haloalkane and the nucleophile. We can conduct similar rate studies for the reaction of 2-bromo-2-methylpropane with water. The results of these experiments show that *the rate of hydrolysis of the bromide is proportional to the concentration of only the starting halide.*

$$\text{Rate} = k[(CH_3)_3CBr] \text{ mol L}^{-1} \text{ s}^{-1}$$

What does this observation mean? First, it is clear that the haloalkane has to undergo some transformation on its own before anything else takes place. Second, because the final product contains a hydroxy group, water (or, in general, any nucleophile) must enter the reaction, but at a later stage and not in a way that will affect the rate law. The only way to explain this behavior is to postulate that any steps that follow the initial reaction of the halide are relatively fast. In other words, *the observed rate is that of the slowest step in the sequence:* the **rate-determining step.** It follows that only those species involved in the transition state of this step enter into the rate expression: in this case, only the starting haloalkane.

For an analogy, think of the rate-determining step as a bottleneck. Imagine a water hose with several attached clamps restricting the flow (Figure 7-1). We can see that the rate at which the water will spew out of the end is controlled by the narrowest constriction. If we were to reverse the direction of flow (to model the reversibility of a reaction), again the rate of flow would be controlled by this point. Such is the case in transformations involving more than one step, for example, solvolysis. What, then, are the steps in our example?

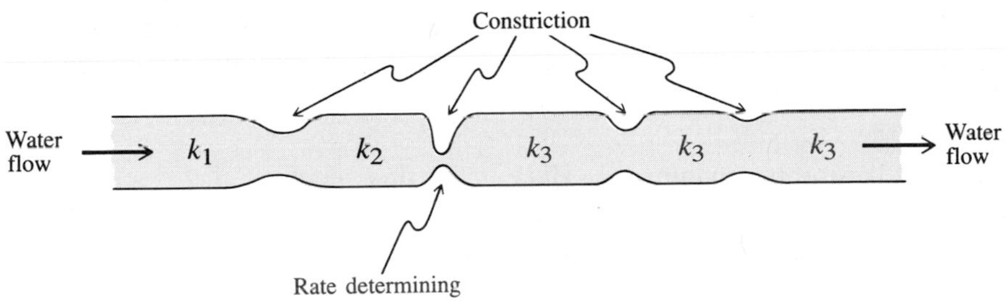

FIGURE 7-1 The rate at which water flows through a hose is controlled by the narrowest constriction. The flow rate at various points is labeled k_1, k_2, and k_3; the overall flow stays constant (k_3) after this constriction.

The mechanism of solvolysis involves carbocation formation

The hydrolysis of 2-bromo-2-methylpropane is said to proceed by **unimolecular nucleophilic substitution,** abbreviated **S_N1.** The number 1 indicates that only one molecule, the haloalkane, participates in the rate-determining step: The rate of the reaction does *not* depend on the concentration of the nucleophile. The mechanism consists of three steps.

STEP 1. The rate-determining step is the dissociation of the haloalkane to an alkyl cation and bromide.

Dissociation of Halide to Form a Carbocation

$$(CH_3)_3C\text{—}Br \underset{}{\overset{\text{Rate determining}}{\rightleftharpoons}} (CH_3)_3C^+ + Br^-$$

1,1-Dimethylethyl cation (*tert*-Butyl cation)

This conversion is an example of heterolytic cleavage. The hydrocarbon product contains a positively charged central carbon atom attached to three other groups and bearing only an electron sextet. Such a structure is called a **carbocation.**

STEP 2. The 1,1-dimethylethyl (*tert*-butyl) cation formed in step 1 is a powerful electrophile and is immediately trapped by the surrounding water. This process can be viewed as a nucleophilic attack by the solvent on the electron-deficient carbon.

Nucleophilic Attack by Water

$$(CH_3)_3C^+ + :\ddot{O}H_2 \overset{\text{Fast}}{\rightleftharpoons} (CH_3)_3C\text{—}\overset{+}{\ddot{O}}H_2$$

An alkyloxonium ion

$$R-\ddot{\underset{..}{O}}-H + H^+ \rightleftharpoons R-\ddot{O}^+(H)(H)$$

Alkyloxonium ion

The resulting species is an example of an **alkyloxonium ion,** the conjugate acid of an alcohol—in this case 2-methyl-2-propanol, the eventual product of the sequence.

STEP 3. Like the hydronium ion, H_3O^+, the first member of the series, all alkyloxonium ions are strong acids. They are therefore readily deprotonated by the water in the reaction medium to furnish the final alcohol.

Deprotonation

$$(CH_3)_3C-\overset{+}{O}H_2 + H_2\ddot{O} \xrightarrow{\text{Fast}} (CH_3)_3C\ddot{O}H + H\overset{+}{\ddot{O}}H_2$$

Alkyloxonium ion (Strongly acidic) — **2-Methyl-2-propanol**

Figure 7-2 compares the potential-energy diagrams for the S_N2 reaction of chloromethane with hydroxide ion and the S_N1 reaction of 2-bromo-2-methylpropane with water. The latter exhibits three transition states, one for each step in the mechanism. The first has the highest energy—and thus is rate determining—because it requires the separation of opposite charges.

EXERCISE 7-2

Using the bond-strength data in Table 3-1, calculate the $\Delta H°$ for the hydrolysis of 2-bromo-2-methylpropane to 2-methyl-2-propanol and hydrogen bromide.

All three steps of the mechanism of solvolysis are reversible. The overall equilibrium can be driven in either direction by the suitable choice of reaction

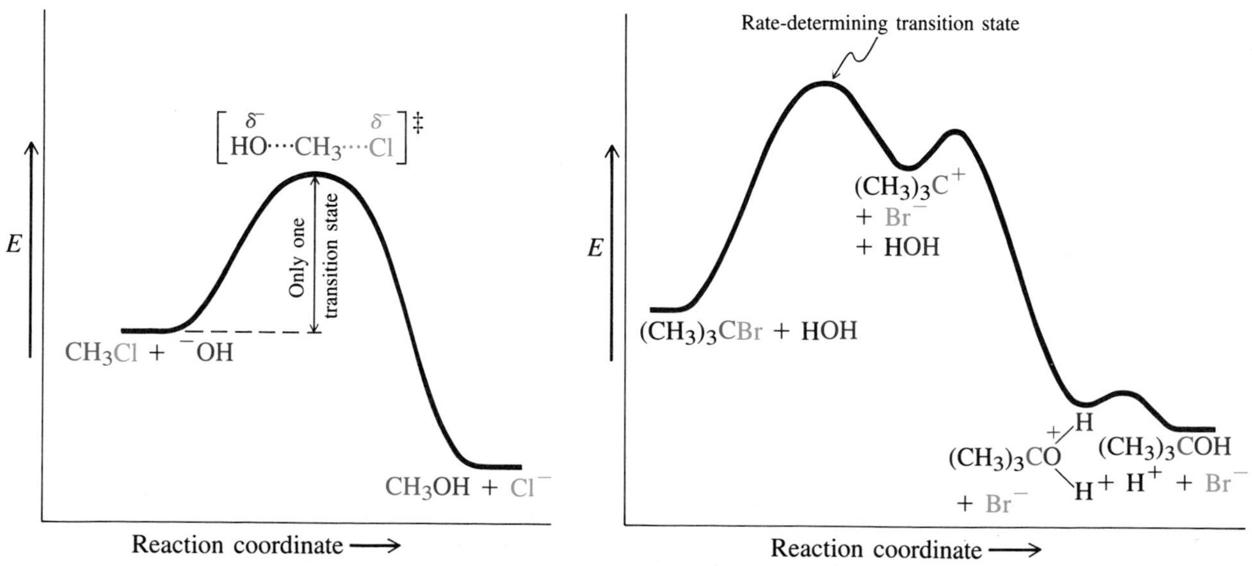

FIGURE 7-2 Potential-energy diagrams for (A) S_N2 reaction of chloromethane with hydroxide and (B) S_N1 hydrolysis of 2-bromo-2-methylpropane. Whereas the S_N2 process occurs in a single step, the S_N1 mechanism involves three distinct events: rate-determining dissociation of the haloalkane into a halide ion and a carbocation, nucleophilic attack by water on the latter to give an alkyloxonium ion, and proton loss to furnish the final product.

conditions. Thus, a large excess of nucleophilic solvent ensures complete solvolysis. In Chapter 9 we shall see how this reaction can be reversed to permit the synthesis of tertiary haloalkanes from alcohols.

In summary, the kinetics of haloalkane solvolysis lead us to a mechanism in which initial dissociation to form a carbocation is the crucial, rate-determining step. Can we back up our mechanistic hypothesis with experimental observations?

7-3 Stereochemical Consequences of S_N1 Reactions

The proposed mechanism of unimolecular nucleophilic substitution has predictable stereochemical consequences, because of the structure of the intermediate carbocation. To minimize electron repulsion, the positively charged carbon assumes trigonal planar geometry, the result of sp^2 hybridization (Sections 1-3 and 1-8). Such an intermediate is therefore achiral. Hence, starting with an optically active tertiary (or secondary) haloalkane in which the stereocenter bears the departing halogen, we should obtain racemic S_N1 products (Figure 7-3). This result is, in fact, observed in many solvolyses.

EXERCISE 7-3

(*R*)-3-Bromo-3-methylhexane loses its optical activity when dissolved in propanone (acetone). Explain.

EXERCISE 7-4

Hydrolysis of molecule A (shown in the margin) gives two alcohols. Explain.

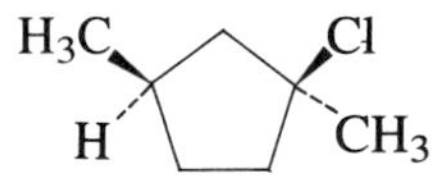

H_3C H Cl CH_3

A

CH_3 $CH_3CH_2CH_2$ C—Br CH_3CH_2 ⟶ Br^- + [CH_3—C $CH_2CH_2CH_3$ CH_2CH_3]$^+$

(*R*)-3-Bromo-3-methylhexane

Planar, achiral

H_2O, $-HBr$ ⟶ OH, CH_3—C—$CH_2CH_2CH_3$, CH_2CH_3 **(*R*)-3-Methyl-3-hexanol**

+

H_2O, $-HBr$ ⟶ CH_3, CH_2CH_3, C—$CH_2CH_2CH_3$, OH **(*S*)-3-Methyl-3-hexanol**

Racemic mixture

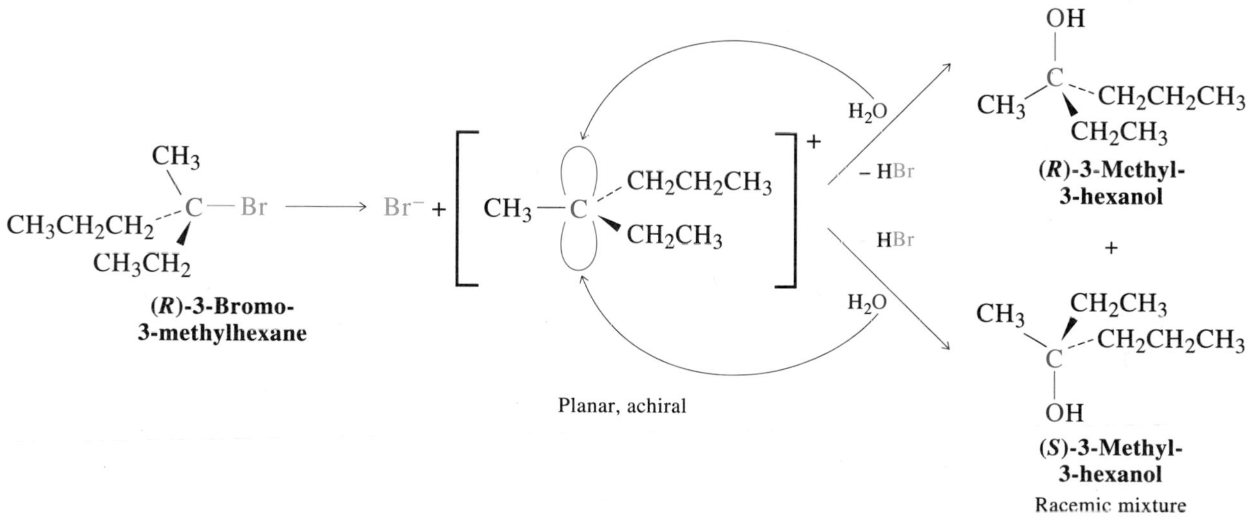

FIGURE 7-3 The mechanism of hydrolysis of (*R*)-3-bromo-3-methylhexane predicts the stereochemistry of the reaction. Initial ionization furnishes a planar, achiral carbocation. This ion, when trapped with water, yields racemic alcohol.

7-4 Effects of Solvent, Leaving Group, and Nucleophile on Unimolecular Substitution

As in S_N2 reactions, varying the solvent, the leaving group, and the nucleophile has a profound effect on unimolecular substitution.

Polar solvents accelerate the S_N1 reaction

Heterolytic cleavage of the C–X bond in the rate-determining step of the S_N1 reaction entails a transition-state structure that is highly polarized (Figure 7-4), eventually leading to two fully charged ions. In contrast, in a typical S_N2 transition state, charges are not created; rather, they are dispersed (Figure 6-4).

Because of this polar transition state, the rate of an S_N1 reaction increases as solvent polarity is increased. The effect is particularly striking when the solvent is changed from aprotic to protic. For example, hydrolysis of 2-bromo-2-methylpropane is much faster in pure water than in a 9:1 mixture of propanone (acetone) and water. The protic solvent accelerates the S_N1 reaction, because it stabilizes the transition state shown in Figure 7-4 by hydrogen bonding with the leaving group. Remember that, in contrast, the S_N2 reaction is accelerated in polar *aprotic* solvents, mainly because of a solvent effect on the reactivity of the nucleophile and *not* of the substrate.

Effect of Solvent on the Rate of an S_N1 Reaction

Reaction	Relative rate
$(CH_3)_3CBr \xrightarrow{100\%\ H_2O} (CH_3)_3COH + HBr$	400,000
$(CH_3)_3CBr \xrightarrow{90\%\ \text{propanone (acetone)},\ 10\%\ H_2O} (CH_3)_3COH + HBr$	1

The S_N1 reaction speeds up with better leaving groups

Because the leaving group departs in the rate-determining step of the S_N1 reaction, it is not surprising that the rate of the reaction increases as the leaving group-ability of the departing group improves. Thus, tertiary iodoalkanes are more readily solvolyzed than the corresponding bromides, and the latter are in turn more reactive than chlorides. Sulfonates are particularly prone to departure.

Relative Rate of Solvolysis of RX (R = Tertiary Alkyl)

$$X = \text{—}OSO_2R' > \text{—}I > \text{—}Br > \text{—}Cl$$

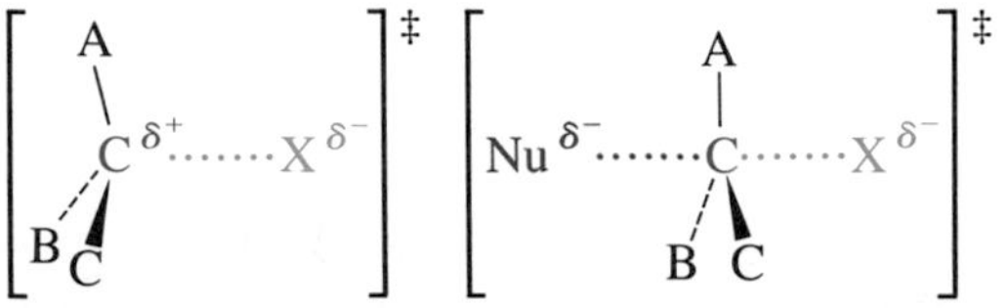

$$\left[\ {}^{A}_{B}\!\!\!>\!C^{\delta+}\cdots\cdots X^{\delta-}\ \right]^{\ddagger} \qquad \left[\ Nu^{\delta-}\cdots\cdots \overset{A}{\underset{B\ C}{C}}\cdots\cdots X^{\delta-}\ \right]^{\ddagger}$$

S_N1 $\qquad$ S_N2

FIGURE 7-4 The respective transition states for the S_N1 and S_N2 reactions explain why the former is strongly accelerated by polar solvents. Heterolytic cleavage involves charge separation, a process aided by polar solvation.

The strength of the nucleophile affects the product distribution but not the reaction rate

Does varying the nature of the incoming nucleophile affect the rate of solvolysis? The answer is no. Recall that, in the S_N2 process, the rate of reaction increases significantly as the nucleophilicity of the attacking species improves. However, because the rate-determining step of unimolecular substitution does *not* include the nucleophile, changing its structure (or concentration) should not alter the rate of disappearance of the haloalkane. Nevertheless, when two or more nucleophiles compete for capture of the intermediate carbocation, their relative strengths and concentrations may profoundly influence the *product distribution*.

For example, solvolysis of a 0.1 M solution of 2-chloro-2-methylpropane in methanol gives the expected 2-methoxy-2-methylpropane, with a rate constant k_1. Quite a different result is obtained when the same experiment is carried out in the presence of an equivalent amount of sodium azide: The product is 1,1-dimethylethyl (*tert*-butyl) azide, still formed at the *same* rate. In this case, the much more powerful nucleophile N_3^- (Table 6-8) wins out in competition with methanol. The rate of disappearance of 2-chloro-2-methylpropane is determined by k_1 (regardless of the product eventually formed), but the relative rates of formation of the *products* depend on the relative reactivities of the competing nucleophiles (k_{CH_3OH} is much smaller than $k_{N_3^-}$).

Competing Nucleophiles in the S_N1 Reaction

$$(CH_3)_3CCl + CH_3OH + NaN_3 \xrightarrow[\text{Rate determining}]{k_1} (CH_3)_3C^+ + Cl^-$$

$$(CH_3)_3C^+ \xrightarrow{k_{CH_3OH}} (CH_3)_3COCH_3 + HCl$$

2-Methoxy-2-methylpropane

$$(CH_3)_3C^+ \xrightarrow{k_{N_3^-}} (CH_3)_3CN_3 + NaCl$$

1,1-Dimethylethyl azide (*tert*-Butyl azide)

EXERCISE 7-5

A solution of 2-methyl-2-propyl methanesulfonate in propanone (acetone) containing excess sodium chloride and sodium bromide produces only 2-bromo-2-methylpropane. Explain.

To summarize, we have seen further evidence supporting the S_N1 mechanism for the reaction of tertiary (and secondary) haloalkanes with certain nucleophiles. The stereochemistry of the process, the influence of the solvent and the leaving-group ability on the rate, and the absence of such an effect when the strength of the nucleophile is varied are consistent with the unimolecular route. The next question to be answered is, Why? What is so special about tertiary haloalkanes that they undergo conversion by the S_N1 pathway, whereas primary systems follow S_N2? How do secondary haloalkanes fit into this scheme?

7-5 Effect of the Alkyl Group on the S_N1 Reaction: Carbocation Stability

Somehow, the degree of substitution at the reacting carbon must control the pathway followed in the reaction of haloalkanes (and related derivatives) with nucleophiles. We shall see that only secondary and tertiary systems can form carbocations. For this reason, tertiary halides, whose steric bulk prevents them from undergoing S_N2 reactions, transform solely by the S_N1 mechanism, primary haloalkanes only by S_N2, and secondary haloalkanes by either route, depending on conditions.

Carbocation stability increases from primary to secondary to tertiary

We have learned that primary haloalkanes undergo *only* direct nucleophilic substitution. In contrast, secondary systems often and tertiary ones always transform through carbocation intermediates. The reasons for this difference are twofold: First, steric hindrance increases along the series, thereby slowing down S_N2; and second, increasing alkyl substitution stabilizes carbocationic centers. Only secondary and tertiary cations are energetically feasible under the conditions of the S_N1 reaction.

Relative Stability of Carbocations

$$\underset{\text{Primary}}{CH_3CH_2CH_2\overset{+}{C}H_2} < \underset{\text{Secondary}}{CH_3CH_2\overset{+}{C}HCH_3} < \underset{\text{Tertiary}}{(CH_3)_3\overset{+}{C}}$$

Now we can see why tertiary haloalkanes solvolyze so readily. Because tertiary carbocations are more stable than their less substituted relatives, they form more easily. But what is the reason for this order of stability?

Hyperconjugation stabilizes positive charge

Note that the order of carbocation stability parallels that of the corresponding radicals. Both trends have their roots in the same phenomenon: *hyperconjugation*. Recall from Section 3-2 that hyperconjugation is the result of overlap of a *p* orbital with a neighboring bonding molecular orbital, such as that of a C–H or a C–C bond. In a radical the *p* orbital is singly filled; in a carbocation it is empty. In both cases the alkyl group donates electron density to and stabilizes the electron-deficient center. Figure 7-5 compares the methyl cation, devoid of hyperconjugation, with the much more stable 1,1-dimethylethyl (*tert*-butyl) cation.

Secondary systems undergo both S_N1 and S_N2 reactions

As you will have gathered from the preceding discussion, secondary haloalkanes exhibit the most varied substitution behavior. Do they prefer a bimolecular substitution pathway or are they more likely to enter into carbocation formation? *Both* are possible: Steric hindrance slows but does not preclude direct nucleophilic attack. At the same time, dissociation becomes competitive because of the relative stability of secondary carbocations. The pathway chosen depends on the reaction conditions: the solvent, the leaving group, and the nucleophile.

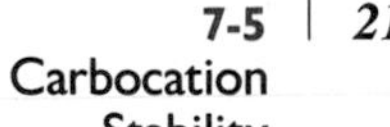

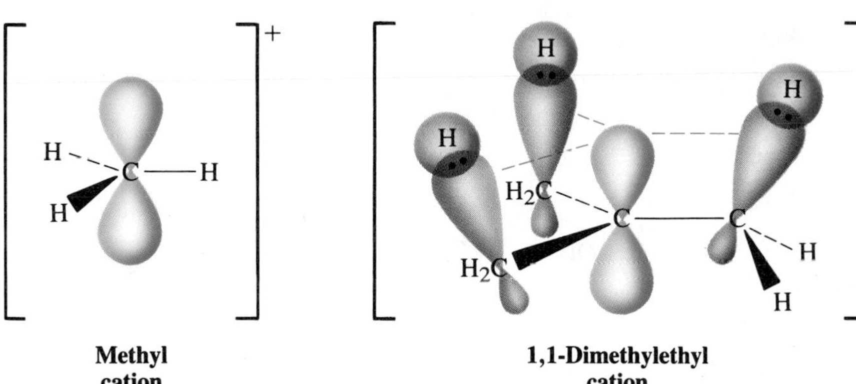

FIGURE 7-5 The methyl cation is not stabilized by hyperconjugation (left), whereas the 1,1-dimethylethyl (*tert*-butyl) cation benefits from three hyperconjugative interactions (right).

If we use a substrate carrying a very good leaving group, a nucleophile that is poor, and a polar protic solvent (S_N1 conditions), *unimolecular* substitution is favored. If we employ a high concentration of a good nucleophile, a polar aprotic solvent, and a haloalkane bearing a reasonable leaving group (S_N2 conditions), *bimolecular* substitution predominates. Table 7-2 summarizes our observations regarding the reactivity of haloalkanes toward nucleophiles.

Substitution of a Secondary Haloalkane under S_N2 Conditions

$$(CH_3)_2CH\text{—}Br + CH_3S^- \xrightarrow{\text{Propanone (acetone)}} CH_3S\text{—}CH(CH_3)_2 + Br^-$$

Substitution of a Secondary Substrate under S_N1 Conditions

$$(CH_3)_2CH\text{—}OSO_2CF_3 \xrightarrow{H_2O} (CH_3)_2CH\text{—}OH + CF_3SO_3H$$

TABLE 7-2 Reactivity of R–X in Nucleophilic Substitutions: $R\text{–}X + Nu^- \longrightarrow R\text{–}Nu + X^-$

R	S_N1	S_N2
CH_3	Not observed in solution (methyl cation too high in energy)	Frequent; fast with good nucleophiles and good leaving groups
Primary	Not observed in solution (primary carbocations too high in energy)[a]	Frequent; fast with good nucleophiles and good leaving groups, slow when branching at C2 is present in R
Secondary	Relatively slow; best with good leaving groups in polar protic solvents	Relatively slow; best with high concentrations of good nucleophiles in polar aprotic solvents
Tertiary	Frequent; particularly fast in polar, protic solvents and with good leaving groups	Extremely slow

a. Exceptions are resonance-stabilized carbocations; see Chapter 14.

EXERCISE 7-6

Explain the following results.

(a) (R)-2-chloro... Cl, H, *R* + CN^- $\xrightarrow{\text{Propanone (acetone)}}$ H, CN, *S*

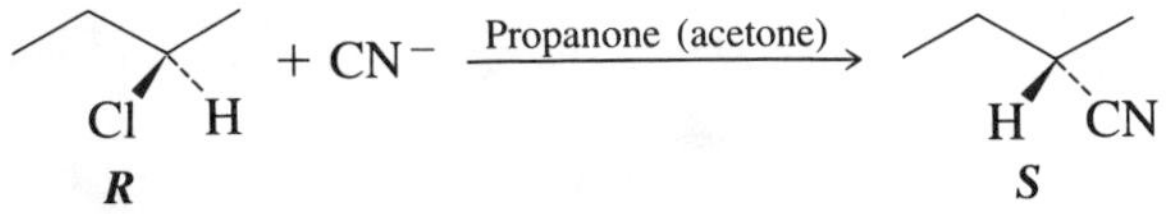

(b) I, H, *R* + CH_3OH ⟶ OCH_3, *R* + *S*

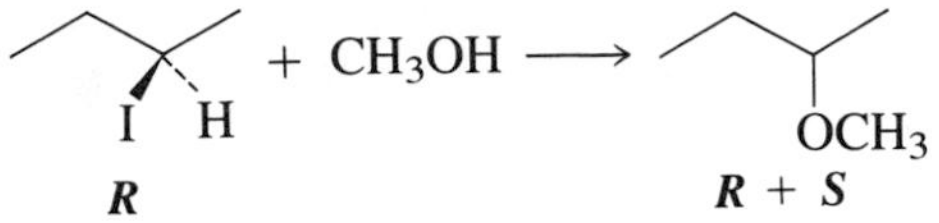

In contrast with S_N2 processes, S_N1 reactions are of limited use in synthesis because the chemistry of carbocations is complex. As we shall see in Chapter 9, these species are prone to rearrangements, frequently resulting in complicated mixtures of products. In addition, carbocations undergo another important reaction, as we shall see next: *loss of a proton* to furnish a double bond.

To summarize, tertiary haloalkanes are reactive in the presence of nucleophiles even though they are too sterically hindered to undergo S_N2 reactions: The tertiary carbocation is readily formed because it is stabilized by hyperconjugation. Subsequent trapping by a nucleophile, such as a solvent (solvolysis), results in the product of nucleophilic substitution. Primary haloalkanes do not react in this manner: The primary cation is too highly energetic (unstable) to be formed in solution. The primary substrate follows the S_N2 route. Secondary systems are converted into substitution products through either pathway, depending on the nature of the leaving group, the solvent, and the nucleophile.

7-6 Unimolecular Elimination: E1

We know that carbocations are readily trapped by nucleophiles through attack at the positively charged carbon. This is not their only mode of reaction, however. An alternative is deprotonation, furnishing a new class of compounds, the alkenes. Starting from a branched haloalkane, the overall transformation constitutes the removal of HX with the simultaneous generation of a double bond. The general term for such a process is **elimination,** abbreviated **E.**

Elimination

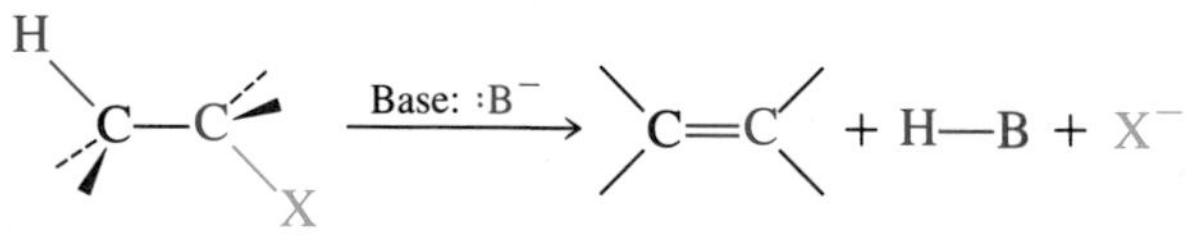

$$\text{H—C—C—X} \xrightarrow{\text{Base: :B}^-} \text{C=C} + \text{H—B} + \text{X}^-$$

Eliminations can occur by several mechanisms. Let us establish the one that is followed during solvolysis.

When 2-bromo-2-methylpropane is dissolved in methanol, it rapidly disappears. As expected, the major product, 2-methoxy-2-methylpropane, arises by methanolysis. However, careful analysis reveals the presence of a significant amount of another compound, 2-methylpropene, the product of elimination. Thus, there is a new mechanism through which the tertiary halide is transformed, competing with solvolysis. What is it? Once again, we turn to a kinetic analysis. The result: The rate of alkene formation depends on the concentration of only the

$$(CH_3)_3CBr \xrightleftharpoons{CH_3OH} H_3C\text{—}\overset{+}{C}(CH_3)_2 + Br^-$$

2-Bromo-2-methylpropane

E1 ↙ ↘ S_N1, CH_3OH

$$H_2C{=}C(CH_3)_2 + H^+ + Br^- \qquad (CH_3)_3COCH_3 + H^+ + Br^-$$

20% **2-Methylpropene** — 80% **2-Methoxy-2-methylpropane**

starting halide; the reaction is first order. Because they are unimolecular, eliminations of this type are labeled **E1.** *The rate-determining step in the E1 process is the same as that in S_N1 reactions: dissociation to a carbocation. This intermediate then has a second pathway at its disposal along with nucleophilic trapping: loss of a proton from a carbon adjacent to the one bearing the positive charge.*

How exactly is the proton lost? Figure 7-6 depicts this process in two ways: first, with orbitals and, second, with electron-pushing arrows. The solvent uses a lone pair on oxygen to attack the hydrogen, removing it as a proton. For simplicity, we often show the protons that evolve in eliminations by using the notation H^+. However, in ordinary organic reactions, there is no such thing as a "free" proton: It is usually removed by some Lewis base. In aqueous solution, water plays this role, giving H_3O^+; here, the proton is carried off by CH_3OH as $CH_3OH_2^+$, an alkyloxonium ion. The carbon left behind rehybridizes from sp^3 to sp^2. As the C–H bond breaks, its electrons shift to overlap in a π fashion with the vacant p orbital at the neighboring cationic center. The result is a hydrocarbon containing a double bond: an alkene.

Any hydrogen positioned on *any* carbon *next to the center bearing the leaving group* can participate in the E1 reaction. The 1,1-dimethylethyl (*tert*-butyl) cation has nine such hydrogens, each of which is equally reactive. In this case, the product is the same regardless of the identity of the proton lost. In other cases, more than one product may be obtained. The pathways involved will be discussed in more detail in Chapter 11.

The E1 Reaction Can Give Product Mixtures

$$(CH_3CH_2)_2CH\text{—}C(CH_3)(Cl)\text{—}CH(CH_3)_2 \xrightarrow[-HCl^*]{CH_3OH,\ \Delta} (CH_3CH_2)_2CH\text{—}C(CH_3)(OCH_3)\text{—}CH(CH_3)_2$$

S_N1 product

$$+ (CH_3CH_2)_2CH\text{—}C(=CH_2)\text{—}CH(CH_3)_2 + (CH_3CH_2)_2CH(CH_3)C{=}C(CH_3)CH_3 + (CH_3CH_2)_2C{=}C(CH_3)CH(CH_3)_2$$

E1 products

*This notation indicates that the elements of the acid have been removed from the starting material. In reality, the proton ends up protonating the base. This system will be used occasionally in other elimination reactions in this book.

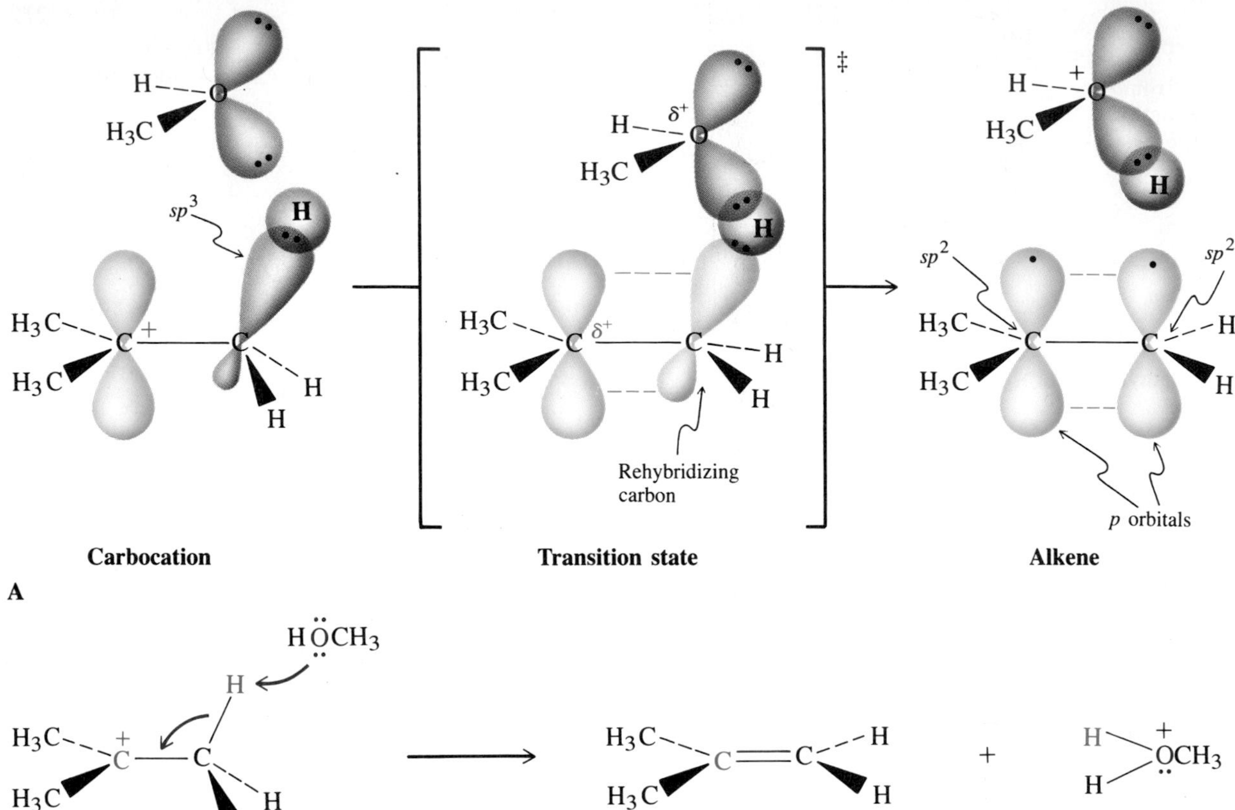

FIGURE 7-6 The alkene-forming step in unimolecular elimination (E1): deprotonation of a 1,1-dimethylethyl (*tert*-butyl) cation by the solvent methanol. (A) An orbital description of proton abstraction. An electron pair on the oxygen atom in the solvent attacks a hydrogen on a carbon adjacent to that bearing the positive charge. The proton is transferred, leaving an electron pair behind. As the carbon rehybridizes from sp^3 to sp^2, these electrons redistribute over the two *p* orbitals of the new double bond. (B) An electron-pushing scheme of the same process. One arrow shows movement of an electron pair from O to H, forming an O–H bond; the other shows the electrons of the breaking C–H linkage shifting between the two carbon atoms, giving rise to the π component of the new double bond.

TABLE 7-3 Ratio of S_N1 to E1 Product in the Hydrolyses of 2-Halo-2-methylpropanes at 25°C

X in $(CH_3)_3CX$	Ratio S_N1:E1
Cl	95:5
Br	95:5
I	96:4

The nature of the leaving group should have no effect on the ratio of substitution to elimination, because the carbocation formed is the same in either case. This is indeed observed qualitatively (Table 7-3). The product ratio may be affected by the addition of a mild base, however, which will accelerate the deprotonation of the carbocation at the expense of nucleophilic attack. For example, adding an equivalent of NaOH to aqueous ethanol containing 2-chloro-2-methylpropane leads to the virtually exclusive formation of 2-methylpropene. In this case, hydroxide, a much stronger base than either water or ethanol, attacks the intermediate carbocation exclusively and rapidly at the hydrogen rather than at the carbon atom. This change, however, is complicated by the fact that strong bases are capable of reacting with haloalkanes by a direct elimination pathway. This reaction will be the subject of the next section.

Addition of Hydroxide Favors E1 over S_N1

$$(CH_3)_3CCl \rightleftharpoons H_3C-\overset{+}{C}(CH_3)_2 + Cl^-$$

E1 Relatively fast, HO^-: $Cl^- + HOH + CH_2{=}C(CH_3)_2$

2-Methylpropene

S_N1 Relatively slow, CH_3CH_2OH, H_2O: $(CH_3)_3COCH_2CH_3 + (CH_3)_3COH + H^+ + Cl^-$

2-Ethoxy-2-methylpropane **2-Methyl-2-propanol**

EXERCISE 7-7

If 2-bromo-2-methylpropane is dissolved in aqueous ethanol at 25°C, a mixture of $(CH_3)_3COCH_2CH_3$ (30%), $(CH_3)_3COH$ (60%), and $(CH_3)_2C{=}CH_2$ (10%) is obtained. Explain.

To summarize, carbocations formed in solvolysis reactions are not only trapped by nucleophiles to give S_N1 products but also deprotonated in an elimination (E1) reaction. In this process, the nucleophile (usually the solvent) acts as a base.

7-7 Bimolecular Elimination: E2

In addition to S_N2, S_N1, and E1 reactions, there is a fourth mode of reactivity of haloalkanes with nucleophiles. In it the nucleophile again acts as a base and again effects elimination, but this time by a bimolecular mechanism.

Strong bases effect bimolecular elimination

The preceding section has taught us that unimolecular elimination may compete with substitution. We also know that the fraction of E1 products but not their rate of formation can be increased by adding dilute base. A dramatic change of the kinetics is observed at higher concentrations of strong base, however. The rate of alkene formation becomes proportional to the concentrations of both the starting halide *and* the base: The kinetics of elimination are now second order and the process is called **bimolecular elimination,** abbreviated **E2.**

Kinetics of the E2 Reaction of 2-Chloro-2-methylpropane

$$(CH_3)_3CCl + Na^{+}\,{}^{-}OH \xrightarrow{k} CH_2{=}C(CH_3)_2 + NaCl + H_2O$$

$$\text{Rate} = k[(CH_3)_3CCl][{}^{-}OH] \text{ mol L}^{-1}\text{ s}^{-1}$$

What causes this change in mechanism? Strong bases can attack haloalkanes before carbocation formation. The target is a hydrogen on a carbon atom next to that carrying the leaving group. This reaction pathway is not restricted to tertiary halides, although in secondary and primary systems it must compete with the S_N2 process.

Competition Between E2 and S_N2 Reactions

$$CH_3CH_2CH_2Br \xrightarrow{CH_3O^-Na^+,\ CH_3OH} CH_3CH_2CH_2OCH_3 + (H)(H_3C)C{=}C(H)(H)$$

92% **1-Methoxypropane** — 8% **Propene**

EXERCISE 7-8

What products do you expect from the reaction of bromocyclohexane with hydroxide ion?

EXERCISE 7-9

Give the products (if any) of the E2 reaction of the following substrates: CH_3CH_2I; CH_3I; $(CH_3)_3CCl$; $(CH_3)_3CCH_2I$.

E2 reactions proceed in one step

The bimolecular elimination mechanism consists of a single step. The bonding changes that occur in its transition state are shown in Figure 7-7, first with electron-pushing arrows, and second with orbitals. Three changes take place:

1. Deprotonation by the base
2. Departure of the leaving group
3. Rehybridization of the reacting carbon centers from sp^3 to sp^2 to furnish the two p orbitals of the emerging double bond

All three take place *simultaneously:* The E2 reaction is a one-step, *concerted* process.

Notice that the E1 (Figure 7-6) and E2 mechanisms are very similar, differing only in the sequence of events. In the bimolecular reaction proton abstraction and leaving-group departure occur at the same time. In the E1 process the halide leaves first, to be followed by an attack by the base. A good way of thinking about the difference is to imagine that the strong base required for the E2 reaction is more aggressive. It does not wait for the tertiary or secondary halide to dissociate but attacks the substrate directly.

Experiments elucidate the detailed structure of the E2 transition state

Relative Reactivity in the E2 Reaction

RCl < RBr < RI

What is the experimental evidence in support of a one-step process with a transition state like that depicted in Figure 7-7? There are four pieces of relevant information. First, the second-order rate law requires that the haloalkane and the base both be involved in the rate-determining step. Second, it is found that better leaving groups result in faster eliminations. This observation implies that the bond to the leaving group is partially broken in the transition state.

EXERCISE 7-10

Explain the result in the reaction shown below.

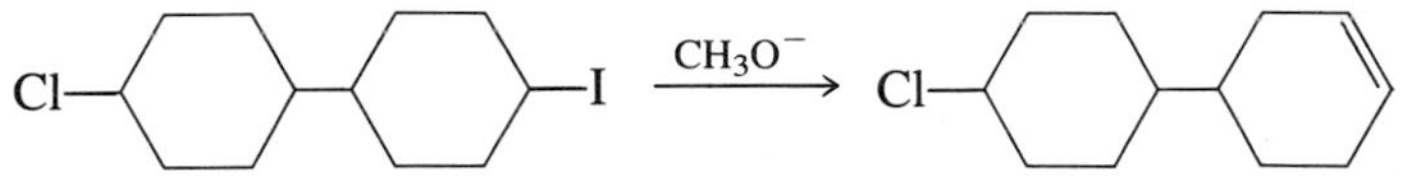

Cl–(cyclohexyl)–(cyclohexyl)–I $\xrightarrow{CH_3O^-}$ Cl–(cyclohexyl)–(cyclohexenyl)

A third type of experiment confirms that the E2 transition state also contains a partially broken C–H bond. It is known that deuterium, the isotope of hydrogen with mass = 2, forms somewhat stronger bonds to carbon than does ordinary hydrogen, an effect resulting in slightly increased activation energies for cleavage of C–D bonds relative to those of C–H linkages. If a C–H bond breaks in the rate-determining step of the E2 reaction, then replacing H by D should lead to a decrease in the observed reaction rate, a result called an **isotope effect.** This is indeed the case.

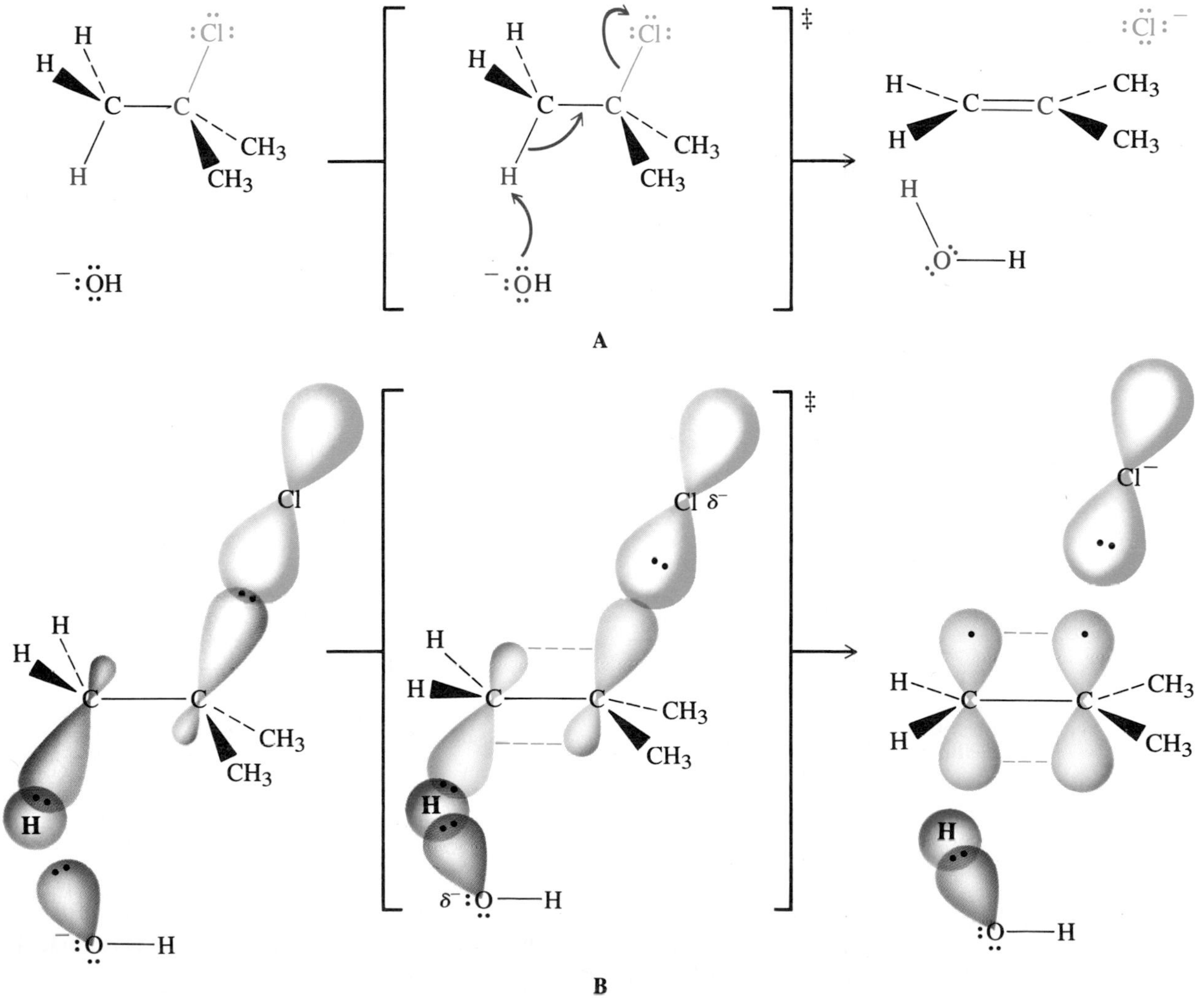

FIGURE 7-7 The E2 reaction of 2-chloro-2-methylpropane with hydroxide ion. (A) Electron-pushing description. (B) Orbital description. Note the *anti* stereochemistry of the transition state.

A

B

$$CH_3\overset{Br}{\underset{H}{C}}CH_3 + CH_3CH_2O^- \xrightarrow[-\ HBr]{CH_3CH_2OH} CH_2{=}C(H)(CH_3) \quad \text{Faster}$$

$$CD_3\overset{Br}{\underset{H}{C}}CD_3 + CH_3CH_2O^- \xrightarrow[-\ DBr]{CH_3CH_2OH} CD_2{=}C(H)(CD_3) \quad \text{Slower}$$

$$\frac{k_{CH_3CHBrCH_3}}{k_{CD_3CHBrCD_3}} = \text{deuterium isotope effect} = 6.7$$

These three sets of experiments establish that the bonding changes in the E2 reaction do indeed occur in a single step. There is also a fourth characteristic feature of the mechanism shown in Figure 7-7: its stereochemistry. The substrate is pictured reacting in a conformation that places the breaking C–H and C–X bonds in an *anti* relationship. How can we establish the structure of the transition state with such precision? For this purpose we can use the principles of conformation and stereochemistry. Treatment of *cis*-1-bromo-4-(1,1-dimethylethyl)-cyclohexane with strong base leads to rapid bimolecular elimination to the corresponding alkene. In contrast, under the same conditions the *trans* isomer reacts only very slowly. Why? When we examine the most stable chair conformation of the *cis* compound, we find that there are two hydrogens located *anti* to the axial bromine substituent. This geometry is very similar to that required by the E2 transition state, and consequently elimination is easy. Conversely, the *trans* system has no C–H bonds aligned *anti* to the equatorial leaving group. E2 elimination in this case would require either ring-flip to a diaxial conformer (see Section 4-4) or removal of a hydrogen *gauche* to the bromine, both energetically costly. The latter is an example of an elimination proceeding through a *syn* transition state (*syn*, Greek, together). We will return to E2 elimination in further detail in Chapter 11.

Anti Elimination Occurs Readily for _cis_- But Not for _trans_-1-Bromo-4-(1,1-dimethylethyl)cyclohexane

$(CH_3)_3C$ (chair, axial Br, two H) $\xrightarrow[\text{Fast}]{Na^+\ ^-OCH_3,\ CH_3OH}$ (cyclohexene with $C(CH_3)_3$) $\xleftarrow[\text{Very slow}]{Na^+\ ^-OCH_3,\ CH_3OH}$ $(CH_3)_3C$ (chair, equatorial Br)

cis-1-Bromo-4-(1,1-dimethylethyl)-cyclohexane
(Two *anti* hydrogens)

trans-1-Bromo-4-(1,1-dimethylethyl)-cyclohexane
(No *anti* hydrogens)

EXERCISE 7-11

The rate of elimination of *cis*-1-bromo-4-(1,1-dimethylethyl)cyclohexane (above) is proportional to the concentration of both substrate and base, but that of the trans isomer is proportional *only* to the concentration of the substrate. Explain.

EXERCISE 7-12

The isomer of 1,2,3,4,5,6-hexachlorocyclohexane shown in the margin undergoes E2 elimination 7000 times *more slowly* than any of its stereoisomers. Explain.

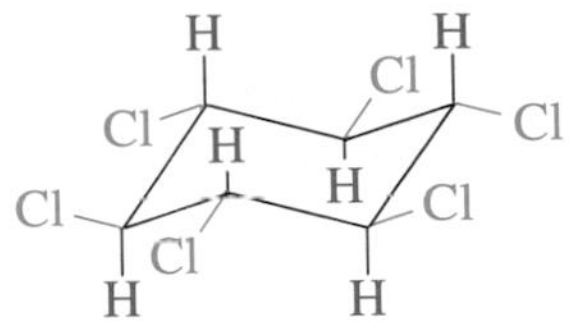

In summary, strong bases react with haloalkanes not only by substitution but also by elimination. The kinetics of these reactions are second order, an observation pointing to a bimolecular mechanism. An *anti* transition state is preferred, in which the base abstracts a proton at the same time as the leaving group departs.

7-8 Competition Between Substitution and Elimination

The multiple reaction pathways that haloalkanes may follow in the presence of nucleophiles may seem confusing: S_N2, S_N1, E2, E1. Given the many parameters that influence the relative importance of these transformations, are there some simple guidelines that might allow us to predict, at least roughly, what the outcome of any particular reaction will be? The answer is a cautious yes. This section will explain how consideration of *base strength* and *steric bulk* of the reacting species can help us decide whether substitution or elimination will predominate.

Weakly basic nucleophiles give substitution

Good nucleophiles that are weak bases give good yields of S_N2 products with primary and secondary halides and of S_N1 products with tertiary substrates. Examples include I^-, Br^-, RS^-, N_3^-, $RCOO^-$, and PR_3. Thus, 2-bromopropane reacts with both iodide and acetate ions cleanly through the S_N2 pathway, with virtually no competing elimination.

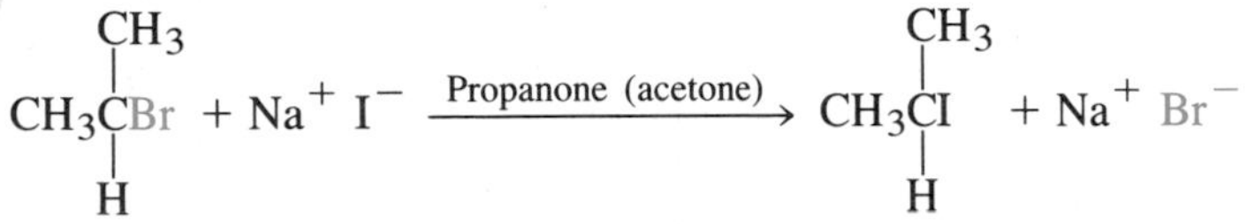

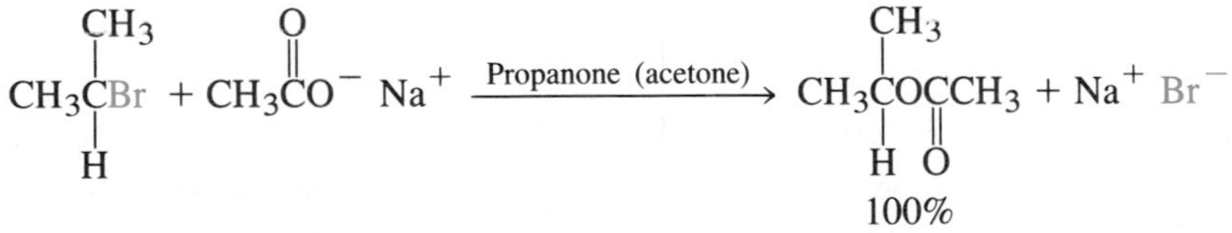

$$CH_3CH(CH_3)Br + Na^+\ I^- \xrightarrow{\text{Propanone (acetone)}} CH_3CH(CH_3)I + Na^+\ Br^-$$

$$CH_3CH(CH_3)Br + CH_3C(=O)O^-\ Na^+ \xrightarrow{\text{Propanone (acetone)}} CH_3CH(CH_3)OC(=O)CH_3 + Na^+\ Br^-$$
100%

Most of these reagents also furnish substitution products in their reactions with tertiary systems, through the S_N1 mechanism.

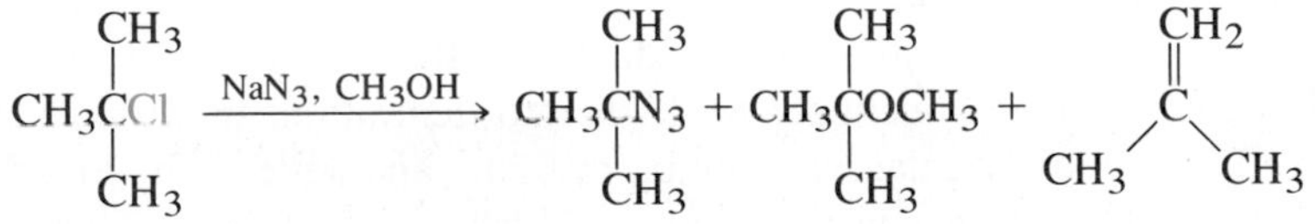

$$(CH_3)_3CCl \xrightarrow{NaN_3,\ CH_3OH} (CH_3)_3CN_3 + (CH_3)_3COCH_3 + CH_2{=}C(CH_3)_2$$

Weak nucleophiles such as water and alcohols react at appreciable rates only with secondary and tertiary halides, substrates capable of following the S_N1 pathway. Unimolecular elimination is usually only a minor side reaction.

$$\underset{\text{Br}}{CH_3CH_2\overset{|}{C}HCH_2CH_3} \xrightarrow{H_2O,\ CH_3OH,\ 80°C} \underset{85\%}{CH_3CH_2\underset{}{\overset{\text{OH}}{\overset{|}{C}}}HCH_2CH_3} + \underset{15\%}{CH_3CH{=}CHCH_2CH_3}$$

Strongly basic nucleophiles give more elimination as steric bulk increases

We have seen (Section 7-7) that strong bases may give rise to elimination through the E2 pathway. Is there some straightforward way to predict how much elimination will occur in competition with substitution in any particular situation? Yes, but other factors need to be considered. Let us examine the reactions of sodium ethoxide, a strong base, with several halides, measuring the relative amounts of ether and alkene produced in each case.

		Ether (Substitution product)		Alkene (Elimination product)
$CH_3CH_2CH_2Br$	$\xrightarrow[-\ HBr]{CH_3CH_2O^-Na^+,\ CH_3CH_2OH}$	$CH_3CH_2CH_2OCH_2CH_3$ 91%	+	$(H_3C)(H)C{=}C(H)(H)$ 9%
$CH_3\overset{CH_3}{\underset{H}{C}}CH_2Br$	$\xrightarrow[-\ HBr]{CH_3CH_2O^-Na^+,\ CH_3CH_2OH}$	$CH_3\overset{CH_3}{\underset{H}{C}}CH_2OCH_2CH_3$ 40%	+	$(H_3C)(H_3C)C{=}C(H)(H)$ 60%
$CH_3\overset{CH_3}{\underset{H}{C}}Br$	$\xrightarrow[-\ HBr]{CH_3CH_2O^-Na^+,\ CH_3CH_2OH}$	$CH_3\overset{CH_3}{\underset{H}{C}}OCH_2CH_3$ 13%	+	$(H)(H)C{=}C(H)(CH_3)$ 87%

Reactions of simple primary halides with strongly basic nucleophiles give mostly S_N2 products. As steric bulk is increased around the carbon bearing the leaving group, substitution is retarded relative to elimination because an attack at carbon is subject to more steric hindrance than an attack on hydrogen. Thus, branched primary substrates give about equal amounts of S_N2 and E2 reaction, and the latter is the major outcome with secondary systems.

The S_N2 mechanism does not operate for tertiary halides. Addition of dilute base suppresses S_N1 substitution by diverting the intermediate carbocation through the E1 pathway (Section 7-6); higher concentrations of strong base give exclusive E2 reaction. Acetate, ammonia, cyanide, and aqueous hydroxide, reagents that are intermediate in their base strength between the halides and the alkoxides, react with primary and secondary haloalkanes primarily through the S_N2 route but furnish mostly E2 products with tertiary substrates.

Sterically hindered basic nucleophiles favor elimination

We have seen that primary haloalkanes react by substitution with good nucleophiles, including strong bases. The situation changes when the steric bulk of the nucleophile hinders attack at the electrophilic carbon. In this case elimination may predominate, even with primary systems, through deprotonation at the less hindered periphery of the molecule.

$$CH_3CH_2CH_2CH_2Br \xrightarrow[-HBr]{(CH_3)_3CO^-K^+,\ (CH_3)_3COH} \underset{85\%}{CH_3CH_2CH{=}CH_2} + \underset{15\%}{CH_3CH_2CH_2CH_2OC(CH_3)_3}$$

Two examples of sterically hindered bases are potassium *tert*-butoxide and lithium diisopropylamide (LDA). When used in elimination reactions, they are frequently dissolved in their conjugate acids, 2-methyl-2-propanol and *N*-(1-methylethyl)-1-methylethanamine (diisopropylamine), respectively.

Sterically Hindered Bases

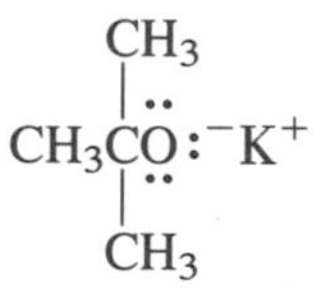

$CH_3C(CH_3)_2\ddot{\underset{..}{O}}{:}^-K^+$

Potassium *tert*-butoxide

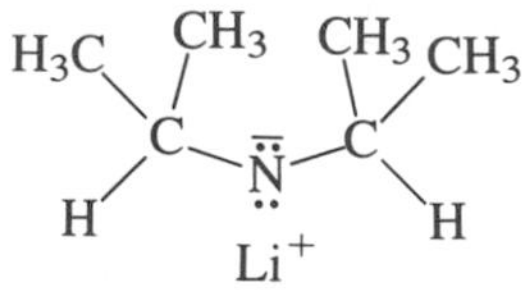

$[(CH_3)_2CH]_2\ddot{N}^-\ Li^+$

Lithium diisopropylamide (LDA)

In summary, we have identified three principal factors that affect the competition between substitution and elimination: basicity of the nucleophile, steric hindrance in the haloalkane, and steric bulk around the nucleophilic (basic) atom.

FACTOR 1. Base strength of the nucleophile

Weak Bases	**Strong Bases**
H_2O*, ROH*, PR_3, halides, RS^-, N_3^-, NC^-, $RCOO^-$	HO^-, RO^-, H_2N^-, R_2N^-
Substitution more likely	Likelihood of elimination increased

FACTOR 2. Steric hindrance around the reacting carbon

Sterically Unhindered	**Sterically Hindered**
Primary haloalkanes	Branched primary, secondary, tertiary haloalkanes
Substitution more likely	Likelihood of elimination increased

FACTOR 3. Steric hindrance in the nucleophile (strong base)

Sterically Unhindered	**Sterically Hindered**
HO^-, CH_3O^-, $CH_3CH_2O^-$, H_2N^-	$(CH_3)_3CO^-$, $[CH(CH_3)_2]_2N^-$
Substitution may occur	Elimination strongly favored

For simple predictive purposes, we assume that their relative importance is equal in determining the ratio of elimination to substitution. Thus, the "majority rules." This method of analysis is quite reliable. Verify that it applies to the examples of this section and the summary section that follows.

*Reacts only with S_N1 substrates; no reaction with simple primary halides.

EXERCISE 7-13

Which nucleophile in each of the following pairs will give a higher elimination : substitution product ratio in reaction with 1-bromo-2-methylpropane?

(a) $N(CH_3)_3$, $P(CH_3)_3$ **(b)** H_2N^-, $(CH_3\overset{\overset{\displaystyle CH_3}{|}}{C}H)_2N^-$ **(c)** I^-, Cl^-

EXERCISE 7-14

In all cases where substitution and elimination compete, it is found that higher reaction temperatures lead to greater proportions of elimination products. Thus, the amount of elimination accompanying hydrolysis of 2-bromo-2-methylpropane doubles as the temperature is raised from 25 to 65°C, and that from reaction of 2-bromopropane with ethoxide rises from 80% to 25°C at nearly 100% at 55°C. Explain.

7-9 Summary of Reactivity of Haloalkanes

Primary, secondary, and tertiary haloalkanes may react with nucleophiles through different pathways. Table 7-4 provides a summary of the most likely mechanisms by which the haloalkanes undergo substitution and elimination.

PRIMARY HALOALKANES. Unhindered primary alkyl substrates always react in a bimolecular way and almost always give predominantly substitution products, except when sterically hindered strong bases, such as potassium *tert*-butoxide, are employed. In these cases, the S_N2 pathway is slowed down sufficiently for steric reasons to allow the E2 mechanism to take over. Another way of reducing substitution is to introduce branching. However, even in these cases, good nucleophiles still furnish predominantly substitution products. Only strong bases, such as alkoxides, RO^-, or amides, R_2N^-, tend to react by elimination.

TABLE 7-4 Likely Mechanisms by Which Haloalkanes React with Nucleophiles (Bases)

	Type of nucleophile (base)			
Type of haloalkane	**Poor nucleophile (e.g., H_2O)**	**Weakly basic, good nucleophile (e.g., I^-)**	**Strongly basic, unhindered nucleophile (e.g., CH_3O^-)**	**Strongly basic, hindered nucleophile (e.g., $(CH_3)_3CO^-$)**
Methyl	No reaction	S_N2	S_N2	S_N2
Primary				
Unhindered	No reaction	S_N2	S_N2	E2
Branched	No reaction	S_N2	E2	E2
Secondary	Slow S_N1, E1	S_N2	E2	E2
Tertiary	S_N1, E1	S_N1, E1	E2	E2

Reactivity of Primary Haloalkanes R–X with Nucleophiles (Bases)

For unhindered primary R–X:

S_N2 with good nucleophiles

$$CH_3CH_2CH_2Br + {}^-CN \xrightarrow{\text{Propanone (acetone)}} CH_3CH_2CH_2CN + Br^-$$

S_N2 with strong base, too

$$CH_3CH_2CH_2Br + CH_3O^- \xrightarrow{CH_3OH} CH_3CH_2CH_2OCH_3 + Br^-$$

But E2 with strong, hindered base

$$CH_3CH_2CH_2Br + CH_3\overset{\overset{\large CH_3}{|}}{\underset{\underset{\large CH_3}{|}}{C}}O^- \xrightarrow[-\,HBr]{(CH_3)_3COH} CH_3CH{=}CH_2$$

No (or exceedingly slow) reaction with poor nucleophiles (CH_3OH)

For branched primary R–X:

S_N2 with good nucleophiles (although slow compared with unhindered R–X)

$$CH_3\overset{\overset{\large CH_3}{|}}{\underset{\underset{\large H}{|}}{C}}CH_2Br + I^- \xrightarrow{\text{Propanone (acetone)}} CH_3\overset{\overset{\large CH_3}{|}}{\underset{\underset{\large H}{|}}{C}}CH_2I + Br^-$$

E2 with strong base (not necessarily hindered)

$$CH_3\overset{\overset{\large CH_3}{|}}{\underset{\underset{\large H}{|}}{C}}CH_2Br + CH_3CH_2O^- \xrightarrow[-\,HBr]{CH_3CH_2OH} CH_3\overset{\overset{\large CH_3}{|}}{C}{=}CH_2$$

No (or exceedingly slow) reaction with poor nucleophiles

SECONDARY HALOALKANES. Secondary alkyl systems undergo, depending on conditions, both eliminations and substitutions by either possible pathway: uni- and bimolecular. Good nucleophiles favor S_N2, bases result in E2, and polar nonnucleophilic media give mainly S_N1 and E1.

Reactivity of Secondary Haloalkanes R–X with Nucleophiles (Bases)

S_N1 and E1 when X is a good leaving group in a highly polar, nonnucleophilic medium

$$CH_3\overset{\overset{\large CH_3}{|}}{\underset{\underset{\large H}{|}}{C}}Br \xrightarrow[-\,HBr]{CH_3CH_2OH} \underset{\text{Major}}{CH_3\overset{\overset{\large CH_3}{|}}{\underset{\underset{\large H}{|}}{C}}OCH_2CH_3} + \underset{\text{Minor}}{CH_3CH{=}CH_2}$$

S_N2 with high concentrations of good, nonbasic nucleophiles

$$CH_3\overset{CH_3}{\underset{H}{\overset{|}{\underset{|}{C}}}}Br + CH_3S^- \xrightarrow{CH_3CH_2OH} CH_3\overset{CH_3}{\underset{H}{\overset{|}{\underset{|}{C}}}}SCH_3 + Br^-$$

E2 with high concentrations of strong base

$$CH_3\overset{CH_3}{\underset{H}{\overset{|}{\underset{|}{C}}}}Br + CH_3CH_2O^- \xrightarrow[-\ HBr]{CH_3CH_2OH} CH_3CH{=}CH_2$$

TERTIARY HALOALKANES. Tertiary systems eliminate (E2) with concentrated strong base and are substituted in nonbasic media (S_N1). Bimolecular substitution is not observed, but elimination by E1 accompanies S_N1.

Reactivity of Tertiary Haloalkanes R–X with Nucleophiles (Bases)

S_N1 and E1 in polar solvents when X is a good leaving group and no base is present

$$CH_3CH_2\overset{CH_3}{\underset{CH_3}{\overset{|}{\underset{|}{C}}}}Br \xrightarrow[-\ HBr]{HOH,\ \text{propanone (acetone)}} CH_3CH_2\overset{CH_3}{\underset{CH_3}{\overset{|}{\underset{|}{C}}}}OH + \text{alkenes}$$

E2 with high concentrations of strong base

$$CH_3CH_2\overset{\overset{CH_3}{|}}{\underset{\underset{CH_3}{|}}{\overset{CH_2}{\underset{CH_2}{\overset{|}{\underset{|}{C}}}}}}Cl \xrightarrow[-\ HCl]{CH_3O^-,\ CH_3OH} CH_3CH_2\overset{\overset{CH_3}{|}}{\overset{CH_2}{\overset{|}{C}}}{=}\underset{H}{\underset{|}{C}}CH_3$$

E1 with added weak base or low concentrations of strong base

$$CH_3\overset{CH_3}{\underset{CH_3}{\overset{|}{\underset{|}{C}}}}Br \xrightarrow[-\ HBr]{CH_3CH_2OH,\ \text{weak base}} CH_2{=}\overset{CH_3}{\overset{|}{C}}CH_3$$

EXERCISE 7-15

Predict which reaction in each of the following pairs will have a higher E2:E1 product ratio and explain why.

(a) $CH_3CH_2CH(CH_3)Br \xrightarrow{CH_3OH}$ $\quad CH_3CH_2CH(CH_3)Br \xrightarrow{CH_3O^-Na^+,\ CH_3OH}$

(b) Iodocyclohexane $\xrightarrow{(CH_3CH(CH_3))_2N^-Li^+,\ (CH_3CH(CH_3))_2NH}$ $\quad$ Iodocyclohexane $\xrightarrow{\text{Propanone (acetone)}}$

NEW REACTIONS

1. BIMOLECULAR SUBSTITUTION—S_N2 (Sections 6-3 through 6-9, 7-5)

Primary and secondary substrates only

$$H_3C(H)(CH_2CH_3)C\text{—}I \xrightarrow{:Nu^-} Nu\text{—}C(CH_3)(H)(CH_2CH_3) + I^-$$

Direct backside displacement with 100% inversion of configuration

2. UNIMOLECULAR SUBSTITUTION—S_N1 (Sections 7-1 through 7-5)

Secondary and tertiary substrates only

$$CH_3CH_2C(CH_3)_2Br \xrightarrow[-\ Br^-]{} CH_3CH_2C^+(CH_3)_2 \xrightarrow{:Nu^-} CH_3CH_2C(CH_3)_2Nu$$

Through carbocation: Chiral systems are racemized

3. UNIMOLECULAR ELIMINATION—E1 (Section 7-6)

Secondary and tertiary substrates only

$$CH_3C(CH_3)(H)\text{—}C(CH_3)(H)Cl \xrightarrow[-\ Cl^-]{} CH_3C(CH_3)(H)\text{—}C^+(CH_3)H \xrightarrow{:B^-} CH_3C(CH_3)(H)\text{—}C(=CH_2)H + (H_3C)_2C{=}C(CH_3)H + BH$$

Through carbocation

4. BIMOLECULAR ELIMINATION–E2 (Section 7-7)

$$CH_3CH_2CH_2I \xrightarrow{:B^-} CH_3CH{=}CH_2 + BH + I^-$$

Simultaneous elimination of leaving group and neighboring proton

IMPORTANT CONCEPTS

1. Secondary haloalkanes undergo slow and tertiary haloalkanes fast unimolecular substitution in polar media. When the solvent serves as the nucleophile, the process is called solvolysis.

2. The slowest, or rate-determining step, in unimolecular substitution is dissociation of the C–X bond to form a carbocation intermediate. Added strong nucleophile changes the product but not the reaction rate.

3. Carbocations are stabilized by hyperconjugation: Tertiary are the most stable, followed by secondary. Primary and methyl cations are too unstable to form in solution.

4. Racemization often results when unimolecular substitution takes place at a chiral carbon.

5. Unimolecular elimination to form an alkene accompanies substitution in secondary and tertiary systems. Elimination is favored by the addition of base.

6. High concentrations of strong base may bring about bimolecular elimination. Expulsion of the leaving group accompanies removal of a hydrogen from the neighboring carbon by the base. The mechanism requires an *anti* conformational arrangement of the hydrogen and the leaving group.

7. Substitution is favored by unhindered substrates and small, less basic nucleophiles.

8. Elimination is favored by hindered substrates and bulky, more basic nucleophiles.

PROBLEMS

1. What is the major substitution product of each of the following solvolysis reactions?

(a) $CH_3C(CH_3)_2Br \xrightarrow{CH_3CH_2OH}$

(b) $(CH_3)_2CBrCH_2CH_3 \xrightarrow{CF_3CH_2OH}$

(c) 1-chloro-1-ethylcyclopentane (CH_3CH_2, Cl on the same ring carbon) $\xrightarrow{CH_3OH}$

(d) cyclohexyl–$C(Br)(CH_3)$–CH_3 $\xrightarrow{CH_3COOH}$

(e) $CH_3C(CH_3)_2Cl \xrightarrow{D_2O}$

(f) $CH_3C(CH_3)_2Cl \xrightarrow{\text{cyclohexyl-OD (H)}}$

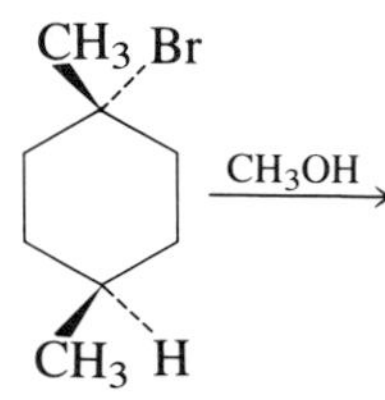

2. Write the two major substitution products of the reaction shown in the margin. **(a)** Write a mechanism to explain the formation of each of them. **(b)** Monitoring the reaction mixture reveals that an *isomer* of the starting material is generated as an intermediate. Draw its structure and explain how it is formed.

3. Give the two major substitution products of the following reaction.

OSO_2CH_3

H_3C C_6H_5 $\xrightarrow{CH_3CH_2OH}$

H_3C C_6H_5

H

4. How would each reaction in Problem 1 be affected by the addition of each of the following substances to the solvolysis mixture?

(a) H_2O **(b)** H_2S **(c)** KI

(d) NaN_3 **(e)** Propanone (acetone) **(f)** $CH_3CH_2OCH_2CH_3$

5. Rank the following carbocations in decreasing order of stability.

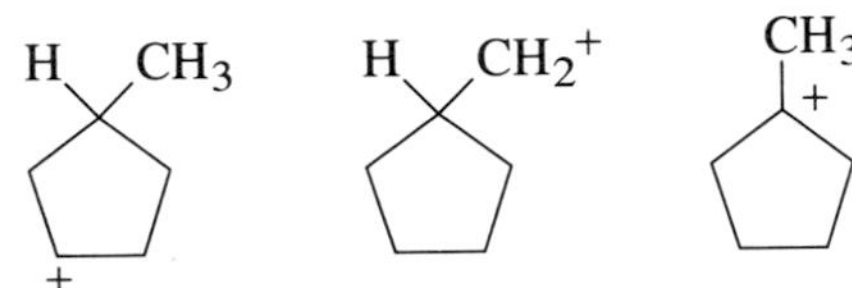

6. Rank the compounds in each group below in order of decreasing rate of solvolysis in aqueous propanone (acetone).

(a) $CH_3CH(CH_3)CH_2CH_2Cl$ $CH_3CH(CH_3)CHClCH_3$ $CH_3CCl(CH_3)CH_2CH_3$

(b) $OCCH_3$ (O) on cyclohexane; Cl on cyclohexane; HO on cyclohexane

(c) Br on cyclohexane; Cl on cyclohexane; CH_3 Cl on cyclohexane

7. Give the products of the following substitution reactions. Indicate whether they arise through the S_N1 or the S_N2 process. Formulate the detailed mechanisms of their generation.

(a) $(CH_3)_2CHOSO_2CF_3 \xrightarrow{CH_3CH_2OH}$ **(b)** cyclopentane with CH_3 and Br $\xrightarrow{CH_3SH,\ CH_3OH}$

(c) $CH_3CH_2CH_2CH_2Br \xrightarrow{(C_6H_5)_3P,\ DMSO}$ **(d)** $CH_3CH_2CHClCH_2CH_3 \xrightarrow{\text{NaI, propanone (acetone)}}$

8. Give the product of each substitution reaction shown below. Which of these transformations should proceed faster in a polar, aprotic solvent (such as propanone or DMSO) than in a polar, protic solvent (such as water or

CH_3OH)? Explain your answer on the basis of the mechanism you expect to be operating in each case.
(a) $CH_3CH_2CH_2Br + Na^+ \ ^-CN \longrightarrow$
(b) $(CH_3)_2CHCH_2I + Na^+N_3^- \longrightarrow$
(c) $(CH_3)_3CBr + HSCH_2CH_3 \longrightarrow$
(d) $(CH_3)_2CHOSO_2CH_3 + HOCH(CH_3)_2 \longrightarrow$

9. Propose a synthesis of (*R*)-$CH_3CHN_3CH_2CH_3$, starting from (*R*)-2-chlorobutane.

10. Two substitution reactions of (*S*)-2-bromobutane are shown below. Show their stereochemical outcomes.

$$(S)\text{-}CH_3CH_2CHBrCH_3 \xrightarrow{\overset{O}{\overset{\|}{HCOH}}}$$

$$(S)\text{-}CH_3CH_2CHBrCH_3 \xrightarrow{\overset{O}{\overset{\|}{HCO^-}}Na^+,\ DMSO}$$

11. Write all possible E1 products of each reaction in Problem 1.

12. Write the products of the following elimination reactions. Specify the predominant mechanism (E1 or E2) and formulate it in detail.

(a) $(CH_3CH_2)_3CBr \xrightarrow{NaNH_2,\ NH_3}$ **(b)** $CH_3CH_2CH_2CH_2Cl \xrightarrow{KOC(CH_3)_3,\ (CH_3)_3COH}$

(c) 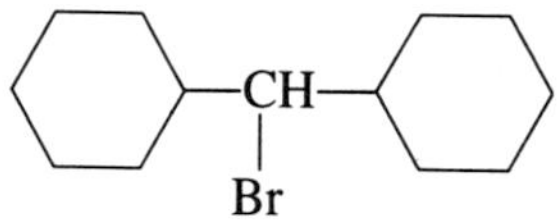 (cyclohexyl)–CH(Br)–(cyclohexyl) $\xrightarrow{\text{Excess KOH, } CH_3CH_2OH}$ **(d)** 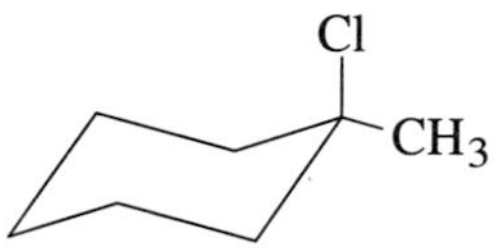 1-chloro-1-methylcyclohexane (chair; Cl, CH_3) $\xrightarrow{NaOCH_3,\ CH_3OH}$

13. Three reactions of 2-chloro-2-methylpropane are shown below. **(a)** Provide the major product of each transformation. **(b)** Compare the rates of the three reactions. Assume identical solution polarities and reactant concentrations. Explain mechanistically.

$$(CH_3)_3CCl \xrightarrow{H_2S,\ CH_3OH}$$

$$(CH_3)_3CCl \xrightarrow{CH_3\overset{O}{\overset{\|}{C}}O^-K^+,\ CH_3OH}$$

$$(CH_3)_3CCl \xrightarrow{CH_3O^-K^+,\ CH_3OH}$$

14. Give the major product(s) of the following reactions. Indicate which of the following mechanism(s) is in operation: S_N1, S_N2, E1, or E2. If no reaction occurs, write "no reaction."

(a) 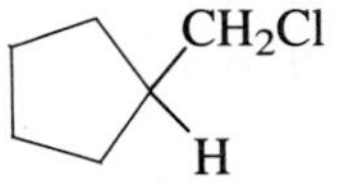 cyclopentane with CH_2Cl and H $\xrightarrow{KOC(CH_3)_3,\ (CH_3)_3COH}$ **(b)** $CH_3\overset{F}{\overset{|}{C}}HCH_2CH_3 \xrightarrow{\text{KBr, propanone (acetone)}}$

(c) H_3C–C(CH_2CH_3)(H)–Br $\xrightarrow{H_2O}$

(d) iodocyclohexane $\xrightarrow{NaNH_2,\ liquid\ NH_3}$

(e) $(CH_3)_2CHCH_2CH_2CH_2Br \xrightarrow{NaOCH_2CH_3,\ CH_3CH_2OH}$

(f) H_3C–C(Br)(CH_2CH_3)–$CH_2CH_2CH_3$ $\xrightarrow{NaI,\ propanone\ (acetone)}$

(g) cyclopentyl(OH)(H) $\xrightarrow{KOH,\ CH_3CH_2OH}$

(h) Cl–cyclohexyl–$CH_2CH_2CH_2Br$ $\xrightarrow{Excess\ KCN,\ CH_3OH}$

(i) (R)-$CH_3CH_2CH(OSO_2C_6H_4CH_3)CH_3$ $\xrightarrow{NaSH,\ CH_3CH_2OH}$

(j) 1-ethyl-1-iodocyclohexane (CH_3CH_2, I) $\xrightarrow{CH_3OH}$

(k) $(CH_3)_3CCHBrCH_3 \xrightarrow{KOH,\ CH_3CH_2OH}$

(l) $CH_3CH_2Cl \xrightarrow{CH_3COH\ (\text{with C=O})}$

15. Consider the reaction shown below. Will it proceed by substitution or by elimination? What factors determine the most likely mechanism? Write the expected product. (Hint: Draw the chair conformation of the substrate.)

Cyclohexane bearing CH_3, Cl, CH_3, and $CH_3\overset{|}{C}HCH_3$ substituents $\xrightarrow{NaOCH_2CH_3,\ CH_3CH_2OH}$

16. Fill the spaces in the table below with the major product(s) of the reaction of each haloalkane with the reagents shown.

	Reagent			
Haloalkane	H_2O	$NaSeCH_3$	$NaOCH_3$	$KOC(CH_3)_3$
CH_3Cl	______	______	______	______
$CH_3CH_2CH_2Cl$	______	______	______	______
$(CH_3)_2CHCl$	______	______	______	______
$(CH_3)_3CCl$	______	______	______	______

17. Indicate the major mechanism(s) (simply specify S_N2, S_N1, E2, or E1) required for the formation of each product that you wrote in Problem 16.

18. For each equation below, indicate whether the reaction would work well, poorly, or not at all. Formulate alternative products, if appropriate.

(a) $CH_3CH_2CH(Br)CH_3 \xrightarrow{\text{NaOH, propanone (acetone)}} CH_3CH_2CH(OH)CH_3$

(b) $CH_3CH(CH_3)CH_2Cl \xrightarrow{CH_3OH} CH_3CH(CH_3)CH_2OCH_3$

(c) 1-chloro-cyclohexane (H, Cl on same carbon) $\xrightarrow{\text{HCN, } CH_3OH}$ cyclohexane with H, CN

(d) $CH_3\text{—}C(CH_3)(CH_3SO_2O)\text{—}CH_2CH_2CH_2CH_2OH \xrightarrow{(CH_3CH_2)_2O}$ six-membered oxygen ring with H_3C and CH_3 on the carbon adjacent to O

(e) cyclopentane bearing CH_3 and CH_2I on the same carbon $\xrightarrow{NaSCH_3, CH_3OH}$ cyclopentane bearing CH_3 and CH_2SCH_3

(f) $CH_3CH_2CH_2Br \xrightarrow{NaN_3, CH_3OH} CH_3CH_2CH_2N_3$

(g) $(CH_3)_3CCl \xrightarrow{\text{NaI, propanone (acetone)}} (CH_3)_3CI$

(h) $(CH_3CH_2)_2O \xrightarrow{CH_3I} (CH_3CH_2)_2\overset{+}{O}CH_3 + I^-$

(i) $CH_3I \xrightarrow{CH_3OH} CH_3OCH_3$

(j) $(CH_3CH_2)_3COCH_3 \xrightarrow{NaBr, CH_3OH} (CH_3CH_2)_3CBr$

(k) $CH_3CH(CH_3)CH_2CH_2Cl \xrightarrow{NaOCH_2CH_3, CH_3CH_2OH} CH_3CH(CH_3)CH{=}CH_2$

(l) $CH_3CH_2CH_2CH_2Cl \xrightarrow{NaOCH_2CH_3, CH_3CH_2OH} CH_3CH_2CH{=}CH_2$

19. Propose syntheses of the molecules below from the indicated starting materials. Make use of any other reagents or solvents that you need. In some cases, there may be no alternative but to employ a reaction that results in a mixture of products. If so, use reagents and conditions that will maximize the yield of the desired material (compare Problem 16 in Chapter 6).

(a) $CH_3CH_2CHICH_3$, from butane
(b) $CH_3CH_2CH_2CH_2I$, from butane
(c) $(CH_3)_3COCH_3$, from methane and 2-methylpropane
(d) Cyclohexene, from cyclohexane
(e) Cyclohexanol, from cyclohexane
(f) 1,3-dithiane (six-membered ring with two S), from 1,3-dibromopropane

20. [(1-Bromo-1-methyl)ethyl]benzene, as shown in the margin, undergoes solvolysis in a unimolecular, strictly first order process. The reaction rate for [RBr] = 0.1 M RBr in 9:1 propanone (acetone):water is measured to be 2×10^{-4} mol L^{-1} s^{-1}. **(a)** Calculate the rate constant k from these data. What is the product of this reaction? **(b)** In the presence of 0.1 M LiCl, the rate is found to increase to 4×10^{-4} mol L^{-1} s^{-1}, although the reaction still remains strictly first order. Calculate the new rate constant k_{LiCl} and suggest an explanation. **(c)** If 0.1 M LiBr is present instead of LiCl, the measured rate *drops* to 1.6×10^{-4} mol L^{-1} s^{-1}. Explain this observation and write the appropriate chemical equations to describe the reactions.

RBr = $C_6H_5\text{—}C(CH_3)_2\text{—}Br$

21. The stabilities of three cyclic cations illustrated below differ greatly. Predict their order of stability and provide a rationalization for your assignments.

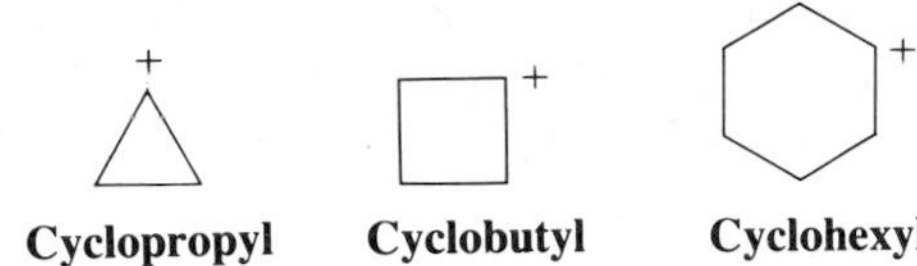

Cyclopropyl **Cyclobutyl** **Cyclohexyl**

22. Match each of the transformations below to the correct reaction profile shown below, and draw the structures of the species present at all points on the energy curves marked by capital letters.

(a) $(CH_3)_3CCl + (C_6H_5)_3P \longrightarrow$ **(b)** $(CH_3)_2CHI + KBr \longrightarrow$

(c) $(CH_3)_3CBr + HOCH_2CH_3 \longrightarrow$ **(d)** $CH_3CH_2Br + NaOCH_2CH_3 \longrightarrow$

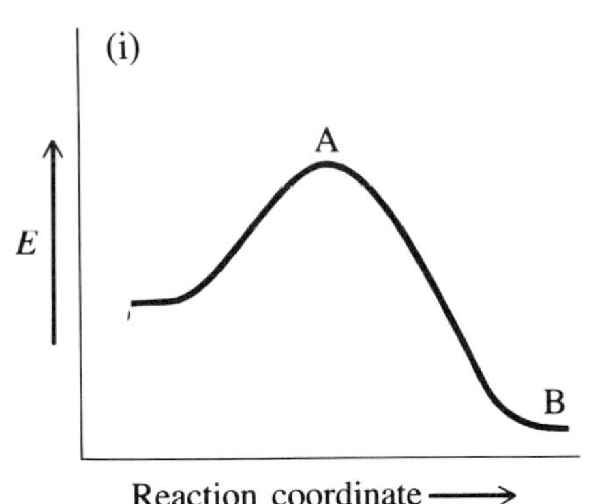

Reaction coordinate ⟶

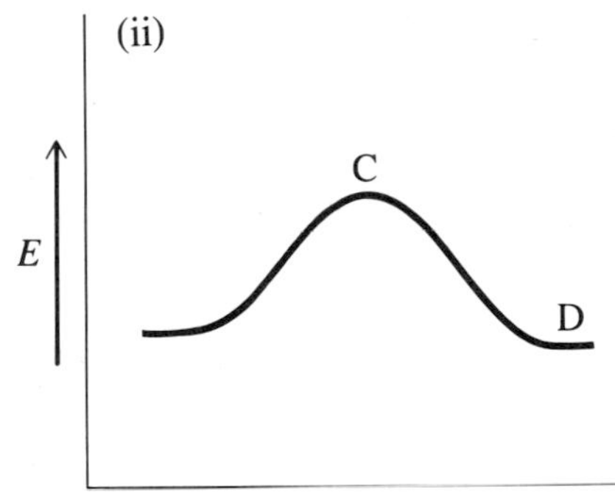

Reaction coordinate ⟶

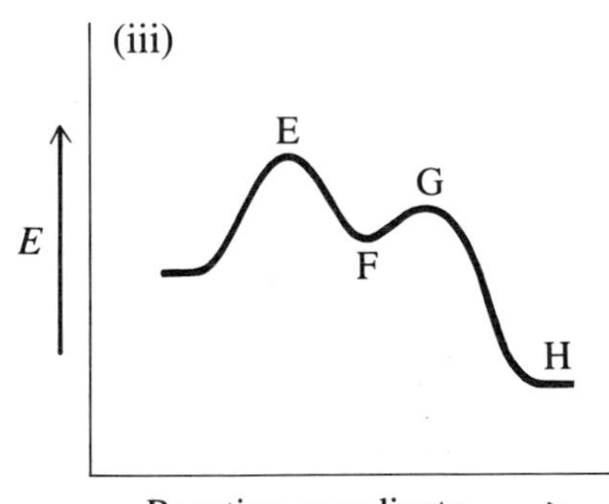

Reaction coordinate ⟶

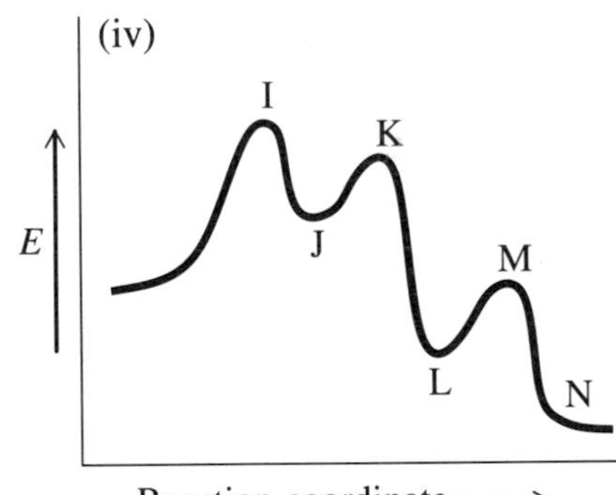

Reaction coordinate ⟶

23. Formulate the structure of the most likely product of the following reaction of 4-chloro-4-methyl-1-pentanol in neutral polar solution.

$$(CH_3)_2\overset{\overset{\displaystyle Cl}{|}}{C}CH_2CH_2CH_2OH \longrightarrow HCl + C_6H_{12}O$$

In *basic* solution the starting material again converts to a molecule with the molecular formula $C_6H_{12}O$, but with a completely different structure. What is it? Explain the difference between the two results.

24. The reaction below can proceed through both E1 and E2 mechanisms.

$$C_6H_5CH_2\overset{\overset{\displaystyle CH_3}{|}}{\underset{\underset{\displaystyle CH_3}{|}}{C}}Cl \xrightarrow{NaOCH_3,\ CH_3OH} C_6H_5CH{=}C(CH_3)_2 + C_6H_5CH_2\overset{\overset{\displaystyle CH_3}{|}}{C}{=}CH_2$$

The E1 rate constant $k_{E1} = 1.4 \times 10^{-4}\ s^{-1}$ and the E2 rate constant $k_{E2} = 1.9 \times 10^{-4}\ L\ mol^{-1}\ s^{-1}$; 0.02 M haloalkane. **(a)** What is the predominant elimination mechanism with 0.5 M $NaOCH_3$? **(b)** What is the predominant elimination mechanism with 2.0 M $NaOCH_3$? **(c)** At what concentration of base does exactly 50% of the starting material react by an E1 route and 50% by an E2 pathway?

$(CH_3)_3COH \xrightarrow{\text{Conc. HBr}} (CH_3)_3CBr$

25. When 2-methyl-2-propanol is shaken with concentrated aqueous HBr, rapid formation of 2-bromo-2-methylpropane occurs (margin), the reverse of S_N1 hydrolysis (Section 7-1). Propose a detailed mechanism for this process.

26. Give the mechanism and major product for the reaction of a secondary haloalkane in a polar aprotic solvent with the nucleophiles below. The pK_a value of the conjugate acid of the nucleophile is given in parentheses.

(a) N_3^- (4.6) **(b)** H_2N^- (35) **(c)** NH_3 (9.5)
(d) HSe^- (3.7) **(e)** F^- (3.2) **(f)** $C_6H_5O^-$ (9.9)
(g) PH_3 (−12) **(h)** NH_2OH (6.0) **(i)** NCS^- (−0.7)

27. Cortisone is an important steroidal anti-inflammatory agent. Cortisone can be synthesized efficiently from the alkene shown below.

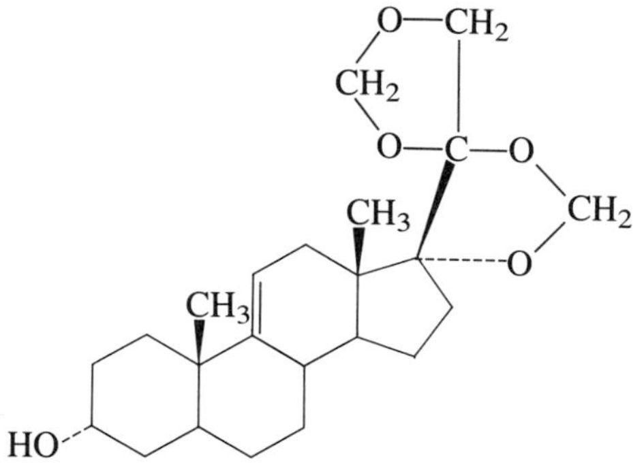

O—CH2 CH2 O—C—O CH3 CH2 O CH3 HO
Alkene

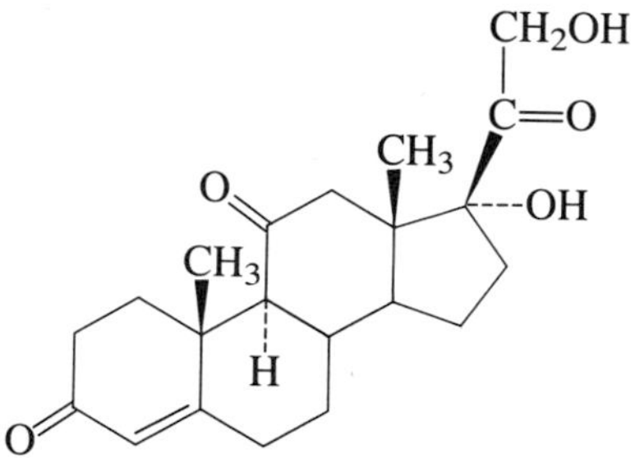

CH2OH C=O CH3 OH O CH3 H O
Cortisone

Of the three chlorinated compounds shown below, two give reasonable yields of the alkene shown above by E2 elimination with base, but one does not. Which one does not work well, and why? What does it give during attempted E2 elimination?

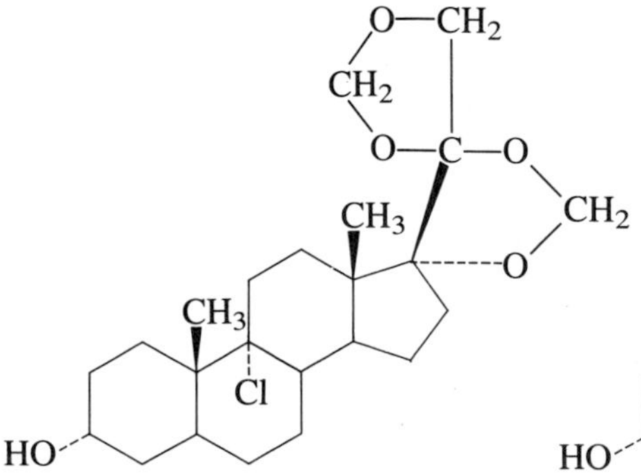

O—CH2 CH2 O—C—O CH3 CH2 O CH3 Cl HO
A

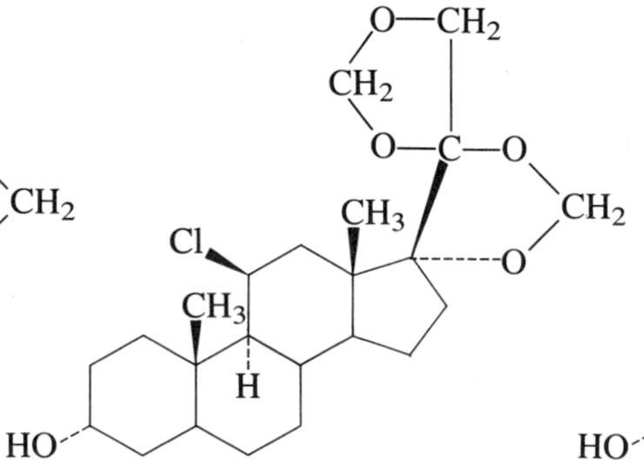

O—CH2 CH2 O—C—O Cl CH3 CH2 O CH3 H HO
B

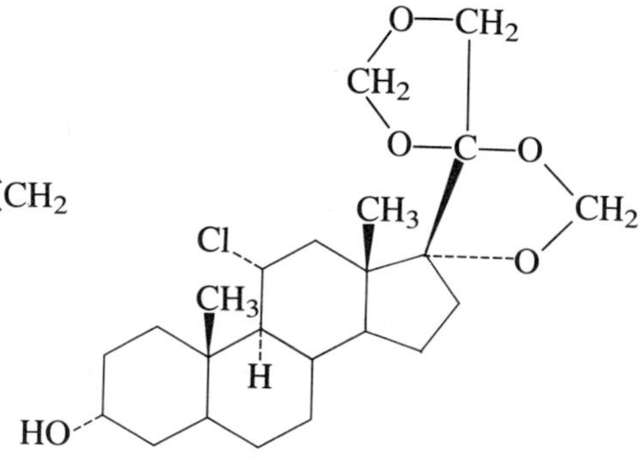

O—CH2 CH2 O—C—O Cl CH3 CH2 O CH3 H HO
C

28. The chemistry of derivatives of *trans*-decalin is of interest because this ring system is part of the structure of steroids. Make models of the brominated systems i and ii to help you answer the following.

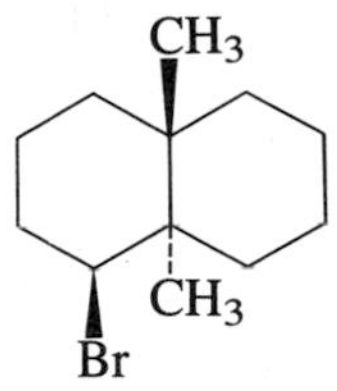

CH_3 CH_3 CH_3 CH_3 Br Br

i 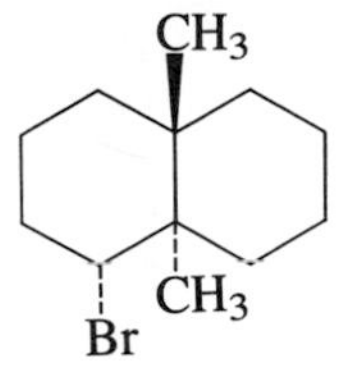 ii

(a) One of the molecules undergoes E2 reaction with $NaOCH_2CH_3$ in CH_3CH_2OH considerably faster than the other. Identify which is which and explain. **(b)** The deuterated analogs of i and ii depicted below react with base to give the products shown.

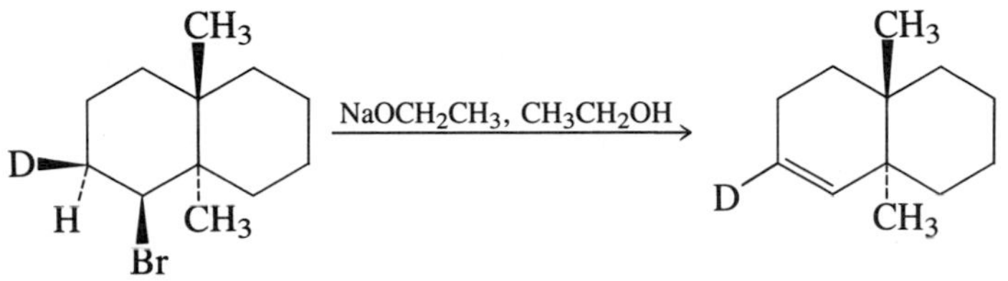

CH_3 D H CH_3 Br **i-deuterated** $\xrightarrow{NaOCH_2CH_3,\ CH_3CH_2OH}$ CH_3 D CH_3 (All D retained)

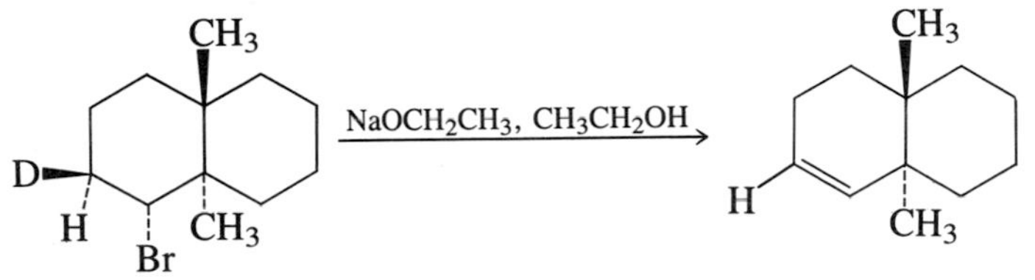

CH_3 D H CH_3 Br **ii-deuterated** $\xrightarrow{NaOCH_2CH_3,\ CH_3CH_2OH}$ CH_3 H CH_3 (All D lost)

Specify whether *anti* or *syn* eliminations have taken place. Draw the conformations that the molecules must adopt for elimination to occur. Does your answer help you in solving (a)? **(c)** Does either reaction in (b) exhibit a deuterium isotope effect? Explain.

8

Hydroxy Functional Group

Properties of the Alcohols and Strategic Syntheses

We now know a great deal about the simplest organic molecules: the alkanes and haloalkanes. We have studied their structures and properties, and we can study reaction mechanisms to help us understand and predict their transformations. Now it is time to explore the oxygen-containing compounds produced in these reactions.

H—O—H
Water

CH_3—O—H
Methanol
(An alcohol)

CH_3—O—CH_3
Methoxymethane
(An ether)

This chapter introduces the chemistry of the alcohols. From Chapter 2 we know that alcohols have carbon backbones bearing the substituent OH, the **hydroxy** group. They may be viewed as derivatives of water in which one hydrogen has been replaced by an alkyl group. Replacement of the second hydrogen gives an **ether** (Chapter 9).

Alcohols are abundant in nature and varied in structure (see, e.g., Section 4-7). Simple alcohols are used as solvents; others aid in the synthesis of more complex molecules. They are a good example of how functional groups shape the properties and applications of organic compounds. Our discussion will begin with the naming of alcohols, followed by a brief description of their structures and other physical properties, particularly in comparison with those of the alkanes and haloalkanes. Finally, their preparation will introduce us to strategies for efficiently synthesizing new organic compounds.

8-1 Naming the Alcohols

Like other compounds, alcohols may have both systematic and common names. Systematic nomenclature treats alcohols as derivatives of alkanes. The ending *-e* of the alkane is replaced by **-ol.** Thus, an alkane is converted into an **alkanol.** For example, the simplest alcohol is derived from methane. It is methanol. Ethanol stems from ethane, propanol from propane, and so on. In more complicated, branched systems, the name of the alcohol is based on the longest chain *containing the OH substituent*—not necessarily the longest chain in the molecule.

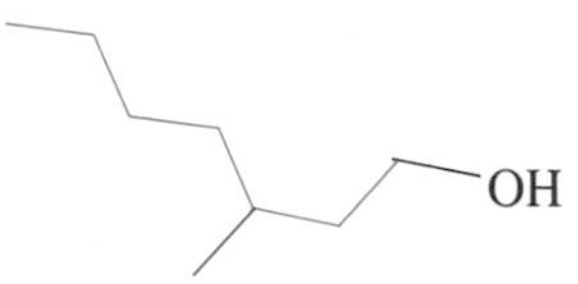

OH OH

A methylheptanol 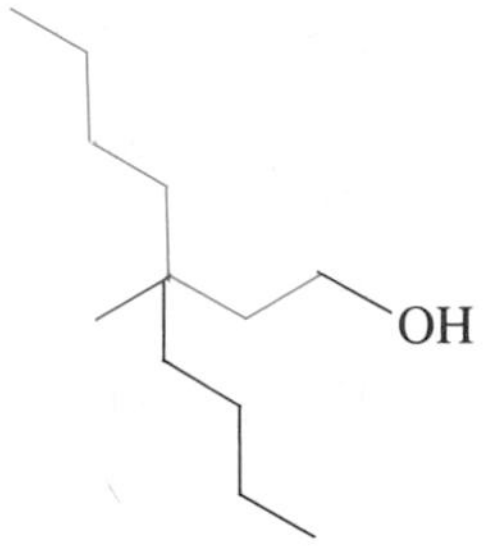 **A butylmethylheptanol**

To locate positions along the chain, number each carbon atom beginning from the end closest to the OH group.

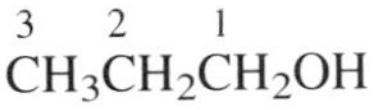

$\overset{3}{C}H_3\overset{2}{C}H_2\overset{1}{C}H_2OH$ — **1-Propanol**

$\overset{1}{C}H_3\overset{2}{C}H(OH)\overset{3}{C}H_3$ — **2-Propanol**

$\overset{1}{C}H_3\overset{2}{C}H(OH)\overset{3}{C}H_2\overset{4}{C}H_2\overset{5}{C}H_3$ — **2-Pentanol**

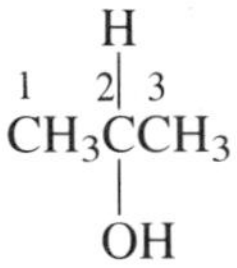

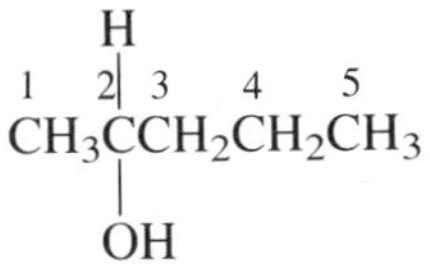

The names of other substituents along the chain can then be added to the alkanol stem as prefixes. Complex alkyl appendages are named according to the IUPAC rules for hydrocarbons (Section 2-3).

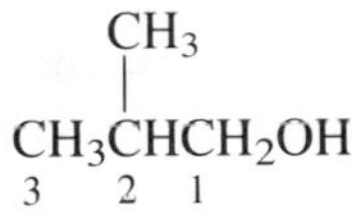

$\underset{3}{C}H_3\underset{2}{C}H(CH_3)\underset{1}{C}H_2OH$ — **2-Methyl-1-propanol**

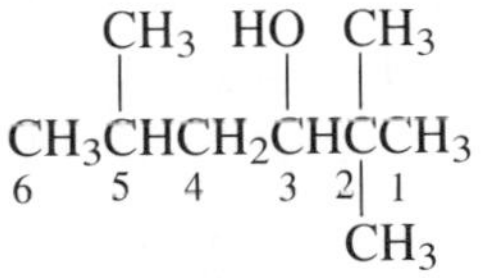

$\underset{6}{C}H_3\underset{5}{C}H(CH_3)\underset{4}{C}H_2\underset{3}{C}H(OH)\underset{2}{C}(CH_3)_2\underset{1}{C}H_3$ — **2,2,5-Trimethyl-3-hexanol**

Cyclic alcohols are called **cycloalkanols.** Here the carbon carrying the functional group automatically receives the number 1.

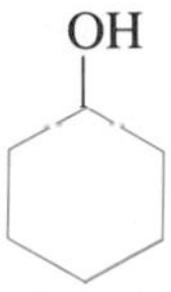

OH HO CH_2CH_3 OH Cl

Cyclohexanol 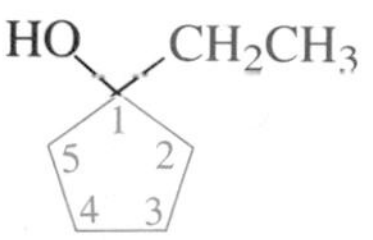 **1-Ethylcyclopentanol** 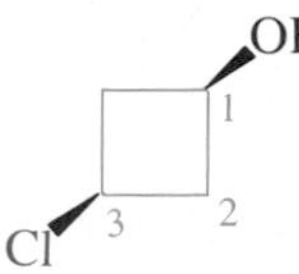 ***cis*-3-Chlorocyclobutanol**

When named as a substituent, the OH group is called *hydroxy*.

Like haloalkanes, alcohols can be classified as primary, secondary, or tertiary.

RCH_2OH	$R\overset{OH}{\underset{H}{C}}R'$	$R\overset{OH}{\underset{R''}{C}}R'$
A primary alcohol	A secondary alcohol	A tertiary alcohol

EXERCISE 8-1

Draw the structures of the following alcohols. **(a)** (*S*)-3-Methyl-3-hexanol; **(b)** *trans*-2-bromocyclopentanol; **(c)** 2,2-dimethyl-1-propanol (neopentyl alcohol).

EXERCISE 8-2

Name the following compounds.

(a) $CH_3\overset{CH_3}{C}HCH_2\overset{OH}{C}HCH_3$ **(b)** CH_3CH_2–(cyclohexane ring)–OH **(c)** $CH_3\overset{Br}{C}H\overset{Cl}{C}HCH_2OH$

In common nomenclature, the name of the alkyl group is followed by the word *alcohol,* written separately. Common names are found in the older literature, and although it is best not to use them, we should be able to recognize them.

CH_3OH	$CH_3\overset{CH_3}{\underset{OH}{C}}H$	$CH_3\overset{CH_3}{\underset{CH_3}{C}}OH$
Methyl **alcohol**	Isopropyl **alcohol**	*tert*-Butyl **alcohol**

In common names a Greek letter indicates the position of a substituent. In alphabetical order, the carbon bearing the hydroxy group is labeled α, the neighboring carbon β, and so on.

In summary, alcohols can be named as alkanols (IUPAC) or alkyl alcohols. In IUPAC nomenclature, the name is derived from the chain bearing the hydroxy group, whose position is given the lowest possible number. Common names incorporate the Greek alphabet to designate the position of additional substituents relative to the OH function.

8-2 Structural and Physical Properties of Alcohols

The hydroxy functional group strongly shapes the physical characteristics of the alcohols. It affects their molecular structure and enhances their ability to enter into hydrogen bonding. As a result, it raises their boiling points and increases their solubilities in water.

The structure of alcohols resembles that of water

Figure 8-1 shows how closely the structure of methanol resembles those of water and of methoxymethane (dimethyl ether). In all three the oxygen atoms are

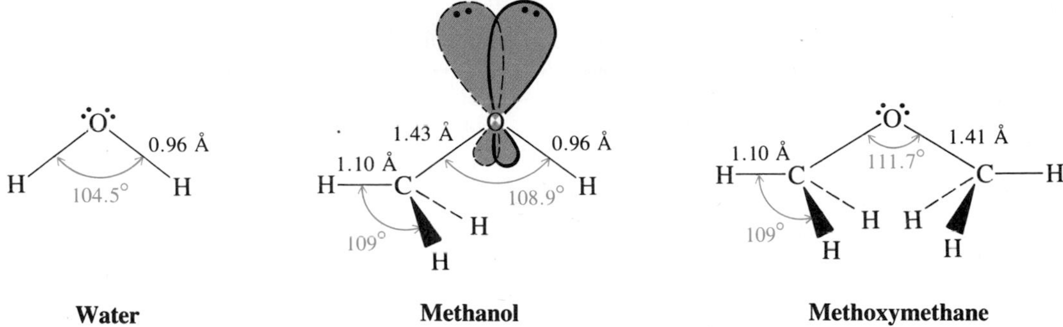

FIGURE 8-1 The similarity in structure of water, methanol, and methoxymethane. The oxygens are approximately sp^3 hybridized, so all three exhibit nearly tetrahedral bond angles around the heteroatom. Remember that the oxygen bears two lone electron pairs in two nonbonding sp^3 hybrid orbitals.

roughly sp^3 hybridized and their bond angles nearly tetrahedral. The minor differences are due to the steric effect produced by replacement of hydrogen atoms by alkyl groups. The O–H bond is considerably shorter than the C–H bond, in part because of the high electronegativity of oxygen relative to that of carbon. Consistent with this bond shortening is the order of bond strengths: $DH^\circ_{O-H} =$ 104 kcal mol^{-1}; $DH^\circ_{C-H} = 98$ kcal mol^{-1}.

The electronegativity of oxygen causes an unsymmetrical distribution of charge in alcohols. This effect polarizes the O–H bond so that the hydrogen has a partial positive charge and gives rise to a molecular dipole (Section 1-3) similar to that observed for water.

Bond and Molecular Dipoles of Water and Methanol

H–O–H Molecular dipole CH_3–O–H Molecular dipole

Hydrogen bonding raises their boiling points and water solubilities

In Section 6-2 we invoked the polarity of the haloalkanes to explain why their boiling points are higher than those of the corresponding nonpolar alkanes. The polarity of alcohols is similar to that of the haloalkanes. Does this mean that the boiling points of haloalkanes and alcohols should correspond? Inspection of Table 8-1 shows that they do not: Alcohols have unusually high boiling points, much higher than those of comparable alkanes and haloalkanes.

The explanation lies in hydrogen bonding. Hydrogen bonds may form between the oxygen atoms of one alcohol molecule and the hydroxy hydrogen atoms of another. Alcohols build up an extensive network of these interactions (Figure 8-2A). Although hydrogen bonds are longer and much weaker ($DH^\circ \sim$ 5 kcal mol^{-1}) than the covalent O–H linkage ($DH^\circ \sim 104$ kcal mol^{-1}), so many of them form that their combined strength makes it difficult for molecules to escape the liquid. The result is a higher boiling point.

TABLE 8-1 Physical Properties of Alcohols and Selected Analogous Haloalkanes and Alkanes

Compound	IUPAC name	Common name	Melting point (°C)	Boiling point (°C)	Solubility in H_2O at 23°C
CH_3OH	Methanol	Methyl alcohol	−97.8	65.0	Infinite
CH_3Cl	Chloromethane	Methyl chloride	−97.7	−24.2	0.74 g/100 mL
CH_4	Methane		−182.5	−161.7	3.5 mL (gas)/100 mL
CH_3CH_2OH	Ethanol	Ethyl alcohol	−114.7	78.5	Infinite
CH_3CH_2Cl	Chloroethane	Ethyl chloride	−136.4	12.3	0.447 g/100 mL
CH_3CH_3	Ethane		−183.3	−88.6	4.7 mL (gas)/100 mL
$CH_3CH_2CH_2OH$	1-Propanol	Propyl alcohol	−126.5	97.4	Infinite
$CH_3CHOHCH_3$	2-Propanol	Isopropyl alcohol	−89.5	82.4	Infinite
$CH_3CHClCH_3$	2-Chloropropane	Isopropyl chloride	−117.2	35.7	0.305 g/100 mL
$CH_3CH_2CH_3$	Propane		−187.7	−42.1	6.5 mL (gas)/100 mL
$CH_3CH_2CH_2CH_2OH$	1-Butanol	Butyl alcohol	−89.5	117.3	8.0 g/100 mL
$(CH_3)_3COH$	2-Methyl-2-propanol	*tert*-Butyl alcohol	25.5	82.2	Infinite
$CH_3(CH_2)_4OH$	1-Pentanol	Pentyl alcohol	−79	138	2.2 g/100 mL
$(CH_3)_3CCH_2OH$	2,2-Dimethyl-1-propanol	Neopentyl alcohol	53	114	Infinite

The effect is even more pronounced in water, which has two hydrogens available for hydrogen bonding (Figure 8-2B). This phenomenon explains why water, with a molecular weight of only 18, has a boiling point of 100°C. Without this property, water would be a gas at ordinary temperatures. Considering the importance of water in all living organisms, imagine how the absence of liquid water would have affected evolution on our planet.

Hydrogen bonding in water and alcohols is responsible for another property: Many alcohols are appreciably water soluble (Table 8-1). This behavior contrasts with that of the nonpolar alkanes, which are poorly solvated by this medium. Because of their characteristic insolubility in water, alkanes are said to be **hydro-**

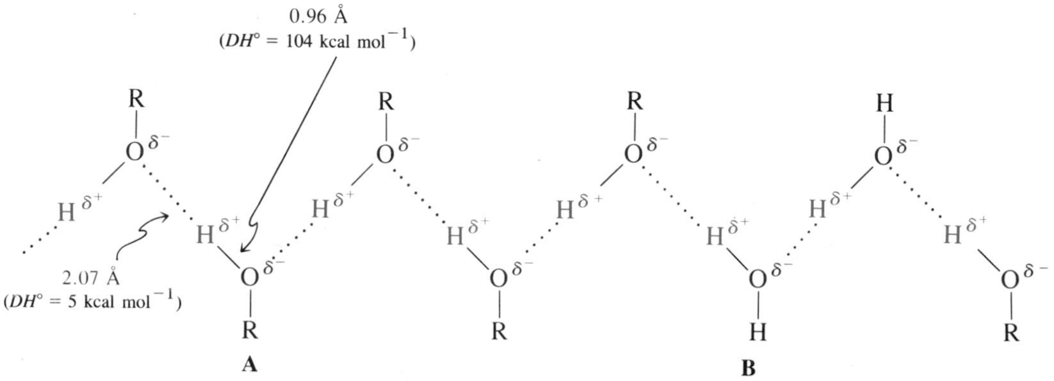

0.96 Å ($DH° = 104$ kcal mol^{-1})

2.07 Å ($DH° = 5$ kcal mol^{-1})

A B

FIGURE 8-2 Hydrogen bonding (A) in an alcohol and (B) between an alcohol and water.

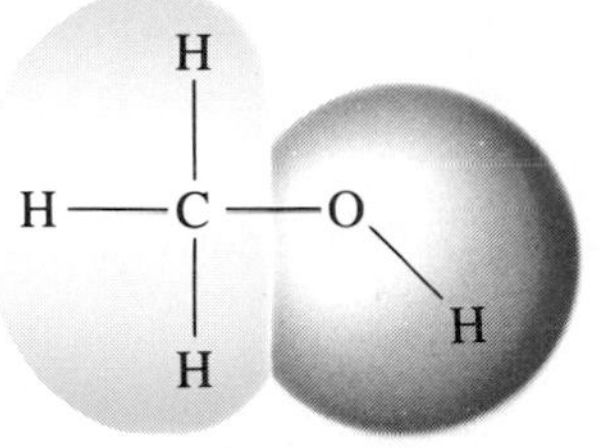

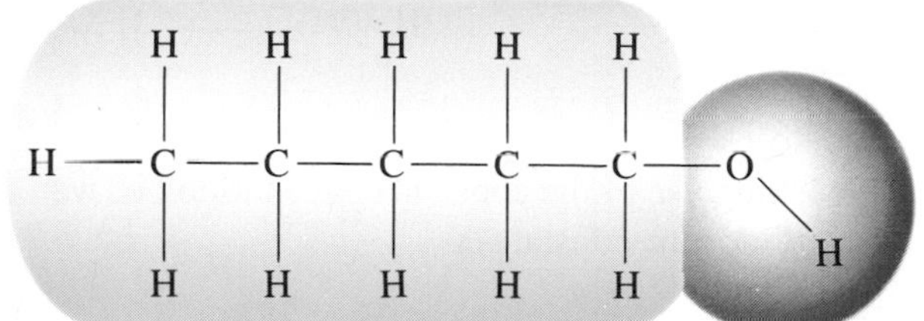

Methanol **1-Pentanol**

FIGURE 8-3 The hydrophobic and hydrophilic parts of methanol and 1-pentanol. The polar functional group dominates the physical properties of methanol: The molecule is completely soluble in water but only partially so in hexane. Conversely, the increased size of the hydrophobic part in the higher alcohol leads to infinite solubility in hexane but reduced solubility in water (Table 8-1).

phobic (*hydro,* Greek, water; *phobos,* Greek, fear). So are most alkyl chains. Conversely, the OH group and other polar substituents, such as COOH and NH_2, are **hydrophilic:** They enhance water solubility.

As the values in Table 8-1 show, the larger the alkyl (hydrophobic) part of an alcohol, the lower its solubility in water. At the same time, the alkyl part increases the solubility of the alcohol in nonpolar solvents (Figure 8-3). (For example, alcohols are effective at removing greasy deposits from the tape heads of cassette decks and VCRs.) The "waterlike" structure of the lower alcohols, particularly methanol and ethanol, also makes them excellent solvents for polar compounds and even salts. It is not surprising, then, that alcohols are popular solvents in the S_N2 reaction (Chapter 6).

In summary, the oxygen in alcohols (and ethers) is tetrahedral and sp^3 hybridized. The covalent O–H bond is shorter and stronger than the C–H bond. Because of the electronegativity of the oxygen, alcohols, like water and ethers, exhibit appreciable molecular polarity. The hydroxy hydrogen enters into hydrogen bonding with other alcohol molecules. These properties lead to a substantial increase in the boiling points and in the solubilities of alcohols in polar solvents relative to those of the alkanes and haloalkanes.

8-3 Alcohols as Acids and Bases

Many applications of the alcohols depend on their ability to act both as acids and bases. (See the review of these concepts in Section 6-7.) Thus, deprotonation gives alkoxide ions. We shall see how structural features affect their pK_a values. The lone electron pairs on oxygen render alcohols basic as well, and protonation results in alkyloxonium ions.

The acidity of alcohols resembles that of water

The acidity of alcohols in water is expressed by the equilibrium constant K.

$$\mathrm{ROH} + \mathrm{H_2O} \overset{K}{\rightleftharpoons} \mathrm{H_3O^+} + \underset{\textbf{Alkoxide ion}}{\mathrm{RO^-}}$$

Making use of the constant concentration of water (Section 6-7), we derive a new equilibrium constant K_a.

$$K_a = K[\mathrm{H_2O}] = \frac{[\mathrm{H_3O^+}][\mathrm{RO^-}]}{[\mathrm{ROH}]} \text{ mol L}^{-1}, \text{ and } \mathrm{p}K_a = -\log K_a$$

Table 8-2 lists the pK_a values of several alcohols and related compounds. A comparison of these values with those given in Table 6-5 for mineral and other strong acids shows that alcohols, like water, are fairly weak acids. Their acidity is far greater, however, than that of alkanes and haloalkanes.

Why are alcohols acidic, whereas alkanes and haloalkanes are not? The answer lies in the relatively strong electronegativity of the oxygen to which the proton is attached, which stabilizes the negative charge of the alkoxide ion.

Strong bases are needed to convert alcohols into their conjugate bases

To drive the equilibrium between alcohol and alkoxide to the side of the conjugate base, it is necessary to use a base *stronger* than the alkoxide formed (i.e., a base derived from a conjugate acid *weaker* than the alcohol). An example is the reaction of sodium amide, $NaNH_2$, with methanol to furnish sodium methoxide and ammonia.

$$\underset{\mathrm{p}K_a = 15.5}{\mathrm{CH_3OH}} + \underset{\textbf{Sodium amide}}{\mathrm{Na^+\ {}^-NH_2}} \overset{K}{\rightleftharpoons} \underset{\textbf{Sodium methoxide}}{\mathrm{Na^+\ {}^-OCH_3}} + \underset{\mathrm{p}K_a = 35}{\mathrm{NH_3}}$$

This equilibrium lies well to the right ($K \sim 10^{35-15.5} = 10^{19.5}$), because methanol is a much stronger acid than ammonia, or, conversely, because amide is a much stronger base than methoxide. The next chapter describes the uses of strong bases in synthetic applications of alkoxides.

TABLE 8-2 pK_a Values of Alcohols and Related Compounds in Water

Compound	pK_a	Compound	pK_a
H_2O	15.7	HOCl	7.53
CH_3OH	15.5	$ClCH_2CH_2OH$	14.3
CH_3CH_2OH	15.9	CF_3CH_2OH	12.4
$(CH_3)_2CHOH$	17.1	$CF_3CH_2CH_2OH$	14.6
$(CH_3)_3COH$	18	$CF_3CH_2CH_2CH_2OH$	15.4
H_2O_2	11.64		

It is often sufficient to generate alkoxides in equilibrium concentrations. For this purpose, we may add an alkali metal hydroxide to the alcohol.

$$\underset{pK_a = 15.9}{CH_3CH_2OH} + Na^+ \ {}^-OH \xrightleftharpoons{K} CH_3CH_2O^-Na^+ + \underset{pK_a = 15.7}{H_2O}$$

With this base present, approximately one-half of the alcohol will exist as the alkoxide (assuming equimolar concentrations of starting materials).

EXERCISE 8-3

Which of the following bases are strong enough to cause essentially complete deprotonation of methanol? The pK_a of the conjugate acid is given in parentheses.
(a) KCN (9.2); **(b)** $CH_3CH_2CH_2CH_2Li$ (50); **(c)** CH_3CO_2Na (4.7); **(d)** $LiN[CH(CH_3)_2]_2$ (LDA, 40); **(e)** KH (38); **(f)** CH_3SNa (10).

Steric hindrance and inductive effects control the acidity of alcohols

Table 8-2 shows a large variation in the pK_a values of the alcohols. A closer look at the first column reveals that the acidity decreases (pK_a increases) from methanol to primary, secondary, and finally tertiary systems.

Relative pK_a Values of Alcohols (in Solution)

$$\underset{\text{Strongest acid}}{CH_3OH} < \text{primary} < \text{secondary} < \underset{\text{Weakest acid}}{\text{tertiary}}$$

This ordering has been ascribed to steric hindrance to solvation and to hydrogen bonding in the alkoxide. Because solvation and hydrogen bonding stabilize the negative charge on oxygen, interference with these processes leads to an increase in pK_a.

The second column in Table 8-2 reveals another contribution to the pK_a of alcohols: The presence of halogens increases acidity. Recall that the carbon of the C–X bond is positively polarized as a result of the high electronegativity of X (Section 1-3). Electron withdrawal by the halogen also causes atoms further away to be slightly positively charged. This phenomenon of transmission of charge through a chain of atoms is called an **inductive effect,** and it stabilizes the negative charge on the alkoxide oxygen by electrostatic attraction. The inductive effect in alcohols increases with the number of electronegative groups but decreases with distance from the oxygen.

EXERCISE 8-4

Rank the following alcohols in order of increasing acidity.

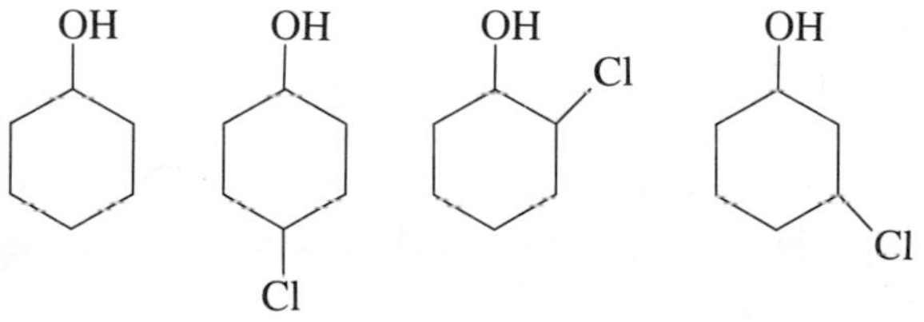

OH OH OH OH
Cl
Cl Cl

EXERCISE 8-5

Which side of the following equilibrium reaction is favored (assuming equimolar concentrations of starting materials)?

$$(CH_3)_3CO^- + CH_3OH \rightleftharpoons (CH_3)_3COH + CH_3O^-$$

The lone electron pairs on oxygen make alcohols basic

Alcohols may also be basic, although weakly so. Very strong acids are required to protonate the OH group, as indicated by the low pK_a values (strong acidity) of their conjugate acids, the alkyloxonium ions (Table 8-3). Molecules that may be both acids and bases are called **amphoteric** (*ampho*, Greek, both).

The amphoteric nature of the hydroxy functional group characterizes the chemical reactivity of alcohols. In strong acids, they exist as alkyloxonium ions, in neutral media as alcohols, and in strong bases as alkoxides.

Alcohols Are Amphoteric

$$R{-}\overset{+}{O}H_2 \underset{\text{Mild base}}{\overset{\text{Strong acid}}{\rightleftharpoons}} ROH \underset{\text{Mild acid}}{\overset{\text{Strong base}}{\rightleftharpoons}} RO^-$$

Alkyloxonium ion | **Alcohol** | **Alkoxide ion**

In summary, alcohols are amphoteric. They are acidic by virtue of the electronegativity of the oxygen and are converted into alkoxides by strong bases. In solution the steric bulk of branching inhibits solvation of the alkoxide, thereby raising the pK_a of the corresponding alcohol. Electron-withdrawing substituents close to the functional group lower the pK_a. Alcohols are also weakly basic and can be protonated by strong acids to furnish alkyloxonium ions.

8-4 Industrial Sources of Alcohols: Carbon Monoxide and Ethene

Let us now turn to the preparation of alcohols. We start in this section with methods of special importance in industry. The sections that follow will take up procedures that are used more generally in the synthetic laboratory to introduce the hydroxy functional group into a wide range of organic molecules.

Methanol is made on a large scale (more than 1 billion gallons, or 4 billion liters, in the United States in 1992) from a pressurized mixture of CO and H_2 called **synthesis gas.** The reaction makes use of a catalyst consisting of copper, zinc oxide, and chromium(III) oxide.

$$CO + 2\,H_2 \xrightarrow{\text{Cu-ZnO-}Cr_2O_3\text{, 250°C, 50–100 atm}} CH_3OH$$

Changing the catalyst to rhodium or ruthenium leads to 1,2-ethanediol (ethylene glycol), an important industrial chemical that is the principal component of automobile antifreeze.

TABLE 8-3 pK_a Values of Four Protonated Alcohols

Compound	pK_a
$CH_3\overset{+}{O}H_2$	−2.2
$CH_3CH_2\overset{+}{O}H_2$	−2.4
$(CH_3)_2CH\overset{+}{O}H_2$	−3.2
$(CH_3)_3C\overset{+}{O}H_2$	−3.8

$$2\ CO + 3\ H_2 \xrightarrow{\text{Rh or Ru, pressure, heat}} \underset{\text{OH}}{CH_2}\text{—}\underset{\text{OH}}{CH_2}$$

1,2-Ethanediol (Ethylene glycol)

Other reactions that would permit the selective formation of a given alcohol from synthesis gas are the focus of much current research, because the gasification of coal in the presence of water provides a ready supply of mixtures of CO and H_2.

$$\text{Coal} \xrightarrow{\text{Air, } H_2O,\ \Delta} x\ CO + y\ H_2$$

The abundant coal supplies in the United States and other parts of the world could then be used as a source of liquid fuels, industrial raw materials, and feedstocks.

Another catalytic reaction of synthesis gas furnishes alcohols only as by-products: the cobalt- or iron-mediated formation of hydrocarbons. This reaction, which was discovered at about the turn of the century, was developed in Germany beginning in the 1920s to provide fuel and oils. It enabled that country to satisfy its energy needs (particularly gasoline) during the Second World War, when its supply of petroleum was virtually shut off. The process is known as the **Fischer-Tropsch reaction.**

$$n\ CO + (2n + 1)\ H_2 \xrightarrow{\text{Co or Fe, pressure, 200–350°C}} C_nH_{2n+2} + n\ H_2O$$

At the height of production, in 1943, more than 500,000 tons (450 million kg) of hydrocarbon and other products (gasoline, diesel fuel, oils, waxes, and detergents) were made in Germany by this process.

Ethanol is prepared in large quantities by fermentation of sugars or by the phosphoric acid-catalyzed hydration of ethene (ethylene). The United States produced about 700 million pounds (320 million kg) of ethanol in 1992.

$$CH_2\text{=}CH_2 + HOH \xrightarrow{H_3PO_4,\ 300°C} \underset{\text{H}}{CH_2}\text{—}\underset{\text{OH}}{CH_2}$$

In summary, the industrial preparation of methanol and 1,2-ethanediol proceeds by reduction of carbon monoxide with hydrogen, that of ethanol by the acid-catalyzed hydration of ethene (ethylene).

8-5 Synthesis of Alcohols by Nucleophilic Substitution

In the laboratory we can prepare alcohols from a wide variety of starting materials. For example, conversions of haloalkanes into alcohols by S_N2 and S_N1 processes were described in Chapters 6 and 7. These methods are not as widely used as one might think, however, because the required halides are often only accessible from the corresponding alcohols (Chapter 9). They also suffer from the usual drawbacks of nucleophilic substitution: Bimolecular elimination can be

BOX 8-1 Synthesis of Alcohols from Acetates

The preparation of alcohols by the reaction of haloalkanes with hydroxide ion can be complicated by competing eliminations. The latter may be minimized in some cases by employing a less basic oxygen nucleophile, such as acetate (Section 7-8). The resulting alkyl acetate can then be converted into the desired alcohol by aqueous hydroxide.

$$\underset{\textbf{Acetic acid}}{H{-}O\overset{O}{\overset{\|}{C}}CH_3} \qquad \underset{\textbf{Alkyl acetate}}{R{-}O\overset{O}{\overset{\|}{C}}CH_3}$$

The second step involves breaking the carbonyl oxygen bond, thereby leaving the alkoxy residue unchanged. We shall discuss this reaction, known as *ester hydrolysis,* in Chapter 20. This approach has been used to make (*R*)-2-hydroxycarboxylic acids, which, unlike their enantiomers, are not readily available from nature but are extremely useful in studies of enzyme behavior. The key step is the displacement of the halide in an (*S*)-2-halocarboxylic ester with inversion.

The starting material is obtained from the naturally abundant (*S*)-2-aminocarboxylic acid.

STEP 1. Acetate formation (S_N2 reaction)

$$\underset{\textbf{1-Bromo-3-methylpentane}}{CH_3CH_2\overset{CH_3}{\overset{|}{C}}HCH_2CH_2Br} + CH_3\overset{O}{\overset{\|}{C}}O^-Na^+ \xrightarrow{\text{DMF, 80°C}} \underset{\substack{95\% \\ \textbf{3-Methylpentyl acetate}}}{CH_3CH_2\overset{CH_3}{\overset{|}{C}}HCH_2CH_2O\overset{O}{\overset{\|}{C}}CH_3} + Na^+Br^-$$

STEP 2. Conversion to the alcohol (ester hydrolysis)

$$CH_3CH_2\overset{CH_3}{\overset{|}{C}}HCH_2CH_2O{\wr}\overset{O}{\overset{\|}{C}}CH_3 + Na^{+-}OH \xrightarrow[-CH_3\overset{O}{\overset{\|}{C}}O^-Na^+]{H_2O} \underset{\substack{85\% \\ \textbf{3-Methyl-1-pentanol}}}{CH_3CH_2\overset{CH_3}{\overset{|}{C}}HCH_2CH_2OH}$$

Synthesis of an Alcohol by a Displacement-Hydrolysis Sequence

$$\underset{2S}{\text{Cl}-\overset{\overset{O=C-OC(CH_3)_3}{|}}{C}-\text{H}}\text{ (}CH_2C_6H_5\text{)} \xrightarrow[S_N2]{R\overset{O}{\overset{\|}{C}}O^{-}\ {}^{+}Cs,\ DMF} \underset{\substack{2R \\ 65\%}}{\text{H}-\overset{\overset{O=C-OC(CH_3)_3}{|}}{C}-O\overset{\|}{\underset{O}{C}}R}\text{ (}CH_2C_6H_5\text{)} \xrightarrow[\text{Ester hydrolysis}]{\text{1. NaOH, }H_2O\text{ 2. HCl, }CH_2Cl_2} \underset{\substack{2R \\ 84\%}}{\text{H}-\overset{\overset{O=C-OH}{|}}{C}-OH}\text{ (}CH_2C_6H_5\text{)}$$

R– = phenyl

(*R*)-2-hydroxy-3-phenylpropanoic acid

a major side reaction of hindered systems, and tertiary halides form carbocations that may undergo E1 reactions. Nevertheless, a number of alcohol syntheses have made use of nucleophilic substitution reactions, two of which are shown here.

Alcohols by Nucleophilic Substitution

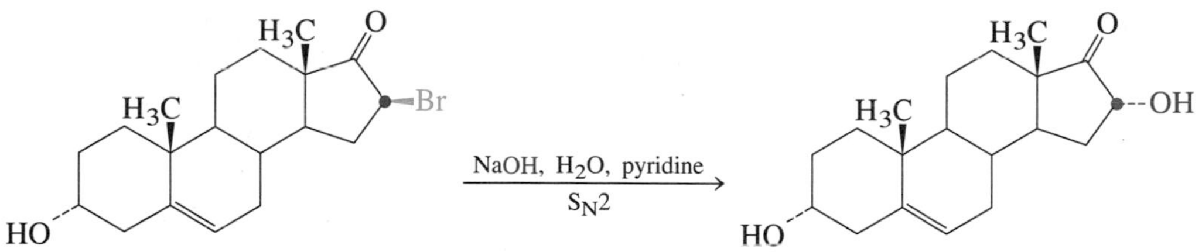

H_3C O H_3C Br HO — NaOH, H_2O, pyridine / S_N2 → H_3C O H_3C OH HO

95%

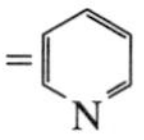

Steroid found in human umbilical cord blood

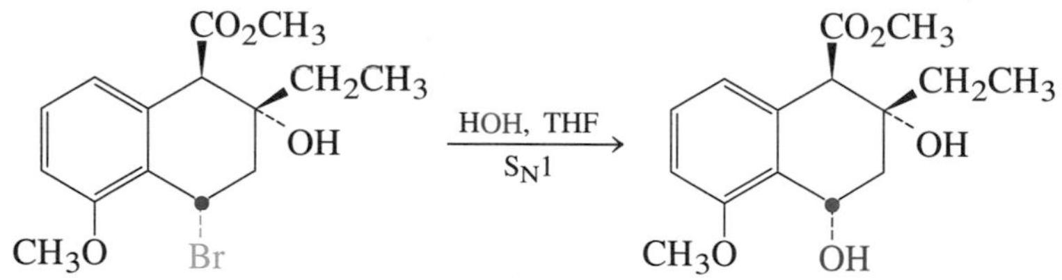

Pyridine =

CO_2CH_3 CH_2CH_3 OH CH_3O Br — HOH, THF / S_N1 → CO_2CH_3 CH_2CH_3 OH CH_3O OH

86%

Precursor for syntheses of antitumor antibiotics

THF = = oxacyclopentane (tetrahydrofuran)

In the first example, stereochemical inversion was achieved by employing S_N2 conditions. The second reaction follows the S_N1 mechanism (solvolysis). Stereoselectivity in this case is observed because the bottom face of the molecule is considerably less sterically hindered than the top. Both nucleophilic substitutions use organic **cosolvents** (pyridine and THF, respectively) to improve the solubility of the substrates.

EXERCISE 8-6

Show how you might convert the following haloalkanes into alcohols.
(a) Bromoethane; **(b)** chlorocyclohexane; **(c)** 3-chloro-3-methylpentane.

In summary, alcohols may be prepared from haloalkanes by nucleophilic substitution, provided the haloalkane is readily available and side reactions such as elimination do not interfere.

8-6 Oxidation–Reduction Relationship Between Alcohols and Carbonyl Compounds

This section describes an important synthesis of alcohols: reduction of aldehydes and ketones. Later we shall see that these compounds also may be converted to alcohols with concomitant formation of a new carbon–carbon bond. Because of the versatility of aldehydes and ketones in synthesis, we shall also illustrate their preparation by alcohol oxidation.

Oxidation and reduction have special meanings in organic chemistry

We can readily recognize inorganic oxidation and reduction processes as the loss and gain of electrons, respectively. With organic compounds, it is often less clear whether electrons are being gained or lost in a reaction. Hence, organic chemists find it more useful to define oxidation and reduction in other terms. A process that adds electronegative atoms such as halogen or oxygen to, or removes hydrogen from, a molecule constitutes an **oxidation;** conversely, the removal of oxygen or the addition of hydrogen is defined as **reduction.**

This definition of an oxidation–reduction relationship allows us to connect alcohols to aldehydes and ketones. Addition of two hydrogen atoms to the double bond of a carbonyl group constitutes reduction to the corresponding alcohol. Aldehydes give primary alcohols, ketones, secondary alcohols. The reverse process, removal of hydrogen to furnish carbonyl compounds, is an example of oxidation.

The Redox Relationship Between Alcohols and Carbonyl Compounds

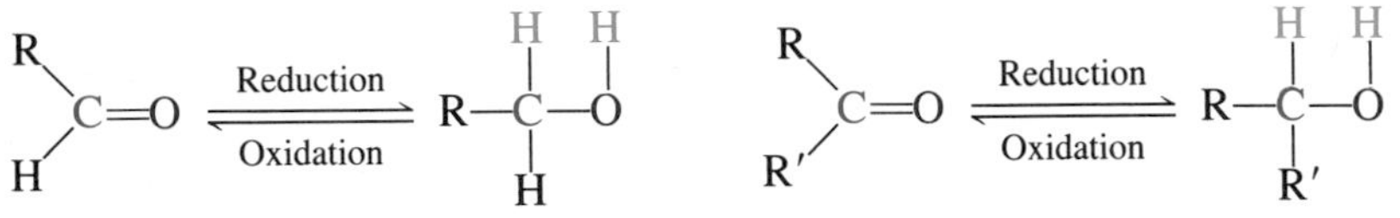

$$\underset{\textbf{Aldehyde}}{RHC{=}O} \underset{\text{Oxidation}}{\overset{\text{Reduction}}{\rightleftharpoons}} \underset{\textbf{Primary alcohol}}{R{-}CH_2{-}O{-}H} \qquad \underset{\textbf{Ketone}}{RR'C{=}O} \underset{\text{Oxidation}}{\overset{\text{Reduction}}{\rightleftharpoons}} \underset{\textbf{Secondary alcohol}}{R{-}CHR'{-}O{-}H}$$

Reduction of carbonyl compounds can be carried out either by addition of molecular hydrogen or by exposure to hydride reagents.

Catalytic hydrogenation converts aldehydes and ketones to alcohols

The addition of gaseous hydrogen to a double bond, called **hydrogenation,** requires a catalyst and, in many cases, must be carried out under high pressure in order to proceed at a useful rate. Most such hydrogenations utilize **heterogeneous catalysts** (*héteros,* Greek, other; *génos,* Greek, kind) that are insoluble in the reaction solvent. The reaction takes place on the surface of the suspended particles, which typically consist of finely divided metals such as platinum, palladium, or nickel, often deposited on a supporting material such as carbon, to maximize the surface area. Catalytic hydrogenation is also an important reaction of alkenes; its mechanism will be discussed in Chapter 12.

Hydrogenation of an Aldehyde

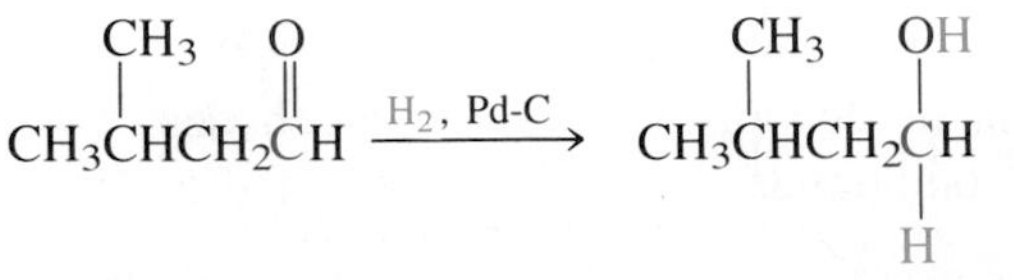

$$\underset{\textbf{3-Methylbutanal}}{CH_3CH(CH_3)CH_2CHO} \xrightarrow{H_2,\ \text{Pd-C}} \underset{\textbf{3-Methyl-1-butanol}}{CH_3CH(CH_3)CH_2CH_2OH}$$

Hydrogenation of a Ketone

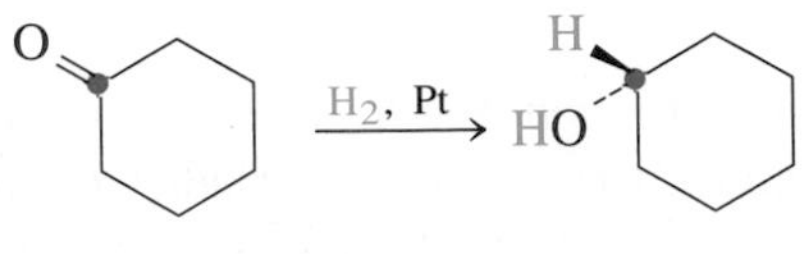

$$\textbf{Cyclohexanone} \xrightarrow{H_2,\ \text{Pt}} \textbf{Cyclohexanol}$$

Hydride reagents also reduce the carbonyl group

The electrons in the **carbonyl group,** C=O, are not distributed evenly between the two component atoms. Because oxygen is more electronegative than carbon, the carbon of a carbonyl group is electrophilic, the oxygen nucleophilic. This polarization can be represented by a charge-separated resonance form.

Polar Character of the Carbonyl Function

$$\overset{\delta^+}{C}{=}\overset{\delta^-}{\ddot{O}} \qquad \left[C{=}\ddot{\underset{..}{O}} \longleftrightarrow \overset{+}{C}{-}\ddot{\underset{..}{O}}{:}^- \right]$$

Because the carbonyl carbon is electrophilic, nucleophilic H^- may be delivered to it by **hydride reagents.** Two such species are sodium borohydride, $NaBH_4$, and lithium aluminum hydride, $LiAlH_4$. These two compounds possess the advantage of higher solubility in common organic solvents than do simpler analogs such as LiH and NaH.

Hydride Reductions of Aldehydes and Ketones

$$RCHO + NaBH_4 \xrightarrow{CH_3CH_2OH} R{-}CH(OH){-}H$$

$$RC(O)R' + LiAlH_4 \xrightarrow{(CH_3CH_2)_2O} \xrightarrow{\text{HOH work-up}} R{-}C(OH)(H){-}R'$$

Hydride reagents arise from the addition of the respective alkali metal hydride to borane, BH_3, or alane, AlH_3, thus giving the central atom an electron octet. Like methane and the ammonium ion, NH_4^+, both BH_4^- and AlH_4^- ions are tetrahedral, with four equivalent hydrogens.

$$H_2B{:}H \xrightarrow{:H^-Na^+} [H{:}BH_2{:}H]^- \; Na^+$$

$$H_2Al{:}H \xrightarrow{:H^-Li^+} [H{:}AlH_2{:}H]^- \; Li^+$$

The chemistry of these reagents is dominated by the hydridic (H^-) character of the hydrogen atoms. Reduction of a carbonyl group is achieved by addition of a hydride to carbon and a proton to oxygen.

Hydride Reductions of Aldehydes and Ketones

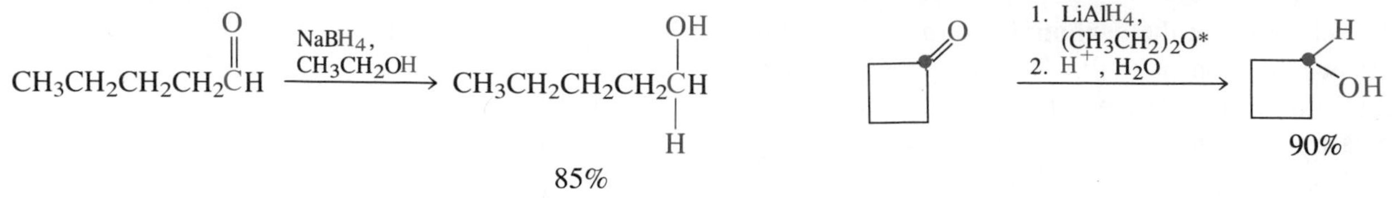

$$\underset{\textbf{Pentanal}}{CH_3CH_2CH_2CH_2\overset{O}{\overset{\|}{C}}H} \xrightarrow[CH_3CH_2OH]{NaBH_4,} \underset{\substack{85\% \\ \textbf{1-Pentanol}}}{CH_3CH_2CH_2CH_2\overset{OH}{\underset{H}{C}}H}$$

$$\textbf{Cyclobutanone} \xrightarrow[\text{2. } H^+,\ H_2O]{\text{1. } LiAlH_4,\ (CH_3CH_2)_2O^*} \underset{90\%}{\textbf{Cyclobutanol}}$$

EXERCISE 8-7

Formulate all of the expected products of $NaBH_4$ reduction of the following compounds. (Hint: Remember the possibility of stereoisomerism.)

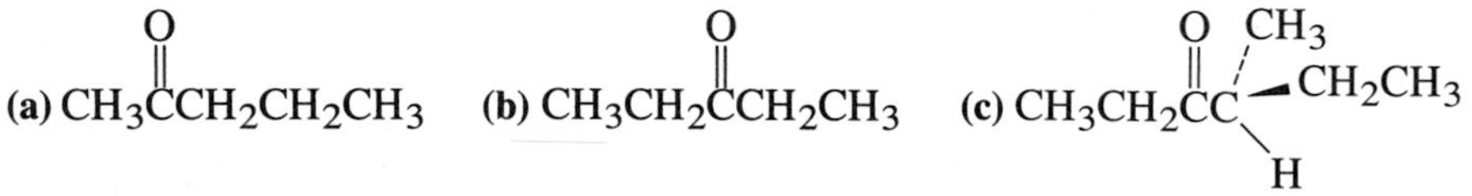

(a) $CH_3\overset{O}{\overset{\|}{C}}CH_2CH_2CH_3$ **(b)** $CH_3CH_2\overset{O}{\overset{\|}{C}}CH_2CH_3$ **(c)** $CH_3CH_2\overset{O}{\overset{\|}{C}}C(CH_3)(H)CH_2CH_3$

EXERCISE 8-8

Hydride reductions are often highly stereoselective, with the delivery of hydrogen from the less hindered side of the substrate molecule. Predict the likely stereochemical outcome of the treatment of compound A with $NaBH_4$.

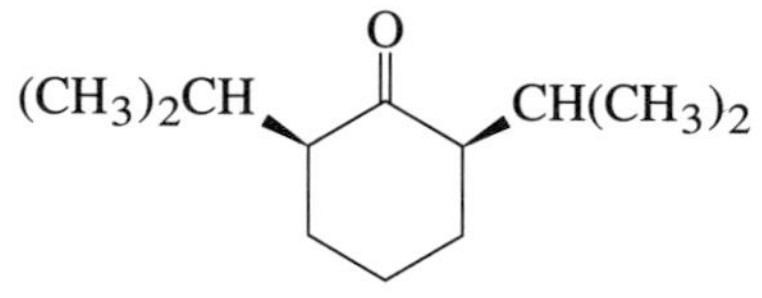

2,6-di[$(CH_3)_2CH$ / $CH(CH_3)_2$]cyclohexanone

A

Although free hydride ion is a powerful base that is immediately protonated by protic solvents (see Exercise 8-3e), attachment to boron in BH_4^- moderates its reactivity considerably, thus allowing $NaBH_4$ to be used in solvents such as ethanol. When an aldehyde or a ketone is exposed to borohydride, it donates an H^- to the carbonyl carbon, with simultaneous protonation of the carbonyl oxygen by the solvent. The ethoxide by-product combines with the remaining boron fragment, giving ethoxyborohydride.

Mechanism of $NaBH_4$ Reduction

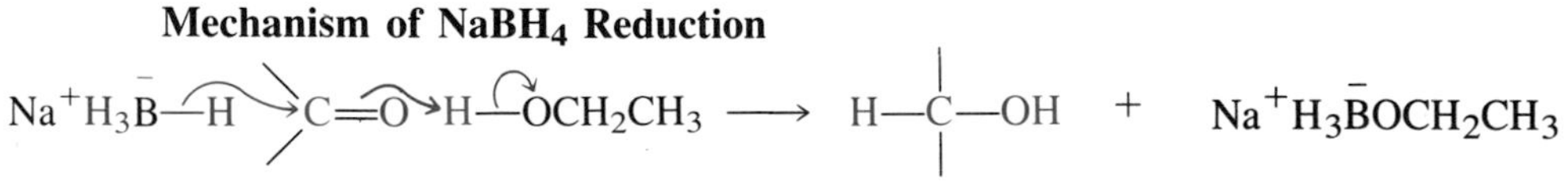

$$Na^+H_3\bar{B}{-}H \quad {>}C{=}O \quad H{-}OCH_2CH_3 \longrightarrow H{-}\overset{|}{\underset{|}{C}}{-}OH \quad + \quad Na^+H_3\bar{B}OCH_2CH_3$$

Ethanol solvent **Product alcohol** **Sodium ethoxyborohydride**

The resulting ethoxyborohydride may attack three more carbonyl substrates before all the hydride atoms of the original reagent have been used up. As a result, one equivalent of borohydride is capable of reducing *four* equivalents of aldehyde

---

*The numbers refer to reagents that are used *sequentially*. Thus, the first reaction involves treatment of the substrate to the left of the arrow with the reagents listed after the number 1. The product of this transformation then undergoes a reaction with the reagents listed after the number 2, and so on. The last reaction gives the final product shown on the right.

or ketone to alcohol. The boron reagent is finally converted into tetra-ethoxyborate, $^{-}B(OCH_2CH_3)_4$.

Lithium aluminum hydride is more reactive than sodium borohydride. Its hydrogens are much more basic and are attacked vigorously by water and alcohols to give hydrogen gas. Reductions utilizing lithium aluminum hydride are therefore carried out in aprotic solvents, such as ethoxyethane (diethyl ether).

Reaction of Lithium Aluminum Hydride with Protic Solvents

$$LiAlH_4 + 4\ CH_3OH \xrightarrow{\text{Fast}} LiAl(OCH_3)_4 + 4\ H—H\uparrow$$

Addition of lithium aluminum hydride to an aldehyde or ketone furnishes initially alkoxyaluminum hydride, which continues to deliver H^- to three more carbonyl groups, thus reducing a total of four equivalents of aldehyde or ketone. Addition of aqueous acid (aqueous, or HOH, work-up) consumes excess reagent, hydrolyzes the tetraalkoxyaluminate, and releases the product alcohol.

Mechanism of $LiAlH_4$ Reduction

$$Li^+\ H_3\bar{A}l—H \quad \rangle C=O \longrightarrow H—\overset{|}{\underset{|}{C}}—O\bar{A}lH_3\ Li^+ \xrightarrow[\text{React with three more } \rangle C=O]{\text{Repeat three times:}}$$

Lithium alkoxyaluminum hydride

$$(H—\overset{|}{\underset{|}{C}}—O)_4Al^-\ Li^+ \xrightarrow{\text{HOH work-up}} 4\ H—\overset{|}{\underset{|}{C}}—OH + Al(OH)_3 + LiOH$$

Lithium tetraalkoxy-aluminate | **Product alcohol**

EXERCISE 8-9

Formulate reductions that would give rise to the following alcohols. **(a)** 1-Decanol; **(b)** 4-methyl-2-pentanol; **(c)** cyclopentylmethanol; **(d)** 1,4-cyclohexanediol.

Chromium reagents oxidize alcohols to carbonyl compounds

We have seen several methods for synthesizing alcohols from aldehydes and ketones by reduction with hydrogen or hydride reagents. The reverse is also possible: Alcohols may be oxidized to produce aldehydes and ketones. A useful reagent for this purpose is a transition metal in a high oxidation state: chromium(VI). In this form, chromium has a yellow-orange color. On exposure to an alcohol, the Cr(VI) species is reduced to the deep green Cr(III). The reagent is usually supplied as a dichromate salt ($K_2Cr_2O_7$ or $Na_2Cr_2O_7$) or as CrO_3. Oxidation of secondary alcohols to ketones is often carried out in aqueous acid solution.

Oxidation of a Secondary Alcohol to a Ketone with Aqueous Cr(VI)

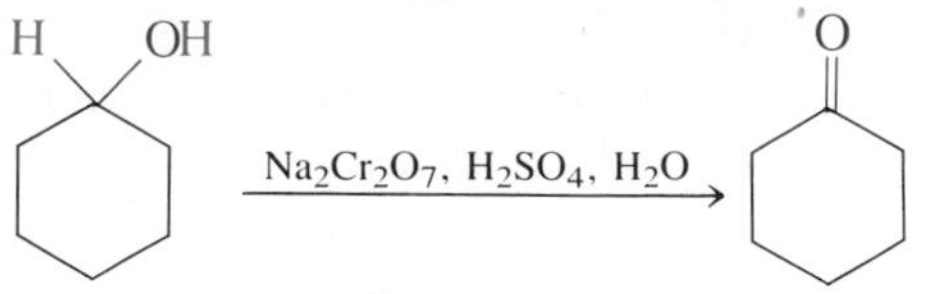

$$\text{cyclohexanol} \xrightarrow{Na_2Cr_2O_7,\ H_2SO_4,\ H_2O} \text{cyclohexanone}\quad 96\%$$

$$CH_3CH_2CH_2OH \xrightarrow{K_2Cr_2O_7,\ H_2SO_4,\ H_2O} CH_3CH_2CH{=}O$$

Propanal

$$\xrightarrow{\text{Overoxidation}} CH_3CH_2C({=}O)OH$$

Propanoic acid

EXERCISE 8-10

Write a balanced equation for the redox process shown above. The inorganic products are $Cr_2(SO_4)_3$ and Na_2SO_4.

Under these conditions, primary alcohols tend to *overoxidize* to a carboxylic acid, as shown in the margin for 1-propanol. In the absence of water, however, aldehydes are not susceptible to overoxidation. Therefore, a water-free form of Cr(VI) has been developed by reaction of CrO_3 with HCl, followed by the addition of the organic base pyridine. The result is the oxidizing agent **pyridinium chlorochromate,** abbreviated as $pyH^+\ CrO_3Cl^-$ or just **PCC** (margin), which gives excellent yields of aldehydes upon exposure to primary alcohols in dichloromethane solvent.

$$\text{C}_5\text{H}_5\overset{+}{\text{N}}\text{H}\ CrO_3Cl^-$$

Pyridinium chlorochromate (PCC or $pyH^+\ CrO_3Cl^-$)

PCC Oxidation of a Primary Alcohol to an Aldehyde

$$CH_3CH_2CH_2CH_2CH_2CH_2CH_2CH_2CH_2CH_2OH \xrightarrow{pyH^+\ CrO_3Cl^-,\ CH_2Cl_2} CH_3CH_2CH_2CH_2CH_2CH_2CH_2CH_2CH_2CH{=}O$$

92%

Chromic esters are intermediates in alcohol oxidation

What is the mechanism of the chromium(VI) oxidation of alcohols? The first step is formation of an intermediate called a **chromic ester;** the oxidation state of chromium stays unchanged in this process.

Chromic Ester Formation from an Alcohol

$$RCH_2\ddot{O}H + H\ddot{O}{-}\overset{:O:}{\underset{:O:}{\overset{\|}{\underset{\|}{Cr^{VI}}}}}{-}\ddot{O}H \rightleftharpoons RCH_2\ddot{O}{-}\overset{:O:}{\underset{:O:}{\overset{\|}{\underset{\|}{Cr^{VI}}}}}{-}\ddot{O}H + H_2\ddot{O}$$

Chromic acid **Chromic ester**

The next step in alcohol oxidation is equivalent to an E2 reaction. Here water (or pyridine, in the case of PCC) acts as a mild base, removing the proton next to the alcohol oxygen; $HCrO_3^-$ functions as a leaving group. The donation of an electron pair to chromium changes its oxidation state by two units, yielding Cr(IV).

Aldehyde Formation from a Chromic Ester

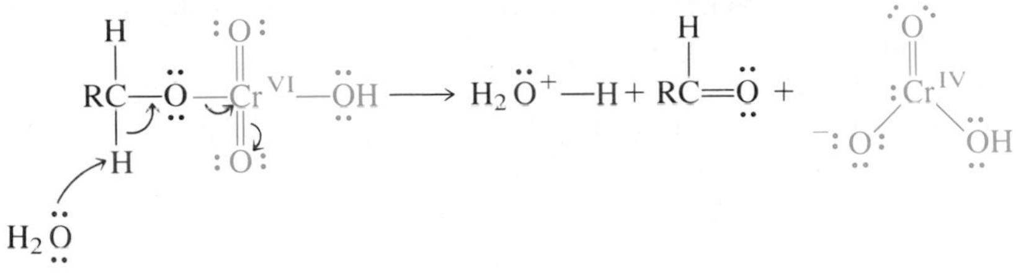

$$\mathrm{RC(H)(H)-\ddot{O}-Cr^{VI}(=O)_2-\ddot{O}H} + H_2\ddot{O} \longrightarrow H_2\ddot{O}^+{-}H + \mathrm{RC(H){=}\ddot{O}} + {:}\mathrm{Cr^{IV}(=O)(O^-)(\ddot{O}H)}$$

In contrast with the kinds of E2 reactions discussed so far, this elimination furnishes a carbon–oxygen instead of a carbon–carbon double bond. The Cr(IV) species formed disproportionates to Cr(III) and Cr(V), which may function as an oxidizing agent independently. Eventually all Cr(VI) is reduced to Cr(III).

Tertiary alcohols cannot be readily oxidized by chromate because they do not carry hydrogens next to the functional group.

EXERCISE 8-11

Formulate a synthesis of each of the following carbonyl compounds from the corresponding alcohol.

(a) $CH_3CH_2\overset{O}{\overset{\|}{C}}CH(CH_3)_2$ **(b)** 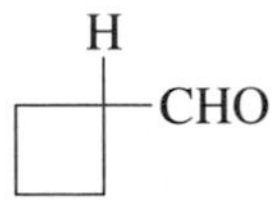 cyclobutane–C(H)–CHO **(c)** CH_3CH_2–cyclohexyl–$\overset{O}{\overset{\|}{C}}CH_3$

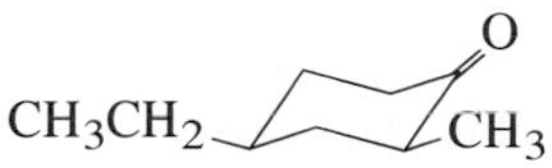

BOX 8-2 Biological Oxidation and Reduction

Alcohols are metabolized by oxidation to carbonyl compounds. In biological systems ethanol is converted to acetaldehyde by the cationic oxidizing agent *nicotinamide adenine dinucleotide* (abbreviated as NAD^+; for its structure see Chapter 25). The process is catalyzed by the enzyme *alcohol dehydrogenase*. (The latter also catalyzes the reverse process, reduction of aldehydes and ketones to alcohols; see Problems 27 and 28 at the end of this chapter.) By subjecting the two enantiomers of 1-deuterioethanol to the enzyme, the biochemical oxidation was found to be stereospecific, the NAD^+ removing only the hydrogen marked below by the solid arrowhead from C1 of the alcohol.

Other alcohols are similarly oxidized biochemically. The relatively high toxicity of methanol (''wood alcohol'') is due largely to its oxidation to formaldehyde. The latter specifically interferes with a system responsible for the transfer of one-carbon fragments between nucleophilic sites in biomolecules.

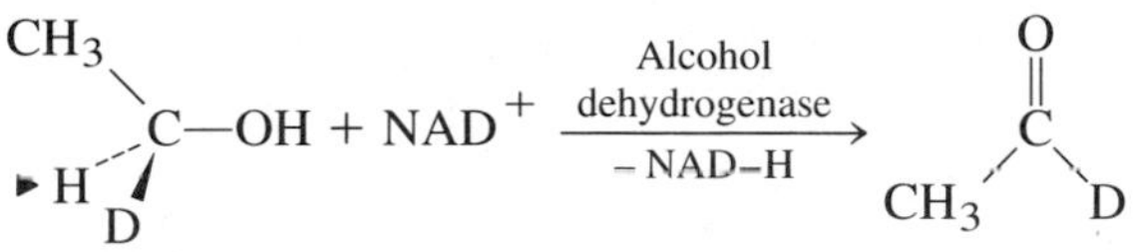

$$\mathrm{CH_3C(H)(D)OH} + NAD^+ \xrightarrow[\text{– NAD–H}]{\text{Alcohol dehydrogenase}} \mathrm{CH_3C(=O)D}$$

(S)-1-Deuterioethanol

$$\mathrm{CH_3C(D)(H)OH} + NAD^+ \xrightarrow[\text{– NAD–D}]{\text{Alcohol dehydrogenase}} \mathrm{CH_3C(=O)H}$$

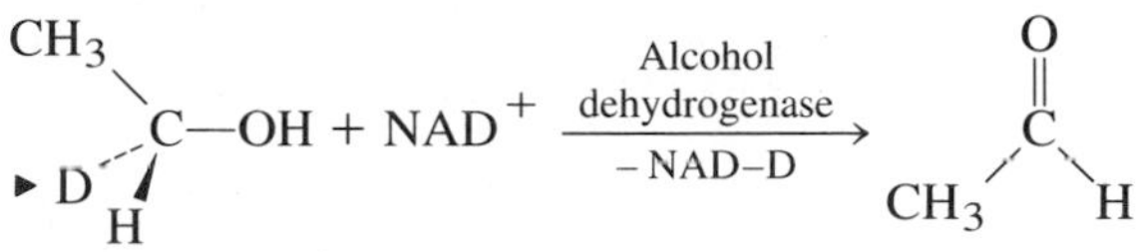

(R)-1-Deuterioethanol

To summarize, reductions of aldehydes and ketones constitute general syntheses of primary and secondary alcohols, respectively. Either catalytic hydrogenation or hydride reagents may be employed in these processes. The reverse reactions, oxidations of primary alcohols to aldehydes and secondary alcohols to ketones, are achieved with chromium(VI) reagents. Use of pyridinium chlorochromate (PCC) prevents overoxidation of aldehydes.

8-7 Organometallic Reagents: Sources of Nucleophilic Carbon

The reduction of aldehydes and ketones with hydride reagents is a useful way of synthesizing alcohols. This approach would be even more powerful if, instead of hydride, we could use a source of *nucleophilic carbon*. Attack by a carbon nucleophile on a carbonyl group would give an alcohol and simultaneously form a carbon–carbon bond. *We would thus have constructed a product with more carbon atoms and, therefore, more complexity than the starting materials.*

To achieve such transformations, we need to find a way of making carbon-based nucleophiles, R^-. This section will describe how this goal can be reached. Metals, particularly lithium and magnesium, act on haloalkanes to generate new compounds, called **organometallic reagents,** in which a carbon atom of an organic group is bound to a metal. These species are strong bases, and, as we shall see in the subsequent section, nucleophiles, and they are extremely useful in organic syntheses.

Alkyllithium and alkylmagnesium reagents are prepared from haloalkanes

Organometallic compounds of lithium and magnesium are most conveniently prepared by direct reaction of a haloalkane with the metal suspended in ethoxyethane (diethyl ether) or oxacyclopentane (tetrahydrofuran, THF). The reactivity of the haloalkanes increases in the order Cl < Br < I; fluorides are not normally used as starting materials in these reactions.

Alkyllithium Synthesis

$$CH_3Br + 2\ Li \xrightarrow{(CH_3CH_2)_2O,\ 0\text{–}10°C} CH_3Li + LiBr$$

Methyllithium

Alkylmagnesium (Grignard) Synthesis

$$(CH_3)_2CHI + Mg \xrightarrow{THF,\ 20°C} (CH_3)_2CHMgI$$

1-Methylethylmagnesium iodide

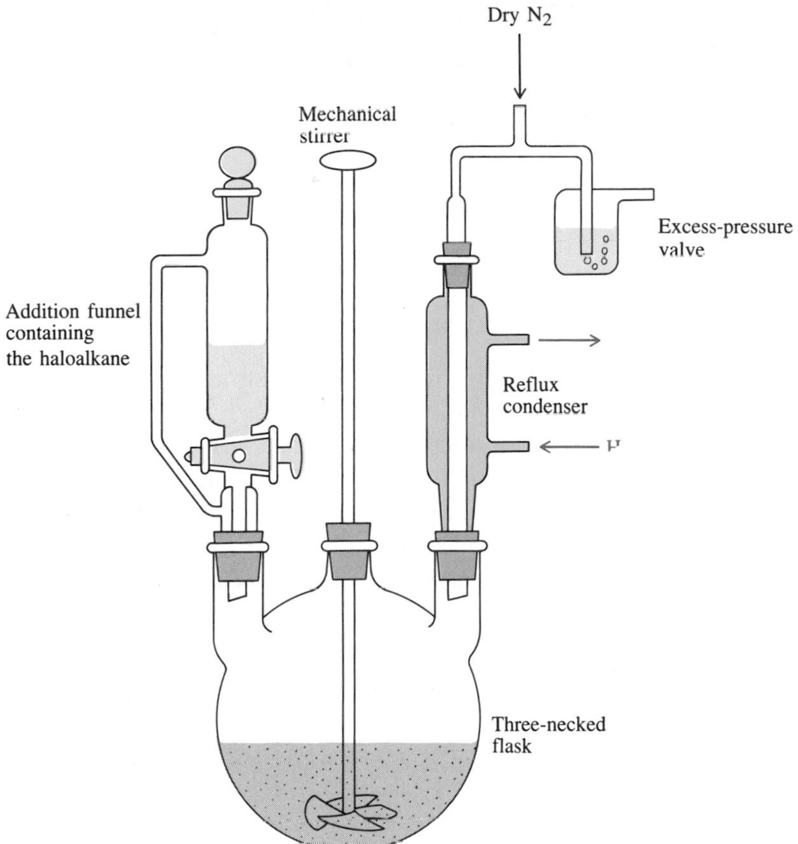

FIGURE 8-4 Apparatus for preparing organometallic reagents by the addition of a haloalkane to suspended lithium or magnesium in an ether solvent.

Organomagnesium compounds, RMgX, are also called **Grignard reagents,** named after their discoverer, F. A. Victor Grignard.* Their formation occurs in reactions starting with primary, secondary, and tertiary haloalkanes (and, as we shall see in later chapters, with haloalkenes and halobenzenes).

A typical experimental apparatus for this preparation of organometallic compounds, a process known as **metallation,** is shown in Figure 8-4. As the reaction proceeds, the suspended metal slowly dissolves. Alkyllithium compounds and Grignard reagents are rarely isolated; they are formed in solution and used immediately in the desired reaction. They are sensitive to air and moisture; thus, they must be prepared and handled under rigorously air-free and dry conditions.

The formulas RLi and RMgX oversimplify the true structures of these reagents. For example, alkylmagnesium halides are stabilized by bonding to two ether molecules, thereby providing the metal with an electron octet. The solvent is said to be **coordinated** to the metal. When the structures of Grignard reagents are written, this coordination is rarely shown. It is crucial, however, because the reaction is very difficult in its absence.

*Professor François Auguste Victor Grignard, 1871–1935, University of Lyon, France, Nobel Prize 1912 (chemistry).

Grignard Reagents Are Coordinated to Solvent

$$\begin{array}{c} CH_3CH_2\ddot{O}CH_2CH_3 \\ \downarrow \\ R{-}X + Mg \xrightarrow{(CH_3CH_2)_2O} R{-}Mg{-}X \\ \uparrow \\ CH_3CH_2\ddot{O}CH_2CH_3 \end{array}$$

The alkylmetal bond is strongly polar

Alkyllithium and alkylmagnesium reagents have strongly polarized carbon–metal bonds; the strongly electropositive metal (Table 1-2) is the positive end of the dipole. The degree of polarization is sometimes referred to as "percentage of ionic bond character." The carbon–lithium bond, for example, has about 40% ionic character and the carbon–magnesium bond 35%. Such systems react chemically as if they contained a negatively charged carbon. To symbolize this behavior, we can show the carbon–metal bond with a resonance form that places the full negative charge on the carbon atom: a **carbanion.**

Carbon–Metal Bond in Alkyllithium and Alkylmagnesium Compounds

$$\left[-\overset{|}{\underset{|}{C}}{}^{\delta^-}{-}M^{\delta^+} \longleftrightarrow -\overset{|}{\underset{|}{C}}{:}^- \; M^+ \right]$$

Polarized (M = metal) **Charge separated**

The preparation of alkylmetals from haloalkanes illustrates an important principle in synthetic organic chemistry: **reverse polarization.** In a haloalkane the presence of the electronegative halogen turns the carbon into an electrophilic center. On treatment with a metal, the $C^{\delta^+}{-}X^{\delta^-}$ unit is converted into $C^{\delta^-}{-}M^{\delta^+}$. In other words, the direction of polarization is reversed. Metallation has turned an electrophilic carbon into a nucleophilic center.

The alkyl group in alkylmetals is strongly basic

$$\underset{\textbf{Organometal}}{\overset{\delta^-}{R}{-}\overset{\delta^+}{M}} + \overset{\delta^+}{H}{-}\overset{\delta^-}{OH} \longrightarrow \underset{\textbf{Alkane}}{R{-}H} + \underset{\textbf{Metal hydroxide}}{M{-}OH}$$

Carbanions are very strong bases. In fact, organometallic reagents are much more basic than amides or alkoxides, because carbon is considerably less electronegative than either nitrogen or oxygen (Table 1-2) and much less capable of supporting a negative charge. Recall (Table 6-5, Section 6-7) that alkanes are *extremely* weak acids: The pK_a of methane is estimated to be 50! It is not surprising, therefore, that carbanions are such strong bases: They are, after all, the *conjugate bases of alkanes*. Their basicity makes organometallic reagents moisture sensitive. In the presence of water, hydrolysis occurs—often violently—to furnish the metal hydroxide and an alkane. The outcome of this transformation is predictable on purely electrostatic grounds (see margin).

Hydrolysis of an Organometallic Reagent

$$\underset{\text{3-Methylpentylmagnesium bromide}}{CH_3CH_2\overset{\overset{CH_3}{|}}{C}HCH_2CH_2MgBr} + HOH \longrightarrow \underset{\substack{100\% \\ \text{3-Methylpentane}}}{CH_3CH_2\overset{\overset{CH_3}{|}}{C}HCH_2CH_2H} + BrMgOH$$

The sequence metallation–hydrolysis affords a means by which a haloalkane can be converted to an alkane. A more direct way of achieving the same goal is the reaction of a haloalkane with the powerful hydride donor lithium aluminum hydride, an S_N2 displacement of halide by H^-.

$$\underset{\text{1-Bromononane}}{CH_3(CH_2)_7CH_2{-}Br} \xrightarrow[-LiBr]{LiAlH_4,\ (CH_3CH_2)_2O} \underset{\text{Nonane}}{CH_3(CH_2)_7CH_2{-}H}$$

A useful application of metallation–hydrolysis is the introduction of hydrogen isotopes, such as deuterium, into a molecule by exposure of the organometallic compound to labeled water.

Introduction of Deuterium by Reaction of an Organometallic Reagent with D_2O

$$(CH_3)_3CCl \xrightarrow{\substack{1.\ Mg \\ 2.\ D_2O}} (CH_3)_3CD$$

EXERCISE 8-12

Show how you would prepare 1-deuteriocyclohexane from cyclohexane.

In summary, haloalkanes can be converted into organometallic compounds of lithium or magnesium (Grignard reagents) by reaction with the respective metals in ether solvents. In these compounds, the alkyl group is negatively polarized, a charge distribution opposite that found in the haloalkane. Although the alkyl–metal bond is to a large extent covalent, the carbon attached to the metal behaves as a strongly basic carbanion, exemplified by its ready protonation.

8-8 Organometallic Reagents in the Synthesis of Alcohols

Among the most useful applications of organometallic reagents of magnesium and lithium are those in which the alkyl group reacts as a nucleophile. Like the hydrides, these reagents can attack the carbonyl group of an aldehyde or ketone to produce an alcohol. The difference is that a new carbon–carbon bond is formed in the process.

Alcohol Syntheses from Aldehydes, Ketones, and Organometallics

$$:\!\ddot{O}\!:^{\delta^-}\!\!=\!C^{\delta^+} + M^{\delta^+}\!-\!R^{\delta^-} \longrightarrow -\overset{:\ddot{O}:^- M^+}{\underset{R}{C}}- \xrightarrow{H^+,\ H_2O} -\overset{OH}{\underset{R}{C}}-$$

The concept of electron pushing can help us understand the reaction. In the first step, the nucleophilic alkyl group in the organometallic compound attacks the carbonyl carbon. As an electron pair from the alkyl group shifts to generate the new carbon–carbon linkage, it ''pushes'' two electrons from the double bond onto the oxygen, thus producing a metal alkoxide. The addition of a dilute aqueous acid furnishes the alcohol by hydrolyzing the metal–oxygen bond, another example of aqueous work-up.

The reaction of organometallic compounds with *formaldehyde* results in *primary alcohols.*

Formation of a Primary Alcohol from a Grignard Reagent and Formaldehyde

$$CH_3CH_2CH_2CH_2MgBr + H_2C{=}O \xrightarrow{(CH_3CH_2)_2O} \xrightarrow{H^+,\ H_2O} CH_3CH_2CH_2CH_2\overset{H}{\underset{H}{C}}OH$$

93%

Butylmagnesium bromide — **Formaldehyde** — **1-Pentanol**

However, *aldehydes* other than formaldehyde convert into *secondary alcohols.*

Formation of a Secondary Alcohol from a Grignard Reagent and an Aldehyde

$$CH_3\overset{MgBr}{C}HCH_3 + CH_3\overset{O}{\overset{\|}{C}}H \xrightarrow{(CH_3CH_2)_2O} \xrightarrow{H^+,\ H_2O} (CH_3)_2CH\overset{OH}{\underset{H}{C}}CH_3$$

54%

1-Methylethyl magnesium bromide — **Acetaldehyde** — **3-Methyl-2-butanol**

Ketones furnish *tertiary alcohols.*

Formation of a Tertiary Alcohol from a Grignard Reagent and a Ketone

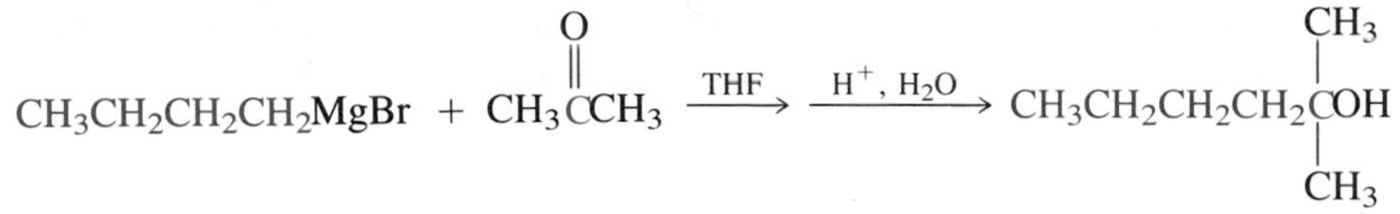

$$CH_3CH_2CH_2CH_2MgBr + CH_3\overset{\overset{\displaystyle O}{\|}}{C}CH_3 \xrightarrow{THF} \xrightarrow{H^+, H_2O} CH_3CH_2CH_2CH_2\underset{\underset{\displaystyle CH_3}{|}}{\overset{\overset{\displaystyle CH_3}{|}}{C}}OH$$

95%

Butylmagnesium bromide **Propanone (Acetone)** **2-Methyl-2-hexanol**

EXERCISE 8-13

Write a synthetic scheme for the conversion of 2-bromopropane, $(CH_3)_2CHBr$, into 2-methyl-1-propanol, $(CH_3)_2CHCH_2OH$.

EXERCISE 8-14

Propose efficient syntheses of the following products from starting materials containing no more than four carbons.

(a) $CH_3(CH_2)_4OH$ **(b)** 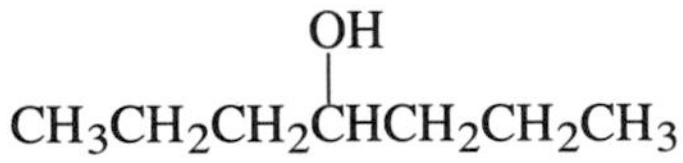 $CH_3CH_2CH_2\overset{\overset{\displaystyle OH}{|}}{C}HCH_2CH_2CH_3$

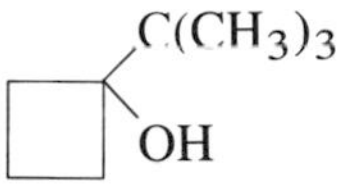

(c) Cyclobutane bearing $C(CH_3)_3$ and OH on the same carbon **(d)** 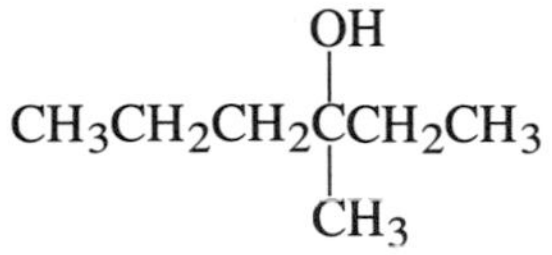 $CH_3CH_2CH_2\underset{\underset{\displaystyle CH_3}{|}}{\overset{\overset{\displaystyle OH}{|}}{C}}CH_2CH_3$

In summary, alkyllithium and alkylmagnesium reagents add to aldehydes and ketones to give alcohols in which the alkyl group of the organometallic reagent has formed a bond to the original carbonyl carbon.

8-9 Complex Alcohols: An Introduction to Synthetic Strategy

The reactions introduced so far are part of the "vocabulary" of organic chemistry, and unless we know the vocabulary, we cannot speak the language of organic chemistry. These reactions allow us to manipulate molecules and interconvert functional groups, so it is important to become familiar with these transformations—their types, the reagents used, the conditions under which they occur (especially when the conditions are crucial to the success of the process), and the limitations of each.

This task may seem monumental, one that will require much memorization. But *it is made easier by an understanding of the reaction mechanisms*. We already know that reactivity can be predicted from a small number of factors, such as electronegativity, coulombic forces, and bond strengths. Let us see how organic chemists apply this understanding to devise useful synthetic strategies.

Let us begin with a few examples in which we predict reactivity on mechanistic grounds. Then we shall turn to synthesis—the making of molecules. How do chemists develop new synthetic methods, and how can we make a "target" molecule as efficiently as possible? These topics are closely related. The second, known as **total synthesis,** usually requires a series of reactions. In studying these tasks, therefore, we will also be reviewing much of the reaction chemistry that we have discussed so far.

Mechanisms help in predicting the outcome of a reaction

First, recall how we predict the outcome of a reaction. What are the factors that let a particular mechanism go forward? Here are three examples.

How to Predict the Outcome of a Reaction on Mechanistic Grounds

$$\underset{\text{Not formed}}{ICH_2CH_2CH_2Br} \xleftarrow[\times]{I^-,\ \text{propanone (acetone)}} FCH_2CH_2CH_2Br \xrightarrow{I^-,\ \text{propanone (acetone)}} FCH_2CH_2CH_2I$$

Explanation: Bromide is a better leaving group than fluoride.

$$\underset{\text{Not formed}}{CH_3C(OCH_3)(H)MgBr} \xleftarrow[\times]{(CH_3CH_2)_2O} H_3C\overset{\delta^+}{C}(=O^{\delta^-})H + \overset{\delta^-}{C}H_3\overset{\delta^+}{Mg}Br \xrightarrow{(CH_3CH_2)_2O} CH_3C(\overset{-}{O}\overset{+}{Mg}Br)(H)CH_3$$

Explanation: The positively polarized carbonyl carbon forms a bond to the negatively polarized alkyl group of the organometallic reagent.

Other bromides + 3-bromo-1-methylcyclohexane (Not formed) + (bromomethyl)cyclohexane $\xleftarrow[\times]{Br_2,\ h\nu}$ methylcyclohexane $\xrightarrow{Br_2,\ h\nu}$ 1-bromo-1-methylcyclohexane

Explanation: The tertiary C–H bond is weaker than a primary or secondary C–H bond, and Br is quite selective in radical halogenation.

EXERCISE 8-15

Predict and explain the outcome of each reaction below on mechanistic grounds.

(a) $ClCH_2CH_2CH_2C(Br)(CH_3)_2 + CH_3CH_2OH \longrightarrow$

(b) $ClCH_2CH_2CH_2C(CH_2Cl)(CH_3)_2 + (CH_3)_3CO^-\ {}^+K \xrightarrow{(CH_3)_3COH}$

(c) $HOCH_2CH_2CH_2C(OH)(CH_3)_2 \xrightarrow{PCC,\ CH_2Cl_2}$

New reactions lead to new synthetic methods

New reactions are found by design or by accident. For example, consider how two different students might discover the reactivity of a Grignard reagent with a ketone to give an alcohol. The first student, knowing about electronegativity and the electronic makeup of ketones, would predict that the nucleophilic alkyl group of the Grignard species should attach itself to the electrophilic carbonyl carbon. This student would be pleased by the successful outcome of the experiment, verifying chemical principles in practice. The second student, with less knowledge, might attempt to dilute a particularly concentrated solution of a Grignard reagent with what he might conceive to be a perfectly good polar solvent: propanone (acetone). A violent reaction would immediately reveal the powerful potential of the reagent in alcohol synthesis.

Once a reaction is discovered, it is important to show its scope and its limitations. For this purpose, many different substrates are tested, side products (if any) noted, new functional groups subjected to the reaction conditions, and mechanistic studies carried out. Should these investigations prove the new reaction to be generally applicable, it will be added as a new synthetic method to the organic chemist's arsenal.

Because a reaction leads to a very specific change in a molecule, it is frequently useful to emphasize this "molecular alteration." A simple example is the addition of a Grignard reagent to formaldehyde. What is the structural change in this transformation? A one-carbon unit is added to an alkyl group. The method is valuable because it allows a straightforward one-carbon extension, also called a *homologation*.

$$\underset{\textbf{Alkyl group}}{R{-}M} + \underset{\textbf{One-carbon unit}}{H_2C{=}O} \longrightarrow R{-}CH_2{-}OH$$

Even though our synthetic vocabulary at this stage is relatively limited, we already have quite a number of molecular alterations at our disposal. For example, bromoalkanes are excellent starting points for numerous transformations.

$$RCH_2CH_2Br \xrightarrow{Nu:^-,\ DMSO} RCH_2CH_2Nu$$

$$RCH_2CH_2Br \xrightarrow{KOC(CH_3)_3,\ (CH_3)_3COH} RCH{=}CH_2$$

$$RCH_2CH_2Br \xrightarrow{Mg,\ (CH_3CH_2)_2O} RCH_2CH_2MgBr$$

$$RCH_2CH_2MgBr \xrightarrow{D_2O} RCH_2CH_2D$$

$$RCH_2CH_2MgBr \xrightarrow{H_2C=O} RCH_2CH_2CH_2{-}OH$$

$$RCH_2CH_2MgBr \xrightarrow{CH_3\overset{O}{\overset{\|}{C}}CH_3} RCH_2CH_2\overset{OH}{\overset{|}{\underset{CH_3}{\underset{|}{C}}}}CH_3$$

$$RCH_2CH_2MgBr \xrightarrow{CH_3\overset{O}{\overset{\|}{C}}H} RCH_2CH_2\overset{OH}{\overset{|}{\underset{H}{\underset{|}{C}}}}CH_3$$

Each one of the products in the scheme can enter into further transformations of its own, thereby leading to more complicated products.

When we ask, "What good is a reaction? What sort of structures can we make by applying it?" we address a problem of *synthetic methodology*. Let us ask a different question. Suppose that we want to prepare a specific target molecule. How would we go about devising an efficient route to it? How do we find suitable starting materials? The problem with which we are dealing now is *total synthesis*.

Organic chemists want to make complex molecules for specific purposes. For example, certain compounds might have valuable medicinal properties but are not readily available from natural sources. Biochemists need a particular isotopically labeled molecule to trace metabolic pathways. Physical organic chemists frequently design novel structures to study. There are many reasons for the total synthesis of organic molecules.

Whatever the final target, a successful synthesis is characterized by brevity and high overall yield. The starting materials should be readily available, preferably commercially, and inexpensive. Moreover, safety and environmental concerns demand that the reagents used be relatively nontoxic and easy to handle.

Retrosynthetic analysis simplifies synthesis problems

Many compounds that are commercially available and inexpensive are also small, containing five or fewer carbon atoms. Therefore, the most frequent task facing the synthetic planner is that of building up a larger, complicated molecule from smaller, simple fragments. The best approach to the preparation of the target is to work its synthesis *backward* on paper, an approach called **retrosynthetic analysis*** (*retro,* Latin, backward). In this analysis, strategic carbon–carbon bonds are ''broken'' at points at which their formation appears possible. For example, a retrosynthetic analysis of the synthesis of 3-hexanol from two three-carbon units would suggest its formation from a propyl organometallic compound and propanal.

Retrosynthetic Analysis of 3-Hexanol Synthesis from Two Three-Carbon Fragments

$$CH_3CH_2CH_2\overset{\overset{\displaystyle OH}{|}}{C}HCH_2CH_3 \Longrightarrow \underset{\textbf{Propylmagnesium bromide}}{CH_3CH_2CH_2MgBr} + \underset{\textbf{Propanal}}{H\overset{\overset{\displaystyle O}{\|}}{C}CH_2CH_3}$$

The double-shafted arrow indicates the so-called **strategic disconnection.** We recognize that the bond ''broken'' in this analysis, that between C3 and C4 in the product, is one we can construct by using a transformation we know, $CH_3CH_2CH_2MgBr + CH_3CH_2CHO$. In this case only one reaction is necessary to achieve the connection; in others, it might require several steps. Two alternate, but inferior retrosyntheses of 3-hexanol are

$$CH_3CH_2CH_2\overset{\overset{\displaystyle OH}{|}}{C}HCH_2CH_3 \Longrightarrow NaBH_4 + CH_3CH_2CH_2\overset{\overset{\displaystyle O}{\|}}{C}CH_2CH_3$$

$$CH_3CH_2CH_2\overset{\overset{\displaystyle OH}{|}}{C}HCH_2CH_3 \Longrightarrow NaO\overset{\overset{\displaystyle O}{\|}}{C}CH_3 + CH_3CH_2CH_2\overset{\overset{\displaystyle Br}{|}}{C}HCH_2CH_3$$

They are not as good as the first because they do not significantly *simplify* the target structure: No carbon–carbon bonds are ''broken.''

*Pioneered by Professor Elias J. Corey, b. 1928, Harvard University, Nobel Prize 1990 (chemistry).

Retrosynthetic analysis aids in alcohol construction

Let us apply retrosynthetic analysis to the preparation of a tertiary alcohol, 4-ethyl-4-nonanol. There are two steps to follow at each stage of the process. First, we identify all possible strategic disconnections, ''breaking'' all bonds that can be formed by reactions we know. Second, we evaluate the relative merits of these disconnections, seeking the one that best simplifies the target structure. The strategic bonds in 4-ethyl-4-nonanol are those around the functional group. There are three disconnections leading to simpler precursors. Path *a* cleaves the ethyl group from C4, suggesting as the starting materials for its construction ethylmagnesium bromide and 4-nonanone. Cleavage *b* is an alternative possibility leading to a propyl Grignard reagent and 3-octanone as precursors. Finally, disconnection *c* reveals a third synthesis route derived from addition of pentylmagnesium bromide to 3-hexanone.

Partial Retrosynthetic Analysis of the Synthesis of 4-Ethyl-4-nonanol

Target: 4-Ethyl-4-nonanol, $CH_3CH_2\text{-}C(OH)(CH_2CH_2CH_2CH_2CH_3)\text{-}CH_2CH_2CH_3$ (strategic bonds around C4 marked as disconnections)

$$\xRightarrow{a} CH_3CH_2MgBr + CH_3CH_2CH_2\overset{\overset{\large O}{\|}}{C}CH_2CH_2CH_2CH_2CH_3$$

Ethylmagnesium bromide + **4-Nonanone**

$$\xRightarrow{b} CH_3CH_2CH_2MgBr + CH_3CH_2\overset{\overset{\large O}{\|}}{C}CH_2CH_2CH_2CH_2CH_3$$

Propylmagnesium bromide + **3-Octanone**

$$\xRightarrow{c} CH_3CH_2CH_2CH_2CH_2MgBr + CH_3CH_2\overset{\overset{\large O}{\|}}{C}CH_2CH_2CH_3$$

Pentylmagnesium bromide + **3-Hexanone**

Evaluation reveals that pathway *c* is best: The necessary building blocks are almost equal in size, containing five and six carbons; thus, this disconnection provides the greatest simplification in structure.

EXERCISE 8-16

Apply retrosynthetic analysis to 4-ethyl-4-nonanol, disconnecting the carbon–*oxygen* bond. Does this lead to an efficient synthesis? Explain.

Can we pursue either of the fragments arising from disconnection by pathway *c* to even simpler starting materials? Yes; recall (Section 8-6) that ketones are obtained from oxidation of secondary alcohols by Cr(VI) reagents. We may therefore envision preparation of 3-hexanone from the corresponding alcohol, 3-hexanol.

$$CH_3CH_2CH_2\overset{\overset{\large O}{\|}}{C}CH_2CH_3 \Longrightarrow Na_2Cr_2O_7 + CH_3CH_2CH_2\overset{\overset{\large OH}{|}}{C}HCH_2CH_3$$

Because we earlier identified an efficient disconnection of the latter into three-carbon fragments, we are now in a position to present our complete synthetic scheme.

Synthesis of 4-Ethyl-4-nonanol

$$CH_3CH_2\overset{O}{\overset{\|}{C}}H \xrightarrow[2.\ H^+,\ H_2O]{1.\ CH_3CH_2CH_2MgBr,\ (CH_3CH_2)_2O} CH_3CH_2\overset{OH}{\overset{|}{C}}HCH_2CH_2CH_3 \xrightarrow{Na_2Cr_2O_7,\ H_2SO_4,\ H_2O}$$

$$CH_3CH_2\overset{O}{\overset{\|}{C}}CH_2CH_2CH_3 \xrightarrow[2.\ H^+,\ H_2O]{1.\ CH_3CH_2CH_2CH_2CH_2MgBr,\ (CH_3CH_2)_2O} CH_3CH_2\overset{OH}{\overset{|}{\underset{CH_2CH_2CH_2CH_2CH_3}{\underset{|}{C}}}}CH_2CH_2CH_3$$

This example illustrates a very powerful general sequence for the construction of complex alcohols: first, Grignard or organolithium addition to an aldehyde to give a secondary alcohol; then oxidation to a ketone; and finally, addition of another organometallic reagent to give a tertiary alcohol.

Utility of Alcohol Oxidations in Synthesis

$$R\overset{O}{\overset{\|}{C}}H \xrightarrow[2.\ H^+,\ H_2O]{1.\ R'MgBr,\ (CH_3CH_2)_2O} R\overset{OH}{\overset{|}{\underset{R'}{\underset{|}{C}}}}H \xrightarrow{CrO_3,\ H^+,\ H_2O} R\overset{O}{\overset{\|}{C}}R' \xrightarrow[2.\ H^+,\ H_2O]{1.\ R''MgBr,\ (CH_3CH_2)_2O} R\overset{OH}{\overset{|}{\underset{R''}{\underset{|}{C}}}}R'$$

EXERCISE 8-17

Write an economical retrosynthetic analysis of 3-cyclobutyl-3-octanol.

EXERCISE 8-18

Show how you would prepare 2-methyl-2-propanol from methane as the only organic starting material.

Watch out for pitfalls in planning syntheses

There are several considerations to keep in mind when practicing synthetic chemistry that will help to avoid designing unsuccessful or low-yielding approaches to a target molecule. First, *try to minimize the total number of transformations required to convert the initial starting material into the desired product*. This point is so important that in some cases it is worthwhile to accept a low-yield step if it allows a significant shortening of the synthetic sequence. For example (assuming all starting materials to be of comparable cost), a seven-step synthesis in which each step has an 85% yield is less productive than a four-step synthesis with three yields at 95% and one at 45%. The overall efficiency in the first sequence comes to $(0.85 \times 0.85 \times 0.85 \times 0.85 \times 0.85 \times 0.85 \times 0.85) \times 100 = 32\%$, whereas the second synthesis, in addition to being three steps shorter, gives $(0.95 \times 0.95 \times 0.95 \times 0.45) \times 100 = 39\%$.

Second, *do not use reagents whose molecules have functional groups that would interfere with the desired reaction*. For example, treating a hydroxyaldehyde with a Grignard reagent leads to an acid–base reaction, destroying the Grignard, and not to carbon–carbon bond formation.

$$\mathrm{HOCH_2CH_2\overset{OH}{\underset{CH_3}{C}}H} \;\not\leftrightarrow\; \mathrm{HOCH_2CH_2\overset{O}{\overset{\|}{C}}H + CH_3MgBr \longrightarrow BrMgOCH_2CH_2\overset{O}{\overset{\|}{C}}H + \overset{H}{\underset{|}{C}}H_3}$$

A possible solution to this problem would be to add two equivalents of Grignard reagent: *one* to react with the acidic hydrogen as above, the *second* to achieve the desired addition to the carbonyl group.

Do not try to make a Grignard reagent from a bromoketone. Such a reagent is not stable and will, as soon as it is formed, decompose by reacting with its own carbonyl group.

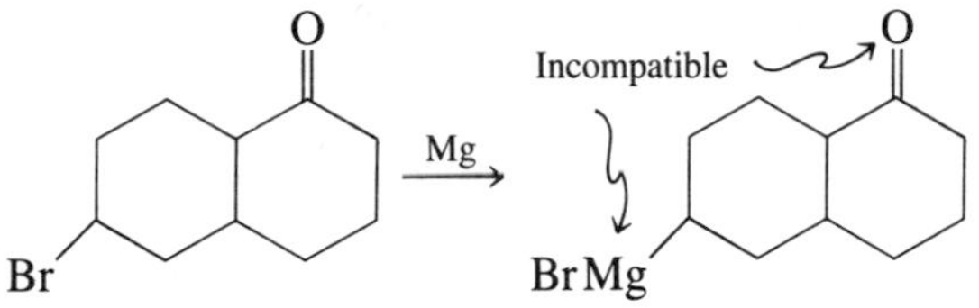

Br (bromo bicyclic ketone) $\xrightarrow{\text{Mg}}$ BrMg (bicyclic ketone) — Incompatible

Third, *take into account any mechanistic and structural constraints affecting the reactions under consideration*. For example, radical brominations are more selective than chlorinations. Keep in mind the structural limitations on nucleophilic reactions, and do not forget the lack of reactivity of the 2,2-dimethyl-1-halopropanes (neopentyl halides). Although sometimes difficult to recognize, many haloalkanes have "neopentyllike" structures and are similarly unreactive. Nevertheless, such systems do form organometallic reagents and may be further functionalized in this manner. For example, treatment of the Grignard reagent made from 1-bromo-2,2-dimethylpropane with formaldehyde leads to the corresponding alcohol.

$$\underset{\textbf{1-Bromo-2,2-dimethylpropane}}{\mathrm{(CH_3)_3CCH_2Br}} \xrightarrow[\text{2. } CH_2=O]{\text{1. Mg}} \underset{\textbf{3,3-Dimethyl-1-butanol}}{\mathrm{(CH_3)_3CCH_2CH_2OH}}$$

Neopentyllike Hindered Haloalkanes

$\mathrm{CH_3CH_2C(CH_3)_2CH_2Br}$

1-(CH_2Cl)-1-CH_3-cyclohexane

bicyclic structure with CH_3 and Br

Tertiary halides, if incorporated into a more complex framework, also are sometimes difficult to recognize. Remember that tertiary halides do not undergo S_N2 reactions but eliminate in the presence of bases.

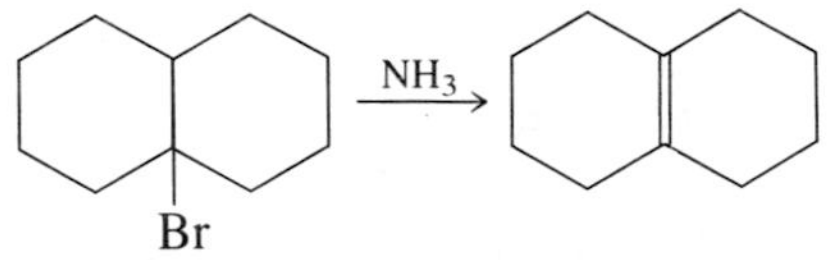

Br (tertiary bromodecalin) $\xrightarrow{NH_3}$ (octalin)

Expertise in synthesis, as in many other aspects of organic chemistry, develops largely from practice. Planning the synthesis of complex molecules requires a review of the reactions and mechanisms covered in earlier sections. The knowledge thus acquired can then be applied to the solution of synthetic problems.

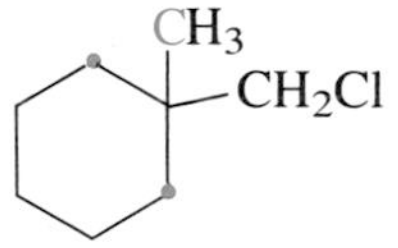

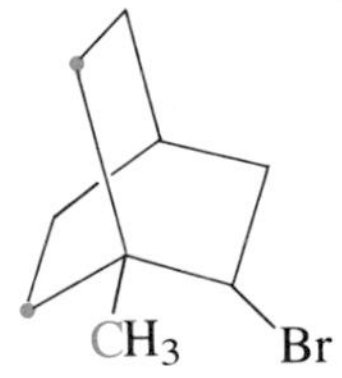

NEW REACTIONS

1. ACID-BASE PROPERTIES OF ALCOHOLS (Section 8-3)

$$R\text{—}\overset{+}{O}H_2 \underset{}{\overset{H^+}{\rightleftharpoons}} ROH \overset{\text{Base }:B^-}{\rightleftharpoons} RO^- + BH$$

Alkyloxonium ion **Alcohol** **Alkoxide**

Acidity: $RO\text{–}H \approx HO\text{–}H > H_2N\text{–}H > H_3C\text{–}H$
Basicity: $RO^- \approx HO^- < H_2N^- < H_3C^-$

Industrial Preparation of Alcohols

2. SYNTHESIS GAS (Section 8-4)

$$\text{Coal} \xrightarrow{\text{Air, } H_2O,\ \Delta} x\,CO + y\,H_2$$

Synthesis gas

3. METHANOL SYNTHESIS FROM SYNTHESIS GAS (Section 8-4)

$$CO + 2\,H_2 \xrightarrow{\text{Cu-ZnO-}Cr_2O_3\text{, 250°C, 50–100 atm}} CH_3OH$$

4. FISCHER-TROPSCH SYNTHESIS (Section 8-4)

$$n\,CO + (2n + 1)\,H_2 \xrightarrow{\text{Co or Fe, pressure, 200–350°C}} C_nH_{2n+2} + n\,H_2O$$

Hydrocarbons

5. ETHANOL BY HYDRATION OF ETHENE (Section 8-4)

$$CH_2{=}CH_2 + HOH \xrightarrow{H_3PO_4\text{, 300°C}} CH_3CH_2OH$$

Laboratory Preparation of Alcohols

6. NUCLEOPHILIC DISPLACEMENT OF HALIDES AND OTHER LEAVING GROUPS BY HYDROXIDE ION (Section 8-5)

$$RCH_2X + HO^- \underset{S_N2}{\xrightarrow{H_2O}} RCH_2OH + X^-$$

X = halide, sulfonate
Primary, secondary (tertiary undergoes elimination)

$$R'\overset{R}{\underset{R''}{C}}X \underset{S_N1}{\xrightarrow{H_2O\text{, propanone (acetone)}}} R'\overset{R}{\underset{R''}{C}}OH$$

Best method for tertiary

7. CATALYTIC HYDROGENATION OF ALDEHYDES AND KETONES (Section 8-6)

$$R\overset{O}{\overset{\|}{C}}H \xrightarrow{H_2\text{, Pt}} RCH_2OH \qquad R\overset{O}{\overset{\|}{C}}R' \xrightarrow{H_2\text{, catalyst}} R\overset{OH}{\underset{H}{C}}R'$$

Aldehyde **Primary alcohol** **Ketone** **Secondary alcohol**

8. REDUCTION OF ALDEHYDES AND KETONES BY HYDRIDES (Section 8-6)

$$\mathrm{R\overset{O}{\overset{\|}{C}}H} \xrightarrow{\mathrm{NaBH_4,\ CH_3CH_2OH}} \mathrm{RCH_2OH} \qquad \mathrm{R\overset{O}{\overset{\|}{C}}R'} \xrightarrow{\mathrm{NaBH_4,\ CH_3CH_2OH}} \mathrm{R\overset{OH}{\overset{|}{\underset{H}{\underset{|}{C}}}}R'}$$

$$\underset{\textbf{Aldehyde}}{\mathrm{R\overset{O}{\overset{\|}{C}}H}} \xrightarrow[\text{2. } \mathrm{H^+,\ H_2O}]{\text{1. } \mathrm{LiAlH_4,\ (CH_3CH_2)_2O}} \underset{\textbf{Primary alcohol}}{\mathrm{RCH_2OH}} \qquad \underset{\textbf{Ketone}}{\mathrm{R\overset{O}{\overset{\|}{C}}R'}} \xrightarrow[\text{2. } \mathrm{H^+,\ H_2O}]{\text{1. } \mathrm{LiAlH_4,\ (CH_3CH_2)_2O}} \underset{\textbf{Secondary alcohol}}{\mathrm{R\overset{OH}{\overset{|}{\underset{H}{\underset{|}{C}}}}R'}}$$

Oxidation of Alcohols

9. CHROMIUM REAGENTS (Section 8-6)

$$\underset{\textbf{Primary alcohol}}{\mathrm{RCH_2OH}} \xrightarrow{\mathrm{PCC,\ CH_2Cl_2}} \underset{\textbf{Aldehyde}}{\mathrm{R\overset{O}{\overset{\|}{C}}H}} \qquad \underset{\textbf{Secondary alcohol}}{\mathrm{R\overset{OH}{\overset{|}{C}}HR}} \xrightarrow{\mathrm{Na_2Cr_2O_7,\ H_2SO_4}} \underset{\textbf{Ketone}}{\mathrm{R\overset{O}{\overset{\|}{C}}R}}$$

Organometallic Reagents

10. REACTION OF METALS WITH HALOALKANES (Section 8-7)

$$\mathrm{RX + Li} \xrightarrow{\mathrm{(CH_3CH_2)_2O}} \underset{\textbf{Alkyllithium reagent}}{\mathrm{RLi}}$$

$$\mathrm{RX + Mg} \xrightarrow{\mathrm{(CH_3CH_2)_2O}} \underset{\textbf{Grignard reagent}}{\mathrm{RMgX}}$$

R cannot contain acidic groups such as O–H or electrophilic groups such as C=O

11. ADDITION OF ORGANOMETALLIC COMPOUNDS TO ALDEHYDES AND KETONES (Section 8-8)

$$\mathrm{RLi\ or\ RMgX} + \underset{\textbf{Formaldehyde}}{\mathrm{CH_2{=}O}} \longrightarrow \underset{\textbf{Primary alcohol}}{\mathrm{RCH_2OH}}$$

$$\mathrm{RLi\ or\ RMgX} + \underset{\textbf{Aldehyde}}{\mathrm{R'\overset{O}{\overset{\|}{C}}H}} \longrightarrow \underset{\textbf{Secondary alcohol}}{\mathrm{R\overset{OH}{\overset{|}{\underset{H}{\underset{|}{C}}}}R'}}$$

$$\mathrm{RLi\ or\ RMgX} + \underset{\textbf{Ketone}}{\mathrm{R'\overset{O}{\overset{\|}{C}}R''}} \longrightarrow \underset{\textbf{Tertiary alcohol}}{\mathrm{R\overset{OH}{\overset{|}{\underset{R''}{\underset{|}{C}}}}R'}}$$

Aldehyde or ketone cannot contain other groups that react with organometallic reagents such as O–H or other C=O groups

12. HYDROLYSIS (Section 8-7)

$$RLi \text{ or } RMgX + H_2O \longrightarrow RH$$
$$RLi \text{ or } RMgX + D_2O \longrightarrow RD$$

13. ALKANES FROM HALOALKANES AND LITHIUM ALUMINUM HYDRIDE (Section 8-7)

$$RX + LiAlH_4 \xrightarrow{(CH_3CH_2)_2O} RH$$

IMPORTANT CONCEPTS

1. Alcohols are alkanols in IUPAC nomenclature. The stem containing the functional group gives the alcohol its name. Alkyl and halo substituents are added as prefixes.
2. Like water, alcohols have a polarized and short O–H bond. The proton is hydrophilic and enters into hydrogen bonding. Consequently, alcohols have unusually high boiling points and, in many cases, appreciable water solubility. The alkyl portion of the molecule is hydrophobic.
3. Again like water, alcohols are amphoteric: They are both acidic and basic. Complete deprotonation to an alkoxide occurs with bases whose conjugate acids are considerably weaker than the alcohol. Protonation gives an alkyloxonium ion. In solution the order of acidity is primary > secondary > tertiary alcohol. Electron-withdrawing substituents increase the acidity (and reduce the basicity).
4. The hydrogenation of aldehydes and ketones furnishes alcohols and requires a catalyst.
5. The conversion of the electrophilic alkyl group in a haloalkane, $C^{\delta^+}–X^{\delta^-}$, into its nucleophilic analog in an organometallic compound, $C^{\delta^-}–M^{\delta^+}$, is an example of reverse polarization.
6. The carbon atom in the carbonyl group, C=O, of an aldehyde or a ketone is electrophilic and therefore subject to attack by nucleophiles, such as hydride in hydride reagents or alkyl in organometallic compounds. Subsequent to aqueous work-up, the products of such transformations are alcohols.
7. The oxidation of alcohols to aldehydes and ketones by chromium(VI) reagents opens up important synthetic possibilities based on further reactions with organometallic reagents.
8. Retrosynthetic analysis aids in planning the synthesis of complex organic molecules by identifying strategic bonds that may be constructed in an efficient sequence of reactions.

PROBLEMS

1. Name the following alcohols according to the IUPAC nomenclature system. Indicate stereochemistry (if any) and label the hydroxy groups as primary, secondary, or tertiary.

(a) $CH_3CH_2CH(OH)CH_3$ (OH on the third carbon)

(b) $CH_3CH(Br)CH_2CH(OH)CH_2CH_3$

(c) $HOCH_2CH(CH_2CH_2CH_3)_2$

(d) H---C(CH$_2$Cl)(CH$_3$)(OH) [stereo drawing: CH_2Cl, H, CH_3, OH]

(e) cyclobutane bearing CH_2CH_3 (wedge) and ---OH (hash)

(f) cyclodecane ring bearing OH (wedge) and ---Br (hash) on adjacent carbons

(g) $C(CH_2OH)_4$

(h) 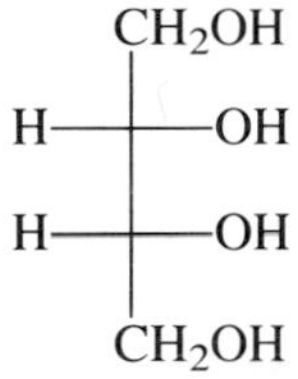 Fischer projection: CH_2OH (top); H—C—OH; H—C—OH; CH_2OH (bottom)

(i) 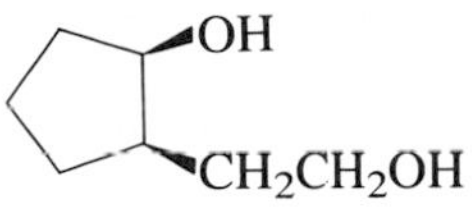 cyclopentane bearing OH and CH_2CH_2OH

(j) 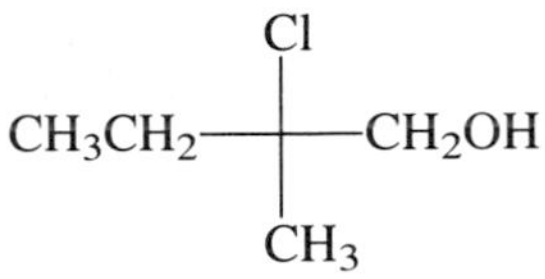 CH_3CH_2—C(Cl)(CH_3)—CH_2OH

2. Draw the structures of the following alcohols. **(a)** 2-(Trimethylsilyl)ethanol; **(b)** 1-methylcyclopropanol; **(c)** 3-(1-methylethyl)-2-hexanol; **(d)** (*R*)-2-pentanol; **(e)** 3,3-dibromocyclohexanol.

3. Rank each group of compounds in order of increasing boiling point. **(a)** Cyclohexane, cyclohexanol, chlorocyclohexane; **(b)** 2,3-dimethyl-2-pentanol, 2-methyl-2-hexanol, 2-heptanol.

4. Explain the order of water solubilities for the compounds in each group below. **(a)** Ethanol > chloroethane > ethane; **(b)** methanol > ethanol > 1-propanol.

5. 1,2-Ethanediol exists to a much greater extent in the *gauche* conformation than does 1,2-dichloroethane. Explain. Would you expect the *gauche*:*anti* conformational ratio of 2-chloroethanol to be similar to that of 1,2-dichloroethane or more like that of 1,2-ethanediol?

6. Rank the compounds in each group in order of decreasing acidity.
(a) $CH_3CHClCH_2OH$, $CH_3CHBrCH_2OH$, $ClCH_2CH_2CH_2OH$
(b) $CH_3CCl_2CH_2OH$, CCl_3CH_2OH, $(CH_3)_2CClCH_2OH$
(c) $(CH_3)_2CHOH$, $(CF_3)_2CHOH$, $(CCl_3)_2CHOH$

7. Write an appropriate equation to show how each of the alcohols below acts as, first, a base, and, second, an acid in solution. How do the base and acid strengths of each compare with those of methanol? **(a)** $(CH_3)_2CHOH$; **(b)** CH_3CHFCH_2OH; **(c)** CCl_3CH_2OH.

8. Given the pK_a values of −2.2 for $CH_3\overset{+}{O}H_2$ and 15.5 for CH_3OH, calculate the pH at which **(a)** methanol will contain exactly equal amounts of $CH_3\overset{+}{O}H_2$ and CH_3O^-; **(b)** 50% CH_3OH and 50% $CH_3\overset{+}{O}H_2$ will be present; **(c)** 50% CH_3OH and 50% CH_3O^- will be present.

9. Do you expect hyperconjugation to be important in the stabilization of alkyloxonium ions (e.g., $R\overset{+}{O}H_2$, $R_2\overset{+}{O}H$)? Explain your answer.

10. Evaluate each of the possible alcohol syntheses below as being good (the desired alcohol is the major or only product), not so good (the desired alcohol is a minor product), or worthless.

(a) $CH_3CH_2CH_2CH_2Cl \xrightarrow{H_2O,\ CH_3\overset{O}{\overset{\|}{C}}CH_3} CH_3CH_2CH_2CH_2OH$

(b) 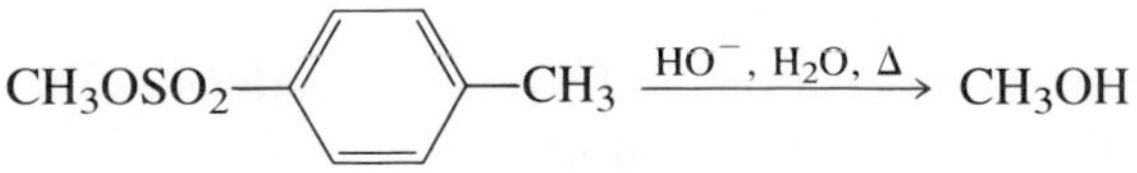 CH_3OSO_2—C_6H_4—$CH_3 \xrightarrow{HO^-,\ H_2O,\ \Delta} CH_3OH$

(c) 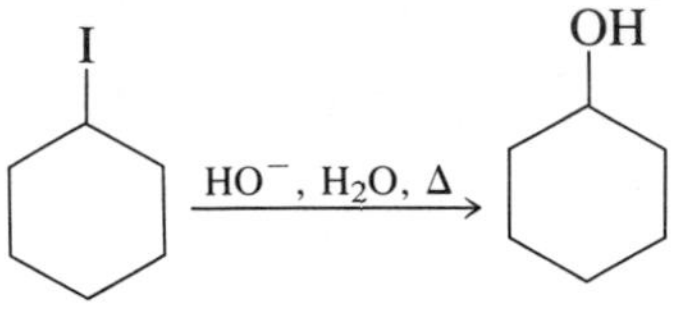 Iodocyclohexane $\xrightarrow{HO^-,\ H_2O,\ \Delta}$ cyclohexanol

(d) $CH_3CH(I)CH_2CH_2CH_3 \xrightarrow{H_2O,\ \Delta} CH_3CH(OH)CH_2CH_2CH_3$

(e) $CH_3CH(CN)CH_3 \xrightarrow{HO^-,\ H_2O,\ \Delta} CH_3CH(OH)CH_3$

(f) $CH_3OCH_3 \xrightarrow{HO^-,\ H_2O,\ \Delta} CH_3OH$

(g) 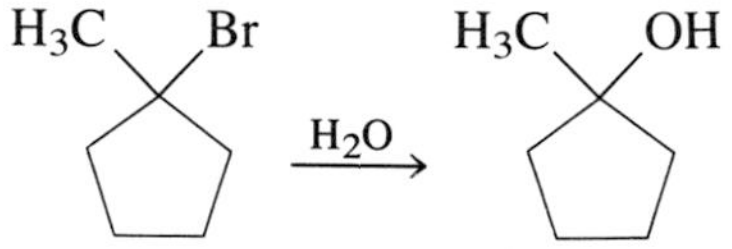 1-bromo-1-methylcyclopentane (H_3C, Br) $\xrightarrow{H_2O}$ 1-methylcyclopentanol (H_3C, OH)

(h) $CH_3CH(CH_3)CH_2Cl \xrightarrow{HO^-,\ H_2O,\ \Delta} CH_3CH(CH_3)CH_2OH$

11. Give the major product(s) of each of the following reactions. Aqueous work-up steps (when necessary) have been omitted.

(a) $CH_3CH{=}CHCH_3 \xrightarrow{H_3PO_4,\ H_2O,\ \Delta}$

(b) $CH_3\overset{O}{\overset{\|}{C}}CH_2CH_2\overset{O}{\overset{\|}{C}}CH_3 \xrightarrow{H_2,\ Pt}$

(c) 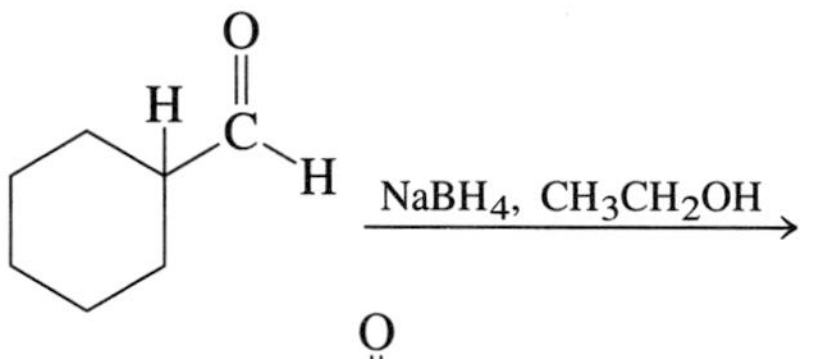 Cyclohexane–CH(H)–C(=O)H $\xrightarrow{NaBH_4,\ CH_3CH_2OH}$

(d) 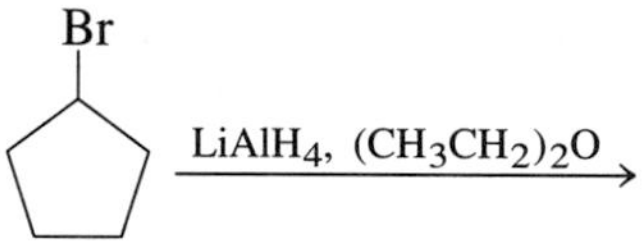 Bromocyclopentane (Br) $\xrightarrow{LiAlH_4,\ (CH_3CH_2)_2O}$

(e) 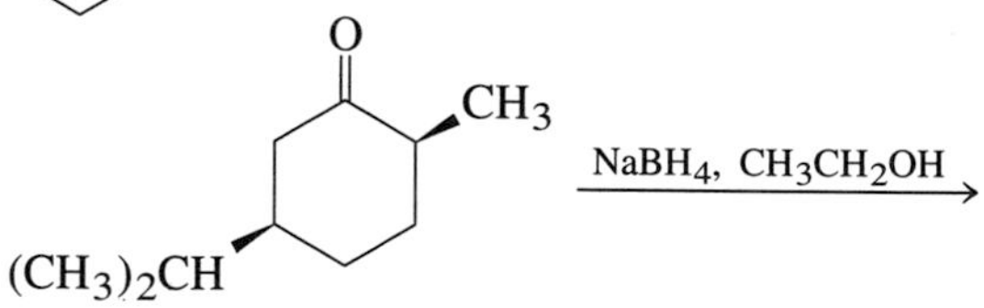 Cyclohexanone with CH_3 (wedge) and $(CH_3)_2CH$ (wedge) substituents $\xrightarrow{NaBH_4,\ CH_3CH_2OH}$

(f) 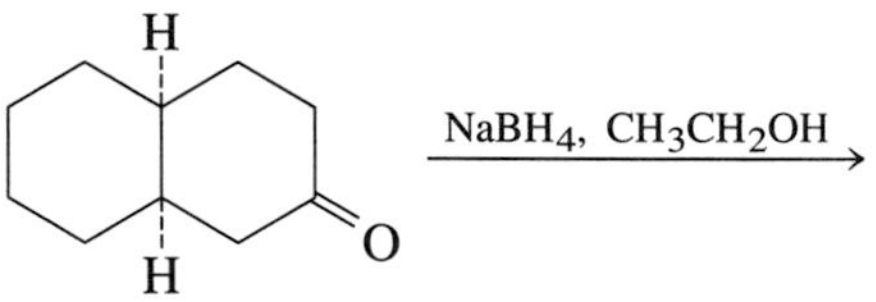 Decalone with ring-fusion H, H and =O $\xrightarrow{NaBH_4,\ CH_3CH_2OH}$

12. What is the direction of the equilibrium below? (Hint: The pK_a for H_2 is about 38.)

$$H^- + H_2O \rightleftharpoons H_2 + HO^-$$

13. Formulate the product of each of the following reactions. The solvent in each case is $(CH_3CH_2)_2O$.

(a) $CH_3\overset{O}{\overset{\|}{C}}H \xrightarrow{1.\ LiAlD_4\quad 2.\ H^+,\ H_2O}$

(b) $CH_3\overset{O}{\overset{\|}{C}}H \xrightarrow{1.\ LiAlH_4\quad 2.\ D^+,\ D_2O}$

(c) $CH_3CH_2I \xrightarrow{LiAlD_4}$

14. Give the major product(s) of each of the following reactions (after work-up with aqueous acid in the cases of d, f, and h).

(a) $CH_3(CH_2)_5CHClCH_3 \xrightarrow{Mg,\ (CH_3CH_2)_2O}$ (b) Product of (a) $\xrightarrow{D_2O}$

(c) 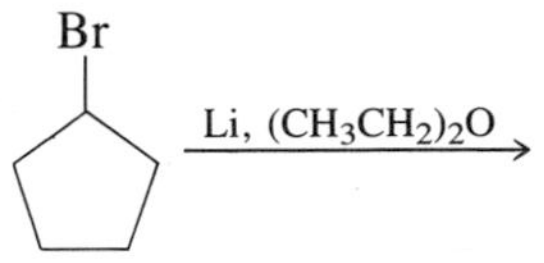 Bromocyclopentane $\xrightarrow{Li,\ (CH_3CH_2)_2O}$ (d) Product of (c) 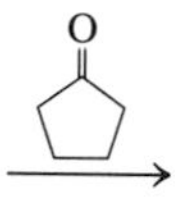 $\xrightarrow{\text{cyclopentanone}}$

(e) $CH_3CH_2CH_2Cl + Mg \xrightarrow{(CH_3CH_2)_2O}$ (f) Product of (e) + $C_6H_5CCH_3$ (O above C) $\longrightarrow$

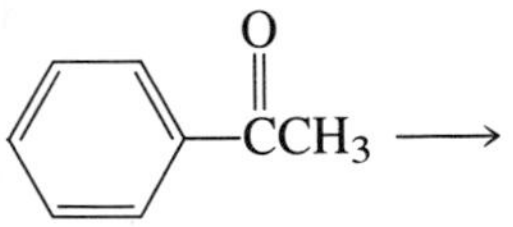

(g) 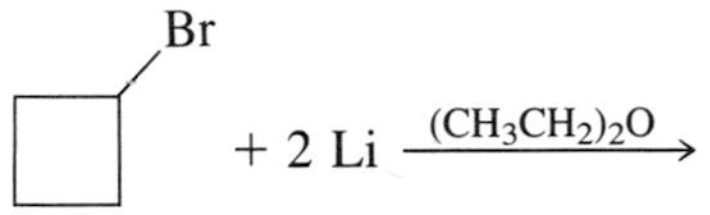 Bromocyclobutane $+ 2\ Li \xrightarrow{(CH_3CH_2)_2O}$ (h) 2 mol product of (g) +1 mol $CH_3\overset{O}{\overset{\|}{C}}CH_2CH_2\overset{O}{\overset{\|}{C}}CH_3 \longrightarrow$

15. Give the major product(s) of each of the following reactions (after aqueous work-up). The solvent in each case is ethoxyethane (diethyl ether).

(a) 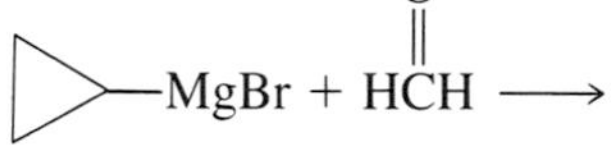 Cyclopropyl–MgBr + $H\overset{O}{\overset{\|}{C}}H \longrightarrow$ (b) $CH_3\overset{CH_3}{\overset{|}{C}}HCH_2MgCl + CH_3\overset{O}{\overset{\|}{C}}H \longrightarrow$

(c) $C_6H_5CH_2Li + C_6H_5\overset{O}{\overset{\|}{C}}H \longrightarrow$ (d) 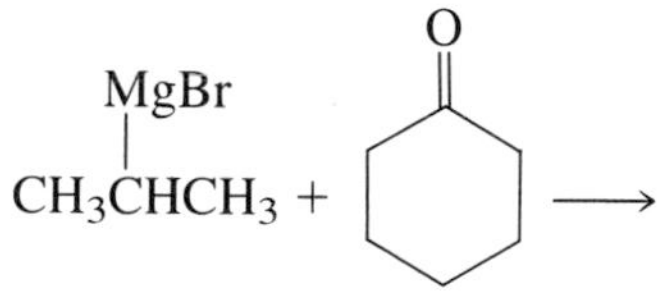 $CH_3\overset{MgBr}{\overset{|}{C}}HCH_3$ + cyclohexanone $\longrightarrow$

(e) 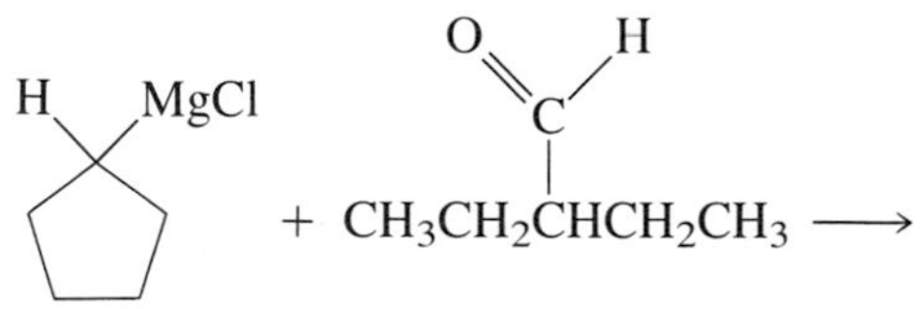 1-(H)(MgCl)cyclopentane + $CH_3CH_2\overset{|}{C}HCH_2CH_3$ (with C(=O)H substituent: O=C–H) $\longrightarrow$

16. Give the expected major product of each reaction shown below. PCC is the abbreviation for pyridinium chlorochromate (Section 8-6).

(a) $CH_3CH_2CH_2OH \xrightarrow{Na_2Cr_2O_7,\ H_2SO_4,\ H_2O}$ (b) $(CH_3)_2CHCH_2OH \xrightarrow{PCC,\ CH_2Cl_2}$

(c) 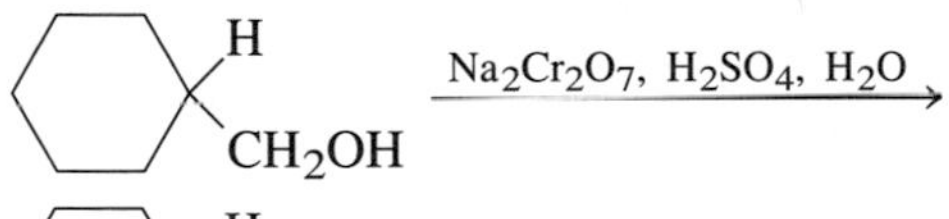 Cyclohexyl(H)$CH_2OH \xrightarrow{Na_2Cr_2O_7,\ H_2SO_4,\ H_2O}$ (d) 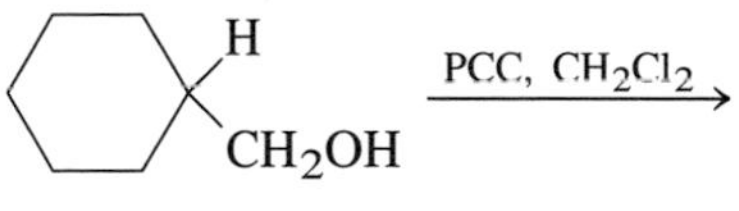 Cyclohexyl(H)$CH_2OH \xrightarrow{PCC,\ CH_2Cl_2}$

(e) 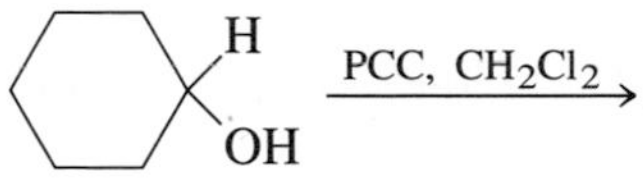 Cyclohexyl(H)OH $\xrightarrow{PCC,\ CH_2Cl_2}$

17. Give the expected major product of each reaction *sequence* shown below. PCC refers to pyridinium chlorochromate.

(a) $(CH_3)_2CHOH$ $\xrightarrow{1.\ CrO_3,\ H_2SO_4,\ H_2O;\ 2.\ CH_3CH_2MgBr,\ (CH_3CH_2)_2O;\ 3.\ H^+,\ H_2O}$

(b) $CH_3CH_2CH_2CH_2Cl \xrightarrow[\text{4. } H^+,\ H_2O]{\text{1. } {}^-OH,\ H_2O \quad \text{2. PCC, } CH_2Cl_2 \quad \text{3. cyclopentyl-Li, } (CH_3CH_2)_2O}$

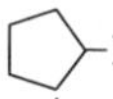

(c) Product of (b) $\xrightarrow[\text{3. } H^+,\ H_2O]{\text{1. } CrO_3,\ H_2SO_4,\ H_2O \quad \text{2. } LiAlD_4,\ (CH_3CH_2)_2O}$

18. Unlike Grignard and organolithium reagents, organometallic compounds of the most electropositive metals (Na, K, etc.) react rapidly with haloalkanes. As a result, attempts to convert RX to RNa or RK by reaction with the corresponding metal lead to alkanes by a reaction called *Wurtz* coupling.

$$2\ RX + 2\ Na \longrightarrow R{-}R + 2\ NaX$$

which is the result of

$$R{-}X + 2\ Na \longrightarrow R{-}Na + NaX$$

followed rapidly by

$$R{-}Na + R{-}X \longrightarrow R{-}R + NaX$$

When it was still in use, the Wurtz coupling reaction was employed mainly for the preparation of alkanes from the coupling of two identical alkyl groups (e.g., equation 1 below). Suggest a reason why Wurtz coupling might not be a useful method for coupling two different alkyl groups (equation 2 below).

$$2\ CH_3CH_2CH_2Cl + 2\ Na \longrightarrow CH_3CH_2CH_2CH_2CH_2CH_3 + 2\ NaCl \quad (1)$$

$$CH_3CH_2Cl + CH_3CH_2CH_2Cl + 2\ Na \longrightarrow CH_3CH_2CH_2CH_2CH_3 + 2\ NaCl \quad (2)$$

19. The reaction of two equivalents of Mg with 1,4-dibromobutane produces compound A. The reaction of A with two equivalents of CH_3CHO (acetaldehyde), followed by work-up with dilute aqueous acid, produces compound B, having the formula $C_8H_{18}O_2$. What are the structures of A and B?

20. Suggest the best synthetic route to each of the simple alcohols below, using in each case a simple alkane as your initial starting molecule. What are some disadvantages of beginning syntheses with alkanes?

(a) Methanol **(b)** Ethanol **(c)** 1-Propanol
(d) 2-Propanol **(e)** 1-Butanol **(f)** 2-Butanol
(g) 2-Methyl-2-propanol

21. For each alcohol in Problem 20, suggest (if possible) a synthetic route that starts with, first, an aldehyde and, second, a ketone.

22. Outline the best method for preparing each of the following compounds from an appropriate alcohol.

(a) 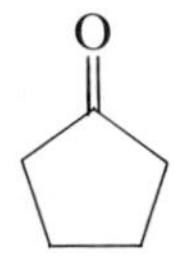 cyclopentanone **(b)** $CH_3CH_2CH_2CH_2COOH$ **(c)** 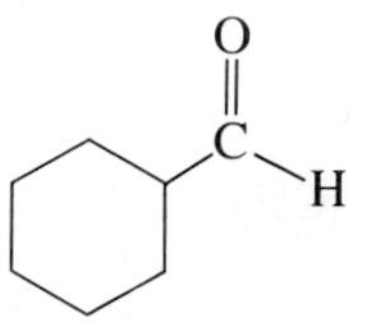 cyclohexanecarbaldehyde (cyclohexyl–C(=O)H)

(d) 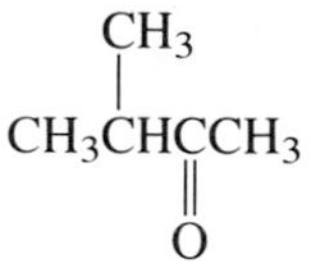 $CH_3CH(CH_3)CCH_3$ (with C=O on the second C from the right: $CH_3CH(CH_3)C(=O)CH_3$) (e) 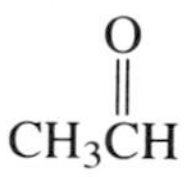 CH_3CH with C=O: $CH_3C(=O)H$

23. Suggest three different syntheses of 2-methyl-2-hexanol. Each route should utilize one of the starting materials listed below. Then use any number of steps and any other reagents needed.

(a) 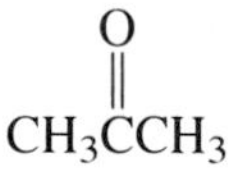 $CH_3C(=O)CH_3$ **(b)** 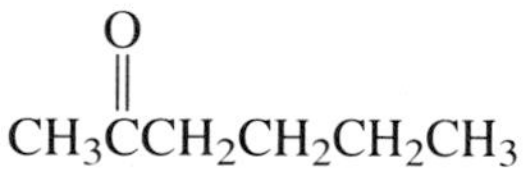 $CH_3C(=O)CH_2CH_2CH_2CH_3$ **(c)** $CH_3CH_2CH_2CH_2CHO$

24. Devise three different syntheses of 3-octanol starting with **(a)** a ketone; **(b)** an aldehyde; **(c)** an aldehyde different from that employed in (b).

25. Propose sensible synthetic schemes for the preparation of each of the following compounds, using only the organic starting material(s) indicated. Use any organic solvents or inorganic reagents necessary.

(a) 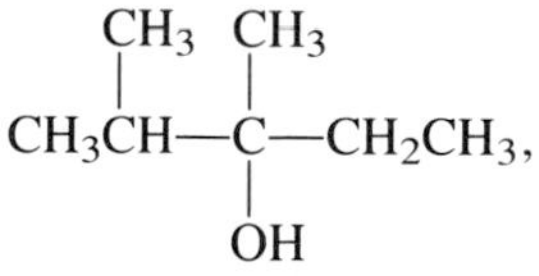 $CH_3CH(CH_3)—C(CH_3)(OH)—CH_2CH_3$, from ethane and propane

(b) 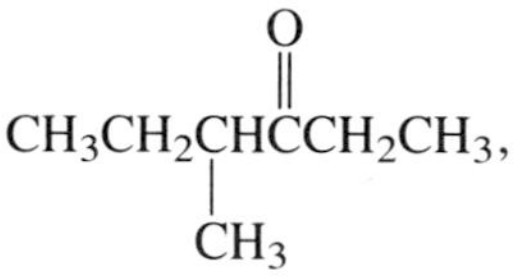 $CH_3CH_2CH(CH_3)C(=O)CH_2CH_3$, from butane, ethane, and formaldehyde $\left(HC(=O)H\right)$

26. Waxes are naturally occurring esters (alkyl alkanoates) containing long, straight alkyl chains. Whale oil contains the wax 1-hexadecyl hexadecanoate, as shown in the margin. How would you synthesize this wax, using an S_N2 reaction?

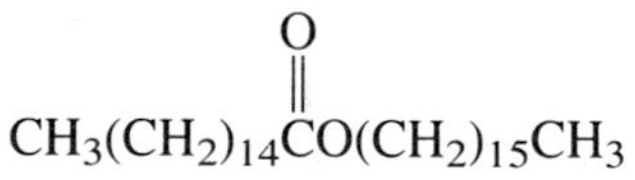

$CH_3(CH_2)_{14}C(=O)O(CH_2)_{15}CH_3$
1-Hexadecyl hexadecanoate

27. The B vitamin commonly known as niacin is used by the body to synthesize the coenzyme nicotinamide adenine dinucleotide (NAD^+; Chapter 25). In the presence of a variety of enzyme catalysts, the reduced form of this substance (NADH) acts as a biological hydride donor, capable of reducing aldehydes and ketones to alcohols, according to the general formula

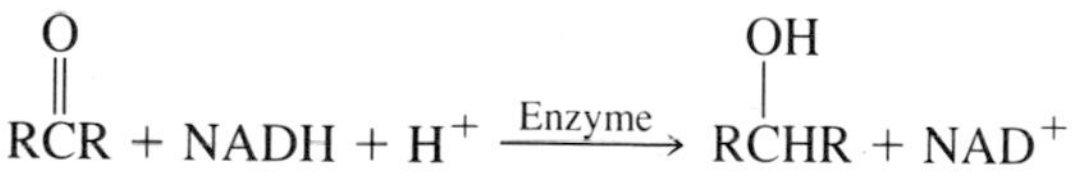

$$RC(=O)R + NADH + H^+ \xrightarrow{\text{Enzyme}} RCH(OH)R + NAD^+$$

The COOH functional group of carboxylic acids is not reduced. Write the products of the NADH reduction of each of the molecules below.

(a)  $CH_3C(=O)H + NADH \xrightarrow{\text{Alcohol dehydrogenase}}$

(b)  $CH_3C(=O)C(=O)OH + NADH \xrightarrow{\text{Lactate dehydrogenase}}$ **Lactic acid**

2-Oxopropanoic acid
(Pyruvic acid)

(c) $\overset{O}{\overset{\|}{HOC}}CH_2\overset{OO}{\overset{\|\|}{CCOH}}$ + NADH $\xrightarrow{\text{Malate dehydrogenase}}$ **Malic acid**

2-Oxobutanedioic acid (Oxaloacetic acid)

28. Reductions by NADH (Problem 27) are stereospecific, with the stereochemistry of the product controlled by an enzyme. The common forms of lactate and malate dehydrogenases produce exclusively the *S* stereoisomers of lactic and malic acids, respectively. Draw these stereoisomers.

29. Chemically modified steroids have become increasingly important in medicine. Give the possible product(s) of the reactions shown below. In each case identify the major stereoisomer formed on the basis of delivery of the attacking reagent from the less hindered side of the substrate molecule. (Hint: Make models and refer to Section 4-7.)

(a) [steroid structure: HO, H, CH$_3$, H, H, H, O, CH$_3$, CH(OH)—CH$_3$] $\xrightarrow[\text{2. } H^+,\ H_2O]{\text{1. Excess } CH_3MgI}$

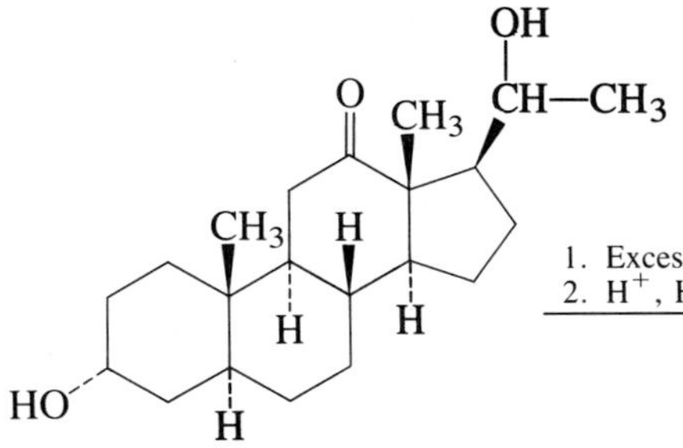

(b) [steroid structure: HO, H, CH$_3$, H, H, H, O, CH$_3$, O] $\xrightarrow[\text{2. } H^+,\ H_2O]{\text{1. Excess } CH_3Li}$

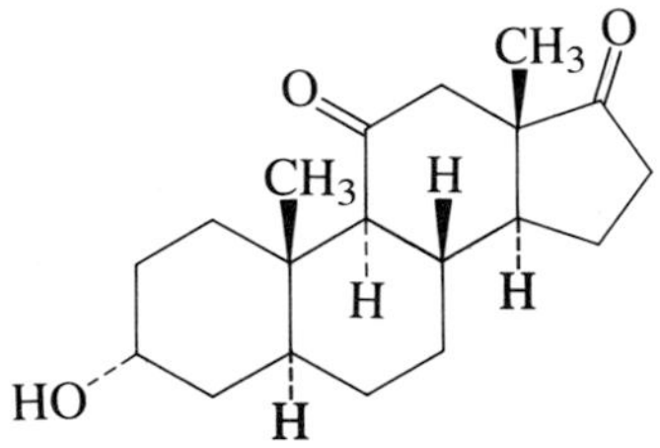

9

Further Reactions of Alcohols and the Chemistry of Ethers

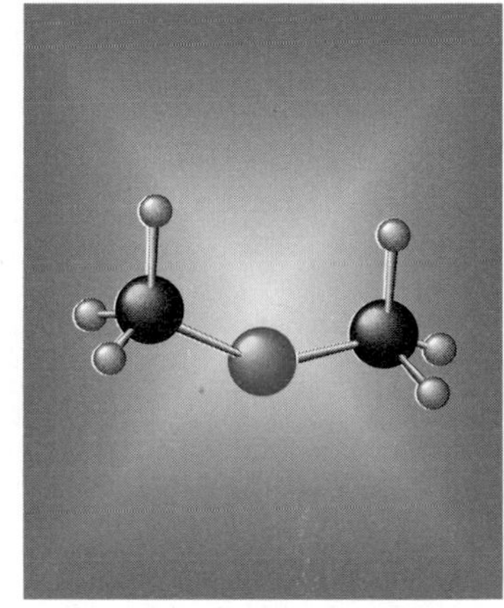

Now that we have been introduced to the properties of alcohols and the methods of their preparation, we shall examine the further transformations of the hydroxy substituent into other functional groups. Figure 9-1 depicts a variety of reaction modes available to alcohols. Usually at least one of the four bonds marked *a*, *b*, *c*, and *d* is cleaved. In Chapter 8 we learned that oxidation to aldehydes and ketones breaks bonds *a* and *d*. We found that the use of this reaction in combination with additions of organometallic reagents provides us with the means to prepare alcohols of considerable structural diversity. To further explore the reactions of alcohols, we shall start by reviewing their acidic and basic properties. Deprotonation at bond *a* furnishes alkoxides, which are valuable both as strong bases and as nucleophiles (Section 7-8). Strong acids transform alcohols into alkyloxonium ions (Section 8-3), converting the OH into a good leaving group. Subsequently, bond *b* may break, thereby leading to substitution; or elimination can take place by

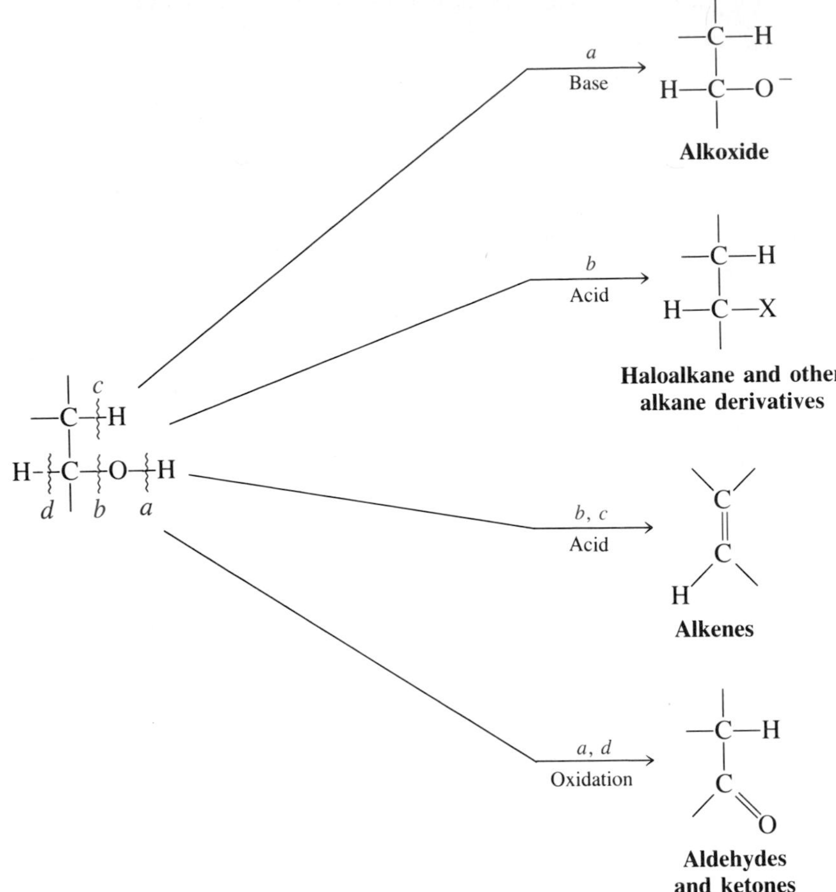

FIGURE 9-1 Four typical reaction modes of alcohols. In each, one or more of the four bonds marked *a–d* are cleaved (wavy line denotes bond cleavage): (*a*) deprotonation by base; (*b*) protonation by acid followed by uni- or bimolecular substitution; (*b*, *c*) elimination; and (*a*, *d*) oxidation.

cleavage of bonds *b* and *c*. We shall see that the carbocations arising from acid treatment of secondary and tertiary alcohols have a varied chemistry. An introduction to the preparation of esters and their applications in synthesis is followed by the chemistry of ethers and sulfur compounds. Alcohols, ethers, and their sulfur analogs occur widely in nature and have numerous applications in industry and medicine.

9-1 Preparation of Alkoxides

As described in Section 8-3, alcohols can be both acids and bases. In this section we shall review the methods by which alcohols are deprotonated to furnish their conjugate bases, the alkoxides.

Strong bases are needed to deprotonate alcohols

To remove a proton from the OH group of an alcohol (Figure 9-1, cleavage of bond *a*), we must use a base stronger than the alkoxide. Examples include lithium diisopropylamide, butyllithium, and alkali metal hydrides such as potassium hydride, KH. The latter are particularly useful because the only by-product of the reaction is hydrogen gas.

Three Ways of Making Methoxide from Methanol

$$\underset{pK_a = 15.5}{CH_3OH} + Li^{+\,-}N(CH(CH_3)_2)CH(CH_3)_2 \overset{K = 10^{24.5}}{\rightleftharpoons} CH_3O^-Li^+ + \underset{pK_a = 40}{HN(CH(CH_3)_2)CH(CH_3)_2}$$

$$\underset{pK_a = 15.5}{CH_3OH} + CH_3CH_2CH_2CH_2Li \overset{K = 10^{34.5}}{\rightleftharpoons} CH_3O^-Li^+ + \underset{pK_a = 50}{CH_3CH_2CH_2CH_2H}$$

$$\underset{pK_a = 15.5}{CH_3OH} + K^+H^- \overset{K = 10^{22.5}}{\rightleftharpoons} CH_3O^-K^+ + \underset{pK_a = 38}{H\text{—}H}$$

EXERCISE 9-1

Considering the pK_a data in Table 6-4, would you use sodium cyanide as a reagent to convert methanol into sodium methoxide? Explain your answer.

Alkali metals also deprotonate alcohols

Another common way to obtain alkoxides is by the reaction of alcohols with alkali metals. Such metals reduce water—in some cases, violently—to yield alkali metal hydroxides and hydrogen gas. When the more reactive metals (sodium, potassium, cesium) are exposed to water in air, the hydrogen generated can ignite spontaneously or even detonate.

$$2\ H\text{—}OH + 2\ M\ (Li, Na, K) \longrightarrow 2\ M^+\ {}^-OH + H_2$$

The alkali metals act similarly on the alcohols to give alkoxides, but the transformation is less vigorous. Here are two examples.

Alkoxides from Alcohols and Alkali Metals

$$2\ CH_3CH_2OH + 2\ Na \longrightarrow 2\ CH_3CH_2O^-Na^+ + H_2$$
$$2\ (CH_3)_3COH + 2\ K \longrightarrow 2\ (CH_3)_3CO^-K^+ + H_2$$

The reactivity of the alcohols used in this process decreases with increasing substitution, methanol being most reactive and tertiary alcohols least reactive.

Relative Reactivity of ROH with Alkali Metals

$$R = CH_3 > \text{primary} > \text{secondary} > \text{tertiary}$$

2-Methyl-2-propanol reacts so slowly that it can be used to safely destroy potassium residues in the laboratory.

What are alkoxides good for? We have already seen that they can be useful reagents in organic synthesis. For example, the reaction of hindered alkoxides with haloalkanes gives elimination.

$$CH_3CH_2CH_2CH_2Br \xrightarrow{(CH_3)_3CO^-K^+,\ (CH_3)_3COH} CH_3CH_2CH{=}CH_2 + (CH_3)_3COH + K^+Br^-$$

Less branched alkoxides attack primary haloalkanes by the S_N2 reaction to give ethers. This method is described in Section 9-6.

In summary, the treatment of an alcohol with a strong base or an alkali metal gives an alkoxide. Let us now inspect the protonation of alcohols with strong acids.

9-2 Preparation of Alkyloxonium Ions: Substitution and Elimination Reactions of Alcohols

We have seen that heterolytic cleavage of the O–H bond in alcohols is readily achieved with strong bases. Can we break the C–O linkage (bond *b*, Figure 9-1) as easily? Yes, but now we need acid. Recall (Section 6-7) that water has a high pK_a: It is a weak acid. Consequently, hydroxide, its conjugate base, is an exceedingly poor leaving group. *For alcohols to undergo substitution or elimination reactions, the OH must first be converted into a better leaving group.*

Water can be a leaving group

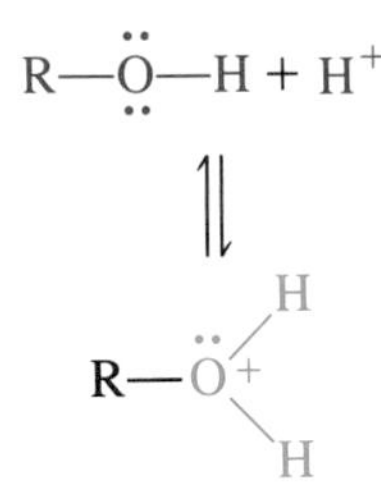

Alkyloxonium ion

The simplest way of turning the hydroxy substituent in alcohols into a good leaving group is by protonation of the oxygen to form an alkyloxonium ion. Recall (Section 8-3) that this process ties up one of the oxygen lone pairs in bonding to a proton. The positive charge therefore resides on the oxygen atom. *Protonation turns OH from a bad leaving group into a good one, neutral water.*

Alkyloxonium ions derived from primary alcohols are subject to nucleophilic attack. For example, the alkyloxonium ion resulting from the treatment of 1-butanol with HBr undergoes displacement by bromide to form 1-bromobutane. Note that in the following scheme color is again used in a functional sense. The originally nucleophilic (red) oxygen is protonated by the electrophilic proton (blue) to give the alkyloxonium ion containing an electrophilic (blue) carbon and H_2O as a leaving group (green). In the subsequent S_N2 reaction, bromide acts as a nucleophile.

Primary Bromoalkane Synthesis from an Alcohol

$$CH_3CH_2CH_2CH_2OH + HBr \longrightarrow CH_3CH_2CH_2CH_2\overset{+}{O}H_2 + Br^- \longrightarrow CH_3CH_2CH_2CH_2Br + H_2O$$

Iodoalkane Synthesis

$$HO(CH_2)_6OH + 2\ HI \longrightarrow I(CH_2)_6I + 2\ H_2O$$

1,6-Hexanediol → **1,6-Diiodohexane** (85%)

Iodoalkanes also can be made in this fashion; however, primary chloroalkanes cannot, because chloride ion is too weak a nucleophile. The conversion of a primary alcohol into the corresponding chloroalkane requires other methods (Section 9-4).

Secondary and tertiary alcohols undergo carbocation reactions with acids

Alkyloxonium ions derived from secondary and tertiary alcohols, in contrast to their primary counterparts, readily lose water to give the corresponding carbocations. Depending on conditions, these ions undergo S_N1 or E1 reactions or both. When nucleophiles are present, we observe S_N1 products. Thus, a common method of preparing a tertiary haloalkane is simply to mix a tertiary alcohol with excess concentrated aqueous hydrogen halide. The product forms in minutes at room temperature. The mechanism, which is precisely the reverse of that of solvolysis (Section 7-2), proceeds by the formation of a carbocation.

Conversion of 2-Methyl-2-propanol to 2-Bromo-2-methylpropane

$$(CH_3)_3COH + \underset{\text{Excess}}{HBr} \rightleftharpoons (CH_3)_3CBr + H_2O$$

Mechanism of the S_N1 Reaction of Tertiary Alcohols with Hydrogen Halides

$$(CH_3)_3C{-}\ddot{O}H + H{-}Br \rightleftharpoons (CH_3)_3C{-}\overset{+}{O}H_2 + Br^-$$

$$\rightleftharpoons H_2O + (CH_3)_3\overset{+}{C} + Br^- \rightleftharpoons H_2O + (CH_3)_3C{-}Br$$

In the absence of good nucleophiles, E1 products are obtained (Section 7-6), especially at elevated temperatures. This reaction, called **dehydration** because it involves the loss of a molecule of water (Figure 9-1, breaking bonds *b* and *c*), is one of the methods for the synthesis of alkenes (Chapter 11).

Alcohol Dehydration by the E1 Mechanism

$$\text{Cyclohexanol} \xrightarrow[-H_2O]{H_2SO_4,\ 130\text{–}140°C} \text{Cyclohexene (87\%)}$$

Cyclohexanol → 87% **Cyclohexene**

In this case the counterion of the acid is the poor nucleophile HSO_4^-, and proton loss from the intermediate carbocation is observed. Dehydration of tertiary alcohols is even easier, often occurring at or only slightly above room temperature.

EXERCISE 9-2

Write the structure of the product you expect from the reaction of 4-methyl-1-pentanol with concentrated aqueous HI. Give the mechanism of its formation.

EXERCISE 9-3

Write the structure of the products you expect from the reaction of 1-methylcyclohexanol with **(a)** concentrated HCl and **(b)** concentrated H_2SO_4. Compare and contrast the mechanisms of the two processes.

In summary, treatment of alcohols with strong acid leads to protonation to give alkyloxonium ions, which are relatively stable if they are primary, undergoing S_N2 reactions in the presence of good nucleophiles. Alkyloxonium ions from secondary or tertiary alcohols convert into carbocations, which furnish products of substitution and elimination (dehydration).

9-3 Carbocation Rearrangements

Carbocations can rearrange by both hydrogen and alkyl shifts to become new carbocations, which can then undergo S_N1 and E1 reactions. Unless there is a thermodynamic driving force toward one product, the result is likely to be a complex mixture.

Hydrogen shifts give new S_N1 products

Treatment of 2-propanol with hydrogen bromide at low temperatures (to prevent elimination) gives 2-bromopropane, as expected. However, exposure of the more highly substituted secondary alcohol, 3-methyl-2-butanol, to the same reaction conditions produces a surprising result. The expected S_N1 product, 2-bromo-3-methylbutane, is only a minor component of the reaction mixture, and the major product is 2-bromo-2-methylbutane.

Normal S_N1 Reaction of an Alcohol (No Rearrangement)

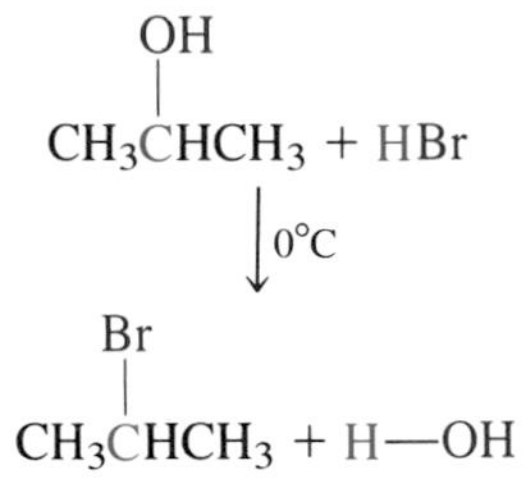

$CH_3CH(OH)CH_3 + HBr \xrightarrow{0°C} CH_3CH(Br)CH_3 + H{-}OH$

Hydrogen Shift in the S_N1 Reaction of an Alcohol with HBr

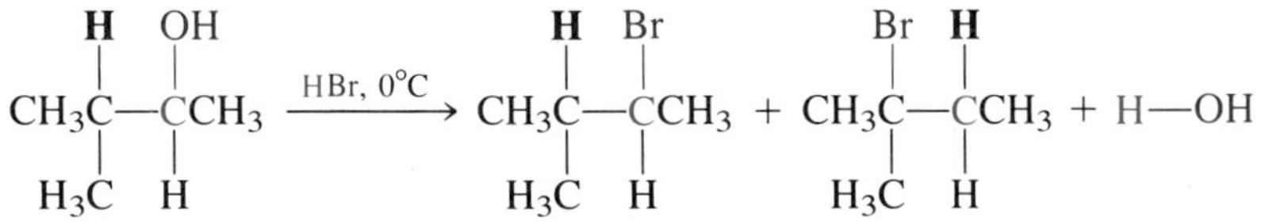

$CH_3C(\mathbf{H})(H_3C){-}C(OH)(H)CH_3 \xrightarrow{HBr,\ 0°C} CH_3C(\mathbf{H})(H_3C){-}C(Br)(H)CH_3 + CH_3C(Br)(H_3C){-}C(\mathbf{H})(H)CH_3 + H{-}OH$

3-Methyl-2-butanol

Minor product: **2-Bromo-3-methylbutane** (Normal product)

Major product: **2-Bromo-2-methylbutane** (Rearranged product)

In this mechanistic scheme and those that follow, color is used to indicate the electrophilic (blue), nucleophilic (red), and leaving-group (green) character of the reacting centers. Therefore, a color may again "switch" from one group or atom to another as the reaction proceeds.

In this reaction scheme and in those that follow, color is used to indicate the origin of groups or atoms in a product. The migrating group or atom is in **bold-face** type.

What is the mechanism of this transformation? The answer is that *carbocations can undergo rearrangement* by **hydrogen shifts,** in which the hydrogen moves with both electrons from its original position to the neighboring carbon atom. Initially, protonation of the alcohol followed by loss of water gives the expected secondary carbocation. A shift of the tertiary hydrogen to the electron-deficient neighbor then generates a tertiary cation, *which is more stable*. This species is finally trapped by bromide ion to give the rearranged S_N1 product. The mechanism for this rearrangement is illustrated below. (Recall that the arrows denote the flow of electron pairs.)

Mechanism of Carbocation Rearrangement

STEP 1. Protonation

$CH_3C(H)(H_3C){-}C(\ddot{O}H)(H)CH_3 + H^+ \rightleftharpoons CH_3C(H)(H_3C){-}C(\overset{+}{O}H_2)(H)CH_3$

STEP 2. Loss of water

$CH_3C(H)(H_3C){-}C(\overset{+}{O}H_2)(H)CH_3 \rightleftharpoons CH_3C(H)(H_3C){-}\overset{+}{C}(H)CH_3 + H_2\ddot{O}$

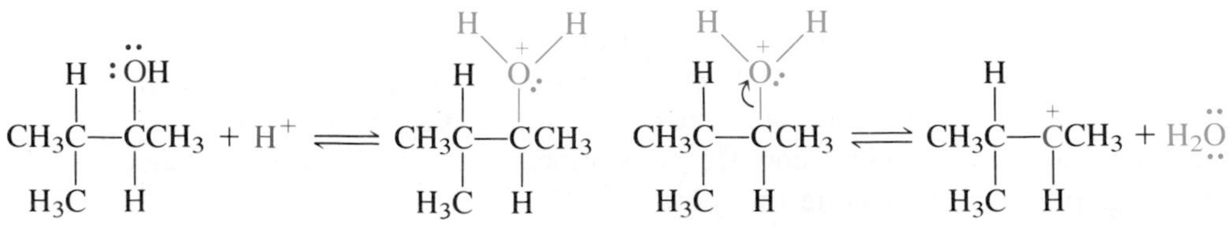

STEP 3. Hydrogen shift

$CH_3\overset{H}{C}—\overset{+}{C}CH_3 \longrightarrow CH_3\overset{+}{C}—\overset{H}{C}CH_3$ (lower substituents: H_3C, H in both)

Secondary cation (Less stable) → **Tertiary cation** (More stable)

STEP 4. Trapping by bromide

$CH_3\overset{+}{C}—\overset{H}{C}CH_3 + :\ddot{Br}:^- \rightleftharpoons CH_3\overset{:\ddot{Br}:}{C}—\overset{H}{C}CH_3$ (lower substituents: H_3C, H)

The details of the transition state of the observed hydrogen shift are shown schematically in Figure 9-2. The process is sometimes called **hydride transfer** (although *free* hydride ions, $H:^-$, are never actually present). A simple rule to remember when executing hydrogen shifts in carbocations is that *the hydrogen and the positive charge exchange places* between the two neighboring carbon atoms participating in the reaction.

Hydrogen shifts are generally very fast. They are particularly favored when the new carbocation is more stable than the original one, as in the example depicted in Figure 9-2.

EXERCISE 9-4

2-Methylcyclohexanol on treatment with HBr gives 1-bromo-1-methylcyclohexane. Explain by a mechanism.

FIGURE 9-2 The rearrangement of a carbocation by a hydrogen shift. (A) Dotted-line notation; (B) orbital picture. Note that the migrating hydrogen and the positive charge exchange places.

A: H, H_3C, H_3C, C, sp^3 — $\overset{+}{C}$, H, CH_3, sp^2 ⇌ [H, δ^+ C, δ^+ C, H_3C, H_3C, H, CH_3, Both carbons are rehybridizing]‡ ⇌ H_3C, H_3C, $\overset{+}{C}$, sp^2 — C, H, H, CH_3, sp^3

B: H, H_3C, H_3C, C, +, C, H, CH_3 ⇌ [H, δ^+, δ^+, C, C, H_3C, H_3C, H, CH_3, *p* back lobe increasing, *p* back lobe decreasing]‡ ⇌ H_3C, H_3C, +, C, C, H, H, CH_3

EXERCISE 9-5

Predict the major product from the following reactions.

(a) 2-Methyl-3-pentanol + H_2SO_4, CH_3OH solvent

(b) 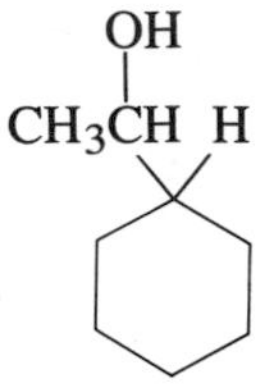 (Cyclohexane with $CH_3CH(OH)$ and H on one carbon) + HCl

Primary carbocations are too unstable to be formed by rearrangement. However, carbocations of comparable stability—for example, secondary–secondary or tertiary–tertiary—equilibrate readily. In this case, any added nucleophile will trap all carbocations present, furnishing mixtures of products.

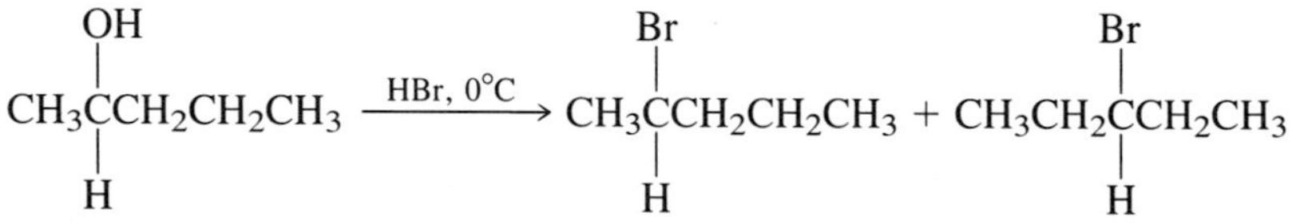

$$CH_3\underset{H}{\overset{OH}{C}}CH_2CH_2CH_3 \xrightarrow{HBr,\ 0^\circ C} CH_3\underset{H}{\overset{Br}{C}}CH_2CH_2CH_3 + CH_3CH_2\underset{H}{\overset{Br}{C}}CH_2CH_3$$

Under conditions favoring dissociation, haloalkanes also can undergo carbocation rearrangements. For example, ethanolysis of 2-bromo-3-ethyl-2-methylpentane gives the two possible tertiary ethers.

Rearrangement During Solvolysis of a Haloalkane

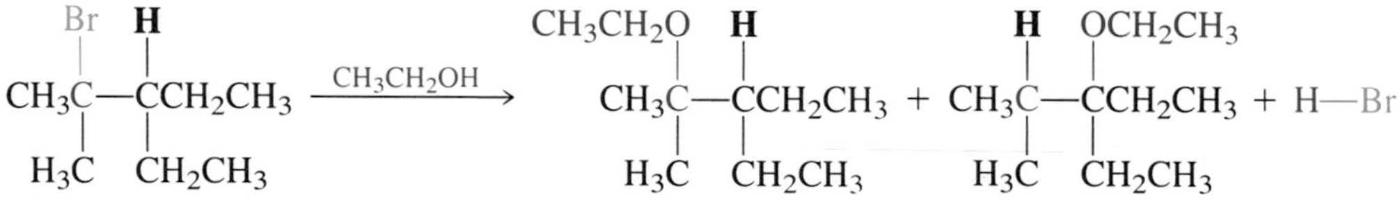

$$CH_3\underset{H_3C}{\overset{Br}{C}}—\underset{CH_2CH_3}{\overset{H}{C}}CH_2CH_3 \xrightarrow{CH_3CH_2OH} CH_3\underset{H_3C}{\overset{CH_3CH_2O}{C}}—\underset{CH_2CH_3}{\overset{H}{C}}CH_2CH_3 + CH_3\underset{H_3C}{\overset{H}{C}}—\underset{CH_2CH_3}{\overset{OCH_2CH_3}{C}}CH_2CH_3 + H—Br$$

2-Bromo-3-ethyl-2-methylpentane → **2-Ethoxy-3-ethyl-2-methylpentane** (Normal product) + **3-Ethoxy-3-ethyl-2-methylpentane** (Rearranged product)

EXERCISE 9-6

Give a mechanism for the reaction shown above. Then predict the outcome of the reaction of 2-chloro-4-methylpentane with methanol.

Carbocation rearrangements also give new E1 products

How does the rearrangement of intermediates affect the outcome of reactions under conditions that favor elimination? At elevated temperatures and in relatively nonnucleophilic media, rearranged carbocations yield alkenes by the E1 mechanism. For example, treatment of 2-methyl-2-pentanol with sulfuric acid at 80°C gives the same major alkene product as that formed when the starting material is 4-methyl-2-pentanol. The conversion of the latter alcohol includes a hydrogen shift of the initial carbocation, followed by deprotonation.

Rearrangement During E1 Elimination

$$\underset{\text{2-Methyl-2-pentanol}}{CH_3C(OH)(CH_3)CH_2CH_2CH_3} \xrightarrow[-\,H_2O]{H_2SO_4,\ 80^{\circ}C} \underset{\substack{\text{Major product}\\ \text{2-Methyl-2-pentene}}}{(H_3C)_2C{=}C(H)CH_2CH_3} \xleftarrow[\substack{-\,H_2O\\ \text{With}\\ \text{rearrangement}}]{H_2SO_4,\ 80^{\circ}C} \underset{\text{4-Methyl-2-pentanol}}{CH_3CH(CH_3)CH_2CH(OH)CH_3}$$

EXERCISE 9-7

(a) Give mechanisms for the reactions shown above. **(b)** Treatment of 4-methylcyclohexanol with hot acid gives 1-methylcyclohexene. Explain by a mechanism.

Other carbocation rearrangements involve alkyl shifts

Carbocations without suitable (secondary and tertiary) hydrogens next to the positively charged carbon can undergo another mode of rearrangement, known as **alkyl group migration,** or **alkyl shift.**

Rearrangement by Alkyl Shift During S_N1 Reaction

$$\underset{\text{3,3-Dimethyl-2-butanol}}{CH_3C(CH_3)_2{-}CH(OH)CH_3} \xrightarrow[-\,HOH]{HBr} \underset{\substack{94\%\\ \text{2-Bromo-2,3-dimethylbutane}}}{CH_3C(Br)(CH_3){-}CH(CH_3)CH_3}$$

As in the hydrogen shift, the migrating group takes its electron pair with it to form a bond to the neighboring carbocation. *The moving alkyl group and the positive charge exchange places.*

$$CH_3C(CH_3)_2{-}CH(OH)CH_3 + H^+ \underset{+\,H_2O}{\overset{-\,H_2O}{\rightleftharpoons}} CH_3C(CH_3)_2{-}\overset{+}{C}HCH_3 \xrightarrow{CH_3\ \text{shift}} CH_3\overset{+}{C}(CH_3){-}CH(CH_3)CH_3 \underset{-\,Br^-}{\overset{+\,Br^-}{\rightleftharpoons}} CH_3C(Br)(CH_3){-}CH(CH_3)CH_3$$

EXERCISE 9-8

At higher temperatures, 3,3-dimethyl-2-butanol gives two products of E1 reaction, one derived from the carbocation present prior to rearrangement, the other from that formed after alkyl shift has taken place. Give the structures of these elimination products.

Primary alcohols may undergo rearrangement

Treatment of a primary alcohol with HBr or HI normally produces the corresponding haloalkane through S_N2 reaction of the alkyloxonium ion (Section 9-2). However, it is possible in some cases to observe alkyl and hydrogen shifts to primary carbons bearing leaving groups, even though primary carbocations are not formed in solution. An example is furnished by 2,2-dimethyl-1-propanol (neopentyl alcohol). Treatment of this compound with strong acid causes rearrangement, despite the fact that a primary carbocation cannot be an intermediate.

Rearrangement in a Primary Substrate

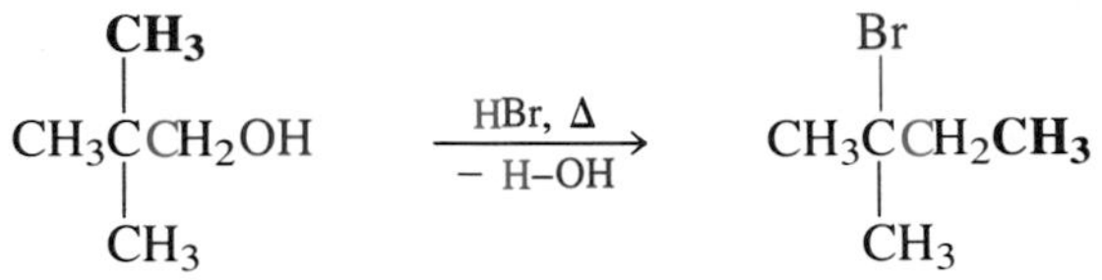

$$CH_3C(CH_3)_2CH_2OH \xrightarrow[- \text{H–OH}]{\text{HBr}, \Delta} CH_3C(Br)(CH_3)CH_2CH_3$$

2,2-Dimethyl-1-propanol (Neopentyl alcohol) → **2-Bromo-2-methylbutane**

In this case, after protonation to form the alkyloxonium ion, steric hindrance interferes with direct displacement by bromide (Section 6-9). Instead, water leaves *at the same time* as a methyl group migrates from the neighboring carbon, thus avoiding formation of a primary carbocation.

Mechanism of Concerted Alkyl Shift

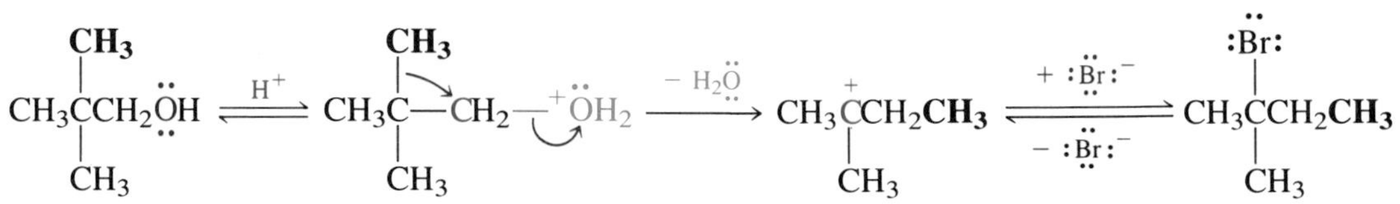

$$CH_3C(CH_3)_2CH_2\ddot{O}H \underset{}{\overset{H^+}{\rightleftharpoons}} CH_3C(CH_3)_2-CH_2-\overset{+}{\ddot{O}}H_2 \xrightarrow{-H_2\ddot{O}} CH_3\overset{+}{C}(CH_3)CH_2CH_3 \underset{-\,:\ddot{Br}:^-}{\overset{+\,:\ddot{Br}:^-}{\rightleftharpoons}} CH_3C(:\ddot{Br}:)(CH_3)CH_2CH_3$$

Nucleophile—red
Electrophile—blue
Leaving group—green

Rearrangements of primary substrates are relatively difficult processes, usually requiring elevated temperatures and long reaction times.

In summary, another mode of reactivity of carbocations, in addition to regular S_N1 and E1 processes, is rearrangement by hydrogen or alkyl shifts. In such rearrangements, the migrating group delivers its bonding electron pair to a positively charged carbon neighbor, exchanging places with the charge. Rearrangement may lead to a more stable cation—as in the conversion of a secondary cation into a tertiary one. Primary alcohols also can undergo rearrangement, but do so by concerted pathways and not through the intermediacy of primary cations.

9-4 Organic and Inorganic Esters from Alcohols

Among the most useful reactions of alcohols is their conversion to esters. This term commonly refers to **organic esters,** also called **carboxylates** or **alkanoates** (Table 2-1). They are formally derived from organic (carboxylic) acids by replacement of the acidic hydrogen with an alkyl group. **Inorganic esters** are the analogous derivatives of inorganic acids.

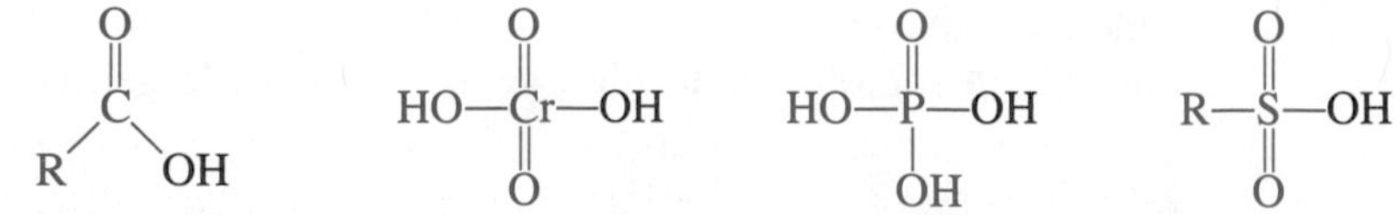

$$RC(=O)OH \qquad HO-Cr(=O)_2-OH \qquad HO-P(=O)(OH)-OH \qquad R-S(=O)_2-OH$$

A carboxylic acid | **Chromic acid** | **Phosphoric acid** | **A sulfonic acid**

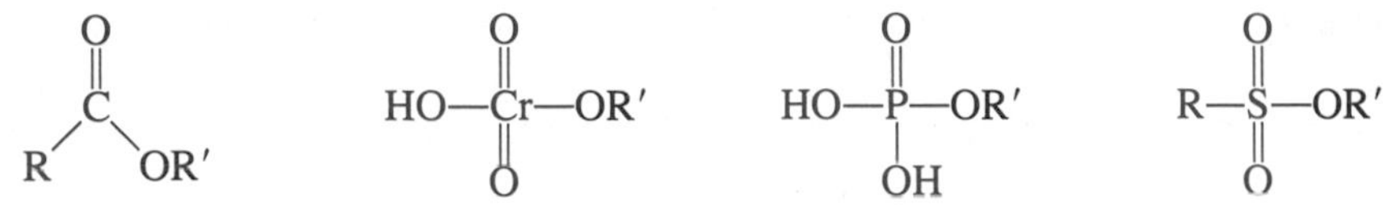

$$\underset{\text{A carboxylate ester (An organic ester)}}{R-\overset{O}{\overset{\|}{C}}-OR'} \qquad \underset{\text{A chromate ester}}{HO-\overset{O}{\overset{\|}{\underset{O}{\underset{\|}{Cr}}}}-OR'} \qquad \underset{\text{A phosphate ester}}{HO-\overset{O}{\overset{\|}{\underset{OH}{\underset{|}{P}}}}-OR'} \qquad \underset{\text{A sulfonate ester}}{R-\overset{O}{\overset{\|}{\underset{O}{\underset{\|}{S}}}}-OR'}$$

A carboxylate ester (An organic ester) **A chromate ester** **A phosphate ester** **A sulfonate ester** (Inorganic esters)

A preparation of organic esters by reaction of haloalkanes with alkanoate ions was presented in Section 7-8. We have also discussed the role chromate esters play in the oxidation of alcohols to aldehydes and ketones (Section 8-6). Living systems make use of alkyl phosphates in many ways, including conversion of alcohols into leaving groups. Others will be outlined in Chapters 20 and 26. Sulfonates were briefly introduced as substrates in nucleophilic displacement reactions in Section 6-7. Here we shall see how these compounds can be made directly by reactions of alcohols with carboxylic acids and a variety of inorganic reagents. Organic and inorganic esters are valuable synthetic intermediates.

Alcohols react with carboxylic acids to give organic esters

Alcohols react with carboxylic acids in the presence of catalytic amounts of mineral acid to give esters and water, a process called **esterification.** Starting materials and products in this transformation form an equilibrium that can be shifted in either direction as shown below. The formation and reactions of organic esters will be presented in detail in Chapters 19 and 20.

Esterification

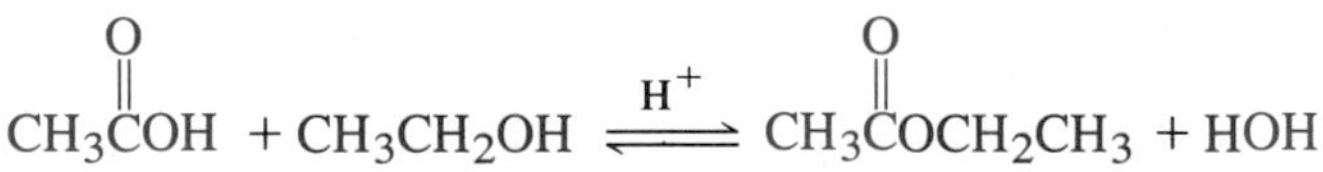

$$\underset{\text{Acetic acid}}{CH_3\overset{O}{\overset{\|}{C}}OH} + \underset{\text{Ethanol solvent}}{CH_3CH_2OH} \overset{H^+}{\rightleftharpoons} \underset{\text{Ethyl acetate}}{CH_3\overset{O}{\overset{\|}{C}}OCH_2CH_3} + HOH$$

Inorganic esters are intermediates in haloalkane synthesis

Primary and secondary alcohols react with phosphorus tribromide, a readily available commercial compound, to give bromoalkanes and phosphorous acid. This method constitutes a general way of making bromoalkanes from alcohols. All three bromine atoms are transferred from phosphorus to alkyl groups.

Bromoalkane Synthesis using PBr_3

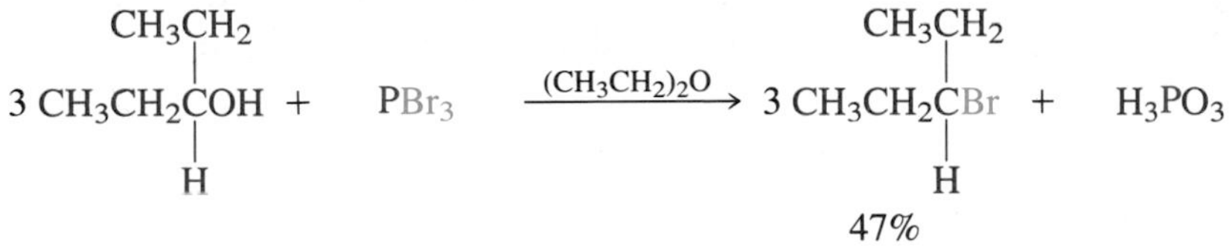

$$\underset{\text{3-Pentanol}}{3\ CH_3CH_2\overset{CH_3CH_2}{\overset{|}{\underset{H}{\underset{|}{C}}}}OH} + \underset{\text{Phosphorus tribromide}}{PBr_3} \xrightarrow{(CH_3CH_2)_2O} \underset{\text{3-Bromopentane}}{\underset{47\%}{3\ CH_3CH_2\overset{CH_3CH_2}{\overset{|}{\underset{H}{\underset{|}{C}}}}Br}} + \underset{\text{Phosphorous acid}}{H_3PO_3}$$

What is the mechanism of action of PBr_3? In the first step the alcohol and the phosphorus reagent form a protonated inorganic ester, a derivative of phosphorous acid.

STEP 1.

$$\mathrm{RCH_2\ddot{O}H} + \mathrm{PBr_3} \longrightarrow \mathrm{RCH_2\overset{+}{O}(H){-}PBr_2} + \mathrm{Br^-}$$

Next, $HOPBr_2$, a good leaving group, is displaced by the bromide generated in step 1, finally producing the haloalkane.

STEP 2.

$$:\ddot{\mathrm{Br}}:^- + \mathrm{RCH_2{-}\overset{+}{O}(H){-}PBr_2} \longrightarrow \mathrm{RCH_2\ddot{Br}:} + \mathrm{H\ddot{O}PBr_2}$$

This method of haloalkane synthesis is especially efficient because $HOPBr_2$ continues to react successively with two more molecules of alcohol, converting them to haloalkane as well.

$$\mathrm{RCH_2\ddot{O}H} + \mathrm{H\ddot{O}PBr_2} \longrightarrow \mathrm{RCH_2\overset{+}{O}(H){-}P(Br)\ddot{O}H} + :\ddot{\mathrm{Br}}:^-$$

$$:\ddot{\mathrm{Br}}:^- + \mathrm{RCH_2{-}\overset{+}{O}(H){-}P(Br)\ddot{O}H} \longrightarrow \mathrm{RCH_2\ddot{Br}:} + \mathrm{H\ddot{O}P(Br)\ddot{O}H}$$

$$\mathrm{RCH_2OH} + \mathrm{(HO)_2PBr} \longrightarrow \longrightarrow \mathrm{RCH_2\ddot{Br}:} + \mathrm{H_3PO_3}$$

Note the need first to convert the hydroxy substituent in the alcohol into a good leaving group, as in the transformation of alcohols by hydrogen bromide. The use of PBr_3 to form bromoalkanes is preferable, however, because the intermediates are less prone to rearrangement than those in reactions involving the strongly acidic HBr.

What if, instead of a bromoalkane, we want the corresponding iodoalkane? The required phosphorus triiodide, PI_3, is best generated in the reaction mixture where it will be used, because it is a reactive species. We do this by adding red elemental phosphorus and elemental iodine to the alcohol (margin). The reagent is consumed as soon as it is formed.

$$\mathrm{CH_3(CH_2)_{14}CH_2OH} \xrightarrow{\mathrm{P,\ I_2,\ \Delta}} \underset{85\%}{\mathrm{CH_3(CH_2)_{14}CH_2I}} + \mathrm{H_3PO_3}$$

A chlorinating agent commonly used to convert alcohols into chloroalkanes is thionyl chloride, $SOCl_2$. Simply warming an alcohol in its presence results in the evolution of SO_2 and HCl and the formation of the chloroalkane.

Chloroalkane Synthesis with $SOCl_2$

$$\mathrm{CH_3CH_2CH_2OH} + \mathrm{SOCl_2} \longrightarrow \underset{91\%}{\mathrm{CH_3CH_2CH_2Cl}} + \mathrm{O{=}S{=}O} + \mathrm{HCl}$$

Mechanistically, the alcohol RCH_2OH again first forms an inorganic ester, RCH_2O_2SCl. The chloride ion created in this process then acts like a nucleophile and attacks the ester, a reaction yielding one molecule each of SO_2 and HCl.

Mechanism of Substitution using Thionyl Chloride

$$RCH_2OH + ClS(=O)Cl \longrightarrow RCH_2OS(=O)Cl + H^+ + Cl^-$$

$$H^+ + :\ddot{Cl}: + CH_2(R)-\ddot{O}-S(=\ddot{O})-\ddot{Cl}: \longrightarrow :\ddot{Cl}CH_2R + \ddot{O}=S=\ddot{O} + H\ddot{Cl}:$$

The reaction works better in the presence of an amine, which neutralizes the hydrogen chloride generated. One such reagent is *N,N*-diethylethanamine (triethylamine), which forms the corresponding ammonium hydrochloride under these conditions.

$$(CH_3CH_2)_3N: + HCl \longrightarrow (CH_3CH_2)_3\overset{+}{N}H\ Cl^-$$

***N,N*-Diethylethanamine (Triethylamine)**

Alkyl sulfonates are versatile substrates for substitution reactions

The inorganic esters in the reactions of $SOCl_2$ are special examples of leaving groups derived from sulfur-based acids. They are related to the sulfonates (Section 6-7). Alkyl sulfonates contain excellent leaving groups and can be readily prepared from the corresponding sulfonyl chlorides and an alcohol. A mild base such as pyridine or another amine is often added to remove the HCl formed.

Synthesis of an Alkyl Sulfonate

$$CH_3CH(CH_3)CH_2OH + CH_3S(=O)_2Cl + C_5H_5N \longrightarrow CH_3CH(CH_3)CH_2OS(=O)_2CH_3 + C_5H_5\overset{+}{N}H\ Cl^-$$

2-Methyl-1-propanol + **Methanesulfonyl chloride (Mesyl chloride)** + **Pyridine** ⟶ **2-Methylpropyl methanesulfonate (2-Methylpropyl mesylate)** + **Pyridinium hydrochloride**

Unlike the inorganic esters derived from phosphorus tribromide and thionyl chloride, alkyl sulfonates are often crystalline solids that can be isolated and purified before further reaction. They then can be used in reactions with a variety of nucleophiles to give the corresponding products of nucleophilic substitution. The displacement of sulfonate groups by halide ions (Cl^-, Br^-, I^-) readily yields the corresponding haloalkanes, particularly with primary and secondary systems, in which S_N2 reactivity is good. In addition, however, alkyl sulfonates allow replacement of the hydroxy group by *any* good nucleophile: They are not limited to halides alone, as was the case with hydrogen, phosphorus, and thionyl halides.

Sulfonate Intermediates in Nucleophilic Displacement of the Hydroxy Group in Alcohols

$$R{-}OH \longrightarrow R{-}OS(=O)_2R' \longrightarrow R{-}Nu$$

Substitution Reactions of Alkyl Sulfonates

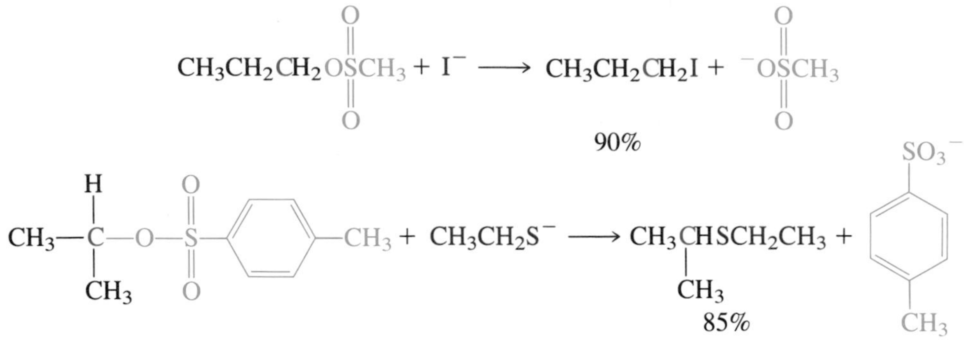

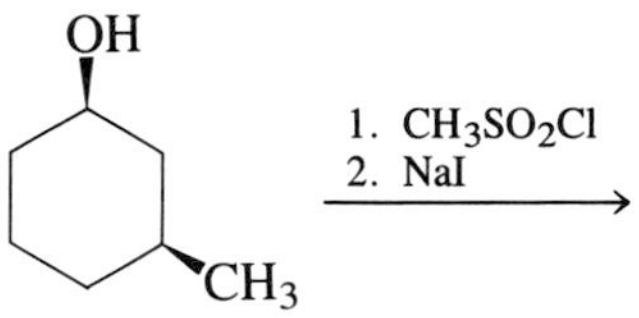

$$CH_3CH_2CH_2OS(=O)_2CH_3 + I^- \longrightarrow CH_3CH_2CH_2I + {}^-OS(=O)_2CH_3$$

90%

$$CH_3CH(CH_3)\text{—O—}SO_2\text{—}C_6H_4\text{—}CH_3 + CH_3CH_2S^- \longrightarrow CH_3CH(CH_3)SCH_2CH_3 + {}^-O_3S\text{—}C_6H_4\text{—}CH_3$$

85%

OH (cyclohexane with CH3), 1. CH_3SO_2Cl 2. NaI →

EXERCISE 9-9

What is the product of the reaction sequence shown in the margin?

EXERCISE 9-10

Supply reagents with which you would prepare the following haloalkanes from the corresponding alcohols.

(a) $I(CH_2)_6I$ **(b)** $(CH_3CH_2)_3CCl$ **(c)** (skeletal structure with Br)

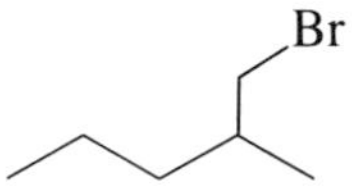

In summary, alcohols react with carboxylic acids by loss of water to furnish organic esters and with inorganic halides, such as PBr_3, PCl_5, $SOCl_2$, and RSO_2Cl, by loss of HX to produce inorganic esters. These inorganic esters contain good leaving groups in nucleophilic substitutions that are, for example, displaced by halide ions to give the corresponding haloalkanes.

9-5 Names and Physical Properties of Ethers

Ethers have been mentioned on several occasions in preceding chapters (see Table 2-1). It is time to introduce this class of compounds more systematically. This section gives the rules for naming ethers and describes some of their physical properties.

In the IUPAC system, ethers are alkoxyalkanes

The IUPAC system for naming **ethers** treats them as alkanes that bear an alkoxy substituent, that is, as alkoxyalkanes. The smaller substituent is considered part of the alkoxy group, and the larger defines the stem.

The alkoxyalkanes may be thought of as derivatives of alcohols in which the hydroxy proton has been replaced by an alkyl. Their common names are based on this picture: The names of the two alkyl groups are followed by the word ether. Hence, CH_3OCH_3 is dimethyl ether, $CH_3OCH_2CH_3$ is ethyl methyl ether, and so forth.

Ethers are generally fairly unreactive (except for strained cyclic derivatives, see Section 9-9) and are therefore frequently used as solvents in organic reactions. Some of these ether solvents are cyclic; they may even contain several ether functions. All have common names.

$CH_3OCH_2CH_3$
Methoxyethane

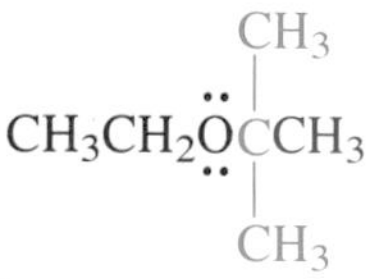

$CH_3CH_2\ddot{O}C(CH_3)_3$
2-Ethoxy-2-methylpropane

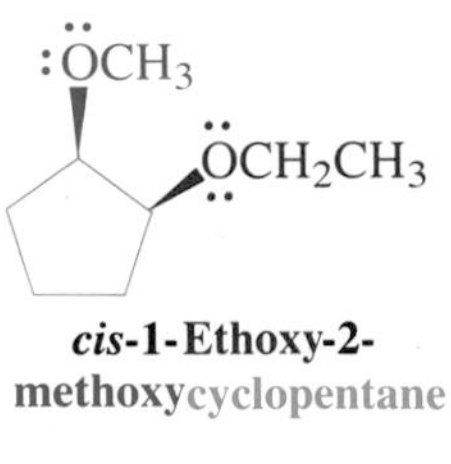

$:\ddot{O}CH_3$ / $\ddot{O}CH_2CH_3$ (cyclopentane)
***cis*-1-Ethoxy-2-methoxycyclopentane**

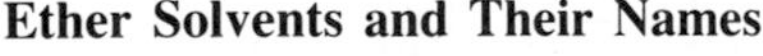

Ether Solvents and Their Names

$CH_3CH_2OCH_2CH_3$		$CH_3OCH_2CH_2OCH_3$	
Ethoxyethane (Diethyl ether)	**1,4-Dioxacyclohexane (1,4-Dioxane)**	**1,2-Dimethoxyethane (Glycol dimethyl ether, glyme)**	**Oxacyclopentane (Tetrahydrofuran, THF)**

Cyclic ethers are members of a class of cycloalkanes in which one or more carbons have been replaced by a *heteroatom*—in this case, oxygen. (A **heteroatom** is defined as any atom except carbon and hydrogen.) Cyclic compounds of this type, called **heterocycles,** are discussed more fully in Chapter 25.

The simplest system for naming cyclic ethers is based on the **oxacycloalkane** stem, in which the prefix *oxa* indicates the replacement of carbon by oxygen in the ring. Thus, three-membered cyclic ethers are oxacyclopropanes (other names used are oxiranes, epoxides, and ethylene oxides), four-membered systems are oxacyclobutanes, and the next two higher homologs are oxacyclopentanes (tetrahydrofurans) and oxacyclohexanes (tetrahydropyrans). The compounds are numbered by starting at the oxygen and proceeding around the ring.

The physical properties of ethers reflect the absence of hydrogen bonding

The molecular formula of simple alkoxyalkanes is $C_nH_{2n+2}O$, identical with that of the alkanols. However, because of the absence of hydrogen bonding, the boiling points of ethers are much lower than those of the corresponding isomeric alcohols (Table 9-1). The two smallest members of the series are water miscible, but ethers become less water soluble as the hydrocarbon residues increase in size. For example, methoxymethane is completely water soluble, whereas ethoxyethane forms only an approximately 10% aqueous solution.

In summary, ethers can be named as alkoxyalkanes or as dialkyl ethers. They have lower boiling points than comparable alcohols because they cannot enter into hydrogen bonding with themselves.

TABLE 9-1 Boiling Points of Ethers and the Isomeric 1-Alkanols

Ether	Name	Boiling point (°C)	1-Alkanol	Boiling point (°C)
CH_3OCH_3	Methoxymethane (Dimethyl ether)	−23	CH_3CH_2OH	78.5
$CH_3OCH_2CH_3$	Methoxyethane (Ethyl methyl ether)	10.8	$CH_3CH_2CH_2OH$	82.4
$CH_3CH_2OCH_2CH_3$	Ethoxyethane (Diethyl ether)	34.5	$CH_3(CH_2)_3OH$	117.3
$(CH_3CH_2CH_2CH_2)_2O$	1-Butoxybutane (Dibutyl ether)	142	$CH_3(CH_2)_7OH$	194.5

9-6 Williamson Ether Synthesis

Alkoxides are excellent nucleophiles. This section describes their use in the most common method for the preparation of ethers.

Ethers are prepared by S_N2 reactions

The simplest way to synthesize an ether is to have an alkoxide react with a primary haloalkane or a sulfonate ester under typical S_N2 conditions (Chapter 6). This approach is known as the **Williamson* ether synthesis.** The alcohol from which the alkoxide is derived can be used as the solvent (if inexpensive), but other polar molecules, such as dimethyl sulfoxide (DMSO) or hexamethylphosphoric triamide (HMPA), are often better (Table 6-6).

Williamson Ether Syntheses

$$CH_3CH_2CH_2CH_2O^-Na^+ + ClCH_2CH_2CH_2CH_3 \xrightarrow[\text{or DMSO, 9.5 h}]{CH_3CH_2CH_2CH_2OH,\ 14\ h} CH_3CH_2CH_2CH_2OCH_2CH_2CH_2CH_3 + Na^+Cl^-$$

60% (butanol solvent)
95% (DMSO solvent)
1-Butoxybutane

$$\text{cyclopentyl}(H)O^-Na^+ + CH_3(CH_2)_{15}CH_2OSO_2CH_3 \xrightarrow{DMSO} \text{cyclopentyl}(H)OCH_2(CH_2)_{15}CH_3 + Na^{+-}O_3SCH_3$$

91%
Cyclopentoxyheptadecane

Nucleophile—red
Electrophile—blue
Leaving group—green

Because alkoxides are strong bases, their use in ether synthesis is restricted to primary unhindered alkylating agents; otherwise a significant amount of E2 product is formed (Section 7-8).

EXERCISE 9-11

Write Williamson syntheses for the following ethers. **(a)** 1-Ethoxybutane (two ways); **(b)** 2-methoxypentane (Are there two ways that work well?); **(c)** propoxycyclohexane; **(d)** 1,4-diethoxybutane.

Cyclic ethers can be prepared by intramolecular Williamson synthesis

The Williamson ether synthesis is also applicable to the preparation of cyclic ethers, starting from halo alcohols. Figure 9-3 depicts the reaction of hydroxide ion with a bromoalcohol. The black curved lines denote the chain of carbon atoms linking the functional groups. The mechanism consists of initial formation

*Professor Alexander W. Williamson, 1824–1904, University College, London.

BOX 9-1 Aromatic Ethers and Williamson Synthesis

Ethers containing one aromatic group (see Table 2-1) attached to oxygen may be prepared readily by Williamson synthesis. Such alkoxyarenes often have pleasant aromas. For example, exposure of 2-naphthalenol (2-naphthol) to KOH deprotonates the aromatic alcohol. Subsequent addition of bromoethane results in 2-ethoxynaphthalene, commonly called Nerolin II. It is used in perfumery as a substitute for essence of orange blossoms.

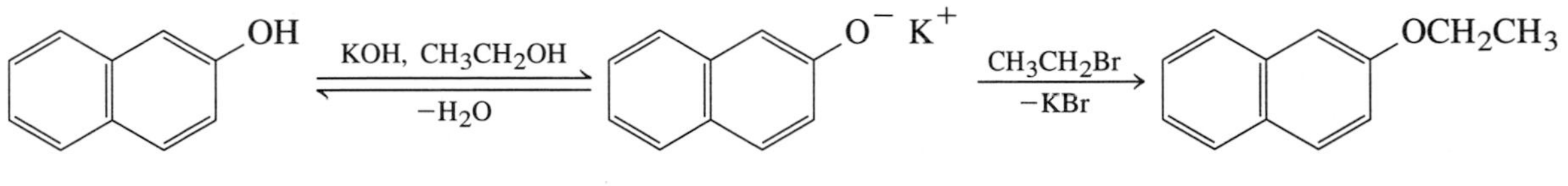

$$\text{2-Naphthalenol-OH} \underset{-H_2O}{\overset{KOH,\ CH_3CH_2OH}{\rightleftharpoons}} \text{ArO}^-\ K^+ \xrightarrow[-KBr]{CH_3CH_2Br} \text{ArOCH}_2\text{CH}_3$$

2-Naphthalenol (2-Naphthol) (An aromatic alcohol) — **2-Ethoxynaphthalene (Nerolin II)** (An alkoxyarene)

of a bromoalkoxide by fast proton transfer to the base, followed by ring closure to furnish the cyclic ether. The latter process is an example of an intramolecular displacement. Cyclic ether formation is much faster than the side reaction shown in the figure, direct displacement of bromide by hydroxide to give a diol.

Intramolecular Williamson synthesis allows the preparation of cyclic ethers of various sizes, including small rings.

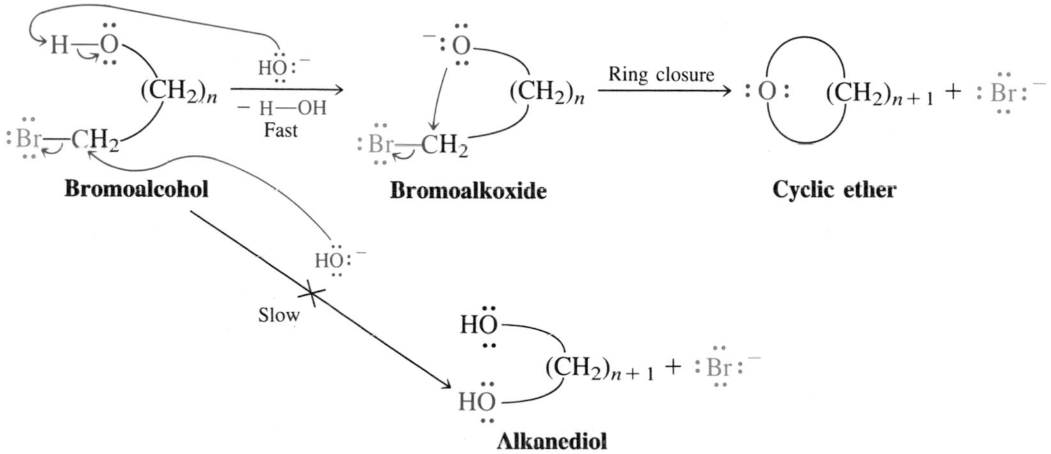

Bromoalcohol: $H-O-(CH_2)_n-CH_2-Br$ $\xrightarrow[-H-OH,\ Fast]{HO:^-}$ Bromoalkoxide: $^-:O-(CH_2)_n-CH_2-Br$ $\xrightarrow{\text{Ring closure}}$ Cyclic ether: $:O:\ (CH_2)_{n+1}$ + $:Br:^-$

Bromoalcohol $\xrightarrow[\text{Slow}]{HO:^-}$ Alkanediol: $HO-(CH_2)_{n+1}-OH$ + $:Br:^-$

FIGURE 9-3 The mechanism of cyclic ether synthesis from a bromoalcohol and hydroxide ion (top). A competing but slower side reaction, direct displacement of bromide by hydroxide, is also shown (below). The curved lines denote a chain of carbon atoms. The Williamson synthesis of cyclic ethers involves an intramolecular nucleophilic substitution.

Cyclic Ether Synthesis

$HOCH_2CH_2Br + HO^- \longrightarrow$ [oxacyclopropane ring: O (1), C2, C3•] $+ Br^- + HOH$

Oxacyclopropane (Oxirane, ethylene oxide)

The blue dot indicates the carbon atom in the product that corresponds to the site of ring closure in the starting material.

$HO(CH_2)_4CH_2Br + HO^- \longrightarrow$ [oxacyclohexane ring: O (1), C2, C3, C4, C5, C6•] $+ Br^- + HOH$

Oxacyclohexane (Tetrahydropyran)

EXERCISE 9-12

Give the product of the reaction of 5-bromo-3,3-dimethyl-1-pentanol with hydroxide and suggest its mechanism.

Ring size controls the speed of cyclic ether formation

A comparison of the relative rates of cyclic ether formation reveals a surprising fact: Three-membered rings form quickly, about as fast as five-membered rings. Six-membered ring systems, four-membered rings, and the larger oxacycloalkanes are generated more slowly. What effects are at work here? The answer includes both entropy factors and ring strain.

Relative Rates of Cyclic Ether Formation

$$k_3 \geq k_5 > k_6 > k_4 \geq k_7 > k_8$$

k_n = reaction rate, n = ring size

The preparation of an oxacyclopropane from a 2-bromoalcohol is favored by entropy, because nucleophile and leaving group are as close to each other as possible. Therefore, even though ring strain is at its worst in this case, the transition-state energy is relatively small, because a favorable entropy contribution allows relatively rapid ring construction. What about four-membered rings? Oxacyclobutanes are generated much more slowly. The entropy factor here is considerably worse, because the two reacting centers are separated by an extra methylene group. The ring strain, however, is about the same. The net result is a very low relative rate of formation. But the synthesis of five-membered ring ethers is easy. Although the reacting centers are even farther apart, the strain is much less. Proceeding to oxacyclohexanes, strain no longer plays a role, but the entropy factor gets worse. This trend continues for the larger rings, their rates of formation suffering, in addition, from eclipsing, *gauche,* and transannular strain.

The intramolecular Williamson synthesis is stereospecific

The Williamson ether synthesis proceeds with inversion of configuration at the carbon bearing the leaving group, in accord with expectations based on an S_N2 mechanism. The attacking nucleophile approaches the electrophilic carbon from

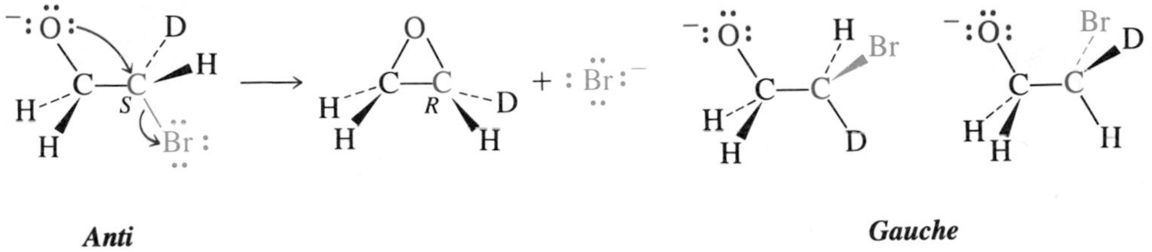

Anti

Gauche

FIGURE 9-4 Only the *anti* conformation of a 2-bromoalkoxide allows for oxacyclopropane formation. The two *gauche* conformers cannot undergo intramolecular backside attack at the bromine-bearing carbon.

the opposite side of the leaving group. Only one conformation of the haloalkoxide can undergo efficient substitution. For example, oxacyclopropane formation requires an *anti* arrangement of the nucleophile and the leaving group. The alternative two *gauche* conformations cannot give the product (Figure 9-4).

EXERCISE 9-13

(1*R*,2*R*)-2-Bromocyclopentanol reacts rapidly with sodium hydroxide to yield an optically inactive product. In contrast, the (1*S*,2*R*) isomer is much less reactive. Explain.

In summary, ethers are prepared by the Williamson synthesis, an S_N2 reaction of an alkoxide with a haloalkane. This reaction works best with primary halides or sulfonates that do not undergo ready elimination. Cyclic ethers are formed by the intramolecular version of this method. The relative rates of ring closure in this case are highest for three- and five-membered rings.

9-7 Ethers from Alcohols and Mineral Acid

An even simpler route to ethers employs strong acid. Protonation of the OH group in one alcohol generates water as a leaving group. Nucleophilic displacement of this leaving group by a second alcohol then results in the corresponding alkoxyalkane.

Alcohols give ethers by both S_N2 and S_N1 mechanisms

We have learned that treatment of primary alcohols with HBr or HI furnishes the corresponding haloalkanes through intermediate alkyloxonium ions (Section 9-2). However, when strong nonnucleophilic acids—such as sulfuric acid—are used at elevated temperatures, the products are ethers.

Ether Synthesis from a Primary Alcohol with Strong Acid

$$2\ CH_3CH_2OH \xrightarrow{H_2SO_4,\ 130°C} CH_3CH_2OCH_2CH_3 + HOH$$

In this reaction, the strongest nucleophile present in solution is the unprotonated starting alcohol. As soon as one alcohol molecule has been protonated, nucleophilic attack begins, the ultimate products being an ether and water.

BOX 9-2 Chemiluminescence of 1,2-Dioxacyclobutanes

$$(CH_3)_2C(OOH)\text{—}C(CH_3)_2Br \xrightarrow[-\,Br^-,\ H\text{–}OH]{HO^-} \text{3,3,4,4-Tetramethyl-1,2-dioxacyclobutane} \xrightarrow{\Delta} 2\ CH_3CCH_3 (=O) + h\nu$$

A 2-bromohydroperoxide — 3,3,4,4-Tetramethyl-1,2-dioxacyclobutane (A 1,2-dioxetane) — Propanone (Acetone)

An intramolecular Williamson-type reaction of a special kind is that in which a 2-bromohydroperoxide is the reactant. The peroxide product is a 1,2-dioxacyclobutane (1,2-dioxetane). This species is unusual because it decomposes to the corresponding carbonyl compounds with emission of light *(chemiluminescence)*. Dioxacyclobutanes seem to be responsible for the *bioluminescence* of certain species in nature. Terrestrial organisms, such as the firefly, the glowworm, and certain click beetles, are well known to emit light. However, most bioluminescent species live in the ocean; they range from microscopic bacteria and plankton to fish. The emitted light serves many purposes and seems to be important in courtship and communication, sex differentiation, finding prey, and hiding from or scaring off predators.

An example of a chemiluminescent molecule in nature is firefly luciferin. The base oxidation of this molecule furnishes a dioxacyclobutanone intermediate that decomposes in a manner analogous to that of 3,3,4,4-tetramethyl-1,2-dioxacyclobutane to give a complex heterocycle, carbon dioxide, and emitted light.

Firefly luciferin $\xrightarrow{\text{Base, } O_2}$ 1,2-Dioxacyclobutanone intermediate $\longrightarrow$ heterocycle $+\ CO_2 + h\nu$

Mechanism of Ether Synthesis

$$CH_3CH_2\ddot{O}H + CH_3CH_2\ddot{O}H \underset{-H^+}{\overset{+H^+}{\rightleftharpoons}} CH_3CH_2\text{—}\overset{+}{O}H_2 \ (+\ CH_3CH_2\ddot{O}H) \rightleftharpoons (CH_3CH_2)_2\overset{+}{O}\text{—}H + H_2\ddot{O} \underset{+H^+}{\overset{-H^+}{\rightleftharpoons}} (CH_3CH_2)_2\ddot{O} + H^+ + H_2\ddot{O}$$

Only symmetric ethers can be prepared by this method.

At even higher temperatures, elimination of water to generate an alkene is observed. This reaction proceeds by an E2 mechanism (Section 7-7), in which the neutral alcohol serves as the base that attacks the alkyloxonium ion.

Alkene Synthesis from an Alcohol and Strong Acid at High Temperature

$$CH_3\overset{H}{\overset{|}{C}}HCH_2OH \xrightarrow{H_2SO_4,\ 180^\circ C} CH_3CH{=}CH_2 + HOH$$

Secondary and tertiary ethers can be made from secondary and tertiary alcohols in the same way. However, in these cases, initial carbocation formation occurs, followed by trapping with alcohol (S_N1), as described in Section 9-2.

$$2\ CH_3\underset{H}{\underset{|}{\overset{OH}{\overset{|}{C}}}}CH_3 \xrightarrow{H^+} (CH_3)_2CHOCH(CH_3)_2 + HOH + H^+$$

2-Propanol → 75% 2-(1-Methylethoxy)propane (Diisopropyl ether)

The major side reaction follows the E1 pathway.

It is harder to synthesize ethers containing two different alkyl groups, because mixing two alcohols in the presence of an acid usually results in mixtures of all three possible products. However, mixed ethers containing one tertiary and one primary or secondary alkyl substituent can be prepared in good yield in the presence of dilute acid. Under these conditions, the much more rapidly formed tertiary carbocation is trapped by the other alcohol.

Synthesis of a Mixed Ether

$$CH_3\underset{CH_3}{\underset{|}{\overset{CH_3}{\overset{|}{C}}}}OH + CH_3CH_2OH \xrightarrow[-\ HOH]{15\%\ \text{aqueous}\ NaHSO_4} CH_3\underset{CH_3}{\underset{|}{\overset{CH_3}{\overset{|}{C}}}}OCH_2CH_3$$

80%

2-Ethoxy-2-methylpropane

EXERCISE 9-14

Write mechanisms for the following two reactions. **(a)** 1,4-Butanediol + $H^+ \rightarrow$ oxacyclopentane (tetrahydrofuran); **(b)** 5-methyl-1,5-hexanediol + $H^+ \rightarrow$ 2,2-dimethyloxacyclohexane (2,2-dimethyltetrahydropyran).

Ethers also form by alcoholysis

As we know, tertiary and secondary ethers may also form by the alcoholysis of the corresponding haloalkanes or alkyl sulfonates (Section 7-1). The starting material is simply dissolved in an alcohol until the S_N1 process is complete.

1-Chloro-1-methylcyclohexane (H_3C, Cl on the same ring carbon)

$\downarrow CH_3CH_2OH$

(H_3C, OCH_2CH_3 on the same ring carbon) $+ H^+ + Cl^-$

86%

1-Ethoxy-1-methylcyclohexane

EXERCISE 9-15

There are several ways of constructing an ether from an alcohol and a haloalkane. Which approach would you choose for the preparation of **(a)** 2-methyl-2-(1-methylethoxy)butane; **(b)** 1-methoxy-2,2-dimethylpropane?

In summary, ethers can be prepared by treatment of alcohols with acid through S_N2 and S_N1 pathways, with alkyloxonium ions or carbocations as intermediates, and by alcoholysis of secondary or tertiary haloalkanes or alkyl sulfonates.

9-8 Reactions of Ethers

As mentioned earlier, ethers are normally rather inert. They do, however, react slowly with oxygen by radical mechanisms to form hydroperoxides and peroxides. Because peroxides can decompose explosively, extreme care should be taken with samples of ethers that have been exposed to air for several days. This section describes a more preparatively useful reaction, cleavage by strong acid.

Peroxides from Ethers

$$2\ \text{RO}\overset{|}{\underset{|}{\text{C}}}\text{H} + \text{O}_2 \longrightarrow 2\ \text{RO}\overset{|}{\underset{|}{\text{C}}}\text{—O—OH} \longrightarrow \text{RO}\overset{|}{\underset{|}{\text{C}}}\text{—O—O—}\overset{|}{\underset{|}{\text{C}}}\text{OR}$$

An ether hydroperoxide   **An ether peroxide**

The oxygen in ethers, like that in alcohols, may be protonated to generate alkyloxonium ions. The subsequent reactivity of these ions depends on the alkyl substituents. With primary groups and strong nucleophilic acids such as HBr, S_N2 displacement occurs.

$$\underset{\textbf{Ethoxyethane}}{CH_3CH_2OCH_2CH_3} \xrightarrow{HBr} \underset{\textbf{Bromoethane}}{CH_3CH_2Br} + \underset{\textbf{Ethanol}}{CH_3CH_2OH}$$

Mechanism of Ether Cleavage

$$CH_3CH_2\ddot{\underset{\cdot\cdot}{O}}CH_2CH_3 \underset{}{\overset{H^+}{\rightleftharpoons}} CH_3CH_2\text{—}\ddot{O}^{+}(H)(CH_2CH_3)$$

Alkyloxonium ion

$$Br^- + CH_3CH_2\text{—}\ddot{O}^{+}(H)(CH_2CH_3) \longrightarrow CH_3CH_2Br + H\ddot{\underset{\cdot\cdot}{O}}CH_2CH_3$$

The alcohol formed as the second product may react to give more of the bromoalkane.

EXERCISE 9-16

Treatment of methoxymethane with hot HI gives iodomethane. Suggest a mechanism.

EXERCISE 9-17

Reaction of oxacyclohexane (tetrahydropyran; shown in the margin) with HI gives 1,5-diiodopentane. Give a mechanism for this reaction.

Oxacyclohexane (Tetrahydropyran)

Ethers containing secondary or tertiary alkyl groups transform even in dilute acid to give intermediate carbocations, which are either trapped by nucleophiles (S_N1) or deprotonated (E1).

EXERCISE 9-18

Show how you would achieve the following interconversion (the dashed arrow indicates that several steps are required).

$$BrCH_2CH_2CH_2OH \dashrightarrow DCH_2CH_2CH_2OH$$

In summary, ethers are cleaved by (strong) acids. Protonation of an ether containing methyl or primary alkyl groups gives an alkyloxonium ion that is subject to S_N2 attack by nucleophiles. Carbocation formation follows protonation when secondary and tertiary groups are present, leading to S_N1 and E1 products.

9-9 Reactions of Oxacyclopropanes

Although ordinary ethers are relatively inert, the strained ring in oxacyclopropane undergoes a variety of ring-opening reactions with nucleophiles. This section presents details of these processes.

Nucleophilic ring opening of oxacyclopropanes is regioselective

Oxacyclopropane is subject to bimolecular ring opening by anionic nucleophiles. The reaction proceeds by nucleophilic attack, with the ether oxygen functioning as an intramolecular leaving group. Because of the symmetry of the substrate, substitution occurs to the same extent at either carbon. For example,

$$\overset{\ddot{O}}{CH_2\!-\!CH_2} + CH_3\ddot{S}:^- \xrightarrow[-\,HO^-]{HOH} H\ddot{O}CH_2CH_2\ddot{S}CH_3$$

This S_N2 transformation is unusual for two reasons. First, alkoxides are usually very bad leaving groups. Second, the leaving group does not actually "leave"; it stays bound to the molecule. The driving force is the release of strain as a result of opening the ring.

BOX 9-3 Protecting Groups in Synthesis

$$\text{ROH} \xrightarrow[-H_2O]{(CH_3)_3COH,\ H^+} \text{ROC(CH}_3)_3 \xrightarrow{\text{Carry out reactions on R using Grignard reagents, oxidizing agents, etc.}} \text{R}'\text{OC(CH}_3)_3 \xrightarrow{H^+,\ H_2O} \text{R}'\text{OH}$$

Protection step — **Protected alcohol** — R changed into R′ — Deprotection

To remove the potentially complicating presence of a reactive functionality, chemists use *protecting groups*. That is, a functional group is temporarily converted into an unreactive form *(protection)* to prevent its interference with transformations to be carried out elsewhere in the molecule. Subsequently, the original unit is regenerated *(deprotection)*. For example, one protected form of an alcohol is a 1,1-dimethylethyl ether, readily obtained from an acid-catalyzed reaction with 2-methyl-2-propanol (*tert*-butyl alcohol) (Section 9-7). The protected functional group is now inert to base, organometallic reagents, and oxidizing and reducing agents. Reactions involving such species may therefore be completed without interference. The protecting group is removed by dilute aqueous acid as shown above.

Another method of alcohol protection is esterification (Section 9-4).

One application of these protecting groups is in the synthesis of the sex hormone testosterone (Section 4-7) from the cholesterol-derived starting material shown. Natural sources of steroid hormones are far too limited to meet the needs of medicine and research; these molecules must be synthesized. In our case, selective reduction of the carbonyl group at C17 and oxidation of the hydroxy function at C3 are required.

Formation of 1,1-dimethylethyl (*tert*-butyl) ether at C3 is followed by reduction at C17. A second protection step at C17 involves esterification (Section 9-4). Esters are stable in dilute acid, which hydrolyzes tertiary ethers. This strategy allows the hydroxy group at C3 to be freed and oxidized to a ketone, while that at C17 remains protected. Exposure to strong acid finally converts the product of the sequence shown here to testosterone.

What is the situation with unsymmetric systems? Consider, for example, the reaction of 2,2-dimethyloxacyclopropane with methoxide. There are *two* possible reaction sites: at the primary carbon (*a*), to give 1-methoxy-2-methyl-2-propanol; and at the tertiary carbon (*b*), to yield 2-methoxy-2-methyl-1-propanol. Evidently this system transforms solely through path *a*.

Nucleophilic Ring Opening of an Unsymmetrically Substituted Oxacyclopropane

$$\underset{\textbf{1-Methoxy-2-methyl-2-propanol}}{CH_3OCH_2C(OH)(CH_3)_2} \xleftarrow[a]{CH_3O^-,\ CH_3OH} \underset{\textbf{2,2-Dimethyloxacyclopropane}}{\text{oxacyclopropane}} \xrightarrow[b]{CH_3O^-,\ CH_3OH\ \times} \underset{\substack{\textbf{2-Methoxy-2-methyl-1-propanol}\\ \text{(Not formed)}}}{HOCH_2C(OCH_3)(CH_3)_2}$$

H_3C O 17 H_3C H H H HO 3

$(CH_3)_3COH, H^+$
Protects C3 OH

H_3C O 17 H_3C H H H $(CH_3)_3CO$ 3

1. $LiAlH_4$, $(CH_3CH_2)_2O$
2. H_2O
Reduces carbonyl at C17 to OH

H_3C OH 17 H_3C H H H $(CH_3)_3CO$ 3

Esterification
Protects C17 OH

H_3C O OCR 17 H_3C H H H $(CH_3)_3CO$ 3

R = phenyl

Dilute H^+, H_2O
Deprotects C3 but not C17 OH

PCC, CH_2Cl_2
Oxidizes C3 OH

H_3C O OCR 17 H_3C H H H O 3

Is this result surprising? No, because, as we know, if there is more than one possibility, S_N2 attack will be at the *less* substituted carbon center (Section 6-9). This selectivity in the nucleophilic opening of substituted oxacyclopropanes is referred to as **regioselectivity** because, of two possible and similar "regions," the nucleophile attacks only one.

In addition, when ring opening occurs at a stereocenter, inversion is observed. Thus, we find that the rules of nucleophilic substitution developed for simple alkyl derivatives also apply to strained cyclic ethers.

Acids catalyze oxacyclopropane ring opening

Ring opening of oxacyclopropanes is also catalyzed by acids. The reaction in this case proceeds through initial cyclic alkyloxonium ion formation followed by ring opening as a result of nucleophilic attack.

Acid-Catalyzed Ring Opening of Oxacyclopropane

$$H_2C\overset{O}{\frown}CH_2 + CH_3OH \xrightarrow{H_2SO_4} HOCH_2CH_2OCH_3$$

2-Methoxyethanol

Mechanism of Acid-Catalyzed Ring Opening

$$\text{oxacyclopropane} \underset{-H^+}{\overset{H^+,\ CH_3\ddot{O}H}{\rightleftharpoons}} \text{Cyclic alkyloxonium ion} + H\ddot{O}CH_3 \longrightarrow HOCH_2CH_2\overset{+}{O}(H)CH_3 \xrightarrow{-H^+} H\ddot{O}CH_2CH_2\ddot{O}CH_3$$

Cyclic alkyloxonium ion

Is this reaction also regioselective and stereospecific, like the anionic nucleophilic opening of oxacyclopropanes discussed first? Yes, but the details are different. Thus, acid-catalyzed methanolysis of 2,2-dimethyloxacyclopropane proceeds by exclusive ring opening at the *more* hindered carbon.

Acid-Catalyzed Ring Opening of 2,2-Dimethyloxacyclopropane

$$\text{2,2-Dimethyloxacyclopropane} \xrightarrow{H_2SO_4,\ CH_3OH} HOCH_2C(CH_3)_2OCH_3$$

2,2-Dimethyloxacyclopropane → **2-Methoxy-2-methyl-1-propanol**

Why is the more hindered position attacked? Protonation at the oxygen of the ether generates a reactive intermediate alkyloxonium ion with substantially polarized oxygen–carbon bonds. This polarization places partial positive charges on the ring carbons. Because alkyl groups act as electron donors (Section 7-5), more positive charge is located on the tertiary than on the primary carbon.

Mechanism of Acid-Catalyzed Ring Opening of 2,2-Dimethyloxacyclopropane by Methanol

$$\text{2,2-Dimethyloxacyclopropane} \xrightarrow{H^+} \text{protonated ion } (O^{\delta+}, C^{\delta+})$$

Primary carbocationlike (attack at primary $C^{\delta+}$) → $-H^+$, CH_3OH → **1-Methoxy-2-methyl-2-propanol** (Not formed)

or

Tertiary carbocationlike (attack at tertiary $C^{\delta+}$) → $-H^+$, CH_3OH → **2-Methoxy-2-methyl-1-propanol**

This uneven charge distribution counteracts steric hindrance: Methanol is attracted by coulombic forces more to the tertiary than to the primary center. Although the result is clear-cut in this example, it is less so in cases in which the two carbons are not quite as different. For example, mixtures of isomeric products are formed by acid-catalyzed ring opening of 2-methyloxacyclopropane.

Why do we not simply write free carbocations as intermediates in the acid-catalyzed ring openings? The reason is that inversion is observed when the reaction occurs at a stereocenter. Like the reaction of oxacyclopropanes with anionic nucleophiles, the acid-catalyzed process involves backside displacement, in this case on a highly polarized cyclic alkyloxonium ion.

Hydride and organometallic reagents convert strained ethers to alcohols

The highly reactive lithium aluminum hydride is able to cause ring opening of oxacyclopropanes, a reaction leading to alcohols. Ordinary ethers, lacking the ring strain of oxacyclopropanes, do not react with $LiAlH_4$. The reaction proceeds by the S_N2 mechanism. Thus, in unsymmetric systems, the hydride attacks the less substituted side.

Ring Opening of an Oxacyclopropane by Lithium Aluminum Hydride

$$\text{(H)(H)C—C(H)(R) (oxacyclopropane)} \xrightarrow[\text{2. } H^+,\ H_2O]{\text{1. } LiAlH_4,\ (CH_3CH_2)_2O} H\text{—}CH_2CHR(OH)$$

Less hindered

EXERCISE 9-19

What oxacyclopropane would give 3-hexanol (Section 8-9) after treatment with $LiAlH_4$ followed by acidic aqueous work-up? Is there more than one possible answer?

Similarly, Grignard reagents and alkyllithium compounds furnish alcohols in which the alkyl portion of the organometallic reagent has been extended by two carbons.

Oxacyclopropane Ring Opening by a Grignard Reagent

$$H_2C\text{—}CH_2 \text{ (with O bridge)} + CH_3CH_2CH_2CH_2MgBr \xrightarrow{THF} \xrightarrow{H^+,\ H_2O} \underset{\substack{62\% \\ \textbf{1-Hexanol}}}{CH_3CH_2CH_2CH_2CH_2CH_2OH}$$

EXERCISE 9-20

Propose an efficient synthesis of 3,3-dimethyl-1-butanol from starting materials containing no more than four carbons.

EXERCISE 9-21

Predict the major product of ring opening of 2,2-dimethyloxacyclopropane on treatment with each of the following. **(a)** $LiAlH_4$, then H^+, H_2O; **(b)** $CH_3CH_2CH_2MgBr$, then H^+, H_2O; **(c)** CH_3SNa in CH_3OH; **(d)** dilute HCl in CH_3CH_2OH; **(e)** concentrated aqueous HBr.

BOX 9-4 Epoxy Resins as Adhesives

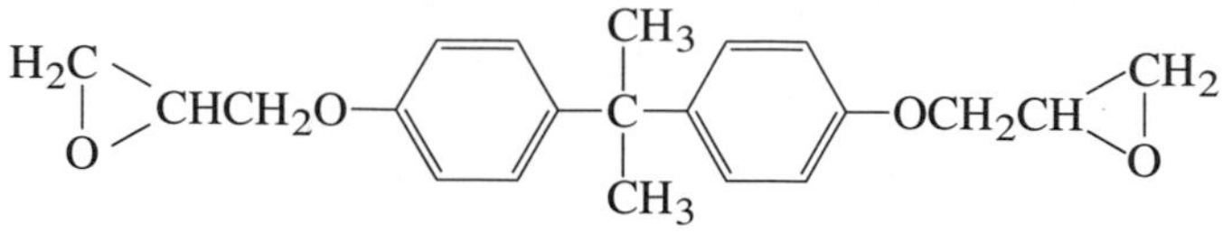

DGEBA epoxy resin
("Diglycidyl ether of bisphenol A")

The reactivity of oxacyclopropanes has been exploited in the development of the *epoxy resins* (a term derived from "epoxide," a common name for oxacyclopropane). A typical example is DGEBA.

Epoxy resins undergo a transformation called *curing* upon heating or exposure to a variety of reagents, usually acids or bases. The result is an extremely hard solid. Most resins contain two oxacyclopropane functional groups. The base-catalyzed curing process begins with ring opening by hydroxide. The resulting alkoxide units then attack other molecules of the resin, a process called *cross-linking*. Each ring opening liberates another alkoxide, which cross-links with another molecule of resin, ultimately providing a material of very high molecular weight. The process may be shown schematically (see below).

When curing takes place in contact with a solid containing surface hydroxy groups (glass is a good example), *surface bonding* occurs through covalent linkages: The resin acts as a strong adhesive.

Curing of an Epoxy Resin

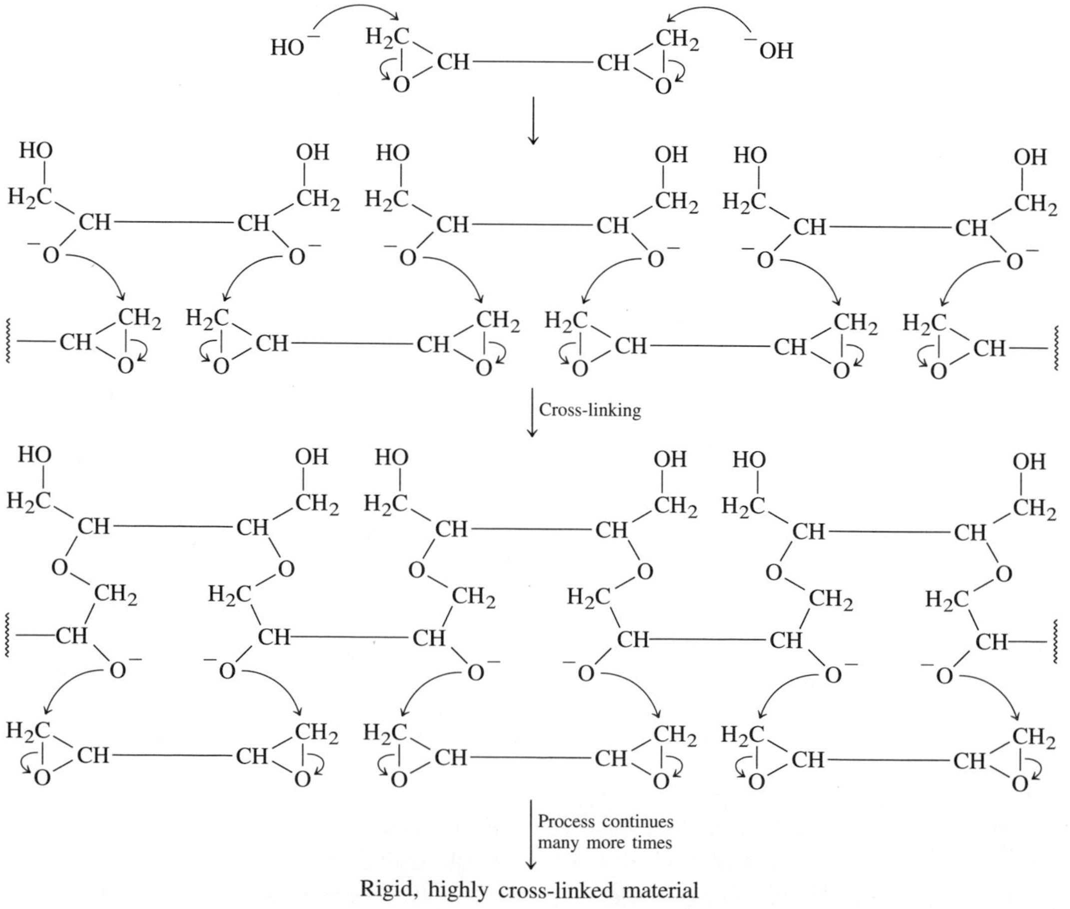

Cross-linking

Process continues many more times

Rigid, highly cross-linked material

In summary, although ordinary ethers are relatively inert, the ring in oxacyclopropanes can be opened both regioselectively and stereospecifically. For anionic nucleophiles, the usual rules of bimolecular nucleophilic substitution hold: Attack is at the less hindered carbon center, which undergoes inversion. Acid catalysis, however, changes the regioselectivity: Attack is at the more hindered center. Hydride and organometallic reagents behave like other anionic nucleophiles, furnishing alcohols by an S_N2 pathway.

9-10 Sulfur Analogs of Alcohols and Ethers

We conclude our study of the chemistry of alcohols and ethers by looking at some of their sulfur analogs. These are compounds in which a sulfur heteroatom replaces the oxygen in the functional group. In this section we shall see what difference this substitution makes.

The sulfur analogs of alcohols and ethers are thiols and sulfides

The sulfur analogs of alcohols, R–SH, are called **thiols** in the IUPAC system (*theion,* Greek, brimstone—an older name for sulfur). The ending *thiol* is added to the alkane stem to yield the alkanethiol name. The SH group is referred to as **mercapto,** and its location is indicated by numbering the longest chain, as in alkanol nomenclature.

$CH_3\ddot{S}H$ — **Methanethiol**

$CH_3CH_2CH(CH_3)CH_2\ddot{S}H$ (C4, C3, C2, C1) — **2-Methyl-1-butanethiol**

$CH_3CH_2CH(\ddot{S}H)CH_2CH_3$ — **3-Pentanethiol**

Cyclohexane–$\ddot{S}H$ — **Cyclohexanethiol**

The sulfur analogs of ethers (common name, thioethers) are called **sulfides,** as in alkyl ether nomenclature. The RS group is named **alkylthio,** the RS^- group **alkanethiolate.**

$CH_3\ddot{S}CH_2CH_3$ — **Ethyl methyl sulfide**

$CH_3C(CH_3)_2\ddot{S}(CH_2)_6CH_3$ — **Heptyl (1,1-dimethyl)-ethyl sulfide**

$CH_3\ddot{S}{:}^-$ — **Methanethiolate ion**

Thiols are less hydrogen bonded and more acidic than alcohols

Sulfur, because of its large size, its diffuse orbitals, and the relatively nonpolarized S–H bond (Table 1-2), does not enter into hydrogen bonding very efficiently. Thus, the boiling points of thiols are not as abnormally high as those of alcohols; rather, their volatilities lie close to those of the analogous haloalkanes (Table 9-2).

TABLE 9-2 Comparison of the Boiling Points of Thiols, Haloalkanes, and Alcohols

Compound	Boiling point (°C)
CH_3SH	6.2
CH_3Br	3.6
CH_3Cl	−24.2
CH_3OH	65.0
CH_3CH_2SH	37
CH_3CH_2Br	38.4
CH_3CH_2Cl	12.3
CH_3CH_2OH	78.5

Partly because of the relatively weak S–H bond, thiols are also more acidic than water, with pK_a values ranging from 9 to 12. They can therefore be more readily deprotonated by hydroxide and alkoxide ions.

Acidity of Thiols

$$\underset{pK_a = 9\text{–}12}{R\ddot{S}H} + H\ddot{O}:^- \rightleftharpoons R\ddot{S}:^- + HOH$$

Thiols and sulfides react much like alcohols and ethers

Many reactions of thiols and sulfides resemble those of their oxygen analogs. The sulfur in these compounds is even more nucleophilic than the oxygen in alcohols and ethers. Therefore, thiols and sulfides are readily made by nucleophilic attack of RS^- or HS^- on haloalkanes. A large excess of the HS^- is used in the preparation of thiols to ensure that the product does not react with the starting halide to give the dialkyl sulfide.

$$CH_3CH(CH_3)Br + \underset{\text{Excess}}{Na^{+-}SH} \xrightarrow{CH_3CH_2OH} \underset{\textbf{2-Propanethiol}}{CH_3CH(CH_3)SH} + Na^+Br^-$$

Sulfides are prepared in an analogous way by alkylation of thiols in the presence of base, such as hydroxide. The base generates the alkanethiolate, which reacts with the haloalkane by an S_N2 process. Because of the strong nucleophilicity of thiolates, there is no competition from hydroxide in this displacement.

Sulfides by Alkylation of Thiols

$$RSH + R'Br \xrightarrow{NaOH} RSR' + NaBr + H_2O$$

The nucleophilicity of sulfur also explains the ability of sulfides to attack haloalkanes to furnish **sulfonium ions.**

$$(H_3C)_2\ddot{S}: + CH_3\text{–}\ddot{I}: \longrightarrow (H_3C)_2\overset{+}{\ddot{S}}\text{–}CH_3 + :\ddot{I}:^-$$

95%

Trimethylsulfonium iodide

Like their alkyloxonium analogs, sulfonium salts are subject to nucleophilic attack at carbon, the sulfide functioning as the leaving group.

$$H\ddot{O}:^- + CH_3\text{–}\overset{+}{\ddot{S}}(CH_3)_2 \longrightarrow H\ddot{O}CH_3 + \ddot{S}(CH_3)_2$$

EXERCISE 9-22

(a) Sulfide A is a powerful poison known as "mustard gas," a devastating chemical-warfare agent used in the First World War and again in the recent war between Iraq and Iran. Propose a synthesis starting with oxacyclopropane. **(b)** Its mechanism of action is believed to involve sulfonium salt B, which is thought to react with nucleophiles in the body. How is B formed and how would it react with nucleophiles?

$ClCH_2CH_2SCH_2CH_2Cl$ (A) $\quad ClCH_2CH_2\overset{+}{S}(CH_2CH_2)\ Cl^-$ (B)

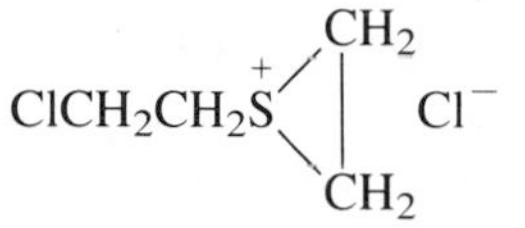

Valence-shell expansion of sulfur accounts for the special reactivity of thiols and sulfides

As a third-row element with *d* orbitals, sulfur's valence shell can expand to accommodate more electrons than are allowed by the octet rule. We have already seen that in some of its compounds sulfur is surrounded by 10 or even 12 valence electrons, and this capacity enables sulfur compounds to undergo reactions inaccessible to the corresponding oxygen analogs. For example, oxidation of thiols with strong oxidizing agents, such as hydrogen peroxide or potassium permanganate, gives the corresponding sulfonic acids. In this way methanethiol is converted into methanesulfonic acid. Sulfonic acids react with PCl_5 to give sulfonyl chlorides. These are used in sulfonate synthesis, as discussed in Section 9-4.

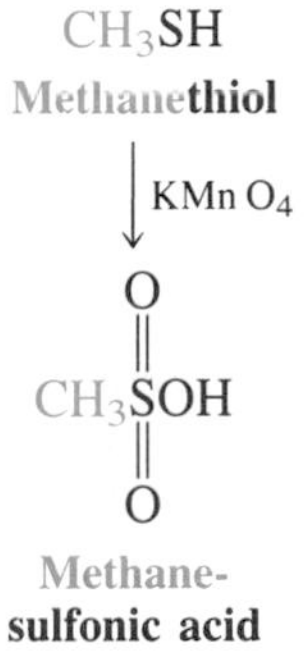

CH_3SH **Methanethiol** $\xrightarrow{KMnO_4}$ CH_3SO_2OH (O=S=O) **Methanesulfonic acid**

Milder oxidation of thiols, by the use of iodine, results in the formation of **disulfides,** which are readily reduced back to thiols by alkali metals.

The Thiol-Disulfide Redox Reaction

Oxidation

$$2\ CH_3CH_2CH_2SH + I_2 \longrightarrow CH_3CH_2CH_2SSCH_2CH_2CH_3 + 2\ HI$$

1-Propanethiol → **Dipropyl disulfide**

Reduction

$$CH_3CH_2CH_2SSCH_2CH_2CH_3 \xrightarrow[2.\ H^+,\ H_2O]{1.\ Li,\ liquid\ NH_3} CH_3CH_2CH_2SH$$

Reversible disulfide formation from thiols is an important biological process. Many proteins and peptides contain free SH groups that form bridging disulfide linkages. Nature exploits this mechanism to link amino acid chains. By thus helping to control the shape of enzymes in three dimensions, the mechanism makes biocatalysis far more efficient and selective.

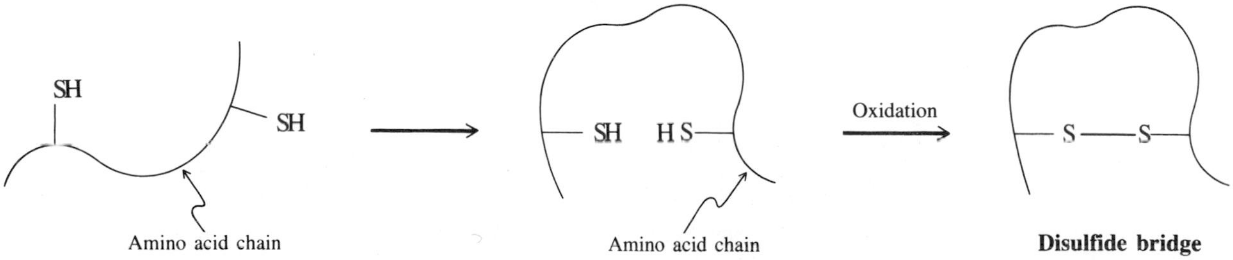

SH SH — Amino acid chain → SH HS — Amino acid chain $\xrightarrow{Oxidation}$ S—S **Disulfide bridge**

$CH_3\ddot{S}CH_3$

Dimethyl sulfide

$\downarrow H_2O_2$

$CH_3S(=O)CH_3$

Dimethyl sulfoxide (DMSO)

$\downarrow H_2O_2$

$CH_3S(=O)_2CH_3$

Dimethyl sulfone

Sulfides are readily oxidized to **sulfones,** a transformation proceeding through the intermediacy of a **sulfoxide.** For example, oxidation of dimethyl sulfide first gives dimethyl sulfoxide, which subsequently furnishes dimethyl sulfone. Dimethyl sulfoxide has already been mentioned as a highly polar nonprotic solvent of great use in organic chemistry, particularly in nucleophilic substitutions (see Section 6-8, Table 6-6).

In summary, the naming of thiols and sulfides is related to the system used for alcohols and ethers. Thiols are more volatile, more acidic, and more nucleophilic than alcohols. Thiols and sulfides can be oxidized, the former to disulfides or sulfonic acids, the latter to sulfoxides and sulfones.

9-11 Physiological Properties and Uses of Alcohols and Ethers

Modern industrial *methanol* synthesis uses the catalytic reduction of carbon monoxide with hydrogen at high pressures and temperatures (Section 8-4). Methanol is sold as a solvent for paint and other materials, as a fuel for camp stoves and soldering torches, and as a synthetic intermediate. It is highly poisonous, and ingestion or chronic exposure may lead to blindness. Death from ingestion of as little as 30 mL has been reported. It is sometimes added to commercial ethanol to render it unfit for consumption (denatured alcohol). The toxicity of methanol is thought to be due to metabolic oxidation to formaldehyde, $CH_2{=}O$, which interferes with the physiochemical processes of vision. Further oxidation to formic acid, HCOOH, causes acidosis, an unusual lowering of the blood pH. This condition disrupts oxygen transport in the blood and leads eventually to coma.

Methanol has been studied as a possible precursor of gasoline. For example, certain zeolite catalysts (Section 3-3) allow the conversion of methanol into a mixture of hydrocarbons, ranging in length from four-carbon chains to ten-carbon ones, with a composition that on distillation yields largely gasoline (see Table 3-3).

$$n\ CH_3OH \xrightarrow{\text{Zeolite, 340–375°C}} \underset{67\%}{C_nH_{2n+2}} + \underset{6\%}{C_nH_{2n}} + \underset{27\%}{\text{aromatics}}$$

Ethanol—diluted by various amounts of flavored water—is an alcoholic beverage. It is classified pharmacologically as a general depressant, because it induces a nonselective, reversible depression of the central nervous system. Approximately 95% of the alcohol ingested is metabolized in the body (usually in the liver) to products that are eventually transformed into carbon dioxide and water. Although high in calories, ethanol has little nutritional value.

The rate of metabolism of most drugs in the liver increases with their concentration, but this is not true for alcohol, which is degraded linearly with time. An adult metabolizes about 10 mL of pure ethanol per hour, roughly the ethanol content of a cocktail, a shot of spirits, or a can of beer. Depending on a person's weight, the ethanol content of the drink, and the speed with which it is consumed, as few as two or three drinks can produce a level of alcohol in the blood that is more than 0.1%, a concentration at or above the legal limit beyond which the operation of a motorized vehicle is prohibited in much of the United States.

Ethanol is poisonous. Its lethal concentration in the bloodstream has been estimated at 0.4%. Its effects include progressive euphoria, disinhibition, disorientation, and decreased judgment (drunkenness), followed by general anesthesia, coma, and death. It dilates the blood vessels, producing a ''warm flush,'' but it actually decreases body temperature. Although long-term ingestion of moderate amounts (the equivalent of about two beers a day) does not appear to be harmful, larger amounts can be the cause of a variety of physical and psychological disorders, usually described by the general term *alcoholism*. These disorders include hallucinations, psychomotor agitation, liver diseases, dementia, gastritis, and psychological dependence.

Interestingly, a near-toxic dose of ethanol is applied in cases of acute methanol or 1,2-ethanediol (ethylene glycol) poisoning. This treatment prevents the metabolism of the more toxic alcohols and allows their excretion before damaging concentrations of secondary products can accumulate.

Ethanol destined for human consumption is prepared by fermentation of sugars and starch (rice, potatoes, corn, wheat, flowers, fruit, etc., Chapter 24). Fermentation is catalyzed by enzymes in a multistep sequence that converts carbohydrates into ethanol and carbon dioxide.

$$\underset{\textbf{Starch}}{(C_6H_{10}O_5)_n} \xrightarrow{\text{Enzymes}} \underset{\textbf{Glucose}}{C_6H_{12}O_6} \xrightarrow{\text{Enzymes}} 2\ CH_3CH_2OH + 2\ CO_2$$

Commercial alcohol not intended as a beverage is made industrially by hydration of ethene (Section 8-4). It is used, for example, as a solvent in perfumes, varnishes, and shellacs and as a synthetic intermediate, as demonstrated in earlier equations. Interest in ethanol production has surged recently because of its potential as a gasoline additive (''gasohol'').

2-Propanol is toxic but (unlike methanol) is not absorbed through the skin. It is popular as a rubbing alcohol and used as a solvent and a cleaning agent.

1,2-Ethanediol (ethylene glycol) is prepared by oxidation of ethene to oxacyclopropane, followed by hydrolysis in quantities exceeding 2 million tons in the United States per year. Its low melting point (−11.5°C), its high boiling point (198°C), and its complete miscibility with water make it a useful antifreeze.

$$\underset{\substack{\textbf{Ethene}\\\textbf{(Ethylene)}}}{CH_2{=}CH_2} \xrightarrow{\text{Oxidation}} \underset{\substack{\textbf{Oxacyclopropane}\\\textbf{(Ethylene oxide)}}}{\overset{O}{\triangle}} \xrightarrow{H_2O} \underset{\substack{\textbf{1,2-Ethanediol}\\\textbf{(Ethylene glycol)}}}{HOCH_2CH_2OH}$$

1,2,3-Propanetriol (glycerol, glycerine), $HOCH_2CHOHCH_2OH$, is a viscous greasy substance, soluble in water, and nontoxic. It is obtained by alkaline hydrolysis of triglycerides, the major component of fatty tissue. The sodium and potassium salts of the long-alkyl-chain acids produced from fats (''fatty acids,'' Chapter 19) are sold as soaps.

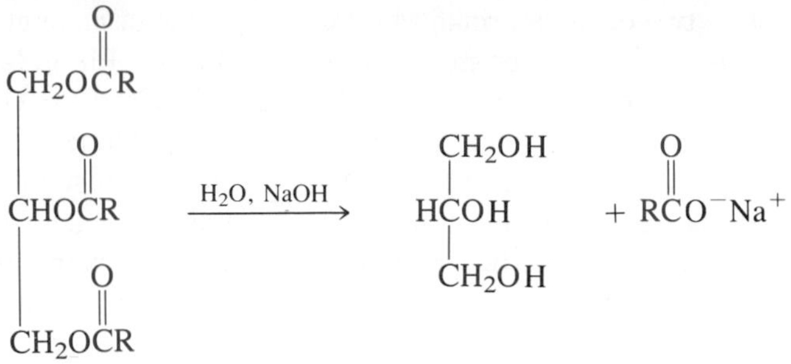

$$\begin{array}{c} \text{CH}_2\text{OCR} \ (\text{C=O}) \\ | \\ \text{CHOCR} \ (\text{C=O}) \\ | \\ \text{CH}_2\text{OCR} \ (\text{C=O}) \end{array} \xrightarrow{\text{H}_2\text{O, NaOH}} \begin{array}{c} \text{CH}_2\text{OH} \\ | \\ \text{HCOH} \\ | \\ \text{CH}_2\text{OH} \end{array} + \text{RCO}^-\text{Na}^+ \ (\text{C=O})$$

Triglyceride **("Fat")** R = long alkyl chain — **1,2,3-Propanetriol** **(Glycerol, glycerine)** — **Soap**

$$\begin{array}{c} \text{CH}_2\text{OH} \\ | \\ \text{CHOH} \\ | \\ \text{CH}_2\text{OH} \end{array} + 3\ \text{HONO}_2 \longrightarrow \begin{array}{c} \text{CH}_2\text{ONO}_2 \\ | \\ \text{CHONO}_2 \\ | \\ \text{CH}_2\text{ONO}_2 \end{array} + 3\ \text{H}_2\text{O}$$

Nitroglycerine

Phosphorus esters of 1,2,3-propanetriols (phosphoglycerides, Section 20-4) are primary components of cell membranes.

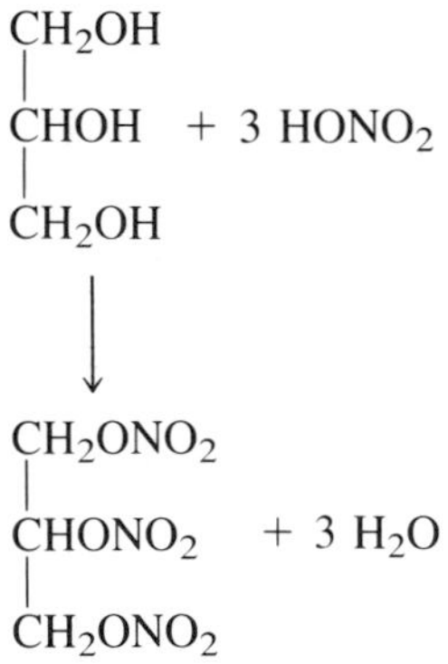

1,2,3-Propanetriol is used in lotions and other cosmetics, as well as in medicinal preparations. Treatment with nitric acid gives a trinitrate ester known as nitroglycerine, an extremely powerful explosive. The explosive potential of this substance results from its shock-induced, highly exothermic decomposition to gaseous products (N_2, CO_2, H_2O gas, O_2), raising temperatures to more than 3000°C and creating pressures higher than 2000 atmospheres in a fraction of a second.

Cholesterol is an important steroid alcohol (Section 4-7).

Ethoxyethane (diethyl ether) was at one time used as a general anesthetic. It produces unconsciousness by depressing central nervous system activity. Because of adverse effects such as irritation of the respiratory tract and extreme nausea, its use has been discontinued, and 1-methoxypropane (methyl propyl ether, "neothyl") and other compounds have replaced it in such applications. Ethoxyethane and other ethers are explosive when mixed with air.

Oxacyclopropane (oxirane, ethylene oxide) is a fumigating agent for seeds and grains. In nature, oxacyclopropane derivatives control insect metamorphosis (see Box 12-2) and are formed in the course of enzyme-catalyzed oxidation of aromatic hydrocarbons, often leading to highly **carcinogenic** (cancer-causing) products (see Section 16-7).

Many *natural products,* some of which are quite physiologically active, contain alcohol and ether groups.

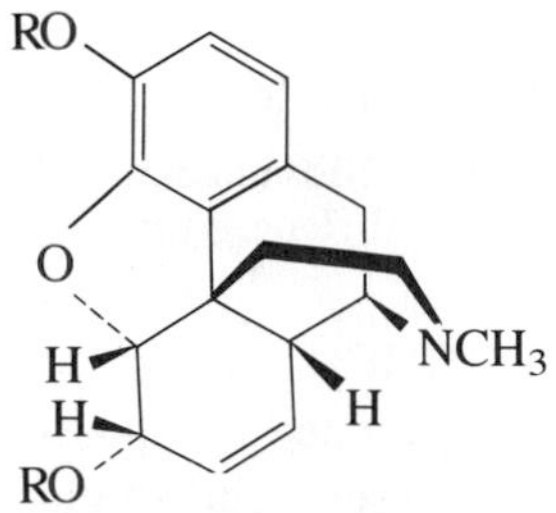

RO, O, H, H, RO, NCH$_3$, H

Morphine
(R = H)
Heroin
$\left(\text{R} = \underset{\text{O}}{\overset{}{\text{CCH}_3}}\right)$

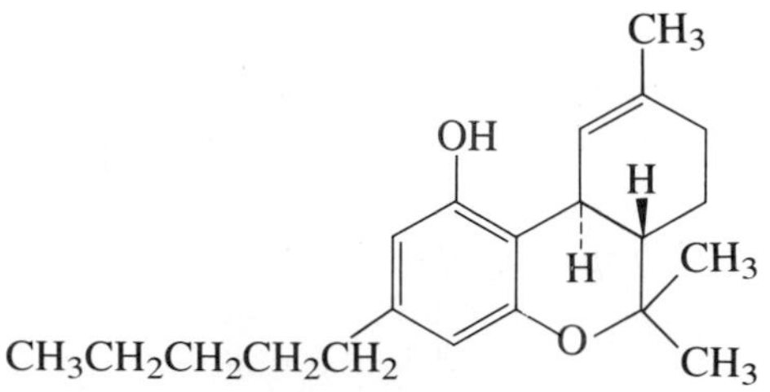

CH$_3$, OH, H, H, CH$_3$, CH$_3$CH$_2$CH$_2$CH$_2$CH$_2$, O, CH$_3$

Tetrahydrocannabinol

The lower thiols and sulfides are most notorious for their foul smell. *Ethanethiol* is detectable by its odor even when diluted in fifty million parts of air. The major volatile components of the skunk's defensive spray are *3-methyl-1-butanethiol,* trans-*2-butene-1-thiol,* and trans-*2-butenyl methyl disulfide*.

$$CH_3CH(CH_3)CH_2CH_2SH$$

3-Methyl-1-butanethiol

$$(H)(H_3C)C{=}C(CH_2SH)(H)$$

***trans*-2-Butene-1-thiol**

$$(H)(H_3C)C{=}C(CH_2SSCH_3)(H)$$

***trans*-2-Butenyl methyl disulfide**

Strangely enough, when highly diluted, sulfur compounds can have a rather pleasant odor. For example, the smell of freshly chopped onions or garlic is due to the presence of small thiols and sulfides. Dimethyl sulfide is a component of the aroma of black tea.

Many beneficial drugs contain sulfur in their molecular framework. Particularly well known are the *sulfonamide,* or *sulfa drugs,* powerful antibacterial agents (Section 15-11):

$$H_2N{-}C_6H_4{-}SO_2NH{-}C_4H_3N_2$$

Sulfadiazine
(An antibacterial drug)

$$H_2N{-}C_6H_4{-}S(=O)_2{-}C_6H_4{-}NH_2$$

Diaminodiphenylsulfone
(An antileprotic drug)

To summarize, alcohols and ethers have various uses, both as chemical raw materials and as medicinal agents. Many of their derivatives can be found in nature; others are readily synthesized.

NEW REACTIONS

1. ALKOXIDES FROM ALCOHOLS (Section 9-1)

Using strong bases

$$ROH \xrightleftharpoons{\text{Strong base}} RO^-$$

Examples of strong bases: $Li^+\ {}^-N[CH(CH_3)_2]_2$; $CH_3CH_2CH_2CH_2Li$; K^+H^-

Using alkali metals

$$ROH + M \longrightarrow RO^-\ M^+ + \tfrac{1}{2}H_2$$

M = Li, Na, K

Substitution Reactions of Alcohols

2. USING HYDROGEN HALIDES (Section 9-2)

$$\text{Primary ROH} \xrightarrow{\text{Conc. HX}} RX$$

X = Br or I (S_N2 mechanism)

$$\text{Secondary or tertiary ROH} \xrightarrow{\text{Conc. HX}} RX$$

X = Cl, Br, or I (S_N1 mechanism)

3. USING PHOSPHORUS REAGENTS (Section 9-4)

$$3\ ROH + PBr_3 \longrightarrow 3\ RBr + H_3PO_3$$
$$6\ ROH + 2\ P + 3\ I_2 \longrightarrow 6\ RI + 2\ H_3PO_3$$

S_N2 mechanism with primary and secondary ROH
Less likelihood of carbocation rearrangements than with HX

4. USING SULFUR REAGENTS (Section 9-4)

$$ROH + SOCl_2 \xrightarrow{N(CH_2CH_3)_3} RCl + SO_2 + (CH_3CH_2)_3\overset{+}{N}H\ Cl^-$$
$$ROH + R'SO_2Cl \longrightarrow \underset{\textbf{Alkyl sulfonate}}{ROSO_2R'} \xrightarrow{Nu^-,\ DMSO} RNu + R'SO_3^-$$

Carbocation Rearrangements in Alcohols

5. CARBOCATION REARRANGEMENTS BY ALKYL AND HYDROGEN SHIFTS (Section 9-3)

$$\text{(R)H–C–C–OH} \xrightarrow[-\ H_2O]{H^+} \text{(R)H–C–}\overset{+}{C}\text{(H)} \longrightarrow \overset{+}{C}\text{–C(H(R))(H)}$$

$$\xrightarrow{-\ H^+,\ E1} \text{C=C–H(R)} \qquad \xrightarrow{S_N1,\ Nu^-} \text{Nu–C–C(H(R))–H}$$

6. CONCERTED ALKYL SHIFTS FROM PRIMARY ALCOHOLS (Section 9-3)

$$R\text{–}C(R')(R'')\text{–}CH_2OH \xrightarrow{H^+} R\text{–}C(R')(R'')\text{–}CH_2\text{–}\overset{+}{O}H_2 \xrightarrow[-\ H_2O]{} (R)(R'')\overset{+}{C}\text{–}CH_2R' \longrightarrow \text{etc.}$$

Elimination Reactions of Alcohols

7. DEHYDRATION WITH STRONG NONNUCLEOPHILIC ACID (Sections 9-2, 9-3, 9-7)

$$\text{H–C–C–OH} \xrightarrow{H_2SO_4,\ heat} \text{C=C} + H_2O$$

Carbocation rearrangements may occur

Temperature required		
Primary ROH:	170–180°	(E2 mechanism)
Secondary ROH:	100–140°	(usually E1)
Tertiary ROH:	25–80°	(E1 mechanism)

Preparation of Ethers

8. WILLIAMSON SYNTHESIS (Section 9-6)

$$\mathrm{ROH} \xrightarrow{\text{NaH, DMSO}} \mathrm{RO^-Na^+} \xrightarrow[\mathrm{S_N2}]{\text{R'X, DMSO}} \mathrm{ROR'}$$

R′ must be methyl or primary
ROH can be primary or secondary (also tertiary if R′ = methyl)
Ease of intramolecular version forming cyclic ethers: $k_3 \geq k_5 > k_6 > k_4 \geq k_7 > k_8$
(k_n = reaction rate, n = ring size)

9. MINERAL ACID METHOD (Section 9-7)

Primary alcohols

$$\mathrm{RCH_2OH} \xrightarrow{\mathrm{H^+}\text{, low temperature}} \mathrm{RCH_2\overset{+}{O}H_2} \xrightarrow[-\mathrm{H_2O}]{\mathrm{RCH_2OH},\ 130\text{–}140°} \mathrm{RCH_2OCH_2R}$$

Secondary alcohols

$$\mathrm{RCH(OH)R} \xrightarrow[-\ \mathrm{H_2O}]{\mathrm{H^+}} \mathrm{R_2CH{-}O{-}CHR_2} + \text{E1 products}$$

Tertiary alcohols

$$\mathrm{R_3COH} + \mathrm{R'OH} \xrightarrow[\mathrm{S_N1},\ -\ \mathrm{H_2O}]{\mathrm{NaHSO_4,\ H_2O}} \mathrm{R_3C{-}OR'} + \text{E1 products}$$

R′ = (mainly) primary

Reactions of Ethers

10. CLEAVAGE BY HYDROGEN HALIDES (Section 9-8)

$$\mathrm{ROR} \xrightarrow{\text{Conc. HX}} \mathrm{RX} + \mathrm{ROH} \xrightarrow{\text{Conc. HX}} 2\ \mathrm{RX}$$

X = Br or I
Primary R: S_N2 mechanism
Secondary or tertiary R: S_N1 mechanism

11. NUCLEOPHILIC OPENING OF OXACYCLOPROPANES (Section 9-9)

Anionic nucleophiles

$$\text{(2,2-dialkyloxacyclopropane)} + \mathrm{Nu^-} \longrightarrow \xrightarrow{\mathrm{H^+,\ H_2O}} \mathrm{NuCH_2C(OH)R_2}$$

Examples of Nu^-: HO^-, RO^-, RS^-

Acid-catalyzed opening

$$\text{(protonated 2,2-dialkyloxacyclopropane)} + \mathrm{Nu} \longrightarrow \mathrm{HOCH_2C(Nu)R_2}$$

Examples of Nu: H_2O, ROH, halide

12. NUCLEOPHILIC OPENING OF OXACYCLOPROPANE BY LITHIUM ALUMINUM HYDRIDE (Section 9-9)

$$\overset{O}{H_2C\!-\!CH_2} \xrightarrow[2.\ H^+,\ H_2O]{1.\ LiAlH_4,\ (CH_3CH_2)_2O} CH_3CH_2OH$$

13. NUCLEOPHILIC OPENING OF OXACYCLOPROPANE BY ORGANOMETALLIC COMPOUNDS (Section 9-9)

$$\text{RLi or RMgX} + \overset{O}{H_2C\!-\!CH_2} \xrightarrow{THF} \xrightarrow{H^+,\ H_2O} RCH_2CH_2OH$$

Sulfur Compounds

14. PREPARATION OF THIOLS AND SULFIDES (Section 9-10)

$$\underset{\text{Excess}}{RX + HS^-} \longrightarrow \underset{\textbf{Thiol}}{RSH}$$

$$RSH + R'X \xrightarrow{\text{Base}} \underset{\textbf{Alkyl sulfide}}{RSR'}$$

15. ACIDITY OF THIOLS (Section 9-10)

$$RSH + HO^- \rightleftharpoons RS^- + H_2O \qquad pK_a(RSH) = 9\text{–}12$$

Acidity of RSH > H_2O ~ ROH

16. NUCLEOPHILICITY OF SULFIDES (Section 9-10)

$$R_2\ddot{\underset{\cdot\cdot}{S}} + R'X \longrightarrow \underset{\textbf{Sulfonium salt}}{R_2\overset{+}{S}R'\ X^-}$$

17. OXIDATION OF THIOLS (Section 9-10)

$$RSH \xrightarrow{KMnO_4\text{ or }H_2O_2} \underset{\textbf{Alkanesulfonic acid}}{RSO_3H}$$

$$RSH \underset{\text{Li, liquid }NH_3}{\overset{I_2}{\rightleftharpoons}} \underset{\textbf{Dialkyl disulfide}}{RS\!-\!SR}$$

18. OXIDATION OF SULFIDES (Section 9-10)

$$RSR' \xrightarrow{H_2O_2} \underset{\textbf{Dialkyl sulfoxide}}{R\overset{O}{\overset{\|}{\underset{\cdot\cdot}{S}}}R'} \xrightarrow{H_2O_2} \underset{\textbf{Dialkyl sulfone}}{R\overset{O}{\overset{\|}{\underset{\underset{O}{\|}}{S}}}R'}$$

IMPORTANT CONCEPTS

1. The reactivity of ROH with alkali metals to give alkoxides and hydrogen follows the order R = CH_3 > primary > secondary > tertiary.
2. In the presence of acid and a nucleophilic counterion, primary alcohols undergo S_N2 reactions. Secondary and tertiary alcohols tend to form carbocations in the presence of acid, capable of E1 and S_N1 product formation, before and after rearrangement.
3. Carbocation rearrangements occur by hydrogen and alkyl group shifts. They usually result in interconversion of secondary carbocations or conversion of a secondary into a tertiary carbocation. Primary alkyloxonium ions can rearrange by a concerted process involving loss of water and simultaneous hydrogen or alkyl shift to give secondary or tertiary carbocations.
4. Synthesis of primary and secondary haloalkanes can be achieved with less risk of rearrangement by using methods involving inorganic esters.
5. Ethers are prepared by either the Williamson ether synthesis or by reaction of alcohols with strong nonnucleophilic acids. The first method is best when S_N2 reactivity is high. In the latter case elimination (dehydration) is a competing process at higher temperatures.
6. Whereas nucleophilic ring opening of oxacyclopropanes by anions is at the less substituted ring carbon according to the rules of the S_N2 reaction, acid-catalyzed opening favors the more substituted carbon, because of charge control of nucleophilic attack.
7. Sulfur has more diffuse orbitals than does oxygen. In thiols, the S–H bond is less polarized than the O–H bond in alcohols, thus leading to diminished hydrogen bonding. Because the S–H bond is also weaker than the O–H bond, the acidity of thiols is greater than that of alcohols.
8. **Note on color use:** Throughout the main portions of the text, beginning in Chapter 6, reacting species in mechanisms and most examples of new transformations are color coded red for nucleophiles, blue for electrophiles, and green for leaving groups. Color coding is *not* used in exercises, summaries of new reactions, or chapter-end problems.

PROBLEMS

1. On which side of the equation do you expect each of the following equilibria to lie (left or right)?

(a) $(CH_3)_3COH + K^{+}\,{}^{-}OH \rightleftharpoons (CH_3)_3CO^{-}K^{+} + H_2O$

(b) $CH_3OH + NH_3 \rightleftharpoons CH_3O^{-} + NH_4^{+}$ ($pK_a = 9.2$)

(c) CH_3CH_2OH + (piperidine ring with $N^{-}Li^{+}$) $\rightleftharpoons CH_3CH_2O^{-}Li^{+}$ + (piperidine ring with N–H) ($pK_a = 40$)

(d) NH_3 ($pK_a = 35$) + $Na^{+}H^{-} \rightleftharpoons Na^{+}\,{}^{-}NH_2 + H_2$ ($pK_a \sim 38$)

2. Give the expected major product of each reaction shown below.

(a) $CH_3CH_2CH_2OH \xrightarrow{\text{Conc. HI}}$

(b) $(CH_3)_2CHCH_2CH_2OH \xrightarrow{\text{Conc. HBr}}$

(c) (cyclohexane ring bearing H and OH) $\xrightarrow{\text{Conc. HI}}$

(d) $(CH_3CH_2)_3COH \xrightarrow{\text{Conc. HCl}}$

3. For each of the following alcohols, write the structure of the alkyloxonium ion produced after protonation by strong acid; if the alkyloxonium ion is capable of losing water readily, write the structure of the resulting carbocation; and if the carbocation obtained is likely to be susceptible to rearrangement, write the structures of all new carbocations that might be reasonably expected to form.

(a) $CH_3CH_2CH_2OH$ **(b)** $CH_3CH(OH)CH_3$ **(c)** $CH_3CH_2CH_2CH_2OH$

(d) $(CH_3)_2CHCH_2OH$ **(e)** $(CH_3)_3CCH_2CH_2OH$ **(f)** 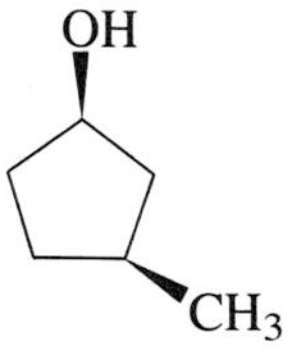 (cyclopentane ring bearing OH and CH$_3$ on wedge bonds)

(g) 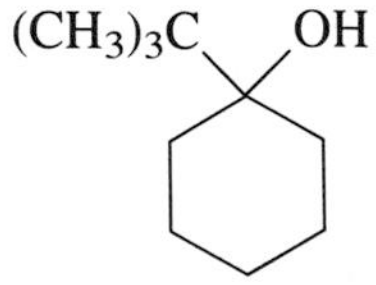 (cyclohexane ring bearing $(CH_3)_3C$ and OH on the same carbon) **(h)** 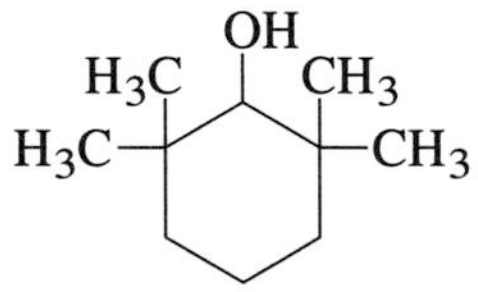 (cyclohexane ring bearing OH, flanked by carbons each bearing two methyl groups: H_3C, H_3C and CH_3, CH_3)

4. Write all products of the reaction of each of the alcohols in Problem 3 with concentrated H_2SO_4 under elimination conditions.

5. Write all sensible products of the reaction of each of the alcohols in Problem 3 with concentrated aqueous HBr.

6. Primary alcohols are often converted into bromides by reaction with NaBr in H_2SO_4. Explain how this transformation works and why it might be considered a superior method to that using concentrated aqueous HBr.

$$CH_3CH_2CH_2CH_2OH \xrightarrow{NaBr,\ H_2SO_4} CH_3CH_2CH_2CH_2Br$$

7. What are the most likely product(s) of each of the following reactions?

(a) 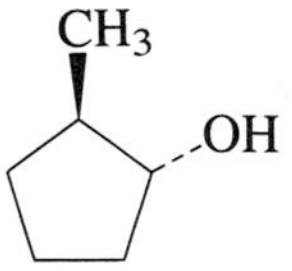 (cyclopentane ring bearing wedged CH_3 and dashed OH on adjacent carbons) $\xrightarrow{CH_3CH_2OH,\ H_2SO_4}$ **(b)** $CH_3C(CH_3)_2CH_2OH \xrightarrow{\text{Conc. HI}}$

(c) 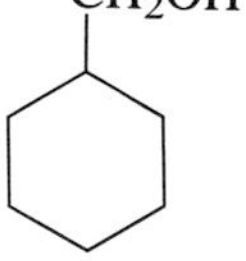 (cyclohexane ring bearing CH_2OH) $\xrightarrow{\text{Conc. } H_2SO_4,\ 180°C}$ **(d)** $CH_3C(CH_3)_2\text{—}CHICH_3 \xrightarrow{H_2O}$

8. Give the expected main product of the reaction of each of the alcohols in Problem 3 with PBr_3. Compare the results with those of Problem 5.

9. Give the expected product(s) of the reaction of 1-pentanol with each of the following reagents.

(a) $K^+\ {}^-OC(CH_3)_3$ **(b)** Sodium metal
(c) CH_3Li **(d)** Concentrated HI
(e) Concentrated HCl **(f)** FSO_3H
(g) Concentrated H_2SO_4 at 130°C **(h)** Concentrated H_2SO_4 at 180°C

(i) CH_3SO_2Cl, $(CH_3CH_2)_3N$
(j) PBr_3
(k) $SOCl_2$
(l) $K_2Cr_2O_7 + H_2SO_4 + H_2O$
(m) PCC, CH_2Cl_2
(n) $(CH_3)_3COH +$ H_2SO_4 (As catalyst)

10. Give the expected product(s) of the reaction of *trans*-3-methylcyclopentanol with each of the reagents in Problem 9.

11. Suggest a good synthetic method for preparing each of the following haloalkanes from the corresponding alcohols.

(a) $CH_3CH_2CH_2Cl$ (b) $CH_3CH_2CH(CH_3)CH_2Br$ (c)

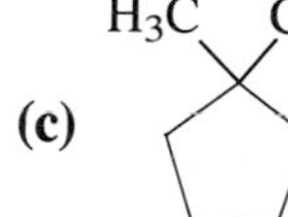

(d) $CH_3CHICH(CH_3)_2$

12. Name each of the following molecules according to IUPAC.

(a) $(CH_3)_2CHOCH_2CH_3$ (b) $CH_3OCH_2CH_2OH$ (c) cyclopentyl–O–cyclopentyl

(d) $(ClCH_2CH_2)_2O$ (e) 1-(H_3C)-1-(OCH_3)cyclopentane (f) CH_3O–cyclohexane–OCH_3

(g) CH_3OCH_2Cl

13. Explain why the boiling points of ethers are lower than those of the isomeric alcohols. Would you expect the relative water solubilities to differ in a similar way?

14. Write the expected major product(s) of each of the following attempted ether syntheses.

(a) $CH_3CH_2CH_2Cl + CH_3CH_2CH(O^-)CH_2CH_3 \xrightarrow{DMSO}$

(b) $CH_3CH_2CH_2O^- + CH_3CH_2CH(Cl)CH_2CH_3 \xrightarrow{HMPA}$

(c) 1-(H_3C)-1-(O^-)cyclohexane $+ CH_3I \xrightarrow{DMSO}$

(d) $(CH_3)_2CHO^- + (CH_3)_2CHCH_2CH_2Br \xrightarrow{(CH_3)_2CHOH}$

(e) cyclohexyl(H)O^- + cyclohexyl(H)Cl $\xrightarrow{\text{Cyclohexanol}}$

(f) $(C_5H_9)_2C(CH_3)$–$O^- + CH_3CH_2I \xrightarrow{DMSO}$

15. For each synthesis proposed in Problem 14 that is not likely to give a good yield of ether product, suggest an alternative synthesis beginning with suitable alcohols or haloalkanes that will give a superior result. (Hint: See Problem 1 of Chapter 7.)

16. Write the product(s) of reaction of each of the following molecules with NaOH in dilute solution in DMSO.

(a) $CH_3\overset{OH}{\overset{|}{C}}HCH_2CH_2\overset{Cl}{\overset{|}{C}}H_2$ **(b)** 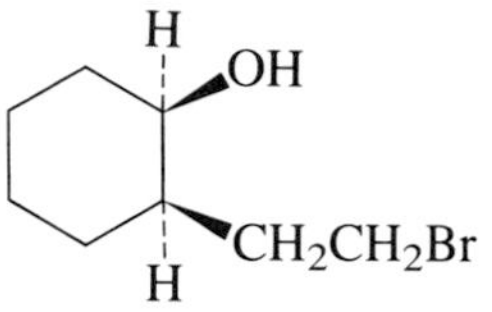 (cyclohexane with H, OH; H, CH_2CH_2Br) **(c)** 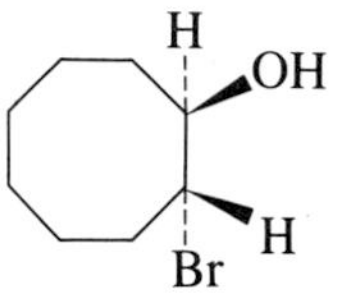 (cyclooctane with H, OH; Br, H)

17. Propose efficient syntheses for each of the following ethers, using haloalkanes or alcohols as starting materials.

(a) $CH_3CH_2\overset{CH_3}{\overset{|}{C}}HOCH_2CH_3$ **(b)** 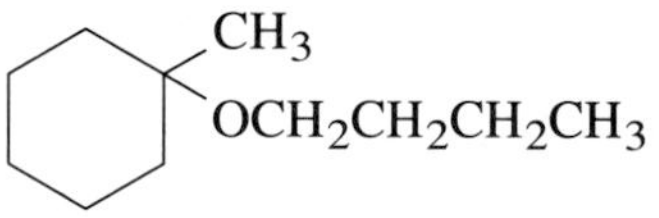 (cyclohexane with CH_3, $OCH_2CH_2CH_2CH_3$)

(c) 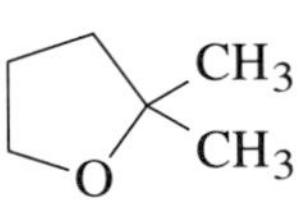 (tetrahydrofuran ring O with CH_3, CH_3) **(d)** 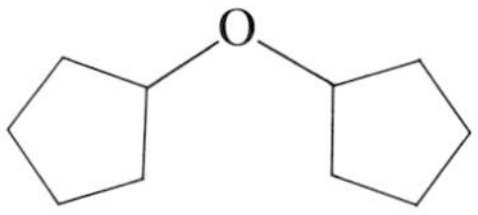 (cyclopentyl–O–cyclopentyl)

18. Give the major product(s) of each of the following reactions.

(a) $CH_3CH_2OCH_2CH_2CH_3 \xrightarrow{\text{Excess conc. HI}}$ **(b)** $CH_3OCH(CH_3)_2 \xrightarrow{\text{Excess conc. HBr}}$

(c) $CH_3OCH_2CH_2OCH_3 \xrightarrow{\text{Excess conc. HI}}$ **(d)** 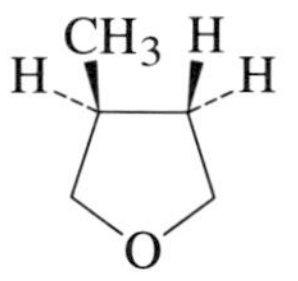 (tetrahydrofuran ring O with CH_3, H; H, H) $\xrightarrow{\text{Excess conc. HBr}}$

(e) 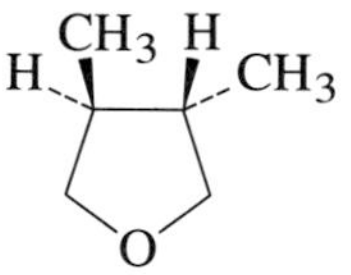 (tetrahydrofuran ring O with CH_3, H; H, CH_3) $\xrightarrow{\text{Excess conc. HBr}}$ **(f)** 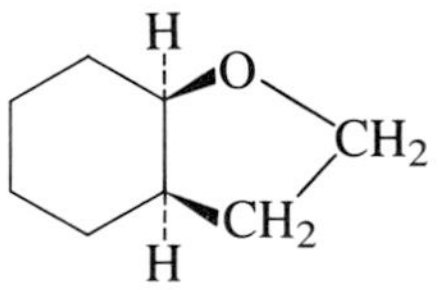 (cyclohexane fused with H, O, CH_2, CH_2, H) $\xrightarrow{\text{Excess conc. HBr}}$

19. Give the major product(s) of each of the following reactions.

(a)  (oxirane) $\xrightarrow{Na^{+}\,^{-}NH_2,\ NH_3}$ **(b)** 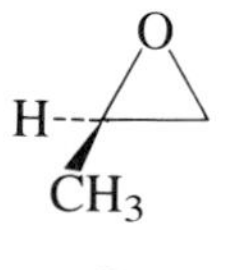 (oxirane with H, CH_3) $\xrightarrow{Na^{+}\,^{-}SCH_2CH_3,\ CH_3CH_2OH}$

(c) 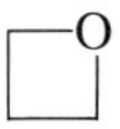 (oxetane) $\xrightarrow{\text{Excess conc. HBr}}$ **(d)** 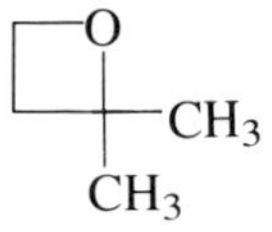 (oxetane with CH_3, CH_3) $\xrightarrow{\text{Dilute HCl in } CH_3OH}$

(e) 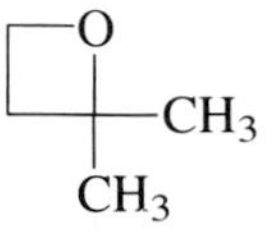 (oxetane with CH_3, CH_3) $\xrightarrow{Na^{+}\,^{-}OCH_3 \text{ in } CH_3OH}$ **(f)** 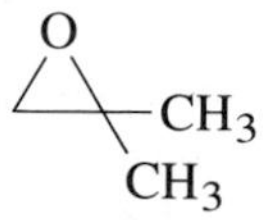 (oxirane with CH_3, CH_3) $\xrightarrow{\text{1. } LiAlD_4,\ (CH_3CH_2)_2O \quad \text{2. } H^+,\ H_2O}$

(g) (oxirane with CH_3, CH_3) $\xrightarrow{\text{1. } (CH_3)_2CHMgCl,\ (CH_3CH_2)_2O \quad \text{2. } H^+,\ H_2O}$ **(h)** (oxirane) $\xrightarrow{\text{1. (cyclopentyl)}-Li,\ (CH_3CH_2)_2O \quad \text{2. } H^+,\ H_2O}$

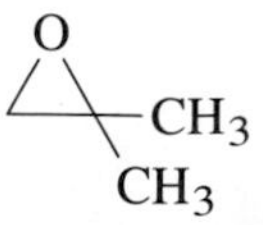

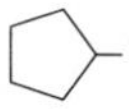

20. For each alcohol in Problem 20 of Chapter 8, suggest a synthetic route that starts with an oxacyclopropane (if possible!).

21. Give the major product(s) expected from each of the reactions shown below. Watch stereochemistry!

(a) 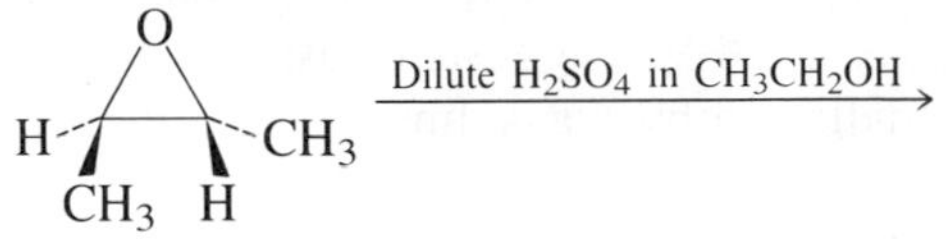 H, CH_3 / CH_3, H oxacyclopropane $\xrightarrow{\text{Dilute } H_2SO_4 \text{ in } CH_3CH_2OH}$ **(b)** 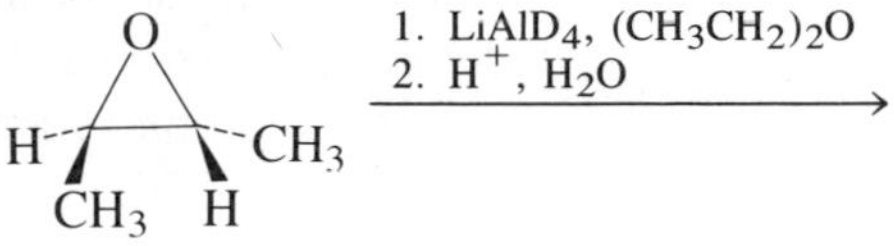 H, CH_3 / CH_3, H oxacyclopropane $\xrightarrow{\text{1. } LiAlD_4, (CH_3CH_2)_2O \quad \text{2. } H^+, H_2O}$

22. Name each of the following compounds according to IUPAC.

(a) 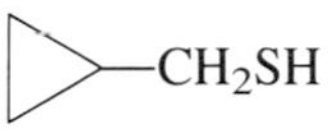 cyclopropyl$-CH_2SH$ **(b)** $CH_3CH_2CH(CH_3)SCH_3$ **(c)** $CH_3CH_2CH_2SO_3H$ **(d)** CF_3SO_2Cl

23. In each of the following pairs of compounds indicate which is the stronger acid and which is the stronger base. **(a)** CH_3SH, CH_3OH; **(b)** HS^-, HO^-; **(c)** H_3S^+, H_2S.

24. Give reasonable products for each of the following reactions.

(a) $ClCH_2CH_2CH_2CH_2Cl \xrightarrow{\text{One equivalent } Na_2S}$

(b) 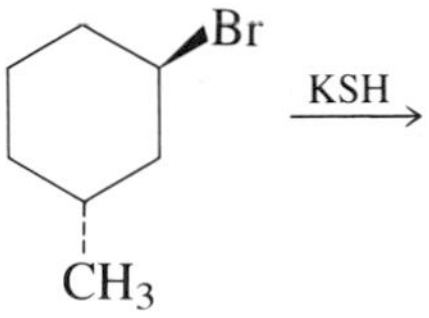 Br, CH_3 cyclohexane $\xrightarrow{KSH}$ **(c)** 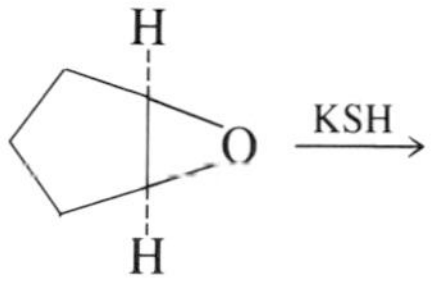 H, H, O (cyclopentane oxide) $\xrightarrow{KSH}$ **(d)** 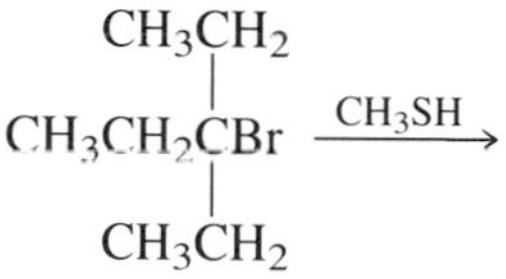 $CH_3CH_2C(Br)(CH_2CH_3)_2$ (CH_3CH_2, CH_3CH_2CBr, CH_3CH_2) $\xrightarrow{CH_3SH}$

(e) 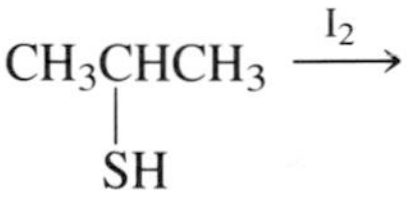 $CH_3CH(SH)CH_3 \xrightarrow{I_2}$ **(f)** 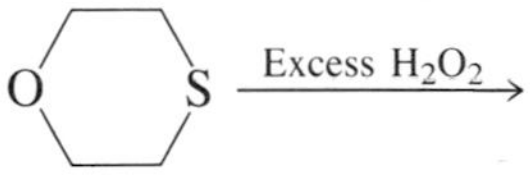 O, S (1,4-oxathiane) $\xrightarrow{\text{Excess } H_2O_2}$

25. Give the structures of compounds A, B, and C (with stereochemistry!) from the information in the scheme below. (Hint: A is acyclic.) To what compound class does the product belong?

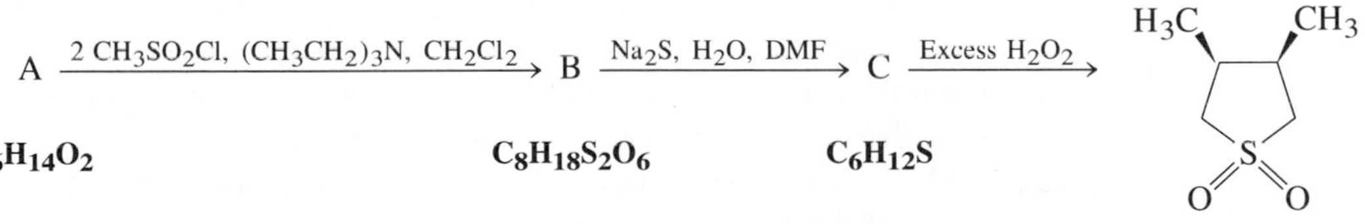

$$\underset{C_6H_{14}O_2}{A} \xrightarrow{2\,CH_3SO_2Cl,\ (CH_3CH_2)_3N,\ CH_2Cl_2} \underset{C_8H_{18}S_2O_6}{B} \xrightarrow{Na_2S,\ H_2O,\ DMF} \underset{C_6H_{12}S}{C} \xrightarrow{\text{Excess } H_2O_2}$$ (product: H_3C, CH_3 substituted ring with S, O, O)

26. In an attempt to make 1-chloro-1-cyclobutylpentane, the following reaction sequence was employed. The actual product isolated, however, was not the desired molecule but an isomer of it. Suggest a structure for the product and give a mechanistic explanation for its formation.

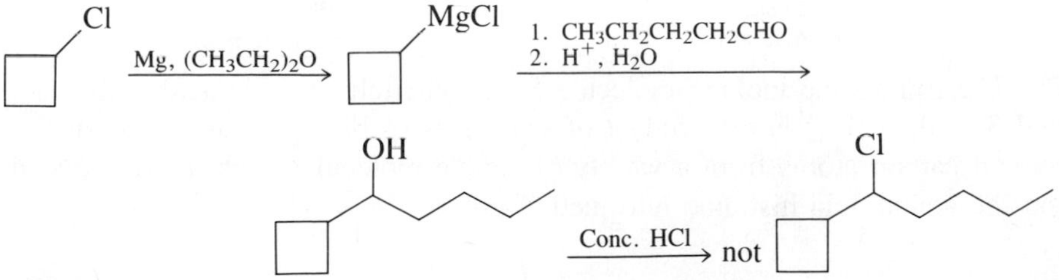

Cl (chloromethylcyclobutane) $\xrightarrow{Mg,\ (CH_3CH_2)_2O}$ MgCl $\xrightarrow{\text{1. } CH_3CH_2CH_2CH_2CHO \quad \text{2. } H^+, H_2O}$

OH $\xrightarrow{\text{Conc. HCl}}$ not Cl

O^- Br

$H\text{---}CCH_2CH_2CH_2C\text{---}H$

CH_3 D

↓ DMSO

27. What is the product of the reaction shown in the margin? (Pay attention to stereochemistry at the reacting centers.) What is the kinetic order of this reaction?

28. Propose syntheses of the following molecules, choosing reasonable starting materials on the basis of the principles of synthetic strategy introduced in preceding chapters, particularly in Section 8-9. Suggested positions for carbon–carbon bond formation are indicated by wavy lines.

(a) $CH_3CH_2CH\text{⌇}CH_2CH_2SO_3H$ (with a cyclopentyl group attached to the CH through a wavy-line bond)

(b) $CH_3CH_2CH_2\text{⌇}C(CH_3)(CH_2CH_3)\text{⌇}CHO$

29. Give efficient syntheses of each of the following compounds, beginning with the indicated starting material.

(a) *trans*-1-Bromo-2-methylcyclopentane, from *cis*-2-methylcyclopentanol

(b) 3-Cyanopentane, from 3-pentanol

(c) 3-Chloro-3-methylhexane, from 3-methyl-2-hexanol

(d) (six-membered ring containing O and S), from 2-bromoethanol (2 mol)

30. Compare the following methods of alkene synthesis from a general primary alcohol. State the advantages and disadvantages of each one.

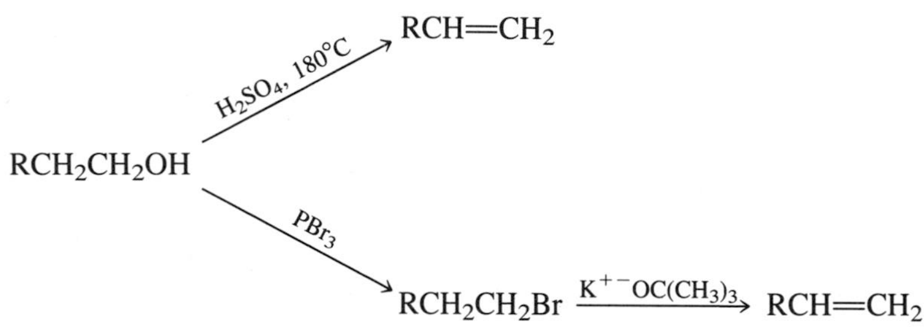

31. Sugars, being polyhydroxylic compounds (Chapter 24), undergo reactions characteristic of alcohols. In one of the later steps in glycolysis (the metabolism of glucose), one of the glucose metabolites with a remaining hydroxy group, 2-phosphoglyceric acid, is converted into 2-phosphoenolpyruvic acid. This reaction is catalyzed by the enzyme enolase in the presence of a Lewis acid such as Mg^{2+}. **(a)** How would you classify this reaction? **(b)** What is the possible role of the Lewis acidic metal ion?

$$HOCH_2\text{—}CH(OPO_3^{2-})\text{—}COOH \xrightarrow{\text{Enolase, } Mg^{2+}} CH_2{=}C(OPO_3^{2-})(CO_2H)$$

2-Phospho-glyceric acid **2-Phosphoenol-pyruvic acid**

32. The formidable looking molecule 5-methyltetrahydrofolic acid (abbreviated 5-methyl-FH_4) is the product of sequences of biological reactions that convert carbon atoms from a variety of simple molecules such as formic acid and the amino acid histidine into methyl groups.

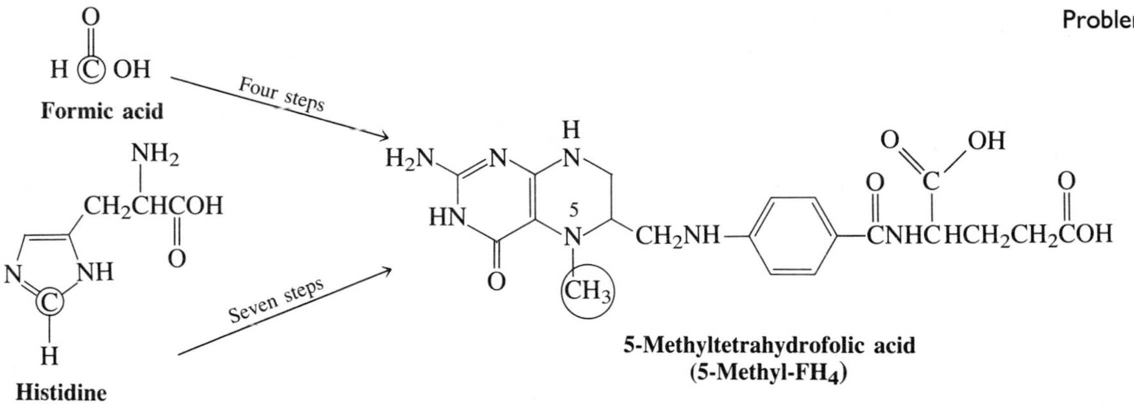

The simplest synthesis of 5-methyltetrahydrofolic acid is from tetrahydrofolic acid (FH_4) and trimethylsulfonium ion, a reaction carried out by microorganisms in the soil.

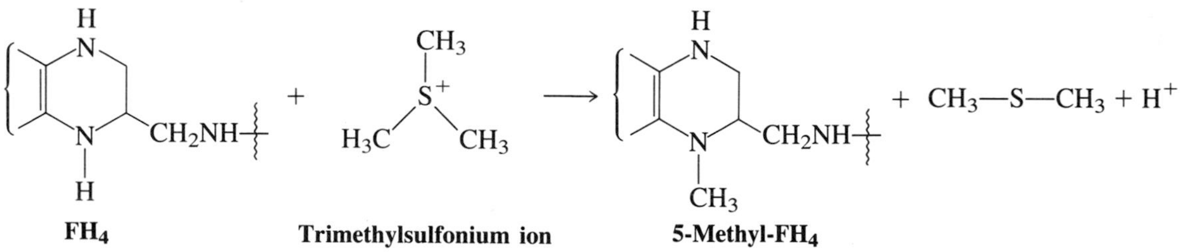

(a) Can this reaction be reasonably assumed to proceed through a nucleophilic substitution mechanism? Write the mechanism using "electron pushing" arrow notation. **(b)** Identify the nucleophile, the nucleophilic and electrophilic atoms participating in the reaction, and the leaving group. **(c)** Based on the concepts presented in Sections 6-8 through 6-10 and in Sections 9-2 and 9-9, are all the groups that you identified in (b) behaving in a reasonable way in this reaction? Does it help to know that species such as H_3S^+ are very strong acids (e.g., pK_a of $CH_3SH_2^+$ is -7)?

33. The role of 5-methyl-FH_4 (Problem 32) in biology is to serve as a donor of methyl groups to small molecules. The synthesis of the amino acid methionine from homocysteine is perhaps the best-known example.

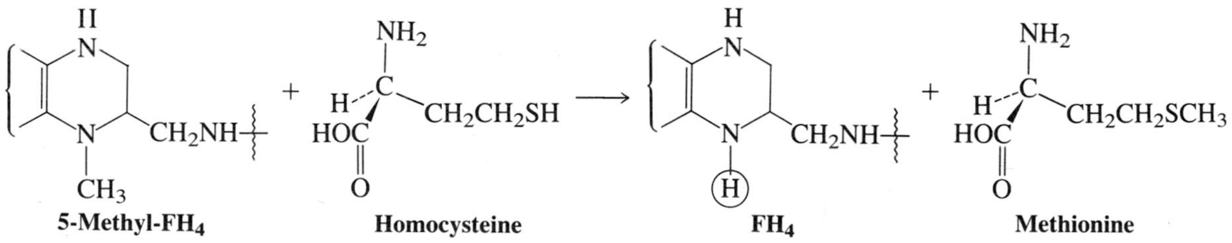

For this problem, answer the same questions that were posed in Problem 32. The pK_a of the circled hydrogen in FH_4 is 5. Does this cause a problem with any feature of your mechanism? In fact, methyl transfer reactions of 5-methyl-FH_4 require a proton source. Review the material in Section 9-2, especially the subsection titled "Water can be a leaving group." Then suggest a useful role for a proton in the reaction above.

34. Epinephrine (adrenalin) is produced in your body in a two-step process that accomplishes the transfer of a methyl group from methionine (Problem 33) to norepinephrine (see reactions 1 and 2 below). **(a)** Explain in detail what is going on mechanistically in these two reactions, and analyze the role played by the molecule of ATP. **(b)** Would you expect methionine to react directly with norepinephrine? Explain. **(c)** Propose a laboratory synthesis of epinephrine from norepinephrine.

REACTION 1.

Methionine + ATP → S-Adenosylmethionine + $H_4P_3O_{10}^-$ (Triphosphate)

Methionine $H_2N\text{-}CH(COOH)\text{-}CH_2CH_2\text{—S—}CH_3$ + **ATP** $HO\text{—}P(O)(OH)\text{—O—}P(O)(OH)\text{—O—}P(O)(OH)\text{—O—}CH_2$-(ribose, OH OH)-adenine (NH_2)

→ **S-Adenosylmethionine** $H_2N\text{-}CH(COOH)\text{-}CH_2CH_2\text{—}\overset{+}{S}(CH_3)\text{—}CH_2$-(ribose, OH OH)-adenine (NH_2) + $H_4P_3O_{10}^-$ **(Triphosphate)**

REACTION 2.

S-Adenosylmethionine + **Norepinephrine** (3,4-$(HO)_2C_6H_3$—C(HO)(H)—CH_2NH_2) →

S-Adenosylhomocysteine $H_2N\text{-}CH(COOH)\text{-}CH_2CH_2\text{—S—}CH_2$-(ribose, OH OH)-adenine (NH_2) + **Epinephrine** (3,4-$(HO)_2C_6H_3$—C(HO)(H)—CH_2NHCH_3) + H^+

35. **(a)** Only the trans isomer of 2-bromocyclohexanol can react with sodium hydroxide to form an oxacyclopropane-containing product. Explain the lack of reactivity of the cis isomer. [Hint: Draw the available conformations of both the cis and trans isomers around the C1–C2 bonds (compare Figure 4-6). Use models if necessary.] **(b)** The synthesis of some oxacyclopropane-containing steroids has been achieved by use of a two-step procedure starting with steroidal bromoketones. Suggest suitable reagents for accomplishing a conversion such as the following one.

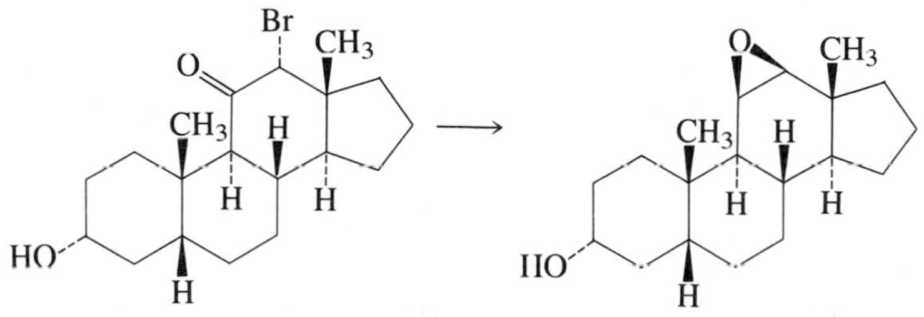

(c) Do any of the steps in your proposed sequence have specific stereochemical requirements for the success of the oxacyclopropane-forming step?

36. Freshly cut garlic contains allicin, a compound responsible for the true garlic odor. Allicin also possesses substantial antibacterial properties and is apparently effective in limiting the increase in cholesterol levels for animals on high-cholesterol diets. Propose a short synthesis of allicin, starting with 3-chloropropene.

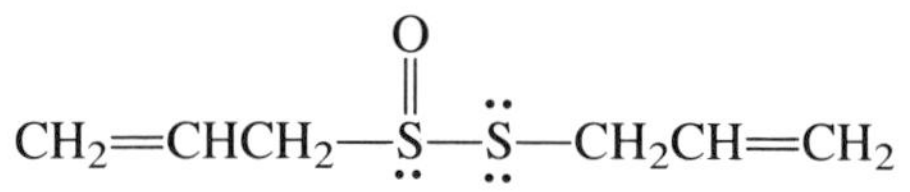

$CH_2{=}CHCH_2{-}S({=}O){-}S{-}CH_2CH{=}CH_2$

Allicin

10

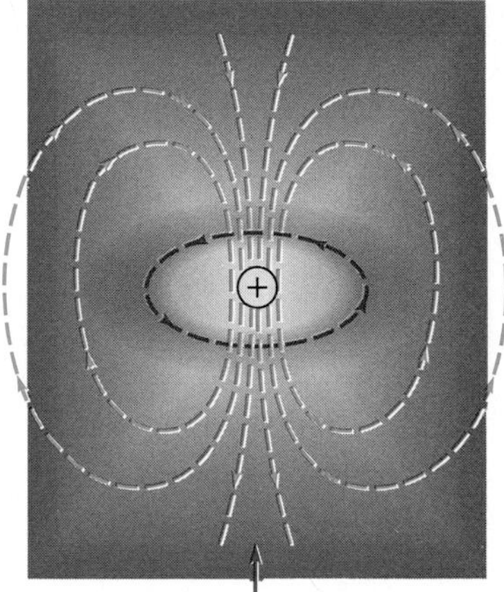

Using Nuclear Magnetic Resonance Spectroscopy to Deduce Structure

Now that we have studied a variety of organic reactions and functional groups, we should be able to plan the synthesis of a reasonably complicated organic molecule in the laboratory. But how can we identify a molecule that we have synthesized, and how can we ascertain its structure? How do we know, for example, that the Grignard reagent we have used has really converted a ketone into the desired alcohol?

This chapter will help answer such questions. We shall see that a tool known as NMR spectroscopy provides a kind of "fingerprint" of a molecule's structure. First, however, consider how physical measurements and chemical tests can elucidate structure.

10-1 Physical and Chemical Tests

To study our sample, we must first purify it—by chromatography, distillation, or recrystallization. We can then compare its melting point, refractive index, and other physical properties with data for known compounds. If our measurements match values in the literature (or appropriate handbooks), we can be reasonably certain of the identity and structure of our molecule. Yet many substances made in the laboratory are new: No published data are available. We need ways to determine their structures *ourselves*.

Elemental analysis will reveal the sample's gross chemical composition. Tests of the chemistry of the compound can then help us to identify its functional groups. For example, we saw in Section 1-9 that we can distinguish between methoxymethane and ethanol on the basis of their physical properties.

The problem becomes considerably more difficult for larger molecules, which vary far more in structure. What if a reaction gave us an alcohol of molecular formula $C_7H_{16}O$? A test with sodium metal would reveal a hydroxy functional group—but not an unambiguous structure. In fact, there are many possibilities, only three of which are shown here.

Three Structural Possibilities for an Alcohol $C_7H_{16}O$

$CH_3(CH_2)_5CH_2OH$ $\quad$ $CH_3C(CH_3)_2CH_2CH_2CH_2OH$ $\quad$ $CH_3C(CH_3)(CH_2CH_3)CH_2CH_2OH$

EXERCISE 10-1

Write the structures of several secondary and tertiary alcohols having the molecular formula $C_7H_{16}O$.

To differentiate between these alternatives, a modern organic chemist makes use of another tool: spectroscopy.

10-2 Defining Spectroscopy

Spectroscopy is a technique for analyzing molecules based on how they absorb radiation. Although there are many types of spectroscopy, four are used most often in organic chemistry: (1) nuclear magnetic resonance (NMR) spectroscopy; (2) infrared (IR) spectroscopy; (3) ultraviolet (UV) spectroscopy; and (4) mass spectroscopy (MS). The first, **NMR spectroscopy,** probes the structure in the vicinity of individual nuclei, particularly hydrogens and carbons, and provides the most detailed information regarding the construction of a molecule.

We begin with a simple overview of spectroscopy as it relates to NMR, IR, and UV. Then we describe how a spectrometer works. Finally, we discuss in more detail the principles and applications of NMR spectroscopy. We shall return to the other major forms of spectroscopy in Chapters 11, 14, and 20.

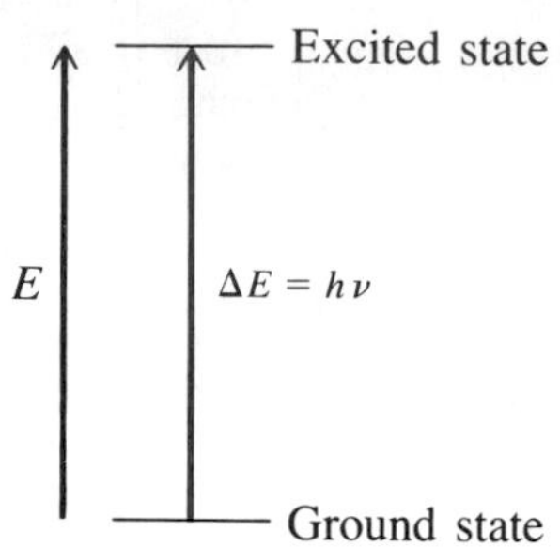

FIGURE 10-1 The energy difference, ΔE, between the ground state and the excited state of a molecule is overcome by incident radiation of frequency ν matched exactly to equal ΔE. [ν, frequency of absorbed radiation; h (Planck's constant) = 6.626×10^{-27} erg s (1 erg = 10^{-7}J).]

Molecules undergo distinctive excitations

Organic molecules absorb electromagnetic radiation in discrete "packets" of energy, or **quanta.** Absorption occurs only when radiation supplying exactly the right packet reaches the compound under investigation. If the frequency of the incident radiation is ν, the packet has energy $\Delta E = h\nu$ (Figure 10-1).

The absorbed energy causes some kind of electronic or mechanical "motion" in the molecule, a process called **excitation.** This motion also is quantized; and because a molecule can undergo many different kinds of excitation, each kind of motion requires its own distinctive energy. X-rays, for example, which are a form of high-energy radiation, can promote electrons in atoms from inner shells to outer ones; this change, called an **electronic transition,** requires energy higher than 300 kcal mol^{-1}. Ultraviolet radiation and visible light, in contrast, excite valence-shell electrons, typically from a filled bonding molecular orbital to an unfilled antibonding one (see Figure 1-12); here the energy needed ranges from 40 to 300 kcal mol^{-1}. Infrared radiation causes vibrational excitation of a compound's bonds (ΔE = 2–10 kcal mol^{-1}), whereas quanta of microwave radiation cause bond rotations to occur ($\Delta E = \sim 10^{-4}$ kcal mol^{-1}). Finally, radio waves can produce transitions *within* an atomic nucleus ($\Delta E = \sim 10^{-6}$ kcal mol^{-1}); we shall see in the next section how this phenomenon is the basis of nuclear magnetic resonance spectroscopy.

Figure 10-2 depicts the various forms of radiation, the energy (ΔE) related to each, and the corresponding wavelengths. Remember that radiation increases in energy with increasing frequency (ν) or wavenumber ($\tilde{\nu}$) but *decreasing* wavelength (λ).*

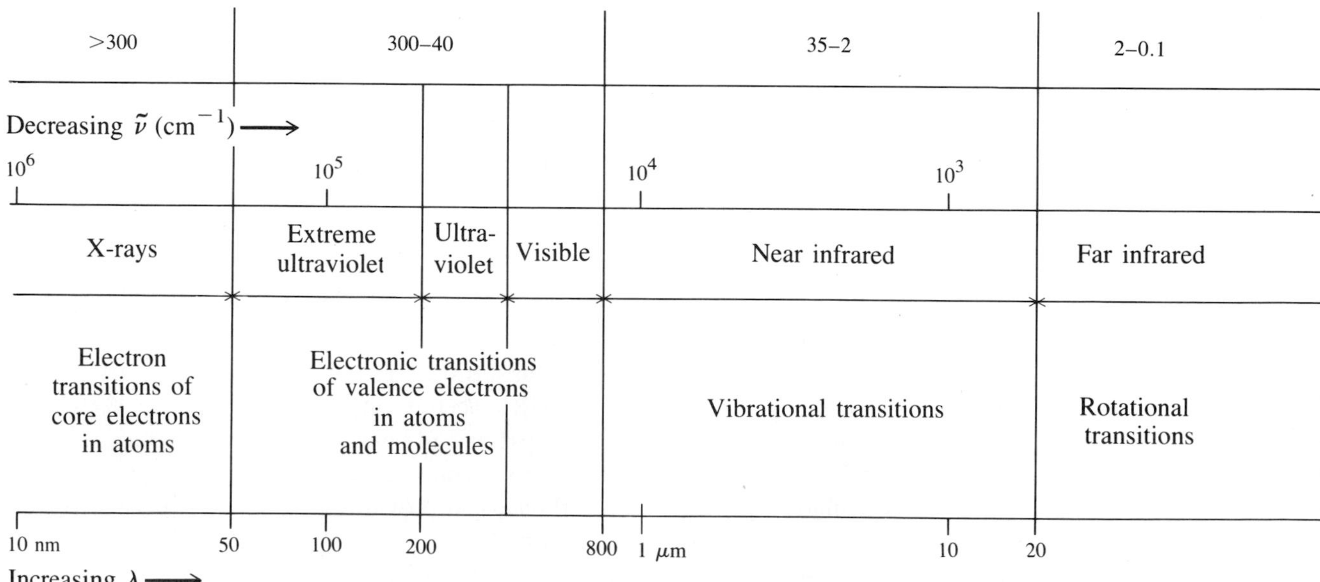

*Wavenumber, defined as $\tilde{\nu} = 1/\lambda$, is the number of waves per centimeter. This quantity is related to (but should not be confused with) frequency, $\nu = c/\lambda$ (in cycles per second, or hertz, named for a German physicist, Heinrich Rudolf Hertz, 1857–1894), in which c = velocity of light = 3×10^{10} cm s^{-1}. A simple conversion between ΔE (kcal mol^{-1}) and λ (nm) is given by the equation $\Delta E = 28{,}600/\lambda$.

EXERCISE 10-2

What type of radiation (in wavelength, λ) would be minimally required to initiate the radical chlorination of methane? (Refer to Section 3-4.)

A spectrometer records the absorption of radiation

As illustrated in Figure 10-1, the absorption of a quantum of radiation by a molecule brings about a transition from its (normal) ground state to an excited state. Spectroscopy is a procedure by which these absorptions can be mapped by instruments called **spectrometers.**

Figure 10-3 shows how the spectrometer operates. It contains a source of electromagnetic radiation with a frequency in the region of interest, such as the infrared or radio. The apparatus is designed so that radiation of one specific wavelength passes through the sample. The frequency of this incident light is changed continuously, and its intensity (relative to a reference beam) is measured at a detector and recorded on calibrated paper. In the absence of absorption, the sweep of radiation appears as a straight line, the **baseline.** Whenever the sample absorbs incident light, however, the resulting change in intensity at the detector registers as a **peak,** or deviation from the baseline. The resulting pattern, or **plot,** is the **spectrum** (Latin, appearance, apparition) of the sample.

In summary, electromagnetic radiation is absorbed in discrete quanta of incident energy measurable by spectroscopy. The four used most often in organic chemistry are nuclear magnetic resonance (NMR) spectroscopy, infrared (IR) spectroscopy, ultraviolet (UV) spectroscopy, and mass spectroscopy (MS).

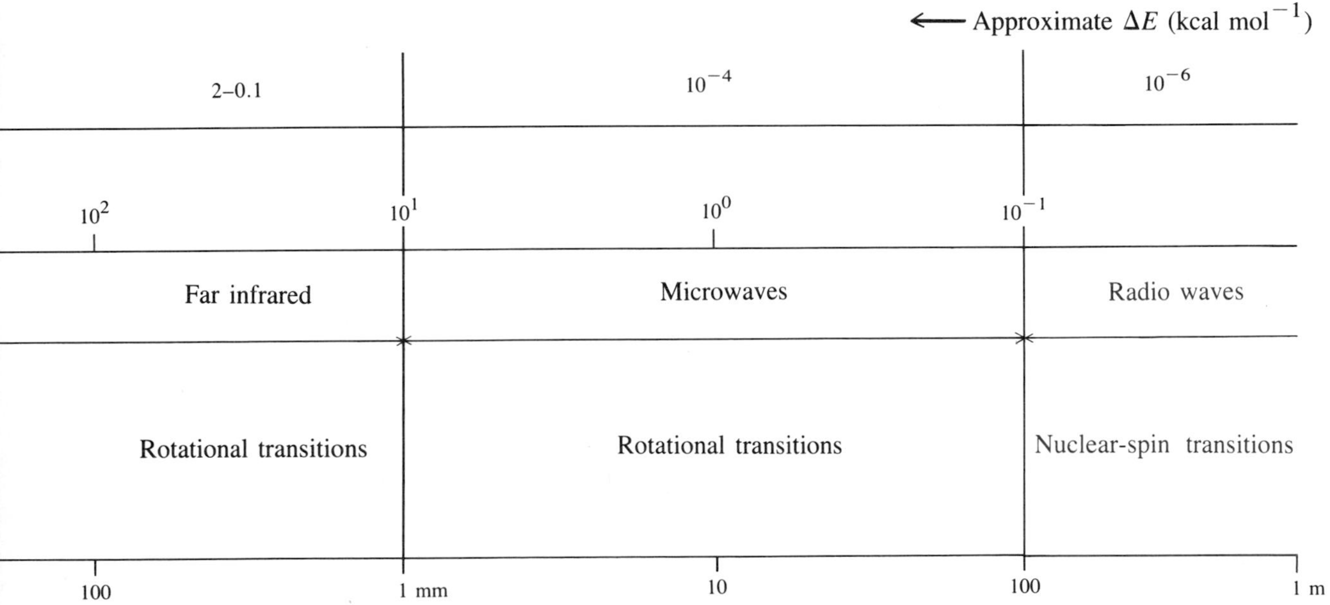

FIGURE 10-2 The spectrum of electromagnetic radiation. The top line is an energy scale, in units of kcal mol^{-1}, increasing from right to left. Below are the corresponding wavenumbers, $\tilde{\nu}$, in units of cm^{-1}. The types of radiation associated with the principal types of spectroscopy and the transitions induced by each are given in the middle. A wavelength scale is at the bottom (λ, in units of nanometers, 1 nm = 10^{-9} m; micrometers, 1 μm = 10^{-6} m; millimeters, mm; and meters, m).

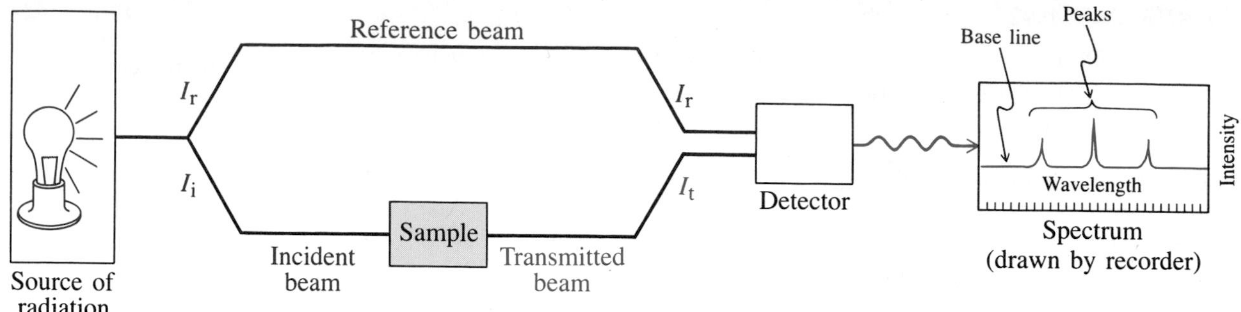

FIGURE 10-3 General schematic diagram of a spectrometer. The source beam is split into reference and incident beams of equal intensity ($I_r = I_i$). The incident beam traverses the sample and emerges as the transmitted beam (I_t). When no absorption occurs, its intensity, I_t, will equal I_i and hence I_r. When $I_t = I_r$, the detector notes no difference and a straight line (zero or baseline) is drawn by the recorder. If absorption occurs, then $I_t \neq I_r$ and a peak is recorded. The resulting diagram is called a spectrum.

10-3 Proton Nuclear Magnetic Resonance

Nuclear magnetic resonance spectroscopy requires low-energy radiation in the radio-frequency range. This section presents the principles behind this technique.

Nuclear spins can be excited by the absorption of radio waves

Many atomic nuclei behave as if they were spinning and are therefore said to have a **nuclear spin.** One of those nuclei is hydrogen, written as ^1H (the hydrogen isotope of mass 1) to differentiate it from other isotopes (deuterium, tritium). Let us consider the simplest form of hydrogen, the proton. Because the proton is positively charged, its spinning motion creates (as does any moving charged particle) a magnetic field. The net result is that a proton may be viewed as a tiny bar magnet floating freely in solution or in space. When the proton is exposed to an *external* magnetic field of strength H_0, it may have one of two orientations: It may be aligned either with H_0, an energetically favorable choice, or (unlike a normal bar magnet) against H_0, a move that costs energy. The two possibilities are designated the α and β **spin states,** respectively (Figure 10-4).

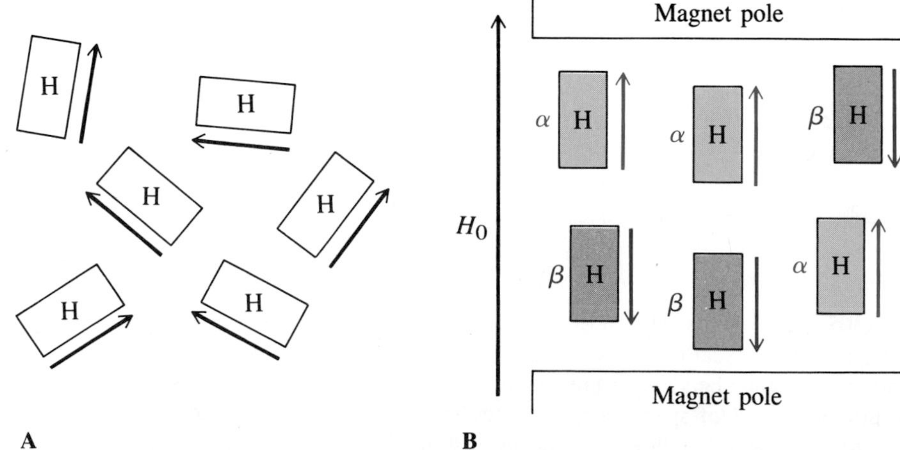

FIGURE 10-4 (A) Single protons (H) act as tiny bar magnets. The direction of the magnetic field created by nuclear spin is indicated by the arrow. (B) In a magnetic field, H_0, the nuclear spins align with (α) or against (β) the field.

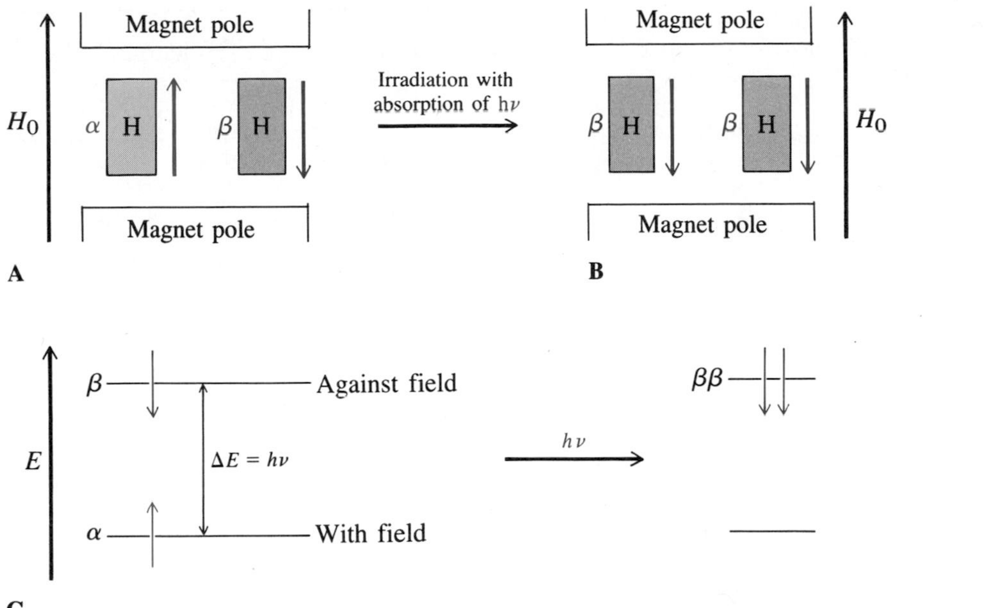

FIGURE 10-5 (A) When protons are exposed to an external magnetic field, two spin states differing in energy by $\Delta E = h\nu$ are generated. (B) Irradiation with energy of the right frequency ν causes absorption, "flipping" the nuclear spin of a proton from the α to the β state (resonance). (C) An energy diagram, showing a proton (denoted by only an arrow) gaining the energy $\Delta E = h\nu$ and undergoing "spin flip" from the α to the β state.

These two energetically different states afford the necessary condition for spectroscopy. Irradiation of the sample with a source of just the right frequency to bridge the difference in energy between the α and β states produces **resonance,** an absorption of energy as an α proton "flips" to the β spin state. This phenomenon is illustrated for a pair of protons in Figure 10-5. After excitation, the nuclei relax and return to their original states by a variety of pathways (which will not be discussed here), all of which release the absorbed energy. At resonance, therefore, there is continuous excitation and relaxation.

The difference in energy, ΔE, between spin states α and β depends directly on the external field strength, H_0. The stronger the external field, the larger the difference in energy: The absorption frequency, ν, is proportional to H_0.

How much energy must be expended for the spin of a proton to flip from α to β? To answer this question, we need to know typical values for H_0. The commercial magnets employed today range in field strength from about 14,100 to 176,250 gauss* (abbreviated G). The corresponding ν values needed to observe resonance lie in the radio-frequency range between 60 and 750 MHz (megahertz, millions of hertz, or cycles s^{-1}). For example, at 21,150 G, hydrogen nuclei require irradiation with radio waves of 90 MHz to cause resonance. Because $\Delta E_{\beta-\alpha} = h\nu$, we can calculate how much energy is being absorbed in this process. The amount is very small, on the order of 9×10^{-6} kcal mol^{-1}. Equili-

*Karl Friedrich Gauss, 1777–1855, German mathematician.

TABLE 10-1 NMR Activity and Natural Abundance of Selected Nuclei

Nucleus	NMR activity	Natural abundance (%)	Nucleus	NMR activity	Natural abundance (%)
^{1}H	Active	99.985	^{16}O	Inactive	99.759
^{2}H (D)	Active	0.015	^{17}O	Active	0.037
^{3}H (T)	Active	0	^{18}O	Inactive	0.204
^{12}C	Inactive	98.89	^{19}F	Active	100
^{13}C	Active	1.11	^{31}P	Active	100
^{14}N	Active	99.63	^{35}Cl	Active	75.53
^{15}N	Active	0.37	^{37}Cl	Active	24.47

bration between the two states is fast, and typically only slightly more than one-half of all proton nuclei in a magnetic field will adopt the α state, the remainder having a β spin.

Many nuclei undergo magnetic resonance

Protons are not the only nuclei capable of magnetic resonance. Table 10-1 lists a number of nuclei responsive to NMR and of importance in organic chemistry, as well as several that lack NMR activity.

Upon exposure to equal magnetic fields, *different NMR-active nuclei will resonate at different values of ν.* For example, if we were to scan a hypothetical spectrum of a sample of deuteriochlorofluoromethane, CHDClF (D = ^{2}H, the hydrogen isotope of mass 2), in a 21,250-G magnet,* we would observe six absorptions corresponding to the six NMR-active nuclei in the molecule: the highly abundant ^{1}H, ^{19}F, ^{35}Cl, and ^{37}Cl, and the much less plentiful ^{13}C (1.11%) and ^{2}H (0.015%), as shown in Figure 10-6. Nuclei with even numbers of both protons and neutrons, such as ^{12}C, lack magnetic properties and do not give rise to NMR signals.

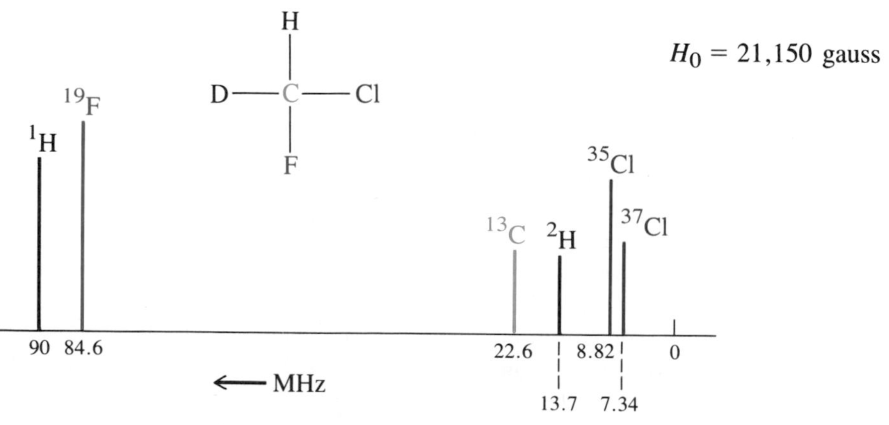

FIGURE 10-6 A hypothetical NMR spectrum of CHDClF at 21,150 G. Because each NMR-active nucleus resonates at a characteristic frequency, six lines are observed. We show them here with similar heights for simplicity. Actually, special techniques are required to observe signals of less abundant nuclei such as ^{2}H and ^{13}C. In addition, it would not normally be possible to observe all these lines on a single instrument: Most NMR spectrometers are not capable of scanning such a wide range of frequencies.

*For comparison, the maximum intensity of the Earth's magnetic field anywhere on its surface is about 0.7 G.

High-resolution NMR spectroscopy can differentiate nuclei of the same element

Consider now the NMR spectrum of chloro(methoxy)methane (chloromethyl methyl ether), $ClCH_2OCH_3$. A sweep at 21,250 G from 0 to 90 MHz would give large peaks for the most abundant nuclei present (Figure 10-7A). However, modern instrumentation enables us to view a small part of the spectrum in the immediate vicinity of a peak in great detail. With this technique, called **high-resolution NMR spectroscopy,** we may study the hydrogen resonance from 90,000,000 to 90,000,900 Hz. We find that what appeared to be only one peak in that region actually consists of two peaks that were not resolved at first (Figure 10-7B). Similarly, the high-resolution ^{13}C spectrum measured in the vicinity of 22.6 MHz shows two peaks (Figure 10-7C). These absorptions reveal the presence of *two* types of hydrogens and carbons. *Because high-resolution NMR spectroscopy distinguishes both hydrogen and carbon atoms in different structural environments, it is a powerful tool for elucidating structures.* The organic chemist uses NMR spectroscopy more often than any other spectroscopic technique.

Proton NMR spectrometers operate between 60 and 750 MHz

NMR spectroscopy has been routinely available since about 1960. The first mass-produced proton NMR spectrometers used magnets at 14,100 G, which set the resonance frequency of hydrogen nuclei at about 60 MHz. Recent advances in

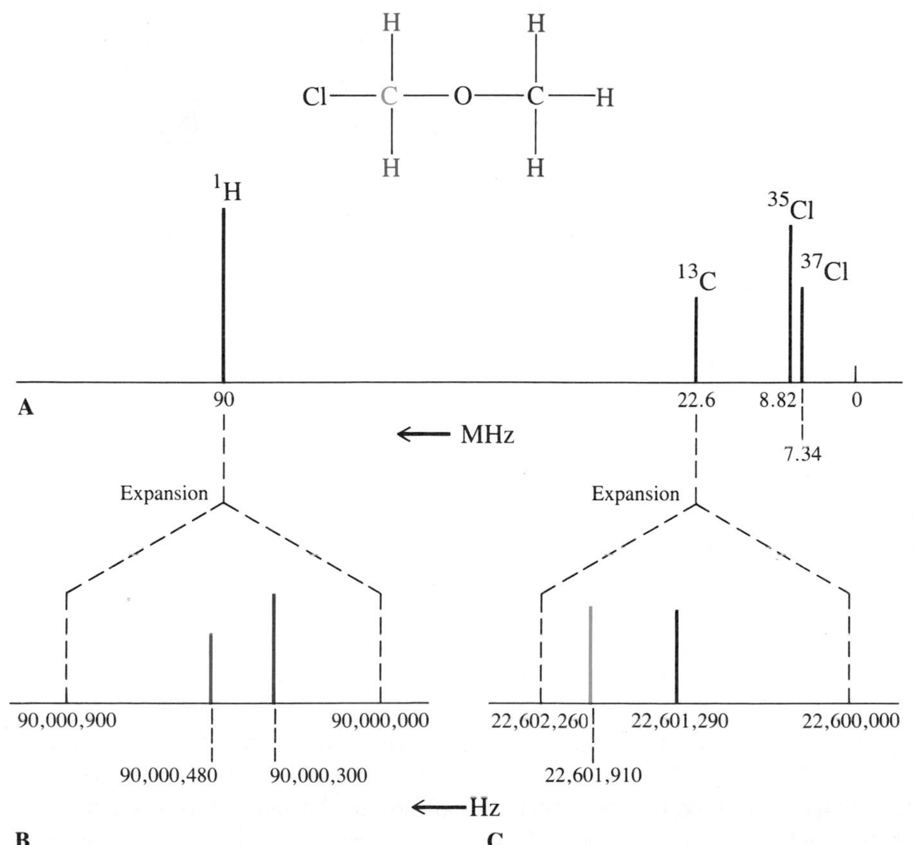

FIGURE 10-7 High resolution can reveal additional peaks in an NMR spectrum. (A) At low resolution, the spectrum of $ClCH_2OCH_3$ at 21,250 G shows four peaks for the four different, most abundant nuclei present in the molecule. Signals are not seen for ^{12}C or ^{16}O, NMR-inactive nuclei (Table 10-1). (B) At high resolution, the hydrogen spectrum shows two peaks for the two sets of hydrogens (one shown in blue in the structure, the other in red). Note that the high-resolution sweep covers only 0.001% of that at low resolution. (C) The high-resolution ^{13}C spectrum (see Section 10-9) shows peaks for the two different carbon atoms in the molecule.

BOX 10-1 Recording an NMR Spectrum

The sample to be studied (a few milligrams) is usually dissolved in a solvent (0.3–0.5 mL) preferably not containing any atoms that themselves absorb in the NMR range under investigation. Typical solvents are tetrachloromethane, CCl_4; or deuterated ones such as trichlorodeuteriomethane (deuteriochloroform), $CDCl_3$; hexadeuteriopropanone (hexadeuterioacetone), CD_3COCD_3; hexadeuteriobenzene, C_6D_6; and octadeuteriooxacyclopentane (octadeuteriotetrahydrofuran), C_4D_8O. The solution is transferred into an NMR sample container, a cylindrical precision-bore glass tube, which is inserted into the magnet. (The photo below, left, shows the magnet housing and the inlet for the sample.) To make sure that all molecules in the sample are rapidly averaged with respect to their position in the magnetic field, the NMR tube is rapidly spun by an air jet. Energy from a radiofrequency (RF) generator is emitted by one of two coils of wire surrounding the sample cavity. The second coil is connected to an RF receiver and detects any energy absorption by the sample, converting it into a plot on a recorder as the spectrum is scanned.

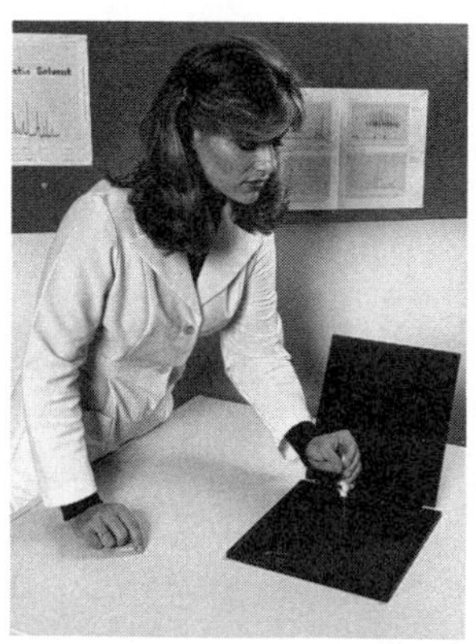

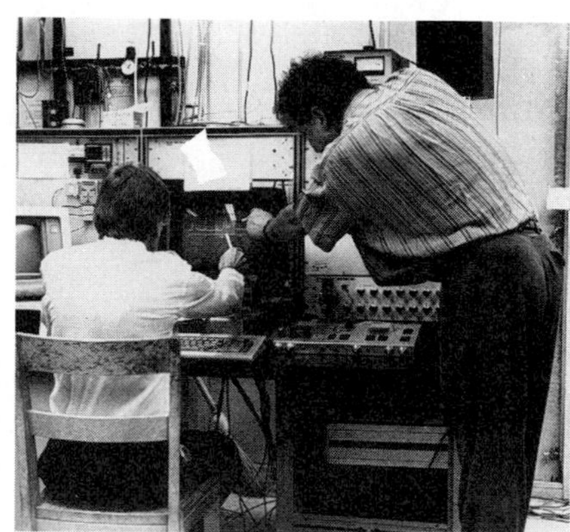

technology have led to the widespread use of NMR spectrometers at higher field strength, which yield hydrogen resonance frequencies as high as 750 MHz and greatly improved resolution as a result. This book will depict most ^1H NMR spectra at 90 MHz, with the occasional use of higher fields for comparison.

NMR spectra may be recorded in either of two ways, depending on the design of the instrument. The magnetic field may be held constant while the radio frequency is scanned, or the field may be varied in the presence of constant radio frequency. The spectra obtained are indistinguishable. For convenience, all are calibrated in hertz (s^{-1}), as if frequency had been varied at constant H_0. An actual ^1H NMR spectrum of $ClCH_2OCH_3$ is shown in Figure 10-8.

In summary, certain nuclei, such as ^1H and ^{13}C, can be viewed as tiny atomic magnets that, when exposed to a magnetic field, can align with it (α) or against it (β). These two states are of unequal energy, a condition giving rise to nuclear magnetic resonance spectroscopy. At resonance, radio-frequency radiation is absorbed by the nucleus to effect α-to-β transitions (excitation). The β state relaxes to the α state by giving off a small amount of heat. The resonance frequency, which is characteristic of the nucleus and its environment, is proportional to the strength of the external magnetic field.

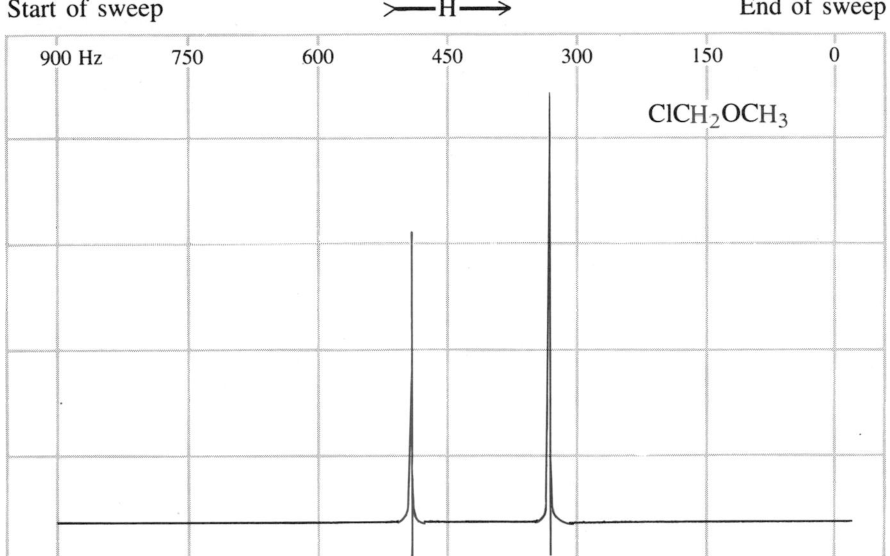

FIGURE 10-8 90-MHz ^{1}H NMR spectrum of chloro(methoxy)methane. The zero-hertz line is set at exactly 90 MHz at the right-hand side of the spectral paper. The strength of the applied magnetic field (H) increases from left to right.

10-4 Using NMR Spectra to Analyze Molecular Structure: The Proton Chemical Shift

Why do the two different groups of hydrogens in chloro(methoxy)methane give rise to distinct NMR peaks? How does the molecular structure affect the position of an NMR signal? This section will answer these questions.

We shall see that the position of an NMR absorption, also called the chemical shift, depends on the electron density around the hydrogen, which in turn is controlled by the molecular structure in the vicinity of the observed nucleus (its "structural environment"). Therefore, *the NMR chemical shifts of the hydrogens in a molecule will be important clues for determining the exact makeup of a compound.*

The position of an NMR signal depends on the nucleus's electronic environment

The high-resolution ^{1}H NMR spectrum of chloro(methoxy)methane depicted in Figure 10-8 reveals that the two kinds of hydrogens give rise to two separate resonance absorptions. What is the origin of this effect? It is the electronic environment of the hydrogen nucleus. A free proton is essentially unperturbed by electrons. Organic molecules, however, contain covalently bonded hydrogen nuclei, *not* free protons, and the electrons in these bonds affect nuclear magnetic resonance absorptions.*

Bound hydrogens are surrounded by electronic shells whose electron density varies, depending on the polarity of the bond, the hybridization of the attached atom, and the presence of electron-donating or -withdrawing groups. When a

*In discussions of NMR, the terms *proton* and *hydrogen* are frequently (albeit incorrectly) interchanged. "Proton NMR" and "protons in molecules" are used even in reference to covalently bound hydrogen.

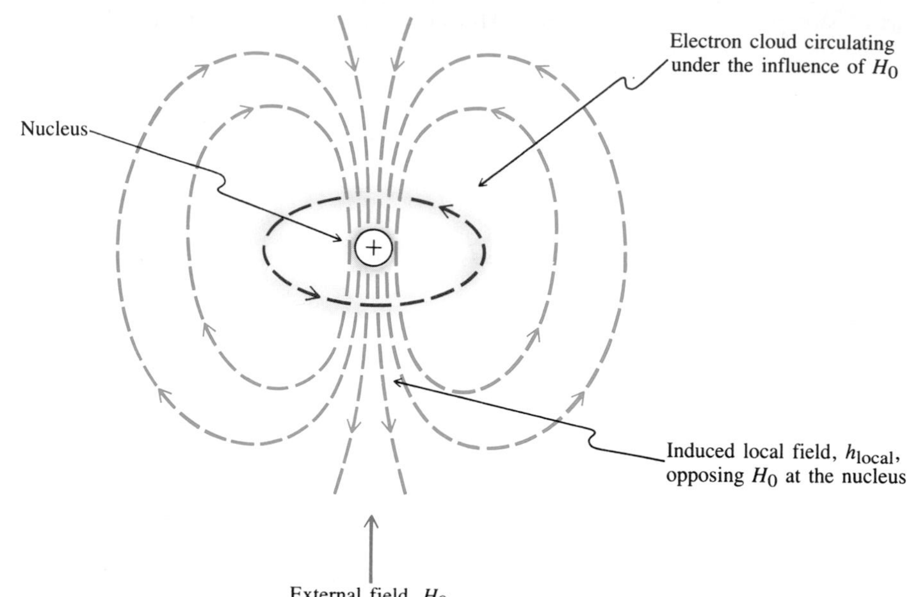

FIGURE 10-9 The external field, H_0, causes an electronic current of the bonding electrons around a hydrogen nucleus, which in turn generates a local magnetic field opposing H_0 (you may recognize this as Lenz's law, named for a Russian physicist, Heinrich Friedrich Emil Lenz, 1804–1865).

nucleus surrounded by electrons is exposed to a magnetic field of strength H_0, these electrons move in such a way as to generate a small **local magnetic field, h_{local},** *opposing* H_0. As a consequence, the total field strength near the hydrogen nucleus is *reduced,* and the nucleus is thus said to be **shielded** from H_0 by its electron cloud (Figure 10-9). The degree of shielding depends on the amount of electron density surrounding the nucleus. Adding electrons increases shielding; their removal causes **deshielding.**

What is the effect of shielding on the relative position of an NMR absorption? Using the currently customary procedure of keeping the radio frequency constant and varying the magnetic field, we find that a higher external field strength is required to overcome the shielding effect and cause resonance. Spectra are recorded with field strength increasing from left to right, so the phrase "a shift upfield" means that the signal (peak) appears farther *to the right* (Figure 10-10).

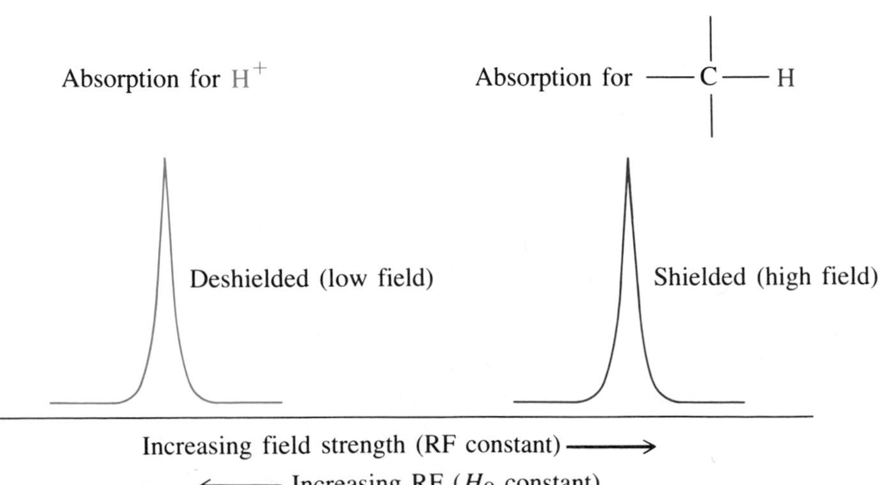

FIGURE 10-10 Effect of shielding on the absorption of a covalently bound proton. At constant RF of energy $h\nu$, the free proton resonates at H_0. Shielding decreases the value of the field around a given nucleus to $H_0 - h_{local}$. Hence, to "match" $h\nu$, the external field must be increased by an amount equal to h_{local}. The net result is a displacement of the corresponding signal to the right-hand side of the spectrum.

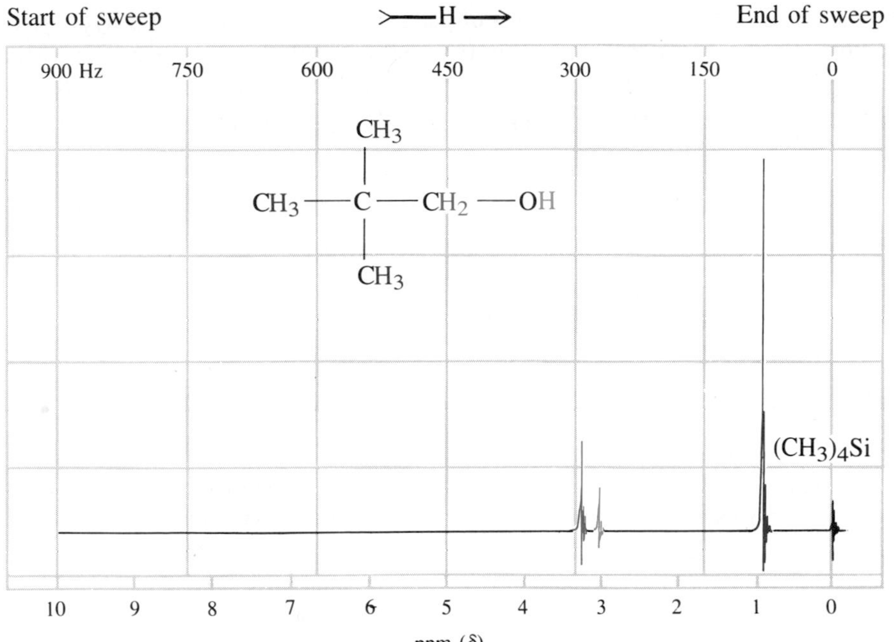

FIGURE 10-11 90-MHz ^1H NMR spectrum of 2,2-dimethyl-1-propanol (containing a little tetramethylsilane) in CCl_4. Three peaks are observed for the three sets of different hydrogens. (The scale at the bottom indicates the chemical shift in δ, a subject discussed in the next subsection.)

Because each chemically distinct hydrogen has a unique electronic environment, it gives rise to a characteristic resonance. Likewise, *chemically equivalent hydrogens show peaks at the same position.* Figure 10-11 shows this for the NMR spectrum of 2,2-dimethyl-1-propanol.

The chemical shift describes the position of an NMR peak

In which manner are spectral data reported? As noted earlier, most hydrogen absorptions in 90-MHz ^1H NMR fall within a range of 900 Hz. Rather than record the exact frequency of each resonance, we measure it relative to an internal standard, the compound tetramethylsilane, $(CH_3)_4Si$ (b.p. 26.5°C). Its 12 equivalent hydrogens are shielded relative to those in most organic molecules, a situation resulting in a resonance line conveniently removed from the usual spectral range. The position of the NMR absorptions of a compound under investigation can then be measured (in hertz) relative to the internal standard. In this way, the signals of, for example, 2,2-dimethyl-1-propanol (Figure 10-11) are reported as being located 78, 258, and 287 Hz downfield from $(CH_3)_4Si$.

A problem with these numbers, however, is that *they vary with the strength of the applied magnetic field.* Because field strength and resonance frequency are directly proportional, doubling or tripling the field strength will double or triple the distance (in hertz) of the observed peaks relative to $(CH_3)_4Si$. To avoid this complication and to be able to compare reported literature spectra at different field strengths, we standardize the measured frequency by dividing the distance to $(CH_3)_4Si$ (in hertz) by the frequency of the spectrometer. This procedure yields a *field-independent* number, the **chemical shift δ.**

The Chemical Shift

$$\delta = \frac{\text{distance of peak from } (CH_3)_4Si \text{ in hertz}}{\text{spectrometer frequency in megahertz}} \text{ ppm}$$

The chemical shift is reported in units of parts per million (ppm)—for hydrogen, usually to three significant figures. For $(CH_3)_4Si$, δ is defined as 0.00. The NMR spectrum of 2,2-dimethyl-1-propanol in Figure 10-11 would then be reported in the following format: ^1H NMR (90 MHz, CCl_4) $\delta = 0.87, 2.87, 3.19$ ppm.

EXERCISE 10-3

The two NMR peaks of 1,2-dimethoxyethane are 288 and 297 Hz away from $(CH_3)_4Si$ at 90 MHz. What is their chemical shift? How many hertz would separate them from $(CH_3)_4Si$ when measured in a 100-MHz spectrometer?

Functional groups have characteristic chemical shifts

A powerful use of NMR spectroscopy derives from the finding that the chemical shift is characteristic of the structural environment around the nucleus being observed. The hydrogen chemical shifts typical of standard organic structural units are listed in Table 10-2. It is important to be familiar with the chemical shift ranges for the structural types discussed so far: alkanes, haloalkanes, ethers, alcohols, aldehydes, and ketones. Others will be discussed in more detail in subsequent chapters.

Cumulative Deshielding in Chloromethanes

CH_3Cl
$\delta = 3.05$ ppm

CH_2Cl_2
$\delta = 5.30$ ppm

$CHCl_3$
$\delta = 7.27$ ppm

Note that the absorptions of the alkane hydrogens occur at relatively high field ($\delta = 0.8$–1.7 ppm). A hydrogen close to an electron-withdrawing group or atom (such as a halogen or oxygen) is shifted to relatively lower field: Such substituents *deshield* their neighbors. Table 10-3 shows how adjacent heteroatoms affect the chemical shifts of a methyl group. The more electronegative the atom, the more deshielded are the methyl hydrogens relative to methane. Several such substituents exert a cumulative effect, as seen in the series of the three chlorinated methanes shown in the margin. The deshielding influence of electron-withdrawing groups diminishes rapidly with distance.

CH_3—CH_2—CH_2—Br
$\delta = 1.06$ 1.81 3.47 ppm

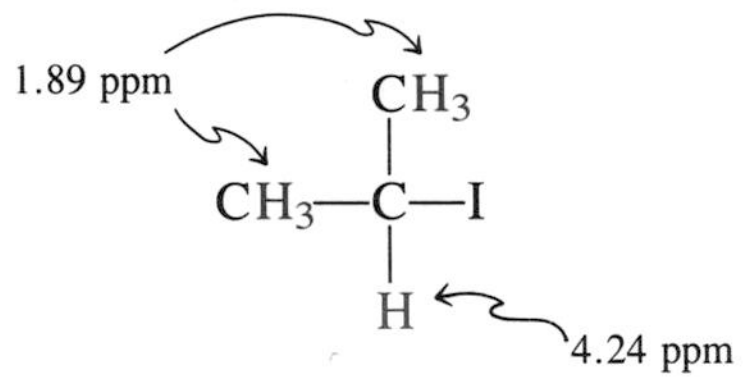

$(CH_3)_2CHI$: CH_3 1.89 ppm; H 4.24 ppm

EXERCISE 10-4

Explain the assignment of the ^1H NMR signals of chloro(methoxy)methane (Figure 10-8).

As noted in Table 10-2, hydroxy (and amine) hydrogens absorb over a range of frequencies. In spectra of such samples, the absorption peak of the OH group is frequently relatively broad. This variability of chemical shift is due to hydrogen bonding and depends on temperature, concentration, and the presence of other hydrogen-bonding species such as water. In simple terms, different degrees of hydrogen bonding alter the electronic environment of the hydrogen nuclei. Similar considerations apply to amines. Because of this line broadening, amine and alcohol hydrogen absorptions are usually readily recognized in NMR spectra.

In summary, the various hydrogen atoms present in an organic molecule can be recognized by their characteristic NMR peaks at certain chemical shifts δ. An

TABLE 10-2 Typical Hydrogen Chemical Shifts in Organic Molecules

Type of hydrogen[a]	Chemical shift δ in ppm	
Primary alkyl, RCH_3	0.8–1.0	Alkane and alkanelike hydrogens
Secondary alkyl, RCH_2R'	1.2–1.4	
Tertiary alkyl, R_3CH	1.4–1.7	
Allylic (next to a double bond), $R_2C{=}C(CH_3)R'$	1.6–1.9	Hydrogens adjacent to unsaturated functional groups
Benzylic (next to a benzene ring), $ArCH_2R$	2.2–2.5	
Ketone, $RC(=O)CH_3$	2.1–2.6	
Alkyne, $RC{\equiv}CH$	1.7–3.1	
Chloroalkane, RCH_2Cl	3.6–3.8	Hydrogens adjacent to electronegative atoms
Bromoalkane, RCH_2Br	3.4–3.6	
Iodoalkane, RCH_2I	3.1–3.3	
Ether, RCH_2OR'	3.3–3.9	
Alcohol, RCH_2OH	3.3–4.0	
Terminal alkene, $R_2C{=}CH_2$	4.6–5.0	Alkene hydrogens
Internal alkene, $R_2C{=}CHR'$	5.2–5.7	
Aromatic, ArH	6.0–9.5	
Aldehyde, $RC(=O)H$	9.5–9.9	
Alcoholic hydroxy, ROH	0.5–5.0 (variable)	
Amine, RNH_2	0.5–5.0 (variable)	

a. R, R′, alkyl groups; Ar, aromatic group (not argon).

TABLE 10-3 The Deshielding Effect of Electronegative Atoms

CH_3X	Electronegativity of X (from Table 1-2)	Chemical shift δ in ppm of CH_3 group
CH_3F	4.0	4.26
CH_3OH	3.4	3.40
CH_3Cl	3.2	3.05
CH_3Br	3.0	2.68
CH_3I	2.7	2.16
CH_3H	2.2	0.23

electron-poor environment is deshielded and leads to low-field (high-δ) absorptions, whereas an electron-rich environment results in the opposite (shielded or high-field peaks). The chemical shift δ is measured in parts per million by dividing the difference between the measured resonance and that of the internal standard, tetramethylsilane, $(CH_3)_4Si$, in hertz by the spectrometer frequency in megahertz. The NMR spectra for the OH groups of alcohols and the NH_2 groups of amines exhibit characteristically broad peaks with concentration- and moisture-dependent δ values.

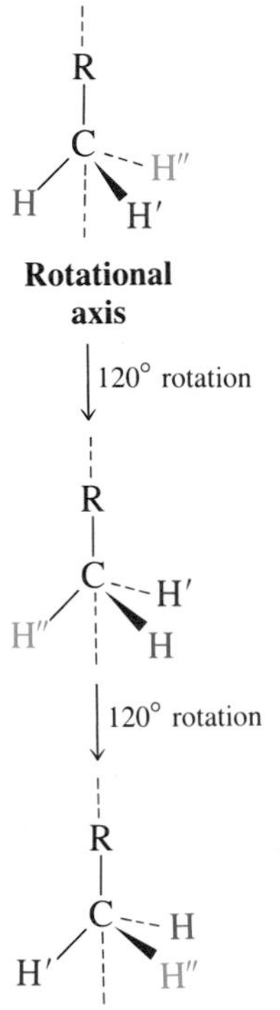

Rotational axis

120° rotation

120° rotation

FIGURE 10-12 Counterclockwise rotation of a methyl group as a test of symmetry.

10-5 Tests for Chemical Equivalence

In the NMR spectra presented so far, two or more hydrogens occupying positions that are chemically equivalent give rise to only *one* NMR absorption. It can be said, in general, that *chemically equivalent protons have the same chemical shift*. However, we shall see that it is not always easy to identify chemically equivalent nuclei. We shall resort to the symmetry operations presented in Chapter 5 to help us decide on the expected NMR spectrum of a specific compound.

Molecular symmetry helps establish chemical equivalence

To establish chemical equivalence, we have to recognize the symmetry of molecules and their substituent groups. As we know, one form of symmetry is the presence of a mirror plane (Section 5-1). Another is rotational equivalence. For example, Figure 10-12 demonstrates how two successive 120° rotations of a methyl group allow each hydrogen to occupy the positions of the other two without effecting any structural change. Thus, in a rapidly rotating methyl group all hydrogens are equivalent and should have the same chemical shift. We shall see shortly that this is indeed the case.

Application of the principles of rotational or mirror symmetry or both allows the assignment of equivalent nuclei in other compounds (Figure 10-13).

EXERCISE 10-5

How many NMR absorptions would you expect for **(a)** 2,2,3,3-tetramethylbutane; **(b)** $CH_3OCH_2CH_2OCH_2CH_2OCH_3$; **(c)** oxacyclopropane?

CH_3CH_3 (*a a*)

$CH_3OCH_2CH_2OCH_3$ (*a b b a*)

FIGURE 10-13 A variety of organic molecules containing chemical-shift-equivalent hydrogens. The colors (and letters) blue (*a*) and red (*b*) differentiate between nuclei giving rise to separate absorptions with distinct chemical shifts. All structures have rotational or mirror symmetry or both.

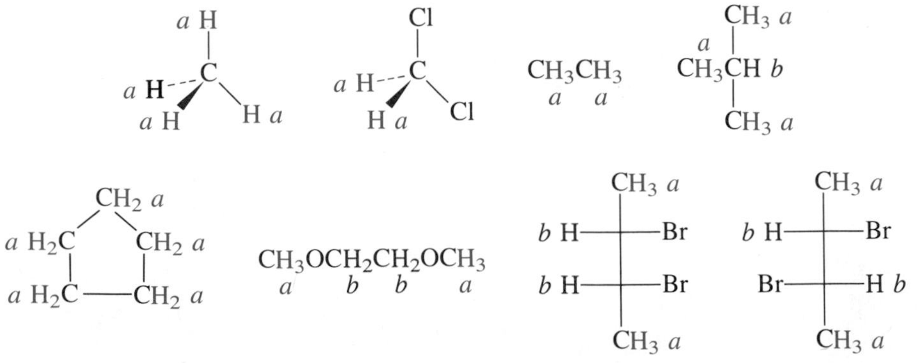

BOX 10-2 Magnetic Resonance Imaging in Medicine

In the mid-1980s a powerful application of NMR in medical diagnosis was demonstrated, *magnetic resonance imaging,* or *MRI*. A patient's entire body is placed within the poles of a large electromagnet, and proton NMR spectra are collected and computer processed to give a series of cross-sectional plots of signal intensity. These are combined to produce a three-dimensional image of tissue proton density, as shown below.

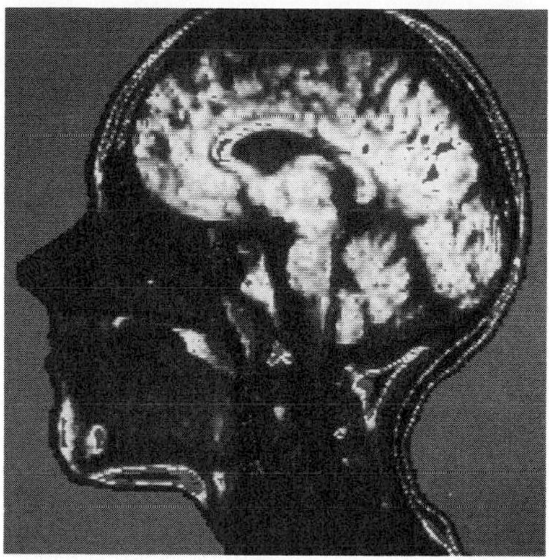

Because most of the signal is due to water, variations from normal water-density patterns may be detected and used in diagnosis. Improvements in the late 1980s shortened the time needed for analysis from minutes to seconds, permitting direct viewing of blood flow, kidney secretion, and other medically relevant phenomena. MRI is particularly helpful in detecting abnormalities not readily found by CAT scans and conventional X-rays. The technique is also *noninvasive,* requiring neither ionizing radiation nor the injection of radioactive substances for visualization, as is the case for other imaging methods.

Conformational interconversion may result in equivalence on the NMR time scale

Let us look more closely at two more examples, chloroethane and cyclohexane. The first should have two NMR peaks because it has the two sets of equivalent hydrogens; the second has twelve chemically equivalent hydrogen nuclei and is expected to show only one absorption. However, are these expectations really justified? Consider the possible conformations of these two molecules (Figure 10-14).

FIGURE 10-14 (A) Newman projections of chloroethane. H_{b_3} is located *anti* to the chlorine substituent and is therefore not in the same environment as H_{b_1} and H_{b_2}. However, fast rotation averages all the methyl hydrogens on the NMR time scale. (B) In any given conformation of cyclohexane, the axial hydrogens are different from the equatorial ones. However, conformational flip is rapid on the NMR time scale, so only one average signal is observed. Colors are used here to distinguish between environments and thus to indicate specific chemical shifts.

CH_3CH_2Cl
b a

a a a a a a

Begin with chloroethane. The most stable conformation is the staggered arrangement, in which one of the methyl hydrogens (H_{b_3}) is located *anti* with respect to the chlorine atom. We expect this particular nucleus to have a chemical shift different from the two *gauche* hydrogens (H_{b_1} and H_{b_2}). In fact, however, the NMR spectrometer cannot resolve that difference, because the fast rotation of the methyl group averages out the signals for H_b. This rotation is said to be "fast on the NMR time scale." The resulting absorption appears at an average δ of the two signals expected for H_b.

In theory, it should be possible to slow the rotation in chloroethane by cooling the sample. In practice, "freezing" the rotation is very difficult to do, because the activation barrier to rotation is only a few kilocalories per mole. We would have to cool the sample to about $-180°C$, at which point most solvents would crystallize—and ordinary NMR spectroscopy would not be possible!

A similar situation is encountered for cyclohexane. Here fast conformational isomerism causes the axial hydrogens to be in equilibrium with the equatorial ones on the NMR time scale (Figure 10-14B), so at room temperature the NMR spectrum shows only one sharp line at $\delta = 1.36$ ppm. However, in contrast with that for chloroethane, the process is slow enough at $-90°C$ that, instead of a single absorption, two are observed: one for the six axial hydrogens at $\delta = 1.12$ ppm; the other for the six equatorial hydrogens at $\delta = 1.60$ ppm. The conformational isomerization in cyclohexane is frozen on the NMR time scale at this temperature because the activation barrier to ring flip is much higher ($E_a = 10.8$ kcal mol^{-1}; Section 4-3) than the barrier to rotation in chloroethane.

In general, the lifetime of a species in such an equilibrium must be on the order of about a second to allow its resolution by NMR. If this period decreases substantially, an average spectrum is obtained. Such time- and temperature-dependent NMR phenomena are in fact used by chemists to estimate the activation parameters of chemical processes.

In summary, the properties of symmetry, particularly mirror images and rotations, help to establish the chemical-shift equivalence or nonequivalence of the hydrogens in organic molecules. Those structures that undergo rapid conformational changes on the NMR time scale show only averaged spectra at room temperature. Examples are rotamers and ring conformers. In some cases, these processes may be "frozen" at low temperatures to allow distinct absorptions to be observed.

10-6 Integration

So far we have looked only at the *position* of NMR peaks. We shall see in this section that another useful feature of NMR spectroscopy is its ability to measure the relative integrated intensity of a signal, which is proportional to the relative number of nuclei giving rise to that absorption.

Integration reveals the number of hydrogens responsible for an NMR peak

The more hydrogens of one kind there are in a molecule, the more intense the corresponding NMR absorption relative to the other signals. By measuring the area under a peak (the "integrated area") and comparing it with the corresponding peak areas of other signals, we can quantitatively estimate the nuclear ratios. For example, in the spectrum of 2,2-dimethyl-1-propanol (Figure 10-11), three signals are observed, with relative areas of 9:2:1.

To simplify this measurement, the spectrometer has an electronic feature called **integration.** At the push of a button, the instrument switches into the *integrator mode,* which allows a rescanning of the spectrum to record the intensity ratio of the observed absorptions. This spectral integration is plotted (for clarity, displaced from the original spectral baseline; Figure 10-15) as a straight, initially horizontal line (baseline). Whenever a point in the scan is reached at which the normal recording mode would draw a peak, the pen in the integrator mode moves vertically upward a distance proportional to the intensity of the peak. It then again moves horizontally until the next peak is reached and so forth. A ruler can be used to measure the distance by which the horizontal line is displaced at every peak. *The relative sizes of these displacements furnish the ratio of hydrogens giving rise to the various signals.* Figure 10-15 depicts the ^1H NMR spectra of 2,2-dimethyl-1-propanol and 1,2-dimethoxyethane, including plots of the integration. Many higher-field instruments also provide digital readouts of integrated peak intensity.

Integration aids in structure elucidation

Let us see how integration is used along with chemical shifts to determine structure. Consider, for example, the three products obtained in the monochlorination of 1-chloropropane, $CH_3CH_2CH_2Cl$. All have the same molecular formula $C_3H_6Cl_2$ and very similar physical properties (such as boiling points).

$$CH_3CH_2CH_2Cl \xrightarrow[-\ HCl]{Cl_2,\ h\nu,\ 100^\circ C} \underset{10\%}{CH_3CH_2CHCl_2} + \underset{27\%}{CH_3CHClCH_2Cl} + \underset{14\%}{ClCH_2CH_2CH_2Cl}$$

1,1-Dichloropropane	**1,2-Dichloropropane**	**1,3-Dichloropropane**
(b.p. 87–90°C)	(b.p. 96°C)	(b.p. 120°C)

NMR spectroscopy clearly distinguishes all three isomers. 1,1-Dichloropropane contains three types of nonequivalent hydrogens, a situation giving rise to three NMR signals in the ratio of 3:2:1. The single hydrogen absorbs at relatively low field ($\delta = 5.93$ ppm) because of the cumulative deshielding effect of the two halogen atoms; the others absorb at relatively high field ($\delta = 1.01$ and 2.34 ppm).

1,2-Dichloropropane also shows three sets of signals associated with CH_3, CH_2, and CH groups. In contrast, however, their chemical shifts are quite different: Each of the latter now bears a halogen atom and gives rise to a low-field signal as a result ($\delta = 3.68$ ppm for the CH_2,* and $\delta = 4.17$ ppm for the CH). Only one signal, shown by integration to represent three hydrogens, and therefore the CH_3 group, occurs at relatively high field ($\delta = 1.70$ ppm).

*This description is somewhat simplified. The two hydrogens of this CH_2 group are not equivalent, because the presence of the adjacent stereocenter at C2 renders the molecule chiral and devoid of mirror symmetry. Indeed, on a high-field instrument with greater resolving power, these hydrogens have measurably different chemical shifts ($\delta = 3.62$ and 3.74).

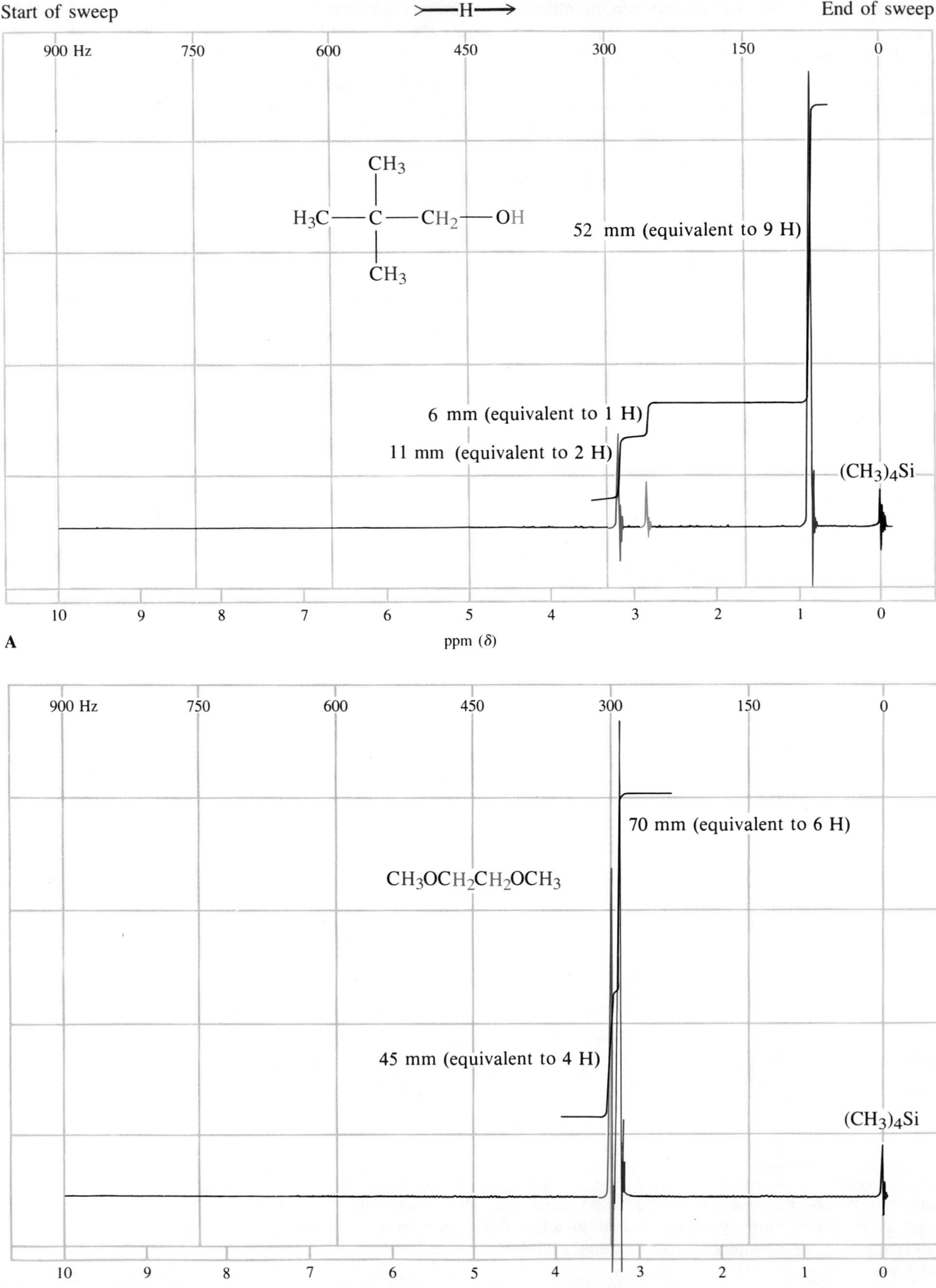

Start of sweep
H
End of sweep
900 Hz
750
600
450
300
150
0
CH3
H3C—C—CH2—OH
CH3
52 mm (equivalent to 9 H)
6 mm (equivalent to 1 H)
11 mm (equivalent to 2 H)
(CH3)4Si
10
9
8
7
6
5
4
3
2
1
0
A
ppm (δ)
900 Hz
750
600
450
300
150
0
70 mm (equivalent to 6 H)
CH3OCH2CH2OCH3
45 mm (equivalent to 4 H)
(CH3)4Si
10
9
8
7
6
5
4
3
2
1
0
B
ppm (δ)

FIGURE 10-15 (opposite) Integrated 90-MHz spectra of (A) 2,2-dimethyl-1-propanol and (B) 1,2-dimethoxyethane, in CCl_4 with added $(CH_3)_4Si$. For example A, the integrated areas measured by a ruler are 11:6:52 (in mm). (Note that the integrator converts units of area into units of distance.) Normalization through division by the smallest number gives a peak ratio of 1.8:1:8.7. The slight deviations from integer numbers are due to small experimental errors: rounding off gives the expected ratio of 2:1:9. Note that the integration gives only *ratios,* not absolute values for the number of hydrogens present in the sample. Thus, in example B, the integrated peak ratio is 3:2, yet the compound contains hydrogens in a ratio of 6:4.

Finally, 1,3-dichloropropane shows only two peaks ($\delta = 3.71$ and 2.25 ppm) in a relative ratio of 2:1, a pattern clearly distinct from those of the other two isomers. By this means, the structures of the three products are readily assigned by a simple measurement.

EXERCISE 10-6

Chlorination of chlorocyclopropane gives three compounds of molecular formula $C_3H_4Cl_2$. Draw their structures and describe how you would differentiate them by ^1H NMR.

To summarize, the NMR spectrometer in the integration mode records the relative areas under the various peaks, values that represent the relative numbers of hydrogens giving rise to these absorptions. This technique can be a powerful aid in elucidating structures.

10-7 Spin–Spin Splitting: The Influence of Nonequivalent Neighboring Hydrogens

The high-resolution NMR spectra presented so far have rather simple line patterns—single sharp peaks, also called **singlets.** The compounds giving rise to these spectra have one feature in common: In each, nonequivalent hydrogens are separated by at least one carbon or oxygen atom. These examples were chosen for good reason, because neighboring hydrogen nuclei can complicate the spectrum as the result of a phenomenon called **spin–spin splitting** or **spin–spin coupling.**

Figure 10-16 shows that the NMR spectrum of 1,1-dichloro-2,2-diethoxyethane has four absorptions, characteristic of four sets of hydrogens (H_a–H_d). Instead of single peaks, they adopt more complex patterns called **multiplets:** two two-peak absorptions (or **doublets**), one of four peaks **(quartet),** and one of three **(triplet).** These multiplets are caused by the number and kind of hydrogen atoms directly adjacent to the nuclei giving rise to the absorption.

In conjunction with chemical shifts and integration, spin–spin splitting frequently helps us arrive at a complete structure for an unknown compound. How can it be understood?

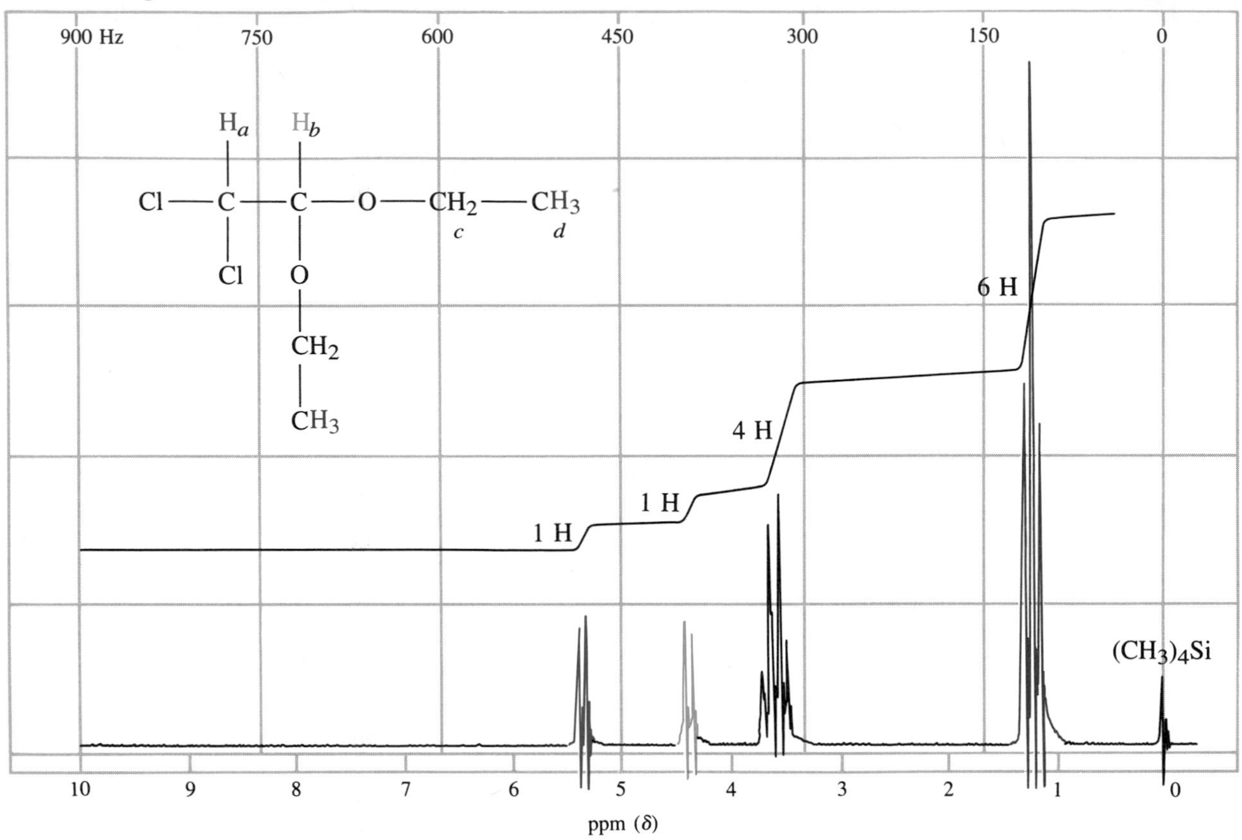

FIGURE 10-16 Spin–spin splitting in the 90-MHz spectrum of 1,1-dichloro-2,2-diethoxyethane in CCl_4. The splitting patterns include two doublets, one triplet, and one quartet for the four types of protons. These multiplets reveal the influence of adjacent hydrogens.

One neighbor splits the signal of a resonating nucleus into a doublet

Let us first consider the two doublets of relative integration 1, assigned to the two single hydrogens H_a and H_b. The splitting of these peaks is explained by the behavior of nuclei in an external magnetic field: They are like tiny magnets aligned with (α) or against (β) the field. The energy difference between the two states is minuscule (see Section 10-3), and at room temperature their populations are nearly equal. In the case under consideration, this means that there are two magnetic types of H_a—approximately half next to an H_b in the α state, the other half with a neighboring H_b in its β state. Conversely, H_b has two types of neighboring H_a—half of them in the α, half in the β state. What are the consequences of this phenomenon in the NMR spectrum?

Those protons of type H_a that have as their neighbor an H_b aligned *with* the field are exposed to a total magnetic field that is strengthened by the addition of that due to the α spin of H_b. To achieve resonance for this type of H_a, a smaller external field strength is required than that necessary for H_a in the absence of a perturbing neighbor. A peak at lower field than that expected is indeed observed (Figure 10-17A). However, this absorption is due to only half the H_a protons. The other half has H_b in its β state as a neighbor. Because H_b in its β state is aligned *against* the external field, the strength of the local field around H_a in this case is *diminished*. To achieve resonance, the external field H_0 has to be increased; an upfield shift is observed (Figure 10-17B).

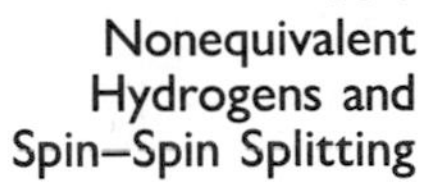

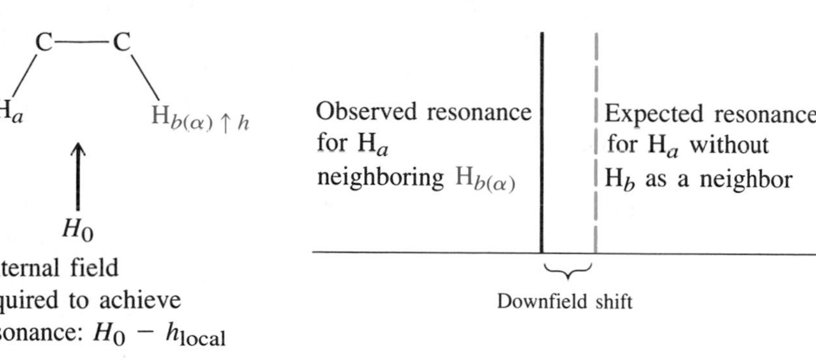

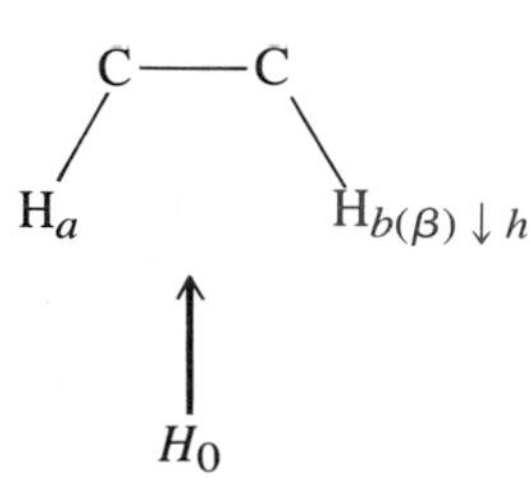

External field required to achieve resonance: $H_0 + h_{local}$

B

Observed resonance for H_a neighboring $H_{b(\beta)}$

Upfield shift

FIGURE 10-17 The effect of a hydrogen nucleus on the chemical shift of its neighbor is an example of spin–spin splitting. Two peaks are generated, because the hydrogen under observation has two types of neighbors. (A) When the neighboring nucleus H_b is in its α state, it contributes a local field h_{local}, so the external field needed for H_a to achieve resonance is reduced from H_0 to $H_0 - h_{local}$, thereby resulting in a downfield shift of the H_a peak. (B) When the neighboring nucleus is in its β state, its local field opposes the external one, and as a result an increase in the external field from H_0 to $H_0 + h_{local}$ is necessary for resonance to occur. The H_a peak is shifted to higher field.

Because the local contribution of H_b to H_0, whether positive or negative, is of the same magnitude, the downfield shift of the hypothetical signal equals the upfield shift. The single absorption expected for a neighbor-free H_a is said to be *split* by H_b into a doublet. Integration of each peak of this doublet shows a contribution of 0.5 hydrogens each. The chemical shift of H_a is reported as the center of the doublet (Figure 10-18).

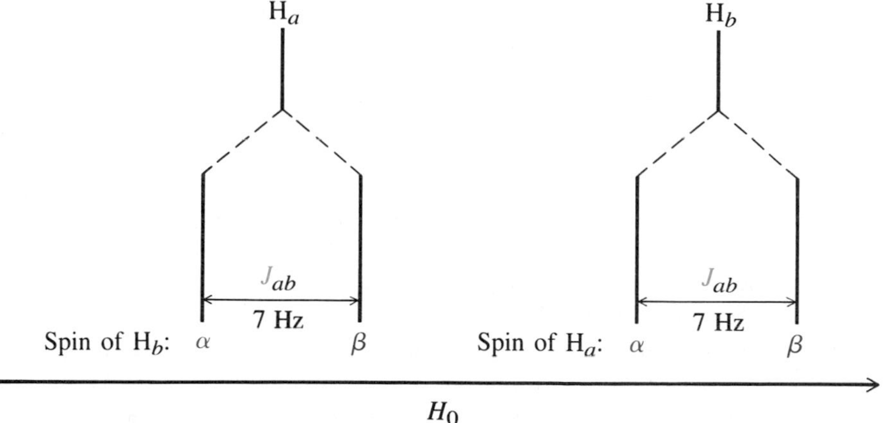

FIGURE 10-18 Spin–spin splitting between H_a and H_b in 1,1-dichloro-2,2-diethoxyethane. The coupling constant J_{ab} is the same for both doublets. The chemical shift is reported as the center of the doublet in the following format: $\delta_{H_a} =$ 5.36 ppm (d, $J = 7$ Hz, 1 H), $\delta_{H_b} = 4.39$ ppm (d, $J = 7$ Hz, 1 H), in which "d" stands for the splitting pattern (doublet), and the last entry refers to the integrated value of the absorption.

The signal for H_b is subject to similar considerations. This hydrogen also has two types of hydrogens as neighbors—$H_{a(\alpha)}$ and $H_{a(\beta)}$. Consequently its absorption lines appear in the form of a doublet. So, in NMR jargon, H_b is also split by H_a. The amount of this mutual splitting is equal; that is, the distance (in hertz) between the individual peaks making up each doublet is identical. This distance is termed the **coupling constant, *J*.** In our example, $J_{ab} = 7$ Hz (Figure 10-18). Because the coupling constant is related only to magnetic field contributions by neighboring nuclei, it is *independent of the external field strength.* Coupling constants *remain unchanged* regardless of the field strength of the NMR instrument being used.

Coupling Between Close-Lying Hydrogens

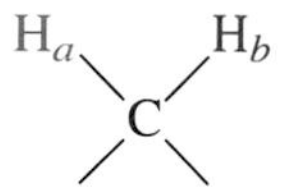

J_{ab}, geminal coupling, variable 0–18 Hz

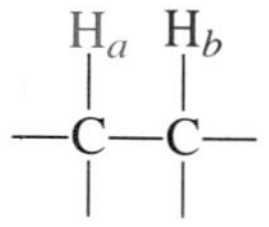

J_{ab}, vicinal coupling, typically 6–8 Hz

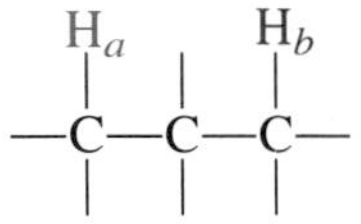

J_{ab}, 1,3-coupling, usually negligible

Spin–spin splitting is generally observed only between hydrogens that are immediate neighbors, bound either to the same carbon atom [**geminal coupling** (*geminus,* Latin, twin)] or to two adjacent carbons [**vicinal coupling** (*vicinus,* Latin, neighbor)]. Hydrogen nuclei separated by a bridge larger than two atoms are usually too far apart to exhibit appreciable coupling.

Note also that *the NMR signals of nuclei with equivalent chemical shifts do not exhibit spin–spin splitting*. Thus, the NMR spectrum of ethane, CH_3–CH_3, consists of a *single line* at $\delta = 0.85$ ppm. Splitting is observed only between nuclei with *different* chemical shifts.

Local-field contributions from more than one hydrogen are additive

How do we handle nuclei with two or more neighboring hydrogens? It turns out that we must consider the influence of each neighbor separately. Let us return to the spectrum of 1,1-dichloro-2,2-diethyoxyethane shown in Figure 10-16. In addition to the two doublets assignable to H_a and H_b, this spectrum records a triplet due to the methyl protons H_d and a quartet assignable to the methylene hydrogens H_c. Because these two nonequivalent sets of nuclei are next to each other, vicinal coupling is observed as expected. However, compared with the peak patterns for H_a and H_b, those for H_c and H_d are considerably more complicated. They can be understood by expanding on the explanation used for the mutual coupling of H_a and H_b.

Consider first the triplet whose chemical shift and integrated value allow it to be assigned to the hydrogens H_d of the two methyl groups. Instead of one peak, we observe three, in the approximate ratio 1:2:1. The splittings must be due to coupling to the adjacent methylene groups—but how?

The three equivalent methyl hydrogens in each ethoxy group have two equivalent methylene hydrogens as their neighbors, and each of these may adopt the α or β spin orientation. Thus, each H_d may "see" its two H_c neighbors as an $\alpha\alpha$, $\alpha\beta$, $\beta\alpha$, or $\beta\beta$ combination. Those methyl hydrogens that are adjacent to the first possibility, $H_{c(\alpha\alpha)}$, are exposed to a twice-strengthened local field and give rise to a lower-field absorption. In the $\alpha\beta$ or $\beta\alpha$ combination, one of the H_c nuclei is aligned with the external field and the other is opposed to it. The net result is no net local-field contribution at H_d. In these cases, a spectral peak should appear at a chemical shift identical with the one expected if there were no coupling between H_c and H_d. Moreover, because *two* equivalent combinations of neighboring H_cs [$H_{c(\alpha\beta)}$ and $H_{c(\beta\alpha)}$] contribute to this signal (instead of only one, as did $H_{c(\alpha\alpha)}$ to the first peak), its approximate height should be double that of the first peak, as observed. Finally, H_d may have the $H_c(\beta\beta)$ combination as its neighbor. In this case, the local field subtracts from the external field, and an upfield

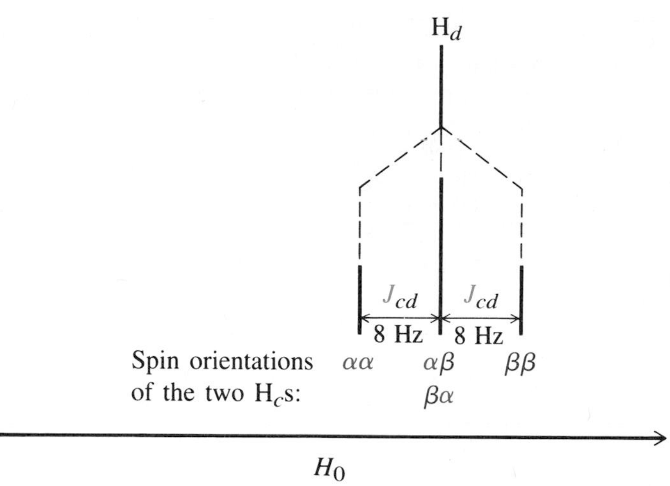

FIGURE 10-19 Nucleus H_d is represented by a three-peak NMR pattern because of the presence of three magnetically non-equivalent neighbor combinations: $H_{c(\alpha\alpha)}$, $H_{c(\alpha\beta \text{ and } \beta\alpha)}$, and $H_{c(\beta\beta)}$. The chemical shift of the absorption is reported as that of the center line of the triplet: $\delta_{H_d} =$ 1.23 ppm (t, $J = 8$ Hz, 6 H), in which "t" stands for triplet.

peak of relative intensity 1 is produced. The resulting pattern for H_d is a *1:2:1 triplet* with a total integration corresponding to six hydrogens (because there are two methyl groups). The splitting of the methyl group by the adjacent methylene group is represented in Figure 10-19. The coupling constant J_{cd}, measured as the distance between each pair of adjacent peaks, is 8 Hz.

The quartet observed in Figure 10-16 for H_c can be analyzed in the same manner. This nucleus is exposed to four different types of H_d proton combinations as neighbors: one in which all protons are aligned with the field [$H_{d(\alpha\alpha\alpha)}$]; three equivalent arrangements in which one H_d is opposed to the external field and the other two are aligned with it [$H_{d(\beta\alpha\alpha,\alpha\beta\alpha,\alpha\alpha\beta)}$]; another set of three equivalent arrangements in which only one proton remains aligned with the field [$H_{d(\beta\beta\alpha,\beta\alpha\beta,\alpha\beta\beta)}$]; and a final possibility in which all H_ds oppose the external magnetic field [$H_{d(\beta\beta\beta)}$]. The resulting spectrum is predicted—and observed—to consist of a *1:3:3:1 quartet* (integrated intensity 4) (Figure 10-20). The coupling constant J_{cd} is identical with that measured in the triplet for H_d (8 Hz).

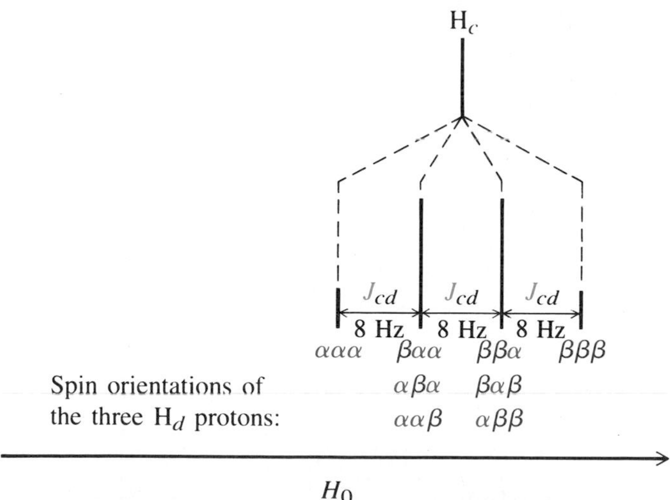

FIGURE 10-20 Splitting of H_c into a quartet by the various spin combinations of H_d. The chemical shift of the quartet is reported as its midpoint: $\delta_{H_c} = 3.63$ ppm (q, $J = 8$ Hz, 4 H), in which "q" stands for quartet.

TABLE 10-4 NMR Splittings of a Set of Hydrogens with *N* Equivalent Neighbors and Their Integrated Ratios (Pascal's Triangle)

Equivalent neighboring (*N*) hydrogens	Number of peaks ($N + 1$)	Name for peak pattern (abbreviation)	Integrated ratios of individual peaks
0	1	Singlet (s)	1
1	2	Doublet (d)	1:1
2	3	Triplet (t)	1:2:1
3	4	Quartet (q)	1:3:3:1
4	5	Quintet (quin)	1:4:6:4:1
5	6	Sextet (sex)	1:5:10:10:5:1
6	7	Septet (sep)	1:6:15:20:15:6:1

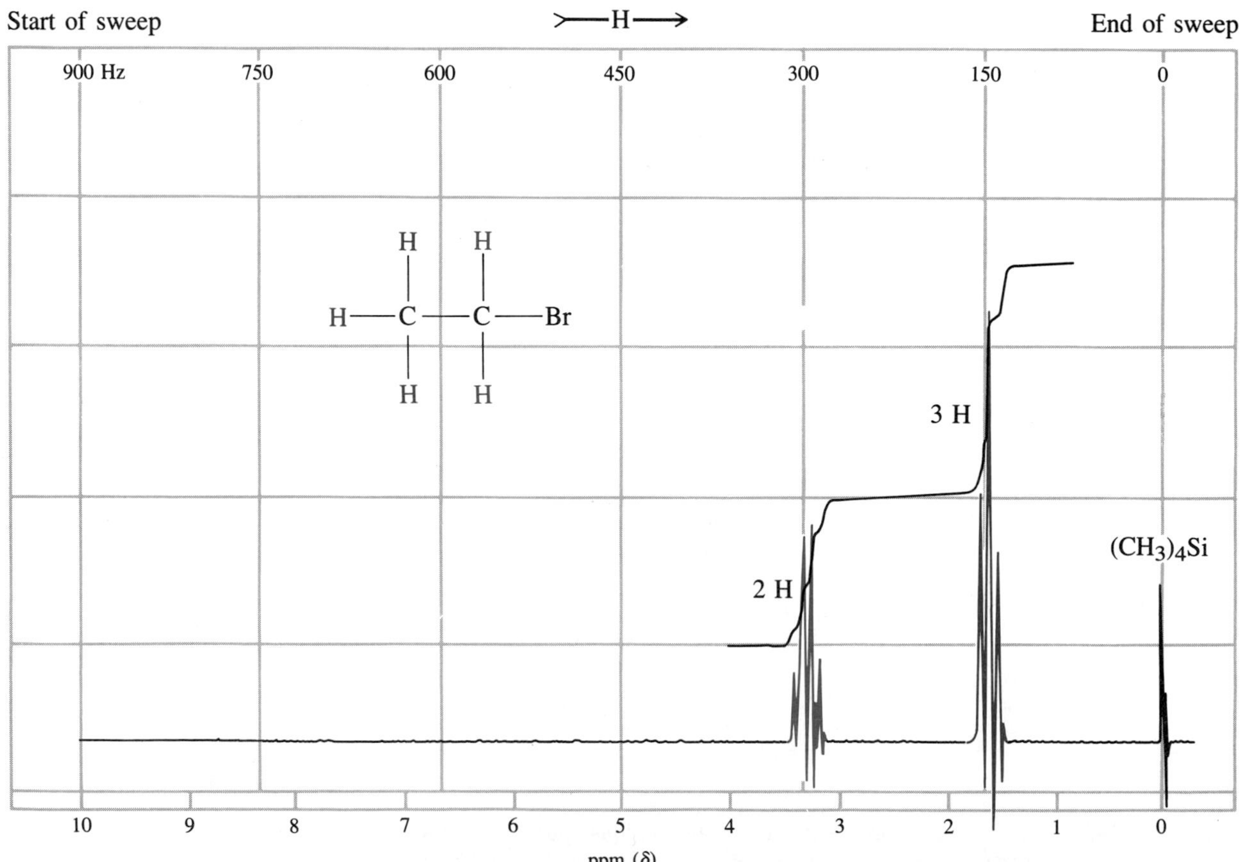

FIGURE 10-21 The 90-MHz NMR spectrum of bromoethane in CCl_4 illustrates the $N + 1$ rule. The methylene group appears as a quartet at $\delta = 3.24$ ppm, $J = 7$ Hz. The methyl hydrogens, on the other hand, absorb as a triplet at $\delta = 1.58$ ppm, $J = 7$ Hz.

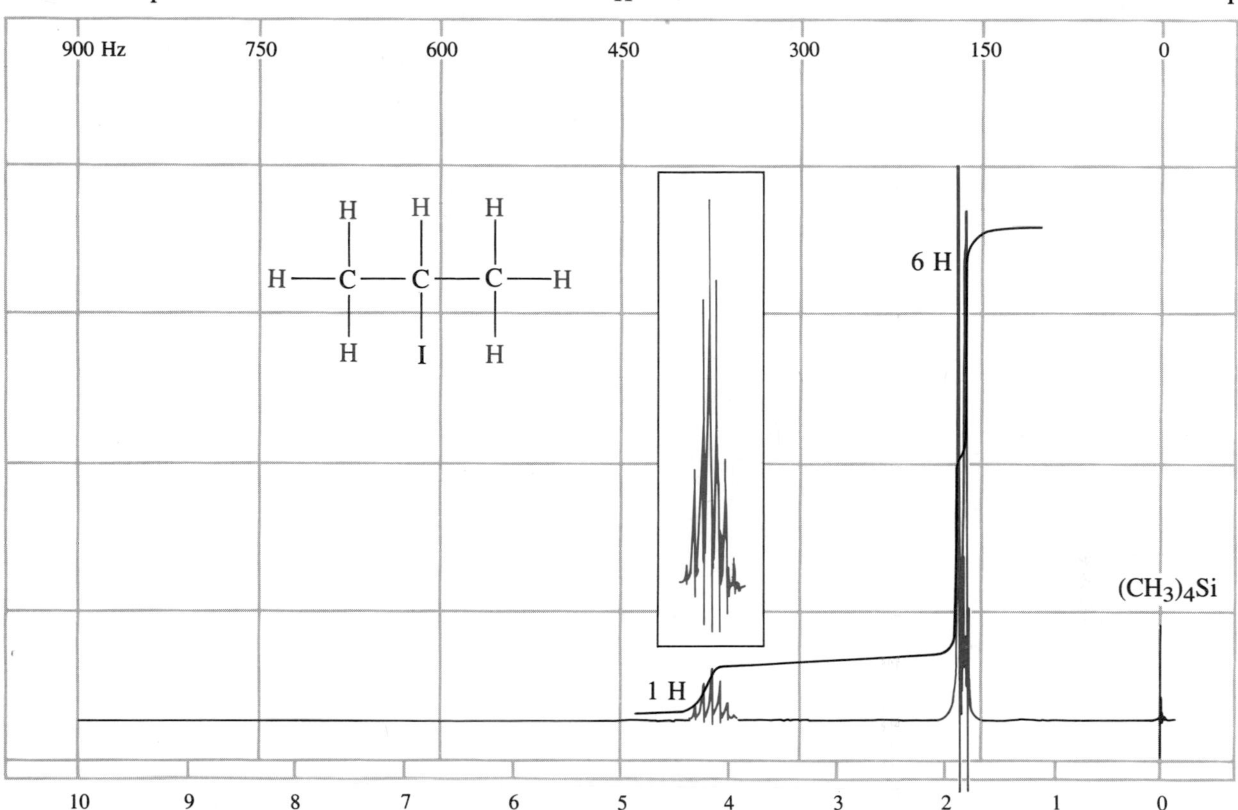

FIGURE 10-22 90-MHz NMR spectrum of 2-iodopropane in CCl_4: $\delta = 4.12$ (sep, $J = 7.5$ Hz, 1 H), 1.82 (d, $J = 7.5$ Hz, 6 H), in which "sep" stands for septet. Note that the outer peaks of the septet are of such small intensity that they are difficult to see in the spectrum recorded on scale. It is therefore frequently advisable to "blow up" split peaks in intensity to clarify some of their features. Such an enlargement is shown in the inset: the septet for the tertiary hydrogen has been rerecorded at higher sensitivity.

In many cases spin–spin splitting is given by the *N* + 1 rule

We can summarize our analysis so far by a set of simple rules:

1. Equivalent nuclei located adjacent to one neighboring hydrogen resonate as a *doublet*.
2. Equivalent nuclei located adjacent to two hydrogens of a second set of equivalent nuclei resonate as a *triplet*.
3. Equivalent nuclei located adjacent to a set of three equivalent hydrogens resonate as a *quartet*.

Table 10-4 shows the expected splitting patterns for nuclei adjacent to N equivalent neighbors. The NMR signals of these nuclei *split into N + 1 peaks,* a result known as the $N + 1$ **rule.** Their relative ratio is given by a mathematical mnemonic device called Pascal's* triangle. Each number in this triangle is the sum of the two numbers closest to it in the line above.

Typical splitting patterns are shown in Figures 10-21 and 10-22. In Figure 10-22 equivalent nuclei on two groups (the two equivalent methyls in 2-iodopropane) give rise to a septet ($N + 1 = 7$) for the central hydrogen. Note that the outer peaks of this septet are of such small intensity that they might easily be

*Blaise Pascal, 1623–1662, French mathematician, physicist, and religious philosopher.

missed. The illustration shows the multiplet recorded at higher instrument "gain" (sensitivity setting) to reveal all seven lines. As expected, the signal for the two methyl groups in this molecule appears as a doublet because of coupling to the tertiary hydrogen. In both spectra, integrations of the multiplets reveal the relative number of hydrogens responsible for each.

It is important to remember that nonequivalent nuclei mutually split one another. In other words, the observation of one split absorption necessitates the presence of another split signal in the spectrum. Moreover, the coupling constants for these patterns must be the same. Some frequently encountered multiplets and the corresponding structural units are shown in Table 10-5.

EXERCISE 10-7

Predict the NMR spectra of **(a)** ethoxyethane (diethyl ether); **(b)** 1,3-dibromopropane; **(c)** 2-methyl-2-butanol; **(d)** 1,1,2-trichloroethane. Specify approximate chemical shifts and multiplicities.

To summarize, spin–spin splitting occurs between vicinal and geminal nonequivalent hydrogens. Usually N equivalent neighbors will split the absorption of the observed hydrogen into $N + 1$ peaks, their relative intensities being in accord with Pascal's triangle. The common alkyl groups give rise to characteristic NMR patterns.

10-8 Spin–Spin Splitting: Some Complications

The rules governing the appearance of split peaks outlined in Section 10-7 are somewhat idealized. There are cases in which, because of a relatively small difference in δ between two absorptions, more complex patterns (multiplets) are observed that are not interpretable without the use of computers. Moreover, the $N + 1$ rule may not be applicable in a direct way if two or more types of neighboring hydrogens are coupled to the resonating nucleus with fairly different coupling constants. Finally, the hydroxy proton may appear as a singlet (see Figure 10-11) even if vicinal hydrogens are present. Let us look in turn at each of these complications.

Close-lying peak patterns may give rise to non-first-order spectra

A careful look at the spectra shown in Figures 10-16 and 10-21 shows that the relative intensities of the splitting patterns do not conform to the idealized peak ratios expected from consideration of Pascal's triangle: The patterns are skewed. The exact intensity ratios dictated by Pascal's triangle and the $N + 1$ rule are observed only when the difference between the chemical shifts of coupled protons is much larger than their coupling constant: $\Delta\delta \gg J$. Under these circumstances, the spectrum is said to be **first order.*** However, when the difference

*This expression derives from the term *first-order theory;* that is, one that takes into account only the most important variables and terms of a system.

TABLE 10-5 Frequently Observed Spin–Spin Splittings in Common Alkyl Groups

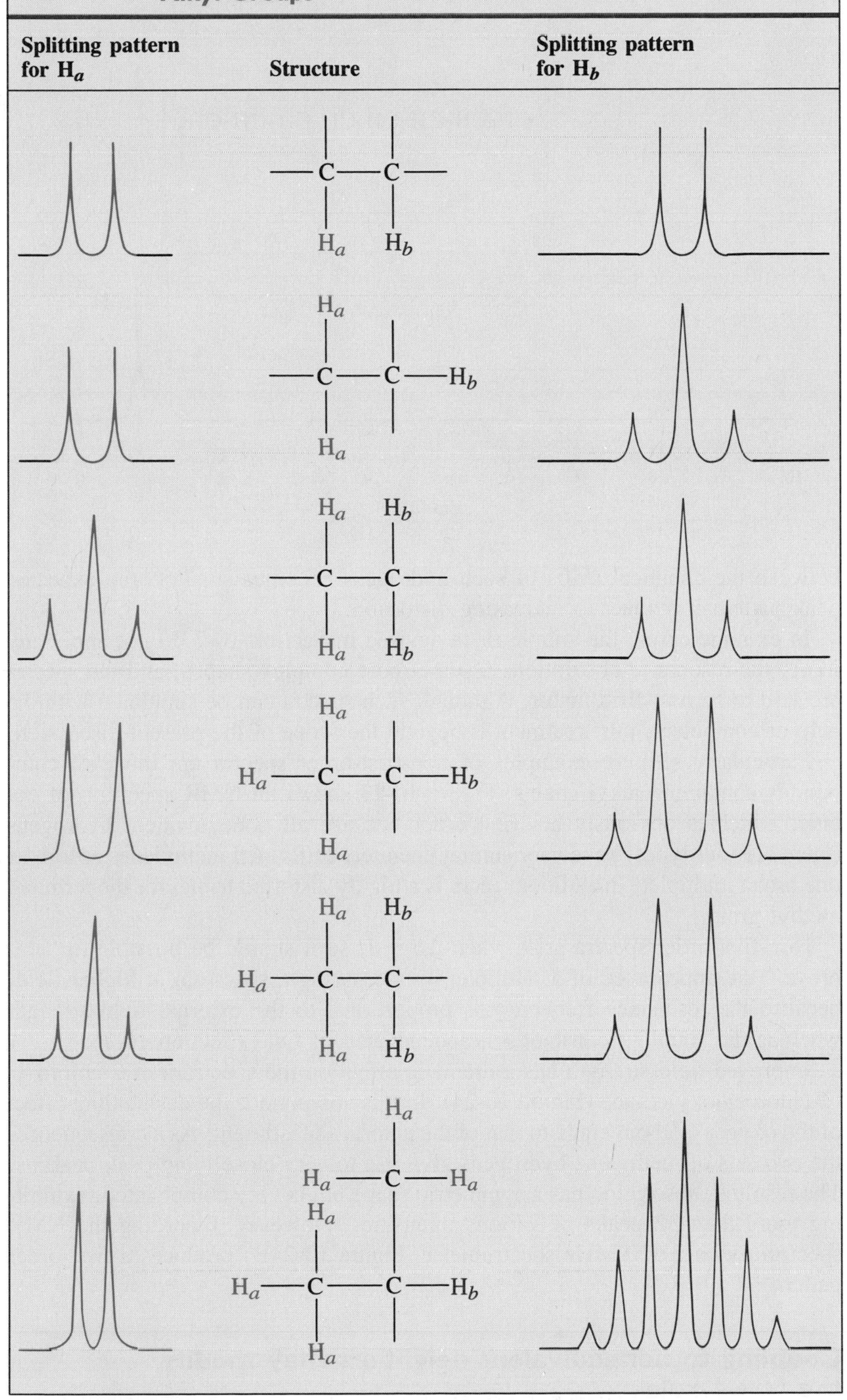

Note: H_a and H_b are assumed to have no other coupled nuclei in their vicinity.

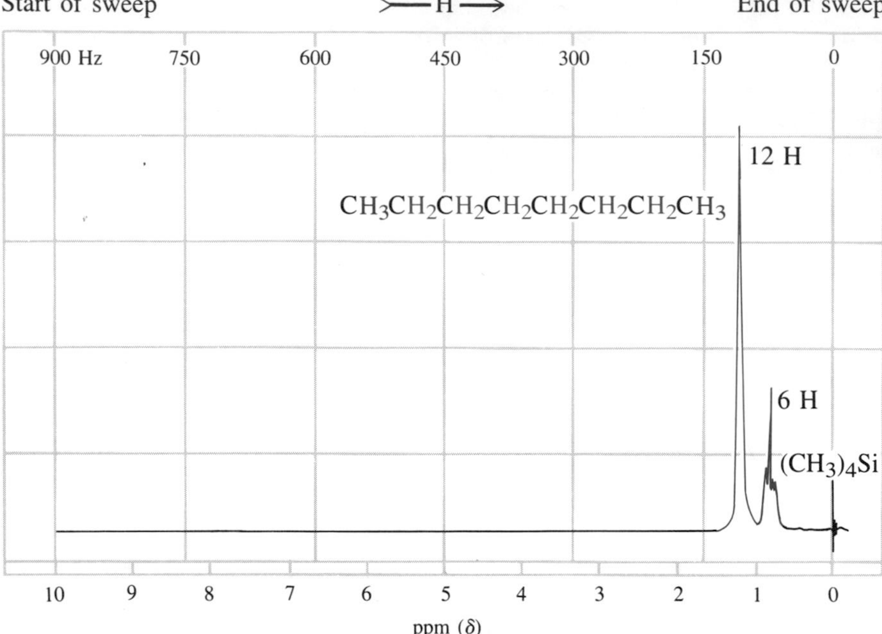

FIGURE 10-23 90-MHz NMR spectrum of octane in CCl_4. Compounds containing alkyl chains often display such non-first-order patterns.

between the chemical shifts of such hydrogens becomes smaller, the expected peak pattern is subject to increasing distortion.

In extreme cases, the simple rules devised in Section 10-7 do not apply any more, the resonance absorptions assume more complex shapes, and the spectra are said to be **non-first order.** Although such spectra can be simulated with the help of computers, this treatment is beyond the scope of the present discussion.

Particularly striking examples of non-first-order spectra are those of compounds containing alkyl chains. Figure 10-23 shows an NMR spectrum of octane, which is obviously not first order because all nonequivalent hydrogens (there are four types) have very similar chemical shifts. All methylenes absorb as one broad multiplet. In addition, there is a highly distorted triplet for the terminal methyl groups.

Non-first-order spectra arise when $\Delta\delta \sim J$, so it should be possible to "improve" the appearance of a multiplet by measuring a spectrum at higher field, because the resonance frequency is proportional to the external field strength whereas the coupling constant J is independent of field (Section 10-7).

Improved field strength has a dramatic effect on the spectrum of 2-chloro-1-(2-chloroethoxy)ethane (Figure 10-24). In this compound, the deshielding effect of the oxygen is about equal to that of the chlorine substituent. As a consequence, the two sets of methylene hydrogens give rise to very close lying peak patterns. The resulting absorption has a symmetric shape but is very complicated, exhibiting more than 32 peaks of various intensities. However, recording the NMR spectrum with a 500-MHz spectrometer (Figure 10-24B) produces a first-order pattern.

Coupling to nonequivalent neighbors may modify the *N* + 1 rule

When hydrogens are coupled to two sets of nonequivalent neighbors, complicated splitting patterns may result. The spectrum of 1,1,2-trichloropropane illus-

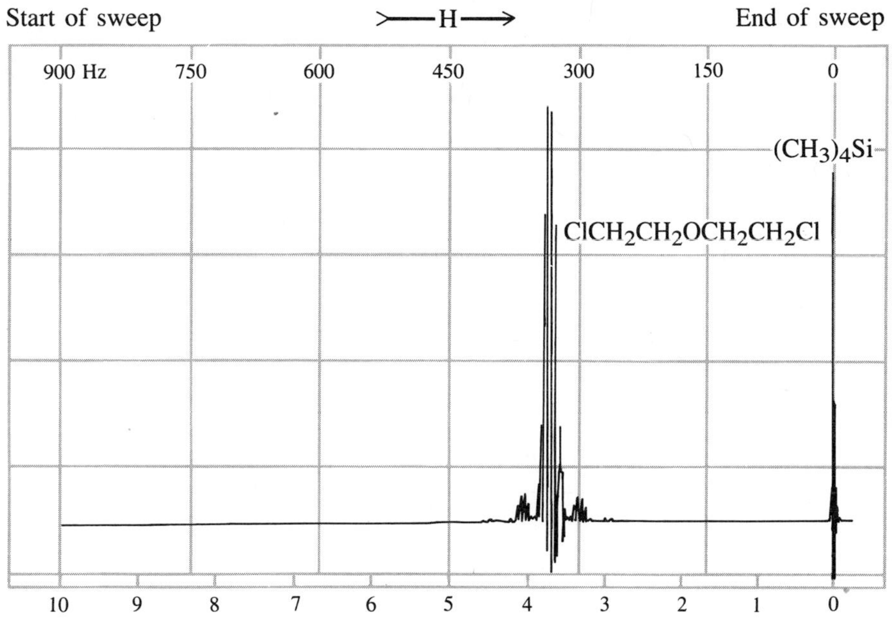

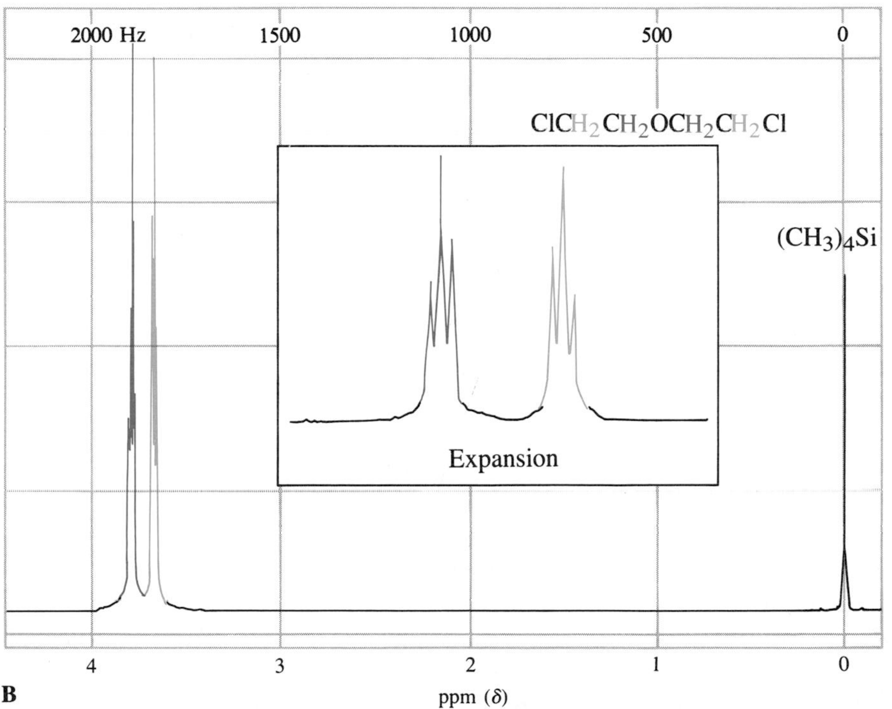

FIGURE 10-24 The effect of increased field strength on a non-first-order NMR spectrum: 2-chloro-1-(2-chloroethoxy)ethane at (A) 90 MHz; (B) 500 MHz in CCl_4. At high field strength, the complex multiplet observed at 90 MHz is simplified into two slightly distorted triplets, as might be expected for two mutually coupled CH_2 groups.

trates this point (Figure 10-25). In this compound, the hydrogen at C2 is located between a methyl and a $CHCl_2$ group, and it is coupled to the hydrogens of each independently.

Let us analyze the spectrum in detail. We first notice two doublets, one at low field ($\delta = 5.69$ ppm, $J = 3.6$ Hz, 1 H) and one at high field ($\delta = 1.64$ ppm, $J = 6.8$ Hz, 3 H). The low-field absorption is assignable to the hydrogen at C1 (H_a), adjacent to two deshielding halogens; and the methyl hydrogens (H_c) reso-

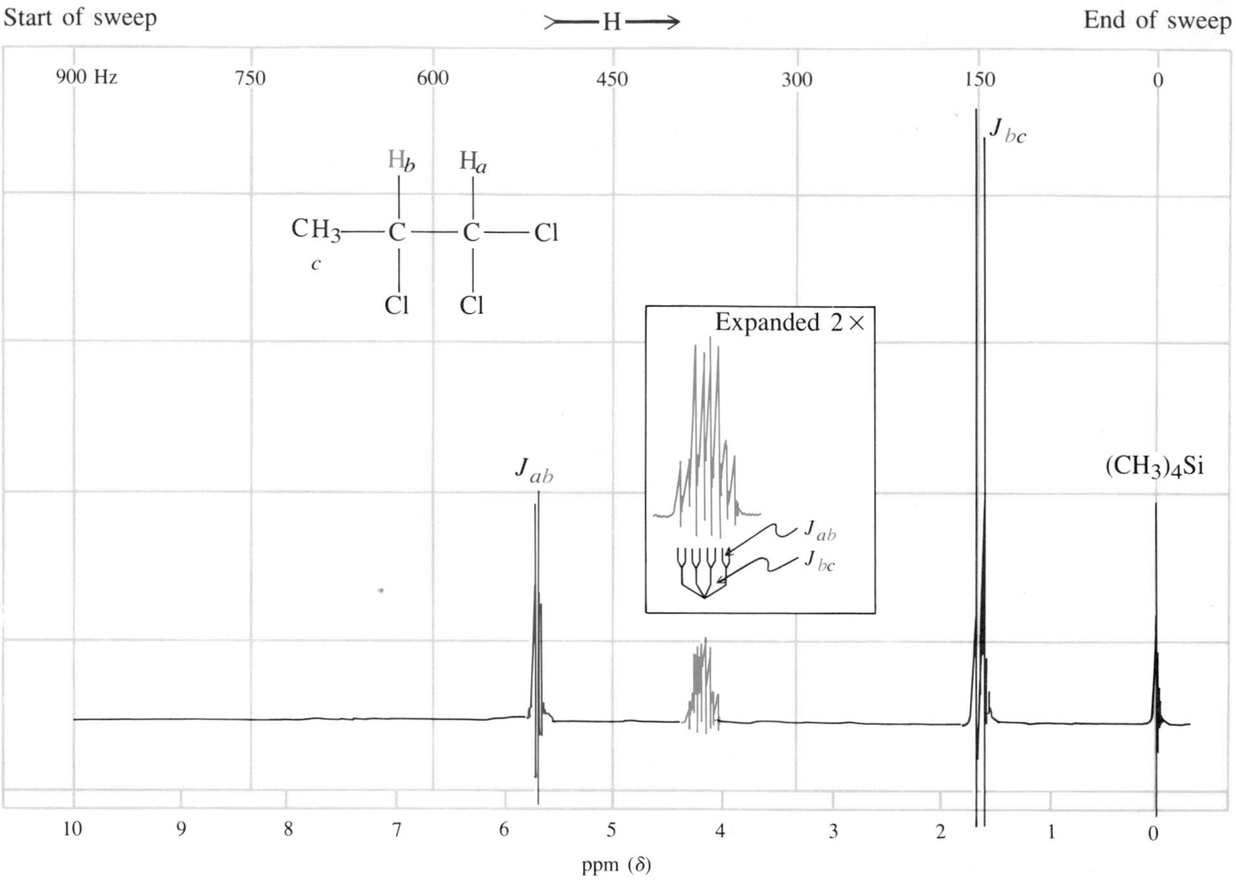

FIGURE 10-25 90-MHz NMR spectrum of 1,1,2-trichloropropane in CCl_4. Nucleus H_b gives rise to a quartet of doublets at $\delta = 4.18$ ppm: eight peaks.

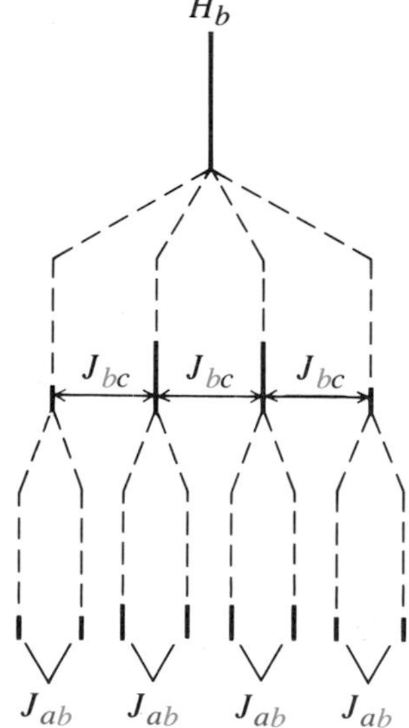

FIGURE 10-26 The splitting pattern for H_b in 1,1,2-trichloropropane. Each of the four lines arising from coupling to the methyl group is further split into a doublet by the hydrogen on C1.

nate as expected at highest field. In accord with the $N + 1$ rule, each signal is split into a doublet because of coupling with the hydrogen at C2 (H_b). The resonance of the latter, however, is quite different in appearance from what we expect. The nucleus giving rise to this absorption has a total of four hydrogens as its neighbors: H_a and three H_c. Application of the $N + 1$ rule suggests that a quintet should be observed. However, the signal for H_b at $\delta = 4.18$ ppm consists of *eight* lines, with relative intensities that do not conform to those expected for ordinary splitting patterns (see Tables 10-4 and 10-5). What is the cause of this complexity?

The $N + 1$ rule applies strictly only to splitting by *equivalent* neighbors. In this molecule we have two sets of different adjacent nuclei that couple to H_b *with different coupling constants*. The effect of these couplings can be understood, however, if we apply the $N + 1$ rule sequentially. The methyl group causes a splitting of the H_b resonance into a quartet, with $J_{bc} = 6.8$ Hz. Then, coupling by the hydrogen at C1 further splits *each peak* in this quartet into a doublet, with $J_{ab} = 3.6$ Hz, both splits resulting in the observed eight-line pattern (Figure 10-26). The hydrogen at C2 is said to be split into a quartet of doublets.

The hydrogens on C2 of 1-bromopropane also couple to two nonequivalent sets of neighbors. In this case, however, the resulting splitting pattern appears to conform with the $N + 1$ rule, and a (slightly distorted) sextet is observed (Figure 10-27). The reason is that the coupling constants to the two different groups are very similar, about 6–7 Hz. Although an analysis similar to that given above for 1,1,2-trichloropropane would lead us to predict as many as 12 lines in this signal (a quartet of triplets), the nearly equal coupling constants cause many of the lines to overlap, thus simplifying the pattern. The hydrogens in many simple alkyl derivatives display similar coupling constants and, therefore, spectra that are in accord with the $N + 1$ rule.

Fast proton exchange decouples hydroxy hydrogens

With our knowledge of vicinal coupling, let us now return to the NMR of alcohols. We note in the NMR spectrum of 2,2-dimethyl-1-propanol (Figure 10-11) that the OH absorption appears as a single peak, devoid of any splitting. This is curious, because the hydrogen is adjacent to two others, which should cause its appearance as a triplet. The CH_2 hydrogens that show up as a singlet should in turn appear as a doublet with the same coupling constant. Why, then, do we not observe spin–spin splitting? It is because the weakly acidic OH hydrogens are rapidly ionizing and transferring both between alcohol molecules and to traces of water under conditions of the NMR measurement. As a consequence of this process, the NMR spectrometer sees only an average signal for the OH hydrogen. No coupling is visible, because the binding time of the proton to the oxygen is too short (about 10^{-5} s). It follows that the CH_2 nuclei are similarly uncoupled, a condition resulting in the observed singlet.

Absorptions of this type are said to be **decoupled** by **fast proton exchange.** The exchange may be slowed by removal of traces of water or acid or by cooling.

FIGURE 10-27 90-MHz NMR spectrum of 1-bromopropane in CCl_4.

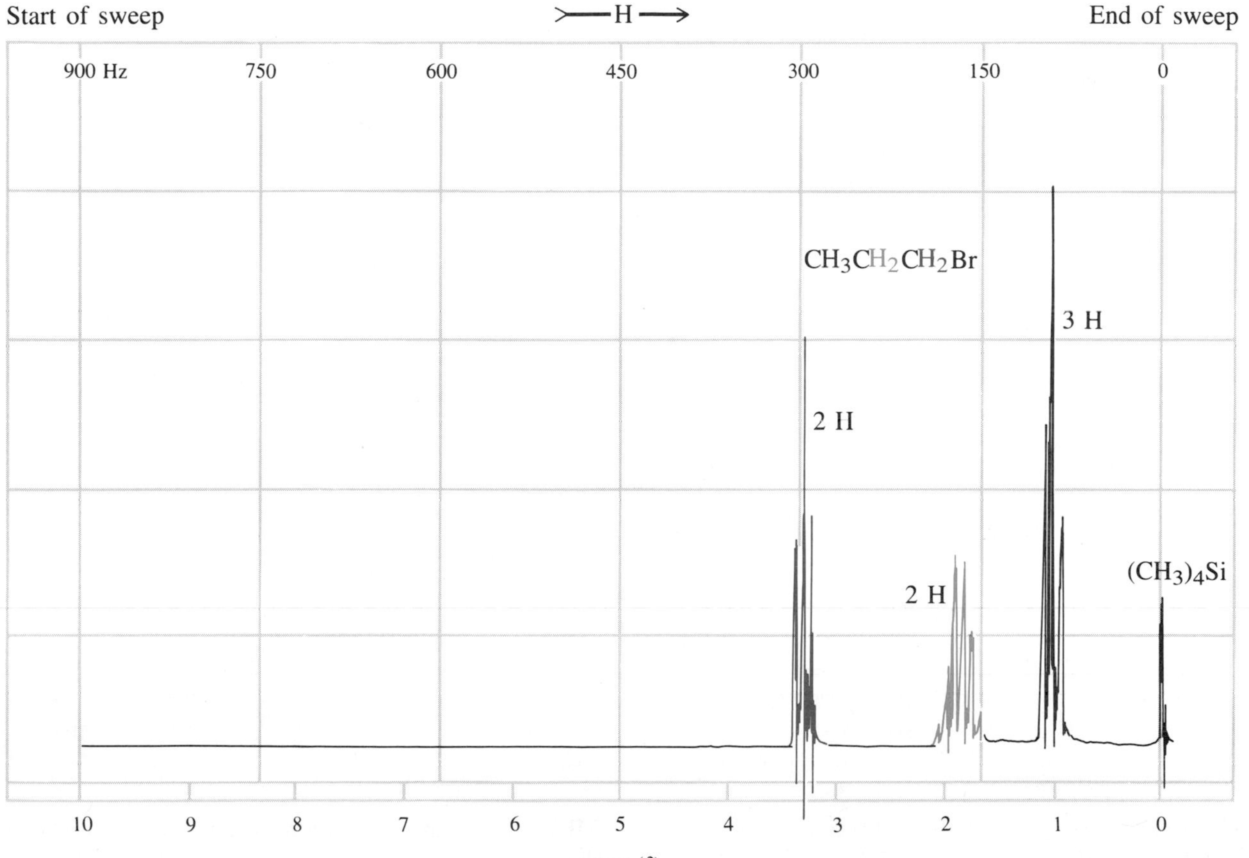

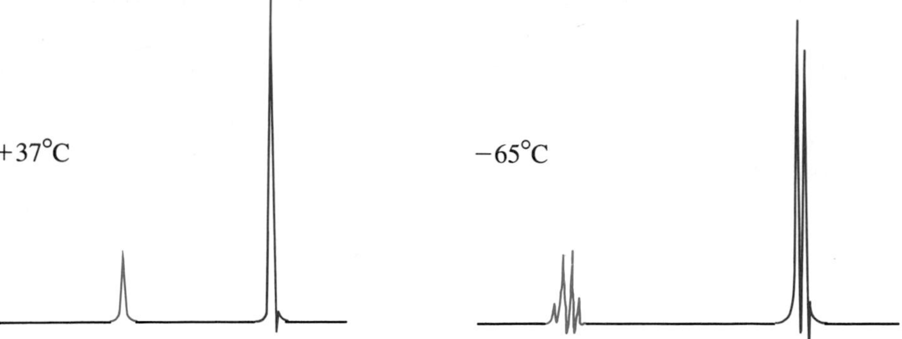

FIGURE 10-28 Temperature dependence of spin–spin splitting in methanol. The singlets at 37°C illustrate the effect of fast proton exchange in alcohols. (After H. Günther, *NMR-Spektroskopie,* Georg Thieme Verlag, Stuttgart, 1973.)

In these cases the OH bond retains its integrity long enough to be observed on the NMR time scale. An example is shown in Figure 10-28 for methanol. At 37°C, two singlets are observed, corresponding to the two types of hydrogens, both devoid of spin–spin splitting. However, at −65°C, the expected coupling pattern is detectable: a quartet and a doublet.

In summary, in many NMR spectra the peak patterns are not first order because the differences between the chemical shifts of nonequivalent hydrogens are close to the values of the corresponding coupling constants. Use of higher-field NMR instruments may improve the appearance of such spectra. Coupling to nonequivalent hydrogen neighbors occurs separately, with different coupling constants. In some cases they are sufficiently dissimilar to allow for an analysis of the multiplets. In many simple alkyl derivatives they are sufficiently similar ($J = 6$–7 Hz) that the spectra observed are simplified to those predicted in accordance with the $N + 1$ rule. Vicinal coupling through the oxygen in alcohols is frequently not observed because of decoupling by fast proton exchange.

10-9 Carbon-13 Nuclear Magnetic Resonance

Proton nuclear magnetic resonance is a powerful method for determining organic structures because most organic compounds contain hydrogens. Of even greater potential utility is NMR spectroscopy of carbon. After all, by definition, *all* organic compounds contain this element. In combination with ^{1}H NMR, it has become the most important analytical tool in the hands of the organic chemist. This section will summarize some of the essential aspects of this technique.

Carbon NMR utilizes an isotope in low natural abundance: ^{13}C

Carbon NMR is possible. However, there is a complication: The most abundant isotope of carbon, carbon-12, is not detectable by NMR. Fortunately, another isotope, carbon-13, is present in nature at a level of about 1.11%. Its behavior in the presence of a magnetic field is the same as that of hydrogen. One might therefore expect it to give spectra very similar to those observed in ^{1}H NMR spectroscopy. This expectation turns out to be only partly correct, because of several important (and very useful) differences between the two types of NMR techniques.

Carbon-13 NMR (^{13}C NMR) spectra are much more difficult to record than are hydrogen spectra, not only because of the low natural abundance of the nucleus under observation, but also because of the much weaker magnetic resonance of ^{13}C. Thus, under comparable conditions, ^{13}C signals are about 1/6000 as strong as those for hydrogen. For these weak signals to be observed, ^{13}C spectra must be scanned up to several thousand times and stored in a computer. The computer then adds up all the signals to produce one composite spectrum with greatly improved signal-to-noise characteristics. To collect data at a reasonable rate requires a special technique called Fourier transform NMR (FT NMR).* Even though the details of this method are beyond the scope of this introduction, it has made ^{13}C NMR as routine a tool in the organic laboratory as ^{1}H NMR.

One advantage of the low abundance of ^{13}C is the absence of carbon–carbon coupling. Just like hydrogens, two adjacent carbons, if magnetically nonequivalent (as they are in, e.g., bromoethane), split each other. In practice, however, such splitting is not observed. Why? Because coupling can occur only if two ^{13}C isotopes come to lie next to each other. With the abundance of ^{13}C in the molecule at 1.11%, this event has a very low probability. Most ^{13}C nuclei are surrounded by only ^{12}C nuclei, which, having no spin, do not give rise to spin–spin splitting. This feature simplifies ^{13}C NMR spectra appreciably, reducing the problem of their analysis to a determination of the coupling patterns to any attached hydrogens.

Figure 10-29 depicts the ^{13}C NMR spectrum of bromoethane. The chemical shift δ is defined as in ^{1}H NMR and is determined relative to an internal standard, normally the carbon absorption in $(CH_3)_4Si$. The chemical-shift range of carbon is much larger than that of hydrogen. For most organic compounds, it covers a distance of about 200 ppm, in contrast with the relatively narrow spectral "window" (10 ppm) of hydrogen. Figure 10-29 reveals the relative complexity of the absorptions caused by extensive ^{13}C–H spin–spin splittings. Not surprisingly, directly bound hydrogens are most strongly coupled (~125–200 Hz). Coupling tapers off rapidly, however, with increasing distance from the ^{13}C nucleus under observation, such that the geminal coupling constant $J_{^{13}C-C-H}$ is in the range of only 0.7–6 Hz.

EXERCISE 10-8

Predict the ^{13}C NMR spectral pattern of 1-bromopropane.

Hydrogen decoupling gives single lines

A technique that completely removes ^{13}C–H coupling is called **broad-band hydrogen** (or **proton**) **decoupling.** This method employs a strong, broad radio-frequency signal that covers the resonance frequencies of all the hydrogens and is applied at the same time as the ^{13}C spectrum is recorded. For example, in a magnetic field of 58,750 G, carbon-13 resonates at 62.8 MHz, hydrogen at 250 MHz. To obtain a proton-decoupled carbon spectrum at this field strength, we irradiate the sample at both frequencies. The first radio-frequency signal is used to produce carbon magnetic resonance. Simultaneous exposure to the second signal causes all the hydrogens to undergo rapid α–β spin flips, fast enough to average their local magnetic field contributions. The net result is the absence

*Invented by Professor Richard R. Ernst, b. 1933, Federal Technical Institute (ETH), Zurich, Switzerland, Nobel Prize 1991 (chemistry); named for Baron (Jean Baptiste) Joseph Fourier, 1768–1830, French mathematician.

FIGURE 10-29 62.8-MHz ^{13}C NMR spectrum of bromoethane, showing the complexity of ^{13}C–H coupling. There is an upfield quartet (δ = 18.3 ppm, J = 126 Hz) and a downfield triplet (δ = 26.6 ppm, J = 151 Hz) resonance for the two carbon atoms. Note the large chemical-shift range. Tetramethylsilane, defined to be located at δ = 0 ppm as in ^{1}H NMR, absorbs as a quartet (J = 118 Hz; the two outside peaks are barely visible) because of coupling of each carbon to three equivalent hydrogens. The inset shows a portion of the spectrum expanded horizontally to reveal the fine splitting of each of the main peaks that is due to coupling of each ^{13}C with protons on the neighboring carbon.

of coupling. Use of this technique simplifies the ^{13}C NMR spectrum of bromoethane to two single lines, as shown in Figure 10-30.

The power of proton decoupling becomes evident when spectra of relatively complex molecules are recorded. *Every magnetically distinct carbon gives only one single peak in the ^{13}C NMR spectrum.* Consider, for example, a hydrocarbon such as methylcyclohexane. Analysis by ^{1}H NMR is made very difficult by the close-lying chemical shifts of the eight different types of hydrogens. However, a proton-decoupled ^{13}C spectrum shows only five peaks, clearly depicting the presence of the five different types of carbons and revealing the twofold symme-

CH3CH2Br

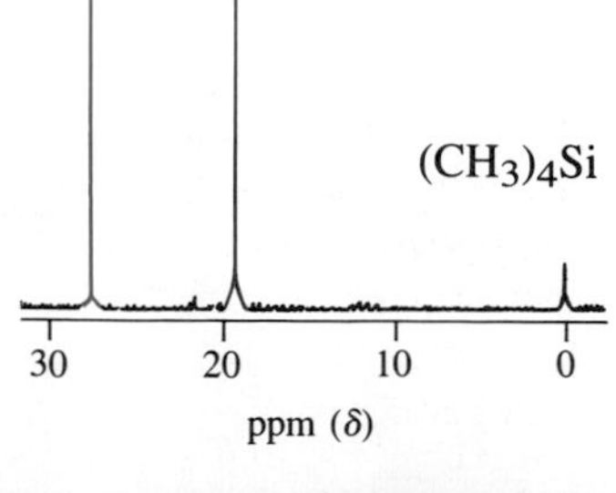

FIGURE 10-30 This 62.8-MHz ^{13}C NMR spectrum of bromoethane has been recorded with broad-band decoupling at 250 MHz. All lines simplify to singlets, including the absorption for $(CH_3)_4Si$.

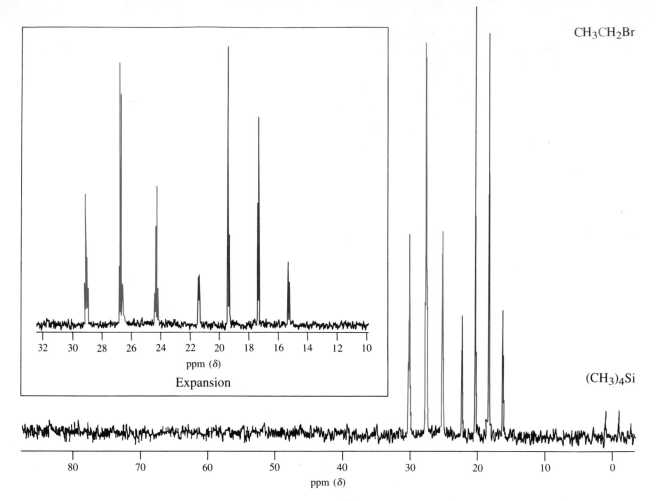

TABLE 10-6 Typical ^{13}C NMR Chemical Shifts

Type of carbon	Chemical shift δ in ppm
Primary alkyl, RCH_3	5–20
Secondary alkyl, RCH_2R'	20–30
Tertiary alkyl, R_3CH	30–50
Quarternary alkyl, R_4C	30–45
Allylic, $R_2C{=}C(R'')CH_2R'$	20–40
Chloroalkane, RCH_2Cl	25–50
Bromoalkane, RCH_2Br	20–40
Ether or alcohol, RCH_2OR' or RCH_2OH	50–90
Aldehyde or ketone, $RC(=O)H$ or $RC(=O)R'$	170–200
Alkene, aromatic, $R_2C{=}CR_2$	100–150
Alkyne, $RC{\equiv}CR$	50–95

try in the structure (Figure 10-31). These spectra also exhibit a limitation in ^{13}C NMR spectroscopy: Integration is not usually possible. As a consequence of the FT NMR method, peak intensities no longer correspond to numbers of nuclei.

Table 10-6 shows that carbon, like hydrogen (Table 10-2), has characteristic chemical shifts depending on its structural environment. As in ^1H NMR, electron-withdrawing groups cause deshielding and the chemical shifts go up in the

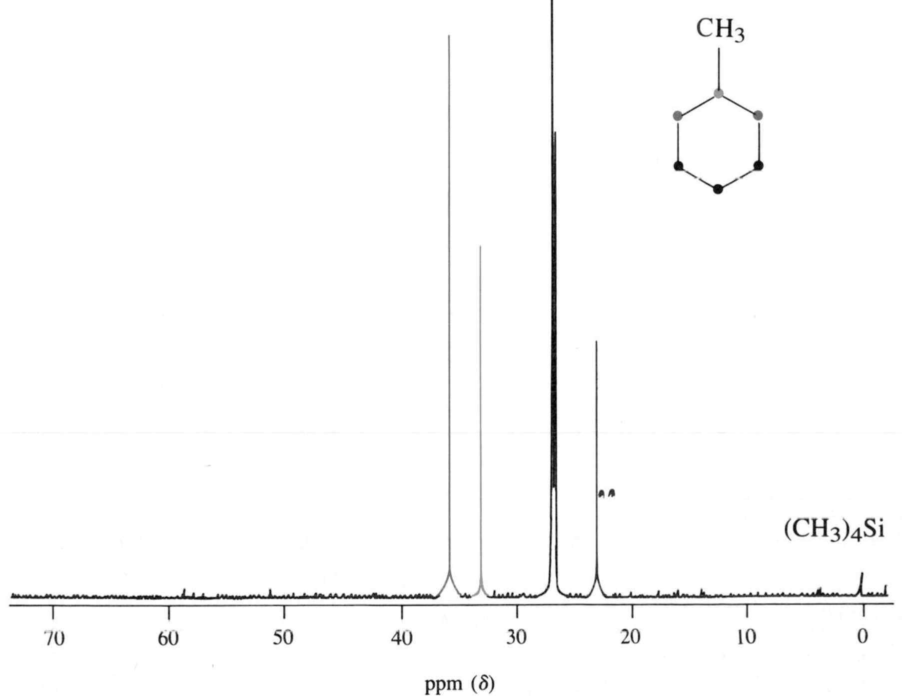

FIGURE 10-31 62.8-MHz ^{13}C NMR spectrum of methylcyclohexane with hydrogen decoupling (in C_6D_6). Each of the five magnetically different types of carbon in this compound gives rise to a distinct peak: $\delta = 23.1$, 26.7, 26.8, 33.1, and 35.8 ppm.

order primary < secondary < tertiary carbon. Apart from the diagnostic usefulness of such δ values, a knowledge of the number of different carbon atoms in the molecule can be an aid to structural identification. Consider, for example, the analytical differentiation of methylcyclohexane from other isomers with the same molecular formula C_7H_{14}. Many of the possibilities have a different number of nonequivalent carbons incorporated in their structure and therefore give distinctly different carbon spectra (find some with the same number of ^{13}C NMR peaks). Notice how much the (lack of) symmetry in a molecule influences the complexity of the carbon spectrum.

Number of ^{13}C Peaks in Some C_7H_{14} Isomers

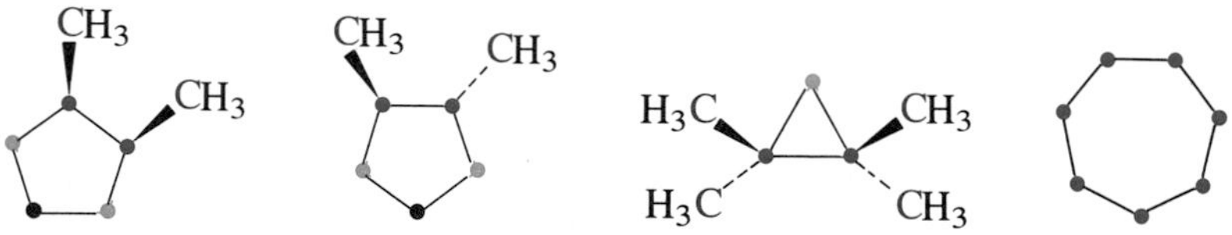

CH_3 CH_3 — Four peaks

CH_3 CH_3 — Four peaks

H_3C CH_3 H_3C CH_3 — Three peaks

One peak

EXERCISE 10-9

How many peaks would you expect in the proton-decoupled ^{13}C NMR spectra of the following compounds? (Hint: Look for symmetry.)

(a) 2,2-dimethyl-1-propanol **(b)** H H

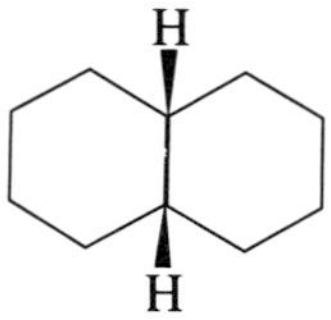

(c) H H H H **(d)**

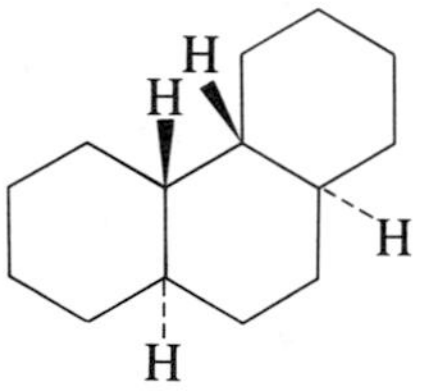

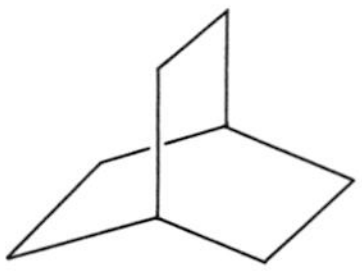

The attached proton test simplifies the interpretation of ^{13}C NMR spectra

The FT technique for the measurement of ^{13}C NMR spectra is extremely versatile, allowing data to be collected and presented in a variety of ways. One of the simplest and most widely used is the attached proton test (APT). In the APT mode, the hydrogen-decoupled ^{13}C NMR spectrum depicts methylene (CH_2) carbons as normal (positive) peaks but methyl and methine (CH) signals in inverted form. Figure 10-32 shows the APT ^{13}C NMR spectrum of methylcyclohexane: The CH_2 groups are immediately identifiable.

EXERCISE 10-10

Are the bicyclic compounds shown below readily distinguished by their proton-decoupled ^{13}C NMR spectra? Would an APT spectrum be of use in solving this problem?

(a) 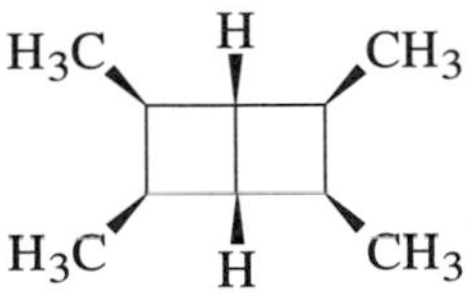 H_3C, H, CH_3, H_3C, H, CH_3 (b) 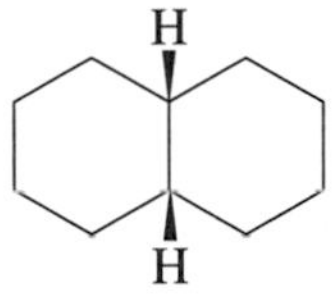 H, H

In summary, ^{13}C NMR is not quite as easy to measure as ^{1}H NMR because of the low natural abundance of the carbon-13 isotope and its intrinsically lower sensitivity in this experiment. Two additional potential complications are ^{13}C–^{1}H and ^{13}C–^{13}C coupling. The first is taken care of by broad-band proton decoupling, and the second never really presents a problem because of the scar-

BOX 10-3 Structural Characterization of Medicinal Agents from Marine Sources

The sea is a rich source of pharmaceutically useful substances. The rare 25-carbon compound manoalide (a *sesterterpene;* see Section 4-7) was isolated from a sponge in 1977, but only 58 mg were obtained. As a result, elemental analysis, chemical tests, and any other procedures that would destroy the tiny amount of available material could not be used in structural elucidation. Instead, a combination of spectroscopic techniques (IR, UV, MS, and NMR) was employed, leading in 1980 to the structure shown below for manoalide, which was later confirmed by total synthesis.

The characterization of manoalide relied in part on ^{1}H and ^{13}C NMR data. For example, five signals for protons on double bonds or on carbons between two oxygens appear between $\delta = 4.8$ and 6.2 ppm (see Table 10-2). The ^{13}C NMR shows the C=O carbon at $\delta = 172.3$ ppm and clearly distinguishes between alkene carbons ($\delta > 115$ ppm), the carbon adjacent to one ether oxygen ($\delta = 63.3$ ppm), the carbons between two oxygens ($\delta = 91.7$ and 99.1 ppm), and the remaining tetrahedral carbons (δ between 15 and 45 ppm; see Table 10-6), all of which were resolved and identified by number of attached hydrogens. While such a structural elucidation may seem extraordinary, improvements in instrumentation since 1980 now allow comparable procedures to be carried out on *micrograms*.

Clinical trials in 1990 revealed manoalide to be a potent drug with outstanding ability to block the release of the enzymes responsible for pain and inflammation; it may become a very useful treatment for the symptoms of arthritis and muscular dystrophy in the 1990s. Additional aspects of the structural determination of manoalide will be presented in later chapters.

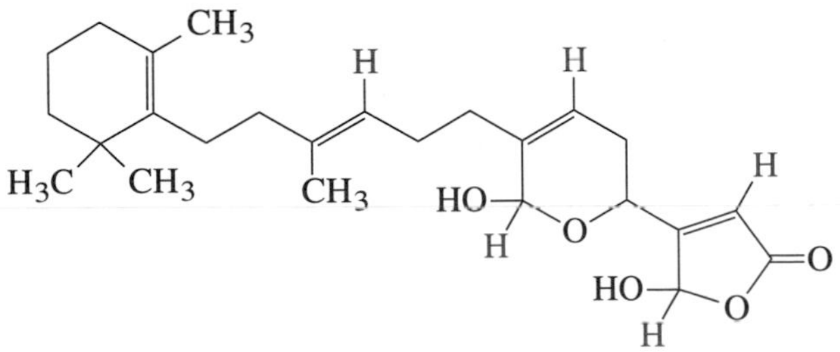

CH_3 H H H H_3C CH_3 CH_3 HO H O O HO H O

Manoalide

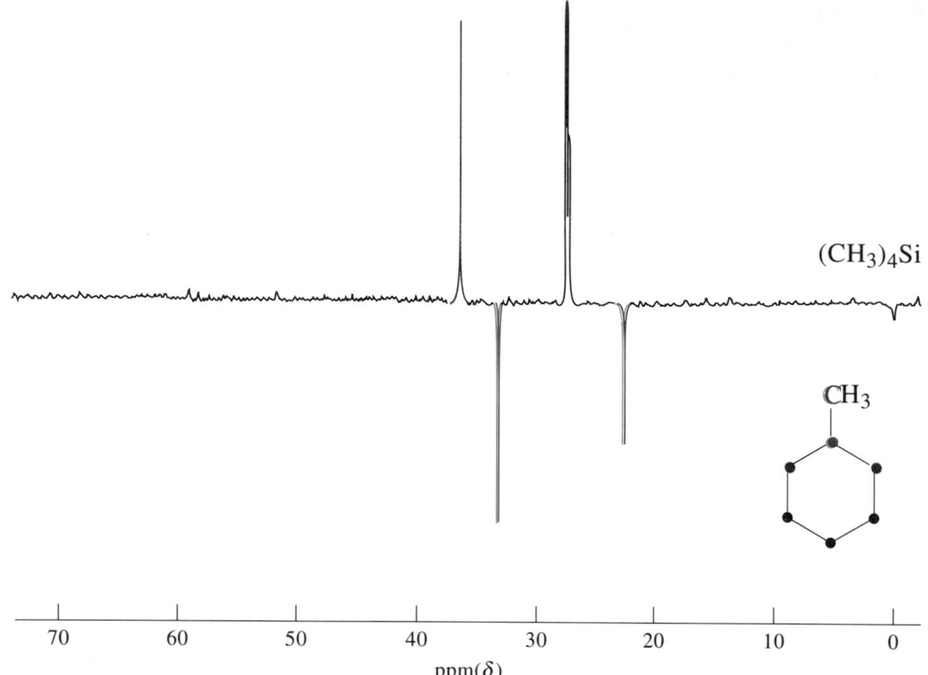

FIGURE 10-32 62.8-MHz ^{13}C NMR spectrum of methylcyclohexane with hydrogen decoupling (in C_6D_6) but recorded in the APT (attached-proton-test) mode. The signals for the methylene carbons are normal, whereas the methyl and methine (CH) carbons give inverted peaks.

city of ^{13}C. The ^{13}C NMR chemical-shift range is large, about 200 ppm in organic molecules. ^{13}C NMR spectra usually cannot be integrated, but the attached proton test readily distinguishes CH_2 groups from CH and CH_3 units.

IMPORTANT CONCEPTS

1. NMR is the most important spectroscopic tool in the elucidation of the structures of organic molecules.
2. Spectroscopy is possible because molecules exist in various energetic forms, those at lower energy being convertible into states of higher energy by absorption of discrete quanta of electromagnetic radiation.
3. NMR is possible because certain nuclei, especially ^{1}H and ^{13}C, when exposed to a strong magnetic field, align with it (α) or against it (β). The α-to-β transition can be effected by radio-frequency radiation, leading to resonance and a spectrum with characteristic absorptions. The higher the external field strength, the higher the resonance frequency. For example, a magnetic field of 21,150 gauss causes hydrogen to absorb at 90 MHz.
4. High-resolution NMR allows for the differentiation of hydrogen and carbon nuclei in different chemical environments. Their characteristic position in the spectrum is measured as the chemical shift, δ, in ppm from an internal standard, tetramethylsilane.
5. The chemical shift is highly dependent on the presence (shielding) or absence (deshielding) of electron density. The former results in relatively high-field [to the right, toward $(CH_3)_4Si$] peaks, the latter in low-field ones. Therefore, electron-donor substituents shield, and electron-withdrawing components deshield. The hydroxy proton of alcohols (and the NH proton in amines) shows a variable chemical shift and a broad absorption because of hydrogen bonding and proton exchange.
6. Chemically equivalent hydrogens and carbons have the same chemical shift. Equivalence is best established by the application of symmetry operations, such as those involving the use of mirror planes and rotations.
7. The number of hydrogens giving rise to a peak is measured by integration.
8. The number of hydrogen neighbors of a nucleus is given by the spin–spin splitting pattern of its

NMR resonance, following the $N + 1$ rule. Equivalent hydrogens show no mutual spin–spin splitting.

9. When the chemical-shift difference between coupled hydrogens is comparable to their coupling constant, non-first-order spectra with complicated patterns are observed.

10. When the constants for coupling to nonequivalent types of neighboring hydrogens are different, the $N + 1$ rule is applied sequentially.

11. Carbon NMR utilizes the low-abundance ^{13}C isotope. Carbon–carbon coupling is not observed in ordinary ^{13}C spectra. Carbon–hydrogen coupling can be removed by proton decoupling, thereby simplifying most ^{13}C spectra to a collection of single peaks.

PROBLEMS

1. Where on the chart presented in Figure 10-2 would the following be located: sound waves ($\nu \sim 1$ kHz $= 10^3$ Hz $= 10^3$ s^{-1}, or cycles s^{-1}); AM radio waves ($\nu \sim 1$ MHz $= 1000$ kHz $= 10^6$ Hz); FM and TV broadcast frequencies ($\nu \sim 100$ MHz $= 10^8$ s^{-1})?

2. Convert each of the following quantities into the specified units. **(a)** 1050 cm^{-1} into λ, in μm; **(b)** 510 nm (green light) into ν, in s^{-1} (cycles s^{-1}, or hertz); **(c)** 6.15 μm into $\tilde{\nu}$, in cm^{-1}; **(d)** 2250 cm^{-1} into ν, in s^{-1} (Hz).

3. Convert each of the following quantities into energies, in kcal mol^{-1}. **(a)** A bond rotation of 750 wavenumbers (cm^{-1}); **(b)** a bond vibration of 2900 wavenumbers (cm^{-1}); **(c)** an electronic transition of 350 nm (ultraviolet light, capable of sunburn); **(d)** a 20-Hz pipe-organ note; **(e)** a 40,000-Hz dog-whistle note; **(f)** the broadcast frequency of the audio signal of TV channel 6 (87.25 MHz); **(g)** a "hard" X-ray with a 0.07-nm wavelength.

4. Calculate to three significant figures the amount of energy absorbed by a hydrogen when it undergoes an α-to-β spin flip in the field of a **(a)** 21,150-G magnet ($\nu = 90$ MHz); **(b)** 117,500-G magnet ($\nu = 500$ MHz).

5. Sketch a hypothetical low-resolution NMR spectrum, showing the positions of the resonance peaks for all magnetic nuclei for each of the following molecules. Assume an external magnetic field of 21,150 G. How would the spectra change if the magnetic field were 84,600 G?

(a) $CHCl_3$ (chloroform) **(b)** $CFCl_3$ (Freon 11) **(c)** $CF_3CHClBr$ (Halothane)

6. If the NMR spectra of the molecules in Problem 5 were recorded by using high resolution for each nucleus, what differences would be observed?

7. The ^1H NMR spectrum of $CH_3COCH_2C(CH_3)_3$, 4,4-dimethyl-2-pentanone, taken at 90 MHz shows signals at the following positions: 92, 185, and 205 Hz, downfield from tetramethylsilane. **(a)** What are the chemical shifts (δ) of these signals? **(b)** What would their positions be in Hz, relative to tetramethylsilane, if the spectrum were recorded at 60 MHz? At 360 MHz? **(c)** Assign each signal to a set of hydrogens in the molecule.

8. Which hydrogens exhibit the more downfield signal relative to $(CH_3)_4Si$ in the proton NMR? Explain.

(a) $(CH_3)_2O$ or $(CH_3)_3N$ (b) $CH_3\overset{O}{\overset{\|}{C}}OCH_3$ (↑ or ↑) (c) $CH_3CH_2CH_2OH$ (↑ or ↑)

(d) $(CH_3)_2S$ or $(CH_3)_2S{=}O$

9. How many signals would be present in the ^1H NMR spectrum of each molecule illustrated below? What would the *approximate* chemical shift be for each of these signals? Ignore spin–spin splitting in this and the next problem.

(a) $CH_3CH_2CH_2CH_3$ (b) $CH_3CH(Br)CH_3$ (c) $HOCH_2C(CH_3)_2Cl$ (d) $CH_3CH(CH_3)CH_2CH_3$

(e) $CH_3C(CH_3)_2NH_2$ (f) $CH_3CH_2CH(CH_2CH_3)_2$ (g) $CH_3OCH_2CH_2CH_3$ (h) $CH_2{-}CH_2$, $CH_2{-}C{=}O$

(i) $CH_3CH_2{-}C(=O)H$ (j) $CH_3CH(OCH_3){-}C(CH_3)_2{-}CH_3$

10. For each compound in each group of isomers shown below, indicate the number of signals in the ^1H NMR spectrum, the *approximate* chemical shift of each signal, and the integration ratios for the signals. Finally, indicate whether all the isomers in each group can be distinguished from each other by means of these three pieces of information alone.

(a) $CH_3C(CH_3)(Br)CH_2CH_3$, $BrCH_2CH(CH_3)CH_2CH_3$, $CH_3CH(CH_3)CH_2CH_2Br$

(b) $ClCH_2CH_2CH_2CH_2OH$, $CH_3CH(CH_2Cl)CH_2OH$, $CH_3C(CH_3)(Cl)CH_2OH$

(c) $ClCH_2C(CH_3)(Br){-}CH(CH_3)CH_3$, $ClCH_2CH(CH_3){-}C(CH_3)(Br)CH_3$, $ClCH_2C(CH_3)_2{-}CH(Br)CH_3$, $ClCH_2CH(Br)C(CH_3)_2CH_3$

11. The 90-MHz ^1H NMR spectra for two haloalkanes are shown on the next page. Propose structures for these compounds that are consistent with the spectra. **(a)** $C_5H_{11}Cl$, spectrum A; **(b)** $C_4H_8Br_2$, spectrum B.

12. The ^1H NMR signals for three molecules with ether functional groups are given below. All the signals are singlets (single, sharp peaks). Propose structures for these compounds. **(a)** $C_3H_8O_2$, $\delta = 3.3$ and 4.4 ppm (ratio 3:1); **(b)** $C_4H_{10}O_3$, $\delta = 3.3$ and 4.9 ppm (ratio 9:1); **(c)** $C_5H_{12}O_2$, $\delta =$

1.2 and 3.1 ppm (ratio 1:1). Compare and contrast these spectra with that of 1,2-dimethyoxyethane (Figure 10-15B).

13. **(a)** The ^1H NMR spectrum of a ketone with the molecular formula $C_6H_{12}O$ has $\delta = 1.2$ and 2.1 ppm (ratio 3:1). Propose a structure for this molecule. **(b)** Each of two isomeric molecules related to the ketone in (a) has the molecular formula $C_6H_{12}O_2$. Their ^1H NMR spectra are described as follows: isomer 1, $\delta = 1.5$ and 2.0 ppm (ratio 3:1); isomer 2, $\delta = 1.2$ and 3.6 ppm (ratio 3:1). All signals in these spectra are singlets. Propose structures for these compounds. To what compound class do they belong?

14. Below are shown three $C_4H_8Cl_2$ isomers on the left and three sets of proton NMR data on the right. Match the structures to the proper spectral data. (Hint: You may find it helpful to sketch the spectra on a piece of scratch paper.)

(a) $CH_3CH_2\overset{Cl}{\overset{|}{C}}H\overset{Cl}{\overset{|}{C}}H_2$ **(i)** $\delta = 1.5$ (d, 6 H) and 4.1 (quin, 2 H) ppm

(b) $CH_3\overset{Cl}{\overset{|}{C}}H\overset{Cl}{\overset{|}{C}}HCH_3$ **(ii)** $\delta = 1.6$ (d, 3 H), 2.1 (q, 2 H), 3.6 (t, 2 H), and 4.2 (sex, 1 H) ppm

(c) $CH_3\overset{Cl}{\overset{|}{C}}HCH_2\overset{Cl}{\overset{|}{C}}H_2$ **(iii)** $\delta = 1.0$ (t, 3 H), 1.9 (quin, 2 H), 3.6 (d, 2 H), and 3.9 (quin, 1 H) ppm

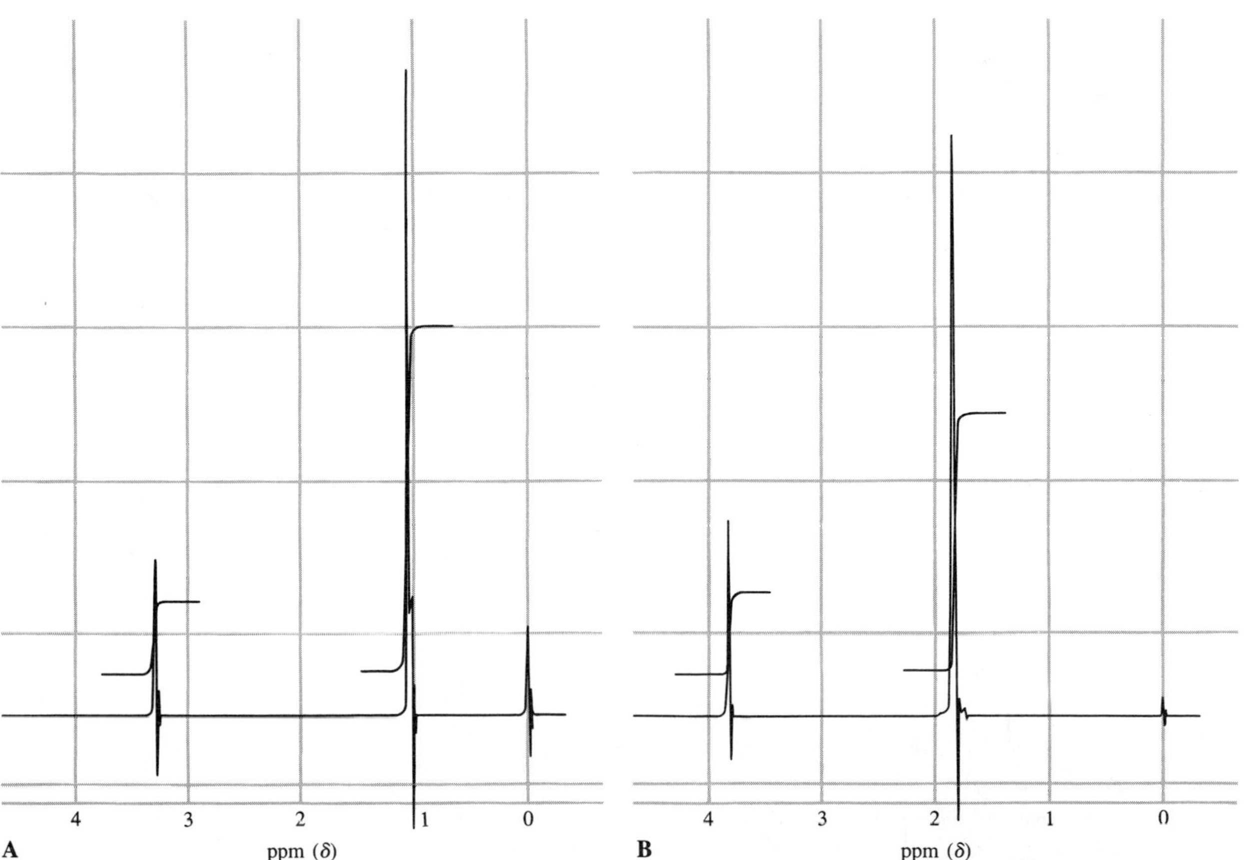

15. ^{1}H NMR spectra C through F correspond to four isomeric alcohols with the molecular formula $C_5H_{12}O$. Try to assign their molecular structures.

16. Sketch proton NMR spectra for the following compounds. Estimate chemical shifts (see Section 10-4) and show the proper multiplets for peaks that exhibit spin–spin coupling. **(a)** $CH_3CH_2OCH_2Br$; **(b)** $CH_3OCH_2CH_2Br$; **(c)** $CH_3CH_2CH_2OCH_2CH_2CH_3$; **(d)** $CH_3CH(OCH_3)_2$.

17. A hydrocarbon with the formula C_6H_{14} gives rise to 90-MHz ^{1}H NMR spectrum G (p. 368). What is its structure? This molecule has a structural feature similar to another molecule whose spectrum is illustrated in this chapter. What molecule is that? Explain similarities and differences in the spectra of the two molecules.

18. Treatment of the alcohol corresponding to NMR spectrum D in Problem 15 with hot concentrated HBr yields a substance with the formula $C_5H_{11}Br$. Its NMR spectrum exhibits signals at $\delta = 1.0$ (t, 3 H), 1.2 (s, 6 H), and 1.6 (q, 2 H) ppm. Explain. (Hint: See NMR spectrum C in Problem 15.)

19. The ^{1}H NMR spectrum of 1-chloropentane (p. 368) is shown at 60 MHz (spectrum H) and 500 MHz (spectrum I). Explain the differences in appearance of the two spectra, and assign the signals to specific hydrogens in the molecule.

20. Can the three isomeric pentanes be distinguished unambiguously from their broad-band proton-decoupled ^{13}C NMR spectra *alone?* How about the five isomeric hexanes?

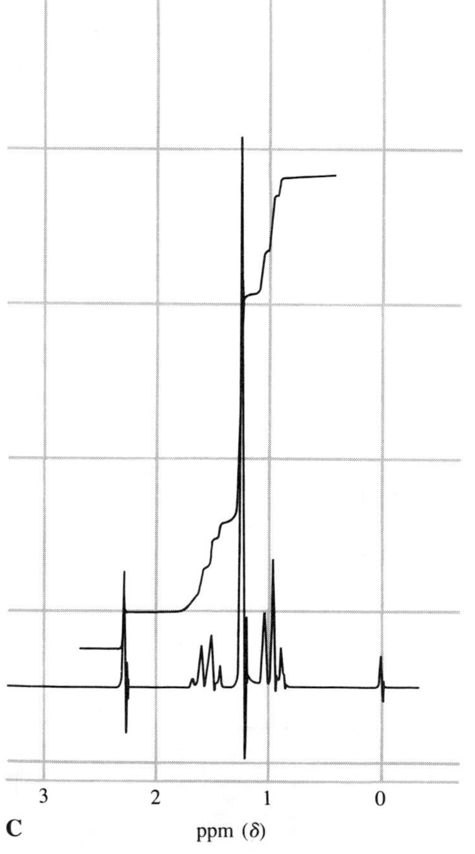

C

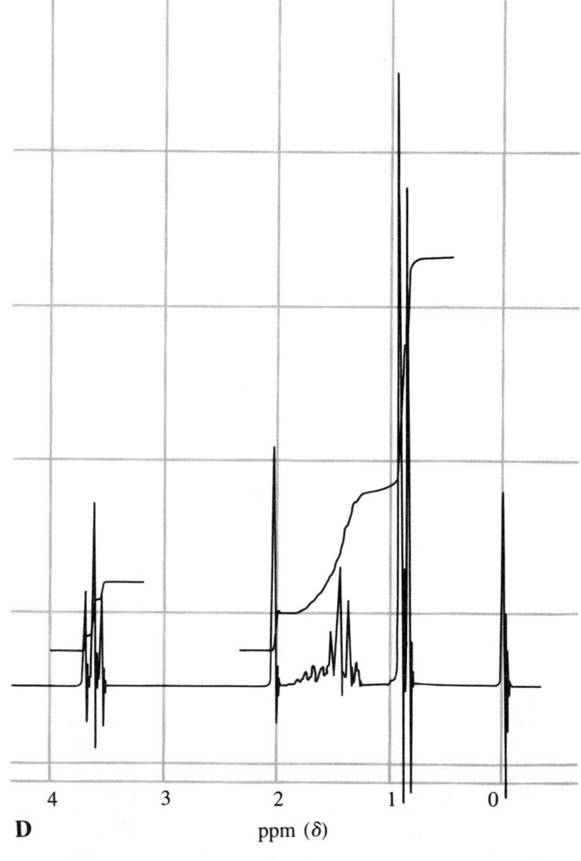

D

21. Predict the ^{13}C NMR spectra of the compounds in Problem 9, with and without proton decoupling.

22. Rework Problem 10 as it pertains to ^{13}C NMR spectroscopy.

23. From each group of three molecules, pick the one whose structure is most consistent with the proton-decoupled ^{13}C NMR data. Explain your choices. **(a)** $CH_3(CH_2)_4CH_3$, $(CH_3)_3CCH_2CH_3$, $(CH_3)_2CHCH(CH_3)_2$; $\delta =$ 19.5 and 33.9 ppm. **(b)** 1-Chlorobutane, 1-chloropentane, 3-chloropentane; $\delta = 13.2$, 20.0, 34.6, and 44.6 ppm. **(c)** Cyclopentanone, cycloheptanone, cyclononanone; $\delta = 24.0$, 30.0, 43.5, and 214.9 ppm.
(d) $ClCH_2CHClCH_2Cl$, $CH_3CCl_2CH_2Cl$, $CH_2{=}CHCH_2Cl$; $\delta = 45.1$, 118.3, and 133.8 ppm.

24. Propose a reasonable structure for each of the following molecules on the basis of the given molecular formula and of the ^1H and proton-decoupled ^{13}C NMR data. **(a)** $C_7H_{16}O$, spectra J and K (p. 369); **(b)** $C_8H_{18}O_2$, spectra L and M (p. 370).

25. The 90-MHz ^1H NMR spectrum of cholesteryl benzoate (see Section 4-7) is shown as spectrum N (p. 370). Although complex, it contains a number of distinguishing features. Analyze as many of the peaks or peak patterns as you can. The inset is a twofold expansion of the signal at $\delta = 4.8$ ppm. Why is it so complex, and can you explain the simplicity by comparison of the absorption at $\delta = 5.4$ ppm?

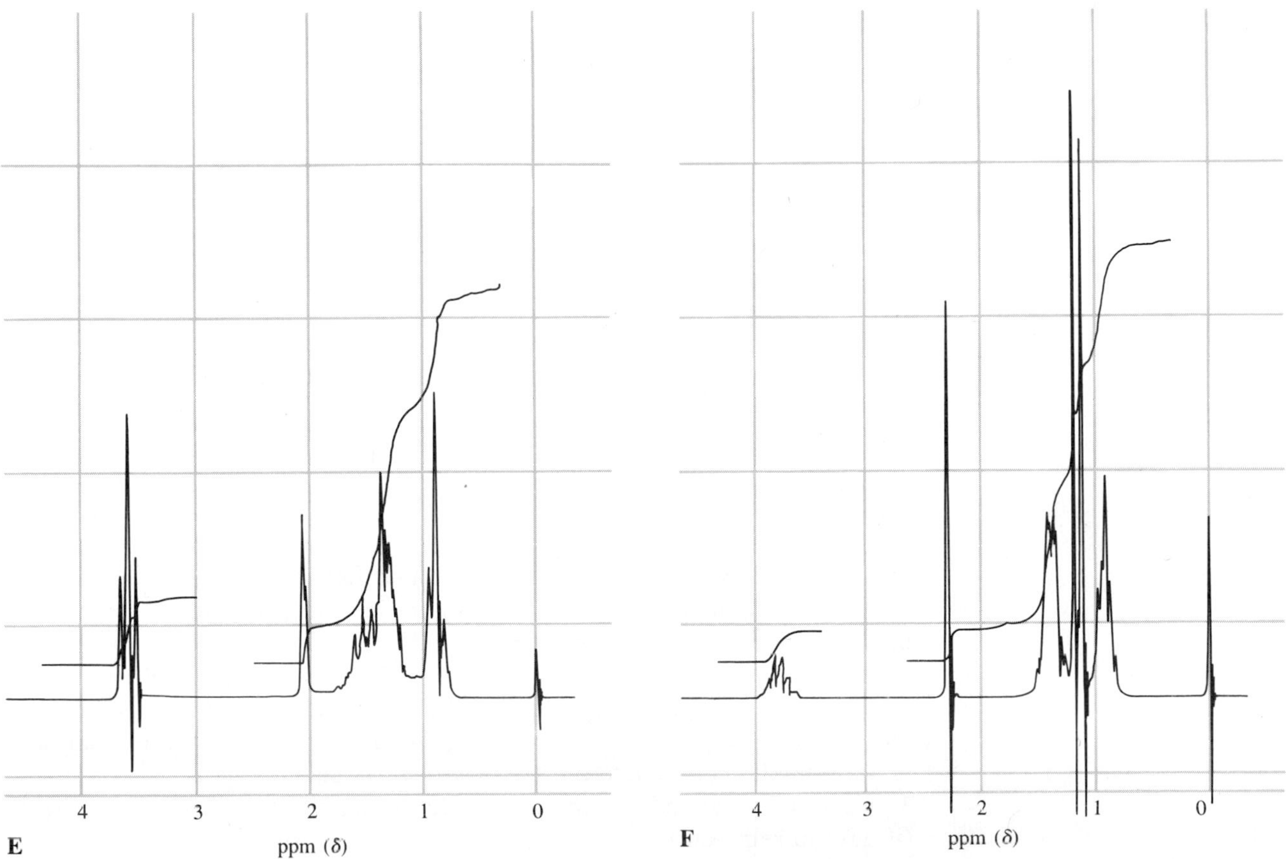

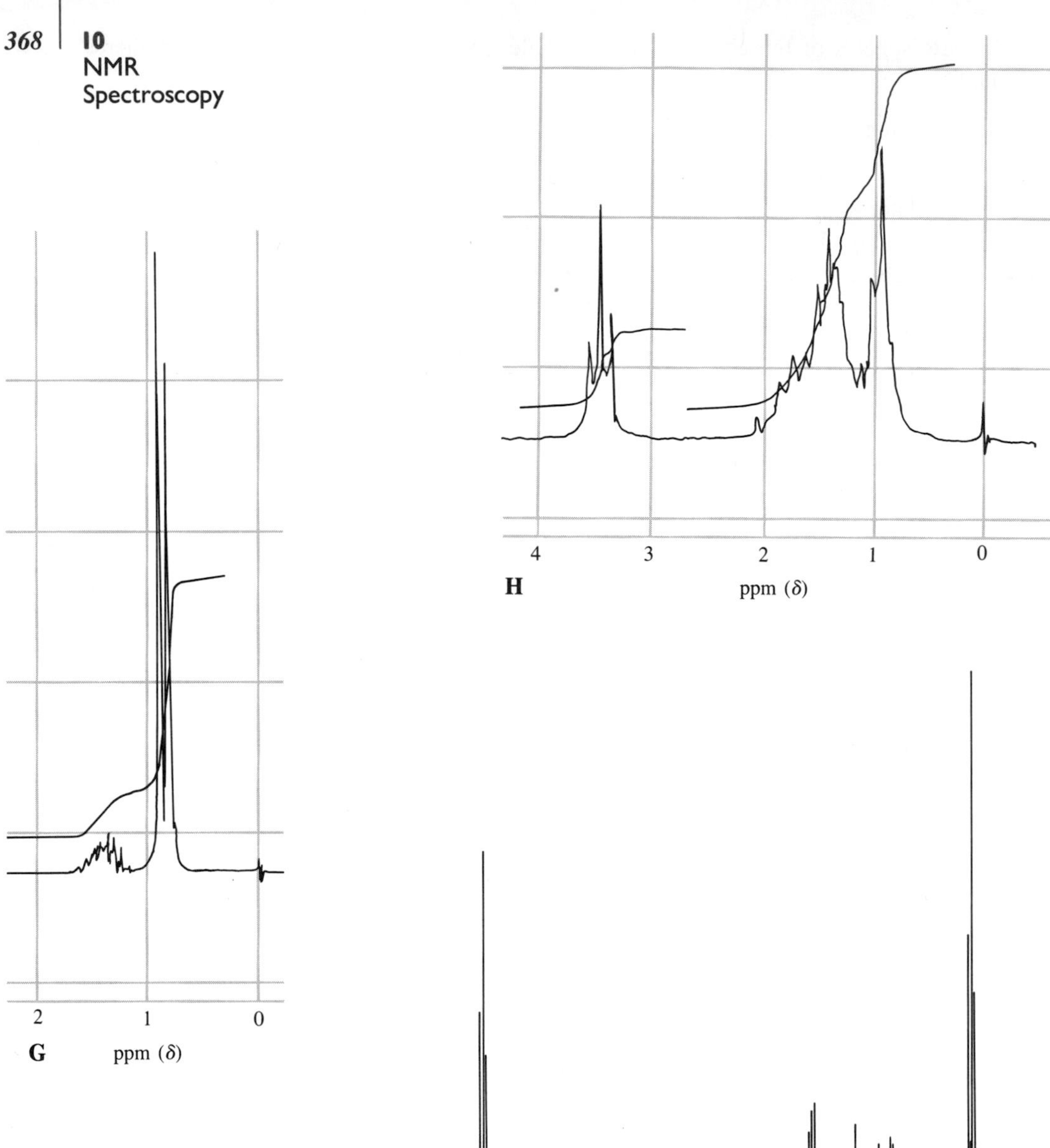

26. The terpene α-terpineol has the molecular formula $C_{10}H_{18}O$ and is a constituent of pine oil. As the -ol ending in the name indicates, it is an alcohol. Use its ^1H NMR spectrum (spectrum O, p. 371) to deduce as much as you can about the structure of α-terpineol. [Hints: (1) α-Terpineol has the 1-methyl-4-(1-methylethyl)cyclohexane framework also found in a number of other terpenes (e.g., carvone, Problem 8 of Chapter 5). (2) In your analysis of spectrum O, concentrate on the most obvious features (peaks at $\delta = 1.1$, 1.6, and 5.3 ppm) and use chemical shifts, integrations, and splitting patterns (if any) to help you.]

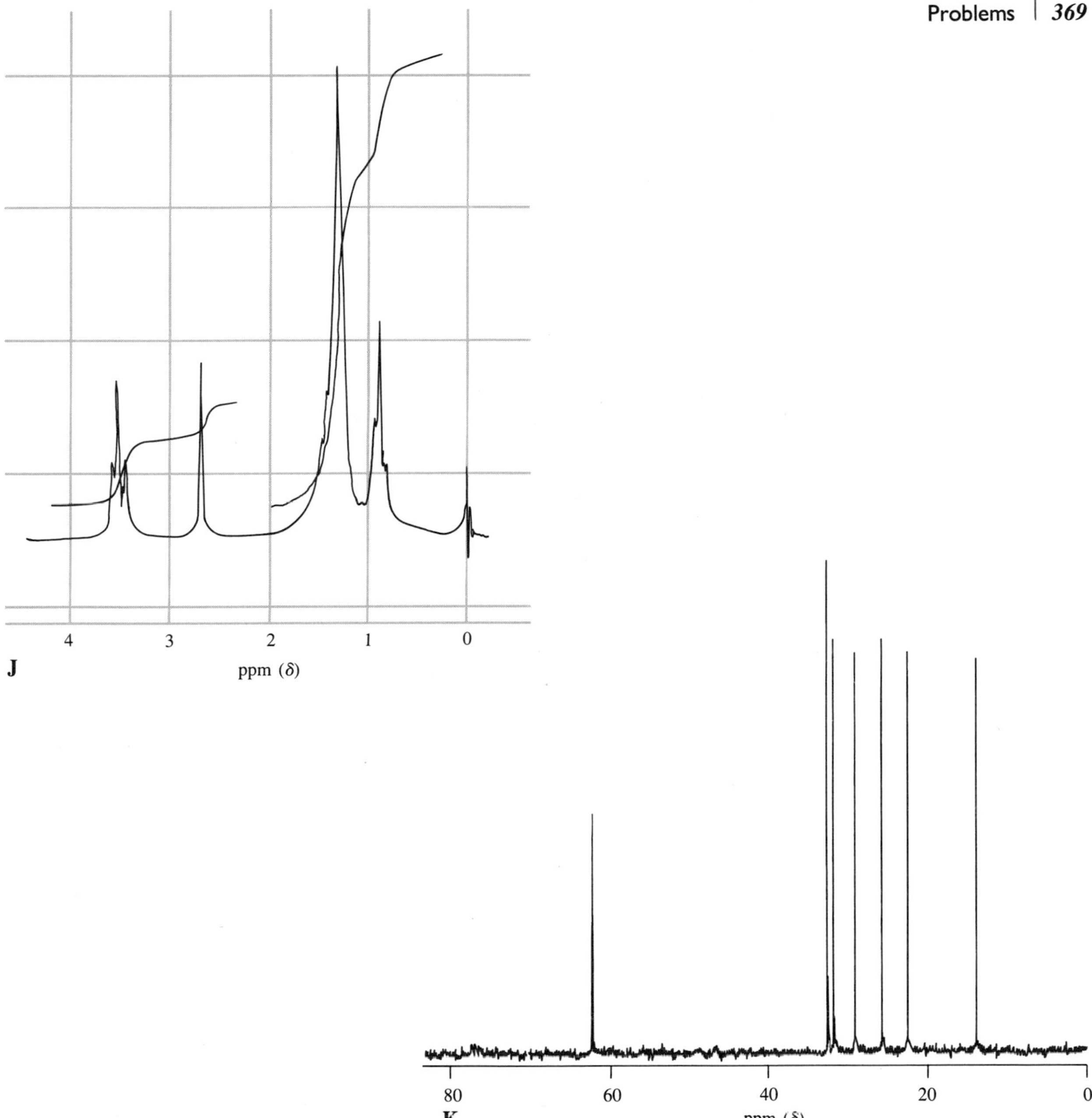

27. Study of the solvolysis of derivatives of menthol [5-methyl-2-(1-methylethyl)cyclohexanol] has greatly enhanced our understanding of these types of reactions. Heating the 4-methylbenzenesulfonate ester of the isomer shown below in 2,2,2-trifluoroethanol (a highly ionizing solvent of low nucleophilicity) leads to two products with the molecular formula $C_{10}H_{18}$. **(a)** The major product displays 10 different signals in its ^{13}C NMR spectrum. Two of them occur at relatively low field, about $\delta = 120$ and 145 ppm, respectively. The ^{1}H NMR spectrum exhibits a multiplet near $\delta = 5$ ppm (1 H); all other signals are upfield of $\delta = 3$ ppm. Identify this compound. **(b)** The minor product gives only seven ^{13}C signals. Again, two are at low field ($\delta \sim 125$

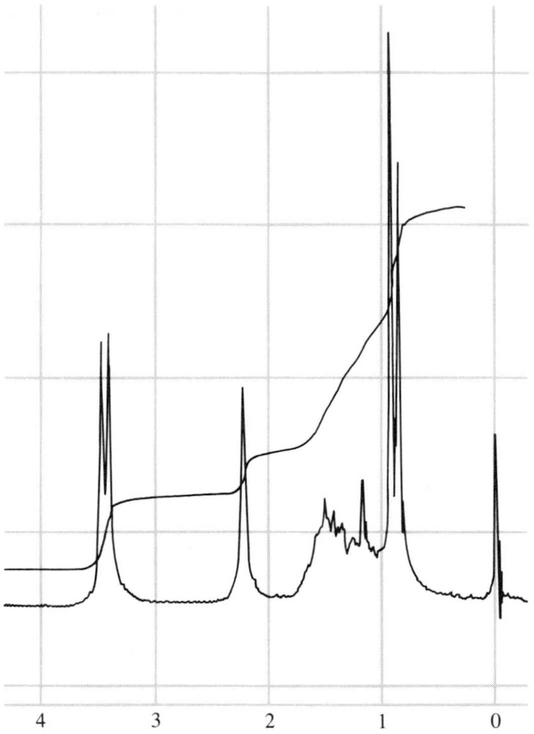

L

ppm (δ)

M

ppm (δ)

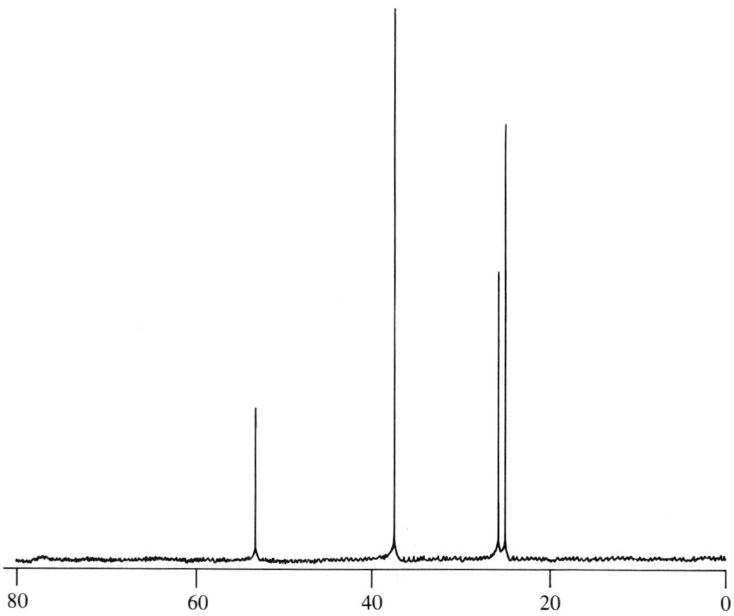

Cholesteryl benzoate

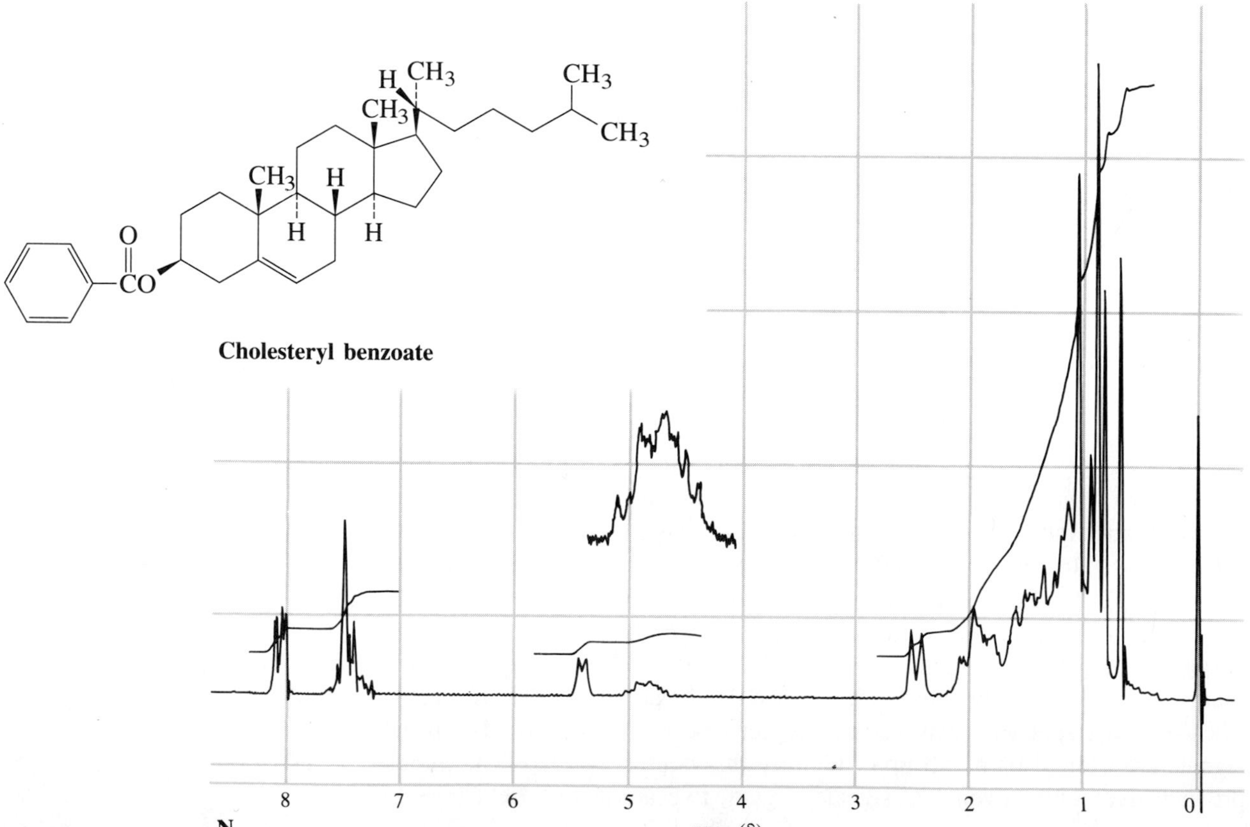

N

ppm (δ)

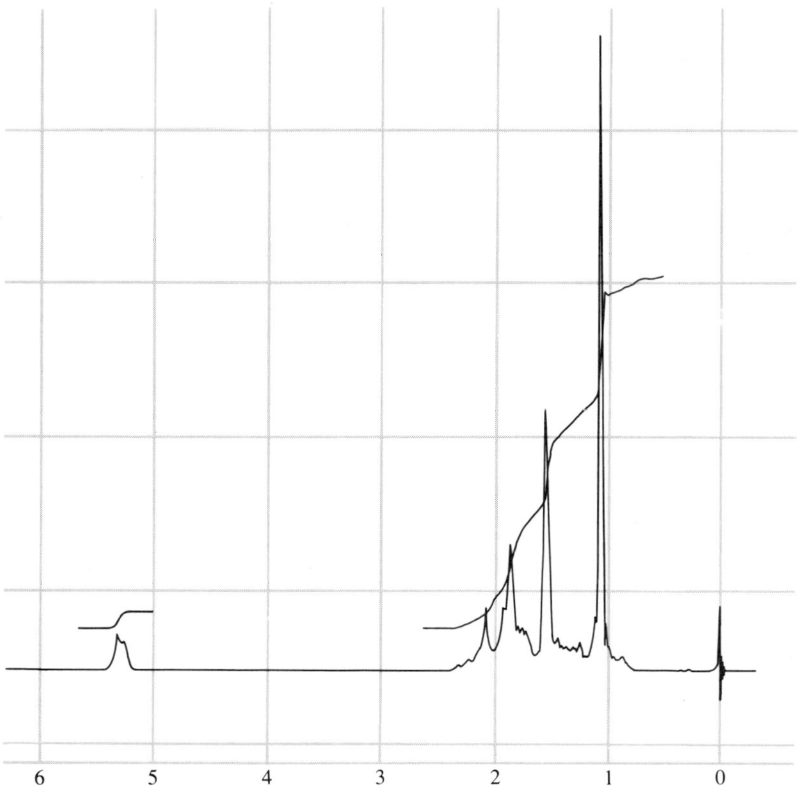

O

and 140 ppm), but, in contrast with the ^1H NMR data on the major isomer, there are no signals at lower field than $\delta = 3$ ppm. Identify this product and explain its formation mechanistically. **(c)** When the solvolysis is carried out starting with the ester labeled with deuterium at C2, the ^1H spectrum of the resulting major product isomer in (a) reveals a significant reduction of the intensity of the signal at $\delta = 5$ ppm, a result indicating the partial (!) incorporation of deuterium at the position associated with this peak. How might this result be explained? [Hint: The answer lies in the mechanism of formation of the minor product, in (b).]

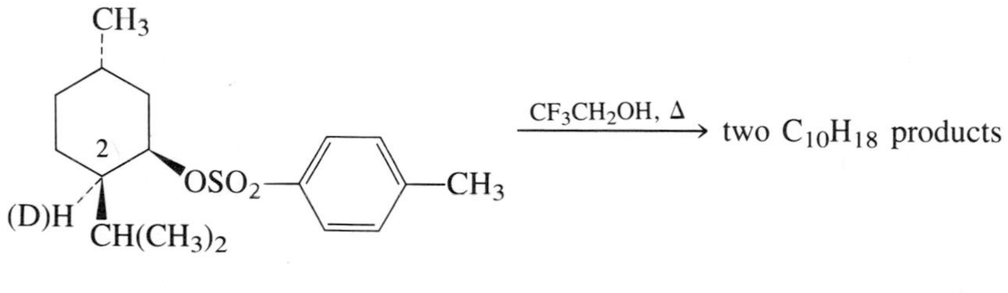

11

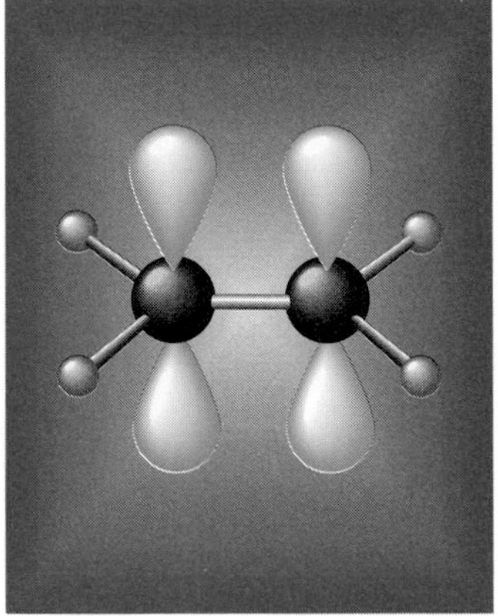

Alkenes and Infrared Spectroscopy

In the preceding five chapters, we have learned about several organic functional groups and begun our study of spectroscopy. For the most part, we have concentrated on molecules containing only single bonds. In this chapter we begin to look closely at the classes of organic compounds that are characterized by the presence of multiple bonds.

The alkenes contain a carbon–carbon double bond. In Chapter 12 we shall see that this bond is the source of considerable chemical reactivity, making alkenes valuable sources of plastics, synthetic fibers, construction materials, and a variety of other industrially useful substances. Many gaseous alkenes give oils by addition reactions to the double bond, which is why this class of compounds used to be called olefins (from *oleum facere,* Latin, to make oil). Because alkenes can undergo addition reactions, they are said to be **unsaturated.** In contrast, alkanes are referred to as **saturated** compounds.

This chapter takes up the names and physical properties of the alkenes, introducing new techniques for evaluating the relative stability of isomers. A discussion of their preparation will afford an opportunity to review and augment what we know about elimination reactions. We shall also present a second form of spectroscopy—infrared (IR) spectroscopy—and show how it is used to help organic chemists determine the presence or absence of characteristic functional groups in a molecule.

11-1 Naming the Alkenes

A carbon–carbon double bond is the characteristic functional group of the alkenes. Their general formula is C_nH_{2n}, the same as that for the cycloalkanes.

Like other organic compounds, some alkenes are still known by their common names, in which the *-ane* ending of the corresponding alkanes is replaced by **-ylene.** Substituent names are added as prefixes.

Common Names of Typical Alkenes

$CH_2{=}CH_2$	$CH_2{=}C(CH_3)H$	$Cl_2C{=}C(Cl)H$
Ethylene	**Propylene**	**Trichloroethylene**
(Fruit-ripening hormone in plants)	(Raw material for plastics)	(Common cleaning solvent)

In IUPAC nomenclature, the simpler ending **-ene** is used instead of -ylene, as in ethene and propene. More complicated systems require adaptations and extensions of the rules for naming alkanes (Section 2-3).

RULE 1. To name the stem, find the longest chain that *includes* the functional group—in this case, *both* carbons making up the double bond. The molecule may have still longer carbon chains, but ignore them.

$CH_2{=}CHCH(CH_3)CH_2CH_3$	$CH_2{=}CHCH(CH_2CH_2CH_3)(CH_2)_4CH_3$	$CH_3CH_2CH_2CH_2C(CH_3){=}C(CH_2CH_3)CH_2CH_2CH_2CH_3$
A methylpentene	**A propyloctene**	**An ethylmethyldecene**
	(Not a hexene or a nonane derivative)	(Not a pentene or a heptene or an octene derivative)

RULE 2. Indicate the location of the double bond in the main chain by number, starting at the end *closer* to the double bond. (Cycloalkenes do not require the numerical prefix, but the carbons making up the double bond are assigned the numbers 1 and 2.) Alkenes that have the same molecular formula but differ in the location of the double bond (like 1-butene and 2-butene) are called **double-bond isomers.** A 1-alkene is also referred to as a **terminal alkene;** the others are labeled **internal.** Note that alkenes are easily depicted in line notation.

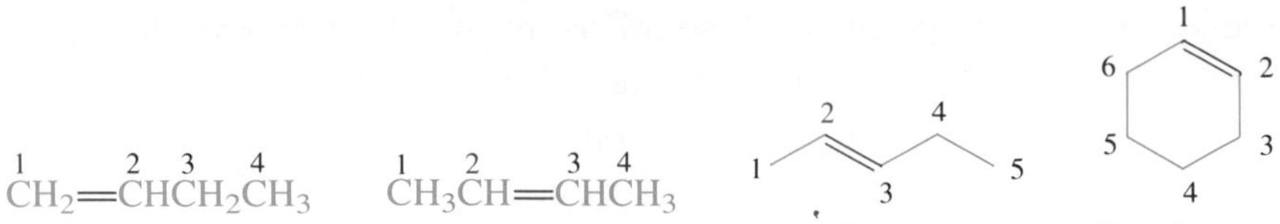

$\overset{1}{C}H_2\!=\!\overset{2}{C}H\overset{3}{C}H_2\overset{4}{C}H_3$
1-Butene
(A terminal alkene; not 3-butene)

$\overset{1}{C}H_3\overset{2}{C}H\!=\!\overset{3}{C}H\overset{4}{C}H_3$
2-Butene
(An internal alkene and a double-bond isomer of 1-butene)

2-Pentene
(Not 3-pentene)

Cyclohexene

RULE 3. Add substituents and their positions to the alkene name as prefixes. If the alkene stem is symmetric, begin from the end that gives the first substituent along the chain the lowest possible number.

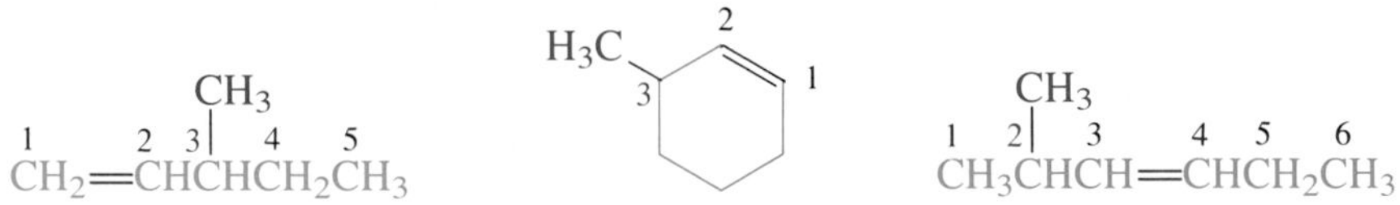

$CH_2\!=\!CHCH(CH_3)CH_2CH_3$
3 Methyl-1-pentene

3 Methylcyclohexene
(Not 6-methylcyclohexene)

$CH_3CH(CH_3)CH\!=\!CHCH_2CH_3$
2-Methyl-3-hexene
(Not 5-methyl-3-hexene)

EXERCISE 11-1

Name the following two alkenes.

(a) 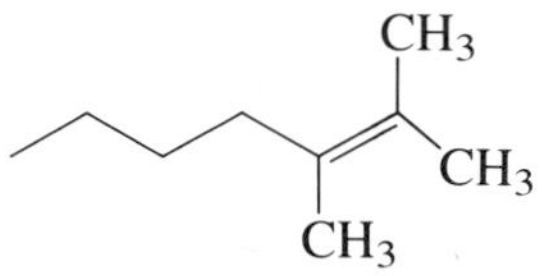 (b)

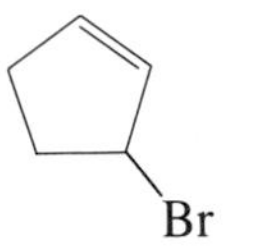

RULE 4. Identify any stereoisomers. In a 1,2-disubstituted ethene, the two substituents may be on the same side of the double bond or on opposite sides. The first stereochemical arrangement is called cis, and the second trans, in analogy to the cis-trans names of the disubstituted cycloalkanes (Section 4-1). Two alkenes of the same molecular formula differing only in their stereochemistry are called cis-trans isomers and are examples of diastereomers: stereoisomers that are not mirror images of each other.

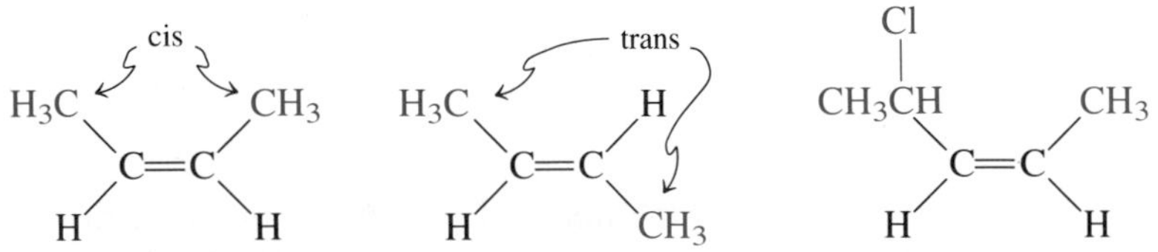

***cis*-2-Butene**

***trans*-2-Butene**

4-Chloro-*cis*-2-pentene

EXERCISE 11-2

Name the following two alkenes.

(a) 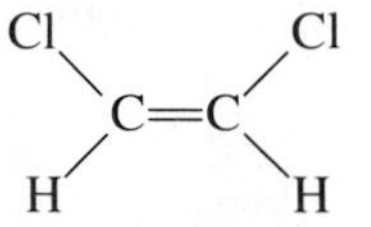 (b)

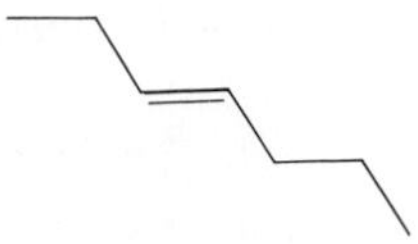

In the smaller substituted cycloalkenes, the double bond can exist only in the cis configuration. The trans arrangement is prohibitively strained (as building a model reveals). However, in larger cycloalkenes, trans isomers are stable.

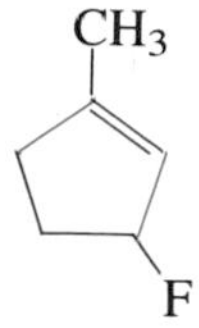

CH_3 F | CH_3 CH_2CH_3 H_3C

3-Fluoro-1-methylcyclopentene 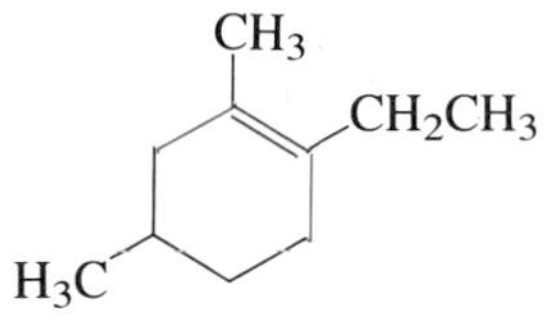 **1-Ethyl-2,4-dimethylcyclohexene** ***trans*-Cyclodecene**

(In both cases, only the cis isomer is stable)

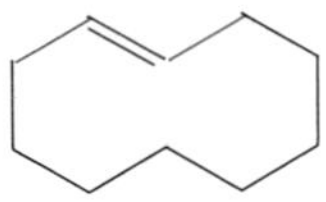

RULE 5. Use a more general method, the *E,Z* system, to label more complex diastereomers. The labels *cis* and *trans* cannot be applied when there are three or more different substituents attached to the double bond carbons. An alternative system for naming such alkenes has been adopted by IUPAC: the ***E,Z* system.** In this convention, the sequence rules devised for establishing priority in *R,S* names (Section 5-3) are applied separately to the two groups on each double-bonded carbon. When the two groups of highest priority are on opposite sides of the double bond, the molecule is of the *E* configuration (E from *entgegen,* German, opposite). When the two substituents of highest priority appear on the same side, the molecule is a *Z* isomer (Z from *zusammen,* German, together).

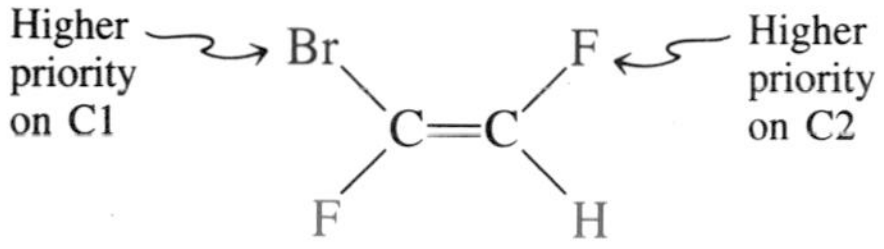

Higher priority on C1 → Br, F / C=C / F, H ← Higher priority on C2

(*Z*)-1-Bromo-1,2-difluoroethene

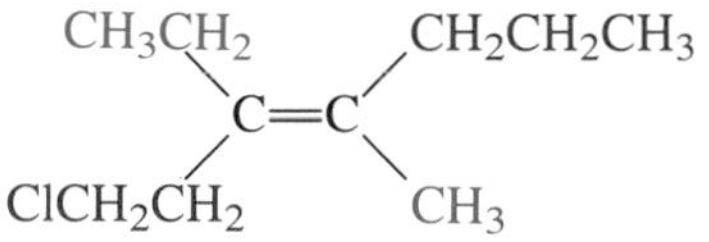

CH_3CH_2, $CH_2CH_2CH_3$ / C=C / $ClCH_2CH_2$, CH_3

(*E*)-1-Chloro-3-ethyl-4-methyl-3-heptene

EXERCISE 11-3

Name the following two alkenes.

(a) D, D / C=C / H_3C, H

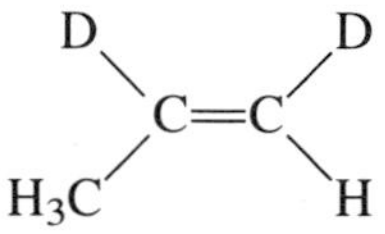

(b) F, OCH_3 / C=C / H_3C, CH_2CH_3

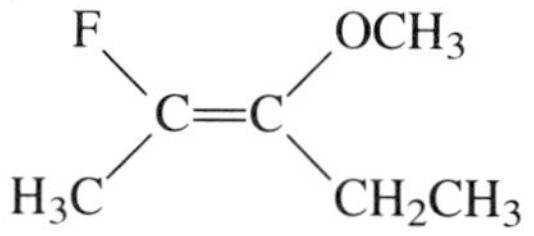

RULE 6. Give the hydroxy functional group precedence over the double bond in numbering a chain. Alcohols containing double bonds are named as **alkenols,** and the stem incorporating both functions is numbered to give the carbon bearing the OH group the lowest possible assignment. Note that the last *e* in alkene is dropped in the naming of alkenols.

$\overset{3}{C}H_2{=}\overset{2}{C}H\overset{1}{C}H_2OH$

2-Propen-1-ol

(Not 1-propen-3-ol)

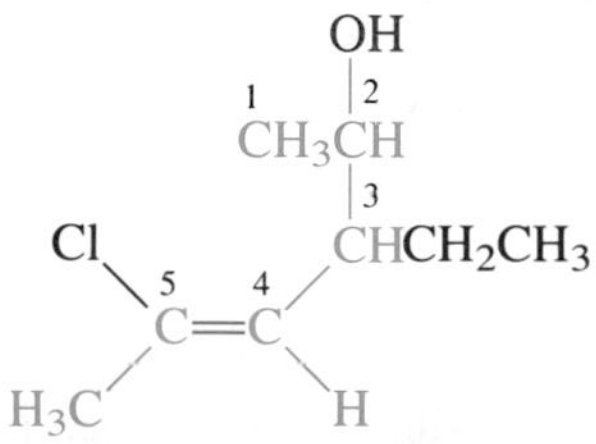

OH / $\overset{1}{C}H_3\overset{2}{C}H$ / $\overset{3}{C}HCH_2CH_3$ / Cl, H_3C — $\overset{5}{C}{=}\overset{4}{C}$ — H

(*Z*)-5-Chloro-3-ethyl-4-hexen-2-ol

(The two stereocenters are unspecified)

EXERCISE 11-4

Draw the structures of the following molecules. **(a)** *trans*-3-Penten-1-ol; **(b)** 3-cyclohexenol.

RULE 7. Substituents containing a double bond are named **alkenyl,** for example, ethenyl (common name, vinyl), 2-propenyl (allyl), and *cis*-1-propenyl.

$CH_2{=}CH{-}$ **Ethenyl (Vinyl)**

$CH_2{=}CH{-}CH_2{-}$ **2-Propenyl (Allyl)**

$\underset{H_3C}{\overset{H}{}}\!C{=}C\overset{H}{}$ ***cis*-1-Propenyl**

As usual, the numbering of a substituent chain begins at the point of attachment to the basic stem.

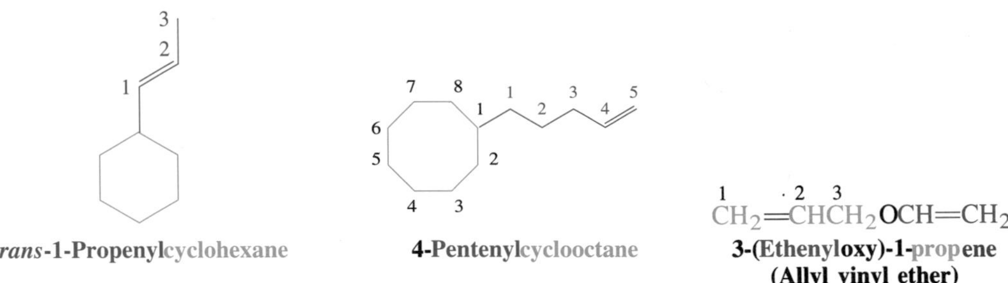

***trans*-1-Propenylcyclohexane** **4-Pentenylcyclooctane** **3-(Ethenyloxy)-1-propene (Allyl vinyl ether)**

EXERCISE 11-5

(a) Draw the structure of *trans*-1,2-diethenylcyclopropane. **(b)** Name the structure shown in the margin.

11-2 Structure and Bonding in Ethene: The Pi Bond

The carbon–carbon double bond in alkenes has special electronic and structural features. This section describes the hybridization of the carbon atoms in this functional group, the nature of its two bonds, defined as σ and π, and their relative strengths. We consider ethene, the simplest of the alkenes.

The double bond consists of sigma and pi components

In ethene both carbon atoms are sp^2 hybridized, as in the methyl radical (Section 3-2). Figure 11-1A summarizes the structure of $CH_3\cdot$. Three sp^2 hybrids overlap with the hydrogen 1*s* orbitals to form three σ bonds (Section 1-7). In addition, one singly occupied perpendicular *p* orbital extends above and below the plane of the atoms. Let us now see how two sp^2-hybridized carbons may combine to form a carbon–carbon double bond.

We begin by removing one hydrogen from each of two methyl radicals, leaving a singly occupied sp^2 hybrid on each. The carbon–carbon σ bond is formed by combining these orbitals. This leaves the *p* orbitals on each carbon in a parallel alignment, but close enough to overlap (Figure 11-1B). This second type of bonding interaction, called a **π bond,** is typical of the double bonds in alkenes.

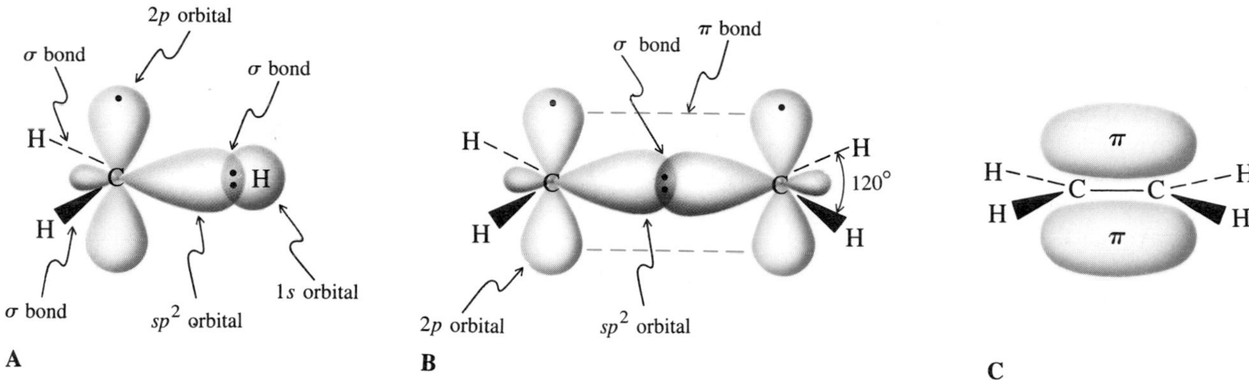

FIGURE 11-1 A molecular-orbital picture of (A) the methyl radical helps us understand (B) the double bond in ethene. The σ carbon–carbon bond is made by sp^2–sp^2 overlap. The pair of p orbitals perpendicular to the ethene molecular plane overlap to form the additional π bond. For clarity, this overlap is indicated in (B) by the dashed green line; the orbital lobes are shown artificially separated. Another way of presenting the π bond is depicted in (C), in which the "π electron cloud" is above and below the molecular plane.

The electrons in a π bond are delocalized over both carbons *above and below* the molecular plane, as indicated in Figure 11-1C.

The pi bond in ethene is relatively weak

As our orbital picture shows, the double bond is made up of two different types of bonds: a σ bond and a π bond. How much does each contribute to the total double-bond strength? We know from Section 1-7 that bonds are made by overlap of orbitals and that their relative strengths depend on the effectiveness of this overlap. Therefore, we can expect overlap in a σ bond to be considerably better than that in a π bond, because the sp^2 orbitals lie along the internuclear axis (Figure 11-1). Thus, a π bond should be weaker than a σ bond. This situation is illustrated in energy-level-interaction diagrams (Figures 11-2 and 11-3) analogous to those used to describe the bonding in the hydrogen molecule (Figures 1-11 and 1-12). Figure 11-4 summarizes our predictions of the relative energies of the molecular orbitals that make up the double bond in ethene.

Thermal isomerization allows us to measure the strength of the pi bond

How do these predictions of the π bond strength compare to experimental values? We can measure the energy required to interconvert the cis form of a substituted alkene—say, 1,2-dideuterioethene—with its trans isomer. In this process, called **thermal isomerization,** the two p orbitals responsible for the π bond are rotated 180°. At the midpoint of this rotation—90°—the π (but not the σ) bond has been broken (Figure 11-5). Thus, the activation energy for the reaction can be roughly equated with the π energy of the double bond.

Experimentally, thermal isomerization requires fairly high temperatures (400–500°C) to take place at measurable rates. The activation energy is 65 kcal mol^{-1}, a value usually assigned to the strength of the π bond. At temperatures below 300°C, most double bonds are configurationally stable; that is, cis stays cis and trans remains trans. The strength of the double bond as a whole in ethene—in

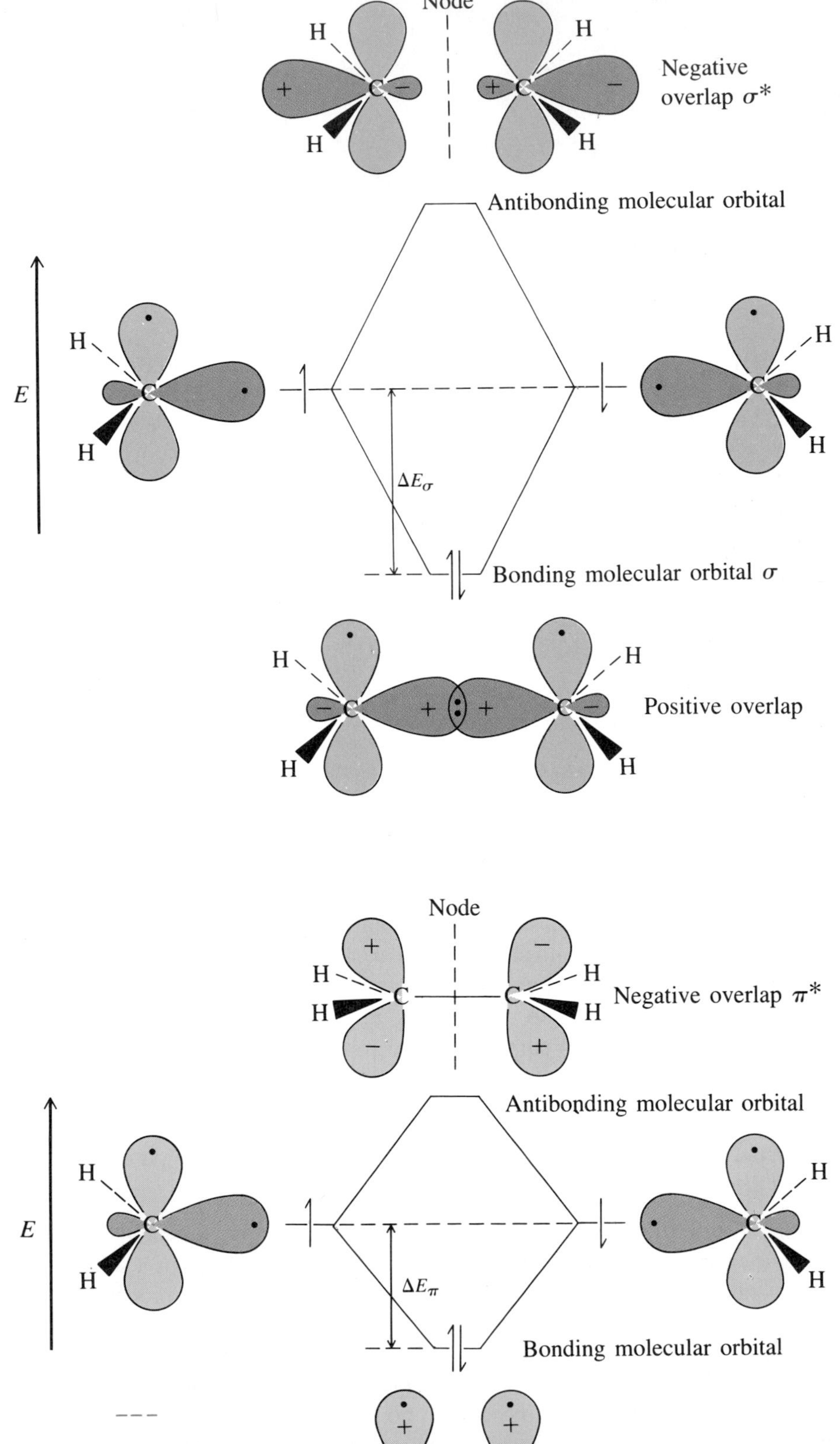

FIGURE 11-2 Overlap between two sp^2 hybrid orbitals (containing one electron each) determines the relative strength of the σ bond of ethene. In-phase interaction between regions of the wave function having the *same* sign reinforces bonding (compare in-phase overlap of waves, Figure 1-4B) and creates a *bonding molecular orbital*. [Recall: These signs do *not* refer to charges; the + designations are arbitrarily chosen (see Figure 1-11).] Both electrons end up occupying this orbital and have a high probability of being located near the internuclear axis. The orbital stabilization energy, ΔE_σ, corresponds to the σ bond strength. The out-of-phase interaction, between regions of *opposite* sign (compare Figure 1-4C), results in an unfilled *antibonding molecular orbital* (designated σ^*) with a node.

FIGURE 11-3 Compare this picture of the formation of the π bond in ethene with Figure 11-2. In-phase interaction between two parallel p orbitals (containing one electron each) results in positive overlap and a filled bonding π orbital. The representation of this orbital reflects the probability of finding the electrons between the carbons above and below the molecular plane. Because π overlap is less effective than σ, the stabilization energy, ΔE_π, is smaller than ΔE_σ. The π bond is therefore weaker than the σ bond.

other words, the energy required for dissociation into two methylene fragments—is estimated to be 173 kcal mol^{-1}. Consequently, the σ bond in this molecule amounts to about 108 kcal mol^{-1} (Figure 11-6). The bond between an alkyl substituent or a hydrogen atom and the alkenyl carbon is also strong in comparison with the analogous bonds in alkanes (Table 3-2). To a large extent, this effect is due to the improved overlap between the relatively compact sp^2 hybrids and the substituent orbitals. As a consequence, radical reactions of alkenes do not occur by abstraction of the strongly bound alkenyl (vinyl) hydrogen. In fact, most of the chemistry of the double bond is characterized by the reactivity of the weaker bond: the π bond (Chapter 12).

E

σ^* orbital

π^* orbital

π orbital

σ orbital

Antibonding orbitals: π^*, σ^*
Bonding orbitals: π, σ

FIGURE 11-4 Energy ordering of the molecular orbitals making up the double bond. The four electrons occupy only bonding orbitals.

Hybridization explains the molecular structure of ethene

The use of sp^2 hybrid orbitals for forming the double bond has significant structural consequences. Ethene is planar, with two trigonal carbon atoms and bond angles close to 120° (Figure 11-7). Therefore, the double bond exhibits no three-dimensional stereochemistry but only that related to cis-trans isomerism.

In summary, the characteristic hybridization scheme for the double bond of an alkene accounts for its physical and electronic features. This hybridization also explains the formation of a strong σ, as well as a weaker π, bond; stable cis and trans isomers; and the strength of the alkenyl–substituent bond. It gives rise to a planar double bond, incorporating trigonal carbon atoms.

11-3 Physical Properties of Alkenes

Although the presence of the carbon–carbon double bond alters many of the physical properties of the alkenes relative to those of the alkanes, the boiling points of alkenes lie very close to those of the corresponding alkanes and reflect their similarities as hydrocarbons. Ethene (b.p. −103.7°C), propene (b.p. −47.4°C), and the butenes are gases at room temperature. The pentenes boil just above room temperature (b.p. 30–40°C). Other physical properties, however, reflect the influence of the functional group. Let us consider polarity and acidity.

Depending on their structure, alkenes may exhibit weak dipolar character. Why? Bonds between alkyl groups and an alkenyl carbon are polarized in the direction of the sp^2 hybridized atom, because the degree of s character in an sp^2

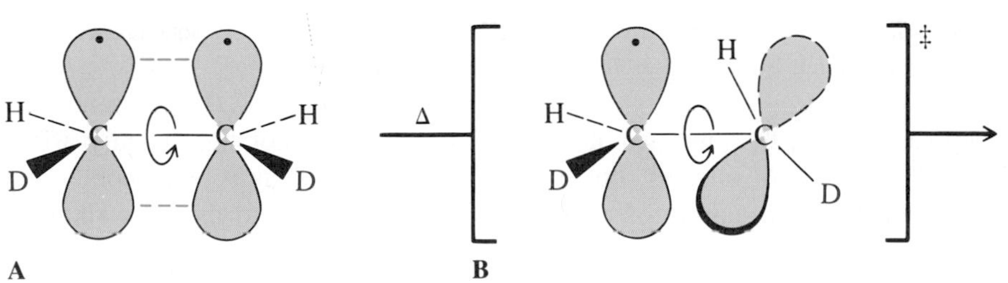

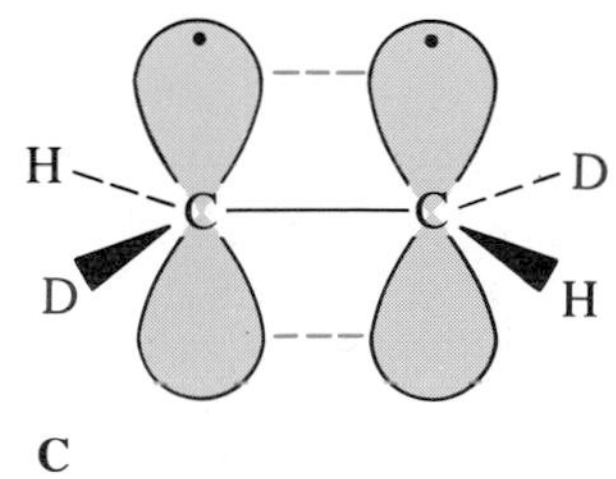

FIGURE 11-5 Thermal isomerization of *cis*-dideuterioethene to the trans isomer requires breaking the π bond. The reaction proceeds from starting material (A) through rotation around the C–C bond until it reaches the point of highest energy, the transition state (B). At this stage, the two p orbitals used to construct the π bond are perpendicular to each other. Further rotation in the same direction results in a product in which the two deuterium atoms are trans (C).

The symbol ‡ in Figure 11-5 denotes a transition state.

hybrid orbital is greater than that in an sp^3. Electrons in orbitals with increased s character are held closer to the nucleus than those in orbitals containing more p character. This effect makes the sp^2 carbon relatively electron withdrawing (although much less so than electronegative atoms such as O and Cl) and creates a dipole along the substituent–alkenyl carbon bond.

Often, particularly in cis-disubstituted alkenes, a net molecular dipole is the result. In trans-disubstituted alkenes, such dipoles are small, because the polarizations of individual local bonds are opposed and tend to cancel each other.

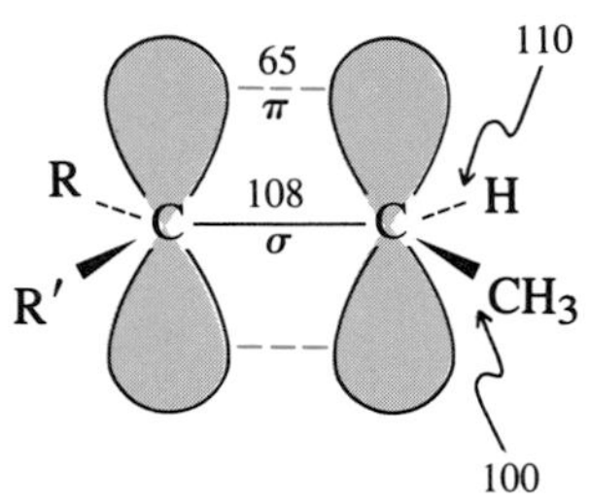

FIGURE 11-6 Approximate bond strengths in an alkene (in kcal mol^{-1}). Note the relative weakness of the π bond.

Polarization in Alkenes

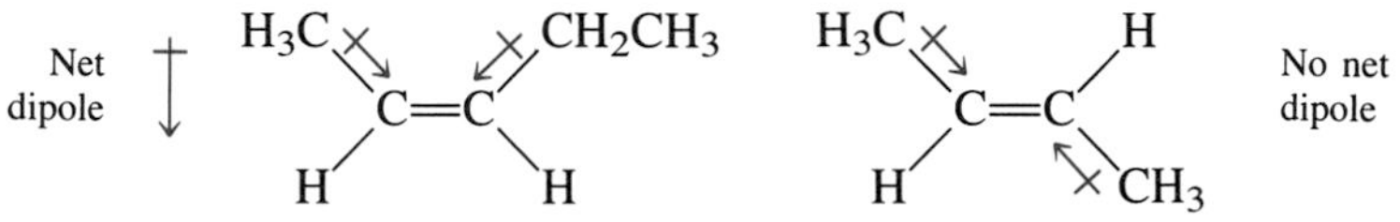

Net dipole — cis: H_3C, CH_2CH_3 / $C{=}C$ / H, H

No net dipole — trans: H_3C, H / $C{=}C$ / H, CH_3

Another consequence of the electron-attracting character of the sp^2 carbon is the increased acidity of the alkenyl hydrogen. Whereas ethane has an approximate pK_a of 50, ethene is somewhat more acidic with a pK_a of 44. Even so, ethene is a very poor source of protons compared with other compounds, such as the carboxylic acids or alcohols.

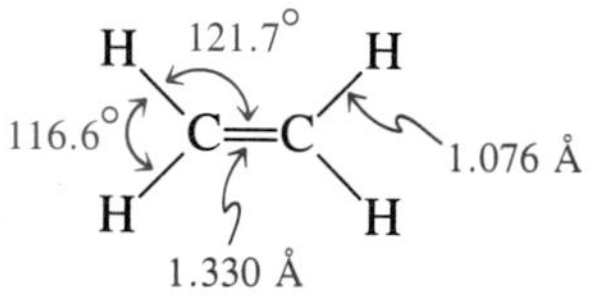

H 121.7° H, 116.6°, C=C, H H, 1.076 Å, 1.330 Å

FIGURE 11-7 Molecular structure of ethene.

Acidity of the Ethenyl Hydrogen

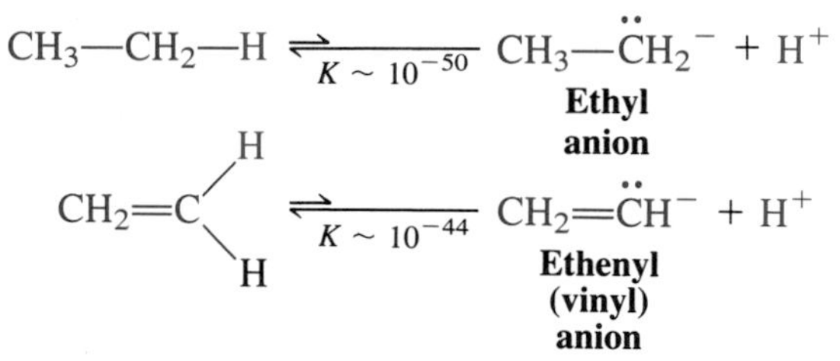

$$CH_3{-}CH_2{-}H \underset{K\,\sim\,10^{-50}}{\rightleftharpoons} CH_3{-}\ddot{C}H_2^{-} + H^+$$

Ethyl anion

$$CH_2{=}C\genfrac{}{}{0pt}{}{H}{H} \underset{K\,\sim\,10^{-44}}{\rightleftharpoons} CH_2{=}\ddot{C}H^{-} + H^+$$

Ethenyl (vinyl) anion

EXERCISE 11-6

Ethenyllithium (vinyllithium) is not generally prepared by direct deprotonation of ethene but rather from chloroethene (vinyl chloride) (Section 8-7).

$$CH_2{=}CHCl + 2\,Li \xrightarrow{(CH_3CH_2)_2O} \underset{60\%}{CH_2{=}CHLi} + LiCl$$

On treatment of ethenyllithium with propanone (acetone) followed by aqueous work-up, a colorless liquid is obtained in 74% yield. Propose a structure.

To summarize, the presence of the double bond does not significantly affect the boiling points of alkenes, but it does give rise to weakly polar bonds and somewhat increased acidity of alkenyl hydrogens relative to those in alkanes.

11-4 Nuclear Magnetic Resonance of Alkenes

The double bond exerts characteristic effects on the ^1H and ^{13}C signals of alkenes (see Tables 10-2 and 10-6). We shall see how to make use of this information in structural assignments.

The pi electrons exert a deshielding effect on alkenyl hydrogens

Figure 11-8 shows the ^{1}H NMR spectrum of *trans*-2,2,5,5-tetramethyl-3-hexene. Only two signals are observed, one for the 18 equivalent methyl hydrogens and one for the 2 alkenyl protons. The absorptions appear as singlets because the methyl hydrogens are too far away from the alkenyl hydrogens to produce detectable coupling. The low-field resonance of the latter ($\delta = 5.30$ ppm) is typical of hydrogen atoms bound to alkenyl carbons. Terminal alkenyl hydrogens ($RR'C{=}CH_2$) appear at $\delta = 4.6$–5.0 ppm, their internal counterparts ($RCH{=}CHR'$) at $\delta = 5.2$–5.7 ppm.

Why is deshielding so pronounced? Although the electron-withdrawing character of the sp^2-hybridized carbon is partly responsible, another phenomenon is more important: *the movement of the electrons in the π bond*. When subjected to an external magnetic field perpendicular to the double-bond axis, these electrons enter into a circular motion. This motion induces a local magnetic field that *reinforces* the external field at the edge of the double bond (Figure 11-9). As a consequence, less external magnetic field strength is required to bring the alkenyl hydrogens into resonance: They are strongly deshielded (Section 10-4).

EXERCISE 11-7

The hydrogens on methyl groups attached to alkenyl carbons resonate at about $\delta = 1.6$ ppm (see Table 10-2). Explain the deshielding of these hydrogens relative to hydrogens on methyl groups in alkanes. (Hint: Try to apply the principles in Figure 11-9.)

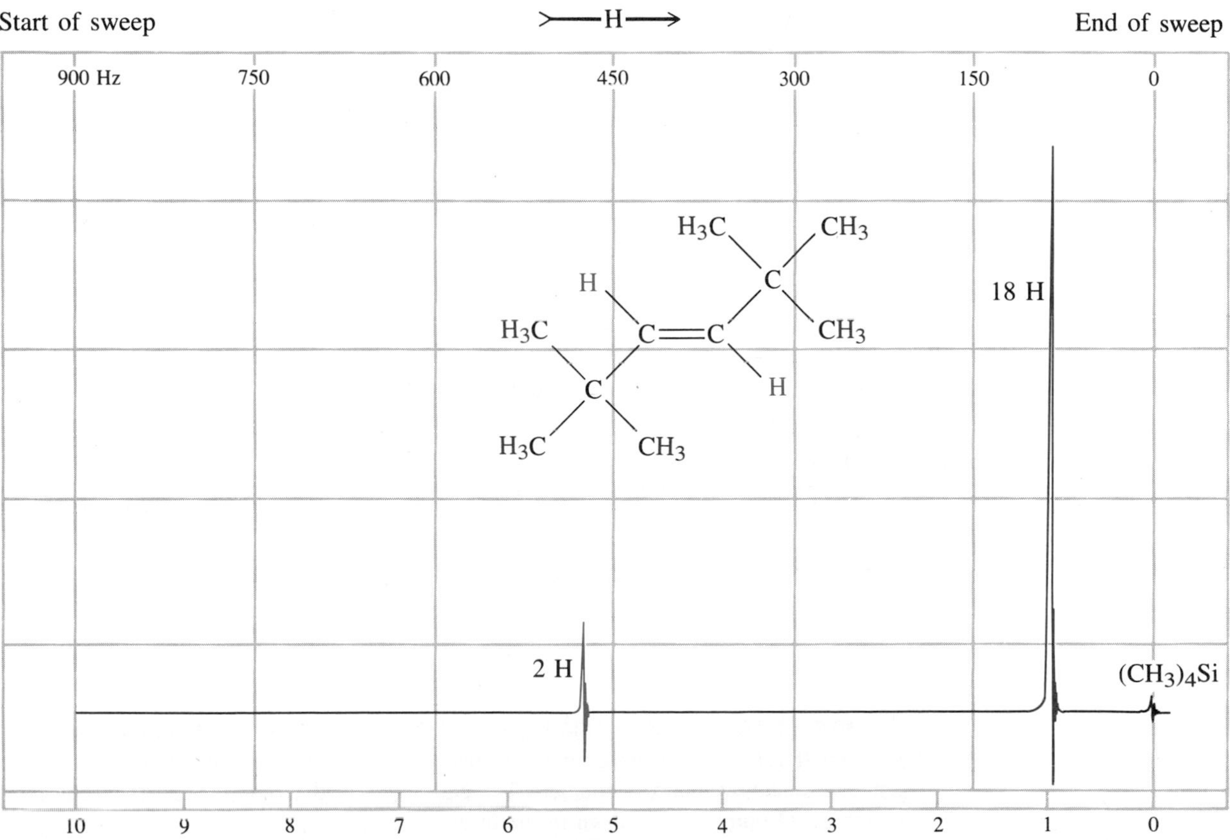

FIGURE 11-8 90-MHz ^{1}H NMR spectrum of *trans*-2,2,5,5-tetramethyl-3-hexene in CCl_4, illustrating the deshielding effect of the π bond in alkenes. It reveals two sharp singlets for two sets of hydrogens: the 18 methyl hydrogens at $\delta =$ 0.97 ppm and 2 highly deshielded alkenyl protons at $\delta = 5.30$ ppm.

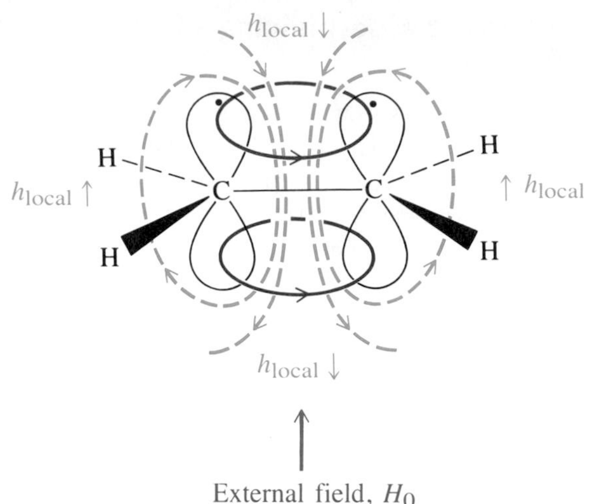

FIGURE 11-9 Movement of electrons in the π bond causes pronounced deshielding of alkenyl hydrogens. An external field, H_0, induces a circular motion of the π electrons (shown in red) above and below the plane of a double bond. This motion in turn induces a local field (shown in green) that opposes H_0 at the center of the double bond but reinforces it in the regions occupied by the alkenyl hydrogens.

Cis coupling through the double bond is different from trans

When a double bond is not symmetrically substituted, the alkenyl hydrogens will be nonequivalent, a situation leading to observable spin–spin coupling such as that shown in the spectra of *cis*- and *trans*-3-chloropropenoic acid (Figure 11-10). Note that the coupling constant for the hydrogens situated cis ($J = 9$ Hz) is different from that for the hydrogens arranged trans ($J = 14$ Hz).

Table 11-1 gives the magnitude of the various possible couplings around a double bond. Although the range of J_{cis} overlaps that of J_{trans}, within a set of isomers the first is always smaller than the second. In this way cis and trans isomers can be readily distinguished.

J_{cis} and J_{trans} are called **vicinal coupling constants.** Coupling between nonequivalent terminal hydrogens is small and labeled **geminal coupling** (Table 11-1). Coupling to neighboring alkyl hydrogens (**allylic,** see Section 11-1) and across the double bond (**1,4** or **long-range**) is also possible, sometimes giving rise to complicated spectral patterns. Thus, the simple rule devised for saturated systems, discounting coupling between hydrogens farther than two intervening atoms apart, does not hold for alkenes.

Further coupling leads to more complex spectra

The spectra of 3,3-dimethyl-1-butene and 1-pentene illustrate the potential complexity of the coupling patterns (Figure 11-11). In both spectra, the alkenyl hy-

FIGURE 11-10 (opposite) 90-MHz ^1H NMR spectra of (A) *cis*-3-chloropropenoic acid and (B) the corresponding trans isomer, each in CCl_4. The two alkenyl hydrogens are nonequivalent and coupled. The carboxylic acid proton ($-CO_2H$) resonates at $\delta = 12.35$ ppm and is shown in the inset.

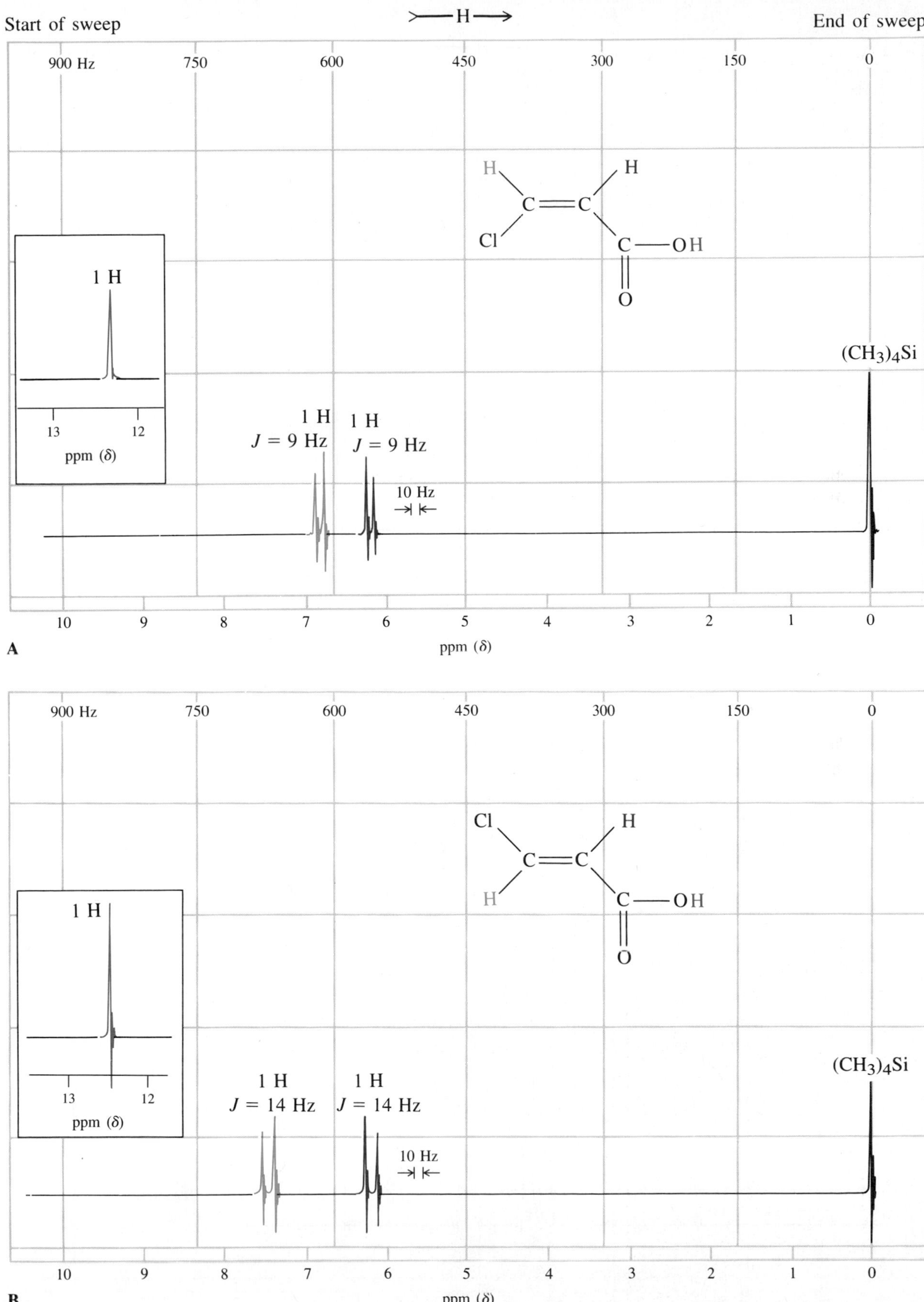

Start of sweep
H
End of sweep
900 Hz
750
600
450
300
150
0
H
H
C=C
Cl
C—OH
O
1 H
13
12
ppm (δ)
(CH3)4Si
1 H
J = 9 Hz
1 H
J = 9 Hz
10 Hz
10
9
8
7
6
5
4
3
2
1
0
A
ppm (δ)
900 Hz
750
600
450
300
150
0
Cl
H
C=C
H
C—OH
O
1 H
13
12
ppm (δ)
(CH3)4Si
1 H
J = 14 Hz
1 H
J = 14 Hz
10 Hz
10
9
8
7
6
5
4
3
2
1
0
B
ppm (δ)

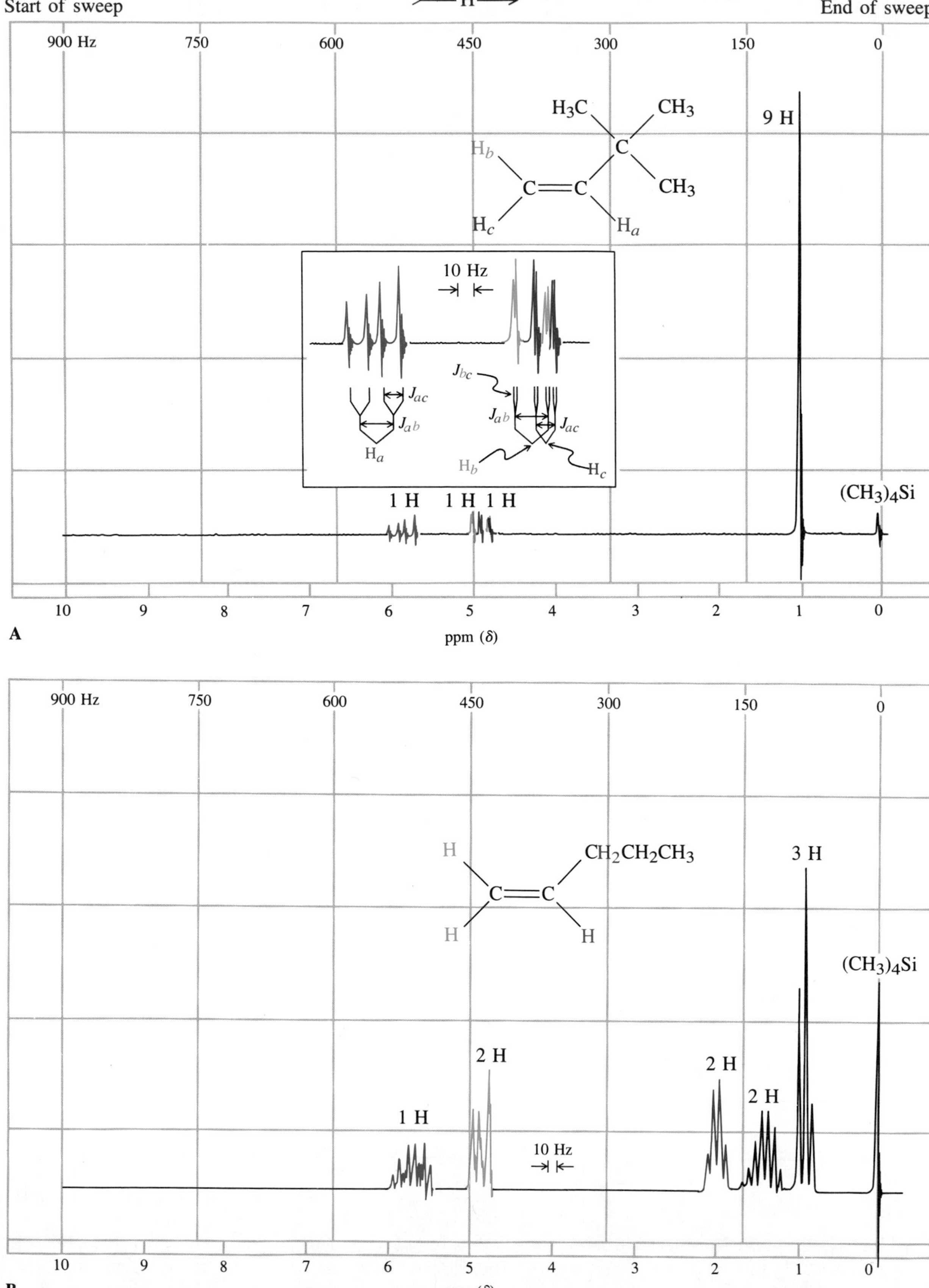
Start of sweep
H
End of sweep
900 Hz
750
600
450
300
150
0
H_3C
CH_3
CH_3
C
H_b
H_c
H_a
C=C
9 H
10 Hz
J_{ac}
J_{ab}
H_a
J_{bc}
J_{ab}
J_{ac}
H_b
H_c
1 H
1 H 1 H
$(CH_3)_4Si$
10
9
8
7
6
5
4
3
2
1
0
A
ppm (δ)
900 Hz
750
600
450
300
150
0
H
H
C=C
$CH_2CH_2CH_3$
H
3 H
$(CH_3)_4Si$
2 H
2 H
2 H
1 H
10 Hz
10
9
8
7
6
5
4
3
2
1
0
B
ppm (δ)

TABLE 11-1 Coupling Constants Around a Double Bond

Type of coupling	Name	J (Hz) Range	Typical
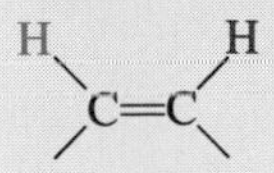 H(C=C)H cis	Vicinal, cis	6–14	10
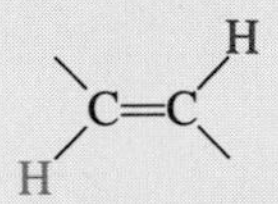 H(C=C)H trans	Vicinal, trans	11–18	16
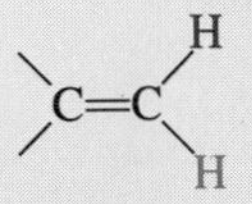 C=CH₂	Geminal	0–3	2
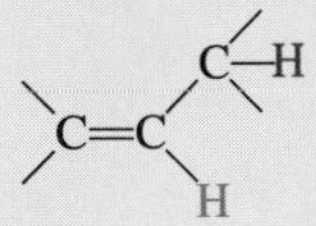 C=C(H)–C–H	None	4–10	6
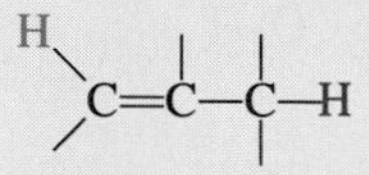 H–C=C–C–H	Allylic, (1,3)-cis or -trans	0.5–3	2
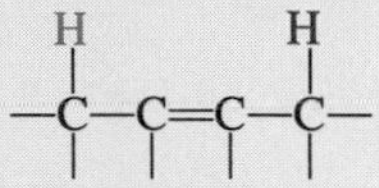 H–C–C=C–C–H	(1,4)- or long-range	0–1.6	1

drogens appear as complex multiplets. In 3,3-dimethyl-1-butene, H_a located on the more highly substituted carbon atom resonates at lower field ($\delta = 5.81$ ppm) and in the form of a double doublet with two relatively large coupling constants ($J_{ab} = 18$ Hz, $J_{ac} = 10.5$ Hz). Both hydrogens H_b and H_c also absorb as double doublets because of their respective coupling to H_a and their mutual coupling ($J_{bc} = 1.5$ Hz). Because of the small chemical-shift difference between them, their signals overlap, but, as shown in the inset (twofold expansion), the coupling pattern can be readily analyzed and both hydrogens assigned. In the spectrum of 1-pentene, additional coupling due to the attached alkyl group (see Table 11-1) creates a situation that is too complex for a first-order analysis. However, the two sets of alkenyl hydrogens (terminal and internal) are clearly differentiated. In addition, the electron-withdrawing effect of the sp^2 carbon causes the directly attached (allylic) CH_2 group to be slightly deshielded, its multiplet appearing at lower field ($\delta = 1.94$ ppm) than that of the other alkyl absorptions.

FIGURE 11-11 (opposite) 90-MHz ^{1}H NMR spectra of (A) 3,3-dimethyl-1-butene and (B) 1-pentene, each in CCl_4.

EXERCISE 11-8

Ethyl 2-butenoate (ethyl crotonate), $CH_3CH{=}CHCO_2CH_2CH_3$, in CCl_4 has the following ^{1}H NMR spectrum: $\delta = 6.95$ (dq, $J = 16$, 6.8 Hz, 1 H), 5.81 (dq, $J = 16$, 1.7 Hz, 1 H), 4.13 (q, $J = 7$ Hz, 2 H), 1.88 (dd, $J = 6.8$, 1.7 Hz, 3 H), and 1.24 (t, $J = 7$ Hz, 3 H) ppm; dd denotes a doublet of doublets, dq a doublet of quartets. Assign the various hydrogens and indicate whether the double bond is substituted cis or trans (consult Table 11-1).

BOX 11-1 Prostaglandins

The prostaglandins (PGs) are a family of extremely potent hormonelike substances with many biological functions, including muscle stimulation, inhibition of platelet aggregation, lowering of blood pressure, enhancement of inflammatory reactions, and induction of labor in childbirth. Indeed, the anti-inflammatory effects of aspirin are due to its ability to suppress prostaglandin biosynthesis. Illustrated here are three members of the PG family, of which the most biologically active is that labeled PGE_2.

The ^{1}H NMR spectra of these are quite complicated, with many overlapping absorptions. In contrast, ^{13}C NMR permits rapid identification of PG derivatives, merely by counting the peaks in three chemical shift ranges. For example, PGE_2 is readily distinguished by the presence of two signals near $\delta = 70$ ppm for two alcohol carbons, four alkene ^{13}C resonances between $\delta = 125$ and 140 ppm, and two carbonyls above $\delta = 170$ ppm.

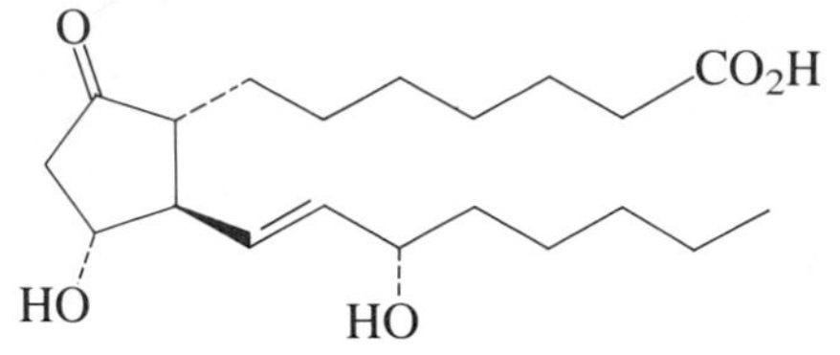

O, CO_2H, HO, HO

PGE_1

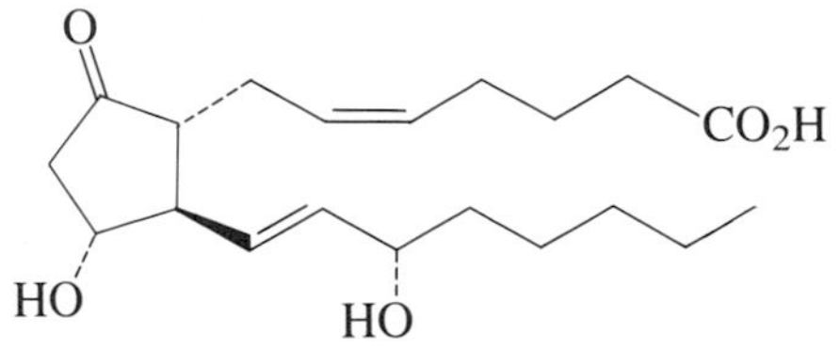

O, CO_2H, HO, HO

PGE_2

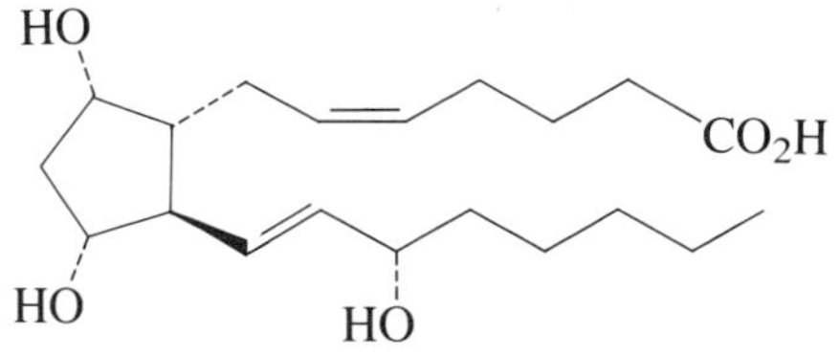

HO, CO_2H, HO, HO

$PGF_{2\alpha}$

Alkenyl carbons are also deshielded in the ^{13}C NMR

The carbon NMR absorptions of the alkenes are also highly revealing. Relative to alkanes, the corresponding alkenyl carbons (with similar substituents) absorb at about 100-ppm lower field (see Table 10-6). Two examples are shown in Table 11-2, in which the carbon chemical shifts of an alkene are compared with those of its saturated counterpart. Recall that, in broad-band decoupled ^{13}C NMR spectroscopy, all magnetically unique carbons absorb as sharp single lines (Section 10-9). It is therefore very easy to determine the presence of sp^2 carbons by this method.

In summary, NMR is highly effective in establishing the presence of double bonds in organic molecules. Alkenyl hydrogens and carbons are strongly deshielded. The order of coupling is $J_{gem} < J_{cis} < J_{trans}$. In addition, various coupling constants are typical for allylic substituents.

11-5 Infrared Spectroscopy

Another method of identifying carbon–carbon double bonds and other functional groups is **infrared (IR) spectroscopy,** which measures the vibrational excitation of atoms around the bonds that connect them. The position of the absorption lines depends on the types of functional groups present, and the IR spectrum as a whole is a unique ''fingerprint'' of the entire molecule.

Absorption of infrared light causes molecular vibrations

In nuclear magnetic resonance, radio waves cause nuclear spins to change their alignment with the magnetic field ($\Delta E \sim 10^{-6}$ kcal mol^{-1}; Section 10-2). Ultraviolet–visible spectroscopy is performed with higher energy light, which induces electronic transitions ($\Delta E \sim$ 40–300 kcal mol^{-1}; Section 14-11). At energies slightly lower than those of visible radiation, molecules absorb light by undergoing a process called **vibrational excitation.** This portion of the electromagnetic spectrum is the infrared region (see Figure 10-2). The intermediate range, or *middle infrared,* is most useful to the organic chemist. IR absorption bands are described by either the wavelength, λ, of the absorbed light in micrometers (10^{-6} m; $\lambda \sim$ 2.5–16.7 μm; see Figure 10-2) or its reciprocal value, called wavenumber, $\tilde{\nu}$ (in units of cm^{-1}; $\tilde{\nu} = 1/\lambda$). Thus, a typical infrared spectrum ranges from 600 to 4000 cm^{-1}, and the energy changes associated with absorption of this radiation range from 1 to 10 kcal mol^{-1}.

Figure 10-3 describes a simple IR spectrometer. Modern systems use sophisticated rapid-scan techniques and are linked with computers. This equipment allows for data storage, spectra manipulation, and computer library searches, so that unknown compounds can be matched with stored spectra.

Vibrational excitation can be envisioned simply by thinking of two atoms A and B linked by a bond as two weights on a spring that stretches and compresses at a certain frequency, ν (Figure 11-12). In this picture, the frequency of the vibrations between two atoms depends both on the strength of the bond between them and on their atomic weights. In fact, it is governed by Hooke's* law, just like the motion of a spring.

Hooke's Law and Vibrational Excitation

$$\tilde{\nu} = k\sqrt{f\frac{(m_1 + m_2)}{m_1 m_2}}$$

$\tilde{\nu}$ = vibrational frequency in wavenumbers (cm^{-1})
k = constant
f = force constant, indicating the strength of the spring (bond)
m_1, m_2 = masses of attached weights (atoms)

This equation might lead us to expect every individual bond in a molecule to show one specific absorption band in the infrared spectrum. However, in practice, an interpretation of the entire infrared spectrum is considerably more complex, because molecules that absorb infrared light undergo not only stretching but also various bending motions (Figure 11-13), and combinations of the two as well. The bending vibrations are mostly of weaker intensity, they overlap with other absorptions, and they may show complicated patterns. The practicing organic chemist can, however, find some good use for IR spectroscopy for two reasons: The vibrational bands of many functional groups appear at characteristic wavenumbers, and the entire infrared spectrum may be used as a unique fingerprint of a compound.

Functional groups have typical infrared absorptions

Table 11-3 lists the characteristic stretching wavenumbers for the bonds (shown in red) of some common organic structural units. We shall show the IR spectra

*Professor Robert Hooke, 1635–1703, physicist at Gresham College, London.

TABLE 11-2 Comparison of ^{13}C NMR Absorptions of Alkenes with the Corresponding Alkane Carbon Chemical Shifts (in ppm)

$(CH_3)_2C{=}C(CH_3)_2$: C=C 122.8; CH_3 18.9

$CH_3CH{=}CHCH_2CH_3$ (H, H cis): C 123.7, C 132.7; CH_3 12.3; CH_2 20.5; CH_3 14.0

Alkenes

$(CH_3)_2CH{-}CH(CH_3)_2$: CH 34.0; CH_3 19.2

$CH_3CH_2CH_2CH_2CH_3$: 13.5, 22.2, 34.1

Alkanes

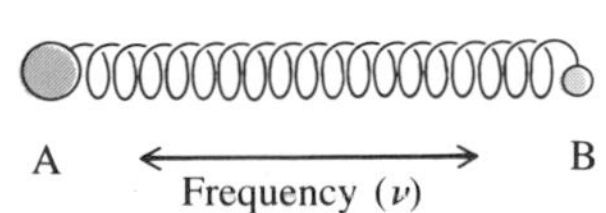

FIGURE 11-12 Two unequal weights on an oscillating ("vibrating") spring: a model for vibrational excitation of a bond.

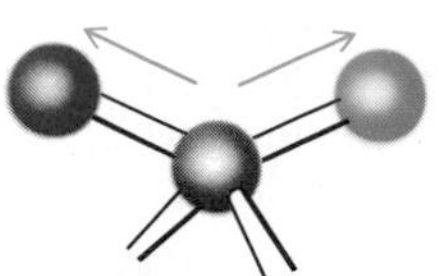

Symmetric stretching (both outside atoms move away from or toward the center)

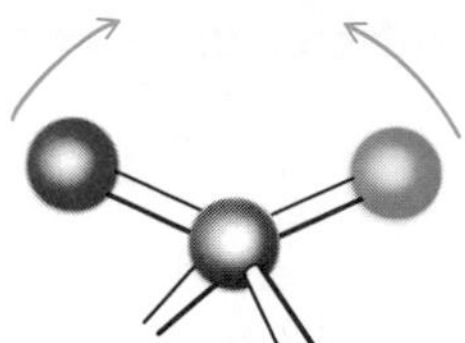

Symmetric bending vibration in a plane (scissoring)

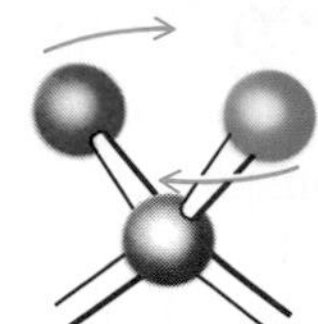

Symmetric bending vibration out of a plane (twisting)

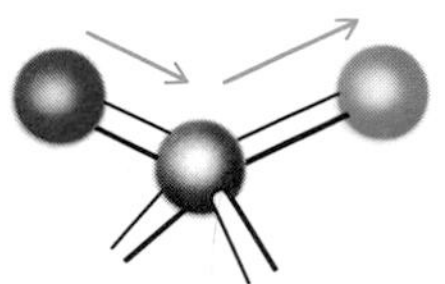

Asymmetric stretching (as one atom moves toward the center, the other moves away)

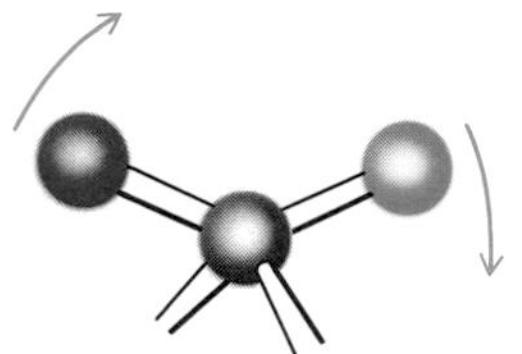

Asymmetric bending vibration in a plane (rocking)

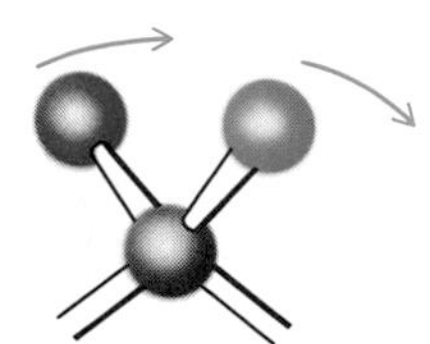

Asymmetric bending vibration out of a plane (wagging)

FIGURE 11-13 Various vibrational modes around tetrahedral carbon. The motions are labeled symmetric and asymmetric stretching or bending, scissoring, rocking, twisting, and wagging.

typical of new functional groups when we introduce each of the corresponding compound classes in subsequent chapters.

Figures 11-14 and 11-15 show the IR spectra of pentane and hexane. Although these spectra have similar main features, their fine structures are different. These differences become clearer at higher recorder sensitivity, particularly in the range between 600 and 1500 cm^{-1}, called the **fingerprint region.** The typical C–H stretching absorptions for the alkanes are seen in the range from 2840 to 3000 cm^{-1}. Three other bands, due to bending motions, stand out at about 1460, 1380, and 730 cm^{-1}. All saturated hydrocarbons (including alkyl-substituted cycloalkanes) show similar absorptions.

TABLE 11-3 Characteristic Infrared Stretching Wavenumber Ranges of Organic Molecules

Bond or functional group	$\tilde{\nu}$ (cm^{-1})	Bond or functional group	$\tilde{\nu}$ (cm^{-1})
RO—H (alcohols)	3200–3650	RC≡N (nitriles)	2220–2260
RC(=O)O—H (carboxylic acids)	2500–3300	RC(=O)H, RC(=O)R′ (aldehydes, ketones)	1690–1750
R_2N—H (amines)	3250–3500	RC(=O)OR′ (esters)	1735–1750
RC≡C—H (alkynes)	3260–3330	RC(=O)OH (carboxylic acids)	1710–1760
>C=C<H (alkenes)	3050–3150	>C=C< (alkenes)	1620–1680
—C—H (alkanes)	2840–3000	RC—OR′ (alcohols, ethers)	1000–1260
RC≡CH (alkynes)	2100–2260		

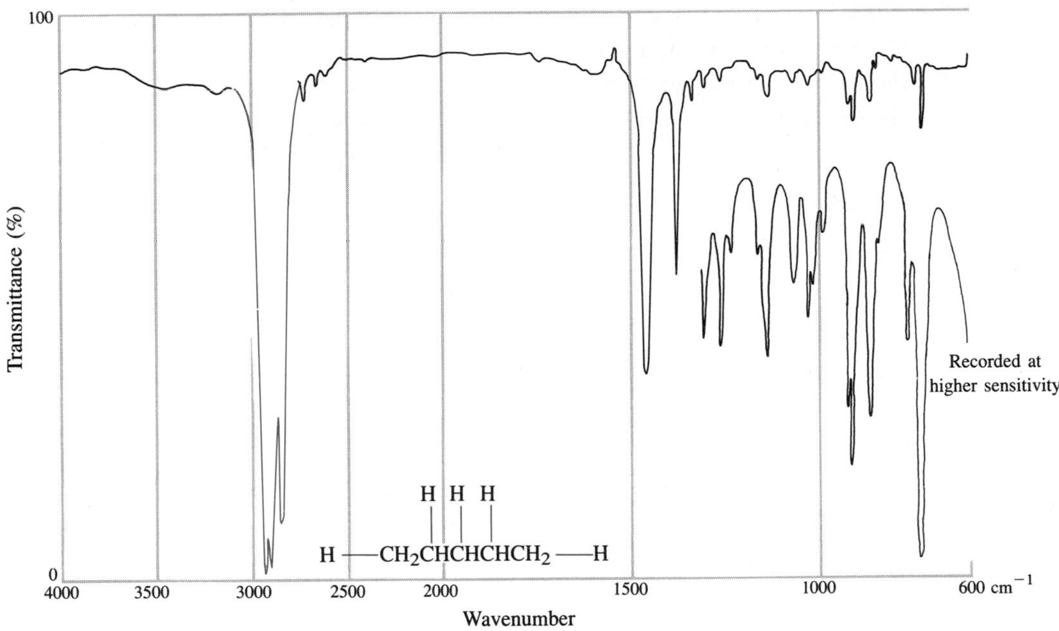

FIGURE 11-14 IR spectrum of pentane. Note the format: Wavenumber is plotted (decreasing from left to right) against percentage of transmittance. A 100% transmittance means *no* absorption; therefore "peaks" in an IR spectrum point *down*. The spectrum shows absorbances at $\tilde{\nu}_{\text{C–H stretch}} = 2960$, 2930, and 2870 cm^{-1}; $\tilde{\nu}_{\text{C–H bend}} = 1460$, 1380, and 730 cm^{-1}. The region between 600 and 1300 cm^{-1} is also shown recorded at higher sensitivity, revealing details of the pattern in the fingerprint region.

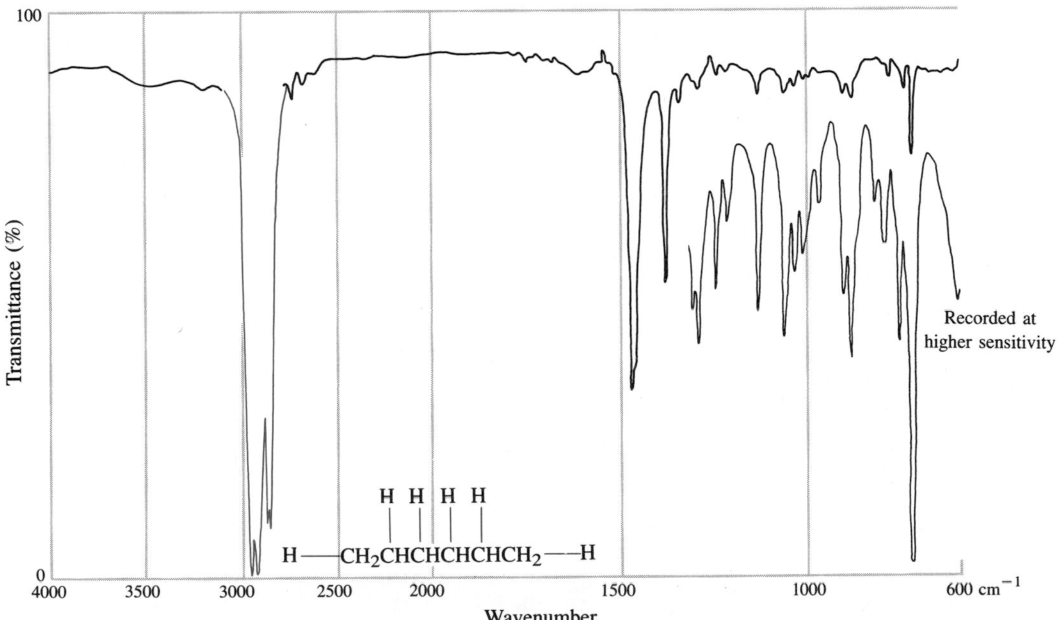

FIGURE 11-15 IR spectrum of hexane. Comparison with that of pentane (Figure 11-14) shows that the location and appearance of the major bands are very similar, but the two fingerprint regions show significant differences at higher recorder sensitivity.

Figure 11-16 shows the IR spectrum of 1-hexene. A characteristic feature of alkenes when compared with alkanes is the stronger C_{sp^2}–H bond, which should therefore have a higher energy peak in the IR spectrum. Indeed, as the figure shows, there is a sharp spike at 3080 cm^{-1} due to this stretching mode, at

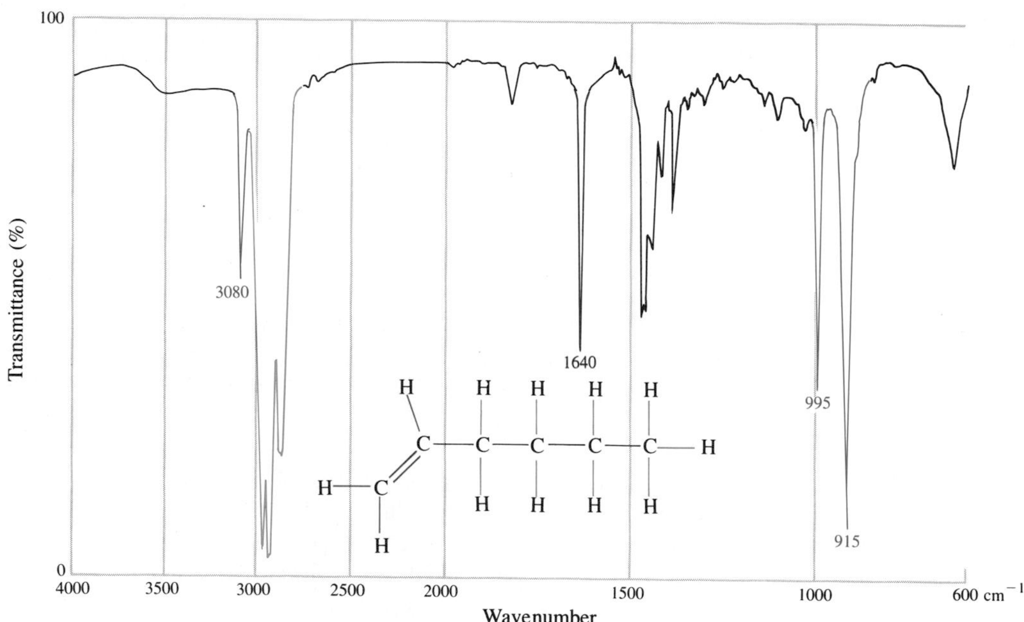

FIGURE 11-16 IR spectrum of 1-hexene: $\tilde{\nu}_{C_{sp^2}\text{–H stretch}} = 3080\ cm^{-1}$; $\tilde{\nu}_{C=C\text{ stretch}} = 1640\ cm^{-1}$; $\tilde{\nu}_{C_{sp^2}\text{–H bend}} = 995$ and $915\ cm^{-1}$.

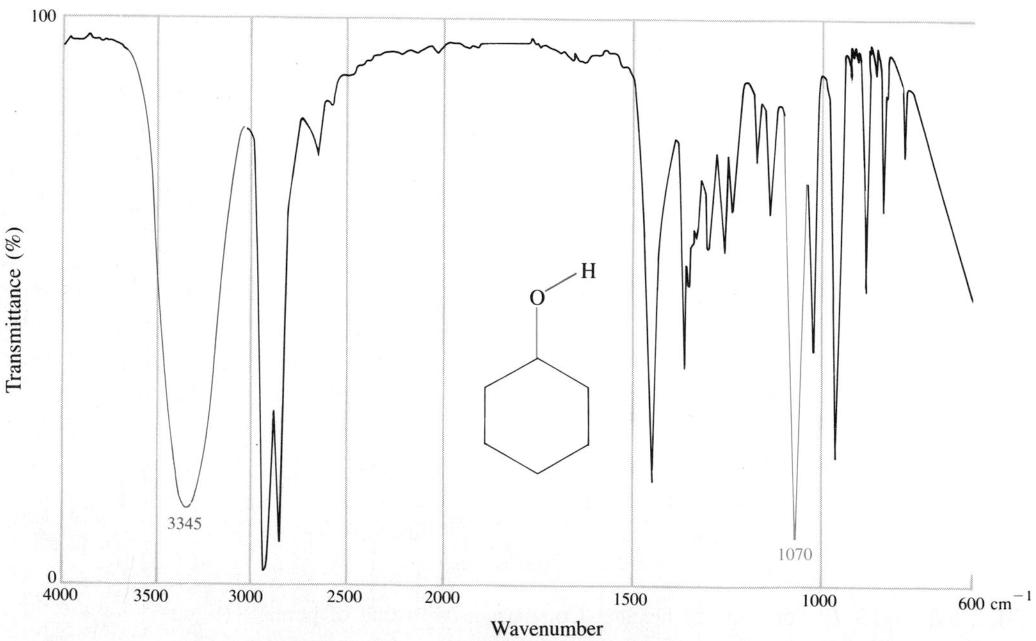

FIGURE 11-17 IR spectrum of cyclohexanol: $\tilde{\nu}_{O\text{–H stretch}} = 3345\ cm^{-1}$; $\tilde{\nu}_{C\text{–O}} = 1070\ cm^{-1}$. Note the broad O–H peak.

slightly higher wavenumber than the remainder of the C–H stretching absorptions. According to Table 11-3, the C=C stretching band should appear between about 1620 and 1680 cm^{-1}. Figure 11-16 shows a relatively strong and sharp band at 1640 cm^{-1} assigned to this vibration. The other strong peaks are the result of bending motions. For example, the two signals at 915 and 995 cm^{-1} are typical of a terminal alkene.

Two other strong bending modes may be used as a diagnostic tool for the substitution pattern in alkenes. One results in a single band at 890 cm^{-1} and is characteristic of 1,1-dialkylethenes; the other gives a sharp peak at 970 cm^{-1} and is produced by the C_{sp^2}–H bending mode of a trans double bond. The presence or absence of such bands is often corroborating evidence for the presence of specifically substituted double bonds. This measurement, in conjunction with NMR (Section 11-4), allows for fairly certain structural assignments.

The O–H stretching absorption is the most characteristic band in the IR spectra of alcohols, appearing as a strong, broad peak over a fairly wide range (3200–3650 cm^{-1}, Figure 11-17). The broadness of this peak is due to hydrogen bonding to other alcohol molecules or to water. Dry alcohols in dilute solution exhibit sharper bands in a narrower range (3620–3650 cm^{-1}).

Approximate Infrared Frequencies of Strong Bending Modes for Alkenes

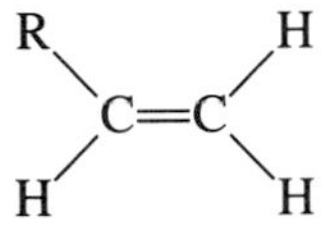

RHC=CH₂ (R, H / H, H): 915, 995 cm^{-1}

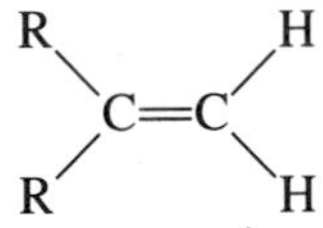

$R_2C=CH_2$ (R, H / R, H): 890 cm^{-1}

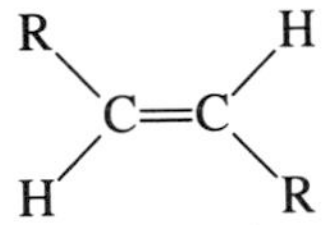

trans RHC=CHR (R, H / H, R): 970 cm^{-1}

EXERCISE 11-9

Three alkenes with the formula C_4H_8 exhibit the following IR absorptions: alkene A, 964 cm^{-1}; alkene B, 908 and 986 cm^{-1}; alkene C, 890 cm^{-1}. Assign structures to each.

BOX 11-2 The Garlic Story: Infrared Spectroscopy in Food Chemistry

You may have noticed that Table 11-3 omits infrared data for some functional groups, such as haloalkanes. This material is omitted because the IR bands for the stretching motions of C–X bonds lie in the fingerprint region of the spectrum, where assignment of individual absorptions is difficult. However, many sulfur-containing functional groups are readily identified by IR. The S=O bond in allicin, the principal volatile compound obtained from crushed garlic (see Problem 36 of Chapter 9), gives rise to an intense band at 1080 cm^{-1}. Allicin is unstable: At room temperature, the 1080-cm^{-1} IR absorption disappears in less than 24 h at room temperature. The major decomposition product, which also smells like garlic, was identified spectroscopically. Two likely candidates, 2-propene-1-thiol and its corresponding disulfide, were synthesized and found to have similar spectra, with alkene C=C and alkenyl C–H bands at 1630 and 3070 cm^{-1}, respectively. However, the thiol also shows an absorption for the S–H group at 2535 cm^{-1}. The absence of this band in the IR of the allicin decomposition product led to its identification as the disulfide, not the thiol.

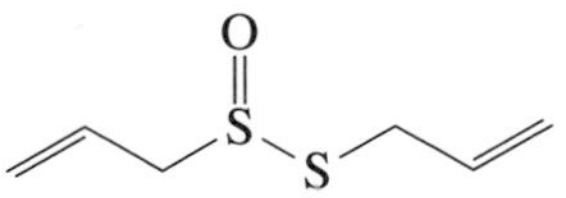

Allicin
IR: 1080 cm^{-1}

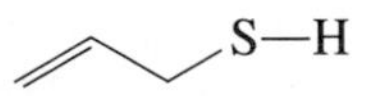

2-Propene-1-thiol
IR: 2535 cm^{-1}

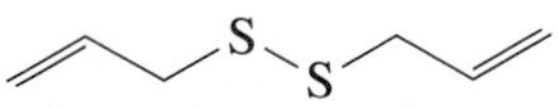

2-Propenyl disulfide
IR: No S=O or S–H bands

In summary, the presence of specific functional groups can be ascertained by infrared spectroscopy. Infrared light causes the vibrational excitation of bonds in molecules. Strong bonds and light atoms vibrate at relatively high stretching frequencies measured in wavenumbers (reciprocal wavelengths). Conversely, weak bonds and heavy atoms absorb at lower wavenumbers, as would be expected from Hooke's law. Because of the variety of stretching and bending modes, infrared spectra usually show complicated patterns. These are, however, diagnostic fingerprints for particular compounds. The presence of variously substituted alkenes may be detected by stretching signals at about 3080 (C–H) and 1640 (C=C) cm^{-1}, and bending modes between 890 and 990 cm^{-1}. Alcohols show a characteristic band for the OH group in the range 3200–3650 cm^{-1}.

11-6 Degree of Unsaturation: Another Aid to Identifying Molecular Structure

Some C_5H_8 Hydrocarbons

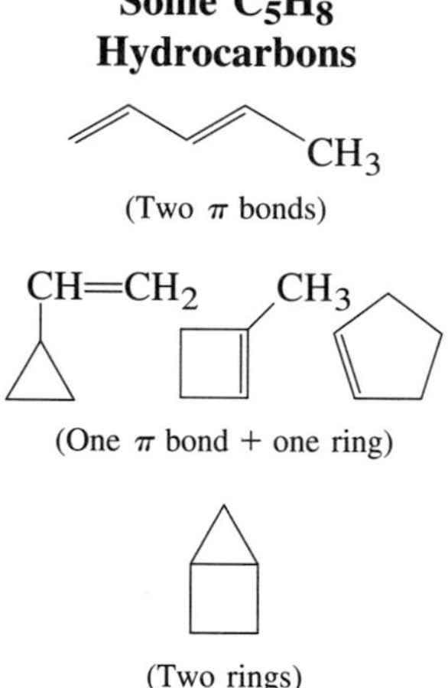

NMR and IR spectroscopy are important tools for determining the structure of an unknown. However, concealed in the molecular formula of every compound is an additional piece of information that can make the job easier. This is the **degree of unsaturation,** defined as the *sum of the numbers of rings and of π bonds* present in the molecule. Table 11-4 illustrates the relationship between molecular formula, structure, and degree of unsaturation for several hydrocarbons.

As Table 11-4 shows, each increase in the degree of unsaturation corresponds to a decrease of *two* hydrogens in the molecular formula. Therefore, starting with the general formula for acyclic alkanes (saturated; degree of unsaturation = 0), C_nH_{2n+2} (Section 2-2), the degree of unsaturation may be determined for any hydrocarbon merely by comparing the actual number of hydrogens present with the number required for the molecule to be saturated, $2n + 2$, where n = the number of carbons atoms present. For example, what is the degree of unsaturation in a hydrocarbon of the formula C_5H_8? A *saturated* compound with five carbons has the formula C_5H_{12} (C_nH_{2n+2}, with $n = 5$). Because C_5H_8 is four hydrogens short of being saturated, the degree of unsaturation is $4/2 = 2$. All

TABLE 11-4 Degree of Unsaturation as a Key to Structure

Formula	Representative structures	Degree of unsaturation
C_6H_{14}		0
C_6H_{12}	(one π bond); (one ring)	1
C_6H_{10}	; ;	2
C_6H_8	; ;	3

molecules with this formula contain a combination of rings and π bonds adding up to two. Several examples are shown in the margin on the previous page.

The presence of heteroatoms may affect the calculation. Let us compare the molecular formulas of several saturated compounds: ethane, C_2H_6, and ethanol, C_2H_6O, have the same number of hydrogen atoms; chloroethane, C_2H_5Cl, has one less, ethanamine, C_2H_7N, one more. The number of hydrogens required for saturation is reduced by the presence of halogen, increased when nitrogen is present, and unaffected by oxygen. We can generalize the procedure for determination of the degree of unsaturation from a molecular formula as follows.

STEP 1. Determine from the number of carbons (n_C), halogens (n_X), and nitrogens (n_N) in the molecular formula the number of hydrogens required for the molecule to be saturated, H_{sat}.

$$H_{sat} = 2n_C + 2 - n_X + n_N \qquad \text{(Oxygen and sulfur are disregarded)}$$

STEP 2. Compare H_{sat} with the actual number of hydrogens in the molecular formula, H_{actual}, to determine the degree of unsaturation.

$$\text{Degree of unsaturation} = (H_{sat} - H_{actual})/2$$

EXERCISE 11-10

Calculate the degree of unsaturation indicated by each of the following molecular formulas. **(a)** C_5H_{10}; **(b)** $C_9H_{12}O$; **(c)** C_8H_7ClO; **(d)** $C_8H_{15}N$; **(e)** $C_4H_8Br_2$.

EXERCISE 11-11

Spectroscopic data for three compounds with the molecular formula C_5H_8 are given below; m denotes a complex multiplet. Assign structures to each. (Hint: One is acyclic; the others each contain one ring.) **(a)** IR 910, 1000, 1650, 3100 cm^{-1}; ^1H NMR δ = 2.79 (t, J = 8 Hz), 4.8–6.2 (m) ppm, integrated intensity ratio of the signals = 1:3. **(b)** IR 900, 995, 1650, 3050 cm^{-1}; ^1H NMR δ = 0.5–1.5 (m), 4.8–6.0 (m) ppm, integrated intensity ratio of the signals = 5:3. **(c)** IR 1611, 3065 cm^{-1}; ^1H NMR δ = 1.5–2.5 (m), 5.7 (m) ppm, integrated intensity ratio of the signals = 3:1. Is there more than one possibility?

In summary, the degree of unsaturation is equal to the sum of the numbers of rings and of π bonds in a molecule. Calculation of this parameter makes solving structure problems from spectroscopic data easier.

11-7 Relative Stability of Double Bonds: Heats of Hydrogenation

We have seen several possible substitution patterns around a double bond, depending on the number of substituents and their positions. Is there a difference in the relative thermodynamic stability of these molecular arrangements? For example, are the three isomeric butenes—1-butene, *cis*-2-butene, and *trans*-2-butene—equally stable? The heat of hydrogenation, an important measure of relative energy content, provides an answer.

The heat of hydrogenation is a measure of stability

How can we establish the relative energy of any compound? One method was presented in Section 3-9: measuring heat of combustion. The greater the energy content of the molecule, the more energy released in this process. Another possibility, particularly applicable to the alkenes, is to measure the heat of another reaction: *hydrogenation of the double bond.*

When an alkene and hydrogen gas are mixed in the presence of a catalyst (usually palladium or platinum), the two hydrogen atoms in H_2 add to the double bond to give the saturated alkane (see Section 12-2), much like the hydrogenation of a carbonyl compound to give an alcohol (Section 8-6). The heat of this reaction, or **heat of hydrogenation,** can be measured accurately and is about -30 kcal mol^{-1} per double bond: The reaction is very exothermic.

Hydrogenation of an Alkene

$$\text{>C=C<} + \text{H—H} \xrightarrow{\text{Pd or Pt}} \text{—}\underset{\text{H}}{\overset{|}{\text{C}}}\text{—}\underset{\text{H}}{\overset{|}{\text{C}}}\text{—}$$

$\Delta H^\circ \sim -30$ kcal mol^{-1}

Hydrogenation of each butene isomer leads to the same product: butane. If their respective energy contents are equal, their heats of hydrogenation should also be equal; however, as shown below, they are not.

Hydrogenation of Three Butenes

$$(CH_3CH_2)(H)C{=}C(H)(H) + H_2 \xrightarrow{Pt} CH_3CH_2CH_2CH_3 \qquad \Delta H^\circ = -30.3 \text{ kcal mol}^{-1}$$

1-Butene **Butane**

$$(H_3C)(H)C{=}C(CH_3)(H) + H_2 \xrightarrow{Pt} CH_3CH_2CH_2CH_3 \qquad \Delta H^\circ = -28.6 \text{ kcal mol}^{-1}$$

***cis*-2-Butene** **Butane**

$$(H_3C)(H)C{=}C(H)(CH_3) + H_2 \xrightarrow{Pt} CH_3CH_2CH_2CH_3 \qquad \Delta H^\circ = -27.6 \text{ kcal mol}^{-1}$$

***trans*-2-Butene** **Butane**

The most heat is evolved by hydrogenation of the terminal double bond; the next most exothermic reaction is that with *cis*-2-butene; and finally the trans isomer gives off the least heat. Therefore, the thermodynamic stability of the butenes must increase in the order 1-butene $<$ *cis*-2-butene $<$ *trans*-2-butene (Figure 11-18).

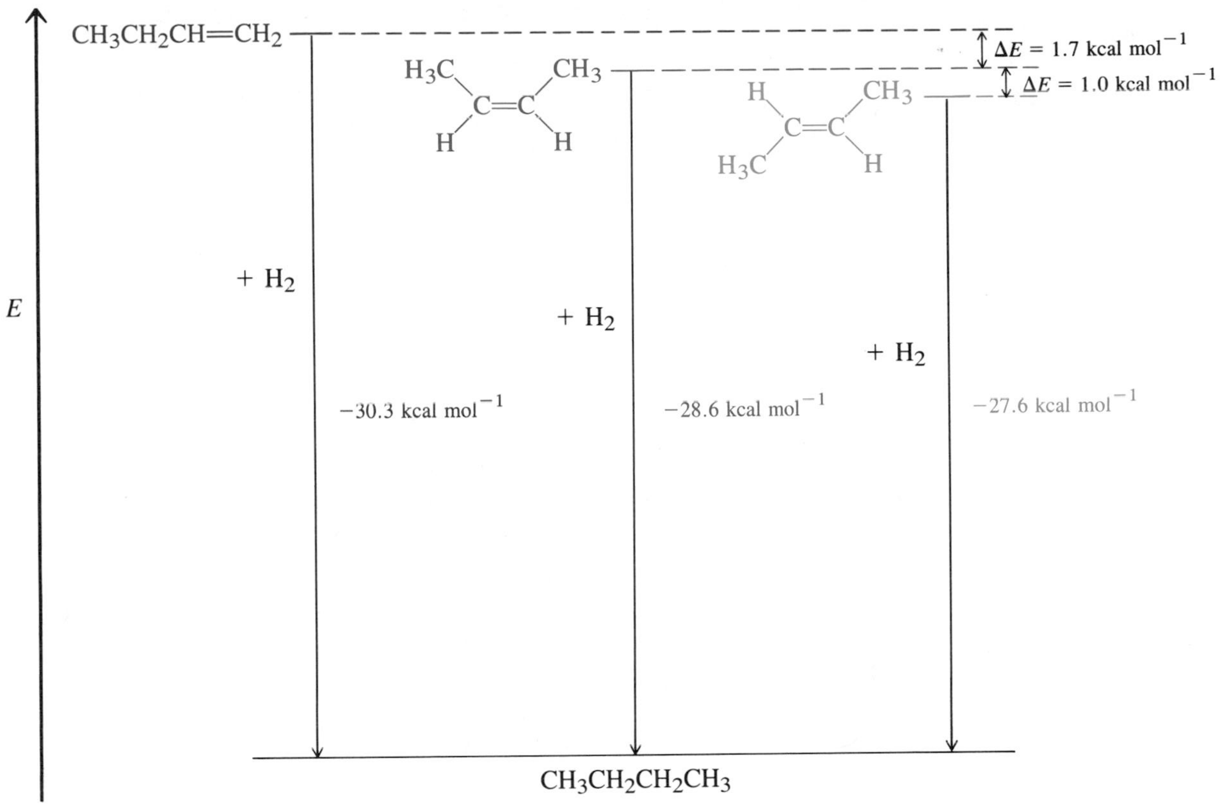

$CH_3CH_2CH{=}CH_2$

$\Delta E = 1.7$ kcal mol^{-1}

$\Delta E = 1.0$ kcal mol^{-1}

(cis-2-butene: H_3C, CH_3 / H, H on C=C)

(trans-2-butene: H, CH_3 / H_3C, H on C=C)

E

$+ H_2$ -30.3 kcal mol^{-1}

$+ H_2$ -28.6 kcal mol^{-1}

$+ H_2$ -27.6 kcal mol^{-1}

$CH_3CH_2CH_2CH_3$

FIGURE 11-18 The relative energy contents of the butene isomers, as measured by their heats of hydrogenation, tell us their relative stabilities.

Highly substituted alkenes are most stable; trans isomers are more stable than cis

These results may be generalized: The relative stability of the alkenes increases with increasing substitution, and trans isomers are usually more stable than their cis counterparts. The first trend cannot be explained in a straightforward way, but it is due in part to hyperconjugation. Just as the stability of a radical increases with increasing alkyl substitution (Section 3-2), the *p* orbitals of a π bond may be stabilized by alkyl substituents. The second finding is more easily understood by looking at molecular models. In cis-disubstituted alkenes, the substituent groups frequently crowd each other.

Relative Stabilities of the Alkenes

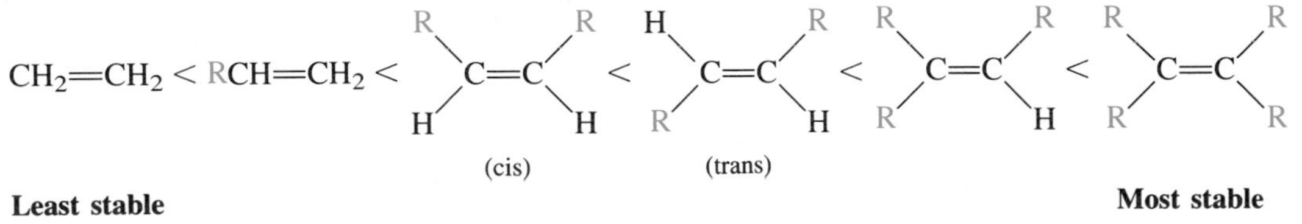

$CH_2{=}CH_2 < RCH{=}CH_2 <$ (R, R / H, H on C=C) (cis) $<$ (H, R / R, H on C=C) (trans) $<$ (R, R / R, H on C=C) $<$ (R, R / R, R on C=C)

Least stable **Most stable**

This steric interference is energetically disadvantageous and absent in the corresponding trans isomers (Figure 11-19).

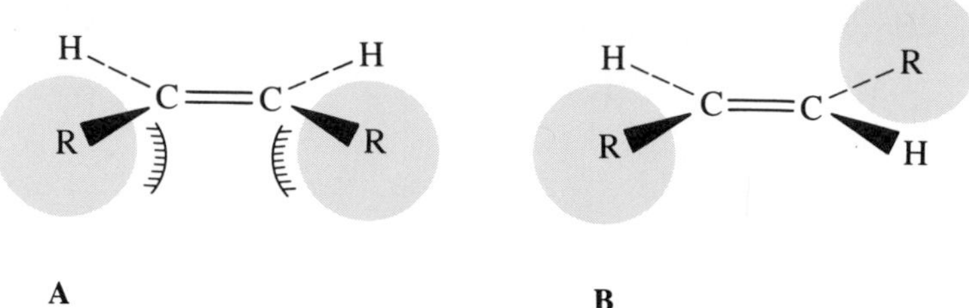

FIGURE 11-19 (A) Steric congestion in cis-disubstituted alkenes and (B) its absence in trans alkenes explain the greater stability of the latter isomer.

EXERCISE 11-12

Rank in order of stability of the double bond to hydrogenation (order of $\Delta H°$ of hydrogenation): 2,3-dimethyl-2-butene, *cis*-3-hexene, *trans*-4-octene, and 1-hexene.

Structure of *trans*-Cyclooctene

Cycloalkenes are exceptions to the generalization that trans alkenes are more stable than their cis isomers. In the small- and medium-ring members of this class of compounds (Section 4-2), the trans isomers are much more strained (Section 11-1). The smallest isolated simple trans cycloalkane is *trans*-cyclooctene. It is 9.2 kcal mol^{-1} less stable than the cis isomer and has a highly twisted structure.

EXERCISE 11-13

Alkene A hydrogenates to B with an estimated release of 65 kcal mol^{-1}, more than double the values of the hydrogenations shown in Figure 11-18. Explain.

A $\xrightarrow{H_2, \text{ catalyst}}$ B (H, H) $\qquad \Delta H° = -65$ kcal mol^{-1}

In summary, the relative energies of isomeric alkenes can be estimated by measuring their heats of hydrogenation. The more energetic alkene has a higher $\Delta H°$ of hydrogenation. Stability increases with increasing substitution because of hyperconjugation. Trans alkenes are more stable than their cis isomers because of steric hindrance. Exceptions are the small- and medium-ring cycloalkenes, in which cis substitution is more stable than trans because of ring strain.

11-8 Preparation of Alkenes from Haloalkanes and Alkyl Sulfonates: Bimolecular Elimination Revisited

With the physical aspects of alkene structure and stability as a background, let us now concern ourselves with the various ways in which alkenes can be made. The most general approach is by *elimination,* in which two adjacent groups on a carbon framework are removed. The E2 reaction (Section 7-7) is the most common laboratory source of alkenes. Another method of alkene synthesis, the dehydration of alcohols, is described in Section 11-9.

Scheme for Elimination

$$-\underset{A}{\overset{|}{C}}-\underset{B}{\overset{|}{C}}- \longrightarrow \;\rangle C{=}C\langle \; + AB$$

Regioselectivity in E2 reactions depends on the base

In Chapter 7 we discussed how haloalkanes (or alkyl sulfonates) in the presence of base can undergo elimination of the elements of HX with simultaneous formation of a carbon–carbon double bond. Strong bases, such as sodium ethoxide, are best. They help prevent the reactants from entering E1 pathways and thus avoid the complications that arise from thc formation of carbocations.

With many substrates, elimination can be in more than one direction, thus giving rise to double-bond isomers. In such cases, can we control which hydrogen is attacked, that is, the *regioselectivity* of the reaction (Section 9-9)? The answer is yes, to a limited extent. A simple example is the dehydrobromination of 2-bromo-2-methylbutane. Reaction with sodium ethoxide in hot ethanol furnishes mainly 2-methyl-2-butene, but also some 2-methyl-1-butene.

E2 Reaction of 2-Bromo-2-methylbutane with Ethoxide

$$CH_3CH_2\underset{Br}{\overset{CH_3}{C}}CH_3 \xrightarrow[-\,HBr]{CH_3CH_2O^-Na^+,\ CH_3CH_2OH,\ 70^\circ C} (H_3C)(H)C{=}C(CH_3)(CH_3) + (CH_3CH_2)(H_3C)C{=}CH_2$$

2-Bromo-2-methylbutane → 70% **2-Methyl-2-butene** + 30% **2-Methyl-1-butene**

In our example, the first alkene contains a trisubstituted double bond, so it is thermodynamically more stable than the second. Indeed, many eliminations are regioselective in this way, with the thermodynamically preferred product predominating. This result can be explained by analysis of the transition state of the reaction (Figure 11-20). Elimination of the elements of hydrogen bromide proceeds through attack by the base on one of the vicinal hydrogens situated *anti* to the leaving group. In the transition state there is partial C–H bond rupture, partial C–C double-bond formation, and partial cleavage at C–Br (compare Figure 7-7). The transition state leading to 2-methyl-2-butene is slightly more stabilized than that generating 2-methyl-1-butene: The more stable product is formed faster because *the structure of the transition state of the reaction resembles that of the products to some extent*. Elimination reactions of this type that lead to the more highly substituted alkene are said to follow thc **Saytzev* rule.**

*Alexander M. Saytzev (also spelled Zaitsev or Saytzeff), 1841–1910, Russian chemist.

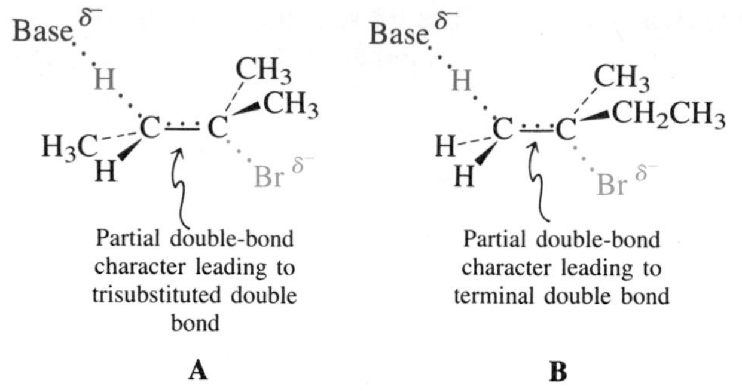

$$\text{Base}^{\delta^-}\cdots H-\overset{H_3C,\ H}{C}\cdots C(CH_3)_2\cdots Br^{\delta^-}$$

Partial double-bond character leading to trisubstituted double bond

A

$$\text{Base}^{\delta^-}\cdots H-\overset{H,\ H}{C}\cdots C(CH_3)(CH_2CH_3)\cdots Br^{\delta^-}$$

Partial double-bond character leading to terminal double bond

B

FIGURE 11-20 The two transition states leading to products in the dehydrobromination of 2-bromo-2-methylbutane. Transition state A is preferred over transition state B because there are more substituents around the partial double bond (Saytzev rule).

A different product distribution is obtained when a more hindered base is used in the same reaction; more of the thermodynamically *less* favored terminal alkene is generated.

E2 Reaction of 2-Bromo-2-methylbutane with *tert*-Butoxide, a Hindered Base

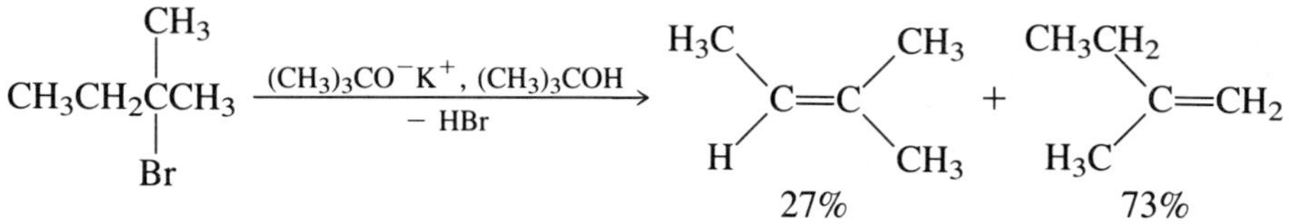

$$CH_3CH_2\overset{CH_3}{\underset{Br}{C}}CH_3 \xrightarrow[-\ HBr]{(CH_3)_3CO^-K^+,\ (CH_3)_3COH} \underset{27\%}{(H_3C)(H)C{=}C(CH_3)_2} + \underset{73\%}{(CH_3CH_2)(H_3C)C{=}CH_2}$$

To see why, we again examine the transition state. Removal of a secondary hydrogen (from C3 in the starting bromide) is sterically more difficult than abstracting one of the more exposed methyl hydrogens. When the base used is bulky, as in our example, the energy of the transition state leading to the more stable product is increased by steric interference relative to that leading to the less substituted isomer; thus, the latter becomes the major product. An E2 reaction that generates the thermodynamically less favored isomer due to such steric constraints is said to follow the **Hofmann* rule.**

EXERCISE 11-14

When the following reaction is carried out with *tert*-butoxide in 2-methyl-2-propanol (*tert*-butyl alcohol), two products, A and B, are formed in the ratio 23:77. When ethoxide in ethanol is used, this ratio changes to 82:18. What are A and B, and how do you explain the difference in their ratios in the two experiments?

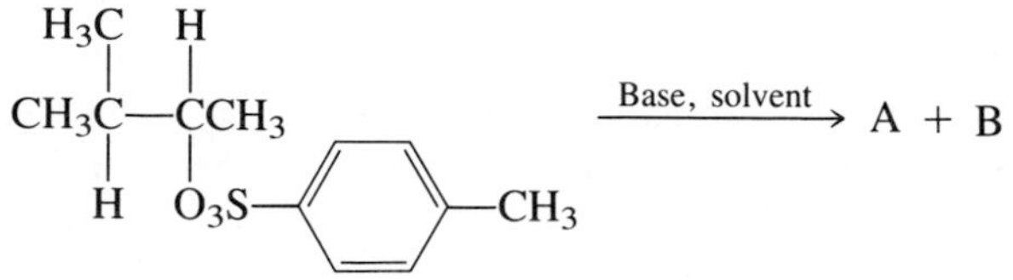

$$CH_3\overset{H_3C}{\underset{H}{C}}-\overset{H}{\underset{O_3S-C_6H_4-CH_3}{C}}CH_3 \xrightarrow{\text{Base, solvent}} A + B$$

*Professor August Wilhelm von Hofmann, 1818–1892, University of Berlin.

E2 reactions often favor trans over cis

Depending on the structure of the alkyl substrate, the E2 reaction can also lead to cis,trans alkene mixtures, in some cases with selectivity. For example, treatment of 2-bromopentane with sodium ethoxide furnishes 51% *trans-* and only 18% *cis*-2-pentene, the remainder of the product being the terminal regioisomer. The outcome of this and related reactions appears to be controlled again to some extent by the relative thermodynamic stabilities of the products, the most stable trans double bond being formed preferentially.

Stereoselective Dehydrobromination of 2-Bromopentane

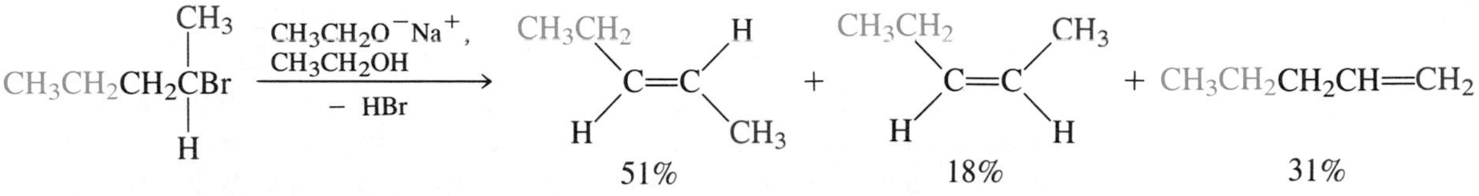

$$CH_3CH_2CH_2CBr(CH_3)(H) \xrightarrow[-\,HBr]{CH_3CH_2O^-Na^+,\ CH_3CH_2OH} \underset{51\%}{(CH_3CH_2)(H)C{=}C(H)(CH_3)} + \underset{18\%}{(CH_3CH_2)(H)C{=}C(CH_3)(H)} + \underset{31\%}{CH_3CH_2CH_2CH{=}CH_2}$$

Unfortunately from a synthetic viewpoint, complete trans selectivity is rare in E2 reactions. Chapter 13 deals with alternative methods for the preparation of stereochemically pure cis and trans alkenes.

Some E2 processes are stereospecific

Recall (Section 7-7) that the preferred transition state of elimination places the proton to be removed and the leaving group *anti* with respect to each other. This fact has additional consequences when reaction may lead to *Z* or *E* stereoisomers. For example, the E2 reaction of the two diastereomers of 2-bromo-3-methylpentane to give 3-methyl-2-pentene is completely stereospecific. Employment of the (*R*,*R*) or the (*S*,*S*) isomer or both leads *exclusively* to the formation of the (*E*) isomer of the alkene. Conversely, reaction of the (*R*,*S*) or (*S*,*R*) diastereomer or both gives only the (*Z*) alkene.

Stereospecificity in the E2 Reaction of 2-Bromo-3-methylpentane

(*R*,*R*) or (*S*,*S*) 2-bromo-3-methylpentane $\xrightarrow[-\,HBr]{Base}$ **(*E*)-3-Methyl-2-pentene**

(*S*,*R*) or (*R*,*S*) 2-bromo-3-methylpentane $\xrightarrow[-\,HBr]{Base}$ **(*Z*)-3-Methyl-2-pentene**

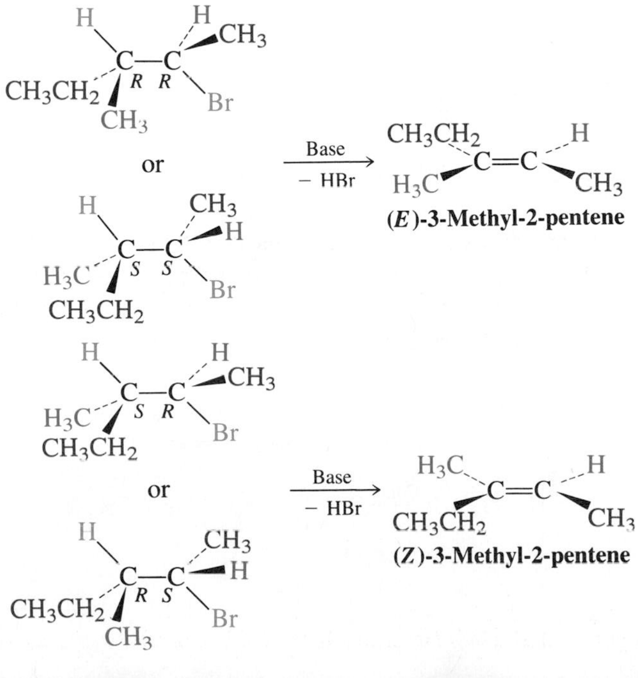

As shown in these dashed-wedged line structures, *anti* elimination of HBr dictates the eventual configuration around the double bond. The reaction is stereospecific, one diastereomer (and its mirror image) producing only one stereoisomeric alkene, the other diastereomer furnishing the opposite configuration.

EXERCISE 11-15

Which diastereomer of 2-bromo-3-deuteriobutane gives (*E*)-2-deuterio-2-butene, and which diastereomer gives the *Z* isomer?

In summary, alkenes are most generally made by E2 reactions. Generally, the thermodynamically more stable internal alkenes are formed faster than the terminal isomers (Saytzev rule). The reaction may be stereoselective, producing greater quantities of trans isomers than their cis counterparts from racemic starting materials. It is also stereospecific, certain haloalkane diastereomers furnishing only one out of the two possible stereoisomeric alkenes. Bulky bases may lead to more of the products with thermodynamically less stable (e.g., terminal) double bonds (Hofmann rule).

11-9 Preparation of Alkenes by Dehydration of Alcohols

Treatment of alcohols with mineral acid at elevated temperatures results in loss of water, a process called **dehydration,** which proceeds by E1 or E2 pathways (Chapters 7 and 9). This section reviews how this reaction can be used to synthesize alkenes.

The usual method employed to dehydrate an alcohol is to heat it in the presence of sulfuric or phosphoric acid at relatively high temperatures (120–170°C).

Acid-Mediated Dehydration of Alcohols

$$\underset{\text{H}\quad\text{OH}}{-\overset{|}{\underset{|}{\text{C}}}-\overset{|}{\underset{|}{\text{C}}}-} \xrightarrow{\text{Acid, }\Delta} \text{>C=C<} + \text{HOH}$$

Another, sometimes more efficient, dehydration procedure requires the use of aluminum oxide (alumina), Al_2O_3, as a Lewis acid catalyst. In this case vapors of the alcohol are passed over alumina powder at a temperature ranging from 350 to 400°C, and the alkene is collected in a trap at the exit of the reaction vessel.

The ease of elimination of water from alcohols increases with increasing substitution of the hydroxy-bearing carbon.

Relative Reactivity of Alcohols (ROH) in Dehydration Reactions

R = primary < secondary < tertiary

$$CH_3CH_2OH \xrightarrow[-\ \text{HOH}]{\text{Conc. } H_2SO_4,\ 170^\circ\text{C}} CH_2{=}CH_2$$

$$\underset{\text{H}\quad\text{H}}{\overset{\text{HO}\quad\text{H}}{\text{CH}_3\text{C}{-}\text{CCH}_3}} \xrightarrow[-\text{ HOH}]{50\%\ \text{H}_2\text{SO}_4,\ 100^\circ\text{C}} \underset{80\%}{\text{CH}_3\text{CH}{=}\text{CHCH}_3} + \underset{\text{Trace}}{\text{CH}_2{=}\text{CHCH}_2\text{CH}_3}$$

$$(\text{CH}_3)_3\text{COH} \xrightarrow[-\text{ HOH}]{\text{Conc. H}_2\text{SO}_4,\ 50^\circ\text{C}} \underset{100\%}{\text{H}_2\text{C}{=}\text{C}(\text{CH}_3)_2}$$

Secondary and tertiary alcohols dehydrate by the unimolecular elimination pathway (E1), discussed in Sections 7-6 and 9-2. Protonation of the weakly basic hydroxy oxygen forms an alkyloxonium ion, now containing water as a good potential leaving group. Loss of H_2O supplies the respective secondary or tertiary carbocations, and deprotonation furnishes the alkene. The reaction is subject to all the side reactions of which carbocations are capable, particularly hydrogen and alkyl shifts (Section 9-3).

Dehydration with Rearrangement

$$\underset{\text{H}\qquad\qquad\ \text{H}}{\overset{\text{CH}_3\qquad\quad \text{OH}}{\text{CH}_3\text{C}{-}\text{CH}_2{-}\text{CCH}_3}} \xrightarrow[-\text{ H}_2\text{O}]{\text{H}_2\text{SO}_4,\ \Delta} \underset{54\%}{(\text{H}_3\text{C})_2\text{C}{=}\text{C}(\text{H})\text{CH}_2\text{CH}_3} + \underset{8\%}{\underset{\text{H}}{\overset{\text{CH}_3}{\text{CH}_3\text{CCH}{=}\text{CHCH}_3}}} + \text{other minor isomers}$$

EXERCISE 11-16

Referring to Sections 7-6 and 9-3, write a mechanism for the preceding reaction.

Typically, the thermodynamically most stable alkene or alkene mixture results from unimolecular dehydration in the presence of acid. Thus, whenever possible, the most highly substituted system is generated; if there is a choice, trans-substituted alkenes predominate over the cis isomers. For example, acid-catalyzed dehydration of 2-butanol furnishes the equilibrium mixture of butenes, consisting of 74% *trans*-2-butene, 23% of the cis isomer, and only 3% 1-butene.

Treatment of primary alcohols with mineral acids at elevated temperatures also leads to alkenes; for example, ethanol gives ethene and 1-propanol yields propene (Section 9-7).

$$\text{CH}_3\text{CH}_2\text{CH}_2\text{OH} \xrightarrow{\text{Conc. H}_2\text{SO}_4,\ 180^\circ\text{C}} \text{CH}_3\text{CH}{=}\text{CH}_2$$

The mechanism of this reaction begins with the initial protonation of oxygen. Then, attack by hydrogen sulfate ion or another alcohol molecule effects bimolecular elimination of the elements of H_3O^+ (see Section 7-7).

BOX 11-3 Acid-Catalyzed Dehydration of α-Terpineol

Acid-catalyzed dehydration is not generally useful for preparative purposes because product mixtures often form. An example is the dehydration of α-terpineol, a naturally occurring, unsaturated terpene alcohol (see below and Section 4-7) with a pleasant lilac fragrance, isolated from pine oil. Fortunately, in this case the product mixture also has a pleasing aroma, due in part to the presence of *limonene,* a major constituent of lemon and orange oils and *α-terpinene,* which also possesses a lemon fragrance. Indeed, most of us have encountered this hydrocarbon mixture—it is used in the manufacture of scented soaps!

OH

α-Terpineol $\xrightarrow[-\ H_2O]{33\%\ H_2SO_4,\ 1\ h,\ 100°C}$ Terpinolene (15%) + Limonene (9%) + α-Terpinene (28.5%) + Isoterpinolene (18.5%) + γ-Terpinene (15%)

EXERCISE 11-17

(a) Propose a mechanism for the formation of propene from 1-propanol on treatment with hot conc. H_2SO_4. **(b)** Propene is also formed when propoxypropane (dipropyl ether) is subjected to the same conditions (below). Explain.

$$CH_3CH_2CH_2OCH_2CH_2CH_3 \xrightarrow{\text{Conc. } H_2SO_4,\ 180°} 2\ CH_3CH{=}CH_2 + H_2O$$

In summary, alkenes can be made by dehydration of alcohols. Secondary and tertiary systems proceed through the intermediacy of carbocations, whereas primary alcohols can undergo E2 reactions from the intermediate alkyloxonium ions. All systems are subject to rearrangement and thus frequently give mixtures.

NEW REACTIONS

1. HYDROGENATION OF ALKENES (Section 11-7)

$$\rangle C{=}C\langle + H_2 \xrightarrow{\text{Pd or Pt}} -\underset{H}{\overset{|}{C}}-\underset{H}{\overset{|}{C}}- \qquad \Delta H° \sim -30\ \text{kcal mol}^{-1}$$

Order of stability of the double bond

$$R_2C{=}CH_2 < \text{RHC=CHR (cis)} < \text{RHC=CHR (trans)} < \text{more substituted alkene}$$

Preparation of Alkenes

2. FROM HALOALKANES, E2 WITH UNHINDERED BASE (Section 11-8)

$$\text{H}-\overset{|}{\underset{|}{\text{C}}}-\overset{|}{\underset{\text{X}}{\text{C}}}-\text{CH}_3 \xrightarrow[-\text{HX}]{\text{CH}_3\text{CH}_2\text{O}^-\text{Na}^+,\ \text{CH}_3\text{CH}_2\text{OH}} \text{C}=\text{C}(\text{CH}_3)$$

Saytzev rule

More substituted (more stable) alkene

3. FROM HALOALKANES, E2 WITH STERICALLY HINDERED BASE (Section 11-8)

$$\text{H}-\overset{|}{\underset{|}{\text{C}}}-\overset{|}{\underset{\text{X}}{\text{C}}}-\text{CH}_3 \xrightarrow[-\text{HX}]{(\text{CH}_3)_3\text{CO}^-\text{K}^+,\ (\text{CH}_3)_3\text{COH}} \text{C(H)}-\text{C}=\text{CH}_2$$

Hofmann rule

Less substituted (less stable) alkene

4. STEREOCHEMISTRY OF E2 REACTION (Section 11-8)

$$\text{H(R')(R)C}-\text{C(R'')(R''')X} \xrightarrow[-\text{HX}]{\text{Base}} \text{R'(R)C}=\text{C(R'')(R''')}$$

Anti elimination

5. DEHYDRATION OF ALCOHOLS (Section 11-9)

$$-\underset{\text{H}}{\overset{|}{\underset{|}{\text{C}}}}-\underset{\text{OH}}{\overset{|}{\underset{|}{\text{C}}}}- \xrightarrow[-\text{H}_2\text{O}]{\text{H}_2\text{SO}_4,\ \Delta} \text{C}=\text{C}$$

Most stable alkene is major product
Secondary, tertiary: E1 mechanism
Carbocations may rearrange

Order of reactivity: primary < secondary < tertiary

IMPORTANT CONCEPTS

1. Alkenes are unsaturated molecules. Their IUPAC names are derived from alkanes, the longest chain incorporating the double bond serving as the stem. Double-bond isomers include terminal, internal, cis, and trans arrangements. Tri- and tetrasubstituted alkenes are named according to the *E,Z* system, in which the *R,S* priority rules apply.
2. The double bond is composed of a σ and a π part. The former is obtained by overlap of the two sp^2 hybrid lobes on carbon, the latter by interaction of the two remaining *p* orbitals. The π bond is weaker ($\sim$65 kcal mol^{-1}) than its σ counterpart ($\sim$108 kcal mol^{-1}) but strong enough to allow for the existence of stable cis and trans isomers.
3. The functional group in the alkenes is flat, sp^2 hybridization being responsible for the possibility of creating dipoles and for the relatively high acidity of the alkenyl hydrogen.
4. Alkenyl hydrogens and carbons appear at low field in ^{1}H NMR (δ = 4.6–5.7 ppm) and ^{13}C NMR (δ = 100–140 ppm) experiments, respectively. J_{trans} is larger than J_{cis}, $J_{geminal}$ is very small, and $J_{allylic}$ variable but small.

5. Infrared spectroscopy measures vibrational excitation. The energy of the incident radiation ranges from about 1 to 10 kcal mol^{-1} ($\lambda \sim$ 2.5–16.7 μm; $\tilde{\nu} \sim$ 600–4000 cm^{-1}). Characteristic peaks are observed for certain functional groups, a consequence of stretching, bending, and other modes of vibration, and their combination. Moreover, each molecule exhibits a characteristic infrared spectral pattern, called a fingerprint.
6. Alkanes show IR bands characteristic of C–H bonds in the range 2840 to 3000 cm^{-1}. The C=C stretching absorption for alkenes occurs in the range 1620 to 1680 cm^{-1}, that for the alkenyl C–H bond around 3100 cm^{-1}. Bending modes sometimes give useful peaks below 1500 cm^{-1}. Alcohols are usually characterized by a broad peak for the O–H stretch between 3200 and 3650 cm^{-1}.
7. Degree of unsaturation (number of rings + number of π bonds) is calculated from the molecular formula by using the equation

$$\text{Degree of unsaturation} = (H_{sat} - H_{actual})/2$$

where $H_{sat} = 2n_C + 2 - n_X + n_N$ (disregard oxygen and sulfur).
8. The relative stability of isomeric alkenes can be established by comparing heats of hydrogenation. It decreases with decreasing substitution; trans isomers are more stable than cis.
9. Elimination of haloalkanes (and other alkyl derivatives) may follow the Saytzev rule (nonbulky base, internal alkene formation) or the Hofmann rule (bulky base, terminal alkene formation). Trans alkenes as products predominate over cis alkenes. Elimination is stereospecific, as dictated by the *anti* transition state.
10. Dehydration of alcohols in the presence of strong acid usually leads to a mixture of products, with the most stable alkene being the major constituent.

PROBLEMS

1. Name each of the following molecules in accord with the IUPAC system of nomenclature.

(a) CH_3CH_2 and H on one carbon, CH_3 and H on the other: $CH_3CH_2(H)C{=}C(CH_3)(H)$

(b) $CH_3CH_2CHCH{=}CH_2$ with CH_3CH_2 on the CH

(c) Cl, H on one carbon; H and $CH_2CH_2CHCH_3$ (with OH) on the other

(d) F, Cl on one carbon; Br, I on the other

(e) $HOCH_2$, CH_3CH_2 on one carbon; $CHCH_3$ (with CF_3), H on the other

(f) CH_3CH_2, H on one carbon; Cl, Cl on the other

(g) CH_3O, H_3C on one carbon; OCH_3, H on the other

(h) CH_3CH (with CH_3), H_3C on one carbon; $CH_2CH_2CH_3$, H on the other

(i) Cyclohexene ring bearing CH_3 and CH_2CH_3

2. Assign the structures of the following molecules on the basis of the indicated ^{1}H NMR spectra. Consider stereochemistry, where applicable.
(a) C_4H_7Cl, NMR spectrum A; **(b)** $C_5H_8O_2$, NMR spectrum B;
(c) $C_6H_{11}I$, NMR spectrum C (p. 406); **(d)** another $C_6H_{11}I$, NMR spectrum D (p. 407); **(e)** $C_3H_4Cl_2$, NMR spectrum E (p. 408).

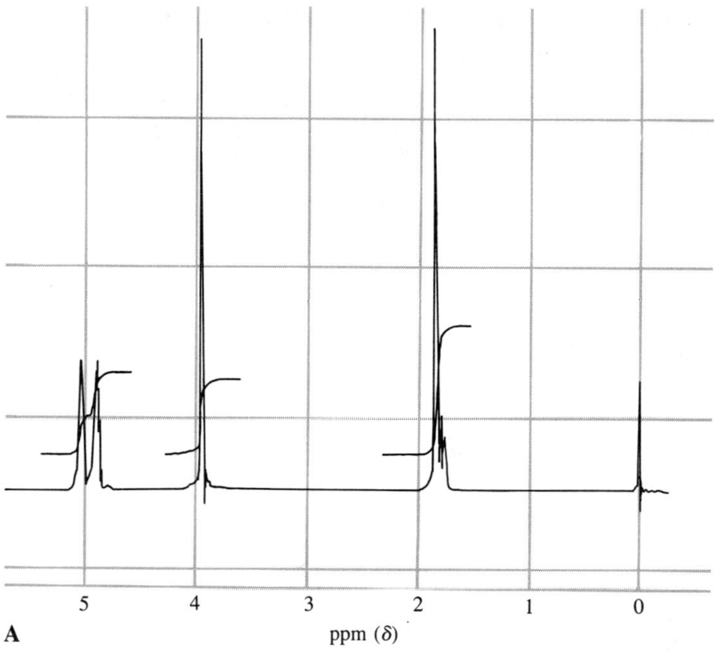
5
4
3
2
1
0
A
ppm (δ)

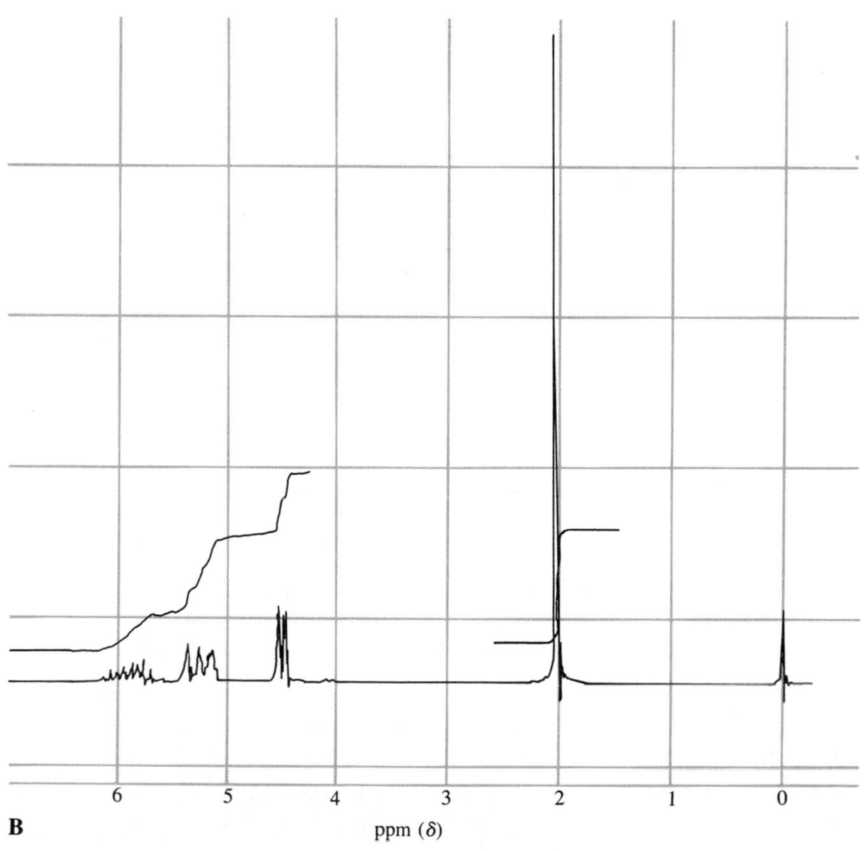
6
5
4
3
2
1
0
B
ppm (δ)

3. Explain the splitting patterns in ^{1}H NMR spectrum D in detail. Inset is $\delta = 5.7$–6.7 region expanded fivefold.

4. For each of the following pairs of alkenes, indicate whether measurements of polarity alone would be sufficient to distinguish the compounds from one another. Where possible, predict which compound would be more polar.

(a) H, CH_3 / C=C / H_3C, H and $CH_3CH_2CH{=}CH_2$

(b) H_3C, $CH_2CH_2CH_3$ / C=C / H, H and CH_3CH_2, CH_2CH_3 / C=C / H, H

(c) H_3C, $CH_2CH_2CH_3$ / C=C / H, H and H, CH_2CH_3 / C=C / CH_3CH_2, H

5. The molecular formulas and ^{13}C NMR data (in ppm) for several compounds are given below. The splitting pattern of each signal, taken from the undecoupled spectrum, is given in parentheses. Deduce structures for each. **(a)** C_4H_6: 30.2 (t), 136.0 (d); **(b)** C_4H_6O: 18.2 (q), 134.9 (d), 153.7 (d),

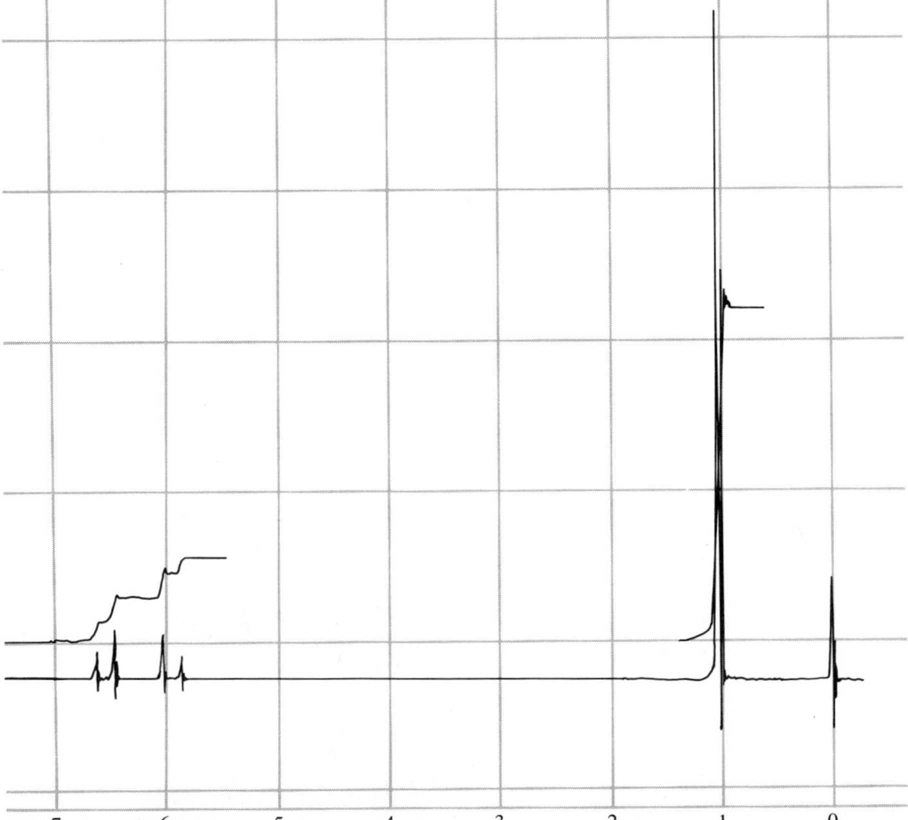

C

193.4 (d); **(c)** C_4H_8: 13.6 (q), 25.8 (t), 112.1 (t), 139.0 (d); **(d)** $C_5H_{10}O$: 17.6 (q), 25.4 (q), 58.8 (t), 125.7 (d), 133.7 (s); **(e)** C_5H_8: 15.8 (t), 31.1 (t), 103.9 (t), 149.2 (s); **(f)** C_7H_{10}: 25.2 (t), 41.9 (d), 48.5 (t), 135.2 (d). (Hint: This one is difficult. The molecule has one double bond. How many rings must it have?)

6. Data from the ^{13}C NMR spectra of several compounds, all with the formula C_5H_{10}, are given below. The splitting of each signal, taken from the undecoupled spectrum, is given in parentheses after the chemical-shift value. Propose structures for each. **(a)** 25.3 (t); **(b)** 13.3 (q), 17.1 (q), 25.5 (q), 118.7 (d), 131.7 (s); **(c)** 12.0 (q), 13.8 (q), 20.3 (t), 122.8 (d), 132.4 (d).

7. From the Hooke's law equation, would you expect the C–X bonds of common haloalkanes (X = Cl, Br, I) to have IR bands at higher or lower wavenumbers than are typical for bonds between carbon and lighter elements (e.g., oxygen)?

8. Convert each of the following IR frequencies into micrometers.
(a) 1720 cm^{-1} (C=O) **(b)** 1650 cm^{-1} (C=C) **(c)** 3300 cm^{-1} (O—H)
(d) 890 cm^{-1} (alkene bend) **(e)** 1100 cm^{-1} (C—O) **(f)** 2260 cm^{-1} (C≡N)

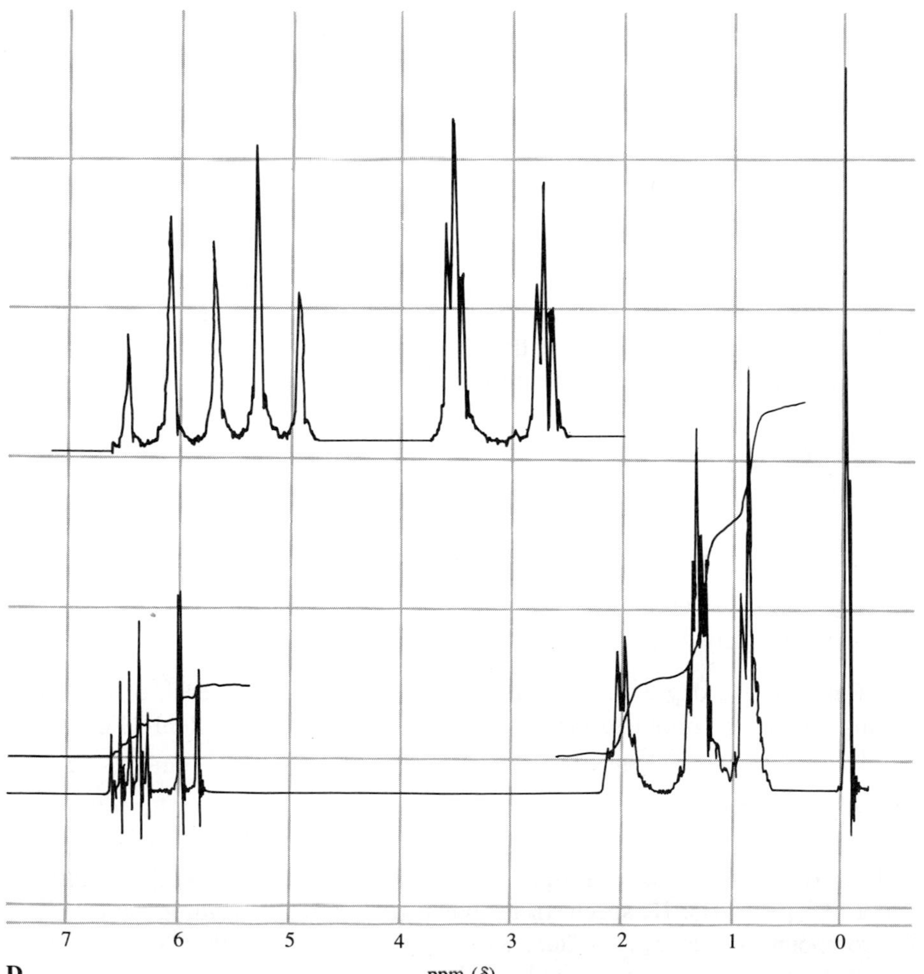

D

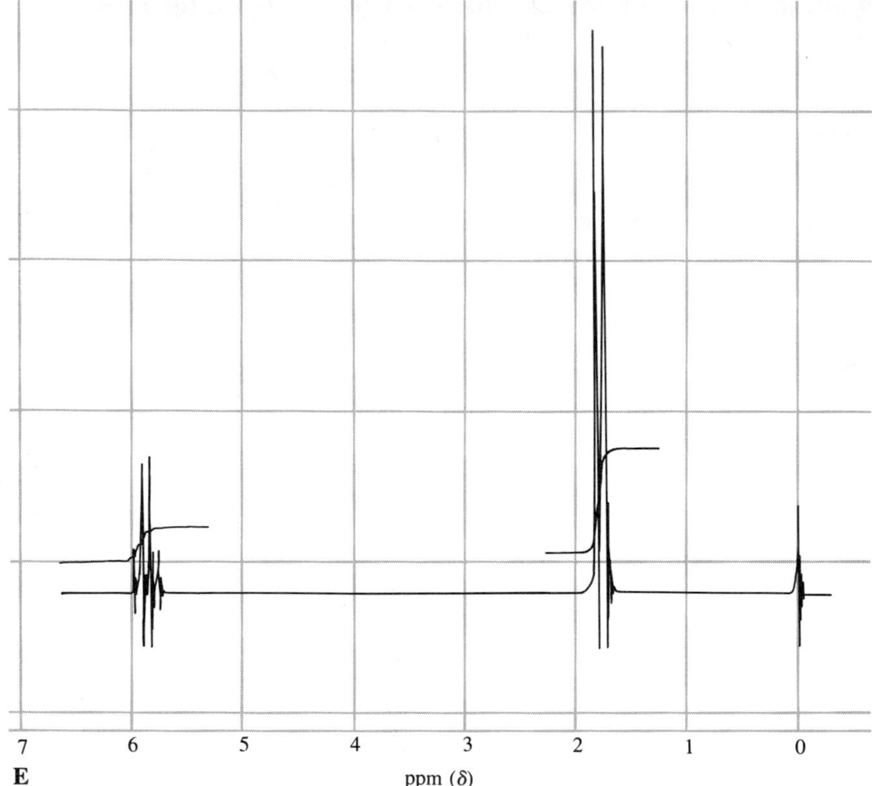

9. You have just entered the chemistry stockroom to look for several isomeric bromopentanes. There are three bottles on the shelf marked $C_5H_{11}Br$, but their labels have fallen off. The NMR machine is broken, so you devise the following experiment in an attempt to determine which isomer is in which bottle: A sample of the contents in each bottle is treated with NaOH in aqueous ethanol, and then the IR spectrum of each product or product mixture determined. Here are the results.

(i) $C_5H_{11}Br$ isomer in bottle A $\xrightarrow{\text{NaOH}}$ IR bands at 1660, 2850–3020, and 3350 cm^{-1}

(ii) $C_5H_{11}Br$ isomer in bottle B $\xrightarrow{\text{NaOH}}$ IR bands at 1670 and 2850–3020 cm^{-1}

(iii) $C_5H_{11}Br$ isomer in bottle C $\xrightarrow{\text{NaOH}}$ IR bands at 2850–2960 and 3350 cm^{-1}

(a) What do the data tell you about each product or product mixture?
(b) Suggest possible structures for the contents of each bottle.

10. Calculate the degree of unsaturation that corresponds to each of the following molecular formulas. **(a)** C_7H_{12} (use the result in Problem 11); **(b)** $C_8H_7NO_2$; **(c)** C_6Cl_6; **(d)** $C_{10}H_{22}O_{11}$; **(e)** $C_6H_{10}S$; **(f)** $C_{18}H_{28}O_2$.

11. A hydrocarbon with the molecular formula C_7H_{12} exhibits the following spectroscopic data: ^{1}H NMR $\delta = 1.3$ (m, 2 H), 1.7 (m, 4 H), 2.2 (m, 4 H), and 4.8 (quin, $J = 3$ Hz, 2 H) ppm; ^{13}C NMR $\delta = 26.8$, 28.7, 35.7, 106.9, and 149.7 ppm. The IR spectrum is shown (spectrum F). Hydrogenation of the compound furnishes a product with the molecular formula C_7H_{14}. Suggest a structure for the compound consistent with these data.

12. The isolation of a new form of molecular carbon, C_{60}, was reported in 1990. The substance has the shape of a soccer ball of carbon atoms and possesses the nickname "buckyball" (you don't want to know the IUPAC name). Hydrogenation produces a hydrocarbon with the molecular formula $C_{60}H_{36}$. How many degrees of unsaturation are present in C_{60}? In $C_{60}H_{36}$? Does the hydrogenation result place limits on the numbers of π bonds and rings in "buckyball"? (More discussion of C_{60} will follow in Chapter 15.)

13. Place the alkenes in each group in order of increasing stability of the double bond and increasing heat of hydrogenation.

(a) $CH_2{=}CH_2$ $\quad (H_3C)_2C{=}C(CH_3)_2 \quad (H_3C)_2C{=}CH_2$

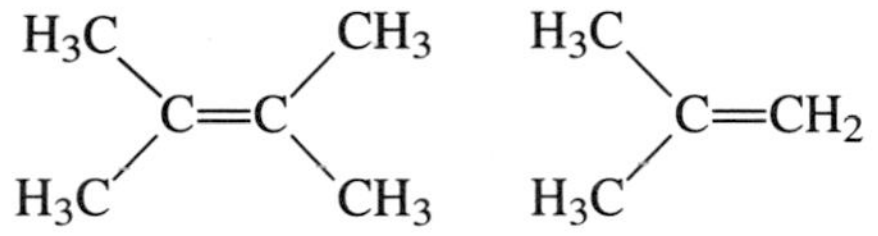

(b) H, H_3C C=C H, $CH(CH_3)_2$ $\quad$ H, H_3C C=C $CH(CH_3)_2$, H $\quad$ H, $(CH_3)_2CH$ C=C H, $CH(CH_3)_2$

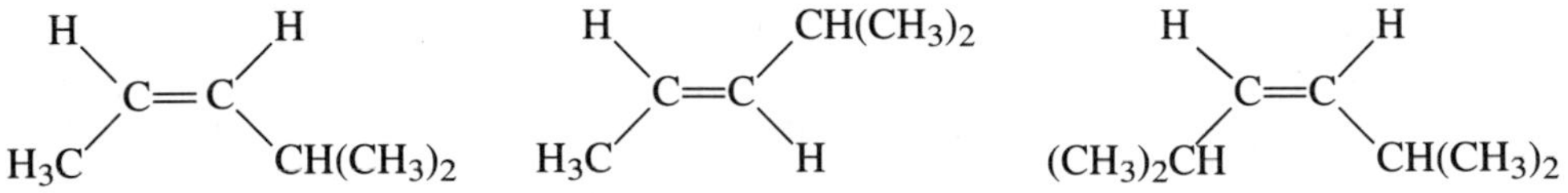

(c)

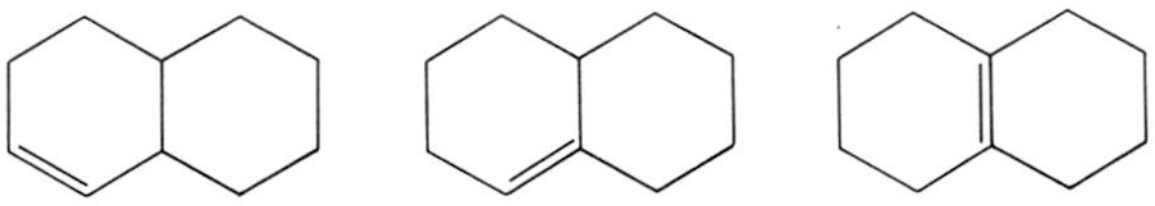

(d) H_3C, CH_3 $\quad$ H_3C, CH_2 $\quad$ H_3C, CH_3

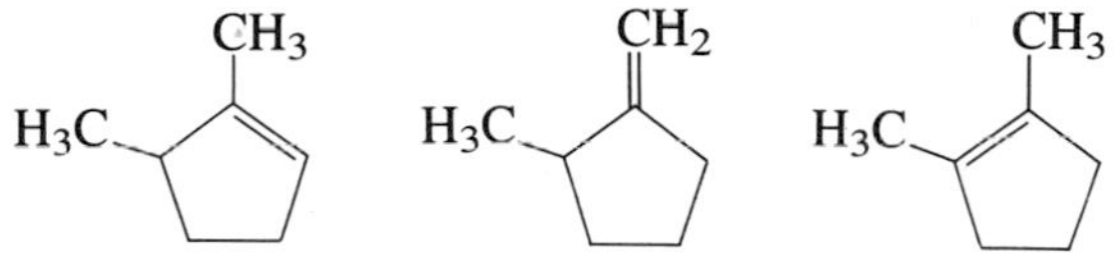

(e)

100

Transmittance (%)

3072

1649

888

0

4000 3500 3000 2500 2000 1500 1000 600 cm^{-1}

Wavenumber

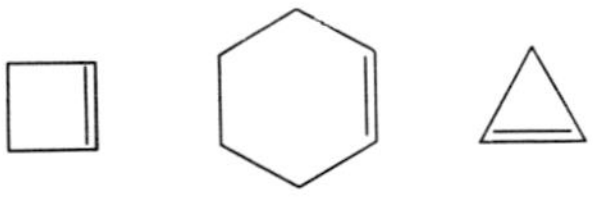

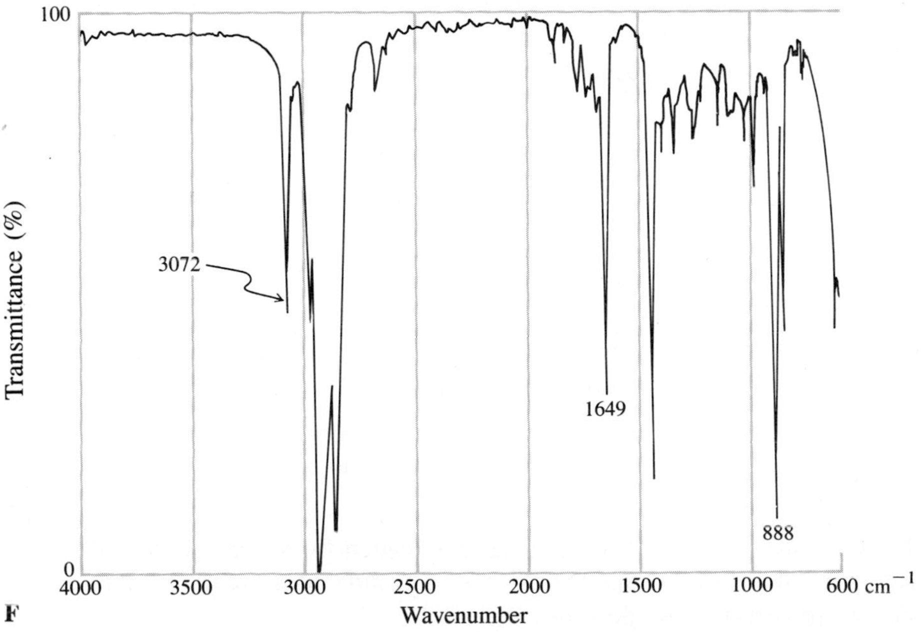

F

14. The reaction between 2-bromobutane and sodium ethoxide in ethanol gives rise to three E2 products. What are they? Predict their relative amounts.

15. What key structural feature distinguishes haloalkanes that give more than one stereoisomer on E2 elimination (e.g., 2-bromobutane, Problem 14) from those that give only a single isomer exclusively (e.g., 2-bromo-3-methylpentane, Section 11-8)?

16. Write the most likely major product(s) of each of the following haloalkanes with sodium ethoxide in ethanol or potassium *tert*-butoxide in 2-methyl-2-propanol (*tert*-butanol). **(a)** Chloromethane; **(b)** 1-bromopentane; **(c)** 2-bromopentane; **(d)** 1-chloro-1-methylcyclohexane; **(e)** (1-bromoethyl)-cyclopentane; **(f)** (2*R*,3*R*)-2-chloro-3-ethylhexane; **(g)** (2*R*,3*S*)-2-chloro-3-ethylhexane; **(h)** (2*S*,3*R*)-2-chloro-3-ethylhexane.

17. Draw Newman projections of the four stereoisomers of 2-bromo-3-methylpentane in the conformations required for E2 elimination. (See the structures labeled "Stereospecificity in the E2 Reaction of 2-Bromo-3-methylpentane" on page 399.) Are the reactive conformations also the most stable conformations? Explain.

18. Referring to the answer to Problem 11 of Chapter 7, predict (qualitatively) the relative amounts of isomeric alkenes that are formed in the elimination reactions shown.

19. Referring to the answers to Problem 4 of Chapter 9, predict (qualitatively) the relative yields of all the alkenes formed in each reaction.

20. Referring to Problem 27 of Chapter 7, write the structure of the alkene that you would expect to be formed as the major product from E2 elimination of each of the chlorinated steroids shown.

21. An enzyme has been discovered in the bacteria *Escherichia coli* that catalyzes the dehydration of a thioester derivative of (−)-3-hydroxydecanoic acid to give a mixture of the corresponding thioester derivatives of *trans*-2-decenoic acid and *cis*-3-decenoic acid:

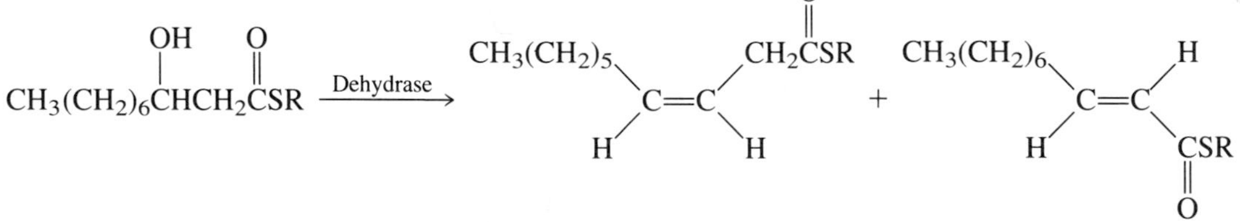

$$CH_3(CH_2)_6\overset{OH}{\overset{|}{C}}HCH_2\overset{O}{\overset{\|}{C}}SR \xrightarrow{\text{Dehydrase}} \underset{H}{\overset{CH_3(CH_2)_5}{}}C{=}C\underset{H}{\overset{CH_2\overset{O}{\overset{\|}{C}}SR}{}} + \underset{H}{\overset{CH_3(CH_2)_6}{}}C{=}C\underset{\underset{O}{\overset{\|}{C}}SR}{\overset{H}{}}$$

How does this result compare with those that can be expected from simple acid-catalyzed dehydration (e.g., H_2SO_4 and heat)?

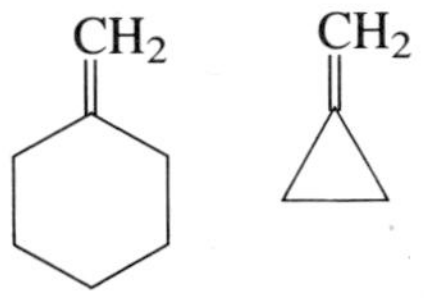

CH_2 / CH_2

A B

22. 1-Methylcyclohexene is more stable than methylenecyclohexane (A, in the margin), but methylenecyclopropane (B) is more stable than 1-methylcyclopropene. Explain.

23. Give the products of bimolecular elimination from each of the following isomeric halogenated compounds. (Note: The structures are depicted as Fischer projections; see Section 5-4.)

(a) Fischer projection: C_6H_5 (top); H—C—Br; H—C—CH_3; C_6H_5 (bottom)

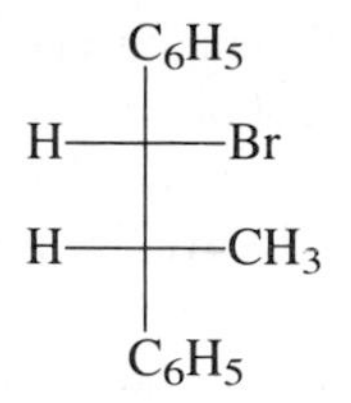

(b) Fischer projection: C_6H_5 (top); H—C—Br; H_3C—C—H; C_6H_5 (bottom)

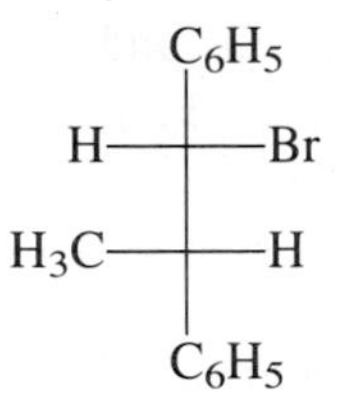

One of these compounds undergoes elimination 50 times faster than the other. Which compound is it? Why? (Hint: See Problem 17.)

24. Explain in detail the differences between the mechanisms giving rise to the following two experimental results.

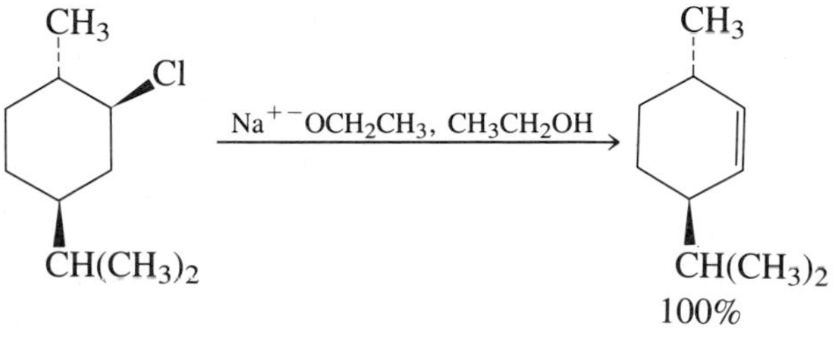

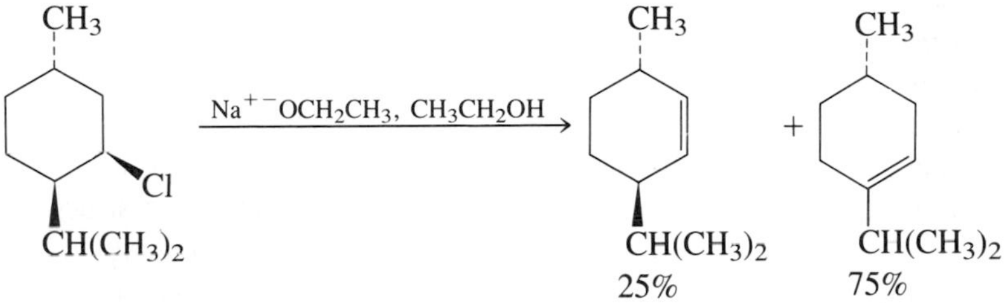

CH_3, Cl, $CH(CH_3)_2$ substituted cyclohexane $\xrightarrow{Na^{+}\,^{-}OCH_2CH_3,\ CH_3CH_2OH}$ CH_3, $CH(CH_3)_2$ cyclohexene
100%

CH_3, Cl, $CH(CH_3)_2$ substituted cyclohexane $\xrightarrow{Na^{+}\,^{-}OCH_2CH_3,\ CH_3CH_2OH}$ CH_3, $CH(CH_3)_2$ cyclohexene + CH_3, $CH(CH_3)_2$ cyclohexene
25% 75%

25. You have just been named president of the famous perfume company, Scents 'R' Us. Searching for a hot new item to market, you run across a bottle labeled only $C_{10}H_{20}O$, which contains a liquid with a wonderfully sweet rose aroma. You want more, so you set out to elucidate its structure. Do so from the following data. (i) ^{1}H NMR: clear signals at $\delta = 0.94$ (d, $J = 7$ Hz, 3 H), 1.63 (s, 3 H), 1.71 (s, 3 H), 3.68 (t, $J = 7$ Hz, 2 H), 5.10 (t, $J = 6$ Hz, 1 H) ppm; the other 8 H have overlapping absorptions in the range $\delta = 1.3$–2.2 ppm. (ii) ^{13}C NMR (^{1}H decoupled): $\delta = 60.7$, 125.0, 130.9 ppm; all other signals are upfield of $\delta = 40$ ppm. (iii) IR: $\tilde{\nu} = 1640$ and 3350 cm^{-1}. (iv) Oxidation with buffered PCC gives a compound with the molecular formula $C_{10}H_{18}O$. Its spectra show the following changes compared with the starting material: ^{1}H NMR: signal at $\delta = 3.68$ ppm is gone, but a new signal is seen at $\delta = 9.64$ ppm; ^{13}C NMR: signal at $\delta =$ 60.7 ppm is gone, replaced by one at $\delta = 202.1$ ppm; IR: loss of signal at $\tilde{\nu} = 3350$ cm^{-1}; new peak at $\tilde{\nu} = 1728$ cm^{-1}. (v) Hydrogenation gives $C_{10}H_{22}O$, identical to that formed on hydrogenation of the natural product geraniol.

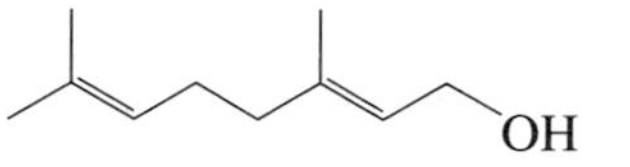

OH

Geraniol

26. Using the information in Table 11-3, match up each set of IR signals below with one of the naturally occurring compounds in the following list: camphor; menthol; chrysanthemic ester; epiandrosterone. You can find the structures of the natural products in Section 4-7. **(a)** 3355 cm^{-1}; **(b)** 1630, 1725, 3030 cm^{-1}; **(c)** 1730, 3410 cm^{-1}; **(d)** 1738 cm^{-1}.

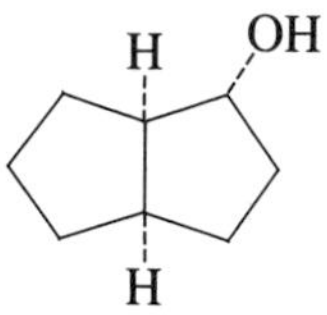

H OH

H

27. Identify compounds A, B, and C from the following information and explain the chemistry that is taking place. Reaction of the alcohol shown in the margin with 4-methylbenzenesulfonyl chloride in pyridine produced A ($C_{15}H_{20}SO_3$). Reaction of A with lithium diisopropylamide (LDA, Section 7-8) produces a single product, B (C_8H_{12}), which displays in its ^1H NMR a two-proton multiplet at about $\delta = 5.6$ ppm. If, however, A is treated with NaI before the reaction with LDA, two products are formed: B and an isomer, C, whose NMR shows a multiplet at $\delta = 5.2$ ppm that integrates as only one proton.

28. The *citric acid cycle* is a series of biological reactions that play a central role in cell metabolism. The cycle includes dehydration reactions of both malic and citric acids, yielding fumaric and aconitic acids, respectively (all common names). Both proceed strictly by enzyme-catalyzed *anti*-elimination mechanisms.

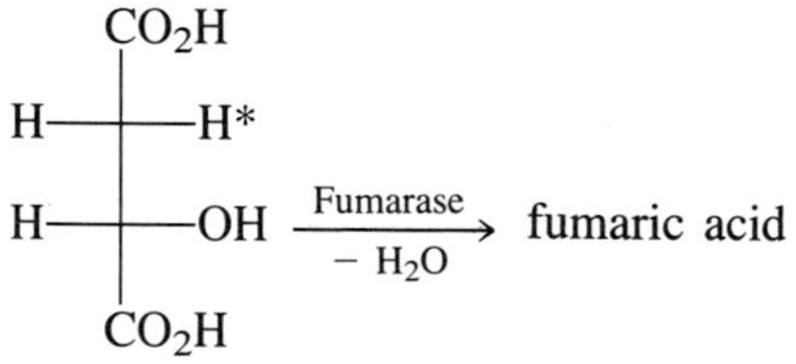

CO_2H
H—|—H*
H—|—OH $\xrightarrow[-\ H_2O]{\text{Fumarase}}$ fumaric acid
CO_2H
Malic acid

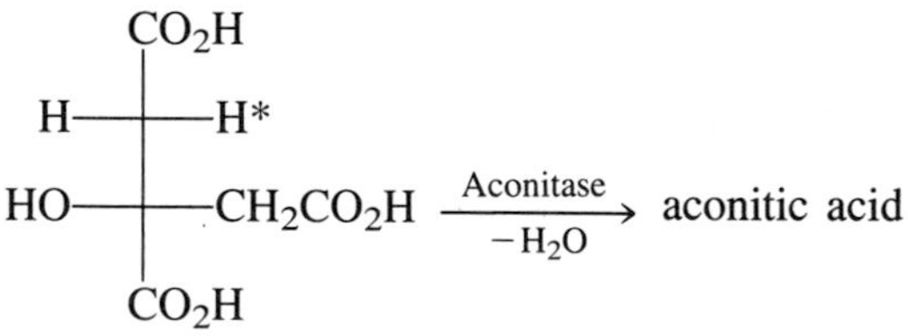

CO_2H
H—|—H*
HO—|—CH_2CO_2H $\xrightarrow[-H_2O]{\text{Aconitase}}$ aconitic acid
CO_2H
Citric acid

(a) In each dehydration, only the hydrogen identified by an asterisk is removed, together with the OH group on the carbon below. Write the structures for fumaric and aconitic acids as they are formed in these reactions. Make sure that the stereochemistry of each product is clearly indicated. **(b)** Specify the stereochemistry of each of these products, using either cis-trans or *E*,*Z* notation, as appropriate. **(c)** Isocitric acid (shown in the margin) is also dehydrated by aconitase. How many stereoisomers can exist for isocitric acid? Remembering that this reaction proceeds through *anti* elimination, write the structure of a stereoisomer of isocitric acid that will give on dehydration the same isomer of aconitic acid that is formed from citric acid. Label the chiral carbons in this isomer of isocitric acid, using *R*,*S* notation.

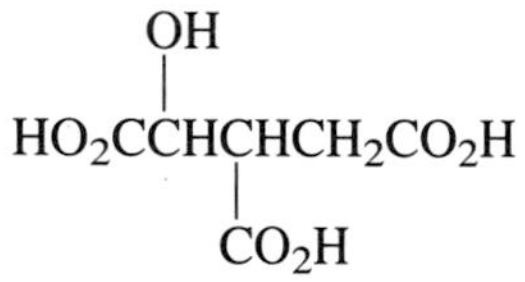

OH
$HO_2CCHCHCH_2CO_2H$
CO_2H
Isocitric acid

12

Reactions of Alkenes

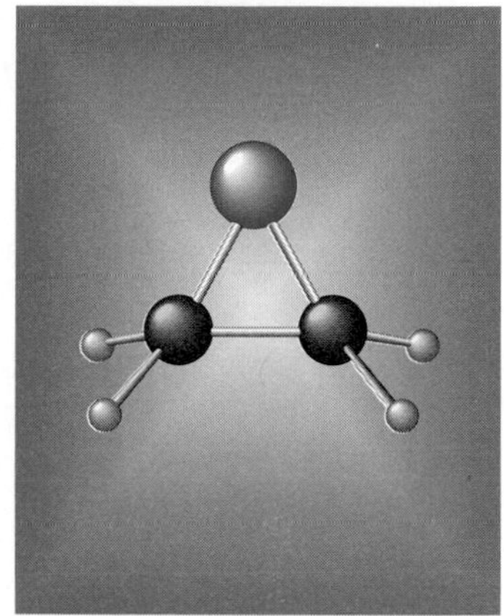

The double bond can undergo a variety of reactions, many of which lead to saturated products by addition. Additions modify the functionality at *both* alkene carbons and can therefore lead to an especially diverse array of useful organic molecules. In this chapter we shall see that additions to the π bond are generally exothermic: They are almost certain to occur if a pathway can be found. We shall discuss several of these pathways, how to control the direction of the addition and the stereochemistry of the products, and their use in synthesis.

A discussion of hydrogenation will describe details of the catalytic activation process. Then we shall present a large class of reactions requiring electrophiles, among which are protons, halogens, and mercuric ions. Other additions treated in this chapter include hydroboration, a reaction giving alkylboranes (which may be further functionalized); oxidation (which can lead to diols or even complete rupture of the double bond); and radical reactions. We conclude with reactions that start with molecules as simple as ethene and are capable of producing polymers of diverse structure, strength, elasticity, and function.

12-1 Why Addition Reactions Proceed: Thermodynamic Feasibility

The carbon–carbon π bond is relatively weak, and the chemistry of the alkenes is largely governed by its reactions. The most common transformation is **addition** of a reagent A–B to give a saturated compound. In the process, the A–B bond is broken, and A and B form single bonds to carbon. Thus, the *thermodynamic feasibility* of this process depends on the strength of the π bond, the dissociation energy DH°_{A-B}, and the strengths of the newly formed bonds of A and B.

Addition to the Alkene Double Bond

$$\text{>C=C<} + \text{A–B} \xrightarrow{\Delta H^\circ = ?} -\overset{\text{A}}{\overset{|}{\underset{|}{\text{C}}}}-\overset{\text{B}}{\overset{|}{\underset{|}{\text{C}}}}-$$

Recall that we can estimate the ΔH° of such reactions by subtracting the combined strength of the bonds made from that of the bonds broken (Section 3-4).

$$\Delta H^\circ = (DH^\circ_{\pi\text{ bond}} + DH^\circ_{A-B}) - (DH^\circ_{C-A} + DH^\circ_{C-B})$$

in which C stands for carbon.

Table 12-1 gives the DH° values (obtained by using the data from Table 3-1 and Section 3-5 and by equating the strength of the π bond to 65 kcal mol^{-1}) and the estimated ΔH° values for various additions to ethene. In all the examples, the combined strength of the bonds formed exceeds, sometimes significantly, that of the bonds broken. Therefore, *if kinetically feasible, additions to alkenes should proceed to products with release of energy*.

EXERCISE 12-1

Calculate the ΔH° for the addition of H_2O_2 to ethene to give 1,2-ethanediol (ethylene glycol) ($DH^\circ_{HO-OH} = 51$ kcal mol^{-1}).

12-2 Catalytic Hydrogenation

The simplest reaction of the double bond is its saturation with hydrogen. As discussed in Section 11-7, this reaction allows us to estimate the relative stability of substituted alkenes from their heats of hydrogenation. The process requires a catalyst, which may be either heterogeneous or homogeneous, that is, either insoluble or soluble in the reaction medium.

Hydrogenation occurs on the surface of heterogeneous catalysts

The hydrogenation of an alkene to an alkane, although exothermic, will not take place even at elevated temperatures. Ethene and hydrogen can be heated in the

TABLE 12-1 Estimated $\Delta H°$ (All Values in kcal mol^{-1}) for Additions to Ethene

$CH_2{=}CH_2$	+	A—B	⟶	H—C(A)—C(B)—H, with H on each C	
$DH°_{\pi\ bond}$		$DH°_{A-B}$		$DH°_{A-C}$ $DH°_{B-C}$	$\sim\Delta H°$
				Hydrogenation	
$CH_2{=}CH_2$	+	H—H	⟶	$H-CH_2-CH_2-H$	−27
65		104		98 98	
				Bromination	
$CH_2{=}CH_2$	+	Br—Br	⟶	H—C(Br)(H)—C(Br)(H)—H	−25
65		46		~68 ~68	
				Hydrochlorination	
$CH_2{=}CH_2$	+	H—Cl	⟶	H—C(H)(H)—C(Cl)(H)—H	−10
65		103		~98 80	
				Hydration	
$CH_2{=}CH_2$	+	H—OH	⟶	H—C(H)(H)—C(OH)(H)—H	−6
65		119		~98 92	

gas phase to 200°C for prolonged periods without any measurable change. However, as soon as a catalyst is added, hydrogenation proceeds even at room temperature at a steady rate. The catalysts frequently are the same as those used for the catalytic hydrogenation of carbonyl compounds to alcohols (Section 8-6): insoluble materials such as palladium (e.g., dispersed on carbon, Pd-C), platinum (Adams's* catalyst, PtO_2, which is converted into colloidal platinum metal in the presence of hydrogen), and nickel (finely dispersed, as in a preparation called Raney† nickel, Ra-Ni).

The major function of the catalyst is the activation of hydrogen to generate metal-bound hydrogen on the catalyst surface (Figure 12-1). Without the metal, thermal cleavage of the strong H–H bond is energetically prohibitive. Solvents commonly used in such hydrogenations are methanol, ethanol, acetic acid, and ethyl acetate, among others, as shown in the following example.

*Professor Roger Adams, 1889–1971, University of Illinois.
†Dr. Murray Raney, 1885–1966, Raney Catalyst Company.

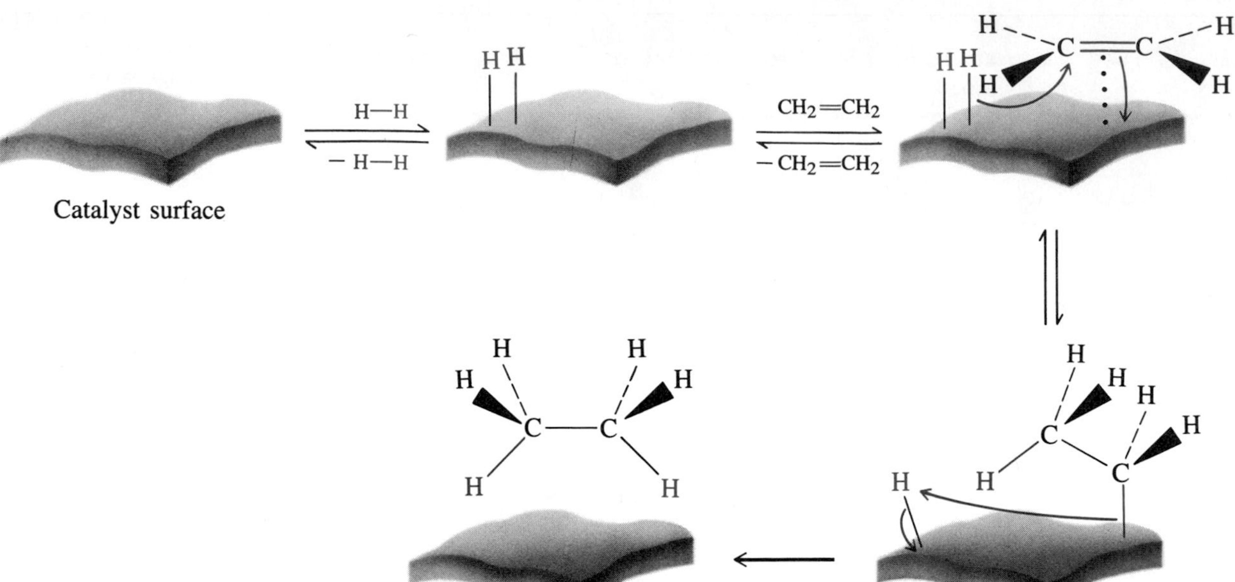

FIGURE 12-1 In the catalytic hydrogenation of ethene to produce ethane, the hydrogens bind to the catalyst surface and are then delivered to the carbons of surface-adsorbed alkene.

$$(CH_3)_2C{=}CHCH_2CH_2CH_3 \xrightarrow{\text{1 atm H–H, }PtO_2\text{, }CH_3OH\text{, 25°C}} CH_3\underset{H}{\overset{CH_3}{C}}H{-}CHCH_2CH_2CH_3 \quad 100\%$$

2-Methyl-2-hexene → **2-Methylhexane**

Hydrogenation is stereospecific

An important feature of catalytic hydrogenation is *stereospecificity*. The two hydrogen atoms are added to the same face of the double bond **(*syn* addition).** For example, 1-ethyl-2-methylcyclohexene is hydrogenated over platinum to give specifically *cis*-1-ethyl-2-methylcyclohexane. Addition of hydrogen can be from above or from below the molecular plane with equal probability. Therefore, each stereocenter is generated as both image and mirror image, and the product is racemic.

$$\text{1-Ethyl-2-methylcyclohexene} \xrightarrow{H_2\text{, }PtO_2\text{, }CH_3CH_2OH\text{, 25°C}} \text{(H, }CH_2CH_3\text{; H, }CH_3\text{)} + \text{(H, }CH_3CH_2\text{; H, }H_3C\text{)} \quad 82\%$$

1-Ethyl-2-methylcyclohexene → ***cis*-1-Ethyl-2-methylcyclohexane (Racemic)**

However, if steric hindrance inhibits hydrogenation on one side of a ring, addition will occur exclusively to the *less hindered* side. Thus, hydrogenation of the bicyclic alkene car-3-ene, a constituent of turpentine, over platinum gives only one saturated product, with the common name *cis*-carane. The prefix cis indicates that the methyl group and the cyclopropane ring are on the same side of the cyclohexane ring. This result shows that hydrogen has been added only from the less hindered face of the double bond, opposite the three-membered ring, thus pushing the methyl group cis to that ring. (Make a model and use a tabletop to represent the catalyst surface, as in Figure 12-1.)

Car-3-ene $\xrightarrow{100\text{ atm } H_2,\ PtO_2,\ CH_3CH_2OH,\ 25°C}$ 98% *cis*-Carane not [other isomer]

Car-3-ene **cis-Carane** 98%

EXERCISE 12-2

Catalytic hydrogenation of (*S*)-2,3-dimethyl-1-pentene gives only one optically active product. Show the product and explain the result.

In summary, hydrogenation of the double bond in alkenes requires a catalyst. This transformation occurs stereospecifically by *syn* addition, and, when there is a choice, from the least hindered side of the molecule.

12-3 Nucleophilic Character of the Pi Bond: Electrophilic Addition of Hydrogen Halides

As noted earlier, the π electrons of a double bond are not as strongly bound as those of a σ bond. The electron cloud above and below the molecular plane of the alkene is polarizable and subject to attack by electron-deficient species, just as are the lone electron pairs in typical Lewis bases. Halogens and the mercuric ion are examples of electrophilic species that attack the π electrons of the double bond. As in hydrogenation, the double bond is altered by *addition,* but by different mechanisms. These transformations, called electrophilic additions, can be regioselective and stereospecific. We begin this section with the simplest of all electrophiles, the proton.

Electrophilic attack by protons gives carbocations

The proton of a strong acid may add to a double bond to yield a carbocation. The transition state for the process is the same as that formulated for the deprotonation step in the E1 reaction (Section 7-6). In the absence of a good nucleophile capable of trapping the carbocation, rearrangement may be observed. However, in the presence of such a nucleophile, particularly at low temperatures, the carbocation is intercepted to give the product of an **electrophilic addition** to the double bond. For example, treatment of alkenes with hydrogen halides leads to the corresponding haloalkanes.

BOX 12-1 Optically Active Amino Acids by Asymmetric Hydrogenation

Amino acids, the building blocks of proteins (see Chapter 26), are synthesized in the laboratory and in industry in optically active form by catalytic hydrogenation of achiral, nitrogen-substituted alkenes called *enamides*. The process uses a homogeneous (soluble) catalyst, consisting of a metal such as rhodium or ruthenium and an optically active *phosphine ligand,* which binds to the metal. Bulky groups on the phosphorus atom direct the hydrogenation to one face of the alkene with high selectivity, thereby furnishing a single enantiomer.

Most of the ligands used in such *asymmetric hydrogenations* are available in both enantiomeric forms; therefore, either enantiomer of a product may be prepared. For example, hydrogenation of (*Z*)-2-acetamido-3-phenylpropenoic acid using a rhodium catalyst containing the ligand "Degphos" produces the (*S*)-phenylalanine derivative shown below with over 99% stereoselectivity.

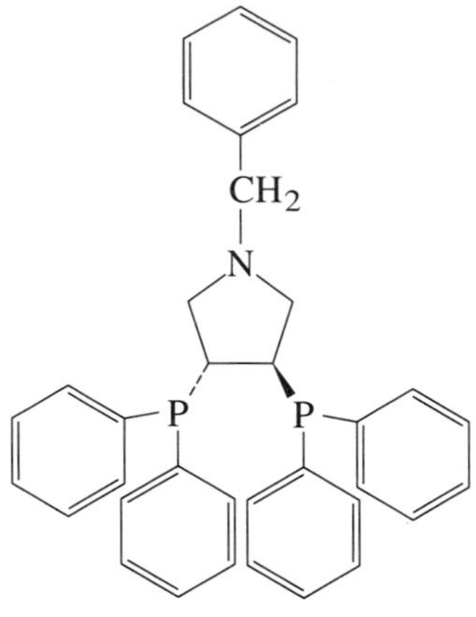

(*R,R*)-Degphos

(A phosphine ligand)

(*S*)-Phenylalanine is used in the manufacture of the artificial sweetener aspartame. Its *R* enantiomer, which occurs in the antibiotic gramicidin S, is equally readily obtained by hydrogenation of the same enamide with Rh in the presence of the enantiomeric ligand, (*S*,*S*)-Degphos.

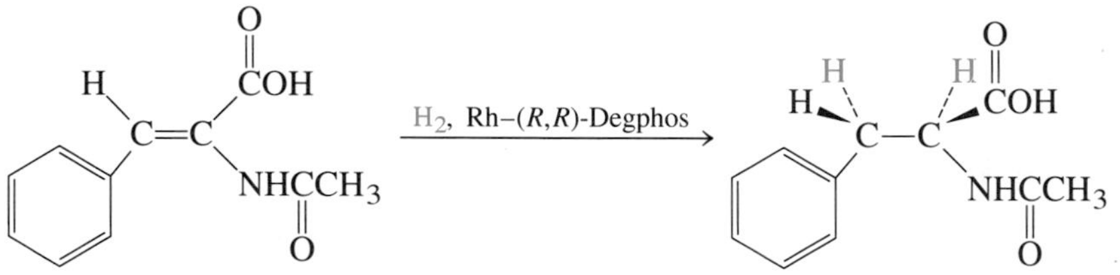

H_2, Rh–(*R,R*)-Degphos

(*Z*)-2-Acetamido-3-phenylpropenoic acid

(An enamide)

(*S*)-*N*-Acetylphenylalanine

Electrophilic Addition of HX to Alkenes

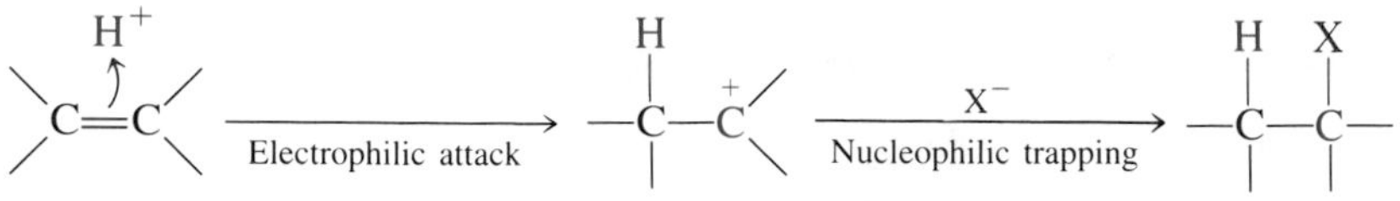

Electrophilic attack

Nucleophilic trapping

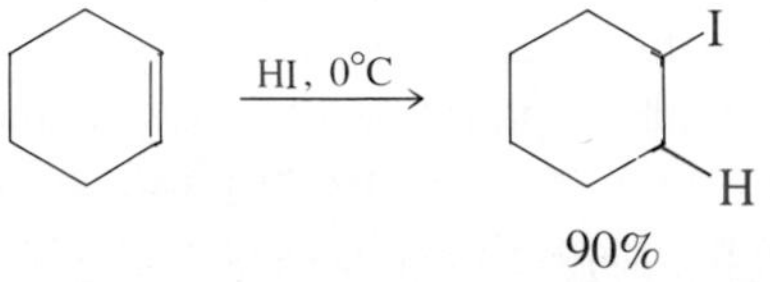

HI, 0°C

90%

Cyclohexene

Iodocyclohexane

In a typical experiment, the gaseous hydrogen halide is bubbled through pure or dissolved alkene. Alternatively, HX can be added in a solvent, such as acetic acid. Aqueous work-up furnishes the haloalkane in high yield. All the hydrogen halides can be used successfully in such addition reactions.

The Markovnikov rule predicts regioselectivity in electrophilic additions

Are additions of HX to unsymmetric alkenes regioselective? To answer this question, let us consider the reaction of propene with hydrogen chloride. Two products are possible: 2-chloropropane and 1-chloropropane. However, the only product observed is 2-chloropropane.

Regioselective Electrophilic Addition to Propene

$$CH_3CH{=}CH_2 \xrightarrow{HCl} CH_3\underset{Cl}{C}H\underset{H}{C}H_2 \quad \text{but no} \quad CH_3\underset{H}{C}H\underset{Cl}{C}H_2$$

Less substituted — **2-Chloropropane** — **1-Chloropropane**

Similarly, reaction of 2-methylpropene with hydrogen bromide gives only 2-bromo-2-methylpropane, and 1-methylcyclohexene combines with HI to furnish only 1-iodo-1-methylcyclohexane.

$$(H_3C)_2C{=}CH_2 \xrightarrow{HBr} CH_3\overset{CH_3}{\underset{Br}{C}}CH_2H$$

Less substituted

1-Methylcyclohexene (CH_3) $\xrightarrow{HI}$ 1-iodo-1-methylcyclohexane (CH_3, I, H)

Less substituted

We can see from these examples that, if the carbon atoms participating in the double bond are not equally substituted, *the proton from the hydrogen halide attaches itself to the less substituted carbon*. As a consequence, the halogen ends up at the more substituted carbon. This phenomenon, referred to as the **Markovnikov* rule,** can be explained in terms of what we know about the mechanism of electrophile additions of protons to alkenes. The key is the relative stability of the resulting carbocations.

Consider the hydrochlorination of propene. The regiochemistry of the reaction is determined in the first step, in which the proton attacks the π system to give an intermediate carbocation. Carbocation generation is rate determining; once it occurs, reaction with chloride proceeds quickly. Let us look at the crucial first step in more detail. The proton may attack either of the two carbon atoms of the double bond. Addition to the internal carbon leads to the primary propyl cation.

Protonation of Propene at C2

$$H_3C(H)C{=}CH_2 + H^+ \longrightarrow \left[H_3C(H)\overset{\delta^+}{C}\cdots C(H)_2 \cdots H^{\delta+} \right]^\ddagger \longrightarrow CH_3CH_2CH_2^+$$

TS-1 — **Primary carbocation** (Not observed)

*Professor Vladimir V. Markovnikov, 1838–1904, formulated his rule in 1869, University of Moscow.

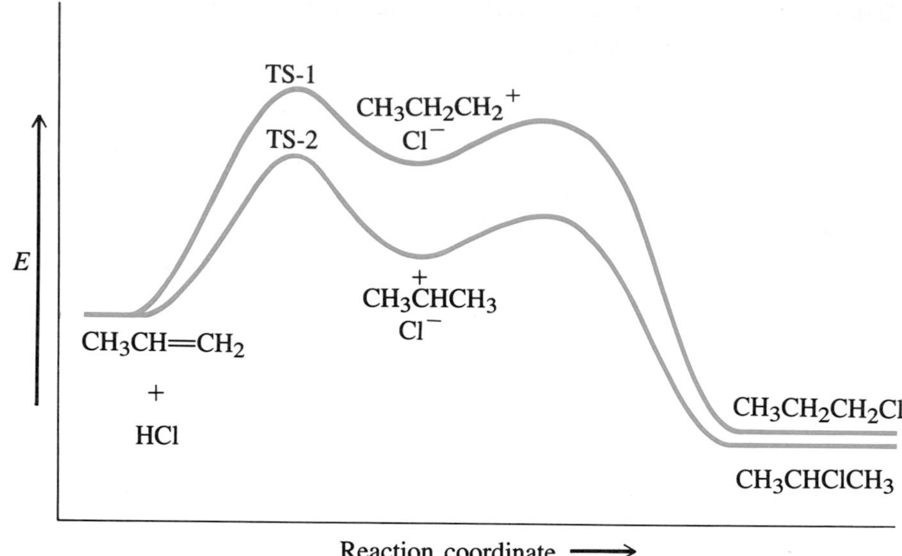

FIGURE 12-2 Potential-energy diagram for the two possible modes of HCl addition to propene. Transition state 1 (TS-1), which leads to the higher-energy primary propyl cation, is less favored than transition state 2 (TS-2), which gives the 1-methylethyl (isopropyl) cation.

In contrast, protonation at the terminal carbon results in formation of the secondary 1-methylethyl (isopropyl) cation.

Protonation of Propene at C1

$$CH_3(H)C{=}CH_2 + H^+ \longrightarrow \left[CH_3(H)\overset{\delta^+}{C}{\cdots}C(H)_2{\cdots}H^{\delta^+} \right]^{\ddagger} \longrightarrow CH_3\overset{+}{C}HCH_3$$

TS-2 **Secondary carbocation** (Favored)

The second species is more stable and, because the structure of the transition state for protonation resembles that of the resulting cation, is formed considerably faster. Figure 12-2 is a potential-energy diagram of this situation.

On the basis of this analysis, we can rephrase the empirical Markovnikov rule: HX adds to unsymmetric alkenes in such a way that *the initial protonation gives the more stable carbocation*. For alkenes that are similarly substituted at both sp^2 carbons, product mixtures are to be expected, because carbocations of comparable stability are formed.

EXERCISE 12-3

Predict the outcome of the addition of HBr to **(a)** 1-hexene; **(b)** *trans*-2-pentene; **(c)** 2-methyl-2-butene; **(d)** 4-methylcyclohexene. How many isomers can be formed in each case?

EXERCISE 12-4

Draw a potential-energy diagram for the reaction in (c) of Exercise 12-3.

In summary, additions of hydrogen halides to alkenes are electrophilic reactions that begin with protonation of the double bond to give a carbocation. Trapping of the carbocation by halide ion gives the final product. The Markovnikov rule predicts the regioselectivity of hydrohalogenation to haloalkanes.

12-4 Alcohol Synthesis by Electrophilic Hydration: Thermodynamic Control

So far, the nucleophilic trapping agents have been the counterions to the protons of the attacking acids. What about other nucleophiles? When an alkene is exposed to an *aqueous* solution of an acid having a poorly nucleophilic counterion, such as sulfuric acid, water plays the role of the trapping nucleophile and intercepts the carbocation formed by the initial protonation. In the overall reaction, which follows the Markovnikov rule, the elements of water are added to the double bond, an **electrophilic hydration.**

The process is the reverse of the acid-induced dehydration of alcohols (Section 11-9), and its mechanism is the same in reverse, as illustrated in the hydration of 2-methylpropene, a reaction of industrial importance leading to 2-methyl-2-propanol (*tert*-butyl alcohol).

Electrophilic Hydration

1-Methylcyclohexene $\xrightarrow{\text{50\% HOH, } H_2SO_4}$ 1-Methylcyclohexanol (50%)

Mechanism of the Hydration of 2-Methylpropene

$$(H_3C)_2C{=}CH_2 \underset{-H^+}{\overset{H^+}{\rightleftharpoons}} (H_3C)_2\overset{+}{C}CH_3 \underset{-\,HOH}{\overset{+\,HOH}{\rightleftharpoons}} (CH_3)_3C{-}\overset{+}{O}H_2 \underset{+H^+}{\overset{-H^+}{\rightleftharpoons}} (CH_3)_3C{-}OH$$

Alkene hydration and alcohol dehydration are equilibrium processes

In the mechanism of alkene hydration, *all the steps are reversible*. The proton acts only as a catalyst and is not consumed in the reaction. Indeed, without the acid, hydration would not occur; alkenes are stable in neutral water. The presence of acid, however, establishes an equilibrium between alkene and alcohol. This equilibrium may be driven toward the alcohol by using low reaction temperatures and a large excess of water. Conversely, we have seen (Section 11-9) that treatment of the alcohol with concentrated acid favors dehydration, especially at elevated temperatures.

Hydration–Dehydration Equilibrium

$$RCH{=}CH_2 + H_2O \underset{}{\overset{\text{Catalytic } H^+}{\rightleftharpoons}} RCH(OH)CH_3$$

$$\text{Alcohol} \underset{H_2SO_4\text{, excess } H_2O\text{, low temperature}}{\overset{\text{Conc. } H_2SO_4\text{, high temperature}}{\rightleftharpoons}} \text{alkene} + H_2O$$

EXERCISE 12-5

Treatment of 2-methylpropene with catalytic deuterated sulfuric acid (D_2SO_4) in D_2O gives $(CD_3)_3COD$. Explain by a mechanism.

The reversibility of alkene protonation leads to alkene equilibration

Section 11-9 explained that the acid-catalyzed dehydration of alcohols gives mixtures of alkenes in which the more stable isomers predominate. Equilibrating carbocation rearrangements are partly responsible for these results. However, another pathway is **reversible protonation** of the initial alkene products arising by E1 pathways.

Recall the dehydration of 2-butanol. Loss of water from the protonated alcohol gives the 2-butyl carbocation. Proton loss from this cation can then give any of three observed products: 1-butene, *cis*-2-butene, or *trans*-2-butene. However, under the strongly acidic conditions, a proton can re-add to the double bond. As noted earlier, this addition will be highly regioselective (Markovnikov rule) and will regenerate the same secondary carbocation that was formed in the initial dehydration. Because this cation may again lose a proton to give any of the same three alkene isomers, the net effect is *interconversion* of the isomers to an *equilibrium mixture,* in which the thermodynamically most stable isomer is the major component. This system is therefore an example of a reaction that is under *thermodynamic control*.

Mechanism of Thermodynamic Control in Acid-Catalyzed Dehydrations

STEP 1. Initial proton loss might give terminal alkene.

$$CH_3CH_2\underset{H}{\overset{OH}{C}}CH_3 \underset{-\,H_2O}{\overset{H^+}{\rightleftharpoons}} CH_3CH_2\underset{H}{\overset{+}{C}}CH_3 \overset{-\,H^+}{\rightleftharpoons} CH_3CH_2CH{=}CH_2$$

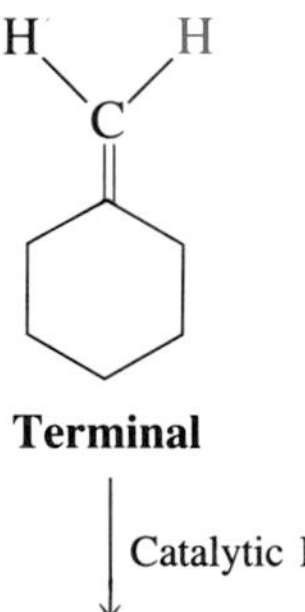

STEP 2. Reprotonation regenerates carbocation, which eventually leads to internal alkene.

$$CH_3CH_2CH{=}CH_2 \overset{H^+}{\rightleftharpoons} CH_3CH_2\underset{H}{\overset{+}{C}}CH_3 \overset{-\,H^+}{\rightleftharpoons} CH_3CH{=}CHCH_3$$

By this procedure, less stable alkenes may be catalytically converted into their more stable isomers.

Acid-Catalyzed Equilibration of Alkenes

$$\underset{\textbf{Cis}}{(CH_3)_3C(H)C{=}C(H)C(CH_3)_3} \xrightarrow{\text{Catalytic } H^+} \underset{\textbf{Trans}}{(CH_3)_3C(H)C{=}C(H)C(CH_3)_3}$$

EXERCISE 12-6

Write a mechanism for the following rearrangement. What is the driving force for the reaction?

H_3C $\xrightarrow{H^+}$ H_3C CH_2CH_3

In summary, the carbocation formed by addition of a proton to an alkene may be trapped by water to give an alcohol, the reverse of alkene synthesis by alcohol dehydration. Reversible protonation equilibrates alkenes in the presence of acid, thereby forming a thermodynamically controlled mixture of isomers.

12-5 Electrophilic Addition of Halogens to Alkenes

Reagents that do not appear to contain electrophilic atoms also can attack double bonds electrophilically. An example is the halogenation of alkenes, which proceeds with addition of two halogen atoms to the double bond to give a vicinal dihalide. The reaction works best for chlorine and bromine. Fluorine reacts too violently (often explosively) with alkenes, whereas diiodide formation is virtually thermoneutral and is not generally observed.

Halogenation of Alkenes

$$\mathrm{C{=}C} \xrightarrow{\mathrm{X{-}X}} \mathrm{XC{-}CX}$$

X = Cl, Br

Vicinal dihalide

EXERCISE 12-7

Calculate (as in Table 12-1) the $\Delta H°$ values for the addition of F_2 and I_2 to ethene. (For $DH°_{X_2}$, see Section 3-5.)

Bromine addition is particularly easy to recognize because bromine solutions immediately change from red to colorless when exposed to an alkene. This phenomenon is sometimes used to test for unsaturation. Saturated systems do not react with bromine unless heat or light initiates a radical reaction (Section 3-5).

Halogenations are best carried out at room temperature or, with cooling, in inert halogenated solvents such as the halomethanes.

Electrophilic Halogen Addition of Br_2 to 1-Hexene

$$\mathrm{CH_3(CH_2)_3CH{=}CH_2} \xrightarrow{\mathrm{Br{-}Br,\ CCl_4}} \mathrm{CH_3(CH_2)_3CHBrCH_2Br}$$

1-Hexene | 90% **1,2-Dibromohexane**

Halogen additions to double bonds may seem to be similar to hydrogenations. However, their mechanism is quite different, as revealed by the stereochemistry of bromination; similar arguments hold for the other halogens.

Bromination occurs through *anti* addition

What is the stereochemistry of bromination? Are the two bromine atoms added from the same side of the double bond (as in catalytic hydrogenation) or from opposite sides? Let us examine the bromination of cyclohexene. Double addition on the same side should give *cis*-1,2-dibromocyclohexane; the alternative would result in *trans*-1,2-dibromocyclohexane. The second alternative is borne out by experiment—only *anti* addition is observed. Because *anti* addition to the two reacting carbon atoms may occur with equal probability in two possible ways—in either case, from both above and below the π bond—the product is racemic.

Anti Bromination of Cyclohexene

Br_2, CCl_4 → Br, Br + Br, Br

83%

Racemic _trans_-1,2-dibromocyclohexane

With acyclic alkenes, the reaction is also cleanly stereospecific. For example, *cis*-2-butene is brominated to furnish a racemic mixture of (2*R*,3*R*)- and (2*S*,3*S*)-2,3-dibromobutane; *trans*-2-butene results in the meso diastereomer.

Stereospecific 2-Butene Bromination

cis-2-Butene —Br_2, CCl_4→ **(2*R*,3*R*)-2,3-Dibromobutane** + **(2*S*,3*S*)-2,3-Dibromobutane**

trans-2-Butene —Br_2, CCl_4→ Identical **_meso_-2,3-Dibromobutane**

Cyclic bromonium ions explain the stereochemistry

What mechanism explains the observed stereochemistry? How does bromine attack the electron-rich double bond even though it does not appear to contain an electrophilic center? The answers lie in the polarizability of the Br–Br bond, which is prone to heterolytic cleavage on reaction with a nucleophile. The π electron cloud of the alkene is nucleophilic and attacks one end of the bromine molecule with simultaneous displacement of the second bromine atom as bromide ion in an S_N2-like process. The resulting intermediate is a cyclic **bromonium ion,** in which the bromine bridges both carbon atoms of the original double bond to form a three-membered ring (Figure 12-3).

The intermediacy of the cyclic bromonium ion explains the stereochemistry of bromination. The structure of this ion is rigid, and it may be attacked only on the side opposite the bromine. This attack is made by the bromide ion generated in

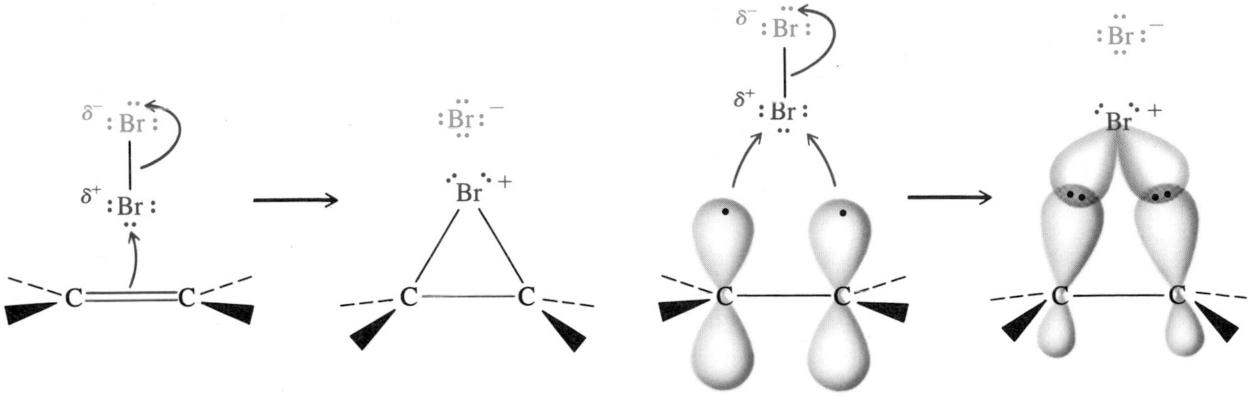

A

B

FIGURE 12-3
(A) Electron-pushing picture of cyclic bromonium ion formation. The alkene (red) acts as a nucleophile to displace bromide ion (green) from bromine. The molecular bromine behaves as if it were strongly polarized, one atom as a bromine cation, the other as an anion. (B) Molecular-orbital picture of bromonium ion formation.

the first step; the three-membered ring is thus opened stereospecifically (compare nucleophilic ring opening of oxacyclopropanes in Section 9-9). The leaving group is the bridging bromine. In symmetric bromonium ions, attack is equally probable at either carbon atom, thereby giving the racemic (or meso) products observed.

Nucleophilic Opening of a Cyclic Bromonium Ion

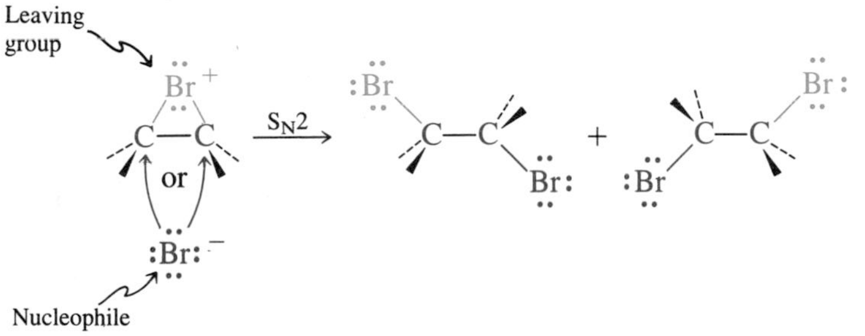

Leaving group

Nucleophile

S_N2

EXERCISE 12-8

Draw the intermediate in the bromination of cyclohexene, using the conformational picture below. Show why the product is racemic. What can you say about the initial conformation of the product?

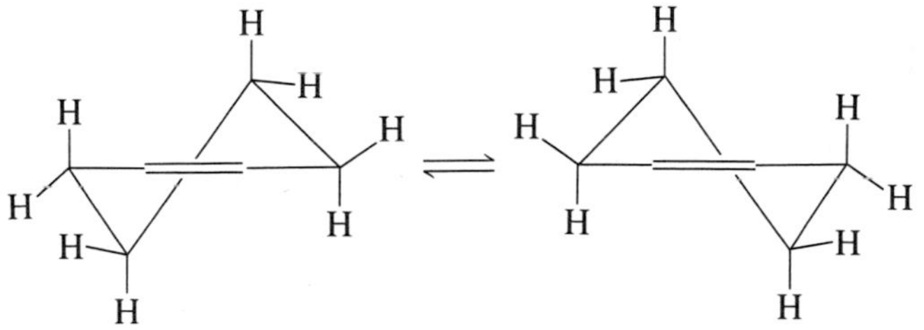

Conformational flip in cyclohexene

In summary, halogens add, as electrophiles, to alkenes to produce intermediate bridged halonium ions. These are opened stereospecifically by the halide ions displaced in the initial step to give overall *anti* addition to the double bond.

12-6 The Generality of Electrophilic Addition

A number of electrophile–nucleophile combinations are capable of addition to double bonds, the reactions resulting in a wide variety of useful products.

The bromonium ion can be trapped by other nucleophiles

The creation of a bromonium ion in alkene brominations suggests that, in the presence of other nucleophiles, competition might be observed in the trapping of the intermediate. Indeed, this is possible. For example, bromination of cyclopentene in the presence of excess chloride ion (added as a salt) gives a mixture of *trans*-1,2-dibromocyclopentane and *trans*-1-bromo-2-chlorocyclopentane.

Competitive Trapping of a Cyclic Bromonium Ion

(Although all products are racemic, only one enantiomer is shown in each case)

Br_2, Cl^-, CCl_4, 0°C

H :Br⁺ H + :Br:⁻ + :Cl:⁻

:Br:⁻ H Br: H Br:

:Cl:⁻ H Br: H Cl:

A large excess of the competing nucleophile may prevent the formation of mixtures of products. For example, bromination of cyclopentene in water as solvent gives the vicinal bromoalcohol (common name, bromohydrin) exclusively. In this case, the bromonium ion is attacked by water. The net transformation is the *anti* addition of the elements of Br–OH to the double bond. The other product formed is HBr. The corresponding chloroalcohols (chlorohydrins) can be made from chlorine in water through the intermediacy of a chloronium ion.

Bromoalcohol (Bromohydrin) Synthesis

Br_2, H_2O, 0°C; $-Br^-$

H :Br⁺ H + $\ddot{O}H_2$ ⟶ H Br: H OH⁺ H ⟶ ($-H^+$) H :Br: H :OH

***trans*-2-Bromocyclopentanol**

EXERCISE 12-9

Show the expected product from the reaction of **(a)** *trans*-2-butene and **(b)** *cis*-2-pentene with aqueous chlorine. Show the stereochemistry clearly.

Vicinal haloalcohols undergo intramolecular ring closure in the presence of base to give oxacyclopropanes (Section 9-9) and are therefore useful intermediates in organic synthesis.

Oxacyclopropane Formation from an Alkene Through the Haloalcohol

Cyclohexene $\xrightarrow{Cl_2,\ H_2O,\ 0°C}$ trans-2-chlorocyclohexanol (Cl, H, OH, H), 70–73% $\xrightarrow[-HCl]{NaOH,\ H_2O,\ 25°C}$ oxacyclopropane (H, O, H), 70–73%

If alcohol is used as a solvent instead of water in these halogenations, the corresponding vicinal haloethers are produced, as shown in the margin.

Vicinal Haloether Synthesis

Cyclohexene $\xrightarrow[-\ HBr]{Br\ Br,\ H\text{–}OCH_3}$ (H, OCH_3, Br, H) 76%

***trans*-1-Bromo-2-methoxycyclohexane**

Halonium ion opening can be regioselective

In contrast with dihalogenations, mixed additions to double bonds can pose regiochemical problems. Is the addition of Br–OR to an unsymmetric double bond selective? The answer is yes. For example, 2-methylpropene is converted by aqueous bromine only into 1-bromo-2-methyl-2-propanol; none of the alternative regioisomer, 2-bromo-2-methyl-1-propanol, is formed.

$$(H_3C)_2C{=}CH_2 \xrightarrow[-\ HBr]{Br\text{–}Br,\ H\text{–}OH} CH_3C(OH)(CH_3)CH_2Br \quad \text{but no} \quad CH_3C(Br)(CH_3)CH_2OH$$

82% yield

1-Bromo-2-methyl-2-propanol **2-Bromo-2-methyl-1-propanol**

The electrophilic halogen in the product always becomes linked to the less substituted carbon of the original double bond, and the subsequently added nucleophile attaches to the more highly substituted center.

How can this be explained? The situation is very similar to the acid-catalyzed nucleophilic ring opening of oxacyclopropanes (Section 9-9), in which the intermediate contains a protonated oxygen in the three-membered ring. In both reactions, the nucleophile attacks the more highly substituted carbon of the ring, because this carbon is more positively polarized than the other.

Recall:
Nucleophile—red
Electrophile—blue
Leaving group—green

Regioselective Opening of the Bromonium Ion Formed from 2-Methylpropene

Greater δ^+ here

$$(CH_3)_2C\overset{Br^+}{\frown}CH_2 + H\text{—}\ddot{O}H \longrightarrow (CH_3)_2C(\text{—}\overset{+}{O}H_2)\text{—}CH_2\ddot{Br}: \xrightarrow{-\ H^+} CH_3C(CH_3)(:\ddot{O}H)\text{—}CH_2\ddot{Br}:$$

Attack at more substituted carbon of bromonium ion

A simple rule of thumb is that electrophilic additions of unsymmetric reagents of this type add in a regiochemical sense that is Markovnikovlike, the electrophilic unit emerging at the less substituted carbon of the double bond. Mixtures are formed only when the two carbons are not sufficiently differentiated [see (b) of Exercise 12-10].

EXERCISE 12-10

What are the product(s) of the following reactions?

(a) $CH_3CH{=}CH_2 \xrightarrow{Cl_2,\ CH_3OH}$ (b) 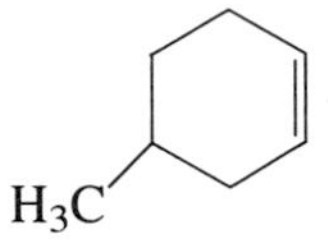 4-methylcyclohexene (H_3C-substituted cyclohexene) $\xrightarrow{Br_2,\ H_2O}$

EXERCISE 12-11

What would be a good alkene precursor for a racemic mixture of (2*R*,3*R*)- and (2*S*,3*S*)-2-bromo-3-methoxypentane? What other isomeric products might you expect to find from the reactions that you propose?

Alkenes in general can undergo stereo- and regiospecific addition reactions with reagents of the type A–B, in which the A–B bond is polarized such that A acts as the electrophile A^+, B as the nucleophile B^-. Table 12-2 shows how such reagents add to 2-methylpropene.

In summary, halonium ions are subject to stereospecific and regioselective ring opening in a manner mechanistically very similar to the nucleophilic opening of protonated oxacyclopropanes. Halonium ions can be trapped by halide ions,

TABLE 12-2 Reagents A–B That Add to Alkenes by Electrophilic Attack

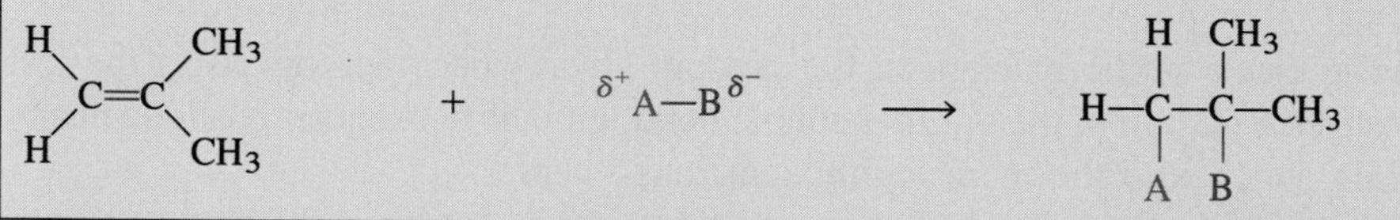

$$(H)_2C{=}C(CH_3)_2 + {}^{\delta^+}A{-}B^{\delta^-} \longrightarrow H{-}\underset{A}{\overset{H}{C}}{-}\underset{B}{\overset{CH_3}{C}}{-}CH_3$$

Name	Structure	Addition product to 2-methylpropene
Bromine chloride	Br—Cl	$BrCH_2C(CH_3)_2$ with Cl on C
Cyanogen bromide	Br—CN	$BrCH_2C(CH_3)_2$ with CN on C
Iodine chloride	I—Cl	$ICH_2C(CH_3)_2$ with Cl on C
Sulfenyl chlorides	RS—Cl	$RSCH_2C(CH_3)_2$ with Cl on C
Mercuric salts	XHg—X[a]	$XHgCH_2C(CH_3)_2$ with X on C

a. X here denotes acetate.

water, and alcohols to give vicinal dihaloalkanes, haloalcohols, and haloethers, respectively. The principle of electrophilic addition can be applied to any reagent A–B containing a polarized or polarizable bond.

12-7 Oxymercuration–Demercuration: A Special Electrophilic Addition

The last example in Table 12-2 constitutes an electrophilic addition of a mercuric salt to an alkene. The reaction is called mercuration; and the resulting compound is an alkylmercury derivative, from which the mercury can be removed in a subsequent step. One particularly useful reaction sequence is **oxymercuration–demercuration,** in which mercuric acetate acts as the reagent. In the first step (oxymercuration), treatment of an alkene with this species in the presence of water leads to the corresponding addition product.

Oxymercuration

CH_3 + $CH_3\overset{O}{\overset{\|}{C}}OHgO\overset{O}{\overset{\|}{C}}CH_3$ + H—OH $\xrightarrow{THF}$ CH_3, OH, H, $HgO\underset{\underset{O}{\|}}{C}CH_3$ + $CH_3\overset{O}{\overset{\|}{C}}OH$

Mercuric acetate **Alkylmercuric acetate**

In the subsequent demercuration, the mercury-containing substituent can be replaced by hydrogen through treatment with sodium borohydride in base. The net result is hydration of the double bond to give an alcohol.

Demercuration

CH_3, OH, H, $HgO\underset{\underset{O}{\|}}{C}CH_3$ $\xrightarrow{NaBH_4,\ NaOH,\ H_2O}$ CH_3, OH, H, H + Hg + $CH_3\underset{\underset{O}{\|}}{C}O^-$

1-Methylcyclopentanol

Oxymercuration is usually *anti* stereospecific; it is also regioselective. This outcome implies a mechanism similar to that for the electrophilic addition reactions discussed so far. The mercury reagent can be thought of as initially dissociating into a cationic mercury species and an acetate ion. The cation then attacks the alkene double bond, furnishing a mercurinium ion, probably with a structure similar to that of a cyclic bromonium ion. The water that is present attacks the more substituted carbon to give the alkylmercuric acetate intermediate, which is reduced in a subsequent step by sodium borohydride. The mechanism of the reduction reaction is complex and not completely understood. After reduction, the resulting product is the same as the product of Markovnikov hydration of the starting material.

Oxymercuration–demercuration is a valuable alternative to acid-catalyzed hydration, because it is not susceptible to the rearrangements that commonly occur under acidic conditions.

BOX 12-2 Synthesis of a Juvenile Hormone Analog

An application of the oxymercuration–demercuration reaction to the synthesis of an analog of juvenile hormone is shown below. Juvenile hormone is a substance that controls the larval metamorphosis in insects. It is produced by the male wild silk moth *Hyalophora cecropia L.*, and its presence prevents the maturation of insect larvae. The compound itself and modified analogs have been proposed as potential agents in insect control. Unfortunately, however, the activity of this analog is only 1/500 that of the natural compound. This example is noteworthy because the reaction can be controlled so that it will take place only at the least hindered electron-rich double bond.

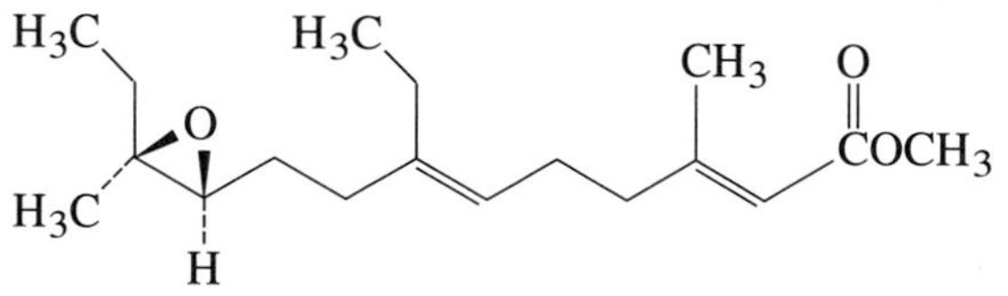

H_3C H_3C CH_3 O O $\overset{\|}{C}OCH_3$ H_3C H

Juvenile hormone

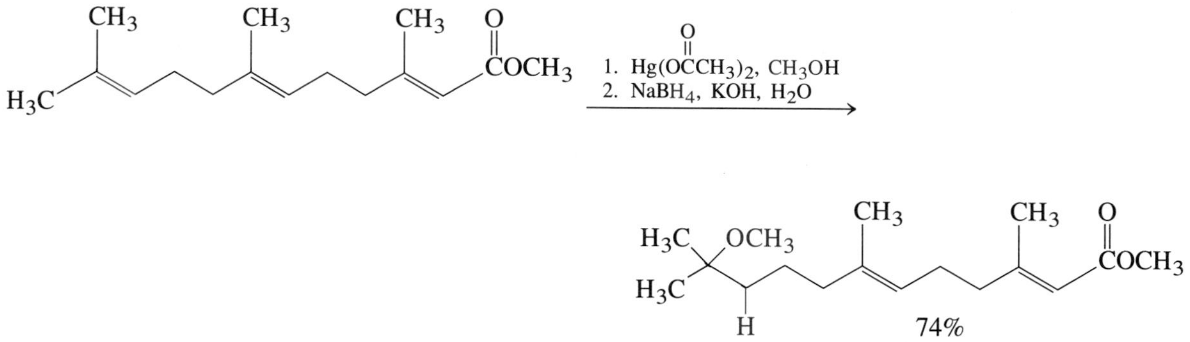

CH_3 CH_3 CH_3 O $\overset{\|}{C}OCH_3$ H_3C

1. $Hg(O\overset{O}{\overset{\|}{C}}CH_3)_2$, CH_3OH
2. $NaBH_4$, KOH, H_2O

H_3C OCH_3 H_3C H CH_3 CH_3 O $\overset{\|}{C}OCH_3$

74%

Analog of juvenile hormone

Mechanism of Oxymercuration–Demercuration

STEP 1. Dissociation

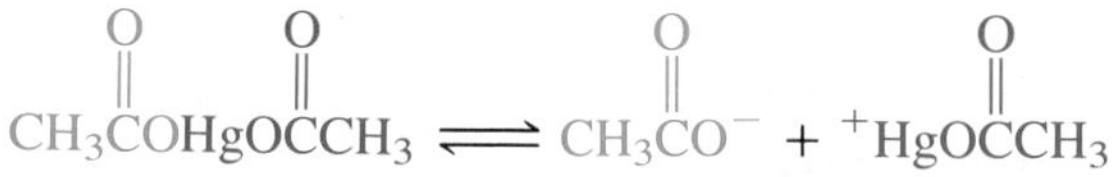

$$CH_3\overset{O}{\overset{\|}{C}}OHgO\overset{O}{\overset{\|}{C}}CH_3 \rightleftharpoons CH_3\overset{O}{\overset{\|}{C}}O^- + {}^+HgO\overset{O}{\overset{\|}{C}}CH_3$$

STEP 2. Electrophilic attack

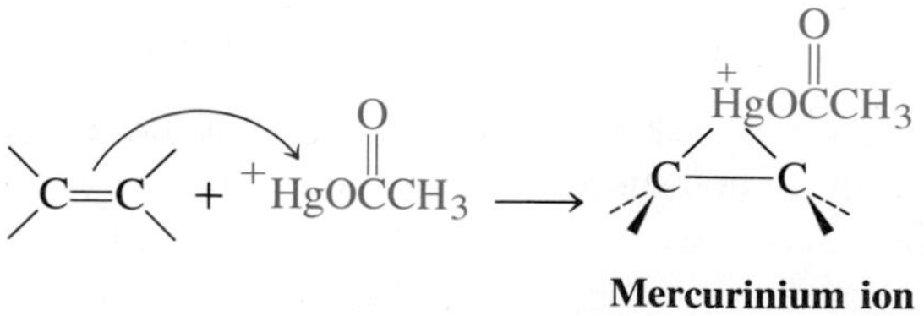

$$\text{>C=C<} + {}^+HgO\overset{O}{\overset{\|}{C}}CH_3 \longrightarrow \text{C—C (bridged by } \overset{+}{Hg}O\overset{O}{\overset{\|}{C}}CH_3\text{)}$$

Mercurinium ion

STEP 3. Nucleophilic opening

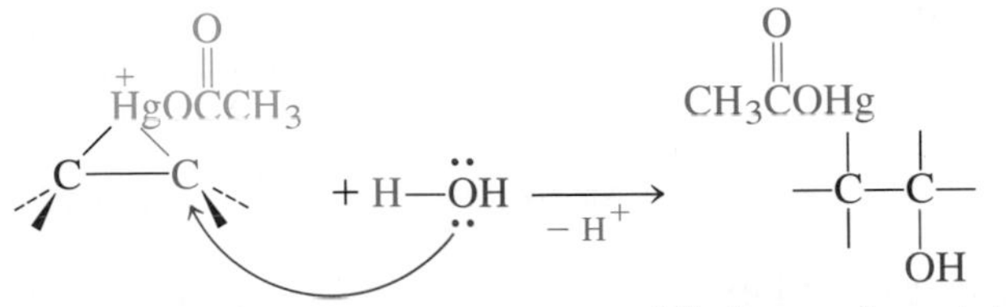

$$\text{(mercurinium ion)} + H{-}\ddot{O}H \xrightarrow{-H^+} CH_3COHg{-}C{-}C{-}OH$$

Alkylmercuric acetate

STEP 4. Reduction

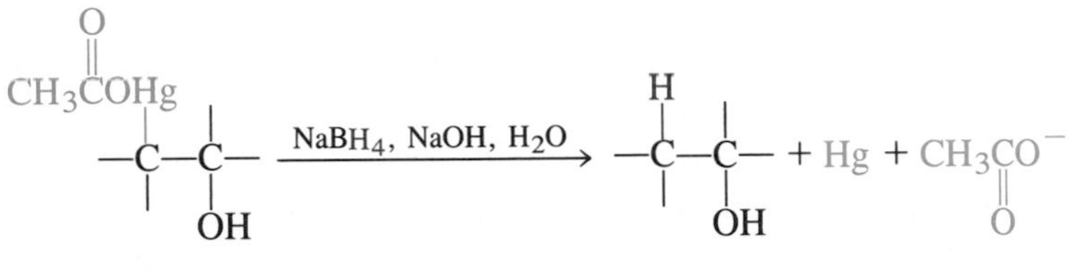

$$CH_3COHg{-}C{-}C{-}OH \xrightarrow{NaBH_4,\ NaOH,\ H_2O} H{-}C{-}C{-}OH + Hg + CH_3CO^-$$

When the oxymercuration of an alkene is executed in an alcohol solvent, demercuration gives an ether, as shown in the margin.

Ether Synthesis by Oxymercuration–Demercuration

1-Hexene

1. $Hg(OCCH_3)_2$, CH_3OH
2. $NaBH_4$, $NaOH$, H_2O

OCH_3

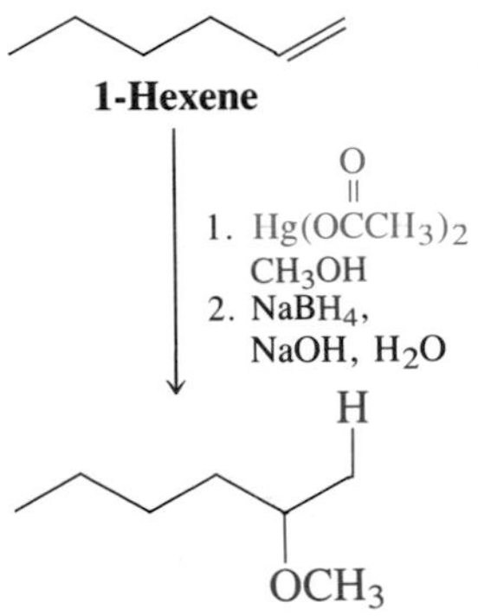

65%

2-Methoxyhexane

EXERCISE 12-12

Explain the result shown below.

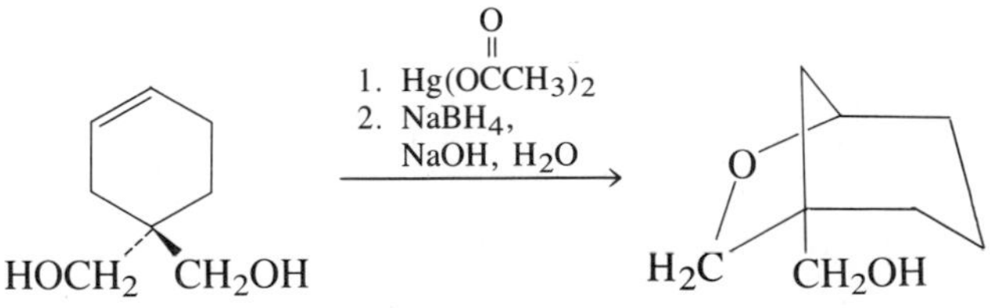

$$\text{HOCH}_2\ \text{CH}_2\text{OH (cyclohexene)} \xrightarrow[\text{2. NaBH}_4,\ \text{NaOH, H}_2\text{O}]{\text{1. Hg(OCCH}_3)_2} \text{H}_2\text{C}\ \ \text{CH}_2\text{OH (bicyclic ether)}$$

42%

In summary, oxymercuration–demercuration is a synthetically useful method for converting alkenes regioselectively (following the Markovnikov rule) into alcohols or ethers. Carbocations are not involved; therefore, rearrangements do not occur.

12-8 Hydroboration–Oxidation: A Stereospecific Anti-Markovnikov Hydration

This section deals with a reaction that seems to lie mechanistically somewhere between hydrogenation and electrophilic addition: the hydroboration of double bonds. The resulting alkylboranes can be oxidized to alcohols.

The boron–hydrogen bond adds across double bonds

Borane, BH_3, adds to double bonds without catalytic activation, a reaction called **hydroboration** by its discoverer, H. C. Brown.*

*Professor Herbert C. Brown, b. 1912, Purdue University, Nobel Prize 1979 (chemistry).

Hydroboration of Alkenes

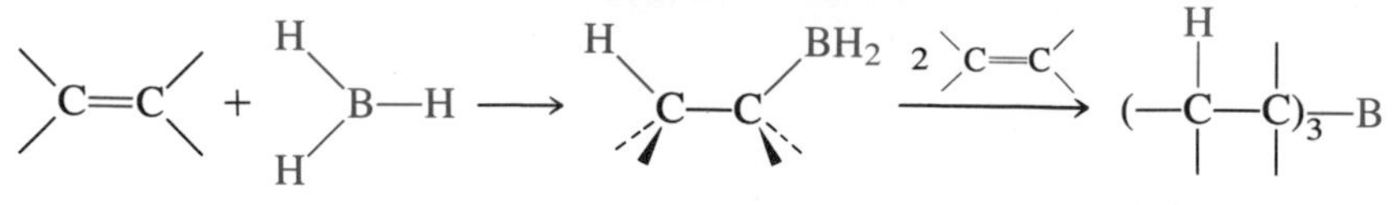

Borane — An alkylborane — A trialkylborane

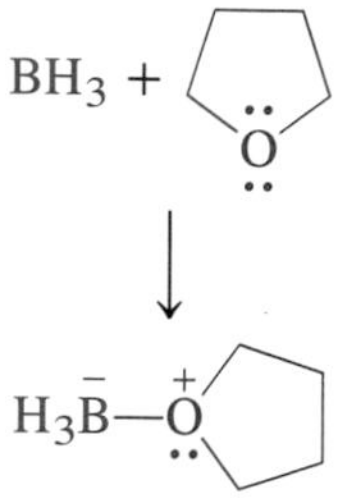

BH_3 + THF → $H_3\bar{B}—\overset{+}{O}$

Borane–THF complex

Borane (which by itself exists as a dimer, B_2H_6) is commercially available in ether and oxacyclopentane (tetrahydrofuran, THF) solutions. In these solutions, borane exists as a Lewis acid-base complex with the ether oxygen, an aggregate that allows the boron to have an electron octet (for the molecular-orbital picture of BH_3, see Figure 1-17).

How does the B–H unit add to the π bond? Because the π bond is electron rich and borane electron poor, it is reasonable to formulate an initial Lewis acid-base complex similar to that of a bromonium ion (Figure 12-3), requiring the participation of the empty p orbital on BH_3. Subsequently, one of the hydrogens is transferred by means of a four-center transition state to one of the alkene carbons, while the boron shifts to the other. The stereochemistry of the addition is *syn*. All three B–H bonds are reactive in this way.

Mechanism of Hydroboration

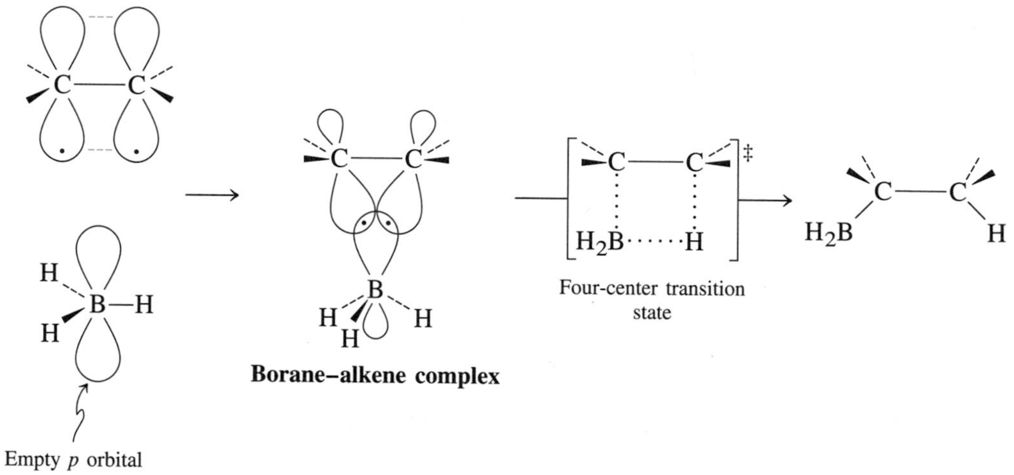

Empty p orbital

Borane–alkene complex

Four-center transition state

Regioselectivity of Hydroboration

$$3\ RCH{=}CH_2 + BH_3 \longrightarrow (RCH_2CH_2)_3B$$

Hydroboration is not only stereospecific (*syn* addition) but also regioselective. Unlike the electrophilic additions described previously, steric and not electronic factors primarily control the regioselectivity: The boron binds to the less hindered (less substituted) carbon. The reactivity of the trialkylboranes resulting from these hydroborations are of special interest to us.

The oxidation of alkylboranes gives alcohols

Trialkylboranes can be oxidized with basic aqueous hydrogen peroxide to furnish alcohols in which the hydroxy function has replaced the boron atom. The net result of the two-step sequence, **hydroboration–oxidation,** is the addition of the elements of water to a double bond. In contrast with the hydrations described in Sections 12-4 and 12-7, those involving borane proceed with the opposite regioselectivity: They are said to be **anti-Markovnikov additions.**

Hydroboration–Oxidation Sequence

$$3\ RCH{=}CHR \xrightarrow{BH_3,\ THF} (RCH_2CHR)_3B \xrightarrow{H_2O_2,\ NaOH,\ H_2O} 3\ RCH_2\overset{R}{\overset{|}{C}}HOH$$

$$\underset{\textbf{4-Methyl-1-pentene}}{(CH_3)_2CHCH_2CH{=}CH_2} \xrightarrow[2.\ H_2O_2,\ NaOH,\ H_2O]{1.\ BH_3,\ THF} \underset{\textbf{4-Methyl-1-pentanol}}{\underset{80\%}{(CH_3)_2CHCH_2CH_2CH_2OH}}$$

In the mechanism of alkylborane oxidation, the nucleophilic hydroperoxide ion attacks the electron-poor boron atom. The resulting species undergoes a rearrangement in which an alkyl group migrates with its electron pair—and with *retention* of configuration—to the neighboring oxygen atom, thus expelling a hydroxide ion in the process.

Mechanism of Alkylborane Oxidation

$$-\overset{/}{\underset{\backslash R}{B}} + \ ^{-}{:}\ddot{O}{-}\ddot{O}H \ (\textbf{Hydroperoxide ion}) \longrightarrow -\overset{|}{\underset{|R}{\overset{-}{B}}}{-}\ddot{O}{-}\ddot{O}H \xrightarrow{-\ HO:^-} -\overset{/}{B}\diagdown\ddot{O}R$$

This process is repeated until all three alkyl groups have migrated to oxygen atoms, finally forming a trialkyl borate $(RO)_3B$. This inorganic ester is then hydrolyzed by base to the alcohol and sodium borate.

$$(RO)_3B + 3\ NaOH \xrightarrow{H_2O} Na_3BO_3 + 3\ ROH$$

Because borane addition to double bonds is so selective, subsequent oxidation allows the stereospecific and regioselective synthesis of alcohols.

A Stereospecific and Regioselective Alcohol Synthesis by Hydroboration–Oxidation

$$\text{1-Methylcyclopentene} \xrightarrow{BH_3,\ THF} \text{(alkylborane: H, } CH_3 \text{, B–, H)} \xrightarrow{H_2O_2,\ NaOH,\ H_2O} \underset{86\%}{\textit{trans}\text{-2-Methylcyclopentanol}}$$

1-Methylcyclopentene → ***trans*-2-Methylcyclopentanol** (86%)

EXERCISE 12-13

Give the products of hydroboration–oxidation of **(a)** propene and **(b)** (*E*)-3-methyl-2-pentene. Show the stereochemistry clearly.

In summary, hydroboration–oxidation constitutes another method for hydrating alkenes. The initial addition is *syn* and regioselective, the boron shifting to the less hindered carbon. Oxidation of alkyl boranes with basic hydrogen peroxide gives anti-Markovnikov alcohols with retention of configuration of the alkyl group.

Peroxycarboxylic Acids

$R{-}\overset{\overset{\displaystyle O}{\|}}{C}{-}\overset{\delta^-}{O}{-}\overset{\delta^+}{O}H$

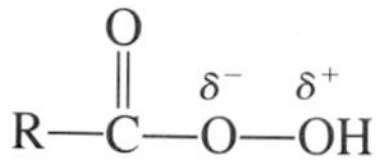

A peroxycarboxylic acid

$CH_3\overset{\overset{\displaystyle O}{\|}}{C}OOH$

Peroxyethanoic (peracetic) acid

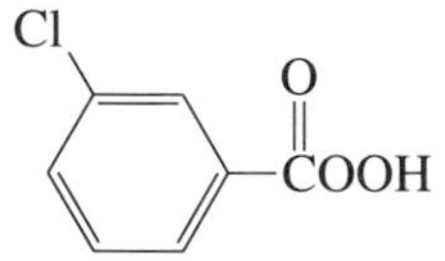

meta-Chloroperoxybenzoic acid (MCPBA)

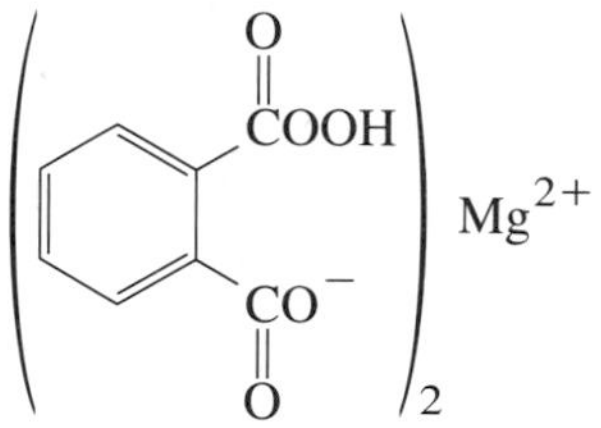

Mg^{2+}

Magnesium monoperoxyphthalate (MMPP)

12-9 Oxacyclopropane Synthesis: Oxidation by Peroxycarboxylic Acids

The next three sections describe how electrophilic oxidizing agents are capable of delivering oxygen atoms to the π bond, thereby producing oxacyclopropanes, vicinal *syn* and *anti* diols, and carbonyl compounds by complete cleavage of the double bond. Let us start with oxacyclopropane formation, a process that can be used to synthesize vicinal *anti* diols.

Peroxycarboxylic acids deliver oxygen atoms to double bonds

The OH group in peroxycarboxylic acids, $R\overset{\overset{\displaystyle O}{\|}}{C}OOH$, contains an electrophilic oxygen. These compounds react with alkenes by adding this oxygen to the double bond to form oxacyclopropanes. The other product of the reaction is a carboxylic acid. The transformation is of value because, as we know, oxacyclopropanes are versatile synthetic intermediates (Section 9-9). It proceeds at room temperature in an inert solvent, such as chloroform, dichloromethane, or benzene. This reaction is commonly referred to as **epoxidation,** a term derived from *epoxide,* one of the older common names for oxacyclopropanes. A popular peroxycarboxylic acid for use in the research laboratory is *meta*-chloroperoxybenzoic acid (MCPBA). For large-scale and industrial purposes, however, the somewhat shock-sensitive MCPBA has been replaced by magnesium monoperoxyphthalate (MMPP).

Oxacyclopropane Formation: Epoxidation of a Double Bond

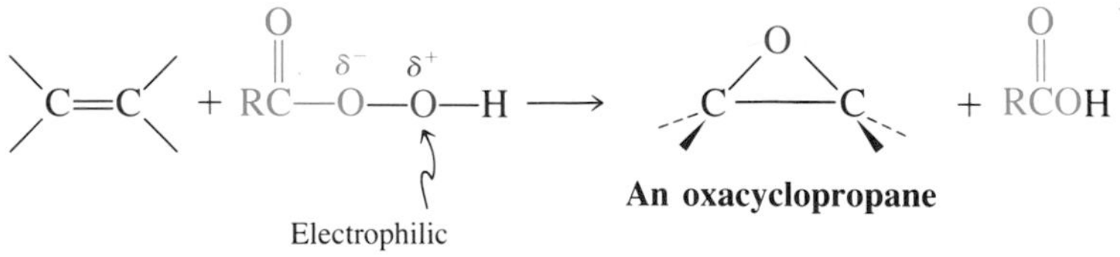

$$>C{=}C< \;+\; R\overset{\overset{\displaystyle O}{\|}}{C}{-}\overset{\delta^-}{O}{-}\overset{\delta^+}{O}{-}H \longrightarrow \text{oxacyclopropane} \;+\; R\overset{\overset{\displaystyle O}{\|}}{C}OH$$

Electrophilic

An oxacyclopropane

$$CH_3CH_2CH{=}CH_2 + \text{MCPBA} \xrightarrow{CHCl_3} CH_3CH_2\overset{O}{CH{-}CH_2} \quad 90\%$$

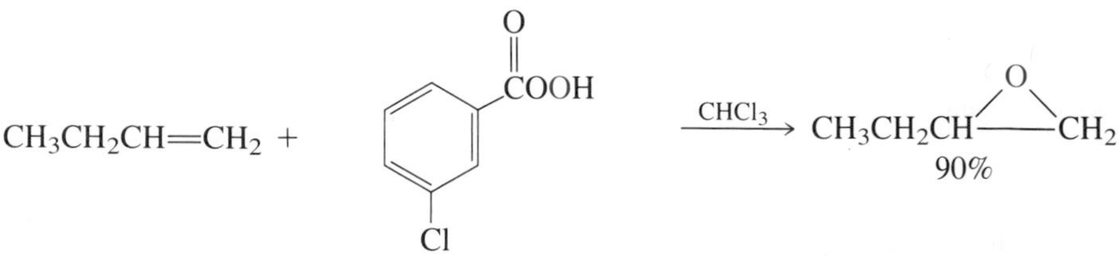

1-Butene *meta*-**Chloroperoxybenzoic acid (MCPBA)** **Ethyloxacyclopropane**

The transfer of oxygen is stereospecifically *syn,* the stereochemistry of the starting alkene being retained in the product. For example, *trans*-2-butene gives *trans*-2,3-dimethyloxacyclopropane; conversely, the *cis*-2-butene yields *cis*-2,3-dimethyloxacyclopropane.

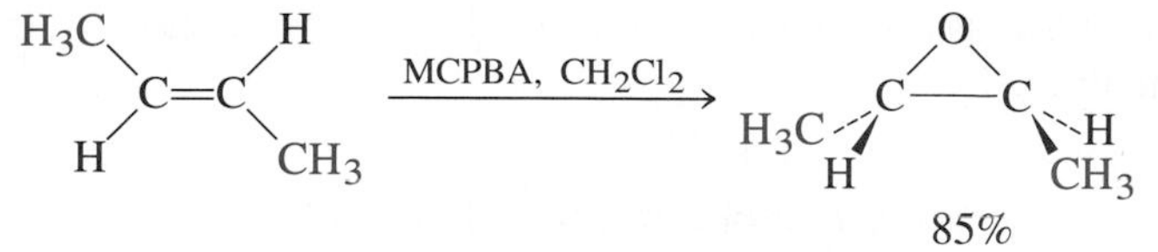

H_3C H C=C H CH_3 $\xrightarrow{\text{MCPBA, } CH_2Cl_2}$ trans-2,3-dimethyloxacyclopropane 85%

***trans*-2-Butene** ***trans*-2,3-Dimethyloxacyclopropane**

What is the mechanism of this oxidation? We can write a cyclic transition state in which the peroxycarboxylic acid proton is transferred to its own carbonyl group at the same time as the electrophilic oxygen is added to the π bond. Formally, the electron pair of the alkene π system participates in the formation of one bond to the oxygen being added, whereas the electron pair responsible for the O–H linkage forms the other bond.

Mechanism of Oxacyclopropane Formation

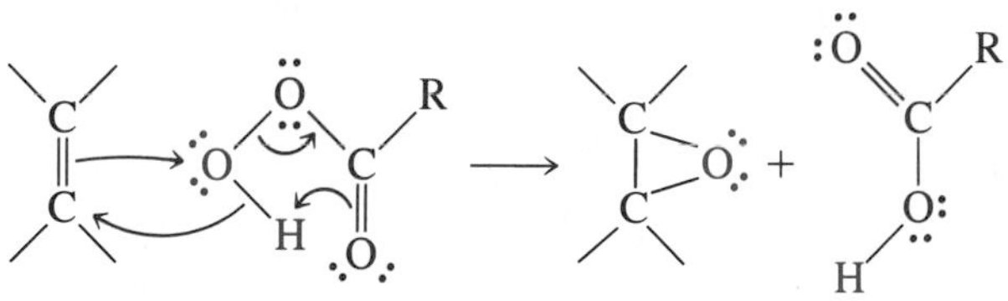

EXERCISE 12-14

Outline a short synthesis of *trans*-2-methylcyclohexanol from cyclohexene. (Hint: Review the reactions of oxacyclopropanes in Section 9-9.)

In accord with the electrophilic mechanism, the reactivity of alkenes to peroxycarboxylic acids increases with alkyl substitution, allowing for selective oxidations.

Relative Rates of Oxacyclopropane Formation (Epoxidation)

$CH_2{=}CH_2$	$<$	$RCH{=}CH_2$	$<$	$RCH{=}CHR$	$\sim$	$R_2C{=}CH_2$	$<$	$R_2C{=}CHR$	$<$	$R_2C{=}CR_2$
1		24		500		500		6500		Very fast

For example,

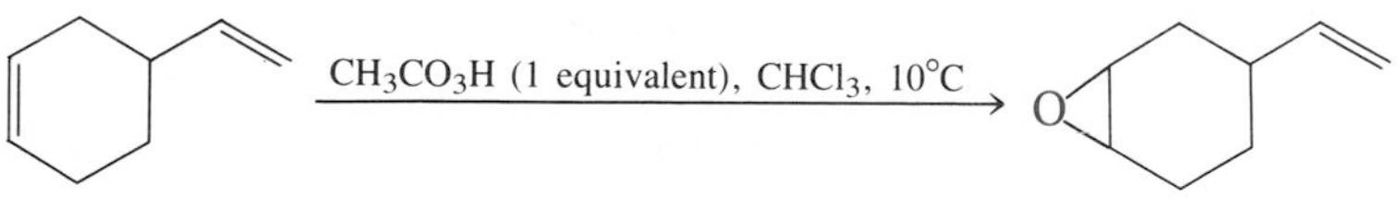

$\xrightarrow{CH_3CO_3H \text{ (1 equivalent), } CHCl_3,\ 10^\circ C}$ 86%

Hydrolysis of oxacyclopropanes furnishes the products of *anti* dihydroxylation of an alkene

Treatment of oxacyclopropanes with water in the presence of catalytic acid or base leads to ring opening (Section 9-9) to the corresponding vicinal diols. In this reaction, water nucleophilically attacks the side opposite the oxygen in the three-membered ring, so the net result of the oxidation–hydrolysis sequence constitutes an ***anti* dihydroxylation** of an alkene. In this way, *trans*-2-butene gives *meso*-

2,3-butanediol, whereas *cis*-2-butene furnishes the racemic mixture of the 2*R*,3*R* and 2*S*,3*S* enantiomers.

Vicinal *Anti* Diol Formation from Alkenes

$$\text{C=C} \xrightarrow[\text{2. H}^+\text{ H}_2\text{O}]{\text{1. MCPBA, CH}_2\text{Cl}_2} \text{HO–C–C–OH}$$

Synthesis of Isomers of 2,3-Butanediol

trans-2-Butene $\xrightarrow[\text{2. H}^+\text{ H}_2\text{O}]{\text{1. MCPBA, CH}_2\text{Cl}_2}$ *meso*-2,3-Butanediol

cis-2-Butene $\xrightarrow[\text{2. H}^+\text{ H}_2\text{O}]{\text{1. MCPBA, CH}_2\text{Cl}_2}$ (2*R*,3*R*)-2,3-Butanediol + (2*S*,3*S*)-2,3-Butanediol

EXERCISE 12-15

Give the products obtained by treating the following alkenes with MCPBA and then aqueous acid. **(a)** 1-Hexene; **(b)** cyclohexene; **(c)** *cis*-2-pentene.

In summary, peroxycarboxylic acids supply oxygen atoms to convert alkenes into oxacyclopropanes (epoxidation). Peroxidation–hydrolysis reactions with peroxycarboxylic acids furnish vicinal *anti*-1,2-diols.

12-10 Synthesis of Vicinal *Syn* Diols by Permanganate Oxidation

Potassium permanganate, in cold aqueous solution, reacts with alkenes under neutral conditions to give the corresponding vicinal *syn* diols. The inorganic product in this reaction is insoluble brown manganese dioxide, MnO_2.

Vicinal *Syn* Dihydroxylation with Permanganate

$$\text{C=C} + \underset{\textbf{Dark purple}}{\text{KMnO}_4} \xrightarrow{0°\text{C, H}_2\text{O, pH = 7}} \text{HO–C–C–OH} + \underset{\textbf{Brown}}{\text{MnO}_2}$$

For example,

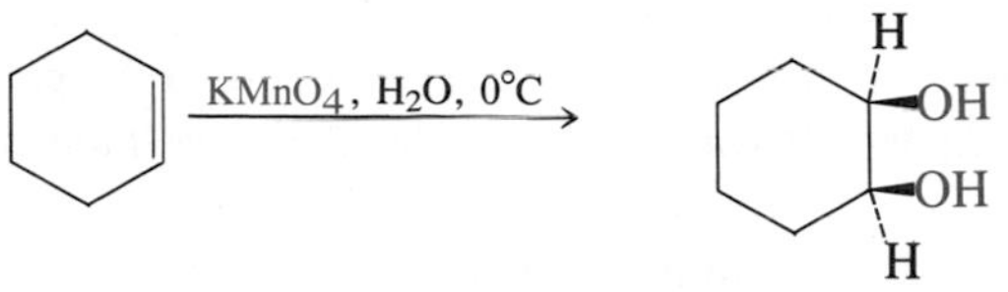

$$\text{Cyclohexene} \xrightarrow{KMnO_4,\ H_2O,\ 0^\circ C} \text{cis-1,2-cyclohexanediol}$$

37%

***cis*-1,2-Cyclohexanediol**

What is the mechanism of this transformation? The initial reaction of the π bond with permanganate constitutes a concerted addition in which three electron pairs move simultaneously to give a cyclic ester containing Mn(V). This process can be viewed as an electrophilic attack on the alkene: Two electrons flow from the alkene onto the metal, which is reduced [Mn(VII) $\rightarrow$ Mn(V)]. For steric reasons, the product can form only in a way that introduces the two oxygen atoms on the *same* face of the double bond—*syn*. This intermediate is reactive, hydrolyzing to give the free diol and an unstable Mn(V) species. Manganese(V) disproportionates into insoluble MnO_2 and (probably) a Mn(VI) species that oxidizes more starting material. Indeed, the decolorization of a purple permanganate solution is a useful test for alkenes.

Mechanism of the Permanganate Oxidation of Alkenes

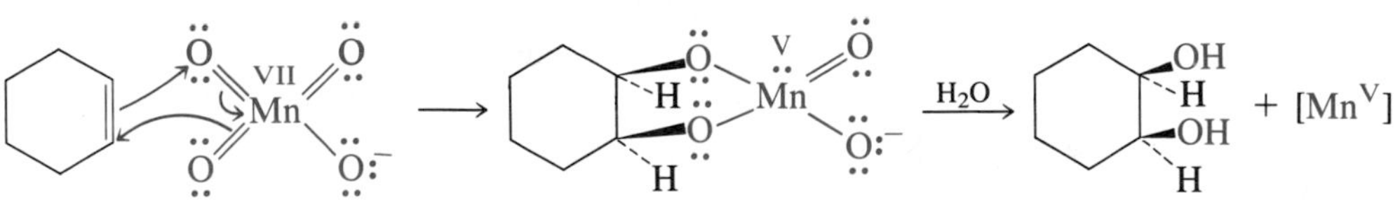

VII Mn → V Mn $\xrightarrow{H_2O}$ + $[Mn^V]$

Syn dihydroxylation can be achieved in better yields by using osmium tetroxide, OsO_4, which is quite similar to permanganate in its mechanism of action. The reagent was originally employed in stoichiometric quantities and led to intermediate isolable cyclic esters. Typically, these intermediates were not isolated but reductively hydrolyzed with H_2S or bisulfite, $NaHSO_3$.

Vicinal *Syn* Dihydroxylation with Osmium Tetroxide

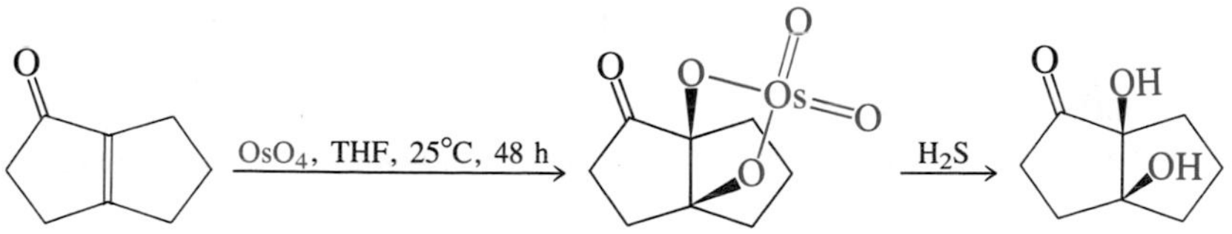

$\xrightarrow{OsO_4,\ THF,\ 25^\circ C,\ 48\ h}$ $\xrightarrow{H_2S}$

90%

However, OsO_4 is expensive and highly toxic; therefore, a newer modification calls for the use of only catalytic quantities of the osmium reagent and stoichiometric amounts of hydrogen peroxide as the oxidizing agent.

EXERCISE 12-16

The stereochemical consequences of the vicinal *syn* dihydroxylation of alkenes are complementary to those of vicinal *anti* dihydroxylation. Show the products (indicate stereochemistry) of the vicinal *syn* dihydroxylation of *cis*- and *trans*-2-butene.

To summarize, cold potassium permanganate oxidizes alkenes to *syn*-1,2-diols. This reaction is accompanied by decolorization of the purple permanganate, a result making it a useful test for the presence of double bonds.

12-11 Oxidative Cleavage: Ozonolysis

Although oxidation of alkenes with cold potassium permanganate or osmium tetroxide breaks only the π bond, other reagents may rupture the σ bond as well. The most general and mildest method of oxidatively cleaving alkenes to carbonyl compounds is through the reaction with ozone, **ozonolysis.**

Ozone, O_3, is produced in the laboratory in an instrument called an ozonator, in which an arc discharge generates 3–4% ozone in a dry oxygen stream. The gas mixture is passed through a solution of the alkene in methanol or dichloromethane. The first isolable intermediate is a species called an **ozonide,** which is reduced directly in a subsequent step by various treatments, such as exposure to zinc in acetic acid, or by reaction with dimethyl sulfide. The net result of the ozonolysis–reduction sequence is the cleavage of the molecule at the carbon–carbon double bond; oxygen becomes attached to each of the carbons that had originally been doubly bonded.

Ozonolysis Reaction of Alkenes

$$\text{C=C} \xrightarrow{O_3} \text{Ozonide} \xrightarrow[-\,[O]]{\text{Reduction}} \text{C=O} + \text{O=C}$$

Ozonide **Carbonyl products**

$$\text{(Z)-}CH_3CH_2(CH_3)C{=}C(H)CH_3 \xrightarrow{\text{1. } O_3,\ CH_2Cl_2 \quad \text{2. Zn, } CH_3COOH} \underset{90\%}{CH_3CH_2\overset{O}{\overset{\|}{C}}CH_3} + CH_3\overset{O}{\overset{\|}{C}}H$$

(Z)-3-Methyl-2-pentene **2-Butanone** **Acetaldehyde**

The mechanism of ozonolysis proceeds through initial electrophilic addition of ozone to the double bond, a transformation that yields the so-called **molozonide.** In this reaction, as in several others already presented, six electrons move in concerted fashion in a cyclic transition state. The molozonide is unstable and breaks apart into a carbonyl and a carbonyl oxide fragment through another cyclic six-electron rearrangement. Recombination of the two fragments as shown yields the ozonide.

Mechanism of Ozonolysis

STEP 1. Molozonide formation and cleavage

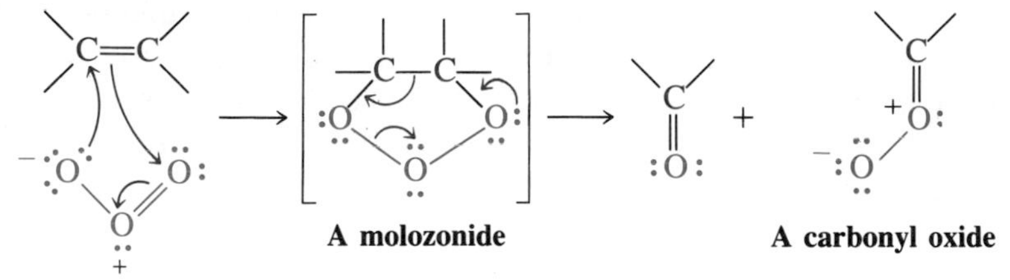

A molozonide

A carbonyl oxide

STEP 2. Ozonide formation and reduction

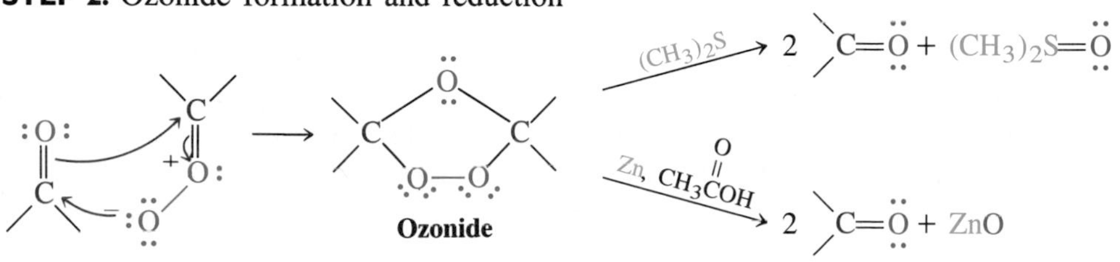

$(CH_3)_2S$ → 2 >C=O + $(CH_3)_2S{=}O$

Ozonide

Zn, CH_3COH (O above C) → 2 >C=O + ZnO

EXERCISE 12-17

An unknown hydrocarbon of the molecular formula $C_{12}H_{20}$ exhibited an ^1H NMR spectrum with a complex multiplet of signals between 1 and 2.2 ppm. Ozonolysis of this compound gave two equivalents of cyclohexanone, whose structure is shown in the margin. What is the structure of the unknown?

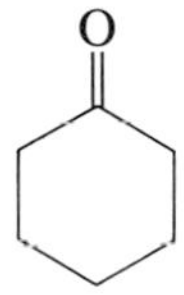

EXERCISE 12-18

Give the products of the following reactions.

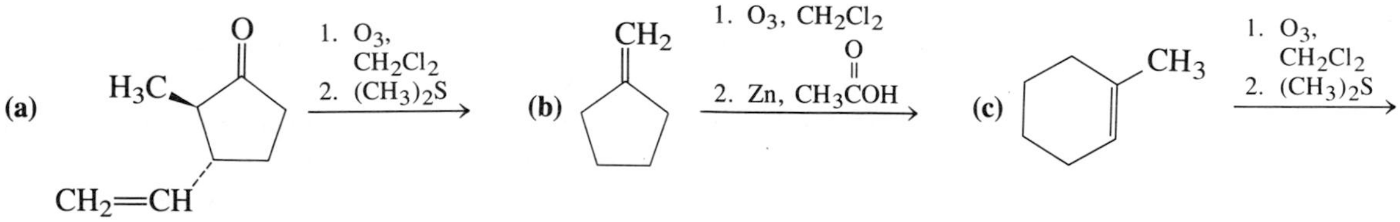

(a) 1. O_3, CH_2Cl_2 2. $(CH_3)_2S$ →

(b) 1. O_3, CH_2Cl_2 2. Zn, CH_3COH →

(c) 1. O_3, CH_2Cl_2 2. $(CH_3)_2S$ →

EXERCISE 12-19

What is the structure of the following starting material?

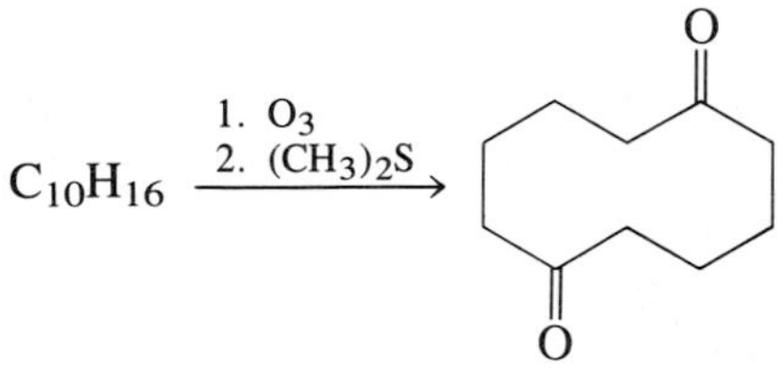

$C_{10}H_{16}$ 1. O_3 2. $(CH_3)_2S$ →

In summary, ozonolysis followed by reduction yields aldehydes and ketones. Mechanistically, the reactions presented in the preceding three sections can be related in that initial attack by an electrophilic oxidizing agent on the π bond leads to its rupture. Unlike the reaction sequences studied in the preceding two sections, however, ozonolysis involves cleavage of both the π and σ bonds.

12-12 Radical Additions: Anti-Markovnikov Product Formation

This section deals with another mode of reactivity of the double bond: radical addition. In contrast with electrophilic reagents, which consume both electrons of the π bond on addition, a radical requires only one electron for bond formation, so that an alkyl radical is formed. The consequences of this difference are anti-Markovnikov products.

Markovnikov Addition of HBr

$CH_3CH_2CH{=}CH_2$ (Freshly distilled) $\xrightarrow{HBr,\ 24\ h}$ $CH_3CH_2CH(Br)CH_2H$

90%

Markovnikov product

Anti-Markovnikov Addition of HBr

$CH_3CH_2CH{=}CH_2$ (Exposed to oxygen) $\xrightarrow{HBr,\ 4\ h}$ $CH_3CH_2CH(H)CH_2Br$

65%

Anti-Markovnikov product

In this section, all radicals and single atoms are shown in green, as in Chapter 3.

Hydrogen bromide can add to alkenes in anti-Markovnikov fashion: a change in mechanism

When freshly distilled 1-butene is exposed to hydrogen bromide, clean Markovnikov addition to give 2-bromobutane is observed. This result is in accord with the ionic mechanism discussed in Section 12-3. Curiously, the same reaction, when carried out with a sample of 1-butene that has been exposed to air, proceeds much faster and gives an entirely different result. In this case, we isolate 1-bromobutane, formed by anti-Markovnikov addition.

This change caused considerable confusion in the early days of alkene chemistry, because one researcher would obtain only one hydrobromination product, whereas another would obtain a different product or mixtures from a seemingly identical reaction. The mystery was solved by Kharasch* in the 1930s, when it was discovered that the culprits responsible for anti-Markovnikov additions were radicals formed from peroxides, ROOR, in alkene samples that had been stored in the presence of air.

The mechanism of the addition reaction under these conditions is not an ionic sequence; rather, it is a much faster **radical chain sequence.** The initiation steps are, first, the homolytic cleavage of the weak RO–OR bond ($DH^\circ \sim$ 35 kcal mol^{-1}) and, then, reaction of the resulting alkoxy radical with hydrogen bromide. The driving force for the second (exothermic) step is the formation of the strong O–H bond. The bromine atom so generated initiates chain propagation by attacking the double bond. One of the π electrons combines with the unpaired electron on the bromine atom to form the carbon–bromine bond. The other π electron remains on carbon, giving rise to a radical.

The halogen atom's attack is *regioselective,* creating the relatively more stable secondary radical rather than the primary one. This result is reminiscent of the ionic additions of hydrogen bromide (Section 12-3), except that the roles of the proton and bromine are reversed. In the ionic mechanism, a proton attacks first to generate the more stable carbocation, which is then trapped by bromide ion. *In the radical mechanism, a bromine atom is the attacking species,* creating the more stable radical center. The latter subsequently reacts with HBr by abstracting a hydrogen and regenerating the chain-carrying bromine atom. Both propagation steps are exothermic, and the reaction proceeds rapidly. As usual, termination is by radical combination or by some other removal of the chain carriers (Section 3-4).

*Professor Morris S. Kharasch, 1895–1957, University of Chicago.

Mechanism of Radical Hydrobromination

INITIATION STEPS.

$$\text{R}\ddot{\text{O}}\text{—}\ddot{\text{O}}\text{R} \xrightarrow{\Delta} 2\,\text{R}\ddot{\text{O}}\cdot \qquad \Delta H^\circ \sim +35 \text{ kcal mol}^{-1}$$

$$\text{R}\ddot{\text{O}}\cdot + \text{H}\ddot{\text{B}}\text{r}: \xrightarrow{\Delta} \text{R}\ddot{\text{O}}\text{H} + :\ddot{\text{B}}\text{r}\cdot \qquad \Delta H^\circ \sim -17 \text{ kcal mol}^{-1}$$

PROPAGATION STEPS.

$$(\text{H})(\text{CH}_3\text{CH}_2)\text{C=CH}_2 + :\ddot{\text{B}}\text{r}\cdot \longrightarrow \text{CH}_3\text{CH}_2\dot{\text{C}}\text{H—CH}_2\ddot{\text{B}}\text{r}: \qquad \Delta H^\circ \sim -3 \text{ kcal mol}^{-1}$$

Secondary radical

$$\text{CH}_3\text{CH}_2\dot{\text{C}}\text{HCH}_2\text{Br} + \text{H}:\ddot{\text{B}}\text{r}: \longrightarrow \text{CH}_3\text{CH}_2\text{CH(H)CH}_2\ddot{\text{B}}\text{r}: + :\ddot{\text{B}}\text{r}\cdot \qquad \Delta H^\circ \sim -7.5 \text{ kcal mol}^{-1}$$

Are radical additions general?

Hydrogen chloride and hydrogen iodide do not give anti-Markovnikov addition products to alkenes; in both cases, one of the propagating steps is endothermic and consequently so slow that the chain reaction terminates. As a result, HBr is the *only* hydrogen halide that adds to an alkene under radical conditions to give anti-Markovnikov products. Additions of HCl and HI proceed by ionic mechanisms only to give normal, Markovnikov products regardless of the presence or absence of radicals. Other reagents, however, do undergo successful radical additions to alkenes. Examples are thiols and some of the halomethanes.

Other Radical Additions to Alkenes

$$\text{CH}_3\text{CH=CH}_2 + \text{CH}_3\text{CH}_2\text{SH} \xrightarrow{\text{ROOR}} \text{CH}_3\text{CH(H)CH}_2\text{SCH}_2\text{CH}_3$$

Ethanethiol → **Ethyl propyl sulfide**

$$\text{CH}_3(\text{CH}_2)_5\text{CH=CH}_2 + \text{HCCl}_3 \xrightarrow{\text{ROOR}} \text{CH}_3(\text{CH}_2)_5\text{CH(H)CH}_2\text{CCl}_3$$

22%

1,1,1-Trichlorononane

In these examples the initiating alkoxy radical abstracts a hydrogen from the substrate to yield a chain carrier, because of the strength of the resulting OH bond. A typical example is trichloromethane (chloroform).

$$\text{RO}\cdot + \text{CHCl}_3 \longrightarrow \text{ROH} + \cdot\text{CCl}_3 \quad \text{not} \quad \text{RO}\cdot + \text{Cl—CHCl}_2 \longrightarrow \text{ROCl} + \cdot\text{CHCl}_2$$

Chain carrier

EXERCISE 12-20

Ultraviolet irradiation of a mixture of 1-octene and diphenylphosphine, $(C_6H_5)_2PH$, furnishes 1-(diphenylphosphino)octane by radical addition. Write a plausible mechanism for this reaction.

$$(C_6H_5)_2PH + H_2C{=}CH(CH_2)_5CH_3 \xrightarrow{h\nu} (C_6H_5)_2P{-}CH_2{-}CH_2(CH_2)_5CH_3$$

Anti-Markovnikov additions are synthetically useful because their products complement those obtained from ionic additions. This type of regiochemical control is an important feature in the development of new synthetic methods.

In summary, radical initiators alter the mechanism of the addition of HBr to alkenes from ionic to radical chain. The consequence of this change is anti-Markovnikov regioselectivity. Other species, most notably thiols and some halomethanes, are capable of undergoing similar reactions.

12-13 Dimerization, Oligomerization, and Polymerization of Alkenes

Is it possible for alkenes to react with each other? Indeed it is, but only in the presence of an appropriate catalyst, for example, an acid, a radical, a base, or a transition metal. In this reaction the unsaturated centers of the alkene monomer (*monos,* Greek, single; *meros,* Greek, part) are linked to form dimers, trimers, **oligomers** (*oligos,* Greek, few, small), and ultimately **polymers** (*polymeres,* Greek, of many parts), substances of great industrial importance.

Polymerization

$$\text{C=C} \quad \text{C=C} \quad \text{C=C} \longrightarrow -\text{C}-\text{C}-\text{C}-\text{C}-\text{C}-\text{C}-$$

Monomers **Polymer**

Carbocations attack pi bonds

Treatment of 2-methylpropene with hot aqueous sulfuric acid gives two dimers: 2,4,4-trimethyl-1-pentene and 2,4,4-trimethyl-2-pentene. This transformation is possible because 2-methylpropene can be protonated under the reaction conditions to furnish the 1,1-dimethylethyl (*tert*-butyl) cation. This species can attack the electron-rich double bond of 2-methylpropene with formation of a new carbon–carbon bond. Electrophilic addition proceeds according to the Markovnikov rule to generate the more stable carbocation. Subsequent deprotonation in each of two directions furnishes a mixture of the two observed products.

Dimerization of 2-Methylpropene

$$CH_2{=}C(CH_3)_2 + CH_2{=}C(CH_3)_2 \xrightarrow{H^+} CH_3\overset{CH_3}{\underset{CH_2}{C}}CH_2\overset{CH_3}{C}{=}CH_2 + CH_3\overset{CH_3}{\underset{CH_3}{C}}CH{=}C(CH_3)_2$$

2,4,4-Trimethyl-1-pentene **2,4,4-Trimethyl-2-pentene**

BOX 12-3 Steroid Synthesis in Nature

A remarkable series of intramolecular alkene couplings occurs in nature as part of the biosynthetic pathway to steroids, including cholesterol and the powerfully biologically active mammalian sex hormones (Section 4-7). In this process, a molecule called *squalene* is first enzymatically oxidized to the oxacyclopropane squalene oxide. Enzymatic acid-catalyzed ring opening occurs, followed by the sequential formation of four carbon–carbon bonds. Each bond-forming step is mechanistically related to alkene oligomerization. Further conversion leads to lanosterol, a biological precursor to cholesterol. Very similar *(biomimetic)* reactions have also been carried out in the laboratory. These processes, which are highly regioselective and stereospecific, are excellent methods for synthesizing many steroids.

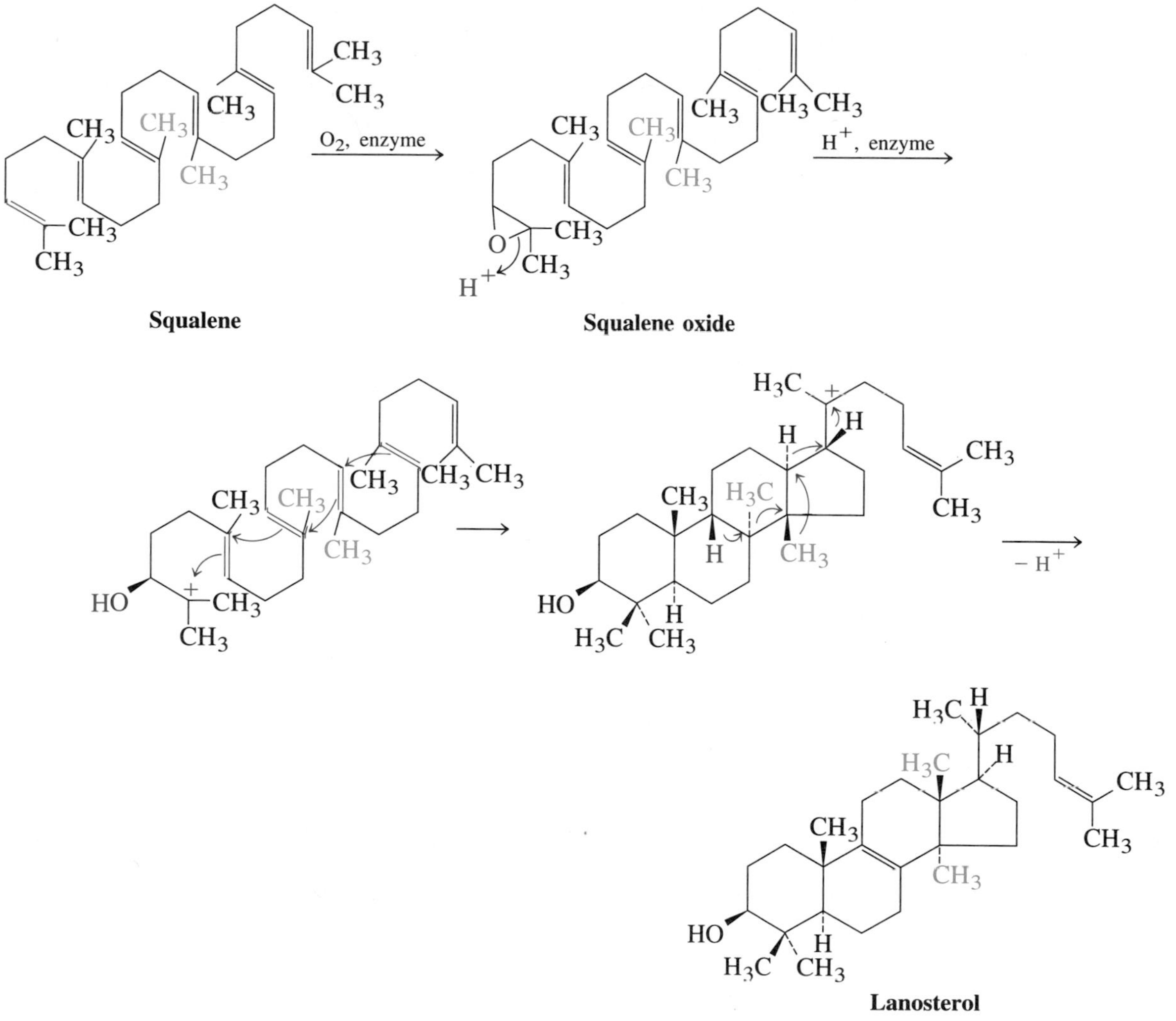

Squalene

O_2, enzyme

Squalene oxide

H^+, enzyme

$-H^+$

Lanosterol

Mechanism of Dimerization of 2-Methylpropene

$$CH_2{=}C(CH_3)_2 \xrightarrow{H^+} CH_3\overset{+}{C}(CH_3)_2 + CH_2{=}C(CH_3)_2 \longrightarrow$$

$$(CH_3)_3C{-}CH_2{-}\overset{+}{C}(CH_3){-}CH_2{-}H \xrightarrow{-H^+} (CH_3)_3CCH_2C(CH_3){=}CH_2 + (CH_3)_3CCH{=}C(CH_3)_2$$

Repeated attack can lead to oligomerization and polymerization

The two dimers of 2-methylpropene tend to react further with the starting alkene. For example, when 2-methylpropene is treated with mineral acid under more stringent conditions, trimers, tetramers, pentamers, and so forth are formed by repeated electrophilic attack of intermediate carbocations on the double bond. This process, which leads to alkane chains of intermediate length, is called **oligomerization.**

Oligomerization of the 2-Methylpropene Dimers

$$(CH_3)_3CCH{=}C(CH_3)_2 + (CH_3)_3CCH_2C(CH_3){=}CH_2 \xrightarrow{H^+} (CH_3)_3CCH_2\overset{+}{C}(CH_3)_2 + CH_2{=}C(CH_3)_2 \longrightarrow$$

$$(CH_3)_3CCH_2C(CH_3)_2CH_2\overset{+}{C}(CH_3)_2 + CH_2{=}C(CH_3)_2 \longrightarrow (CH_3)_3CCH_2C(CH_3)_2CH_2C(CH_3)_2CH_2\overset{+}{C}(CH_3)_2 \quad \text{etc.}$$

At higher temperatures, the oligomerization of alkenes continues to give polymers containing many subunits.

Polymerization of 2-Methylpropene

$$n\ CH_2{=}C(CH_3)_2 \xrightarrow{H^+,\ 200°C} H{-}(CH_2{-}C(CH_3)_2)_{n-1}{-}CH_2C(CH_3){=}CH_2$$

Poly-2-methylpropene (Polyisobutylene)

In summary, catalytic acid causes alkene–alkene additions to occur, a process forming dimers, trimers, oligomers containing several components, and finally polymers, which are composed of a great many alkene subunits.

12-14 Synthesis of Polymers

Many alkenes are suitable monomers for polymerization. Although polymerization can be an unwanted side reaction in their chemistry, it is exceedingly important in the chemical industry, because many polymers have desirable properties, such as durability, inertness to many chemicals, elasticity, transparency, and electrical and thermal resistance.

Although the production of polymers has contributed to pollution—many of them are not biodegradable—they have varied uses as synthetic fibers, films, pipes, coatings, and molded articles. Names such as polyethylene, poly(vinyl chloride) (PVC), Teflon, polystyrene, Orlon, and Plexiglas (Table 12-3) have become household words.

Acid-catalyzed polymerizations, such as that described for poly(2-methylpropene), are carried out with H_2SO_4, HF, or BF_3 as the initiators. Because they proceed through the intermediacy of carbocations, they are also called *cationic polymerizations*. Other mechanisms of polymerizations are *radical, anionic,* and *metal catalyzed.*

Radical polymerizations lead to commercially useful materials

An example of **radical polymerization** is that of ethene in the presence of an organic peroxide at high pressures and temperatures. The reaction proceeds by a mechanism that, in its initial stages, resembles that of the radical addition to alkenes (Section 12-12). The peroxide initiators cleave into alkoxy radicals, which begin polymerization by addition to the double bond of ethene. The alkyl radical thus created attacks the double bond of another ethene molecule, furnishing another radical center, and so on. Termination of the polymerization can be by dimerization, disproportionation of the radical, or other radical-trapping reactions.

Radical Polymerization of Ethene

$$n\ CH_2{=}CH_2 \xrightarrow[>100^{\circ}C]{ROOR,\ 1000\ atm} {-}(CH_2{-}CH_2)_n{-}$$

Polyethene (Polyethylene)

BOX 12-4 Polymers in the Cleanup of Oil Spills

Poly(2-methylpropene) (polyisobutylene) is the principal ingredient in a product called Elastol, which has been demonstrated to be an effective agent in the cleanup of oil spills. When Elastol is sprayed on an oil slick, its polymer chains, which normally wrap tightly about each other, unravel and mix with the oil. The oil is bound by the polymer into a viscous mat that can be skimmed off the water surface. An especially valuable feature of this cleanup method is that the oil may be recovered from the polymer by running the mixture through a special type of pump. The process was successfully demonstrated on an oil spill in New Haven harbor, Connecticut, during 1990.

Poly(2-methylpropene) (Polyisobutylene)

TABLE 12-3 Common Polymers and Their Monomers

Monomer	Structure	Polymer (common name)	Structure	Uses
Ethene	$H_2C{=}CH_2$	Polyethylene	$-\!\!(CH_2CH_2)\!\!-_n$	Food storage bags, containers
Chloroethene (vinyl chloride)	$H_2C{=}CHCl$	Poly(vinyl chloride) (PVC)	$-\!\!(CH_2CH(Cl))\!\!-_n$	Pipes, vinyl fabrics
Tetrafluoroethene	$F_2C{=}CF_2$	Teflon	$-\!\!(CF_2CF_2)\!\!-_n$	Nonstick cookware
Ethenylbenzene (styrene)	$C_6H_5CH{=}CH_2$	Polystyrene	$-\!\!(CH_2CH(C_6H_5))\!\!-_n$	Foam packing material
Propenenitrile (acrylonitrile)	$H_2C{=}C(H)C{\equiv}N$	Orlon	$-\!\!(CH_2CH(CN))\!\!-_n$	Clothing, synthetic fabrics
Methyl 2-methylpropenoate (methyl methacrylate)	$H_2C{=}C(CH_3)COCH_3$ (C=O)	Plexiglas	$-\!\!(CH_2C(CH_3)(CO_2CH_3))\!\!-_n$	Impact-resistant paneling
2-Methylpropene (isobutylene)	$H_2C{=}C(CH_3)CH_3$	Elastol	$-\!\!(CH_2C(CH_3)(CH_3))\!\!-_n$	Oil-spill cleanup

Mechanism of Radical Polymerization of Ethene

INITIATION STEPS.

$$RO{-}OR \longrightarrow RO\cdot$$

$$RO\cdot + CH_2{=}CH_2 \longrightarrow ROCH_2{-}\dot{C}H_2$$

PROPAGATION STEPS.

$$ROCH_2CH_2\cdot + CH_2{=}CH_2 \longrightarrow ROCH_2CH_2CH_2CH_2\cdot$$

$$ROCH_2CH_2CH_2CH_2\cdot \xrightarrow{(n-1)\ CH_2=CH_2} RO{-}(CH_2CH_2)_n{-}CH_2CH_2\cdot$$

Branching in Polyethene (Polyethylene)

$$\sim\sim CH_2\overset{H}{C}(CH_2CH_2\sim)CH_2CH_2\sim\sim$$

Polyethene (polyethylene) produced in this way does not have the expected linear structure. *Branching* occurs by abstraction of a hydrogen along the growing chain by another radical center followed by chain growth originating from the new radical. The average molecular weight of polyethene is almost 1 million.

Polychloroethene [poly(vinyl chloride), PVC] is made by similar radical polymerization. Interestingly, the reaction is regioselective. The peroxide initiator and the intermediate chain radicals add only to the unsubstituted end of the monomer, because the radical center formed next to chlorine is relatively stable. Thus, PVC has a very regular *head-to-tail structure* of molecular weight in excess of 1.5 million. Although PVC itself is fairly hard and brittle, it can be softened by addition of carboxylic acid esters (Section 20-4), called **plasticizers** (*plastikos,* Greek, to form). The resulting elastic material is used in "vinyl leather," plastic covers, and garden hoses.

$$CH_2{=}CHCl \xrightarrow{ROOR} -(CH_2\underset{\displaystyle Cl}{\underset{|}{C}}H)_n-$$

Polychloroethene
[Poly(vinyl chloride)]

Exposure to chloroethene (vinyl chloride) has been linked to the incidence of a rare form of liver cancer (angiocarcinoma). The Occupational Safety and Health Administration (OSHA) has set limits to human exposure of less than an average of 1 ppm per 8-hr working day per worker.

An iron compound, $FeSO_4$, in the presence of hydrogen peroxide promotes the radical polymerization of propenenitrile (acrylonitrile). **Polypropenenitrile** (polyacrylonitrile), $-(CH_2CHCN)_n-$, also known as Orlon, is used to make fibers. Similar polymerizations of other monomers furnish Teflon and Plexiglas.

EXERCISE 12-21

Saran Wrap is made by radical copolymerization of 1,1-dichloroethene and chloroethene. Propose a structure.

Anionic polymerizations require initiation by bases

Anionic polymerizations are initiated by strong bases such as alkyllithiums, amides, alkoxides, and hydroxide. For example, methyl 2-cyanopropenoate (methyl α-cyanoacrylate) polymerizes rapidly in the presence of even small traces of hydroxide. When spread between two surfaces, it forms a tough solid film that cements the surfaces together. For this reason, commercial preparations of this monomer are marketed as Super Glue.

What accounts for this ease of polymerization? When the base attacks the methylene group of α-cyanoacrylate, it generates a carbanion whose negative charge is located next to the nitrile and ester groups, both of which are strongly electron withdrawing. The anion is stabilized because the nitrogen and oxygen atoms polarize their multiple bonds in the sense $^{\delta^+}C{\equiv}N^{\delta^-}$ and $^{\delta^+}C{=}O^{\delta^-}$, and because the charge can be delocalized by resonance.

Anionic Polymerization of Super Glue (Methyl α-Cyanoacrylate)

$$H\ddot{O}:^- + CH_2{=}C(COCH_3{=}O)(C{\equiv}N:) \longrightarrow \left[HOCH_2C{=}C(O^-)OCH_3(C{\equiv}N:) \leftrightarrow HOCH_2\bar{C}(COCH_3)(C{\equiv}N:) \leftrightarrow HOCH_2C(COCH_3){=}C{=}\ddot{N}:^- \right] \xrightarrow{CH_2=C(COCH_3)(C\equiv N:)} HOCH_2C(COOCH_3)(CN:)-\bar{C}(COCH_3)(CN:) \xrightarrow{etc.}$$

Methyl 2-Cyanopropenoate

(α-Cyanoacrylate, Super Glue)

Metal-catalyzed polymerizations produce highly regular chains

An important **metal-catalyzed polymerization** is that initiated by Ziegler-Natta* catalysts. They are typically made from titanium tetrachloride and a trialkylaluminum, such as triethylaluminum, $Al(CH_2CH_3)_3$. The system polymerizes alkenes, particularly ethene, at relatively low pressures with remarkable ease and efficiency.

Although we shall not discuss the mechanism here, two features of Ziegler-Natta polymerization are the regularity with which substituted alkane chains are constructed from substituted alkenes, such as propene, and the high linearity of the chains. The polymers that result possess higher density and much greater strength than those obtained from radical polymerization. An example of this contrast is found in the properties of polyethene (polyethylene) prepared by the two methods. The chain branching that occurs during radical polymerization of ethene results in a flexible, transparent material *(low-density polyethylene)* used for food storage bags, whereas the Ziegler-Natta method produces a tough, chemically resistant plastic *(high-density polyethylene)* that may be molded into containers.

In summary, alkenes are subject to attack by carbocations, radicals, anions, and transition metals to give polymers. In principle, any alkene can function as a monomer. The intermediates are usually formed according to the rules that govern the stability of charges and radical centers.

*Professor Karl Ziegler, 1898–1973, Max Planck Institute for Coal Research, Mülheim, Germany, Nobel Prize 1963 (chemistry); Professor Giulio Natta, 1903–1979, Polytechnic Institute of Milan, Nobel Prize 1963 (chemistry).

12-15 Ethene: An Important Industrial Feedstock

Ethene serves as a case study of the significance of alkenes in industrial chemistry. This monomer is important for polyethene production. More than 20 billion pounds (10 million tons) of this polymer are produced in the United States annually. Currently, the major source of ethene is the pyrolysis of petroleum or hydrocarbons, such as ethane, propane, other alkanes, and cycloalkanes, derived from natural gas. Temperatures range from 750 to 900°C and the yields of ethene from 20 to 30%. Cracking of larger alkanes typically proceeds through C–C bond breaking to alkyl radicals, the further fragmentation of which eliminates ethene (Section 3-3). In 1992, 40 billion pounds (20 million tons) of ethene were made in this way. This amount is equivalent to about 17% of the total production of organic chemicals.

Apart from its direct use as a monomer, ethene is the starting material for many other industrial chemicals, some of which are themselves valuable monomers. For example, ethenyl acetate (vinyl acetate) is obtained in the reaction of ethene with acetic acid in the presence of a palladium(II) catalyst, air, and $CuCl_2$.

$$CH_2{=}CH_2 \xrightarrow{CH_3\overset{\overset{O}{\|}}{C}OH,\ O_2,\ \text{catalytic } PdCl_2 \text{ and } CuCl_2} CH_2{=}CHO\overset{\overset{O}{\|}}{C}CH_3 + H_2O$$

Ethenyl acetate
(Vinyl acetate)

A similar reaction, in which water is used instead of acetic acid, leads to the intermediate ethenol (vinyl alcohol). This species is unstable and spontaneously rearranges to acetaldehyde (see Chapters 13 and 18). The catalytic conversion of ethene into acetaldehyde is also known as the *Wacker process*.

The Wacker Process

$$CH_2{=}CH_2 \xrightarrow{H_2O,\ O_2,\ \text{catalytic } PdCl_2 \text{ and } CuCl_2} CH_2{=}CHOH \longrightarrow CH_3\overset{\overset{O}{\|}}{C}H$$

Ethenol **(Vinyl alcohol)** (Unstable) — **Acetaldehyde**

Chloroethene (vinyl chloride) is made from ethene by a chlorination–dehydrochlorination sequence. Because chlorine is relatively expensive, an indirect process has been developed that uses HCl in the presence of oxygen and $CuCl_2$. These conditions lead to the same intermediate, 1,2-dichloroethane, which is converted into the desired product by elimination of HCl.

Chloroethene (Vinyl Chloride) Synthesis

$$CH_2{=}CH_2 \xrightarrow{Cl_2} \underset{\underset{Cl}{|}}{C}H_2{-}\underset{\underset{Cl}{|}}{C}H_2 \xrightarrow[-\ HCl]{\Delta} CH_2{=}CHCl$$

1,2-Dichloroethane — **Chloroethene** **(Vinyl chloride)**

TABLE 12-4 Chemicals Made from Ethene

Chemical	Tons
1,2-Dichloroethane (ethylene dichloride)	7969
Ethenylbenzene (styrene)	4471
Ethylbenzene	5495
Chloroethene (vinyl chloride)	6613
Oxacyclopropane (ethylene oxide)	2780
1,2-Ethanediol (ethylene glycol)	2562
Acetic acid[a]	1800
Ethanol	351
Ethenyl acetate (vinyl acetate)	1329

a. Also made by other processes.

Oxidation of ethene with oxygen in the presence of silver furnishes oxacyclopropane (ethylene oxide), the hydrolysis of which gives 1,2-ethanediol (ethylene glycol). Hydration of ethene gives ethanol.

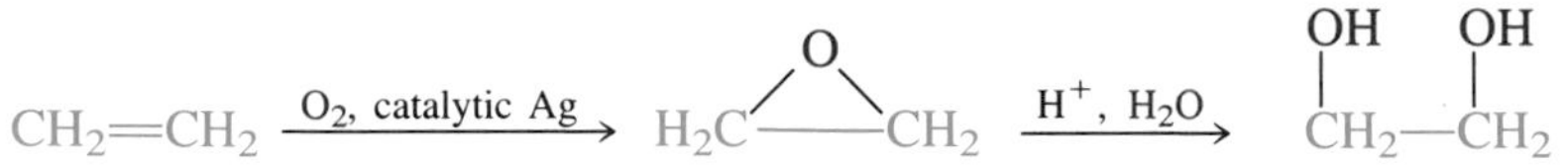

Oxacyclopropane (Ethylene oxide) **1,2-Ethanediol (Ethylene glycol)**

Table 12-4 gives an idea of the sizable amount of ethene-derived raw materials produced in the United States in 1992.

In summary, ethene is a valuable source of various industrial raw materials, particularly monomers, ethanol, and 1,2-ethanediol (ethylene glycol).

12-16 Alkenes in Nature: Insect Pheromones

Many natural products contain π bonds; several were mentioned in Sections 4-7 and 9-11. This section describes in more detail a specific group of naturally occurring alkenes, the **insect pheromones** (*pherein,* Greek, to bear; *hormon,* Greek, to stimulate).

Insect Pheromones

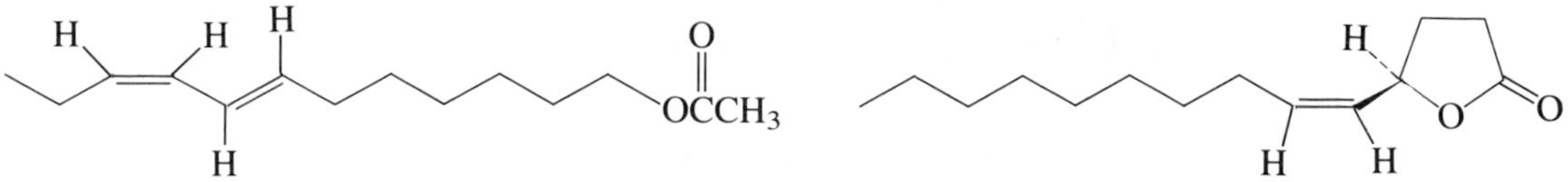

European vine moth **Japanese beetle**

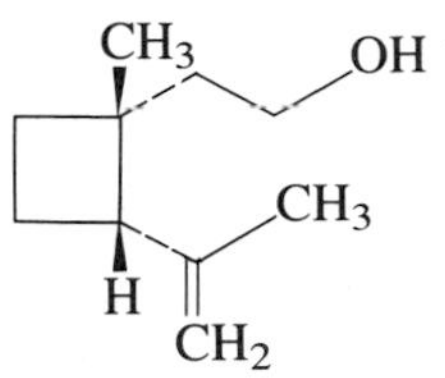

Male boll weevil

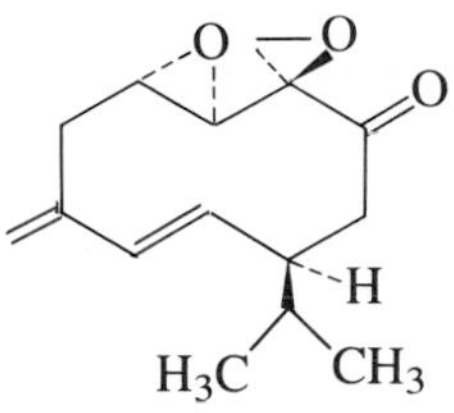

American cockroach

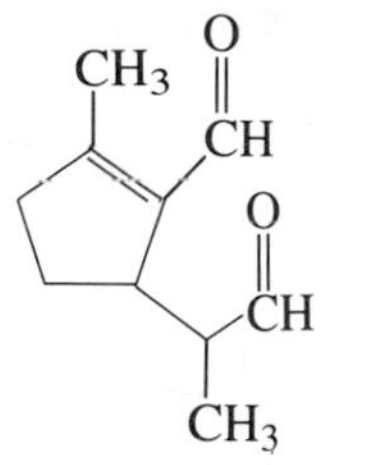

Defense pheromone of larvae of chrysomelid beetle

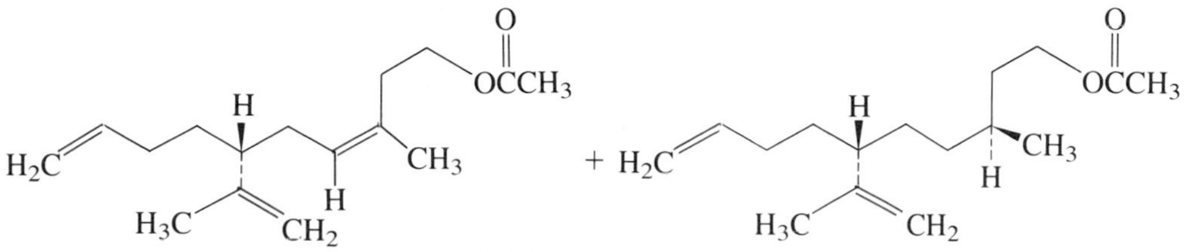

California red scale

Pheromones are chemical substances used for communication within a living species. There are sex, trail, alarm, and defense pheromones, to mention a few. Many insect pheromones are simple alkenes; they are isolated by extraction of certain parts of the insect and separation of the resulting product mixture by chromatographic techniques. Often only minute quantities of the bioactive compound can be isolated, in which case the synthetic organic chemist can play a very important role in the design and execution of total syntheses. Interestingly, the specific activity of a pheromone frequently depends on the configuration around the double bond (e.g., E or Z), as well as on the absolute configuration of any chiral centers present (R,S) *and* the composition of isomer mixtures. For example, the sex attractant for the male silkworm moth, 10-*trans*-12-*cis*-hexadecadien-1-ol (known as bombykol), is 10 billion times more active in eliciting a response than is the 10-*cis*-12-*trans* isomer, and 10 *trillion* times more so than the trans, trans compound.

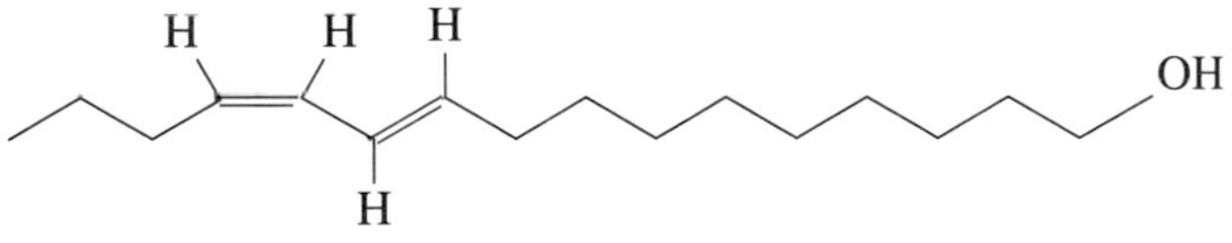

Bombykol

Pheromone research affords an important opportunity for achieving pest control. Minute quantities of sex pheromones can be used per acre of land to confuse male insects about the location of their female partners. These pheromones can thus serve as lures in traps to effectively remove insects without spraying crops with large amounts of other chemicals. It is clear that organic chemists in collaboration with insect biologists will make important contributions in this area in the years to come.

NEW REACTIONS

1. GENERAL ADDITION TO ALKENES (Section 12-1)

$$\text{>C=C<} + \text{A—B} \longrightarrow \text{—}\underset{|}{\overset{\text{A}}{\overset{|}{\text{C}}}}\text{—}\underset{|}{\overset{\text{B}}{\overset{|}{\text{C}}}}\text{—}$$

2. HYDROGENATION (Section 12-2)

$$\text{>C=C<} \xrightarrow{H_2,\ \text{catalyst}} \text{H—C—C—H}$$

***Syn* addition**

Typical catalysts: PtO_2, Pd-C, Ra-Ni

Electrophilic Additions

3. HYDROHALOGENATION (Section 12-3)

$$\text{>C=C<} \xrightarrow{HX} \text{—}\underset{|}{\overset{\text{H}}{\overset{|}{\text{C}}}}\text{—}\underset{|}{\overset{\text{X}}{\overset{|}{\text{C}}}}\text{—}$$

$$\text{(R)(H)C=CH}_2 \xrightarrow{HX} \text{H—}\underset{\text{X}}{\underset{|}{\overset{\text{R}}{\overset{|}{\text{C}}}}}\text{—CH}_3$$

Regiospecific (Markovnikov rule)

Through more stable carbocation

4. HYDRATION (Section 12-4)

$$\text{>C=C<} \xrightarrow{H^+,\ H_2O} \text{—}\underset{|}{\overset{\text{H}}{\overset{|}{\text{C}}}}\text{—}\underset{|}{\overset{\text{OH}}{\overset{|}{\text{C}}}}\text{—}$$

Through more stable carbocation

5. HALOGENATION (Section 12-5)

$$\text{>C=C<} \xrightarrow{X_2,\ CCl_4} \text{X—C—C—X}$$

Stereospecific *(anti)*

$X_2 = Cl_2$ or Br_2, but not I_2

6. VICINAL HALOALCOHOL SYNTHESIS (Section 12-6)

$$\text{>C=C<} \xrightarrow{X_2,\ H_2O} \text{X—C—C—OH}$$

OH attaches to more substituted carbon

7. VICINAL HALOETHER SYNTHESIS (Section 12-6)

$$\text{C=C} \xrightarrow{X_2,\ \text{ROH}} \text{X–C–C–OR}$$

OR attaches to more substituted carbon

8. GENERAL ELECTROPHILIC ADDITIONS (Section 12-6, Table 12-2)

$$\text{C=C} \xrightarrow{\text{AB}} \text{(cyclic A}^{+}\text{ ion)} \xrightarrow{B^-} \text{A–C–C–B}$$

A = electropositive, B = electronegative

B attaches to more substituted carbon

9. OXYMERCURATION–DEMERCURATION (Section 12-7)

$$\text{C=C} \xrightarrow[\text{2. NaBH}_4,\ \text{NaOH, H}_2\text{O}]{\text{1. Hg(OCCH}_3)_2,\ \text{H}_2\text{O}} \text{–C(H)–C(OH)–}$$

Initial addition is *anti,* through mercurinium ion

$$\text{C=C} \xrightarrow[\text{2. NaBH}_4,\ \text{NaOH, H}_2\text{O}]{\text{1. Hg(OCCH}_3)_2,\ \text{ROH}} \text{–C(H)–C(OR)–}$$

OH or OR attaches to more substituted carbon

10. HYDROBORATION (Section 12-8)

$$\text{C=C} + \text{BH}_3 \xrightarrow{\text{THF}} (\text{–C(H)–C})_3\text{–B}$$

$$\text{R(H)C=CH}_2 + \text{BH}_3 \xrightarrow{\text{THF}} (\text{RCH}_2\text{CH}_2)_3\text{–B}$$

Regiospecific

B attaches to less substituted carbon

$$\text{1,2-dimethylcyclohexene} + \text{BH}_3 \xrightarrow{\text{THF}} \text{(H and B added syn)}$$

Stereospecific (*syn*)
and anti-Markovnikov

11. HYDROBORATION–OXIDATION (Section 12-8)

$$>C{=}C< \xrightarrow[\text{2. } H_2O_2,\ HO^-]{\text{1. } BH_3,\ THF} -\overset{H}{\underset{|}{\overset{|}{C}}}-\overset{OH}{\underset{|}{\overset{|}{C}}}-$$

Stereospecific *(syn)* and anti-Markovnikov

OH attaches to less substituted carbon

Oxidation

12. OXACYCLOPROPANE FORMATION (Section 12-9)

$$>C{=}C< \xrightarrow{RCOOH,\ CH_2Cl_2} \text{(oxacyclopropane)} + RCOH$$

Stereospecific *(syn)*

13. VICINAL ANTI DIHYDROXYLATION (Section 12-9)

$$>C{=}C< \xrightarrow[\text{2. } H^+,\ H_2O]{\text{1. } RCOOH,\ CH_2Cl_2} HO-C-C-OH \ (\text{anti}) + RCOH$$

14. VICINAL SYN DIHYDROXYLATION (Section 12-10)

$$>C{=}C< \xrightarrow{KMnO_4,\ H_2O,\ 0°C,\ pH\ 7} HO-C-C-OH \ (\text{syn})$$

$$>C{=}C< \xrightarrow{OsO_4,\ H_2S;\ \text{or catalytic } OsO_4,\ H_2O_2} HO-C-C-OH \ (\text{syn})$$

Through cyclic intermediates

15. OZONOLYSIS (Section 12-11)

$$>C{=}C< \xrightarrow[\text{2. } (CH_3)_2S;\ \text{or Zn, } CH_3COH]{\text{1. } O_3,\ CH_3OH} >C{=}O + O{=}C<$$

Through molozonide and ozonide intermediates

Radical Additions

16. RADICAL HYDROBROMINATION (Section 12-12)

$$>C{=}CH_2 \xrightarrow{HBr,\ ROOR} -\overset{H}{\underset{|}{\overset{|}{C}}}-\overset{Br}{\underset{H}{\overset{|}{C}}}-H$$

Anti-Markovnikov

Does not occur with HCl or HI

17. OTHER RADICAL ADDITIONS (Section 12-12)

$$\text{>C=C<} \xrightarrow{\text{RSH, ROOR}} -\overset{\text{H}}{\underset{|}{\overset{|}{\text{C}}}}-\overset{\text{SR}}{\underset{|}{\overset{|}{\text{C}}}}-$$

$$\text{>C=C<} \xrightarrow{\text{HCX}_3\text{, ROOR}} -\overset{\text{H}}{\underset{|}{\overset{|}{\text{C}}}}-\overset{\text{CX}_3}{\underset{|}{\overset{|}{\text{C}}}}-$$

Anti-Markovnikov

Monomers and Polymers

18. DIMERIZATION, OLIGOMERIZATION, AND POLYMERIZATION (Sections 12-13 and 12-14)

$$n\ \text{>C=C<} \xrightarrow{\text{H}^+ \text{ or RO}\cdot \text{ or B}^-} -(\underset{|}{\overset{|}{\text{C}}}-\underset{|}{\overset{|}{\text{C}}})_n-$$

Ziegler-Natta polymerization

$$n\ \text{CH}_2\text{=CH}_2 \xrightarrow{\text{TiCl}_4\text{, AlR}_3} -(\text{CH}_2\text{CH}_2)_n-$$

19. ETHENE IN INDUSTRIAL PROCESSES (Section 12-15)

Synthesis by cracking

$$\text{R—CH}_2\text{—CH}_2\text{—R} \xrightarrow{\Delta} \text{CH}_2\text{=CH}_2 + \text{R—R} + \text{other hydrocarbons}$$

Ethenyl acetate (vinyl acetate) synthesis

$$\text{CH}_2\text{=CH}_2 \xrightarrow{\text{CH}_3\overset{\text{O}}{\overset{\|}{\text{C}}}\text{OH, O}_2\text{, catalytic PdCl}_2\text{ and CuCl}_2} \text{CH}_2\text{=C}\begin{matrix}\text{H}\\ \text{O}\overset{}{\underset{\underset{\text{O}}{\|}}{\text{C}}}\text{CH}_3\end{matrix}$$

Wacker process

$$\text{CH}_2\text{=CH}_2 \xrightarrow{\text{H}_2\text{O, O}_2\text{, catalytic PdCl}_2\text{ and CuCl}_2} \text{CH}_3\overset{\text{O}}{\overset{\|}{\text{C}}}\text{H}$$

Chloroethene (vinyl chloride) synthesis

$$\text{CH}_2\text{=CH}_2 \xrightarrow{\text{Cl}_2} \underset{\text{Cl}}{\underset{|}{\text{CH}_2}}-\underset{\text{Cl}}{\underset{|}{\text{CH}_2}} \xrightarrow[-\text{ HCl}]{} \text{CH}_2\text{=CHCl}$$

Oxacyclopropane (ethylene oxide) and 1,2-ethanediol (ethylene glycol) synthesis

$$\text{CH}_2\text{=CH}_2 \xrightarrow{\text{O}_2\text{, catalytic Ag}} \text{H}_2\text{C}\overset{\text{O}}{\frown}\text{CH}_2 \xrightarrow{\text{H}^+\text{, H}_2\text{O}} \overset{\text{OH}}{\overset{|}{\text{CH}_2}}-\overset{\text{OH}}{\overset{|}{\text{CH}_2}}$$

IMPORTANT CONCEPTS

1. The reactivity of the double bond manifests itself in exothermic addition reactions leading to saturated products.
2. The hydrogenation of alkenes is immeasurably slow unless a catalyst capable of splitting the strong H–H bond is used. Possible catalysts are palladium on carbon, platinum (as PtO_2), and Raney nickel. Addition of hydrogen is subject to steric control, the least hindered face of the least substituted double bond frequently being attacked preferentially.
3. As a Lewis base, the π bond is subject to attack by acid and electrophiles, such as H^+, X_2, and Hg^{2+}. If the initial intermediate is a free carbocation, the more highly substituted carbocation is formed. Alternatively, a cyclic onium ion is generated subject to nucleophilic ring opening at the more substituted carbon. The former case leads to control of regiochemistry (Markovnikov rule), the latter to control of both regio- and stereochemistry.
4. Mechanistically, hydroboration seems to lie between hydrogenation and electrophilic addition. The first step is π complexation to the electron-deficient boron, whereas the second is a concerted transfer of the hydrogen to carbon. Hydroboration–oxidation results in the anti-Markovnikov hydration of alkenes.
5. Peroxycarboxylic acids may be thought of as containing an electrophilic oxygen atom, transferable to alkenes to give oxacyclopropanes. The process is often called epoxidation.
6. Permanganate and osmium tetroxide act as electrophilic oxidants of alkenes; in the course of the reaction, the oxidation state of the metal is reduced by two units. Addition takes place in a concerted manner through cyclic six-electron transition states to give vicinal *syn* diols.
7. Ozonolysis followed by reduction yields carbonyl compounds derived by cleavage of the double bond.
8. In radical chain additions to alkenes, the chain carrier adds to the π bond to create the more highly substituted radical. This method allows for the anti-Markovnikov hydrobromination of alkenes, as well as the addition of thiols and some halomethanes.
9. Alkenes react with themselves through initiation by charged species, radicals, or some transition metals. The initial attack at the double bond yields a reactive intermediate that perpetuates carbon–carbon bond formation.

PROBLEMS

1. With the help of the $DH°$ values given in Table 3-1 and Section 3-5, calculate the $\Delta H°$ values for addition of each of the following molecules to ethene, using 65 kcal mol^{-1} for the carbon–carbon π bond strength.

(a) Cl_2
(b) IF ($DH° = 67$ kcal mol^{-1})
(c) IBr ($DH° = 43$ kcal mol^{-1})
(d) HF
(e) HI
(f) HO—Cl ($DH° = 60$ kcal mol^{-1})
(g) Br—CN ($DH° = 83$ kcal mol^{-1}; $DH°$ for —C—CN = 124 kcal mol^{-1})
(h) CH_3S—H ($DH° = 88$ kcal mol^{-1}; $DH°$ for —C—S = 60 kcal mol^{-1})

2. Give the expected major product of catalytic hydrogenation of each of the following alkenes. Clearly show and explain the stereochemistry of the resulting molecules.

(a) 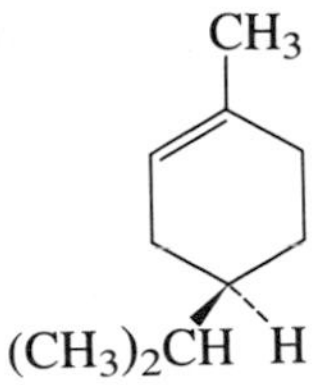 CH_3 ... $(CH_3)_2CH$ H (b) 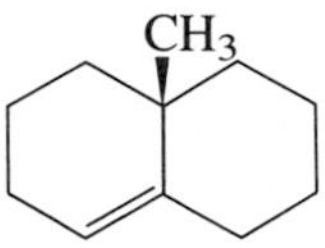 CH_3 (c) 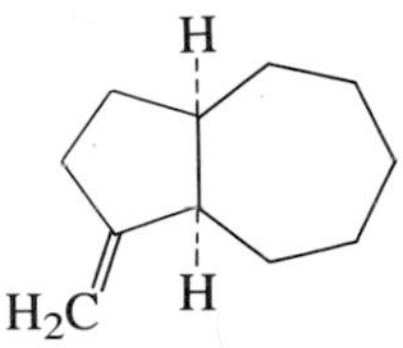 H ... H_2C H

3. Would you expect the catalytic hydrogenation of a small-ring cyclic alkene such as cyclobutene to be more or less exothermic than that of cyclohexene?

4. Give the expected major product from the reaction of each alkene with (i) peroxide-free HBr and (ii) HBr in the presence of peroxides. **(a)** 1-Hexene; **(b)** 2-methyl-1-pentene; **(c)** 2-methyl-2-pentene; **(d)** (Z)-3-hexene; **(e)** cyclohexene.

5. Give the product of addition of Br_2 to each alkene in Problem 4. Pay attention to stereochemistry.

6. What alcohol would be obtained from treatment of each alkene in Problem 4 with aqueous sulfuric acid? Would any of these alkenes give a different product upon oxymercuration–demercuration? With hydroboration–oxidation?

7. Give the reagents and conditions necessary for each of the following transformations, and comment on the thermodynamics of each. **(a)** Cyclohexanol → cyclohexene; **(b)** cyclohexene → cyclohexanol; **(c)** chlorocyclopentane → cyclopentene; **(d)** cyclopentene → chlorocyclopentane.

8. **(a)** Give the product of addition of methylselenyl chloride, CH_3SeCl (polarized $^{\delta+}$Se–Cl$^{\delta-}$), to propene. Explain the regiochemistry. **(b)** Reaction of CH_3SeCl with cyclohexene is a stereospecific *anti* addition. Explain mechanistically.

9. Formulate the product(s) that you would expect from each of the following reactions. Show stereochemistry clearly.

(a) 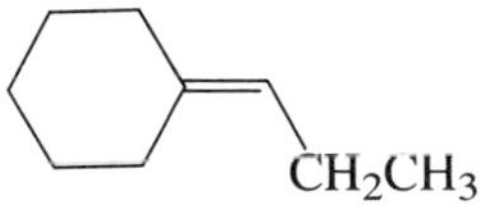 CH_2CH_3 $\xrightarrow{HCl}$

(b) *trans*-3-Heptene $\xrightarrow{Cl_2}$

(c) 1-Ethylcyclohexene $\xrightarrow{Br_2,\ H_2O}$

(d) Product of (c) $\xrightarrow{NaOH,\ H_2O}$

(e) 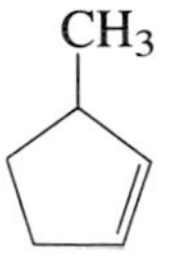 CH_3 $\xrightarrow{\text{1. } Hg(OCCH_3)_2,\ CH_3OH \text{ (with C=O)};\ \text{2. } NaBH_4,\ CH_3OH}$

(f) 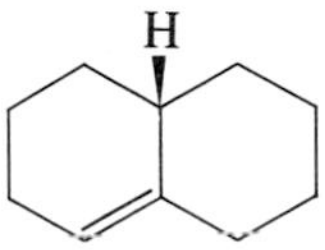 H $\xrightarrow{Br_2,\ \text{excess } Na^+N_3^-}$

(g) 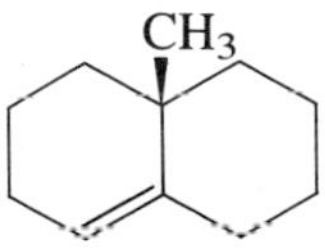 CH_3 $\xrightarrow{\text{1. } BH_3,\ THF;\ \text{2. } H_2O_2,\ NaOH,\ H_2O}$

10. Show how you would synthesize each of the following molecules from an alkene of appropriate structure (your choice).

(a) $(CH_3)_2CH\underset{}{C}HCH_3$ with OH on C-2 (i.e., $(CH_3)_2CHCH(OH)CH_3$) **(b)** $ClCH_2CH(CH_3)OCH(CH_3)_2$

(c) *meso*-$CH_3CH_2CH_2CH(Br)CH(Br)CH_2CH_2CH_3$ (i.e., the 4*R*,5*S* stereoisomer)

(d) $CH_3CH_2CH_2CH(Br)CH(Br)CH_2CH_2CH_3$ (racemic mixture of 4*R*,5*R* and 4*S*,5*S*)

(e) 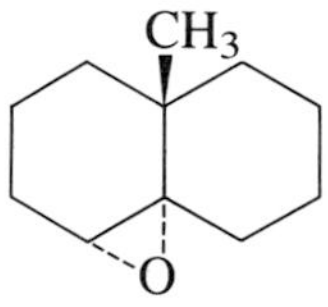 [bicyclic epoxide with angular CH_3] **(f)** 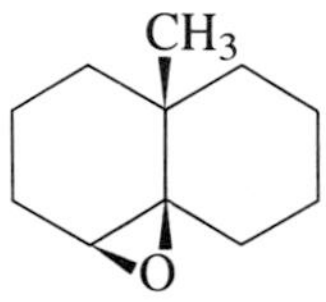 [bicyclic epoxide with angular CH_3] (More challenging. Hint: See Section 12-6.)

11. Propose efficient methods for accomplishing each of the following transformations. Most will require more than one step.

(a) $CH_3CH_2CH(Br)CH_3 \longrightarrow CH_3CH_2CH_2CH_2I$

(b) $CH_3CH(OH)CH_2CH_3 \longrightarrow$ *meso*-$CH_3CH(OH)CH(OH)CH_3$ (i.e., 2*R*,3*S*)

(c) $CH_3CH(OH)CH_2CH_3 \longrightarrow CH_3CH(OH)CH(OH)CH_3$ (1:1 mixture of 2*R*,3*R* and 2*S*,3*S*)

(d) $(CH_3)_2C{=}CHCH_2CH_2CH{=}CH_2 \longrightarrow (CH_3)_2C\overset{O}{—}CHCH_2CH_2CH_2\overset{O}{\overset{\|}{C}}H$ (epoxide at C-2,C-3; aldehyde)

12. Give the expected product of reaction of 2-methyl-1-pentene with each of the following reagents.

(a) H_2, PtO_2, CH_3CH_2OH

(b) D_2, Pd-C, CH_3CH_2OH

(c) BH_3, THF then NaOH + H_2O_2

(d) HCl

(e) HBr

(f) HBr + peroxides

(g) HI + peroxides

(h) $H_2SO_4 + H_2O$

(i) Cl_2

(j) ICl

(k) $Br_2 + CH_3CH_2OH$

(l) CH_3SH + peroxides

(m) MCPBA, CH_2Cl_2

(n) $KMnO_4$, H_2O, 0°C

(o) O_3, then Zn + $CH_3\overset{O}{\overset{\|}{C}}OH$

(p) $Hg(O\overset{O}{\overset{\|}{C}}CH_3)_2 + H_2O$, then $NaBH_4$

(q) $CHBr_3$ + peroxides

(r) Catalytic H_2SO_4 + heat

13. What are the products of reaction of (*E*)-3-methyl-3-hexene with each of the reagents in Problem 12?

14. Write the expected products of reaction of 1-ethylcyclopentene with each of the reagents in Problem 12.

15. Give the expected major product from reaction of 3-methyl-1-butene with each of the following. Explain any differences in the products mechanistically.

(a) 50% aqueous H_2SO_4 **(b)** $Hg(OCCH_3)_2$ (with C=O) in H_2O, followed by $NaBH_4$
(c) BH_3 in THF, followed by NaOH and H_2O_2

16. Answer the question posed in Problem 15 for ethenylcyclohexane (vinylcyclohexane).

17. ^{1}H NMR spectrum A corresponds to a molecule with the formula C_3H_5Cl. The compound shows significant IR bands at 730 (see Problem 7 of Chapter 11), 930, 980, 1630, and 3090 cm^{-1}. **(a)** Deduce the structure of the molecule. **(b)** Assign each NMR signal to a hydrogen or group of hydrogens. **(c)** The "doublet" at $\delta = 4$ ppm has $J = 6$ Hz. Is this in accord with your assignment in (b)? **(d)** This "doublet," on fivefold expansion, becomes a doublet of triplets (inset, spectrum A), with $J \sim 1$ Hz for the triplet splittings. What is the origin of this triplet splitting? Is it reasonable in light of your assignment in (b)?

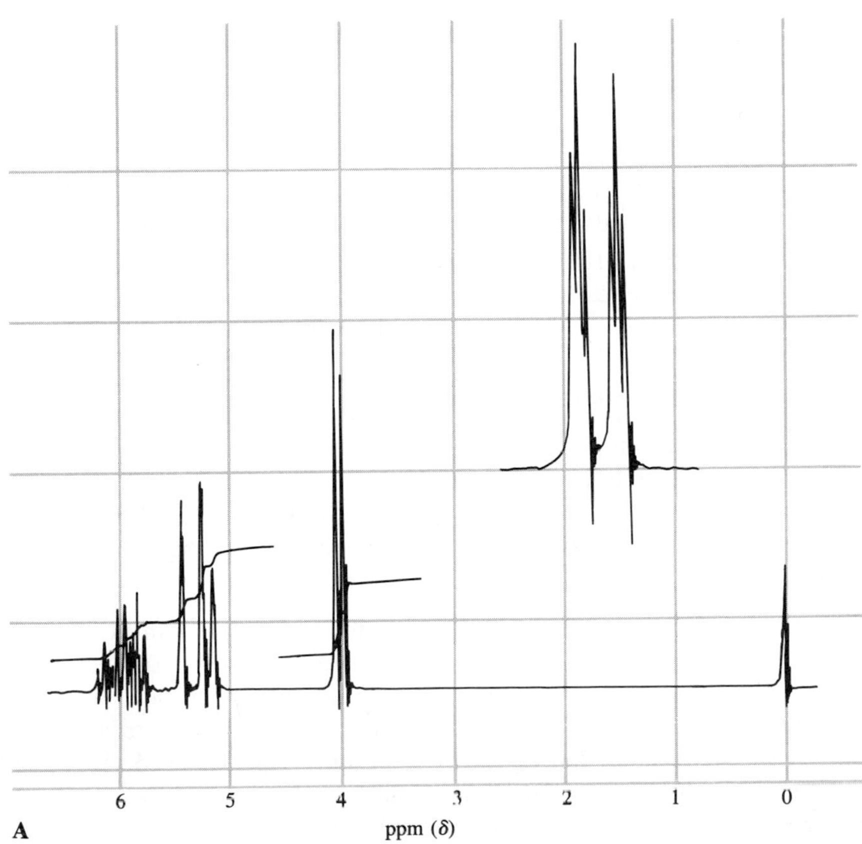

A

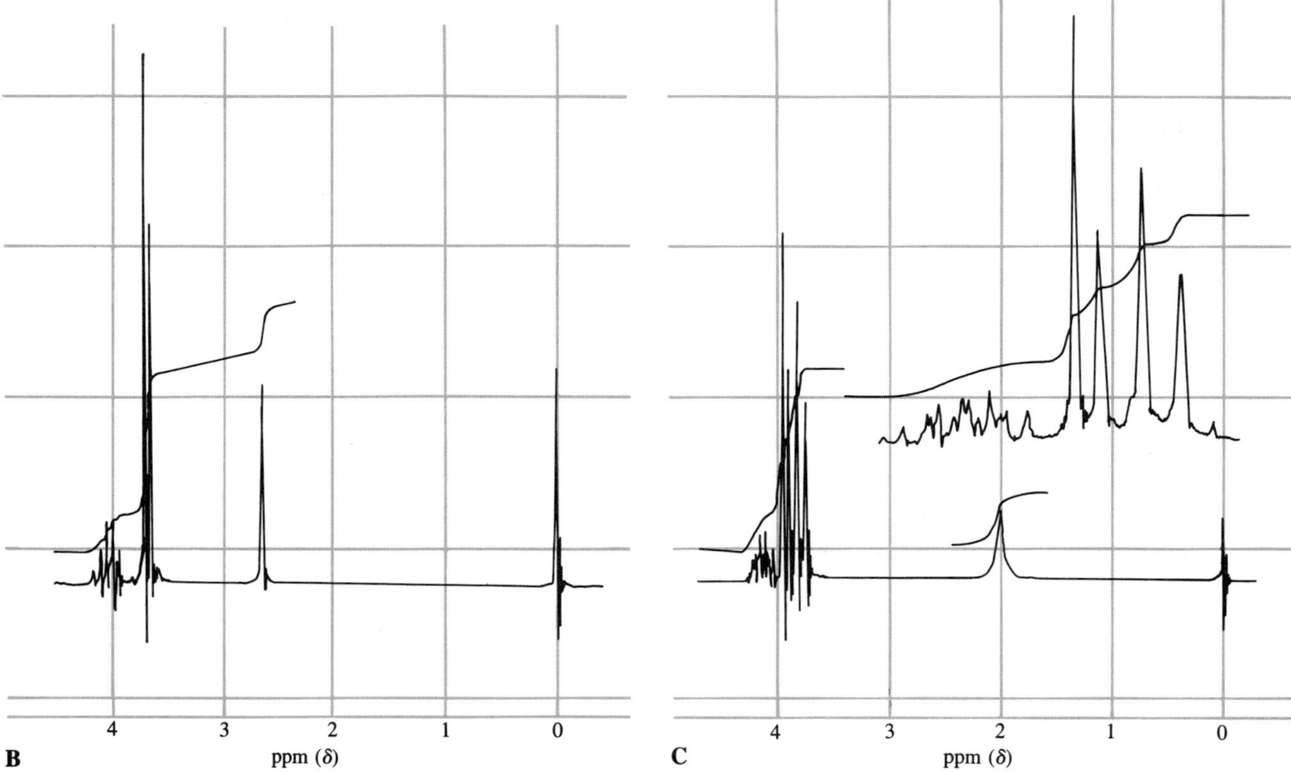

18. Reaction of C_3H_5Cl (Problem 17, spectrum A) with Cl_2 in H_2O gives rise to two products, both $C_3H_6Cl_2O$, whose spectra are shown in B and C. Reaction of either of these with KOH yields the same molecule C_3H_5ClO (spectrum D). The insets show expansions of some of the multiplets. The IR spectrum D reveals bands at 720 and 1260 cm^{-1} and the absence of any signals between 1600 and 1800 cm^{-1} and between 3200 and 3700 cm^{-1}. **(a)** Deduce the structures of the compounds giving rise to spectra B, C, and D. **(b)** Why does reaction of the starting chloride compound with Cl_2 in H_2O give two isomeric products? **(c)** Write mechanisms for the formation of the product C_3H_5ClO from both isomers of $C_3H_6Cl_2O$.

19. ^1H NMR spectrum E corresponds to a molecule with the formula C_4H_8O. Its IR spectrum has important bands at 945, 1015, 1665, 3095, and 3360 cm^{-1}. **(a)** Determine the structure of the unknown. **(b)** Assign each NMR and IR signal. **(c)** Explain the splitting patterns for the signals at δ = 1.2, 4.2, and 5.8 ppm (see inset for fivefold expansion).

20. Reaction of the compound associated with spectrum E with $SOCl_2$ produces a chloroalkane, C_4H_7Cl, whose NMR spectrum is almost identical with E, except that the signal at δ = 3.5 ppm is absent. Treatment with H_2 over PtO_2 results in C_4H_9Cl (spectrum F, p. 462). The IR spectrum of compound E shows bands at 700 (Problem 7, Chapter 11), 925, 985, 1640, and 3090 cm^{-1}. In the IR spectrum of compound F, all but the band near 700 cm^{-1} are gone. Identify these two molecules.

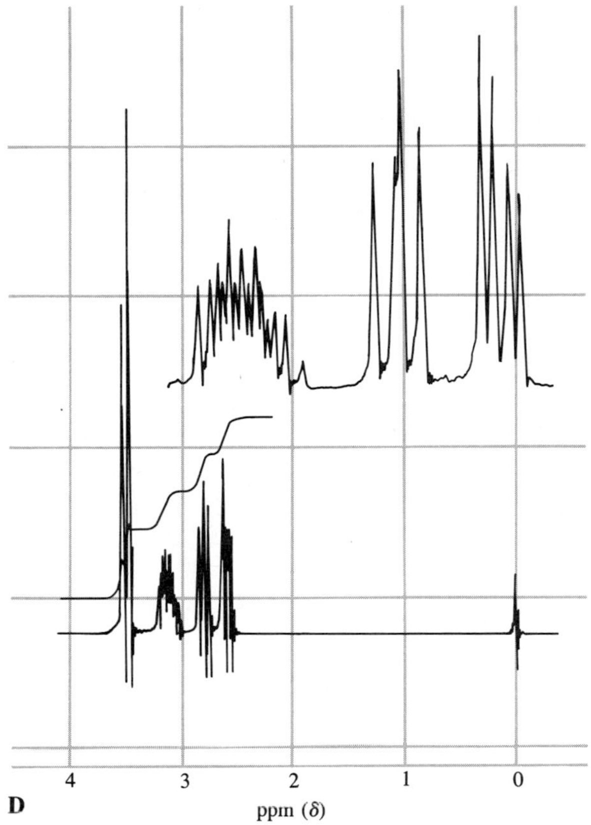
4
3
2
1
0
D
ppm (δ)

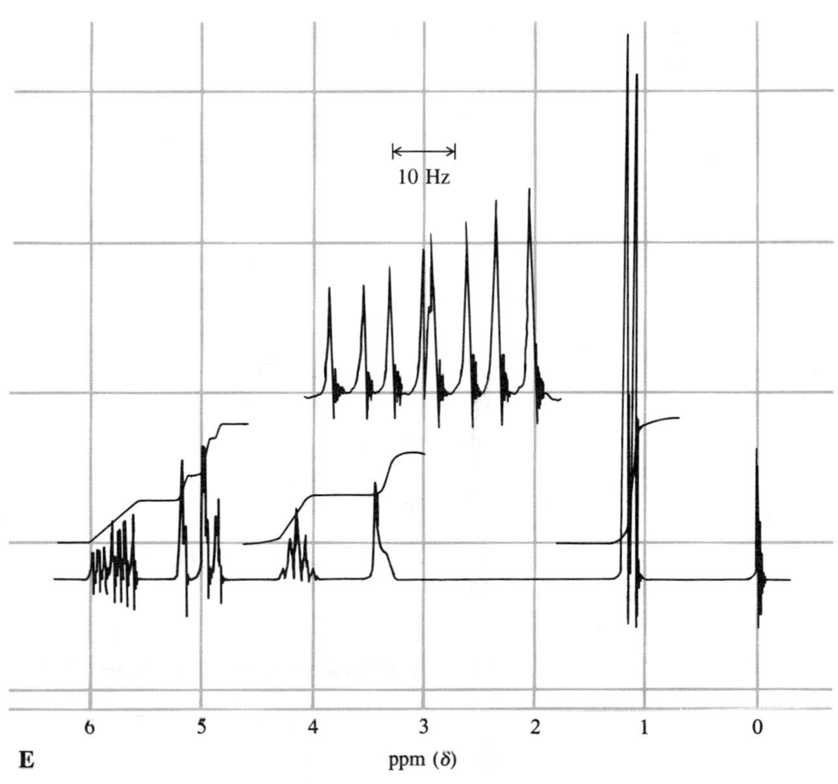
10 Hz
6
5
4
3
2
1
0
E
ppm (δ)

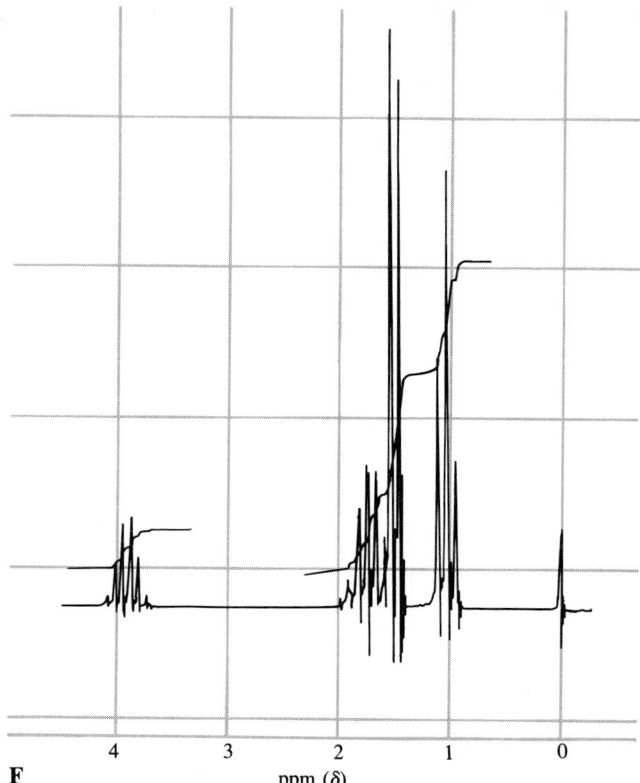

21. Give the structure of an alkene that will give the following carbonyl compounds upon ozonolysis followed by reduction with $(CH_3)_2S$.

(a) CH_3CHO only **(b)** CH_3CHO and CH_3CH_2CHO

(c) $(CH_3)_2C{=}O$ and $H_2C{=}O$ **(d)** $CH_3CH_2\overset{\overset{\displaystyle O}{\|}}{C}CH_3$ and CH_3CHO

(e) Cyclopentanone and CH_3CH_2CHO

22. Plan syntheses of each of the following compounds, utilizing retrosynthetic-analysis techniques. Starting compounds are given in parentheses. However, other simple alkanes or alkenes also may be used, as long as you include at least one carbon–carbon bond-forming step in each synthesis.

(a) $CH_3CH_2\overset{\overset{\displaystyle O}{\|}}{C}\underset{\underset{\displaystyle CH_3}{|}}{C}HCH_3$ (propene)

(b) 2-methylcyclohexanol drawn as a ring, with H_3C on a hashed bond and OH on a wedged bond (cyclohexene)

(c) $CH_3CH_2CH_2\underset{\underset{\displaystyle Cl}{|}}{C}HCH_2CH_2CH_3$ (propene, again)

23. Show how you would convert cyclopentane into each of the following molecules.

(a) *cis*-1,2-Dideuteriocyclopentane **(b)** *trans*-1,2-Dideuteriocyclopentane

(c) 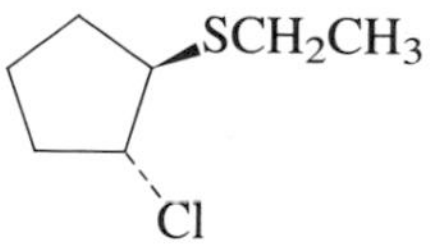 [cyclopentane with SCH_2CH_3 (wedge) and Cl (dash)]

(d) 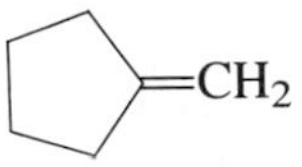 [cyclopentane =CH_2]

(e) 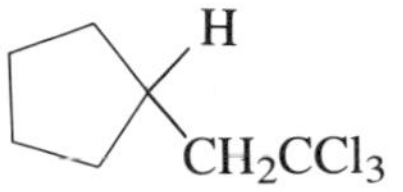 [cyclopentane with H and CH_2CCl_3]

(f) 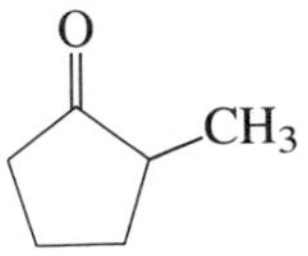 [cyclopentanone with CH_3]

(g) 1,2-Dimethylcyclopentene

(h) *trans*-1,2-Dimethyl-1,2-cyclopentanediol

24. Give the expected major product(s) of each of the following reactions.

(a) $CH_3OCH_2CH_2CH{=}CH_2$ $\xrightarrow{\text{1. Hg(OCCH}_3)_2\text{, CH}_3\text{OH; 2. NaBH}_4\text{, CH}_3\text{OH}}$ (O above the C of $OCCH_3$)

(b) 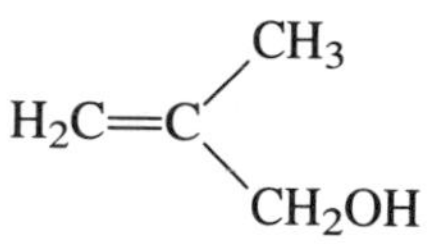 $H_2C{=}C(CH_3)CH_2OH$ $\xrightarrow{\text{1. CH}_3\text{COOH, CH}_2\text{Cl}_2\text{; 2. H}^+\text{, H}_2\text{O}}$ (O above the first C of CH_3COOH)

(c) 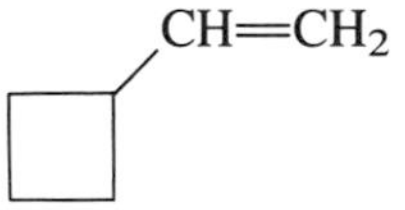 [cyclobutane with $CH{=}CH_2$] $\xrightarrow{\text{Conc. HI}}$

(d) 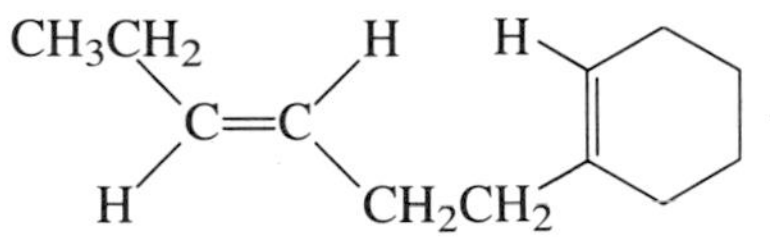 $CH_3CH_2(H)C{=}C(H)CH_2CH_2$–[cyclohexene bearing H] $\xrightarrow{\text{1. Excess O}_3\text{, CH}_2\text{Cl}_2\text{; 2. (CH}_3)_2\text{S}}$

(e) 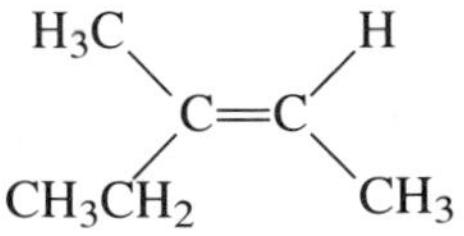 $(H_3C)(CH_3CH_2)C{=}C(H)(CH_3)$ $\xrightarrow{\text{BrCN}}$

(f)  [cyclopentene with Cl (wedge)] $\xrightarrow{\text{Cold KMnO}_4\text{, H}_2\text{O}}$

(g) $CH_3CH{=}CH_2$ $\xrightarrow{\text{Catalytic HF}}$

(h) $CH_2{=}CHNO_2$ $\xrightarrow{\text{Catalytic KOH}}$
(Hint: Draw Lewis structures for the NO_2 group.)

25. (*E*)-5-Hepten-1-ol reacts with the following reagents to give products with the indicated formulas. Determine their structures and explain their formation by detailed mechanisms. **(a)** HCl, $C_7H_{14}O$ (no Cl!); **(b)** Cl_2, $C_7H_{13}ClO$ (IR: 740 cm^{-1}; nothing between 1600 and 1800 cm^{-1} and between 3200 and 3700 cm^{-1}).

26. When a cis alkene is mixed with a small amount of I_2 in the presence of heat or light, it isomerizes to some trans alkene. Propose a detailed mechanism to account for this observation.

27. Treatment of α-terpineol (Chapter 10, Problem 26) with aqueous mercuric acetate followed by sodium borohydride reduction leads predominantly to an isomer of the starting compound ($C_{10}H_{18}O$) instead of a hydration product. This isomer is the chief component in oil of eucalyptus and, appropriately enough, is called eucalyptol. It is popularly used as a flavoring for otherwise foul-tasting medicines because of its pleasant spicy taste and aroma. Deduce a structure for eucalyptol on the basis of sensible mechanistic chemistry and the following proton-decoupled ^{13}C NMR data. (Hint: IR shows nothing between 1600 and 1800 cm^{-1} or between 3200 and 3700 cm^{-1}!)

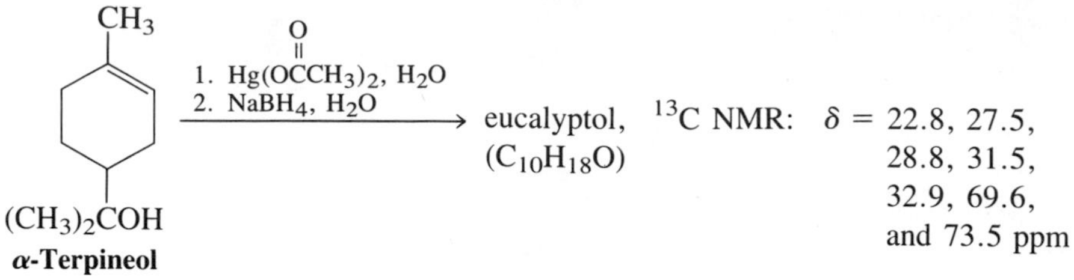

$$\alpha\text{-Terpineol} \xrightarrow[\text{2. } NaBH_4,\ H_2O]{\text{1. } Hg(OCCH_3)_2,\ H_2O} \text{eucalyptol, } (C_{10}H_{18}O)$$

$(CH_3)_2COH$

α-Terpineol

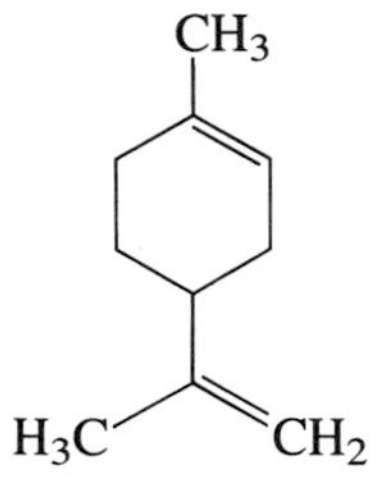

eucalyptol, ^{13}C NMR: δ = 22.8, 27.5, 28.8, 31.5, 32.9, 69.6, and 73.5 ppm

CH_3

H_3C CH_2

Limonene

28. Both borane and MCPBA react highly selectively with molecules such as limonene that contain double bonds in very different environments. Predict the products of reaction of limonene with **(a)** one equivalent of BH_3 in THF, followed by basic aqueous H_2O_2, and **(b)** one equivalent of MCPBA in CH_2Cl_2. Explain your answers.

29. Oil of marjoram contains a pleasant, lemon-scented substance, $C_{10}H_{16}$ (compound G). Upon ozonolysis, G forms two products. One of them, H, has the formula $C_8H_{14}O_2$, and can be independently synthesized in the following way.

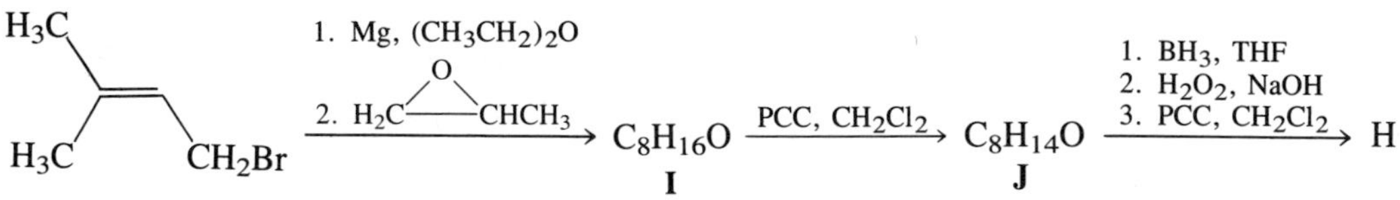

$$(H_3C)_2C{=}CHCH_2Br \xrightarrow[\text{2. } H_2C\overset{O}{—}CHCH_3]{\text{1. Mg, } (CH_3CH_2)_2O} \underset{\textbf{I}}{C_8H_{16}O} \xrightarrow{PCC,\ CH_2Cl_2} \underset{\textbf{J}}{C_8H_{14}O} \xrightarrow[\text{3. PCC, } CH_2Cl_2]{\text{1. } BH_3,\text{ THF; 2. } H_2O_2,\text{ NaOH}} H$$

From this information, propose reasonable structures for compounds G through J.

30. Humulene and α-caryophyllene alcohol are terpene constituents of carnation extracts. The former is converted into the latter by acid-catalyzed hydration in one step. Write a mechanism. Hint: Follow the labeled carbon atoms retrosynthetically. The mechanism includes carbocation-induced cyclizations and hydrogen and alkyl-group migrations. This is difficult.

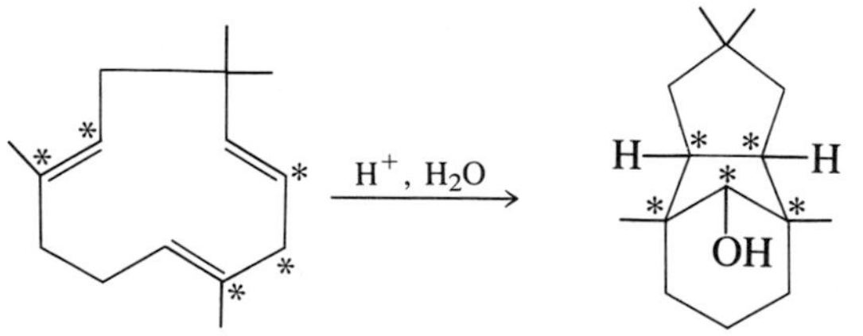

$$\text{Humulene} \xrightarrow{H^+,\ H_2O} \alpha\text{-Caryophyllene alcohol}$$

H * * H * * * OH

Humulene **α-Caryophyllene alcohol**

31. Caryophyllene ($C_{15}H_{24}$) is an unusual sesquiterpene familiar to you as a major cause of the odor of cloves. Determine its structure from the information below. (Caution: The structure is totally different from that of α-caryophyllene alcohol in Problem 30.)

REACTION I.

$$\text{Caryophyllene} \xrightarrow{H_2,\ Pd\text{–}C} C_{15}H_{28}$$

REACTION 2.

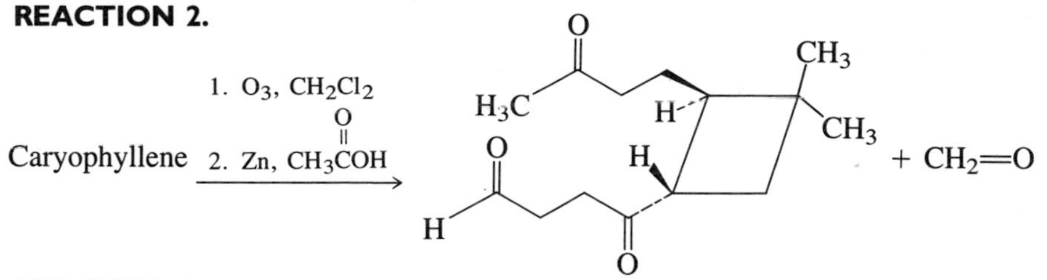

Caryophyllene $\xrightarrow[\text{2. Zn, } CH_3\overset{O}{\overset{\|}{C}}OH]{\text{1. } O_3, CH_2Cl_2}$ [product structure] + $CH_2{=}O$

H₃C, O, CH₃, H, CH₃, H, H, O, O

REACTION 3.

Caryophyllene $\xrightarrow[\text{2. } H_2O_2, \text{ NaOH, } H_2O]{\text{1. One equivalent of } BH_3, \text{ THF}}$ $C_{15}H_{26}O$ $\xrightarrow[\text{2. Zn, } CH_3\overset{O}{\overset{\|}{C}}OH]{\text{1. } O_3, CH_2Cl_2}$ [product structure]

O, CH₃, H₃C, H, CH₃, O, H, H, HOCH₂ H

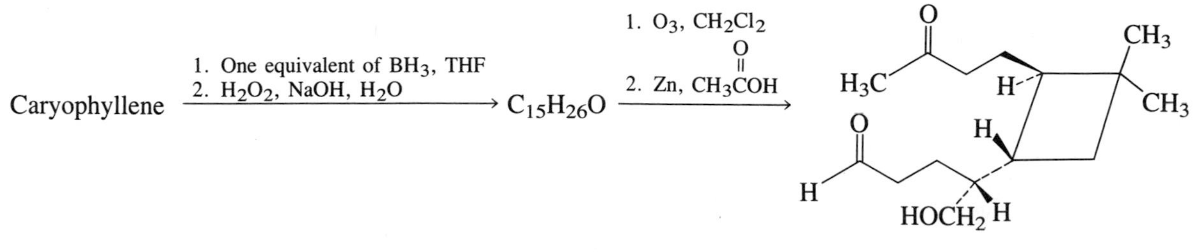

An isomer, isocaryophyllene, gives the same products as caryophyllene upon hydrogenation and ozonolysis. Hydroboration–oxidation of isocaryophyllene gives a $C_{15}H_{26}O$ product isomeric to the one shown in reaction 3; however, ozonolysis converts this $C_{15}H_{26}O$ compound into the same final product shown. In what way do caryophyllene and its isomer differ?

32. Juvenile hormone (JH, Section 12-7) itself has been synthesized in several ways. Two molecules from which it has been synthesized are shown below. Propose completions for syntheses of JH that start with each of them. Your synthesis for (a) should be stereospecific. Also for (a), note that the double bond between C10 and C11 is the most reactive toward electrophilic reagents (compare the synthesis of the JH analog in Section 12-7).

(a) [structure: H₃C, H₃C, H₃C, CH₃, O, COCH₃]

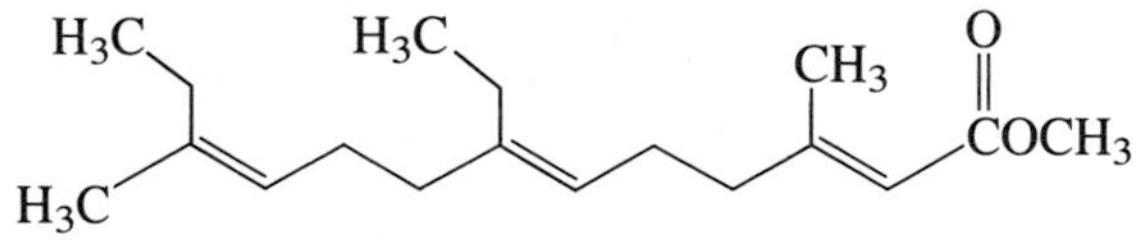

(b) [structure: H₃C, H₃C, CH₃, O, COCH₃, O, Cl]

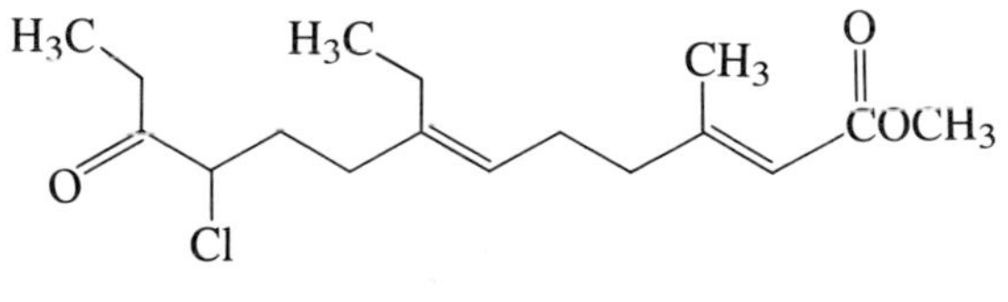

13

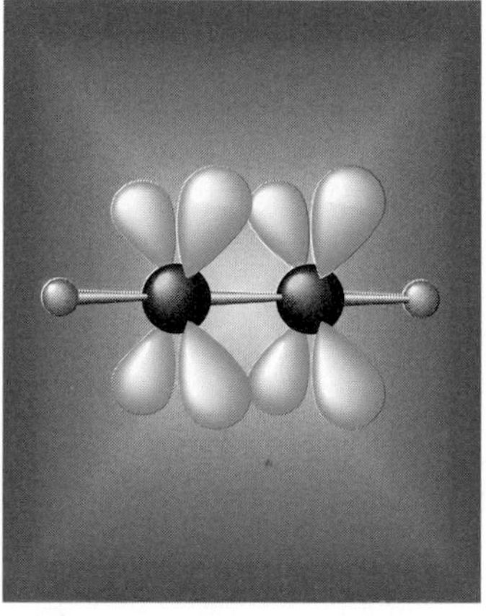

Alkynes

The Carbon–Carbon Triple Bond

We turn now to the structure, preparation, and reactions of the alkynes, compounds characterized by the presence of a carbon–carbon triple bond. Because the $-C\equiv C-$ functional group contains two π linkages (which are mutually perpendicular; recall Figure 1-21B), its reactivity is much like that of the double bond. For example, like alkenes, alkynes are electron rich and subject to attack by electrophiles. They also can be prepared by elimination reactions, and they are most stable when the multiple bond is internal rather than terminal. However, in contrast with their alkenyl counterparts, alkynyl hydrogens are much more acidic, a property allowing easy deprotonation by strong bases. The resulting alkynyl anions are valuable nucleophilic reagents in synthesis.

The smallest alkyne, $HC\equiv CH$ (ethyne or acetylene), is a precursor to a number of industrially important alkene monomers (Chapters 11 and 12). The alkyne functional group is also found in several physiologically active compounds, some of them of natural origin.

We begin with discussions of the naming and structural characteristics of the alkynes. Subsequent sections will cover the spectroscopic and chemical properties of members of this compound class.

13-1 Naming the Alkynes

A carbon–carbon triple bond is the functional group characteristic of the **alkynes.** The general formula for the alkynes is C_nH_{2n-2}, the same as that for the cycloalkenes. The common names for many alkynes are still in use. These include *acetylene,* the common name of the smallest alkyne, C_2H_2. Other alkynes are treated as its derivatives, for example, the alkylacetylenes.

Common Names for Alkynes

$HC{\equiv}CH$	$CH_3C{\equiv}CCH_3$	$CH_3CH_2CH_2C{\equiv}CH$
Acetylene	**Dimethylacetylene**	**Propylacetylene**

The IUPAC rules for naming alkenes (Section 11-1) also apply to alkynes, the ending **-yne** replacing *-ene*. A number indicates the position of the triple bond in the main chain.

$HC{\equiv}CH$	$CH_3C{\equiv}CCH_3$	$\overset{1}{C}H_3\overset{2}{C}{\equiv}\overset{3}{C}\overset{4}{C}H(Br)\overset{5}{C}H_2\overset{6}{C}H_3$	$\overset{4}{C}H_3\overset{3}{C}(CH_3)_2\overset{2}{C}{\equiv}\overset{1}{C}H$
Ethyne	**2-Butyne**	**4-Bromo-2-hexyne**	**3,3-Dimethyl-1-butyne**

Alkynes having the general structure RC≡CH are **terminal,** whereas those with the structure of RC≡CR′ are **internal.**

Substituents bearing a triple bond are called **alkynyl** groups. Thus, the substituent with the structure –C≡CH is named **ethynyl;** its homolog $–CH_2C{\equiv}CH$ is **2-propynyl** (propargyl). Like alkanes and alkenes, alkynes can be depicted in straight-line notation.

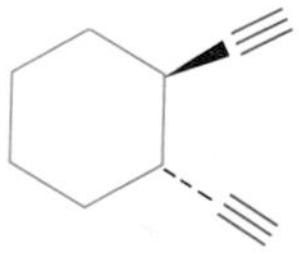

***trans*-1,2-Diethynylcyclohexane**

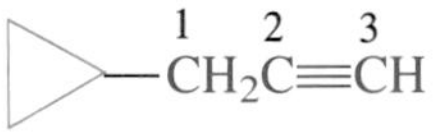

2-Propynylcyclopropane (Propargylcyclopropane)

$HC{\equiv}CCH_2OH$

2-Propyn-1-ol (Propargyl alcohol)

In IUPAC nomenclature, a hydrocarbon containing both double and triple bonds is called an **alkenyne.** The chain is numbered starting from the end closest to either of the functional groups. When a double bond and a triple bond are at equidistant positions from either terminus, the *double* bond is given the lower number. Alkynes incorporating the hydroxy function are named **alkynols.** Note the omission of the final *e* of -ene in -enyne and of -yne in -ynol. The OH group takes precedence over both double and triple bonds when numbering a chain.

$\overset{6}{C}H_3\overset{5}{C}H_2\overset{4}{C}H{=}\overset{3}{C}H\overset{2}{C}{\equiv}\overset{1}{C}H$	$\overset{1}{C}H_2{=}\overset{2}{C}H\overset{3}{C}H_2\overset{4}{C}{\equiv}\overset{5}{C}H$	$H\overset{6}{C}{\equiv}\overset{5}{C}\overset{4}{C}H_2\overset{3}{C}H_2\overset{2}{C}H(OH)\overset{1}{C}H_3$
3-Hexen-1-yne	**1-Penten-4-yne**	**5-Hexyn-2-ol**
(Not 3-hexen-5-yne)	(Not 4-penten-1-yne)	(Not 1-hexyn-5-ol)

EXERCISE 13-1

Give the IUPAC names for **(a)** all the alkynes of composition C_6H_{10};

(b)

```
H3C
   \
    C—C≡CH
H'/
    CH=CH2
```

(c) all butynols. Remember to include and designate stereoisomers.

13-2 Properties and Bonding in the Alkynes

The nature of the triple bond helps explain the physical and chemical properties of the alkynes. In molecular-orbital terms, we shall see that the carbons are *sp* hybridized, and the four singly filled *p* orbitals form two perpendicular π bonds.

Alkynes are relatively nonpolar

Alkynes have boiling points very similar to those of the corresponding alkenes and alkanes. Ethyne is unusual in that it has no boiling point at atmospheric pressure; rather it sublimes at $-84°C$. Propyne (b.p. $-23.2°C$) and 1-butyne (b.p. 8.1°C) are gases, whereas 2-butyne is barely a liquid (b.p. 27°C) at room temperature. The medium-sized alkynes are distillable liquids. Care must be taken in the handling of alkynes: They polymerize very easily—frequently with violence. Ethyne explodes under pressure but can be shipped in pressurized gas cylinders that contain propanone (acetone) and a porous filler such as pumice.

Ethyne is linear and has strong, short bonds

In ethyne, the two carbons are *sp* hybridized (Figure 13-1A). One of the hybrid orbitals on each carbon overlaps with hydrogen, and a σ bond between the two carbon atoms results from mutual overlap of the remaining *sp* hybrids. The two perpendicular *p* orbitals on each carbon contain one electron each. These two sets overlap to form two perpendicular π bonds (Figure 13-1B). Because π bonds are diffuse, the distribution of electrons in the triple bond resembles a cylindrical

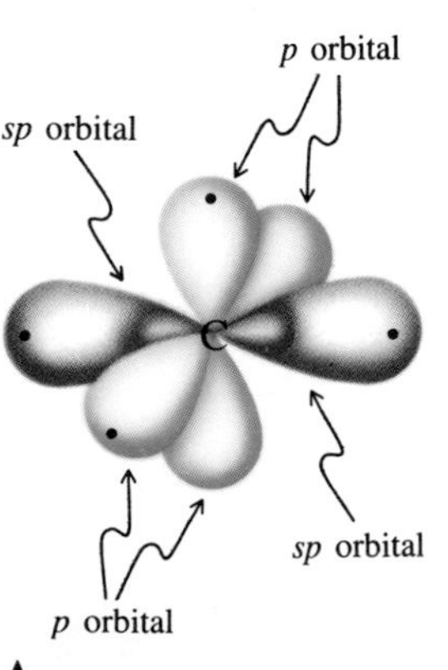

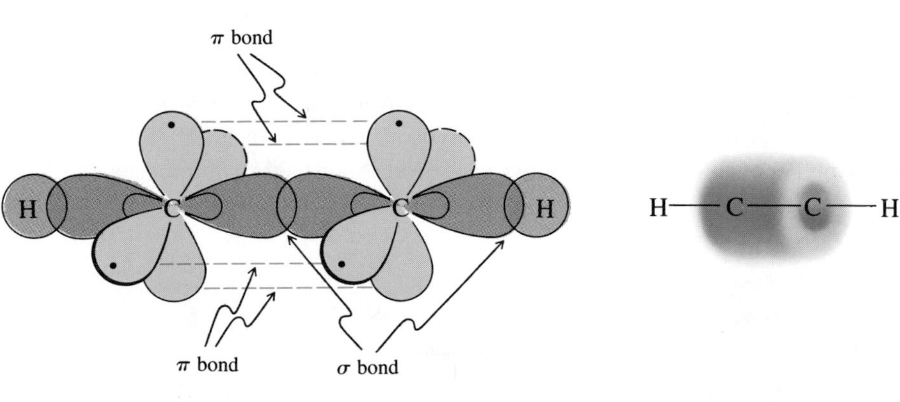

FIGURE 13-1
(A) Molecular-orbital picture of *sp*-hybridized carbon, showing the two perpendicular *p* orbitals. (B) The triple bond in ethyne: The orbitals of two *sp*-hybridized CH fragments overlap to create a σ bond and two π bonds. (C) The two π bonds produce a cylindrical electron cloud around the molecular axis of ethyne.

cloud (Figure 13-1C). As a consequence of hybridization and the two π interactions, the strength of the triple bond is high, about 229 kcal mol^{-1}. The C–H bond-dissociation energy of terminal alkynes is also substantial: 131 kcal mol^{-1}.

Because both carbon atoms in ethyne are *sp* hybridized, its structure is linear (Figure 13-2). The carbon–carbon bond length is 1.203 Å, shorter than that of a double bond (1.33 Å, Figure 11-7). The carbon–hydrogen bond also is short, again because of the relatively large degree of *s* character in the *sp* hybrids used for bonding to hydrogen. The electrons in these orbitals (and in the bonds that they form by overlapping with other orbitals) reside relatively close to the nucleus and produce shorter (and stronger) bonds.

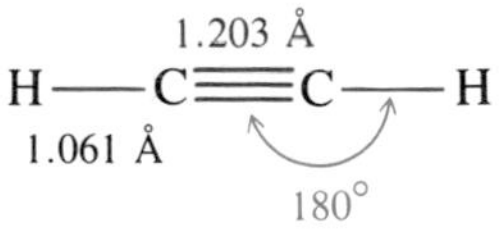

$$H{-}C{\equiv}C{-}H \quad (C{\equiv}C\ 1.203\ \text{Å};\ C{-}H\ 1.061\ \text{Å};\ 180^\circ)$$

FIGURE 13-2 Molecular structure of ethyne.

Terminal alkynes are remarkably acidic

The relatively high *s* character in the carbon hybrid orbitals of terminal alkynes makes them more acidic than alkanes and alkenes. The pK_a of ethyne, for example, is 25, remarkably low compared with that of ethene and ethane.

Relative Acidities of Alkanes, Alkenes, and Alkynes

	$H_3C{-}CH_3$	< $H_2C{=}CH_2$	< $HC{\equiv}CH$
Hybridization:	sp^3	sp^2	sp
pK_a:	50	44	25

This property is useful, because strong bases such as sodium amide in liquid ammonia, alkyllithiums, and Grignard reagents can deprotonate terminal alkynes to the corresponding **alkynyl anions.** These species react as bases and nucleophiles, much like other carbanions (Section 13-6).

Deprotonation of a Terminal Alkyne

$$CH_3CH_2C{\equiv}CH + CH_3CH_2CH_2CH_2Li \xrightarrow{(CH_3CH_2)_2O} CH_3CH_2C{\equiv}CLi + CH_3CH_2CH_2CH_2{-}H$$

EXERCISE 13-2

Strong bases other than those mentioned here for the deprotonation of alkynes were introduced earlier. Two examples are potassium *tert*-butoxide and lithium diisopropylamide (LDA). Would either (or both) of these compounds be suitable for making ethynyl anion from ethyne? Explain, in terms of their pK_a values.

Deprotonation of 1-Alkynes

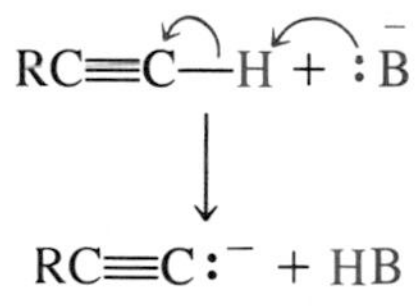

$$RC{\equiv}C{-}H + :B^- \longrightarrow RC{\equiv}C:^- + HB$$

In summary, the characteristic hybridization scheme for the triple bond of an alkyne controls its physical and electronic features. It is responsible for strong bonds, the linear structure, and the relatively acidic alkynyl hydrogen.

13-3 Spectroscopy of the Alkynes

Alkenyl hydrogens are deshielded and give rise to relatively low-field NMR signals compared with those in saturated alkanes (Section 11-4). In contrast,

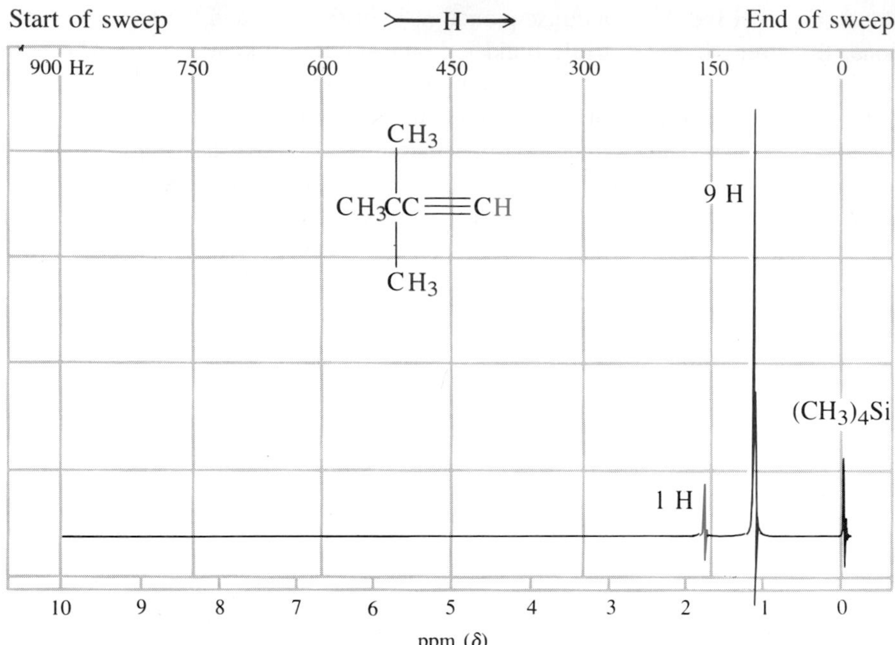

FIGURE 13-3 90-MHz ^{1}H NMR spectrum of 3,3-dimethyl-1-butyne in CCl_4, showing the high-field position (δ = 1.74 ppm) of the signal due to the alkynyl hydrogen.

alkynyl hydrogens have smaller chemical shifts, much closer to those in alkanes. Terminal alkynes are readily identified by IR spectroscopy.

The NMR absorptions of alkyne hydrogens show a characteristic shielding

Unlike alkenyl hydrogens, which are deshielded and give ^{1}H NMR absorptions at δ = 4.6–5.7 ppm, protons bound to *sp*-hybridized carbon atoms are found at δ = 1.7–3.1 ppm (Table 10-2). For example, in the NMR spectrum of 3,3-dimethyl-1-butyne, the alkynyl hydrogen absorbs at δ = 1.74 ppm (Figure 13-3).

Why is the terminal alkyne hydrogen so shielded? Like the π electrons of an alkene, those in the triple bond enter into a circular motion when an alkyne is subjected to an external magnetic field (Figure 13-4). Unlike the case for alkenes, however, the major direction of this motion generates a local magnetic field that *opposes* H_0 in the vicinity of the alkyne hydrogen. The result is a strong *shielding* effect that cancels the deshielding tendency of the electron-withdrawing *sp*-hybridized carbon and gives rise to a relatively small (high-field) chemical shift.

The triple bond transmits spin–spin coupling

Long-Range Coupling in Alkynes

$$-\overset{\text{H}}{\underset{|}{\overset{|}{\text{C}}}}-\text{C}\equiv\text{C}-\text{H}$$

J = 2–4 Hz

The alkyne functional group transmits coupling so well that the terminal hydrogen is split by the hydrogens across the triple bond, even though they are separated from them by three carbons. This result is an example of long-range coupling. The coupling constants are small and range from about 2 to 4 Hz. Figure 13-5 shows the NMR spectrum of 1-pentyne. The alkynyl hydrogen signal at δ = 1.71 ppm is a triplet (J = 2.5 Hz) because of coupling to the two equivalent hydrogens at C3, which appear at δ = 2.07 ppm. The latter, in turn, give rise to a doublet of triplets, reflecting coupling to the two hydrogens at C4 (J = 6 Hz) as well as that at C1 (J = 2.5 Hz).

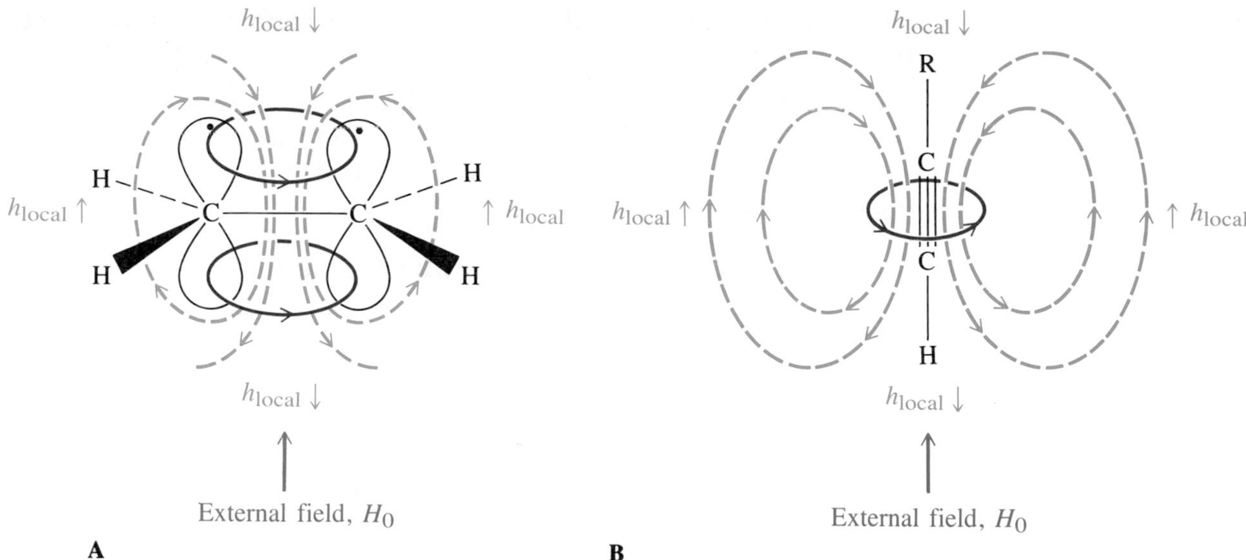

FIGURE 13-4 Electron circulation in the presence of an external magnetic field generates local magnetic fields that cause the characteristic chemical shifts of alkenyl and alkynyl hydrogens. (A) Alkenyl hydrogens are located in a region of space where h_{local} reinforces H_0. Therefore, a smaller applied field is necessary for resonance and these protons are relatively deshielded. (B) Electron circulation in an alkyne generates a local field that opposes H_0 in the vicinity of the alkynyl hydrogen. A higher field is required for resonance because of this shielding effect.

EXERCISE 13-3

Predict the first-order splitting pattern in the ^{1}H NMR spectrum of 3-methyl-1-butyne.

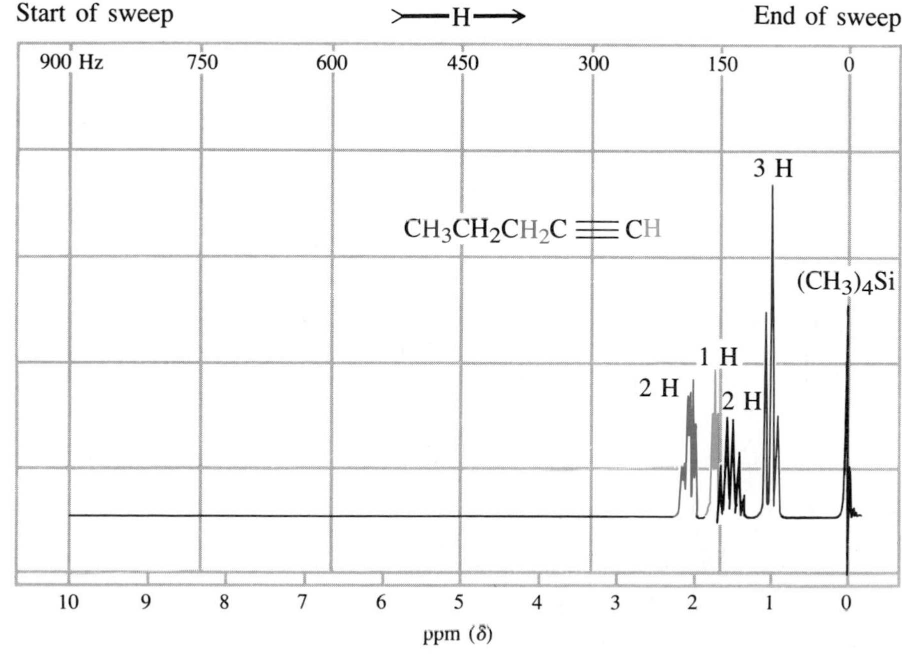

FIGURE 13-5 90-MHz ^{1}H NMR spectrum of 1-pentyne in CCl_4, showing coupling between the alkynyl (green) and propargylic (blue) hydrogens.

The ^{13}C NMR chemical shifts of alkyne carbons are distinct from those of the alkanes and alkenes

Carbon-13 NMR spectroscopy is also useful in deducing the structure of alkynes. For example, the triple-bonded carbons in alkyl-substituted alkynes absorb in the narrow range of δ = 65–85 ppm, quite separate from the chemical shifts of analogous alkane and alkene carbon atoms.

Typical Alkyne ^{13}C NMR Chemical Shifts

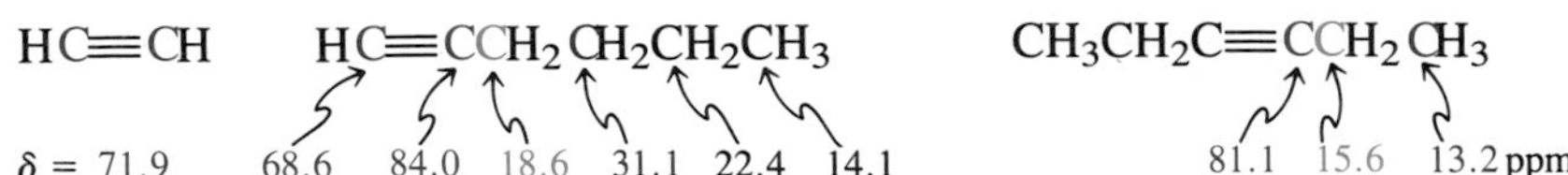

Terminal alkynes give rise to two characteristic infrared absorptions

Infrared spectroscopy is helpful in identifying terminal alkynes. Characteristic stretching bands appear for the alkynyl hydrogen at 3260–3330 cm^{-1} and for the C≡C triple bond at 2100–2260 cm^{-1} (see Figure 13-6). Such data are especially useful when ^{1}H NMR spectra are complex and difficult to interpret. However, the band for the C≡C stretching vibration in internal alkynes is often weak, thus reducing the value of IR spectroscopy for characterizing these systems.

In summary, the cylindrical π cloud around the carbon–carbon triple bond induces local magnetic fields that lead to NMR chemical shifts for alkynyl hydro-

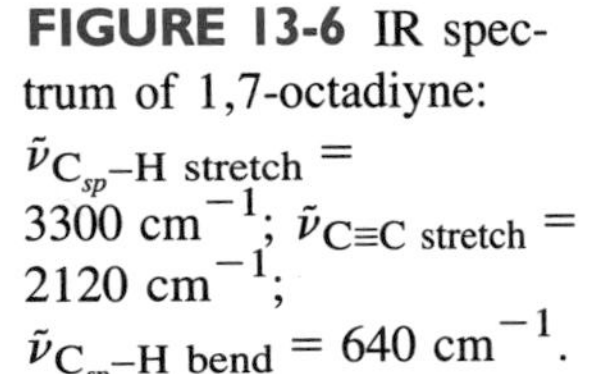

FIGURE 13-6 IR spectrum of 1,7-octadiyne: $\tilde{\nu}_{C_{sp}\text{–H stretch}}$ = 3300 cm^{-1}; $\tilde{\nu}_{C\equiv C\text{ stretch}}$ = 2120 cm^{-1}; $\tilde{\nu}_{C_{sp}\text{–H bend}}$ = 640 cm^{-1}.

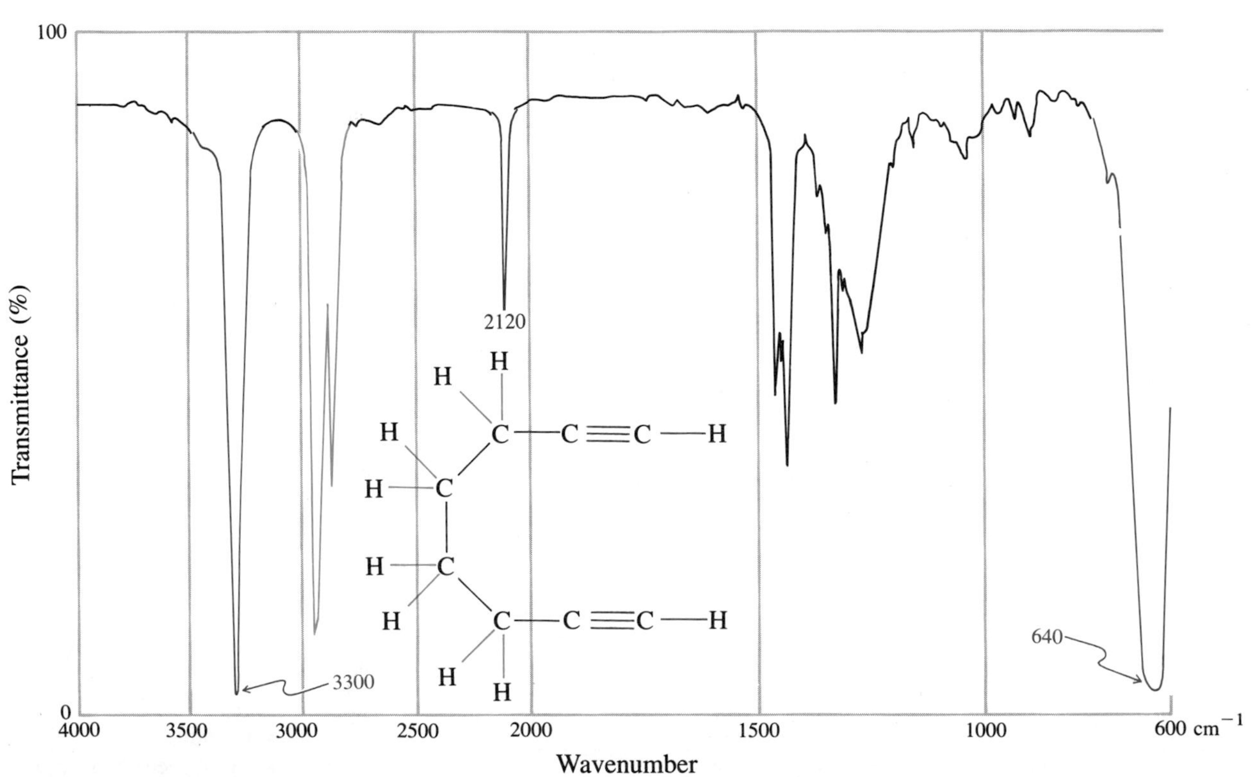

gens at higher field than those of alkenyl protons. Long-range coupling is observed through the C≡C linkage. Infrared spectroscopy provides a useful complement to NMR data, displaying characteristic bands for the C≡C and ≡C–H bonds of terminal alkynes.

13-4 Stability of the Triple Bond

Does the heat content of the triple bond depend on the substituents, as it does for alkenes? Which type of alkyne is more stable, a terminal or an internal isomer? These questions can be addressed by studying the comparative heats of combustion and hydrogenation.

Alkynes are high-energy compounds

Alkynes have a high energy content. Because of this property, alkynes react in many cases with considerable release of energy. For example, when shocked by high pressure or exposed to catalytic amounts of copper, ethyne explosively decomposes to carbon and hydrogen. On combustion, it releases 317 kcal mol^{-1}. This energy is put to practical use in acetylene torches employed in welding, which requires very hot flames (more than 2500°C).

Combustion of Ethyne

$$HC{\equiv}CH + 2.5\ O_2 \longrightarrow 2\ CO_2 + H_2O \qquad \Delta H° = -317\ \text{kcal mol}^{-1}$$

As discussed in Problem 21 of Chapter 3, the flame temperature is related to both the heat of combustion and the quantity of exhaust gas produced. In the combustion of ethyne the heat released is distributed among only three molecules of product gas per molecule of ethyne consumed. Therefore, each product molecule is heated to a higher temperature than in the case of propane, for instance, the chief constituent of natural gas. The heat of combustion of the latter is 530.6 kcal mol^{-1}, but it is distributed among *seven* molecules of product gas (3 CO_2 and 4 H_2O) per molecule of propane, a situation resulting in a cooler flame.

Internal alkynes are more stable than terminal alkynes

Although heats of combustion could be used to determine relative stabilities of alkyne isomers, it is more convenient to compare their heats of hydrogenation, just as we did in the case of alkenes. In the presence of the usual catalysts (platinum or palladium on charcoal) hydrogenation of alkynes proceeds by addition of two molar equivalents of hydrogen to produce alkanes. Thus, both butyne isomers hydrogenate to furnish butane, but on doing so release different amounts of heat.

$$CH_3CH_2C{\equiv}CH + 2\ H_2 \xrightarrow{\text{Catalyst}} CH_3CH_2CH_2CH_3 \qquad \Delta H° = -69.9\ \text{kcal mol}^{-1}$$

$$CH_3C{\equiv}CCH_3 + 2\ H_2 \xrightarrow{\text{Catalyst}} CH_3CH_2CH_2CH_3 \qquad \Delta H° = -65.1\ \text{kcal mol}^{-1}$$

Relative Stabilities of the Alkynes

$RC\equiv CH < RC\equiv CR'$

The results show that the internal π system is more stable than the terminal one. As with the alkenes, the stabilizing influence is hyperconjugation (Sections 3-2 and 11-7).

EXERCISE 13-4

Are the heats of hydrogenation of the butynes consistent with the notion that alkynes are high-energy compounds? Explain. (Hint: Compare these values with the heats of hydrogenation of alkene double bonds.)

In summary, alkynes are highly energetic compounds. Internal isomers are more stable than terminal ones, as shown by the relative heats of hydrogenation.

13-5 Preparation of Alkynes by Double Elimination

The two basic methods used to prepare alkynes are double elimination from 1,2-dihaloalkanes and alkylation of alkynyl anions. This section will discuss the first of these, which provides a synthetic route to alkynes, from alkenes.

Alkynes are prepared from dihaloalkanes by elimination

As discussed in Section 11-8, alkenes can be prepared by E2 reactions of haloalkanes. Application of this principle to alkyne synthesis suggests that treatment of vicinal dihaloalkanes with strong base should result in double elimination to furnish a triple bond.

Double Elimination from Dihaloalkanes to Give Alkynes

$$\text{—}\underset{\substack{|\\ H}}{\overset{\substack{X\\ |}}{C}}\text{—}\underset{\substack{|\\ H}}{\overset{\substack{X\\ |}}{C}}\text{—} \xrightarrow[-\,2\,HX]{\text{Base}} \text{—}C\equiv C\text{—}$$

Vicinal dihaloalkane

Indeed, addition of 1,2-dibromohexane (prepared by bromination of 1-hexene) to sodium amide in liquid ammonia followed by evaporation of solvent and aqueous work-up gives 1-hexyne.

$$CH_3CH_2CH_2CH_2\underset{\substack{|\\ Br}}{CH}\text{—}CH_2Br \xrightarrow[-\,2\,HBr]{1.\ NaNH_2,\ \text{liquid}\ NH_3\quad 2.\ H_2O} CH_3CH_2CH_2CH_2C\equiv CH$$

The other products in this reaction are the protonated base (i.e., NH_3), which is evaporated, and sodium bromide, which is removed in the work-up. Eliminations in liquid ammonia are usually carried out at its boiling point, $-33°C$.

Because vicinal dihaloalkanes are readily available from alkenes by halogenation, this sequence, called **halogenation–double dehydrohalogenation,** is a ready means of converting alkenes into the corresponding alkynes.

A Halogenation–Double Dehydrohalogenation Used in Alkyne Synthesis

$$CH_2{=}CHCH_2CH_2CH{=}CH_2 \xrightarrow[\text{3. } H_2O]{\text{1. } Br_2,\ CCl_4 \quad \text{2. } NaNH_2,\ \text{liquid } NH_3} HC{\equiv}CCH_2CH_2C{\equiv}CH \quad 53\%$$

1,5-Hexadiene **1,5-Hexadiyne**

EXERCISE 13-5

Illustrate the use of halogenation–double dehydrohalogenation in the synthesis of the alkynes **(a)** 2-pentyne; **(b)** 1-octyne; **(c)** 2-methyl-3-hexyne.

Haloalkenes are intermediates in alkyne synthesis by elimination

Dehydrohalogenation of dihaloalkanes proceeds through the intermediacy of haloalkenes, also called **alkenyl halides.** Although mixtures of *E*- and *Z*-haloalkenes are in principle possible, with diastereomerically pure vicinal dihaloalkanes only one product is formed because elimination proceeds stereospecifically *anti* (Section 11-8).

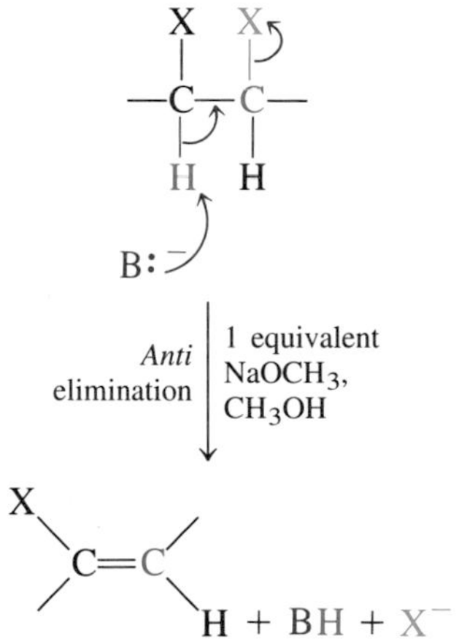

X X —C—C— H H B: ; *Anti* elimination; 1 equivalent $NaOCH_3$, CH_3OH → X–C=C–H + BH + X^-

An alkenyl halide

EXERCISE 13-6

Give the structure of the bromoalkene involved in the bromination–double dehydrobromination of *cis*-2-butene to 2-butyne. Do the same for the trans isomer. (Hint: Refer to Section 12-5 for useful information, and use models.)

The stereochemistry of the intermediate haloalkene is of no consequence when the sequence is used for alkyne synthesis. Both *E*- and *Z*-haloalkenes eliminate with base to give the same alkyne.

In summary, alkynes are made from vicinal dihaloalkanes by double elimination. Alkenyl halides are intermediates, being formed stereospecifically in the first elimination.

13-6 Preparation of Alkynes by Alkylation of Alkynyl Anions

Alkynes can also be prepared from other alkynes. The reaction of terminal alkynyl anions with alkylating agents, such as primary haloalkanes, oxacyclopropanes, aldehydes, or ketones, results in carbon–carbon bond formation. As we know, such anions are readily prepared from terminal alkynes by deprotonation with strong bases (mostly alkyllithium reagents, sodium amide in liquid ammonia, or Grignard reagents). Alkylation with primary haloalkanes is typically done in liquid ammonia or in ether solvents. These solvents sometimes contain 1,2-ethanediamine, $H_2NCH_2CH_2NH_2$; *N,N,N′,N′*-tetramethyl-1,2-ethanediamine, $(CH_3)_2NCH_2CH_2N(CH_3)_2$; or HMPA as cosolvents to increase the nucleophilic power of the anion (Section 6-8). The process is unusual, because ordinary alkyl organometallic compounds are unreactive in the presence of haloalkanes. Alkynyl anions are an exception, however (Section 13-10).

Alkylation of an Alkynyl Anion

$$\text{C}_6\text{H}_{11}\text{C}\equiv\text{CH} \xrightarrow[\text{THF, } H_2NCH_2CH_2NH_2]{CH_3CH_2CH_2CH_2Li,} \text{C}_6\text{H}_{11}\text{C}\equiv\text{C}:^{-}\,{}^{+}\text{Li} \xrightarrow[-\,LiBr]{CH_3CH_2CH_2Br} \text{C}_6\text{H}_{11}\text{C}\equiv\text{CCH}_2\text{CH}_2\text{CH}_3$$

85%

1-Pentynylcyclohexane

Attempted alkylation of alkynyl anions with secondary and tertiary halides leads to E2 products because of the strongly basic character of the nucleophile (recall Section 7-8). Ethyne itself may be alkylated in a series of steps through the selective formation of the monoanion to give mono- and dialkyl-derivatives. The ethynyllithium-1,2-ethanediamine complex is also commercially available.

Reactions with oxacyclopropanes or carbonyl compounds proceed in the same manner as similar transformations of other organometallic reagents (Sections 8-8 and 9-9).

Reactions of Alkynyl Anions

$$\text{HC}\equiv\text{CH} \xrightarrow[-\,NH_2H]{LiNH_2\text{ (1 equivalent), liquid }NH_3} \text{HC}\equiv\text{CLi} \xrightarrow[-\,LiOH]{1.\ H_2C\overset{O}{-}CH_2\quad 2.\ HOH} \text{HC}\equiv\text{CCH}_2\text{CH}_2\text{OH}$$

92%

3-Butyn-1-ol

$$\text{CH}_3\text{C}\equiv\text{CH} \xrightarrow[-\,CH_3CH_2H]{CH_3CH_2MgBr,\ (CH_3CH_2)_2O,\ 20^\circ C} \text{CH}_3\text{C}\equiv\text{CMgBr} \xrightarrow{1.\ \text{cyclopentanone}\quad 2.\ H_2O} \text{1-(1-propynyl)cyclopentanol (OH, C}\equiv\text{CCH}_3)$$

66%

1-(1-Propynyl)cyclopentanol

EXERCISE 13-7

Suggest efficient and short syntheses of the following compounds. (Hint: Review Section 8-9.)

(a) $CH_3(CH_2)_3C\equiv C\overset{OH}{\overset{|}{C}}HC\equiv C(CH_2)_3CH_3$ from $CH_3(CH_2)_3C\equiv CH$

(b) $CH_3CH_2CH_2C\equiv C\overset{OH}{\overset{|}{C}}HCH_2CH_3$ from ethyne

In summary, alkynes can be prepared from other alkynes by alkylation with primary haloalkanes, oxacyclopropanes, or carbonyl compounds. Ethyne itself can be alkylated in a series of steps.

13-7 Reduction of Alkynes: The Relative Reactivity of the Two Pi Bonds

Now we turn from the preparation of alkynes to the characteristic reactions of the triple bond. In many respects, alkynes are like alkenes, except for the availability of two π bonds, one generally being more reactive than the other. Thus, alkynes can undergo additions, such as hydrogenation and electrophilic attacks.

Addition of Reagents A–B to Alkynes

$$R{-}C{\equiv}C{-}R \xrightarrow{A\text{–}B} \underset{A}{\overset{R}{}}{>}C{=}C{<}\underset{B}{\overset{R}{}} \text{ or } \underset{A}{\overset{R}{}}{>}C{=}C{<}\underset{R}{\overset{B}{}} \xrightarrow{A\text{–}B} A{-}\underset{A}{\overset{R}{C}}{-}\underset{B}{\overset{R}{C}}{-}B \text{ or } A{-}\underset{B}{\overset{R}{C}}{-}\underset{A}{\overset{R}{C}}{-}B$$

In this section we introduce two new reactions: step-by-step hydrogenation and one-electron reduction by sodium to give cis and trans alkenes, respectively.

Cis alkenes can be synthesized by catalytic hydrogenation

Alkynes can be hydrogenated under the same conditions used to hydrogenate alkenes. Typically, platinum or palladium on charcoal is suspended in a solution containing the alkyne, and the mixture is exposed to a hydrogen atmosphere. Under these conditions, the triple bond is saturated completely.

Complete Hydrogenation of Alkynes

$$\underset{\textbf{3-Heptyne}}{CH_3CH_2CH_2C{\equiv}CCH_2CH_3} \xrightarrow{H_2,\ Pt} \underset{\textbf{Heptane}}{\underset{100\%}{CH_3CH_2CH_2CH_2CH_2CH_2CH_3}}$$

Hydrogenation is a step-by-step process that may be stopped at the intermediate alkene stage by the use of modified catalysts, such as the **Lindlar* catalyst.** This catalyst is palladium that has been precipitated on calcium carbonate and treated with lead acetate and quinoline. The surface of the metal rearranges to a less active configuration than that of palladium on carbon so that only the more reactive first π bond of the alkyne is hydrogenated. Because the addition of hydrogen is *syn*, this method affords a stereoselective synthesis of cis alkenes.

Hydrogenations with Lindlar Catalyst

Lindlar catalyst: 5% Pd-$CaCO_3$, $Pb(OCCH_3)_2$ (C=O), quinoline

Quinoline

$$CH_3CH_2CH_2C{\equiv}CCH_2CH_3 \xrightarrow{H_2,\ \text{Lindlar catalyst},\ 25°C} \underset{CH_3CH_2CH_2}{\overset{H}{}}{>}C{=}C{<}\underset{CH_2CH_3}{\overset{H}{}}$$

100%

***cis*-3-Heptene**

*Dr. Herbert H. M. Lindlar, b. 1909, Hoffman La Roche and Co. A. G., Basel.

EXERCISE 13-8

Write the structure of the product expected from the following reaction.

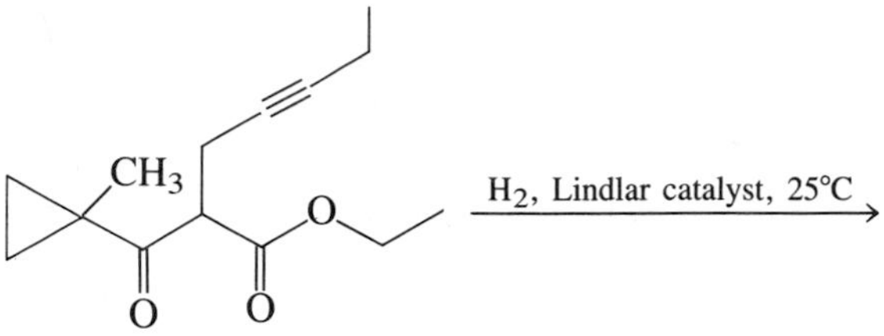

CH_3 $\xrightarrow{H_2,\ \text{Lindlar catalyst, } 25°C}$

O O

EXERCISE 13-9

The perfume industry makes considerable use of naturally occurring substances such as those obtained from rose and jasmine extracts. In many cases the quantities of fragrant oils available by natural product isolation are so small that it is necessary to synthesize them. Examples are found in the olfactory components of violets, which include *trans*-2-*cis*-6-nonadien-1-ol and the corresponding aldehyde. An important intermediate in their large-scale synthesis is *cis*-3-hexen-1-ol, whose industrial preparation is described as "a closely guarded secret." Using the methods in this and the previous sections, propose a synthesis from 1-butyne.

With a method for the stereoselective construction of cis alkenes at our disposal, we might ask: Can we modify the reduction of alkynes to give only trans alkenes? The answer is yes, with a different reducing agent and through a different mechanism.

Sequential one-electron reductions of alkynes produce trans alkenes

When, instead of catalytically activated hydrogen, we use *sodium metal* dissolved in liquid ammonia as the reagent for the reduction of alkynes, we obtain trans alkenes as the products. For example, 3-heptyne is reduced to *trans*-3-heptene in this way. Unlike sodium amide in liquid ammonia, which functions as a strong base, elemental sodium in liquid ammonia acts as a powerful electron donor (i.e., a reducing agent).

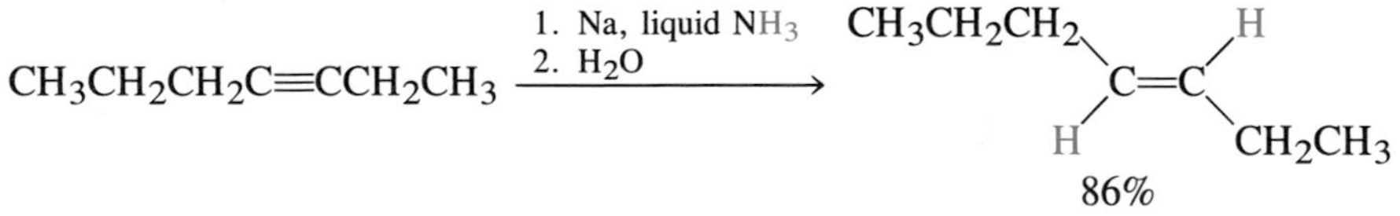

$$CH_3CH_2CH_2C{\equiv}CCH_2CH_3 \xrightarrow[2.\ H_2O]{1.\ Na,\ \text{liquid } NH_3} \underset{H}{\overset{CH_3CH_2CH_2}{}}\!C{=}C\!\underset{CH_2CH_3}{\overset{H}{}}$$

86%

3-Heptyne **trans-3-Heptene**

In the first step of the mechanism of this reduction, the π framework of the triple bond accepts one electron to give a radical anion. This anion is protonated by the ammonia solvent to give an alkenyl radical that is further reduced by accepting another electron to give an alkenyl anion. This species is again protonated to give the product alkene, which is stable to further reduction. The trans stereochemistry of the final alkene is due to the rapid equilibration of the intermediate alkenyl radical between the two possible cis and trans forms. The equilibrium lies on the side of the more stable trans species.

Mechanism of the Reduction of Alkynes by Sodium in Liquid Ammonia

STEP 1. One-electron transfer

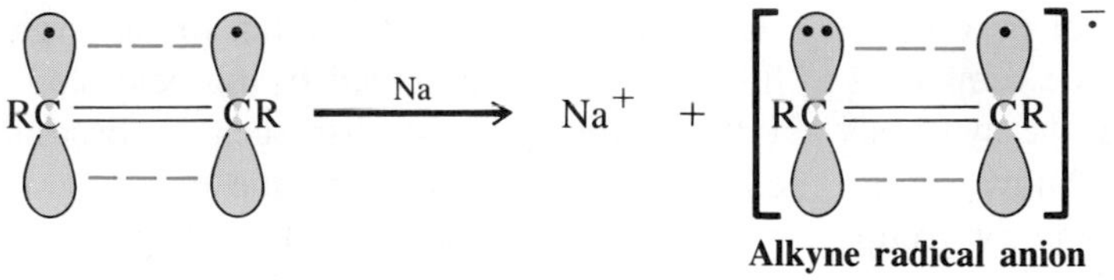

$$RC{\equiv}CR \xrightarrow{Na} Na^+ + [RC{\equiv}CR]^{\overline{\cdot}}$$

Alkyne radical anion

STEP 2. First protonation

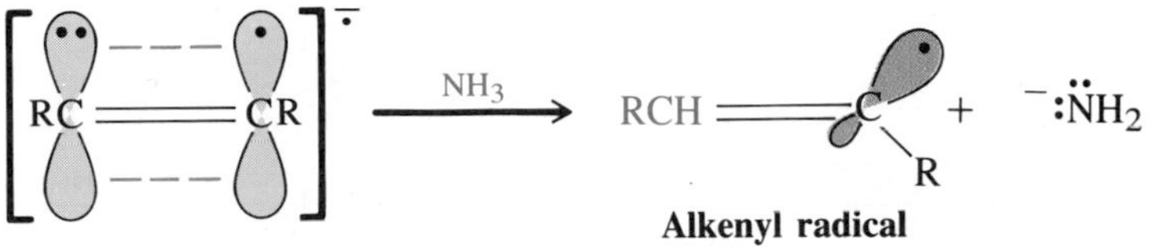

$$[RC{\equiv}CR]^{\overline{\cdot}} \xrightarrow{NH_3} RCH{=}\dot{C}R + {}^-{:}\ddot{N}H_2$$

Alkenyl radical

STEP 3. Alkenyl radical equilibration

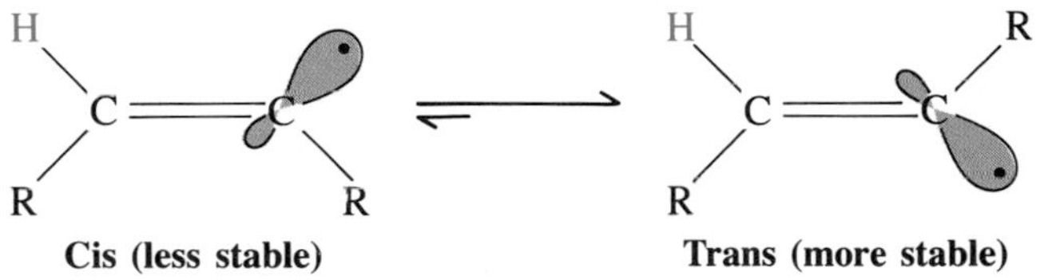

Cis (less stable) ⇌ **Trans (more stable)**

STEP 4. Second one-electron transfer

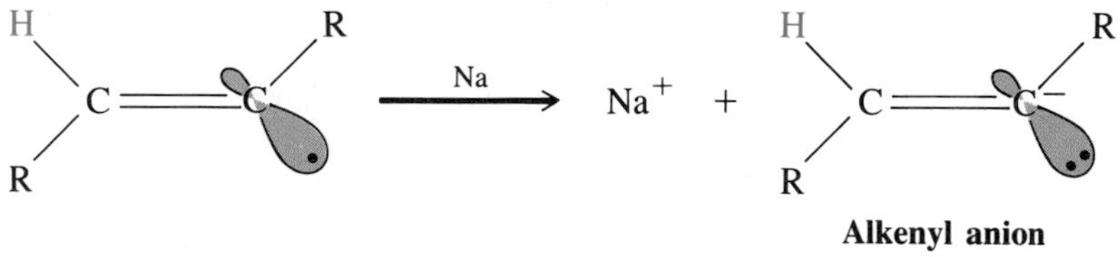

$$\xrightarrow{Na} Na^+ +$$

Alkenyl anion

STEP 5. Second protonation

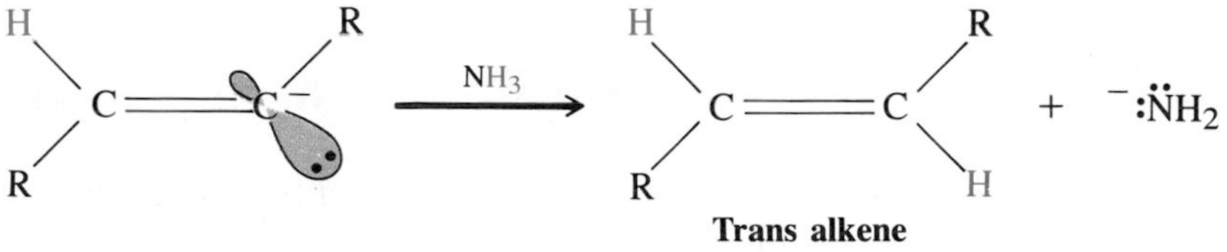

$$\xrightarrow{NH_3} + {}^-{:}\ddot{N}H_2$$

Trans alkene

EXERCISE 13-10

When 1,7-undecadiyne (11 carbons) was treated with a mixture of sodium *and* sodium amide in liquid ammonia, only the internal bond was reduced to give *trans*-7-undecen-1-yne. Explain.

BOX 13-1 Synthesis of a Sex Pheromone

We have seen that pheromones may be used as chemical lures to aid in eradication of pests (Section 12-16). An application of the sequential one-electron reduction of alkynes to the synthesis of the sex pheromone of the spruce budworm is shown in the scheme below. The hydroxy group in the starting material, 10-bromo-1-decanol, is initially protected as an ether by acid-catalyzed addition to 2-methylpropene. This step is necessary to prevent the protonation of the organolithium reagent employed in the next step. 1-Butynyllithium then displaces the bromine, the oxygen is deprotected, and the resulting alkynol reduced to the trans alkenol. Oxidation of the alcohol function gives the pheromone.

$$\underset{\textbf{10-Bromo-1-decanol}}{HO(CH_2)_{10}Br} \xrightarrow[\text{Protection (Sections 9-7 and 12-4)}]{(CH_3)_2C{=}CH_2,\ H^+} (CH_3)_3CO(CH_2)_{10}Br \xrightarrow[-\ LiBr]{LiC{\equiv}CCH_2CH_3,\ THF,\ HMPA}$$

$$(CH_3)_3CO(CH_2)_{10}C{\equiv}CCH_2CH_3 \xrightarrow[\substack{-\ (CH_3)_2C{=}CH_2 \\ \text{Deprotection} \\ \text{(Section 9-8)}}]{H^+,\ H_2O} \underset{\textbf{11-Tetradecyn-1-ol}}{HO(CH_2)_{10}C{\equiv}CCH_2CH_3} \xrightarrow[\text{Reduction}]{\text{Na, liquid } NH_3}$$

$$\underset{\textit{trans}\textbf{-11-Tetradecen-1-ol}}{HO(CH_2)_{10}(H)C{=}C(H)CH_2CH_3} \xrightarrow[\substack{\text{Oxidation} \\ \text{(Section 8-6)}}]{PCC,\ CH_2Cl_2} \underset{\textbf{Sex pheromone of the spruce budworm}}{HC(=O)(CH_2)_9(H)C{=}C(H)CH_2CH_3}$$

In summary, alkynes are very similar in reactivity to alkenes, except that they have two π bonds, both of which may be saturated by addition reactions. Hydrogenation of the first π bond, which gives cis alkenes, is best achieved by using the Lindlar catalyst. Alkynes are converted into trans alkenes by treatment with sodium in liquid ammonia, a process that involves two successive one-electron reductions.

13-8 Electrophilic Addition Reactions of Alkynes

As a center of high electron density, the triple bond is readily attacked by electrophiles. This section will describe the results of three such processes: addition of hydrogen halides, reaction with halogens, and hydration. The hydration is catalyzed by mercury(II) ions. The Markovnikov rule is followed in transformations of terminal alkynes.

Addition of hydrogen halides forms haloalkenes and geminal dihaloalkanes

The addition of hydrogen bromide to 2-butyne yields the (*Z*)-2-bromobutene.

Addition of a Hydrogen Halide to an Internal Alkyne

$$CH_3C{\equiv}CCH_3 \xrightarrow{HBr,\ Br^-} \text{(H)(H}_3\text{C)C}{=}\text{C(CH}_3\text{)(Br)}$$

60%

(Z)-2-Bromobutene

The stereochemistry of this type of addition is frequently (but not always) *anti*, particularly in the presence of excess bromide ion. Another molecule of hydrogen bromide added to the bromoalkene gives the geminal dihaloalkane with regioselectivity in accord with the Markovnikov rule.

$$\text{(H)(H}_3\text{C)C}{=}\text{C(CH}_3\text{)(Br)} \xrightarrow{HBr} CH_3CH_2CBr_2CH_3$$

90%

2,2-Dibromobutane

The addition of hydrogen halides to terminal alkynes also proceeds in accord with the Markovnikov rule.

Addition to a Terminal Alkyne

$$CH_3C{\equiv}CH \xrightarrow{HI,\ -70^\circ C} \text{(I)(H}_3\text{C)C}{=}\text{CH}_2 + CH_3CI_2CH_3$$

35%   65%

It is usually difficult to limit such reactions to addition of a single molecule of HX.

EXERCISE 13-11

Write a step-by-step mechanism for the addition of HBr twice to 2-butyne to give 2,2-dibromobutane. Show clearly the structure of the intermediate formed in each step.

Halogenation also occurs once or twice

Electrophilic addition of halogen to alkynes proceeds through the intermediacy of isolable vicinal dihaloalkenes, the products of a single *anti* addition. Reaction with additional halogen gives tetrahaloalkanes. For example, halogenation of 3-hexyne gives the expected (*E*)-dihaloalkene and the tetrahaloalkane.

$$CH_3CH_2C{\equiv}CCH_2CH_3 \xrightarrow{Br_2,\ CH_3COOH,\ LiBr} \text{(CH}_3\text{CH}_2\text{)(Br)C}{=}\text{C(Br)(CH}_2\text{CH}_3\text{)} \xrightarrow{Br_2,\ CCl_4} CH_3CH_2CBr_2CBr_2CH_2CH_3$$

3-Hexyne   99% **(E)-3,4-Dibromo-3-hexene**   95% **3,3,4,4-Tetrabromohexane**

EXERCISE 13-12

Give the products of addition of one and two molecules of Cl_2 to 1-butyne.

Mercuric ion-catalyzed hydration of alkynes furnishes ketones

In a process analogous to the hydration of alkenes, water can be added to alkynes in a Markovnikov sense to give alcohols—in this case **enols,** in which the hydroxy group is attached to a double bond carbon. As mentioned in Section 12-15, enols spontaneously rearrange to the isomeric carbonyl compounds. This process, called **tautomerism,** interconverts two isomers by simultaneous proton and double bond shifts. The enol is said to **tautomerize** to the carbonyl compound, and the two species are referred to as **tautomers** (*tauto,* Greek, the same; *meros,* Greek, part). In this way, alkynes are ultimately converted into ketones. The reaction is catalyzed by mercuric ions.

Hydration of Alkynes

$$\text{RC}\equiv\text{CR} \xrightarrow{\text{HOH, H}^+\text{, HgSO}_4} \underset{\textbf{Enol}}{\text{RCH}=\overset{\text{OH}}{\overset{|}{\text{C}}}\text{R}} \longrightarrow \underset{\textbf{Ketone}}{\text{R}\underset{\text{H}}{\underset{|}{\overset{\text{H}}{\overset{|}{\text{C}}}}}-\overset{\text{O}}{\overset{\|}{\text{C}}}\text{R}}$$

Hydration follows Markovnikov's rule: Terminal alkynes give methyl ketones.

Hydration of a Terminal Alkyne

$$\text{1-ethynylcyclohexanol (OH)} \xrightarrow{\text{H}_2\text{SO}_4\text{, H}_2\text{O, HgSO}_4} \text{1-acetylcyclohexanol (OH)}\quad 91\%$$

Symmetric internal alkynes give a single carbonyl compound; unsymmetric systems lead to a mixture of ketones.

Hydration of Internal Alkynes

$$\text{CH}_3\text{CH}_2\text{C}\equiv\text{CCH}_2\text{CH}_3 \xrightarrow{\text{H}_2\text{SO}_4\text{, H}_2\text{O, HgSO}_4} \underset{80\%}{\text{CH}_3\text{CH}_2\overset{\text{O}}{\overset{\|}{\text{C}}}\text{CH}_2\text{CH}_2\text{CH}_3}$$

Only possible product

$$\text{CH}_3\text{CH}_2\text{CH}_2\text{C}\equiv\text{CCH}_3 \xrightarrow{\text{H}_2\text{SO}_4\text{, H}_2\text{O, HgSO}_4} \underset{50\%}{\text{CH}_3\text{CH}_2\text{CH}_2\overset{\text{O}}{\overset{\|}{\text{C}}}\text{CH}_2\text{CH}_3} + \underset{50\%}{\text{CH}_3\text{CH}_2\text{CH}_2\text{CH}_2\overset{\text{O}}{\overset{\|}{\text{C}}}\text{CH}_3}$$

EXERCISE 13-13

Give the products of mercuric ion-catalyzed hydration of **(a)** ethyne; **(b)** propyne; **(c)** 1-butyne; **(d)** 2-butyne; **(e)** 2-methyl-3-hexyne.

EXERCISE 13-14

Propose a synthetic scheme that will convert compound A into B.

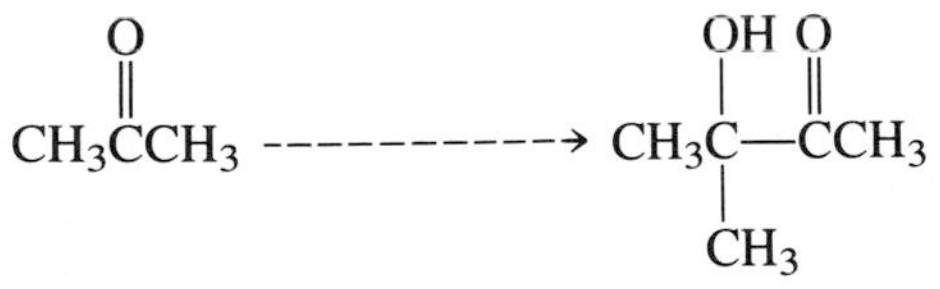

$$\underset{\mathbf{A}}{CH_3\overset{O}{\overset{\|}{C}}CH_3} \dashrightarrow \underset{\mathbf{B}}{CH_3\overset{OH}{\overset{|}{\underset{CH_3}{\underset{|}{C}}}}\text{—}\overset{O}{\overset{\|}{C}}CH_3}$$

To summarize, alkynes can react with electrophiles such as hydrogen halides and halogens either once or twice. Terminal alkynes transform in accord with the Markovnikov rule. Mercuric ion-catalyzed hydration furnishes enols, which convert to ketones by a process called tautomerism.

13-9 Anti-Markovnikov Additions to Triple Bonds

This section describes two methods by which addition to terminal alkynes can be carried out in an anti-Markovnikov manner.

Radical addition of HBr gives 1-bromoalkenes

As in the case of alkenes, hydrogen bromide can add to triple bonds by a radical mechanism in an anti-Markovnikov fashion if light or other radical initiators are present. Both *syn* and *anti* additions are observed.

$$\underset{\textbf{1-Hexyne}}{CH_3(CH_2)_3C{\equiv}CH} \xrightarrow{\text{HBr, ROOR}} \underset{\textit{cis-}\textbf{ and }\textit{trans-}\textbf{1-Bromo-1-hexene}}{\underset{74\%}{CH_3(CH_2)_3CH{=}CHBr}}$$

Aldehydes result from hydroboration–oxidation of terminal alkynes

Terminal alkynes are hydroborated in a regioselective, anti-Markovnikov fashion, the boron attacking the less hindered carbon. However, with borane itself, this reaction leads to hydroboration of both π bonds. To stop at the alkenylborane stage, less reactive bulky borane reagents, such as dicyclohexylborane, are used.

Hydroboration of a Terminal Alkyne

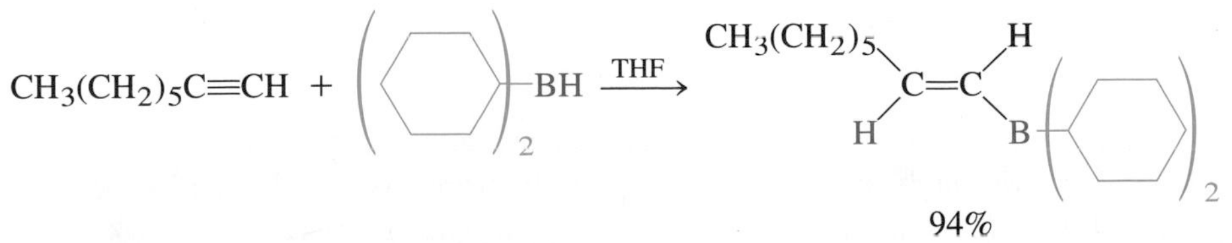

$$\underset{\textbf{1-Octyne}}{CH_3(CH_2)_5C{\equiv}CH} + \underset{\textbf{Dicyclohexylborane}}{(C_6H_{11})_2BH} \xrightarrow{\text{THF}} \underset{94\%}{\overset{CH_3(CH_2)_5\quad\quad H}{\underset{H\quad\quad\quad B(C_6H_{11})_2}{C{=}C}}}$$

EXERCISE 13-15

Dicyclohexylborane is prepared by a hydroboration reaction. What are the starting materials for its preparation?

Like alkylboranes (Section 12-8), alkenylboranes can be oxidized to the corresponding alcohols—in this case, to terminal enols that spontaneously rearrange to aldehydes.

$$CH_3(CH_2)_5C\equiv CH \xrightarrow[\text{2. } H_2O_2,\ HO^-]{\text{1. Dicyclohexylborane}} \left[\begin{matrix} CH_3(CH_2)_5 & & H \\ & C=C & \\ H & & OH \end{matrix} \right] \xrightarrow{\text{Tautomerism}} CH_3(CH_2)_5\overset{H}{\underset{H}{C}}-\overset{O}{\overset{\|}{C}}H$$

Enol — Octanal, 70%

EXERCISE 13-16

Give the products of hydroboration–oxidation of **(a)** ethyne; **(b)** 1-propyne; **(c)** 1-butyne.

EXERCISE 13-17

Outline a synthesis of the molecule shown below from 3,3-dimethyl-1-butyne.

$$(CH_3)_3CCH_2\overset{O}{\overset{\|}{C}}H$$

In summary, HBr in the presence of peroxides undergoes anti-Markovnikov addition to terminal alkynes to give 1-bromoalkenes. Hydroboration–oxidation with sterically hindered boranes furnishes intermediate enols that tautomerize to the final product aldehydes.

13-10 Chemistry of Alkenyl Halides and Cuprate Reagents

We have encountered haloalkenes as intermediates in both the preparation of alkynes by double dehydrohalogenation and the addition of hydrogen halides to triple bonds. This section will present some chemistry of these systems and introduce a new type of organometallic reagent for carbon–carbon bond formation, the organocuprates.

Alkenyl halides do not undergo S_N2 or S_N1 reactions

Compared with haloalkanes, alkenyl halides are relatively unreactive toward nucleophiles. Although we have seen that, with strong bases, alkenyl halides undergo elimination reactions to give alkynes, they do not react with weak bases and relatively nonbasic nucleophiles, such as iodide. Similarly, S_N1 reactions do not normally take place because the intermediate alkenyl cations are species of high energy.

$$CH_2{=}C(H)(Br) \xrightarrow{I^-}\!\!\!\!\!\!\times\;\; CH_2{=}C(H)(I) + Br^-$$

Does not occur

$$CH_2{=}C(H)(Br) \;\not\longrightarrow\; CH_2{=}\overset{+}{C}{-}H + Br^-$$

Ethenyl (vinyl) cation

Does not occur

Alkenyl halides, however, can react through the intermediate formation of alkenyl organometallics (see Exercise 11-6). These species allow access to a variety of specifically substituted alkenes.

Alkenyl Organometallics in Synthesis

$$CH_2{=}C(Br)(H) + Mg \xrightarrow{THF} CH_2{=}C(MgBr)(H) \xrightarrow[2.\ H^+,\ H_2O]{1.\ CH_3\overset{O}{\overset{\|}{C}}CH_3} CH_2{=}C(H){-}C(OH)(CH_3)_2$$

1-Bromoethene (Vinyl bromide)

90% **Ethenylmagnesium bromide (A vinyl Grignard reagent)**

65% **2-Methyl-3-buten-2-ol**

Organocuprates form carbon–carbon bonds

Many important molecules contain double bonds that bear simple alkyl groups (Section 12-9). In principle, these compounds may be derived by the direct alkylation of organometallic reagents. Unfortunately, however, RLi and RMgX molecules do not generally react with haloalkanes (except when R is an alkynyl group; Section 13-6). Fortunately, another type of organometallic reagent, an organocuprate, is very effective. **Organocuprates** have the empirical formula R_2CuLi, and they may be prepared directly by adding two equivalents of an organolithium reagent to one of cuprous iodide, CuI.

Direct Cuprate Synthesis

$$2\ RLi + CuI \longrightarrow \underset{\text{An organocuprate}}{R_2CuLi} + LiI$$

For example,

$$\underset{\textbf{Ethenyllithium}}{2\ CH_2{=}CHLi} + CuI \longrightarrow \underset{\substack{\textbf{Lithium diethenylcuprate}\\ \text{(An organocuprate reagent)}}}{(CH_2{=}CH)_2CuLi} + LiI$$

Lithium alkenylcuprates undergo so-called **coupling reactions** with haloalkanes to give substituted alkene products. Primary iodides are best, although other primary and secondary haloalkanes also give satisfactory results.

Reaction of Organocuprates with Haloalkanes

$$R_2CuLi + R'X \longrightarrow R—R' + LiX$$

$$\underset{\text{(trans-1-chloropropene)}}{CH_3CH=CHCl} \xrightarrow[\text{2. CuI}]{\text{1. Li, }(CH_3CH_2)_2O} (CH_3CH=CH)_2CuLi \xrightarrow[\text{HMPA}]{CH_3(CH_2)_6CH_2I,} CH_3CH=CHCH_2(CH_2)_6CH_3 \quad 90\%$$

Lithium bis(*trans*-1-Propenyl)cuprate

***trans*-2-Undecene**

Such couplings are also accomplished by using alkylcuprate species.

$$CH_3I \xrightarrow[\text{2. CuI}]{\text{1. Li, }(CH_3CH_2)_2O} (CH_3)_2CuLi \xrightarrow[(CH_3CH_2)_2O,\ 0°C]{CH_3(CH_2)_8CH_2I,} CH_3(CH_2)_8CH_2CH_3 \quad 90\%$$

Lithium dimethylcuprate

Undecane

EXERCISE 13-18

Show how you would prepare (*Z*)-3-methyl-2-heptene from *cis*-2-butene.

In summary, alkenyl halides are unreactive in nucleophilic substitutions but can be converted to alkenyllithium or alkenyl Grignard reagents. Organocuprates are species that undergo efficient coupling reactions with haloalkanes.

13-11 Ethyne as an Industrial Starting Material

Ethyne was once one of the four or five major starting materials in the chemical industry for two reasons: Addition reactions to one of the π bonds produce useful alkene monomers (Section 12-14), and it has a high heat content. Its industrial use has declined because of the availability of cheap ethene, propene, butadiene, and other hydrocarbons through oil-based technology. However, during the twenty-first century, oil reserves are expected to dwindle to the point that other sources of energy will have to be developed. One such source is coal. There are currently no known processes for converting coal directly into the alkenes mentioned above; ethyne, however, can be produced from coal and hydrogen or from coke (a coal residue obtained after removal of volatiles) and limestone through the formation of calcium carbide. Consequently, it may once again become an important industrial raw material.

Production of ethyne from coal requires high temperatures

The high energy content of ethyne requires the use of production methods that are costly in energy. One process for making ethyne from coal uses hydrogen in an arc reactor at temperatures as high as several thousand degrees.

$$\text{Coal} + H_2 \xrightarrow{\Delta} \underset{33\%\ \text{conversion}}{HC{\equiv}CH} + \text{nonvolatile salts}$$

The oldest large-scale preparation of ethyne proceeds through calcium carbide. Limestone (calcium oxide) and coke are heated to about 2000°C, which results in the desired product and carbon monoxide.

$$\underset{\textbf{Coke}}{3\ C} + \underset{\textbf{Lime}}{CaO} \xrightarrow{2000^{\circ}C} \underset{\textbf{Calcium carbide}}{CaC_2} + CO$$

The calcium carbide is then treated with water at ambient temperatures to give ethyne and calcium hydroxide.

$$CaC_2 + 2\ H_2O \longrightarrow HC{\equiv}CH + Ca(OH)_2$$

Ethyne is a source of valuable monomers for industry

Ethyne chemistry underwent important commercial development in the 1930s and 1940s in the laboratories of the Badische Anilin and Sodafabriken (BASF) in Ludwigshafen, Germany. Ethyne under pressure was brought into reaction with carbon monoxide, carbonyl compounds, alcohols, and acids in the presence of catalysts to give a multitude of valuable raw materials to be used in further transformations. For example, nickel carbonyl catalyzes the addition of carbon monoxide and water to ethyne to give propenoic (acrylic) acid. Similar exposure to alcohols or amines instead of water results in the corresponding acid derivatives. All of the products are valuable monomers (see Section 12-14).

Industrial Chemistry of Ethyne

$$HC{\equiv}CH + CO + H_2O \xrightarrow{Ni(CO)_4,\ 100\ \text{atm},\ >250^{\circ}C} H_2C{=}CHCOOH$$

Propenoic acid
(Acrylic acid)

$$HC{\equiv}CH + CO + CH_3OH \xrightarrow{Ni(CO)_4,\ \Delta} H_2C{=}CH\overset{O}{\overset{\|}{C}}OCH_3$$

80%

Methyl propenoate
(Methyl acrylate)

Polymerization of propenoic (acrylic) acid and its derivatives produces materials of considerable utility. The polymeric esters **(polyacrylates)** are tough, resilient, and flexible polymers that have replaced natural rubber (see Chapter 14) in many applications. Poly(ethyl acrylate) is used for O-rings, valve seals, and related purposes in automobiles. Other polyacrylates are found in biomedical and dental appliances, such as dentures.

The addition of formaldehyde to ethyne is achieved with high efficiency by using copper acetylide as a catalyst.

$$HC\equiv CH + CH_2{=}O \xrightarrow{Cu_2C_2\text{–}SiO_2,\ 125°C,\ 5\ atm} \underset{\substack{\textbf{2-Propyn-1-ol}\\ \textbf{(Propargyl alcohol)}}}{HC\equiv CCH_2OH} \quad \text{or} \quad \underset{\textbf{2-Butyne-1,4-diol}}{HOCH_2C\equiv CCH_2OH}$$

The resulting alcohols are useful synthetic intermediates. For example, 2-butyne-1,4-diol is a precursor for the production of oxacyclopentane (tetrahydrofuran, one of the solvents most frequently employed for Grignard and organolithium reagents) by hydrogenation, followed by acid-catalyzed dehydration.

Oxacyclopentane (Tetrahydrofuran) Synthesis

$$HOCH_2C\equiv CCH_2OH \xrightarrow{\text{Catalyst},\ H_2} HO(CH_2)_4OH \xrightarrow[260\text{–}280°C,\ 90\text{–}100\ atm]{H_3PO_4,\ pH\ 2,} \underset{\substack{99\%\\ \textbf{Oxacyclopentane}\\ \textbf{(Tetrahydrofuran, THF)}}}{\text{(five-membered ring with O)}} + H_2O$$

A number of technical processes have been developed in which reagents $\overset{\delta^+}{A}\text{–}\overset{\delta^-}{B}$ in the presence of a catalyst add to the triple bond. For example, the catalyzed addition of hydrogen chloride gives chloroethene (vinyl chloride), and addition of hydrogen cyanide produces propenenitrile (acrylonitrile).

Addition Reactions of Ethyne

$$HC\equiv CH + HCl \xrightarrow{Hg^{2+},\ 100\text{–}200°C} \underset{\substack{\textbf{Chloroethene}\\ \textbf{(Vinyl chloride)}}}{H_2C{=}CHCl}$$

$$HC\equiv CH + HCN \xrightarrow{Cu^+,\ NH_4Cl,\ 70\text{–}90°C,\ 1.3\ atm} \underset{\substack{80\text{–}90\%\\ \textbf{Propenenitrile}\\ \textbf{(Acrylonitrile)}}}{H_2C{=}CHCN}$$

Polymers containing at least 85% propenenitrile (acrylonitrile) are called **acrylic fibers.** Their applications include clothing (Orlon), carpets, and insulating materials. Copolymers of acrylonitrile and 10–15% vinyl chloride have fire-retardant properties and are used in children's sleepwear.

In summary, ethyne was once, and may again be in the future, a valuable industrial feedstock because of its ability to react with a large number of substrates to yield useful monomers and other compounds having functional groups. It can be made from coal and H_2 at high temperatures or it can be liberated from calcium carbide by hydrolysis. Some of the industrial reactions that it undergoes are carbonylation, addition of formaldehyde, and addition reactions with HX.

13-12 Naturally Occurring and Physiologically Active Alkynes

Although alkynes are not very abundant in nature, they do exist in some plants and other organisms. The first such substance to be isolated, in 1826, was dehydromatricaria ester.

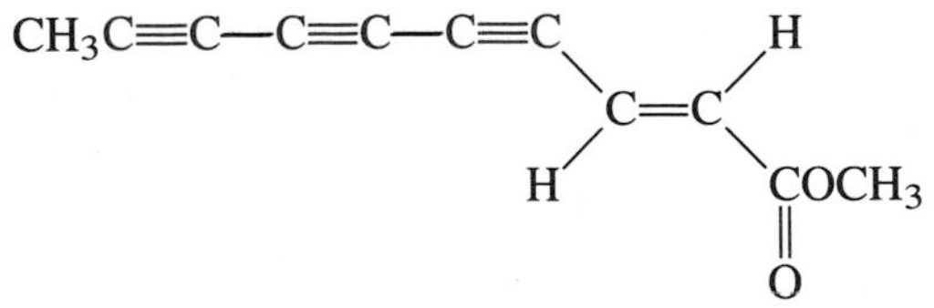

$CH_3C{\equiv}C{-}C{\equiv}C{-}C{\equiv}C$ — C=C (H, H) — $\underset{\underset{O}{\|}}{C}OCH_3$

Dehydromatricaria ester

More than a thousand such compounds are now known, and some of them are physiologically active. For example, some naturally occurring ethynylketones, such as capillin, have fungicidal activity.

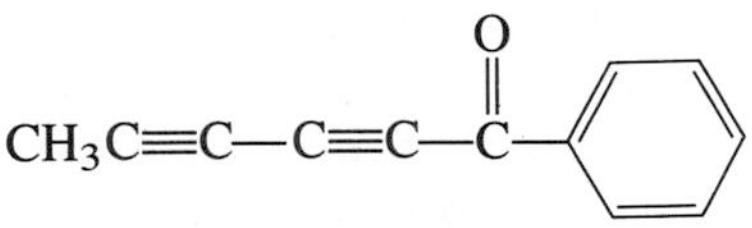

$CH_3C{\equiv}C{-}C{\equiv}C{-}\overset{\overset{O}{\|}}{C}{-}C_6H_5$

Capillin
(Active against skin fungi)

The alkyne ichthyothereol is the active ingredient of a poisonous substance used by the Indians of the Lower Amazon Basin in arrowheads. It causes convulsions in mammals. Two enyne functional groups are incorporated in the compound hystrionicotoxin. It is one of the substances isolated from the skin of "arrow poison frog," a highly colorful species of the genus *Dendrobates*. The frog secretes this compound and similar ones as defensive venoms and mucosal-tissue irritants against both mammals and reptiles. How and why the alkyne units are constructed biosynthetically is not clear.

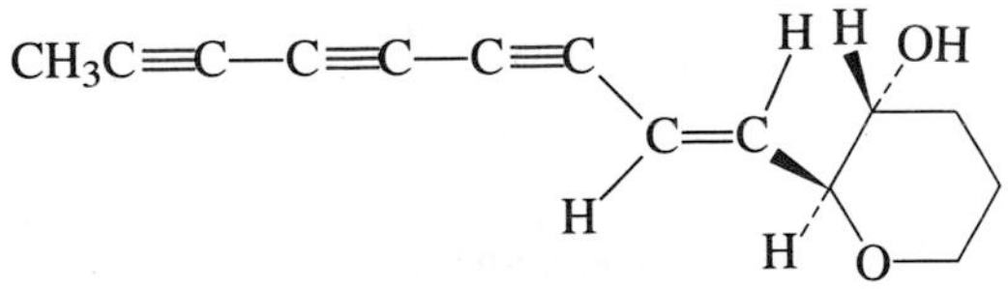

$CH_3C{\equiv}C{-}C{\equiv}C{-}C{\equiv}C$ — C=C — (H, H, OH, H, O ring)

Ichthyothereol
(A convulsant)

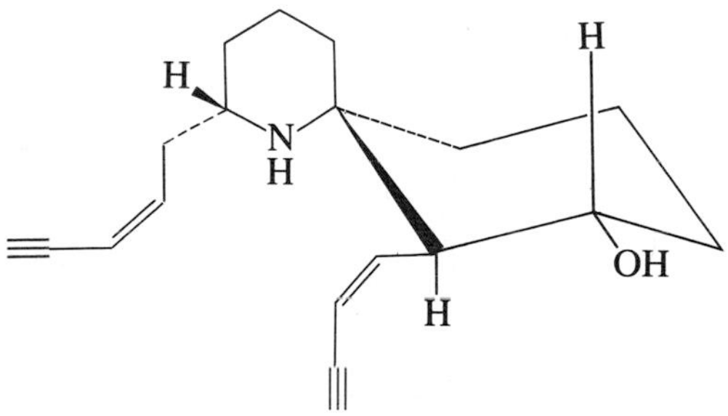

(H, N–H, H, OH, H)

Hystrionicotoxin

Many drugs have been modified by synthesis to contain alkyne substituents, because such compounds are frequently more readily absorbed by the body, less toxic, and more active than the corresponding alkenes or alkanes. For example,

3-methyl-1-pentyn-3-ol is available as a nonprescription hypnotic, and several other alkynols are similarly effective.

Highly reactive enediyne (–C≡C–CH=CH–C≡C–) and trisulfide (RSSSR) functional groups characterize a class of naturally occurring antibiotic–antitumor agents discovered in the late 1980s, such as calicheamicin and esperamicin.

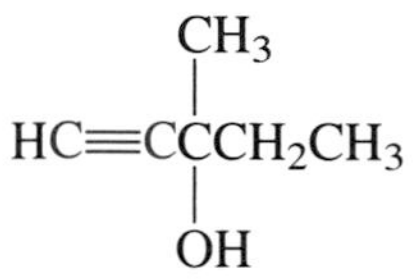

CH_3
HC≡CCCH$_2$CH$_3$
OH

3-Methyl-1-pentyn-3-ol
(Hypnotic)

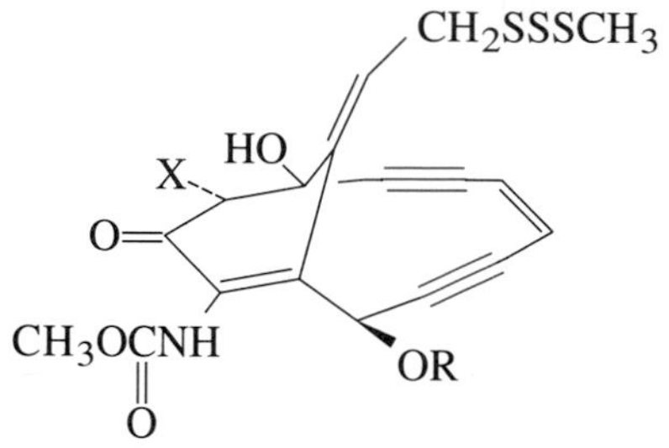

CH_2SSSCH_3, HO, X, O=, CH_3OCNH, O, OR

Calicheamicin (X = H)
Esperamicin (X = OR′)
R and R′ = sugars (Chapter 24)

Ethynyl estrogens, such as 17-ethynylestradiol, are considerably more potent birth control agents than are the naturally occurring hormones (see Section 4-7). The diaminoalkyne tremorine induces symptoms characteristic of Parkinson's disease: spasms of uncontrolled movement. Interestingly, a simple cyclic homolog of tremorine acts as a muscle relaxant and counteracts the effect of tremorine. Compounds that cancel the physiological effects of other compounds are called **antagonists** (*antagonizesthai,* Greek, to struggle against). Finally, ethynyl analogs of amphetamine have been prepared in a search for alternative, more active, more specific, and less addictive central nervous system stimulants.

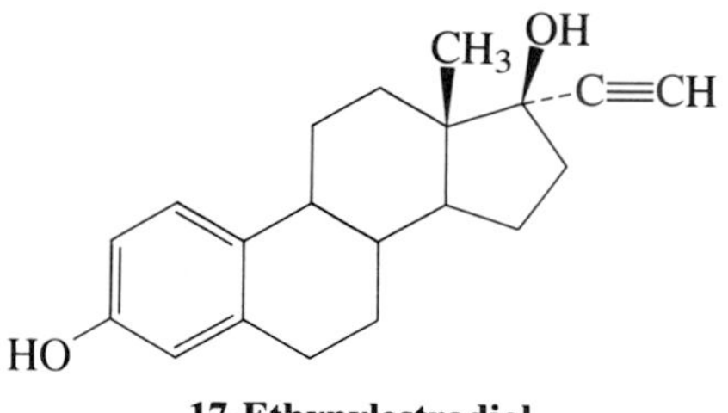

CH_3 OH, C≡CH, HO

17-Ethynylestradiol

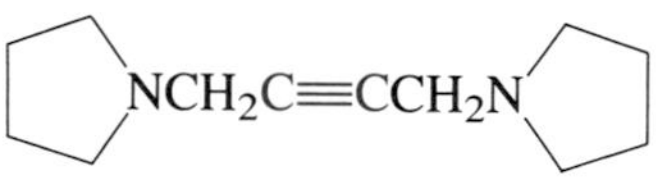

NCH$_2$C≡CCH$_2$N

Tremorine

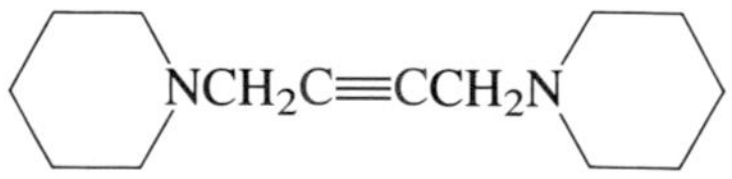

NCH$_2$C≡CCH$_2$N

Tremorine antagonist

NH$_2$
–CH$_2$CHC≡CH

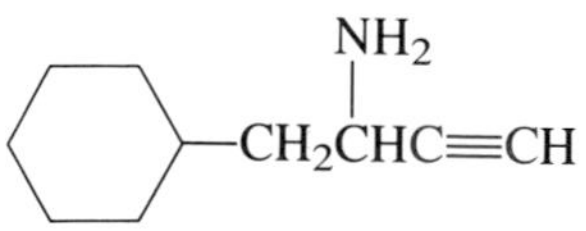

An amphetamine analog
(Active in the central nervous system)

In summary, the alkyne unit is present in a number of physiologically active natural and synthetic compounds.

NEW REACTIONS

1. ACIDITY OF 1-ALKYNES (Section 13-2)

$$RC{\equiv}CH + :B^- \rightleftharpoons RC{\equiv}C:^- + BH$$

$pK_a \sim 25$

Base (B): $NaNH_2$-liquid NH_3; RLi-$(CH_3CH_2)_2O$; RMgX-THF

Preparation of Alkynes

2. DOUBLE ELIMINATION FROM DIHALOALKANES (Section 13-5)

$$\underset{\text{Vicinal dihaloalkane}}{\mathrm{RC(X)(H)\!-\!C(X)(H)R}} \xrightarrow[-\ 2\ \mathrm{HX}]{\mathrm{NaNH_2,\ liquid\ NH_3}} \mathrm{RC{\equiv}CR}$$

Vicinal dihaloalkane

3. FROM ALKENES BY HALOGENATION–DEHYDROHALOGENATION (Section 13-5)

$$\mathrm{RCH{=}CHR} \xrightarrow{\text{1. } \mathrm{X_2, CCl_4}\quad \text{2. } \mathrm{NaNH_2}\text{, liquid } \mathrm{NH_3}} \mathrm{RCH{=}C(R)X} \xrightarrow{\mathrm{NaNH_2}\text{, liquid } \mathrm{NH_3}} \mathrm{RC{\equiv}CR}$$

Alkenyl halide intermediate

Conversion of Alkynes into Other Alkynes

4. ALKYLATION OF ALKYNYL ANIONS (Section 13-6)

$$\mathrm{RC{\equiv}CH} \xrightarrow{\text{1. } \mathrm{NaNH_2}\text{, liquid } \mathrm{NH_3}\quad \text{2. } \mathrm{R'X}} \mathrm{RC{\equiv}CR'}$$

S_N2 reaction: R′ must be primary

5. ALKYLATION WITH OXACYCLOPROPANE (Section 13-6)

$$\mathrm{RC{\equiv}CH} \xrightarrow{\text{1. } \mathrm{CH_3CH_2CH_2CH_2Li, THF}\quad \text{2. } \mathrm{H_2C\overset{O}{-}CH_2}\quad \text{3. } \mathrm{H^+, H_2O}} \mathrm{RC{\equiv}CCH_2CH_2OH}$$

Attack occurs at less substituted carbon in unsymmetric oxacyclopropanes

6. ALKYLATION WITH CARBONYL COMPOUNDS (Section 13-6)

$$\mathrm{RC{\equiv}CH} \xrightarrow{\text{1. } \mathrm{CH_3CH_2CH_2CH_2Li, THF}\quad \text{2. } \mathrm{R'\overset{O}{\overset{\|}{C}}R''}\quad \text{3. } \mathrm{H^+, H_2O}} \mathrm{RC{\equiv}C\underset{R''}{\overset{OH}{C}}R'}$$

Reactions of Alkynes

7. HYDROGENATION (Section 13-7)

$$\mathrm{RC{\equiv}CR} \xrightarrow{\text{Catalyst, } \mathrm{H_2}} \mathrm{RCH_2CH_2R} \qquad \Delta H^\circ \sim -70 \text{ kcal mol}^{-1}$$

Catalysts: Pt, Pd–C

$$\mathrm{RC{\equiv}CR} \xrightarrow{\mathrm{H_2}\text{, Lindlar catalyst}} \mathrm{(H)(R)C{=}C(H)(R)} \qquad \Delta H^\circ \sim -40 \text{ kcal mol}^{-1}$$

Cis alkene

8. REDUCTION WITH SODIUM IN LIQUID AMMONIA (Section 13-7)

$$RC\equiv CR \xrightarrow[2.\ H^+,\ H_2O]{1.\ Na,\ liquid\ NH_3} \underset{\text{Trans alkene}}{(H)(R)C{=}C(R)(H)}$$

9. ELECTROPHILIC (AND MARKOVNIKOV) ADDITIONS: HYDROHALOGENATION, HALOGENATION, AND HYDRATION (Section 13-8)

$$RC\equiv CR \xrightarrow{HX} RCH{=}CXR \xrightarrow{HX} \underset{\text{Geminal dihaloalkane}}{RCH_2CX_2R}$$

$$RC\equiv CH \xrightarrow{2\ HX} RCX_2CH_3$$

$$RC\equiv CR \xrightarrow{Br_2,\ Br^-} \underset{\text{Mainly trans}}{(R)(Br)C{=}C(Br)(R)} \xrightarrow{Br_2} RCBr_2CBr_2R$$

$$RC\equiv CR \xrightarrow{Hg^{2+},\ H_2O} RCH_2\overset{O}{\overset{\|}{C}}R$$

10. RADICAL ADDITION OF HYDROGEN BROMIDE (Section 13-9)

$$RC\equiv CH \xrightarrow{HBr,\ ROOR} \underset{\text{Anti-Markovnikov}}{RCH{=}CHBr}$$

Br attaches to less substituted carbon

11. HYDROBORATION (Section 13-9)

$$RC\equiv CH \xrightarrow{R'_2BH,\ THF} (R)(H)C{=}C(H)(BR'_2)$$

Anti-Markovnikov and stereospecific *(syn)* addition

B attaches to less substituted carbon

Dicyclohexylborane (R′ = cyclohexyl)

12. OXIDATION OF ALKENYLBORANES (Section 13-9)

$$(H)(R)C{=}C(B{<})(H) \xrightarrow{H_2O_2,\ HO^-} \underset{\text{Enol}}{\left[(H)(R)C{=}C(OH)(H)\right]} \xrightarrow[\text{Tautomerism}]{} RCH_2\overset{O}{\overset{\|}{C}}H$$

Organometallic Reagents

13. ALKENYL ORGANOMETALLICS (Section 13-10)

$$(R)(R')C{=}C(X)(R'') \xrightarrow{Mg,\ THF} (R)(R')C{=}C(MgX)(R'')$$

14. COUPLING REACTIONS OF ORGANOCUPRATES (Section 13-10)

$$R\text{—}X \xrightarrow[\text{2. CuI}]{\text{1. Li, }(CH_3CH_2)_2O} \underset{\textbf{Lithium diorganocuprate}}{R_2CuLi} \xrightarrow{R'\text{–}X,\ (CH_3CH_2)_2O} R\text{—}R'$$

R = alkyl or alkenyl

15. INDUSTRIAL PREPARATION AND USES OF ETHYNE (Sections 13-4 and 13-11)

Preparation

$$\text{Directly from coal} + H_2,\ \Delta;\ \text{or from coke} + CaO \longrightarrow CaC_2 \xrightarrow{H_2O} HC{\equiv}CH$$

Combustion (acetylene torch)

$$HC{\equiv}CH + 2.5\ O_2 \longrightarrow 2\ CO_2 + H_2O \qquad \Delta H^\circ = -317\ \text{kcal mol}^{-1}$$

Industrial chemistry

$$C_2H_2 + CO + H_2O \xrightarrow{Ni(CO)_4} CH_2{=}CHCO_2H$$

$$C_2H_2 + CH_2O \xrightarrow{Cu_2C_2} HOCH_2C{\equiv}CH + HOCH_2C{\equiv}CCH_2OH$$

Additions

$$C_2H_2 + HX \longrightarrow CH_2{=}CHX$$

$$C_2H_2 + HCN \longrightarrow CH_2{=}CHCN$$

Catalyzed by transition metal cations

IMPORTANT CONCEPTS

1. The rules for naming alkynes are essentially the same as those formulated for alkenes. Molecules with both double and triple bonds are called alkenynes, the double bond receiving the lower number if both are at equivalent positions. Hydroxy groups are given precedence in numbering alkynyl alcohols (alkynols).
2. The electronic structure of the triple bond reveals two π bonds, perpendicular to each other, and a σ bond, formed by two overlapping *sp* hybrid orbitals. The strength of the triple bond is about 229 kcal mol^{-1}; that of the alkynyl C–H bond is 131 kcal mol^{-1}. Triple bonds form linear structures with respect to other attached atoms, with short C–C (1.20 Å) and C–H (1.06 Å) bonds.
3. The high *s* character of the terminal alkyne carbon makes the bound hydrogen relatively acidic ($pK_a \sim 25$).
4. The chemical shift of the alkynyl hydrogen is low (δ = 1.7–3.1 ppm) compared with that of alkenyl hydrogens because of the shielding effect of an induced electron current around the molecular axis caused by the external magnetic field. The triple bond allows for long-range coupling. IR spectroscopy indicates the presence of the C≡C and ≡C–H bonds of terminal alkynes through bands at 2100–2260 cm^{-1} and 3260–3330 cm^{-1}, respectively.
5. Internal alkynes are more stable than the isomeric terminal alkynes by about 4–5 kcal mol^{-1}.
6. The elimination reaction with vicinal dihaloalkanes proceeds regioselectively and stereospecifically to give alkenyl halides.
7. Selective cis dihydrogenation of alkynes is possible with Lindlar catalyst, the surface of which is less active than palladium on carbon and there-

fore not capable of hydrogenating alkenes. Selective trans hydrogenation is possible with sodium metal dissolved in liquid ammonia because simple alkenes cannot be reduced by one-electron transfer. The stereochemistry is set by the greater stability of a trans disubstituted alkenyl radical intermediate.

8. To stop the hydroboration of terminal alkynes at the alkenylboron intermediate stage, modified dialkylboranes—particularly dicyclohexylborane—are used. Oxidation of the resulting alkenylboranes produces enols that are unstable with respect to rearrangement to carbonyl compounds (tautomerism).

9. Organocuprate reagents undergo coupling reactions with haloalkanes.

PROBLEMS

1. Name each of the following compounds, using the IUPAC system of nomenclature.

(a) $H_3C—C(CH_3)(Cl)—C{\equiv}CH$

(b) $H_3C—C(CH_3)(HO)—C{\equiv}CH$

(c) $CH_3CH_2CH_2CHCH_2CH_2CH_2OH$ with C≡CH on the CH

(d) H_3C, H / C=C \ H, C≡CH

(e) H_3C, CH_3CH_2 / C=C \ C≡CCH$_3$, $CHCH_2CH_2CH_3$ (with CH_3)

(f) cyclopentane with C≡CH and CH=CH$_2$

2. Compare C–H bond strengths in ethane, ethene, and ethyne. Reconcile these data with hybridization, bond polarity, and acidity of the hydrogen.

3. Compare the C2–C3 bonds in propane, propene, and propyne. Should they be any different with respect to either bond length or bond strength? If so, how should they vary?

4. The heats of combustion for three compounds with the molecular formula C_5H_8 are as follows: cyclopentene, $\Delta H_{comb} = -1027$ kcal mol^{-1}; 1,4-pentadiene, $\Delta H_{comb} = -1042$ kcal mol^{-1}; and 1-pentyne, $\Delta H_{comb} = -1052$ kcal mol^{-1}. Explain in terms of relative stability and bond strengths.

5. Rank in order of decreasing stability.
(a) 1-Heptyne and 3-heptyne

(b) cyclopentyl–C≡CCH$_3$, cyclopentyl–CH$_2$C≡CH, and cyclooctyne

(Hint: Make a model of the third structure. Is there anything unusual about its triple bond?)

6. Deduce structures for each of the following. **(a)** Molecular formula C_6H_{10}; NMR spectrum A; no strong IR bands between 2100 and 2300 or 3250 and 3350 cm^{-1}. **(b)** Molecular formula C_7H_{12}; NMR spectrum B; IR bands at about 2120 and 3330 cm^{-1}. **(c)** Molecular formula C_5H_8O; NMR and IR spectra C (p. 496). The NMR spectrum has been recorded at 300 MHz to provide better resolution of the signals between 1.6 and 2.4 ppm (see inset).

7. The IR spectrum of 1,4-nonadiyne displays a strong, sharp band at 3300 cm^{-1}. What is the origin of this absorption? Treatment of 1,4-nonadiyne with $NaNH_2$, then with D_2O, leads to the incorporation of two deuterium atoms, leaving the molecule unchanged otherwise. The IR spectrum reveals that the peak at 3300 cm^{-1} has disappeared, but a new one is present at 2580 cm^{-1}. **(a)** What is the product of this reaction? **(b)** What new bond is responsible for the IR absorption at 2580 cm^{-1}? **(c)** Using Hooke's law, calculate the approximate expected position of this new band from the structure of the original molecule and its IR spectrum. Assume that *k* and *f* have not changed.

8. Write the expected product(s) of each of the following reactions.

(a) $CH_3CH_2C(CH_3)(Cl)CHCH_2Cl$ $\xrightarrow{\text{2 NaNH}_2\text{, liquid NH}_3}$ (CH₃ and Cl on C3)

(b) $CH_3OCH_2CH_2CH_2CH(Br)CH(Br)CH_3$ $\xrightarrow{\text{2 NaNH}_2\text{, liquid NH}_3}$

(c) *meso*-$CH_3CH(CH_3)CH_2CH(Cl)CH(Cl)CH_2CH(CH_3)CH_3$ $\xrightarrow{\text{NaOCH}_3\text{ (1 equivalent), CH}_3\text{OH}}$

(d) (4*R*,5*R*)-$CH_3CH(CH_3)CH_2CH(Cl)CH(Cl)CH_2CH(CH_3)CH_3$ $\xrightarrow{\text{NaOCH}_3\text{ (1 equivalent), CH}_3\text{OH}}$

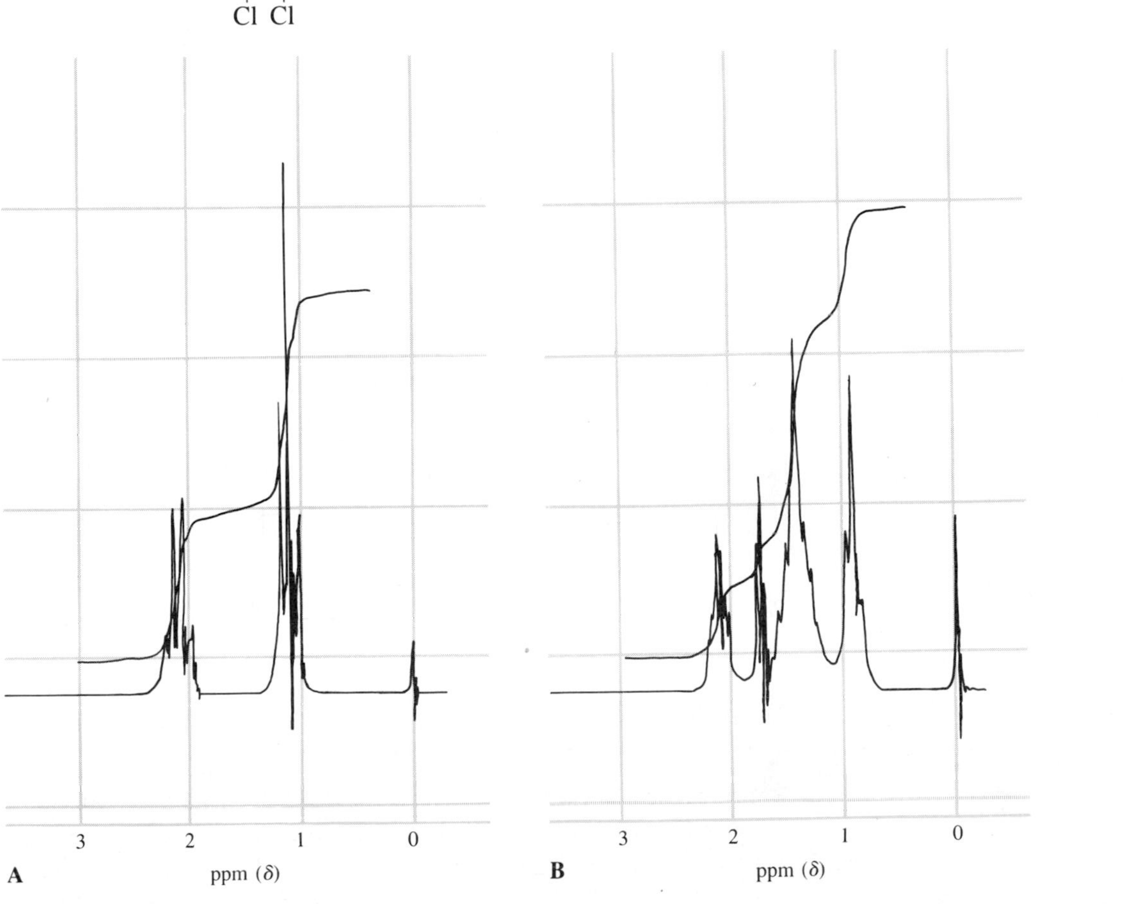

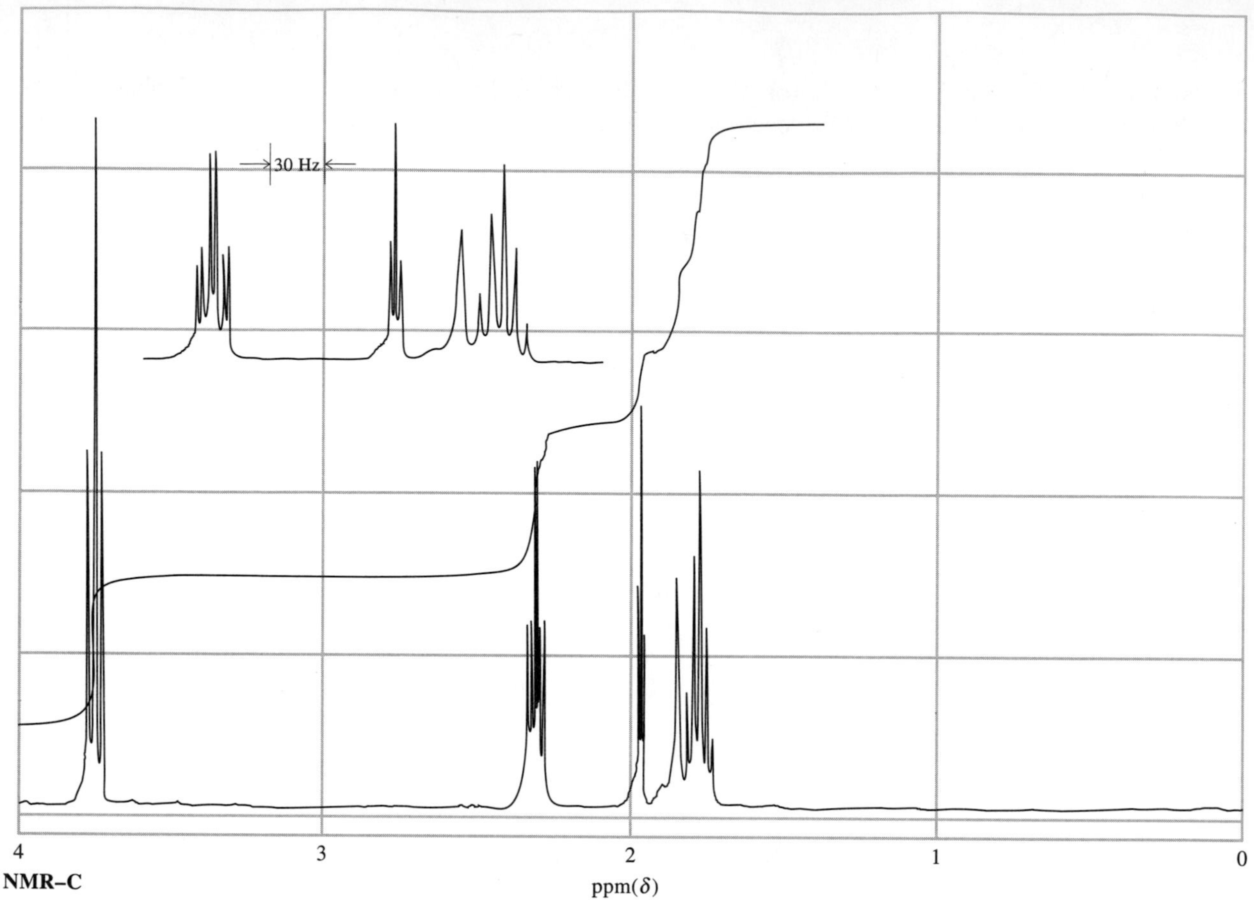

30 Hz

4 3 2 1 0

NMR–C

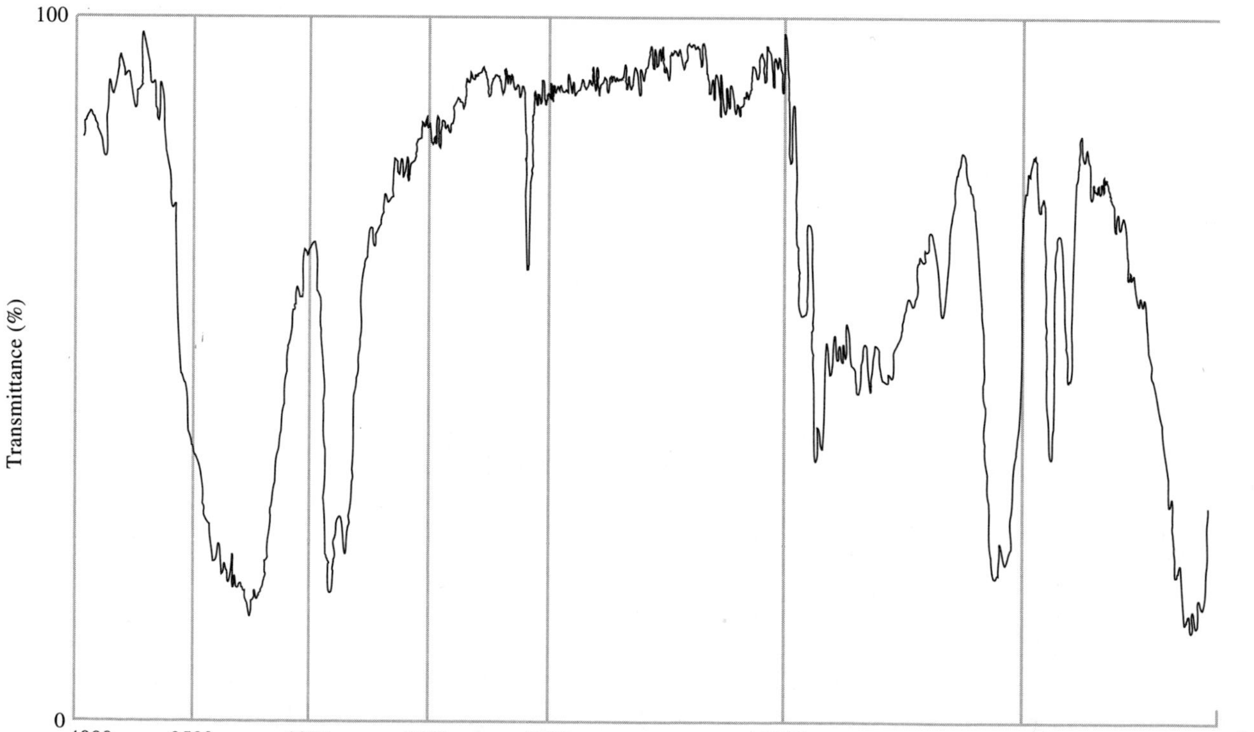

ppm(δ)

100

Transmittance (%)

0

4000 3500 3000 2500 2000 1500 1000 600 cm^{-1}

IR–C

Wavenumber

9. Write the expected major product of reaction of 1-propynyllithium, $CH_3C{\equiv}C^-Li^+$, with each of the following molecules in THF.

(a) CH_3CH_2Br **(b)** $CH_3CH(CH_3)CHClCH(CH_3)_2$

(c) Cyclohexanone **(d)** 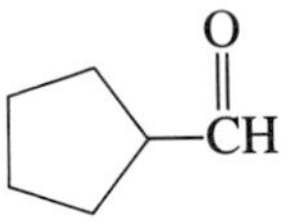 cyclopentyl–CHO (cyclopentanecarbaldehyde)

(e) 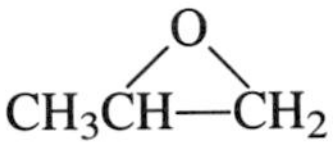 $CH_3CH\text{—}CH_2$ (epoxide, O bridging the two carbons) **(f)** 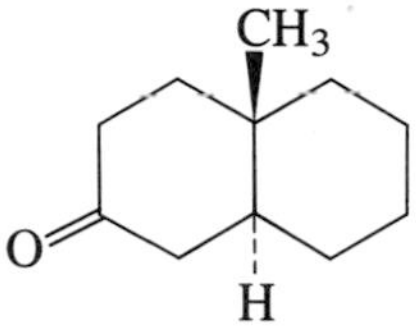 bicyclic ketone with angular CH_3 (wedge) and H (dashed)

10. Propose reasonable syntheses of each of the following alkynes, using the principles of retrosynthetic analysis. Each alkyne functional group in your synthetic target should come from a *separate* molecule, which may be any two-carbon compound (e.g., ethyne, ethene, ethanal).

(a) $CH_3CH_2C{\equiv}CCH_2CH_2CH_3$ **(b)** $CH_3CH_2C(CH_3)(OH)C{\equiv}CH$

(c) $HC{\equiv}CCH_2CH(OH)CH_3$ **(d)** $(CH_3)_3CC{\equiv}CH$ (Be careful! What is wrong with $(CH_3)_3CCl + {}^-{:}C{\equiv}CH$?)

11. Draw the structure of (*R*)-4-deuterio-2-hexyne. Propose an efficient strategy for the synthesis of this compound.

12. Give the expected major product of the reaction of propyne with each of the following reagents. **(a)** D_2, Pd-$BaSO_4$, quinoline; **(b)** Na, ND_3; **(c)** 1 equivalent HI; **(d)** 2 equivalents HI; **(e)** 1 equivalent Br_2; **(f)** 1 equivalent ICl; **(g)** 2 equivalents ICl; **(h)** H_2O, $HgSO_4$, H_2SO_4; **(i)** dicyclohexylborane, then NaOH, H_2O_2.

13. What are the products of the reactions of dicyclohexylethyne with the reagents in Problem 12?

14. Give the products of the reactions of your first two answers to Problem 13 with each of the following reagents. **(a)** H_2, Pd-C, CH_3CH_2OH; **(b)** Br_2, CCl_4; **(c)** BH_3, THF, then NaOH, H_2O_2; **(d)** MCPBA, CH_2Cl_2; **(e)** cold $KMnO_4$, H_2O.

15. Give the products of each reaction or reaction sequence.

(a) $(CH_3)_2CHBr + 2\ Li \xrightarrow{(CH_3CH_2)_2O}$

(b) Product of (a) (2 equivalents) + CuI $\longrightarrow$

(c) Product of (b) + $CH_3CH(CH_3)CH_2CH_2Br \longrightarrow$

(d) $CH_3CH_2CH_2C{\equiv}CCH_2CH_2CH_3$ + HBr (1 equivalent) + Br^- (excess) $\longrightarrow$

(e) Product of (d) $\xrightarrow[\text{2. CuI (0.5 equivalents)}]{\text{1. Li, }(CH_3CH_2)_2O}$

(f) Product of (e) + $H_2C{=}CHCH_2Cl \longrightarrow$

16. Propose reasonable syntheses of each of the following molecules, using an alkyne at least once in each synthesis.

(a) $CH_3CH_2CBr(Cl)CH_3$

(b) $CH_3CH_2CH_2CH_2CI_2CH_3$

(c) *meso*-2,3-Dibromobutane

(d) Racemic mixture of (2*R*,3*R*)- and (2*S*,3*S*)-2,3-dibromobutane

(e) Fischer projection: top CH_3; Br—C—Cl; H—C—Cl; bottom CH_3

(f) $CH_3(CH_2)_2C(=O)(CH_2)_3CH_3$

(g) $HOCH_2CH_2CH(OH)CH_3$

(h) cyclopentyl—$CH_2C(=O)$—H

(i) 1-vinylcyclohexene

(j) vinylcyclohexane

17. Propose a reasonable structure for calcium carbide, CaC_2, on the basis of its chemical reactivity (Section 13-11). What might be a more systematic name for it?

18. Propose *two different* syntheses of linalool, a terpene found in cinnamon, sassafras, and orange flower oils. Start with the eight-carbon ketone shown below and use ethyne as your source of the necessary additional two carbons in both syntheses.

(6-methyl-5-hepten-2-one) $\xrightarrow{?}$ Linalool (OH)

Linalool

19. The synthesis of chamaecynone, the essential oil of the Benihi tree, requires the conversion of a chloroalcohol into an alkynyl ketone. Propose a synthetic strategy to accomplish this task.

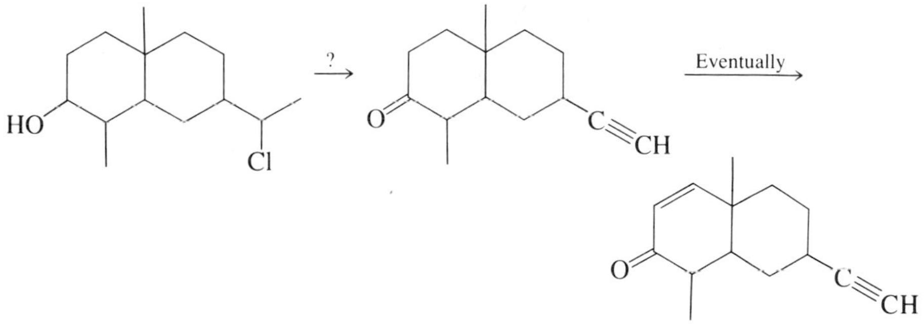

HO Cl ? O C≡CH Eventually O C≡CH

Chamaecynone

20. Synthesis of the sesquiterpene bergamotene proceeds from the alcohol shown below. Suggest a sequence to complete the synthesis.

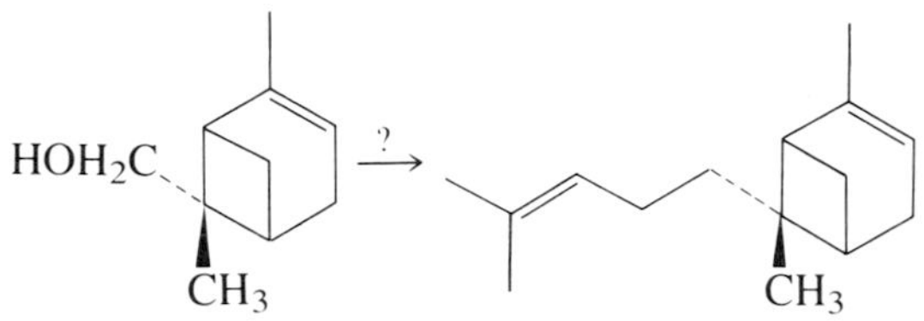

HOH_2C CH_3 ? CH_3

Bergamotene

21. An unknown molecule displays ^{1}H NMR and IR spectra D (on the next page). Reaction with H_2 in the presence of the Lindlar catalyst gives a compound that, after ozonolysis and treatment with Zn in aqueous acid, gives rise to one equivalent of $CH_3\overset{\overset{O}{\|}}{C}\overset{\overset{O}{\|}}{C}H$ and two of $H\overset{\overset{O}{\|}}{C}H$. What was the structure of the original molecule?

22. Formulate a plausible mechanism for the hydration of ethyne in the presence of mercuric chloride. (Hint: Review the hydration of alkenes catalyzed by mercuric ion, Section 12-7.)

23. A synthesis of the sesquiterpene farnesol requires the conversion of a dichloro compound into an alkynol. Suggest a way of achieving this transformation. (Hint: Devise a conversion of the starting compound into a terminal alkyne.)

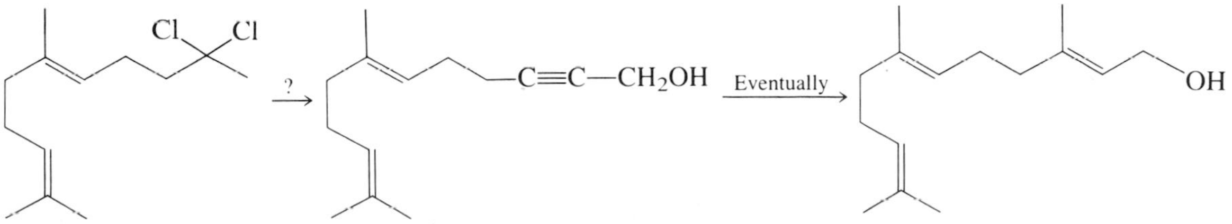

Cl Cl ? C≡C—CH_2OH Eventually OH

Farnesol

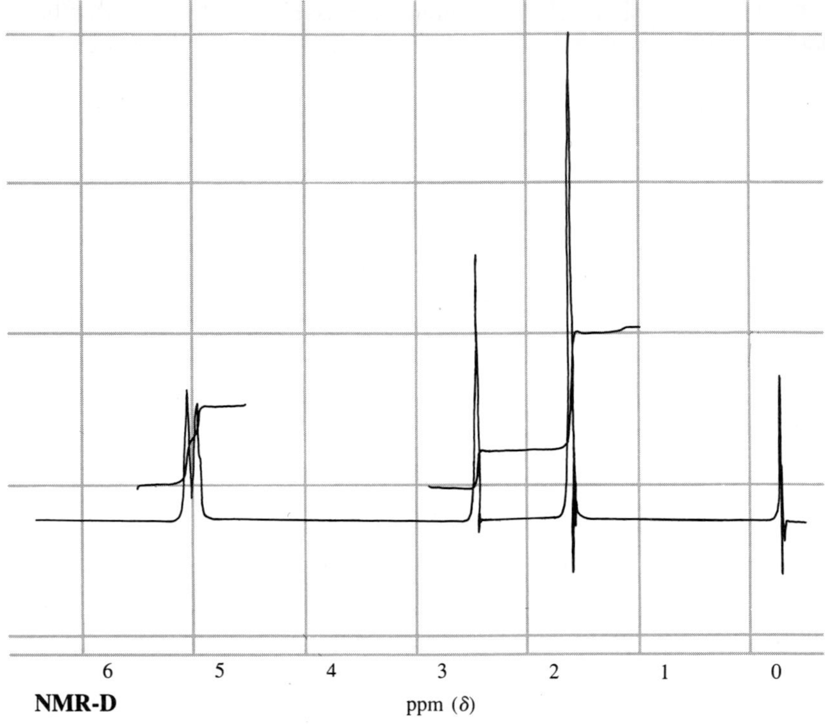

100

Transmittance (%)

0

4000 3500 3000 2500 2000 1500 1000 600 cm^{-1}

IR–D

Wavenumber

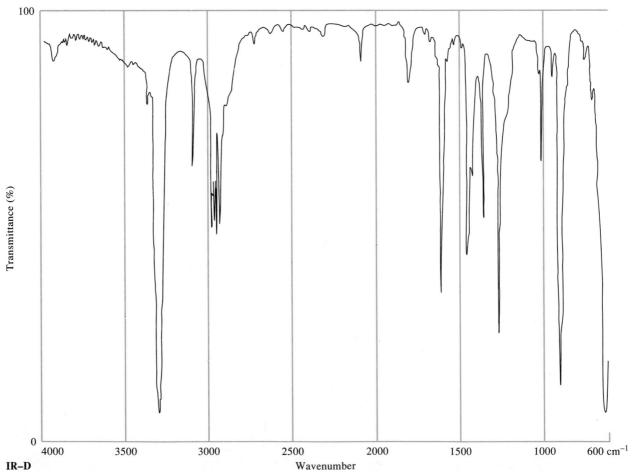

14

Delocalized Pi Systems

Investigation by Ultraviolet and Visible Spectroscopy

The last three chapters pointed out the importance of the overlap between two adjacent, parallel *p* orbitals. This interaction releases energy and gives rise to a π bond, a new center of reactivity. If two overlapping *p* orbitals are energetically advantageous, are *three or more* such interactions better? This chapter provides an answer: Compounds with three or more contiguous *p* orbitals may be stabilized by extended overlap. The *p* electrons are shared freely by several atomic centers: They are **delocalized.**

Our discussion begins with the 2-propenyl system—also called allyl—containing three interacting *p* orbitals. We then proceed to compounds that contain *several* adjacent double bonds, including both dienes and more extended π systems. These molecules undergo not only the usual transformations of alkenes, modified by the special situation of multiple overlap, but also unique thermal and photochemical cycloadditions and ring closures. Next we discuss the cyclic triene *benzene,* which is unusually stable and unreactive and is at the heart of much organic and bioorganic chemistry. (We shall learn much more about benzene's chemistry in later chapters.) Finally, we introduce the technique of ultraviolet and visible spectroscopy for probing the extent of π delocalization.

14-1 Overlap of Three Adjacent *p* Orbitals: Resonance in the 2-Propenyl (Allyl) System

What is the effect of a neighboring double bond on the reactivity of a carbon center? Three key observations answer this question.

Dissociation Energies of Various C–H Bonds

$CH_2{=}CHCH_2{-}H$
$DH° = 87$ kcal mol^{-1}

$(CH_3)_3C{-}H$
$DH° = 93$ kcal mol^{-1}

$(CH_3)_2CH{-}H$
$DH° = 94.5$ kcal mol^{-1}

$CH_3CH_2{-}H$
$DH° = 98$ kcal mol^{-1}

OBSERVATION 1. The primary carbon–hydrogen bond in propene is relatively weak, only 87 kcal mol^{-1}.

$$H_2C{=}C(H)CH_2H \longrightarrow H_2C{=}C(H)CH_2\cdot + H\cdot \qquad DH° = 87 \text{ kcal mol}^{-1}$$

Propene → **2-Propenyl radical**

A comparison with the values found for other hydrocarbons (see margin) shows that it is even weaker than a tertiary C–H bond. *Evidently, the 2-propenyl radical enjoys some type of special stability.*

OBSERVATION 2. 3-Chloropropene dissociates relatively fast under S_N1 (solvolysis) conditions and undergoes rapid unimolecular substitution through a carbocation intermediate.

$$H_2C{=}C(H)CH_2Cl \xrightarrow[-\,Cl^-]{CH_3OH,\ \Delta} H_2C{=}C(H)CH_2^+ \xrightarrow[-\,H^+]{CH_3OH} H_2C{=}C(H)CH_2OCH_3$$

3-Chloropropene → **2-Propenyl cation** → **3-Methoxypropene** (S_N1 product)

This finding clearly contradicts our expectations (recall Section 7-5). *It appears that the cation derived from 3-chloropropene is somehow more stable than other primary carbocations.* By how much? The ease of formation of the 2-propenyl cation in solvolysis reactions has been found to be roughly equal to that of a secondary carbocation.

OBSERVATION 3. The pK_a of propene is about 40.

$$H_2C{=}C(H)CH_2H \underset{K\,\sim\,10^{-40}}{\rightleftharpoons} H_2C{=}C(H)CH_2^- + H^+$$

2-Propenyl anion

Thus, propene is considerably more acidic than propane (p$K_a \sim 50$), and *the formation of the propenyl anion by deprotonation appears unusually favored.*

How can we explain these three observations?

Resonance stabilizes 2-propenyl (allyl) intermediates

Each of the preceding three processes generates a reactive carbon center—a radical, a carbocation, or a carbanion, respectively—that is adjacent to the π framework of a double bond. This arrangement seems to impart special stability.

Why? The reason is resonance and the resulting electron delocalization: Each species may be described by a pair of equivalent contributing resonance structures. These three-carbon intermediates have been given the name **allyl** (followed by the appropriate term: radical, cation, or anion). The activated carbon is labeled **allylic.**

Remember that resonance forms are *not* isomers but partial molecular representations. The true structure (the resonance hybrid) is derived by their superposition, better represented by the dotted-line drawings at the right of the classical picture.

Resonance in the 2-Propenyl (Allyl) System

$$[CH_2{=}CH{-}\dot{C}H_2 \longleftrightarrow \dot{C}H_2{-}CH{=}CH_2] \quad \text{or} \quad H_2C\overset{\overset{H}{|}}{\cdots C \cdots}CH_2 \;(\cdot)$$

Radical

$$[CH_2{=}CH{-}\overset{+}{C}H_2 \longleftrightarrow \overset{+}{C}H_2{-}CH{=}CH_2] \quad \text{or} \quad H_2C\overset{\overset{H}{|}}{\cdots C \cdots}CH_2 \;(+)$$

Cation

$$[CH_2{=}CH{-}\ddot{C}H_2^{-} \longleftrightarrow {}^{-}\ddot{C}H_2{-}CH{=}CH_2] \quad \text{or} \quad H_2C\overset{\overset{H}{|}}{\cdots C \cdots}CH_2 \;(-)$$

Anion

The 2-propenyl (allyl) pi system is represented by three molecular orbitals

The stabilization of the 2-propenyl (allyl) system by resonance can also be described in terms of molecular orbitals. Each of the three carbons is sp^2 hybridized and bears a *p* orbital perpendicular to the molecular plane (Figure 14-1). *Make a model: The structure is symmetric, with equal C–C bond lengths.*

Ignoring the σ framework, we can combine the three *p* orbitals mathematically to give three π molecular orbitals. This process is analogous to mixing two atomic orbitals to give two molecular orbitals describing a π bond (Figures 11-1 and 11-3), except that there is now a third atomic orbital. Of the three resulting molecular orbitals, one (π_1) is bonding and has no nodes, one (π_2) is *nonbonding* (in other words, it has the same energy as a noninteracting *p* orbital) and has one node, and one (π_3) is antibonding, with two nodes, as shown in Figure 14-2.

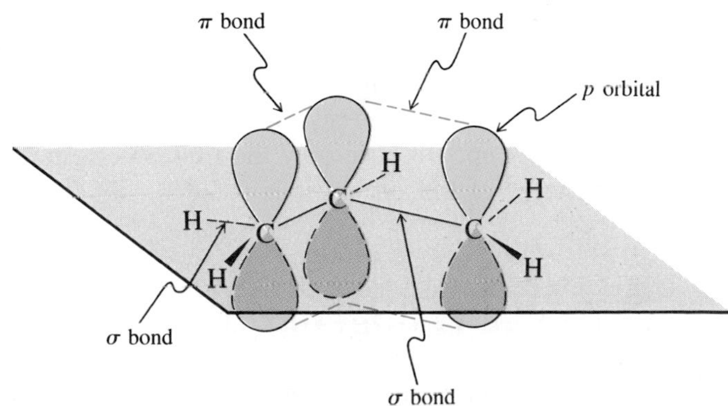

FIGURE 14-1 The three *p* orbitals in the 2-propenyl (allyl) group overlap, giving a symmetric structure with delocalized electrons. The σ framework is shown as black lines.

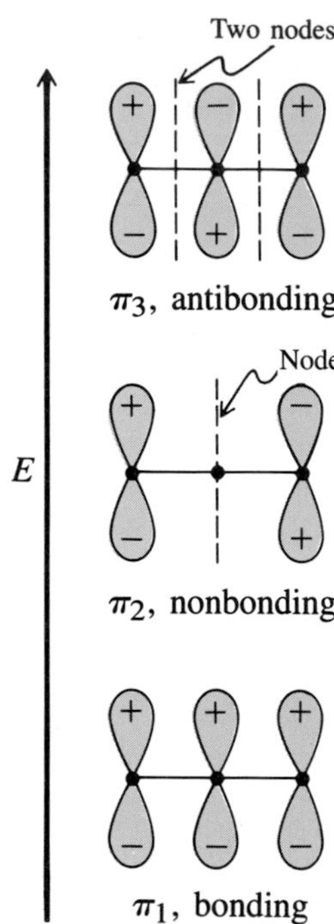

FIGURE 14-2 The three π molecular orbitals of 2-propenyl (allyl), obtained by combining the three adjacent atomic p orbitals.

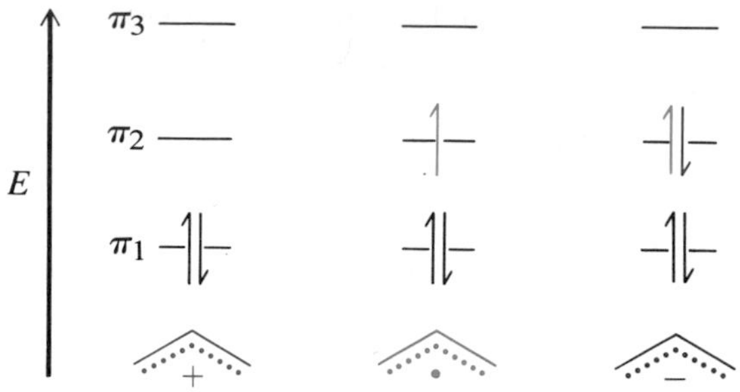

FIGURE 14-3 The Aufbau principle is used to fill up the π molecular orbitals of 2-propenyl (allyl) cation, radical, and anion. In each case the total energy of the π electrons is lower than that of three noninteracting p orbitals. Partial cation, radical, or anion character is present at the end carbons in these systems, a result of the location of the lobes in the π_2 molecular orbital.

We can use the Aufbau principle to fill in the appropriate number of π electrons (Figure 14-3). The cation, with a total of two, contains only one filled orbital, π_1. For the radical and the anion, we place one or two electrons, respectively, into the second molecular orbital, π_2. In all cases, the total π-electron energy of the system is lower (more favorable) than that expected from three noninteracting p orbitals—essentially because π_1 is greatly stabilized and filled in all cases, whereas the antibonding level, π_3, stays empty throughout.

The resonance formulations for the three 2-propenyl species indicate that it is mainly the two *terminal* carbons that accommodate the charges in the ions or the odd electron in the radical. The molecular-orbital picture is consistent with this view: The three structures differ only in the number of electrons present in molecular orbital π_2, which possesses a node passing through the central carbon; therefore, very little of the electron excess or deficiency will show up at this position.

Partial Electron Density Distribution in the 2-Propenyl (Allyl) System

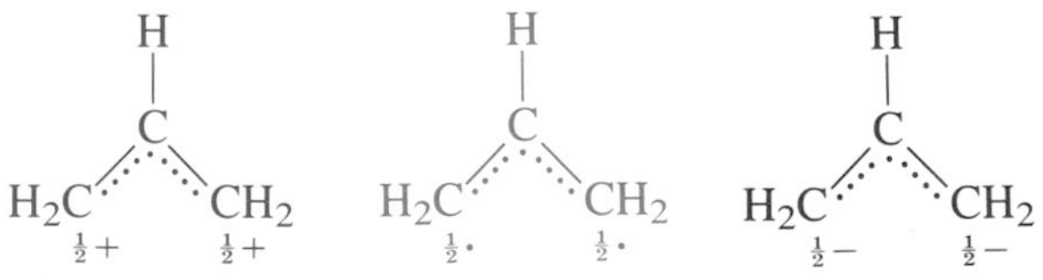

In summary, allylic radicals, cations, and anions are unusually stable. In Lewis terms, this stabilization is readily explained by resonance. In a molecular-orbital description, the three interacting p orbitals form three new molecular orbitals: One is considerably lower in energy than the p level, another one stays the same, and a third moves up. Because only the first two are populated with electrons, the total π energy of the system is lowered.

14-2 Radical Allylic Halogenation

A consequence of delocalization is that resonance-stabilized allylic intermediates can readily participate in reactions of unsaturated molecules. For example, although halogens can add to alkenes to give the corresponding vicinal dihalides (Section 12-5), the course of this reaction is changed when the halogen is present only in low concentrations. These conditions favor radical chain mechanisms and lead to **radical allylic substitution.***

Radical Allylic Halogenation

$$CH_2{=}CHCH_3 \xrightarrow{X_2 \text{ (low conc.)}} CH_2{=}CHCH_2X + HX$$

A reagent frequently used in allylic brominations in the laboratory is *N*-bromobutanimide (*N*-bromosuccinimide, NBS, Section 3-8) suspended in tetrachloromethane. This species is nearly insoluble in CCl_4 and is a steady source of very small amounts of bromine formed by reaction with traces of HBr.

NBS as a Source of Bromine

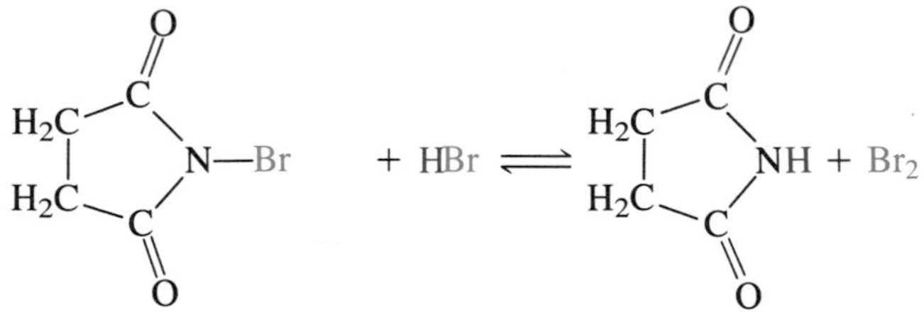

$$\text{NBS} + HBr \rightleftharpoons \text{butanimide (NH)} + Br_2$$

***N*-Bromobutanimide (*N*-Bromosuccinimide, NBS)**

Butanimide

Example:

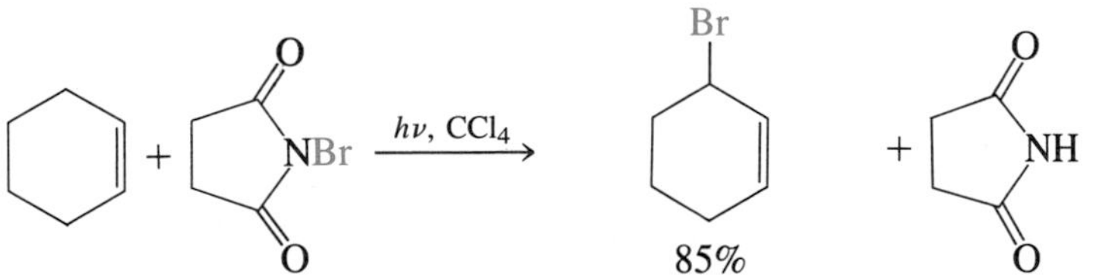

$$\text{cyclohexene} + \text{NBS} \xrightarrow{h\nu,\ CCl_4} \text{3-bromocyclohexene} + \text{butanimide}$$

85%

3-Bromocyclohexene

The bromine reacts with the alkene by a radical chain mechanism (Section 3-4). The process is initiated by light or by traces of radical initiators that cause dissociation of Br_2 into bromine atoms. Propagation of the chain involves abstraction of a weakly bound allylic hydrogen by Br·.

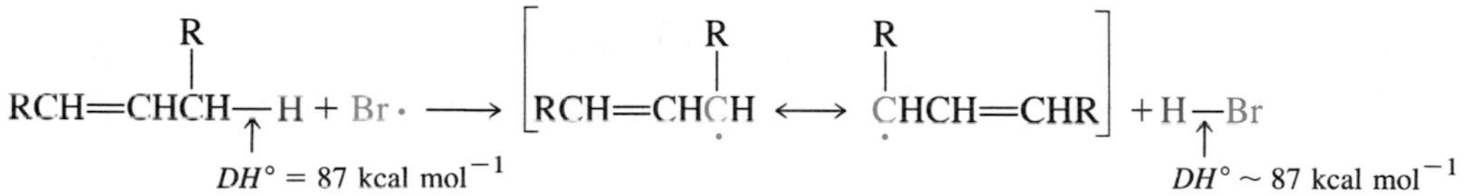

$$RCH{=}CHCH(R){-}H + Br\cdot \longrightarrow \left[RCH{=}CH\dot{C}HR \longleftrightarrow R\dot{C}HCH{=}CHR\right] + H{-}Br$$

$DH^\circ = 87\ \text{kcal mol}^{-1}$ (C–H bond); $DH^\circ \sim 87\ \text{kcal mol}^{-1}$ (H–Br bond)

The resonance-stabilized radical may then react with Br_2 at either end of the allylic system to furnish an allylic bromide and regenerate Br·, which continues the chain. Thus, alkenes that form unsymmetric allylic radicals can give mixtures of products upon treatment with NBS. For example,

*The explanation for this change requires a detailed kinetic analysis that is beyond the scope of this book. Suffice it to say that at low bromine concentrations the competing addition processes are reversible, and substitution wins out.

$$CH_2{=}CH(CH_2)_5CH_3 \xrightarrow[-\ HBr]{NBS,\ h\nu} [CH_2{=}CH\dot{C}H(CH_2)_4CH_3 \longleftrightarrow \cdot CH_2CH{=}CH(CH_2)_4CH_3]$$

1-Octene

$$\longrightarrow \underset{\substack{17\% \\ \textbf{3-Bromo-1-octene}}}{CH_2{=}CHCH(Br)(CH_2)_4CH_3} + \underset{\substack{44\% \\ \textbf{1-Bromo-2-octene}}}{BrCH_2CH{=}CH(CH_2)_4CH_3}$$

EXERCISE 14-1

Ignoring stereochemistry, give all the isomeric bromoheptenes that are possible from NBS treatment of *trans*-2-heptene.

Allylic chlorinations are important in industry because chlorine is relatively cheap. For example, 3-chloropropene (allyl chloride) is made commercially by the gas-phase chlorination of propene at 400°C.

$$CH_3CH{=}CH_2 + Cl_2 \xrightarrow{400°C} \underset{\substack{\textbf{3-Chloropropene} \\ \textbf{(Allyl chloride)}}}{ClCH_2CH{=}CH_2} + HCl$$

EXERCISE 14-2

Predict the outcome of the allylic bromination of the following substrates with NBS (1 equivalent).

(a) Cyclohexene **(b)** [bicyclic structure: two fused cyclohexane rings sharing a central double bond] **(c)** 1-Methylcyclohexene

The biochemical degradation of unsaturated molecules frequently involves radical abstraction of allylic hydrogens by oxygen-containing species. Such processes will be discussed in Chapter 22.

To summarize, under radical conditions, alkenes containing allylic bonds enter into allylic halogenation. A particularly good reagent for allylic bromination is *N*-bromobutanimide (*N*-bromosuccinimide, NBS).

14-3 Nucleophilic Substitution of Allylic Halides: Kinetic and Thermodynamic Control

As our example of 3-chloropropene in Section 14-1 shows, allylic halides dissociate readily to produce allylic cations. These can be trapped by nucleophiles at either end in S_N1 reactions. Allylic halides also readily undergo S_N2 transformations.

Allylic halides undergo S_N1 reactions

The ready dissociation of allylic halides has important chemical consequences. Different allylic halides may give identical products upon solvolysis if they dissociate to the same allylic cation. For example, the hydrolysis of either 1-chloro-

2-butene or 3-chloro-1-butene results in the same alcohol mixture. The reason is the intermediacy of the same allylic cation.

Hydrolysis of Isomeric Allylic Chlorides

$$CH_3CH{=}CHCH_2Cl \xrightarrow[-Cl^-]{} \left[\overset{4}{C}H_3\overset{3}{C}H{=}\overset{2}{C}H\overset{1}{C}H_2^{+} \longleftrightarrow \overset{4}{C}H_3\overset{3}{\overset{+}{C}}H\overset{2}{C}H{=}\overset{1}{C}H_2\right] \xleftarrow[-Cl^-]{} CH_3\underset{}{\overset{Cl}{C}}HCH{=}CH_2$$

1-Chloro-2-butene — **Allylic cation** — **3-Chloro-1-butene**

$$\downarrow HOH$$

$$CH_3CH{=}CHCH_2OH + CH_3\overset{OH}{C}HCH{=}CH_2 + H^+$$

Minor — Major

2-Buten-1-ol — **3-Buten-2-ol**

EXERCISE 14-3

Hydrolysis of (*R*)-3-chloro-1-butene gives exclusively racemic alcohols. Explain.

Interestingly, the major product from this hydrolysis reaction is 3-buten-2-ol. The terminal double bond is created, even though it is thermodynamically *less* favored. Why? The less stable isomer must be formed faster. We say that the reaction is under **kinetic control:** For short reaction times or at low temperatures, the outcome is determined kinetically.

At higher temperatures, the more stable product, 2-buten-1-ol, starts to predominate. The difference can be demonstrated by studying the equilibration of the two isomers. The one that is kinetically favored forms first, but its formation is reversible. If the acidic mixture is now heated, the alcohols start to regenerate the allylic cation, and the slower nucleophilic reaction leading to the thermodynamically more stable isomer has a chance to succeed. The reaction is under **thermodynamic control.**

Kinetic Compared with Thermodynamic Control

$$CH_3\overset{OH}{C}HCH{=}CH_2 + H^+ \underset{}{\overset{\text{Kinetic control (fast, irreversible at low temperature)}}{\rightleftharpoons}} \left[CH_3CH{=}CHCH_2^{+} \longleftrightarrow CH_3\overset{+}{C}HCH{=}CH_2\right] + HOH \overset{\text{Thermodynamic control (slow)}}{\rightleftharpoons} CH_3CH{=}CHCH_2OH + H^+$$

Less stable product, predominates when reaction time is short or at low temperatures

More stable product, predominates when reaction time is long or at higher temperatures

The potential-energy diagram in Figure 14-4 shows the lower activation barrier associated with the formation of the less stable product. At higher temperatures and longer reaction times, the (relatively slow) generation of the thermodynamically favored isomer becomes competitive.

Why does the formation of the less stable product have a lower activation barrier? The reason is that the allylic cation in this system is unsymmetric; its

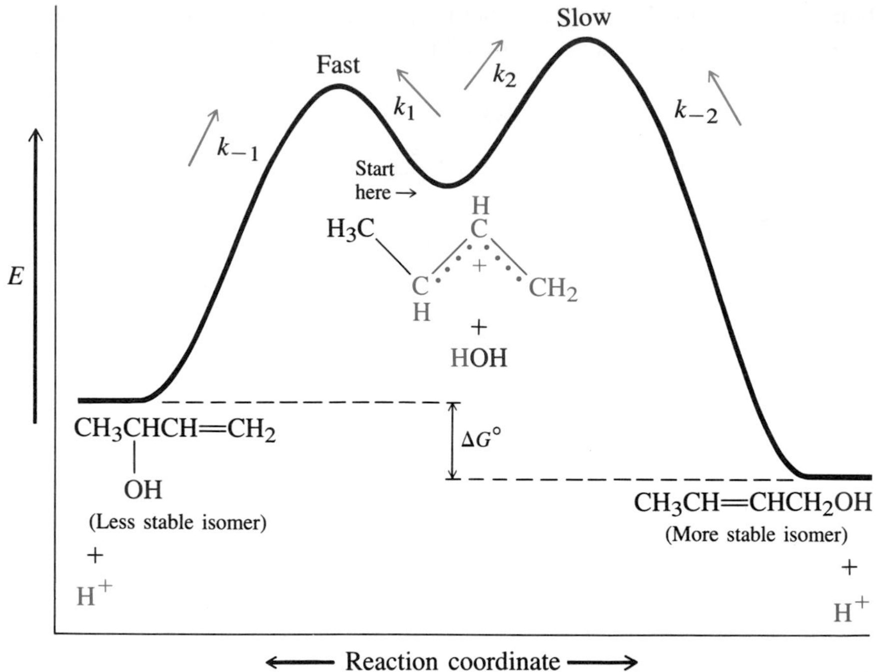

FIGURE 14-4 Kinetic control compared with thermodynamic control in the reaction of 1-methyl-2-propenyl cation with water. At low temperatures, the lower activation energy barrier leading to the formation of 3-buten-2-ol results in predominant formation of this isomer (i.e., $k_1 > k_2$). At higher temperatures, the reverse process (governed by k_{-1}) becomes competitive, regenerating the cation, which slowly forms more and more of the thermodynamic product, 2-buten-1-ol. (Alkyloxonium ion intermediates have been omitted from the diagram for clarity.)

positive charge is *unequally* distributed between carbons 1 and 3. More positive charge resides at C3, the more substituted carbon, a condition leading to faster attack by the nucleophile (in this case, water) at this position.

EXERCISE 14-4

Treatment of 3-buten-2-ol with cold hydrogen bromide gives 1-bromo-2-butene and 3-bromo-1-butene in a 15:85 ratio. After heating, this ratio changes to give mainly 1-bromo-2-butene. Explain.

Allylic halides can also undergo S_N2 reactions

S_N2 reactions of allylic halides are faster than those of the corresponding saturated haloalkanes. Overlap between the double bond and the *p* orbital in the transition state of the displacement (see Figure 6-4) stabilizes the latter, thereby producing a relatively low activation barrier.

The S_N2 Reactions of 3-Chloro-1-propene and 1-Chloropropane

Reaction	Relative rate
$CH_2{=}CHCH_2Cl + I^- \xrightarrow{\text{Propanone (acetone), 50°C}} CH_2{=}CHCH_2I + Cl^-$	73
$CH_3CH_2CH_2Cl + I^- \xrightarrow{\text{Propanone (acetone), 50°C}} CH_3CH_2CH_2I + Cl^-$	1

EXERCISE 14-5

The solvolysis of 3-chloro-3-methyl-1-butene in acetic acid at 25°C gives initially a mixture containing mostly the structurally isomeric chloride with some of the acetate. After a longer period of time, no allylic chloride remains, and the acetate is the only product present. Explain this result.

$$CH_3-\underset{CH_3}{\overset{Cl}{C}}-CH{=}CH_2 \xrightarrow{CH_3\overset{O}{\overset{\|}{C}}OH,\ CH_3\overset{O}{\overset{\|}{C}}O^{-}\,{}^{+}K} (CH_3)_2C{=}CH-CH_2-Cl + (CH_3)_2C{=}CH-CH_2-O\overset{O}{\overset{\|}{C}}CH_3$$

3-Chloro-3-methyl-1-butene → **1-Chloro-3-methyl-2-butene** + **3-Methyl-2-butenyl acetate**

In summary, allylic halides undergo both S_N1 and S_N2 reactions. At low temperatures (conditions favoring kinetic control), unimolecular substitution in unsymmetric systems results in nucleophilic attack at the more substituted allylic position, an outcome reflecting the greater presence of partial positive charge in the intermediate carbocation. Prolonged heating and longer reaction times lead to a predominance of the more stable substitution product (thermodynamic control). With good nucleophiles, allylic halides undergo S_N2 reaction more rapidly than the corresponding saturated substrates.

14-4 Allylic Organometallic Reagents: Useful Three-Carbon Nucleophiles

Propene is appreciably more acidic than propane because of the relative stability of the conjugated carbanion that results from deprotonation. Therefore, allylic lithium reagents can be made from propene derivatives by proton abstraction by an alkyllithium. The process is facilitated by *N,N,N′,N′*-tetramethylethane-1,2-diamine (tetramethylethylenediamine, TMEDA), a good solvating agent.

Allylic Deprotonation

$$CH_3CH_2CH_2CH_2Li + H_2C{=}C(CH_3)_2 \xrightarrow{(CH_3)_2NCH_2CH_2N(CH_3)_2\ (TMEDA)} H_2C{=}C(CH_3)CH_2Li + CH_3CH_2CH_2CH_2-H$$

An even more straightforward way of producing an allylic organometallic is Grignard formation. For example,

$$CH_2{=}CHCH_2Br \xrightarrow{Mg,\ THF,\ 0°C} CH_2{=}CHCH_2MgBr$$

3-Bromo-1-propene → **2-Propenylmagnesium bromide**

Like their alkyl counterparts, allylic lithium and Grignard reagents can function as nucleophiles.

$$CH_2{=}CHCH_2MgBr + CH_3\overset{\overset{\displaystyle O}{\|}}{C}CH_3 \xrightarrow{(CH_3CH_2)_2O} \xrightarrow{H^+,\ H_2O} CH_2{=}CHCH_2\underset{\underset{\displaystyle CH_3}{|}}{\overset{\overset{\displaystyle OH}{|}}{C}}CH_3$$

85%

2-Methyl-4-penten-2-ol

EXERCISE 14-6

Show how to accomplish the following conversion in as few steps as possible.

Cyclohexanone ---→ 1-hydroxycyclohexyl-CH_2CHO (HO, CH_2CHO)

In summary, alkenes tend to be deprotonated at the allylic position, a reaction resulting in the corresponding delocalized anions. Allylic Grignard reagents are made from the corresponding halides. Like their alkyl analogs, allyl organometallics function as nucleophiles.

14-5 Two Neighboring Double Bonds: Conjugated Dienes

Now that we have caught a glimpse of the consequences of delocalization over three atoms, it should be interesting to learn what happens if we go one step further. Let us consider the addition of a fourth *p* orbital, which results in two linked double bonds: a **conjugated diene** (*conjugatio,* Latin, union). In these compounds, delocalization again results in stabilization, as measured by heats of hydrogenation. It is also reflected in their molecular and electronic structures and in their chemistry.

The Simplest Conjugated and Nonconjugated Dienes

$CH_2{=}CH{-}CH{=}CH_2$

1,3-Butadiene

$CH_2{=}CHCH_2CH{=}CH_2$

1,4-Pentadiene

$CH_2{=}C{=}CH_2$

1,2-Propadiene (Allene)

Hydrocarbons with two double bonds are named dienes

Conjugated dienes have to be contrasted with their **nonconjugated** isomers, in which the two double bonds are separated by saturated carbons, and the **allenes** (or **cumulated** dienes), in which the π bonds share a single carbon (Figure 14-5).

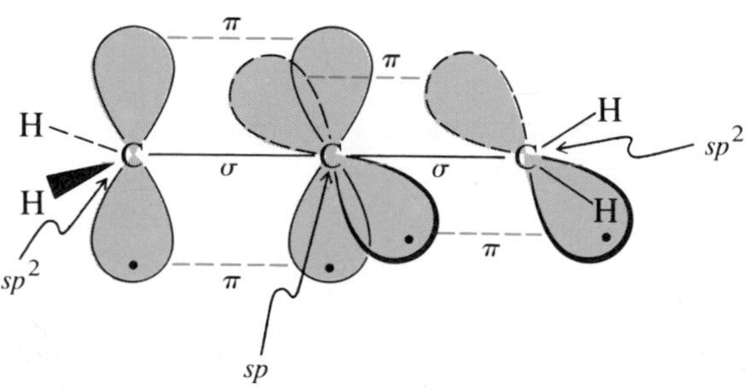

FIGURE 14-5 The two π bonds of an allene share a single carbon and are perpendicular to each other.

The names of conjugated and nonconjugated dienes are derived from those of the alkenes in a straightforward manner. The longest chain incorporating both double bonds is found, then numbered to indicate the positions of the functional and substituent groups. If necessary, cis-trans or E,Z prefixes indicate the geometry around the double bonds. Cyclic dienes are named accordingly.

***trans*-1,3-Pentadiene**

***cis*-2-*trans*-4-Heptadiene**

(Z)-4-Bromo-1,3-pentadiene

***cis*-1,4-Heptadiene**
(A nonconjugated diene)

1,3-Cyclohexadiene

1,4-Cycloheptadiene
(A nonconjugated cyclic diene)

EXERCISE 14-7

Suggest names or draw structures, as appropriate, for the following compounds.

(c) *cis*-3,6-Dimethyl-1,4-cyclohexadiene **(d)** *cis,cis*-1,4-Dibromo-1,3-butadiene

Conjugated dienes are more stable than nonconjugated dienes

The preceding sections noted that delocalization of electrons makes the allylic system especially stable. Does a conjugated diene have the same property? If so, that stability should be manifest in its heat of hydrogenation. We know that the heat of hydrogenation of a terminal alkene such as 1-hexene is about -30 kcal mol^{-1} (see Section 11-7). A compound containing two *noninteracting* terminal double bonds should exhibit a heat of hydrogenation roughly twice this value, about -60 kcal mol^{-1}. Indeed, catalytic hydrogenation of either 1,5-hexadiene or 1,4-pentadiene releases just about that amount of energy.

Heats of Hydrogenation of Nonconjugated Alkenes

$$CH_3CH_2CH{=}CH_2 + H_2 \xrightarrow{Pt} CH_3CH_2CH_2CH_3 \qquad \Delta H^\circ = -30.3 \text{ kcal mol}^{-1}$$

$$CH_2{=}CHCH_2CH_2CH{=}CH_2 + 2\,H_2 \xrightarrow{Pt} CH_3(CH_2)_4CH_3 \qquad \Delta H^\circ = -60.5 \text{ kcal mol}^{-1}$$

$$CH_2{=}CHCH_2CH{=}CH_2 + 2\,H_2 \xrightarrow{Pt} CH_3(CH_2)_3CH_3 \qquad \Delta H^\circ = -60.8 \text{ kcal mol}^{-1}$$

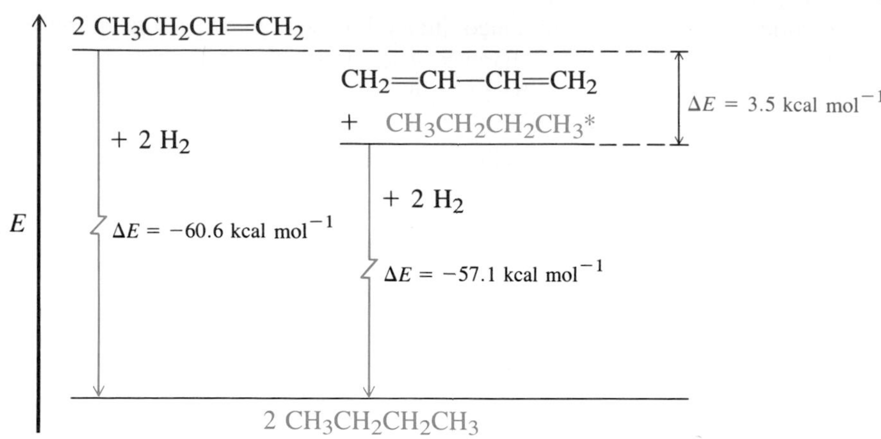

FIGURE 14-6 The heats of hydrogenation reveal the relative stability of two molecules of 1-butene (a monoalkene) and one of 1,3-butadiene (a conjugated diene). The difference, about 3.5 kcal mol^{-1}, reflects the stabilization of the latter due to conjugation. [*To make the two processes energetically comparable, the heat content of one molecule of butane has to be added to that of the diene.]

When the same experiment is carried out with the conjugated diene 1,3-butadiene, *less* energy is produced.

Heat of Hydrogenation of 1,3-Butadiene

$$CH_2{=}CH{-}CH{=}CH_2 + 2\ H_2 \xrightarrow{Pt} CH_3CH_2CH_2CH_3 \qquad \Delta H^\circ = -57.1 \text{ kcal mol}^{-1}$$

The difference, about 3.5 kcal mol^{-1}, is due to a stabilizing interaction between the two double bonds, as illustrated in Figure 14-6. It is referred to as the **resonance energy** of 1,3-butadiene.

EXERCISE 14-8

The heat of hydrogenation of *trans*-1,3-pentadiene is 54.2 kcal mol^{-1}, 6.6 kcal mol^{-1} less than that of 1,4-pentadiene, even lower than that expected from the resonance energy of 1,3-butadiene. Explain. (Hint: See Section 11-7.)

Conjugation in 1,3-butadiene results from overlap of the pi bonds

How do the two double bonds in 1,3-butadiene interact? The answer lies in the alignment of their π systems, an arrangement that permits the *p* orbitals on C2 and C3 to overlap (Figure 14-7A). The π interaction that results is weak but nevertheless amounts to a few kilocalories per mole.

Besides adding stability to the diene, this π interaction also raises the barrier to rotation about the single bond. An inspection of models shows that the molecule can adopt two possible extreme coplanar conformations. In one, designated **s-cis,** the two π bonds lie on the same side of the C2–C3 axis; in the other, called **s-trans,** the π bonds are on opposite sides (Figure 14-7B). The prefix *s* refers to the fact that the bridge between C2 and C3 constitutes a *single* bond. The *s*-cis

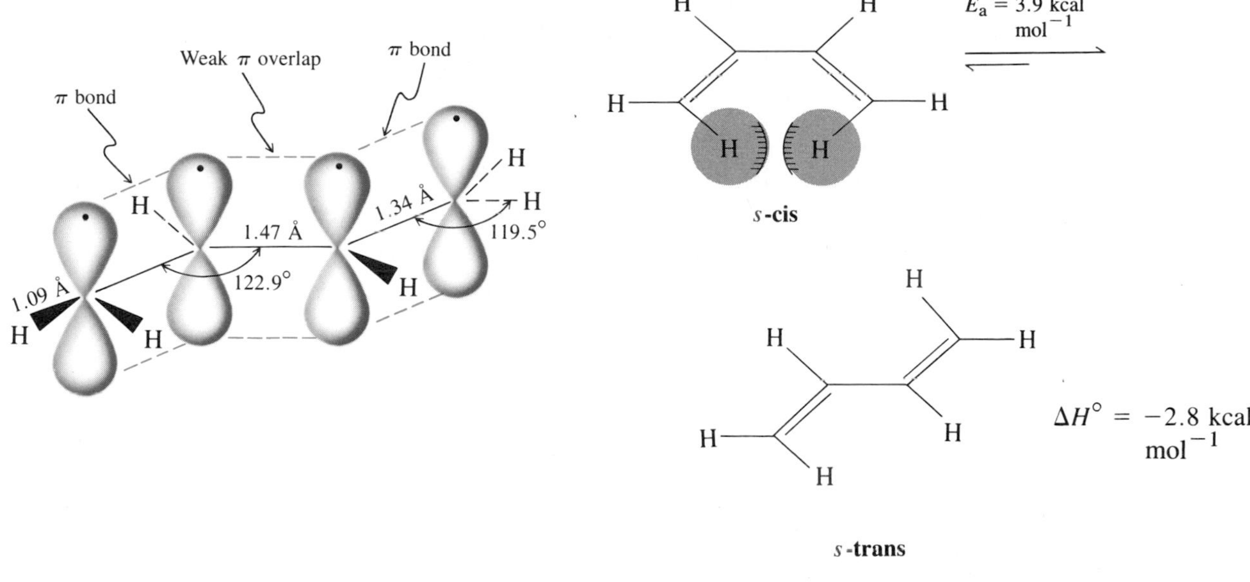

FIGURE 14-7 (A) The structure of 1,3-butadiene. The central bond is shorter than that in an alkane (1.54 Å for the central C–C bond in butane). The *p* orbitals aligned perpendicular to the molecular plane form a contiguous interacting array. (B) 1,3-Butadiene can exist in two planar conformations.

form is almost 3 kcal mol^{-1} less stable than the *s*-trans conformation because of the steric interference between the two hydrogens on the inside of the diene unit.*

EXERCISE 14-9

The dissociation energy of the central C–H bond in 1,4-pentadiene is only 71 kcal mol^{-1}. Explain.

The π-electronic structure of 1,3-butadiene may be described by constructing four molecular orbitals out of the four *p* atomic orbitals (Figure 14-8). Because the four π electrons occupy solely the two bonding levels, the energy of the system is lower than that of four noninteracting *p* orbitals.

In summary, dienes are named according to the rules formulated for ordinary alkenes. Conjugated dienes are more stable than dienes containing two isolated double bonds, as measured by their heats of hydrogenation; the difference is their resonance energy. Conjugation is reflected in the molecular structure of 1,3-butadiene, revealing a relatively short central carbon–carbon bond with a small barrier to rotation of about 4 kcal mol^{-1}. The two rotamers *s*-trans and *s*-cis differ in energy by about 3 kcal mol^{-1}. The molecular-orbital picture of the π system in 1,3-butadiene shows two bonding and two antibonding orbitals. The four electrons are placed in the first two bonding levels.

---

*The *s*-cis conformation is very close in energy to a nonplanar conformation in which the two double bonds are *gauche* (Section 2-7). Whether the *s*-cis or the *gauche* conformation is more stable remains a subject of controversy.

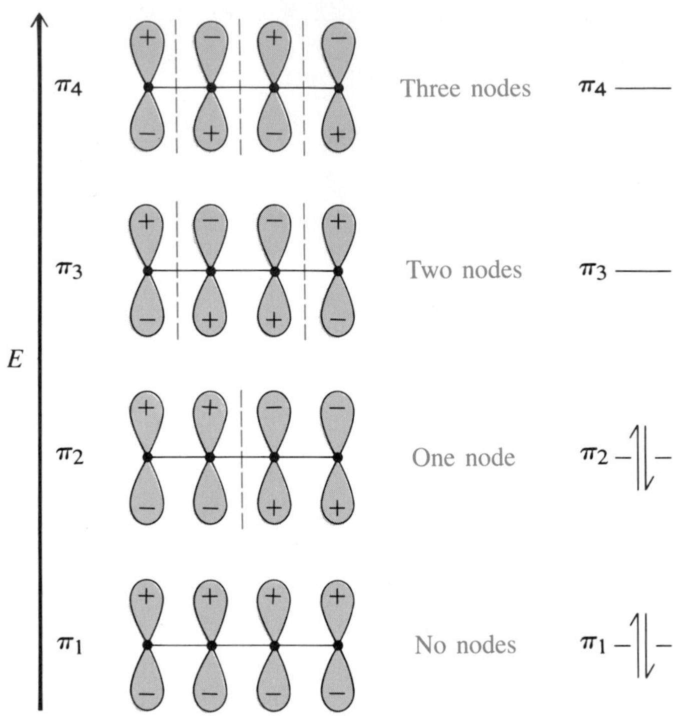

FIGURE 14-8 π-Molecular-orbital description of 1,3-butadiene. Its four electrons are placed into the two lowest π (bonding) orbitals, π_1 and π_2.

14-6 Electrophilic Attack on Conjugated Dienes

Does the structure of conjugated dienes affect their reactivity? These molecules are centers of high electron density because of the presence of the π electrons. In fact, although more stable thermodynamically than dienes with isolated double bonds, conjugated dienes are actually *more reactive* kinetically in the presence of electrophiles and other reagents. 1,3-Butadiene, for example, readily absorbs one mole of gaseous hydrogen chloride. Two isomeric addition products are formed: 3-chloro-1-butene and 1-chloro-2-butene.

$$CH_2{=}CH{-}CH{=}CH_2 + HCl \xrightarrow{25^\circ C} \underset{\substack{80\% \\ \textbf{3-Chloro-1-butene}}}{CH_2{=}CH{-}\overset{\overset{Cl}{|}}{C}H{-}CH_2H} + \underset{\substack{20\% \\ \textbf{1-Chloro-2-butene}}}{HCH_2{-}CH{=}CH{-}CH_2Cl}$$

The generation of the first product is readily understood in terms of ordinary alkene chemistry. It is the result of a Markovnikov addition to one of the double bonds. But what about the second product?

The presence of 1-chloro-2-butene is explained by the reaction mechanism. Initial protonation at C1 gives the thermodynamically most favored allylic cation.

Protonation of 1,3-Butadiene

$$\overset{+}{C}H_2-\underset{|}{\overset{H}{C}}H-CH=CH_2 \xleftarrow[\text{Attack at C2}]{H^+ \;\times} \overset{1}{C}H_2=\overset{2}{C}H-\overset{3}{C}H=\overset{4}{C}H_2 \xrightarrow[\text{Attack at C1}]{H^+}$$

Primary nondelocalized cation not formed

$$\left[CH_3-\overset{+}{C}H-CH=CH_2 \longleftrightarrow CH_3-CH=CH-\overset{+}{C}H_2\right]$$

Delocalized allylic cation formed exclusively

This cation can be trapped by chloride in two possible ways, to form the two observed products: At the terminal carbon, it yields 1-chloro-2-butene; and at the internal carbon, it furnishes 3-chloro-1-butene. The 1-chloro-2-butene is said to form by *1,4-addition* of hydrogen chloride to butadiene, because reaction has occurred at C1 and C4 of the original diene. The other product arises by the normal 1,2-addition. Many electrophilic additions to dienes give rise to product mixtures by both modes of addition.

Nucleophilic Trapping of the Allylic Cation Formed on Protonation of 1,3-Butadiene

$$\underset{\text{1,2-Addition}}{CH_3\underset{|}{\underset{Cl}{C}}HCH=CH_2} \xleftarrow[\text{Attack at internal carbon}]{Cl^-} CH_3CH\overset{\overset{H}{C}}{\cdots +\cdots}CH_2 \xrightarrow[\text{Attack at terminal carbon}]{Cl^-} \underset{\text{1,4-Addition}}{CH_3CH=CH-CH_2Cl}$$

Example:

$$CH_2=CH-CH=CH_2 + Br-Br \xrightarrow{CCl_4} \underset{\substack{54\% \\ \text{3,4-Dibromo-1-butene}}}{\underset{Br}{\underset{|}{C}}H_2-\underset{Br}{\underset{|}{C}}H-CH=CH_2} + \underset{\substack{46\% \\ \text{1,4-Dibromo-2-butene}}}{\underset{Br}{\underset{|}{C}}H_2-CH=CH-\underset{Br}{\underset{|}{C}}H_2}$$

Conjugated dienes also function as monomers in polymerizations induced by electrophiles, radicals, and other initiators (see Sections 12-13 and 12-14).

EXERCISE 14-10

Conjugated dienes can be made by the methods applied to the preparation of ordinary alkenes. Propose syntheses of **(a)** 2,3-dimethylbutadiene from 2,3-dimethyl-1,4-butanediol; **(b)** 1,3-cyclohexadiene from cyclohexane.

EXERCISE 14-11

Write the products of 1,2-addition and 1,4-addition of **(a)** HBr and **(b)** DBr to 1,3-cyclohexadiene. What is unusual about the products of 1,2- and 1,4-addition of HX to simple cyclic 1,3-dienes?

In summary, conjugated dienes are electron rich and are attacked by electrophiles to give intermediate allylic cations on the way to 1,2- or 1,4-addition products.

14-7 Delocalization among More Than Two Pi Bonds: Extended Conjugation and Benzene

What happens if a molecule contains more than two conjugated double bonds? Are cyclic conjugated systems different from their acyclic analogs? This section will begin to answer these questions.

Extended pi systems are thermodynamically stable but kinetically reactive

When more than two double bonds are in conjugation, the molecule is called an **extended π system.** An example is 1,3,5-hexatriene, the next higher double-bond homolog of 1,3-butadiene. This compound is quite reactive and readily polymerizes, particularly in the presence of electrophiles. Despite its reactivity as a delocalized π system, it is also relatively stable thermodynamically.

The increased reactivity of this extended π system is due to the low activation barriers for electrophilic additions, which proceed through highly delocalized carbocations. For example, the bromination of 1,3,5-hexatriene produces a substituted pentadienyl cation intermediate that can be described by three resonance structures.

Bromination of 1,3,5-Hexatriene

$$\underset{\textbf{1,3,5-Hexatriene}}{CH_2{=}CH{-}CH{=}CH{-}CH{=}CH_2} \xrightarrow{Br_2} \left[\begin{array}{c} BrCH_2{-}\overset{+}{C}H{-}CH{=}CH{-}CH{=}CH_2 \\ \updownarrow \\ BrCH_2{-}CH{=}CH{-}\overset{+}{C}H{-}CH{=}CH_2 \\ \updownarrow \\ BrCH_2{-}CH{=}CH{-}CH{=}CH{-}\overset{+}{C}H_2 \end{array}\right] + Br^-$$

$$\downarrow$$

$$\underset{\substack{\textbf{5,6-Dibromo-1,3-hexadiene} \\ \text{(A 1,2-addition product)}}}{BrCH_2\overset{Br}{\overset{|}{C}}HCH{=}CHCH{=}CH_2} + \underset{\substack{\textbf{3,6-Dibromo-1,4-hexadiene} \\ \text{(A 1,4-addition product)}}}{BrCH_2CH{=}CH\overset{Br}{\overset{|}{C}}HCH{=}CH_2} + \underset{\substack{\textbf{1,6-Dibromo-2,4-hexadiene} \\ \text{(A 1,6-addition product)}}}{BrCH_2CH{=}CHCH{=}CHCH_2Br}$$

The final mixture is the result of 1,2-, 1,4-, and 1,6-additions, the last product being the most favored thermodynamically, because it retains an internal conjugated diene system.

EXERCISE 14-12

On treatment with two equivalents of bromine, 1,3,5-hexatriene has been reported to give moderate amounts of 1,2,5,6-tetrabromo-3-hexene. Write a mechanism for the formation of this product.

Some highly extended π systems are found in nature. An example is β-carotene, the orange coloring agent in carrots, and its biological degradation product, vitamin A.

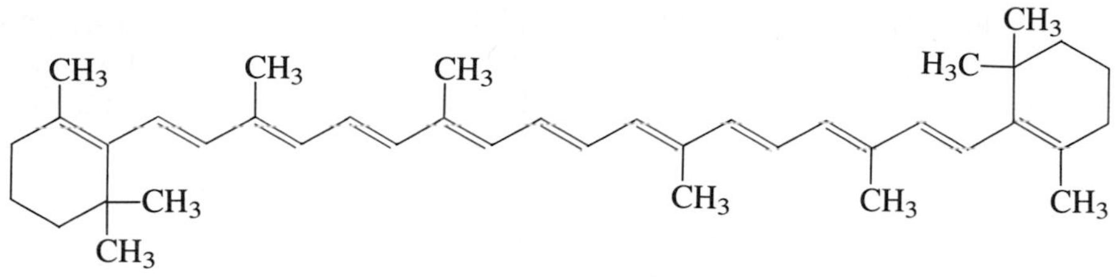

β-Carotene

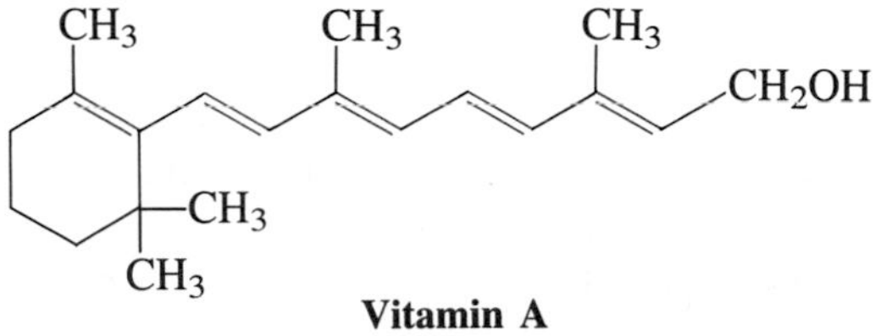

Vitamin A

Compounds of this type can be very reactive because there are many potential sites for attack by reagents that add to double bonds. In contrast, some cyclic conjugated systems may be considerably less reactive, depending on the number of π electrons (see Chapter 15). The most striking example of this effect is benzene, the cyclic analog of 1,3,5-hexatriene.

Benzene, a cyclic triene, is unusually stable

Cyclic conjugated systems are special cases. The most common examples are the cyclic triene C_6H_6, better known as benzene, and its derivatives (Chapters 15, 16, and 22). In contrast with hexatriene, benzene is unusually stable both thermodynamically and kinetically, because of its special electronic makeup (see Chapter 15). That benzene is unusual can be seen by drawing its resonance structures: There are two *equally* contributing forms. Benzene does not readily undergo addition reactions typical of unsaturated systems, such as catalytic hydrogenation, hydration, halogenation, and oxidation. In fact, because of its low reactivity, benzene can be used as a solvent in organic reactions.

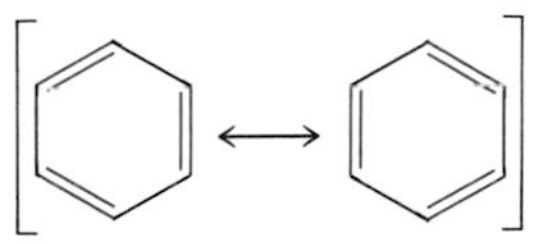

Benzene Is Unusually Unreactive

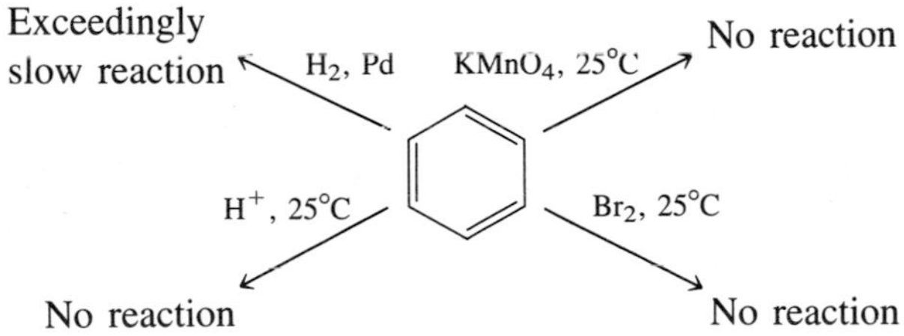

Exceedingly slow reaction ← H_2, Pd — Benzene — $KMnO_4$, 25°C → No reaction

No reaction ← H^+, 25°C — Benzene — Br_2, 25°C → No reaction

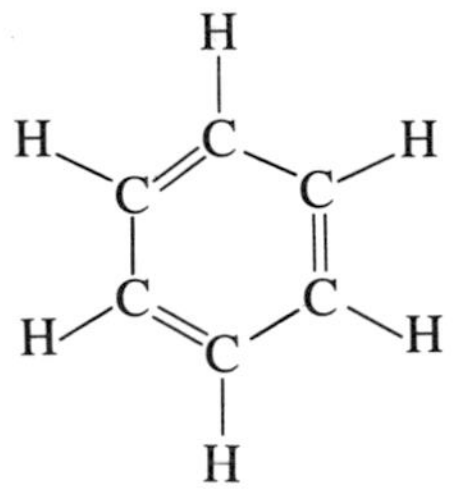

In the chapters that follow, we shall see that the unusual lack of reactivity of benzene is related to the number of π electrons present in its cyclic conjugation, namely, six. The next section introduces a reaction that is made possible only because its transition state benefits from six-electron cyclic overlap.

Benzene and Its Resonance Structures

Benzene

To summarize, acyclic extended conjugated systems show increasing reactivity because of the many sites open to attack by reagents and the ease of formation of delocalized intermediates. In contrast, the cyclohexatriene benzene is unusually unreactive.

14-8 A Special Transformation of Conjugated Dienes: Diels-Alder Cycloaddition

Conjugated double bonds participate in more than just the reactions typical of the alkenes, such as electrophilic addition. This section describes a process in which conjugated dienes and alkenes combine to give substituted cyclohexenes. In this transformation, known as Diels-Alder cycloaddition, the nuclei at the ends of the diene add to the alkene double bond, thereby closing a ring. The new bonds form simultaneously and stereospecifically.

The cycloaddition of dienes to alkenes gives cyclohexenes

When a mixture of 1,3-butadiene and ethene is heated in the gas phase, a remarkable reaction takes place in which cyclohexene is formed by the simultaneous generation of two new carbon–carbon bonds. This is the simplest example of the **Diels-Alder* reaction,** in which a conjugated diene adds to an alkene to yield cyclohexene derivatives. The Diels-Alder reaction is in turn a special case of the more general class of **cycloaddition reactions** between π systems. In the Diels-Alder reaction, an assembly of four conjugated atoms containing four π electrons reacts with a double bond containing two π electrons. For that reason, the reaction is also called a [4 + 2]cycloaddition.

Diels-Alder Cycloaddition of Ethene and 1,3-Butadiene

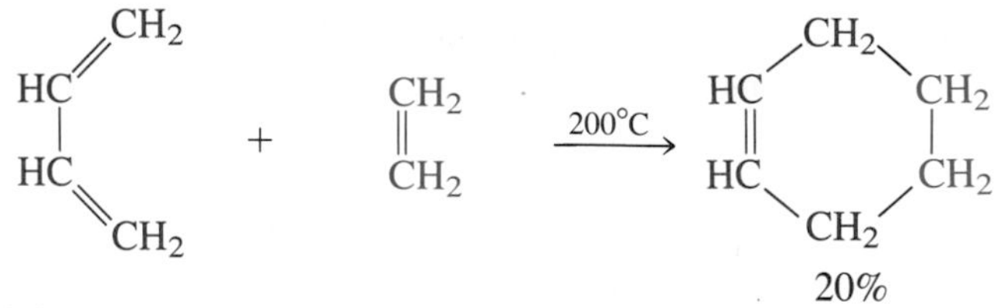

$CH_2{=}CH{-}CH{=}CH_2$ + $CH_2{=}CH_2$ $\xrightarrow{200°C}$ cyclohexene (20%)

1,3-Butadiene (Four π electrons) + **Ethene** (Two π electrons) → **Cyclohexene** (A cycloadduct), 20%

$F_3C(H)C{=}CH_2$

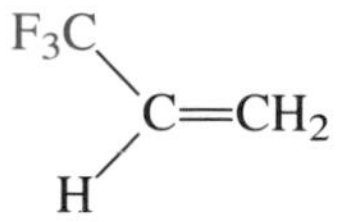

An electron-poor alkene

$H_2C{=}C(CH_3){-}C(CH_3){=}CH_2$

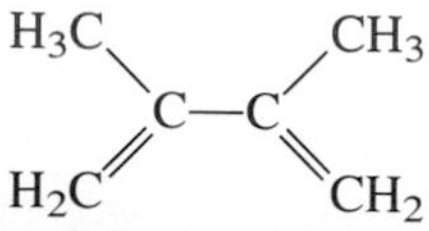

An electron-rich diene

The prototype reaction of butadiene and ethene actually does not work very well and gives only low yields of cyclohexene. It is much better to use an *electron-poor alkene* with an *electron-rich diene*. Substitution of the alkene with electron-attracting groups and of the diene with electron-donating groups therefore creates excellent reaction partners.

The trifluoromethyl group, for example, is inductively (Section 8-3) electron attracting due to its highly electronegative fluorine atoms. The presence of such a

*Professor Otto P. H. Diels, 1876–1954, University of Kiel, Germany, Nobel Prize 1950 (chemistry); Professor Kurt Alder, 1902–1958, University of Köln, Germany, Nobel Prize 1950 (chemistry).

substituent enhances the Diels-Alder reactivity of an alkene. Conversely, alkyl groups are electron donating because of hyperconjugation (Section 7-5); their presence increases electron density and is beneficial to dienes in the Diels-Alder reaction.

Other substituents may interact with double bonds by resonance. For example, carbonyl-containing groups and nitriles are good electron acceptors. Double bonds containing such substituents are electron poor because of the contribution of resonance forms that place a positive charge on an alkene carbon atom.

Groups That Are Electron Withdrawing by Resonance

$$\left[H_2C{=}CH{-}C(R){=}\ddot{O}: \longleftrightarrow H_2C{=}CH{-}\overset{+}{C}(R){-}\ddot{O}:^- \longleftrightarrow H_2\overset{+}{C}{-}CH{=}C(R){-}\ddot{O}:^- \right]$$

$$\left[H_2C{=}CH{-}C{\equiv}N: \longleftrightarrow H_2C{=}CH{-}\overset{+}{C}{=}\ddot{N}:^- \longleftrightarrow H_2\overset{+}{C}{-}CH{=}C{=}\ddot{N}:^- \right]$$

EXERCISE 14-13

Classify each of the alkenes below as electron poor or electron rich, relative to ethene. Explain your assignments.

(a) $H_2C{=}CHCH_2CH_3$ (b) cyclohexene (c) maleic anhydride (d) 3,3,4,4-tetrafluorocyclobutene

EXERCISE 14-14

The double bond in nitroethene, $H_2C{=}CHNO_2$, is electron poor, and that in methoxyethene, $H_2C{=}CHOCH_3$, is electron rich. Explain, using resonance structures.

An example of reaction partners that undergo efficient Diels-Alder cycloaddition are 2,3-dimethyl-1,3-butadiene and propenal (acrolein).

$$\underset{\textbf{2,3-Dimethyl-1,3-butadiene}}{H_2C{=}C(CH_3){-}C(CH_3){=}CH_2} + \underset{\textbf{Propenal}}{H_2C{=}CH{-}CHO} \xrightarrow{100^\circ C,\ 3\ h} \underset{\textbf{Diels-Alder adduct}}{\text{3,4-dimethylcyclohex-3-enecarbaldehyde}}\ \ 90\%$$

The carbon–carbon double bond in the cycloaddition product is electron rich and sterically hindered. It therefore does not compete with propenal in further Diels-Alder reaction with additional diene.

The parent 1,3-butadiene, without additional substituents, is electron rich enough to undergo cycloadditions with electron-poor alkenes.

$$\text{1,3-butadiene} + CH_2{=}CH\overset{O}{\overset{\|}{C}}OCH_2CH_3 \xrightarrow{160^\circ C,\ 15\ h} \text{ethyl 3-cyclohexenecarboxylate}$$

Ethyl propenoate (Ethyl acrylate) 94%

In cycloadditions, the substituted ethene is frequently called the **dienophile** (''diene loving''), to contrast it with the diene. Many typical dienes and dienophiles have common names, reflecting their widespread use in Diels-Alder syntheses of substances of industrial and pharmaceutical importance (Table 14-1).

EXERCISE 14-15

Formulate the products of [4 + 2]cycloaddition of tetracyanoethene with **(a)** 1,3-butadiene; **(b)** cyclopentadiene; **(c)** 1,2-dimethylenecyclohexane.

The Diels-Alder reaction is concerted

The Diels-Alder reaction takes place in one step. Both new carbon–carbon single bonds and the new π bond form simultaneously, just as the three π bonds in the starting materials break. As mentioned earlier (Section 6-4), one-step reactions, in which bond breaking happens at the same time as bond making, are *concerted.* The concerted nature of this transformation can be depicted in either of two ways: a dotted circle, representing the six delocalized π electrons, or by electron-pushing arrows. Just as six-electron cyclic overlap stabilizes benzene (Section 14-7), the Diels-Alder process benefits from the presence of such an array in its transition state.

Two Pictures of the Transition State of the Diels-Alder Reaction

or ‡

Dotted-line picture **Electron-pushing picture**

A molecular-orbital representation (Figure 14-9) clearly shows bond formation by overlap of the p orbitals of the dienophile with the terminal p orbitals of the diene. While these four carbons rehybridize to sp^3, the remaining two internal diene p orbitals give rise to the new π bond.

The Diels-Alder reaction is stereospecific

As a consequence of the concerted mechanism, the Diels-Alder reaction is *stereospecific*. For example, reaction of 1,3-butadiene with dimethyl *cis*-2-butenedioate (dimethyl maleate, a cis alkene) gives dimethyl *cis*-4-cyclohexene-1,2-dicarboxylate. *The stereochemistry at the original double bond of the dienophile is retained in the product*. In the complementary reaction, dimethyl *trans*-2-butenedioate (dimethyl fumarate, a trans alkene) gives the trans adduct.

TABLE 14-1 Typical Dienes and Dienophiles in the Diels-Alder Reaction

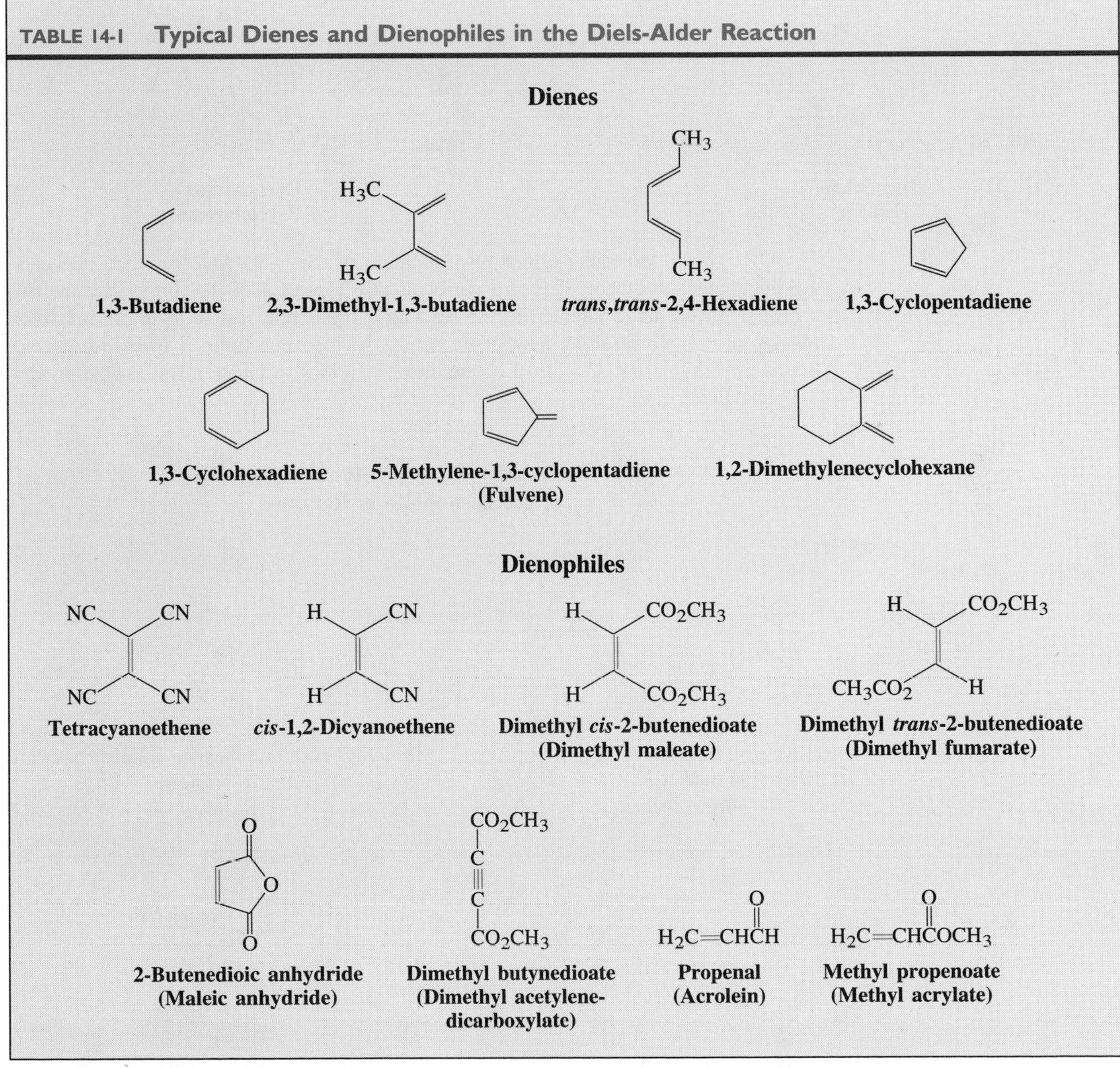

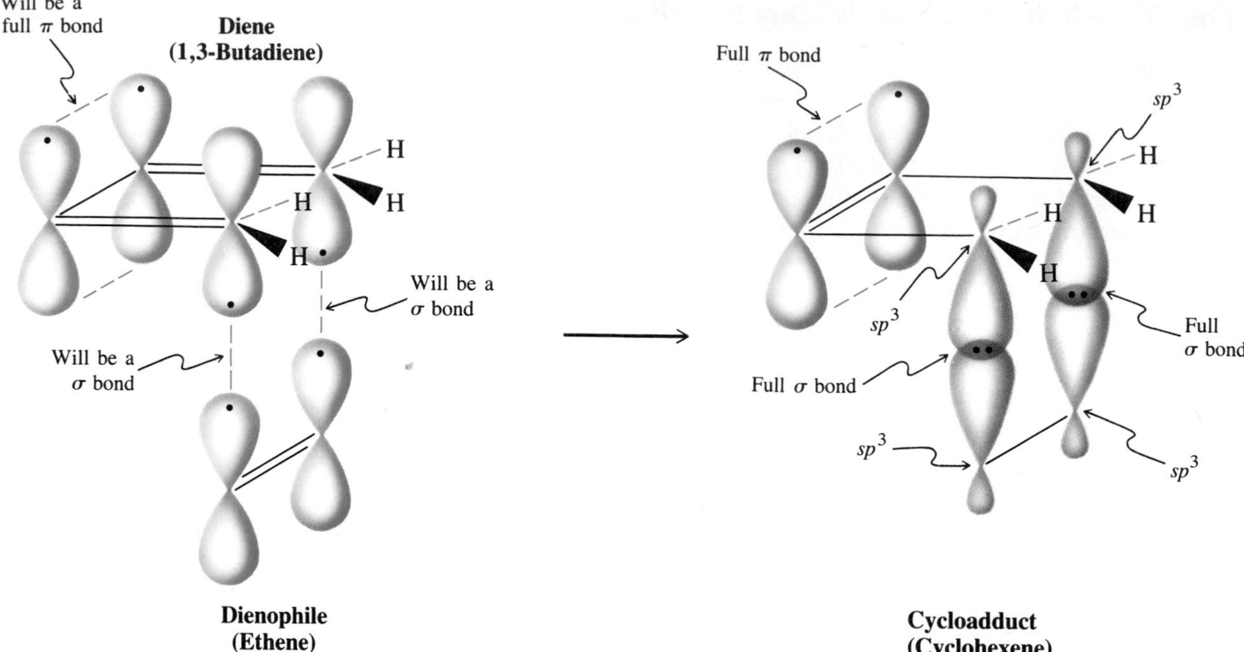

FIGURE 14-9 Molecular-orbital representation of the Diels-Alder reaction between 1,3-butadiene and ethene. The two *p* orbitals at C1 and C4 of the former and the two *p* orbitals of the latter interact, as the reacting carbons rehybridize to sp^3 to maximize overlap in the two resulting new single bonds. At the same time, π overlap between the two *p* orbitals on C2 and C3 of the diene increases to create a full double bond.

In the Diels-Alder Reaction, the Stereochemistry of the Dienophile Is Retained

+ (H)(O=COCH₃)C=C(H)(COCH₃=O) —150–160°C, 20 h→ cyclohexene with CO_2CH_3, H, CO_2CH_3, H

Dimethyl *cis*-2-butenedioate
(Dimethyl maleate)
(Cis starting material)

68%
Dimethyl *cis*-4-cyclohexene-1,2-dicarboxylate
(Cis product)

+ (H)(COCH₃)C=C(H_3COC)(H) —200–205°C, 3.5 h→ cyclohexene with CO_2CH_3, H, H, CO_2CH_3

Dimethyl *trans*-2-butenedioate
(Dimethyl fumarate)
(Trans starting material)

95%
Dimethyl *trans*-4-cyclohexene-1,2-dicarboxylate
(Trans product)

Similarly, *the stereochemistry of the diene is also retained.*

In the Diels-Alder Reaction, the Stereochemistry of the Diene Is Retained

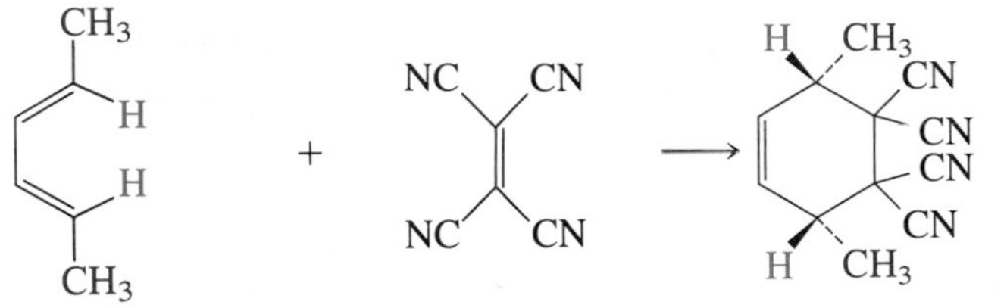

CH_3 H H CH_3 + NC CN NC CN → H CH_3 CN CN CN CN H CH_3

***trans,trans*-2,4-Hexadiene** (Both methyls "outside")

Tetracyanoethene

(Methyls end up cis)

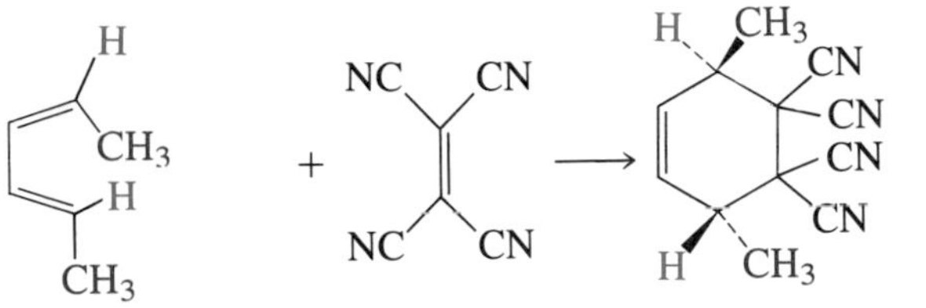

H CH_3 H CH_3 + NC CN NC CN → H CH_3 CN CN CN CN H CH_3

***cis,trans*-2,4-Hexadiene** (One methyl "inside"; one methyl "outside")

(Methyls end up trans)

EXERCISE 14-16

Add the missing products or starting materials to the following schemes.

(a) 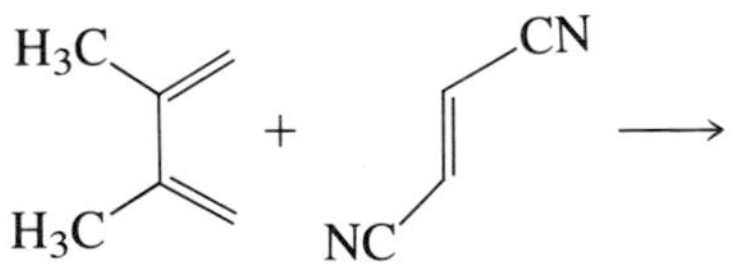 H_3C, H_3C (2,3-dimethyl-1,3-butadiene) + NC–CH=CH–CN →

(b) 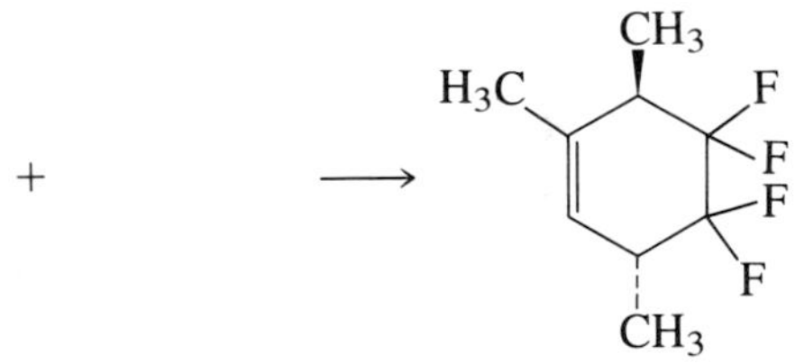 + → CH_3, H_3C, F, F, F, F, CH_3

EXERCISE 14-17

cis-trans-2,4-Hexadiene reacts very sluggishly in [4 + 2]cycloadditions; the trans,trans isomer does so much more rapidly. Explain. [Hint: The Diels-Alder reaction requires the *s*-cis arrangement of the diene (Figure 14-9, and see Figure 14-7).]

Diels-Alder cycloadditions follow the endo rule

The Diels-Alder reaction is stereospecific, not only with respect to the substitution pattern of the original double bonds, but also with respect to the orientation of the starting materials relative to each other. Consider the reaction of 1,3-cyclopentadiene with dimethyl *cis*-2-butenedioate. Two products are conceivable, one in which the two ester substituents on the bicyclic frame are on the same side (cis) as the bridge, the other in which they are on the side opposite (trans) from the bridge. The first is called the **exo adduct,** the second the **endo adduct** (*exo*, Greek, outside; *endo*, Greek, within). The terms refer to the position of groups in bridged systems. Exo substituents are placed cis with respect to the shorter bridge; endo substituents are positioned trans to this bridge.

Exo and Endo Cycloadditions to Cyclopentadiene

The Diels-Alder reaction usually proceeds with *endo selectivity; only the endo product is formed,* a result referred to as the **endo rule.** This preference appears to originate from an attractive interaction between the π system of the diene and that of the unsaturated substituents on the dienophile.

The Endo Rule

EXERCISE 14-18

Predict the products of the following reactions (show the stereochemistry clearly). **(a)** *trans,trans*-2,4-Hexadiene with methyl propenoate; **(b)** *trans*-1,3-pentadiene with 2-butenedioic anhydride (maleic anhydride); **(c)** 1,3-cyclopentadiene with dimethyl *trans*-2-butenedioate (dimethyl fumarate).

EXERCISE 14-19

The Diels-Alder reaction can also occur in an intramolecular fashion. Draw the two transition states leading to products in the following reaction.

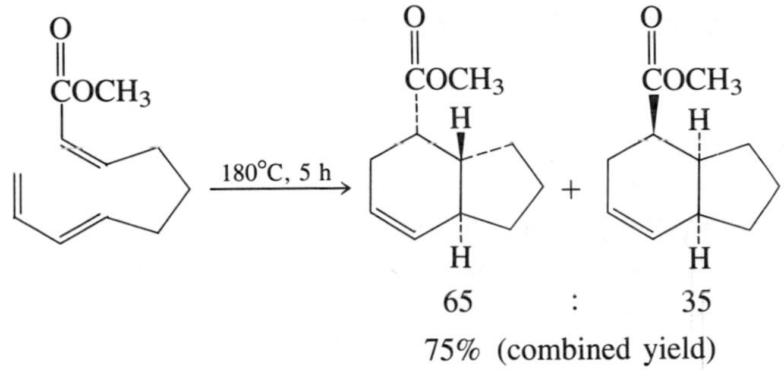

180°C, 5 h

65 : 35

75% (combined yield)

In summary, the Diels-Alder reaction is a cycloaddition that proceeds best between electron-rich 1,3-dienes and electron-poor dienophiles to furnish cyclohexenes. It is stereospecific with respect to the stereochemistry of the double bonds and with respect to the arrangements of the substituents on diene and dienophile: It follows the endo rule.

BOX 14-1 A Cycloaddition Reaction Promoted by Light

Photochemical [2+2]Cycloaddition

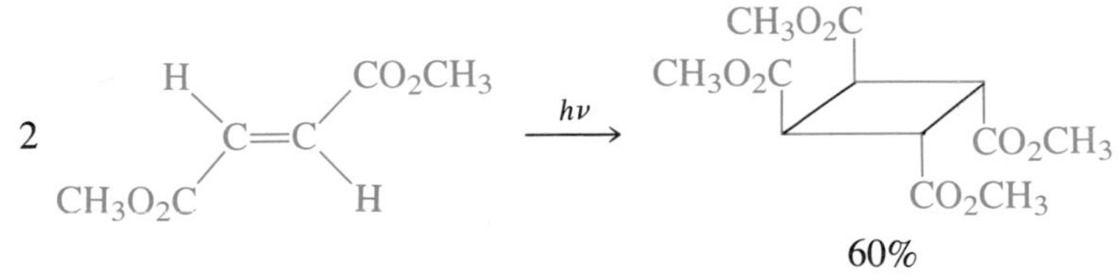

$$2\ \text{H}(\text{CH}_3\text{O}_2\text{C})\text{C}{=}\text{C}(\text{CO}_2\text{CH}_3)\text{H} \xrightarrow{h\nu} \text{cyclobutane-1,2,3,4-tetracarboxylate (}\text{CO}_2\text{CH}_3\text{)}_4$$

60%

Although isolated double bonds do not normally undergo cycloadditions to each other when heated, they do so in the presence of light to give four-membered rings. Such reactions are called [2 + 2]cycloadditions.

An unusual example is the intramolecular photochemical conversion of norbornadiene into quadricyclane (common names) shown below.

The reaction is thermally unfavorable, and light energy is needed to drive it to the strained product. The latter rapidly reverses to starting material in the presence of metal catalysts, releasing 26 kcal mol^{-1} of strain energy. A system of this type may become important in efforts to convert the energy of sunlight into a chemically storable and transportable form.

A Photochemical-Energy Storage System

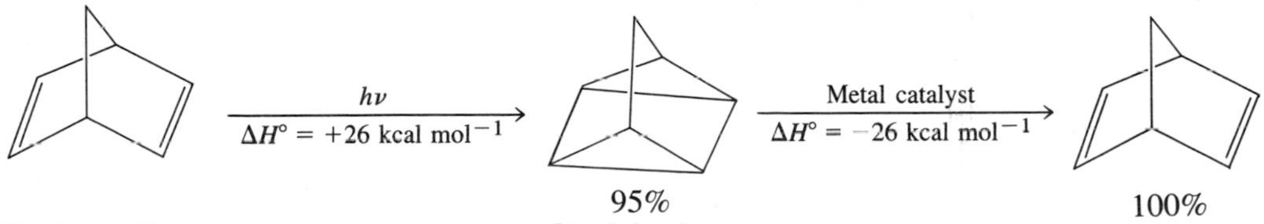

$$\text{Norbornadiene} \xrightarrow[\Delta H^\circ = +26\ \text{kcal mol}^{-1}]{h\nu} \text{Quadricyclane} \xrightarrow[\Delta H^\circ = -26\ \text{kcal mol}^{-1}]{\text{Metal catalyst}} \text{Norbornadiene}$$

Norbornadiene — **Quadricyclane** 95% — 100%

14-9 Electrocyclic Reactions

The Diels-Alder reaction couples the ends of two separate π systems. Can rings be formed by the linkage of the termini of a *single* conjugated di-, tri-, or polyene? Yes, and this section will describe the conditions under which such ring closures, called **electrocyclic reactions,** take place. Cycloadditions and electrocyclic reactions belong to a class of transformations called **pericyclic** (*peri,* Greek, around) because they exhibit transition states with a cyclic array of nuclei and electrons.

Electrocyclic transformations are driven by heat or light

Let us consider first the conversion of 1,3-butadiene to cyclobutene. This process is endothermic because of ring strain. Indeed, the reverse reaction, ring *opening* of cyclobutene, occurs readily upon heating. However, ring *closure* of *cis*-1,3,5-hexatriene to 1,3-cyclohexadiene is exothermic and takes place thermally. Is it possible to drive these transformations in the thermally disfavored directions? Again, the answer is yes. Irradiation with ultraviolet light **(photochemical conditions)** provides sufficient energy to reverse both processes.

Electrocyclic Reactions

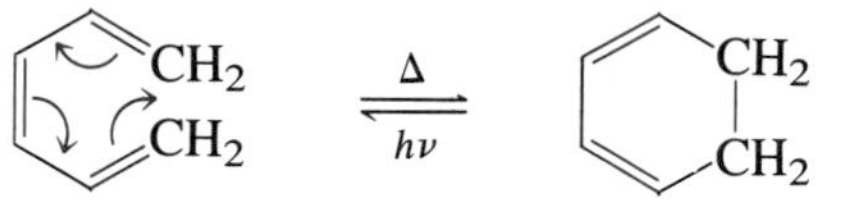

CH_2 / CH_2 $\underset{h\nu}{\overset{\Delta}{\rightleftharpoons}}$ CH_2 / CH_2 $\Delta H° = -14.5$ kcal mol^{-1}

***cis*-1,3,5-Hexatriene** **1,3-Cyclohexadiene**

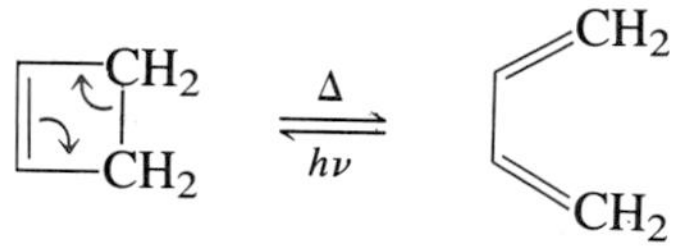

CH_2 / CH_2 $\underset{h\nu}{\overset{\Delta}{\rightleftharpoons}}$ CH_2 / CH_2 $\Delta H° = -9.7$ kcal mol^{-1}

Cyclobutene **1,3-Butadiene**

EXERCISE 14-20

When heated, benzocyclobutene A in the presence of dimethyl-*trans*-2-butenedioate B gives C. Explain. (Hint: Combine an electrocyclic with a Diels-Alder reaction.)

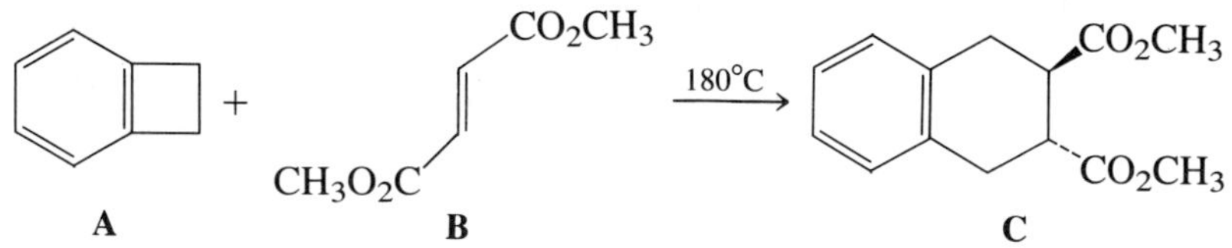

A + CO_2CH_3 / CH_3O_2C (B) $\xrightarrow{180°C}$ CO_2CH_3 / CO_2CH_3 (C)

A **B** **C**

Electrocyclic reactions are concerted and stereospecific

Like the Diels-Alder cycloaddition, electrocyclic reactions are concerted and stereospecific. Thus, the thermal isomerization of *cis*-3,4-dimethylcyclobutene gives only *cis,trans*-2,4-hexadiene.

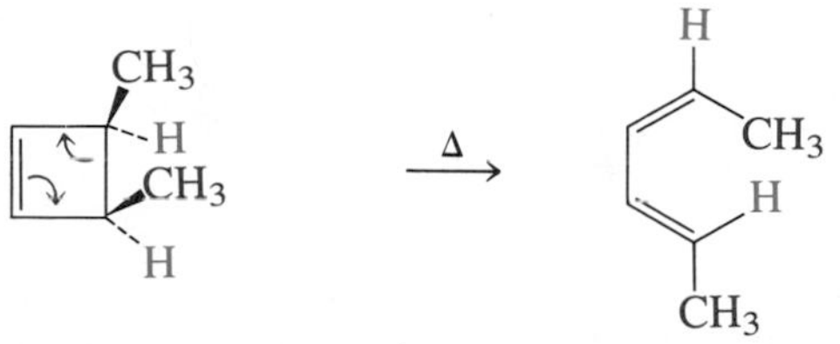

cis-3,4-Dimethylcyclobutene → (Δ) cis,trans-2,4-Hexadiene

cis-3,4-Dimethylcyclobutene **cis,trans-2,4-Hexadiene**

Heated *trans*-3,4-dimethylcyclobutene, however, opens to *trans,trans*-2,4-hexadiene.

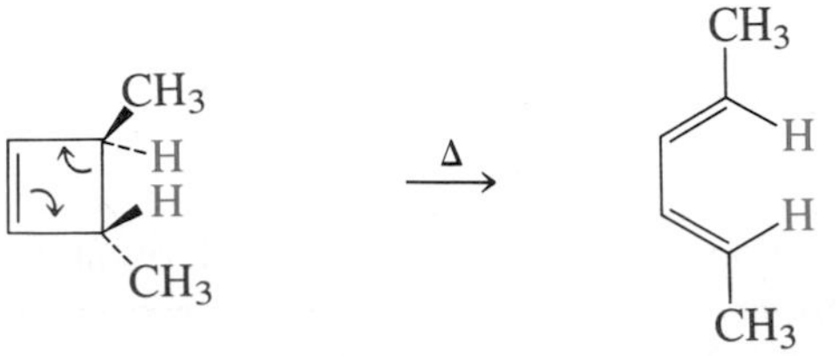

trans-3,4-Dimethylcyclobutene **trans,trans-2,4-Hexadiene**

Figure 14-10 takes a closer look at these processes. As the bond between carbons C3 and C4 in the cyclobutene is broken, these carbon atoms must rehybridize from sp^3 to sp^2 and rotate to permit overlap between the emerging p orbitals and those originally present. In thermal cyclobutene ring opening, the carbon atoms are found to rotate *in the same direction,* either both clockwise or both counterclockwise. This mode of reaction is called a **conrotatory** process.

Fascinatingly, the photochemical closure **(photocyclization)** of butadiene to cyclobutene proceeds with stereochemistry exactly *opposite* to that observed in the thermal opening. In this case, the products arise by rotation of the two reacting carbons in opposite directions. In other words, if one rotates clockwise, the other does so counterclockwise. This mode of movement is called **disrotatory** (Figure 14-11).

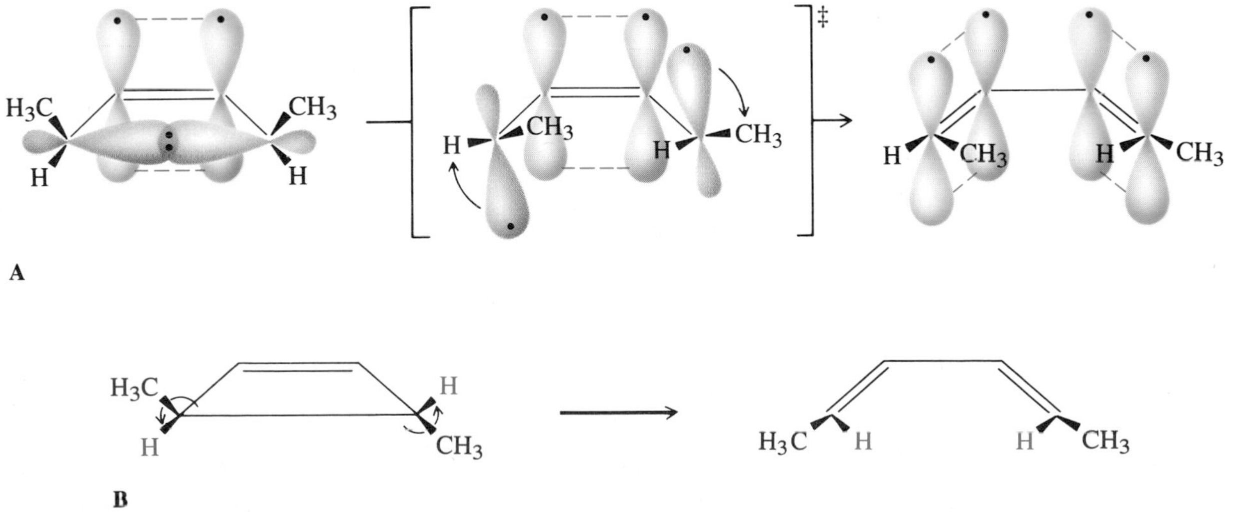

FIGURE 14-10 (A) Conrotatory ring opening of *cis*-3,4-dimethylcyclobutene. Both the reacting carbons rotate clockwise. The sp^3 hybrid lobes in the ring change into p orbitals, the carbons becoming sp^2 hybridized. Overlap of these p orbitals with those already present in the cyclobutene starting material creates the two double bonds of the cis,trans diene. (B) Similar conrotatory opening of *trans*-3,4-dimethylcyclobutene proceeds to the trans,trans diene.

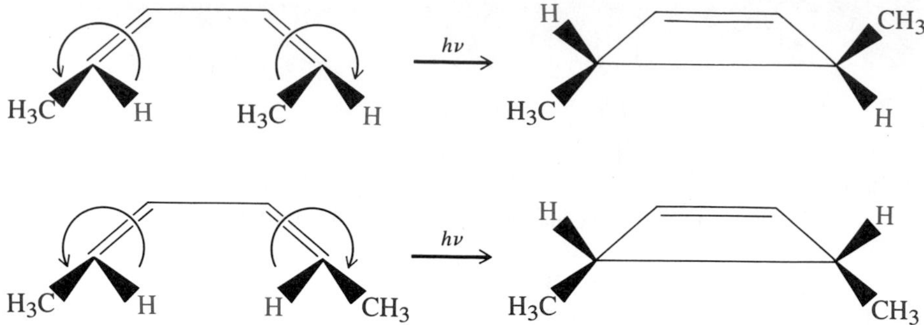

FIGURE 14-11 Disrotatory photochemical ring closure of *cis,trans*- and *trans,trans*-2,4-hexadiene. In the disrotatory mode, one carbon rotates clockwise, the other counterclockwise.

Can these observations be generalized? Let us look at the stereochemistry of the *cis*-1,3,5-hexatriene–cyclohexadiene interconversion. Surprisingly, the six-membered ring is formed thermally by the disrotatory mode, as can be shown by using derivatives. For example, heated *trans,cis,trans*-2,4,6-octatriene gives *cis*-5,6-dimethyl-1,3-cyclohexadiene, and *cis,cis,trans*-2,4,6-octatriene converts to *trans*-5,6-dimethyl-1,3-cyclohexadiene, both disrotatory closures.

Stereochemistry of the Thermal 1,3,5-Hexatriene Ring Closure

CH_3 CH_3 —Δ, Disrotatory→ CH_3 CH_3

***trans,cis,trans*-2,4,6-Octatriene** → ***cis*-5,6-Dimethyl-1,3-cyclohexadiene**

CH_3 CH_3 —Δ, Disrotatory→ CH_3 CH_3

***cis,cis,trans*-2,4,6-Octatriene** → ***trans*-5,6-Dimethyl-1,3-cyclohexadiene**

In contrast, the corresponding photochemical reactions occur in conrotatory fashion.

Stereochemistry of the Photochemical 1,3,5-Hexatriene Ring Closure

CH_3 CH_3 —$h\nu$→ CH_3 CH_3

This stereocontrol is observed in many other electrocylic transformations and is governed by the symmetry properties of the relevant π molecular orbitals

BOX 14-2 An Extraordinary Electrocyclic Reaction of Anticancer Agents

Heating of 3-hexene-1,5-diyne to 200° induces an electrocyclization (Bergman* reaction) related to the ring closure of 1,3,5-hexatrienes. However, the extra electrons in the two triple bonds mean that, instead of a cyclohexadiene, a reactive intermediate known as a 1,4-benzene diradical is generated.

Electrocyclization of *cis*-3-Hexene-1,5-diyne

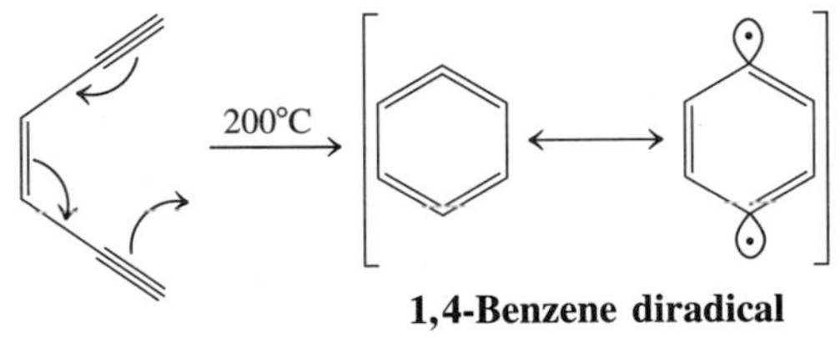

200°C

1,4-Benzene diradical

The enediyne-containing antitumor antibiotics calicheamicin and esperamicin (Section 13-12) exhibit impressive biological activity (4000 times greater than the clinical drug adriamycin). A sequence of enzyme-catalyzed steps (shown for calicheamicin below) cleaves an S–S bond and adds sulfur to the double bond at C9. The change in hybridization from sp^2 to sp^3 at C9 "pinches" the molecule, bringing C2 and C7 within bonding distance. (Make a model!) Bergman-type electrocyclization ensues (below −10°C!), a reaction leading to a benzene diradical.

It is these radical centers that are thought to decompose the DNA (Chapter 26) of tumor cells by hydrogen abstraction. Ironically, the enzyme system that activates the antitumor molecule usually *protects* cells from damage by radicals (see Chapter 25). Thus, these anticancer agents destroy tumor cells *with their own defense systems!* Although the enediyne-to-benzene diradical transformation was discovered in the laboratory in 1973, the finding that such a remarkable process occurs in nature was, to say the least, unexpected.

*Professor Robert G. Bergman, b. 1942, University of California, Berkeley.

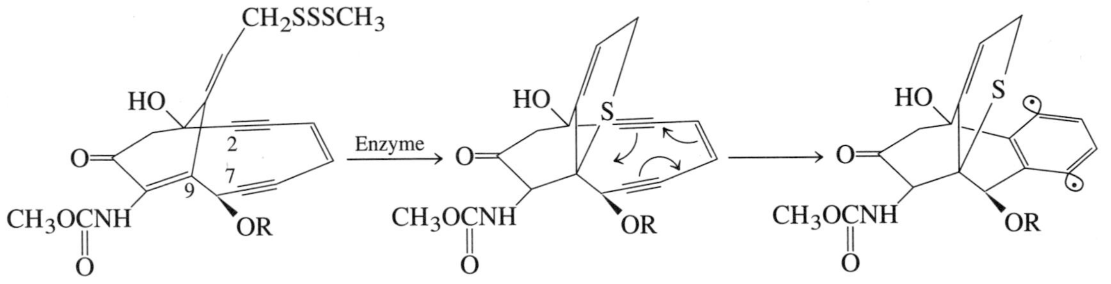

CH_2SSSCH_3 HO O= 2 7 9 CH_3OCNH O OR — Enzyme → HO O= S CH_3OCNH O OR → HO S O= CH_3OCNH O OR

involved. The **Woodward-Hoffmann* rules** describe these interactions and predict the stereochemical outcome of all electrocyclic processes, a subject left to an advanced treatment of organic chemistry.

*Professor Robert B. Woodward, 1917–1979, Harvard University, Nobel Prize 1965 (chemistry); Professor Roald Hoffmann, b. 1937, Cornell University, Nobel Prize 1981 (chemistry).

EXERCISE 14-21

Photolysis of ergosterol gives provitamin D_2, a precursor of vitamin D_2 (an antirachitic agent, shown in the margin). Is the ring opening conrotatory or disrotatory?

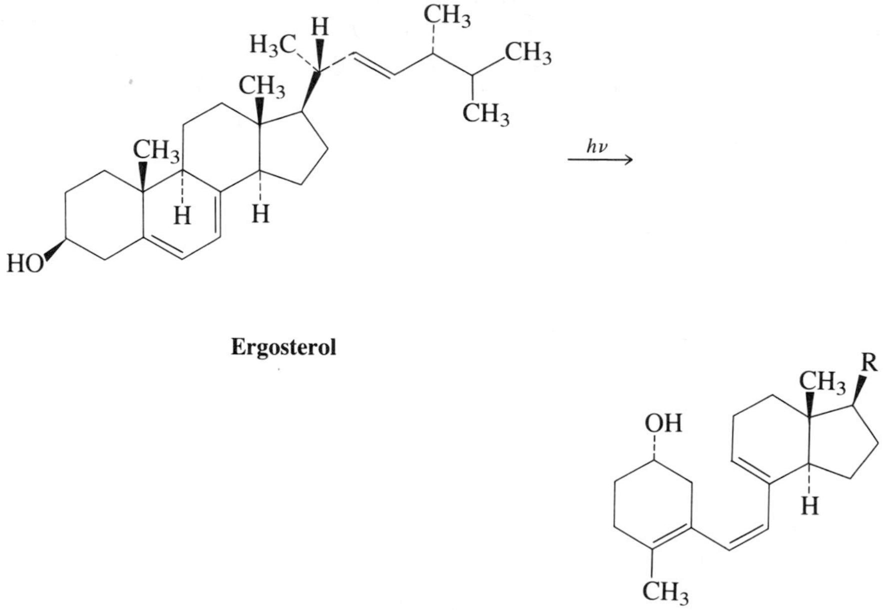

Ergosterol

Provitamin D_2

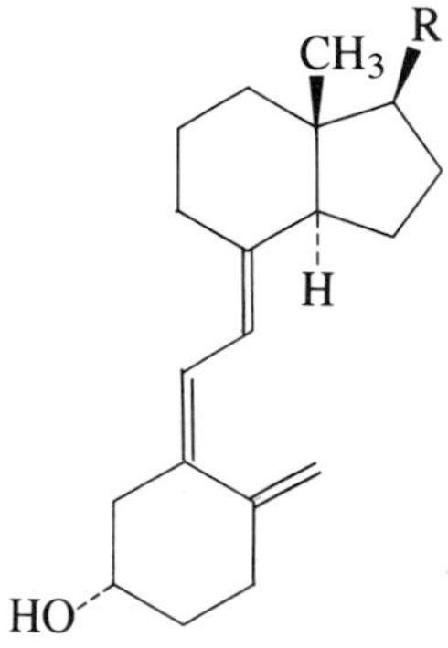

Vitamin D_2

In summary, conjugated dienes and hexatrienes are capable of (reversible) electrocyclic ring closures to cyclobutenes and 1,3-cyclohexadienes, respectively. The diene–cyclobutene system prefers thermal conrotatory and photochemical disrotatory modes. The triene–cyclohexadiene system reacts in the opposite way, proceeding through thermal disrotatory and photochemical conrotatory rearrangements. The stereochemistry of such electrocyclic reactions is governed by the Woodward-Hoffmann rules.

14-10 Polymerization of Conjugated Dienes: Rubber

Like simple alkenes (Section 12-14), conjugated dienes can be polymerized. The elasticity of the resulting materials has led to their use as synthetic rubbers. The biochemical pathway to natural rubber features an activated form of the five-carbon unit 2-methyl-1,3-butadiene (isoprene, see Section 4-7). The latter is an important building block in nature.

1,3-Butadiene can form cross-linked polymers

When 1,3-butadiene is polymerized at C1 and C2, it yields a polyethenylethene (polyvinylethylene).

1,2-Polymerization of 1,3-Butadiene

$$2n\ CH_2{=}CH{-}CH{=}CH_2 \xrightarrow{\text{Initiator}} {-}(\underset{\underset{CH_2}{\|}}{\underset{CH}{|}}CH{-}CH_2{-}\underset{\underset{CH_2}{\|}}{\underset{CH}{|}}CH{-}CH_2)_n{-}$$

Alternatively, polymerization at C1 and C4 gives either *trans*-polybutadiene, *cis*-polybutadiene, or a mixed polymer.

1,4-Polymerization of 1,3-Butadiene

$$n\ CH_2{=}CH{-}CH{=}CH_2 \xrightarrow{\text{Initiator}} {-}(CH_2{-}CH{=}CH{-}CH_2)_n{-}$$

***cis*- or *trans*-Polybutadiene**

Butadiene polymerization is unique in that the product itself may be unsaturated. The double bonds in this initial polymer may be linked by further treatment with added chemicals, such as radical initiators, or by radiation. In this way, **cross-linked polymers** arise, in which individual chains have been connected into a more rigid framework (Figure 14-12). Cross-linking generally increases the density and hardness of such materials. It also greatly affects a property characteristic of butadiene polymers. The individual chains in *most* polymers can be moved past each other, a process allowing for their molding and shaping. In cross-linked systems, however, such deformations are rapidly reversible: The chain snaps back to its original shape (more or less). Such **elasticity** is characteristic of rubbers.

Synthetic rubbers are derived from poly-1,3-dienes

Polymerization of 2-methyl-1,3-butadiene (isoprene, Section 4-7) by a Ziegler-Natta catalyst (Section 12-14) results in a synthetic rubber *(polyisoprene)* of almost 100% *Z* configuration. Similarly, 2-chloro-1,3-butadiene furnishes an elastic, heat- and oxygen-resistant polymer called neoprene, again almost exclusively in the *Z* form. Over 2 million metric tons of synthetic rubber are produced in the United States every year.

$$n\ H_2C{=}\overset{\overset{CH_3}{|}}{C}{-}CH{=}CH_2 \xrightarrow{TiCl_4,\ AlR_3} \begin{matrix} H_3C & & H \\ & C{=}C & \\ {-}(H_2C & & CH_2)_n{-} \end{matrix}$$

2-Methyl-1,3-butadiene **(*Z*)-Polyisoprene**

$$n\ H_2C{=}\overset{\overset{Cl}{|}}{C}{-}CH{=}CH_2 \xrightarrow{TiCl_4,\ AlR_3} \begin{matrix} Cl & & CH_2)_n{-} \\ & C{=}C & \\ {-}(H_2C & & H \end{matrix}$$

2-Chloro-1,3-butadiene **Neoprene**

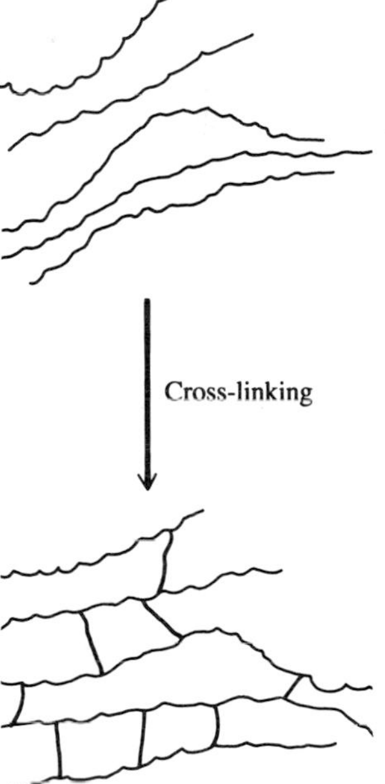

FIGURE 14-12
Cross-linking underlies the elasticity of polybutadiene chains in rubber.

Natural *Hevea* rubber is a 1,4-polymerized (Z)-poly(2-methyl-1,3-butadiene), similar in structure to polyisoprene. To increase its elasticity, it is treated with hot elemental sulfur in a process called **vulcanization** (*Vulcanus,* Latin, the Roman god of fire), which creates sulfur cross-links. This reaction was discovered by Goodyear* in 1839. One of the earliest and most successful uses of the product, "Vulcanite," was the manufacture of dentures that could be molded for good fit. Prior to the 1860s, false teeth were embedded in animal bone, ivory, or metal. The swollen appearance of George Washington's lips (visible on U.S. currency) is attributed to his ill-fitting ivory dentures. Nowadays, such appliances are made from acrylics (Section 13-11).

Polyisoprene is the basis of natural rubber

How is rubber made in nature? Plants construct the polyisoprene framework of natural rubber by using as a building block 3-methyl-3-butenyl pyrophosphate (isopentenyl pyrophosphate). This molecule is an ester of pyrophosphoric acid and 3-methyl-3-buten-1-ol. An enzyme equilibrates a small amount of this material with the 2-butenyl isomer, an allylic pyrophosphate.

Biosynthesis of the Two Isomers of 3-Methylbutenyl Pyrophosphate

$H_2C{=}C(CH_3)CH_2CH_2OH$ **3-Methyl-3-buten-1-ol**

$HO{-}P(=O)(OH){-}OH + HO{-}P(=O)(OH){-}OH$ **Phosphoric acid**

$\xrightarrow{-\,HOH}$ $HO{-}P(=O)(OH){-}O{-}P(=O)(OH){-}OH$ **Pyrophosphoric acid**

3-Methyl-3-buten-1-ol + pyrophosphoric acid $\xrightarrow{-\,HOH}$

$$H_2C{=}C(CH_3)CH_2CH_2O{-}P(=O)(OH){-}O{-}P(=O)(OH){-}OH \underset{}{\overset{\text{Enzyme}}{\rightleftharpoons}} (H_3C)_2C{=}CHCH_2O{-}P(=O)(OH){-}O{-}P(=O)(OH){-}OH$$

3-Methyl-3-butenyl pyrophosphate — **3-Methyl-2-butenyl pyrophosphate**

Although the subsequent processes are enzymatically controlled, they can be formulated simply in terms of familiar mechanisms (OPP = pyrophosphate).

*Charles Goodyear, 1800–1860, American inventor.

Mechanism of Natural Rubber Synthesis

STEP 1. Ionization to stabilized (allylic) cation

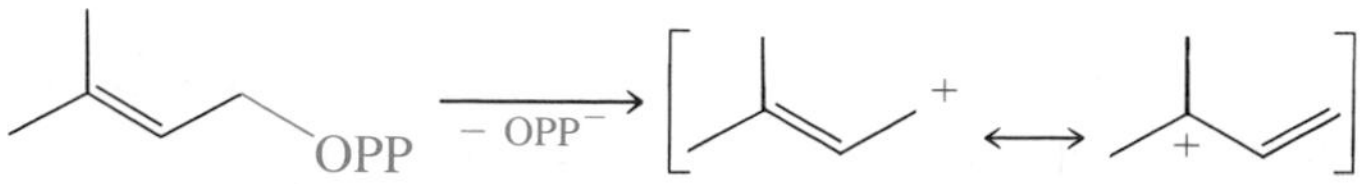

OPP —(− OPP⁻)→ [+ ↔ +]

STEP 2. Electrophilic attack

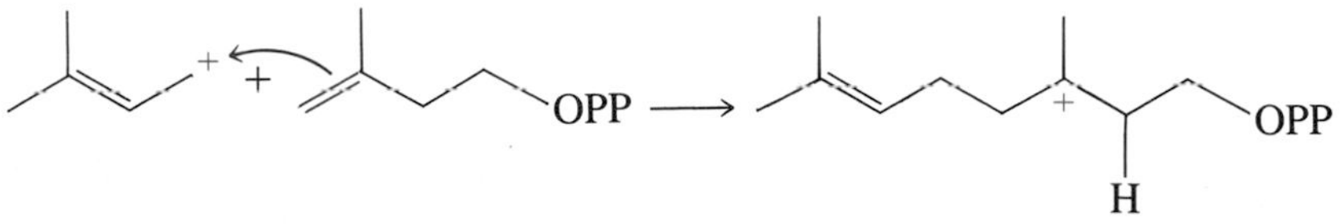

OPP → OPP H

STEP 3. Proton loss

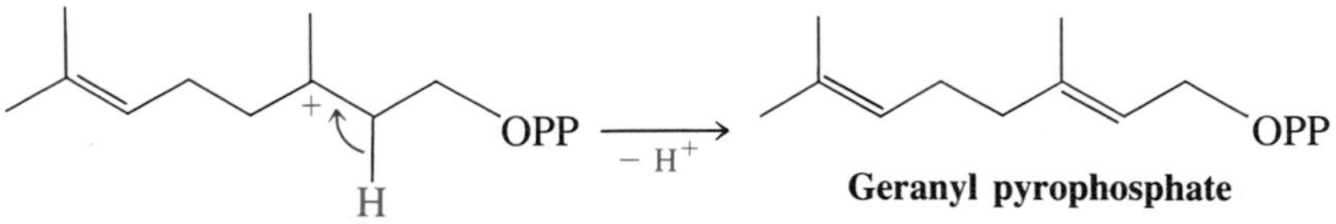

H OPP —(− H^+)→ OPP

Geranyl pyrophosphate

STEP 4. Second oligomerization

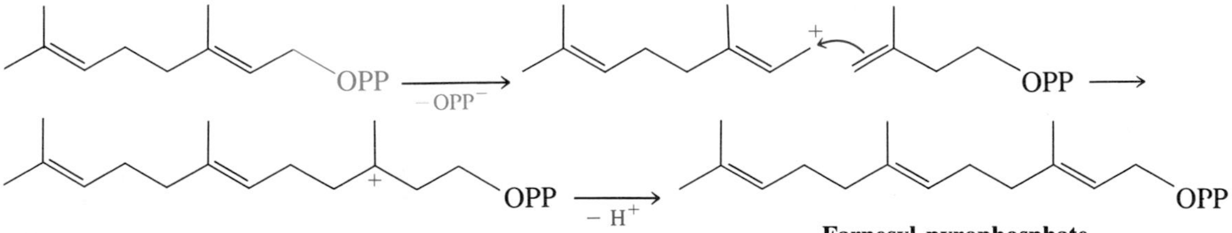

OPP —(−OPP⁻)→ OPP →

OPP —(− H^+)→ OPP

Farnesyl pyrophosphate

In the first step, ionization of the allylic pyrophosphate gives an allylic cation. Attack by a molecule of 3-methyl-3-butenyl pyrophosphate, followed by proton loss, yields a dimer called geranyl pyrophosphate. Repetition of this process leads to natural rubber.

Many natural products are composed of 2-methyl-1,3-butadiene (isoprene) units

Many natural products are derived from 3-methyl-3-butenyl pyrophosphate, including the terpenes first discussed in Section 4-7. Indeed, the structures of terpenes can be dissected into five-carbon units connected as in 2-methyl-1,3-butadiene. Their structural diversity can be attributed to the multiple ways in which 3-methyl-3-butenyl pyrophosphate can couple. The monoterpene geraniol and the sesquiterpene farnesol, two of the most widely distributed substances in the plant kingdom, form by hydrolysis of their corresponding pyrophosphates.

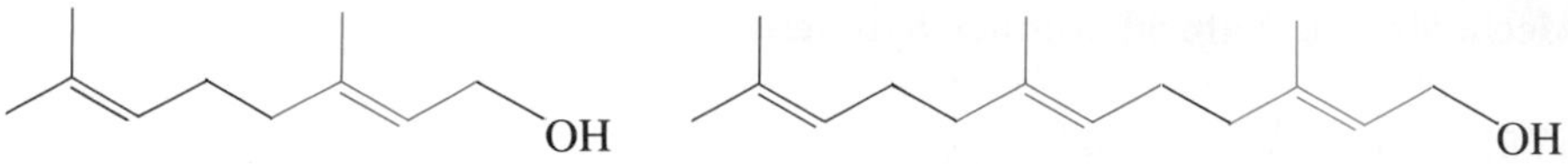

OH OH

Geraniol Farnesol

Coupling of two molecules of farnesyl pyrophosphate leads to squalene, a biosynthetic precursor of the steroid nucleus (Section 12-13).

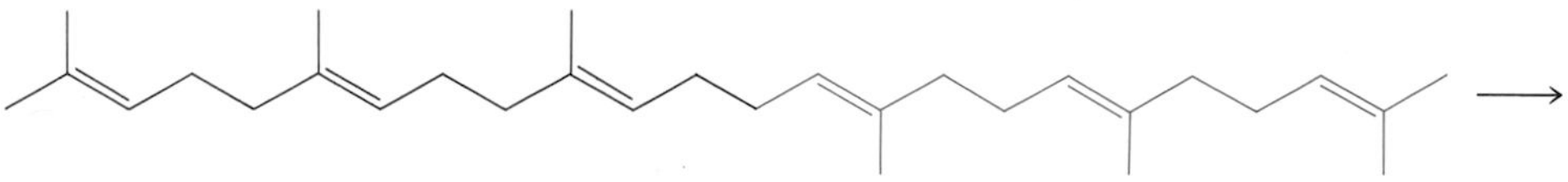

⟶ steroids

Squalene

Bicyclic substances, such as camphor, a chemical used in mothballs, nasal sprays, and muscle rubs, are built up from geranyl pyrophosphate by enzymatically controlled electrophilic carbon–carbon bond formations.

Camphor Biosynthesis from Geranyl Pyrophosphate

OPP — Cis, trans isomerization ⟶ OPP ⟶ $-\ OPP^-$ + ⟶

Geranyl pyrophosphate

[+ is the same as +] ⟶ + ⟶ ⟶ O

Camphor

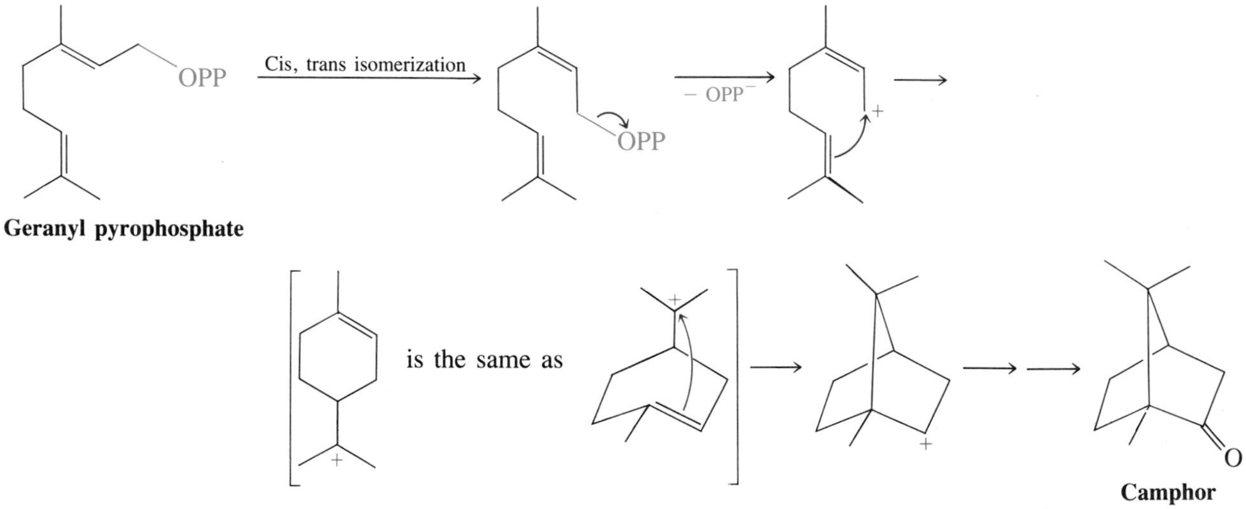

Other higher terpenes are constructed by similar cyclization reactions.

To summarize, 1,3-butadiene polymerizes in a 1,2 or 1,4 manner to give polybutadienes with various amounts of cross-linking and therefore variable elasticity. Synthetic rubber can be made from 2-methyl-1,3-butadiene and contains various amounts of *E* and *Z* double bonds. Natural rubber is constructed by isomerization of 3-methyl-3-butenyl pyrophosphate to the 2-butenyl system, ionization, and electrophilic (step-by-step) polymerization. Similar mechanisms account for the incorporation of 2-methyl-1,3-butadiene (isoprene) units into the polycyclic structure of terpenes.

14-11 Electronic Spectra: Ultraviolet and Visible Spectroscopy

Section 10-2 explained that organic molecules may absorb radiation at various wavelengths. Spectroscopy is possible because the phenomenon is restricted to

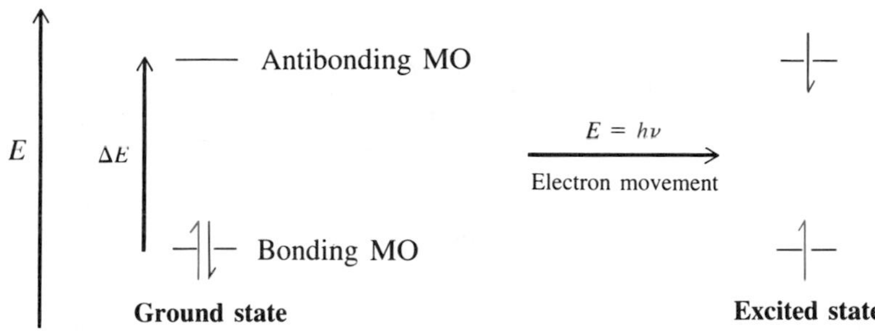

FIGURE 14-13 Electronic excitation by transfer of an electron from a bonding to an antibonding orbital converts a molecule from its ground electronic state to an excited state.

quanta of defined energies, $h\nu$, to effect specific excitations with energy change ΔE.

$$\Delta E = h\nu = \frac{hc}{\lambda} \qquad (c = \text{velocity of light})$$

This section will discuss a form of spectroscopy that requires electromagnetic radiation of relatively high energy, within the wavelength ranges of 200 to 400 nm, called **ultraviolet spectroscopy,** and 400 to 800 nm, **visible spectroscopy** (see Figure 10-2). Both are useful for investigating the electronic structures of unsaturated molecules and for measuring the extent of their conjugation.

A UV-visible spectrometer is constructed according to the general scheme in Figure 10-3. As in NMR, samples are usually dissolved in solvents that do not absorb in the spectral region under scrutiny. Examples are ethanol, methanol, and cyclohexane, none of which have absorption peaks above 200 nm. The events triggered by electromagnetic radiation at the UV and visible wavelengths, the excitation of electrons from filled bonding (and nonbonding) to unfilled antibonding molecular orbitals, are recorded as **electronic spectra.**

Ultraviolet and visible light give rise to electronic excitations

Consider the bonds in an average molecule: We can safely assume that, except for lone pairs, all electrons occupy bonding molecular orbitals. The compound is said to be in its **ground electronic state.** Electronic spectroscopy is possible because ultraviolet radiation and visible light have sufficient energy to transfer many such electrons to antibonding orbitals, thereby creating an **excited electronic state** (Figure 14-13). Dissipation of the absorbed energy may occur in the form of a chemical reaction (cf. Section 14-9), as emission of light (fluorescence, phosphorescence), or simply as emission of heat.

Organic σ bonds have a large gap between bonding and antibonding orbitals. To excite electrons in such bonds requires wavelengths much below the practical range (<200 nm). As a result, the technique has found its major uses in the study of π systems, in which filled and unfilled orbitals are much closer in energy. Excitation of such electrons gives rise to $\boldsymbol{\pi \rightarrow \pi^*}$ **transitions.** Nonbonding (n) electrons are even more readily promoted through $\boldsymbol{n \rightarrow \pi^*}$ **transitions** (Figure 14-14). Because the number of π molecular orbitals is equal to that of the component p orbitals, the simple picture presented in Figure 14-14 is rapidly compli-

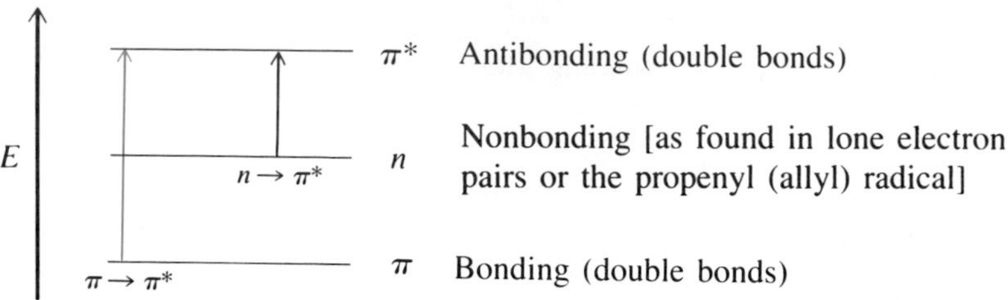

FIGURE 14-14 Electronic transitions in a simple π system. The wavelength of radiation required to cause them is revealed as a peak in the ultraviolet or visible spectrum.

cated by extending conjugation: The number of possible transitions skyrockets and with it the complexity of the spectra.

A typical UV spectrum is that of 2-methyl-1,3-butadiene (isoprene), shown in Figure 14-15. The position of a peak is defined by the wavelength at its maximum, a λ_{max} value (in nanometers). Its height is reported as the **molar extinction coefficient** or **molar absorptivity,** ϵ, which is characteristic of the molecule. The value of ϵ is calculated by dividing the measured peak height (absorbance, A) by the molar concentration, C, of the sample (assuming a standard cell length of 1 cm).

$$\epsilon = \frac{A}{C}$$

The size of ϵ can range from less than a hundred to several hundred thousand. It provides a good estimate of the efficiency of light absorption. Electronic spectral peaks are frequently broad, as in Figure 14-16, not the sharp lines typical of many NMR spectra.

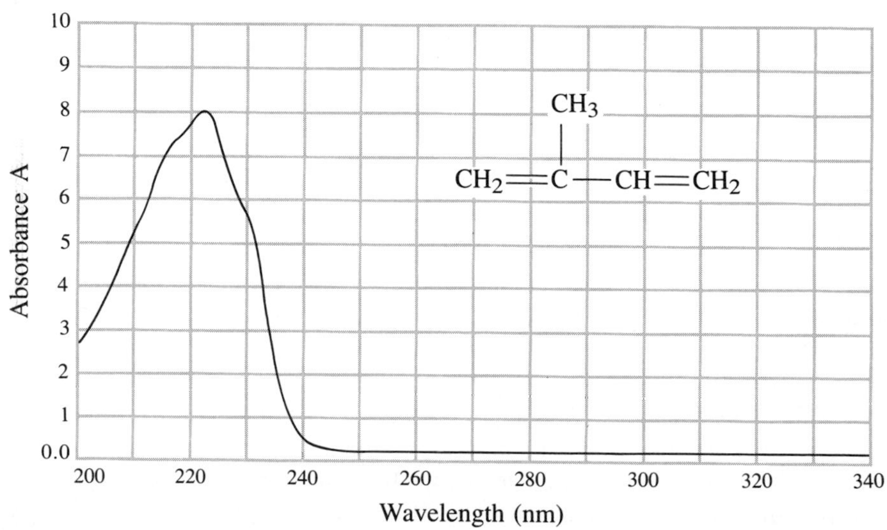

FIGURE 14-15 Ultraviolet spectrum of 2-methyl-1,3-butadiene in methanol, λ_{max} = 222.5 nm (ϵ = 10,800). The two indentations at the sides of the main peak are called shoulders.

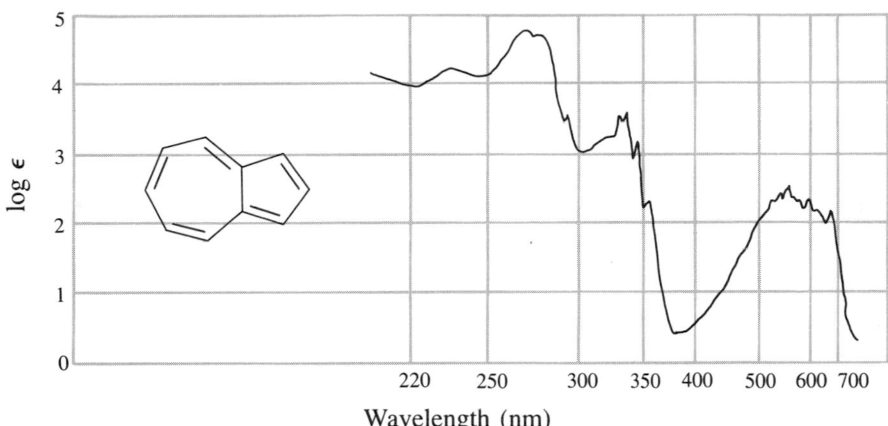

FIGURE 14-16 UV-visible spectrum of azulene in cyclohexane. The absorbance is plotted as log ϵ to compress the scale. The horizontal axis, representing wavelength, also is nonlinear.

Electronic spectra tell us the extent of delocalization

Electronic spectra often indicate the size and degree of delocalization in an extended π system. The more double bonds there are in conjugation, the longer is the wavelength for the lowest energy excitation (and the more peaks will appear in the spectrum). For example, ethene absorbs at $\lambda_{max} = 171$ nm, and an unconjugated diene, such as 1,4-pentadiene, at $\lambda_{max} = 178$ nm. A conjugated diene, such as 1,3-butadiene, absorbs at much lower energy ($\lambda_{max} = 217$ nm). Further extension of the conjugated system leads to corresponding incremental increases in the λ_{max} values, as shown in Table 14-2. The hyperconjugation of alkyl groups and the improved π overlap in rigid, planar cyclic systems both seem to contribute. Beyond 400 nm (in the visible range), molecules become colored, first yellow, then orange, red, violet, and finally blue-green. For example, the contiguous array of 11 double bonds in β-carotene is responsible for its characteristic intense orange appearance. Finally, conjugation in cyclopolyenes is governed by a totally separate set of rules, to be introduced in the next two chapters. Just compare the electronic spectra of benzene (Figure 15-6) with that of azulene (Figure 14-16), and then both with the data in Table 14-2.

Why should larger conjugated π systems have more readily accessible and lower energy excited states? The answer is pictured in Figure 14-17. As the overlapping p-orbital array gets longer, the energy gap between filled and unfilled orbitals gets smaller, and more bonding and antibonding orbitals are available to give rise to additional electronic excitations.

EXERCISE 14-22

Each substitution of a hydrogen by an alkyl group at an sp^2 carbon in a conjugated system causes the wavelength associated with the lowest energy $\pi \rightarrow \pi^*$ transition to increase by 5 nm. Use this fact and the measured value of λ_{max} for 1,3-butadiene to calculate λ_{max} for 2-methylbutadiene and 2,5-dimethyl-2,4-hexadiene. Compare your results with the measured values in Table 14-2.

TABLE 14-2 λ_{max} **Values for the Lowest Energy Transitions in Ethene and Conjugated Pi Systems**

Alkene structure	Name	λ_{max} (nm)	ϵ
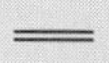	Ethene	171	15,500
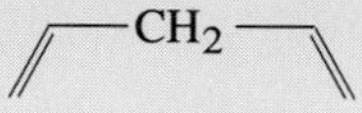 CH_2	1,4-Pentadiene	178	Not measured
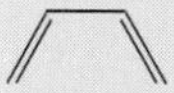	1,3-Butadiene	217	21,000
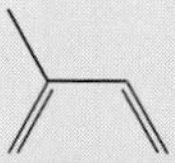	2-Methyl-1,3-butadiene	222.5	10,800
	*trans*-1,3,5-Hexatriene	268	36,300
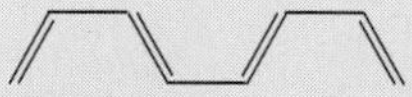	*trans,trans*-1,3,5,7-Octatetraene	330	Not measured
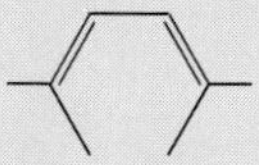	2,5-Dimethyl-2,4-hexadiene	241.5	13,100
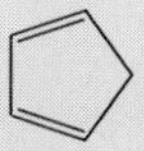	1,3-Cyclopentadiene	239	4,200
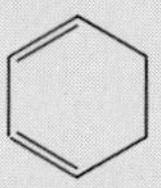	1,3-Cyclohexadiene	259	10,000
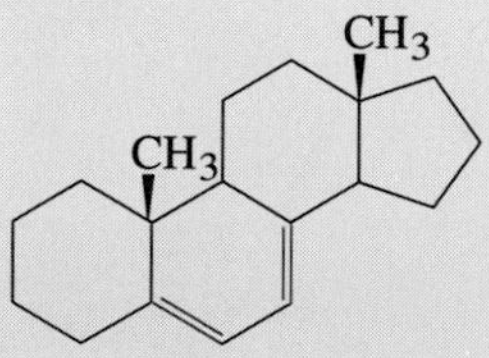 CH_3, CH_3	A steroid diene	282	Not measured
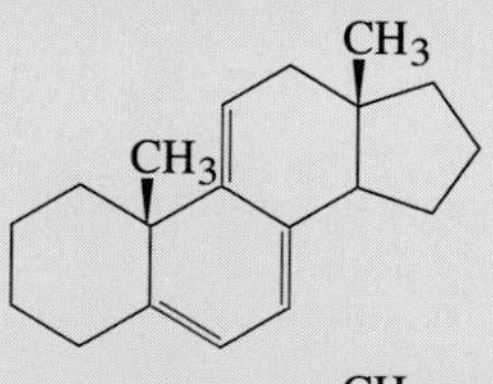 CH_3, CH_3	A steroid triene	324	Not measured
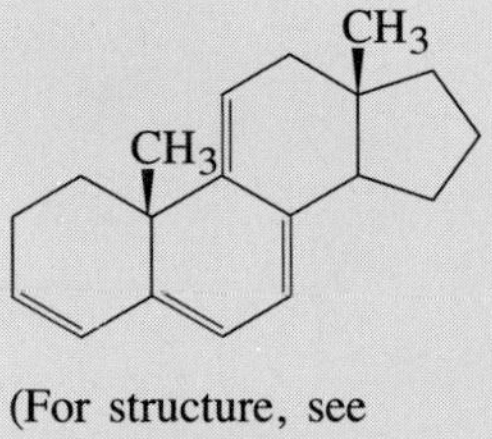 CH_3, CH_3	A steroid tetraene	355	Not measured
(For structure, see Section 14-7)	*β*-Carotene (Vitamin A precursor)	497 (orange)	133,000
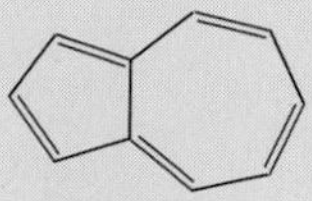	Azulene, a cyclic conjugated hydrocarbon	696 (blue-violet)	150

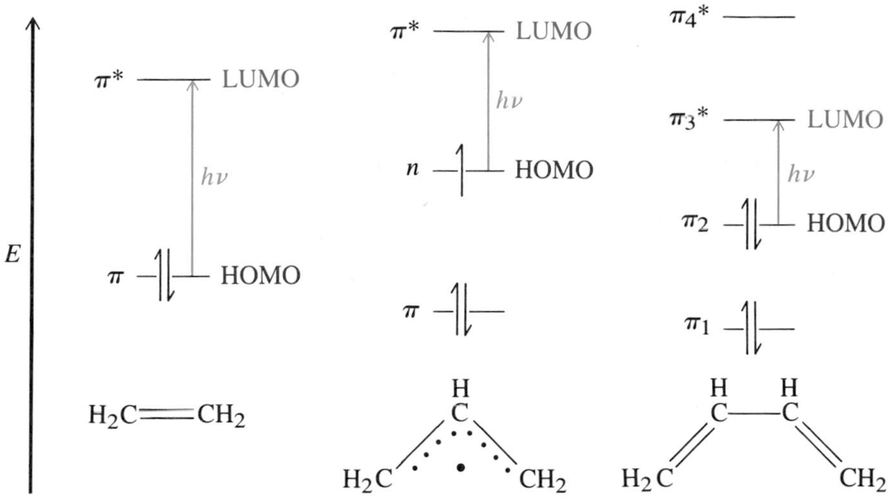

E

π^* — LUMO; $h\nu$; π — HOMO; $H_2C{=}CH_2$

π^* — LUMO; $h\nu$; n — HOMO; π; allyl radical $H_2C\cdots CH\cdots CH_2$

π_4^*; π_3^* — LUMO; $h\nu$; π_2 — HOMO; π_1; $H_2C{=}CH{-}CH{=}CH_2$

FIGURE 14-17 The energy gap between the highest occupied and lowest unoccupied molecular orbitals (HOMO and LUMO) decreases along the series ethene, the 2-propenyl (allyl) radical, and butadiene. Excitation therefore requires less energy and is observed at longer wavelengths.

In summary, UV and visible spectroscopy can be used to detect electronic excitations in conjugated molecules. With an increasing number of molecular orbitals, there is an increasing variety of possible transitions and hence of absorption bands. The band of longest wavelength is typically associated with the movement of an electron from the highest occupied to the lowest unoccupied molecular orbital. Its energy decreases with increasing conjugation.

NEW REACTIONS

1. RADICAL ALLYLIC HALOGENATIONS (Section 14-2)

$$RCH_2CH{=}CH_2 \xrightarrow{\text{NBS, }CCl_4,\ h\nu} R\underset{}{\overset{Br}{C}}HCH{=}CH_2$$

$DH°$ of allylic C–H bond ~ 87 kcal mol^{-1}

2. THERMODYNAMIC COMPARED WITH KINETIC CONTROL IN S_N1 REACTIONS OF ALLYLIC DERIVATIVES (Section 14-3)

$$CH_3CH{=}CHCH_2X \xleftarrow{\text{Slow}} CH_3CH{=}CHCH_2^+ + X^- \xrightleftharpoons{\text{Fast}} CH_3\overset{X}{C}HCH{=}CH_2$$

More stable product

Stability of primary allylic cation ~ ordinary secondary cation

Less stable product

(Formation reversible at higher temperature)

3. S_N2 REACTIVITY OF ALLYLIC HALIDES (Section 14-3)

$$CH_2{=}CHCH_2X + Nu\!:^- \xrightarrow{\text{Propanone or DMSO}} CH_2{=}CHCH_2Nu + X^-$$

Faster than ordinary primary halides

4. ALLYLIC GRIGNARD REAGENTS (Section 14-4)

$$CH_2{=}CHCH_2Br \xrightarrow{Mg,\ (CH_3CH_2)_2O} CH_2{=}CHCH_2MgBr$$

Can be used in additions to carbonyl compounds

5. ALLYLLITHIUM REAGENTS (Section 14-4)

$$RCH_2CH{=}CH_2 \xrightarrow{CH_3CH_2CH_2CH_2Li,\ TMEDA} R\ddot{C}HCH{=}CH_2\ Li^+$$

pK_a of allylic C–H bond ~ 40

6. HYDROGENATION OF CONJUGATED DIENES (Section 14-5)

$$CH_2{=}CH{-}CH{=}CH_2 \xrightarrow{H_2,\ Pd/C,\ CH_3CH_2OH} CH_3CH_2CH_2CH_3 \qquad \Delta H° = -57.1\ \text{kcal mol}^{-1}$$

but compare

$$CH_2{=}CH{-}CH_2{-}CH{=}CH_2 \xrightarrow{H_2,\ Pd/C,\ CH_3CH_2OH} CH_3(CH_2)_3CH_3 \qquad \Delta H° = -60.8\ \text{kcal mol}^{-1}$$

7. ELECTROPHILIC REACTIONS OF 1,3-DIENES: 1,2- AND 1,4-ADDITION (Section 14-6)

$$CH_2{=}CH{-}CH{=}CH_2 \xrightarrow{HX} CH_2{=}CH\overset{X}{\overset{|}{C}}HCH_3 + XCH_2CH{=}CHCH_3$$

$$CH_2{=}CH{-}CH{=}CH_2 \xrightarrow{X_2} CH_2{=}CH\overset{X}{\overset{|}{C}}HCH_2X + XCH_2CH{=}CHCH_2X$$

8. DIELS-ALDER REACTION (CONCERTED AND STEREOSPECIFIC, ENDO RULE) (Section 14-8)

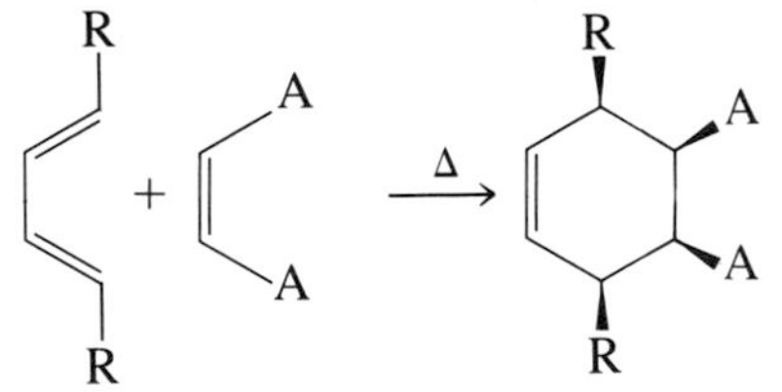

R R A A Δ R R A A

A = electron acceptor

Requires *s*-cis diene; better with electron-poor dienophile

9. ELECTROCYCLIC REACTIONS (Section 14-9)

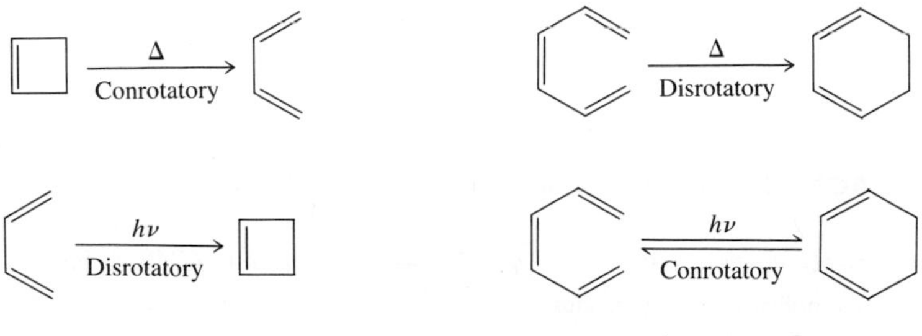

Δ Conrotatory

Δ Disrotatory

$h\nu$ Disrotatory

$h\nu$ Conrotatory

10. POLYMERIZATION OF 1,3-DIENES (Section 14-10)

1,2-Polymerization

$$2n\ CH_2{=}CH{-}CH{=}CH_2 \xrightarrow{\text{Initiator}} {-}(\underset{\underset{\underset{CH_2}{\|}}{CH}}{CH}{-}CH_2{-}\underset{\underset{\underset{CH_2}{\|}}{CH}}{CH}{-}CH_2)_n{-}$$

1,4-Polymerization

$$n\ CH_2{=}CH{-}CH{=}CH_2 \xrightarrow{\text{Initiator}} {-}(CH_2{-}CH{=}CH{-}CH_2)_n{-}$$

Cis or trans

11. 3-METHYL-3-BUTENYL PYROPHOSPHATE AS A BIOCHEMICAL BUILDING BLOCK (Section 14-10)

$$CH_2{=}\overset{\overset{CH_3}{|}}{C}{-}CH_2CH_2OPP \underset{}{\overset{\text{Enzyme}}{\rightleftharpoons}} (CH_3)_2C{=}CHCH_2OPP \longrightarrow (CH_3)_2C{=}CHCH_2{}^+ + {}^-OPP$$

3-Methyl-3-butenyl pyrophosphate — **Allylic cation** — **Pyrophosphate ion**

C–C bond formation

$$\text{(allylic cation)} + \text{(3-methyl-3-butenyl)}\text{OPP} \longrightarrow \text{(tertiary cation)}\text{OPP} \xrightarrow[-\ H^+]{}$$

$$\text{OPP} \xrightarrow[-\ ({}^-OPP)]{} \text{(allylic cation)} \longrightarrow \text{etc.}$$

IMPORTANT CONCEPTS

1. The 2-propenyl (allyl) system is stabilized by resonance. Its molecular-orbital description shows the presence of three π molecular levels: one bonding, one nonbonding, and one antibonding. Its structure is symmetric, any charges or odd electrons being equally distributed between the two end carbons.
2. The chemistry of the 2-propenyl (allyl) cation is subject to both thermodynamic and kinetic control. Nuclcophilic trapping may occur more rapidly at an internal carbon that bears relatively more positive charge, giving the thermodynamically less stable product. The kinetic product may rearrange to its thermodynamic isomer by dissociation followed by eventual thermodynamic trapping.
3. The stability of allylic radicals allows radical halogenations of alkenes at the allylic position.
4. The S_N2 reaction of allylic halides is accelerated by orbital overlap in the transition state.
5. The special stability of allylic anions allows allylic deprotonation by a strong base, such as butyllithium–TMEDA.
6. 1,3-Dienes reveal the effects of conjugation by their resonance energy and a relatively short internal bond (1.47 Å).
7. Electrophilic attack on 1,3-dienes leads to the preferential formation of allylic cations.
8. Extended conjugated systems are reactive because they have many sites for possible attack and the resulting intermediates are stabilized by resonance.

9. Benzene has special stability because of cyclic delocalization.

10. The Diels-Alder reaction is a concerted stereospecific cycloaddition reaction of an *s*-cis diene to a dienophile; it leads to cyclohexene or 1,4-cyclohexadiene derivatives. It follows the endo rule.

11. Conjugated dienes and trienes equilibrate with their respective cyclic isomers by concerted and stereospecific electrocyclic reactions.

12. Polymerization of 1,3-dienes results in 1,2- or 1,4-additions to give polymers that are capable of further cross-linking. Synthetic rubbers can be synthesized in this way. Natural rubber is made by electrophilic carbon–carbon bond formation involving biosynthetic five-carbon cations derived from 3-methyl-3-butenyl pyrophosphate.

13. Ultraviolet and visible spectroscopy gives a way of estimating the extent of conjugation in a molecule. Peaks in electronic spectra are usually broad and are reported as λ_{max} (nm). Their relative intensity is given by the molar absorptivity (extinction coefficient) ϵ.

PROBLEMS

1. Draw all resonance forms and a representation of the appropriate resonance hybrid for each of the following species.

(a) 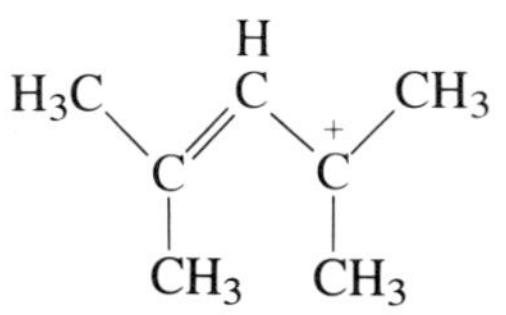 H$_3$C, H, C, CH$_3$, C, C, +, CH$_3$, CH$_3$

(b) H$_3$C, H, C, +, CH$_3$, C, CH, CH$_2$, CH$_3$

(c) CH$_3$

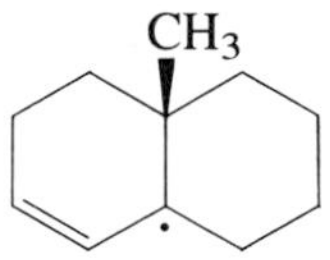

(d)

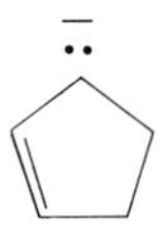

(e)

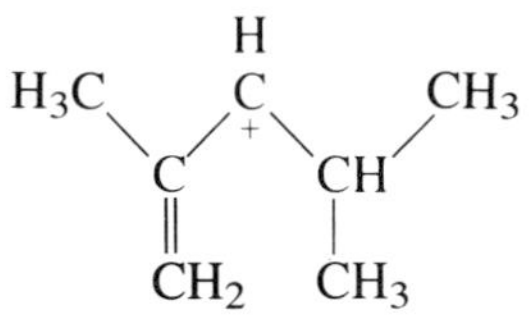

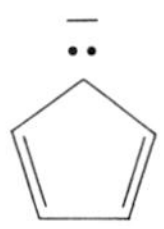

2. Illustrate by means of appropriate structures (including all relevant resonance forms) the initial species formed by **(a)** breaking the weakest C–H bond in 1-butene; **(b)** treating 4-methylcyclohexene with a powerful base (e.g., butyllithium–TMEDA); **(c)** heating a solution of 3-chloro-1-methylcyclopentene in aqueous ethanol.

3. Rank primary, secondary, tertiary, and allylic radicals in order of decreasing stability. Do the same for the corresponding carbocations. Do the results indicate something about the relative ability of hyperconjugation and resonance to stabilize radical and cationic centers?

4. Give the major product(s) of each of the following reactions. If more than one product forms, indicate which is kinetic (i.e., formed fastest at low temperature and short reaction time) and which is thermodynamic (i.e., formed in highest yield at higher temperature after longer reaction times).

(a) 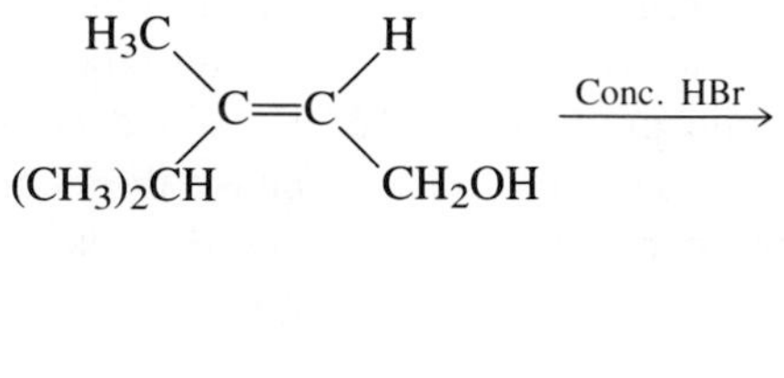 H$_3$C, H, C=C, (CH$_3$)$_2$CH, CH$_2$OH $\xrightarrow{\text{Conc. HBr}}$

(b) 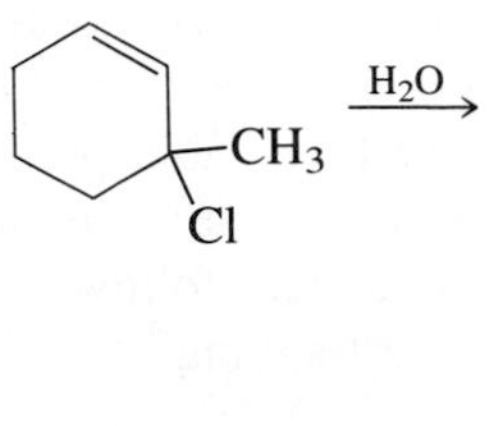 CH$_3$, Cl $\xrightarrow{H_2O}$

(c) 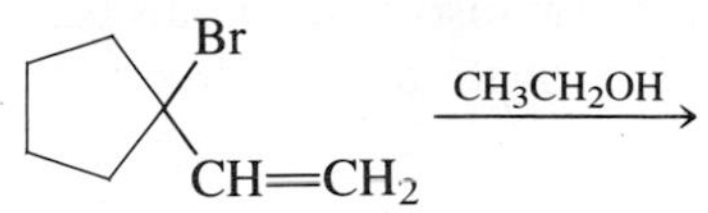 [1-bromo-1-vinylcyclopentane: cyclopentane bearing Br and CH=CH$_2$] $\xrightarrow{CH_3CH_2OH}$

(d) 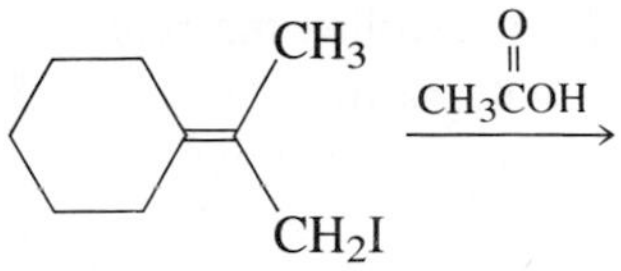 [cyclohexylidene=C(CH$_3$)(CH$_2$I)] $\xrightarrow{CH_3\overset{O}{\overset{\|}{C}}OH}$

(e) 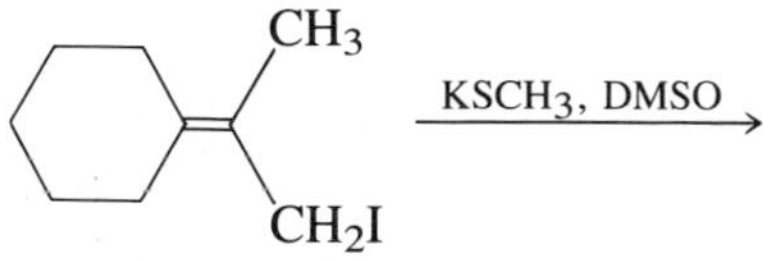 [cyclohexylidene=C(CH$_3$)(CH$_2$I)] $\xrightarrow{KSCH_3,\ DMSO}$

(f) 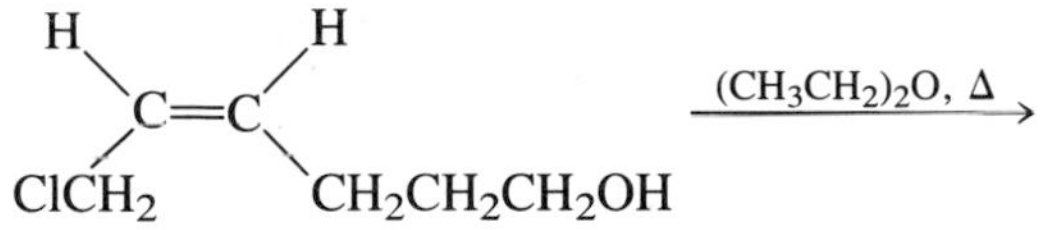 H(ClCH$_2$)C=C(H)(CH$_2$CH$_2$CH$_2$OH) $\xrightarrow{(CH_3CH_2)_2O,\ \Delta}$

5. Formulate detailed mechanisms for the reactions in Problem 4(a, c, e, f).

6. Rank primary, secondary, tertiary, and (primary) allylic chlorides in approximate order of **(a)** decreasing S_N1 reactivity; **(b)** decreasing S_N2 reactivity.

7. Rank the following six molecules in approximate order of decreasing S_N1 reactivity and decreasing S_N2 reactivity.

(a) $CH_3\underset{}{\overset{Cl}{\overset{|}{C}}}HCH{=}CH_2$

(b) 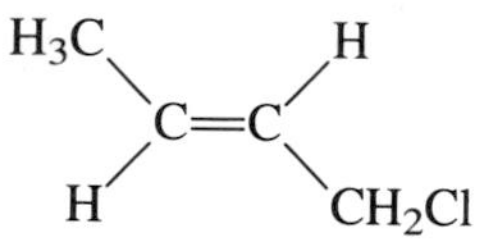 (H$_3$C)(H)C=C(H)(CH$_2$Cl)

(c) (H$_3$C)(H)C=C(CH$_3$)(CH$_2$Cl)

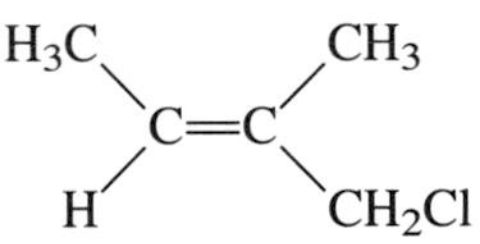

(d) 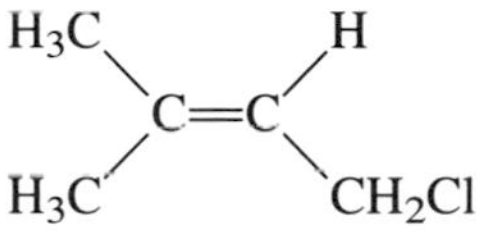 (H$_3$C)(H$_3$C)C=C(H)(CH$_2$Cl)

(e) $(CH_3)_2\overset{Cl}{\overset{|}{C}}CH{=}CH_2$

(f) $CH_2{=}CHCH_2Cl$

8. Give the major product(s) of each of the following reactions.

(a) 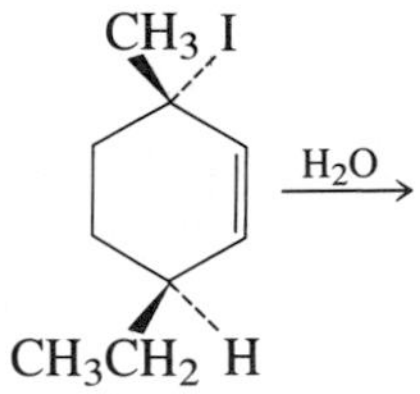 [cyclohexene bearing CH$_3$ and I on one carbon, CH$_3$CH$_2$ and H on the opposite carbon] $\xrightarrow{H_2O}$

(b) 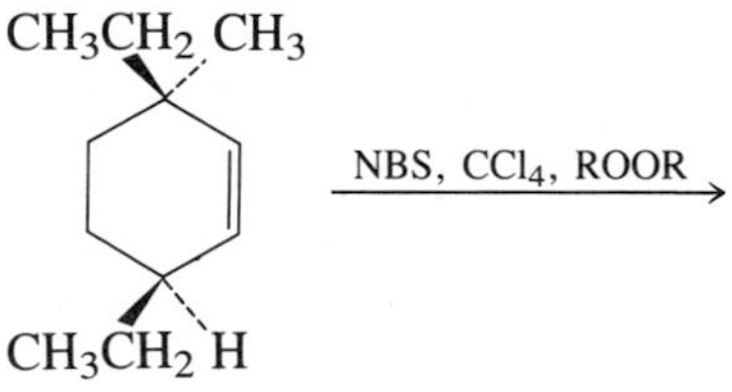 [cyclohexene bearing CH$_3$CH$_2$ and CH$_3$ on one carbon, CH$_3$CH$_2$ and H on the opposite carbon] $\xrightarrow{NBS,\ CCl_4,\ ROOR}$

(c) (*S*)-$CH_3CH_2\overset{CH_3}{\overset{|}{C}}H{-}CH{=}CH_2$ $\xrightarrow{NBS,\ CCl_4,\ ROOR}$

(d) 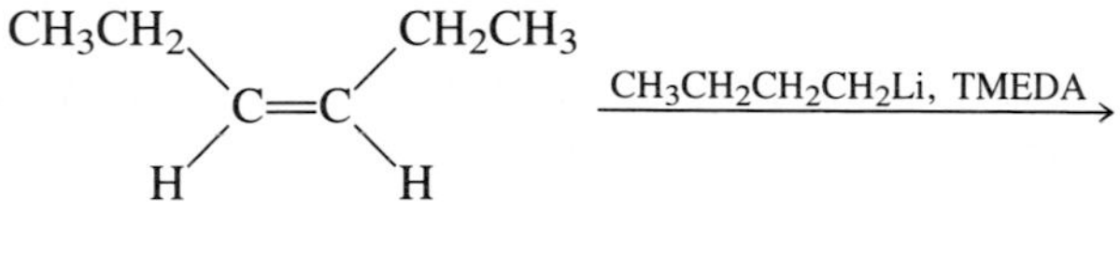 (CH$_3$CH$_2$)(H)C=C(CH$_2$CH$_3$)(H) $\xrightarrow{CH_3CH_2CH_2CH_2Li,\ TMEDA}$

(e) Product of (d) $\xrightarrow[2.\ H^+,\ H_2O]{1.\ CH_3\overset{O}{\overset{\|}{C}}H,\ THF}$

(f) 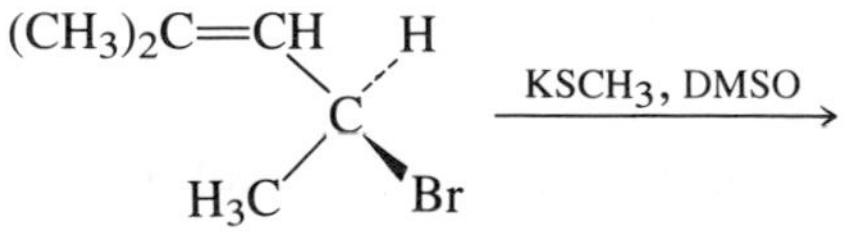 $(CH_3)_2C{=}CH$–C(H)(H$_3$C)(Br) $\xrightarrow{KSCH_3,\ DMSO}$

9. The following reaction sequence gives rise to two isomeric products. What are they? Explain the mechanism of their formation.

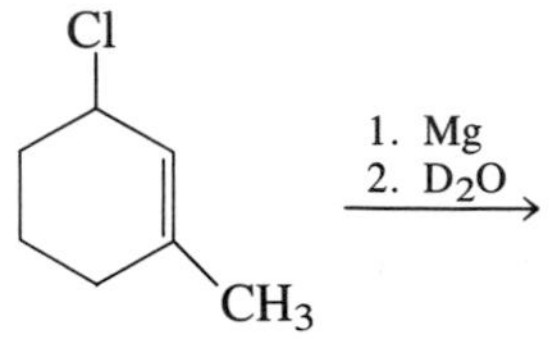

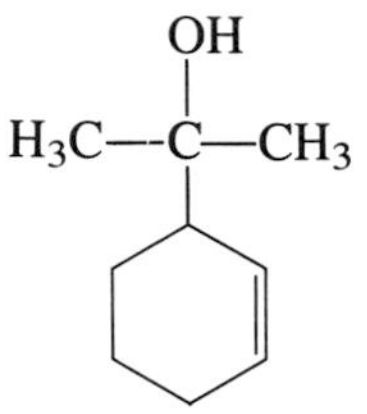

Cl, CH_3; 1. Mg 2. D_2O →

OH
$H_3C—C—CH_3$

10. Starting with cyclohexene, propose a reasonable synthesis of the cyclohexene derivative shown in the margin.

11. Give a systematic name to each of the following molecules.

(a) 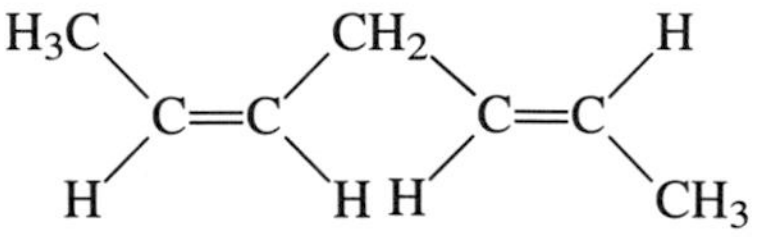 H_3C, CH_2, H; C=C, C=C; H, H H, CH_3

(b) $CH_2{=}CH—CH{=}CH—CH_2OH$

(c) 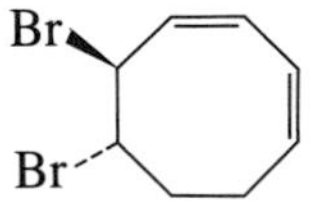 Br, Br

(d)

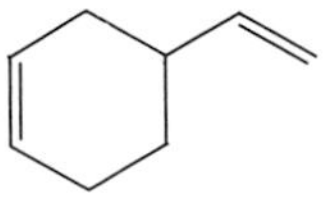

12. Compare the allylic bromination reactions of 1,3-pentadiene and 1,4-pentadiene. Which should be faster? Which is more energetically favorable? How do the product mixtures compare?

$$CH_2{=}CH—CH{=}CH—CH_3 \xrightarrow{\text{NBS, ROOR, CCl}_4}$$
$$CH_2{=}CH—CH_2—CH{=}CH_2 \xrightarrow{\text{NBS, ROOR, CCl}_4}$$

13. Compare the addition of H^+ to 1,3-pentadiene and 1,4-pentadiene (see Problem 12). Write the structures of the products. Draw a qualitative reaction profile showing both dienes and both proton addition products on the same graph. Which diene adds the proton faster? Which one gives the more stable product?

14. What products would you expect from the electrophilic addition of each of the following reagents to 1,3-cycloheptadiene? **(a)** HI; **(b)** Br_2 in H_2O; **(c)** IN_3; **(d)** H_2SO_4 in CH_3CH_2OH.

15. Give the products of the reaction of *trans*-1,3-pentadiene with each of the reagents in Problem 14.

16. What are the products of reaction of 2-methyl-1,3-pentadiene with each of the reagents in Problem 14?

17. Arrange the following carbocations in order of decreasing stability. Draw all possible resonance forms for each of them.

(a) $CH_2{=}CH—\overset{+}{C}H_2$ **(b)** $CH_2{=}CH^+$ **(c)** $CH_3CH_2{}^+$

(d) $CH_3—CH{=}CH—\overset{+}{C}H—CH_3$ **(e)** $CH_2{=}CH—CH{=}CH—CH_2{}^+$

18. Sketch the molecular orbitals for the pentadienyl system in order of ascending energy (see Figures 14-2 and 14-8). Indicate how many electrons are present, and in which orbitals, for **(a)** the free radical, **(b)** the cation, and

(c) the anion (see Figures 14-3 and 14-8). Draw all reasonable resonance forms for any one of these three species.

19. Dienes may be prepared by elimination reactions of substituted allylic compounds. For example,

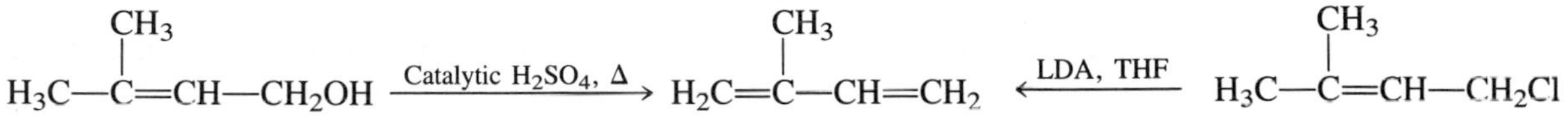

$$H_3C{-}\overset{CH_3}{\overset{|}{C}}{=}CH{-}CH_2OH \xrightarrow{\text{Catalytic } H_2SO_4,\ \Delta} H_2C{=}\overset{CH_3}{\overset{|}{C}}{-}CH{=}CH_2 \xleftarrow{\text{LDA, THF}} H_3C{-}\overset{CH_3}{\overset{|}{C}}{=}CH{-}CH_2Cl$$

Propose detailed mechanisms for each of these 2-methyl-1,3-butadiene (isoprene) syntheses.

20. Give the structures of all possible products of the acid-catalyzed dehydration of vitamin A (Section 14-7).

21. Propose a synthesis of each of the following molecules by Diels-Alder reactions.

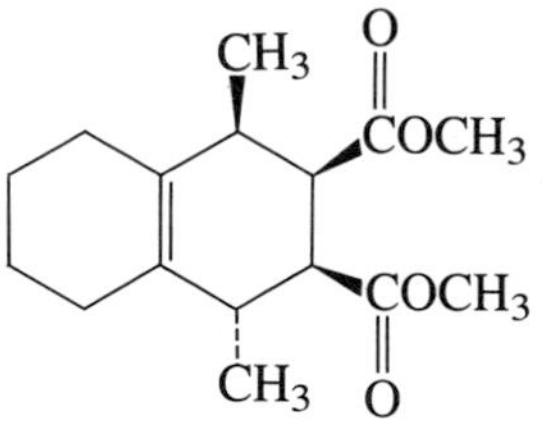

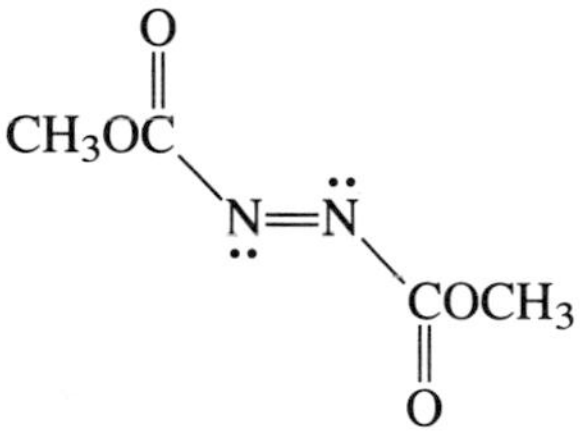

(a) 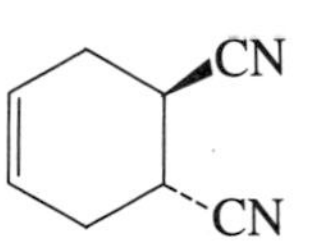 CN, CN (b) 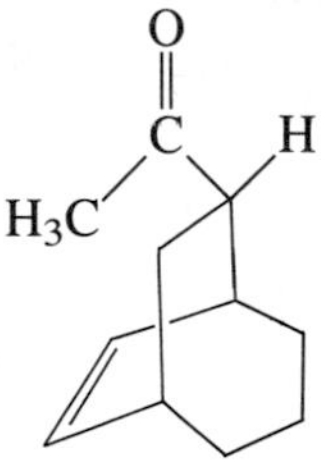 H_3C, C, O, H (c) CH_3, O, $COCH_3$, $COCH_3$, CH_3, O

22. Dimethyl azodicarboxylate takes part in the Diels-Alder reaction as a dienophile. Write the structure of the product of cycloaddition of this molecule with each of the following dienes. **(a)** 1,3-Butadiene; **(b)** *trans,trans*-2,4-hexadiene; **(c)** 5,5-dimethoxycyclopentadiene; **(d)** 1,2-dimethylenecyclohexane.

$CH_3OC(=O){-}\ddot{N}{=}\ddot{N}{-}C(=O)OCH_3$

Dimethyl azodicarboxylate

23. Bicyclic diene A reacts readily with appropriate alkenes by the Diels-Alder reaction, whereas diene B is totally unreactive. Explain.

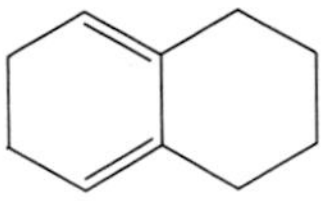

A 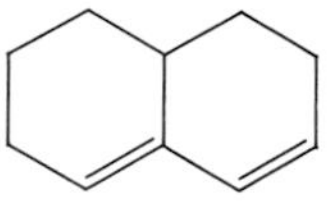 B

24. Formulate the expected product of each of the following reactions.

(a) 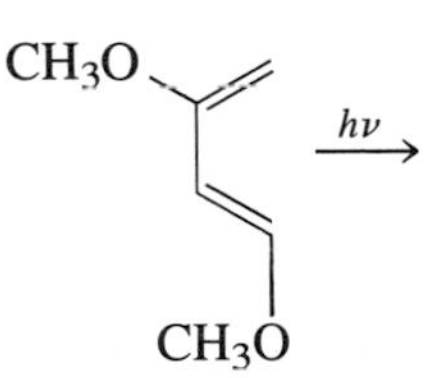 CH_3O, CH_3O $\xrightarrow{h\nu}$ (b) 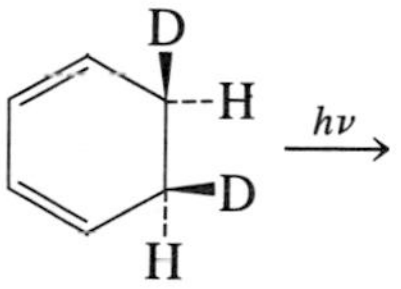 D, H, D, H $\xrightarrow{h\nu}$

(c) 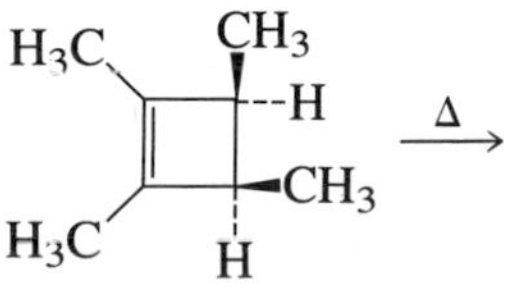 H_3C, CH_3, H, CH_3, H_3C, H $\xrightarrow{\Delta}$ (d) 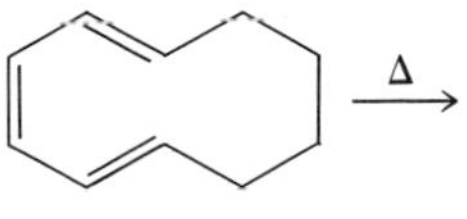 $\xrightarrow{\Delta}$

25. Give abbreviated structures of each of the following. **(a)** (*E*)-1,4-Poly-2-methyl-1,3-butadiene [(*E*)-1,4-polyisoprene]; **(b)** 1,2-poly-2-methyl-1,3-butadiene (1,2-polyisoprene); **(c)** 3,4-poly-2-methyl-1,3-butadiene (3,4-polyisoprene); **(d)** copolymer of 1,3-butadiene and ethenylbenzene (styrene, $C_6H_5CH{=}CH_2$, SBR, used in automobile tires); **(e)** copolymer of 1,3-butadiene and propenenitrile (acrylonitrile, $CH_2{=}CHCN$, latex); **(f)** copolymer of 2-methyl-1,3-butadiene (isoprene) and 2-methylpropene (butyl rubber, for inner tubes).

CH_3 H_3C C CH_2

Limonene

26. The structure of the terpene limonene is shown in the margin. Identify the two 2-methyl-1,3-butadiene (isoprene) units in limonene. **(a)** Treatment of isoprene with catalytic amounts of acid leads to a variety of oligomeric products, one of which is limonene. Devise a detailed mechanism for the acid-catalyzed conversion of two molecules of isoprene into limonene. Take care to use sensible intermediates in each step. **(b)** Two molecules of isoprene may also be converted into limonene by a completely different mechanism, which takes place in the strict absence of catalysts of any kind. Describe this mechanism. What is the name of this reaction?

27. The carbocation derived from geranyl pyrophosphate (Section 14-10) is the biosynthetic precursor of not only camphor but also limonene (Problem 26) and α-pinene (Chapter 4, Problem 18). Formulate mechanisms for the formation of the latter two.

28. What is the longest wavelength electronic transition in each of the following species? Use molecular-orbital designations such as $n \rightarrow \pi^*$, $\pi_1 \rightarrow \pi_2$, in your answer. (Hint: Prepare a molecular-orbital energy diagram such as that in Figure 14-17 for each.) **(a)** 2-Propenyl (allyl) cation; **(b)** 2-propenyl (allyl) radical; **(c)** formaldehyde, $H_2C{=}O$; **(d)** N_2; **(e)** pentadienyl anion (Problem 18); **(f)** 1,3,5-hexatriene.

29. Ethanol, methanol, and cyclohexane are commonly used solvents for UV spectroscopy because they do not absorb radiation of wavelength longer than 200 nm. Why not?

30. The ultraviolet spectrum of a 2×10^{-4} M solution of 3-penten-2-one exhibits a $\pi \rightarrow \pi^*$ absorption at 224 nm with $A = 1.95$ and an $n \rightarrow \pi^*$ band at 314 nm with $A = 0.008$. Calculate the molar absorptivities (extinction coefficients) for these bands.

31. In a published synthetic procedure, propanone (acetone) is treated with ethenyl (vinyl) magnesium bromide, and the reaction mixture is then neutralized with strong aqueous acid. The product exhibits ^{1}H NMR spectrum A. What is its structure?

If the reaction mixture is (improperly) allowed to remain in contact with aqueous acid for too long, a mixture of products is obtained. This mixture gives rise to the ^{1}H NMR spectrum B. New signals are present at $\delta = 1.77$ (several lines), 4.10 (a doublet with $J = 8$ Hz), and 5.45 (a broad triplet with $J = 8$ Hz) ppm with relative intensities 6:2:1. What is the structure of the new compound? How did it get there?

32. Farnesol is a molecule that makes flowers smell nice (lilacs, for instance). Treatment with hot concentrated H_2SO_4 converts farnesol first into bisabolene and finally into cadinene, a compound of the essential oils of junipers and cedars. Propose detailed mechanisms for these conversions.

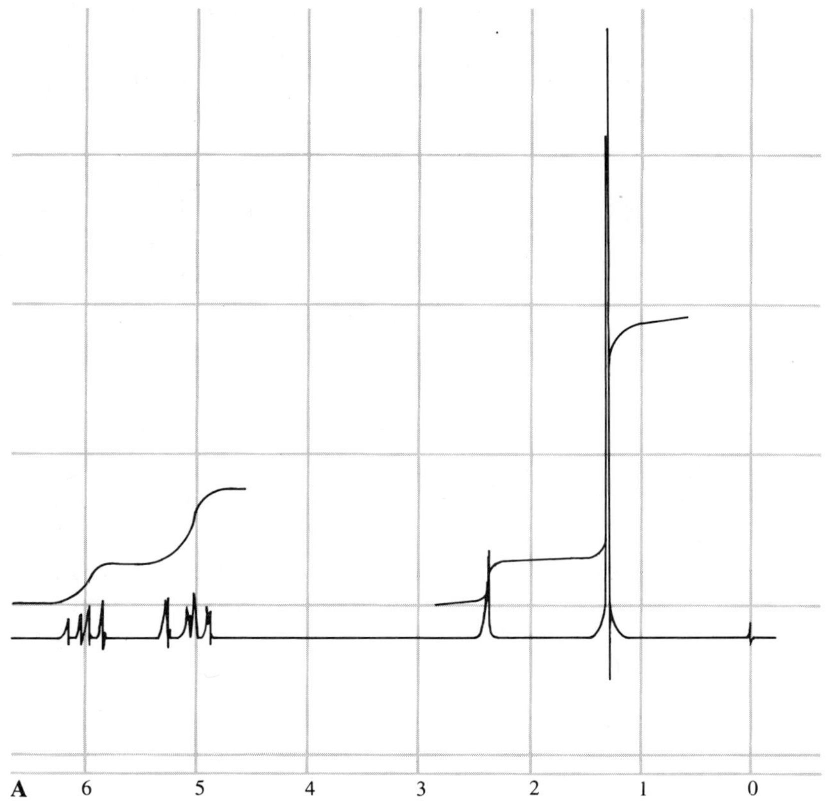
A
6
5
4
3
2
1
0

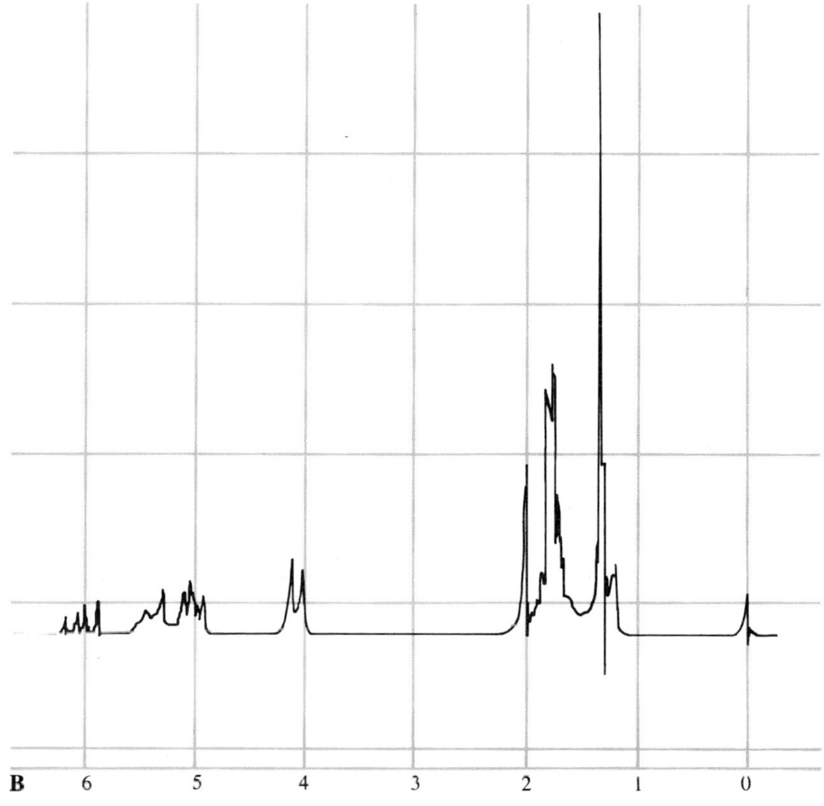
B
6
5
4
3
2
1
0

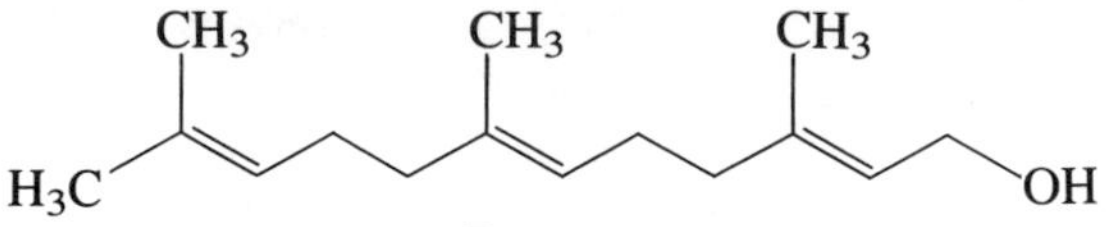

CH3 CH3 CH3
H3C OH
Farnesol

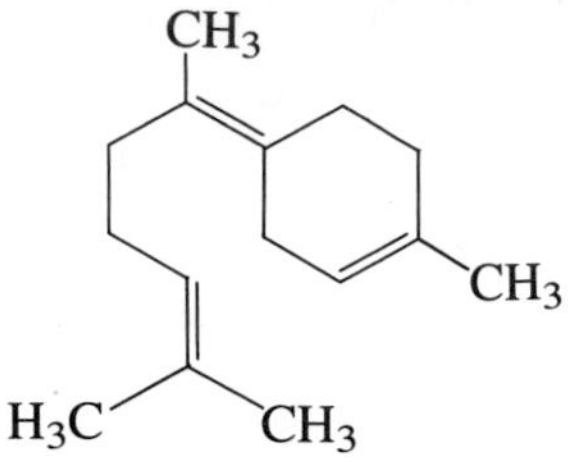

CH3
CH3
H3C CH3
Bisabolene

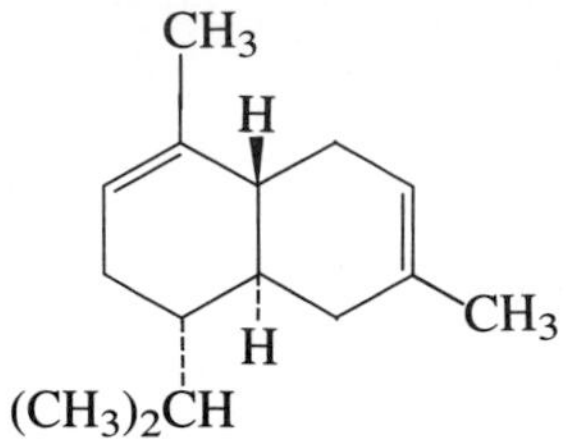

CH3
H
H
CH3
$(CH_3)_2CH$
Cadinene

33. The ratio of 1,2- to 1,4-addition of Br_2 to 1,3-butadiene (Section 14-6) is temperature dependent. Identify the kinetic and thermodynamic products, and explain your choices.

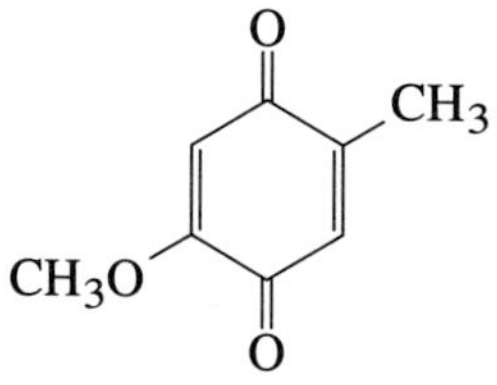

34. Diels-Alder cycloaddition of 1,3-butadiene with the cyclic dienophile shown in the margin takes place at only one of the two carbon–carbon double bonds in the latter to give a single product. Give its structure and explain your answer. Watch stereochemistry.

O
CH3
CH3O
O

This transformation was the initial step in the total synthesis of cholesterol (Section 4-7), completed by R. B. Woodward (see Section 14-9) in 1951. This achievement, monumental for its time, revolutionized synthetic organic chemistry.

15

Unusual Stability of the Cyclic Electron Sextet

Benzene, Other Cyclic Polyenes, and Electrophilic Aromatic Substitution

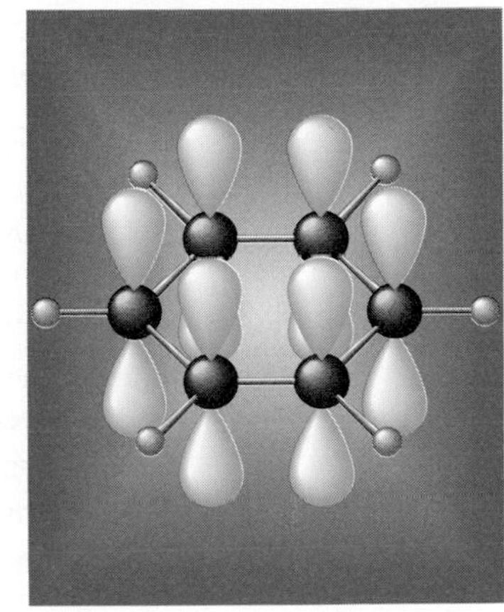

In 1825 the English scientist Faraday* pyrolyzed whale oil to obtain a colorless liquid (b.p. 80.1°C, m.p. 5.5°C) that had the empirical formula CH. This compound posed a problem for the theory that all carbons had to have four valences to other atoms, and it was of particular interest because of its unusual stability and chemical inertness. The molecule was named **benzene** (Section 14-7), and it was eventually shown to have the molecular formula C_6H_6. Various investigators proposed many incorrect structures, such as Dewar benzene, Claus benzene, Ladenburg[†] prismane, and benzvalene. Since then, Dewar benzene, prismane, and benzvalene (but not Claus benzene) have in fact been synthesized. These compounds are unstable and isomerize to benzene in very exothermic transformations. It was Kekulé[‡] who suggested, in 1865, that benzene should be viewed as a set of equilibrating

*Professor Michael Faraday, 1791–1867, Royal Institution of Chemistry, London.

[†]Named for Professor James Dewar, 1842–1923, Cambridge University, England; Professor Adolf Carl Ludwig Claus, 1838–1900, University of Freiburg, Germany; and Professor Albert Ladenburg, 1842–1911, University of Breslau, Germany, respectively.

[‡]Professor F. August Kekulé, see Section 1-4.

Proposed Benzene Structures

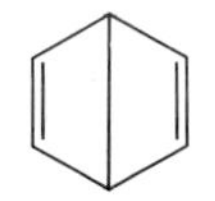

Dewar benzene

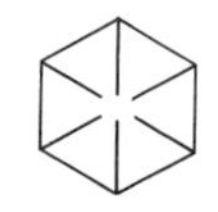

Claus benzene

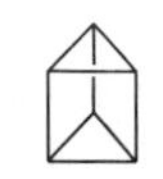

Ladenburg prismane

Benzvalene

cyclohexatriene *isomers*. We now know that this notion was not quite right. In terms of modern electronic theory, benzene is best described by two equivalent cyclohexatrienic *resonance* structures.

This chapter begins by introducing the physical and chemical properties of benzene. We look first at the system of naming substituted benzenes, second at the electronic and molecular structure of the parent molecule, and third at the evidence for an unusual stabilizing energy, the *aromaticity* of benzene. This aromaticity and the special structure of benzene affect its spectral properties and reactivity. Subsequently, we shall see what happens when two or more benzene rings are fused to give more extended π systems. Do these compounds also enjoy the special stability and properties of benzene? Is aromaticity unique to six-membered rings, or are there other cyclic polyenes that possess this property? This chapter answers these questions as well. Finally, the chapter introduces electrophilic aromatic substitution, a class of reactions that affords a ready synthetic path to substituted benzenes.

15-1 Naming the Benzenes

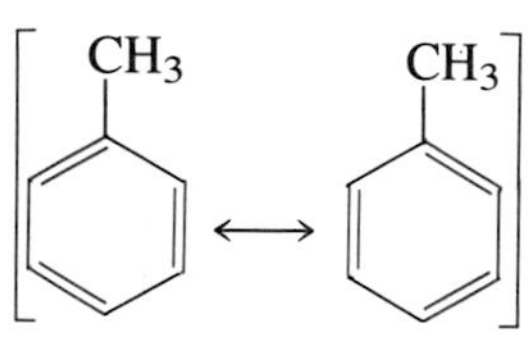

is the same as

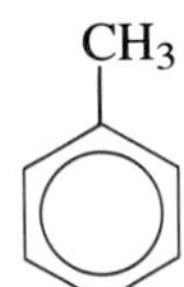

Because of their strong aroma, many derivatives of benzene are called **aromatic compounds.** Benzene, even though its odor is not particularly pleasant, is viewed as the ''parent'' aromatic molecule. Whenever the symbol for the benzene ring with its three double bonds is written, it should be understood to represent only one of a pair of contributing resonance structures. Alternatively, the ring is sometimes drawn as a regular hexagon with an inscribed circle.

Many monosubstituted benzenes are named by adding a substituent prefix to the word benzene.

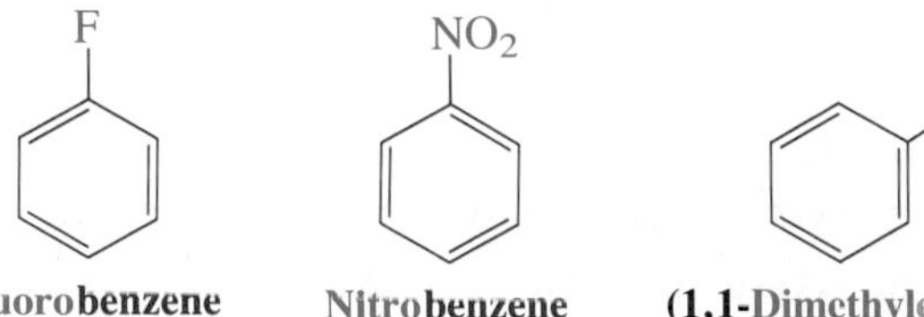

Fluorobenzene **Nitrobenzene** **(1,1-Dimethylethyl)benzene (*tert*-Butylbenzene)**

There are three possible arrangements of disubstituted benzenes. These are designated by the prefixes **1,2- (ortho-,** or ***o*-,** Greek, straight) for adjacent substituents, **1,3 (meta-,** or ***m*-,** Greek, transposed) for 1,3-disubstitution, and **1,4 (para-,** or ***p*-,** Greek, beyond) for 1,4-disubstitution. The substituents are listed in alphabetical order.

1,2-Dichlorobenzene (*o*-Dichlorobenzene) **1-Bromo-3-nitrobenzene (*m*-Bromonitrobenzene)** **1-Ethyl-4-(1-methylethyl)benzene (*p*-Ethylisopropylbenzene)**

To name tri- and more highly substituted derivatives, the six carbons of the ring are numbered with the lowest set of locants, and the substituents are labeled accordingly, as in cyclohexane nomenclature.

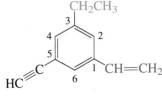

1-Bromo-2,3-dimethylbenzene **1,2,4-Trinitrobenzene** **1-Ethenyl-3-ethyl-5-ethynylbenzene**

The following benzene derivatives will be encountered in this book.

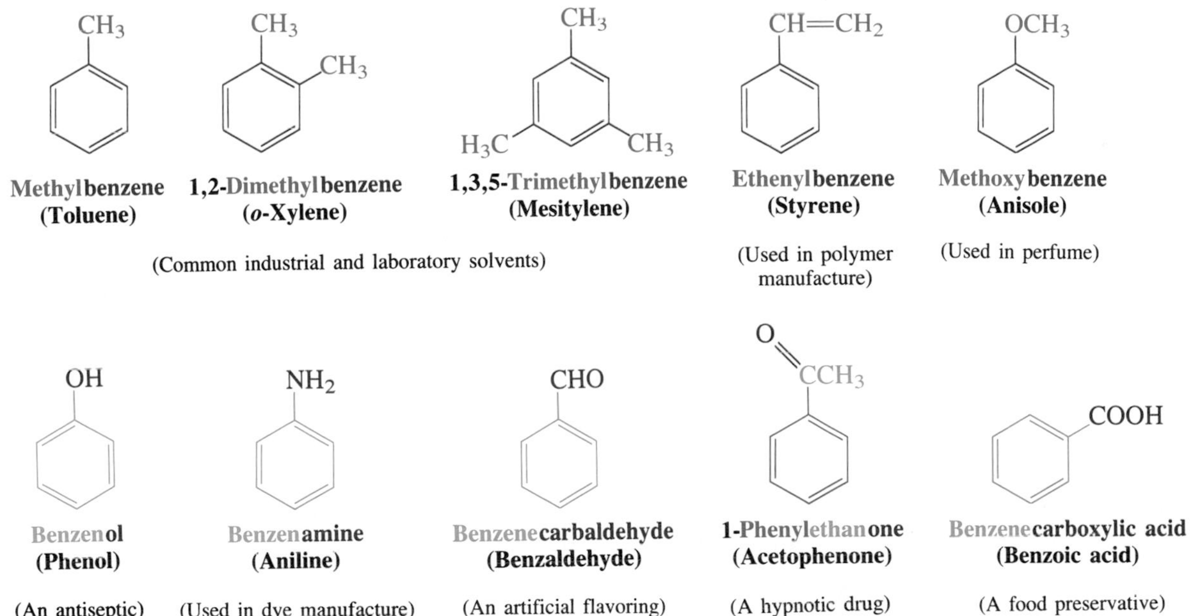

Methylbenzene (Toluene) **1,2-Dimethylbenzene (*o*-Xylene)** **1,3,5-Trimethylbenzene (Mesitylene)**
(Common industrial and laboratory solvents)

Ethenylbenzene (Styrene) (Used in polymer manufacture)

Methoxybenzene (Anisole) (Used in perfume)

Benzenol (Phenol) (An antiseptic)

Benzenamine (Aniline) (Used in dye manufacture)

Benzenecarbaldehyde (Benzaldehyde) (An artificial flavoring)

1-Phenylethanone (Acetophenone) (A hypnotic drug)

Benzenecarboxylic acid (Benzoic acid) (A food preservative)

We shall employ IUPAC nomenclature for all but three systems. Following the indexing preferences of *Chemical Abstracts,* the common names phenol, benzaldehyde, and benzoic acid will be used in place of their systematic counterparts.

Ring-substituted derivatives of such compounds are named by numbering the ring positions or by using the prefixes *o*-, *m*-, and *p*-. The substituent that gives the compound its base name is placed at carbon 1.

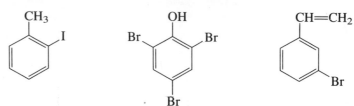

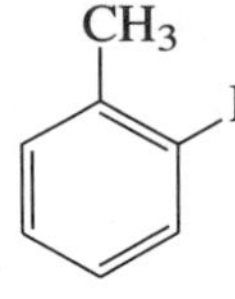

1-Iodo-2-methylbenzene (o-Iodotoluene) **2,4,6-Tribromophenol** **1-Bromo-3-ethenylbenzene (m-Bromostyrene)**

A number of aromatic compounds have common names, many of which refer to their fragrance and natural sources. Several such names have been accepted by IUPAC. As before, a consistent logical naming of these compounds will be adhered to as much as possible, with common names mentioned in parentheses.

Aromatic Flavoring Agents

Methyl 2-hydroxybenzoate (Methyl salicylate, oil of wintergreen flavor) **4-Hydroxy-3-methoxybenzaldehyde (Vanillin, vanilla flavor)** **5-Methyl-2-(1-methylethyl)phenol (Thymol, thyme flavor)**

Phenylmethanol (Benzyl alcohol)

***trans*-1-(4-Bromophenyl)-2-methylcyclohexane**

The generic term for substituted benzenes is **arene.** An arene as a substituent is referred to as an **aryl group,** abbreviated **Ar.** The parent aryl substituent is **phenyl,** C_6H_5. The $C_6H_5CH_2$– group, which is related to the 2-propenyl (allyl) substituent (Section 22-1), is called **phenylmethyl (benzyl).**

EXERCISE 15-1

Write systematic and common names of the following substituted benzenes.

(a) (b) (c)

EXERCISE 15-2

Draw the structures of **(a)** (1-methylbutyl)benzene; **(b)** 1-ethenyl-4-nitrobenzene (*p*-nitrostyrene); **(c)** 2-methyl-1,3,5-trinitrobenzene (2,4,6-trinitrotoluene—the explosive TNT).

EXERCISE 15-3

The following names are incorrect. Write the correct form. **(a)** 3,5-Dichlorobenzene; **(b)** *o*-aminophenyl fluoride; **(c)** *p*-fluorobromobenzene.

In summary, simple singly substituted benzenes are named by placing the substituent name before the word benzene. For more highly substituted systems, 1,2-, 1,3-, and 1,4- or ortho, meta, and para prefixes indicate the positions of disubstitution or the ring is numbered and the so-labeled substituents are named in alphabetical order. Many simple substituted benzenes have common names.

15-2 Structure and Resonance Energy of Benzene: A First Look at Aromaticity

Benzene is unusually unreactive: At room temperature, benzene is inert to acids, H_2, Br_2, and $KMnO_4$, reagents that readily add to conjugated alkenes (Section 14-6). This section explains why that is so: The cyclic six-electron arrangement imparts a special stability in the form of a large resonance energy (Section 14-7). We shall first review the evidence for the structure of benzene and then estimate its resonance energy by comparing its heat of hydrogenation with those of model systems that lack cyclic conjugation, such as 1,3-cyclohexadiene.

The benzene ring contains six equally overlapping *p* orbitals

The electronic structure of the benzene ring is shown in Figure 15-1. All carbons are sp^2 hybridized, and each p orbital overlaps to an equal extent with its two neighbors. The consequently delocalized electrons form a *π cloud* above and below the ring.

According to this picture, the benzene molecule should be a completely symmetric hexagon with equal C–C bond lengths. Such is in fact the experimentally determined structure (Figure 15-2), which reveals the absence of alternation between single and double bonds. This type of alternation would have been expected if benzene were a conjugated triene, a "cyclohexatriene." The C–C bond length in benzene is 1.39 Å, *between* the values found for the single (1.47 Å) and the double bond (1.34 Å) in 1,3-butadiene (Figure 14-7).

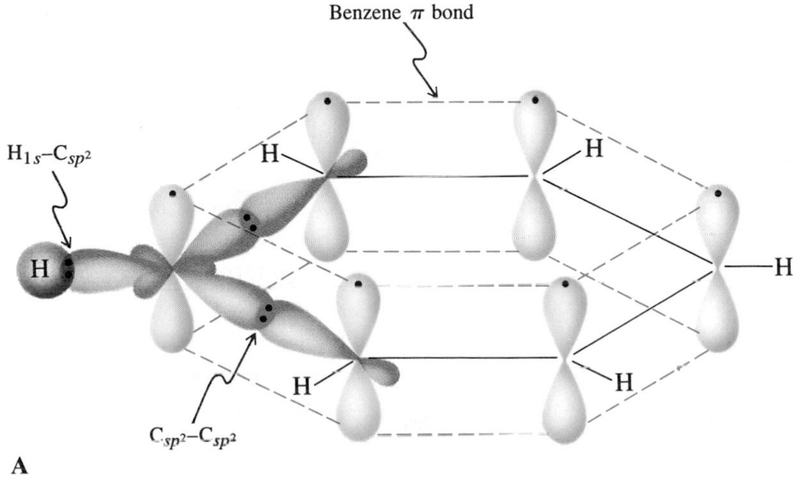

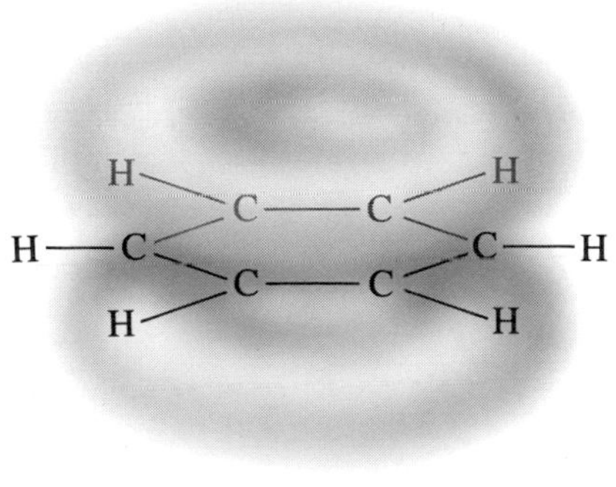

FIGURE 15-1 Orbital picture of the bonding in benzene. (A) The σ framework is depicted as straight lines except for the bonding to one carbon, in which the p orbital and the sp^2 hybrids are shown explicitly. (B) The six overlapping p orbitals in benzene form a π electron cloud located above and below the molecular plane.

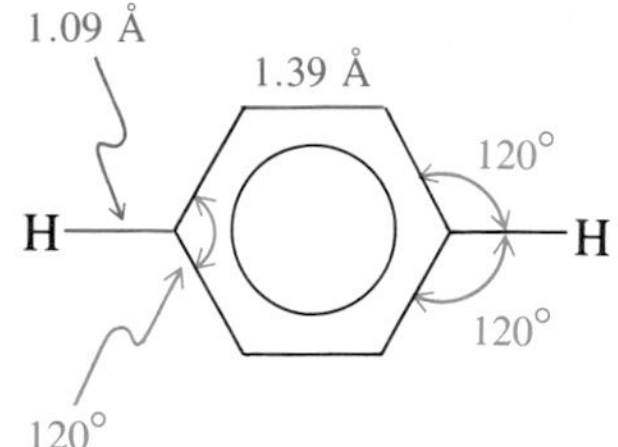

FIGURE 15-2 The molecular structure of benzene. All six C–C bonds are equal; all bond angles are 120°.

Benzene is especially stable: heats of hydrogenation

A way to establish the relative stability of a series of alkenes is to measure their heats of hydrogenation (Sections 11-7 and 14-5). We may carry out a similar experiment with benzene, relating its heat of hydrogenation to those of 1,3-cyclohexadiene and cyclohexene. These molecules are conveniently compared because hydrogenation changes all three into cyclohexane.

The hydrogenation of cyclohexene is exothermic by -28.6 kcal mol^{-1}, a value expected for the hydrogenation of a cis double bond (Section 11-7). The heat of hydrogenation of 1,3-cyclohexadiene ($\Delta H° = -54.9$ kcal mol^{-1}) is slightly less than double that of cyclohexene, because of the resonance stabilization in a conjugated diene (Section 14-5); the energy of that stabilization is $2 \times (28.6) - 54.9 = 2.3$ kcal mol^{-1}.

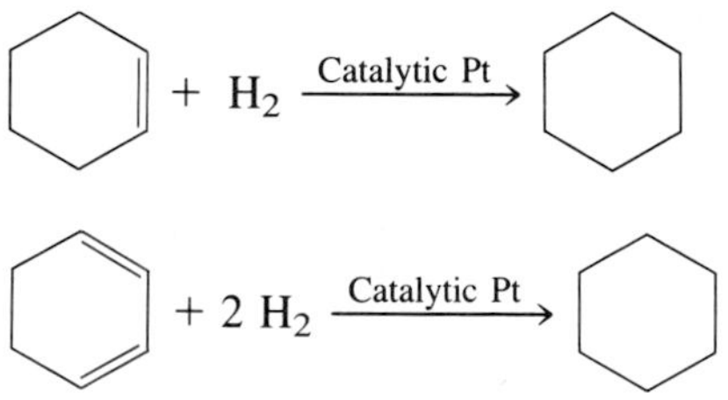

$$+ \; H_2 \xrightarrow{\text{Catalytic Pt}} \qquad \Delta H° = -28.6 \text{ kcal mol}^{-1}$$

$$+ \; 2\, H_2 \xrightarrow{\text{Catalytic Pt}} \qquad \Delta H° = -54.9 \text{ kcal mol}^{-1}$$

Armed with these numbers, we can calculate the expected heat of hydrogenation of benzene, as though it were simply composed of three double bonds like that of cyclohexene.

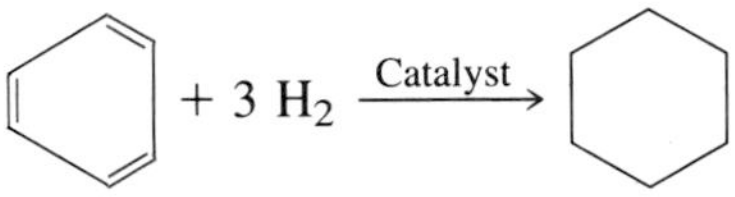

$$+ \; 3\, H_2 \xrightarrow{\text{Catalyst}} \qquad \Delta H° = ?$$

$$\begin{aligned}\Delta H° &= 3\,(\Delta H° \text{ of hydrogenation of } \bigcirc) + 3\,(\text{resonance correction in } \bigcirc)\\ &= (3 \times -28.6) + (3 \times 2.3) \text{ kcal mol}^{-1}\\ &= -85.8 + 6.9 \text{ kcal mol}^{-1}\\ &= -78.9 \text{ kcal mol}^{-1}\end{aligned}$$

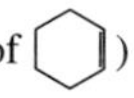

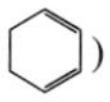

Now let us look at the experimental data. Although benzene is hydrogenated only with difficulty (Section 14-7), special catalysts carry out this reaction, so the heat of hydrogenation of benzene can be measured: $\Delta H° = -49.3$ kcal mol^{-1}, much less than the -78.9 kcal mol^{-1} predicted.

Figure 15-3 summarizes these results. It is immediately apparent that benzene is *much* more stable than a cyclic triene containing alternating single and double bonds. The difference is the **resonance energy** of benzene, about 30 kcal mol^{-1}. Other terms used to describe this quantity are *delocalization energy, aromatic stabilization,* or simply the **aromaticity** of benzene. The original meaning of the word *aromatic* has changed with time, now referring to a thermodynamic property rather than to odor.

To summarize, the structure of benzene is a regular hexagon made up of six sp^2-hybridized carbons. The C–C bond length is between those of a single and a double bond. The electrons occupying the p orbitals form a π cloud above and below the plane of the ring. The structure of benzene can be represented by two

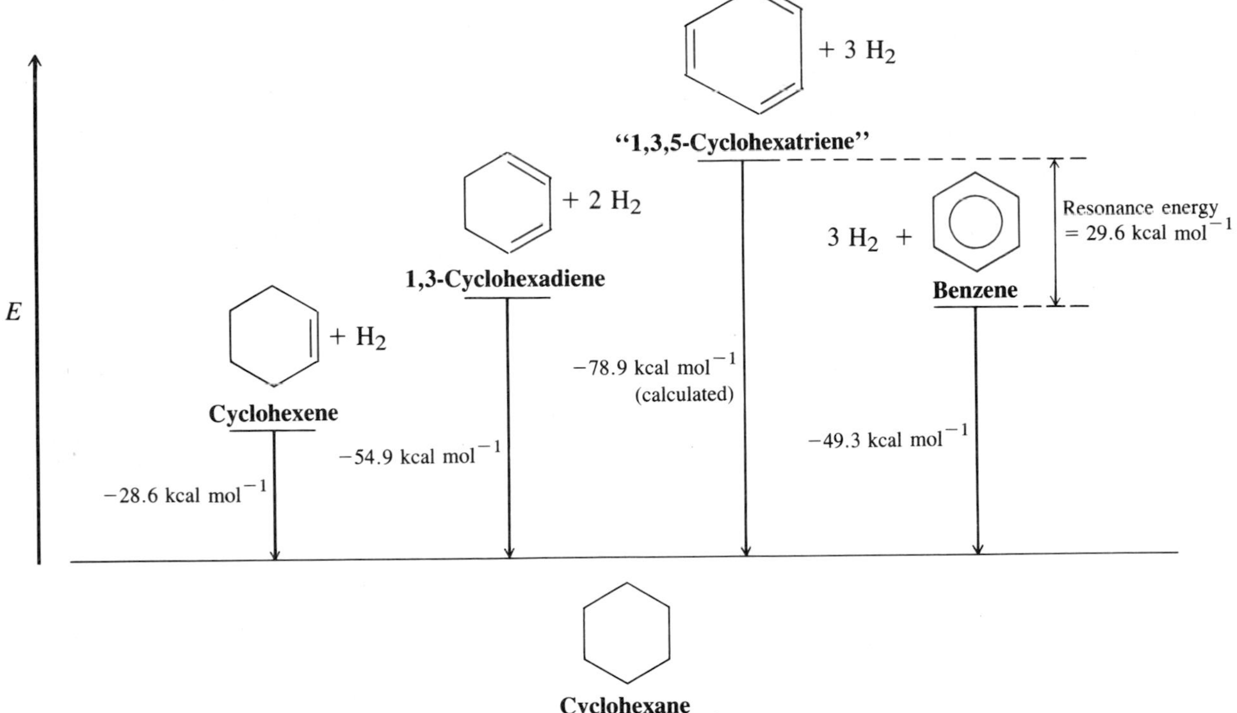

FIGURE 15-3 Heats of hydrogenation provide a measure of benzene's unusual stability. Experimental values for cyclohexene and 1,3-cyclohexadiene allow us to estimate the heat of hydrogenation for the hypothetical "1,3,5-cyclohexatriene." Comparison with the experimental $\Delta H°$ for benzene gives a value of approximately 29.6 kcal mol^{-1} for the aromatic resonance energy.

equally contributing cyclohexatriene resonance forms. Hydrogenation to cyclohexane releases about 30 kcal mol^{-1} less energy than is expected on the basis of nonaromatic models. This difference is the resonance energy of benzene.

15-3 Pi Molecular Orbitals of Benzene

This section compares the six π molecular orbitals of benzene with those of 1,3,5-hexatriene, the open-chain analog. Both sets are the result of the contiguous overlap of six p orbitals, yet the cyclic system differs considerably from the acyclic one. A comparison of the energies of the bonding orbitals in these two compounds shows that cyclic conjugation of three double bonds is better than acyclic conjugation.

Cyclic overlap modifies the energy of benzene's molecular orbitals

Figure 15-4 compares the π molecular orbitals of benzene with those of 1,3,5-hexatriene. The acyclic triene follows a pattern similar to that of 1,3-butadiene (Figure 14-8), but with two more molecular orbitals: The orbitals all have different energies, the number of nodes increasing as we go up. The picture for benzene is different in all respects: different orbital energies, two sets of degenerate (equal energy) orbitals, and completely different nodal patterns.

Is the cyclic π system more stable than the acyclic one? To answer this question, we have to compare the combined energies of the three filled bonding

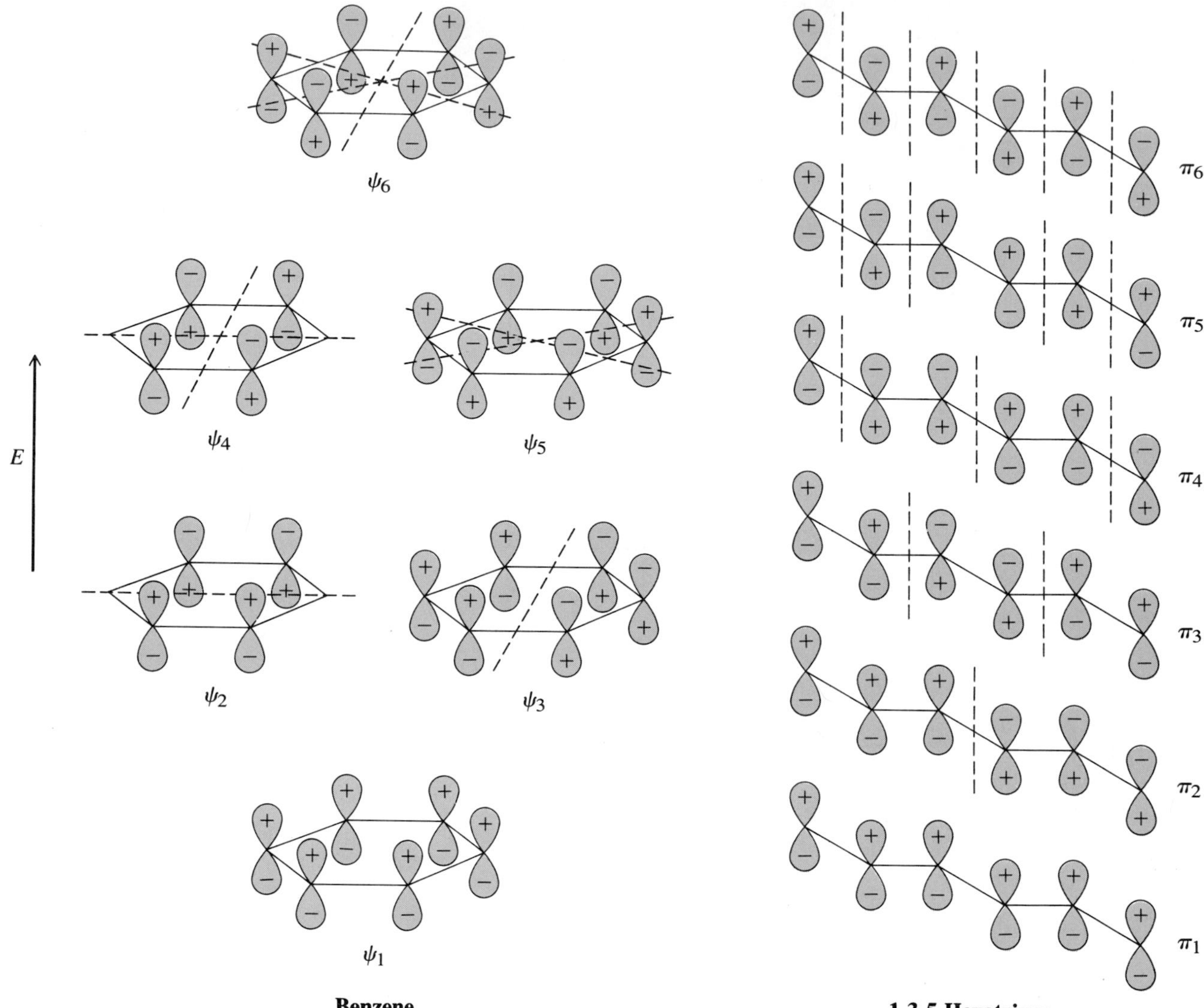

FIGURE 15-4 Pi molecular orbitals of benzene compared with those of 1,3,5-hexatriene. The orbitals are shown at equal size for simplicity. Favorable overlap (bonding) takes place between orbital lobes of equal sign. A sign change is indicated by a node (dashed line). As the number of nodes increases, so does the energy of the orbitals. Note that benzene has two sets of degenerate orbitals, the lower energy set occupied (ψ_2, ψ_3), the other not (ψ_4, ψ_5).

orbitals in both. Figure 15-5 makes the answer obvious: *The cyclic π system is far more stable than the acyclic one*. When proceeding from 1,3,5-hexatriene to benzene, two of the bonding orbitals (π_1 and π_3) are lowered in energy, one is raised; the effect of the latter is more than offset by the former.

Some reactions have aromatic transition states

The structure of benzene accounts in a simple way for several reactions that proceed by what has seemed to be a complicated, concerted movement of three electron pairs: the Diels-Alder reaction (Section 14-8), permanganate addition to

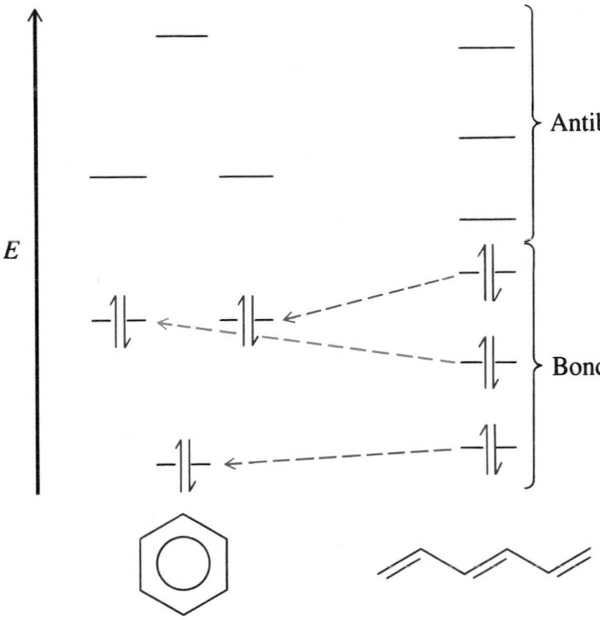

Antibonding molecular orbitals

E

Bonding molecular orbitals

FIGURE 15-5 Energy levels of the π molecular orbitals in benzene and 1,3,5-hexatriene. In both, the six π electrons fill the three bonding molecular orbitals. In benzene, two of them are lower in energy and one is higher than the corresponding orbitals in 1,3,5-hexatriene. The net result is a more stable system of six π electrons.

alkenes (Section 12-10), and the first step of ozonolysis (Section 12-11). In all three processes there is a transition state with cyclic overlap of six electrons in π orbitals (or orbitals with π character). This electronic arrangement is similar to that in benzene and is energetically more favored than the alternative, sequential bond breaking and bond making. Such transition states are called aromatic.

Aromatic Transition States

VII Mn

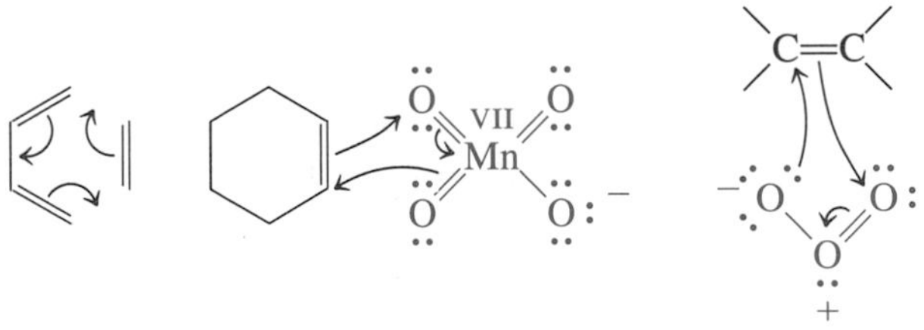

Diels-Alder reaction **Permanganate addition** **Ozonolysis**

EXERCISE 15-4

If benzene were a cyclohexatriene, 1,2-dichloro- and 1,2,4-trichlorobenzene should each exist as two isomers. Draw them.

EXERCISE 15-5

The thermal ring opening of cyclobutene to 1,3-butadiene is exothermic by about 10 kcal mol^{-1} (Section 14-9). Conversely, the same reaction for benzocyclobutene, A, to B (shown in the margin) is *endothermic* by the same amount. Explain.

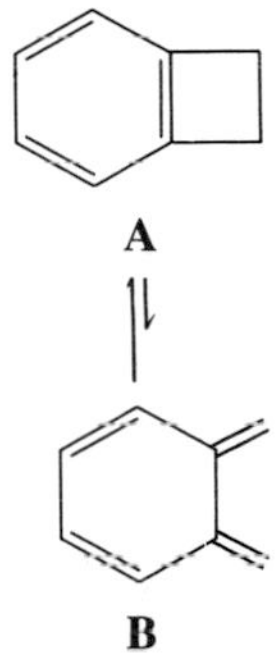

A

B

$\Delta H° \sim +10$ kcal mol^{-1}

To summarize, two of the filled π molecular orbitals of benzene are lower in energy than are those in 1,3,5-hexatriene. Benzene therefore is stabilized by considerably more resonance energy than is its acyclic analog. A similar orbital structure also stabilizes aromatic transition states.

15-4 Spectral Characteristics of the Benzene Ring

Is the unique nature of benzene and its derivatives reflected in their spectra? This section shows that the electronic arrangement in these molecules gives rise to characteristic ultraviolet spectra. The hexagonal structure is also manifest in specific infrared bands. What is most striking, cyclic delocalization causes induced ring currents in NMR spectroscopy, thereby resulting in unusual deshielding of hydrogens attached to the aromatic ring. Moreover, the different coupling constants of 1,2- (ortho), 1,3- (meta), and 1,4- (para) hydrogens in substituted benzenes reveal the substitution pattern.

The UV-visible spectrum of benzene reflects its electronic structure

Cyclic delocalization in benzene gives rise to a characteristic arrangement of energy levels for its molecular orbitals (Figure 15-4). In particular, the energetic gap between bonding and antibonding orbitals is relatively large (Figure 15-5). Is this reflected in its electronic spectrum? As shown in Section 14-11, the answer is yes: Compared to the spectra of the acyclic trienes, we expect smaller λ_{max} values for benzene and its derivatives. You can verify this effect in Figure 15-6: The highest wavelength absorption for benzene occurs at 261 nm, closer to that of 1,3-cyclohexadiene (259 nm, Table 14-2) than of 1,3,5-hexatriene (268 nm).

The ultraviolet and visible spectra of aromatic compounds vary with the introduction of substituents; this phenomenon has been exploited in the tailored synthesis of dyes (Section 22-11). Simple substituted benzenes absorb between 250 and 290 nm. For example, the water-soluble 4-aminobenzoic acid (*p*-aminobenzoic acid, or PABA), 4-H_2N-C_6H_4-COOH, has a λ_{max} at 289 nm, with a rather high extinction coefficient of 18,600. Because of this property (and because it is virtually nontoxic), it is used in suntan lotions, in which it filters out the dangerous ultraviolet radiation emanating from the sun in this wavelength region.

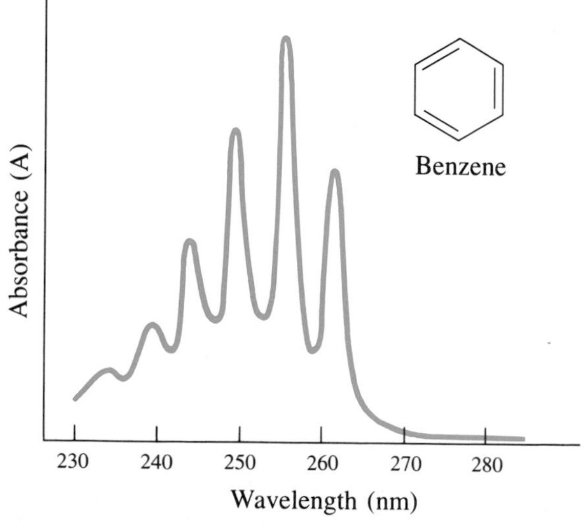

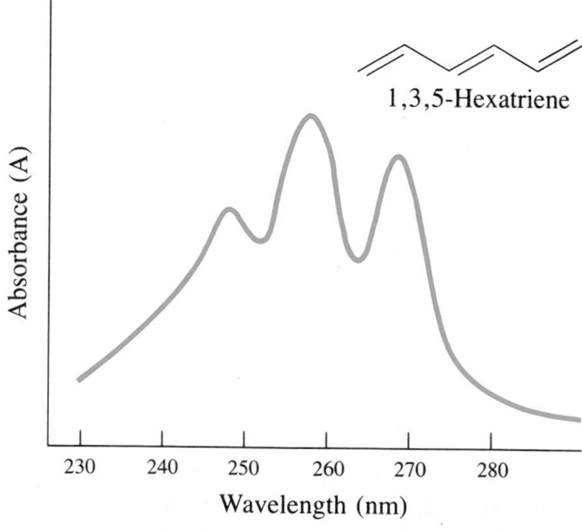

FIGURE 15-6 The distinctive ultraviolet spectra of benzene: $\lambda_{max}(\epsilon)$ = 234(30), 238(50), 243(100), 249(190), 255(220), 261(150) nm; and 1,3,5-hexatriene: $\lambda_{max}(\epsilon)$ = 247(33,900), 258(43,700), 268(36,300) nm. The extinction coefficients of the absorptions of 1,3,5-hexatriene are very much larger than those of benzene; therefore the spectrum at the right was taken at lower concentration.

The infrared spectrum reveals substitution patterns in benzene derivatives

The infrared spectra of benzene and its derivatives show characteristic bands in three regions. The first is at 3030 cm^{-1} for the phenyl–hydrogen stretching mode. The second ranges from 1500 to 2000 cm^{-1} and includes aromatic ring C–C stretching vibrations. Finally, a useful set of bands due to C–H out-of-plane bending motions is found between 650 and 1000 cm^{-1}.

Typical Infrared C–H Out-of-plane Bending Vibrations for Substituted Benzenes (cm^{-1})

Monosubstituted	1,2-Disubstituted	1,3-Disubstituted	1,4-Disubstituted
690–710 730–770	735–770	690–710 750–810	790–840

Their precise location indicates the specific substitution pattern. For example, 1,2-dimethylbenzene (*o*-xylene) has this band at 738 cm^{-1}, the 1,4 isomer at 793 cm^{-1}, and the 1,3 isomer (Figure 15-7) shows two absorptions in this range, at 690 and 765 cm^{-1}.

The NMR spectra of benzene derivatives show the effects of an electronic ring current

A powerful spectroscopic technique for the identification of benzene and its derivatives is based on ^{1}H NMR. The cyclic delocalization of the aromatic ring gives rise to unusual deshielding, which causes the ring hydrogens to resonate at very low field ($\delta \sim 6.5$–8.5 ppm), even lower than the already rather deshielded alkenyl hydrogens ($\delta \sim 4.6$–5.7 ppm, see Section 11-4).

The ^{1}H NMR spectrum of benzene, for example, exhibits a sharp singlet for the six equivalent hydrogens at $\delta = 7.27$ ppm. How can this strong deshielding be explained? In a simplified picture, the cyclic π system with its delocalized electrons may be compared to a loop of conducting metal. When such a loop is exposed to a perpendicular magnetic field (H_0), an electric current (called a **ring current**) flows in the loop, in turn generating a new local magnetic field (h_{local}). This induced field opposes H_0 on the inside of the loop (Figure 15-8), but it reinforces H_0 on the outside—just where the hydrogens are located. This reinforcement results in a local field in their vicinity equal to $H_0 + h_{local}$. So, to cause resonance at constant radio frequency ν, the applied field strength has to be reduced to ($H_0 - h_{local}$): The nuclei are deshielded. This effect is strongest close to the ring and diminishes rapidly with increasing distance from it. Thus, benzylic nuclei are deshielded only about 0.4 to 0.8 ppm more than their allylic counterparts, and hydrogens further away from the π system have chemical shifts that do not differ much from each other and are similar to those in the alkanes.

Chemical Shifts of Allylic and Benzylic Hydrogens

$CH_2{=}CH{-}CH_3$

Allylic: δ 1.68 ppm

$C_6H_5{-}CH_3$

Benzylic: δ 2.35 ppm

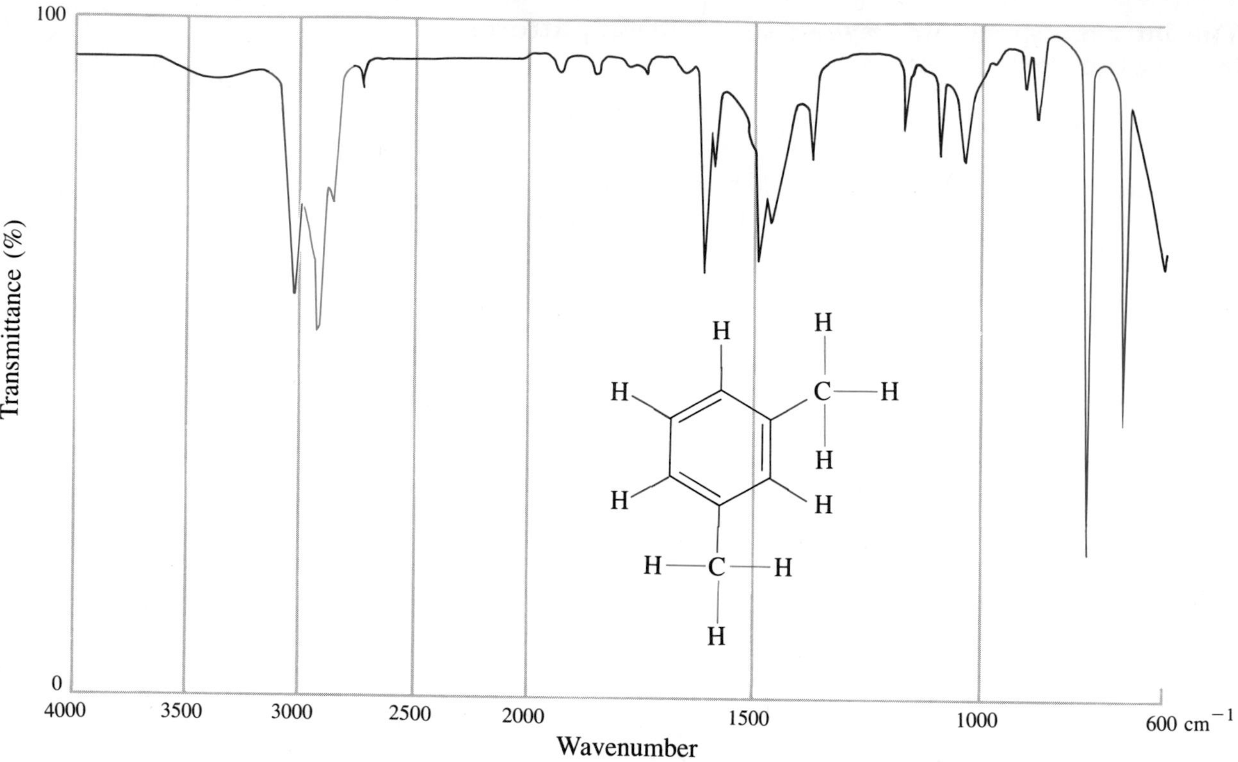

FIGURE 15-7 The infrared spectrum of 1,3-dimethylbenzene (*m*-xylene). There are two C–H stretching absorptions, one due to the aromatic bonds (3030 cm^{-1}), the other to saturated C–H bonds (2920 cm^{-1}). The two bands at 690 and 765 cm^{-1} are typical of 1,3-disubstituted benzenes.

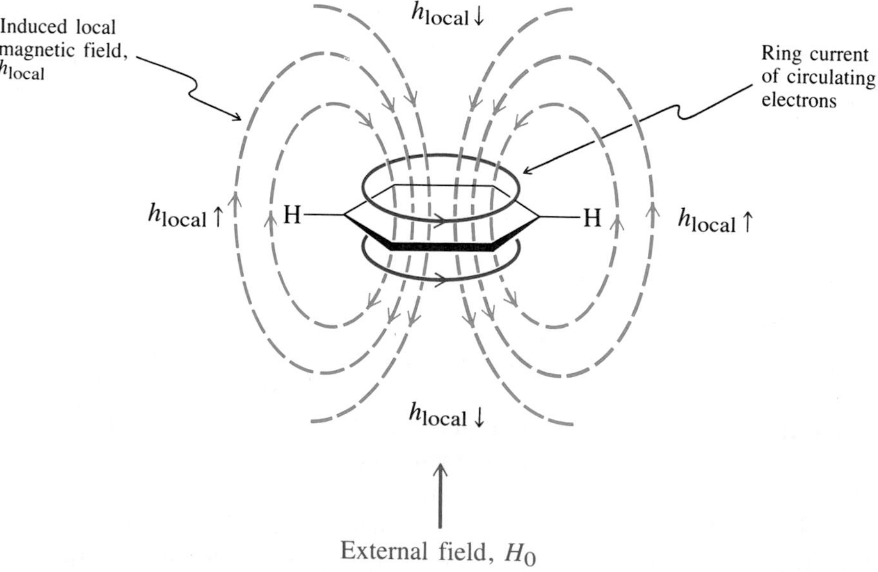

FIGURE 15-8 The π electrons of benzene may be compared to those in a loop of conducting metal. Exposure of this loop of electrons to an external magnetic field H_0 causes them to circulate like an electric current. This ''ring current'' generates a local field reinforcing H_0 on the outside of the ring. Thus, the hydrogens resonate at a lower applied field strength.

FIGURE 15-9 A portion of the 90-MHz ^{1}H NMR spectrum of bromobenzene in CCl_4, a non–first-order spectrum.

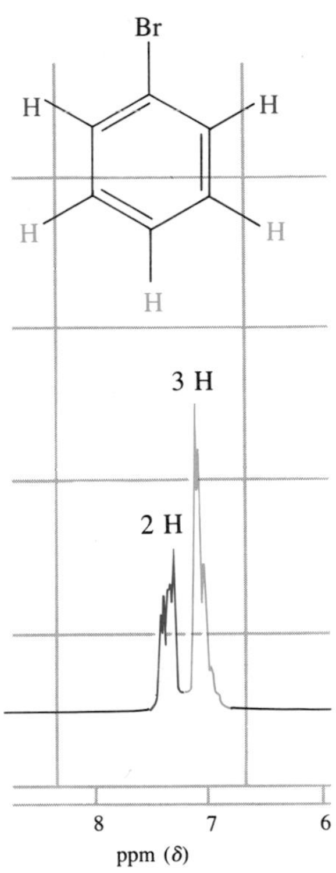

Whereas benzene exhibits a sharp singlet in its NMR spectrum, substituted derivatives may have more complicated patterns. For example, introduction of one substituent renders the hydrogens positioned ortho, meta, and para nonequivalent and subject to mutual coupling. An example is the NMR spectrum of bromobenzene, in which the ortho hydrogens are shifted downfield and the meta and para hydrogens slightly upfield, relative to the chemical shift of benzene. Moreover, all the protons are coupled to one another, thus giving rise to a complicated spectral pattern (Figure 15-9).

Figure 15-10 shows the NMR spectrum of 4-(*N,N*-dimethylamino)benzaldehyde. The large chemical shift difference between the two sets of hydrogens on the ring results in a first-order pattern of two doublets. The observed coupling constant is 9 Hz, a typical splitting between ortho protons.

A first-order spectrum revealing all three types of coupling is shown in Figure 15-11. 1-Methoxy-2,4-dinitrobenzene (2,4-dinitroanisole) bears three ring hydrogens with different chemical shifts and distinct splittings. The hydrogen ortho to the methoxy group appears at $\delta = 7.30$ ppm as a doublet with a 9-Hz ortho coupling. The hydrogen flanked by the two nitro groups (at $\delta = 8.73$ ppm) also appears as a doublet, with a small (3-Hz) meta coupling. Finally, we find the remaining ring hydrogen absorbing at $\delta = 8.50$ ppm as a double doublet because

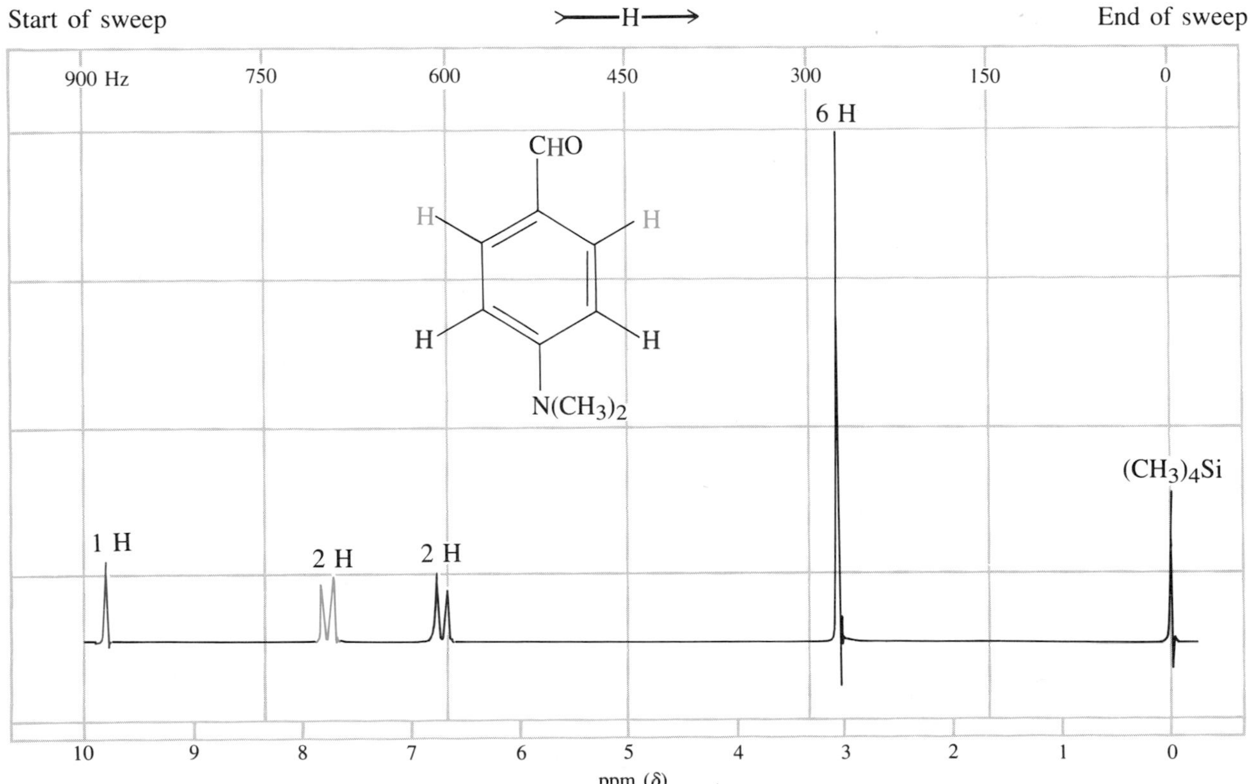

FIGURE 15-10 90-MHz ^{1}H NMR spectrum of 4-(*N,N*-dimethylamino)benzaldehyde (*p*-dimethylaminobenzaldehyde) in $CDCl_3$. In addition to the two aromatic doublets ($J = 9$ Hz), there are singlets for the methyl ($\delta = 3.07$ ppm) and the formyl hydrogens ($\delta = 9.80$ ppm).

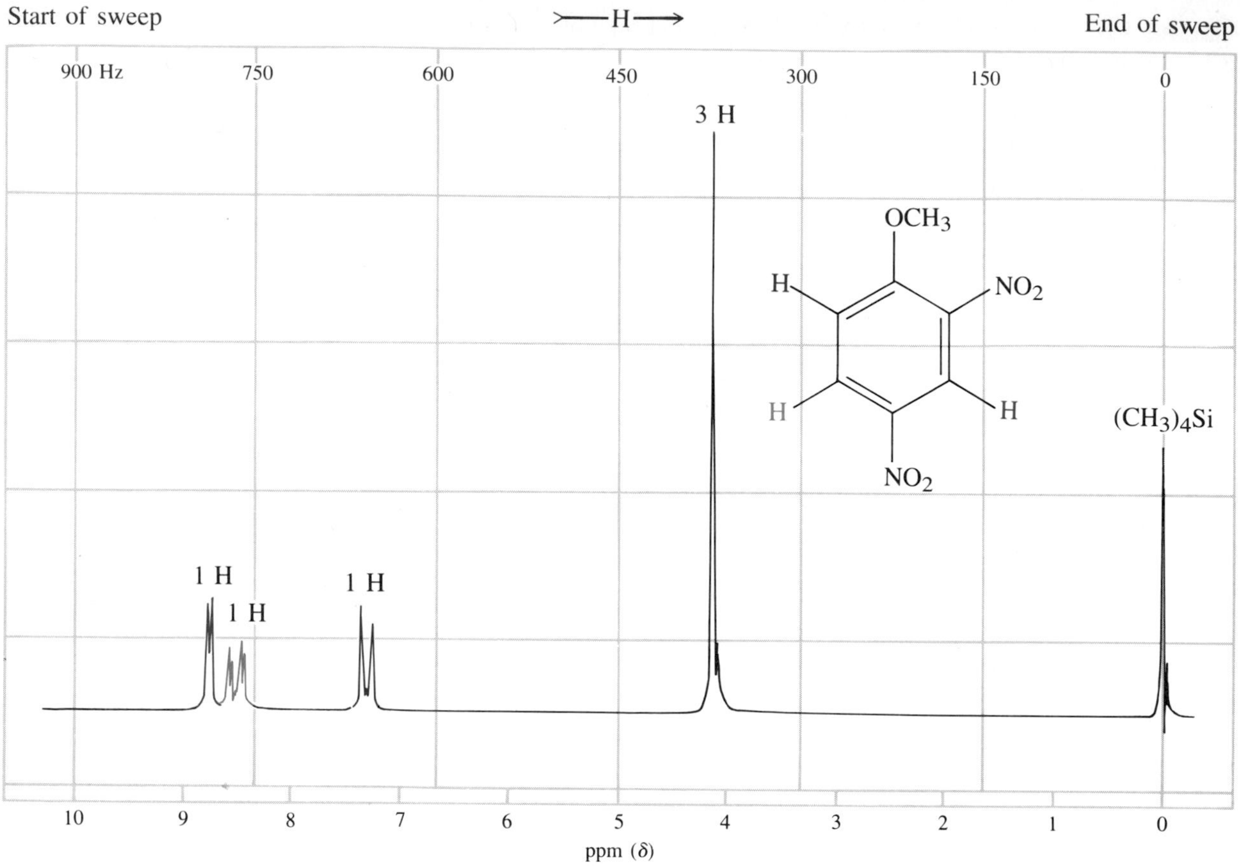

FIGURE 15-11 ^{1}H NMR spectrum of 1-methoxy-2,4-dinitrobenzene (2,4-dinitroanisole) in $CDCl_3$.

of simultaneous coupling to the two other ring protons. Para coupling between the hydrogens at C3 and C6 is too small (<1 Hz) to be resolved; it is evident as a slight broadening of the resonances of these protons.

In contrast with ^{1}H NMR, ^{13}C NMR chemical shifts of carbons in benzene derivatives are dominated by hybridization and substituent effects. As a result of the large chemical shift range for carbon nuclei (200 ppm), the contribution from the ring current (only a few parts per million) becomes relatively less significant. As a result, benzene carbons exhibit chemical shifts similar to those in alkenes, between 120 and 135 ppm when unsubstituted.

EXERCISE 15-6

Can the three isomeric trimethylbenzenes be distinguished solely on the basis of the number of peaks in their proton-decoupled ^{13}C NMR spectra? Explain.

EXERCISE 15-7

A hydrocarbon has the molecular formula $C_{10}H_{14}$. The spectral data for this compound are as follows: ^{1}H NMR (90 MHz) $\delta = 7.02$ (brs, 4 H), 2.82 (septet, $J = 7.0$ Hz, 1 H), 2.28 (s, 3 H), and 1.22 ppm (d, $J = 7.0$ Hz, 6 H); ^{13}C NMR $\delta = 21.3$, 24.2, 38.9, 126.6, 128.6, 134.8, and 145.7 ppm; IR $\tilde{\nu} = 3030$, 2970, 2880, 1515, 1465, and 813 cm^{-1}; UV $\lambda_{max}(\epsilon) = 265(450)$. What is its structure?

In summary, benzene and its derivatives can be recognized and structurally characterized by their spectral data. Electronic absorptions take place between 250 and 290 nm. The infrared vibrational bands are found at 3030 cm^{-1}

($C_{aromatic}$–H), from 1500 to 2000 cm^{-1} (C–C), and from 650 to 1000 cm^{-1} (C–H out-of-plane bending). Most informative is NMR, with low-field resonances for the aromatic hydrogens and carbons. Coupling is largest between the ortho hydrogens, smaller in their meta and para counterparts.

15-5 Polycyclic Benzenoid Hydrocarbons

What happens if several benzene rings are fused to give a more extended π system? Molecules of this class are called **polycyclic benzenoid** or **polycyclic aromatic hydrocarbons (PAHs).** In these structures two or more benzene rings share two or more carbon atoms. Do these compounds also enjoy the special stability of benzene? The next two sections will show that they largely do.

There is no simple system for naming these structures; so we shall use their common names. There is only one way to fuse one benzene ring to another. The resulting compound is called naphthalene. Further fusion can occur in a linear manner to give anthracene, tetracene, pentacene, and so on, a series called the **acenes. Angular fusion,** also called **peri fusion,** results in phenanthrene, which can be further annulated to a variety of other benzenoid polycycles.

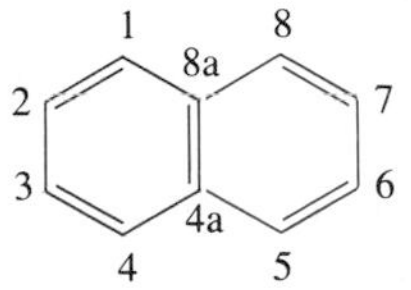

Naphthalene 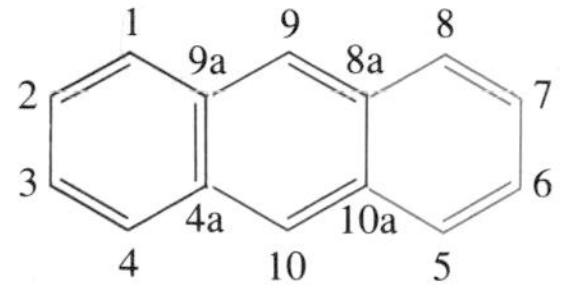 **Anthracene** 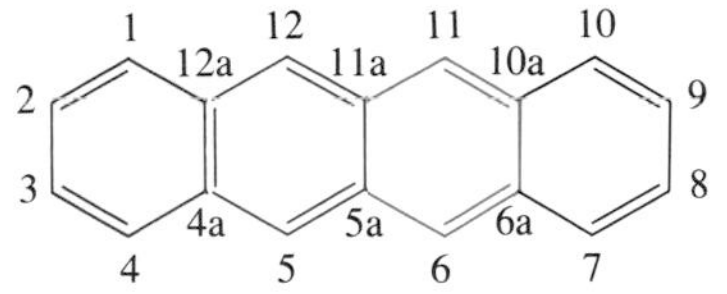 **Tetracene (Naphthacene)** **Phenanthrene**

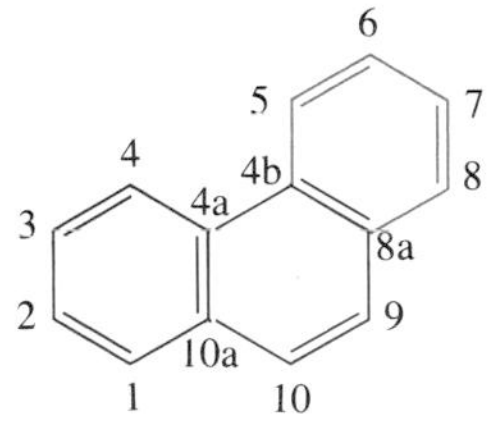

Each has its own numbering system around the periphery. A quaternary carbon is given the number of the preceding carbon in the sequence followed by the letters *a*, *b*, and so on, depending on how close it is to that carbon.

EXERCISE 15-8

Name the following compounds or draw their structures.

(a) 2,6-Dimethylnaphthalene **(b)** 1-Bromo-6-nitrophenanthrene **(c)** 9,10-Diphenylanthracene

(d) Br

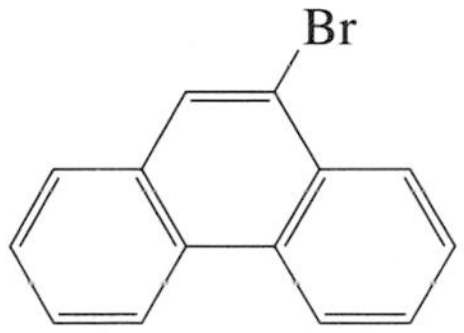

(e) NO_2 HO_3S

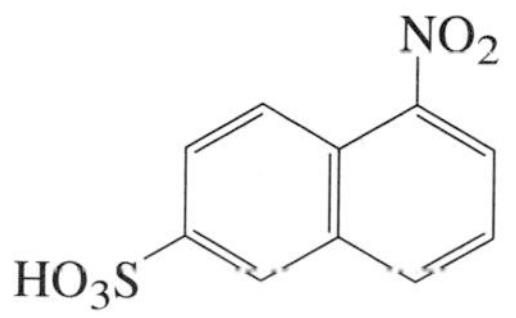

15-6 Fused Benzenoid Hydrocarbons: Naphthalene and the Tricyclic Systems

In contrast with benzene, which is a liquid, naphthalene is a colorless crystalline material with a melting point of 80°C. It is probably best known as a moth repellent and insecticide, although in these capacities it has been partly replaced by chlorinated compounds such as 1,4-dichlorobenzene (*p*-dichlorobenzene).

Is naphthalene still aromatic? Does it share benzene's delocalized electronic structure and thermodynamic stability? Its spectral properties suggest strongly that it does. Particularly revealing are the ultraviolet and NMR spectra.

Naphthalene is aromatic: a look at spectra

The ultraviolet spectrum of naphthalene (Figure 15-12) shows a pattern typical for an extended conjugated system, with peaks at wavelengths as long as 320 nm. On the basis of this observation, the electrons are delocalized more extensively than in benzene (Section 15-2 and Figure 15-6). Thus, it appears that the added four π electrons enter into efficient overlap with those of the attached benzene ring. In fact, it is possible to draw several resonance structures.

Resonance Structures in Naphthalene

Alternatively, the continuous overlap of the 10 *p* orbitals can be shown as in Figure 15-13.

According to these representations, the structure of naphthalene should be symmetric, with planar and almost hexagonal benzene rings and two perpendicular mirror planes bisecting the molecule. X-ray crystallographic measurements confirm this prediction (Figure 15-14). The C–C bonds deviate only slightly in length from those in benzene (1.39 Å), and they are clearly different from pure single (1.54 Å) and double bonds (1.33 Å).

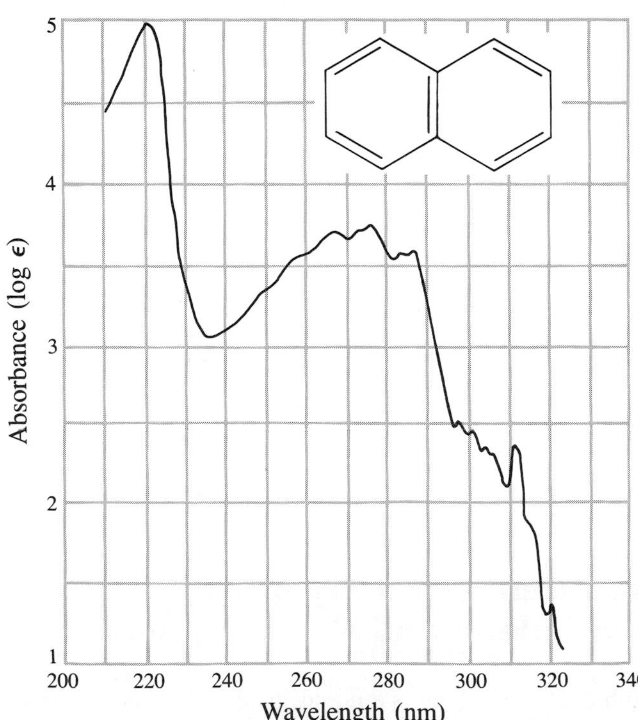

FIGURE 15-12 Extended π conjugation in naphthalene is manifest in its UV spectrum (measured in 95% ethanol). The complexity and location of the absorptions are typical of extended π systems.

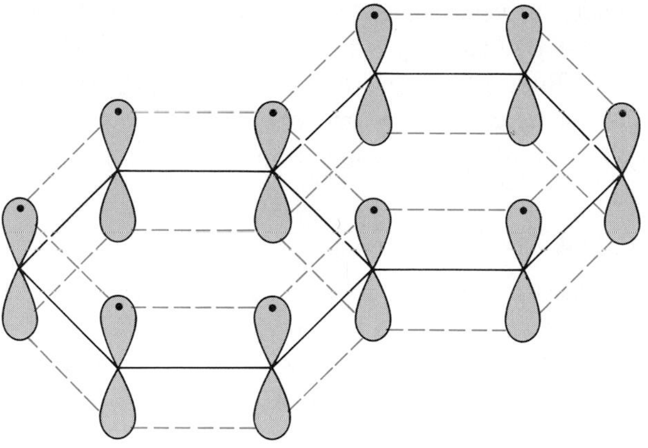

FIGURE 15-13 Orbital picture of naphthalene, showing its extended overlap of *p* orbitals.

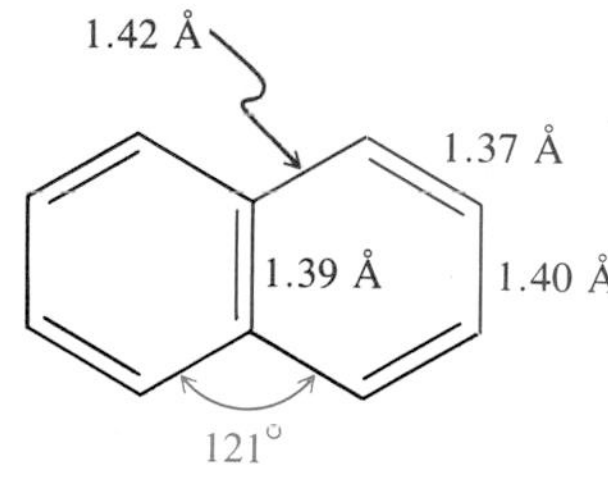

FIGURE 15-14 The molecular structure of naphthalene. The bond angles within the rings are 120°.

Further evidence of aromaticity is found in the ^{1}H NMR spectrum of naphthalene (Figure 15-15). Two symmetric multiplets can be observed at $\delta = 7.40$ and 7.77 ppm, characteristic of aromatic hydrogens deshielded by the ring-current effect of the π electron loop (see Section 15-4, Figure 15-8). Coupling in the

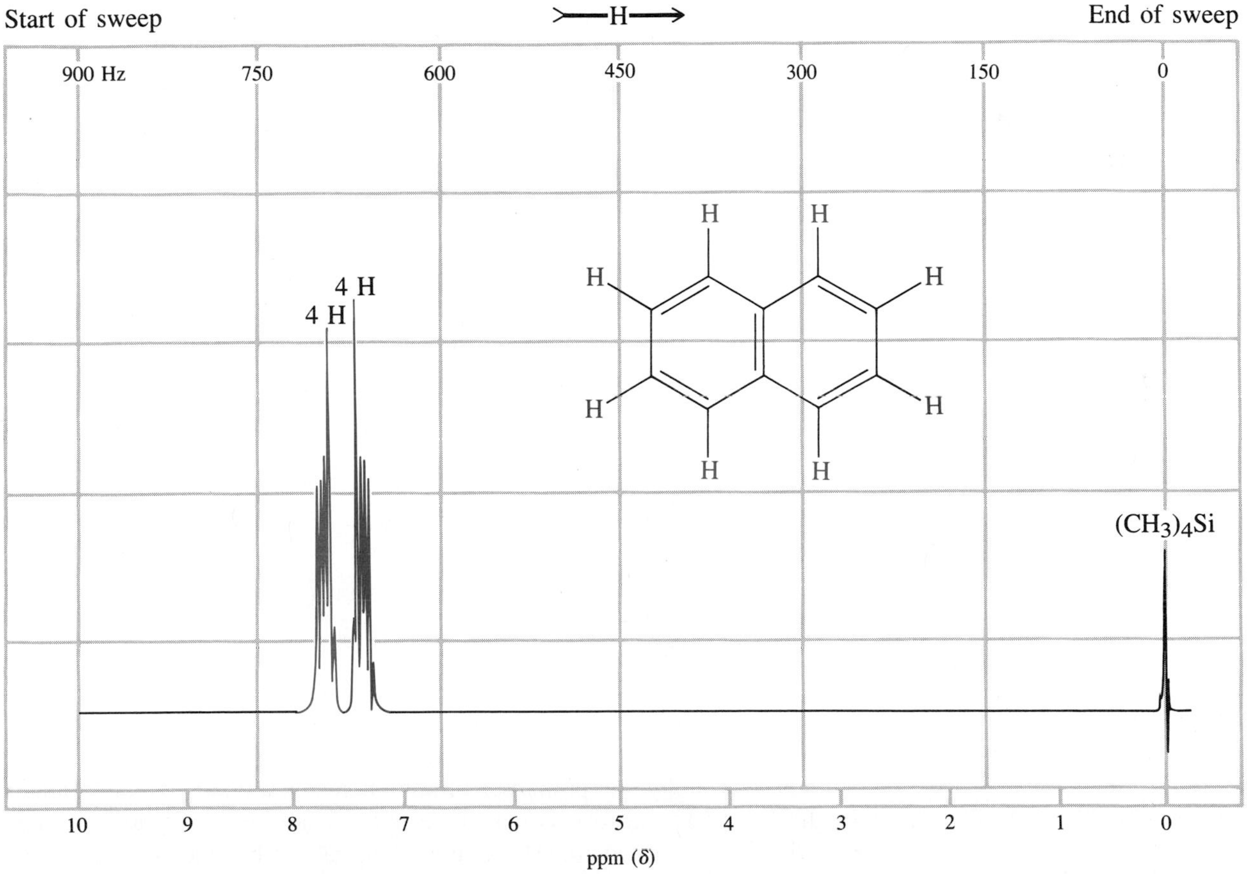

FIGURE 15-15 90-MHz ^{1}H NMR spectrum of naphthalene in CCl_4 reveals the characteristic deshielding due to a ring of π electrons.

naphthalene nucleus is very similar to that in substituted benzenes: J_{ortho} = 7.5 Hz, J_{meta} = 1.4 Hz, and J_{para} = 0.7 Hz. The ^{13}C NMR spectrum shows three lines at δ = 126.5, 128.5, and 134.4 ppm (quaternary carbons), chemical shifts that are in the range of other benzene derivatives. Thus, on the basis of structural and spectral criteria, naphthalene is aromatic.

EXERCISE 15-9

A substituted naphthalene $C_{10}H_8O_2$ gave the following spectral data: ^1H NMR δ = 5.20 (broad s, 2 H), 6.92 (dd, J = 7.5 Hz and 1.4 Hz, 2 H), 7.00 (d, J = 1.4 Hz, 2 H), and 7.60 (d, J = 7.5 Hz, 2 H) ppm; ^{13}C NMR δ = 107.5, 115.3, 123.0, 129.3, 136.8, and 155.8 ppm; IR $\tilde{\nu}$ = 3100 cm^{-1}. What is its structure?

Most fused benzenoid hydrocarbons are aromatic

These properties of naphthalene hold for most of the other polycyclic benzenoid hydrocarbons. It appears that the cyclic delocalization in the individual benzene rings is not significantly perturbed by the fact that they have to share at least one π bond. Linear and angular fusion of a third benzene ring onto naphthalene results in the systems anthracene and phenanthrene. Although isomeric and seemingly very similar, they have different thermodynamic stabilities: Anthracene is about 6 kcal mol^{-1} less stable than phenanthrene even though both are aromatic. Enumeration of the various resonance structures of the molecules explains why. Anthracene has only four, and only two contain two fully aromatic benzene rings (red in the structures below). Phenanthrene has five, three of which incorporate two aromatic benzenes, one even three.

Resonance in Anthracene

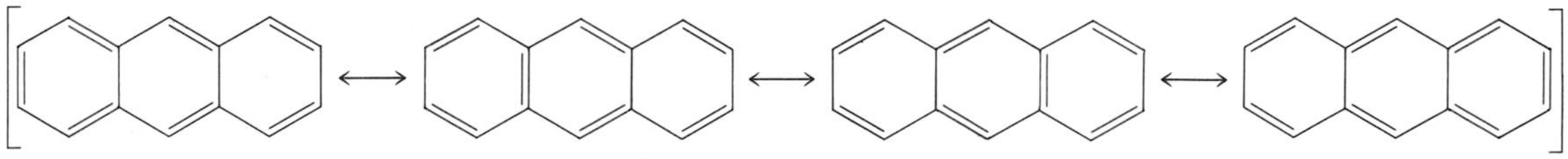

Resonance in Phenanthrene

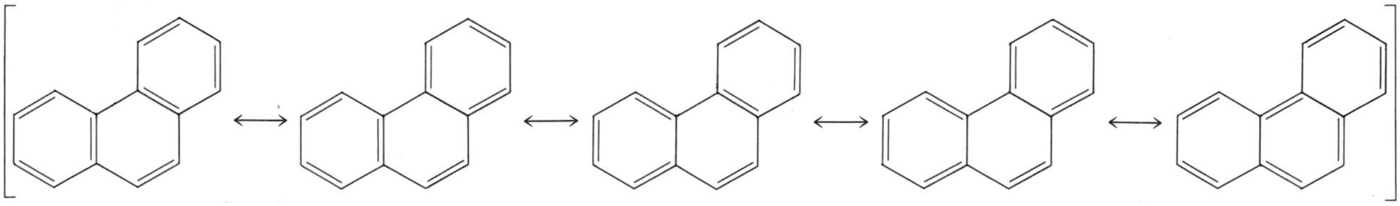

EXERCISE 15-10

Draw all the possible resonance forms of tetracene (naphthacene, Section 15-5). What is the maximum number of completely aromatic benzene rings in these structures?

In summary, the physical properties of naphthalene are typical of an aromatic system. Its UV spectrum reveals that there is extensive delocalization of all π electrons, its molecular structure shows bond lengths and bond angles very simi-

BOX 15-1 Buckyball Looks Like It Should Be Aromatic, But It Isn't

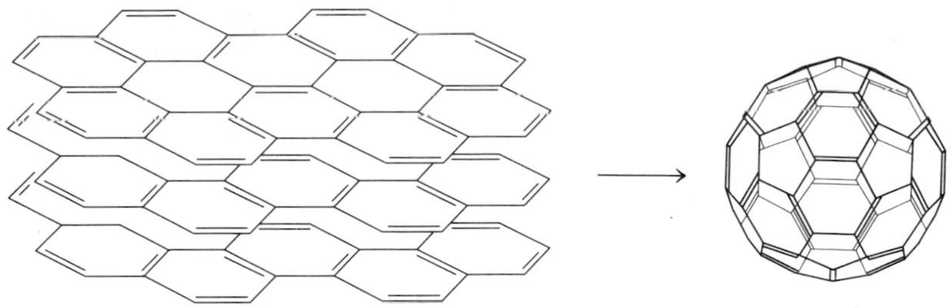

Graphite **C_{60}, Buckminsterfullerene (Buckyball)**

For many years scientists have wondered whether it might be possible to induce rings of carbons to connect to form a surface that closes on itself like a ball. The answer was found in 1990: When graphite is vaporized in a high-amperage arc under a helium atmosphere, the soot formed contains novel all-carbon molecules known as *fullerenes*. The most prevalent fullerene contains 60 carbon atoms equally spaced on the surface of a sphere (see above). C_{60} is called buckminsterfullerene because its soccer-ball shape is reminiscent of the geodesic domes popularized by Buckminster Fuller,* thus the nickname "buckyball." It is a completely conjugated system containing five- and six-membered rings.

The formation of fullerenes is believed to begin with the linkage of individual carbon atoms in the arc to give short chains. These chains combine and close to form large rings, which can connect to give bi- and tricyclic structures. When the composition of such a system reaches or exceeds C_{30}, transannular bond formation ensues, leading to the network of fused five- and six-membered rings characteristic of the fullerenes.

Although chemists initially expected C_{60} to be aromatic because of its many benzenelike rings, studies of its reactivity revealed alkenelike tendencies: Buckyball undergoes addition reactions across the double bonds at the fusion between six-membered rings. Indeed, the X-ray structure of the product of one such reaction was used to prove the soccer-ball shape of C_{60}. What accounts for the lack of aromaticity? Evidently the enforced curvature of the rings reduces the overlap of the π bonds and hence the stabilization of the delocalized molecular orbitals.

An exciting new discovery is the ability of the fullerenes to encapsulate metal atoms and ions, which may lead to substances with useful electrical and magnetic properties such as superconductivity. Amazingly, C_{60} occurs *in nature:* In mid-1992 it was identified in 500-million-year-old carbon-rich mineral deposits in the Russian Karelian Republic.

*Richard Buckminster Fuller, 1895–1983, American inventor and philosopher.

lar to those in benzene, and its ^{1}H NMR spectrum reveals deshielded ring hydrogens indicative of an aromatic ring current. Other polycyclic benzenoid hydrocarbons have similar properties and are also considered aromatic.

15-7 Other Cyclic Polyenes: Hückel's Rule

Is the special stability and reactivity associated with cyclic delocalization unique to benzene and polycyclic benzenoids, or do other cyclic π systems have similar properties? Indeed, other cyclic conjugated polyenes are aromatic if they contain $4n + 2$ π electrons ($n = 0, 1, 2, 3, \ldots$). This pattern is known as **Hückel's* rule.**

1,3-Cyclobutadiene, the smallest cyclic polyene, is not aromatic

1,3-Cyclobutadiene is a highly strained and extremely reactive molecule, devoid of any aromatic properties. It can be prepared and observed only at very low temperatures. The reactivity of cyclobutadiene can be seen in its rapid Diels-Alder reactions, in which it can act as both diene and dienophile.

Reactions of Cyclobutadiene

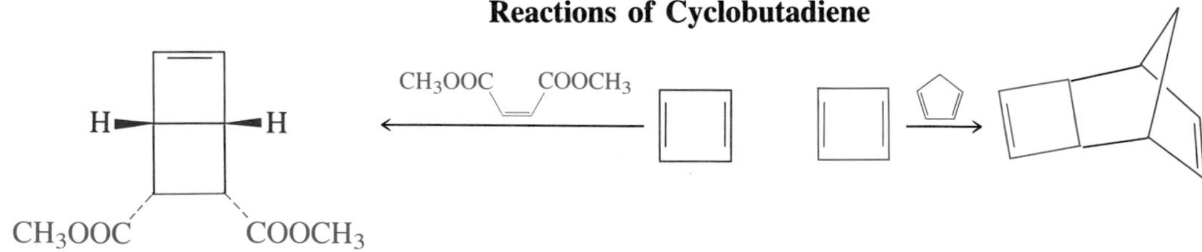

EXERCISE 15-11

1,3-Cyclobutadiene dimerizes at temperatures as low as $-200°C$ to give the two products shown. Explain mechanistically.

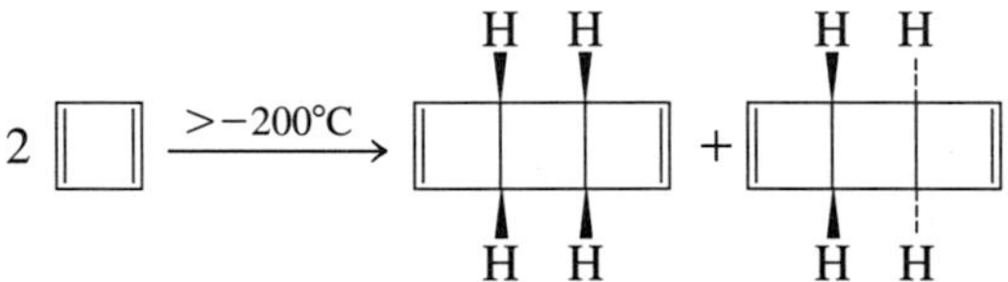

Substituted cyclobutadienes are less reactive, particularly if the substituents are bulky; they have been used to probe the spectroscopic features of the cyclic system of four π electrons. Particularly interesting is the ^1H NMR spectrum of 1,2,3-tris(1,1-dimethylethyl)cyclobutadiene (1,2,3-tri-*tert*-butylcyclobutadiene), in which the ring hydrogen resonates at $\delta = 5.38$ ppm, at much higher field than expected for an aromatic system. This and other properties of cyclobutadiene show that it is quite unlike benzene.

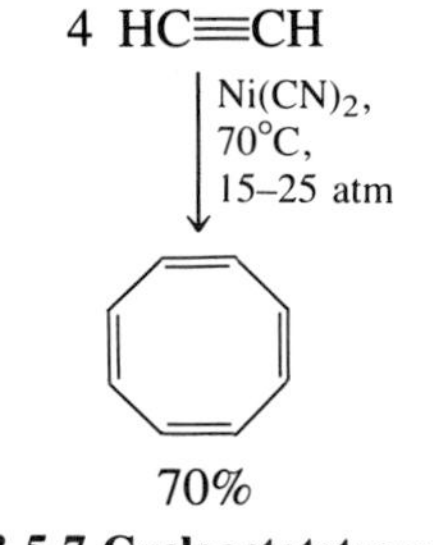

1,3,5,7-Cyclooctatetraene

1,3,5,7-Cyclooctatetraene is nonplanar and nonaromatic

Let us now examine the properties of the next higher cyclic polyene analog of benzene, 1,3,5,7-cyclooctatetraene. First prepared in 1911 by Willstätter,† this substance is now readily available from a special reaction, the nickel-catalyzed cyclotetramerization of ethyne. It is a yellow liquid, b.p. 152°C, that is stable if

*Professor Erich Hückel, 1896–1984, University of Marburg, Germany.

†Professor Richard Willstätter, 1872–1942, Technical University, Munich, Germany, Nobel Prize 1915 (chemistry).

kept cold but readily polymerizes when heated. It is oxidized by air, readily hydrogenated to cyclooctene and cyclooctane, and subject to electrophilic additions and to cycloadditions. This chemical reactivity is again not what would be expected if the molecule were aromatic.

Spectral and structural data confirm the lack of aromaticity. Thus, the ^1H NMR spectrum shows a sharp singlet at $\delta = 5.68$ ppm, typical of an alkene. The molecular-structure determination reveals that cyclooctatetraene is actually *nonplanar* and tub-shaped (Figure 15-16). The double bonds are nearly orthogonal, and they alternate with single bonds.

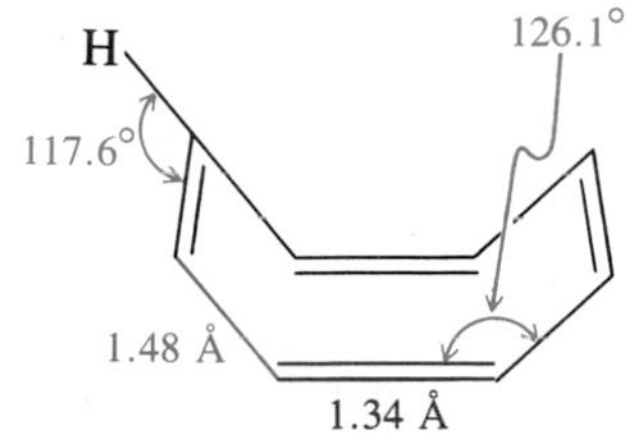

FIGURE 15-16 The molecular structure of 1,3,5,7-cyclooctatetraene. Note the alternating single and double bonds of this nonplanar, nonaromatic molecule.

EXERCISE 15-12

On the basis of the molecular structure of 1,3,5,7-cyclooctatetraene, would you describe its double bonds as conjugated (i.e., does it exhibit extended π overlap)? Would it be correct to draw two resonance forms for this molecule, as we do for benzene?

EXERCISE 15-13

Formulate the product of the reaction of 1,3,5,7-cyclooctatetraene with 1 mol of Br_2 (there is only one, and it forms in 100% yield). Watch stereochemistry.

Only cyclic conjugated polyenes containing $4n + 2$ pi electrons are aromatic

Unlike cyclobutadiene and cyclooctatetraene, certain higher cyclic conjugated polyenes are aromatic. All of them have one property in common: They contain $4n + 2$ π electrons.

The first such system was prepared in 1956 by Sondheimer;* it was 1,3,5,7,9,11,13,15,17-cyclooctadecanonaene, containing 18 π electrons. To avoid such cumbersome names, Sondheimer introduced a simpler system of naming cyclic conjugated polyenes. He named completely conjugated monocyclic hydrocarbons $(CH)_N$ as **[*N*]annulenes,** in which *N* denotes the ring size. Thus, cyclobutadiene would be called [4]annulene; benzene, [6]annulene; cyclooctatetraene, [8]annulene. The first unstrained aromatic system in the series after benzene is [18]annulene.

[18]Annulene
(1,3,5,7,9,11,13,15,17-Cyclooctadecanonaene)

---

*Professor Franz Sondheimer, 1926–1981, University College, London.

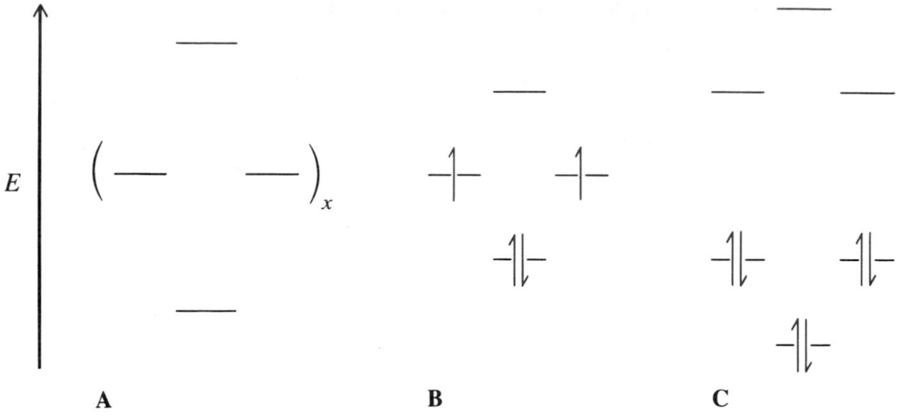

FIGURE 15-17 (A) Hückel's $4n + 2$ rule is based on the regular pattern of the π molecular orbitals in cyclic conjugated polyenes. The energy levels are equally spaced, and only the highest and lowest ones are nondegenerate. (B) Molecular-orbital levels in 1,3-cyclobutadiene. Four π electrons are not enough to result in a closed shell, so the molecule is not aromatic. (C) The six π electrons in benzene produce a closed-shell configuration.

[18]Annulene contains delocalized electrons, is fairly planar, and shows little alternation of the single and double bonds. Like benzene, therefore, it can be described by a set of two equal resonance structures. In accord with its aromatic character, the molecule is relatively stable and undergoes electrophilic aromatic substitution. It also exhibits a benzenelike ring-current effect in ^{1}H NMR.

Since the preparation of [18]annulene, many other annulenes have been made: Those with $4n$ π electrons, such as cyclobutadiene and cyclooctatetraene, are not aromatic, but those with $4n + 2$ electrons, such as benzene and [18]annulene, are aromatic. This behavior had been predicted earlier by the theoretical chemist Hückel, who formulated this $4n + 2$ rule in 1938.

Hückel's rule expresses the regular molecular-orbital patterns in cyclic conjugated polyenes. The p orbitals mix to give an equal number of π molecular orbitals as shown in Figure 15-17. All levels are composed of degenerate pairs, except for the lowest bonding and highest antibonding orbitals. A closed-shell system is possible only if all bonding molecular orbitals are occupied (see Section 1-7), that is, only if there are $4n + 2$ π electrons.

EXERCISE 15-14

On the basis of Hückel's rule, label the following molecules as aromatic or nonaromatic. **(a)** [30]Annulene; **(b)** [16]annulene.

In summary, cyclic conjugated polyenes are aromatic if their π electron count is $4n + 2$. This number corresponds to a completely filled set of bonding molecular orbitals. Conversely, $4n$ π systems have open-shell structures that are unstable, reactive, and lack aromatic ring-current effects in ^{1}H NMR.

15-8 Hückel's Rule and Charged Molecules

Hückel's rule also applies to charged molecules, as long as cyclic delocalization can occur. This section shows how charged aromatic systems can be prepared.

The cyclopentadienyl anion and the cycloheptatrienyl cation are aromatic

1,3-Cyclopentadiene is unusually acidic ($pK_a \sim 16$) because the cyclopentadienyl anion resulting from deprotonation contains a delocalized, aromatic system of six π electrons. The negative charge is equally distributed over all five carbon atoms. For comparison, the pK_a of propene is 40.

Aromatic Cyclopentadienyl Anion

H H

$\rightleftharpoons$ H^+ + [$\longleftrightarrow$ $\longleftrightarrow$ $\longleftrightarrow$ etc.] or

$pK_a = 16$

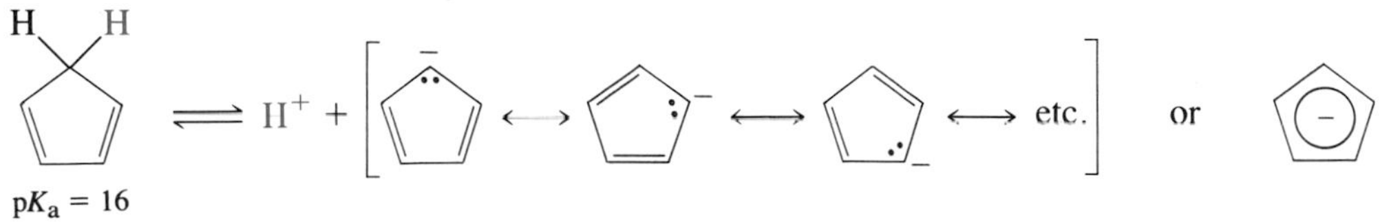

In contrast, the cyclopentadienyl cation, a system of four π electrons, can be produced only at low temperature and is extremely reactive.

When 1,3,5-cycloheptatriene is treated with bromine, a stable salt is formed, cycloheptatrienyl bromide. In this molecule, the organic cation contains six delocalized π electrons, and the positive charge is equally distributed over seven carbons. Even though a carbocation, the system is remarkably unreactive, as expected for an aromatic system.

Aromatic Cycloheptatrienyl Cation

H H

$\xrightarrow[-\ HBr]{Br–Br,\ \Delta}$ Br^- [$\longleftrightarrow$ $\longleftrightarrow$ $\longleftrightarrow$ etc.] or

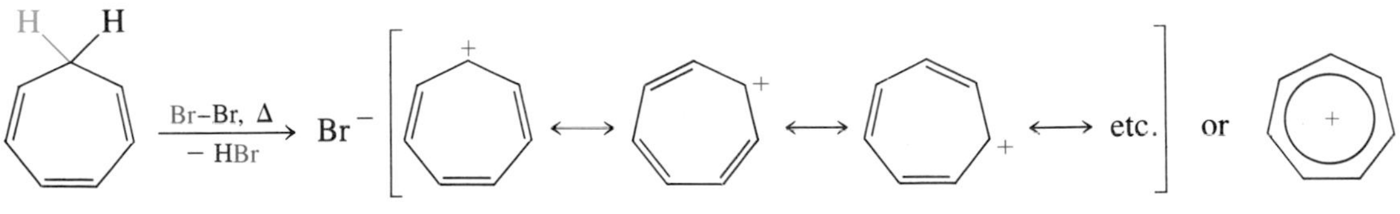

EXERCISE 15-15

Draw an orbital picture of **(a)** the cyclopentadienyl anion and **(b)** cycloheptatrienyl cation (consult Figure 15-1).

EXERCISE 15-16

On the basis of Hückel's rule, label the following molecules aromatic or nonaromatic. **(a)** Cyclopropenyl cation; **(b)** cyclononatetraenyl anion.

Nonaromatic cyclic polyenes can form aromatic dianions and dications

Cyclic systems of $4n$ π electrons can be converted into their aromatic counterparts by two-electron oxidations and reductions. For example, cyclooctatetraene is reduced by alkali metals to the corresponding aromatic dianion. This species is planar, contains fully delocalized electrons, and is relatively stable. It also exhibits an aromatic ring current in ^1H NMR.

Nonaromatic Cyclooctatetraene Forms an Aromatic Dianion

K, THF → + 2 K^+

Eight π electrons, nonaromatic

Ten π electrons, aromatic

Similarly, [16]annulene can be either reduced to its dianion or oxidized to its dication, both products being aromatic. On formation of the dication, the configuration of the molecule changes.

Aromatic [16]Annulene Dication and Dianion from Nonaromatic [16]Annulene

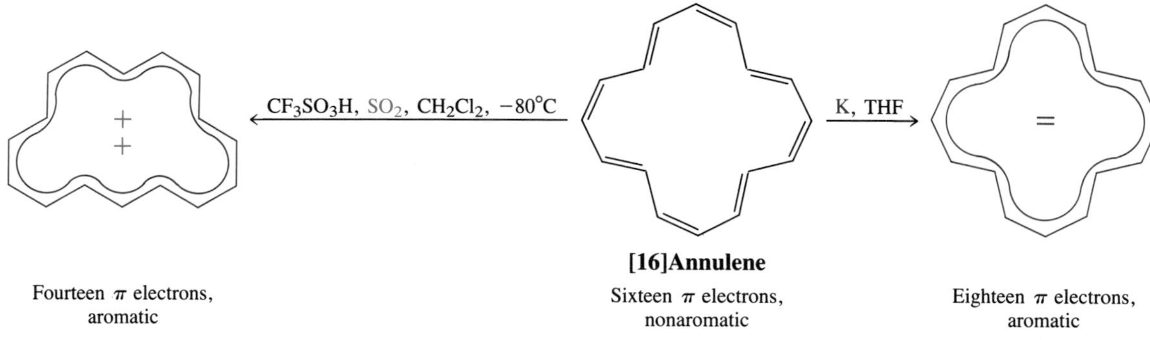

Fourteen π electrons, aromatic

[16]Annulene
Sixteen π electrons, nonaromatic

Eighteen π electrons, aromatic

EXERCISE 15-17

The triene A can be readily deprotonated twice to give the stable dianion B. However, the neutral analog of B, the tetraene C (pentalene), is extremely unstable. Explain.

H H H H Base →

A **B** **C**

EXERCISE 15-18

Azulene is a deep blue (see Figure 14-16) hydrocarbon. Is it aromatic?

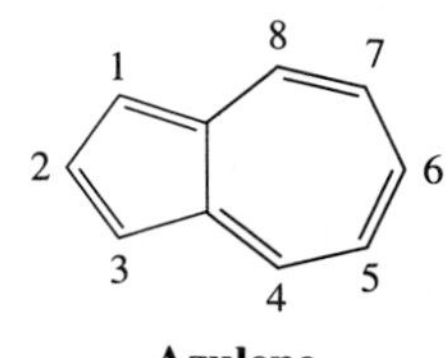

Azulene

In summary, charged species may be aromatic, provided they exhibit cyclic delocalization and obey the $4n + 2$ rule.

15-9 Synthesis of Benzene Derivatives: Electrophilic Aromatic Substitution

In this section we shall begin to explore the reactivity of benzene, the prototype aromatic compound. The stability of benzene makes it relatively unreactive. As a result, its chemical transformations require special conditions and proceed through new pathways.

Benzene undergoes substitution reactions with electrophiles

Benzene can be attacked by electrophiles. In contrast with the corresponding reactions of alkenes, this reaction results in *substitution* of hydrogens—**electrophilic aromatic substitution**—*not addition* to the ring.

Electrophilic Aromatic Substitution

$$C_6H_6 + \underset{\text{Electrophile}}{E^+X^-} \longrightarrow C_6H_5E + H^+X^-$$

Under the conditions employed for these transformations, nonaromatic conjugated polyenes would rapidly polymerize. However, the stability of the benzene ring allows it to survive. Let us begin with the general mechanism of electrophilic aromatic substitution.

Electrophilic aromatic substitution in benzene proceeds by addition of the electrophile and then by proton loss

The mechanism of electrophilic aromatic substitution has two steps. First, the electrophile E^+ attacks the benzene nucleus, much as it would attack an ordinary double bond. The cationic intermediate thus formed then loses a proton to regenerate the aromatic ring.

Mechanism of Electrophilic Aromatic Substitution

STEP 1. Electrophilic attack

$$C_6H_6 + E^+ \rightleftharpoons [\text{three resonance forms of the cation } C_6H_6E^+]$$

STEP 2. Proton loss

$$[\text{three resonance forms of the cation } C_6H_6E^+] \rightleftharpoons C_6H_5E + H^+$$

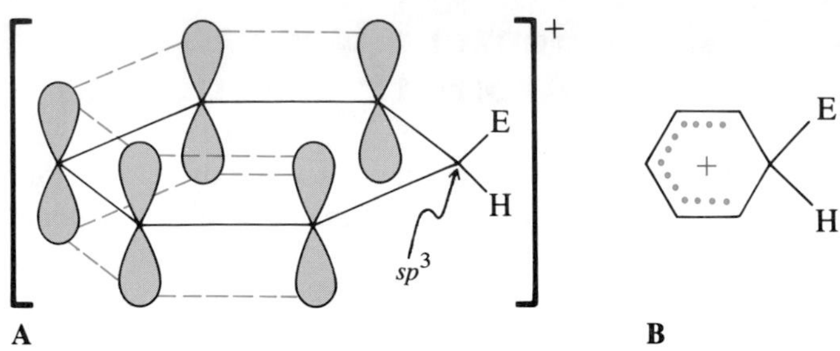

FIGURE 15-18
(A) Orbital picture of the cationic intermediate resulting from attack by an electrophile on the benzene ring. Aromaticity is lost because cyclic conjugation is interrupted by the sp^3-hybridized carbon. The four electrons in the π system are not shown.
(B) Dotted-line notation to indicate delocalized nature of the charge in the cation.

The first step is not favored thermodynamically. Although charge is delocalized in the cationic intermediate, the formation of the C–E bond results in an sp^3-hybridized carbon in the ring, which interrupts cyclic conjugation: *The intermediate is not aromatic* (Figure 15-18). However, the next step, loss of the proton at the sp^3-hybridized carbon, regenerates the aromatic ring. This process is more favored than nucleophilic trapping by the anion that accompanies E^+. The latter would give a nonaromatic addition product. The overall substitution is exothermic, the new bonds being stronger than the old ones.

A potential-energy diagram (Figure 15-19) shows that the first step in this reaction is rate determining. Proton loss is much faster, because it leads to the aromatic product in an exothermic step, which furnishes the driving force for the overall sequence.

The details of this mechanism depend on the electrophile. The following sections will look more closely at the most common reagents employed in this transformation.

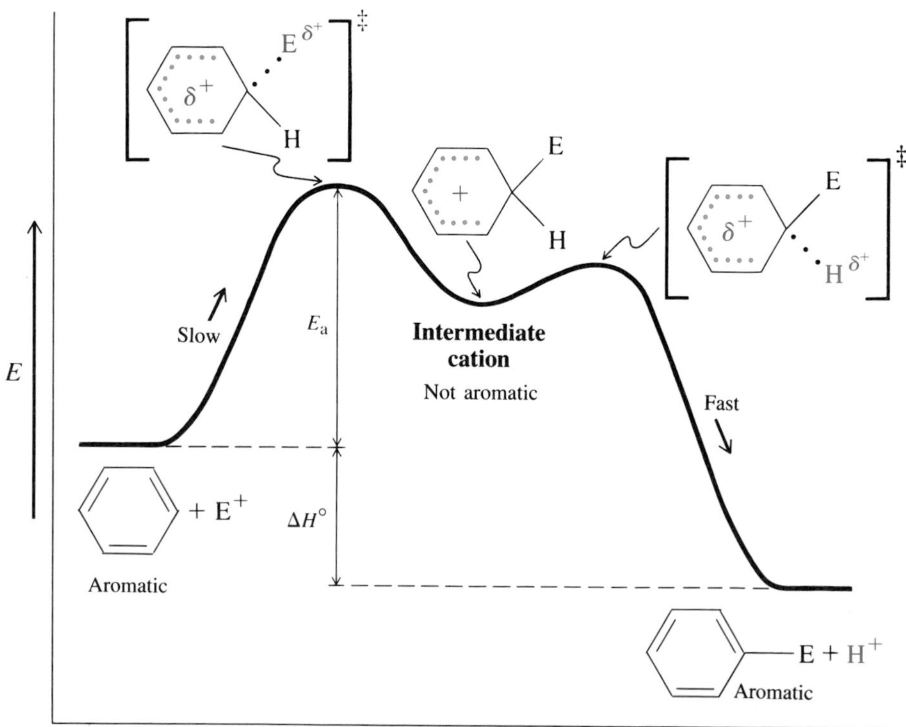

FIGURE 15-19
Potential-energy diagram describing the course of the reaction of benzene with an electrophile. The first transition state is rate determining. Proton loss is relatively fast. The overall rate of the reaction is determined by E_a, the exothermicity by $\Delta H°$.

EXERCISE 15-19

Review Section 12-1 and explain why addition reactions to benzene do not occur. (Calculate the $\Delta H°$ for some additions.)

In summary, the general mechanism of electrophilic aromatic substitution begins with electrophilic attack by E^+ to give an intermediate, charge-delocalized but nonaromatic cation in a rate-determining step. Subsequent fast proton loss regenerates the (now substituted) aromatic ring.

15-10 Halogenation of Benzene: The Need for a Catalyst

An example of electrophilic aromatic substitution is halogenation. Benzene is normally unreactive in the presence of halogens, because halogens are not electrophilic enough to disrupt its aromaticity. However, the halogen may be activated by Lewis acids, such as ferric halides, FeX_3, or aluminum halides, AlX_3, to become a much more powerful electrophile.

Bromination of Benzene

Br_2, $FeBr_3$

Bromobenzene

How does this activation work? The characteristic of Lewis acids is their ability to accept electron pairs. When a halogen such as bromine is exposed to $FeBr_3$, the two molecules combine as Lewis acid and base.

Activation of Bromine by the Lewis Acid $FeBr_3$

$$:\ddot{Br}-\ddot{Br}: \curvearrowright FeBr_3 \longrightarrow \left[:\ddot{Br}-\overset{+}{\ddot{Br}}-\overset{-}{FeBr_3} \longleftrightarrow \overset{+}{\ddot{Br}}=\ddot{Br}-\overset{-}{FeBr_3} \right]$$

This complex serves to polarize the Br–Br bond, thereby imparting electrophilic character to one of the bromine atoms. Electrophilic attack on benzene occurs as described in the preceding section.

Electrophilic Attack on Benzene by Activated Bromine

$$:\ddot{Br}-\overset{+}{\ddot{Br}}-\overset{-}{FeBr_3} \longrightarrow \text{(cyclohexadienyl cation, H, Br)} + FeBr_4^-$$

The $FeBr_4^-$ formed in this step now functions as a base, abstracting a proton from the hexadienyl cation intermediate. This transformation not only furnishes the two products of the reaction, bromobenzene and hydrogen bromide, but also regenerates the $FeBr_3$ catalyst.

Bromobenzene Formation

$$:\ddot{Br}-\overset{-}{FeBr_3} \longrightarrow \text{(bromobenzene)} + HBr + FeBr_3$$

A quick calculation confirms the exothermicity of the electrophilic bromination of benzene. A phenyl–hydrogen bond (approximately 111 kcal mol^{-1},

TABLE 15-1 Strengths ($DH°$) of Bonds A–B ($kcal\ mol^{-1}$)

A	B	$DH°$
F	F	37
Cl	Cl	58
Br	Br	46
I	I	36
F	C_6H_5	126
Cl	C_6H_5	96
Br	C_6H_5	81
I	C_6H_5	65
C_6H_5	H	111
F	H	135.8
Cl	H	103.2
Br	H	87.5
I	H	71.3

Table 15-1) and a bromine molecule (46 $kcal\ mol^{-1}$) are lost in the process. Counterbalancing this, a phenyl–bromine bond ($DH° = 81\ kcal\ mol^{-1}$) and an H–Br bond ($DH° = 87.5\ kcal\ mol^{-1}$) are formed. Thus, the overall reaction is exothermic by $168.5 - 157 = 11.5\ kcal\ mol^{-1}$.

As in the radical halogenation of alkanes (Section 3-7), the exothermicity of aromatic halogenation decreases down the periodic table. Fluorination is so exothermic that direct reaction of fluorine with benzene is explosive. Chlorination, on the other hand, is controllable but requires the presence of an activating catalyst, such as aluminum chloride or ferric chloride. The mechanism of this reaction is identical with that of bromination. Finally, electrophilic iodination with iodine is endothermic and thus not normally possible.

EXERCISE 15-20

When benzene is dissolved in D_2SO_4, its ^1H NMR absorption at $\delta = 7.27$ ppm disappears and a new compound is formed having a molecular weight of 84. What is it? Propose a mechanism for its formation.

EXERCISE 15-21

Professor G. Olah* and his colleagues exposed benzene to the especially strong acid system HF-SbF_5 in an NMR tube and observed a new ^1H NMR spectrum with absorptions at $\delta = 5.69$ (2 H), 8.22 (2 H), 9.42 (1 H), and 9.58 (2 H) ppm. Propose a structure for this species.

In summary, the halogenation of benzene increases in exothermicity from I_2 (endothermic) to F_2 (explosive). Chlorinations and brominations are achieved with the help of Lewis acid catalysts, which polarize the X–X bond and activate the halogen by increasing its electrophilic power.

15-11 Nitration and Sulfonation of Benzene

Nitration of Benzene

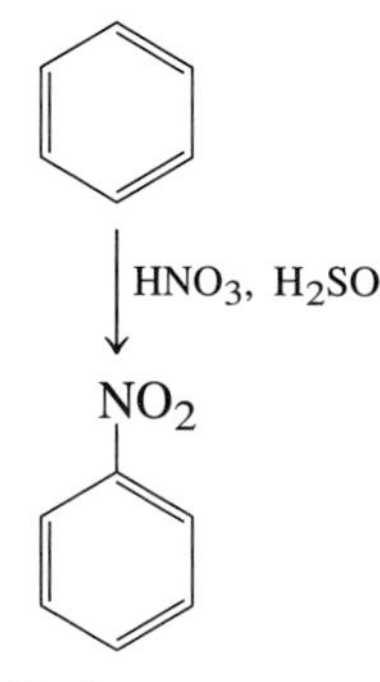

Nitrobenzene

In two other typical electrophilic substitutions of benzene, the electrophiles are the nitronium ion, NO_2^+, leading to nitrobenzene, and sulfur trioxide, SO_3, giving benzenesulfonic acid.

Benzene is subject to electrophilic attack by the nitronium ion

To bring about nitration of the ring at moderate temperatures, it is not sufficient just to treat benzene with concentrated nitric acid. Because the nitrogen in the nitrate group of HNO_3 has no electrophilic power, it must somehow be activated. Addition of concentrated sulfuric acid serves this purpose, protonating the nitric acid. Loss of water then yields the **nitronium ion,** NO_2^+, a strong electrophile.

*Professor George A. Olah, b. 1927, University of Southern California, Los Angeles, Nobel Prize 1994 (chemistry).

Activation of Nitric Acid by Sulfuric Acid

$$HO{-}\overset{+}{N}(=O){-}O^- + H{-}OSO_3H \rightleftharpoons HO^+(H){-}\overset{+}{N}(=O){-}O^- + HSO_4^-$$

Nitric acid

$$H\overset{+}{O}(H){-}\overset{+}{N}(=O){-}O^- \longrightarrow H_2O: + O{=}\overset{+}{N}{=}O$$

Nitronium ion

Electrophilic attack on benzene proceeds by a simple variation of the general mechanism discussed earlier.

Mechanism of Aromatic Nitration

$$C_6H_6 + :O{=}\overset{+}{N}{=}O: \longrightarrow [C_6H_6NO_2]^+ \xrightarrow{^-:OSO_3H} C_6H_5NO_2 + HOSO_3H$$

Nitrobenzene

Sulfonation is reversible

Concentrated sulfuric acid does not sulfonate benzene at room temperature. However, a more reactive form, called *fuming sulfuric acid,* permits electrophilic attack by SO_3. Commercial fuming sulfuric acid is made by adding about 8% of sulfur trioxide, SO_3, to the concentrated acid. Because of the strong electron-withdrawing effect of the three oxygens, the sulfur in SO_3 is electrophilic enough to attack benzene directly. Subsequent proton transfer results in the sulfonated product, benzenesulfonic acid.

Sulfonation of Benzene

$$C_6H_6 \xrightarrow{SO_3,\ H_2SO_4} C_6H_5SO_3H$$

Benzenesulfonic acid

Mechanism of Aromatic Sulfonation

$$C_6H_6 + SO_3 \rightleftharpoons [C_6H_6SO_3]^{+/-} \rightleftharpoons C_6H_5S(=O)_2OH$$

Benzenesulfonic acid

Aromatic sulfonation is easily reversible. The reaction of sulfur trioxide with water to give sulfuric acid is so exothermic that heating benzenesulfonic acid in dilute aqueous acid completely reverses sulfonation.

Hydration of SO_3

$$SO_3 + H\ddot{O}H \longrightarrow O{=}S(=O)(OH){-}OH + \text{heat}$$

Reverse Sulfonation

$$C_6H_5SO_3H \xrightarrow{H_2O,\ \text{catalytic } H_2SO_4,\ 100°C} C_6H_5H + HOSO_3H$$

The reversibility of sulfonation may be used to control further aromatic substitution processes. The ring carbon containing the substituent is blocked from attack, and electrophiles are directed to other positions. Thus, the sulfonic acid group can be introduced to serve as a *directing blocking group* and then removed by reverse sulfonation. Synthetic applications of this strategy will be discussed in Section 16-5.

Benzenesulfonic acids have important uses

Sulfonation is also important in the synthesis of *detergents*. Until recently, long-chain branched alkylbenzenes were sulfonated to the corresponding sulfonic acids, then converted into their sodium salts. Because such detergents are not readily biodegradable, they have been replaced by more environmentally acceptable alternatives. We shall examine this class of compounds in Chapter 19.

Aromatic Detergent Synthesis

$$R{-}C_6H_5 \xrightarrow{SO_3,\ H_2SO_4} R{-}C_6H_4{-}SO_3H \xrightarrow[-\ H_2O]{NaOH} R{-}C_6H_4{-}SO_3^-Na^+$$

R = branched alkyl group

Another application of sulfonation is to the manufacture of dyes, because the sulfonic acid group imparts water solubility (Chapter 22).

Also useful are the **sulfonyl chlorides,** the acid chlorides of sulfonic acids (see Section 9-4). They are usually prepared by reaction of the sodium salt of the acid with PCl_5 or $SOCl_2$.

Preparation of Benzenesulfonyl Chloride

$$2\ C_6H_5SO_2O^-Na^+ + PCl_5 \longrightarrow 2\ C_6H_5SO_2Cl + POCl_3 + 2\ Na^+Cl^-$$

Sulfonyl chlorides are frequently employed in synthesis. For example, recall that the hydroxy group of an alcohol may be turned into a good leaving group by conversion of the alcohol into the 4-methylbenzenesulfonate (*p*-toluenesulfonate, tosylate; Sections 6-7 and 9-4).

Sulfonyl chlorides are important precursors of **sulfonamides,** many of which are chemotherapeutic agents, such as the *sulfa drugs* discovered in 1932 (Section 9-11). Sulfonamides are derived from the reaction of a sulfonyl chloride with an amine. Sulfa drugs specifically contain the 4-aminobenzenesulfonamide (sulfanilamide) function. Their mode of action is to interfere with the bacterial enzymes that help to synthesize folic acid (Box 25-5).

Sulfa Drugs

RHN—C₆H₄—SO_2NHR'

General structure

H_2N—C₆H₄—SO_2NH—(5-methylisoxazol-3-yl)

Sulfamethoxazole (Gantanol)
(Antibacterial, used to treat urinary infections)

H_2N—C₆H₄—SO_2NH—(3-methoxypyrazin-2-yl)

Sulfalene (Kelfizina)
(Antilepral)

H_2N—C₆H₄—SO_2NH—(pyrimidin-2-yl)

Sulfadiazine
(Antimalarial)

About 15,000 sulfa derivatives have been synthesized and screened for antibacterial activity; some have become new drugs. With the advent of antibiotics, the medicinal use of sulfa drugs has greatly diminished, but their discovery was a milestone in the systematic development of medicinal chemistry.

EXERCISE 15-22

Formulate mechanisms for **(a)** the reverse of sulfonation; **(b)** the hydration of SO_3.

In summary, nitration of benzene requires the generation of the nitronium ion, NO_2^+, which functions as the active electrophile. The nitronium ion is formed by the loss of water from protonated nitric acid. Sulfonation is achieved with fuming sulfuric acid, in which sulfur trioxide, SO_3, is the electrophile. Sulfonation is reversed by hot aqueous acid. Benzenesulfonic acids are used in the preparation of detergents, dyes, compounds containing leaving groups, and sulfa drugs.

15-12 Friedel-Crafts Alkylation

None of the electrophilic substitutions mentioned so far have led to carbon–carbon bond formation, one of the primary challenges in organic chemistry. In principle, such reactions could be carried out with benzene in the presence of a sufficiently electrophilic carbon-based electrophile. This section introduces the first of two such transformations, the **Friedel-Crafts* reactions.** The secret to

*Professor Charles Friedel, 1832–1899, Sorbonne, Paris; Professor James M. Crafts, 1839–1917, Massachusetts Institute of Technology, Boston.

the success of both processes is the use of a Lewis acid, usually aluminum chloride. In the presence of this reagent, haloalkanes attack benzene to form alkylbenzenes.

In 1877, Friedel and Crafts discovered that a haloalkane reacts with benzene in the presence of an aluminum halide. The resulting products are the alkylbenzene and hydrogen halide. This reaction, which can be carried out in the presence of other Lewis acid catalysts, is called the **Friedel-Crafts alkylation** of benzene.

Friedel-Crafts Alkylation

$$C_6H_5\text{–}H + RX \xrightarrow{AlX_3} C_6H_5\text{–}R + HX$$

The reactivity of the haloalkane increases in the order RI < RBr < RCl < RF. Typical Lewis acids are (in the order of increasing activity) BF_3, $SbCl_5$, $FeCl_3$, $AlCl_3$, and $AlBr_3$.

Friedel-Crafts Alkylation of Benzene with Chloroethane

$$CH_3CH_2Cl + C_6H_5\text{–}H \xrightarrow{AlCl_3,\ 25^\circ C} C_6H_5\text{–}CH_2CH_3 + HCl$$

28%
Ethylbenzene

With primary halides, the reaction begins with coordination of the Lewis acid to the halogen of the haloalkane, much as in the activation of halogens in electrophilic halogenation. This coordination places a partial positive charge on the halogen-bearing carbon, rendering it more electrophilic. Attack on the benzene ring is followed by proton loss in the usual manner, giving the observed product.

Mechanism of Friedel-Crafts Alkylation with Primary Haloalkanes

STEP 1. Haloalkane activation

$$RCH_2\text{—}\ddot{\underset{\cdot\cdot}{X}}\!: \curvearrowright AlX_3 \rightleftharpoons \overset{\delta^+}{RCH_2}\!:\!\overset{+}{\underset{\cdot\cdot}{\ddot{X}}}\!:\!\overset{-}{AlX_3}$$

STEP 2. Electrophilic attack

$$C_6H_6 + \overset{\delta^+}{RH_2C}\cdots\overset{+}{X}\text{—}\overset{-}{AlX_3} \longrightarrow \overset{+}{C_6H_6}(H)(CH_2R) + AlX_4^-$$

STEP 3. Proton loss

$$\overset{+}{C_6H_6}(H)(CH_2R) + :\ddot{X}\text{—}\overset{-}{AlX_3} \longrightarrow C_6H_5\text{–}CH_2R + HX + AlX_3$$

With secondary and tertiary halides, free carbocations are usually formed as intermediates; these attack the benzene ring in the same way as the cation NO_2^+.

EXERCISE 15-23

Write a mechanism for the formation of (1,1-dimethylethyl)benzene (*tert*-butylbenzene) from 2-chloro-2-methylpropane (*tert*-butyl chloride), benzene, and catalytic $AlCl_3$.

Intramolecular Friedel-Crafts alkylations can be used to fuse a new ring onto the benzene nucleus.

An Intramolecular Friedel-Crafts Alkylation

Cl, H

$AlCl_3$, CS_2 and CH_3NO_2 (solvents), 25°C, 72 h; − HCl

31%

Tetralin

(Common name)

Friedel-Crafts alkylations can be carried out with any starting material, such as an alcohol or alkene, that functions as a precursor to a carbocation (Sections 9-2 and 12-3).

Friedel-Crafts Alkylations Using Other Carbocation Precursors

H + $CH_3CH_2CHCH_3$ (OH) —BF_3, 60°C, 9 h; − HOH→ $CH_3CH_2CHCH_3$

36%

(1-Methylpropyl)benzene

H + (cyclohexene) —HF, 0°C→ H

62%

Cyclohexylbenzene

EXERCISE 15-24

In 1991 more than 4.3 billion pounds of (1-methylethyl)benzene (isopropylbenzene or cumene), an important industrial intermediate in the manufacture of phenol (Chapter 22), was synthesized in the United States from propene and benzene in the presence of phosphoric acid. Write a mechanism for its formation in this reaction.

In summary, the Friedel-Crafts alkylation produces carbocations (or their equivalents) capable of electrophilic aromatic substitution by formation of aryl–carbon bonds. Haloalkanes, alkenes, and alcohols can be used to achieve aromatic alkylation in the presence of a Lewis or mineral acid.

15-13 Limitations of Friedel-Crafts Alkylations

The alkylation of benzenes under Friedel-Crafts conditions is accompanied by two important limiting reactions: one is *polyalkylation;* the second, *carbocation rearrangements*. Both cause the yield of the desired products to diminish and lead to mixtures that are difficult to separate.

Consider first polyalkylation. Benzene reacts with 2-bromopropane in the presence of $FeBr_3$ as a catalyst to give products of both single and double substitution. The yields are low because of the formation of many by-products.

$$C_6H_6 + (CH_3)_2CHBr \xrightarrow[-\ HBr]{FeBr_3} C_6H_5CH(CH_3)_2 + p\text{-}C_6H_4[CH(CH_3)_2]_2$$

25% **(1-Methylethyl)benzene (Isopropylbenzene)**

15% **1,4-Bis(1-methylethyl)benzene (*p*-Diisopropylbenzene)**

The electrophilic aromatic substitutions we studied in Sections 15-10 and 15-11 can be stopped at the monosubstitution stage. Why do Friedel-Crafts alkylations have the problem of multiple electrophilic substitution? It is because the substituents differ in electronic structure. Bromination, nitration, and sulfonation introduce an electron-withdrawing group into the benzene ring, which renders the product *less* susceptible than the starting material to electrophilic attack. In contrast, an alkylated benzene is more electron rich than unsubstituted benzene and thus *more* susceptible to electrophilic attack.

EXERCISE 15-25

Treatment of benzene with chloromethane in the presence of aluminum chloride results in a complex mixture of tri-, tetra-, and pentamethylbenzenes. One of the components in this mixture crystallizes out selectively: m.p. = 80°C; molecular formula = $C_{10}H_{14}$; ^{1}H NMR $\delta = 2.27$ (s, 12 H) and 7.15 (s, 2 H) ppm; ^{13}C NMR $\delta = 19.2$, 131.2, and 133.8 ppm. Draw a structure for this product.

The second complication in aromatic alkylation is skeletal rearrangement (Section 9-3). For example, the attempted propylation of benzene with 1-bromopropane and $AlCl_3$ produces (1-methylethyl)benzene.

$$C_6H_6 + CH_3CH_2CH_2Br \xrightarrow[-\ HBr]{AlCl_3} C_6H_5CH(CH_3)_2$$

The starting haloalkane rearranges to the thermodynamically favored 1-methylethyl (isopropyl) cation in the presence of the Lewis acid.

Rearrangement of 1-Bromopropane to 1-Methylethyl (Isopropyl) Cation

$$CH_3\overset{H}{\overset{|}{C}}H{-}CH_2{-}Br + AlCl_3 \longrightarrow CH_3\overset{+}{C}HCH_3 + Br\overset{-}{Al}Cl_3$$

1-Methylethyl (isopropyl) cation

EXERCISE 15-26

Attempted alkylation of benzene with 1-chlorobutane in the presence of $AlCl_3$ gave not only the expected butylbenzene, but, as a major product, (1-methylpropyl)benzene. Write a mechanism for this reaction.

Because of these limitations, Friedel-Crafts alkylations are used rarely in synthetic chemistry. Is there a way to improve this process? It would require an electrophilic carbon species that cannot rearrange and that would, moreover, deactivate the ring to prevent further substitution. There is such a species—an acylium cation—and it is used in the second Friedel-Crafts reaction, the topic of the next section.

In summary, Friedel-Crafts alkylation suffers from overalkylation and skeletal rearrangements by both hydrogen and alkyl shifts.

15-14 Friedel-Crafts Alkanoylation (Acylation)

The second electrophilic aromatic substitution that forms carbon–carbon bonds is **Friedel-Crafts alkanoylation,** or **acylation.** This reaction proceeds through the intermediacy of **acylium cations,** with the general structure $RC{\equiv}O\!:^+$. This section will describe how these ions readily attack benzene to form ketones.

Friedel-Crafts Alkanoylation

$$C_6H_6 \xrightarrow[-\,HX]{R\overset{O}{\overset{\|}{C}}X,\ AlX_3} C_6H_5\overset{O}{\overset{\|}{C}}R$$

Friedel-Crafts alkanoylation employs alkanoyl chlorides

Benzene reacts with alkanoyl (acyl) halides in the presence of an aluminum halide to give 1-phenylalkanones (phenyl ketones). An example is the preparation of 1-phenylethanone (acetophenone) from benzene and acetyl chloride, by using aluminum chloride as the Lewis acid.

Friedel-Crafts Alkanoylation of Benzene with Acetyl Chloride

$$\text{C}_6\text{H}_5\text{H} + \text{CH}_3\overset{\overset{\text{O}}{\|}}{\text{C}}\text{Cl} \xrightarrow[\text{2. H}_2\text{O, H}^+]{\text{1. AlCl}_3} \text{C}_6\text{H}_5\overset{\overset{\text{O}}{\|}}{\text{C}}\text{—CH}_3 + \text{HCl}$$

61%
1-Phenylethanone (Acetophenone)

Alkanoyl (acyl) chlorides are reactive derivatives of carboxylic acids. They are readily formed from the acids by reaction with $SOCl_2$. (We shall explore this process in detail in Chapter 19.)

Preparation of an Alkanoyl (Acyl) Chloride

$$\text{R}\overset{\overset{\text{O}}{\|}}{\text{C}}\text{OH} + \text{SOCl}_2 \longrightarrow \text{R}\overset{\overset{\text{O}}{\|}}{\text{C}}\text{Cl} + \text{SO}_2 + \text{HCl}$$

Alkanoyl halides react with Lewis acids to produce acylium ions

Acylium cations are formed by the reaction of alkanoyl halides with aluminum chloride. The Lewis acid forms a complex by coordination with the carbonyl oxygen. This complex is in equilibrium with another species in which the aluminum chloride is bound to the halogen. Dissociation then produces the acylium ion, which is stabilized by resonance.

Acylium Ion Generation

$$\text{R}\overset{\overset{:\text{O}:}{\|}}{\text{C}}\text{—}\ddot{\text{X}}: + \text{AlCl}_3 \rightleftharpoons \text{R}\overset{\overset{^{+}:\text{O—}\overset{-}{\text{AlCl}_3}}{\|}}{\text{C}}\text{—}\ddot{\text{X}}: \rightleftharpoons \text{R}\overset{\overset{:\text{O}:}{\|}}{\text{C}}\text{—}\overset{+}{\ddot{\text{X}}}\text{—}\overset{-}{\text{AlCl}_3} \rightleftharpoons [\text{R}\overset{+}{\text{C}}\equiv\text{O}: \longleftrightarrow \text{R}\overset{+}{\text{C}}=\ddot{\text{O}}] + :\ddot{\text{X}}\overset{-}{\text{AlCl}_3}$$

Acylium ion

Sometimes carboxylic anhydrides are used in alkanoylation in place of the halides. These molecules react with Lewis acids in a similar way.

Acylium Ions from Carboxylic Anhydrides

$$\text{R}\overset{\overset{:\text{O}:}{\|}}{\text{C}}\text{—}\ddot{\text{O}}\text{—}\overset{\overset{:\text{O}:}{\|}}{\text{C}}\text{R} + \text{AlCl}_3 \rightleftharpoons \text{R}\overset{\overset{^{+}:\text{O—}\overset{-}{\text{AlCl}_3}}{\|}}{\text{C}}\text{—}\ddot{\text{O}}\text{—}\overset{\overset{:\text{O}:}{\|}}{\text{C}}\text{R} \rightleftharpoons \text{R}\overset{\overset{:\text{O}:}{\|}}{\text{C}}\text{—}\underset{\underset{^{-}\text{AlCl}_3}{|}}{\overset{+}{\ddot{\text{O}}}}\text{—}\overset{\overset{:\text{O}:}{\|}}{\text{C}}\text{R} \rightleftharpoons [\text{R}\overset{+}{\text{C}}\equiv\text{O}: \longleftrightarrow \text{R}\overset{+}{\text{C}}=\ddot{\text{O}}] + \text{Cl}_3\overset{-}{\text{Al}}\ddot{\text{O}}\overset{\overset{:\text{O}:}{\|}}{\text{C}}\text{R}$$

Acylium ions undergo electrophilic aromatic substitution

The acylium ion is sufficiently electrophilic to attack benzene by the usual aromatic substitution mechanism.

Electrophilic Alkanoylation

Because the newly introduced alkanoyl substituent is electron withdrawing (see Section 14-8), it deactivates the ring and protects it from further substitution. The effect is accentuated by the formation of a strong complex between the aluminum chloride catalyst and the carbonyl function of the product ketone.

Lewis Acid Complexation with 1-Phenylalkanones

This complexation removes the $AlCl_3$ from the reaction mixture and necessitates the use of *at least one full equivalent* of the Lewis acid to allow the reaction to go to completion. Aqueous work-up is necessary to liberate the ketone from its aluminum chloride complex, as illustrated by the following examples.

1. $AlCl_3$ (1.7 equivalents)
2. H_2O, H^+
− HCl

84%

1-Phenyl-1-propanone
(Propiophenone)

1. $AlCl_3$ (2.4 equivalents)
2. H_2O, H^+
− CH_3COOH

85%

1-Phenylethanone
(Acetophenone)

EXERCISE 15-27

The simplest alkanoyl chloride, formyl chloride, H–C(=O)–Cl, is unstable, decomposing to HCl and CO upon attempted preparation. Therefore, direct Friedel-Crafts formylation of benzene is impossible. An alternative process, the Gattermann-Koch reaction, enables the introduction of the formyl group, –CHO, into the benzene ring by treatment with CO under pressure, in the presence of HCl and Lewis acid catalysts. For example, methylbenzene (toluene) can be formylated at the para position in this way in 51%

yield. Write a mechanism for this reaction. (Hints: Start with the Lewis structure for CO and proceed by considering the species that may arise in the presence of acid.)

$$CH_3C_6H_5 + CO + HCl \xrightarrow{AlCl_3,\ CuCl} p\text{-}CH_3C_6H_4CHO$$

In summary, the problems of Friedel-Crafts alkylation are avoided in Friedel-Crafts alkanoylations, in which an alkanoyl halide or carboxylic acid anhydride is the reaction partner, in the presence of a Lewis acid. The intermediate acylium cations undergo electrophilic aromatic substitution to yield the corresponding aromatic ketones.

NEW REACTIONS

1. HYDROGENATION OF BENZENE (Section 15-2)

$$C_6H_6 \xrightarrow{H_2,\ \text{catalyst}} C_6H_{12}$$

$\Delta H° = -49.3$ kcal mol^{-1}
Resonance energy: ~ -30 kcal mol^{-1}

Electrophilic Aromatic Substitution

2. CHLORINATION, BROMINATION, NITRATION, AND SULFONATION (Sections 15-10 and 15-11)

$$C_6H_6 \xrightarrow{X_2,\ FeX_3} C_6H_5X + HX \qquad X = Cl, Br$$

$$C_6H_6 \xrightarrow{HNO_3,\ H_2SO_4} C_6H_5NO_2 + H_2O$$

$$C_6H_6 \underset{H_2SO_4,\ H_2O,\ \Delta}{\overset{SO_3,\ H_2SO_4}{\rightleftharpoons}} C_6H_5SO_3H \qquad \text{Reversible}$$

3. BENZENESULFONYL CHLORIDES (Section 15-11)

$$C_6H_5SO_3Na + PCl_5 \longrightarrow C_6H_5SO_2Cl + POCl_3 + NaCl$$

4. FRIEDEL-CRAFTS ALKYLATION (Section 15-12)

$$C_6H_6 + RX \xrightarrow{AlCl_3} C_6H_5R + HX + \text{overalkylated product}$$

R^+ is subject to carbocation rearrangements

Intramolecular

$$C_6H_5(CH_2)_4Cl \xrightarrow{AlCl_3} \text{tetralin} + HCl$$

Alcohols and alkenes as substrates

$$C_6H_6 + \underset{}{\overset{\text{OH}}{\underset{|}{}}}\text{RCHR}' \xrightarrow[-\,H_2O]{BF_3,\ 60^\circ C} C_6H_5\overset{\text{R}'}{\text{CHR}}$$

$$C_6H_6 + RCH{=}CH_2 \xrightarrow{HF,\ 0^\circ C} C_6H_5\overset{\text{R}}{\text{CHCH}_3}$$

5. FRIEDEL-CRAFTS ALKANOYLATION (Section 15-14)

$$C_6H_6 + R\overset{O}{\overset{\|}{C}}Cl \xrightarrow{1.\ AlCl_3\quad 2.\ H_2O} C_6H_5\overset{O}{\overset{\|}{C}}R + HCl$$

Requires at least one full equivalent of Lewis acid

Anhydrides

$$C_6H_6 + CH_3\overset{O}{\overset{\|}{C}}O\overset{O}{\overset{\|}{C}}CH_3 \xrightarrow{1.\ AlCl_3\quad 2.\ H_2O} C_6H_5\overset{O}{\overset{\|}{C}}CH_3 + CH_3COOH$$

IMPORTANT CONCEPTS

1. Substituted benzenes are named by adding prefixes or suffixes to the word *benzene*. Disubstituted systems are labeled as 1,2-, 1,3-, and 1,4-, or ortho, meta, and para, depending on the location of the substituents. Many benzene derivatives have common names, sometimes used as bases for naming their substituted analogs. As a substituent, an aromatic system is called aryl; the parent aryl substituent, C_6H_5, is called phenyl; its homolog $C_6H_5CH_2$ is named phenylmethyl (benzyl).
2. Benzene is not a cyclohexatriene but a delocalized cyclic system of six π electrons. It is a regular hexagon of six sp^2-hybridized carbons. All six p orbitals overlap equally with their neighbors. Its unusually low heat of hydrogenation indicates a resonance energy of about 30 kcal mol^{-1}. The stability imparted by aromatic delocalization is also evident in the transition state of some reactions, such as the Diels-Alder cycloaddition and ozonolysis.
3. The special structure of benzene gives rise to unusual UV, IR, and NMR spectral data. ^1H NMR spectroscopy is particularly diagnostic because of the unusual deshielding of aromatic hydrogens by an induced ring current. Moreover, the substitution pattern is revealed by examination of the o, m, and p coupling constants.
4. The polycyclic benzenoid hydrocarbons are composed of linearly or angularly fused benzene rings. The simplest members of this class of compounds are naphthalene, anthracene, and phenanthrene.
5. In these molecules, benzene rings share two (or more) carbon atoms, whose π electrons are delocalized over the entire ring system. Thus, naphthalene shows some of the properties characteristic of the aromatic ring in benzene: The electronic spectra reveal extended conjugation, ^1H NMR exhibits deshielding ring-current effects, and there is little bond alternation.
6. Benzene is the smallest member of the class of aromatic cyclic polyenes following Hückel's $4n + 2$ rule. Most of the $4n$ π systems are relatively reactive species devoid of aromatic properties. Hückel's rule also extends to aromatic

charged systems, such as the cyclopentadienyl anion, cycloheptatrienyl cation, and cyclooctatetraene dianion.

7. The most important reaction of benzene is electrophilic aromatic substitution. The rate-determining step is addition by the electrophile to give a delocalized hexadienyl cation in which the aromatic character of the original benzene ring has been lost. Fast deprotonation restores the aromaticity of the (now-substituted) benzene ring. Exothermic substitution is preferred over endothermic addition. The reaction can lead to halo- and nitrobenzenes, benzenesulfonic acids, and alkylated and alkanoylated derivatives. When necessary, Lewis acid (chlorination, bromination, Friedel-Crafts reaction) or mineral acid (nitration, sulfonation) catalysts are applied. These enhance the electrophilic power of the reagents (chlorination, bromination, primary alkylation, sulfonation), or generate strong, positively charged electrophiles (nitration, NO_2^+; alkylation, R^+; alkanoylation, RCO^+).

8. Sulfonation of benzene is a reversible process. The sulfonic acid group is removed by heating with dilute aqueous acid.

9. Benzenesulfonic acids are precursors of benzenesulfonyl chlorides. The chlorides react with alcohols to form sulfonic esters containing useful leaving groups and with amines to give sulfonamides, some of which are medicinally important (sulfa drugs).

10. In contrast with other electrophilic substitutions, Friedel-Crafts alkylations activate the aromatic ring to further electrophilic substitution, leading to product mixtures.

PROBLEMS

1. Name each of the following compounds using the IUPAC system and, if possible, by a reasonable common alternative. (Hint: If necessary, refer to the examples on page 552.)

(a) COOH, Cl

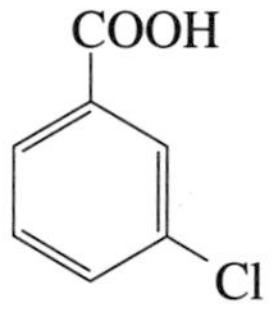

(b) OCH_3, NO_2

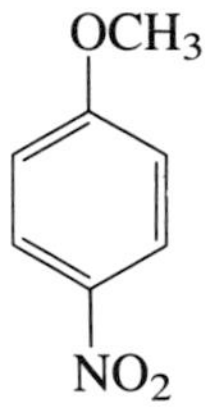

(c) OH, CHO

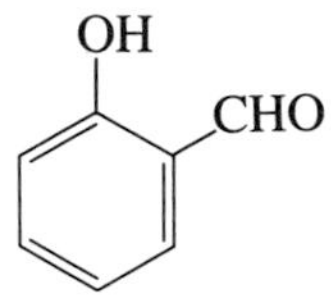

(d) NH_2, COOH

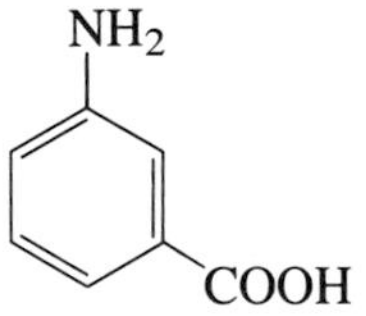

(e) NH_2, CH_3, CH_2CH_3

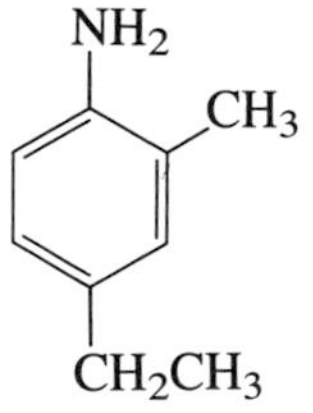

(f) CH_3, CH_3, Br

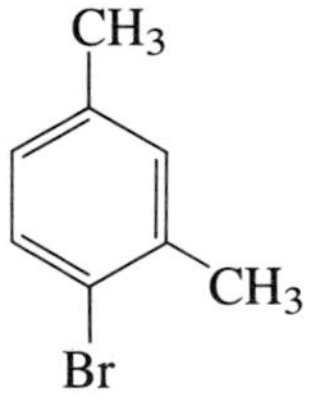

(g) OH, CH_3O, OCH_3, Br

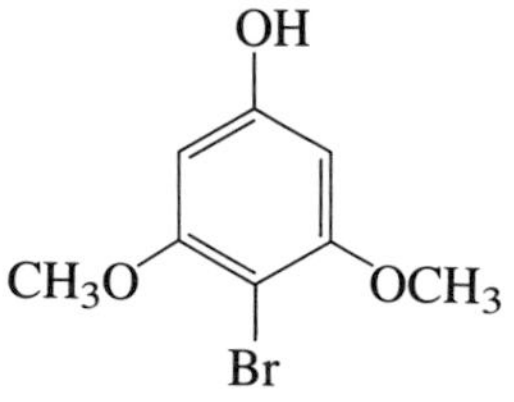

(h) $-CH_2CH_2OH$

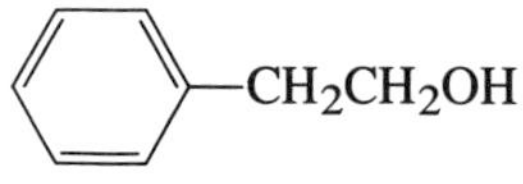

(i) CH_3, O

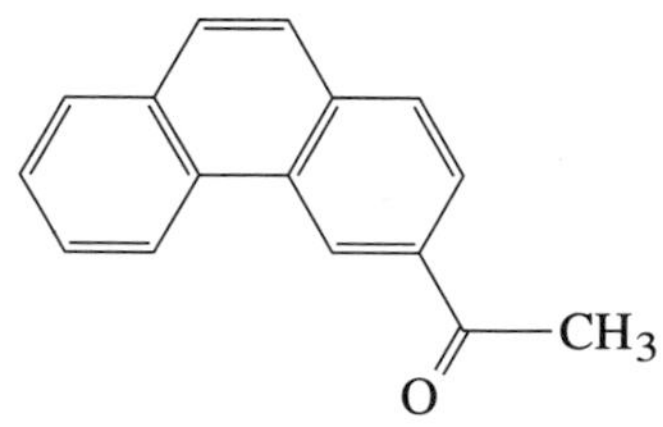

2. Give a proper IUPAC name for each of the following commonly named substances.

(a) CH_3, CH_3, H_3C, CH_3
Durene

(b) OH, OH, $CH_2(CH_2)_4CH_3$
Hexylresorcinol

(c) OH, OCH_3, $CH_2CH{=}CH_2$
Eugenol

3. Draw the structure of each of the following compounds. If the name itself is incorrect, give a correct systematic alternative. **(a)** *o*-Chlorobenzaldehyde; **(b)** 2,4,6-trihydroxybenzene; **(c)** 4-nitro-*o*-xylene; **(d)** *m*-isopropylbenzoic acid; **(e)** 4,5-dibromoaniline; **(f)** *p*-methoxy-*m*-nitroacetophenone.

4. The complete combustion of benzene is exothermic by approximately -789 kcal mol^{-1}. What would this value be if benzene lacked aromatic stabilization?

5. The ^{1}H NMR spectrum of naphthalene shows two multiplets (Figure 15-15). The upfield absorption ($\delta = 7.40$ ppm) is due to the hydrogens at C2, C3, C6, and C7, and the downfield multiplet ($\delta = 7.77$ ppm) is due to the hydrogens at C1, C4, C5, and C8. Why do you suppose the latter hydrogens are more deshielded than the former?

6. Complete hydrogenation of 1,3,5,7-cyclooctatetraene is exothermic by -101 kcal mol^{-1}. Hydrogenation of cyclooctene proceeds with $\Delta H° =$ -23 kcal mol^{-1}. Are these data consistent with the description of cyclooctatetraene presented in the chapter?

7. Which of the following structures qualify as being aromatic, according to Hückel's rule?

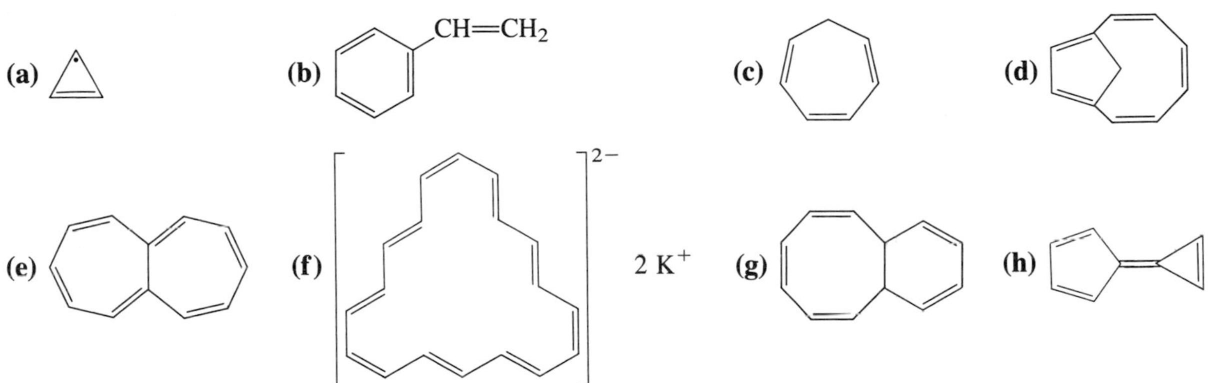

8. Following are spectroscopic and other data for several compounds. Propose a structure for each of them. **(a)** Molecular formula = $C_6H_4Br_2$. ^{1}H NMR spectrum A (p. 590). ^{13}C NMR: 3 peaks. IR: $\tilde{\nu} = 745$ (s, broad) cm^{-1}. UV: $\lambda_{max}(\epsilon) = 263(150)$, $270(250)$, and $278(180)$ nm. **(b)** Molecular formula = C_7H_7BrO. ^{1}H NMR spectrum B (p. 590). ^{13}C NMR: 7 peaks. IR: $\tilde{\nu} = 765$(s) and 680(s) cm^{-1}. **(c)** Molecular formula = $C_9H_{11}Br$. ^{1}H NMR spectrum C (p. 591). ^{13}C NMR: $\delta = 20.6$(q), 23.6(q), 124.2(s), 129.0(d), 136.0(s), and 137.7(s) ppm.

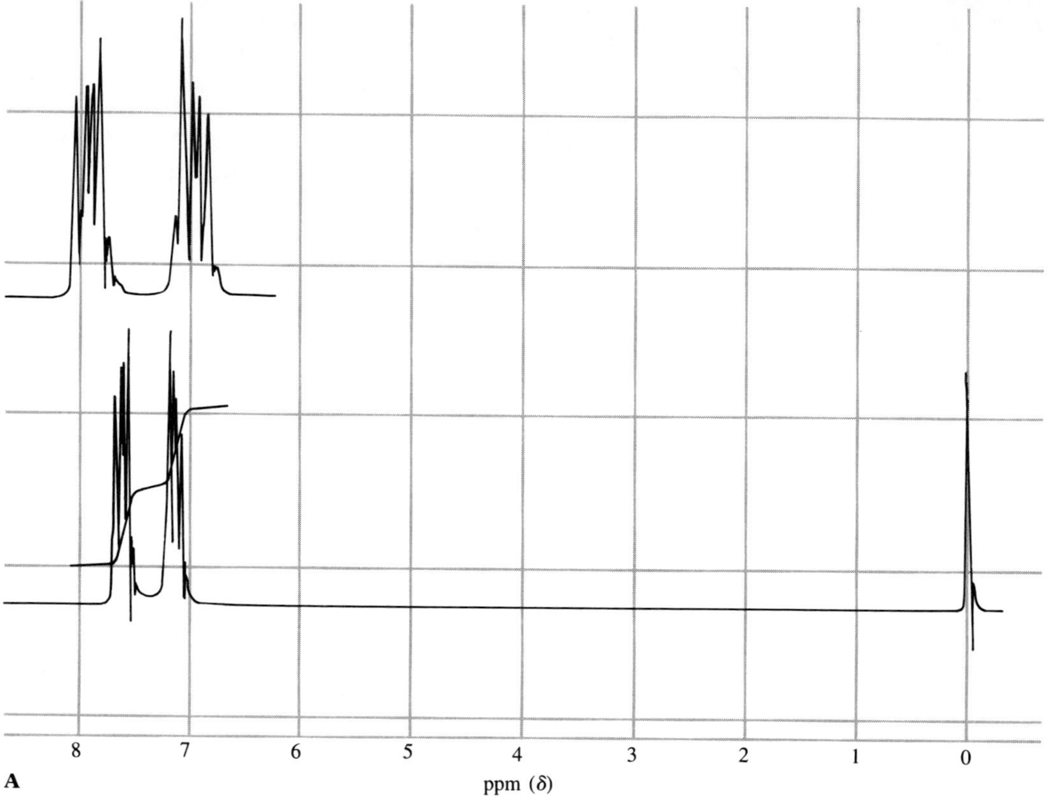

A

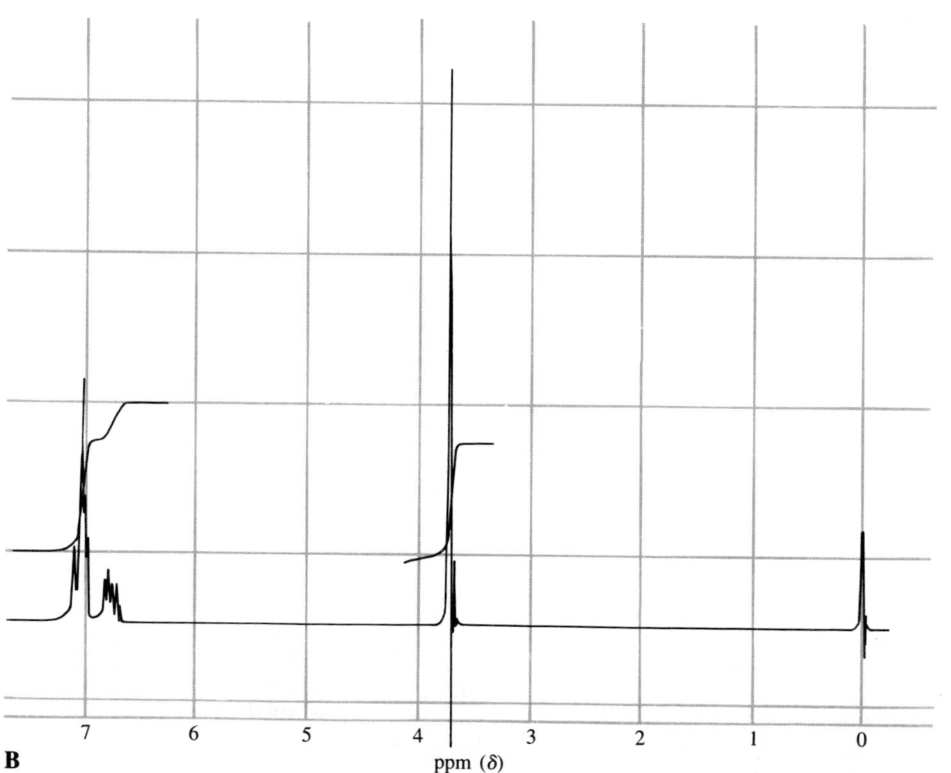

B

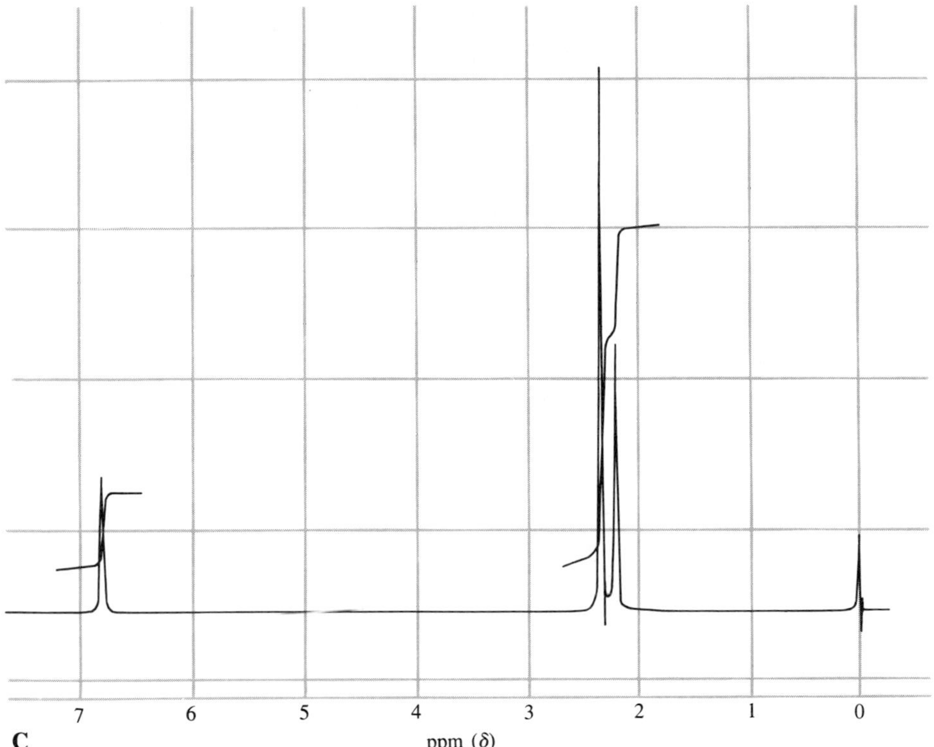

9. The species resulting from the addition of benzene to HF-SbF_5 (Exercise 15-21) shows the following ^{13}C NMR absorptions: $\delta = 52.2$(t), 136.9(d), 178.1(d), and 186.6(d) ppm. The signals at $\delta = 136.9$ and $\delta = 186.6$ are twice the intensity of the other signals. Assign the signals in this spectrum.

10. Write the expected major product that should form with the addition to benzene of each of the following reagent mixtures.

(a) $Cl_2 + AlCl_3$

(b) $T_2O + T_2SO_4$ (T = tritium, ^3H)

(c) $ICl + FeCl_3$ (Careful! $DH^\circ_{ICl} = 50$ kcal mol^{-1}. Is this reaction exothermic?)

(d) N_2O_5 (which tends to dissociate into NO_2^+ and NO_3^-)

(e) $(CH_3)_2C{=}CH_2 + H_3PO_4$

(f) $(CH_3)_3CCH_2CH_2Cl + AlCl_3$

(g) $(CH_3)_2\overset{\displaystyle Br}{\overset{|}{C}}CH_2CH_2\overset{\displaystyle Br}{\overset{|}{C}}(CH_3)_2 + AlBr_3$

(h) H_3C—⟨benzene ring⟩—$COCl + AlCl_3$

11. Write mechanisms for reactions (c) and (f) in Problem 10.

12. Propose a mechanism for the direct chlorosulfonylation of benzene (in the margin), an alternative synthesis of benzenesulfonyl chloride.

[benzene ring] + 2 $ClSO_3H$ → [benzene ring]—SO_2Cl + HCl + H_2SO_4

13. The text states that alkylated benzenes are more susceptible to electrophilic attack than benzene itself. Draw a graph like Figure 15-19 to show how the energy profile of electrophilic substitution of methylbenzene (toluene) would differ quantitatively from that of benzene.

14. Like haloalkanes, haloarenes are readily converted into organometallic reagents, which are sources of nucleophilic carbon.

C_6H_5—Br $\xrightarrow{Mg,\ (CH_3CH_2)_2O,\ 25°C}$ $C_6H_5^{\delta-}$—$^{\delta+}MgBr$

Phenylmagnesium bromide

C_6H_5—Cl $\xrightarrow{Mg,\ THF,\ 50°C}$ $C_6H_5^{\delta-}$—$^{\delta+}MgCl$

Phenylmagnesium chloride

Grignard reagents

C_6H_5—Br $\xrightarrow{Li,\ (CH_3CH_2)_2O,\ 25°C}$ $C_6H_5^{\delta-}$—$^{\delta+}Li$ $\xrightarrow{CuI,\ (CH_3CH_2)_2O,\ 0°C}$ $(C_6H_5)_2CuLi$

Phenyllithium **Lithium diphenylcuprate**

The chemical behavior of these reagents is very similar to that of their alkyl counterparts. Write the main product of each of the following sequences.

(a) C_6H_5Br $\xrightarrow{1.\ Li,\ (CH_3CH_2)_2O;\ 2.\ CH_3CHO;\ 3.\ H^+,\ H_2O}$

(b) C_6H_5Cl $\xrightarrow{1.\ Mg,\ THF;\ 2.\ CH_2\text{—}CH_2\ (\text{bridged by O});\ 3.\ H^+,\ H_2O}$

(c) C_6H_5Br $\xrightarrow{1.\ Li,\ (CH_3CH_2)_2O;\ 2.\ CuI,\ (CH_3CH_2)_2O;\ 3.\ (CH_3)_2CHCH_2CH_2I}$

15. Give efficient syntheses of the following, beginning with benzene. **(a)** 1-Phenyl-1-heptanol; **(b)** 2-phenyl-2-butanol; **(c)** 1-phenyloctane. (Hint: Use a method from Problem 14. Why won't Friedel-Crafts alkylation work?)

16. Because of cyclic delocalization, structures A and B shown below for *o*-dimethylbenzene (*o*-xylene) are simply two resonance forms for the same molecule. Can the same be said for the two dimethylcyclooctatetraene structures C and D? Explain.

CH_3 CH_3 **A** | CH_3 CH_3 **B** | CH_3 CH_3 **C** | CH_3 CH_3 **D**

17. The energy levels of the 2-propenyl (allyl) and cyclopropenyl π systems are compared qualitatively in the diagram on the next page. **(a)** Draw the three molecular orbitals of each system, using plus and minus signs and dotted lines to indicate bonding overlap and nodes, as in Figure 15-4. Does either of these systems possess degenerate molecular orbitals? **(b)** How many π electrons would give rise to the maximum stabilization of the cyclopropenyl system, relative to 2-propenyl (allyl)? (Compare Figure 15-5, for benzene.) Draw Lewis structures for both systems with this number of π electrons and any appropriate atomic charges. **(c)** Could the cyclopropenyl system drawn in (b) qualify as "aromatic"? Explain.

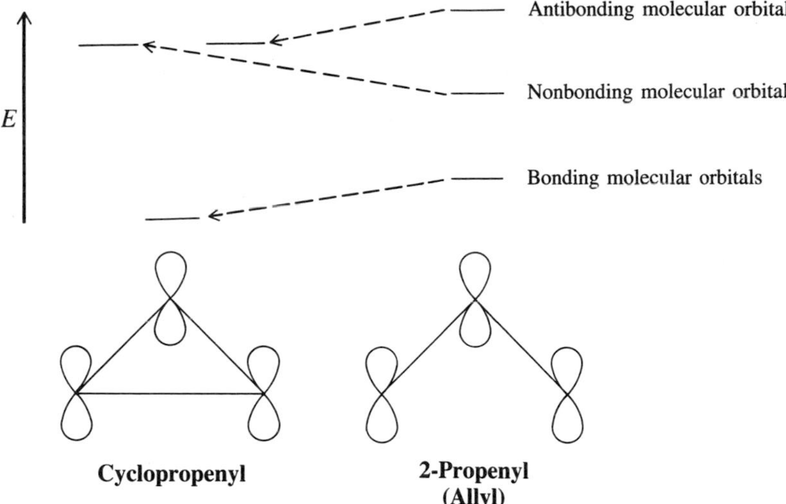

18. **(a)** The ^{1}H NMR spectrum of [18]annulene shows two signals, at δ = 9.28 (12 H) and −2.99 (6 H) ppm. The negative chemical shift value refers to a resonance *upfield* (to the *right*) of $(CH_3)_4Si$. Explain this spectrum. (Hint: Consult Figure 15-8.) **(b)** The unusual molecule 1,6-methano[10]annulene (shown in the margin) exhibits two sets of signals in the ^{1}H NMR spectrum at δ = 7.10 (8 H) and −0.50 (2 H) ppm. Is this result a sign of aromatic character?

H H

1,6-Methano[10]annulene

19. The ^{1}H NMR spectrum of the most stable isomer of [14]annulene shows two signals, at δ = −0.61 (4 H) and 7.88 (10 H) ppm. Two possible structures for [14]annulene are shown below. How do they differ? Which one corresponds to the NMR spectrum described?

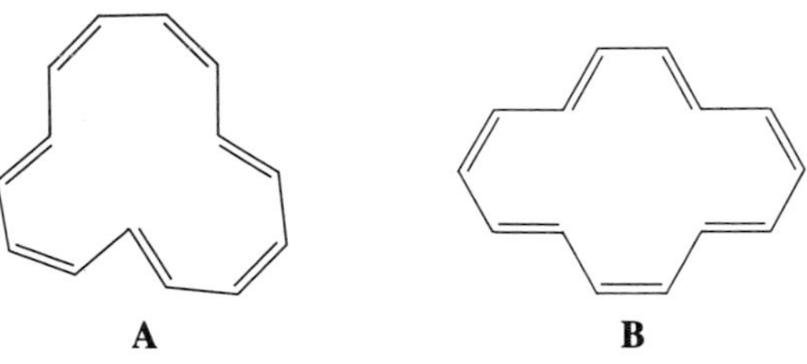

20. Explain the following reaction and the indicated stereochemical result mechanistically.

Benzene + H_2C—O—C(H)---CH_3 (epoxide) $\xrightarrow[\text{2. } H^+,\ H_2O]{\text{1. } AlCl_3}$ $HOCH_2$, H, CH_3 on C attached to benzene ring

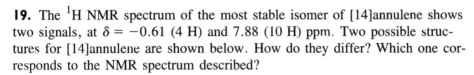

21. Metal-substituted benzenes have a long history of use in medicine. Before antibiotics were discovered, phenylarsenic derivatives were the only treatment for a number of diseases. Phenylmercury compounds continue to be used as fungicides and antimicrobial agents to the present day. Based on the general principles explained in this chapter and your knowledge of the characteristics of compounds of Hg^{2+} (see Section 12-7), propose a sensible synthesis of phenylmercury acetate (shown in the margin).

$HgOCCH_3$ (O above the C)

Phenylmercury acetate

16

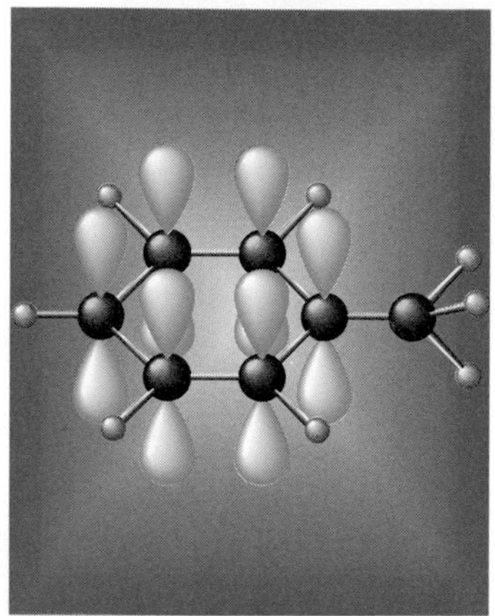

Electrophilic Attack on Derivatives of Benzene

Substituents Control Regioselectivity

Chapter 15 introduced a new transformation: the electrophilic replacement of a hydrogen on a benzene ring. With most electrophiles, this process can be stopped at monosubstitution because the newly introduced group deactivates the ring, thereby slowing further reaction. We also saw that alkyl groups have the opposite effect. As a result, in Friedel-Crafts alkylation, higher substitution is difficult to avoid. What are the factors that contribute to the activating or deactivating nature of substituents on a benzene ring? Furthermore, does subsequent substitution occur at preferred positions around the ring? This chapter will answer these questions and explore strategies for the selective synthesis of highly substituted benzene derivatives.

16-1 Directing Inductive Effects of Alkyl Groups

What makes a monosubstituted benzene more or less susceptible to further electrophilic attack? What controls the position where an electrophile will attack next? It is the *substituent* on the ring, which has either an activating effect, by donating electron density, or a deactivating effect, by withdrawing it, and which directs incoming electrophiles to specific positions. Let us explore the electrophilic substitution reactions of alkyl-substituted benzenes. We shall begin with methylbenzene (toluene), in which the methyl group on the benzene ring is electron donating because of hyperconjugation (Section 7-5).

Groups that donate electrons by hyperconjugation are activating and direct ortho and para

Electrophilic bromination of methylbenzene (toluene) occurs considerably faster than the bromination of benzene itself. The reaction also is regioselective: It results mainly in para (60%) and ortho (40%) substitutions, with virtually no meta product.

Electrophilic Bromination of Methylbenzene (Toluene) Gives Ortho and Para Substitution

CH_3 (toluene) $\xrightarrow[-\ HBr]{Br{-}Br,\ FeBr_3,\ CCl_4}$ (ortho: CH_3, Br) + (meta: CH_3, Br) + (para: CH_3, Br)

40%	<1%	60%
1-Bromo-2-methylbenzene (*o*-Bromotoluene)	**1-Bromo-3-methylbenzene (*m*-Bromotoluene)**	**1-Bromo-4-methylbenzene (*p*-Bromotoluene)**

Is bromination a special case? The answer is no; nitration and sulfonation of the alkylbenzene give similar results—mainly ortho and para substitutions. Evidently, the nature of the attacking electrophile has little influence on the distribution of isomers; it is the methyl group that matters. Because each of these reactions takes place more rapidly than the corresponding reaction of benzene, we say that the methyl substituent is **activating.** Because there is virtually no meta product, we also say that the methyl substituent is **ortho and para directing.**

Electrophilic Nitration and Sulfonation of Methylbenzene (Toluene)

CH_3 (toluene) $\xrightarrow[-\ H_2O]{HNO_3,\ H_2SO_4,\ 30°C}$ (ortho: CH_3, NO_2) + (meta: CH_3, NO_2) + (para: CH_3, NO_2)

60%	5%	35%
1-Methyl-2-nitrobenzene (*o*-Nitrotoluene)	**1-Methyl-3-nitrobenzene (*m*-Nitrotoluene)**	**1-Methyl-4-nitrobenzene (*p*-Nitrotoluene)**

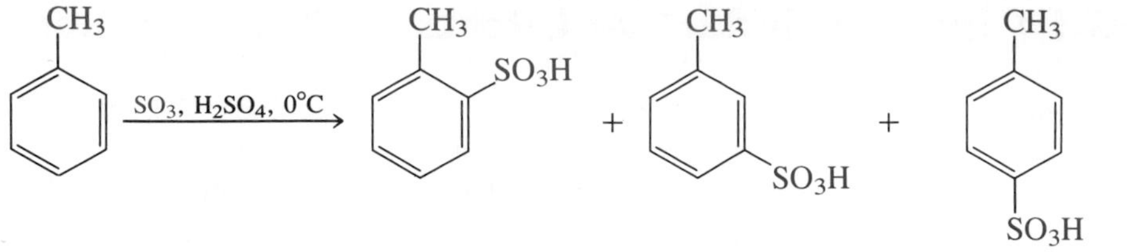

CH_3-benzene $\xrightarrow{SO_3,\ H_2SO_4,\ 0°C}$ 2-Methylbenzenesulfonic acid + 3-Methylbenzenesulfonic acid + 4-Methylbenzenesulfonic acid

43%	4%	53%
2-Methylbenzene-sulfonic acid (*o*-Toluene-sulfonic acid)	**3-Methylbenzene-sulfonic acid (*m*-Toluene-sulfonic acid)**	**4-Methylbenzene-sulfonic acid (*p*-Toluene-sulfonic acid)**

Can we explain this selectivity by a mechanism? Let us inspect the possible resonance structures of the cations formed after the electrophile, E^+, has attacked the ring in the first, and rate-determining, step.

Ortho, Meta, and Para Attack on Methylbenzene (Toluene)

Ortho attack (E^+ = electrophile)

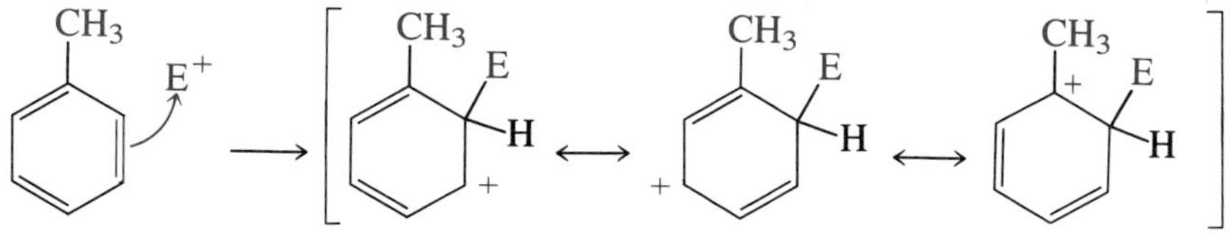

Most significant resonance structure

More stable cation

Meta attack

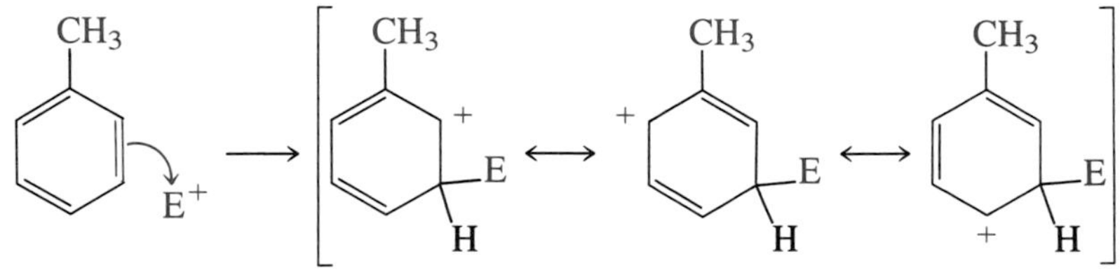

Less stable cation

Para attack

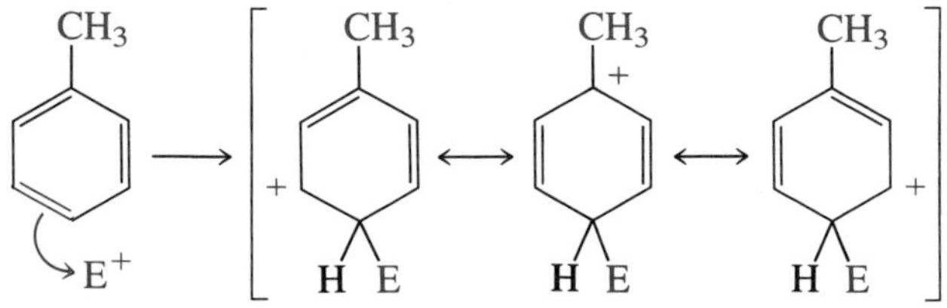

Most significant resonance structure

More stable cation

The alkyl group is electron donating by hyperconjugative (Section 7-5) and inductive effects (Section 8-3). Attacks at the ortho and para positions produce an intermediate carbocation in which one of the resonance forms places the positive charge *next* to the alkyl substituent. Because the alkyl group can donate electron density to stabilize the positive charge, that resonance form is a more important contributor to the resonance hybrid than the others, in which the positive charge is at an unsubstituted carbon. Meta attack, however, produces an intermediate in which *none* of the resonance structures benefits from such direct stabilization. Thus, electrophilic attack on a carbon located ortho or para to the methyl (or another alkyl group) leads to a cationic intermediate that is more stable than the one derived from attack at the meta carbon. The transition state leading to the more stable intermediate is of relatively low energy, and it therefore forms relatively rapidly.

By comparison, electrophilic substitution of benzene itself produces carbocations lacking stabilizing substituents (Section 15-9). Thus, the transition state energies for ortho and para attack on methylbenzene (and, in general, alkylbenzenes) are also lower than that for attack on benzene, an effect leading to faster substitution reactions for methylbenzene.

Groups that withdraw electrons inductively are deactivating and meta directing

The strongly electronegative fluorine atoms in (trifluoromethyl)benzene render the trifluoromethyl group inductively electron withdrawing. In this case reaction is very sluggish. Under stringent conditions, substitution does take place—but *only* at the meta positions: The trifluoromethyl group is **deactivating** and **meta directing.**

Electrophilic Nitration of (Trifluoromethyl)benzene Gives Meta Substitution

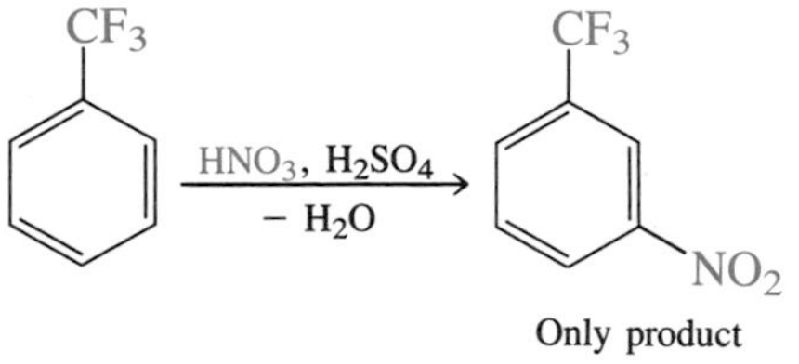

CF_3 $\xrightarrow[-\ H_2O]{HNO_3,\ H_2SO_4}$ CF_3 NO_2

Only product

Once again, the explanation lies in the various resonance structures for the cation produced by ortho, meta, and para attack. The presence of the substituent *de*stabilizes the carbocations resulting from electrophilic attack at all positions in the ring.

Ortho, Meta, and Para Attack on (Trifluoromethyl)benzene

Ortho attack

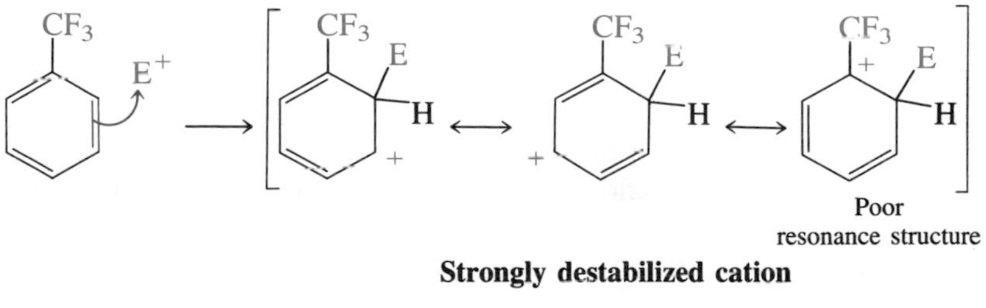

CF_3 E^+ ⟶ [CF_3 E H + ⟷ CF_3 E H + ⟷ CF_3 + E H]

Poor resonance structure

Strongly destabilized cation

Meta attack

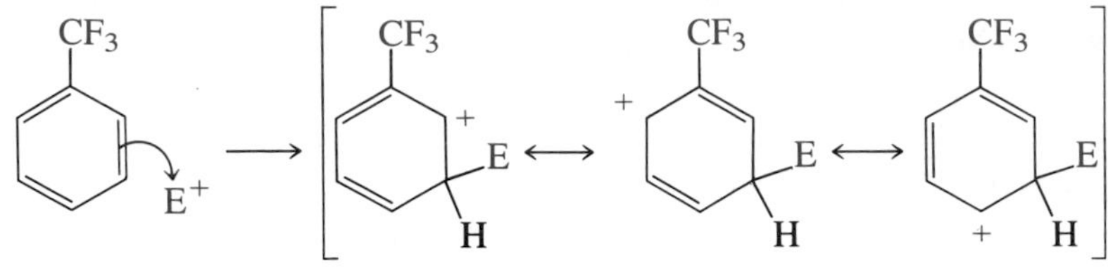

CF_3 E^+ → [CF_3 + E H ⟷ + CF_3 E H ⟷ CF_3 E + H]

Less destabilized cation

Para attack

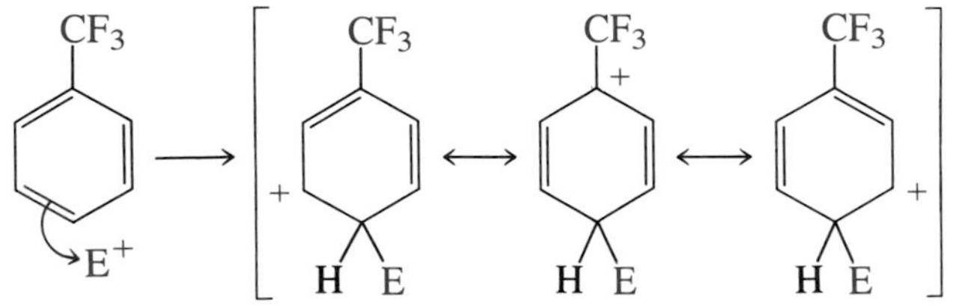

CF_3 E^+ → [CF_3 + H E ⟷ CF_3 + H E ⟷ CF_3 + H E]

Poor resonance structure

Strongly destabilized cation

Ortho and para attack are relatively unfavored for the same reasons that they are favored with methylbenzene (toluene): In each case, one of the resonance structures in the intermediate cation places the positive charge next to the substituent. This structure is stabilized by an electron-donating group, but it is *destabilized* by an *electron-withdrawing* substituent—removing electron density from a positively charged center is energetically unfavored. Meta attack avoids this situation. The destabilizing inductive effect is still felt in the meta intermediate, but to a lesser extent. Therefore, the trifluoromethyl group directs substitution meta, or, more accurately, *away* from the ortho and para carbons.

EXERCISE 16-1

Rank the following compounds in order of decreasing activity in electrophilic substitution.

(a) 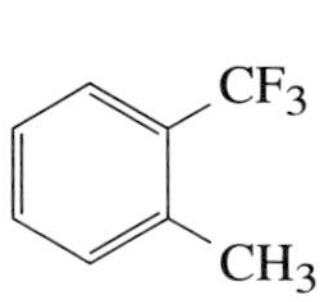 CF_3, CH_3 **(b)** 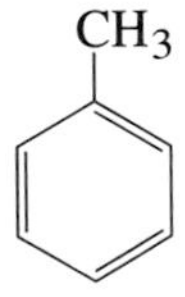 CH_3 **(c)** 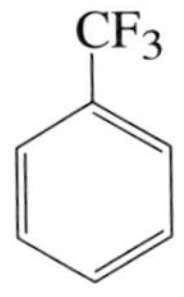 CF_3 **(d)** 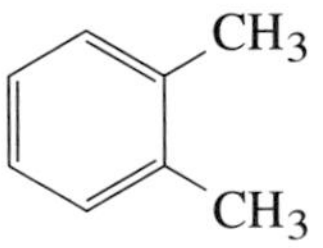 CH_3, CH_3

EXERCISE 16-2

Electrophilic bromination of an equimolar mixture of methylbenzene (toluene) and (trifluoromethyl)benzene with one equivalent of bromine gives only 1-bromo-2-methylbenzene and 1-bromo-4-methylbenzene. Explain.

In summary, electron-donating substituents inductively activate the benzene ring and direct electrophiles ortho and para; their electron-accepting counterparts deactivate the benzene ring and direct electrophiles to the meta positions.

16-2 Effects of Substituents in Resonance with the Benzene Ring

What is the influence of substituents whose electrons are in resonance with those of the benzene ring? This section answers this question by again comparing the resonance structures of the intermediates formed by the various modes of electrophilic attack.

Groups that donate electrons by resonance activate and direct ortho and para

Benzene rings bearing the groups NH_2 and OH are strongly activated. For example, halogenations of benzenamine (aniline) and phenol not only need no catalysts but also are difficult to stop at single substitution. The reactions proceed very rapidly to furnish exclusively ortho- and para-substituted products.

Electrophilic Brominations of Benzenamine (Aniline) and Phenol Give Ortho and Para Substitution

NH_2 — 3 Br–Br, H_2O / − 3 HBr → NH_2, Br, Br, Br; 100%

OH — 3 Br–Br, H_2O / − 3 HBr → OH, Br, Br, Br; 100%

Benzenamine (Aniline) **2,4,6-Tribromobenzenamine (2,4,6-Tribromoaniline)** **Phenol** **2,4,6-Tribromophenol**

Better control of substitution is attained with less activated derivatives of these substrates, such as *N*-phenylacetamide (acetanilide) and methoxybenzene (anisole). These related substituents are, again, ortho and para directing.

Electrophilic Nitration of *N*-Phenylacetamide (Acetanilide) and Methoxybenzene (Anisole)

HNCCH$_3$ (with O) — HNO_3, H_2SO_4, 20°C / − H_2O → HNCCH$_3$, NO_2 (ortho) + HNCCH$_3$, NO_2 (meta) + HNCCH$_3$, NO_2 (para)

N-Phenylacetamide (Acetanilide)

21% **N-(2-Nitrophenyl)-acetamide (o-Nitroacetanilide)**

Trace **N-(3-Nitrophenyl)-acetamide (m-Nitroacetanilide)**

79% **N-(4-Nitrophenyl)-acetamide (p-Nitroacetanilide)**

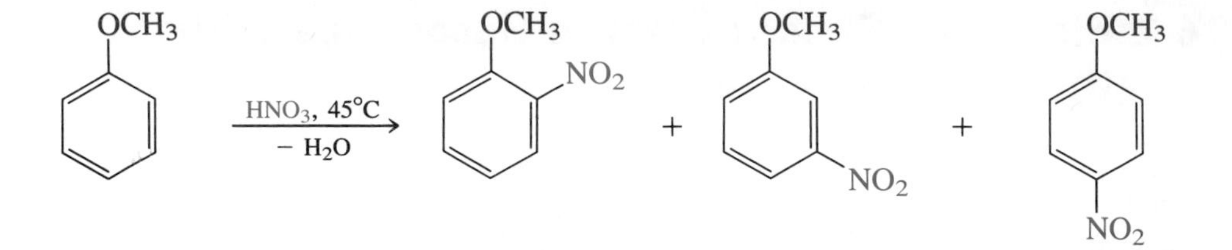

OCH_3 → (HNO_3, 45°C; $- H_2O$) → 40% + 2% + 58%

Methoxybenzene (Anisole)

1-Methoxy-2-nitrobenzene (*o*-Nitroanisole) 40%

1-Methoxy-3-nitrobenzene (*m*-Nitroanisole) 2%

1-Methoxy-4-nitrobenzene (*p*-Nitroanisole) 58%

Both the activated nature of these compounds and the observed regioselectivity upon electrophilic substitution can be explained by writing resonance forms for the various intermediate cations.

Ortho, Meta, and Para Attack on Benzenamine (Aniline)

Ortho attack

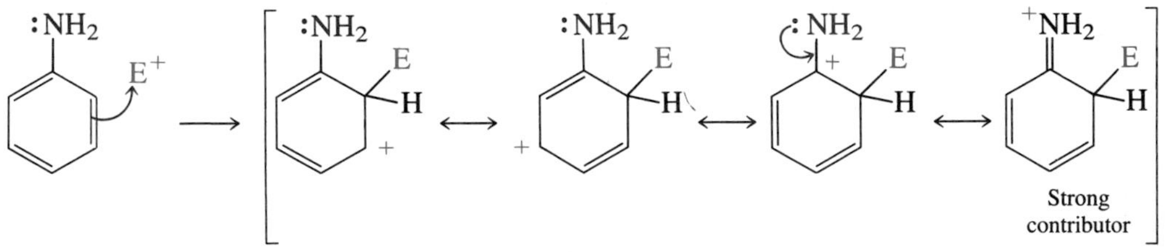

$:NH_2$, E^+

Strong contributor

Strongly stabilized cation

Meta attack

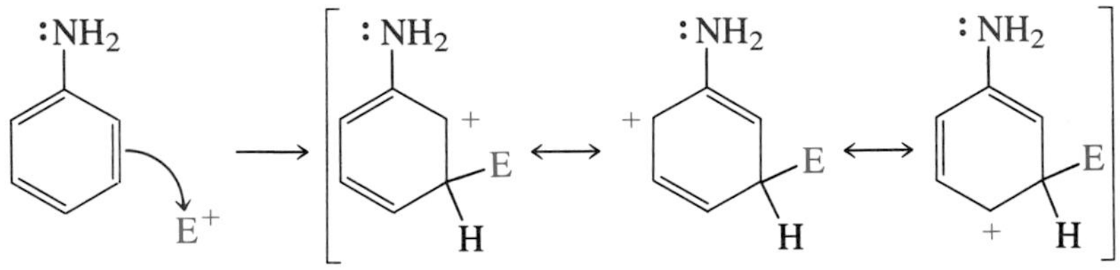

$:NH_2$, E^+

Para attack

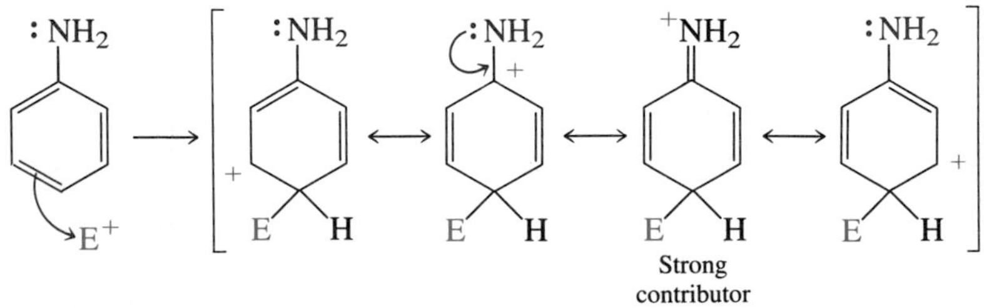

$:NH_2$, E^+

Strong contributor

Strongly stabilized cation

Because nitrogen is more electronegative than carbon, the amino group in benzenamine (aniline) is inductively electron withdrawing. However, the lone electron pair on the nitrogen atom may participate in resonance, thereby stabilizing the intermediate cations resulting from ortho and para (but not meta) substitutions. This resonance contribution by far outweighs the inductive effect. The result is a much reduced activation barrier for ortho or para attack. Consequently,

BOX 16-1 Nitration of Phenol Produces a Powerful Explosive, Picric Acid

Successive treatment of phenol with concentrated sulfuric and nitric acids furnishes 2,4,6-trinitrophenol, also known as *picric acid*. This yellow crystalline material detonates powerfully when shocked or heated and has a long history of use as both a military and an industrial explosive. The use of picric acid in artillery shells during World War I created an unprecedented demand from the chemical manufacturing industry for phenol.

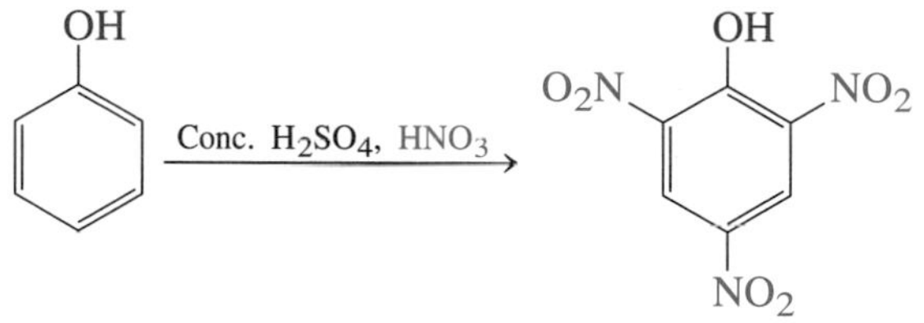

OH

Conc. H_2SO_4, HNO_3 →

OH, O_2N, NO_2, NO_2

65%

2,4,6-Trinitrophenol (Picric acid)

benzenamine (aniline) is strongly activated toward electrophilic substitution relative to benzene itself, and the reaction is highly regioselective as well.

EXERCISE 16-3

Formulate resonance structures for the various modes of electrophilic attack on methoxybenzene (anisole).

EXERCISE 16-4

In strongly acidic solution, benzenamine (aniline) becomes quite unreactive to electrophilic attack, and increased meta substitution is observed. Explain. (Hint: The nitrogen atom in benzenamine may behave as a base.)

Groups that withdraw electrons by resonance deactivate and direct meta

Many groups *deactivate* the benzene ring by *both* inductive and resonance effects. A good example is the carboxy group in benzoic acid, C_6H_5COOH. Nitration of benzoic acid takes place at only about 1/1000th the rate of benzene and gives predominantly meta substitution. The COOH group is deactivating and meta directing. Other substituents showing similar behavior include SO_3H, C≡N, and other carbonyl-containing groups such as CHO and $COOCH_3$, and NO_2.

Electrophilic Nitration of Benzoic Acid

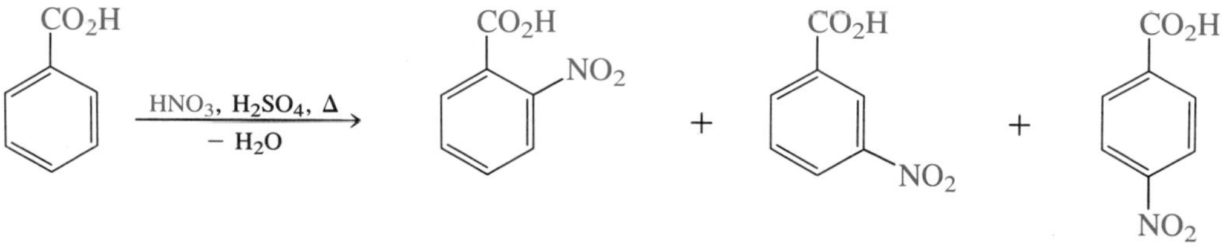

CO_2H —[HNO_3, H_2SO_4, Δ; $- H_2O$]→

CO_2H, NO_2 (ortho)	+	CO_2H, NO_2 (meta)	+	CO_2H, NO_2 (para)
18.5%		80%		1.5%
2-Nitrobenzoic acid (*o*-Nitrobenzoic acid)		**3-Nitrobenzoic acid (*m*-Nitrobenzoic acid)**		**4-Nitrobenzoic acid (*p*-Nitrobenzoic acid)**

The C–O double bond in the COOH group is strongly polarized as a result of both the electronegativity difference between carbon and oxygen (Section 8-6) and the contribution from a resonance form that places a positive charge on the former.

Polarization in the C–O Double Bond

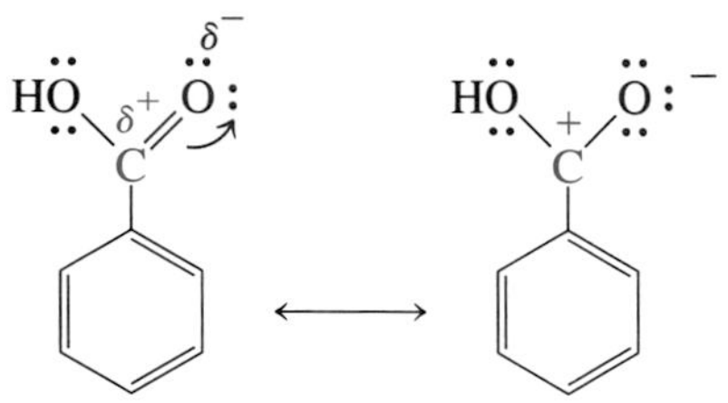

Let us see how this polarization affects the resonance structures of the cations resulting from electrophilic attack.

Ortho, Meta, and Para Attack on Benzoic Acid

Ortho attack

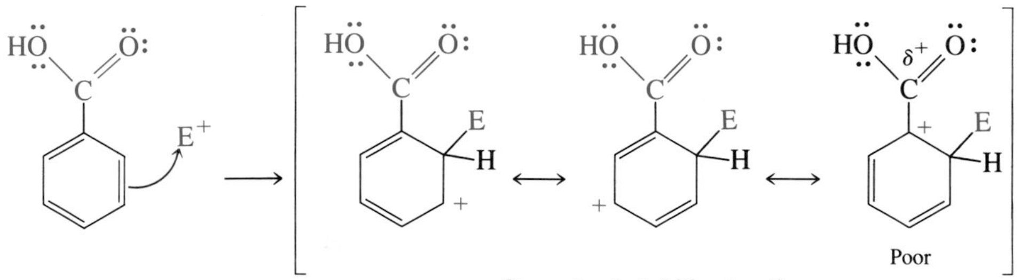

Poor

Strongly destabilized cation

Meta attack

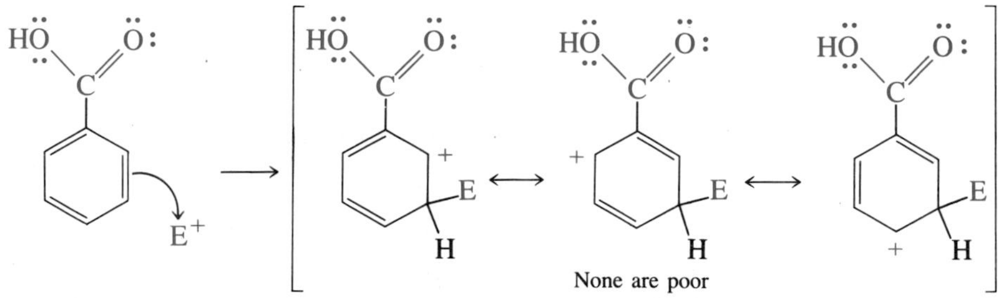

None are poor

Less destabilized cation

Para attack

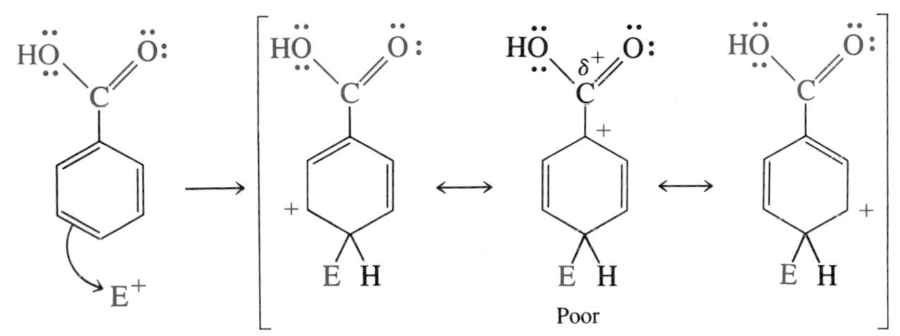

Strongly destabilized cation

Attack at the meta position avoids placing the positive charge next to the electron-withdrawing carboxy group, whereas ortho and para attacks necessitate the formulation of poor resonance structures. The carbocation resulting from meta substitution is still destabilized by the substituent, however, and substitution is slower than in the case of benzene itself.

Hence, it appears that deactivating groups, whether operating by induction or resonance, direct incoming electrophiles to the meta position, whereas activating groups direct to the ortho and para carbons. This statement is true for all classes of substituents except one—the halogens.

There is always an exception: halogen substituents, although deactivating, direct ortho and para

Halogen substituents inductively withdraw electron density (Section 8-3); however, they are donors by resonance. On balance the inductive effect wins out, rendering haloarenes *deactivated*. Nevertheless, the electrophilic substitution that does take place occurs mainly at the ortho and para positions.

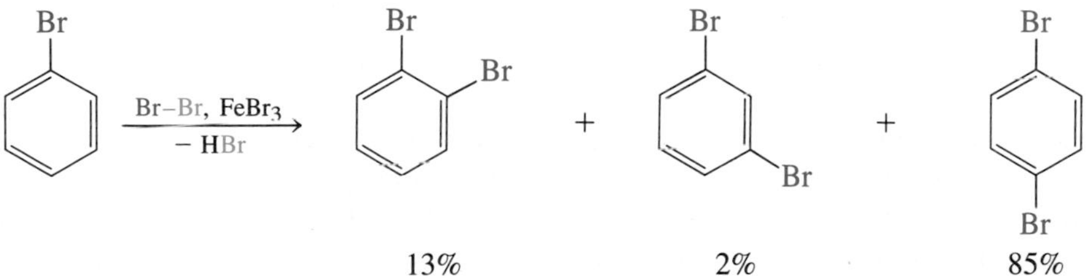

13% **1,2-Dibromobenzene (*o*-Dibromobenzene)**

2% **1,3-Dibromobenzene (*m*-Dibromobenzene)**

85% **1,4-Dibromobenzene (*p*-Dibromobenzene)**

The competition between resonance and inductive effects explains this seemingly contradictory reactivity. Again, we must examine the resonance structures for the various possible intermediates.

Ortho, Meta, and Para Attack on a Halobenzene

Ortho attack

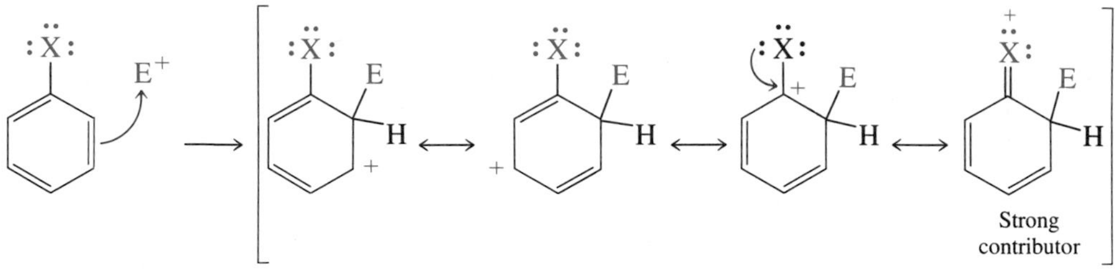

Strong contributor

More stable cation

Meta attack

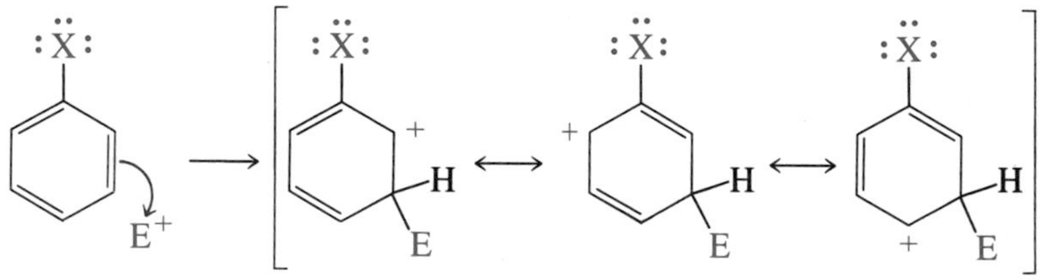

Less stable cation

Para attack

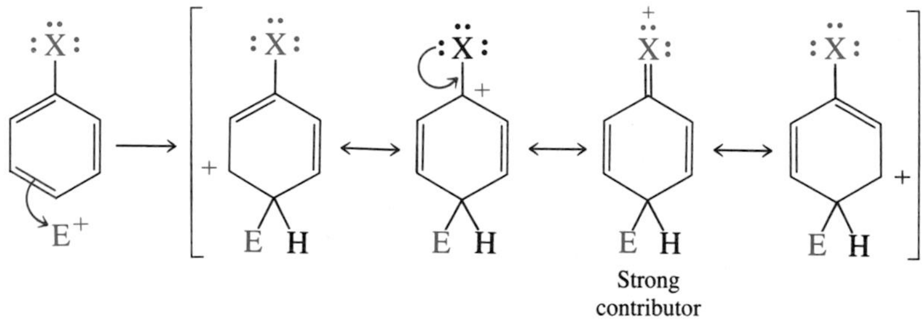

Strong contributor

More stable cation

Note that ortho and para attack leads to resonance structures in which the positive charge is placed next to the halogen substituent. Although this might be expected to be unfavorable, because the halogen is inductively electron withdrawing, resonance with the lone electron pairs allows the charge to be delocalized. Therefore, ortho and para substitutions become the preferred modes of reaction. The inductive effect of the halogen is still strong enough to make all three possible cations less stable than the one derived from benzene itself. Therefore, we have the unusual result that halogens are ***ortho and para directing,*** but ***deactivating.***

This section completes the survey of the regioselectivity of electrophilic attack on monosubstituted benzenes, summarized in Table 16-1. Table 16-2 ranks various substituents by their activating power and lists the product distributions obtained on electrophilic nitration of the benzene ring.

TABLE 16-1 Effects of Substituents in Electrophilic Aromatic Substitution

Ortho and para directors	Meta directors
Moderate and strong activators	Strong deactivators
$-\ddot{N}H_2$ $-\ddot{N}HR$ $-\ddot{N}R_2$ $-\ddot{N}HCR$ (C=O with :O:)	$-NO_2$ $-CF_3$ $-\overset{+}{N}R_3$ $-\overset{O}{\overset{\|}{C}}OH$
$-\ddot{\underset{..}{O}}H$ $-\ddot{\underset{..}{O}}R$	$-\overset{O}{\overset{\|}{C}}OR$ $-\overset{O}{\overset{\|}{C}}R$ $-SO_3H$ $-C\equiv N$
Weak activators	
Alkyl, phenyl	
Weak deactivators	
$-\ddot{\underset{..}{F}}:$ $-\ddot{\underset{..}{Cl}}:$ $-\ddot{\underset{..}{Br}}:$ $-\ddot{\underset{..}{I}}:$	

EXERCISE 16-5

Specify whether the benzene rings in the following compounds are activated or deactivated.

(a) 1,4-di(CH_2CH_3)benzene (b) benzene with $-\overset{O}{\overset{\|}{C}}-CH_3$ and ortho CH_3 (c) benzene with COOH and para CF_3 (d) benzene with OCH_3 and para $N(CH_3)_2$

TABLE 16-2 Relative Rates and Orientational Preferences in the Nitration of Some Monosubstituted Benzenes, RC_6H_5

		Percentage of isomer		
R	Relative rate	Ortho	Meta	Para
OH	1000	40	<2	58
CH_3	25	58	4	38
H	1			
CH_2Cl	0.71	32	15.5	52.5
I	0.18	41	<0.2	59
Cl	0.033	31	<0.2	69
$CO_2CH_2CH_3$	0.0037	24	72	4
CF_3	2.6×10^{-5}	6	91	3
NO_2	6×10^{-8}	5	93	2
$\overset{+}{N}(CH_3)_3$	1.2×10^{-8}	0	89	11

EXERCISE 16-6

Explain why **(a)** NO_2, **(b)** $\overset{+}{N}R_3$, and **(c)** SO_3H are strong deactivators. **(d)** Why should phenyl be activating and ortho and para directing? (Hint: Draw resonance structures for the appropriate cationic intermediates.)

16-3 Electrophilic Attack on Disubstituted Benzenes

Do the rules developed so far in this chapter predict the reactivity and regioselectivity of still higher substitution? We shall see that they do, provided we take into account the individual effect of each substituent. Let us investigate the reactions of disubstituted benzenes with electrophiles.

Substituent effects are additive

The effects of two substituents on the relative rate and orientation of electrophilic substitution of the benzene ring are additive. For example, we know that alkyl groups are activating and direct substitutions ortho and para. We indeed find that the dimethylbenzenes (xylenes) are several times as reactive as toluene. All positions that are either ortho or para to a methyl group will be the subject of attack, regardless of whether they are also meta to a substituent. (Remember that the effect of an alkyl group on a meta carbon is negligible.) However, steric effects may hinder substitution between two groups situated meta to each other, particularly if the groups are bulky.

In general, a doubly activated position (e.g., ortho to one, para to a second alkyl group) is more reactive than a singly activated one (Figure 16-1). Sulfonation of 1,3-dimethylbenzene (*m*-xylene) occurs only in the 2- and 4- (which is the same as 6-) positions, both being doubly (ortho and para) activated, whereas C5 is meta to both substituents and hence is not as reactive. In contrast, 1,2- and 1,4-dimethylbenzene (*o*- and *p*-xylene) are susceptible to attack at all positions, albeit more slowly, because each carbon is only singly (either ortho or para) activated.

Similar arguments apply to other activating substituents. Resonance activators usually override the effect of an inductively acting donor, such as an alkyl group. For example, under conditions that limit strongly activated benzenes to monobromination, the disinfectant 4-methylphenol (*p*-cresol) reacts to give mainly 2-bromo-4-methylphenol (see margin).

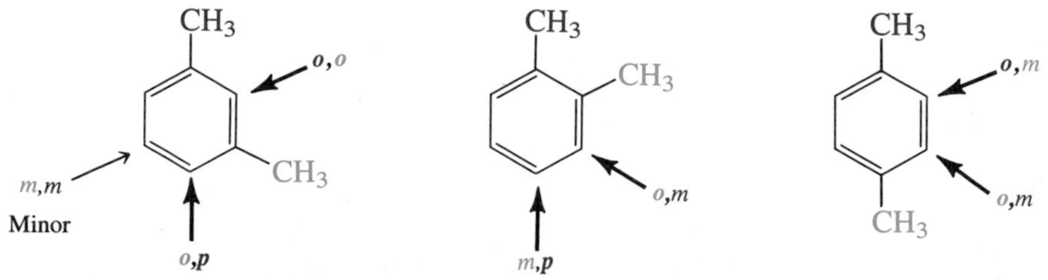

Conditions for Monohalogenation of a Strongly Activated Benzene

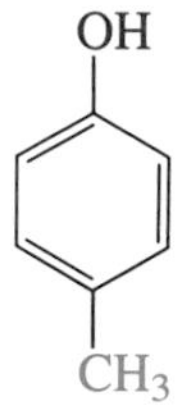

OH, CH_3
4-Methylphenol (*p*-Cresol)

Br–Br, $CHCl_3$, 0°C; − HBr

OH, Br, CH_3
80%
2-Bromo-4-methylphenol

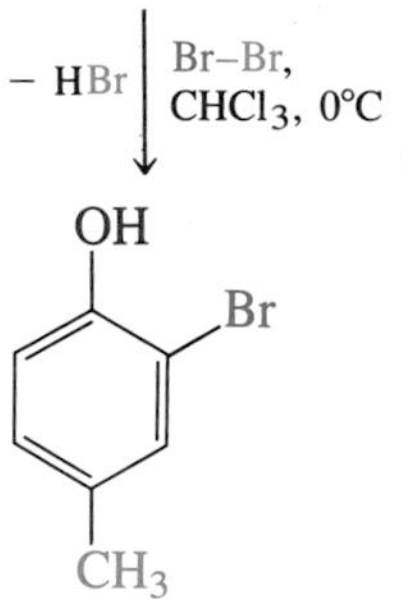

CH_3, CH_3: *o,o*; *m,m* Minor; *o,p*
CH_3, CH_3: *o,m*; *m,p*
CH_3, CH_3: *o,m*; *o,m*

FIGURE 16-1 Electrophilic attack on the dimethylbenzenes (xylenes). All positions ortho or para to a methyl group are activated.

EXERCISE 16-7

Predict the result of the mononitration of

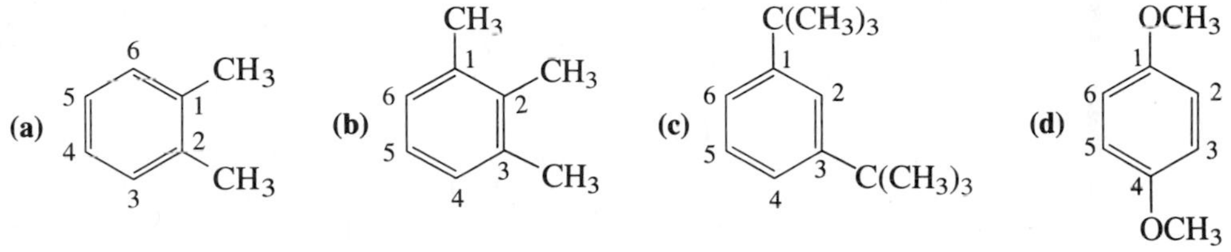

(a) 1,2-dimethylbenzene (CH_3 at 1 and 2) (b) 1,2,3-trimethylbenzene (CH_3 at 1, 2, 3) (c) 1,3-bis($C(CH_3)_3$)benzene ($C(CH_3)_3$ at 1 and 3) (d) 1,4-dimethoxybenzene (OCH_3 at 1 and 4)

EXERCISE 16-8

The food preservative BHT (*tert*-butylated hydroxytoluene) has the structure shown below. Suggest a synthesis starting from 4-methylphenol (*p*-cresol).

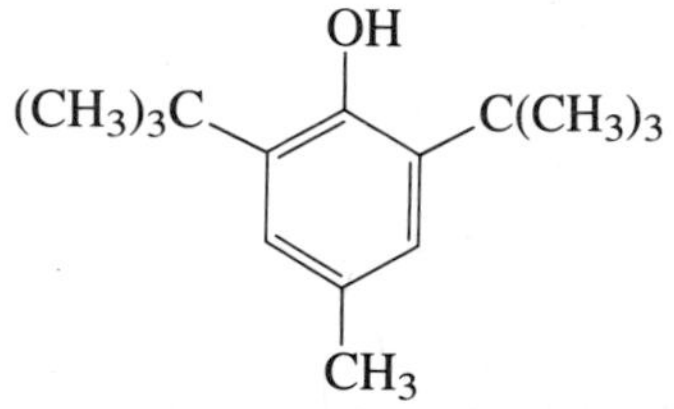

4-Methyl-2,6-bis(1,1-dimethylethyl)phenol
(2,6-Di-*tert*-butyl-4-methylphenol)

When all substituents are deactivating, reaction is sluggish, and attack is directed to the position that is meta to most of them. This effect can be seen in the nitration of 1,3-benzenedicarboxylic (isophthalic) and 1,2-benzenedicarboxylic (phthalic) acids.

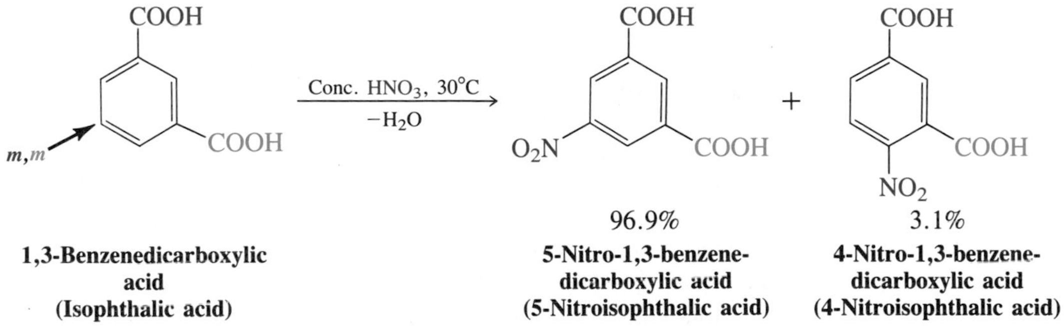

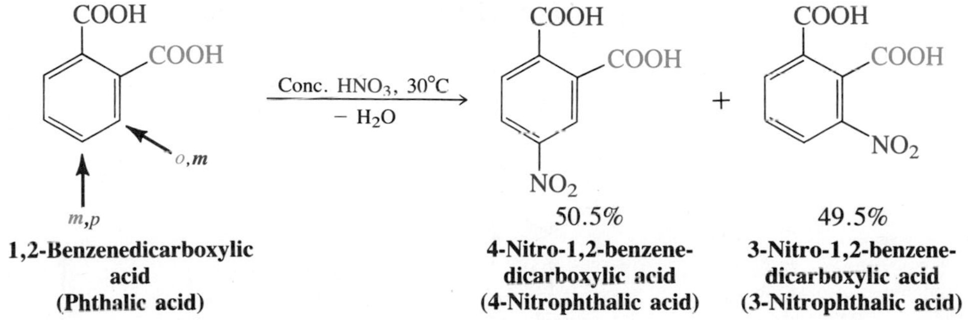

$$\text{1,3-Benzenedicarboxylic acid} \xrightarrow[-H_2O]{\text{Conc. } HNO_3,\ 30°C} \text{5-Nitro (96.9\%)} + \text{4-Nitro (3.1\%)}$$

m,m

1,3-Benzenedicarboxylic acid (Isophthalic acid)

96.9% **5-Nitro-1,3-benzene-dicarboxylic acid (5-Nitroisophthalic acid)**

3.1% **4-Nitro-1,3-benzene-dicarboxylic acid (4-Nitroisophthalic acid)**

$$\text{1,2-Benzenedicarboxylic acid} \xrightarrow[-H_2O]{\text{Conc. } HNO_3,\ 30°C} \text{4-Nitro (50.5\%)} + \text{3-Nitro (49.5\%)}$$

o,m *m,p*

1,2-Benzenedicarboxylic acid (Phthalic acid)

50.5% **4-Nitro-1,2-benzene-dicarboxylic acid (4-Nitrophthalic acid)**

49.5% **3-Nitro-1,2-benzene-dicarboxylic acid (3-Nitrophthalic acid)**

EXERCISE 16-9

Predict the result of the mononitration of

(a) 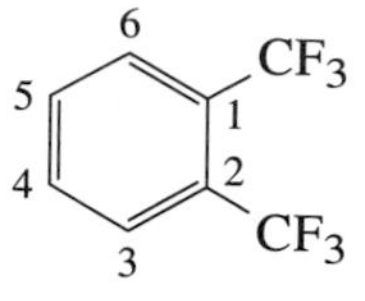 1,2-bis(CF_3)benzene (ring positions numbered 1–6; CF_3 at 1 and 2)

(b) 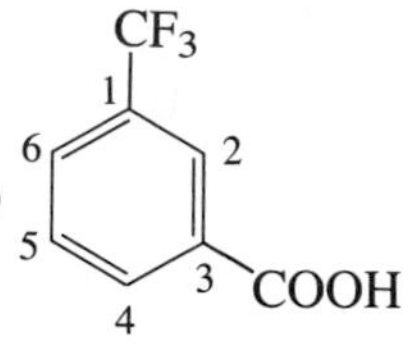 ring with CF_3 at 1 and COOH at 3 (positions numbered 1–6)

(c) 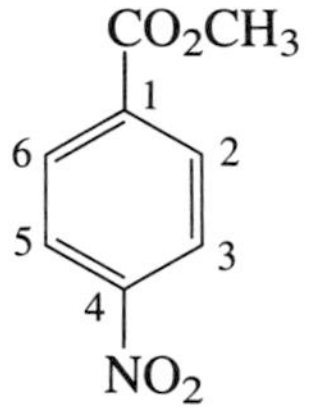 ring with CO_2CH_3 at 1 and NO_2 at 4 (positions numbered 1–6)

(d) 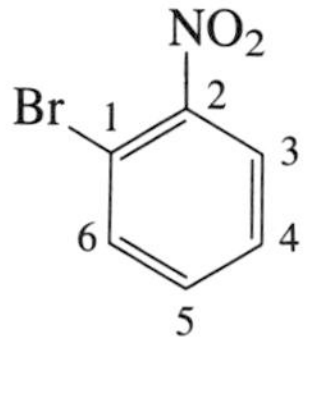 ring with Br at 1 and NO_2 at 2 (positions numbered 1–6)

The activating substituent wins out

The situation becomes more complicated when one group directs ortho and para while another directs meta. Usually each substituent acts independently. If the groups reinforce each other in directing attack, relatively clean results may be obtained.

Electrophilic Substitution with Reinforcing *o*, *m*, and *p* Directors

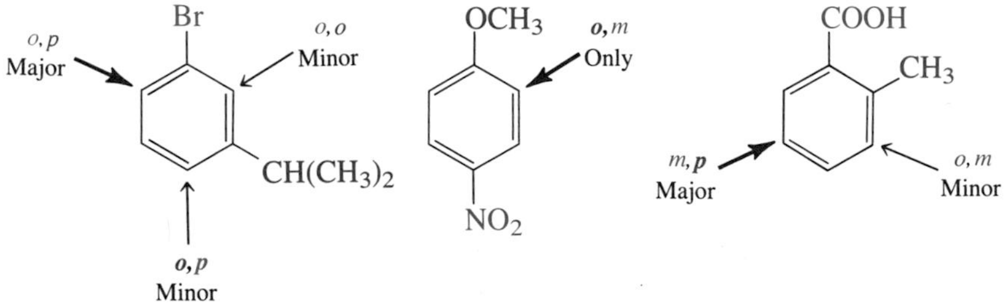

Br, $CH(CH_3)_2$: *o,p* Major; *o,o* Minor; *o,p* Minor

OCH_3, NO_2: *o,m* Only

COOH, CH_3: *m,p* Major; *o,m* Minor

If the substituents compete with each other in controlling the site of reaction, *the stronger activator wins out*. For example, if there is competition between two ortho, para directors, the better electron donor generally has the upper hand.

Regiocontrol by the Stronger Activator (Red) in Electrophilic Aromatic Substitution

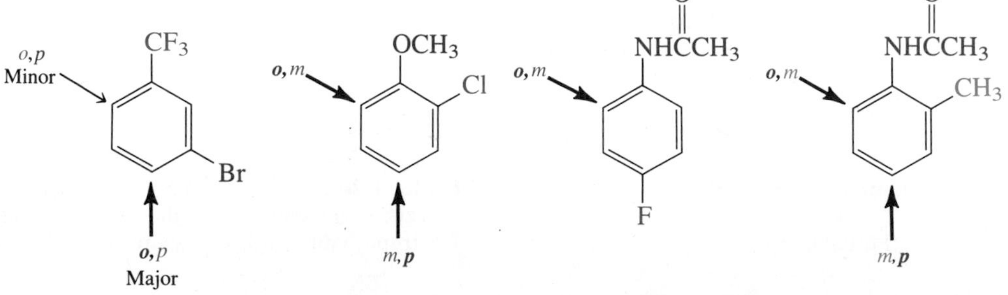

CF_3, Br: *o,p* Minor; *o,p* Major

OCH_3, Cl: *o,m*; *m,p*

$NHCCH_3$ (C=O), F: *o,m*

$NHCCH_3$ (C=O), CH_3: *o,m*; *m,p*

EXERCISE 16-10

Predict the site of electrophilic aromatic substitution in

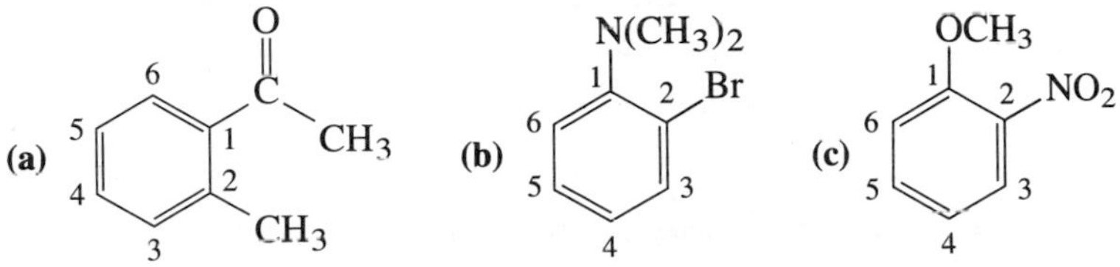

(a) 1-(2-methylphenyl)ethanone ($C(=O)CH_3$ at C1, CH_3 at C2; ring positions 1–6) (b) $N(CH_3)_2$ at C1, Br at C2 (ring positions 1–6) (c) OCH_3 at C1, NO_2 at C2 (ring positions 1–6)

In summary, electrophilic aromatic substitution of multiply substituted benzenes is controlled by the strongest activator and, to a certain extent, by steric effects. The greatest product selectivity appears if there is only one activator and all other groups deactivate or if all groups cooperate in their directing effects.

16-4 Sources of Benzene and Other Aromatic Hydrocarbons

This section examines industrial benzene production and laboratory syntheses of aromatics. Many benzenoid hydrocarbons, including benzene, are derived from coal. Heating coal in the absence of air produces coal gas containing methane and other gaseous products. The remainder of the distillate is *coal tar,* which can be fractionally distilled to give benzene, methylbenzene (toluene), the dimethylbenzenes (xylenes), phenol, naphthalene, and higher polycyclic aromatic hydrocarbons, as well as heterocyclic compounds (Chapter 25). The residue is *coke,* used in large quantities for the smelting of iron ore to steel.

Coal is not simply carbon, but an amorphous polymer consisting of layers of weakly linked polycyclic aromatic and hydroaromatic compounds. On heating, the primary coal structure degrades to fragments of molecular weight between 300 and 1000, a large proportion of which are soluble in organic solvents. Coal solubilization and the conversion of coal into liquid fuels are of current interest in efforts to use coal as a source of new industrial chemical feedstocks.

Industrial preparations of benzenes begin with cyclohexanes

Benzene is obtained commercially from aromatic distillation fractions in oil refineries, the steam cracking of alkenes, the so-called hydrodealkylation of methylbenzene (toluene), and the pyrolysis of coal. The United States produces about 1.6 billion gallons of benzene yearly, mainly by catalytic reforming of gasoline fractions (see Section 3-3), which is also a source of methylbenzene (toluene) and the dimethylbenzenes (xylenes). In this process, called **platforming,** C_{6-8} hydrocarbons are dehydrogenated to simple aromatic compounds.

Industrial Preparation of a Substituted Aromatic Hydrocarbon

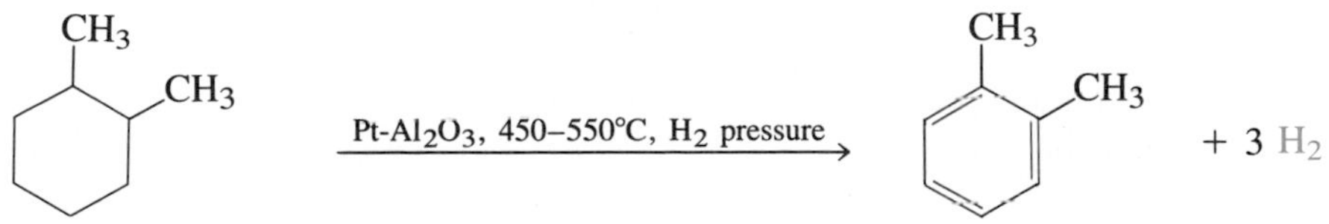

$$\text{1,2-Dimethylcyclohexane} \xrightarrow{\text{Pt-}Al_2O_3,\ 450\text{–}550^\circ C,\ H_2\ \text{pressure}} \text{1,2-Dimethylbenzene} + 3\,H_2$$

1,2-Dimethylcyclohexane → **1,2-Dimethylbenzene (*o*-Xylene)**

If hydrogen is produced in these reactions, why is more added? The answer is, to suppress the formation of high-molecular-weight carbonaceous material (ultimately, coke), which tends to clog and deactivate the catalyst surface.

Benzene, methylbenzene (toluene), and the dimethylbenzenes (xylenes) are added to gasoline because they improve engine performance.

Dehydrogenation is also used to produce aromatic compounds in the laboratory

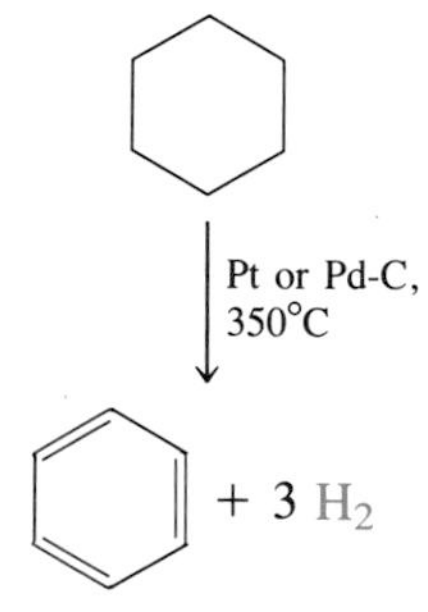

Dehydrogenation of cyclohexane, cyclohexene, and cyclohexadiene derivatives **(hydroaromatic compounds)** is a laboratory method for the preparation of substituted benzenes. The transformation is usually carried out at elevated temperatures with platinum or palladium metal as a catalyst, either as a fine powder or as a deposit on activated charcoal. The mechanism of the reaction is probably the reverse of the mechanism of hydrogenation of double bonds (Section 12-2). Dehydrogenations of this type have found some use in the synthesis of fused benzenes such as naphthalene (Section 15-5).

Pd-C, 300°C $\longrightarrow$ + 4 H_2

82%

Naphthalene

In summary, benzene is obtained commercially mainly by reforming of petroleum distillates. In the laboratory, hydroaromatics can be aromatized by Pd-C.

16-5 Synthetic Strategies

The synthesis of a specific benzene derivative requires planning to ensure that the desired substitution pattern is consistent with the directing power of the first group introduced. This section will present several useful reactions that reverse the directing power of a substituent when the target pattern is "wrong." Strategies to block ring positions and to moderate the activating power of hydroxy and amine groups also will be discussed.

Specifically substituted benzenes can be constructed by carefully planned electrophilic substitutions

The interconversion of the nitro group in nitrobenzenes (meta director) and an amino substituent (ortho, para director) illustrates how we can control the substitution pattern by using simple reagents. Reduction of NO_2 to NH_2 is effected by either catalytic hydrogenation or exposure to acid in the presence of active metals such as iron or zinc amalgam. The reverse oxidation reaction employs trifluoroperacetic acid.

Reversible Conversion of a Meta into an Ortho, Para Director

NO_2-benzene $\underset{CF_3COOH \text{ (with C=O)}}{\overset{Zn(Hg),\ HCl,\ \text{or } H_2,\ Ni,\ \text{or } Fe,\ HCl}{\rightleftharpoons}}$ NH_2-benzene

For an example of the application of this strategy, consider the preparation of 3-bromobenzenamine. Direct bromination of benzenamine (aniline) leads to complete ortho and para substitution (Section 16-2) and is therefore useless. However, bromination of nitrobenzene allows preparation of 3-bromonitrobenzene, which can be converted to the required target molecule by one of the reduction methods introduced above.

NO_2-benzene $\xrightarrow{Br_2,\ FeBr_3}$ 3-Br-NO_2-benzene $\xrightarrow{Fe,\ HCl}$ 3-Br-NH_2-benzene

Notice that the outcome is a benzene in which two ortho, para directors emerge positioned meta to each other.

EXERCISE 16-11

Would nitration of bromobenzene be a useful alternative way to begin a synthesis of 3-bromobenzenamine?

EXERCISE 16-12

Propose a synthesis from benzene of 3-aminobenzenesulfonic acid (metanilic acid, used in the synthesis of azo dyes, such as "Metanil yellow," and certain sulfa drugs).

EXERCISE 16-13

Use the methods presented above to devise a synthesis of 4-nitrobenzenesulfonic acid from benzene.

Another example of conversion of one substituent into another with different directing ability is found in connection with the Friedel-Crafts reactions. Friedel-Crafts alkanoylation (acylation) introduces a meta-directing carbonyl substituent to the benzene ring. Treatment with zinc amalgam and concentrated HCl effects the conversion of the C=O group to CH_2, an ortho, para director. This process, known as the **Clemmensen* reduction,** also provides a way to synthesize alkylbenzenes without the complication of alkyl group rearrangement. For example, butylbenzene may be synthesized by the sequence of Friedel-Crafts alkanoylation with butanoyl chloride, followed by Clemmensen reduction.

---

*E. C. Clemmensen, 1876–1941, President of Clemmensen Chemical Corporation, Newark, N.J.

Synthesis of an Alkylbenzene Without Rearrangement

Benzene $\xrightarrow{CH_3CH_2CH_2CCl\ (C{=}O),\ AlCl_3}$ $C_6H_5C(=O)CH_2CH_2CH_3$ (51%) $\xrightarrow{HCl,\ Zn(Hg),\ \Delta}$ $C_6H_5CH_2CH_2CH_2CH_3$ (59%)

Attempted synthesis of the same alkylbenzene by Friedel-Crafts alkylation fails because of the interference of polyalkylation and the formation of the rearranged product (1-methylpropyl)benzene (*sec*-butylbenzene; Section 15-13).

EXERCISE 16-14

Give an efficient synthesis of (2-methylpropyl)benzene (isobutylbenzene), starting from benzene. What would you expect as the major monosubstitution product of Friedel-Crafts alkylation of benzene with 1-chloro-2-methylpropane (isobutyl chloride)?

Friedel-Crafts electrophiles do not attack strongly deactivated benzene rings

Let us examine possible syntheses of 1-(3-nitrophenyl)ethanone (*m*-nitroacetophenone). Because both groups are meta directors, two possibilities appear available: nitration of 1-phenylethanone or Friedel-Crafts acetylation of nitrobenzene. However, in practice, only the first route succeeds.

Successful and Unsuccessful Syntheses of 1-(3-Nitrophenyl)ethanone (*m*-Nitroacetophenone)

Benzene $\xrightarrow{1.\ CH_3COCl,\ AlCl_3;\ 2.\ H^+,\ H_2O;\ 3.\ HNO_3,\ H_2SO_4}$ 3-O_2N-$C_6H_4COCH_3$ $\xleftarrow[\times]{1.\ HNO_3,\ H_2SO_4;\ 2.\ CH_3COCl,\ AlCl_3;\ 3.\ H^+,\ H_2O}$ Benzene

The failure of the second route results from a combination of factors. One is the extreme deactivation of the nitrobenzene ring. Another is the relatively low electrophilicity of the acylium ion, at least compared with other electrophiles in aromatic substitution. As a general rule, neither Friedel-Crafts alkylations nor alkanoylations take place with benzene derivatives strongly deactivated by meta-directing groups. Friedel-Crafts reactions are also not possible on a benzene ring bearing an NH_2 or an NHR group because reactions of the nitrogen group, as a Lewis base, interfere.

EXERCISE 16-15

Propose a synthesis of 2-bromo-4-ethylbenzenamine, starting from benzene. Consider carefully the order in which you introduce the groups.

Reversible sulfonation allows the efficient synthesis of ortho-disubstituted benzenes

A problem of another sort arises in the attempt to prepare an *o*-disubstituted benzene, even when one of the groups is an ortho, para director. Although appreciable amounts of ortho isomers may form in electrophilic substitutions of benzenes containing such groups, the para isomer is the major product in most such cases (Section 16-1). Suppose you required an efficient synthesis of 1-(1,1-dimethylethyl)-2-nitrobenzene [*o*-(*t*-butyl)nitrobenzene]? Direct nitration of (1,1-dimethylethyl)benzene (*t*-butylbenzene) is unsatisfactory.

A Poor Synthesis of 1-(1,1-Dimethylethyl)-2-nitrobenzene [*o*-(*t*-Butyl)nitrobenzene]

$C(CH_3)_3$ benzene $\xrightarrow[-\ H_2O]{HNO_3,\ H_2SO_4,\ 30°C}$ ortho-NO_2 + meta-NO_2 + para-NO_2

16%	11%	73%
1-(1,1-Dimethylethyl)-2-nitrobenzene [*o*-(*t*-Butyl)nitrobenzene]	**1-(1,1-Dimethylethyl)-3-nitrobenzene [*m*-(*t*-Butyl)nitrobenzene]**	**1-(1,1-Dimethylethyl)-4-nitrobenzene [*p*-(*t*-Butyl)nitrobenzene]**

A clever solution makes use of reversible sulfonation (Section 15-11) as a blocking procedure. For steric reasons, (1,1-dimethylethyl)benzene is sulfonated almost entirely para, blocking this carbon from further electrophilic attack. Nitration now can occur only ortho to the alkyl group. Heating in aqueous acid removes the blocking group and completes the synthesis.

Reversible Sulfonation as a Blocking Procedure

$C(CH_3)_3$ benzene $\xrightarrow{SO_3,\ \text{conc. } H_2SO_4}$ 4-SO_3H $\xrightarrow{HNO_3,\ H_2SO_4}$ 2-NO_2, 4-SO_3H $\xrightarrow{H^+,\ H_2O,\ \Delta}$ 2-NO_2 product

EXERCISE 16-16

Suggest a synthetic route from benzene to 1,3-dibromo-2-nitrobenzene.

Blocking strategies moderate the activating power of amine and hydroxy groups

We noted in Section 16-2 that electrophilic substitutions of benzenamine (aniline) and phenol are sometimes difficult to control. The derivatives *N*-phenylacetamide (acetanilide) and methoxybenzene (anisole) are more satisfactory substrates, especially for halogenation, nitration, and the Friedel-Crafts reactions. The synthesis of 2-nitrobenzenamine (*o*-nitroaniline) below illustrates the reversible acetylation of the nitrogen in benzenamine as well as the use of sulfonation to block the para position.

A Synthesis of 2-Nitrobenzenamine (*o*-Nitroaniline)

NH_2 —[$CH_3\overset{O}{\overset{\|}{C}}Cl$, pyridine]→ $HN\overset{O}{\overset{\|}{C}}CH_3$ —[SO_3, conc. H_2SO_4]→ $HN\overset{O}{\overset{\|}{C}}CH_3$ / SO_3H —[HNO_3]→ $HN\overset{O}{\overset{\|}{C}}CH_3$ / NO_2 / SO_3H —[1. H^+, H_2O, Δ 2. ^-OH, H_2O]→ NH_2 / NO_2

Blocking the Oxygen Atom in Phenol

OH

Conc. HI ⇅ NaOH, CH_3I

OCH_3

Finally, control of substitution on the ring in phenol is obtained by methylation of the oxygen atom (by Williamson ether synthesis; Section 9-6). Ether cleavage by concentrated HI (Section 9-8) regenerates the hydroxy group (margin).

EXERCISE 16-17

Apply the strategy discussed above to a synthesis of 2-chloro-4-acetylphenol, starting with phenol.

In summary, by careful choice of the sequence in which new groups are introduced, it is possible to devise specific syntheses of multiply substituted benzenes. Such strategies can change the directing power of substituents, modify their activating ability, and reversibly block positions on the ring.

16-6 Reactivity of Polycyclic Benzenoid Hydrocarbons

This section illustrates the use of resonance forms to predict the regioselectivity and reactivity of polycyclic aromatic molecules, using naphthalene as an example. Some biological implications of the reactivity of these substances will be explored in Section 16-7.

Naphthalene is activated toward electrophilic substitution

The aromatic character of naphthalene is manifest in its reactivity: It undergoes electrophilic substitution rather than addition. For example, treatment with bromine, even in the absence of a catalyst, results in smooth conversion into 1-bromonaphthalene. The mild conditions required for this process reveal that naphthalene is activated with respect to electrophilic aromatic substitution.

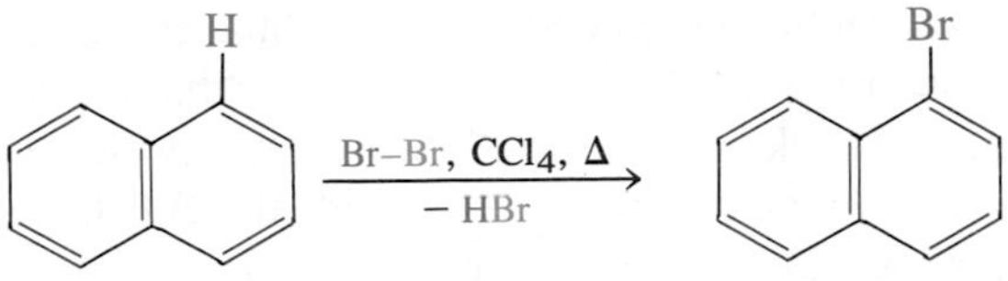

H — Br–Br, CCl_4, Δ / – HBr → Br

75%

1-Bromonaphthalene

Other electrophilic substitutions are also readily achieved and, again, are highly selective for reaction at C1. For example,

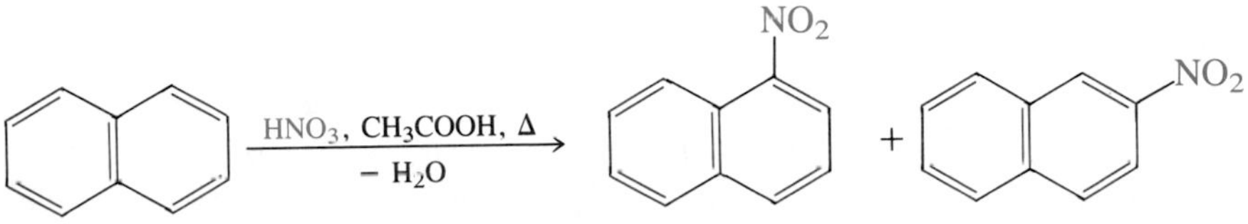

HNO_3, CH_3COOH, Δ / – H_2O → NO_2 + NO_2

Major — **1-Nitronaphthalene**

Minor — **2-Nitronaphthalene**

The highly delocalized nature of the intermediate explains the ease of attack. The cation can be nicely pictured as a hybrid of five resonance structures.

Electrophilic Reactivity of Naphthalene: Attack at C1

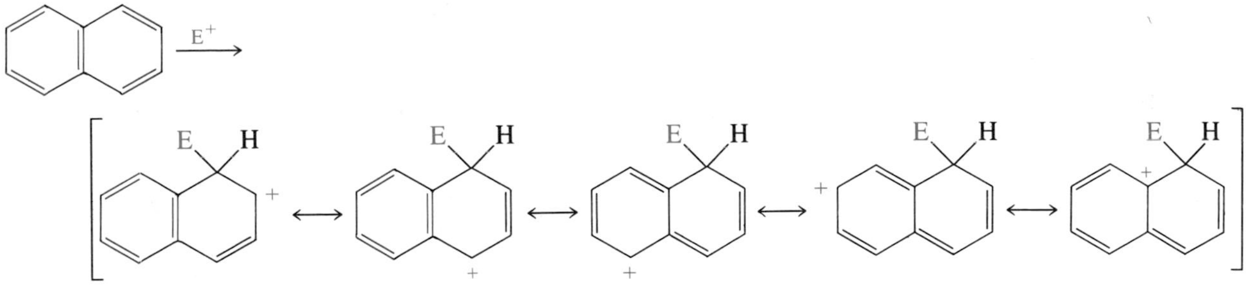

E^+ →

[E H + ⟷ E H + ⟷ E H + ⟷ E H + ⟷ E H +]

However, attack of an electrophile at C2 also produces a cation that may be described by five contributing resonance forms.

Electrophilic Attack on Naphthalene at C2

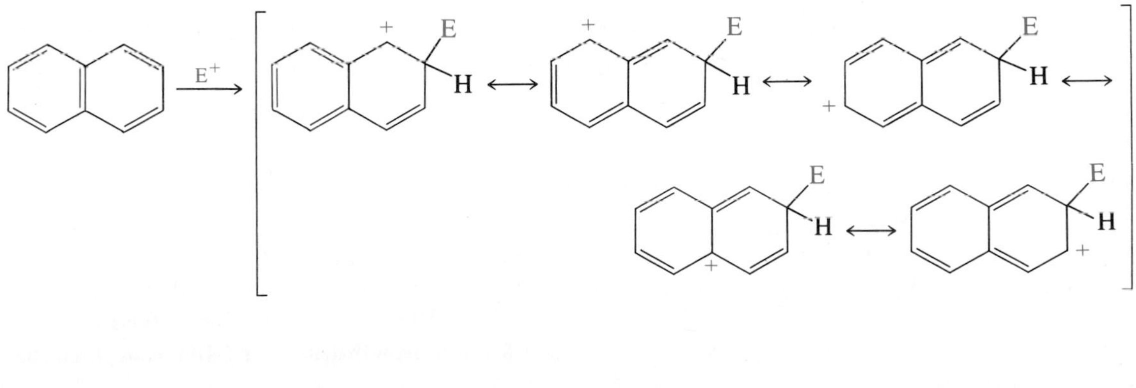

E^+ → [+ E H ⟷ + E H ⟷ E H + ⟷ E H + ⟷ E H +]

Why, then, do electrophiles prefer to attack naphthalene at C1 rather than at C2? Closer inspection of the resonance contributors for the two cations reveals an important difference: Attack at C1 allows *two* of the resonance forms of the intermediate to keep an intact benzene ring, with the full benefit of aromatic cyclic delocalization. Attack at C2 allows only *one* such structure, so the resulting carbocation is less stable, and the transition state leading to it less energetically favorable. Because the first step in electrophilic aromatic substitution is rate determining, attack at C1 is therefore faster than at C2.

Electrophiles attack substituted naphthalenes regioselectively

The rules of orientation in electrophilic attack on monosubstituted benzenes extend easily to naphthalenes. *The substituted ring is the one most affected by the substituent already present:* An activating group usually directs the incoming electrophile to the same ring, a deactivating group directs it away. For example, 1-naphthalenol (1-naphthol) undergoes electrophilic nitration at C2 and C4.

Nitration of 1-Naphthalenol (1-Naphthol)

OH; HNO_3, H_2SO_4; $- H_2O$; ^+O–H; H, NO_2; and; NO_2, H

Para attack

Ortho attack

$- H^+$

OH, NO_2; +; OH, NO_2

Major
4-Nitro-1-naphthalenol (4-Nitro-1-naphthol)

Minor
2-Nitro-1-naphthalenol (2-Nitro-1-naphthol)

Deactivating groups in one ring usually direct electrophilic substitutions to the other ring and preferentially in the positions C5 and C8.

NO_2; 8; 5; HNO_3, H_2SO_4, 0°C; $- H_2O$; NO_2 NO_2; +; NO_2; NO_2

30%
1,8-Dinitronaphthalene

60%
1,5-Dinitronaphthalene

EXERCISE 16-18

On the basis of the relative viability of the sets of resonance structures arising from initial electrophilic attack, predict the position of electrophilic aromatic nitration in **(a)** 1-ethylnaphthalene; **(b)** 2-nitronaphthalene; and **(c)** 5-methoxy-1-nitronaphthalene.

In summary, naphthalene is activated with respect to electrophilic aromatic substitution; favored attack takes place at C1. Electrophilic attack on a substituted naphthalene takes place on an activated ring and away from a deactivated ring, with regioselectivity in accordance with the general rules developed for electrophilic aromatic substitution of benzene derivatives.

16-7 Polycyclic Aromatic Hydrocarbons and Cancer

Many polycyclic benzenoid hydrocarbons are carcinogenic. The first observation of human cancer caused by such compounds was made in 1775 by Sir Percival Pott, a surgeon at London's St. Bartholomew's hospital, who recognized that chimney sweeps were prone to scrotal cancer. Since then, a great deal of research has gone into identifying which polycyclic benzenoid hydrocarbons have this physiological property and how their structure correlates with activity. A particularly well studied molecule is benz[*a*]pyrene, a widely distributed environmental pollutant. It is frequently produced in the combustion of organic matter, such as automobile fuel and oil (for domestic heating and industrial power generation), in incineration of refuse, in forest fires, in burning cigarettes and cigars, and even in roasting meats. The annual release into the atmosphere in the United States alone has been estimated at 1300 tons.

Carcinogenic Benzenoid Hydrocarbons

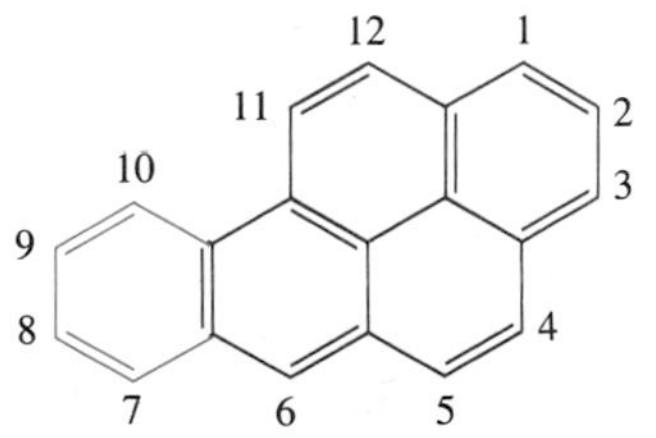

Benz[*a*]pyrene 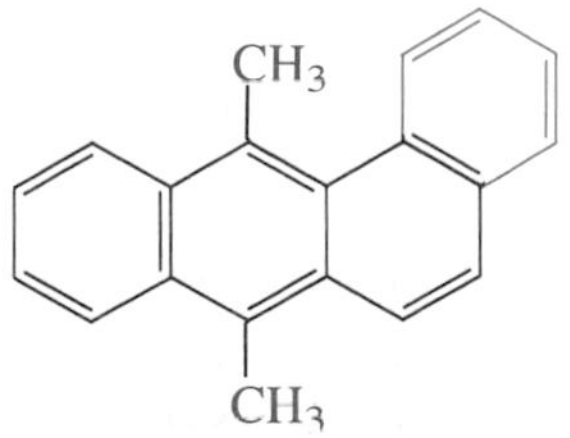 **7,12-Dimethylbenz[*a*]anthracene** 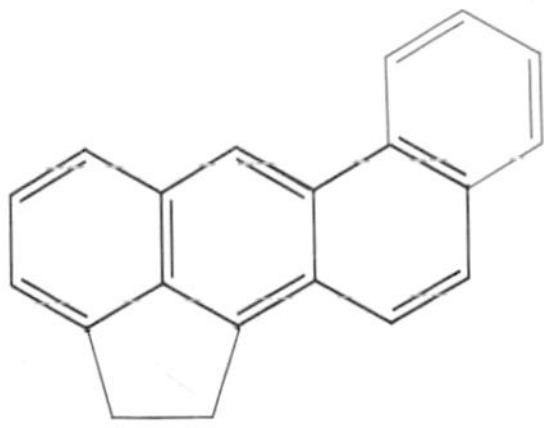 **Cholanthrene**

What is the mechanism of carcinogenic action of benz[*a*]pyrene? An oxidizing enzyme (an *oxidase*) of the liver converts the hydrocarbon into the oxacyclopropane at C7 and C8. Another enzyme *(epoxide hydratase)* catalyzes the hydration of the product to the trans diol. Further oxidation then results in the ultimate carcinogen, a new oxacyclopropane at C9 and C10.

Enzymatic Conversion of Benz[*a*]pyrene into the Ultimate Carcinogen

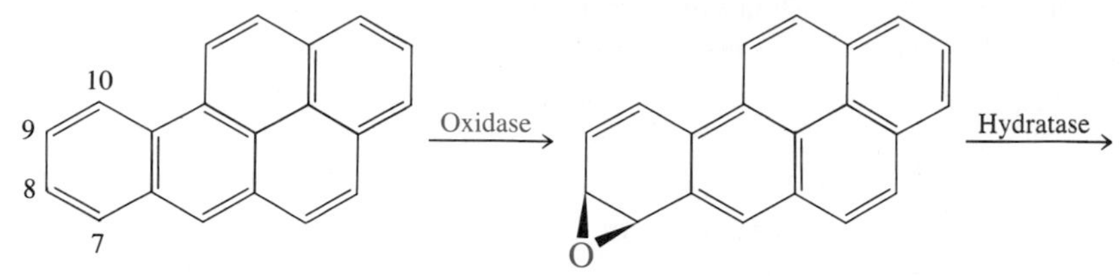

10
9
8
7
Oxidase
Hydratase

Benz[*a*]pyrene oxacyclopropane

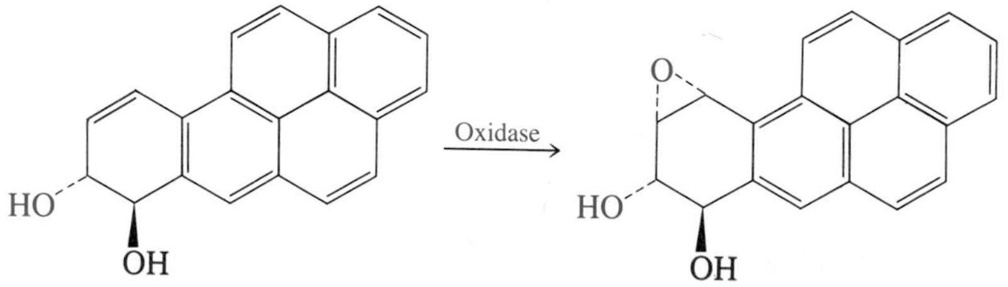

HO
OH
Oxidase
HO
OH

7,8-Dihydrobenz[*a*]pyrene-*trans*-7,8-diol

Carcinogen

What makes the compound carcinogenic? It is believed that the amine nitrogen in guanine, one of the bases in DNA (see Chapter 26), attacks the three-membered ring as a nucleophile. This reaction significantly alters the structure of one of the base pairs in DNA, leading to a mismatch during DNA replication.

The Carcinogenic Event

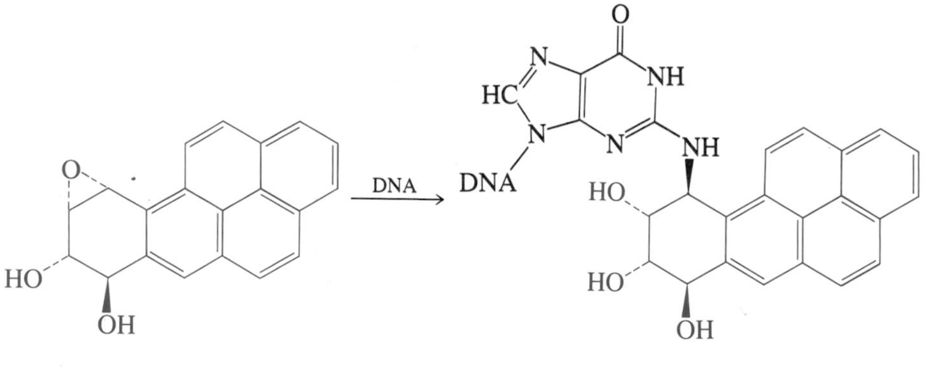

DNA

This change can lead to an alteration (mutation) of the genetic code, which may then generate a line of rapidly and indiscriminately proliferating cells typical of cancer. Not all mutations are carcinogenic; in fact, most of them lead to the destruction of only the one affected cell. Exposure to the carcinogen simply increases the likelihood of a carcinogenic event.

Notice that the carcinogen acts as an alkylating agent on DNA. This observation implies that other alkylating agents could also be carcinogenic, and indeed that is found to be the case. The Occupational Safety and Health Administration (OSHA) has published a list of carcinogens that includes simple alkylating agents such as 1,2-dibromoethane and ethyl methanesulfonate (recall Problem 16 of Chapter 1).

Carcinogenic Alkylating Agents and Sites of Reactivity

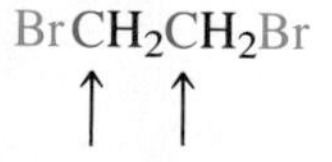

$BrCH_2CH_2Br$

1,2-Dibromoethane

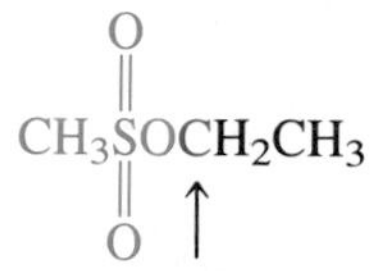

$CH_3SO_2OCH_2CH_3$

Ethyl methanesulfonate

The discovery of carcinogenicity in a number of organic compounds necessitated their replacement in synthetic applications. Both 1- and 2-naphthalenamines (naphthyl amines) were once widely used in the synthesis of dyes because of the brilliant colors of many of their derivatives (azo dyes; see Section 22-11). These substances were discovered to be carcinogens many years ago, a finding leading to the development of synthetic routes that avoided their use as intermediates as well as of new dyes with completely unrelated structures. A more recent example is chloro(methoxy)methane ($ClCH_2OCH_3$, chloromethyl methyl ether), once a commonly used reagent for the blocking of alcohols by ether formation. The discovery of carcinogenicity in this alkylating agent in the 1970s resulted in the development of several less hazardous reagents.

NEW REACTIONS

Electrophilic Substitution of Substituted Benzenes

1. ORTHO- AND PARA-DIRECTING GROUPS (Sections 16-1 and 16-2)

G-benzene $\xrightarrow{E^+}$ ortho-E-G-benzene + para-E-G-benzene

Ortho isomer **Para isomer**
(Usually predominates)

G = NH_2, OH; strongly activating
= NHCOR, OR; moderately activating
= alkyl, aryl; weakly activating
= halogen; weakly deactivating

2. META-DIRECTING GROUPS (Sections 16-1 and 16-2)

G-benzene $\xrightarrow{E^+}$ meta-E-G-benzene

Meta isomer

G = $\overset{+}{N}(CH_3)_3$, NO_2, CF_3, C≡N, SO_3H; very strongly deactivating
= CHO, COR, COOH, COOR, $CONH_2$; strongly deactivating

Synthesis of Benzene and Derivatives

3. INDUSTRIAL BENZENE SYNTHESIS BY PLATFORMING (Section 16-4)

$$\text{cyclohexane} \xrightarrow{\text{Pt-Al}_2\text{O}_3,\ 450\text{–}550^\circ\text{C},\ \text{H}_2\ \text{pressure}} \text{benzene} + 3\ H_2$$

4. DEHYDROGENATION OF HYDROAROMATICS (Section 16-4)

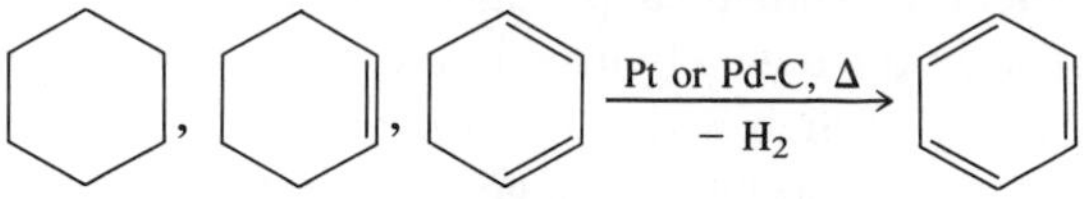

Cyclohexane, cyclohexene, 1,3-cyclohexadiene $\xrightarrow[-\ H_2]{\text{Pt or Pd-C, }\Delta}$ benzene

Synthetic Planning: Switching and Blocking of Directing Power

5. INTERCONVERSION OF NITRO AND AMINO GROUPS (Section 16-5)

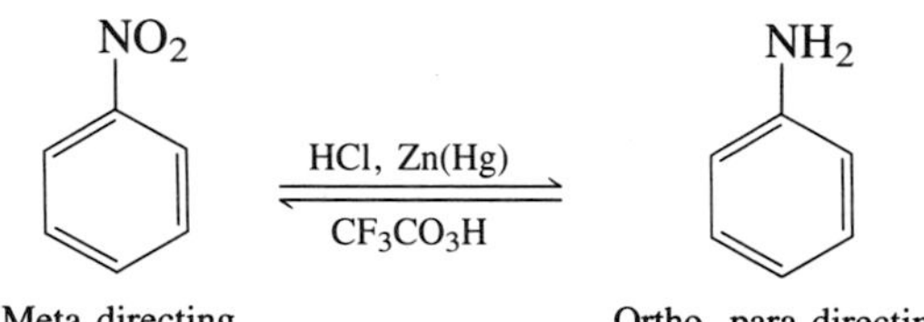

6. CONVERSION OF ALKANOYL INTO ALKYL (Section 16-5)

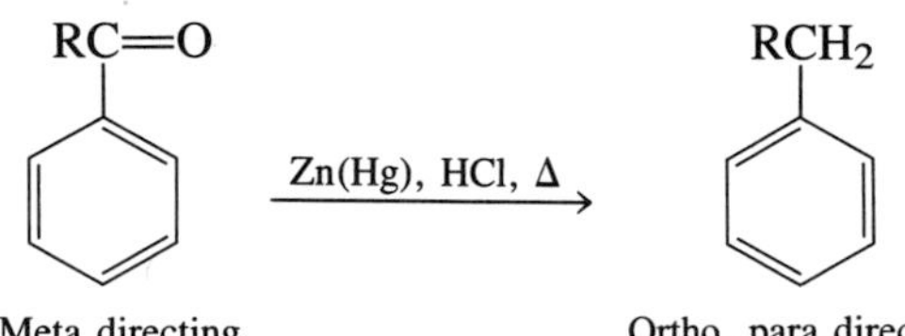

Clemmensen reduction

7. BLOCKING BY SULFONATION (Section 16-5)

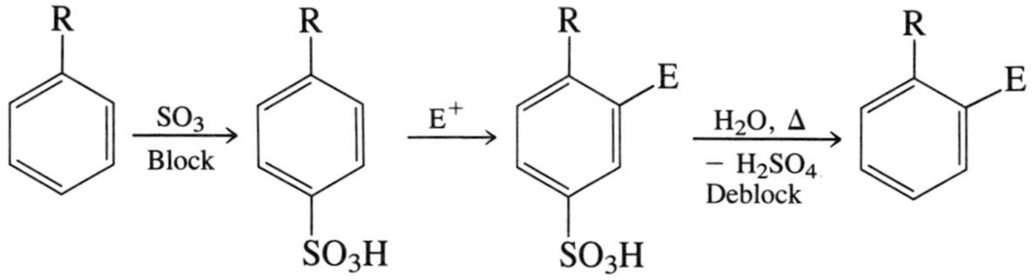

R-benzene $\xrightarrow[\text{Block}]{SO_3}$ 4-R-benzenesulfonic acid (SO_3H) $\xrightarrow{E^+}$ 4-R-3-E-benzenesulfonic acid (SO_3H) $\xrightarrow[-\ H_2SO_4]{H_2O,\ \Delta}$ 1-R-2-E-benzene

Deblock

8. MODERATION OF STRONG ACTIVATORS (Section 16-5)

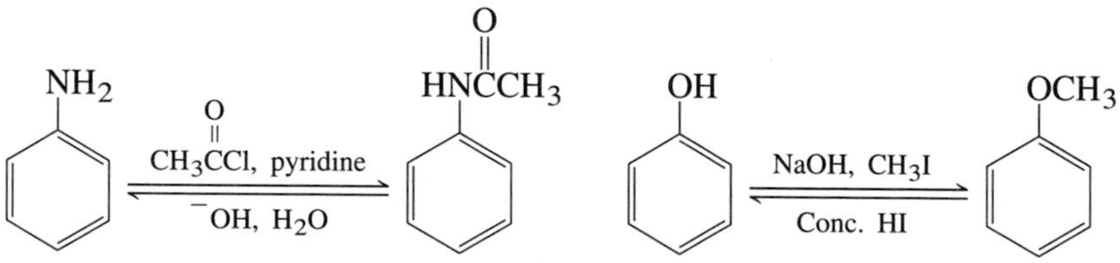

NH_2 (Strongly activated) $\underset{^{-}OH,\ H_2O}{\overset{\overset{O}{\|}}{CH_3CCl},\text{ pyridine}}{\rightleftharpoons}$ $\overset{O}{\overset{\|}{HNCCH_3}}$ (Moderately activated)

OH (Strongly activated) $\underset{\text{Conc. HI}}{\overset{NaOH,\ CH_3I}{\rightleftharpoons}}$ OCH_3 (Moderately activated)

9. ELECTROPHILIC AROMATIC SUBSTITUTION OF NAPHTHALENE (Section 16-6)

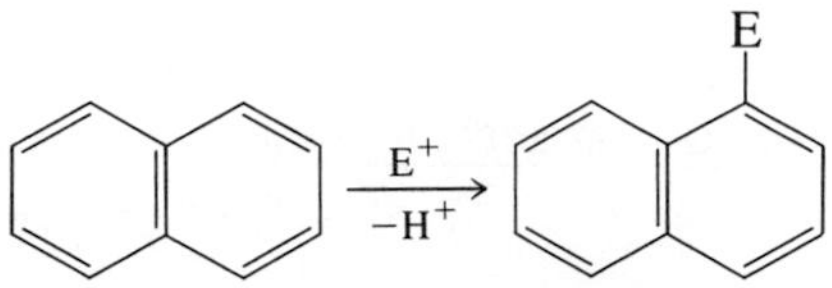

Naphthalene $\xrightarrow[-H^+]{E^+}$ 1-E-naphthalene

IMPORTANT CONCEPTS

1. Substituents on the benzene ring can be divided into two classes: those that activate the ring by electron donation and those that deactivate it by electron withdrawal. The mechanisms of donation and withdrawal are based on either hyperconjugation, induction, or resonance. These effects may operate simultaneously to either reinforce or oppose each other. Amino and alkoxy substituents are strongly activating, alkyl and phenyl groups weakly so; nitro, trifluoromethyl, sulfonic and carboxylic acid, nitrile, and cationic groups are strongly deactivating, whereas halogens are weakly so.
2. Activators direct electrophiles ortho and para; deactivators direct meta, although at a much lower rate. The exceptions are the halogens, which direct ortho and para.
3. If there are several substituents, electrophilic aromatic substitution is governed by the activating power of each group. Generally, the stronger activator (or weaker deactivator) controls the regioselectivity of attack, increasing in the following order: strongly deactivating meta director < less strongly deactivating meta director < deactivating ortho, para director < weakly activating ortho, para director < strongly activating ortho, para director.

$$NO_2 < CHO < Br < CH_3 < OH$$

4. Strategies for the synthesis of highly substituted benzenes rely on the directing power of the substituents, the synthetic ability to change the sense of direction of these substituents by chemical manipulation, and the use of blocking groups.
5. Naphthalene undergoes favored electrophilic substitution at C1 because of the relative stability of the intermediate carbocation.
6. Electron-donating substituents on one of the naphthalene rings direct electrophiles to the same ring, ortho and para. Electron-withdrawing substituents direct electrophiles away from that ring; substitution is mainly at C5 and C8.
7. The actual carcinogen derived from benz[*a*]pyrene appears to be an oxacyclopropanediol in which C7 and C8 bear hydroxy groups and C9 and C10 are bridged by oxygen. This molecule alkylates one of the nitrogens of one of the DNA bases, thus causing mutations.

PROBLEMS

1. Rank the compounds in each of the following groups in order of decreasing reactivity toward electrophilic substitution. Explain your rankings.

(a) $C_6H_5CCl_3$ $C_6H_5CH_3$ $C_6H_5CHCl_2$ $C_6H_5CH_2Cl$

(b) $C_6H_5CH_2CH_3$ $C_6H_5CH_2CCl_3$ $C_6H_5CH_2CF_3$ $C_6H_5CF_2CH_3$

(c) $C_6H_5OCH_3$ $C_6H_5O^-Na^+$ $C_6H_5OCCH_3$ (with C=O)

2. Specify whether you expect the benzene rings in the following compounds to be activated or deactivated.

(a) 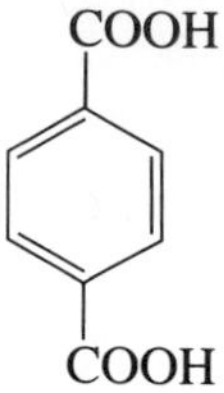 COOH / COOH (b) 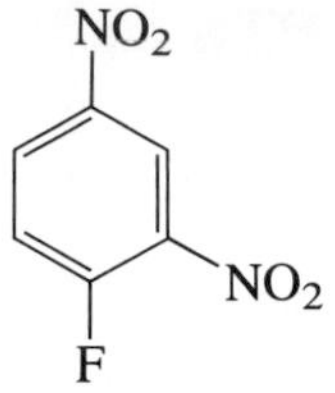 NO_2 / NO_2 / F (c) 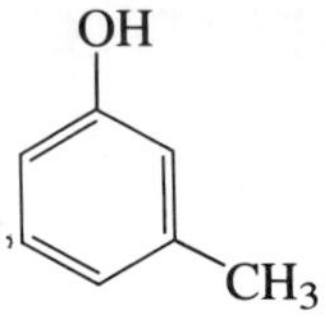 OH / CH_3

(d) 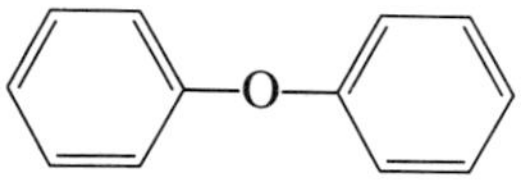 O (e) 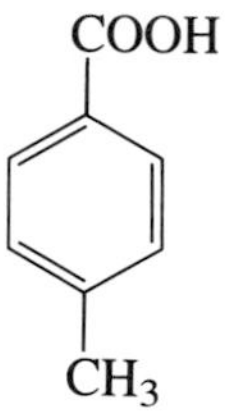 COOH / CH_3 (f) NH_2 / OH

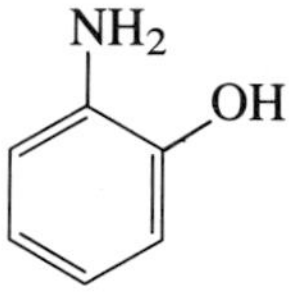

(g) 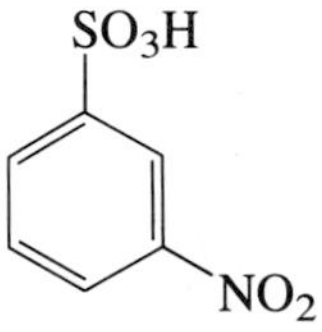 SO_3H / NO_2 (h) 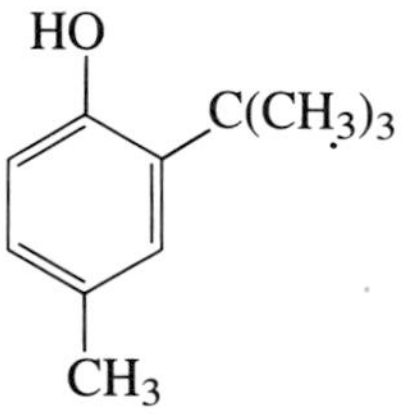 HO / $C(CH_3)_3$ / CH_3 (i) 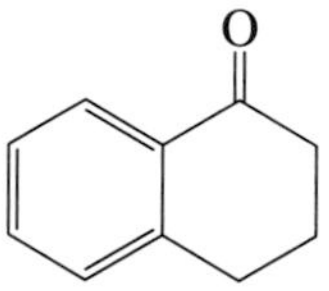 O

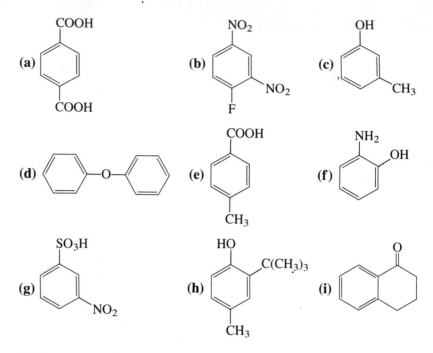

3. Draw appropriate resonance forms to explain the deactivating meta-directing character of the SO_3H group in benzenesulfonic acid.

4. Do you agree with the following statement? "Strongly electron-withdrawing substituents on benzene rings are meta directing because they deactivate the meta positions less than they deactivate the ortho and para positions." Explain your answer.

5. Give the expected major product(s) of each of the following electrophilic substitution reactions.

(a) 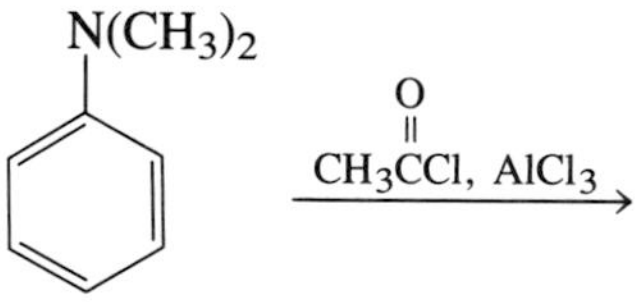 $N(CH_3)_2$ $\xrightarrow{CH_3\overset{O}{\overset{\|}{C}}Cl,\ AlCl_3}$ (b) Cl $\xrightarrow{Br_2,\ FeBr_3}$ (c) $\overset{O}{\overset{\|}{C}}CH_2CH_3$ $\xrightarrow{HNO_3,\ H_2SO_4}$

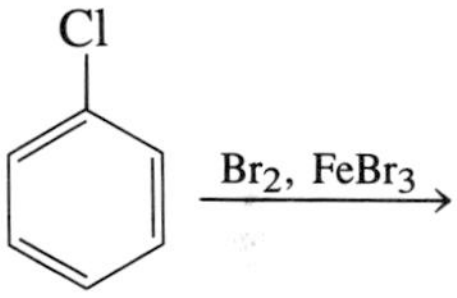

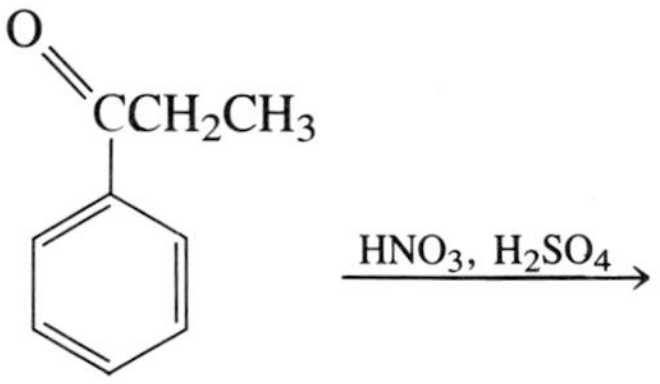

(d) 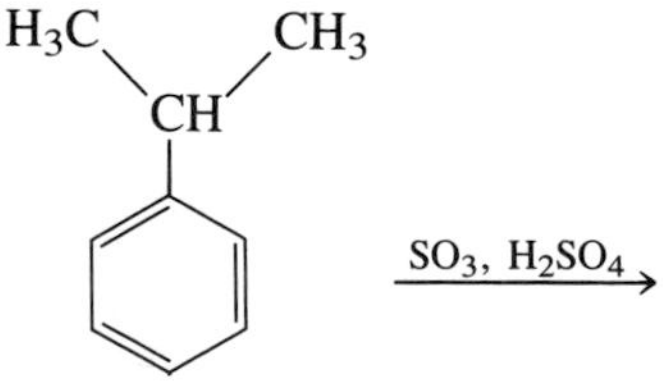 H_3C CH CH_3 $\xrightarrow{SO_3,\ H_2SO_4}$ (e) 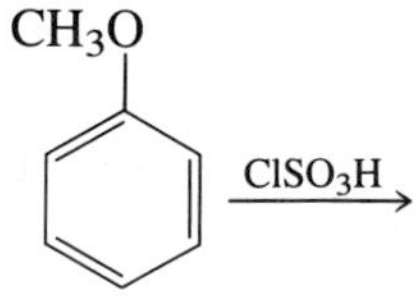 CH_3O $\xrightarrow{ClSO_3H}$ (f) 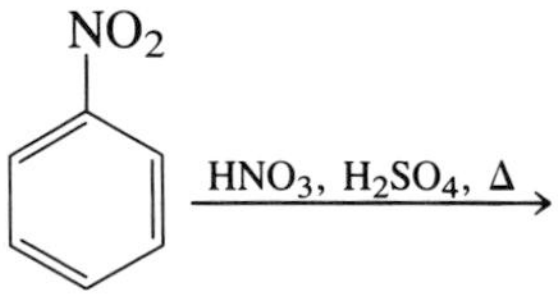 NO_2 $\xrightarrow{HNO_3,\ H_2SO_4,\ \Delta}$

6. Give the expected major product(s) of each of the following reactions.

(a) 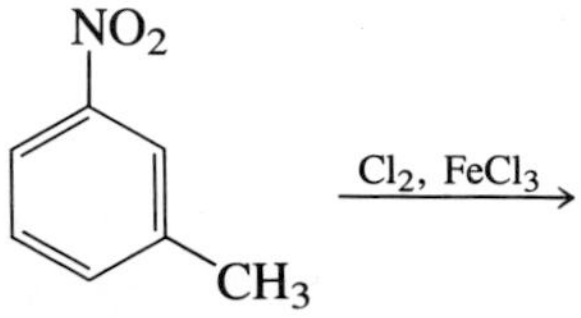 NO_2 / CH_3 $\xrightarrow{Cl_2,\ FeCl_3}$ (b) 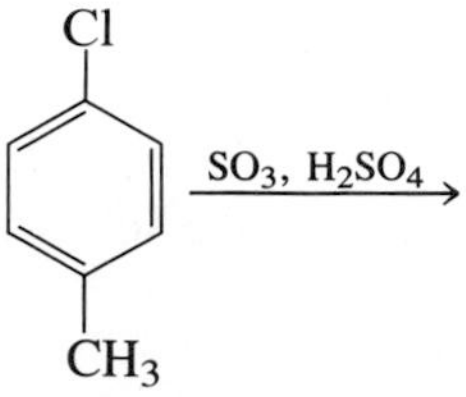 Cl / CH_3 $\xrightarrow{SO_3,\ H_2SO_4}$ (c) 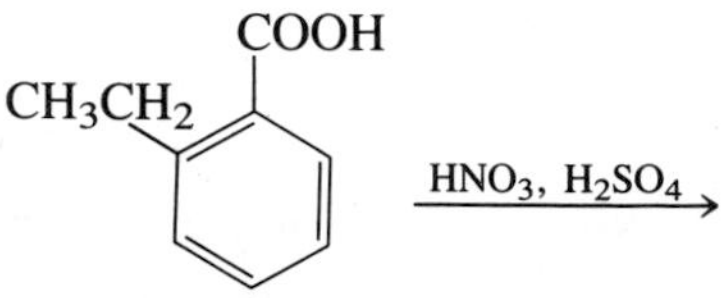 COOH / CH_3CH_2 $\xrightarrow{HNO_3,\ H_2SO_4}$

(d) 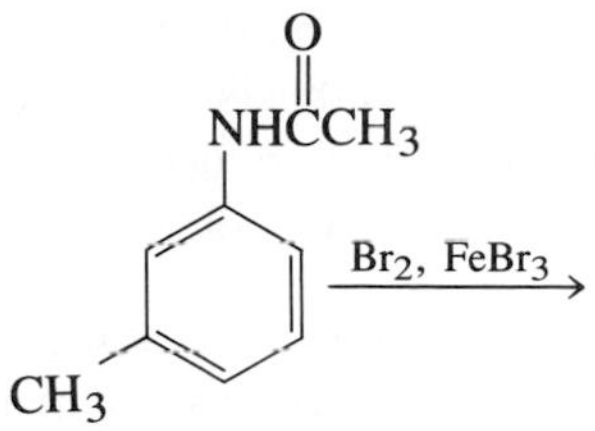 [m-methylacetanilide: NHCCH$_3$ (C=O), CH$_3$] $\xrightarrow{Br_2,\ FeBr_3}$

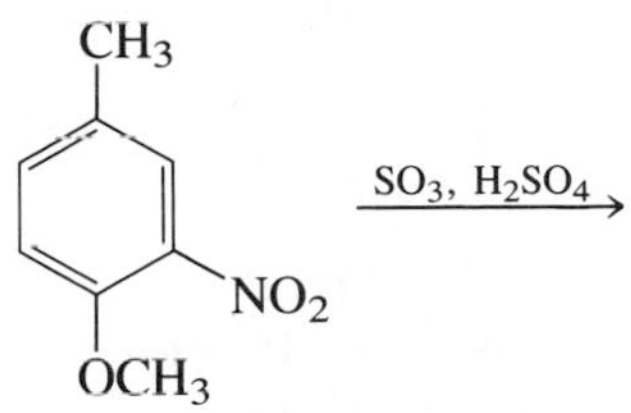

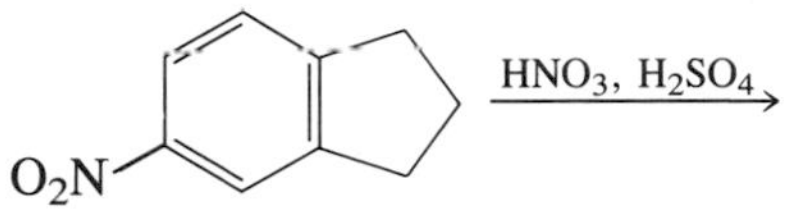

(e) 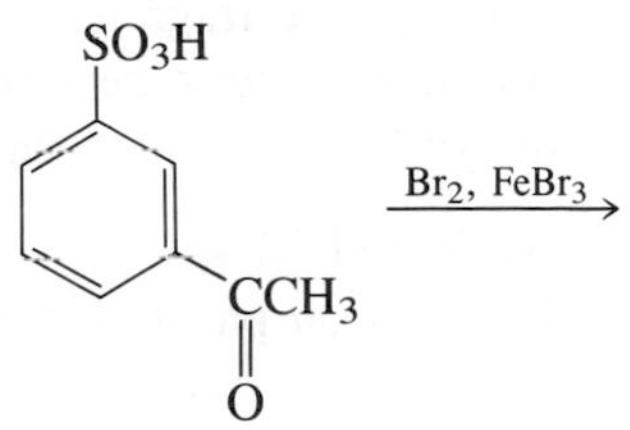 [SO$_3$H, CCH$_3$ (C=O) meta] $\xrightarrow{Br_2,\ FeBr_3}$

(f) [CH$_3$, NO$_2$, OCH$_3$] $\xrightarrow{SO_3,\ H_2SO_4}$

(g) [O$_2$N-indane] $\xrightarrow{HNO_3,\ H_2SO_4}$

(h) [NHCCH$_3$ (C=O), NO$_2$ para] $\xrightarrow{Cl_2,\ FeCl_3}$

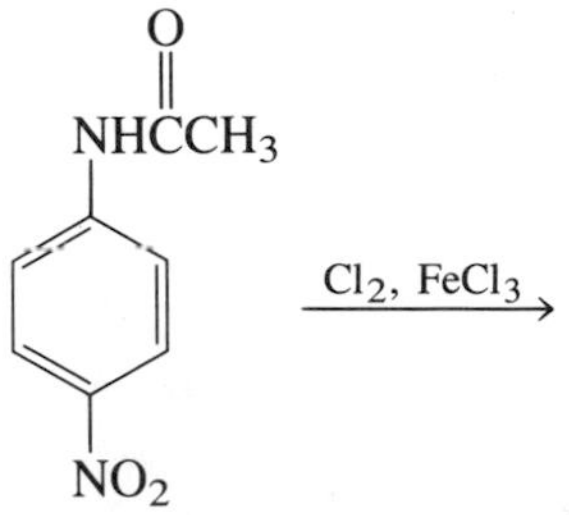

(i) 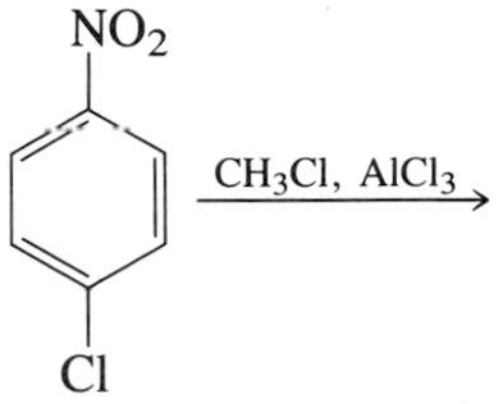 [NO$_2$, Cl para] $\xrightarrow{CH_3Cl,\ AlCl_3}$

7. **(a)** When a mixture containing one mole each of the three dimethylbenzenes (*o*-, *m*-, and *p*-xylene) is treated with one mole of chlorine in the presence of a Lewis acid catalyst, one of the three hydrocarbons is monochlorinated in 100% yield while the other two remain completely unreacted. Which isomer reacts? Explain the differences in reactivity. **(b)** The same experiment carried out on a mixture of the three trimethylbenzenes shown below gives a similar outcome. Answer the questions posed in (a) for this mixture of compounds.

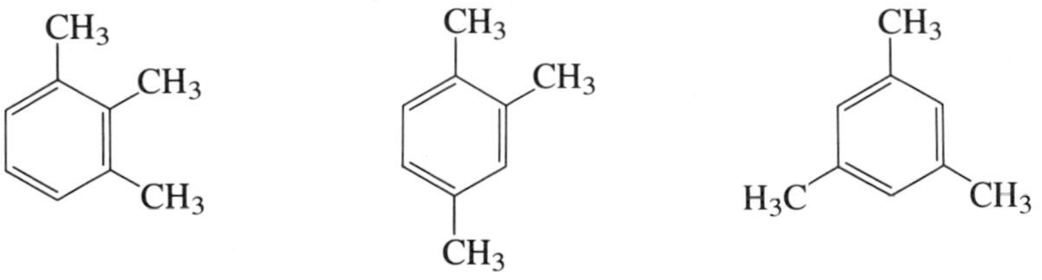

1,2,3-Trimethylbenzene **1,2,4-Trimethylbenzene** **1,3,5-Trimethylbenzene**

8. Propose a reasonable synthesis of each of the following multiply substituted arenes from benzene.

(a) 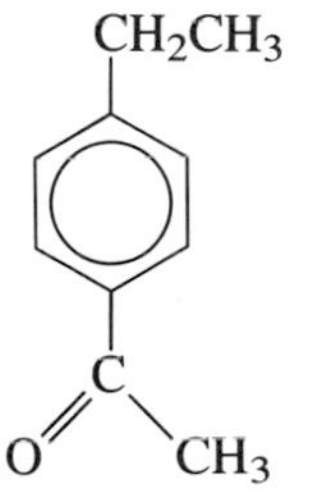 [CH$_2$CH$_3$, C(=O)CH$_3$ para] (b) [NO$_2$, Cl meta] (c) [C(=O)CH$_3$, SO$_3$H meta] (d) [NO$_2$, SO$_3$H para]

(e) [Cl, NO$_2$, NO$_2$] (f) [Cl, Br, O$_2$N] (g) [Br, Cl] (h) [CH$_3$, Br, Br]

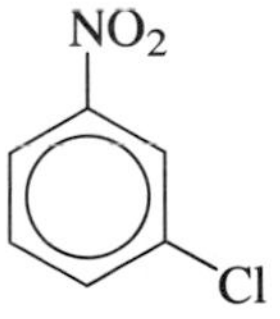

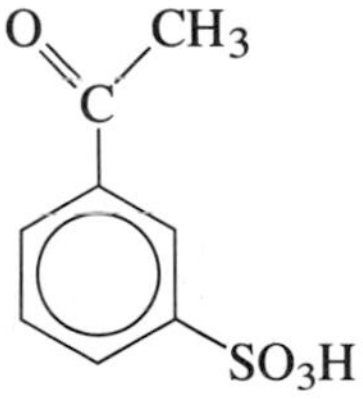

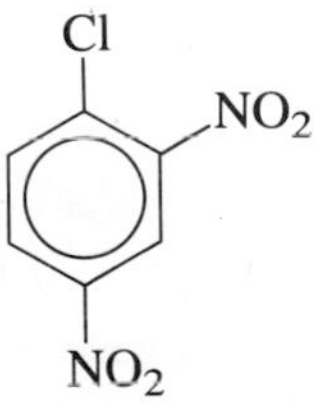

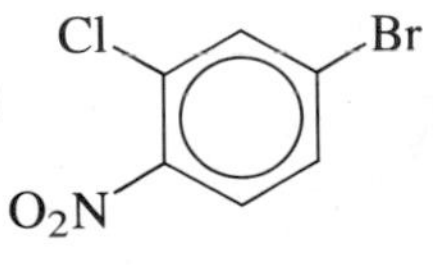

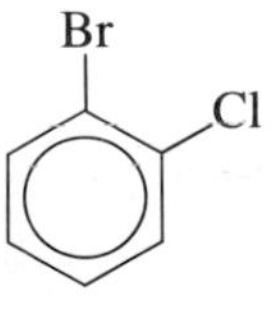

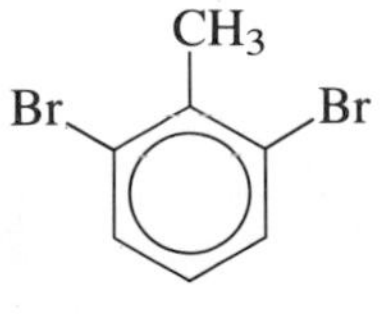

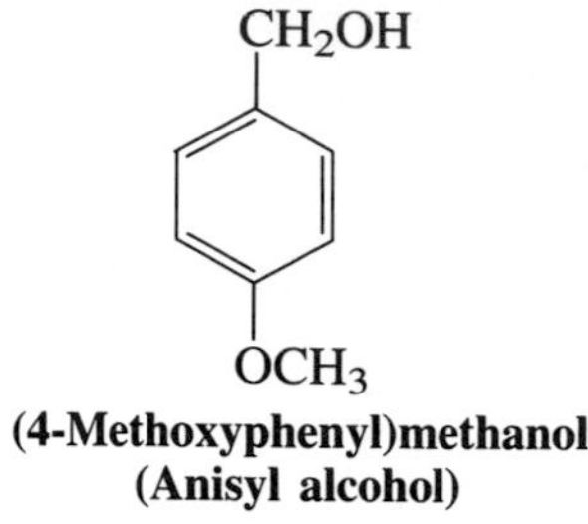

(4-Methoxyphenyl)methanol (Anisyl alcohol)

9. (4-Methoxyphenyl)methanol (anisyl alcohol) contributes both to the flavor of licorice and to the fragrance of lavender. Propose a synthesis of this compound from methoxybenzene (anisole). (Hint: Consider your range of options for alcohol synthesis. If necessary, refer to Problem 14 of Chapter 15.)

10. The NMR and IR spectra for four unknown compounds A–D are presented below and on pages 624–627. Possible empirical formulas for them (not in any particular order) are C_6H_5Br, C_6H_6BrN, and $C_6H_5Br_2N$ (one of

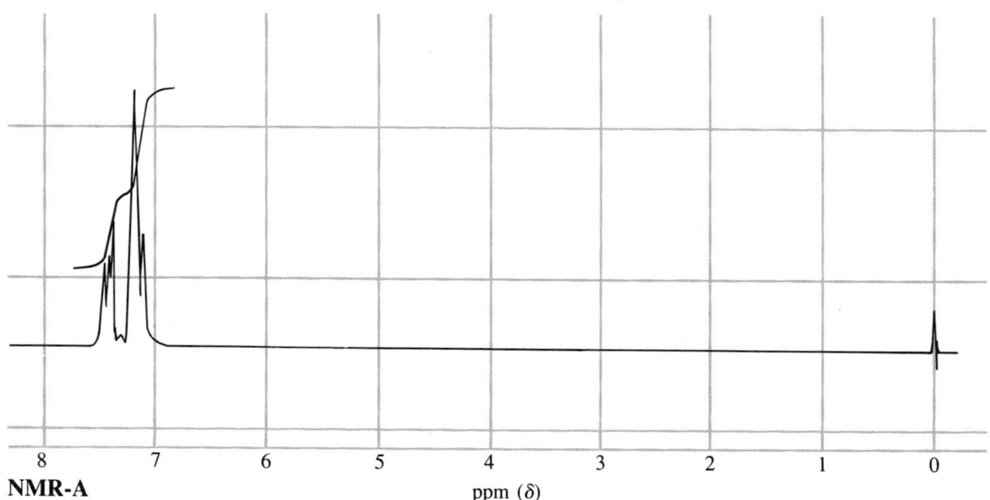

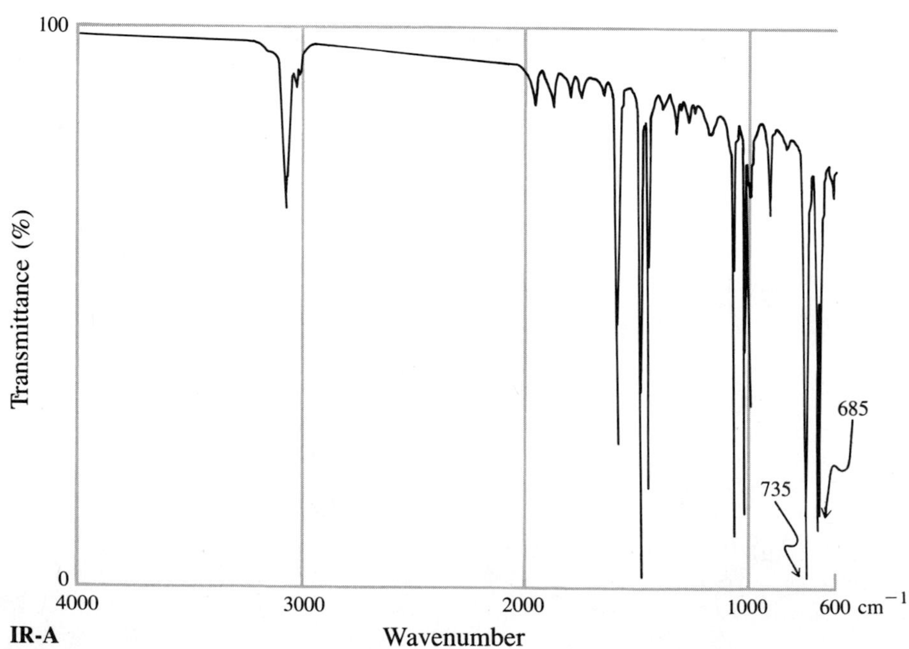

these is used twice—two of the unknowns are isomers). Propose a structure and suggest a synthesis of each unknown from benzene.

11. Catalytic hydrogenation of naphthalene over Pd-C results in rapid addition of two moles of H_2. Propose a structure for this product.

12. Predict the major mononitration product of each of the following disubstituted naphthalenes. **(a)** 1,3-Dimethylnaphthalene; **(b)** 1-chloro-5-methoxynaphthalene; **(c)** 1,7-dinitronaphthalene; **(d)** 1,6-dichloronaphthalene.

13. Write the expected product(s) of each of the following reactions (p. 626).

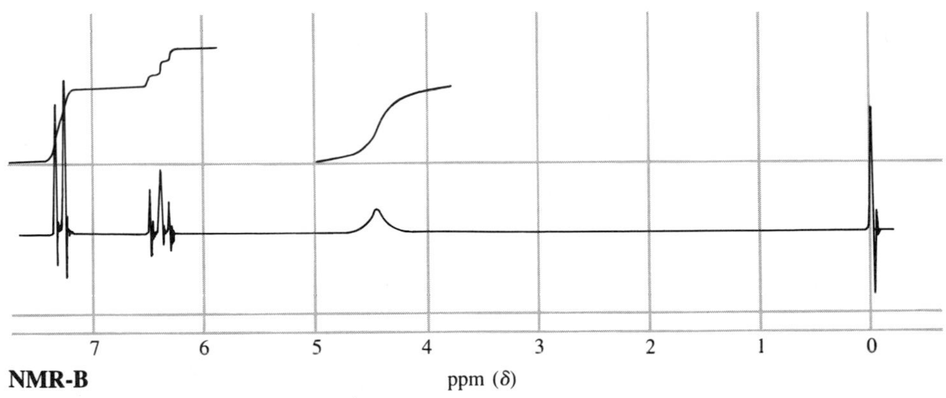

NMR-B

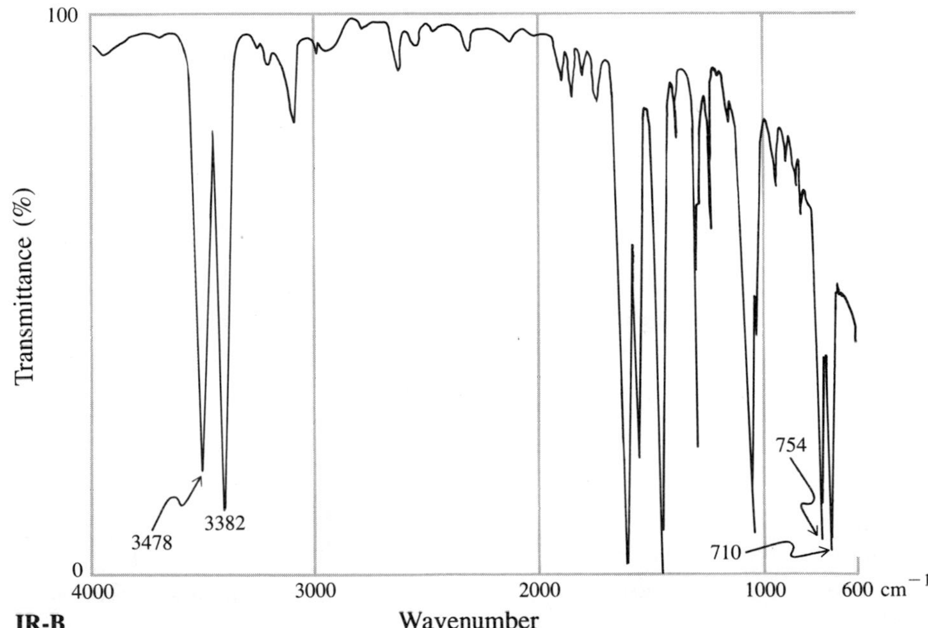

IR-B

(a)

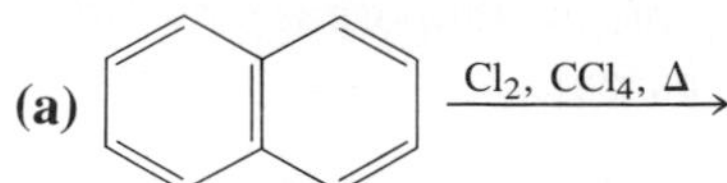

(b) OCH_3 $\xrightarrow{HNO_3}$

(c) CH_3 $\xrightarrow{\text{Conc. } H_2SO_4,\ \Delta}$

(d) $\xrightarrow{CH_3\overset{O}{\overset{\|}{C}}Cl,\ AlCl_3,\ CS_2}$

(e) NO_2 $\xrightarrow{Br_2,\ FeBr_3}$

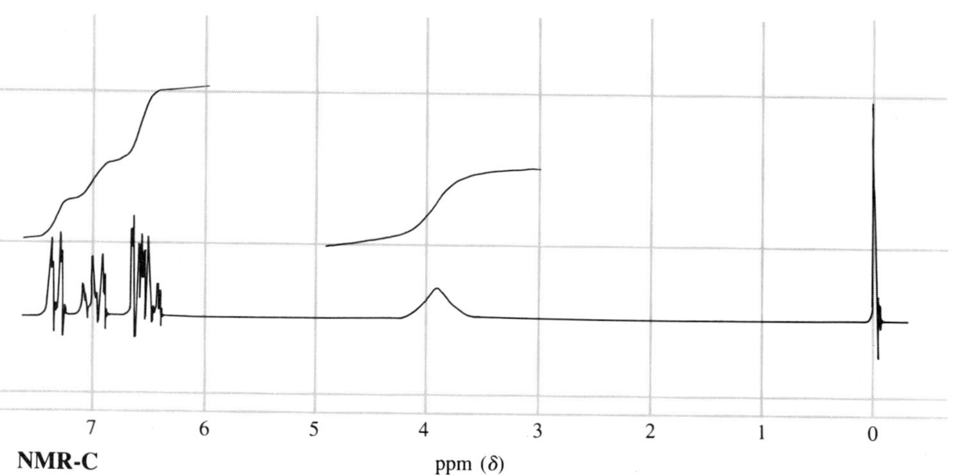

NMR-C

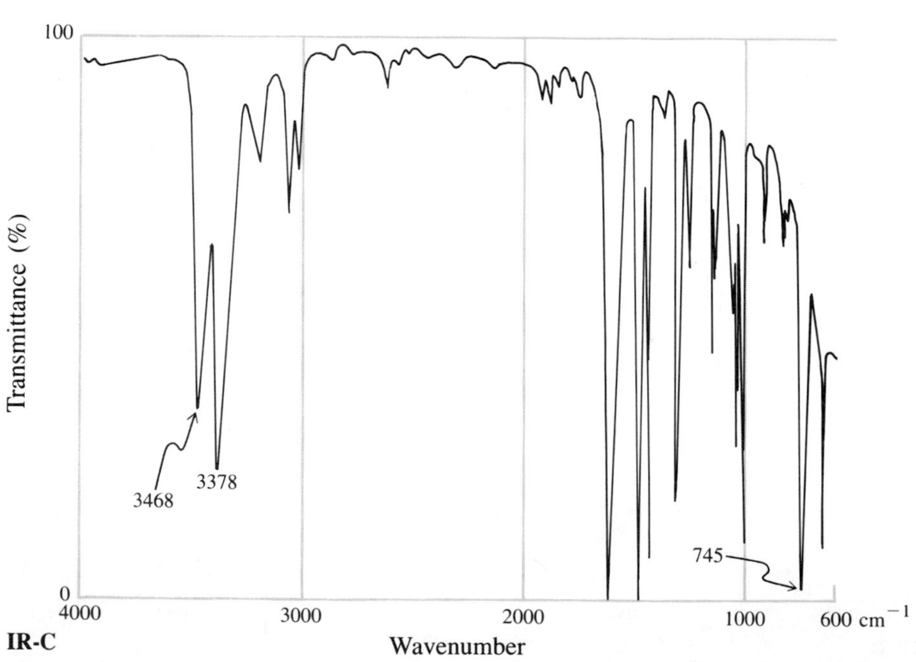

IR-C

14. Electrophilic substitution on the benzene ring of benzenethiol (thiophenol, C_6H_5SH) is not possible. Why? What do you think happens when benzenethiol is allowed to react with an electrophile? (Hint: Recall Chapter 9 regarding sulfur compounds.)

15. Although methoxy is a strongly activating (and ortho, para-directing) group, the meta positions in methoxybenzene (anisole) are actually slightly *deactivated* toward electrophilic substitution relative to benzene. Explain.

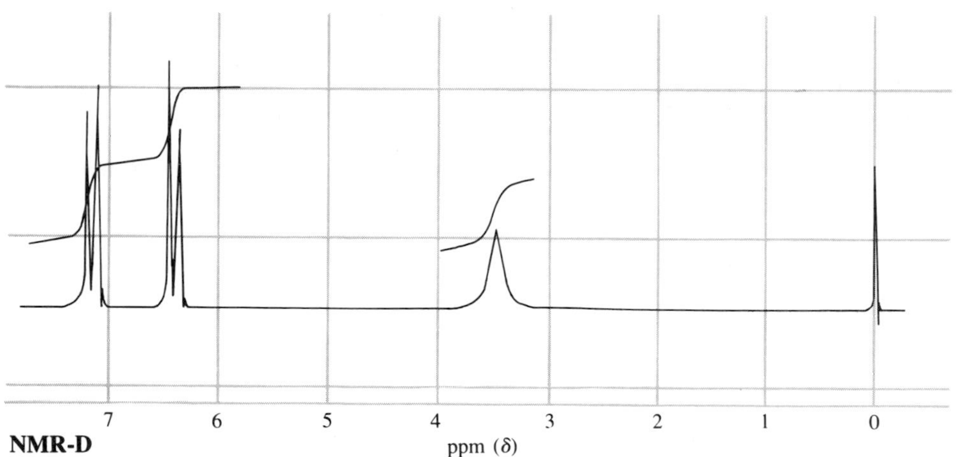

NMR-D

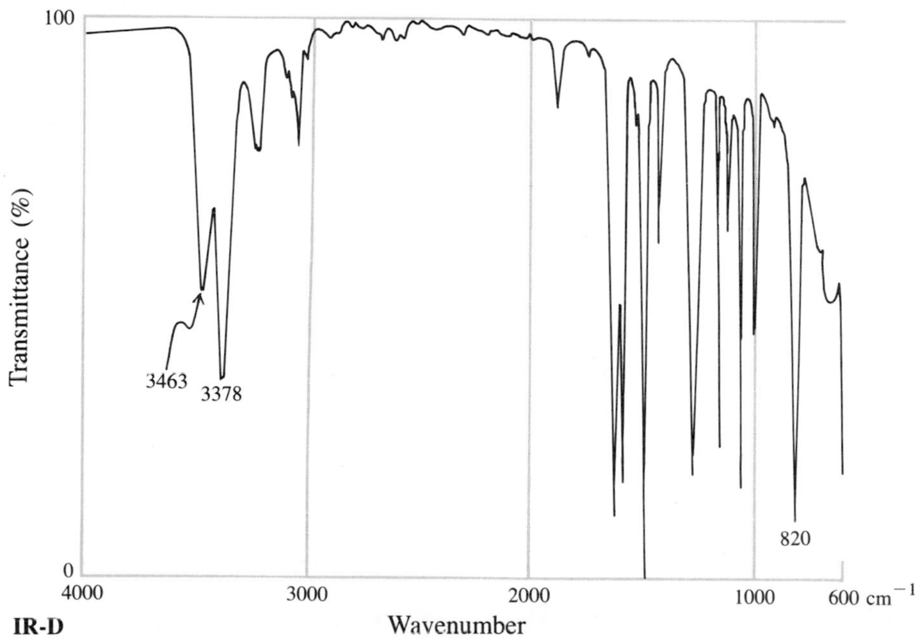

IR-D

16. Predict the result of mononitration of

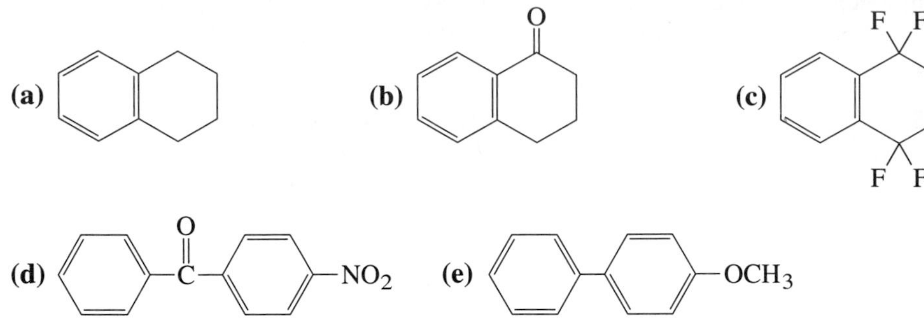

17. The *nitroso* group, NO, as a substituent on a benzene ring acts as an ortho, para-directing group but is deactivating. Explain this finding by the Lewis structure of the nitroso group and its inductive and resonance interactions with the benzene ring. (Hint: Consider possible similarities to another type of substituent that is ortho, para-directing but deactivating.)

18. Typical conditions for nitrosation are illustrated in the equation below. Propose a detailed mechanism for this reaction.

OH (phenol) $\xrightarrow{NaNO_2,\ HCl,\ H_2O}$ OH / NO (ortho) + OH / NO (para)

17

Aldehydes and Ketones

The Carbonyl Group

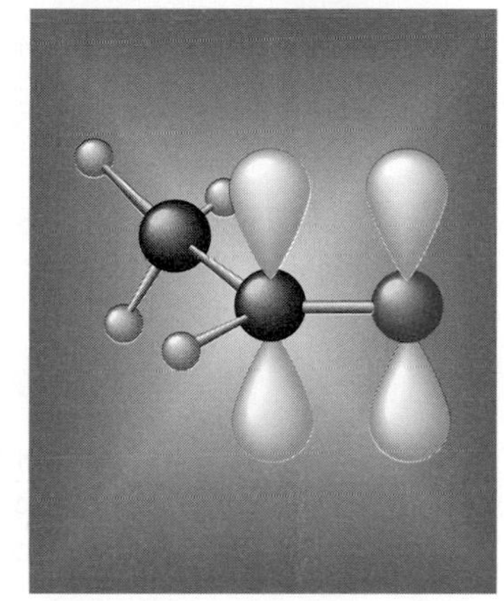

We turn now to the chemistry of the carbon–oxygen double bond, the **carbonyl group.** In this and the next chapter we focus on two classes of carbonyl compounds: **aldehydes,** in which the carbon atom of the carbonyl group is bound to at least one hydrogen atom, and **ketones,** in which it is bound to two carbons. These compounds occur widely in nature, contributing to the flavors and aromas of many foods and assisting in the biological functions of a number of enzymes. In addition, industry makes considerable use of aldehydes and ketones, both as reagents and as solvents in synthesis. Indeed, the carbonyl function is frequently considered to be the most important in organic chemistry.

Carbonyl group

An aldehyde

A ketone

After explaining how to name aldehydes and ketones, this chapter will look at their structures and physical properties. Like alcohols, carbonyl groups are weak Lewis bases, because the oxygen bears two lone electron pairs. Also, the carbon–oxygen double bond is polarized, thereby making the carbonyl carbon electrophilic. The remainder of the chapter will show how these properties shape the chemistry of this versatile functional group.

17-1 Naming the Aldehydes and Ketones

For historical reasons, the simpler aldehydes often retain their common names. These are derived from the common name of the corresponding carboxylic acid, with the ending *-ic acid* replaced by **-aldehyde.**

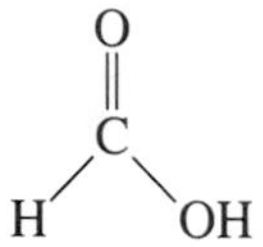

Formic acid 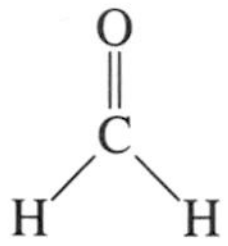 **Formaldehyde** 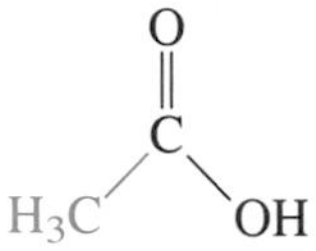 **Acetic acid** 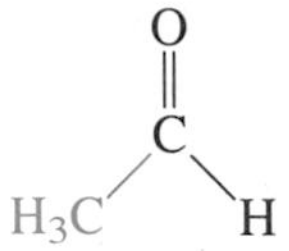 **Acetaldehyde** 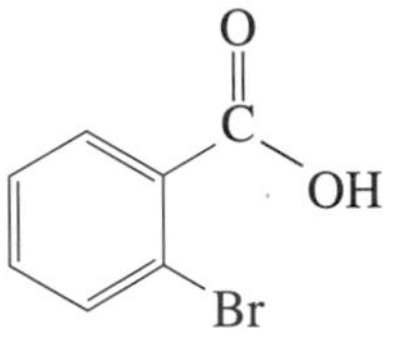 ***o*-Bromo-benzoic acid** 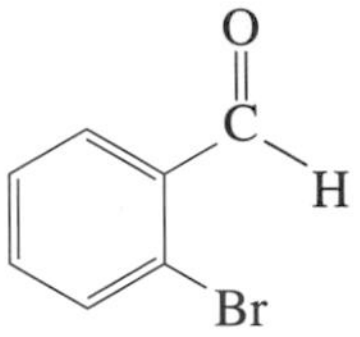 ***o*-Bromo-benzaldehyde**

Many ketones also have common names that consist of the names of the substituent groups followed by the word *ketone*. Dimethyl ketone, the simplest example, is a common solvent best known as **acetone.** Phenyl ketones have common names ending in **-phenone.**

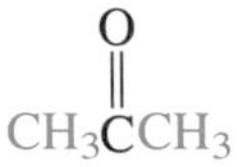

CH_3COCH_3 Dimethyl **ketone** **(Acetone)**

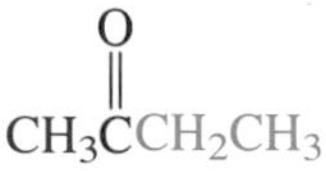

$CH_3COCH_2CH_3$ Ethyl **methyl ketone**

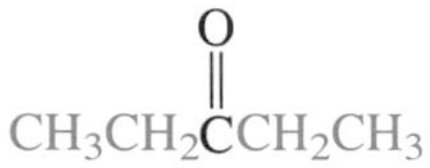

$CH_3CH_2COCH_2CH_3$ Diethyl **ketone**

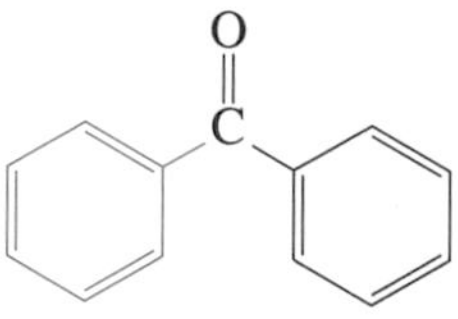

Benzo**phenone**

IUPAC names treat aldehydes as derivatives of the alkanes, with the ending *-e* replaced by **-al.** An alkane thus becomes an **alkanal.** Methanal, the systematic name of the simplest aldehyde, is thus derived from methane, ethanal from ethane, propanal from propane, and so on. However, *Chemical Abstracts* retains the common names for the first two, and so shall we. We number the substituent chain beginning with the carbonyl carbon.

$HCHO$ Methan**al**

CH_3CHO Ethan**al**

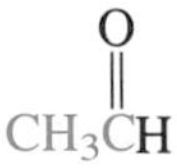

CH_3CH_2CHO Propan**al**

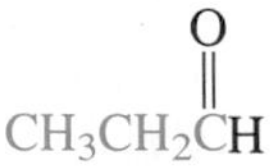

$ClCH_2CH_2CH_2CHO$ **4-Chloro**butan**al**

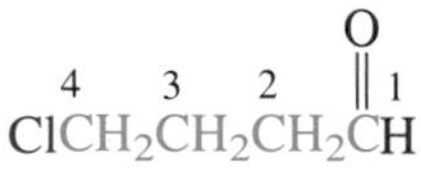

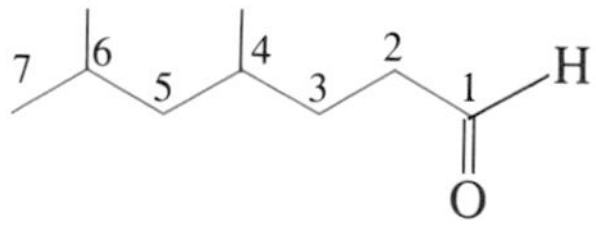

4,6-Dimethylheptan**al**

Notice that the names parallel those of the 1-alkanols (Section 8-1), except that the position of the aldehyde carbonyl group does not have to be specified; *its carbon is defined as C1.*

Aldehydes not readily named after alkanes are instead called **carbaldehydes.** The parent aromatic aldehyde, for instance, is benzenecarbaldehyde, although its common name, benzaldehyde, is still widely used and accepted by *Chemical Abstracts*.

Cyclohexane-carbaldehyde

4-Hydroxy-3-methoxy-benzenecarbaldehyde (4-Hydroxy-3-methoxy-benzaldehyde)

Ketones are called **alkanones,** the ending *-e* of the alkane replaced with **-one.** We can see, for example, why acetone should be called propanone. *The carbonyl carbon is assigned the lowest possible number in the chain, regardless of the presence of other substituents or the OH, C=C, or C≡C functional groups.* Aromatic ketones are named as aryl-substituted alkanones. Ketones, unlike aldehydes, may also be part of a ring, an arrangement giving compounds called **cycloalkanones.**

2-Pentanone | **4-Chloro-6-methyl-3-heptanone** | **2,2-Dimethylcyclopentanone** | **1-Phenylethanone (Acetophenone)**

Notice that we assign the number 1 to the carbonyl carbon when it is in a ring.

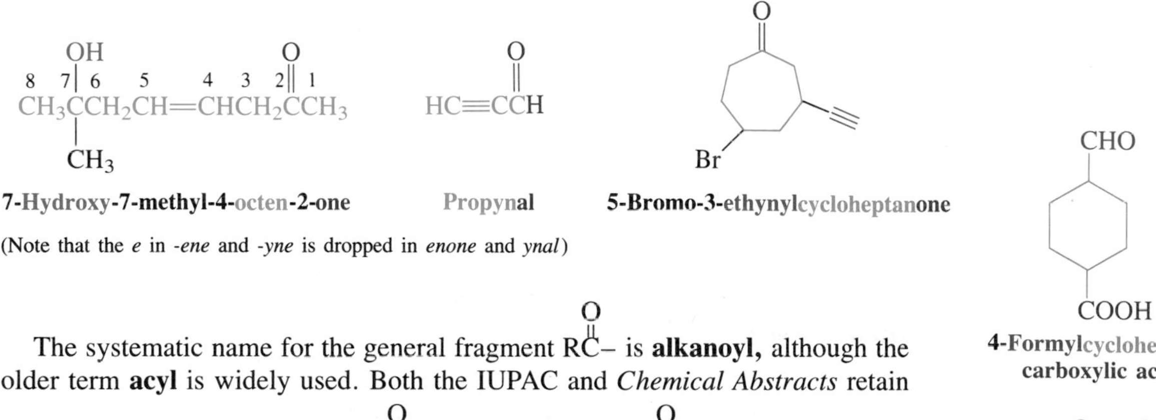

Aldehydes and Ketones with Other Functional Groups

7-Hydroxy-7-methyl-4-octen-2-one | **Propynal** | **5-Bromo-3-ethynylcycloheptanone**

(Note that the *e* in *-ene* and *-yne* is dropped in *enone* and *ynal*)

4-Formylcyclohexane-carboxylic acid

3-Oxobutanal

The systematic name for the general fragment $R\overset{O}{\overset{\|}{C}}-$ is **alkanoyl,** although the older term **acyl** is widely used. Both the IUPAC and *Chemical Abstracts* retain the common names **formyl** for $H\overset{O}{\overset{\|}{C}}-$ and **acetyl** for $CH_3\overset{O}{\overset{\|}{C}}-$. The term **oxo** denotes the location of a ketone carbonyl group when it is present together with an aldehyde function.

EXERCISE 17-1

Name or draw the structures of the following compounds.

(a) 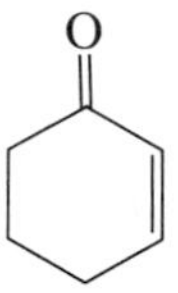 [structure: cyclohexenone]   (b) 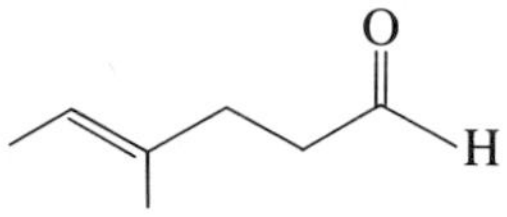 [structure: zigzag aldehyde]   (c) 4-Octyn-3-one

(d) 3-Hydroxybutanal   (e) 4-Bromocyclohexanone

There are various ways of drawing aldehydes and ketones. As usual, condensed formulas or the zigzag notation may be used. Note that the condensed formulas for aldehydes are written as RCHO, and *never* as RCOH, to prevent confusion with the hydroxy group of alcohols.

Various Ways of Writing Aldehyde and Ketone Structures

Butanal: $CH_3CH_2CH_2\overset{O}{\overset{\|}{C}}H$ $CH_3CH_2CH_2CHO$ (Not a hydroxy group)

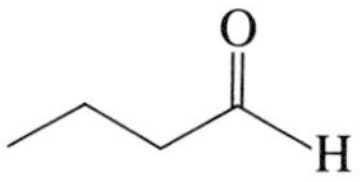

Butanone: $CH_3CH_2\overset{O}{\overset{\|}{C}}CH_3$ $CH_3CH_2COCH_3$

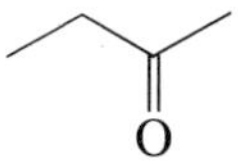

In summary, aldehydes and ketones are named systematically as alkanals and alkanones. The carbonyl group takes precedence over the hydroxy function and C–C double and triple bonds in numbering. With these rules, the usual guidelines for numbering the stem and labeling the substituents are followed.

17-2 Structure of the Carbonyl Group

If we think of the carbonyl group as an oxygen analog of the alkene functional group, we can correctly predict its molecular-orbital description, the structures of aldehydes and ketones, and some of their physical properties. However, the alkene and carbonyl double bonds do differ considerably in reactivity because of the electronegativity of oxygen and its two lone pairs of electrons.

The carbonyl group contains a short, strong, and very polar bond

Both the carbon and the oxygen of the carbonyl group are sp^2 hybridized. They therefore lie in the same plane as the two additional groups on carbon, with bond angles approximating 120°. Perpendicular to the molecular frame are two p orbitals, one on carbon and one on oxygen, making up the π bond (Figure 17-1).

Figure 17-2 shows some of the structural features of acetaldehyde. As expected, the molecule is planar, with a trigonal carbonyl carbon and a short

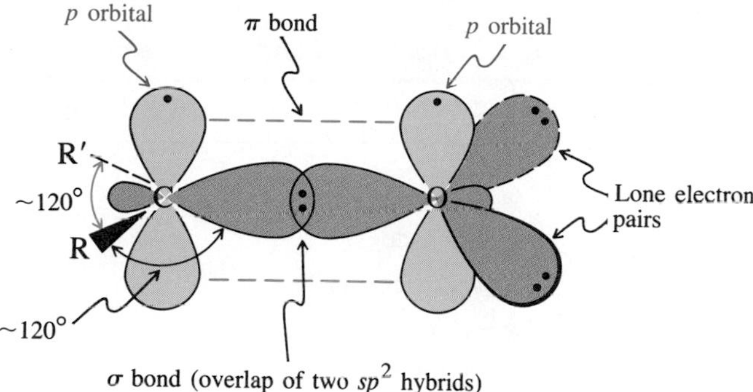

FIGURE 17-1 Molecular-orbital picture of the carbonyl group. The sp^2 hybridization and the orbital arrangement are similar to that of ethene (Figure 11-1). However, both the two lone electron pairs and the electronegativity of the oxygen atom modify the properties of the functional group.

carbon–oxygen bond, indicative of its double bond character. Not surprisingly, this bond is also rather strong, ranging from 175 to 180 kcal mol^{-1}.

Comparison with the electronic structure of an alkene double bond reveals two important differences. First, the oxygen atom bears two lone electron pairs located in two sp^2 hybrid orbitals. Second, oxygen is more electronegative than carbon. This property causes an appreciable polarization of the carbon–oxygen double bond, with a partial positive charge on carbon and an equal amount of negative charge on oxygen. Thus, the carbon is electrophilic, the oxygen nucleophilic and slightly basic. This polarization can be described either by a polar resonance structure for the carbonyl moiety or by partial charges. As we have seen (Section 16-2), the partial positive charge on the carbonyl carbon renders alkanoyl groups electron withdrawing.

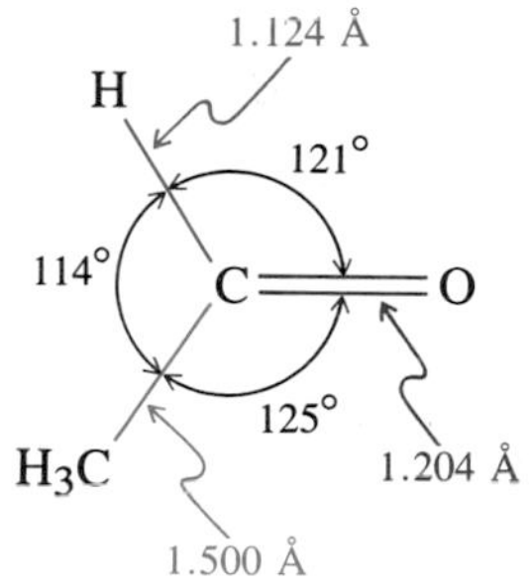

H, 1.124 Å, 121°, 114°, C=O, 125°, H_3C, 1.204 Å, 1.500 Å

FIGURE 17-2 Molecular structure of acetaldehyde.

Descriptions of a Carbonyl Group

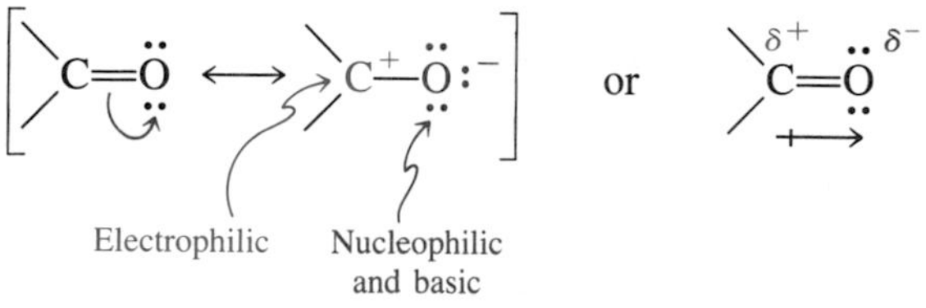

$$\left[\rangle C{=}\ddot{O} \longleftrightarrow \rangle \overset{+}{C}{-}\ddot{O}{:}^{-} \right] \quad \text{or} \quad \rangle \overset{\delta+}{C}{=}\overset{\delta-}{\ddot{O}}$$

Electrophilic — Nucleophilic and basic

Polarization alters the physical constants of aldehydes and ketones

The polarization of the carbonyl functional group makes the boiling points of aldehydes and ketones higher than those of hydrocarbons of similar molecular weight (Table 17-1). Because of their polarity, the smaller carbonyl compounds are soluble in water. For example, acetaldehyde and propanone (acetone) are completely miscible with water. As the hydrophobic hydrocarbon part of the molecule increases in size, however, water solubility decreases. Carbonyl compounds with more than six carbons are rather insoluble in water.

TABLE 17-1 Boiling Points of Aldehydes and Ketones

Formula	Name	Boiling point (°C)
HCHO	Formaldehyde	−21
CH_3CHO	Acetaldehyde	21
CH_3CH_2CHO	Propanal (propionaldehyde)	49
CH_3COCH_3	Propanone (acetone)	56
$CH_3CH_2CH_2CHO$	Butanal (butyraldehyde)	76
$CH_3CH_2COCH_3$	Butanone (ethyl methyl ketone)	80
$CH_3CH_2CH_2CH_2CHO$	Pentanal	102
$CH_3COCH_2CH_2CH_3$	2-Pentanone	102
$CH_3CH_2COCH_2CH_3$	3-Pentanone	102

To summarize, the carbonyl group in aldehydes and ketones is an oxygen analog of the carbon–carbon double bond. However, the electronegativity of oxygen polarizes the π bond, thereby rendering the alkanoyl substituent electron withdrawing. The arrangement of bonds around the carbon and oxygen is planar, a consequence of sp^2 hybridization.

17-3 Spectroscopic Properties of Aldehydes and Ketones

The carbonyl group gives rise to characteristic spectra. In ^1H NMR spectroscopy, the formyl hydrogen of the aldehydes is very strongly deshielded, appearing between 9 and 10 ppm, a chemical shift that is unique for this class of compounds. The reason for this effect is twofold. First, the movement of the π electrons, like that in alkenes (Section 11-4), causes a local magnetic field strengthening the external field. Second, the charge on the positively polarized carbon exerts an additional deshielding effect. Figure 17-3 shows the ^1H NMR spectrum of propanal with the formyl hydrogen resonating at $\delta = 9.89$ ppm, split into a triplet ($J = 2$ Hz) because of a small coupling to the methylene hydrogens on the other side of the carbonyl function. The latter are also slightly deshielded relative to alkane hydrogens because of the electron-withdrawing character of the functional group. This effect is also seen in the ^1H NMR spectra of ketones: These hydrogens normally appear in the region $\delta = 2.0$–2.8 ppm.

^1H NMR Deshielding in Aldehydes and Ketones

$RCH_2C(=O)H$: $\delta \sim 2.5$ (CH_2), ~ 9.8 ppm (CHO)

$RCH(R')C(=O)CH_3$: $\delta \sim 2.6$ (CH), ~ 2.0 ppm (CH_3)

Carbon-13 NMR spectra are diagnostic of both aldehydes *and* ketones, because of the characteristic chemical shift of the carbonyl carbon. Partly because of the electronegativity of the directly bound oxygen, the carbonyl carbons in aldehydes and ketones appear at even lower field (~200 ppm) than do the sp^2-hybridized carbon atoms of alkenes (Section 11-4). The carbons next to the carbonyl group are also deshielded relative to those located farther away. The ^{13}C NMR spectrum of cyclohexanone is shown in Figure 17-4.

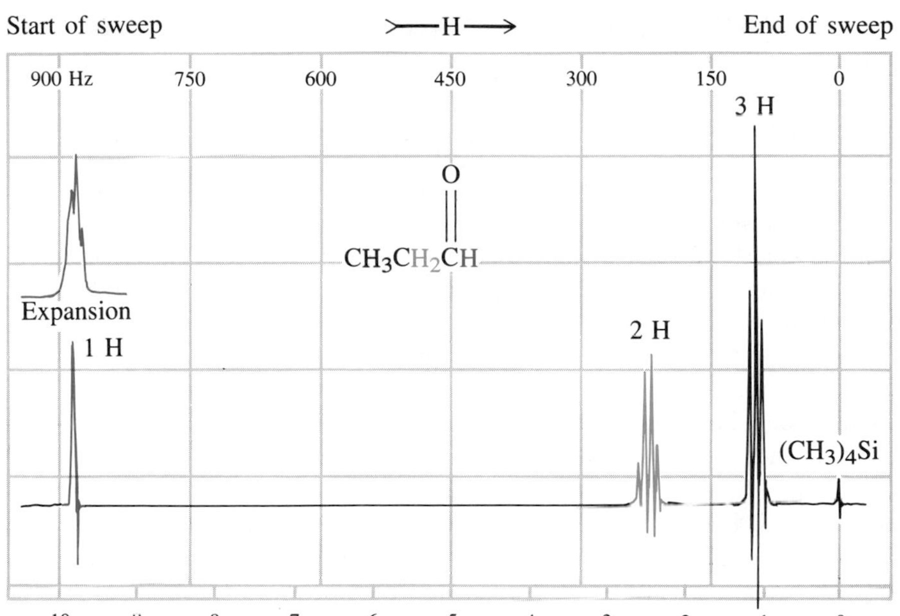

FIGURE 17-3 90-MHz ^1H NMR spectrum of propanal in CCl_4. The formyl hydrogen (at δ = 9.8 ppm) is strongly deshielded.

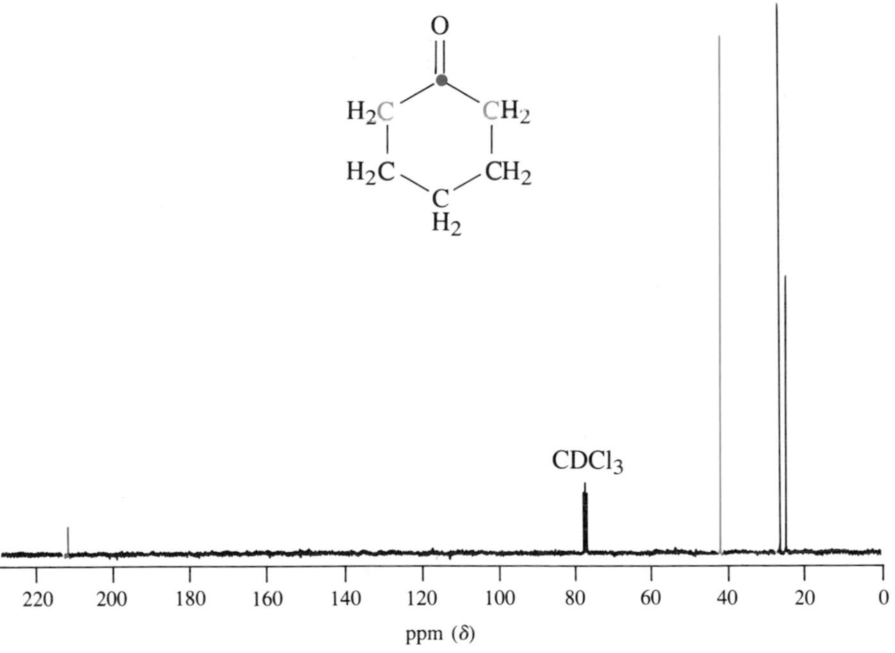

FIGURE 17-4 ^{13}C NMR spectrum of cyclohexanone at 75.5 MHz in $CDCl_3$. The carbonyl carbon at 211.8 ppm is strongly deshielded relative to the other carbons. Because of symmetry, the molecule exhibits only four peaks; the three methylene carbon resonances absorb at increasingly lower field the closer they are to the carbonyl group. The triplet at 77 ppm is due to the carbon in the solvent, $CDCl_3$, split by deuterium. (The rules for spin–spin splitting by deuterium are different from those for hydrogen and will not be discussed here.)

^{13}C NMR Chemical Shifts of Typical Aldehydes and Ketones

$CH_3—\overset{O}{\overset{\|}{C}}H$	$CH_3—CH_2—\overset{O}{\overset{\|}{C}}H$	$CH_3\overset{O}{\overset{\|}{C}}CH_3$	$CH_3\overset{O}{\overset{\|}{C}}—CH_2—CH_2—CH_3$
δ = 31.2 199.6 ppm	δ = 5.2 36.7 201.8 ppm	δ = 30.2 205.1 ppm	δ = 29.3 206.6 45.2 17.5 13.5 ppm

Infrared spectroscopy is a useful way of directly detecting the presence of a carbonyl group. The C=O stretching frequency gives rise to an intense band that typically appears in a relatively narrow range (1690–1750 cm^{-1}; Figure 17-5). The carbonyl absorption for aldehydes typically appears at about 1735 cm^{-1}. Those for acyclic alkanones and cyclohexanone are found at about 1715 cm^{-1}. Conjugation with either alkene or benzene π systems reduces the carbonyl infrared frequency by about 30–40 cm^{-1}; thus 1-phenylethanone (acetophenone) exhibits an IR band at 1680 cm^{-1}. Conversely, the stretching frequency increases for carbonyl groups in rings with fewer than six atoms: Cyclopentanone absorbs at 1745 cm^{-1}, cyclobutanone at 1780 cm^{-1}.

Carbonyl groups also exhibit characteristic electronic spectra, because the nonbonding lone electron pairs on the oxygen atom undergo low-energy $n \rightarrow \pi^*$ and $\pi \rightarrow \pi^*$ transitions (Figure 17-6). For example, propanone (acetone) shows an $n \rightarrow \pi^*$ band at 280 nm (ϵ = 15) in hexane. The corresponding $\pi \rightarrow \pi^*$ transition appears at about 190 nm (ϵ = 1100). Conjugation with a carbon–carbon double bond shifts absorptions to longer wavelengths. For example, the electronic spectrum of 3-buten-2-one, CH_2=$CHCOCH_3$, has peaks at 324 nm (ϵ = 24, $n \rightarrow \pi^*$) and 219 nm (ϵ = 3600, $\pi \rightarrow \pi^*$).

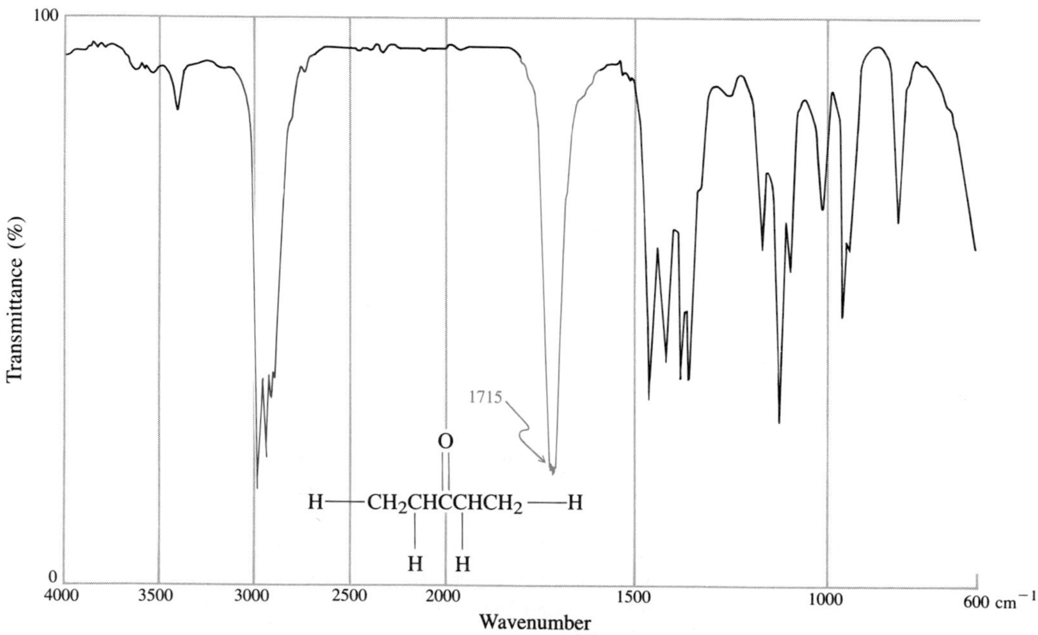

FIGURE 17-5 IR spectrum of 3-pentanone; $\tilde{\nu}_{C=O\ stretch}$ = 1715 cm^{-1}.

Electronic Transitions of Propanone (Acetone) and 3-Buten-2-one

$CH_3\overset{O}{\overset{\|}{C}}CH_3$ **Propanone (Acetone)**		$CH_2{=}CH\overset{O}{\overset{\|}{C}}CH_3$ **3-Buten-2-one**	
$\lambda_{max}(\epsilon) = 280(15)$	$n \longrightarrow \pi^*$	$\lambda_{max}(\epsilon) = 324(24)$	$n \longrightarrow \pi^*$
$190(1100)$	$\pi \longrightarrow \pi^*$	$219(3600)$	$\pi \longrightarrow \pi^*$

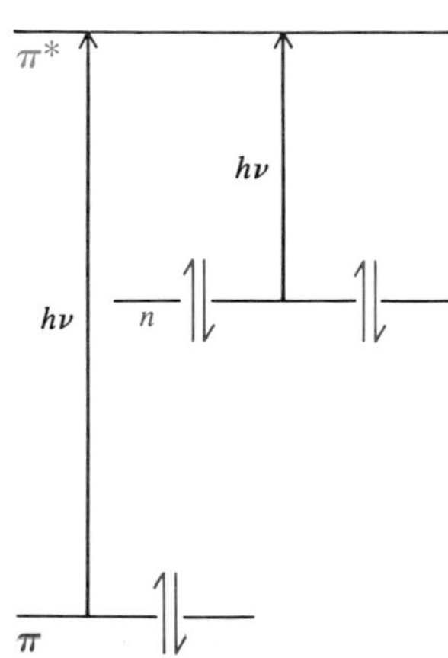

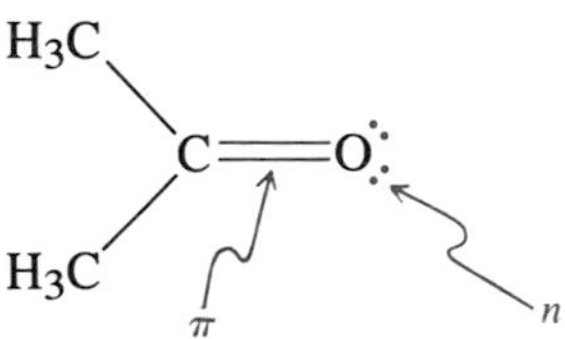

FIGURE 17-6 The $\pi \rightarrow \pi^*$ and $n \rightarrow \pi^*$ transitions in propanone (acetone).

EXERCISE 17-2

An unknown C_4H_6O exhibited the following spectral data: ^{1}H NMR (CCl_4) $\delta = 2.03$ (dd, $J = 6.7$, 1.6 Hz, 3 H), 6.06 (ddq, $J = 16.1$, 7.7, 1.6 Hz, 1 H), 6.88 (dq, $J = 16.1$, 6.7 Hz, 1 H), 9.47 (d, $J = 7.7$ Hz, 1 H) ppm; ^{13}C NMR (CCl_4) $\delta = 18.4$, 132.8, 152.1, 191.4 ppm; UV $\lambda_{max}(\epsilon) = 220(15{,}000)$ and 314(32) nm. Suggest a structure.

To summarize the spectroscopic features of aldehydes and ketones, the NMR spectra of the formyl hydrogens and the carbonyl carbons show strong deshielding. The carbon–oxygen double bond gives rise to a strong infrared band at about 1715 cm^{-1}, which is shifted to lower frequency by conjugation and to higher frequency in small rings. Finally, the ability of nonbonding electrons to be excited into the π^* molecular orbitals causes the carbonyl group to exhibit characteristic, relatively long wavelength UV absorptions.

17-4 Preparation of Aldehydes and Ketones

The most important ways to prepare aldehydes and ketones have already been described in connection with the chemistry of other functional groups. This section introduces some industrial preparations and reviews the methods we have studied, pointing out special features and additional examples. Other routes to aldehydes and ketones will be described in later chapters.

Formaldehyde and propanone are important industrial carbonyl compounds

The most important aldehyde in industry is formaldehyde; the most important ketone is propanone (acetone). About 6 billion pounds of formaldehyde is made yearly in the United States by oxidation of methanol.

$$CH_3OH \xrightarrow{O_2,\ 600\text{–}650°C,\ \text{catalytic Ag}} CH_2{=}O$$

In aqueous solution (formalin), it has applications as a disinfectant and a fungicide. Its greatest use is in the preparation of phenolic resins (Section 22-6).

Propanone (acetone) is a valuable by-product of the cumene hydroperoxide process (Section 22-4) and is sold as a solvent and starting material for the production of other industrial materials. Annual production in the United States alone is about 2 billion pounds.

Butanal is made by a process called hydroformylation, in which propene is exposed to synthesis gas ($CO + H_2$; Section 8-4) in the presence of a cobalt or rhodium catalyst. The reaction can also be used to prepare other aldehydes.

$$CH_3CH{=}CH_2 + CO + H_2 \xrightarrow{\text{Co or Rh, }\Delta\text{, pressure}} CH_3CH_2CH_2CHO$$

Four methods allow laboratory synthesis of aldehydes and ketones

Table 17-2 summarizes four approaches to synthesizing aldehydes and ketones. First, we have seen (Section 8-6) that *oxidation* of alcohols by chromium (VI) reagents gives carbonyl compounds. Secondary alcohols give ketones and primary alcohols give aldehydes, but, in the latter case, only in the absence of water, to avoid overoxidation to carboxylic acids. The chromium oxidant is selective even in the presence of alkene and alkyne units.

Selective Alcohol Oxidation

$$\underset{\textbf{3-Octyn-2-ol}}{CH_3\overset{OH}{\overset{|}{C}}HC{\equiv}C(CH_2)_3CH_3} \xrightarrow{CrO_3,\ H_2SO_4,\ \text{propanone (acetone), } 0^\circ C} \underset{\substack{80\% \\ \textbf{3-Octyn-2-one}}}{CH_3\overset{O}{\overset{\|}{C}}C{\equiv}C(CH_2)_3CH_3}$$

Use of PCC (CrO_3 + Pyridine + HCl) to Oxidize a Primary Alcohol to an Aldehyde

$$\text{(2,2-dimethyl-1,3-dioxolan-4-yl)}\text{–}CH(CH_3)CH_2OH \xrightarrow{PCC,\ CH_2Cl_2,\ Na^{+\,-}OOCCH_3} \text{(2,2-dimethyl-1,3-dioxolan-4-yl)}\text{–}CH(CH_3)\overset{O}{\overset{\|}{C}}H \quad 85\%$$

TABLE 17-2 Syntheses of Aldehydes and Ketones

Reaction	Illustration
1. Oxidation of alcohols	$—CH_2OH \xrightarrow{PCC,\ CH_2Cl_2} —\overset{O}{\overset{\|}{C}}H$
2. Ozonolysis of alkenes	$\rangle C{=}C\langle \xrightarrow{1.\ O_3,\ CH_2Cl_2 \quad 2.\ (CH_3)_2S} \rangle C{=}O + O{=}C\langle$
3. Hydration of alkynes	$—C{\equiv}C— \xrightarrow{H_2O,\ H^+,\ Hg^{2+}} —\overset{O}{\overset{\|}{C}}—CH_2—$
4. Friedel-Crafts alkanoylation	$C_6H_6 \xrightarrow{1.\ RCOCl,\ AlCl_3,\ CS_2 \quad 2.\ H^+,\ H_2O} C_6H_5—\overset{O}{\overset{\|}{C}}—R$

Overoxidation of aldehydes in the presence of water is due to hydration to a 1,1-diol (Section 17-6). Oxidation of this diol leads to the carboxylic acid.

Water Causes the Overoxidation of Primary Alcohols

$$\text{RCH}_2\text{OH} \xrightarrow{\text{Cr(VI), H}^+} \text{RCH}{=}\text{O} \xrightarrow{\text{H}_2\text{O}} \text{RC(OH)}_2\text{H} \xrightarrow{\text{Cr(VI)}} \text{RC(=O)OH}$$

Another mild reagent that specifically oxidizes allylic alcohols is manganese dioxide. Ordinary alcohols are not attacked at room temperature, as shown in the selective oxidation to form a steroid found in the adrenal gland.

Selective Allylic Oxidations with Manganese Dioxide

Not oxidized (not allylic)

MnO_2, $CHCl_3$, 25°C

62%

EXERCISE 17-3

Design a synthesis of cyclohexyl 1-propynyl ketone starting from cyclohexane. You may use any other reagents.

Ozonolysis

1. O_3
2. Reducing agent

$$\text{CH}_3\text{C(=O)(CH}_2)_4\text{CH=O}$$

85%

The second method of preparation that we studied is the oxidative cleavage of carbon–carbon double bonds—*ozonolysis* (Section 12-11). Exposure to ozone followed by treatment with a mild reducing agent, such as catalytically activated hydrogen or dimethyl sulfide, cleaves alkenes to give aldehydes and ketones.

Third, *hydration* of the carbon–carbon triple bond yields enols that tautomerize to carbonyl compounds (Sections 13-8 and 13-9). In the presence of mercuric ion, addition of water follows Markovnikov's rule to furnish ketones.

Markovnikov Hydration of Alkynes

$$\text{RC}{\equiv}\text{CH} \xrightarrow{\text{HOH, H}^+\text{, Hg}^{2+}} \left[\text{(HO)(R)C}{=}\text{CH}_2\right] \longrightarrow \text{RC(=O)CH}_3$$

Anti-Markovnikov addition is observed in hydroboration–oxidation.

Anti-Markovnikov Hydration of Alkynes

$$\text{RC}{\equiv}\text{CH} \xrightarrow{(\text{C}_6\text{H}_{11})_2\text{BH}} \text{(R)(H)C}{=}\text{C(H)B(C}_6\text{H}_{11})_2 \xrightarrow{\text{H}_2\text{O}_2\text{, HO}^-} \left[\text{(R)(H)C}{=}\text{C(H)(OH)}\right] \longrightarrow \text{RCH}_2\text{CH=O}$$

Finally, Section 15-14 discussed the synthesis of aryl ketones by *Friedel-Crafts alkanoylation,* a form of electrophilic aromatic substitution. The example shown below furnishes an industrially useful perfume additive.

Friedel-Crafts Alkanoylation (Acylation)

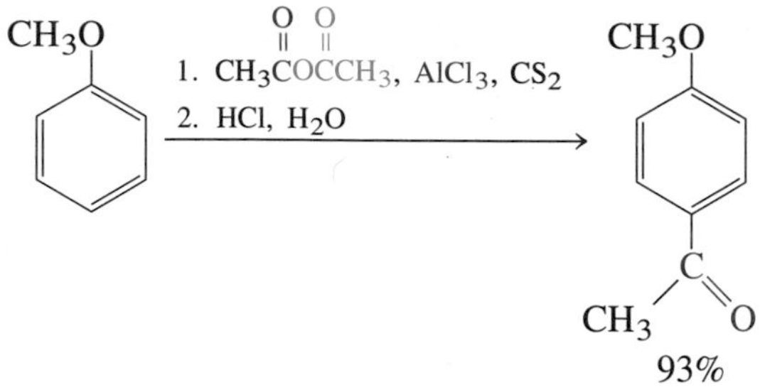

In summary, four methods of synthesizing aldehydes and ketones are oxidation of alcohols, oxidative cleavage of alkenes, hydration of alkynes, and Friedel-Crafts alkanoylation. Many other approaches will be discussed in later chapters.

17-5 Reactivity of the Carbonyl Group: Mechanisms of Addition

Our knowledge of the carbonyl group's structure helps us understand its characteristic reactions. This section begins the discussion of the chemistry of the carbonyl group in aldehydes and ketones. Like the π bond in the alkenes, the carbon–oxygen double bond is prone to additions. For example, catalytic hydrogenation yields alcohols. Electrophiles attack at oxygen, nucleophiles at carbon.

There are three regions of reactivity in aldehydes and ketones

Aldehydes and ketones contain three regions at which most reactions occur: the Lewis basic oxygen, the electrophilic carbonyl carbon, and the adjacent carbon.

Regions of Reactivity in Aldehydes and Ketones

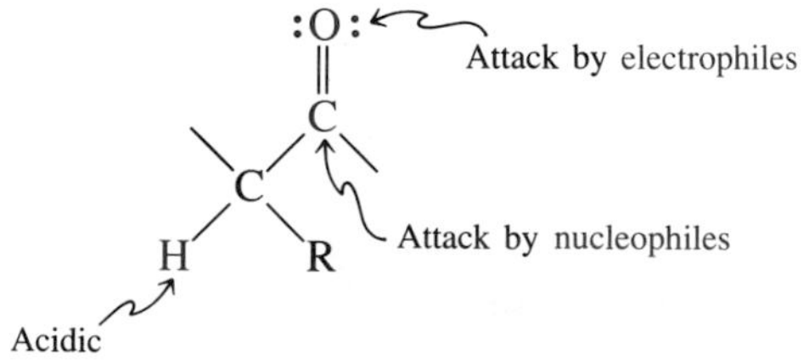

The remainder of this chapter is concerned with the first two areas of reactivity, both of which lead to addition to the carbonyl π bond. This section illustrates two distinct ways in which this is achieved: catalytic hydrogenation and ionic addition. Other reactions, centered instead on the acidic hydrogen of the adjacent carbon, will be the subject of Chapter 18.

The carbonyl pi bond undergoes hydrogenation

Like carbon–carbon π bonds, the carbonyl group is susceptible to catalytic hydrogenation (Section 8-6), which results in the formation of alcohols. Aldehydes and ketones are usually more sluggish than alkenes in this reaction, requiring pressure or elevated temperature to proceed at a useful rate.

$$CH_3\overset{O}{\overset{\|}{C}}CH_2CH_3 \xrightarrow{H_2,\ \text{Raney Ni},\ 80°C,\ 5\ \text{atm}} CH_3\underset{H}{\underset{|}{\overset{OH}{\overset{|}{C}}}}CH_2CH_3$$

This difference in reactivity can be exploited in selective hydrogenations of the carbon–carbon π bond in unsaturated aldehydes (enals) or ketones (enones).

The catalytic hydrogenations of aldehydes and ketones are addition reactions that proceed through surface-bound intermediates. Additions to the carbon–oxygen double bond may also occur by ionic mechanisms that exploit the dipolar nature of the functional group.

Selective Enone Hydrogenation

H_2, Pt, 1 atm, 25°C

100%

The carbonyl group undergoes ionic additions

Polar reagents add to the dipolar carbonyl group according to Coulomb's law (Section 1-2). Nucleophiles bond to the carbon, electrophiles to the oxygen. Sections 8-6 and 8-8 described a number of such additions by organometallic and hydride reagents to give alcohols. Table 17-3 reviews these processes.

Ionic Additions to the Carbonyl Group

$$\overset{\delta^-}{O}=\overset{\delta^+}{C}\langle \;+\; \overset{\delta^+}{X}—\overset{\delta^-}{Y} \longrightarrow —\underset{Y}{\underset{|}{\overset{OX}{\overset{|}{C}}}}—$$

TABLE 17-3 Additions of Hydride and Organometallic Reagents to Aldehydes and Ketones

Reaction	Illustration	
1. Aldehyde + hydride reagent	$RCHO \xrightarrow{NaBH_4,\ CH_3CH_2OH} RCH_2OH$	Primary alcohol
2. Ketone + hydride reagent	$R_2CO \xrightarrow{NaBH_4,\ CH_3CH_2OH} R_2CHOH$	Secondary alcohol
3. Formaldehyde + Grignard reagent	$H_2CO \xrightarrow{R'MgX,\ (CH_3CH_2)_2O} R'CH_2OH^a$	Primary alcohol
4. Aldehyde + Grignard reagent	$RCHO \xrightarrow{R'MgX,\ (CH_3CH_2)_2O} R'RCHOH^a$	Secondary alcohol
5. Ketone + Grignard reagent	$R_2CO \xrightarrow{R'MgX,\ (CH_3CH_2)_2O} R'R_2COH^a$	Tertiary alcohol

a. After aqueous work-up.

The hydride reagents $NaBH_4$ and $LiAlH_4$ reduce carbonyl groups but not carbon–carbon double bonds. These reagents therefore convert unsaturated aldehydes and ketones into unsaturated alcohols, with a selectivity opposite that exhibited in catalytic hydrogenation.

$$C_6H_5\text{—CH=CH—CHO} \xrightarrow[\text{2. } H^+,\ H_2O]{\text{1. } LiAlH_4,\ (CH_3CH_2)_2O,\ -10^\circ C} C_6H_5\text{—CH=CH—CH}_2\text{OH}$$

90%

Because the nucleophilic reagents illustrated in Table 17-3 are very strong bases, their additions are irreversible. This and the following sections consider ionic additions of less basic nucleophiles Nu–H, such as water, alcohols, thiols, and amines. These processes are not strongly exothermic but instead establish equilibria that can be pushed in either direction by the appropriate choice of reaction conditions.

What is the mechanism of the ionic addition of these milder reagents to the carbon–oxygen double bond? Two pathways can be formulated: nucleophilic addition–protonation and electrophilic protonation–addition. The first, which begins with nucleophilic attack, occurs under neutral or basic conditions. As the nucleophile approaches the electrophilic carbon, the latter rehybridizes, and the electron pair of the π bond moves over to the oxygen, thereby producing an alkoxide ion. Subsequent protonation, usually from a protic solvent such as water or alcohol, yields the final addition product.

Nucleophilic Addition–Protonation

$$\overset{\delta^+}{C}=\overset{\delta^-}{\ddot{O}}: + Nu:^- \rightleftharpoons Nu\text{—C—}\ddot{O}:^- \underset{-\ H_2\ddot{O}}{\overset{H\ddot{O}H}{\rightleftharpoons}} Nu\text{—C—}\ddot{O}H + H\ddot{O}:^-$$

Alkoxide ion

Note that the new Nu–C bond is made up entirely of the electron pair of the nucleophile. The entire transformation is reminiscent of an S_N2 reaction. In that process a leaving group is displaced. Here it is an electron pair that is displaced from a shared position between carbon and oxygen to one solely on the oxygen atom. Additions of strongly basic nucleophiles to carbonyl groups typically follow the nucleophilic addition–protonation pathway.

The second mechanism predominates under acidic conditions and begins with electrophilic attack. In the first step, protonation of the carbonyl oxygen occurs, facilitated by the polarization of the C=O bond and the presence of lone electron pairs on the oxygen atom. The latter is only weakly basic, as shown by the pK_a of the conjugate acid, which ranges from about -7 to -8. Thus, in a dilute acidic medium, in which many carbonyl addition reactions are carried out, most of the carbonyl compound will stay unprotonated. However, the small amount of protonated material behaves like a very reactive carbon electrophile. Nucleophilic attack by the nucleophile completes the addition process and shifts the first, unfavorable equilibrium.

Electrophilic Protonation–Addition

$$\overset{\delta^+}{>}C=\overset{\delta^-}{\ddot{O}} + H^+ \rightleftharpoons \left[>\overset{+}{C}-\ddot{O}H \longleftrightarrow >C=\overset{+}{O}H \right] \underset{-:Nu}{\overset{:Nu^-}{\rightleftharpoons}} Nu-C-\ddot{O}H$$

Protonated carbonyl group

$pK_a \sim -8$

The electrophilic protonation–addition mechanism is best suited for reactions of relatively weakly basic nucleophiles. The acidic conditions are incompatible with strongly basic nucleophiles, because protonation of the nucleophile occurs.

In summary, there are three regions of reactivity in aldehydes and ketones. The first two are the two atoms of the carbonyl group and are the subject of the remainder of this chapter. The third is the adjacent carbon. The reactivity of the carbonyl group is governed by addition processes. Catalytic hydrogenation gives alcohols. Ionic additions of NuH (Nu = OH, OR, SR, NR_2) are reversible and may begin with nucleophilic attack at the carbonyl carbon, followed by electrophilic trapping of the alkoxide anion so generated. Alternatively, in acidic media, protonation precedes the addition of the nucleophile.

17-6 Addition of Water to Form Hydrates

The next three sections introduce the reactions of aldehydes and ketones with water and alcohols. These compounds attack the carbonyl group through both mechanisms just outlined, involving both acid and base catalysis.

Water hydrates the carbonyl group

Water is capable of attacking the carbonyl function of aldehydes and ketones. The transformation can be catalyzed by either acid or base and leads to equilibration with the corresponding **geminal diols,** $RC(OH)_2R'$, also called **carbonyl hydrates.**

Hydration of the Carbonyl Group

$$>C=O + HOH \underset{K}{\overset{H^+ \text{ or } HO^-}{\rightleftharpoons}} HO-C-OH$$

Geminal diol

In the base-catalyzed mechanism, the hydroxide functions as the nucleophile. Water then protonates the intermediate adduct, a hydroxy alkoxide, to give the product diol and to regenerate the catalyst.

Mechanism of Base-Catalyzed Hydration

$$\text{C=}\ddot{\text{O}} + {}^{-}\text{:}\ddot{\text{O}}\text{H} \rightleftharpoons \text{HO--C--}\ddot{\text{O}}\text{:}^{-} \xrightleftharpoons{\text{H}\ddot{\text{O}}\text{H}} \text{HO--C--}\ddot{\text{O}}\text{H} + \text{H}\ddot{\text{O}}\text{:}^{-}$$

Hydroxy alkoxide **Geminal diol**

In the acid-catalyzed mechanism, the sequence of events is reversed. Here, initial protonation facilitates nucleophilic attack by the weak nucleophile water. Subsequently, the catalytic proton is lost to reenter the catalytic cycle.

Mechanism of Acid-Catalyzed Hydration

$$\text{C=}\ddot{\text{O}} + \text{H}^{+} \rightleftharpoons \text{C=}\ddot{\text{O}}^{+}\text{--H} \xrightleftharpoons{\text{H}_2\ddot{\text{O}}} \text{H}_2\overset{+}{\text{O}}\text{--C--}\ddot{\text{O}}\text{H} \rightleftharpoons \text{HO--C--}\ddot{\text{O}}\text{H} + \text{H}^{+}$$

Protonated carbonyl **Geminal diol**

Hydration is reversible

$$\left[\text{C=}\ddot{\text{O}} \longleftrightarrow \overset{+}{\text{C}}\text{--}\ddot{\text{O}}\text{:}^{-}\right]$$

Electrophilic

As these equations indicate, hydrations of aldehydes and ketones are reversible. The equilibrium lies to the left for ketones and to the right for formaldehyde and aldehydes bearing electron-withdrawing substituents. For ordinary aldehydes, the equilibrium constant approaches unity.

How can these trends be explained? Substituents that make the carbonyl carbon more electrophilic favor hydration. Look again at the resonance structures of the carbonyl group, as described in Section 17-2 and repeated here in the margin. In the dipolar resonance form, the carbon possesses carbocation character. Substituents that stabilize carbocations reduce the electrophilic character of the carbonyl group. Therefore *the more alkylated carbonyl carbon is the more stable,* and the favorability of hydration thus *decreases* along the series formaldehyde, acetaldehyde, propanone. Conversely, *electron-withdrawing substituents destabilize the positively polarized carbon,* favoring the addition of nucleophiles. We shall see that the same trends govern other addition reactions.

Equilibrium Constants K for the Hydration of Typical Carbonyl Compounds

Cl_3CCH (C=O)	$K > 10^4$
$H_2C{=}O$	$K > 10^3$
CH_3CH (C=O)	$K \sim 1$
CH_3CCH_3 (C=O)	$K < 10^{-2}$

Relative Reactivities of Carbonyl Groups

$$Cl_3C\overset{\overset{O^{\delta-}}{\|}}{C^{\delta+}}H > H\overset{\overset{O^{\delta-}}{\|}}{C^{\delta+}}H > H_3C\overset{\overset{O^{\delta-}}{\|}}{C^{\delta+}}H > H_3C\overset{\overset{O^{\delta-}}{\|}}{C^{\delta+}}CH_3$$

Electron-withdrawing CCl_3 group generates additional positive charge at carbonyl carbon

Electron-donating CH_3 groups reduce positive charge at carbonyl carbon

EXERCISE 17-4

(a) Rank in order of increasing favorability of hydration: $Cl_3C\overset{O}{\overset{\|}{C}}H$, $Cl_3C\overset{O}{\overset{\|}{C}}CH_3$, $Cl_3C\overset{O}{\overset{\|}{C}}CCl_3$. **(b)** Treatment of propanone (acetone) with $H_2{}^{18}O$ results in the formation of labeled propanone, $CH_3\overset{^{18}O}{\overset{\|}{C}}CH_3$. Explain.

In summary, the carbonyl group of the aldehydes and ketones is hydrated by water. Aldehydes are more reactive than ketones. Electron-withdrawing substituents render the carbonyl carbon more electrophilic. Hydration is an equilibrium process that may be catalyzed by both acids and bases.

17-7 Addition of Alcohols to Form Hemiacetals and Acetals

In this section we shall see that alcohols add to carbonyl groups in much the same manner as water does. Both acids and bases catalyze the process, but acids catalyze the further transformation of the initial addition product to acetals by replacement of the hydroxy group with an alkoxy substituent.

Aldehydes and ketones form hemiacetals reversibly

Not surprisingly, alcohols also undergo addition to aldehydes and ketones, by a mechanism virtually identical with that outlined for water. The adducts formed are called **hemiacetals** (*hemi,* Greek, half), because they are intermediates on the way to acetals.

Hemiacetal Formation

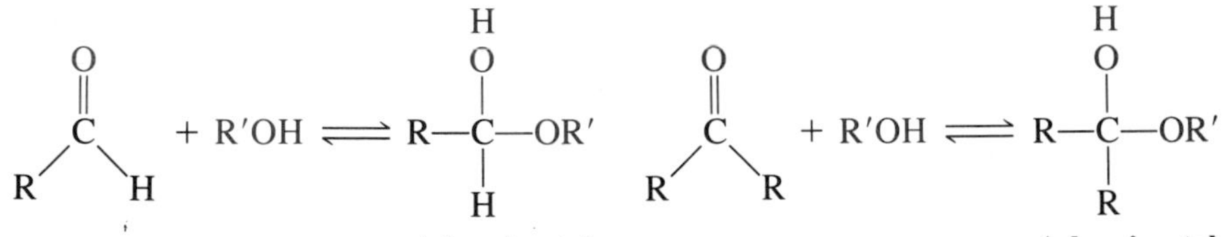

$$RCHO + R'OH \rightleftharpoons R\text{—}C(OH)(H)\text{—}OR'$$

A hemiacetal

$$RCOR + R'OH \rightleftharpoons R\text{—}C(OH)(R)\text{—}OR'$$

A hemiacetal

Like hydration, these addition reactions are governed by equilibria that usually favor the starting carbonyl compound. *Hemiacetals, like hydrates, are therefore usually not isolable.* Exceptions are those formed from reactive carbonyl compounds such as formaldehyde or 2,2,2-trichloroacetaldehyde. Hemiacetals are also isolable from hydroxy aldehydes and ketones when cyclization leads to the formation of relatively strain-free five- and six-membered rings.

Intramolecular Hemiacetal Formation

$$HOCH_2CH_2CH_2CH_2CHO \xrightarrow{H^+} \text{cyclic hemiacetal}$$

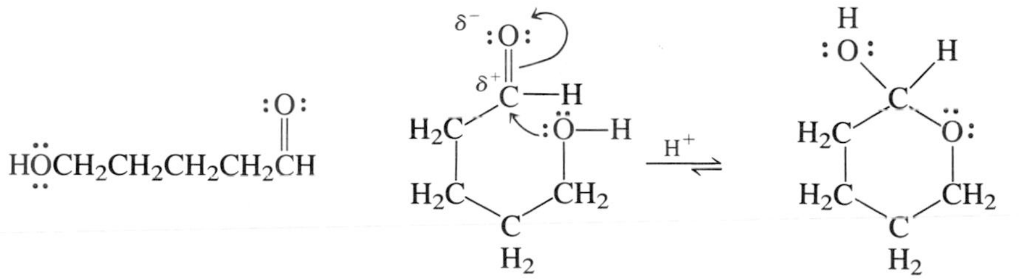

5-Hydroxypentanal — **A cyclic hemiacetal, stable**

Intramolecular hemiacetal formation is common in sugar chemistry (Chapter 24). For example, glucose, the most common simple sugar in nature, exists as an equilibrium mixture of an acyclic pentahydroxyaldehyde and two stereoisomeric cyclic hemiacetals. The latter forms constitute more than 99% of the mixture in aqueous solution.

Glucose

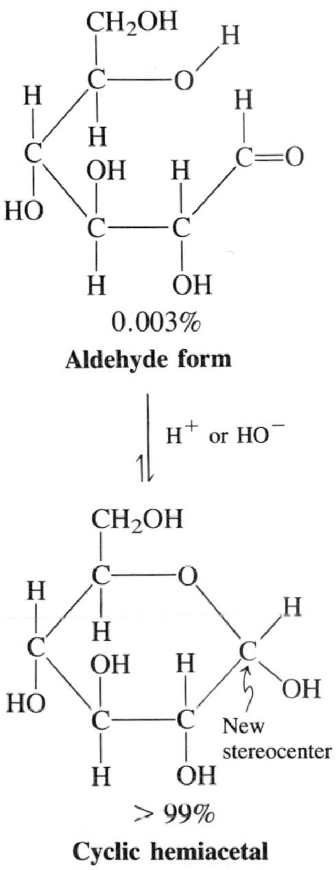

0.003%
Aldehyde form

H^+ or HO^-

New stereocenter

> 99%
Cyclic hemiacetal
(Two stereoisomers)

Acids catalyze acetal formation

In the presence of excess alcohol, the *acid*-catalyzed reaction of aldehydes and ketones proceeds beyond the hemiacetal stage. Under these conditions, the hydroxy function of the initial adduct is replaced by another alkoxy unit derived from the alcohol. The resulting compounds are called **acetals.** (**Ketal** is an older term for an acetal derived from a ketone.)

Acetal Synthesis

$$\mathrm{RC(=O)R} + 2\,\mathrm{R'OH} \underset{}{\overset{\mathrm{H^+}}{\rightleftharpoons}} \mathrm{R{-}C(OR')(R){-}OR'} + \mathrm{H_2O}$$

An acetal

The net change is the replacement of the carbonyl oxygen by two alkoxy groups and the formation of one equivalent of water.

Let us examine the mechanism of this transformation for an aldehyde. The initial reaction is the ordinary acid-catalyzed addition of the first molecule of alcohol. The resulting hemiacetal can be protonated at the hydroxy group, changing this substituent into water, a good leaving group. On loss of water, the resulting carbocation is stabilized by resonance with a lone electron pair on oxygen. A second molecule of alcohol now adds to the electrophilic carbon, initially giving the protonated acetal, which is then deprotonated to the final product.

Mechanism of Acetal Formation

STEP 1. Hemiacetal generation

$$\mathrm{RCHO} \underset{-\mathrm{H^+}}{\overset{+\mathrm{H^+}}{\rightleftharpoons}} \left[\mathrm{RCH{=}\overset{+}{O}H} \longleftrightarrow \mathrm{R\overset{+}{C}H{-}OH}\right] \underset{-\mathrm{R'OH}}{\overset{\mathrm{R'OH}}{\rightleftharpoons}} \mathrm{R{-}CH(OH){-}\overset{+}{O}(H)R'} \underset{+\mathrm{H^+}}{\overset{-\mathrm{H^+}}{\rightleftharpoons}} \mathrm{R{-}CH(OH){-}OR'}$$

Hemiacetal

STEP 2. Acetal generation

$$\mathrm{R{-}CH(OH){-}OR'} \underset{-\mathrm{H^+}}{\overset{+\mathrm{H^+}}{\rightleftharpoons}} \mathrm{R{-}CH(\overset{+}{O}H_2){-}OR'} \underset{+\mathrm{H_2O}}{\overset{-\mathrm{H_2O}}{\rightleftharpoons}} \left[\mathrm{R\overset{+}{C}H{-}OR'} \longleftrightarrow \mathrm{RCH{=}\overset{+}{O}R'}\right] \underset{-\mathrm{R'OH}}{\overset{\mathrm{R'OH}}{\rightleftharpoons}}$$

$$\mathrm{R{-}CH(OR'){-}\overset{+}{O}(H)R'} \underset{+\mathrm{H^+}}{\overset{-\mathrm{H^+}}{\rightleftharpoons}} \mathrm{R{-}CH(OR'){-}OR'}$$

Acetal

Each step is reversible; the entire sequence, starting from the carbonyl compound and ending with the acetal, is an equilibrium process. The overall equilibrium may be shifted in either direction by manipulating the reaction conditions: toward acetal, by using excess alcohol or by continually removing water from the reaction medium; toward aldehyde or ketone by using excess water. The latter process is called **acetal hydrolysis.**

In summary, the reaction of alcohols with aldehydes and ketones results in hemiacetal formation. This process, like hydration, is reversible and is catalyzed by both acids and bases. Hemiacetals are converted into relatively stable acetals by acid and excess alcohol.

17-8 Acetals as Protecting Groups

Is it possible to carry out a reaction exclusively at one electrophilic site in a molecule when several are present? For example, is there a way to selectively transform a carbon–halogen bond in the presence of an aldehyde or ketone function? This section will describe strategies for *masking* or *protecting* carbonyl groups for this purpose.

Cyclic acetals protect carbonyl groups from attack by nucleophiles

1,2-Ethanediol and related diols react with aldehydes and ketones in the presence of catalytic acid to form cyclic acetals. The products readily revert to aldehydes and ketones in the presence of excess acidic water, yet they are also relatively inert. Thus, they are not attacked by many basic, organometallic, and hydride reagents. These properties make cyclic acetals useful as a kind of masked aldehyde or ketone; the acetal acts as a **protecting group** for the carbonyl function.

An example is the alkylation of an alkynyl anion with 3-iodopropanal 1,2-ethanediol acetal.

$CH_3\overset{O}{\overset{\|}{C}}H + HOCH_2CH_2OH \xrightarrow[-\,H_2O]{H^+}$ cyclic acetal (H_2C—CH_2 ring with two O, C bearing H_3C and H)

73%

A cyclic acetal

Use of a Protected Aldehyde in Synthesis

$CH_3(CH_2)_3C{\equiv}C^-Li^+ + ICH_2CH_2{-}C(H)(OCH_2CH_2O) \xrightarrow{-\,LiI}$

1-Hexynyllithium **3-Iodopropanal 1,2-ethanediol acetal**

$CH_3(CH_2)_3{-}{\equiv}{-}$ (1,3-dioxolane) $\xrightarrow[-\,HOCH_2CH_2OH]{H^+,\,H_2O}$ $CH_3(CH_2)_3{-}{\equiv}{-}$CHO

Deprotection

70% **4-Nonynal 1,2-ethanediol acetal**

90% **4-Nonynal**

If the same alkylation is attempted with unprotected 3-iodopropanal, the alkynyl anion attacks the carbonyl group.

EXERCISE 17-5

Suggest a convenient way of converting compound A into B.

$$\underset{\textbf{A}}{\mathrm{CH_3\overset{\overset{\displaystyle O}{\|}}{C}(CH_2)_4Br}} \longrightarrow \underset{\textbf{B}}{\mathrm{CH_3\overset{\overset{\displaystyle O}{\|}}{C}(CH_2)_4CH_2OH}}$$

If a diol may be used to protect a carbonyl function, then a carbonyl compound should be able to protect a diol. This is indeed the case. For example, propanone (acetone) blocks the acidic sites in vicinal diols by forming an acetal with them. The protection of diols as propanone (acetone) acetals is an important reaction in sugar chemistry (Chapter 24).

Protection of a Vicinal Diol as the Propanone (Acetone) Acetal

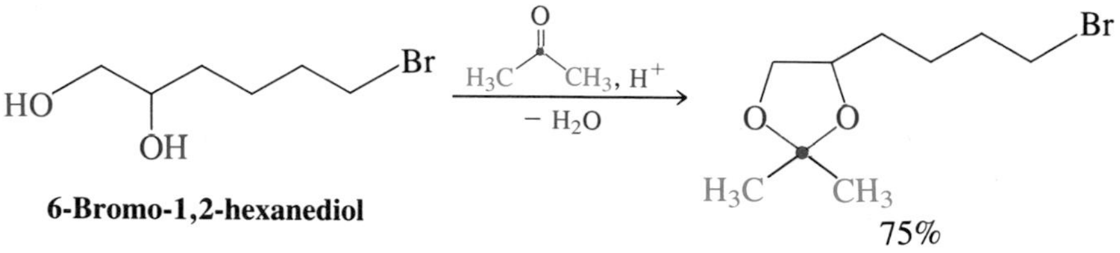

6-Bromo-1,2-hexanediol

Thiols react with the carbonyl group to form thioacetals

Thiols, the sulfur analogs of alcohols (see Section 9-10), react with aldehydes and ketones by a mechanism identical with the one described for alcohols. Instead of a proton catalyst, a Lewis acid, such as BF_3 or $ZnCl_2$, is often used. The reaction produces the sulfur analogs of acetals, **thioacetals.**

Cyclic Thioacetal Formation from a Ketone

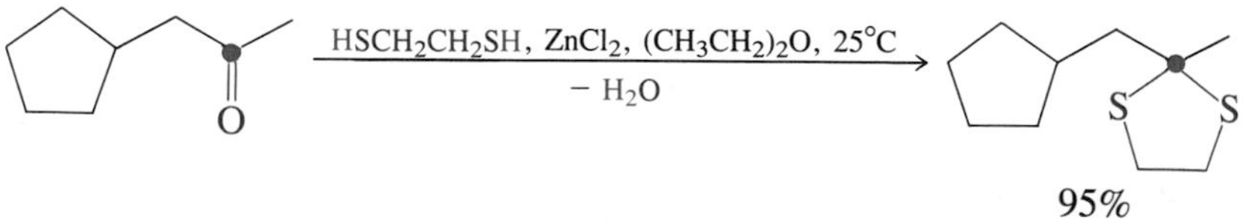

95%

A cyclic thioacetal

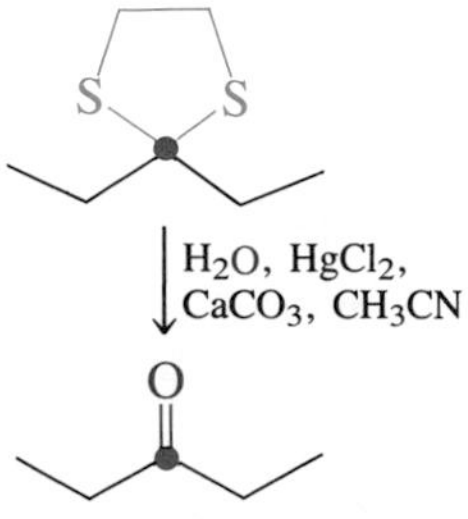

These sulfur derivatives are stable in aqueous acid, a medium that hydrolyzes ordinary acetals. The difference in reactivity may be useful in synthesis if it is necessary to differentiate two different carbonyl groups in the same molecule. Hydrolysis of thioacetals is carried out by using mercuric chloride in aqueous acetonitrile. The driving force is the formation of insoluble mercuric sulfides.

Thioacetals are **desulfurized** to the corresponding hydrocarbon by treatment with Raney nickel (Section 12-2). Thioacetal generation followed by desulfurization is used to convert a carbonyl into a methylene group.

BOX 17-1 Protecting Groups in Vitamin C Synthesis

HO H OH H 1 CH_2OH $HOCH_2$ H OH O

Sorbose

$\xrightarrow{H^+,\ 2\ CH_3CCH_3}$ (O above C)

H O CH3 O 2 H O H2C 4 3 CH2OH 6 O O H 1 H3C CH3 CH3 + 2 H2O

Virtually all vitamin C sold is synthetic material. At an early stage of the synthesis, the primary alcohol group at C1 of the sugar sorbose must be selectively oxidized to a carboxylic acid function, without affecting the OH groups at carbons 3 through 6. Sorbose, like glucose (Section 17-7), exists mostly as a cyclic hemiacetal. Upon acid-catalyzed reaction with excess propanone, the hydroxy groups at C2 and C3 are blocked as a five-membered cyclic acetal and those at C4 and C6 are protected as a six-membered acetal ring. Potassium permanganate then converts the unprotected CH_2OH into CO_2H.

OH HO C C C O H—C—O HO—C—H CH_2OH

Vitamin C

S S $\xrightarrow{\text{Raney Ni, } H_2}$ H H

65%

EXERCISE 17-6

Suggest a possible synthesis of cyclodecane from [bicyclic alkene structure].

In summary, cyclic acetals are good protecting groups for the carbonyl function and for diols. Thiols show similar reactivity. The formation of thioacetals is usually catalyzed by Lewis acids. Their hydrolysis requires the presence of mercuric salts. Thioacetals are reduced by Raney nickel to hydrocarbons.

17-9 Nucleophilic Addition of Ammonia and Its Derivatives

Ammonia and the amines may be regarded as nitrogen analogs of water and alcohols. We might therefore expect them to add just as effectively to aldehydes and ketones; and in fact they do, giving products corresponding to those we have

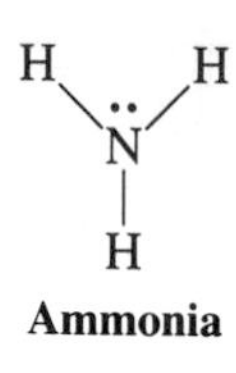

Ammonia

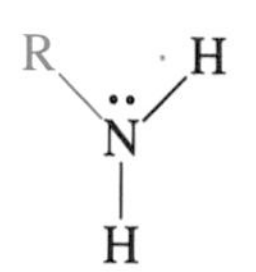

Primary amine

just studied. We shall see one important difference, however. The products of addition of amines and their derivatives lose water, furnishing either of two new derivatives of the original carbonyl compounds, imines and enamines.

Ammonia and primary amines form imines

On exposure to an amine, aldehydes and ketones form **hemiaminals,** the nitrogen analogs of hemiacetals. Hemiaminals of primary amines readily lose water to form a carbon–nitrogen double bond. This function is called an **imine** (an older name is **Schiff* base**) and is the nitrogen analog of a carbonyl group.

Imine Formation from Amines and Aldehydes or Ketones

R—NH₂ + C=O —Addition⇌ [H, H, R–N⁺–C–O⁻] —Deprontonation at N, reprotonation at O⇌ [R, H–N–C–OH] A hemiaminal —(− H_2O / + H_2O, Elimination of water)⇌ C=N–R An imine

The mechanism of the elimination of water from the hemiaminal is the same as that for the decomposition of a hemiacetal to the carbonyl compound and alcohol. It begins with protonation of the hydroxy group. (Protonation of the more basic nitrogen just leads back to the carbonyl compound.) Dehydration follows, and then deprotonation of the intermediate **iminium ion.**

Mechanism of Hemiaminal Dehydration

Hemiaminal (R, H–N–C–OH) —($+ H^+$ / $- H^+$)⇌ (R, H–N–C–$\overset{+}{O}H_2$) —(− HOH / + HOH)⇌ [R, H–N–C⁺ ⟷ R, H–N⁺=C] Iminium ion —($- H^+$ / $+ H^+$)⇌ R–N=C Imine

Processes such as imine formation from a primary amine and an aldehyde or ketone, in which two molecules are joined with the elimination of water, are called **condensations.** Imine formation is reversible, and the usual measures have to be employed to shift the equilibrium in the desired direction.

Condensation of a Ketone with a Primary Amine

$$RNH_2 + O{=}C(R')(R'') \rightleftharpoons RN{=}C(R')(R'') + H_2O$$

---

*Professor Hugo Schiff, 1834–1915, University of Florence, Italy.

Examples:

$$CH_3\overset{O}{\overset{\|}{C}}H + H_2NCH_2CH_2CH_2CH_3 \xrightarrow{KOH} \underset{83\%}{CH_3CH{=}NCH_2CH_2CH_2CH_3} + H_2O$$

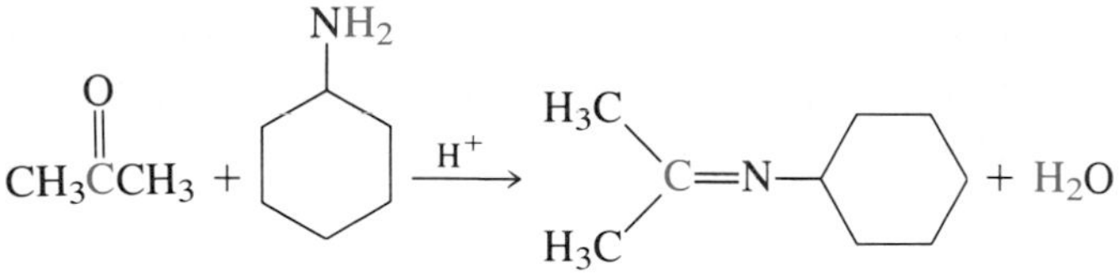

$$CH_3\overset{O}{\overset{\|}{C}}CH_3 + \text{(cyclohexylamine, } C_6H_{11}NH_2\text{)} \xrightarrow{H^+} \underset{95\%}{(H_3C)_2C{=}N{-}C_6H_{11}} + H_2O$$

EXERCISE 17-7

Reagent A has been used with aldehydes to prepare crystalline imidazolidine derivatives such as B, for the purpose of their isolation and structural identification. Write a mechanism for the formation of B.

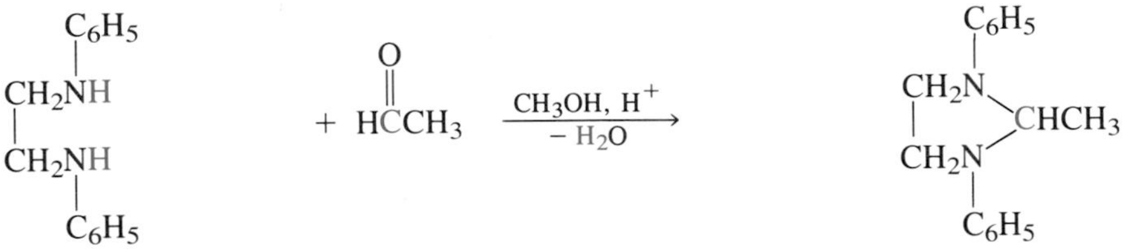

$$\underset{\mathbf{A}}{C_6H_5NHCH_2CH_2NHC_6H_5} + H\overset{O}{\overset{\|}{C}}CH_3 \xrightarrow[-\,H_2O]{CH_3OH,\ H^+} \underset{\mathbf{B}}{\text{(cyclic } CH_2N(C_6H_5){-}CHCH_3{-}N(C_6H_5)CH_2\text{)}}$$

A
***N,N'*-Diphenyl-1,2-ethanediamine**

B
2-Methyl-1,3-diphenyl-1,3-diazacyclopentane
(2-Methyl-1,3-diphenylimidazolidine)
(m.p. 102°C)

Special imines aid in identification of aldehydes and ketones

Several amine derivatives condense with aldehydes and ketones to form imine products that are highly crystalline and often have sharp melting points. For example, *hydroxylamine,* H_2NOH, in the form of its hydrochloride, condenses with aldehydes to **oximes.**

A Typical Oxime Synthesis

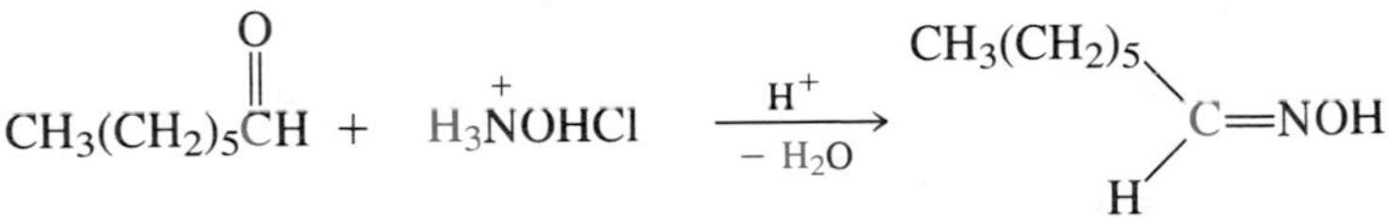

$$CH_3(CH_2)_5\overset{O}{\overset{\|}{C}}H + H_3\overset{+}{N}OHCl^- \xrightarrow[-\,H_2O]{H^+} \underset{93\%}{\overset{CH_3(CH_2)_5}{\underset{H}{}}\!\!\!C{=}NOH}$$

Hydroxylamine hydrochloride

Heptanal oxime
(m.p. 57°C)

Derivatives of *hydrazine,* H_2NNH_2—in particular, *phenylhydrazine,* $C_6H_5NHNH_2$, and 2,4-*dinitrophenylhydrazine,* 2,4-$(NO_2)_2C_6H_3NHNH_2$—condense to the corresponding **hydrazones.**

BOX 17-2 Biological Transformations Involving Imines

The related molecules pyridoxal and pyridoxamine assist in the interconversion of carbonyl groups and primary amine functions in biology. Pyridoxamine undergoes enzyme-catalyzed condensation with the carbonyl group of 2-oxocarboxylic acids to produce an imine. Rearrangement furnishes a new imine, which hydrolyzes to pyridoxal and a 2-aminocarboxylic acid (an *amino acid*). In the forward direction this process synthesizes several of the naturally occurring amino acids; in the reverse mode it aids in their metabolism.

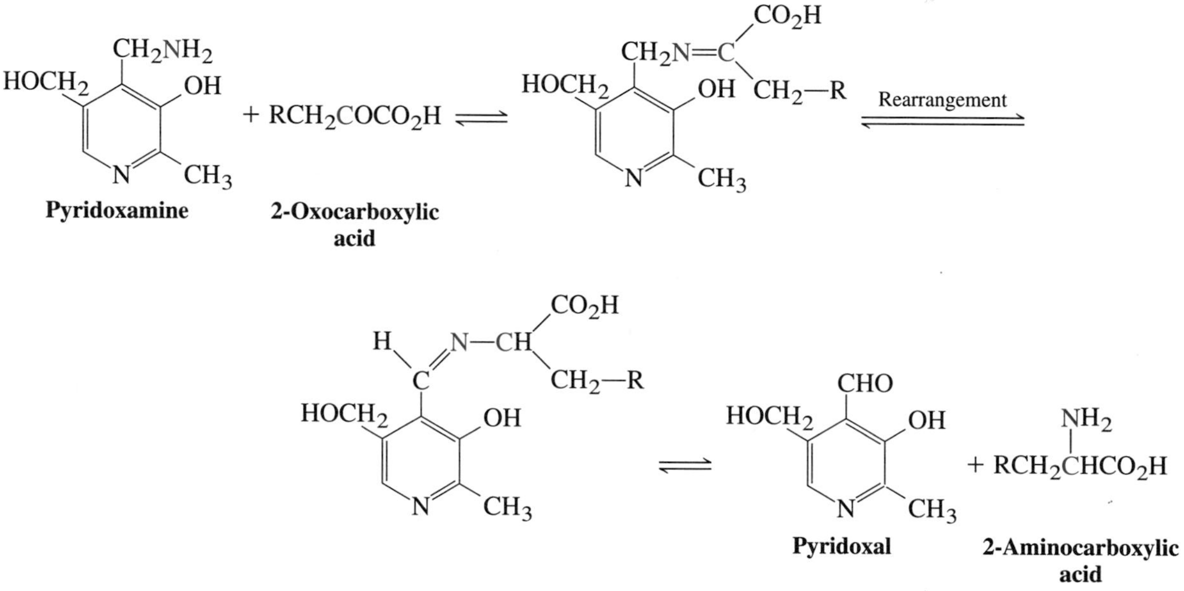

CH_2NH_2 HOCH$_2$ OH N CH$_3$ + RCH_2COCO_2H ⇌ CO_2H $CH_2N{=}C$ HOCH$_2$ OH $CH_2{-}R$ N CH$_3$ Rearrangement ⇌

Pyridoxamine **2-Oxocarboxylic acid**

CO_2H H N—CH C $CH_2{-}R$ HOCH$_2$ OH N CH$_3$ ⇌ CHO HOCH$_2$ OH N CH$_3$ + $RCH_2CH(NH_2)CO_2H$

Pyridoxal **2-Aminocarboxylic acid**

Formation of a 2,4-Dinitrophenylhydrazone from a Ketone

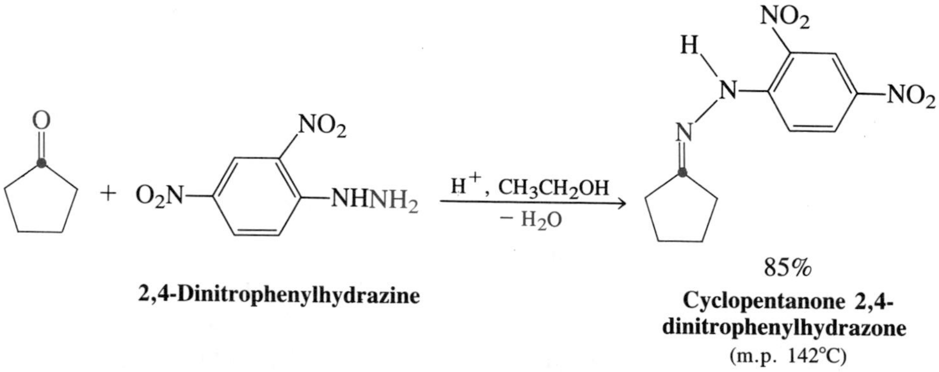

Cyclopentanone + O_2N–$C_6H_3(NO_2)$–$NHNH_2$ $\xrightarrow[-\ H_2O]{H^+,\ CH_3CH_2OH}$ cyclopentylidene=N–NH–$C_6H_3(NO_2)_2$

2,4-Dinitrophenylhydrazine

85%
Cyclopentanone 2,4-dinitrophenylhydrazone
(m.p. 142°C)

Finally, the hydrazine derivative *semicarbazide*, reacts with aldehydes and ketones to give **semicarbazones.**

A Semicarbazone Synthesis

$$\text{C}_6\text{H}_{11}\text{CHO} + \text{H}_2\text{N—NHCNH}_2 \xrightarrow[-\,\text{H}_2\text{O}]{\text{H}^+,\ \text{CH}_3\text{CH}_2\text{OH}} \text{C}_6\text{H}_{11}\text{CH=N—NHCNH}_2$$

Semicarbazide

90%

Cyclohexanecarbaldehyde semicarbazone

Melting points have been tabulated for imine derivatives of thousands of aldehydes and ketones. Before spectroscopic methods became routinely available, chemists used comparisons of these melting points to determine whether an aldehyde or ketone of unknown structure was identical to a previously characterized compound. For example, the distillation of castor oil yields a compound with the formula $C_7H_{14}O$. Its boiling point of about 150°C is similar to those recorded for heptanal, 2-heptanone, and 4-heptanone. The unknown is readily identified as heptanal by preparing and determining the melting points of its semicarbazone (109°C) and 2,4-dinitrophenylhydrazone (108°C). These values match those tabulated for the derivatives of the aldehyde and are quite different from those recorded for the two ketones: The respective derivatives of 2-heptanone exhibit melting points of 127 and 89°C, and those of 4-heptanone, 133 and 75°C.

Condensations with secondary amines give enamines

The condensations of amines described so far are possible only for primary derivatives, because the nitrogen of the amine has to supply both of the protons necessary to form water. Reaction with a secondary amine therefore takes a different course. After the initial addition, water is eliminated by deprotonation at *carbon* to produce an **enamine.** This functional group incorporates both the *ene* function of an alkene and the *amino* group of an amine.

$$\text{RR'NH}$$

Secondary amine

Enamine Formation

$$\text{CH}_3\text{CH}_2\overset{\text{O}}{\overset{\|}{\text{C}}}\text{CH}_2\text{CH}_3 + \text{pyrrolidine} \rightleftharpoons \text{CH}_3\text{CH}_2\text{—}\overset{\text{OH}}{\overset{|}{\underset{\text{CH}_2\text{CH}_3}{\underset{|}{\text{C}}}}}\text{—N(pyrrolidinyl)} \underset{+\,\text{H}_2\text{O}}{\overset{-\,\text{HOH}}{\rightleftharpoons}} \text{CH}_3\text{CH=C(N-pyrrolidinyl)CH}_2\text{CH}_3$$

90%

An enamine

Enamine formation is reversible, and hydrolysis occurs readily in the presence of acidic water. Enamines are useful substrates in alkylations (Section 18-4).

EXERCISE 17-8

Write the products of the following reactions.

(a) 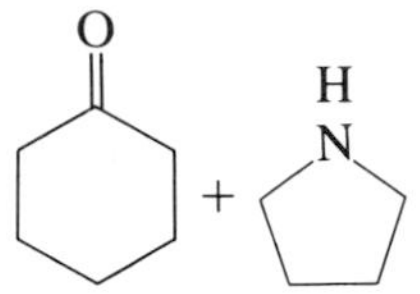 cyclohexanone + pyrrolidine (b) $CH_3CH(NH_2)CH(NH_2)CH_3 + CH_3CH_2\overset{O}{\overset{\|}{C}}-\overset{O}{\overset{\|}{C}}CH_3$

(c) 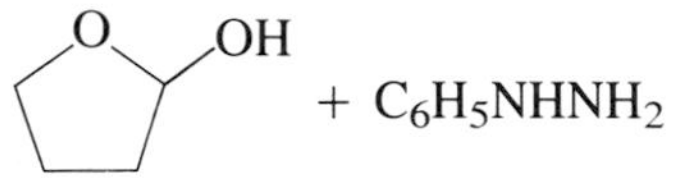 2-hydroxytetrahydrofuran $+ C_6H_5NHNH_2$

(Hint: See Section 17-7)

In summary, amines attack aldehydes and ketones to form imines by condensation. Hydroxylamine gives oximes, hydrazines lead to hydrazones, and semicarbazide results in semicarbazones. Secondary amines react with aldehydes and ketones to give enamines.

17-10 Deoxygenation of the Carbonyl Group

Synthesis of a Hydrazone

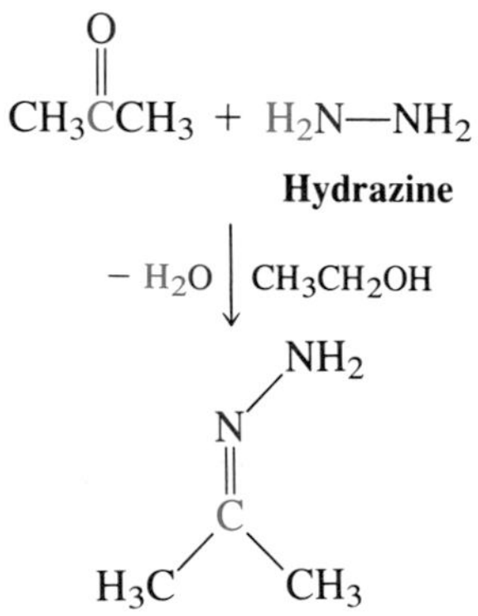

$$CH_3\overset{O}{\overset{\|}{C}}CH_3 + \underset{\textbf{Hydrazine}}{H_2N{-}NH_2} \xrightarrow[-\,H_2O]{CH_3CH_2OH} (CH_3)_2C{=}N{-}NH_2$$

Propanone (acetone) hydrazone

In Section 17-5 we reviewed methods by which carbonyl compounds are reduced to alcohols. Reduction of the C=O group to CH_2 **(deoxygenation)** is also possible. Two ways this may be achieved are Clemmensen reduction (Section 16-5) and thioacetal formation followed by desulfurization (Section 17-8). This section presents a third method for deoxygenating aldehydes and ketones—the Wolff-Kishner reduction.

Strong base converts simple hydrazones to hydrocarbons

Condensation of hydrazine itself with aldehydes and ketones produces simple hydrazones. These derivatives undergo decomposition with evolution of nitrogen when treated with base at elevated temperatures. The product of this reaction, called the **Wolff-Kishner* reduction,** is the corresponding hydrocarbon.

Wolff-Kishner Reduction

$$R\overset{N-NH_2}{\overset{\|}{C}}R' + NaOH \xrightarrow{(HOCH_2CH_2)_2O,\ 180\text{–}200°C} RCH_2R' + N_2$$

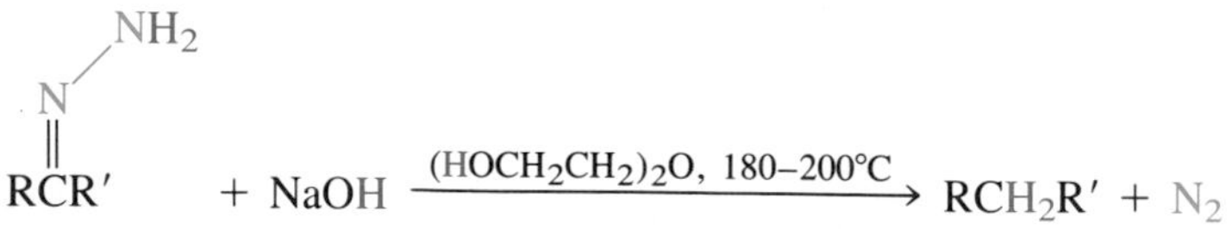

The mechanism of nitrogen elimination includes a sequence of base-mediated hydrogen shifts. The base first removes a proton from the hydrazone to give the corresponding delocalized anion. Reprotonation may occur on nitrogen, regenerating the starting material, or on carbon, thereby furnishing the product. The base

*Professor Ludwig Wolff, 1857–1919, University of Jena, Germany; Professor N. M. Kishner, 1867–1935, University of Moscow.

removes another proton from nitrogen on the new intermediate to generate a new anion, which rapidly decomposes irreversibly, with the extrusion of nitrogen gas. The alkyl anion so formed is rapidly protonated to the hydrocarbon.

Mechanism of Nitrogen Elimination in the Wolff-Kishner Reduction

$R_2C{=}N{-}\ddot{N}H_2 + HO^- \underset{+H_2O}{\overset{-HOH}{\rightleftharpoons}}$ (Deprotonation) $\left[R_2C{=}N{-}\ddot{N}H^- \longleftrightarrow R_2\ddot{C}^-{-}N{=}NH\right] \underset{}{\overset{HOH}{\rightleftharpoons}}$ (Reprotonation)

$RCHR'{-}N{=}NH + HO^- \underset{+H_2O}{\overset{-HOH}{\rightleftharpoons}}$ (Deprotonation) $RCHR'{-}N{=}N{:}^- \xrightarrow{\text{Irreversible loss of } N_2} R\ddot{C}^-HR' \xrightarrow{H_2O,\ \text{protonation}} RCH_2R' + HO^-$

In practice, the Wolff-Kishner reduction is carried out without isolating the intermediate hydrazone. An 85% aqueous solution of hydrazine (hydrazine hydrate) is added to the carbonyl compound in the high-boiling alcohol $HOCH_2CH_2OCH_2CH_2OH$ (b.p. 245°C), containing sodium hydroxide, and the mixture is heated. Aqueous work-up yields the hydrocarbon. For example,

$\xrightarrow[\text{2. } H_2O]{\text{1. } H_2NNH_2,\ H_2O,\ (HOCH_2CH_2)_2O,\ NaOH,\ \Delta}$

CH3 (diketone) → CH3 (hydrocarbon), 69%

Wolff-Kishner reduction complements the Clemmensen and thioacetal desulfurization methods of deoxygenating aldehydes and ketones. Thus, the Clemmensen reduction is unsuitable for compounds containing acid-sensitive groups, and hydrogenation of multiple bonds can accompany desulfurization with hydrogen and Raney nickel. Such functional groups are generally not affected by Wolff-Kishner conditions.

Wolff-Kishner reduction aids in alkylbenzene synthesis

We have already seen that the products of Friedel-Crafts alkanoylations may be converted to alkyl benzenes by using Clemmensen reduction. Wolff-Kishner deoxygenation is also frequently employed for this purpose and is particularly useful for acid-sensitive, base-stable substrates.

Wolff-Kishner Reduction of a Friedel-Crafts Alkanoylation Product

$$C_6H_5CCH_3(=O) \xrightarrow{NH_2NH_2,\ H_2O,\ KOH,\ HOCH_2CH_2OCH_2CH_2OCH_2CH_2OH,\ \Delta} C_6H_5CH_2CH_3$$

95%

EXERCISE 17-9

Propose a synthesis of hexylbenzene from hexanoic acid.

In summary, the Wolff-Kishner reduction is the decomposition of a hydrazone by base, the second part of a method of deoxygenating aldehydes and ketones. It complements Clemmensen and thioacetal desulfurization procedures.

17-11 Addition of Hydrogen Cyanide to Give Cyanohydrins

Besides alcohols and amines, several other nucleophilic reagents may attack the carbonyl group. Particularly important are carbon nucleophiles, because new carbon–carbon bonds can be made in this way. Section 8-8 explained that organometallic compounds, such as Grignard and alkyllithium reagents, add to aldehydes and ketones to produce alcohols. In contrast with additions by alcohols and amines, these are normally *not* reversible. The next two sections deal with the behavior of carbon nucleophiles that are not organometallic reagents—the additions of cyanide ion and of a new class of compounds called ylides.

Hydrogen cyanide adds to the carbonyl group to form hydroxy alkanenitrile adducts commonly called **cyanohydrins.** This process is reversible because the negative charge on the cyanide ion is fairly well stabilized (Table 6-5). The equilibrium may be shifted toward the adduct by the use of liquid HCN as solvent. However, it is dangerous to use such large amounts of HCN, which is volatile and highly toxic. Alternatively, HCN may be generated in situ from a cyanide salt by the slow addition of an acid.

Cyanohydrin Formations

$$CH_3CH(=O) + HCN \rightleftharpoons CH_3C(OH)(H)CN$$

70%

2-Hydroxypropanenitrile (Acetaldehyde cyanohydrin)

$$\text{Cyclohexanone} + Na^{+}\ {}^{-}CN \xrightarrow[-\ NaCl]{\text{Conc. HCl}} \text{1-(HO)(CN)cyclohexane}$$

60%

1-Hydroxycyclohexanecarbonitrile (Cyclohexanone cyanohydrin)

The mechanism of cyanohydrin formation begins with nucleophilic attack by cyanide ion and ends with protonation at oxygen.

Mechanism of Cyanohydrin Formation

$$\rangle C{=}\ddot{O}: + {}^{-}:C{\equiv}N: \rightleftharpoons (NC)C{-}\ddot{O}:^{-} \underset{-H^+}{\overset{+H^+}{\rightleftharpoons}} (NC)C{-}\ddot{O}H$$

The reaction is readily reversed by the addition of base, which shifts the equilibrium to the free cyanide side by removing protons from the equation.

$$CH_3CH(OH)CN \rightleftharpoons CH_3CH{=}O + HCN \xrightarrow[\text{Shifts equilibrium}]{HO^-} CH_3CH{=}O + HOH + {}^{-}CN$$

We will see in subsequent chapters that cyanohydrins are useful intermediates because the nitrile group can be modified by further reaction.

In summary, the carbonyl group in aldehydes and ketones can be attacked by carbon-based nucleophiles. Organometallic reagents give alcohols and cyanide gives cyanohydrins.

17-12 Addition of Phosphorus Ylides: The Wittig Reaction

Another useful reagent in nucleophilic additions contains a carbanion that is stabilized by an adjacent, positively charged phosphorus group. Such a species is called a **phosphorus ylide,** and its attack on aldehydes and ketones is the **Wittig* reaction.** In the most commonly used ylides, the other substituents on phosphorus are phenyl groups.

$$RHC^{-}{-}\overset{+}{P}(C_6H_5)_3$$

Ylide

Deprotonation of phosphonium salts gives phosphorus ylides

Phosphorus ylides are most conveniently prepared from haloalkanes by a two-step sequence; the first step is the nucleophilic displacement of halide by triphenylphosphine to furnish an alkyltriphenylphosphonium salt.

Phosphonium Salt Synthesis

$$(C_6H_5)_3P: + RCH_2{-}\ddot{X}: \xrightarrow{C_6H_6} RCH_2\overset{+}{P}(C_6H_5)_3 + :\ddot{X}:^{-}$$

Triphenylphosphine — **An alkyltriphenylphosphonium halide**

*Professor Georg Wittig, 1897–1987, University of Heidelberg, Germany, Nobel Prize 1979 (chemistry).

The positively charged phosphorus atom renders any neighboring proton acidic. In the second step, deprotonation by bases, such as alkoxides, sodium hydride, or butyllithium, gives the ylide. Ylides can be isolated, but they are usually produced and used as reagents immediately.

Ylide Formation

$$RCH_2\overset{+}{P}(C_6H_5)_3X^- + CH_3CH_2CH_2CH_2Li \xrightarrow{THF} \left[R\ddot{\overset{-}{C}}H-\overset{+}{P}(C_6H_5)_3 \longleftrightarrow RCH{=}P(C_6H_5)_3 \right] + CH_3CH_2CH_2CH_2H + LiX$$

Ylide

Notice that we may formulate a second resonance structure for the ylide by delocalizing the negative charge onto the phosphorus. In this form the valence shell on phosphorus has been expanded and a carbon–phosphorus double bond is present.

The Wittig reaction forms carbon–carbon double bonds

When an ylide is exposed to an aldehyde or ketone, their reaction ultimately produces an alkene by coupling the ylide carbon with that of the carbonyl. The other product of this reaction is triphenylphosphine oxide.

The Wittig Reaction

$$>C{=}P(C_6H_5)_3 + O{=}C< \longrightarrow >C{=}C< + (C_6H_5)_3P{=}O$$

Ylide | **Aldehyde or ketone** | **Alkene** | **Triphenylphosphine oxide**

$$CH_3CH_2CH_2\overset{O}{\overset{\|}{C}}H + CH_3CH_2\overset{CH_3}{\overset{|}{C}}{=}P(C_6H_5)_3 \xrightarrow[(C_6H_5)_3PO]{(CH_3CH_2)_2O,\ 10^\circ C} CH_3CH_2CH_2CH{=}\overset{CH_3}{\overset{|}{C}}CH_2CH_3$$

66%

3-Methyl-3-heptene

The Wittig reaction is a valuable addition to our synthetic arsenal because it forms carbon–carbon double bonds. In contrast with eliminations (Sections 11-8 and 11-9), it gives rise to alkenes in which the position of the newly formed double bond is unambiguous. Compare, for example, two syntheses of 2-ethyl-1-butene, one by the Wittig reaction, the other by elimination.

Comparison of Two Syntheses of 2-Ethyl-1-butene

By Wittig reaction

$$CH_3CH_2\overset{O}{\overset{\|}{C}}CH_2CH_3 + CH_2{=}P(C_6H_5)_3 \longrightarrow CH_3CH_2\overset{CH_2}{\overset{\|}{C}}CH_2CH_3 + (C_6H_5)_3P{=}O$$

(Only one isomer)

By elimination

$$\underset{\underset{Br}{|}}{CH_3CH_2\overset{\overset{CH_3}{|}}{C}CH_2CH_3} \xrightarrow{Base} CH_3CH_2\overset{\overset{CH_2}{\|}}{C}CH_2CH_3 + CH_3CH_2\overset{\overset{CH_3}{|}}{C}{=}CHCH_3$$

(Mixture of isomers)

What is the mechanism of the Wittig reaction? The negatively polarized carbon in the ylide is nucleophilic and can attack the carbonyl group. The result is a **phosphorus betaine,*** a dipolar species of the kind called a *zwitterion* (*Zwitter,* German, hybrid). The betaine is short lived and rapidly forms a neutral **oxaphosphacyclobutane (oxaphosphetane),** characterized by a four-membered ring containing phosphorus and oxygen. This substance then decomposes to the product alkene and triphenylphosphine oxide. The driving force for the last step is the formation of the very strong phosphorus–oxygen double bond.

BOX 17-3 The Wittig Reaction in Synthesis

$$CH_3CH_2CH_2C{\equiv}CCH_2Br \xrightarrow{P(C_6H_5)_3} CH_3CH_2CH_2C{\equiv}CCH_2\overset{+}{P}(C_6H_5)_3Br^- \xrightarrow{CH_3CH_2O^-Na^+}$$

$$CH_3CH_2CH_2C{\equiv}CCH{=}P(C_6H_5)_3 \xrightarrow{CH_3CH_2O_2C(CH_2)_8CHO} CH_3CH_2CH_2C{\equiv}CCH{=}CH(CH_2)_8CO_2CH_2CH_3$$

Cis and trans isomers

The Wittig reaction has been employed extensively in total synthesis. For example, the total synthesis of the pheromone (see Section 12-16) bombykol, the sex attractant of the silkworm moth, employs a Wittig reaction to construct the backbone of the molecule, as shown above.

The Wittig reaction has also found extensive application in industry. The chemical company Badische Anilin und Soda Fabriken (BASF) in Germany synthesizes vitamin A_1 (Section 14-7) by a Wittig reaction in the crucial step. In this case, the reaction is stereoselective, giving only the trans alkene.

BASF Vitamin A_1 Synthesis

$\overset{+}{P}(C_6H_5)_3X^-$ $\xrightarrow{CH_3O^-,\ CH_3OH}$ $P(C_6H_5)_3$ $\xrightarrow{OHC\text{–}CH{=}C(CH_3)CH_2OCCH_3}$ $\overset{O}{\|}{OCCH_3}$ $\xrightarrow[\text{Ester hydrolysis}]{HO^-,\ H_2O}$ OH

Vitamin A_1

---

*Betaine is the name of an amino acid, $(CH_3)_3\overset{+}{N}CH_2COO^-$, which is found in beet sugar (*beta,* Latin, beet) and exists as a zwitterion.

Mechanism of the Wittig Reaction

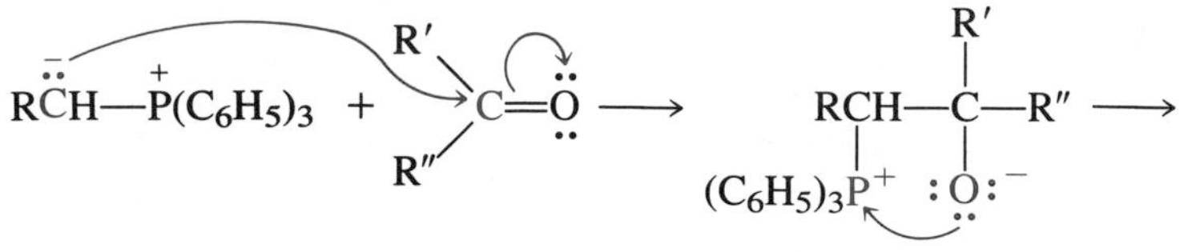

$$R\ddot{\bar{C}}H-\overset{+}{P}(C_6H_5)_3 + \underset{R''}{\overset{R'}{>}}C=\ddot{O} \longrightarrow RCH-\underset{\underset{:\ddot{O}:^-}{|}}{\overset{\overset{R'}{|}}{C}}-R'' \longrightarrow$$
$$(C_6H_5)_3P^+$$

A phosphorus betaine

$$\begin{array}{cc} RCH-C\overset{R'}{\diagup} & \\ | \quad | \diagdown R'' & \longrightarrow RCH=C\underset{R''}{\overset{R'}{<}} + (C_6H_5)_3P=O \\ (C_6H_5)_3P-\ddot{O}: & \end{array}$$

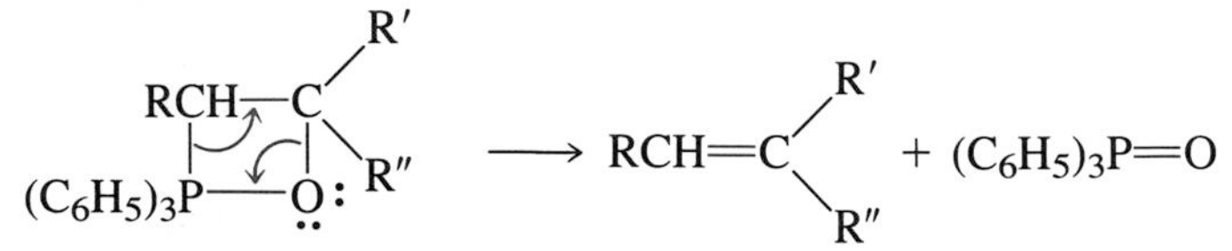

An oxaphosphacyclobutane (Oxaphosphetane)

Wittig reactions can be carried out in the presence of ether, ester, halogen, alkene, and alkyne functions. However, they are only sometimes stereoselective, and mixtures of Z and E alkenes may form.

$$CH_3CH_2CH_2CH=P(C_6H_5)_3 + CH_3(CH_2)_4CHO \xrightarrow[-\ (C_6H_5)_3PO]{THF} \underset{\substack{70\% \\ \text{(cis:trans ratio = 6:1)}}}{CH_3(CH_2)_2CH=CH(CH_2)_4CH_3}$$

EXERCISE 17-10

Propose syntheses of 3-methylenecyclohexene from **(a)** 2-cyclohexenone and **(b)** 3-bromocyclohexene, using Wittig reactions.

EXERCISE 17-11

Propose a synthesis of the dienone shown below from the indicated starting materials. [Hint: Make use of a protecting group (Section 17-8).]

$$CH_3\overset{O}{\overset{\|}{C}}CH_2CH_2CH=CHCH=CH_2$$

$$\text{from } CH_3\overset{O}{\overset{\|}{C}}CH_2CH_2CH_2Br \text{ and } H\overset{O}{\overset{\|}{C}}CH=CH_2$$

EXERCISE 17-12

Develop concise synthetic routes from starting material to product. You may use any material in addition to the given compound (more than one step will be required).

(a) 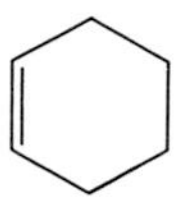 [cyclohexene] $\dashrightarrow$ $CH_2=CH(CH_2)_4CH=CH_2$ (Hint: Section 12-11)

(b) (Hint: Section 14-8)

To summarize, phosphorus ylides add to aldehydes and ketones to give betaines that decompose by forming carbon–carbon double bonds. The Wittig reaction affords a means of synthesizing alkenes from carbonyl compounds and haloalkanes by way of the corresponding phosphonium salts.

17-13 Oxidation by Peroxycarboxylic Acids: The Baeyer-Villiger Oxidation

This section shows how the hydroperoxy group of peroxycarboxylic acids adds to the carbonyl group of ketones to eventually give esters by rearrangement.

Addition of a peroxycarboxylic acid gives the peroxide analog of a hemiacetal. This unstable adduct decomposes through a cyclic transition state in which an alkyl group shifts from the original carbonyl carbon to oxygen to give an ester. This transformation is called the **Baeyer-Villiger* oxidation.**

Baeyer-Villiger Oxidation

$$R''\overset{O}{\overset{\|}{C}}R' + R\overset{O}{\overset{\|}{C}}OOH \longrightarrow [\text{adduct}] \longrightarrow R''\overset{O}{\overset{\|}{C}}OR' + HO\overset{O}{\overset{\|}{C}}R$$

Ketone **Peroxycarboxylic acid** **Ester**

$$CH_3\overset{O}{\overset{\|}{C}}CH_2CH_3 \xrightarrow{CF_3\overset{O}{\overset{\|}{C}}OOH,\ CH_2Cl_2} CH_3\overset{O}{\overset{\|}{C}}OCH_2CH_3 \quad 72\%$$

2-Butanone **Ethyl acetate**

*Professor Johann Friedrich Wilhelm Adolf von Baeyer, 1835–1917, University of Munich, Nobel Prize 1905 (chemistry); Victor Villiger, 1868–1934, BASF, Ludwigshafen, Germany.

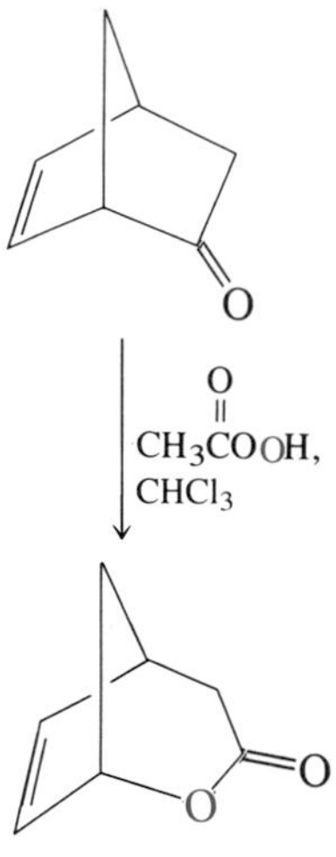

O

$\overset{O}{\overset{\|}{CH_3COOH}}$,
$CHCl_3$

56%

Cyclic ketones are converted into cyclic esters (margin). Attack is at the carbonyl rather than at the carbon–carbon double bond.

Unsymmetric ketones can in principle lead to two different esters. Why is only one observed? The answer is that some substituents migrate more easily than others. Experiments have established their relative ease of migration, or **migratory aptitude.** The ordering suggests that the migrating carbon possesses carbocationic character in the transition state for rearrangement.

Migratory Aptitudes in the Baeyer-Villiger Reaction

Methyl < primary < phenyl ~ secondary < cyclohexyl < tertiary

EXERCISE 17-13

Predict the outcome of the following oxidations with a peroxycarboxylic acid.

(a) $CH_2{=}CHCH_2CH_2\overset{O}{\overset{\|}{C}}CH_3$ **(b)** 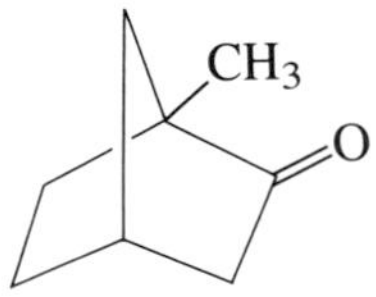 CH_3, O **(c)** $(CH_3)_3C\overset{O}{\overset{\|}{C}}CH_2CH_3$

In summary, ketones can be oxidized with peroxycarboxylic acids to give esters; with unsymmetric ketones, the esters can be formed selectively, by migration of only one of the substituents.

17-14 Oxidative Chemical Tests for Aldehydes

Although the advent of NMR and other spectroscopy has made chemical tests for functional groups a rarity, they are still used in special cases in which other analytical tests may fail. Two characteristic simple tests for aldehydes will again turn up in the discussion of sugar chemistry in Chapter 24; they make use of the ready oxidation of aldehydes to carboxylic acids. The first is **Fehling's* test,** in which cupric ion is the oxidant. In a basic medium, the precipitation of red cuprous oxide indicates the presence of an aldehyde function.

Fehling's Test

$$R\overset{O}{\overset{\|}{C}}H + Cu^{2+} \xrightarrow{\text{NaOH, tartrate, } H_2O} \underset{\text{Brick-red}}{Cu_2O} + R\overset{O}{\overset{\|}{C}}OH$$

The second is **Tollens's† test,** in which a solution of silver ion precipitates a silver mirror on exposure to an aldehyde.

*Professor Hermann C. von Fehling, 1812–1885, Polytechnic School of Stuttgart, Germany.

†Professor Bernhard C. G. Tollens, 1841–1918, University of Göttingen, Germany.

Tollens's Test

$$\mathrm{R\overset{\overset{O}{\|}}{C}H + Ag^+ \xrightarrow{NH_3,\ H_2O} \underset{Mirror}{Ag} + R\overset{\overset{O}{\|}}{C}OH}$$

The Fehling's and Tollen's tests are not commonly used in large-scale syntheses of carboxylic acids from aldehydes because the yields are often low. However, these tests are very convenient and visually dramatic ways to detect the CHO functional group.

NEW REACTIONS

Synthesis of Aldehydes and Ketones

1. OXIDATION OF ALCOHOLS (Section 17-4)

$$\mathrm{R\overset{\overset{R'}{|}}{C}HOH \xrightarrow{CrO_3,\ H_2SO_4} R\overset{\overset{O}{\|}}{C}R'}$$

$$\mathrm{RCH_2OH \xrightarrow{PCC,\ CH_2Cl_2} \underset{\text{Stable to oxidizing agent}}{R\overset{\overset{O}{\|}}{C}H}}$$

Allylic oxidation

$$\mathrm{>C{=}C<\!\!\!\!-\overset{H\ \ OH}{C}\!\!< \xrightarrow{MnO_2,\ CHCl_3} >C{=}C<\!\!\!\!-\overset{\overset{O}{\|}}{C}-}$$

2. OZONOLYSIS OF ALKENES (Section 17-4)

$$\mathrm{>C{=}C< \xrightarrow[\text{2. Zn, } CH_3CO_2H]{\text{1. } O_3,\ CH_2Cl_2} >C{=}O + O{=}C<}$$

3. HYDRATION OF ALKYNES (Section 17-4)

$$\mathrm{RC{\equiv}CH \xrightarrow{H_2O,\ Hg^{2+},\ H_2SO_4} R\overset{\overset{O}{\|}}{C}CH_3}$$

4. FRIEDEL-CRAFTS ALKANOYLATION (Section 17-4)

$$\mathrm{C_6H_6 + R\overset{\overset{O}{\|}}{C}Cl \xrightarrow[\text{2. } H^+,\ H_2O]{\text{1. } AlCl_3} C_6H_5\overset{\overset{O}{\|}}{C}R + HCl}$$

Reactions of Aldehydes and Ketones

5. HYDROGENATION (Section 17-5)

$$\text{R}\overset{\text{O}}{\overset{\|}{\text{C}}}\text{R}' \xrightarrow{\text{H}_2\text{, Raney Ni, pressure}} \text{R}\underset{\text{H}}{\overset{\text{OH}}{\text{C}}}\text{R}'$$

Selectivity

$$\text{RCH}{=}\text{CHCH}_2\text{CH}_2\overset{\text{O}}{\overset{\|}{\text{C}}}\text{R}' \xrightarrow{\text{H}_2\text{, Pt}} \text{R(CH}_2)_4\overset{\text{O}}{\overset{\|}{\text{C}}}\text{R}'$$

6. REDUCTION BY HYDRIDES (Section 17-5)

$$\text{R}\overset{\text{O}}{\overset{\|}{\text{C}}}\text{H} \xrightarrow{\text{NaBH}_4\text{, CH}_3\text{CH}_2\text{OH}} \text{RCH}_2\text{OH} \qquad \text{R}\overset{\text{O}}{\overset{\|}{\text{C}}}\text{R}' \xrightarrow[\text{2. H}^+\text{, H}_2\text{O}]{\text{1. LiAlH}_4\text{, (CH}_3\text{CH}_2)_2\text{O}} \text{R}\underset{\text{H}}{\overset{\text{OH}}{\text{C}}}\text{R}'$$

Selectivity

$$\text{RCH}{=}\text{CHCH}_2\text{CH}_2\overset{\text{O}}{\overset{\|}{\text{C}}}\text{R}' \xrightarrow[\text{2. H}^+\text{, H}_2\text{O}]{\text{1. LiAlH}_4\text{, (CH}_3\text{CH}_2)_2\text{O}} \text{RCH}{=}\text{CHCH}_2\text{CH}_2\underset{\text{H}}{\overset{\text{OH}}{\text{C}}}\text{R}'$$

7. ADDITION OF ORGANOMETALLIC COMPOUNDS (Section 17-5)

$$\text{RLi or RMgX} + \underset{\textbf{Formaldehyde}}{\text{CH}_2{=}\text{O}} \xrightarrow{\text{THF}} \underset{\textbf{Primary alcohol}}{\text{RCH}_2\text{OH}}$$

$$\text{RLi or RMgX} + \underset{\textbf{Aldehyde}}{\text{R}'\overset{\text{O}}{\overset{\|}{\text{C}}}\text{H}} \xrightarrow{\text{THF}} \underset{\textbf{Secondary alcohol}}{\text{R}\underset{\text{H}}{\overset{\text{OH}}{\text{C}}}\text{R}'}$$

$$\text{RLi or RMgX} + \underset{\textbf{Ketone}}{\text{R}'\overset{\text{O}}{\overset{\|}{\text{C}}}\text{R}''} \xrightarrow{\text{THF}} \underset{\textbf{Tertiary alcohol}}{\text{R}\underset{\text{R}''}{\overset{\text{OH}}{\text{C}}}\text{R}'}$$

8. ADDITION OF WATER AND ALCOHOLS—HEMIACETALS (Sections 17-6 and 17-7)

$$\text{R}\overset{\text{O}}{\overset{\|}{\text{C}}}\text{R}' \underset{}{\overset{\text{H}_2\text{O, H}^+\text{ or HO}^-}{\rightleftharpoons}} \text{R}\underset{\text{OH}}{\overset{\text{OH}}{\text{C}}}\text{R}'$$

Carbonyl hydrate

(A geminal diol)

$$\mathrm{RCR'} \xrightleftharpoons{\mathrm{R''OH,\ H^+\ or\ HO^-}} \mathrm{RC(OH)(OR'')R'}$$

Hemiacetal

K_{eq}: $R\text{–}\overset{O}{\overset{\|}{C}}\text{—}R < R\text{–}\overset{O}{\overset{\|}{C}}\text{–}H < H_2C{=}O$

Intramolecular addition

$$\mathrm{HO(CH_2)_4COR} \xrightarrow{\mathrm{H^+\ or\ HO^-}}$$

Cyclic hemiacetal

9. ACID-CATALYZED ADDITION OF ALCOHOLS—ACETALS (Sections 17-7 and 17-8)

$$\mathrm{RCR'} + 2\,\mathrm{R''OH} \xrightleftharpoons{\mathrm{H^+}} \mathrm{RC(OR'')_2R'} + \mathrm{H_2O}$$

Acetal

Cyclic acetals

$$\mathrm{RCR'} + \mathrm{HOCH_2CH_2OH} \xrightleftharpoons{\mathrm{H^+}} + \mathrm{H_2O}$$

Ketone

Ketone, protected as cyclic acetal

(Stable to base, $LiAlH_4$, RMgX)

$$\text{Diol} + \mathrm{CH_3CCH_3} \xrightleftharpoons{\mathrm{H^+}} + \mathrm{H_2O}$$

Diol

Diol, protected as propanone (acetone) acetal

10. THIOACETALS (Section 17-8)

Formation

$$\mathrm{RCR'} + 2\,\mathrm{R''SH} \xrightarrow{\mathrm{BF_3\ or\ ZnCl_2,\ (CH_3CH_2)_2O}} \mathrm{RC(SR'')_2R'}$$

(Stable to aqueous acid, base, $LiAlH_4$, RMgX)

Hydrolysis

$$\text{R''S(SR'')C(R)(R')} \xrightarrow{\text{H}_2\text{O, HgCl}_2\text{, CaCO}_3\text{, CH}_3\text{CN}} \text{RC(=O)R'}$$

11. RANEY NICKEL DESULFURIZATION (Section 17-8)

$$\text{(dithiolane) RC(R')} \xrightarrow{\text{Raney Ni, H}_2} \text{RCH}_2\text{R'}$$

12. ADDITION OF AMINE DERIVATIVES (Section 17-9)

$$\text{RC(=O)R'} \xrightarrow{\text{R''NH}_2\text{, H}^+} \underset{\textbf{Imine}}{\text{R(R')C=NR''}} + \text{H}_2\text{O}$$

$$\text{RC(=O)R'} \xrightarrow[\text{Hydroxylamine}]{\text{H}_2\text{NOH, H}^+} \underset{\textbf{Oxime}}{\text{R(R')C=NOH}} + \text{H}_2\text{O}$$

$$\text{RC(=O)R'} \xrightarrow[\text{2,4-Dinitrophenylhydrazine}]{\text{2,4-(O}_2\text{N)}_2\text{C}_6\text{H}_3\text{NHNH}_2\text{, H}^+} \underset{\textbf{2,4-Dinitrophenylhydrazone}}{\text{R(R')C=N-NH-C}_6\text{H}_3\text{(NO}_2\text{)}_2} + \text{H}_2\text{O}$$

$$\text{RC(=O)R'} \xrightarrow[\text{Semicarbazide}]{\text{H}_2\text{NNHC(=O)NH}_2\text{, H}^+} \underset{\textbf{Semicarbazone}}{\text{R(R')C=N-NHC(=O)NH}_2} + \text{H}_2\text{O}$$

13. ENAMINES (Section 17-9)

$$\text{RCH}_2\text{C(=O)R'} + \underset{\textbf{Secondary amine}}{\text{R''R'''NH}} \rightleftharpoons \underset{\textbf{Enamine}}{\text{RCH=C(R')N(R'')R'''}} + \text{H}_2\text{O}$$

14. WOLFF-KISHNER REDUCTION (Section 17-10)

$$\mathrm{RC(=O)R'} \xrightarrow{\mathrm{H_2NNH_2,\ H_2O,\ HO^-,\ \Delta}} \mathrm{RCH_2R'}$$

15. CYANOHYDRINS (Section 17-11)

$$\mathrm{R\overset{O}{\overset{\|}{C}}R'} + \mathrm{HCN} \rightleftharpoons \mathrm{RR'C(OH)CN}$$

Cyanohydrin

16. WITTIG REACTION (Section 17-12)

$$\mathrm{R''CH_2X} + \underset{\textbf{Triphenylphosphine}}{\mathrm{P(C_6H_5)_3}} \xrightarrow{\mathrm{C_6H_6}} \underset{\textbf{Phosphonium halide}}{\mathrm{R''CH_2\overset{+}{P}(C_6H_5)_3\ X^-}}$$

Works with primary or secondary haloalkanes

$$\mathrm{R''CH_2\overset{+}{P}(C_6H_5)_3\ X^-} \xrightarrow{\text{Base}} \underset{\textbf{Ylide}}{\mathrm{R''CH{=}P(C_6H_5)_3}}$$

$$\mathrm{R\overset{O}{\overset{\|}{C}}R'} + \mathrm{R''CH{=}P(C_6H_5)_3} \xrightarrow{\mathrm{THF}} \mathrm{RR'C{=}CHR''} + \mathrm{(C_6H_5)_3P{=}O}$$

(Not always stereoselective)

17. BAEYER-VILLIGER OXIDATION (Section 17-13)

$$\underset{\textbf{Ketone}}{\mathrm{R\overset{O}{\overset{\|}{C}}R'}} + \mathrm{R''\overset{O}{\overset{\|}{C}}OOH} \xrightarrow{\mathrm{CH_2Cl_2}} \underset{\textbf{Ester}}{\mathrm{R\overset{O}{\overset{\|}{C}}OR'}} + \mathrm{R''\overset{O}{\overset{\|}{C}}OH}$$

Migratory aptitudes in Baeyer-Villiger oxidation

Methyl < primary < phenyl ~ secondary < cyclohexyl < tertiary

18. TEST FOR ALDEHYDES (Section 17-14)

$$\mathrm{R\overset{O}{\overset{\|}{C}}H} \xrightarrow{\mathrm{Cu^{2+}\ or\ Ag^+}} \mathrm{R\overset{O}{\overset{\|}{C}}OH} + \underset{\textbf{Red precipitate}}{\mathrm{Cu_2O}} \text{ or } \underset{\textbf{Mirror}}{\mathrm{Ag}}$$

IMPORTANT CONCEPTS

1. The carbonyl group is the functional group of the aldehydes (alkanals) and ketones (alkanones). It has precedence over the hydroxy, alkenyl, and alkynyl groups in the naming of molecules.
2. The carbon–oxygen double bond and its two attached nuclei in aldehydes and ketones form a plane. The C=O unit is polarized, with a partial negative charge on oxygen and a partial positive charge on carbon.
3. The ^{1}H NMR spectra of aldehydes exhibit a peak at $\delta \sim 9.8$ ppm. The carbonyl carbon absorbs at ~200 ppm. Aldehydes and ketones exhibit strong infrared bands in the region 1690–1750 cm^{-1}; this absorption is due to the stretching of the C=O bond. Because of the availability of low-energy $n \rightarrow \pi^*$ transitions, the electronic spectra of aldehydes and ketones have relatively long wavelength bands.
4. The carbon–oxygen double bond undergoes catalytic hydrogenation and ionic additions. The catalysts for the former are heterogeneous transition metal surfaces; for the latter, acid or base.
5. The reactivity of the carbonyl group increases with increasing electrophilic character of the carbonyl carbon. Therefore, aldehydes are more reactive than ketones.
6. Primary amines undergo condensation reactions with aldehydes and ketones to imines; secondary amines condense to enamines.
7. The combination of Friedel-Crafts alkanoylation and Wolff-Kishner or Clemmensen reduction allows synthesis of alkylbenzenes free of the limitations of Friedel-Crafts alkylation.
8. The Wittig reaction is an important carbon–carbon bond-forming reaction that produces alkenes directly from aldehydes and ketones.
9. The reaction of peroxycarboxylic acids with the carbonyl group of ketones produces esters.

PROBLEMS

1. Name or draw the structure of each of the following compounds.

(a)  $(CH_3)_2CH\overset{O}{\overset{\|}{C}}CH(CH_3)_2$

(b) 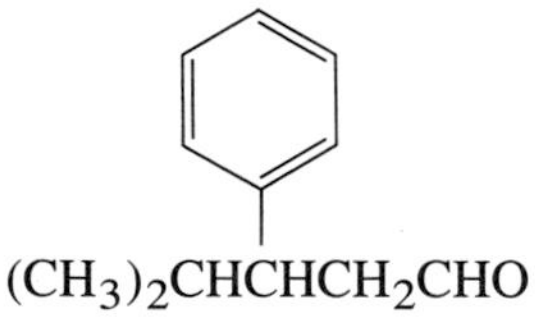 $(CH_3)_2CHCHCH_2CHO$ (with C_6H_5 on the third carbon)

(c) $CH_3\overset{O}{\overset{\|}{C}}CH{=}CH_2$

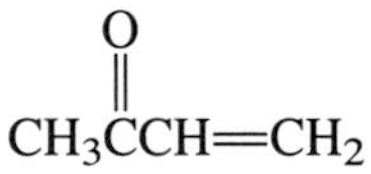

(d) 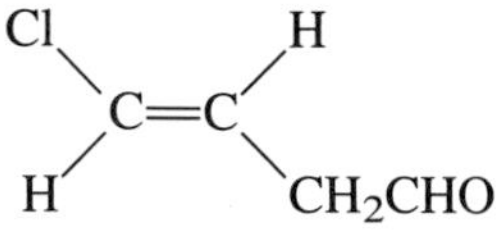 Cl(H)C=C(H)CH$_2$CHO (Cl and H on the first carbon; H and CH_2CHO on the second)

(e) 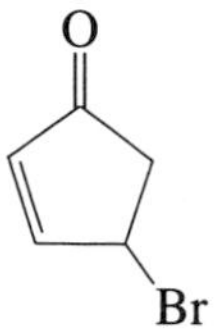 (cyclopentenone structure with Br)

(f) 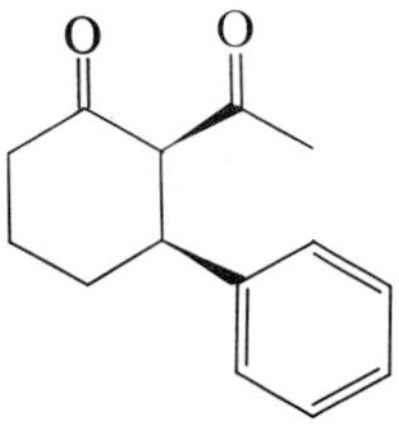 (cyclohexanone structure with acetyl and phenyl substituents)

(g) (*Z*)-2-Acetyl-2-butenal

(h) *trans*-3-Chlorocyclobutanecarbaldehyde

2. Spectroscopic data for two carbonyl compounds with the formula $C_8H_{12}O$ are given below. Suggest a structure for each. The letter m stands for the appearance of this particular part of the spectrum as an uninterpretable multiplet. **(a)** ^{1}H NMR: $\delta = 1.6$ (m, 4 H), 2.15 (s, 3 H), 2.19 (m, 4 H), and 6.78 (t, 1 H) ppm. ^{13}C NMR: $\delta = 21.8$, 22.2, 23.2, 25.0, 26.2, 139.8, 140.7, and 198.6 ppm. **(b)** ^{1}H NMR: $\delta = 0.94$ (t, 3 H), 1.48 (sex, 2 H), 2.21 (q, 2 H), 5.8–7.1 (m, 4 H), and 9.56 (d, 1 H) ppm. ^{13}C NMR: $\delta =$ 13.6, 21.9, 35.2, 129.0, 135.2, 146.7, 152.5, and 193.2 ppm.

3. The compounds described in Problem 2 have very different ultraviolet spectra. One has $\lambda_{max}(\epsilon) = 232(13{,}000)$ and 308(1450) nm, whereas the other has $\lambda_{max}(\epsilon) = 272(35{,}000)$ nm and a weaker absorption near 320 nm (this value is hard to determine accurately because of the intensity of the stronger absorption). Using the structures that you determined in Problem 2, match the compounds to these UV spectral data. Explain the spectra in terms of the structures.

4. Indicate which reagent or combination of reagents is best suited for each of the following reactions.

(a) 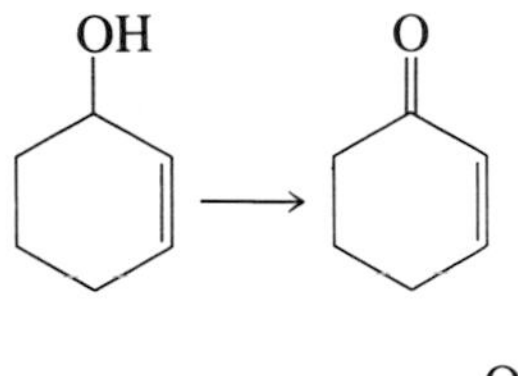 2-cyclohexen-1-ol (OH) $\longrightarrow$ 2-cyclohexen-1-one (O)

(b) $CH_3CH_2CH(CH_2OH)CH_2CH_3 \longrightarrow CH_3CH_2CH(CHO)CH_2CH_3$

(c) 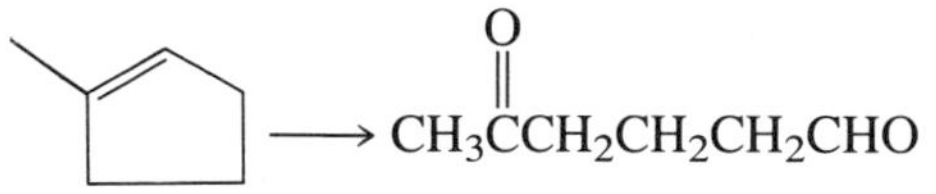 1-methylcyclopentene $\longrightarrow CH_3\overset{O}{\overset{\|}{C}}CH_2CH_2CH_2CHO$

(d) 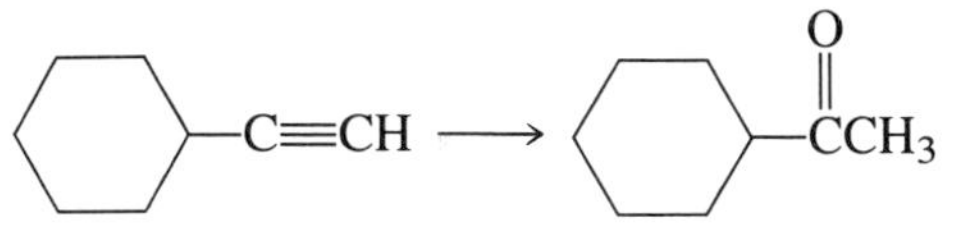 cyclohexyl–$C{\equiv}CH \longrightarrow$ cyclohexyl–$\overset{O}{\overset{\|}{C}}CH_3$

(e) $(CH_3)_2CHC{\equiv}CCH(CH_3)_2 \longrightarrow (CH_3)_2CH\overset{O}{\overset{\|}{C}}CH_2CH(CH_3)_2$

(f) 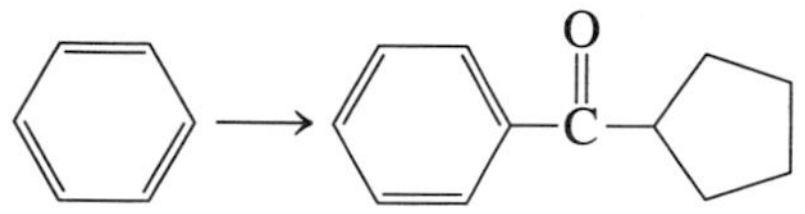 benzene $\longrightarrow$ C_6H_5–$\overset{O}{\overset{\|}{C}}$–cyclopentyl

5. Write the expected products of ozonolysis (followed by mild reduction—e.g., by Zn) of each of the following molecules.

(a) $CH_3CH_2CH_2CH{=}CH_2$ **(b)** 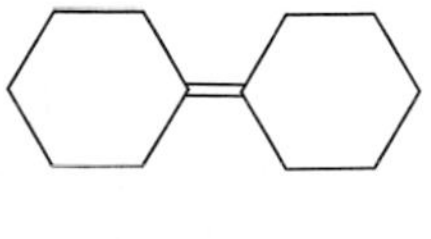 cyclohexylidenecyclohexane **(c)** 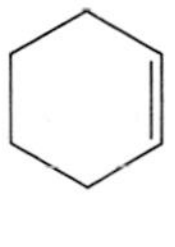 cyclohexene **(d)** 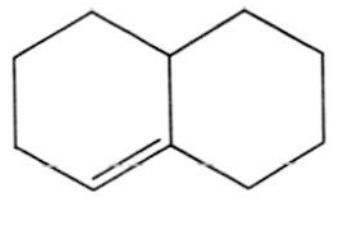 bicyclic octahydronaphthalene (ring-fusion double bond structure as drawn)

6. For each of the following groups, rank the molecules in decreasing order of reactivity toward addition of a nucleophile to the most electrophilic sp^2-hybridized carbon.

(a) $(CH_3)_2C{=}O$, $(CH_3)_2C{=}NH$, $(CH_3)_2C{=}\overset{+}{O}H$ **(b)** $CH_3\overset{O}{\overset{\|}{C}}CH_3$, $CH_3\overset{O}{\overset{\|}{C}}\overset{O}{\overset{\|}{C}}CH_3$, $CH_3\overset{O}{\overset{\|}{C}}\overset{O}{\overset{\|}{C}}\overset{O}{\overset{\|}{C}}CH_3$

(c) $BrCH_2COCH_3$, CH_3COCH_3, CH_3CHO, $BrCH_2CHO$

7. Give the expected products of reaction of 4-acetylcyclohexene with each of the following reagents.
(a) H_2 (1 equivalent), Pd, CH_3CH_2OH
(b) $LiAlH_4$, $(CH_3CH_2)_2O$, then H^+, H_2O
(c) CH_3CH_2MgBr, $(CH_3CH_2)_2O$, then H^+, H_2O

8. Give the expected product(s) of each of the following reactions.

(a) 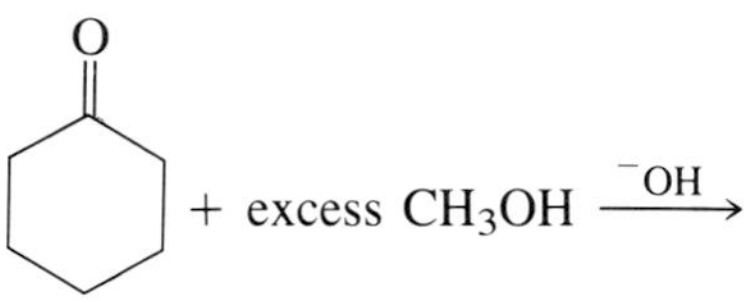 cyclohexanone + excess $CH_3OH \xrightarrow{^-OH}$

(b) 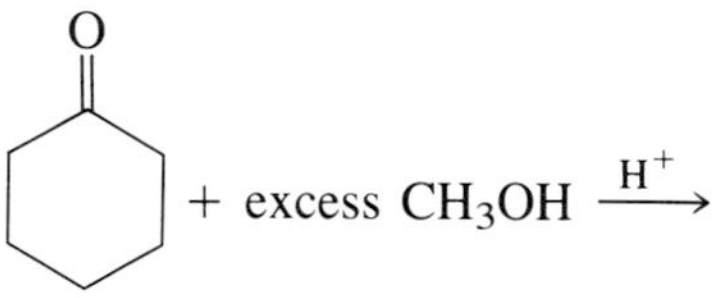 cyclohexanone + excess $CH_3OH \xrightarrow{H^+}$

(c) 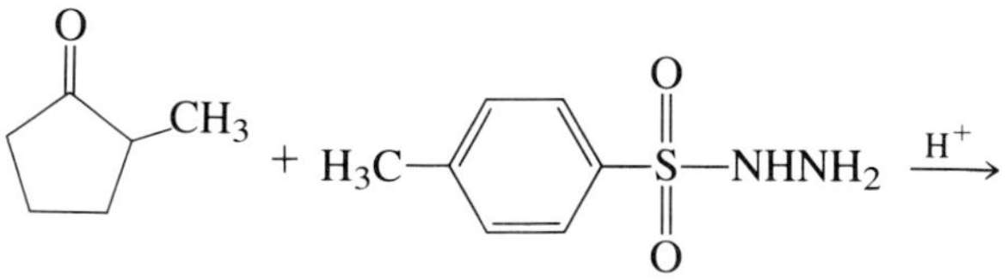 2-methylcyclopentanone (CH_3) + $H_3C-C_6H_4-SO_2-NHNH_2 \xrightarrow{H^+}$

(d) 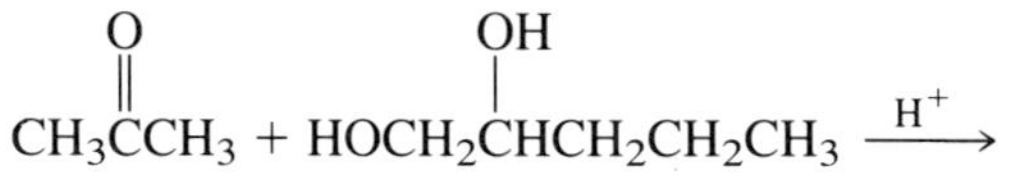 $CH_3\overset{O}{\overset{\|}{C}}CH_3 + HOCH_2\overset{OH}{\overset{|}{C}}HCH_2CH_2CH_3 \xrightarrow{H^+}$

(e) 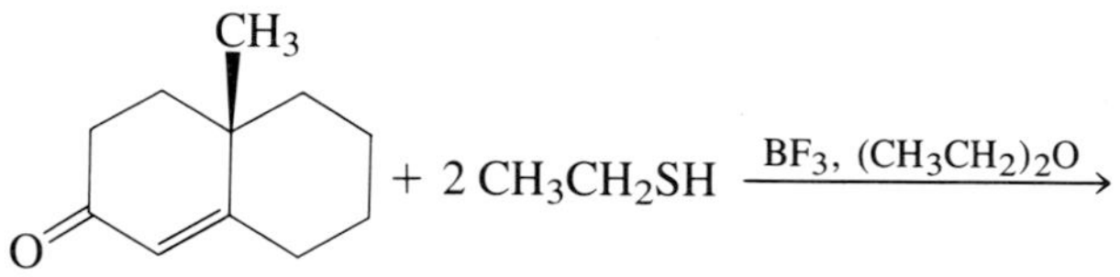 (bicyclic enone with angular CH_3, O) + 2 $CH_3CH_2SH \xrightarrow{BF_3,\ (CH_3CH_2)_2O}$

(f) 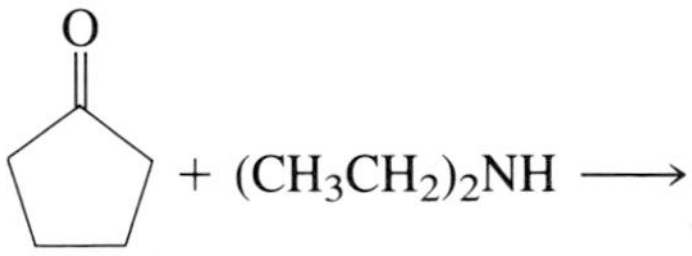 cyclopentanone + $(CH_3CH_2)_2NH \longrightarrow$

9. Formulate the mechanism of the BF_3-catalyzed reaction of CH_3SH with butanal (Section 17-8).

10. Overoxidation of primary alcohols to carboxylic acids is caused by the water present in the usual aqueous acidic Cr(VI) reagents. The water adds to the initial aldehyde product to form a hydrate, which is further oxidized (Section 17-6). In view of these facts, explain the following two observations. **(a)** Water adds to ketones to form hydrates, but no overoxidation follows the conversion of a secondary alcohol into a ketone. **(b)** Successful oxidation of primary alcohols to aldehydes by the water-free PCC reagent requires that the alcohol be added slowly to the Cr(VI) reagent. If, instead, the PCC is added *to the alcohol,* a new side reaction forms an ester. This is illustrated (on the following page) for 1-butanol.

$$CH_3CH_2CH_2CH_2OH \xrightarrow{PCC,\ CH_2Cl_2} CH_3CH_2CH_2\overset{O}{\overset{\|}{C}}OCH_2CH_2CH_2CH_3$$

(c) Give the products expected from reaction of 3-phenyl-1-propanol and water-free CrO_3 (1) when the alcohol is added to the oxidizing agent and (2) when the oxidant is added to the alcohol.

11. Explain the results of the reactions below by means of mechanisms.

(a) :ÖH :O: $\underset{\rightleftharpoons}{H^+}$ Ö: :ÖH

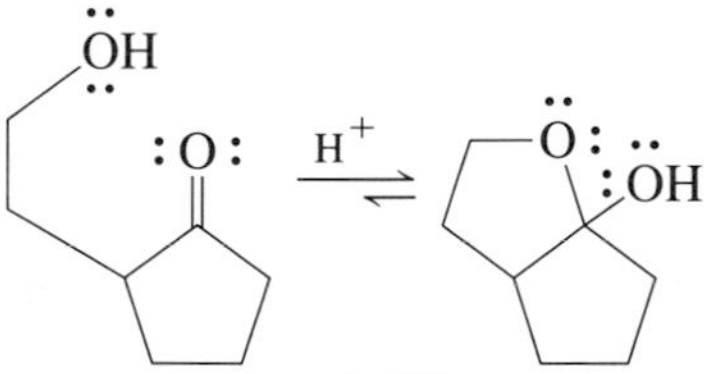

(b) HO ... H, O $\xrightarrow{CrO_3}$ O, O

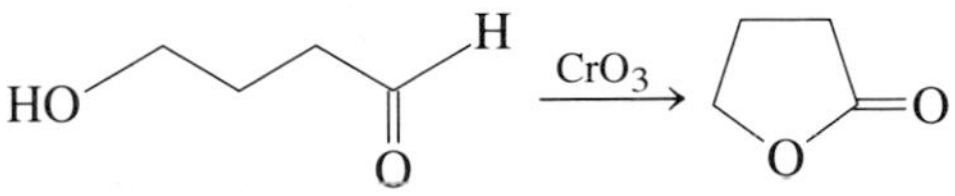

(c) HO ... O O $\xrightarrow{H^+}$ O O OH

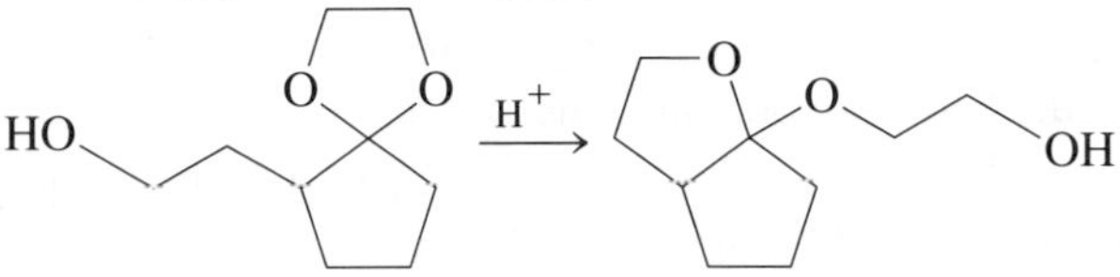

(d) Discuss why hemiacetal formation may be catalyzed by either acid or base, but acetal formation is catalyzed by only acid, not base.

12. Formulate a plausible mechanism for the reaction below. The product is a precursor of *mediquox* (shown in the margin), an agent used to treat respiratory infections in chickens (no, we are not making this up).

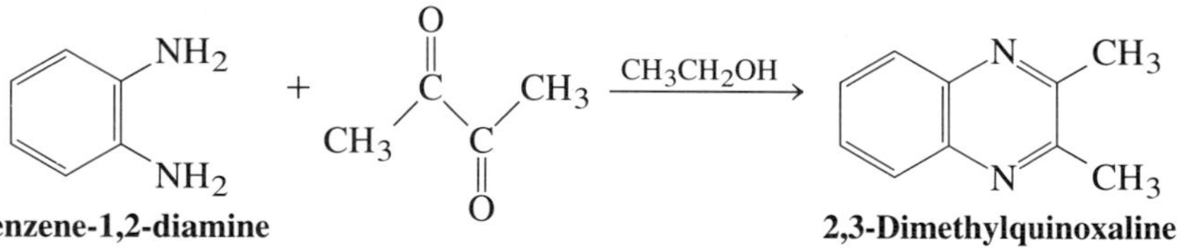

NH_2, NH_2 + CH_3C(=O)C(=O)CH_3 $\xrightarrow{CH_3CH_2OH}$ N, N, CH_3, CH_3

Benzene-1,2-diamine **2,3-Dimethylquinoxaline**

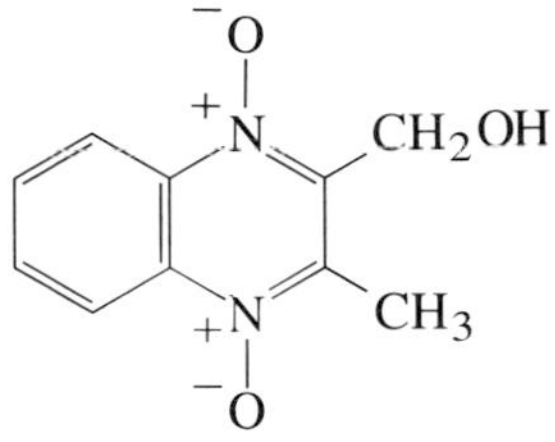

$^-$O, $^+$N, CH_2OH, $^+$N, CH_3, $^-$O

Mediquox

13. The formation of imines, oximes, hydrazones, and related derivatives from carbonyl compounds is reversible. Write a detailed mechanism for the acid-catalyzed hydrolysis of cyclohexanone semicarbazone to cyclohexanone and semicarbazide.

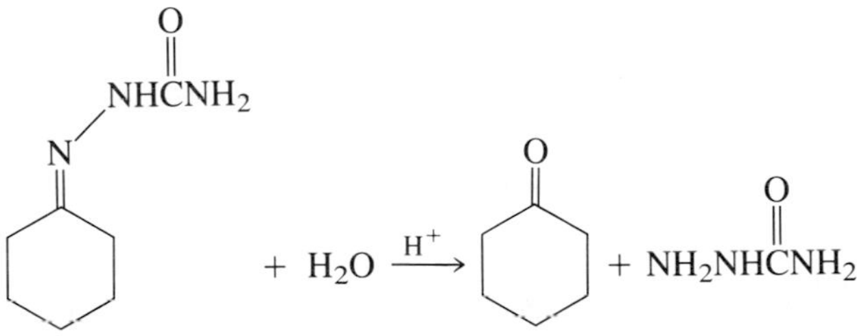

N–$NH\overset{O}{\overset{\|}{C}}NH_2$ + H_2O $\xrightarrow{H^+}$ O + $NH_2NH\overset{O}{\overset{\|}{C}}NH_2$

14. Propose reasonable syntheses of each of the following molecules, beginning with the indicated starting material.

(a) $HOCH_2\overset{OH}{\overset{|}{C}}HCH_2CH_2CH_2\underset{CH_3}{\underset{|}{\overset{OH}{\overset{|}{C}}}}CH_3$ from $HOCH_2\overset{OH}{\overset{|}{C}}HCH_2CH_2CH_2OH$

(b) $C_6H_5N{=}C(CH_2CH_3)_2$ from 3-pentanol

(c) 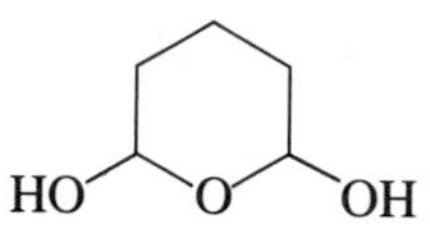 [structure: 2,6-bis(hydroxy)tetrahydropyran, HO, O, OH] from 1,5-pentanediol

(d) 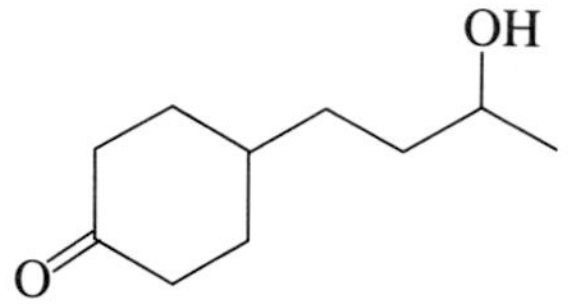 [structure with OH and O] from [structure with O]

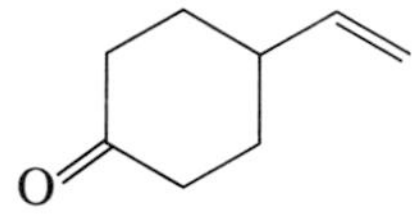

15. The UV absorptions and colors of 2,4-dinitrophenylhydrazone derivatives of aldehydes and ketones depend sensitively on the structure of the carbonyl compound. Suppose that you are asked to identify the contents of three bottles whose labels have fallen off. The labels indicate that one bottle contained butanal, one contained *trans*-2-butenal, and one contained *trans*-3-phenyl-2-propenal. The 2,4-dinitrophenylhydrazones prepared from the contents of the bottles have the following characteristics.

Bottle 1: m.p. 187–188°C; λ_{max} = 377 nm; orange color
Bottle 2: m.p. 121–122°C; λ_{max} = 358 nm; yellow color
Bottle 3: m.p. 252–253°C; λ_{max} = 394 nm; red color

Match up the hydrazones with the aldehydes (*without* first looking up the melting points of these derivatives), and explain your choices. (Hint: See Section 14-11.)

16. Indicate the reagent(s) best suited to effect these transformations.

(a) 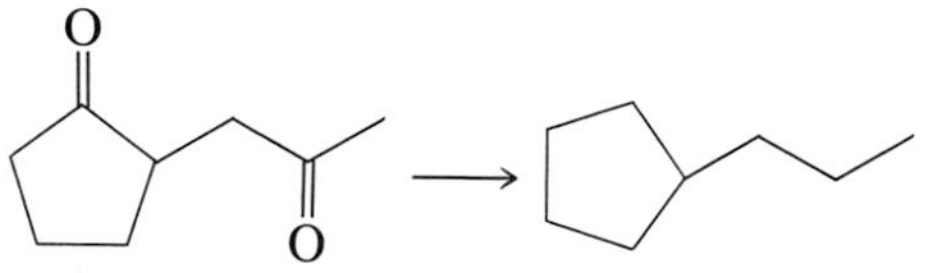 [structure with O, O] $\longrightarrow$ [structure]

(b) $CH_3CH{=}CHCH_2CH_2\overset{O}{\overset{\|}{C}}H \longrightarrow CH_3CH_2CH_2CH_2CH_2\overset{O}{\overset{\|}{C}}H$

(c) $CH_3CH{=}CHCH_2CH_2\overset{O}{\overset{\|}{C}}H \longrightarrow CH_3CH{=}CHCH_2CH_2CH_2OH$

(d) 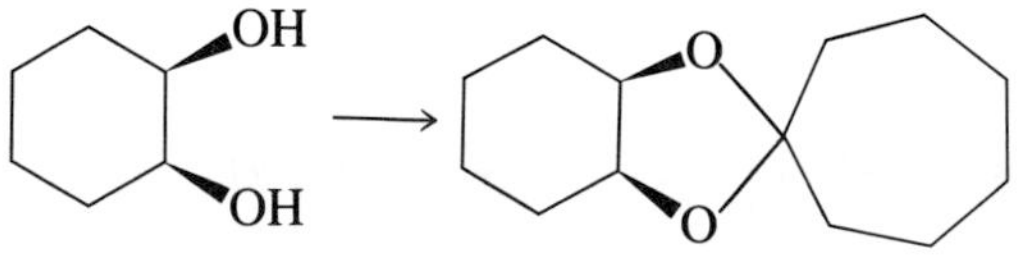 [structure with OH, OH] $\longrightarrow$ [structure with O, O]

17. For each of the following molecules, propose *two* methods of synthesis, from the different precursor molecules indicated.

(a) $CH_3CH{=}CHCH_2CH(CH_3)_2$ from (1) an aldehyde and (2) a different aldehyde

(b) 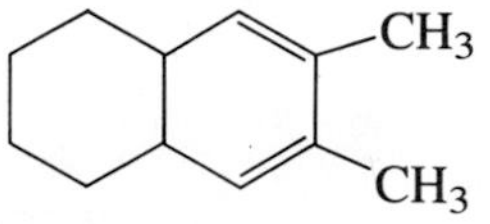 [structure with CH_3, CH_3] from (1) a dialdehyde and (2) a diketone

18. Three isomeric ketones with the molecular formula $C_7H_{14}O$ are converted into heptane by Clemmensen reduction. Compound A gives a single product on Baeyer-Villiger oxidation; compound B gives two different products in very different yields; compound C gives two different products in virtually a 1:1 ratio. Identify A, B, and C.

19. Give the product(s) of reaction of hexanal with each of the following reagents.

(a) $HOCH_2CH_2OH, H^+$ **(b)** $LiAlH_4$, then H^+, H_2O **(c)** NH_2OH, H^+

(d) NH_2NH_2, KOH, heat **(e)** $(CH_3)_2CHCH_2CH{=}P(C_6H_5)_3$ **(f)** pyrrolidine (N–H), H^+

(g) Ag^+, NH_3 **(h)** CrO_3, H_2SO_4, H_2O **(i)** HCN

20. Give the product(s) of reaction of cycloheptanone with each of the reagents in Problem 19.

21. Formulate a detailed mechanism for the Baeyer-Villiger oxidation of the ketone shown in the margin (refer to Exercise 17-13).

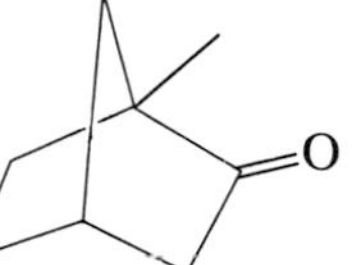

22. Give the two theoretically possible Baeyer-Villiger products from each of the following compounds. Indicate which one is preferentially formed.

(a) cyclohexyl–C(=O)–CH_3 (ring carbon bearing H) **(b)** 2-methylcyclopentanone (CH_3) **(c)** $(CH_3)_2CH\overset{O}{\overset{\|}{C}}CH_2CH(CH_3)_2$

(d) 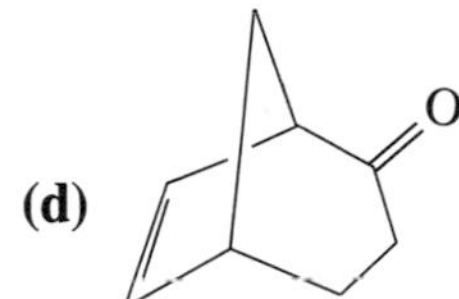

(e) $C_6H_5\overset{O}{\overset{\|}{C}}CH_3$

23. Propose efficient syntheses of each of the following molecules, beginning with the indicated starting materials.

(a) CHO-substituted cyclohexane with H_3C and OH on C-4 from CHO-substituted cyclohexane with H and OH on C-4

(b) cyclohexyl–C(=O)–CH_3 from CHO-substituted cyclohexane with H and OH on C-4

(c) 2-methyltetrahydrofuran-2-ol (OH, CH_3) from $ClCH_2CH_2CH_2OH$

24. The rate of the reaction of NH_2OH with aldehydes and ketones is very sensitive to pH. It is very slow in solutions more acidic than pH 2 or more basic than pH 7. It is fastest in moderately acidic solution (pH ~ 4). Suggest explanations for these observations.

25. Compound D, formula $C_8H_{14}O$, is converted by $CH_2{=}P(C_6H_5)_3$ into E, C_9H_{16}. Treatment of D with $LiAlH_4$ yields *two* isomeric products F and G, both $C_8H_{16}O$, in unequal yield. Heating either F or G with concentrated H_2SO_4 produces H, with the formula C_8H_{14}. Ozonolysis of H produces a keto aldehyde after $Zn–H^+$, H_2O treatment. Oxidation of this keto aldehyde with aqueous Cr(VI) produces

$$CH_3\overset{\overset{\large O}{\|}}{C}CH_2CH_2CH_2\overset{\overset{\large CH_3}{|}}{C}HCO_2H$$

Identify compounds D through H. Pay particular attention to the stereochemistry of D.

26. In 1862, it was discovered that cholesterol (for structure, see Section 4-7) is converted into a new substance named coprostanol by the action of bacteria in the human digestive tract. Make use of the following information to deduce the structure of coprostanol. Identify the structures of unknowns J through M as well. (i) Coprostanol, on treatment with Cr(VI) reagents, gives compound J, UV $\lambda_{max}(\epsilon) = 281(22)$ nm and IR $= 1710\ cm^{-1}$. (ii) Treatment of cholesterol with H_2 over Pt gives compound K, a stereoisomer of coprostanol. Treatment of K with the Cr(VI) reagent gives compound L, which has a UV peak very similar to that of J, $\lambda_{max}(\epsilon) = 258(23)$ nm, and turns out to be a stereoisomer of J. (iii) Careful treatment of cholesterol with Cr(VI) reagent produces M: UV $\lambda_{max}(\epsilon) = 286(109)$ nm. Treatment of M with H_2 over Pt gives compound L described above.

27. Three reactions involving compound M (see Problem 26) are described below. Answer the questions that follow the descriptions. **(a)** Treatment of M with catalytic amounts of acid in ethanol solvent causes isomerization to N: UV $\lambda_{max}(\epsilon) = 241(17{,}500)$ and $310(72)$ nm. Propose a structure for N. **(b)** Hydrogenation of N (H_2-Pd, ether solvent) produces J (Problem 26). Is this the result that you would have predicted, or is there anything unusual about it? **(c)** Wolff-Kishner reduction of N (H_2NNH_2, H_2O, HO^-, Δ) produces 3-cholestene. Propose a mechanism for this transformation.

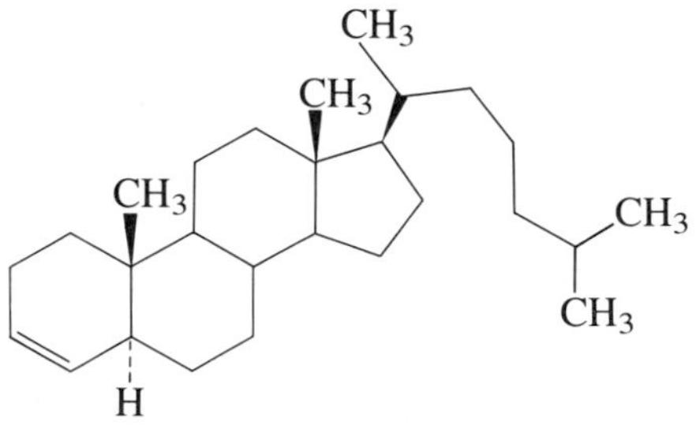

CH_3 CH_3 CH_3 CH_3 CH_3 H

3-Cholestene

18

Enols and Enones

α,β-Unsaturated Alcohols, Aldehydes, and Ketones

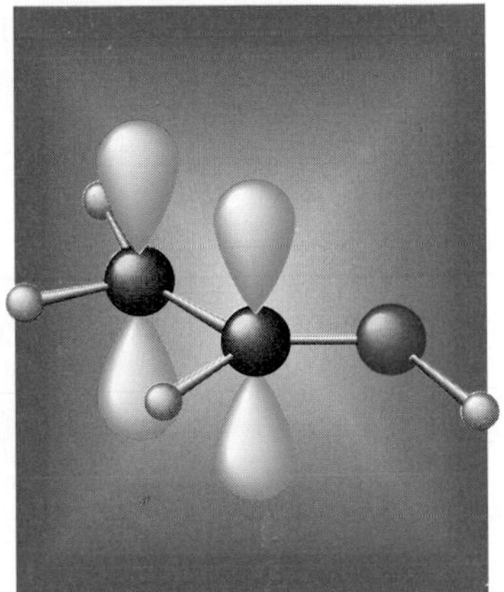

The last chapter examined two sites of reactivity in aldehydes and ketones. We saw in Sections 17-5 through 17-12 that the carbonyl oxygen is easily attacked by electrophiles (usually protons), its carbon by nucleophiles. We turn now to a third site of reactivity, the carbon *next to* the carbonyl group, also called the ***α*-carbon.** Hydrogens on α-carbons are rendered acidic by the neighboring carbonyl group; loss of such ***α*-hydrogens** leads to either of two electron-rich species: unsaturated alcohols, called enols, or their corresponding anions, known as enolate ions. Both species are important nucleophilic intermediates in reactions of carbonyl compounds. We shall see that they are attacked by electrophiles such as protons, alkylating agents, halogens, and even other carbonyl carbons.

α-carbon

α-hydrogen

We begin by introducing the chemistry of enolates and enols. Especially important will be a reaction between enolate ions and carbonyl compounds, the aldol condensation. This process is widely used to form carbon–carbon bonds both in the laboratory and in nature. Among the possible products of aldol condensation are α,β-unsaturated aldehydes and ketones, which contain conjugated carbon–carbon and carbon–oxygen π bonds. As expected, electrophilic additions may take place at the carbon–carbon double bond. More significantly, α,β-unsaturated carbonyl compounds are also subject to nucleophilic attack, which may occur at the carbonyl carbon or may involve the entire conjugated system.

18-1 Acidity of Aldehydes and Ketones: Enolate Ions

The pK_a values of aldehyde and ketone α-hydrogens range from 19 to 21, much lower than the pK_a values of ethene (44) or ethyne (25), but higher than those of alcohols (15–18). Strong bases can therefore remove an α-hydrogen. The anions that result are known as **enolate ions** or simply **enolates.**

Deprotonation of a Carbonyl Compound

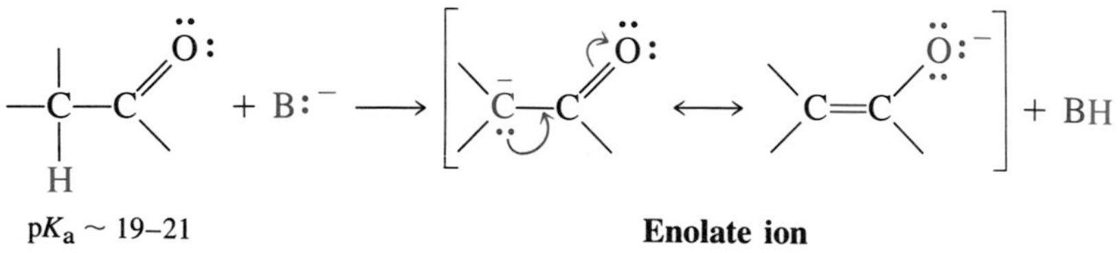

pK_a ~ 19–21

+ B:$^-$ ⟶ [⟷] + BH

Enolate ion

Why are aldehydes and ketones relatively acidic? We know that acid strength is enhanced by stabilization of the conjugate base (Section 6-7). Enolate ions are strongly stabilized by resonance, which delocalizes the negative charge onto an electronegative oxygen. An example of enolate formation is the deprotonation of cyclohexanone by lithium diisopropylamide (LDA; Section 7-8).

Enolate Preparation

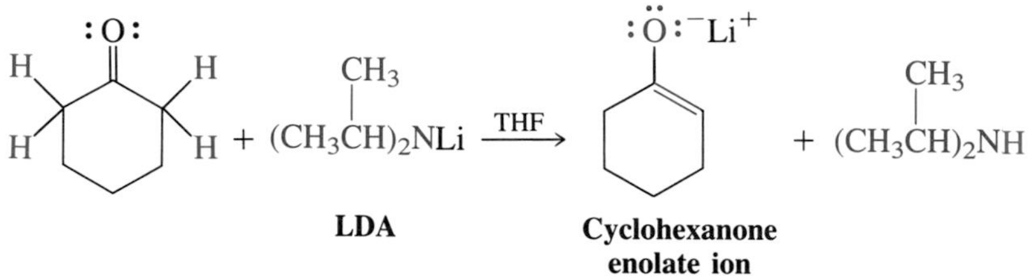

$$+ (CH_3CH)_2NLi \xrightarrow{THF} + (CH_3CH)_2NH$$

LDA **Cyclohexanone enolate ion**

EXERCISE 18-1

Identify the most acidic hydrogens in each of the following molecules. Give the structure of the enolate ion arising from deprotonation. **(a)** Acetaldehyde; **(b)** propanal; **(c)** propanone; **(d)** 4-heptanone; **(e)** cyclopentanone.

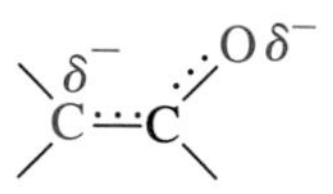

Resonance hybrid

Both resonance forms contribute to the characteristics of the enolate ion and thus to the chemistry of carbonyl compounds. The resonance hybrid possesses partial negative charges on both carbon and oxygen; as a result, electrophiles can attack at either position, although in most cases the carbon is attacked. A species that can react at two different sites to give two different products is called **ambident** ("two fanged": from *ambi,* Latin, both; *dens,* Latin, tooth). The enolate ion is thus an ambident anion. For example, alkylation of cyclohexanone enolate with 3-chloropropene occurs at carbon (C-alkylation), but protonation takes place at oxygen (O-protonation). The product of protonation is an unsaturated alcohol, called an **alkenol,** or **enol** for short. Enols are unstable and rapidly isomerize back to the original ketone (recall Section 13-8).

Ambident Behavior of Cyclohexanone Enolate Ion

:O: CH$_2$CH=CH$_2$ ← CH$_2$=CHCH$_2$Cl, THF / C-alkylation, − Cl$^-$ [:Ö:$^-$ ⟷ :O:]$^-$ H$^+$, H$_2$O / O-protonation → :ÖH → :O: H

62%

2-(2-Propenyl)cyclohexanone | Cyclohexanone enolate ion | Cyclohexanone enol | Cyclohexanone

EXERCISE 18-2

Give the products of reaction of cyclohexanone enolate with **(a)** iodoethane (reacts by C-alkylation) and **(b)** chlorotrimethylsilane, $(CH_3)_3Si–Cl$ (reacts by O-silylation).

In summary, the hydrogens on the carbon next to the carbonyl group in aldehydes and ketones are acidic, with pK_a values ranging from 19 to 21. Deprotonation leads to the corresponding enolate ions, which may be attacked by electrophilic reagents at either oxygen or carbon. Protonation at oxygen gives enols.

18-2 Keto–Enol Equilibria

We have seen that protonation of an enolate at oxygen leads to an enol. The enol, an unstable isomer of an aldehyde or ketone, rapidly converts to the carbonyl system: It **tautomerizes** (Section 13-8). These isomers are called **enol** and **keto tautomers.** This section begins by discussing factors influencing their equilibria, in which the keto form usually predominates. It then describes the mechanism of tautomerism and its chemical consequences.

An enol equilibrates with its keto form in acidic or basic solution

Enol–keto tautomerism proceeds by either acid or base catalysis. Base simply removes the proton from the enol oxygen, reversing the initial protonation. Subsequent (and slower) C-protonation furnishes the thermodynamically more stable keto form.

Base-Catalyzed Enol–Keto Equilibration

C=C–ÖH + :B$^-$ ⇌ [C=C–Ö:$^-$ ⟷ $\ddot{C}^-$–C=Ö] + BH ⇌ H–C–C=Ö + :B$^-$

Enol form | **Enolate ion** | **Keto form**

In the acid-catalyzed process, the enol form is protonated at the double bond to give the resonance-stabilized carbocation next to oxygen. This species is simply a protonated carbonyl function. Its deprotonation then gives the keto form.

Acid-Catalyzed Enol–Keto Equilibration

$$\text{>C=C(}\ddot{\text{O}}\text{H)} + H^+ \rightleftharpoons \left[-\overset{H}{\underset{|}{C}}-\overset{+}{C}(\ddot{\text{O}}\text{H}) \longleftrightarrow -\overset{H}{\underset{|}{C}}-C{=}\overset{+}{\ddot{\text{O}}}\text{H} \right] \rightleftharpoons -\overset{H}{\underset{|}{C}}-C{=}\ddot{\text{O}}\text{:} + H^+$$

Enol form **Protonated carbonyl system** **Keto form**

Both the acid- and base-catalyzed enol–keto interconversions occur relatively fast in solution, whenever there are traces of the required catalysts.

Substituents can shift the keto–enol equilibrium

The equilibrium constants for the conversion of the keto into the enol forms are very small for ordinary aldehydes and ketones, only traces of enol being present. However, relative to its keto form, the enol of acetaldehyde is about a hundred times more stable than the enol of propanone (acetone), because the less substituted aldehyde group is less stable than the more substituted ketone group.

Keto–Enol Equilibria

$$\underset{\textbf{Acetaldehyde}}{H{-}CH_2\overset{O}{\overset{\|}{C}}H} \underset{K = 6 \times 10^{-7}}{\rightleftharpoons} \underset{\textbf{Ethenol (Vinyl alcohol)}}{H_2C{=}C(OH)H} \qquad \Delta G° \sim +8.5 \text{ kcal mol}^{-1}$$

$$\underset{\textbf{Propanone (Acetone)}}{H{-}CH_2\overset{O}{\overset{\|}{C}}CH_3} \underset{K = 5 \times 10^{-9}}{\rightleftharpoons} \underset{\textbf{2-Propenol}}{H_2C{=}C(OH)CH_3} \qquad \Delta G° \sim +11.3 \text{ kcal mol}^{-1}$$

Enol formation leads to deuterium exchange and stereoisomerization

What are some consequences of enol formation by tautomerism? One is that treatment of a ketone with traces of acid or base in D_2O solvent leads to the complete exchange of *all* the α-hydrogens.

Hydrogen–Deuterium Exchange of Enolizable Hydrogens

$$\underset{\textbf{2-Butanone}}{CH_3\overset{O}{\overset{\|}{C}}CH_2CH_3} \xrightarrow{D_2O,\ DO^-} \underset{\textbf{1,1,1,3,3-Pentadeuterio-2-butanone}}{CD_3\overset{O}{\overset{\|}{C}}CD_2CH_3}$$

This reaction can be conveniently followed by ^{1}H NMR, because the signal for these hydrogens slowly disappears as each hydrogen is sequentially replaced by deuterium. In this way the number of α-hydrogens present in a molecule can be readily determined.

EXERCISE 18-3

Formulate mechanisms for the base- and acid-catalyzed replacement of a single α-hydrogen in propanone by deuterium from D_2O.

EXERCISE 18-4

Write the products (if any) of deuterium incorporation by the treatment of the following compounds with D_2O–NaOD.

(a) Cycloheptanone **(b)** 2,2-Dimethylpropanal
(c) 3,3-Dimethyl-2-butanone **(d)** O= [bicyclic ketone] CHO

Another consequence of enol formation, or **enolization,** is the ease with which stereoisomers at α-carbons interconvert. For example, treatment of cis-2,3-disubstituted cyclopentanones with mild base furnishes the corresponding trans isomers. The latter are more stable for steric reasons.

Base-Catalyzed Isomerism of an α-Substituted Ketone

O, $CH_2CH{=}CH_2$, CH_3 —10% KOH, CH_3CH_2OH⇌ O^-K^+, $CH_2CH{=}CH_2$, CH_3 ⇌ O, $CH_2CH{=}CH_2$, CH_3 >95%

The reaction proceeds through the enolate ion, in which the α-carbon is planar and, hence, no longer a stereocenter. Reprotonation from the side cis to the 3-methyl group results in the trans diastereomer (Section 5-5).

Another consequence of enolization is the difficulty of maintaining optical activity in a compound whose only stereocenter is an α-carbon. Why? As the (achiral) enol converts back to the keto form, a racemic mixture of *R* and *S* enantiomers is produced. For example, at room temperature, optically active 3-phenyl-2-butanone racemizes with a half-life of minutes in basic ethanol.

Racemization of Optically Active 3-Phenyl-2-butanone

C_6H_5, H, H_3C C—$\overset{O}{\overset{\|}{C}}CH_3$ —$CH_3CH_2O^-$, CH_3CH_2OH⇌ C_6H_5, H_3C C=C O^-, CH_3 —$CH_3CH_2O^-$, CH_3CH_2OH⇌ C_6H_5, H_3C, H C—$\overset{O}{\overset{\|}{C}}CH_3$

(*S*)-3-Phenyl-2-butanone Achiral **(*R*)-3-Phenyl-2-butanone**

In summary, aldehydes and ketones are in equilibrium with their enol forms, which are roughly 10 kcal mol^{-1} less stable. Keto–enol equilibration is catalyzed by acid or base. Enolization allows for easy H–D exchange in D_2O and causes stereoisomerization at stereocenters next to the functional group.

18-3 Halogenation of Aldehydes and Ketones

This section examines a reaction of the carbonyl group that can proceed through the intermediacy of either enols or enolate ions—halogenation. Aldehydes and ketones react with halogens at the carbon next to the carbonyl group. In contrast with deuteration, the extent of halogenation depends on whether acid or base catalysis has been used.

In the presence of acid, halogenation usually stops after the first halogen has been introduced, as shown in the following examples.

Acid-Catalyzed α-Halogenation of Ketones

$$H{-}CH_2\overset{\overset{\displaystyle O}{\|}}{C}CH_3 \xrightarrow{Br{-}Br,\ CH_3CO_2H,\ H_2O,\ 70^\circ C} BrCH_2\overset{\overset{\displaystyle O}{\|}}{C}CH_3 + HBr$$

44%

Bromopropanone (Bromoacetone)

2-Methylcyclohexanone $\xrightarrow{Cl{-}Cl,\ H^+,\ CCl_4}$ 2-Chloro-2-methylcyclohexanone + HCl

85%

2-Methylcyclohexanone **2-Chloro-2-methylcyclohexanone**

The rate of the acid-catalyzed halogenation is independent of the halogen concentration, an observation suggesting a rate-determining first step involving the carbonyl substrate. This step is enolization. The halogen then attacks the double bond to give an intermediate oxygen-stabilized halocarbocation. Subsequent deprotonation of this species furnishes the product.

Mechanism of the Acid-Catalyzed Bromination of Propanone (Acetone)

STEP 1. Enolization

$$CH_3\overset{\overset{\displaystyle :O:}{\|}}{C}CH_3 \underset{}{\overset{H^+}{\rightleftharpoons}} H_2C{=}C(\ddot{O}H)(CH_3)$$

STEP 2. Halogen attack

$$H_2C{=}C(\ddot{O}H)(CH_3) + Br{-}Br \longrightarrow \left[BrH_2C{-}\overset{+}{C}(\ddot{O}H)(CH_3) \longleftrightarrow BrH_2C{-}C({=}\overset{+}{\ddot{O}}H)(CH_3) \right] + Br^-$$

STEP 3. Deprotonation

$$BrCH_2\overset{\overset{+:O-H}{\|}}{C}CH_3 \longrightarrow BrCH_2\overset{\overset{:O:}{\|}}{C}CH_3 + H^+$$

Why is halogenation slower after the first halogen has been introduced? The answer lies in the requirement for enolization. To repeat halogenation, the halo carbonyl compound must enolize again by the usual acid-catalyzed mechanism. However, the electron-withdrawing power of the halogen makes protonation, the initial step in enolization, *more difficult* than in the original carbonyl compound.

Halogenation Slows Down Enolization

$$BrCH_2\overset{\overset{:O:}{\|}}{C}CH_3 \overset{H^+}{\rightleftharpoons} BrCH_2\overset{\overset{+:O-H}{\|}}{C}CH_3$$

Less basic than unsubstituted ketone (arrow to :O:)

Electron withdrawing (arrow under $BrCH_2$)

BOX 18-1 Haloform Reaction: A Test for Methyl Ketones

The base-mediated halogenation of methyl ketones can proceed beyond complete halogenation of the methyl group. The trihalomethyl substituent can function as a leaving group, giving rise eventually to a carboxylic acid and a trihalomethane. The process is called the *haloform reaction,* after the common name of the product.

Mechanism of the Haloform Reaction

$$R\overset{\overset{:O:}{\|}}{C}CBr_3 + {}^-:\ddot{O}H \longrightarrow R\overset{\overset{:\ddot{O}:^-}{|}}{\underset{\underset{:\ddot{O}H}{|}}{C}}-CBr_3 \longrightarrow$$

$$R\overset{\overset{:O:}{\|}}{C}\ddot{O}H + {}^-:CBr_3 \longrightarrow R\overset{\overset{:O:}{\|}}{C}\ddot{O}:^- + HCBr_3 \xrightarrow[H_2O]{H^+,} R\overset{\overset{O}{\|}}{C}OH + HCBr_3$$

When the halogen is iodine, triiodomethane (iodoform) precipitates as a yellow solid. Its formation, called the *iodoform reaction,* is a qualitative test for the $R\overset{\overset{O}{\|}}{C}CH_3$ structural unit.

An Iodoform Reaction

$$C_6H_5-\overset{\overset{O}{\|}}{C}-CH_3 \xrightarrow[2.\ H^+,\ H_2O]{1.\ I_2,\ NaOH} C_6H_5-\overset{\overset{O}{\|}}{C}-OH + \underset{\text{Yellow precipitate}}{CHI_3}$$

Iodoform, by the way, is a topical disinfectant. It was responsible for the "smell of a doctor's office" familiar to patients of past generations because of its widespread use and characteristic odor.

Mechanism of Halogenation of an Enolate Ion

$$^{-}\!:\ddot{O}\!-\!C(R)\!=\!CH_2 + Br\!-\!Br \longrightarrow R\overset{\overset{\displaystyle :O:}{\|}}{C}CH_2Br + Br^-$$

Therefore, the singly halogenated product is not attacked by additional halogen until the starting aldehyde or ketone has been used up.

Base-mediated halogenation is entirely different. It proceeds instead by the formation of an enolate ion, which then attacks the halogen. Here the reaction continues until it *completely* halogenates the α-carbon. Why is base-catalyzed halogenation so difficult to stop at the stage of monohalogenation? The electron-withdrawing power of the halogen increases the acidity of the remaining α-hydrogens, accelerating further enolate formation and hence further halogenation.

EXERCISE 18-5

Write the products of the acid- and base-catalyzed bromination of acetylcyclopentane.

In summary, halogenation in acid can proceed selectively to the monohalocarbonyl compounds. In base, complete replacement of all α-hydrogens occurs.

18-4 Alkylation of Aldehydes and Ketones

Alkylation of an enolate provides a general way to introduce an alkyl substituent next to the carbonyl group of a ketone. This section presents examples of this process and compares it with alkylation of another intermediate, an enamine.

Alkylation of enolates can be difficult to control

Most alkylations occur at carbon. When a ketone possesses only a *single* α-hydrogen, high yields of the alkylation product can be obtained. (Aldehydes cannot be alkylated in this way.)

Alkylation of a Ketone

$$C_6H_5\overset{\overset{\displaystyle O}{\|}}{C}CH(CH_3)_2 \xrightarrow[-\,H\text{–}H,\ -\,NaBr]{1.\ NaH,\ benzene,\ \Delta\quad 2.\ (CH_3)_2C{=}CHCH_2Br} C_6H_5\overset{\overset{\displaystyle O}{\|}}{C}C(CH_3)_2CH_2CH{=}C(CH_3)_2 \quad 88\%$$

2-Methyl-1-phenyl-1-propanone → **2,2,5-Trimethyl-1-phenyl-4-hexen-1-one**

In other cases, prevention of dialkylation is a problem. Under the reaction conditions, a monoalkylated ketone might become deprotonated by the starting enolate and could therefore undergo further alkylation.

Single and Double Alkylations of a Ketone

Bicyclic ketone (α-CH$_2$: H, H) $\xrightarrow[-\,H\text{–}H,\ -\,NaI]{1.\ NaH,\ CH_3OCH_2CH_2OCH_3\quad 2.\ CH_3I}$ monomethylated ketone (H, CH_3) 27% + dimethylated ketone (CH_3, CH_3) 38%

Another complication arises in the alkylation of unsymmetric ketones: Both α positions are subject to electrophilic attack.

Alkylation of an Unsymmetric Ketone

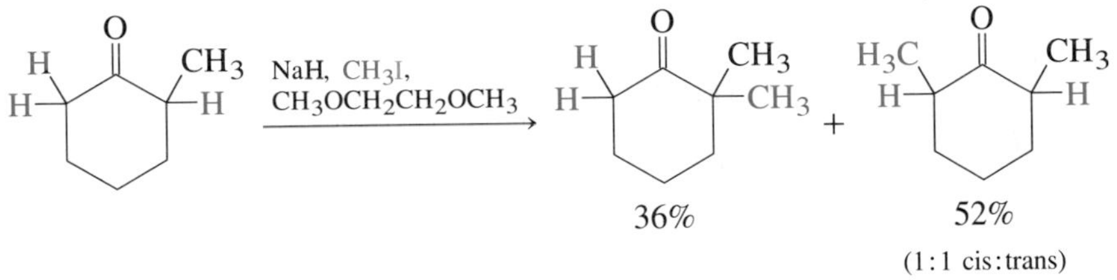

2-Methylcyclohexanone $\xrightarrow{\text{NaH, } CH_3I,\ CH_3OCH_2CH_2OCH_3}$ 2,2-dimethylcyclohexanone (36%) + 2,6-dimethylcyclohexanone (52%, 1:1 cis:trans)

EXERCISE 18-6

What is the mechanism of the C-alkylation of cyclohexanone enolate with 3-chloropropene (Section 18-1)? (Hint: See Chapter 6.) What product(s) would you expect from reaction of cyclohexanone enolate with **(a)** 2-bromopropane and **(b)** 2-bromo-2-methylpropane? (Hint: See Chapter 7.)

EXERCISE 18-7

The reaction of the compound shown in the margin with base gives three isomeric products $C_8H_{12}O$. What are they? (Hint: Try intramolecular alkylations.)

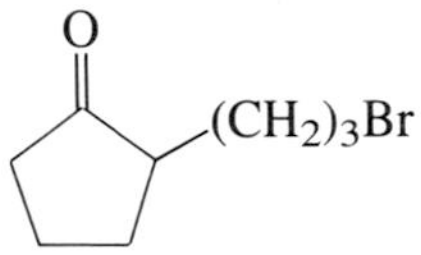

2-[$(CH_2)_3Br$]cyclopentanone

Enamines afford an alternative route for the alkylation of aldehydes and ketones

Section 17-9 showed that the reaction of secondary amines with aldehydes or ketones produces enamines. As the resonance forms shown below indicate, the nitrogen substituent renders the enamine carbon–carbon double bond electron rich. Furthermore, the dipolar resonance contribution gives rise to significant nucleophilicity at carbon, even though the enamine is neutral. As a result, electrophiles may attack at this position. Let us see how this attack may be used to synthesize alkylated aldehydes and ketones.

Resonance in Enamines

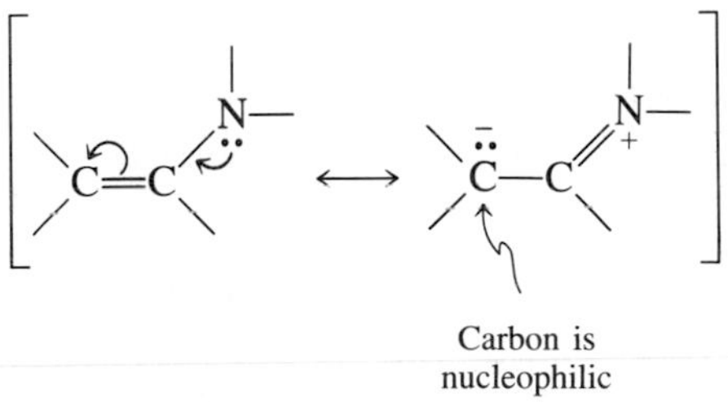

$$\left[>C{=}C{<}\ddot{N}{<} \longleftrightarrow >\overset{-}{\ddot{C}}{-}C{=}\overset{+}{N}{<} \right]$$

Carbon is nucleophilic

Exposure of enamines to haloalkanes results in alkylation at carbon to produce an iminium salt. On aqueous work-up, iminium salts hydrolyze by a mechanism that is the reverse of the one formulated for imine formation in Section 17-9. The results are a new alkylated aldehyde or ketone and the original secondary amine.

Alkylation of an Enamine

$$CH_3CH_2\overset{O}{\overset{\|}{C}}CH_2CH_3 + \text{Azacyclopentane (Pyrrolidine)} \xrightarrow[-\,H_2O]{H^+} \text{(pyrrolidinyl)}(CH_3CH_2)C{=}CHCH_3 \xrightarrow[-\,I^-]{CH_3{-}I}$$

3-Pentanone **Azacyclopentane (Pyrrolidine)**

Alkylated at carbon

$$\xrightarrow[-\,H^+]{HOH} CH_3CH_2\overset{O}{\overset{\|}{C}}CH(CH_3)_2$$

An iminium salt − pyrrolidine **2-Methyl-3-pentanone**

EXERCISE 18-8

Formulate the mechanism for the final step of the sequence shown above: hydrolysis of the iminium salt.

How does this sequence compare with alkylation of the enolate? It is far superior, because it minimizes double or multiple alkylation. It has the additional advantage that it can be used to prepare alkylated aldehydes, as shown below.

$$(CH_3)_2CH\overset{O}{\overset{\|}{C}}H \xrightarrow{\text{1. pyrrolidine (N–H)};\ 2.\ CH_3CH_2Br;\ 3.\ H^+,\ H_2O} (CH_3)_2\underset{CH_3CH_2}{C}\overset{O}{\overset{\|}{C}}H$$

67%

2-Methylpropanal **2,2-Dimethylbutanal**

EXERCISE 18-9

Alkylations of the enolate of ketone A are very difficult to stop before dialkylation occurs, as illustrated below. Show how you would use an enamine to prepare monoalkylated ketone B (see top of next page).

$$\text{A} \xrightarrow{1.\ NaH,\ C_6H_6;\ 2.\ BrCH_2CO_2C_2H_5} \text{(1,1-bis)}\ C_2H_5O_2CCH_2,\ CH_2CO_2C_2H_5$$

94%

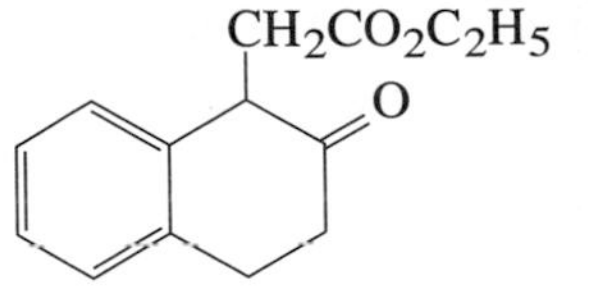

$$\text{B: 1-}(CH_2CO_2C_2H_5)\text{-2-tetralone}$$

B

In summary, enolates give rise to alkylated derivatives upon exposure to haloalkanes. In these reactions, control of the extent and the position of alkylation, when there is a choice, may be a problem. Enamines derived from aldehydes and ketones undergo alkylation to the corresponding iminium salts, which can hydrolyze to the alkylated carbonyl compounds.

18-5 Attack by Enolates on the Carbonyl Function: Aldol Condensation

The next three sections describe reactions of aldehydes and ketones that involve the attack of an enolate ion at a carbonyl carbon. The product of this process is a hydroxy carbonyl compound. Subsequent elimination of water leads to α,β-unsaturated aldehydes and ketones, the overall two-step sequence constituting a condensation reaction.

Aldehydes undergo base-catalyzed condensations

Addition of a small amount of dilute aqueous sodium hydroxide to acetaldehyde at low temperature initiates the conversion of the aldehyde to a dimer, 3-hydroxybutanal. Upon heating, this hydroxyaldehyde transforms into the final condensation product, the α,β-unsaturated aldehyde *trans*-2-butenal (crotonaldehyde). This reaction is an example of the **aldol condensation.**

Aldol Condensation of Acetaldehyde

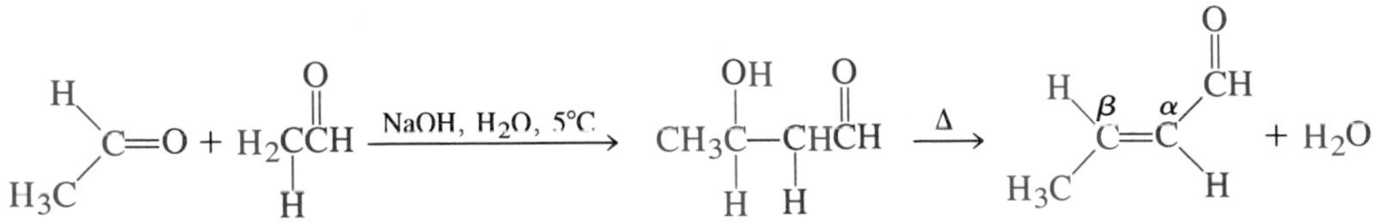

$$(H)(H_3C)C{=}O + H_2C(H)\overset{O}{\overset{\|}{C}}H \xrightarrow{NaOH,\ H_2O,\ 5°C} CH_3\underset{H}{\overset{OH}{C}}{-}\underset{H}{C}H\overset{O}{\overset{\|}{C}}H \xrightarrow{\Delta} \underset{H_3C}{\overset{H}{}}\overset{\beta}{C}{=}\overset{\alpha}{C}\underset{H}{\overset{\overset{O}{\|}}{CH}} + H_2O$$

3-Hydroxybutanal — ***trans*-2-Butenal (Crotonaldehyde)** (An α,β-unsaturated aldehyde)

The aldol condensation is general for aldehydes and, as we shall see, sometimes succeeds with ketones as well. We shall first describe its mechanism before turning to its uses in synthesis.

The aldol condensation highlights the two most important facets of carbonyl group reactivity: enolate formation and nucleophilic attack at a carbonyl carbon. The base sets up an equilibrium between a small amount of the aldehyde and its corresponding enolate ion. Because the ion is surrounded by excess aldehyde, its nucleophilic α-carbon can then attack the carbonyl group of another molecule of aldehyde. Protonation of the resulting alkoxide furnishes 3-hydroxybutanal, which has the common name *aldol* (from *ald*ehyde alcoh*ol*).

Mechanism of Aldol Formation

STEP 1. Enolate generation

$$HC(=O)-CH_2-H + {}^-:\ddot{O}H \rightleftharpoons H_2C{=}C(-\ddot{O}:^-)H + H\ddot{O}H$$

Small equilibrium concentration of enolate

STEP 2. Nucleophilic attack

$$CH_3CH(=O) \quad CH_2{=}C(-\ddot{O}:^-)H \rightleftharpoons CH_3CH(-\ddot{O}:^-)CH_2CH(=O)$$

STEP 3. Protonation

$$CH_3CH(-\ddot{O}:^-)CH_2CH(=O) + HOH \rightleftharpoons CH_3CH(-\ddot{O}H)CH_2CH(=O) + HO^-$$

50–60%

3-Hydroxybutanal (Aldol)

Note that hydroxide ion functions as a catalyst in this reaction. The last two steps of the sequence drive the initially unfavorable equilibrium toward product, but the overall reaction is not very exothermic and therefore is quite reversible. The aldol is formed in 50–60% yield and does not react further if its preparation is carried out at low temperature (5°C).

EXERCISE 18-10

Give the structure of the hydroxyaldehyde product of aldol condensation at 5°C of each of the following aldehydes. **(a)** Propanal; **(b)** butanal; **(c)** 2-phenylacetaldehyde; and **(d)** 3-phenylpropanal.

EXERCISE 18-11

Can benzaldehyde undergo aldol condensation? Why or why not?

At elevated temperature, the aldol is converted into its enolate ion, which eliminates hydroxide ion to yield the final product. The net result of this second sequence is a hydroxide-catalyzed dehydration of the aldol.

Mechanism of Dehydration

$$CH_3\underset{H}{\overset{OH}{C}}-\underset{H}{\overset{H}{C}}-\overset{O}{\overset{\|}{C}}H + {}^{-}\!:\ddot{O}H \rightleftharpoons CH_3\underset{H}{\overset{OH}{C}}-\overset{H}{C}=C\begin{matrix}\ddot{O}:^{-} \\ H\end{matrix} + HOH$$

$$CH_3\underset{H}{\overset{:\ddot{O}H}{C}}-\overset{H}{C}=C\begin{matrix}\ddot{O}:^{-} \\ H\end{matrix} \longrightarrow CH_3CH=CH\overset{:O:}{\overset{\|}{C}}H + H\ddot{O}:^{-}$$

What makes the aldol condensation synthetically useful? Depending on the temperature, it leads either to a hydroxycarbonyl compound or to the formation of a new carbon–carbon double bond, the latter characterizing an α,β-unsaturated carbonyl compound. For example, at low temperature,

$$\underset{\substack{\textbf{2-Methylpropanal}}}{CH_3\underset{CH_3}{C}H\overset{O}{\overset{\|}{C}}H} \xrightarrow{NaOH,\ H_2O,\ 5°C} \underset{\substack{85\% \\ \textbf{3-Hydroxy-2,2,4-trimethylpentanal}}}{CH_3\overset{CH_3}{C}H-\underset{H}{\overset{OH}{C}}-\underset{CH_3}{\overset{CH_3}{C}}CHO}$$

At higher temperature, however,

$$\underset{\textbf{Heptanal}}{CH_3(CH_2)_5CHO} \xrightarrow{K_2CO_3,\ H_2O,\ \Delta} \underset{\substack{80\% \\ \textbf{Z-2-Pentyl-2-nonenal}}}{CH_3(CH_2)_5\overset{\beta}{C}H=\overset{\alpha}{C}(CHO)(CH_2)_4CH_3}$$

EXERCISE 18-12

Give the structure of the α,β-unsaturated aldehyde product of aldol condensation at higher temperature of each of the aldehydes in Exercise 18-10.

Ketones can be substrates in the aldol condensation

So far, only aldehydes have been discussed as substrates in the aldol condensation. What about ketones? Treatment of propanone (acetone) with base does indeed lead to some 4-hydroxy-4-methyl-2-pentanone, but the yield is very low and the product is in equilibrium with the starting material.

BOX 18-2 Aldol Condensations in Nature

Aldol condensations occur in natural systems. For example, collagen fibers are strengthened by chemical cross-linking of aldehyde units through aldol condensations. Collagen is the most abundant fibrous protein in mammals, being the major fibrous component of skin, bone, tendon, cartilage, and teeth. One of its functions is to hold cells together in discrete units. Its structure is basically a staggered array of *tropocollagen* molecules, which consist of triply stranded helical polypeptide chains (Chapter 26). Cross-linking (see Sections 12-14 and 14-10 for cross-linking of polymers) of these chains is by aldol condensations catalyzed by enzymes. First, lysine residues (Section 26-1) in the chain are enzymatically oxidized to aldehyde derivatives. Then, an aldol condensation cross-links two chains.

The extent of cross-linking depends on the function of the tissue. For example, the collagen in the Achilles' tendon of rats is highly cross-linked, but that of the more flexible tail tendon is less so.

$$\text{H-N} \qquad\qquad \text{N-H}$$
$$\text{H-C-(CH}_2)_2\text{-CH}_2\text{-CH}_2\text{-}\overset{+}{\text{N}}\text{H}_3 \quad \text{H}_3\overset{+}{\text{N}}\text{-CH}_2\text{-CH}_2\text{-(CH}_2)_2\text{-C-H}$$
$$\text{O=C} \qquad\qquad \text{C=O}$$

Lysine residues

Oxidation

$$\text{H-C-(CH}_2)_2\text{-CH}_2\text{-C(=O)H} \quad \text{HC(=O)-CH}_2\text{-(CH}_2)_2\text{-C-H}$$

Aldehyde derivatives

$- H_2O$

$$\text{H-C-(CH}_2)_2\text{-CH}_2\text{-CH=C(CHO)-(CH}_2)_2\text{-C-H}$$

Aldol cross-link

Aldol Formation from Propanone (Acetone)

$$\underset{94\%}{CH_3\overset{\overset{\displaystyle O}{\|}}{C}CH_3} \underset{}{\overset{HO^-}{\rightleftharpoons}} \underset{6\%}{CH_3\overset{\overset{\displaystyle OH}{|}}{\underset{\underset{\displaystyle CH_3}{|}}{C}}CH_2\overset{\overset{\displaystyle O}{\|}}{C}CH_3}$$

4-Hydroxy-4-methyl-2-pentanone

What explains the lesser driving force of the aldol reaction of ketones? Because their carbonyl bond is somewhat stronger (by about 3 kcal mol^{-1}) than that in aldehydes, aldol addition of ketones is endothermic. To drive the reaction forward, we can extract the product alcohol continuously from the reaction mixture as it is formed. Alternatively, under more vigorous conditions, dehydration and removal of water move the equilibrium toward the α,β-unsaturated ketone, as shown in the margin.

$$CH_3\overset{\overset{\displaystyle OH}{|}}{\underset{\underset{\displaystyle CH_3}{|}}{C}}CH_2\overset{\overset{\displaystyle O}{\|}}{C}CH_3 \xrightarrow{NaOH,\ H_2O,\ \Delta} (H_3C)_2C{=}CH\overset{\overset{\displaystyle O}{\|}}{C}CH_3$$

80%

4-Methyl-3-penten-2-one

+

H_2O (removed)

EXERCISE 18-13

Formulate the mechanism for the aldol addition of propanone. This process is reversible. Propose a mechanism for the conversion of 4-hydroxy-4-methyl-2-pentanone to propanone in the presence of ^-OH, an example of a *retro-aldol reaction.*

In summary, treatment of enolizable aldehydes with catalytic base leads to hydroxy aldehydes at low temperature, to α,β-unsaturated aldehydes upon heating. The reaction proceeds by enolate attack on the carbonyl function. Aldol addition to a ketone carbonyl group is energetically unfavorable. To drive the aldol condensation of ketones to product, special conditions have to be used, such as removal of the water or the aldol formed in the reaction.

18-6 Crossed Aldol Condensation

A drawback of the aldol reaction is the lack of selectivity when two different aldehydes are present. In such a situation, called **crossed aldol condensation,** enolates of both aldehydes are present and may react with the carbonyl groups of either starting compound. For example, a 1:1 mixture of acetaldehyde and propanal gives the four possible aldol addition products in comparable amounts.

Nonselective Crossed Aldol Reaction of Acetaldehyde and Propanal

(All four reactions occur simultaneously)

$$CH_3\overset{\overset{\displaystyle O}{\|}}{C}H + CH_3\overset{\overset{\displaystyle O}{\|}}{C}H + \underset{\text{Not involved}}{CH_3CH_2\overset{\overset{\displaystyle O}{\|}}{C}H} \longrightarrow CH_3\overset{\overset{\displaystyle OH}{|}}{\underset{\underset{\displaystyle H}{|}}{C}}{-}CH_2\overset{\overset{\displaystyle O}{\|}}{C}H$$

3-Hydroxybutanal

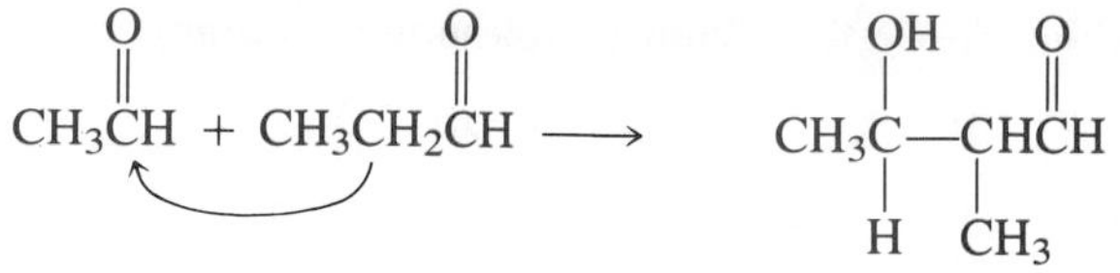

$$\mathrm{CH_3\overset{O}{\overset{\|}{C}}H + CH_3CH_2\overset{O}{\overset{\|}{C}}H \longrightarrow CH_3\underset{H}{\overset{OH}{C}}-\underset{CH_3}{C}H\overset{O}{\overset{\|}{C}}H}$$

3-Hydroxy-2-methylbutanal

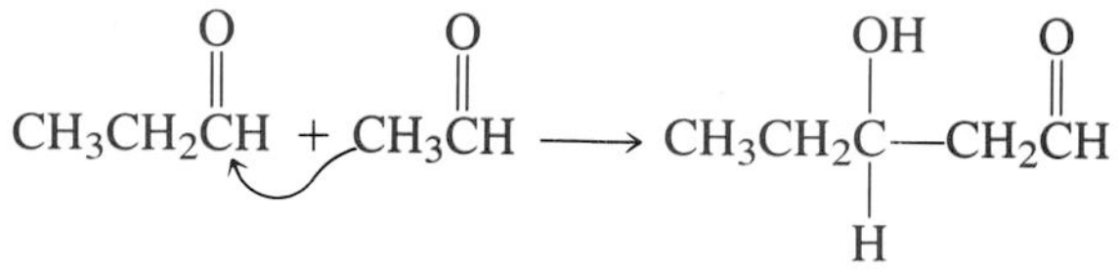

$$\mathrm{CH_3CH_2\overset{O}{\overset{\|}{C}}H + CH_3\overset{O}{\overset{\|}{C}}H \longrightarrow CH_3CH_2\underset{H}{\overset{OH}{C}}-CH_2\overset{O}{\overset{\|}{C}}H}$$

3-Hydroxypentanal

$$\mathrm{CH_3CH_2\overset{O}{\overset{\|}{C}}H + CH_3CH_2\overset{O}{\overset{\|}{C}}H + \underset{\text{Not involved}}{CH_3\overset{O}{\overset{\|}{C}}H} \longrightarrow CH_3CH_2\underset{H}{\overset{OH}{C}}-\underset{CH_3}{C}H\overset{O}{\overset{\|}{C}}H}$$

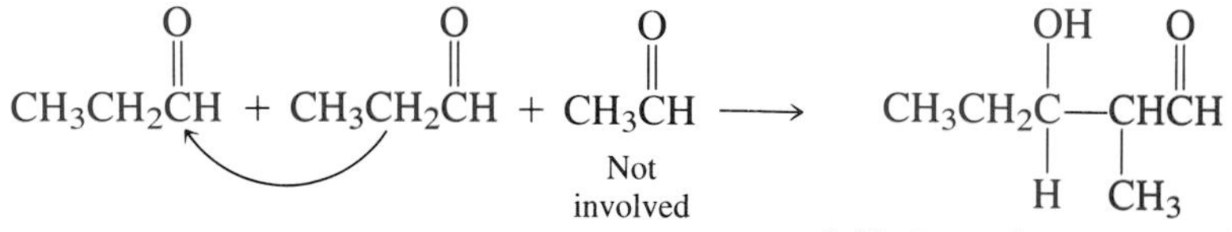

3-Hydroxy-2-methylpentanal

Can we ever efficiently synthesize a single aldol product from the reaction of two different aldehydes? We can, when one of the aldehydes has *no enolizable hydrogens,* because two of the four possible condensation products cannot form. We add the enolizable aldehyde slowly to an excess of the nonenolizable reactant. As soon as the enolate ion forms, it reacts preferentially with the other aldehyde.

A Successful Crossed Aldol Condensation

$$\mathrm{CH_3\underset{CH_3}{\overset{CH_3}{C}}CHO + \underset{\text{Added slowly}}{CH_3CH_2CHO} \xrightarrow{NaOH,\ H_2O,\ \Delta} CH_3\underset{CH_3}{\overset{CH_3}{C}}CH{=}\overset{CH_3}{C}CHO + H_2O}$$

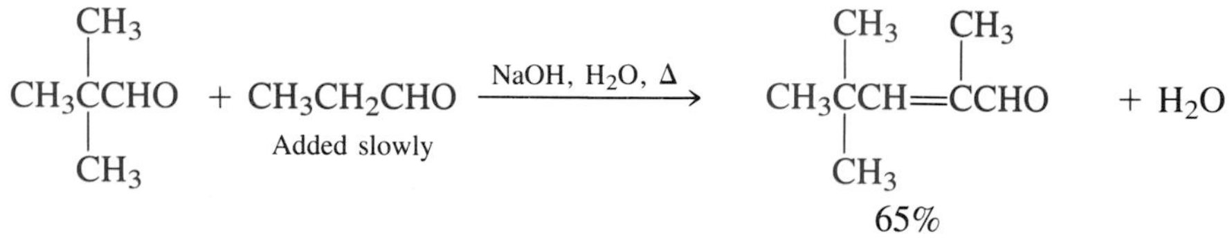

2,2-Dimethylpropanal   **Propanal**   65% **2,4,4-Trimethyl-2-pentenal**

EXERCISE 18-14

Show the likely products of the following aldol condensations.

(a) 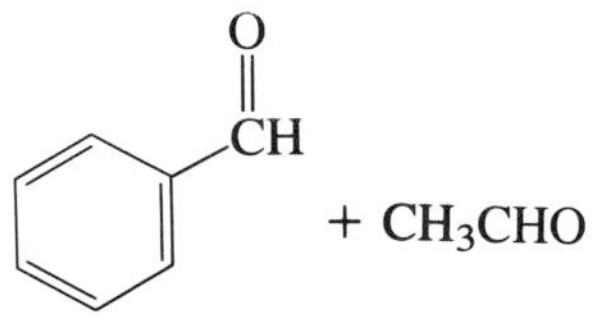 $\mathrm{C_6H_5\overset{O}{\overset{\|}{C}}H + CH_3CHO}$ **(b)** 2 cyclohexane-CHO (reacts with itself)

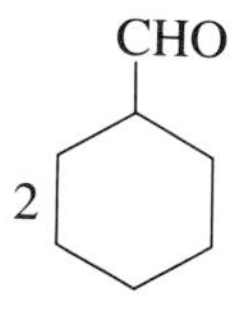

(c) $\mathrm{CH_2{=}CHCHO + CH_3CH_2CHO}$

In summary, crossed aldol condensations furnish product mixtures unless one of the reaction partners cannot enolize.

BOX 18-3 Enzymes in Crossed Aldol Condensations

The biosynthesis of sugars uses crossed aldol condensations (catalyzed by an enzyme appropriately named *aldolase*) to construct carbon–carbon bonds. A protein-bound primary amine (a substituent on the amino acid lysine) first condenses with the carbonyl group of 1,3-dihydroxypropanone (1,3-dihydroxyacetone) monophosphate to form an iminium salt. Enzyme-catalyzed deprotonation gives an enamine, whose nucleophilic carbon attacks the carbonyl group of 2,3-dihydroxypropanal (glyceraldehyde) 3-phosphate, in a nitrogen analog of a crossed aldol condensation. Hydrolysis of the resulting iminium salt furnishes the six-carbon sugar *fructose*.

Enzyme—NH_2 + O=C($CH_2OPO_3^{2-}$)(CHH–OH) $\xrightarrow{-H_2O}$ Enzyme—$\overset{+}{N}$(H)=C($CH_2OPO_3^{2-}$)(C---H, HO, H) :Base ⇌

1,3-Dihydroxyacetone monophosphate

Enzyme—$\ddot{N}$(H)—C($CH_2OPO_3^{2-}$)=C(H)(HO) + H–C(=O)–C(H)(HO)($CH_2OPO_3^{2-}$) ⟶ Enzyme—$\overset{+}{N}$(H)=C[$CH_2OPO_3^{2-}$]–C(HO, H)–C(H, OH)–C(H, OH)–$CH_2OPO_3^{2-}$ $\xrightarrow{H_2O}$ O=C[$CH_2OPO_3^{2-}$]–C(HO, H)–C(H, OH)–C(H, OH)–$CH_2OPO_3^{2-}$

Enamine

(*R*)-Glyceraldehyde-3-phosphate

Fructose-1,6-diphosphate

18-7 Intramolecular Aldol Condensation

It is possible to carry out aldol condensation between an enolate ion and a carbonyl group *in the same molecule;* such a reaction is called an **intramolecular aldol condensation.** This section describes the utility of this type of reaction in the synthesis of cyclic compounds.

Treatment of a dilute solution of hexanedial with aqueous base results in formation of a cyclic product. As soon as an enolate ion is generated, it attacks the carbonyl carbon at the opposite end of the molecule.

$$\underset{\textbf{Hexanedial}}{\overset{O}{\overset{\|}{H C}}CH_2CH_2CH_2CH_2\overset{O}{\overset{\|}{C}}H} \xrightarrow{KOH,\ H_2O} \underset{\textbf{1-Cyclopentenecarbaldehyde}}{\underset{62\%}{\text{(cyclopentene with }\overset{O}{\overset{\|}{C}}H\text{ and H)}}} + H_2O$$

Attack at the carbonyl carbon of a different molecule (*intermolecular* aldol condensation) is minimized by the low concentration of the dialdehyde and the favorable kinetics of five-membered ring formation.

Intramolecular ketone condensations are a ready source of cyclic and bicyclic enones. Usually the least strained ring is formed, typically one that is five or six membered. Thus, reaction of 2,5-hexanedione results in condensation between an enolate at C1 and the C5 carbonyl group. An alternative pathway, attack of a C3 enolate on C5 to give a strained three-membered ring, is not observed.

Intramolecular Aldol Condensation of a Dione

$$\underset{\textbf{2,5-Hexanedione}}{CH_3\overset{O}{\overset{\|}{\underset{5}{C}}}CH_2\underset{3}{C}H_2\overset{O}{\overset{\|}{\underset{1}{C}}}CH_3} \xrightarrow{NaOH,\ H_2O} \underset{\textbf{3-Methyl-2-cyclopentenone}}{\underset{42\%}{\text{(cyclopentenone with }CH_3)}} + H_2O$$

Why is intramolecular aldol condensation of ketones so readily achieved when the intermolecular analog is so difficult? Part of the answer lies in the thermodynamics of the process. We know that reaction of two ketone molecules suffers from an unfavorable enthalpy contribution because of the loss of a strong carbonyl bond (Section 18-5). In addition, the process must overcome the loss of entropy arising from the conversion of two molecules into one as the initial hydroxyketone product forms. In contrast, the intramolecular transformation merely converts an acyclic substrate into a cyclic product, a process "costing" much less in entropic terms and therefore more favorable thermodynamically.

EXERCISE 18-15

Predict outcome for intramolecular aldol condensations of the following compounds.

(a) Cyclodecane-1,5-dione **(b)** $C_6H_5\overset{O}{\overset{\|}{C}}(CH_2)_2\overset{O}{\overset{\|}{C}}CH_3$

(c) (cyclopentanone bearing) $CH_2\underset{\underset{O}{\|}}{C}(CH_2)_3CH_3$ **(d)** 2,7-Octanedione

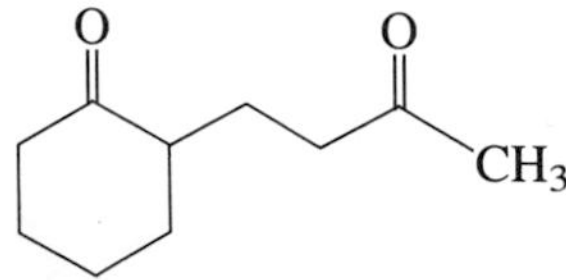

2-(3-Oxobutyl)cyclohexanone

EXERCISE 18-16

The intramolecular aldol condensation of 2-(3-oxobutyl)cyclohexanone (margin) can in principle lead to four different compounds (ignoring stereochemistry). Draw them and suggest which one would be the most likely to form.

EXERCISE 18-17

Prepare the following compounds from any starting material, using aldol reactions in the crucial step. (Hint: The last preparation requires a double aldol addition.)

(a) 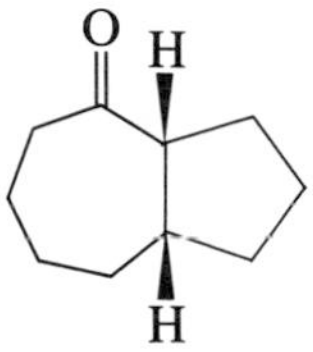 (b) 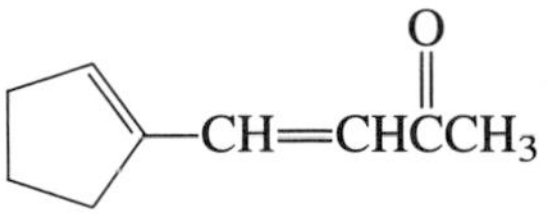 $-CH{=}CHCCH_3$ (c)

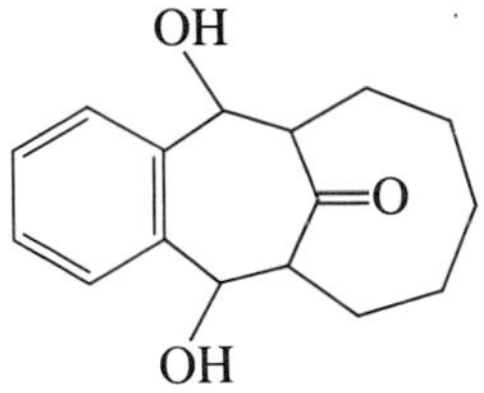

In summary, intramolecular aldol condensation succeeds with both aldehydes and ketones. It can be highly selective and gives the least strained cycloalkenones.

18-8 Other Preparations of α,β-Unsaturated Aldehydes and Ketones

The remainder of the chapter is about α,β-unsaturated aldehydes and ketones. We shall find that their chemistry either may be a simple composite of the individual behavior of their two types of double bonds or may involve the α,β-unsaturated carbonyl, or **enone,** functional group as a whole. As later chapters will indicate, this complex reactivity is quite typical of molecules with two functional groups, or **difunctional compounds.**

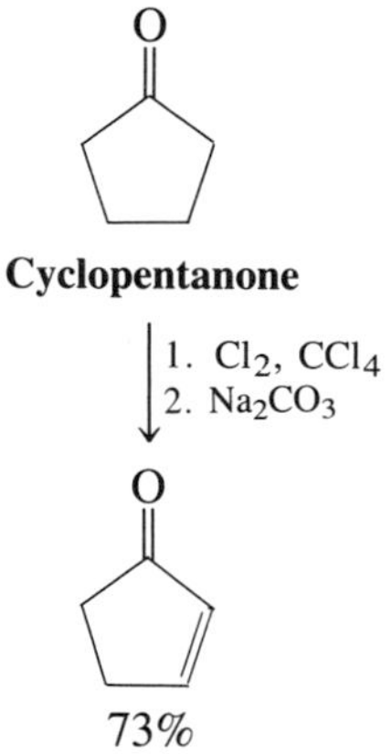

Sections 18-5 through 18-7 showed how α,β-unsaturated aldehydes and ketones are prepared by the aldol condensation. There are other synthetic routes to this class of molecules. For example, the carbon–carbon double bond can be introduced next to the carbonyl function by halogenation (Section 18-3), followed by base-mediated dehydrohalogenation, as shown in the margin.

Cyclopentanone

1. Cl_2, CCl_4
2. Na_2CO_3

73%

2-Cyclopentenone

Another way of introducing the double bond is by a Wittig reaction (Section 17-12) of carbonyl ylides, which are stabilized by resonance.

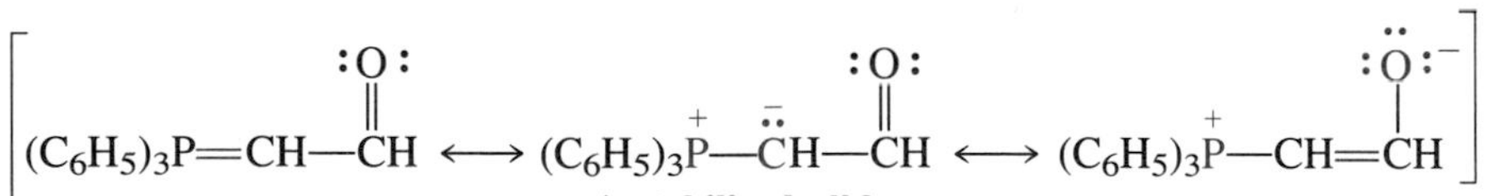

$$\left[(C_6H_5)_3P{=}CH{-}\overset{:O:}{\overset{\|}{C}}H \longleftrightarrow (C_6H_5)_3\overset{+}{P}{-}\overset{..}{\overset{-}{C}}H{-}\overset{:O:}{\overset{\|}{C}}H \longleftrightarrow (C_6H_5)_3\overset{+}{P}{-}CH{=}\overset{:\ddot{O}:^-}{\overset{|}{C}}H\right]$$

A stabilized ylide

Such **stabilized ylides** are comparatively unreactive and can be readily isolated and stored. However, they do react with aldehydes to form the corresponding α,β-unsaturated aldehydes, as the following example shows.

Wittig Synthesis of an α,β-Unsaturated Aldehyde

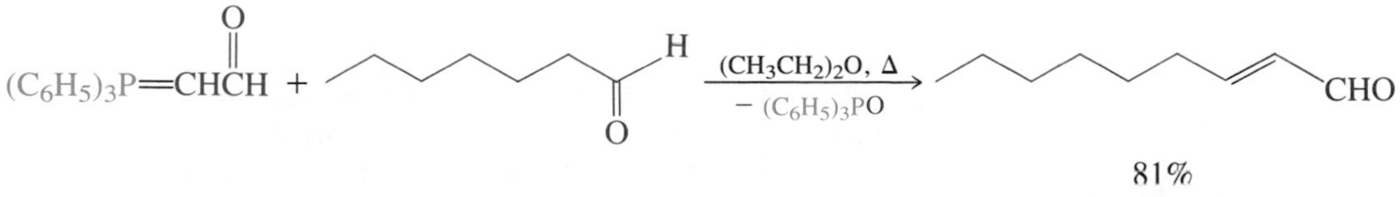

$$(C_6H_5)_3P{=}CH\overset{O}{\overset{\|}{C}}H + \text{Heptanal} \xrightarrow[-\,(C_6H_5)_3PO]{(CH_3CH_2)_2O,\ \Delta} \text{trans-2-Nonenal (CHO)}$$

Heptanal

81%

***trans*-2-Nonenal**

A fourth way to prepare α,β-unsaturated aldehydes and ketones is by oxidation of allylic alcohols with manganese dioxide, MnO_2 (Section 17-4). Vitamin A, for example, can be oxidized in this way to all-*trans*-retinal, a molecule of importance in the chemistry of vision (see Box 18-4).

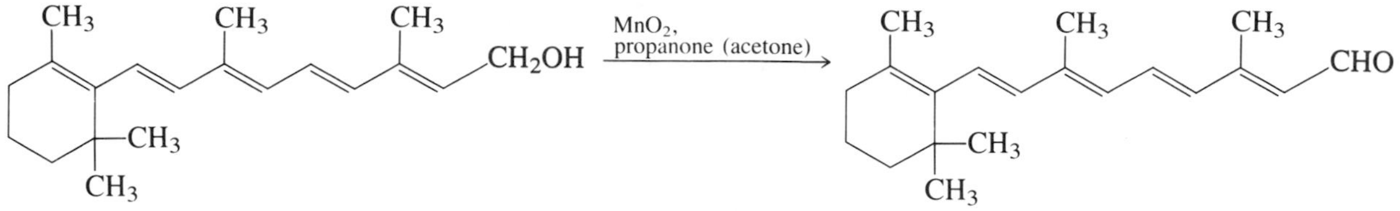

$$\text{Vitamin A} \xrightarrow{MnO_2,\ \text{propanone (acetone)}} \text{All-}trans\text{-retinal} \quad 80\%$$

Vitamin A | **All-*trans*-retinal**

In summary, this section reviewed synthetic methods of preparing α,β-unsaturated aldehydes and ketones. These are aldol condensations, halogenation–dehydrohalogenation of saturated aldehydes and ketones, Wittig reactions with stabilized ylides, and MnO_2 oxidations of allylic alcohols.

18-9 Properties of α,β-Unsaturated Aldehydes and Ketones

Conjugation lends additional stability to α,β-unsaturated aldehydes and ketones relative to their nonconjugated counterparts. This section presents consequences of this stabilization and describes reactions of enone systems characteristic of their component functional groups.

Conjugated unsaturated aldehydes and ketones are more stable than their unconjugated isomers

Like conjugated dienes (Section 14-5), α,β-unsaturated aldehydes and ketones are stabilized by resonance.

Resonance in 2-Butenal

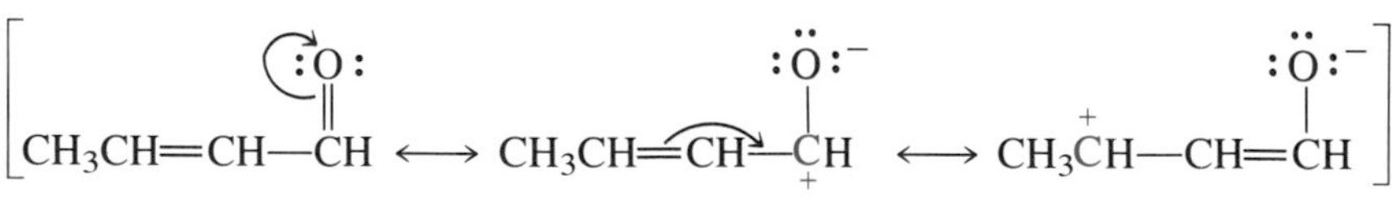

$$\left[CH_3CH{=}CH{-}\overset{\overset{\displaystyle :\ddot{O}:}{\|}}{C}H \longleftrightarrow CH_3CH{=}CH{-}\overset{\overset{\displaystyle :\ddot{O}:^-}{|}}{\underset{+}{C}}H \longleftrightarrow CH_3\overset{+}{C}H{-}CH{=}\overset{\overset{\displaystyle :\ddot{O}:^-}{|}}{C}H \right]$$

Thus, unconjugated β,γ-unsaturated carbonyl compounds rearrange readily to their conjugated isomers. The carbon–carbon double bond is said to ''move into conjugation'' with the carbonyl group, as the following examples show.

Isomerization of β,γ-Unsaturated Carbonyl Compounds to Conjugated Systems

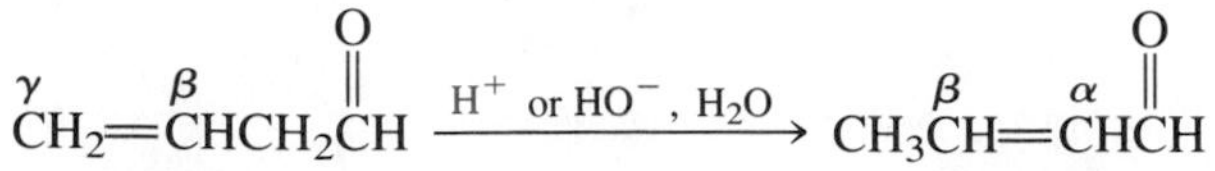

$$\overset{\gamma}{C}H_2{=}\overset{\beta}{C}HCH_2\overset{\overset{\displaystyle O}{\|}}{C}H \xrightarrow{H^+ \text{ or } HO^-,\ H_2O} CH_3\overset{\beta}{C}H{=}\overset{\alpha}{C}H\overset{\overset{\displaystyle O}{\|}}{C}H$$

3-Butenal | **2-Butenal**

$$\text{3-Cyclohexenone} \xrightarrow{H^+ \text{ or } HO^-,\ H_2O} \text{2-Cyclohexenone}$$

3-Cyclohexenone **2-Cyclohexenone**

The isomerization can be acid or base catalyzed. In the base-catalyzed reaction, the intermediate is the conjugated dienolate ion, which is reprotonated at the carbon terminus.

Mechanism of Base-Mediated Isomerization of β,γ-Unsaturated Carbonyl Compounds

$$CH_2{=}CHCH_2\overset{\overset{:O:}{\|}}{C}H + H\ddot{O}:^- \rightleftharpoons$$

Protonation here gives conjugated aldehyde

$$\left[CH_2{=}CH{-}\ddot{\bar{C}}H{-}\overset{\overset{:O:}{\|}}{C}H \longleftrightarrow CH_2{=}CH{-}CH{=}\overset{\overset{:\ddot{O}:^-}{|}}{C}H \longleftrightarrow :\bar{C}H_2{-}CH{=}CH{-}\overset{\overset{:O:}{\|}}{C}H\right] + H\ddot{O}H \rightleftharpoons$$

Dienolate ion

$$CH_3CH{=}CH\overset{\overset{:O:}{\|}}{C}H + H\ddot{O}:^-$$

EXERCISE 18-18

Propose a mechanism for the *acid*-catalyzed isomerization of 3-butenal to 2-butenal. (Hint: An intermediate is the *dienol* shown in the margin.)

$$CH_2{=}CHCH{=}C(\ddot{O}H)H$$

Dienol

α,β-Unsaturated aldehydes and ketones undergo the reactions typical of their component functional groups

α,β-Unsaturated aldehydes and ketones undergo many reactions that are perfectly predictable from the known chemistry of the carbon–carbon and carbon–oxygen double bonds. For example, hydrogenation by palladium on carbon gives the saturated carbonyl compound.

$$\xrightarrow{H_2,\ \text{Pd-C},\ CH_3CO_2CH_2CH_3\ \text{(ethyl acetate solvent)}} \quad 95\%$$

The double bond of some conjugated enones and enals can undergo ''conjugate reduction'' by a reducing system described earlier in the conversion of alkynes into trans alkenes: an alkali metal in liquid ammonia (Section 13-7). The mechanisms of these reductions are similar and include two one-electron transfers and

BOX 18-4 Reactions of Unsaturated Aldehydes in Nature: The Chemistry of Vision

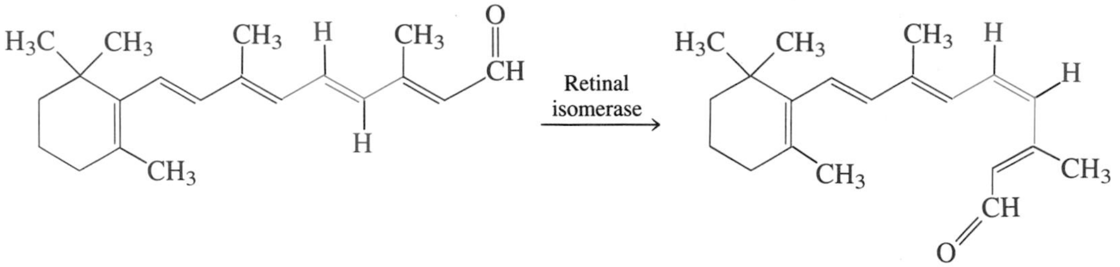

H_3C CH_3 CH_3 H CH_3 O CH H CH_3 —Retinal isomerase→ H_3C CH_3 CH_3 H H CH_3 CH_3 CH O

***trans*-Retinal** **cis-Retinal**

Vitamin A (retinol) is an important nutritional factor in vision. Vitamin A deficiency causes night blindness. Living organisms use an enzyme called *retinol dehydrogenase* to oxidize the vitamin to *trans*-retinal. This molecule is present in the light receptor cells of the human eye, but before it can fulfill its biological function it has to be isomerized by another enzyme, *retinal isomerase,* to give *cis*-retinal. This molecule fits well into the active site of a protein called *opsin* (approximate molecular weight 38,000). As shown below, *cis*-retinal reacts with one of the amine substituents of opsin to form the imine *rhodopsin,* the light-sensitive chemical unit in the eye. The electronic spectrum of rhodopsin, with a λ_{max} at 506 nm (ϵ = 40,000), has been interpreted as being indicative of the presence of a protonated imine group.

When a photon strikes rhodopsin, the *cis*-retinal part isomerizes extremely rapidly, in only picoseconds (10^{-12} s), to the trans isomer. This isomerization induces a tremendous geometric change, which appears to severely disrupt the snug fit of the original molecule in the protein cavity. Within nanoseconds (10^{-9} s), a series of new intermediates form from this photoproduct, accompanied by conformational changes in the protein structure, followed by eventual hydrolysis of the ill-fitting retinal unit. This sequence initiates a nerve impulse perceived by us as light. The *trans*-retinal is then reisomerized to the cis form by retinal isomerase and reforms rhodopsin, ready for another photon. What is extraordinary about this mechanism is its sensitivity, which allows the eye to register as little as one photon impinging on the retina. Curiously, all known visual systems in nature, even though they might have a completely different evolutionary history, use the retinal system for visual excitation. Evidently, this molecule offers an optimal solution to the problem of vision.

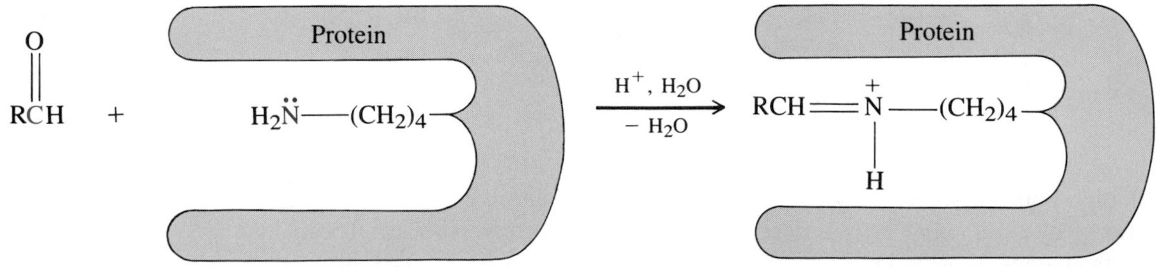

$$\mathrm{RCH{=}O} \; + \; \mathrm{H_2\ddot{N}{-}(CH_2)_4{-}Protein} \xrightarrow[-\,H_2O]{H^+,\,H_2O} \mathrm{RCH{=}\overset{+}{N}(H){-}(CH_2)_4{-}Protein}$$

***cis*-Retinal** **Opsin** **Rhodopsin**

two protonations. The method allows for the selective reduction of the conjugated double bond in the presence of unconjugated ones.

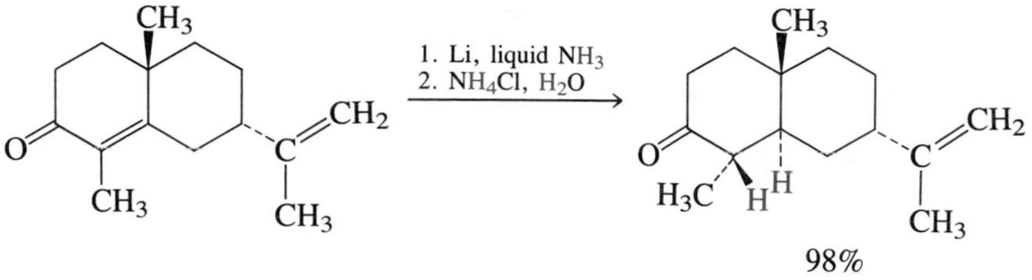

$$\xrightarrow[\text{2. } NH_4Cl,\ H_2O]{\text{1. Li, liquid } NH_3}$$

98%

Electrophilic attack is at the carbon–carbon π system. For example, bromination furnishes a dibromocarbonyl compound.

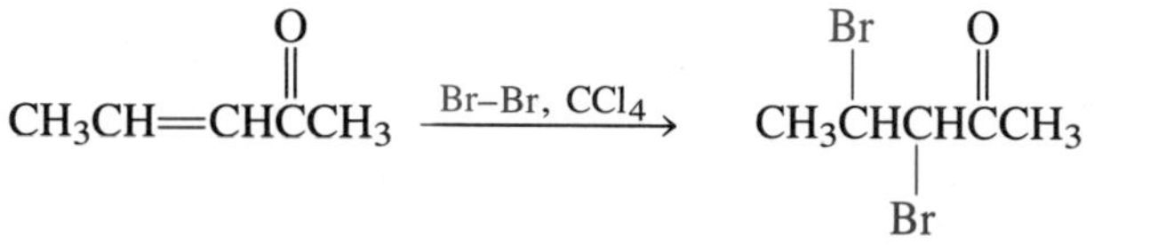

$$CH_3CH{=}CHCCH_3 \xrightarrow{Br{-}Br,\ CCl_4} CH_3CHCHCCH_3$$

60%

3-Penten-2-one **3,4-Dibromo-2-pentanone**

The carbonyl function undergoes the usual addition reactions (Section 17-9). Thus, nucleophilic addition of amines results in the expected condensation products (see, however, the next section).

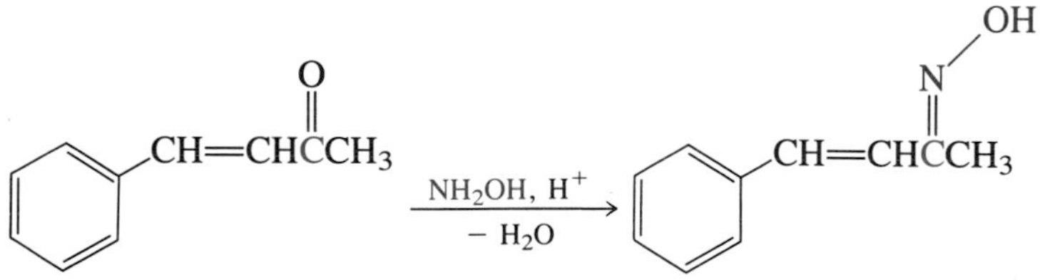

$$C_6H_5CH{=}CHCCH_3 \xrightarrow[-\ H_2O]{NH_2OH,\ H^+} C_6H_5CH{=}CHC(=NOH)CH_3$$

4-Phenylbut-3-en-2-one **Oxime** (m.p. 115°C)

EXERCISE 18-19

Propose a synthesis of 1-pentanol starting from propanal.

In summary, α,β-unsaturated aldehydes and ketones are more stable than their nonconjugated counterparts. Either base or acid catalyzes interconversion of the isomeric systems. Reactions typical of alkenes and carbonyl compounds are also characteristic of α,β-unsaturated aldehydes and ketones.

18-10 Conjugate Additions to α,β-Unsaturated Aldehydes and Ketones

We now show how the conjugated carbonyl group of α,β-unsaturated aldehydes and ketones can enter into reactions that involve the entire functional system. These are 1,4-additions of the type encountered with 1,3-butadiene (Section 14-6). Depending on the reagents, the reactions proceed by acid-catalyzed, radical, or nucleophilic addition mechanisms.

1,4-Additions involve the entire conjugated system

Addition reactions involving only one of the π bonds of a conjugated system are classified as 1,2-additions (compare Section 14-6). Examples are the additions of Br_2 to the carbon–carbon double bond and NH_2OH to the carbon–oxygen double bond of α,β-unsaturated aldehydes and ketones discussed in the previous section.

1,2-Addition of a Polar Reagent A–B to a Conjugated Enone

$$>C{=}C{<}\text{–}C({=}O)\text{–} \xrightarrow{A\text{–}B} \text{–}C(B)\text{–}C(A)\text{–}C({=}O)\text{–} \quad \text{or} \quad >C{=}C{<}\text{–}C(OA)(B)\text{–}$$

However, several reagents add to the conjugated π system in a 1,4-manner, a result also called **conjugate addition.** In these transformations, the nucleophilic part of a reagent attaches itself to the β-carbon, and the electrophilic part (commonly, a proton) binds to the carbonyl oxygen. The initial product is an enol, which subsequently rearranges to its keto form.

1,4-Addition of a Polar Reagent A–B to a Conjugated Enone

$$>\overset{\beta}{C}{=}\overset{\alpha}{C}{<}\text{–}C({=}O)\text{–} \xrightarrow{A\text{–}B} \text{–}C(B)\text{–}C{=}C(OA)\text{–} \xrightarrow{A\,=\,H} \text{–}C(B)\text{–}C(H)\text{–}C({=}O)\text{–}$$

New bond to carbon

Oxygen and nitrogen nucleophiles undergo conjugate additions

Water, alcohols, and amines can be induced to undergo 1,4-additions, as the following examples show. Although these reactions can be catalyzed by acid or base, the products are usually formed faster and in higher yields with base.

$$CH_2{=}CH\overset{O}{\overset{\|}{C}}CH_3 \xrightleftharpoons{HOH,\ Ca(OH)_2,\ -5°C} HOCH_2\underset{H}{\underset{|}{C}}H\overset{O}{\overset{\|}{C}}CH_3$$

3-Buten-2-one **4-Hydroxy-2-butanone**

$$CH_3CH{=}CH\overset{O}{\overset{\|}{C}}H \xrightleftharpoons{CH_3OH,\ CH_3O^-K^+} CH_3\overset{CH_3O}{\overset{|}{C}}\underset{H}{\underset{|}{C}}H\overset{O}{\overset{\|}{C}}H$$

50%

2-Butenal **3-Methoxybutanal**

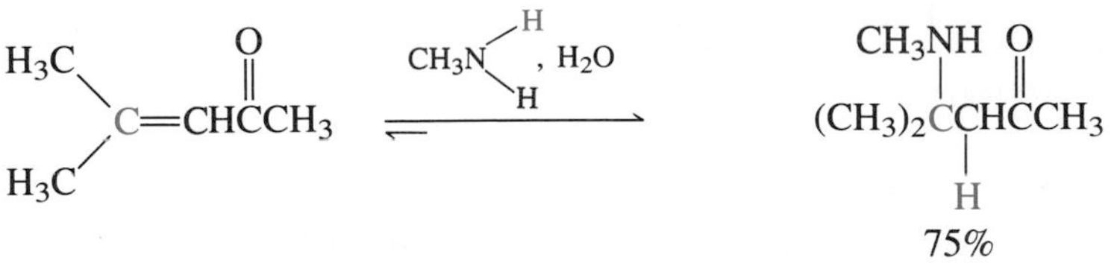

$$(CH_3)_2C{=}CHCCH_3\ (\text{C}{=}\text{O}) \xrightleftharpoons{CH_3NH_2,\ H_2O} (CH_3)_2C(NHCH_3)CH_2CCH_3\ (\text{C}{=}\text{O})$$

4-Methyl-3-penten-2-one → 75% **4-Methyl-4-(methylamino)-2-pentanone**

Note that the hydration of an α,β-unsaturated carbonyl compound is the reverse of the second step of the aldol condensation. Indeed, at elevated temperatures, the 1,4-addition becomes reversible, and other products may be formed—for example, derived from aldol or amine condensation reactions (Section 18-5).

The mechanism of the base-catalyzed addition to conjugated aldehydes and ketones is direct nucleophilic attack at the β-carbon to give the enolate ion, which is subsequently protonated.

Mechanism of Base-Catalyzed Hydration of α,β-Unsaturated Aldehydes and Ketones

$$HO^- + \ C{=}C{-}C{=}O \ \rightleftharpoons\ HOC{-}C{=}C{-}O^- \ \xrightleftharpoons{HOH}\ HOC{-}CHC{=}O \ + \ ^-OH$$

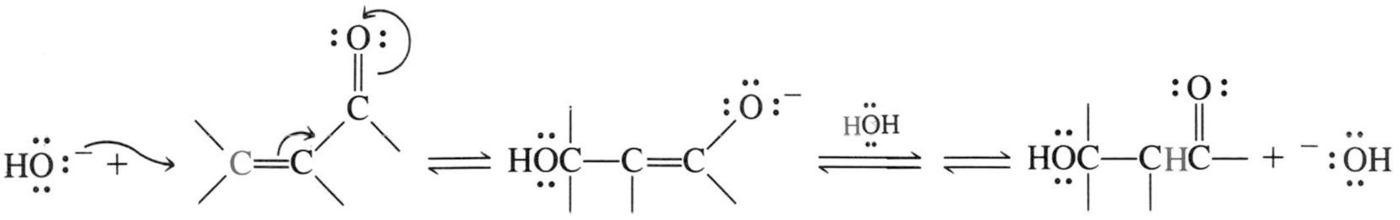

EXERCISE 18-20

Treatment of 3-chloro-2-cyclohexenone with sodium methoxide in methanol gave 3-methoxy-2-cyclohexenone. Write the mechanism of this reaction.

Hydrogen cyanide also undergoes conjugate addition

Treatment of a conjugated aldehyde or ketone with cyanide in the presence of acid may result in attack by cyanide at the β-carbon, in contrast with cyanohydrin formation (Section 17-11). Although this transformation appears to give a 1,2-adduct to the carbon–carbon double bond, it proceeds through a 1,4-addition pathway. The reaction involves protonation of the oxygen, then nucleophilic β-attack, and finally enol–keto tautomerization.

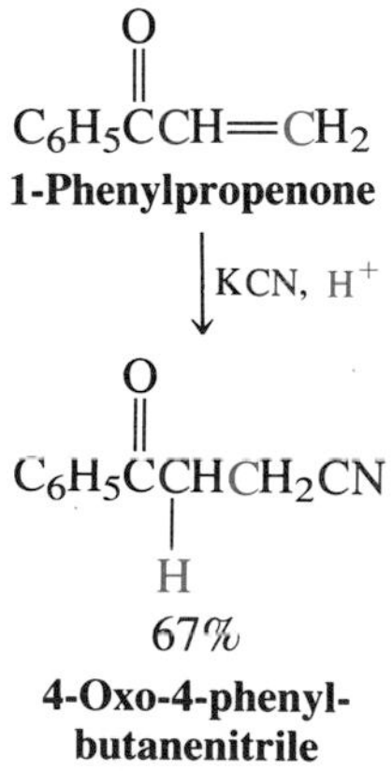

$$C_6H_5CCH{=}CH_2\ (\text{C}{=}\text{O})$$

1-Phenylpropenone

$\downarrow$ KCN, H^+

$$C_6H_5CCH_2CH_2CN\ (\text{C}{=}\text{O})$$

67%

4-Oxo-4-phenyl-butanenitrile

EXERCISE 18-21

Formulate the mechanism of the acid-catalyzed 1,4-addition of cyanide to 1-phenylpropenone.

In summary, α,β-unsaturated aldehydes and ketones are synthetically useful building blocks in organic synthesis because of their ability to undergo 1,4-additions. Hydrogen cyanide addition leads to β-cyano carbonyl compounds; oxygen and nitrogen nucleophiles can also add to the β-carbon.

18-11 1,2- and 1,4-Additions of Organometallic Reagents

Organometallic reagents may add to the α,β-unsaturated carbonyl function in either 1,2 or 1,4 fashion. Organolithium reagents, for example, react preferentially by direct nucleophilic attack at the carbonyl carbon.

Exclusive 1,2-Addition of an Organolithium

$$(H_3C)_2C{=}CH\overset{O}{\overset{\|}{C}}CH_3 \xrightarrow[2.\ H^+,\ H_2O]{1.\ CH_3Li,\ (CH_3CH_2)_2O} (H_3C)_2C{=}CH\underset{CH_3}{\underset{|}{\overset{OH}{\overset{|}{C}}}}CH_3$$

81%

4-Methyl-3-penten-2-one → **2,4-Dimethyl-3-penten-2-ol**

Conversely, cuprates give only products of conjugate addition.

Exclusive 1,4-Addition of a Lithium Organocuprate

$$CH_3(CH_2)_5CH{=}\underset{CH_3}{\underset{|}{C}}\overset{O}{\overset{\|}{C}}H \xrightarrow[2.\ H^+,\ H_2O]{1.\ (CH_3)_2CuLi,\ THF,\ -78^\circ C,\ 4\ h} CH_3(CH_2)_5\overset{CH_3}{\overset{|}{C}}H\underset{CH_3}{\underset{|}{C}}H\overset{O}{\overset{\|}{C}}H$$

40%

2-Methyl-2-nonenal → **2,3-Dimethylnonanal**

The copper-mediated 1,4-addition reactions are thought to proceed through complex electron transfer mechanisms. The first isolable intermediate is an enolate ion, which can be trapped by alkylating species, as shown in Section 18-4. Conjugate addition followed by alkylation constitutes a useful sequence for α,β-dialkylation of unsaturated aldehydes and ketones.

α,β-Dialkylation of Unsaturated Carbonyl Compounds

$$>C{=}C<\overset{O}{\overset{\|}{C}}- \xrightarrow[2.\ R'X]{1.\ R_2CuLi} -\underset{R}{\underset{|}{\overset{|}{C}}}-\underset{R'}{\underset{|}{\overset{|}{C}}}-C(=O)-$$

The following example illustrates this reaction.

$$\text{2-cyclohexenone} \xrightarrow[2.\ CH_3I]{1.\ (CH_3CH_2CH_2CH_2)_2CuLi,\ THF} \text{trans-3-butyl-2-methylcyclohexanone } (CH_3,\ CH_2CH_2CH_2CH_3) + \text{cis isomer}$$

84%, 4:1

***trans-* and *cis*-3-Butyl-2-methylcyclohexanone**

Reactions of ordinary Grignard reagents with α,β-unsaturated aldehydes and ketones give 1,2-addition, 1,4-addition, or both, depending on the structures of the reacting species and conditions.

BOX 18-5 Prostaglandins: α,β-Dialkylation in Synthesis

The α,β-dialkylation procedure has been exploited in the total synthesis of *prostaglandins,* powerful physiologically active compounds (see Section 19-14). They appear to regulate a remarkable variety of bodily functions, including those of the endocrine, reproductive, nervous, digestive, hemostatic, respiratory, cardiovascular, and renal systems. Because of these properties, they are potential drugs in the treatment of hypertension, asthma, fever, inflammations, and ulcers. One of the commercially available prostaglandins induces labor in pregnant women. Others have applications in animal breeding by controlling the day in which the animal goes into heat.

A synthesis of prostaglandin $PGF_{2\alpha}$, developed by Stork,* includes two conjugate additions as well as an aldol reaction.

*Professor Gilbert Stork, b. 1921, Columbia University.

A Prostaglandin Synthesis

(R, R′, R″ are protecting groups; C_5H_{11} = pentyl)

RO, O, + LiCu, C_5H_{11}, OR′, 2; $(CH_3CH_2)_2O$, −78°C; 1,4-Addition

O^-Li^+, RO, C_5H_{11}, OR′; 1. $CH_2{=}O$, −50°C; 2. H^+, H_2O; Aldol condensation

O, CH_2, RO, C_5H_{11}, OR′; 1. $[R''O(CH_2)_4 \ldots]_2CuLi$; 2. H^+, H_2O; 1,4-Addition

O, CH_2, $(CH_2)_4OR''$, RO, C_5H_{11}, OR′ → →

OH, CH_2, $(CH_2)_3CO_2H$, HO, C_5H_{11}, OH

$PGF_{2\alpha}$

EXERCISE 18-22

Show how you might synthesize the following compounds from 3-methyl-2-cyclohexenone. (Hints: Work backward; and the last step in (b) is an intramolecular aldol condensation.)

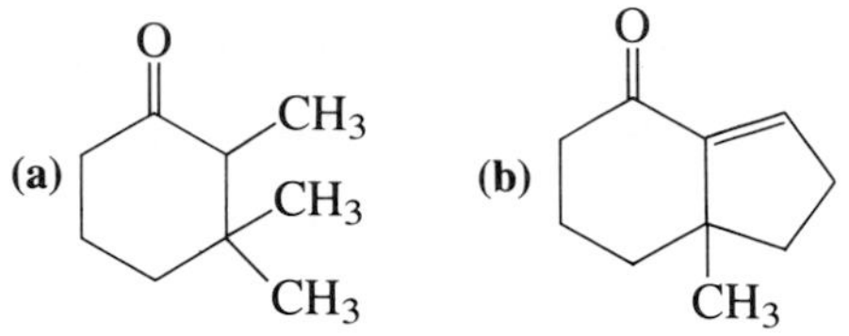

(a) 2,3,3-trimethylcyclohexanone [structure: cyclohexanone ring with CH_3, CH_3, CH_3 substituents]

(b) bicyclic enone [structure: cyclohexanone fused to cyclopentene ring, with angular CH_3]

In summary, organolithium reagents add 1,2 and organocuprate reagents add 1,4 to α,β-unsaturated carbonyl systems. The latter process initially gives rise to a β-substituted enolate, which upon exposure to haloalkanes gives rise to α,β-dialkylated aldehydes and ketones.

18-12 Conjugate Additions of Enolate Ions: Michael Addition and Robinson Annulation

Like other nucleophiles, enolate ions undergo conjugate additions to α,β-unsaturated aldehydes and ketones, in a reaction known as the **Michael* addition.** This transformation works best with enolates derived from β-dicarbonyl compounds (Chapter 21), but it also works with simpler systems, as the following examples show.

Michael Addition

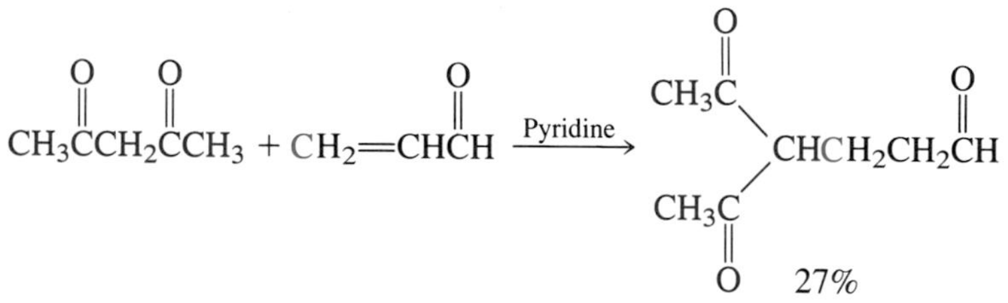

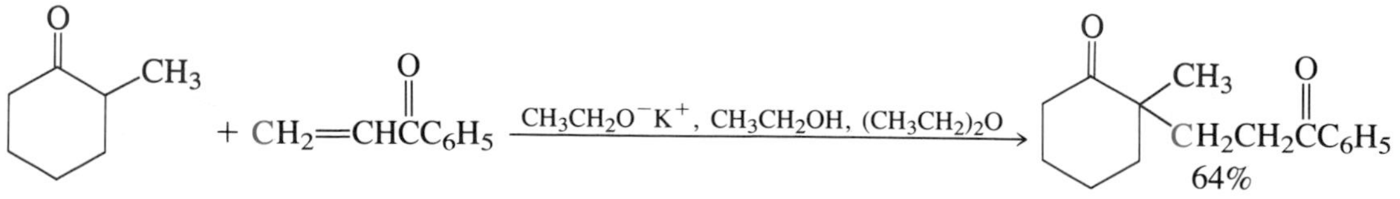

$$CH_3\overset{O}{\overset{\|}{C}}CH_2\overset{O}{\overset{\|}{C}}CH_3 + CH_2{=}CH\overset{O}{\overset{\|}{C}}H \xrightarrow{\text{Pyridine}} (CH_3\overset{O}{\overset{\|}{C}})_2CHCH_2CH_2\overset{O}{\overset{\|}{C}}H \quad 27\%$$

$$\text{2-methylcyclohexanone} + CH_2{=}CH\overset{O}{\overset{\|}{C}}C_6H_5 \xrightarrow{CH_3CH_2O^-K^+,\ CH_3CH_2OH,\ (CH_3CH_2)_2O} \text{2-methyl-2-}(CH_2CH_2\overset{O}{\overset{\|}{C}}C_6H_5)\text{cyclohexanone} \quad 64\%$$

The mechanism of the Michael addition includes nucleophilic attack by the enolate ion at the β-carbon of the unsaturated carbonyl compound (the Michael "acceptor"), followed by protonation of the resulting enolate.

---

*Professor Arthur Michael, 1853–1942, Harvard University.

Mechanism of the Michael Addition

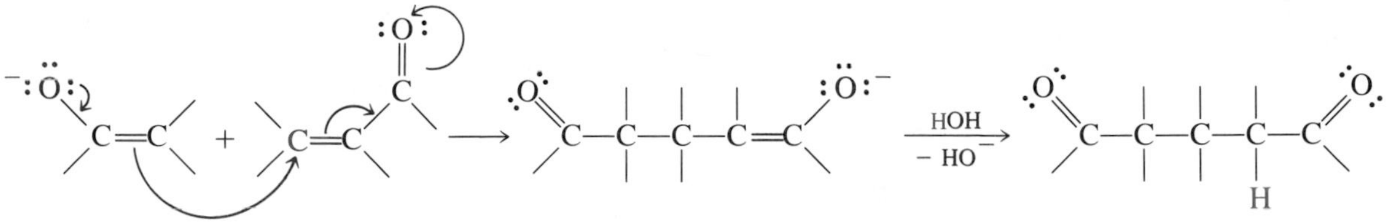

$$^{-}\!:\ddot{O}: \;\; C{=}C \; + \; C{=}C{-}C{=}O \longrightarrow O{=}C{-}C{-}C{-}C{=}C{-}\ddot{O}:^{-} \xrightarrow[-\,HO^{-}]{HOH} O{=}C{-}C{-}C{-}C(H){-}C{=}O$$

As the mechanism indicates, the reaction works because of the nucleophilic potential of the α-carbon of an enolate and the electrophilic reactivity of the β-carbon of an α,β-unsaturated carbonyl compound.

With some Michael acceptors, such as 3-buten-2-one, the products of the initial addition are capable of a subsequent intramolecular aldol condensation, which creates a new ring.

Michael Addition Followed by Intramolecular Aldol Condensation

2-methylcyclohexanone (H_3C, O) + $CH_2{=}CH\overset{O}{\overset{\|}{C}}CH_3$ (**3-Buten-2-one**) $\xrightarrow[\text{Michael addition}]{CH_3CH_2O^-Na^+,\ CH_3CH_2OH,\ (CH_3CH_2)_2O,\ -10°C}$

(CH_3, O, O, CH_3) $\xrightarrow[\text{Intramolecular aldol condensation}]{}$ (CH_3, O, OH) 54% $\xrightarrow[-\,HOH]{\Delta,\ HO^-}$ (CH_3, O) 86%

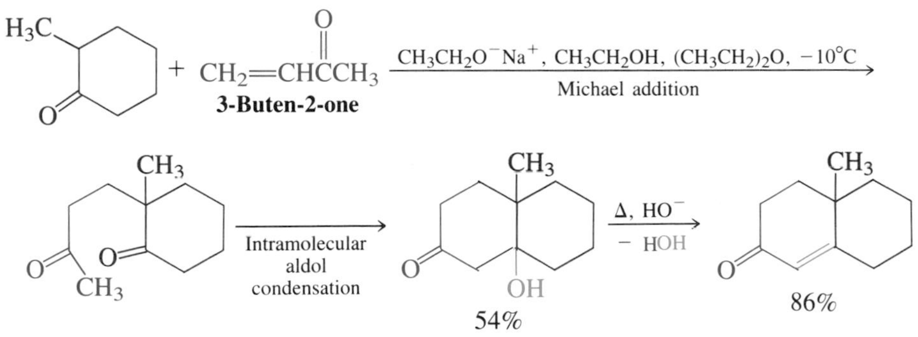

The synthetic sequence of a Michael addition followed by an intramolecular aldol condensation is also called a **Robinson* annulation.**

Robinson Annulation

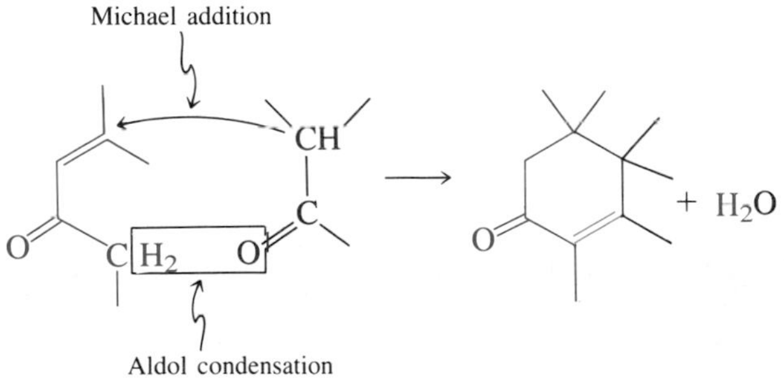

Michael addition

O, CH, C, C$\boxed{H_2 \quad O}$ $\longrightarrow$ O + H_2O

Aldol condensation

The Robinson annulation has found extensive use in the synthesis of polycyclic ring systems, including steroids.

---

*Sir Robert Robinson, 1886–1975, Oxford University, Nobel Prize 1947 (chemistry).

Steroid Synthesis by Robinson Annulation

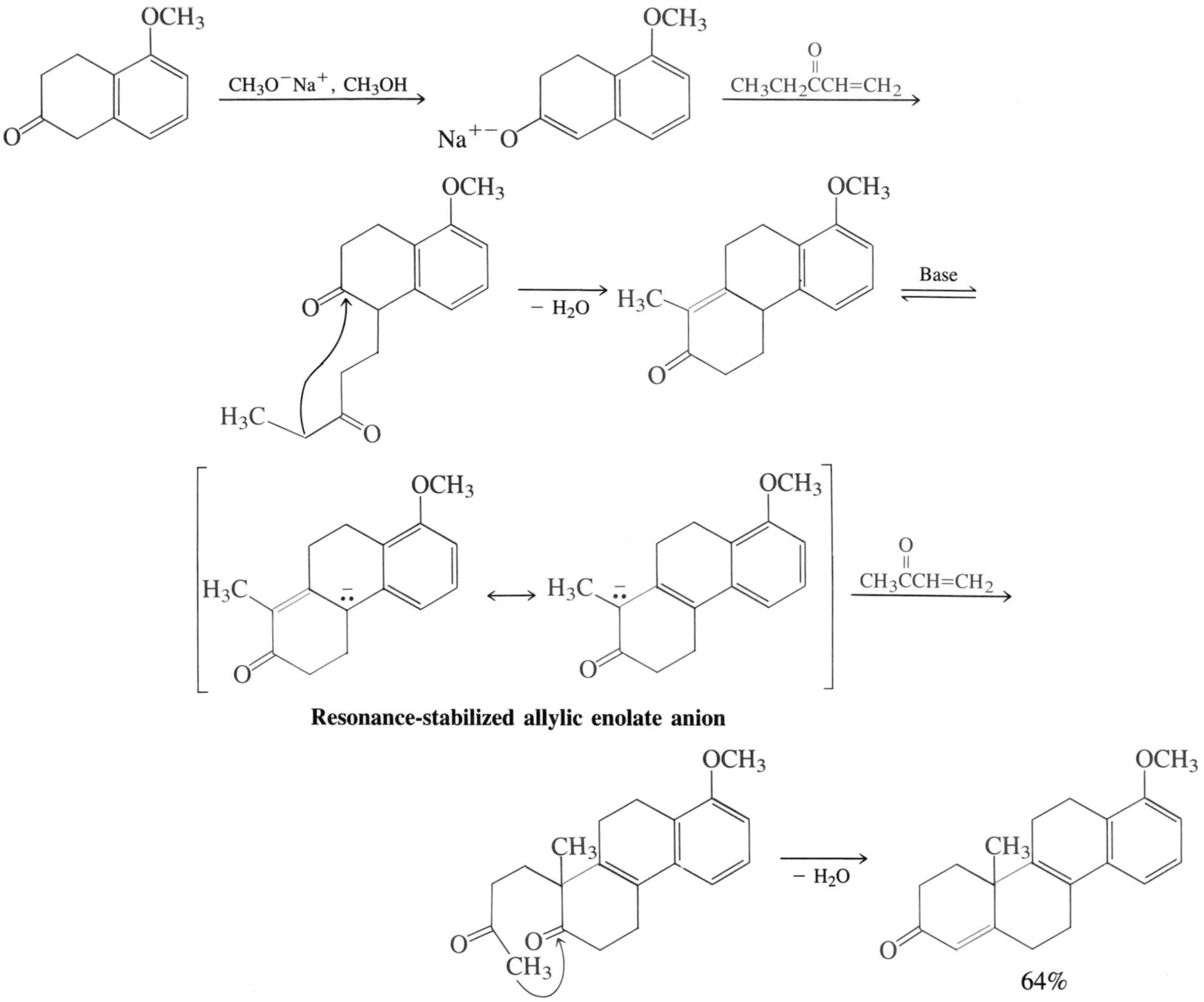

OCH_3 $\xrightarrow{CH_3O^-Na^+,\ CH_3OH}$ $Na^{+-}O$ OCH_3 $\xrightarrow{CH_3CH_2\overset{O}{\overset{\|}{C}}CH{=}CH_2}$

OCH_3 O H_3C O $\xrightarrow[-\ H_2O]{}$ H_3C OCH_3 O $\xrightleftharpoons{\text{Base}}$

H_3C OCH_3 O $\longleftrightarrow$ H_3C OCH_3 O $\xrightarrow{CH_3\overset{O}{\overset{\|}{C}}CH{=}CH_2}$

Resonance-stabilized allylic enolate anion

OCH_3 CH_3 O O CH_3 $\xrightarrow[-\ H_2O]{}$ OCH_3 CH_3 O

64%

EXERCISE 18-23

Propose syntheses of the following compounds by Michael or Robinson reactions.

(a) $CH_3\overset{O}{\overset{\|}{C}}CH_2CH_2CH_2\overset{O}{\overset{\|}{C}}CH_3$ **(b)** 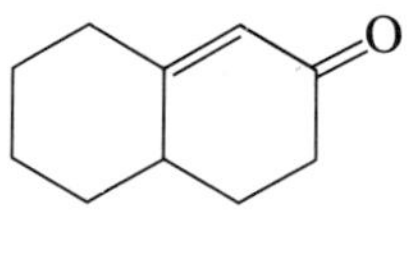 O **(c)** 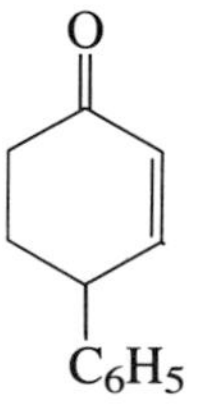 O, C_6H_5

In summary, the Michael addition results in the conjugate addition of an enolate ion to give dicarbonyl compounds. The Robinson annulation reaction combines a Michael addition with a subsequent intramolecular aldol condensation to give new cyclic enones.

NEW REACTIONS

Synthesis and Reactions of Enolates and Enols

1. ENOLATE IONS (Section 18-1)

$$\mathrm{RCH_2\overset{O}{\overset{\|}{C}}R'} \xrightarrow[\text{or other strong base, } -78^\circ\text{C}]{\text{LDA or KH or } (CH_3)_3CO^-K^+} \mathrm{RCH{=}C(O^-)R'}$$

Enolate ion

2. KETO–ENOL EQUILIBRIA (Section 18-2)

$$\mathrm{RCH_2\overset{O}{\overset{\|}{C}}R'} \underset{\text{Tautomerism}}{\overset{\text{Catalytic } H^+ \text{ or } HO^-}{\rightleftharpoons}} \mathrm{RCH{=}C(OH)R'}$$

3. HYDROGEN–DEUTERIUM EXCHANGE (Section 18-2)

$$\mathrm{RCH_2\overset{O}{\overset{\|}{C}}R'} \xrightarrow{D_2O,\ DO^- \text{ or } D^+} \mathrm{RCD_2\overset{O}{\overset{\|}{C}}R'}$$

4. STEREOISOMERIZATION (Section 18-2)

$$\mathrm{(R)(H)(R')C{-}\overset{O}{\overset{\|}{C}}R''} \overset{H^+ \text{ or } HO^-}{\rightleftharpoons} \mathrm{(R)(R')(H)C{-}\overset{O}{\overset{\|}{C}}R''}$$

5. HALOGENATION (Section 18-3)

$$\mathrm{RCH_2\overset{O}{\overset{\|}{C}}R'} \xrightarrow[-\ HX]{X_2,\ H^+} \mathrm{R\underset{X}{\underset{|}{C}}H\overset{O}{\overset{\|}{C}}R'}$$

6. ENOLATE ALKYLATION (Section 18-4)

$$\mathrm{RCH{=}C(O^-)R'} \xrightarrow[-\ X^-]{R''X} \mathrm{R\underset{R''}{\underset{|}{C}}H\overset{O}{\overset{\|}{C}}R'}$$

S_N2 reaction: R″X must be methyl or primary halide

7. ENAMINE ALKYLATION (Section 18-4)

$$\underset{}{>C{=}C{<}}\!\!-\!N(R)-R \xrightarrow{R'X} -\underset{R'}{\overset{|}{C}}-C{=}\overset{+}{N}(R)-R\ X^- \xrightarrow{H^+,\ H_2O} -\underset{R'}{\overset{|}{C}}-C(=O)- \; + \; R_2NH$$

S_N2 reaction: R′X must be methyl, primary, or secondary halide

8. ALDOL CONDENSATIONS (Sections 18-5 through 18-7)

$$2\ RCH_2\overset{O}{\overset{\|}{C}}H \underset{}{\overset{HO^-}{\rightleftharpoons}} RCH_2\underset{H}{\overset{OH}{C}}-\underset{R}{C}H\overset{O}{\overset{\|}{C}}H \overset{HO^-,\ \Delta}{\rightleftharpoons} RCH_2CH{=}C(CHO)(R) + H_2O$$

Aldol adduct **Condensation product**

Crossed aldol condensation (one aldehyde not enolizable)

$$R\overset{O}{\overset{\|}{C}}H + R'CH_2\overset{O}{\overset{\|}{C}}H \xrightarrow[-\ H_2O]{HO^-,\ \Delta} RCH{=}C(R')(CHO)$$

Ketones

$$R\overset{O}{\overset{\|}{C}}CH_2R' \overset{HO^-}{\rightleftharpoons} R\underset{CH_2R'}{\overset{OH}{C}}-\overset{R'}{C}H-\overset{O}{\overset{\|}{C}}R \xrightarrow[\text{Drive equilibrium}]{\Delta,\ -\ H_2O} R\underset{CH_2R'}{C}{=}\overset{R'}{C}-\overset{O}{\overset{\|}{C}}R$$

Intramolecular aldol condensation

$$(CH_2)_n\text{-ring with } C(=O)CH_2R \text{ and } C(=O)R' \xrightarrow[-\ H_2O]{HO^-} \text{cyclic enone: } (CH_2)_n\text{ ring, } C(=O)-C(R){=}C(R')$$

Unstrained rings preferred

9. SYNTHESIS OF α,β-UNSATURATED ALDEHYDES AND KETONES (Sections 18-8 and 18-9)

Aldol condensation (see preceding reactions)

Bromination–dehydrobromination of aldehydes and ketones

$$RCH_2CH_2\overset{O}{\overset{\|}{C}}R' \xrightarrow[2.\ Base]{1.\ Br_2,\ H^+} RCH{=}CH\overset{O}{\overset{\|}{C}}R'$$

Wittig reaction with stabilized ylides

$$\text{RCH(=O)} + (C_6H_5)_3P{=}CHC(=O)R' \xrightarrow{\text{THF}} (R)(H)C{=}CHC(=O)R' + (C_6H_5)_3P{=}O$$

Oxidation of allylic alcohols

$$\text{>C=C(-)–C(OH)(H)–} \xrightarrow{\text{MnO}_2\text{, propanone}} \text{>C=C(-)–C(=O)–}$$

Isomerization of β,γ-unsaturated aldehydes and ketones to conjugated carbonyl compounds

$$RCH{=}CHCH_2CH(=O) \xrightarrow{H^+ \text{ or } HO^-,\ H_2O} RCH_2CH{=}CHCH(=O)$$

Reactions of α,β-Unsaturated Aldehydes and Ketones

10. REDUCTIONS (Section 18-9)

Hydrogenation

$$\text{>C=C(-)–C(=O)–} \xrightarrow{H_2,\ \text{Pd},\ CH_3CH_2OH} \text{–C(H)–C(H)–C(=O)–}$$

One-electron transfer reduction

$$\text{>C=C(-)–C(=O)–} \xrightarrow{\text{Li, liquid } NH_3} \text{–C(H)–C(H)–C(=O)–}$$

11. ADDITION OF HALOGEN (Section 18-9)

$$\text{>C=C(-)–C(=O)–} \xrightarrow{X_2,\ CCl_4} \text{–C(X)–C(X)–C(=O)–}$$

12. CONDENSATIONS WITH AMINE DERIVATIVES (Section 18-9)

$$\begin{array}{c} O \\ \| \\ >C{=}C{-}C{-} \end{array} \xrightarrow{ZNH_2} \begin{array}{c} N{-}Z \\ \| \\ >C{=}C{-}C{-} \end{array}$$

Z = OH, RNH, R, etc.

Conjugate Additions to α,β-Unsaturated Aldehydes and Ketones

13. HYDROGEN CYANIDE ADDITION (Section 18-10)

$$>C{=}C{-}\overset{O}{\overset{\|}{C}}{-} \xrightarrow{KCN,\ H^+} {-}\underset{CN}{\overset{|}{\underset{|}{C}}}{-}\underset{H}{\overset{|}{\underset{|}{C}}}{-}\overset{O}{\overset{\|}{C}}{-}$$

14. WATER, ALCOHOLS, AMINES (Section 18-10)

$$>C{=}C{-}\overset{O}{\overset{\|}{C}}{-} \xrightarrow{H_2O} {-}\underset{OH}{\overset{|}{\underset{|}{C}}}{-}\underset{H}{\overset{|}{\underset{|}{C}}}{-}\overset{O}{\overset{\|}{C}}{-}$$

$$>C{=}C{-}\overset{O}{\overset{\|}{C}}{-} \xrightarrow{ROH} {-}\underset{OR}{\overset{|}{\underset{|}{C}}}{-}\underset{H}{\overset{|}{\underset{|}{C}}}{-}\overset{O}{\overset{\|}{C}}{-}$$

$$>C{=}C{-}\overset{O}{\overset{\|}{C}}{-} \xrightarrow{RNH_2} {-}\underset{RNH}{\overset{|}{\underset{|}{C}}}{-}\underset{H}{\overset{|}{\underset{|}{C}}}{-}\overset{O}{\overset{\|}{C}}{-}$$

15. ORGANOMETALLIC REAGENTS (Section 18-11)

$$>C{=}C{-}\overset{O}{\overset{\|}{C}}{-} \xrightarrow[2.\ H^+,\ H_2O]{1.\ RLi,\ (CH_3CH_2)_2O} >C{=}C{-}C(OH)(R){-}$$

1,2-Addition

$$>C{=}C{-}\overset{O}{\overset{\|}{C}}{-} \xrightarrow[2.\ H^+,\ H_2O]{1.\ R_2CuLi,\ THF} {-}\underset{R}{\overset{|}{\underset{|}{C}}}{-}\underset{H}{\overset{|}{\underset{|}{C}}}{-}\overset{O}{\overset{\|}{C}}{-}$$

1,4-Addition

Additions of RMgX may be 1,2 or 1,4, depending on reagent and substrate structure

Cuprate additions followed by enolate alkylations

$$\text{C=C–C(=O)–} \xrightarrow[\text{2. R'X}]{\text{1. } R_2CuLi\text{, THF}} \text{–C(R)–C(R')–C(=O)–}$$

16. MICHAEL ADDITION (Section 18-12)

$$^{-}\text{O(R)C=C} + \text{C=C–C(=O)R'} \longrightarrow \xrightarrow{H^+} \text{RC(=O)–C–C–C(H)–C(=O)R'}$$

17. ROBINSON ANNULATION (Section 18-12)

$$\text{enone (O=C–CH}_2\text{–)} + \text{O=C–CH} \xrightarrow{HO^-} \text{cyclohexenone}$$

IMPORTANT CONCEPTS

1. Hydrogens next to the carbonyl group (α-hydrogens) are acidic because of the electron-withdrawing nature of the functional group and because the resulting enolate ion is resonance stabilized.
2. Electrophilic attack on enolates can occur at both the α-carbon and the oxygen. Haloalkanes usually prefer the former. Protonation of the latter leads to enols.
3. Enamines are neutral analogs of enolates. They can be β-alkylated to give iminium cations that hydrolyze to aldehydes and ketones on aqueous work-up.
4. Aldehydes and ketones are in equilibrium with their tautomeric enol forms; the enol–keto conversion is catalyzed by acid or base. This equilibrium allows for facile α-deuteration and stereochemical equilibration.
5. α-Halogenation of carbonyl compounds may be acid or base catalyzed. With acid, the enol is halogenated by attack at the double bond; subsequent renewed enolization is slowed down by the halogen substituent. With base, the enolate is attacked at carbon, and subsequent enolate formation is accelerated by the halogens introduced.
6. Enolates are nucleophilic and reversibly attack the carbonyl carbon of aldehydes and ketones in the aldol condensation, and the β-carbon of α,β-unsaturated carbonyl compounds in the Michael addition.
7. α,β-Unsaturated aldehydes and ketones show the normal chemistry of each individual double bond, but the entire conjugated system may react as a whole, as revealed by the ability of these compounds to undergo acid- and base-mediated 1,4-additions. Cuprates add to the β-position.

PROBLEMS

1. Write the structures of every enol and enolate ion that can arise from each of the following carbonyl compounds.

(a) $CH_3CH_2\overset{O}{\overset{\|}{C}}CH_2CH_3$ **(b)** $CH_3\overset{O}{\overset{\|}{C}}CH(CH_3)_2$ **(c)** 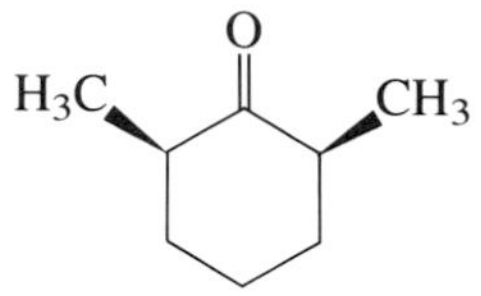 [2,6-dimethylcyclohexanone, H_3C and CH_3 wedged]

(d) 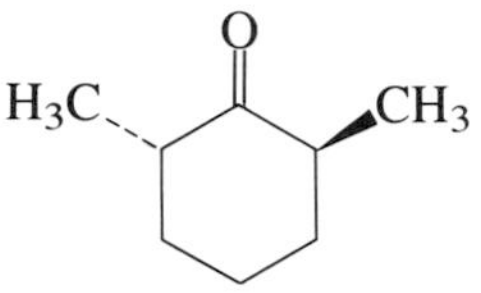 [2,6-dimethylcyclohexanone, H_3C hashed, CH_3 wedged] **(e)** 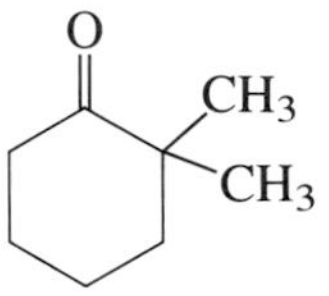 [2,2-dimethylcyclohexanone, CH_3, CH_3] **(f)** 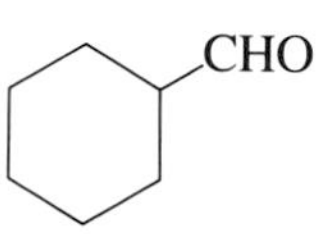 [cyclohexane-CHO]

(g) $(CH_3)_3C\overset{O}{\overset{\|}{C}}H$ **(h)** $(CH_3)_3CCH_2\overset{O}{\overset{\|}{C}}H$

2. What product(s) would form if each carbonyl compound in Problem 1 were treated with **(a)** alkaline D_2O; **(b)** 1 equivalent of Br_2 in acetic acid; **(c)** excess Cl_2 in aqueous base?

3. Describe the experimental conditions that would be best suited for the efficient synthesis of each of the following compounds from the corresponding nonhalogenated ketone.

(a) $C_6H_5\overset{Br}{\overset{|}{C}}H\overset{O}{\overset{\|}{C}}CH_3$ **(b)** $CH_3\underset{Cl}{\overset{Cl}{C}}\overset{O}{\overset{\|}{C}}\underset{Cl}{\overset{Cl}{C}}CH_3$ **(c)** $CH_3\overset{O}{\overset{\|}{C}}CH_2Cl$

4. Give the product(s) that would be expected on reaction of 3-pentanone with 1 equivalent of LDA, followed by addition of 1 equivalent of

(a) CH_3CH_2Br **(b)** $(CH_3)_2CHCl$ **(c)** 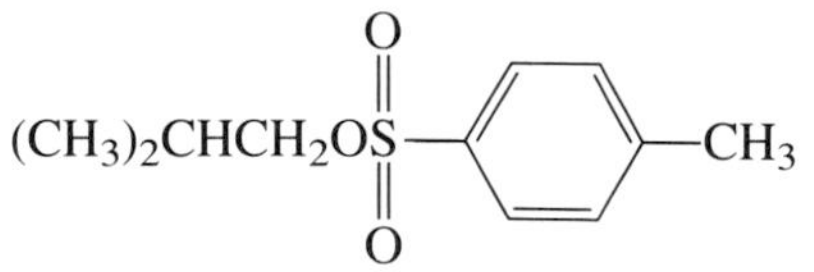 $(CH_3)_2CHCH_2O\overset{O}{\underset{O}{\overset{\|}{\underset{\|}{S}}}}$–C$_6H_4$–$CH_3$ (p-tolyl) **(d)** $(CH_3)_3CCl$

5. Give the product(s) of the following reaction sequences.

(a) 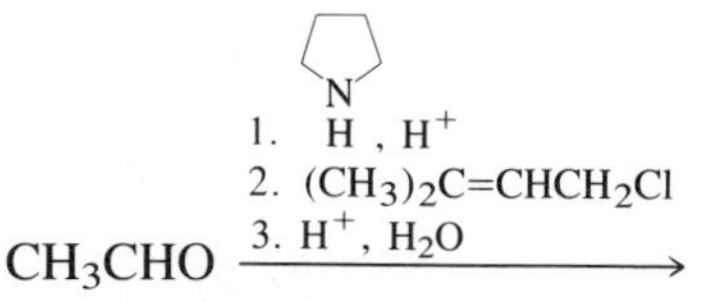 $CH_3CHO \xrightarrow[\text{3. } H^+,\ H_2O]{\text{1. pyrrolidine (N–H)},\ H^+;\ \text{2. } (CH_3)_2C{=}CHCH_2Cl}$

(b) 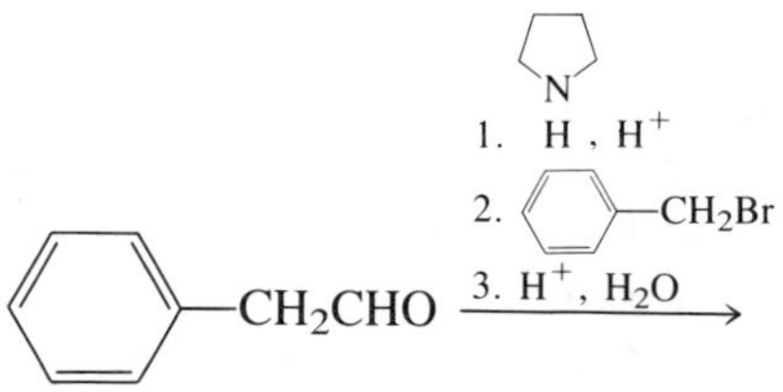 $C_6H_5CH_2CHO \xrightarrow[\text{3. } H^+,\ H_2O]{\text{1. pyrrolidine (N–H)},\ H^+;\ \text{2. } C_6H_5CH_2Br}$

6. The problem of double compared with single alkylation of ketones by iodomethane and base is mentioned in Section 18-4. Write a detailed mechanism showing how some double alkylation occurs even when only 1 equivalent each of the iodide and base is used. Suggest a reason why the use of the enamine alkylation procedure solves this problem.

7. Would the use of an enamine instead of an enolate improve the likelihood of successful alkylation of a ketone by a secondary haloalkane?

8. Formulate a mechanism for the acid-catalyzed hydrolysis of the pyrrolidine enamine of cyclohexanone (shown in the margin).

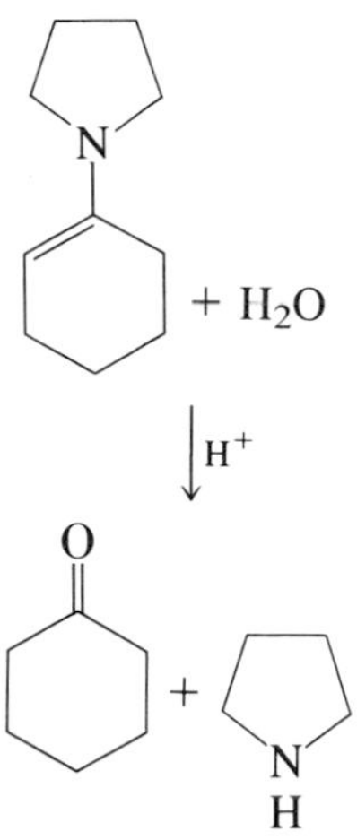

9. Give the likely products of each of the following aldol addition reactions.

(a)  2 $C_6H_5CH_2CHO \xrightarrow{NaOH,\ H_2O}$

(b) 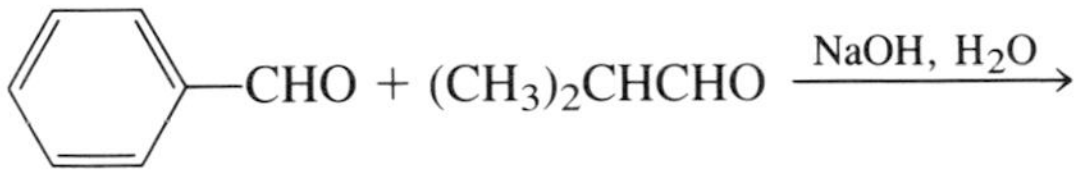 $C_6H_5CHO + (CH_3)_2CHCHO \xrightarrow{NaOH,\ H_2O}$

(c) $HC(=O)C(OCH_3)(CH_3)CH_2CH_2CH_2C(=O)CH_3 \xrightarrow{NaOH,\ H_2O}$

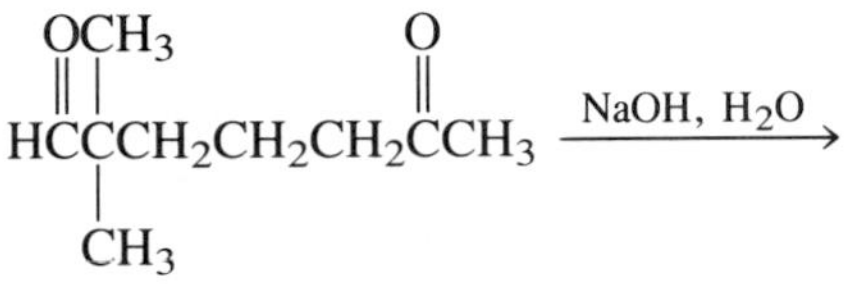

(d) 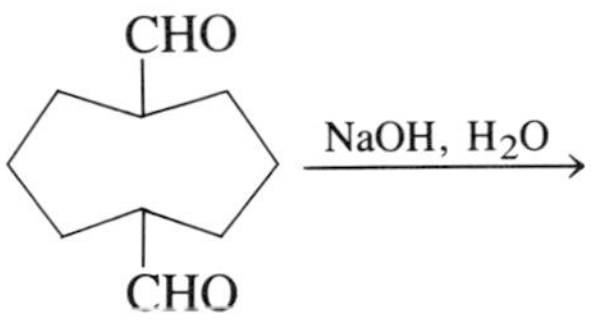 cyclooctane-1,5-dicarbaldehyde (CHO, CHO) $\xrightarrow{NaOH,\ H_2O}$

Margin: 1-(1-cyclohexenyl)pyrrolidine + H_2O $\xrightarrow{H^+}$ cyclohexanone + pyrrolidine (N–H)

10. Describe how you would prepare each of the following compounds, using an aldol condensation.

(a) 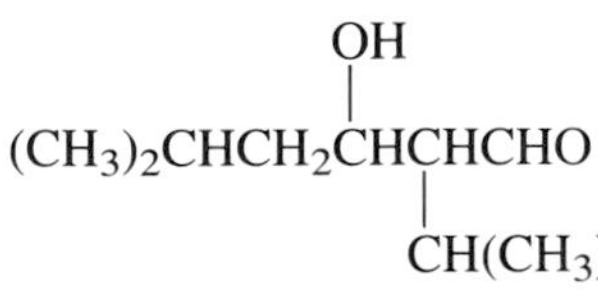 $(CH_3)_2CHCH_2CH(OH)CH(CH(CH_3)_2)CHO$

(b) 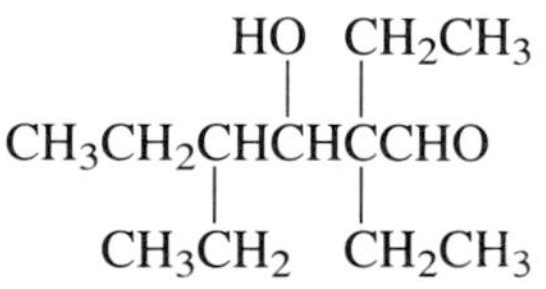 $CH_3CH_2CH(OH)CH(CH_2CH_3)C(CH_2CH_3)_2CHO$

(c) $(CH_3)_3CCH(OH)CH(CH_2CH_2CH_2CH_3)CHO$

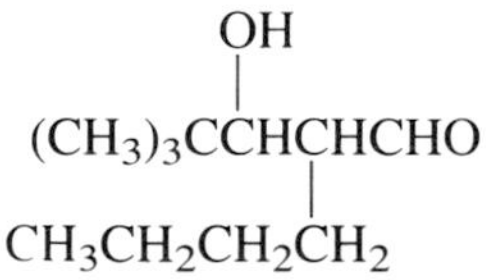

(d) 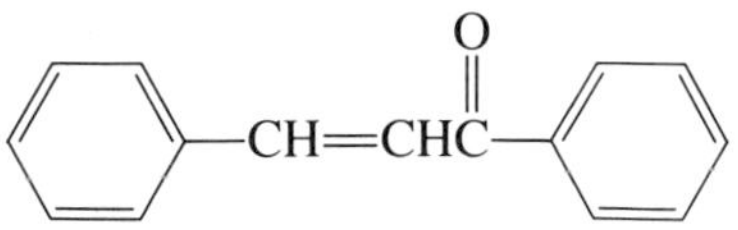 $C_6H_5CH{=}CHC(=O)C_6H_5$

(e) 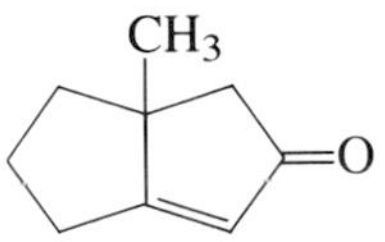 bicyclic enone with CH_3 at ring junction, =O

(f) 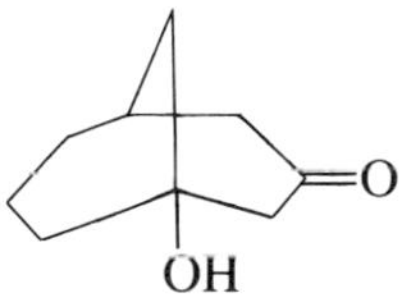 bridged bicyclic ketone with OH, =O

11. Aldol condensations may be catalyzed by acids. Suggest a role for H^+ in the acid-catalyzed version. (Hint: Consider what kind of nucleophile might exist in acid solution, where enolate ions are *unlikely* to be present.)

12. A fresh salad may contain the following compounds: 2-hexenal (aroma of tomatoes), 3-octen-2-one (flavor of mushrooms), and 2-nonenal and 2,4-nonadienal (flavor and aroma of cucumbers). In Section 18-8, the synthesis of 2-nonenal was illustrated. Write similar sequences for the preparation of these other naturally occurring compounds from saturated aldehydes.

13. Four general synthetic routes to α,β-unsaturated aldehydes and ketones are described in Section 18-8. For each of the following three compounds, select the route that, you feel, might be especially useful and practical.

(a) 2-cyclohexenone (b) $CH_3CH{=}C(CH_2CH_2CH_3)CH{=}O$ (c) $H_2C{=}CHC(=O)CH_2CH_2CH_2CH_2CH_3$

14. For each carbonyl compound listed in Problem 13, write the expected major product of reaction with each of the following reagents.

(a) H_2, Pd, CH_3CH_2OH

(b) $LiAlH_4$, $(CH_3CH_2)_2O$

(c) Cl_2, CCl_4

(d) KCN, H^+, H_2O

(e) CH_3Li, $(CH_3CH_2)_2O$

(f) $(CH_3CH_2CH_2CH_2)_2CuLi$, THF

(g) $NH_2NHC(=O)NH_2$, CH_3CH_2OH

(h) $(CH_3CH_2CH_2CH_2)_2CuLi$, followed by treatment with $CH_2{=}CHCH_2Cl$ in THF

15. Give the expected product(s) of each of the following reactions.

(a) $C_6H_5C(=O)CH_2CH_2CH_3$ $\xrightarrow{\text{1. LDA, THF; 2. } CH_3CH_2Br\text{, HMPA}}$

(b) cis-fused bicyclic ketone (ring-fusion H, H; =O) $\xrightarrow{\text{1. NaH, THF; 2. } BrCH_2C(=O)CH_3}$

(c) 4,4-dimethyl-2-cyclohexenone (H_3C, CH_3; O) $\xrightarrow{\text{1. } (CH_3)_2CuLi\text{, THF; 2. } C_6H_5CH_2Cl}$

(d) cyclohexanone bearing CH_3 and $(CH_2)_4Br$ $\xrightarrow{\text{LDA, THF}}$

16. Write the products of each of the following reactions after aqueous work-up.

(a) $C_6H_5C(=O)CH_3 + CH_2{=}CHC(=O)C_6H_5$ $\xrightarrow{\text{LDA, THF}}$

(b) cyclohexanone $+ (CH_3)_2C{=}CHC(=O)H$ $\xrightarrow{\text{NaOH, } H_2O}$

(c) 2-cyclopentenone $\xrightarrow{\text{1. } (CH_2{=}CH)_2CuLi\text{, THF; 2. } CH_2{=}CHC(=O)CH_3}$

(d) bicyclic enone (CH_3; O) $\xrightarrow{\text{1. } (CH_3)_2CuLi\text{, THF; 2. } (CH_3)_2C{=}CHC(=O)CH_3}$

(e) Write the results that you expect from base treatment of the products of reactions (c) and (d).

17. Write the final products of the following reaction sequences.

(a) tricyclic ketone (O) $+ CH_2{=}CHC(=O)CH_3$ $\xrightarrow{NaOCH_3\text{, } CH_3OH\text{, } \Delta}$

(b) 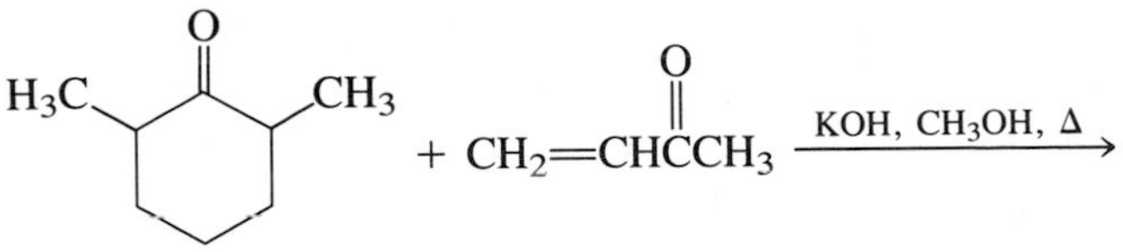 [structure: 2,6-dimethylcyclohexanone] + $CH_2{=}CH\overset{O}{\overset{\|}{C}}CH_3$ $\xrightarrow{\text{KOH, CH}_3\text{OH, }\Delta}$

(c) [structure: cyclohexanone] $\xrightarrow[]{\text{1. NaH, }(CH_3CH_2)_2O \quad \text{2. } HC{\equiv}C\overset{O}{\overset{\|}{C}}CH_3}$

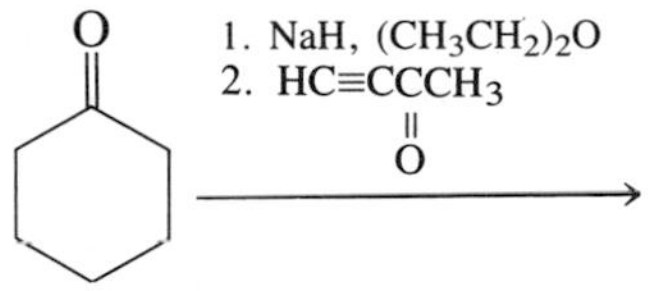

(d) Write a detailed mechanism for reaction sequence (c).

18. Propose syntheses of the following compounds using Michael additions followed by aldol condensations (i.e., Robinson annulation). Each of the compounds shown has been instrumental in one or more total syntheses of steroidal hormones.

(a) 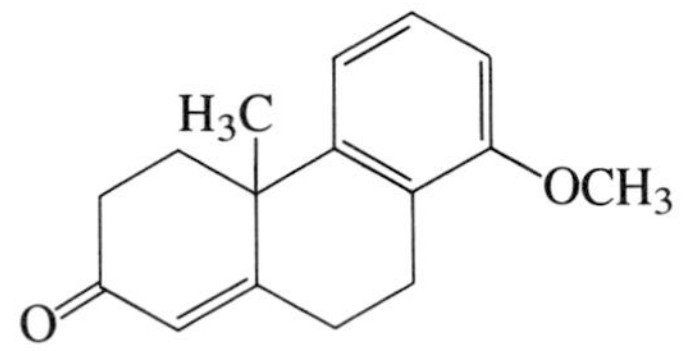 [structure: H_3C, OCH_3, O]

(b) 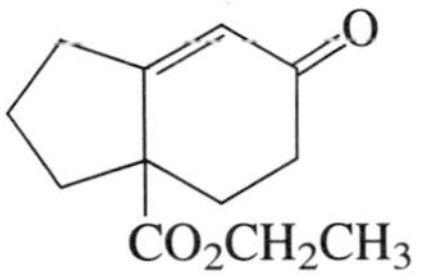 [structure: O, $CO_2CH_2CH_3$]

(c) 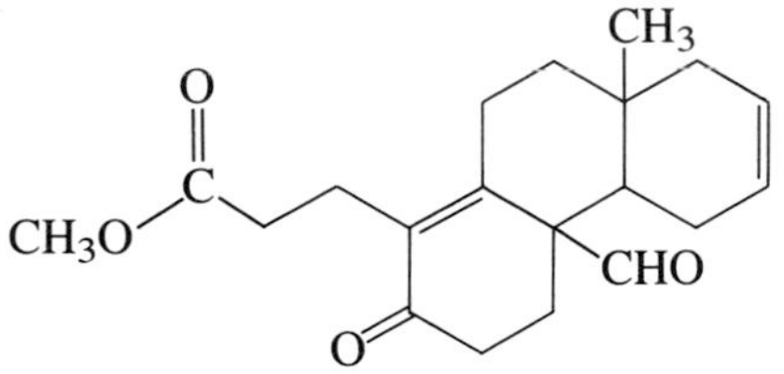 [structure: CH_3O–C(=O)–, CH_3, CHO, O]

(d) 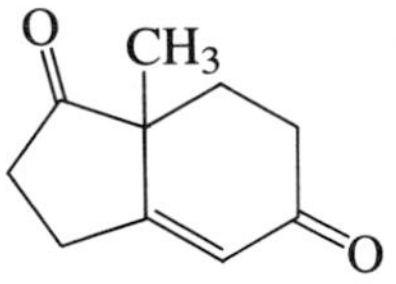 [structure: O, CH_3, O]

19. Would you expect addition of HCl to the double bond of 3-buten-2-one (shown in the margin) to follow Markovnikov's rule in orientation of addition? Explain your answer by a mechanistic argument.

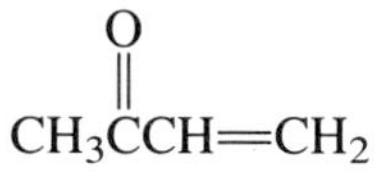

$CH_3\overset{O}{\overset{\|}{C}}CH{=}CH_2$

3-Buten-2-one

20. A very clever synthesis of steroids in the cortisone family includes the sequence of compounds shown below. Describe how (reagents, reaction conditions) each of the three transformations (a, b, c) shown below might be carried out.

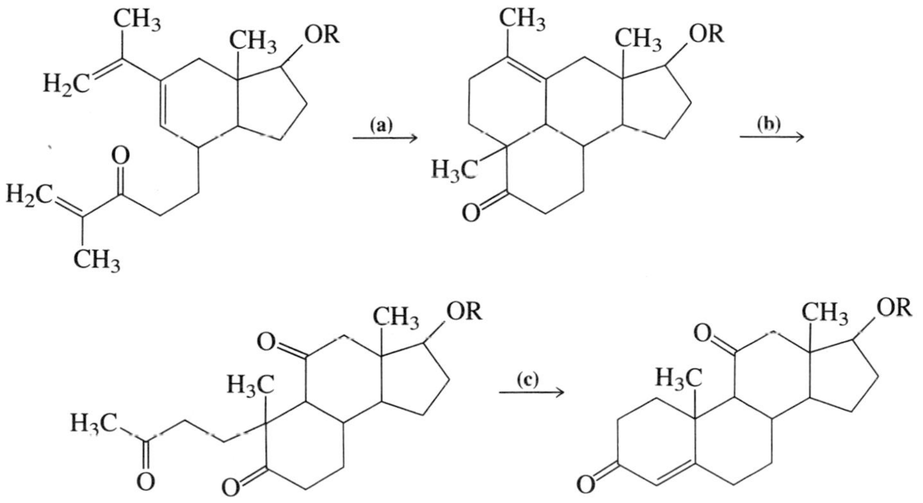

[structures: CH_3, H_2C, CH_3, OR, O, H_2C, CH_3 $\xrightarrow{(a)}$ CH_3, CH_3, OR, H_3C, O $\xrightarrow{(b)}$ O, CH_3, OR, H_3C, H_3C, O, O $\xrightarrow{(c)}$ O, CH_3, OR, H_3C, O]

$R = Si(CH_3)_2C(CH_3)_3$, an alcohol-protecting group

The last reaction of the sequence (c) can give rise to an undesirable isomeric product. Suggest a structure for this isomer, write a mechanism for its formation, and suggest a reason why the above structure is the actual product.

21. Using the information given below, propose structures for each of these compounds. **(a)** $C_5H_{10}O$, NMR spectrum A, UV $\lambda_{max}(\epsilon) = 280(18)$ nm; **(b)** C_5H_8O, NMR spectrum B, UV $\lambda_{max}(\epsilon) = 220(13{,}200), 310(40)$ nm; **(c)** C_6H_{12}, NMR spectrum C (p. 716), UV $\lambda_{max}(\epsilon) = 189(8{,}000)$ nm; **(d)** $C_6H_{12}O$, NMR spectrum D (p. 716), UV $\lambda_{max}(\epsilon) = 282(25)$ nm.

Next, for each of the following reactions, name an appropriate reagent for the indicated interconversion. (The letters refer to the compounds giving rise to NMR spectra A through D. **(e)** A → C; **(f)** B → D; **(g)** B → A.

22. Treatment of cyclopentane-1,3-dione with iodomethane in the presence of base leads mainly to a mixture of three products.

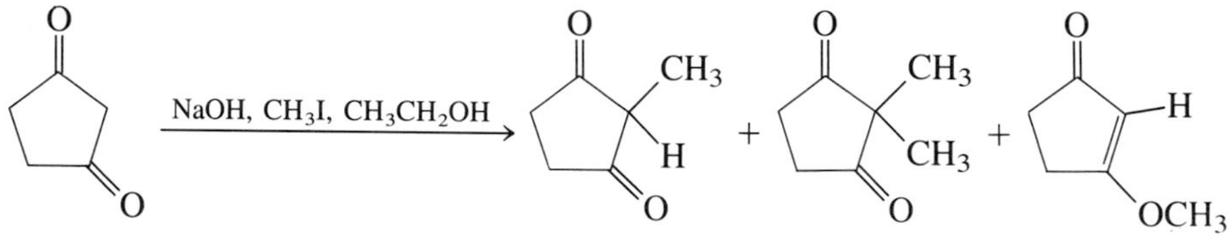

NaOH, CH_3I, CH_3CH_2OH → A + B + C

(a) Give a mechanistic description of how these three products are formed. **(b)** Reaction of product (C) with a cuprate reagent results in loss of the methoxy group. For example,

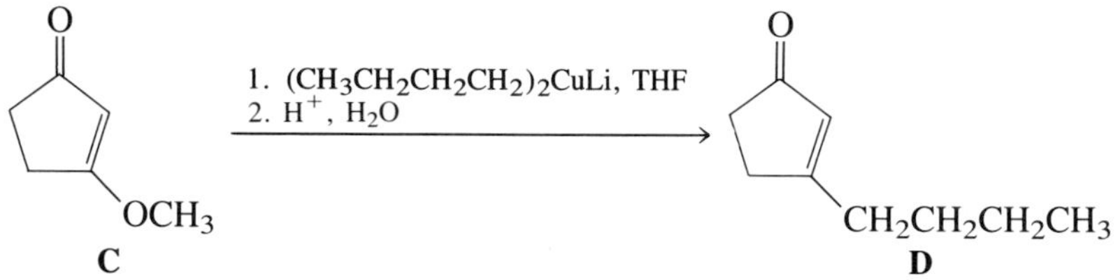

C → 1. $(CH_3CH_2CH_2CH_2)_2CuLi$, THF 2. H^+, H_2O → D

Suggest a mechanism for this reaction, which is another synthetic route to enones substituted at the β-carbon. (Hint: See Exercise 18-20.)

23. A somewhat unusual synthesis of cortisone-related steroids includes the following two reactions.

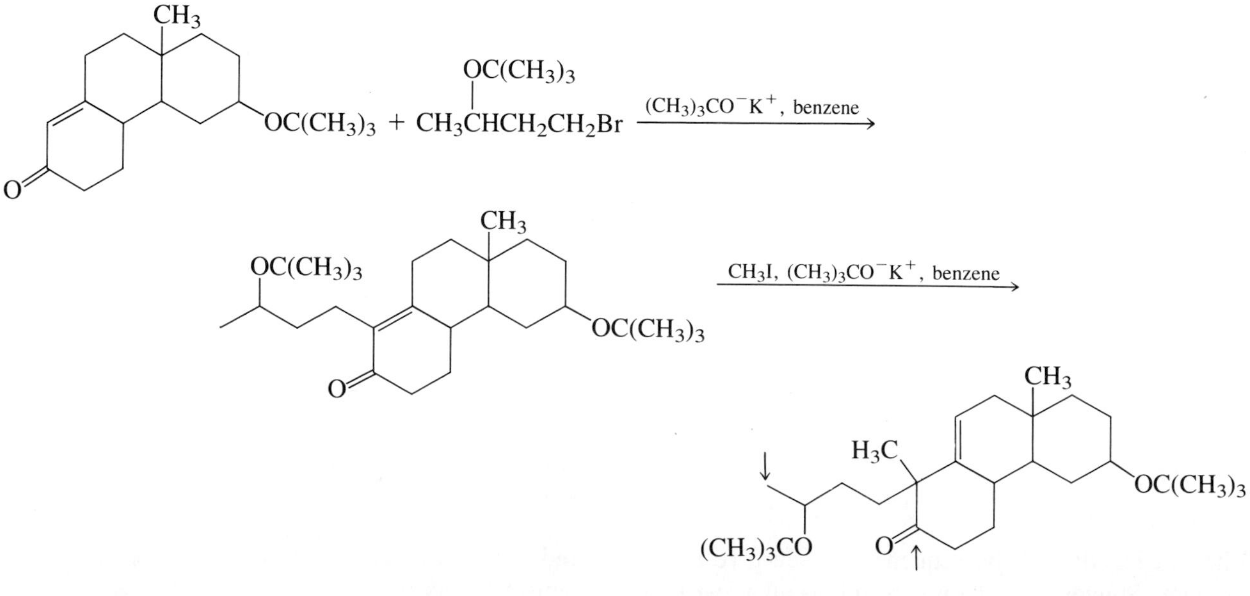

$+ CH_3CHCH_2CH_2Br$ (with $OC(CH_3)_3$ on CH) $\xrightarrow{(CH_3)_3CO^-K^+,\ \text{benzene}}$

$\xrightarrow{CH_3I,\ (CH_3)_3CO^-K^+,\ \text{benzene}}$

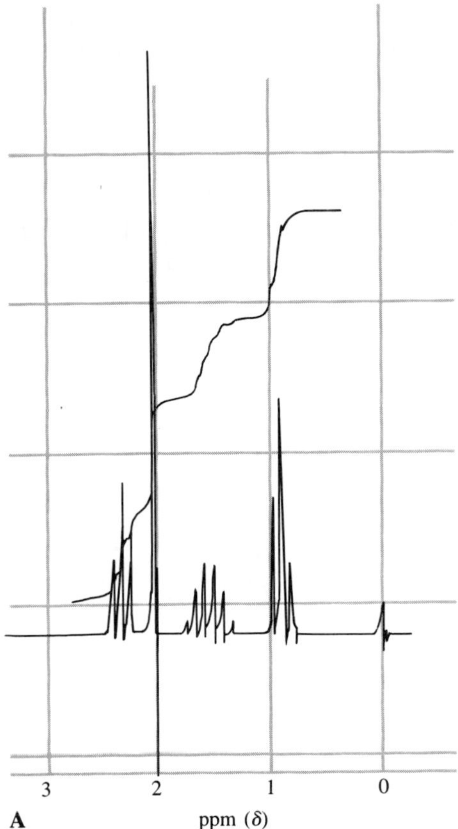
3
2
1
0
A
ppm (δ)

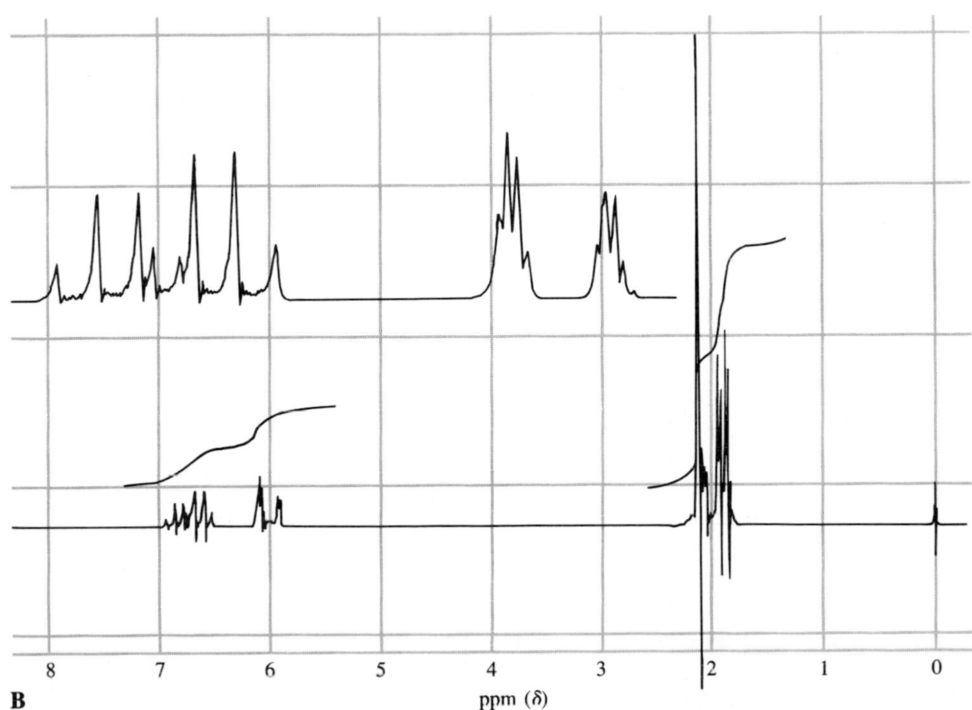
8
7
6
5
4
3
2
1
0
B
ppm (δ)

(a) Propose mechanisms for these two transformations. Be careful in choosing the initial site of deprotonation in the starting enone. The alkenyl hydrogen, in particular, is *not* the one initially removed by base in this reaction. **(b)** Propose a sequence of reactions that will connect the carbons marked by arrows in the structure above to form another six-membered ring.

24. The following steroid synthesis contains modified versions of two key types of reactions presented in this chapter. Identify these reaction types and give detailed mechanisms for each of the transformations shown.

$\xrightarrow{K^{+\,-}OC(CH_3)_3,\ THF}$

$\xrightarrow{K_2CO_3,\ CH_3OH}$

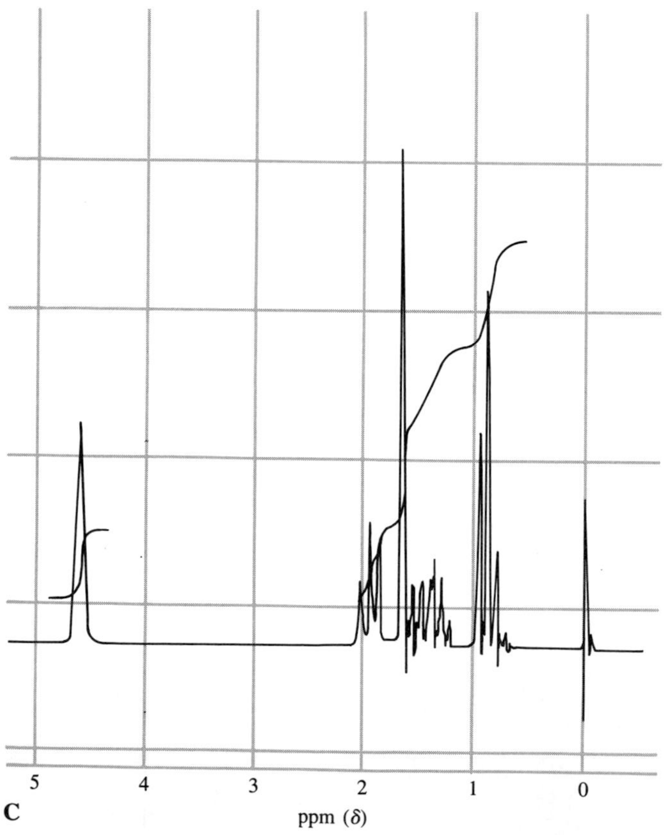

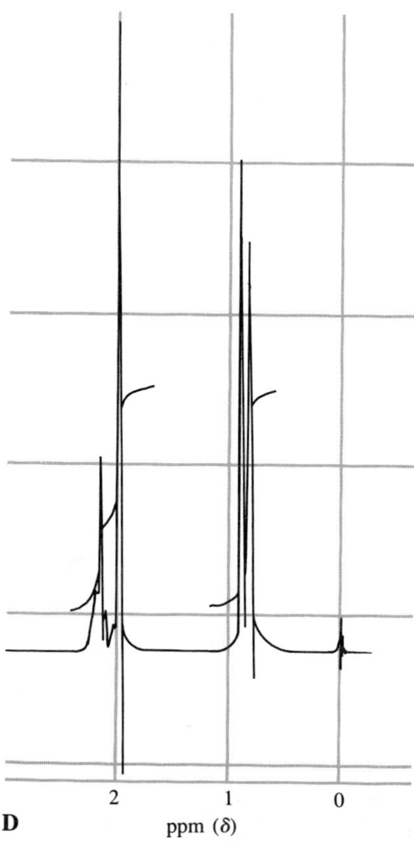

25. Devise reasonable plans for carrying out the following syntheses. Ignore stereochemistry in your strategies.

(a) 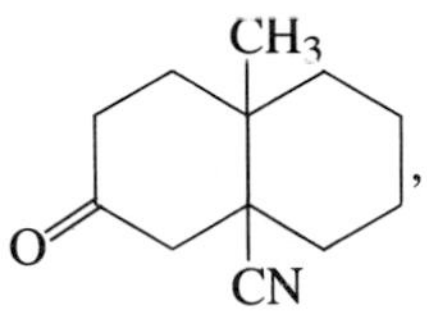 CH_3 / CN [structure], starting from cyclohexanone

(b) 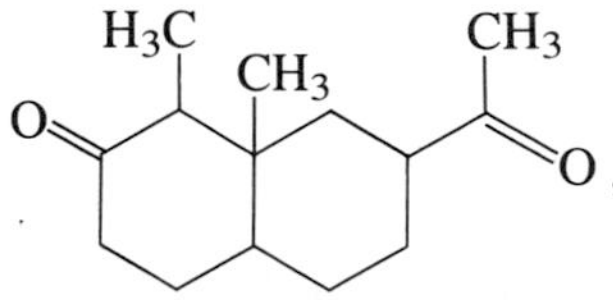 H_3C / CH_3 / CH_3 / O [structure] O, starting from 2-cyclohexenone

(Hint: Prepare O CH_3 CH_2 in your first step.)

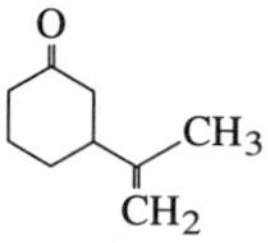

26. Write reagents (a, b, c, d, e) where they have been omitted from the following synthetic sequence. Each letter may correspond to one or more reaction steps. This sequence is the beginning of a synthesis of germanicol, a naturally occurring triterpene (see p. 718). The diol used in the step between (a) and (b) provides selective protection of the more reactive carbonyl group. [Hint: See Problem 23 when formulating (b).]

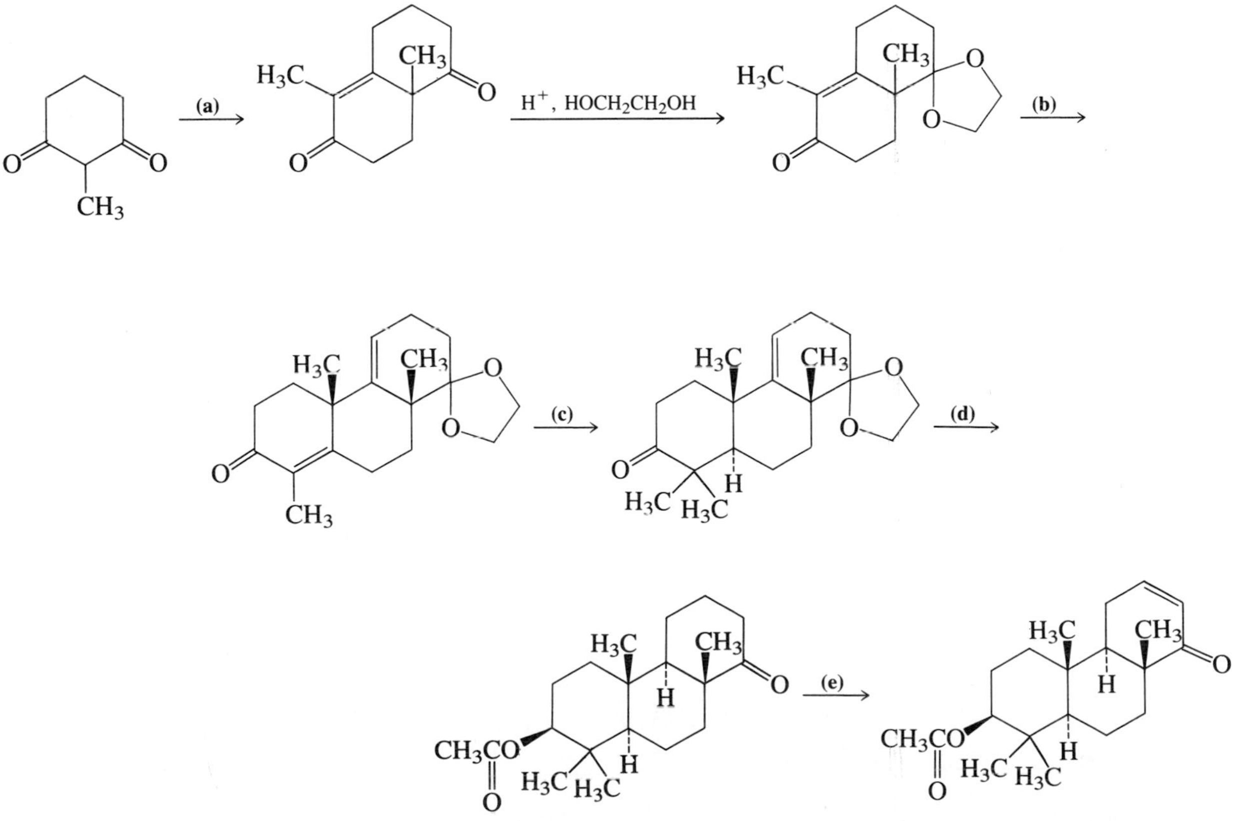

[Structure: O, O, CH_3] $\xrightarrow{\text{(a)}}$ [Structure: H_3C, CH_3, O, O] $\xrightarrow{H^+,\ HOCH_2CH_2OH}$ [Structure: H_3C, CH_3, O, O, O] $\xrightarrow{\text{(b)}}$

[Structure: H_3C, CH_3, O, O, O, CH_3] $\xrightarrow{\text{(c)}}$ [Structure: H_3C, CH_3, O, O, O, H, H_3C, H_3C] $\xrightarrow{\text{(d)}}$

[Structure: H_3C, CH_3, H, O, CH_3CO, O, H, H_3C, H_3C] $\xrightarrow{\text{(e)}}$ [Structure: H_3C, CH_3, H, O, CH_3CO, O, H, H_3C, H_3C]

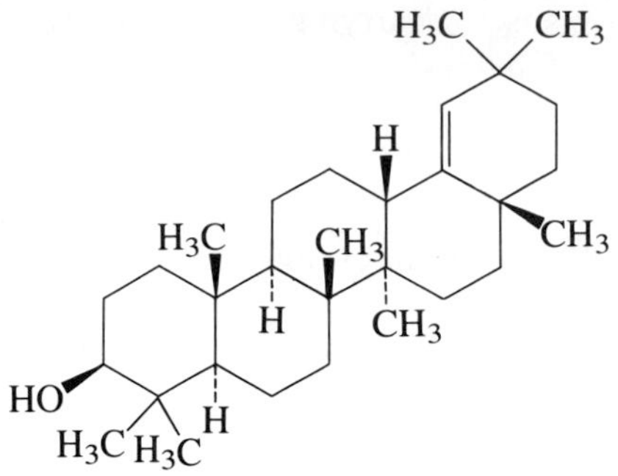

H_3C CH_3 H H_3C CH_3 CH_3 H CH_3 HO H H_3C H_3C

Germanicol

27. **(a)** The enzymatic oxidation of a lysine group to give an aldehyde is described in Section 18-5 (Box 18-2). What sort of intermediate(s) might appear in this oxidation? (Hints: Refer to Problem 22 of Chapter 4 and to Sections 3-9 and 17-9.) **(b)** A similar enzyme-catalyzed oxidation is the first step in the metabolism of amphetamine, which takes place in the endoplasmic reticulum of the liver. Write the structures of both the final product of this oxidation and the intermediate that immediately precedes its formation.

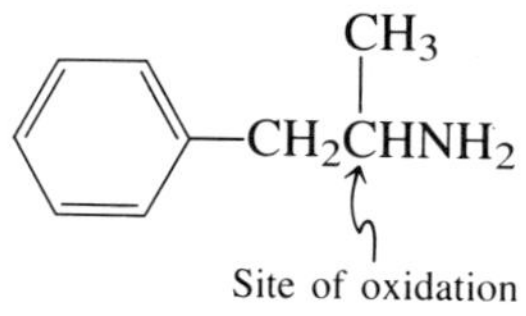

C_6H_5–$CH_2CH(CH_3)NH_2$

Site of oxidation

Amphetamine

19

Carboxylic Acids

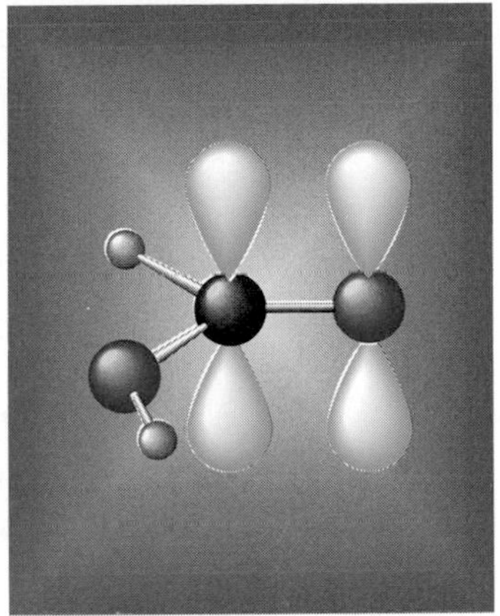

When a hydroxy group is attached to the carbonyl function, the **carboxy group,** a new functional group characteristic of **carboxylic acids,** is formed. This substituent is usually written COOH or CO_2H, and both of these conventions will be used in the following sections.

O
‖
—COH
Carboxy group

Carboxylic acids occur widely in nature; they are also important industrial chemicals. For example, acetic acid not only is the most important building block in the assembly of complex molecules in biology but also is a commodity produced industrially in large quantities. We all know it well: Vinegar derives its characteristic flavor and aroma from acetic acid.

Much of the reactivity of carboxylic acids can be anticipated if we view them as hydroxy carbonyl derivatives. Thus, the hydrogen is acidic, the oxygens basic and nucleophilic, and the carbonyl carbon is subject to nucleophilic attack.

Electrophilic → C (δ^+); Basic → O (δ^-) and OH oxygen; Acidic → H

This chapter first introduces the system of naming carboxylic acids and some of their physical and spectroscopic characteristics. We then examine their acidity and basicity, two properties that are influenced strongly by the interaction between the electron-withdrawing carbonyl group and the hydroxy function. Methods for the preparation of the carboxy group are discussed next, followed by a survey of its reactivity. The latter will feature a new substitution mechanism, *addition–elimination,* for the replacement of the hydroxy group by other nucleophiles, such as halide, alkoxide, and amide. The chemistry of the carboxylic acid derivatives, which result from these transformations, is the subject of Chapter 20.

19-1 Naming the Carboxylic Acids

Like other organic compounds, many carboxylic acids have common names that are used frequently in the literature. These names often indicate the natural sources from which the acids were first isolated (Table 19-1). *Chemical Abstracts* retains the common names for the two simplest, **formic** and **acetic** acids.

The IUPAC system derives the names of the carboxylic acids by replacing the ending *-e* in the name of the alkane by **-oic acid.** The alkanoic acid stem is numbered by assigning the number 1 to the carboxy carbon, and labeling any substituents along the longest chain incorporating the CO_2H group accordingly.

$CH_3CH(Br)COOH$
2-Bromopropanoic acid
(α-Bromopropionic acid)

$CH_2{=}CHCOOH$
Propenoic acid
(Acrylic acid)

$CH_3CH_2CH(CH_3)CH(CH_3)COOH$
2,3-Dimethylpentanoic acid
(α,β-Dimethylvaleric acid)

TABLE 19-1 Names and Natural Sources of Carboxylic Acids

Structure	IUPAC name	Common name	Natural source
$HCOOH$	Methanoic acid	Formic acid[a]	From the "destructive distillation" of ants (*formica,* Latin, ant)
CH_3COOH	Ethanoic acid	Acetic acid[a]	Vinegar (*acetum,* Latin, vinegar)
CH_3CH_2COOH	Propanoic acid	Propionic acid	Dairy products (*pion,* Greek, fat)
$CH_3CH_2CH_2COOH$	Butanoic acid	Butyric acid	Butter (particularly if rancid) (*butyrum,* Latin, butter)
$CH_3(CH_2)_3COOH$	Pentanoic acid	Valeric acid	Valerian root
$CH_3(CH_2)_4COOH$	Hexanoic acid	Caproic acid	Odor of goats (*caper,* Latin, goat)

a. Used by *Chemical Abstracts*.

The carboxy function takes precedence over any other discussed so far. In multiply functionalized carboxylic acids, the main chain is chosen to include other functional groups as much as possible. Saturated cyclic acids are named as **cycloalkanecarboxylic acids.** Their aromatic counterparts are the **benzoic acids.** In these compounds, the carbon attached to the carboxy functional group is C1.

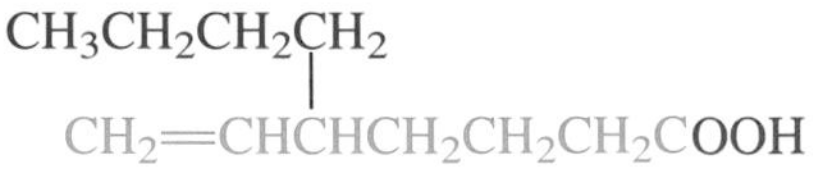

$CH_3CH_2CH_2CH_2$ (on C5 of) $CH_2{=}CHCHCH_2CH_2CH_2COOH$
5-Butyl-6-heptenoic acid
(Better than 5-ethenylnonanoic acid)

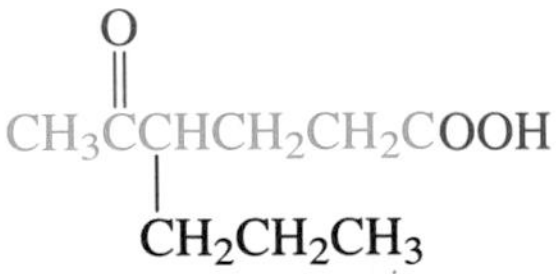

$CH_3C(=O)CHCH_2CH_2COOH$ with $CH_2CH_2CH_3$ on C4
5-Oxo-4-propylhexanoic acid

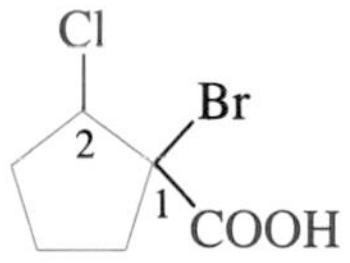

(Cyclopentane with Cl on C2, Br and COOH on C1)
1-Bromo-2-chloro-cyclopentane-carboxylic acid

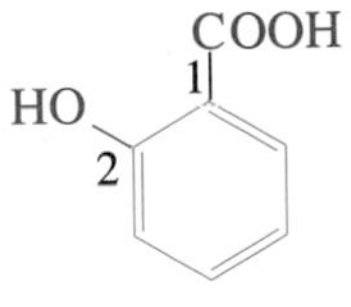

(Benzene ring with COOH on C1, HO on C2)
2-Hydroxybenzoic acid (*o*-Hydroxybenzoic acid, salicylic acid)

Dicarboxylic acids should be referred to as **dioic acids.**

$HOC(=O)C(=O)OH$
Ethanedioic acid
(Oxalic acid)

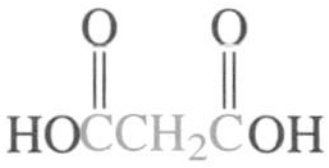

$HOC(=O)CH_2C(=O)OH$
Propanedioic acid
(Malonic acid)

$HOOCCH_2CH_2COOH$
Butanedioic acid
(Succinic acid)

$HOOC(CH_2)_3COOH$
Pentanedioic acid
(Glutaric acid)

$HOOC(CH_2)_4COOH$
Hexanedioic acid
(Adipic acid)

$HO_2CCH{=}CHCO_2H$
***cis*-2-Butenedioic acid**
(Maleic acid)
or
***trans*-2-Butenedioic acid**
(Fumaric acid)

Their common names reflect their natural origins. For example, butanedioic (succinic) acid was discovered in the distillate from amber (*succinum,* Latin, amber), 2-hydroxybenzoic (salicylic) acid is a widely known analgesic (painkilling) folk remedy that occurs in willow bark (*salix,* Latin, willow), and *trans*-2-butenedioic (fumaric) acid is found in the plant *Fumaria,* which was burned in ancient times to ward off evil spirits (*fumus,* Latin, smoke).

EXERCISE 19-1

Give systematic names or write the structure, as appropriate, of the following compounds.

(a) 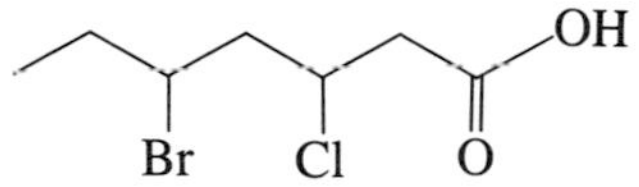 (skeletal chain with Br, Cl, and C(=O)OH)

(b) (cyclohexane with COOH and ring C=O at opposite position)

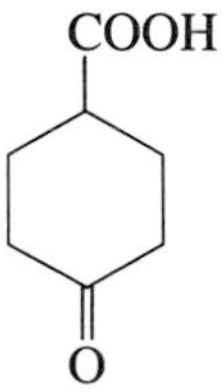

(c) (benzene ring with COOH, CH_3O, and NO_2)

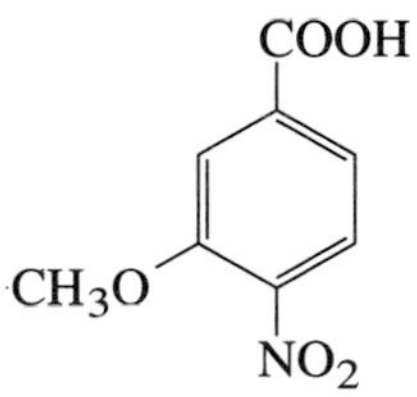

(d) 2,2-Dibromohexanedioic acid **(e)** 4-Hydroxypentanoic acid **(f)** 4-(1,1-Dimethylethyl)benzoic acid

In summary, the systematic naming of the carboxylic acids is based on the alkanoic acid stem. Cyclic derivatives are called cycloalkanecarboxylic acids, their aromatic counterparts benzoic acids, and dicarboxylic systems are labeled alkanedioic acids.

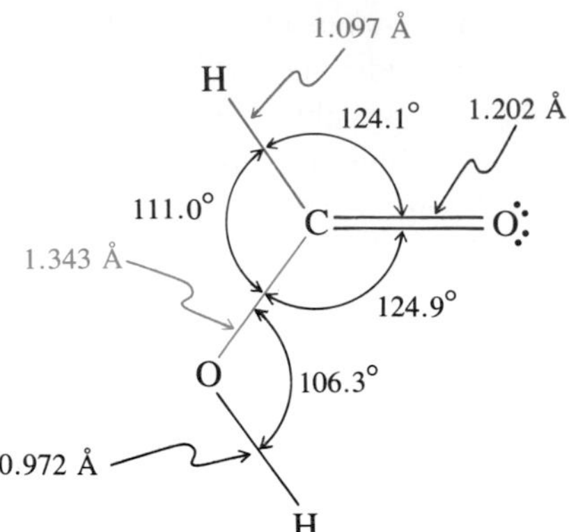

FIGURE 19-1 Molecular structure of formic acid. Like the other carboxylic acids, it is planar, with a roughly equilateral trigonal arrangement about the carbonyl carbon.

19-2 Structural and Physical Properties of Carboxylic Acids

What is the structure of a typical carboxylic acid? Do carboxylic acids have characteristic physical properties? This section answers these questions, beginning with the structure of formic acid. We shall see that carboxylic acids exist mainly as hydrogen-bonded dimers.

Formic acid is planar

The molecular structure of formic acid is shown in Figure 19-1. It is roughly planar, as expected for a ''hydroxyformaldehyde,'' with an approximately trigonal carbonyl carbon. (See the structure of methanol, Figure 8-1, and that of acetaldehyde, Figure 17-2.) These structural characteristics are found in carboxylic acids in general.

The carboxy group is polar and forms hydrogen-bonded dimers

The carboxy function is strongly polar because of the polarizable carbonyl double bond and the hydroxy group, which forms hydrogen bonds to other polarized molecules, such as water, alcohols, and other carboxylic acids. Not surprisingly, therefore, the lower carboxylic acids (up to butanoic acid) are completely soluble in water. As neat liquids and even in fairly dilute solutions (in aprotic solvents), carboxylic acids exist to a large extent as hydrogen-bonded dimers, each O–H···O interaction ranging in strength from about 6 to 8 kcal mol^{-1}.

Carboxylic Acids Form Dimers Readily

$$2\ RCOOH \longrightarrow R{-}C\begin{matrix} {/\!\!/}O\cdots H{-}O\backslash \\ \backslash O{-}H\cdots O{/\!\!/} \end{matrix}C{-}R$$

Two hydrogen bonds

TABLE 19-2 Melting and Boiling Points of Functional Alkane Derivatives with Various Chain Lengths

Derivative	Melting point (°C)	Boiling point (°C)
CH_4	−182.5	−161.7
CH_3Cl	−97.7	−24.2
CH_3OH	−97.8	65.0
HCHO	−92.0	−21.0
HCOOH	8.4	100.6
CH_3CH_3	−183.3	−88.6
CH_3CH_2Cl	−136.4	12.3
CH_3CH_2OH	−114.7	78.5
CH_3CHO	−121.0	20.8
CH_3COOH	16.7	118.2
$CH_3CH_2CH_3$	−187.7	−42.1
$CH_3CH_2CH_2Cl$	−122.8	46.6
$CH_3CH_2CH_2OH$	−126.5	97.4
CH_3COCH_3	−95.0	56.5
CH_3CH_2CHO	−81.0	48.8
CH_3CH_2COOH	−20.8	141.8

Table 19-2 shows that carboxylic acids have relatively high melting and boiling points, because they form hydrogen bonds in the solid as well as the liquid state.

Carboxylic acids, especially those possessing relatively low molecular weights and correspondingly high volatility, exhibit characteristically strong odors. For example, the presence of butanoic acid contributes to the characteristic strong flavor of many cheeses, and *E*-3-methyl-2-hexenoic acid was identified in 1991 as the principal compound responsible for the smell of human sweat.

O
OH

***E*-3-Methyl-2-hexenoic acid**

In summary, the carboxy function is planar and contains a polarizable carbonyl group. Carboxylic acids exist as hydrogen-bonded dimers and exhibit unusually high melting and boiling points.

19-3 NMR and IR Spectroscopy of Carboxylic Acids

The polarizable double bond and the hydroxy group also shape the spectra of carboxylic acids. In this section, we show how both NMR and IR methods are used to characterize the carboxy group.

The carboxy hydrogen and carbon are deshielded

As in aldehydes and ketones, the hydrogens positioned on the carbon next to the carbonyl group are slightly deshielded in ^1H NMR spectra. The effect diminishes

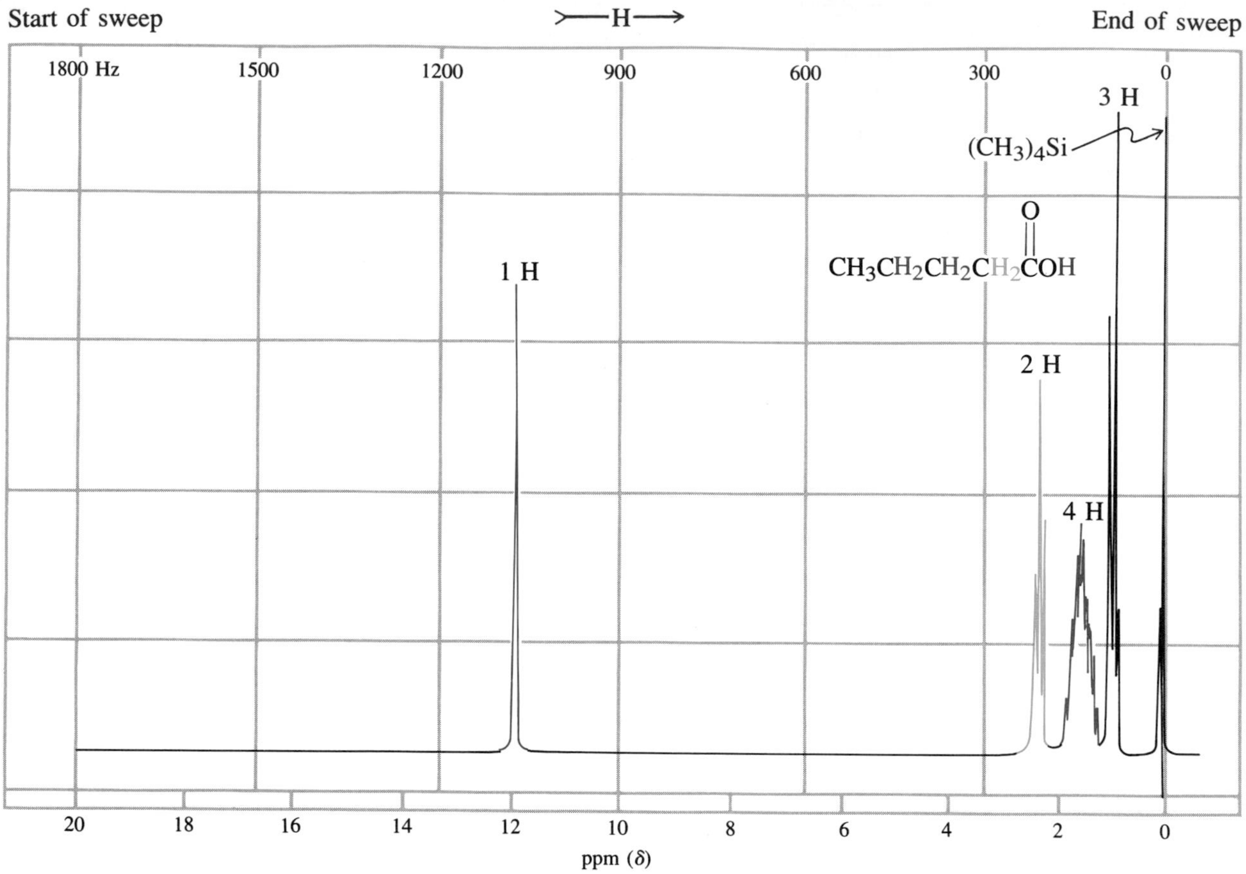

FIGURE 19-2 90-MHz ^{1}H NMR spectrum of pentanoic acid in CCl_4. The scale has been expanded to 20 ppm, to allow the signal for the acid proton at $\delta = 11.83$ ppm to be shown. The methylene hydrogens at C2 absorb at the next lowest field as a triplet ($\delta = 2.25$ ppm, $J = 7$ Hz), followed by a four-hydrogen multiplet for the next two sets of methylenes. The methyl group appears as a distorted triplet at highest field ($\delta = 0.90$ ppm, $J = 6$ Hz).

rapidly with increasing distance from the functional group. The hydroxy proton resonates at very low field ($\delta = 10$–13 ppm). As in the NMR spectra of alcohols, its chemical shift varies strongly with concentration, solvent, and temperature, because of the strong ability of the OH group to enter into hydrogen bonding. The ^{1}H NMR spectrum of pentanoic acid is shown in Figure 19-2.

^{1}H NMR Chemical Shifts of Alkanoic Acids

CH_3COOH	CH_3CH_2COOH	$(CH_3)_2CHCOOH$	$HCOOH$
$\delta = 2.08$	1.16 2.36	1.21 2.56	8.08 ppm

The ^{13}C NMR chemical shifts of carboxylic acids are also similar to those of the aldehydes and ketones, with moderately deshielded carbons next to the carbonyl group and the typically low-field carbonyl absorptions. However, the amount of deshielding is smaller, because the positive polarization of the carboxy carbon is somewhat attenuated by the presence of the extra OH group.

Typical ^{13}C NMR Chemical Shifts of Alkanoic Acids

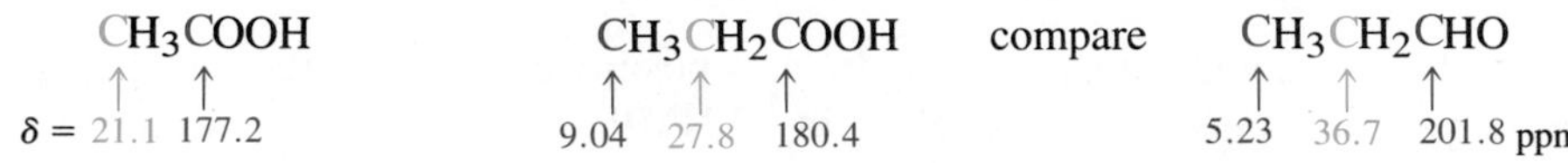

This attenuation is best understood by writing resonance structures. Recall from Section 17-2 that aldehydes and ketones are described by two resonance forms. The dipolar structure reflects the polarization of the C=O bond. Even though this form is a minor contributor (because of the lack of an octet on carbon), it explains the deshielding of the carbonyl carbon. In carboxylic acids, however, the corresponding structure contributes much less to the resonance hybrid: The hydroxy oxygen can donate an electron pair to give a third form in which the carbon and both oxygen atoms have octets. The degree of positive charge on the carbonyl carbon and therefore its deshielding are greatly reduced.

Resonance in Aldehydes and Ketones

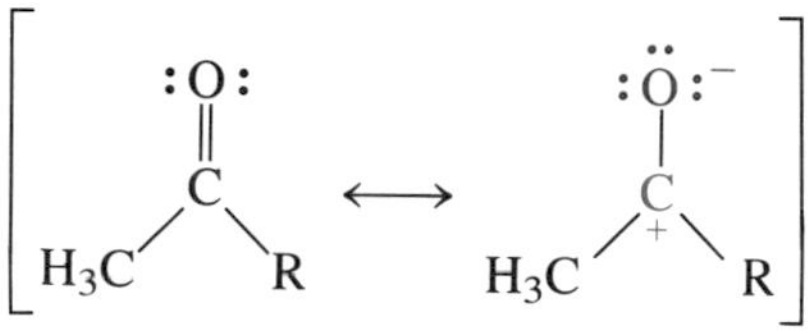

The contribution of the second resonance structure, though minor, explains the strong deshielding of the carbonyl and adjacent carbons

Resonance in Carboxylic Acids

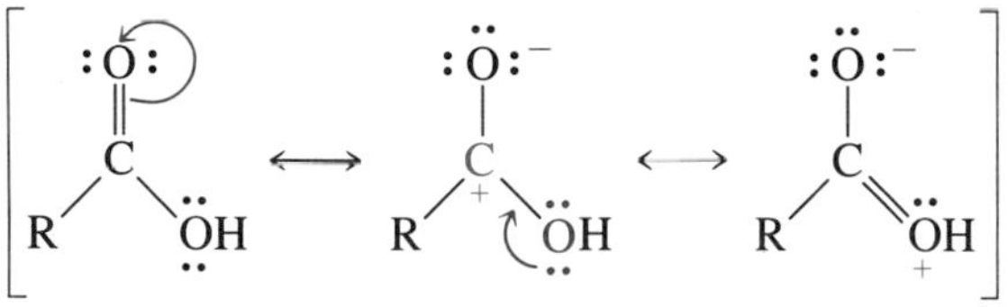

The third resonance structure explains the attenuated deshielding of the carbonyl carbon relative to that of aldehydes and ketones

EXERCISE 19-2

A foul-smelling carboxylic acid with b.p. 164°C gave the following NMR data: ^{1}H NMR (CCl_4) $\delta = 1.00$ (t, $J = 7.4$ Hz, 3 H), 1.65 (sextet, $J = 7.5$ Hz, 2 H), 2.31 (t, $J = 7.4$ Hz, 2 H), and 11.68 (s, 1 H) ppm; ^{13}C NMR (CS_2) $\delta = 13.4$, 18.5, 36.3, and 179.6 ppm. Assign a structure to it.

The carboxy group shows two important IR bands

The carboxy group consists of a carbonyl group and an attached hydroxy substituent. Consequently, both characteristic stretching frequencies are seen in the infrared spectrum. The O–H bond gives rise to a very broad band at lower wavenumber (2500–3300 cm^{-1}) than is observed for alcohols, largely because of strong hydrogen bonding. The IR spectrum of propanoic acid is shown in Figure 19-3.

In summary, the NMR signals for carboxylic acids reveal highly deshielded acid protons and carbonyl carbons and moderately deshielded nuclei next to the functional group. The infrared spectrum shows characteristic bands at about 3000 (O–H) and 1710 (C=O) cm^{-1}.

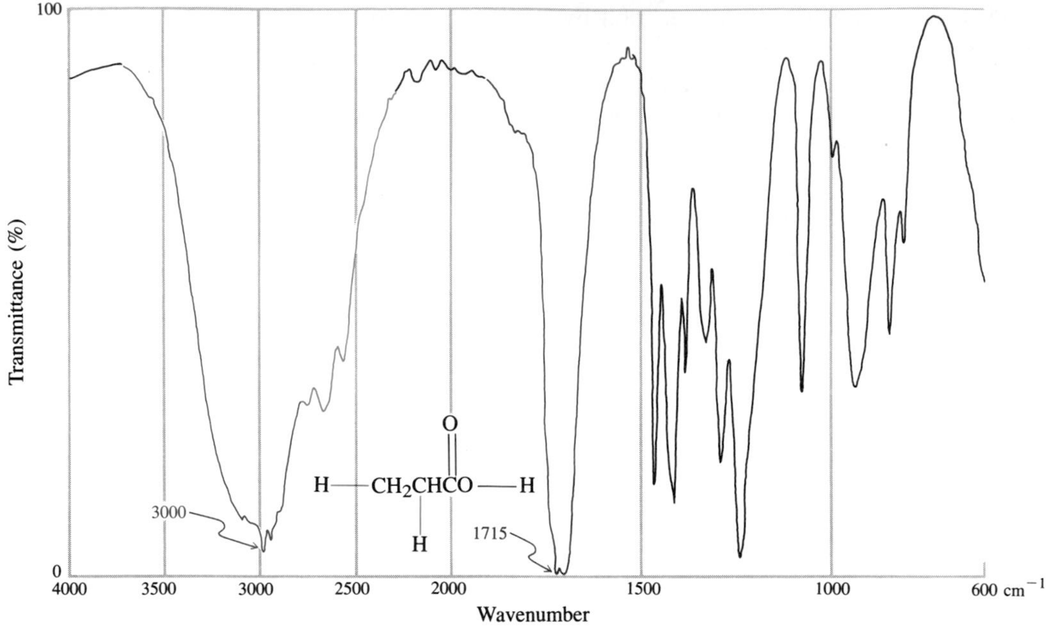

FIGURE 19-3 IR spectrum of propanoic acid: $\tilde{\nu}_{\text{O–H stretch}} = 3000\ \text{cm}^{-1}$; $\tilde{\nu}_{\text{C=O stretch}} = 1715\ \text{cm}^{-1}$. The peaks associated with these stretching vibrations are broad because of hydrogen bonding.

19-4 Acidic and Basic Character of Carboxylic Acids

Like alcohols (Section 8-3), carboxylic acids exhibit both acidic and basic character: Deprotonation to carboxylate ions is relatively easy, protonation more difficult.

Carboxylic acids are relatively strong acids

Carboxylic acids have much lower pK_a values than alcohols do, even though the hydrogen involved in each case is that of a hydroxy group.

Carboxylic Acids Dissociate Readily

$$\mathrm{R\overset{\displaystyle O}{\overset{\|}{C}}\ddot{O}H} + \mathrm{H_2\ddot{O}} \rightleftharpoons \mathrm{R\overset{\displaystyle O}{\overset{\|}{C}}\ddot{O}\!:^-} + \mathrm{H\ddot{O}H_2^+}$$

$K_a \sim 10^{-4}$–10^{-5}, p$K_a \sim 4$–5 **Carboxylate ion**

Why do carboxylic acids dissociate to a greater extent than do alcohols? The difference is that their hydroxy substituent is attached to a carbonyl group, whose positively polarized carbonyl carbon exerts a powerful electron-withdrawing inductive effect. In addition, the carboxylate ion is stabilized strongly by resonance, much as is the enolate ion formed by deprotonation of the α-carbon in aldehydes and ketones (Section 18-1).

Resonance in Carboxylate and Enolate Ions

Carboxylate ion (B = base)

$$\text{B:}^- + \text{RCOH} \rightleftharpoons \text{BH} + \left[\text{RC(=O)}-\ddot{\text{O}}:^- \longleftrightarrow \text{RC}(-\text{O}:^-)=\text{O}\right] \qquad \text{p}K_a \sim 4\text{–}5$$

Enolate ion

$$\text{B:}^- + \text{R}'\text{CCH}_2\text{R} \rightleftharpoons \text{BH} + \left[\text{R}'\text{C(=O)}-\ddot{\text{C}}\text{HR} \longleftrightarrow \text{R}'\text{C}(-\text{O}:^-)=\text{CHR}\right] \qquad \text{p}K_a \sim 19\text{–}21$$

In contrast with enolates, the two resonance structures in carboxylate ions are equivalent (Section 1-5). As a result, carboxylates are symmetric, with equal carbon–oxygen bond lengths (1.26 Å), in between the lengths typical of the carbon–oxygen double (1.20 Å) and single (1.34 Å) bonds in the corresponding acids (Figure 19-1).

Electron-withdrawing substituents increase the acidity of carboxylic acids

As with alcohols, the inductive effects of electron-withdrawing groups close to the carboxy function increase the acidity. Table 19-3 shows the pK_a values of selected carboxylic acids. Note that two or three such groups on the α-carbon can result in carboxylic acids that are nearly as strong as typical mineral acids.

TABLE 19-3 pK_a Values of Various Carboxylic and Other Acids

Compound	pK_a	Compound	pK_a
Alkanoic acids		**Benzoic acids**	
CH_3COOH	4.74	4-$CH_3C_6H_4COOH$	4.36
$ClCH_2COOH$	2.86	C_6H_5COOH	4.19
$Cl_2CHCOOH$	1.26	4-ClC_6H_4COOH	3.98
Cl_3CCOOH	0.64	**Other acids**	
F_3CCOOH	0.23	H_3PO_4	2.15 (first pK_a)
$CH_3CH_2CH_2COOH$	4.9	HNO_3	−1.3
$CH_3CH_2CHClCOOH$	3.8	HCl	−2.2
$CH_3CHClCH_2COOH$	4.1	H_2SO_4	−5.2 (first pK_a)
$ClCH_2CH_2CH_2COOH$	4.5	H_2O	15.7
		CH_3OH	15.5
Dioic acids			
$HOOCCOOH$	2.77, 5.81		
$HOOCCH_2COOH$	3.15, 6.30		
$HOOCCH_2CH_2COOH$	5.84, 6.34		
$HOOC(CH_2)_4COOH$	5.57, 6.38		

EXERCISE 19-3

In the following sets of acids, rank the components in order of *decreasing* acidity.

(a)  CH_3CH_2COOH $CH_3CH(Br)COOH$ CH_3CBr_2COOH

(b) 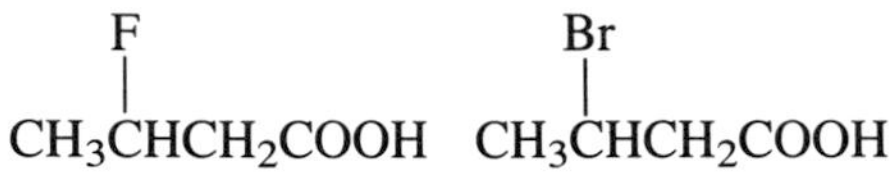 $CH_3CH(F)CH_2COOH$ $CH_3CH(Br)CH_2COOH$

(c) 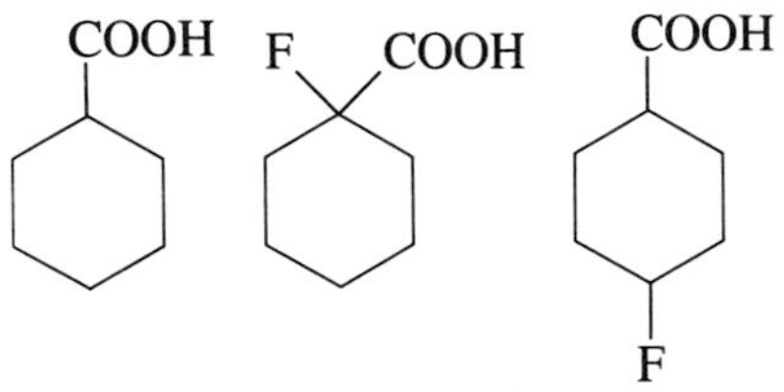 Cyclohexanecarboxylic acid; 1-fluorocyclohexanecarboxylic acid (F and COOH on the same carbon); 4-fluorocyclohexanecarboxylic acid (COOH at top, F at bottom)

The dioic acids have two pK_a values, one for each CO_2H group. In ethanedioic (oxalic) and propanedioic (malonic) acids, the first pK_a is much lower than the second, because of the electron-withdrawing effects of the two carboxy groups on each other. In the higher dioic acids, the two pK_a values are very close.

The relatively strong acidity of carboxylic acids means that their corresponding **carboxylate salts** are readily made by treatment of the acid with base, such as NaOH, Na_2CO_3, or $NaHCO_3$. These salts are named by specifying the metal and replacing the ending *-ic acid* in the acid name by **-ate.** Thus, $HCOO^-Na^+$ is called sodium formate; $CH_3COO^-Li^+$ is named lithium acetate; and so on. Carboxylate salts are much more water soluble than the corresponding acids, because the polar anionic group is readily solvated.

Carboxylate Salt Formation

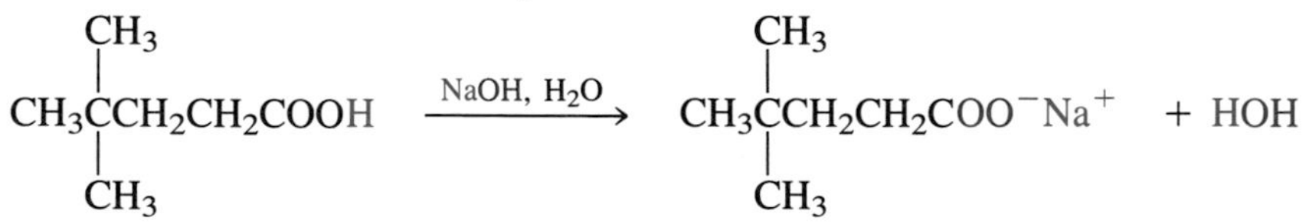

$$CH_3C(CH_3)_2CH_2CH_2COOH \xrightarrow{NaOH,\ H_2O} CH_3C(CH_3)_2CH_2CH_2COO^-Na^+ + HOH$$

4,4-Dimethylpentanoic acid (Slightly water soluble) → **Sodium 4,4-dimethylpentanoate** (Water soluble)

Carboxylic acids may be protonated on the carbonyl oxygen

The lone electron pairs on both of the oxygen atoms in the carboxy group can, in principle, be protonated, just as alcohols are protonated by strong acids to give alkyloxonium ions (Section 8-3). The available evidence indicates that it is the carbonyl oxygen that is more basic. Why? Protonation at the carbonyl oxygen gives a species whose positive charge is delocalized by resonance. The alternative possibility, resulting from protonation at the hydroxy group, does not benefit from such stabilization.

BOX 19-1 Soaps from Long-Chain Carboxylates

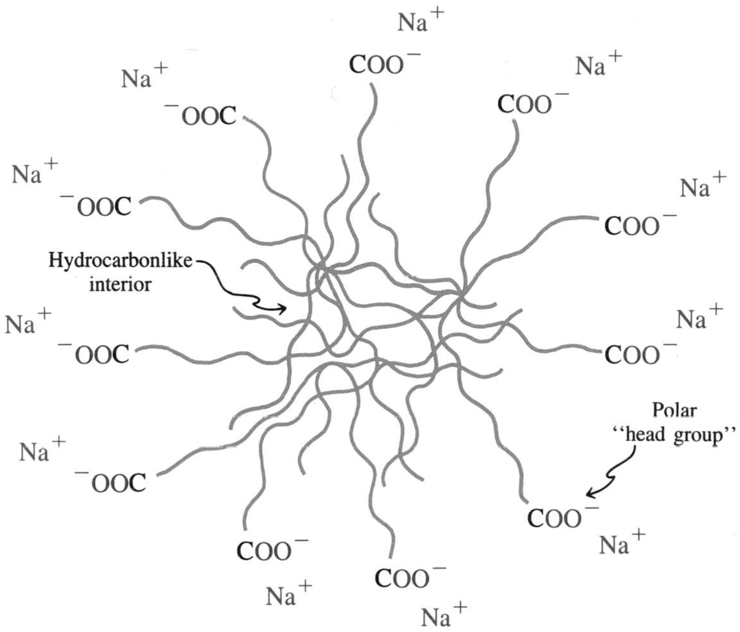

The sodium and potassium salts of long-chain carboxylic acids have the interesting property of aggregating as spherical clusters called *micelles* in aqueous solution. In such aggregates, all the hydrophobic alkyl chains of the acids attempt to occupy the same region in space because of their attraction to one another by London forces and their tendency to avoid exposure to polar water as much as possible. As shown above, the polar, water-solvated carboxylate "head groups" form a spherical wall around the hydrocarbonlike center.

Because these carboxylate salts also create films on aqueous surfaces, they act as soaps. The polar groups stick into the water while the alkyl chains assemble into a hydrophobic layer. This construction reduces the surface tension of the water, permitting it to permeate cloth and other fabrics and giving rise to the foaming typical of soaps. Cleansing is accomplished by dissolving ordinarily water-insoluble materials (oils, fats) in the hydrocarbon interior of the micelles.

One problem with long-chain carboxylate soaps is that they form curdlike precipitates with the ions present in hard water (e.g., Mg^{2+} and Ca^{2+}). Detergents based on alkane sulfonates, $RSO_3^-\ Na^+$, and alkyl sulfates, $ROSO_3^-\ Na^+$, avoid this drawback but have caused serious pollution problems in lakes and streams because branches in their alkyl chains made early versions *nonbiodegradable*. The microorganisms associated with normal sewage treatment procedures are capable of breaking down only straight-chain systems.

Certain steroid *bile acids* such as *cholic acid* (Section 4-7) also have surfactant or detergent-like properties and are found in the bile duct. These substances are released into the upper intestinal tract where they emulsify water-insoluble fats through the formation of micelles. Hydrolytic enzymes can then digest the dispersed fat molecules.

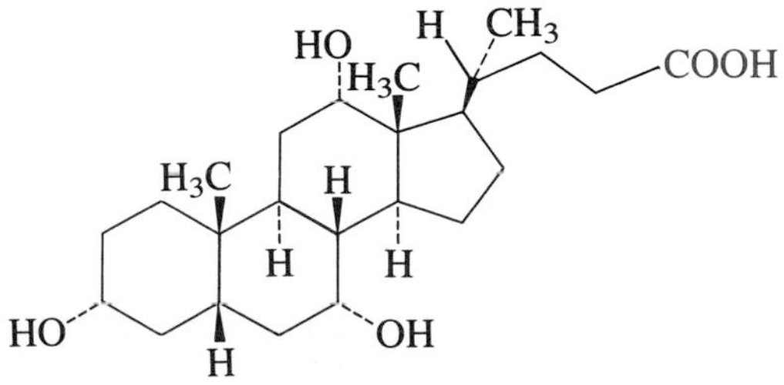

Cholic acid

Protonation of a Carboxylic Acid

$$\text{RC}(=\ddot{\text{O}}{:})\overset{+}{\text{O}}\text{H}_2 \xleftarrow[\text{Not observed}]{\times} \text{RC}(=\ddot{\text{O}}{:})\ddot{\text{O}}\text{H} + \text{H}^+ \underset{K\,\sim\,10^{-6}}{\rightleftharpoons} \left[\text{RC}(=\overset{+}{\ddot{\text{O}}}\text{H})\ddot{\text{O}}\text{H} \longleftrightarrow \text{RC}(\ddot{\text{O}}\text{H})=\overset{+}{\ddot{\text{O}}}\text{H}\right]$$

$pK_a \sim -6$

Note that protonation is nevertheless very difficult, as shown by the very high acidity ($pK_a \sim -6$) of the conjugate acid. We shall see, however, that such protonations are important in many reactions of the carboxylic acids and their derivatives.

EXERCISE 19-4

The pK_a of protonated propanone (acetone) is -7.2 and that of protonated acetic acid is -6.1. Explain.

In summary, carboxylic acids are acidic because the polarized carbonyl carbon is strongly electron-withdrawing, and deprotonation gives resonance-stabilized anions. Electron-withdrawing groups increase acidity, although this effect wears off rapidly with increasing distance from the carboxy group. Protonation is difficult, but possible, and occurs on the carbonyl oxygen to give a resonance-stabilized cation.

19-5 Carboxylic Acid Synthesis in Industry

Carboxylic acids are useful reagents and synthetic precursors. The two simplest ones are manufactured on a large scale by the chemical industry. Formic acid is prepared conveniently by reaction of powdered sodium hydroxide with carbon monoxide under pressure. This transformation proceeds by nucleophilic addition followed by protonation.

Formic Acid Synthesis

$$\text{NaOH} + \text{CO} \xrightarrow{150°\text{C, 100 psi}} \text{HCOO}^-\text{Na}^+ \xrightarrow{\text{H}^+,\ \text{H}_2\text{O}} \text{HCOOH}$$

There are three important industrial preparations of acetic acid: ethene oxidation through acetaldehyde (Section 12-15); air oxidation of butane; and carbonylation of methanol. The mechanisms of these reactions are complex.

Acetic Acid by Oxidation of Ethene

$$\text{CH}_2{=}\text{CH}_2 \xrightarrow[\text{Wacker process}]{\text{O}_2,\ \text{H}_2\text{O, catalytic PdCl}_2\text{ and CuCl}_2} \text{CH}_3\text{CHO} \xrightarrow{\text{O}_2,\ \text{catalytic Co}^{3+}} \text{CH}_3\text{COOH}$$

Acetic Acid by Oxidation of Butane

$$CH_3CH_2CH_2CH_3 \xrightarrow{O_2,\ \text{catalytic } Co^{3+},\ 15\text{–}20\ \text{atm},\ 180°C} CH_3COOH$$

Acetic Acid by Carbonylation of Methanol

$$CH_3OH \xrightarrow[\text{Monsanto process}]{CO,\ \text{catalytic } Rh^{3+},\ I_2,\ 30\text{–}40\ \text{atm},\ 180°C} CH_3COOH$$

Annual production of acetic acid in the United States alone approaches 2 million tons (2×10^9 kg). It is used to manufacture monomers for polymerization, such as ethenyl acetate (vinyl acetate; Section 12-15), as well as pharmaceuticals, dyes, and pesticides. Two dicarboxylic acids in large-scale chemical production are hexanedioic (adipic) acid, used in the manufacture of nylon (see Section 21-12), and 1,4-benzenedicarboxylic (terephthalic) acid, whose polymeric esters with diols are fashioned into plastic sheets and films.

COOH–C₆H₄–COOH

1,4-Benzenedicarboxylic acid (Terephthalic acid)

19-6 Methods for Introducing the Carboxy Functional Group

The oxidation of primary alcohols and aldehydes to carboxylic acids by aqueous Cr(VI) has been described previously (Sections 8-6 and 17-4). This section presents two additional reagents suitable for this purpose. It is also possible to introduce the carboxy function by adding a carbon atom to a molecule. This transformation may be achieved in either of two ways: the carbonation of organometallic reagents and the hydrolysis of nitriles.

Oxidation of primary alcohols and of aldehydes furnishes carboxylic acids

Primary alcohols oxidize to aldehydes, which in turn may readily oxidize further to the corresponding carboxylic acids.

Carboxylic Acids by Oxidation

$$RCH_2OH \xrightarrow{\text{Oxidation}} R\overset{O}{\overset{\|}{C}}H \xrightarrow{\text{Oxidation}} R\overset{O}{\overset{\|}{C}}OH$$

In addition to aqueous CrO_3, $KMnO_4$ and nitric acid, HNO_3, are frequently used in this process. Because it is one of the cheapest strong oxidants, nitric acid is often chosen for large-scale and industrial applications.

$$2\ HNO_3 + ClCH_2CH_2\overset{O}{\overset{\|}{C}}H \xrightarrow{25°C} ClCH_2CH_2\overset{O}{\overset{\|}{C}}OH + 2\ NO_2 + H_2O$$

3-Chloropropanal → **3-Chloropropanoic acid** (79%)

EXERCISE 19-5

Give the products of nitric acid oxidation of **(a)** pentanal; **(b)** 1,6-hexanediol; **(c)** 4-(hydroxymethyl)cyclohexanecarbaldehyde.

Organometallic reagents react with carbon dioxide to give carboxylic acids

Organometallic reagents attack carbon dioxide (usually in the solid form known as "Dry Ice") much as they would attack aldehydes or ketones. The product of this **carbonation** process is a carboxylate salt, which upon protonation by aqueous acid yields the carboxylic acid.

Carbonation of Organometallics

$$R^{\delta^-}\text{—}Mg^{\delta^+}X + {}^{\delta^+}C(=O^{\delta^-})_2 \xrightarrow{\text{THF}} R\text{—}C(=O)\text{—}\ddot{O}{:}^{-}\ {}^{+}MgX \xrightarrow[-\ XMgOH]{H^+,\ HOH} RCOOH$$

$$RLi + CO_2 \xrightarrow{\text{THF}} RCOO^-Li^+ \xrightarrow[-LiOH]{H^+,\ HOH} RCOOH$$

Recall that the organometallic reagent is usually made from the corresponding haloalkane: $RX + Mg \rightarrow RMgX$. Hence, carbonation of the latter allows the two-step transformation of RX into RCOOH, the carboxylic acid with one more carbon. For example,

$$CH_3CH_2CH(Cl)CH_3 + Mg \xrightarrow{\text{THF}} CH_3CH_2CH(MgCl)CH_3 \xrightarrow{1.\ CO_2\quad 2.\ H^+,\ H_2O} CH_3CH_2CH(COOH)CH_3$$

2-Chlorobutane → 86% **2-Methylbutanoic acid**

Nitriles hydrolyze to carboxylic acids

Another method for converting a haloalkane into a carboxylic acid with an additional carbon atom is through the preparation and hydrolysis of a nitrile, RC≡N. Recall (Section 6-3) that cyanide ion, $^-$:C≡N:, is a good nucleophile and may be used to synthesize nitriles through S_N2 reactions. Hydrolysis of the nitrile in hot acid or base furnishes the corresponding carboxylic acid (and ammonia).

Carboxylic Acids from Haloalkanes Through Nitriles

$$RX \xrightarrow[-\ X^-]{^-CN} RC{\equiv}N \xrightarrow{1.\ HO^-\quad 2.\ H^+,\ H_2O} RCOOH + NH_3$$

The mechanism of this reaction will be discussed in detail in Section 20-8.

Carboxylic acid synthesis using nitrile hydrolysis is preferable to Grignard carbonation when the substrate contains groups that react with organometallic reagents, such as the hydroxy, carbonyl, and nitro functionalities.

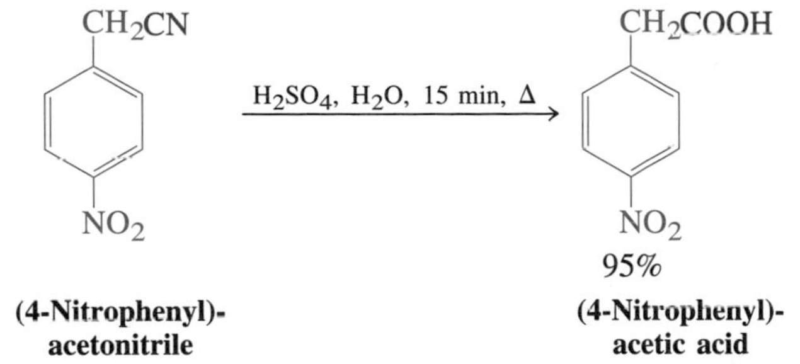
CH2CN
H2SO4, H2O, 15 min, Δ
CH2COOH
NO2
NO2
95%
(4-Nitrophenyl)-
acetonitrile
(4-Nitrophenyl)-
acetic acid

Hydrolysis of the nitrile group in a cyanohydrin, prepared by addition of HCN to an aldehyde or a ketone (Section 17-11), is a general route to 2-hydroxycarboxylic acids.

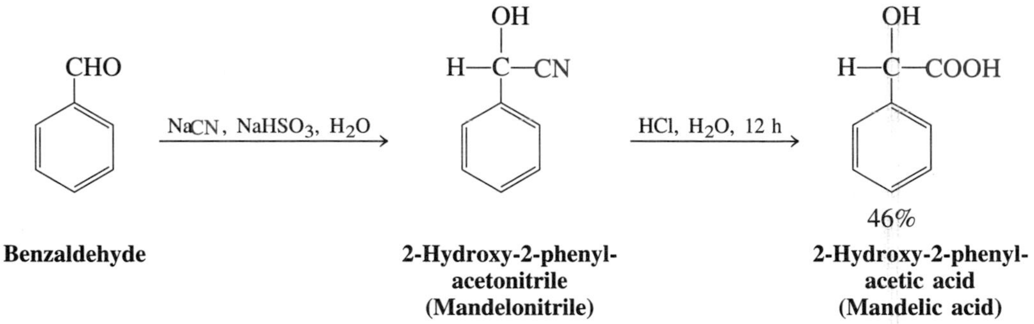
CHO
NaCN, NaHSO3, H2O
OH
H—C—CN
HCl, H2O, 12 h
OH
H—C—COOH
46%
Benzaldehyde
2-Hydroxy-2-phenyl-
acetonitrile
(Mandelonitrile)
2-Hydroxy-2-phenyl-
acetic acid
(Mandelic acid)

EXERCISE 19-6

Suggest ways to effect the following conversions (more than one step will be required).

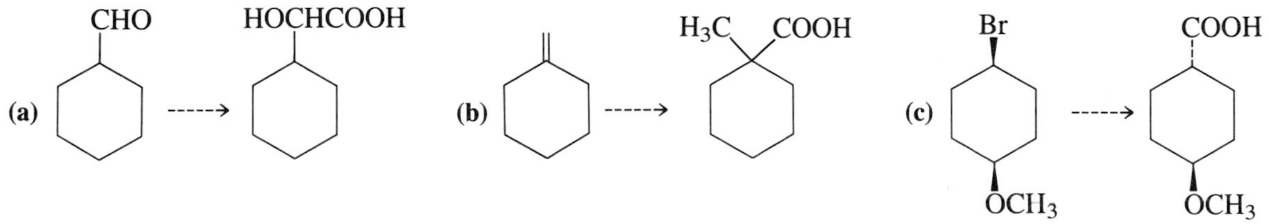
(a)
CHO
HOCHCOOH
(b)
H3C
COOH
(c)
Br
OCH3
COOH
OCH3

In summary, several reagents oxidize primary alcohols and aldehydes to carboxylic acids. Haloalkanes are transformed into the carboxylic acid containing one additional carbon atom by either conversion to an organometallic reagent and carbonation or displacement of halide by cyanide followed by nitrile hydrolysis.

19-7 Substitution at the Carboxy Carbon: The Addition–Elimination Mechanism

Carboxylic acids show reactivity similar to that of aldehydes and ketones (Section 17-5): The carbonyl carbon is subject to nucleophilic attack and the oxygen is the target of electrophiles. However, the course of nucleophilic addition is different from that for aldehydes and ketones. The carboxy OH may be converted into a leaving group, thereby giving rise to new carbonyl derivatives. This section introduces the general mechanisms by which this substitution process occurs for both carboxylic acids and their derivatives.

The carbonyl carbon is attacked by nucleophiles

$$\underset{\text{R}\diagup\quad\diagdown\text{L}}{\overset{:\ddot{\text{O}}:}{\overset{\|}{\text{C}}}}$$

Carboxylic acid derivative

Carbonyl carbons are electrophilic and potentially can be attacked by nucleophiles. This type of reactivity is observed in the carboxylic acids and the **carboxylic acid derivatives,** substances with the general formula RCOL (L stands for leaving group).

Carboxylic Acid Derivatives

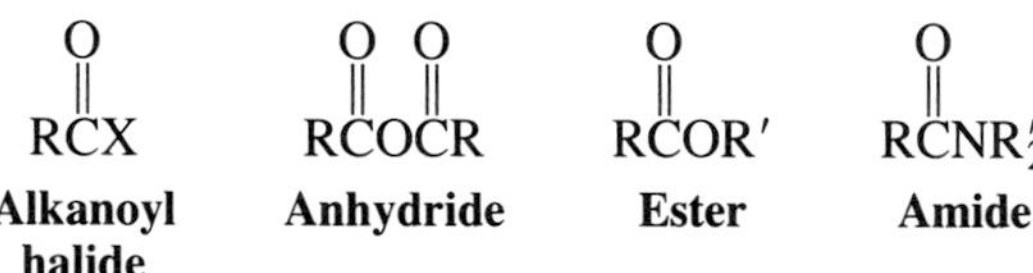

Alkanoyl halide **Anhydride** **Ester** **Amide**

In contrast to the addition products of aldehydes and ketones (Sections 17-5 through 17-7), the intermediate formed upon attack of a nucleophile on a carboxy carbon can decompose *by eliminating a leaving group*. The overall result is substitution of the nucleophile for the leaving group through a process called **addition–elimination.** The species formed first in this transformation contains (in contrast with both starting material and product) a tetrahedral carbon center. It is therefore called the **tetrahedral intermediate.**

Color code reminder:
Nucleophile—red
Electrophile—blue
Leaving group—green

Nucleophilic Substitution by Addition–Elimination

$$\text{R}\overset{\text{O}}{\overset{\|}{\text{C}}}\text{L} + \text{H—Nu} \rightleftharpoons \text{R—}\underset{\text{Nu}}{\overset{\text{O—H}}{\overset{|}{\underset{|}{\text{C}}}}}\text{—L} \rightleftharpoons \text{R}\overset{\text{O}}{\overset{\|}{\text{C}}}\text{Nu} + \text{L—H}$$

Tetrahedral intermediate

Substitution by addition–elimination is the most important pathway for the formation of carboxylic acid derivatives as well as for their interconversion, that is, $\text{R}\overset{\text{O}}{\overset{\|}{\text{C}}}\text{L} \rightarrow \text{R}\overset{\text{O}}{\overset{\|}{\text{C}}}\text{L}'$. The remainder of this section and those that follow will describe how such derivatives are prepared from carboxylic acids, and Chapter 20 will explore their properties and chemistry.

Addition–elimination is catalyzed by acid or base

Addition–elimination reactions benefit from acid catalysis. The acid functions in two ways: First, it protonates the carbonyl oxygen, activating the carbonyl group toward nucleophilic attack (Section 17-5). Second, protonation of L facilitates elimination by generating a better leaving group (recall Sections 6-7 and 9-2).

Mechanism of Acid-Catalyzed Addition–Elimination

STEP 1. Protonation

$$RC(=O)L + H^+ \rightleftharpoons RC(=\overset{+}{O}H)L$$

STEP 2. Addition–elimination

$$RC(=\overset{+}{O}H)L + :NuH \rightleftharpoons R{-}C(OH)(L){-}\overset{+}{Nu}H \rightleftharpoons R{-}C(OH)(Nu){-}\overset{+}{L}H \rightleftharpoons RC(=\overset{+}{O}H)Nu + LH$$

STEP 3. Deprotonation

$$RC(=\overset{+}{O}H)Nu \rightleftharpoons RC(=O)Nu + H^+$$

In base-catalyzed addition–elimination, the base (denoted as $:B^-$) ensures the maximum concentration of negatively charged (deprotonated) nucleophile (such as HO^- or RO^-) when it is the attacking species.

Mechanism of Base-Catalyzed Addition–Elimination

STEP 1. Deprotonation of NuH

$$Nu{-}H + {}^-{:}B \rightleftharpoons {}^-{:}Nu + BH$$

STEP 2. Addition–elimination

$$RC(=O)L + {}^-{:}Nu \rightleftharpoons R{-}C(O^-)(L){-}Nu \rightleftharpoons RC(=O)Nu + {}^-{:}L$$

STEP 3. Regeneration of catalyst

$${}^-{:}L + H{-}B \rightleftharpoons LH + {}^-{:}B$$

(Alternatively, $^-$:L may act as a base in step 1)

Substitution in carboxylic acids is complicated by a poor leaving group and the acidic proton

Let us now see how the general addition–elimination process may be applied to the conversion of carboxylic acids to their derivatives. Two problems are immediately evident. First, the hydroxide ion is a very poor leaving group. Second, recall that the hydroxy proton is acidic and that most nucleophiles are bases. Therefore, an acid-base reaction can interfere with nucleophilic attack.

Competing Reactions of a Carboxylic Acid with a Nucleophile

$$\mathrm{R{-}C(O^-)(Nu){-}OH} \underset{\substack{\text{Reversible}\\ \text{nucleophilic}\\ \text{addition}\\ (\text{path } a)}}{\overset{\mathrm{Nu:^-}}{\rightleftharpoons}} \mathrm{RC({=}O){-}OH} \underset{\substack{\text{Reversible}\\ \text{acid-base}\\ \text{reaction}\\ (\text{path } b)}}{\overset{\mathrm{:Nu^-}}{\rightleftharpoons}} \mathrm{RC({=}O){-}O^-} + \mathrm{NuH}$$

Esterification

$$\mathrm{RCOOH + R'OH} \xrightarrow{H^+} \mathrm{RCOOR' + H_2O}$$

If a nucleophile is very basic (e.g., an alkoxide), essentially irreversible formation of the carboxylate ion will interfere with attack on the carbonyl carbon. However, if the nucleophile is less basic, carboxylate formation is reversible, and nucleophilic addition may become competitive. A typical example of addition–elimination of a carboxylic acid is **esterification,** which is defined as the condensation of an alcohol and a carboxylic acid to yield an ester and water. The nucleophile, an alcohol, is a weak base, and acid is present both to activate the carbonyl toward addition and to convert the carboxy OH into a better leaving group. The sections that follow will examine this and other carboxy substitutions in detail, illustrating how the problems associated with both the addition and the elimination steps are solved.

In summary, nucleophilic attack on the carbonyl group of carboxylic acid derivatives is a key step in substitution by addition–elimination. Either acid or base catalysis may be observed. In the case of carboxylic acids, the process is complicated by the poor leaving group (hydroxide) and competitive deprotonation of the acid by the nucleophile, acting as a base. With less basic nucleophiles, addition can occur.

19-8 Carboxylic Acid Derivatives: Alkanoyl (Acyl) Halides and Anhydrides

With this section we begin a survey of the preparation of carboxylic acid derivatives. Replacement of the hydroxy group in RCOOH by halide gives rise to **alkanoyl (acyl) halides;** substitution by alkanoate ($RCOO^-$) furnishes anhydrides. *Both processes first require transformation of the hydroxy functionality into a better leaving group.*

Alkanoyl (acyl) halides are formed by using inorganic derivatives of carboxylic acids

$$\mathrm{RC({=}O)X}$$
Alkanoyl (acyl) halide

The conversion of carboxylic acids to alkanoyl (acyl) halides employs the same reagents used in the synthesis of haloalkanes from alcohols (Section 9-4), $SOCl_2$

and PBr_3. The problem to be solved in both cases is identical—changing a poor leaving group (OH) into a good one.

Alkanoyl (Acyl) Halide Synthesis

$$CH_3CH_2CH_2\overset{O}{\overset{\|}{C}}OH \xrightarrow{Cl\overset{O}{\overset{\|}{S}}Cl,\ \text{reflux}} CH_3CH_2CH_2\overset{O}{\overset{\|}{C}}Cl + O{=}S{=}O + HCl$$

Butanoic acid → **Butanoyl chloride** (85%)

$$3\ \text{C}_6\text{H}_{11}\overset{O}{\overset{\|}{C}}OH \xrightarrow{PBr_3} 3\ \text{C}_6\text{H}_{11}\overset{O}{\overset{\|}{C}}Br + H_3PO_3$$

90%

(These reactions fail with formic acid, HCOOH, because formyl chloride, HCOCl, and formyl bromide, HCOBr, are unstable.)

The mechanisms governing these transformations are reminiscent of those in the reactions of alcohols. Initially, an inorganic derivative of the acid is generated. The newly formed substituent is a good leaving group, facilitating acid-catalyzed addition–elimination, with halide as the nucleophile.

Mechanism of Alkanoyl (Acyl) Chloride Formation with Thionyl Chloride ($SOCl_2$)

STEP 1. Activation

$$R\overset{O}{\overset{\|}{C}}OH + Cl\overset{O}{\overset{\|}{S}}Cl \longrightarrow R\overset{O}{\overset{\|}{C}}{}^{\delta+}O\overset{O}{\overset{\|}{S}}Cl + HCl$$

STEP 2. Addition

$$R\overset{O}{\overset{\|}{C}}O\overset{O}{\overset{\|}{S}}Cl + H^+ + :Cl:^- \longrightarrow R\overset{^+OH}{\overset{\|}{C}}O\overset{O}{\overset{\|}{S}}Cl + :Cl:^- \longrightarrow R{-}\underset{Cl}{\overset{OH}{C}}{-}O\overset{O}{\overset{\|}{S}}Cl$$

Tetrahedral intermediate

STEP 3. Elimination

$$R{-}\underset{Cl}{\overset{OH}{C}}{-}O{-}\overset{O}{\overset{\|}{S}}{-}Cl \longrightarrow R\overset{O}{\overset{\|}{C}}Cl + O{=}S{=}O + H^+ + :Cl:^-$$

The mechanism of alkanoyl bromide formation using phosphorus tribromide (PBr_3) is similar (see Section 9-4).

Acids combine with alkanoyl halides to produce anhydrides

$$\mathrm{RC(=O)OC(=O)R}$$

Carboxylic anhydride

The electronegative power of the halogens in alkanoyl halides activates the carbonyl function to attack by other, even weak, nucleophiles (Chapter 20). For example, treatment of alkanoyl halides with carboxylic acids results in **carboxylic anhydrides.**

$$\underset{\textbf{Butanoic acid}}{CH_3CH_2CH_2\overset{O}{\overset{\|}{C}}OH} + \underset{\textbf{Butanoyl chloride}}{Cl\overset{O}{\overset{\|}{C}}CH_2CH_2CH_3} \xrightarrow{\Delta,\ 8\ h} \underset{\textbf{Butanoic anhydride}}{\underset{85\%}{CH_3CH_2CH_2\overset{O}{\overset{\|}{C}}O\overset{O}{\overset{\|}{C}}CH_2CH_2CH_3}} + HCl$$

As the name indicates, carboxylic anhydrides are formally derived from the corresponding acids by loss of water. Although acid dehydration is not a general method for anhydride synthesis, cyclic examples may be prepared from dicarboxylic acids. A condition for the success of this transformation is that the ring closure lead to a five- or six-membered ring product.

Cyclic Anhydride Formation

$$\underset{\substack{\textbf{Butanedioic acid}\\\textbf{(Succinic acid)}}}{HOOC{-}CH_2{-}CH_2{-}COOH} \xrightarrow{300^{\circ}C} \underset{\substack{95\%\\\textbf{Butanedioic anhydride}\\\textbf{(Succinic anhydride)}}}{\text{cyclic } H_2C{-}C(=O){-}O{-}C(=O){-}CH_2} + H_2O$$

Because the halogen in the alkanoyl halide and the RCO_2 substituent in the anhydride are good leaving groups, and because they activate the adjacent carbonyl function, these carboxylic acid derivatives are useful synthetic intermediates for the preparation of other compounds, a topic to be discussed in Sections 20-2 and 20-3.

EXERCISE 19-7

Suggest two preparations, starting from carboxylic acids or their derivatives, for each of the following compounds.

(a) $CH_3\overset{O}{\overset{\|}{C}}O\overset{O}{\overset{\|}{C}}CH_2CH_3$ **(b)** $CH_3\overset{H_3C}{\overset{|}{C}}H\overset{O}{\overset{\|}{C}}Cl$

EXERCISE 19-8

Propose a mechanism for the thermal formation of butanedioic anhydride from the dioic acid.

In summary, the hydroxy group in COOH can be replaced by halogen by using the same reagents employed in the conversion of alcohols into haloalkanes—$SOCl_2$ and PBr_3. The resulting alkanoyl (acyl) halides are sufficiently reactive to be attacked by carboxylic acids to generate carboxylic anhydrides. Cyclic examples of the latter may be made from dicarboxylic acids by thermal dehydration.

19-9 Carboxylic Acid Derivatives: Esters

Esters have the general formula $RC(=O)OR'$. They are perhaps the most important of the carboxylic acid derivatives. This section describes two methods by which esters are made—the mineral acid-catalyzed reaction of carboxylic acids with alcohols and the synthesis of methyl esters using diazomethane.

Carboxylic acids react with alcohols to form esters

When a carboxylic acid and an alcohol are mixed together, no reaction takes place. However, upon addition of catalytic amounts of a mineral acid, such as sulfuric acid or HCl, the two components combine in an equilibrium process to give an ester and water (Section 9-4).

Acid-Catalyzed Esterification

$$\underset{\textbf{Carboxylic acid}}{RC(=O)OH} + \underset{\textbf{Alcohol}}{R'OH} \underset{}{\overset{H^+}{\rightleftharpoons}} \underset{\textbf{Ester}}{RC(=O)OR'} + H_2O$$

Esterification is not very exothermic. How can the equilibrium be shifted toward the ester product? One way is to use an excess of either of the two starting materials; another is to remove the ester or the water product from the reaction mixture. In practice, esterifications are most often accomplished by using the alcohol as a solvent.

$$\underset{\textbf{Acetic acid}}{CH_3C(=O)OH} + \underset{\textbf{Solvent}}{CH_3OH} \overset{H_2SO_4,\ \Delta}{\rightleftharpoons} \underset{\substack{85\% \\ \textbf{Methyl acetate}}}{CH_3C(=O)OCH_3} + H_2O$$

The opposite of esterification is **ester hydrolysis.** This reaction is carried out under the same conditions as esterification, but, to shift the equilibrium, an excess of water is used in a water-miscible solvent.

$$\underset{\textbf{Ethyl 2,2-dimethylhexanoate}}{CH_3CH_2CH_2CH_2C(CH_3)_2COOCH_2CH_3} \xrightarrow{H_2SO_4,\ HOH,\ \text{propanone (acetone)},\ \Delta} \underset{\substack{85\% \\ \textbf{2,2-Dimethylhexanoic acid}}}{CH_3CH_2CH_2CH_2C(CH_3)_2COOH} + CH_3CH_2OH$$

Esterification proceeds through acid-catalyzed addition–elimination

The presence of the acid catalyst features prominently in the mechanism of ester formation: It causes the carbonyl function to undergo nucleophilic attack by the alcohol and the hydroxy group to leave as water.

Mechanism of Acid-Catalyzed Esterification and Ester Hydrolysis

STEP 1. Protonation of the carboxy group

$$\mathrm{RC(=O)OH} \underset{-\,H^+}{\overset{+\,H^+}{\rightleftharpoons}} \left[\mathrm{RC(={}^+OH)OH} \longleftrightarrow \mathrm{R\overset{+}{C}(OH)OH} \longleftrightarrow \mathrm{RC(OH)={}^+OH}\right]$$

STEP 2. Attack by methanol

$$\mathrm{RC(={}^+OH)OH} + \mathrm{CH_3OH} \rightleftharpoons \mathrm{RC(OH)(OH)\overset{+}{O}(H)CH_3} \underset{+\,H^+}{\overset{-\,H^+}{\rightleftharpoons}} \mathrm{RC(OH)(OH)OCH_3}$$

Tetrahedral intermediate

Relay point:
← Can go back to starting material
Can go forward to product →

STEP 3. Elimination of water

$$\mathrm{RC(OH)(OH)OCH_3} \underset{-\,H^+}{\overset{+\,H^+}{\rightleftharpoons}} \mathrm{RC(OH)(\overset{+}{O}H_2)OCH_3} \underset{+\,H_2O}{\overset{-\,H_2O}{\rightleftharpoons}}$$

$$\left[\mathrm{RC(={}^+OH)OCH_3} \longleftrightarrow \mathrm{R\overset{+}{C}(OH)OCH_3}\right] \underset{+\,H^+}{\overset{-\,H^+}{\rightleftharpoons}} \mathrm{RC(=O)OCH_3}$$

Initially, protonation of the carbonyl oxygen gives a delocalized carbocation (step 1). Now the carbonyl carbon is susceptible to nucleophilic attack by methanol. Proton loss from the initial adduct furnishes the tetrahedral intermediate (step 2). This species is a crucial relay point, because it can react in either of *two* ways in the presence of the mineral acid catalyst. First, it can lose methanol by the reverse of steps 1 and 2. This process, beginning with protonation of the methoxy oxygen, leads back to the carboxylic acid. The second possibility, however, is protonation at either hydroxy oxygen, leading to elimination of water and to the ester (step 3). All the steps are reversible; therefore, either addition of excess alcohol or removal of water favors esterification by shifting the equilibria in steps 2 and 3, respectively. Ester hydrolysis proceeds by the reverse of the sequence and is favored by aqueous conditions.

EXERCISE 19-9

Give the mechanism of esterification of a general carboxylic acid RCOOH with methanol in which the alcohol oxygen is labeled with the ^{18}O isotope ($CH_3{}^{18}OH$). Does the labeled oxygen appear in the ester or the water product?

Hydroxy acids may undergo intramolecular esterification to lactones

When hydroxy acids are treated with catalytic amounts of mineral acid, cyclic esters—or **lactones**—may form. This process is called **intramolecular esterification** and is favorable for formation of five- and six-membered rings.

Formation of a Lactone

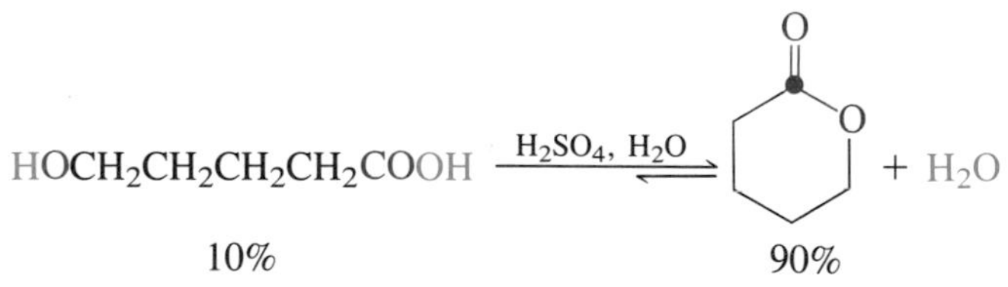

$$HOCH_2CH_2CH_2CH_2COOH \xrightleftharpoons{H_2SO_4,\ H_2O} \text{(six-membered lactone, C=O, ring O)} + H_2O$$

10% 90%

EXERCISE 19-10

Explain the following result by a mechanism. (Hint: See Section 17-7.)

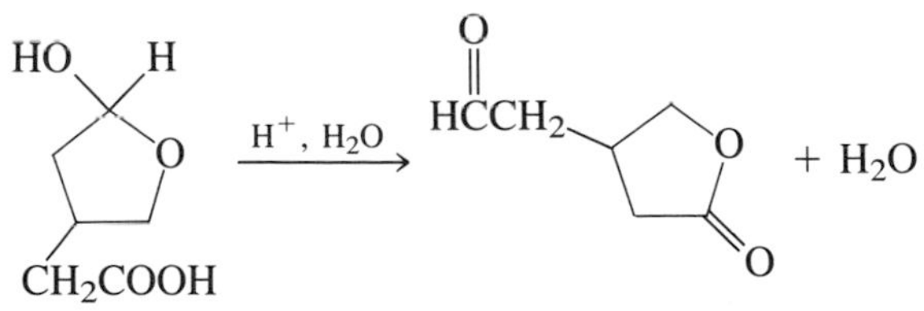

(Cyclic five-membered hemiacetal bearing HO and H on the carbon next to ring O, with CH_2COOH substituent) $\xrightarrow{H^+,\ H_2O}$ (five-membered lactone bearing $HCCH_2$ with C=O, i.e. $\overset{O}{\overset{\|}{H C}}CH_2$ substituent) $+ H_2O$

Carboxylic acids form esters by other processes

In addition to acid-catalyzed esterification, other reactions can transform carboxylic acids into esters. Section 7-8 described one, the nucleophilic substitution of haloalkanes with carboxylate ions. Another method is used only on a small scale for the specific conversion of an acid into its methyl ester. This transformation employs **diazomethane,** CH_2N_2, a highly reactive, toxic, and explosive gas, and is driven by the production of gaseous nitrogen.

BOX 19-2 Macrocyclic Lactones

Commercial Synthesis of a Macrocyclic Musk

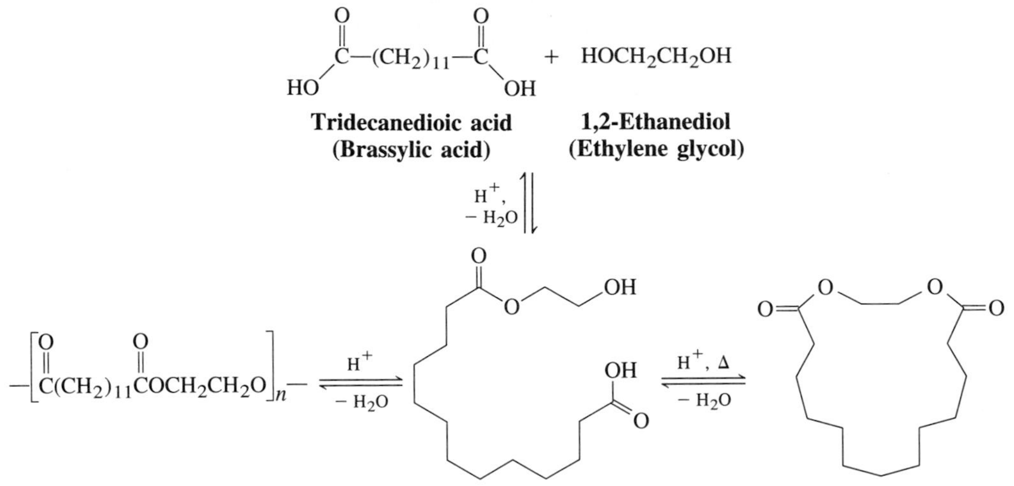

$HOOC-(CH_2)_{11}-COOH$ + $HOCH_2CH_2OH$

Tridecanedioic acid (Brassylic acid) **1,2-Ethanediol (Ethylene glycol)**

H^+, $-H_2O$

$-[C(=O)(CH_2)_{11}C(=O)OCH_2CH_2O]_n-$ $\underset{-H_2O}{\overset{H^+}{\rightleftharpoons}}$ (monoester) $\underset{-H_2O}{\overset{H^+, \Delta}{\rightleftharpoons}}$ (ethylene brassylate)

Polymer **Ethylene brassylate**

Lactones with rings of more than six members may be synthesized if ring strain and transannular interactions are minimized (see Sections 4-2 through 4-5 and 9-6). The preparation of large-ring (macrocyclic) lactones requires high-dilution conditions to ensure that esterification is intramolecular. Polymerization is still a problem. For example, one of the most important commercial musk fragrances is a dilactone, ethylene brassylate. Acid-catalyzed esterification of tridecanedioic acid with 1,2-ethanediol initially gives a monoester, which then polymerizes. Strong heating reverses polymer formation and gives the macrocycle, which is distilled from the reaction mixture, thereby shifting the equilibrium in favor of the volatile 17-membered ring product.

Macrocyclic lactone synthesis is also a topic of considerable interest to the pharmaceutical industry because this function constitutes the basic framework of many medicinally valuable compounds. Examples include erythromycin A, a widely used *macrolide antibiotic,* and FK-506, a powerful *immunosuppressant* that shows great promise in controlling the rejection of transplanted organs in human patients.

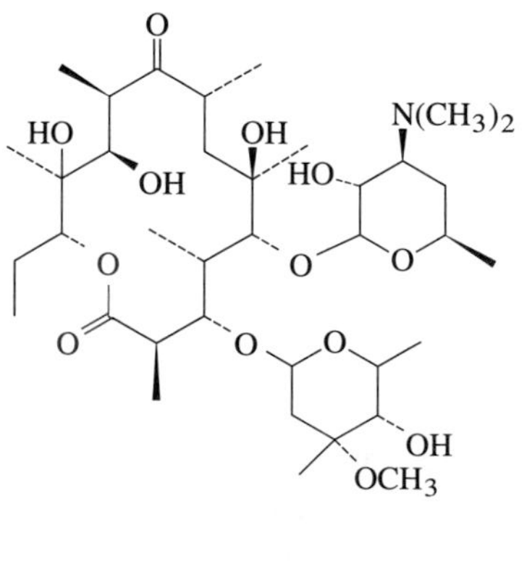

Erythromycin A 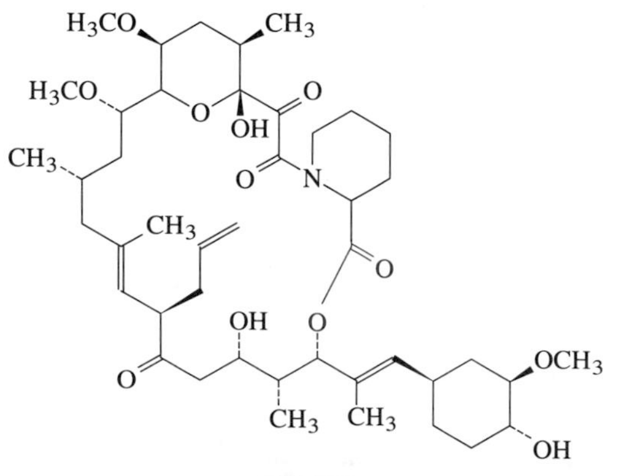 **FK-506**

$$CH_3C{\equiv}CCOOH + CH_2N_2 \xrightarrow{(CH_3CH_2)_2O} CH_3C{\equiv}CCOOCH_2H + N_2$$

80%

2-Butynoic acid **Methyl 2-butynoate**

In summary, carboxylic acids react with alcohols to form esters, as long as a mineral acid catalyst is present. This reaction is only slightly exothermic, and its equilibrium may be shifted in either direction by the choice of reaction conditions. The reverse of ester formation is ester hydrolysis. The mechanism of esterification is acid-catalyzed addition of alcohol to the carbonyl group followed by acid-catalyzed dehydration. Intramolecular ester formation results in lactones, favored only when five- or six-membered rings are produced. Esters can be formed from carboxylic acids by other mechanisms, for example, the reaction of carboxylate ions with (primary) haloalkanes, and, for methyl alkanoates, of carboxylic acids with diazomethane.

19-10 Carboxylic Acid Derivatives: Amides

This section shows that amines are also capable of attacking the carbonyl function in carboxylic acids to form another class of derivatives, **carboxylic amides.** The mechanism of this reaction is again addition–elimination, but complicated by competing deprotonation.

$$\overset{O}{\overset{\|}{RCNR'_2}}$$

Carboxylic amide

Amines react with carboxylic acids as bases and as nucleophiles

Because nitrogen is not as electronegative as oxygen, amines are both more basic and more nucleophilic than alcohols (Chapter 21). They can react in either mode with carboxylic acids. Thus, exposure of the acid to the amine initially leads to the rapid formation of ammonium carboxylates, in which the negative charge protects the carbonyl carbon from nucleophilic attack.

Ammonium Salts from Carboxylic Acids

$$\overset{:O:}{\overset{\|}{RC\ddot{O}}}{-}H + :NH_3 \rightleftharpoons \overset{:O:}{\overset{\|}{RC\ddot{O}}}{:}^{-}\ H\overset{+}{N}H_3$$

Ammonia **An ammonium alkanoate**

Upon heating, however, salt formation is reversed and a slower but thermodynamically more favored process takes over. In this second mode of reaction, the nitrogen acts as a nucleophile. Addition–elimination leads to the amide.*

*Remember not to confuse the names of carboxylic amides with those of the alkali salts of amines, also called amides (e.g., lithium amide, $LiNH_2$).

Mechanism of Amide Formation

$$RC(=O)OH + :NH_3 \rightleftharpoons R{-}C(O^-)(OH)({}^+NH_3) \xrightleftharpoons{\text{Proton transfer}} R{-}C(O^-)({}^+OH_2)(NH_2) \rightleftharpoons H_2O: + RC(=O)NH_2$$

Tetrahedral intermediates **An amide**

Formation of an Amide from an Amine and a Carboxylic Acid

$$CH_3CH_2CH_2COOH + (CH_3)_2NH \xrightarrow{155°C} CH_3CH_2CH_2\overset{O}{\overset{\|}{C}}N(CH_3)_2 + HOH$$

84%

***N,N*-Dimethyl butanamide**

Amide formation is reversible. Thus, treatment of amides with hot acidic or basic water regenerates the component carboxylic acids and amines.

Dicarboxylic acids react with amines to give imides

Dicarboxylic acids may react twice with the amine nitrogen of ammonia or of a primary amine. This sequence gives rise to **imides,** the nitrogen analogs of cyclic anhydrides.

$$\begin{matrix} CH_2COOH \\ | \\ CH_2COOH \end{matrix} \xrightarrow{NH_3} \begin{matrix} CH_2COO^-NH_4^+ \\ | \\ CH_2COO^-NH_4^+ \end{matrix} \xrightarrow{290°C} \text{succinimide (cyclic } H_2C{-}C(=O){-}\ddot{N}H{-}C(=O){-}CH_2) + 2\ H_2O + NH_3$$

Butanedioic acid

83%

Butanimide (Succinimide)

Recall the use of *N*-halobutanimides in halogenations (Sections 3-8 and 14-2).

Amino acids cyclize to lactams

In analogy to hydroxycarboxylic acids, some amino acids undergo cyclization to the corresponding cyclic amides, called **lactams** (Section 20-7).

$$H_2\overset{+}{N}(H)CH_2CH_2CH_2\overset{O}{\overset{\|}{C}}O^- \rightleftharpoons H_2\ddot{N}CH_2CH_2CH_2\overset{O}{\overset{\|}{C}}OH \xrightarrow{\Delta} \text{(five-membered ring lactam, }C(=O){-}\ddot{N}H) + HOH$$

86%

A lactam

EXERCISE 19-11

Formulate a detailed mechanism for the formation of butanimide from butanedioic acid and ammonia.

In summary, amines react with carboxylic acids to form amides by an addition–elimination process that begins with nucleophilic attack by the amine on the carboxy carbon. Amide formation is complicated by reversible deprotonation of the carboxylic acid by the basic amine to give an ammonium salt.

19-11 Reduction of Carboxylic Acids by Lithium Aluminum Hydride

Carboxylate salts are generally unreactive toward attack by nucleophiles. However, lithium aluminum hydride is capable of reducing carboxylic acids all the way to the corresponding primary alcohols, which are obtained upon aqueous acidic work-up.

Reduction of a Carboxylic Acid

$$\text{RCOOH} \xrightarrow[\text{2. } H^+, H_2O]{\text{1. } LiAlH_4, \text{ THF}} \text{RCH}_2\text{OH}$$

$$\text{(bicyclic)–COOH} \xrightarrow[\text{2. } H^+, H_2O]{\text{1. } LiAlH_4, \text{ THF}} \text{(bicyclic)–CH}_2\text{OH} \quad 65\%$$

Although the exact mechanism of this transformation is not completely understood, it is clear that the hydride reagent first acts as a base, forming the lithium salt of the acid and hydrogen gas. Despite the negative charge, lithium aluminum hydride is so reactive that it is capable of donating two hydrides to the carbonyl function of the carboxylate. The product of this sequence, a simple alkoxide, gives the alcohol after protonation.

EXERCISE 19-12

Propose synthetic schemes that produce compound B from compound A.

(a) $CH_3CH_2CH_2CN$ (**A**) $\quad CH_3CH_2CH_2CH_2OH$ (**B**)

(b) cyclopropyl–CH_2COOH (**A**) $\quad$ cyclopropyl–CH_2CD_2OH (**B**)

In summary, the nucleophilic reactivity of lithium aluminum hydride is sufficiently great to effect the reduction of carboxylates to primary alcohols.

19-12 Bromination Next to the Carboxy Group: The Hell-Volhard-Zelinsky Reaction

Like aldehydes and ketones, alkanoic acids can be brominated at the α-carbon by exposure to Br_2. The addition of a trace amount of PBr_3 is necessary to get the reaction started. Because the highly corrosive PBr_3 is difficult to handle, it is often generated in the reaction flask (in situ). This is achieved by the addition of a little elemental (red) phosphorus to the mixture of starting materials; it is converted into PBr_3 instantaneously by the bromine present.

A Hell-Volhard-Zelinsky* Reaction

$$CH_3CH_2CH_2CH_2COOH \xrightarrow{\text{Br–Br, trace P}} CH_3CH_2CH_2\underset{}{\overset{\text{Br}}{\overset{|}{C}}}HCOOH + HBr$$

80%

2-Bromopentanoic acid

The alkanoyl bromide, formed by the reaction of PBr_3 with the carboxylic acid (Section 19-8), is subject to rapid acid-catalyzed enolization. The enol is brominated to give the 2-bromoalkanoyl bromide. The latter then undergoes an exchange reaction with unreacted acid to furnish the product bromoacid and another molecule of alkanoyl bromide, which reenters the reaction cycle.

Mechanism of the Hell-Volhard-Zelinsky Reaction

STEP 1. Alkanoyl bromide formation

$$3\ RCH_2\overset{\overset{\displaystyle O}{\|}}{C}OH + PBr_3 \longrightarrow 3\ RCH_2\overset{\overset{\displaystyle O}{\|}}{C}Br + H_3PO_3$$

Alkanoyl bromide

STEP 2. Enolization

$$RCH_2\overset{\overset{\displaystyle O}{\|}}{C}Br \underset{}{\overset{H^+}{\rightleftharpoons}} RCH{=}C\begin{matrix} OH \\ Br \end{matrix}$$

Enol

STEP 3. Bromination

$$RCH{=}C\begin{matrix} OH \\ Br \end{matrix} \xrightarrow{\text{Br–Br}} R\underset{\underset{\displaystyle Br}{|}}{C}H\overset{\overset{\displaystyle O}{\|}}{C}Br + HBr$$

*Professor Carl M. Hell, 1849–1926, University of Stuttgart, Germany; Professor Jacob Volhard, 1834–1910, University of Halle, Germany; Professor Nicolai D. Zelinsky, 1861–1953, University of Moscow.

STEP 4. Exchange

$$\underset{\substack{|\\ \mathrm{Br}}}{\mathrm{RCH}}\overset{\mathrm{O}}{\overset{\|}{\mathrm{C}}}\mathrm{Br} + \mathrm{RCH_2}\overset{\mathrm{O}}{\overset{\|}{\mathrm{C}}}\mathrm{OH} \rightleftharpoons \underset{\substack{|\\ \mathrm{Br}}}{\mathrm{RCH}}\overset{\mathrm{O}}{\overset{\|}{\mathrm{C}}}\mathrm{OH} + \underset{\text{Reenters step 2}}{\mathrm{RCH_2}\overset{\mathrm{O}}{\overset{\|}{\mathrm{C}}}\mathrm{Br}}$$

EXERCISE 19-13

Formulate detailed mechanisms for steps 2 and 3 of the Hell-Volhard-Zelinsky reaction. (Hints: Review Sections 18-2 for step 2 and 18-3 for step 3.)

The bromocarboxylic acids formed in the Hell-Volhard-Zelinsky reaction can be converted into other 2-substituted derivatives. For example, treatment with aqueous base gives 2-hydroxy acids, whereas amines yield α-amino acids (Chapter 26). An example of the latter is the synthesis of racemic 2-aminohexanoic acid (norleucine, a naturally occurring but rare amino acid) shown below.

$$\underset{\textbf{Hexanoic acid}}{\mathrm{CH_3(CH_2)_4COOH}} \xrightarrow[\text{70–100°C, 4 h}]{\mathrm{Br_2},\ \text{trace}\ \mathrm{PCl_3},} \underset{\textbf{2-Bromohexanoic acid}}{\underset{86\%}{\mathrm{CH_3(CH_2)_3}\overset{\mathrm{Br}}{\overset{|}{\mathrm{C}}}\mathrm{HCOOH}}} \xrightarrow[\text{50°C, 30 h}]{\mathrm{NH_3},\ \mathrm{H_2O},} \underset{\textbf{2-Aminohexanoic acid (Norleucine)}}{\underset{64\%}{\mathrm{CH_3(CH_2)_3}\overset{\mathrm{NH_2}}{\overset{|}{\mathrm{C}}}\mathrm{HCOOH}}}$$

In summary, with trace amounts of phosphorus (or phosphorus tribromide), carboxylic acids are brominated at C2 (the Hell-Volhard-Zelinsky reaction). The transformation proceeds through 2-bromoalkanoyl bromide intermediates.

19-13 Decarboxylation of Carboxylic Acids by One-Electron Transfer: The Hunsdiecker Reaction

We conclude our survey of reactions of the carboxylic acids by describing the results of oxidation of the carboxylate ion. Removal of one electron from this ion produces a very unstable RCOO · radical, which decomposes to an alkyl radical by loss of carbon dioxide **(decarboxylation).**

Generation and Decomposition of an RCOO · Radical

$$\mathrm{R}\overset{:\mathrm{O}:}{\overset{\|}{\mathrm{C}}}\ddot{\underset{\cdot\cdot}{\mathrm{O}}}:^- \xrightarrow[\text{Oxidation}]{-1e} \mathrm{R}\overset{:\mathrm{O}:}{\overset{\|}{\mathrm{C}}}\ddot{\underset{\cdot\cdot}{\mathrm{O}}}\cdot \xrightarrow{\text{Decarboxylation}} \mathrm{R}\cdot + \ddot{\mathrm{O}}{=}\mathrm{C}{=}\ddot{\mathrm{O}}$$

In the **Hunsdiecker* reaction,** the oxidation is performed chemically: A silver salt of the carboxylic acid is treated with a halogen, usually bromine. Silver bromide precipitates, carbon dioxide is evolved, and a bromoalkane is formed in which the bromine has taken the position of the carboxy function.

Hunsdiecker Reaction

$$RCOO^-Ag^+ + Br{-}Br \longrightarrow RBr + CO_2 + AgBr$$

The reaction allows for the conversion of a carboxylate salt into a bromoalkane containing one less carbon atom.

$$\underset{\textbf{Dodecanoic acid}}{CH_3(CH_2)_{10}COOH} \xrightarrow[2.\ Br-Br,\ CCl_4]{1.\ AgNO_3,\ KOH,\ H_2O} \underset{\textbf{1-Bromoundecane}}{\underset{67\%}{CH_3(CH_2)_{10}Br}}$$

The mechanism of this degradation reaction involves an alkanoyl hypobromite, RCOOBr, which decomposes to the RCOO· radical. After decarboxylation, the alkyl radical abstracts bromine from another molecule of hypobromite to yield the bromoalkane and another RCOO· radical.

Mechanism of the Hunsdiecker Reaction

STEP 1. Hypobromite formation

$$RC(=\ddot{O}\colon)\ddot{O}\colon^-Ag^+ \xrightarrow{Br-Br} \underset{\textbf{Hypobromite}}{RC(=\ddot{O}\colon)\ddot{O}Br} + AgBr$$

STEP 2. RCOO· radical formation

$$RC(=\ddot{O}\colon)\ddot{O}{-}\ddot{Br}\colon \longrightarrow RC(=\ddot{O}\colon)\ddot{O}\cdot + \colon\ddot{Br}\cdot$$

STEP 3. RCOO· radical decomposition

$$R{-}C(=\ddot{O}\colon)\ddot{O}\cdot \longrightarrow R\cdot + \ddot{O}{=}C{=}\ddot{O}$$

STEP 4. Haloalkane and RCOO· formation

$$R\cdot + RC(=\ddot{O}\colon)\ddot{O}Br \longrightarrow RBr + RC(=\ddot{O}\colon)\ddot{O}\cdot$$

Other examples of decarboxylation will appear in Chapters 23 and 24.

*Dr. Heinz Hunsdiecker, b. 1904, Vogt and Co., Köln-Braunsfeld, Germany.

EXERCISE 19-14

Give the product of Hunsdiecker reaction of **(a)** pentanoic acid; **(b)** 2-ethylbutanoic acid; **(c)** benzoic acid.

In summary, oxidation of a silver carboxylate by a halogen produces an unstable RCOO· radical, which decomposes to carbon dioxide and an alkyl radical. The latter reacts with a halogen source to give the corresponding haloalkane.

19-14 Biological Activity of Carboxylic Acids

Considering the variety of reactions that carboxylic acids can undergo, it is no wonder that they are very important, not only as synthetic intermediates in the laboratory, but also in biological systems. This section will provide a glimpse of the enormous structural and functional diversity of natural carboxylic acids. A discussion of amino acids will be deferred to Chapter 26.

As Table 19-1 indicates, even the simplest carboxylic acids are abundant in nature. Formic acid is present not only in ants, where it functions as an alarm pheromone, but also in plants. For example, one reason why human skin hurts after it touches the stinging nettle is that formic acid is deposited in the wounds.

Acetic acid is formed through the enzymatic oxidation of ethanol produced by fermentation. Vinegar (*vin,* French, wine; *aigre,* French, sour) is the term given to the dilute (ca. 4–12%) aqueous solution thus generated in ciders, wines, and malt extracts. Louis Pasteur in 1864 established the involvement of bacteria in the oxidation stage of this ancient process.

Fatty acids are derived from coupling of acetic acid

Acetic acid exhibits diverse biological activities, ranging from a defense pheromone in some ants and scorpions to the primary building block for the biosynthesis of more naturally occurring organic compounds than any other single precursor substance. For example, 3-methyl-3-butenyl pyrophosphate, the crucial precursor in the buildup of the terpenes (Section 14-10), is made by the enzymatic conversion of three molecules of CH_3COOH to an intermediate called mevalonic acid. Further reactions degrade the system to the five-carbon (isoprene) unit of the product.

$$3\ CH_3{-}COOH \xrightarrow{\text{Enzymes}} CH_3{-}\underset{\displaystyle CH_2{-}CH_2OH}{\overset{\displaystyle CH_2{-}COOH}{C}}OH \xrightarrow{\text{Enzymes}} CH_3{-}\underset{\displaystyle CH_2{-}CH_2O{-}\underset{\displaystyle OH}{\overset{\displaystyle O}{\overset{\|}{P}}}{-}O{-}\underset{\displaystyle OH}{\overset{\displaystyle O}{\overset{\|}{P}}}{-}OH}{\overset{\displaystyle CH_2}{\overset{\|}{C}}}$$

Mevalonic acid **3-Methyl-3-butenyl pyrophosphate**

A conceptually more straightforward mode of multiple coupling is found in the biosynthesis of **fatty acids.** This class of compounds derives its name from its source, the natural **fats,** which are esters of long-chain carboxylic acids (see Section 20-4). Hydrolysis or **saponification** (so called because the corresponding salts form soaps—see Box 19-1; *sapo,* Latin, soap) yields the corresponding fatty acids. The most important of these are between 12 and 22 carbons long and may be unsaturated.

Fatty Acids

$CH_3(CH_2)_{14}COOH$

Hexadecanoic acid
(Palmitic acid)

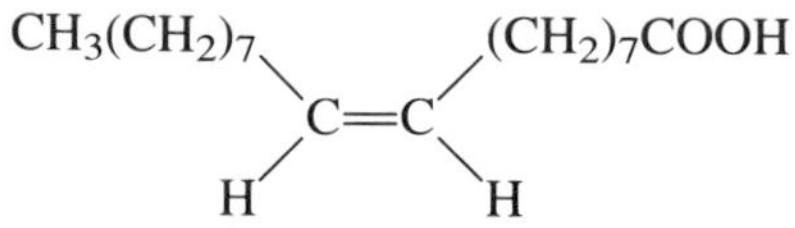

$CH_3(CH_2)_7CH{=}CH(CH_2)_7COOH$ (cis, H and H on the same side)

***cis*-9-Octadecenoic acid**
(Oleic acid)

In accord with their biosynthetic origin, fatty acids consist mostly of even-numbered carbon chains. A very elegant experiment demonstrated that linear coupling occurred in a highly regular fashion. Here, singly labeled radioactive (^{14}C) acetic acid was fed to several organisms. The resulting fatty acids were labeled only at every other carbon atom.

$$CH_3{}^{14}COOH \xrightarrow{\text{Organism}}$$

$$CH_3{}^{14}CH_2CH_2{}^{14}CH_2CH_2{}^{14}CH_2CH_2{}^{14}CH_2CH_2{}^{14}CH_2CH_2{}^{14}CH_2CH_2{}^{14}CH_2CH_2{}^{14}COOH$$

Labeled hexadecanoic (palmitic) acid

The mechanism of chain formation is very complex, but the following scheme provides a general idea of the process. A key player is the mercapto group of an important biological relay compound called coenzyme A (abbreviated HSCoA; Figure 19-4). This function serves to bind acetic acid in the form of a **thioester** called acetyl CoA. Carboxylation transforms some of this thioester into malonyl CoA. The two acetyl groups are then transferred to two molecules of **acyl carrier protein.** Coupling occurs with loss of CO_2 to furnish a 3-oxobutanoic thioester.

Mechanism of Coupling of Acetic Acid Units

STEP 1. Thioester formation

$$\underset{\textbf{Acetic acid}}{CH_3COOH} + \underset{\textbf{Coenzyme A}}{HSCoA} \longrightarrow \underset{\textbf{Acetyl coenzyme A}}{CH_3\overset{O}{\overset{\|}{C}}SCoA} + HOH$$

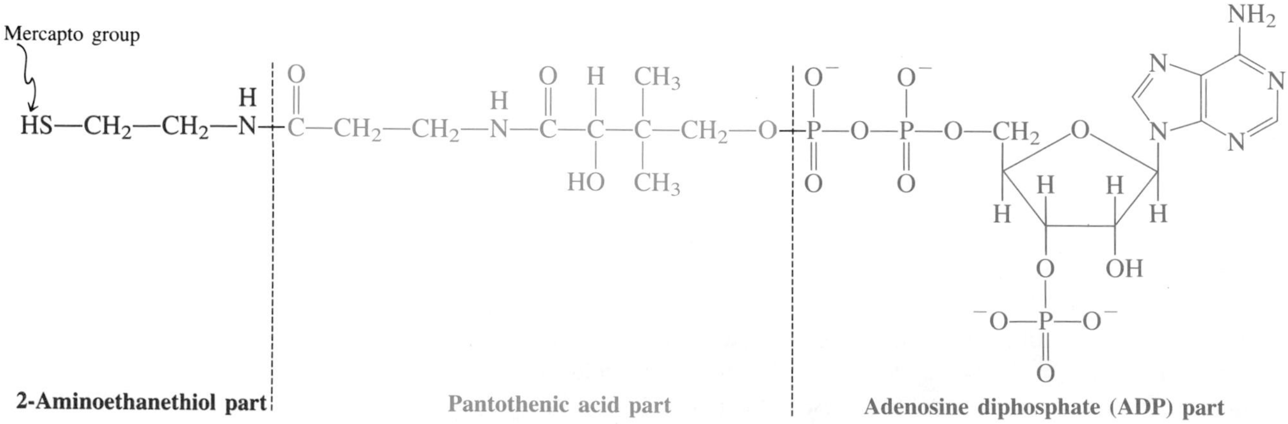

FIGURE 19-4 Structure of coenzyme A. For this discussion, the important part is the mercapto function. For convenience, the molecule may be abbreviated HSCoA.

STEP 2. Carboxylation

$$CH_3\overset{\overset{O}{\|}}{C}SCoA + CO_2 \xrightarrow{\text{Acetyl CoA carboxylase}} HO\overset{\overset{O}{\|}}{C}CH_2\overset{\overset{O}{\|}}{C}SCoA$$

Malonyl CoA

STEP 3. Acetyl group transfer

$$CH_3\overset{\overset{O}{\|}}{C}SCoA + HS\text{—}\boxed{\text{protein}} \longrightarrow CH_3\overset{\overset{O}{\|}}{C}S\text{—}\boxed{\text{protein}} + HSCoA$$

Acyl carrier protein

$$HO\overset{\overset{O}{\|}}{C}CH_2\overset{\overset{O}{\|}}{C}SCoA + HS\text{—}\boxed{\text{protein}} \longrightarrow HO\overset{\overset{O}{\|}}{C}CH_2\overset{\overset{O}{\|}}{C}S\text{—}\boxed{\text{protein}} + HSCoA$$

Acyl carrier protein

STEP 4. Coupling

$$HO\overset{\overset{O}{\|}}{C}CH_2\overset{\overset{O}{\|}}{C}S\text{—}\boxed{\text{protein}} \xrightarrow[-CO_2]{CH_3\overset{\overset{O}{\|}}{C}S\text{—}\boxed{\text{protein}}} CH_3\overset{\overset{O}{\|}}{C}CH_2\overset{\overset{O}{\|}}{C}S\text{—}\boxed{\text{protein}}$$

A 3-oxobutanoic thioester

Reduction of the ketone function to a methylene group follows. The resulting butanoic thioester repeatedly undergoes a similar sequence of reactions elongating the chain, always by two carbons. The eventual product is a long-chain alkanoyl group, which is removed from the protein by hydrolysis.

$$CH_3CH_2\text{+}CH_2CH_2\text{+}_nCH_2\overset{\overset{O}{\|}}{C}\text{—}S\text{—}\boxed{\text{protein}} \xrightarrow{H_2O} CH_3CH_2\text{+}CH_2CH_2\text{+}_nCH_2\overset{\overset{O}{\|}}{C}\text{—}OH + HS\text{—}\boxed{\text{protein}}$$

Fatty acid

Arachidonic acid is a biologically important unsaturated fatty acid

Naturally occurring *unsaturated* fatty acids are capable of undergoing further transformations leading to a variety of unusual structures. An example is arachidonic acid, which is the biological precursor to a multitude of important chemicals in the human body, such as prostaglandins, thromboxanes, prostacyclins, and leukotrienes.

COOH

Arachidonic acid

HO COOH HO OH

Prostaglandin $F_{2\alpha}$

(Induces labor, abortion, menstruation)

Thromboxane A_2
(Contracts smooth muscles; aggregates blood platelets)

Prostacyclin I_2, sodium salt
(The most potent natural inhibitor of platelet aggregation; vasodilator, used in heart-bypass operations, in kidney patients, and in others)

Leukotriene B_4
(Potent chemotactic factor; e.g., causes cell migrations)

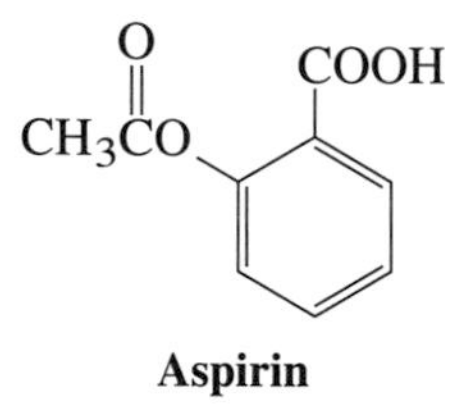

Aspirin

Each of these compounds possesses powerful biological properties. For example, some of the prostaglandins are responsible for the tissue inflammation associated with rheumatoid arthritis. The salicylate derivative aspirin is capable of combating such symptoms by inhibiting the conversion of arachidonic acid to prostaglandins. Other anti-inflammatory agents, such as the **corticosteroids** (Section 4-7), function instead by blocking the biosynthesis of arachidonic acid itself.

EXERCISE 19-15

Identify the arachidonic acid backbone in each of its four derivatives shown above.

Nature also produces complex polycyclic carboxylic acids

Many biologically active natural products that have carboxy groups as substituents of complex polycyclic frames have physiological potential that derives from other sites in the molecule. In these compounds the function of the carboxy group may be to impart water solubility, to allow for salt formation or ion transport, and to enable micellar-type aggregation. Two examples are gibberellic acid, one of a group of plant growth-promoting substances manufactured by fermentation, and lysergic acid, a major product of hydrolyzed extracts of ergot, a fungal parasite that lives on grasses, including rye. Many lysergic acid derivatives possess powerful psychotomimetic activity. In the Middle Ages, thousands who ate rye bread contaminated by ergot experienced the poisonous effects characteristic of these compounds: hallucinations, convulsions, delirium, epilepsy, and death. The synthetic lysergic acid diethylamide (LSD) is one of the most powerful hallucinogens known. The effective oral dose for humans is only about 0.05 mg.

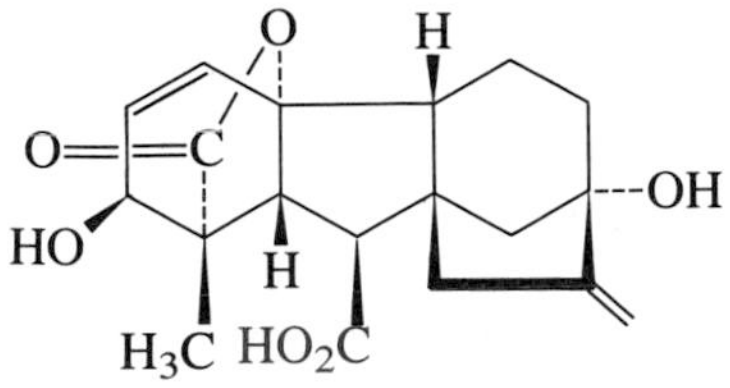

Gibberellic acid

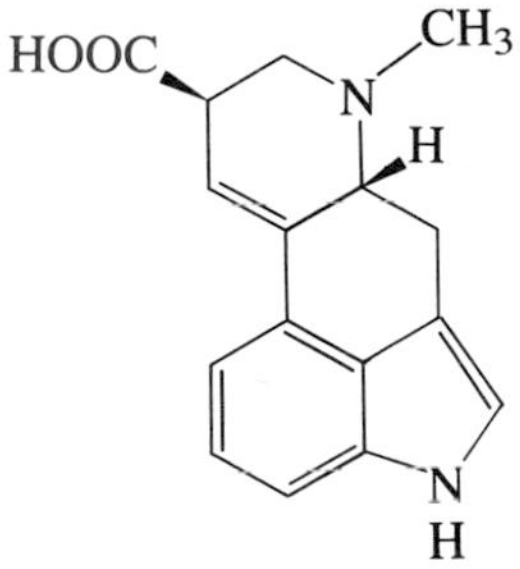

Lysergic acid

In summary, the numerous naturally occurring carboxylic acids are structurally and functionally diverse.

NEW REACTIONS

1. ACIDITY OF CARBOXYLIC ACIDS (Section 19-4)

$$RCOOH + H_2\ddot{O} \rightleftharpoons \left[RC(=O)\ddot{O}:^- \longleftrightarrow {}^-:\ddot{O}C(R)=\ddot{O} \right] + H_3\ddot{O}^+$$

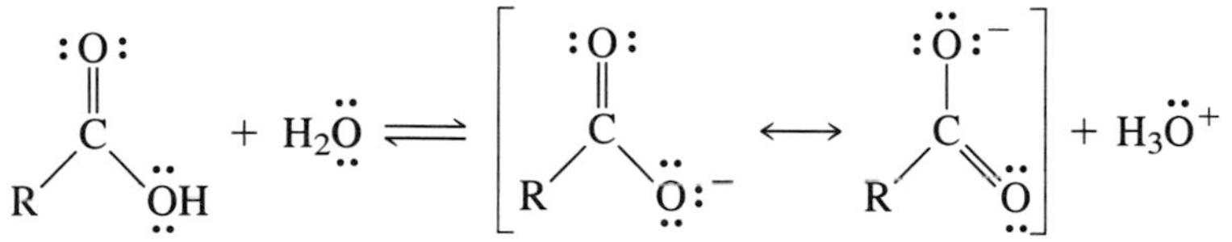

Resonance-stabilized carboxylate ion

$K_a = 10^{-4}–10^{-5}$, $pK_a \sim 4–5$

Salt formation

$$RCOOH + NaOH \longrightarrow RCOO^-Na^+ + H_2O$$

Also with Na_2CO_3, $NaHCO_3$

2. BASICITY OF CARBOXYLIC ACIDS (Section 19-4)

$$R\text{—}C(=\ddot{O}:)\ddot{O}H + H^+ \rightleftharpoons \left[R\text{—}C(=\overset{+}{O}\text{—}H)\ddot{O}\text{—}H \longleftrightarrow R\text{—}C(\ddot{O}\text{—}H)=\overset{+}{O}\text{—}H \right]$$

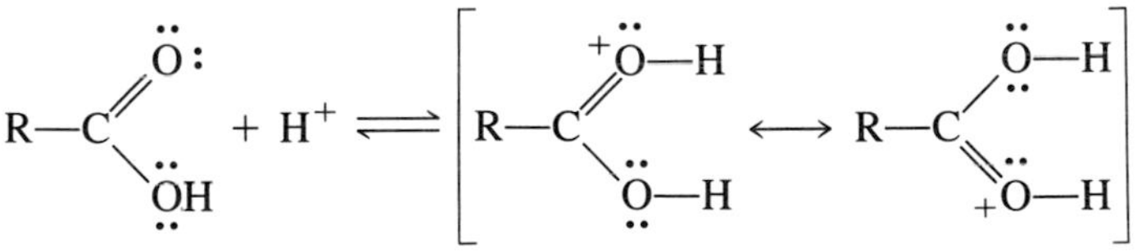

Resonance-stabilized protonated carboxylic acid

Preparation of Carboxylic Acids

3. FORMIC ACID (Section 19-5)

$$CO + NaOH \xrightarrow{\Delta,\ 100\ psi} HCOO^-Na^+ \xrightarrow{H^+,\ H_2O} HCOOH$$

4. ACETIC ACID (Section 19-5)

Ethene oxidation

$$CH_2{=}CH_2 \xrightarrow{O_2,\ H_2O,\ \text{catalytic } Pd^{2+} \text{ and } Cu^{2+}} CH_3CHO \xrightarrow{O_2,\ \text{catalytic } Co^{3+}} CH_3COOH$$

Butane oxidation

$$CH_3CH_2CH_2CH_3 \xrightarrow{O_2,\ \text{catalytic } Co^{3+},\ \Delta,\ \text{pressure}} CH_3COOH$$

Carbonylation of methanol

$$CH_3OH \xrightarrow{CO,\ \text{catalytic } Rh^{3+},\ I_2,\ \Delta,\ \text{pressure}} CH_3COOH$$

5. OXIDATION OF PRIMARY ALCOHOLS AND ALDEHYDES (Section 19-6)

$$RCH_2OH \xrightarrow{\text{Oxidizing agent}} RCOOH$$

Oxidizing agents: aqueous CrO_3, $KMnO_4$, HNO_3

$$RCHO \xrightarrow{\text{Oxidizing agent}} RCOOH$$

Oxidizing agents: aqueous CrO_3, $KMnO_4$, Ag^+, H_2O_2, HNO_3

6. CARBONATION OF ORGANOMETALLIC REAGENTS (Section 19-6)

$$RMgX + CO_2 \xrightarrow{THF} RCOO^{-}\,{}^{+}MgX \xrightarrow{H^+,\ H_2O} RCOOH$$

$$RLi + CO_2 \xrightarrow{THF} RCOO^{-}Li^{+} \xrightarrow{H^+,\ H_2O} RCOOH$$

7. HYDROLYSIS OF NITRILES (Section 19-6)

$$RC{\equiv}N \xrightarrow{H_2O,\ \Delta,\ H^+ \text{ or } HO^-} RCOOH + NH_3$$

Reactions of Carboxylic Acids

8. NUCLEOPHILIC ATTACK AT THE CARBONYL GROUP (Section 19-7)

Base-catalyzed addition–elimination

$$R\overset{\overset{\displaystyle O}{\|}}{C}L + :Nu^- \xrightarrow[\text{Addition}]{} R{-}\underset{\underset{\displaystyle Nu}{|}}{\overset{\overset{\displaystyle O^-}{|}}{C}}{-}L \xrightarrow[\text{Elimination}]{} R\overset{\overset{\displaystyle O}{\|}}{C}Nu + L^-$$

L = leaving group

Tetrahedral intermediate

Acid-catalyzed addition–elimination

$$\mathrm{R\overset{O}{\overset{\|}{C}}L \underset{}{\overset{H^+}{\rightleftharpoons}} R\overset{{}^{+}OH}{\overset{\|}{C}}L \overset{:NuH}{\rightleftharpoons} R{-}\underset{{}^{+}NuH}{\overset{OH}{\overset{|}{\underset{|}{C}}}}{-}L \overset{-H^+}{\rightleftharpoons}}$$

$$\mathrm{\underset{\textbf{Tetrahedral intermediate}}{R{-}\underset{Nu}{\overset{OH}{\overset{|}{\underset{|}{C}}}}{-}L} \overset{H^+}{\rightleftharpoons} R{-}\underset{Nu}{\overset{OH}{\overset{|}{\underset{|}{C}}}}{-}\overset{+}{L}{-}H \overset{-HL}{\rightleftharpoons} R\overset{{}^{+}OH}{\overset{\|}{C}}Nu \overset{-H^+}{\rightleftharpoons} R\overset{O}{\overset{\|}{C}}Nu}$$

Derivatives of Carboxylic Acids

9. ALKANOYL HALIDES (Section 19-8)

$$\mathrm{RCOOH + SOCl_2 \longrightarrow \underset{\textbf{Alkanoyl chloride}}{R\overset{O}{\overset{\|}{C}}Cl} + SO_2 + HCl}$$

$$\mathrm{3\ RCOOH + PBr_3 \longrightarrow 3\ \underset{\textbf{Alkanoyl bromide}}{R\overset{O}{\overset{\|}{C}}Br} + H_3PO_3}$$

10. CARBOXYLIC ANHYDRIDES (Section 19-8)

$$\mathrm{RCOOH + R\overset{O}{\overset{\|}{C}}Cl \xrightarrow{\Delta} \underset{\textbf{Anhydride}}{R\overset{O}{\overset{\|}{C}}O\overset{O}{\overset{\|}{C}}R} + HCl}$$

Cyclic anhydrides

$$\mathrm{(CH_2)_n(COOH)_2 \xrightarrow[-H_2O]{\Delta} (CH_2)_n(C{=}O)_2O}$$

Best for five- or six-membered rings

11. ESTERS (Section 19-9)

Acid-catalyzed esterification

$$RCO_2H + R'OH \underset{K \sim 1}{\overset{H^+}{\rightleftharpoons}} RC(=O)OR' + H_2O$$

Cyclic esters (lactones)

$$(CH_2)_n(CH_2OH)(COOH) \overset{H^+}{\rightleftharpoons} (CH_2)_n[CH_2-O-C(=O)] + H_2O$$

Lactone

$K > 1$ for five- and six-membered rings

Nucleophilic displacements with carboxylate ions

$$RCO_2^- \; Na^+ + R'X \xrightarrow{HMPA} RCO_2R' + NaX$$

Diazomethane reaction

$$RCOOH + CH_2N_2 \xrightarrow{(CH_3CH_2)_2O} RCOOCH_3 + N_2$$

12. CARBOXYLIC AMIDES (Section 19-10)

$$RCOOH + R'NH_2 \rightarrow RCOO^- + R'NH_3^+ \xrightarrow{\Delta} RC(=O)NHR' + H_2O$$

Imides

$$(CH_2)_n(COOH)(COOH) + R'NH_2 \xrightarrow{\Delta} (CH_2)_n[C(=O)-NR'-C(=O)] + 2\,H_2O$$

Cyclic amides (lactams)

$$(CH_2)_n(CH_2NH_2)(COOH) \xrightarrow{\Delta} (CH_2)_n[CH_2-NH-C(=O)] + H_2O$$

Lactam

13. REDUCTION WITH LITHIUM ALUMINUM HYDRIDE (Section 19-11)

$$\text{RCOOH} \xrightarrow[\text{2. } H^+,\ H_2O]{\text{1. } LiAlH_4,\ (CH_3CH_2)_2O} \text{RCH}_2\text{OH}$$

14. BROMINATION: HELL-VOLHARD-ZELINSKY REACTION (Section 19-12)

$$\text{RCH}_2\text{COOH} \xrightarrow{Br_2,\ \text{trace P}} \text{R}\underset{}{\overset{\text{Br}}{\underset{|}{\text{C}}}}\text{HCOOH}$$

One-Carbon Degradation of Carboxylic Acids

15. HUNSDIECKER REACTION (Section 19-13)

$$\text{RCOO}^-\text{Ag}^+ \xrightarrow{Br_2} \text{RBr} + CO_2 + \text{AgBr}$$

IMPORTANT CONCEPTS

1. Carboxylic acids are named as alkanoic acids. The carbonyl carbon is numbered 1 in the longest chain incorporating the carboxy group. Dicarboxylic acids are called alkanedioic acids. Cyclic and aromatic systems are called cycloalkanecarboxylic and benzoic acids, respectively. In these systems the ring carbon bearing the carboxy group is assigned the number 1.
2. The carboxy group is approximately trigonally planar. Except in very dilute solution, carboxylic acids form dimers by hydrogen bonding.
3. The carboxylic acid proton chemical shift is variable but relatively high ($\delta = 10$–13), because of hydrogen bonding. The carbonyl carbon is also relatively deshielded, but not as much as in aldehydes and ketones because of the resonance contribution of the hydroxy group. The carboxy function shows two important infrared bands, one at about 1710 cm^{-1} for the C=O bond, and a very broad band centered around 3000 cm^{-1} for the O–H group.
4. The carbonyl group in carboxylic acids is subject to addition by nucleophiles to give an unstable tetrahedral intermediate. This intermediate may decompose by elimination of the hydroxy group to give a carboxylic acid derivative.
5. Alkanoyl halides are formed by the action of inorganic halides ($SOCl_2$, PBr_3) on carboxylic acids.
6. Lithium aluminum hydride is a strong enough nucleophile to add to the carbonyl group of carboxylate ions. This process allows the reduction of carboxylic acids to primary alcohols.

PROBLEMS

1. Name (IUPAC or *Chemical Abstracts* system) or draw the structure of each of the following compounds.

(a) $CH_3\overset{CH_3}{\overset{|}{C}}HCH_2\overset{Cl}{\overset{|}{C}}HCO_2H$ **(b)** $CH_3CH_2\underset{H_2C=CH}{\underset{|}{C}}HCO_2H$ **(c)** $(H_3C)((CH_3)_2CH)C=C(Br)(CO_2H)$

(d) 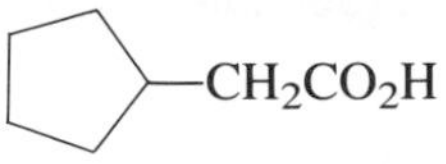 [cyclopentyl]–CH_2CO_2H (e) 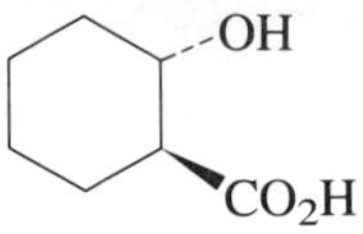 [cyclohexane with OH and CO_2H] (f) 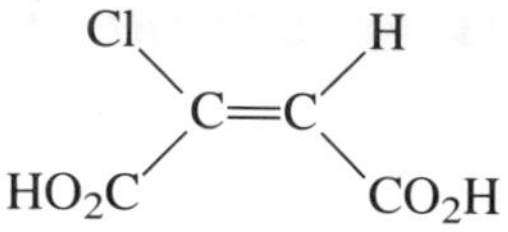 Cl(HO$_2$C)C=C(H)CO_2H

(g) 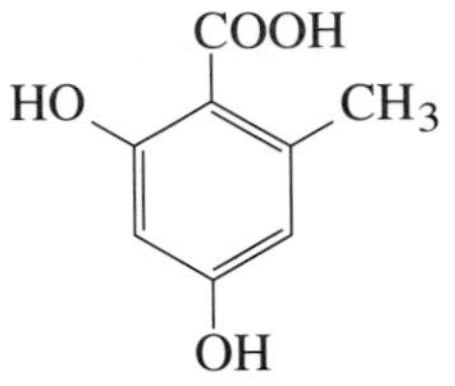 [benzene ring with COOH, HO, CH_3, OH]
o-Orsellinic acid

(h) 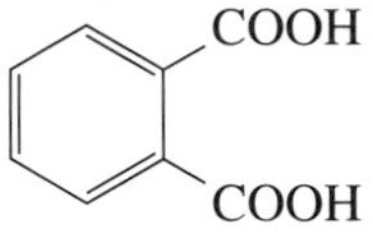 [benzene ring with two COOH]
Phthalic acid

(i) 4-aminobutanoic acid (also known as GABA, a critical participant in brain biochemistry); **(j)** *meso*-2,3-dimethylbutanedioic acid; **(k)** 2-oxopropanoic acid (pyruvic acid); **(l)** *trans*-2-formylcyclohexanecarboxylic acid; **(m)** (*Z*)-3-phenyl-2-butenoic acid; **(n)** 1,8-naphthalenedicarboxylic acid.

2. Rank the group of molecules below in decreasing order of boiling point and solubility in water. Explain your answers.

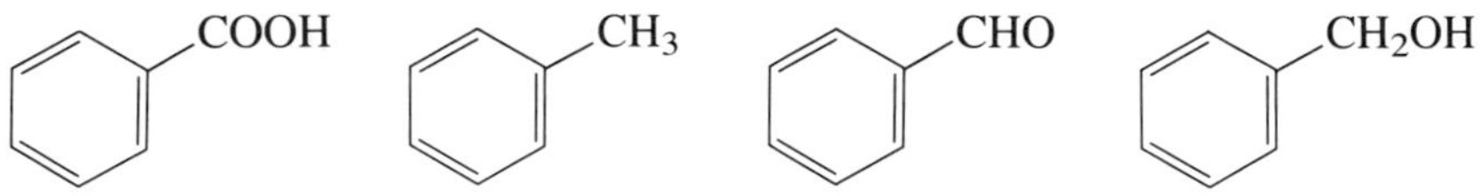

[benzene]–COOH [benzene]–CH_3 [benzene]–CHO [benzene]–CH_2OH

3. Rank each of the following groups of organic compounds in order of decreasing acidity.

(a) $CH_3CH_2CO_2H$, $CH_3\overset{O}{\overset{\|}{C}}CH_2OH$, $CH_3CH_2CH_2OH$

(b) $BrCH_2CO_2H$, $ClCH_2CO_2H$, FCH_2CO_2H

(c) $CH_3\overset{Cl}{\overset{|}{C}}HCH_2CO_2H$, $ClCH_2CH_2CH_2CO_2H$, $CH_3CH_2\overset{Cl}{\overset{|}{C}}HCO_2H$

(d) CF_3CO_2H, CBr_3CO_2H, $(CH_3)_3CCO_2H$

(e) 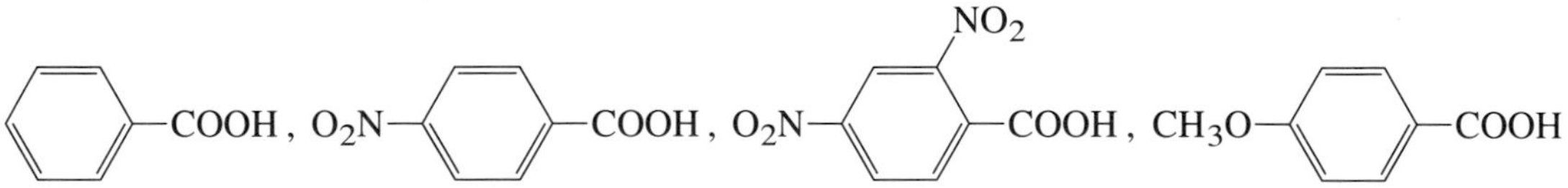 [benzene]–COOH, O_2N–[benzene]–COOH, O_2N–[benzene with NO_2]–COOH, CH_3O–[benzene]–COOH

4. (a) An unknown compound A has the formula $C_7H_{12}O_2$ and the infrared spectrum A (p. 759). To which class does this compound belong? **(b)** Use the other spectra (NMR-B, p. 760, and F, p. 762; IR-D, E, and F, pp. 761–762) and spectroscopic and chemical information in the reaction sequence to determine the structures of compound A and the other unknown substances B through F. References are made to relevant sections of earlier chapters, but do not look them up before you have tried to solve the problem without the extra help. **(c)** Another unknown compound, G, has the formula $C_8H_{14}O_4$ and the NMR and IR spectra labeled G (pp. 762–763). Propose a structure for this compound. **(d)** Compound G may be readily synthesized from B.

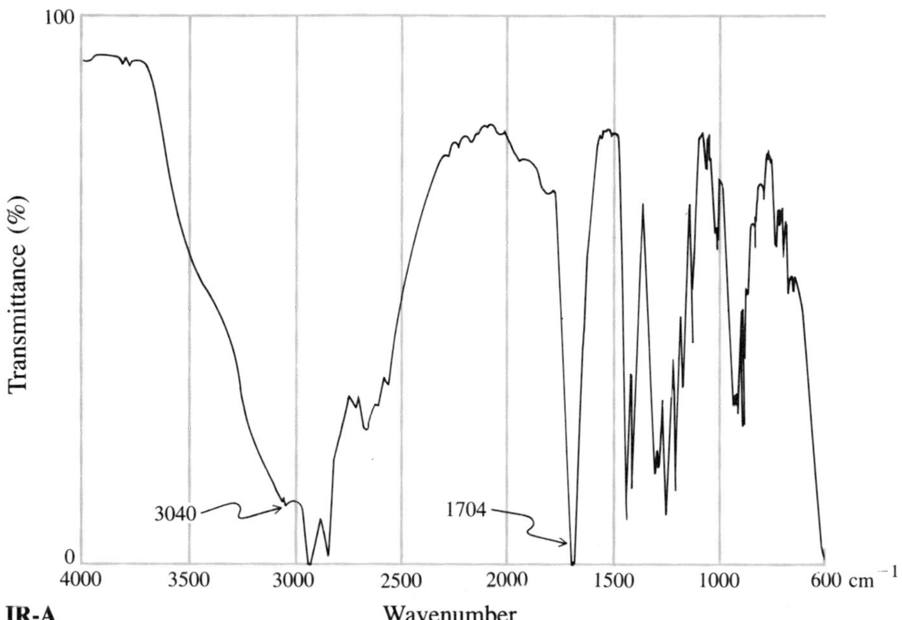

Propose a sequence that accomplishes this efficiently. **(e)** Propose a completely different sequence from that shown in (b) for the conversion of C into A. **(f)** Finally, construct a synthetic scheme that is the reverse of that shown in (b); namely, the conversion of A into B.

C_6H_{10} **B** $\xrightarrow[\text{Section 12-7}]{\text{1. } Hg(OCCH_3)_2,\ H_2O \text{ (C=O)};\ \text{2. } NaBH_4}$ $C_6H_{12}O$ **C** $\xrightarrow[\text{Section 8-6}]{CrO_3,\ H_2SO_4,\ \text{propanone (acetone), } 0°C}$

B: ^{13}C NMR: δ = 22.1, 24.5, 126.2 ppm; ^1H NMR-B

C: ^{13}C NMR: δ = 24.4, 25.9, 35.5, 69.5 ppm

$C_6H_{10}O$ **D** $\xrightarrow[\text{Section 17-12}]{CH_2{=}P(C_6H_5)_3}$ C_7H_{12} **E** $\xrightarrow[\text{Section 12-8}]{\text{1. } BH_3,\ THF;\ \text{2. } HO^-,\ H_2O_2}$ $C_7H_{14}O$ **F** $\xrightarrow[\text{Section 8-6}]{Na_2Cr_2O_7,\ H_2O,\ H_2SO_4}$ **A**

D: ^{13}C NMR: δ = 23.8, 26.5, 40.4, 208.5 ppm; IR-D

E: IR-E

F: IR-F; ^1H NMR-F

5. Give the products of each reaction shown below.

(a) $(CH_3)_2CH_2CH_2CO_2H + SOCl_2 \longrightarrow$

(b) $(CH_3)_2CH_2CH_2CO_2H + CH_3COBr \longrightarrow$

(c) (cyclopentane)–COOH $+ CH_3CH_2OH \xrightarrow{H^+}$

(d) CH_3O–(benzene ring)–$COOH + NH_3 \longrightarrow$

(e) Product of (d), heated strongly

(f) Phthalic acid (Problem 1h), heated strongly

6. Fill in suitable reagents to carry out the following transformations.

(a) $(CH_3)_2CHCH_2CHO \longrightarrow (CH_3)_2CHCH_2CO_2H$

(b) cyclopentyl–CHO $\longrightarrow$ cyclopentyl–CH(OH)CO_2H

(c) 2-bromodecalin (Br) $\longrightarrow$ decalin–CO_2H

(d) $CH_3CH(OH)CH_2CH_2Cl \longrightarrow CH_3CH(OH)CH_2CH_2CO_2H$

(e) $CH_3CH_2CH(CO_2H)CH_3 \longrightarrow CH_3CH_2CH(CH_3)\text{—}\overset{O}{\overset{\|}{C}}O\overset{O}{\overset{\|}{C}}\text{—}CH(CH_3)CH_2CH_3$

(f) $(CH_3)_3CCO_2H \longrightarrow (CH_3)_3CCO_2CH(CH_3)_2$

(g) $C_6H_5NHCH_3 \longrightarrow C_6H_5N(COCH_3)CH_3$

(h) cyclopropyl–CH_2CO_2H $\longrightarrow$ cyclopropyl–CH_2Br

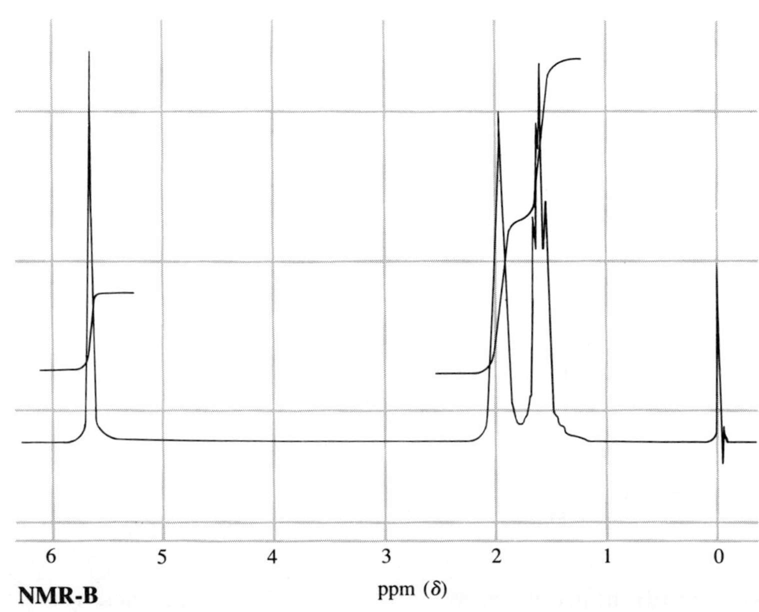

NMR-B

7. Propose syntheses of each of the following carboxylic acids that employ at least one reaction that forms a carbon–carbon bond.

(**a**) $CH_3CH_2CH_2CH_2CH_2CH_2CO_2H$ (**b**) $CH_3CH(OH)CH_2CO_2H$ (**c**) $H_3C{-}C(CH_3)_2{-}CO_2H$

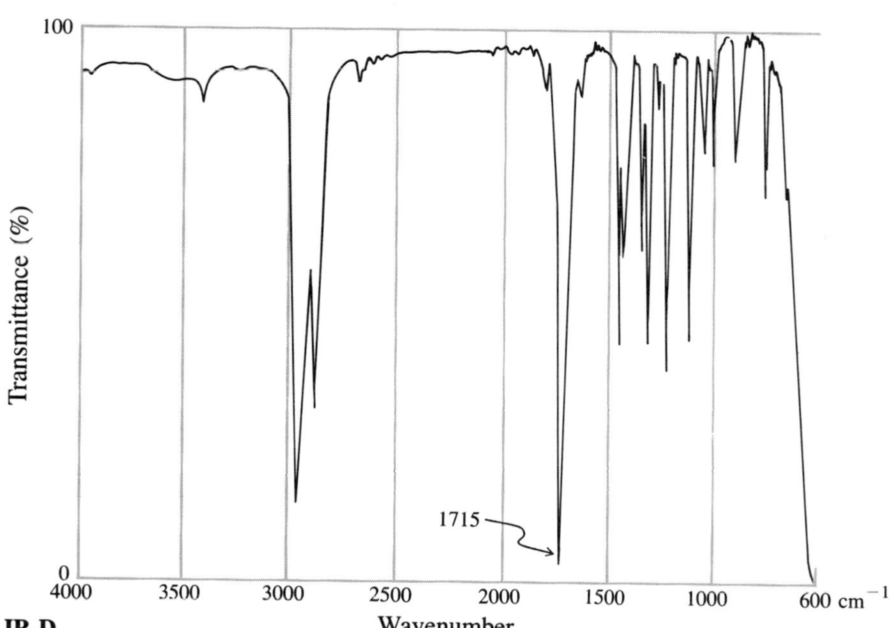

IR-D

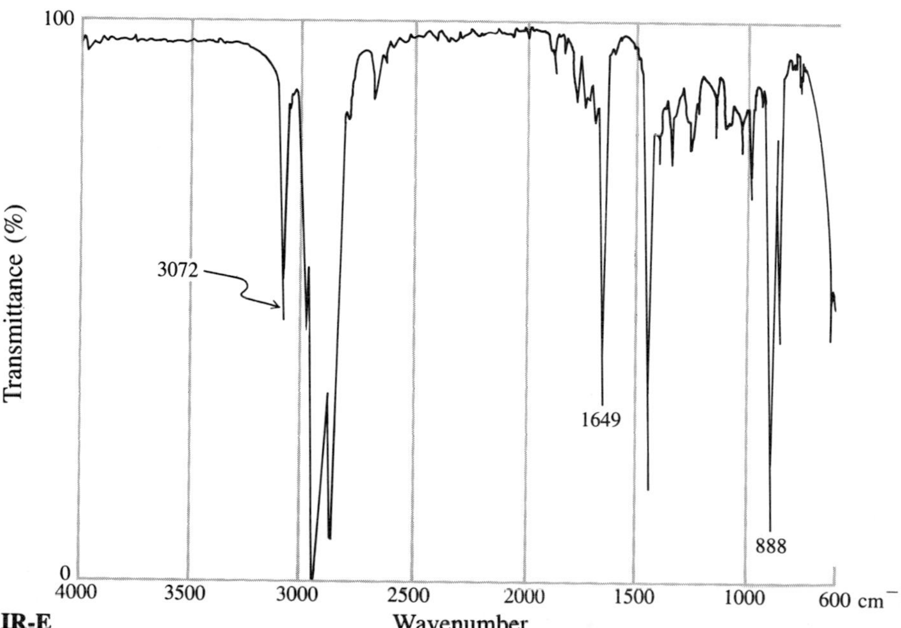

IR-E

8. (a) Write a mechanism for the esterification of propanoic acid with ^{18}O-labeled ethanol. Show clearly the fate of the ^{18}O label. **(b)** Acid-catalyzed hydrolysis of an unlabeled ester with ^{18}O-labeled water ($H_2{}^{18}O$) leads to incorporation of some ^{18}O into *both* oxygens of the carboxylic acid product. Explain by a mechanism. (Hint: You must use the fact that all steps in the mechanism are reversible.)

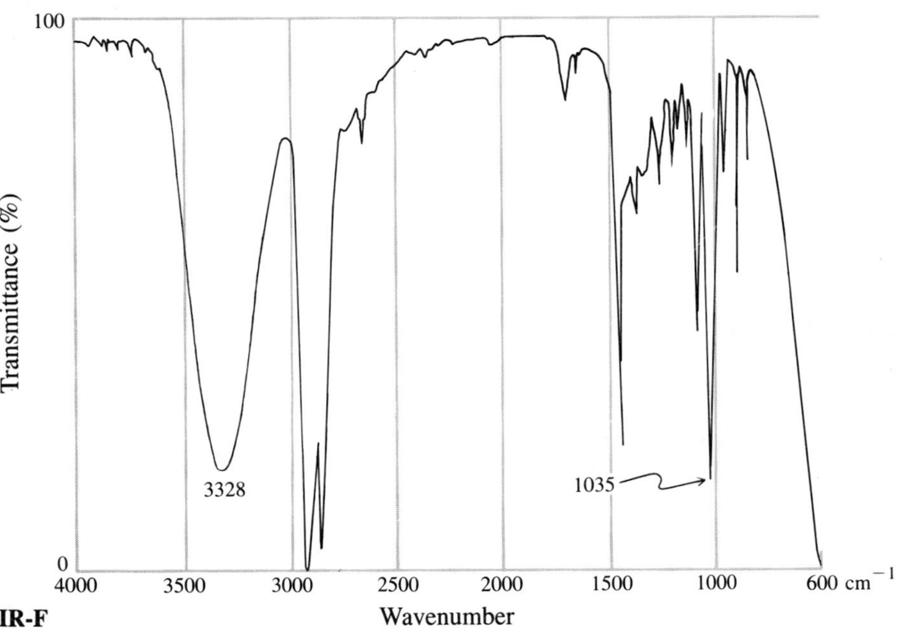

IR-F

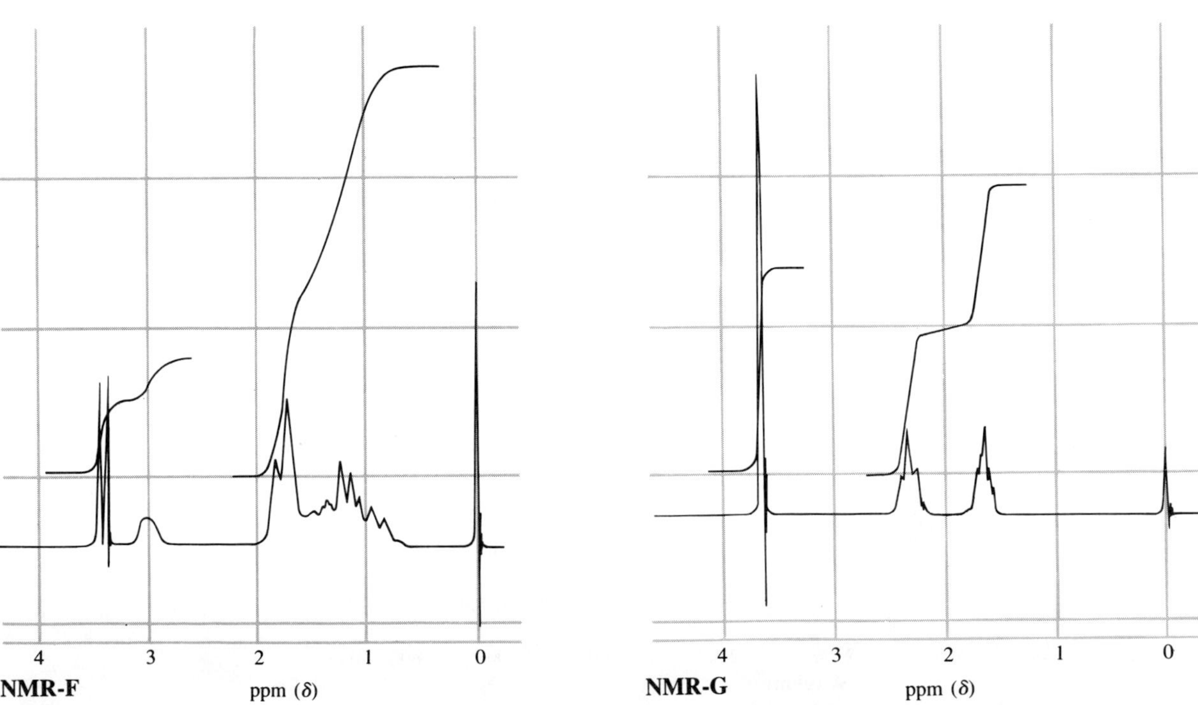

NMR-F **NMR-G**

9. Give the products of reaction of propanoic acid with each of the following reagents.

(a) $SOCl_2$ **(b)** PBr_3 **(c)** CH_3CH_2COBr + pyridine
(d) $(CH_3)_2CHOH + HCl$ **(e)** CH_2N_2, $(CH_3CH_2)_2O$ **(f)** KOH, CH_3CH_2I, DMSO
(g) C_6H_5–CH_2NH_2 **(h)** Product of (g), heated strongly **(i)** $LiAlH_4$, $(CH_3CH_2)_2O$, then H^+, H_2O
(j) Br_2, P **(k)** $AgNO_3$, KOH, H_2O, then Br_2, CCl_4

10. Give the product of reaction of cyclopentanecarboxylic acid with each of the reagents in Problem 9.

11. Suggest a preparation of hexanoic acid from pentanoic acid.

12. Give reagents and reaction conditions that would allow efficient conversion of 2-methylbutanoic acid into **(a)** the corresponding alkanoyl chloride; **(b)** the corresponding methyl ester; **(c)** the corresponding ester with 2-butanol; **(d)** the anhydride; **(e)** the *N*-methylamide; **(f)** 2-bromobutane; **(g)** $CH_3CH_2CH(CH_3)CH_2OH$ **(h)** $CH_3CH_2CBr(CH_3)CO_2H$

13. Show how the Hell-Volhard-Zelinsky reaction might be used in the synthesis of each of the compounds at the top of the next page, beginning in each case with a simple monocarboxylic acid. Write detailed mechanisms for all the reactions in *one* of your syntheses.

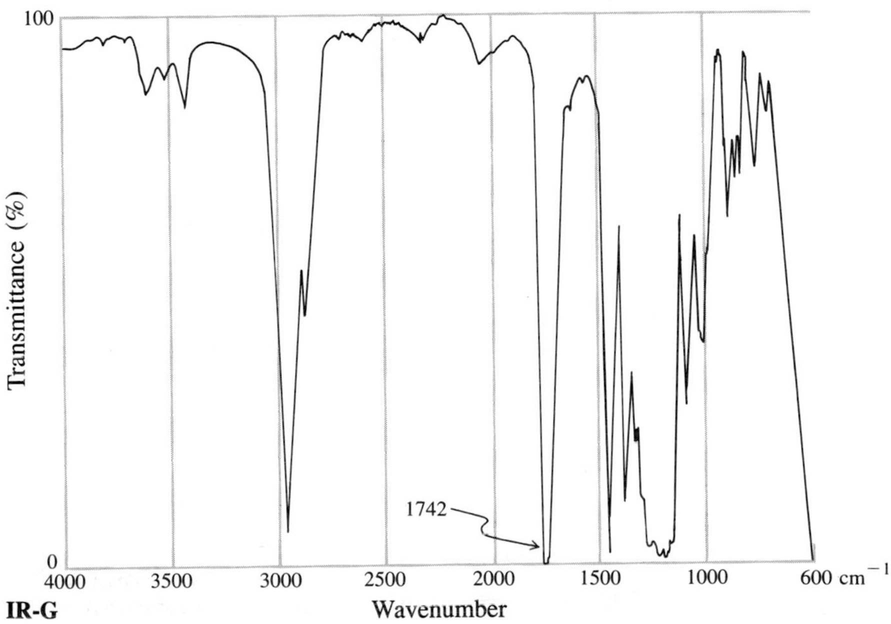

IR-G

(a) $CH_3CH_2CH(NH_2)CO_2H$ (b) 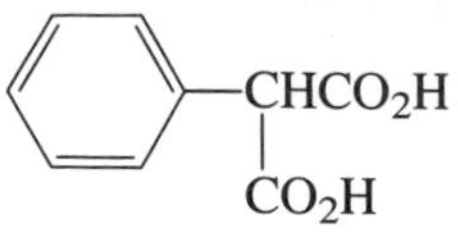 $C_6H_5CH(CO_2H)CO_2H$ (c) $CH_3CH_2CH(CH_3)CH_2CH(OH)CO_2H$

(d) $HO_2CCH_2SSCH_2CO_2H$ (e) $(CH_3CH_2)_2NCH_2CO_2H$ (f) $(C_6H_5)_3\overset{+}{P}CH(CH_3)CO_2H\ Br^-$

14. Although the original Hell-Volhard-Zelinsky reaction is limited to the synthesis of bromocarboxylic acids, the chloro and iodo analogs may be prepared by modifications of the procedure. Thus, alkanoyl chlorides may be converted to their α-chloro and α-bromo derivatives by reaction with *N*-chloro and *N*-bromobutanimide (*N*-chloro- and *N*-bromosuccinimide, NCS and NBS: Sections 3-8 and 14-2), respectively. Reaction with I_2 gives α-iodo compounds. Suggest a mechanism for any one of these processes.

$$C_6H_5CH_2CH_2COCl \xrightarrow{NCS,\ HCl,\ SOCl_2,\ 70°} C_6H_5CH_2CHClCOCl \quad 84\%$$

$$C_6H_5CH_2CH_2COCl \xrightarrow{NBS,\ HBr,\ SOCl_2,\ 70°} C_6H_5CH_2CHBrCOCl \quad 71\%$$

$$C_6H_5CH_2CH_2COCl \xrightarrow{I_2,\ HI,\ SOCl_2,\ 85°} C_6H_5CH_2CHICOCl \quad 75\%$$

15. Propose a mechanism for the exchange between an alkanoyl bromide and a carboxylic acid, as occurs in step 4 of the Hell-Volhard-Zelinsky reaction. (Hint: Refer to the mechanisms presented in Section 19-8.)

16. Hunsdiecker decarboxylation of cyclopropanecarboxylic acids is a method of choice for synthesis of bromocyclopropanes. Explain why this is so by describing the problems with other methods for the synthesis of haloalkanes as applied to cyclopropyl compounds.

$CH_3\overset{O}{\overset{\|}{C}}NH_2$

Acetamide

17. How would you expect the acidity of acetamide to compare with that of acetic acid? With that of propanone (acetone)? Which protons in acetamide are the most acidic? Where would you expect acetamide to be protonated by very strong acid?

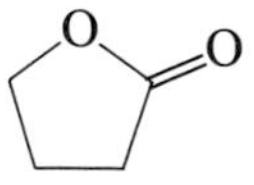

γ-Butyrolactone

18. Attempted CrO_3 oxidation of 1,4-butanediol to butanedioic acid results in significant yields of ‘‘γ-butyrolactone.’’ Explain mechanistically.

19. Following the general mechanistic scheme given in Section 19-7, write detailed mechanisms for each of the following substitution reactions. (Note: These transformations are part of Chapter 20, but you should be able to solve the problem without looking ahead.)

(a) 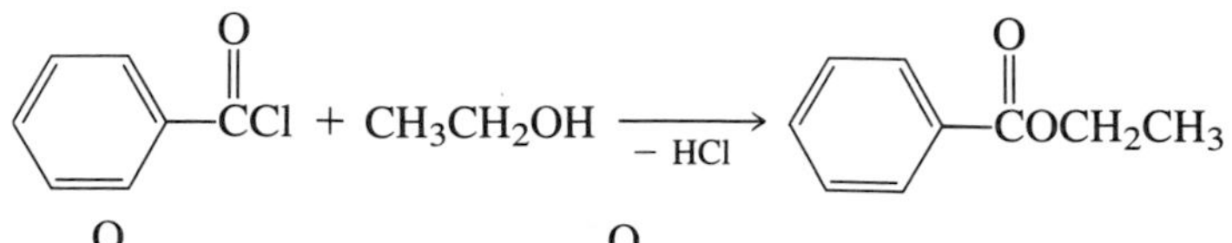 $C_6H_5\overset{O}{\overset{\|}{C}}Cl + CH_3CH_2OH \xrightarrow[-\ HCl]{} C_6H_5\overset{O}{\overset{\|}{C}}OCH_2CH_3$

(b) $CH_3\overset{O}{\overset{\|}{C}}NH_2 + H_2O \xrightarrow{Acid} CH_3\overset{O}{\overset{\|}{C}}OH + \overset{+}{N}H_4$

20. Suggest structures for the products of each reaction in the following synthetic sequence.

(starting bicyclic ketone, IR: 1745 cm^{-1}) $\xrightarrow[\text{3. } H^+,\ H_2O]{\text{1. pyrrolidine, } H^+ \quad \text{2. } (CH_3)_2C{=}CHCH_2Br}$ $C_{14}H_{22}O$ **H** (IR: 1675 and 1745 cm^{-1}) $\xrightarrow{HOCH_2CH_2OH,\ H^+}$ $C_{16}H_{26}O_2$ **I** (IR: 1670 cm^{-1}) $\xrightarrow{KMnO_4,\ NaHCO_3,\ \text{propanone (acetone)}}$

$C_{13}H_{20}O_4$ **J** (IR: 1715 and 3000 (broad) cm^{-1}) $\xrightarrow[\text{2. } NaBH_4]{\text{1. } H^+,\ H_2O}$ $C_{11}H_{18}O_3$ **K** (IR: 1715, 3000 (broad), and 3350 cm^{-1}) $\xrightarrow{\text{Catalytic } H^+,\ \Delta}$ $C_{11}H_{16}O_2$ **L** (IR: 1770 cm^{-1})

21. S_N2 reactions of simple carboxylate ions with haloalkanes in aqueous solution generally do not give good yields of esters. **(a)** Explain why this is so. **(b)** Reaction of 1-iodobutane with sodium acetate gives an excellent yield of ester if carried out in acetic acid (below). Why is the latter a better solvent for this process than water?

$$CH_3CH_2CH_2CH_2I + CH_3\overset{O}{\overset{\|}{C}}O^-Na^+ \xrightarrow{CH_3CO_2H,\ 100°C} CH_3CH_2CH_2CH_2O\overset{O}{\overset{\|}{C}}CH_3 + Na^+I^-$$

Iodobutane — **Sodium acetate** — **Butylacetate** (95%)

(c) The reaction of 1-iodobutane with sodium dodecanoate proceeds surprisingly well in aqueous solution, much better than the reaction with sodium acetate (see the following equation). Explain this observation. (Hint: Sodium dodecanoate is a soap and forms micelles in water. See Box 19-1.)

$$CH_3CH_2CH_2CH_2I + CH_3(CH_2)_{10}CO_2^-Na^+ \xrightarrow{H_2O} CH_3(CH_2)_{10}\overset{O}{\overset{\|}{C}}OCH_2CH_2CH_2CH_3$$

22. (a) Diazomethane, CH_2N_2, is usually represented as a resonance hybrid of two contributing Lewis structures. Write them. **(b)** Propose a mechanism for the formation of a methyl ester from diazomethane and a carboxylic acid.

23. The *iridoids* are a class of monoterpenes with powerful and varied biological activities. They include insecticides, agents of defense against predatory insects, and animal attractants. On the next page is a synthesis of neonepetalactone, one of the nepetalactones, which are primary constituents of catnip. Use the information given to deduce the structures that have been left out, including that of neonepetalactone itself.

$C_{10}H_{16}O_2$ **M** $\xrightarrow{\text{Base}}$ [cyclopentene aldehyde structure] –CHO $\xrightarrow{CrO_3, H_2SO_4, 0°C}$ $C_{10}H_{14}O_2$ **N** $\xrightarrow{CH_2N_2, (CH_3CH_2)_2O}$

IR: 890, 1645, 1725 (very strong), and 1705 cm^{-1}

IR: 890, 1630, 1640, 1720, and 3000 (very broad) cm^{-1}

$C_{11}H_{16}O_2$ **O** $\xrightarrow{\text{1. Dicyclohexylborane, THF; 2. } HO^-, H_2O_2}$ $C_{11}H_{18}O_3$ **P** $\xrightarrow{H^+, H_2O, \Delta}$ $C_{10}H_{14}O_2$ **Neonepetalactone**

IR: 890, 1630, 1640, and 1720 cm^{-1}

IR: 1630, 1720, and 3335 cm^{-1}

IR: 1645 and 1710 cm^{-1}
UV λ_{max} = 241 nm

24. Propose *two* possible mechanisms for the following reaction. (Hint: Consider the possible sites of protonation in the molecule and the mechanistic consequences of each.) Devise an isotope-labeling experiment that might distinguish your two mechanisms.

[Ph–CH(OH)–C$_6$H$_4$–COOH] $\xrightarrow{HCl, \Delta}$ [lactone: Ph–CH–O–C(=O) fused to benzene ring]

25. Propose a short synthesis of 2-butynoic acid, $CH_3C{\equiv}CCO_2H$, starting from propyne. (Hint: Review Section 13-2.)

26. The benzene rings of many compounds in nature are prepared by a biosynthetic pathway similar to that operating in fatty acid synthesis. Acetyl units are coupled, but the ketone functions are not reduced. The result is a *polyketide thioester,* which forms rings by intramolecular aldol condensation.

$CH_3\overset{O}{\overset{\|}{C}}CH_2\overset{O}{\overset{\|}{C}}CH_2\overset{O}{\overset{\|}{C}}CH_2\overset{O}{\overset{\|}{C}}{-}S{-}\boxed{\text{protein}}$

Polyketide thioester

o-Orsellinic acid [for structure, see Problem 1(g)] is a derivative of salicylic acid that is prepared biosynthetically from the polyketide thioester shown. Explain how this transformation might take place. Hydrolysis of the thioester to give the free carboxylic acid is the last step.

20

Carboxylic Acid Derivatives and Mass Spectroscopy

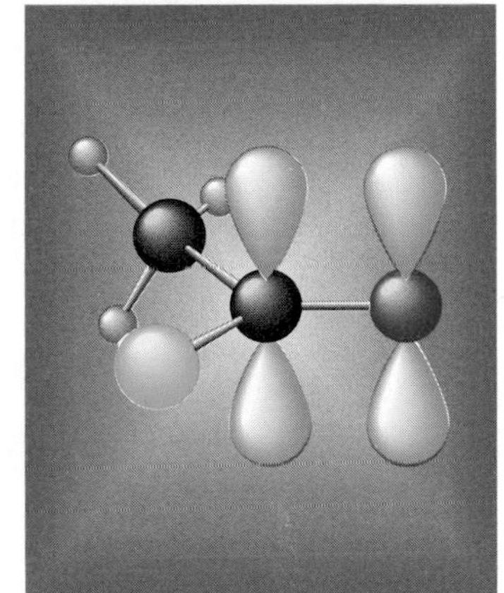

The previous chapter showed how chemists prepare the four principal carboxylic acid derivatives: halides, anhydrides, esters, and amides. This chapter deals with their chemistry, with particular emphasis on nucleophilic substitutions and functional-group interconversions. Each has a substituent, L, which can function as a leaving group in substitution reactions. We already know, for instance, that displacement of halide in $R\overset{O}{\overset{\|}{C}}X$ by a carboxylic acid leads to anhydrides.

$$R-\overset{:O:}{\overset{\|}{C}}-L$$

$L = X,\ \overset{O}{\overset{\|}{C}}R\text{ (as }O\overset{O}{\overset{\|}{C}}R),\ OR,\ NR_2$

Carboxylic acid derivatives

We begin with a comparison of their structures, properties, and relative reactivities. We then explore the chemistry of each type of compound. Halides and anhydrides are valuable reagents in the synthesis of other carbonyl compounds, and esters and amides possess enormous importance in nature. For example, the esters include common flavoring agents, waxes, fats, and oils; urea and penicillin both are amides. The alkanenitriles, RC≡N, are also treated here, because they undergo similar reactions. Finally, the chapter introduces another physical technique of great value to the organic chemist—mass spectroscopy.

20-1 Relative Reactivities, Structures, and Spectra of Carboxylic Acid Derivatives

Carboxylic acid derivatives undergo substitution reactions with nucleophiles, such as water, organometallic compounds, and hydride reducing agents. These transformations proceed through the familiar addition–elimination sequence (Section 19-7).

Addition–Elimination in Carboxylic Acid Derivatives

$$\mathrm{RC(=O)L} + :\mathrm{Nu}^- \longrightarrow \mathrm{R{-}C(O^-)(Nu){-}L} \longrightarrow \mathrm{RC(=O)Nu} + :\mathrm{L}^-$$

The relative reactivities of the substrates follow a consistent order: Alkanoyl (acyl) halides are most reactive, followed by anhydrides, then esters, and finally the amides, which are least reactive.

Relative Reactivities of Carboxylic Acid Derivatives in Nucleophilic Addition–Elimination

$$\underset{\text{Amide}}{\mathrm{RCNR'_2}} < \underset{\text{Ester}}{\mathrm{RCOR'}} < \underset{\text{Anhydride}}{\mathrm{RCOCR}} < \underset{\text{Alkanoyl halide}}{\mathrm{RCX}}$$

(each with C=O)

Nucleophilic Addition–Elimination of Carboxylic Acid Derivatives

$$\mathrm{RCX} + \mathrm{HOH} \xrightarrow[\text{Fast}]{20^\circ\mathrm{C}} \mathrm{RCOOH} + \mathrm{HX}$$

$$\mathrm{RCOCR} + \mathrm{HOH} \xrightarrow[\text{Slow}]{20^\circ\mathrm{C}} \mathrm{RCOOH} + \mathrm{RCOOH}$$

$$\mathrm{RCOR'} + \mathrm{HOH} \xrightarrow{\text{Very slow: Needs catalyst and heat}} \mathrm{RCOOH} + \mathrm{R'OH}$$

$$\mathrm{RCNR'_2} + \mathrm{HOH} \xrightarrow{\text{Exceedingly slow: Needs catalyst and prolonged heating}} \mathrm{RCOOH} + \mathrm{HNR'_2}$$

This order of reactivities depends directly on the extent to which the lone electron pairs on the leaving group, L, delocalize into the carbonyl function. The order depends also on the inductive effect of L on the carbonyl carbon.

In the carboxylic acid derivatives, just as in the acids themselves, the carbonyl group and any lone pairs on the adjacent substituent enter into resonance.

Resonance in Carboxylic Acid Derivatives

$$\left[R\overset{\overset{:\ddot{O}:}{\|}}{C}\ddot{L} \longleftrightarrow R\overset{\overset{:\ddot{O}:^-}{|}}{C}{=}L^+ \right]$$

The resulting dipolar resonance form is most important when L is NR_2, because nitrogen is the least electronegative atom in the series. Because the contribution of this resonance form reduces the partial positive charge on the carbonyl carbon (Section 19-3), it is not surprising that amides are the least reactive of the carboxylic acid derivatives. In esters the importance of resonance is decreased as a result of the greater electronegativity of oxygen; their reactivity is correspondingly higher than that of the amides. Anhydrides are more reactive than esters, because the lone electron pairs on the central oxygen are shared by two carbonyl groups. Finally, resonance is least important in the alkanoyl (acyl) chlorides and bromides for two reasons: The halogens are very electronegative, and their larger *p* orbitals overlap relatively poorly with the 2*p* lobes on the carbonyl carbon.

The greater the resonance, the shorter the C–L bond

The extent of resonance can be observed directly in the structures of carboxylic acid derivatives. As we go from the most reactive systems (alkanoyl halides) to the least reactive (esters and amides), the C–L bond becomes progressively shorter, reflecting increased double-bond character (Table 20-1). The NMR spectra of amides also reveal that rotation about this bond has become restricted. For example, *N*,*N*-dimethylformamide at room temperature exhibits *two* singlets for the two methyl groups, because rotation around the C–N bond is so slow on the NMR time scale. The evidence points to considerable π overlap between the lone pair on nitrogen and the carbonyl carbon, as a result of the increased importance of the dipolar resonance form in amides. The measured barrier to this rotation is about 21 kcal mol^{-1}.

TABLE 20-1 C–L Bond Lengths in $R\overset{O}{\overset{\|}{C}}L$ Compared with R–L Single Bond Distances

L	Bond length (Å) in R–L	Bond length (Å) in $R\overset{O}{\overset{\|}{C}}$–L
Cl	1.78	1.79 (not shorter)
OCH_3	1.43	1.36 (shorter by 0.07 Å)
NH_2	1.47	1.36 (shorter by 0.11 Å)

TABLE 20-2 Carbonyl Stretching Frequencies of Carboxylic Acid Derivatives RCL (with C=O)

L	$\tilde{\nu}_{C=O}$ (cm^{-1})	
Cl	1790–1815	
OCR	1740–1790 1800–1850	Two bands are observed, corresponding to asymmetric and symmetric stretching motions
OR	1735–1750	
NR'_2	1650–1690	

Slow Rotation in *N*,*N*-Dimethylformamide

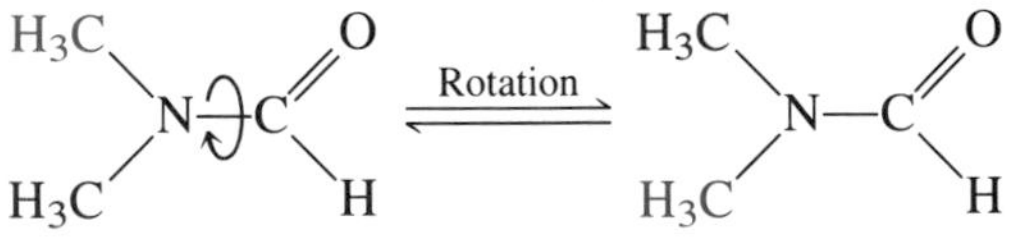

$$(CH_3)_2N{-}CH{=}O \underset{}{\overset{\text{Rotation}}{\rightleftharpoons}} (CH_3)_2N{-}CH{=}O$$

$E_a = 21$ kcal mol^{-1}

$CH_3CNHNHC_6H_5$ (with C=O)

EXERCISE 20-1

The methyl group in the ^1H NMR spectrum of 1-acetyl 2-phenylhydrazide, shown in the margin, exhibits two singlets at $\delta = 2.02$ and 2.10 ppm at room temperature. On heating to 100°C in the NMR probe, the same compound gives rise to only one signal in that region. Explain.

Infrared spectroscopy can also be used to probe resonance in carboxylic acid derivatives. The dipolar resonance structure weakens the C=O bond and causes a corresponding decrease in the carbonyl stretching frequency (Table 20-2).

The ^{13}C NMR signals of the carbonyl carbons in carboxylic acid derivatives are less sensitive to polarity differences and fall in a narrow range near 170 ppm.

^{13}C NMR Chemical Shifts of the Carbonyl Carbon in Carboxylic Acid Derivatives

CH_3CCl	CH_3COCCH_3	CH_3COH	CH_3COCH_3	CH_3CNH_2
$\delta = 170.3$	166.9	177.2	170.7	172.6 ppm

Carboxylic acid derivatives are basic and acidic

The extent of resonance in carboxylic acid derivatives is also seen in their basicity (protonation at the carbonyl oxygen) and acidity (enolate formation). In all cases, protonation requires strong acid, but it gets easier as the electron-donating ability of the L group increases.

BOX 20-1 Spectroscopic Characterization of the Lactone Ring in Manoalide

The identification of the lactone ring in the anti-inflammatory natural product manoalide, isolated from a marine sponge (Box 10-3), depended largely on IR and UV spectroscopy.

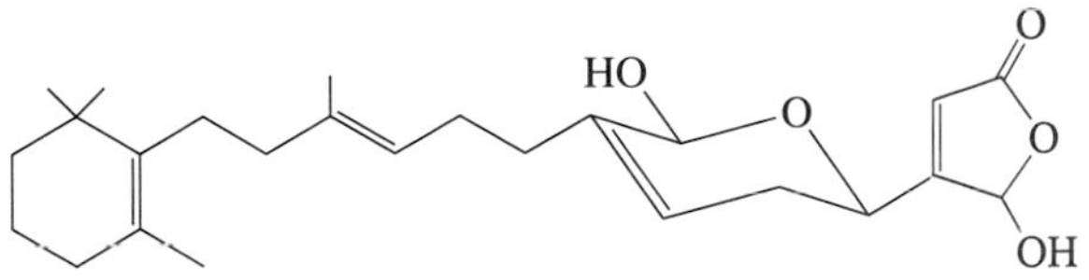

HO O O O OH

Manoalide

The signal at $\delta = 172.3$ ppm in the ^{13}C NMR spectrum revealed that a carboxylic acid derivative was present but not which type. The IR spectrum showed a carbonyl stretching frequency at 1765 cm^{-1}. Although higher than usual for an ester group, recall that the incorporation of a carbonyl group in a ring containing fewer than six atoms raises the value of $\tilde{\nu}_{C=O}$ (Section 17-3). This value is in the range observed for γ-butyrolactone and its derivatives.

The presence of unsaturation in the lactone ring of manoalide was suggested by the UV spectrum. Saturated esters normally exhibit λ_{max} values in the range 200–210 nm, but conjugation with a double bond shifts this band into the range 215–250 nm (see Section 14-11). The λ_{max} value for manoalide is 227 nm.

Support for the presence of an α,β-unsaturated five-membered lactone ring with an alkyl substituent at C3 and a hydroxy group at C4 relied on comparison with the spectra of known substances, such as the simpler lactone shown below, which displays an IR band at 1760 cm^{-1}.

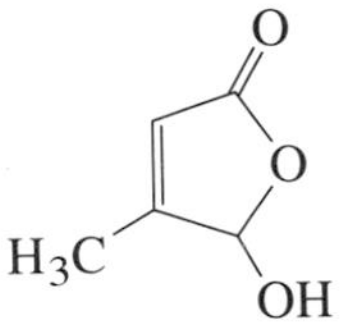

O O H_3C OH

Protonation of a Carboxylic Acid Derivative

$$\underset{R\quad\ddot{L}}{\overset{:O:}{\underset{\|}{C}}} \xrightarrow{H^+} \left[\underset{R\quad\ddot{L}}{\overset{+\ddot{O}-H}{\underset{\|}{C}}} \longleftrightarrow \underset{R\quad\ddot{L}}{\overset{:\ddot{O}-H}{\underset{|}{C^+}}} \longleftrightarrow \underset{R\quad L^+}{\overset{:\ddot{O}-H}{\underset{|}{C}}} \right]$$

A relatively strong contribution of this resonance structure stabilizes the protonated species

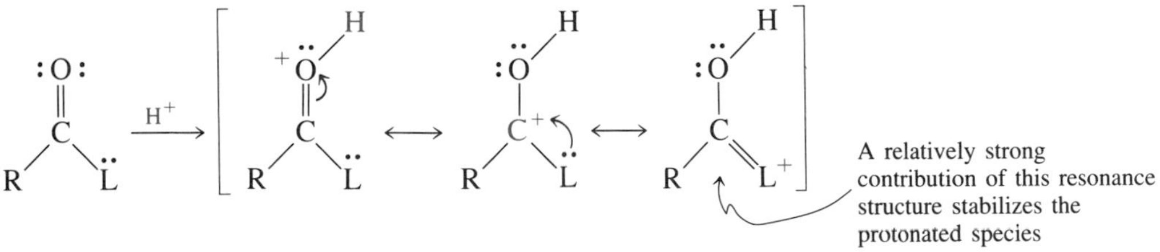

EXERCISE 20-2

Acetyl chloride is a much weaker base than acetamide. Explain, using resonance structures.

For related reasons, the acidity of the hydrogens next to the carbonyl group increases along the series. The acidity of a ketone lies between those of an alkanoyl chloride and an ester.

Acidities of α-Hydrogens in Carboxylic Acid Derivatives in Comparison with Propanone (Acetone)

$$\underset{\sim 30}{CH_3\overset{O}{\overset{\|}{C}}N(CH_3)_2} < \underset{\sim 25}{CH_3\overset{O}{\overset{\|}{C}}OCH_3} < \underset{\sim 20}{CH_3\overset{O}{\overset{\|}{C}}CH_3} < \underset{\sim 16}{CH_3\overset{O}{\overset{\|}{C}}Cl}$$

pK_a: ~30, ~25, ~20, ~16

EXERCISE 20-3

Can you think of a reaction from a previous chapter that takes advantage of the relatively high acidity of the α-hydrogens in alkanoyl (acyl) halides?

In summary, the relative electronegativity (and, for the halogens, the size) of L in $\mathrm{R\overset{O}{\overset{\|}{C}}L}$ controls the extent of resonance of the lone electron pair(s) and therefore the relative reactivity of a carboxylic acid derivative in nucleophilic addition–elimination reactions. This effect manifests itself in structural and spectroscopic measurements, as well as in the relative acidity and basicity of the α-hydrogen and the carbonyl oxygen, respectively.

20-2 Chemistry of Alkanoyl Halides

The alkanoyl **halides,** $\mathrm{R\overset{O}{\overset{\|}{C}}X}$, are named after the alkano*ic acid* from which they are derived. The halides of cycloalkane*carboxylic acids* are called cycloalkane**carbonyl halides.**

$\mathrm{CH_3\overset{O}{\overset{\|}{C}}Cl}$
Acetyl chloride

$\mathrm{CH_3\underset{}{\overset{CH_3}{\overset{|}{C}}}HCH_2\overset{O}{\overset{\|}{C}}Br}$
3-Methylbutanoyl bromide

$\mathrm{CH_3CH_2CH_2CH_2\overset{O}{\overset{\|}{C}}F}$
Pentanoyl fluoride

$\mathrm{C_6H_{11}\overset{O}{\overset{\|}{C}}Cl}$
Cyclohexanecarbonyl chloride

Alkanoyl halides undergo addition–elimination reactions in which nucleophiles displace the halide leaving group.

Addition–Elimination Reactions of Alkanoyl Halides

$$\mathrm{R\overset{:O:}{\overset{\|}{C}}\ddot{X}: + :Nu^- \longrightarrow R{-}\underset{Nu}{\underset{|}{\overset{:\ddot{O}:^-}{\overset{|}{C}}}}{-}\ddot{X}: \longrightarrow R\overset{:O:}{\overset{\|}{C}}Nu + :\ddot{X}:^-}$$

Figure 20-1 shows the variety of nucleophilic reagents and the corresponding products. It is because of this wide range of reactivity that alkanoyl halides are useful synthetic intermediates.

Let us consider these transformations one by one (except for anhydride formation, which was covered in Section 19-8). Examples will be restricted to alkanoyl

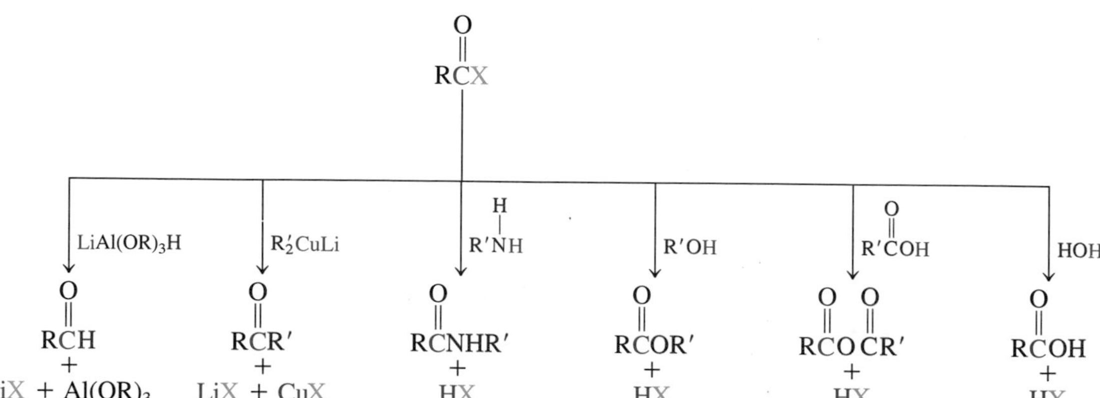

FIGURE 20-1 Nucleophilic addition–elimination reactions of alkanoyl halides.

chlorides, which are the most readily accessible, but their transformations can be generalized to a considerable extent to the other alkanoyl halides.

Water hydrolyzes alkanoyl chlorides to carboxylic acids

Alkanoyl chlorides react with water, often violently, to give the corresponding carboxylic acids and hydrogen chloride. This transformation is a simple example of addition–elimination.

Alkanoyl Chloride Hydrolysis

$$\underset{\text{Propanoyl chloride}}{CH_3CH_2\overset{O}{\overset{\|}{C}}Cl} + HOH \longrightarrow \underset{\substack{100\% \\ \text{Propanoic acid}}}{CH_3CH_2\overset{O}{\overset{\|}{C}}OH} + HCl$$

Mechanism of Alkanoyl Chloride Hydrolysis

$$CH_3CH_2\overset{:O:}{\overset{\|}{C}}\ddot{C}l: + H_2\ddot{O} \longrightarrow CH_3CH_2-\underset{:\overset{+}{O}H_2}{\overset{:\ddot{O}:^-}{C}}-\ddot{C}l: \xrightarrow{-:\ddot{C}l:^-} CH_3CH_2\overset{:O:}{\overset{\|}{C}}\overset{+}{\ddot{O}}H_2 \xrightarrow{-H^+} CH_3CH_2\overset{:O:}{\overset{\|}{C}}\ddot{O}H$$

Alcohols convert alkanoyl chlorides into esters

The analogous reaction of alkanoyl chlorides with alcohols is a highly effective way of producing esters. A base such as an alkali metal hydroxide, pyridine, or a tertiary amine is usually added to neutralize the HCl by-product. Because alkanoyl chlorides are readily made from the corresponding carboxylic acids (Section 19-8), the sequence $RCOOH \rightarrow RCOCl \rightarrow RCOOR'$ is a good method for esterification. By maintaining neutral conditions, this preparation avoids the equilibrium problem of acid-catalyzed ester formation (Section 19-9).

Ester Synthesis from Carboxylic Acids Through Alkanoyl Chlorides

$$R\overset{O}{\overset{\|}{C}}OH \xrightarrow[-\,HCl]{SOCl_2} R\overset{O}{\overset{\|}{C}}Cl \xrightarrow[-\,HCl]{R'OH,\ base} R\overset{O}{\overset{\|}{C}}OR'$$

$$\underset{\text{Acetyl chloride}}{CH_3\overset{O}{\overset{\|}{C}}Cl} + \underset{\text{1-Propanol}}{HOCH_2CH_2CH_3} \xrightarrow{N(CH_2CH_3)_3} \underset{\substack{75\% \\ \text{Propyl acetate}}}{CH_3\overset{O}{\overset{\|}{C}}OCH_2CH_2CH_3} + \underset{\text{Triethylammonium chloride}}{H\overset{+}{N}(CH_2CH_3)_3Cl^-}$$

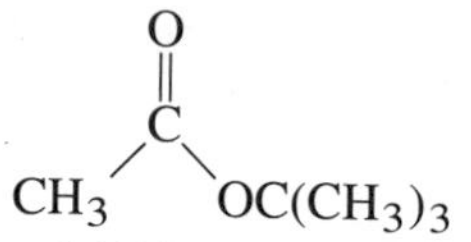

$$\mathrm{CH_3\overset{O}{\overset{\|}{C}}OC(CH_3)_3}$$

1,1-Dimethylethyl acetate (*tert*-Butyl acetate)

EXERCISE 20-4

You have learned that 2-methyl-2-propanol (*tert*-butyl alcohol) dehydrates in the presence of acid (Section 9-2). Suggest a synthesis of 1,1-dimethylethyl acetate (*tert*-butyl acetate, shown in the margin) from acetic acid. Avoid conditions that might dehydrate the alcohol.

Amines convert alkanoyl chlorides into amides

Secondary and primary amines, as well as ammonia, convert alkanoyl chlorides into amides. Again, the hydrogen chloride formed is neutralized by added base (which can be excess amine).

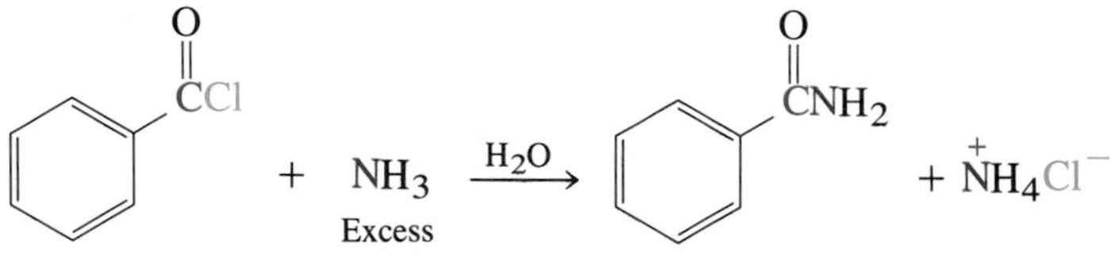

$$\mathrm{C_6H_5\overset{O}{\overset{\|}{C}}Cl} + \underset{\text{Excess}}{\mathrm{NH_3}} \xrightarrow{H_2O} \mathrm{C_6H_5\overset{O}{\overset{\|}{C}}NH_2} + \mathrm{\overset{+}{N}H_4Cl^-}$$

Benzoyl chloride → 86% Benzamide

$$\mathrm{CH_2{=}CH\overset{O}{\overset{\|}{C}}Cl} + 2\,\mathrm{CH_3NH_2} \xrightarrow{\text{Benzene, } 5^\circ\text{C}} \mathrm{CH_2{=}CH\overset{O}{\overset{\|}{C}}NHCH_3} + \mathrm{CH_3\overset{+}{N}H_3Cl^-}$$

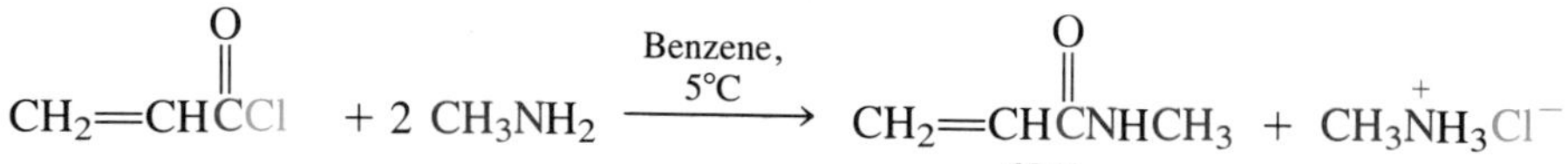

Propenoyl chloride → 68% *N*-Methylpropenamide

The mechanism of this transformation is, again, addition–elimination, beginning with attack of the nucleophilic amine nitrogen at the carbonyl carbon.

Mechanism of Amide Formation from Alkanoyl Chlorides

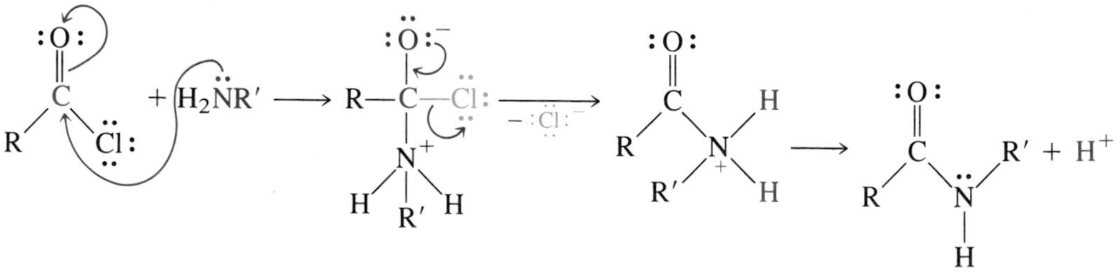

$$\mathrm{R\overset{:O:}{\overset{\|}{C}}\ddot{C}l:} + \mathrm{H_2\ddot{N}R'} \longrightarrow \mathrm{R{-}\underset{H-\overset{+}{N}(R')-H}{\overset{:\ddot{O}:^-}{C}}{-}\ddot{C}l:} \xrightarrow{-:\ddot{C}l:^-} \mathrm{R\overset{:O:}{\overset{\|}{C}}\overset{+}{N}(R')H_2} \longrightarrow \mathrm{R\overset{:O:}{\overset{\|}{C}}\ddot{N}(H)R'} + \mathrm{H^+}$$

Note that, in the last step, a proton must be lost from nitrogen to give the amide. Consequently, tertiary amines (which have no hydrogens on nitrogen) cannot form amides.

EXERCISE 20-5

Some amide preparations from alkanoyl halides require a primary or secondary amine that is too expensive to use also as the base to neutralize the hydrogen halide. Suggest a solution to this problem.

Organometallic reagents convert alkanoyl chlorides into ketones

Organometallic reagents attack the carbonyl group of alkanoyl chlorides to give the corresponding ketones. However, the latter are themselves prone to further attack by the relatively unselective organolithium and Grignard reagents to give alcohols (see Section 8-8). Ketone formation is best achieved using diorganocuprates (Section 13-10), which are more selective than RLi or RMgX and do not add to the product ketone.

$$2\ C_6H_{11}CCl(=O) + [(CH_3)_2C{=}CH]_2CuLi \longrightarrow 2\ C_6H_{11}C(=O)CH{=}C(CH_3)_2 + LiCl + CuCl$$

70%

Reduction of alkanoyl chlorides results in aldehydes

We can convert an alkanoyl chloride into an aldehyde by hydride reduction. In this transformation, we again face a selectivity problem: Sodium borohydride and lithium aluminum hydride convert aldehydes into alcohols. To prevent such overreduction, we must modify $LiAlH_4$ by letting it react first with three molecules of 2-methyl-2-propanol (*tert*-butyl alcohol; see Section 8-6). This treatment neutralizes three of the reactive hydride atoms, leaving one behind that is nucleophilic enough to attack an alkanoyl chloride but not the resulting aldehyde.

Reductions by Modified Lithium Aluminum Hydride

Preparation of reagent

$$LiAlH_4 + 3\ (CH_3)_3COH \longrightarrow LiAl[OC(CH_3)_3]_3H + 3\ H{-}H$$

Lithium tri(*tert*-butoxy)aluminum hydride

Reduction

$$R\overset{O}{\overset{\|}{C}}Cl + LiAl[OC(CH_3)_3]_3H \xrightarrow[2.\ H^+,\ H_2O]{1.\ \text{Ether solvent}} R\overset{O}{\overset{\|}{C}}H + LiCl + Al[OC(CH_3)_3]_3$$

$$CH_3CH{=}CH\overset{O}{\overset{\|}{C}}Cl \xrightarrow[2.\ H^+,\ H_2O]{1.\ LiAl[OC(CH_3)_3]_3H,\ (CH_3OCH_2CH_2)_2O,\ -78^\circ C} CH_3CH{=}CH\overset{O}{\overset{\|}{C}}H$$

48%

2-Butenoyl chloride → **2-Butenal**

EXERCISE 20-6

Prepare the following compounds from butanoyl chloride.

(a) $CH_3CH_2CH_2\overset{O}{\overset{\|}{C}}OH$ (b) 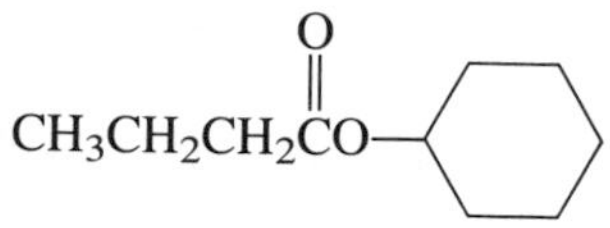 $CH_3CH_2CH_2\overset{O}{\overset{\|}{C}}O$–cyclohexyl

(c) $CH_3CH_2CH_2\overset{O}{\overset{\|}{C}}N(CH_3)_2$ (d) $CH_3CH_2CH_2\overset{O}{\overset{\|}{C}}CH_2CH_3$

(e) $CH_3CH_2CH_2\overset{O}{\overset{\|}{C}}H$

In summary, alkanoyl chlorides are attacked by a variety of nucleophiles, the reactions leading to new carboxylic acid derivatives, ketones, and aldehydes by addition–elimination mechanisms. Aldehydes can also be made from alkanoyl chlorides by reduction. The reactivity of alkanoyl halides makes them useful synthetic relay points on the way to other carbonyl derivatives.

20-3 Chemistry of Carboxylic Anhydrides

Carboxylic anhydrides, $R\overset{O}{\overset{\|}{C}}O\overset{O}{\overset{\|}{C}}R$, are named by simply adding the term *anhydride* to the acid name (or names, in the cases of mixed anhydrides). This method also applies to cyclic derivatives.

$CH_3\overset{O}{\overset{\|}{C}}O\overset{O}{\overset{\|}{C}}CH_3$
Acetic anhydride

$CH_3\overset{O}{\overset{\|}{C}}O\overset{O}{\overset{\|}{C}}CH_2CH_3$
Acetic propanoic anhydride
(A mixed anhydride)

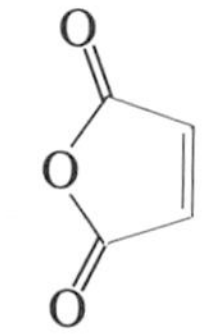

2-Butenedioic anhydride (Maleic anhydride)

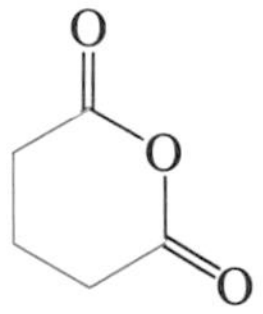

Pentanedioic anhydride (Glutaric anhydride)

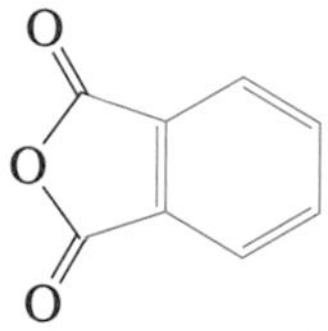

1,2-Benzenedicarboxylic anhydride (Phthalic anhydride)

The reactions of carboxylic anhydrides with nucleophiles, although less vigorous, are completely analogous to those of the alkanoyl halides. The leaving group is a carboxylate instead of a halide ion.

Nucleophilic Addition–Elimination of Anhydrides

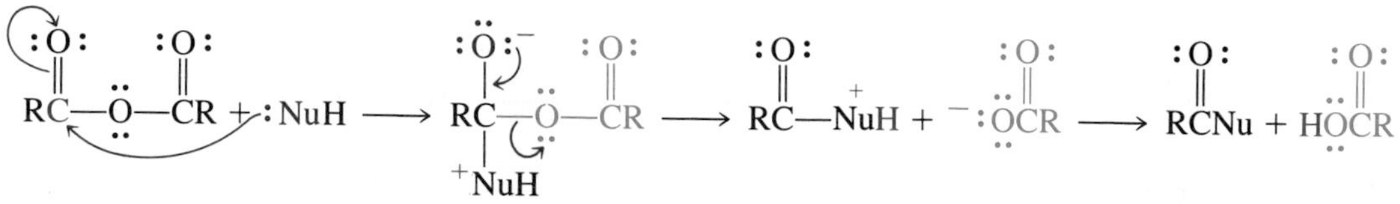

$$R\overset{:O:}{\overset{\|}{C}}-\ddot{O}-\overset{:O:}{\overset{\|}{C}}R + :NuH \longrightarrow R\overset{:\ddot{O}:^-}{\overset{|}{\underset{^+NuH}{\underset{|}{C}}}}-\ddot{O}-\overset{:O:}{\overset{\|}{C}}R \longrightarrow R\overset{:O:}{\overset{\|}{C}}-\overset{+}{Nu}H + {}^-:\ddot{O}\overset{:O:}{\overset{\|}{C}}R \longrightarrow R\overset{:O:}{\overset{\|}{C}}Nu + H\ddot{O}\overset{:O:}{\overset{\|}{C}}R$$

$$\mathrm{CH_3\overset{O}{\overset{\|}{C}}O\overset{O}{\overset{\|}{C}}CH_3 \xrightarrow{HOH} CH_3\overset{O}{\overset{\|}{C}}OH + HO\overset{O}{\overset{\|}{C}}CH_3}$$

100%

Acetic anhydride **Acetic acid**

$$\mathrm{CH_3CH_2\overset{O}{\overset{\|}{C}}O\overset{O}{\overset{\|}{C}}CH_2CH_3 \xrightarrow{CH_3OH} CH_3CH_2\overset{O}{\overset{\|}{C}}OCH_3 + HO\overset{O}{\overset{\|}{C}}CH_2CH_3}$$

83%

Propanoic anhydride **Methyl propanoate** **Propanoic acid**

In every addition–elimination reaction except hydrolysis, the carboxylic acid side product is usually undesired and is removed by work-up with basic water. Cyclic anhydrides undergo similar nucleophilic addition–elimination reactions that lead to ring opening.

Nucleophilic Ring Opening of Cyclic Anhydrides

$$\xrightarrow{CH_3OH,\ 100°C} \mathrm{HO\overset{O}{\overset{\|}{C}}CH_2CH_2\overset{O}{\overset{\|}{C}}OCH_3}$$

96%

Butanedioic (succinic) anhydride

EXERCISE 20-7

Treatment of butanedioic (succinic) anhydride with ammonia at elevated temperatures leads to a compound $C_4H_5NO_2$. What is its structure?

EXERCISE 20-8

Formulate the mechanism for the reaction of acetic anhydride with methanol in the presence of sulfuric acid.

In summary, anhydrides react with nucleophiles in the same way as alkanoyl halides do, except that the leaving group is a carboxylate ion. Cyclic anhydrides furnish dicarboxylic acid derivatives.

20-4 Names and Properties of Esters

As mentioned in Section 19-9, esters, $\mathrm{R\overset{O}{\overset{\|}{C}}OR'}$, constitute the most important class of carboxylic acid derivatives. They are particularly widespread in nature. Many have characteristically pleasant odors and are important contributors to natural and artificial fruit flavors.

Esters are alkyl alkanoates

Esters are named as alkyl alkanoates. The ester grouping, $-\overset{\text{O}}{\overset{\|}{\text{C}}}\text{OR}$, as a substituent is called **alkoxycarbonyl.** A cyclic ester is called a **lactone** (the common name, Section 19-9); a systematic name would be **oxa-2-cycloalkanone** (Section 25-1). Depending on ring size, its name is preceded by α, β, γ, δ, and so on.

$CH_3\overset{O}{\overset{\|}{C}}OCH_3$
Methyl acetate

$CH_3CH_2\overset{O}{\overset{\|}{C}}OCH_2CH_3$
Ethyl propanoate
(Ethyl propionate)

$CH_3\overset{O}{\overset{\|}{C}}OCH_2CH_2\overset{CH_3}{\overset{|}{C}}HCH_3$
3-Methylbutyl acetate
(Isopentyl acetate)
(A component of banana flavor)

$CH_3CH_2\overset{O}{\overset{\|}{C}}OCH_2\overset{CH_3}{\overset{|}{C}}HCH_3$
2-Methylpropyl propanoate
(Isobutyl propionate)
(A component of rum flavor)

$CH_3CH_2CH_2CH_2\overset{O}{\overset{\|}{C}}OCH_2CH_2\overset{CH_3}{\overset{|}{C}}HCH_3$
3-Methylbutyl pentanoate
(Isopentyl valerate)
(A component of apple flavor)

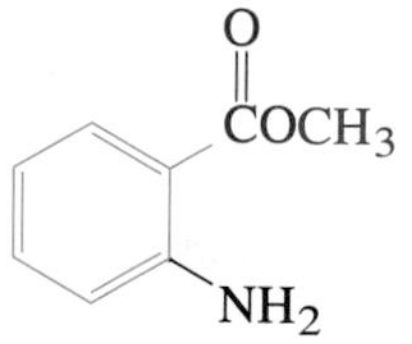

$\overset{O}{\overset{\|}{C}}OCH_3$, NH_2
Methyl 2-aminobenzoate
(Methyl anthranilate)
(A component of grape flavor)

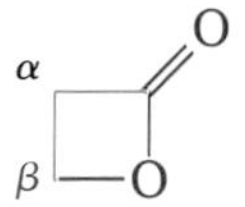

α, β, O, O
***β*-Propiolactone**

(This compound is a carcinogen and would be systematically called oxa-2-cyclobutanone; see Section 25-1)

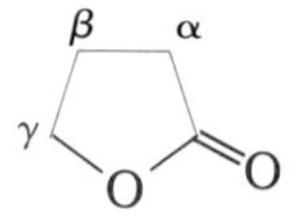

β, α, γ, O, O
***γ*-Butyrolactone**
(Better: oxa-2-cyclopentanone)

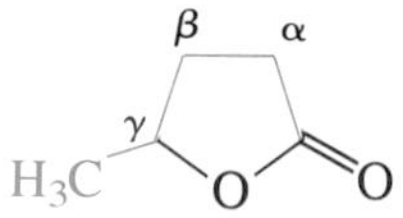

β, α, γ, H_3C, O, O
***γ*-Valerolactone**
(Better: 5-methyloxa-2-cyclopentanone)

EXERCISE 20-9

Name the following esters.

(a) $CH_3CH_2\overset{O}{\overset{\|}{C}}OCH_2CH_2CH_3$ **(b)** $CH_3O\overset{O}{\overset{\|}{C}}CH_2CH_2\overset{O}{\overset{\|}{C}}OCH_3$

(c) $CH_2{=}CHCO_2CH_3$

Lower esters, such as ethyl acetate (b.p. 77°C) and butyl acetate (b.p. 127°C), are used as solvents. For example, butyl butanoate has recently replaced the ozone-depleting trichloroethane as a cleaning solvent in the manufacture of electronic components such as computer chips. Higher nonvolatile esters are used as softeners (called plasticizers; see Section 12-14) for brittle polymers—in flexible tubing (e.g., Tygon tubing), rubber pipes, and upholstery.

Some waxes, oils, and fats are esters

Esters made up of long-chain carboxylic acids and long-chain alcohols are the main constituents of animal- and plant-derived **waxes.**

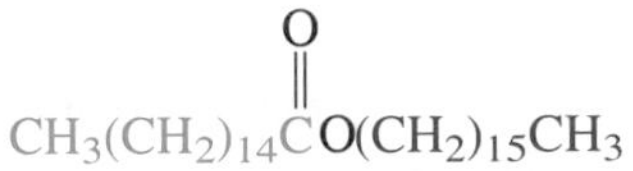

$$CH_3(CH_2)_{14}\overset{O}{\overset{\|}{C}}O(CH_2)_{15}CH_3$$

Hexadecyl hexadecanoate
(Cetyl palmitate)
(Wax from the sperm whale)

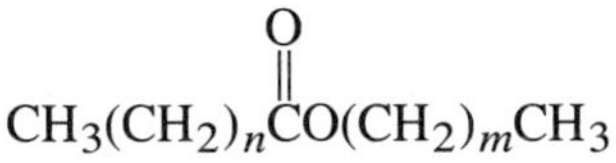

$$CH_3(CH_2)_n\overset{O}{\overset{\|}{C}}O(CH_2)_mCH_3$$

$n = 24, 26;\ m = 29, 31$

Beeswax

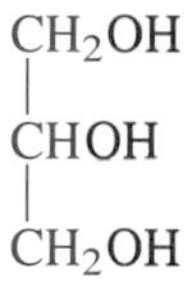

$$\begin{array}{c} CH_2OH \\ | \\ CHOH \\ | \\ CH_2OH \end{array}$$

1,2,3-Propanetriol
(Glycerol)

$$\begin{array}{l} CH_2O\overset{O}{\overset{\|}{C}}R \\ | \\ HCO\overset{O}{\overset{\|}{C}}R' \\ | \\ CH_2O\overset{O}{\overset{\|}{C}}R'' \end{array}$$

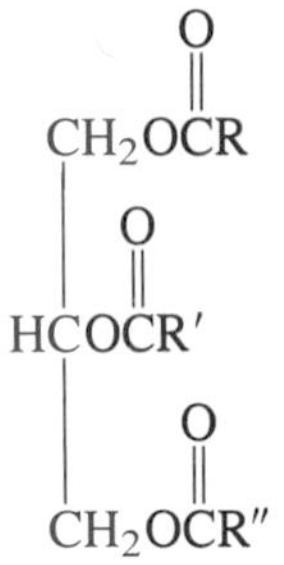

1,2,3-Propanetriol triester
(Triglyceride)

Triesters of 1,2,3-propanetriol (glycerol) constitute the **oils** and **fats** found in plants and animals (see Section 19-14). They are also called **triglycerides.** The carboxylic acid parts of these esters are **fatty acids.**

There is no essential chemical difference between fats and oils; fats just happen to be solids at room temperature. However, oils are usually more unsaturated. They may be converted into solid fats by catalytic hydrogenation. A variety of cooking fats and margarines are produced in this way. Saturated fats have been implicated as a dietary factor in atherosclerosis (hardening of the arteries). Thus, for considerations of health, vegetable oils, which are highly unsaturated, have become increasingly popular. Biologically, fats are used as a source of energy, their metabolism leading ultimately to CO_2 and water.

Waxes and fats are constituents of **lipids,** which are defined as water-insoluble biomolecules highly soluble in organic solvents such as chloroform. They serve as molecular ''fuel'' and energy stores, as well as components of cell membranes. An important class of membrane lipids are the **phospholipids,** which are di- and triesters derived from carboxylic acids and phosphoric acid. In the **phosphoglycerides,** glycerol is esterified by two adjacent fatty acids and a phosphate unit, which bears another ester substituent, such as that derived from choline, $[HOCH_2CH_2N(CH_3)_3]^+HO^-$.

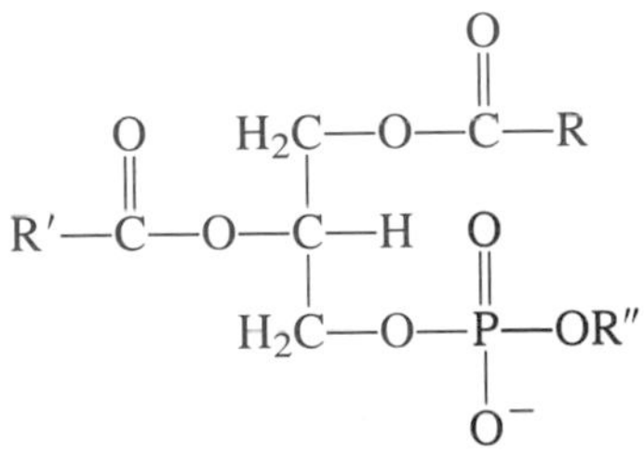

$$\begin{array}{l} \qquad\qquad\quad H_2C{-}O{-}\overset{O}{\overset{\|}{C}}{-}R \\ \qquad\qquad\qquad | \\ R'{-}\overset{O}{\overset{\|}{C}}{-}O{-}C{-}H \\ \qquad\qquad\qquad | \\ \qquad\qquad\quad H_2C{-}O{-}\overset{O}{\overset{\|}{P}}{-}OR'' \\ \qquad\qquad\qquad\qquad\quad | \\ \qquad\qquad\qquad\qquad\quad O^- \end{array}$$

A phosphoglyceride

Hexadecanoic (palmitic) acid unit

$$\begin{array}{l} \qquad\qquad\qquad\qquad\qquad\qquad\qquad CH_2O\overset{O}{\overset{\|}{C}}(CH_2)_{14}CH_3 \\ \qquad\qquad\qquad\qquad\qquad\qquad\qquad | \\ CH_3(CH_2)_7CH\overset{cis}{=}CH(CH_2)_7\overset{O}{\overset{\|}{C}}OCH \\ \qquad\qquad\qquad\qquad\qquad\qquad\qquad | \\ \qquad\qquad\qquad\qquad\qquad\qquad\qquad CH_2O\overset{O}{\overset{\|}{P}}O(CH_2)_2\overset{+}{N}(CH_3)_3 \\ \qquad\qquad\qquad\qquad\qquad\qquad\qquad\qquad\quad | \\ \qquad\qquad\qquad\qquad\qquad\qquad\qquad\qquad\quad O^- \end{array}$$

cis-9-Octadecenoic (oleic) acid unit

Choline unit

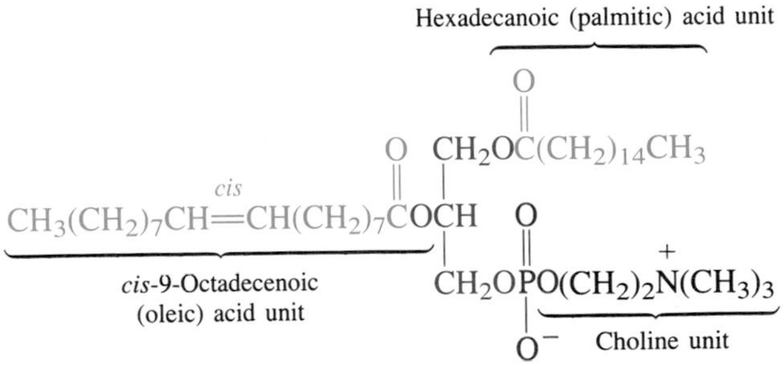

Palmitoyloleoylphosphatidyl choline

Because these molecules carry two long hydrophobic fatty acid chains and a polar head group (the phosphate and choline substituent), they are capable of forming micelles in aqueous solution (see Section 19-4). In the micelles, the phosphate unit is solubilized by water, and the ester chains are clustered inside the hydrophobic micellar sphere (Figure 20-2A).

Phosphoglycerides can also aggregate in a different way: They may form a unimolecular sheet called a **lipid bilayer** (Figure 20-2B). This capability is significant because, whereas micelles usually are limited in size (<200 Å in diameter), bilayers may be as much as 1 mm (10^7 Å) in length. This property makes them ideal constituents of cell membranes, which act as permeability barriers regulating molecular transport into and out of the cell. Lipid bilayers are relatively stable molecular assemblies. The forces that drive their formation are similar to those at work in micelles: London interactions between the hydrophobic alkane chains, and coulombic and solvation forces among the polar head groups between each other and water.

In summary, esters are named as alkyl alkanoates. Many of them have a pleasant odor and occur in nature as fragrant agents, waxes, oils, and fats.

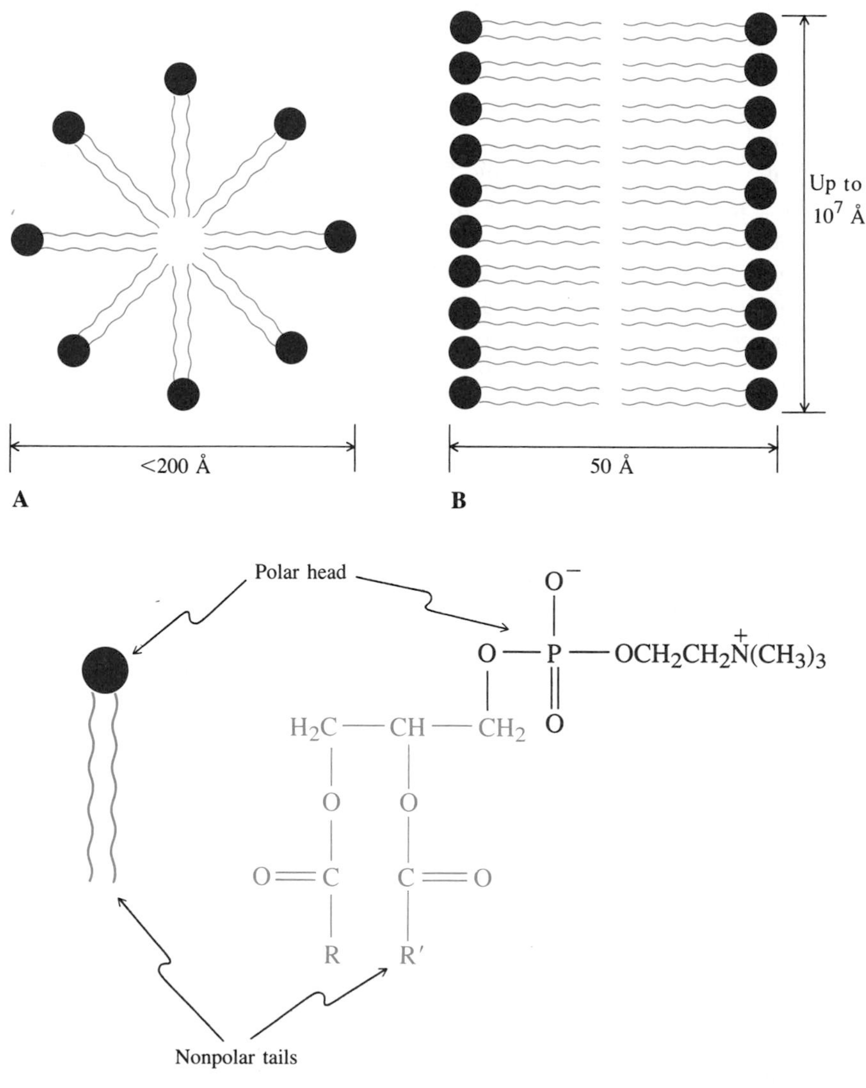

FIGURE 20-2 Phospholipids are substituted esters essential to the structures of cell membranes. These molecules aggregate to form (A) a micelle or (B) a lipid bilayer. (C) The polar head groups and nonpolar tails in phospholipids drive these aggregations. (After *Biochemistry,* 3d ed., by Lubert Stryer. W. H. Freeman and Company. Copyright © 1975, 1981, 1988.)

20-5 Chemistry of Esters

Esters display extensive carbonyl chemistry but reduced reactivity relative to that of either alkanoyl halides or carboxylic anhydrides. This section discusses the transformations that esters undergo with a variety of nucleophilic reagents.

Esters hydrolyze to carboxylic acids

When esters are heated in the presence of excess water and strong acids, they hydrolyze to carboxylic acids and alcohols. The mechanism of this reaction is the reverse of acid-catalyzed esterification (Section 19-9).

EXERCISE 20-10

Formulate a mechanism for the acid-catalyzed hydrolysis of γ-butyrolactone.

Strong *bases* also promote ester hydrolysis through an addition–elimination mechanism (Section 19-7). The base (B) converts the poor nucleophile water into the negatively charged and more highly nucleophilic hydroxide ion.

$$B:^- + H—OH \longrightarrow {}^-:OH + B—H$$

Ester hydrolysis is frequently achieved by using hydroxide itself, in at least stoichiometric amounts, as the base.

Mechanism of Base-Mediated Ester Hydrolysis

$$RCOCH_3 \text{ (C=O)} + {}^-:\ddot{O}H \rightleftharpoons R—C(:\ddot{O}:^-)(:\ddot{O}H)—\ddot{O}CH_3 \rightleftharpoons RC(=O)\ddot{O}—H + {}^-:\ddot{O}CH_3 \longrightarrow RC(=O)\ddot{O}:^- + H\ddot{O}CH_3$$

Carboxylate ion

Unlike acid-catalyzed hydrolysis, which is reversible, the base-mediated process (saponification; see Section 19-14) is driven essentially to completion by the last step, in which the stoichiometric base converts the acid into a carboxylate ion. The carboxylic acid product is obtained by acidic aqueous work-up.

$$CH_3CH(CH_3)CH_2COCH_3 \text{ (C=O)} \xrightarrow[2.\ H^+,\ H_2O]{1.\ KOH,\ H_2O,\ CH_3OH,\ \Delta} CH_3CH(CH_3)CH_2COH \text{ (C=O)} + CH_3OH$$

Methyl 3-methylbutanoate — **3-Methylbutanoic acid** (100%)

EXERCISE 20-11

Formulate a mechanism for the base-mediated hydrolysis of γ-butyrolactone.

Transesterification takes place with alcohols

Esters react with alcohols in an acid- or base-catalyzed transformation called **transesterification.** It allows for the direct conversion of one ester into another

Transesterification

$$\mathrm{R\overset{O}{\overset{\|}{C}}OR' + R''OH \underset{}{\overset{H^+ \text{ or } ^-OR''}{\rightleftharpoons}} R\overset{O}{\overset{\|}{C}}OR'' + R'OH}$$

without proceeding through the free acid. Transesterification is a reversible reaction. To shift the equilibrium, a large excess of the alcohol is usually employed, sometimes in the form of solvent.

$$\mathrm{C_{17}H_{35}\overset{O}{\overset{\|}{C}}OCH_2CH_3 + CH_3OH \xrightarrow{H^+ \text{ or } ^-OCH_3} \underset{90\%}{C_{17}H_{35}\overset{O}{\overset{\|}{C}}OCH_3} + CH_3CH_2OH}$$

Ethyl octadecanoate **Solvent** **Methyl octadecanoate**

Lactones are opened to hydroxy esters by transesterification.

$$\gamma\text{-Butyrolactone} + \mathrm{BrCH_2CH_2CH_2OH} \xrightarrow{H^+} \mathrm{HOCH_2CH_2CH_2\overset{O}{\overset{\|}{C}}OCH_2CH_2CH_2Br}\ (80\%)$$

γ-Butyrolactone **3-Bromopropanol** **3-Bromopropyl 4-hydroxybutanoate**

The mechanisms of transesterification by acid and base are straightforward extensions of the mechanisms of the corresponding hydrolyses to the carboxylic acids. Thus, acid-catalyzed transesterification begins with protonation of the carbonyl oxygen, followed by nucleophilic attack by alcohol on the carbonyl carbon. In contrast, in base the alcohol is first deprotonated, and the resulting alkoxide ion then adds to the ester carbonyl group.

EXERCISE 20-12

Formulate mechanisms for the acid- and base-catalyzed transesterifications of γ-butyrolactone by 3-bromopropanol.

Amines convert esters into amides

Amines, which are more nucleophilic than alcohols, readily transform esters into amides. No catalyst is needed, but heating is required.

Amide Formation from Methyl Esters

$$\mathrm{R\overset{O}{\overset{\|}{C}}OCH_3 + R'\overset{H}{N}H \xrightarrow{\Delta} R\overset{O}{\overset{\|}{C}}NHR' + CH_3OH}$$

$$\mathrm{CH_3(CH_2)_7CH{=}CH(CH_2)_7\overset{O}{\overset{\|}{C}}OCH_3 + CH_3(CH_2)_{11}\overset{H}{N}H \xrightarrow{230°C}}$$

Methyl 9-octadecenoate **1-Dodecanamine**

$$\mathrm{\underset{69\%}{CH_3(CH_2)_7CH{=}CH(CH_2)_7\overset{O}{\overset{\|}{C}}NH(CH_2)_{11}CH_3} + CH_3OH}$$

***N*-Dodecyl-9-octadecenamide**

Grignard reagents transform esters into alcohols

Esters can be converted into alcohols by using *two* equivalents of a Grignard reagent. In this way, ordinary esters are transformed into tertiary alcohols, whereas formic esters furnish secondary alcohols.

Alcohols from Esters and Grignard Reagents

$$\underset{\text{Ethyl propanoate}}{CH_3CH_2\overset{O}{\overset{\|}{C}}OCH_2CH_3} + \underset{\text{Propylmagnesium bromide}}{2\ CH_3CH_2CH_2MgBr} \xrightarrow[-\ CH_3CH_2OH]{1.\ (CH_3CH_2)_2O \quad 2.\ H^+,\ H_2O} \underset{\substack{69\% \\ \text{4-Ethyl-4-heptanol}}}{CH_3CH_2\overset{OH}{\overset{|}{\underset{CH_2CH_2CH_3}{\underset{|}{C}}}}CH_2CH_2CH_3}$$

$$\underset{\text{Methyl formate}}{H\overset{O}{\overset{\|}{C}}OCH_3} + \underset{\text{Butylmagnesium bromide}}{2\ CH_3CH_2CH_2CH_2MgBr} \xrightarrow[-\ CH_3OH]{1.\ (CH_3CH_2)_2O \quad 2.\ H^+,\ H_2O} \underset{\substack{85\% \\ \text{5-Nonanol}}}{H\overset{OH}{\overset{|}{\underset{CH_2CH_2CH_2CH_3}{\underset{|}{C}}}}CH_2CH_2CH_2CH_3}$$

The reaction probably begins with addition of the organometallic to the carbonyl function in the usual manner to give the magnesium salt of a hemiacetal (Section 17-7). At room temperature, rapid elimination results in the formation of an intermediate ketone (or aldehyde, from formates). The resulting carbonyl group then immediately adds a second equivalent of Grignard reagent. Subsequent acidic aqueous work-up leads to the observed alcohol.

Mechanism of the Alcohol Synthesis from Esters and Grignard Reagents

$$R\overset{O}{\overset{\|}{C}}OCH_3 + R'MgBr \longrightarrow R{-}\overset{O-MgBr}{\overset{|}{\underset{R'}{\underset{|}{C}}}}{-}OCH_3 \xrightarrow[-\ CH_3OMgBr]{}$$

$$\underset{\substack{\text{(Cannot be stopped} \\ \text{at this stage)}}}{R\overset{O}{\overset{\|}{C}}R'} \xrightarrow{R'MgBr} R{-}\overset{OMgBr}{\overset{|}{\underset{R'}{\underset{|}{C}}}}{-}R' \xrightarrow{H^+,\ H_2O} R{-}\overset{OH}{\overset{|}{\underset{R'}{\underset{|}{C}}}}{-}R'$$

EXERCISE 20-13

Propose a synthesis of triphenylmethanol, $(C_6H_5)_3COH$, beginning with methyl benzoate (shown in the margin) and bromobenzene.

$$C_6H_5\overset{O}{\overset{\|}{C}}OCH_3$$

Esters are reduced by hydride reagents to give alcohols or aldehydes

The reduction of esters to alcohols by $LiAlH_4$ requires 0.5 equivalent of the hydride, because only two hydrogens are needed per ester function.

Reduction of an Ester to an Alcohol

$$\text{(pyrrolidin-1-yl)}\underset{\underset{CH_3}{|}}{N}CH\overset{\overset{O}{\|}}{C}OCH_2CH_3 \xrightarrow[-\,CH_3CH_2OH]{1.\ LiAlH_4\ (0.5\ \text{equivalent}),\ (CH_3CH_2)_2O;\ 2.\ H^+,\ H_2O} \text{(pyrrolidin-1-yl)}\underset{\underset{CH_3}{|}}{N}CHCH_2OH\quad 90\%$$

A milder reducing agent allows the reaction to be stopped at the aldehyde oxidation stage. Such a reagent is bis(2-methylpropyl)aluminum hydride (diisobutylaluminum hydride), when used at low temperatures in toluene.

Reduction of an Ester to an Aldehyde

$$CH_3\overset{\overset{H_3C}{|}}{C}H\overset{\overset{O}{\|}}{C}OCH_2CH_3 + (CH_3\overset{\overset{CH_3}{|}}{C}HCH_2)_2AlH \xrightarrow[-\,CH_3CH_2OH]{1.\ \text{Toluene},\ -60°C;\ 2.\ H^+,\ H_2O} CH_3\overset{\overset{CH_3}{|}}{C}HCHO$$

Ethyl 2-Methylpropanoate + **Bis(2-methylpropyl)aluminum hydride (Diisobutylaluminum hydride)** → **2-Methylpropanal**

Esters form enolates that can be alkylated

The acidity of the α-hydrogens in esters is sufficiently high that **ester enolates** are formed by treatment of esters with strong base at low temperatures. Ester enolates react like ketone enolates, undergoing alkylations.

Alkylation of an Ester Enolate

$$\underset{pK_a \sim 25}{CH_3CO_2CH_2CH_3} \xrightarrow{\text{LDA, THF, } -78°C} CH_2{=}C(O^-Li^+)(OCH_2CH_3) \xrightarrow[-\,LiBr]{CH_2{=}CHCH_2Br,\ \text{HMPA}} CH_2{=}CHCH_2CH_2\overset{\overset{O}{\|}}{C}OCH_2CH_3\quad 97\%$$

Ethyl acetate enolate ion → **Ethyl 4-pentenoate**

The pK_a of esters is about 25. Consequently, ester enolates exhibit the typical side reactions of strong bases: E2 processes (especially with secondary, tertiary, and branched halides) and deprotonations. The most characteristic reaction of ester enolates is the Claisen condensation, in which the enolate attacks the carbonyl carbon of another ester. This process will be discussed in Chapter 23.

EXERCISE 20-14

Give the products of the reaction of ethyl cyclohexanecarboxylate with the following compounds or under the following conditions (and followed by acidic aqueous work-up, if necessary). **(a)** H^+, H_2O; **(b)** HO^-, H_2O; **(c)** CH_3O^-, CH_3OH; **(d)** NH_3, Δ; **(e)** 2 CH_3MgBr; **(f)** $LiAlH_4$; **(g)** 1. LDA, 2. CH_3I.

In summary, with acidic or basic water, esters hydrolyze to the corresponding carboxylic acids or carboxylates; with alcohols, they undergo transesterification; and with amines at elevated temperatures, they furnish amides. Grignard reagents add twice to give tertiary alcohols (or secondary alcohols, from formates). Lithium aluminum hydride reduces esters all the way to the alcohols, whereas bis(2-methylpropyl)aluminum (diisobutylaluminum) hydride allows the process to be stopped at the aldehyde stage. With LDA, it is possible to form ester enolates, which can be alkylated by electrophiles.

20-6 Amides: The Least Reactive Carboxylic Acid Derivatives

Among all carboxylic acid derivatives, the amides, $R\overset{O}{\overset{\|}{C}}NR'_2$, are the least susceptible to nucleophilic attack. After a brief introduction to amide naming, this section describes their reactions.

Amides are named alkanamides, cyclic amides are lactams

Amides are called **alkanamides,** the ending *-e* of the alkane stem having been replaced by **-amide.** In common names, the ending *-ic* of the acid name is replaced by the **-amide** suffix. In cyclic systems the ending *-carboxylic acid* is replaced by **-carboxamide.** Substituents on the nitrogen are indicated by the prefix *N*-, or *N,N*-, depending on the number of groups. Accordingly, there are primary, secondary, and tertiary amides.

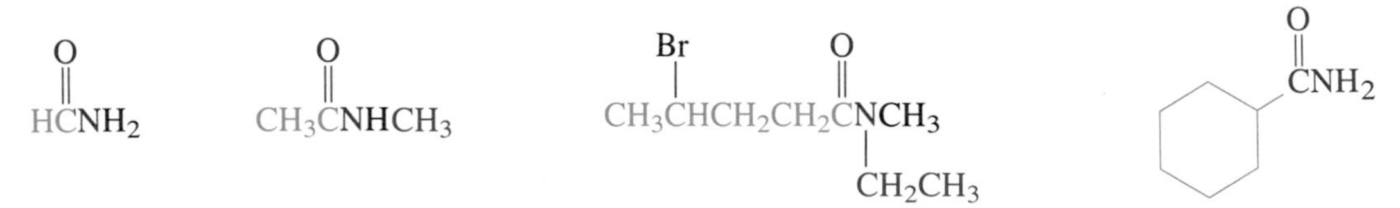

$H\overset{O}{\overset{\|}{C}}NH_2$ **Formamide** (A primary amide)

$CH_3\overset{O}{\overset{\|}{C}}NHCH_3$ ***N*-Methylacetamide** (A secondary amide)

$CH_3\overset{Br}{\overset{|}{C}}HCH_2CH_2\overset{O}{\overset{\|}{C}}N(CH_2CH_3)CH_3$ **4-Bromo-*N*-ethyl-*N*-methylpentanamide** (A tertiary amide)

$C_6H_{11}\overset{O}{\overset{\|}{C}}NH_2$ **Cyclohexanecarboxamide**

There are several amide derivatives of carbonic acid, H_2CO_3: ureas, carbamic acids, and carbamic esters (urethanes).

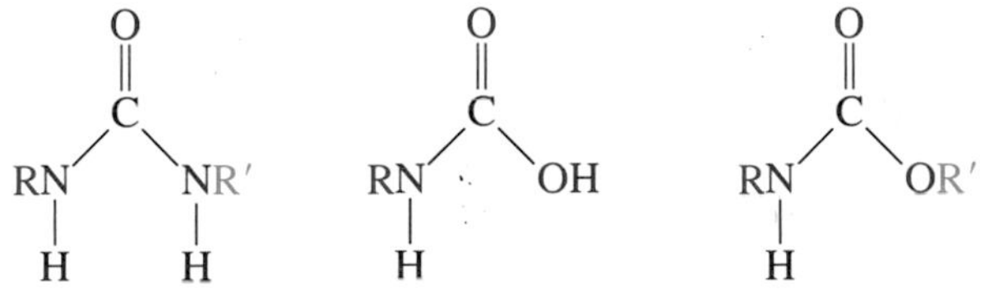

$RNH-\overset{O}{\overset{\|}{C}}-NHR'$ **A urea**

$RNH-\overset{O}{\overset{\|}{C}}-OH$ **A carbamic acid**

$RNH-\overset{O}{\overset{\|}{C}}-OR'$ **A carbamic ester (Urethane)**

Cyclic amides are called **lactams** (Section 19-10)—a systematic name would be aza-2-cycloalkanones (Section 25-1)—and the rules for naming them follow those used for lactones. The penicillins are annulated β-lactams.

BOX 20-2 Penicillin: An Antibiotic Containing a β-Lactam Ring

The discovery of penicillin as a powerful broad-spectrum antibiotic was one of the milestones in medicinal chemistry. As with many such advances, serendipity played a major role. In 1928 the British bacteriologist Alexander Fleming* noted that several *Staphylococcus* cultures set aside on a laboratory bench had been contaminated by microorganisms from the laboratory air. A green mold, *Penicillium notatum,* had grown in some places, and the *Staphylococcus* in its vicinity was disintegrating. The substance causing this antibiotic activity was called penicillin.

Penicillin became available in pure form only about 10 years later. Many different penicillins have subsequently been synthesized with different R groups. *Penicillin G* has a phenylmethyl (benzyl, $C_6H_5CH_2$) group attached to the amide function; *ampicillin* has a phenylaminomethyl ($C_6H_5CHNH_2$) substituent. Structurally and biologically related are the *cephalosporins,* important antibiotics that are frequently active when the penicillins are not.

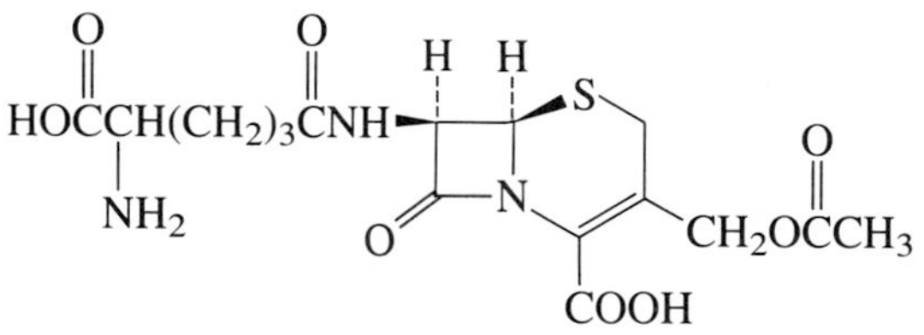

Cephalosporin C

The strained β-lactam ring is responsible for the antibiotic activity of these drugs. Because ring strain is relieved on opening, β-lactams are unusually reactive compared with ordinary amides. The enzyme *transpeptidase,* which catalyzes a crucial cross-linking reaction in the biosynthesis of bacterial cell walls, accepts penicillin as a substrate. The penicillin then alkanoylates a nucleophilic oxygen of the enzyme, rendering it inactive. Cell wall construction stops, and the organism soon dies. The reaction is the reverse of amide formation from esters (Section 20-5).

Penicillin in Action

Transpeptidase + penicillin

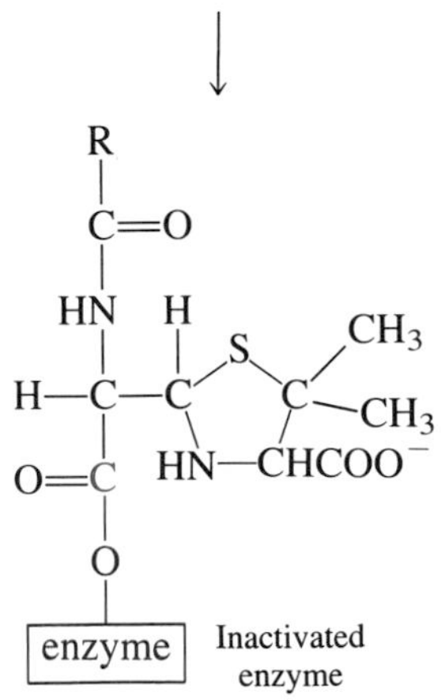

Inactivated enzyme

Some bacteria are resistant to penicillin because they produce an enzyme, *penicillinase,* that hydrolyzes the β-lactam ring before it can attach itself to transpeptidase. The rate of this hydrolysis depends on the structure of the β-lactam. Cephalosporins are not affected by penicillinase. Nevertheless, the continual emergence of new, antibiotic-resistant bacterial strains, resulting from the frequently indiscriminate prescription of penicillin and other antibiotics, has spurred intensive ongoing efforts to discover novel, more active, and more selective systems.

*Sir Alexander Fleming, 1881–1955, Royal College of Surgeons, England.

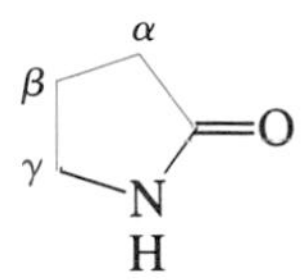

γ-Butyrolactam (Systematic name: aza-2-cyclopentanone)

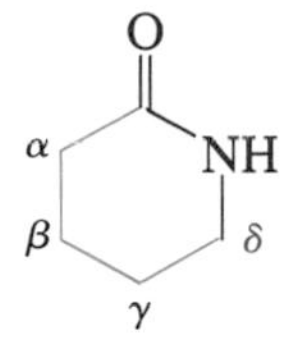

δ-Valerolactam (Systematic name: aza-2-cyclohexanone)

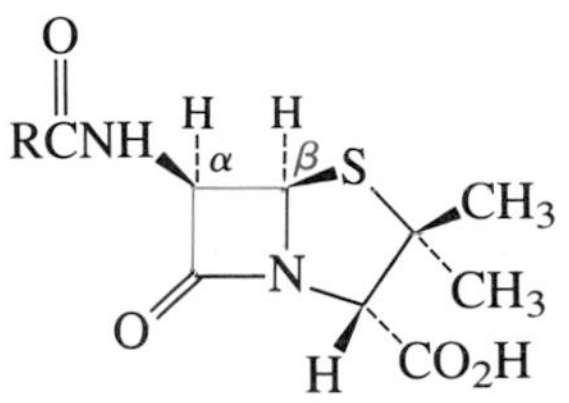

Penicillin (A β lactam derivative)

Amide hydrolysis requires strong heating in concentrated acid or base

The amides are the least reactive of the carboxylic acid derivatives, mainly because of the extra resonance capability of the nitrogen lone electron pair. As a consequence, their nucleophilic addition–eliminations require relatively harsh conditions. For example, hydrolysis occurs only on prolonged heating in strongly acidic or basic water. Acid hydrolysis liberates the amine in the form of the corresponding ammonium salt.

Acid Hydrolysis of an Amide

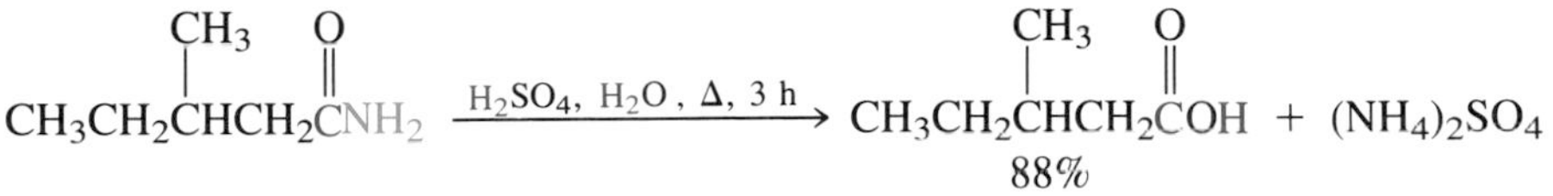

$$CH_3CH_2CH(CH_3)CH_2CONH_2 \xrightarrow{H_2SO_4,\ H_2O,\ \Delta,\ 3\ h} CH_3CH_2CH(CH_3)CH_2COOH + (NH_4)_2SO_4$$

3-Methylpentanamide → **3-Methylpentanoic acid** (88%)

Mechanism of Hydrolysis of Amides by Aqueous Acid

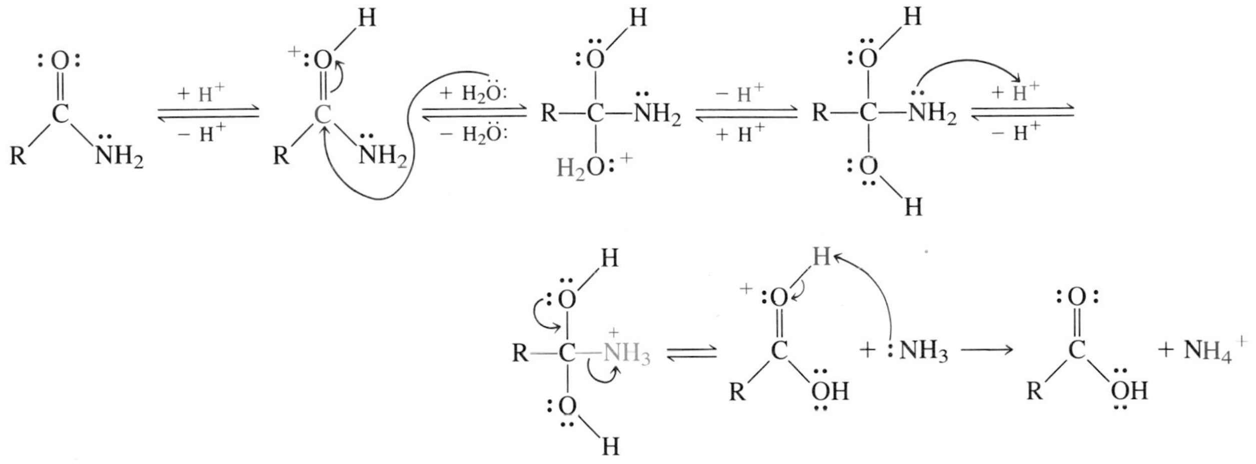

$$RC(=O)NH_2 \underset{-H^+}{\overset{+H^+}{\rightleftharpoons}} RC(=OH^+)NH_2 \underset{-H_2O}{\overset{+H_2O}{\rightleftharpoons}} R\text{—}C(OH)(NH_2)(OH_2^+) \underset{+H^+}{\overset{-H^+}{\rightleftharpoons}} R\text{—}C(OH)_2NH_2 \underset{-H^+}{\overset{+H^+}{\rightleftharpoons}}$$

$$R\text{—}C(OH)_2NH_3^+ \rightleftharpoons RC(=OH^+)OH + :NH_3 \longrightarrow RC(=O)OH + NH_4^+$$

Base hydrolysis initially furnishes the carboxylate salt and the amine. Acidic aqueous work-up then produces the acid.

Base Hydrolysis of an Amide

$$CH_3CH_2CH_2\overset{O}{\overset{\|}{C}}\ddot{N}HCH_3 \xrightarrow{H\ddot{O}:^-,\ H_2O,\ \Delta} CH_3CH_2CH_2\overset{O}{\overset{\|}{C}}\ddot{O}:^- + CH_3\ddot{N}H_2 \xrightarrow{H^+,\ H_2O} \underset{87\%}{CH_3CH_2CH_2\overset{O}{\overset{\|}{C}}\ddot{O}H} + CH_3\overset{+}{N}H_3$$

N-Methylbutanamide — **Butanoic acid**

Mechanism of Hydrolysis of Amides by Aqueous Base

$$RC(=\ddot{O})\ddot{N}H_2 \underset{}{\overset{^-:\ddot{O}H}{\rightleftharpoons}} R{-}C(:\ddot{O}:^-)(:\ddot{O}H){-}\ddot{N}H_2 \rightleftharpoons RC(=\ddot{O})\ddot{O}H + :\ddot{N}H_2^- \longrightarrow RC(=\ddot{O})\ddot{O}:^- + H\ddot{N}H_2$$

Amides can be reduced to amines or aldehydes

In contrast with carboxylic acids and esters, the reaction of amides with lithium aluminum hydride produces amines instead of alcohols.

Reduction of an Amide to an Amine

$$(CH_3)_2CHCH_2CH_2\overset{O}{\overset{\|}{C}}N(CH_2CH_3)_2 \xrightarrow[2.\ H^+,\ H_2O]{1.\ LiAlH_4,\ (CH_3CH_2)_2O} \underset{85\%}{(CH_3)_2CHCH_2CH_2CH_2N(CH_2CH_3)_2}$$

***N,N*-Diethyl-4-methylpentanamide** — ***N,N*-Diethyl-4-methylpentanamine**

H_3C, H_3C, N, H, O (5,5-dimethyl-2-pyrrolidinone structure in margin)

EXERCISE 20-15

What product would you expect from $LiAlH_4$ reduction of the compound depicted in the margin?

The mechanism of reduction begins with hydride addition, which gives a tetrahedral intermediate. Elimination of an aluminum alkoxide leads to an **iminium ion.** Addition of a second hydride gives the final amine product.

Mechanism of Amide Reduction by Hydride

$$R\overset{:O:}{\overset{\|}{C}}\ddot{N}R'_2 \xrightarrow{LiAlH_4} R{-}\underset{H}{\underset{|}{C}}(\ddot{O}{-}Al{-}){-}\ddot{N}R'_2 \xrightarrow{-\,-Al\ddot{O}:^-} \underset{\textbf{Iminium ion}}{R\overset{\overset{+}{N}R'_2}{\overset{\|}{C}}H} \xrightarrow{LiAlH_4} R\underset{H}{\underset{|}{\overset{\ddot{N}R'_2}{\overset{|}{C}}}}H$$

Reduction of amides by bis(2-methylpropyl)aluminum hydride (diisobutylaluminum hydride) furnishes aldehydes. Recall that esters are also converted to aldehydes by this reagent (Section 20-5).

Reduction of an Amide to an Aldehyde

$$CH_3(CH_2)_3\overset{O}{\overset{\|}{C}}N(CH_3)_2 \xrightarrow[2.\ H^+,\ H_2O]{1.\ (CH_3\overset{CH_3}{\overset{|}{C}}HCH_2)_2AlH,\ (CH_3CH_2)_2O} CH_3(CH_2)_3CHO$$

N,N-Dimethylpentanamide — **Pentanal** 92%

EXERCISE 20-16

Treatment of amide A with $LiAlH_4$, followed by acidic aqueous work-up, gave B. Explain. (Hint: Review Sections 17-8 and 17-9.)

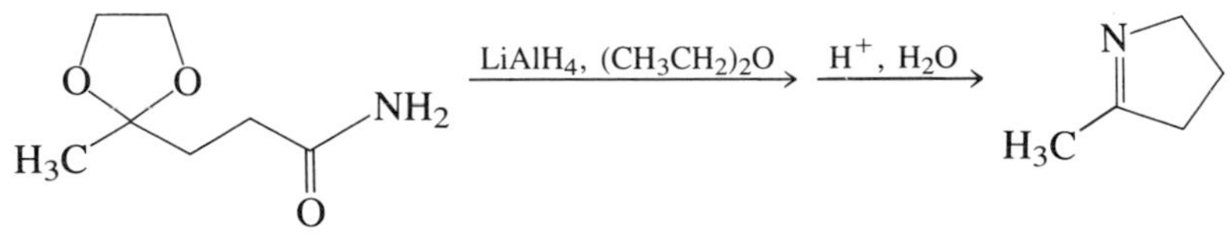

A (2-methyl-1,3-dioxolan-2-yl propanamide; H_3C, O, O, NH_2, O) $\xrightarrow{LiAlH_4,\ (CH_3CH_2)_2O}$ $\xrightarrow{H^+,\ H_2O}$ B (N, H_3C)

A **B**

In summary, carboxylic amides are named as alkanamides, or lactams if cyclic. They can be hydrolyzed to carboxylic acids by acid or base and reduced to amines by lithium aluminum hydride. Reduction by bis(2-methylpropyl)aluminum hydride (diisobutylaluminum hydride) stops at the aldehyde stage.

20-7 Amidates and Their Halogenation: The Hofmann Rearrangement

Hydrogens on both the carbon and nitrogen atoms next to the amide carbonyl group are acidic. Removal of the NH hydrogen, which has a pK_a of about 15, leads to an **amidate ion.** The CH proton is less acidic, with a pK_a of about 30 (Section 20-1); therefore, deprotonation of the α-carbon, leading to an **amide enolate,** is more difficult.

$$R\ddot{\bar{C}}H\overset{O}{\overset{\|}{C}}\ddot{N}H_2 + H^+ \not\rightleftharpoons RCH_2\overset{O}{\overset{\|}{C}}\ddot{N}H_2 \rightleftharpoons RCH_2\overset{O}{\overset{\|}{C}}\ddot{N}H^- + H^+$$

Amide enolate ion — p$K_a \sim 30$ (CH), p$K_a \sim 15$ (NH) — **Amidate ion**

Practically speaking, therefore, a proton may be removed from carbon only with tertiary amides, in which the nitrogen is blocked.

The amidate ion formed by deprotonation of a primary amide is a synthetically useful nucleophile. This section will focus on one of its reactions, the Hofmann rearrangement.

EXERCISE 20-17

The pK_a of 1,2-benzenedicarboxamide (phthalimide, A) is 8.3, considerably lower than the pK_a of benzamide (B). Why?

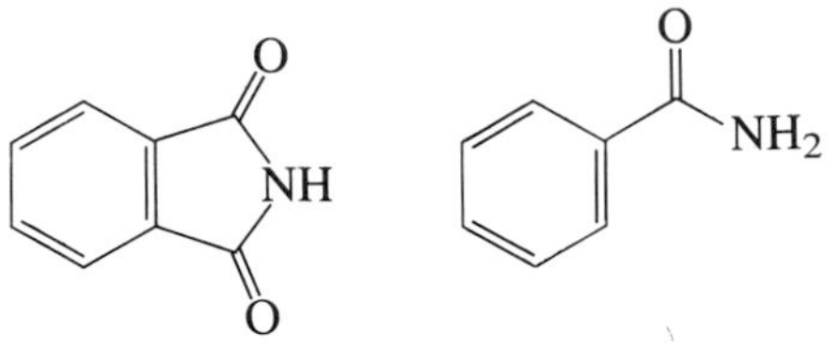

A B

In the presence of base, primary amides undergo a special halogenation reaction, the **Hofmann* rearrangement.** In it the carbonyl group is expelled from the molecule to give a primary amine with one carbon less in the chain.

Hofmann Rearrangement

$$\text{RCNH}_2 \xrightarrow{X_2,\ \text{NaOH},\ H_2O} \text{RNH}_2 + \text{O=C=O}$$

$$\underset{\textbf{Nonanamide}}{CH_3(CH_2)_6CH_2CONH_2} \xrightarrow{Cl_2,\ \text{NaOH}} \underset{\substack{66\% \\ \textbf{Octanamine}}}{CH_3(CH_2)_6CH_2NH_2} + \text{O=C=O}$$

The Hofmann rearrangement begins with deprotonation of nitrogen to form an amidate ion. Halogenation of the nitrogen follows, much like the α-halogenation of aldehyde and ketone enolates (Section 18-3). Subsequently, the second proton on the nitrogen is abstracted by additional base to give an *N*-haloamidate, which spontaneously eliminates halide. The species formed contains an uncharged nitrogen atom surrounded by only an electron sextet. Such intermediates, called **nitrenes,** are highly reactive and short lived. In the Hofmann rearrangement the acyl nitrene undergoes a 1,2-shift of an alkyl group to give an **isocyanate,** R–N=C=O, a nitrogen analog of carbon dioxide, O=C=O. The *sp*-hybridized carbonyl carbon in the isocyanate is highly electrophilic and is attacked by water to produce an unstable **carbamic acid.** Finally, the carbamic acid decomposes to carbon dioxide and the amine.

Mechanism of Hofmann Rearrangement

STEP 1. Amidate formation

$$\text{R}\overset{\overset{:\text{O}:}{\|}}{\text{C}}\ddot{\text{N}}\text{H}_2 + {}^{-}:\ddot{\underset{\cdot\cdot}{\text{O}}}\text{H} \rightleftharpoons \text{R}\overset{\overset{:\text{O}:}{\|}}{\text{C}}\ddot{\underset{\cdot\cdot}{\text{N}}}\text{H}^{-} + \text{H}\ddot{\underset{\cdot\cdot}{\text{O}}}\text{H}$$

*This is the Hofmann of the Hofmann rule of E2 reactions (Section 11-8).

STEP 2. Halogenation

$$\mathrm{RC(=O)NH^-} + :\!\ddot{X}\!-\!\ddot{X}\!: \longrightarrow \mathrm{RC(=O)NH(X)} + :\!\ddot{X}\!:^-$$

STEP 3. *N*-Haloamidate formation

$$\mathrm{RC(=O)NH(X)} + {}^-:\!\ddot{O}H \rightleftharpoons \mathrm{RC(=O)\ddot{N}^-\!-\!\ddot{X}}: + \mathrm{H\ddot{O}H}$$

An *N*-halo-amidate

STEP 4. Halide elimination

$$\mathrm{RC(=O)\ddot{N}^-\!-\!\ddot{X}}: \longrightarrow \mathrm{RC(=O)\ddot{N}} + :\!\ddot{X}\!:^-$$

An acyl nitrene

STEP 5. Rearrangement

$$\mathrm{R\!-\!C(=O)\!-\!\ddot{N}}: \longrightarrow \ddot{O}\!=\!C\!=\!\ddot{N}\!-\!R$$

An isocyanate

STEP 6. Hydrolysis to the carbamic acid and decomposition

$$\mathrm{R\!-\!\ddot{N}\!=\!C\!=\!\ddot{O}} \xrightarrow{H_2O} \mathrm{R(H)N\!-\!C(=O)\!-\!\ddot{O}H} \longrightarrow \mathrm{R\ddot{N}H_2} + \mathrm{CO_2}$$

A carbamic acid

EXERCISE 20-18

Write a detailed mechanism for the addition of water to an isocyanate under basic conditions and for the decarboxylation of the resulting carbamic acid.

EXERCISE 20-19

Suggest a sequence by which you could convert ester A into amine B.

$$\underset{\mathbf{A}}{\mathrm{CH_3CH(COOCH_3)CH_2CH_2CH_3}} \dashrightarrow \underset{\mathbf{B}}{\mathrm{CH_3CH(NH_2)CH_2CH_2CH_3}}$$

In summary, treatment of primary and secondary amides with base leads to deprotonation at nitrogen, giving amidate ions. Bases abstract protons from the α-carbon of tertiary amides. In the Hofmann rearrangement, amides react with halogens in base to furnish amines with one less carbon. The reaction involves an alkyl shift, which converts an acyl nitrene into an isocyanate.

BOX 20-3 Methyl Isocyanate and the Bhopal Tragedy

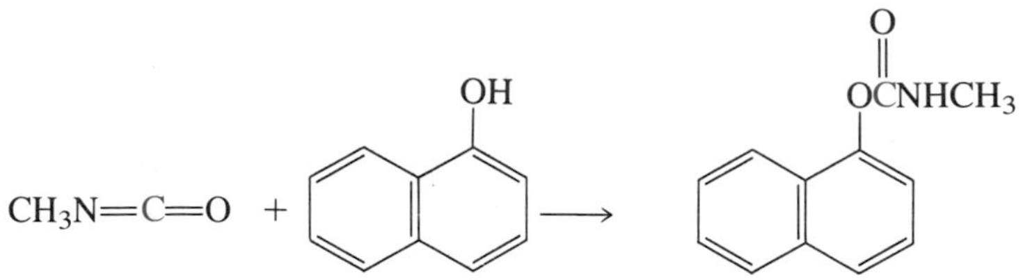

$$CH_3N{=}C{=}O \; + \; \text{1-Naphthalenol} \longrightarrow \text{1-Naphthyl } N\text{-methylcarbamate}$$

Methyl isocyanate **1-Naphthalenol (1-Naphthol)** **1-Naphthyl *N*-methylcarbamate (Sevin, an insecticide)**

The reaction of methyl isocyanate with various alcohols and amines is used in the industrial preparation of several powerful herbicides and insecticides.

Consumption of methyl isocyanate in the United States has been estimated at 30 to 35 million pounds per year. In late 1984 in the city of Bhopal, India, a massive leak of this substance, used in the preparation of the insecticide Sevin, resulted in the deaths of more than 2000 people; at least 300,000 more were exposed to it. This catastrophe, the worst chemical industrial accident in history, led to a complete reappraisal of the safety measures for the handling of large quantities of toxic chemicals.

The toxicity of the isocyanate function derives from its rapid reaction with nucleophilic sites in biological molecules. Indiscriminate attack on the hydroxy, amino, and thiol groups in, for example, peptides and proteins, inactivates them with respect to their biological functions. Other substances, which would be similarly affected by such attack, include small molecules involved in transmission of nerve impulses and various aspects of cell regulation.

20-8 Alkanenitriles: A Special Class of Carboxylic Acid Derivatives

$CH_3CH(CH_3)CH_2C{\equiv}N$
3-Methylbutanenitrile

$CH_3CH_2C{\equiv}N$
Propanenitrile
(Propionitrile)

$CH_3C{\equiv}N$
Acetonitrile

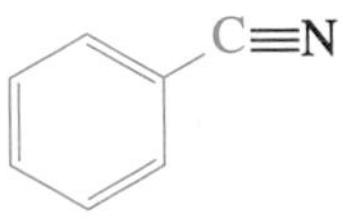

$C_6H_5C{\equiv}N$
Benzonitrile

Nitriles, RC≡N, are considered derivatives of carboxylic acids because the nitrile carbon is in the same oxidation state as the carboxy carbon and because nitriles are readily interconverted with other carboxylic acid derivatives. This section describes the rules for naming nitriles, the structure and bonding in the nitrile group, and some of its spectral characteristics. Then it compares the chemistry of the nitrile group with that of other carboxylic acid derivatives.

In IUPAC nomenclature, nitriles are named from alkanes

A systematic way of naming this class of compounds is as **alkanenitriles.** In common names the ending *-ic acid* of the carboxylic acid is usually replaced with **-nitrile.** The chain is numbered like those of carboxylic acids. Similar rules apply to dinitriles derived from dicarboxylic acids. The substituent CN is called **cyano.** Cyanocycloalkanes are labeled cycloalkane**carbonitriles.**

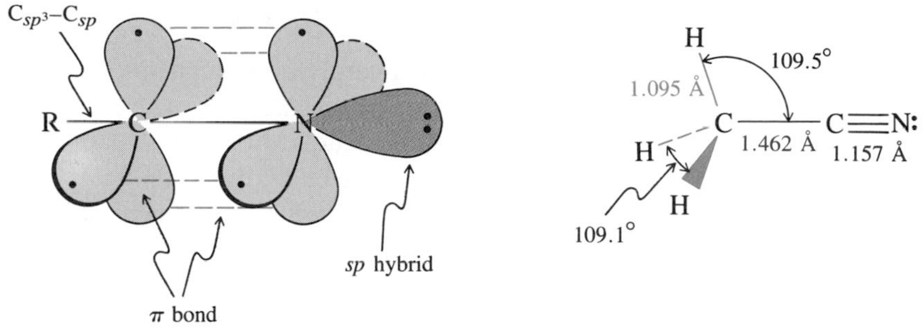

A B

FIGURE 20-3 (A) Molecular-orbital picture of the nitrile group, showing the *sp* hybridization of both atoms in the C≡N function. (B) Molecular structure of acetonitrile, which is similar to that of the corresponding alkyne.

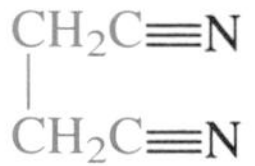

Butanedinitrile (Succinonitrile) — $CH_2C{\equiv}N$ / $CH_2C{\equiv}N$

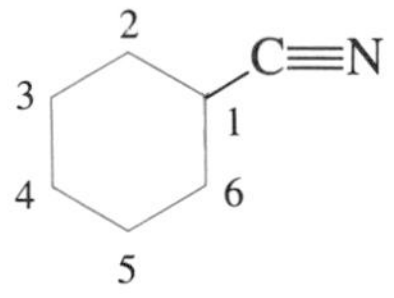

Cyclohexanecarbonitrile

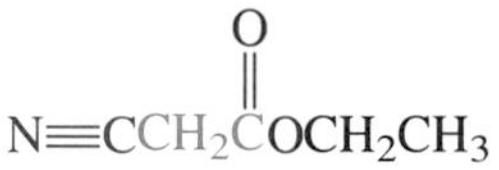

Ethyl cyanoacetate — $N{\equiv}CCH_2\overset{O}{\overset{\|}{C}}OCH_2CH_3$

The C≡N bond in nitriles resembles the C≡C bond in alkynes

In the nitriles both atoms in the functional group are *sp* hybridized, and there is a lone electron pair on nitrogen occupying an *sp* hybrid orbital pointing away from the molecule along the C–N axis. The hybridization and structure of the nitrile functional group very much resemble those of the alkynes (Figure 20-3; see also Figures 13-1 and 13-2).

In the infrared spectrum, the C≡N stretching vibration appears at about 2250 cm^{-1}, in the same range as the C≡C absorption but much more intense. The ^1H NMR spectra of nitriles indicate that protons near the nitrile group are deshielded about as much as those in other carboxylic acid and alkyne derivatives (Table 20-3).

The ^{13}C NMR absorption for the nitrile carbon appears at lower field ($\delta \sim$ 112–126 ppm) than that of the alkynes ($\delta \sim$ 65–85 ppm), because nitrogen is more electronegative than carbon.

TABLE 20-3 ^1H NMR Chemical Shifts of Substituted Methanes CH_3X

X	δ_{CH_3} (ppm)
—H	0.23
—Cl	3.06
—OH	3.39
—CH(=O)	2.18
—COOH	2.08
—$CONH_2$	2.02
—C≡N	1.98
—C≡CH	1.80

EXERCISE 20-20

1,3-Dibromopropane was treated with sodium cyanide in dimethyl sulfoxide-d_6 and the mixture monitored by ^{13}C NMR. After a few minutes, four new intermediate peaks appeared, one of which was located well downfield from the others at δ = 117.6 ppm. Subsequently, another three peaks began growing at δ = 119.1, 22.6, and 17.6 ppm, at the expense of the signals of starting material and the intermediate. Explain.

Nitriles undergo hydrolysis to carboxylic acids

As mentioned in Section 19-6, nitriles can be hydrolyzed by aqueous acid or base to give the corresponding carboxylic acids. The mechanisms of these reactions proceed through the intermediate amide and include addition–elimination steps.

In the acid-catalyzed process, initial protonation on nitrogen facilitates nucleophilic attack by water. Loss of a proton from oxygen furnishes a neutral intermediate, which is a tautomer of an amide. A second protonation on the nitrogen occurs, to be followed again by deprotonation of oxygen and formation of the amide. Hydrolysis of the latter proceeds as described in Section 20-6.

Mechanism of the Acid-Catalyzed Hydrolysis of Nitriles

$$R{-}C{\equiv}N: \underset{-H^+}{\overset{+H^+}{\rightleftharpoons}} \left[R{-}C{\equiv}\overset{+}{N}{-}H \longleftrightarrow R{-}\overset{+}{C}{=}\ddot{N}{-}H\right] \underset{-H_2\ddot{O}}{\overset{+H_2\ddot{O}}{\rightleftharpoons}} RC(=\ddot{N}H)\overset{+}{\ddot{O}}H_2 \underset{+H^+}{\overset{-H^+}{\rightleftharpoons}}$$

$$\underset{\text{Tautomer of amide}}{RC(=\ddot{N}H)\ddot{O}H} \underset{-H^+}{\overset{+H^+}{\rightleftharpoons}} \left[RC(=\overset{+}{N}H_2)\ddot{O}H \longleftrightarrow RC(\ddot{N}H_2)=\overset{+}{\ddot{O}}H\right] \underset{+H^+}{\overset{-H^+}{\rightleftharpoons}} RC(\ddot{N}H_2)=\ddot{O}: \xrightarrow{H^+} \xrightarrow{H_2O} RCOOH + NH_4^+$$

In base-catalyzed nitrile hydrolysis, direct attack of hydroxide gives the anion of the amide tautomer, which protonates on nitrogen. The remaining proton on the oxygen is then removed by base, and a second *N*-protonation gives the amide. Hydrolysis is completed as described in Section 20-6.

Mechanism of the Base-Catalyzed Hydrolysis of Nitriles

$$R{-}C{\equiv}N: \overset{^-:\ddot{O}H}{\rightleftharpoons} RC(=\ddot{N}:^-)\ddot{O}H \overset{HOH}{\rightleftharpoons} RC(=\ddot{N}H)\ddot{O}{-}H + {}^-:\ddot{O}H \underset{+H\ddot{O}H}{\overset{-H\ddot{O}H}{\rightleftharpoons}}$$

$$\left[RC(=\ddot{N}H)\ddot{O}:^- \longleftrightarrow RC(\ddot{N}H^-)=\ddot{O}:\right] \overset{HOH}{\rightleftharpoons} RC(\ddot{N}H_2)=\ddot{O}: + {}^-:\ddot{O}H \longrightarrow \xrightarrow{H^+} R\overset{O}{\overset{\|}{C}}OH + NH_4^+$$

The conditions for the hydrolysis of nitriles are usually stringent, requiring concentrated acid or base at high temperatures.

$$\underset{\substack{\textbf{Hexanedinitrile}\\ \textbf{(Adiponitrile)}}}{N{\equiv}C(CH_2)_4C{\equiv}N} \xrightarrow{H^+,\ H_2O,\ 300°C} \underset{\substack{97\%\\ \textbf{Hexanedioic acid}\\ \textbf{(Adipic acid)}}}{HOOC(CH_2)_4COOH}$$

Organometallic reagents attack nitriles to give ketones

Strong nucleophiles, such as organometallic reagents, add to nitriles to give anionic imine salts. Work-up with acidic water gives the neutral imine, which is rapidly hydrolyzed to the ketone (Section 17-9).

Ketone Synthesis from Nitriles

$$R{-}C{\equiv}N + R'M \longrightarrow \underset{R\quad R'}{\overset{:\ddot{N}^{-}M^{+}}{\overset{\|}{C}}} \xrightarrow[-\,MOH]{H^+,\ HOH} \underset{R\quad R'}{\overset{\ddot{N}H}{\overset{\|}{C}}} \xrightarrow{H^+,\ H_2O} \underset{R\quad R'}{\overset{O}{\overset{\|}{C}}} + NH_4^{+}$$

(M = metal)

$$CH_3CN \xrightarrow[2.\ H^+,\ H_2O]{1.\ CH_3(CH_2)_3CH_2MgBr,\ THF} CH_3\overset{O}{\overset{\|}{C}}(CH_2)_4CH_3$$

44%

Acetonitrile → **2-Heptanone**

Reduction of nitriles by hydride reagents leads to aldehydes and amines

As in its reactions with esters and amides, bis(2-methylpropyl)aluminum (diisobutylaluminum) hydride adds to nitriles only once to give an imine derivative. Aqueous hydrolysis then produces aldehydes.

Aldehyde Synthesis from Nitriles

$$R{-}C{\equiv}N + R'_2AlH \longrightarrow R{-}C(H){=}N{-}AlR'_2 \xrightarrow{H^+,\ H_2O} \underset{R\quad H}{\overset{O}{\overset{\|}{C}}}$$

1-(1,3-benzodioxol-5-yl)cyclopropanecarbonitrile (–CN) $\xrightarrow[2.\ H^+,\ H_2O]{1.\ (CH_3CH(CH_3)CH_2)_2AlH}$ 1-(1,3-benzodioxol-5-yl)cyclopropanecarbaldehyde (–CHO)

85%

Treatment of nitriles with strong hydride reducing agents results in double hydride addition, giving the amine on aqueous work-up. The best reagent for this purpose is lithium aluminum hydride.

$$CH_3CH_2CH_2CN \xrightarrow[2.\ H^+,\ H_2O]{1.\ LiAlH_4} CH_3CH_2CH_2CH_2NH_2$$

86%

Butanenitrile → **Butanamine**

EXERCISE 20-21

The reduction of a nitrile by $LiAlH_4$ to give an amine adds four hydrogen atoms to the C–N triple bond: two from the reducing agent and two from the water in the aqueous work-up. Formulate a mechanism for this transformation.

Like the triple bond of alkynes (Section 13-7), the nitrile group is hydrogenated by catalytically activated hydrogen. The result is the same as that with reduction by lithium aluminum hydride—amine formation. All four hydrogens are from the hydrogen gas.

$$\underset{\textbf{Butanenitrile}}{CH_3CH_2CH_2C{\equiv}N} \xrightarrow{H_2,\ PtO_2,\ CH_3CH_2OH,\ CHCl_3} \underset{\textbf{Butanamine}}{\underset{96\%}{CH_3CH_2CH_2CH_2NH_2}}$$

EXERCISE 20-22

Show how you would prepare the following compounds from pentanenitrile.

(a) $CH_3(CH_2)_3COOH$ **(b)** $CH_3(CH_2)_3\overset{O}{\overset{\|}{C}}(CH_2)_3CH_3$

(c) $CH_3(CH_2)_3\overset{O}{\overset{\|}{C}}H$ **(d)** $CH_3(CH_2)_3CD_2ND_2$

In summary, nitriles are named as alkanenitriles. Both atoms making up the C–N unit are *sp* hybridized, the nitrogen bearing a lone pair in an *sp* hybrid orbital. The nitrile stretching vibration appears at 2250 cm^{-1}, the ^{13}C NMR absorption at about 120 ppm. Acid- or base-catalyzed hydrolysis of nitriles gives carboxylic acids, and organometallic reagents (RLi, RMgBr) add to give ketones after hydrolysis. With bis(2-methylpropyl)aluminum hydride (diisobutylaluminum hydride), addition and hydrolysis furnishes aldehydes, whereas $LiAlH_4$ or catalytically activated hydrogen converts the nitrile function to the amine.

20-9 Measuring the Molecular Weight of Organic Compounds: Mass Spectroscopy

This section introduces one last important physical technique used by organic chemists to characterize organic molecules: **mass spectroscopy,** which is employed to measure molecular weights. The section begins with a description of the apparatus used and the physical principles on which it is based. Organic molecules fragment under the conditions necessary for measuring molecular weights, giving rise to characteristic recorded patterns called **mass spectra.**

The mass spectrometer separates molecular ions by weight

Mass spectroscopy is not spectroscopy in the conventional sense, because no radiation is absorbed (Section 10-2). Instead, charged particles travel through a magnetic field, which deflects them from a linear path. Because lighter species are deflected more than heavier ones, ions of different mass can be separated.

The sample is introduced into an inlet chamber (Figure 20-4, upper right). It is vaporized and a small quantity is then allowed to leak into the source chamber of the spectrometer. Here the neutral molecules (M) pass through a beam of electrons, usually accelerated to 70 eV (nearly 1600 kcal mol^{-1}). Upon electron impact, the molecules eject an electron to form a radical cation, $M^{+\cdot}$, called the **molecular ion.** Most organic molecules undergo only a single ionization.

Ionization of a Molecule on Electron Impact

$$\underset{\textbf{Neutral molecule}}{M} + \underset{\textbf{Ionizing beam}}{e\ (70\ \text{eV})} \longrightarrow \underset{\textbf{Radical cation (Molecular ion)}}{M^{+\cdot}} + 2e$$

As a charged particle, the molecular ion can be accelerated by an electric field set up within the source chamber. The accelerated $M^{+\cdot}$ now enters a magnetic field, which deflects it into a circular path. (Neutral molecules are not accelerated; they remain in the source chamber, eventually to be pumped out.)

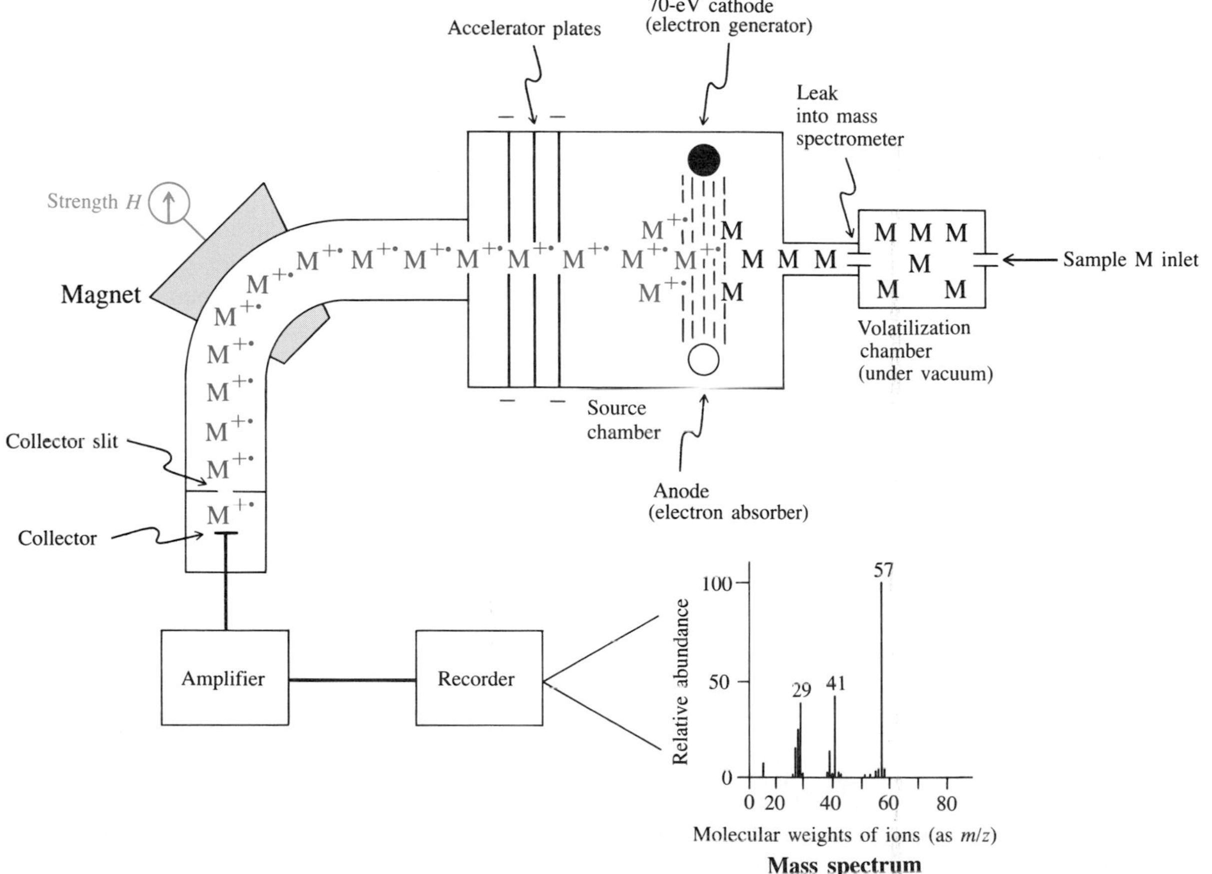

FIGURE 20-4 Schematic diagram of a mass spectrometer.

Molecular Weights of Organic Molecules

CH_4
$m/z = 16$

CH_3OH
$m/z = 32$

$CH_3\overset{\overset{\displaystyle O}{\|}}{C}OCH_3$
$m/z = 74$

The radius of the curved path depends on both the mass of the ion and the strength of the magnetic field, which can be varied in much the same way as that of an NMR spectrometer (Section 10-3). Thus, as the field strength changes, so does the path of the ions, *and only certain ions can pass through the collector slit*. Finally, the arrival of ions at the collector is translated electronically into a signal and recorded as a peak on a chart.

The mass spectrometer distinguishes between the different ions formed from organic molecules according to their ratios of mass to charge, m/z. If only singly charged species are considered, the m/z value equals the mass of the ion in question. Mass spectra are plotted as m/z values (on the abscissa) versus peak height (on the ordinate), the latter being a measure of the relative number of ions with this molecular weight.

EXERCISE 20-23

Three unknown compounds containing only C, H, and O gave rise to the following molecular weights. Draw as many reasonable structures as you can. **(a)** $m/z = 46$; **(b)** $m/z = 30$; **(c)** $m/z = 56$.

Molecular ions undergo fragmentation

Fragmentation of Methane in the Mass Spectrometer

$CH_4^{+\cdot}$
$-H_2$ → $CH_2^{+\cdot}$; $-H\cdot$ → CH_3^{+}
$CH_2^{+\cdot}$ $-H_2$ → $C^{+\cdot}$; CH_3^{+} $-H_2$ → CH^{+}

Odd-electron (radical) cations | Even-electron cations

Mass spectroscopy gives information not only about the molecular ion but also about its component structural units. Because the energy of the ionizing beam far exceeds that required to break typical organic bonds, some of the ionized molecules break apart into virtually all possible pairs of fragments. This **fragmentation** gives rise to a number of additional mass-spectral peaks, *all of lower mass* than the molecular ion (also called the **parent ion**) from which they are derived. The spectrum that results is called the **mass-spectral fragmentation pattern.** The most intense peak in the spectrum is called the **base peak.** Its intensity is defined to be 100.

For example, the mass spectrum of methane contains, in addition to the parent ion peak, lines for CH_3^{+}, $CH_2^{+\cdot}$, CH^{+}, and $C^{+\cdot}$ (Figure 20-5). These are formed by the processes shown in the margin. The relative abundance of these species, as indicated by the height of the peaks, gives a useful indication of the relative ease of their formation. It can be seen that the first C–H bond is cleaved

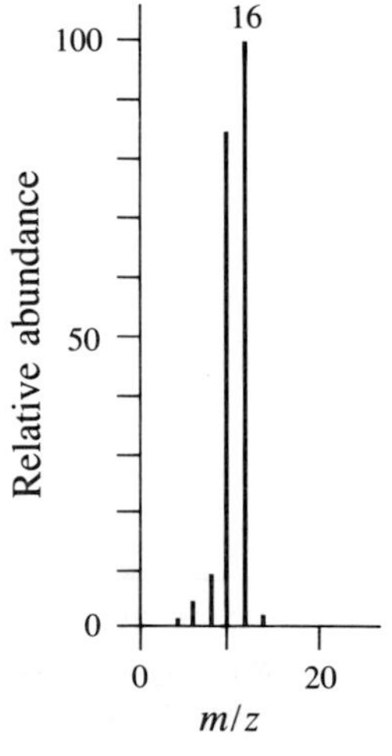

Tabulated Spectrum

m/z	Relative abundance (%)	Molecular or fragment ion
17	1.1	$(M + 1)^{+\cdot}$
16	100.0 (base peak)	$M^{+\cdot}$ (parent ion)
15	85.0	$(M - 1)^{+}$
14	9.2	$(M - 2)^{+\cdot}$
13	3.9	$(M - 3)^{+}$
12	1.0	$(M - 4)^{+\cdot}$

FIGURE 20-5 Mass spectrum of methane. At the left is the spectrum actually recorded; at the right is the tabulated form, the largest peak **(base peak)** being defined as 100%. For methane, the base peak at $m/z = 16$ is due to the parent ion. Fragmentation gives rise to peaks of lower mass.

readily, the $m/z = 15$ peak reaching 85% of the abundance of the parent ion. Breaking two, three, or four C–H bonds is more difficult, and the corresponding ions have lower relative abundance.

Mass spectra reveal the presence of isotopes

An unusual feature in the mass spectrum of methane is a small (1.1%) peak at $m/z = 17$; it is designated $(M + 1)^{+\cdot}$. How is it possible to have an ion present that has an extra mass unit? The answer lies in the fact that carbon is not isotopically pure. About 1.1% of natural carbon is the ^{13}C isotope (see Table 10-1), giving rise to the additional peak. In the mass spectrum of ethane, the height of the $(M + 1)^{+\cdot}$ peak, at $m/z = 31$, is about 2.2% that of the parent ion. The reason for this finding is statistical. The chance of finding a ^{13}C atom in a compound containing two carbons is double that expected of a one-carbon molecule. For a three-carbon moiety, it would be threefold, and so on. A mass spectrum of the eighteen-carbon steroid estrone (see Section 4-7) is shown in Figure 20-6. The height of the $(M + 1)^{+\cdot}$ peak is 18 times 1.1% of the height of $M^{+\cdot}$. There is a simple rule of thumb for organic compounds that lack isotopically impure heteroatoms: The $(M + 1)^{+\cdot}$ peak has a height (relative to $M^{+\cdot}$) of n times 1.1%, where n is the number of carbon atoms in the molecule.

Other elements, too, have naturally occurring higher isotopes: Hydrogen (deuterium, ^{2}H, about 0.015% abundance), nitrogen (0.366% ^{15}N), and oxygen (negligible ^{17}O, 0.204% ^{18}O) are examples. These isotopes also contribute to the intensity of peaks at masses higher than $M^{+\cdot}$.

Fluorine and iodine are isotopically pure. However, chlorine (75.53% ^{35}Cl; 24.47% ^{37}Cl) and bromine (50.54% ^{79}Br; 49.46% ^{81}Br) each exist as a mixture of two isotopes and give rise to readily identifiable isotopic patterns. For example, the mass spectrum of 1-bromopropane (Figure 20-7) shows two peaks of nearly equal intensity at $m/z = 122$ and 124, and only a very small peak at 123, which is the sum of the average atomic weights of the elements in the periodic table. Why? The true isotopic composition of the molecule is a nearly 1:1 mixture of $CH_3CH_2CH_2{}^{79}Br$ and $CH_3CH_2CH_2{}^{81}Br$. Similarly, the spectra of monochloroalkanes exhibit ions two mass units apart in a 3:1 intensity ratio, because of the presence of about 75% $R^{35}Cl$ and 25% $R^{37}Cl$. Peak patterns such as these are useful in revealing the presence of chlorine or bromine.

EXERCISE 20-24

What peak pattern do you expect for the molecular ion of dibromomethane?

EXERCISE 20-25

Nonradical compounds containing C, H, and O have even molecular weights, those containing C, H, O and an odd number of N atoms have odd molecular weights, but those with an even number of N atoms are even again. Explain.

In summary, molecules can be ionized by an electron beam at 70 eV to give radical cations that are accelerated by an electric field and then separated by the different deflections that they undergo in a magnetic field. In a mass spectrometer, this effect is used to measure the molecular weights of molecules. The molec-

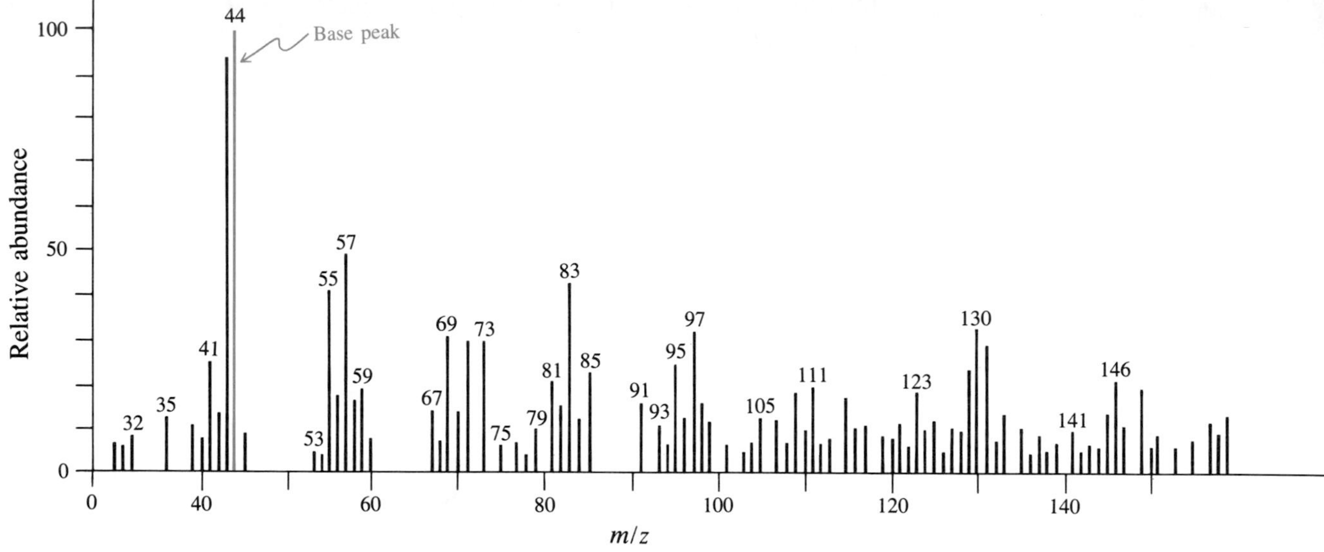

FIGURE 20-6 Mass spectrum (above and on the facing page) of the female sex hormone estrone. The molecule contains eighteen carbon atoms and thus the $(M + 1)^{+\cdot}$ peak height is predicted to be approximately 18 times 1.1% of the intensity of the $M^{+\cdot}$ peak, a value close to that observed ($18 \times 1.1\% \times 62\% = 12.3\%$). Note the extensive and complex fragmentation pattern. The base peak is found at $m/z = 44$.

ular ion is usually accompanied by less massive fragments and isotopic ''satellites'' due to the presence of less abundant isotopes. In some cases, such as with Cl and Br, more than one isotope may be present in substantial quantities.

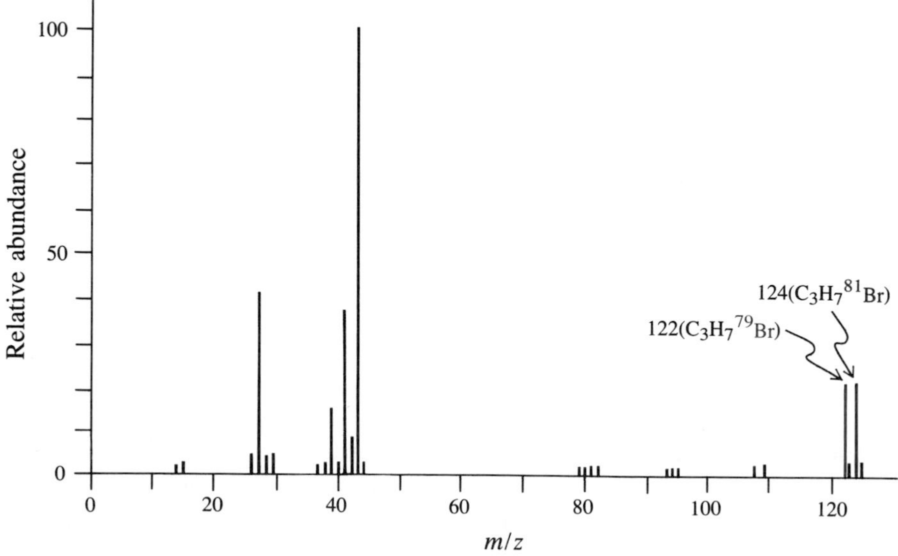

FIGURE 20-7 Mass spectrum of 1-bromopropane. Note the nearly equal heights of the peaks at $m/z = 122$ and 124, due to the almost equal abundances of the two bromine isotopes.

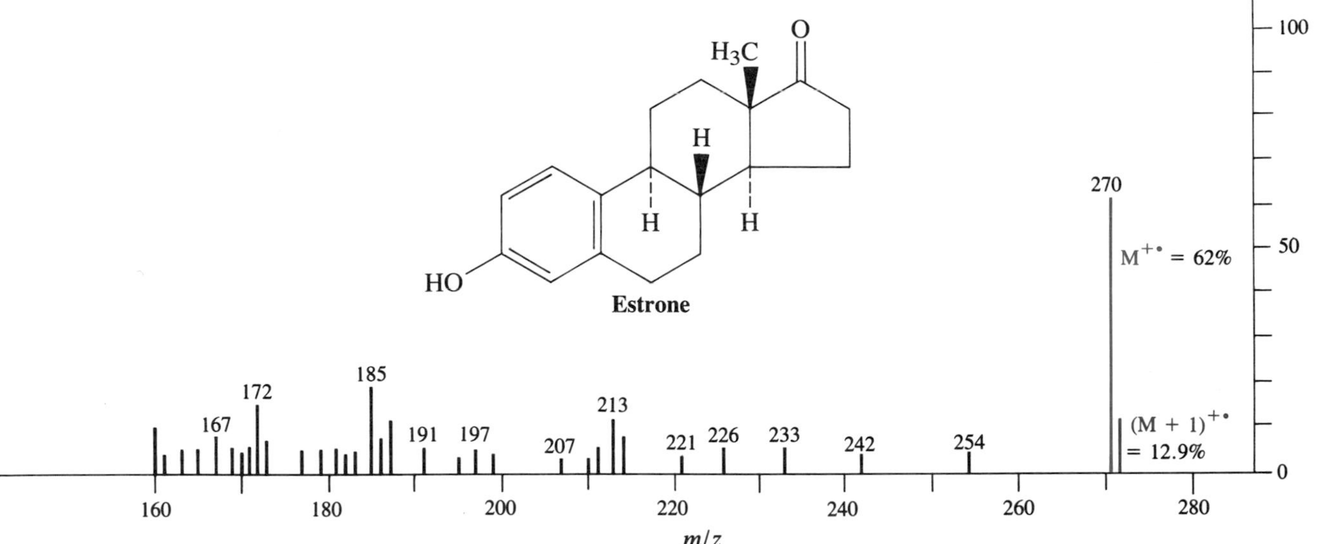

20-10 Fragmentation Patterns of Organic Molecules

On electron impact, molecules fragment at weaker bonds first and then at the stronger ones. The resulting fragments may themselves fall apart into smaller pieces. The detection of this fragmentation pattern by the mass spectrometer provides clues to the structure of the parent molecule.

Fragmentation is more likely at a highly substituted center

The relative ease of bonding dissociation can be seen in the mass spectra of the isomeric hydrocarbons pentane, 2-methylbutane, and 2,2-dimethylpropane (Figures 20-8, 20-9, and 20-10). In each case, the molecular ion ($m/z = 72$) produces a relatively small peak, but the spectra are otherwise very different for the three compounds. Pentane fragments by more or less indiscriminate C–C bond breaking (Figure 20-8; see margin). Thus, in addition to the molecular ion signal, there is a line at $m/z = 57$ $(M - CH_3)^+$ followed by peaks indicating progressive loss of CH_2 units: $m/z = 43$ $(M - CH_2CH_3)^+$, and $m/z = 29$ $(CH_3CH_2)^+$. These fragment peaks are surrounded by clusters of smaller lines because of the presence of ^{13}C $(M + 1)^{+\cdot}$ and the loss of hydrogens [$(M - 1)^+$, $(M - 2)^+$, etc.].

The mass spectrum of 2-methylbutane (Figure 20-9) shows a pattern similar to that of pentane; however, the relative intensities of the various peaks differ. Thus, there is a larger $(M - 1)^+$ peak at $m/z = 71$, and strong $(M - \text{alkyl})^+$ signals at $m/z = 57$ and 43, because of the relative stability of the cations produced by preferred fragmentation at the more highly substituted center at C2.

Fragment Ions from Pentane

CH_3^+ ($m/z = 15$)

$C_3H_7^+$ ($m/z = 43$)

$CH_3–CH_2–CH_2–CH_2–CH_3^{+\cdot}$ ($m/z = 72$)

$C_2H_5^+$ ($m/z = 29$)

$C_4H_9^+$ ($m/z = 57$)

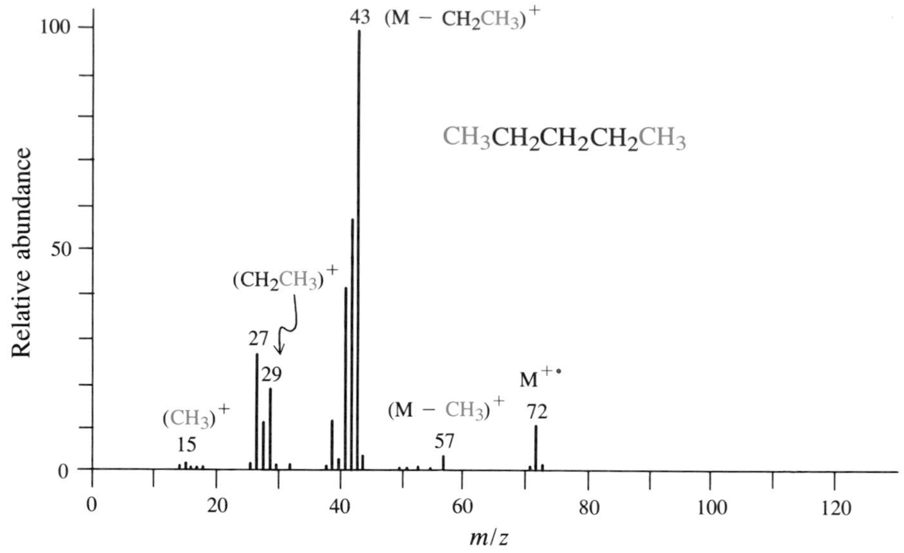

Relative abundance

100 50 0

43 $(M - CH_2CH_3)^+$

$CH_3CH_2CH_2CH_2CH_3$

$(CH_2CH_3)^+$

27 29

$M^{+\bullet}$

$(CH_3)^+$

15

$(M - CH_3)^+$ 57

72

0 20 40 60 80 100 120

m/z

FIGURE 20-8 Mass spectrum of pentane, revealing that all C–C bonds in the chain have been ruptured.

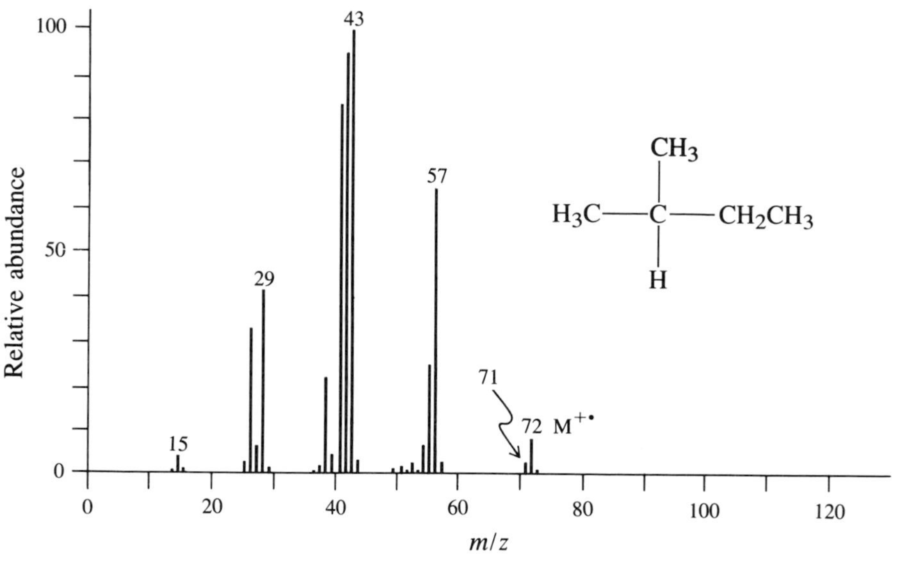

FIGURE 20-9 Mass spectrum of 2-methylbutane. The peaks at $m/z = 43$ and 57 result from preferred fragmentation around C2 to give secondary carbocations.

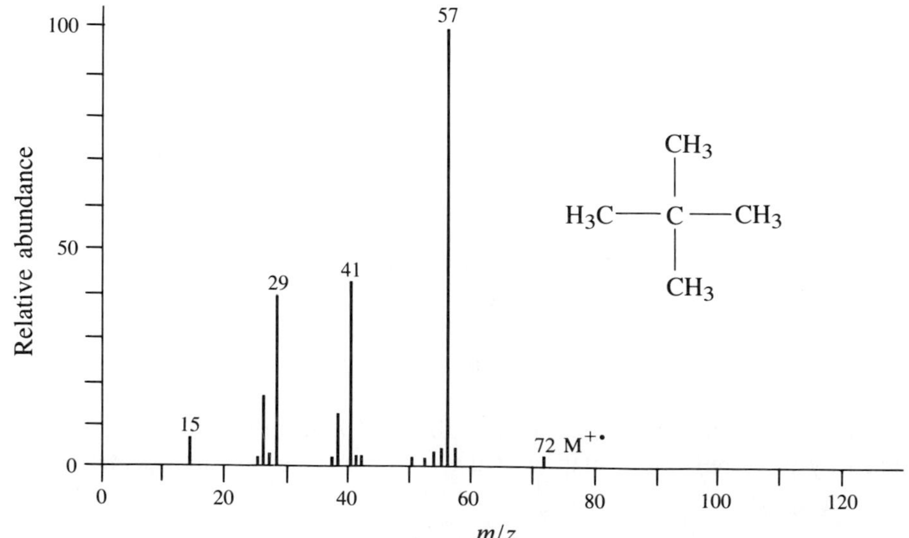

FIGURE 20-10 Mass spectrum of 2,2-dimethylpropane. Only a very weak molecular ion peak is seen, because the fragmentation to give a tertiary cation is favored.

Preferred Fragmentation of 2-Methylbutane

$$\left[\mathrm{H_3C{-}\underset{H}{\overset{CH_3}{C}}{\nmid}CH_2CH_3}\right]^{+\cdot} \xrightarrow{-\ \mathrm{C_2H_5}\cdot} \mathrm{H_3C{-}\overset{+}{C}(CH_3)H}$$

$m/z = 72$ $\qquad$ $m/z = 43$

The preference for fragmentation at a highly substituted center is even more pronounced in the mass spectrum of 2,2-dimethylpropane (Figure 20-10). Here, loss of a methyl radical from the molecular ion produces the 1,1-dimethylethyl (*tert*-butyl) cation as the base peak at $m/z = 57$. This fragmentation is so easy that the molecular ion is barely visible. The spectrum also reveals peaks at $m/z = 41$ and 29, the result of complex structural reorganizations, like the carbocation rearrangements discussed in Section 9-3.

$$\left[\mathrm{H_3C{-}\underset{CH_3}{\overset{CH_3}{C}}{-}CH_3}\right]^{+\cdot}$$

$m/z = 72$

$$\downarrow$$

$$\mathrm{H_3C{-}\overset{+}{C}(CH_3)CH_3}$$

$m/z = 57$

Fragmentations also help identify functional groups

Particularly easy fragmentation of relatively weak bonds is also seen in the mass spectra of the haloalkanes. The fragment ion $(M - X)^+$ is frequently the base peak in these spectra. A similar phenomenon is observed in the mass spectra of alcohols, which eliminate water to give a large $(M - H_2O)^{+\cdot}$ peak 18 mass units below the parent ion (Figure 20-11). Thc bonds to the C–OH group also readily dissociate in a process called ***α* cleavage,** leading to resonance-stabilized hydroxycarbocations, as shown below.

$$\mathrm{>\overset{+}{C}{-}\ddot{O}H} \rightleftharpoons \mathrm{>C{=}\overset{+}{O}H}$$

The strong peak at $m/z = 31$ in the mass spectrum of 1-butanol is due to the hydroxymethyl cation, $^+CH_2OH$, which arises from α cleavage.

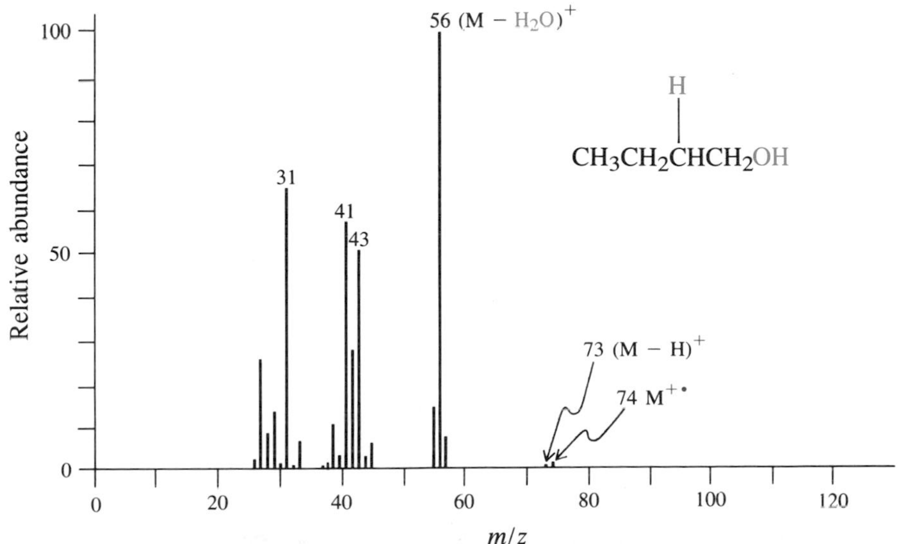

FIGURE 20-11 Mass spectrum of 1-butanol. The molecular ion, at $m/z = 74$, gives rise to a small peak because of ready loss of water to give the ion at $m/z = 56$. Other fragment ions are probably propyl ($m/z = 43$), 2-propenyl (allyl) ($m/z = 41$), and hydroxymethyl ($m/z = 31$).

Alcohol Fragmentation by Dehydration and α Cleavage

$$\left[\begin{array}{c} \text{HO}\quad\text{H} \\ \text{R—C—CHR}' \\ \text{H} \end{array}\right]^{+\cdot} \longrightarrow [\text{RCH=CHR}']^{+\cdot} + H_2O$$

M$^{+\cdot}$ **(M − 18)$^{+\cdot}$**

− R· → $[\text{HO–C(H)(CH}_2\text{R}')]^+$; − H· → $[\text{HO–C(R)(CH}_2\text{R}')]^+$; − R′CH$_2$· → $[\text{HO–C(R)(H)}]^+$

EXERCISE 20-26

Try to predict the appearance of the mass spectrum of 3-methyl-3-heptanol.

The fragmentation patterns of carbonyl compounds are often useful in structural identifications. For example, the mass spectra of the isomeric ketones 2-pentanone, 3-pentanone, and 3-methyl-2-butanone (Figure 20-12) reveal very clean and distinct fragment ions. The predominant decomposition pathway is α cleavage, which severs an alkyl bond to the carbonyl function to give the corresponding **acylium cation** and an alkyl radical.

$$\left[\text{R}-\overset{+}{\text{C}}=\ddot{\text{O}} \longleftrightarrow \text{R}-\text{C}\equiv\overset{+}{\text{O}}:\right]$$

Acylium cation

α Cleavage of Carbonyl Compounds

$$:\overset{+}{\text{O}}\equiv\text{CR}' \xleftarrow[-\text{R}\cdot]{\alpha\text{ cleavage}} \left[\text{R}-\overset{\overset{\Large :\text{O}:}{\|}}{\text{C}}-\text{R}'\right]^{+\cdot} \xrightarrow[-\text{R}'\cdot]{\alpha\text{ cleavage}} \text{RC}\equiv\overset{+}{\text{O}}:$$

The acylium cation forms easily because of resonance stabilization. These fragment ions allow the gross composition of the two alkyl groups in a ketone to be read from the spectrum. In this way, 2-pentanone is readily differentiated from 3-pentanone: α cleavage of 2-pentanone gives two acylium ions, at $m/z = 43$ and 71, but 3-pentanone gives only one, at $m/z = 57$. (The $m/z = 29$ peak in the mass spectrum of the latter is due partly to $CH_3CH_2^+$ and partly to $HC{\equiv}O^+$, which arises by loss of C_2H_4 from $CH_3CH_2C{\equiv}O^+$.)

α Cleavage in 2-Pentanone

$$\underset{m/z\,=\,71}{:\overset{+}{\text{O}}\equiv\text{CCH}_2\text{CH}_2\text{CH}_3} \longleftarrow \underset{m/z\,=\,86}{\text{H}_3\text{C}-\overset{\overset{:\text{O}:}{\|}}{\text{C}}-\text{CH}_2\text{CH}_2\text{CH}_3} \longrightarrow \underset{m/z\,=\,43}{\text{CH}_3\text{C}\equiv\overset{+}{\text{O}}:}$$

2-Pentanone

α Cleavage in 3-Pentanone

$$\underset{m/z\,=\,86}{\text{CH}_3\text{CH}_2-\overset{\overset{:\text{O}:}{\|}}{\text{C}}-\text{CH}_2\text{CH}_3} \longrightarrow \underset{m/z\,=\,57}{\text{CH}_3\text{CH}_2\text{C}\equiv\overset{+}{\text{O}}:}$$

3-Pentanone

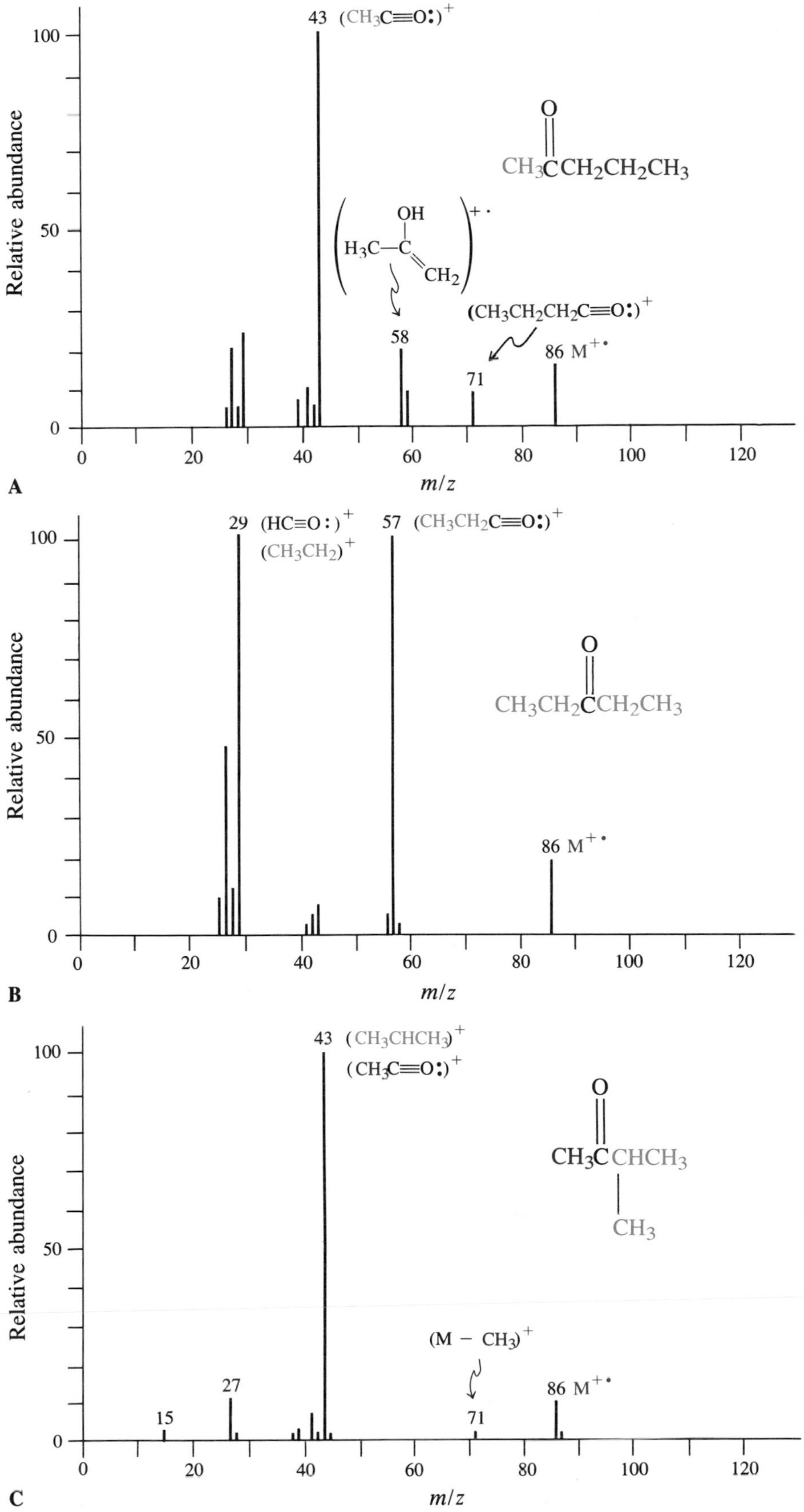

FIGURE 20-12 Mass spectra of (A) 2-pentanone, showing two peaks for α cleavage and one for McLafferty rearrangement; (B) 3-pentanone, showing only a single α cleavage peak due to symmetry; and (C) 3-methyl-2-butanone, showing two α cleavages.

Can 2-pentanone be distinguished from 3-methyl-2-butanone? Not by the observation of α cleavage—in both molecules, the substituent groups are CH_3 and C_3H_7. However, comparison of the mass spectra of the two compounds (Figure 20-12A and C) reveals an additional prominent peak for 2-pentanone at $m/z = 58$, signifying the loss of a molecular fragment of weight $m/z = 28$. This fragment is absent from the spectra of both other isomers and is characteristic of the presence of hydrogens located gamma to the carbonyl group. Compounds with this structural feature and with sufficient flexibility to allow the γ-hydrogen to be close to the carbonyl oxygen decompose by the **McLafferty* rearrangement.** In this reaction, the molecular ion of the starting ketone splits into two pieces (a neutral fragment and a radical cation) in a unimolecular process.

McLafferty Rearrangement

$$\left[\begin{array}{c} \gamma\ \text{RHC}(\text{H})-\beta\ \text{H}_2\text{C}-\underset{\alpha}{\text{CH}_2}-\text{C}(=\text{O})-\text{R}' \end{array}\right]^{+\cdot} \longrightarrow \text{RCH}{=}\text{CH}_2 + \left[\text{H}_2\text{C}{=}\text{C}(\text{OH})-\text{R}'\right]^{+\cdot}$$

The McLafferty rearrangement yields an alkene and the enol form of a new ketone. In the case of 2-pentanone, ethene and the enol of 2-propanone are produced; the radical cation of the latter is observed with $m/z = 58$.

$$\left[\text{CH}_3\overset{\text{O}}{\overset{\|}{\text{C}}}\underset{\alpha}{\text{CH}_2}\underset{\beta}{\text{CH}_2}\underset{\gamma}{\text{CH}_2}\text{H}\right]^{+\cdot} \longrightarrow \left[\text{H}_3\text{C}-\text{C}(\text{OH}){=}\text{CH}_2\right]^{+\cdot} + \text{CH}_2{=}\text{CH}_2$$

$m/z = 86$ $\qquad$ $m/z = 58$

Neither 3-pentanone nor 3-methyl-2-butanone possesses a γ-hydrogen; therefore, neither is able to undergo McLafferty rearrangement.

EXERCISE 20-27

How would you tell the difference between **(a)** 3-methyl-2-pentanone and 4-methyl-2-pentanone, and **(b)** 2-ethylcyclohexanone and 3-ethylcyclohexanone, using only mass spectroscopy?

Similar rearrangements and α cleavages can also be seen in the mass spectra of aldehydes and carboxylic acid derivatives.

EXERCISE 20-28

Interpret the labeled peaks in the mass spectra of pentanal, pentanoic acid, and methyl pentanoate shown in Figure 20-13.

*Professor Fred W. McLafferty, b. 1923, Cornell University, Ithaca, New York.

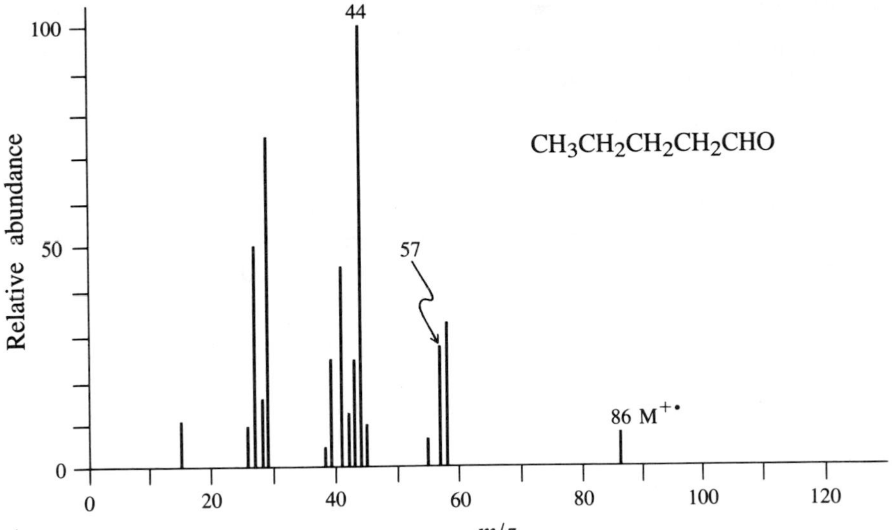

Relative abundance

$CH_3CH_2CH_2CH_2CHO$

44

57

86 $M^{+\bullet}$

m/z

A

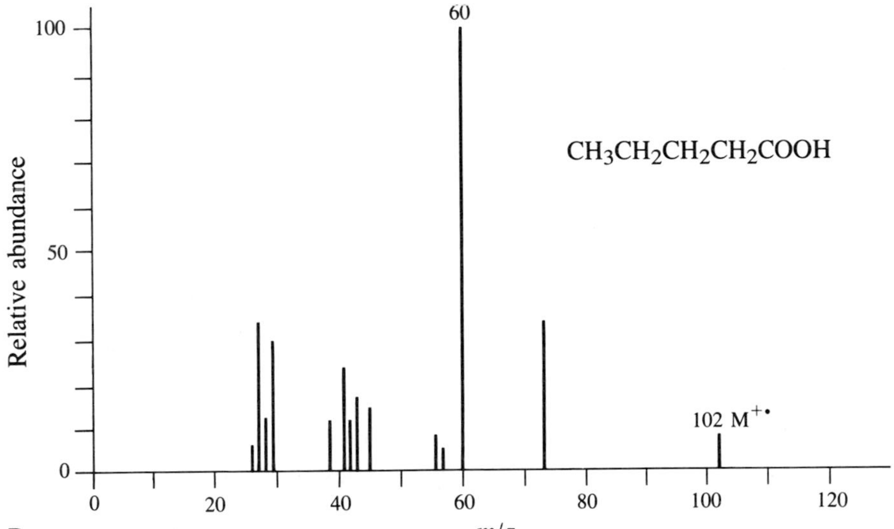

Relative abundance

$CH_3CH_2CH_2CH_2COOH$

60

102 $M^{+\bullet}$

m/z

B

Relative abundance

$$CH_3CH_2CH_2CH_2\overset{\displaystyle O}{\overset{\|}{C}}OCH_3$$

57

74

85

116 $M^{+\bullet}$

m/z

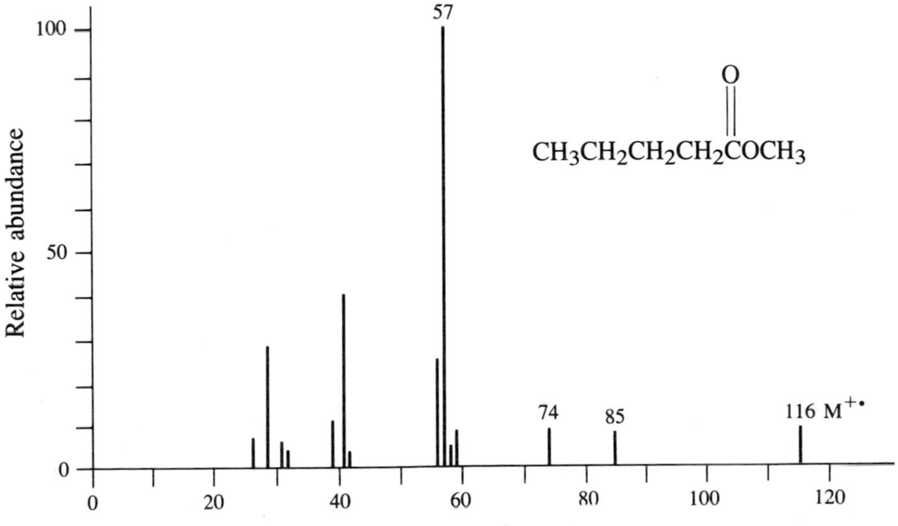

C

FIGURE 20-13 Mass spectra of (A) pentanal; (B) pentanoic acid; and (C) methyl pentanoate. (See Exercise 20-28.)

BOX 20-4 The Contribution of Mass Spectroscopy to the Characterization of Manoalide

Completion of the structural elucidation of manoalide (Boxes 10-3 and 20-1) required that information from several sources, including mass spectroscopy, be reconciled. The parent ion in the mass spectrum was too weak for exact mass determination with the instruments of the late 1970s (such limitations no longer exist). However, a strong peak at M − 18, corresponding to loss of water, was found to have exact m/z = 398.2459. The composition fitting this mass is $C_{25}H_{34}O_4$ (m/z = 398.2457); extrapolation gives a formula of $C_{25}H_{36}O_5$ for manoalide itself. The base peak at m/z = 137 was identified by analogy with similar compounds as the $C_{10}H_{17}$ fragment arising from cleavage of an allylic C–C bond.

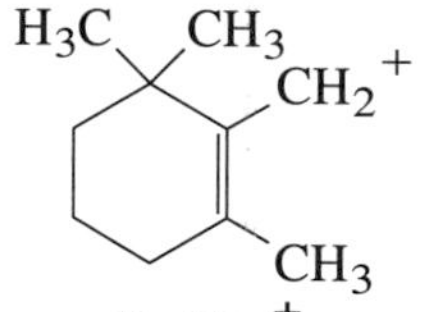

H_3C CH_3 CH_2^+ CH_3

$\mathbf{C_{10}H_{17}}^+$

NMR and mass spectroscopic information also identified an acyclic C_5H_8 (isoprene) unit (see Section 14-10). The remaining fragment was identified as follows. Selective protection of one OH group followed by PCC oxidation of the other (Section 8-6) gave a product with new bands in the IR spectrum at 1725 cm^{-1} and in the UV spectrum with λ_{max} = 211 nm. Only an α,β-unsaturated, six-membered lactone is consistent with these data, identifying the final piece of the puzzle as an unsaturated cyclic hemiacetal (see below).

The potential therapeutic value of manoalide arises from a rare mode of biological activity: It is one of only a tiny number of nonsteroidal substances that inhibits the release of arachidonic acid (Section 19-14). Manoalide inactivates the enzyme that cleaves arachidonic acid from the phospholipid ester (Section 20-4) by which it is transported through cell membranes. Recall that arachidonic acid is the biochemical precursor to both the prostaglandins, which contribute to inflammation, and the leukotrienes, the overproduction of which is associated with uncontrolled cell division. When these substances are present to excess in the outer layers of the skin, they give rise to abnormalities such as the lesions due to psoriasis. Other nonsteroidal anti-inflammatory drugs ("NSAIDs") such as aspirin are effective as prostaglandin but not leukotriene inhibitors. By blocking the release of arachidonic acid, manoalide impedes both prostaglandin *and* leukotriene biosynthesis, without the undesirable side effects of the usual steroid treatments.

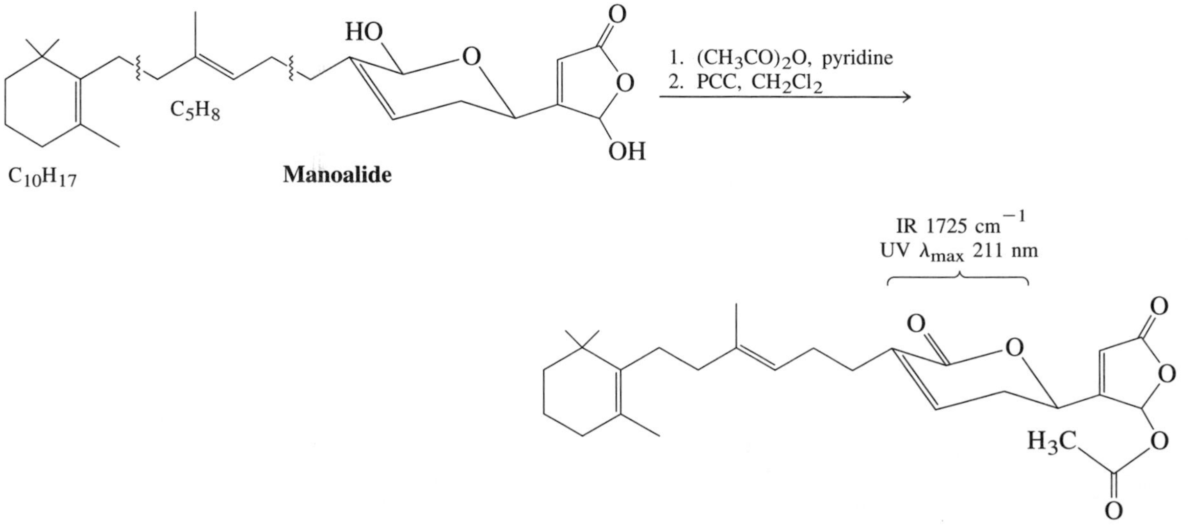

HO O O O OH

$C_{10}H_{17}$ C_5H_8

Manoalide

1. $(CH_3CO)_2O$, pyridine
2. PCC, CH_2Cl_2

IR 1725 cm^{-1}
UV λ_{max} 211 nm

O O O O H_3C O O

In summary, fragmentation patterns can be interpreted for structural elucidation. For example, the radical cations of alkanes cleave to form the most stable positively charged fragments, haloalkanes fragment by rupture of the carbon–halogen bond, alcohols readily dehydrate and undergo α cleavage, and carbonyl compounds decompose by both α cleavage and McLafferty rearrangement.

20-11 High-Resolution Mass Spectroscopy

Consider substances with the following molecular formulas: C_7H_{14}, $C_6H_{10}O$, $C_5H_6O_2$, and $C_5H_{10}N_2$. They all possess the same **integral mass;** that is, to the nearest integer, all four would be expected to exhibit a parent ion at $m/z = 98$. However, the atomic weights of the elements are composites of the masses of their naturally occurring isotopes, *which are not integers.* Thus, if we use the atomic masses for the most abundant isotopes of C, H, O, and N (Table 20-4) to calculate the **exact mass** corresponding to each of the molecular formulas above, we see significant differences.

TABLE 20-4 Exact Masses of Several Common Isotopes

Isotope	Mass
^{1}H	1.00783
^{12}C	12.00000
^{14}N	14.0031
^{16}O	15.9949
^{32}S	31.9721
^{35}Cl	34.9689
^{37}Cl	36.9659
^{79}Br	78.9183
^{81}Br	80.9163

Exact Masses of Four Compounds with $m/z = 98$

C_7H_{14}	$C_6H_{10}O$	$C_5H_6O_2$	$C_5H_{10}N_2$
98.1096	98.0732	98.0368	98.0845

Can we use mass spectroscopy to differentiate between these species? Yes. Modern **high-resolution mass spectrometers** are capable of distinguishing between ions that differ in mass by as little as a few thousandths of a mass unit. We can therefore measure the exact mass of any parent or fragment ion. By comparing this experimentally determined value with that calculated for each species possessing the same integral mass, we can assign a molecular formula to the unknown ion.

Modern high-resolution instruments possess computer programs designed to match exact masses with possible formulas. As a result, high-resolution mass spectroscopy is now the most widely used method for determining the molecular formulas of unknowns.

EXERCISE 20-29

Choose the molecular formula that matches the exact mass. **(a)** $m/z = 112.0888$, C_8H_{16}, $C_7H_{12}O$, or $C_6H_8O_2$; **(b)** $m/z = 86.1096$, C_6H_{14}, $C_4H_6O_2$, or $C_4H_{10}N_2$.

NEW REACTIONS

1. ORDER OF REACTIVITY OF CARBOXYLIC ACID DERIVATIVES (Section 20-1)

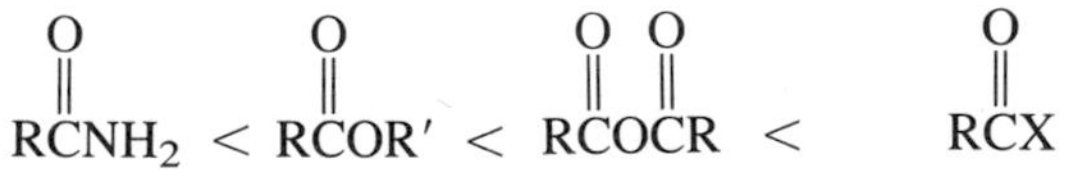

$$\underset{\textbf{Amide}}{\overset{O}{\overset{\|}{R}}\!CNH_2} < \underset{\textbf{Ester}}{\overset{O}{\overset{\|}{R}}\!COR'} < \underset{\textbf{Anhydride}}{R\overset{O}{\overset{\|}{C}}O\overset{O}{\overset{\|}{C}}R} < \underset{\textbf{Alkanoyl halide}}{R\overset{O}{\overset{\|}{C}}X}$$

Esters and amides require acid or base catalysts to react with weak nucleophiles

2. BASICITY OF THE CARBONYL OXYGEN (Section 20-1)

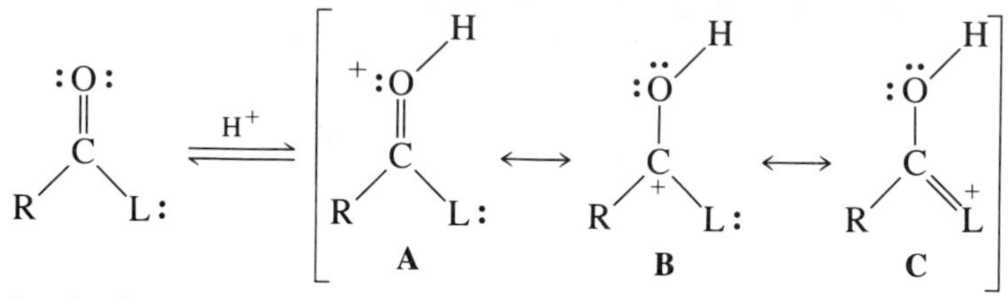

$$\mathrm{RC(=\ddot{O}{:})L{:}} \underset{}{\overset{H^+}{\rightleftharpoons}} \left[\underset{\mathbf{A}}{\mathrm{RC(={}^{+}{:}O{-}H)L{:}}} \longleftrightarrow \underset{\mathbf{B}}{\mathrm{R\overset{+}{C}(-\ddot{O}{-}H)L{:}}} \longleftrightarrow \underset{\mathbf{C}}{\mathrm{RC(-\ddot{O}{-}H)(=\overset{+}{L})}} \right]$$

L = leaving group
Basicity increases with increasing contribution of resonance structure C

3. ENOLATE FORMATION (Section 20-1)

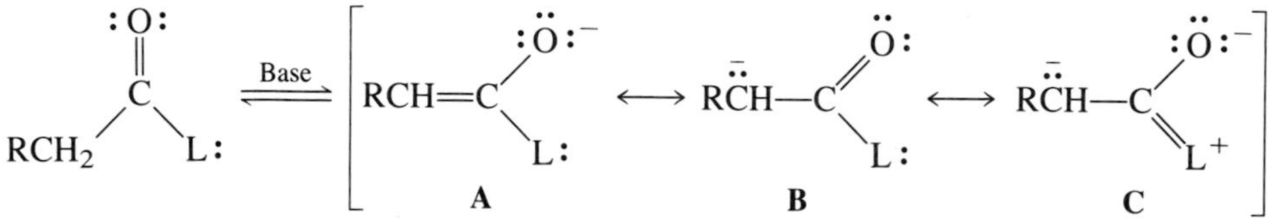

$$\mathrm{RCH_2C(=\ddot{O}{:})L{:}} \overset{\text{Base}}{\rightleftharpoons} \left[\underset{\mathbf{A}}{\mathrm{RCH{=}C(-\ddot{O}{:}^-)L{:}}} \longleftrightarrow \underset{\mathbf{B}}{\mathrm{R\bar{\ddot{C}}H{-}C(=\ddot{O}{:})L{:}}} \longleftrightarrow \underset{\mathbf{C}}{\mathrm{R\bar{\ddot{C}}H{-}C(-\ddot{O}{:}^-)(=L^+)}} \right]$$

Acidity of the neutral derivative generally increases with decreasing contribution of resonance structure C in the anion

Reactions of Alkanoyl Halides

4. WATER (Section 20-2)

$$\mathrm{R\overset{O}{\overset{\|}{C}}X + H_2O \longrightarrow \underset{\textbf{Carboxylic acid}}{R\overset{O}{\overset{\|}{C}}OH} + HX}$$ Very reactive

5. CARBOXYLIC ACIDS (Sections 19-8 and 20-2)

$$\mathrm{R\overset{O}{\overset{\|}{C}}X + R'CO_2H \longrightarrow \underset{\textbf{Carboxylic anhydride}}{R\overset{O}{\overset{\|}{C}}O\overset{O}{\overset{\|}{C}}R'} + HX}$$

6. ALCOHOLS (Section 20-2)

$$\mathrm{R\overset{O}{\overset{\|}{C}}X + R'OH \longrightarrow \underset{\textbf{Ester}}{R\overset{O}{\overset{\|}{C}}OR'} + HX}$$

(Removed with pyridine, triethylamine, or other base)

7. AMINES (Section 20-2)

$$\mathrm{R\overset{O}{\overset{\|}{C}}X + R'NH_2 \longrightarrow \underset{\textbf{Amide}}{R\overset{O}{\overset{\|}{C}}NHR'} + HX}$$

(Removed with pyridine, triethylamine, excess $R'NH_2$, or other base)

8. CUPRATE REAGENTS (Section 20-2)

$$\mathrm{R\overset{O}{\overset{\|}{C}}X \xrightarrow[2.\ H^+,\ H_2O]{1.\ R'_2CuLi,\ THF} \underset{\textbf{Ketone}}{R\overset{O}{\overset{\|}{C}}R'} + CuX + LiX}$$

9. HYDRIDES (Section 20-2)

$$\mathrm{R\overset{O}{\overset{\|}{C}}X \xrightarrow[2.\ H^+,\ H_2O]{1.\ LiAl[OC(CH_3)_3]_3H,\ (CH_3CH_2)_2O} \underset{\textbf{Aldehyde}}{R\overset{O}{\overset{\|}{C}}H} + LiX + Al[OC(CH_3)_3]_3}$$

Reactions of Carboxylic Acid Anhydrides

10. WATER (Section 20-3)

$$\mathrm{R\overset{O}{\overset{\|}{C}}O\overset{O}{\overset{\|}{C}}R + H_2O \longrightarrow 2\ \underset{\textbf{Carboxylic acid}}{R\overset{O}{\overset{\|}{C}}OH}}$$

11. ALCOHOLS (Section 20-3)

$$\mathrm{R\overset{O}{\overset{\|}{C}}O\overset{O}{\overset{\|}{C}}R + R'OH \longrightarrow \underset{\textbf{Ester}}{R\overset{O}{\overset{\|}{C}}OR'} + R\overset{O}{\overset{\|}{C}}OH}$$

12. AMINES (Section 20-3)

$$\mathrm{R\overset{O}{\overset{\|}{C}}O\overset{O}{\overset{\|}{C}}R + R'NH_2 \longrightarrow \underset{\textbf{Amide}}{R\overset{O}{\overset{\|}{C}}NHR'} + R\overset{O}{\overset{\|}{C}}OH}$$

Reactions of Esters

13. WATER (ESTER HYDROLYSIS) (Sections 19-9 and 20-5)

Acid catalysis

$$\mathrm{R\overset{O}{\overset{\|}{C}}OR' + H_2O \xrightarrow{\text{Catalytic } H^+} \underset{\textbf{Carboxylic acid}}{R\overset{O}{\overset{\|}{C}}OH} + R'OH}$$

Base catalysis

$$\mathrm{R\overset{O}{\overset{\|}{C}}OR' + \underset{\text{1 equivalent}}{{}^-OH} \xrightarrow{H_2O} \underset{\textbf{Carboxylate ion}}{R\overset{O}{\overset{\|}{C}}O^-} + R'OH}$$

14. ALCOHOLS (TRANSESTERIFICATION) AND AMINES (Section 20-5)

$$\mathrm{R\overset{O}{\overset{\|}{C}}OR' + R''OH \xrightarrow{H^+ \text{ or } {}^-OR''} \underset{\textbf{Ester}}{R\overset{O}{\overset{\|}{C}}OR''} + R'OH}$$

$$\mathrm{R\overset{O}{\overset{\|}{C}}OR' + R''NH_2 \xrightarrow{\text{Heat}} \underset{\textbf{Amide}}{R\overset{O}{\overset{\|}{C}}NHR''} + R'OH}$$

15. ORGANOMETALLIC REAGENTS (Section 20-5)

$$\mathrm{RCOR''} \xrightarrow[\text{2. } H^+,\ H_2O]{\text{1. } 2\ R'MgX,\ (CH_3CH_2)_2O} \mathrm{R{-}C(OH)(R'){-}R'} + \mathrm{R''OH}$$

(RCOR″ with C=O; product: **Tertiary alcohol**)

Methyl formate

$$\mathrm{HCOCH_3} \xrightarrow[\text{2. } H^+,\ H_2O]{\text{1. } 2\ R'MgX,\ (CH_3CH_2)_2O} \mathrm{H{-}C(OH)(R'){-}R'} + \mathrm{CH_3OH}$$

Secondary alcohol

16. HYDRIDES (Section 20-5)

$$\mathrm{RCOR'} \xrightarrow[\text{2. } H^+,\ H_2O]{\text{1. } LiAlH_4,\ (CH_3CH_2)_2O} \mathrm{RCH_2OH}$$

$$\mathrm{RCOR'} \xrightarrow[\text{2. } H^+,\ H_2O]{\text{1. } (CH_3CH(CH_3)CH_2)_2AlH,\ \text{toluene},\ -60^\circ C} \mathrm{RCH}(=O)$$

17. ENOLATES (PRODUCTS OBTAINED AFTER ACIDIC WORK-UP) (Section 20-5)

$$\mathrm{RCH_2COR'} \xrightarrow{\text{LDA, THF}} \left[\mathrm{R\ddot{C}H^-{-}C({=}O)OR'} \longleftrightarrow \mathrm{RCH{=}C(O^-)OR'}\right] \xrightarrow{R''X} \mathrm{RCH(R'')COR'}$$

Ester enolate ion

Reactions of Amides

18. WATER (Section 20-6)

$$\mathrm{RCNHR'} + \mathrm{H_2O} \xrightarrow{H^+,\ \Delta} \mathrm{RCOH} + \mathrm{R'\overset{+}{N}H_3}$$

Carboxylic acid

$$\mathrm{RCNHR'} + \mathrm{H_2O} \xrightarrow{HO^-,\ \Delta} \mathrm{RCO^-} + \mathrm{R'NH_2}$$

19. HYDRIDES (Section 20-6)

$$\mathrm{RCNHR'} \xrightarrow[\text{2. } H^+,\ H_2O]{\text{1. } LiAlH_4,\ (CH_3CH_2)_2O} \mathrm{RCH_2NH_2}$$

Amine

$$\mathrm{RCNHR'} \xrightarrow[\text{2. } H^+,\ H_2O]{\text{1. } (CH_3CH(CH_3)CH_2)_2AlH,\ (CH_3CH_2)_2O} \mathrm{RCH}(=O)$$

Aldehyde

20. ENOLATES AND AMIDATES (Section 20-7)

$$\underset{\uparrow \atop \mathrm{p}K_a \sim 30}{\mathrm{RCH_2}}\overset{:\mathrm{O}:}{\overset{\|}{\mathrm{C}}}\mathrm{NR'_2} \xrightarrow{\text{Base}} \mathrm{RCH{=}C}\begin{matrix}\diagup \ddot{\mathrm{O}}:^- \\ \diagdown \mathrm{NR'_2}\end{matrix}$$

Amide enolate ion

$$\mathrm{RCH_2}\overset{:\mathrm{O}:}{\overset{\|}{\mathrm{C}}}\underset{\uparrow \atop \mathrm{p}K_a \sim 15}{\ddot{\mathrm{N}}\mathrm{H}}\mathrm{R'} \xrightarrow{\text{Base}} \mathrm{RCH_2}\overset{:\ddot{\mathrm{O}}:^-}{\overset{|}{\mathrm{C}}}{=}\ddot{\mathrm{N}}\mathrm{R'}$$

Amidate ion

21. HOFMANN REARRANGEMENT (Section 20-7)

$$\mathrm{R}\overset{\mathrm{O}}{\overset{\|}{\mathrm{C}}}\mathrm{NH_2} \xrightarrow{\mathrm{Br_2,\ NaOH,\ H_2O,\ 75^\circ C}} \underset{\textbf{Amine}}{\mathrm{RNH_2}} + \mathrm{CO_2}$$

Reactions of Nitriles

22. WATER (Section 20-8)

$$\mathrm{RC{\equiv}N} + \mathrm{H_2O} \xrightarrow{\mathrm{H^+\ or\ HO^-,\ \Delta}} \underset{\textbf{Amide}}{\mathrm{R}\overset{\mathrm{O}}{\overset{\|}{\mathrm{C}}}\mathrm{NH_2}} \xrightarrow{\mathrm{H^+\ or\ HO^-,\ \Delta}} \underset{\textbf{Carboxylic acid}}{\mathrm{R}\overset{\mathrm{O}}{\overset{\|}{\mathrm{C}}}\mathrm{OH}}$$

23. ORGANOMETALLIC REAGENTS (Section 20-8)

$$\mathrm{RC{\equiv}N} \xrightarrow[\mathrm{2.\ H^+,\ H_2O}]{\mathrm{1.\ R'MgX\ or\ R'Li}} \underset{\textbf{Ketone}}{\mathrm{R}\overset{\mathrm{O}}{\overset{\|}{\mathrm{C}}}\mathrm{R'}}$$

24. HYDRIDES (Section 20-8)

$$\mathrm{RC{\equiv}N} \xrightarrow[\mathrm{2.\ H^+,\ H_2O}]{\mathrm{1.\ LiAlH_4}} \underset{\textbf{Amine}}{\mathrm{RCH_2NH_2}}$$

$$\mathrm{RC{\equiv}N} \xrightarrow[\mathrm{2.\ H^+,\ H_2O}]{\mathrm{1.\ (CH_3\overset{CH_3}{\overset{|}{C}}HCH_2)_2AlH}} \underset{\textbf{Aldehyde}}{\mathrm{R}\overset{\mathrm{O}}{\overset{\|}{\mathrm{C}}}\mathrm{H}}$$

25. CATALYTIC HYDROGENATION (Section 20-8)

$$\mathrm{RC{\equiv}N} \xrightarrow{\mathrm{H_2,\ PtO_2,\ CH_3CH_2OH}} \underset{\textbf{Amine}}{\mathrm{RCH_2NH_2}}$$

IMPORTANT CONCEPTS

1. The electrophilic reactivity of the carbonyl carbon in carboxylic acid derivatives is weakened by good electron-donating substituents. This effect, measurable by IR spectroscopy, is responsible not only for the decrease in the reactivity with nucleophiles and acid but also for the increased basicity along the series: alkanoyl halides–anhydrides–esters–amides. Electron donation by resonance from the nitrogen in amides is so pronounced that there is hindered rotation around the amide bond on the NMR time scale.
2. Carboxylic acid derivatives are named as alkanoyl halides, carboxylic anhydrides, alkyl alkanoates, alkanamides, and alkanenitriles, depending on the functional group.
3. Carbonyl stretching frequencies in the IR spectra are diagnostic of the carboxylic acid derivatives: Alkanoyl chlorides absorb at 1790–1815 cm^{-1}, anhydrides at 1740–1790 and 1800–1850 cm^{-1}, esters at 1735–1750 cm^{-1}, and amides at 1650–1690 cm^{-1}.
4. Carboxylic acid derivatives generally react with water (under acid or base catalysis) to hydrolyze to the corresponding carboxylic acid; they combine with alcohols to give esters, and with amines to furnish amides. With Grignard and other organometallic reagents, they form ketones; esters may react further to form the corresponding alcohols. Reduction by hydrides gives products in various oxidation states: aldehydes, alcohols, or amines.
5. Long-chain esters are the constituents of animal and plant waxes. Triesters of glycerol are contained in natural oils and fats. Their hydrolysis gives soaps. Triglycerides containing phosphoric acid ester subunits belong to the class of phospholipids. Because they carry a highly polar head group and hydrophobic tails, they form micelles and lipid bilayers.
6. Transesterification can be used to convert one ester into another.
7. The functional group of nitriles has some similarity to that of the alkynes. The two component atoms are *sp* hybridized. The IR stretching vibration appears at about 2250 cm^{-1}. The hydrogens next to the cyano group are deshielded in ^1H NMR. The ^{13}C NMR absorptions for nitrile carbons are at relatively low field ($\delta \sim$ 112–126 ppm), a consequence of the electronegativity of nitrogen.
8. Mass spectroscopy is a technique for ionizing molecules and separating the resulting ions magnetically by molecular weight. Because the ionizing beam has high energy, the ionized molecules also fragment to smaller particles, all of which are separated and recorded as the mass spectrum of a compound. The presence of certain elements (such as Cl, Br) can be detected by their isotopic patterns. The presence of fragment-ion signals in mass spectra can be used to deduce the structure of a molecule.
9. High-resolution mass spectral data allow determination of molecular formulas from exact mass values.

PROBLEMS

1. Name (IUPAC system) or draw the structure of each of the following.

(a) $(CH_3)_2CHCH_2\overset{O}{\overset{\|}{C}}Cl$

(b) cyclopentane ring bearing COCl and CH_3 on the same carbon

(c) $CF_3\overset{O}{\overset{\|}{C}}O\overset{O}{\overset{\|}{C}}CF_3$

(d) $C_6H_5\!-\!\overset{O}{\overset{\|}{C}}O\overset{O}{\overset{\|}{C}}CH_2CH_3$

(e) $(CH_3)_3C\overset{O}{\overset{\|}{C}}OCH_2CH_3$

(f) $CH_3\overset{O}{\overset{\|}{C}}NH\!-\!C_6H_5$

(g) Propyl butanoate

(h) Butyl propanoate

(i) 2-Chloroethyl benzoate

(j) *N,N*-Dimethylbenzamide

(k) 2-Methylhexanenitrile

(l) Cyclopentanecarbonitrile

2. Use resonance structures to explain in detail the relative order of acidity of carboxylic acid derivatives, as presented in Section 20-1.

3. In each of the following pairs of compounds, decide which possesses the indicated property to the greater degree. **(a)** Length of C–X bond: acetyl fluoride or acetyl chloride. **(b)** Acidity of the boldface H: $\mathbf{CH_2}(COCH_3)_2$ or $\mathbf{CH_2}(COOCH_3)_2$. **(c)** Reactivity toward addition of a nucleophile: (i) an amide or (ii) an imide (as shown in the margin). **(d)** High-energy infrared carbonyl stretching frequency: ethyl acetate or ethenyl acetate.

$CH_3\overset{O}{\overset{\|}{C}}N(CH_3)_2$

i

$CH_3\overset{O}{\overset{\|}{C}}\underset{CH_3}{\underset{|}{N}}\overset{O}{\overset{\|}{C}}CH_3$

ii

4. Give the product(s) of each of the following reactions.

(a) 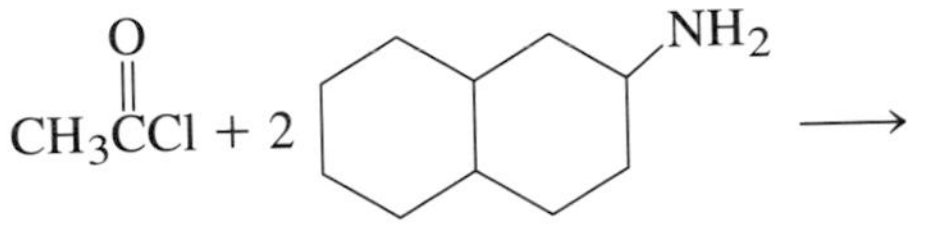 $CH_3\overset{O}{\overset{\|}{C}}Cl + 2$ (decahydronaphthalene with NH_2 substituent) $\longrightarrow$

(b) $\left(C_6H_5\right)_2CuLi + CH_3CH_2CH_2CH_2CH_2\overset{O}{\overset{\|}{C}}Cl \xrightarrow{THF}$

(c) $(H_3C)_3C\overset{O}{\overset{\|}{C}}Cl \xrightarrow{LiAl[OC(CH_3)_3]_3H,\ THF,\ -78°C}$

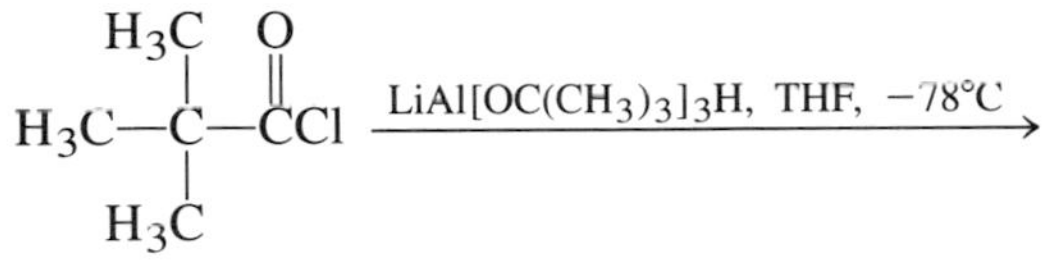

(d) 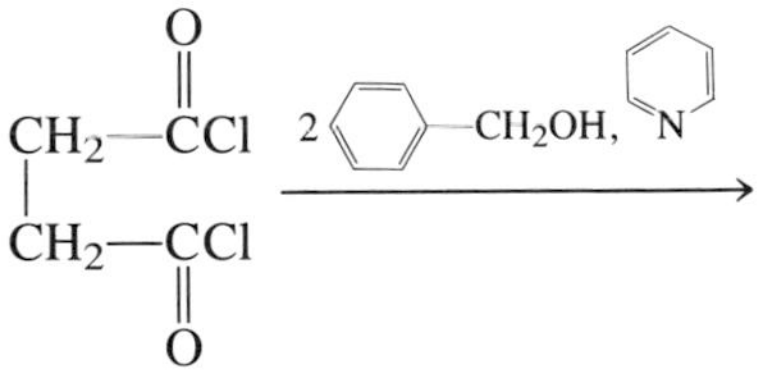 $\underset{CH_2-\overset{\|}{\underset{O}{C}}Cl}{\overset{CH_2-\overset{O}{\overset{\|}{C}}Cl}{|}} \xrightarrow{2\ C_6H_5CH_2OH,\ \text{pyridine}}$

(e) 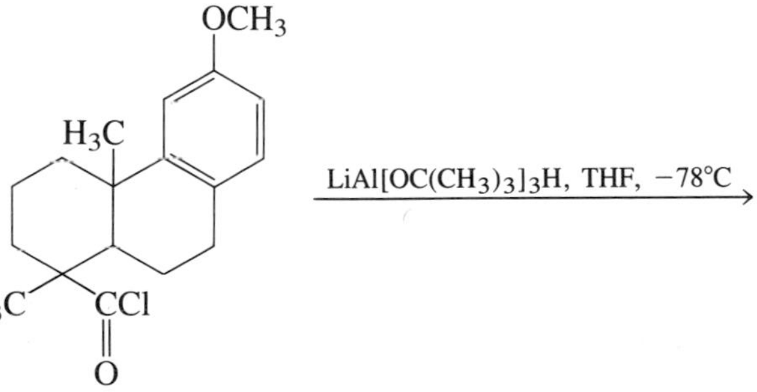 (tricyclic compound with OCH_3 on aromatic ring, H_3C at ring junction, H_3C and $\overset{O}{\overset{\|}{C}}Cl$ substituents) $\xrightarrow{LiAl[OC(CH_3)_3]_3H,\ THF,\ -78°C}$

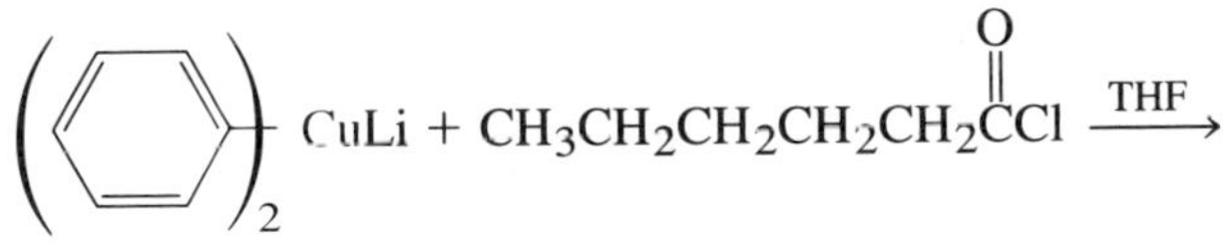

5. Give the product(s) of the reactions of acetic anhydride with each of the reagents below. Assume in all cases that the reagent is present in excess.

(a) $(CH_3)_2CHOH$ **(b)** NH_3

(c) 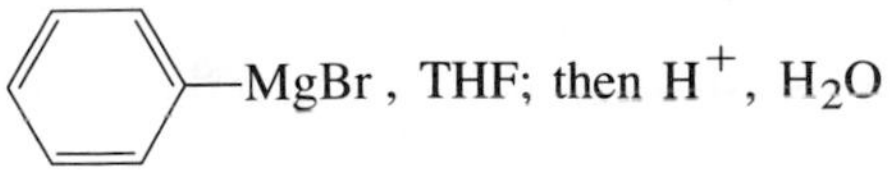 C_6H_5–MgBr, THF; then H^+, H_2O **(d)** $LiAlH_4$, $(CH_3CH_2)_2O$; then H^+, H_2O

6. Give the product(s) of the reaction of butanedioic (succinic) anhydride with each of the reagents in Problem 5.

7. Give the products of reaction of methyl pentanoate with each of the following.

(a) NaOH, H_2O, heat; then H^+, H_2O
(b) $(CH_3)_2CHCH_2CH_2OH$ (excess), H^+
(c) $(CH_3CH_2)_2NH$, heat
(d) CH_3MgI (excess), $(CH_3CH_2)_2O$; then H^+, H_2O
(e) $LiAlH_4$, $(CH_3CH_2)_2O$; then H^+, H_2O
(f) $[(CH_3)_2CHCH_2]_2AlH$, toluene, low temperature; then H^+, H_2O

8. Give the products of reaction of γ-valerolactone (5-methyloxa-2-cyclopentanone, Section 20-4) with each of the reagents in Problem 7.

9. Draw the structure of each of the following. **(a)** β-Butyrolactone; **(b)** β-valerolactone; **(c)** δ-valerolactone; **(d)** β-propiolactam; **(e)** α-methyl-δ-valerolactam; **(f)** N-methyl-γ-butyrolactam.

10. Formulate a mechanism for the acid-catalyzed transesterification of ethyl-2-methylpropanoate (ethyl isobutyrate) into the corresponding methyl ester. Your mechanism should clearly illustrate the catalytic role of the proton.

11. Give the product of each of the following reactions.

(a) $COOCH_3$, $COOCH_3$ (bicyclic diester) $\xrightarrow[\text{2. } H^+,\ H_2O]{\text{1. KOH, } H_2O}$

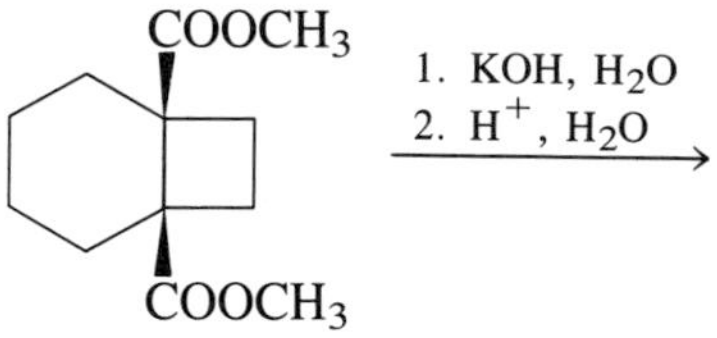

(b) H_3C, CH_3CH_2 (substituted lactone, =O) $\xrightarrow{(CH_3)_2CHNH_2,\ CH_3OH,\ \Delta}$

(c) $CH_3\overset{O}{\overset{\|}{C}}OCH_3$ + excess (cyclopentyl)MgBr $\xrightarrow[\text{2. } H^+,\ H_2O]{\text{1. } (CH_3CH_2)_2O,\ 20°C}$

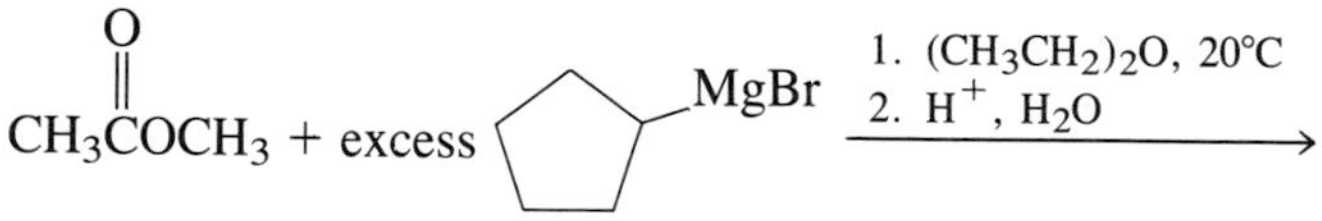

(d) (phenyl)–CH(CH_3)($COOCH_2CH_3$) $\xrightarrow{\text{1. LDA, THF, } -78°C;\ \text{2. } CH_3I,\ \text{HMPA};\ \text{3. } H^+,\ H_2O}$

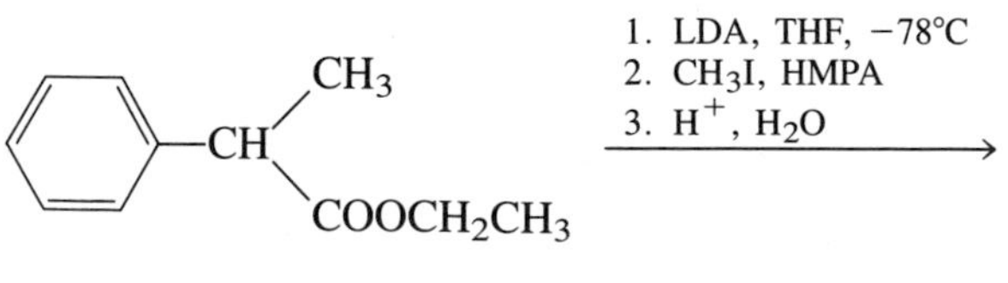

(e) (cyclohexyl)$COOCH_3$ $\xrightarrow[\text{2. } H^+,\ H_2O]{\text{1. } (CH_3\overset{CH_3}{\overset{|}{C}}HCH_2)_2AlH,\ \text{toluene, } -60°C}$

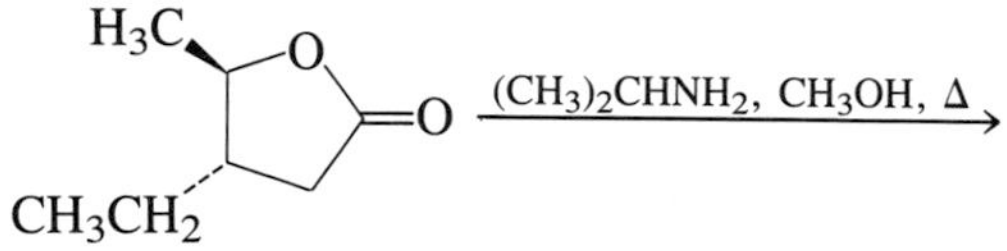

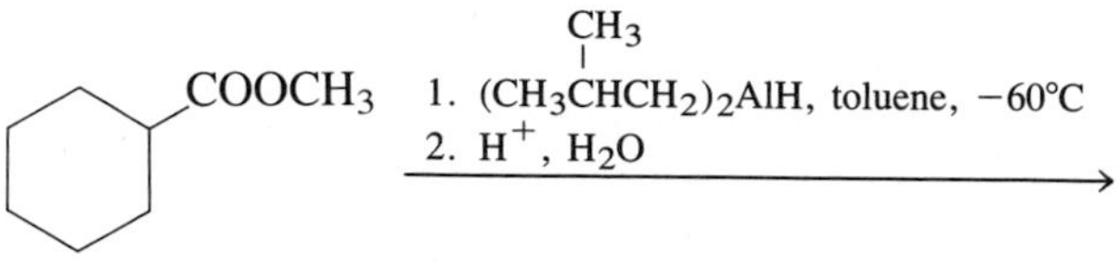

12. A useful synthesis of certain types of diols includes the reaction of a "bis-Grignard" reagent with a lactone:

1. $BrMgCH_2CH_2CH_2CH_2MgBr$, THF
2. H^+, H_2O

(a) Formulate a mechanism for this transformation. (b) Show how you would apply this general method to the synthesis of diols A and B.

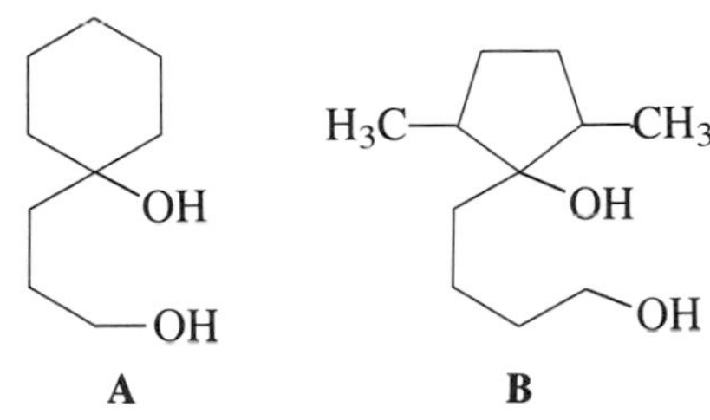

13. Formulate a mechanism for the formation of acetamide, $CH_3\overset{O}{\overset{\|}{C}}NH_2$, from methyl acetate and ammonia.

14. Give the products of the reactions of pentanamide with the reagents given in Problem 7(a, e, f). Repeat for *N,N*-dimethylpentanamide.

15. What reagents would be necessary to carry out the following transformations? **(a)** Cyclohexanecarbonyl chloride → pentanoylcyclohexane; **(b)** 2-butenedioic (maleic) anhydride → *Z*-butene-1,4-diol; **(c)** 3-methylbutanoyl bromide → 3-methylbutanal; **(d)** benzamide → 1-phenylmethanamine; **(e)** propanenitrile → 3-hexanone; **(f)** methyl propanoate → 4-ethyl-4-heptanol.

16. On treatment with strong base followed by protonation, compounds A and B undergo cis-trans isomerization, but compound C does not. Explain.

$COOCH_3$ / $COOCH_3$ **A**; $CON(CH_3)_2$ / $CON(CH_3)_2$ **B**; $CONH_2$ / $CONH_2$ **C**

17. 2-Aminobenzoic (anthranilic) acid is prepared from 1,2-benzenedicarboxylic anhydride (phthalic anhydride) by using the two reactions shown below. Explain these processes mechanistically.

NH_3, 300°C → ; 1. NaOH, Br_2, 80°C 2. H^+, H_2O →

1,2-Benzenedicarboxylic anhydride (Phthalic anhydride) → **1,2-Benzenedicarboximide (Phthalimide)** → **2-Aminobenzoic acid (Anthranilic acid)**

18. Show how you might synthesize chlorpheniramine, a powerful antihistamine used in several decongestants, from each of carboxylic acids A and B shown on the next page. Use a different carboxylic amide in each synthesis.

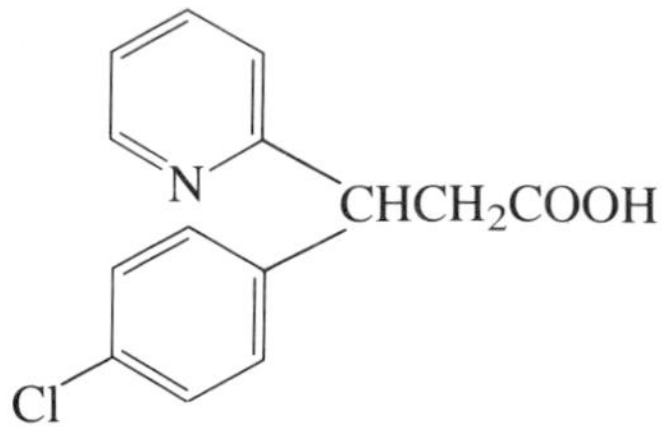

CHCH$_2$COOH

Cl

A

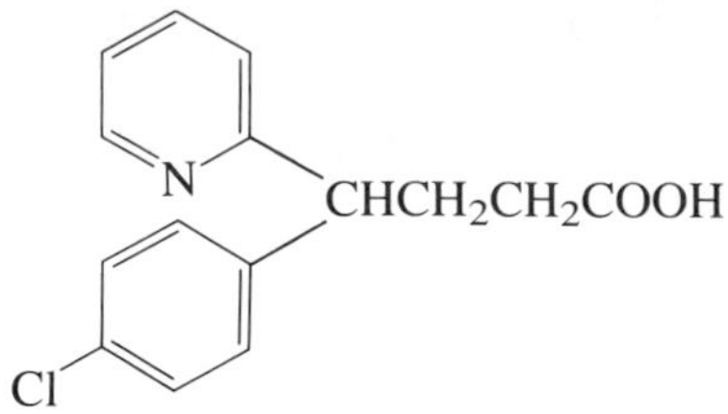

CHCH$_2$CH$_2$COOH

Cl

B

CHCH$_2$CH$_2$N(CH$_3$)$_2$

Cl

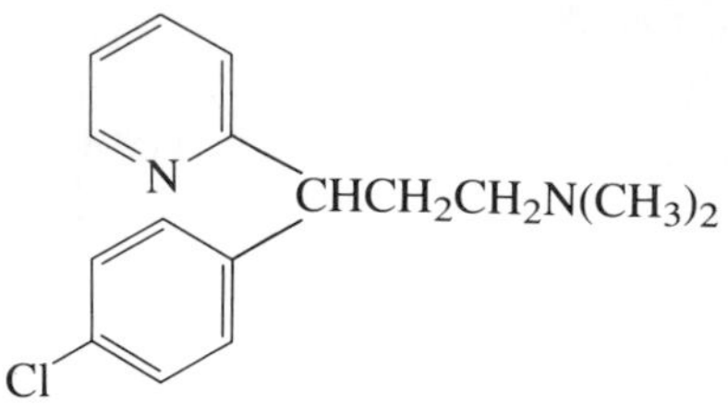

Chlorpheniramine

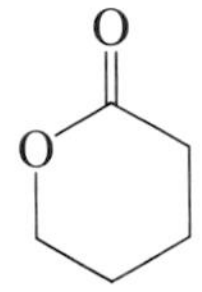

1735 cm^{-1} 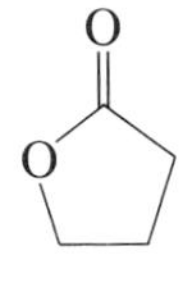 1770 cm^{-1} 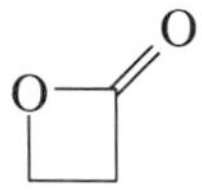 1840 cm^{-1}

19. Although esters typically have carbonyl stretching frequencies at about 1740 cm^{-1} in the infrared spectrum, the corresponding band for lactones can vary greatly with ring size. Three examples are shown in the margin. Propose an explanation for the IR bands of these smaller-ring lactones.

20. On completing a synthetic procedure, every chemist is faced with the job of cleaning glassware. Because the compounds present may be dangerous in some way or have unpleasant properties, a little serious chemical thinking is often beneficial before ''doing the dishes.'' Suppose that you have just completed a synthesis of hexanoyl chloride, perhaps to carry out the reaction in Problem 4(b); first, however, you must clean the glassware contaminated with this alkanoyl halide. *Both hexanoyl chloride and hexanoic acid have terrible odors.* **(a)** Would cleansing the glassware with soap and water be a good idea? Explain. **(b)** Suggest a more pleasant alternative, based on the chemistry of alkanoyl halides and the physical properties (particularly the odors) of the various carboxylic acid derivatives.

21. Show how you would carry out the following transformation, in which the ester function at the lower left of the molecule is converted into a hydroxy group but that at the upper right is preserved. (Hint: Do not try ester hydrolysis. Look carefully at how the ester groups are linked to the steroid and consider an approach based upon transesterification.)

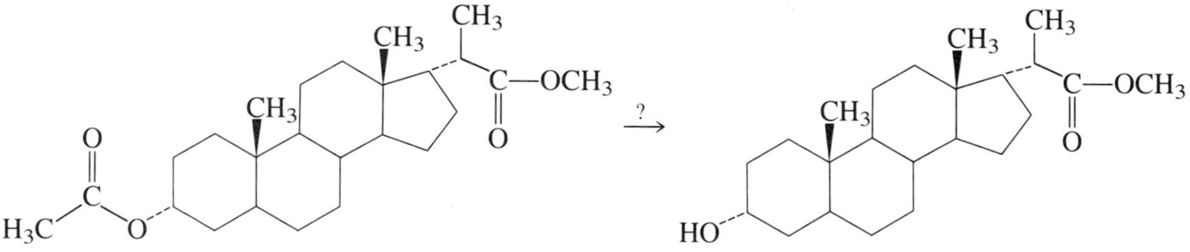

$$\xrightarrow{?}$$

22. The removal of the C17 side chain of certain steroids is a critical element in the synthesis of a number of hormones, such as testosterone, from steroids in the more readily available pregnane family.

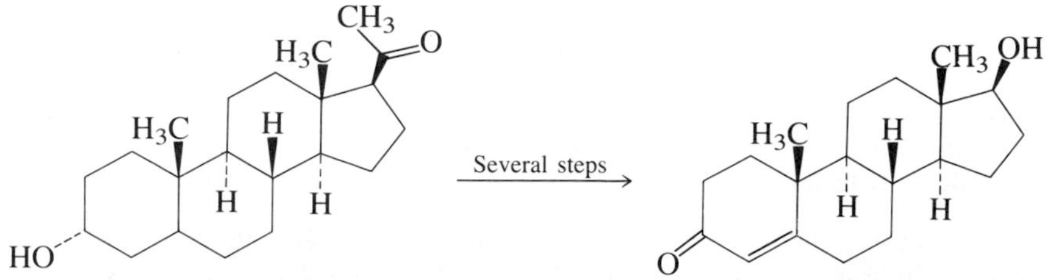

$$\xrightarrow{\text{Several steps}}$$

Pregnan-3α-ol-20-one

Testosterone

How would you carry out the comparable transformation, shown in the margin, of acetylcyclopentane into cyclopentanol? (Note: In this and subsequent synthetic problems you may need to use reactions from several areas of carbonyl chemistry discussed in Chapters 17–20.)

H_3C O ? OH

23. Propose a synthetic sequence to convert carboxylic acid A, shown below, into the naturally occurring sesquiterpene α-curcumene.

CH_3 CO_2H CH_3 ? CH_3 CH_3 H_3C CH_3

A **α-Curcumene**

24. Propose a synthetic scheme for the conversion of lactone A into amine B, a precursor to the naturally occurring monoterpene C.

CH_3 CH_3 H_3C O O CH_3 ? H_3C H_3C HO CH_3 CH_3 $N(CH_3)_2$ → H_3C H_3C HO H_2C CH_3 CH_3

A **B** **C**

25. Propose a synthesis of β-selinene, a member of a very common family of sesquiterpenes, beginning with the alcohol shown here. Use a nitrile in your synthesis. Inspection of a model may help you choose a way to obtain the desired stereochemistry. (Is the isopropenyl group axial or equatorial?)

CH_3 OH H H_2C ? CH_3 CH_2 H H_2C CH_3

β-Selinene

26. Give the structure of the product of the first of the following reactions, and then propose a scheme that will ultimately convert it into the methyl-substituted ketone at the end of the scheme. This example illustrates a common method for introduction of "angular methyl groups" into synthetically prepared steroids. (Hint: It will be necessary to protect the ketone carbonyl.)

O H $\xrightarrow{HCN}$ $C_{11}H_{15}NO$ IR: 1715, 2250 cm^{-1} ? CH_3 O H

27. Assign as many peaks as you can in the mass spectrum of 1-bromopropane (Figure 20-7).

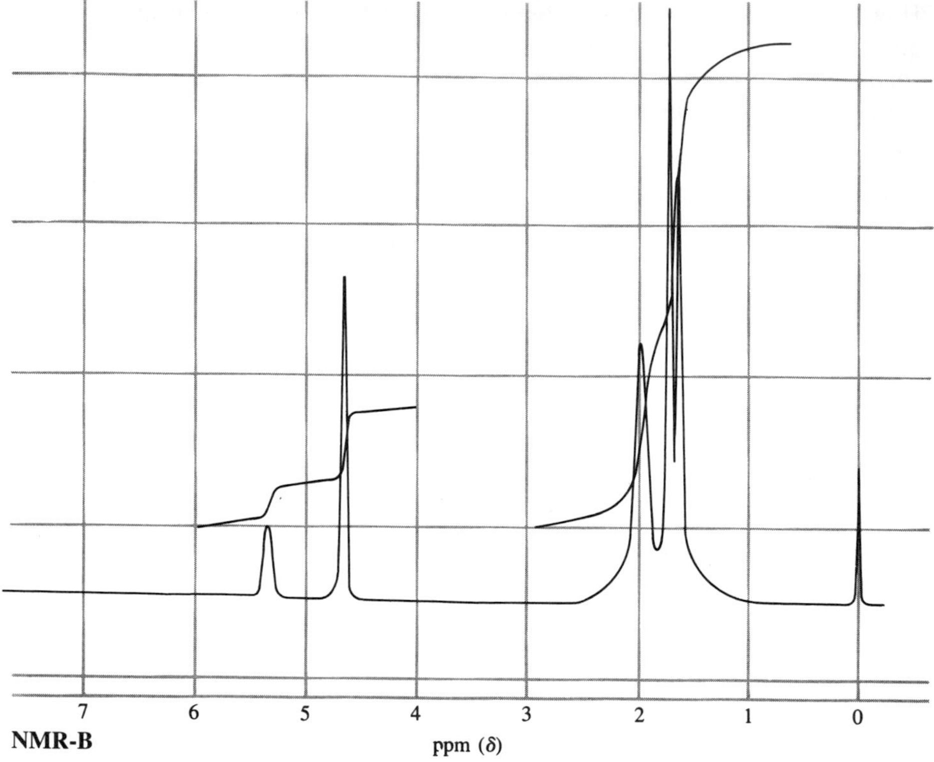

28. The following table lists selected mass-spectral data for three isomeric alcohols with the formula $C_5H_{12}O$. On the basis of the peak positions and intensities, suggest structures for each of the three isomers. A dash means that the peak is very weak or absent entirely.

Relative Peak Intensities

m/z	Isomer A	Isomer B	Isomer C
88 M^+	—	—	—
87 $(M-1)^+$	2	2	—
73 $(M-15)^+$	—	7	55
70 $(M-18)^+$	38	3	3
59 $(M-29)^+$	—	—	100
55 $(M-15-18)^+$	60	17	33
45 $(M-43)^+$	5	100	10
42 $(M-18-28)^+$	100	4	6

29. Following are spectroscopic and analytical characteristics of two unknown compounds. Propose a structure for each. **(a)** Empirical formula: $C_8H_{16}O$. ^{1}H NMR: $\delta = 0.90$ (t, 3 H), 1.0–1.6 (m, 8 H), 2.05 (s, 3 H), and 2.25 (t, 2 H) ppm. IR: 1715 cm^{-1}. UV: $\lambda_{max}\epsilon = 280(15)$ nm. MS: *m/z* for $M^{+\cdot}$ is 128; intensity of $(M+1)^+$ peak is 9% of $M^{+\cdot}$ peak; important fragments are at $m/z = 113$ $(M-15)^+$, $m/z = 85$ $(M-43)^+$, $m/z = 71$ $(M-57)^+$, $m/z = 58$ $(M-70)^+$ (the second largest peak), and $m/z = 43$

$(M - 79)^+$ (the base peak). **(b)** Empirical formula: C_5H_8. ^1H NMR: spectrum B. ^{13}C NMR (undecoupled from H): $\delta = 20.5$ (q), 23.8 (q), 28.0 (t), 30.6 (t), 30.9 (t), 41.2 (d), 108.4 (t), 120.8 (d), 133.2 (s), and 149.7 (s) ppm. IR: significant bands at 3060 (medium), 3010 (medium), 1680 (weak), 1646 (medium), and 880 (very strong) cm^{-1}. UV: $\lambda_{max} < 200$ nm. MS: m/z for $M^{+\cdot}$ is 136; intensity of $(M + 1)^+$ peak is 11% of $M^{+\cdot}$ peak; important fragments are at $m/z = 121$ $(M - 15)^+$, $m/z = 95$ $(M - 41)^+$, $m/z = 68$ $(M - 68)^+$ (the base peak), and $m/z = 41$ $(M - 95)^+$.

30. Spectroscopic data for two carboxylic acid derivatives are given below. Identify these compounds, which may contain C, H, O, N, Cl, and Br, but no other elements. **(a)** ^1H NMR: spectrum C (one signal has been amplified to reveal all peaks in the multiplet). IR: 1728 cm^{-1}. High-resolution mass spectrum: m/z for the molecular ion is 116.0837. See table (margin) for important MS fragmentation peaks. **(b)** ^1H NMR: spectrum D (p. 822). IR: 1739 cm^{-1}. High-resolution mass spectrum: The intact molecule gives two peaks with almost equal intensity $m/z = 179.9786$ and 181.9766. See table (margin, p. 822) for important MS fragmentation peaks.

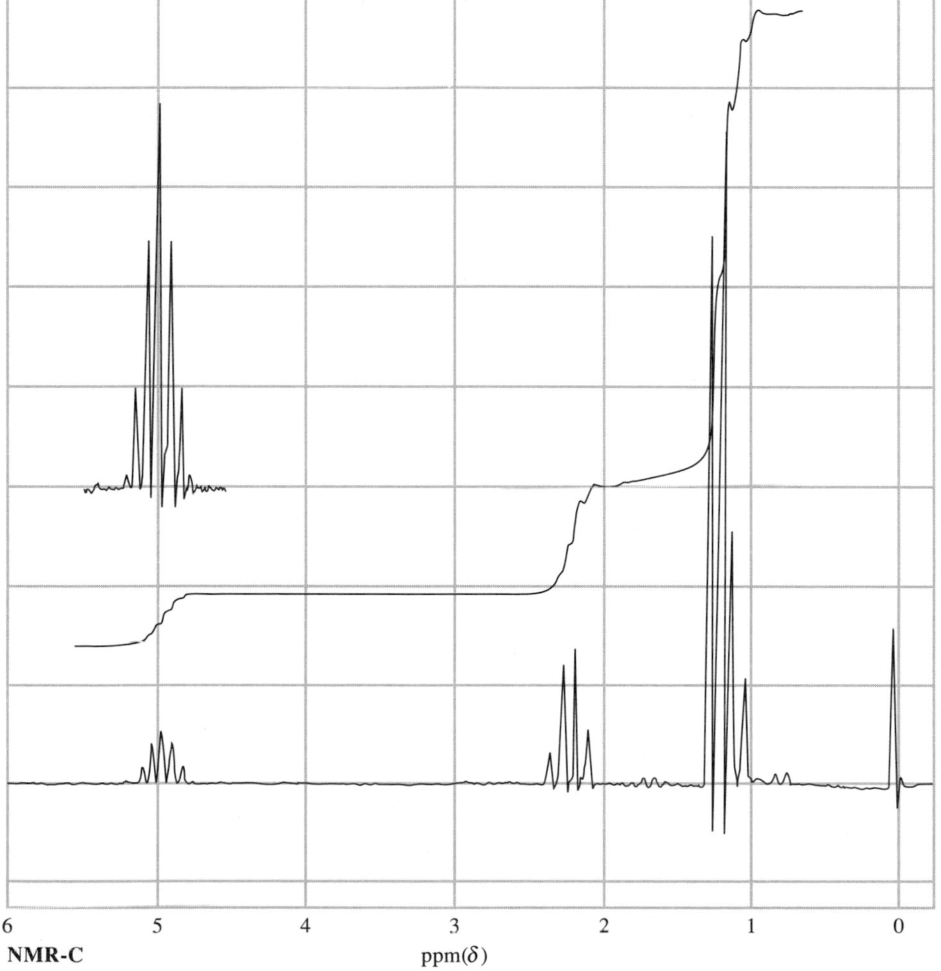

Mass Spectrum of Unknown C

m/z	Intensity relative to base peak (%)
116	0.5
101	12
75	26
57	100
43	66
29	34

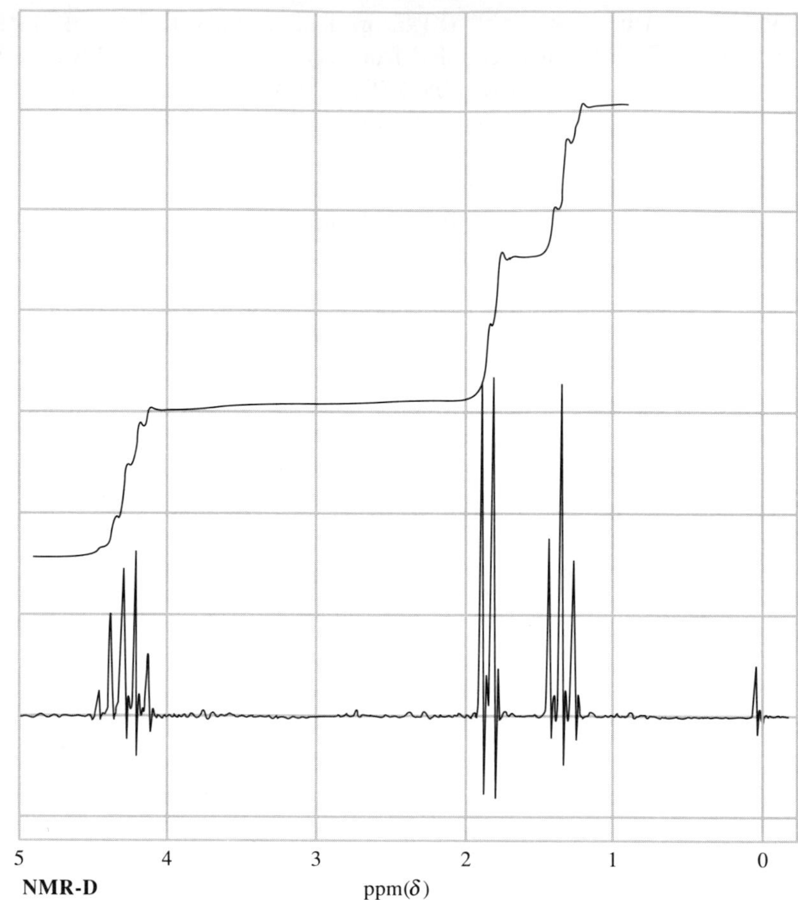

Mass Spectrum of Unknown D

m/z	Intensity relative to base peak (%)
182	13
180	13
109	78
107	77
101	3
29	100

21

Amines and Their Derivatives

Functional Groups Containing Nitrogen

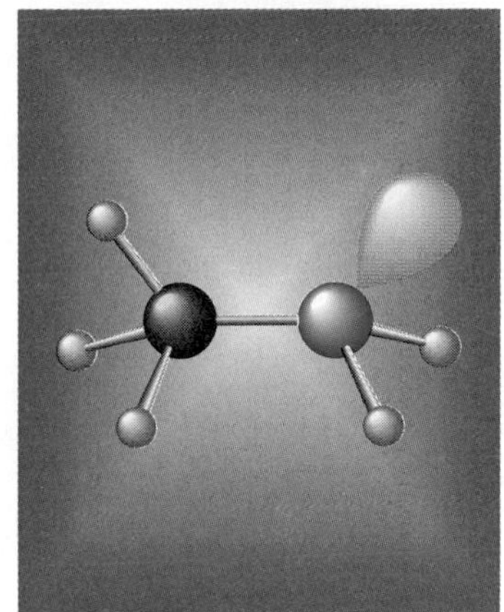

Amines and other nitrogen-bearing compounds are among the most abundant organic molecules. As components of the amino acids, peptides, proteins, and alkaloids, they are essential to biochemistry. Many, such as the neurotransmitters, possess powerful physiological activity; related substances have found medicinal uses as decongestants, anesthetics, sedatives, and stimulants.

In many respects, the chemistry of the amines is analogous to that of the alcohols and ethers (Chapters 8 and 9). For example, all amines are basic (although primary and secondary ones can also behave as acids), they form hydrogen bonds, and they act as nucleophiles in substitution reactions. However, there are some differences in reactivity, because nitrogen is less electronegative than oxygen. Thus, primary and secondary amines are less acidic and form weaker hydrogen bonds than alcohols and ethers, and they are more basic and more nucleophilic. This chapter will show that these properties underlie their physical and chemical characteristics and give us a variety of ways to synthesize amines.

21-1 Naming the Amines

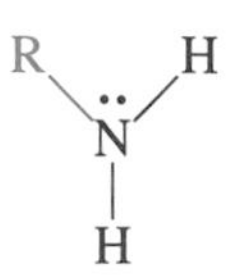

Ammonia

Primary amine

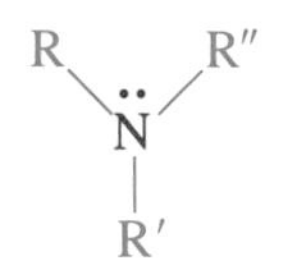

Secondary amine

Tertiary amine

Amines are derivatives of ammonia, in which one (primary), two (secondary), or three (tertiary) of the hydrogens have been replaced by alkyl or aryl groups. Therefore, amines are related to ammonia in the same sense as ethers and alcohols are related to water. The system for naming amines is confused by the variety of common names in the literature. Probably the best way to name aliphatic amines is that used by *Chemical Abstracts*—that is, as **alkanamines,** in which the name of the alkane stem is modified by replacing the ending *-e* by **-amine.** The position of the functional group is indicated by a prefix designating the carbon atom to which it is attached, as in the alcohols (Section 8-1).

CH_3NH_2
Methanamine

$CH_3CH(CH_3)CH_2NH_2$
2-Methyl-1-propanamine

$CH_3CH(NH_2)CH{=}CHCH_3$
3-Penten-2-amine

Substances with two amine functions are **diamines,** an example of which is 1,4-butanediamine. Its contribution to the smell of dead fish and rotting flesh leads to its singularly appropriate common name, putrescine.

$H_2NCH_2CH_2CH_2CH_2NH_2$
1,4-Butanediamine
(Putrescine)

The aromatic amines, or anilines, are called **benzenamines** (Section 15-1). For secondary and tertiary amines, the largest alkyl substituent on nitrogen is chosen as the alkanamine stem, and the other groups are named by using the letter *N*-, followed by the name of the additional substituent(s).

$C_6H_5NH_2$
Benzenamine
(Aniline)

$CH_3N(H)CH_2CH_3$
N-Methylethanamine

$CH_3N(CH_3)CH_2CH_2CH_3$
N,N-Dimethyl-1-propanamine

CH_3NH_2
Methylamine

$(CH_3)_3N$
Trimethylamine

$CH_3N(C_6H_{11})CH_2C_6H_5$
Benzylcyclohexylmethylamine

An alternative way to name amines treats the functional group, called **amino-,** as a substituent of the alkane stem. This procedure is analogous to naming alcohols as hydroxyalkanes.

$CH_3CH_2NH_2$
Aminoethane

$(CH_3)_2NCH_2CH_2CH_3$
N,N-Dimethylaminopropane

$FCH_2CH_2CH(CH_3)N(H)CH_2CH_3$
3-(Ethylamino)-1-fluorobutane

Many common names are based on the label **alkylamine** (see margin), as in the naming of alkyl alcohols.

EXERCISE 21-1

Name each of the following molecules twice, first as an alkanamine, then as an alkyl amine.

(a) 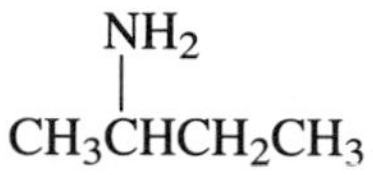 $CH_3CH(NH_2)CH_2CH_3$ (b) $C_6H_5N(CH_3)_2$ (c) $BrCH_2CH_2CH_2CH_2CH(CH_3)NH_2$

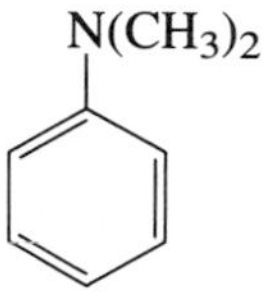

BOX 21-1 Physiologically Active Amines

A large number of physiologically active compounds contain nitrogen. Several simple amines are used as drugs. Examples are epinephrine, propylhexedrine, hexamethylenetetramine, amphetamine, and mescaline. A recurring pattern in many (but not all) of these compounds is the 2-phenylethanamine (β-phenethylamine) unit. Although the mechanism of action of these substances is not understood, it seems that such a structural feature is required for their binding to a receptor site.

Not only are the molecules potent central nervous system stimulants, but they also increase cardiovascular activity, raise body temperature, and cause loss of appetite. The last function is one reason for their use in diets and the treatment of obesity. Unfortunately, they can lead to chemical dependency and are potentially dangerous, particularly when taken without discretion.

Many amines in which the nitrogen is part of a ring (nitrogen heterocycles, Chapter 25) also have powerful physiological effects.

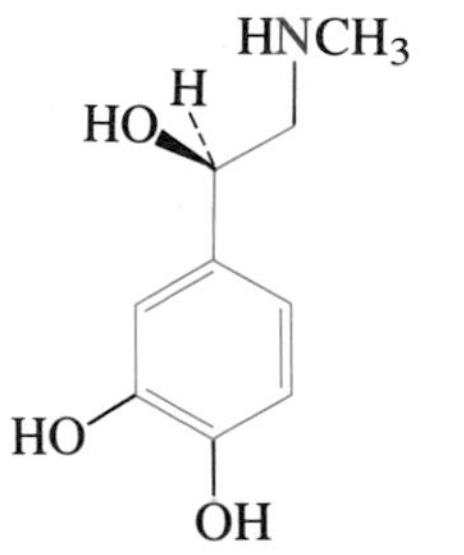

Epinephrine (Adrenaline)
(Adrenergic stimulant)

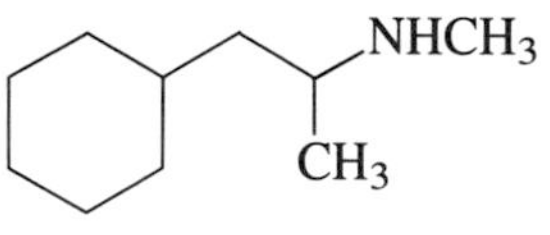

Propylhexedrine (Benzedrex)
(Nasal decongestant)

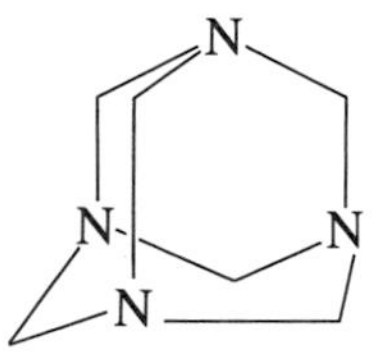

Hexamethylenetetramine (Urotropine)
(Antibacterial agent)

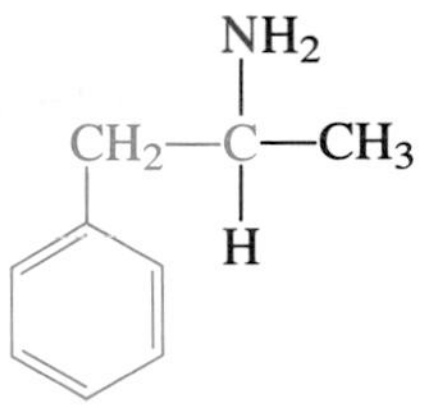

Amphetamine
(Antidepressant, central nervous system stimulant)

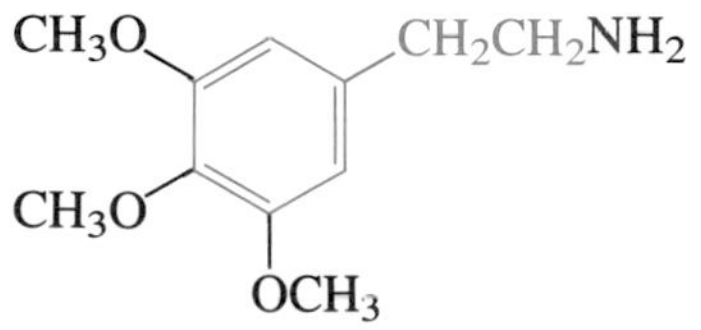

Mescaline
(Hallucinogen)

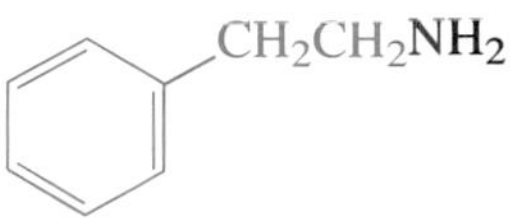

2-Phenylethanamine (β-Phenethylamine)

EXERCISE 21-2

Draw structures for the following compounds (common name in parentheses). **(a)** 2-Propynamine (propargylamine); **(b)** (*N*-2-propenyl)phenylmethanamine (*N*-allylbenzylamine); **(c)** *N*,2-dimethyl-2-propanamine (*tert*-butylmethylamine).

In summary, there are several systems for naming amines. *Chemical Abstracts* uses names of the type *alkanamine* and *benzenamine*. Alternatives are the labels *aminoalkane, aniline,* and *alkylamine*.

21-2 Structural and Physical Properties of Amines

Let us now look at some of the structural and physical characteristics of simple amines. Amines adopt a tetrahedral geometry around the heteroatom, but this arrangement is not rigid because of a rapid isomerization process called inversion. This section also compares the polarity and hydrogen-bonding ability of amines with that of alcohols.

The alkanamine molecule is tetrahedral

The nitrogen orbitals in amines are very nearly sp^3 hybridized (see Section 1-8), forming an approximately tetrahedral arrangement. Three vertices of the tetrahedron are occupied by the three substituents; the fourth is the center of the lone electron pair. The term **pyramidal** is often used to describe the geometry adopted by the nitrogen and its three substituents. Figure 21-1 depicts the structure of methanamine (methylamine).

EXERCISE 21-3

The bonds to nitrogen in methanamine (methylamine; see Figure 21-1) are slightly longer than the bonds to oxygen in the structure of methanol (Figure 8-1). Explain.

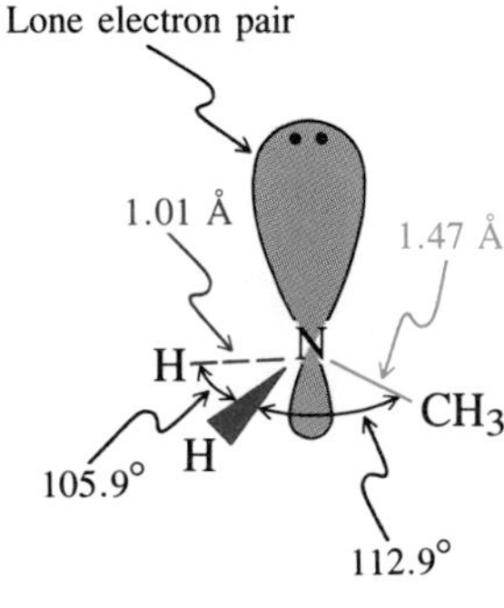

FIGURE 21-1 Nearly tetrahedral structure of methanamine (methylamine).

The tetrahedral geometry around an amine nitrogen suggests that it should be chiral if it bears three different substituents, the lone electron pair serving as the fourth. The image and mirror image of such a compound are not superimposable, by analogy with carbon-based stereocenters (Section 5-1). This is illustrated with the simple chiral alkanamine *N*-methylethanamine (ethylmethylamine).

Image and Mirror Image of *N*-Methylethanamine (Ethylmethylamine)

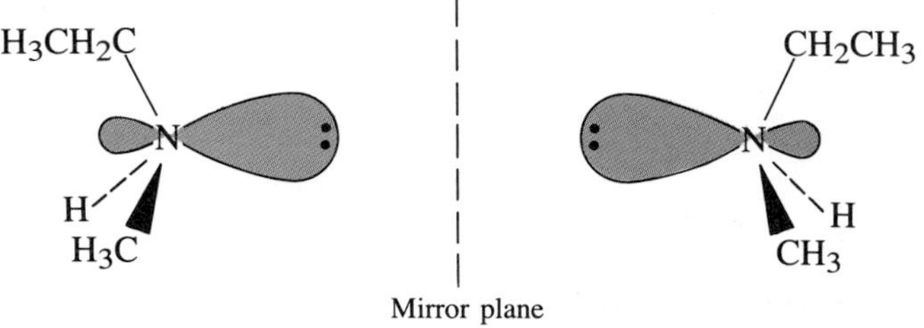

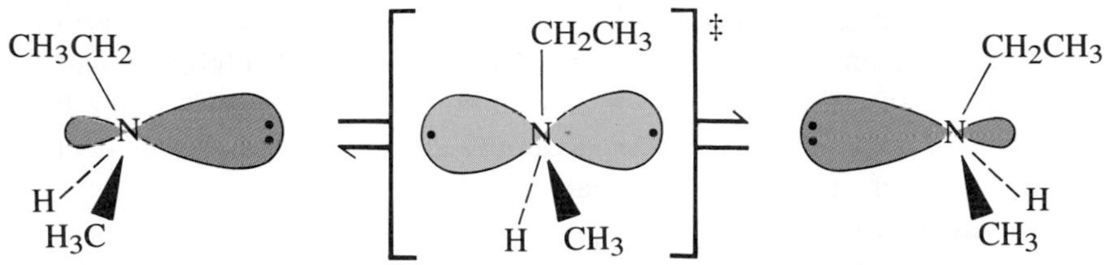

FIGURE 21-2 Inversion at nitrogen rapidly interconverts the two enantiomers of *N*-methylethanamine (ethylmethylamine). Thus, the compound exhibits no optical activity.

However, samples of the amine prove *not* to be optically active. Why? Amines are not configurationally stable at nitrogen, because of rapid isomerization by a process called **inversion.** The molecule passes through a transition state incorporating an sp^2-hybridized nitrogen atom, as illustrated in Figure 21-2. The barrier to this motion in ordinary small amines has been measured by spectroscopic techniques and found to be between 5 and 7 kcal mol^{-1}. It is therefore impossible to keep an enantiomerically pure, simple di- or trialkylamine from racemizing at room temperature when the nitrogen atom is the only stereocenter.

Amines form weaker hydrogen bonds than alcohols do

Because alcohols readily form hydrogen bonds (Section 8-2), they have unusually high boiling points. In principle, so should amines, and indeed Table 21-1 bears out this expectation. However, because amines form weaker hydrogen bonds* than alcohols do, their boiling points are lower and their solubility in

TABLE 21-1 Physical Properties of Amines, Alcohols, and Alkanes

Compound	Melting point (°C)	Boiling point (°C)	Compound	Melting point (°C)	Boiling point (°C)
CH_4	−182.5	−161.7	$(CH_3)_2NH$	−93	7.4
CH_3NH_2	−93.5	−6.3	$(CH_3)_3N$	−117.2	2.9
CH_3OH	−97.5	65.0			
			$(CH_3CH_2)_2NH$	−48	56.3
CH_3CH_3	−183.3	−88.6	$(CH_3CH_2)_3N$	−114.7	89.3
$CH_3CH_2NH_2$	−81	16.6			
CH_3CH_2OH	−114.1	78.5	$(CH_3CH_2CH_2)_2NH$	−40	110
			$(CH_3CH_2CH_2)_3N$	−94	155
$CH_3CH_2CH_3$	−187.7	−42.1			
$CH_3CH_2CH_2NH_2$	−83	47.8	NH_3	−77.7	−33.4
$CH_3CH_2CH_2OH$	−126.2	97.4	H_2O	0	100

*Whereas *all* amines can act as proton acceptors in hydrogen bonding, only primary and secondary amines can function as proton donors, because tertiary amines lack such protons.

water less. In general, the boiling points of the amines lie between those of the corresponding alkanes and alcohols. The smaller amines are soluble in water and in alcohols because they can form hydrogen bonds to the solvent. If the hydrophobic part of an amine exceeds six carbons, the solubility in water decreases rapidly; the larger amines are essentially insoluble in water.

To summarize, amines adopt an approximately tetrahedral structure in which the lone electron pair occupies one vertex of the tetrahedron. They can, in principle, be chiral at nitrogen but are difficult to maintain in enantiomerically pure form because of fast nitrogen inversion. Amines have boiling points higher than those of alkanes of similar size. Boiling points are lower than those of the analogous alcohols because of weaker hydrogen bonding, and their water solubility is between that of comparable alkanes and alcohols.

21-3 Spectroscopy of the Amine Group

Primary and secondary amines can be recognized by infrared spectroscopy because they exhibit a characteristic broad N–H stretching absorption in the range between 3250 and 3500 cm^{-1}. Primary amines show two strong peaks in this range, whereas secondary amines give rise to only a very weak single line. Primary amines also show a band near 1600 cm^{-1} that is due to a scissoring motion of the NH_2 group (Section 11-5, Figure 11-3). Tertiary amines do not give rise to such signals because they do not have a hydrogen that is bound to nitrogen. Figure 21-3 shows the infrared spectrum of cyclohexanamine.

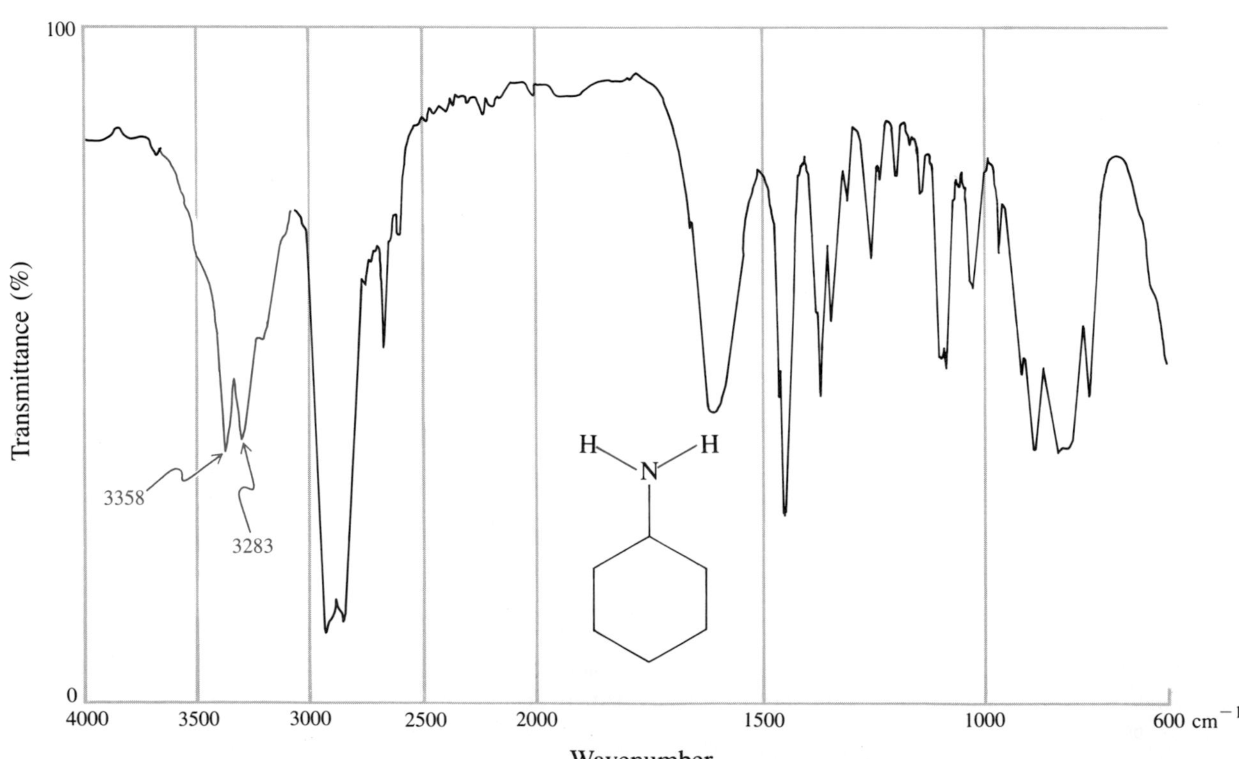

FIGURE 21-3 Infrared spectrum of cyclohexanamine. The amine exhibits two strong peaks between 3250 and 3500 cm^{-1}, characteristic of the N–H stretching absorptions of the primary amine functional group. Note also the broad band near 1600 cm^{-1}, which results from scissoring motions of the N–H bonds.

Nuclear magnetic resonance spectroscopy is also useful for detecting the presence of amino groups. Amine hydrogens resonate to give (sometimes broadened) peaks almost anywhere in the normal hydrogen range, like the OH signal in the NMR spectra of alcohols. Their chemical shift depends mainly on the rate of exchange of protons with water in the solvent and the degree of hydrogen bonding. Figure 21-4 shows the ^1H NMR spectrum of azacyclohexane (piperidine), a cyclic secondary amine. The amine hydrogen appears at $\delta = 1.29$ ppm, and there are two other sets of signals, at $\delta = 1.52$ and 2.73 ppm. The absorption at lowest field can be assigned to the hydrogens neighboring the nitrogen, which are deshielded by the nearby electronegative nitrogen atom.

EXERCISE 21-4

Would you expect the hydrogens next to the heteroatom in an amine, RCH_2NH_2, to be more or less deshielded than those in an alcohol, RCH_2OH? Explain.

The ^{13}C NMR spectra of amines show a similar trend: Carbons directly bound to nitrogen resonate at considerably lower field than do the carbon atoms in

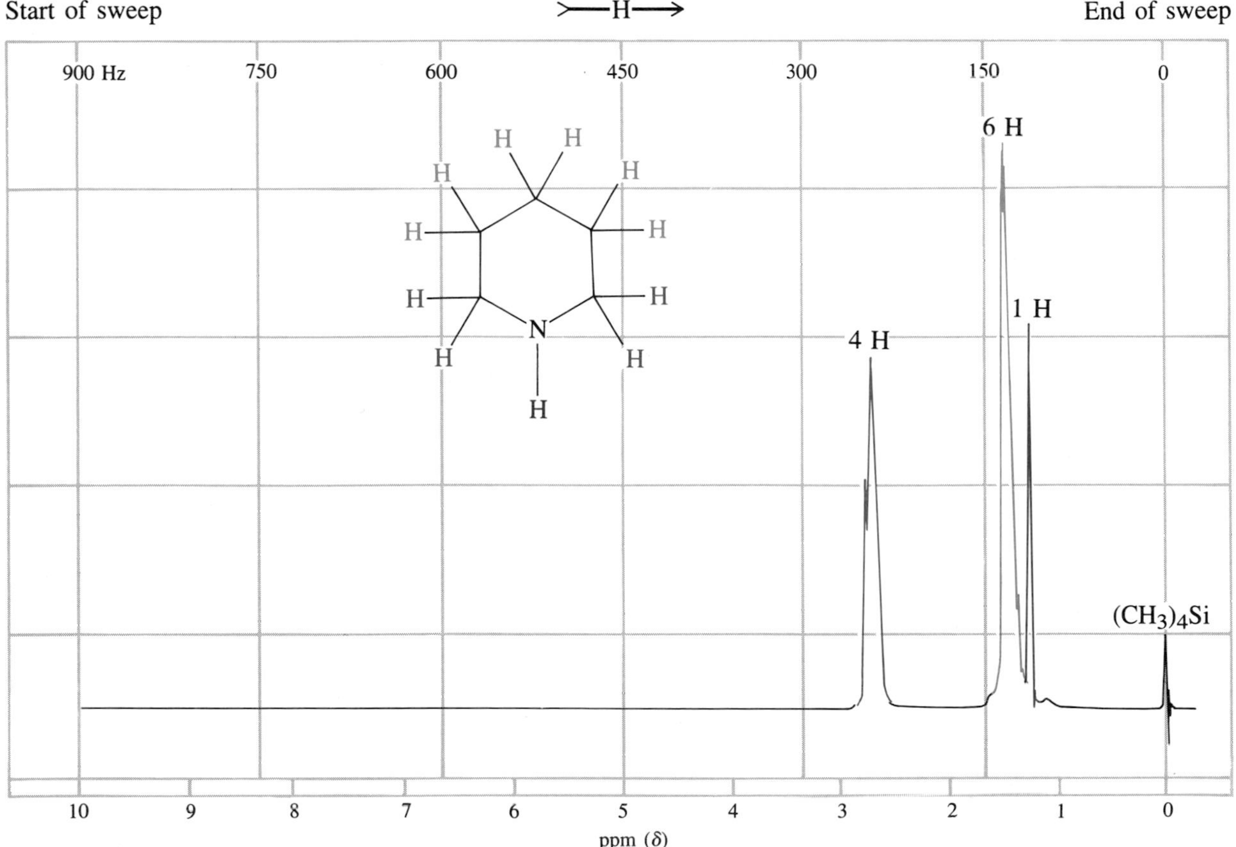

FIGURE 21-4 90 MHz ^1H NMR spectrum of azacyclohexane (piperidine) in dry CCl_4. Like the OH hydrogen signal in alcohols, the amine NH peak may appear almost anywhere in the normal hydrogen chemical-shift range. Here $\delta = 1.29$ ppm, and the NH absorption is sharp, because of the use of dry solvent.

alkanes. However, as in the hydrogen spectra (Exercise 21-4), nitrogen is less deshielding than oxygen is.

^{13}C Chemical Shifts in Various Amines (ppm)

Piperidine: 25.5, 27.2, 47.5 (N–H) compared with tetrahydropyran: 23.6, 26.6, 68.0 (O); cyclohexanamine: 25.1, 25.7, 36.7, 50.4 (NH_2); $(CH_3CH_2)_2NH$: 15.4, 44.2; $CH_3N(H)CH(CH_3)_2$: 33.6, 50.9, 22.6

Mass spectroscopy also can be used to establish the presence of nitrogen in an organic compound. Unlike carbon, which is tetravalent, nitrogen is trivalent. Because of these valence requirements and because nitrogen has an even atomic weight (14), molecules incorporating one nitrogen (or any odd number of nitrogens) have an *odd* molecular weight (recall Exercise 20-25). For example, the mass spectrum of *N,N*-diethylethanamine (triethylamine) shows the peak of a molecular ion at $m/z = 101$ (Figure 21-5). The base peak, at $m/z = 86$, is caused by the loss of a methyl group. Such a fragmentation is favored because it results in a resonance-stabilized **iminium ion.**

Mass-Spectral Fragmentation of *N,N*-Diethylethanamine

$$[(CH_3CH_2)_2\ddot{N}CH_2 \text{–}\!\!\!\!\wr\, CH_3]^{+\cdot} \longrightarrow CH_3\cdot + [(CH_3CH_2)_2\ddot{N}\text{—}CH_2^{+} \longleftrightarrow (CH_3CH_2)_2\overset{+}{N}{=}CH_2]$$

_N,N_-Diethylethanamine (Triethylamine) **Iminium ion**

The rupture of the C–C bond next to nitrogen is frequently so easy that the molecular ion cannot be observed. For example, in the mass spectrum of 1-hexanamine, the molecular ion ($m/z = 101$) is barely visible; the dominating peak corresponds to the methyleneiminium fragment $[CH_2{=}NH_2]^+$ ($m/z = 30$).

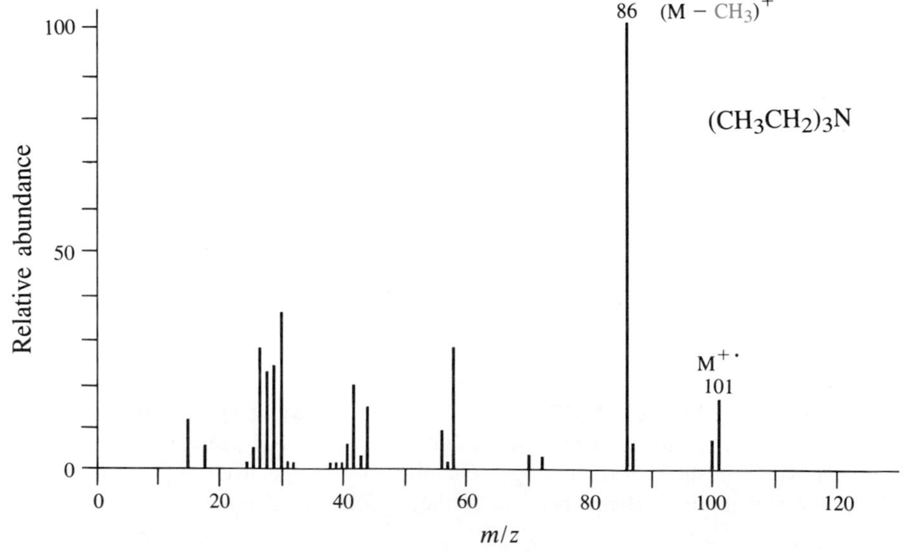

FIGURE 21-5 Mass spectrum of *N,N*-diethylethanamine (triethylamine), showing a molecular ion peak at $m/z = 101$. In general, molecules incorporating one nitrogen atom have an odd molecular weight. The base peak is due to loss of a methyl group, resulting in an iminium ion with $m/z = 86$.

EXERCISE 21-5

What approximate spectral data (IR, NMR, *m/z*) would you expect for *N*-ethyl-2,2-dimethylpropanamine, shown below?

$$CH_3\overset{\displaystyle CH_3}{\underset{\displaystyle CH_3}{\overset{|}{\underset{|}{C}}}}CH_2NHCH_2CH_3$$

***N*-Ethyl-2,2-dimethylpropanamine**

In summary, the IR stretching absorption of the N–H bond ranges between 3250 and 3500 cm^{-1}; the corresponding ^1H NMR peak can be found at variable δ. The electron-withdrawing nitrogen deshields neighboring carbons and hydrogens, although to a lesser extent than in alcohols and ethers. The mass spectra of simple alkanamines that contain only one nitrogen atom have odd-numbered molecular ion peaks, because of the trivalent character of nitrogen. Fragmentation occurs in such a way as to produce resonance-stabilized iminium ions.

21-4 Acidity and Basicity of Amines

Like the alcohols (Section 8-3), amines are both basic and acidic. Because nitrogen is less electronegative than oxygen, the acidity of amines is about 20 orders of magnitude less than that of comparable alcohols. Conversely, the lone pair is much more available for protonation, thereby causing amines to be better bases.

Acidity and Basicity of Amines

Amine acting as an acid: $$R\ddot{N}H(H) + {}^{-}{:}B \underset{}{\overset{K_a}{\rightleftharpoons}} R\ddot{\underset{\cdot\cdot}{N}}H^{-} + HB$$

Amine acting as a base: $$R\ddot{N}H_2 + HA \overset{K_b}{\rightleftharpoons} R\overset{+}{N}H_2(H) + {}^{-}{:}A$$

Amines are very weak acids

We have seen evidence that amines are much less acidic than alcohols: Amide ions, R_2N^-, are used to deprotonate alcohols (Section 9-1). The equilibrium of this proton transfer is strongly shifted to the side of the alkoxide ion. The high value of the equilibrium constant, about 10^{20}, is due to the strong basicity of amide ions, which is reflected, in turn, in the low acidity of amines. The pK_a of ammonia and alkanamines is on the order of 35.

Acidity of Amines

$$R\ddot{N}H(H) + H_2\ddot{O} \overset{K_a}{\rightleftharpoons} R\ddot{\underset{\cdot\cdot}{N}}H^{-} + H_2\ddot{O}H^{+} \qquad K_a = \frac{[R\ddot{N}H^-][H_2\ddot{O}H^+]}{[R\ddot{N}H(H)]} \sim 10^{-35}$$

Amine (Weak acid) **Amide ion** (Strong base) $pK_a \sim 35$

The deprotonation of amines requires exceedingly strong bases, such as alkyllithium reagents. For example, lithium diisopropylamide, the special sterically hindered base used in some bimolecular elimination reactions (Section 7-8), is made in the laboratory by treatment of *N*-(1-methylethyl)-2-propanamine (diisopropylamine) with butyllithium.

Preparation of LDA

$$\underset{\substack{\textbf{N-(1-Methylethyl)-}\\ \textbf{2-propanamine}\\ \textbf{(Diisopropylamine)}}}{CH_3\overset{CH_3}{\overset{|}{C}}H\ddot{N}H\overset{CH_3}{\overset{|}{C}}HCH_3} \xrightarrow[-\ CH_3CH_2CH_2CH_2H]{CH_3CH_2CH_2CH_2Li} \underset{\substack{\textbf{Lithium}\\ \textbf{diisopropylamide, LDA}}}{(CH_3\overset{CH_3}{\overset{|}{C}}H)_2\ddot{\underset{..}{N}}^{-}Li^{+}}$$

An alternative synthesis of amide ions is the treatment of amines with alkali metals. Alkali metals dissolve in amines (albeit relatively slowly) with the evolution of hydrogen and the formation of amine salts. For example, sodium amide can be made in liquid ammonia from sodium metal in the presence of catalytic amounts of Fe^{3+}, which facilitates electron transfer to the amine. In the absence of such a catalyst, sodium simply dissolves in ammonia to form a strongly reducing solution (Section 13-7).

Preparation of Sodium Amide

$$2\ Na + 2\ NH_3 \xrightarrow{\text{Catalytic } Fe^{3+}} 2\ NaNH_2 + H_2$$

Amines are moderately basic

Amines deprotonate water to a *small* extent to form ammonium and hydroxide ions. Thus, amines are more strongly basic than alcohols, but not nearly as basic as alkoxides; their pK_b values (Section 6-7) are about 4.

Basicity of Amines

$$\underset{\textbf{Amine}}{R\ddot{N}H_2} + H\ddot{\underset{..}{O}}H \overset{K_b}{\rightleftharpoons} \underset{\textbf{Ammonium ion}}{R\overset{H}{\overset{|+}{N}}H_2} + H\ddot{\underset{..}{O}}:^{-} \qquad K_b = \frac{[R\overset{H}{\overset{|+}{N}}H_2][H\ddot{\underset{..}{O}}:^{-}]}{[R\ddot{N}H_2]} \sim 10^{-4}$$

$pK_b \sim 4$

Alkanamines are slightly more basic than ammonia but less basic than hydroxide ion ($pK_b = -1.7$).

pK_b Values of a Series of Simple Amines

	NH_3	CH_3NH_2	$(CH_3)_2NH$	$(CH_3)_3N$
$pK_b =$	4.76	3.38	3.27	4.21

The protonation of amines gives ammonium salts. Depending on the number of substituents on nitrogen, these salts can be primary, secondary, or tertiary.

$$R\ddot{N}H_2 + H^+Cl^- \longrightarrow RNH_3{}^+Cl^-$$
Primary ammonium chloride

$$R_2\ddot{N}H + H^+Cl^- \longrightarrow R_2NH_2{}^+Cl^-$$
Secondary ammonium chloride

$$R_3N\!: + H^+I^- \longrightarrow R_3NH^+I^-$$
Tertiary ammonium iodide

Ammonium salts are named by attaching the substituent names to the ending *-ammonium* followed by the name of the anion.

$CH_3NH_3{}^+Cl^-$
Methylammonium chloride

Ammonium salts are weakly acidic

It is useful to view the basicity of the amines as a reflection of the acidity of their conjugate acids (Section 6-7), the ammonium ions. These species are stronger acids than water or alcohols but much weaker than carboxylic acids.

Acidity of Ammonium Ions

$$R\overset{+}{N}H_3 + H_2\ddot{O} \overset{K_a}{\rightleftharpoons} R\ddot{N}H_2 + H_2\overset{+}{\ddot{O}}H \qquad K_a = \frac{[R\ddot{N}H_2][H_2\overset{+}{\ddot{O}}H]}{[R\overset{+}{N}H_3]} \sim 10^{-10}$$

$$pK_a \sim 10$$

The pK_b value is easily converted into the pK_a of the conjugate acid by using the relation $pK_a + pK_b = 14$ (Section 6-7).

pK_a Values of a Series of Simple Ammonium Salts*

	$\overset{+}{N}H_4$	$CH_3\overset{+}{N}H_3$	$(CH_3)_2\overset{+}{N}H_2$	$(CH_3)_3\overset{+}{N}H$
$pK_a =$	9.24	10.62	10.73	9.79

Optically active amines are resolving agents

Some naturally occurring, optically active amines, particularly alkaloids (Chapter 25), are useful in the resolution of enantiomers (Section 5-8). For example, upon acid-base reaction with a racemic mixture of a chiral carboxylic acid, they form diastereomeric ammonium salts that can readily be crystallized. Fractional crystallization may then be followed by acidification, to separate the pure enantiomers of the acid and allow the recovery of the amine for further use.

*A confusing practice in the literature is to refer to the pK_a value of an ammonium salt as being that of the neutral amine. In the statement "the pK_a of methanamine is 10.62," what is given is the pK_a of the methylammonium ion. The pK_a of methanamine is actually 35.

BOX 21-2 Separation of Organic Compounds by Using Acid-Base Properties

Organic chemicals with different acid or base strengths may frequently be separated by using extraction techniques. Such procedures work best when the substances in the original mixture are much less soluble in water than in solvents such as hexane or ethoxyethane (diethyl ether).

First, a solution of the mixture in either solvent is treated with mild aqueous base, such as $NaHCO_3$, thereby converting any carboxylic acid that may be present into its sodium salt. This salt, being water soluble, moves from the organic into the aqueous layer. After the layers are separated, the carboxylic acid may be brought out of the water solution by acidification with strong mineral acid, such as HCl. Next, the organic solution that remains is treated with acid, thus converting any amine present into the corresponding water-soluble ammonium salt, which moves into the water phase. Separation of the layers gives an organic solution, which contains neutral (that is, only very weakly acidic or basic) substances. The amine is removed from water solution by treatment with strong base, such as NaOH.

A flow chart that describes such a separation procedure schematically is shown below.

Separation of Acidic and Basic Organic Compounds from Neutral Substances

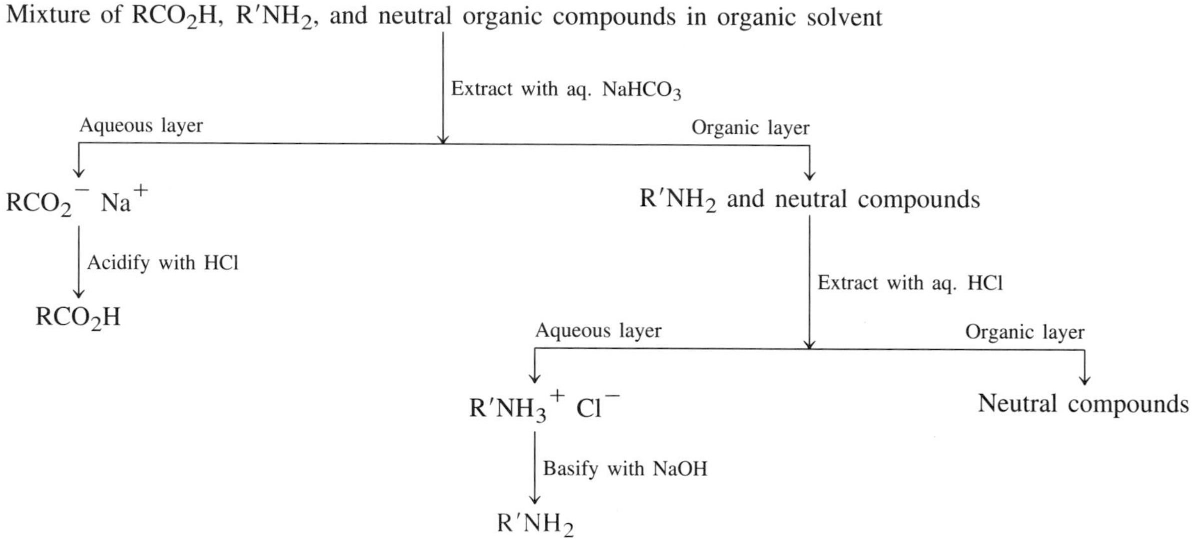

Mixture of RCO_2H, $R'NH_2$, and neutral organic compounds in organic solvent

↓ Extract with aq. $NaHCO_3$

Aqueous layer: $RCO_2^- \ Na^+$ → Acidify with HCl → RCO_2H

Organic layer: $R'NH_2$ and neutral compounds → Extract with aq. HCl

- Aqueous layer: $R'NH_3^+ \ Cl^-$ → Basify with NaOH → $R'NH_2$
- Organic layer: Neutral compounds

In summary, amines are poor acids and require alkyllithium reagents or alkali metal treatment to form amide ions. In contrast, they are good bases, although they are weaker than hydroxide ion.

21-5 Synthesis of Amines by Alkylation

Amines can be synthesized by alkylating nitrogen atoms. Several such procedures take advantage of an important property of the nitrogen in many compounds: It is *nucleophilic*.

Amines can be derived from other amines

As nucleophiles, amines react with haloalkanes to give ammonium salts (Section 6-3). Unfortunately, this reaction is not clean, because the resulting amine product usually undergoes further alkylation. How does this complication arise?

Consider the alkylation of ammonia with bromomethane. When this transformation is carried out with equimolar quantities of starting materials, the product (methylammonium bromide), as soon as it is formed, exchanges a proton with the starting ammonia. The small quantities of methanamine generated in this way then compete effectively with the ammonia for the alkylating agent, and this further methylation generates a dimethylammonium salt.

The process does not stop there, either. This salt can donate a proton to either of the other two nitrogen bases present, furnishing *N*-methylmethanamine (dimethylamine). This compound constitutes yet another nucleophile competing for bromomethane; its further reaction leads to *N*,*N*-dimethylmethanamine (trimethylamine) and, eventually, to tetramethylammonium bromide, a *quaternary* ammonium salt. *The final outcome is a mixture of alkylammonium salts and alkanamines.*

Methylation of Ammonia

FIRST ALKYLATION. Gives primary amine

$$H_3N\!: + CH_3Br \longrightarrow CH_3\overset{+}{N}H_3\ Br^-$$

Methylammonium bromide

$$CH_3\overset{+}{N}H_2(H)\ Br^- + :NH_3 \rightleftharpoons CH_3\ddot{N}H_2 + H\overset{+}{N}H_3\ Br^-$$

Methanamine

SECOND ALKYLATION. Gives secondary amine

$$CH_3\ddot{N}H_2 + CH_3Br \longrightarrow (CH_3)_2\overset{+}{N}H_2\ Br^-$$

Dimethylammonium bromide

$$(CH_3)_2\overset{+}{N}H(H)\ Br^- + :NH_3 \text{ or } CH_3\ddot{N}H_2 \rightleftharpoons (CH_3)_2\ddot{N}H + H\overset{+}{N}H_3\ Br^- \text{ or } CH_3\overset{+}{N}H_2(H)\ Br^-$$

***N*-Methylmethanamine (Dimethylamine)**

THIRD ALKYLATION. Gives tertiary amine

$$(CH_3)_2\ddot{N}H + CH_3Br \longrightarrow (CH_3)_3\overset{+}{N}H\ Br^-$$

Trimethylammonium bromide

$$(CH_3)_3\overset{+}{N}H\ Br^- + \text{amine} \rightleftharpoons (CH_3)_3N\!: + \text{amine hydrobromide}$$

***N*,*N*-Dimethylmethanamine (Trimethylamine)**

FOURTH ALKYLATION. Gives quaternary ammonium salt

$$(CH_3)_3N\!: + CH_3Br \longrightarrow (CH_3)_4N^+Br^-$$

Tetramethylammonium bromide

The mixture of products obtained upon treatment of haloalkanes with ammonia or amines is a serious drawback that limits the usefulness of direct alkylation in synthesis. As a result, indirect alkylation methods are frequently applied.

EXERCISE 21-6

Like other amines, benzenamine (aniline) can be benzylated with chloromethylbenzene (benzyl chloride), $C_6H_5CH_2Cl$. In contrast with the reaction with alkanamines, which proceeds at room temperature, this transformation requires heating to between 90 and 95°C. Explain. (Hint: Review Section 16-2.)

Indirect alkylation is more effective

Controlled alkylation of amines requires a nitrogen-containing nucleophile that will undergo reaction *only once*. For example, cyanide ion, ^-CN, turns primary and secondary haloalkanes into nitriles, which can be reduced subsequently to the corresponding amines (Section 20-8). This sequence allows the conversion $RX \rightarrow RCH_2NH_2$.

Conversion of a Haloalkane into the Homologous Amine by Cyanide Displacement–Reduction

$$RX + {}^-CN \longrightarrow RC{\equiv}N + X^-$$

$$RC{\equiv}N \xrightarrow{\text{LiAlH}_4 \text{ or } H_2\text{, metal catalyst}} RCH_2NH_2$$

$$\underset{\textbf{1,8-Dibromooctane}}{Br(CH_2)_8Br} + NaCN \xrightarrow[-\ 2NaBr]{DMSO} \underset{\textbf{Decanedinitrile}}{\underset{93\%}{NC(CH_2)_8CN}} \xrightarrow{H_2\text{, Raney Ni, 100 atm}} \underset{\textbf{1,10-Decanediamine}}{\underset{80\%}{H_2NCH_2(CH_2)_8CH_2NH_2}}$$

$R{-}N_3$
Alkyl azide

To introduce an amino group without additional carbons requires a modified nitrogen nucleophile, which should be unreactive after the first alkylation. Such a nucleophile is the **azide ion,** N_3^-, which reacts with haloalkanes to furnish **alkyl azides.** These in turn are reduced by catalytic hydrogenation (Pd-C) or by lithium aluminum hydride to the corresponding primary amines.

Azide Displacement–Reduction

$$\text{(3-cyclopentylpropyl)}{-}Br + Na^+N_3^- \xrightarrow[\substack{-\ Na^+Br^- \\ S_N2 \text{ reaction}}]{CH_3CH_2OH}$$

$$\underset{\textbf{3-Cyclopentylpropyl azide}}{\underset{91\%}{\text{(3-cyclopentylpropyl)}{-}N{=}\overset{+}{N}{=}N^-}} \xrightarrow{\substack{1.\ LiAlH_4,\ (CH_3CH_2)_2O \\ 2.\ H^+,\ H_2O}} \underset{\textbf{3-Cyclopentylpropanamine}}{\underset{89\%}{\text{(3-cyclopentylpropyl)}{-}NH_2}}$$

A nonreductive approach to synthesizing primary amines uses the anion of 1,2-benzenedicarboximide (phthalimide), the imide of 1,2-benzenedicarboxylic (phthalic) acid. This process is also known as the **Gabriel* synthesis.** Because

*Professor Siegmund Gabriel, 1851–1924, University of Berlin.

the nitrogen in this system is adjacent to two carbonyl functions, the acidity of the NH group ($pK_a = 8.3$) is much greater than that of an ordinary amide ($pK_a = 15$). Deprotonation can therefore be achieved with as mild a base as carbonate ion, and the resulting anion monoalkylated in good yield. The amine subsequently can be liberated by acidic hydrolysis, initially as the ammonium salt. Base treatment of the salt then produces the free amine.

Gabriel Synthesis of a Primary Amine

$\xrightarrow[-\ H_2O]{:NH_3,\ 300°}$ (97%) $\xrightarrow{K_2CO_3,\ H_2O}$:N:⁻K⁺ $\xrightarrow[-\ KBr]{HC\equiv CCH_2Br,\ DMF,\ 100°C}$

1,2-Benzenedicarboxylic acid (Phthalic acid)

97%
1,2-Benzenedicarboxímide (Phthalimide)

:NCH₂C≡CH $\xrightarrow{H_2SO_4,\ H_2O,\ 120°C}$ (phthalic acid) + $HC\equiv CCH_2\overset{+}{N}H_3HSO_4^-$ $\xrightarrow{NaOH,\ H_2O}$

93%
***N*-2-Propynyl-1,2-benzenedicarboximide (*N*-Propargylphthalimide)**

$HC\equiv CCH_2\ddot{N}H_2$ + (disodium 1,2-benzenedicarboxylate, O^-Na^+, O^-Na^+)

73%
2-Propynamine (Propargylamine)

Removed in aqueous work-up

EXERCISE 21-7

The cleavage of an *N*-alkyl-1,2-benzenedicarboximide (*N*-alkyl phthalimide) is frequently carried out with base or with hydrazine, H_2NNH_2. The respective products of these two treatments are the 1,2-benzenedicarboxylate A or hydrazide B. Write mechanisms for these two transformations.

A (COO^-, COO^-) $\xleftarrow[-\ RNH_2]{^-OH}$ NR $\xrightarrow[-\ RNH_2]{H_2NNH_2}$ B (NH–NH)

EXERCISE 21-8

Show how you would apply the Gabriel method to the synthesis of each of the following amines. **(a)** 1-Hexanamine; **(b)** 3-methylpentanamine; **(c)** cyclohexanamine; **(d)** $H_2NCH_2CO_2H$, the amino acid glycine. (Note: During the last of these syntheses, the carboxylic acid group should be protected as an ester. Can you see why?) For each of these four syntheses, would the azide displacement–reduction method be equally good, better, or worse?

To summarize, amines can be made from ammonia or other amines by simple alkylation, but this method gives mixtures and poor yields. It is better to use indirect methods that employ nitrile and azide groups, or protected systems, such as 1,2-benzenedicarboxylic imide (phthalimide) in the Gabriel synthesis.

21-6 Synthesis of Amines by Reductive Amination

Another method of amine synthesis, called **reductive amination,** begins by the condensation of amines with carbonyl compounds to produce imines (Section 17-9). Like the carbon–oxygen double bond in aldehydes and ketones, the carbon–nitrogen double bond in imines can then be reduced by catalytic hydrogenation or by hydride reagents. Reductive amination may be performed on either aldehydes or ketones.

Reductive Amination of a Ketone

$$RR'C{=}O + H_2NR'' \underset{\text{Condensation}}{\rightleftharpoons} RR'C{=}NR'' + H_2O$$

$$RR'C{=}NR'' \xrightarrow[\text{Reduction}]{} R'{-}\overset{R}{\underset{H}{C}}{-}NHR''$$

This reaction succeeds because of the selectivity of the reducing agents: catalytically activated hydrogen or sodium cyanoborohydride, $Na^{+}\,{}^{-}BH_3CN$. Both react faster with the imine double bond than with the carbonyl group under the conditions employed. In a typical procedure, the carbonyl component and the amine are allowed to equilibrate with the imine and water in the presence of the reducing agent.

Amine Synthesis by Reductive Amination

$$C_6H_5{-}CHO + NH_3 \underset{+\,H_2O}{\overset{-\,H_2O}{\rightleftharpoons}} C_6H_5{-}CH{=}NH \xrightarrow{H_2,\ Ni,\ CH_3CH_2OH,\ 70°C,\ 90\ atm} C_6H_5{-}\overset{H}{C}H\overset{H}{N}H$$

Benzaldehyde — Not isolated — 89% **Phenylmethanamine (Benzylamine)**

$$\text{Cyclohexanone} + NH_3 \underset{+H_2O}{\overset{-H_2O}{\rightleftharpoons}} \text{(imine, NH; not isolated)} \xrightarrow{NaBH_3CN,\ CH_3CH_2OH,\ pH = 2\text{–}3} \text{Cyclohexanamine (H, }NH_2\text{)}\ 61\%$$

Cyclohexanone Not isolated **Cyclohexanamine**

Reductive aminations with secondary amines give the corresponding *N*,*N*-dialkylamino derivatives.

$$\text{Ketone } (H_3C,\ H_3C,\ CH_3) \xrightarrow{(CH_3)_2NH,\ NaBH_3CN,\ CH_3OH} \text{Amine } (H,\ N(CH_3)_2;\ H_3C,\ H_3C,\ CH_3)\ 89\%$$

The reaction proceeds through the intermediacy of an iminium ion, which is reduced by addition of H^- from cyanoborohydride.

$$CH_2{=}O + \text{C}_5\text{H}_9\text{–}\underset{H}{N}CH_2C_6H_5 \longrightarrow \text{C}_5\text{H}_9\text{–}\overset{+}{N}(=CH_2)CH_2C_6H_5 \xrightarrow{NaBH_3CN,\ CH_3OH} \text{C}_5\text{H}_9\text{–}N(HCH_2)CH_2C_6H_5\ 100\%$$

_N_-(Phenylmethyl)-cyclopentanamine (Benzylcyclopentylamine) **Iminium ion** **_N_-Methyl-_N_-(phenylmethyl)-cyclopentanamine (Benzylcyclopentylmethylamine)**

EXERCISE 21-9

Formulate a mechanism for the reductive amination with the secondary amine shown in the example above.

EXERCISE 21-10

Explain the following transformation by a mechanism.

$$\overset{O}{\overset{\|}{H C}}CH_2CH_2\overset{NH_2}{\overset{|}{C}H}CH_2CH_2\overset{O}{\overset{\|}{C}}H \xrightarrow{NaBH_3CN,\ CH_3OH} \text{(pyrrolizidine, N)}\ 35\%$$

In summary, reductive amination furnishes alkanamines by reductive condensation of amines with aldehydes and ketones.

21-7 Synthesis of Amines from Carboxylic Amides

Carboxylic amides can be versatile precursors of amines (Section 20-6). Recall that amides are readily available by reaction of alkanoyl halides with amines. Reduction with lithium aluminum hydride then converts them into amines.

Utility of Amides in Amine Synthesis

$$\mathrm{R\overset{O}{\overset{\|}{C}}Cl + H_2\ddot{N}R' \xrightarrow[-\,HCl]{Base} R\overset{O}{\overset{\|}{C}}\ddot{N}HR' \xrightarrow{LiAlH_4,\ (CH_3CH_2)_2O} RCH_2\ddot{N}HR'}$$

Primary amides can be turned into amines also by oxidation with bromine or chlorine in the presence of sodium hydroxide, in other words, by the Hofmann rearrangement (Section 20-7). Recall that in this transformation the carbonyl group is extruded as carbon dioxide, so the resulting amine bears one less carbon than the starting material.

Amines by Hofmann Rearrangement

$$\mathrm{R\overset{O}{\overset{\|}{C}}NH_2 \xrightarrow{Br_2,\ NaOH,\ H_2O} RNH_2 + O{=}C{=}O}$$

EXERCISE 21-11

Suggest synthetic methods for the preparation of *N*-methylhexanamine from hexanamine (two syntheses) and from *N*-hexylmethanamide.

In summary, amides can be reduced to amines by treatment with lithium aluminum hydrides. The Hofmann rearrangement converts amides to amines with loss of the carboxy carbon atom.

21-8 Quaternary Ammonium Salts: Hofmann Elimination

The alkylation of tertiary amines with haloalkanes furnishes quaternary ammonium salts (Section 21-5). These species are reactive, giving alkenes by elimination of a proton and a tertiary amine upon heating with base. This transformation, called **Hofmann* elimination,** has been especially useful in determining the structure of naturally occurring amines.

Nucleophilic attack of amines on haloalkanes results in ammonium ions. If these are quaternary, further alkylation is impossible because there are no more replaceable protons on the nitrogen. Quaternary ammonium salts are nevertheless unstable in the presence of strong base, because of a bimolecular elimination reaction that furnishes alkenes (Section 7-7). The base attacks the hydrogen in the β-position with respect to the nitrogen, and a trialkylamine departs as a neutral leaving group.

Bimolecular Elimination of Quaternary Ammonium Ions

$$\mathrm{H{-}C{-}C{-}\overset{+}{N}R_3 \;\; {}^{-}\!:\ddot{\underset{..}{O}}H \longrightarrow C{=}C + HOH + :NR_3}$$

Alkene

*This is the Hofmann of the Hofmann rule for E2 reactions (Section 11-8) and the Hofmann rearrangement (Section 20-7).

In the procedure of Hofmann elimination, the amine is first completely methylated with excess iodomethane **(exhaustive methylation)** and then treated with wet silver oxide (a source of HO^-) to produce the ammonium hydroxide. Heating degrades this salt to the alkene. When more than one regioisomer is possible, Hofmann elimination, in contrast with most E2 processes, tends to give less substituted alkenes as the major products.

Hofmann Elimination of Butanamine

$$\underset{\textbf{Butanamine}}{CH_3CH_2CH_2CH_2NH_2} \xrightarrow{\text{Excess } CH_3I,\ K_2CO_3,\ H_2O} \underset{\textbf{Butyltrimethylammonium iodide}}{CH_3CH_2CH_2CH_2\overset{+}{N}(CH_3)_3I^-} \xrightarrow[-\ AgI]{Ag_2O,\ H_2O}$$

$$\underset{\textbf{Butyltrimethylammonium hydroxide}}{CH_3CH_2\overset{\overset{H}{|}}{C}HCH_2\overset{+}{N}(CH_3)_3\ HO^-} \xrightarrow{\Delta} \underset{\textbf{1-Butene}}{CH_3CH_2CH{=}CH_2} + N(CH_3)_3 + HOH$$

EXERCISE 21-12

Give the structures of the possible alkene products of the Hofmann elimination of **(a)** *N*-ethylpropanamine (ethylpropylamine) and **(b)** 2-butanamine.

The Hofmann elimination of amines has been used to elucidate the structure of nitrogen-containing natural products, such as alkaloids (Section 25-8). Each sequence of exhaustive methylation and Hofmann elimination cleaves one C–N bond. Repeated cycles allow the location of the heteroatom to be pinpointed, particularly if the nitrogen atom is part of a ring. In this case the first carbon–nitrogen bond cleavage opens the ring.

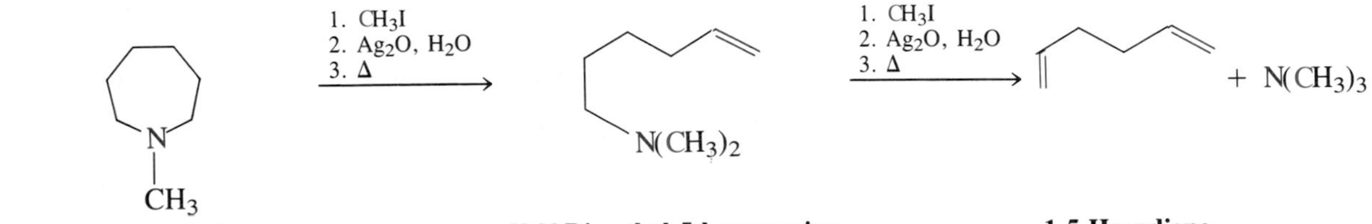

***N*-Methylazacycloheptane** ***N,N*-Dimethyl-5-hexenamine** **1,5-Hexadiene**

EXERCISE 21-13

Why is exhaustive *methyl*ation and not, say, ethylation used in Hofmann eliminations for structure elucidation? (Hint: Look for other possible elimination pathways.)

EXERCISE 21-14

An unknown amine of the molecular formula $C_7H_{13}N$ has a ^{13}C NMR spectrum containing only three lines of $\delta = 21.0$, 26.8, and 47.8 ppm. Three cycles of Hofmann elimination are required to form 3-ethenyl-1,4-pentadiene (trivinylmethane) and its double-bond isomers (as side products arising from base-catalyzed isomerization). Propose a structure for the unknown.

In summary, quaternary ammonium salts, synthesized by amine alkylation, undergo bimolecular elimination in the presence of base to give alkenes.

21-9 Mannich Reaction: Alkylation of Enols by Iminium Ions

The **Mannich* reaction** forms carbon–carbon bonds by using iminium ions (Section 21-3) to alkylate enols. First, the ion is prepared by condensation of formaldehyde with ammonia or a primary or secondary amine. The iminium ion, a strong electrophile, then reacts with the α-carbon of enolizable aldehydes and ketones. The result is a β-aminocarbonyl compound or its ammonium salt.

Typically, an aldehyde or ketone is heated in the presence of formaldehyde, the amine, and HCl in alcohol solvent. This process gives the hydrochloride salt of the product. The free amine, called a **Mannich base,** can be obtained upon treatment with base.

Mannich Reaction

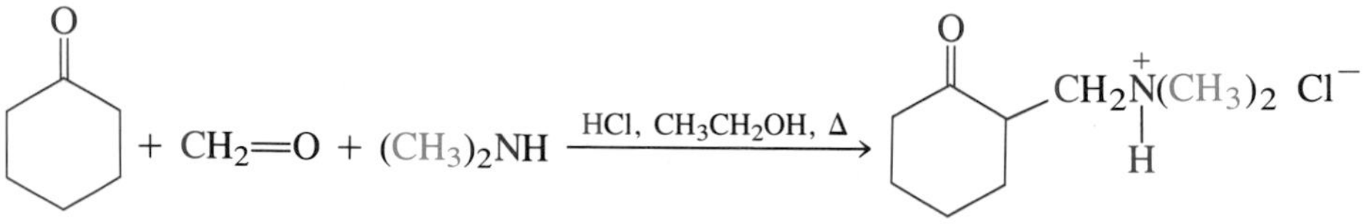

$$\text{cyclohexanone} + CH_2{=}O + (CH_3)_2NH \xrightarrow{HCl,\ CH_3CH_2OH,\ \Delta} \text{2-}(CH_2\overset{+}{N}H(CH_3)_2\ Cl^-)\text{cyclohexanone}$$

85%
Salt of Mannich base

$$\underset{\textbf{2-Methylpropanal}}{CH_3CH(CH_3)CH{=}O} + CH_2{=}O + CH_3NH_2 \xrightarrow[2.\ HO^-,\ H_2O]{1.\ HCl,\ CH_3CH_2OH,\ \Delta} CH_3C(CH_3)(CH_2NHCH_3){-}CH{=}O$$

70%
2-Methyl-(2-*N*-methyl-aminomethyl)propanal Mannich base

This process is closely related to another reaction of enolizable carbonyl compounds, the aldol condensation (Section 18-5). In the Mannich reaction, however, a nucleophilic enol attacks a carbon–nitrogen double bond and not a carbonyl group.

Mechanism of the Mannich Reaction

STEP 1. Iminium ion formation

$$CH_2{=}O + (CH_3)_2\overset{+}{N}H_2\ Cl^- \longrightarrow CH_2{=}\overset{+}{N}(CH_3)_2\ Cl^- + H_2O$$

STEP 2. Enolization

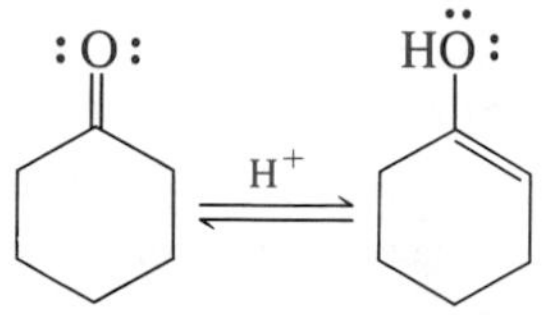

$$\text{cyclohexanone } (:\!O\!:) \underset{}{\overset{H^+}{\rightleftharpoons}} \text{cyclohexenol } (H\ddot{O}\!:)$$

*Professor Carl Mannich, 1877–1947, University of Berlin.

STEP 3. Carbon–carbon bond formation

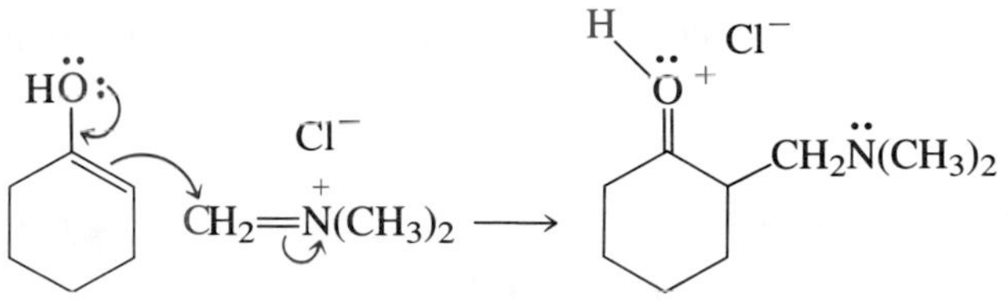

Cl^- $CH_2=\overset{+}{N}(CH_3)_2$ ⟶ $CH_2\ddot{N}(CH_3)_2$ Cl^-

STEP 4. Hydrochloride salt formation

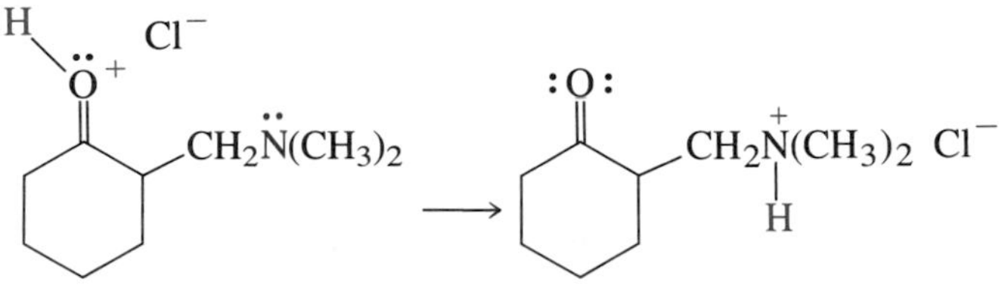

$CH_2\ddot{N}(CH_3)_2$ Cl^- ⟶ $CH_2\overset{+}{N}H(CH_3)_2\ Cl^-$

Salt of Mannich base

An example of the Mannich reaction in natural product synthesis is shown below. In this instance one ring is formed by condensation of the amine with one aldehyde group. Mannich reaction of the resulting iminium salt with the enol of the other aldehyde follows. The product has the framework of retronecine, an alkaloid that occurs in many shrubs and is hepatotoxic (causes liver damage) to grazing livestock.

Mannich Reaction in Synthesis

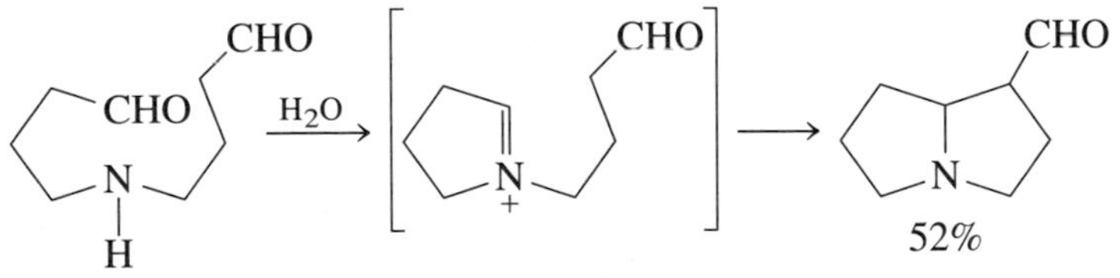

CHO CHO N H $\xrightarrow{H_2O}$ [CHO N+] ⟶ CHO N 52%

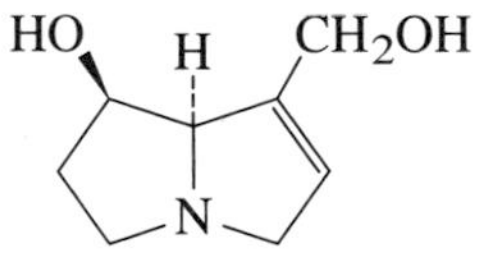

HO H CH_2OH N

Retronecine

EXERCISE 21-15

Write the products of each of the following Mannich reactions. **(a)** Ammonia + formaldehyde + cyclopentanone; **(b)** 1-hexanamine + formaldehyde + 2-methylpropanal; **(c)** *N*-methylmethanamine + formaldehyde + propanone; **(d)** cyclohexanamine + formaldehyde + cyclohexanone.

EXERCISE 21-16

β-Dialkylamino alcohols and their esters are useful local anesthetics. Suggest a synthesis of the anesthetic Tutocaine hydrochloride, beginning with 2-butanone.

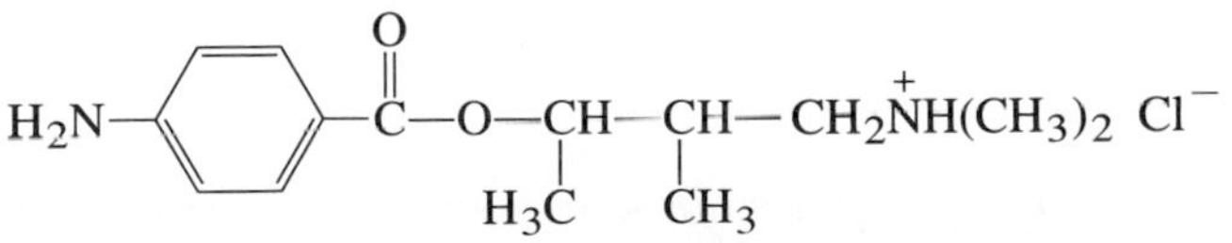

$H_2N-C_6H_4-\overset{O}{\overset{\|}{C}}-O-\underset{H_3C}{CH}-\underset{CH_3}{CH}-CH_2\overset{+}{N}H(CH_3)_2\ Cl^-$

Tutocaine hydrochloride

In summary, condensation of formaldehyde with amines furnishes iminium ions, which are electrophilic and may attack the enols of aldehydes and ketones (Mannich reaction). The products are β-aminocarbonyl compounds.

21-10 Nitrosation of Amines: *N*-Nitrosamines and Diazonium Ions

Amines react with nitrous acid, through nucleophilic attack on (electrophilic attack *by*) the **nitrosyl cation,** NO^+. The product depends very much on whether the reactant is an alkanamine or a benzenamine (aniline), and whether it is primary, secondary, or tertiary. This section deals with alkanamines; aromatic amines will be discussed in the next chapter.

To generate NO^+, we must first prepare the unstable nitrous acid by the treatment of sodium nitrite with aqueous HCl. In such an acid solution, an equilibrium is established with the nitrosyl cation. (Compare this sequence with the preparation of the nitronium cation from nitric acid; Section 15-11.)

Nitrosyl Cation from Nitrous Acid

$$Na^{+}\,{}^{-}\!:\ddot{\underset{..}{O}}-\ddot{N}=\ddot{\underset{..}{O}} \xrightarrow[-\,NaCl]{HCl} H\ddot{\underset{..}{O}}-\ddot{N}=\ddot{\underset{..}{O}} \overset{H^+}{\rightleftharpoons} H_2\overset{+}{\ddot{O}}-\ddot{N}=\ddot{\underset{..}{O}} \rightleftharpoons \left[{}^{+}\!:N=\ddot{\underset{..}{O}} \longleftrightarrow :N\equiv\overset{+}{O}:\right] + H_2\ddot{\underset{..}{O}}$$

Sodium nitrite **Nitrous acid** **Nitrosyl cation**

The nitrosyl cation is electrophilic and is attacked by amines to form an *N*-nitrosammonium salt.

$$-\overset{|}{N}: + {}^{+}\ddot{N}=\ddot{\underset{..}{O}} \longrightarrow -\overset{|}{\underset{|}{N}}\overset{+}{-}\ddot{N}=\ddot{\underset{..}{O}}$$

***N*-Nitrosammonium salt**

$R_2N-N=O$
***N*-Nitrosamine**

The course of the reaction now depends on whether the amine nitrogen bears none, one, or two hydrogens. *Tertiary N*-nitrosammonium salts are stable at low temperatures but decompose on heating to give a mixture of compounds. *Secondary N*-nitrosammonium salts are simply deprotonated to furnish the relatively stable ***N*-nitrosamines** as the major products.

$$(CH_3)_2NH \xrightarrow{NaNO_2,\ HCl,\ H_2O,\ 0°C} (CH_3)_2\overset{H}{\overset{|}{N}}\overset{+}{-}N=O\ Cl^- \xrightarrow[-\,HCl]{} (H_3C)_2N-N=O$$

88–90%
***N*-Nitrosodimethylamine**

Similar treatment of *primary* amines initially gives the analogous monoalkyl-*N*-nitrosamines. However, these products are unstable because of the remaining proton on the nitrogen, and they rapidly decompose to complex mixtures. By a series of hydrogen shifts, these compounds first rearrange to the corresponding diazohydroxides. Then protonation followed by loss of water gives highly reactive **diazonium ions,** $R{-}N_2^{\ +}$. When R is a secondary or a tertiary alkyl group, these ions lose molecular nitrogen, N_2, and form the corresponding carbocations, which may rearrange, deprotonate, or undergo nucleophilic trapping (Section 9-3) to yield the observed mixtures of compounds.

Mechanism of Decomposition of Primary *N*-Nitrosamines

STEP 1. Rearrangement to a diazohydroxide

$$\text{R(H)N}{-}\text{N}{=}\ddot{\text{O}}\ \underset{-\,\text{H}^+}{\overset{+\,\text{H}^+}{\rightleftharpoons}}\ \left[\text{R(H)}\ddot{\text{N}}{-}\ddot{\text{N}}{=}\overset{+}{\ddot{\text{O}}}\text{H} \longleftrightarrow \text{R(H)}\overset{+}{\text{N}}{=}\ddot{\text{N}}{-}\ddot{\text{O}}\text{H}\right]\ \underset{+\,\text{H}^+}{\overset{-\,\text{H}^+}{\rightleftharpoons}}\ \text{R}{-}\ddot{\text{N}}{=}\ddot{\text{N}}{-}\ddot{\text{O}}\text{H}$$

Diazohydroxide

STEP 2. Loss of water to give a diazonium ion

$$\text{R}{-}\ddot{\text{N}}{=}\ddot{\text{N}}{-}\ddot{\text{O}}\text{H}\ \underset{-\,\text{H}^+}{\overset{+\,\text{H}^+}{\rightleftharpoons}}\ \text{R}{-}\ddot{\text{N}}{=}\ddot{\text{N}}{-}\overset{+}{\ddot{\text{O}}}\text{H}_2\ \underset{+\,\text{H}_2\text{O}}{\overset{-\,\text{H}_2\text{O}}{\rightleftharpoons}}\ \text{R}{-}\overset{+}{\text{N}}{\equiv}\text{N}{:}$$

Diazonium cation

BOX 21-3 *N*-Nitrosodialkanamines Are Carcinogens

N-Nitrosodialkanamines are notoriously potent carcinogens in a variety of animals. Although there is no direct evidence, they are suspected of causing cancer in humans as well. Most nitrosamines appear to cause liver cancer, but certain of them are very organ specific in their carcinogenic potential (bladder, lungs, esophagus, nasal cavity, etc.).

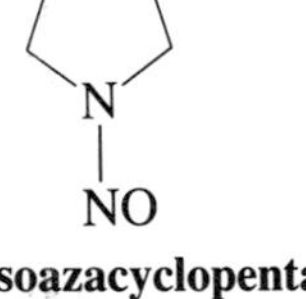

_N_-Nitrosoazacyclopentane (_N_-Nitrosopyrrolidine)

Their mode of carcinogenic action appears to involve initial enzymatic oxidation of one of the α positions, which allows eventual formation of a monoalkyl-*N*-nitrosamine. This compound then decomposes to a carbocation that, as a powerful electrophile, is thought to attack one of the bases in DNA to inflict the kind of genetic damage that seems to lead to cancerous cell behavior.

Nitrosamines have been detected in a variety of cured meats, such as smoked fish, frankfurters (*N*-nitrosodimethylamine), and fried bacon [*N*-nitrosoazacyclopentane (*N*-nitrosopyrrolidine), see Exercise 25-6]. Moreover, they can be formed from natural amines and added nitrite ion at physiological pH in the stomach of test animals. The level of nitrite salts in the body depends on environmental factors, such as water supply and consumption of natural foods as well as food additives. For example, fresh spinach contains about 5 mg kg^{-1} of nitrite. On storing the vegetable in a refrigerator, its nitrite level may increase to 300 mg kg^{-1}. Nitrite has been widely used to preserve meats and enhance their color and flavor. In 1976, the FDA limited the amount of nitrite permissible for such purposes to between 50 and 125 ppm. Subsequently, this rule was relaxed somewhat because of the lack of data pointing to nitrite as a carcinogenic additive.

STEP 3. Nitrogen loss to give a carbocation

$$R{-}\overset{+}{N}{\equiv}N: \xrightarrow[-\ N_2]{} R^+ \longrightarrow \text{product mixtures}$$

EXERCISE 21-17

The following result was reported in 1991. What does this observation suggest regarding the applicability of the mechanism shown above to diazonium ions with *primary* R groups?

$$CH_3CH_2CH_2{-}C(H)(D)(NH_2) \xrightarrow{NaNO_2,\ HCl,\ H_2O} (D)(H)(HO)C{-}CH_2CH_2CH_3$$

Pure *R* enantiomer → 100% **Pure *S* enantiomer**

In summary, nitrous acid attacks amines, thereby causing *N*-nitrosation. Secondary amines give *N*-nitrosamines, which are notorious for their carcinogenicity. *N*-Nitrosamines derived from primary amines decompose through carbocations to a variety of products.

21-11 Diazomethane, Carbenes, and Cyclopropane Synthesis

The nitrosyl cation also attacks the nitrogen of *N*-methylamides. The products, *N*-methyl-*N*-nitrosamides, are precursors to useful synthetic intermediates.

Nitrosation of an *N*-Methylamide

$$R\overset{O}{\overset{\|}{C}}NHCH_3 \xrightarrow[-\ H^+]{NO^+} R\overset{O}{\overset{\|}{C}}N(NO)CH_3$$

***N*-Methylamide** → ***N*-Methyl-*N*-nitrosamide**

Base treatment of *N*-methyl-*N*-nitrosamides gives diazomethane

N-Methyl-*N*-nitrosamides are converted to **diazomethane,** CH_2N_2, upon treatment with aqueous base.

Making Diazomethane

$$CH_3N(N{=}O)\overset{O}{\overset{\|}{C}}NH_2 \xrightarrow{KOH,\ H_2O,\ (CH_3CH_2)_2O,\ 0^\circ C} CH_2{=}\overset{+}{N}{=}\ddot{N}:^- + NH_3 + K_2CO_3 + H_2O$$

***N*-Methyl-*N*-nitrosourea** → **Diazomethane**

Diazomethane is used in the synthesis of methyl esters from carboxylic acids (Section 19-9). However, it is exceedingly toxic and highly explosive in the gaseous state (b.p. −24°C) and in concentrated solutions. It is therefore usually generated in dilute ether solution and immediately allowed to react with the acid. This method is very mild and permits esterification of molecules possessing acid- and base-sensitive functional groups, as shown in the following example.

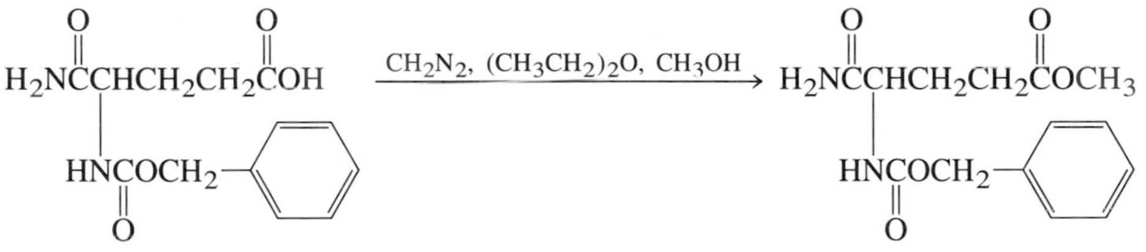

$$H_2NC(=O)CH(NHCOCH_2C_6H_5)CH_2CH_2COOH \xrightarrow{CH_2N_2,\ (CH_3CH_2)_2O,\ CH_3OH} H_2NC(=O)CH(NHCOCH_2C_6H_5)CH_2CH_2COOCH_3$$

75%

Diazomethane forms methylene, which converts alkenes to cyclopropanes

On exposure to light, heat, or catalytic copper metal, diazomethane extrudes N_2 to give the highly reactive species **methylene,** $H_2C:$, the simplest **carbene.**

$R_2C:$
Carbene

$$H_2\overset{-}{\ddot{C}}-\overset{+}{N}\equiv N: \xrightarrow{h\nu \text{ or } \Delta \text{ or Cu}} H_2C: \quad + :N\equiv N:$$

Methylene

When methylene is generated in the presence of compounds containing double bonds, addition takes place to furnish cyclopropanes, usually stereospecifically.

Methylene Additions to Double Bonds

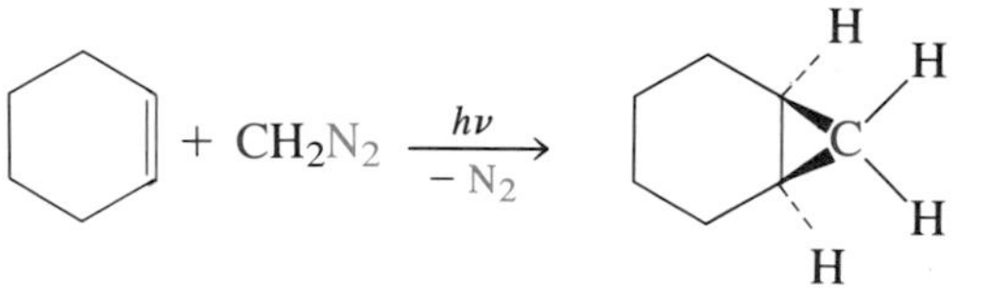

$$\text{Cyclohexene} + CH_2N_2 \xrightarrow[-N_2]{h\nu} \text{Bicyclo[4.1.0]heptane}$$

40%
Bicyclo[4.1.0]heptane

$$cis\text{-}CH_3CH_2CH{=}CHCH_2CH_3 + CH_2N_2 \xrightarrow[-N_2]{Cu,\ \Delta} cis\text{-Diethylcyclopropane}$$

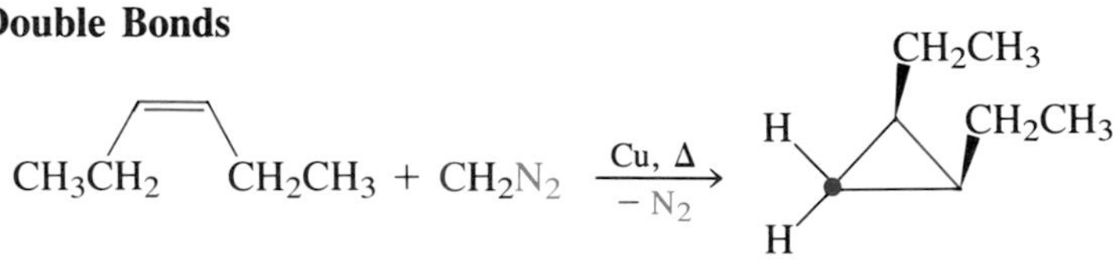

50–70%
***cis*-Diethylcyclopropane**

EXERCISE 21-18

Diazomethane is the simplest member of the class of compounds called *diazoalkanes* or *diazo compounds,* $R_2C{=}N_2$. When the diazo compound A is irradiated in heptane solution at −78°C, it gives a hydrocarbon C_4H_6, exhibiting three signals in ^1H NMR and two signals in ^{13}C NMR spectroscopy, all in the aliphatic region. Suggest a structure for this molecule.

$$CH_2{=}CHCH_2CH{=}\overset{+}{N}{=}\ddot{N}:^-$$

A

Halogenated carbenes and carbenoids also give cyclopropanes

Cyclopropanes also may be synthesized from halogenated carbenes, which are prepared from halomethanes. For example, treatment of trichloromethane (chloroform) with a strong base causes an unusual elimination reaction in which both the proton and the leaving group are removed from the same carbon. The product is dichlorocarbene, which gives cyclopropanes when generated in the presence of alkenes.

Dichlorocarbene from Chloroform and Its Trapping by Cyclohexene

$$HCCl_3 + (CH_3)_3CO^- \xrightarrow[-\ (CH_3)_3COH]{} {}^-{:}CCl_2(Cl) \longrightarrow {:}CCl_2 + Cl^-$$

Dichlorocarbene

$$\text{Cyclohexene} + {:}CCl_2 \longrightarrow \text{7,7-dichlorobicyclo[4.1.0]heptane (59\%)}$$

In another route to cyclopropanes, diiodomethane is treated with zinc powder (usually activated with copper) to generate ICH_2ZnI, called the **Simmons-Smith* reagent.** This species is an example of a **carbenoid,** or carbenelike substance, because, like carbenes, it also converts alkenes to cyclopropanes stereospecifically. Use of the Simmons-Smith reagent in cyclopropane synthesis avoids the hazards associated with diazomethane preparation.

Simmons-Smith Reagent in Cyclopropane Synthesis

$$\text{trans-}H(H_3C)C{=}C(CH_3)H + CH_2I_2 \xrightarrow[-\ \text{Metal iodide}]{\text{Zn-Cu, }(CH_3CH_2)_2O} \text{trans-1,2-dimethylcyclopropane}$$

To summarize, *N*-methylnitrosamides release diazomethane on treatment with hydroxide. Diazomethane is a useful synthetic intermediate in the methyl esterification of carboxylic acids and as a methylene source for forming cyclopropanes from alkenes. Halogenated carbenes, which are formed by dehydrohalogenation of halomethanes, and the Simmons-Smith reagent, a carbenoid arising from the reaction of diiodomethane with zinc, also convert alkenes to cyclopropanes.

21-12 Amines in Industry: Nylon

In addition to their significance in medicine (Box 21-1) and in the resolution of enantiomers (Section 21-4), the amines have numerous industrial applications. This section deals with one commercially important amine, 1,6-hexanediamine (hexamethylenediamine, HMDA), needed in the manufacture of nylon. This

*Dr. Howard E. Simmons, b. 1929, and Dr. Ronald D. Smith, b. 1930, both with E. I. du Pont de Nemours and Company, Wilmington, Delaware.

compound is copolymerized with hexanedioic (adipic) acid to produce nylon 6,6, out of which hosiery, gears, and millions of tons of textile fiber are made.

Copolymerization of Adipic Acid with HMDA

$$\mathrm{HO\overset{O}{\overset{\|}{C}}(CH_2)_4\overset{O}{\overset{\|}{C}}OH} + \mathrm{H_2N(CH_2)_6NH_2} \longrightarrow \begin{matrix}\mathrm{^{-}O\overset{O}{\overset{\|}{C}}(CH_2)_4\overset{O}{\overset{\|}{C}}O^{-}} \\ \mathrm{H_2\overset{+}{N}(CH_2)_6\overset{+}{N}H_2} \\ \mathrm{|\quad\quad\quad |} \\ \mathrm{H\quad\quad\quad H}\end{matrix} \xrightarrow[\text{Polymerization}]{270°\text{C, 250 psi} \atop -\ H_2O}$$

Hexanedioic acid (Adipic acid) — **1,6-Hexanediamine (Hexamethylenediamine)** — **Double salt**

$$-\!\!\left[\mathrm{NH(CH_2)_6NH\overset{O}{\overset{\|}{C}}(CH_2)_4\overset{O}{\overset{\|}{C}}NH(CH_2)_6NH\overset{O}{\overset{\|}{C}}(CH_2)_4\overset{O}{\overset{\|}{C}}}\right]_n\!\!-$$

Nylon 6,6

Nylon 6,6 is a polyamide formed by condensation of the acid with the diamine under pressure. The high demand for nylon has stimulated the development of several ingenious cheap syntheses of the monomeric precursors. Originally, du Pont made the diamine from hexanedioic (adipic) acid. The diacid was turned into hexanedinitrile (adiponitrile) by treatment with ammonia. Finally, catalytic hydrogenation furnished the diamine.

$$\mathrm{HO\overset{O}{\overset{\|}{C}}(CH_2)_4\overset{O}{\overset{\|}{C}}OH} \xrightarrow[-\ 4\ H_2O]{NH_3,\ \Delta} \mathrm{N{\equiv}C(CH_2)_4C{\equiv}N} \xrightarrow{H_2,\ Ni,\ 130°C,\ 2000\ psi} \mathrm{H_2N(CH_2)_6NH_2}$$

Hexanedinitrile (Adiponitrile) — **1,6-Hexanediamine (Hexamethylene-diamine)**

Du Pont later discovered a still shorter hexanedinitrile synthesis, using 1,3-butadiene as a starting material. Chlorination of butadiene furnished a mixture of 1,2- and 1,4-dichlorobutene (Section 14-6). This mixture can be directly converted into the dinitrile with sodium cyanide in the presence of cuprous cyanide. Selective hydrogenation then furnishes the desired product.

Hexanedinitrile (Adiponitrile) from 1,3-Butadiene

$$\mathrm{CH_2{=}CH{-}CH{=}CH_2} \xrightarrow{Cl-Cl} \mathrm{ClCH_2CH{=}CHCH_2Cl} + \mathrm{ClCH_2\overset{Cl}{\overset{|}{C}}HCH{=}CH_2} \xrightarrow[-\ NaCl]{CuCN,\ NaCN}$$

$$\mathrm{NCCH_2CH{=}CHCH_2CN} \xrightarrow{H_2,\ catalyst} \mathrm{NC(CH_2)_4CN}$$

In the mid-1960s, Monsanto developed a process that uses a more expensive starting material, but takes just one step: the electrolytic hydrodimerization of propenenitrile (acrylonitrile).

Electrolytic Hydrodimerization of Propenenitrile (Acrylonitrile)

$$2\ \mathrm{CH_2{=}CHC{\equiv}N} + 2\ e + 2\ \mathrm{H^+} \longrightarrow \mathrm{N{\equiv}CCH_2CH_2{-}CH_2CH_2{\equiv}N}$$

To counter Monsanto's challenge, du Pont devised yet another synthesis, again starting with 1,3-butadiene, but now avoiding the consumption of chlorine,

Hydrogen Cyanide Addition to 1,3-Butadiene

$$CH_2{=}CHCH{=}CH_2 + 2\,HCN \xrightarrow{\text{Catalyst}} NC(CH_2)_4CN$$

removing the toxic-waste problems of the disposal of copper salts, and using cheaper hydrogen cyanide rather than sodium cyanide. The synthesis uses the conceptually simplest approach: direct regioselective addition of two molecules of hydrogen cyanide to butadiene. A transition metal catalyst is needed, such as iron, cobalt, or nickel. Typically required also are Lewis acids and phosphines, usually triphenylphosphine, $P(C_6H_5)_3$.

In summary, amines and their salts have uses as drugs, resolving agents, and industrial chemicals, the last being exemplified by 1,6-hexanediamine.

NEW REACTIONS

1. ACIDITY OF AMINES AND AMIDE FORMATION (Section 21-4)

$$RNH_2 + H_2O \rightleftharpoons R\bar{N}H + H_3O^+ \qquad K_a \sim 10^{-35}$$

$$R_2NH + CH_3CH_2CH_2CH_2Li \rightleftharpoons \underset{\textbf{Lithium dialkylamide}}{R_2N^-Li^+} + CH_3CH_2CH_2CH_3$$

$$2\,NH_3 + 2\,Na \xrightarrow{\text{Catalytic } Fe^{3+}} 2\,NaNH_2 + H_2$$

2. BASICITY OF AMINES (Section 21-4)

$$RNH_2 + H_2O \rightleftharpoons R\overset{+}{N}H_3 + HO^- \qquad K_b \sim 10^{-4}$$

$$R\overset{+}{N}H_3 + H_2O \rightleftharpoons RNH_2 + H_3O^+ \qquad K_a \sim 10^{-10}$$

Salt formation

$$RNH_2 + HCl \longrightarrow \underset{\textbf{Alkylammonium chloride}}{R\overset{+}{N}H_3Cl^-}$$

General for primary, secondary, and tertiary amines

Preparation of Amines

3. AMINES BY ALKYLATION (Section 21-5)

$$R\ddot{N}H_2 + R'X \longrightarrow R\overset{+}{N}(R')H_2\,X^-$$

General for primary, secondary, and tertiary amines

Drawback: Multiple alkylation

$$R\overset{+}{N}(R')H_2\,X^- + R'X \longrightarrow \longrightarrow \longrightarrow R\overset{+}{N}R'_3\,X^-$$

4. PRIMARY AMINES FROM NITRILES (Section 21-5)

$$RX + {}^-CN \xrightarrow[-X^-]{\text{DMSO, } S_N2} RCN \xrightarrow{\text{1. LiAlH}_4,\ (CH_3CH_2)_2O \text{ 2. } H_2O \text{ or } H_2\text{, catalyst}} RCH_2NH_2$$

R limited to methyl, primary, and secondary alkyl groups

5. PRIMARY AMINES FROM AZIDES (Section 21-5)

$$RX + N_3^- \xrightarrow[-\,X^-]{CH_3CH_2OH,\ S_N2} RN_3 \xrightarrow{1.\ LiAlH_4,\ (CH_3CH_2)_2O;\ 2.\ H_2O} RNH_2$$

R limited to methyl, primary, and secondary alkyl groups

6. PRIMARY AMINES BY GABRIEL SYNTHESIS (Section 21-5)

$$\text{Phthalimide (C}_6\text{H}_4(\text{CO})_2\text{NH)} \xrightarrow{1.\ K_2CO_3,\ H_2O;\ 2.\ RX,\ DMF} \text{C}_6\text{H}_4(\text{CO})_2\text{N—R} \xrightarrow{1.\ H_2SO_4,\ H_2O,\ 120°C;\ 2.\ NaOH,\ H_2O} RNH_2$$

R limited to methyl, primary, and secondary alkyl groups

7. AMINES BY REDUCTIVE AMINATION (Section 21-6)

$$RC(=O)R' \xrightarrow{NH_3,\ NaBH_3CN,\ H_2O\ \text{or alcohol}} R\text{—}CH(NH_2)\text{—}R'$$

Reductive methylation with formaldehyde

$$R_2NH + CH_2{=}O \xrightarrow{NaBH_3CN,\ CH_3OH} R_2NCH_3$$

8. AMINES FROM CARBOXYLIC AMIDES (Section 21-7)

$$RC(=O)NR'R'' \xrightarrow{1.\ LiAlH_4,\ (CH_3CH_2)_2O;\ 2.\ H_2O} RCH_2NR'R''$$

9. HOFMANN REARRANGEMENT (Section 21-7)

$$RC(=O)NH_2 \xrightarrow{Br_2,\ NaOH,\ H_2O} RNH_2 + CO_2$$

Reactions of Amines

10. HOFMANN ELIMINATION (Section 21-8)

$$RCH_2CH_2NH_2 \xrightarrow{\text{Excess } CH_3I,\ K_2CO_3} RCH_2CH_2\overset{+}{N}(CH_3)_3\ I^- \xrightarrow[-\,AgI]{Ag_2O,\ H_2O}$$

$$RCH_2CH_2\overset{+}{N}(CH_3)_3\ {}^-OH \xrightarrow{\Delta} RCH{=}CH_2 + N(CH_3)_3 + H_2O$$

11. MANNICH REACTION (Section 21-9)

$$RC(=O)CH_2R' + CH_2{=}O + (CH_3)_2NH \xrightarrow{1.\ HCl;\ 2.\ HO^-} RC(=O)CHR'(CH_2N(CH_3)_2)$$

12. NITROSATION OF AMINES (Section 21-10)

Tertiary amines

$$R_3N \xrightarrow{NaNO_2,\ H^+X^-} R_3\overset{+}{N}\text{—}NO\ X^-$$

Tertiary *N*-nitrosammonium salt

Secondary amines

$$RR'NH \xrightarrow{NaNO_2,\ H^+} RR'N\text{—}N\text{=}O$$

***N*-Nitrosamine**

Primary amines

$$RNH_2 \xrightarrow{NaNO_2,\ H^+} RN\text{=}NOH \xrightarrow[-\,H_2O]{H^+} RN_2^+ \xrightarrow[-\,N_2]{} R^+ \longrightarrow \text{mixture of products}$$

13. DIAZOMETHANE (Section 21-11)

$$CH_3N(N\text{=}O)\overset{O}{\overset{\|}{C}}R \xrightarrow{KOH} CH_2\text{=}\overset{+}{N}\text{=}\ddot{N}:^-$$

Reactions of diazomethane

$$R\overset{O}{\overset{\|}{C}}OH + CH_2N_2 \longrightarrow R\overset{O}{\overset{\|}{C}}OCH_3 + N_2$$

$$RCH\text{=}CHR' + CH_2N_2 \xrightarrow{h\nu \text{ or } \Delta \text{ or Cu}} \text{cyclopropane (R, R')}$$

Stereospecific

Other sources of carbenes or carbenoids

$$CHCl_3 \xrightarrow{Base} :CCl_2$$

$$CH_2I_2 \xrightarrow{Zn\text{-}Cu} ICH_2ZnI$$

14. NYLON 6,6 (Section 21-12)

$$HO\overset{O}{\overset{\|}{C}}(CH_2)_4\overset{O}{\overset{\|}{C}}OH + H_2N(CH_2)_6NH_2 \xrightarrow[-\,H_2O]{} \text{—}\left[NH(CH_2)_6NH\overset{O}{\overset{\|}{C}}(CH_2)_4\overset{O}{\overset{\|}{C}}\right]_n\text{—}$$

Hexanedioic acid (Adipic acid) — **1,6-Hexanediamine (Hexamethylenediamine)** — **Nylon 6,6**

IMPORTANT CONCEPTS

1. Amines can be viewed as derivatives of ammonia, just as ethers and alcohols can be regarded as derivatives of water.
2. *Chemical Abstracts* names amines as alkanamines (and benzenamines), alkyl substituents on the nitrogen being designated as *N*-alkyl. Another system is based on the label aminoalkane. Common names are based on the label alkylamine.
3. The nitrogen in amines is sp^3 hybridized, the nonbonding electron pair functioning as the equivalent of a substituent. This tetrahedral arrangement inverts rapidly through a planar transition state.
4. The lone electron pair in amines is less tightly held than in alcohols and ethers, because nitrogen is less electronegative than oxygen. The consequences are a diminished capability for hydrogen bonding, higher basicity and nucleophilicity, and lower acidity.
5. Infrared spectroscopy helps to differentiate between primary and secondary amines. Nuclear magnetic resonance spectroscopy indicates the presence of nitrogen-bound hydrogens; both hydrogen and carbon atoms are deshielded in the vicinity of the nitrogen. Mass spectra are characterized by iminium ion fragments.
6. Indirect methods, such as displacements with azide or cyanide, or reductive amination, are superior to direct alkylation of ammonia for the synthesis of amines.
7. The NR_3 group in a quaternary amine, $R'-\overset{+}{N}R_3$, is a good leaving group in E2 reactions; this enables the Hofmann elimination to take place.
8. The nucleophilic reactivity of amines manifests itself in reactions with electrophilic carbon, as in haloalkanes, aldehydes, and ketones, and carboxylic acids and their derivatives.
9. Carbenes and carbenoids are useful for the synthesis of cyclopropanes from alkenes.

PROBLEMS

1. Give at least two names for each of the following amines.

(a) $CH_3CH_2CH_2\underset{}{C}HNH_2$ (with CH_3CH_2 on the CH carbon)

(b) $(H_3C)_2CHNHCH_3$

(c) Benzene ring bearing $-NH_2$ with Cl at the adjacent (ortho) position

(d) Benzene ring bearing $-N(CH_3)CH_2CH_2CH_3$

(e) $(CH_3)_3N$

(f) $CH_3\overset{O}{\overset{\|}{C}}CH_2CH_2N(CH_3)_2$

(g) Cyclopentyl$-N(CH_3)CH_2CH_2CH_2CH_2CH(CH_3)CH_2Cl$

(h) $(CH_3CH_2)_2NCH_2CH{=}CH_2$

2. Give structures that correspond to each of the following names. **(a)** *N,N*-Dimethyl-3-cyclohexenamine; **(b)** *N*-ethyl-2-phenylethylamine; **(c)** 2-aminoethanol; **(d)** *m*-chloroaniline.

3. As mentioned in Section 21-2, the inversion of nitrogen requires a change of hybridization. **(a)** What is the approximate energy difference between pyramidal nitrogen (sp^3 hybridized) and trigonal planar nitrogen (sp^2 hybridized) in ammonia and simple amines? **(b)** Compare the nitrogen atom in ammonia with the carbon atom in each of the following: methyl cation, methyl radical, and methyl anion. Compare the most stable geometries and the hybridizations of each of these species. Using fundamental notions of orbital energies and bond strengths, explain the similarities and differences among them.

4. Use the following NMR- and mass-spectral data to identify the structures of the two unknown compounds, A and B.

A: ^{1}H NMR $\delta = 0.92$ (t, $J = 6$ Hz, 3 H), 1.32 (broad s, 12 H), 2.28 (broad s, 2 H), and 2.69 (t, $J = 7$ Hz, 2 H) ppm
Mass spectrum m/z (relative intensity) = 129(0.6) and 30(100)

B: ^{1}H NMR $\delta = 1.00$ (s, 9 H), 1.17 (s, 6 H), 1.28 (s, 2 H), and 1.42 (s, 2 H) ppm
Mass spectrum m/z (relative intensity) = 129(0.05), 114(3), 72(4), and 58(100)

5. Spectroscopic data (^{13}C NMR and IR) are presented below for several isomeric amines of the formula $C_6H_{15}N$. Propose a structure for each compound. **(a)** ^{13}C NMR: $\delta = 23.7$ (q) and 45.3 (d) ppm. IR: 3300 cm^{-1}. **(b)** ^{13}C NMR: $\delta = 12.6$ (q) and 46.9 (t) ppm. IR: no bands in 3250–3500 cm^{-1} range. **(c)** ^{13}C NMR: $\delta = 12.0$ (q), 23.9 (t), and 52.3 (t) ppm. IR: 3280 cm^{-1}. **(d)** ^{13}C NMR: $\delta = 14.2$ (q), 23.2 (t), 27.1 (t), 32.3 (t), 34.6 (t), and 42.7 (t) ppm. IR: 1600 (broad), 3280, and 3365 cm^{-1}. **(e)** ^{13}C NMR: $\delta = 25.6$ (q), 38.7 (q), and 53.2 (s) ppm. IR: no bands in 3250–3500 cm^{-1} region.

6. Mass-spectral data for two of the compounds in Problem 5 are given below. Match the mass spectrum with a compound. **(a)** m/z (relative intensity) = 101(8), 86(11), 72(79), 58(10), 44(40), and 30(100). **(b)** m/z (relative intensity) = 101(3), 86(30), 58(14), and 44(100).

7. Is a molecule with a high pK_b value a stronger or weaker base than a molecule with a low pK_b value? Explain by using a general equilibrium equation.

8. How would you expect the following classes of compounds to compare with simple primary amines as bases and acids?

(a) Carboxamides; for example, CH_3CONH_2 **(b)** Imides; for example, $CH_3CONHCOCH_3$

(c) Enamines; for example, $CH_2{=}CHN(CH_3)_2$ **(d)** Benzenamines; for example, $C_6H_5{-}NH_2$

9. Several functional groups containing nitrogen are considerably stronger bases than are ordinary amines. One is the amidine group found in DBN and DBU, both of which are widely used as bases in a variety of organic reactions.

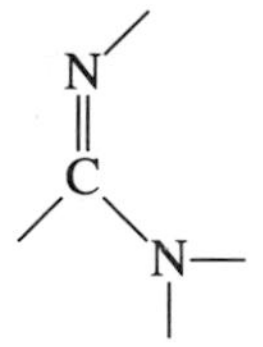

Amidine group

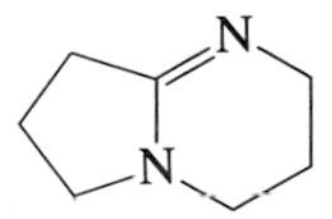

1,5-Diazabicyclo[4.3.0]non-5-ene (DBN)

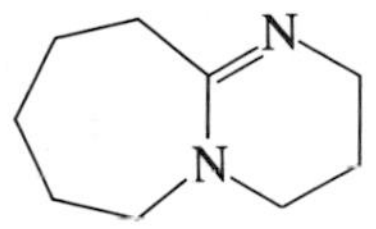

1,8-Diazabicyclo[5.4.0]undec-7-ene (DBU)

Another unusually strong organic base is guanidine, $H_2N\overset{NH}{\overset{\|}{C}}NH_2$. Indicate which nitrogen in each of these bases is the one most likely to be protonated and explain the enhanced strength of these bases relative to simple amines.

10. The following are proposed syntheses of amines. In each case, indicate whether the synthesis will work well, poorly, or not at all. If a synthesis will not work well, explain why.

(a) $CH_3CH_2CH_2CH_2Cl \xrightarrow[\text{2. LiAlH}_4\text{, (CH}_3\text{CH}_2)_2\text{O}]{\text{1. KCN, CH}_3\text{CH}_2\text{OH}} CH_3CH_2CH_2CH_2NH_2$

(b) $(CH_3)_3CCl \xrightarrow[\text{2. LiAlH}_4\text{, (CH}_3\text{CH}_2)_2\text{O}]{\text{1. NaN}_3\text{, DMSO}} (CH_3)_3CNH_2$

(c) Cyclohexyl–$CONH_2$ $\xrightarrow{\text{Br}_2\text{, NaOH, H}_2\text{O}}$ Cyclohexyl–NH_2

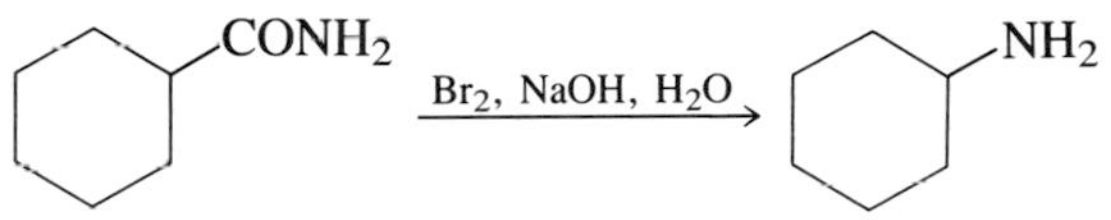

(d) Cyclopentyl–$CH_2CH_2CH_2CH_2$–Br $\xrightarrow{\text{CH}_3\text{NH}_2}$ Cyclopentyl–$CH_2CH_2CH_2CH_2$–$NHCH_3$

(e) Decalin (H, H)–CH_2Br (CH_3) $\xrightarrow[\text{2. H}^+\text{, H}_2\text{O, }\Delta]{\text{1. potassium phthalimide (N}^-\text{K}^+\text{), DMF}}$ Decalin (H, H)–CH_2NH_2 (CH_3)

(f) C_6H_5COCl $\xrightarrow[\text{2. LiAlH}_4\text{, (CH}_3\text{CH}_2)_2\text{O}]{\text{1. C}_6\text{H}_5\text{CH}_2\text{NH}_2\text{, HO}^-}$ $(C_6H_5CH_2)_2NH$

(g) Cyclopentyl–CHO $\xrightarrow[\text{2. NaBH}_3\text{CN, CH}_3\text{CH}_2\text{OH}]{\text{1. (CH}_3)_3\text{CNH}_2}$ Cyclopentyl–$CH_2NHC(CH_3)_3$

(h) $NH_2CH_2CH_2CHO \xrightarrow{\text{NaBH}_3\text{CN, CH}_3\text{CH}_2\text{OH}}$ Azetidine (NH)

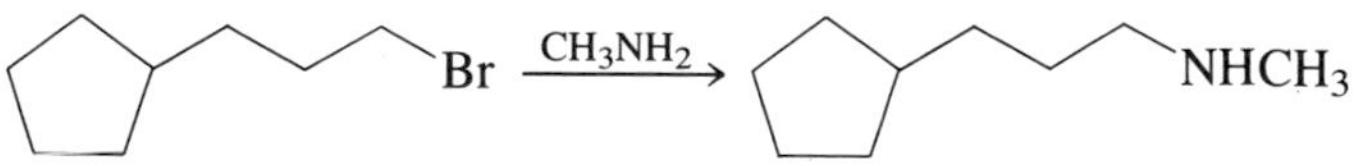

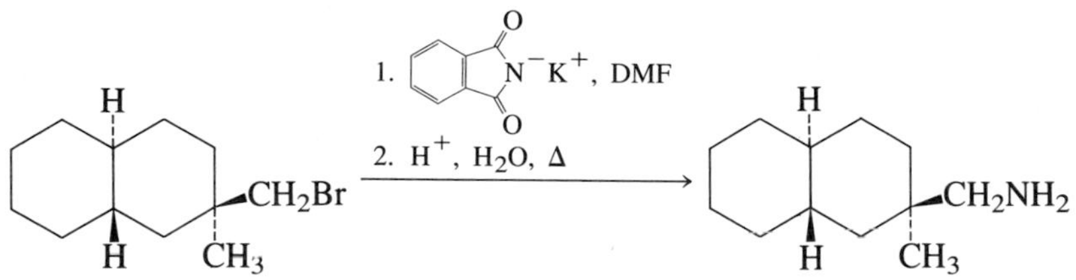

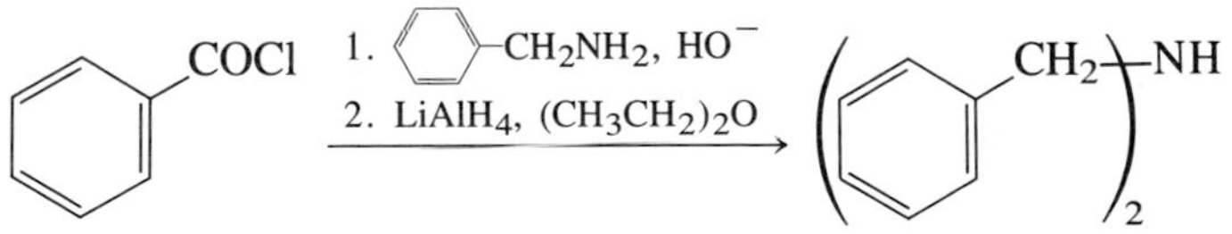

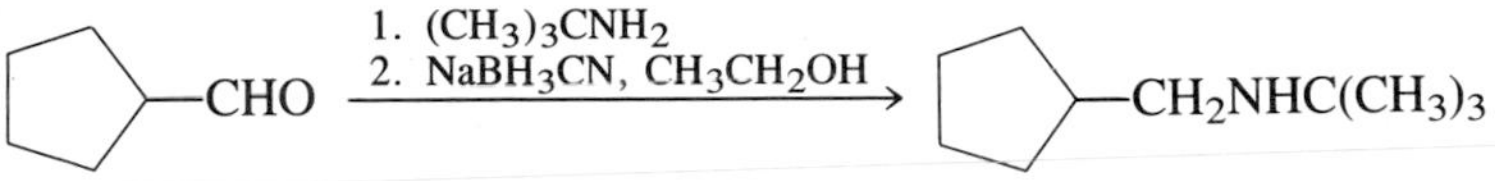

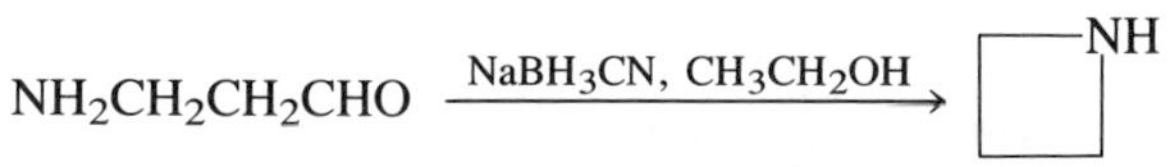

(i) Br-cyclohexane → [1. HNO_3, H_2SO_4; 2. Fe, H^+] → 4-bromocyclohexylamine (Br, NH_2)

(j) azetidinone (NH, O) → [1. NaH, THF; 2. CH_3I; 3. $LiAlH_4$, $(CH_3CH_2)_2O$] → N-methylazetidine (N–CH_3)

11. For each synthesis in Problem 10 that does not work well, propose an alternative synthesis of the final amine, starting either with the same material or with a material of similar structure and functionality.

12. In the past several years, pseudoephedrine has gradually replaced phenylpropanolamine as the favored decongestant in over-the-counter cold remedies. (Pseudoephedrine is less likely to cause drowsiness.)

C_6H_5–CH(OH)–CH(NH_2)–CH_3

Phenylpropanolamine

C_6H_5–CH(OH)–CH($NHCH_3$)–CH_3

Pseudoephedrine

Suppose that you are the director of a major pharmaceutical laboratory with a huge stock of phenylpropanolamine on hand and the president of the company issues the order, ''Pseudoephedrine from now on!'' Analyze all your options, and propose the best solution that you can find for the problem.

C_6H_5–CH_2–CH(N(H)CH_2CH_3)–CH_3

Apetinil

13. Apetinil, an appetite suppressant (i.e., diet pill), has the structure shown in the margin. Is it a primary, a secondary, or a tertiary amine? Propose an efficient synthesis of Apetinil from each of the following starting materials. Try to use a variety of methods.

(a) $C_6H_5CH_2COCH_3$ **(b)** $C_6H_5CH_2CH(Br)CH_3$ **(c)** $C_6H_5CH_2CH(CH_3)COOH$

14. Give the structures of the possible alkene products of Hofmann elimination of each of the following amines. If a compound can be cycled through multiple eliminations, give the products of each cycle.

(a) C_6H_5–$CH(NH_2)CH_2CH_3$ **(b)** 1-methylcyclohexylamine (H_3C, NH_2) **(c)** H_3C–(N-CH_3 pyrrolidine)

(d) indoline (N–H) **(e)** indolizidine (N)

15. Reaction of the tertiary amine tropinone with (bromomethyl)benzene (benzyl bromide) gives not one but two quaternary ammonium salts, A and B.

Tropinone ($C_8H_{13}NO$) + $C_6H_5CH_2Br$ ⟶ A + B $[C_{15}H_{20}NO]^+ Br^-$

Tropinone

($C_8H_{13}NO$)

Compounds A and B are stereoisomers that are interconverted by base; that is, base treatment of either pure isomer leads to an equilibrium mixture of the two. **(a)** Propose structures for A and B. **(b)** What kind of stereoisomers are A and B? **(c)** Suggest a mechanism for the equilibration of A and B by base.

16. Attempted Hofmann elimination of an amine containing a hydroxy group on the β-carbon gives an oxacyclopropane product instead of an alkene.

$HOCH_2$—CH_2NH_2 $\xrightarrow{\text{1. Excess } CH_3I \text{; 2. } Ag_2O, H_2O \text{; 3. } \Delta}$ H_2C—CH_2 (oxacyclopropane) + $(CH_3)_3N$

(a) Propose a sensible mechanism for this transformation. **(b)** Pseudoephedrine (see Problem 12) and ephedrine are closely related, naturally occurring compounds, as the similar names imply. In fact, they are stereoisomers. From the results of the following reactions, deduce the precise stereochemistries of ephedrine and pseudoephedrine.

Ephedrine $\xrightarrow{\text{1. } CH_3I \text{; 2. } Ag_2O, H_2O \text{; 3. } \Delta}$ oxacyclopropane with phenyl, H, H, CH_3

Pseudoephedrine $\xrightarrow{\text{1. } CH_3I \text{; 2. } Ag_2O, H_2O \text{; 3. } \Delta}$ oxacyclopropane with H, phenyl, H, CH_3

17. Show how each of the following molecules might be synthesized by Mannich or Mannichlike reactions. (Hint: Work backward, identifying the bond made in the Mannich reaction.)

(a) $CH_3COCH_2CH_2N(CH_2CH_3)_2$

(b) 1-[(dimethylamino)methyl]-2-indanone ($-N(CH_3)_2$, $=O$)

(c) H_3C—$CH(NH_2)$—CN

(d) $CH_3CH_2CH_2COCH(CH_2CH_3)CH_2N(CH_3)_2$

(e) $CH_3COCH_2CH_2N(CH_3)CH_2CH_2COCH_3$

18. Tropinone (Problem 15) was first synthesized by Sir Robert Robinson (famous for the Robinson annulation reaction; Section 18-12), in 1917, by the following reaction. Show a mechanism for this transformation.

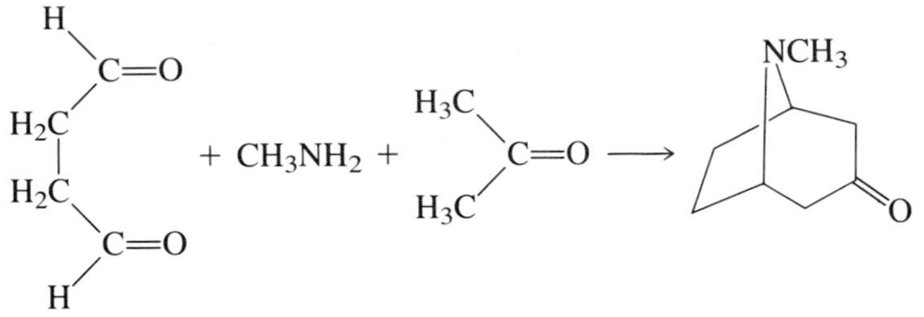

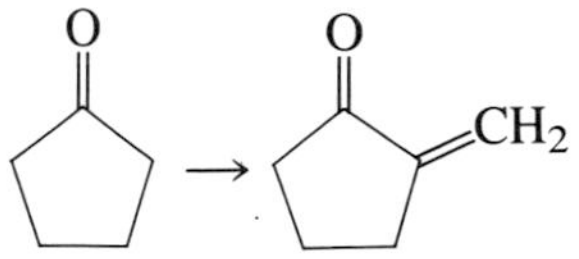

H, C=O, H_2C, H_2C, C=O, H + CH_3NH_2 + H_3C, C=O, H_3C → NCH_3, O

O → O, CH_2

19. Illustrate a method for achieving the transformation shown in the margin, using combinations of reactions presented in Sections 21-8 and 21-9.

20. Give the expected product(s) of each of the following reactions.

(a) 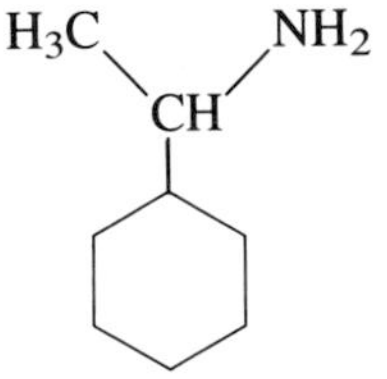 H_3C, NH_2, CH (cyclohexyl) $\xrightarrow{\text{NaNO}_2\text{, HCl, H}_2\text{O}}$ **(b)** 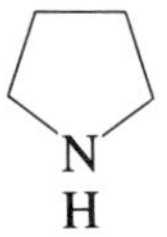 N H (pyrrolidine) $\xrightarrow{\text{NaNO}_2\text{, HCl, 0°C}}$

21. Write the expected products of each of the following reactions.

(a) *E*-2-pentene + $CHCl_3$ $\xrightarrow{\text{KOC(CH}_3)_3\text{, (CH}_3)_3\text{COH}}$

(b) 1-Methylcyclohexene + CH_2I_2 $\xrightarrow{\text{Zn-Cu, (CH}_3\text{CH}_2)_2\text{O}}$

(c) Propene + CH_2N_2 $\xrightarrow{\text{Cu, }\Delta}$

(d) *Z*-1,2-diphenylethene + $CHBr_3$ $\xrightarrow{\text{KOC(CH}_3)_3\text{, (CH}_3)_3\text{COH}}$

(e) *E*-1,3-pentadiene + 2 CH_2I_2 $\xrightarrow{\text{Zn-Cu, (CH}_3\text{CH}_2)_2\text{O}}$

(f) $CH_2{=}CHCH_2CH_2CH_2CHN_2$ $\xrightarrow{h\nu}$

22. Reductive amination of *excess* formaldehyde with a primary amine leads to the formation of a *di*methylated tertiary amine as the product (see the example below). Propose an explanation.

$$(CH_3)_3CCH_2NH_2 + 2\ CH_2{=}O \xrightarrow{\text{NaBH}_3\text{CN, CH}_3\text{OH}} (CH_3)_3CCH_2N(CH_3)_2$$

84%

2,2-Dimethylpropanamine **_N,N_,2,2-Tetramethylpropanamine**

23. Several of the natural amino acids are synthesized from 2-keto carboxylic acids by an enzyme-catalyzed reaction with a special coenzyme called pyridoxamine. Use electron-pushing arrows to describe each step in the following synthesis of phenylalanine from phenylpyruvic acid.

Pyridoxamine + Phenylpyruvic acid ($C_6H_5CH_2COCO_2H$) ⟶ (imine: $HOCH_2$, OH, CH_3 pyridine with $CH_2N{=}C(CO_2H)CH_2C_6H_5$) ⟶

(N–H dihydropyridine tautomer with H–$C{=}$, $N{=}C(CO_2H)CH_2C_6H_5$) ⟶ (pyridine with H–$C{=}N$–$CH(CO_2H)CH_2C_6H_5$) $\xrightarrow{H_2O}$

Pyridoxal + Phenylalanine ($C_6H_5CH_2CH(NH_2)CO_2H$)

24. From the following information, deduce the structure of coniine, an amine found in poison hemlock, which, deservedly, has a very bad reputation. IR: 3330 cm^{-1}. ^1H NMR: $\delta = 0.91$ (t, $J = 7$ Hz, 3 H), 1.33 (s, 1 H), 1.52 (m, 10 H), 2.70 (t, $J = 6$ Hz, 2 H), and 3.0 (m, 1 H) ppm. Mass spectrum: molecular ion m/z = 127; also, m/z = 84(100) and 56(20).

$$\text{Coniine} \xrightarrow[3.\ \Delta]{1.\ CH_3I\quad 2.\ Ag_2O,\ H_2O} \text{mixture of three compounds} \xrightarrow[3.\ \Delta]{1.\ CH_3I\quad 2.\ Ag_2O,\ H_2O} (CH_3)_3N + \text{mixture of 1,4-octadiene and 1,5-octadiene}$$

25. Pethidine, the active ingredient in the narcotic analgesic Demerol, was subjected to two successive exhaustive methylations with Hofmann eliminations, and then ozonolysis, with the following results. **(a)** Propose a structure for pethidine based on this information:

$$\underset{\textbf{Pethidine}}{C_{15}H_{21}NO_2} \xrightarrow[3.\ \Delta]{1.\ CH_3I\quad 2.\ Ag_2O,\ H_2O} C_{16}H_{23}NO_2 \xrightarrow[3.\ \Delta]{1.\ CH_3I\quad 2.\ Ag_2O,\ H_2O}$$

$$(CH_3)_3N + C_{14}H_{16}O_2 \xrightarrow[2.\ Zn,\ H_2O]{1.\ O_3,\ CH_2Cl_2} 2\ CH_2O + C_6H_5C(CO_2CH_2CH_3)(CHO)_2$$

$\mathrm{OHCCH_2}$–C(C_6H_5)($CO_2CH_2CH_3$)–CH_2CHO

(b) Propose a synthesis of pethidine that begins with ethyl phenylacetate and *cis*-1,4-dibromo-2-butene. (Hint: First prepare the dialdehyde ester shown in the margin, and then convert it into pethidine.)

26. Skytanthine is a monoterpene alkaloid with the following properties. Analysis: $C_{11}H_{21}N$. ^1H NMR: two CH_3 doublets ($J = 7$ Hz) at $\delta = 1.20$ and 1.33 ppm; one CH_3 singlet at $\delta = 2.32$ ppm; other hydrogens give rise to broad signals at $\delta = 1.3$–2.7 ppm. IR: no bands $\geq 3100\ \mathrm{cm}^{-1}$. Deduce the structures of skytanthine and degradation products A, B, and C from this information.

Skytanthine $\xrightarrow{\text{1. } CH_3I;\ \text{2. } Ag_2O,\ H_2O;\ \text{3. } \Delta}$ $C_{12}H_{23}N$ **A** (IR: $\tilde{\nu} = 1646\ \mathrm{cm}^{-1}$) $\xrightarrow{\text{1. } O_3,\ CH_2Cl_2;\ \text{2. Zn, } H_2O}$ $CH_2{=}O$ + $C_{11}H_{21}NO$ **B** (IR: $\tilde{\nu} = 1715\ \mathrm{cm}^{-1}$) $\xrightarrow{\text{1. } m\text{-}ClC_6H_4CO_3H,\ CH_2Cl_2;\ \text{2. KOH, } H_2O}$ CH_3COOH + $C_9H_{19}NO$ **C** (IR: $\tilde{\nu} = 3620\ \mathrm{cm}^{-1}$) $\xrightarrow{\text{Careful oxidation}}$ 3-methyl-2-($CH_2N(CH_3)_2$)cyclopentanone (O, CH_3, $CH_2N(CH_3)_2$) (IR: $\tilde{\nu} = 1745\ \mathrm{cm}^{-1}$)

27. Many alkaloids are synthesized in nature from a precursor molecule called norlaudanosoline which, in turn, appears to be derived from the condensation of amine (i) with aldehyde (ii). Formulate a mechanism for this transformation. Note that a carbon–carbon bond is formed in the process. Name a reaction presented in this chapter that is closely related to this carbon–carbon bond formation.

(HO)$_2C_6H_3CH_2CH_2NH_2$ (**i**) + (HO)$_2C_6H_3CH_2CHO$ (**ii**) $\longrightarrow$ **Norlaudanosoline** (HO, HO, NH, OH, OH) $\longrightarrow$ alkaloids

22

Chemistry of Benzene Substituents

Alkylbenzenes, Phenols, and Benzenamines

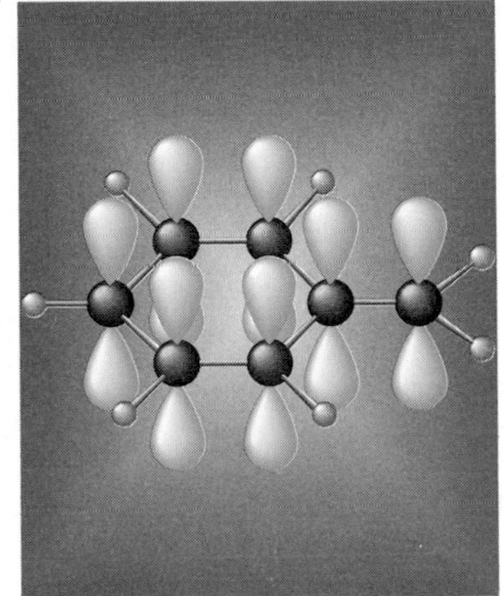

Chapters 15 and 16 introduced the chemistry of benzene. We know that the special stability of the aromatic ring makes it relatively unreactive. How does the benzene ring modify the behavior of neighboring reactive centers? This chapter takes a closer look at the effect exerted by the ring on the reactivity of carbon, oxygen, and nitrogen substituents. We shall see that the interaction of the ring and the substituent allows the formation of compounds that are stabilized by resonance. After discussing the special reactivity of aryl-substituted (benzylic) carbon atoms, we consider the preparation and reactions of phenols and benzenamines (anilines). These compounds occur widely in nature and are used in synthetic procedures as precursors to substances such as aspirin, dyes, and vitamins.

Phenylmethyl (benzyl) system

2-Propenyl (allyl) system

22-1 Reactivity at the Phenylmethyl (Benzyl) Carbon: Benzylic Resonance Stabilization

The **phenylmethyl (benzyl)** group, $C_6H_5CH_2$, may be viewed as a benzene ring whose π system overlaps with an extra p orbital. This interaction, called **benzylic resonance,** stabilizes adjacent radical, cationic, and anionic centers in much the same way that overlap of a π bond and a third p orbital stabilizes 2-propenyl (allyl) intermediates (Section 14-1).

Benzylic radicals are reactive intermediates in the halogenation of alkylbenzenes

We have seen that benzene will not react with chlorine or bromine unless a Lewis acid is added. The acid catalyzes halogenation of the ring (Section 15-10).

$$\text{No reaction} \xleftarrow{\text{Br–Br}} \text{C}_6\text{H}_6 \xrightarrow[-\text{HBr}]{\text{Br–Br, FeBr}_3} \text{C}_6\text{H}_5\text{Br}$$

In contrast, heat or light allows attack by chlorine or bromine on methylbenzene (toluene) even in the absence of a catalyst. Analysis of the products shows that reaction takes place at the methyl group, *not* at the aromatic ring, and that excess halogen leads to multiple substitution.

$$\text{C}_6\text{H}_5\text{CH}_2\text{H} + \text{Br}_2 \xrightarrow{\Delta} \text{C}_6\text{H}_5\text{CH}_2\text{Br} + \text{HBr}$$

(Bromomethyl)-benzene

$$\text{C}_6\text{H}_5\text{CH}_3 \xrightarrow[-\text{HCl}]{\text{Cl}_2,\ h\nu} \text{C}_6\text{H}_5\text{CH}_2\text{Cl} \xrightarrow[-\text{HCl}]{\text{Cl}_2,\ h\nu} \text{C}_6\text{H}_5\text{CHCl}_2 \xrightarrow[-\text{HCl}]{\text{Cl}_2,\ h\nu} \text{C}_6\text{H}_5\text{CCl}_3$$

(Chloromethyl)-benzene **(Dichloromethyl)-benzene** **(Trichloromethyl)-benzene**

Each substitution yields one molecule of hydrogen halide as a by-product.

As in the halogenation of alkanes (Sections 3-4 through 3-6) and the allylic halogenation of alkenes (Section 14-2), the mechanism of benzylic halogenation proceeds through radical intermediates. Heat or light induces dissociation of the halogen molecule into atoms. One of them abstracts a benzylic hydrogen, a reaction giving HX and a phenylmethyl (benzyl) radical. This intermediate reacts with another molecule of halogen to give the product, a (halomethyl)benzene, and another halogen atom, which propagates the chain process.

Mechanism of Benzylic Halogenation

$$\text{X}\cdot + \text{C}_6\text{H}_5\text{CH}_2\text{H} \xrightarrow[-\text{HX}]{} \text{C}_6\text{H}_5\dot{\text{C}}\text{H}_2 \xrightarrow{\text{X}_2} \text{C}_6\text{H}_5\text{CH}_2\text{X} + \text{X}\cdot$$

Phenylmethyl (benzyl) radical

$CH_2\cdot$ ⟷ CH_2 ⟷ CH_2 ⟷ CH_2 or CH_2

A B

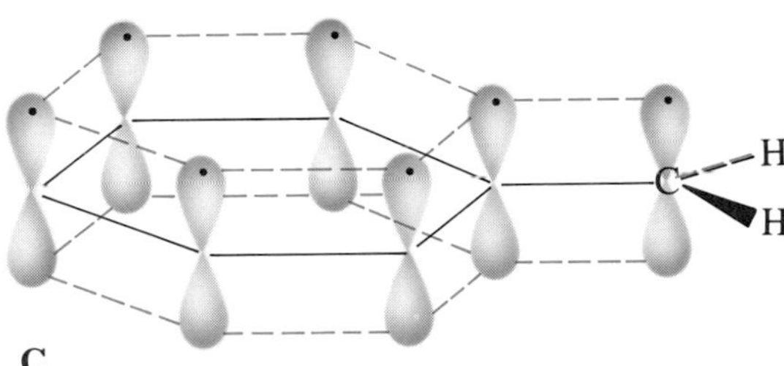

FIGURE 22-1 The benzene π system of the phenylmethyl (benzyl) radical enters into resonance with the adjacent radical center. The extent of delocalization may be depicted by (A) resonance structures, (B) dotted lines, or (C) orbitals.

What explains the ease of benzylic halogenation? The answer lies in the stabilization of the phenylmethyl (benzyl) radical by the phenomenon called benzylic resonance (Figure 22-1). As a consequence, the benzylic C–H bond is relatively weak ($DH° = 87$ kcal mol^{-1}); its cleavage is thermodynamically relatively favorable and proceeds with a low activation energy.

Inspection of the resonance structures in Figure 22-1 reveals why the halogen attacks only the *benzylic* position and not an aromatic carbon: Reaction at any but the benzylic carbon would destroy the aromatic character of the benzene ring.

EXERCISE 22-1

For each of the compounds named below, draw the structure and indicate where radical halogenation is most likely to occur upon heating in the presence of Br_2. Then rank the compounds in approximate descending order of reactivity under bromination conditions. **(a)** Ethylbenzene; **(b)** 1,2-diphenylethane; **(c)** 1,3-diphenylpropane; **(d)** diphenylmethane; **(e)** (1-methylethyl)benzene.

Benzylic cations delocalize the positive charge

Benzylic resonance can also have a profound effect on the reactivity of benzylic halides and sulfonates. Thus, the 4-methylbenzenesulfonate (tosylate) of phenylmethanol (benzyl alcohol) reacts with ethanol much more rapidly than does the corresponding ethyl sulfonate: $k_2:k_1 = 100:1$.

$$CH_3CH_2OS(=O)_2\text{–}C_6H_4\text{–}CH_3 + CH_3CH_2OH \xrightarrow{k_1} CH_3CH_2OCH_2CH_3 + HO_3S\text{–}C_6H_4\text{–}CH_3$$

Ethyl 4-methylbenzenesulfonate
(A simple primary tosylate)

$$C_6H_5CH_2OSO_2C_6H_4CH_3 + CH_3CH_2OH \xrightarrow{k_2} C_6H_5CH_2OCH_2CH_3 + HO_3SC_6H_4CH_3$$

Phenylmethyl 4-methylbenzenesulfonate
(A benzylic tosylate)

(Ethoxymethyl)benzene

The reaction rates are different because the mechanisms are different. Ethyl 4-methylbenzenesulfonate reacts with ethanol by an S_N2 process. This transformation is slow because ethanol is a poor nucleophile. The benzylic derivative, however, like allylic halides and sulfonates (Section 14-3), can dissociate readily to form a delocalized carbocation and therefore can follow an S_N1 mechanism.

$$C_6H_5CH_2OSO_2C_6H_4CH_3 \longrightarrow \left[C_6H_5\overset{+}{C}H_2 \longleftrightarrow \cdots \longleftrightarrow \cdots \longleftrightarrow \cdots \right]$$

Benzylic cation

EXERCISE 22-2

Which one of the two chlorides will solvolyze more rapidly: (chloromethyl)benzene, $C_6H_5CH_2Cl$, or chlorodiphenylmethane, $(C_6H_5)_2CHCl$? Explain your answer.

EXERCISE 22-3

Phenylmethanol (benzyl alcohol) is converted into (chloromethyl)benzene in the presence of hydrogen chloride much more rapidly than ethanol is converted into chloroethane. Explain.

Benzylic halides and sulfonates, like their allylic counterparts, also show enhanced S_N2 reactivity. The transition state is stabilized by overlap with the benzene π system (Figure 22-2).

$$C_6H_5CH_2Br + {}^-CN \xrightarrow{S_N2} C_6H_5CH_2CN + Br^-$$

81%

(Bromomethyl)benzene (Benzyl bromide)
($\sim 10^2$ times faster than S_N2 reactions of primary bromoalkanes)

Phenylacetonitrile

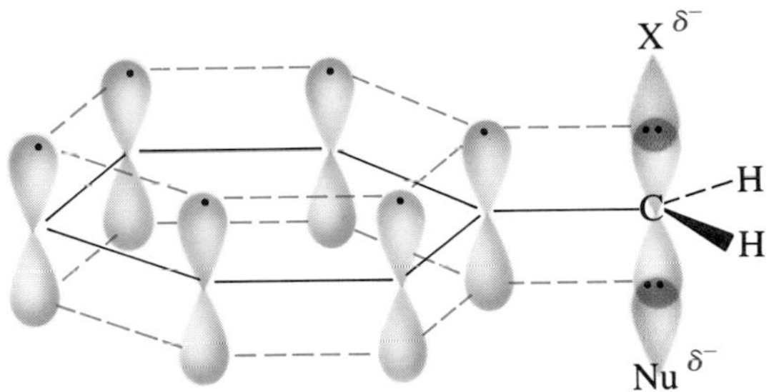

FIGURE 22-2 The benzene π system overlaps with the orbitals of the S_N2 transition state at a benzylic center. As a result, the transition state is stabilized, thereby lowering the activation barrier toward S_N2 reactions of (halomethyl)benzenes.

Resonance in benzylic anions makes benzylic hydrogens relatively acidic

A negative charge adjacent to a benzene ring, as in phenylmethyl (benzyl) anion, is stabilized by conjugation in much the same way that the corresponding radical and cation are stabilized.

Resonance in Benzylic Anions

CH_3 (pK_a ~ 41) ⇌ [$^-{:}CH_2$ ⟷ CH_2 ⟷ CH_2 ⟷ CH_2] + H^+

The acidity of methylbenzene (toluene; pK_a ~ 41) is therefore considerably greater than that of ethane (pK_a ~ 50) and comparable to that of propene (pK_a ~ 40), which is deprotonated to produce the resonance-stabilized 2-propenyl (allyl) anion (Section 14-4). Consequently, methylbenzene (toluene) can be deprotonated by butyllithium. The corresponding Grignard reagent is made in the usual way from a (halomethyl)benzene and magnesium.

Deprotonation of Methylbenzene

CH_3 + $CH_3CH_2CH_2CH_2Li$ $\xrightarrow{(CH_3)_2NCH_2CH_2N(CH_3)_2,\ THF,\ \Delta}$ CH_2Li + $CH_3CH_2CH_2CH_2H$

Methylbenzene (Toluene) — **Phenylmethyllithium (Benzyllithium)**

EXERCISE 22-4

Which molecule in each of the following pairs is more reactive with the indicated reagents, and why?

(a) $(C_6H_5)_2CH_2$ or $C_6H_5CH_3$, with $CH_3CH_2CH_2CH_2Li$

(b) p-$CH_3OC_6H_4CH_2Br$ or p-$CH_3OC_6H_4CH_2Cl$, with $NaOCH_3$ in CH_3OH

(c) $C_6H_5CH_2OH$ or p-$O_2NC_6H_4CH_2OH$, with HCl

In summary, benzylic radicals, cations, and anions are stabilized by resonance with the benzene ring. This effect allows for relatively easy radical halogenations, S_N1 and S_N2 reactions, and benzylic anion formation.

22-2 Benzylic Oxidations and Reductions

Because it is aromatic, the benzene ring is quite unreactive. It does undergo certain transformations—in particular, electrophilic aromatic substitution (Chapters 15 and 16)—but oxidations and reductions of the ring are difficult to achieve. However, *benzylic* positions are more susceptible to such transformations. This section describes how certain reagents oxidize and reduce alkyl substituents on the benzene ring.

Oxidation of substituted benzenes leads to benzoic acids

Reagents such as $KMnO_4$ and $Na_2Cr_2O_7$ oxidize alkylbenzenes to benzoic acids. Benzylic carbon–carbon bonds are cleaved in this process.

Benzylic Oxidations of Alkyl Chains

$$p\text{-}H_3C\text{-}C_6H_4\text{-}CH_2CH_2CH_2CH_3 \xrightarrow[2.\ H^+,\ H_2O]{1.\ KMnO_4,\ HO^-,\ \Delta} p\text{-}HOOC\text{-}C_6H_4\text{-}COOH \quad 80\%$$

1-Butyl-4-methylbenzene → **1,4-Benzenedicarboxylic acid (Terephthalic acid)**

The special reactivity of the benzylic position is also seen in the mild conditions required for the oxidation of benzylic alcohols to the corresponding carbonyl compounds. For example, manganese dioxide, MnO_2, performs this oxidation selectively in the presence of other (nonbenzylic) hydroxy groups. (Recall that MnO_2 was used in the conversion of allylic alcohols into α,β-unsaturated aldehydes and ketones; see Section 18-8.)

Selective Oxidation of a Benzylic Alcohol with Manganese Dioxide

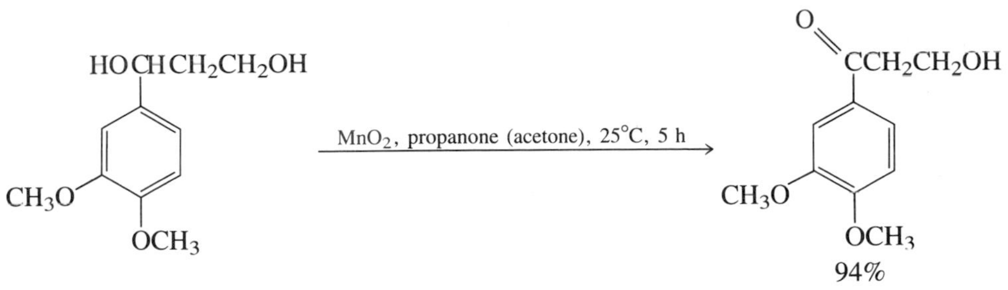

HOCHCH$_2$CH$_2$OH (3,4-dimethoxyphenyl; CH$_3$O, OCH$_3$) $\xrightarrow{MnO_2,\ \text{propanone (acetone)},\ 25°C,\ 5\ h}$ CCH$_2$CH$_2$OH (O) (CH$_3$O, OCH$_3$) 94%

Benzylic ethers are cleaved by hydrogenolysis

Exposure of benzylic alcohols or ethers to hydrogen in the presence of metal catalysts results in rupture of the reactive benzylic carbon–oxygen bond. This transformation is an example of **hydrogenolysis,** cleavage of a σ bond by catalytically activated hydrogen.

Cleavage of Benzylic Ethers by Hydrogenolysis

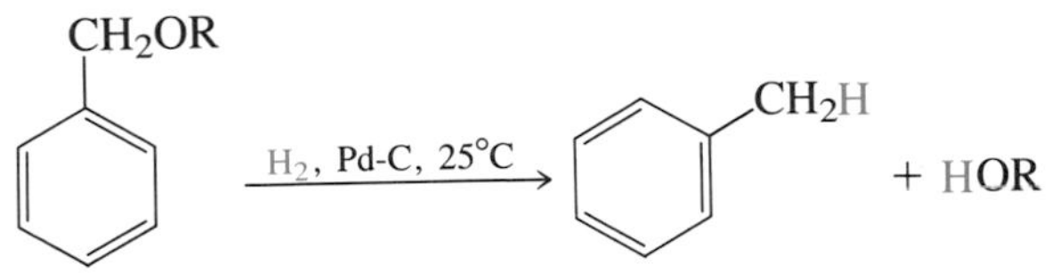

C$_6$H$_5$CH$_2$OR $\xrightarrow{H_2,\ \text{Pd-C},\ 25°C}$ C$_6$H$_5$CH$_2$H + HOR

EXERCISE 22-5

Write synthetic schemes that would connect the following starting materials with their products.

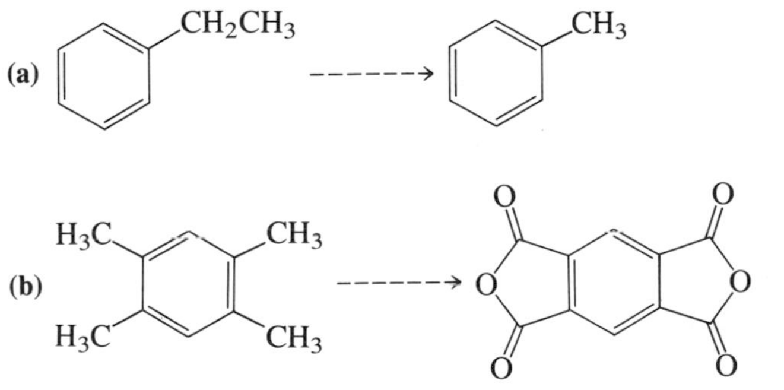

(a) C$_6$H$_5$CH$_2$CH$_3$ ------→ C$_6$H$_5$CH$_3$

(b) H$_3$C, H$_3$C, CH$_3$, CH$_3$ (1,2,4,5-tetramethylbenzene) ------→ (benzene with two fused anhydride rings: O, O, O, O, O, O)

Hydrogenolysis is not possible for ordinary alcohols and ethers. Therefore, the phenylmethyl (benzyl) substituent is a valuable protecting group for hydroxy functions. The following scheme shows its use in part of a synthesis of a compound in the eudesmane class of essential oils, which includes substances of importance in both medicine and perfumery.

Because the hydrogenolysis of the phenylmethyl (benzyl) ether in the final step occurs under neutral conditions, the tertiary alcohol function survives untouched. A tertiary butyl ether would have been a worse choice as a protecting group, because cleavage of its carbon–oxygen bond would have required strong acid (Section 9-8), which may cause dehydration (Section 9-2).

HO CH$_3$ → 1. NaH, THF 2. $C_6H_5CH_2Br$, C_6H_6, DMSO; Protection of OH (Section 9-6) → RO CH$_3$ 80% (R = $C_6H_5CH_2$) → CH_3COOH, H_2O; Deprotection of carbonyl group (Section 17-8) →

RO CH$_3$ O 93% → $CH_3CH{=}P(C_6H_5)_3$, DMSO; Wittig reaction (Section 17-12) → RO CH$_3$ CHCH$_3$ 94% → 1. BH_3, THF 2. Oxidation; Hydroboration–oxidation (Section 12-8) →

RO CH$_3$ COCH$_3$ 99% → 1. CH_3Li, $(CH_3CH_2)_2O$ 2. H^+, H_2O (Section 8-8) → RO CH$_3$ CH$_3$ CCH$_3$ OH 98% → H_2, Pd-C, CH_3CH_2OH; Deprotection of OH → HO CH$_3$ CH$_3$ CCH$_3$ OH 98%

To summarize, benzylic oxidations of alkyl groups occur in the presence of permanganate or chromate; benzylic alcohols are converted into the corresponding ketones by manganese dioxide. The benzylic ether function can be cleaved by hydrogenolysis in a transformation that allows the phenylmethyl (benzyl) substituent to be used as a protecting group for the hydroxy function in alcohols.

22-3 Names and Properties of Phenols

Arenes substituted by hydroxy groups are called **phenols** (IUPAC name: benzenols; Section 15-1). The π system of the benzene ring overlaps with an occupied p orbital on the oxygen atom, a situation resulting in delocalization similar to that found in benzylic anions (Section 22-1). As one result of this extended conjugation, phenols possess an unusual, enolic structure. Recall that enols are usually unstable: They tautomerize easily to the corresponding ketones because of the relatively strong carbonyl bond (Section 18-2). Phenols, however, prefer the enol to the keto form because the aromatic character of the benzene ring is preserved.

Keto and Enol Forms of Phenol

O H H ⇌ [:ÖH ⟷ :ÖH ⟷ $^+$ÖH ⟷ $^+$ÖH ⟷ $^+$ÖH]

2,4-Cyclohexadienone **Phenol**

Phenols and their ethers are ubiquitous in nature; some derivatives have medicinal and herbicidal applications, whereas others are important industrial materials. This section first explains the names of these compounds. It then describes an important difference between phenols and alkanols—phenols are stronger acids because of the neighboring aromatic ring.

Phenols are hydroxyarenes

Phenol itself was formerly known as carbolic acid. It forms colorless needles (m.p. 41°C), has a characteristic odor, and is somewhat soluble in water. Aqueous solutions of it (or its methyl-substituted derivatives) are used as disinfectants, but its main use is for the preparation of polymers (phenolic resins). Total U.S. production of phenol in 1992 was 1.9 million tons. Pure phenol causes severe skin burns and is poisonous; deaths have been reported from the ingestion of as little as 1 g. Fatal poisoning may also result from absorption through the skin.

Substituted phenols are named as phenols, benzenediols, or benzenetriols, according to the system described in Section 15-1. These substances find uses in the photography, dyeing, and tanning industries. Compounds containing carboxylic acid or sulfonic acid functionalities are called **hydroxybenzoic** or **hydroxybenzenesulfonic acids,** respectively. Many have common names.

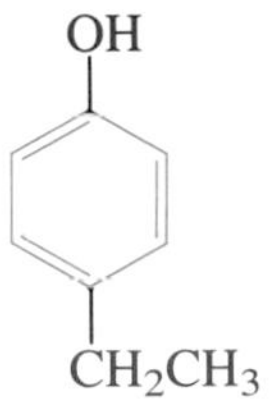

OH, CH_2CH_3 — **4-Ethylphenol (*p*-Ethylphenol)**

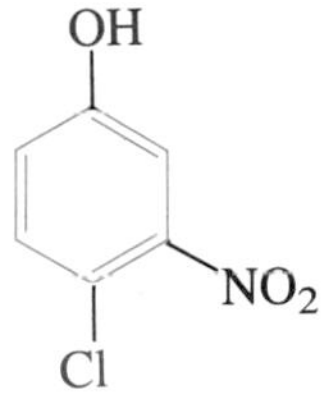

OH, NO_2, Cl — **4-Chloro-3-nitrophenol**

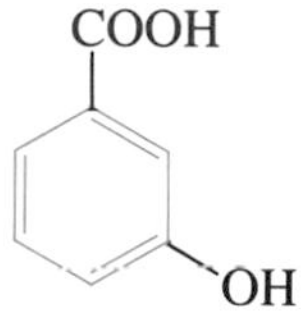

COOH, OH — **3-Hydroxybenzoic acid (*m*-Hydroxybenzoic acid)**

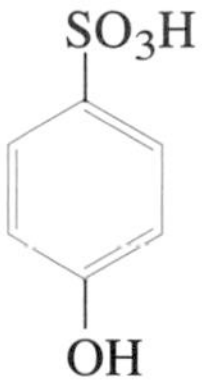

SO_3H, OH — **4-Hydroxybenzenesulfonic acid**

OH, CH_3 — **4-Methylphenol (*p*-Cresol)**

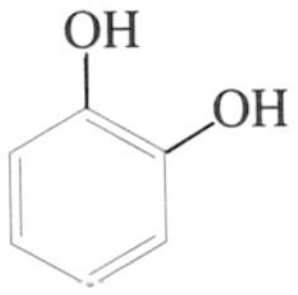

1,2-Benzenediol (Catechol)

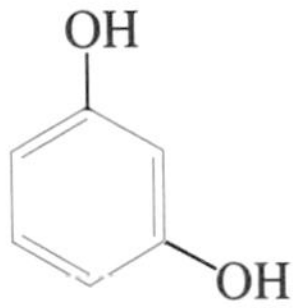

1,3-Benzenediol (Resorcinol)

1,4-Benzenediol (Hydroquinone)

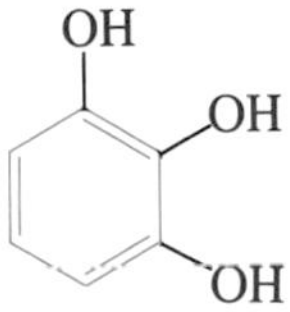

1,2,3-Benzenetriol (Pyrogallol)

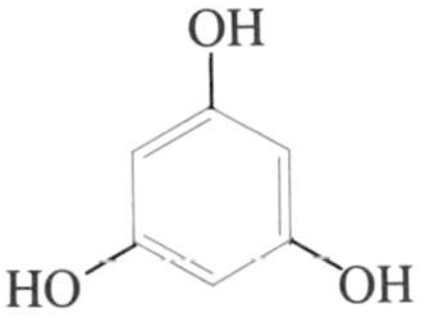

1,3,5-Benzenetriol (Phloroglucinol)

Phenyl ethers are named as **alkoxybenzenes.** As a substituent, C_6H_5O is called **phenoxy.**

Phenols are unusually acidic

Phenols have pK_a values that range from 8 to 10. Even though they are less acidic than the carboxylic acids (pK_a = 3–5), they are stronger than the alkanols (pK_a = 16–18). The reason is resonance: The negative charge in the conjugate base, called the **phenoxide ion,** is stabilized by delocalization into the ring.

Acidity of Phenol

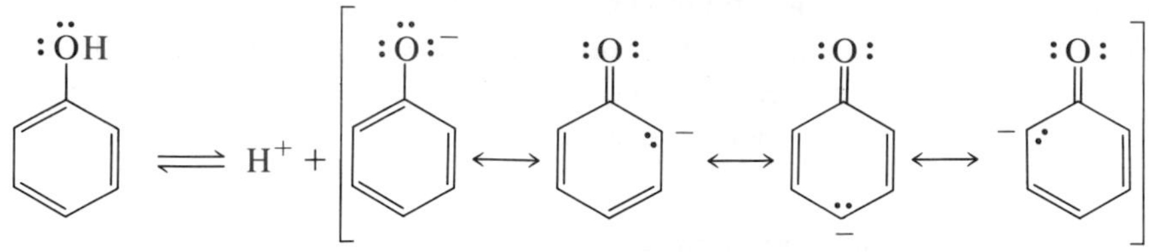

:ÖH H⁺ + :Ö:⁻ :O: :O: :O:

$pK_a \sim 10$ **Phenoxide ion**

The acidity of phenols is greatly affected by substituents that are capable of resonance. 4-Nitrophenol (*p*-nitrophenol), for example, has a pK_a of 7.15.

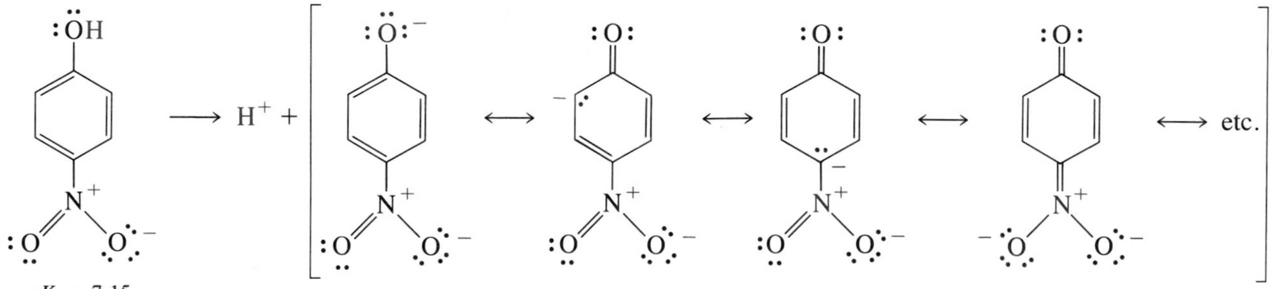

$pK_a = 7.15$ etc.

The 2-isomer has similar acidity ($pK_a = 7.22$), whereas nitrosubstitution at C3 results in a pK_a of 8.39. Multiple nitration increases the acidity to that of carboxylic or even mineral acids. Electron-donating substituents have the opposite effect, raising the pK_a.

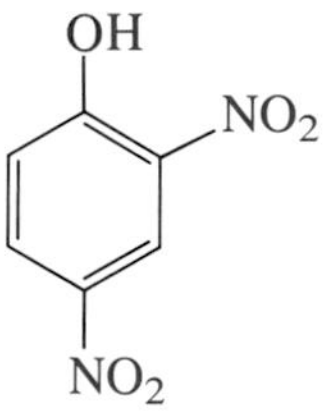

2,4-Dinitrophenol $pK_a = 4.09$

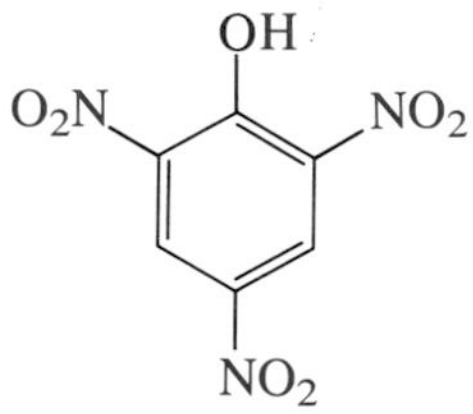

2,4,6-Trinitrophenol (Picric acid) $pK_a = 0.25$

4-Methylphenol (*p*-Cresol) $pK_a = 10.26$

As Section 22-5 will show, the oxygen in phenol and its ethers is also weakly basic, and its protonation gives rise to ether cleavage.

EXERCISE 22-6

Why is 3-nitrophenol (*m*-nitrophenol) less acidic than the other two isomers but more acidic than phenol itself?

EXERCISE 22-7

Rank in order of increasing acidity: phenol, A; 3,4-dimethylphenol, B; 3-hydroxybenzoic (*m*-hydroxybenzoic) acid, C; 4-(fluoromethyl)phenol [*p*-(fluoromethyl)phenol], D.

In summary, phenols exist in the enol form because of aromatic stabilization. They are named according to the rules for naming aromatic compounds explained in Section 15-1. Those derivatives bearing carboxy or sulfonic acid groups on the ring are called hydroxybenzoic acids or hydroxybenzenesulfonic acids. Phenols are acidic because the corresponding anions are resonance stabilized.

22-4 Preparation of Phenols: Nucleophilic Aromatic Substitution

Phenols are synthesized quite differently from the way ordinary alcohols are made because of the aromatic ring. Direct *electrophilic* addition of OH to arenes is difficult, because of the scarcity of reagents that generate an electrophilic hydroxy group, such as HO^+. Instead, phenols are prepared by *nucleophilic* displacement of a leaving group from the arene ring by hydroxide, HO^-. This section discusses the ways in which this transformation may be achieved.

Nucleophilic aromatic substitution may follow an addition–elimination pathway

Treatment of 1-chloro-2,4-dinitrobenzene with hydroxide replaces the halogen with the nucleophile, furnishing the corresponding substituted phenol. Other nucleophiles such as ammonia may be similarly employed, forming benzenamines. Processes such as these, in which a group other than hydrogen is displaced from an aromatic ring, are called **ipso substitutions** (*ipso,* Latin, on itself). The products of these reactions are intermediates in the manufacture of useful dyes.

Nucleophilic Aromatic Ipso Substitution

Cl, NO_2, NO_2 (benzene ring) + :Nu^- ⟶ Nu, NO_2, NO_2 (benzene ring) + Cl^-

Cl, NO_2, NO_2 (benzene ring) $\xrightarrow{Na_2CO_3,\ HOH,\ 100°C}$ OH, NO_2, NO_2 (benzene ring) + NaCl

90%

1-Chloro-2,4-dinitrobenzene **2,4-Dinitrophenol**

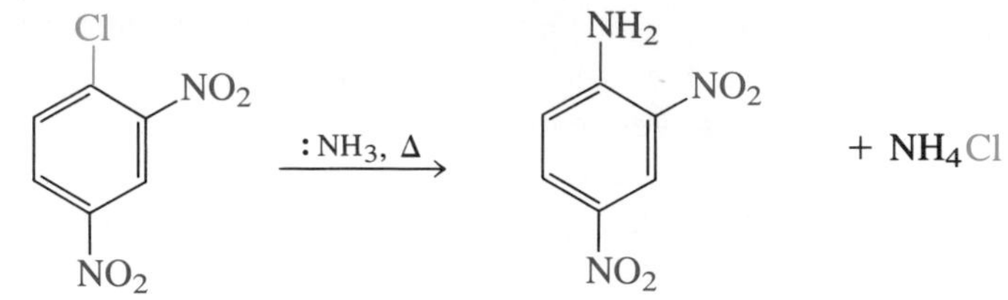

$$\text{1-chloro-2,4-dinitrobenzene} \xrightarrow{:NH_3,\ \Delta} \text{2,4-dinitrobenzenamine} + NH_4Cl$$

85%

2,4-Dinitrobenzenamine
(2,4-Dinitroaniline)

The transformation is called **nucleophilic aromatic substitution.** The key to its success is the presence of one or more strongly electron-withdrawing groups on the benzene ring located ortho or para to the leaving group. Such substituents, first, decrease the electron density in the ring, thereby making it more favorable for nucleophilic attack; second, they stabilize an intermediate anion by resonance. In contrast with the S_N2 reaction of haloalkanes, substitution in these reactions takes place by a *two-step mechanism,* an *addition–elimination sequence* similar to the mechanism of substitution of carboxylic acid derivatives (Chapter 20).

Mechanism of Nucleophilic Aromatic Substitution

STEP 1. Addition (facilitated by resonance stabilization)

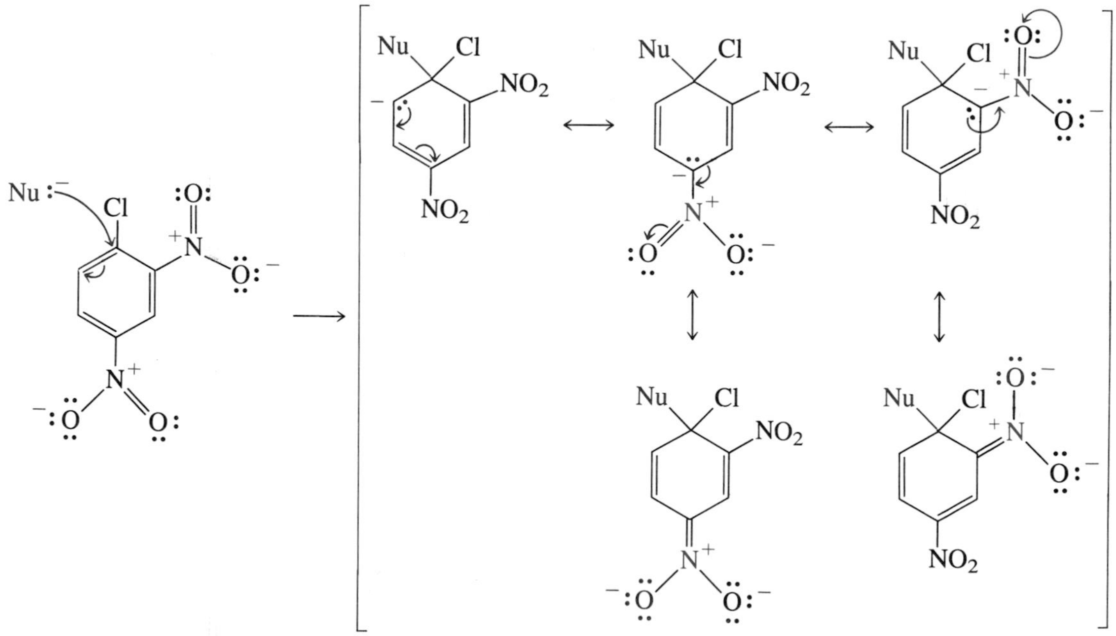

The negative charge is strongly stabilized by resonance involving the ortho- and para-NO_2 groups

STEP 2. Elimination (only one resonance structure is shown)

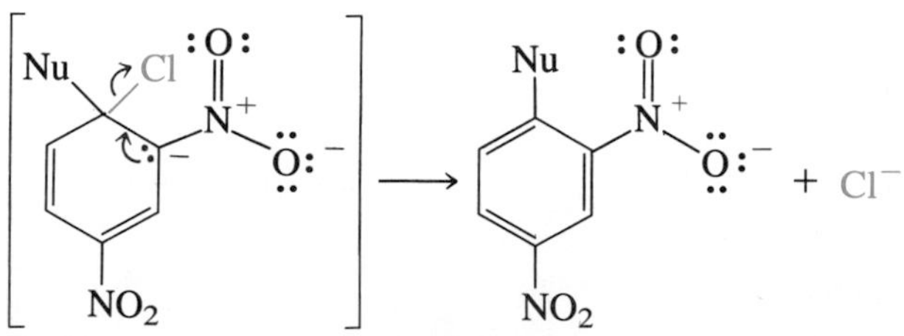

+ Cl^-

In the first step, ipso attack by the nucleophile produces an anion with a highly delocalized charge, for which several resonance structures may be written; five are shown. Note the ability of the negative charge to be delocalized into the electron-withdrawing groups. In contrast, such delocalization is *not* possible in 1-chloro-3,5-dinitrobenzene, in which these groups are located meta; so this compound does *not* undergo ipso substitution under the conditions employed.

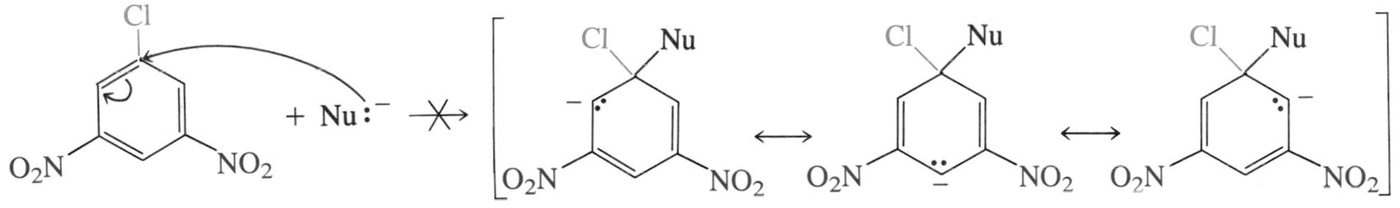

Meta-NO_2 groups provide *no* resonance stabilization for the negative charge

In the second step, the leaving group is expelled to regenerate the aromatic ring. The reactivity of haloarenes in nucleophilic substitutions increases with the nucleophilicity of the reagent and the number of electron-withdrawing groups on the ring, particularly if they are in the ortho and para positions.

EXERCISE 22-8

Write the expected product of reaction of 1-chloro-2,4-dinitrobenzene with $NaOCH_3$ in refluxing CH_3OH.

Another example of nucleophilic aromatic substitution is the conversion of arenesulfonic acids into phenols by heating in molten NaOH. Phenol itself was at one time manufactured from sodium benzenesulfonate in this manner.

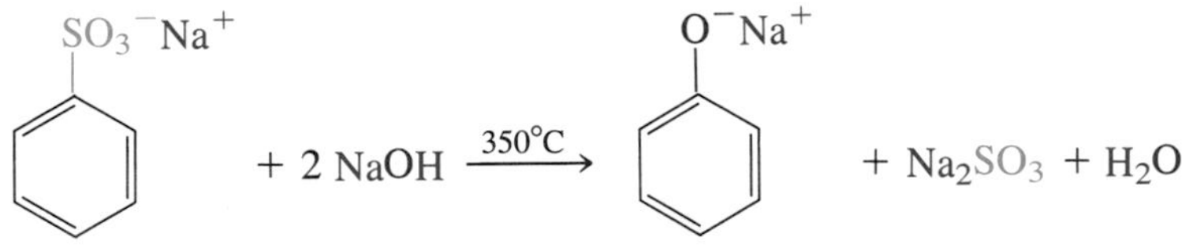

$$C_6H_5SO_3^-Na^+ + 2\ NaOH \xrightarrow{350°C} C_6H_5O^-Na^+ + Na_2SO_3 + H_2O$$

Sodium salt of phenol

The reaction initially produces the phenoxide salt, which is subsequently protonated with HCl.

EXERCISE 22-9

Propose a mechanism for the following conversion. Considering that the first step is rate determining, draw a potential-energy diagram depicting the progress of the reaction.

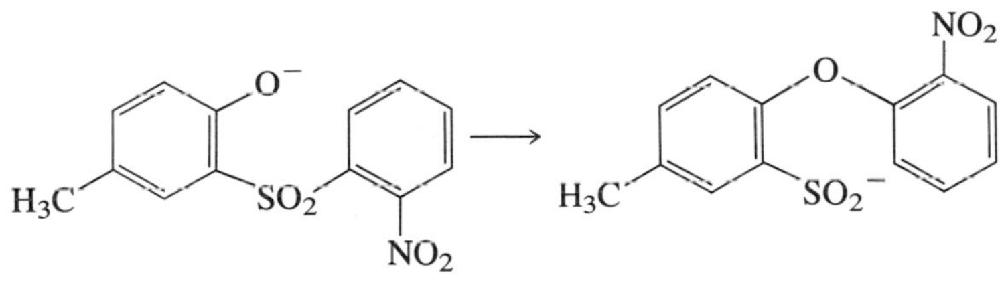

BOX 22-1 Toxicity of Chlorophenols

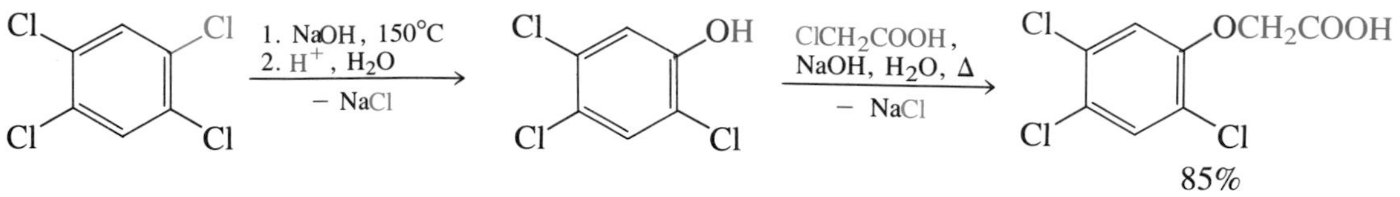

1,2,4,5-Tetrachlorobenzene $\xrightarrow[-\ NaCl]{1.\ NaOH,\ 150°C;\ 2.\ H^+,\ H_2O}$ 2,4,5-Trichlorophenol $\xrightarrow[-\ NaCl]{ClCH_2COOH,\ NaOH,\ H_2O,\ \Delta}$ 2,4,5-Trichlorophenoxyacetic acid (85%)

2,4,5-Trichlorophenol (2,4,5-TCP)

2,4,5-Trichlorophenoxyacetic acid (2,4,5-T)

The direct nucleophilic substitution of chloride in chloroarenes is a synthetic pathway to a number of herbicides, pesticides, and antibacterials. For example, as shown above, hydroxylation of 1,2,4,5-tetrachlorobenzene gives 2,4,5-trichlorophenol (2,4,5-TCP), an intermediate in the synthesis of 2,4,5-trichlorophenoxyacetic acid (2,4,5-T). This acid is a powerful herbicide of particular value in brush control. A 1:1 mixture of the butyl esters of 2,4,5-T and its 2,4-dichloro analog (2,4-D) was used in large amounts (estimated at more than 10 million gallons from 1965 to 1970) as a defoliant (code name, Agent Orange) during the Vietnam war.

These chemicals are toxic irritants. 2,4,5-T is notorious because of a much more toxic impurity that forms in small quantities during its preparation: 2,3,7,8-tetrachlorodibenzo-*p*-dioxin (TCDD or, popularly but incorrectly, *dioxin*). Heating 2,4,5-T to between 500 and 600°C has been shown to produce TCDD, and thus extreme care has to be taken to control the reaction temperatures in the preparation of 2,4,5-T. TCDD can also be made directly from 2,4,5-trichlorophenol by coupling through double dehydrochlorination (below). The toxicity of TCDD (lethal dose, for test animals, in moles per kilogram of body weight) is about 500 times that of strychnine and more than 100,000 times that of sodium cyanide. It is embryotoxic, teratogenic (causing deformations of the fetus), and a suspected carcinogen in humans. In smaller than lethal concentrations, it causes severe skin rashes and lesions *(chloracne)*. In 1976, a runaway reaction in a chemical plant in Seveso, Italy, led to the accidental release of a cloud of overheated 2,4,5-trichlorophenol contaminated with TCDD. It is estimated that more than 130 pounds of the poison was vaporized, causing numerous deaths among animals and severe skin irritations in many humans.

Other examples of chlorinated phenols with physiological activity are 2,3,4,5,6-pentachlorophenol, a fungicide, and hexachlorophene (common name), a skin germicide once used in soaps and other toiletry products. It was banned when it was discovered to cause brain damage.

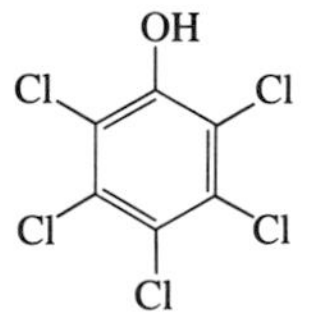

2,3,4,5,6-Pentachlorophenol
(A fungicide)

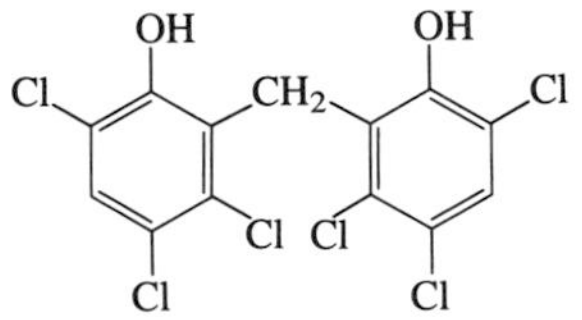

Hexachlorophene
(A skin germicide)

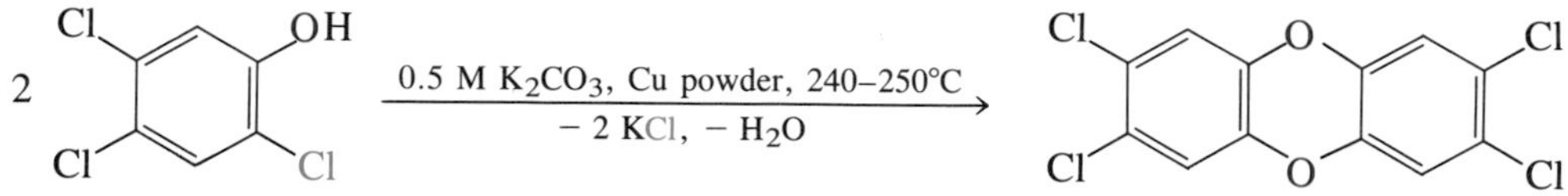

2 2,4,5-Trichlorophenol $\xrightarrow[-\ 2\ KCl,\ -\ H_2O]{0.5\ M\ K_2CO_3,\ Cu\ powder,\ 240–250°C}$

2,3,7,8-Tetrachlorodibenzo-*p*-dioxin (Dioxin)

Haloarenes may react through benzyne intermediates

Haloarenes do not undergo simple S_N2 or S_N1 reactions. However, at highly elevated temperatures and pressures, it is possible to effect nucleophilic substitution. For example, if exposed to hot sodium hydroxide followed by neutralizing work-up, chlorobenzene furnishes phenol.

$$\text{Chlorobenzene} \xrightarrow[\text{2. } H^+, H_2O]{\text{1. NaOH, } H_2O, 340°C, 150 \text{ atm}} \text{Phenol} + \text{NaCl}$$

Similar treatment with potassium amide results in benzenamine (aniline).

$$\text{Chlorobenzene} \xrightarrow[\text{2. } H^+, H_2O]{\text{1. } KNH_2, \text{ liquid } NH_3} \text{Benzenamine (Aniline)} + \text{KCl}$$

It is tempting to assume that these substitutions follow mechanisms similar to those formulated for the haloalkanes. However, when the last reaction is performed with radioactively labeled chlorobenzene (^{14}C at C1), a very curious result is obtained: Only half of the product is substituted at the labeled carbon; in the other half, the nitrogen is located at the *neighboring* position.

$$\text{Chlorobenzene-1-}^{14}\text{C} \xrightarrow[- \text{KCl}]{KNH_2, \text{ liquid } NH_3} \underset{\text{Benzenamine-1-}^{14}\text{C}}{50\%} + \underset{\text{Benzenamine-2-}^{14}\text{C}}{50\%}$$

Direct substitution mechanisms do not seem to operate these reactions. What, then, is the answer to this puzzle? A clue is the attachment of the incoming nucleophile *only* at the ipso or at the ortho position relative to the leaving group. This observation can be accounted for by an initial base-induced elimination of HX from the benzene ring, a process reminiscent of the dehydrohalogenation of alkenyl halides to give alkynes (Section 13-5). In the present case, elimination through a phenyl anion intermediate gives a highly strained and reactive species called **benzyne,** or **1,2-dehydrobenzene.**

Mechanism of Nucleophilic Substitution of Simple Haloarenes

STEP 1. Elimination

$$\text{Haloarene (X, H)} \xrightarrow[- H\ddot{N}H_2]{^-:\ddot{N}H_2} \text{Phenyl anion} \longrightarrow \text{Benzyne (Reactive intermediate)} + X^-$$

STEP 2. Addition

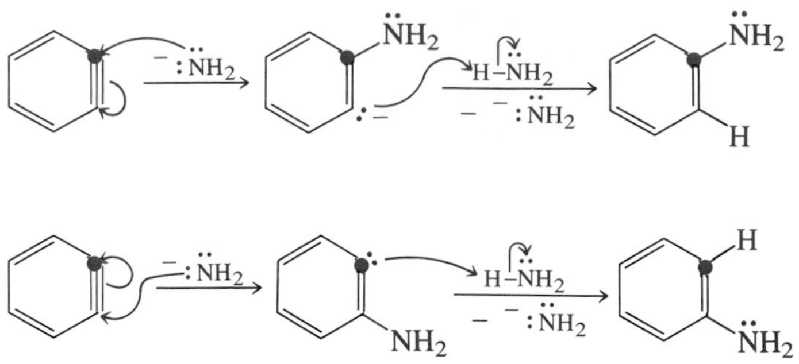

Why is benzyne so reactive? Recall that alkynes normally adopt a linear structure, a consequence of the *sp* hybridization of the carbons making up the triple bond (Section 13-2). Because of benzyne's cyclic structure, its triple bond is forced to be bent. The molecule has only a fleeting life and is rapidly attacked by any nucleophile present. For example, amide ion or even ammonia solvent can add to furnish the product benzenamine (aniline). Because the two ends of the triple bond are equally reactive, addition can occur at either carbon. These two possibilities explain the product mixture observed from labeled chlorobenzene.

EXERCISE 22-10

1-Chloro-4-methylbenzene (*p*-chlorotoluene) is not a good starting material for the preparation of 4-methylphenol (*p*-cresol) by direct reaction with hot NaOH because it forms a mixture of two products. Why does it do so, and what are the two products? Propose a synthesis from methylbenzene (toluene).

EXERCISE 22-11

Explain the regioselectivity observed in the following reaction. (Hint: Consider the effect of the methoxy group on the selectivity of attack on the benzyne by amide ion.)

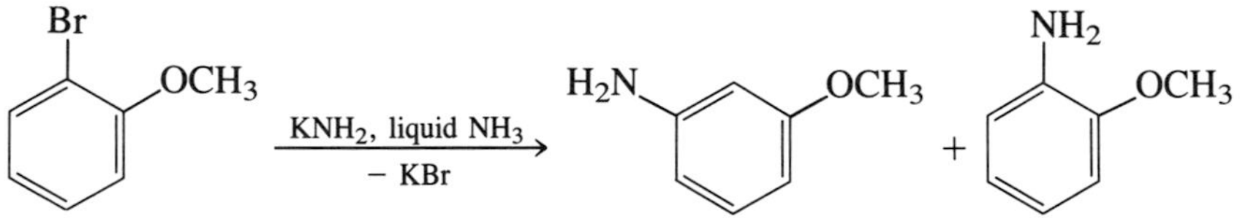

KNH_2, liquid NH_3; $-$ KBr

Major

Minor

Generation of Benzyne, a Reactive Intermediate

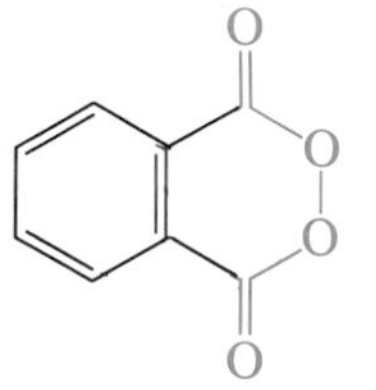

1,2-Benzenedicarbonyl peroxide (Phthaloyl peroxide)

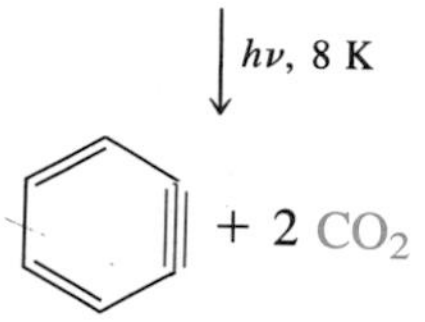

$h\nu$, 8 K

$+ \; 2\,CO_2$

Benzyne is a strained cycloalkyne

Although benzyne is too reactive to be isolated and stored in a bottle, it can be observed spectroscopically under special conditions. Photolysis of 1,2-benzenedicarbonyl (phthaloyl) peroxide at 8 K ($-265°C$) in frozen argon (m.p. $= -189°C$) produces a species whose IR and UV spectra are assignable to benzyne, formed by expulsion of two molecules of CO_2.

Although benzyne is represented as a cycloalkyne (Figure 22-3), its triple bond exhibits an IR stretching frequency of 1846 cm^{-1}, intermediate between the values for normal double and triple bonds. The bond is weakened considerably by poor orbital overlap. The normally symmetric benzene ring is also distorted and destabilized by introduction of the strained triple bond.

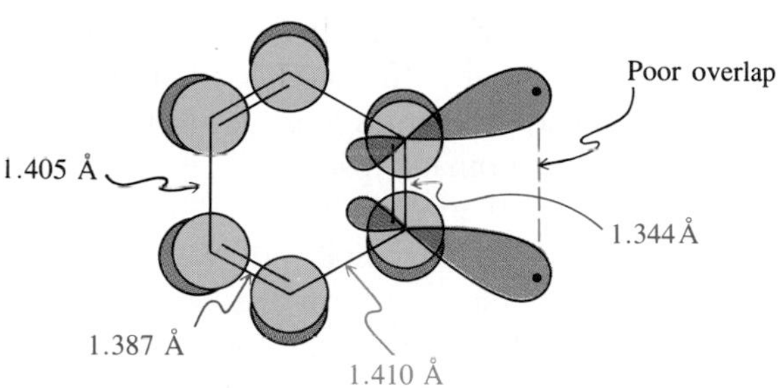

FIGURE 22-3 Orbital picture of benzyne reveals that the six aromatic π electrons are located in orbitals that are perpendicular to the two additional hybrid orbitals making up the distorted triple bond. The latter overlap only poorly; therefore, benzyne is highly reactive. Its molecular structure is quite deformed from regular hexagonal symmetry because of the triple bond.

Phenols are produced from arenediazonium salts

The most general laboratory procedure for making phenols is through **arenediazonium salts,** $ArN_2{}^+X^-$. Recall that primary alkanamines can be *N*-nitrosated but that the resulting species rearrange to diazonium ions, which are unstable—they lose nitrogen to give carbocations (Section 21-10). In contrast, primary benzenamines (anilines) are attacked by cold nitrous acid, in a reaction called **diazotization,** to give relatively stable arenediazonium salts.

Diazotization

NH_2 ... R $\xrightarrow{NaNO_2,\ H^+,\ H_2O,\ 0°C}$ $N \equiv N^+$ ... R

Arenediazonium ion

Phenyl cation

The reactions of these species will be discussed in detail in Section 22-10. When arenediazonium ions are heated, nitrogen is evolved and reactive **aryl cations** are produced. These ions are trapped by water to give the desired phenols.

Decomposition of Arenediazonium Salts

$N \equiv N^+$ ... R $\xrightarrow[-\ N_2]{\Delta}$ + ... R $\xrightarrow[-\ H^+]{HOH}$ OH ... R

Aryl cation

EXERCISE 22-12

Propose a synthesis of (4-phenylmethyl)phenol (*p*-benzylphenol) from benzene.

In summary, if a benzene ring bears enough strongly electron-withdrawing substituents and a leaving group, nucleophilic addition to give an intermediate anion with delocalized charge becomes feasible, followed by elimination of the leaving group (nucleophilic aromatic ipso substitution). Phenols result when the nucleophile is hydroxide ion, benzenamines (anilines) when it is ammonia, and alkoxybenzenes when alkoxides are employed. Very strong bases are capable of eliminating HX from haloarenes to form the reactive intermediate benzyne, which is subject to nucleophilic attack to give substitution products. Finally, phenols may be prepared by decomposition of arenediazonium salts in water.

BOX 22-2 Economics of Industrial Synthesis of Phenol: The Cumene Hydroperoxide Process

$$CH_3CH{=}CH_2 + C_6H_6 + O_2 \longrightarrow CH_3\overset{O}{\overset{\|}{C}}CH_3 + C_6H_5OH$$

Another industrial preparation of phenol highlights the economic constraints on any process that has commercial significance. In this approach, called the cumene hydroperoxide process (shown above), benzene and propene are oxidized in a series of steps by air to phenol and propanone (acetone). Although the goal of the sequence is to make the former product, it is the sales potential of the ketone by-product that makes the process economically feasible.

The synthesis proceeds through several separate reactions. In the first, benzene is converted into 1-methylethylbenzene (isopropylbenzene, or cumene) by Friedel-Crafts alkylation with propene under acidic conditions (Section 15-12).

In the second reaction (shown below), the alkyl benzene is oxidized by air to the corresponding hydroperoxide. The ease with which this transformation occurs is due to the ready initiation of a radical chain process through the tertiary benzylic radical. In the final step, the hydroperoxide is treated with dilute acid to give the two products, phenol and propanone (acetone), by acid-catalyzed rearrangement.

$$C_6H_5H + CH_3CH{=}CH_2 \xrightarrow{H^+,\ \Delta} C_6H_5CH(CH_3)_2 \xrightarrow[\text{Initiator}]{O_2}$$

1-Methylethylbenzene (Isopropylbenzene, or cumene)

$$C_6H_5C(CH_3)_2{-}O{-}O{-}H \xrightarrow{10\%\ H_2SO_4,\ H_2O,\ \Delta} C_6H_5OH + CH_3\overset{O}{\overset{\|}{C}}CH_3$$

1-Methyl-1-phenylethyl hydroperoxide (Cumene hydroperoxide)

22-5 Alcohol Chemistry of Phenols

This section shows that the phenol hydroxy group undergoes several of the reactions of alcohols (Chapter 9), such as protonation, Williamson ether synthesis, and esterification.

The oxygen in phenols is only weakly basic

We have seen that phenols are unusually acidic. They (and their ethers) can also be protonated by strong acids to give the corresponding **phenyloxonium ions.** Thus, as with the alkanols, the hydroxy group imparts amphoteric character (Section 8-3). However, the basicity of phenol is even less than that of the alkanols, because the lone electron pairs on the oxygen are delocalized into the benzene ring (Section 16-2). The pK_a values for phenyloxonium ions are, therefore, lower than those of alkyloxonium ions.

pK_a Values of Methyl- and Phenyloxonium Ion

$$CH_3\overset{+}{O}H_2 \rightleftharpoons CH_3\ddot{O}H + H^+ \qquad C_6H_5\overset{+}{O}H_2 \rightleftharpoons C_6H_5\ddot{O}H + H^+$$

$pK_a = -2.2$ $\qquad$ $pK_a = -6.7$

Unlike secondary and tertiary alkyloxonium ions derived from alcohols, phenyloxonium derivatives do not dissociate to form phenyl cations, because such ions have too high an energy content (see Section 22-10). The phenyl–oxygen bond in phenols is very difficult to break. However, after protonation of alkoxybenzenes, the bond between the *alkyl* group and oxygen is readily cleaved in the presence of nucleophiles such as Br^- or I^- (e.g., from HBr or HI) to give phenol and the corresponding haloalkane.

$$\text{3-}CH_3OC_6H_4COOH \xrightarrow{HBr,\ \Delta} \text{3-}HOC_6H_4COOH + CH_3Br$$

90%

3-Methoxybenzoic acid (*m*-Methoxybenzoic acid) $\qquad$ **3-Hydroxybenzoic acid (*m*-Hydroxybenzoic acid)**

EXERCISE 22-13

Why does cleavage of an alkoxybenzene by acid not produce a halobenzene and the alkanol?

Alkoxybenzenes are prepared by Williamson ether synthesis

The Williamson ether synthesis (Section 9-6) permits easy preparation of many alkoxybenzenes. The phenoxide ions obtained by deprotonation of phenols (Section 22-3) are good nucleophiles. They can displace the leaving groups from haloalkanes and alkyl sulfonates.

$$\text{3-Chlorophenol (m-Chlorophenol)} + CH_3CH_2CH_2Br \xrightarrow[-\,NaBr,\ -\,HOH]{NaOH,\ H_2O} \text{1-Chloro-3-propoxybenzene (m-Chlorophenyl propyl ether)}\ (63\%)$$

3-Chlorophenol
(*m*-Chlorophenol)

$OCH_2CH_2CH_3$ / Cl — 63%

1-Chloro-3-propoxybenzene
(*m*-Chlorophenyl propyl ether)

Esterification leads to phenyl alkanoates

The reaction of a carboxylic acid with a phenol (Section 19-9) to form a phenyl ester is endothermic. Therefore, esterification requires an activated carboxylic acid derivative, such as an alkanoyl halide or a carboxylic anhydride.

BOX 22-3 Aspirin: A Physiologically Active Phenyl Alkanoate

$$\text{2-Hydroxybenzoic acid} \xrightarrow[-\,CH_3COOH]{CH_3\overset{O}{\overset{\|}{C}}O\overset{O}{\overset{\|}{C}}CH_3,\ H^+,\ \Delta} \text{2-Acetyloxybenzoic acid}$$

2-Hydroxybenzoic acid
(*o*-Hydroxybenzoic acid,
salicylic acid)

2-Acetyloxybenzoic acid
(*o*-Acetoxybenzoic acid,
acetylsalicylic acid, aspirin)

Acetylation of 2-hydroxybenzoic (*o*-hydroxybenzoic) acid, also known as salicylic acid, results in the formation of *aspirin,* probably the most consumed drug in the world. Production capacity in the United States alone is about 22 million pounds (10 million kg) per year.

Aspirin is used as an analgesic, antipyretic, and antirheumatic (anti-inflammatory, particularly for arthritis). It interferes with the synthesis of prostaglandins (Section 19-14), thereby reducing fever, pain, and inflammation. Even though it is a popular medicine, it has some unwelcome side effects—particularly, gastric irritation and even bleeding. Recently, it has been linked to some childhood illnesses—notably, Reye's syndrome, a brain condition. Because of some of these drawbacks, there has been a gradual increase in the sales of other analgesics, such as Tylenol, prepared from 4-aminophenol (*p*-aminophenol) by acetylation.

$$\text{4-Aminophenol} \xrightarrow[-\,CH_3COOH]{CH_3\overset{O}{\overset{\|}{C}}O\overset{O}{\overset{\|}{C}}CH_3,\ CH_3COOH} \text{N-(4-Hydroxyphenyl)acetamide}\ (HN\overset{O}{\overset{\|}{C}}CH_3)$$

4-Aminophenol
(*p*-Aminophenol)

***N*-(4-Hydroxyphenyl)acetamide**
[*N*-(*p*-Hydroxyphenyl)acetamide, Tylenol]

$$\text{4-Methylphenol } (p\text{-Cresol}) + CH_3CH_2\overset{\overset{\displaystyle O}{\|}}{C}Cl \xrightarrow[-\,NaCl,\ -\,HOH]{NaOH,\ H_2O} \text{4-Methylphenyl propanoate } (p\text{-Methylphenyl propanoate})$$

4-Methylphenol (*p*-Cresol) — **Propanoyl chloride** — **4-Methylphenyl propanoate (*p*-Methylphenyl propanoate)**

EXERCISE 22-14

Explain why, in the preparation of Tylenol, the amide is formed rather than the ester.

In summary, the oxygen in phenols and alkoxybenzenes may be protonated even though it is less basic than the oxygen in the alkanols and alkoxyalkanes. Protonated phenols and their derivatives do not ionize to phenyl cations, but the ethers can be cleaved to phenols and haloalkanes by HX. Alkoxybenzenes are made by Williamson ether synthesis, aryl alkanoates by alkanoylation.

22-6 Electrophilic Substitution of Phenols

The aromatic ring in phenols is also a center of reactivity. The interaction between the OH group and the ring strongly activates the ortho and para positions toward electrophilic substitution (Section 16-2). For example, even dilute nitric acid causes nitration.

$$C_6H_5OH \xrightarrow{HNO_3,\ CHCl_3,\ 15^\circ C} \text{2-}O_2NC_6H_4OH + \text{4-}O_2NC_6H_4OH$$

26% **2-Nitrophenol (*o*-Nitrophenol)** + 61% **4-Nitrophenol (*p*-Nitrophenol)**

Friedel-Crafts alkanoylation (acylation) of phenols is complicated by ester formation and is better carried out on ether derivatives of phenol.

$$C_6H_5OCH_3 + CH_3\overset{\overset{\displaystyle O}{\|}}{C}Cl \xrightarrow[-\,HCl]{AlCl_3,\ CS_2} \text{4-}CH_3OC_6H_4COCH_3$$

Methoxybenzene (Anisole) — 70% **1-(4-Methoxyphenyl)ethanone (*p*-Methoxyacetophenone)**

Phenols are halogenated so readily that no catalyst is required, and multiple halogenations are frequently observed (Sections 16-2 and 16-3). As shown below, tribromination occurs in water at 20°C, but the reaction can be controlled to produce the monohalogenation product through the use of a lower temperature and a less polar solvent.

Halogenation in Water at 20°C

Halogenation in $CHCl_3$ at 0°C

OH, 3 Br–Br, H_2O, 20°C, − 3 HBr → Br, OH, Br, Br, 100%

Phenol **2,4,6-Tribromophenol**

but

OH, CH_3, Br–Br, $CHCl_3$, 0°C, − HBr → OH, Br, CH_3, 80%

4-Methylphenol (*p*-Cresol) **2-Bromo-4-methylphenol**

Electrophilic attack at the para position is frequently predominant because of steric effects. However, it is normal to obtain mixtures resulting from both ortho and para substitutions, and their compositions are highly dependent on reagents and reaction conditions.

EXERCISE 22-15

Friedel-Crafts methylation of methoxybenzene (anisole) with chloromethane in the presence of $AlCl_3$ gives a 2:1 ratio of ortho:para products. Treatment of methoxybenzene with 2-chloro-2-methylpropane (*tert*-butyl chloride) under the same conditions furnishes only 1-methoxy-4-(1,1-dimethylethyl)benzene (*p-tert*-butylanisole). Explain.

Under basic conditions, phenols can undergo electrophilic substitution, even with very mild electrophiles, through intermediate phenoxide ions. An industrially important application is the reaction with formaldehyde, which leads to *o*- and *p*-hydroxymethylation. Mechanistically, these processes may be considered enolate condensations, much like the aldol reaction (Section 18-5).

Hydroxymethylation of Phenol

OH, HO^- ⇌ [O^- ⟷ O ⟷ O] + HOH, $CH_2{=}O$ →

O, H, CH_2O^- + O, $^-OCH_2$, H, HOH, $- HO^-$ → OH, CH_2OH + OH, CH_2OH

The initial aldol products are unstable: They dehydrate on heating, giving reactive intermediates called **quinomethanes.**

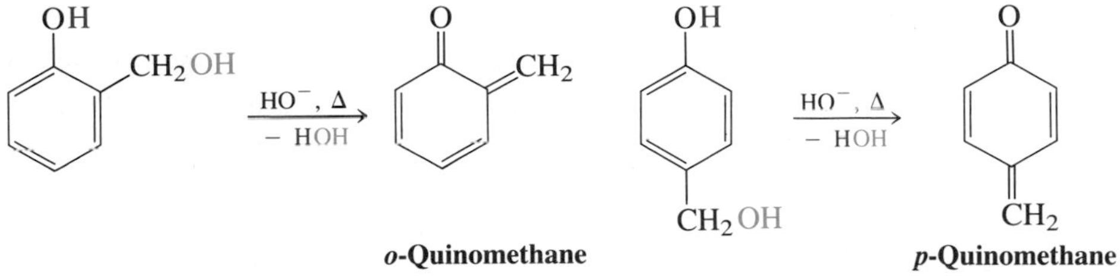

o-Quinomethane *p*-Quinomethane

Because quinomethanes are α,β-unsaturated carbonyl compounds, they may undergo Michael additions (Section 18-12) with excess phenoxide ion. The resulting phenols can be hydroxymethylated again and the entire process repeated. Eventually, a complex phenol–formaldehyde copolymer, also called a **phenolic resin,** is formed. Total production of these resins in the United States in 1992 exceeded 1.4 million tons. Their major uses are in plywood (45%), insulation (14%), molding compounds (9%), fibrous and granulated wood (9%), laminates (8%), and foundry core binders (5%).

Phenolic Resin Synthesis

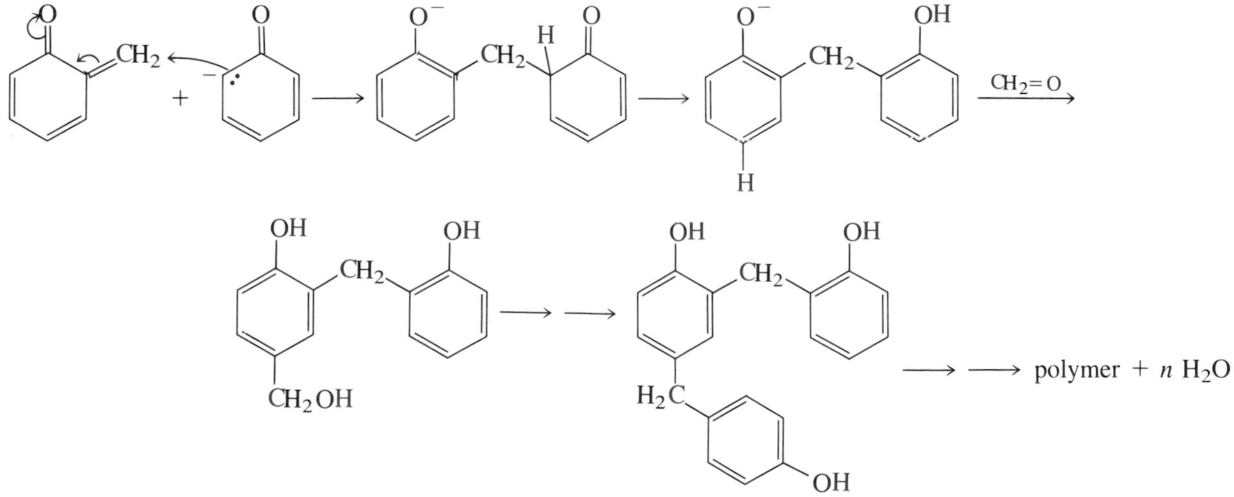

$\longrightarrow \longrightarrow$ polymer + $n\ H_2O$

In the **Kolbe* reaction,** phenoxide attacks carbon dioxide to furnish the salt of 2-hydroxybenzoic acid (*o*-hydroxybenzoic acid, salicylic acid, precursor to aspirin; see Box 22-3).

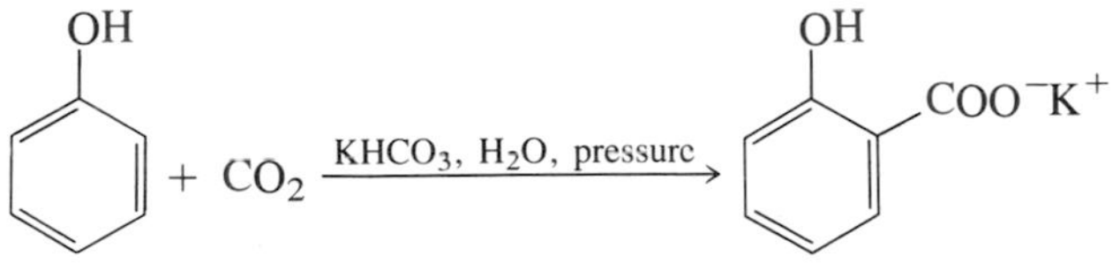

$$\text{C}_6\text{H}_5\text{OH} + CO_2 \xrightarrow{KHCO_3,\ H_2O,\ \text{pressure}} o\text{-HOC}_6\text{H}_4\text{COO}^-K^+$$

EXERCISE 22-16

Formulate a mechanism for the Kolbe reaction.

*Professor Adolph Wilhelm Hermann Kolbe, 1818–1884, University of Leipzig, Germany.

EXERCISE 22-17

Hexachlorophene (see Box 22-1) is prepared in one step from 2,4,5-trichlorophenol and formaldehyde in the presence of sulfuric acid. How does this reaction proceed? (Hint: Formulate an acid-catalyzed hydroxymethylation for the first step.)

In summary, the benzene ring in phenols is subject to electrophilic aromatic substitution, particularly under basic conditions. Phenoxide ions can be hydroxymethylated and carbonated.

22-7 Claisen and Cope Rearrangements

At 200°C, 2-propenyloxybenzene (allyl phenyl ether) undergoes an unusual reaction that is made possible by the direct phenyl–oxygen linkage: The starting material rearranges to 2-(2-propenyl)phenol (*o*-allylphenol).

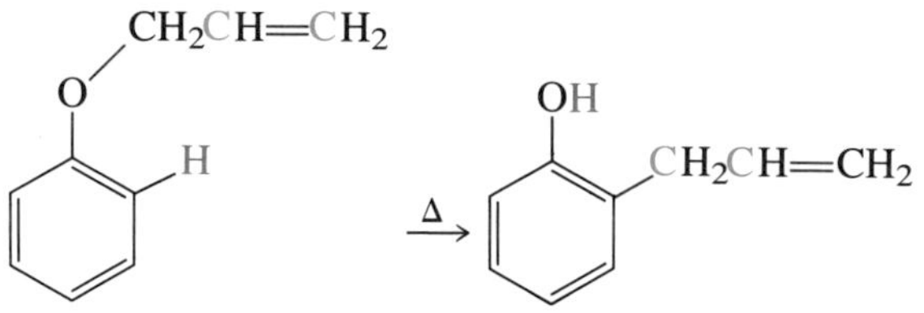

$CH_2CH{=}CH_2$ O H $\xrightarrow{\Delta}$ OH $CH_2CH{=}CH_2$

75%

2-Propenyloxybenzene (Allyl phenyl ether) → **2-(2-Propenyl)phenol (*o*-Allylphenol)**

This transformation, called the **Claisen* rearrangement,** is another concerted reaction with a transition state that accommodates the movement of six electrons. The initial intermediate is a high-energy isomer, 6-(2-propenyl)-2,4-cyclohexadienone, which enolizes to the final product.

Mechanism of the Claisen Rearrangement

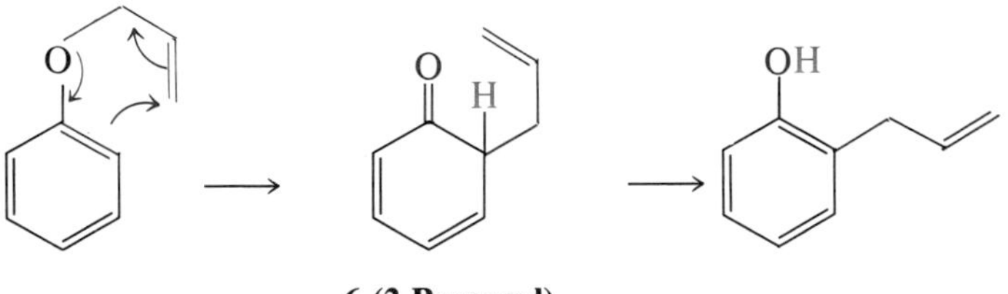

O → O H → OH

6-(2-Propenyl)-2,4-cyclohexadienone

With the nonaromatic 1-ethenyloxy-2-propene (allyl vinyl ether), the Claisen rearrangement stops at the initial stage because the carbonyl group in this molecule is stable. This is called the **aliphatic Claisen rearrangement.**

---

*Professor Ludwig Claisen, 1851–1930, University of Berlin.

Aliphatic Claisen Rearrangement

$$CH_2{=}CH{-}O{-}CH_2{-}CH{=}CH_2 \xrightarrow{255^\circ C} CH_2{=}CH{-}CH_2{-}CH_2{-}CH{=}O$$

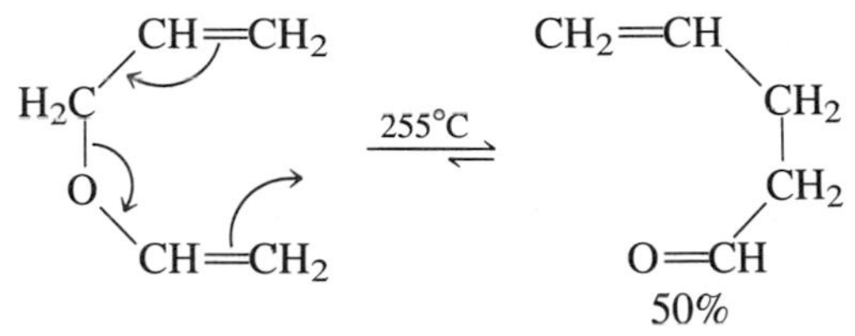

1-Ethenyloxy-2-propene (Allyl vinyl ether) → **4-Pentenal** (50%)

The carbon analog of the Claisen rearrangement is called the **Cope* rearrangement;** it takes place in compounds containing 1,5-diene units.

Cope Rearrangement

H_5C_6 (178°C) 72%

3-Phenyl-1,5-hexadiene → ***trans*-1-Phenyl-1,5-hexadiene**

EXERCISE 22-18

Explain the following transformation by a mechanism.

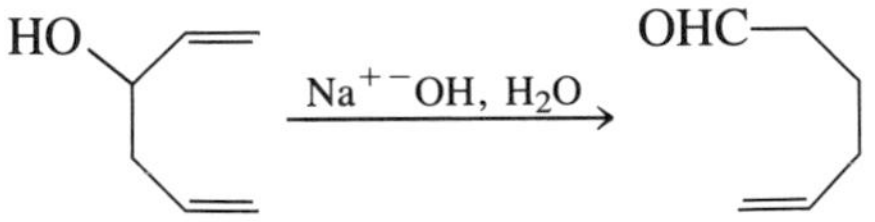

HO → ($Na^{+-}OH, H_2O$) → OHC

In summary, 2-propenyloxybenzene rearranges to 2-(2-propenyl)phenol (*o*-allylphenol) by an electrocyclic mechanism that moves six electrons (Claisen rearrangement). Similar concerted reactions are undergone by aliphatic unsaturated ethers (aliphatic Claisen rearrangement) and by hydrocarbons containing 1,5-diene units (Cope rearrangement).

22-8 Oxidation of Phenols: Cyclohexadienediones (Benzoquinones)

Phenols can be oxidized to carbonyl derivatives. Unlike the oxidation of alkanols, that of the benzene ring does not stop until it reaches the stage of a new class of cyclic diketones, called **cyclohexadienediones (benzoquinones).**

Cyclohexadienediones (benzoquinones) and benzenediols (hydroquinones and catechols) are redox couples

1,2- and **1,4-Benzenediols** (**catechols** and **hydroquinones**) are oxidized to the corresponding diketones, 3,5-cyclohexadiene-1,2-diones and 2,5-cyclohexa-

*Professor Arthur C. Cope, 1909–1966, Massachusetts Institute of Technology.

diene-1,4-diones (*o*-benzoquinones and *p*-benzoquinones), by a variety of oxidizing agents, such as sodium dichromate or silver oxide. Yields can be variable if the resulting diones are reactive, like the 1,2-dione, which partly decomposes under the conditions of its formation.

Cyclohexadienediones (Benzoquinones) from Oxidation of 1,2- and 1,4-Benzenediols

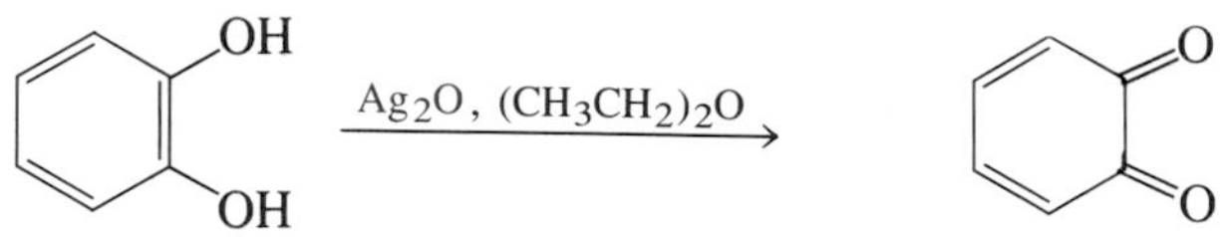

Ag_2O, $(CH_3CH_2)_2O$ →

Low yield

1,2-Benzenediol (Catechol) → **3,5-Cyclohexadiene-1,2-dione (*o*-Benzoquinone)**

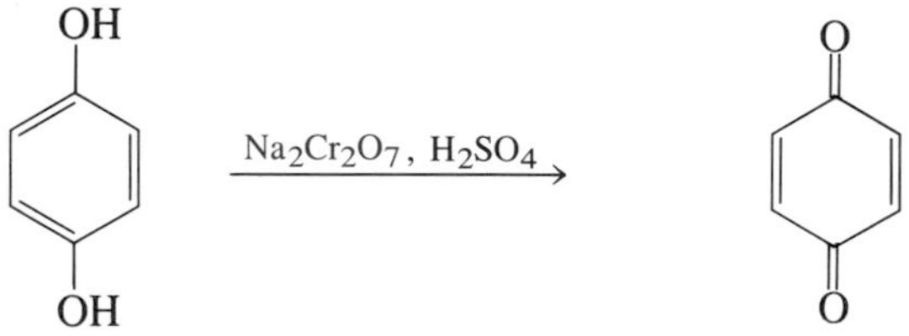

$Na_2Cr_2O_7$, H_2SO_4 →

92%

1,4-Benzenediol (Hydroquinone) → **2,5-Cyclohexadiene-1,4-dione (*p*-Benzoquinone)**

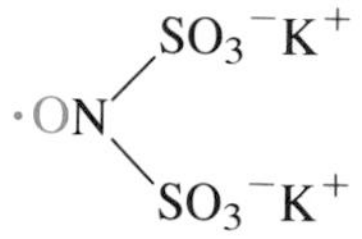

$\cdot ON(SO_3^-K^+)_2$

Potassium nitrosodisulfonate (Frémy's salt)

Even simple phenols can be oxidized to such diones. The second oxygen is introduced at C4 (para) by several oxidants. A special reagent that accomplishes this transformation is potassium nitrosodisulfonate, $\cdot ON(SO_3^-K^+)_2$, a radical species also known as **Frémy's* salt.**

Oxidation of Phenols by Frémy's Salt

$\cdot ON(SO_3K)_2$ →

82%

2-Methylphenol (*o*-Cresol) → **2-Methyl-2,5-cyclohexadiene-1,4-dione (*o*-Toluquinone)**

The redox process that interconverts 1,4-benzenediol (hydroquinone) and 2,5-cyclohexadiene-1,4-dione (*p*-benzoquinone) can be visualized as a sequence of proton and electron transfers. Initial deprotonation gives a phenoxide ion, which is transformed into a **phenoxy radical** by one-electron oxidation. Dissociation of the remaining OH group furnishes a **semiquinone radical anion,** and a second one-electron oxidation step leads to the benzoquinone. All of the intermediate

*Professor Edmond Frémy, 1814–1894, École Polytechnique, Paris.

species involved in this sequence benefit from considerable resonance stabilization (two forms are shown for the semiquinone). We shall see in Section 22-9 that redox processes similar to that shown here occur widely in nature.

Redox Relation Between 2,5-Cyclohexadiene-1,4-dione (*p*-Benzoquinone) and 1,4-Benzenediol (Hydroquinone)

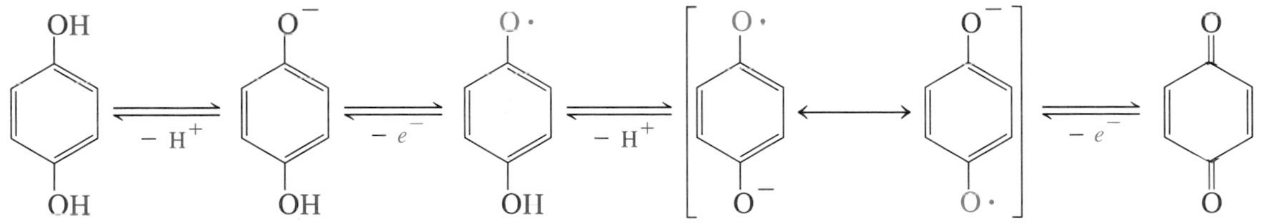

OH, OH; $-H^+$; O⁻, OH; $-e^-$; O·, OH; $-H^+$; [O·, O⁻ ⟷ O⁻, O·]; $-e^-$; O, O

Phenoxide ion — Phenoxy radical — Semiquinone radical anion

EXERCISE 22-19

Give a minimum of two additional resonance forms each for the phenoxide ion, phenoxy radical, and semiquinone radical anion shown in the scheme above.

The enone units in 2,5-cyclohexadiene-1,4-diones (*p*-benzoquinones) undergo conjugate and Diels–Alder additions

2,5-Cyclohexadiene-1,4-diones (*p*-benzoquinones) function as reactive α,β-unsaturated ketones in conjugate additions (see Section 18-10). For example, hydrogen chloride adds to give an intermediate hydroxy dienone that enolizes to the aromatic 2-chloro-1,4-benzenediol.

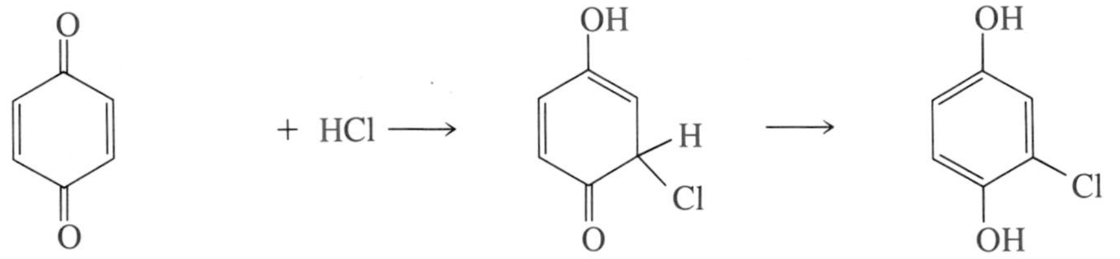

+ HCl ⟶ ⟶

2,5-Cyclohexadiene-1,4-dione (*p*-Benzoquinone) — **6-Chloro-4-hydroxy-2,4-cyclohexadienone** — **2-Chloro-1,4-benzenediol**

The double bonds also undergo cycloadditions to dienes (Section 14-8). The initial cycloadduct to 1,3-butadiene tautomerizes on heating with acid to the aromatic system.

Diels–Alder Reactions of 2,5-Cyclohexadiene-1,4-dione (*p*-Benzoquinone)

+ ⟶ (C_6H_6, 20°C, 48 h) ⟶ (HCl, Δ)

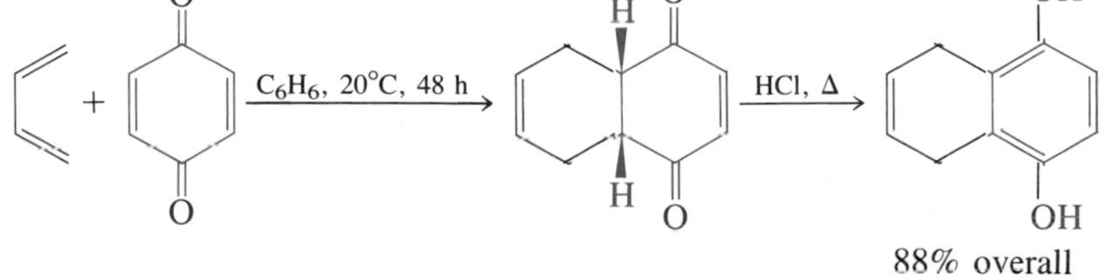

88% overall

EXERCISE 22-20

Explain the following result by a mechanism.

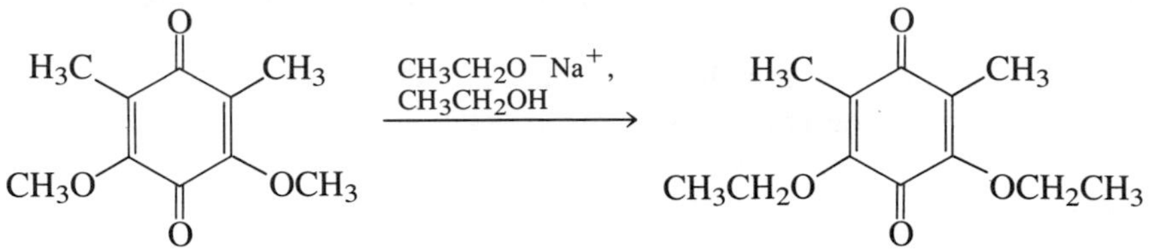

$$\text{2,6-dimethoxy-3,5-dimethyl-}p\text{-benzoquinone (}H_3C, CH_3, CH_3O, OCH_3\text{)} \xrightarrow[CH_3CH_2OH]{CH_3CH_2O^-Na^+,} \text{2,6-diethoxy-3,5-dimethyl-}p\text{-benzoquinone (}H_3C, CH_3, CH_3CH_2O, OCH_2CH_3\text{)}$$

In summary, phenols are oxidized to the corresponding diones (benzoquinones). The diones enter into reversible redox reactions that yield the corresponding diols. They also undergo conjugate additions and Diels-Alder additions to the double bonds.

22-9 Oxidation–Reduction Processes in Nature

This section describes some chemical processes involving 1,4-benzenediols (hydroquinones) and 2,5-cyclohexadiene-1,4-diones (*p*-benzoquinones) that occur in nature. We begin with an introduction to the biochemical reduction of O_2. Oxygen also engages in reactions that can cause damage to biomolecules. Naturally occurring **antioxidants** inhibit these transformations. The section concludes with a discussion of the properties of synthetic preservatives.

Ubiquinones mediate the biological reduction of oxygen to water

Nature makes use of the benzoquinone–hydroquinone redox couple in reversible oxidation reactions. These processes are part of the complicated cascade by which oxygen is used in biochemical degradations. An important series of compounds used for this purpose are the **ubiquinones** (a name coined to indicate their ubiquitous presence in nature), also collectively called **coenzyme Q (CoQ,** or simply **Q**). The ubiquinones are substituted 2,5-cyclohexadiene-1,4-dione (*p*-benzoquinone) derivatives bearing a side chain made up of 2-methylbutadiene units (isoprene; Sections 4-7 and 14-10). An enzyme system that utilizes NADH (Chapter 25) converts CoQ to its reduced form (QH_2).

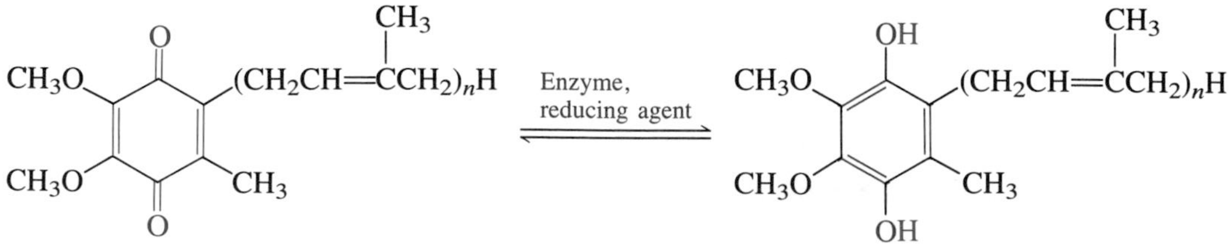

$$\text{Ubiquinone: } CH_3O, CH_3O, CH_3, (CH_2CH{=}C(CH_3)CH_2)_nH \;\; \underset{\text{reducing agent}}{\overset{\text{Enzyme,}}{\rightleftharpoons}} \;\; \text{Hydroquinone (OH, OH): } CH_3O, CH_3O, CH_3, (CH_2CH{=}C(CH_3)CH_2)_nH$$

Ubiquinones (n = 6, 8, 10)
(Coenzyme Q)

Reduced form of coenzyme Q
(Reduced Q, or QH_2)

QH_2 participates in a chain of redox reactions with electron-transporting iron-containing proteins called **cytochromes.** The reduction of Fe^{3+} to Fe^{2+} in cytochrome *b* by QH_2 begins a sequence of electron transfers involving six different proteins. The chain ends with reduction of O_2 to water by addition of four electrons and four protons.

$$O_2 + 4\,H^+ + 4\,e^- \longrightarrow 2\,H_2O$$

Phenol derivatives protect cell membranes from oxidative damage

The biochemical conversion of oxygen to water involves several intermediates, including **superoxide,** $O_2^{-\cdot}$, the product of one-electron reduction, and **hydroxy radical,** $\cdot OH$, which arises from cleavage of H_2O_2. Both are highly reactive species capable of initiating reactions that damage organic molecules of biological importance. An example is the phosphoglyceride shown below, a component of cell membranes derived from the unsaturated fatty acid *cis,cis*-octadeca-9,12-dienoic acid (linoleic acid).

INITIATION STEP.

$CH_3(CH_2)_4$ (R) –CH=CH–CH_2(C11, H H)–CH=CH– $CH_2(CH_2)_6CO$–CH(CH_2O–C(=O)$(CH_2)_{14}CH_3$)–CH_2O–PO(O^-)(=O)$(CH_2)_2N(CH_3)_3^+$ (R′) $\xrightarrow[-\ HOH]{\cdot OH}$ [three resonance structures] **Pentadienyl radical**

The doubly allylic hydrogens at C11 are readily abstracted by radicals such as $\cdot OH$ (Chapter 14).

PROPAGATION STEP 1.

Pentadienyl radical (R, R′) $\xrightarrow{O_2}$ R–CH(O–O·)–CH=CH–CH=CH–R′

Peroxy radical

The resonance-stabilized pentadienyl radical combines rapidly with O_2 in the first of two propagation steps. Reaction occurs at either C9 or C13 (shown here), giving either of two peroxy radicals containing conjugated diene systems.

PROPAGATION STEP 2.

Peroxy radical (R, R′) + R–CH=CH–CH₂(C11, H H)–CH=CH–R′ ⟶ R–CH(O–OH)–CH=CH–CH=CH–R′ + pentadienyl radical (R, R′)

Lipid hydroperoxide

In the second propagation step this species removes a hydrogen atom from C11 of another molecule of lipid, thereby forming a new dienyl radical and a molecule of **lipid hydroperoxide.** The dienyl radical may then reenter propagation step 1. In this way a large number of lipid molecules may be oxidized following just a single initiation event.

Numerous studies have confirmed that lipid hydroperoxides are toxic, their products of decomposition even more so. For example, loss of ·OH by cleavage of the relatively weak O–O bond gives rise to an alkoxy radical, which may decompose by breaking a neighboring C–C bond (β-scission), thereby forming an unsaturated aldehyde.

β-Scission of a Lipid Alkoxy Radical

$\xrightarrow{-\ \cdot OH}$ $\xrightarrow{-\ \cdot R'}$

Alkoxy radical

$CH_3(CH_2)_4$

***trans*-4-Hydroxy-2-nonenal**

Propanedial (Malondialdehyde)

Through related but more complex mechanisms, certain lipid hydroperoxides decompose to give unsaturated hydroxyaldehydes such as *trans*-4-hydroxy-2-nonenal as well as the dialdehyde propanedial (malondialdehyde). Molecules of these general types are partly responsible for the smell of rancid fats.

Both propanedial and the α,β-unsaturated aldehydes are extremely toxic, because they are highly reactive toward the proteins that are present in close proximity to the lipids in cell membranes. For example, both dials and enals are capable of reacting with nucleophilic amino and mercapto groups from two different portions of one protein, or from two different protein molecules; and these reactions produce cross-linking (Section 14-10). Cross-linking severely inhibits protein molecules from carrying out their biological functions (Chapter 26).

Cross-linking of Proteins by Reaction with Unsaturated Aldehydes

$\xrightarrow{\text{Protein}_1\text{–SH}}$ $\text{protein}_1\text{—S}$ $\xrightarrow{\text{Protein}_2\text{–NH}_2}$ $\text{protein}_1\text{—S}$ N—protein_2

Processes such as these are thought by many to contribute to the development of emphysema, atherosclerosis (the underlying cause of several forms of heart disease and stroke), certain chronic inflammatory and autoimmune diseases, cancer, and, possibly, the process of aging itself.

Does nature provide the means for biological systems to protect themselves from such damage? A variety of naturally occurring antioxidant systems defend lipid molecules inside cell membranes from oxidative destruction. The most important is **vitamin E,** a reducing agent that possesses a long hydrocarbon chain (see Problem 13 of Chapter 2), a feature making it lipid soluble. Vitamin E contains a structure similar to 1,4-benzenediol (Section 22-8). The corresponding phenoxide ion is an excellent electron donor. The protective qualities of vitamin E rest on its ability to break the propagation chain of lipid oxidation by the reduction of radical intermediates.

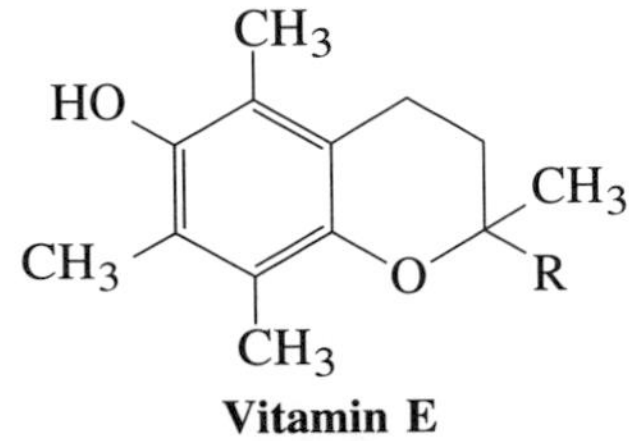

Vitamin E
(α-Tocopherol)
R = branched $C_{16}H_{33}$ chain

Reactions of Vitamin E with Lipid Hydroperoxy and Alkoxy Radicals

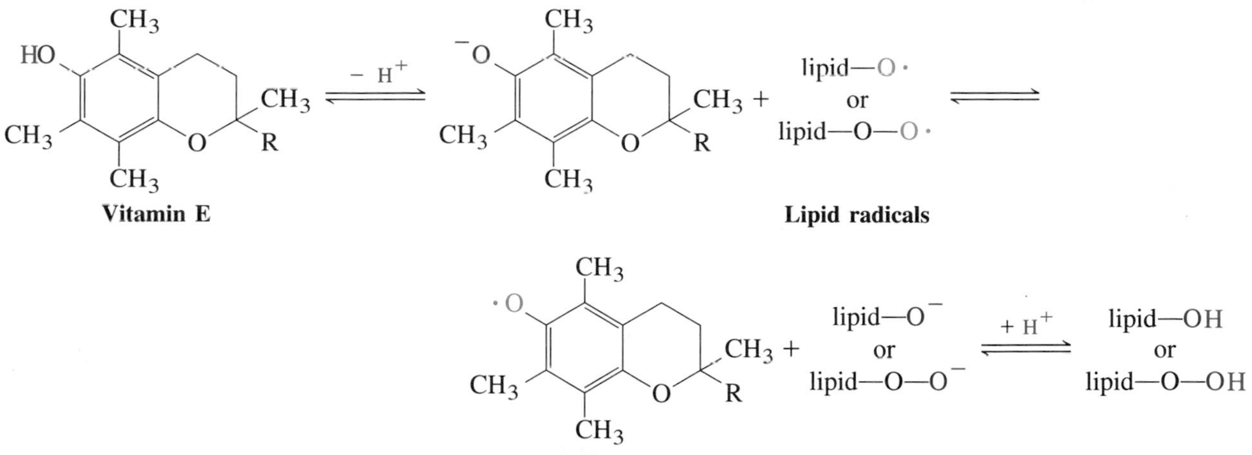

Vitamin E $\underset{}{\overset{-\ H^+}{\rightleftharpoons}}$ (anion) CH_3 + lipid—O· or lipid—O—O· (Lipid radicals) $\rightleftharpoons$

α-Tocopheroxy radical + lipid—O^- or lipid—O—O^- $\overset{+\ H^+}{\rightleftharpoons}$ lipid—OH or lipid—O—OH

In this process lipid radicals are reduced and protonated. Vitamin E is oxidized to an α-tocopheroxy radical, which is relatively unreactive because of extensive delocalization and the steric hindrance of the methyl substituents. Vitamin E is regenerated at the membrane surface by reaction with water-soluble reducing agents such as **vitamin C.**

Regeneration of Vitamin E by Vitamin C

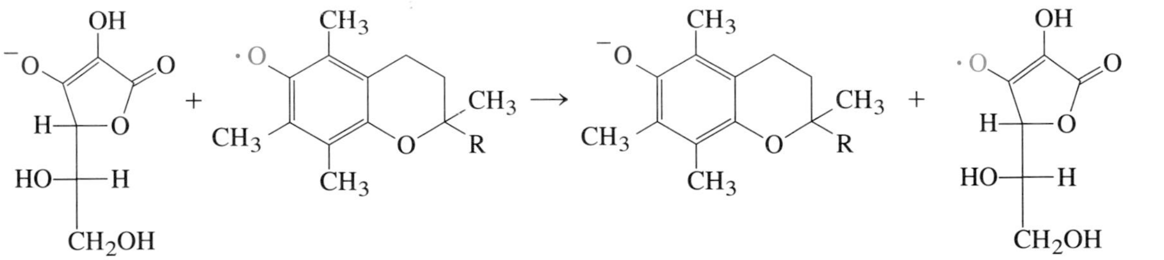

Vitamin C + α-tocopheroxy radical $\longrightarrow$ vitamin E anion + Semidehydroascorbic acid

The product of vitamin C oxidation eventually decomposes to lower molecular weight water-soluble compounds, which are excreted by the body.

EXERCISE 22-21

Vitamin C is an effective antioxidant because its oxidation product is stabilized by resonance. Give another resonance form for this species.

Benzoquinones consume glutathione, an intracellular reducing agent

Virtually all living cells contain the substance **glutathione,** a peptide that incorporates a mercapto functional group (see Chapters 9 and 26). The latter serves to reduce disulfide linkages in proteins to SH groups and to maintain the iron in hemoglobin in the 2+ oxidation state (Chapter 26). Glutathione also participates in the reduction of oxidants that may be present in the interior of the cell, such as hydrogen peroxide, H_2O_2.

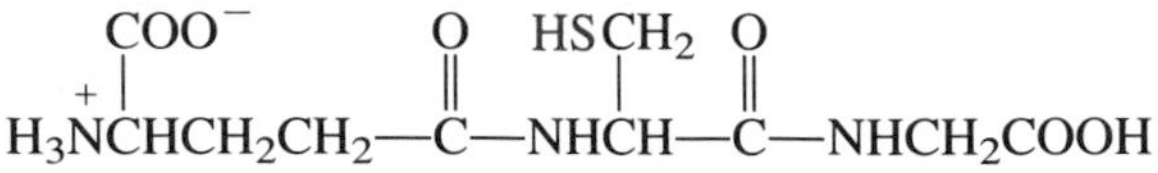

$$\underset{\textbf{Glutathione}}{\overset{+}{H_3N}\underset{|}{\overset{COO^-}{C}}HCH_2CH_2-\overset{O}{\overset{\|}{C}}-NH\overset{HSCH_2}{\overset{|}{C}}H-\overset{O}{\overset{\|}{C}}-NHCH_2COOH}$$

Glutathione is converted into a disulfide (Section 9-10) during this process but is regenerated by an enzyme-mediated reduction.

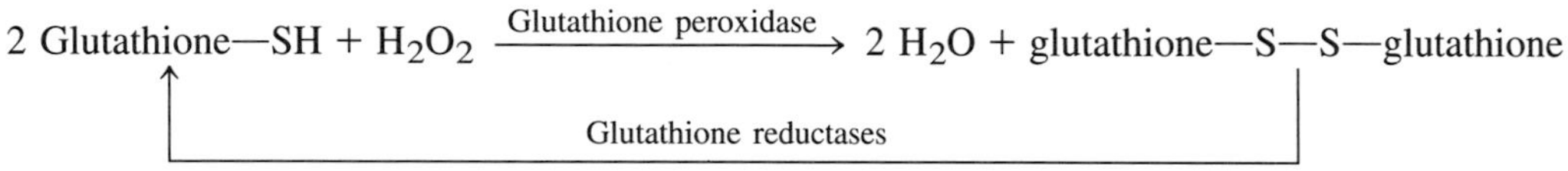

$$2\ \text{Glutathione}-SH + H_2O_2 \xrightarrow{\text{Glutathione peroxidase}} 2\ H_2O + \text{glutathione}-S-S-\text{glutathione}$$

Glutathione reductases

Cyclohexadienediones (benzoquinones) and related compounds react irreversibly with glutathione in the liver by conjugate addition. Cell death results if glutathione destruction is extensive. Tylenol is an example of a substance that exhibits such liver toxicity at very high doses. Cytochrome P-450, a redox enzyme system in the liver, oxidizes Tylenol to an imine derivative of a cyclohexadienedione, which in turn consumes glutathione. Vitamin C is capable of reversing the oxidation.

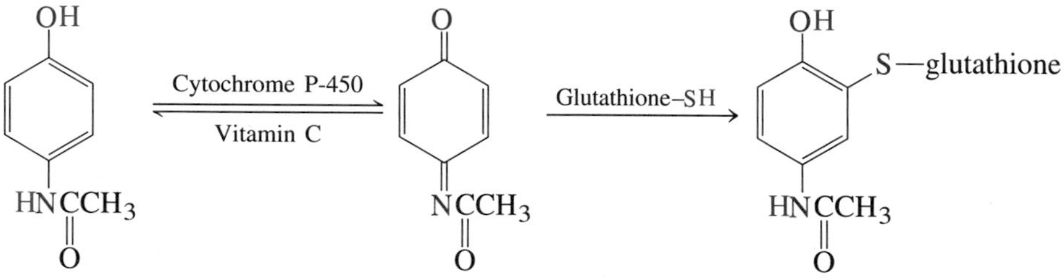

OH; HNCCH3 (C=O); Cytochrome P-450 ⇌ Vitamin C; O; NCCH3 (C=O); Glutathione–SH →; OH; S—glutathione; HNCCH3 (C=O)

Tylenol

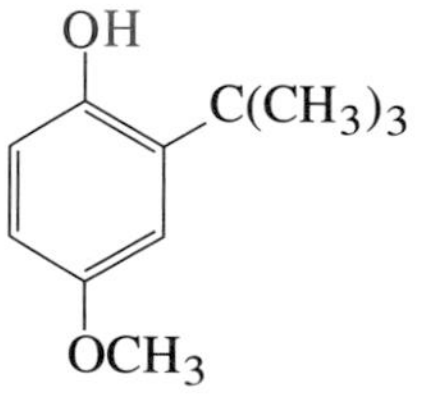

OH; C(CH3)3; OCH3

2-(1,1-Dimethylethyl)-4-methoxyphenol (BHA)

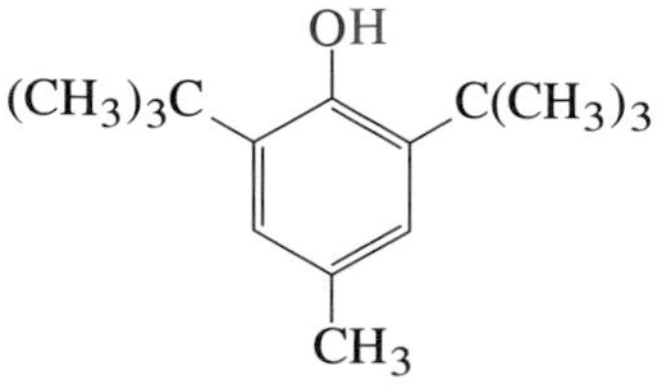

OH; (CH3)3C; C(CH3)3; CH3

2,6-Bis(1,1-dimethylethyl)-4-methylphenol (BHT)

Synthetic analogs of vitamin E are preservatives

Synthetic phenol derivatives are widely used as antioxidants and preservatives in the food industry. Perhaps two of the most familiar are 2-(1,1-dimethylethyl)-4-methoxyphenol (butylated hydroxyanisole or **BHA**) and 2,6-bis(1,1-dimethylethyl)-4-methylphenol (butylated hydroxytoluene or **BHT**). For example, addition of BHA to butter increases its storage life from months to years. Both BHA and BHT function like vitamin E, reducing oxygen radicals and interrupting the propagation of oxidation processes.

In summary, oxygen-derived radicals are capable of initiating radical chain reactions in lipids, thereby leading to toxic decomposition products. Vitamin E is a naturally occurring phenol derivative, which functions as an antioxidant to inhibit these processes within membrane lipids. Vitamin C and glutathione are biological reducing agents located in the intra- and extracellular aqueous environments. High concentrations of cyclohexadienediones (benzoquinones) can bring about cell death by consumption of glutathione; vitamin C can protect the cell by reduction of the benzoquinone. Synthetic food preservatives are structurally designed to mimic the antioxidant behavior of vitamin E.

22-10 Arenediazonium Salts

The final two sections of this chapter conclude our examination of functional group chemistry that has been modified by conjugation with a benzene ring. We focus now on nitrogen substituents and describe the chemistry of arenediazonium salts. As mentioned in Section 22-4, *N*-nitrosation of primary benzenamines (anilines) furnishes these salts, which can be used in the synthesis of phenols. Arenediazonium salts are stabilized by resonance and are converted to haloarenes, arenecarbonitriles, and other aromatic derivatives through replacement of nitrogen by the appropriate nucleophile.

Arenediazonium salts are stabilized by resonance

The reason for the relative stability of arenediazonium salts is resonance and the high energy of the aryl cations formed by loss of nitrogen. One of the electron pairs making up the aromatic π system can be delocalized into the functional group, which results in charge-separated resonance structures containing a double bond between the benzene ring and the attached nitrogen.

Resonance in the Benzenediazonium Cation

Only at high temperatures (>50°C) can nitrogen extrusion form the very reactive phenyl cation. When this is done in aqueous solution, phenols are produced (Section 22-4).

Why is the phenyl cation so reactive? After all, it is a carbocation that is part of a benzene ring. Should it not be resonance stabilized, like the phenylmethyl (benzyl) cation? The answer is no, as may be seen in the molecular-orbital picture of the phenyl cation (Figure 22-4). The empty orbital associated with the positive charge is one of the sp^2 hybrids aligned *perpendicular* to the π framework that normally produces aromatic resonance stabilization. Hence, this orbital cannot overlap with the π bonds, and the positive charge cannot be delocalized.

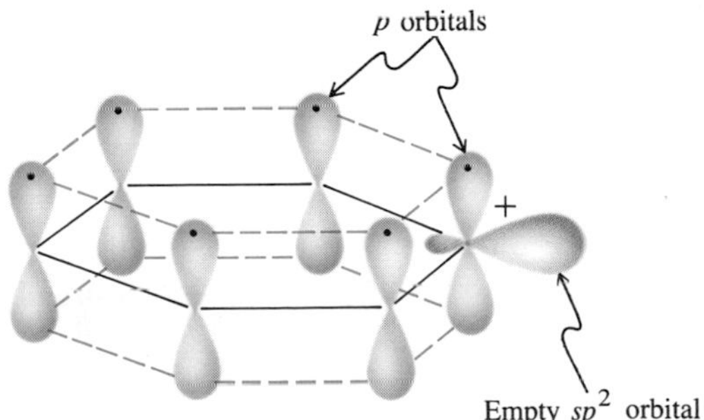

FIGURE 22-4 Orbital picture of the phenyl cation. Its empty sp^2 orbital is aligned perpendicular to the six-π-electron framework of the aromatic ring. As a result, the positive charge is not stabilized by resonance.

Arenediazonium salts can be converted into other substituted benzenes

When arenediazonium salts are decomposed in the presence of nucleophiles other than water, the corresponding substituted benzenes are formed. For example, diazotization of arenamines (anilines) in the presence of hydrogen iodide results in the corresponding iodoarenes.

$$\text{(3,4-methylenedioxy-6-aminobenzaldehyde: O, O, CHO, } NH_2\text{)} \xrightarrow{CH_3COOH,\ HI,\ NaNO_2} \text{(O, O, CHO, I)} + N_2$$

53%

Attempts to obtain other haloarenes in this way are frequently complicated by side reactions. One solution to this problem is the **Sandmeyer* reaction,** which makes use of the fact that the exchange of the nitrogen substituent for halogen is considerably facilitated by the presence of cuprous [Cu(I)] salts. The detailed mechanism of this process is complex and involves radicals.

Sandmeyer Reactions

$$\text{(}CH_3\text{, }NH_2\text{)} \xrightarrow{HCl,\ NaNO_2,\ 0^\circ C} \text{(}CH_3\text{, }N_2^+Cl^-\text{)} \xrightarrow{CuCl,\ 60^\circ C} \text{(}CH_3\text{, Cl)} + N_2$$

2-Methylbenzenamine
(*o*-Methylaniline)

79% overall
1-Chloro-2-methylbenzene
(*o*-Chlorotoluene)

$$\text{(Cl, }NH_2\text{)} \xrightarrow{1.\ HBr,\ NaNO_2,\ 0^\circ C\quad 2.\ CuBr,\ 100^\circ C} \text{(Cl, Br)}$$

2-Chlorobenzenamine
(*o*-Chloroaniline)

73%
1-Bromo-2-chlorobenzene
(*o*-Bromochlorobenzene)

Cuprous ion catalysis is also used for the preparation of aromatic nitriles from arenamines (anilines). For this purpose, cuprous cyanide, CuCN, is added to the diazonium salt, in the presence of excess potassium cyanide.

$$\text{(}CH_3\text{, }NH_2\text{)} \xrightarrow{1.\ HCl,\ NaNO_2,\ 0^\circ C\quad 2.\ CuCN,\ KCN,\ 50^\circ C} \text{(}CH_3\text{, CN)}$$

70%
4-Methylbenzonitrile
(*p*-Tolunitrile)

*Dr. Traugott Sandmeyer, 1854–1922, Geigy Company, Basel.

Thermal decomposition of diazonium tetrafluoroborates yields the corresponding fluoroarenes. The transformation is known as the **Schiemann* reaction,** and no additional catalysts are required. The reaction is important, because direct electrophilic fluorination of benzenes with fluorine is difficult to control, owing to the extremely exothermic character of the reaction.

Schiemann Reaction

COOH, $N_2^+Cl^-$ $\xrightarrow{HBF_4}$ COOH, $N_2^+BF_4^-$ $\xrightarrow{120°C}$ COOH, F $+ N_2 + BF_3$

78%

A diazonium tetrafluoroborate

2-Fluorobenzoic acid (*o*-Fluorobenzoic acid)

The diazonium group can be removed reductively by reducing agents. The sequence diazotization–reduction is a way to replace the amino group in arenamines (anilines) with hydrogen. The reducing agent employed is aqueous hypophosphorous acid, H_3PO_2. This method is especially useful in syntheses in which an amino group is used as a removable directing substituent in electrophilic aromatic substitution.

Reductive Removal of a Diazonium Group

CH_3, Br, NH_2 $\xrightarrow{NaNO_2,\ H^+,\ H_2O}$ CH_3, Br, N_2^+ $\xrightarrow{H_3PO_2,\ H_2O,\ 25°C}$ CH_3, Br, H

85%

1-Bromo-3-methylbenzene (*m*-Bromotoluene)

Another application of diazotization in synthetic strategy is illustrated in the synthesis of 1,3-dibromobenzene (*m*-dibromobenzene). Direct electrophilic bromination of benzene is not feasible for this purpose; after the first bromine has been introduced, the second will attack ortho or para. What is required is a meta-directing substituent, which can be transformed eventually into bromine. The nitro group is such a substituent. Double nitration of benzene furnishes 1,3-dinitrobenzene (*m*-dinitrobenzene). Reduction leads to the benzenediamine, which is then converted into the dihalo derivative.

Synthesis of 1,3-Dibromobenzene, Using a Diazotization Strategy

$\xrightarrow{HNO_3,\ H_2SO_4,\ \Delta}$ NO_2, NO_2 $\xrightarrow{H_2,\ Pd}$ NH_2, NH_2 $\xrightarrow{1.\ NaNO_2,\ H^+,\ H_2O;\ 2.\ CuBr,\ 100°C}$ Br, Br

*Professor Günther Schiemann, 1899–1969, Technical University of Hannover, Germany.

EXERCISE 22-22

Propose a synthesis of 1,3,5-tribromobenzene from benzene.

To summarize, arenediazonium salts, which are more stable than alkanediazonium salts because of resonance, are starting materials not only for phenols, but also for haloarenes, arenecarbonitriles, and reduced aromatics by displacement of nitrogen gas. The intermediates in some of these reactions may be aryl cations, highly reactive because of the absence of any electronically stabilizing features, but other, more complicated mechanisms may be followed. The ability to transform arenediazonium salts in this way gives considerable scope to the regioselective construction of substituted benzenes.

22-11 Electrophilic Substitution with Arenediazonium Salts: Diazo Coupling

Being positively charged, arenediazonium ions are electrophilic. Although the salts are not very reactive, they can accomplish electrophilic aromatic substitution if the substrate is an activated arene, such as phenol or benzenamine (aniline). This reaction, called **diazo coupling,** leads to highly colored compounds called **azo dyes.** For example, reaction of *N,N*-dimethylbenzenamine (*N,N*-dimethylaniline) with benzenediazonium chloride gives the brilliant orange dye Butter Yellow. This compound was once used as a food coloring agent but has been declared a suspect carcinogen by the Food and Drug Administration.

Diazo Coupling

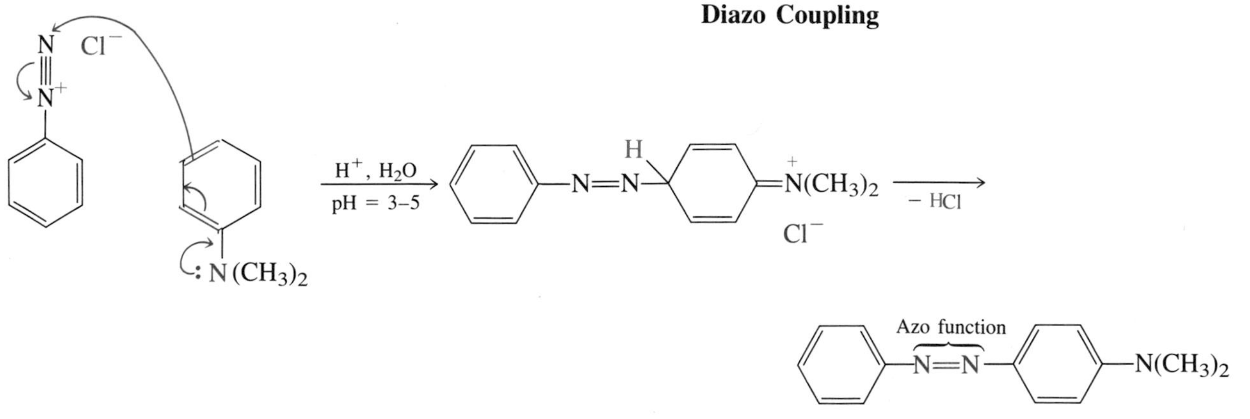

$N \equiv N^+$ Cl^- ; $:N(CH_3)_2$; H^+, H_2O, pH = 3–5 ; H ; $-N{=}N-$; $\overset{+}{=}N(CH_3)_2$; Cl^- ; $-$ HCl

Azo function

$-N{=}N-$; $-N(CH_3)_2$

4-Dimethylaminoazobenzene
(*p*-Dimethylaminoazobenzene, Butter Yellow)

Dyes used in the clothing industry usually contain sulfonic acid groups that impart water solubility and allow the dye molecule to attach itself ionically to charged sites on the polymer framework of the textile.

Industrial Dyes

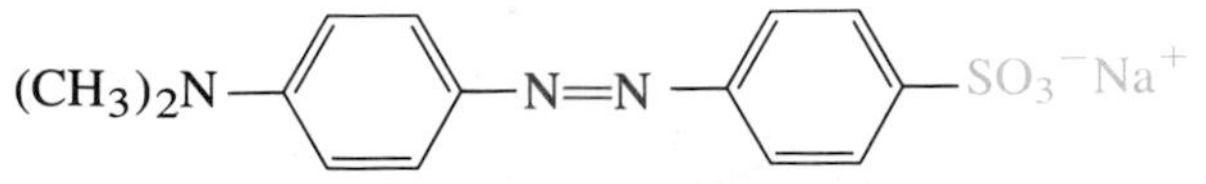

$(CH_3)_2N-$; $-N{=}N-$; $-SO_3^-\ Na^+$

Methyl Orange

pH = 3.1, red
pH = 4.4, yellow

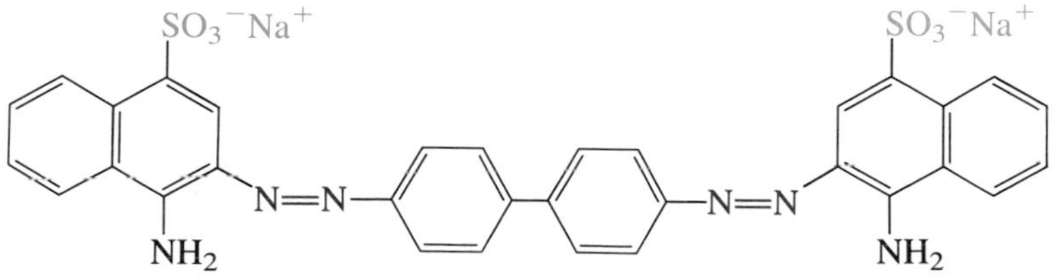

$SO_3^-Na^+$ $SO_3^-Na^+$

N=N N=N

NH_2 NH_2

Congo Red

pH = 3.0, bluc-violet
pH = 5.0, red

EXERCISE 22-23

Write the products of diazo coupling of benzenediazonium chloride with each of the following molecules. **(a)** Methoxybenzene; **(b)** 1-chloro-3-methoxybenzene; **(c)** 1-(dimethylamino)-4-methylbenzene.

In summary, arenediazonium cations attack activated benzene rings by diazo coupling, a process that furnishes azobenzenes, which are often highly colored.

NEW REACTIONS

Benzylic Resonance

1. RADICAL HALOGENATION (Section 22-1)

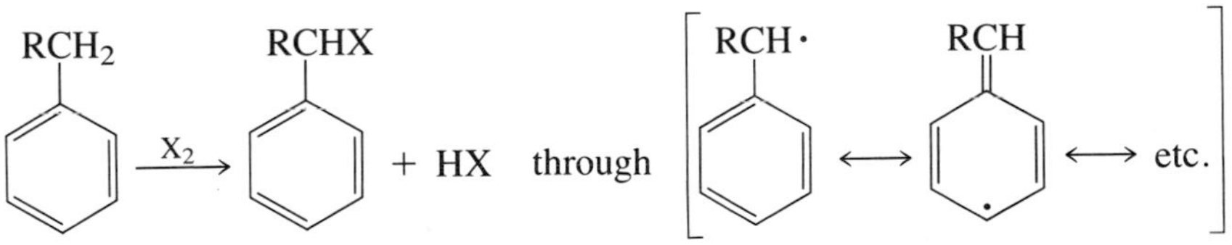

RCH_2 —X_2→ RCHX + HX through [RCH· ⟷ RCH ⟷ etc.]

Requires heat, light, or a radical initiator

Benzylic radical

2. SOLVOLYSIS (Section 22-1)

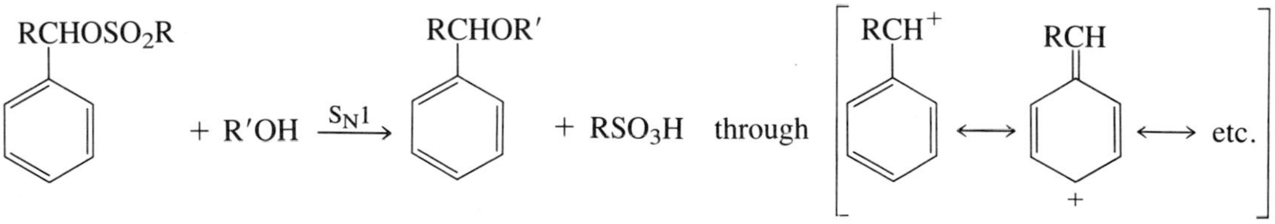

$RCHOSO_2R$ + R′OH —S_N1→ RCHOR′ + RSO_3H through [RCH^+ ⟷ RCH (+) ⟷ etc.]

Benzylic cation

3. S_N2 REACTIONS OF (HALOMETHYL)BENZENES (Section 22-1)

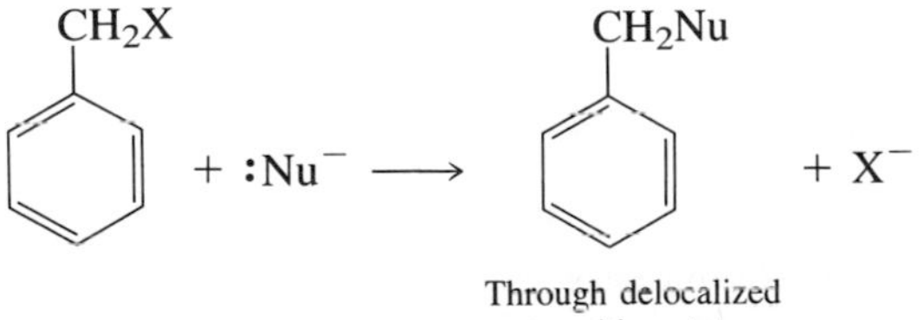

CH_2X + $:Nu^-$ ⟶ CH_2Nu + X^-

Through delocalized transition state

4. BENZYLIC DEPROTONATION (Section 22-1)

$C_6H_5CH_3$ + RLi ⟶ $C_6H_5CH_2Li$ + RH

$pK_a \sim 41$

Phenylmethyllithium (Benzyllithium)

Oxidation and Reduction Reactions on Aromatic Side Chains

5. OXIDATION (Section 22-2)

$C_6H_5CH_2R$ —(1. $KMnO_4$, HO^-, Δ; 2. H^+, H_2O)⟶ C_6H_5COOH

Benzylic alcohols

C_6H_5CHOHR —(MnO_2, propanone (acetone))⟶ $C_6H_5C(=O)R$

6. REDUCTION BY HYDROGENOLYSIS (Section 22-2)

$C_6H_5CH_2OR$ —(H_2, Pd-C, ethanol)⟶ $C_6H_5CH_3$ + ROH

$C_6H_5CH_2$ is a protecting group for ROH

Phenols and Ipso Substitution

7. ACIDITY (Section 22-3)

C_6H_5OH ⇌ H^+ + [$C_6H_5\ddot{O}:^-$ ⟷ (cyclohexadienone carbanion form) ⟷ etc.]

$pK_a \sim 10$

Much stronger acid than simple alkanols

Phenoxide ion

8. NUCLEOPHILIC AROMATIC SUBSTITUTION (Section 22-4)

1-Cl-2,4-$(NO_2)_2C_6H_3$ —($Nu:^-$)⟶ [ipso adduct: Cl, Nu on same carbon; NO_2, NO_2; delocalized negative charge] ⟶ 1-Nu-2,4-$(NO_2)_2C_6H_3$ + Cl^-

Nucleophile attacks at ipso position

9. AROMATIC SUBSTITUTION THROUGH BENZYNE INTERMEDIATES (Section 22-4)

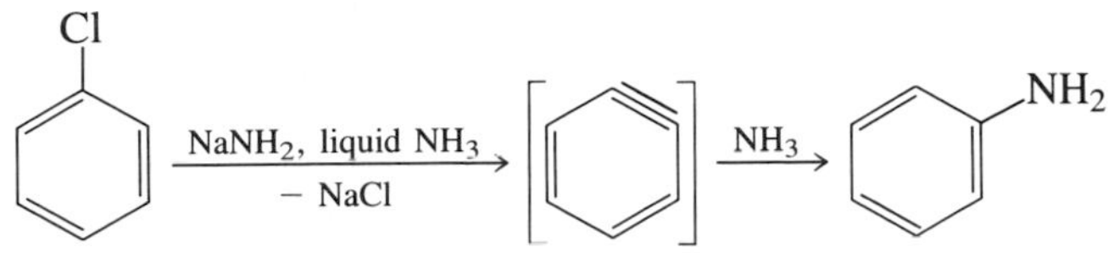

Cl

$NaNH_2$, liquid NH_3; $-$ NaCl; NH_3

NH_2

Nucleophile attacks at either ipso or ortho position

10. PREPARATION OF PHENOLS BY NUCLEOPHILIC AROMATIC SUBSTITUTION (Section 22-4)

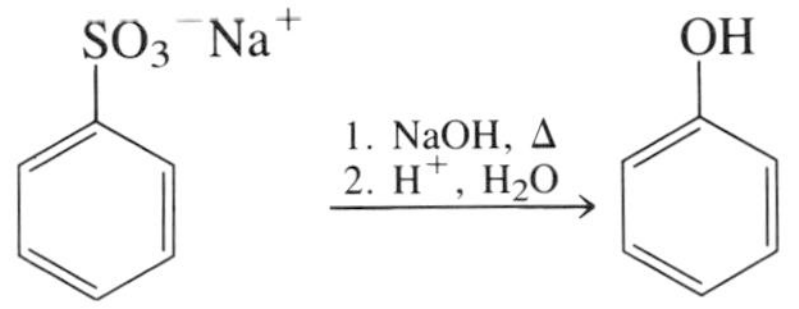

$SO_3^- Na^+$

1. NaOH, Δ
2. H^+, H_2O

OH

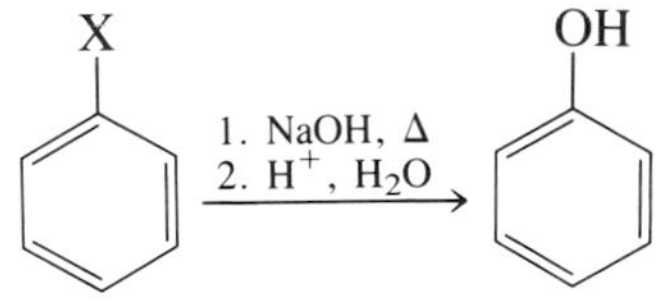

X

1. NaOH, Δ
2. H^+, H_2O

OH

11. ARENEDIAZONIUM SALT HYDROLYSIS (Section 22-4)

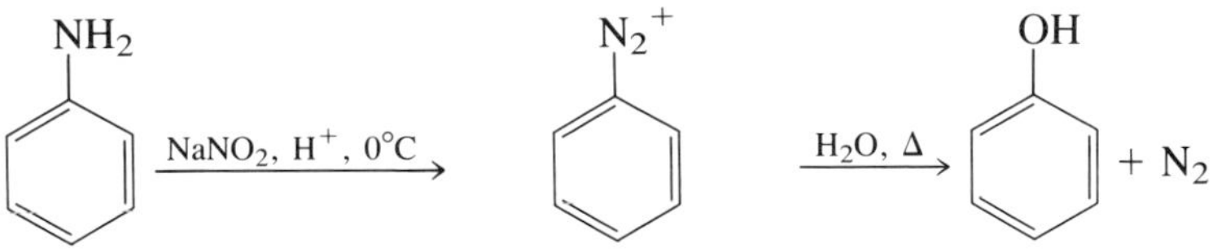

NH_2

$NaNO_2$, H^+, 0°C

N_2^+

H_2O, Δ

OH

$+ N_2$

Benzenediazonium cation

Reactions of Phenols and Alkoxybenzenes

12. ETHER CLEAVAGE (Section 22-5)

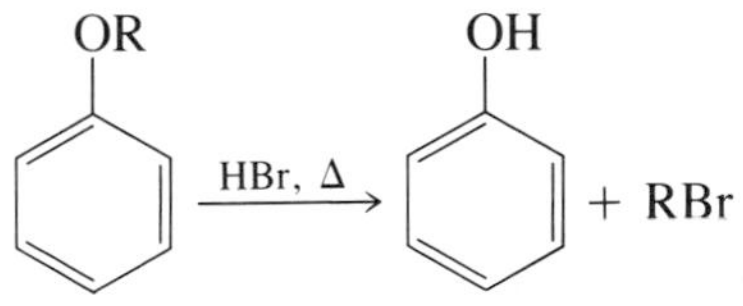

OR

HBr, Δ

OH

+ RBr

Aryl C–O bond is not cleaved

13. ETHER FORMATION (Section 22-5)

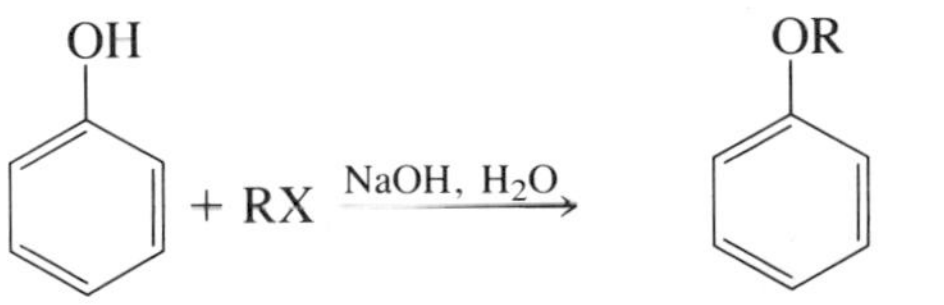

OH

+ RX

NaOH, H_2O

OR

Alkoxybenzene

Williamson method (Section 9-6)

14. ESTERIFICATION (Section 22-5)

$$C_6H_5OH + RCOCl \xrightarrow{\text{Base}} C_6H_5OCOR$$

Phenyl alkanoate

15. ELECTROPHILIC AROMATIC SUBSTITUTION (Section 22-6)

$$C_6H_5OCH_3 + E^+ \longrightarrow o\text{-}E\text{-}C_6H_4OCH_3 + p\text{-}E\text{-}C_6H_4OCH_3 + H^+$$

16. PHENOLIC RESINS (Section 22-6)

$$C_6H_5OH + CH_2{=}O \xrightarrow{\text{NaOH}} o\text{-}HOCH_2C_6H_4OH \xrightarrow[-\,H_2O]{HO^-}$$

$$o\text{-quinone methide} \xrightarrow{C_6H_5OH,\ \text{NaOH}} \text{bis(2-hydroxyphenyl)methane} \longrightarrow \longrightarrow \text{polymer}$$

17. KOLBE REACTION (Section 22-6)

$$C_6H_5OH + CO_2 \xrightarrow[2.\ H^+,\ H_2O]{1.\ \text{NaOH, pressure}} o\text{-}HOC_6H_4COOH$$

18. CLAISEN REARRANGEMENT (Section 22-7)

Aromatic Claisen rearrangement

$$\text{allyl phenyl ether} \xrightarrow{\Delta} \text{6-allyl-2,4-cyclohexadienone (H)} \longrightarrow \text{2-allylphenol (OH)}$$

Aliphatic Claisen rearrangement

$$\text{allyl vinyl ether} \longrightarrow \text{4-pentenal } (O{=}CH\text{–}, \text{H})$$

19. COPE REARRANGEMENT (Section 22-7)

R R

20. OXIDATION (Section 22-8)

OH OH O OH

$Na_2Cr_2O_7$, H^+

$\cdot ON(SO_3)_2K_2$

2,5-Cyclohexadiene-1,4-dione
(*p*-Benzoquinone)

21. CONJUGATE ADDITIONS TO 2,5-CYCLOHEXADIENE-1,4-DIONES (*p*-BENZOQUINONES) (Section 22-8)

O O + HCl ⟶ OH OH Cl

22. DIELS–ALDER CYCLOADDITIONS TO 2,5-CYCLOHEXADIENE-1,4-DIONES (*p*-BENZOQUINONES) (Section 22-8)

O O + ⟶ O O H H

23. LIPID PEROXIDATION (Section 22-9)

H H C R R′ —O_2, radical chain reaction→ H OOH C R R′ —Via fragmentation to alkoxy radicals, followed by β-scission→ toxic substances such as 4-hydroxy-2-alkenals

24. INHIBITION BY ANTIOXIDANTS (Section 22-9)

H O· C R R′ or H OO· C R R′ + HO CH_3 CH_3 CH_3 O CH_3 R ⟶ H OH C R R′ or H OOH C R R′ + ·O CH_3 CH_3 CH_3 O CH_3 R

Vitamin E
(or BHA or BHT)

25. VITAMIN C AS AN ANTIOXIDANT (Section 22-9)

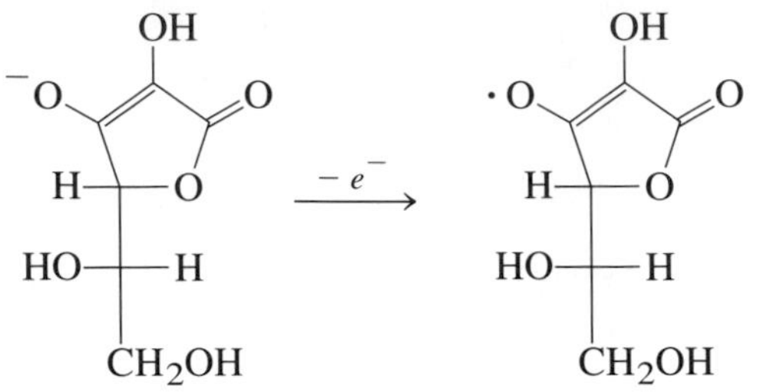

$\xrightarrow{-e^-}$

Semidehydroascorbic acid

Arenediazonium Salts

26. SANDMEYER REACTIONS (Section 22-10)

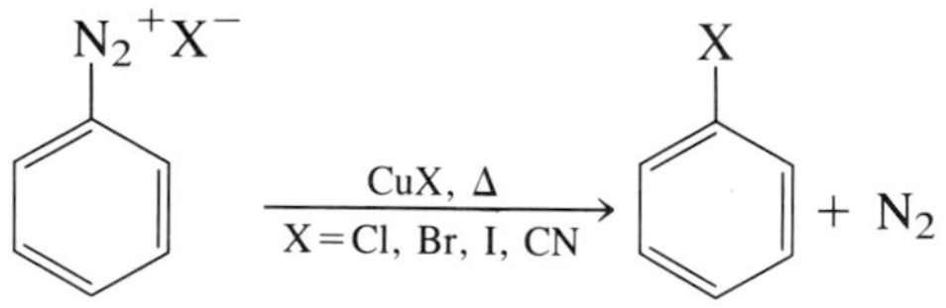

$N_2{}^+X^-$ $\xrightarrow[X = Cl, Br, I, CN]{CuX,\ \Delta}$ X $+ N_2$

27. SCHIEMANN REACTION (Section 22-10)

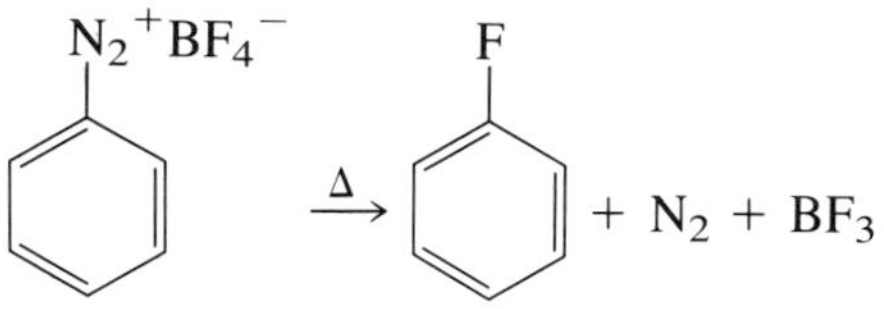

$N_2{}^+BF_4{}^-$ $\xrightarrow{\Delta}$ F $+ N_2 + BF_3$

28. REDUCTION (Section 22-10)

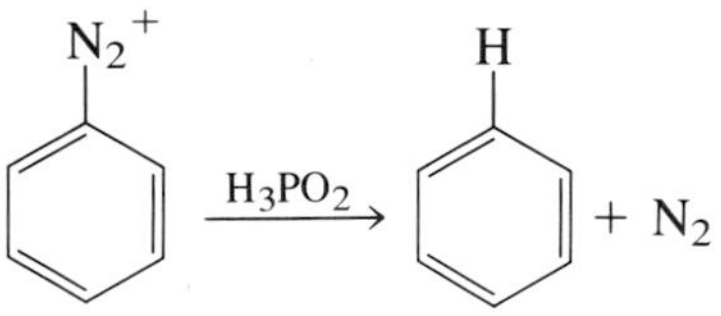

$N_2{}^+$ $\xrightarrow{H_3PO_2}$ H $+ N_2$

29. DIAZO COUPLING (Section 22-11)

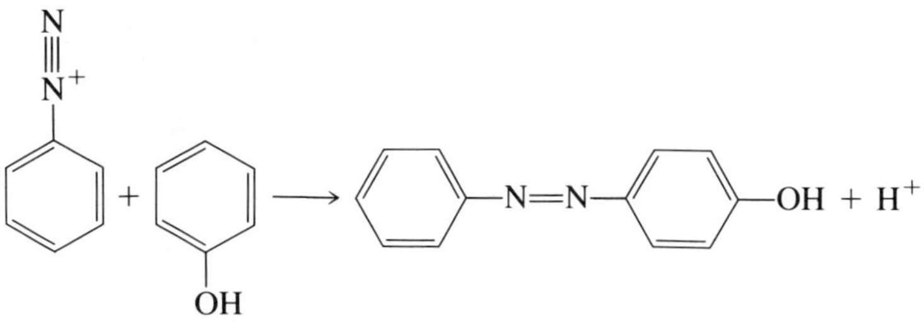

$N\equiv N^+$ + OH $\longrightarrow$ —N=N— —OH + H^+

Occurs only with strongly activated rings

Azo compound

IMPORTANT CONCEPTS

1. Phenylmethyl and other benzylic radicals, cations, and anions are reactive intermediates stabilized by resonance of the resulting centers with a benzene π system.
2. Nucleophilic aromatic ipso substitution accelerates with the nucleophilicity of the attacking species and with the number of electron-withdrawing groups on the ring, particularly if they are located ortho or para to the point of attack.
3. Benzyne is destabilized by the strain imposed by the two *sp*-hybridized carbons forming the triple bond.
4. Phenols are aromatic enols, undergoing reactions typical of the hydroxy function and the aromatic ring.
5. Cyclohexadienediones and benzenediols function as redox couples in the laboratory and in nature.
6. Vitamin E and the highly substituted phenol derivatives BHA and BHT function as inhibitors of the radical-chain oxidation of lipids. Vitamin C is also an antioxidant, capable of regenerating vitamin E at the surface of cell membranes.
7. Arenediazonium ions are stabilized by resonance but furnish reactive aryl cations in which the positive charge cannot be delocalized into the aromatic ring.
8. The amino group can be used to direct electrophilic aromatic substitution, after which it is removable by diazotization and reduction.

PROBLEMS

1. Give the expected major product(s) of each of the following reactions.

(a) [ethylbenzene, $C_6H_5CH_2CH_3$] $\xrightarrow{Cl_2\text{(1 equivalent)}, h\nu}$

(b) [tetralin] $\xrightarrow{\text{NBS (1 equivalent)}, h\nu}$

2. By drawing appropriate resonance structures, illustrate why halogen atom attachment at the para position of phenylmethyl (benzyl) radical is unfavored compared with attachment at the benzylic position.

3. Triphenylmethyl radical, $(C_6H_5)_3C\cdot$, is stable at room temperature in dilute solution in an inert solvent, and salts of triphenylmethyl cation, $(C_6H_5)_3C^+$, may be isolated as stable crystalline solids. Propose explanations for the unusual stabilities of these species.

4. Give the expected products of the following reactions or reaction sequences.

(a) $BrCH_2CH_2CH_2$–[benzene ring, para]–CH_2Br $\xrightarrow{H_2O, \Delta}$

(b) [benzene ring with CH_2Cl] $\xrightarrow{\text{1. KCN, DMSO; 2. } H^+, H_2O, \Delta}$

(c) [indane] $\xrightarrow{\text{1. } CH_3CH_2CH_2CH_2Li, (CH_3)_2NCH_2CH_2N(CH_3)_2, \text{THF; 2. } C_6H_5CHO\text{; 3. } H^+, H_2O, \Delta}$ $C_{16}H_{14}$

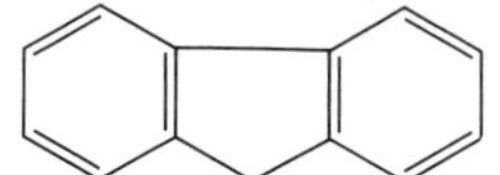
Fluorene

5. The hydrocarbon with the common name fluorene is acidic enough ($pK_a \sim 23$) to be a useful indicator in deprotonation reactions of compounds of greater acidity. Indicate the most acidic hydrogen(s) in fluorene. Draw resonance structures to explain the relative stability of its conjugate base.

6. Outline a straightforward, practical, and efficient synthesis of each of the following compounds. Start with benzene or methylbenzene. Assume that the para isomer (but *not* the ortho isomer) may be separated efficiently from any mixtures of ortho and para substitution products.

(a) CH_2CH_2Br **(b)** $CONH_2$, Cl **(c)** O, $COOCH_3$ **(d)** COOH, Br, Br

7. Rank the following compounds in descending order of reactivity toward aqueous base.

Br, NO_2 — Br, NO_2 — Br, NO_2 — Br, NO_2, NO_2 — Br, NO_2, NO_2

8. Predict the main product(s) of the following reactions. In each case, describe the mechanism(s) in operation.

(a) Cl, NO_2, NO_2 $\xrightarrow{NH_2NH_2}$

(b) Cl, Cl, O_2N, NO_2 $\xrightarrow{NaOCH_3,\ CH_3OH}$

(c) Cl, CH_3 $\xrightarrow{LiN(CH_2CH_3)_2,\ (CH_3CH_2)_2NH}$

9. Starting with benzenamine, propose a synthesis of aklomide, an agent used to treat certain exotic fungal and protozoal infections in veterinary medicine. Several intermediates are shown to give you the general route. Fill in the blanks that remain; each requires up to three sequential reactions.

NH_2 $\xrightarrow{(a)}$ NH_2, Br $\xrightarrow{(b)}$ NO_2, Cl, Br $\xrightarrow{(c)}$ NO_2, Cl, CN $\xrightarrow{(d)}$ NO_2, Cl, $CONH_2$

Aklomide

10. Explain the mechanism of the following synthetic transformation. (Hint: Two equivalents of butyllithium are used.)

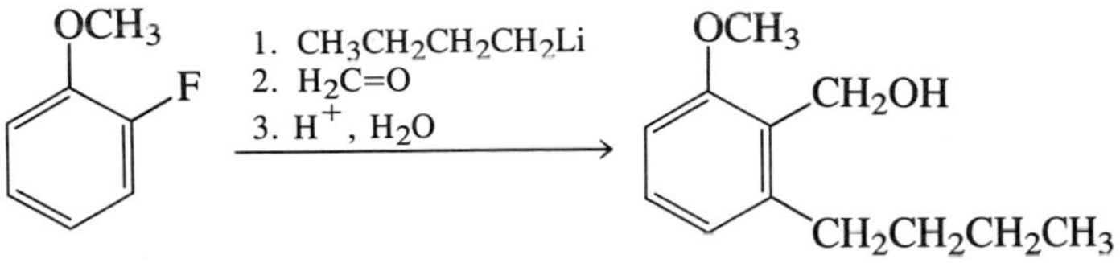

OCH_3, F (1. $CH_3CH_2CH_2CH_2Li$; 2. $H_2C{=}O$; 3. H^+, H_2O) → OCH_3, CH_2OH, $CH_2CH_2CH_2CH_3$

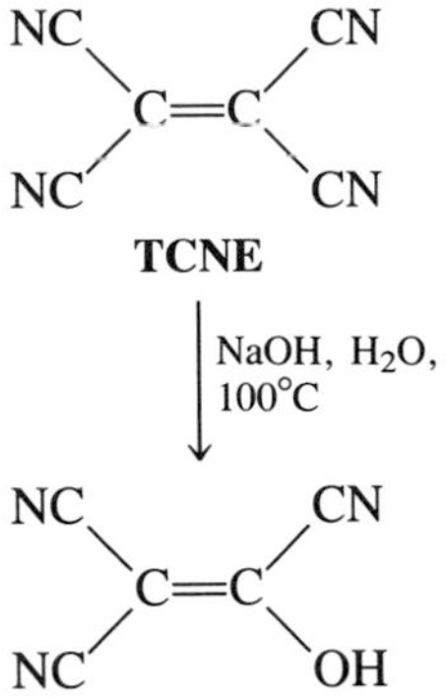

11. In a rather unusual reaction, tetracyanoethene (TCNE) is converted by boiling in aqueous base into tricyanoethenol (whose enol form is stabilized by the three nitrile groups). Suggest a mechanism for this transformation. (Hint: This reaction occurs for the same reason that one form of nucleophilic aromatic substitution takes place.)

$(NC)_2C{=}C(CN)_2$ **TCNE** (NaOH, H_2O, 100°C) → $(NC)_2C{=}C(CN)OH$

12. Give the expected major product(s) of each of the following reactions and reaction sequences.

(a) 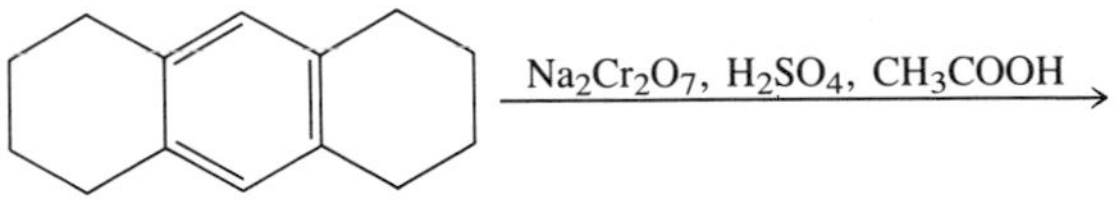 → $Na_2Cr_2O_7$, H_2SO_4, CH_3COOH

(b) 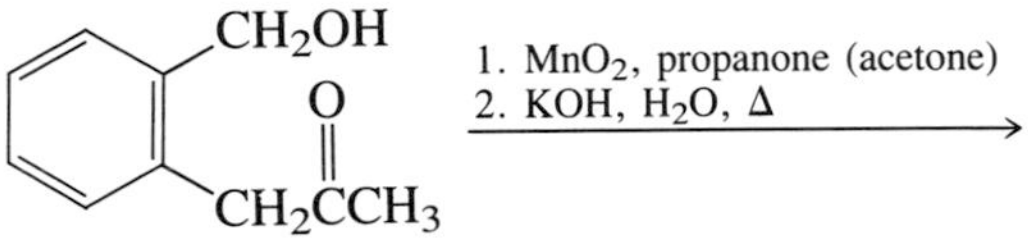 CH_2OH, CH_2CCH_3 (C=O) → 1. MnO_2, propanone (acetone); 2. KOH, H_2O, Δ

(c) CH_3 → 1. $(CH_3)_2CHCl$, $AlCl_3$; 2. HNO_3, H_2SO_4; 3. $KMnO_4$, NaOH, Δ; 4. H^+, H_2O

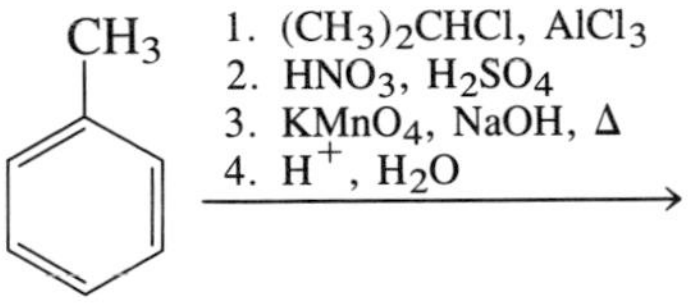

13. Rank the following compounds in order of descending acidity.

(a) CH_3OH **(b)** CH_3COOH

(c) 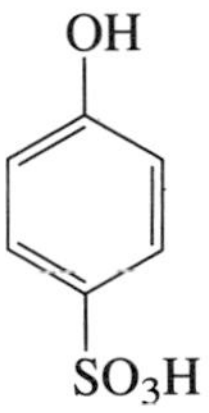 OH, SO_3H **(d)** 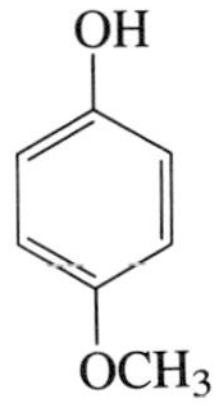 OH, OCH_3 **(e)**  OH, CF_3 **(f)** 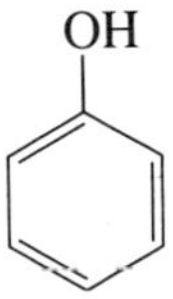 OH

14. Design a synthesis of each of the following phenols, starting with either benzene or any monosubstituted benzene derivative.

(a)  OH, CH_3 **(b)** 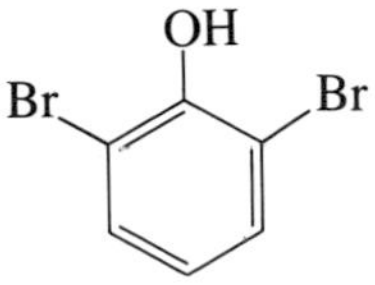 OH, Br, Br **(c)** The three benzenediols **(d)** 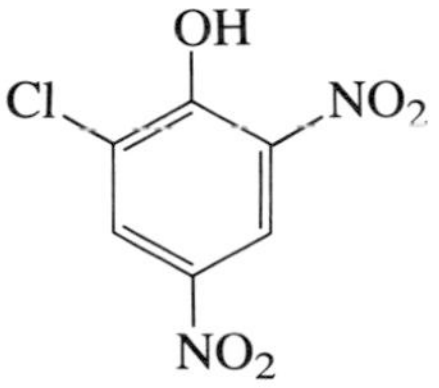 OH, Cl, NO_2, NO_2

15. Starting with benzene, propose syntheses of each of the following phenol derivatives.

(a) 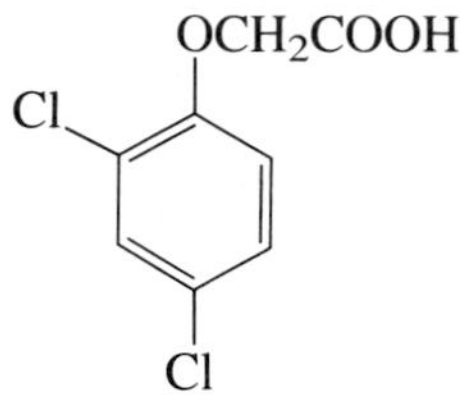 OCH_2COOH, Cl, Cl
The herbicide 2,4-D

(b) 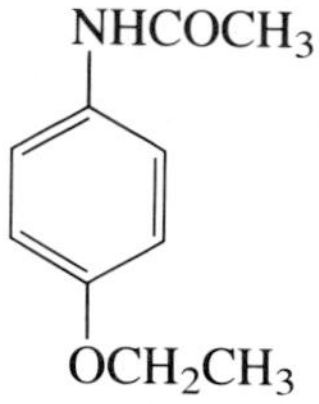 $NHCOCH_3$, OCH_2CH_3
Phenacetin
(The active ingredient in Midol)

(c) 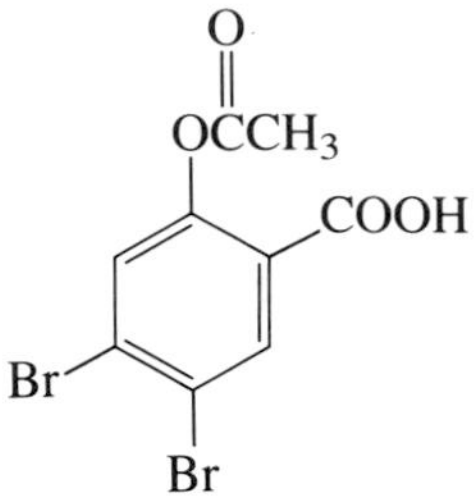 $OCCH_3$ (O), COOH, Br, Br
Dibromoaspirin
(An experimental drug for the treatment of sickle-cell anemia)

16. Name each compound shown below.

(a) 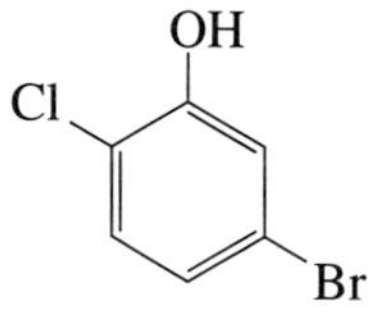 OH, Cl, Br

(b) 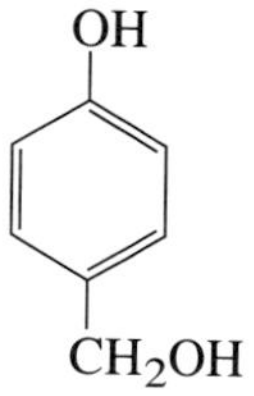 OH, CH_2OH

(c) 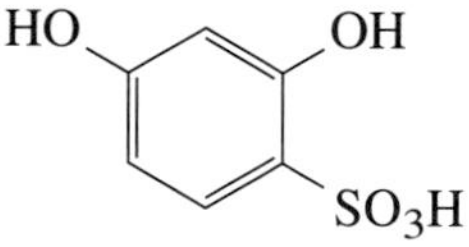 HO, OH, SO_3H

(d) 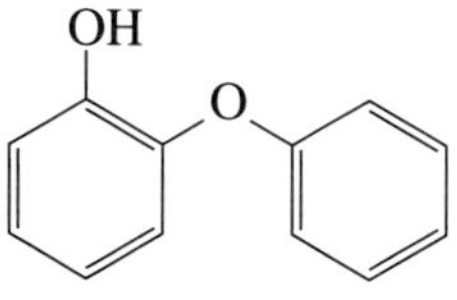 OH, O

(e) 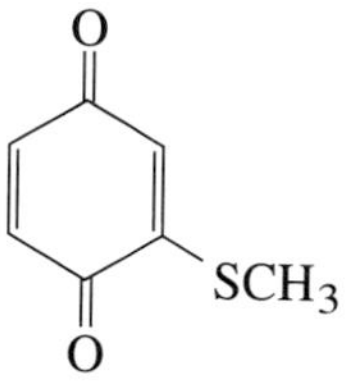 O, O, SCH_3

17. Give the expected product(s) of each of the following reaction sequences.

(a) 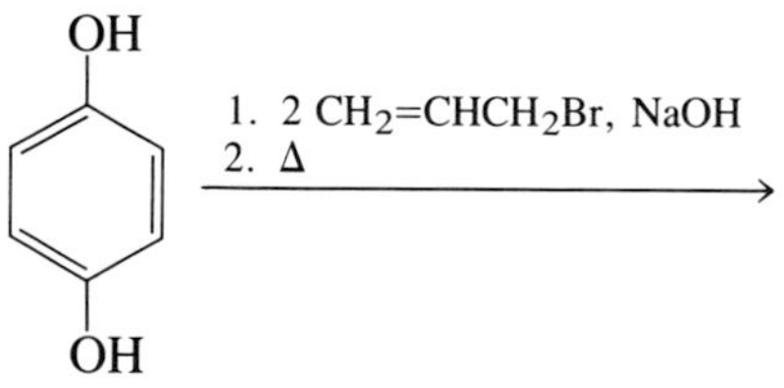 OH, OH $\xrightarrow{\text{1. } 2\ CH_2{=}CHCH_2Br,\ NaOH \quad \text{2. } \Delta}$

(b) 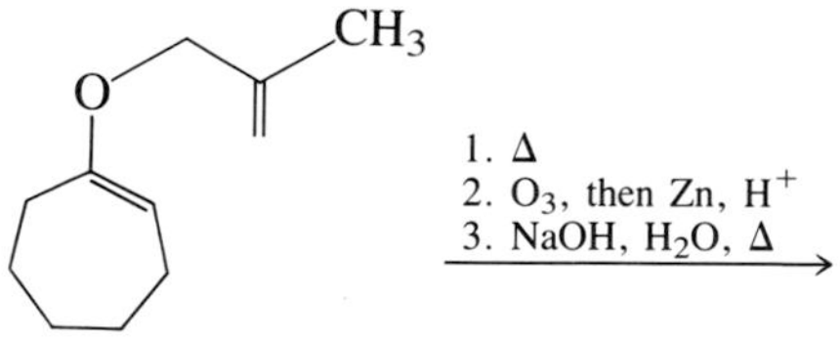 O, CH_3 $\xrightarrow{\text{1. } \Delta \quad \text{2. } O_3, \text{ then Zn, } H^+ \quad \text{3. } NaOH,\ H_2O,\ \Delta}$

(c) 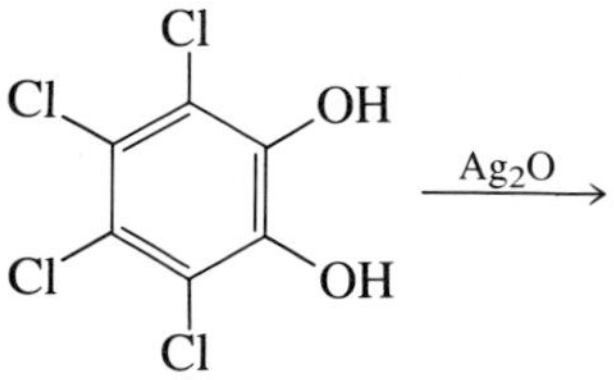 Cl, Cl, Cl, Cl, OH, OH $\xrightarrow{Ag_2O}$

(d) 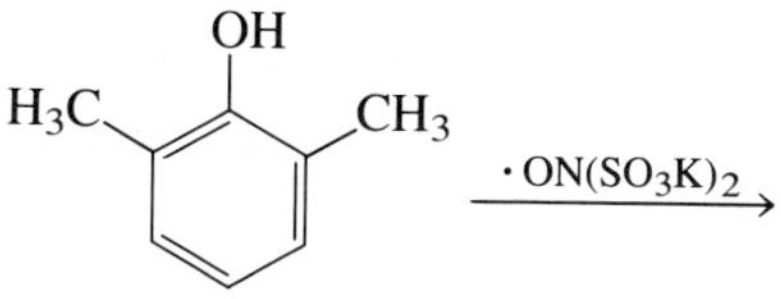 OH, H_3C, CH_3 $\xrightarrow{\cdot ON(SO_3K)_2}$

(e) 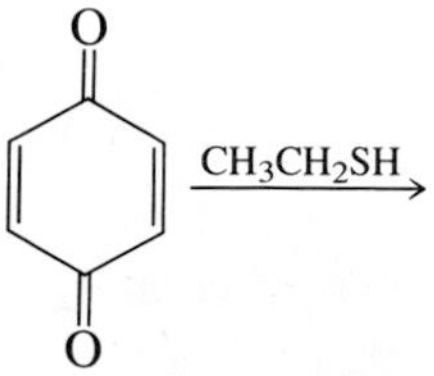 O, O $\xrightarrow{CH_3CH_2SH}$ (two possibilities)

(f) 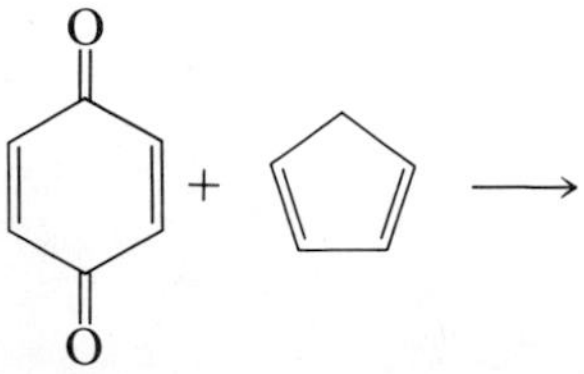 O, O + $\longrightarrow$

18. As a children's medicine, Tylenol has a major marketing advantage over aspirin: *Liquid Tylenol* preparations (essentially, Tylenol dissolved in flavored water) are stable, whereas comparable aspirin solutions are not. Explain.

19. Black-and-white photographic film contains tiny crystals of silver bromide. On exposure to light, the silver bromide becomes photoactivated. In this form, it is readily converted to particles of black silver metal by a reducing agent called a *developer,* an example of which is 4-(*N*-methylamino)phenol (Metol). After exposure and development, the film is washed with a *fixer,* such as ammonium thiosulfate, which dissolves and removes unreduced silver bromide. Give the structure of Metol and of the product of its oxidation by silver bromide. [Hint: The oxidation of Metol is closely related to that of 1,4-benzenediol (hydroquinone).]

20. Biochemical oxidation of aromatic rings is catalyzed by a group of liver enzymes called aryl hydroxylases. Part of this chemical process is the conversion of toxic aromatic hydrocarbons such as benzene into water-soluble phenols, which can be easily excreted. However, the primary purpose of the enzyme is to enable the synthesis of biologically useful compounds, such as the amino acid tyrosine.

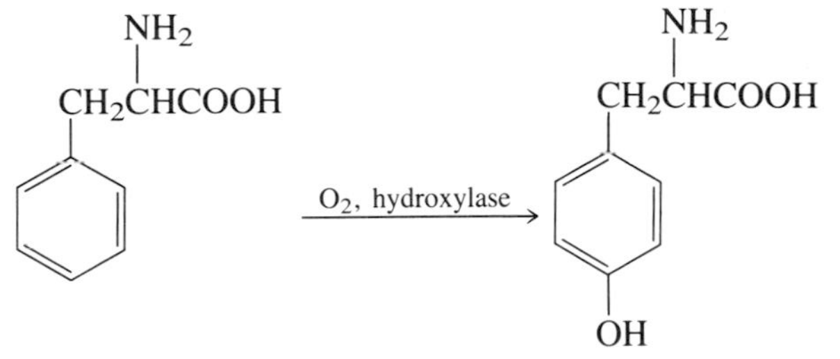

NH_2 | $CH_2CHCOOH$ (on benzene ring) $\xrightarrow{O_2,\ \text{hydroxylase}}$ NH_2 | $CH_2CHCOOH$ (on benzene ring with para OH)

Phenylalanine **Tyrosine**

(a) Extrapolating from your knowledge of benzene chemistry, which of the three following possibilities seems most reasonable: The oxygen is introduced by electrophilic attack on the ring; the oxygen is introduced by free-radical attack on the ring; or the oxygen is introduced by nucleophilic attack on the ring? **(b)** It is widely suspected that oxacyclopropanes play a role in arene hydroxylation. Part of the evidence is the following observation: If the site to be hydroxylated is initially labeled with deuterium, a substantial proportion of the product still contains deuterium atoms, which have apparently migrated to the position ortho to the site of hydroxylation.

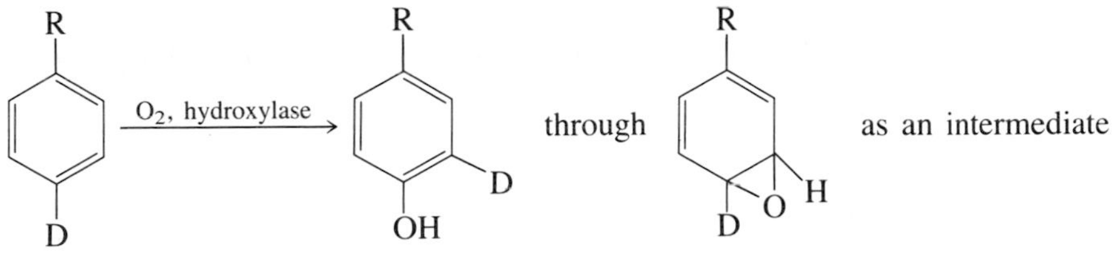

R–C₆H₄–D $\xrightarrow{O_2,\ \text{hydroxylase}}$ R, OH, D-substituted benzene through [R-substituted ring with D, H and epoxide O] as an intermediate

Suggest a plausible mechanism for the formation of the oxacyclopropane intermediate and its conversion into the observed product. (Hint: Hydroxylase converts O_2 into hydrogen peroxide, HO–OH.) Assume the availability of catalytic amounts of acids and bases, as necessary.

Note: In victims of the genetically transmitted disorder called phenylketonuria (PKU), the hydroxylase enzyme system described here does not function properly. Instead, phenylalanine in the brain is converted into 2-phenyl-2-oxopropanoic (phenylpyruvic) acid, the reverse of the process shown in Problem 23 of Chapter 21. The buildup of this compound in the brain can lead to severe retardation; thus people with PKU (which can be diagnosed at birth) must be restricted to diets low in phenylalanine.

21. A common application of the Cope rearrangement is in ring-enlargement sequences. Fill in the reagents and products missing from the following scheme, which illustrates the construction of a 10-membered ring.

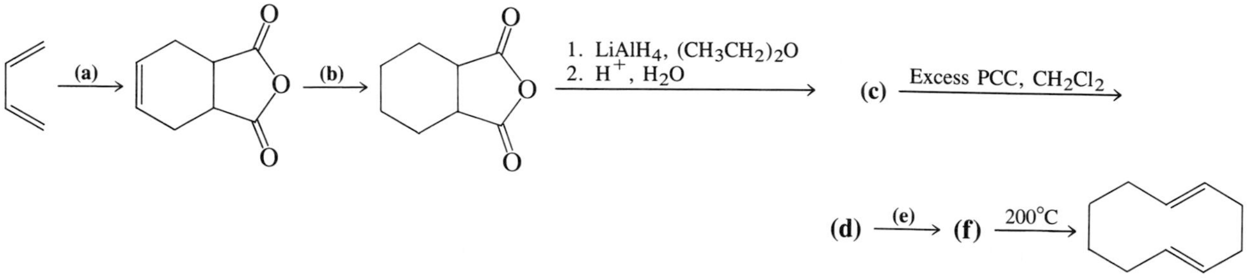

(a) → (b) → 1. $LiAlH_4$, $(CH_3CH_2)_2O$ 2. H^+, H_2O → (c) Excess PCC, CH_2Cl_2 →

(d) (e) → (f) 200°C →

22. Formulate a complete mechanism for the diazotization of benzenamine (aniline) in the presence of HCl and $NaNO_2$ and a plausible mechanism (based on what you have learned in Section 22-10) for its conversion into iodobenzene by treatment with aqueous iodide ion (e.g., from K^+I^-).

23. Devise a synthesis of each of the following substituted benzene derivatives. Start each synthesis with benzene.

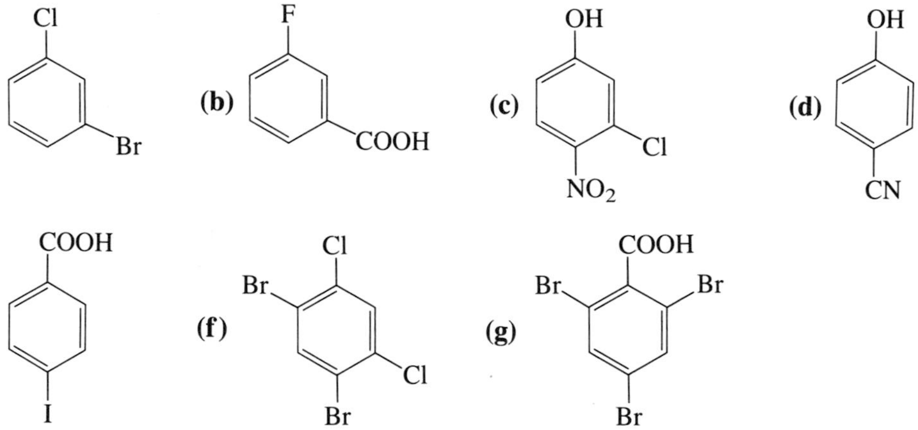

(a) Cl, Br (b) F, COOH (c) OH, Cl, NO_2 (d) OH, CN

(e) COOH, I (f) Cl, Br, Cl, Br (g) COOH, Br, Br, Br

24. Write the most reasonable structure of the product of each of the following reaction sequences.

(a) NH_2, SO_3H — 1. $NaNO_2$, HCl, 5°C 2. HO⁻, phenol (OH) → Golden Yellow

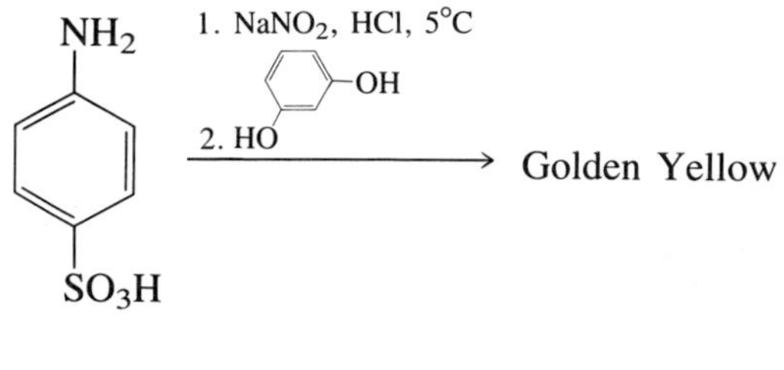

(b) NH_2, SO_3H — 1. $NaNO_2$, HCl, 5°C 2. diphenylamine (NH) → Metanil Yellow

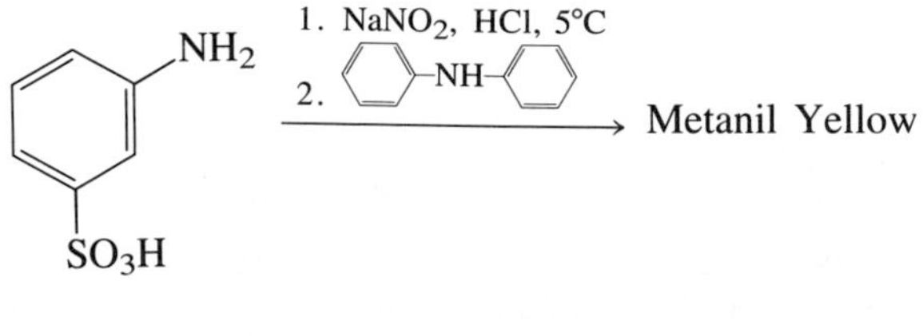

For the following reaction, assume that electrophilic substitution is preferentially on the most activated ring.

(c) 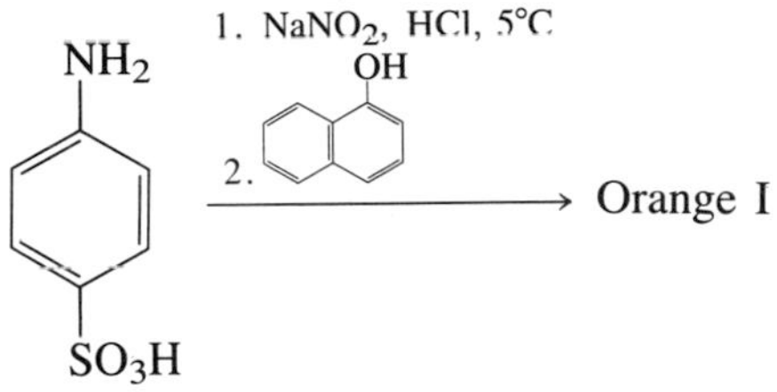 p-H_2N–C_6H_4–SO_3H $\xrightarrow{\text{1. NaNO}_2\text{, HCl, 5°C; 2. 1-naphthol (OH)}}$ Orange I

25. Show the reagents that would be necessary for the synthesis by diazo coupling of each of the three compounds listed below. For structures, see Section 22-11.

(a) Methyl Orange **(b)** Congo Red

(c) Prontosil, H_2N–$C_6H_3(NH_2)$–N=N–C_6H_4–SO_2NH_2, which is converted microbially into sulfanilamide, H_2N–C_6H_4–SO_2NH_2

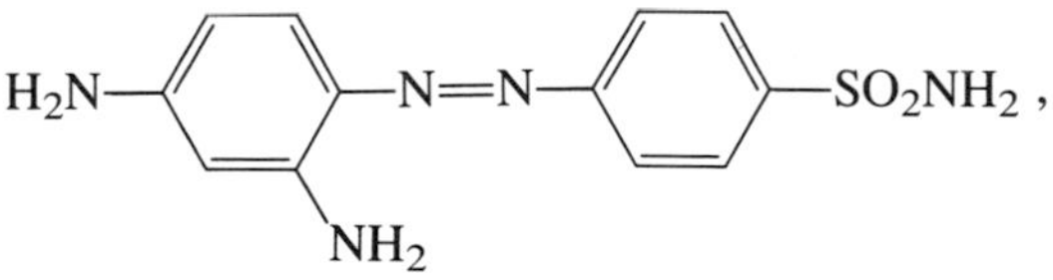

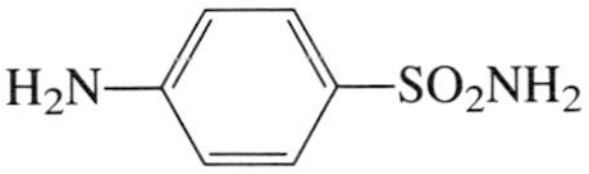

(The accidental discovery of the antibacterial properties of prontosil in the 1930s led indirectly to the development of sulfa drugs as antibiotics in the 1940s.)

26. **(a)** Give the key reaction that illustrates the inhibition of fat oxidation by the preservative BHT. **(b)** The extent to which fat oxidation occurs in the body may be determined by measuring the amount of *pentane* exhaled in the breath. Increasing the amount of vitamin E in the diet decreases the amount of pentane exhaled. Examine the processes described in Section 22-9 and identify one that could produce pentane. You will have to do some extrapolating from the specific reactions shown in the section.

27. The urushiols are the irritants in poison ivy and poison oak that give you rashes and make you itch after touching them. Use the following information to determine the structures of urushiols I ($C_{21}H_{36}O_2$) and II ($C_{21}H_{34}O_2$), the two major members of this family of unpleasant compounds.

Urushiol II $\xrightarrow{H_2\text{, Pd-C, }CH_3CH_2OH}$ urushiol I

Urushiol II $\xrightarrow{\text{Excess }CH_3I\text{, NaOH}}$ $C_{23}H_{38}O_2$ **Dimethylurushiol II** $\xrightarrow{\text{1. }O_3\text{, }CH_2Cl_2\text{; 2. Zn, }H_2O}$

$CH_3CH_2CH_2CH_2CH_2CH_2CHO$ + $C_{16}H_{24}O_3$ **Aldehyde A**

Synthesis of aldehyde A

OCH_3-benzene (anisole) $\xrightarrow{\text{1. } SO_3,\ H_2SO_4;\ \text{2. } HNO_3,\ H_2SO_4}$ $C_7H_7NSO_6$ (**B**) $\xrightarrow{H^+,\ H_2O,\ \Delta}$ $C_7H_7NO_3$ (**C**) $\xrightarrow{\text{1. } H_2,\ Pd,\ CH_3CH_2OH;\ \text{2. } NaNO_2,\ H^+,\ H_2O;\ \text{3. } H_2O,\ \Delta}$

$C_7H_8O_2$ (**D**) $\xrightarrow{\text{1. } CO_2,\ \text{pressure},\ KHCO_3,\ H_2O;\ \text{2. } NaOH,\ CH_3I;\ \text{3. } H^+,\ H_2O}$ $C_9H_{10}O_4$ (**E**) $\xrightarrow{\text{1. } LiAlH_4,\ (CH_3CH_2)_2O;\ \text{2. } H^+,\ H_2O;\ \text{3. } MnO_2,\ \text{propanone (acetone)}}$

$C_9H_{10}O_3$ (**F**) $\xrightarrow{\text{1. } C_6H_5CH_2O(CH_2)_6CH{=}P(C_6H_5)_3;\ \text{2. Excess } H_2,\ Pd\text{-}C,\ CH_3CH_2OH;\ \text{3. PCC},\ CH_2Cl_2}$ aldehyde A

28. Is the site of reaction in the biosynthesis of norepinephrine from dopamine (see Chapter 5, Problem 26) consistent with the principles outlined in this chapter? Would it be easier or more difficult to duplicate this transformation nonenzymatically? Explain.

23

Dicarbonyl Compounds

We tend to view organic molecules in terms of the chemistry of their functional groups. What happens if there are *several* functional units in the same molecule? We know that in some instances they react independently. For example, 5-hexenoic acid may be hydrogenated in the presence of a catalyst (as an alkene) or deprotonated with base (as a carboxylic acid).

$$\underset{\textbf{Hexanoic acid}}{CH_3(CH_2)_4\overset{O}{\overset{\|}{C}}OH} \xleftarrow{H_2,\ Pt} \underset{\textbf{5-Hexenoic acid}}{CH_2{=}CH(CH_2)_3\overset{O}{\overset{\|}{C}}OH} \xrightarrow[-\ HOH]{NaOH} \underset{\textbf{Sodium 5-hexenoate}}{CH_2{=}CH(CH_2)_3\overset{O}{\overset{\|}{C}}O^-\,Na^+}$$

In other transformations, however, the chemical behavior typical of one functional group is drastically altered by the presence of another. This situation is particularly likely if the groups are in close proximity. For example, α,β-unsaturated carbonyl compounds may undergo nucleophilic addition to the C–C double bond (1,4-addition; Sections 18-9 through 18-12).

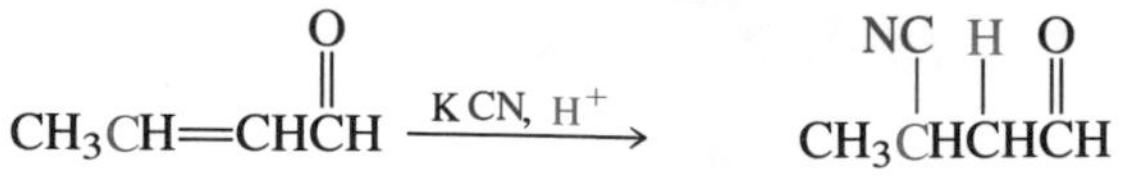

$$\underset{\textbf{2-Butenal}}{CH_3CH{=}CH\overset{O}{\overset{\|}{C}}H} \xrightarrow{KCN,\ H^+} \underset{\textbf{2-Methyl-4-oxobutanenitrile (3-Cyanobutanal)}}{CH_3\overset{NC}{\overset{|}{C}}H\overset{H}{\overset{|}{C}}H\overset{O}{\overset{\|}{C}}H}$$

In sharp contrast, the nonconjugated analogs undergo 1,2-addition to the C–O function (Chapter 17); the C–C double bond is immune to nucleophilic attack.

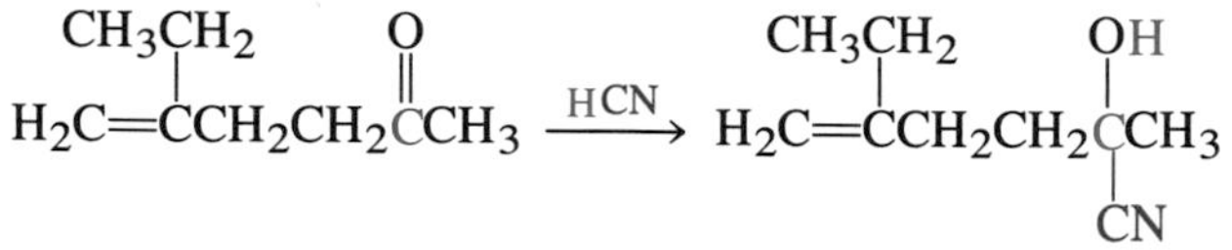

$$\underset{\textbf{5-Ethyl-5-hexen-2-one}}{H_2C{=}\overset{CH_3CH_2}{\overset{|}{C}}CH_2CH_2\overset{O}{\overset{\|}{C}}CH_3} \xrightarrow{HCN} \underset{\textbf{5-Ethyl-2-hydroxy-2-methyl-5-hexenenitrile}}{H_2C{=}\overset{CH_3CH_2}{\overset{|}{C}}CH_2CH_2\overset{OH}{\overset{|}{\underset{CN}{\underset{|}{C}}}}CH_3}$$

This chapter focuses on **dicarbonyl compounds**—molecules bearing two carbonyl groups. Several of these compounds have common names, in which Greek letters designate whether the two carbonyl carbons are adjacent or separated by one or more atoms.

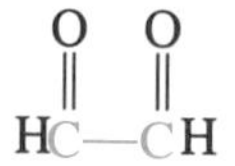

$\overset{O}{\overset{\|}{HC}}-\overset{O}{\overset{\|}{CH}}$
Ethanedial (Glyoxal)
(An α-dicarbonyl compound)

$C_6H_5-\overset{O}{\overset{\|}{C}}-\overset{O}{\overset{\|}{C}}-C_6H_5$
Diphenylethanedione (Benzil)
(An α-dicarbonyl compound)

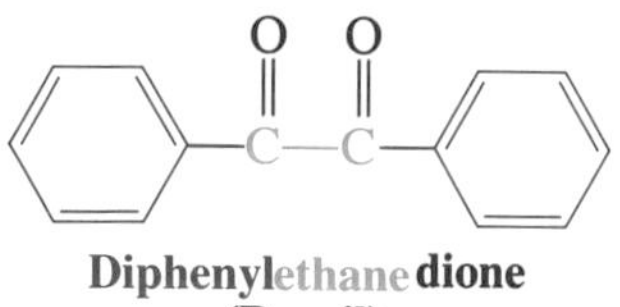

$HO\overset{O}{\overset{\|}{C}}-\overset{O}{\overset{\|}{C}}OH$
Ethanedioic acid (Oxalic acid)
(An α-dicarbonyl compound)

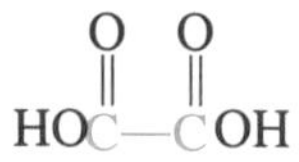

$CH_3\overset{O}{\overset{\|}{C}}CH_2\overset{O}{\overset{\|}{C}}CH_3$
2,4-Pentanedione (Acetylacetone)
(A β-dicarbonyl compound)

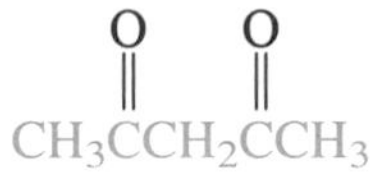

$CH_3\overset{O}{\overset{\|}{C}}CH_2\overset{O}{\overset{\|}{C}}OCH_3$
Methyl 3-oxobutanoate (Methyl acetoacetate)
(A β-dicarbonyl compound)

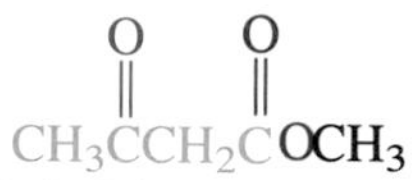

$HO\overset{O}{\overset{\|}{C}}CH_2\overset{O}{\overset{\|}{C}}OH$
Propanedioic acid (Malonic acid)
(A β-dicarbonyl compound)

$CH_3\overset{O}{\overset{\|}{C}}CH_2CH_2\overset{O}{\overset{\|}{C}}H$
4-Oxopentanal
(A γ-dicarbonyl compound)

If the two carbonyl functions are separated by *more than one carbon,* as in 4-oxopentanal, their chemistry can be independent. However, the α- and β-difunctional systems exhibit novel behavior. We begin with a description of the Claisen condensation, a route to the β-dicarbonyl compounds, which are important for their versatility in synthesis. We then discuss α-dicarbonyl compounds and the preparation of their α-hydroxycarbonyl precursors.

23-1 β-Dicarbonyl Compounds: Claisen Condensations

Ester enolates undergo addition–elimination reactions with ester functions, furnishing β-keto esters. These transformations, known as **Claisen* condensations,** are the ester analogs of the aldol reaction (Section 18-5).

*This is the Claisen of the Claisen rearrangement (Section 22-7).

Claisen condensations form β-dicarbonyl compounds

Ethyl acetate reacts with a stoichiometric amount of sodium ethoxide to give ethyl 3-oxobutanoate (ethyl acetoacetate).

Claisen Condensation of Ethyl Acetate

$$\underset{\textbf{Ethyl acetate}}{CH_3\overset{\overset{\displaystyle O}{\|}}{C}OCH_2CH_3 + CH_3\overset{\overset{\displaystyle O}{\|}}{C}OCH_2CH_3} \xrightarrow[-\ CH_3CH_2OH]{Na^{+}\,{}^{-}OCH_2CH_3,\ CH_3CH_2OH} \underset{\substack{75\% \\ \textbf{Ethyl 3-oxobutanoate} \\ \textbf{(Ethyl acetoacetate)}}}{CH_3\overset{\overset{\displaystyle O}{\|}}{C}CH_2\overset{\overset{\displaystyle O}{\|}}{C}OCH_2CH_3}$$

The Claisen condensation begins with formation of the ester enolate ion. Addition–elimination of this species to the carbonyl group of another ester molecule furnishes a 3-ketoester. These steps are reversible, and the overall process is endothermic to this stage. Indeed, the initial deprotonation of the ester ($pK_a \sim 25$) by ethoxide is particularly unfavorable (pK_a of ethanol = 15.9). Nonetheless, the equilibrium shifts to the product, because the base deprotonates C2 of the 3-ketoester ($pK_a \sim 11$) essentially irreversibly. Work-up with aqueous acid protonates the conjugate base of the latter, completing the process.

Mechanism of the Claisen Condensation

STEP 1. Ester enolate formation

$$CH_3\overset{\overset{\displaystyle O}{\|}}{C}OCH_2CH_3 \xrightleftharpoons{Na^{+}\,{}^{-}OCH_2CH_3} Na^{+}\left[{}^{-}\!:CH_2\!-\!C(=\ddot{O}:)(\ddot{O}CH_2CH_3) \longleftrightarrow CH_2\!=\!C(-\ddot{O}:^{-})(\ddot{O}CH_2CH_3)\right] + CH_3CH_2OH$$

STEP 2. Nucleophilic addition

$$CH_3\overset{\overset{\displaystyle :\ddot{O}:}{\|}}{C}OCH_2CH_3 + {}^{-}\!:CH_2\overset{\overset{\displaystyle O}{\|}}{C}OCH_2CH_3 \rightleftharpoons CH_3\!-\!\underset{\underset{\displaystyle CH_2\overset{}{C}(=O)OCH_2CH_3}{|}}{\overset{\overset{\displaystyle :\ddot{O}:^{-}}{|}}{C}}\!-\!OCH_2CH_3$$

STEP 3. Elimination

$$CH_3\underset{\underset{\displaystyle CH_2C(=O)OCH_2CH_3}{|}}{\overset{\overset{\displaystyle :\ddot{O}:^{-}}{|}}{C}}\!-\!OCH_2CH_3 \rightleftharpoons \underset{\textbf{3-Ketoester}}{CH_3\overset{\overset{\displaystyle O}{\|}}{C}CH_2\overset{\overset{\displaystyle O}{\|}}{C}OCH_2CH_3} + {}^{-}\!:\ddot{O}CH_2CH_3$$

STEP 4. Deprotonation

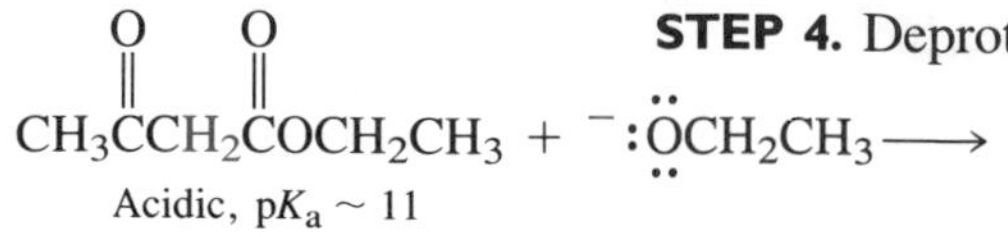

$$\mathrm{CH_3\overset{O}{\overset{\|}{C}}CH_2\overset{O}{\overset{\|}{C}}OCH_2CH_3} + {}^{-}\!:\!\ddot{\underset{..}{O}}CH_2CH_3 \longrightarrow$$

Acidic, pK_a ~ 11

$$\left[\mathrm{CH_3\overset{O}{\overset{\|}{C}}-\overset{..}{\bar{C}}H-\overset{O}{\overset{\|}{C}}OCH_2CH_3} \longleftrightarrow \mathrm{CH_3\overset{:\ddot{O}:^-}{\overset{|}{C}}=CH-\overset{O}{\overset{\|}{C}}OCH_2CH_3} \longleftrightarrow \mathrm{CH_3\overset{O}{\overset{\|}{C}}-CH=\overset{:\ddot{O}:^-}{\overset{|}{C}}OCH_2CH_3}\right] + \mathrm{CH_3CH_2OH}$$

STEP 5. Protonation on aqueous work-up

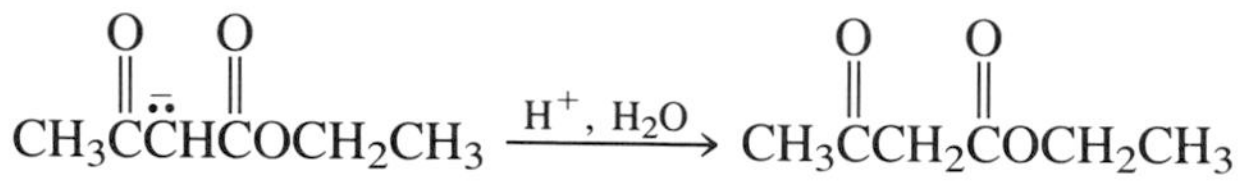

$$\mathrm{CH_3\overset{O}{\overset{\|}{C}}\overset{..}{\bar{C}}H\overset{O}{\overset{\|}{C}}OCH_2CH_3} \xrightarrow{H^+,\ H_2O} \mathrm{CH_3\overset{O}{\overset{\|}{C}}CH_2\overset{O}{\overset{\|}{C}}OCH_2CH_3}$$

To prevent transesterification, both the alkoxide and the ester should be derived from the same alcohol.

EXERCISE 23-1

Give the products of Claisen condensation of **(a)** ethyl propanoate; **(b)** ethyl 3-methylbutanoate; **(c)** ethyl pentanoate. For each, the base is sodium ethoxide, the solvent ethanol.

Protons flanked by two carbonyl groups are acidic

Why is deprotonation of the 3-ketoester so favorable? The acidity of the hydrogens flanked by the two carbonyl groups is much enhanced by resonance stabilization of the corresponding anion. Table 23-1 lists the pK_a values of several β-dicarbonyl compounds and related systems, such as methyl cyanoacetate and propanedinitrile (malonodinitrile).

The importance of the final deprotonation step in the Claisen condensation is clearly apparent if the ester bears only one α-hydrogen. The product of a reaction with such an ester would be a 2,2-disubstituted 3-ketoester lacking any of the acidic protons necessary to drive the equilibrium. Hence, no Claisen condensation product is observed.

TABLE 23-1 pK_a Values for β-Dicarbonyl and Related Compounds

Name	Structure	pK_a
2,4-Pentanedione (Acetylacetone)	$\mathrm{CH_3\overset{O}{\overset{\|}{C}}CH_2\overset{O}{\overset{\|}{C}}CH_3}$	9
Methyl 2-cyanoacetate	$\mathrm{NCCH_2\overset{O}{\overset{\|}{C}}OCH_3}$	9
Ethyl 3-oxobutanoate (Ethyl acetoacetate)	$\mathrm{CH_3\overset{O}{\overset{\|}{C}}CH_2\overset{O}{\overset{\|}{C}}OCH_2CH_3}$	11
Propanedinitrile (Malonodinitrile)	$\mathrm{NCCH_2CN}$	13
Diethyl propanedioate (Diethyl malonate)	$\mathrm{CH_3CH_2O\overset{O}{\overset{\|}{C}}CH_2\overset{O}{\overset{\|}{C}}OCH_2CH_3}$	13

Failure of a Claisen Condensation

$$2\ (CH_3)_2CH\overset{\overset{O}{\|}}{C}OCH_2CH_3 \xrightleftharpoons{Na^{+}\,{}^{-}OCH_2CH_3,\ CH_3CH_2OH} (CH_3)_2CH\overset{\overset{O}{\|}}{C}-\underset{\underset{CH_3}{|}}{\overset{\overset{CH_3}{|}}{C}}-\overset{\overset{O}{\|}}{C}OCH_2CH_3 + CH_3CH_2OH$$

Ethyl 2-methylpropanoate **Ethyl 2,2,4-trimethyl-3-oxopentanoate**

That this result is due to an unfavorable equilibrium may be demonstrated by treating a 2,2-disubstituted 3-ketoester with base: A **retro-Claisen condensation** takes place, proceeding through a mechanism that is the exact reverse of the forward reaction.

Retro-Claisen Condensation

$$(CH_3)_2CH\overset{\overset{:O:}{\|}}{C}-\underset{\underset{CH_3}{|}}{\overset{\overset{CH_3}{|}}{C}}-\overset{\overset{:O:}{\|}}{C}\ddot{O}CH_2CH_3 + CH_3CH_2\ddot{O}:^- \xrightarrow{CH_3CH_2\ddot{O}H} (CH_3)_2CH\underset{\underset{CH_3CH_2\ddot{O}:}{|}}{\overset{\overset{:\ddot{O}:^-}{|}}{C}}-\underset{\underset{CH_3}{|}}{\overset{\overset{CH_3}{|}}{C}}-\overset{\overset{:\ddot{O}:}{\|}}{C}OCH_2CH_3 \longrightarrow$$

$$(CH_3)_2CH\overset{\overset{:O:}{\|}}{C}\ddot{O}CH_2CH_3 + (CH_3)_2C{=}\overset{\overset{:\ddot{O}:^-}{|}}{C}\ddot{O}CH_2CH_3 \xrightleftharpoons{CH_3CH_2\ddot{O}H} (CH_3)_2CH\overset{\overset{:O:}{\|}}{C}\ddot{O}CH_2CH_3 + CH_3CH_2\ddot{O}:^-$$

EXERCISE 23-2

Explain the following observation.

$$2\ CH_3\overset{\overset{O}{\|}}{C}-\underset{\underset{CH_3}{|}}{\overset{\overset{CH_3}{|}}{C}}-COOCH_3 \xrightarrow[2.\ H^+,\ H_2O]{1.\ CH_3O^-Na^+,\ CH_3OH} CH_3\overset{\overset{O}{\|}}{C}CH_2COOCH_3 + 2\ (CH_3)_2CHCOOCH_3$$

Claisen condensations can involve two different esters

Mixed Claisen condensations involve two different esters. Like crossed aldol condensations (Section 18-6), they are typically unselective and furnish product mixtures. However, a selective mixed condensation is possible when one of the reacting partners has no α-hydrogens, as in ethyl formate or ethyl benzoate.

Selective Mixed Claisen Condensations

$$H\overset{\overset{O}{\|}}{C}OCH_2CH_3 + CH_3\overset{\overset{O}{\|}}{C}OCH_2CH_3 \xrightarrow[2.\ H^+,\ H_2O]{1.\ CH_3CH_2O^-Na^+,\ CH_3CH_2OH} H\overset{\overset{O}{\|}}{C}CH_2\overset{\overset{O}{\|}}{C}OCH_2CH_3$$

Ethyl formate 80% **Ethyl 3-oxopropanoate**

$$C_6H_5CO_2CH_2CH_3 + CH_3CH_2\overset{\overset{O}{\|}}{C}OCH_2CH_3 \xrightarrow[2.\ H^+,\ H_2O]{1.\ CH_3CH_2O^-Na^+,\ CH_3CH_2OH} C_6H_5\overset{\overset{O}{\|}}{C}\underset{\underset{CH_3}{|}}{C}H\overset{\overset{O}{\|}}{C}OCH_2CH_3$$

71%

Ethyl benzoate **Ethyl 2-methyl-3-oxo-3-phenylpropanoate**

EXERCISE 23-3

Give all the Claisen condensation products that would result from treatment of a mixture of ethyl acetate and ethyl propanoate with sodium ethoxide in ethanol.

Intramolecular and double Claisen condensations result in cyclic compounds

The intramolecular version of the Claisen reaction, called the **Dieckmann* condensation,** produces cyclic 3-ketoesters. As expected (Section 9-6), it works best for the formation of five- and six-membered rings.

$$CH_3CH_2O\overset{O}{\overset{\|}{C}}CH_2CH_2CH_2CH_2CH_2\overset{O}{\overset{\|}{C}}OCH_2CH_3 \xrightarrow[2.\ H^+,\ H_2O]{1.\ CH_3CH_2O^-Na^+,\ CH_3CH_2OH} \text{(cyclohexanone with } CO_2CH_2CH_3\text{)}$$

Diethyl heptanedioate → **Ethyl 2-oxocyclohexanecarboxylate** (60%)

Cyclic compounds may also be obtained by (intermolecular followed by intramolecular) **double Claisen condensations** with diesters such as diethyl ethanedioate (diethyl oxalate).

$$CH_3CH_2O\overset{OO}{\overset{\|\|}{CC}}OCH_2CH_3 + CH_3CH_2O\overset{O}{\overset{\|}{C}}(CH_2)_3\overset{O}{\overset{\|}{C}}OCH_2CH_3 \xrightarrow[2.\ H^+,\ H_2O]{1.\ CH_3CH_2O^-Na^+,\ CH_3CH_2OH} \text{(cyclopentanedione with two } CO_2CH_2CH_3\text{)}$$

Diethyl ethanedioate (Diethyl oxalate) + **Diethyl pentanedioate** → **Diethyl 4,5-dioxo-1,3-cyclopentanedicarboxylate** (80%)

EXERCISE 23-4

Formulate a mechanism for the reaction of diethyl ethanedioate with diethyl pentanedioate.

EXERCISE 23-5

Formulate a mechanism for the following reaction.

$$\text{Diethyl 1,2-benzenedicarboxylate} + CH_3CO_2CH_2CH_3 \xrightarrow[2.\ H^+,\ H_2O]{1.\ CH_3CH_2O^-Na^+,\ CH_3CH_2OH} \text{(1,3-indandione with } CO_2CH_2CH_3\text{ at C-2)}$$

Diethyl 1,2-benzenedicarboxylate (Diethyl phthalate) (benzene ring bearing two $CO_2CH_2CH_3$ groups)

60–80%

*Professor Walter Dieckmann, 1869–1925, University of Munich, Germany.

Ketones undergo mixed Claisen reactions

Ketones can participate in the Claisen condensation. Because they are more acidic than esters, they are deprotonated before the ester has a chance to undergo self-condensation. The products (after acidic work-up) may be β-diketones, β-ketoaldehydes, or other β-dicarbonyl compounds. The reaction can be carried out with a variety of ketones and esters both inter- and intramolecularly.

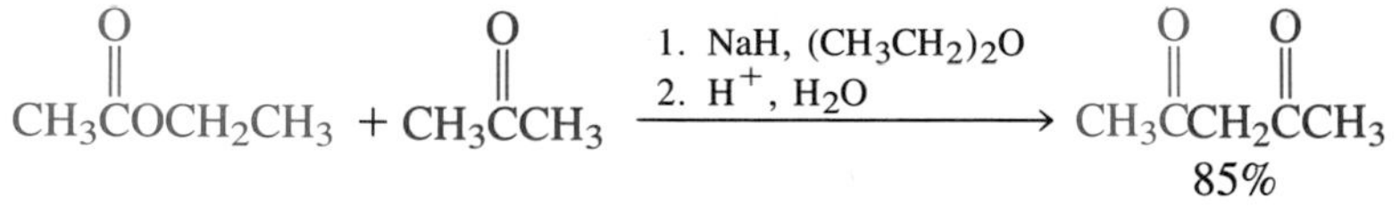

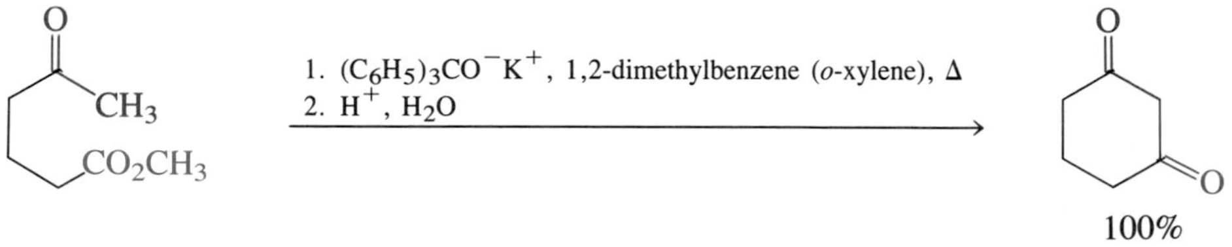

$$CH_3\overset{O}{\overset{\|}{C}}OCH_2CH_3 + CH_3\overset{O}{\overset{\|}{C}}CH_3 \xrightarrow[2.\ H^+,\ H_2O]{1.\ NaH,\ (CH_3CH_2)_2O} \underset{85\%}{CH_3\overset{O}{\overset{\|}{C}}CH_2\overset{O}{\overset{\|}{C}}CH_3}$$

1. $(C_6H_5)_3CO^-K^+$, 1,2-dimethylbenzene (*o*-xylene), Δ
2. H^+, H_2O

Methyl 5-oxohexanoate → 100% **1,3-Cyclohexanedione**

BOX 23-1 Claisen Condensations in Biochemistry

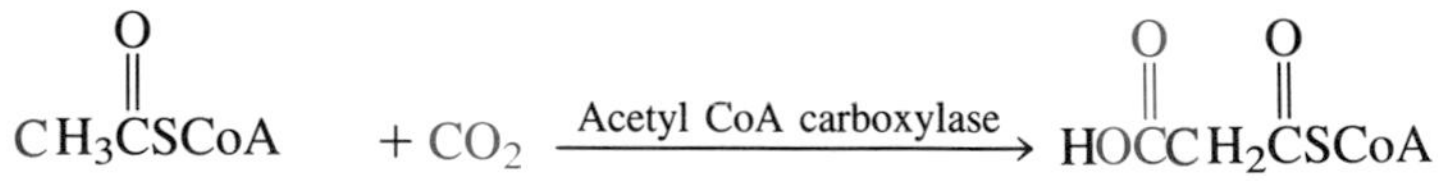

$$\underset{\textbf{Acetyl coenzyme A}}{CH_3\overset{O}{\overset{\|}{C}}SCoA} + CO_2 \xrightarrow{\text{Acetyl CoA carboxylase}} \underset{\textbf{Malonyl CoA}}{HO\overset{O}{\overset{\|}{C}}CH_2\overset{O}{\overset{\|}{C}}SCoA}$$

The coupling processes that build fatty acid chains from thioesters of coenzyme A (Section 19-14) are forms of Claisen condensations. The carboxylation of acetyl CoA into malonyl CoA (shown above) is a variant in which the carbon of CO_2 rather than that of an ester carbonyl group is the site of nucleophilic attack.

The methylene group in the carboxylated species is much more reactive than the methyl group in acetyl thioesters and participates in a wide variety of Claisenlike condensations. Although these processes require enzyme catalysis, they may be formulated in simplified form, shown below.

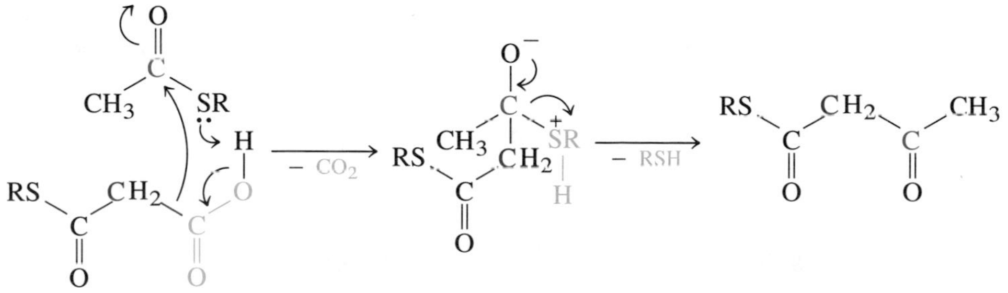

$$\xrightarrow{-\,CO_2} \qquad \xrightarrow{-\,RSH} \quad RS\overset{O}{\overset{\|}{C}}CH_2\overset{O}{\overset{\|}{C}}CH_3$$

(RSH = acyl carrier protein; see Section 19-14)

Formates lead to 3-ketoaldehydes.

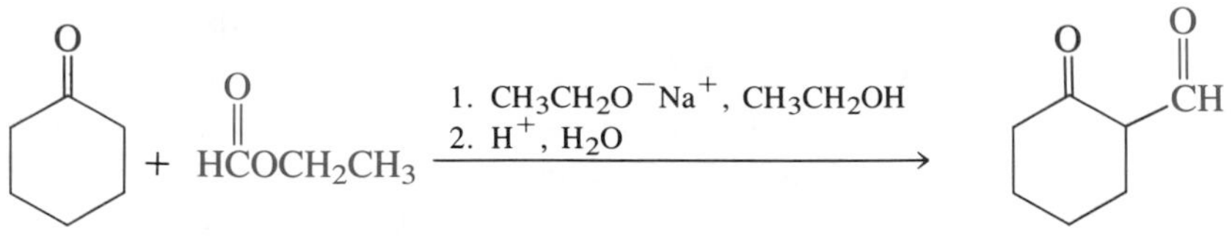

$$\text{Cyclohexanone} + \underset{\textbf{Ethyl formate}}{\overset{O}{\overset{\|}{H C}}OCH_2CH_3} \xrightarrow[\text{2. } H^+,\ H_2O]{\text{1. } CH_3CH_2O^-Na^+,\ CH_3CH_2OH} \underset{\textbf{2-Formylcyclohexanone}}{\underset{75\%}{\text{2-(}\overset{O}{\overset{\|}{C}}H\text{)cyclohexanone}}}$$

EXERCISE 23-6

Suggest syntheses of the following molecules by Claisen or Dieckmann condensations.

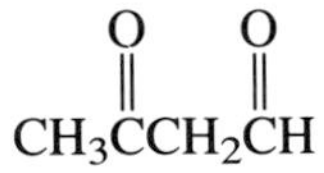

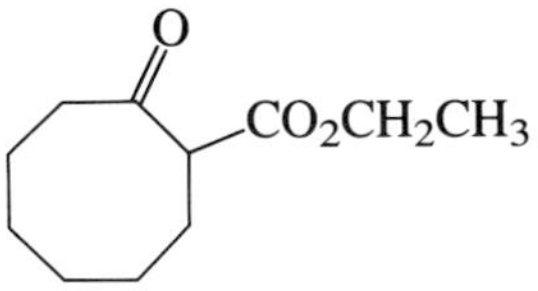

(a) 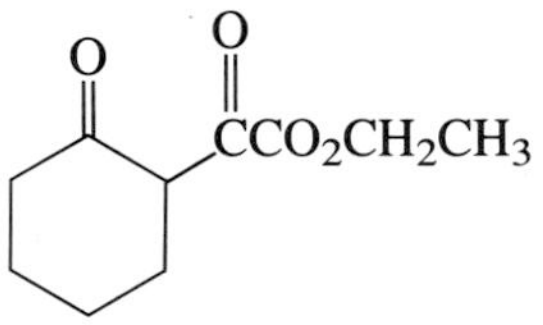 2-oxocyclohexyl–$\overset{O}{\overset{\|}{C}}CO_2CH_2CH_3$ **(b)** $CH_3\overset{O}{\overset{\|}{C}}CH_2\overset{O}{\overset{\|}{C}}H$

(c) 2-oxocyclooctyl–$CO_2CH_2CH_3$ **(d)** 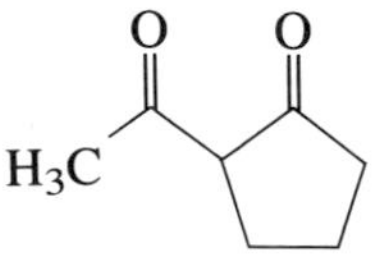 H_3C–$\overset{O}{\overset{\|}{C}}$–(2-oxocyclopentyl)

In summary, Claisen condensations are endothermic and therefore would not occur without a stoichiometric amount of base strong enough to deprotonate the resulting 3-ketoester. Mixed Claisen condensations between two esters are non-selective, unless they are intramolecular (Dieckmann condensation) or one of the components is devoid of α-hydrogens. Ketones also participate in selective mixed Claisen reactions because they are more acidic than esters.

23-2 β-Dicarbonyl Compounds as Synthetic Intermediates

Having seen how to prepare β-dicarbonyl compounds, let us explore their synthetic utility. This section will show that the corresponding anions are readily alkylated and that 3-ketoesters are hydrolyzed to the corresponding acids, which can be decarboxylated to give ketones or new carboxylic acids. These transformations open up versatile synthetic routes to other functionalized molecules.

β-Dicarbonyl anions are nucleophilic

The unusual acidity of β-ketocarbonyl compounds may be used to synthetic advantage, because the enolate ions obtained by deprotonation can be alkylated to give substituted derivatives. For example, in this way ethyl 3-oxobutanoate is readily converted into alkylated analogs.

β-Ketoester Alkylations

$$\underset{\text{Ethyl 3-oxobutanoate}}{CH_3\overset{O}{\overset{\|}{C}}\underset{H}{\underset{|}{C}}H\overset{O}{\overset{\|}{C}}OCH_2CH_3} \xrightarrow[\substack{-\,CH_3CH_2OH\\ -\,NaI}]{\substack{1.\ Na^{+-}OCH_2CH_3,\ CH_3CH_2OH\\ 2.\ CH_3I}} \underset{\substack{65\%\\ \text{Ethyl 2-methyl-3-oxobutanoate}}}{CH_3\overset{O}{\overset{\|}{C}}\underset{CH_3}{\underset{|}{C}}H\overset{O}{\overset{\|}{C}}OCH_2CH_3} \xrightarrow[\substack{-(CH_3)_3COH\\ -\,KBr}]{\substack{1.\ K^{+-}OC(CH_3)_3,\ (CH_3)_3COH\\ 2.\ C_6H_5CH_2Br}} \underset{\substack{77\%\\ \text{Ethyl 2-methyl-2-(phenylmethyl)-3-oxobutanoate}}}{CH_3\overset{O}{\overset{\|}{C}}-\overset{CH_3}{\overset{|}{\underset{CH_2C_6H_5}{\underset{|}{C}}}}-\overset{O}{\overset{\|}{C}}OCH_2CH_3}$$

Other β-dicarbonyl compounds undergo similar reactions.

$$\underset{\text{2,4-Pentanedione}}{CH_3\overset{O}{\overset{\|}{C}}\underset{H}{\underset{|}{C}}H\overset{O}{\overset{\|}{C}}CH_3} \xrightarrow[\substack{-\,KHCO_3\\ -\,KI}]{K_2CO_3,\ CH_3I,\ \text{propanone (acetone)},\ \Delta} \underset{\substack{77\%\\ \text{3-Methyl-2,4-pentanedione}}}{CH_3\overset{O}{\overset{\|}{C}}\underset{CH_3}{\underset{|}{C}}H\overset{O}{\overset{\|}{C}}CH_3}$$

$$\underset{\text{Diethyl propanedioate}}{CH_3CH_2O\overset{O}{\overset{\|}{C}}\underset{H}{\underset{|}{C}}H\overset{O}{\overset{\|}{C}}OCH_2CH_3} \xrightarrow[\substack{-\,CH_3CH_2OH\\ -\,NaBr}]{\substack{1.\ Na^{+-}OCH_2CH_3,\ CH_3CH_2OH\\ 2.\ CH_3CH_2CH(CH_3)Br}} \underset{\substack{84\%\\ \text{Diethyl 2-(1-methylpropyl)propanedioate}}}{CH_3CH_2O\overset{O}{\overset{\|}{C}}\underset{CH_3CH_2CHCH_3}{\underset{|}{C}}H\overset{O}{\overset{\|}{C}}OCH_2CH_3}$$

EXERCISE 23-7

Give a synthesis of 2,2-dimethyl-1,3-cyclohexanedione from methyl 5-oxohexanoate.

3-Ketoacids readily undergo decarboxylation

Hydrolysis of 3-ketoesters furnishes 3-ketoacids, which in turn readily undergo decarboxylation under mild conditions. The products, ketones and carboxylic acids, contain the alkyl groups introduced in prior alkylation steps.

Decarboxylation of 3-Ketoacids

$$R\overset{O}{\overset{\|}{C}}\underset{R'}{\underset{|}{C}}H\overset{O}{\overset{\|}{C}}OCH_2CH_3 \xrightarrow[\text{Hydrolysis}]{} \underset{\text{3-Ketoacid}}{R\overset{O}{\overset{\|}{C}}\underset{R'}{\underset{|}{C}}H\overset{O}{\overset{\|}{C}}OH} \xrightarrow{\Delta} \underset{\text{Ketone}}{R\overset{O}{\overset{\|}{C}}\underset{H}{\underset{|}{C}}HR'} + CO_2$$

$$\text{CH}_3\text{CH}_2\text{O}\overset{\text{O}}{\overset{\|}{\text{C}}}\underset{\text{R}}{\underset{|}{\text{C}}}\text{H}\overset{\text{O}}{\overset{\|}{\text{C}}}\text{OCH}_2\text{CH}_3 \xrightarrow[\text{Hydrolysis}]{} \text{HO}\overset{\text{O}}{\overset{\|}{\text{C}}}\underset{\text{R}}{\underset{|}{\text{C}}}\text{H}\overset{\text{O}}{\overset{\|}{\text{C}}}\text{OH} \xrightarrow{\Delta} \text{R}\underset{\text{H}}{\underset{|}{\text{C}}}\text{H}\overset{\text{O}}{\overset{\|}{\text{C}}}\text{OH} + \text{CO}_2$$

Carboxylic acid

$$\text{CH}_3\overset{\text{O}}{\overset{\|}{\text{C}}}\underset{(\text{CH}_2)_3\text{CH}_3}{\underset{|}{\text{C}}}\text{H}\overset{\text{O}}{\overset{\|}{\text{C}}}\text{OCH}_2\text{CH}_3 \xrightarrow[\substack{-\ \text{CH}_3\text{CH}_2\text{OH} \\ -\ \text{CO}_2}]{\substack{1.\ \text{NaOH, H}_2\text{O} \\ 2.\ \text{H}_2\text{SO}_4\text{, H}_2\text{O, 100°C}}} \text{CH}_3\overset{\text{O}}{\overset{\|}{\text{C}}}\underset{\text{H}}{\underset{|}{\text{C}}}\text{H}(\text{CH}_2)_3\text{CH}_3$$

60%

Ethyl 2-butyl-3-oxobutanoate **2-Heptanone**

$$\text{CH}_3\text{CH}_2\text{O}\overset{\text{O}}{\overset{\|}{\text{C}}}\underset{\text{CH}_3\text{CH}_2\text{CHCH}_3}{\underset{|}{\text{C}}}\text{H}\overset{\text{O}}{\overset{\|}{\text{C}}}\text{OCH}_2\text{CH}_3 \xrightarrow[\substack{-\ \text{CH}_3\text{CH}_2\text{OH} \\ -\ \text{CO}_2}]{\text{H}_2\text{SO}_4\text{, H}_2\text{O, }\Delta} \text{CH}_3\text{CH}_2\overset{\text{CH}_3}{\overset{|}{\text{C}}}\text{H}\underset{\text{H}}{\underset{|}{\text{C}}}\text{HCOOH}$$

65%

Diethyl 2-(1-methylpropyl)propanedioate **3-Methylpentanoic acid**

Decarboxylation has a concerted mechanism with a cyclic transition state, somewhat like that of the McLafferty rearrangement in mass spectroscopy (Section 20-10).

Mechanism of Decarboxylation of 3-Ketoacids

$$\text{H}_3\text{C}\overset{:\text{O}:}{\overset{\|}{\text{C}}}\text{CH}_2\text{C}(=\text{O})\text{OH} \longrightarrow \text{H}_3\text{C}\overset{:\ddot{\text{O}}\text{H}}{\overset{|}{\text{C}}}{=}\text{CH}_2 + \text{O}{=}\text{C}{=}\text{O} \longrightarrow \text{H}_3\text{C}\overset{\text{O}}{\overset{\|}{\text{C}}}\text{CH}_3 + \text{CO}_2$$

Loss of CO_2 can occur readily only from the free carboxylic acid. If the ester is hydrolyzed under basic conditions, the resulting carboxylate salt is usually neutralized with acid to enable subsequent decarboxylation. Decarboxylation of substituted propanedioic (malonic) acids follows the same mechanism.

EXERCISE 23-8

Formulate a detailed mechanism for the decarboxylation of $CH_3CH(COOH)_2$ (methylmalonic acid).

The acetoacetic ester synthesis leads to methyl ketones

The combination of alkylation followed by ester hydrolysis and finally decarboxylation allows ethyl 3-oxobutanoate (ethyl acetoacetate) to be converted ultimately into 3-substituted or 3,3-disubstituted methyl ketones. This strategy is called the **acetoacetic ester synthesis.**

Acetoacetic Ester Synthesis

$$CH_3\overset{O}{\overset{\|}{C}}CH_2\overset{O}{\overset{\|}{C}}OCH_2CH_3 \dashrightarrow CH_3\overset{O}{\overset{\|}{C}}-\underset{R'}{\overset{R}{C}}-\overset{O}{\overset{\|}{C}}OCH_2CH_3 \dashrightarrow CH_3\overset{O}{\overset{\|}{C}}CH\begin{matrix}R\\R'\end{matrix}$$

3,3-Disubstituted methyl ketone

Methyl ketones with either one or two substituent groups on C3 can be synthesized by using the acetoacetic sequence.

Syntheses of Substituted Methyl Ketones

$$CH_3\overset{O}{\overset{\|}{C}}CH_2\overset{O}{\overset{\|}{C}}OCH_2CH_3 \xrightarrow[\text{2. } CH_3CH_2CH_2CH_2Br]{\text{1. } NaOCH_2CH_3,\ CH_3CH_2OH} \underset{72\%}{CH_3\overset{O}{\overset{\|}{C}}\underset{CH_2CH_2CH_2CH_3}{CH}\overset{O}{\overset{\|}{C}}OCH_2CH_3} \xrightarrow[\text{2. } H_2SO_4,\ H_2O,\ 100°C]{\text{1. } NaOH,\ H_2O} \underset{60\%}{CH_3\overset{O}{\overset{\|}{C}}CH_2CH_2CH_2CH_2CH_3}$$

2-Heptanone

1. $KOC(CH_3)_3$, $(CH_3)_3COH$
2. $CH_3CH_2CH_2CH_2I$

$$\underset{80\%}{CH_3\overset{O}{\overset{\|}{C}}\underset{CH_3CH_2CH_2CH_2\quad CH_2CH_2CH_2CH_3}{C}\overset{O}{\overset{\|}{C}}OCH_2CH_3} \xrightarrow[\text{2. } HCl,\ H_2O,\ 100°C]{\text{1. } KOH,\ H_2O,\ 100°C} \underset{64\%}{CH_3\overset{O}{\overset{\|}{C}}-\overset{CH_2CH_2CH_2CH_3}{CH}CH_2CH_2CH_2CH_3}$$

3-Butyl-2-heptanone

EXERCISE 23-9

Propose syntheses of the following ketones, beginning with ethyl 3-oxobutanoate (ethyl acetoacetate). **(a)** 2-Hexanone; **(b)** 2-octanone; **(c)** 3-ethyl-2-pentanone; **(d)** 4-phenyl-2-butanone.

The malonic ester synthesis furnishes carboxylic acids

Diethyl propanedioate (malonic ester) is a good starting material for preparing 2-alkylated and 2,2-dialkylated acetic acids, a method called the **malonic ester synthesis.**

Malonic Ester Synthesis

$$CH_3CH_2O\overset{O}{\overset{\|}{C}}CH_2\overset{O}{\overset{\|}{C}}OCH_2CH_3 \dashrightarrow CH_3CH_2O\overset{O}{\overset{\|}{C}}-\underset{R'}{\overset{R}{C}}-\overset{O}{\overset{\|}{C}}OCH_2CH_3 \dashrightarrow H-\underset{R'}{\overset{R}{C}}-COOH$$

2,2-Dialkylated acetic acid

Like the acetoacetic ester route to ketones, the malonic ester synthesis can lead to carboxylic acids with either one or two substituents at C2.

Synthesis of a 2,2-Dialkylated Acetic Acid

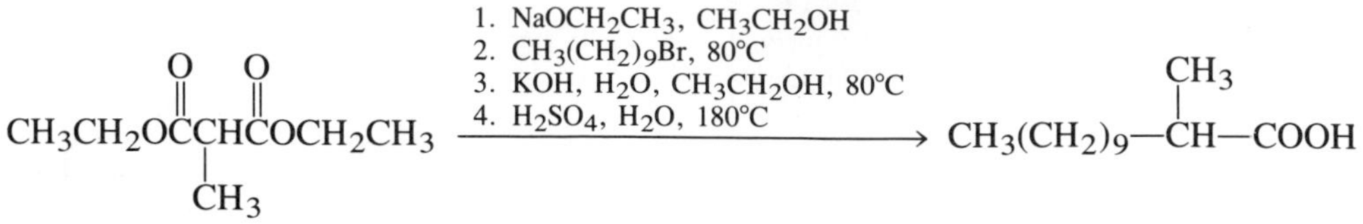

$$\underset{\text{Diethyl 2-methylpropanedioate (Diethyl methylmalonate)}}{CH_3CH_2O\overset{O}{\overset{\|}{C}}\underset{\underset{CH_3}{|}}{C}H\overset{O}{\overset{\|}{C}}OCH_2CH_3} \xrightarrow[\text{4. } H_2SO_4,\ H_2O,\ 180°C]{\text{1. } NaOCH_2CH_3,\ CH_3CH_2OH \quad \text{2. } CH_3(CH_2)_9Br,\ 80°C \quad \text{3. } KOH,\ H_2O,\ CH_3CH_2OH,\ 80°C} \underset{\text{2-Methyldodecanoic acid}}{\underset{74\%}{CH_3(CH_2)_9{-}\overset{CH_3}{\overset{|}{C}}H{-}COOH}}$$

Diethyl 2-methylpropanedioate (Diethyl methylmalonate)

2-Methyldodecanoic acid

EXERCISE 23-10

(a) Give the structure of the product formed after each of the first three steps in the synthesis of 2-methyldodecanoic acid shown above. **(b)** Propose a synthesis of the starting material for this synthesis, diethyl 2-methylpropanedioate.

The rules and limitations governing S_N2 reactions apply to the alkylation steps. Thus, tertiary haloalkanes exposed to β-dicarbonyl anions give mainly elimination products. However, the anions can be successfully attacked by alkanoyl halides, α-bromoesters, α-bromoketones, and oxacyclopropanes.

EXERCISE 23-11

The first-mentioned compound in each of the following parts is treated with the subsequent series of reagents. Give the final products.

(a) $CH_3CH_2O_2C(CH_2)_5CO_2CH_2CH_3$: $NaOCH_2CH_3$; $CH_3(CH_2)_3I$; NaOH; and H^+, H_2O, Δ

(b) $CH_3CH_2O_2CCH_2CO_2CH_2CH_3$: $NaOCH_2CH_3$; CH_3I; KOH; and H^+, H_2O, Δ

(c) $CH_3\overset{O}{\overset{\|}{C}}\underset{\underset{CH_3}{|}}{C}HCO_2CH_3$: NaH, C_6H_6; $C_6H_5\overset{O}{\overset{\|}{C}}Cl$; and H^+, H_2O, Δ

(d) $CH_3\overset{O}{\overset{\|}{C}}CH_2CO_2CH_2CH_3$: $NaOCH_2CH_3$; $BrCH_2CO_2CH_2CH_3$; NaOH; and H^+, H_2O, Δ

(e) $CH_3CH_2CH(CO_2CH_2CH_3)_2$: $NaOCH_2CH_3$; $BrCH_2CO_2CH_2CH_3$; and H^+, H_2O, Δ

(f) $CH_3\overset{O}{\overset{\|}{C}}CH_2CO_2CH_2CH_3$: $NaOCH_2CH_3$; $BrCH_2\overset{O}{\overset{\|}{C}}CH_3$; and H^+, H_2O, Δ

EXERCISE 23-12

Propose a synthesis of cyclohexanecarboxylic acid from diethyl propanedioate (malonate), $CH_2(CO_2CH_2CH_3)_2$, and 1-bromo-5-chloropentane, $Br(CH_2)_5Cl$.

In summary, β-dicarbonyl compounds such as ethyl 3-oxobutanoate (acetoacetate) and diethyl propanedioate (malonate) are versatile synthetic building blocks for elaborating more complex molecules. Their unusual acidity makes it easy to form the corresponding anions, which can be used in nucleophilic displacement reactions with a wide variety of substrates. Their hydrolysis produces 3-ketoacids that are unstable and undergo decarboxylation on heating.

23-3 β-Dicarbonyl Anion Chemistry: Michael Additions

Reaction of the stabilized anions derived from β-dicarbonyl compounds and related analogs (Table 23-1) with α,β-unsaturated carbonyl compounds leads to 1,4-additions. This transformation, an example of **Michael addition** (Section 18-12), requires only a catalytic amount of base and works with α,β-unsaturated ketones, aldehydes, nitriles, and carboxylic acid derivatives, all of which are termed **Michael acceptors.**

Michael Addition

$$CH_2(CO_2CH_2CH_3)_2 + CH_2{=}CH\overset{O}{\overset{\|}{C}}CH_3 \xrightarrow[CH_3CH_2OH,\ -10\text{ to }25^\circ C]{\text{Catalytic } CH_3CH_2O^-Na^+,} (CH_3CH_2O_2C)_2CHCH_2CH_2\overset{O}{\overset{\|}{C}}CH_3$$

3-Buten-2-one (Methyl vinyl ketone) (Michael acceptor)

71% **Diethyl 2-(3-oxobutyl)propanedioate**

Why do stabilized anions undergo 1,4- rather than 1,2-addition to Michael acceptors? The latter process occurs but is reversible with relatively stable anionic nucleophiles, because it leads to a relatively high-energy alkoxide intermediate. Conjugate addition is favored thermodynamically, because it produces a resonance-stabilized enolate ion.

EXERCISE 23-13

Formulate a detailed mechanism for the Michael addition process depicted above. Why is the base required in only catalytic amounts?

EXERCISE 23-14

Give the products of the following Michael additions [base in square brackets].

(a) $CH_3CH_2CH(CO_2CH_2CH_3)_2 + CH_2{=}CH\overset{O}{\overset{\|}{C}}H\ [Na^{+\,-}OCH_2CH_3]$

(b) (5,5-dimethyl-1,3-cyclohexanedione) $+\ CH_2{=}CHC{\equiv}N\ [Na^{+\,-}OCH_3]$

(c) 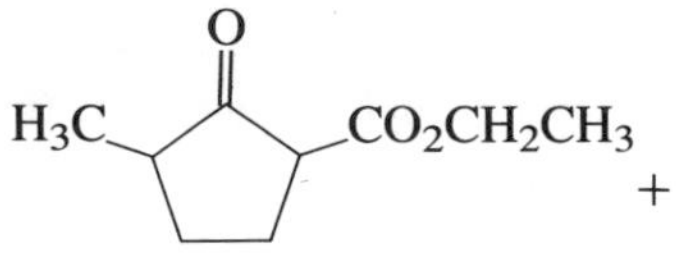 H_3C–(2-oxocyclopentyl)–$CO_2CH_2CH_3$ + $CH_3CH{=}CHCO_2CH_2CH_3$ [$K^{+-}OCH_2CH_3$]

EXERCISE 23-15

Explain the following observation. (Hint: Consider proton transfer in the first Michael adduct.)

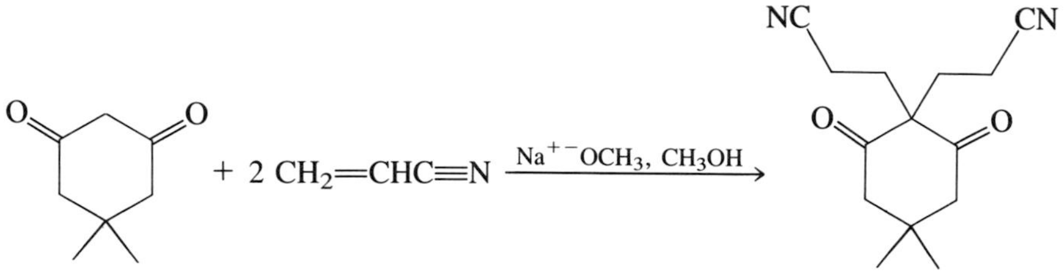

$$\text{5,5-dimethylcyclohexane-1,3-dione} + 2\ CH_2{=}CHC{\equiv}N \xrightarrow{Na^{+-}OCH_3,\ CH_3OH} \text{2,2-bis(2-cyanoethyl)-5,5-dimethylcyclohexane-1,3-dione}\ (81\%)$$

A useful synthetic application of Michael addition of anions of β-ketoesters to α,β-unsaturated ketones is shown below. The process leads to a diketone in which the enolate of one is positioned to form a six-membered ring upon aldol condensation with the carbonyl group of the other. Recall (Section 18-12) that the synthesis of six-membered rings by Michael addition followed by aldol condensation is called **Robinson annulation.**

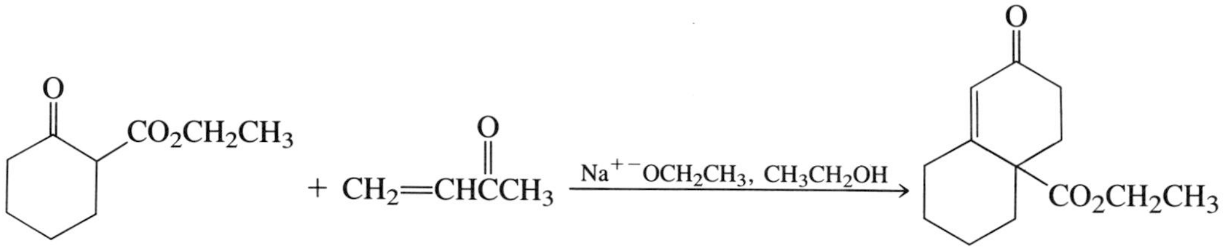

$$\text{2-}(CO_2CH_2CH_3)\text{cyclohexanone} + CH_2{=}CH\overset{O}{\overset{\|}{C}}CH_3 \xrightarrow{Na^{+-}OCH_2CH_3,\ CH_3CH_2OH} \text{bicyclic enone with }CO_2CH_2CH_3\ (70\%)$$

EXERCISE 23-16

Formulate a detailed mechanism for the transformation shown above.

In summary, β-dicarbonyl anions, like ordinary enolate anions, undergo Michael additions to α,β-unsaturated carbonyl compounds. Addition of a β-ketoester to enones gives a diketone, which can generate six-membered rings by intramolecular aldol condensation (Robinson annulation).

23-4 α-Dicarbonyl Compounds

This section discusses the preparation and reactions of α-dicarbonyl compounds, molecules in which the C=O groups strongly influence each other's reactivity.

α-Diketones and α-ketoaldehydes are made by oxidation of α-hydroxycarbonyl compounds

α-Diketones and α-ketoaldehydes are frequently obtained by oxidation of α-hydroxycarbonyl precursors. Because of the sensitivity of the products, special reagents and reaction conditions are required to prevent side reactions. A relatively simple example is $KMnO_4$ in acetic anhydride solvent, which converts α-hydroxyketones into the desired products as shown in the conversion of 2-hydroxy-1,2-diphenylethanone (benzoin) into diphenylethanedione (benzil).

$$\xrightarrow{KMnO_4,\ CH_3COCCH_3\ (solvent),\ 5°C}$$

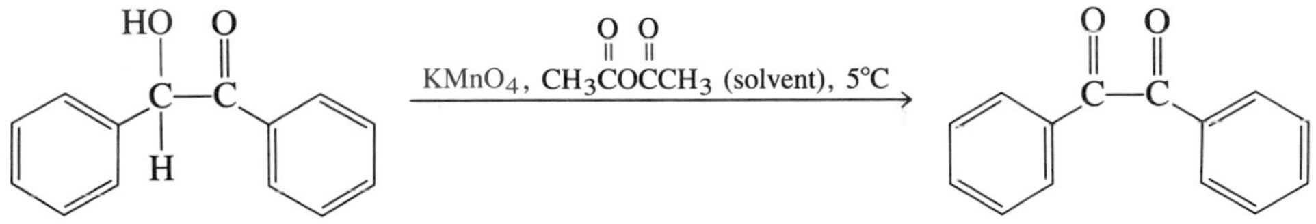

73%

2-Hydroxy-1,2-diphenylethanone (Benzoin) → **Diphenylethanedione (Benzil)**

A milder oxidizing agent is cupric acetate in aqueous acetic acid. The mechanism of its action is complex and includes the transfer of electrons to the metal.

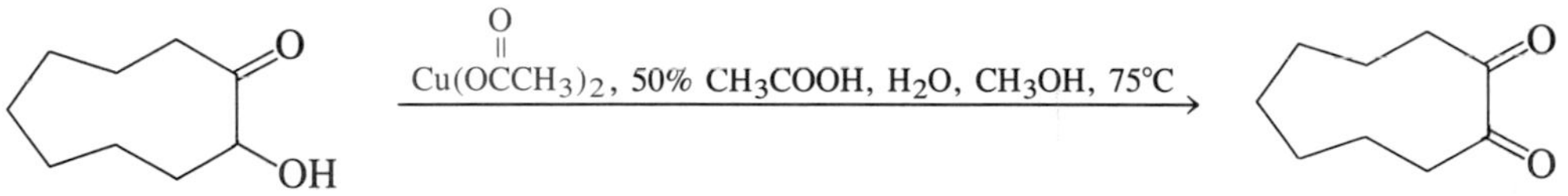

$$\xrightarrow{Cu(OCCH_3)_2,\ 50\%\ CH_3COOH,\ H_2O,\ CH_3OH,\ 75°C}$$

75%

2-Hydroxycyclononanone → **1,2-Cyclononanedione**

α-Dicarbonyl compounds are unusually reactive

The reactivity of α-dicarbonyl compounds derives from the proximity of the two carbonyl double bonds. Each enhances the reactivity of the other toward nucleophilic attack. Ethanedial (glyoxal), for example, is readily hydrated, and it is quite difficult to completely dehydrate this compound, unlike normal aldehydes and ketones. Treatment with sodium hydroxide induces a rearrangement that produces the sodium salt of hydroxyacetic acid (glycolic acid).

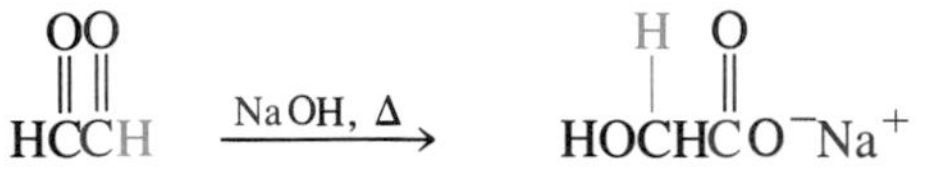

$$HCCH \xrightarrow{NaOH,\ \Delta} HOCHCO^-Na^+$$

Ethanedial (Glyoxal) → **Sodium hydroxyacetate (Glycolic acid sodium salt)**

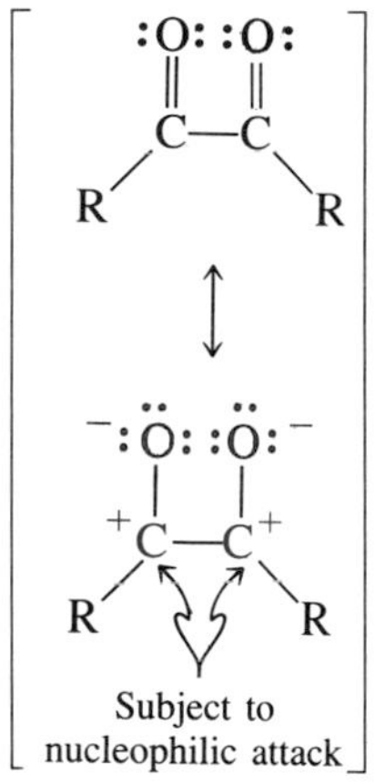

This reaction is general for other α-dicarbonyl derivatives, such as α-ketoaldehydes and α-diketones. It has been named the **benzilic acid rearrangement,** because diphenylethanedione (benzil) is converted into diphenylhydroxyacetic (benzilic) acid under these conditions.

Activation of Dicarbonyl Compounds

Subject to nucleophilic attack

Benzilic Acid Rearrangement

$$\xrightarrow{KOH,\ H_2O,\ CH_3CH_2OH,\ 100°C}$$

OH — C — COO⁻K⁺

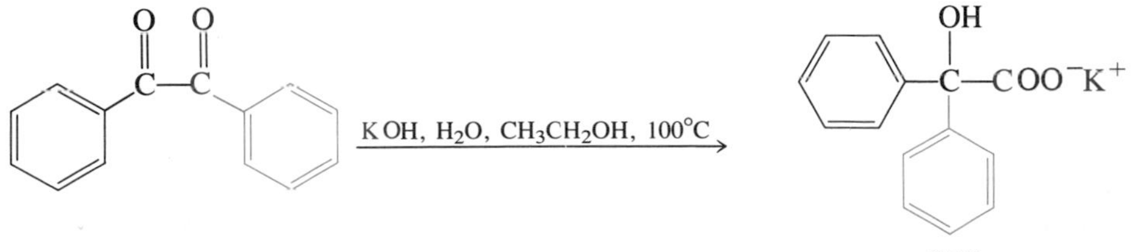

95%

Diphenylethanedione (Benzil) → **Potassium diphenylhydroxyacetate (Benzilic acid)**

How does this rearrangement proceed? The mechanism of the benzilic acid synthesis begins with nucleophilic addition of hydroxide ion to one of the activated carbonyl carbons. Subsequently, the molecule reorganizes by migration of the substituent on the alkoxide carbon to the neighboring carbonyl function. The migrating group takes with it its electron pair, which is donated to the neighboring carbonyl double bond. Proton transfer completes the sequence.

Mechanism of the Benzilic Acid Rearrangement

STEP 1. Addition of hydroxide ion

$$\mathrm{RC(=O){-}C(=O)R} + {}^{-}\!:\!\ddot{\mathrm{O}}\mathrm{H} \rightleftharpoons \mathrm{HOC(R)(O^-){-}C(=O)R}$$

STEP 2. Rearrangement

$$\mathrm{HOC(R)(O^-){-}C(=O)R} \longrightarrow \mathrm{HOC(=O){-}C(R)(O^-)R}$$

STEP 3. Proton transfer

$$\mathrm{HOC(=O){-}C(R)(O^-)R} \rightleftharpoons {}^{-}\!:\!\mathrm{OC(=O){-}C(R)(OH)R}$$

The second step in the benzilic acid rearrangement is, in principle, reversible. However, it is thermodynamically favorable in the direction shown, because the carboxylic acid function generated enjoys greater stabilization due to resonance than does the simple carbonyl group present initially (see Section 19-3). The final proton transfer drives the entire process to completion.

When cyclic α-diketones undergo this transformation, a ring carbon migrates and leads to **ring contraction.**

Benzilic Acid Rearrangement of 1,2-Cyclohexanedione

1,2-Cyclohexanedione $\xrightarrow{\text{NaOH, H}_2\text{O, 250°C}}$ [intermediate: $:\ddot{\mathrm{O}}:^-$, OH] $\longrightarrow$ $\xrightarrow{\text{H}^+,\ \text{H}_2\text{O}}$ 1-Hydroxycyclopentanecarboxylic acid (COOH, OH) 80%

1,2-Cyclohexanedione

1-Hydroxycyclopentanecarboxylic acid

EXERCISE 23-17

Give the products of benzilic acid rearrangement of the following.

(a) 2,3-Butanedione **(b)** 1,2-Cycloheptanedione **(c)** [structure: phenanthrene-9,10-dione]

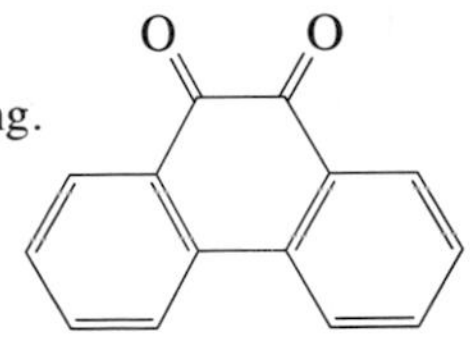

In summary, oxidation of α-hydroxyketones furnishes α-diketones. The α-diketo function is more reactive than an isolated carbonyl group. Hydration occurs readily, and base causes the benzilic acid rearrangement.

23-5 Preparation of α-Hydroxyketones: Alkanoyl (Acyl) Anion Equivalents

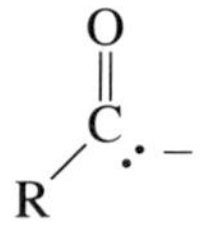

How are the α-hydroxyketones required for the synthesis of α-diketones prepared? One might try addition of an **alkanoyl (acyl) anion** to an aldehyde or ketone, as follows.

$R\overset{O}{\overset{\|}{C}}:^-$

Alkanoyl (acyl) anion

A Plausible (but Unfeasible) Synthesis of α-Hydroxyketones

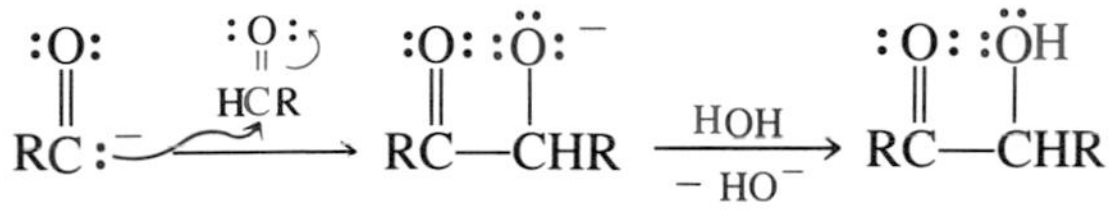

$$R\overset{:O:}{\overset{\|}{C}}:^- \xrightarrow{HCR(=O)} R\overset{:O:}{\overset{\|}{C}}-\overset{:\ddot{O}:^-}{\overset{|}{C}}HR \xrightarrow[-\ HO^-]{HOH} R\overset{:O:}{\overset{\|}{C}}-\overset{:\ddot{O}H}{\overset{|}{C}}HR$$

BOX 23-2 2-Oxopropanoic (Pyruvic) Acid, a Natural α-Ketoacid

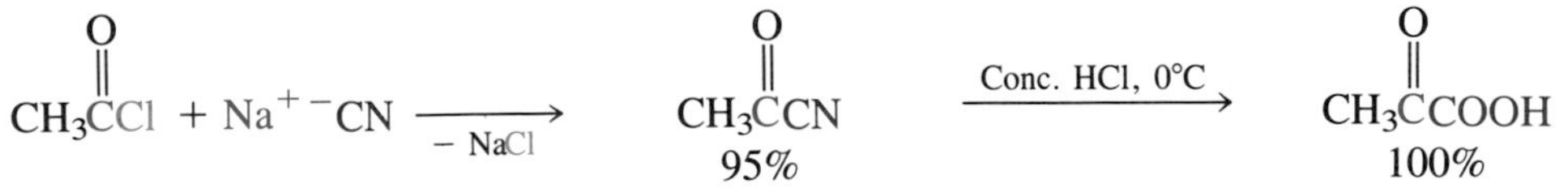

$$CH_3\overset{O}{\overset{\|}{C}}Cl + Na^{+\,-}CN \xrightarrow[-\ NaCl]{} CH_3\overset{O}{\overset{\|}{C}}CN \xrightarrow{\text{Conc. HCl, 0°C}} CH_3\overset{O}{\overset{\|}{C}}COOH$$

95% **2-Oxopropanenitrile**; 100% **2-Oxopropanoic acid (Pyruvic acid)**

2-Oxopropanoic (pyruvic) acid is an α-ketoacid. It can be prepared by the hydrolysis of 2-oxopropanenitrile, which is available by the treatment of acetyl chloride with sodium cyanide.

The two molecules 2-hydroxypropanoic (lactic) and 2-oxopropanoic (pyruvic) acid are interconverted in the body. This task is carried out by an enzyme in the muscle, *lactic acid dehydrogenase,* which reduces pyruvic to lactic acid during physical exercise. The enzyme reverses the process when muscles rest. The reduction is highly stereoselective, giving only the (*S*)-(+) acid. In nature, 2-oxopropanoic (pyruvic) acid is made by the enzymatic degradation of carbohydrates (Chapter 24).

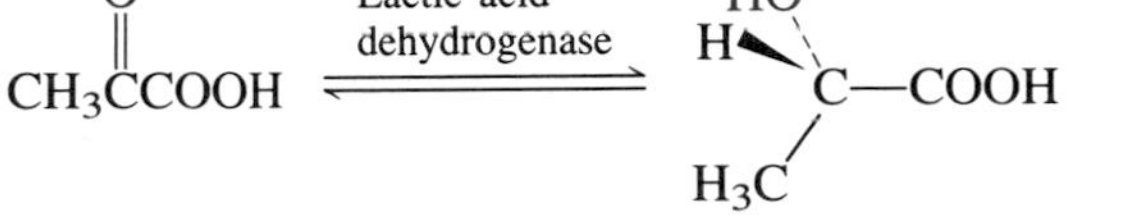

$$CH_3\overset{O}{\overset{\|}{C}}COOH \xrightleftharpoons{\text{Lactic acid dehydrogenase}} (HO)(H)(H_3C)C-COOH$$

(*S*)-(+)-2-Hydroxypropanoic (lactic) acid

Unfortunately, alkanoyl anions cannot readily be generated for synthetic applications such as this. As a result, special nucleophiles called **alkanoyl (acyl) anion equivalents** or **masked alkanoyl (acyl) anions** have been developed. These species contain negatively charged carbon atoms that can undergo addition reactions and later be transformed into carbonyl groups. This section describes two such reagents and illustrates their use in synthesis.

EXERCISE 23-18

Why are alkanoyl anions not formed by reaction of a base with an aldehyde? (Hint: See Sections 17-5 and 18-1.)

Cyclic dithioacetals are masked alkanoyl anion precursors

The hydrogens on the methylene group positioned between the two sulfur atoms in 1,3-dithiacyclohexane (1,3-dithiane) are relatively acidic ($pK_a = 31$). The negative charge on the corresponding anion is inductively stabilized by the highly polarizable sulfur atoms.

Deprotonation of 1,3-Dithiacyclohexane

$pK_a = 31.1$ H H — $CH_3CH_2CH_2CH_2Li$, THF → (anion) H Li^+ + $CH_3CH_2CH_2CH_2$–H

These anions add to aldehydes and ketones, thereby furnishing alcohols with an adjacent thioacetal function.

(dithiane anion) + $C_6H_5CH=O$ → H^+, H_2O → 2-substituted dithiane, H, CH(OH)C_6H_5

91%

2-(1-Hydroxyphenylmethyl)-1,3-dithiacyclohexane

The thioacetals that result from such addition may be hydrolyzed by using mercuric salts (Section 17-8), this hydrolysis furnishing the corresponding carbonyl compounds.

The synthesis of 1-acetyl-2-cyclohexen-1-ol illustrates the formation of a substituted 1,3-dithiane from an aldehyde. Addition of the dithiane anion to a ketone is followed by hydrolysis of the resulting thioacetal, to give the product.

$$CH_3CHO \xrightarrow[\text{(Section 17-8)}]{HS(CH_2)_3SH,\ CHCl_3,\ HCl,\ -H_2O}$$ 2-methyl-1,3-dithiane (91%) $$\xrightarrow{1.\ CH_3CH_2CH_2CH_2Li,\ THF;\ 2.\ \text{2-cyclohexenone};\ 3.\ H^+,\ H_2O}$$ (70%) $$\xrightarrow{H_2O,\ HgCl_2,\ CaCO_3,\ CH_3CN}$$ (93%)

1-Acetyl-2-cyclohexen-1-ol

In this synthesis the electrophilic carbonyl carbon of the starting aldehyde is transformed into a *nucleophilic* atom, thc ncgatively charged C2 of a 1,3-dithiane anion. After the latter is added to the ketone, hydrolysis of the thioacetal function regenerates the original electrophilic carbonyl group. The sequence therefore employs the *reversal of the polarization* of this carbon atom to form the carbon–carbon bond. Reagents exhibiting reverse polarization greatly increase the strategies available to chemists in planning syntheses.

EXERCISE 23-19

Formulate a synthesis of 2-hydroxy-2,4-dimethyl-3-pentanone beginning with simple aldehydes and ketones and using a 1,3-dithiane anion.

Thiazolium ions catalyze aldehyde coupling

Masked alkanoyl anions may be generated catalytically in the reaction of aldehydes with **thiazolium salts.** These salts are derived from thiazoles by alkylation at nitrogen. Thiazole is a heteroaromatic compound containing sulfur and nitrogen. Thiazolium salts have an unusual feature—a relatively acidic proton located between the two heteroatoms (at C2).

Thiazole

Thiazolium salt

Thiazolium Cations Are Acidic

Thiazolium cation ($pK_a < 20$) $+ \ {}^-{:}\ddot{O}H \rightleftharpoons$ [resonance structures of the ylide] $+$ HOH

Thiazolium cation

In the presence of thiazolium salts, aldehydes undergo conversion into α-hydroxyketones. An example of this process is the conversion of two molecules of butanal into 5-hydroxy-4-octanone. The catalyst is *N*-dodecylthiazolium bromide, which contains a long-chain alkyl substituent to improve its solubility in organic solvents.

Aldehyde Coupling

$$2\ CH_3CH_2CH_2\overset{O}{\overset{\|}{C}}H \xrightarrow[\text{N-Dodecylthiazolium bromide}]{} CH_3CH_2CH_2\overset{O}{\overset{\|}{C}}-\overset{OH}{\overset{|}{C}}HCH_2CH_2CH_3$$

76%

Butanal **5-Hydroxy-4-octanone**

The mechanism of this reaction begins with reversible addition of C2 in the deprotonated thiazolium salt to the carbonyl function of an aldehyde.

Mechanism of Thiazolium Ion Catalysis in Aldehyde Coupling

STEP 1. Deprotonation of thiazolium ion
STEP 2. Nucleophilic attack by catalyst

HOH

$+\ ^-OH$

STEP 3. Masked alkanoyl anion formation

HO^-

$pK_a \sim 25$

Alkanoyl anion equivalent

$+\ HOH$

BOX 23-3 Total Synthesis of the Sex Attractant of the Bark Beetle

1,3-Dithiacyclohexane anions participate in S_N2 alkylations with suitable haloalkanes. Hydrolysis of the resulting thioacetals furnishes ketones. An example is the laboratory preparation of a sex attractant of the bark beetle *Ips confusus*. 2-(2-Methylpropyl)-1,3-dithiacyclohexane anion is first alkylated with 2-(bromomethyl)-1,3-butadiene. Hydrolysis of the thioacetal group in the product and reduction of the carbonyl function then gives the natural product.

$CH_2CH(CH_3)_2$

2-(2-Methylpropyl)-1,3-dithiane anion

$BrCH_2$ / $-\ Br^-$ → $(CH_3)_2CHCH_2$, CH_2 — 51%

$HgCl_2$ → 59%

$NaBH_4$ → 66%

Sex attractant of the bark beetle

STEP 4. Nucleophilic attack on initial aldehyde

STEP 5. Liberation of α-hydroxyketone

The product alcohol of step 2 is unique in that the thiazolium unit is a substituent. This group is electron withdrawing and increases the acidity of the adjacent proton. Deprotonation leads to an unusually stable masked alkanoyl anion. Nucleophilic attack by this anion on another molecule of aldehyde, followed by loss of the thiazolium substituent, liberates the α-hydroxyketone.

Comparison of the thiazolium method for synthesis of α-hydroxyketones with the use of dithiane anions is instructive. Thiazolium salts have the advantage in that they are needed in only catalytic amounts. However, their use is limited to the synthesis of molecules $R{-}\overset{O}{\overset{\|}{C}}{-}\overset{OH}{\overset{|}{C}H}{-}R$ in which the two R groups are identical. The dithiane method is more versatile and can be used to prepare a much wider variety of substituted α-hydroxyketones.

EXERCISE 23-20

Which of the following compounds can be prepared by using thiazolium ion catalysts, and which are only accessible from 1,3-dithiane anions? Formulate syntheses of at least two of these substances, one by each route.

(a) $CH_3CH_2{-}\overset{O}{\overset{\|}{C}}{-}\overset{OH}{\overset{|}{C}H}{-}CH_2CH_3$ **(b)** $CH_3(CH_2)_3{-}\overset{O}{\overset{\|}{C}}{-}\overset{OH}{\overset{|}{C}H}{-}CH_2CH_3$ **(c)** $CH_3CH_2\underset{CH_3}{\underset{|}{C}H}{-}\overset{O}{\overset{\|}{C}}{-}\underset{CH_3}{\underset{|}{\overset{OH}{\overset{|}{C}}}}CH_2CH_3$

(d) $C_6H_5{-}\overset{O}{\overset{\|}{C}}{-}\overset{OH}{\overset{|}{C}H}{-}C_6H_5$ **(e)** $(CH_3)_2CH{-}\overset{O}{\overset{\|}{C}}{-}\overset{OH}{\overset{|}{C}H}{-}CH(CH_3)_2$

In summary, α-hydroxyketones are available from addition of masked alkanoyl (acyl) anions to aldehydes and ketones. The conversion of aldehydes into the anions of the corresponding 1,3-dithiacyclohexanes (1,3-dithianes) illustrates the method of reverse polarization. The electrophilic carbon changes into a nucleophilic center, thereby allowing addition to an aldehyde or ketone carbonyl group. Thiazolium ions catalyze the dimerization of aldehydes, again through the transformation of the carbonyl carbon into a nucleophilic atom.

BOX 23-4 Thiamine: A Naturally Catalytically Active Thiazolium Ion

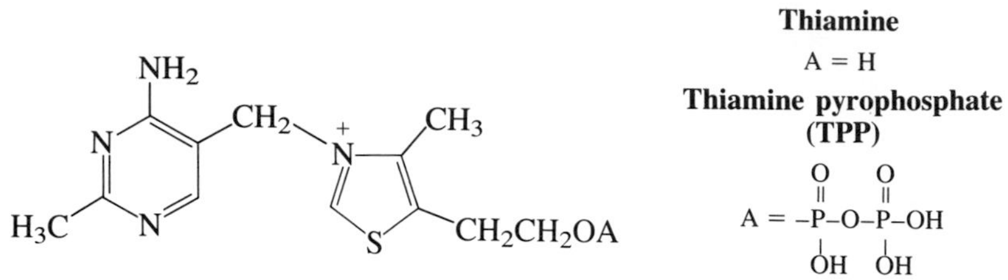

Thiamine

A = H

Thiamine pyrophosphate (TPP)

$$A = -\overset{O}{\overset{\|}{\underset{OH}{\underset{|}{P}}}}-O-\overset{O}{\overset{\|}{\underset{OH}{\underset{|}{P}}}}-OH$$

The catalytic activity of thiazolium salts in aldehyde dimerization has an analogy in nature, the action of *thiamine,* or vitamin B_1. Thiamine, in the form of its pyrophosphate, is a coenzyme for several biochemical transformations that include intermediates of the type appearing in the synthesis of α-hydroxyketones. As shown below, the enzyme *transketolase* catalyzes the transfer of (hydroxymethyl)carbonyl units from sugars (Chapter 24) to aldehydes, thereby producing new sugars in the process. The active site of the enzyme (Chapter 26) contains a molecule of thiamine pyrophosphate (TPP).

Mechanistically (facing page), the deprotonated thiazolium ion first attacks the carbonyl group of the donor molecule to form an addition product, in a way completely analogous to addition to aldehydes. Because the donor sugar contains a hydroxy group next to the reaction site, this initial product can decompose by the reverse of the addition process to an aldehyde and a new thiamine intermediate. This species attacks another aldehyde to produce a new addition product. The catalyst then dissociates as thiamine pyrophosphate, releasing a new sugar molecule.

Action of Transketolase

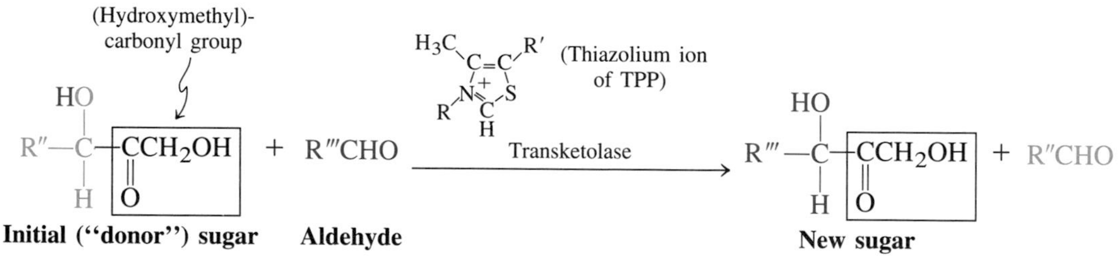

(Hydroxymethyl)-carbonyl group

$$R''-\underset{H}{\overset{HO}{C}}-\underset{O}{\overset{}{\boxed{\underset{\|}{C}CH_2OH}}} + R'''CHO \xrightarrow[\text{Transketolase}]{\text{(Thiazolium ion of TPP)}} R'''-\underset{H}{\overset{HO}{C}}-\boxed{\underset{\|}{C}CH_2OH} + R''CHO$$

Initial ("donor") sugar — Aldehyde — New sugar

NEW REACTIONS

Synthesis of β-Dicarbonyl Compounds

I. CLAISEN AND RETRO-CLAISEN CONDENSATIONS (Section 23-1)

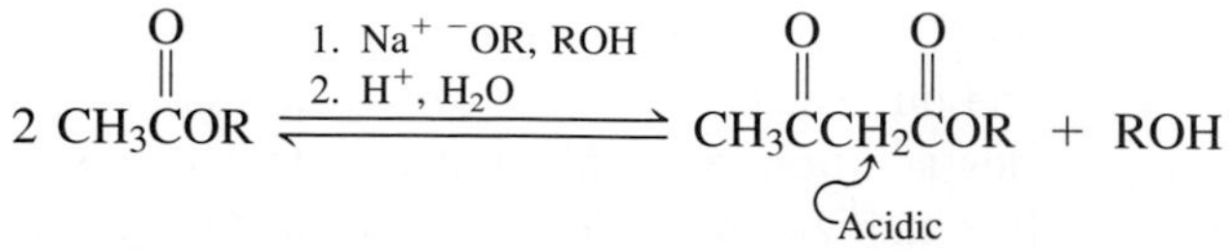

$$2\ CH_3\overset{O}{\overset{\|}{C}}OR \underset{}{\overset{1.\ Na^+\ ^-OR,\ ROH \quad 2.\ H^+,\ H_2O}{\rightleftharpoons}} CH_3\overset{O}{\overset{\|}{C}}CH_2\overset{O}{\overset{\|}{C}}OR + ROH$$

Acidic

Endothermic reaction; equilibrium driven to anion of product by excess base

Sugar Activation

$$\text{Deprotonated thiamine pyrophosphate} + R''\text{—CH(OH)—C(=O)CH}_2\text{OH} \xrightleftharpoons{H^+} \text{Addition compound}$$

Deprotonated thiamine pyrophosphate

Donor sugar

Addition compound

Removal of Old Aldehyde

$$\rightleftharpoons \left[\quad \longleftrightarrow \quad \right] + R''\overset{O}{\overset{\|}{C}}H + H^+$$

Introduction of New Aldehyde

$$H^+ + R'''\overset{O}{\overset{\|}{C}}H + \text{TPP–}\ddot{C}(OH)CH_2OH \rightleftharpoons HO\text{—CH(R''')—C(OH)(TPP)—CH}_2\text{OH} \rightleftharpoons HO\text{—CH(R''')—C(=O)—CH}_2\text{OH} + \text{TPP}$$

New sugar

2. DIECKMANN CONDENSATION (Section 23-1)

$$(CH_2)_n\begin{cases}CO_2R\\CH_2CO_2R\end{cases} \xrightarrow[\text{2. } H^+,\ H_2O]{\text{1. } Na^{+\,-}OR',\ R'OH} (CH_2)_n\text{(ring: } C{=}O\text{, }CHCO_2R) + ROH$$

3. β-DIKETONE SYNTHESIS (Section 23-1)

$$\mathrm{R\overset{O}{\overset{\|}{C}}OR' + CH_3\overset{O}{\overset{\|}{C}}CH_3 \xrightarrow[2.\ H^+,\ H_2O]{1.\ Na^{+\,-}OR',\ R'OH} R\overset{O}{\overset{\|}{C}}CH_2\overset{O}{\overset{\|}{C}}CH_3 + R'OH}$$

Intramolecular

$$\mathrm{(CH_2)_n\text{ ring precursor bearing } \overset{O}{\overset{\|}{C}}CH_3 \text{ and } CO_2R \xrightarrow[2.\ H^+,\ H_2O]{1.\ Na^{+\,-}OR,\ ROH} (CH_2)_n \text{ ring with } C(=O)CH_2C(=O) + ROH}$$

3-Ketoesters as Synthetic Building Blocks

4. ENOLATE ALKYLATION (Section 23-2)

$$\mathrm{R\overset{O}{\overset{\|}{C}}CH_2CO_2R' \xrightarrow[2.\ R''X]{1.\ Na^{+\,-}OR',\ R'OH} R\overset{O}{\overset{\|}{C}}\underset{R''}{\underset{|}{C}H}CO_2R'}$$

5. 3-KETOACID DECARBOXYLATION (Section 23-2)

$$\mathrm{R\overset{O}{\overset{\|}{C}}CH_2\overset{O}{\overset{\|}{C}}OR' \xrightarrow[-\ R'O^-]{HO^-} R\overset{O}{\overset{\|}{C}}CH_2\overset{O}{\overset{\|}{C}}O^- \xrightarrow{H^+} R\overset{O}{\overset{\|}{C}}CH_2\overset{O}{\overset{\|}{C}}OH \xrightarrow[-\ CO_2]{\Delta} R\overset{O}{\overset{\|}{C}}CH_3}$$

6. ACETOACETIC ESTER SYNTHESIS OF METHYL KETONES (Section 23-2)

$$\mathrm{CH_3\overset{O}{\overset{\|}{C}}CH_2\overset{O}{\overset{\|}{C}}OR' \xrightarrow{1.\ NaOR',\ R'OH;\ 2.\ R''X;\ 3.\ HO^-;\ 4.\ H^+,\ \Delta} CH_3\overset{O}{\overset{\|}{C}}CH_2R''}$$

$\mathrm{R''}$ = alkyl, alkanoyl (acyl), $\mathrm{CH_2\overset{O}{\overset{\|}{C}}OR'''}$, $\mathrm{CH_2\overset{O}{\overset{\|}{C}}R'''}$
$\mathrm{R''X}$ = oxacyclopropane

7. MALONIC ESTER SYNTHESIS OF CARBOXYLIC ACIDS (Section 23-2)

$$\mathrm{RO\overset{O}{\overset{\|}{C}}CH_2\overset{O}{\overset{\|}{C}}OR \xrightarrow{1.\ NaOR,\ ROH;\ 2.\ R'X;\ 3.\ HO^-;\ 4.\ H^+,\ \Delta} R'CH_2\overset{O}{\overset{\|}{C}}OH}$$

$\mathrm{R'}$ = alkyl, alkanoyl (acyl), $\mathrm{CH_2\overset{O}{\overset{\|}{C}}OR''}$, $\mathrm{CH_2\overset{O}{\overset{\|}{C}}R''}$
$\mathrm{R'X}$ = oxacyclopropane

8. MICHAEL ADDITION (Section 23-3)

$$\mathrm{CH_2{=}CH\overset{O}{\overset{\|}{C}}R + H_2C(CO_2CH_3)_2 \xrightarrow{Na^{+\,-}OCH_3,\ CH_3OH} R\overset{O}{\overset{\|}{C}}CH_2CH_2CH(CO_2CH_3)_2}$$

Michael acceptor

α-Dicarbonyl and α-Hydroxycarbonyl Compounds

9. PREPARATION OF α-DIKETONES (Section 23-4)

$$\mathrm{R{-}\overset{OH}{\underset{H}{C}}{-}\overset{O}{\overset{\|}{C}}R'} \xrightarrow{\mathrm{KMnO_4,\ CH_3\overset{O}{\overset{\|}{C}}O\overset{O}{\overset{\|}{C}}CH_3\ \ or\ \ Cu(O\overset{O}{\overset{\|}{C}}CH_3)_2,\ CH_3COOH}} \mathrm{R\overset{O}{\overset{\|}{C}}{-}\overset{O}{\overset{\|}{C}}R'}$$

10. BENZILIC ACID REARRANGEMENT (Section 23-4)

$$\mathrm{R\overset{O}{\overset{\|}{C}}{-}\overset{O}{\overset{\|}{C}}R} \xrightarrow{\mathrm{NaOH}} \mathrm{R{-}\overset{R}{\underset{OH}{C}}{-}\overset{O}{\overset{\|}{C}}O^-Na^+}$$

11. 1,3-DITHIACYCLOHEXANE (1,3-DITHIANE) ANIONS AS ALKANOYL (ACYL) ANION EQUIVALENTS (Section 23-5)

$$\mathrm{R{-}\overset{A'}{\underset{A}{C}}{:}^-} \quad \text{synthetic equivalent of} \quad \mathrm{R{-}\overset{O}{\overset{\|}{C}}{:}^-}$$

A = electron-withdrawing, conjugating, or polarizable group

$$\text{1,3-Dithiane} \xrightarrow[\text{2. } \mathrm{R\overset{O}{\overset{\|}{C}}R'}]{\text{1. } \mathrm{CH_3CH_2CH_2CH_2Li,\ THF}} \text{2-(HO–C(R)(R'))-2-H-1,3-dithiane} \xrightarrow{\mathrm{HgCl_2,\ HgO,\ CH_3OH}} \mathrm{R\overset{HO}{\underset{R'}{C}}{-}\overset{O}{\overset{\|}{C}}H}$$

12. THIAZOLIUM SALTS IN ALDEHYDE COUPLING (Section 23-5)

$$\mathrm{2\ R\overset{O}{\overset{\|}{C}}H} \xrightarrow[\textit{N}\text{-Dodecylthiazolium bromide}]{\text{thiazolium } \mathrm{N^+{-}(CH_2)_{11}CH_3\ Br^-}} \mathrm{R\overset{HO}{\underset{H}{C}}{-}\overset{O}{\overset{\|}{C}}R}$$

IMPORTANT CONCEPTS

1. The Claisen condensation is driven by the stoichiometric generation of a stable β-dicarbonyl anion in the presence of excess base.
2. β-Dicarbonyl compounds contain acidic hydrogens at the carbon between the two carbonyl groups because of the inductive electron-withdrawing effect of the two neighboring carbonyl functions and because the anions resulting from deprotonation are resonance stabilized.
3. Although mixed Claisen condensations between esters are usually not selective, they can be so with certain substrates (nonenolizable esters, intramolecular versions, ketones).
4. 3-Ketoacids are unstable; they decarboxylate in a concerted process through an aromatic transition state. This property, in conjunction with the nucleophilic reactivity of 3-ketoester anions, allows the synthesis of substituted ketones and acids.
5. Because alkanoyl (acyl) anions are not directly available by deprotonation of aldehydes, they have to be made as masked reactive intermediates or stoichiometric reagents by transformations of functional groups.

PROBLEMS

1. When an apparently ordinary reaction is attempted on one functional group in a difunctional compound, interference by the second group may cause it to fail. For each of the following equations, indicate whether the reaction is likely to proceed as written. If not, suggest what might occur instead.

(a) [alkenyl ketone] $\xrightarrow{Br_2,\ CCl_4}$ [dibromo ketone]: Br, Br, O

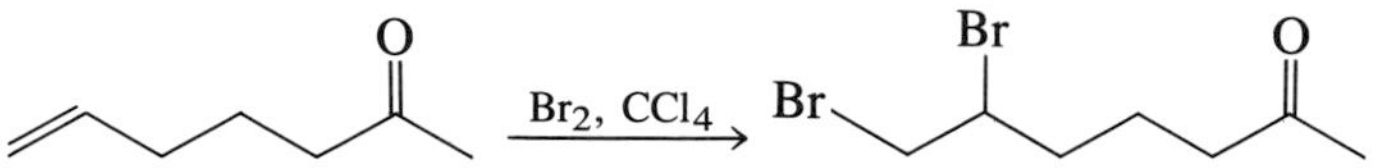

(b) [cyclopentenone with bromoalkyl chain] Br $\xrightarrow{KCN,\ CH_3CH_2OH}$ [cyclopentenone with cyanoalkyl chain] CN

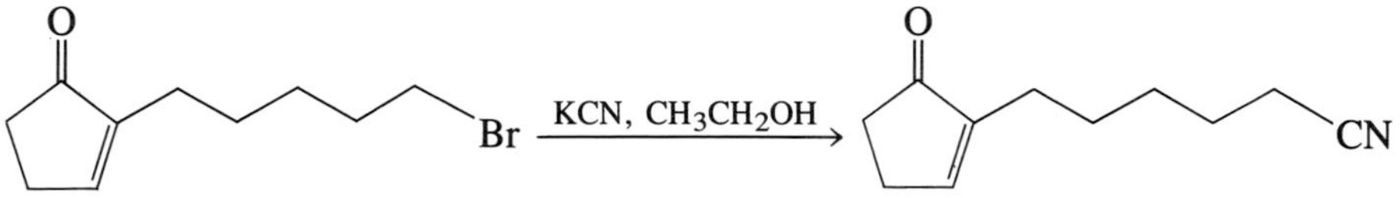

(c) [bicyclic alkene] COOH $\xrightarrow{LiAlH_4,\ (CH_3CH_2)_2O}$ [bicyclic alkene] CH_2OH

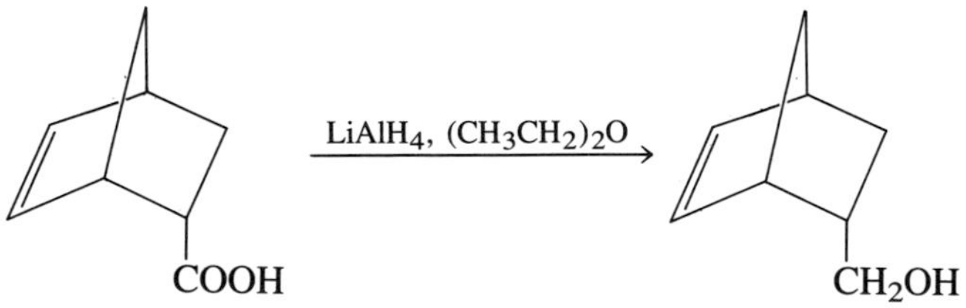

(d) OH [tertiary alcohol with phenyl chain] $\xrightarrow{\text{Conc. HCl}}$ Cl [tertiary chloride with phenyl chain]

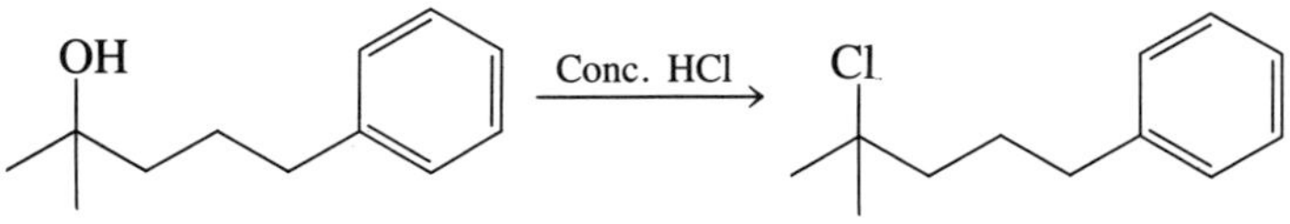

(e) [cyclohexanone with iodoalkyl chain] I $\xrightarrow{LDA,\ THF}$ [cyclohexanone with alkenyl chain]

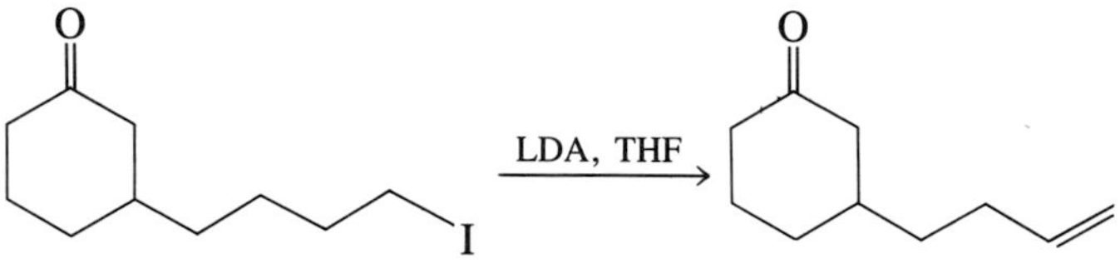

(f) [pyrrolidine, N–H] COOH $\xrightarrow{1.\ SOCl_2\ \ 2.\ (CH_3)_2CHOH}$ [pyrrolidine, N–H, isopropyl ester]

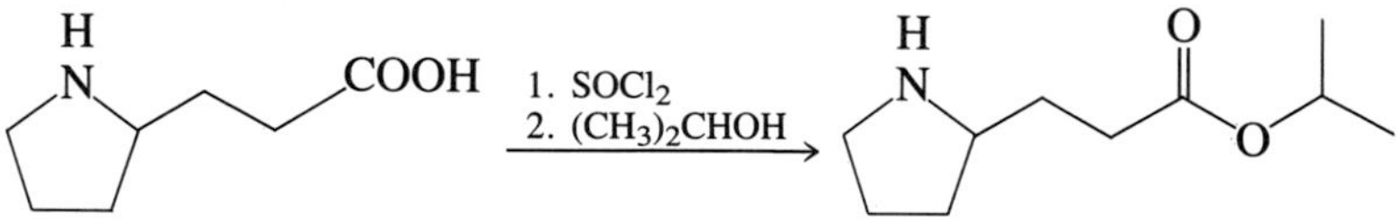

2. Give the expected results of the reaction of each of the following molecules (or combinations of molecules) with excess $NaOCH_2CH_3$ in CH_3CH_2OH, followed by aqueous acid work-up.

(a) $CH_3CH_2CH_2COOCH_2CH_3$

(b) $C_6H_5CH(CH_3)CH_2COOCH_2CH_3$

(c) $C_6H_5CH_2CH(CH_3)COOCH_2CH_3$

(d) $CH_3CH_2OC(=O)(CH_2)_4C(=O)OCH_2CH_3$

(e) $CH_3CH_2OC(=O)CH(CH_3)(CH_2)_4C(=O)OCH_2CH_3$

(f) $C_6H_5CH_2CO_2CH_2CH_3 + HCO_2CH_2CH_3$

(g) $C_6H_5CO_2CH_2CH_3 + CH_3CH_2CH_2CO_2CH_2CH_3$

(h) 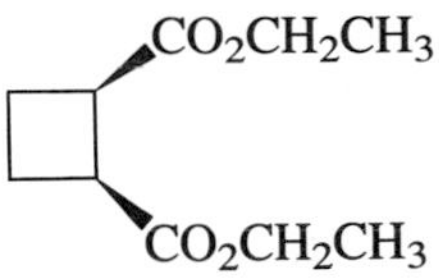 (cyclobutane with two $CO_2CH_2CH_3$ groups) $+ CH_3CH_2O\overset{O}{\overset{\|}{C}}CH_2CH_2\overset{O}{\overset{\|}{C}}OCH_2CH_3$

(i) 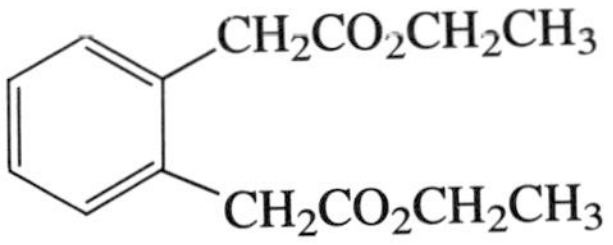 (benzene with two $CH_2CO_2CH_2CH_3$ groups) $+ CH_3CH_2O\overset{O}{\overset{\|}{C}}-\overset{O}{\overset{\|}{C}}OCH_2CH_3$

3. The following mixed Claisen condensation works best when one of the starting materials is present in large excess. Which of the two starting materials should be present in excess? Why? What side reaction will compete if the reagents are present in comparable amounts?

$$CH_3CH_2\overset{O}{\overset{\|}{C}}OCH_3 + (CH_3)_2CH\overset{O}{\overset{\|}{C}}OCH_3 \xrightarrow{NaOCH_3,\ CH_3OH} (CH_3)_2CH\overset{O}{\overset{\|}{C}}\underset{CH_3}{\underset{|}{C}}H\overset{O}{\overset{\|}{C}}OCH_3$$

4. Suggest a synthesis of each of the following β-dicarbonyl compounds by Claisen or Dieckmann condensations.

(a) 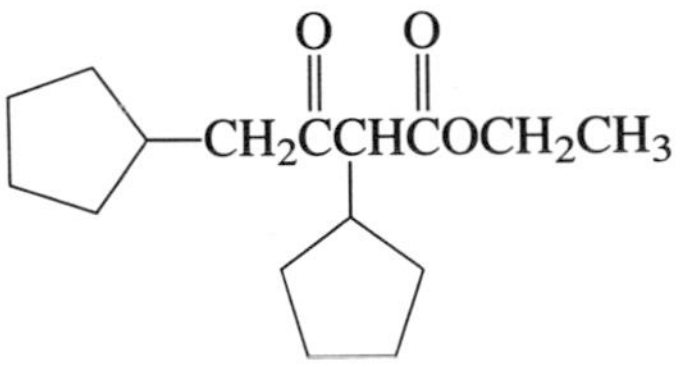 (cyclopentyl)$-CH_2\overset{O}{\overset{\|}{C}}\underset{\text{cyclopentyl}}{\underset{|}{C}}H\overset{O}{\overset{\|}{C}}OCH_2CH_3$

(b) $C_6H_5\overset{O}{\overset{\|}{C}}\underset{C_6H_5}{\underset{|}{C}}H\overset{O}{\overset{\|}{C}}OCH_2CH_3$

(c) 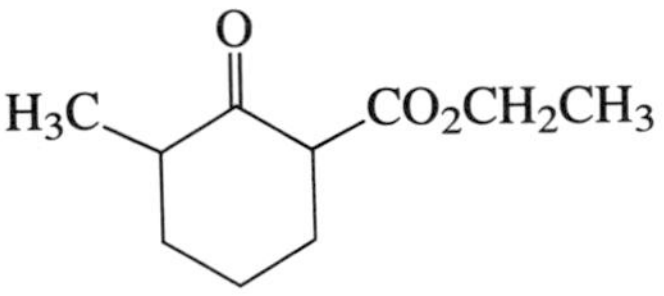 (cyclohexanone with H_3C and $CO_2CH_2CH_3$ on the carbons flanking C=O)

(d) 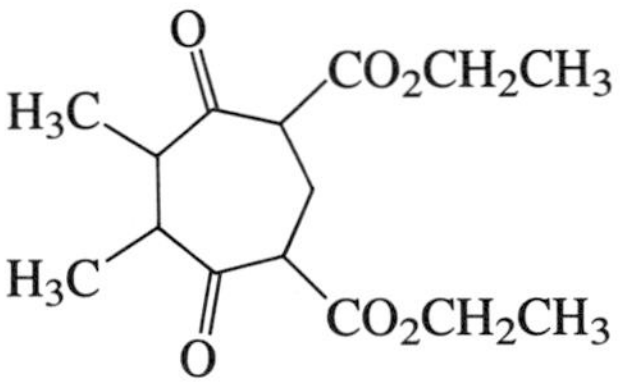 (cycloheptane-1,4-dione with two H_3C groups and two $CO_2CH_2CH_3$ groups)

(e) $H\overset{O}{\overset{\|}{C}}-\overset{O}{\overset{\|}{C}}CH_2\overset{O}{\overset{\|}{C}}OCH_2CH_3$

(f) $C_6H_5\overset{O}{\overset{\|}{C}}CH_2\overset{O}{\overset{\|}{C}}C_6H_5$

(g) $CH_3CH_2O\overset{O}{\overset{\|}{C}}CH_2\overset{O}{\overset{\|}{C}}OCH_2CH_3$

(h) 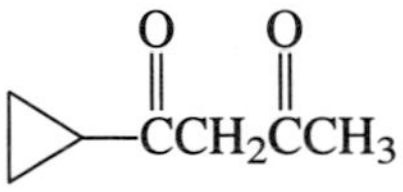 (cyclopropyl)$-\overset{O}{\overset{\|}{C}}CH_2\overset{O}{\overset{\|}{C}}CH_3$

5. Do you think that propanedial, shown in the margin, can be easily prepared by a simple Claisen condensation? Why or why not?

$H\overset{O}{\overset{\|}{C}}CH_2\overset{O}{\overset{\|}{C}}H$
Propanedial

6. Devise a preparation of each of the following ketones, using the acetoacetic ester synthesis.

(a) $CH_3\overset{O}{\overset{\|}{C}}CH_2CH_2\overset{CH_3}{\overset{|}{C}}HCH_3$

(b) 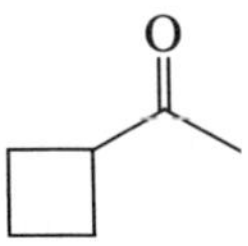 (cyclobutyl methyl ketone)

(c) 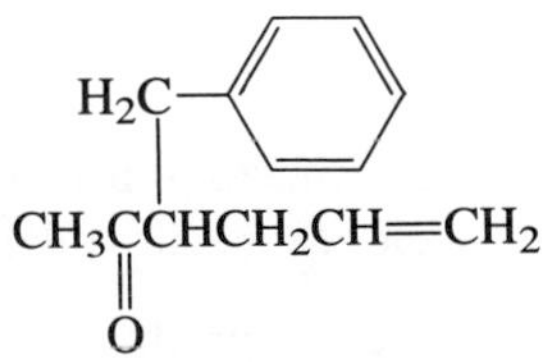 $CH_3\underset{O}{\underset{\|}{C}}\overset{H_2C-C_6H_5}{\overset{|}{C}}HCH_2CH{=}CH_2$

(d) $CH_3\overset{O}{\overset{\|}{C}}\underset{CH_2\underset{O}{\underset{\|}{C}}OCH_2CH_3}{\underset{|}{C}}HCH_2CH_3$

7. Devise a synthesis for each of the four following compounds, using the malonic ester synthesis.

(a) 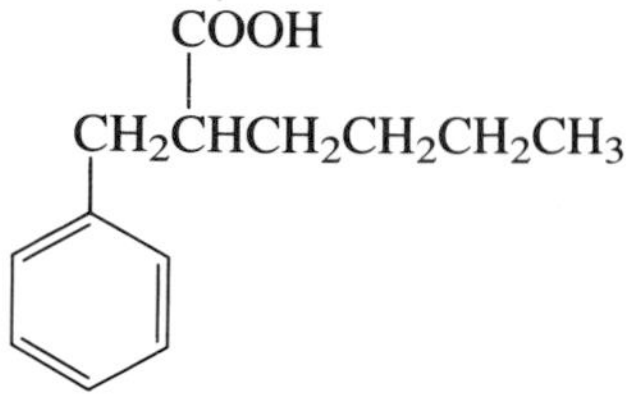 $C_6H_5CH_2CH(COOH)CH_2CH_2CH_2CH_3$ (b) $CH_3CH(CH_3)CH_2CH(CH_3)COOH$

(c) $HOOC{-}CH_2{-}CH_2{-}COOH$ (d) 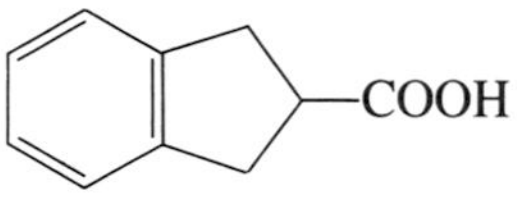 [indane-2-carboxylic acid] —COOH

8. Use the methods described in Section 23-3, with other reactions if necessary, to synthesize each of the following compounds. In each case, your starting materials should include one aldehyde or ketone and one β-dicarbonyl compound.

(a) 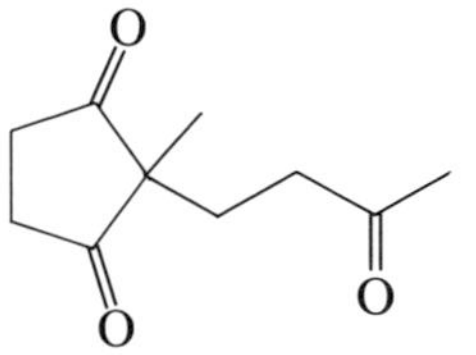 (b) 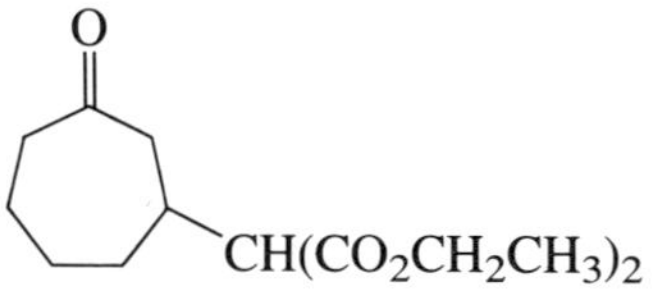 $CH(CO_2CH_2CH_3)_2$ (c)

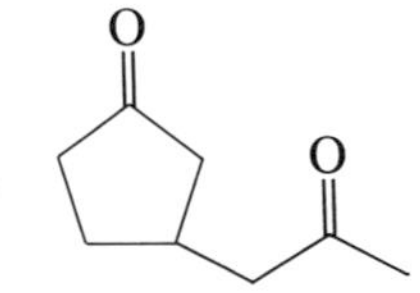

(Hint: A decarboxylation is necessary)

9. Write out, in full detail, the mechanism of the Michael addition of malonic ester to 3-buten-2-one in the presence of ethoxide ion. Be sure to indicate all steps that are reversible. Does the overall reaction appear to be exo- or endothermic? Explain why only a catalytic amount of base is necessary.

10. Using the methods described in this chapter, design a multistep synthesis of each of the following molecules, making use of the indicated building blocks as the sources of all the carbon atoms in your final product.

(a) 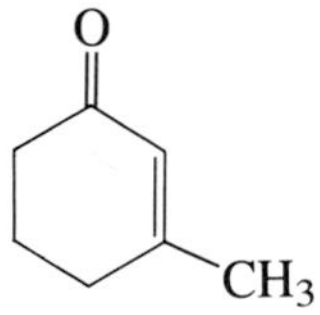 CH_3, from $CH_3CO_2CH_2CH_3$ and $CH_3COCH{=}CH_2$

(b) 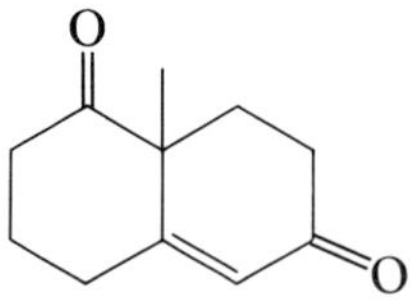 , from CH_3I, $CH_2(CO_2CH_2CH_3)_2$, and $CH_3\overset{O}{\overset{\|}{C}}CH{=}CH_2$ (Hint: First make .)

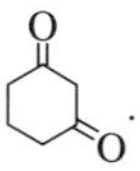

(c) 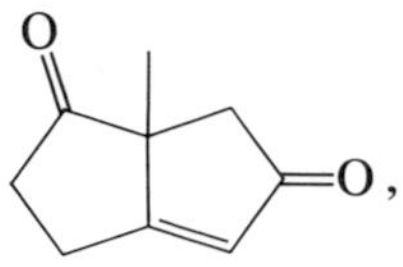 =O, from CH_3I, $CH_2(CO_2CH_2CH_3)_2$, and $BrCH_2\overset{O}{\overset{\|}{C}}CH_3$ (Hint: First make 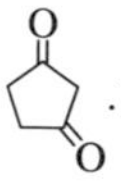 .)

11. Give the products of reaction of the following aldehydes with catalytic *N*-dodecylthiazolium bromide. **(a)** $(CH_3)_2CHCHO$; **(b)** C_6H_5CHO; **(c)** cyclohexanecarbaldehyde; **(d)** $C_6H_5CH_2CHO$.

12. Give the products of the following reactions.

(a) $C_6H_5CHO + HS(CH_2)_3SH \xrightarrow{BF_3}$ **(b)** Product of (a) + $CH_3CH_2CH_2CH_2Li \xrightarrow{THF}$

What are the results of reaction of the substance formed in (b) with each aldehyde in Problem 11, followed by hydrolysis in the presence of $HgCl_2$?

13. What compounds result from treatment of each product formed in Problem 11 with cupric acetate, $Cu(O\overset{O}{\overset{\|}{C}}CH_3)_2$, in aqueous acetic acid? What is the outcome of subsequent exposure of each to hot concentrated aqueous KOH?

14. (a) On the basis of the following data, identify the unknowns A, found in fresh cream prior to churning, and B, possessor of the characteristic yellow color and buttery odor of butter.

A: MS m/z (relative abundance) = 88($M^{+\cdot}$, weak), 45(100), and 43(80)
^{1}H NMR δ = 1.36 (d, J = 7 Hz, 3 H), 2.18 (s, 3 H), 3.73 (broad s, 1 H), 4.22 (q, J = 7 Hz, 1 H) ppm
IR $\tilde{\nu}$ = 1718 and 3430 cm^{-1}

B: MS m/z (relative abundance) = 86(17) and 43(100)
^{1}H NMR δ = 2.29 (s) ppm
IR $\tilde{\nu}$ = 1708 cm^{-1}

(b) What kind of reaction is the conversion of A into B? Does it make sense that this should take place in the churning of cream to make butter? Explain. **(c)** Outline laboratory syntheses of A and B, starting only with compounds containing two carbons. **(d)** The UV spectrum of A has a λ_{max} of 271 nm, whereas that of B has a λ_{max} of 290 nm. (Extension of the latter absorption into the violet region of the visible spectrum is responsible for the yellow color of B.) Explain the difference in λ_{max}.

15. Write chemical equations to illustrate all primary reaction steps that can occur between a base such as ethoxide ion and a carbonyl compound such as acetaldehyde. Explain why the carbonyl carbon is not deprotonated to any appreciable extent in this system.

16. Nootkatone is found in grapefruit. Fill in the necessary steps in the following scheme to make nootkatone from 4-(1-methylethenyl)cyclohexanone.

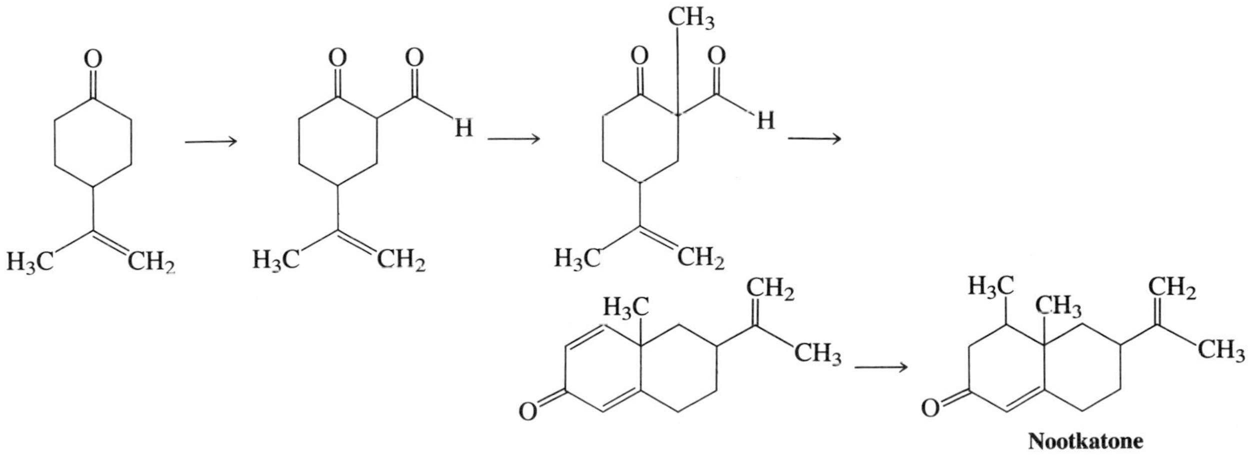

O H CH$_3$ H
H$_3$C CH$_2$ H$_3$C CH$_2$ H$_3$C CH$_2$
H$_3$C CH$_2$ CH$_3$ O → H$_3$C CH$_3$ CH$_2$ CH$_3$ O

Nootkatone

17. The following ketones cannot be synthesized by the acetoacetic ester method (why?), but they can be prepared by a modified version of it. The modification includes the preparation (by Claisen condensation) and use of an appropriate 3-ketoester, $RCCH_2COCH_2CH_3$ (with C=O at both C), containing an R group that appears in the final product. Synthesize each of the following ketones. For each, show the structure and synthesis of the necessary 3-ketoester as well.

(a) $CH_3CH_2\overset{O}{\overset{\|}{C}}CH_2CH_3$

(b) 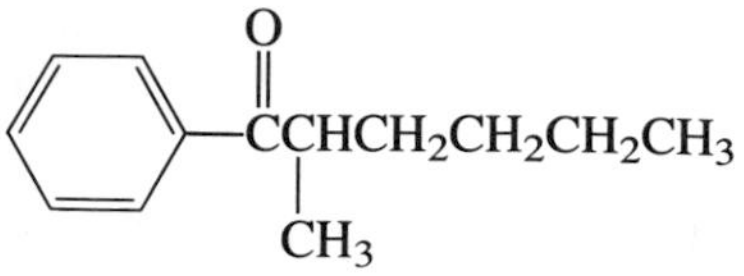 $C_6H_5\overset{O}{\overset{\|}{C}}CHCH_2CH_2CH_2CH_3$ (with CH_3 on CH)

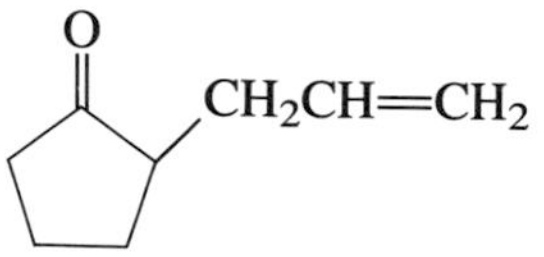

(c) 2-($CH_2CH{=}CH_2$)cyclopentanone

(Hint: Use a Dieckmann condensation.)

(d) 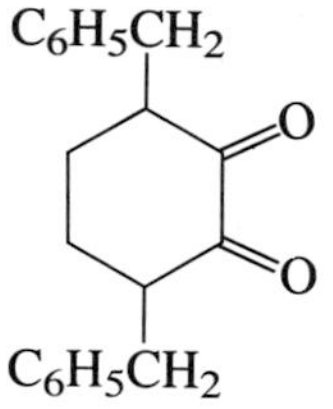 2,5-bis($C_6H_5CH_2$)cyclohexane-1,3... dione ($C_6H_5CH_2$, O, O, $C_6H_5CH_2$)

(Hint: Use a double Claisen condensation.)

18. Some of the most important building blocks for synthesis are very simple molecules. Although cyclopentanone and cyclohexanone are readily commercially available, an understanding of how they might be made from simpler molecules is instructive. The following are possible retrosynthetic analyses (Section 8-9) for both of these ketones. Using them as a guide, write out a synthesis of each ketone from the indicated starting materials.

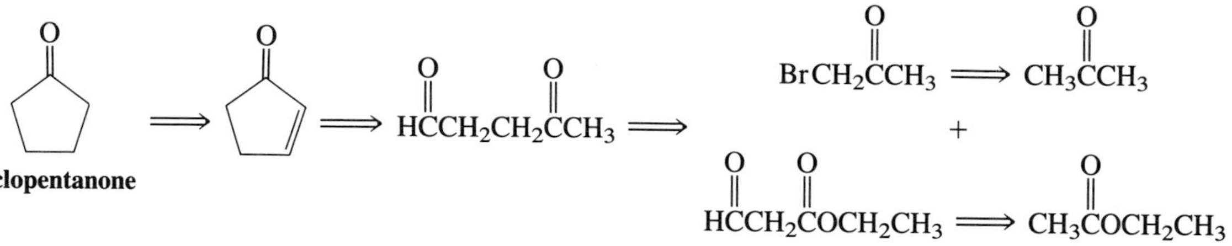

Cyclopentanone $\Longrightarrow$ cyclopentenone $\Longrightarrow$ $H\overset{O}{\overset{\|}{C}}CH_2CH_2\overset{O}{\overset{\|}{C}}CH_3$ $\Longrightarrow$ $BrCH_2\overset{O}{\overset{\|}{C}}CH_3 \Longrightarrow CH_3\overset{O}{\overset{\|}{C}}CH_3$ + $H\overset{O}{\overset{\|}{C}}CH_2\overset{O}{\overset{\|}{C}}OCH_2CH_3 \Longrightarrow CH_3\overset{O}{\overset{\|}{C}}OCH_2CH_3$

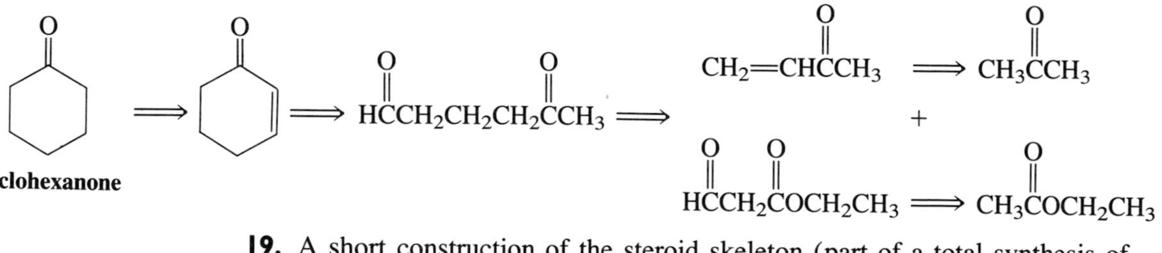

Cyclohexanone $\Longrightarrow$ cyclohexenone $\Longrightarrow$ $H\overset{O}{\overset{\|}{C}}CH_2CH_2CH_2\overset{O}{\overset{\|}{C}}CH_3$ $\Longrightarrow$ $CH_2{=}CH\overset{O}{\overset{\|}{C}}CH_3 \Longrightarrow CH_3\overset{O}{\overset{\|}{C}}CH_3$ + $H\overset{O}{\overset{\|}{C}}CH_2\overset{O}{\overset{\|}{C}}OCH_2CH_3 \Longrightarrow CH_3\overset{O}{\overset{\|}{C}}OCH_2CH_3$

19. A short construction of the steroid skeleton (part of a total synthesis of the hormone estrone) is shown here. Formulate mechanisms for each of the steps. (Hint: A process similar to that taking place in the second step is presented in Problem 11 of Chapter 18.)

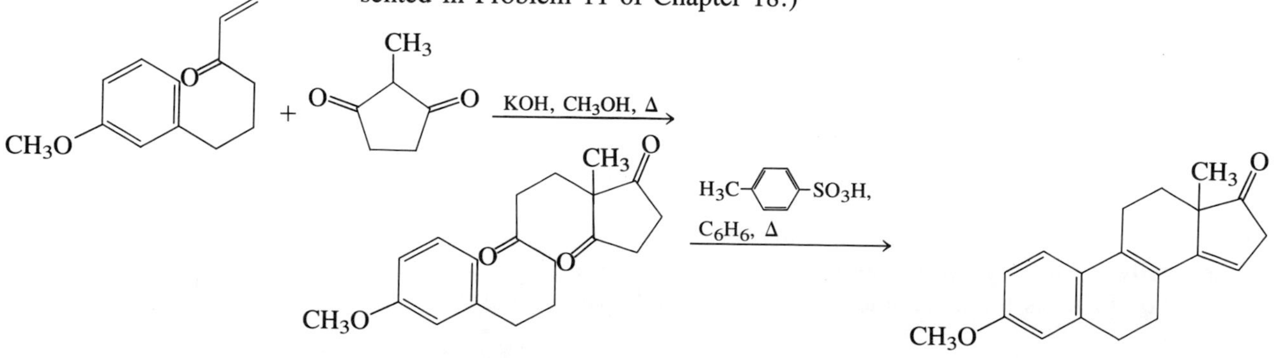

CH_3O-substituted bicyclic vinyl ketone + 2-CH_3-cyclopentane-1,3-dione $\xrightarrow{KOH,\ CH_3OH,\ \Delta}$ triketone (CH_3O, CH_3, O, O, O) $\xrightarrow{H_3C\text{-}C_6H_4\text{-}SO_3H,\ C_6H_6,\ \Delta}$ tetracyclic steroid diene (CH_3O, CH_3, O)

20. Propose short syntheses of each of the following molecules, starting with the material indicated and making use of the benzilic acid rearrangement.

(a) 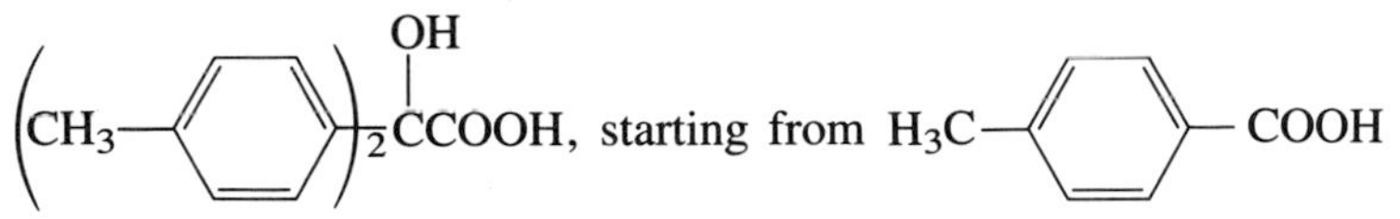 $(CH_3-C_6H_4)_2C(OH)COOH$, starting from $H_3C-C_6H_4-COOH$

(b) $C_6H_5-CH(OH)-COOH$, starting from C_6H_5CHO

(c) $C_6H_5CH_2-C(OH)(C_6H_5)-COOH$, starting from C_6H_5CHO

21. Using methods described in Section 23-5 (i.e., reverse polarization), propose a simple synthesis of each of the following molecules.

(a) $CH_2{=}CHCH(OH)C(O)CH_2C_6H_5$ **(b)** 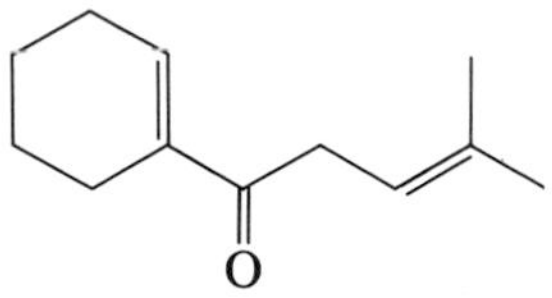 1-cyclohexenyl–C(=O)–CH$_2$CH=C(CH$_3$)$_2$ **(c)** $CH_3CH(OH)CH(CH_3)CHO$

22. Propose a synthesis of ketone iii, which was central in attempts to synthesize several antitumor agents. Start with aldehyde i, lactone ii, and anything else you need.

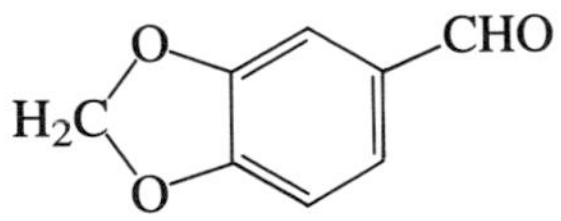

i: 3,4-methylenedioxybenzaldehyde (H_2C, O, O, CHO) ii: unsaturated five-membered lactone (O, O) iii: 3,4-methylenedioxyphenyl ketone of a γ-butyrolactone (H_2C, O, O, O, O, O)

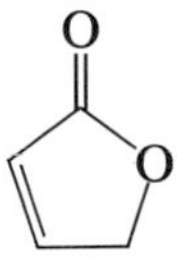

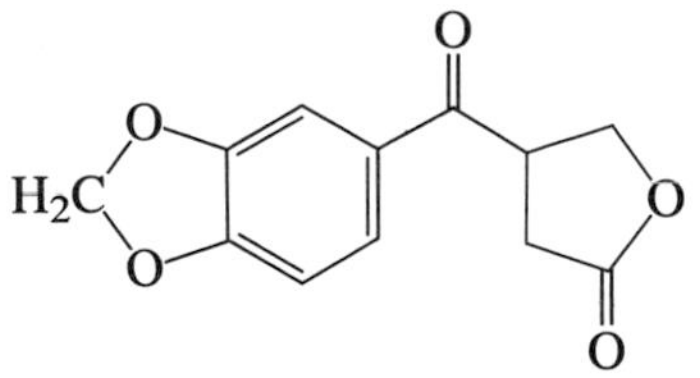

23. Urea reacts with diphenylethanedione (benzil) in the presence of NaOH and heat to give Dilantin, a drug useful for its powerful antiepileptic and anticonvulsive properties. Propose a mechanism for this transformation. (Hint: Begin by condensing one amino group of urea with one carbonyl group of the dione, and continue by *starting* a second such condensation.) What reaction given in this chapter does this process resemble mechanistically?

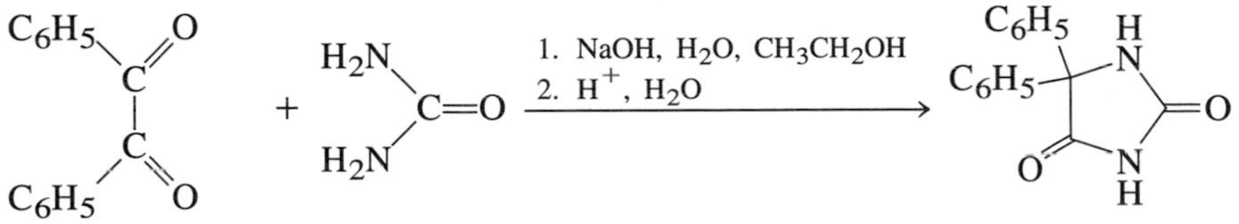

$$C_6H_5C(O)C(O)C_6H_5 + (H_2N)_2C{=}O \xrightarrow[\text{2. } H^+,\ H_2O]{\text{1. NaOH, } H_2O,\ CH_3CH_2OH} \text{Dilantin}$$

Diphenylethanedione (Benzil) **Urea** **Dilantin**

24

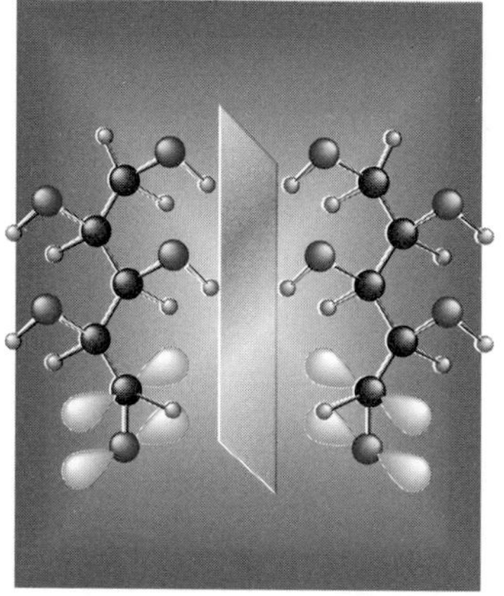

Carbohydrates

Polyfunctional Compounds in Nature

Carbohydrates, a very important class of naturally occurring chemicals, give structure to plants, flowers, vegetables, and trees. In addition, carbohydrates function as chemical energy-storage systems; they are metabolized to water, carbon dioxide, and heat or other energy. As such, they constitute an important source of food. Finally, they serve as the building units of fats (Sections 19-14 and 20-4) and nucleic acids (Section 26-9). Cellulose, starch, and ordinary table sugar are typical carbohydrates. All are said to be **polyfunctional,** because they possess multiple functional groups. Like glucose, $C_6(H_2O)_6$, many of the simple building blocks of complex carbohydrates have the general formula $C_n(H_2O)_n$.

We shall first consider the structure and naming of the simplest carbohydrates, the sugars. We then discuss their chemistry, which is governed by the presence of carbonyl and hydroxy functions along carbon chains of various lengths. Several methods for sugar synthesis and their structural analysis are then introduced, procedures that allow chains to be lengthened or shortened. Finally, we describe the various types of carbohydrates in nature.

24-1 Names and Structures of Carbohydrates

The simplest carbohydrates are the sugars, or **saccharides.** As chain length increases, the increasing number of carbon-based stereocenters gives rise to a multitude of diastereomers. Fortunately for chemists, nature deals mainly with only one of the possible series of enantiomers. Sugars are polyhydroxycarbonyl compounds, so they form stable cyclic hemiacetals, which affords additional structural and chemical variety.

Sugars are classified as aldoses and ketoses

Carbohydrate is the general name for the monomeric (monosaccharides), dimeric (disaccharides), trimeric (trisaccharides), oligomeric (oligosaccharides), and polymeric (polysaccharides) forms of sugar (*saccharum,* Latin, sugar). A **monosaccharide,** or **simple sugar,** is an aldehyde or ketone containing at least two additional hydroxy groups. Thus, the two simplest members of this class of compounds are 2,3-dihydroxypropanal (glyceraldehyde) and 1,3-dihydroxypropanone (1,3-dihydroxyacetone). **Complex sugars** (Section 24-11) are those formed by linking simple sugars, with the elimination of water.

Aldehydic sugars are classified as **aldoses;** those with a ketone function are called **ketoses.** On the basis of their chain length, we call sugars **trioses** (3 carbons), **tetroses** (4 carbons), **pentoses** (5 carbons), **hexoses** (6 carbons), and so on. Therefore, 2,3-dihydroxypropanal (glyceraldehyde) is an aldotriose, whereas 1,3-dihydroxypropanone is a ketotriose.

Glucose, also known as dextrose, blood sugar, or grape sugar (*glykys,* Greek, sweet), is a pentahydroxyhexanal and hence belongs to the class of aldohexoses. It occurs naturally in many fruits and plants and in concentrations ranging from 0.08 to 0.1% in human blood. A corresponding isomeric ketohexose is **fructose,** the sweetest natural sugar (some synthetic sugars are sweeter), which also occurs in many fruits (*fructus,* Latin, fruit) and in honey. Another important natural sugar is the aldopentose **ribose,** which constitutes a building block of the ribonucleic acids (Section 26-9). The empirical formula for all these sugars is $C_n(H_2O)_n$, which is equivalent to hydrated carbon. This is one of the reasons why the compounds in this class are called carbohydrates.

CHO
|
H—C—OH
|
CH_2OH

2,3-Dihydroxypropanal (Glyceraldehyde)
(An aldotriose)

CH_2OH
|
C=O
|
CH_2OH

1,3-Dihydroxypropanone (1,3-Dihydroxyacetone)
(A ketotriose)

Glucose	Fructose	Ribose
CHO	CH_2OH	
H—C—OH	C=O	CHO
HO—C—H	HO—C—H	H—C—OH
H—C—OH	H—C—OH	H—C—OH
H—C—OH	H—C—OH	H—C—OH
CH_2OH	CH_2OH	CH_2OH
Glucose (An aldohexose)	**Fructose** (A ketohexose)	**Ribose** (An aldopentose)

EXERCISE 24-1

To which class of sugars do the following monosaccharides belong?

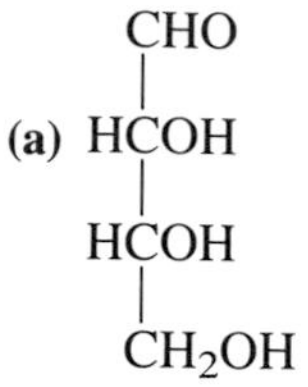

(a) CHO–HCOH–HCOH–CH_2OH **Erythrose**

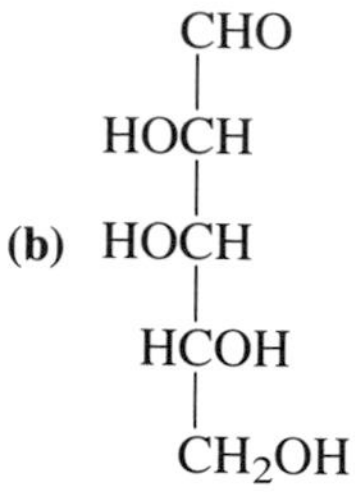

(b) CHO–HOCH–HOCH–HCOH–CH_2OH **Lyxose**

(c) CH_2OH–C=O–HOCH–HCOH–CH_2OH **Xylulose**

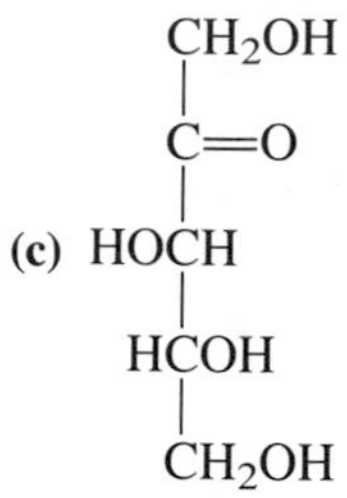

A **disaccharide** is derived from two monosaccharides by formation of an ether (usually, acetal) bridge (Section 17-7). Hydrolysis regenerates the monosaccharides. Ether formation between a mono- and a disaccharide results in a trisaccharide, and repetition of this process eventually produces a natural polymer (polysaccharide). Such polymeric carbohydrates constitute the basic framework of cellulose and starch (Section 24-12).

Most sugars are chiral and optically active

With the exception of 1,3-dihydroxypropanone, all the sugars mentioned so far contain at least one stereocenter. The simplest chiral sugar is 2,3-dihydroxypropanal (glyceraldehyde), with one asymmetric carbon. Its dextrorotatory form is found to be *R*; the levorotatory enantiomer, *S*, as shown in the Fischer projections of the molecule. (See Section 5-4 for a discussion of this notation, which is used extensively to represent sugars.)

Fischer Projections of the Two Enantiomers of 2,3-Dihydroxypropanal (Glyceraldehyde)

CHO / H–C–OH / CH_2OH (Fischer cross) is the same as CHO / H▶C◀OH / CH_2OH

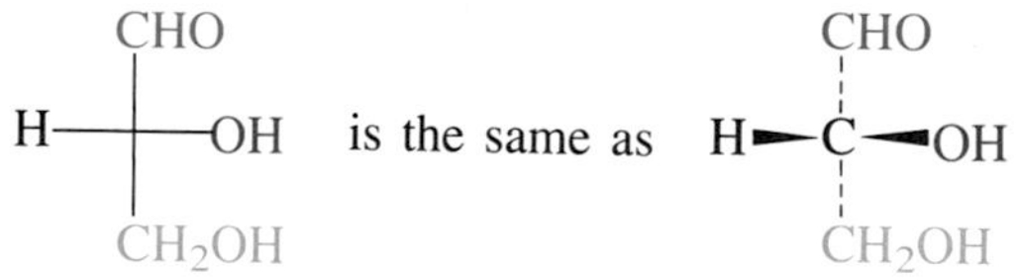

(*R*)-(+)-2,3-Dihydroxypropanal
[D-(+)-Glyceraldehyde]
($[\alpha]_D^{25°C} = +8.7°$)

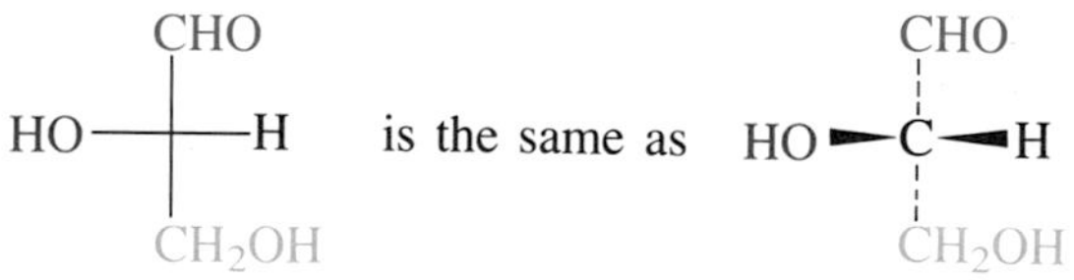

CHO / HO–C–H / CH_2OH (Fischer cross) is the same as CHO / HO▶C◀H / CH_2OH

(*S*)-(−)-2,3-Dihydroxypropanal
[L-(−)-Glyceraldehyde]
($[\alpha]_D^{25°C} = -8.7°$)

Even though *R* and *S* nomenclature is perfectly satisfactory for naming sugars, an older system is still in general use. It was developed before the absolute configuration of sugars were established, and it relates all sugars to 2,3-dihydroxypropanal (glyceraldehyde). Instead of *R* and *S*, it uses the prefixes D for the (+) enantiomer of glyceraldehyde and L for the (−) enantiomer (Section 5-3). Those monosaccharides whose *highest-numbered stereocenter* (i.e., the one farthest from the aldehyde or keto group) has the same absolute configuration as that of D-(+)-2,3-dihydroxypropanal [D-(+)-glyceraldehyde] are then labeled D; those with the opposite configuration at that stereocenter are named L. Two diastereomers that differ *only at one stereocenter* are also called **epimers.**

Designation of a D and an L Sugar

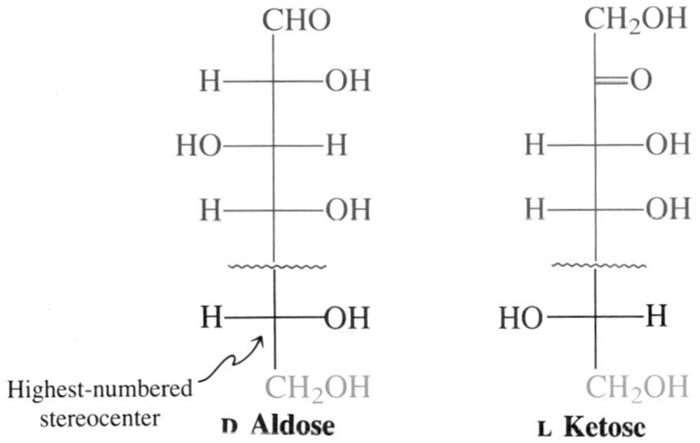

The D,L nomenclature divides the sugars into two groups. As the number of stereocenters increases, so does the number of stereoisomers. For example, the aldotetrose 2,3,4-trihydroxybutanal has two stereocenters and hence may exist as four stereoisomers: two diastereomers, each as a pair of enantiomers.

Like many natural products, these diastereomers have common names that are often used, mainly because the complexity of these molecules leads to long systematic names. This chapter will therefore deviate from our usual procedure of labeling molecules systematically. The isomer of 2,3,4-trihydroxybutanal with 2*R*,3*R* configuration is called erythrose; its diastereomer, threose. Note that each of these has two enantiomers, one belonging to the family of the D sugars, its mirror image to the L sugars. The sign of the optical rotation is not correlated with the D and L label (just as in the *R*,*S* notation; see Section 5-3). For example, D-glyceraldehyde is dextrorotatory, but D-erythrose is levorotatory.

Diastereomeric 2,3,4-Trihydroxybutanals: Erythrose and Threose

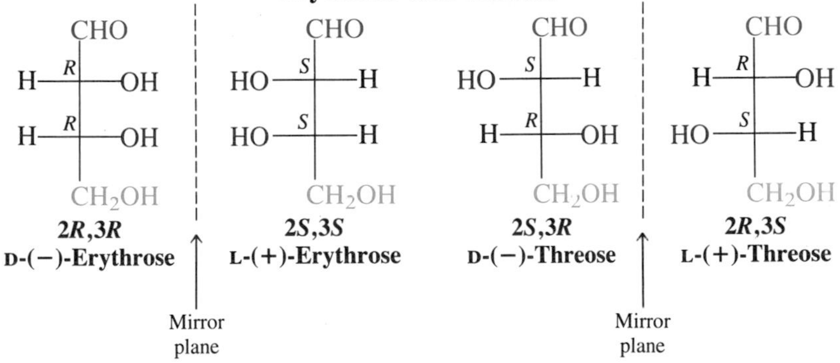

An aldopentose has three stereocenters and hence $2^3 = 8$ stereoisomers. There are $2^4 = 16$ such isomers in the group of aldohexoses. Why then use the D,L nomenclature even though it designates the absolute configuration of only one chiral center? Probably because *almost all naturally occurring sugars have the D configuration*. Evidently somewhere in the structural evolution of the sugar molecules, nature "chose" only one configuration for one end of the chain. The amino acids are another example of such selectivity (Chapter 26).

Figure 24-1 shows the series of D-aldoses up to the aldohexoses, using Fischer projections. To avoid confusion, chemists have adopted a standard way to draw

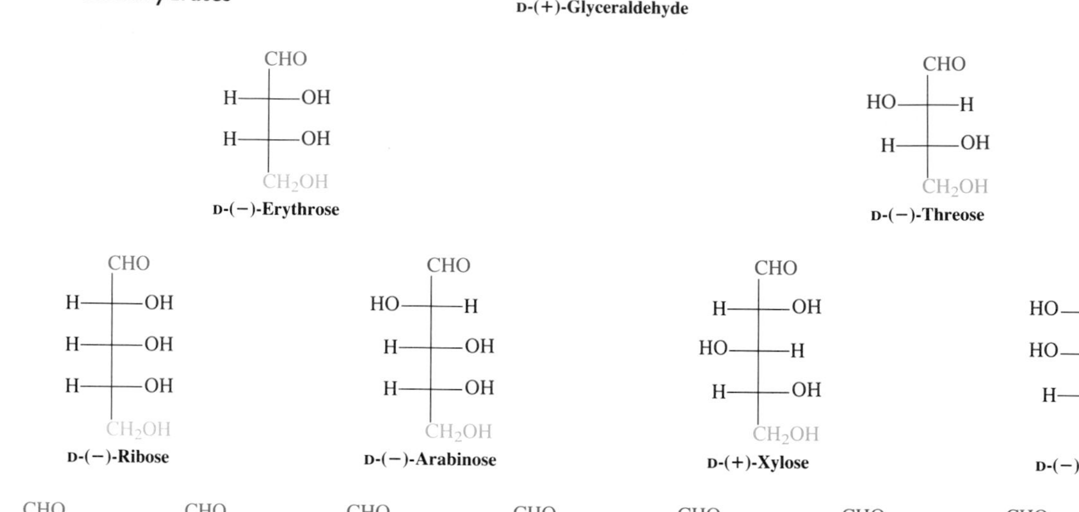

D-(+)-Allose D-(+)-Altrose D-(+)-Glucose D-(+)-Mannose D-(−)-Gulose D-(−)-Idose D-(+)-Galactose D-(+)-Talose

FIGURE 24-1 D-Aldoses (up to the aldohexoses), their signs of rotation, and their common names.

1,3-Dihydroxypropanone

D-(−)-Erythrulose

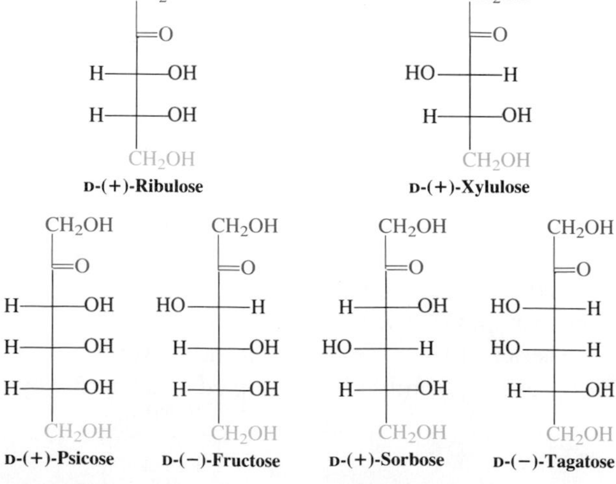

FIGURE 24-2 D-Ketoses (up to the ketohexoses), their signs of rotation, and their common names.

these projections: The carbon chain extends vertically, and the aldehyde terminus is placed at the top. In this convention, the hydroxy group at the highest-numbered stereocenter (at the bottom) points to the right in all D sugars. Figure 24-2 shows the analogous series of ketoses.

EXERCISE 24-2

Give a systematic name for **(a)** D-(−)-ribose and **(b)** D-(+) glucose. Remember to assign the *R* and *S* configuration at each stereocenter.

EXERCISE 24-3

Redraw the dashed-wedged line structure of sugar A (shown in the margin) as a Fischer projection and find its common name in Figure 24-1.

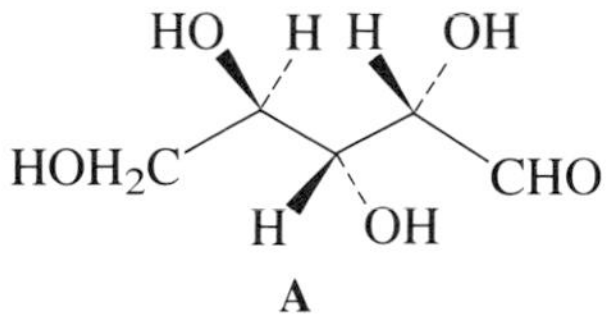

In summary, the simplest carbohydrates are sugars, which are polyhydroxy aldehydes (aldoses) and ketones (ketoses). They are classified as D when the highest-numbered stereocenter is *R*, L when it is *S*. Sugars related to each other by inversion at one stereocenter are called epimers. Most of the naturally occurring sugars belong to the D family.

24-2 Conformations and Cyclic Forms of Sugars

So far, the structures of monosaccharides have been shown in the abstract form of Fischer projections. Dashed-wedged line structures are more realistic, but we must be very careful when we convert a Fischer projection into the correct dashed-wedged line representation (and vice versa). This section describes this procedure and then continues with a discussion of the cyclic isomers that exist in solutions of simple sugars.

Fischer projections depict all-eclipsed conformations

Recall (Section 5-4) that the Fischer projection represents the molecule in an *all-eclipsed arrangement*. It can be translated into an all-eclipsed dashed-wedged line picture.

Fischer Projection and Dashed-Wedged Line Structures for D-(+)-Glucose

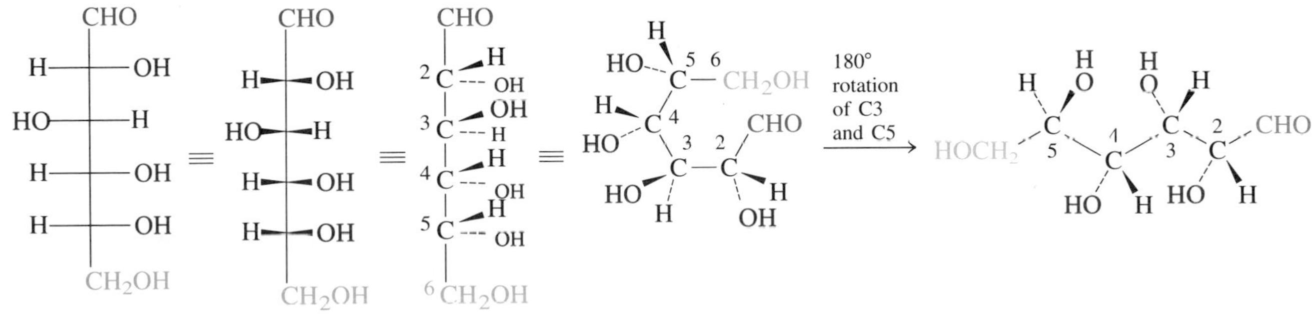

Fischer projection

All-eclipsed dashed-wedged line structure

All-staggered dashed-wedged line structure

A molecular model will help you see that the all-eclipsed form actually possesses a roughly semicircular shape. Notice that the groups on the *right* of the carbon chain in the original Fischer projection now project *downward* in the dashed-wedged line structure. Rotation of two alternate internal carbons by 180° (C3 and C5 in this example) gives the all-staggered conformation.

Sugars form intramolecular hemiacetals

Sugars are hydroxycarbonyl compounds that should be capable of intramolecular hemiacetal formation (see Section 17-7). Indeed, glucose and the other hexoses and pentoses exist as an equilibrium mixture with their cyclic hemiacetal isomers, in which the latter strongly predominate. In principle, any one of the five hydroxy groups could add to the carbonyl group of the aldehyde. However, although five-membered ring formation is known, six-membered rings are usually the preferred product.

To correctly depict a sugar in cyclic form, redraw the dashed-wedged line representation of the all-eclipsed structure with the C2–C3 bond raised above the plane of the paper, as shown at the left in the display below. Rotation of C5 places its hydroxyl group in position to form a six-membered cyclic hemiacetal by addition to the C1 aldehyde carbon. Similarly, a five-membered ring can be made by rotation of C4 to place its OH group in position to bond to C1. This procedure is general for all sugars in the D series.

Cyclic Hemiacetal Formation by Glucose

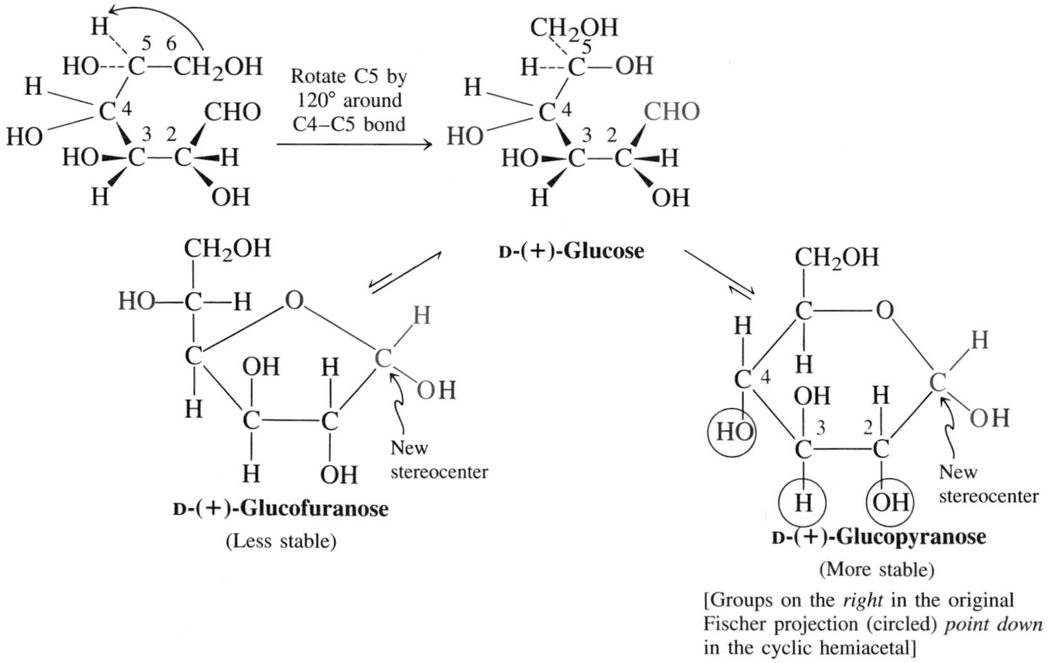

Rotate C5 by 120° around C4–C5 bond

New stereocenter

D-(+)-Glucose

D-(+)-Glucofuranose

(Less stable)

D-(+)-Glucopyranose

(More stable)

[Groups on the *right* in the original Fischer projection (circled) *point down* in the cyclic hemiacetal]

EXERCISE 24-4

Draw the Fischer projection of L-(−)-glucose and illustrate its transformation into the corresponding six-membered cyclic hemiacetal.

The six-membered ring structure of a monosaccharide is called a **pyranose,** a name derived from *pyran,* a six-membered cyclic ether (see Sections 9-6 and 25-1). Sugars in the five-membered ring form are called **furanoses,** from *furan.* In contrast with glucose which exists primarily as the pyranose, fructosc forms both fructopyranose and fructofuranose, in a rapidly equilibrating 70:30 mixture.

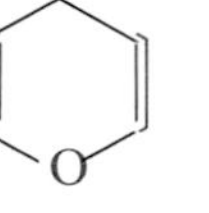

Pyran

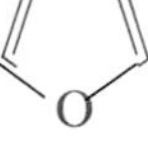

Furan

Cyclic Hemiacetal Formation by Fructose

D-Fructose

New stereocenter

30%

D-(−)-Fructofuranose

New stereocenter

70%

D-(−)-Fructopyranose

Note that, on cyclization, the carbonyl carbon turns into a new stereocenter. As a consequence, hemiacetal formation leads to *two* new compounds, two diastereomers differing in the configuration of the hemiacetal group. If that configuration is S, the sugar is labeled α; when it is R, it is called β. Hence, for example, glucose may form α- or β-glucopyranose or -furanose. Because this type of diastereomer formation is unique to sugars, such isomers have been given a separate name: **anomers.** The new stereocenter is called the **anomeric carbon.**

EXERCISE 24-5

The anomers α- and β-glucopyranose should form in equal amounts because they are enantiomers. True or false? Explain your answer.

Fischer, Haworth, and chair cyclohexane projections help depict cyclic sugars

How can we represent the stereochemistry of the cyclic forms of sugars? One approach uses Fischer projections. We simply draw elongated lines to indicate the bonds formed on cyclization, preserving thc basic "grid" of the original formula.

Adapted Fischer Projections of Glucopyranoses

In the Fischer projection of the α form the anomeric OH points toward the *right*.
In the Fischer projection of the β form the anomeric OH is on the *left*.

O=C—H; H—C—OH; HO—C—H; H—C—OH; H—C—OH; CH_2OH —Cyclization→

S: H—C—OH; H—C—OH; HO—C—H; H—C—OH; H—C—(O); CH_2OH

α-D-(+)-Glucopyranose
(m.p. 146°C)

R: HO—C—H; H—C—OH; HO—C—H; H—C—OH; H—C—(O); CH_2OH

β-D-(+)-Glucopyranose
(m.p. 150°C)

Haworth* projections more accurately represent the real three-dimensional structure of the sugar molecule. The cyclic ether is written in line notation as a pentagon or a hexagon, the anomeric carbon placed on the right, and the ether oxygen put on top. The substituents located above or below the ring are attached to vertical lines. In relating the Haworth projection to a three-dimensional structure, the ring bond at the bottom (between C2 and C3) is understood to be *in front* of the plane of the paper, and the ring bonds containing the oxygen are understood to be in back.

Haworth Projections

Groups on the *right* in the Fischer projection point *down* in the Haworth formula.

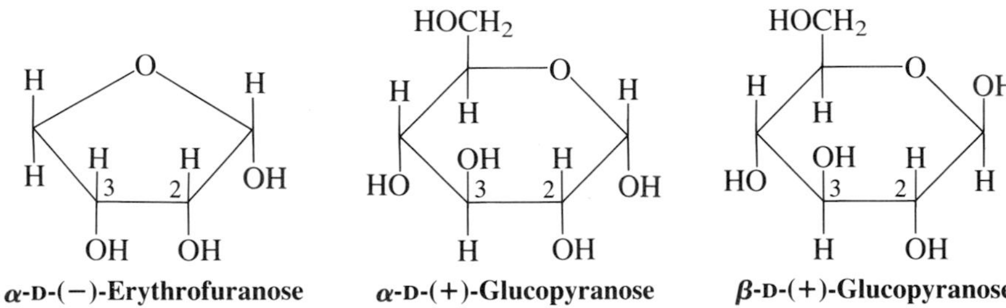

In a Haworth projection, the α anomer has the OH group at the anomeric carbon pointing down, whereas the β anomer has it pointing up.

EXERCISE 24-6

Draw the structure of **(a)** α-D-fructofuranose; **(b)** β-D-glucofuranose; and **(c)** β-D-arabinopyranose.

Haworth projections are used extensively in the literature, but here, to make use of our knowledge of conformation (Sections 4-3 and 4-4), the cyclic forms of sugars will be presented as envelope (for furanoses) or chair (for pyranoses) conformations. As in Haworth notation, the ether oxygen usually will be placed top right, and the anomeric carbon at the right vertex of the envelope or chair.

*Sir W. Norman Haworth, 1883–1950, University of Birmingham, England, Nobel Prize 1937 (chemistry).

Conformational Pictures of Glucofuranose and -pyranose

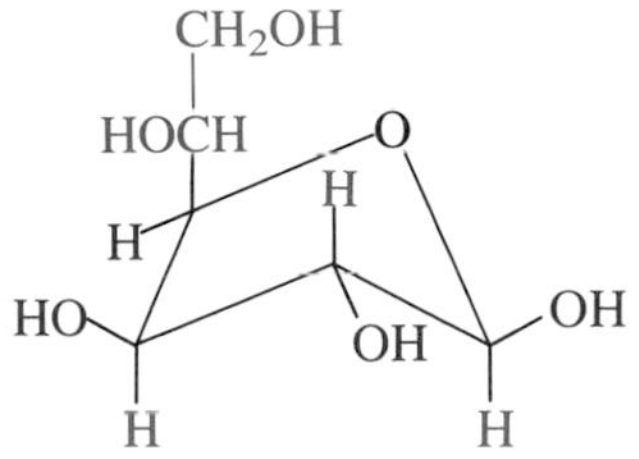

β-D-Glucofuranose

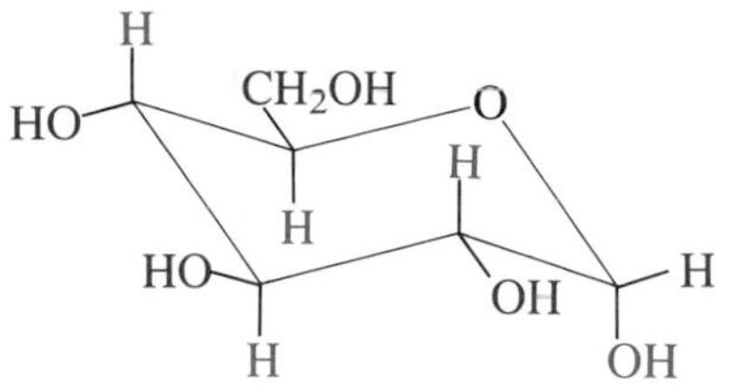

α-D-Glucopyranose

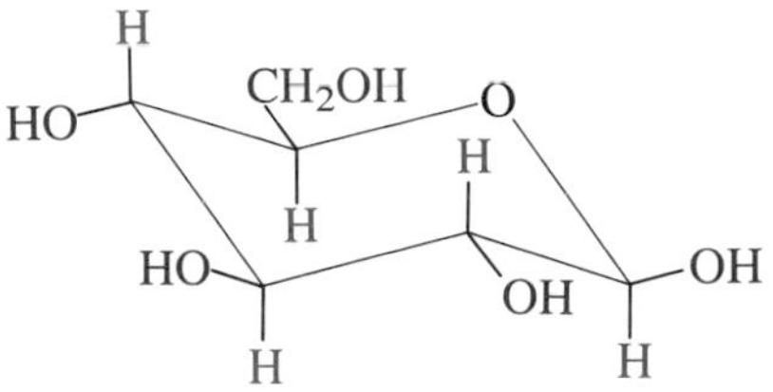

β-D-Glucopyranose

Although there are exceptions, *most aldohexoses adopt the chair conformation* that places the bulky hydroxymethyl group at the C5 terminus *in the equatorial position*. For glucose, this preference means that, in the α form, four of the five substituents can be equatorial, and one is forced to lie axial; in the β form, *all* substituents can be equatorial. This situation is unique for glucose; the other seven D aldohexoses (see Figure 24-1) contain one or more axial substituents.

EXERCISE 24-7

Using the values in Table 4-3, estimate the difference in free energy between the all-equatorial conformer of β-D-glucopyranose and that obtained by ring flip (assume that $\Delta G^\circ_{CH_2OH} = \Delta G^\circ_{CH_3} = 1.7$ kcal mol^{-1} and that the ring oxygen mimics a CH_2 group).

In summary, the hexoses and pentoses can take the form of five- or six-membered cyclic hemiacetals. These structures rapidly interconvert through the open-chain polyhydroxyaldehyde or ketone, with the equilibrium usually favoring the six-membered (pyranose) ring.

24-3 Anomers of Simple Sugars: Mutarotation of Glucose

Glucose precipitates from concentrated solutions at room temperature to give crystals that melt at 146°C. Structural analysis by X-ray diffraction reveals that these crystals contain only the α-D-(+)-glucopyranose anomer (Figure 24-3). If crystalline α-D-(+)-glucoypranose is dissolved in water and its optical rotation measured immediately, a value $[\alpha]_D^{25°C} = +112°$ is obtained. Curiously, this value decreases with time until it reaches a constant $+52.7°$. This change is accelerated by both acids and bases. Evidently, some chemical change alters the initial specific rotation of the sample. The process that gives rise to this effect is the *interconversion* of the α and β anomers.

In solution, the α-pyranose rapidly establishes an equilibrium (in a process that is catalyzed by acid and base; see Section 17-7) with a small amount of the open-chain aldehyde isomer, which in turn undergoes renewed and reversible ring closure to the β anomer.

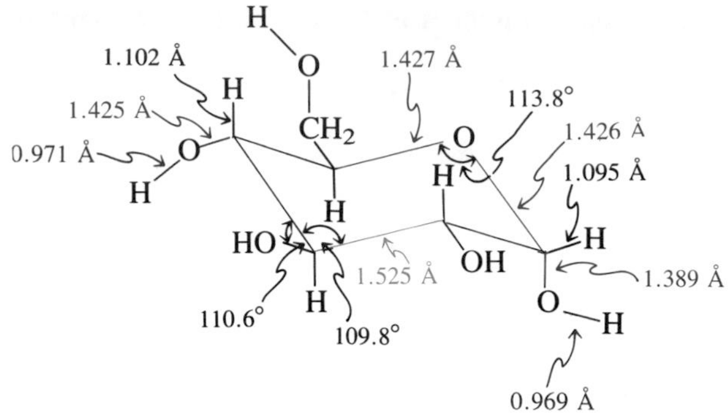

FIGURE 24-3 Structure of α-D-(+)-glucopyranose, with selected bond lengths and angles.

Interconversion of Open-Chain and Pyranose Forms of D-Glucose

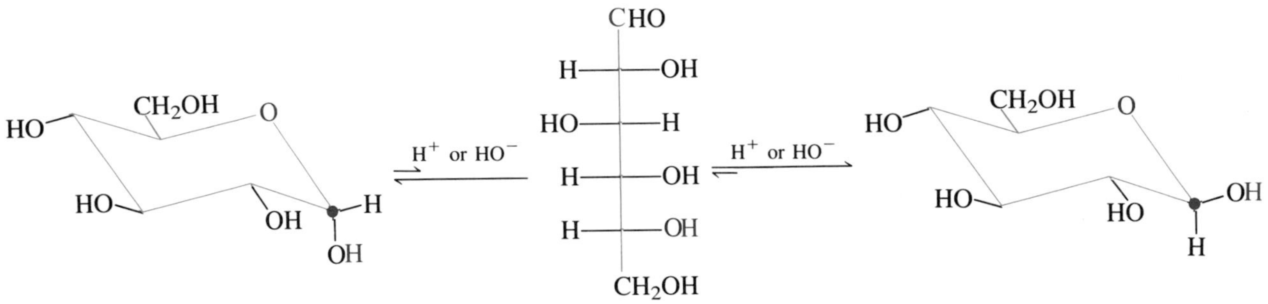

36.4%
α-D-(+)-Glucopyranose
($[\alpha]_D^{25°C} = +112°$)

0.003%
Aldehyde form

63.6%
β-D-(+)-Glucopyranose
($[\alpha]_D^{25°C} = +18.7°$)

The β form has a considerably lower specific rotation (+18.7°) than its anomer; therefore, the observed α value in solution decreases. Similarly, a solution of the pure β anomer [m.p. 150°C, obtainable by crystallizing glucose from acetic acid] gradually increases its specific rotation from +18.7° to +52.7°. At this point, a final equilibrium has been reached, with 36.4% of the α anomer and 63.6% of the β anomer. The change in optical rotation observed when a sugar equilibrates with its anomer is called **mutarotation** (*mutare,* Latin, to change). Interconversion of α and β anomers is a general property of sugars. This includes all monosaccharides capable of existing as cyclic hemiacetals.

EXERCISE 24-8

An alternative mechanism for mutarotation bypasses the aldehyde intermediate and proceeds through oxonium ions. Formulate it.

EXERCISE 24-9

Calculate the equilibrium ratio of α- and β-glucopyranose (that has been given in the text) from the specific rotations of the pure anomers and the observed specific rotation at mutarotational equilibrium.

EXERCISE 24-10

By using Table 4-3, estimate the difference in energy between α- and β-glucopyranose at room temperature (25°C). Then calculate it by using the equilibrium percentage.

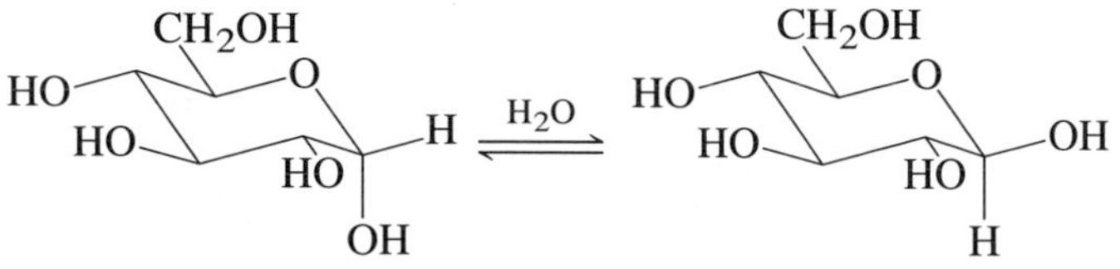

α-D-Glucopyranose $\quad\overset{H_2O}{\rightleftharpoons}\quad$ **β-D-Glucopyranose**

In summary, the hemiacetal carbon (anomeric carbon) can have two configurations: α or β. In solution, the α and β forms of the sugars are in equilibrium with each other. The equilibration can be followed by starting with a pure anomer and observing the changes in specific rotation, a phenomenon also called mutarotation.

24-4 Polyfunctional Chemistry of Sugars: Oxidation to Carboxylic Acids

Simple sugars exist as various isomers: the open-chain carbonyl compound and the α and β anomers of various cyclic forms. Because all of these are rapidly equilibrated, the relative rates of their individual reactions with various reagents will determine the product distribution of a particular transformation. We can therefore divide the reactions of sugars into two groups, those of the linear form and those of the cyclic forms, although sometimes both react competitively. This section describes reactions of aldoses with oxidizing agents, which predominantly attack the open-chain form.

Fehling's and Tollens's tests detect reducing sugars

Because they are polyfunctional compounds, the open-chain monosaccharides undergo the reactions typical of each of their functional groups. For example, aldoses contain the oxidizable aldehyde group and therefore respond to the standard oxidation tests such as exposure to Fehling's or Tollens's solutions (Section 17-14). The α-hydroxy substituent in ketoses is similarly oxidized.

Results of Fehling's and Tollens's Tests on Aldoses and Ketoses

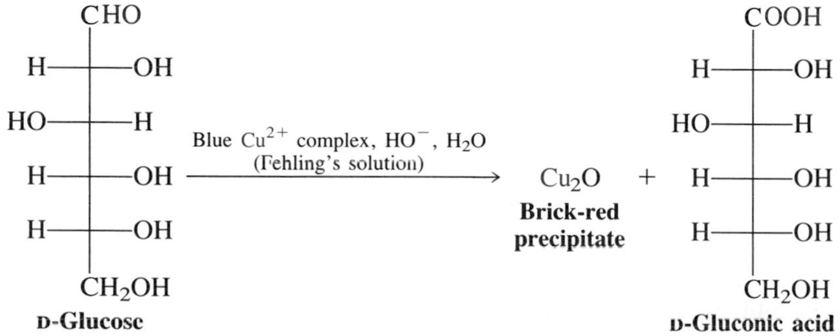

D-Glucose $\xrightarrow[\text{(Fehling's solution)}]{\text{Blue } Cu^{2+} \text{ complex, } HO^-,\ H_2O}$ Cu_2O **Brick-red precipitate** + **D-Gluconic acid** (An aldonic acid)

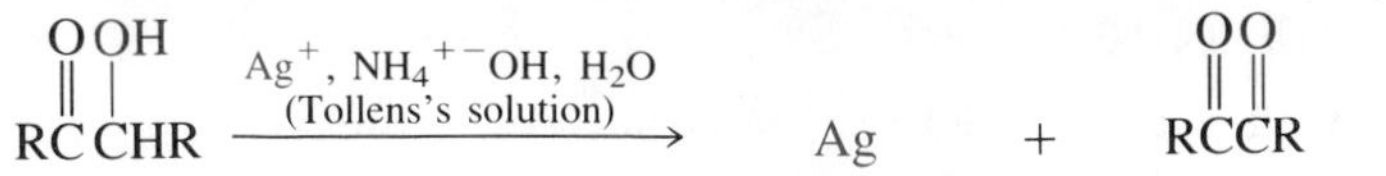

$$\underset{\text{Ketose}}{\text{RC(=O)CH(OH)R}} \xrightarrow[\text{(Tollens's solution)}]{Ag^+,\ NH_4{}^{+\,-}OH,\ H_2O} \underset{\text{Silver mirror}}{Ag} + \underset{\alpha\text{-Dicarbonyl compound}}{\text{RC(=O)C(=O)R}}$$

In these reactions, the aldoses are transformed into **aldonic acids,** ketoses into α-dicarbonyl compounds. Sugars that respond positively to these tests are called **reducing sugars.** All ordinary monosaccharides are reducing sugars.

Oxidation of aldoses can give mono- or dicarboxylic acids

Aldonic acids are made on a preparative scale by oxidation of aldoses with bromine in buffered aqueous solution (pH = 5–6). For example, D-mannose yields D-mannonic acid in this way.

Aldonic Acid Preparation

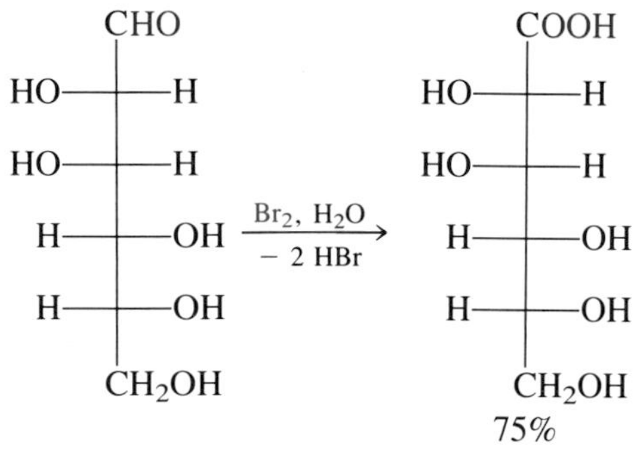

D-Mannose (Fischer projection: CHO; HO–C–H; HO–C–H; H–C–OH; H–C–OH; CH_2OH) $\xrightarrow[-\ 2\ HBr]{Br_2,\ H_2O}$ D-Mannonic acid (COOH; HO–C–H; HO–C–H; H–C–OH; H–C–OH; CH_2OH), 75%

On evaporation of solvent from the aqueous solution of an aldonic acid, the γ-lactone (Section 20-4) forms spontaneously.

Dehydration of an Aldonic Acid to Give an Aldonolactone

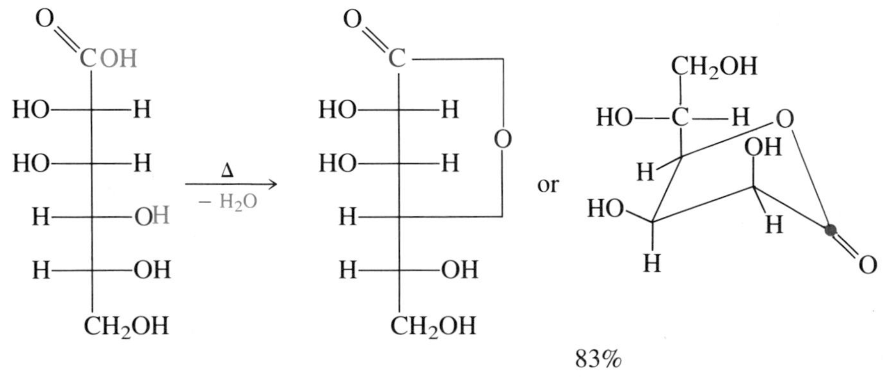

D-Mannonic acid (O=COH; HO–C–H; HO–C–H; H–C–OH; H–C–OH; CH_2OH) $\xrightarrow[-\ H_2O]{\Delta}$ D-Mannono-γ-lactone (O=C; HO–C–H; HO–C–H; H–C–O (ring to C); H–C–OH; CH_2OH) or (ring form: CH_2OH, HO—C—H, O, OH, H, HO, H, H, O), 83%

More vigorous oxidation of an aldose leads to attack at the primary alcohol function as well as at the aldehyde group. The resulting dicarboxylic acid is called an **aldaric,** or **saccharic, acid.** This oxidation can be achieved with warm dilute aqueous nitric acid (see Section 19-6). For example, D-mannose is converted into D-mannaric acid under these conditions.

Preparation of an Aldaric Acid

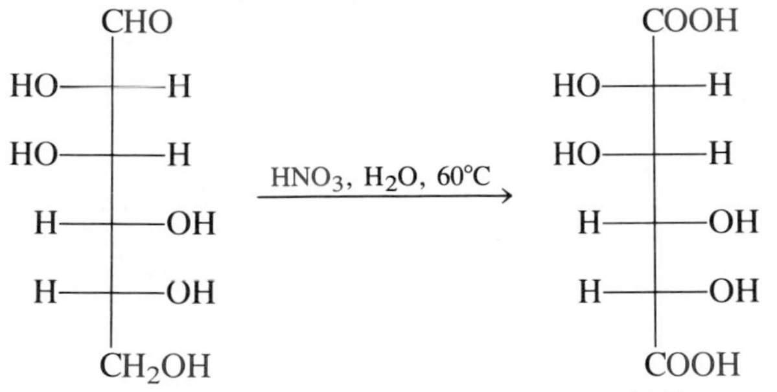

D-Mannose (CHO; HO–H; HO–H; H–OH; H–OH; CH_2OH) $\xrightarrow{HNO_3,\ H_2O,\ 60°C}$ D-Mannaric acid (COOH; HO–H; HO–H; H–OH; H–OH; COOH), 44%

D-Mannose **D-Mannaric acid**

EXERCISE 24-11

The two sugars D-allose and D-glucose (Figure 24-1) differ in configuration only at C3. If you did not know which was which and you had samples of both, a polarimeter, and nitric acid at your disposal, how could you distinguish the two? (Hint: Write the products of oxidation.)

In summary, the chemistry of the sugars is largely that expected for carbonyl compounds containing several hydroxy substituents. Oxidation (by Br_2) of the aldehyde group of aldoses gives aldonic acids; more vigorous oxidation (by HNO_3) converts sugars into aldaric acids.

24-5 Oxidative Cleavage of Sugars

The methods for oxidation of sugars discussed so far leave the basic skeleton intact. A reagent that leads to C–C bonds rupture is periodic acid, HIO_4. This compound oxidatively degrades vicinal diols to give carbonyl compounds.

Oxidative Cleavage of Vicinal Diols with Periodic Acid

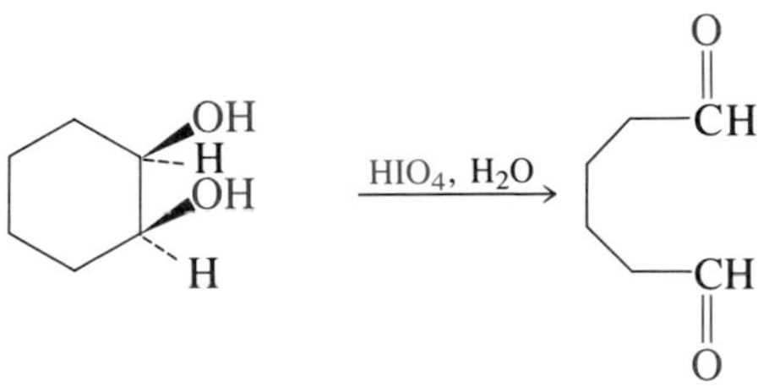

cis-1,2-Cyclohexanediol $\xrightarrow{HIO_4,\ H_2O}$ Hexanedial ($OHC(CH_2)_4CHO$), 77%

***cis*-1,2-Cyclohexanediol** **Hexanedial**

The mechanism of this transformation proceeds through a cyclic **periodate ester,** which decomposes to give two carbonyl groups.

Mechanism of Periodic Acid Cleavage of Vicinal Diols

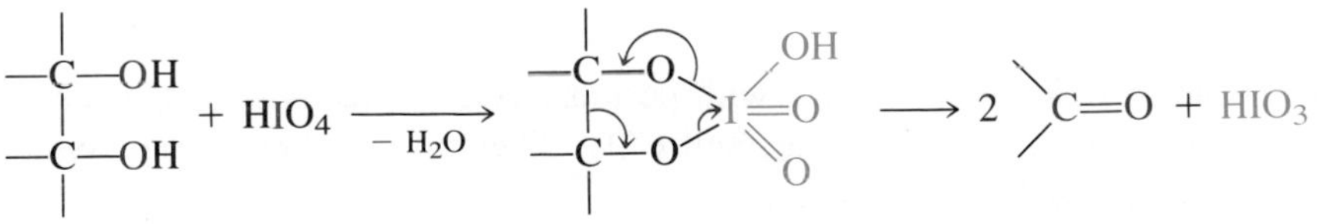

$$-\overset{|}{\underset{|}{C}}-OH,\ -\overset{|}{\underset{|}{C}}-OH + HIO_4 \xrightarrow[-\ H_2O]{} \text{cyclic periodate ester } (C\text{–}O\text{–}I(OH)(=O)_2\text{–}O\text{–}C) \longrightarrow 2\ \rangle C{=}O + HIO_3$$

Cyclic periodate ester

Because most sugars contain several pairs of vicinal diols, oxidation with HIO_4 can give complex mixtures. Sufficient oxidizing agent causes complete degradation of the chain to one-carbon compounds, a technique that has been applied in the structural elucidation of sugars. For example, treatment of glucose with five equivalents of HIO_4 results in the formation of five equivalents of formic acid and one of formaldehyde. Similar degradation of the isomeric fructose consumes an equal amount of oxidizing agent, but the products are three equivalents of the acid, two of the aldehyde, and one of carbon dioxide.

Periodic Acid Degradation of Sugars

D-Glucose (1 CHO; 2 H–C–OH; 3 HO–C–H; 4 H–C–OH; 5 H–C–OH; 6 CH_2OH) $\xrightarrow{5\ HIO_4}$ 5 HCOOH (From C1–C5) + HCHO (From C6)

$$\text{D-Glucose} \xrightarrow{5\,HIO_4} 5\,HCOH\,(=O) + HCH\,(=O)$$

D-Fructose (1 CH_2OH; 2 C=O; 3 HO–C–H; 4 H–C–OH; 5 H–C–OH; 6 CH_2OH) $\xrightarrow{5\ HIO_4}$ 3 HCOOH (From C3–C5) + 2 HCHO (From C1 and C6) + CO_2 (From C2)

D-Glucose **D-Fructose**

It is found that (1) the breaking of each C–C bond in the sugar consumes one molecule of HIO_4; (2) each aldehyde and secondary alcohol unit furnishes an equivalent of formic acid; and (3) the primary hydroxy function gives formaldehyde. The carbonyl group in ketoses gives CO_2. The number of equivalents of HIO_4 consumed reveals the size of the sugar molecule, and the ratios of products are important clues to the number and arrangement of hydroxy and carbonyl functions. In particular, notice that after degradation each carbon fragment retains the same number of attached hydrogen atoms as were present in the original sugar.

EXERCISE 24-12

Write the expected products (and their ratios), if any, of the treatment of the following compounds with HIO_4. **(a)** 1,2-Ethanediol (ethylene glycol); **(b)** 1,2-propanediol; **(c)** 1,2,3-propanetriol; **(d)** 1,3-propanediol; **(e)** 2,4-dihydroxy-3,3-dimethylcyclobutanone; **(f)** D-threose.

EXERCISE 24-13

Would degradation with HIO_4 permit the following sugars to be distinguished? Explain. (For structures, see Figures 24-1 and 24-2.) **(a)** D-Arabinose and D-glucose; **(b)** D-erythrose and D-erythrulose; **(c)** D-glucose and D-mannose.

In summary, oxidative cleavage with periodic acid degrades the sugar backbone to formic acid, formaldehyde, and CO_2. The ratio of these products depends on the structure of the sugar.

24-6 Reduction of Monosaccharides to Alditols

Aldoses and ketoses are reduced by the same types of reducing agents that convert aldehydes and ketones into alcohols. The resulting polyhydroxy compounds are called **alditols.** For example, D-glucose gives D-glucitol (older name, D-sorbitol) when treated with sodium borohydride. The hydride reducing agent traps the small amount of the open-chain form of the sugar, in this way shifting the equilibrium from the unreactive cyclic hemiacetal to the product.

Preparation of an Alditol

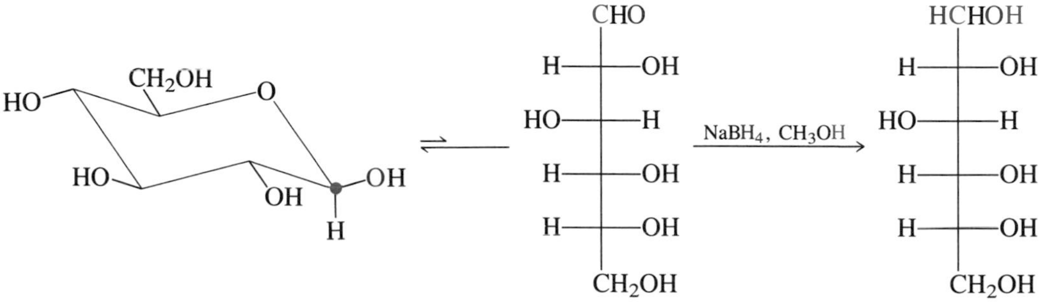

D-Glucose (cyclic) ⇌ open-chain form (CHO; H–OH; HO–H; H–OH; H–OH; CH_2OH) $\xrightarrow{NaBH_4,\ CH_3OH}$ HCHOH; H–OH; HO–H; H–OH; H–OH; CH_2OH

D-Glucose

D-Glucitol
(D-Sorbitol)

Many alditols occur in nature. D-Glucitol is found in red seaweed in concentrations as high as 14%, also in many berries (but not in grapes), in cherries, in plums, in pears, and in apples. It is prepared commercially by high-pressure hydrogenation of D-glucose or by electrochemical reduction.

EXERCISE 24-14

(a) Reduction of D-ribose with $NaBH_4$ gives a product without optical activity. Explain. **(b)** Similar reduction of D-fructose gives two optically active products. Explain.

BOX 24-1 Glucitol (Sorbitol) as a Sweetening Agent

D-Glucitol (D-sorbitol) is about 60% as sweet as common table sugar (sucrose; see Section 24-11) and is commonly encountered in products such as hard candy and chewing gums. You may be surprised to discover that this substance is *not* a noncaloric sweetener. Indeed, it gives rise to virtually the same number of calories upon human ingestion as does an equal weight of sucrose! However, the bacteria responsible for the initiation of tooth decay are unable to consume glucitol. In addition, its metabolism furnishes mostly CO_2. As a result, pharmaceutical preparations for diabetics often use glucitol in place of sucrose, which is metabolized to glucose.

Additional uses of this substance rely on the properties of its aqueous solutions. Such mixtures are viscous, have very low freezing points, and do not lose moisture readily. They are thus used in antifreeze preparations. Glucitol is also employed to improve the flow and drying properties of ink formulations.

In summary, reduction of the carbonyl function in aldoses and ketoses (by $NaBH_4$) furnishes alditols.

24-7 Carbonyl Condensations with Amine Derivatives

As might be expected, the carbonyl function in aldoses and ketoses undergoes condensation reactions with amine derivatives (Section 17-9). For example, treatment of D-mannose with phenylhydrazine gives the corresponding **hydrazone,** D-mannose phenylhydrazone. Surprisingly, the reaction does not stop at this stage but can be induced to continue with additional phenylhydrazine (two extra equivalents). The final product is a double phenylhydrazone, also called an **osazone** (here, phenylosazone). In addition, one equivalent each of benzenamine (aniline), ammonia, and water is generated.

Phenylhydrazone and Phenylosazone Formation

D-Mannose		D-Mannose phenylhydrazone		A phenylosazone
H–C=O		H–C=N—NHC_6H_5		H–C=N—NHC_6H_5
HO—C—H	$C_6H_5NHNH_2$, CH_3CH_3OH, Δ, 30 min → $- H_2O$	HO—C—H	2 $C_6H_5NHNH_2$, CH_3CH_2OH, Δ → $- C_6H_5NH_2$, $- NH_3$, $- H_2O$	C=N—NHC_6H_5
HO—C—H		HO—C—H		HO—C—H
H—C—OH		H—C—OH		H—C—OH
H—C—OH		H—C—OH		H—C—OH
CH_2OH		CH_2OH		CH_2OH
		75%		95%

The mechanism of osazone synthesis is thought to involve tautomerism to a 2-keto amine, followed by a complex sequence of eliminations and condensations. Once formed, the osazones do not continue to react with excess phenylhydrazine but are stable under the conditions of the reaction.

Historically, the discovery of osazone formation marked a significant advance in the practical aspects of sugar chemistry. Sugars, like many other polyhydroxy compounds, are well known for their reluctance to crystallize from syrups. Their osazones, however, readily form yellow crystals with sharp melting points, thus simplifying the isolation and characterization of many sugars, particularly if they have been formed as mixtures or are impure.

EXERCISE 24-15

Compare the structures of the phenylosazones of D-glucose, D-mannose, and D-fructose. Do you notice anything unusual?

In summary, one equivalent of phenylhydrazine converts a sugar into the corresponding phenylhydrazone. Additional hydrazine reagent causes oxidation of the center adjacent to the hydrazone function to furnish the osazone.

24-8 Ester and Ether Formation: Glycosides

Because of their multiple hydroxy groups, sugars can be converted into several alcohol derivatives. This section explores the formation of simple esters and ethers of monosaccharides as well as reactions that take place selectively at the anomeric carbon.

Sugars can be esterified and methylated

Esters can be prepared from monosaccharides by standard techniques (Sections 19-9, 20-2, and 20-3). Excess reagent will completely convert all hydroxy groups, including the hemiacetal function. For example, acetic anhydride transforms β-D-glucopyranose into the pentaacetate.

Complete Esterification of Glucose

CH_2OH, HO, HO, OH, OH, H, O

5 $CH_3\overset{O}{\overset{\|}{C}}O\overset{O}{\overset{\|}{C}}CH_3$, pyridine, 0°C, 24 h; $- 5\ CH_3COOH$

$CH_2O\overset{O}{\overset{\|}{C}}CH_3$, $CH_3\overset{O}{\overset{\|}{C}}O$, $CH_3\overset{}{C}O$ (=O), $O\overset{O}{\overset{\|}{C}}CH_3$, O=C–$CH_3$, H

91%

β-D-Glucopyranose → **β-D-Glucopyranose pentaacetate**

Williamson ether synthesis (Section 9-6) allows complete methylation.

Complete Methylation of a Pyranose

HO, OH, OH, OH, H, O

4 CH_3OSOCH_3 (with two S=O), NaOH; $- 4\ CH_3OSO_3Na$, $- 4$ HOH

CH_3O, OCH_3, OCH_3, OCH_3, H, O

70%

β-D-Ribopyranose → **β-D-Ribopyranose tetramethyl ether**

Notice that the hemiacetal function at C1 is converted into an acetal group. The latter can be selectively hydrolyzed back to the hemiacetal (see Section 17-7).

Selective Hydrolysis of an Acetal in the Presence of Ether Groups

CH_3O, OCH_3, OCH_3, OCH_3, H, O

8% HCl, HOH, Δ

CH_3O, OCH_3, OCH_3, OH, O $+ CH_3OH$

67%

D-Ribopyranose trimethyl ether
(Mixture of α and β forms)

It is also possible to convert the hemiacetal unit of a sugar selectively into the acetal. For example, treatment of D-glucose with acidic methanol leads to the formation of the two methyl acetals. Sugar acetals are called **glycosides.** Thus, glucose forms **glucosides.**

Selective Preparation of a Glycoside (Sugar Acetal)

α- or β-D-Glucopyranose

$\xrightarrow[-\,HOH]{CH_3OH,\ 0.25\%\ HCl,\ H_2O}$

Methyl α-D-glucopyranoside
(m.p. 166°C, $[\alpha]_D^{25°C} = +158°$)

+

Methyl β-D-glucopyranoside
(m.p. 105°C, $[\alpha]_D^{25°C} = -33°$)

Because glycosides contain a blocked anomeric carbon atom, they do not show mutarotation in the absence of acid, they test negatively to Fehling's and Tollens's reagents (they are *non*reducing sugars), and they are unreactive toward reagents that attack carbonyl groups. Such protection can be useful in synthesis and in structural analysis (see Exercise 24-16).

EXERCISE 24-16

The same mixture of glucosides is formed in the methylation of D-glucose with acidic methanol, regardless of whether you start with the α or β form. Why?

EXERCISE 24-17

Draw the structure of methyl α-D-arabinofuranoside.

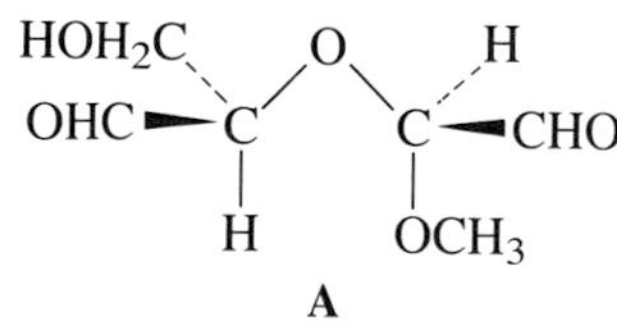

EXERCISE 24-18

Methyl α-D-glucopyranoside consumes two equivalents of HIO_4 to give one equivalent each of formic acid and dialdehyde A (shown in the margin). An unknown aldopentose methyl furanose reacted with one equivalent of HIO_4 to give A, but no formic acid. Suggest a structure for the unknown. Is there more than one solution to this problem?

Neighboring hydroxy groups in sugars can be linked as cyclic ethers

The presence of neighboring pairs of hydroxy groups in the sugars allows for the formation of cyclic ether derivatives. For example, it is possible to synthesize five- or six-membered cyclic sugar acetals from the vicinal (and also β-diol) units by treating them with carbonyl compounds (Section 17-8).

Cyclic Acetal Formation from Vicinal Diols

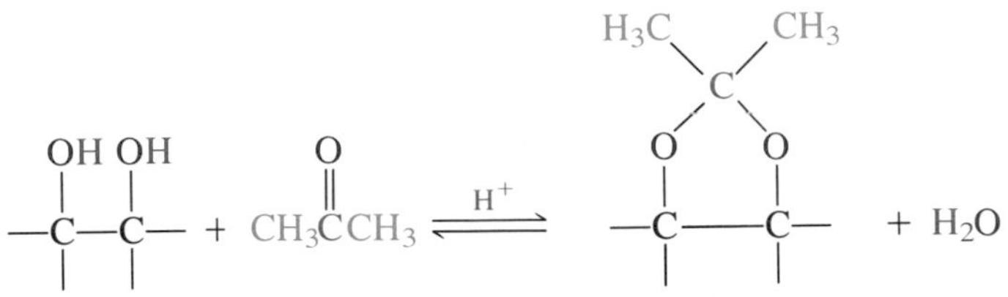

Propanone (acetone) acetal

Such processes work best if the two OH groups are positioned cis to allow a relatively unrestrained ring to form. For example, excess acidic propanone (acetone) converts β-D-arabinopyranose into the double acetal.

Conversion of *cis*-Diols into Cyclic Acetals

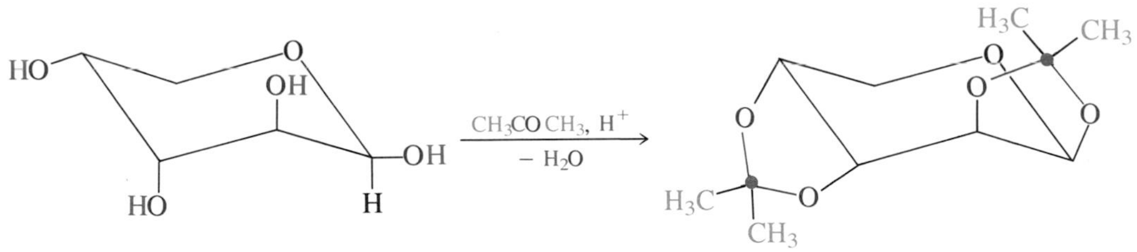

β-D-Arabinopyranose

β-D-Arabinopyranose double acetal

Cyclic acetal and ester formation is often employed to protect selected alcohol functions. The remaining hydroxy groups can then be oxidized to carbonyl compounds, converted into leaving groups, or transformed by elimination.

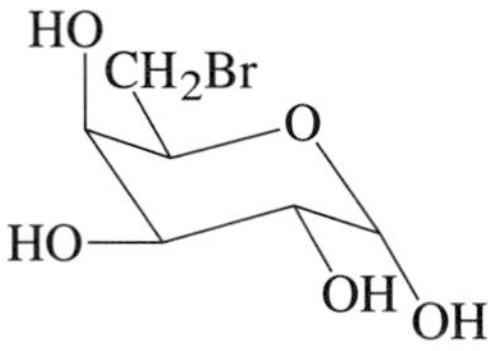

EXERCISE 24-19

Suggest a synthesis of the compound shown in the margin from D-galactose. (Hint: Consider a strategy involving protecting groups.)

In summary, the various hydroxy groups of sugars can be esterified or converted into ethers. The hemiacetal unit can be selectively protected as the acetal, also called a glycoside. Finally, the various diol units in the sugar backbone can be linked as cyclic acetals, depending on steric requirements.

24-9 Step-by-Step Buildup and Degradation of Sugars

Larger sugars can be made from smaller ones and vice versa, by chain lengthening and chain shortening. These transformations can also be used to structurally correlate various sugars, a procedure applied by Fischer to prove the relative configuration of all the stereocenters in the aldoses shown in Figure 24-1.

BOX 24-2 Biosynthesis of Sugars

In nature carbohydrates are produced primarily by a reaction sequence called *photosynthesis*. In this process sunlight impinging on the chlorophyll of green plants is absorbed, and the photochemical energy thus obtained is used to convert carbon dioxide and water into oxygen and the polyfunctional structure of carbohydrates.

Photosynthesis of Glucose in Green Plants

$$6\ CO_2 + 6\ H_2O \xrightarrow[\text{Released metabolic energy}]{\text{Sunlight, chlorophyll}} \underset{\textbf{Glucose}}{C_6(H_2O)_6} + 6\ O_2$$

The detailed mechanism of this transformation is complicated and takes many steps, the first of which is the absorption of one quantum of light by the extended π system (Chapter 14) of chlorophyll, shown below. The mechanism of enzymatic degradation of carbohydrates is still the subject of much research. The cycle of photosynthesis and carbohydrate metabolism is a beautiful example of how nature reuses its resources. First, CO_2 and H_2O are consumed to convert solar energy into chemical energy and O_2. When an organism needs some of the stored energy, it is generated by conversion of carbohydrate into CO_2 and H_2O, using up roughly the same amount of O_2 originally liberated.

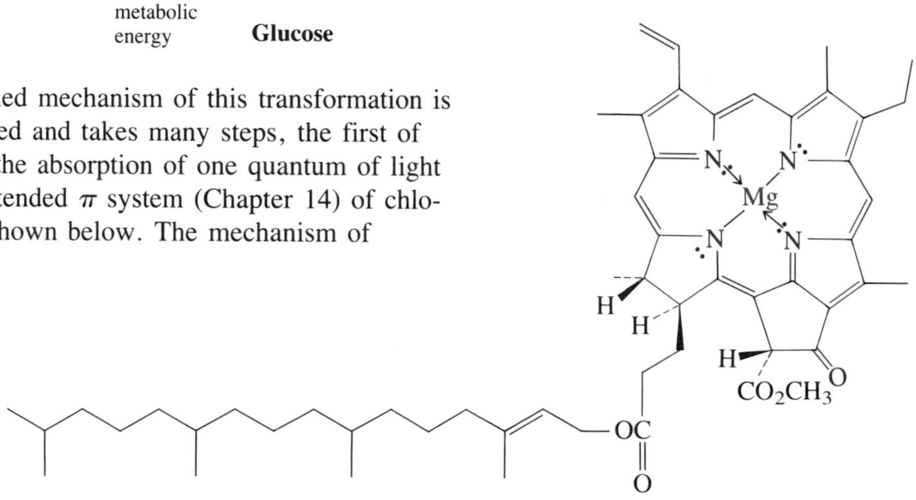

Chlorophyll a

Kiliani-Fischer synthesis lengthens the chain

In the (modified) **Kiliani-Fischer* synthesis** of sugars, an aldose is first treated with HCN to give the corresponding cyanohydrin. Because this transformation forms a new stereocenter, two diastereomers appear. Separation of the diastereomers and partial reduction of the nitrile group by catalytic hydrogenation in aqueous acid then gives the aldehyde groups of the chain-extended sugars.

*Professor Heinrich Kiliani, 1855–1945, University of Freiburg, Germany; Professor Emil Fischer, see Section 5-4.

Kiliani-Fischer Synthesis of Sugars

STEP 1. Cyanohydrin formation

New stereocenter

$$\begin{array}{c} HC{=}O \\ H-\!\!\!-\!\!\!-OH \\ HO-\!\!\!-\!\!\!-H \end{array} \xrightarrow{HCN} \begin{array}{c} CN \\ H-\!\!\!\bullet\!\!\!-OH \\ H-\!\!\!-\!\!\!-OH \\ HO-\!\!\!-\!\!\!-H \end{array} + \begin{array}{c} CN \\ HO-\!\!\!\bullet\!\!\!-H \\ H-\!\!\!-\!\!\!-OH \\ HO-\!\!\!-\!\!\!-H \end{array}$$

Two new diastereomeric nitriles

STEP 2. Reduction and hydrolysis (only one diastereomer is shown)

$$\begin{array}{c} C{\equiv}N \\ H-\!\!\!-\!\!\!-OH \\ H-\!\!\!-\!\!\!-OH \\ HO-\!\!\!-\!\!\!-H \end{array} \xrightarrow{H_2,\ Pd\text{-}BaSO_4,\ H^+,\ H_2O,\ 4\ atm} \begin{array}{c} HC{=}NH \\ H-\!\!\!-\!\!\!-OH \\ H-\!\!\!-\!\!\!-OH \\ HO-\!\!\!-\!\!\!-H \end{array} \xrightarrow[-\ NH_2H]{H_2O} \begin{array}{c} HC{=}O \\ H-\!\!\!-\!\!\!-OH \\ H-\!\!\!-\!\!\!-OH \\ HO-\!\!\!-\!\!\!-H \end{array}$$

Imine **Extended sugar**

In this hydrogenation, a modified palladium catalyst (similar to the Lindlar catalyst, Section 13-7) allows selective reduction of the nitrile to the imine, which hydrolyzes under the reaction conditions. The special catalyst is necessary to prevent the reduction from proceeding all the way to the amine (Section 20-8).

EXERCISE 24-20

What are the products of Kiliani-Fischer chain extension of **(a)** D-erythrose and **(b)** D-arabinose?

Ruff degradation shortens the chain

Whereas the Kiliani-Fischer approach synthesizes higher sugars, complementary strategies degrade higher to lower sugars one carbon at a time. One of these is the **Ruff* degradation.** This procedure removes the carbonyl group of an aldose and converts the neighboring carbon into the aldehyde function of the new sugar.

The Ruff degradation is an oxidative decarboxylation. The sugar is first oxidized to the aldonic acid by aqueous bromine. Exposure to hydrogen peroxide in the presence of ferric ion then leads to the loss of the carboxy group and oxidation of the new terminus to the aldehyde function of the lower aldose. The mechanism of this decarboxylation is related to the one discussed in Section 19-13.

*Professor Otto Ruff, 1871–1939, University of Danzig, Germany.

Ruff Degradation of Sugars

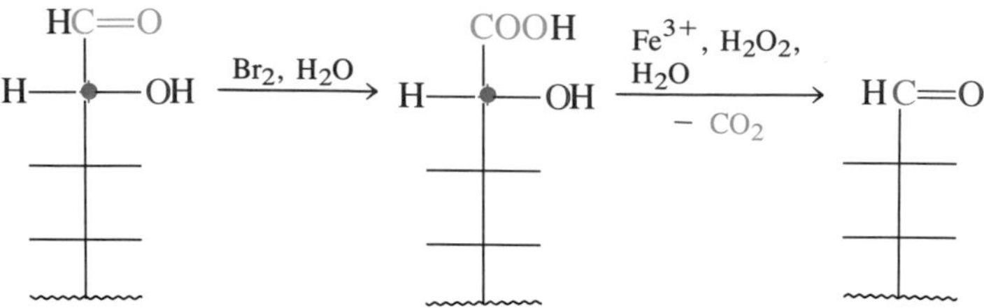

Ruff degradation gives low yields because of the sensitivity of the products to the reaction conditions. Nevertheless, the procedure is useful in structural elucidations (Exercise 24-21). Fischer originally carried out such studies to establish the relative configurations of the monosaccharides (the **Fischer proof**). The next section will describe some of the logic behind his approach to the problem.

EXERCISE 24-21

Ruff degradation of two D pentoses A and B gave two new sugars C and D. Oxidation of C with HNO_3 gave *meso*-2,3-dihydroxybutanedioic (tartaric) acid, that of D resulted in an optically active acid. Oxidation of either A or B with HNO_3 furnished an optically active aldaric acid. What are compounds A, B, C, and D?

In summary, sugars can be made from other sugars by step-by-step one-carbon chain lengthening (Kiliani-Fischer synthesis) or shortening (Ruff degradation).

24-10 Relative Configurations of the Aldoses: An Exercise in Structure Determination

Imagine that we have been presented with 14 jars, each filled with one of the tetroses, pentoses, and hexoses in Figure 24-1, each labeled with a name, but *no structural formula*. How would we establish the structure of each substance?

This was essentially the challenge faced by Fischer in the late nineteenth century, when chemists had no modern spectrometers at their disposal. Fischer showed that this problem can be solved by interpreting the results of a combination of synthetic manipulations. Only one assumption had to be made: The dextrorotatory enantiomer of 2,3-dihydroxypropanal (glyceraldehyde) has the D (and not the L) configuration. This assumption was actually verified only by X-ray structural analysis in 1950, long after Fischer's days. Fischer guessed this absolute configuration and was lucky to have been right; otherwise, all the structures of compounds in Figure 24-1 would have had to be changed into their mirror images. However, at the time it was more important to establish the *relative* configuration of all the stereocenters—that is, to associate each unique sugar with a unique sequence of such centers in its backbone.

The structures of the four- and five-carbon aldoses can be determined from the optical activity of the corresponding aldaric acids

Given the structure of (*R*)-2,3-dihydroxypropanal (D-glyceraldehyde), we will now set out to unambiguously prove the structure of all the higher D aldoses.

(Consult Figure 24-1 as required while following these arguments.) First, we perform a Kiliani-Fischer chain lengthening on D-glyceraldehyde, which gives two new isomeric sugars. Separation and oxidation with nitric acid leads to *meso*-2,3-dihydroxybutanedioic (tartaric) acid from one product and an optically active acid from the other. Therefore, the former must be D-erythrose; the latter, D-threose. Note that the absolute configuration at the next-to-last carbon is the same in both sugars and in our starting material, (*R*)-2,3-dihydroxypropanal (D-glyceraldeyde). Recall that this is a common property of all the D sugars. The difference is at C2: D-Erythrose is 2*R*; D-threose, 2*S*.

CHO; H–C(*R*)–OH; H–C–OH; CH_2OH
D-Erythrose
↓ HNO_3
meso-Tartaric acid

CHO; HO–C(*S*)–H; H–C–OH; CH_2OH
D-Threose
↓ HNO_3
S,*S*-Tartaric acid

Let us now use D-erythrose as our new starting material—because we know its structure—and lengthen the chain further.

CHO; H–C–OH; H–C–OH; CH_2OH
D-Erythrose
Kiliani-Fischer extension

Meso: COOH; H–C–OH; H–C–OH; H–C–OH; COOH ← HNO_3 — **D-Ribose**: CHO; H–C–OH; H–C–OH; H–C–OH; CH_2OH

D-Arabinose: CHO; HO–C–H; H–C–OH; H–C–OH; CH_2OH — HNO_3 → **Optically active**: COOH; HO–C–H; H–C–OH; H–C–OH; COOH

Again, two new sugars ensue (because we have again added a new stereocenter), two pentoses. We know their configuration at C3 and C4 (the same as that in C2 and C3 of their starting material) but not that at C2. Their oxidation again produces one optically inactive dicarboxylic acid and one that is active. The former must therefore have the structure of D-ribose, the latter that of D-arabinose.

A very similar train of thought leads to the unambiguous assignment of the structures of D-xylose (oxidized to a meso dioic acid) and D-lyxose (oxidized to an optically active dioic acid), derived synthetically from D-threose, whose structure we ascertained at the very beginning.

Meso: COOH; H–C–OH; HO–C–H; H–C–OH; COOH ← HNO_3 — **D-Xylose**: CHO; H–C–OH; HO–C–H; H–C–OH; CH_2OH

D-Lyxose: CHO; HO–C–H; HO–C–H; H–C–OH; CH_2OH — HNO_3 → **Optically active**: COOH; HO–C–H; HO–C–H; H–C–OH; COOH

Symmetry properties also define the structures of the six-carbon aldoses

We now know the structures of the four aldopentoses and can extend the chain of each of them. This process gives us four pairs of aldohexoses, each pair distinguished from the other by the unique sequence of stereocenters at C3, C4, and C5. The members of each pair differ only in their configuration at C2.

The structural assignment for the four sugars obtained from D-ribose and D-lyxose, respectively, is again accomplished by oxidation to the corresponding aldaric acids. Both D-allose and D-galactose give optically inactive oxidation products, in contrast with their counterparts D-altrose and D-talose, which give optically active dicarboxylic acids.

D-Allose	**D-Galactose**	**D-Altrose**	**D-Talose**
CHO	CHO	CHO	CHO
H—OH	H—OH	HO—H	HO—H
H—OH	HO—H	H—OH	HO—H
H—OH	HO—H	H—OH	HO—H
H—OH	H—OH	H—OH	H—OH
CH_2OH	CH_2OH	CH_2OH	CH_2OH
(From D-ribose)	(From D-lyxose)	(From D-ribose)	(From D-lyxose)

(Both give meso dicarboxylic acids) (Both give optically active dicarboxylic acids)

The structural assignment of the four remaining sugars cannot be based on the approach taken thus far, because all give optically active diacids on oxidation.

1	**2**	**3**	**4**
CHO	CHO	CHO	CHO
H—OH	HO—H	H—OH	HO—H
HO—H	HO—H	H—OH	H—OH
H—OH	H—OH	HO—H	HO—H
H—OH	H—OH	H—OH	H—OH
CH_2OH	CH_2OH	CH_2OH	CH_2OH

↓ HNO_3

COOH	COOH	COOH	COOH
H—OH	HO—H	H—OH	HO—H
HO—H	HO—H	H—OH	H—OH
H—OH	H—OH	HO—H	HO—H
H—OH	H—OH	H—OH	H—OH
COOH	COOH	COOH	COOH

Enantiomers (1 and 3)

(All optically active)

However, it is found that the two carboxylic acids derived from sugars 1 and 3 are enantiomers—that is, mirror images of each other. This result is possible only if 1 and 3 have the structures of D-glucose or D-gulose. This relation of the two aldaric acids can be verified by building molecular models.

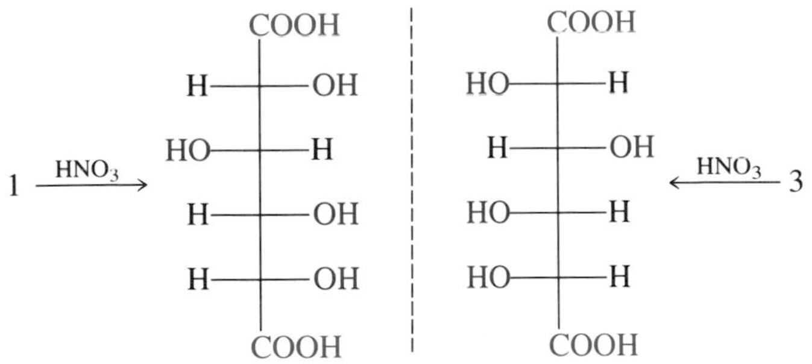

		Mirror plane		
$1 \xrightarrow{HNO_3}$	COOH H—C—OH HO—C—H H—C—OH H—C—OH COOH	¦	COOH HO—C—H H—C—OH HO—C—H HO—C—H COOH	$\xleftarrow{HNO_3} 3$

Mirror plane

We now proceed experimentally as follows. D-Arabinose is converted into a pair of new sugars, 1 and 2, by Kiliani-Fischer chain extension; D-xylose furnishes sugars 3 and 4. With these results in hand, the structural assignments fall into place. Sugar 1 must have the structure of D-glucose, and sugar 3 must have the structure of D-gulose. Therefore, 2 is assigned the structure of D-mannose and 4 that of D-idose.

EXERCISE 24-22

In the preceding discussion, we assigned the structures of D-ribose and D-arabinose by virtue of the fact that on oxidation the first gives a meso dioic acid, the second an optically active isomer. Could you arrive at the same result by ^{13}C NMR spectroscopy?

In summary, step-by-step one-carbon chain lengthening or shortening, in conjunction with the symmetry properties of the various aldaric acids, allows the stereochemical assignments of the aldoses.

24-11 Complex Sugars in Nature: Disaccharides

A substantial fraction of the natural sugars occurs in dimeric, trimeric, higher oligomeric (between 2 and 10 sugar units), and polymeric forms. The sugar most familiar to us is a dimer.

Sucrose is a disaccharide derived from glucose and fructose

Sucrose, ordinary table sugar, is one of the few natural chemicals consumed in unmodified form (water and NaCl are others). Its average yearly annual consumption in the United States is about 150 pounds per person. Sugar is isolated from sugar cane and sugar beets, in which it is particularly abundant (about 14–20% by weight), although it is present in many plants in smaller concentrations. World production is about 100 billion tons a year, and there are countries (e.g., Cuba) whose entire economy depends on the world price of sucrose.

Sucrose has not been discussed in this chapter so far, because it is not a simple monosaccharide but a disaccharide composed of two units, glucose and fructose.

The structure of sucrose can be deduced from its chemical behavior: Acidic hydrolysis splits it into glucose and fructose. It is a nonreducing sugar. It does not form an osazone. It does not undergo mutarotation. These findings suggest that the component monosaccharide units are linked by an acetal bridge connecting the two anomeric carbons; in this way, the two cyclic hemiacetal functions protect each other. X-ray structural analysis confirms this hypothesis: Sucrose is a disaccharide in which the α-D-glucopyranose form of glucose is attached to β-D-fructofuranose in this way.

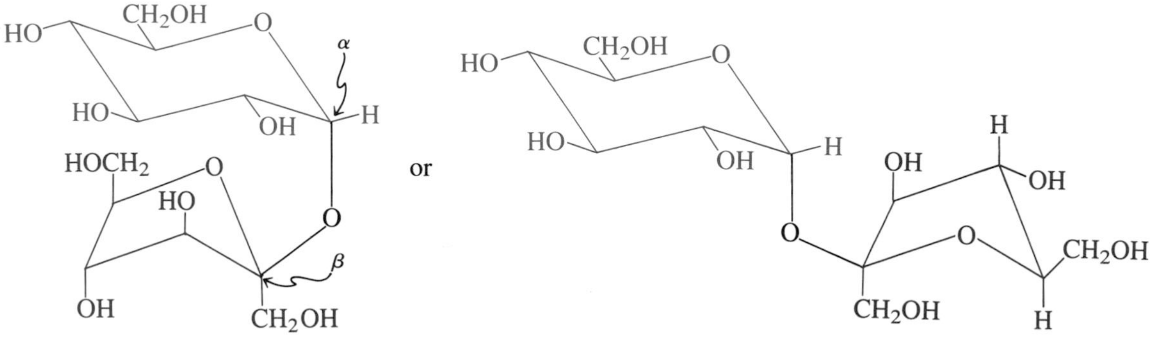

Sucrose, an α-D-glucopyranosyl-β-D-fructofuranose

Two representations of the molecule are shown. At the left, both cyclic sugars are drawn in the usual way: the anomeric carbon on the right, the acetal oxygen on top. Another structure with more favorable steric interactions is shown at the right, a rotamer in which the two sugar units point away from each other.

Sucrose has a specific rotation of +66.5°. Treatment with aqueous acid decreases the rotation until it reaches a value of −20.0°. The same effect is observed with the enzyme invertase. The phenomenon, known as the **inversion of sucrose,** is related to mutarotation of monosaccharides. It includes three separate reactions: hydrolysis of the disaccharide to the component monosaccharides α-D-glucopyranose and β-D-fructofuranose; mutarotation of α-D-glucopyranose to the equilibrium mixture with the β form; and mutarotation of β-D-fructofuranose to the slightly more stable β-D-fructopyranose. Because the value for the specific rotation of fructose (−92°) is more negative than the value for glucose (+52.7°) is positive, the resulting mixture, sometimes called **invert sugar,** has a net negative rotation, *inverted* from that of the original sucrose solution.

Inversion of Sucrose

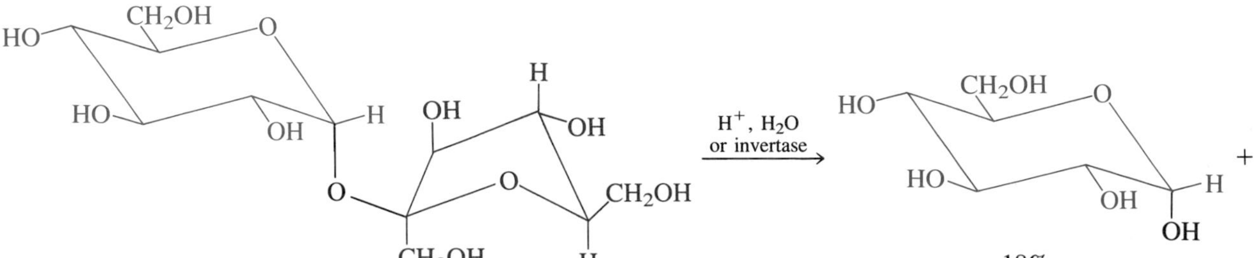

$\xrightarrow{H^+,\ H_2O \text{ or invertase}}$ +

Sucrose

18%

α-D-Glucopyranose

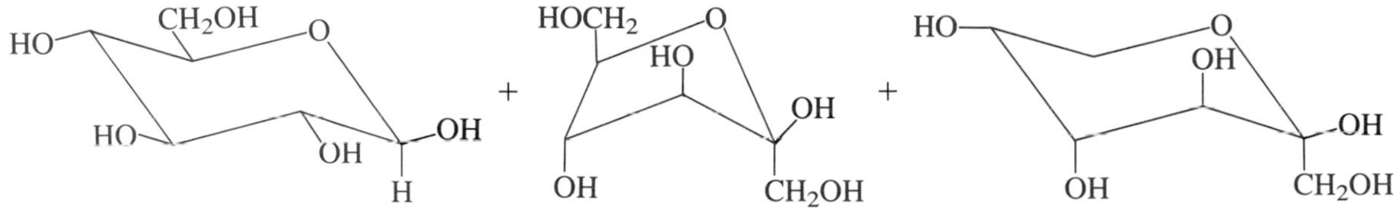

32%
β-D-Glucopyranose

16%
β-D-Fructofuranose

34%
β-D-Fructopyranose

EXERCISE 24-23

Write the products (if any) of the reaction of sucrose with **(a)** excess $(CH_3)_2SO_4$, NaOH; **(b)** 1. H^+, H_2O, 2. $NaBH_4$; and **(c)** NH_2OH.

Acetals link the components of complex sugars

Sucrose contains an acetal linkage between the anomeric carbons of the component sugars. One could imagine other acetal linkages with other hydroxy groups. Indeed, **maltose** (malt sugar), which is obtained in 80% yield by enzymatic (amylase) degradation of starch (to be discussed later in this section), is a dimer of glucose in which the hemiacetal oxygen of one glucose molecule (in the α anomeric form) is bound to C4 of the second.

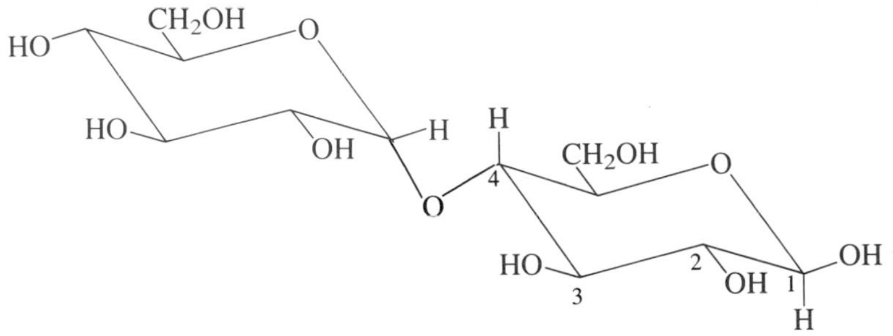

β-Maltose, an α-D-glucopyranosyl-β-D-glucopyranose

In this arrangement, one glucose unit maintains its unprotected hemiacetal structure, with its distinctive chemistry. For example, maltose is a reducing sugar; it forms osazones, and it undergoes mutarotation. Maltose is hydrolyzed to two molecules of glucose by aqueous acid or by the enzyme maltase. It is about one-third as sweet as sucrose.

EXERCISE 24-24

Draw the structure of the initial product of β-maltose when it is subjected to **(a)** Br_2 oxidation; **(b)** phenylhydrazine (3 equivalents); **(c)** conditions that effect mutarotation.

Another common disaccharide is **cellobiose,** obtained by the hydrolysis of cellulose (to be discussed later in this section). Its chemical properties are almost identical with those of maltose, and so is its structure, which is the same except for the stereochemistry at the acetal linkage—β instead of α.

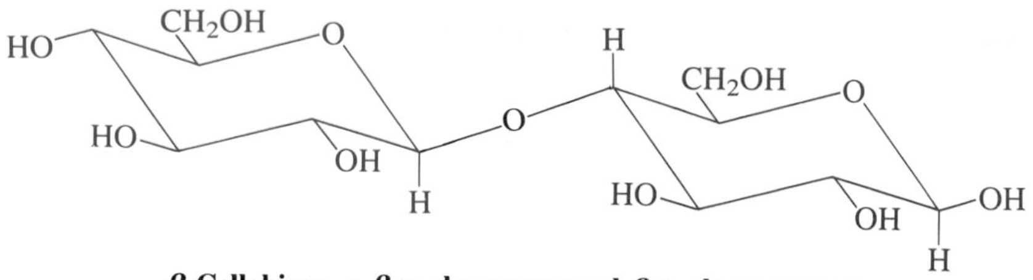

β-Cellobiose, a β-D-glucopyranosyl-β-D-glucopyranose

Aqueous acid cleaves cellobiose into two glucose molecules just as efficiently as it hydrolyzes maltose. However, enzymatic hydrolysis requires a different enzyme, β-glucosidase, which specifically attacks only the β-acetal bridge. In contrast, maltase is specific for α-acetal units of the type found in maltose.

After sucrose, the most abundant natural disaccharide is **lactose** (milk sugar). It is found in human and most animal milk (about 5% solution), constituting more than one-third of the solid residue remaining on evaporation of all volatiles. Its structure is made up of galactose and glucose units, connected in the form of a β-D-galactopyranosyl-D-glucopyranose. Crystallization from water furnishes only the α anomer.

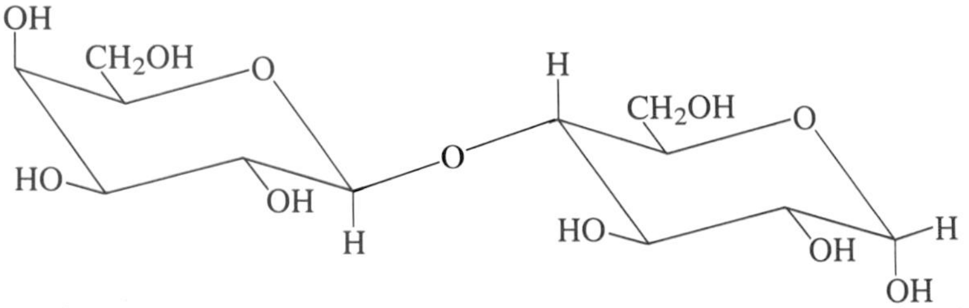

Crystalline α-lactose, a β-D-galactopyranosyl-α-D-glucopyranose

In summary, sucrose is a dimer derived from linking α-D-glucopyranose with β-D-fructofuranose at the anomeric centers. It shows inversion of its optical rotation on hydrolysis to its mutarotating component sugars. The disaccharide maltose is a glucose dimer in which the components are linked by a carbon–oxygen bond between an α anomeric carbon of a glucose molecule and C4 of the second. Cellobiose is almost identical structurally with maltose but has a β configuration at the acetal carbon. Lactose has a β-D-galactose linked to glucose in the same manner as in cellobiose.

24-12 Polysaccharides and Other Sugars in Nature

Polysaccharides are the polymers of monosaccharides. Their possible structural diversity is comparable to that of alkene polymers (Sections 12-13 and 12-14), particularly in variations of chain length and branching. Nature, however, has been remarkably conservative in its construction of such polymers. The three most abundant natural polysaccharides, cellulose, starch, and glycogen, are derived from the same monomer—glucose.

Cellulose and starch are unbranched polymers

Cellulose is a poly-β-glucopyranoside linked at C4, containing about 3000 monomeric units and having a molecular weight of about 500,000. It is largely linear.

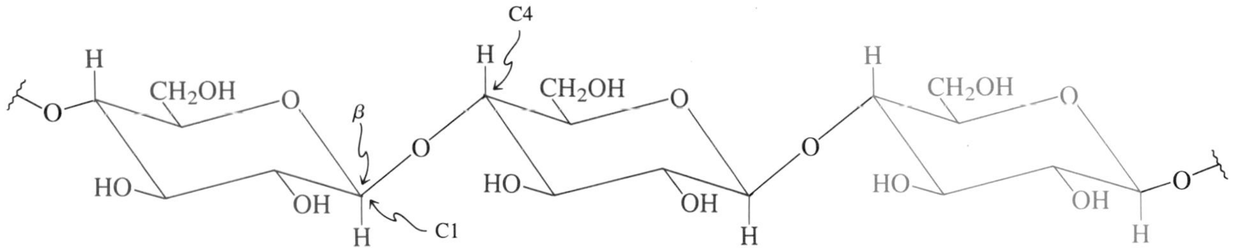

Cellulose

Individual strands of cellulose tend to align with one another and are connected by multiple hydrogen bonds. The development of so many hydrogen bonds is responsible for the highly rigid structure of cellulose and its effective use as the cell wall material in organisms. Thus, cellulose is abundant in trees and other plants. Cotton fiber is almost pure cellulose, as is filter paper. Wood and straw contain about 50% of the polysaccharide.

Several derivatives of cellulose have commercial uses. Conversion of the free hydroxy groups into nitrate esters with nitric acid results in **nitrocellulose.** If the nitrate content is high, this material is explosive and is used in smokeless gunpowder. A lower nitrate content gives a polymer that was important as one of the first commercial plastics—celluloid. For a long time, nitrocellulose was used extensively in the photographic and film industries. Unfortunately, it is highly flammable and gradually decomposes; it is now used only rarely.

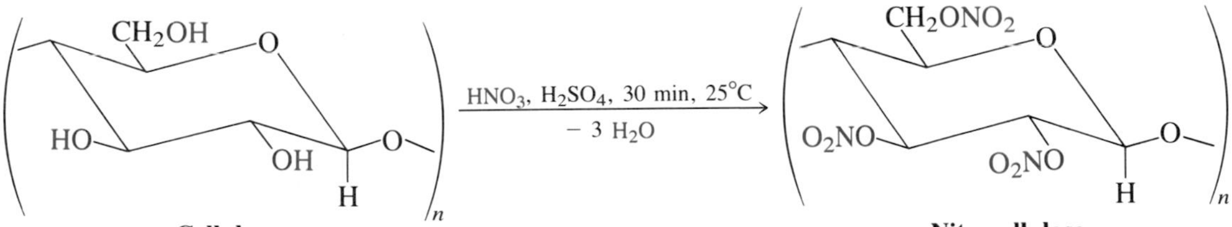

$\xrightarrow[-\ 3\ H_2O]{HNO_3,\ H_2SO_4,\ 30\ min,\ 25^\circ C}$

Cellulose **Nitrocellulose**

Cellulose, which is insoluble in almost all solvents, can be made soluble by blocking the hydroxy groups as adducts to carbon disulfide, the sulfur analog of CO_2. The resulting functional group is called a **xanthate.** Subsequent treatment with acid regenerates the insoluble polymer; this process may be controlled to give fibers (rayon) or sheets (cellophane).

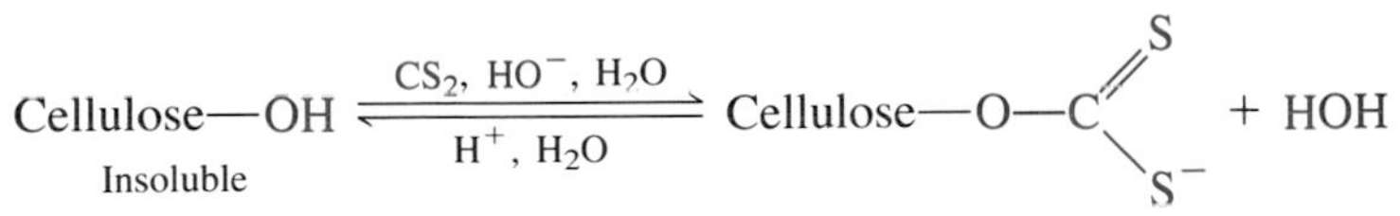

$$\underset{\text{Insoluble}}{\text{Cellulose—OH}} \underset{H^+,\ H_2O}{\overset{CS_2,\ HO^-,\ H_2O}{\rightleftharpoons}} \text{Cellulose—O—C}(=S)S^- + \text{HOH}$$

Cellulose xanthate

(Water soluble)

Like cellulose, **starch** is a polyglucose, but its subunits are connected by α-acetal linkages. It functions as a food reserve in plants and (like cellulose) is readily cleaved by aqueous acid into glucose. Major sources of starch are corn, potatoes, wheat, and rice. Hot water swells granular starch and allows the separation of the two major components: **amylose** (~20%) and **amylopectin** (~80%).

Both are soluble in hot water, but the former is less soluble in cold water. Amylose contains a few hundred glucose units per molecule (molecular weight, 150,000–600,000). Its structure is different from that of cellulose, even though both polymers are unbranched. The difference in the stereochemistry at the anomeric carbons leads to the strong tendency of amylose to form a helical polymer arrangement (not the straight chain shown in the formula). Note that the disaccharide units in amylose are the same as those in maltose.

H
CH_2OH
O
α
C4
HO
H H
HO
CH_2OH
O
C1
O
HO
Maltose unit
HO
H H
CH_2OH
O
HO
H
HO
O

Amylose

In contrast with amylose, amylopectin is branched, mainly at C6, about once every 20 to 25 glucose units. Its molecular weight runs into the millions.

H
CH_2OH
O
HO
H
HO
H
CH_2OH
O
O
Branching point
HO
H H
HO
CH_2
O
HO
H H
HO
CH_2OH
O
HO
H
HO
O

Amylopectin

Glycogen is a source of energy

Another polysaccharide similar to amylopectin but with greater branching (1 per 10 glucose units) and of much larger size (as much as 100 million molecular weight) is **glycogen.** This compound is of considerable biological importance because it is one of the major energy-storage polysaccharides in humans and animals and because it provides an immediate source of glucose between meals and during (strenuous) physical activity. It is accumulated in the liver and in rested skeletal muscle in relatively large amounts. The manner in which cells make use of this energy storage is a fascinating story in biochemistry.

A special enzyme, phosphorylase, first breaks glycogen down to give a derivative of glucose, α-D-glucopyranosyl 1-phosphate. This transformation takes place at one of the nonreducing terminal sugar groups of the glycogen molecule and proceeds step by step—one glucose molecule at a time. Because glycogen is so highly branched, there are many such end groups at which the enzyme can ''nibble'' away, making sure that, at a time of high energy requirements, a sufficient amount of glucose becomes quickly available.

Phosphorylase cannot break α-1,6-glycosidic bonds. As soon as it gets close to such a branching point (in fact, as soon as it reaches a terminal residue four units away from that point), it stops (Figure 24-4). At this stage, a different enzyme comes into play, transferase, which can shift blocks of three terminal glucosyl residues from one branch to another. One glucose substituent remains at the branching point. Now a third enzyme is required to remove this last obstacle to obtaining a new straight chain. This enzyme is specific for the kind of bond at

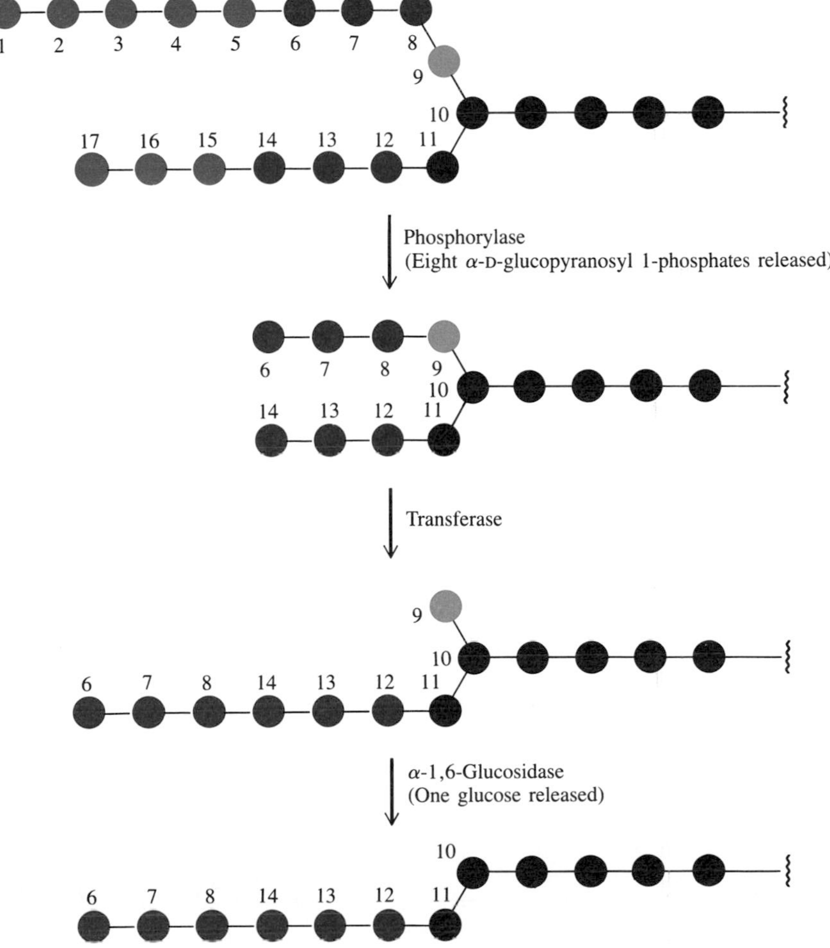

FIGURE 24-4 Steps in the degradation of a glycogen side chain. Initially, phosphorylase removes glucose units 1 through 5 and 15 through 17 step by step. The enzyme is now four sugar units away from a branching point (10). Transferase moves units 6 through 8 in one block and attaches them to unit 14. A third enzyme, α-1,6-glucosidase, debranches the system at glucose unit 10, by removing glucose 9. A straight chain has been formed and phosphorylase can continue its degradation job.

which cleavage is needed; it is α-1,6-glucosidase, also known as the debranching enzyme. Once this enzyme has completed its task, phosphorylase can continue degrading the glucose chain until it reaches another branch, and so on.

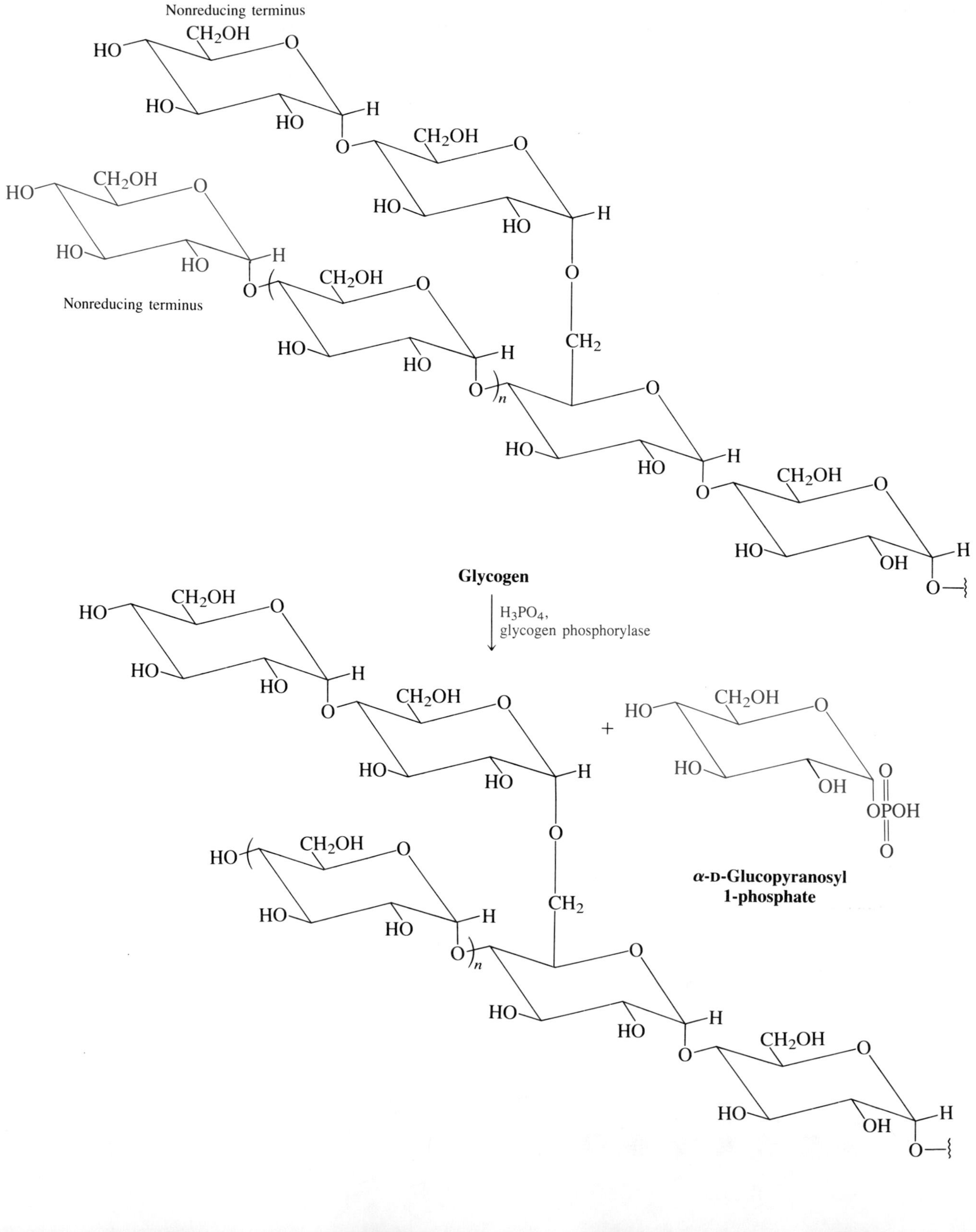

Nonreducing terminus

Nonreducing terminus

Glycogen

H_3PO_4, glycogen phosphorylase

+

α-D-Glucopyranosyl 1-phosphate

The glucose liberated from glycogen is converted into 2-oxopropanoic (pyruvic) acid by a complex pathway (glycolysis) that includes several enzymes. This acid then gives rise to various products, depending on the circumstances.

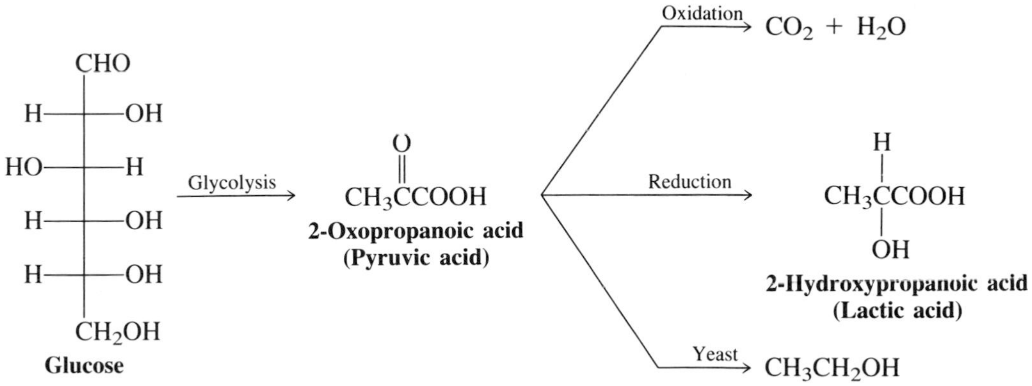

Glucose $\xrightarrow{\text{Glycolysis}}$ $CH_3\overset{O}{\overset{\|}{C}}COOH$ **2-Oxopropanoic acid (Pyruvic acid)**

$\xrightarrow{\text{Oxidation}} CO_2 + H_2O$

$\xrightarrow{\text{Reduction}} CH_3CH(OH)COOH$ **2-Hydroxypropanoic acid (Lactic acid)**

$\xrightarrow{\text{Yeast}} CH_3CH_2OH$

In an aerobic (oxygen-rich) environment, further oxidation results in CO_2 and H_2O, with a maximum gain in energy. If there is a poor supply of oxygen, as, for example, in an actively contracting muscle, incomplete reduction gives 2-hydroxypropanoic (lactic) acid. Some anaerobic organisms, such as yeast, convert 2-oxopropanoic (pyruvic) acid into ethanol.

Modified sugars may contain nitrogen

Many of the naturally occurring sugars have a modified structure or are attached to some other organic molecule. There is a large class of sugars in which at least one of the hydroxy groups has been replaced by an amine function. They are called **glycosylamines** if the nitrogen is attached to the anomeric carbon and **amino deoxy sugars** if it is located elsewhere.

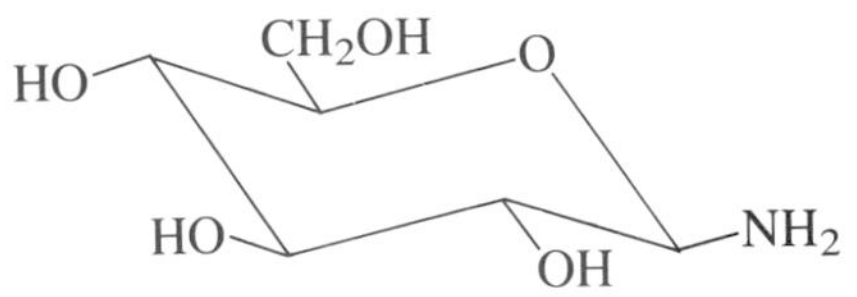

β-D-Glucopyranosylamine
(A glycosylamine)

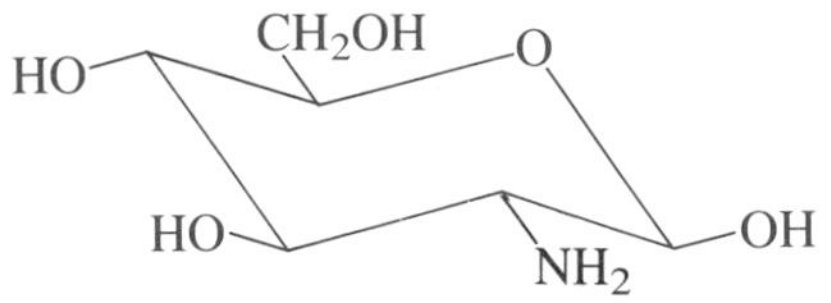

2-Amino-2-deoxy-D-glucopyranose
(An amino deoxy sugar)

Glycosylamines are present in other important biological polymers: the nucleic acids (Section 26-9), which contain the genetic code and are responsible for protein biosynthesis. Ribonucleic acid is a polymer made of subunits called nucleotides, which are substituted glycosylamines. An example is uridylic acid.

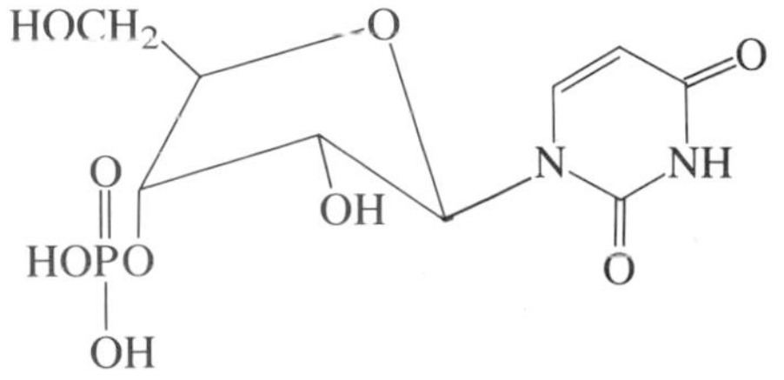

Uridylic acid

If a sugar is attached by its anomeric carbon to the hydroxy group of another complex residue, it is called a **glycosyl group.** The remainder of the molecule (or the product after removal of the sugar by hydrolysis) is called the **aglycon.** Examples are amygdalin and adriamycin. Amygdalin, a derivative of the β-1,6-glycoside gentiobiose, is isolated from apricot pits, along with its monosaccharide analog Laetrile.

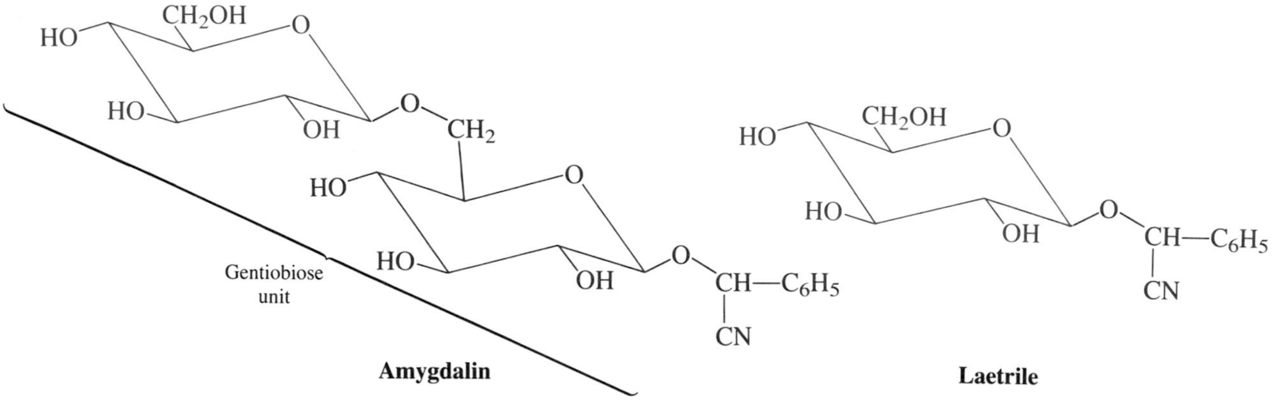

Gentiobiose unit

Amygdalin

Laetrile

Laetrile was briefly notorious for its controversial use in the treatment of cancer. Proponents claimed that the molecule interacts with the enzyme β-glucosidase to release cyanide ion, which destroys cancer cells. Noncancerous cells were said to be able to enzymatically deactivate cyanide to cyanate ion. As a consequence of these claims, many patients, particularly those for whom conventional therapy had failed, sought treatment with Laetrile. Its curative effects were disputed by the medical profession, and the National Institutes of Health conducted a study that indicated the inefficacy of the drug.

Adriamycin is a member of the anthracycline family of antibiotics. Adriamycin and its deoxy analog daunomycin have been remarkably effective in the treatment of a wide variety of human cancers. They now constitute a cornerstone of combination cancer chemotherapy. The aglycon part of these systems is a linear tetracyclic framework incorporating an anthraquinone moiety (derived from anthracene; see Section 15-6). The amino sugar is called daunosamine.

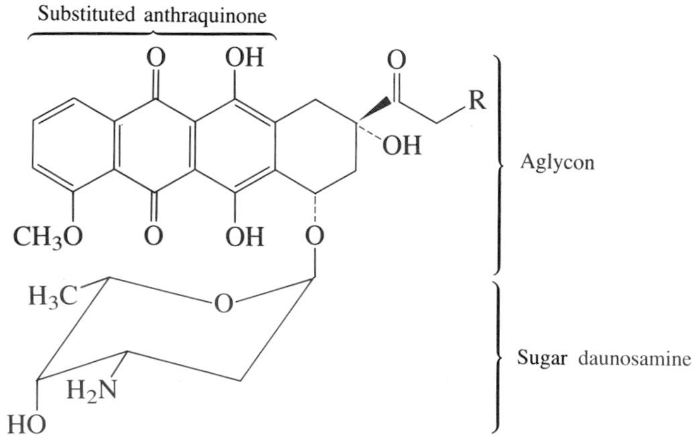

Substituted anthraquinone

Aglycon

Sugar daunosamine

Adriamycin (R = OH)
Daunomycin (R = H)

An unusual group of antibiotics, the aminoglycoside antibiotics, is based almost exclusively on oligosaccharide structures. Of particular therapeutic importance is streptomycin (an antituberculosis agent), isolated in 1944 from cultures of the mold *Streptomyces griseus*.

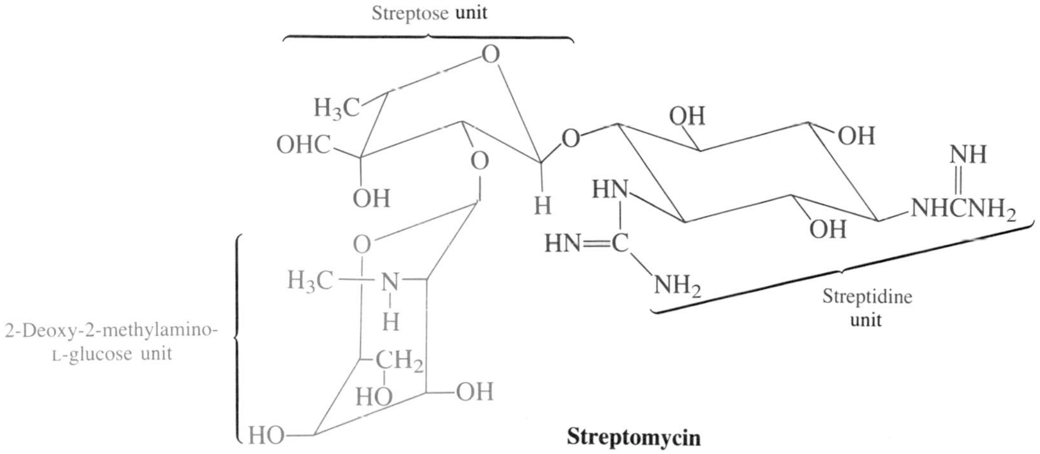

Streptose unit

2-Deoxy-2-methylamino-L-glucose unit

Streptidine unit

Streptomycin

The molecule consists of three subunits: the furanose streptose, the glucose derivative 2-deoxy-2-methylamino-L-glucose (an example of the rare L form), and streptidine, which is actually a hexasubstituted cyclohexane.

In summary, the polysaccharides cellulose, starch, and glycogen are all polyglucosides. Cellulose consists of repeating dimeric cellobiose units. Starch may be regarded as a polymaltose derivative. Its occasional branching poses a challenge to enzymatic degradation, as does glycogen. Metabolism of these polymers first gives monomeric glucose, which is then oxidized to 2-oxopropanoic (pyruvic) acid. Depending on the organism and physiological conditions, this molecule may then be oxidized further to CO_2 and H_2O or reduced to 2-hydroxypropanoic (lactic) acid or ethanol. Finally, many sugars occur in nature in modified form or as simple appendages to other structures. Examples include amino sugars, glycosylamines, amygdalin, and adriamycin. The aminoglycoside antibiotics consist entirely of saccharide molecules, modified and unmodified.

NEW REACTIONS

1. CYCLIC HEMIACETAL FORMATION IN SUGARS (Section 24-2)

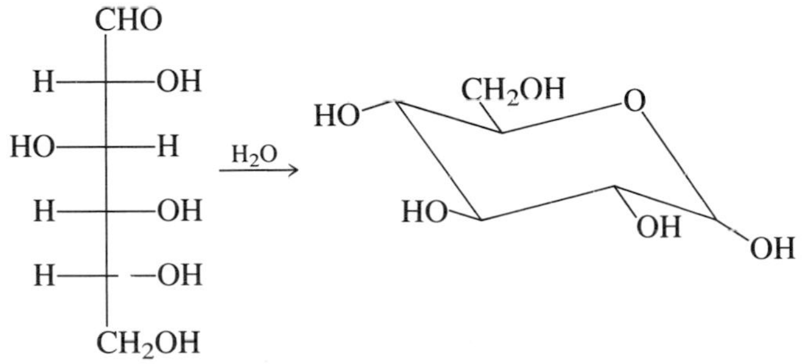

$\xrightarrow{H_2O}$

α- and β-glucopyranoses

2. MUTAROTATION (Section 24-3)

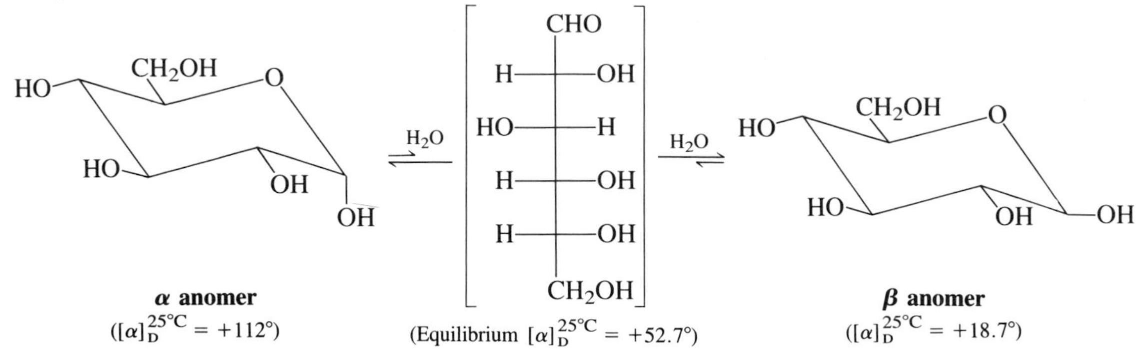

α anomer ($[\alpha]_D^{25°C} = +112°$)

(Equilibrium $[\alpha]_D^{25°C} = +52.7°$)

β anomer ($[\alpha]_D^{25°C} = +18.7°$)

3. OXIDATION (Section 24-4)

Tests for reducing sugars

CHO / CH2OH (aldose) $\xrightarrow[\text{or } Ag^+,\ NH_4OH,\ H_2O \text{ (Tollens's solution)}]{Cu^{2+},\ OH^-,\ H_2O \text{ (Fehling's solution)}}$ COOH / CH2OH + Cu_2O (Red) or Ag (Silver mirror)

Aldonic acid synthesis

CHO, H—OH, CH_2OH $\xrightarrow{Br_2,\ H_2O}$ COOH, H—OH, CH_2OH (**Aldonic acid**) $\xrightarrow{-\ H_2O}$ C(=O)—O—, H, CH_2OH (**γ-Lactone**)

Aldaric acid synthesis

CHO, HO—H, H—OH, CH_2OH $\xrightarrow{HNO_3,\ H_2O}$ COOH, HO—H, H—OH, COOH (**Aldaric acid**)

4. SUGAR DEGRADATION (Section 24-5)

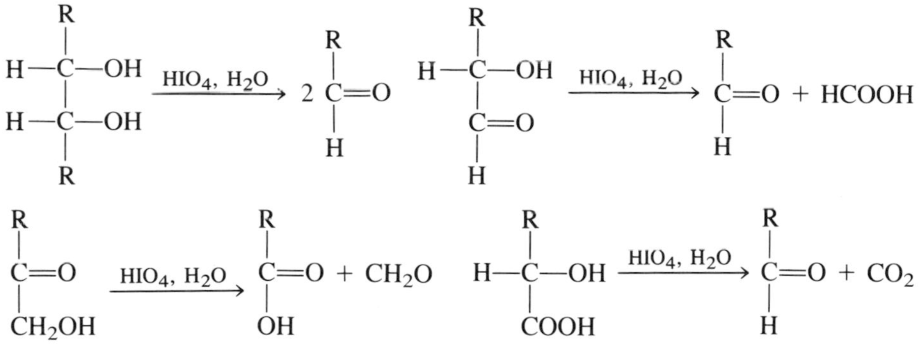

$$\begin{array}{c} R \\ | \\ H-C-OH \\ | \\ H-C-OH \\ | \\ R \end{array} \xrightarrow{HIO_4,\ H_2O} 2\ \begin{array}{c} R \\ | \\ C=O \\ | \\ H \end{array} \qquad \begin{array}{c} R \\ | \\ H-C-OH \\ | \\ C=O \\ | \\ H \end{array} \xrightarrow{HIO_4,\ H_2O} \begin{array}{c} R \\ | \\ C=O \\ | \\ H \end{array} + HCOOH$$

$$\begin{array}{c} R \\ | \\ C=O \\ | \\ CH_2OH \end{array} \xrightarrow{HIO_4,\ H_2O} \begin{array}{c} R \\ | \\ C=O \\ | \\ OH \end{array} + CH_2O \qquad \begin{array}{c} R \\ | \\ H-C-OH \\ | \\ COOH \end{array} \xrightarrow{HIO_4,\ H_2O} \begin{array}{c} R \\ | \\ C=O \\ | \\ H \end{array} + CO_2$$

5. REDUCTION (Section 24-6)

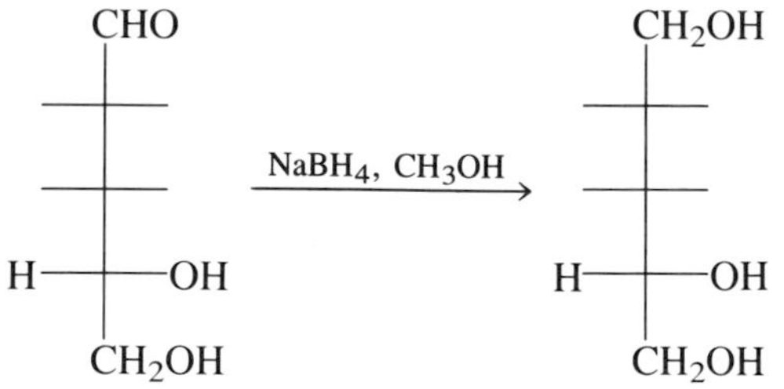

$$\begin{array}{c} CHO \\ \text{—|—} \\ \text{—|—} \\ H\text{—|—}OH \\ CH_2OH \end{array} \xrightarrow{NaBH_4,\ CH_3OH} \begin{array}{c} CH_2OH \\ \text{—|—} \\ \text{—|—} \\ H\text{—|—}OH \\ CH_2OH \end{array}$$

Alditol

6. HYDRAZONES AND OSAZONES (Section 24-7)

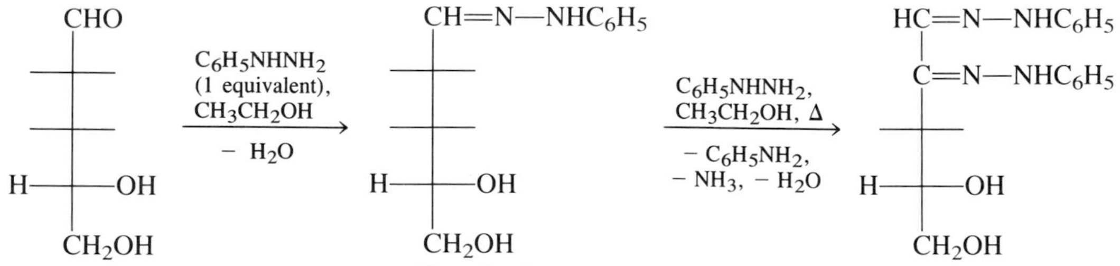

$$\begin{array}{c} CHO \\ \text{—|—} \\ \text{—|—} \\ H\text{—|—}OH \\ CH_2OH \end{array} \xrightarrow[-\ H_2O]{C_6H_5NHNH_2\ (1\ \text{equivalent}),\ CH_3CH_2OH} \begin{array}{c} CH=N-NHC_6H_5 \\ \text{—|—} \\ \text{—|—} \\ H\text{—|—}OH \\ CH_2OH \end{array} \xrightarrow[-\ C_6H_5NH_2,\ -\ NH_3,\ -\ H_2O]{C_6H_5NHNH_2,\ CH_3CH_2OH,\ \Delta} \begin{array}{c} HC=N-NHC_6H_5 \\ | \\ C=N-NHC_6H_5 \\ \text{—|—} \\ H\text{—|—}OH \\ CH_2OH \end{array}$$

Phenylhydrazone **Osazone**

7. ESTERS (Section 24-8)

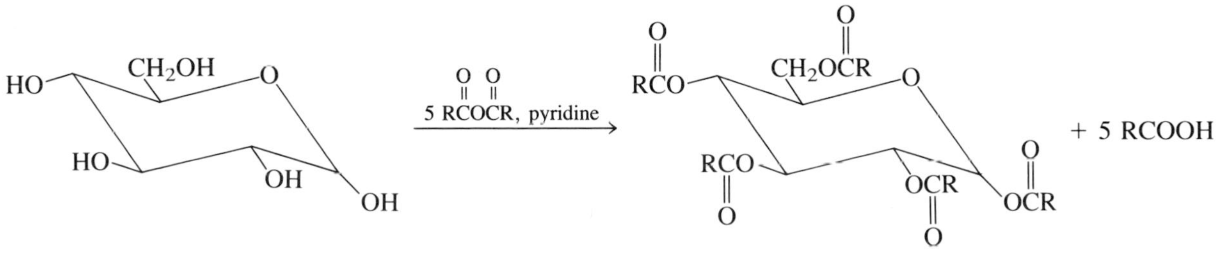

Pyranose (HO, CH_2OH, HO, OH, OH substituents) $\xrightarrow{5\ RCOCR\ (\text{O}\ \text{O}),\ \text{pyridine}}$ pentaester (RCO(=O)O at each position, CH_2OCR(=O)) $+\ 5\ RCOOH$

α and β anomers **α and β anomers**

8. GLYCOSIDES (Section 24-8)

HO, CH_2OH, O, HO, OH, OH — $\xrightleftharpoons[H_2O,\ H^+]{CH_3OH,\ H^+}$ — HO, CH_2OH, O, HO, OH, OCH_3 + H_2O

α and β anomers → **α and β anomers**

9. ETHERS (Section 24-8)

HO, CH_2OH, O, HO, OH, OH — $\xrightarrow[-\ Na_2SO_4]{5\ (CH_3)_2SO_4,\ Na^{+\ -}OH}$ — CH_3O, CH_2OCH_3, O, CH_3O, OCH_3, OCH_3

α and β anomers → **α and β anomers**

10. CYCLIC ACETALS (Section 24-8)

R, R′, O, OH, OH, OH, H — $\xrightarrow[-\ H_2O]{CH_3\overset{O}{\overset{\|}{C}}CH_3,\ H^+}$ — R, R′, O, O, O, OH, H, CH_3, CH_3

11. KILIANI-FISCHER SYNTHESIS (Section 24-9)

CHO, H—OH, CH_2OH (**Sugar**) $\xrightarrow{HCN}$ CN, H—C—OH, H—OH, CH_2OH (**Cyanohydrin**) $\xrightarrow{H_2,\ Pd\text{-}BaSO_4,\ H^+,\ H_2O}$ CHO, H—C—OH, H—OH, CH_2OH (**Extended sugar**)

12. RUFF DEGRADATION (Section 24-9)

CHO, H—OH, H—OH, CH_2OH $\xrightarrow{Br_2,\ H_2O}$ COOH, H—OH, H—OH, CH_2OH $\xrightarrow[-\ CO_2]{Fe^{3+},\ H_2O_2}$ CHO, H—OH, CH_2OH

IMPORTANT CONCEPTS

1. Carbohydrates are naturally occurring polyhydroxy carbonyl compounds that can exist as monomers, dimers, oligomers, and polymers.
2. Monosaccharides are called aldoses if they are aldehydes and ketoses if they are ketones. The chain length is indicated by the prefix tri-, tetr-, pent-, hex-, and so on.
3. Most natural carbohydrates belong to the D family; that is, the stereocenter farthest from the carbonyl group has the same configuration as that in (*R*)-(+)-2,3-dihydroxypropanal [D-(+)-glyceraldehyde].
4. The keto forms of carbohydrates exist in equilibrium with the corresponding five-membered (furanoses) or six-membered (pyranoses) cyclic hemiacetals. The new stereocenter formed by cyclization is called the anomeric carbon, and the two anomers are designated α and β.
5. Haworth projections of D sugars depict the cyclic ether in line notation as a pentagon or a hexagon, the anomeric carbon placed on the right, and the ether oxygen at the top. The substituents located above or below the ring are attached to vertical lines. The ring bond at the bottom (between C2 and C3) is understood to be in front of the plane of the paper, and the ring bonds containing the oxygen are understood to be in back. The α anomer has the OH group at the anomeric carbon pointing down, whereas the β anomer has it pointing up.
6. Equilibration between anomers in solution gives rise to changes in the measured optical rotation called mutarotation.
7. The reactions of the saccharides are characteristic of carbonyl, alcohol, and hemiacetal groups. They include oxidation of the aldehyde to the carboxy function of aldonic acids, double oxidation to aldaric acids, oxidative cleavage of vicinal diol units, reduction to alditols, condensations, esterifications, and acetal formations.
8. Sugars containing hemiacetal functions are called reducing sugars, because they readily reduce Tollens's and Fehling's solutions. Sugars with only acetal groups are nonreducing.
9. The synthesis of higher sugars is based on Kiliani-Fischer chain lengthening, the new carbon being introduced by cyanide ion. The synthesis of lower sugars relies on Ruff chain shortening, a terminal carbon being expelled as CO_2.
10. The Fischer proof uses the techniques of chain lengthening and shortening together with the symmetry properties of aldaric acids to determine the structures of the aldoses.
11. Di- and higher saccharides are formed by ether formation between monomers; the ether bridge usually includes at least one hemiacetal hydroxy group.
12. The change in optical rotation observed in aqueous solutions of sucrose, called the inversion of sucrose, is due to the equilibration of the starting sugar with the various cyclic and anomeric forms of its component monomers.
13. Many sugars contain modified backbones. Amino groups may have replaced hydroxy groups, there may be substituents of various complexity (aglycons), the backbone carbon atoms of a sugar may lack oxygens, and (rarely) the sugar may adopt the L configuration.

PROBLEMS

1. The designations D and L as applied to sugars refer to the configuration of the highest-numbered stereocenter. If the configuration of the highest-numbered stereocenter of D-ribose (Figure 24-1) is switched from D to L, is the product L-ribose? If not, what is the product? How is it related to D-ribose (i.e., what kind of isomers are they)?

2. To which classes of sugars do the following monosaccharides belong? Which are D and which are L?

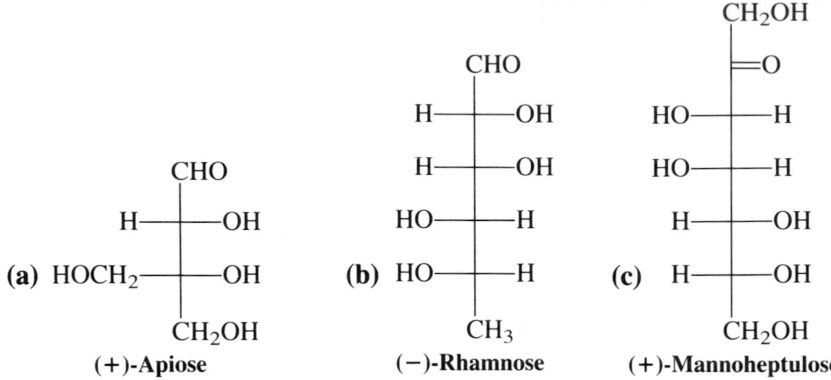

3. Draw open-chain (Fischer-projection) structures for L-(+)-ribose and L-(−)-glucose (see Exercise 24-2). What are their systematic names?

4. Identify the following sugars, which are represented by unconventionally drawn Fischer projections. (Hint: It will be necessary to convert these into more conventional representations *without* inverting any of the stereocenters.)

(a) H; $HOCH_2$ — CHO; OH

(b) CH_2OH; =O; $HOCH_2$ — H; OH

(c) CH_2OH; HO — H; HO — H; H — OH; HO — H; CHO

(d) HO; H — CHO; $HOCH_2$ — H; OH

(e) H; OHC — OH; H — OH; $HOCH_2$ — OH; H

5. Redraw each of the following sugars in open-chain form as a Fischer projection, and find its common name.

(a) OHC; H, OH, H, OH; CH_2OH; H, OH, HO, H

(b) $HOCH_2$; O; H; H; H; OH; OH; HO; CH_2OH

(c) 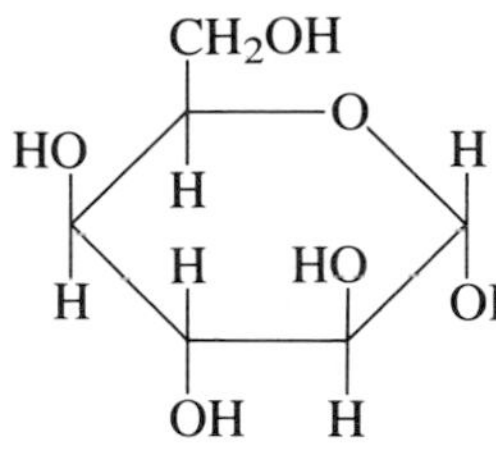 [chemical structure]

(d) 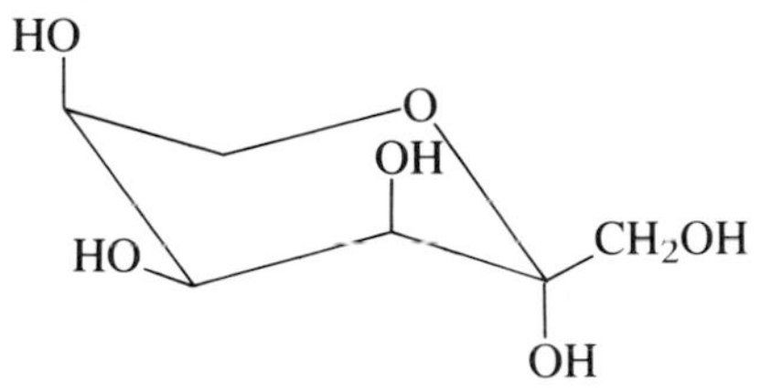 [chemical structure]

6. For each of the following sugars, draw all reasonable cyclic structures, using either Haworth or conformational formulas; indicate which structures are pyranoses and which are furanoses; and label α and β anomers. **(a)** (−)-Threose; **(b)** (−)-allose; **(c)** (−)-ribulose; **(d)** (+)-sorbose; **(e)** (+)-mannoheptulose (Problem 1).

7. Are any of the sugars in Problem 6 incapable of mutarotation? Why?

8. Draw the most stable conformation of each of the following sugars in its pyranose form. **(a)** α-D-Arabinose; **(b)** β-D-galactose; **(c)** β-D-mannose; **(d)** α-D-idose.

9. Write the expected products of the reaction of each of the following sugars with (i) Br_2, H_2O; (ii) HNO_3, H_2O, 60°C; (iii) $NaBH_4$, CH_3OH; and (iv) excess $C_6H_5NHNH_2$, CH_3CH_2OH, Δ. Find the common names of all the products. **(a)** D-(−)-Threose; **(b)** D-(+)-xylose; **(c)** D-(+)-galactose.

10. Draw the Fischer projection of an aldohexose that will give the same osazone as **(a)** D-(−)-idose and **(b)** L-(−)-altrose.

11. **(a)** Which of the aldopentoses (Figure 24-1) would give optically active alditols upon reduction with $NaBH_4$? **(b)** Using D-fructose, illustrate the results of $NaBH_4$ reduction of a ketose. Is the situation more complicated than reduction of an aldose? Explain.

12. Which of the following glucoses and glucose derivatives are capable of undergoing mutarotation? **(a)** α-D-Glucopyranose; **(b)** methyl α-D-glucopyranoside; **(c)** methyl α-2,3,4,6-tetra-*O*-methyl-D-glucopyranoside (i.e., the tetramethyl ether at carbon 2, 3, 4, and 6); **(d)** α-2,3,4,6-tetra-*O*-methyl-D-glucopyranose; **(e)** α-D-glucopyranose 1,2-monopropanone acetal.

13. **(a)** Explain why the oxygen at C1 of an aldopyranose can be methylated so much more easily than the other oxygens in the molecule. **(b)** Explain why the methyl ether unit at C1 of a fully methylated aldopyranose can be hydrolyzed so much more easily than the other methyl ether functions in the molecule. **(c)** Write the expected product(s) of the following reaction.

$$\text{D-Fructose} \xrightarrow{CH_3OH,\ 0.25\%\ HCl,\ H_2O}$$

14. Of the four aldopentoses, two readily form double acetals when treated with excess acidic propanone (acetone), but the other two form only monoacetals. Explain.

15. D-Sedoheptulose is a sugar that plays a role in a metabolic cycle (the *pentose oxidation cycle*) that converts glucose into 2,3-dihydroxypropanal (glyceraldehyde) plus three equivalents of CO_2. Determine the structure of D-sedoheptulose from the following information.

$$\text{D-Sedoheptulose} \xrightarrow{6\ HIO_4} 4\ \overset{\overset{\displaystyle O}{\|}}{H C}OH + 2\ H\overset{\overset{\displaystyle O}{\|}}{C}H + CO_2$$

$$\text{D-Sedoheptulose} \xrightarrow{C_6H_5NHNH_2} \text{an osazone identical with that formed by another sugar, aldoheptose A}$$

$$\text{Aldoheptose A} \xrightarrow{\text{Ruff degradation}} \text{aldohexose B}$$

$$\text{Aldohexose B} \xrightarrow{HNO_3,\ H_2O,\ \Delta} \text{an optically active product}$$

$$\text{Aldohexose B} \xrightarrow{\text{Ruff degradation}} \text{D-ribose}$$

16. Illustrate the results of Kiliani-Fischer chain elongation of D-talose. How many products are formed? Draw them. After treatment with warm HNO_3, do the product(s) give optically active or inactive dicarboxylic acids?

17. Write a plausible mechanism for the decarboxylation step in the Ruff degradation.

18. **(a)** Write a detailed mechanism for the isomerization of β-D-fructofuranose from the hydrolysis of sucrose into an equilibrium mixture of the β-pyranose and β-furanose forms. **(b)** Although fructose usually appears as a furanose when it is part of a polysaccharide, in the pure crystalline form, fructose adopts a β-pyranose structure. Draw β-D-fructopyranose in its most stable conformation. In water at 20°C, the equilibrium mixture contains about 80% pyranose and 20% furanose. **(c)** What is the free-energy difference between the pyranose and furanose forms at this temperature? **(d)** Pure β-D-fructopyranose has $[\alpha]_D^{20°C} = -132°$. The equilibrium pyranose–furanose mixture has $[\alpha]_D^{20°C} = -92°$. Calculate $[\alpha]_D^{20°C}$ for pure β-D-fructofuranose.

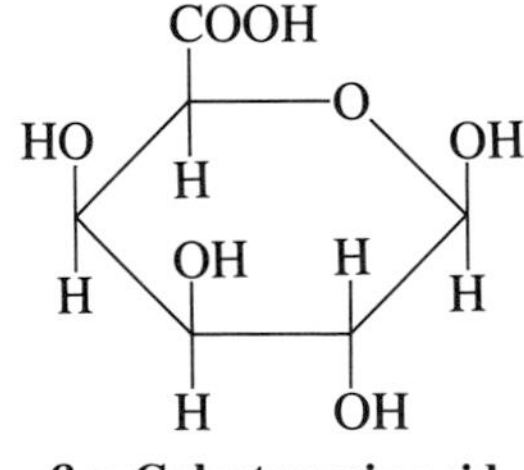

β-D-Galacturonic acid

19. Classify each of the following sugars and sugar derivatives as either reducing or nonreducing. **(a)** D-Glyceraldehyde; **(b)** D-arabinose; **(c)** β-D-arabinopyranose 3,4-monopropanone acetal; **(d)** β-D-arabinopyranose double propanone acetal; **(e)** D-ribulose; **(f)** D-galactose; **(g)** methyl β-D-galactopyranoside; **(h)** β-D-galacturonic acid (as shown in the margin); **(i)** β-cellobiose; **(j)** α-lactose.

20. Is α-lactose capable of mutarotation? Write an equation to illustrate.

21. Trehalose, sophorose, and turanose are disaccharides. Trehalose is found in the cocoons of some insects, sophorose turns up in a few bean varieties, and turanose is an ingredient in low-grade honey made by bees with indigestion from a diet of pine tree sap. Identify among the following structures those that correspond to trehalose, sophorose, and turanose on the basis of the following information: (i) Turanose and sophorose are reducing sugars. Trehalose is a nonreducing sugar. (ii) On hydrolysis, sophorose and trehalose give two molecules each of aldoses. Turanose gives one molecule of an aldose and one molecule of a ketose. (iii) The two aldoses that comprise sophorose are anomers of each other.

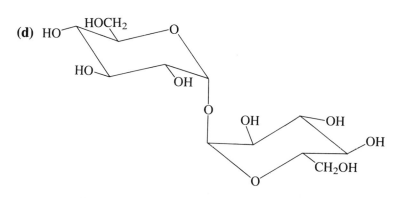

(a)

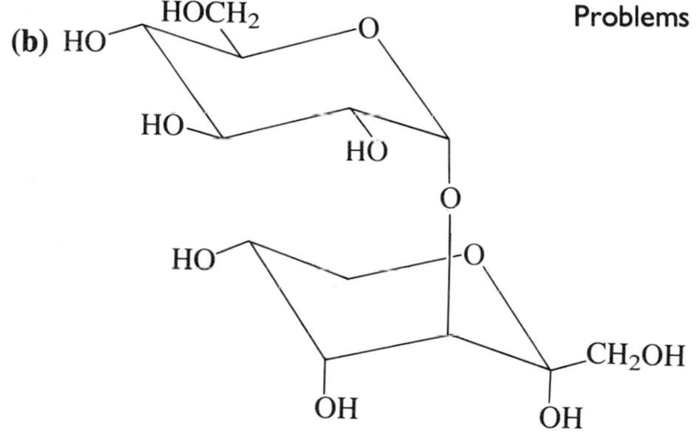

(c)

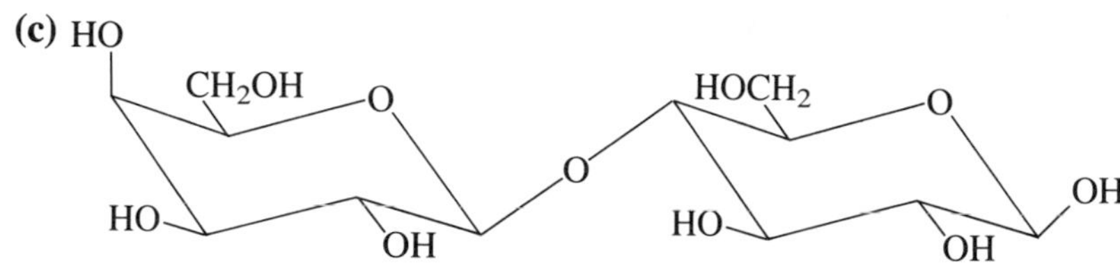

(d)

22. **(a)** A mixture of (*R*)-2,3-dihydroxypropanal (D-glyceraldehyde) and 1,3-dihydroxypropanone (1,3-dihydroxyacetone) that is treated with aqueous NaOH rapidly yields a mixture of three sugars: D-fructose, D-sorbose, and racemic dendroketose (only one enantiomer is shown here). Explain this result by means of a detailed mechanism. **(b)** The same product mixture is also obtained if either the aldehyde or the ketone *alone* is treated with base. Explain. [Hint: Closely examine the intermediates in your answer to (a).]

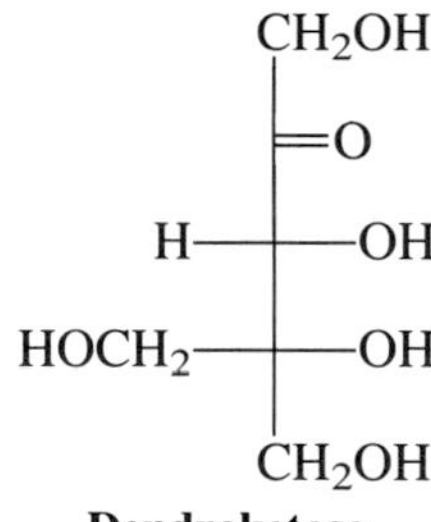

Dendroketose

23. Write or draw the missing reagents and structures a through g. What is the common name of g?

D-(+)-Xylose $\xrightarrow{(a)}$ **(b)** **D-Xylonic acid** $\xrightarrow{(c)}$ **(d)** **Methyl D-xylonate** $\xrightarrow{NH_3,\ \Delta}$ $C_5H_{11}NO_5$ **(e)** $\xrightarrow{Br_2,\ NaOH}$

$CO_2 + C_4H_{11}NO_4$ **(f)** $\xrightarrow{\Delta}$ $NH_3 + C_4H_8O_4$ **(g)**

The preceding sequence (called the *Weerman degradation*) achieves the same end as what procedure described in this chapter?

24. Fischer's solution to the problem of sugar structures was actually much more difficult to achieve experimentally than Section 24-10 implies. For one thing, the only sugars that he could readily obtain from natural sources were glucose, mannose, and arabinose. (Erythrose and threose were, in fact, not then available at all, either naturally or synthetically.) His ingenious solution required a source of gulose so that he could make the critical comparison of dicarboxylic acids described at the end of the section. Unfortunately, gulose does not occur in nature; so Fischer had to make it. His synthesis, from glucose, was difficult because at a key point he got a troublesome mixture of products. Nowadays the following synthesis might be used.

Write or draw the missing reagents and structures a through g. Use Fischer projections for all structures. Follow the instructions and hints in parentheses.

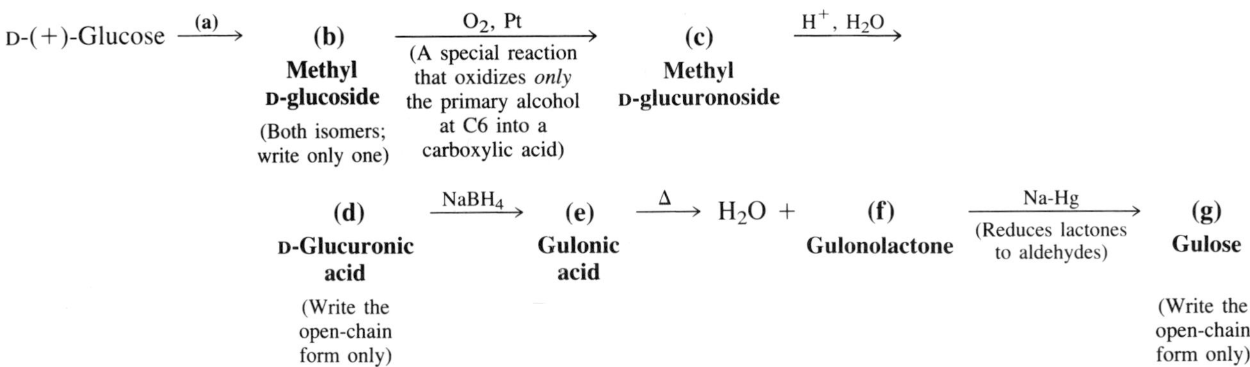

Is the gulose that Fischer synthesized from D-glucose an L sugar or a D sugar? (Be careful. Fischer himself got the wrong answer at first, and that confused *everybody* for *years*.)

L-(−)-Mannose
1. $HSCH_2CH_2SH$, $ZnCl_2$
2. Raney Ni (Section 17-7)
3. O_2, Pt (see Problem 24)
4. Δ ($-H_2O$)
5. Na–Hg (Problem 24 again)

Sugar C

25. Rutinose is a reducing sugar that is part of several bioflavonoids, a group of compounds found in many plants. They have therapeutic value in maintaining the strength of blood-vessel walls. One rutinose-containing bioflavonoid is hesperidin, which is present in lemons and oranges.

On acid hydrolysis, rutinose gives one equivalent each of D-glucose and a sugar C with the formula $C_6H_{12}O_5$. Sugar C reacts with four equivalents of HIO_4 to give four equivalents of formic acid and one equivalent of acetaldehyde. Sugar C can be synthesized as shown in the margin. **(a)** What is the structure of sugar C?

Complete methylation of rutinose (by excess dimethyl sulfate) followed by acid hydrolysis gives one equivalent of methyl 2,3,4-tri-*O*-methyl-D-glucoside and one equivalent of the 2,3,4-tri-*O*-methyl derivative of sugar C. **(b)** What possible structure(s) of rutinose are consistent with these data?

26. Vitamin C (ascorbic acid, Section 22-9) occurs almost universally in the plant and animal kingdoms. (According to Linus Pauling, mountain goats biosynthesize from 12 to 14 g of it per day.) Animals produce it from D-glucose in the liver by the four-step sequence D-glucose → D-glucuronic acid (see Problem 24) → D-glucuronic acid γ-lactone → L-gulonic acid γ-lactone → vitamin C.

CH_2OH — H—C—OH — C—H — C—OH ‖ C—OH — C=O (ring through O) is the same as [HO–C=C(OH)–C(=O)–O–C(H)–, HO—C—H, CH_2OH]

Vitamin C

The enzyme that catalyzes the last reaction, L-gulonolactone oxidase, is absent from humans, some monkeys, guinea pigs, and birds, presumably because of a defective gene resulting from a mutation that may have occurred some 60 million years ago. As a result, we have to get our vitamin C from food or make it synthetically. In fact, the ascorbic acid in almost all vitamin supplements is synthetic. An outline of one of the major commercial syntheses follows. Draw the missing reagents and products a through f.

D-Glucose $\xrightarrow{\text{(a)}}$ D-glucitol $\xrightarrow[\text{(by } \textit{Gluconobacter oxydans})]{\text{Microbial oxidation at C5}}$

(b) **L-Sorbose** (Open chain) ⇌ **(c)** **L-Sorbofuranose** $\xrightarrow[\text{(Two steps)}]{\text{(d)}}$ [diacetonide with CH_3, CH_3, H_3C, CH_3, O, COOH] $\xrightarrow{\text{(e)}}$

CH_2OH — H—C—OH — HO—C—H — H—C—OH — C=O — COOH
2-Keto-L-gulonic acid

$\xrightarrow{\text{(f)}}$

CH_2OH — H—C—OH — C—H — H—C—OH — C=O — C=O (ring through O)
Keto form of vitamin C

⇌ vitamin C

25

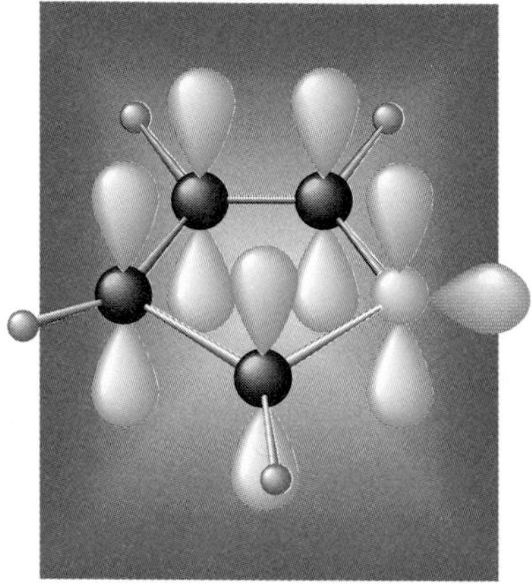

Heterocycles

Heteroatoms in Cyclic Organic Compounds

Carbocyclic compounds are cyclic molecules in which the rings are made up of only carbon atoms. In contrast, **heterocycles** are their analogs in which one or more ring carbons have been replaced by a **heteroatom,** such as nitrogen, oxygen, sulfur, phosphorus, silicon, a metal, and so on. The most common heterocyclic systems contain nitrogen or oxygen or both. Several of their derivatives appeared in the discussion of cyclic ethers—for example, oxacyclohexane (tetrahydropyran) (Section 9-6), cyclic acetals (Sections 17-8 and 24-8), cyclic dicarboxylic acid derivatives (Sections 19-8 and 20-3), halonium ions (Section 12-5), and 1,3-dithiacyclohexanes (dithianes; Section 23-5).

Approximately two-thirds of all published chemical studies deal in one way or another with heterocyclic systems. More than half of the known natural compounds are heterocyclic, and a high percentage of drugs contain heterocycles. Earlier chapters have described many representatives of these natural products. Some additional examples follow on the next page.

This chapter describes the naming, syntheses, and reactions of some saturated and aromatic heterocyclic compounds in order of increasing ring size, starting with the heterocyclopropanes. Some of this chemistry is a simple extension of transformations discussed earlier for carbocycles. However, the heteroatom often causes heterocyclic compounds to exhibit special chemical behavior.

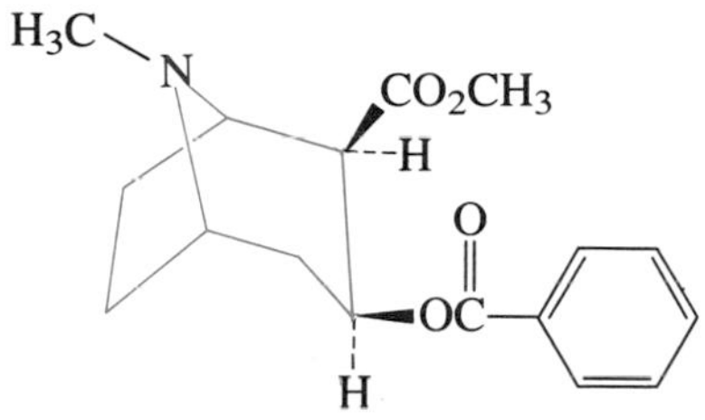

Cocaine

(Stimulant, topical anesthetic; found in coca leaves)

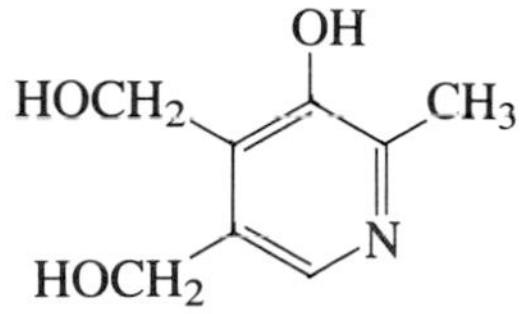

Pyridoxine, vitamin B_6

(Enzyme cofactor vitamin)

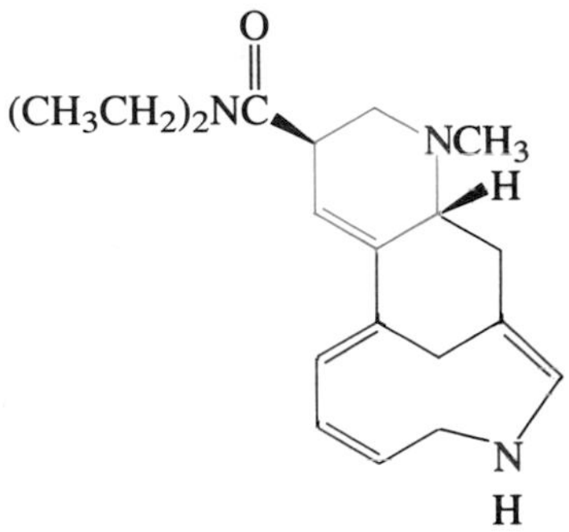

Lysergic acid diethylamide, LSD

(Psychotomimetic)

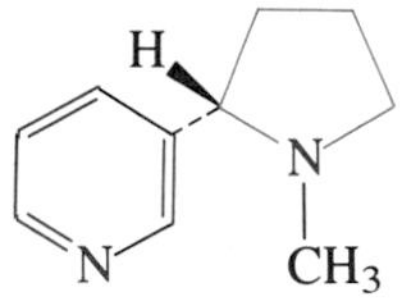

Nicotine

(Found in dried tobacco leaves in 2–8% concentration)

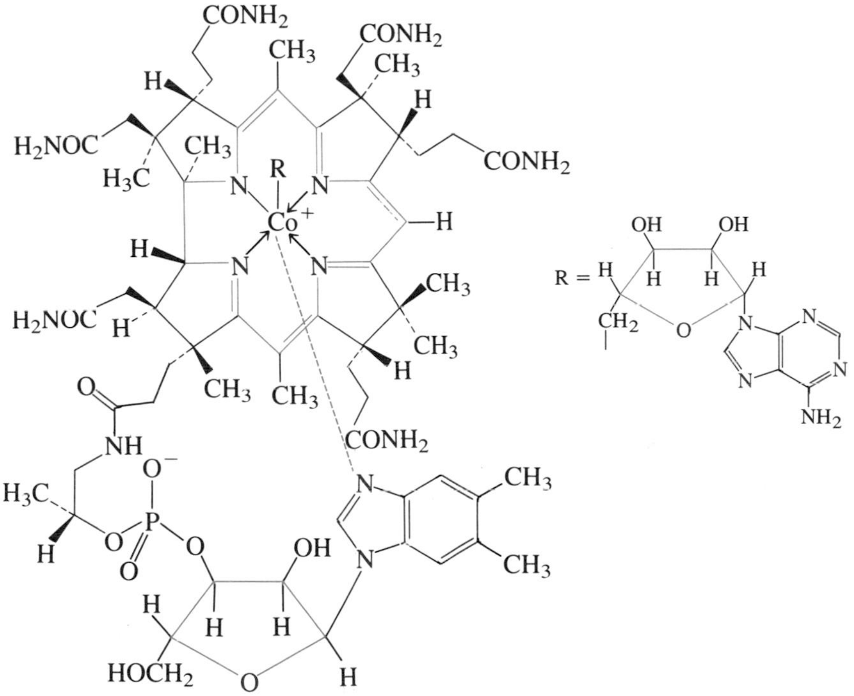

Vitamin B_{12}
(Cobalamin)

(Catalyzes biological rearrangements and methylations)

25-1 Naming the Heterocycles

Like all the other classes of compounds, this one contains many members bearing common names. Moreover, there are several competing systems for naming heterocycles that are sometimes confusing. We will adhere to the simplest system. We regard saturated heterocycles as derivatives of the related carbocycles and use a prefix to denote the presence and identity of the heteroatom: **aza-** for nitrogen, **oxa-** for oxygen, **thia-** for sulfur, **phospha-** for phosphorus, and so on. Other widely used names will be given in parentheses. The location of substituents is indicated by numbering the ring atoms, starting with the heteroatom.

Oxacyclopropane
(Oxirane, ethylene oxide)

***N*-Methylazacyclopropane**
(*N*-Methylaziridine)

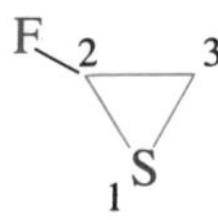

2-Fluorothiacyclopropane
(2-Fluorothiirane)

Oxacyclobutane
(Oxetane)

3-Ethylazacyclobutane
(3-Ethylazetidine)

2,2-Dimethylthiacyclobutane
(2,2-Dimethylthietane)

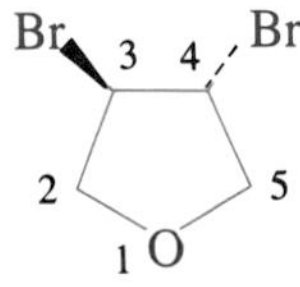

***trans*-3,4-Dibromooxacyclopentane**
(*trans*-3,4-Dibromotetrahydrofuran)

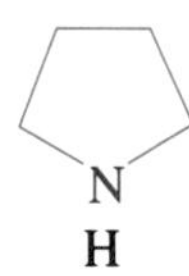

Azacyclopentane
(Pyrrolidine)

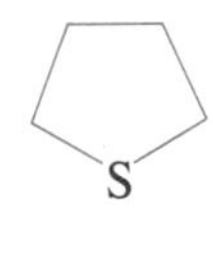

Thiacyclopentane
(Tetrahydrothiophene)

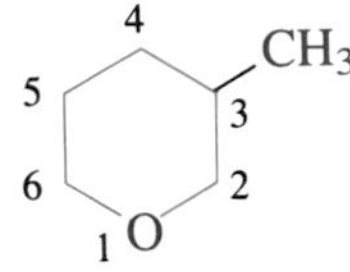

3-Methyloxacyclohexane
(3-Methyltetrahydropyran)

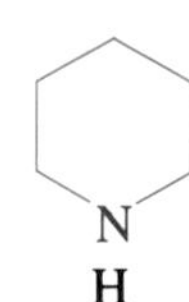

Azacyclohexane
(Piperidine)

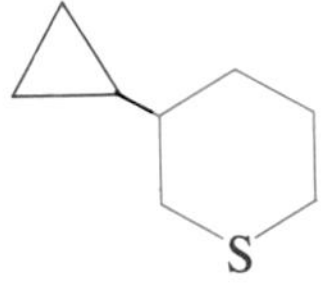

3-Cyclopropylthiacyclohexane
(3-Cyclopropyltetrahydrothiopyran)

The common names of unsaturated heterocycles are so firmly entrenched in the literature that we shall use them here.

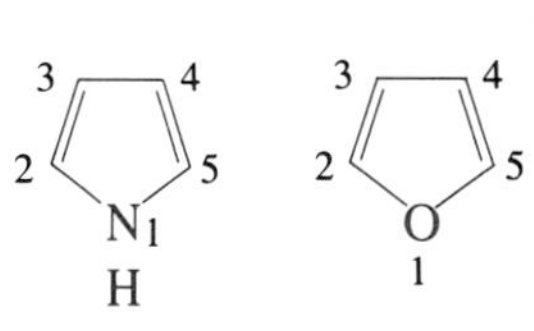

Pyrrole **Furan**

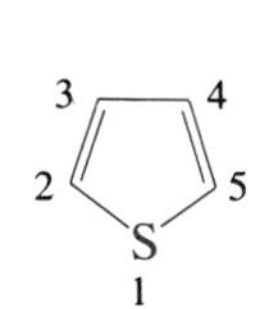

Thiophene

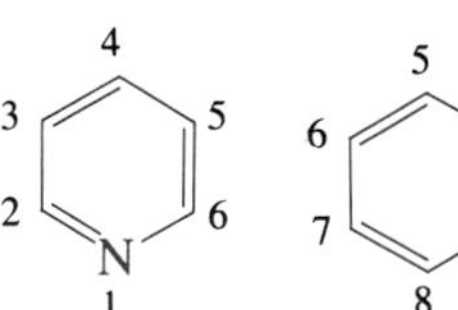

Pyridine

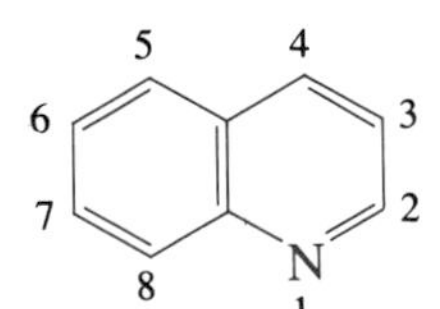

Quinoline

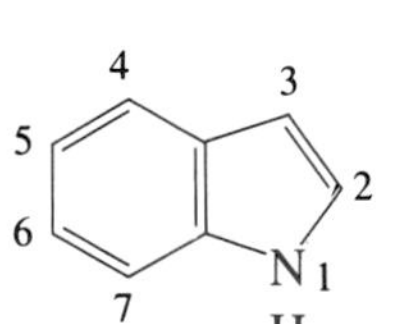

Indole

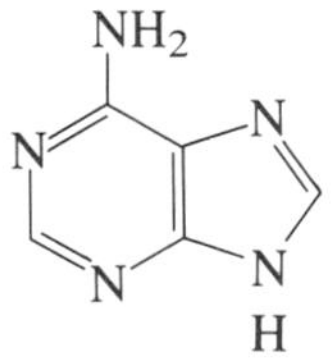

Adenine
(See Section 26-9)

EXERCISE 25-1

Name or draw the following compounds.

(a) *trans*-2,4-Dimethyloxacyclopentane (*trans*-2,4-dimethyltetrahydrofuran)
(b) *N*-Ethylazacyclopropane

(c) 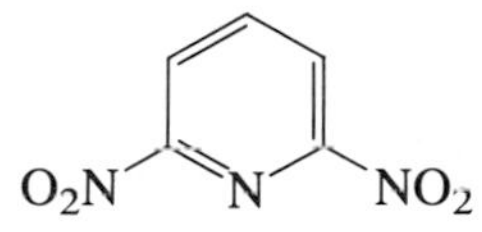 O_2N–(pyridine)–NO_2 (2,6-dinitropyridine)

(d) 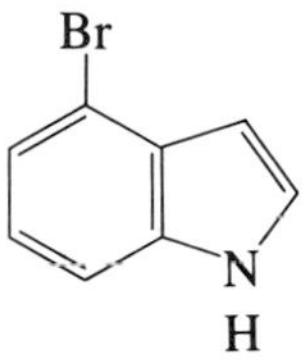 (4-bromoindole, Br, N–H)

25-2 Nonaromatic Heterocycles

As illustrated by the extensive chemistry of the oxacyclopropanes (Section 9-9), ring strain allows the three- and four-membered heterocycles to undergo nucleophilic ring opening readily. In contrast, the larger, unstrained systems are relatively inert to attack.

Ring strain makes heterocyclopropanes and heterocyclobutanes reactive

Heterocyclopropanes are relatively reactive, because ring strain is released by nucleophilic ring opening. Under basic conditions, this process gives rise to inversion at the less substituted center (Section 9-9).

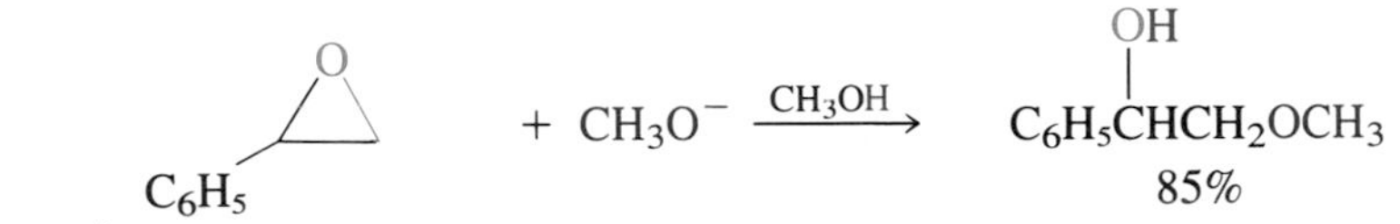

$$C_6H_5\text{-oxacyclopropane} + CH_3O^- \xrightarrow{CH_3OH} C_6H_5CH(OH)CH_2OCH_3 \quad 85\%$$

2-Phenyloxacyclopropane → **2-Methoxy-1-phenylethanol**

$$\text{H}_3\text{C, H, CH}_3\text{, H, N–CH}_2\text{CH}_3 \xrightarrow{70\%\ CH_3CH_2NH_2,\ H_2O,\ 120^\circ C,\ 16\ \text{days}} H_3C(H)(HNCH_2CH_3)C\text{—}C(H)(CH_3)(NHCH_2CH_3) \quad 55\%$$

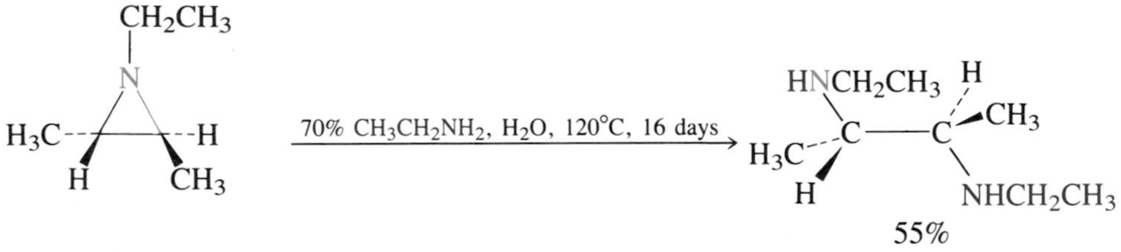

***N*-Ethyl-(2*S*,3*S*)-*trans*-2,3-dimethylazacyclopropane** → ***meso-N,N'*-Diethyl-2,3-butanediamine**

EXERCISE 25-2

Explain the following result by a mechanism. (Hint: Try a ring opening catalyzed by the Lewis acid and consider the options available to the resulting intermediate.)

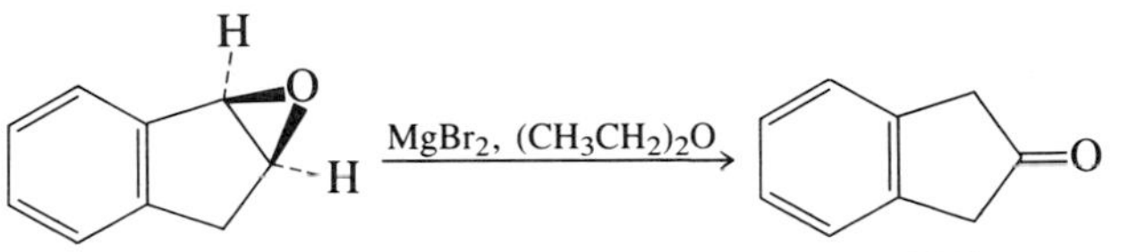

$$\text{(indene oxide)} \xrightarrow{MgBr_2,\ (CH_3CH_2)_2O} \text{(2-indanone)} \quad 100\%$$

EXERCISE 25-3

(2-Chloromethyl)oxacyclopropane reacts with hydrogen sulfide ion (HS^-) to give thiacyclobutan-3-ol. Explain by a mechanism.

The reactivity of the four-membered heterocycloalkanes bears out expectations based on ring strain: They undergo ring opening, as do their three-membered cyclic counterparts, but more stringent reaction conditions are usually required. The reaction of oxacyclobutane with CH_3NH_2 is typical.

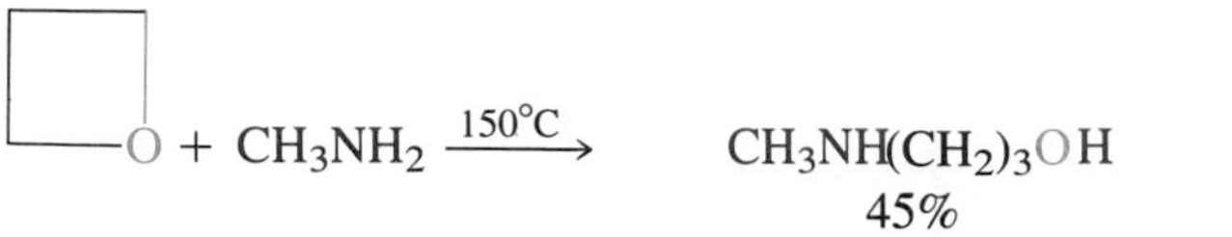

$$\text{(oxacyclobutane)} + CH_3NH_2 \xrightarrow{150°C} CH_3NH(CH_2)_3OH$$

45%

N-Methyl-3-amino-1-propanol

BOX 25-1 Azacyclopropene Antibiotics

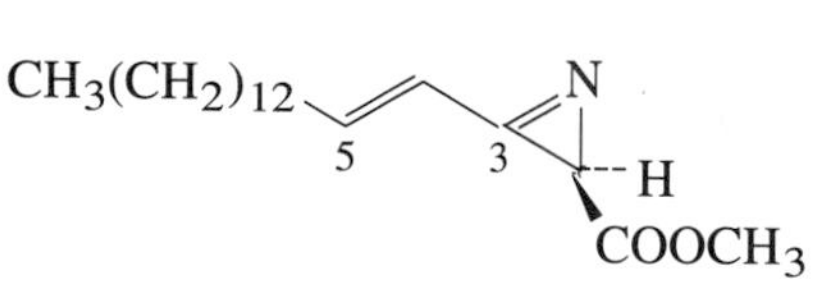

$CH_3(CH_2)_{12}$–CH=CH–(C3=N ring, C–H, $COOCH_3$)

Dysidazirine

$CH_3(CH_2)_{12}$–CH=CH–CH(OH)–CH(NH_2)–CH_2OH

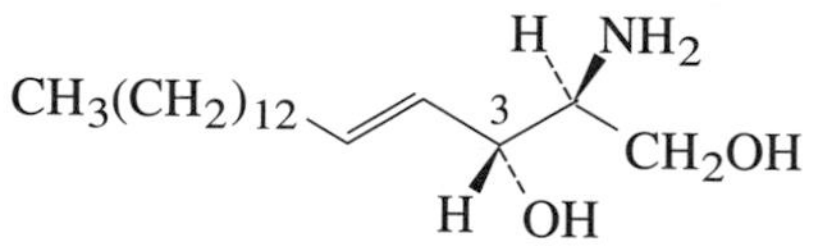

D-Sphingosine

The sea is an abundant source of highly biologically active substances. One of the most unusual is dysidazirine, a natural product that was discovered in 1988 in a South Pacific species of sponge and contains an azacyclopropene ring. This substance is toxic toward certain strains of cancer cells and inhibits the growth of Gram negative bacteria.

It is not known how dysidazirine is synthesized in nature, but one possible precursor is the amino alcohol D-sphingosine, which is a component of cell membranes. Enzyme-mediated oxidation of the secondary alcohol group at C3 of D-sphingosine to a ketone function could be a plausible first step in the transformation to dysidazirine. Intramolecular imine formation between the carbonyl and amino groups would lead to the three-membered heterocyclic ring. Finally, oxidation and esterification at C1 would complete the synthesis.

Although the mechanism of antibiotic action of dysidazirine is not yet known, the chemical reactivity of its C–N double bond is greatly increased by the strain of the three-membered ring. Nucleophiles rapidly attack C3 and also undergo conjugate addition to C5. Ring opening occurs under very mild conditions as well: Catalytic hydrogenation gives a mixture of the corresponding azacyclopropane and acyclic amino esters, as shown below.

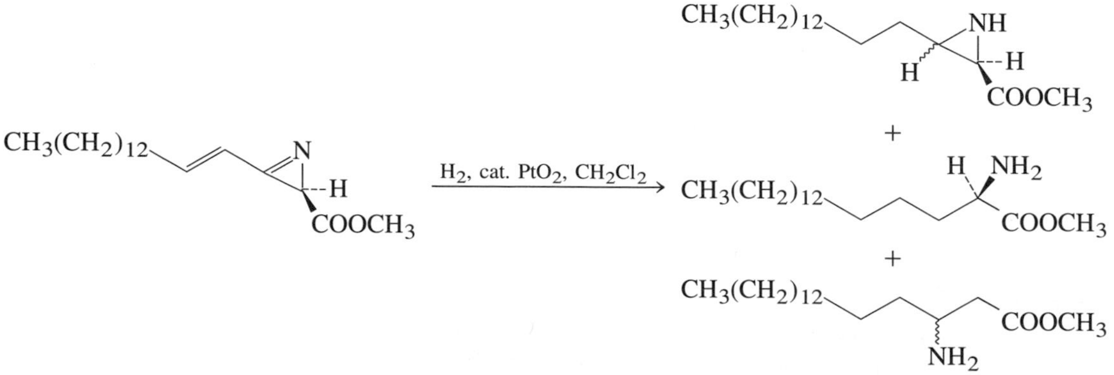

$$\text{Dysidazirine} \xrightarrow{H_2,\ \text{cat. } PtO_2,\ CH_2Cl_2} CH_3(CH_2)_{12}CH_2CH_2\text{-azacyclopropane}(H)(COOCH_3) + CH_3(CH_2)_{12}CH_2CH_2CH_2CH(NH_2)COOCH_3 + CH_3(CH_2)_{12}CH_2CH_2CH(NH_2)CH_2COOCH_3$$

The β-lactam antibiotics (penicillins and cephalosporins) function through related ring-opening processes (see Box 20-2).

EXERCISE 25-4

Treatment of thiacyclobutane with chlorine in $CHCl_3$ at $-70°C$ gives $ClCH_2CH_2CH_2SCl$ in 30% yield. Suggest a mechanism for this transformation. [Hint: The sulfur in sulfides is nucleophilic (Section 9-10).]

EXERCISE 25-5

2-Methyloxacyclobutane reacts with HCl to give two products. Write their structures.

Heterocyclopentanes and heterocyclohexanes are relatively unreactive

The unstrained heterocyclopentanes and heterocyclohexanes are relatively inert. Recall that oxacyclopentane (tetrahydrofuran, THF) is used as a solvent. However, the heteroatoms in five- and six-membered aza- and thiacycloalkanes allow these species to undergo their own characteristic transformations (see Sections 9-10, 17-8, 18-4, and 21-8). In general, ring-opening by cleavage of a bond to the heteroatom does not occur unless the latter is first converted into a good leaving group.

EXERCISE 25-6

Treatment of azacyclopentane (pyrrolidine) with sodium nitrite in acetic acid gives a liquid, b.p. 99–100°C (15 mm Hg), which has the composition $C_4H_8N_2O$. Propose a structure for this compound.

BOX 25-2 Taxol and Cancer

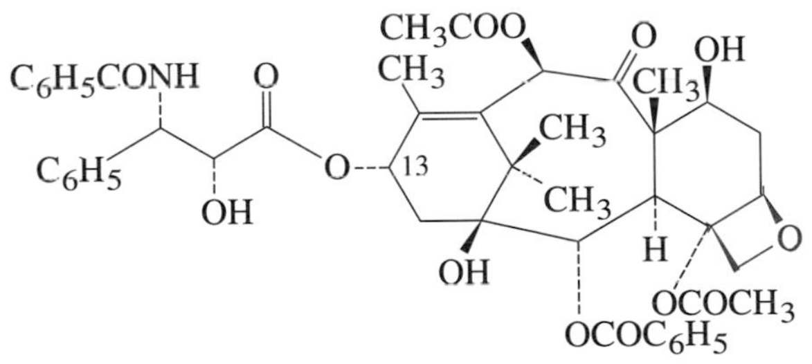

C_6H_5CONH O CH_3COO O OH CH_3 CH_3 C_6H_5 O 13 CH_3 CH_3 OH H O OH $OCOCH_3$ $OCOC_6H_5$

Taxol

Taxol is a highly promising, naturally occurring anticancer agent. Extensive clinical trials have revealed activity against leukemia and several tumor systems, including at least one type of lung tumors. Taxol has been approved by the U.S. Food and Drug Administration as a treatment for ovarian cancer. The availability of taxol from natural sources is extremely limited, however: It is present only to the extent of about 0.02% in the stem bark of the western yew tree. As a result, widespread efforts have been made to develop syntheses of this substance, two of which were completed in 1994.

Although taxol possesses an oxacyclobutane ring, the role that this portion of the molecule may play in anticancer activity is unclear. Several other reactive functions are present, including the allylic ester at C13. Hydrolysis of the four ester groups substantially reduces taxol's anticancer activity, a strong indication of their importance in the molecule's function.

BOX 25-3 Nicotine and Cancer

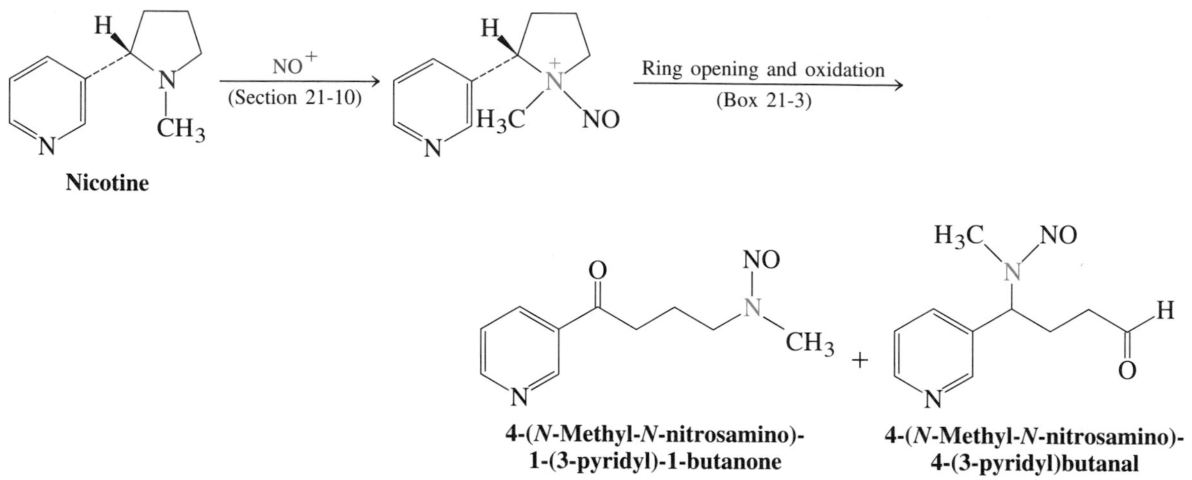

Nicotine

$\xrightarrow[\text{(Section 21-10)}]{NO^+}$ $\xrightarrow[\text{(Box 21-3)}]{\text{Ring opening and oxidation}}$

4-(*N*-Methyl-*N*-nitrosamino)-1-(3-pyridyl)-1-butanone + **4-(*N*-Methyl-*N*-nitrosamino)-4-(3-pyridyl)butanal**

The mechanism by which nicotine in cigarette smoke is converted into highly carcinogenic species is gradually becoming more clearly understood. The initial step appears to be *N*-nitrosation of the azacyclopentane (pyrrolidine) nitrogen. Oxidation and ring opening (compare Box 21-3) occur, giving a mixture of two *N*-nitrosodialkanamines (*N*-nitrosamines), each of which is a powerful carcinogen.

Upon protonation of the oxygen in a nitroso group, these substances become reactive alkylating agents, capable of transferring methyl groups to nucleophilic sites in biological molecules such as DNA, as shown below.

The diazohydroxide that remains decomposes via a diazonium ion to a carbocation, which may inflict additional molecular damage (Section 21-10).

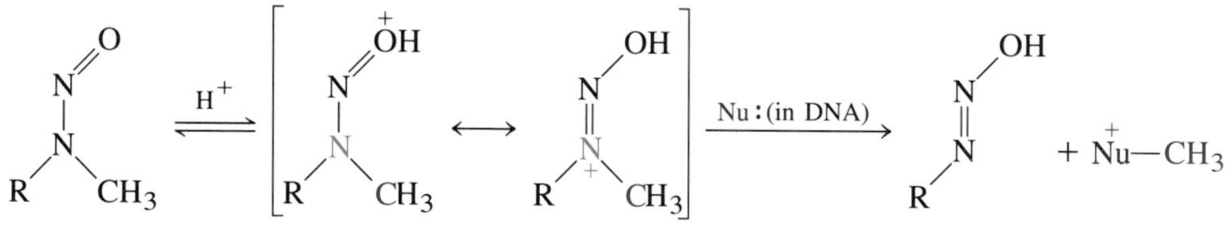

$$R(CH_3)N{-}N{=}O \overset{H^+}{\rightleftharpoons} \left[R(CH_3)N{-}N{=}\overset{+}{O}H \longleftrightarrow R(CH_3)\overset{+}{N}{=}N{-}OH \right] \xrightarrow{\text{Nu:(in DNA)}} R{-}N{=}N{-}OH + \overset{+}{Nu}{-}CH_3$$

In summary, the reactivity of heterocyclopropanes and heterocyclobutanes results from the release of strain by ring opening. The five- and six-membered heterocycloalkanes are less reactive than their smaller-ring counterparts.

Aromatic Heterocyclopentadienes

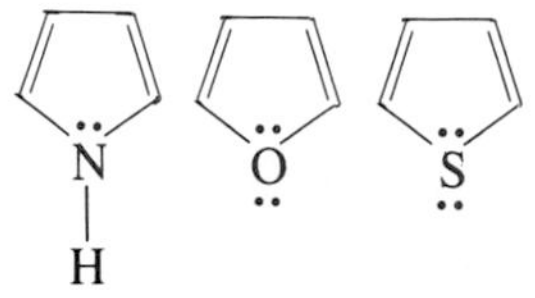

25-3 Structure and Properties of Aromatic Heterocyclopentadienes

Pyrrole, furan, and **thiophene** are 1-hetero-2,4-cyclopentadienes. Each contains a butadiene unit bridged by a heteroatom bearing lone electron pairs. These systems turn out to contain delocalized π electrons in an aromatic six-electron framework. This section will discuss their structure and review the methods used to prepare these compounds.

Pyrrole, furan, and thiophene contain delocalized lone electron pairs

The electronic structure of the three heterocycles pyrrole, furan, and thiophene is similar to that of the cyclopentadienyl anion (Section 15-8). The cyclopentadienyl anion may be viewed as a butadiene bridged by a negatively charged carbon whose electron pair is delocalized over the other four carbons. The heterocyclic analogs contain a neutral atom in that place, again bearing lone electron pairs. One of these pairs is similarly delocalized, furnishing the two electrons needed to satisfy the $4n + 2$ rule (Section 15-7). To maximize overlap, the heteroatoms are hybridized sp^2 (Figure 25-1), the delocalized electron pair being assigned to the remaining p orbital. In pyrrole, the sp^2-hybridized nitrogen bears a hydrogen substituent in the plane of the molecule. For furan and thiophene, the second lone electron pair is placed into one of the sp^2 hybrid orbitals, again in the plane and therefore with no opportunity to achieve overlap. This arrangement is much like that in the phenyl anion (Section 22-4).

Cyclopentadienyl Anion

The delocalization of the lone pair in the 1-hetero-2,4-cyclopentadienes can be described by charge-separated resonance structures, as shown for pyrrole.

Resonance Structures of Pyrrole

Notice that there are four dipolar structures in which a positive charge is placed on the heteroatom and a negative charge successively on each of the carbons. This picture suggests that the heteroatom should be relatively electron poor and that the carbons should be relatively electron rich. Indeed, as we shall see, the reactivity of these compounds bears out that expectation.

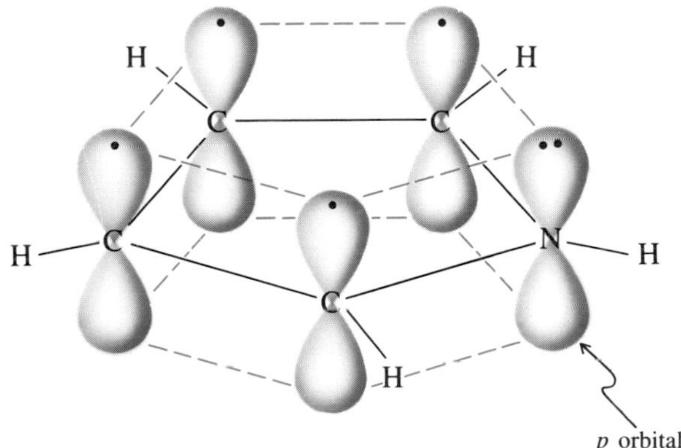

Pyrrole

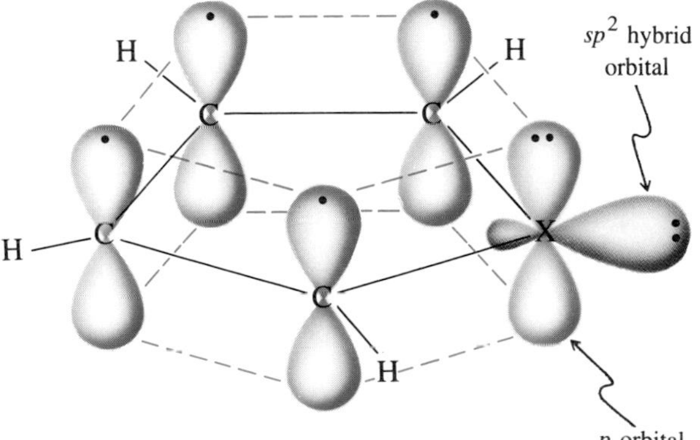

Furan (X = O)
Thiophene (X = S)

FIGURE 25-1 Molecular-orbital pictures of pyrrole, furan (X = O), and thiophene (X = S). The heteroatom in each is sp^2 hybridized and bears one delocalized lone electron pair.

EXERCISE 25-7

Azacyclopentane and pyrrole are both polar molecules. However, the dipole vectors in the two molecules point in opposite directions. What is the sense of direction of this vector in each structure? Explain your answer.

Pyrroles, furans, and thiophenes are prepared from γ-dicarbonyl compounds

Syntheses of the heterocyclopentadienes use a variety of cyclization strategies. A general approach is the **Paal-Knorr* synthesis** (for pyrroles) and its variations (for the other heterocycles). The target molecule is approached from an enolizable γ-dicarbonyl compound that is treated with an amine derivative (for pyrroles), or P_2O_5 (for furans), or P_2S_5 (for thiophenes).

Cyclization of a γ-Dicarbonyl Compound to a 1-Hetero-2,4-Cyclopentadiene

R, O, O, R; $R'NH_2$, or P_2O_5, or P_2S_5; R, X, R

$X = NR', O, S$

$CH_3CCH_2CH_2CCH_3$ (O, O) + $(CH_3)_2CHNH_2$ $\xrightarrow{CH_3COOH,\ \Delta,\ 17\ h}$ H_3C, N, CH_3, $(CH_3)_2CH$

70%

***N*-1-Methylethyl-2,5-dimethylpyrrole**

O, C_6H_5, C_6H_5, O $\xrightarrow{P_2O_5,\ 150°C}$ C_6H_5, O, C_6H_5

62%

$CH_3CCH_2CH_2CCH_3$ (O, O) $\xrightarrow{P_2S_5,\ 140\text{–}150°C}$ H_3C, S, CH_3

60%

2,5-Dimethylthiophene

*Professor Karl Paal, 1860–1935, University of Erlangen, Germany; Professor Ludwig Knorr, 1859–1921, University of Jena, Germany.

EXERCISE 25-8

The following equation is an example of another synthesis of pyrroles. Write a mechanism for this transformation. (Hint: Refer to Section 17-9.)

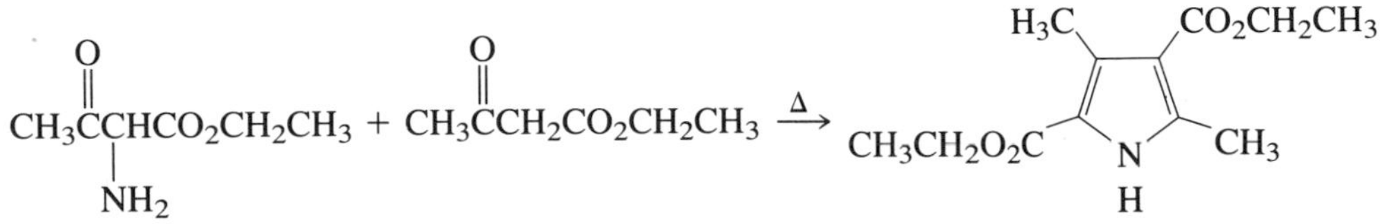

$$CH_3\overset{O}{\overset{\|}{C}}\underset{NH_2}{\underset{|}{C}}HCO_2CH_2CH_3 + CH_3\overset{O}{\overset{\|}{C}}CH_2CO_2CH_2CH_3 \xrightarrow{\Delta}$$

H3C, CO2CH2CH3, CH3CH2O2C, N, H, CH3

Ethyl 2-amino-3-oxobutanoate **Ethyl 3-oxobutanoate** **Diethyl 3,5-dimethylpyrrole-2,4-dicarboxylate**

In summary, pyrrole, furan, and thiophene contain delocalized aromatic π systems analogous to that of the cyclopentadienyl anion. A general method for the preparation of 1-hetero-2,4-cyclopentadienes is based on the cyclization of enolizable 1,4-dicarbonyl compounds.

25-4 Reactions of the Aromatic Heterocyclopentadienes

The reactivity of pyrrole, furan, and thiophene and their derivatives is largely governed by their aromaticity and based on the chemistry of benzene. This section describes some of their reactions, particularly electrophilic aromatic substitution, and introduces indole, a ring-fused analog of pyrrole.

Pyrroles, furans, and thiophenes undergo electrophilic aromatic substitution

As expected for aromatic systems, the 1-hetero-2,4-cyclopentadienes undergo electrophilic substitution. There are two sites of possible attack—at C2 and at C3. Which one should be more reactive? An answer can be found by the same procedure used to predict the regioselectivity of electrophilic aromatic substitution of substituted benzenes (Chapter 16): enumeration of all the possible resonance structures for the two modes of reaction.

Consequences of Electrophilic Attack at C2 and C3 in the Aromatic Heterocyclopentadienes

Attack at C2

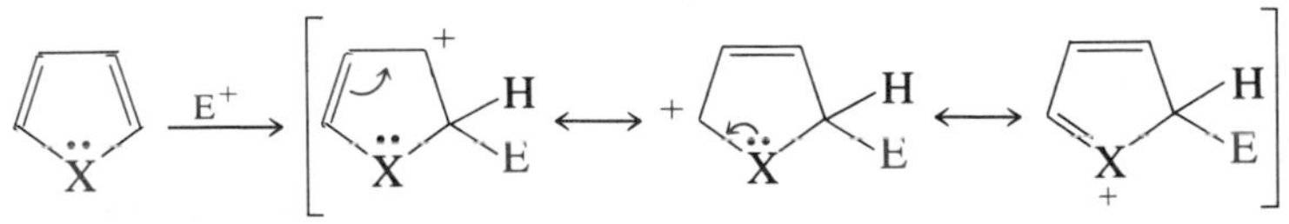

X, E^+, X, H, E, X, H, E, X, H, E, +

Attack at C3

Both modes benefit from the presence of the resonance-contributing heteroatom, but attack at C2 leads to an intermediate with an additional resonance structure, thus indicating this position to be the preferred center of substitution. Indeed, such selectivity is generally observed. However, because C3 also is activated to electrophilic attack, mixtures of products can form, depending on conditions, substrates, and electrophiles.

Electrophilic Aromatic Substitution of Pyrrole, Furan, and Thiophene

CH_3CONO_2, −10°C; − CH_3COOH

50% **2-Nitropyrrole**; 13% **3-Nitropyrrole**

Cl−Cl, CH_2Cl_2; − HCl

64% **2-Chlorofuran**

CH_3CCl, $SnCl_4$; − HCl

64% **2-Acetyl-5-methylthiophene**

The relative nucleophilic reactivity of benzene and the three heterocycles increases in the order benzene ≪ thiophene < furan < pyrrole.

EXERCISE 25-9

The monobromination of thiophene-3-carboxylic acid gives only one product. What is its structure and why is it the only product formed?

Because the lone electron pair on nitrogen is tied up by conjugation, pyrrole is extremely nonbasic. Very strong acid is required to effect protonation, which occurs not on nitrogen but on C2.

Protonation of Pyrrole

$$\text{Pyrrole} + H^+ \rightleftharpoons \text{2H-pyrrolium ion} \qquad pK_a = -4.4$$

EXERCISE 25-10

Explain why pyrrole is protonated on an α-carbon rather than on the nitrogen.

1-Hetero-2,4-cyclopentadienes can undergo ring opening and cycloaddition reactions

Furans can be hydrolyzed under mild conditions to γ-dicarbonyl compounds. The reaction may be viewed as the reverse of the Paal-Knorr-type synthesis of furans. Pyrrole polymerizes under these reaction conditions, whereas thiophene is stable.

Hydrolysis of a Furan to a γ-Dicarbonyl Compound

$$\text{2,5-dimethylfuran} \xrightarrow{CH_3COOH,\ H_2SO_4,\ H_2O,\ \Delta} CH_3\overset{O}{\overset{\|}{C}}CH_2CH_2\overset{O}{\overset{\|}{C}}CH_3$$

90%

2,5-Hexanedione

Raney nickel desulfurization (Section 17-8) of thiophene derivatives results in sulfur-free acyclic saturated compounds.

$$\text{2-thienyl}{-}CH(OCH_2CH_3)_2 \xrightarrow[-\ NiS]{\text{Raney Ni},\ (CH_3CH_2)_2O,\ \Delta} CH_3(CH_2)_3CH(OCH_2CH_3)_2$$

50%

The π system of furan (but not of pyrrole or thiophene) possesses sufficient diene character to undergo Diels-Alder cycloadditions (Section 14-8).

$$\text{Furan} + \text{maleimide} \xrightarrow{(CH_3CH_2)_2O,\ 25°C} \text{oxa-bridged bicyclic imide adduct}$$

95%

Indole is a benzpyrrole

Indole is the most important *benzannulated* (fused-ring) derivative of the 1-hetero-2,4-cyclopentadienes. It forms part of many natural products, including the amino acid tryptophan (Section 26-1).

Tryptophan

Indole is related to pyrrole in the same way that naphthalene is related to benzene. Its electronic makeup is indicated by the various possible resonance structures that can be formulated for the molecule. Although those resonance structures that disturb the cyclic six-π-electron system of the fused benzene ring are less important, they indicate the electron-donating effect of the heteroatom.

Resonance in Indole

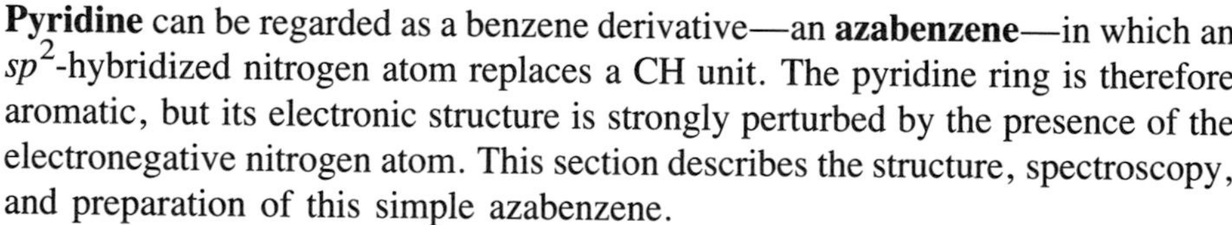

EXERCISE 25-11

Predict the preferred site of electrophilic aromatic substitution in indole. Explain your choice.

In summary, the donation of the lone electron pair on the heteroatom to the diene unit in pyrrole, furan, and thiophene makes the carbon atoms in these systems electron rich and therefore more susceptible to electrophilic aromatic substitution than those in benzene. Electrophilic attack is frequently favored at C2, but substitution at C3 is also observed, depending on conditions, substrates, and electrophiles. Some rings can be opened by hydrolysis or by desulfurization (for thiophenes). The diene unit in furan is reactive enough to undergo Diels-Alder cycloadditions. Indole is a benzpyrrole containing a delocalized π system.

25-5 Structure and Preparation of Pyridine: An Azabenzene

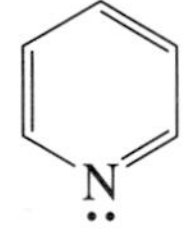

Pyridine

Pyridine can be regarded as a benzene derivative—an **azabenzene**—in which an sp^2-hybridized nitrogen atom replaces a CH unit. The pyridine ring is therefore aromatic, but its electronic structure is strongly perturbed by the presence of the electronegative nitrogen atom. This section describes the structure, spectroscopy, and preparation of this simple azabenzene.

Pyridine is a cyclic aromatic imine

Pyridine contains an sp^2-hybridized nitrogen atom like that in an imine (Section 17-9). In contrast with pyrrole, there is only one electron in the p orbital that completes the aromatic π-electron arrangement of the aromatic ring; as in the phenyl anion, the lone electron pair is located in one of the sp^2 hybrid atomic orbitals in the molecular plane (Figure 25-2). Therefore, in pyridine, the nitrogen does not donate excess electron density to the remainder of the molecule. Quite the contrary: Because nitrogen is more electronegative than carbon (Table 1-2), it withdraws electron density from the ring, both inductively and by resonance.

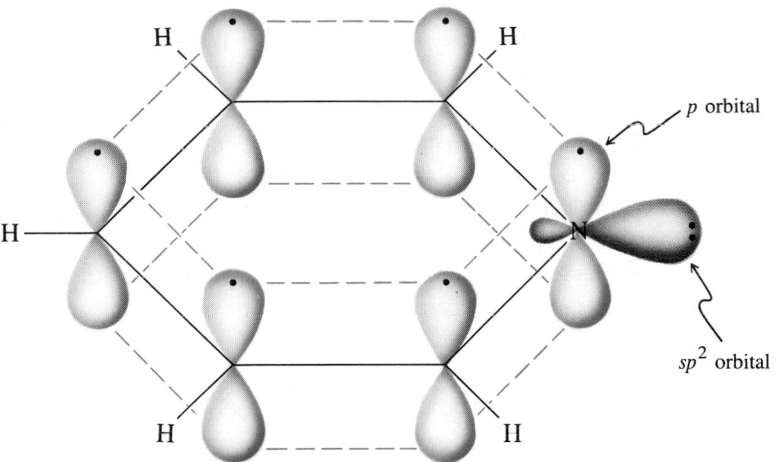

FIGURE 25-2 Molecular-orbital picture of pyridine. The lone electron pair on nitrogen is in an sp^2-hybridized orbital and is *not* part of the aromatic π system.

Resonance in Pyridine

EXERCISE 25-12

Azacyclohexane (piperidine) is a polar molecule. In which direction does its dipole vector point? Answer the same question for pyridine. Explain your answer.

Aromatic delocalization in pyridine is evident in the ^1H NMR spectrum, which reveals the presence of a ring current. The electron-withdrawing capability of the nitrogen is manifest in larger chemical shifts (more deshielding) at C2 and C4, as expected from the resonance picture.

^1H NMR Chemical Shifts (ppm) in Pyridine and Benzene

H 7.27

H 7.46, H 7.06, H 8.50

Pyridine Is a Weak Base

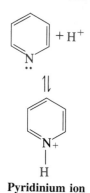

Pyridinium ion

$pK_a = 5.29$

Because the lone pair on nitrogen is not tied up by conjugation (as it is in pyrrole, Exercise 25-10), pyridine is a weak base. (It is used as such in numerous organic transformations.) Compared with alkanamines (pK_a of ammonium salts ~ 10), the pyridinium ion has a low pK_a, because the nitrogen is sp^2 and not sp^3 hybridized (see Section 11-3 for the effect of hybridization on acidity).

Pyridine is the simplest azabenzene. Some of its higher aza analogs are shown here. They behave like pyridine but show the increasing effect of aza substitution—in particular, increasing electron deficiency. Minute quantities of several 1,4-diazabenzene (pyrazine) derivatives are responsible for the characteristic odors of many vegetables. One drop of 2-methoxy-3-(1-methylethyl)-1,4-diazabenzene (2-isopropyl-3-methoxypyrazine) in a large swimming pool would be more than adequate to give the entire pool the odor of raw potatoes.

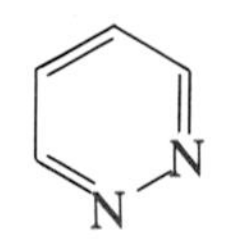

1,2-Diazabenzene (Pyridazine)

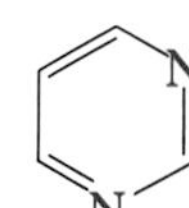

1,3-Diazabenzene (Pyrimidine)

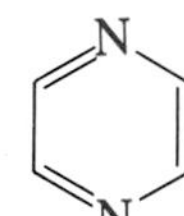

1,4-Diazabenzene (Pyrazine)

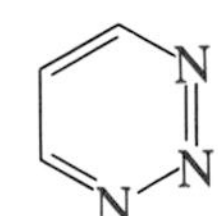

1,2,3-Triazabenzene (1,2,3-Triazine)

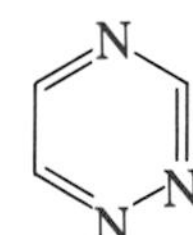

1,2,4-Triazabenzene (1,2,4-Triazine)

1,3,5-Triazabenzene (1,3,5-Triazine)

1,2,4,5-Tetraazabenzene (1,2,4,5-Tetrazine)

2-Methoxy-3-(1-methylethyl)-1,4-diazabenzene

(Potatoes)

2-Methoxy-3-(2-methylpropyl)-1,4-diazabenzene

(Green peppers)

Pyridines are made by condensation reactions

Pyridine and simple alkylpyridines are obtained from coal tar. Many of the more highly substituted pyridines are in turn made by both electrophilic and nucleophilic substitution of the simpler derivatives.

Pyridines can be made by condensation reactions of acyclic starting materials such as carbonyl compounds with ammonia. The most general of these methods is the **Hantzsch* pyridine synthesis.** In this reaction, two molecules of a β-dicarbonyl compound, an aldehyde, and ammonia combine in several steps to give a substituted dihydropyridine, which is readily oxidized by nitric acid to the aromatic system.

*Professor Arthur Hantzsch, 1857–1935, University of Leipzig, Germany.

Hantzsch Synthesis of 2,6-Dimethylpyridine

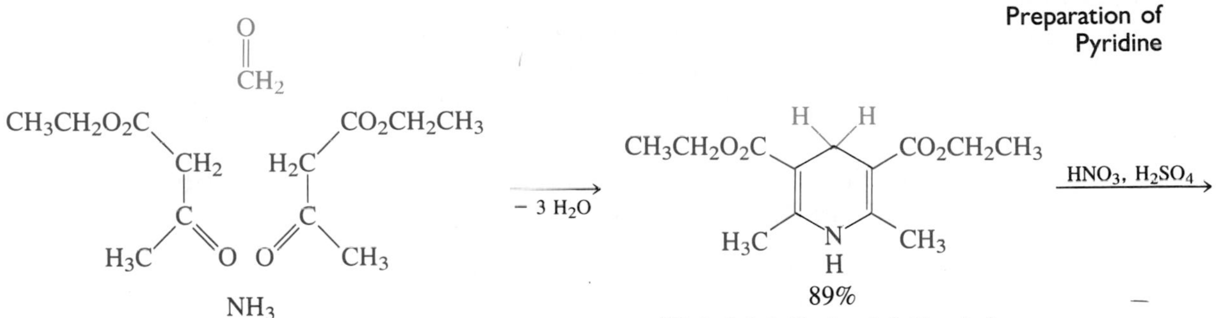

$CH_2=O$ + 2 $CH_3COCH_2CO_2CH_2CH_3$ + NH_3 $\xrightarrow{-\,3\,H_2O}$ Diethyl 1,4-dihydro-2,6-dimethyl-3,5-pyridinedicarboxylate (89%) $\xrightarrow{HNO_3,\ H_2SO_4}$

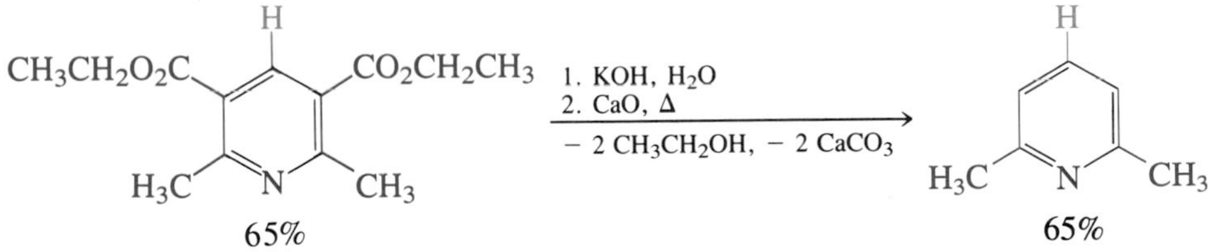

Diethyl 2,6-dimethyl-3,5-pyridinedicarboxylate (65%) $\xrightarrow[-\,2\,CH_3CH_2OH,\ -\,2\,CaCO_3]{1.\ KOH,\ H_2O\quad 2.\ CaO,\ \Delta}$ 2,6-Dimethylpyridine (65%)

If the β-dicarbonyl compound is a 3-keto ester, the resulting product is a 3,5-pyridinedicarboxylic ester. Hydrolysis followed by pyrolysis of the calcium salt of the acid causes decarboxylation.

EXERCISE 25-13

What starting materials would you use in the Hantzsch synthesis of the following pyridines?

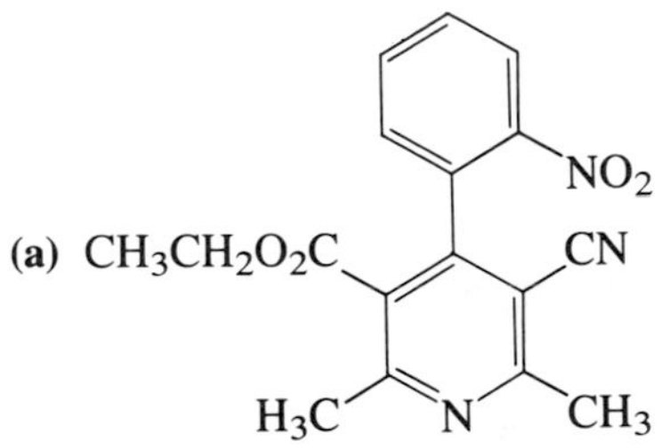

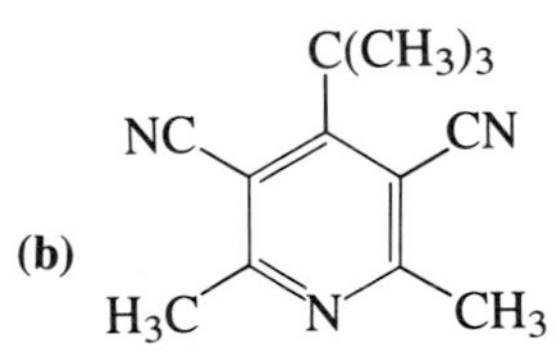

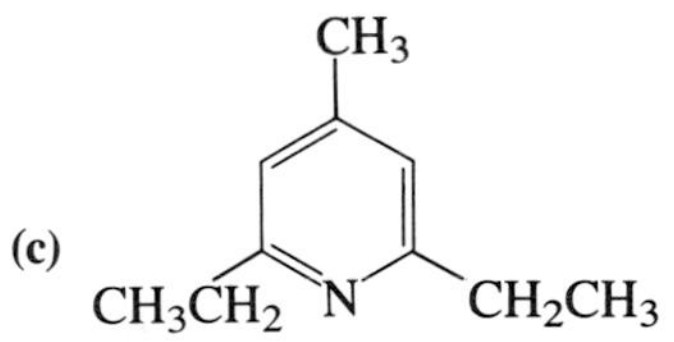

(a) Ethyl 5-cyano-2,6-dimethyl-4-(2-nitrophenyl)pyridine-3-carboxylate ($CH_3CH_2O_2C$, CN, NO_2, H_3C, CH_3)
(b) Pyridine with NC, $C(CH_3)_3$, CN, H_3C, CH_3 substituents
(c) Pyridine with CH_3, CH_3CH_2, CH_2CH_3 substituents

In summary, pyridines are aromatic but electron poor. The lone pair on nitrogen makes the heterocycle weakly basic. Pyridines are prepared by condensation of a β-dicarbonyl compound with ammonia and an aldehyde.

25-6 Reactions of Pyridine

The reactivity of pyridine derives from its dual nature as both an aromatic molecule and a cyclic imine. Both electrophilic and nucleophilic substitution processes may occur, leading to a variety of substituted derivatives.

Pyridine undergoes electrophilic aromatic substitution only under extreme conditions

Because the pyridine ring is electron poor, the system undergoes electrophilic aromatic substitution only with great difficulty, several orders of magnitude more slowly than benzene, and at C3 (see Section 15-9).

Electrophilic Aromatic Substitution of Pyridine

Pyridine $\xrightarrow[- H_2O]{NaNO_3,\ \text{fuming}\ H_2SO_4,\ 300°C}$ 3-Nitropyridine, 4.5%

3-Nitropyridine

Pyridine $\xrightarrow[- HBr]{Br{-}Br,\ H_2SO_4,\ SO_3}$ 3-Bromopyridine, 86%

3-Bromopyridine

EXERCISE 25-14

Explain why electrophilic aromatic substitution of pyridine is at C3.

Activating substituents allow for milder conditions or improved yields.

2,6-Dimethylpyridine $\xrightarrow[- H_2O]{KNO_3,\ \text{fuming}\ H_2SO_4,\ 100°C}$ 2,6-Dimethyl-3-nitropyridine, 81%

2,6-Dimethylpyridine **2,6-Dimethyl-3-nitropyridine**

2-Aminopyridine $\xrightarrow[- HBr]{Br{-}Br,\ CH_3COOH,\ 20°C}$ 2-Amino-5-bromopyridine, 90%

2-Aminopyridine **2-Amino-5-bromopyridine**

Pyridine undergoes nucleophilic substitution

Because the pyridine ring is relatively electron deficient, it undergoes nucleophilic substitution much more readily than does benzene (Section 22-4). Attack at C2 and C4 is preferred because it leads to intermediates in which the negative charge is on the nitrogen. An example of nucleophilic substitution of pyridine is the **Chichibabin* reaction,** in which the heterocycle is converted into 2-aminopyridine by treatment with sodium amide in liquid ammonia.

Chichibabin Reaction

1. $NaNH_2$, liquid NH_3
2. H^+, H_2O

+ H—H

70%

2-Aminopyridine

This reaction proceeds by the addition–elimination mechanism. The first step is attack by $^-:\ddot{N}H_2$ at C2, a process that resembles 1,2-addition to an imine function. Expulsion of a hydride ion, $H:^-$, from C2 occurs, followed by deprotonation of the amine nitrogen to give H_2 and a resonance-stabilized 2-pyridineamide ion. Protonation of the latter by aqueous work-up furnishes the final product. Note the contrast with *electrophilic* substitutions, which involve *proton* loss, not elimination of hydride.

Transformations related to the Chichibabin reaction occur when pyridines are treated with Grignard or organolithium reagents.

Methylbenzene (toluene), 110°C, 8 h

− LiH

49%

2-Phenylpyridine

In most nucleophilic substitutions of pyridines, halides are leaving groups, the 2- and 4-halopyridines being particularly reactive.

$Na^{+-}OCH_3$, CH_3OH

− NaCl

CH_3O

75%

4-Methoxypyridine

*Professor Alexei E. Chichibabin, 1871–1945, University of Moscow.

BOX 25-4 Pyridinium Salts in Nature: Nicotinamide Adenine Dinucleotide

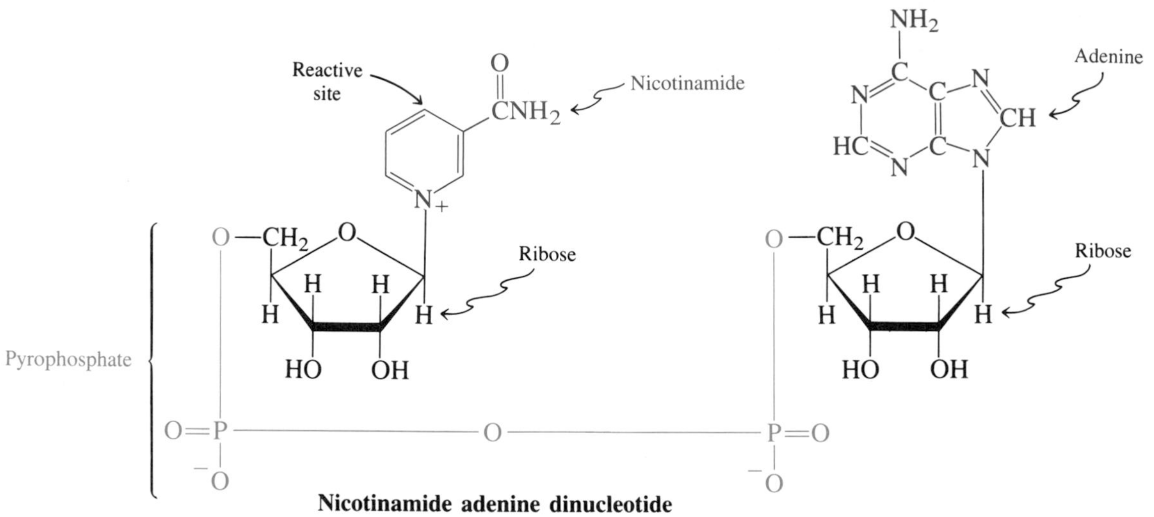

Nicotinamide adenine dinucleotide

A complex pyridinium derivative, *nicotinamide adenine dinucleotide* (NAD^+) is an important biological oxidizing agent. The structure consists of a pyridine ring [derived from 3-pyridinecarboxylic (nicotinic) acid], two ribose molecules (Section 24-1) linked by a pyrophosphate bridge, and the heterocycle adenine (Section 26-9).

Most organisms derive their energy from the oxidation (removal of electrons) of fuel molecules, such as glucose or fatty acids; the ultimate oxidant (electron acceptor) is oxygen, which gives water. Such biological oxidations proceed through a cascade of electron-transfer reactions requiring the intermediacy of special redox reagents. NAD^+ is one such molecule. In the oxidation of a substrate, the pyridinium ring in NAD^+ undergoes a two-electron reduction with simultaneous protonation.

Reduction of NAD^+

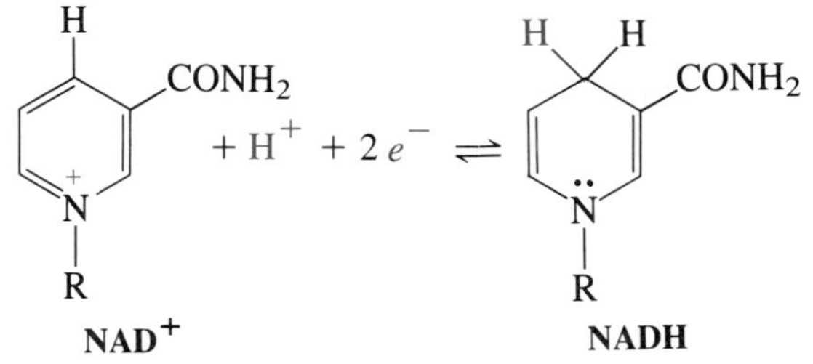

$$NAD^+ + H^+ + 2e^- \rightleftharpoons NADH$$

NAD^+ is the electron acceptor in many biological oxidations of alcohols to aldehydes (including the conversion of vitamin A into retinal, Section 18-9). This reaction can be seen as a transfer of hydride from C1 of the alcohol to the pyridinium nucleus with simultaneous deprotonation.

Oxidation of Alcohols by NAD^+

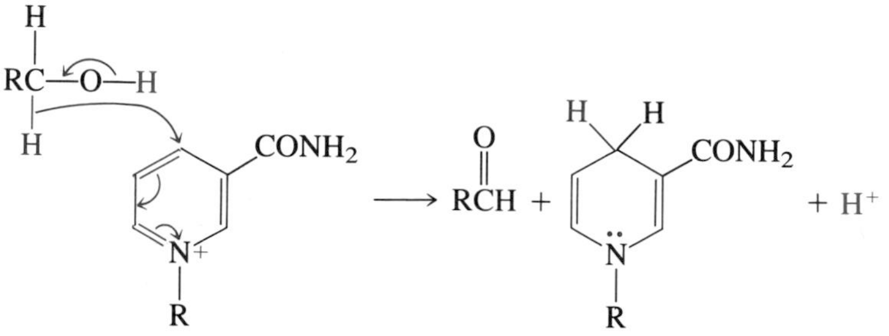

$$RCH_2OH + NAD^+ \longrightarrow RCHO + NADH + H^+$$

EXERCISE 25-15

The relative rates of the reactions of 2-, 3-, and 4-chloropyridine with sodium methoxide in methanol are 3000:1:81,000. Explain.

EXERCISE 25-16

Propose a mechanism for the reaction of 4-chloropyridine with methoxide (see Exercise 25-15). [Hint: Think of the pyridine ring as containing an α,β-unsaturated imine function (see Sections 17-9 and 18-10).]

In summary, pyridine undergoes slow electrophilic aromatic substitution preferentially at C3. Nucleophilic substitution reactions occur more readily to expel hydride or another leaving group from either C2 or C4.

25-7 Quinoline and Isoquinoline: The Benzpyridines

We can imagine the fusion of a benzene ring to pyridine in either of two ways, giving us **quinoline** and **isoquinoline** (1- and 2-azanaphthalene, according to our systematic nomenclature). Both are liquids with high boiling points. Many of their derivatives are found in nature or have been synthesized in the search for physiological activity. Like pyridine, quinoline and isoquinoline are readily available from coal tar.

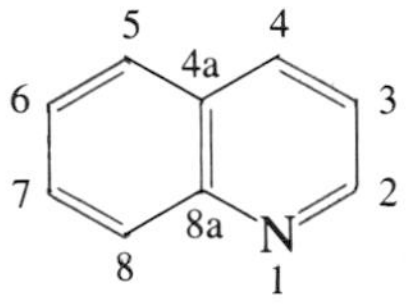

5 4a 4 6 3 7 2 8a N 8 1

Quinoline

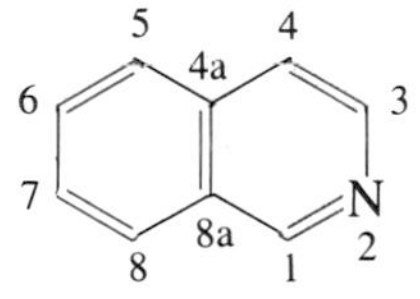

5 4a 4 6 3 7 N 8a 2 8 1

Isoquinoline

As might be expected, because pyridine is electron poor compared with benzene, electrophilic substitutions on quinoline and isoquinoline take place at the *benzene* ring. As with naphthalene, substitution at the carbons next to the ring fusion predominates.

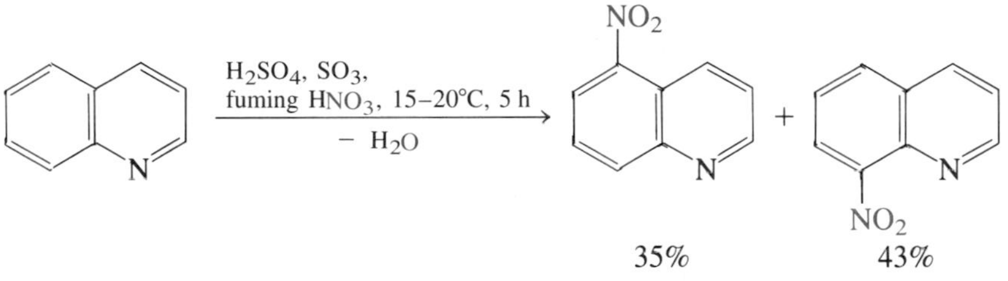

N —[H_2SO_4, SO_3, fuming HNO_3, 15–20°C, 5 h; $- H_2O$]→ NO_2 N + N NO_2

35% **5-Nitroquinoline** 43% **8-Nitroquinoline**

N —[H_2SO_4, HNO_3, 0°C, 30 min; $- H_2O$]→ NO_2 N + N NO_2

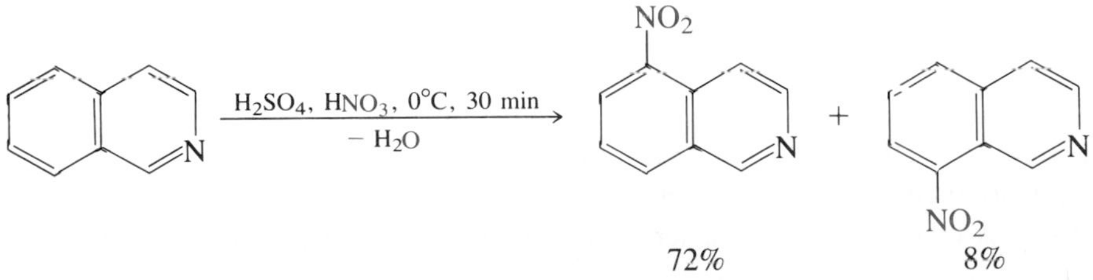

72% **5-Nitroisoquinoline** 8% **8-Nitroisoquinoline**

In contrast with electrophiles, nucleophiles prefer reaction at the electron-poor *pyridine* nucleus. These reactions are quite analogous to those with pyridine.

Chichibabin Reaction of Quinoline and Isoquinoline

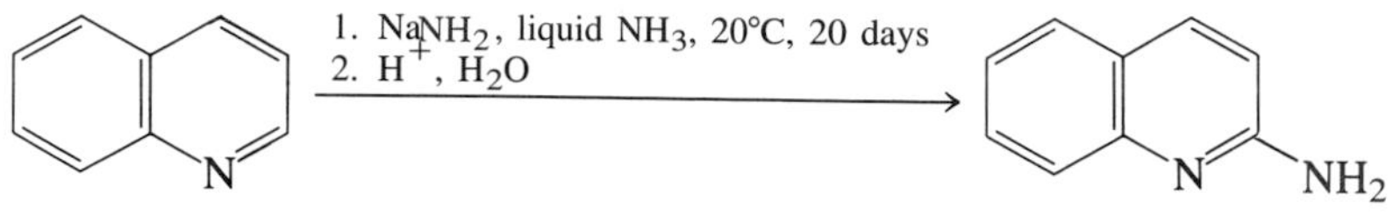

1. $NaNH_2$, liquid NH_3, 20°C, 20 days
2. H^+, H_2O

NH_2

80%

2-Aminoquinoline

BOX 25-5 Azanaphthalenes in Nature

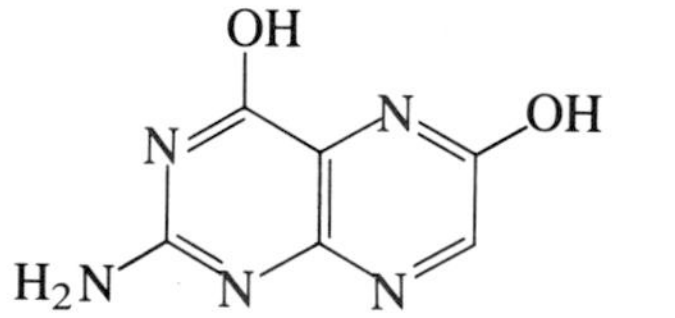

Xanthopterin
(Yellow butterfly and other insect pigment)

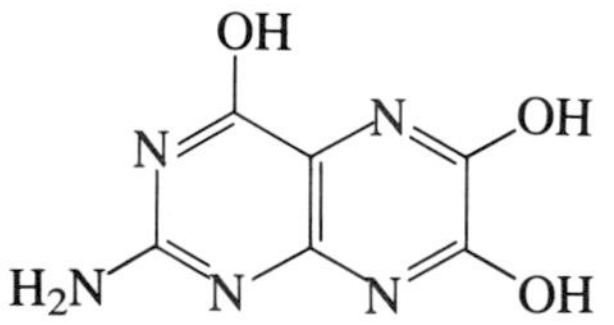

Leucopterin
(Colorless substance found in white butterfly wings)

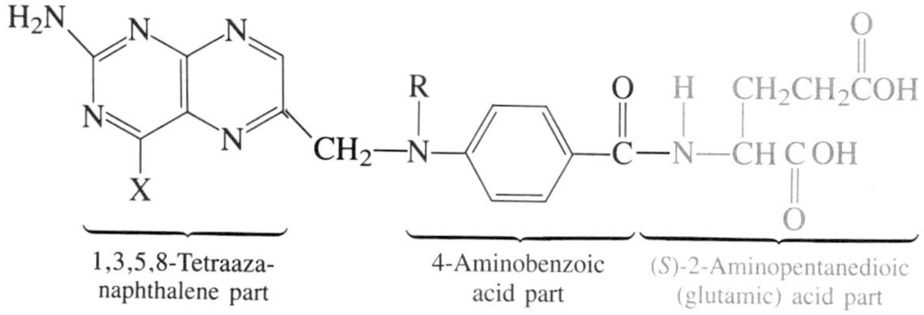

1,3,5,8-Tetraaza-naphthalene part

4-Aminobenzoic acid part

(*S*)-2-Aminopentanedioic (glutamic) acid part

Folic acid (X = OH, R = H)
Methotrexate (X = NH_2, R = CH_3)

The 1,3,5,8-tetraazanaphthalene (pteridine) ring system is present in a number of interesting natural products. Xanthopterin and leucopterin are insect pigments. Folic acid (Section 15-11) is a biologically important molecule incorporating a 1,3,5,8-tetraazanaphthalene (pteridine) ring, 4-aminobenzoic acid, and (*S*)-2-aminopentanedioic (glutamic) acid (Section 26-1). Folic acid is critical to the proper development of the nervous system in very early stages of pregnancy. A deficiency of this substance, which must be obtained from the diet, is associated with crippling and often fatal birth defects such as spina bifida (''open spine'') and anencephaly (a failure of the brain to develop normally). The United States Public Health Service recommends that all women of childbearing age take 0.4 mg of folic acid daily.

Tetrahydrofolic acid functions as a biological carrier of one-carbon units. The reactive part of

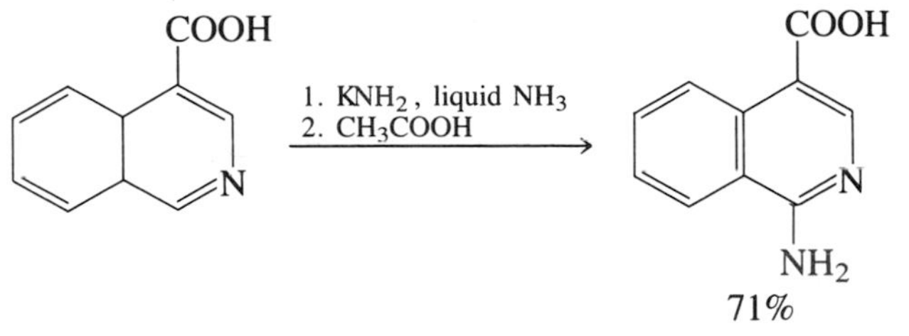

COOH · 1. KNH_2, liquid NH_3 2. CH_3COOH → COOH, N, NH_2

71%

1-Aminoisoquinoline-4-carboxylic acid

the molecule is at nitrogens N5 and N10, as shown below.

A derivative of folic acid, methotrexate, is sufficiently similar structurally that it can enter into some of the reactions of folic acid. It also acts as an inhibitor in some of the processes of cell division that are mediated by folic acid. As a result, it is a useful drug in cancer chemotherapy. Because cancer cells divide much more rapidly than normal cells, they are strongly affected by the presence of this compound.

Riboflavin (vitamin B_2) is a benzannulated analog of 1,3,5,8-tetraazanaphthalene (pteridine) bearing a ribose unit; it is found in animal and plant tissues.

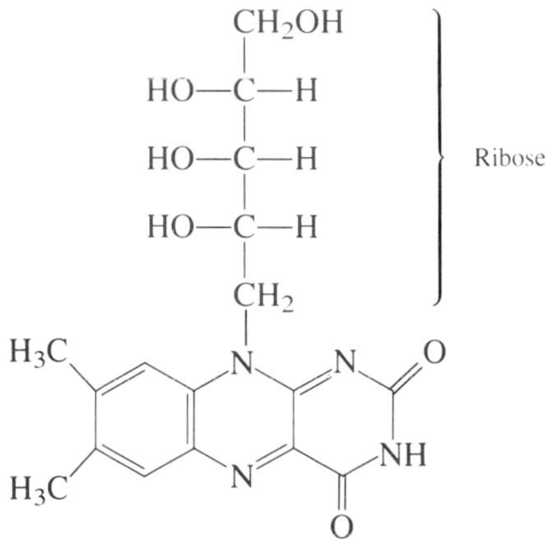

CH_2OH, HO—C—H, HO—C—H, HO—C—H, CH_2 } Ribose

H_3C, H_3C, N, N, O, NH, N, O

Riboflavin (Vitamin B_2)

Tetrahydrofolic Acid as a Carrier of One-Carbon Units

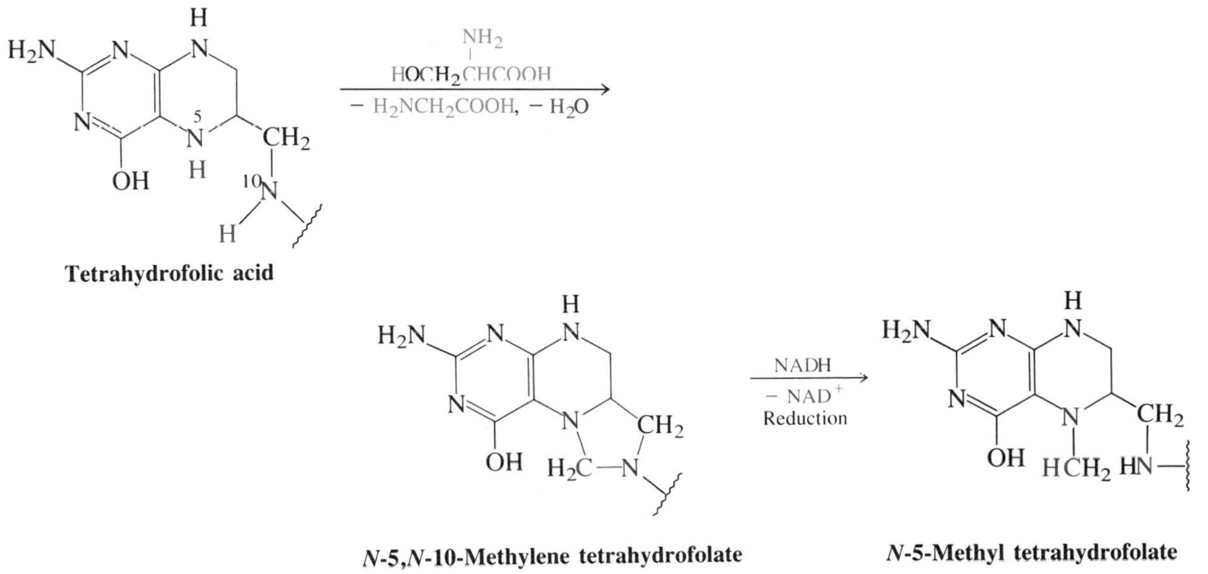

H_2N, N, H, N, N, 5, N, H, CH_2, OH, 10 N, H

Tetrahydrofolic acid

$HOCH_2CH(NH_2)COOH$, $- H_2NCH_2COOH$, $- H_2O$ →

H_2N, N, H, N, N, N, CH_2, OH, H_2C—N

N-5,N-10-Methylene tetrahydrofolate

NADH, $- NAD^+$, Reduction →

H_2N, N, H, N, N, N, CH_2, OH, HCH_2, HN—

N-5-Methyl tetrahydrofolate

EXERCISE 25-17

Quinoline and isoquinoline react with organometallic reagents exactly as pyridine does (Section 25-6). Give the products of their reaction with 2-propenylmagnesium bromide (allylmagnesium bromide).

The structures below are representative of higher aza analogs of naphthalene.

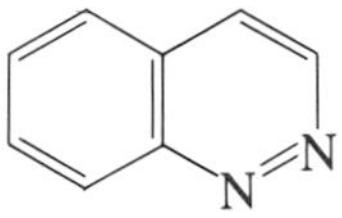

1,2-Diazanaphthalene (Cinnoline)

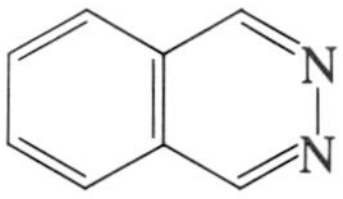

2,3-Diazanaphthalene (Phthalazine)

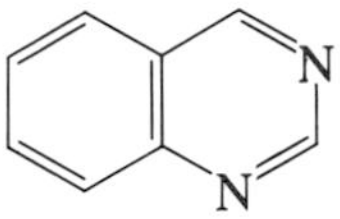

1,3-Diazanaphthalene (Quinazoline)

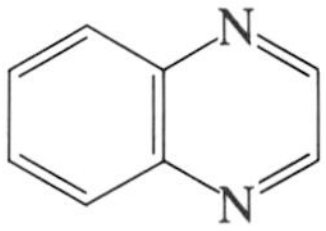

1,4-Diazanaphthalene (Quinoxaline)

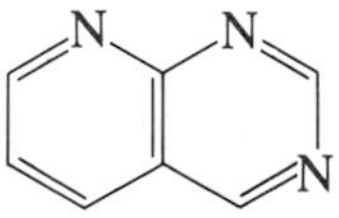

1,3,8-Triazanaphthalene (Pyrido[2,3-*d*]pyrimidine)

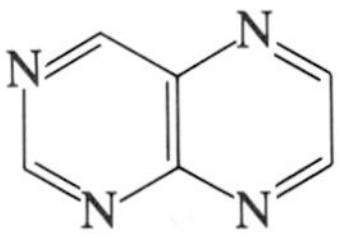

1,3,5,8-Tetraazanaphthalene (Pteridine)

In summary, the azanaphthalenes quinoline and isoquinoline may be regarded as benzpyridines. Electrophiles attack the benzene ring of azanaphthalenes; nucleophiles, the pyridine ring.

25-8 Alkaloids: Physiologically Potent Nitrogen Heterocycles in Nature

The **alkaloids** are natural nitrogen-containing compounds found particularly in plants. The name is derived from their characteristic basic properties (alkali-like), which are induced by the lone electron pair of nitrogen. A variety of alkaloids have been described in the earlier sections.

Many alkaloids have potent pharmacological properties. One of the most abused alkaloids is heroin, the acetyl derivative of morphine (Section 9-11). Morphine and related alkaloids are responsible for the physiological effect of opium poppies. The danger of these drugs is their addictiveness.

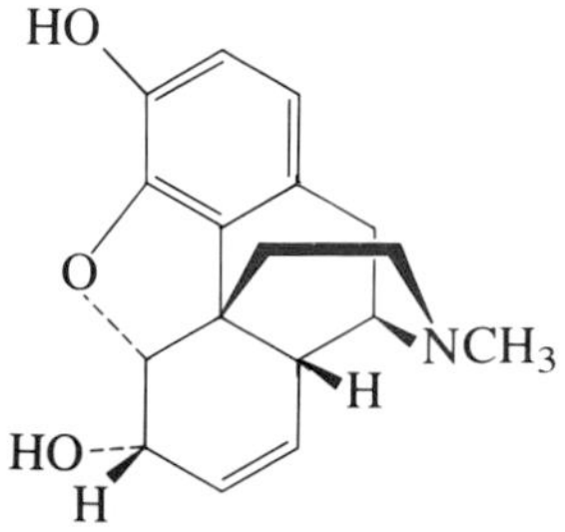

Morphine

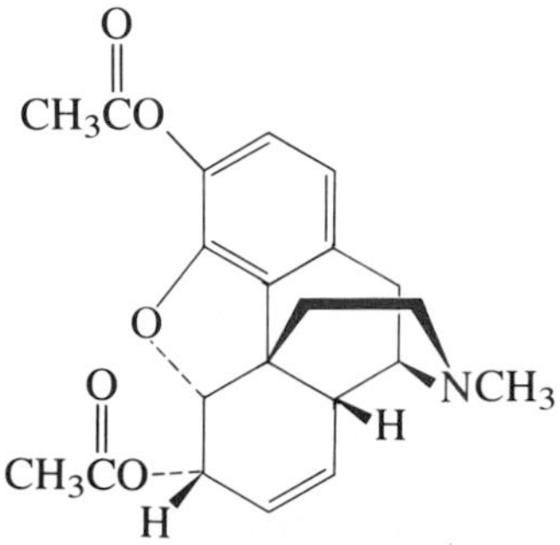

Heroin

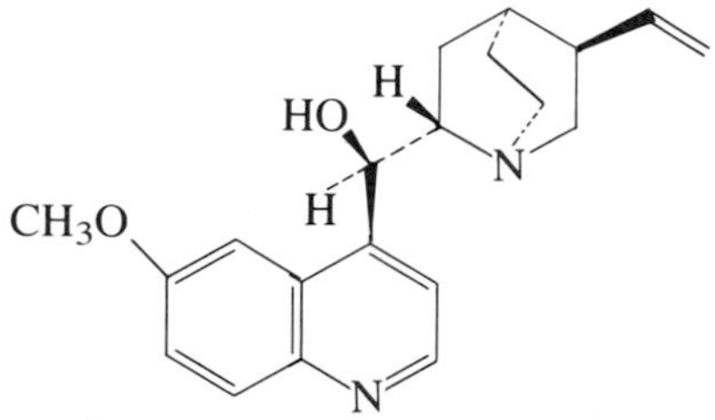

Quinine

Quinine, isolated from cinchona bark (as much as 8% concentration) is the oldest known effective antimalarial agent. A malaria attack consists of a chill accompanied or followed by a fever, which terminates in a sweating stage. Such attacks may recur regularly. The name malaria is derived from the Italian *malo*, bad, and *aria*, air, referring to the old theory that the disease is caused by noxious effluent gases from marshland. However, malaria is actually caused by a protozoan parasite (*Plasmodium* species) transmitted by the bite of an infected female mosquito of the genus *Anopheles*. Millions of people are affected by this disease, which is responsible for a quarter of the deaths of African children between one and four years of age.

Strychnine and brucine are powerful poisons (the lethal dose in animals is approximately 5–8 mg kg^{-1}), the lethal ingredients of many a detective novel.

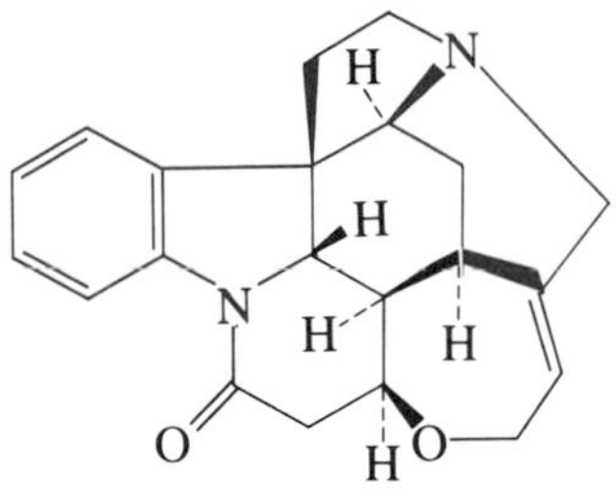

Strychnine 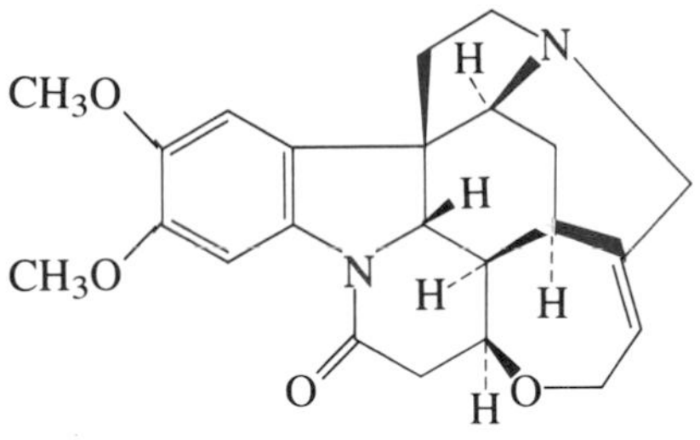 **Brucine**

The isoquinoline and 1,2,3,4-tetrahydroisoquinoline nuclei are abundant among the alkaloids, and their derivatives are physiologically active, for example, as hallucinogens, central nervous system agents (depressants and stimulants), and hypotensives.

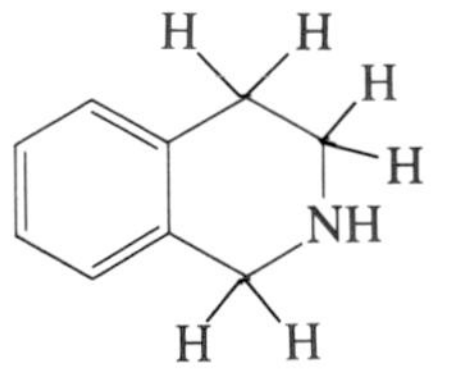

1,2,3,4-Tetrahydroisoquinoline

In summary, many natural nitrogen-containing compounds are alkaloids, which are physiologically active in various ways.

NEW REACTIONS

1. REACTIONS OF HETEROCYCLOPROPANES (Section 25-2)

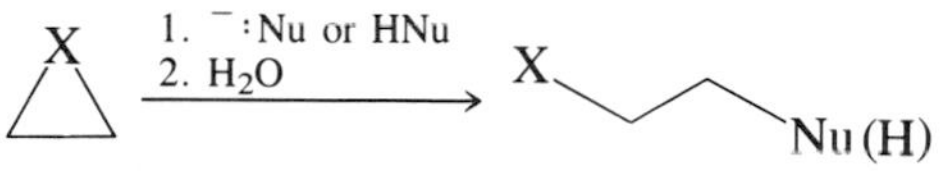

1. $^{-}$:Nu or HNu
2. H_2O

2. RING OPENING OF HETEROCYCLOBUTANES (Section 25-2)

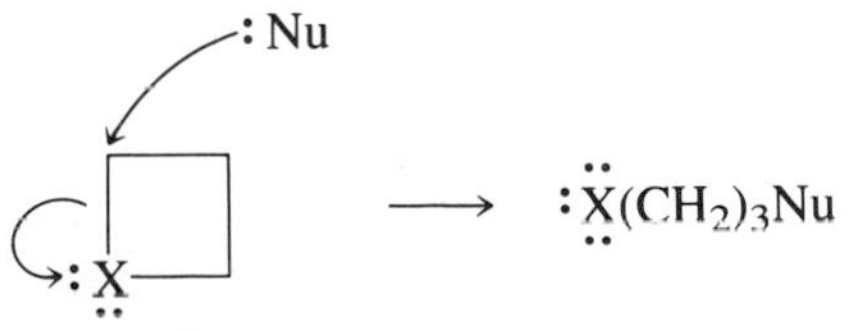

$:\ddot{X}(CH_2)_3Nu$

Less reactive than heterocyclopropanes

3. PAAL-KNORR SYNTHESIS OF 1-HETERO-2,4-CYCLOPENTADIENES (Section 25-3)

R–CO–CH$_2$CH$_2$–CO–R $\xrightarrow{P_2O_5,\ \Delta}$ 2,5-R$_2$-furan

R–CO–CH$_2$CH$_2$–CO–R $\xrightarrow{P_2S_5,\ \Delta}$ 2,5-R$_2$-thiophene

R–CO–CH$_2$CH$_2$–CO–R $\xrightarrow{(NH_4)_2CO_3,\ \Delta}$ 2,5-R$_2$-pyrrole (N–H)

4. REACTIONS OF 1-HETERO-2,4-CYCLOPENTADIENES (Section 25-4)

Electrophilic substitution

$C_4H_4X \xrightarrow{E^+}$ 2-E-substituted ring (X) through cationic intermediate (E, H at C-2; X^+)

Main product

benzene ≪ thiophene (S) < furan (O) < pyrrole (NH)

Relative reactivity

Ring opening

H_3C–(2,5-dimethylfuran)–CH_3 $\xrightarrow{H^+,\ H_2O}$ H_3C–CO–CH$_2$CH$_2$–CO–CH_3

Cycloaddition

furan + maleimide (O, NH, O) ⟶ oxabicyclic adduct (O bridge; H, H; =O; N–H; O)

5. HANTZSCH SYNTHESIS OF PYRIDINES (Section 25-5)

ROC(=O)–CH$_2$–C(=O)–R″ + R′CHO + H$_2$C(–C(=O)OR)–C(=O)–R″ + NH$_3$ ⟶ dihydropyridine (R′; RO$_2$C, CO$_2$R; R″, N–H, R″) $\xrightarrow{HNO_3}$

pyridine (R′; RO$_2$C, CO$_2$R; R″, N, R″) $\xrightarrow{1.\ KOH;\ 2.\ CaO,\ \Delta}$ pyridine (R′; R″, N, R″)

6. REACTIONS OF PYRIDINE (Sections 25-5 and 25-6)

Protonation (Section 25-5)

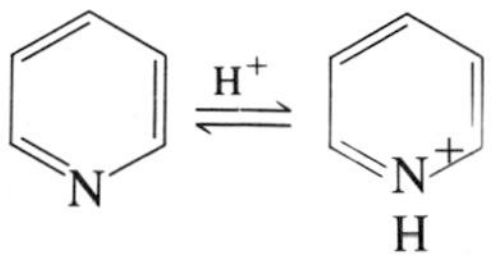
H+
N
N+
H

Pyridinium ion

$pK_a = 5.29$

Electrophilic substitution (Section 25-6)

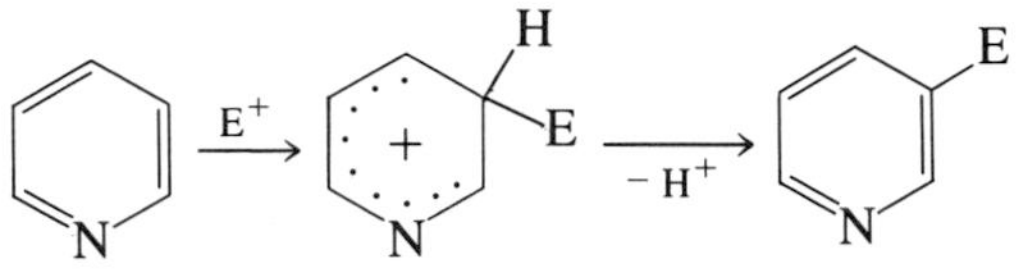
E+
H
E
−H+
N

Ring is deactivated relative to benzene

Nucleophilic substitution (Section 25-6)

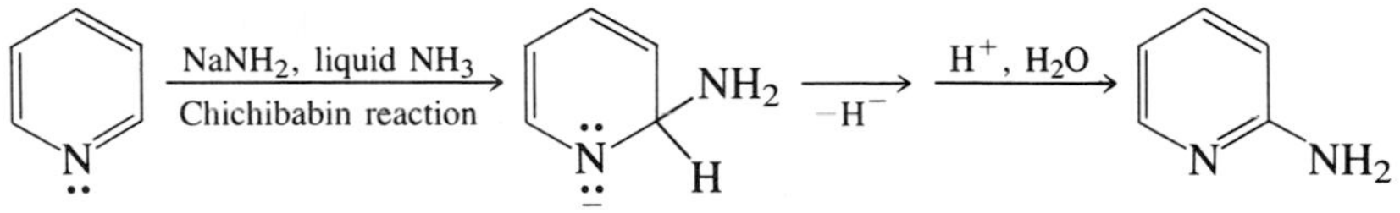
NaNH2, liquid NH3
Chichibabin reaction
NH2
H
−H−
H+, H2O
N
NH2

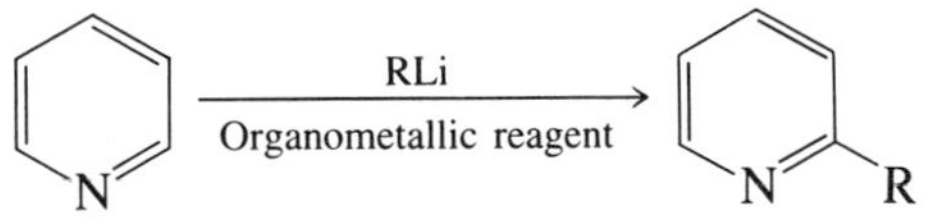
RLi
Organometallic reagent
N
R

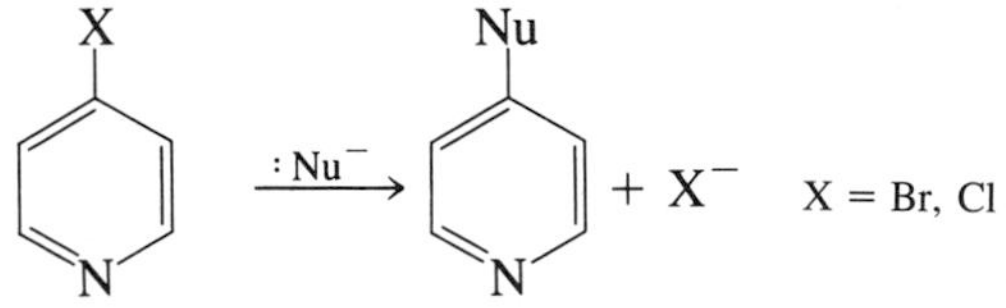
X
:Nu−
Nu
+ X−
X = Br, Cl
N

Halopyridine

7. REACTIONS OF QUINOLINE AND ISOQUINOLINE (Section 25-7)

Electrophilic substitution

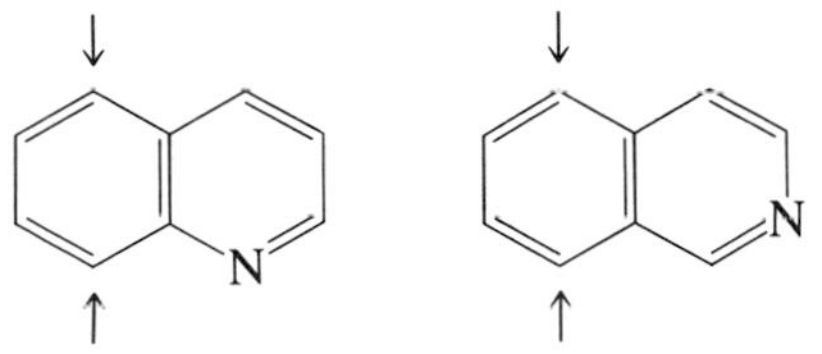
N
N

Nucleophilic substitution

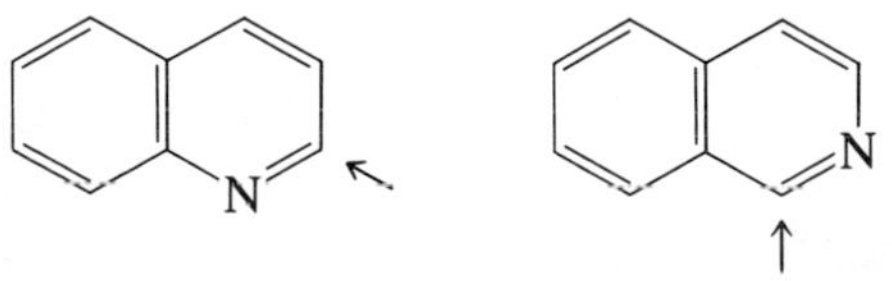
N
N

IMPORTANT CONCEPTS

1. The heterocycloalkanes can be named by using cycloalkane nomenclature. The prefix aza- for nitrogen, oxa- for oxygen, thia- for sulfur, and so on indicates the heteroatom. Other systematic and common names abound in the literature, particularly for the aromatic heterocycles.

2. The strained three- and four-membered heterocycloalkanes easily undergo ring opening with nucleophiles.

3. The 1-hetero-2,4-cyclopentadienes are aromatic and have an arrangement of six π electrons similar to that in the cyclopentadienyl anion. The heteroatom is sp^2 hybridized, the p orbital contributing two electrons to the π system. As a consequence, the diene unit is electron rich and reactive in electrophilic aromatic substitutions.

4. Replacement of one (or more) of the CH units in benzene by an sp^2-hybridized nitrogen gives rise to pyridine (and other azabenzenes). The p orbital on the heteroatom contributes one electron to the π system; the lone electron pair is located in an sp^2 hybrid atomic orbital in the molecular plane. Azabenzenes are electron poor, because the electronegative nitrogen withdraws electron density from the ring by induction and by resonance. Electrophilic aromatic substitution of azabenzenes is sluggish. Conversely, nucleophilic aromatic substitution occurs readily; this is shown by the Chichibabin reaction, substitutions by organometallic reagents next to the nitrogen, and the displacement of halide ion from halopyridines by nucleophiles.

5. The azanaphthalenes (benzpyridines) quinoline and isoquinoline contain an electron-poor pyridine ring, susceptible to nucleophilic attack, and an electron-rich benzene ring that enters into electrophilic aromatic substitution reactions, usually at the positions closest to the heterocyclic unit.

PROBLEMS

1. Name or draw the following compounds. **(a)** *cis*-2,3-Diphenyloxacyclopropane; **(b)** 3-azacyclobutanone; **(c)** 1,3-oxathiacyclopentane; **(d)** 2-butanoyl-1,3-dithiacyclohexane;

(e) 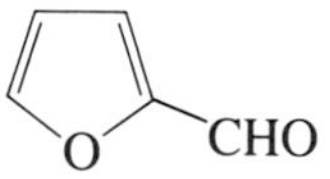 [furan ring with CHO at C2; labels: O, CHO] **(f)** 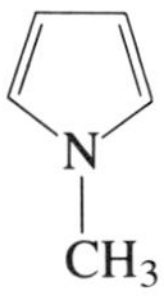 [pyrrole ring with N–CH₃; labels: N, CH_3] **(g)** 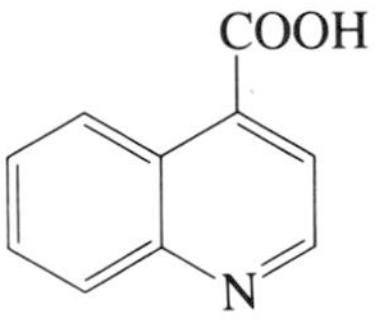 [quinoline with COOH at C4; labels: COOH, N] **(h)** 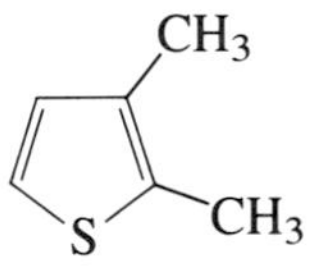 [thiophene with CH₃ at C2 and C3; labels: CH_3, S, CH_3]

2. Give the expected product of each of the following reaction sequences.

(a) 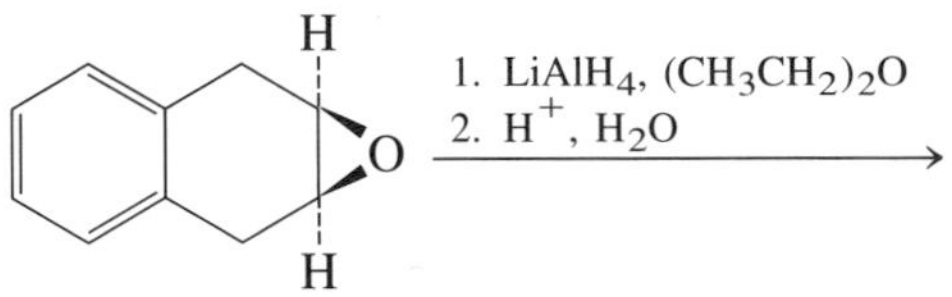 [tetrahydronaphthalene fused epoxide with H, H; label O] $\xrightarrow{\text{1. LiAlH}_4\text{, (CH}_3\text{CH}_2)_2\text{O} \quad \text{2. H}^+\text{, H}_2\text{O}}$

(b) 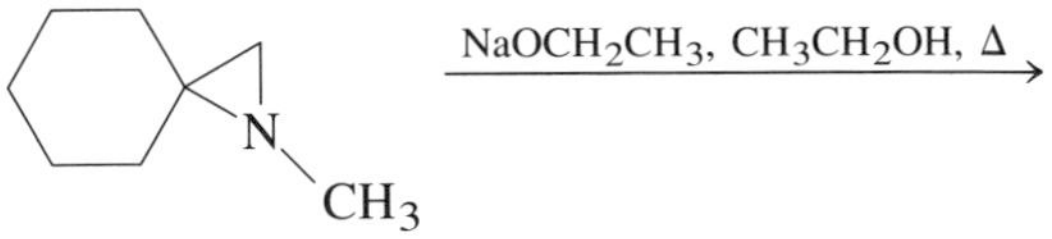 [spiro cyclohexane–aziridine, N–CH_3] $\xrightarrow{\text{NaOCH}_2\text{CH}_3\text{, CH}_3\text{CH}_2\text{OH, }\Delta}$

(c) 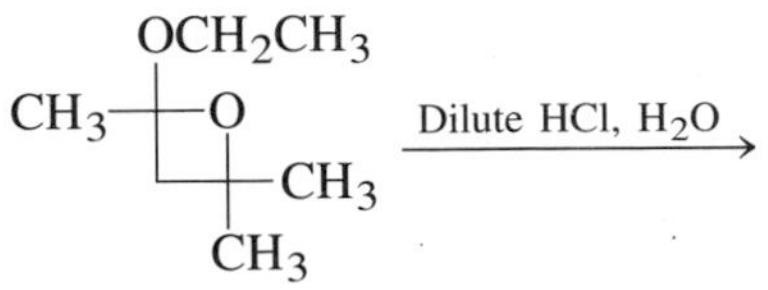 [oxacyclobutane with CH_3 and OCH_2CH_3 at C2, two CH_3 at C3] $\xrightarrow{\text{Dilute HCl, H}_2\text{O}}$

3. The penicillins are a class of antibiotics containing two heterocyclic rings that interfere with the construction of cell walls by bacteria. The interference results from reaction of the penicillin with an amino group of a protein that closes gaps that develop during construction of the cell wall. The insides of the cell leak out, and the organism dies. **(a)** Suggest a reasonable product for the reactions of penicillin G with the amino group of a protein (Protein-NH_2). (Hint: First identify the most reactive electrophilic site in penicillin.)

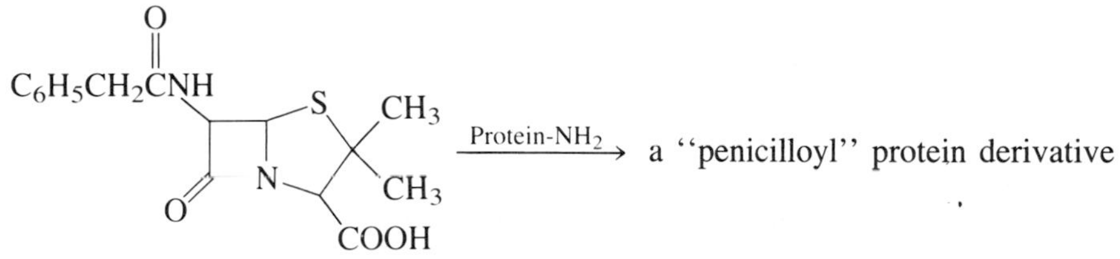

$C_6H_5CH_2\overset{O}{\overset{\|}{C}}NH$ ... CH_3, CH_3, COOH $\xrightarrow{\text{Protein-NH}_2}$ a "penicilloyl" protein derivative

Penicillin G

(b) Penicillin-resistant bacteria secrete an enzyme (penicillinase) that catalyzes hydrolysis of the antibiotic faster than the antibiotic can attack the cell-wall proteins. Propose a structure for the product of this hydrolysis and suggest a reason why hydrolysis destroys the antibiotic properties of penicillin.

Penicillin G $\xrightarrow{H_2O,\ \text{penicillinase}}$ penicilloic acid
(Hydrolysis product; no antibiotic activity)

4. Propose reasonable mechanisms for the transformations shown below.

(a) $\xrightarrow[\text{2. } H^+,\ H_2O]{\text{1. } SnCl_4 \text{ (a Lewis acid), } CH_2Cl_2}$ CH_2OH

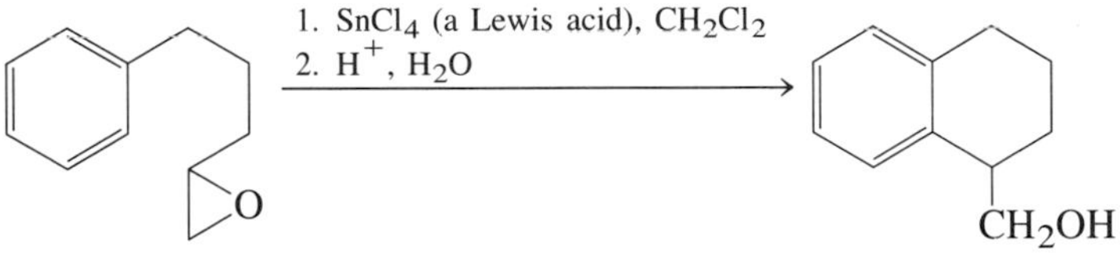

(b) C_6H_5 $\xrightarrow[\text{2. } H^+,\ H_2O]{\text{1. } CH_3CH_2CH_2CH_2Li,\ BF_3 \cdot O(CH_2CH_3)_2,\ \text{THF}}$ C_6H_5, OH

(c) CH_3, H, O, H, CH_3 $\xrightarrow[\text{(Hint: See Section 15-14)}]{MgBr_2,\ CH_3\overset{O}{\overset{\|}{C}}O\overset{O}{\overset{\|}{C}}CH_3}$ $CH_3\underset{O}{\underset{\|}{C}}O$, H, CH_3, Br, H_3C, H

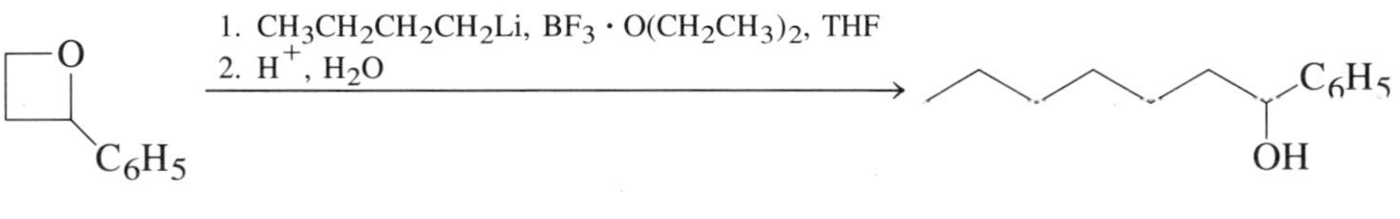

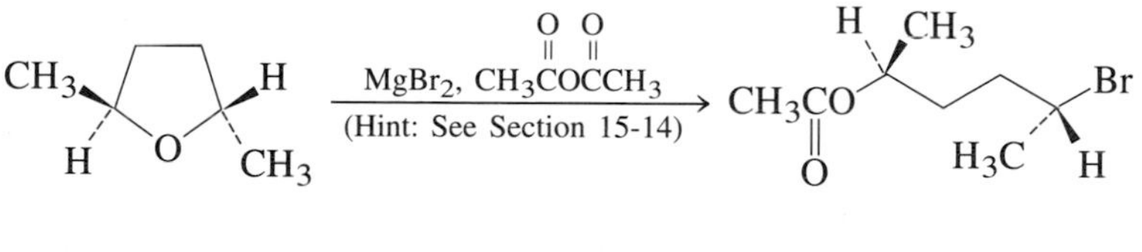

5. Rank the following compounds in increasing order of basicity: water, hydroxide, pyridine, pyrrole, ammonia.

6. The following heterocyclopentadienes contain more than one heteroatom. For each one, identify the orbitals occupied by all lone electron pairs on the heteroatoms and determine whether the molecule qualifies as aromatic. Are any of these heterocycles a stronger base than pyrrole?

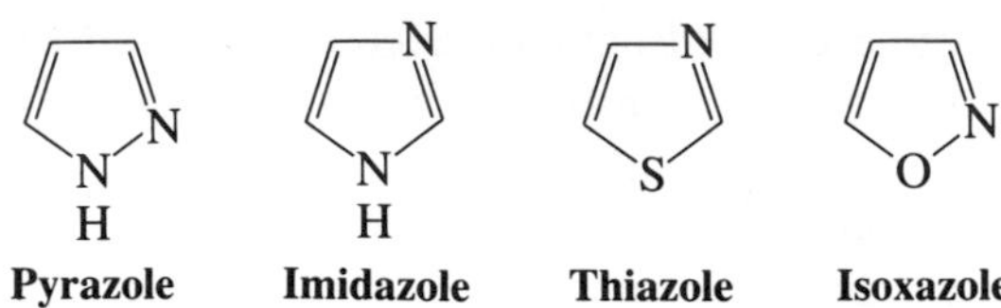

7. Give the product of each of the following reactions.

(a) [2-(2-oxocyclohexyl)cyclohexanone] $\xrightarrow{CH_3NH_2}$ (b) [cyclododecane-1,4-dione] $\xrightarrow{P_2O_5,\ \Delta}$

8. 1-Hetero-2,4-cyclopentadienes can be prepared by condensation of an α-dicarbonyl compound and certain heteroatom-containing diesters. Propose a mechanism for the following pyrrole synthesis.

$$C_6H_5COCOC_6H_5 + CH_3OCOCH_2N(COCH_3)CH_2COOCH_3 \xrightarrow{NaOCH_3,\ CH_3OH,\ \Delta}$$ 3,4-(C_6H_5)$_2$-2,5-(CH_3OOC)$_2$-N-($COCH_3$)pyrrole

How would you use a similar approach to synthesize 2,5-thiophenedicarboxylic acid?

9. Give the expected major product(s) of each of the following reactions. Explain how you chose the position of substitution in each case.

(a) methyl 2-furoate ($COOCH_3$) $\xrightarrow{Cl_2}$ (b) 2-methylthiophene (CH_3) $\xrightarrow{HNO_3,\ H_2SO_4}$ (c) 2-acetylthiophene ($C(=O)CH_3$) $\xrightarrow{CH_3CHClCH_3,\ AlCl_3}$

(d) 3-bromothiophene (Br) $\xrightarrow{Br_2}$ (e) imidazole (N, NH) $\xrightarrow{C_6H_5N_2^+Cl^-,\ NaOH,\ H_2O}$

10. Give the products expected of each of the following reactions.

(a) 4-(H_3C)N(CH_3)-pyridine $\xrightarrow{\text{Fuming } H_2SO_4,\ 270°C}$ (b) furan + S-containing bicyclic anhydride $\xrightarrow{\Delta,\ \text{pressure}}$ (c) 2-bromopyridine (N, Br) $\xrightarrow{KSH,\ CH_3OH,\ \Delta}$

(d) thiophene (S) $\xrightarrow{1.\ C_6H_5COCl,\ SnCl_4\ \ 2.\ \text{Raney Ni},\ \Delta}$ (e) pyridine (N) $\xrightarrow{(CH_3)_3CLi,\ THF,\ \Delta}$

11. Propose a synthesis of each of the following substituted heterocycles, using synthetic sequences presented in this chapter.

(a) CH_3, CH_3, CH_3, O, CH_3 (b) H_5C_6, N, C_6H_5 (c) H_3C, CH_3, N, H (d) CH_3, CH_3, S

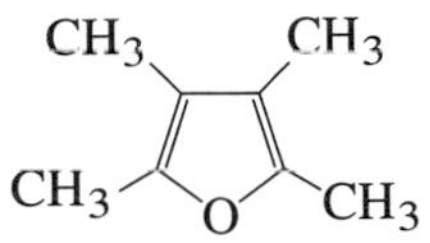

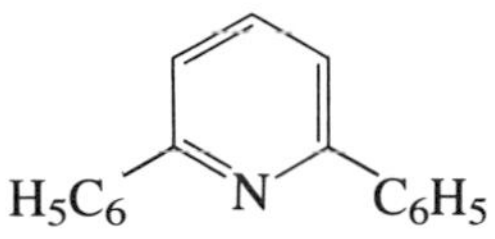

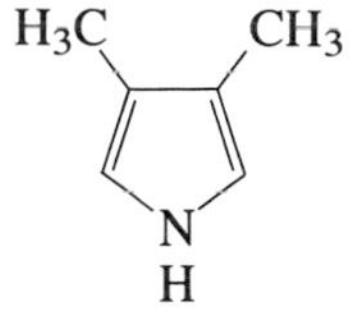

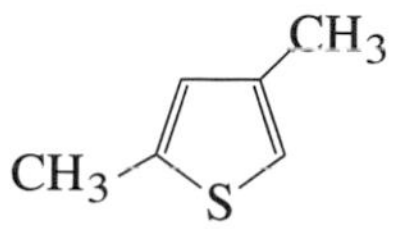

12. Chelidonic acid, a 4-oxacyclohexanone (common name, γ-pyrone), is found in a number of plants and is synthesized from propanone (acetone) and diethyl ethanedioate. Formulate a mechanism for this transformation.

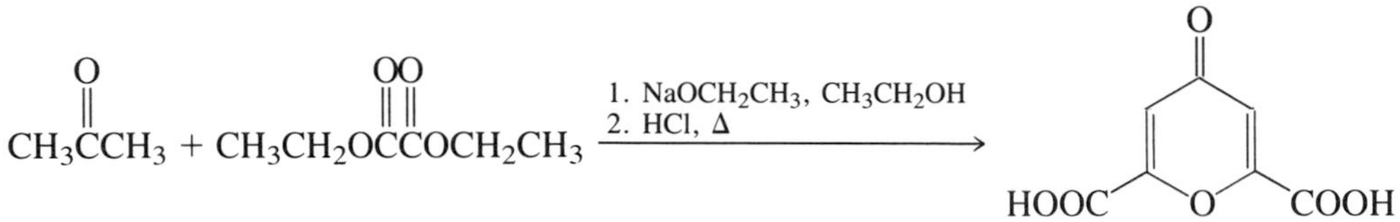

$$CH_3\overset{O}{\overset{\|}{C}}CH_3 + CH_3CH_2O\overset{O}{\overset{\|}{C}}\overset{O}{\overset{\|}{C}}OCH_2CH_3 \xrightarrow[2.\ HCl,\ \Delta]{1.\ NaOCH_2CH_3,\ CH_3CH_2OH}$$ HOOC, O, COOH

Chelidonic acid

13. Reserpine is a naturally occurring indole alkaloid with powerful tranquilizing and antihypertensive activity. Many such compounds possess a characteristic structural feature: one nitrogen atom at a ring fusion separated by two carbons from another nitrogen atom.

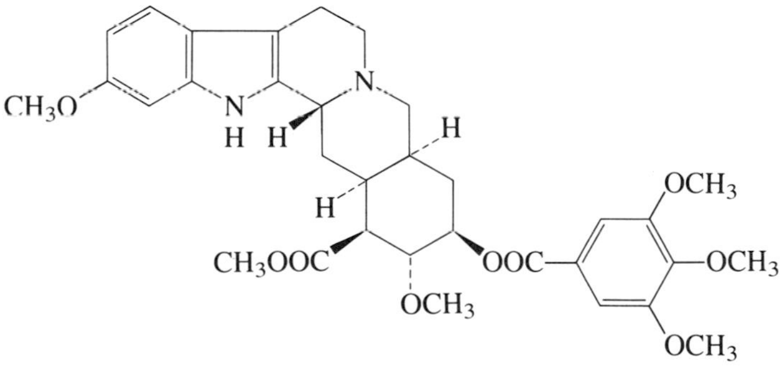

CH_3O, N, H, H, N, H, H, CH_3OOC, OCH_3, OOC, OCH_3, OCH_3, OCH_3

Reserpine

A series of compounds with modified versions of this structural feature have been synthesized and shown also to have antihypertensive activity, as well as antifibrillatory properties. One such synthesis is shown here. Name or draw the missing reagents and products a through c.

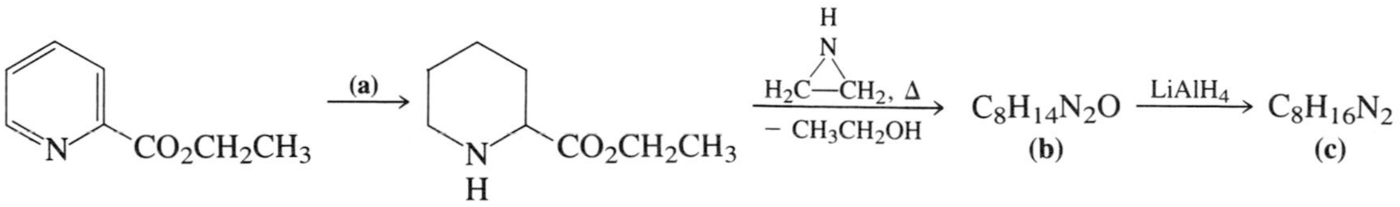

N, $CO_2CH_2CH_3$ $\xrightarrow{(a)}$ N, H, $CO_2CH_2CH_3$ $\xrightarrow[-\ CH_3CH_2OH]{H_2C\text{—}CH_2\ (NH),\ \Delta}$ $C_8H_{14}N_2O$ **(b)** $\xrightarrow{LiAlH_4}$ $C_8H_{16}N_2$ **(c)**

14. Starting with benzenamine (aniline) and pyridine, propose a synthesis for the antimicrobial sulfa drug sulfapyridine.

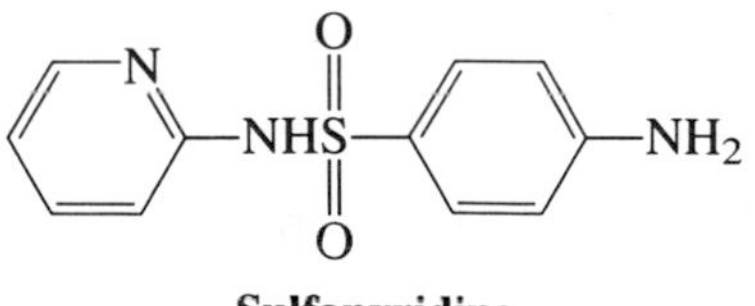

N, —NHS(=O)(=O)—, —NH_2

Sulfapyridine

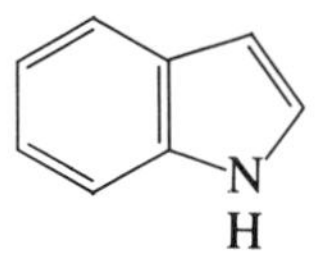

Indole

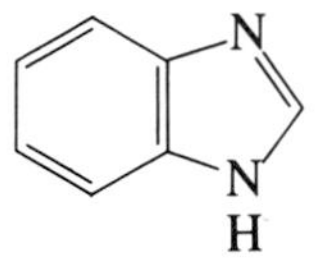

Benzimidazole

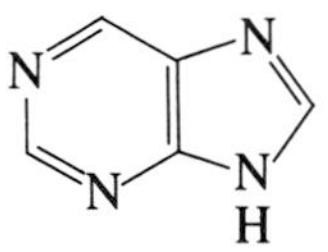

Purine

15. Derivatives of benzimidazole possess biological activity somewhat like that of indoles and purines (of which adenine, Section 25-1, is an example). Benzimidazoles are commonly prepared from benzene-1,2-diamine. Devise a short synthesis of 2-methylbenzimidazole from benzene-1,2-diamine.

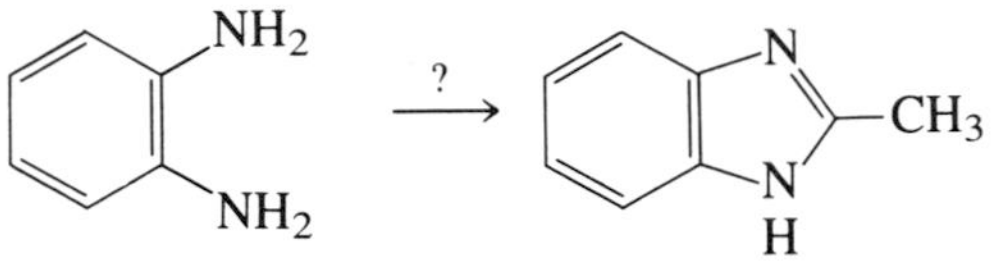

Benzene-1,2-diamine $\xrightarrow{?}$ **2-Methylbenzimidazole**

16. Benzene-1,2-diamine (Problem 15) is also the precursor of choice for the synthesis of quinoxaline derivatives. Propose a simple method for the conversion of benzene-1,2-diamine into 2,3-dimethylquinoxaline, a precursor of mediquox, an agent used to treat respiratory infections in chickens (no, we are not making this up).

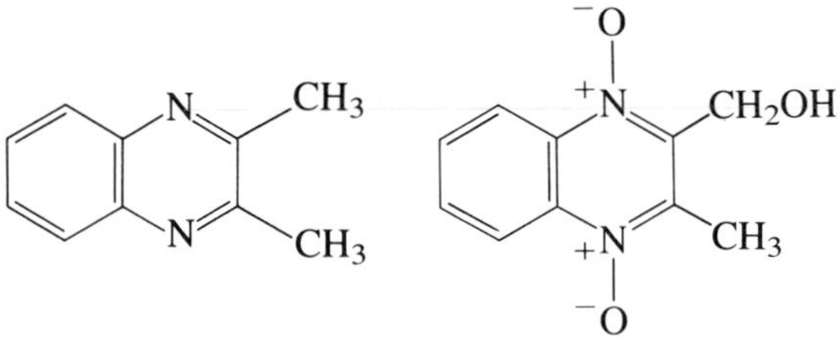

2,3-Dimethylquinoxaline **Mediquox**

17. The Darzens condensation is one of the older methods (1904) for the synthesis of three-membered heterocycles. It is most commonly the reaction of a 2-halo ester with a carbonyl derivative in the presence of base. The following examples of the Darzens condensation show how it is applied to the synthesis of oxacyclopropane and azacyclopropane. Suggest a reasonable mechanism for each of these reactions.

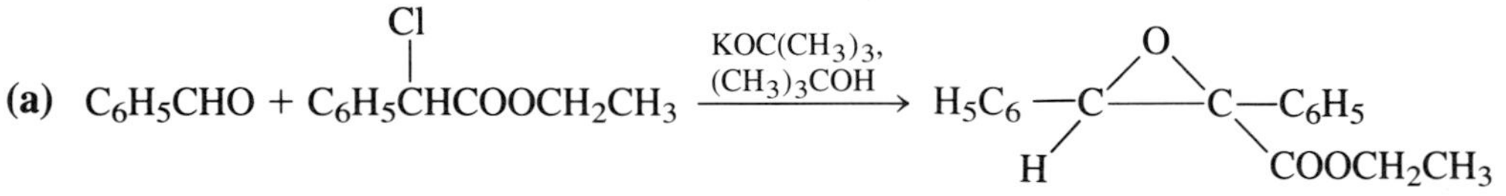

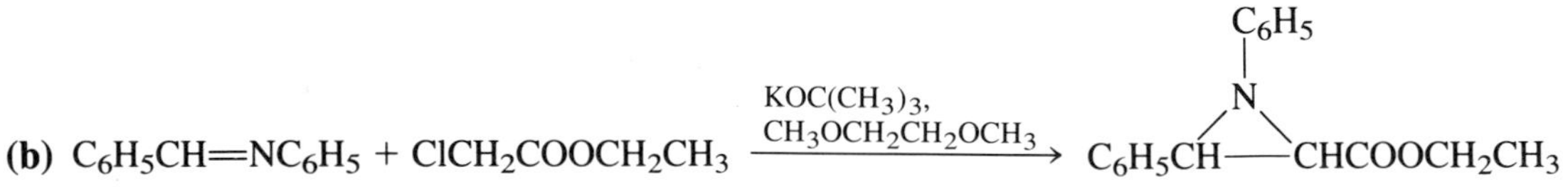

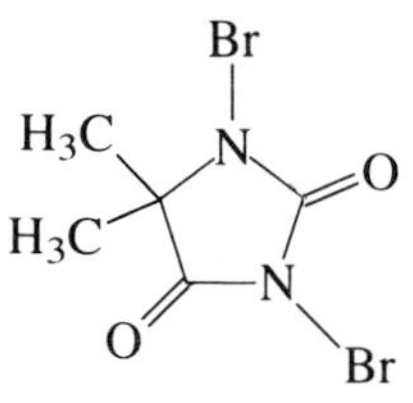

(a) $C_6H_5CHO + C_6H_5CH(Cl)COOCH_2CH_3 \xrightarrow[(CH_3)_3COH]{KOC(CH_3)_3,}$ H_5C_6—C(H)—O—C(C_6H_5)($COOCH_2CH_3$) (oxacyclopropane)

(b) $C_6H_5CH{=}NC_6H_5 + ClCH_2COOCH_2CH_3 \xrightarrow[CH_3OCH_2CH_2OCH_3]{KOC(CH_3)_3,}$ C_6H_5CH—N(C_6H_5)—$CHCOOCH_2CH_3$ (azacyclopropane)

18. (a) The compound shown in the margin, with the common name 1,3-dibromo-5,5-dimethylhydantoin, is useful as a source of electrophilic bromine (Br^+) for addition reactions. Give a more systematic name for this heterocyclic compound. **(b)** An even more remarkable heterocyclic compound (ii) is prepared by the following reaction sequence. Using the given information, deduce structures for i and ii, and name the latter.

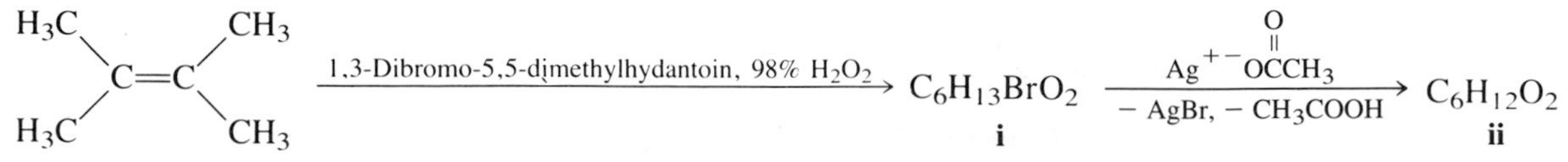

$(CH_3)_2C{=}C(CH_3)_2 \xrightarrow{\text{1,3-Dibromo-5,5-dimethylhydantoin, 98\% } H_2O_2} \underset{\textbf{i}}{C_6H_{13}BrO_2} \xrightarrow[-\,AgBr,\ -\,CH_3COOH]{Ag^{+}\ {}^{-}OC(O)CH_3} \underset{\textbf{ii}}{C_6H_{12}O_2}$

Heterocycle ii is a yellow, crystalline, sweet-smelling compound that decomposes on gentle heating to two molecules of propanone (acetone), one of which is formed directly in its $n \rightarrow \pi^*$ excited state (Sections 14-11 and 17-3). This electronically excited product is chemiluminescent.

$$\text{ii} \longrightarrow CH_3\overset{\overset{\large O}{\|}}{C}CH_3 + \left[CH_3\overset{\overset{\large O}{\|}}{C}CH_3\right]^{n \rightarrow \pi^*} \longrightarrow h\nu + 2\, CH_3\overset{\overset{\large O}{\|}}{C}CH_3$$

Heterocycles similar to ii are responsible for the chemiluminescence produced by a number of species (e.g., fireflies and several deep-sea fish); they also serve as the energy sources in commercial chemiluminescent products.

19. Azacyclohexanes (piperidines) can be synthesized by reaction of ammonia with *cross-conjugated dienones:* ketones conjugated on both sides with double bonds. Propose a mechanism for the following synthesis of 2,2,6,6-tetramethylaza-4-cyclohexanone.

$$(CH_3)_2C{=}CH\overset{\overset{\large O}{\|}}{C}CH{=}C(CH_3)_2 \xrightarrow{NH_3}$$ (2,2,6,6-tetramethyl-4-piperidinone: ring C=O at top; H_3C, CH_3 on each carbon adjacent to N–H)

20. Compound A, C_8H_8O, exhibits ^1H NMR spectrum A (p. 1020). On treatment with concentrated aqueous HCl, it is converted almost instantaneously into a compound that exhibits spectrum B (p. 1020). Approximate integrated intensities of the signals in spectrum B are as follows: $\delta = 7.1$–7.4 (5 H), 4.8 (1 H), 4.2 (2 H), and 3.8 (2 H) ppm. What is compound A, and what is the product of its treatment with aqueous acid?

21. Heterocycle C, C_5H_6O, exhibits ^1H NMR spectrum C and is converted by H_2 and Raney nickel into D, $C_5H_{10}O$, with spectrum D (p. 1021). Identify compounds C and D. (Note: The coupling constants of the compounds in this problem and the next one are rather small; they are therefore not nearly as useful in structure elucidation as those around a benzene ring.)

22. The commercial synthesis of a useful heterocyclic derivative requires treatment of a mixture of aldopentoses (derived from corncobs, straw, etc.) with hot acid under dehydrating conditions. The product, E, has ^1H NMR spectrum E (p. 1022), shows a strong IR band at 1670 cm^{-1}, and is formed in nearly quantitative yield. Identify E and formulate a mechanism for its formation.

$$\text{Aldopentoses} \xrightarrow{H^+,\ \Delta} \underset{\textbf{E}}{C_5H_4O_2}$$

Compound E is a valuable synthetic starting material. The following sequence converts it into furethonium, which is useful in the treatment of glaucoma. What is the structure of furethonium?

$$E \xrightarrow[\text{2. Excess } CH_3I,\ (CH_3CH_2)_2O]{\text{1. } NH_3,\ NaBH_3CN} \text{furethonium}$$

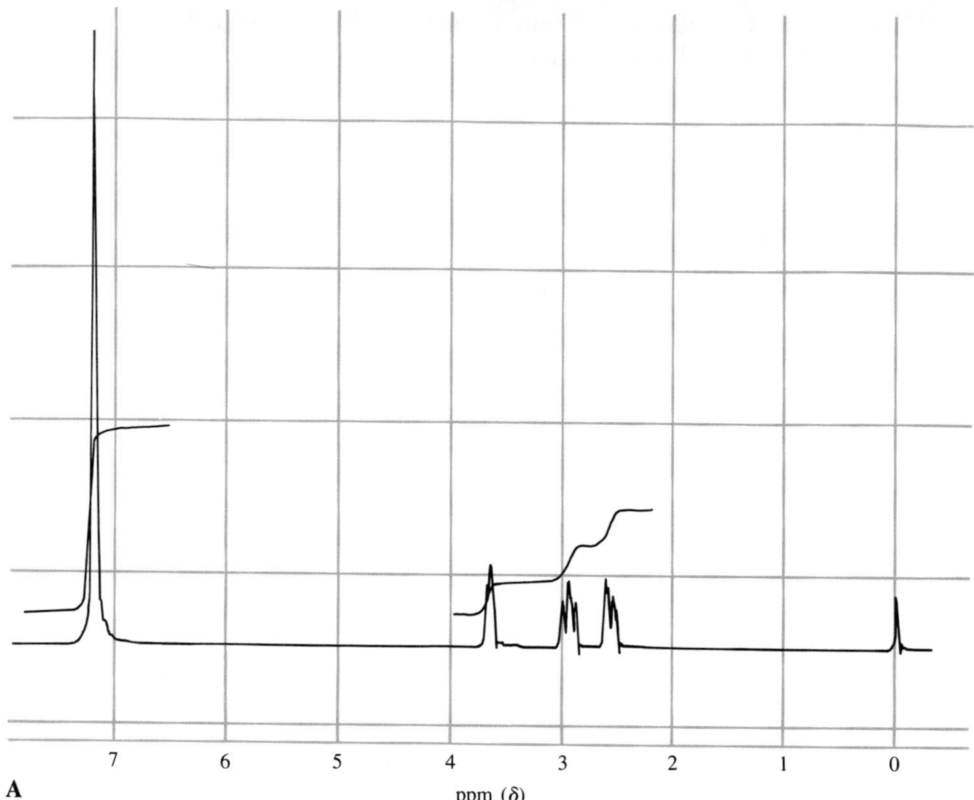

A

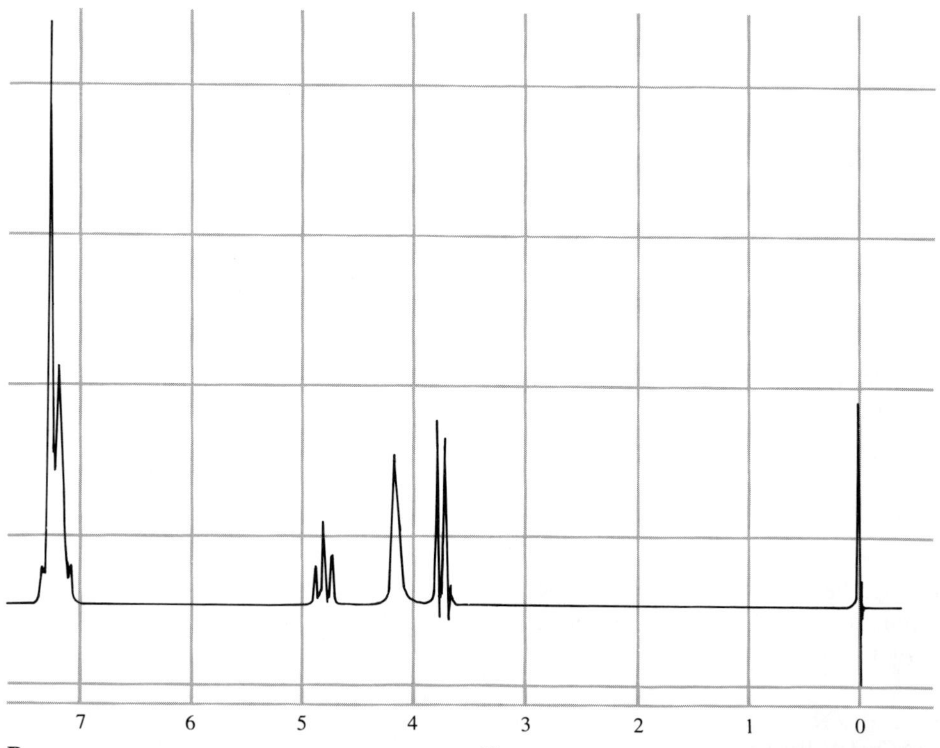

B

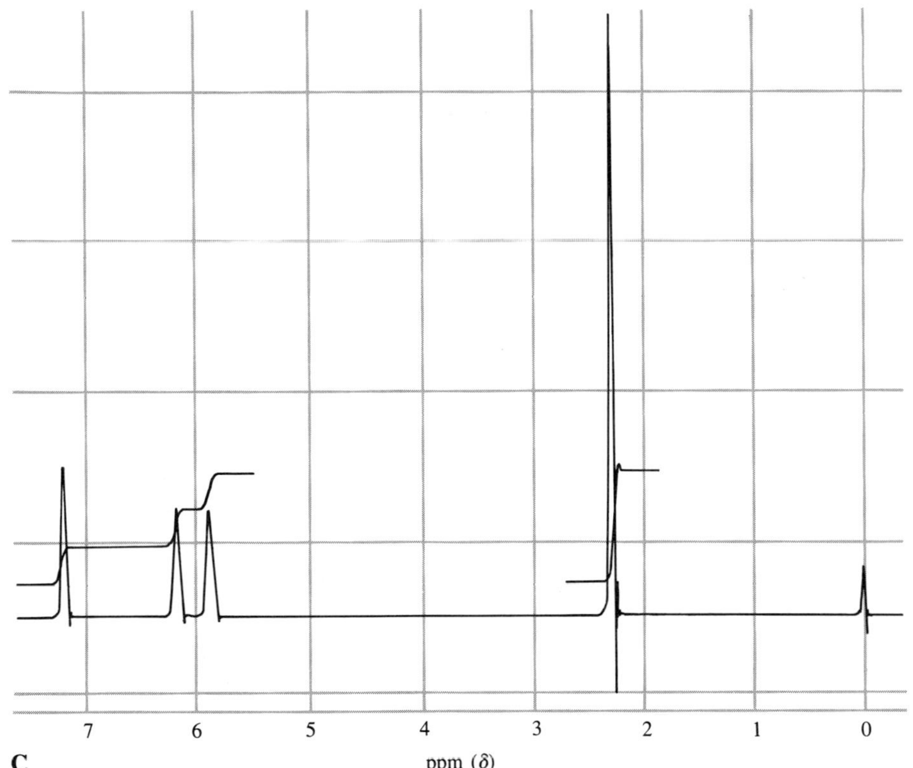
7
6
5
4
3
2
1
0
C
ppm (δ)

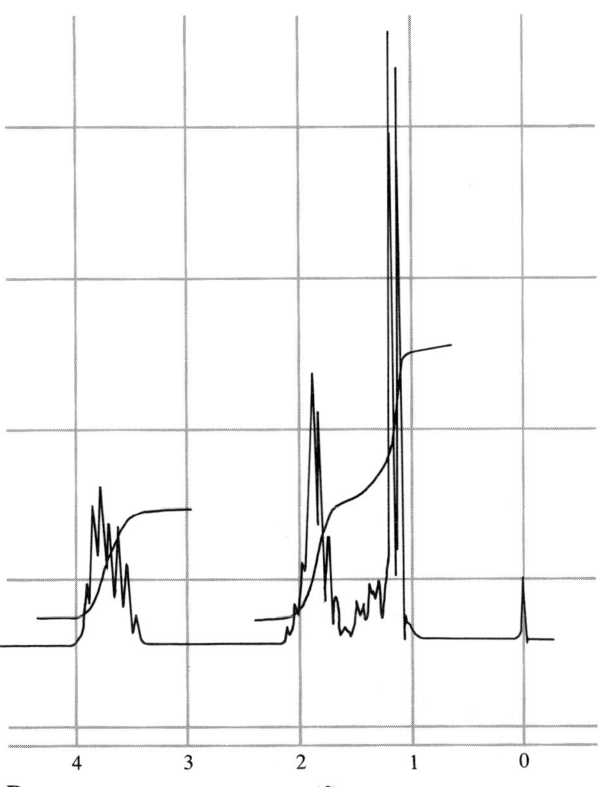
4
3
2
1
0
D
ppm (δ)

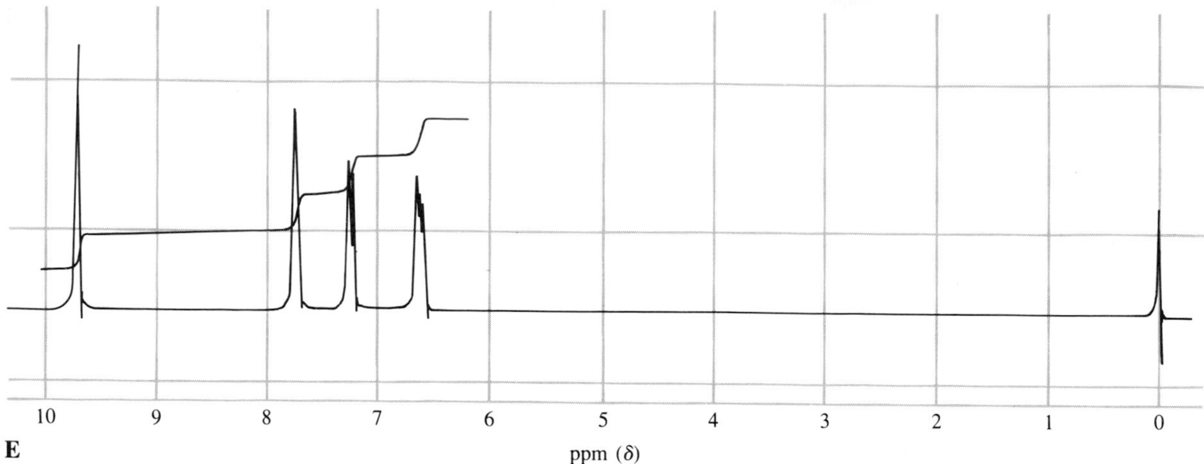

E

23. Treatment of a 3-alkanoylindole with $LiAlH_4$ in $(CH_3CH_2)_2O$ reduces the carbonyl all the way to a CH_2 group. Explain by a plausible mechanism. (Note: Direct S_N2 displacement of alkoxide by hydride is *not* plausible.)

3-(RC(=O))-indole (N–H) $\xrightarrow{\text{Excess } LiAlH_4,\ (CH_3CH_2)_2O,\ \Delta}$ 3-(CH_2R)-indole (N–H)

(Compare C_6H_5–C(=O)R $\xrightarrow{LiAlH_4}$ C_6H_5–CH(OH)–R)

24. The following is a rapid synthesis of one of the heterocycles in this chapter. Draw the structure of the product, which has ^{1}H NMR spectrum F.

Indene $\xrightarrow{\text{1. } O_3,\ CH_2Cl_2;\ \text{2. } (CH_3)_2S;\ \text{3. } NH_3}$

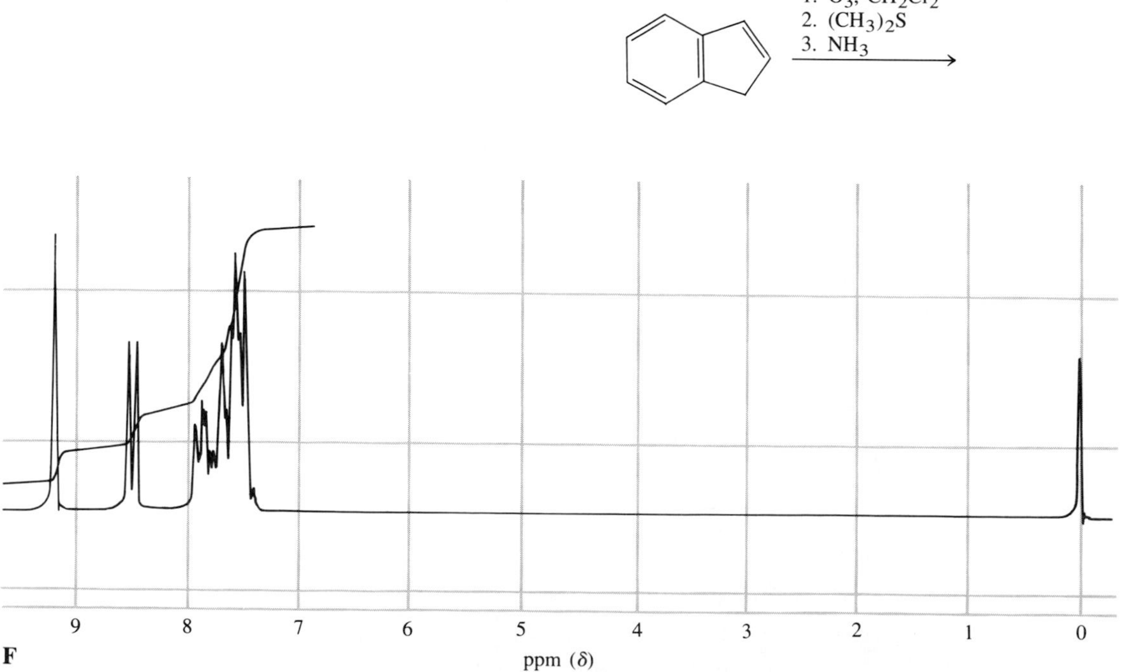

F

26

Amino Acids, Peptides, and Proteins

Nitrogen-Containing Polymers in Nature

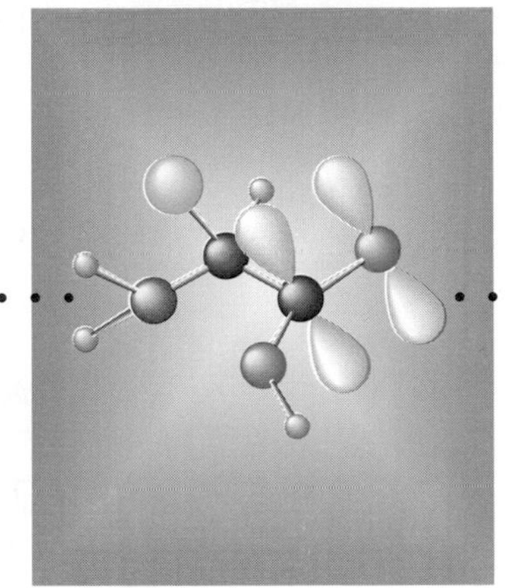

Chapter 24 showed that monosaccharides form polymers by making ether bonds to other monosaccharides. Nature uses the resulting polysaccharides for energy storage and for building cell walls. Now we consider a second type of biological polymer, the polypeptides—in particular, the large natural polypeptides called **proteins.**

Proteins have an astounding diversity of functions in living systems. As **enzymes,** they catalyze transformations ranging in complexity from the simple hydration of carbon dioxide to the replication of entire chromosomes—great coiled strands of DNA, the genetic material in living cells. Enzymes can accelerate certain reactions many millionfold.

We have already encountered the protein rhodopsin, the photoreceptor that generates and transmits nerve impulses in retinal cells (Box 18-4, Section 18-9). Other proteins serve for transport and storage. Thus, hemoglobin carries oxygen; iron is transported in the blood by transferrin and stored in the liver by ferritin. Proteins play a crucial role in coordinated motion, such as muscle contraction. They give mechanical support to skin and bone; they are the antibodies responsible for our immune protection; and they control growth and differentiation, that is, what part of the information stored in DNA is to be used at any given time.

This chapter begins with the structure and preparation of the 20 most common amino acids, the building blocks of proteins. We then show how amino acids are linked by peptide bonds in the three-dimensional structure of hemoglobin and other polypeptides. Some proteins contain thousands of amino acids, but we shall see how to determine the sequence of amino acids in many polypeptides and synthesize these molecules in the laboratory. Finally, we discuss how other polymers, the nucleic acids DNA and RNA, direct the synthesis of proteins in nature.

26-1 Structure and Properties of Amino Acids

Amino acids are carboxylic acids bearing an amine group. The most common of these in nature are the **2-amino acids,** or ***α*-amino acids,** which have the general formula $RCH(NH_2)COOH$; that is, the amino function is located at C2, the α-carbon. The R group can be alkyl or aryl, and it can contain hydroxy, amino, mercapto, sulfide, and carboxy groups. Because of the presence of both amino and carboxy functions, amino acids are both acidic and basic.

The stereocenter of common 2-amino acids has the *S* configuration

More than 500 amino acids occur in nature, but the proteins in all species, from bacteria to humans, consist mainly of only 20. Adult humans can synthesize all but eight, which are called the **essential amino acids** because they must be included in our diet. Although amino acids can be named in a systematic manner, they rarely are; so we shall use their common names. Table 26-1 lists the 20 most common amino acids, along with their structures, their pK_a values, and the three-letter codes that abbreviate their names. We shall see later how to use these codes to describe peptides conveniently.

Amino acids may be depicted by either dashed-wedged line structures or by Fischer projections.

How to Draw (2*S*)-Amino Acids and Their Relation to the L-Sugars

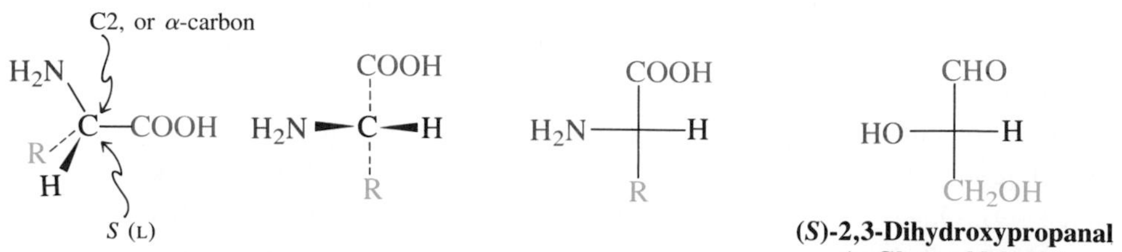

C2, or α-carbon

H_2N C—COOH R H

S (L)

COOH H_2N—C—H R

Dashed-wedged line structures

COOH H_2N—+—H R

CHO HO—+—H CH_2OH

(*S*)-2,3-Dihydroxypropanal (L-Glyceraldehyde)

Fischer projections

TABLE 26-1 Natural (2S)-Amino Acids

$$H_2N\text{—}\overset{COOH}{\underset{R}{C}}\text{—}H$$

R	Name	Three-letter code	pK_a of COOH	pK_a of $^+NH_3$	pK_a of acidic function in R
H	Glycine	Gly	2.4	9.8	—
Alkyl group					
CH_3	Alanine	Ala	2.4	9.9	—
$CH(CH_3)_2$	Valine[a]	Val	2.3	9.7	—
$CH_2CH(CH_3)_2$	Leucine[a]	Leu	2.3	9.7	—
$CH(CH_3)CH_2CH_3$ (S)	Isoleucine[a]	Ile	2.3	9.7	—
$CH_2C_6H_5$	Phenylalanine[a]	Phe	2.6	9.2	—
(cyclic: HN—C(COOH)(H)—CH_2, ring closed via CH_2CH_2)[b]	Proline	Pro	2.0	10.6	—
Hydroxy-containing					
CH_2OH	Serine	Ser	2.2	9.4	—
$CH(OH)CH_3$	Threonine[a]	Thr	2.1	9.1	—
CH_2—C_6H_4—OH	Tyrosine	Tyr	2.2	9.1	10.1
Amino-containing					
CH_2CONH_2	Asparagine	Asn	2.0	8.8	—
$CH_2CH_2CONH_2$	Glutamine	Gln	2.2	9.1	—
$(CH_2)_4NH_2$	Lysine[a]	Lys	2.2	9.2	10.8[c]
$(CH_2)_3NHC(=NH)NH_2$	Arginine	Arg	1.8	9.0	13.2[c]
CH_2-(3-indolyl, N–H)	Tryptophan[a]	Trp	2.4	9.4	—

(continued)

TABLE 26-1 (continued)

R	Name	Three-letter code	pK_a of COOH	pK_a of $^{+}NH_3$	pK_a of acidic function in R
Amino-containing *(continued)*					
H_2C NH N	Histidine	His	1.8	9.2	6.1[c]
Mercapto- or sulfide-containing					
CH_2SH	Cysteine[d]	Cys	1.9	10.3	8.4
$CH_2CH_2SCH_3$	Methionine[a]	Met	2.2	9.3	—
Carboxy-containing					
CH_2COOH	Aspartic acid	Asp	2.0	10.0	3.9
CH_2CH_2COOH	Glutamic acid	Glu	2.1	10.0	4.3

a. Essential amino acids. b. Entire structure. c. pK_a of conjugate acid.
d. The stereocenter is *R*, because the CH_2SH substituent has higher priority than the COOH group.

In all but glycine, the simplest of the amino acids, C2 is a stereocenter and usually adopts the *S* configuration.

Like the names of the sugars (Section 24-1), an older amino acid nomenclature uses the prefixes D and L, which relate all the (2*S*)-amino acids to (*S*)-2,3-dihydroxypropanal (L-glyceraldehyde). As emphasized in the discussion of the natural D-sugars, a molecule belonging to the L family is not necessarily levorotatory. For example, both valine ($[\alpha]_D^{25°C} = +13.9°$) and isoleucine ($[\alpha]_D^{25°C} = +11.9°$) are dextrorotatory.

EXERCISE 26-1

Give the systematic names of alanine, valine, leucine, isoleucine, phenylalanine, serine, tyrosine, lysine, cysteine, methionine, aspartic acid, and glutamic acid.

EXERCISE 26-2

Give dashed-wedged line structures for (*S*)-alanine, (*S*)-phenylalanine, (*R*)-phenylalanine, and (*S*)-proline.

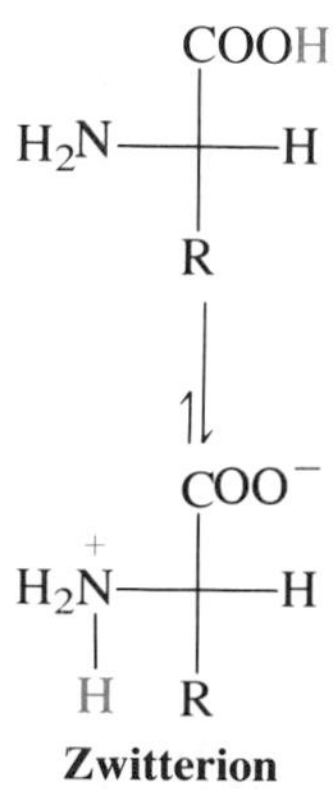

Zwitterion

Amino acids are acidic and basic: zwitterions

Because of their two functional groups, the amino acids are both acidic and basic; that is, they are **amphoteric** (Section 8-3). In the solid state the carboxylic acid group protonates the amine function, thus forming a **zwitterion**—also called a **dipolar ion** or an **inner salt.** This ammonium carboxylate form is favored because an ammonium ion is much less acidic (p$K_a \sim 10$–11) than a carboxylic acid (p$K_a \sim 2$–5). The highly polar zwitterionic structure allows amino acids to form particularly strong crystal lattices. Most of them therefore are fairly insoluble in organic solvents, and they decompose rather than melt when heated.

The structure of an amino acid in aqueous solution depends on the pH. Consider, for example, the simplest member of the series, glycine. The major form in neutral solution is the zwitterion. However, in strong acid ($pH < 1$) glycine exists predominantly as the cationic ammonium carboxylic acid, whereas strongly basic solutions ($pH > 13$) contain mainly the deprotonated 2-aminocarboxylate ion. These forms interconvert by acid-base equilibria (Section 6-7).

$$H_2\overset{+}{N}(H)CH_2COOH \underset{H^+}{\overset{HO^-}{\rightleftharpoons}} H_2\overset{+}{N}(H)CH_2COO^- \underset{H^+}{\overset{HO^-}{\rightleftharpoons}} H_2NCH_2COO^-$$

Predominates at pH < 1 — Predominates at pH ~ 6 — Predominates at pH > 13

Table 26-1 records pK_a values for each functional group of the amino acids. For glycine, the first value (2.4) refers to the equilibrium

$$H_3\overset{+}{N}CH_2COOH + H_2O \rightleftharpoons H_3\overset{+}{N}CH_2COO^- + H_2\overset{+}{O}H$$

$pK_a = 2.4$

$$K_1 = \frac{[H_3\overset{+}{N}CH_2COO^-][H_2\overset{+}{O}H]}{[H_3\overset{+}{N}CH_2COOH]} = 10^{-2.4}$$

Note that this pK_a is more than two units lower than that of an ordinary carboxylic acid (pK_a $CH_3COOH = 4.74$). This difference is a consequence of the electron-withdrawing effect of the protonated amino group. The second pK_a value (9.8) describes the second deprotonation step.

$$H_3\overset{+}{N}CH_2COO^- + H_2O \rightleftharpoons H_2NCH_2COO^- + H_2\overset{+}{O}H$$

$pK_a = 9.8$

$$K_2 = \frac{[H_2NCH_2COO^-][H_2\overset{+}{O}H]}{[H_3\overset{+}{N}CH_2COO^-]} = 10^{-9.8}$$

At the isoelectric point, the charges are neutralized

The pH at which the extent of protonation equals that of deprotonation is called the **isoelectric pH** or the **isoelectric point (p*I*)**. At this pH, the concentration of the charge-neutralized zwitterionic form is at its greatest. The value of p*I* is calculated from the expressions for K_1 and K_2, by setting $[H_3\overset{+}{N}CH_2COOH] = [H_2NCH_2COO^-]$. As can be verified readily, *the isoelectric point is the average of the two pK_a values* of the amino acid.

$$pI = \frac{pK_{COOH} + pK_{\overset{+}{N}H_2H}}{2}$$

The situation is a little more complicated if the side chain of the acid bears an additional acidic or basic function. Table 26-1 shows seven entries in which this is the case. One is tyrosine, bearing an acidic 4-hydroxyphenylmethyl substituent with $pK_a = 10.1$, typical of phenols (Section 22-3).

Assignment of pK_a Values in Selected Amino Acids

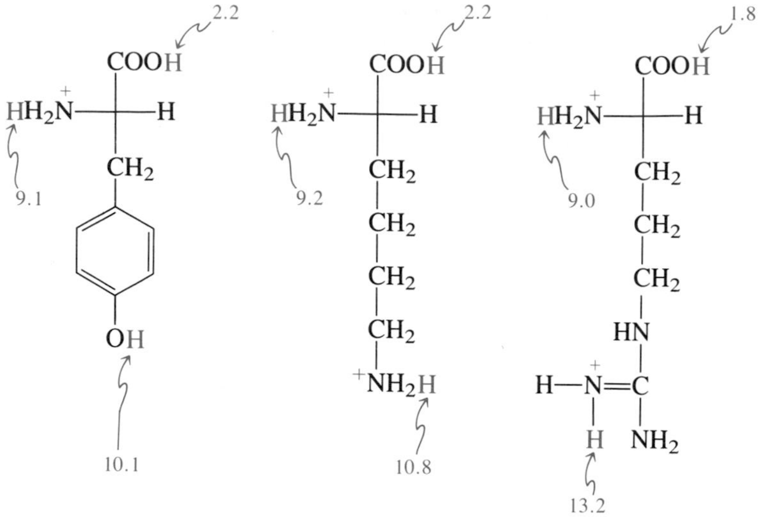

Tyrosine: COOH 2.2; $\overset{+}{H}H_2N$ 9.1; OH 10.1
Lysine: COOH 2.2; $\overset{+}{H}H_2N$ 9.2; $^{+}NH_2H$ 10.8
Arginine: COOH 1.8; $\overset{+}{H}H_2N$ 9.0; $H-\overset{+}{N}=C$ H 13.2

Tyrosine **Lysine** **Arginine**

In such cases, the isoelectric point is the average of the pK_a values for the two equilibria *associated with the charge-neutralized form* of the amino acid. For example, tyrosine may possess any of four structures in aqueous solution, depending on the pH.

Structures of Tyrosine at Different pH Values

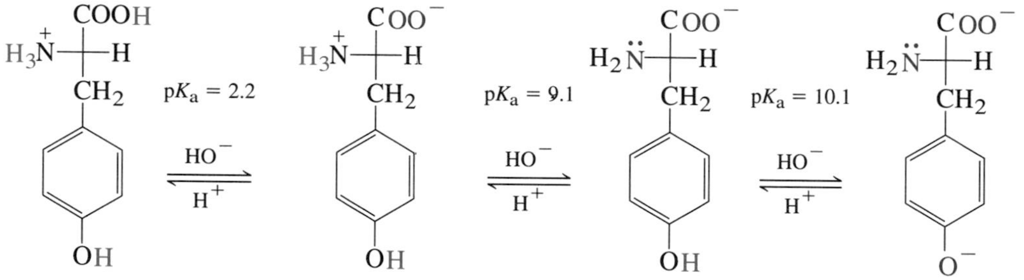

Monocation ⇌ (pK_a = 2.2; HO^- / H^+) Charge-neutralized zwitterion ⇌ (pK_a = 9.1; HO^- / H^+) Monoanion ⇌ (pK_a = 10.1; HO^- / H^+) Dianion

Monocation **Charge-neutralized zwitterion** **Monoanion** **Dianion**

The isoelectric point pI = (2.2 + 9.1)/2 = 5.7.

Lysine has an additional amino group that can be protonated in a strongly acidic medium to furnish a dication. When the pH of the solution is raised, deprotonation of the carboxy group occurs first, to be followed by proton loss from the nitrogen at C2, and finally from the remote ammonium function. The isoelectric point is located halfway between the last two pK_a values, at pI = 10.

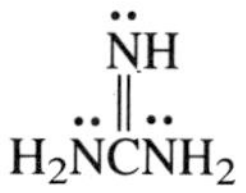

Arginine bears a substituent new to us: the relatively basic **guanidino** group, $-NH\overset{\overset{NH}{\|}}{C}NH_2$ (pK_a of conjugate acid ~ 13), derived from the molecule **guanidine.**

$H_2N\overset{\overset{NH}{\|}}{C}NH_2$
Guanidine

EXERCISE 26-3

Guanidine is found in turnip juice, mushrooms, corn germ, rice hulls, mussels, and earthworms. Its basicity is due to the formation of a highly resonance-stabilized protonated form. Draw its resonance structures. (Hint: Review Section 20-1.)

Histidine contains another new substituent, the **imidazole** ring (see Problem 6 of Chapter 25). In this aromatic heterocycle, one of the nitrogen atoms is hybridized as in pyridine, and the other is hybridized as in pyrrole.

Imidazole

BOX 26-1 Arginine and Nitric Oxide in Biochemistry and Medicine

$$^{-}OOC-C(H)(NH_3^+)-(CH_2)_3NHC(=NH_2^+)NH_2 \xrightarrow[\text{synthase}]{\frac{1}{2}O_2,\ \text{nitric oxide}} {}^{-}OOC-C(H)(NH_3^+)-(CH_2)_3NHC(=NH_2^+)NHOH \xrightarrow[-H_2O]{\frac{1}{2}O_2}$$

L-Arginine **N^G-Hydroxy-L-arginine**

$$^{-}OOC-C(H)(NH_3^+)-(CH_2)_3N(H)-C(=NH_2^+)-N{=}O \xrightarrow[-H^+]{} [H-N{=}O] + \left[{}^{-}OOC-C(H)(NH_3^+)-(CH_2)_3N{=}C{=}NH\right] \xrightarrow{H_2O} {}^{-}OOC-C(H)(NH_3^+)-(CH_2)_3NHC(=O)NH_2$$

N^G-Oxo-L-arginine

$$[H-N{=}O] \rightarrow H\cdot + \cdot N{=}O$$

Nitric oxide **L-Citrulline**

In the late 1980s and early 1990s scientists made a series of startling discoveries. The simple but highly reactive and exceedingly toxic molecule nitric oxide, $:\dot{N}{=}\ddot{O}:$, is synthesized in a wide variety of cells in mammals, including humans, and performs several critical biological functions. Macrophages (cells associated with the body's immune system) destroy bacteria and tumor cells by exposing them to nitric oxide, which is synthesized by the enzyme-catalyzed oxidation of arginine, as shown above.

Nitric oxide is released by cells on the inner walls of blood vessels and causes adjacent muscle fibers to relax. This 1987 discovery explains the effectiveness of nitroglycerin and other organic nitrates as treatments for angina and heart attacks, a nearly century-old mystery: These substances are metabolically converted into NO, which dilates the blood vessels. More recent studies reveal a role for nitric oxide as a neurotransmitter in the brain, perhaps the creator of memory itself! Paradoxically, NO is also a powerful neurotoxin whose uncontrolled release may be responsible for the extensive cell destruction associated with vascular strokes and brain disorders such as Alzheimer's and Huntington's diseases. Indeed, after initiating strokes in mice, administration of N^2-nitro-L-arginine, an inhibitor of NO synthesis, greatly reduces neuronal damage. Such studies are of great interest in the quest for effective therapies for these conditions.

$$^{-}OOC-C(H)(NH-NO_2)-(CH_2)_3NHC(=NH_2^+)NH_2$$

N^2-Nitro-L-arginine

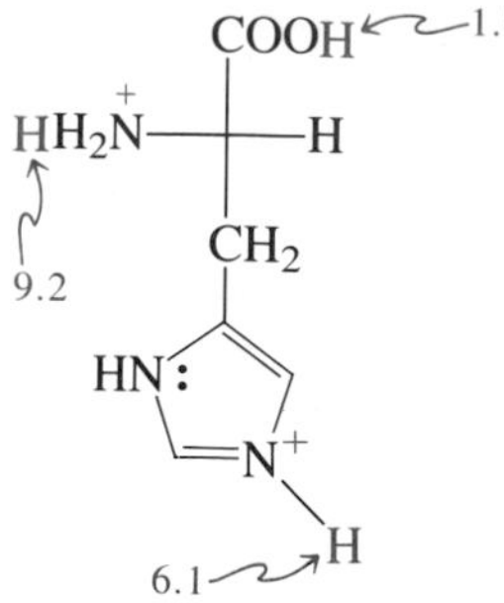

COOH ← 1.8

$H\overset{+}{H_2N}$—C—H

9.2

CH_2

HN:

N^+—H

6.1

Histidine

EXERCISE 26-4

Draw an orbital picture of imidazole. (Hint: Use Figure 25-1 as a model.)

The imidazole ring is relatively basic because the protonated species is stabilized by resonance and can be described by two equivalent resonance forms.

Resonance in Protonated Imidazole

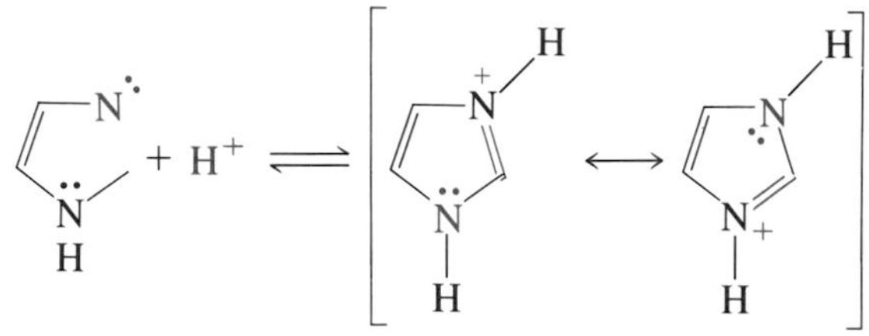

$p K_a = 7.0$

This resonance stabilization is related to that in amides (Section 20-1). Imidazole is significantly protonated at physiological pH. It can therefore function as a proton acceptor and donor at the active site of a variety of enzymes.

The amino acid cysteine bears a mercapto substituent. Recall that, apart from their acidic character, thiols can be oxidized to disulfides under mild conditions (Section 9-10). In nature, various enzymes are capable of oxidatively coupling the mercapto groups in the cysteines of proteins and peptides, thereby reversibly linking peptide strands (Section 9-10).

Aspartic acid and glutamic acid are amino dicarboxylic acids. At physiological pH, both of the carboxy functions are deprotonated, and the molecules exist as the zwitterionic anions aspartate and glutamate. Monosodium glutamate (MSG) is used as a flavor enhancer in various foods.

In summary, there are 20 elementary (2*S*)-amino acids, all of which have common names. Unless there are additional acid-base functions in the chain, their acid-base behavior is governed by two pK_a values, the lower one describing the deprotonation of the carboxy group. At the isoelectric point, which is the average of the two pK_a values, the number of amino acid molecules with net zero charge is maximized. Some amino acids contain additional acidic or basic functions, such as hydroxy, amino, guanidino, imidazolyl, mercapto, and carboxy.

26-2 Synthesis of Amino Acids: A Combination of Amine and Carboxylic Acid Chemistry

Chapter 21 treated the chemistry of amines, Chapters 19 and 20 that of carboxylic acids and their derivatives. We use both in preparing 2-amino acids.

Hell-Volhard-Zelinsky bromination followed by amination converts carboxylic acids to 2-amino acids

What would be the quickest way of introducing a 2-amino substituent into a carboxylic acid? Section 19-12 pointed out that simple 2-functionalization of an acid is possible by the Hell-Volhard-Zelinsky bromination. Furthermore, the bromine in the product can be displaced by nucleophiles, such as ammonia. In these two steps, propanoic acid can be converted into racemic alanine.

$$\underset{\textbf{Propanoic acid}}{CH_3CH_2COOH} \xrightarrow[-\ HBr]{Br\text{–}Br,\ \text{catalytic } PBr_3} \underset{\textbf{2-Bromopropanoic acid}}{\underset{80\%}{CH_3CH(Br)COOH}} \xrightarrow[-\ HBr]{NH_3,\ H_2O,\ 25^{\circ}C,\ 4\ \text{days}} \underset{\textbf{(R,S)-Alanine}}{\underset{56\%}{CH_3CH(\overset{+}{N}H_3)COO^-}}$$

Unfortunately, this approach frequently suffers from relatively low yields. A better synthesis utilizes Gabriel's procedure for the preparation of primary amines (Section 21-5).

The Gabriel synthesis can be adapted to produce amino acids

Recall that *N*-alkylation of 1,2-benzenedicarboxylic imide (phthalimide) anion followed by acid hydrolysis furnishes amines (Section 21-5). To prepare an amino acid instead, we can use diethyl 2-bromopropanedioate (diethyl 2-bromomalonate) in the first step of the reaction sequence. This alkylating agent is readily available from the bromination of diethyl propanedioate (malonate). Now the alkylation product can be hydrolyzed and decarboxylated. Hydrolysis of the imide group then furnishes an amino acid.

Gabriel Synthesis of Glycine

$$HCBr(COOCH_2CH_3)_2 + C_6H_4(CO)_2\ddot{N}{:}^-\ K^+ \xrightarrow[-\ KBr]{} C_6H_4(CO)_2N\text{—}CH(COOCH_2CH_3)_2 \xrightarrow[-\ 2\ CH_3CH_2OH]{H^+,\ H_2O,\ \Delta}$$

Diethyl 2-bromopropanedioate (Diethyl 2-bromomalonate) — **Potassium 1,2-benzenedicarboxylic imide (Potassium phthalimide)** — 85%

$$C_6H_4(CO)_2N\text{—}CH(COOH)_2 \xrightarrow[-\ CO_2]{} \xrightarrow[-\ C_6H_4(COOH)_2]{H^+,\ H_2O} \underset{\textbf{Glycine}}{\underset{85\%}{H_3\overset{+}{N}CH_2COO^-}}$$

One of the advantages of this approach is the versatility of the initially formed 2-substituted propanedioate. This product can itself be alkylated, thus allowing for the preparation of a variety of substituted amino acids.

$$C_6H_4(CO)_2N\text{—}CH(CO_2CH_2CH_3)_2 \xrightarrow{\text{1. } CH_3CH_2O^-Na^+,\ CH_3CH_2OH \quad \text{2. } RX \quad \text{3. } H^+,\ H_2O,\ \Delta} RCH(\overset{+}{N}H_3)COO^-$$

EXERCISE 26-5

Propose Gabriel syntheses of methionine, aspartic acid, and glutamic acid.

Amino acids are prepared from aldehydes by the Strecker synthesis

The crucial step in the **Strecker* synthesis** is a variation of the cyanohydrin formation from aldehydes and hydrogen cyanide (Section 17-11).

$$\mathrm{R\overset{O}{\overset{\|}{C}}H + HCN \rightleftharpoons R{-}\underset{H}{\underset{|}{\overset{OH}{\overset{|}{C}}}}{-}CN}$$

Cyanohydrin

When the same reaction is carried out in the presence of ammonia, it is the intermediate imine that undergoes addition of hydrogen cyanide, to furnish the corresponding 2-amino nitriles. Subsequent acidic or basic hydrolysis results in the desired amino acids.

Strecker Synthesis of Alanine

$$\underset{\textbf{Ethanal (Acetaldehyde)}}{\mathrm{CH_3\overset{O}{\overset{\|}{C}}H}} \xrightarrow[-\,H_2O]{NH_3} \underset{\textbf{Imine}}{\mathrm{CH_3\overset{NH}{\overset{\|}{C}}H}} \xrightarrow{HCN} \underset{\textbf{2-Aminopropanenitrile}}{\mathrm{H_3C{-}\underset{H}{\underset{|}{\overset{NH_2}{\overset{|}{C}}}}{-}CN}} \xrightarrow{H^+,\ H_2O,\ \Delta} \underset{\textbf{Alanine}}{\underset{55\%}{\mathrm{CH_3\overset{^+NH_3}{\overset{|}{C}}HCOO^-}}}$$

EXERCISE 26-6

Propose Strecker syntheses of glycine and methionine.

In summary, racemic amino acids are made by the amination of 2-bromocarboxylic acids, applications of the Gabriel synthesis of amines, and the Strecker synthesis, which proceeds through an imine variation of the preparation of cyanohydrins, followed by hydrolysis.

26-3 Synthesis of Enantiomerically Pure Amino Acids

All the methods of the preceding section produce amino acids in racemic form. However, we noted that most of the amino acids in natural polypeptides have the *S* configuration. Thus, many synthetic procedures—in particular, peptide and protein syntheses—require enantiomerically pure compounds. To meet this requirement, either the racemic amino acids must be resolved (Section 5-8) or a single enantiomer must be prepared by enantioselective reactions.

*Professor Adolf Strecker, 1822–1871, University of Würzburg, Germany.

A conceptually straightforward approach to the preparation of pure enantiomers of amino acids would be resolution of their diastereomeric salts. Typically, the amine group is first protected as an amide, and the resulting product then treated with an optically active amine (Section 21-4), such as the alkaloid brucine (Section 25-8). The two diastereomers formed can be separated by fractional crystallization. Unfortunately, in practice, this method can be tedious and can suffer from poor yields.

Resolution of Racemic Valine

$$\underset{\textbf{(R,S)-Valine}}{(CH_3)_2CH\overset{\overset{+}{N}H_3}{\overset{|}{C}}HCOO^-} + HCOOH \xrightarrow[\text{Protection}]{} \underset{\substack{80\% \\ \textbf{(R,S)-N-Formylvaline}}}{(CH_3)_2CH\overset{HN\overset{O}{\overset{\|}{C}}H}{\overset{|}{C}}HCOOH} + HOH$$

Brucine (abbreviated B), CH_3OH, 0°C

(S) salt: $COO^{-}\,{}^{+}BH$ at top; $HC(=O)NH$ on left, H on right; $CH(CH_3)_2$ at bottom (S)

(R) salt: $HB^{+}\,{}^{-}OOC$ at top; H on left, $NHC(=O)H$ on right; $(CH_3)_2HC$ at bottom (R)

Separate by fractional crystallization

NaOH, H_2O, 0°C (each branch)

Remove brucine and hydrolyze amide

(S)-Valine, 70%: COO^- at top; $H_3\overset{+}{N}$ on left, H on right; $CH(CH_3)_2$ at bottom

(R)-Valine, 70%: ^{-}OOC at top; H on left, $\overset{+}{N}H_3$ on right; $(CH_3)_2HC$ at bottom

In an alternative approach, the stereocenter at C2 is formed enantioselectively. Nature makes use of this strategy in the biosynthesis of amino acids. Thus, the enzyme glutamate dehydrogenase converts the carbonyl group in 2-oxopentanedioic acid into the amine substituent in (*S*)-glutamic acid by a biological reductive amination (for chemical reductive aminations, see Section 21-6). The reducing agent is NADH (Box 25-4, Section 25-6).

$$\underset{\textbf{2-Oxopentanedioic acid}}{HOOCCH_2CH_2\overset{O}{\overset{\|}{C}}COOH} + NH_3 + H^+ \xrightarrow[-\ NAD^+]{\text{NADH, glutamate dehydrogenase}} \underset{\textbf{(S)-Glutamic acid}}{HOOCCH_2CH_2\overset{\overset{+}{N}H_3}{\overset{|}{C}}HCOO^-} + H_2O$$

(*S*)-Glutamic acid is the biosynthetic precursor of glutamine, proline, and arginine. Moreover, it functions to aminate other 2-oxo acids, with the help of another enzyme, transaminase, making additional amino acids available.

$$\mathrm{H{-}\overset{\overset{+}{N}H_3}{\underset{R}{C}}{-}COO^-} + \mathrm{R'\overset{O}{\overset{\|}{C}}COO^-} \xrightleftharpoons{\text{Transaminase}} \mathrm{R\overset{O}{\overset{\|}{C}}COO^-} + \mathrm{H{-}\overset{\overset{+}{N}H_3}{\underset{R'}{C}}{-}COO^-}$$

In summary, optically pure amino acids can be obtained by resolution of a racemic mixture or by enantioselective formation of the C2 stereocenter.

BOX 26-2 Synthesis of Optically Pure Amino Acids

Catalytic asymmetric hydrogenation (Box 12-1, Section 12-2) is only one of many methods that have been developed to prepare optically pure amino acids. Another approach is the alkylation of glycine in the presence of optically active bases. Optically pure D-(4-chlorophenyl)alanine has been obtained by using the naturally occurring alkaloid cinchonine according to the scheme shown below.

Only a catalytic quantity of the alkaloid base is required for the first step of this process, which is carried out in a rapidly stirred mixture of dichloromethane and water. During alkylation, which occurs in the CH_2Cl_2 layer, the base is converted into its conjugate acid, which is ionic and migrates into the water phase. The NaOH dissolved in the latter regenerates the neutral alkaloid, which shifts back into the CH_2Cl_2 phase, thereby leading to alkylation of another molecule of substrate, and so on. This cycle of reactions, which employs a catalyst that repeatedly shuttles back and forth between two immiscible solvents, is an example of *phase-transfer catalysis*. Alkylation gives an *optically enriched* intermediate, from which the pure major enantiomer is separated from residual racemic material by recrystallization.

$$(C_6H_5)_2C{=}N{-}CH_2{-}COOC(CH_3)_3 + BrCH_2{-}C_6H_4{-}Cl \xrightarrow{\text{Cinchonine, NaOH, } H_2O,\ CH_2Cl_2}$$

1,1-Dimethylethyl *N*-(diphenylmethylidene)glycinate

$$(C_6H_5)_2C{=}N{-}CH(CH_2C_6H_4Cl){-}COOC(CH_3)_3 \xrightarrow{\text{1. Recrystallize; 2. } H_3O^+,\ \Delta} H_3\overset{+}{N}{-}CH(CH_2C_6H_4Cl){-}COO^-$$

95% yield; 82% *R*, 18% *S*

100% optically pure

(*R*)-(4-Chlorophenyl)alanine

26-4 Peptides and Proteins: Amino Acid Oligomers and Polymers

Amino acids are very versatile biologically because they can be polymerized. This section will describe the structure and properties of such **polypeptide** chains. These chains twist and fold in three dimensions to form the highly biologically active **proteins.**

Amino acids form peptide bonds

2-Amino acids are the monomer units in polypeptides. The polymer forms by repeated reaction of the carboxylic acid function of one amino acid with the amine group in another to make a chain of amides (Section 20-6). The amide linkage joining amino acids is also called a **peptide bond.**

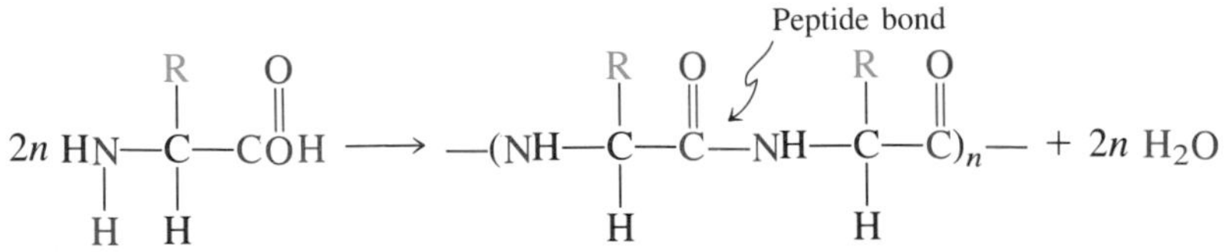

$$2n\ \mathrm{HN(H)-C(R)(H)-C(=O)OH} \longrightarrow -(\mathrm{NH-C(R)(H)-C(=O)-NH-C(R)(H)-C(=O)})_n- + 2n\ H_2O$$

Peptide bond

2-Amino acid (α-Amino acid) — **Polyamino acid (Polypeptide)**

The oligomers formed by linking amino acids in this way are called **peptides.** For example, two amino acids give rise to a **dipeptide,** three to a **tripeptide,** and so on. The individual amino acid units forming the peptide are referred to as **residues.** In some proteins, two or more polypeptide chains are linked by disulfide bridges (Sections 9-10 and 26-1).

Polypeptides are characterized by their sequence of amino acid residues

In drawing a polypeptide chain, the **amino end** or ***N*-terminal amino acid** is placed at the left, the **carboxy end** or ***C*-terminal amino acid** at the right. The configuration at the C2 stereocenters is usually presumed to be *S*.

How to Draw the Structure of a Tripeptide

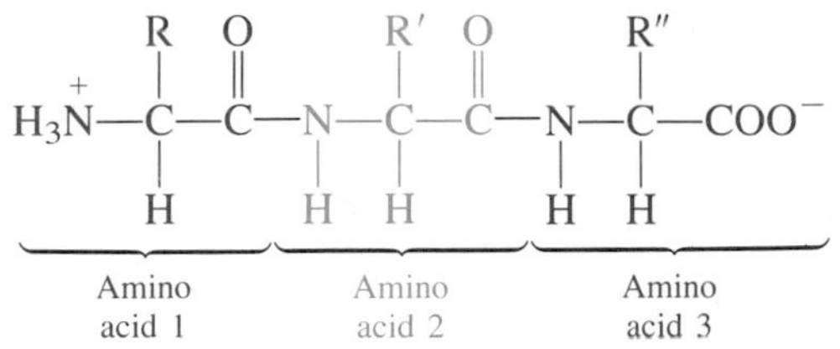

$$\overset{+}{H_3N}-\underbrace{\mathrm{C(R)(H)-C(=O)}}_{\text{Amino acid 1}}-\underbrace{\mathrm{N(H)-C(R')(H)-C(=O)}}_{\text{Amino acid 2}}-\underbrace{\mathrm{N(H)-C(R'')(H)-COO^-}}_{\text{Amino acid 3}}$$

The chain incorporating the amide (peptide) bonds is called the **main chain,** the substituents R, R′, and so on are the **side chains.**

The naming of peptides is straightforward. Starting from the amino end, the names of the individual residues are simply connected in sequence, each regarded as a substituent to the next amino acid, ending with the *C*-terminal residue. Because this procedure rapidly becomes cumbersome, the abbreviations listed in Table 26-1 are used for larger peptides.

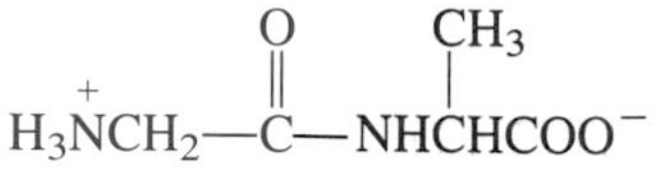

$$H_3\overset{+}{N}CH_2-\overset{O}{\overset{\|}{C}}-NH\overset{CH_3}{\overset{|}{C}}HCOO^-$$

Glycylalanine
Gly-Ala

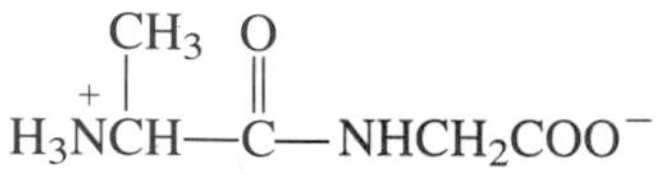

$$H_3\overset{+}{N}\overset{CH_3}{\overset{|}{C}}H-\overset{O}{\overset{\|}{C}}-NHCH_2COO^-$$

Alanylglycine
Ala-Gly

$$H_3\overset{+}{N}\overset{C_6H_5CH_2}{\overset{|}{C}}H-\overset{O}{\overset{\|}{C}}-NH\overset{(CH_3)_2CHCH_2}{\overset{|}{C}}H-\overset{O}{\overset{\|}{C}}-NH\overset{CH_3CHOH}{\overset{|}{C}}HCOO^-$$

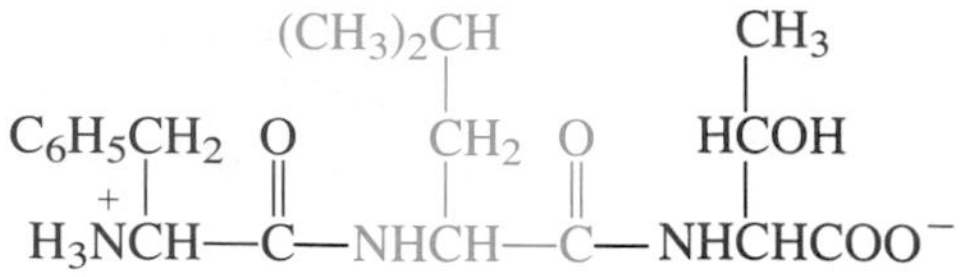

Phenylalanylleucylthreonine
Phe-Leu-Thr

Let us look at some examples of peptides and their structural variety. A dipeptide ester, aspartame, is a low-calorie artificial sweetener (Nutrasweet). In the three-letter notation, the ester end is denoted by OCH_3. Glutathione, a tripeptide, is found in all living cells, and in particularly high concentrations in the lens of the eye. It is unusual in that its glutamic acid residue is linked at the γ-carboxy group (denoted γ-Glu) to the rest of the peptide.

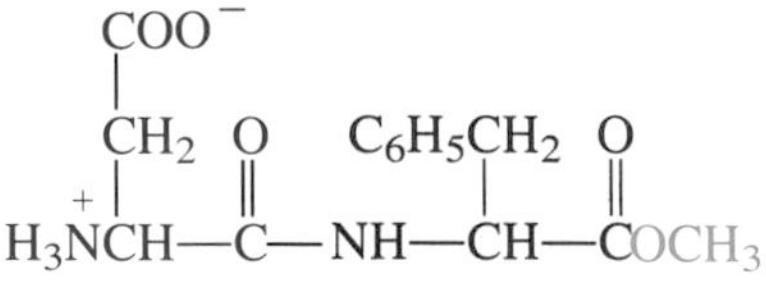

$$H_3\overset{+}{N}\overset{COO^-CH_2}{\overset{|}{C}}H-\overset{O}{\overset{\|}{C}}-NH-\overset{C_6H_5CH_2}{\overset{|}{C}}H-\overset{O}{\overset{\|}{C}}OCH_3$$

Aspartylphenylalanine methyl ester
Asp-Phe-OCH_3
(Aspartame)

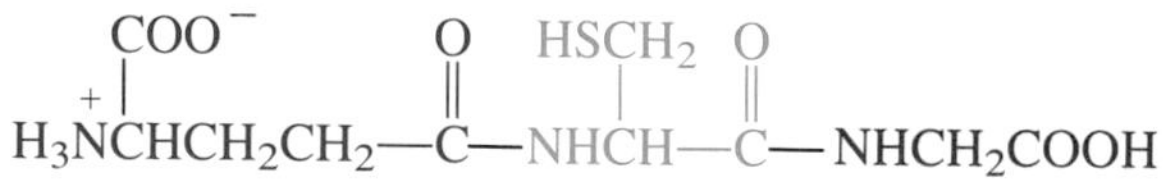

$$H_3\overset{+}{N}\overset{COO^-}{\overset{|}{C}}HCH_2CH_2-\overset{O}{\overset{\|}{C}}-NH\overset{HSCH_2}{\overset{|}{C}}H-\overset{O}{\overset{\|}{C}}-NHCH_2COOH$$

γ-Glutamylcysteinylglycine
γ-Glu-Cys-Gly
(Glutathione)

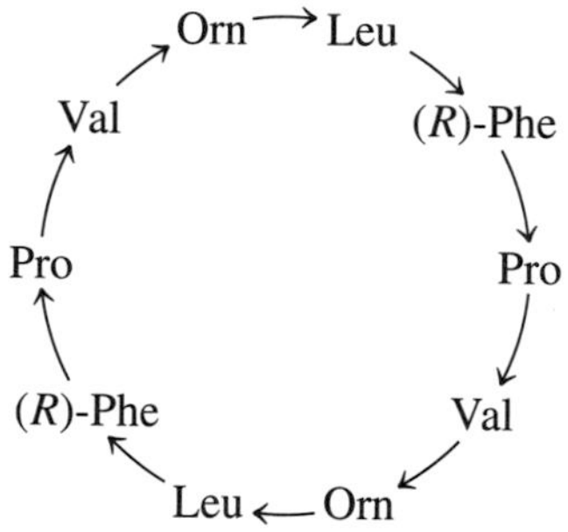

It functions as a biological reducing agent by being readily oxidized enzymatically at the cysteine mercapto unit to the disulfide-bridged dimer.

Gramicidin S is a cyclic peptide antibiotic constructed out of two identical pentapeptides that have been joined head to tail. It contains phenylalanine in the *R* configuration and a rare amino acid, ornithine [Orn, a lower homolog (one less CH_2 group) of lysine]. In the short notation in which gramicidin S is shown in the margin, the sense in which the amino acids are linked (amino to carboxy direction) is indicated by arrows.

Orn → Leu → (*R*)-Phe → Pro → Val → Orn → Leu → (*R*)-Phe → Pro → Val → (back to Orn)

Gramicidin S

Although short notations are practical, they do not reveal some of the noncovalent bonding interactions in peptides. A dashed-wedged line picture of gramicidin S shows several important hydrogen bonds contributing substantially to its stereochemical rigidity.

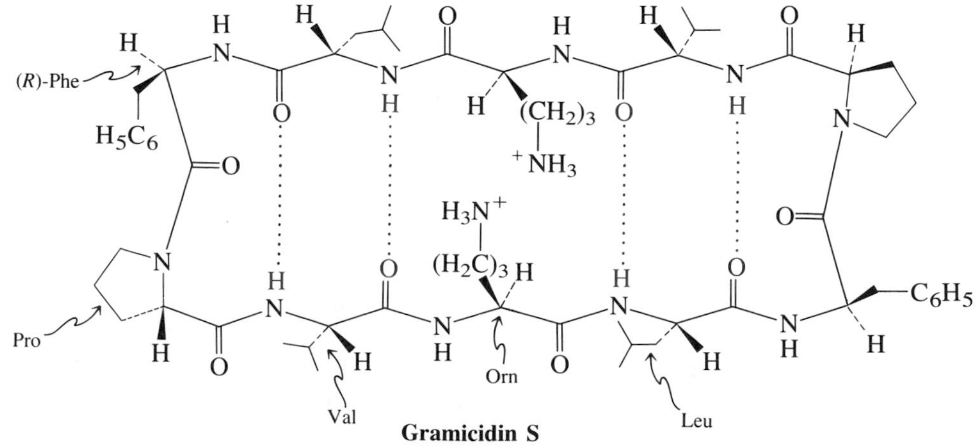

(*R*)-Phe, Pro, Val, Orn, Leu; H_5C_6, $(CH_2)_3$, $^+NH_3$, H_3N^+, $(H_2C)_3$, C_6H_5

Gramicidin S
(Dashed-wedged line notation)

A second feature inducing rigidity is the strongly dipolar nature of the amide functional group (Section 20-1), which causes an appreciable barrier to rotation around the relatively short carbonyl–nitrogen bond.

Insulin provides a good illustration of the three-dimensional structure adopted by a complex sequence of amino acids (Figure 26-1). This protein hormone is a drug important in the treatment of diabetes because of its ability to regulate glucose metabolism.

Insulin contains 51 amino acid residues incorporated into two chains, labeled A and B. The chains are connected by two disulfide bridges, and there is an additional such linkage connecting the cysteine residues at positions 6 and 11 of the A chain, causing it to loop. Both chains fold up in a way that minimizes steric interference and maximizes electrostatic, London, and hydrogen-bonding attractions. These forces give rise to a fairly condensed three-dimensional structure (Figure 26-2).

BOX 26-3 Glutathione and the Toxicity of Methyl Isocyanate: The Bhopal Tragedy

Examination of victims of the Bhopal tragedy (Box 20-3, Section 20-7) revealed that many of these individuals suffered extensive damage to *internal* tissues and organs, characteristic of exposure to methyl isocyanate. This finding was unexpected, because the latter is so chemically reactive that none of it should have been capable of passing through the lungs and circulatory system to more remote locations in the body, such as the liver.

The solution to this mystery was found in 1992 when glutathione, one of whose roles is to protect cells from damage by toxic agents (see Section 22-9), was found to be capable of carrying and releasing methyl isocyanate throughout the body. This insidious effect derives from the ability of methyl isocyanate to react *reversibly* with the mercapto group of the peptide (as shown below).

Intermediates such as *S*-(*N*-methylcarbamoyl)glutathione are currently under investigation as possible mediators of the toxic effects of several other small nitrogen-containing organic compounds as well.

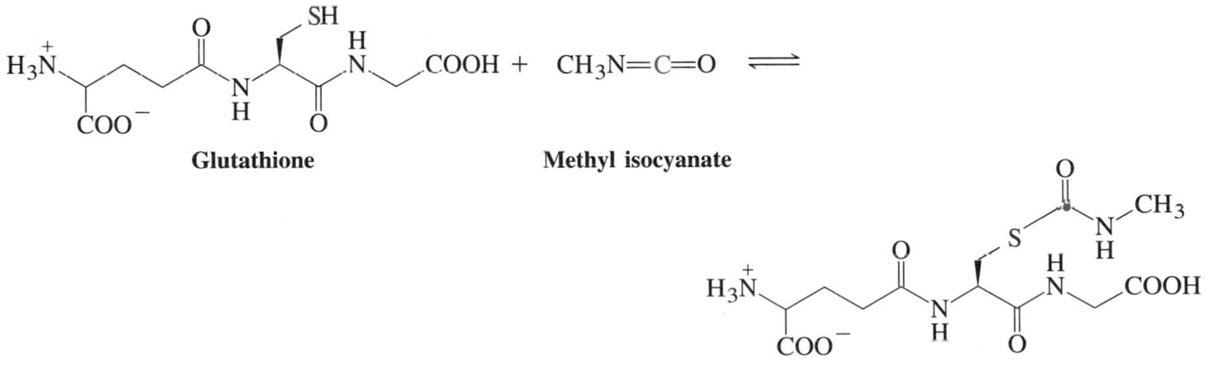

Glutathione **Methyl isocyanate**

***S*-(*N*-Methylcarbamoyl)glutathione**

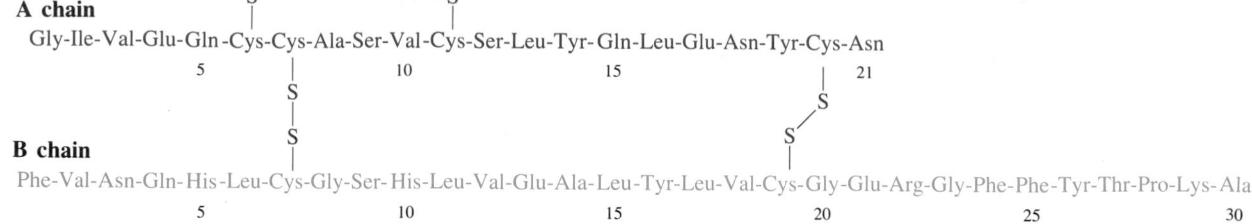

FIGURE 26-1 Bovine (cattle) insulin is actually made up of two amino acid chains, linked by disulfide bridges. The amino (*N*-terminal) end is at the left in both chains.

Because most synthetic methods give only low yields, a major source of insulin, until recently, has been the pancreas of slaughtered animals. An exciting new development is the efficient preparation of the protein by genetic-engineering methods. Huge vats of bacteria carrying the human insulin gene now generate enough material to treat thousands of diabetics around the world.

EXERCISE 26-7

Vasopressin, also known as antidiuretic hormone, controls the excretion of water from the body. Write its full structure. Note that there is an intramolecular disulfide bridge between the two cysteine molecules.

S——————S

Cys-Tyr-Phe-Gln-Asn-Cys-Pro-Arg-Gly-NH_2

Vasopressin

Proteins fold into pleated sheets and helices: secondary and tertiary structure

Insulin and other polypeptide chains adopt well-defined three-dimensional structures. Whereas the sequence of amino acids in the chain defines the **primary structure,** the folding pattern of the chain gives rise to the **secondary structure** of the polypeptide. The secondary structure results mainly from the rigidity of the amide bond and from hydrogen and other noncovalent bonding along the chain(s). Two important arrangements are the pleated sheet, or β configuration, and the α helix.

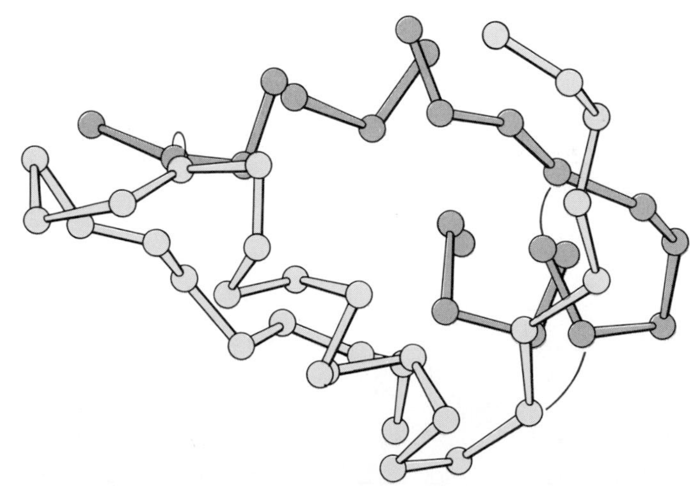

FIGURE 26-2 Three-dimensional structure of insulin. Residues on chain A are blue, those on B green. The disulfide bridges are indicated in red. (After *Biochemistry,* 3d Ed., by Lubert Stryer, W. H. Freeman and Company, Copyright © 1975, 1981, 1988.)

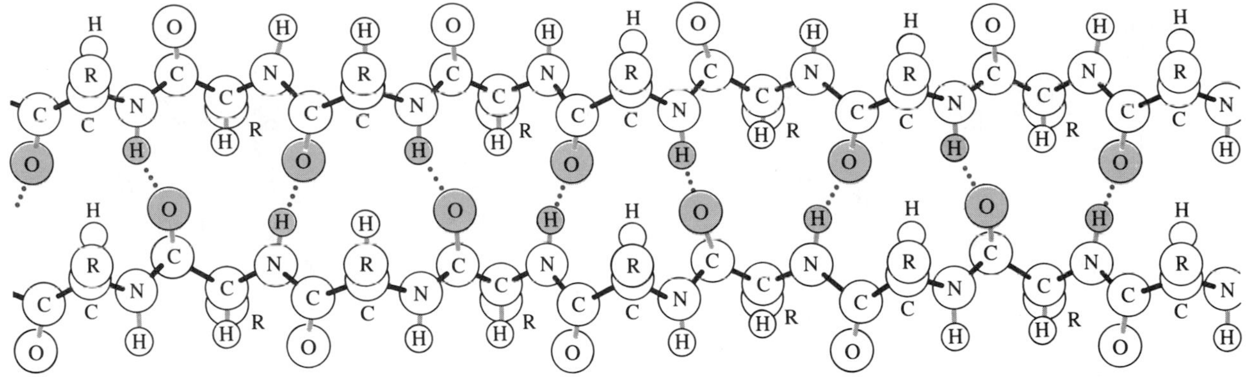

A

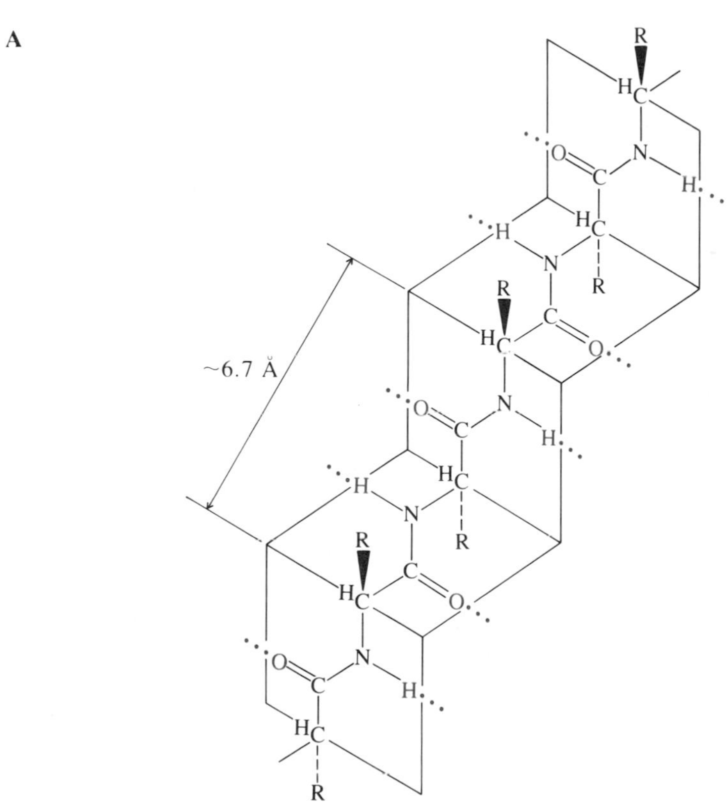

B

FIGURE 26-3 (A) The pleated sheet, or β configuration, which is held in place by hydrogen bonds (dotted lines) between two polypeptide strands. (After "Proteins," by Paul Doty, *Scientific American,* September 1957, Copyright © 1957, Scientific American, Inc.) (B) The colored peptide bonds define the individual pleats; the positions of the side chains, R, are alternately above and below the planes of the sheets. The dotted lines indicate hydrogen bonds to a neighboring chain or to water.

In the **pleated sheet** (Figure 26-3), two chains line up with the amino groups of one peptide opposite the carbonyl groups of a second, thereby allowing hydrogen bonds to form. Such bonds can also develop within a single chain if it loops back on itself. Multiple hydrogen bonding of this type can impart considerable rigidity to a system. The planes of adjacent amide linkages form a specific angle, a geometry that produces the observed pleated-sheet structure.

The ***α* helix** (Figure 26-4) allows for intramolecular hydrogen bonding between nearby amino acids in the chain. There are 3.6 amino acids per turn of the helix, two equivalent points in neighboring turns being about 5.4 Å apart.

Not all polypeptides adopt idealized structures such as these. If too much charge of the same kind builds up along the chain, charge repulsion will enforce a more random orientation. In addition, the bulky proline, because its amino nitrogen is also part of the substituent ring, can cause a kink or bend in an α helix.

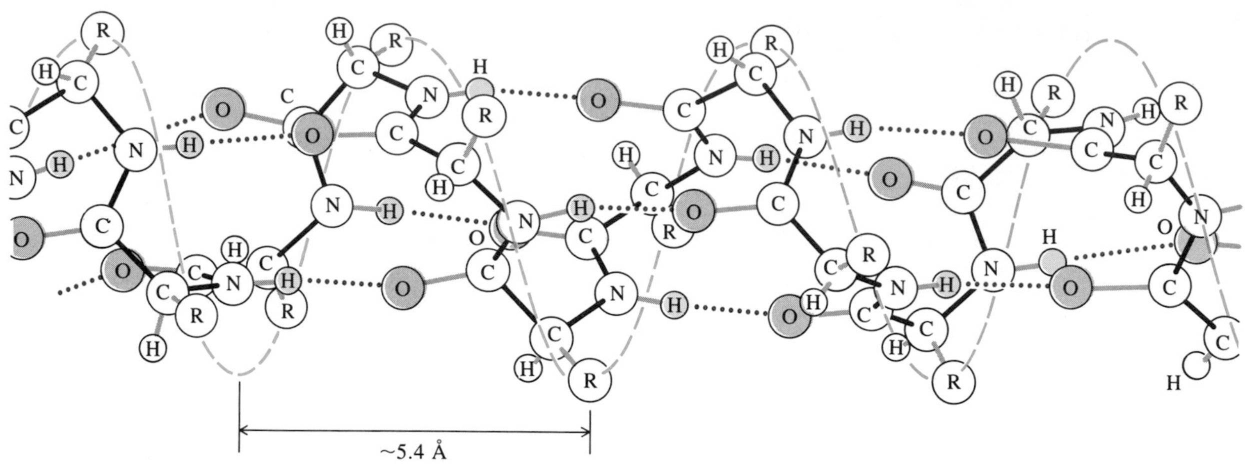

FIGURE 26-4 The α helix, in which the polymer chain is arranged as a right-handed spiral held rigidly in shape by intramolecular hydrogen bonds. (After "Proteins," by Paul Doty, *Scientific American,* September 1957, Copyright © 1957, Scientific American, Inc.)

Further folding, coiling, and aggregation of polypeptides give rise to their **tertiary structure.** A variety of forces, all arising from the R group, come into play to stabilize such molecules, including disulfide bridges, hydrogen bonds, London forces, and electrostatic attraction and repulsion. There are also **micellar effects** (Section 20-4): The polymer adopts a structure that maximizes exposure of polar groups to the aqueous environment, while minimizing exposure of hydrophobic groups (e.g., alkyl and phenyl). Pronounced folding is observed in the **globular proteins,** many of which perform chemical transport and catalysis (e.g., myoglobin and hemoglobin, Section 26-8). In the **fibrous proteins,** such as myosin (in muscle) and α-keratin (in hair, nails, and wool), several α helices are coiled to produce a **superhelix** (Figure 26-5).

The tertiary structure of enzymes and transport proteins (proteins that carry molecules from place to place) usually gives rise to three-dimensional pockets, called **active sites.** The size and shape of the active site provide a highly specific "fit" for the **substrate,** the molecule on which the protein carries out its intended function. The inner surface of the pocket typically contains a specific arrangement of the side chains of polar amino acids that attracts functional groups in the substrate by hydrogen bonding or ionic interactions. In enzymes, the active site aligns functional groups and additional molecules in a way that promotes their reactions with the substrate. **Denaturation,** or breakdown of the tertiary structure of a protein, usually causes precipitation of the protein and destroys its catalytic activity. Denaturation is caused by exposure to excessive heat or extreme pH values. Think, for example, of what happens to clear egg white when it is poured into a hot frying pan or to milk when it is added to lemon tea.

Some molecules, such as hemoglobin (Section 26-8), also adopt a **quaternary structure,** in which two or more amino acid chains, each with its own tertiary structure, combine to form a larger assembly.

In summary, polypeptides are polymers of amino acids linked by amide bonds. Their amino acid sequence can be described in a shorthand notation using the three-letter abbreviations compiled in Table 26-1. The amino end group is placed at the left, the carboxy end at the right. Polypeptides can be cyclic and can also be linked by disulfide and hydrogen bonds. The sequence of amino acids is the primary structure of a polypeptide, folding gives rise to its secondary structure, and further folding, coiling, and aggregation to its tertiary structure.

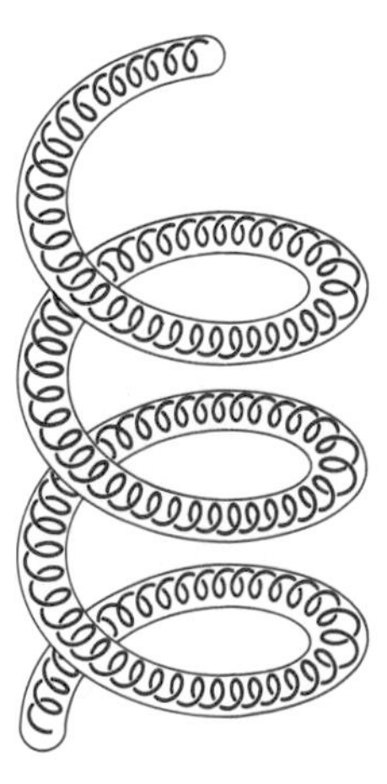

FIGURE 26-5 Idealized picture of a superhelix, a helix within a helix.

26-5 Determination of Primary Structure: Amino Acid Sequencing

Biological function in polypeptides and proteins requires a specific three-dimensional shape and arrangement of functional groups, which, in turn, requires a definite amino acid sequence. One "monkey wrench" residue in an otherwise normal protein can completely alter its behavior. For example, sickle-cell anemia, a potentially lethal condition, is the result of changing a *single* amino acid in hemoglobin (Section 26-8). The determination of the primary structure of a protein, called amino acid or polypeptide **sequencing,** can help us to understand its mechanism of action.

In the late 1950s and early 1960s, it was discovered that amino acid sequences are predetermined by DNA, the molecule containing our hereditary information (Section 26-9). Thus, through a knowledge of protein primary structure, we can learn how genetic material expresses itself. Functionally similar proteins in related species should, and do, have similar primary structures. The closer their sequences of amino acids, the more closely the species are related. Polypeptide sequencing therefore strikes at the heart of the question of the evolution of life itself. This section shows how chemical means, together with analytical techniques, allow us to obtain this information.

First, break the disulfide bridges and purify

Many polypeptides are made up of two or more chains linked by disulfide bridges. The first step in the analysis of such structures is to break these bonds and separate the resulting subunits. One way of doing this is to oxidize the disulfide bridges to sulfonic acids with performic acid.

$$\text{Chain A}—S—S—\text{Chain B} \xrightarrow{HCOOH\ (\text{with } C{=}O)} \text{Chain A}—SO_3H \quad HO_3S—\text{Chain B}$$

The problem of purification is an enormous one, and attempts at its solution consume many hours in the laboratory. Several techniques can separate polypeptides on the basis of size, solubility in a particular solvent, charge, or ability to bind to a support. Although detailed discussions are beyond the scope of this book, we shall briefly describe some of the more widely used methods.

In **dialysis,** the polypeptide is separated from smaller fragments by filtration through a semipermeable membrane. A second method, **gel-filtration chromatography,** uses a carbohydrate polymer in the form of a column of beads as a support. Smaller molecules diffuse more easily into the beads, spending a longer time on the column than large ones do; thus they emerge from the column later than the large molecules. In **ion-exchange chromatography,** a charged support separates molecules according to the amount of charge they carry. Another method based on electric charge is **electrophoresis.** A spot of the mixture to be separated is placed on a piece of moistened absorbing paper that is attached to two electrodes. When the voltage is turned on, positively charged species (e.g., polypeptides rich in protonated amine groups) migrate toward the cathode, negatively charged species (carboxy-rich peptides) toward the anode. The separating power of this technique is extraordinary. More than a thousand different proteins from one species of bacterium have been resolved in a single experiment.

Finally, **affinity chromatography** exploits the tendency of polypeptides to bind very specifically to certain supports by hydrogen bonds and other attractive forces. Peptides of differing sizes and shapes have differing retention times in a column containing such a support.

Second, determine which amino acids are present

Once the individual polypeptide strands are purified, the next step in structural analysis is to establish their composition. To determine which amino acids and how much of each is present in the polypeptide, the entire chain is degraded by amide hydrolysis (6 N HCl, 110°C, 24 h) to give a mixture of the free amino acids. The mixture is then separated and its composition recorded by an automated **amino acid analyzer.**

This instrument consists of a column bearing a negatively charged support, usually containing carboxylate or sulfonate ions. The amino acids pass through the column in slightly acidic solution. Depending on their structure, they are protonated to a greater or lesser degree and therefore are more or less retained on the column. This differential retention separates the amino acids, and they come off the column in a specific order, beginning with the most acidic and ending with the most basic. At the end of the column is a reservoir containing a special indicator. Each amino acid produces a violet color whose intensity is proportional to the amount of that acid present and is recorded in a chromatogram (Figure 26-6). The area under each peak is a measure of the relative amount of a specific amino acid in the mixture.

The amino acid analyzer can readily establish the composition of a polypeptide. For example, the chromatogram of hydrolyzed glutathione (Section 26-4) gives three equal-sized peaks, corresponding to Glu, Gly, and Cys.

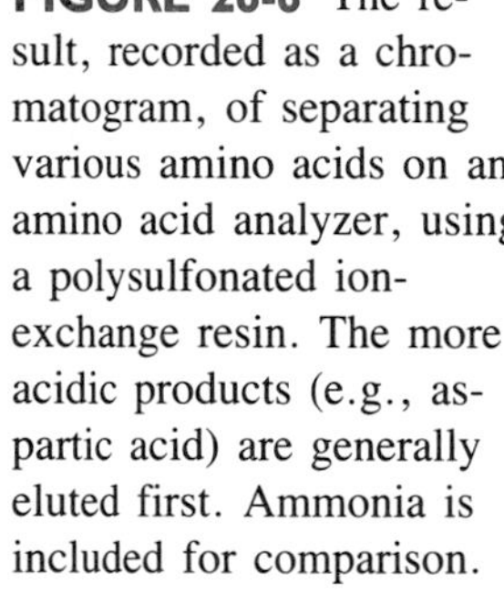

FIGURE 26-6 The result, recorded as a chromatogram, of separating various amino acids on an amino acid analyzer, using a polysulfonated ion-exchange resin. The more acidic products (e.g., aspartic acid) are generally eluted first. Ammonia is included for comparison.

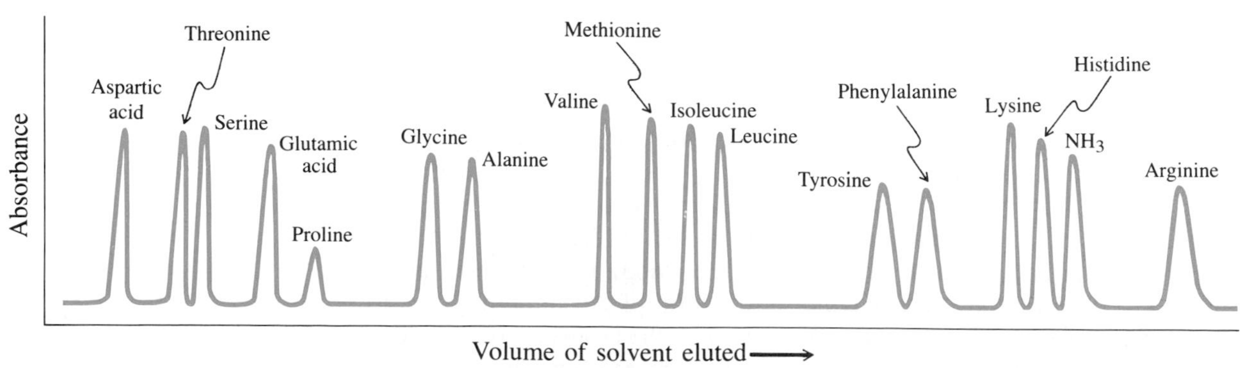

EXERCISE 26-8

Give the expected results of the amino acid analysis of the A chain in insulin (Figure 26-1).

Third, sequence the peptide from the amino (*N*-terminal) end

Once we know the gross makeup of a polypeptide, we must determine the order in which the individual amino acids are bound to each other—the amino acid sequence.

Several different methods can reveal the identity of the residue at the amino end. They all exploit the uniqueness of the free amino substituent, which may enter into specific chemical reactions that serve to "tag" the *N*-terminal amino acid. One such procedure is the **Sanger* degradation.** In this process, the peptide is first exposed to 1-fluoro-2,4-dinitrobenzene. The amino group of the *N*-terminal residue attacks this reagent, picking up a dinitrophenyl tag by nucleophilic aromatic substitution (Section 22-4). After complete hydrolysis of the polypeptide, the tagged amino acid is readily identified in the mixture by its chromatographic behavior.

Sanger Degradation of the A Chain of Insulin

O_2N–$C_6H_3(NO_2)$–F + $H_2\ddot{N}CH_2C(=O)$–NH–$CH(CH(CH_3)CH_2CH_3)$–C(=O)–NH–〰 the A chain of insulin

↓ $-$ HF

O_2N–$C_6H_3(NO_2)$–$NHCH_2C(=O)$–NH–$CH(CH(CH_3)CH_2CH_3)$–C(=O)–NH–〰 the A chain of insulin

↓ 6 N HCl, H_2O

O_2N–$C_6H_3(NO_2)$–$NHCH_2COOH$ + Ile, Ala, Val, Glu, Gln, Ser, Leu, Tyr, Asn, Cys

Tagged Gly

The major drawback is that the Sanger method completely degrades the polypeptide after it has been tagged; information regarding the rest of the amino acid sequence is therefore lost. It would be more useful if a reagent permitted the *selective removal* of the *N*-terminal residue and left intact the remainder of the chain, which could subsequently be exposed to this reagent again. In this way, it might be possible to identify each amino acid in sequence. Such a reagent is phenylisothiocyanate, $C_6H_5N{=}C{=}S$, used in the **Edman† degradation.**

---

*Professor Frederick Sanger, b. 1918, Cambridge University, England, Nobel Prize 1958 and 1980 (chemistry).

†Professor Pehr Edman, b. 1916, University of Lund, Sweden.

Recall (Section 20-7) that isocyanates are very reactive with respect to nucleophilic attack, and the same is true of their sulfur analogs. In the Edman degradation, the terminal amino group adds to the isothiocyanate reagent to give a thiourea derivative. Treatment with mild acid causes the molecule to extrude the tagged amino acid as a phenylthiohydantoin, leaving the remainder of the polypeptide unchanged. The phenylthiohydantoins of all amino acids are well known, so the *N*-terminal end of the original polypeptide can readily be identified. The new chain, carrying a new terminal amino acid, is now ready for another Edman degradation to tag the next residue, and so on. The entire procedure has been automated to allow the routine identification of polypeptides containing 20 or more amino acids.

Edman Degradation of the A Chain of Insulin

$C_6H_5\ddot{N}{=}C{=}S + H_2\ddot{N}CH_2C(O)-NH-CH(CH(CH_3)CH_2CH_3)-C(O)-NH-CH(CH(CH_3)CH_3)-C(O)-NH-$ ~~~ the A chain of insulin

$\downarrow H_2O$

Thiourea part

$C_6H_5NHC(S)NHCH_2C(O)-NH-CH(CH(CH_3)CH_2CH_3)-C(O)-NH-CH(CH(CH_3)CH_3)-C(O)-NH-$ ~~~ the A chain of insulin

$\downarrow H^+, H_2O$

Phenylthiohydantoin (S, NH, CH_2, O, N–C_6H_5) $+ H_2NCH(CH(CH_3)CH_2CH_3)-C(O)-NH-CH(CH(CH_3)CH_3)-C(O)-NH-$ ~~~ the A chain of insulin

Phenylthiohydantoin derived from glycine

$\downarrow$ 1. $C_6H_5N{=}C{=}S$ 2. H^+, H_2O

Phenylthiohydantoin (S, NH, $CHCHCH_3$ with CH_2CH_3, O, N–C_6H_5) $+ H_2NCH(CH(CH_3)CH_3)-C(O)-NH-$ ~~~ the A chain of insulin

Phenylthiohydantoin derived from isoleucine

$\downarrow$

etc.

EXERCISE 26-9

Write a mechanism for the formation of glycine phenylthiohydantoin from the reaction of phenylisothiocyanate with glycine amide, $H_2NCH_2CONH_2$.

Finally, cleave long chains at specific sites

The Edman procedure allows for the ready sequencing of only short polypeptides. For longer ones, the method becomes impractical because of the buildup of impurities. It is therefore necessary to cleave the larger chains into shorter fragments in a selective and predictable manner. These cleavage methods rely mostly on hydrolytic enzymes. For example, trypsin, a digestive enzyme of intestinal liquids, cleaves polypeptides only at the carboxy end of arginine and lysine.

Selective Hydrolysis of the B Chain of Insulin by Trypsin

Phe-Val-Asn-Gln-His-Leu-Cys-Gly-Ser-His-Leu-Val-Glu-Ala-Leu-Tyr-Leu-Val-Cys-Gly-Glu-Arg-Gly-Phe-Phe-Tyr-Thr-Pro-Lys-Ala
5 10 15 20 25 30

↓ Trypsin, H_2O

Phe-Val-Asn-Gln-His-Leu-Cys-Gly-Ser-His-Leu-Val-Glu-Ala-Leu-Tyr-Leu-Val-Cys-Gly-Glu-Arg + Gly-Phe-Phe-Tyr-Thr-Pro-Lys + Ala

A more selective enzyme is clostripain, which cleaves only at the carboxy end of arginine. In contrast, chymotrypsin, which like trypsin is found in mammalian intestines, is less selective and cleaves at the carboxy end of phenylalanine, tryptophan, and tyrosine. Other enzymes have similar selectivity (Table 26-2). In this way, a longer polypeptide is broken down into several shorter ones, which may then be sequenced by the Edman procedure.

EXERCISE 26-10

A polypeptide containing 21 amino acids was hydrolyzed by thermolysin. The products of this treatment were Gly, Ile, Val-Cys-Ser, Leu-Tyr-Gln, Val-Glu-Gln-Cys-Cys-Ala-Ser, and Leu-Glu-Asn-Tyr-Cys-Asn. When the same polypeptide was hydrolyzed by chymotrypsin, the products were Cys-Asn, Gln-Leu-Glu-Asn-Tyr, and Gly-Ile-Val-Glu-Gln-Cys-Cys-Ala-Ser-Val-Cys-Ser-Leu-Tyr. Give the amino acid sequence of this molecule.

In summary, the structure of polypeptides is established by various degradation schemes. First, disulfide bridges are broken, and then the kind and relative abundance of the component amino acids in each polypeptide are determined by complete hydrolysis and amino acid analysis. The *N*-terminal residues can be identified by Sanger or Edman degradation. Repeated Edman degradation gives the sequence of shorter polypeptides. Such shorter pieces are made from longer polypeptides by specific enzymatic hydrolysis.

TABLE 26-2 Specificity of Hydrolytic Enzymes in Polypeptide Cleavage

Enzyme	Site of cleavage
Trypsin	Lys, Arg, carboxy end
Clostripain	Arg, carboxy end
Chymotrypsin	Phe, Trp, Tyr, carboxy end
Pepsin	Asp, Glu, Leu, Phe, Trp, Tyr, carboxy end
Thermolysin	Leu, Ile, Val, amino end

26-6 Synthesis of Polypeptides: A Challenge in the Application of Protecting Groups

In a sense, the topic of peptide synthesis is a trivial one: Only one type of bond, the amide linkage, has to be made. The formation of this linkage was described in Section 19-10. Why discuss it further? This section shows that, in fact, achieving selectivity poses great problems, for which specific solutions have to be found.

Consider even as simple a target as the dipeptide glycylalanine. Just heating glycine and alanine to make the peptide bond by dehydration would result in a complex mixture of di-, tri-, and higher peptides with random sequences. Because the two starting materials can form bonds either to their own kind or to each other, there is no way to prevent random oligomerization.

An Attempt at the Synthesis of Glycylalanine by Thermal Dehydration

$$\text{Gly} + \text{Ala} \xrightarrow[-\,H_2O]{\Delta} \text{Gly-Gly} + \text{Ala-Gly} + \underbrace{\text{Gly-Ala}}_{\text{Desired product}} + \text{Ala-Ala} + \text{Gly-Gly-Ala} + \text{Ala-Gly-Ala} \text{ etc.}$$

Selective peptide synthesis requires protecting groups

To form peptide bonds selectively, the functional groups of the amino acids have to be protected. There are both amino- and carboxy-protecting groups.

The amino group is frequently blocked by a **phenylmethoxycarbonyl group** (abbreviated **carbobenzoxy** or **Cbz**), introduced by reaction of an amino acid with phenylmethyl chloroformate (benzyl chloroformate).

Protection of the Amino Group in Glycine

$$\underset{\textbf{Glycine}}{H_3\overset{+}{N}CH_2COO^-} + \underset{\substack{\textbf{Phenylmethyl chloroformate}\\\textbf{(Benzyl chloroformate)}}}{C_6H_5CH_2O\overset{O}{\overset{\|}{C}}Cl} \xrightarrow[\substack{-\,NaCl\\-\,HOH}]{NaOH} \underset{\substack{80\%\\\textbf{Phenylmethoxycarbonylglycine}\\\textbf{(Carbobenzoxyglycine, Cbz-Gly)}}}{C_6H_5CH_2O\overset{O}{\overset{\|}{C}}NHCH_2COOH}$$

The amino group is deprotected by hydrogenolysis (Section 22-2), which initially furnishes the carbamic acid as a reactive intermediate (Section 20-7). Decarboxylation occurs instantly to restore the amino function.

Deprotection of the Amino Group in Glycine

$$C_6H_5CH_2O\overset{O}{\overset{\|}{C}}NHCH_2COOH \xrightarrow{H_2,\ Pd\text{-}C} C_6H_5CH_2H + \underbrace{HO\overset{O}{\overset{\|}{C}}NH}_{\substack{\text{Carbamic acid}\\\text{function}}}CH_2COOH \longrightarrow CO_2 + \underset{95\%}{H_3\overset{+}{N}CH_2COO^-}$$

Another amino-protecting group is **1,1-dimethylethoxycarbonyl (*tert*-butoxycarbonyl, Boc),** introduced by reaction with bis(1,1-dimethylethyl) dicarbonate (di-*tert*-butyl dicarbonate).

Protection of the Amino Group in Amino Acids as the Boc Derivative

$$\overset{+}{H_3N}\overset{R}{\overset{|}{C}}HCOO^- + (CH_3)_3CO\overset{O}{\overset{\|}{C}}O\overset{O}{\overset{\|}{C}}OC(CH_3)_3 \xrightarrow[-\ CO_2,\ -\ (CH_3)_3COH]{(CH_3CH_2)_3N} (CH_3)_3CO\overset{O}{\overset{\|}{C}}NH\overset{R}{\overset{|}{C}}HCOOH \quad 70\text{–}100\%$$

Bis(1,1-dimethylethyl) dicarbonate (Di-*tert*-butyl dicarbonate)

1,1-Dimethylethoxy-carbonylamino acid (*tert*-Butoxycarbonylamino acid, Boc-amino acid)

Deprotection in this case is achieved by treatment with acid under conditions mild enough to leave other peptide bonds untouched.

Deprotection of Boc-Amino Acids

$$(CH_3)_3CO\overset{O}{\overset{\|}{C}}NH\overset{R}{\overset{|}{C}}HCOOH \xrightarrow{HCl\ \text{or}\ CF_3COOH,\ 25^{\circ}C} \overset{+}{H_3N}\overset{R}{\overset{|}{C}}HCOO^- + CO_2 + CH_2{=}C(CH_3)_2$$

EXERCISE 26-11

The mechanism of the deprotection of Boc-amino acids is different from that of the normal ester hydrolysis (Section 20-5): It proceeds through the intermediate 1,1-dimethylethyl (*tert*-butyl) cation. Formulate this mechanism.

The carboxy terminus of an amino acid is protected by the formation of a simple ester, such as methyl or ethyl. Deprotection results from treatment with base. Phenylmethyl (benzyl) esters can be cleaved by hydrogenolysis under neutral conditions (Section 22-2).

Peptide bonds are formed by using carboxy activation

With the ability to protect either end of the amino acid, we can achieve selective peptide synthesis by coupling an amino-protected unit with a carboxy-protected one. Because the protecting groups are sensitive to acid and base, the peptide bond must be formed under the mildest possible conditions. Special carboxy-activating reagents are used.

Perhaps the most general of these reagents is **dicyclohexylcarbodiimide (DCC).** The electrophilic reactivity of this molecule is similar to that of an isocyanate; it is ultimately hydrated to *N,N′*-dicyclohexylurea.

Peptide Bond Formation with Dicyclohexylcarbodiimide

$$RCOOH + R'NH_2 + C_6H_{11}N{=}C{=}NC_6H_{11} \longrightarrow R\overset{O}{\overset{\|}{C}}NHR' + O{=}C(NHC_6H_{11})_2$$

Dicyclohexyl-carbodiimide **N,N′-Dicyclohexylurea**

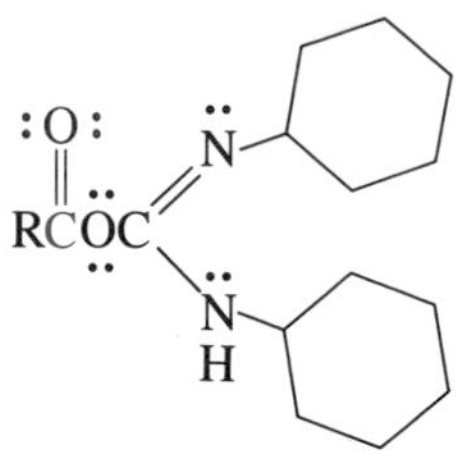

$$\begin{array}{c} :\ddot{O}: \\ \| \\ RC\ddot{O}C(=\ddot{N}\text{-}C_6H_{11})\ddot{N}H\text{-}C_6H_{11} \end{array}$$

An *O*-acyl isourea

The role of DCC is to activate the carbonyl group of the acid to nucleophilic attack by the amine. This activation arises from the formation of an ***O*-acyl isourea,** in which the carbonyl group possesses reactivity similar to that found in an anhydride (Section 20-3).

Armed with this knowledge, let us return to the problem of the synthesis of glycylalanine. It is solved by adding amino-protected glycine to an alanyl ester in the presence of DCC. The resulting product is then deprotected to give the desired dipeptide.

Preparation of Gly-Ala

$$\underset{\text{Boc-Gly}}{(CH_3)_3COCNHCH_2COOH} + \underset{\text{Ala-}OCH_2C_6H_5}{H_2NCH(CH_3)COCH_2C_6H_5} \xrightarrow{\text{DCC}} \underset{\text{Boc-Gly-Ala-}OCH_2C_6H_5}{(CH_3)_3COC(O)NHCH_2C(O)NHCH(CH_3)C(O)OCH_2C_6H_5} \xrightarrow[2.\ H_2,\ \text{Pd-C}]{1.\ H^+,\ H_2O}$$

$$\underset{\text{Gly-Ala}}{H_3\overset{+}{N}CH_2C(O)NHCH(CH_3)COO^-} + C_6H_5CH_3 + CO_2 + CH_2{=}C(CH_3)_2$$

For the preparation of a higher peptide, deprotection of only one end is required, followed by renewed coupling, and so on.

EXERCISE 26-12

Propose a synthesis of Leu-Ala-Val from the component amino acids.

In summary, polypeptides are made by coupling an amino-protected amino acid with another in which the carboxy end is protected. Typical protecting groups are readily cleaved esters and related functions. Coupling proceeds under mild conditions with dicyclohexylcarbodiimide as a dehydrating agent.

26-7 Merrifield Solid-Phase Peptide Synthesis

Polypeptide synthesis has been automated. This ingenious method, known as the **Merrifield* solid-phase peptide synthesis,** uses a solid support of polystyrene to anchor a peptide chain.

Polystyrene is a polymer (Section 12-14) whose subunits are derived from ethenylbenzene (styrene). Although beads of polystyrene are insoluble and rigid when dry, they *swell* considerably in certain organic solvents, such as dichloromethane. The swollen material allows reagents to move in and out of the polymer matrix easily. Thus, its phenyl groups may be functionalized by electrophilic aromatic substitution. For peptide synthesis, a form of Friedel-Crafts alkylation is used to chloromethylate a few percent of the phenyl rings in the polymer.

*Professor Robert Bruce Merrifield, b. 1921, Rockefeller University, New York, Nobel Prize 1984 (chemistry).

Electrophilic Chloromethylation of Polystyrene

$$\text{Polystyrene} \xrightarrow[-\,CH_3CH_2OH]{ClCH_2OCH_2CH_3,\ SnCl_4} \text{Functionalized polystyrene } (p\text{-}CH_2Cl)$$

Polystyrene

Functionalized polystyrene

EXERCISE 26-13

Formulate a plausible mechanism for the chloromethylation of the benzene rings in polystyrene.

A dipeptide synthesis on chloromethylated polystyrene proceeds as follows.

Solid-Phase Synthesis of a Dipeptide

$$(CH_3)_3COC(=O)NHCH(R)COO^- + ClCH_2\text{–C}_6\text{H}_4\text{–polystyrene chain}$$

1. Attachment of protected amino acid ($-\,Cl^-$)

$$(CH_3)_3COC(=O)NHCH(R)C(=O)OCH_2\text{–C}_6\text{H}_4\text{–polystyrene chain}$$

2. Deprotection of amino terminus (CF_3CO_2H, CH_2Cl_2)

$$H_2NCH(R)C(=O)OCH_2\text{–C}_6\text{H}_4\text{–polystyrene chain}$$

3. Coupling to protected second amino acid ($(CH_3)_3COC(=O)NHCH(R')COOH$, DCC)

$$(CH_3)_3COC(=O)NHCH(R')C(=O)NHCH(R)C(=O)OCH_2\text{–C}_6\text{H}_4\text{–polystyrene chain}$$

4. Deprotection of amino terminus (CF_3CO_2H, CH_2Cl_2)

$$H_2NCH(R')C(=O)NHCH(R)C(=O)OCH_2\text{–C}_6\text{H}_4\text{–polystyrene chain}$$

5. Disconnection of dipeptide from polymer (HF)

$$H_3\overset{+}{N}CH(R')C(=O)NHCH(R)COO^- + FCH_2\text{–C}_6\text{H}_4\text{–polystyrene chain}$$

Dipeptide

First an amino-protected amino acid is anchored on the polystyrene by nucleophilic substitution of the benzylic chloride by carboxylate. Deprotection is then followed by coupling with a second amino-protected amino acid. Renewed deprotection and final removal of the dipeptide by treatment with hydrogen fluoride complete the sequence.

The great advantage of solid-phase synthesis is the ease with which products may be isolated. Because all the intermediates are immobilized on the polymer, they can be purified by simple filtration and washing.

Obviously, it is not necessary to stop at the dipeptide stage. Repetition of the deprotection-coupling sequence leads to larger and larger peptides. Merrifield designed a machine that would carry out the required series of manipulations automatically, each cycle requiring only a few hours. In this way, the first total synthesis of the protein insulin was accomplished. More than 5000 separate operations were required to assemble the 51 amino acids in the two separate chains; thanks to the automated procedure, this took only several days.

Automated protein synthesis has opened up exciting possibilities. First, it is used to confirm the structure of polypeptides that have been analyzed by chain degradation and sequencing. Second, it can be used to construct unnatural proteins that might be more active and more specific than natural ones. Such proteins could be invaluable in the treatment of disease or in the understanding of biological function and activity.

In summary, solid-phase synthesis is an automated procedure in which a carboxy-anchored peptide chain is built up from amino-protected monomers by cycles of coupling and deprotection.

26-8 Polypeptides in Nature: Oxygen Transport by the Proteins Myoglobin and Hemoglobin

Two natural polypeptides function as oxygen carriers in vertebrates: the proteins myoglobin and hemoglobin. Myoglobin is active in the muscle, where it stores oxygen and releases it when needed. Hemoglobin is contained in red blood cells and facilitates oxygen transport. Without its presence, blood would be able to absorb only a fraction (about 2%) of the oxygen needed by the body.

How is the oxygen bound in these proteins? The secret of the oxygen-carrying ability of myoglobin and hemoglobin is a special nonpolypeptide unit, called a **heme group,** attached to the protein. Heme is a cyclic organic molecule (called a **porphyrin**) made of four linked, substituted pyrrole units surrounding an iron atom (Figure 26-7). The complex is red, giving blood its characteristic color.

The iron in the heme is attached to four nitrogens but can accommodate two additional groups above and below the plane of the porphyrin ring. In myoglobin, one of these groups is the imidazole ring of a histidine unit attached to one of the α-helical segments of the protein (Figure 26-8A). The other is most important for the protein's function—binding oxygen. Close to the oxygen-binding site is a second imidazole of a histidine unit, which protects this side of the heme by steric hindrance. For example, carbon monoxide, which also binds to the iron in the heme group, and thus blocks oxygen transport, is prevented from binding as strongly as it normally would because of the presence of the second imidazole group. Consequently, CO poisoning can be reversed by administering oxygen to

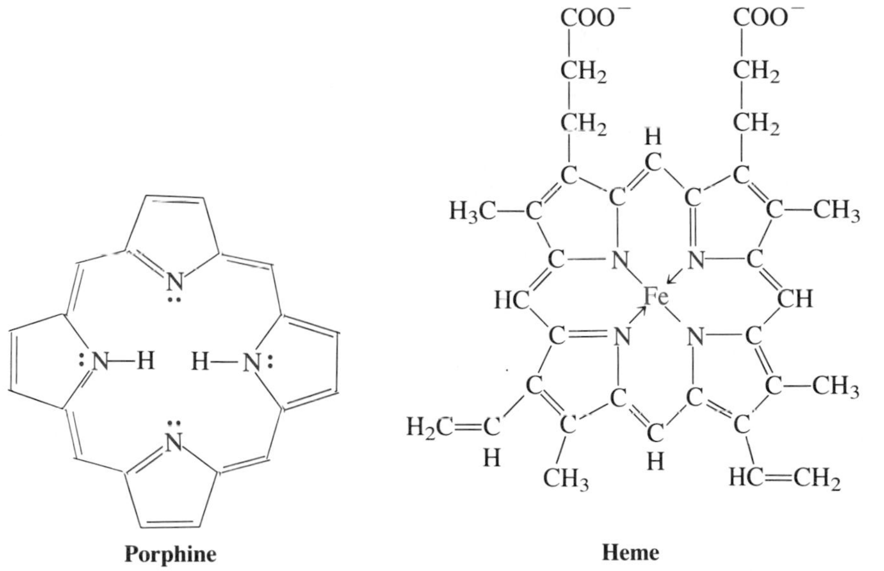

Porphine

Heme

FIGURE 26-7 Porphine is the simplest porphyrin. Note that the system forms an aromatic ring of 18 delocalized π electrons. A biologically important porphyrin is the heme group, responsible for binding oxygen. Two of the bonds to iron are dative (coordinate covalent), indicated by arrows.

a person who has been exposed to the gas. The two imidazole substituents in the neighborhood of the iron atom in the heme group are brought into close proximity by the unique folding pattern of the protein. The rest of the polypeptide chain serves as a mantle, shielding and protecting the active site from unwanted intruders and controlling the kinetics of its action (Figures 26-8B and C).

Myoglobin and hemoglobin offer excellent examples of the four structural levels in proteins. The primary structure of myoglobin consists of 153 amino acid residues of known sequence. Myoglobin has eight α-helical segments that consti-

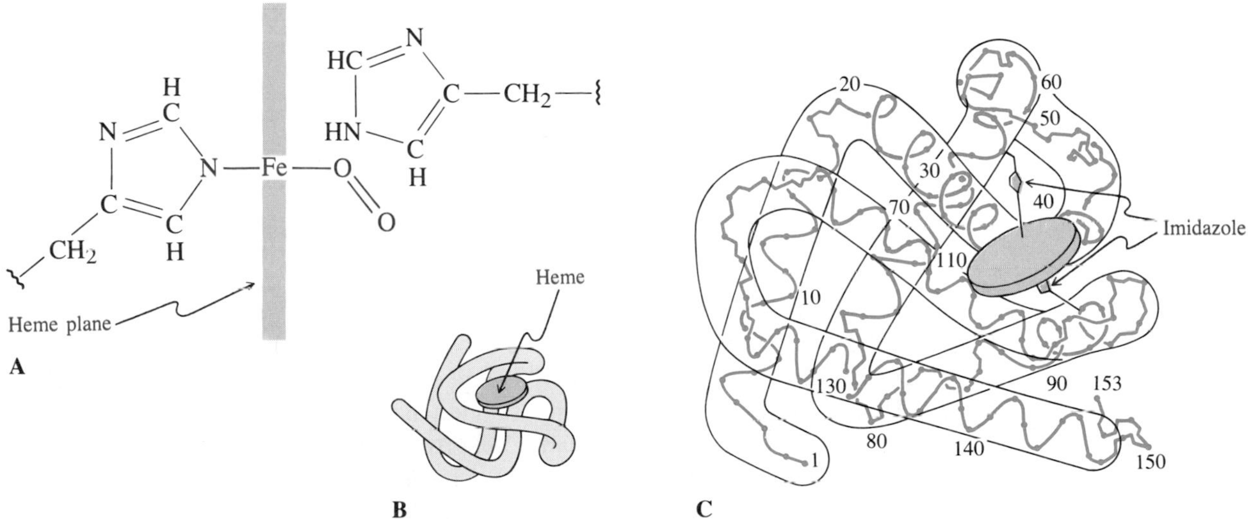

FIGURE 26-8 (A) Schematic representation of the active site in myoglobin, showing the iron atom in the heme plane bound to a molecule of oxygen and to the imidazole nitrogen atom of one histidine residue. (B) Schematic representation of the tertiary structures of myoglobin and its heme. (C) Secondary and tertiary structure of myoglobin. (After "The Hemoglobin Molecule," by M. F. Perutz, *Scientific American,* November 1964, Copyright © 1964, Scientific American, Inc.)

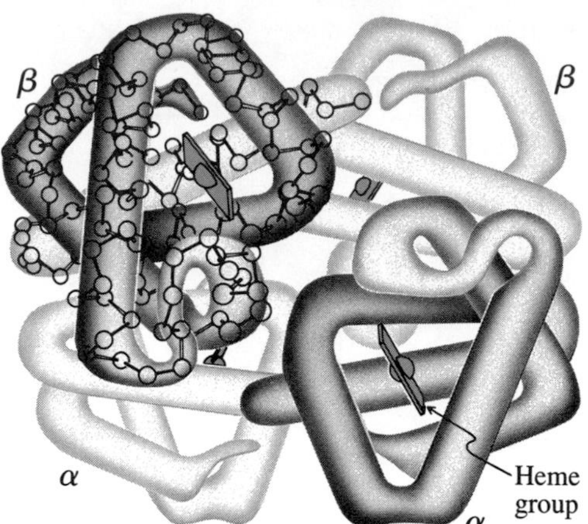

FIGURE 26-9 The quaternary structure of hemoglobin. Each α and β chain has its own heme group. (After R. E. Dickerson and I. Geis, 1969, *The Structure and Action of Proteins,* Benjamin-Cummings, p. 56, Copyright © 1969 by Irving Geis.)

tute its secondary structure, the longest having 23 residues. The tertiary structure has the bends that give myoglobin its three-dimensional shape.

Hemoglobin contains four protein chains: two α *chains* of 141 residues each, and two β *chains* of 146 residues each. Each chain has its own heme group and a tertiary structure similar to that of myoglobin. Although there is little interaction between the two α chains or the two β chains, there are many contacts between them. Furthermore, α_1 is closely attached to β_1, as is α_2 to β_2. These interactions give hemoglobin its quaternary structure (Figure 26-9).

The folding of the hemoglobin and myoglobin of several living species is strikingly similar even though the amino acid sequences differ. This finding implies that this particular tertiary structure is an optimal configuration around the heme group. The folding allows the heme to absorb oxygen as it is introduced through the lung, hang on to it as long as necessary for safe transport, and release it when required.

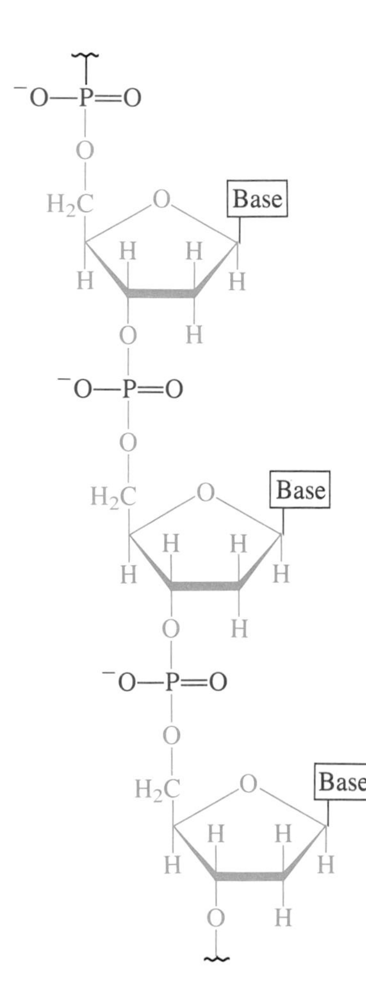

FIGURE 26-10 Part of a DNA chain. The base is a nitrogen heterocycle. The sugar is 2-deoxyribose.

26-9 Biosynthesis of Proteins: Nucleic Acids

How does nature assemble proteins? The answer to this question is based on one of the most exciting discoveries in science, the nature and workings of the genetic code. All hereditary information is embedded in the **deoxyribonucleic acids (DNA).** The expression of this information in the synthesis of the many enzymes necessary for cell function is carried out by the **ribonucleic acids (RNA).** After the carbohydrates and polypeptides, the nucleic acids are the third major type of biological polymer. This section describes their structure and function.

Four heterocycles define the structure of nucleic acids

Considering the structural diversity of natural products, the structures of DNA and RNA are simple. All their components, called **nucleotides,** are polyfunctional, and it is one of the wonders of nature that evolution has eliminated all but a few specific combinations. Nucleic acids are polymers in which phosphate units link sugars, which bear various heterocyclic nitrogen **bases** (Figure 26-10).

In DNA, the sugar units are 2-deoxyriboses, and only four bases are present: **cytosine (C), thymine (T), adenine (A),** and **guanine (G).** The sugar characteristic of RNA is ribose, and again there are four bases, but the nucleic acid incorporates **uracil (U)** instead of thymine.

Nucleic Acid Sugars and Bases

2-Deoxyribose **Ribose**

Cytosine (C) **Thymine (T)** **Adenine (A)** **Guanine (G)** **Uracil (U)**

We construct a nucleotide from three components. First, we replace the hydroxy group at C1 in the sugar with one of the base nitrogens. This combination is called a **nucleoside.** Second, a phosphate substituent is introduced at C5. In this way we obtain the four nucleotides of both DNA and RNA. The positions on the sugars in nucleosides and nucleotides are designated 1′, 2′, and so on, to distinguish them from the carbon atoms in the nitrogen heterocycles.

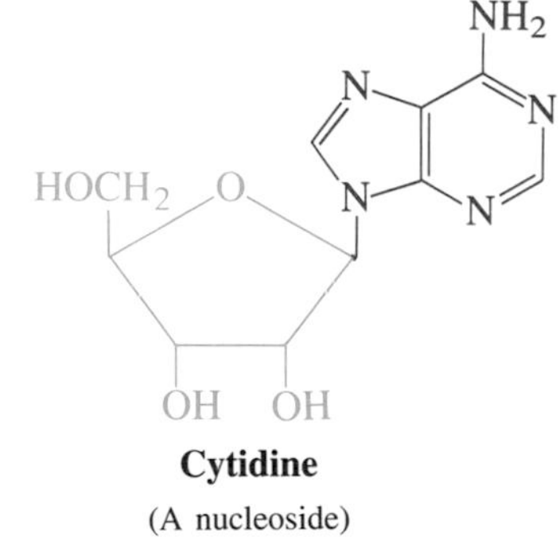

Cytidine

(A nucleoside)

Nucleotides of DNA

2′-Deoxyadenylic acid **2′-Deoxyguanidylic acid**

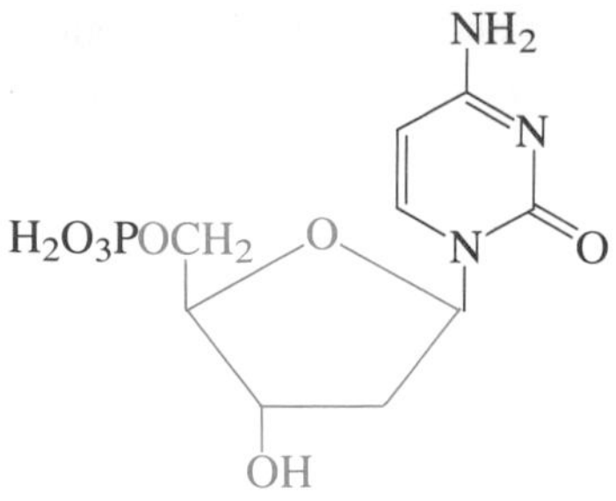

2′-Deoxycytidylic acid

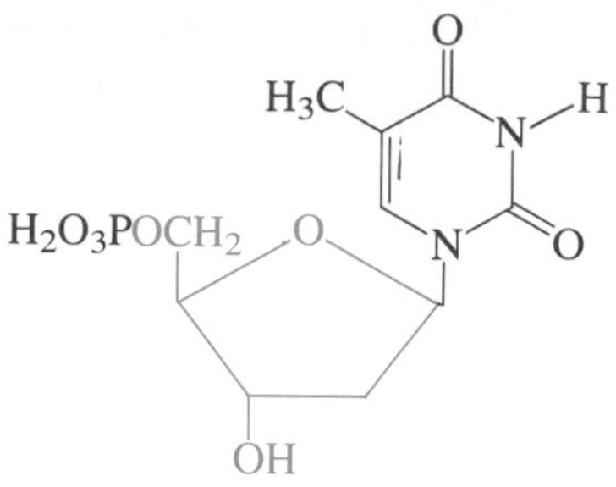

2′-Deoxythymidylic acid

Nucleotides of RNA

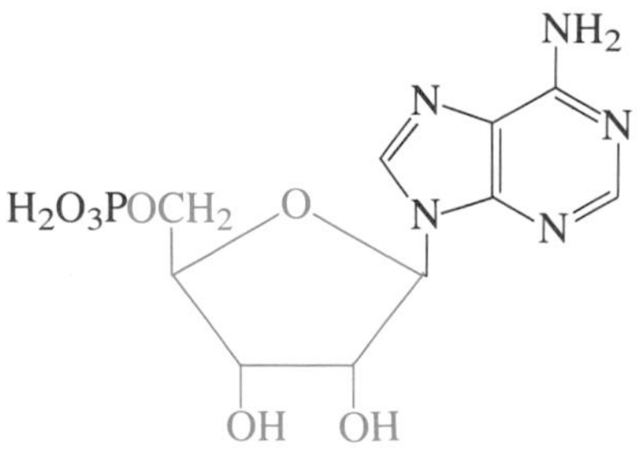

Adenylic acid

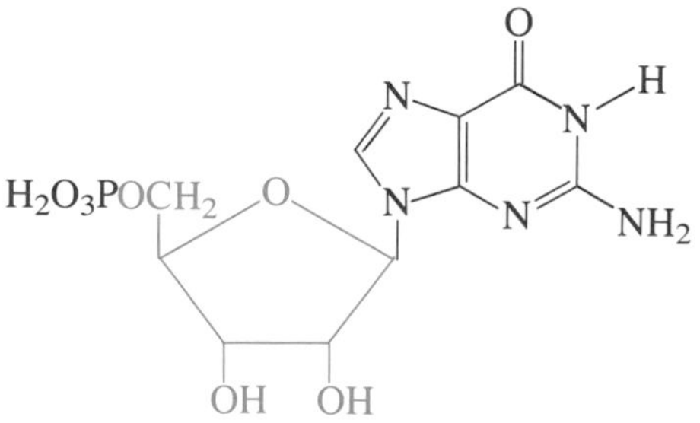

Guanidylic acid

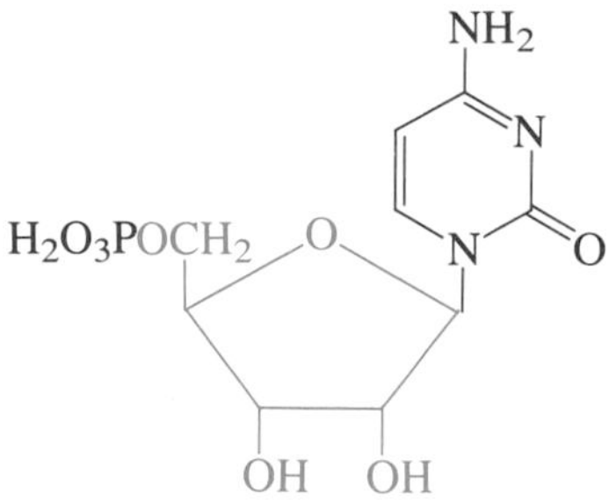

Cytidylic acid

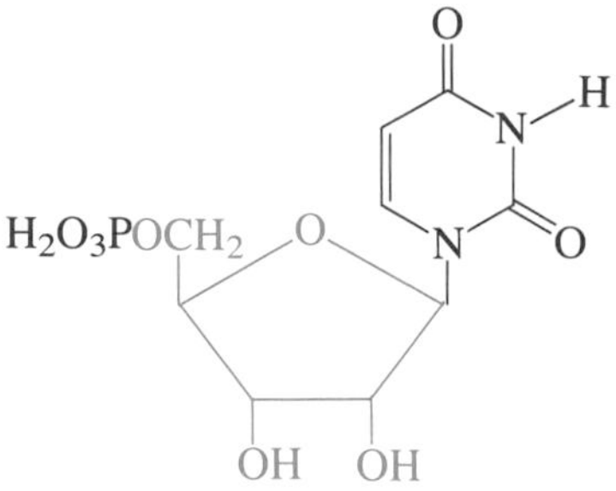

Uridylic acid

The polymeric chain shown in Figure 26-10 is then readily derived by repeatedly forming a phosphate ester bridge from C5′ (called the **5′ end**) of the sugar unit of one nucleotide to C3′ (the **3′ end**) of another.

In this polymer, the bases adopt the same role as that of the 2-substituent in the amino acids of a polypeptide: Their sequence varies from one nucleic acid to another and determines the fundamental biological properties of the system.

Nucleic acids form a double helix

Nucleic acids can form extraordinarily long chains with molecular weights in the billions. Like proteins, they adopt secondary and tertiary structures. In 1953, Watson and Crick* made their well-known proposal that DNA is a double helix composed of two strands with complementary base sequences. A crucial piece of

*Professor James D. Watson, b. 1928, Harvard University, Nobel Prize 1962 (medicine); Professor Francis H. F. C. Crick, b. 1916, Cambridge University, England, Nobel Prize 1962 (medicine).

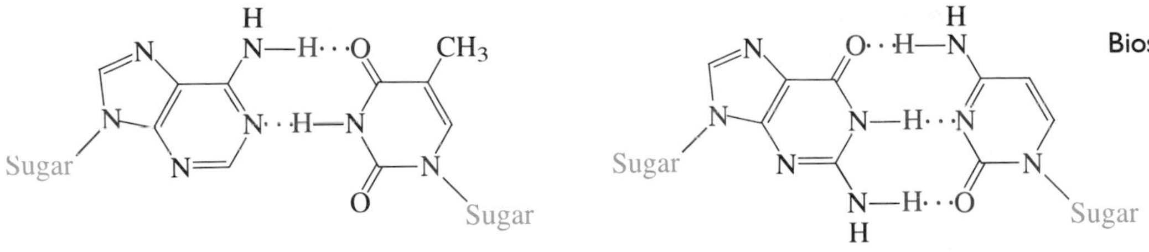

Adenine–thymine

Guanine–cytosine

FIGURE 26-11
Hydrogen-bonding between the base pairs adenine–thymine and guanine–cytosine. The components of each pair are always present in equal amounts.

information was that, in the DNA of various species, the ratio of adenine to thymine, like that of guanine to cytosine, is always one to one. Watson and Crick concluded that two chains are held together by hydrogen bonding in such a way that adenine and guanine in one chain always face thymine and cytosine in the other (Figure 26-11). Thus, if a piece of DNA in one strand has the sequence -A-G-C-T-A-C-G-A-T-C-, this entire segment is hydrogen bonded to a complementary strand -T-C-G-A-T-G-C-T-A-G-, as shown.

~~~—A—G—C—T—A—C—G—A—T—C—~~~
~~~—T—C—G—A—T—G—C—T—A—G—~~~

Because of other structural constraints, the arrangement that maximizes hydrogen bonding and minimizes steric repulsion is the double helix (Figure 26-12).

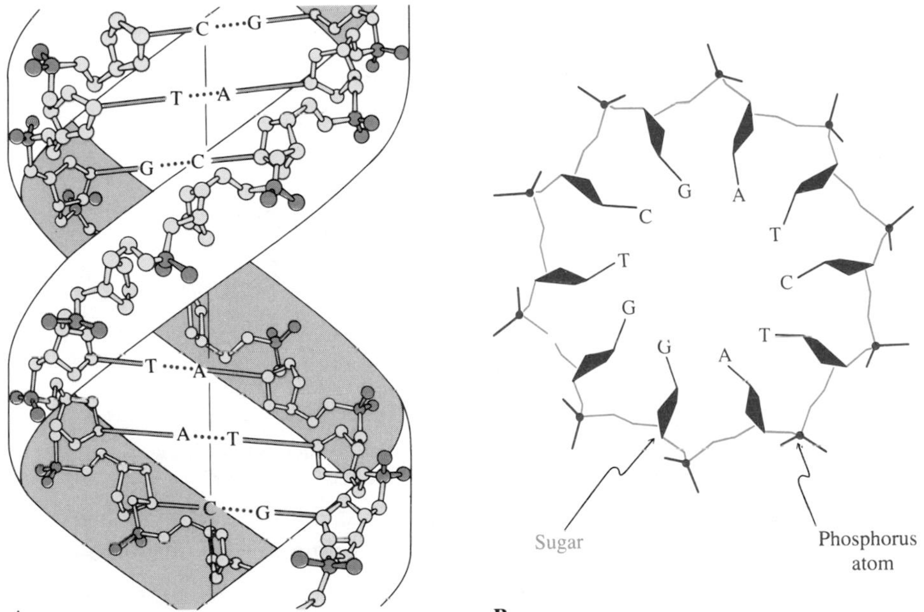

A

B

FIGURE 26-12 (A) The two nucleic acid strands of a DNA double helix are held together by hydrogen bonding between the complementary sets of bases. Note: The two chains run in opposite directions and all the bases are on the inside of the double helix. The diameter of the helix is 20 Å; base–base separation across the strands is ~3.4 Å; the helical turn repeats every 34 Å. (B) One of the strands of a DNA double helix, in a view down the axis of the molecule. (After *Biochemistry,* 3d Ed., by Lubert Stryer, W. H. Freeman and Company, Copyright © 1975, 1981, 1988.)

DNA replicates by unwinding and assembling new complementary strands

There is no restriction on the variety of sequences of the bases in the nucleic acids. Watson and Crick proposed that the specific base sequence of a particular DNA contained all genetic information necessary for the duplication of a cell and, indeed, the growth and development of the organism as a whole. Moreover, the double-helical structure suggested a way in which DNA might **replicate**—make exact copies of itself—and so pass on the genetic code. In this mechanism, each of the two strands of DNA functions as a template. The double helix partly unwinds, and enzymes then begin to assemble the new DNA by coupling nucleotides to each other in a sequence complementary to that in the template, always juxtaposing C to G and A to T (Figure 26-13). Eventually, two complete double helices are produced from the original. This process is at work throughout the entire human genetic material or **genome**—some 3 billion base pairs!

In summary, the nucleic acids DNA and RNA are polymers containing monomeric units called nucleotides. There are four nucleotides for each, varying only in the structure of the base: cytosine (C), thymine (T), adenine (A), and guanine (G) for DNA; cytosine, uracil (U), adenine, and guanine for RNA. The two nucleic acids differ also in the identity of the sugar unit: deoxyribose for DNA, ribose for RNA. DNA replication and RNA synthesis from DNA is facilitated by the complementary character of the base pairs A–T, G–C, and A–U. The double helix partly unwinds and functions as a template for replication.

BOX 26-4 Synthetic Nucleic Acid Bases and Nucleosides in Medicine

5-Fluorouracil (Fluracil)

9-[(2-Hydroxyethoxy)methyl]guanosine (Acyclovir)

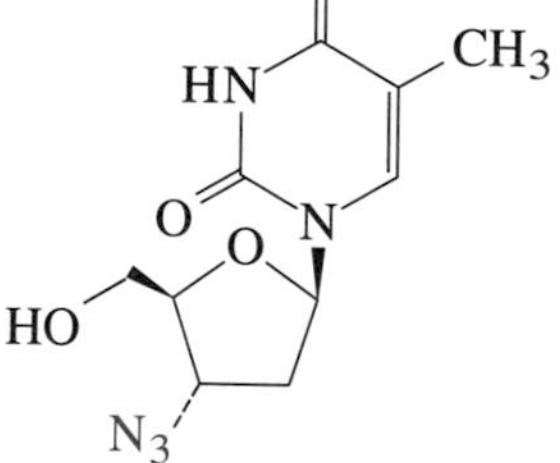

3′-Azido-3′-deoxythymidine (Zidovudine or AZT)

The central role played by nucleic acid replication in biology has been exploited in medicine. Many hundreds of synthetically modified bases and nucleosides have been prepared and their effects on nucleic acid synthesis investigated. Some of these in clinical use include 5-fluorouracil *(fluracil),* an anticancer agent, 9-[(2-hydroxyethoxy)methyl]guanosine *(acyclovir),* which is active against two strains of herpes simplex virus, and 3′-azido-3′-deoxythymidine *(zidovudine* or *AZT),* a drug that combats the AIDS virus.

Substances such as these may interfere with nucleic acid replication by masquerading as legitimate nucleic acid building blocks. The enzymes associated with this process are fooled into incorporating the drug molecule, and synthesis of the biological polymer cannot continue.

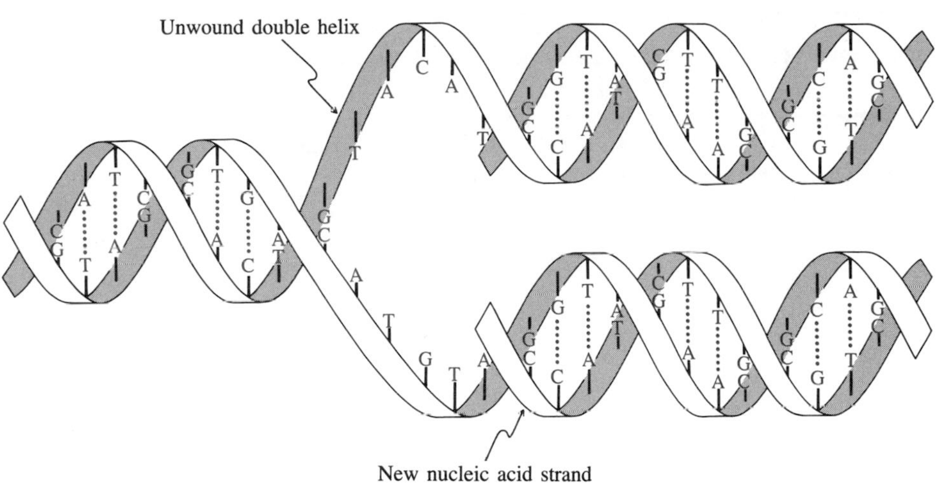

FIGURE 26-13 Model for DNA replication. The double helix initially unwinds to two single strands, each of which is used as a template for reconstruction of the complementary nucleic acid sequence.

26-10 Protein Synthesis Through RNA

DNA gives rise to RNA by a replication process that is very similar to that described for DNA. In RNA, however, ribose takes the place of deoxyribose as the repeating sugar unit, and uracil is incorporated instead of thymine. The resulting nucleic acid is called **messenger RNA (mRNA;** Figure 26-14). Its chain is much shorter than that of DNA, and it does not stay bound to the template but breaks away as its synthesis is finished.

The mRNA is the template responsible for the correct sequencing of the amino acid units in proteins. Each sequence of three bases, called a **codon,** specifies a particular amino acid (Table 26-3). Simple permutation of this three-base code with a total of four bases gives $4^3 = 64$ possible distinct sequences. That number is more than enough, because only 20 different amino acids are needed for protein synthesis. This might seem like overkill, but consider that the next lower alternative—namely, a two-base code—would give only $4^2 = 16$ combinations, too few for the number of different amino acids found in natural proteins.

Codons do not overlap; in other words, the three bases specifying one amino acid are not part of another preceding or succeeding codon. Moreover, the "reading" of the base sequence is consecutive; each codon immediately follows the next, uninterrupted by genetic "commas" or "hyphens." Nature also makes full use of all 64 codons, by allowing for several of them to describe the same amino acid (Table 26-3). Only tryptophan and methionine are characterized by single three-base codes. Some codons act as signals to initiate or terminate production of a polypeptide chain. Note that the initiator codon (AUG) is also the codon for methionine. Thus, if the codon AUG appears *after* a chain has been initiated,

```
~~~~—A—G—C—T—A—C—~~~~~~—A—C—~~~~
     :  :  :  :  :  :         :  :
     U—C—G—A—U—G—~~~~~~—T—G
              mRNA
```

FIGURE 26-14 Simplified picture of messenger RNA synthesis.

TABLE 26-3 Three-Base Code for the Common Amino Acids Used in Protein Synthesis

Amino acid	Base sequence	Amino acid	Base sequence	Amino acid	Base sequence
Ala	GCA	His	CAC	Ser	AGC
	GCC		CAU		AGU
	GCG	Ile	AUA		UCA
	GCU		AUC		UCG
Arg	AGA		AUU		UCC
	AGG	Leu	CUA		UCU
	CGA		CUC	Thr	ACA
	CGC		CUG		ACC
	CGG		CUU		ACG
	CGU		UUA		ACU
Asn	AAC		UUG	Trp	UGG
	AAU	Lys	AAA	Tyr	UAC
Asp	GAC		AAG		UAU
	GAU	Met	AUG	Val	GUA
Cys	UGC	Phe	UUU		GUG
	UGU		UUC		GUC
Gln	CAA	Pro	CCA		GUU
	CAG		CCC	Chain initiation	AUG
Glu	GAA		CCG	Chain termination	UGA
	GAG		CCU		UAA
Gly	GGA				UAG
	GGC				
	GGG				
	GGU				

methionine will be produced. The complete base sequence of the DNA in a cell defines its **genetic code.**

Mutations in the base sequence of DNA can be caused by physical (radiation) or chemical (carcinogens; see, e.g., Section 16-7) interference. Mutations can either replace one base with another or add or delete one base or more. Here is some of the potential value of redundant codons. If, for example, the RNA sequence CCG (proline) were changed as a result of a DNA mutation to the RNA sequence CCC, proline would still be correctly synthesized.

EXERCISE 26-14

Given the following RNA sequence, what amino acid sequence would be produced? Remember that the chain must have both an initiating and a terminating codon. What would happen if the first U in the sequence were eliminated by radiation?

G-G-A-U-G-A-A-G-U-A-U-G-C-A-U-C-A-U-G-C-U-U-A-A-G-C-U-A-G-C-A-A-U

Proteins are synthesized along the mRNA template with the help of a set of other important nucleic acids, called **transfer ribonucleic acids (tRNAs).** These are molecules of relatively low molecular weight, containing from 70 to about 90 nucleotides. Each tRNA is specifically designed to carry one of the 20 amino acids to the mRNA in the course of protein buildup. As the polypeptide chain grows longer, it begins to develop its characteristic secondary and tertiary structure (α helix, pleated sheets, etc.), helped by enzymes that form the necessary disulfide bridges. All of this happens with remarkable speed. It is estimated that a protein made up of approximately 150 amino acid residues can be biosynthesized in less than one minute. Clearly, nature still has the edge over the synthetic organic chemist, at least in this domain.

In summary, RNA is responsible for protein biosynthesis; each three-base sequence, or codon, specifies a particular amino acid. Codons do not overlap, and more than one can specify the same amino acid.

NEW REACTIONS

1. ACIDITY OF AMINO ACIDS (Section 26-1)

$$\overset{+}{H_3N}\underset{}{\overset{R}{\overset{|}{C}}}HCOOH \quad pK_a \sim 2\text{–}3 \qquad \overset{+}{H_3N}\overset{R}{\overset{|}{C}}HCOO^- \quad pK_a \sim 9\text{–}10$$

$$\text{Isoelectric point } pI = \frac{pK_{COOH} + pK_{\overset{+}{N}H_3}}{2}$$

2. STRONGLY BASIC GUANIDINO GROUP IN ARGININE (Section 26-1)

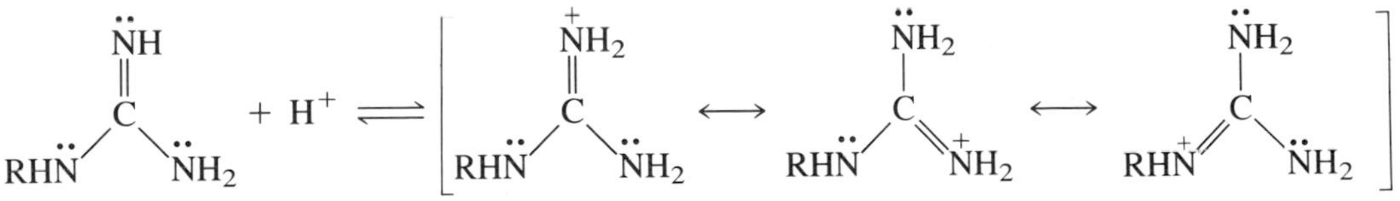

$$RHN\text{–}C(=NH)\text{–}NH_2 + H^+ \rightleftharpoons \left[RHN\text{–}C(=\overset{+}{N}H_2)\text{–}NH_2 \longleftrightarrow RHN\text{–}C(NH_2)=\overset{+}{N}H_2 \longleftrightarrow R\overset{+}{H}N=C(NH_2)\text{–}NH_2 \right]$$

$pK_a \sim 13$

3. BASICITY OF IMIDAZOLE IN HISTIDINE (Section 26-1)

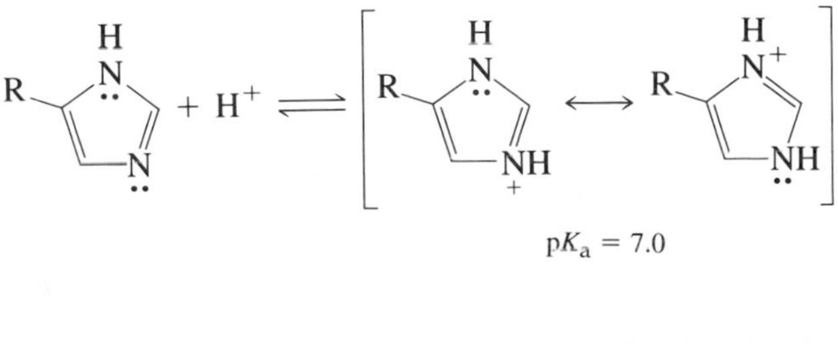

$$\text{R-imidazole} + H^+ \rightleftharpoons \left[\text{R-imidazolium (}\overset{+}{N}H\text{)} \longleftrightarrow \text{R-imidazolium (}\overset{+}{N}\text{–}H\text{)} \right]$$

$pK_a = 7.0$

Preparation of Amino Acids

4. HELL-VOLHARD-ZELINSKY BROMINATION FOLLOWED BY AMINATION (Section 26-2)

$$\mathrm{RCH_2COOH} \xrightarrow[\text{2. } NH_3,\ H_2O]{\text{1. } Br_2,\ \text{catalytic } PBr_3} \mathrm{R\underset{|}{C}HCOO^-}\ (\text{with } \overset{+}{N}H_3 \text{ on the } \alpha\text{-carbon})$$

5. GABRIEL SYNTHESIS (Section 26-2)

$$\text{Phthalimide } N^-K^+ + \mathrm{BrCH(CO_2CH_3)_2} \xrightarrow[-\ KBr]{} \text{Phthalimide-}\mathrm{NCH(CO_2CH_3)_2} \xrightarrow{\text{1. } RO^-Na^+;\ \text{2. } R'X;\ \text{3. } H^+,\ H_2O,\ \Delta} \mathrm{R'\overset{\overset{+}{N}H_3}{C}HCOO^-}$$

6. STRECKER SYNTHESIS (Section 26-2)

$$\mathrm{R\overset{O}{\overset{\|}{C}}H} \xrightarrow{HCN,\ NH_3} \mathrm{R\overset{NH_2}{C}HCN} \xrightarrow{H^+,\ H_2O,\ \Delta} \mathrm{R\overset{\overset{+}{N}H_3}{C}HCOO^-}$$

Polypeptide Sequencing

7. DISULFIDE CLEAVAGE (Section 26-5)

$$\sim S\text{—}S \sim \xrightarrow{HCO_3H} \sim SO_3H \quad SO_3H \sim$$

8. HYDROLYSIS (Section 26-5)

$$\text{Peptide} \xrightarrow{6\ N\ HCl,\ 110°C,\ 24\ h} \text{amino acids}$$

9. SANGER DEGRADATION (Section 26-5)

$$\text{2,4-}(O_2N)_2C_6H_3\text{—F} + \mathrm{H_2N\overset{R}{C}H\overset{O}{\overset{\|}{C}}\text{—}\underset{H}{N}\sim} \xrightarrow[-\ HF]{} \xrightarrow{6\ N\ HCl,\ \Delta} \text{2,4-}(O_2N)_2C_6H_3\text{—}\mathrm{NH\overset{R}{C}HCO_2H} + \text{amino acids}$$

Amino terminus of peptide

10. EDMAN DEGRADATION (Section 26-5)

$$C_6H_5\text{—N=C=S} + \mathrm{H_2N\overset{R}{C}H\overset{O}{\overset{\|}{C}}\text{—}\underset{H}{N}\sim} \xrightarrow{H_2O} \xrightarrow{H^+,\ H_2O} \text{Phenylthiohydantoin} + \mathrm{H_2N}\sim$$

Phenylthiohydantoin — Lower polypeptide

Preparation of Polypeptides

11. PROTECTING GROUPS (Section 26-6)

$$\overset{+}{N}H_3\text{—}RCHCOO^- + C_6H_5CH_2OCCl\ (\text{C=O}) \xrightarrow[-\ NaCl]{NaOH} C_6H_5CH_2OC(=O)NHCH(R)COOH \xrightarrow[\text{Deprotection};\ -\ C_6H_5CH_3;\ -\ CO_2]{H_2,\ Pd\text{-}C} RCH(\overset{+}{N}H_3)COO^-$$

Phenylmethyl chloroformate (Benzyl chloroformate) — **Cbz-protected amino acid**

$$RCH(\overset{+}{N}H_3)COO^- + (CH_3)_3COC(=O)OC(=O)OC(CH_3)_3 \xrightarrow{(CH_3CH_2)_3N} (CH_3)_3COC(=O)NHCH(R)COOH \xrightarrow[\text{Deprotection};\ -\ CO_2;\ -\ CH_2{=}C(CH_3)_2]{H^+,\ H_2O} RCH(\overset{+}{N}H_3)COO^-$$

Bis(1,1-Dimethylethyl) dicarbonate (Di-*tert*-butyl dicarbonate) — **Boc-protected amino acid**

12. PEPTIDE-BOND FORMATION WITH DICYCLOHEXYLCARBODIIMIDE (Section 26-6)

$$\text{Cbz-Gly} + \text{Ala-OCH}_2C_6H_5 + \underset{\textbf{DCC}}{C_6H_{11}N{=}C{=}NC_6H_{11}} \longrightarrow$$

$$\text{Cbz-Gly- Ala-OCH}_2C_6H_5 + C_6H_{11}NHC(=O)NHC_6H_{11}$$

13. MERRIFIELD SOLID-PHASE SYNTHESIS (Section 26-7)

$$\text{(P)} \xrightarrow[-\ CH_3CH_2OH]{ClCH_2OCH_2CH_3,\ SnCl_4} \text{(P)}\text{—}CH_2Cl \xrightarrow[2.\ H^+,\ H_2O]{1.\ (CH_3)_3COC(=O)NHCH(R)COO^-} \text{(P)}\text{—}CH_2OC(=O)\text{—}CH(R)NH_2 \xrightarrow[2.\ H^+,\ H_2O]{1.\ (CH_3)_3COC(=O)NHCH(R')COOH,\ DCC}$$

(P) = polystyrene

$$\text{(P)}\text{—}CH_2OC(=O)\text{—}CH(R)NHC(=O)\text{—}CH(R')NH_2 \xrightarrow{HF} \text{(P)}\text{—}CH_2F + H_3\overset{+}{N}CH(R')C(=O)NHCH(R)COO^-$$

IMPORTANT CONCEPTS

1. Polypeptides are poly(amino acids) linked by amide bonds. Most natural polypeptides are made from only 19 different (2*S*)-amino acids and glycine, all of which have common names and three-letter abbreviations.
2. Amino acids are amphoteric; they can be protonated and deprotonated.
3. Besides fractional crystallization of diastereomers, one way of obtaining enantiomerically pure amino acids is by resolution of a racemic mixture by enzymes.
4. The structures of polypeptides are varied; they can be linear, cyclic, disulfide bridged, pleated sheet, α helical or superhelical, or disordered, depending on size, composition, hydrogen bonding, and electrostatic and London forces.
5. Amino acids are separated mainly by virtue of their pH-dependent differences in ability to bind to solid supports.
6. Polypeptide sequencing entails a combination of selective chain cleavage and amino acid analysis of the resulting shorter polypeptide fragments.
7. Polypeptide synthesis requires end-protected amino acids that are coupled by dicyclohexylcarbodiimide. The product can be selectively deprotected at either end to allow for further ex-

tension of the chain. The use of solid supports, as in the Merrifield synthesis, can be automated.

8. The proteins myoglobin and hemoglobin are polypeptides in which the amino acid chain envelops the active site, heme. The heme contains an iron atom that reversibly binds oxygen, allowing for oxygen uptake, transport, and delivery.

9. The nucleic acids are biological polymers made of phosphate-linked base-bearing sugars. Only four different bases and one sugar are used for DNA and RNA. Because the base pairs adenine–thymine, guanine–cytosine, and adenine–uracil pair up by particularly favorable hydrogen bonding, a nucleic acid can adopt a dimeric helical structure containing complementary base sequences. In DNA, this arrangement unwinds and functions as a template during DNA replication and RNA synthesis. In protein synthesis, each amino acid is specified by a set of three consecutive RNA bases, called a codon. Thus, the base sequence (genetic code) in a strand of RNA translates into a specific amino acid sequence in a protein.

PROBLEMS

1. Draw stereochemically correct structural formulas for isoleucine and threonine (Table 26-1). What is a systematic name for threonine?

2. The abbreviation *allo* means *diastereomer* in amino acid terms. Draw allo-L-isoleucine and give it a systematic name.

3. Draw the structure that each of the following amino acids would have in aqueous solution at the indicated pH values. Calculate the isoelectric point for each amino acid. **(a)** Alanine at pH = 1, 7, and 12; **(b)** serine at pH = 1, 7, and 12; **(c)** lysine at pH = 1, 7, 9.5, and 12; **(d)** histidine at pH = 1, 5, 7, and 12; **(e)** cysteine at pH = 1, 7, 9, and 12; **(f)** aspartic acid at pH = 1, 3, 7, and 12; **(g)** arginine at pH = 1, 7, 12, and 14.

4. Group the amino acids in Problem 3 according to whether they are **(a)** positively charged, **(b)** neutral, or **(c)** negatively charged at pH = 7.

5. Using either one of the methods in Section 26-2 or a route of your own devising, propose a reasonable synthesis of each of the following amino acids in racemic form. **(a)** Val; **(b)** Leu; **(c)** Pro; **(d)** Thr; **(e)** Lys.

6. **(a)** Illustrate the Strecker synthesis of phenylalanine. Is the product chiral? Does it exhibit optical activity? **(b)** It has been found that replacement of NH_3 by an optically active amine in the Strecker synthesis of phenylalanine leads to an excess of one enantiomer of the product. Assign the *R* or *S* configuration to each stereocenter in the following structures, and explain why the use of a chiral amine causes preferential formation of one stereoisomer of the final product.

$C_6H_5CH_2CHO$ —[$C_6H_5CH(CH_3)NH_2$, HCN; $-H_2O$]→ $C_6H_5CH_2$–CH(CN)–NH–CH(CH_3)C_6H_5 (Mainly) —[1. H^+, H_2O; 2. H_2, Pd-C]→ $C_6H_5CH_2$–CH($\overset{+}{N}H_3$)COO^- (Mainly)

7. The antibacterial agent in garlic, allicin (recall Problem 36 of Chapter 9), is synthesized from the unusual amino acid alliin by the action of the enzyme allinase. Because allinase is an extracellular enzyme, this process takes place only when garlic cells are crushed. Propose a reasonable synthesis for the amino acid alliin. (Hint: Begin by designing a synthesis of an amino acid from Table 26-1 that is structurally related to alliin.)

$$H_2C{=}CHCH_2\overset{\overset{\displaystyle O}{\|}}{S}CH_2\overset{\overset{\displaystyle \overset{+}{N}H_3}{|}}{C}HCOO^-$$

Alliin

8. Devise a procedure for separating a mixture of the four stereoisomers of isoleucine into its four components: (+)-isoleucine, (−)-isoleucine, (+)-alloisoleucine, and (−)-alloisoleucine (Problem 2). (Note: Alloisoleucine is much more soluble in 80% ethanol at all temperatures than is isoleucine.)

9. Identify each of the following structures as a dipeptide, tripeptide, and so on, and point out all the peptide bonds.

(a) $H_3\overset{+}{N}-CH[(CH_3)_2CH]-C(=O)-NH-CH[CH_3]-C(=O)-NH-CH[HSCH_2]-COO^-$

(b) $H_3\overset{+}{N}-CH[HOCH_2]-C(=O)-NH-CH[CH_2COO^-]-COO^-$

(c) $H_3\overset{+}{N}-CH[CH_2\text{-imidazolyl}]-C(=O)-NH-CH[CH_3CH(OH)]-C(=O)-N\langle CH_2CH_2CH_2 \text{ ring}\rangle CH-C(=O)-NH-CH[CH_2(CH_2)_3\overset{+}{N}H_3]-COO^-$

(d) $H_3\overset{+}{N}-CH[CH_2C_6H_4OH]-C(=O)-NH-CH_2-C(=O)-NH-CH_2-C(=O)-NH-CH[CH_2C_6H_5]-C(=O)-NH-CH[CH_2CH(CH_3)_2]-COO^-$

10. Using the standard three-letter abbreviations for amino acids, write the peptide structures in Problem 9 in short notation.

11. Indicate which of the amino acids in Problem 3 and the peptides in Problem 9 would migrate in an electrophoresis apparatus at pH = 7 **(a)** toward the anode or **(b)** toward the cathode.

12. Silk consists of β sheets whose polypeptide chains consist of the repeating sequence Gly-Ser-Gly-Ala-Gly-Ala. What characteristics of amino acid side chains appear to favor the β-sheet configuration? Do the illustrations of β-sheet structures (Figure 26-3) suggest an explanation for this preference?

13. Identify as many stretches of α helix as you can in the structure of myoglobin (Figure 26-8C). Prolines are located in myoglobin at positions 37, 88, 100, and 120. How does each of these prolines affect the tertiary structure of the molecule?

14. Of the 153 amino acids in myoglobin, 78 contain polar side chains (i.e., Arg, Asn, Asp, Gln, Glu, His, Lys, Ser, Thr, Trp, and Tyr). When myoglobin adopts its natural folded conformation, 76 of these 78 polar side chains (all but those of two histidines) project outward from its surface. Meanwhile, in addition to the two histidines, the interior of myoglobin contains only Gly, Val, Leu, Ala, Ile, Phe, Pro, and Met. Explain.

15. Explain the following three observations. **(a)** Silk, like most polypeptides with sheet structures, is water insoluble. **(b)** Globular proteins like myoglobin generally dissolve readily in water. **(c)** Disruption of the tertiary structure of a globular protein (denaturation) leads to precipitation from aqueous solution.

16. In your own words, outline the procedure that might have been followed by the researchers who determined which amino acids were present in vasopressin (Exercise 26-7).

17. Write the products of Sanger degradation of the peptides in Problem 9.

18. What would be the outcome of reaction of gramicidin S with 1-fluoro-2,4-dinitrobenzene (Sanger's reagent)? With phenyl isothiocyanate (Edman degradation)?

19. The polypeptide bradykinin is a tissue hormone that can function as a potent pain-producing agent. Treatment of bradykinin with 1-fluoro-2,4-dinitrobenzene followed by complete acid hydrolysis produces *N*-(2,4-dinitrophenyl)arginine together with free Arg, Gly, Phe, Pro, and Ser. Incomplete acid hydrolysis causes random cleavage of many bradykinin molecules into an assortment of peptide fragments that includes Arg-Pro-Pro-Gly, Phe-Arg, Ser-Pro-Phe, and Gly-Phe-Ser. Complete hydrolysis followed by amino acid analysis indicates a ratio of 3 Pro, 2 Phe, 2 Arg, and one each of Gly and Ser. Deduce the amino acid sequence of bradykinin.

20. Somatostatin is a polypeptide hormone with several functions, including regulation of the secretion of insulin by the pancreas. It is useful in the treatment of certain kinds of diabetes. Somatostatin contains one disulfide linkage. After its cleavage by HCO_3H, just a single polypeptide chain is present, which has a molecular weight 98 mass units higher than somatostatin itself. **(a)** What does this tell you?

Treatment of this polypeptide chain with trypsin yields three peptides: Ala-Gly-Cys(SO_3H)-Lys, Thr-Phe-Thr-Ser-Cys(SO_3H), and Asn-Phe-Phe-Trp-Lys. **(b)** Now what do you know about the structure of somatostatin?

Hydrolysis of the polypeptide by chymotrypsin leads to Lys-Thr-Phe, Thr-Ser-Cys(SO_3H), Ala-Gly-Cys(SO_3H)-Lys-Asn-Phe, free Phe, and free Trp. **(c)** Write the entire amino acid sequence of somatostatin.

21. The amino acid sequence of met-enkephalin, a brain peptide with powerful opiatelike biological activity, is Tyr-Gly-Gly-Phe-Met. What would be the products of step-by-step Edman degradation of met-enkephalin?

The peptide shown in Problem 9(d) is leu-enkephalin, a relative of met-enkephalin with similar properties. How would the results of Edman degradation of leu-enkephalin differ from those of met-enkephalin?

22. Secreted by the pituitary gland, corticotropin is a hormone that stimulates the adrenal cortex. Determine its primary structure from the following information. (i) Hydrolysis by chymotrypsin produces six peptides: Arg-Trp, Ser-Tyr, Pro-Leu-Glu-Phe, Ser-Met-Glu-His-Phe, Pro-Asp-Ala-Gly-Glu-Asp-Gln-Ser-Ala-Glu-Ala-Phe, and Gly-Lys-Pro-Val-Gly-Lys-Lys-Arg-Arg-Pro-Val-Lys-Val-Tyr. (ii) Hydrolysis by trypsin produces free lysine, free arginine, and the following five peptides: Trp-Gly-Lys, Pro-Val-Lys, Pro-Val-Gly-Lys, Ser-Tyr-Ser-Met-Glu-His-Phe-Arg, and Val-Tyr-Pro-Asp-Ala-Gly-Glu-Asp-Gln-Ser-Ala-Glu-Ala-Phe-Pro-Leu-Glu-Phe.

23. Glucagon is a pancreatic hormone whose function opposes that of insulin: It causes an increase in glucose levels in the blood. It consists of a polypeptide chain with 29 amino acid units. Treatment of glucagon with thermolysin produces four fragments, including the tripeptide Val-Gln-Tyr, the tetrapeptide Leu-Met-Asn-Thr, a nine-amino-acid peptide A and a 13-amino-acid peptide B. Sanger degradation of A yields *N*-(2,4-dinitrophenyl)leucine, and Sanger degradation of B yields *N*-(2,4-dinitrophenyl)histidine.

Peptide A is not cleaved by chymotrypsin, but clostripain breaks it down into Leu-Asp-Ser-Arg, Ala-Gln-Asp-Phe, and a free Arg. Chymotripsin cleaves peptide B into Ser-Lys-Tyr, Thr-Ser-Asp-Tyr, and His-Ser-Gln-Gly-Thr-Phe. **(a)** At this stage, how much do you know for certain about the structure of glucagon? What uncertainties still remain? **(b)** One of the products of trypsin hydrolysis of the intact glucagon molecule is the peptide Tyr-Leu-Asp-Ser-Arg. Does this help? **(c)** One product of chymotrypsin hydrolysis of the intact hormone is Leu-Met-Asn-Thr, the same tetrapeptide released by thermolysin. Now can you piece together the entire molecule?

24. Propose a synthesis of leu-enkephalin [see Problem 9(d)] from the component amino acids.

25. The following molecule is thyrotropin-releasing hormone (TRH). It is secreted by the hypothalamus, causing the release of thyrotropin from the pituitary gland, which, in turn, stimulates the thyroid gland. The thyroid produces hormones, such as thyroxine, that control metabolism in general.

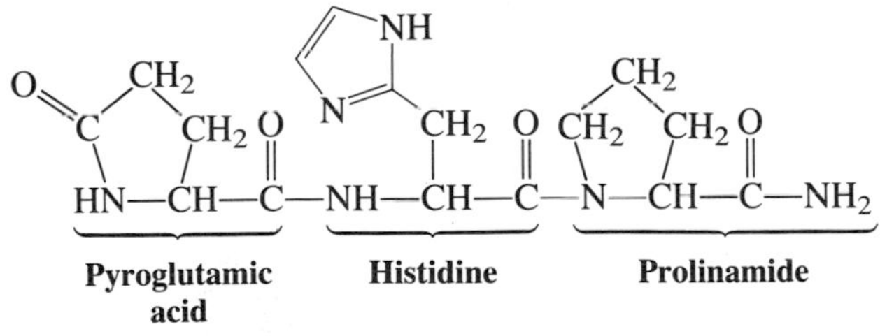

HN—CH—C(=O)—NH—CH—C(=O)—N—CH—C(=O)—NH_2

Pyroglutamic acid **Histidine** **Prolinamide**

The initial isolation of TRH required the processing of four tons of hypothalamic tissue, from which 1 mg of the hormone was obtained. Needless to say, it is a bit more convenient to synthesize TRH in the laboratory than to extract it from natural sources. Devise a synthesis of TRH from Glu, His, and Pro. Note that pyroglutamic acid is just the lactam of Glu and may be readily obtained by heating Glu to between 135 and 140°C.

26. Consider the synthesis of aspartame (Section 26-4). Is there a structural feature in one of its component amino acids that might make the synthesis difficult? What other amino acids contain groups that might cause problems in synthesizing peptides containing them?

27. **(a)** The structures illustrated for the four DNA bases (Section 26-9) represent only the most stable tautomers. Draw one or more alternative tautomers for each of these heterocycles (review tautomerism, Sections 13-8 and 18-2). **(b)** In certain cases, the presence of a small amount of one of these less stable tautomers can lead to an error in DNA replication or mRNA synthesis due to faulty base pairing. One example is the imine tautomer of adenine, which pairs with cytosine instead of thymine. Draw a possible structure for this hydrogen-bonded base pair (see Figure 26-11). **(c)** Using Table 26-3, derive a possible nucleic acid sequence for an mRNA that would code for the five amino acids in met-enkephalin (see Problem 21). If the mispairing described in (b) were at the first possible position in the synthesis of this mRNA sequence, what would be the consequence in the amino acid sequence of the peptide? (Ignore the initiation codon.)

28. Factor VIII is one of the proteins participating in the formation of blood clots. A defect in the gene whose DNA sequence codes for Factor VIII is responsible for classic hemophilia. Factor VIII contains 2332 amino acids. How many nucleotides are needed to code for its synthesis?

29. Hydroxyproline (Hyp), like many other amino acids that are not ''officially'' classified as essential, is nonetheless a very necessary biological substance. It constitutes about 14% of the amino acid content of the protein collagen. Collagen is the main constituent of skin and connective tissue. It is also present, together with inorganic substances, in nails, bones, and teeth. **(a)** The systematic name for hydroxyproline is (2*S*,4*R*)-4-hydroxyazacyclopentane-2-carboxylic acid. Draw a stereochemically correct structural formula for this amino acid. **(b)** Hyp is synthesized in the body in peptide-bound form from peptide-bound proline and O_2, in an enzyme-catalyzed process that requires vitamin C. In the absence of vitamin C, only a defective, Hyp-deficient collagen can be produced. Vitamin C deficiency causes scurvy, a condition characterized by bleeding of the skin and swollen, bleeding gums.

In the following reaction sequence, an efficient laboratory synthesis of hydroxyproline, fill in the necessary reagents (i) and (ii), and formulate detailed mechanisms for the steps marked with an asterisk.

$ClCH_2CH$—CH_2 (epoxide, O) $\xrightarrow{i}$ (phthalimide) N—CH_2CH—CH_2 (epoxide, O) $\xrightarrow{Na^{+}\,{}^{-}CH(COOCH_2CH_3)_2{}^{*}}$

(phthalimide) NCH_2— (lactone, O, O) $COOCH_2CH_3$ $\xrightarrow{SO_2Cl_2}$ (phthalimide) NCH_2— (lactone, O, O) Cl, $COOCH_2CH_3$ $\xrightarrow{ii^{*}}$ hydroxyproline

(c) Gelatin, which is partly hydrolyzed collagen, is rich in hydroxyproline and, as a result, is often touted as a remedy for split or brittle nails. Like most proteins, however, gelatin is almost completely broken down into individual amino acids in the stomach and small intestine before absorption. Is the free hydroxyproline thus introduced into the bloodstream of any use to the body in the synthesis of collagen? (Hint: Does Table 26-3 list a three-base code for hydroxyproline?)

30. Sickle-cell anemia is an often fatal genetic condition caused by a single error in the DNA gene that codes for the β chain of hemoglobin. The correct nucleic acid sequence (read from the mRNA template) begins with AUG-GUGCACCUGACUCCUG**A**GGAGAAG . . . , and so forth. **(a)** Translate this into the corresponding amino acid sequence of the protein. **(b)** The mutation that gives rise to the sickle-cell condition is replacement of the boldface A in the preceding sequence by U. What is the consequence of this error in the corresponding amino acid sequence? **(c)** This amino acid substitution alters the properties of the hemoglobin molecule—in particular, its polarity and its shape. Suggest reasons for both these effects. (Refer to Table 26-1 for amino acid structures and to Figure 26-8C for the structure of myoglobin, which is similar to that of hemoglobin. Note the location of the amino acid substitution in the tertiary structure of the protein.)

Answers to Exercises

CHAPTER 1

1-1

(a)

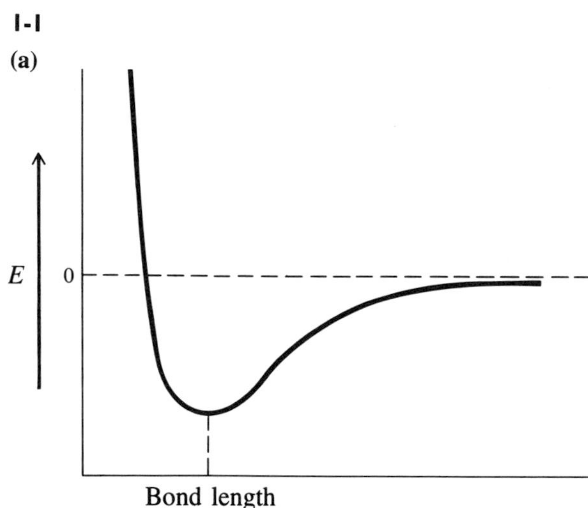

(b) Self-explanatory

1-2

$Li^+ :Br:^-$ $[Na]_2^+ :O:^{2-}$ $Be^{2+}[:F:]_2^-$

$Al^{3+}[:Cl:]_3^-$ $Mg^{2+} :S:^{2-}$

1-3

:F:F: :F:C:F: (with :F: above and below C) H:C:H (with :Cl: above and below C) H:P:H (with H below P)

:Br:I: $^-$:O:H H:N:H (with $^-$ above N) $^-$:C:H (with H above and below C)

1-4

H→O←H SC→O S→O I→Br H→C←H (with H→C from above and H→C from below)

Cl←C→Cl (with H→C from above, C→Cl below) H→C→Cl (with H→C from above, C→Cl below) H→C←H (with H→C from above, C→Cl below)

1-5

You can view NH_3 as being isoelectronic with H_3C^-, H_2O with H_2C^{2-}. Electron repulsion by the free electron pairs causes the bonding electrons to "bend away," giving rise to the respective pyramidal and bent structures.

1-6

H:I: H:C:C:C:H (H above and below each C) H:C:O:H (H above and below C) H:S:S:H

O::Si::O O::O S::C::S

1-7

S::O :F:O:F: H:O:Cl:O: (+ on Cl, − on terminal O) :F:B:N:H (with :F: above and below B, H above and below N; − on B, + on N)

H:C:O:H (H above and below C, H below O, + on O) :Cl:C::O:Cl: (Cl above and below C) $^-$:C:::N: $^-$:C:::C:$^-$

1-8

It should be close to trigonal (counting the lone electron pair), with equal N–O bond lengths and one-half of a negative charge on each oxygen atom.

[:O=N–O:$^-$ ⟷ $^-$:O–N=O:]

1-9

(a) [$^-$:C≡$\overset{+}{N}$–O:$^-$ ⟷ $^{2-}$C=$\overset{+}{N}$=O]

The left-hand structure is preferred, because the charges are more evenly distributed and the negative charge resides on the relatively electronegative oxygen.

(b) 

The left-hand structure is preferred, because the right-hand one has no octet on nitrogen.

1-10

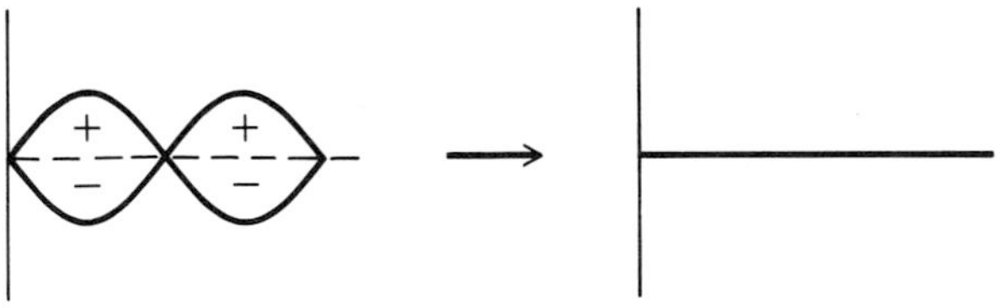

1-11

S $(1s)^2(2s)^2(2p)^6(3s)^2(3p)^4$; P $(1s)^2(2s)^2(2p)^6(3s)^2(3p)^3$

1-12

The molecular-orbital picture is similar to that shown for the bonding in H_2 (Figures 1-11 and 1-12). However, the presence of only one antibonding but two bonding electrons results in net bonding.

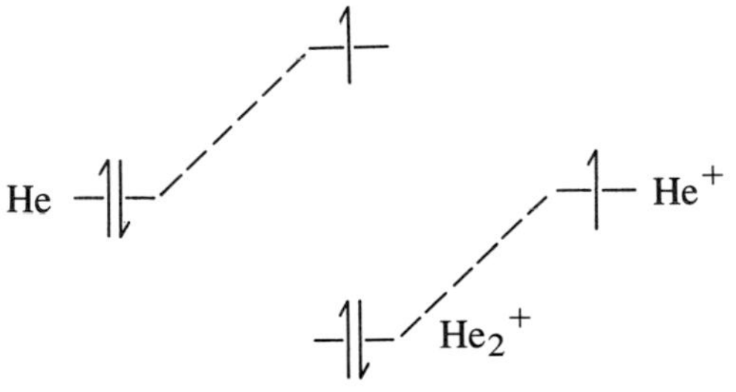

1-13

No octet

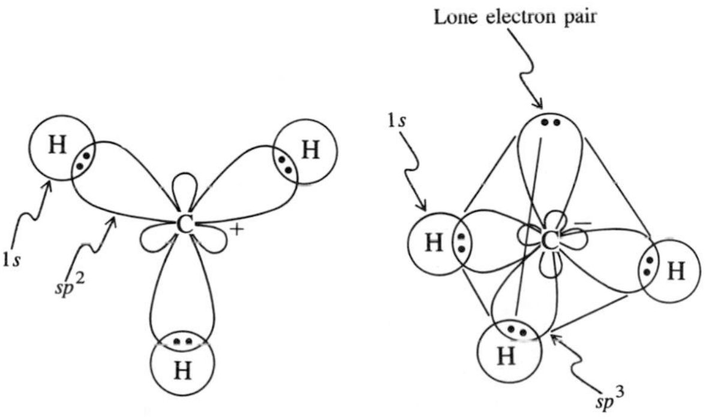

Trigonal, sp^2 hybridized, electron deficient like BH_3

Tetrahedral, sp^3 hybridized, closed shell

1-14

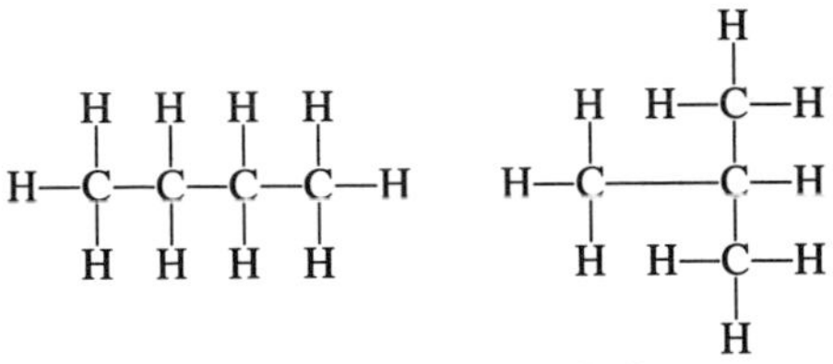

Butane **Isobutane**

1-15

Self-explanatory. Note that the molecules are flexible and can adopt a variety of arrangements in space.

1-16

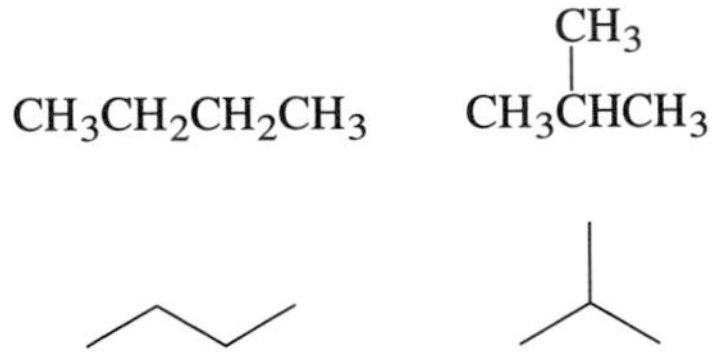

1-17

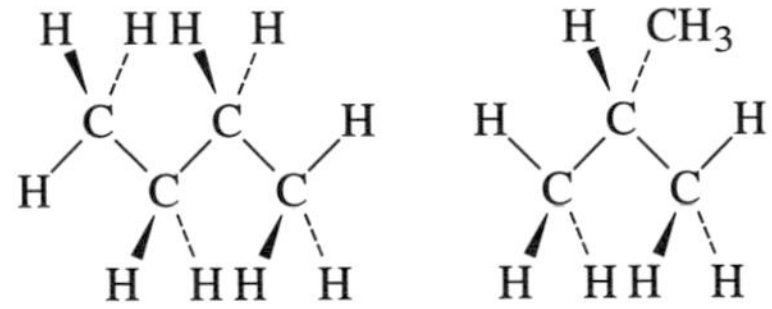

CHAPTER 2

2-1

(a) 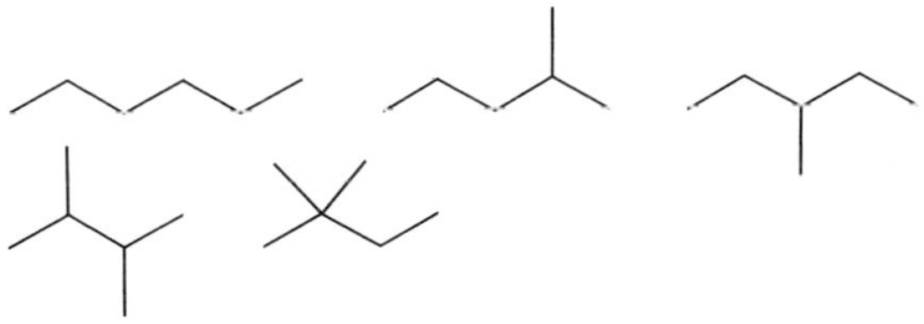

(b) Higher homologs:

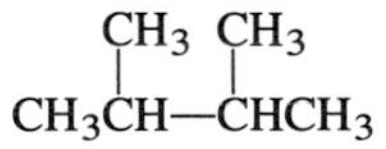

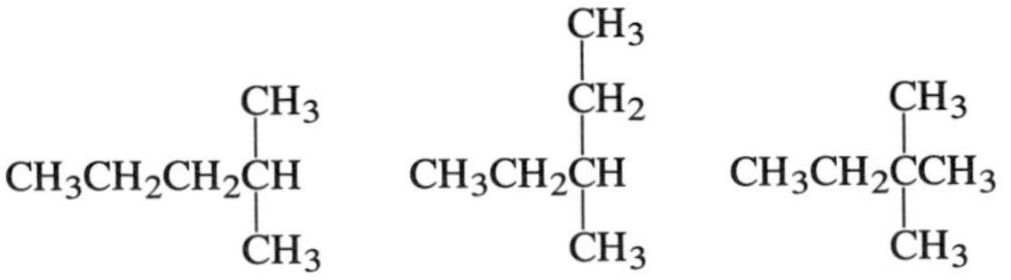

Lower homologs:

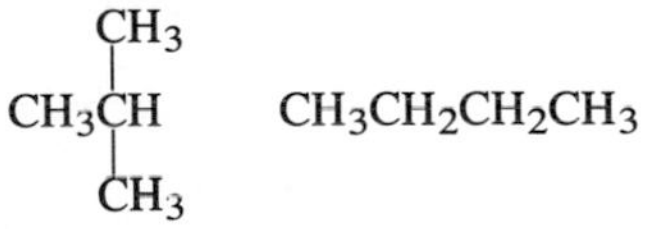

2-2

CH_3
$CH_3CHCH_2CH_2CH_3$

Isohexane

CH_3
CH_3CCH_3
CH_3

Neopentane

2-3

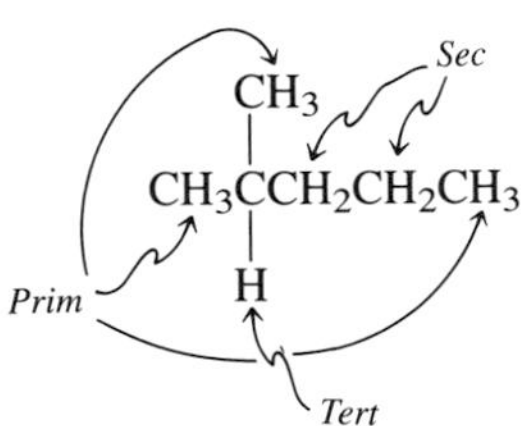

2-4
Self-explanatory

2-5

H_3C H

2-Methylbutane

H CH_3
H_3C H

2,3-Dimethylbutane

2-6
In this example (graphed below), the energy difference between the two staggered conformers turns out to be quite small.

2-7
$0.9 = -1.36 \log K$ kcal mol^{-1} at 25°C
$K = 0.219$; *anti*:*gauche* = 82:18
$0.9 = -RT \ln K = -1.71 \log K$ kcal mol^{-1} at 100°C
$K = 0.297$; *anti*:*gauche* = 77:23

2-8
$$\Delta G^\circ = \Delta H^\circ - T\Delta S^\circ$$
$$= 22.4 \text{ kcal mol}^{-1} - (298 \text{ deg} \times 33.3 \text{ cal deg}^{-1} \text{ mol}^{-1})$$
$$= 12.5 \text{ kcal mol}^{-1}$$

At higher temperatures, ΔG° is less positive, eventually becoming negative. The crossover point, $\Delta G^\circ = 0$, is reached at 400°C, when $\Delta H^\circ = T\Delta S^\circ$.

2-9
$$\Delta G^\circ = \Delta H^\circ - T\Delta S^\circ$$
$$= -15.5 \text{ kcal mol}^{-1} - (298 \text{ deg} \times -31.3 \text{ cal deg}^{-1} \text{ mol}^{-1})$$
$$= -6.17 \text{ kcal mol}^{-1}$$

The entropy is negative because two molecules are converted into one in this reaction.

2-10
After 50% conversion, only 1/2 molar concentration of starting materials is present. Hence, for first order: rate = k[A]. At 50% conversion, rate will be 1/2 initial rate. For second order: rate = k[A][B]. At 50% conversion, rate = k(1/2)[A](1/2)[B] = 1/4 initial rate.

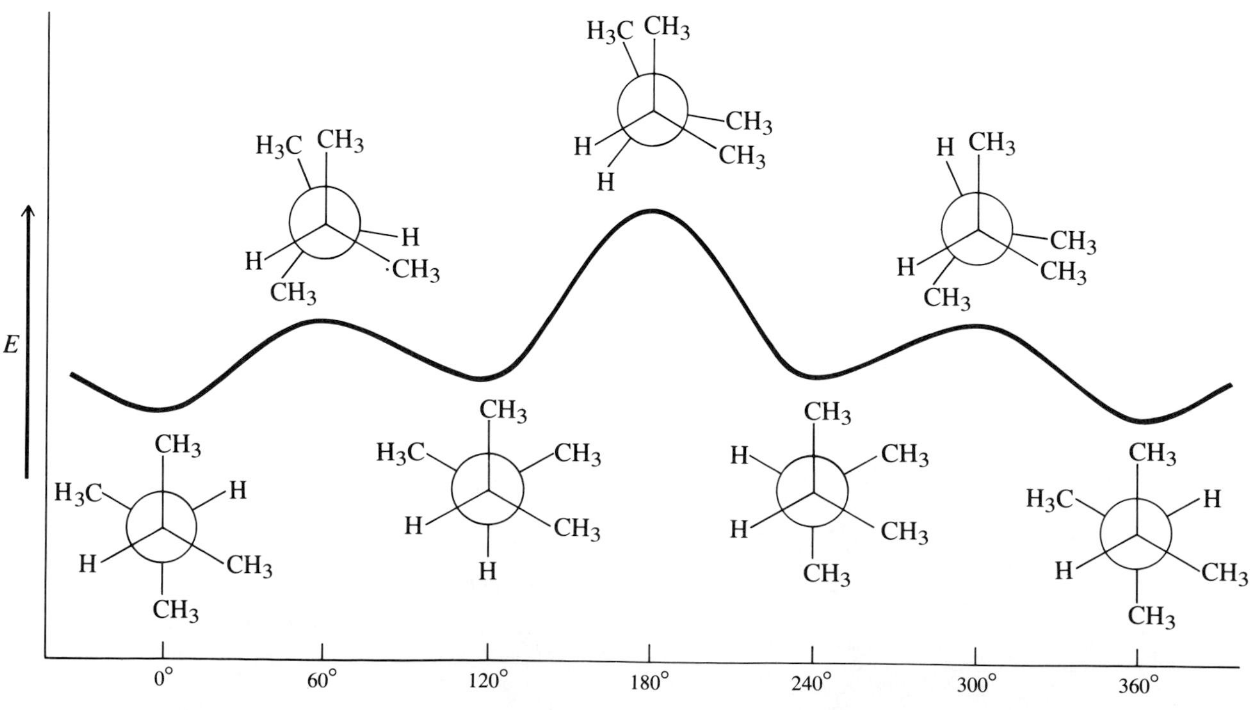

2-11

(a) $+6.17 \text{ kcal mol}^{-1}$

(b) $\Delta G° = 15.5 - (0.773 \times 31.3)$
$= -8.69 \text{ kcal mol}^{-1}$

Hence, the dissociation equilibrium lies on the side of ethene and HCl at this high temperature, where the entropy factor overrides the $\Delta H°$ term.

2-12

$k = 10^{14}e^{-58.4/1.53} = 3.03 \times 10^{-3} \text{ s}^{-1}$

CHAPTER 3

3-1

A simple answer would be that the strength of a bond depends not only on the size and energies of the orbitals, but also on coulombic contributions. Thus, in going from N to O to F, nuclear charge increases in the core, allowing for more nuclear-core–electronic attraction in binding to CH_3. The bond in question becomes more and more polar along the series.

3-2

First: CH_3–$C(CH_3)_3$ $DH° = 84 \text{ kcal mol}^{-1}$
Second: CH_3–CH_3 $DH° = 90 \text{ kcal mol}^{-1}$

3-3

$$CH_3CH_3 + Cl_2 \xrightarrow{h\nu} CH_3CH_2Cl + HCl$$

$$\Delta H° = 98 + 58 - 80 - 103 = -27 \text{ kcal mol}^{-1}$$

Mechanism

Initiation

$$Cl_2 \xrightarrow{h\nu} 2\,:\ddot{Cl}\cdot \qquad \Delta H° = +58 \text{ kcal mol}^{-1}$$

Propagation

$$CH_3CH_3 + :\ddot{Cl}\cdot \rightarrow CH_3CH_2\cdot + H\ddot{Cl}: \qquad \Delta H° = -5 \text{ kcal mol}^{-1}$$

$$CH_3CH_2\cdot + Cl_2 \rightarrow CH_3CH_2\ddot{Cl}: + :\ddot{Cl}\cdot \qquad \Delta H° = -22 \text{ kcal mol}^{-1}$$

Termination

$$:\ddot{Cl}\cdot + :\ddot{Cl}\cdot \rightarrow Cl_2 \qquad \Delta H° = -58 \text{ kcal mol}^{-1}$$

$$CH_3CH_2\cdot + :\ddot{Cl}\cdot \rightarrow CH_3CH_2\ddot{Cl}: \qquad \Delta H° = -80 \text{ kcal mol}^{-1}$$

$$CH_3CH_2\cdot + \cdot CH_2CH_3 \rightarrow CH_3CH_2CH_2CH_3 \qquad \Delta H° = -82 \text{ kcal mol}^{-1}$$

3-4

$$CH_4 + Cl_2 + Br_2 \longrightarrow CH_3Cl + CH_4 + Cl_2 + Br_2 + HCl$$

Cl_2 is more reactive than Br_2.

3-5

$$CH_3CH_2CH_2CH_3 + Cl_2 \xrightarrow{h\nu} CH_3CH_2CH_2CH_2Cl + CH_3CH_2CH(Cl)CH_3 + HCl$$

Ratio of primary to secondary product:

$$(6 \times 1):(4 \times 4) = 6:16 = 3:8$$

In other words, 2-chlorobutane:1-chlorobutane = 8:3.

3-6

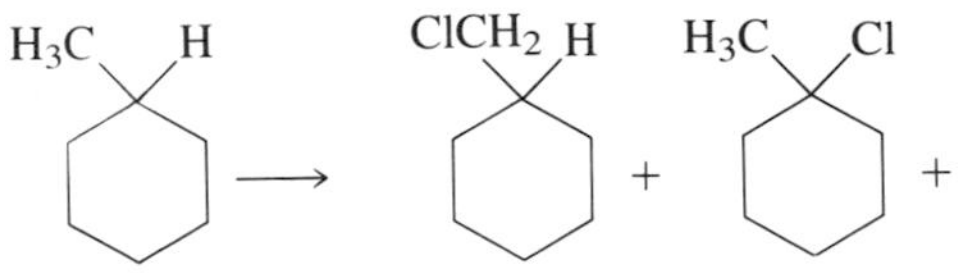

1-methylcyclohexane (H₃C, H) ⟶ A (ClCH₂, H) + B (H₃C, Cl) + C (H₃C, H; Cl on C-2) + D (H₃C, H; Cl on C-3) + E (H₃C, H; Cl on C-4)

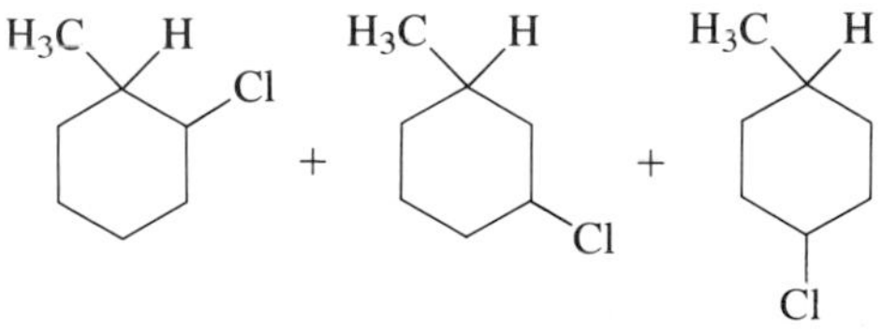

3 primary, 3 types (4, 4, and 2) secondary, 1 tertiary hydrogen. Relative amounts of A, B, C, D, E:

$$A:B:C:D:E = (3 \times 1):(1 \times 5):(4 \times 4):(4 \times 4):(2 \times 4)$$
$$= 3:5:16:16:8$$

This problem is actually more complicated because of the possibility of forming cis and trans isomers (see Section 4-1).

3-7

$CH_3CH_2CH_3$
↑ ↑

Will give mixture, because of competitive primary and secondary chlorination (indicated by arrows).

$$H_3C{-}C(CH_3)_2{-}CH_3 \longrightarrow (CH_3)_3CCH_2Cl$$

Only one type of C–H; so 2,2-dimethylpropane should give good selectivity.

Cl

Same situation as in 2,2-dimethylpropane.

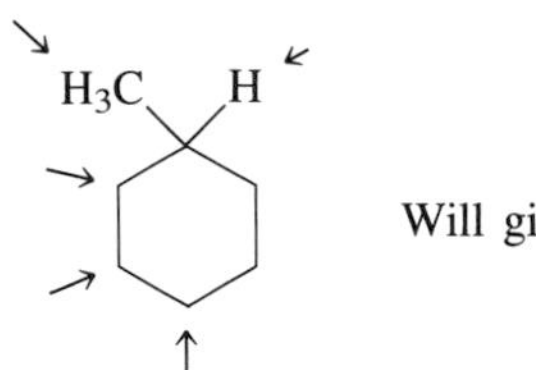

Will give a bad mixture.

3-8

In this isomerization, a secondary hydrogen and a terminal methyl group in butane switch positions:

$CH_3CHCH_2{-}CH_3$ (H)

Hence,

$$\begin{aligned}\Delta H^\circ &= \text{(sum of the strengths of the bonds broken)}\\ &\quad - \text{(sum of the strengths of the bonds made)}\\ &= (94.5 + 87) - (86 + 98)\\ &= -2.5 \text{ kcal mol}^{-1}\end{aligned}$$

3-9

Combustion of the butanes yields four molecules of CO_2 and five of H_2O (standard state: liquid). For the latter, $\Delta H_f^\circ = -[(4 \times 94.1) + (5 \times 68.3)] = -717.9$ kcal mol^{-1}. Subtracting from this value the respective ΔH°_{comb} of the butanes provides their respective heats of formation. (Note that the last significant figure of the values obtained by using the data of Table 3-7 is one unit higher than the numbers quoted in Figure 3-12, because of rounding off.)

3-10

(a) $CH_4 \longrightarrow \cdot CH_3 + \cdot H$, $\Delta H^\circ = 105$ kcal mol^{-1}. Using the formula given in the problem, we have $105 = (52.1 + x) - (-17.9)$ kcal mol^{-1}. Hence, $x = 35$ kcal mol^{-1} $= \Delta H_f(\cdot CH_3)$

(b) ΔH_f of $Br\cdot = \frac{1}{2}DH^\circ(Br_2) = 23$ kcal mol^{-1}. ΔH_f of CH_3Br is obtained from $\cdot CH_3 + \cdot Br \longrightarrow CH_3Br$, $\Delta H^\circ = -71$ kcal mol^{-1} (the DH° of CH_3–Br), and therefore $\Delta H_f(CH_3Br) = (35 + 23) - 71 = -13$ kcal mol^{-1}. For $\cdot CH_3 + Br_2 \longrightarrow CH_3Br + Br\cdot$, $\Delta H^\circ = (-13 + 23) - 35 = -25$ kcal mol^{-1}.

The agreement with the value quoted in Table 3-4 is excellent, although somewhat fortuitous for reasons that require a more advanced treatment of the problem.

CHAPTER 4

4-1

Aspects of ring strain and conformational analysis are discussed in Sections 4-2 through 4-5.

Note that the cycloalkanes are much less flexible than the straight-chain alkanes and thus have less conformational freedom. Cyclopropane must be flat and all hydrogens eclipsed. The higher cycloalkanes have increasingly more flexibility, hydrogens being in staggered positions and the carbon atoms of the ring eventually being able to adopt *anti* conformations.

4-2

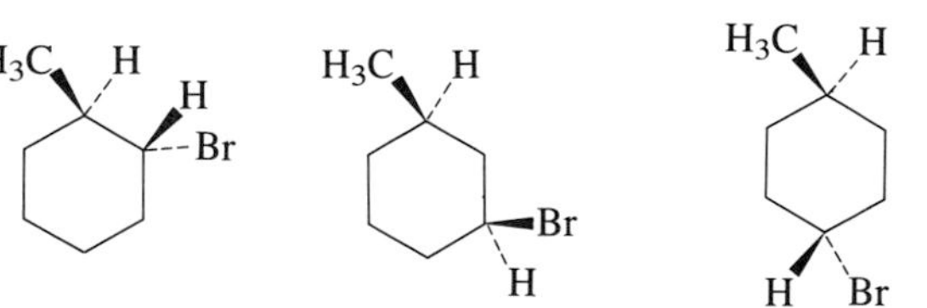

***trans*-1-Bromo-2-methylcyclohexane** | ***cis*-1-Bromo-3-methylcyclohexane** | ***trans*-1-Bromo-4-methylcyclohexane**

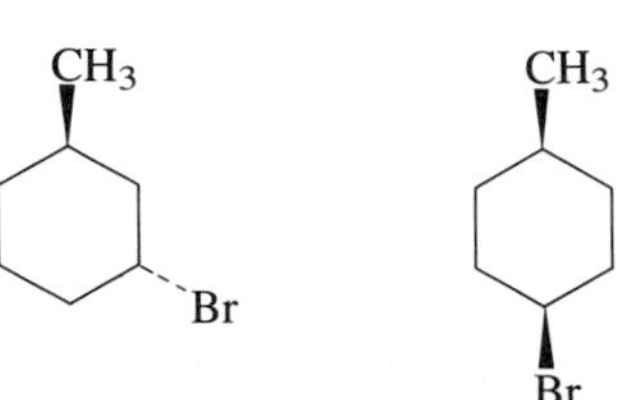

***trans*-1-Bromo-3-methylcyclohexane** | ***cis*-1-Bromo-4-methylcyclohexane**

4-3

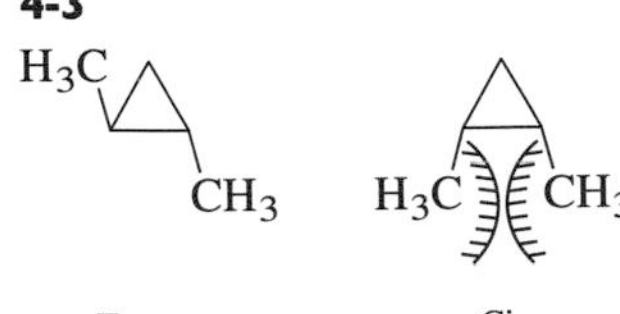

Trans Cis

The cis isomer has a larger heat of combustion (about 1 kcal mol^{-1}).

4-4

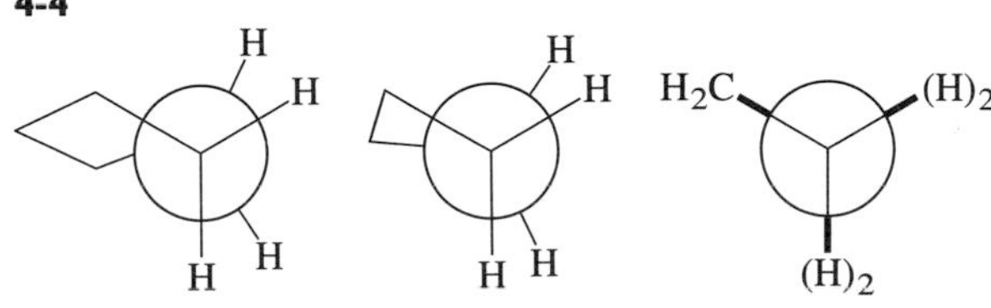

Cyclopentane **Cyclobutane** **Cyclopropane**

The respective C–H torsional angles are roughly 40°, 20°, and 0°.

4-5

$\log K = -1.7/1.36 = -1.25$

$K = 0.056$. Cf. $K = 5/95 = 0.053$

4-6

(a) $\Delta G°$ = energy difference between an axial methyl and axial ethyl group: $1.75 - 1.7$ = about 0.05 kcal mol^{-1}; that is, very small.

(b) Same as (a).

(c) $1.75 + 1.7 = 3.45$ kcal mol^{-1}

4-7

(a)

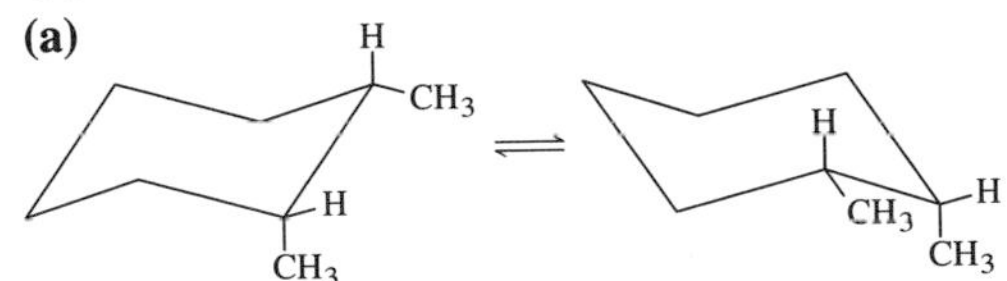

Both axial–equatorial

(b)

Diequatorial Diaxial

(c)

Diequatorial Diaxial

(d)

Both axial–equatorial

4-8

trans-Decalin is fairly rigid. Full chair–chair conformational "flipping" is not possible. In contrast, the axial and equatorial positions in the cis isomer can be interchanged by conformational isomerization of both rings. The barrier to this exchange is small (E_a = 14 kcal mol^{-1}). Because one of the appended bonds is always axial, the cis isomer is less stable than the trans isomer by 2 kcal mol^{-1} (as measured by combustion experiments).

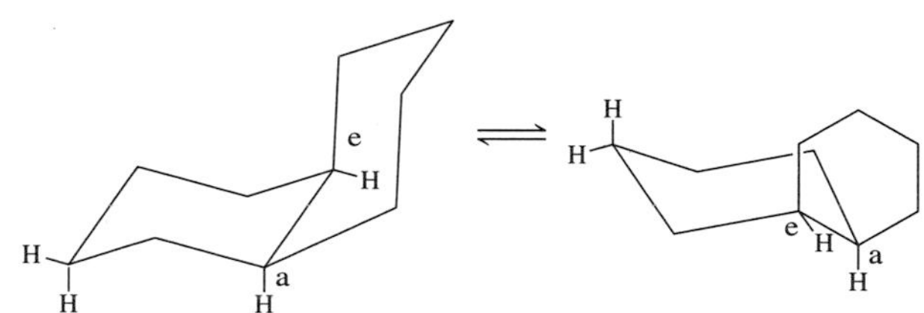

Ring flip in *cis*-decalin

4-9

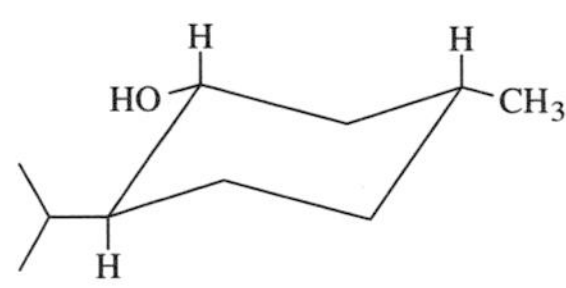

All equatorial

4-10

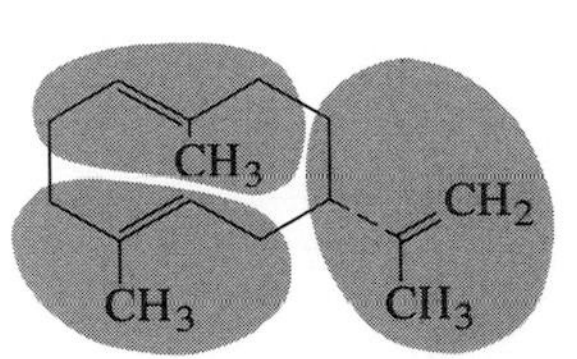

Sesquiterpene

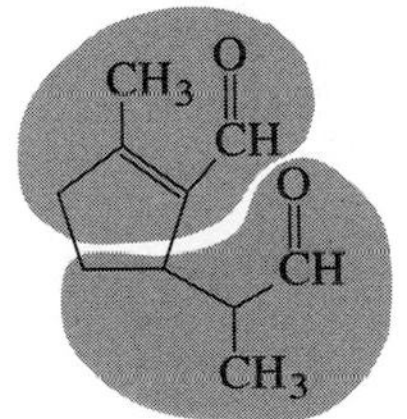

Monoterpene

4-11

Chrysanthemic acid: alkene, carboxylic acid, ester.
Grandisol: alkene, alcohol.
Menthol: alcohol.
Camphor: ketone.
β-Cadinene: alkene.

CHAPTER 5

5-1

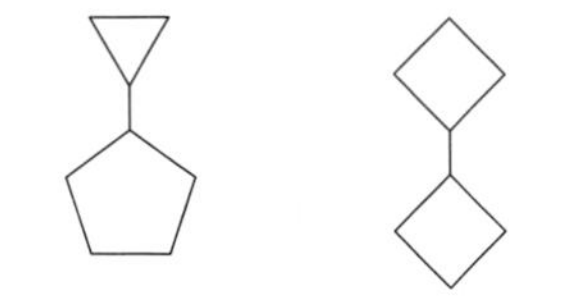

Cyclopropyl-cyclopentane **Cyclobutyl-cyclobutane**

Both hydrocarbons have the same molecular formula: C_8H_{14}. Therefore, they are (constitutional) isomers.

5-2

There are several boat and twist-boat forms of methylcyclohexane, some of which are shown:

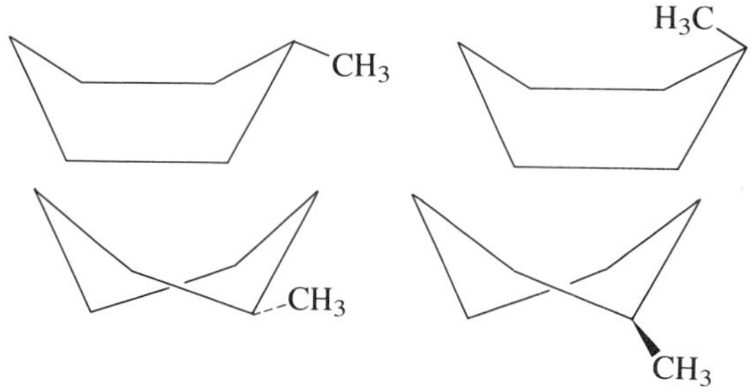

5-3

All are chiral. Note, however, that 2-methylbutadiene (isoprene) itself is achiral. Number of stereocenters: chrysanthemic acid, 2; grandisol, 2; menthol, 3; camphor, 2; β-cadinene, 3; cholesterol, 8; cholic acid, 11; cortisone, 6; testosterone, 6; estradiol, 5; progesterone, 6; norethynodrel, 5; mestranol, 5; RU-486, 5.

5-4

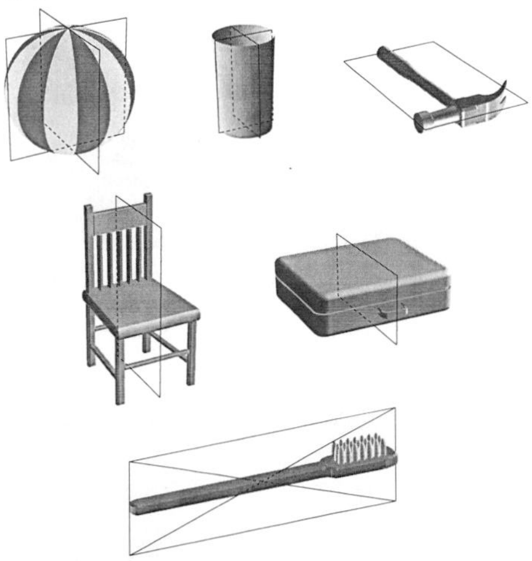

(Illustration courtesy Marie Sat.)

5-5

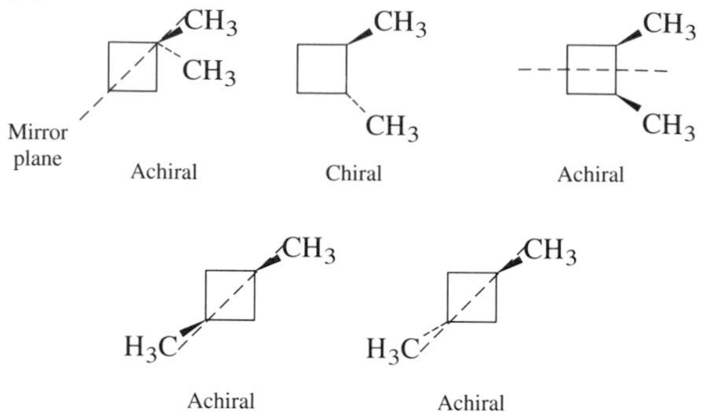

5-6

$$[\alpha] = \frac{6.65}{1 \times 0.1} = 66.5°$$

The enantiomer of natural sucrose has $[\alpha] = -66.5°$.

5-7

Optical purity (%)	Ratio (+/−)	$[\alpha]_{obs.}$
75	87.5/12.5	+17.3°
50	75/25	+11.6°
25	62.5/37.5	+5.8°

5-8

(a) $-CH_2Br > -CCl_3 > -CH_2CH_3 > -CH_3$

(b) 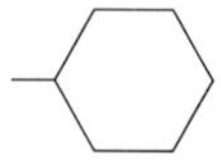 $> -\overset{CH_3}{\overset{|}{C}}HCH_3 > -CH_2\overset{CH_3}{\overset{|}{C}}HCH_3$

(c) $-C(CH_3)_3 > -\overset{CH_3}{\overset{|}{C}}HCH_2CH_3 > -CH_2\overset{CH_3}{\overset{|}{C}}HCH_3 >$
$-CH_2CH_2CH_2CH_3$

(d) $-\overset{Br}{\overset{|}{C}}HCH_3 > -\overset{Cl}{\overset{|}{C}}HCH_3 > -CH_2CH_2Br > -CH_2CH_3$

5-9

(−)-2-Bromobutane: *R*
(+)-2-Bromobutane: *S*
(+)-2-Aminopropanoic acid: *S*
(−)-2-Hydroxypropanoic acid: *R*

5-10

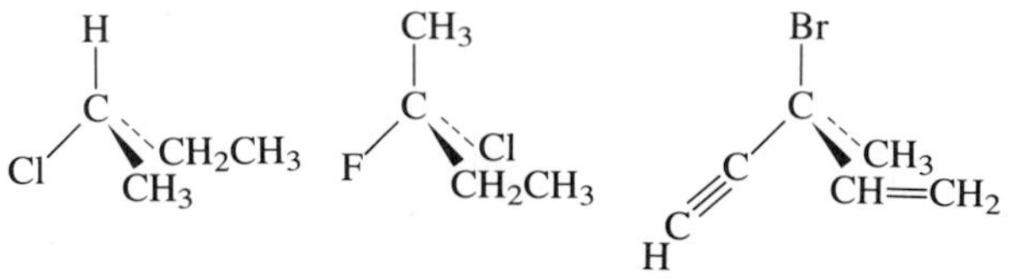

S ***R*** ***S***

5-11

CH_2CH_3 / H—+—Br / CH_3 CH_2CH_3 / Br—+—H / CH_3 H / H_2N—+—CH_3 / COOH

H / HO—+—COOH / CH_3 H / H_3C—+—CH_2CH_3 / Cl CH_3 / CH_3CH_2—+—Cl / F

Br / CH_2=CH—+—CH_3 / C≡CH

5-12

H, CH_3, Cl, 120° H_3C, Cl, H

5-13

a Br / *d* H—+—D *c* / CH_3 *b* → *d* / *a*—+—*b* / *c* **R**

c Cl / *d* F—+—Br *b* / I *a* → *d* / *c*—+—*a* / *b* **R**

c CH_3 / *a* H_2N—+—COOH *b* / H *d* → *d* / *b*—+—*a* / *c* **S**

5-14

CH_3 *b*, Br *a*, H—C (*d*), D *c*

Cl *c*, Br *b*, F—C (*d*), I *a*

CH_3 *c*, H—C (*d*), COOH *b*, NH_2 *a*

Placing the lowest priority substituent *d* at the top of a Fischer projection means that it is located behind the plane of the page, the place required for the correct assignment of absolute configuration by visual inspection.

5-15

	Isoleucine			Alloisoleucine	
	CO_2H			CO_2H	
H_2N	—	H	H_2N	—	H
H_3C	—	H	H	—	CH_3
	CH_2CH_3			CH_2CH_3	

Isoleucine   **Alloisoleucine**

They are diastereomers.

5-16

1: (2*S*,3*S*)-2-Fluoro-3-methylpentane.
2: (2*R*,3*S*)-2-Fluoro-3-methylpentane.
3: (2*R*,3*R*)-2-Fluoro-3-methylpentane.
4: (2*S*,3*S*)-2-Fluoro-3-methylpentane.
1 and 2 are diastereomers; 1 and 3 are enantiomers; 1 and 4 are identical; 2 and 3 and 2 and 4 are diastereomers; 3 and 4 are enantiomers.
Including the mirror image of 2, there are four stereoisomers.

5-17

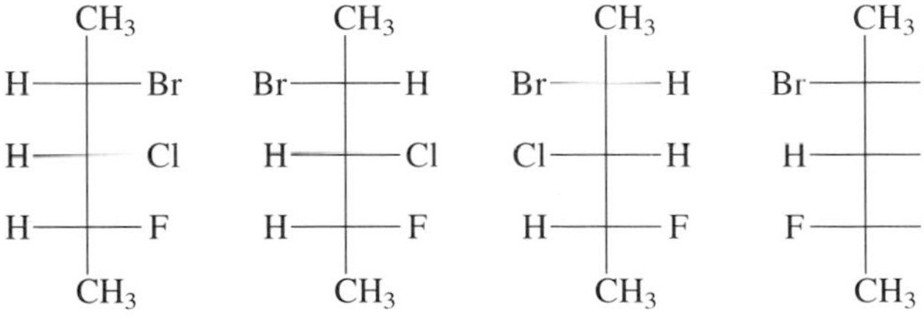

Including the four mirror images, there are four enantiomeric pairs of diastereomers.

5-18

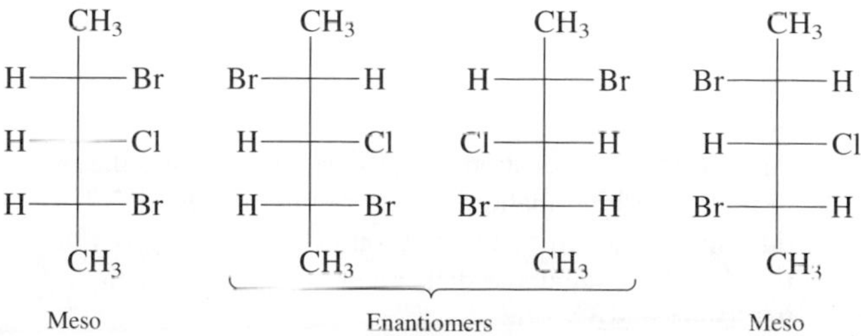

5-19

Mirror plane

(a)

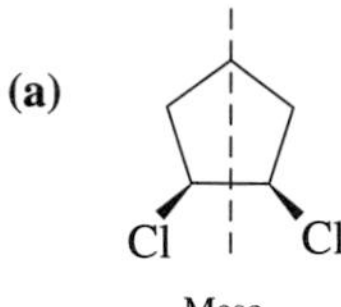

Meso

(b)

Cl Cl

Chiral

(c)

Cl Cl

Meso

(d)

Cl Cl

Chiral

(e)

Cl Cl

Meso

(f)

Cl Cl

Chiral

(g)

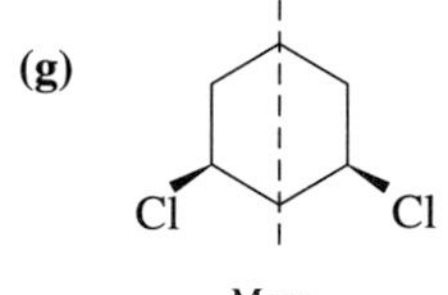

Meso

(h)

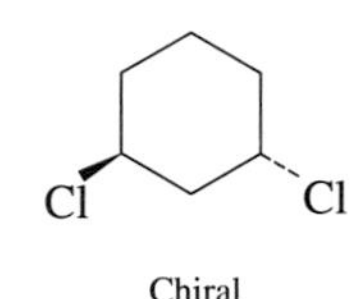

Chiral

5-20

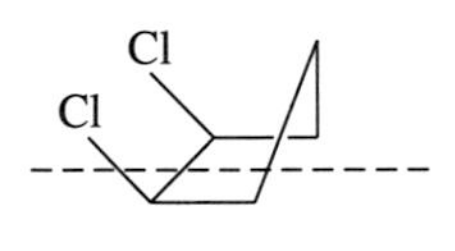

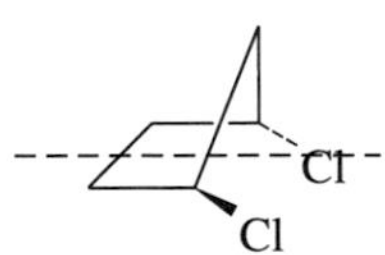

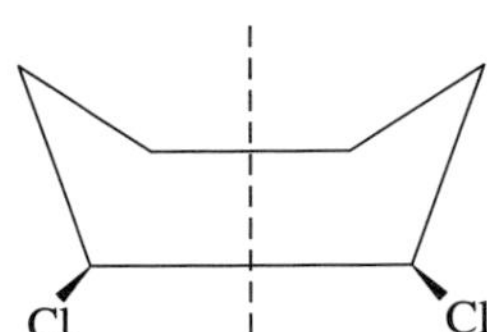

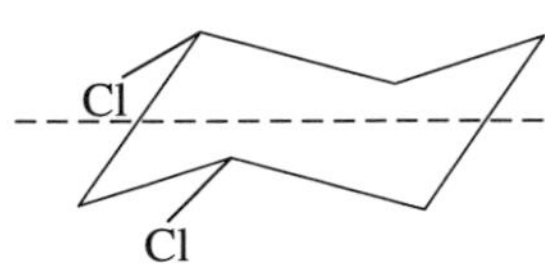

5-21

Almost any halogenation at C2 gives a racemate; the exception is bromination, which results in achiral 2,2-dibromobutane. In addition, bromination at C3 gives the two 2,3-dibromobutane diastereomers, one of which, 2*R*,3*S*, is meso.

5-22

Attack at C1:

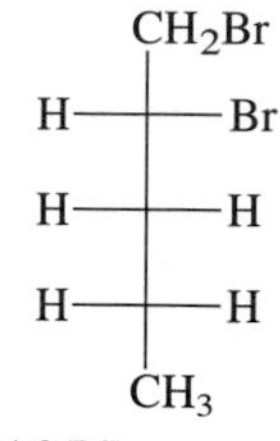

(*R*)-1,2-Dibromopentane
Chiral, optically active

Attack at C2:

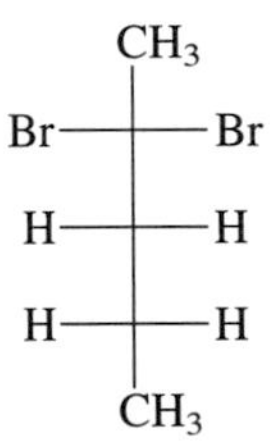

2,2-Dibromopentane
Achiral

Attack at C3:

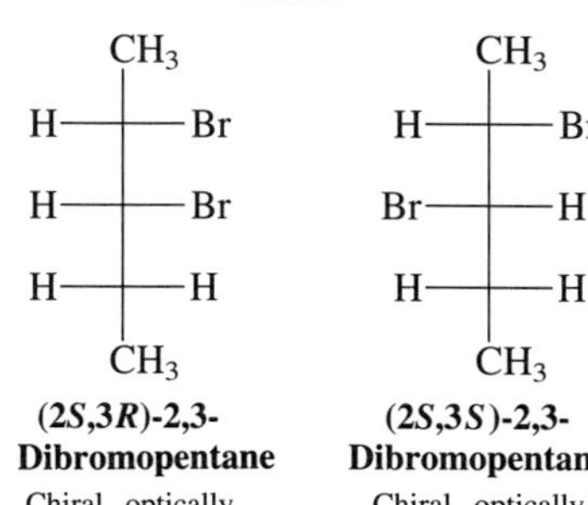

(2*S*,3*R*)-2,3-Dibromopentane
Chiral, optically active

(2*S*,3*S*)-2,3-Dibromopentane
Chiral, optically active

Diastereomers, formed in unequal amounts

Attack at C4:

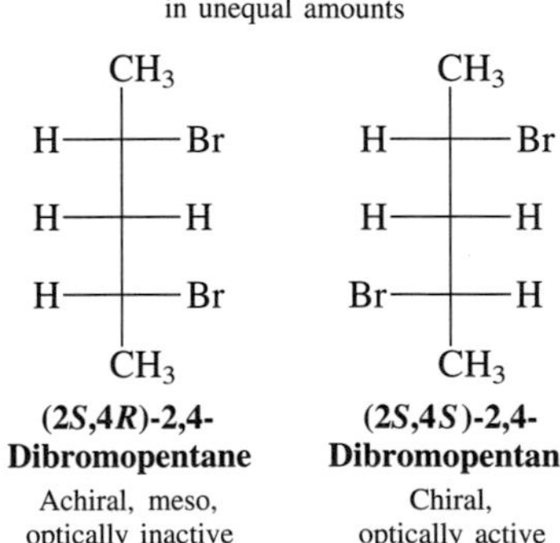

(2*S*,4*R*)-2,4-Dibromopentane
Achiral, meso, optically inactive

(2*S*,4*S*)-2,4-Dibromopentane
Chiral, optically active

Diastereomers, formed in unequal amounts

Attack at C5:

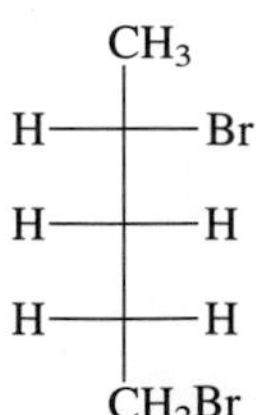

(*S*)-1,4-Dibromopentane
Chiral, optically active

CHAPTER 6

6-1

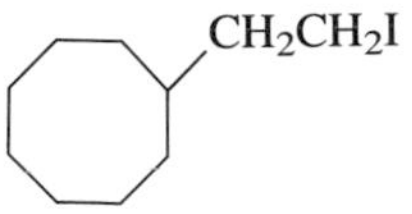

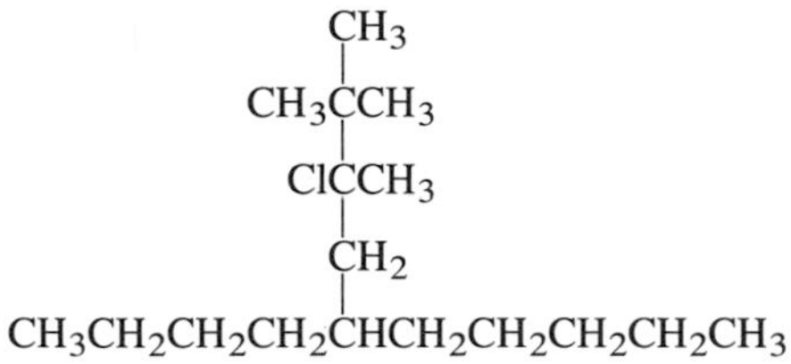

CH_2CH_2I (on cyclooctane ring)

$$
\begin{array}{c}
CH_3 \\
|\\
CH_3CCH_3 \\
|\\
ClCCH_3 \\
|\\
CH_2 \\
|\\
CH_3CH_2CH_2CH_2CHCH_2CH_2CH_2CH_2CH_3
\end{array}
$$

Note the similarity of this structure to that of 6-(2-chloro-2,3,3-trimethylbutyl)undecane. Why is it named so differently?

6-2

(a) $CH_3CH_2CH_2CH_2\ddot{\underset{..}{I}}:$

(b) $CH_3CH_2CH_2CH_2\ddot{\underset{..}{O}}CH_2CH_3$

(c) $CH_3CH_2CH_2CH_2\ddot{N}{=}\overset{+}{N}{=}\ddot{N}:^-$

(d) 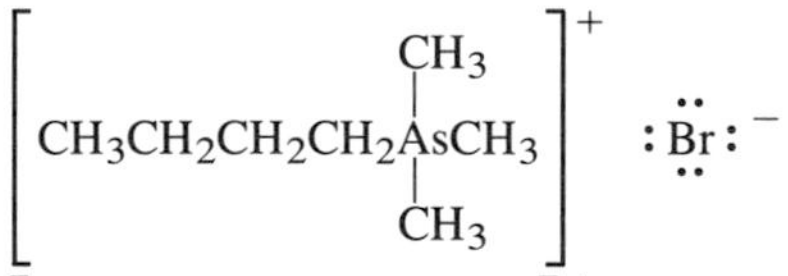 $\left[CH_3CH_2CH_2CH_2\underset{CH_3}{\overset{CH_3}{As}}CH_3\right]^+ \quad :\ddot{\underset{..}{Br}}:^-$

(e) 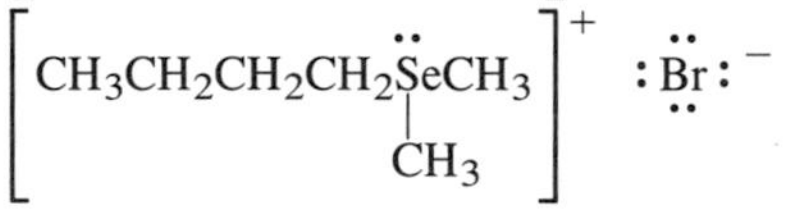 $\left[CH_3CH_2CH_2CH_2\underset{CH_3}{\ddot{Se}}CH_3\right]^+ \quad :\ddot{\underset{..}{Br}}:^-$

6-3

(a) $CH_3I + :N(CH_3)_3$

(b) There are two approaches:

$CH_3\ddot{\underset{..}{S}}:^- + CH_3CH_2\ddot{\underset{..}{I}}:$ or $CH_3\ddot{\underset{..}{I}}: + CH_3CH_2\ddot{\underset{..}{S}}:^-$

6-4

$CH_3I + {}^-:\ddot{N}{=}\overset{+}{N}{=}\ddot{N}:^- \longrightarrow CH_3{-}\ddot{N}{=}\overset{+}{N}{=}\ddot{N}:^- + I^-$

$k = 3 \times 10^{-10}/10^{-4} = 3 \times 10^{-6}\ \text{mol}^{-1}\ \text{L s}^{-1}$

(a) Rate $= k[CH_3I][N_3^-] = (3 \times 10^{-6}) \times (2 \times 10^{-4})$
$= 6 \times 10^{-10}\ \text{mol L}^{-1}\ \text{s}^{-1}$

(b) $1.2 \times 10^{-9}\ \text{mol L}^{-1}\ \text{s}^{-1}$

(c) $2.7 \times 10^{-9}\ \text{mol L}^{-1}\ \text{s}^{-1}$

6-5

Frontside displacement

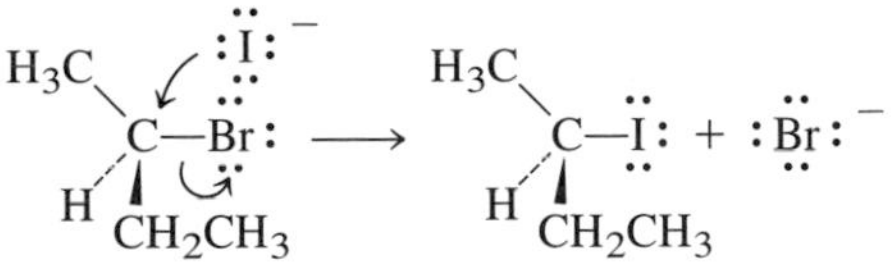

$:\ddot{I}:^-$ + $(H_3C)(H)(CH_2CH_3)C{-}\ddot{Br}: \longrightarrow (H_3C)(H)(CH_2CH_3)C{-}\ddot{\underset{..}{I}}: + :\ddot{\underset{..}{Br}}:^-$

Backside displacement

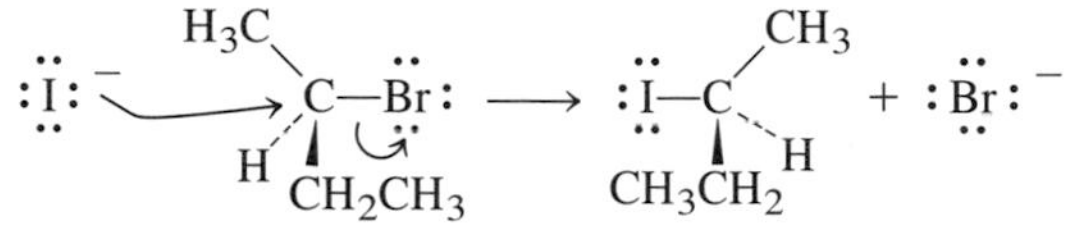

$:\ddot{\underset{..}{I}}:^- + (H_3C)(H)(CH_2CH_3)C{-}\ddot{Br}: \longrightarrow :\ddot{\underset{..}{I}}{-}C(CH_3)(H)(CH_3CH_2) + :\ddot{\underset{..}{Br}}:^-$

6-6

(a) 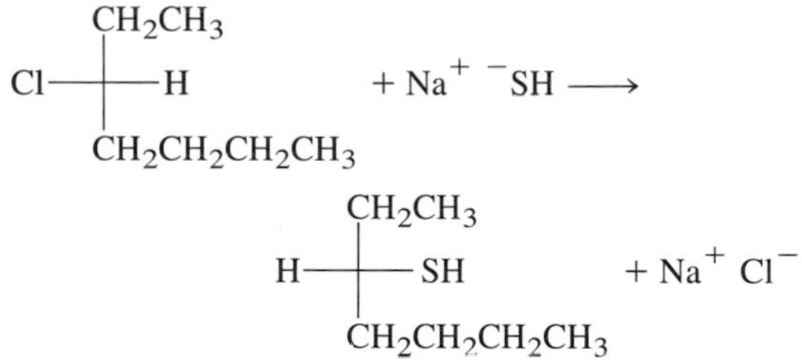 Cl—C(CH₂CH₃)(H)(CH₂CH₂CH₂CH₃) $+ Na^+\ {}^-SH \longrightarrow$ H—C(CH₂CH₃)(SH)(CH₂CH₂CH₂CH₃) $+ Na^+\ Cl^-$

(b) H, Br (skeletal structure) $+ :N(CH_3)_3 \longrightarrow$ $(CH_3)_3\overset{+}{N}$, H (skeletal structure) $+ Br^-$

(c) H, I; H, CH₃ (skeletal structure) $+ K^+\ {}^-SeCH_3 \longrightarrow$ CH_3Se, H; H, CH₃ (skeletal structure) $+ K^+\ I^-$

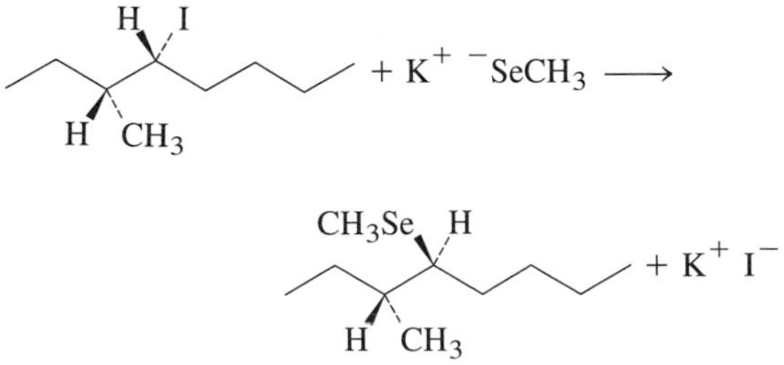

6-7

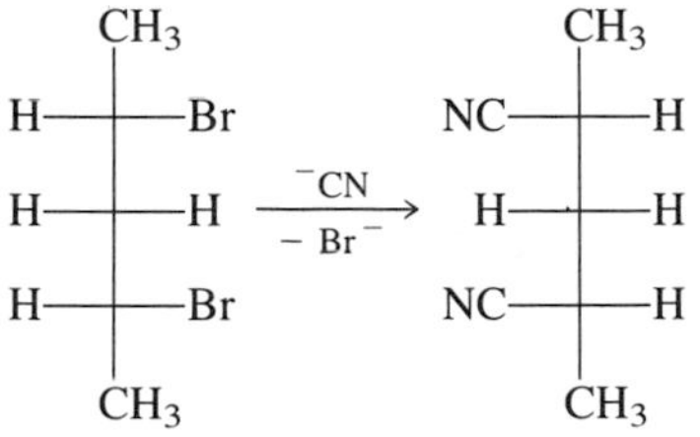

$$
\begin{array}{c}
CH_3 \\
H{-}\!\!\!\!\!\!\quad Br \\
H{-}\quad H \\
H{-}\quad Br \\
CH_3
\end{array}
\xrightarrow[-\ Br^-]{^-CN}
\begin{array}{c}
CH_3 \\
NC{-}\quad H \\
H{-}\quad H \\
NC{-}\quad H \\
CH_3
\end{array}
$$

Meso Meso

CH_3 / I (cyclohexane) $\xrightarrow[-\ I^-]{^-CN}$ CH_3 / CN (cyclohexane)

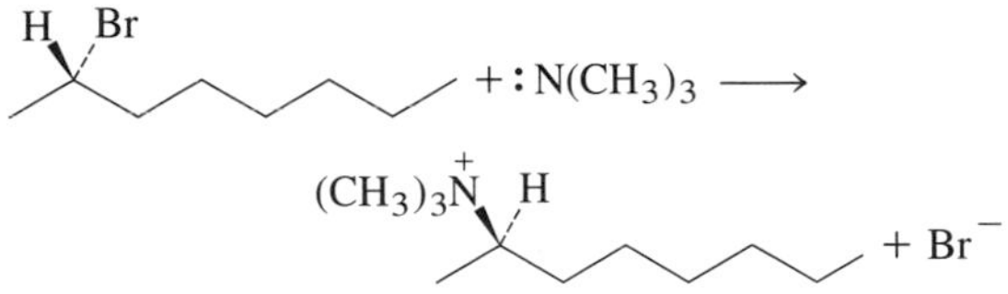

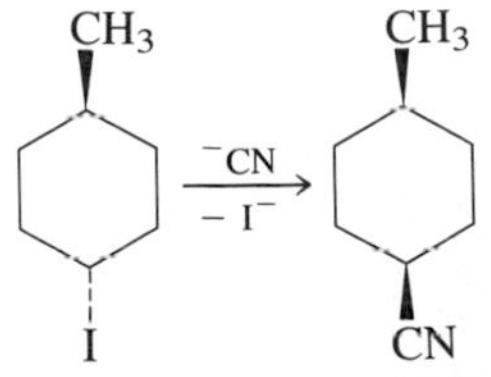

Trans Cis

6-8

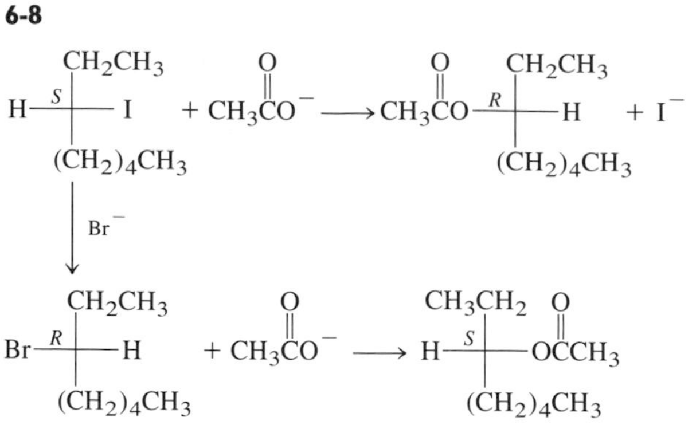

$$\text{H}\overset{S}{-}\text{C}(\text{CH}_2\text{CH}_3)((\text{CH}_2)_4\text{CH}_3)-\text{I} + \text{CH}_3\overset{\text{O}}{\overset{\|}{\text{C}}}\text{O}^- \longrightarrow \text{CH}_3\overset{\text{O}}{\overset{\|}{\text{C}}}\text{O}\overset{R}{-}\text{C}(\text{CH}_2\text{CH}_3)((\text{CH}_2)_4\text{CH}_3)-\text{H} + \text{I}^-$$

$\downarrow$ Br^-

$$\text{Br}\overset{R}{-}\text{C}(\text{CH}_2\text{CH}_3)((\text{CH}_2)_4\text{CH}_3)-\text{H} + \text{CH}_3\overset{\text{O}}{\overset{\|}{\text{C}}}\text{O}^- \longrightarrow \text{H}\overset{S}{-}\text{C}(\text{CH}_3\text{CH}_2)((\text{CH}_2)_4\text{CH}_3)-\text{O}\overset{\text{O}}{\overset{\|}{\text{C}}}\text{CH}_3$$

6-9

(S)-2-Iodooctane $\overset{I^-}{\rightleftharpoons}$ (R)-2-iodooctane, i.e., racemization.

6-10

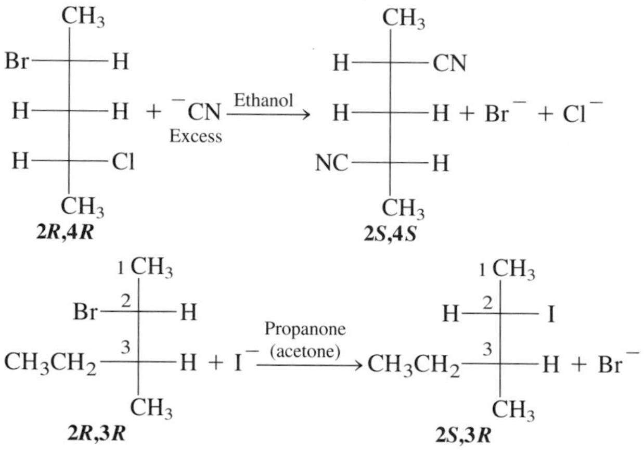

$$\begin{array}{c} \text{CH}_3 \\ \text{Br}-\text{C}-\text{H} \\ \text{H}-\text{C}-\text{H} \\ \text{H}-\text{C}-\text{Cl} \\ \text{CH}_3 \\ \mathbf{2R,4R} \end{array} + {}^-\text{CN} \xrightarrow[\text{Excess}]{\text{Ethanol}} \begin{array}{c} \text{CH}_3 \\ \text{H}-\text{C}-\text{CN} \\ \text{H}-\text{C}-\text{H} \\ \text{NC}-\text{C}-\text{H} \\ \text{CH}_3 \\ \mathbf{2S,4S} \end{array} + \text{Br}^- + \text{Cl}^-$$

$$\begin{array}{c} {}_1\text{CH}_3 \\ \text{Br}-\overset{2}{\text{C}}-\text{H} \\ \text{CH}_3\text{CH}_2-\overset{3}{\text{C}}-\text{H} \\ \text{CH}_3 \\ \mathbf{2R,3R} \end{array} + \text{I}^- \xrightarrow[\text{(acetone)}]{\text{Propanone}} \begin{array}{c} {}_1\text{CH}_3 \\ \text{H}-\overset{2}{\text{C}}-\text{I} \\ \text{CH}_3\text{CH}_2-\overset{3}{\text{C}}-\text{H} \\ \text{CH}_3 \\ \mathbf{2S,3R} \end{array} + \text{Br}^-$$

All four components are diastereomers of their counterparts on p. 186.

6-11

I^- is a better leaving group than Cl^-. Hence the product is $Cl(CH_2)_6SeCH_3$.

6-12

(a) ${}^-OH > {}^-SH$ **(b)** ${}^-PH_2 > {}^-SH$

(c) ${}^-SeH > I^-$ **(d)** $CH_3O\overset{O}{\overset{\|}{C}}O^- > CH_3\overset{O}{\overset{\|}{C}}O^-$

The CH_3O group donates electron density by resonance, destabilizing the negative charge.

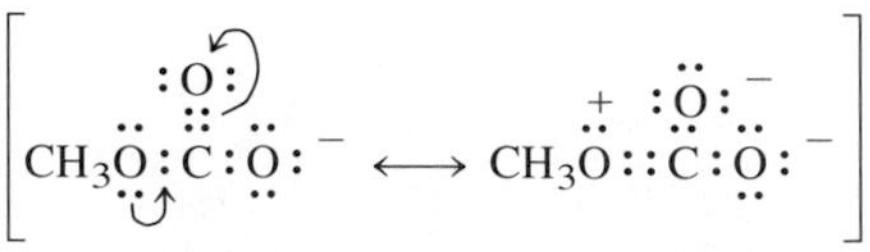

$$\left[\text{CH}_3\ddot{\text{O}}\!:\!\overset{:\text{O}:}{\overset{::}{\text{C}}}\!:\!\ddot{\text{O}}\!:^- \longleftrightarrow \text{CH}_3\overset{+}{\ddot{\text{O}}}\!::\!\overset{:\ddot{\text{O}}:^-}{\text{C}}\!:\!\ddot{\text{O}}\!:^- \right]$$

(e) $HOS\overset{O}{\overset{\|}{}}O^- > HO\overset{O}{\overset{\|}{\underset{\|}{\underset{O}{S}}}}O^-$

The relative acidities of the respective conjugate acids follow the inverse order.

6-13

(a) $HS^- > H_2S$ **(b)** $CH_3S^- > CH_3SH$
(c) $CH_3NH^- > CH_3NH_2$ **(d)** $HSe^- > H_2Se$

6-14

(a) $CH_3S^- > Cl^-$ **(b)** $P(CH_3)_3 > S(CH_3)_2$
(c) $CH_3CH_2Se^- > Br^-$ **(d)** $H_2O > HF$

6-15

(a) $CH_3SeH > CH_3SH$
(b) $(CH_3)_2PH > (CH_3)_2NH$

6-16

(a) CH_3S^- **(b)** $(CH_3)_2NH$

6-17

The more reactive substrates are **(a)** 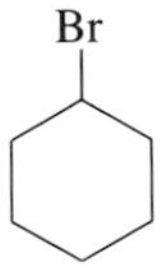 bromocyclohexane (cyclohexane ring bearing Br) and **(b)** $CH_3CH_2CH_2Br$

6-18

(Cyclohexylmethyl bromide, cyclohexane–CH_2Br) > (1-methylcyclohexyl)methyl bromide (H_3C and CH_2Br on the same ring carbon)

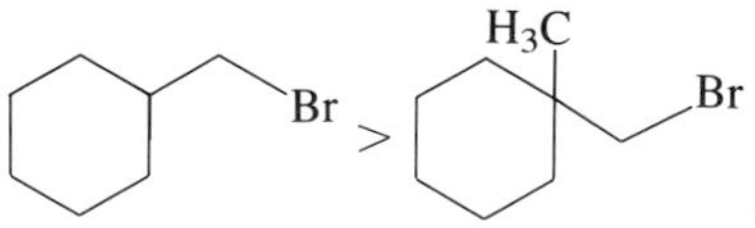

CHAPTER 7

7-1

Compound A is a 2,2-dialkyl-1-halopropane (neopentyl halide) derivative. The carbon bearing the potential leaving group is primary but very hindered and therefore very unreactive with respect to any substitution reactions. Compound B is a 1,1-dialkyl-1-haloethane (*tert*-alkyl halide) derivative and undergoes solvolysis.

7-2

Bonds broken: $67 + 119 = 186$ kcal mol^{-1}
Bonds made: $93 + 87 = 180$ kcal mol^{-1}
$\Delta H° = +6$ kcal mol^{-1}

By this calculation, the reaction should actually be endothermic. It still proceeds because of the excess water employed and the favorable solvation energies of the products.

7-3

H$_3$C Br; *R* ⇌ Achiral carbocation + Br$^-$ ⇌ Br CH$_3$; *S*

The molecule dissociates to the achiral tertiary carbocation. Recombination gives a 1:1 mixture of *R* and *S* product.

7-4

A ⇌ H$_3$C, H, CH$_3$ (Planar carbocation); H$_2$O, $-H^+$ → H$_3$C, H, CH$_3$, OH; H$_2$O, $-H^+$ → H$_3$C, H, OH, CH$_3$

7-5

$(CH_3)_3COSCH_3$ (with two S=O) $\xrightleftharpoons{\text{Propanone}}$ $(CH_3)_3C^+ + CH_3SO_3^-$

Br^- (Faster) → $(CH_3)_3CBr$; Cl^- (Slower) → $(CH_3)_3CCl$

7-6

(a) This is an S_N2 reaction that occurs with inversion.
(b) In a weakly nucleophilic protic solvent, mainly solvolysis occurs through the intermediacy of an achiral carbocation.

7-7

$(CH_3)_3CBr \rightleftharpoons (CH_3)_3C^+ + Br^-$

CH_3CH_2OH, S_N1 → $(CH_3)_3COCH_2CH_3$ + H^+
H_2O, S_N1 → $(CH_3)_3COH$ + H^+
E1 → $H_2C{=}C(CH_3)_2$ + H^+

7-8

Bromocyclohexane $\xrightarrow{HO^-}$ cyclohexanol (S_N2) + cyclohexene (E2)

7-9

$CH_2{=}CH_2$; no E2 possible; $CH_2{=}C(CH_3)_2$; no E2 possible.

7-10

I^- is a better leaving group, thus allowing for selective elimination of HI.

7-11

The cis isomer undergoes E2; the trans compound, E1.

7-12

All chlorines are equatorial, lacking *anti* hydrogens.

7-13

(a) $N(CH_3)_3$, stronger base, worse nucleophile
(b) $(CH_3\overset{CH_3}{\overset{|}{C}}H)_2N^-$, more hindered base
(c) Cl^-, stronger base, worse nucleophile

7-14

Eliminations are usually favored by entropy, and the entropy term in $\Delta G° = \Delta H° - T\Delta S°$ is temperature dependent.

7-15

(a) The second reaction will give more E2 product, because a stronger base is present.
(b) The first reaction will give E2 product, mainly because of the presence of a strong, hindered base.

CHAPTER 8

8-1

(a) 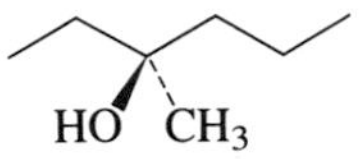 HO CH3 (b) 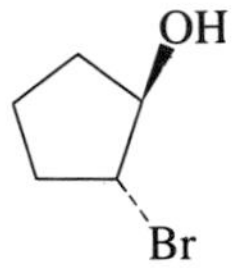 OH Br

(c) $(CH_3)_3CCH_2OH$

8-2

(a) 4-Methyl-2-pentanol
(b) *cis*-4-Ethylcyclohexanol
(c) 3-Bromo-2-chloro-1-butanol

8-3

All the bases whose conjugate acids have pK_a values $\gg$ 15.5—i.e., $CH_3CH_2CH_2CH_2Li$, LDA, and KH.

8-4

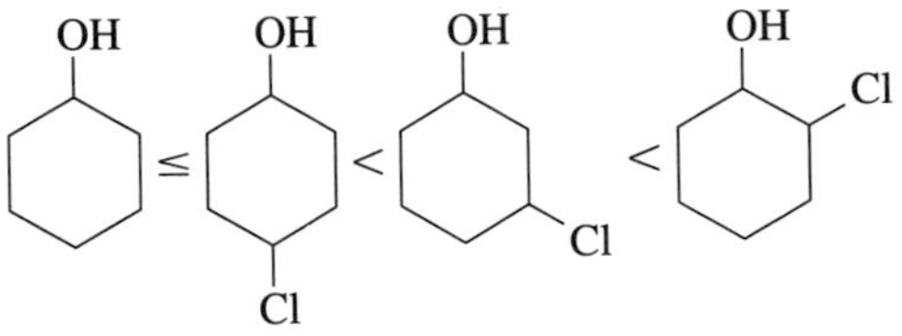

OH ≤ OH, Cl < OH, Cl < OH, Cl

8-5

In solution, $(CH_3)_3COH$ is a weaker acid than CH_3OH. The equilibrium lies to the right.

8-6

(a) NaOH, H_2O
(b) 1. NaO_2CCH_3, 2. NaOH, H_2O
(c) H_2O

8-7

(a) 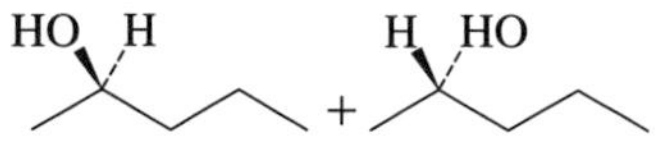 HO H + H HO

(b) $CH_3CH_2\overset{OH}{C}HCH_2CH_3$

(c) 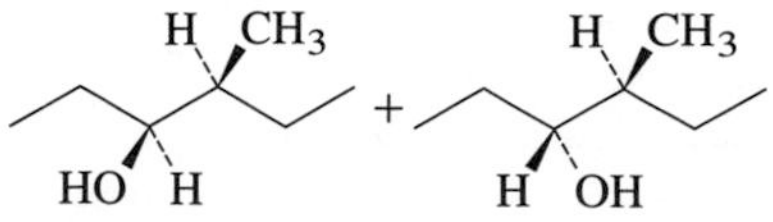 H CH3, HO H + H CH3, H OH

8-8

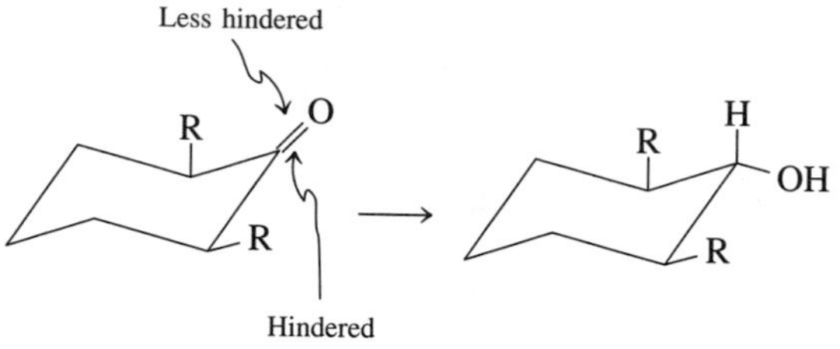

Less hindered, R, O, R, Hindered ⟶ R, H, OH, R

8-9

(a) $CH_3(CH_2)_8CHO + NaBH_4$ (b) 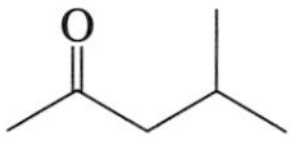 O + $NaBH_4$

(c) 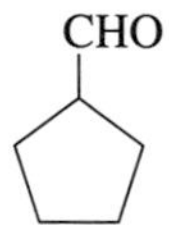 CHO + $NaBH_4$ (d) 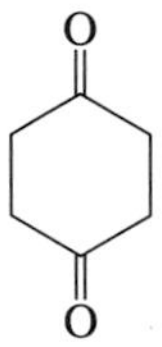 O, O + $NaBH_4$

8-10

1. 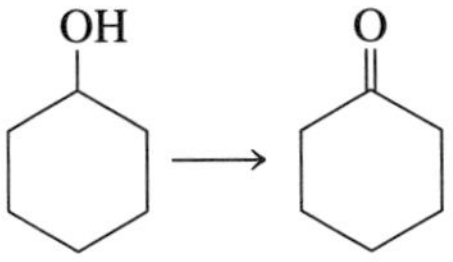 OH ⟶ O $+ 2\,H^+ + 2\,e$

2. $Cr_2O_7^{2-} + 14\,H^+ + 6\,e \longrightarrow 2\,Cr^{3+} + 7\,H_2O$

Addition of (1) and (2) and balancing the electron count gives

3 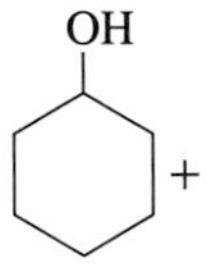 OH $+ Cr_2O_7^{2-} + 8\,H^+ \longrightarrow$

3 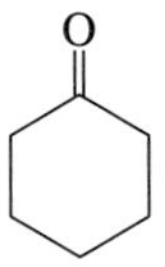 O $+ 2\,Cr^{3+} + 7\,H_2O$

Adding the counterions results in

3 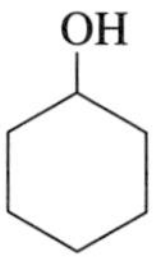 OH $+ Na_2Cr_2O_7 + 4\,H_2SO_4 \longrightarrow$

3 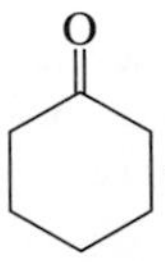 O $+ Cr_2(SO_4)_3 + Na_2SO_4 + 7\,H_2O$

8-11

(a) $CH_3CH_2CH(OH)CH(CH_3)_2 + Na_2Cr_2O_7$

(b) 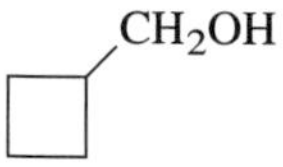 Cyclobutyl-CH_2OH + PCC

(c) 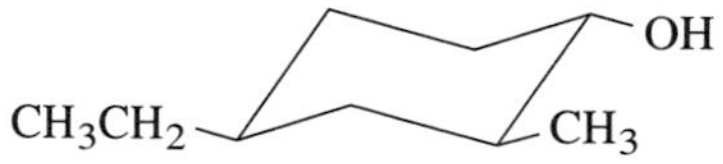 CH_3CH_2-cyclohexane(OH)(CH_3) + $Na_2Cr_2O_7$

8-12

Cyclohexane $\xrightarrow{Br_2,\ h\nu}$ bromocyclohexane $\xrightarrow{Mg}$ cyclohexyl-MgBr $\xrightarrow{D_2O}$ cyclohexane-D

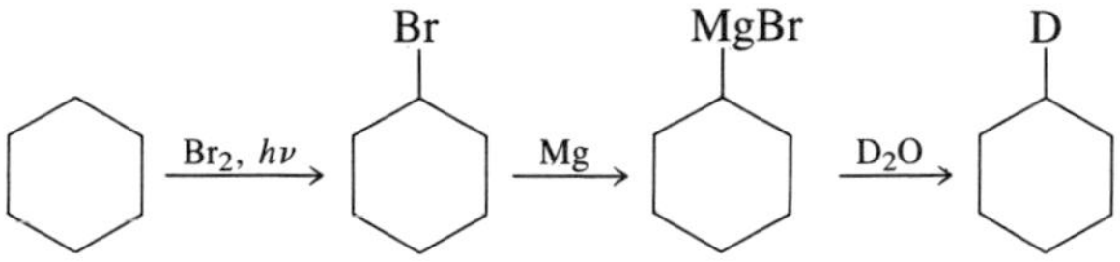

8-13

$(CH_3)_2CHBr \xrightarrow{Mg}$

$(CH_3)_2CHMgBr \xrightarrow{CH_2{=}O} (CH_3)_2CHCH_2OH$

8-14

(a) $CH_3CH_2CH_2CH_2Li + CH_2{=}O$

(b) $CH_3CH_2CH_2MgBr + CH_3CH_2CH_2CHO$

(c) $(CH_3)_3CLi$ + cyclobutanone

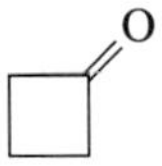

(d) $CH_3CH_2CH_2MgBr + CH_3CCH_2CH_3$ (with C=O)

8-15

(a) Product:

$ClCH_2CH_2CH_2C(CH_3)_2(OCH_2CH_3)$ By S_N1

(b) Product:

$CH_2{=}CHCH_2C(CH_3)_2(CH_2Cl)$ By E2 (hindered base)

The second chlorine is in a neopentyl position.

(c) Product:

$(CH_3)_2C(OH)CH_2CH_2CHO$

The second hydroxy function is tertiary.

8-16

The desired alcohol is tertiary and is therefore readily made from 4-ethylnonane by 1. Br_2, $h\nu$, 2. hydrolysis (S_N1). However, the starting hydrocarbon is itself complex and would require an elaborate synthesis. Thus, the retrosynthetic analysis by C–O disconnection is poor.

8-17

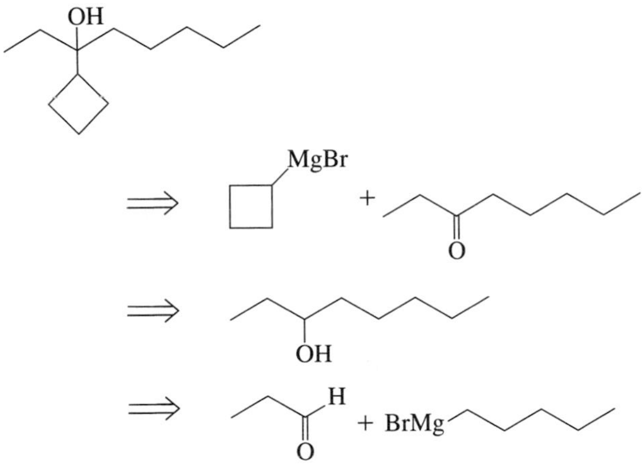

Tertiary alcohol (OH, cyclobutyl) $\Rightarrow$ cyclobutyl-MgBr + ketone (O)

$\Rightarrow$ secondary alcohol (OH)

$\Rightarrow$ aldehyde (H, O) + BrMg-pentyl

8-18

$CH_4 \xrightarrow{Br_2,\ h\nu} CH_3Br \xrightarrow{Mg} CH_3MgBr$

CH_3Br → (1. NaOH, 2. PCC) → $CH_2{=}O$

CH_3MgBr → (1. $CH_2{=}O$, 2. PCC) → CH_3CHO

$CH_3CHO \xrightarrow{1.\ CH_3MgBr,\ 2.\ Na_2Cr_2O_7} CH_3CCH_3$ (with C=O) $\xrightarrow{CH_3MgBr} (CH_3)_3COH$

CHAPTER 9

9-1

$$\underset{pK_a = 15.5}{CH_3OH} + {}^-CN \overset{K = 10^{-6.3}}{\rightleftharpoons} CH_3O^- + \underset{pK = 9.2}{HCN}$$

Answer: No.

9-2

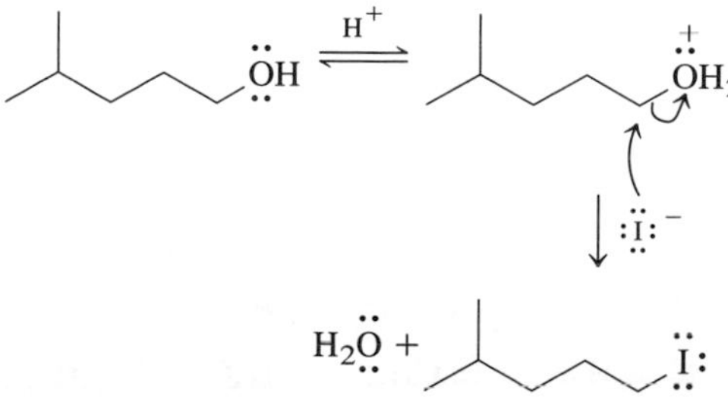

Alcohol (ÖH) $\overset{H^+}{\rightleftharpoons}$ $\overset{+}{O}H_2$; $:\ddot{I}:^-$ ↓

$H_2\ddot{O}$ + alkyl–$\ddot{I}:$

9-3

1-Methylcyclohexanol (HO, CH_3) $\underset{}{\overset{H^+}{\rightleftharpoons}}$ protonated alcohol ($H_2\overset{+}{O}$, CH_3) $\underset{}{\overset{-H_2O}{\rightleftharpoons}}$ tertiary carbocation (CH_3, +)

Carbocation $\xrightarrow{Cl^-}$ 1-chloro-1-methylcyclohexane (Cl, CH_3) **(a)**

Carbocation $\xrightarrow{-H^+ \mid SO_4^{2-}}$ 1-methylcyclohexene **(b)**

The tertiary carbocation is either trapped by the nucleophile (Cl^-) or undergoes E1. (SO_4^{2-} is a poor nucleophile.)

9-4

2-Methylcyclohexanol (CH_3, $\ddot{O}H$) $\overset{H^+}{\rightleftharpoons}$ (CH_3, $\overset{+}{O}H_2$) $\overset{-H_2\ddot{O}:}{\rightleftharpoons}$

(H_3C, H, +) $\longrightarrow$ (CH_3, +, H, H) $\xrightarrow{:\ddot{Br}:^-}$ (H_3C, $\ddot{Br}:$)

Secondary carbocation — Tertiary carbocation

9-5

(a) $CH_3\underset{CH_3}{\overset{OCH_3}{C}}CH_2CH_2CH_3$ **(b)** 1-chloro-1-ethylcyclohexane (CH_3CH_2, Cl)

9-6

$CH_3\underset{H_3C}{\overset{Br}{C}}-\underset{CH_2CH_3}{\overset{H}{C}}CH_2CH_3 \xrightarrow{-Br^-}$

$CH_3\underset{H_3C}{\overset{+}{C}}-\underset{CH_2CH_3}{\overset{H}{C}}CH_2CH_3 \rightleftharpoons CH_3\underset{H_3C}{\overset{H}{C}}-\underset{CH_2CH_3}{\overset{+}{C}}CH_2CH_3 \xrightarrow[-H^+]{CH_3CH_2OH}$

$CH_3\underset{H_3C}{\overset{CH_3CH_2O}{C}}-\underset{CH_2CH_3}{\overset{H}{C}}CH_2CH_3 + CH_3\underset{H_3C}{\overset{H}{C}}-\underset{CH_2CH_3}{\overset{OCH_2CH_3}{C}}CH_2CH_3$

Similarly,

$CH_3\underset{H}{\overset{CH_3}{C}}CH_2\underset{H}{\overset{Cl}{C}}CH_3 \overset{-Cl^-}{\rightleftharpoons} CH_3\underset{H}{\overset{H}{C}}CH_2\underset{H}{\overset{+}{C}}CH_3 \overset{\text{H shift}}{\rightleftharpoons}$

Secondary carbocation

$CH_3\underset{H}{\overset{CH_3}{C}}\overset{+}{C}HCH_2CH_3 \xrightarrow{\text{Second H shift}}$

Secondary carbocation

$CH_3\underset{+}{\overset{CH_3}{C}}CH_2CH_2CH_3 \xrightarrow[-H^+]{CH_3OH} CH_3\underset{CH_3O}{\overset{CH_3}{C}}CH_2CH_2CH_3$

Tertiary carbocation

9-7

(a) $CH_3\underset{CH_3}{\overset{OH}{C}}CH_2CH_2CH_3 \xrightarrow[-H_2O]{\text{Straight E1}} (H_3C)_2C{=}C(H)CH_2CH_3$

$CH_3\underset{H}{\overset{CH_3}{C}}CH_2\underset{H}{\overset{OH}{C}}CH_3 + H^- \underset{+H_2O}{\overset{-H_2O}{\rightleftharpoons}} CH_3\underset{H}{\overset{H_3C}{C}}-\underset{H}{\overset{\mathbf{H}}{C}}H\overset{+}{C}CH_3 \overset{\textbf{H shift}}{\rightleftharpoons}$

$CH_3\underset{H}{\overset{H_3C}{C}}-\overset{+}{C}H\underset{H}{\overset{\mathbf{H}}{C}}CH_3 \rightleftharpoons (H_3C)_2C{=}C(H)CH_2CH_3 + H^+$

(b) 4-Methylcyclohexanol (CH_3, OH) $\xrightarrow[-H_2O]{H^+}$ secondary carbocation (CH_3, +) $\xrightarrow{\text{H shifts}}$ (dashed)

tertiary carbocation (CH_3, +) $\xrightarrow{-H^+}$ 1-methylcyclohexene (CH_3)

9-8

$(CH_3)_3CCH{=}CH_2$ and $(CH_3)_2C{=}C(CH_3)_2$

9-9

$$\text{OH-cyclohexane-CH}_3 \xrightarrow[-\ HCl]{CH_3SO_2Cl} \text{CH}_3\text{SO}_2\text{O-cyclohexane-CH}_3 \xrightarrow[-\ CH_3SO_3Na]{NaI} \text{I-cyclohexane-CH}_3$$

9-10

(a) 1. CH_3SO_2Cl, 2. NaI
(b) HCl
(c) PBr_3

9-11

(a) 1. $CH_3CH_2I + CH_3CH_2CH_2CH_2O^-Na^+$, 2. $CH_3CH_2O^-Na^+ + CH_3CH_2CH_2CH_2I$

(b) Best is O^-Na^+ (2-pentoxide) $+ CH_3I$

The alternative, $CH_3O^-Na^+$ + (2-iodopentane) suffers from competing E2.

(c) O^-Na^+ (cyclohexoxide) $+ CH_3CH_2CH_2Br$

(d) $Na^+\ ^-O$~~~~$O^-\ Na^+$

$+ CH_3CH_2OSO_2CH_3$

9-12

Br~~~(C(CH$_3$)$_2$)~~$\ddot{O}$—H $\xrightleftharpoons{^-:\ddot{O}H,\ -\ H_2\ddot{O}}$

$:\ddot{Br}$~~~~$\ddot{O}:^-$ $\longrightarrow$ (4,4-dimethyltetrahydropyran) $+ :\ddot{Br}:^-$

9-13

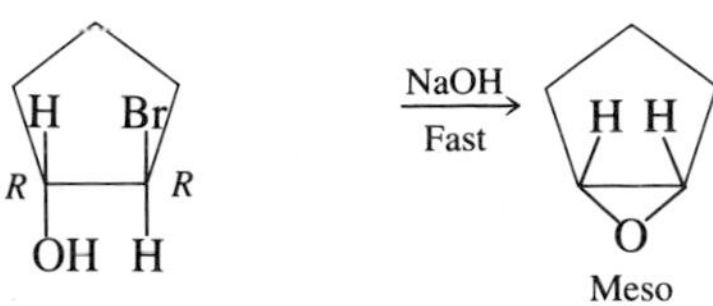

(1*R*,2*R*)-2-Bromocyclopentanol
The nucleophilic oxygen and the leaving group are trans (*anti*)

NaOH, Fast → Meso

$\xrightarrow{NaOH}$ no epoxide formation, relatively slow E2 and S_N2

(1*S*,2*R*)-2-Bromocyclopentanol
Here nucleophile and leaving group are cis (*syn*)

9-14

(a) $H\ddot{O}CH_2CH_2CH_2CH_2\ddot{O}H + H^+ \rightleftharpoons$

$H_2C(CH_2)CH_2\overset{+}{O}H_2$ … $\ddot{O}$:—H $\xrightarrow{-\ H_2\ddot{O}}$ (tetrahydrofuran, O–H$^+$) $\rightleftharpoons$

(tetrahydrofuran) $+ H^+$

(b) $CH_3\overset{CH_3}{\underset{:\ddot{O}H}{C}}CH_2CH_2CH_2CH_2\ddot{O}H + H^+ \xrightleftharpoons[+\ H_2\ddot{O}]{-\ H_2\ddot{O}}$

$(CH_3)_2\overset{+}{C}$—CH_2—CH_2—CH_2—CH_2—$\ddot{O}$—H $\longrightarrow$ (2,2-dimethyltetrahydropyran, $\overset{+}{O}$–H, CH_3, CH_3) $\rightleftharpoons$

(2,2-dimethyltetrahydropyran, CH_3, CH_3) $+ H^+$

9-15

(a) This ether is best synthesized by solvolysis:

$$CH_3CH_2C(CH_3)_2Br + CH_3CH(CH_3)OH \longrightarrow CH_3CH_2C(CH_3)_2{-}O{-}CH(CH_3)CH_3$$

Solvent — **2-Methyl-2-(1-methylethoxy)butane**

The alternative, an S_N2 reaction, would give elimination:

$$CH_3CH_2C(CH_3)_2O^- + CH_3CH(CH_3)Br \longrightarrow$$

$$CH_3CH{=}CH_2 + CH_3CH_2C(CH_3)_2OH$$

(b) This target is best prepared by an S_N2 reaction with a halomethane, because such an alkylating agent cannot undergo elimination. The alternative would be nucleophilic substitution of a 1-halo-2,2-dimethylpropane, a reaction that is normally too slow.

$$CH_3C(CH_3)_2CH_2O^- + CH_3Cl \longrightarrow CH_3C(CH_3)_2CH_2OCH_3 + Cl^-$$

1-Methoxy-2,2-dimethylpropane

$$CH_3C(CH_3)_2CH_2Br + CH_3O^- \longrightarrow \text{slow reaction, impractical}$$

9-16

$$CH_3OCH_3 + 2\ HI \xrightarrow{\Delta} 2\ CH_3I + H_2O$$

Mechanism

$$CH_3\ddot{O}CH_3 + H\ddot{I}: \rightleftharpoons CH_3\overset{+}{O}(H)CH_3 + :\ddot{I}:^-$$

$$:\ddot{I}:^- \ CH_3{-}\overset{+}{O}(H){-}CH_3 \longrightarrow CH_3\ddot{I}: + H\ddot{O}CH_3$$

$$CH_3\ddot{O}H + H\ddot{I}: \rightleftharpoons CH_3\overset{+}{O}(H)H + :\ddot{I}:^-$$

$$:\ddot{I}:^- \ CH_3{-}\overset{+}{O}(H){-}H \longrightarrow CH_3\ddot{I}: + H_2\ddot{O}$$

9-17

Oxacyclohexane (tetrahydropyran) $+ H^+ \longrightarrow$ protonated oxacyclohexane (attacked by $:\ddot{I}:^-$) $\longrightarrow$

$$HO(CH_2)_5I \xrightarrow{H^+}$$

$$H_2\overset{+}{O}(CH_2)_5I \ \ (:\ddot{I}:^-) \longrightarrow$$

$$H_2O + I(CH_2)_5I$$

9-18

$$BrCH_2CH_2CH_2OH \xrightarrow[\text{4. } H^+,\ H_2O]{\text{1. } (CH_3)_3COH,\ H^+ \quad \text{2. Mg} \quad \text{3. } D_2O} DCH_2CH_2CH_2OH$$

9-19

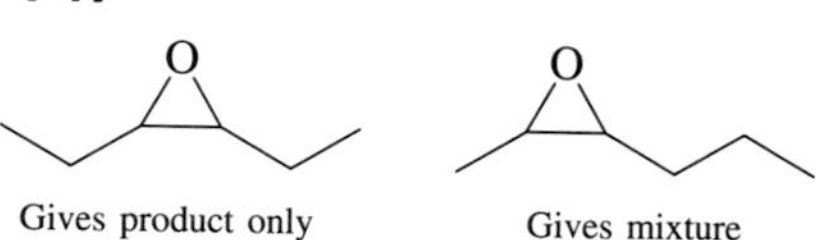

Gives product only — Gives mixture

9-20

$(CH_3)_3CLi$ + oxacyclopropane

9-21

(a) $(CH_3)_3COH$ **(b)** $CH_3CH_2CH_2CH_2C(CH_3)_2OH$

(c) $CH_3SCH_2C(CH_3)_2OH$

(d) $HOCH_2C(CH_3)_2OCH_2CH_3$

(e) $HOCH_2C(CH_3)_2Br$

9-22

(a) Oxacyclopropane $+ HS^- \longrightarrow HOCH_2CH_2S^- \xrightarrow{\text{oxacyclopropane}}$

$$\xrightarrow{H^+,\ H_2O} HOCH_2CH_2SCH_2CH_2OH \xrightarrow{SOCl_2} ClCH_2CH_2SCH_2CH_2Cl$$

(b) Intramolecular sulfonium salt formation

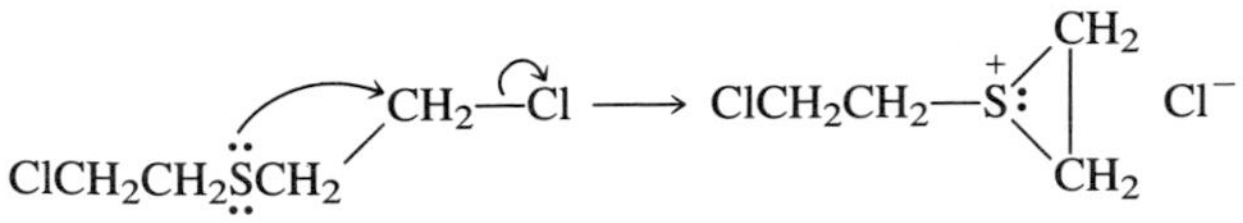

$$ClCH_2CH_2\ddot{S}CH_2\text{—}CH_2\text{—}Cl \longrightarrow ClCH_2CH_2\text{—}\overset{+}{S}(CH_2)_2 \quad Cl^-$$

Nucleophiles attack by ring opening

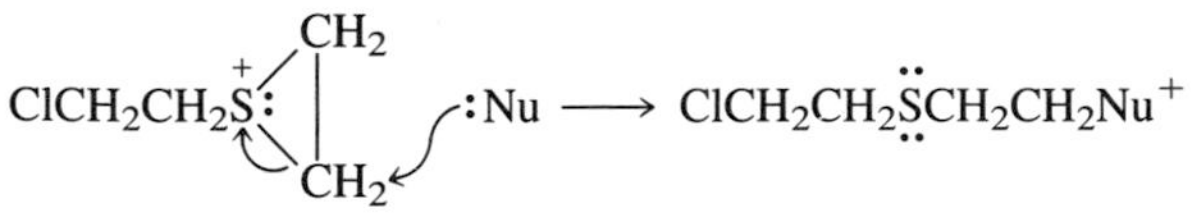

$$ClCH_2CH_2\overset{+}{S}(CH_2)_2 + :Nu \longrightarrow ClCH_2CH_2\ddot{S}CH_2CH_2Nu^+$$

CHAPTER 10

10-1

There are quite a number of isomers, e.g., one substituted propanol, several butanols, pentanols, hexanols, and heptanols. Examples include

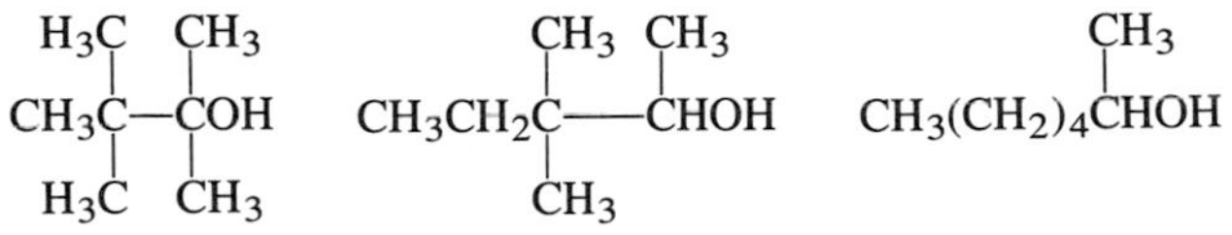

$$(CH_3)_3C\text{—}C(CH_3)_2OH \qquad CH_3CH_2C(CH_3)_2\text{—}CH(CH_3)OH \qquad CH_3(CH_2)_4CH(CH_3)OH$$

10-2

$DH°_{Cl_2} = 58$ kcal mol^{-1}
$\Delta E = 28{,}600/\lambda$
$\lambda = 28{,}600/\Delta E = 490$ nm, in the ultraviolet-visible range

10-3

$\delta = 288/90 = 3.20$ ppm and $\delta = 297/90 = 3.30$ ppm. In a 100-MHz spectrometer, the signals would appear 320 and 330 Hz downfield from $(CH_3)_4Si$.

10-4

The methyl group resonates at higher field; the methylene hydrogens are relatively deshielded because of the cumulative electron-withdrawing effect of the two heteroatoms.

10-5

(a) 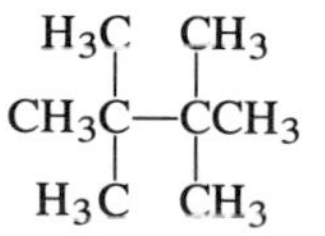 $(CH_3)_3C\text{—}C(CH_3)_3$ One peak

(b) $CH_3OCH_2CH_2OCH_2CH_2OCH_3$ Three peaks

(c) 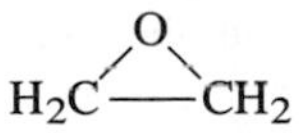 $H_2C\text{—}CH_2$ (bridged by O) One peak

10-6

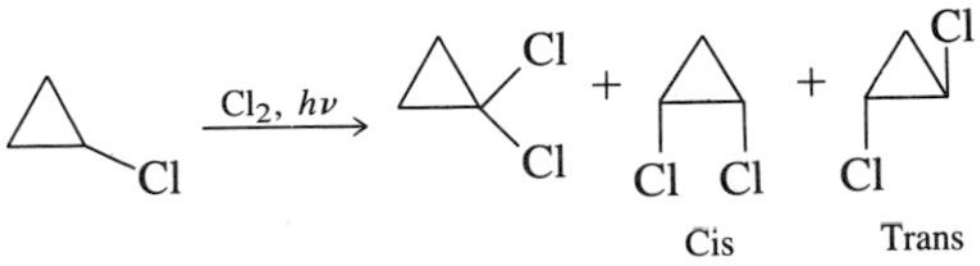

$$\text{chlorocyclopropane} \xrightarrow{Cl_2,\ h\nu} \text{1,1-dichlorocyclopropane} + \text{1,2-dichlorocyclopropane (Cis)} + \text{1,2-dichlorocyclopropane (Trans)}$$

1,1-Dichlorocyclopropane shows only one signal whereas *cis*-1,2-dichlorocyclopropane exhibits three in the ratio 2:1:1. In the cis isomer the lowest field absorption is due to the two equivalent hydrogens next to the chlorine atoms at C1 and C2. The two hydrogens at C3 are not equivalent: One lies cis to the chlorine atoms, the other trans. In contrast, in the trans isomer, the hydrogens at C3 *are* equivalent, as shown by a 180° rotational symmetry operation:

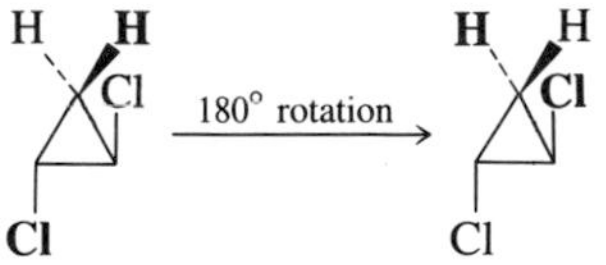

H H Cl Cl —180° rotation→ H H Cl Cl

Therefore, this isomer reveals only two signals (integration ratio, 1:1).

10-7

The following δ values were recorded in CCl_4 solution.

(a) $\delta = 3.38$ (q, $J = 7.1$ Hz, 4 H) and 1.12 (t, $J = 7.1$ Hz, 6 H) ppm

(b) $\delta = 3.53$ (t, $J = 6.2$ Hz, 4 H) and 2.34 (quin, $J = 6.2$ Hz, 2 H) ppm

(c) $\delta = 3.19$ (s, 1 H), 1.48 (q, $J = 6.7$ Hz, 2 H), 1.14 (s, 6 H), and 0.90 (t, $J = 6.7$ Hz, 3 H) ppm

(d) $\delta = 5.58$ (t, $J = 7$ Hz, 1 H) and 3.71 (d, $J = 7$ Hz, 2 H) ppm

10-8

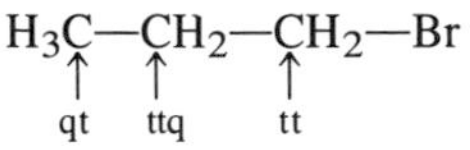

$H_3C\text{—}CH_2\text{—}CH_2\text{—}Br$
qt ttq tt

10-9

(a) 3; **(b)** 3; **(c)** 7; **(d)** 2

10-10

(a) Three lines, one of them at relatively high field (CH_3); all of them are inverted in APT ^{13}C NMR.

(b) Three lines, no CH_3 absorption; the lowest field absorption (methine carbons) is inverted in APT ^{13}C NMR.

CHAPTER 11

11-1

(a) 2,3-Dimethyl-2-heptene
(b) 3-Bromocyclopentene

11-2

(a) *cis*-1,2-Dichloroethene
(b) *trans*-3-Heptene

11-3

(a) (*E*)-1,2-Dideuterio-1-propene
(b) (*Z*)-2-Fluoro-3-methoxy-2-pentene

11-4

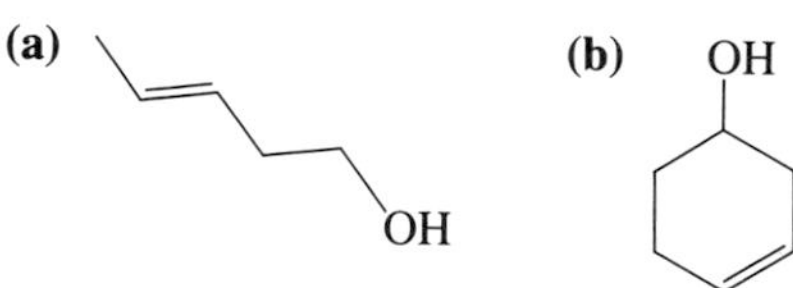

11-5

(b) (1-Methylethenyl)cyclopentane or (1-methylvinyl)cyclopentane

11-6

$$CH_2{=}CHLi + CH_3\overset{\displaystyle O}{\overset{\|}{C}}CH_3 \longrightarrow CH_2{=}CH\underset{\displaystyle CH_3}{\overset{\displaystyle OH}{C}}CH_3$$

The reaction of ethenyllithium (vinyllithium) with carbonyl compounds is like that of other alkyllithium organometallics.

11-7

The induced local magnetic field strengthens H_0 in the region occupied by the methyl hydrogens.

11-8

H_3C (1.88), H (6.95) on one alkene carbon; H (5.81) and $C(=O){-}O{-}CH_2{-}CH_3$ (4.13, 1.24) on the other.

The trans coupling constant is 16.0 Hz. The couplings to the methyl group on the double bond conform with the values in Table 11-1.

11-9

Alkene A: $H_3C(H)C{=}C(H)CH_3$ (trans)

B: $CH_3CH_2CH{=}CH_2$

C: $CH_2{=}C(CH_3)_2$

11-10

(a) $H_{sat} = 12$, degree of unsaturation = 1
(b) $H_{sat} = 20$; degree of unsaturation = 4
(c) $H_{sat} = 17$; degree of unsaturation = 5
(d) $H_{sat} = 19$; degree of unsaturation = 2
(e) $H_{sat} = 8$; degree of unsaturation = 0

11-11

(a) (b) (c)

Another possibility is H, CH_2CH_3 (cyclopropene)

which should, however, exhibit a distinct methyl triplet signal as part of the high-field multiplet absorption.

11-12

1-Hexene < *cis*-3-hexene < *trans*-4-octene < 2,3-dimethyl-2-butene.

11-13

If you can make a model of alkene A (without breaking your plastic sticks), you will notice its extremely strained nature, much of which is released on hydrogenation. You can estimate the excess strain in A (relative to B) by subtracting the $\Delta H°$ of the hydrogenation of a "normal" tetrasubstituted double-bond (~ -27 kcal mol^{-1}) from the $\Delta H°$ of the A-to-B transformation: 38 kcal mol^{-1}.

11-14

$(CH_3)_2C{=}CHCH_3$	$(CH_3)_2CHCH{=}CH_2$
A	**B**

Product B results from abstraction of the more accessible methyl group hydrogen at C1, favored with the more hindered base.

11-15

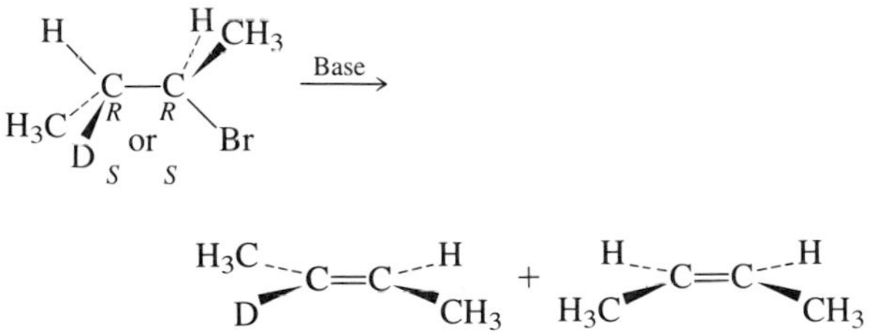

[Structure: (R,R) or (S,S) 2-bromo-3-deuteriobutane, H, CH_3 / H, H_3C, D, Br] $\xrightarrow{\text{Base}}$

[$H_3C(D)C{=}C(H)CH_3$] + [$H(H_3C)C{=}C(H)CH_3$]

E Cis (*Z*)

[Structure: (S,R) or (R,S) 2-bromo-3-deuteriobutane] $\xrightarrow{\text{Base}}$

[$D(H_3C)C{=}C(H)CH_3$] + [$H_3C(H)C{=}C(H)CH_3$]

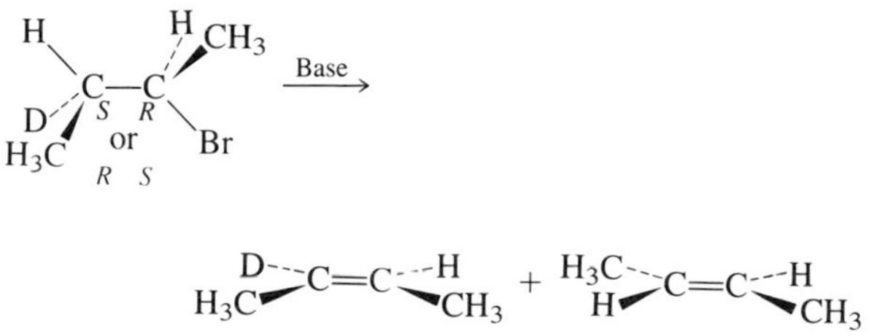

Z Trans (*E*)

Note that in the first case a pair of isomers is formed with the configuration *opposite* that generated in the second. The *E* and *Z* isomers of 2-deuterio-2-butene are isotopically pure in each case; none of the protic 2-butene with the same configuration is generated. The protio-2-butenes are also pure, devoid of any deuterium.

11-16

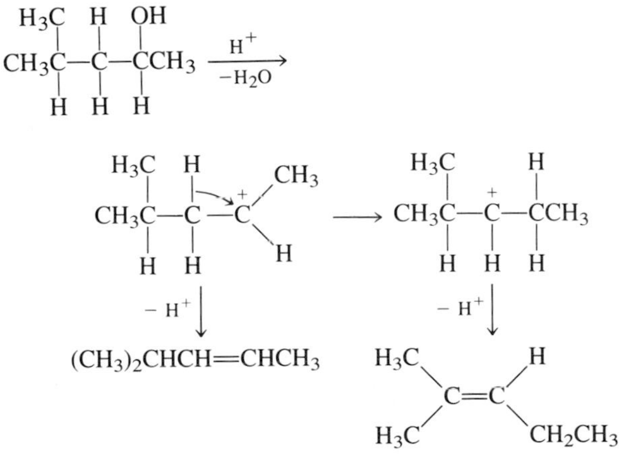

$CH_3C(CH_3)(H)-CH(H)-C(OH)(H)CH_3 \xrightarrow[-H_2O]{H^+}$

$CH_3C(CH_3)(H)-C(H)(H)-\overset{+}{C}(CH_3)(H) \longrightarrow CH_3C(CH_3)-\overset{+}{C}(H)-C(H)(H)CH_3$

$\downarrow -H^+$ $\downarrow -H^+$

$(CH_3)_2CHCH{=}CHCH_3$ $(H_3C)_2C{=}C(H)CH_2CH_3$

11-17

(a) $CH_3CH_2CH_2\ddot{O}H \underset{-H^+}{\overset{+H^+}{\rightleftharpoons}}$

$CH_3CH(H)CH_2\overset{+}{O}H_2 \xrightarrow[-H_2SO_4]{-HOSO_3^-} CH_3CH{=}CH_2 + H_2\ddot{O}$

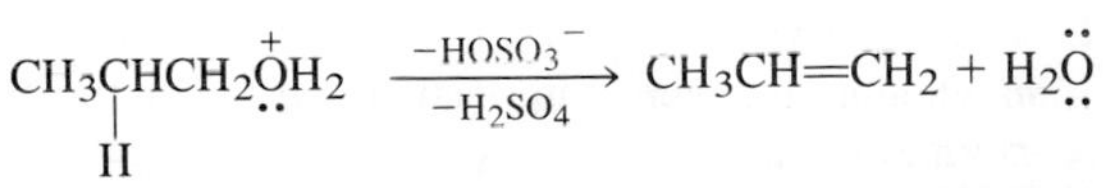

(b) $CH_3CH_2CH_2OCH_2CH_2CH_3 \overset{H^+}{\rightleftharpoons}$ $CH_3CH{=}CH_2 + CH_3CH_2CH_2OH$ in analogy to (a). The propanol may then be dehydrated as in (a).

CHAPTER 12

12-1

Estimating the strengths of the bonds broken and the bonds made

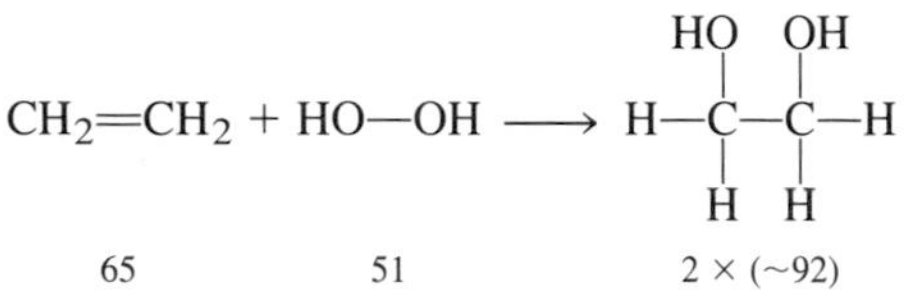

$$CH_2{=}CH_2 + HO{-}OH \longrightarrow H{-}\underset{H}{\overset{HO}{C}}{-}\underset{H}{\overset{OH}{C}}{-}H$$

65 51 2 × (~92)

gives $\Delta H^\circ = -68$ kcal mol^{-1}. Even though very exothermic, this reaction requires a catalyst.

12-2

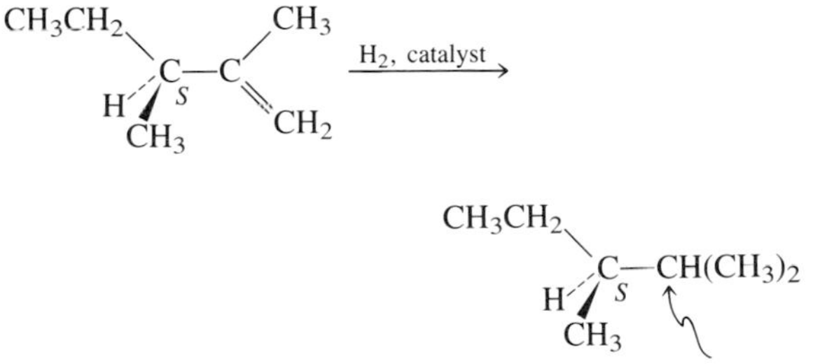

[Structure: CH_3CH_2, H, CH_3 on *S* carbon attached to $C(CH_3){=}CH_2$] $\xrightarrow{H_2,\ \text{catalyst}}$

[Structure: CH_3CH_2, H, CH_3 on *S* carbon attached to $CH(CH_3)_2$]

Not a stereocenter

12-3

(a) 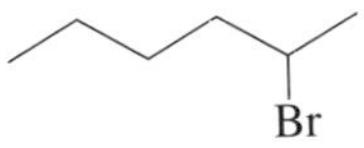 [2-bromohexane skeletal structure] Both enantiomers

Br

(b) 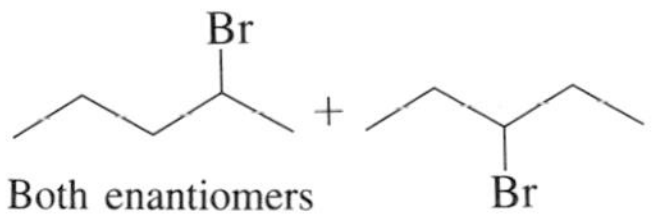 [3-bromohexane] + [3-bromohexane]

Br

Both enantiomers Br

(c) 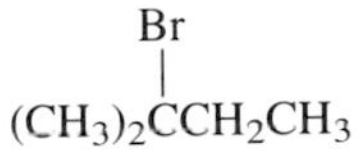 $(CH_3)_2C(Br)CH_2CH_3$

(d) 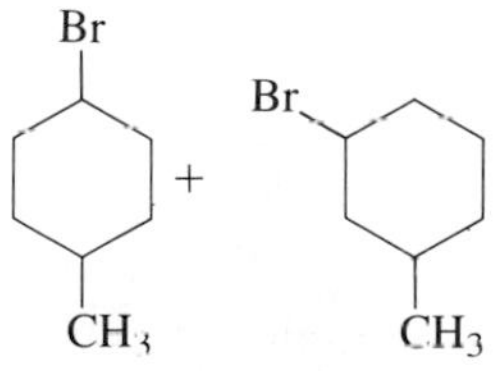 [1-bromo-4-methylcyclohexane] + [1-bromo-3-methylcyclohexane]

Br Br

CH_3 CH_3

Both cis and trans

12-4

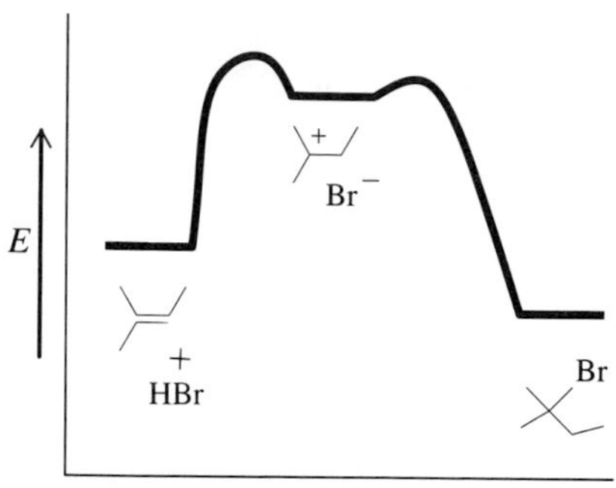

12-5

Protonation to the 1,1-dimethylethyl (*tert*-butyl) cation is reversible. With D^+, fast exchange of all hydrogens for deuterium will occur.

$$CH_2{=}C(CH_3)_2 \underset{-D^+}{\overset{+D^+}{\rightleftharpoons}} DCH_2\overset{+}{C}(CH_3)_2 \underset{+H^+}{\overset{-H^+}{\rightleftharpoons}}$$

$$DCH{=}C(CH_3)_2 \underset{-D^+}{\overset{+D^+}{\rightleftharpoons}} D_2CH\overset{+}{C}(CH_3)_2 \underset{+H^+}{\overset{-H^+}{\rightleftharpoons}}$$

$$D_2C{=}C(CH_3)_2 \underset{-D^+}{\overset{+D^+}{\rightleftharpoons}} D_3C\overset{+}{C}(CH_3)_2 \underset{+H^+}{\overset{-H^+}{\rightleftharpoons}}$$

$$(D_3C)(H_3C)C{=}CH_2 \underset{-D^+}{\overset{+D^+}{\rightleftharpoons}} \text{and so on} \dashrightarrow$$

$$(CD_3)_3C^+ \underset{-D_2O}{\overset{D_2O}{\rightleftharpoons}} (CD_3)_3COD + D^+$$

12-6

H_3C $\underset{-H^+}{\overset{+H^+}{\rightleftharpoons}}$ H_3C, CH_3 $\underset{+H^+}{\overset{-H^+}{\rightleftharpoons}}$

H_3C, CH_3 $\underset{-H^+}{\overset{+H^+}{\rightleftharpoons}}$ H_3C, CH_2CH_3 $\underset{+H^+}{\overset{-H^+}{\rightleftharpoons}}$ H_3C, CH_2CH_3

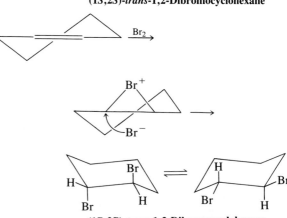

Tetrasubstituted alkene, most stable

12-7

$$\underset{65}{CH_2{=}CH_2} + \underset{37}{F{-}F} \longrightarrow \underset{2 \times (\sim 107)\ \text{kcal mol}^{-1}}{FCH_2{-}CH_2F}$$

$$\Delta H^\circ = -112\ \text{kcal mol}^{-1}$$

$$\underset{65}{CH_2{=}CH_2} + \underset{36}{I{-}I} \longrightarrow \underset{2 \times (\sim 53)\ \text{kcal mol}^{-1}}{ICH_2{-}CH_2I}$$

$$\Delta H^\circ = -5\ \text{kcal mol}^{-1}$$

12-8

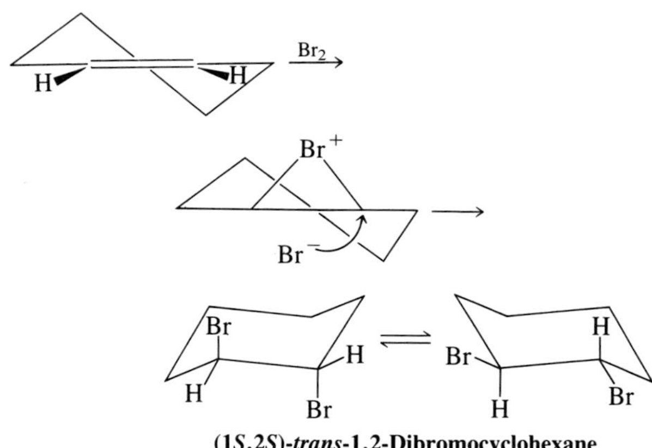

(1*S*,2*S*)-*trans*-1,2-Dibromocyclohexane

Br_2, Br^+, Br^-

H, Br, Br, H ⇌ H, Br, Br, H

(1*R*,2*R*)-*trans*-1,2-Dibromocyclohexane

Anti addition to either conformation gives the trans-diaxial conformer initially.

12-9

(a) Only one diastereomer is formed (as a racemate):

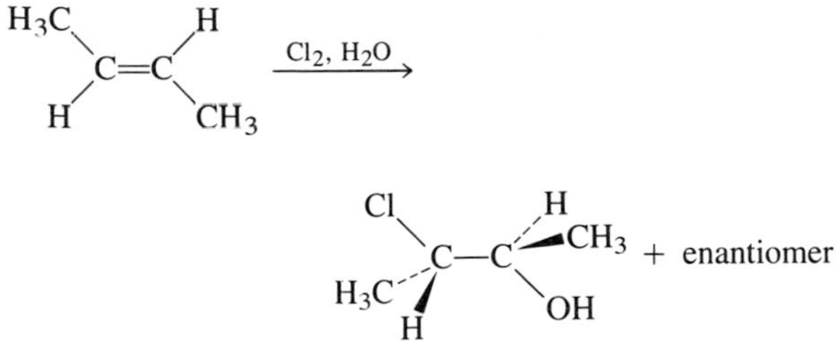

$H_3C(H)C{=}C(H)CH_3$ $\xrightarrow{Cl_2,\ H_2O}$

Cl, H_3C, H–C–C–H, CH_3, OH + enantiomer

(b) Two isomers are formed, but only one diastereomer of each (as racemates):

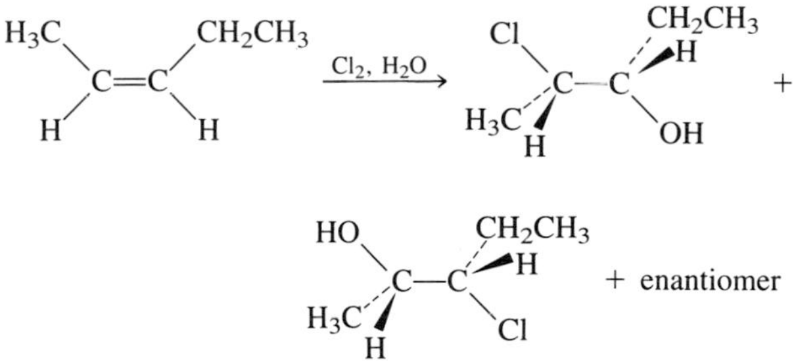

$H_3C(H)C{=}C(H)CH_2CH_3$ $\xrightarrow{Cl_2,\ H_2O}$ Cl, H_3C, H–C–C–CH_2CH_3, H, OH +

HO, H_3C, H–C–C–CH_2CH_3, H, Cl + enantiomer

12-10

(a) $CH_3CH(OCH_3)CH_2Cl$ (both enantiomers)

(b) [cyclohexane structures: H_3C, Br, OH] + [H_3C, OH, Br] + [H_3C, Br, OH] + [H_3C, OH, Br]

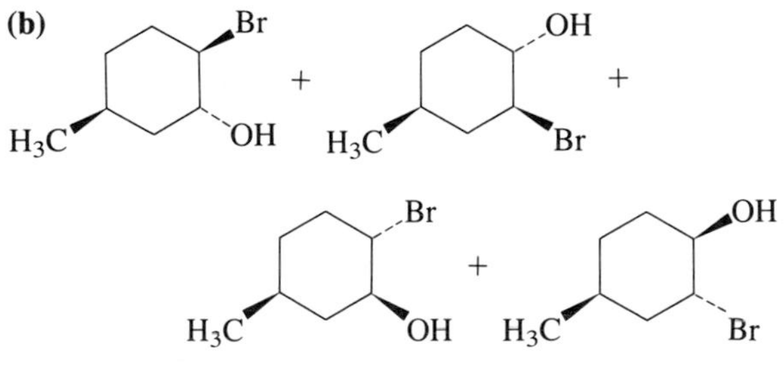

+ all enantiomers

12-11

$H_3C(H)C{=}C(H)CH_2CH_3$ $\xrightarrow{Br_2,\ CH_3OH}$

***cis*-2-Pentene**

Br, H, H_3C, C–C (*R*, *R*), H, CH_2CH_3, OCH_3 + enantiomer

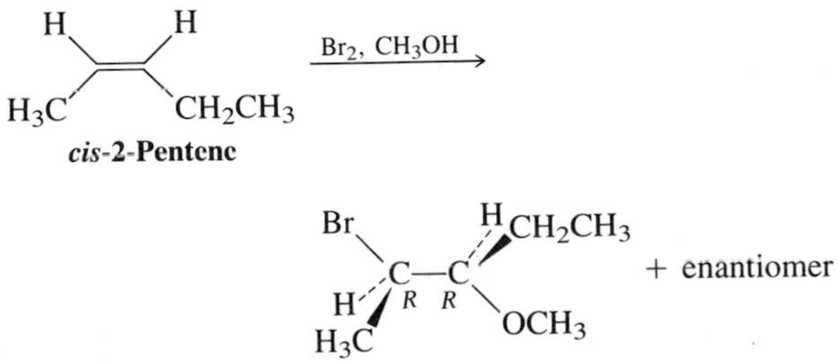

Opening of the bromonium ion can also give (3*R*,2*R*)- and (3*S*,2*S*)-3-bromo-2-methoxypentane.

12-12

Mercuration is followed by *intramolecular* trapping of the mercurinium ion by one of the hydroxy groups.

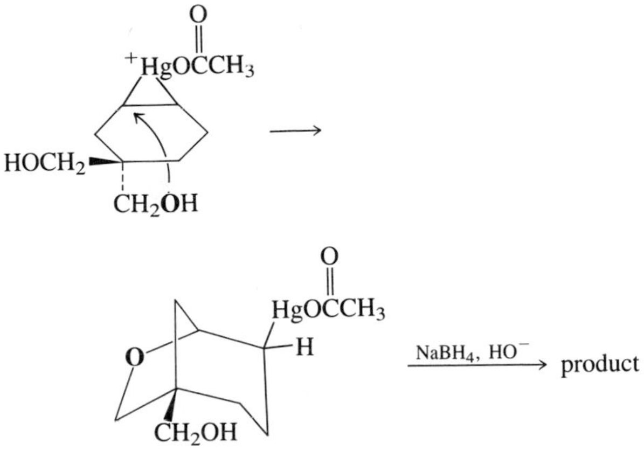

[structure: $^{+}HgOCCH_3$ (C=O), $HOCH_2$, CH_2OH] $\longrightarrow$

[structure: O, $HgOCCH_3$ (C=O), H, CH_2OH] $\xrightarrow{NaBH_4,\ HO^-}$ product

12-13

(a) $CH_3CH_2CH_2OH$

(b) H, H_3C, CH_3CH_2, C–C (*S*, *R*), OH, CH_3, H + enantiomer

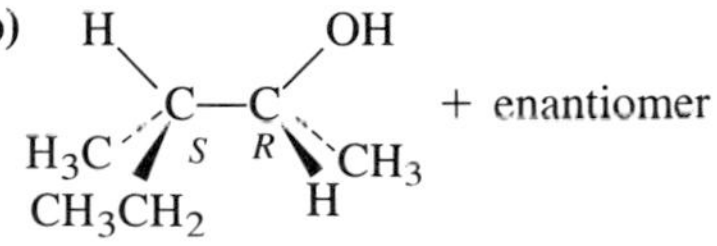

12-14

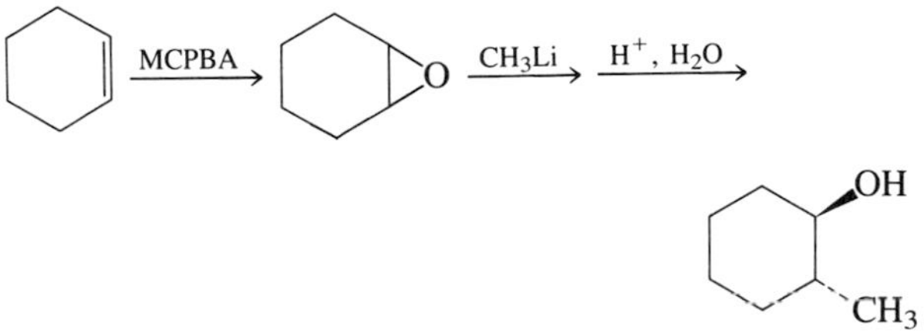

[cyclohexene] $\xrightarrow{MCPBA}$ [cyclohexene oxide, O] $\xrightarrow{CH_3Li}$ $\xrightarrow{H^+,\ H_2O}$

[cyclohexane with OH and CH_3]

12-15

(a) [chain structure] OH + enantiomer (HO, H)

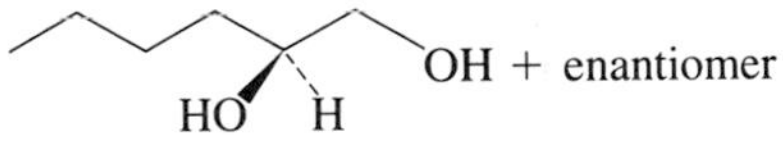

(b) [cyclohexane with OH, OH] + enantiomer

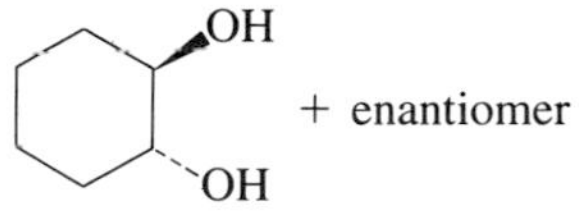

70%

(c) HO, H, H_3C, C–C, H, CH_2CH_3, OH + enantiomer

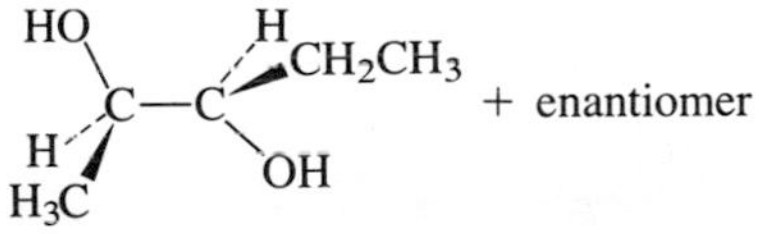

12-16

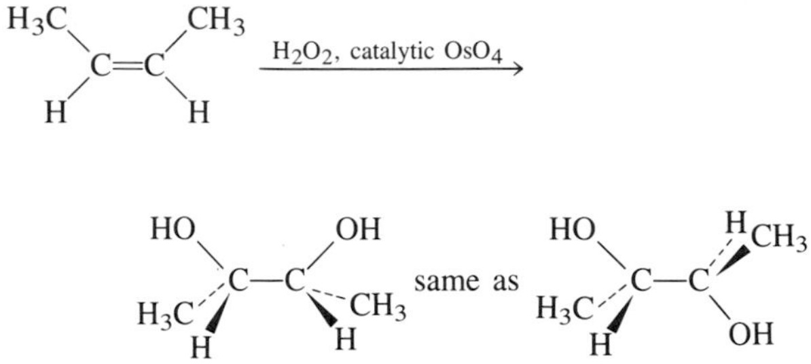

H$_3$C(H)C=C(H)CH$_3$ (cis) $\xrightarrow{H_2O_2,\ \text{catalytic } OsO_4}$

HO, H$_3$C, H–C—C–OH, CH$_3$, H (Eclipsed) same as HO, H$_3$C, H–C—C–H, CH$_3$, OH (Staggered)

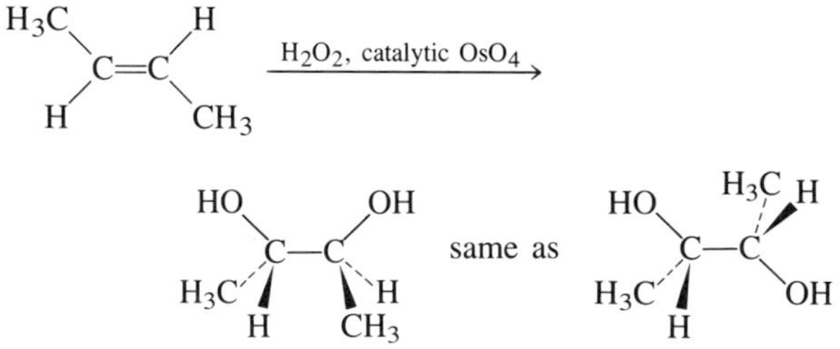

Meso

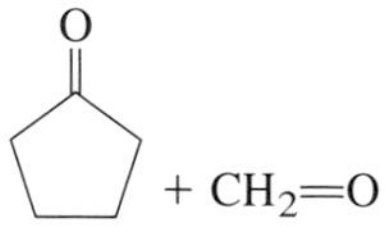

H$_3$C(H)C=C(H)CH$_3$ (trans) $\xrightarrow{H_2O_2,\ \text{catalytic } OsO_4}$

HO, H$_3$C, H–C—C–OH, H, CH$_3$ (Eclipsed) same as HO, H$_3$C, H–C—C–H$_3$C, H, OH (Staggered)

(*R,R*), (*S,S*)

12-17

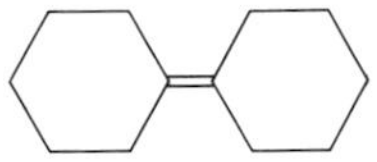

$C_{12}H_{20}$

12-18

(a) 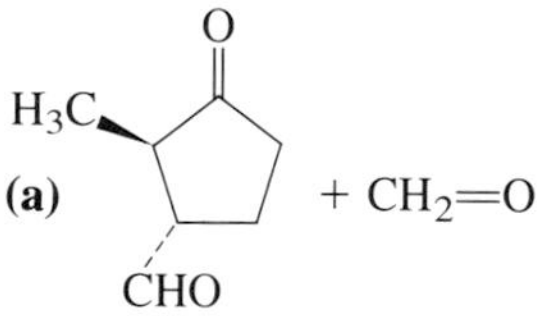 2-methyl-3-formylcyclopentanone (H$_3$C, O, CHO) + $CH_2{=}O$ (b) cyclopentanone (O) + $CH_2{=}O$

(c) O, H, O

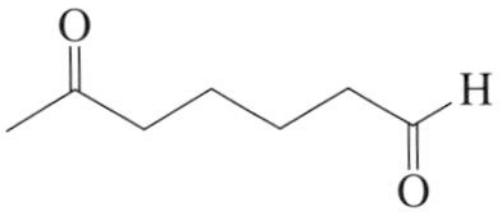

12-19

Do not be fooled by the way structures are drawn.

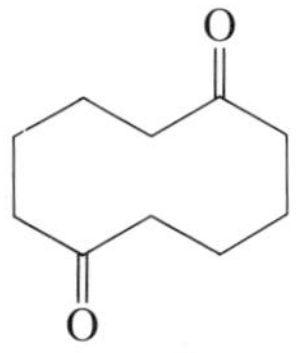

cyclodecane-1,6-dione (O, O) is the same as (O, O) Therefore

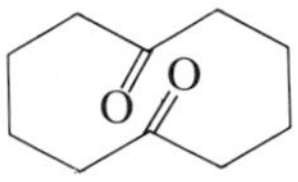

the starting material is

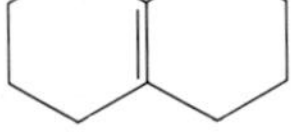

12-20

Initiation

$$(C_6H_5)_2PH \xrightarrow{h\nu} (C_6H_5)_2P\cdot + H\cdot$$

Chain carrier

Propagation

$$CH_3(CH_2)_5CH{=}CH_2 + (C_6H_5)_2P\cdot \rightarrow CH_3(CH_2)_5\dot{C}HCH_2P(C_6H_5)_2$$

More stable radical

$$CH_3(CH_2)_5\dot{C}HCH_2P(C_6H_5)_2 + (C_6H_5)_2PH \rightarrow CH_3(CH_2)_5CH_2CH_2P(C_6H_5)_2 + (C_6H_5)_2P\cdot$$

12-21

This is an irregular copolymer with both monomers incorporated in random numbers but regioselectively along the chain. Write a mechanism for its formation.

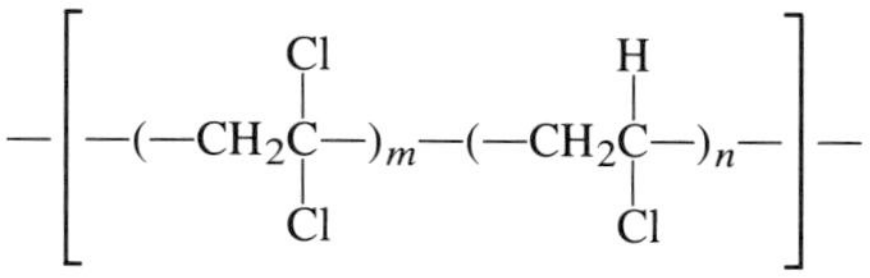

$$-\left[-(-CH_2\overset{Cl}{\underset{Cl}{C}}-)_m-(-CH_2\overset{H}{\underset{Cl}{C}}-)_n-\right]-$$

CHAPTER 13

13-1

(a) 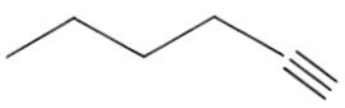 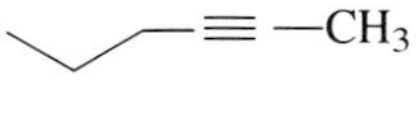 **1-Hexyne** **2-Hexyne** (≡—CH$_3$)

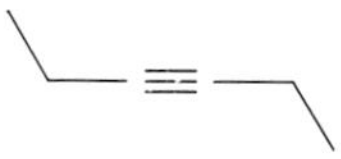

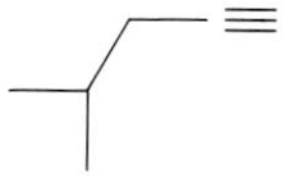

3-Hexyne **4-Methyl-1-pentyne**

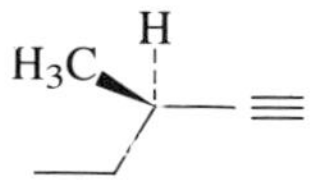

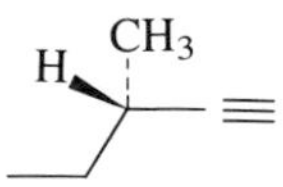

(*R*)-3-Methyl-1-pentyne (H$_3$C, H) **(*S*)-3-Methyl-1-pentyne** (H, CH$_3$)

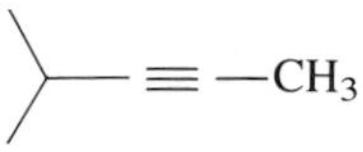

≡—CH$_3$ **4-Methyl-2-pentyne** $(CH_3)_3C$—≡ **3,3-Dimethyl-1-butyne**

(b) (*R*)-3-Methyl-1-penten-4-yne

(c)

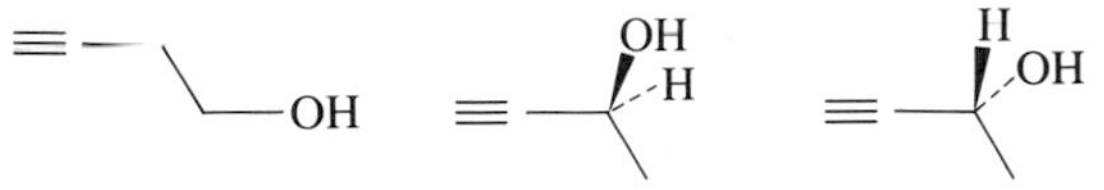

3-Butyn-1-ol (*S*)-3-Butyn-2-ol (*R*)-3-Butyn-2-ol

2-Butyn-1-ol 1-Butyn-1-ol

(This compound is highly unstable and does not exist in solution)

13-2

Only those bases whose conjugate acids have a pK_a higher than that of ethyne ($pK_a = 25$) will deprotonate it: $(CH_3)_3COH$ has a $pK_a \sim 18$, so $(CH_3)_3CO^-$ is too weak; but $[(CH_3)_2CH]_2NH$ has a $pK_a \sim 40$, therefore LDA is a suitable base.

13-3

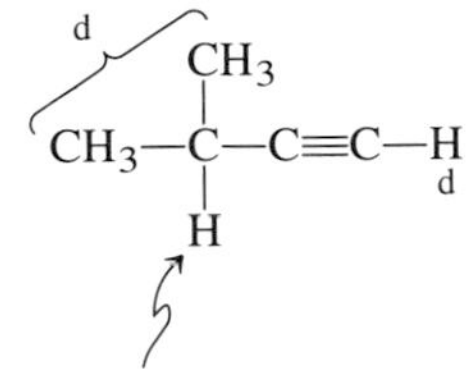

CH_3–C(CH_3)(H)–C≡C–H

Doublet of septets (or septet of doublets)

13-4

From the data in Section 11-7, we can calculate the heat of hydrogenation of the first π bond in the butynes.

$CH_3CH_2C{\equiv}CH + H_2 \rightarrow CH_3CH_2CH{=}CH_2$
$\Delta H° = -(69.9 - 30.3) = -39.6 \text{ kcal mol}^{-1}$

$CH_3C{\equiv}CCH_3 + H_2 \rightarrow$ cis-$CH_3CH{=}CHCH_3$

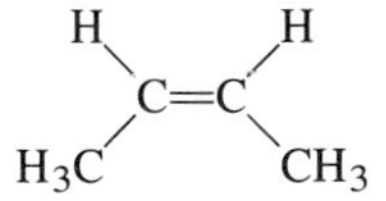

$\Delta H° = -(65.1 - 28.6) = -36.5 \text{ kcal mol}^{-1}$

In both cases more heat is released than expected for a simple C—C double bond.

13-5

The starting materials in each case can be

(a) 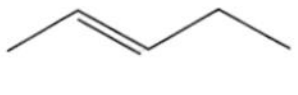 (b) 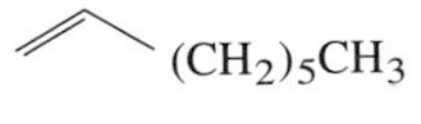 $(CH_2)_5CH_3$ (c)

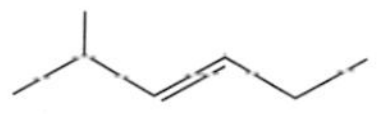

13-6

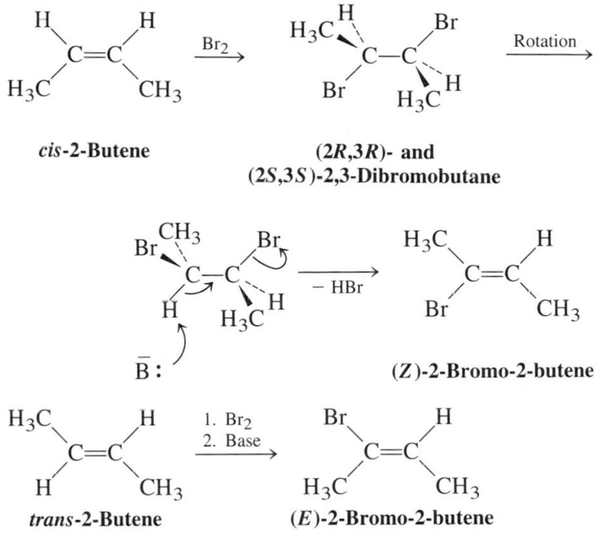

cis-2-Butene $\xrightarrow{Br_2}$ (2*R*,3*R*)- and (2*S*,3*S*)-2,3-Dibromobutane $\xrightarrow{\text{Rotation}}$ $\xrightarrow{-\,HBr}$ (*Z*)-2-Bromo-2-butene

trans-2-Butene $\xrightarrow{\text{1. } Br_2 \text{ 2. Base}}$ (*E*)-2-Bromo-2-butene

13-7

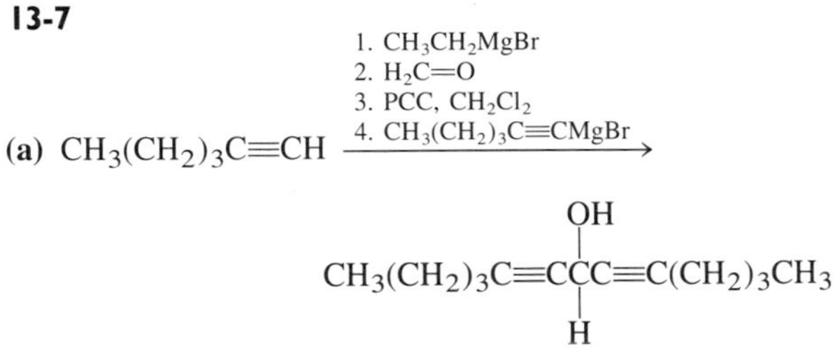

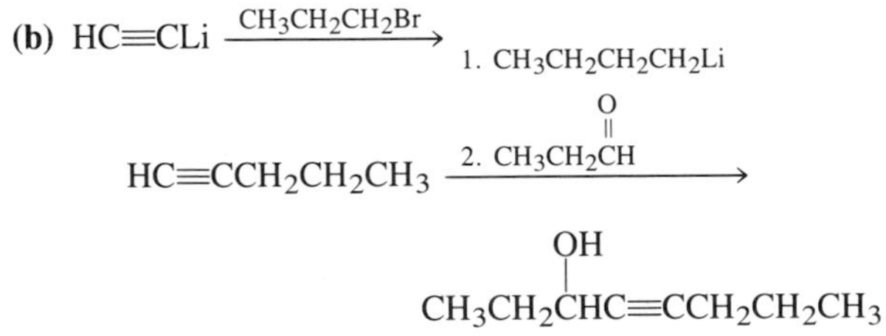

(a) $CH_3(CH_2)_3C{\equiv}CH$ → 1. CH_3CH_2MgBr 2. $H_2C{=}O$ 3. PCC, CH_2Cl_2 4. $CH_3(CH_2)_3C{\equiv}CMgBr$ → $CH_3(CH_2)_3C{\equiv}CCH(OH)C{\equiv}C(CH_2)_3CH_3$

(b) $HC{\equiv}CLi \xrightarrow{CH_3CH_2CH_2Br} HC{\equiv}CCH_2CH_2CH_3$ → 1. $CH_3CH_2CH_2CH_2Li$ 2. CH_3CH_2CHO → $CH_3CH_2CH(OH)C{\equiv}CCH_2CH_2CH_3$

13-8

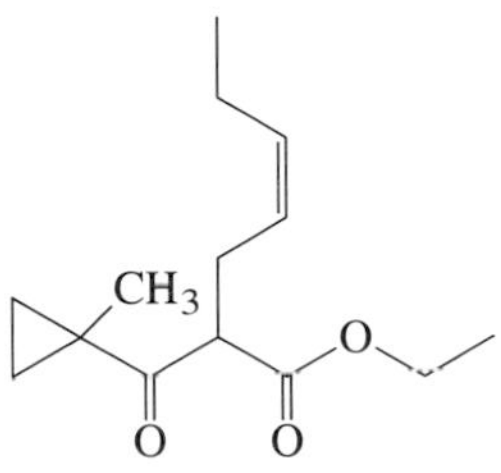

CH_3

13-9

$CH_3CH_2{\equiv}CH$ → 1. $CH_3CH_2CH_2CH_2$ Li 2. (oxirane) 3. H_2, Lindlar cat. → HO

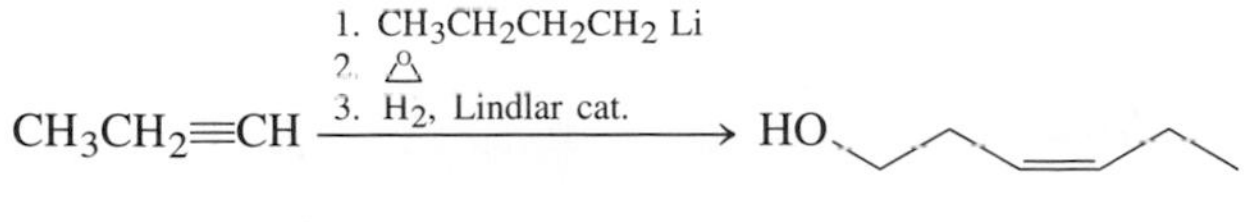

13-10

In the presence of sodium amide, the terminal alkyne unit is deprotonated. Electron transfer to a negatively charged alkynyl group is not favored.

$$CH_3(CH_2)_2C{\equiv}C(CH_2)_4C{\equiv}CH \xrightarrow{NaNH_2,\ \text{liquid } NH_3}$$

$$CH_3(CH_2)_2C{\equiv}C(CH_2)_4C{\equiv}C:^- \xrightarrow{Na,\ \text{liquid } NH_3} \xrightarrow{H^+,\ H_2O}$$

$$CH_3(CH_2)_2CH{=}CH(CH_2)_4C{\equiv}CH$$

75%

13-11

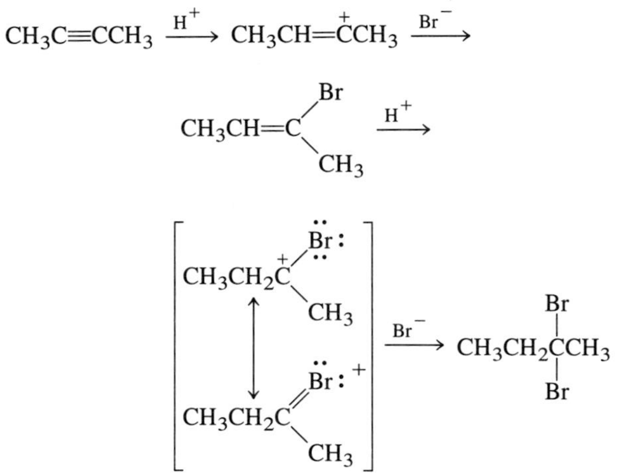

13-12

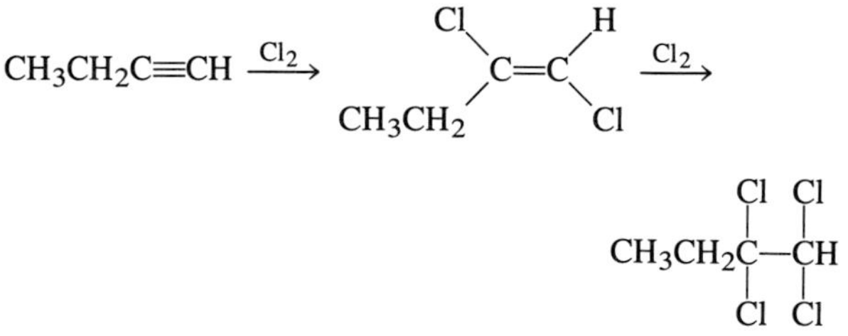

13-13

(a) CH_3CHO (b) $CH_3\overset{O}{\overset{\|}{C}}CH_3$ (c) $CH_3CH_2\overset{O}{\overset{\|}{C}}CH_3$

(d) $CH_3CH_2\overset{O}{\overset{\|}{C}}CH_3$

(e) 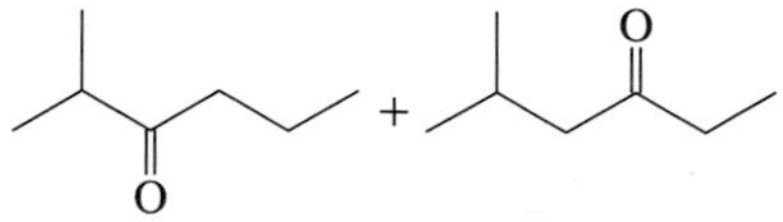

13-14

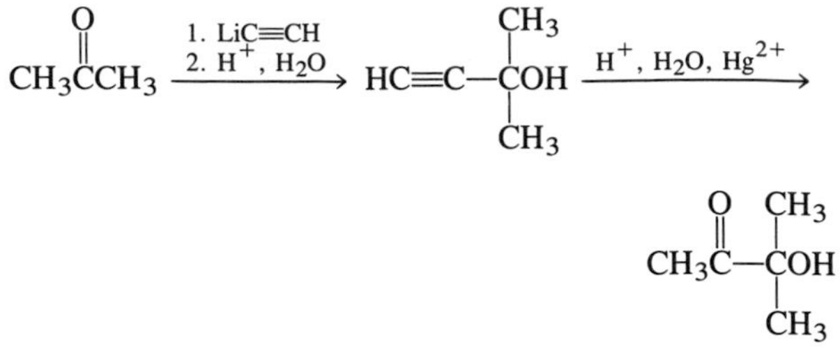

13-15

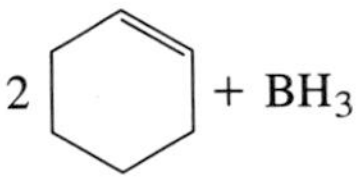

13-16

(a) CH_3CHO (b) CH_3CH_2CHO
(c) $CH_3CH_2CH_2CHO$

13-17

$$(CH_3)_3CC{\equiv}CH \xrightarrow[\text{2. } H_2O_2,\ HO^-]{\text{1. Dicyclohexylborane}} (CH_3)_3CCH_2\overset{O}{\overset{\|}{C}}H$$

13-18

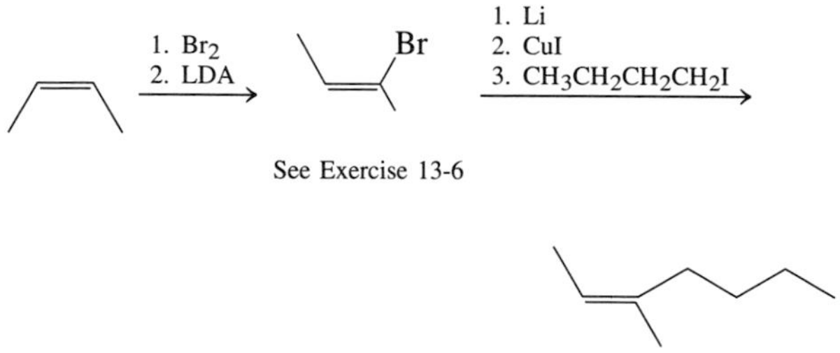

CHAPTER 14

14-1

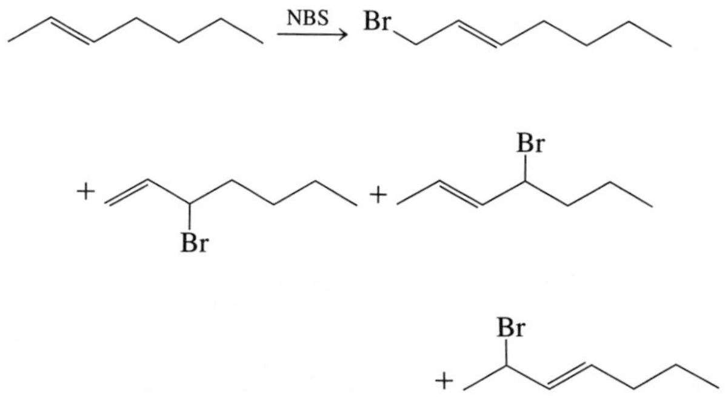

14-2

(a) 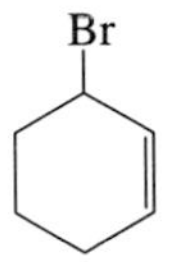 (b)

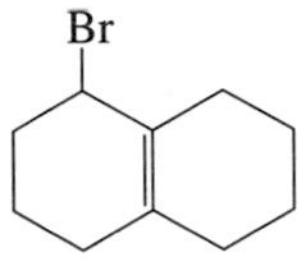

Br Br

(c)

Br + Br

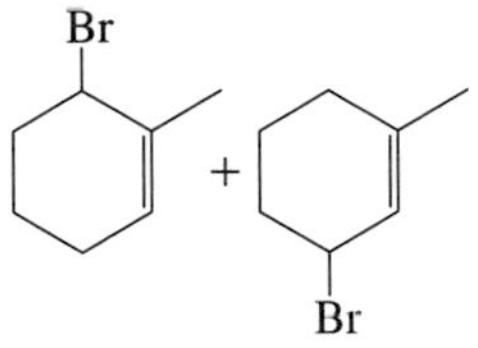

Bromination at the primary allylic position is too slow.

14-3

The intermediate allylic cation is achiral.

14-4

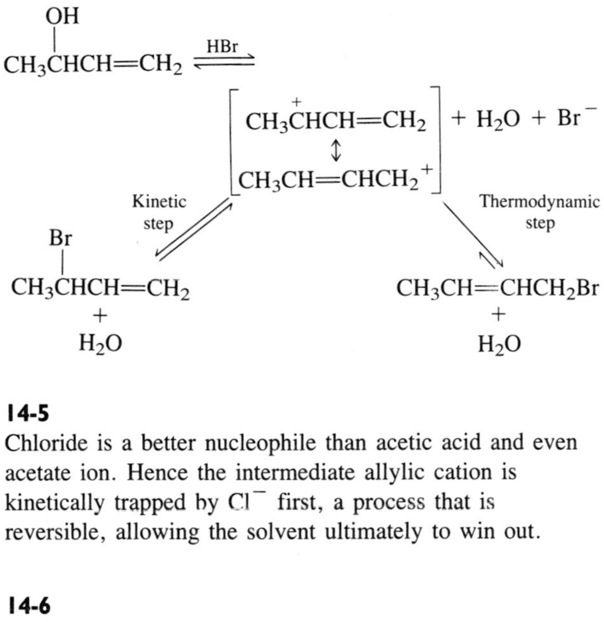

$$CH_3\overset{OH}{\overset{|}{C}}HCH{=}CH_2 \xrightleftharpoons{HBr} \left[\begin{array}{c} CH_3\overset{+}{C}HCH{=}CH_2 \\ \updownarrow \\ CH_3CH{=}CHCH_2^+ \end{array}\right] + H_2O + Br^-$$

Kinetic step: $CH_3\overset{Br}{\overset{|}{C}}HCH{=}CH_2 + H_2O$

Thermodynamic step: $CH_3CH{=}CHCH_2Br + H_2O$

14-5

Chloride is a better nucleophile than acetic acid and even acetate ion. Hence the intermediate allylic cation is kinetically trapped by Cl^- first, a process that is reversible, allowing the solvent ultimately to win out.

14-6

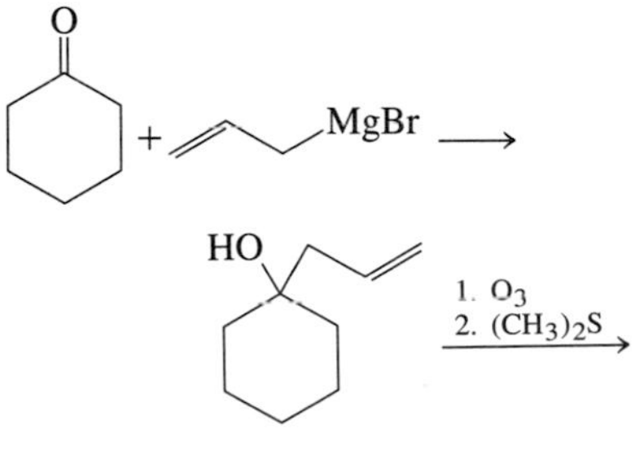

O + MgBr ⟶

HO ... $\xrightarrow{1.\ O_3\quad 2.\ (CH_3)_2S}$ HO ... H O

14-7

(a) 5-Bromo-1,3-cycloheptadiene

(b) (*E*)-2,3-Dimethyl-1,3-pentadiene

(c) 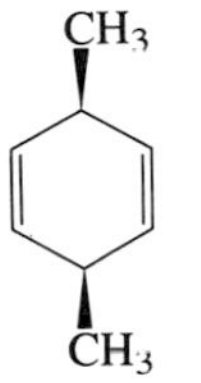 CH_3 ... CH_3 **(d)**  Br Br

14-8

An internal trans double bond is more stable than a terminal double bond by about 2.7 kcal mol^{-1} (see Figure 11-18). This difference plus the expected resonance energy of 3.5 kcal mol^{-1} add up to 6.2 kcal mol^{-1}, pretty close to the observed value.

14-9

The effect of the two allylic double bonds is roughly additive. The $DH°$ of the central methylene bond in 1,4-pentadiene may be estimated by taking the $DH°$ of a secondary C–H bond (95 kcal mol^{-1}) and subtracting twice the amount of allylic stabilization (in this case a little less than expected, about 2 × 12 kcal mol^{-1}).

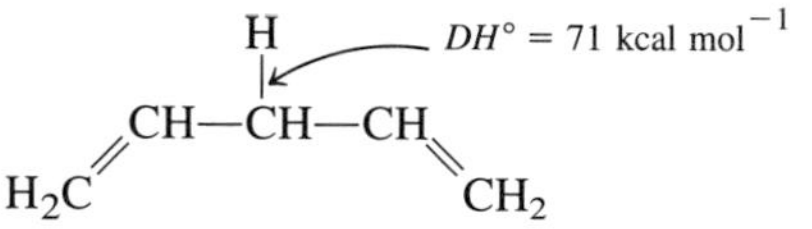

$H_2C{=}CH{-}\overset{H}{\overset{|}{C}}H{-}CH{=}CH_2$ $DH° = 71$ kcal mol^{-1}

14-10

(a) 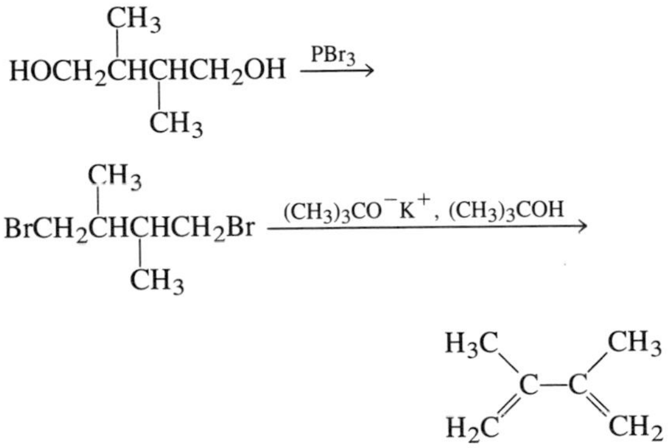 $HOCH_2\overset{CH_3}{\overset{|}{C}}H\underset{CH_3}{\underset{|}{C}}HCH_2OH \xrightarrow{PBr_3}$

$BrCH_2\overset{CH_3}{\overset{|}{C}}H\underset{CH_3}{\underset{|}{C}}HCH_2Br \xrightarrow{(CH_3)_3CO^-K^+,\ (CH_3)_3COH}$ $H_2C{=}C(CH_3){-}C(CH_3){=}CH_2$

(b)

$\xrightarrow[-\ HBr]{Br_2,\ h\nu}$ Br $\xrightarrow[-\ CH_3OH,\ -\ NaBr]{CH_3O^-Na^+}$ $\xrightarrow[-\ HBr]{NBS}$

Br $\xrightarrow[-\ (CH_3)_3COH,\quad KBr]{(CH_3)_3CO^-K^+,\ (CH_3)_3COH}$

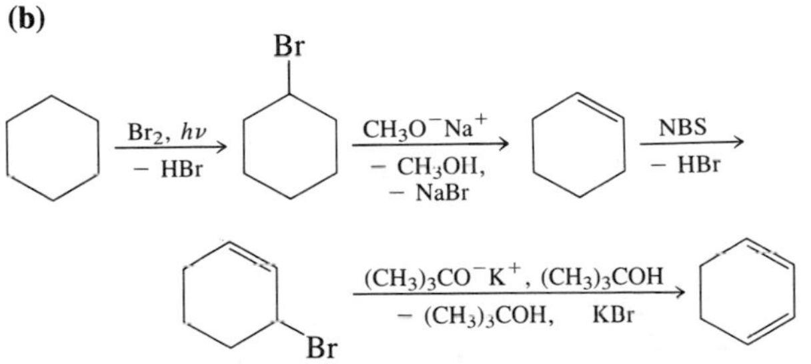

14-11

(a) 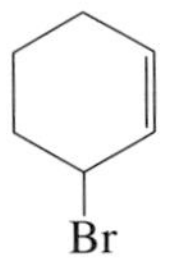 Same product for both modes of addition

Br

(b) 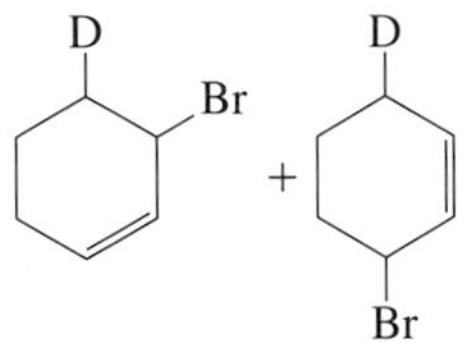 D Br + D Br Both cis and trans

HX adds to unsubstituted cycloalka-1,3-dienes in 1,2- and 1,4- manner to give the same product.

14-12

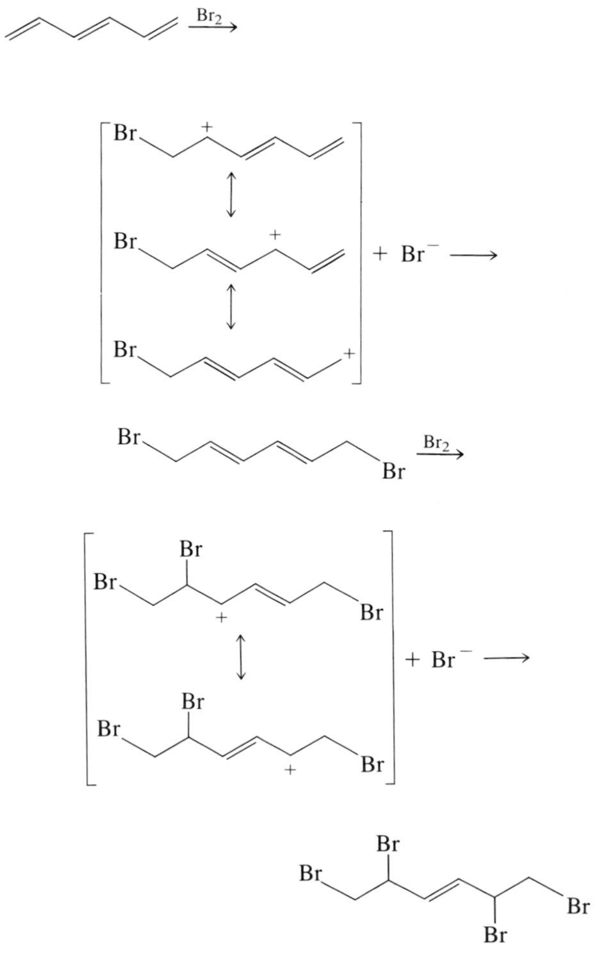

$\xrightarrow{Br_2}$

Br + Br + Br + + Br^- ⟶

Br Br $\xrightarrow{Br_2}$

Br Br + Br Br Br + Br + Br^- ⟶

Br Br Br Br

14-13

(a), (b) Electron rich, because alkyl groups are electron donors.

(c), (d) Electron poor, because the carbonyl group is electron withdrawing by resonance, whereas the fluoroalkyl group is so by induction.

14-14

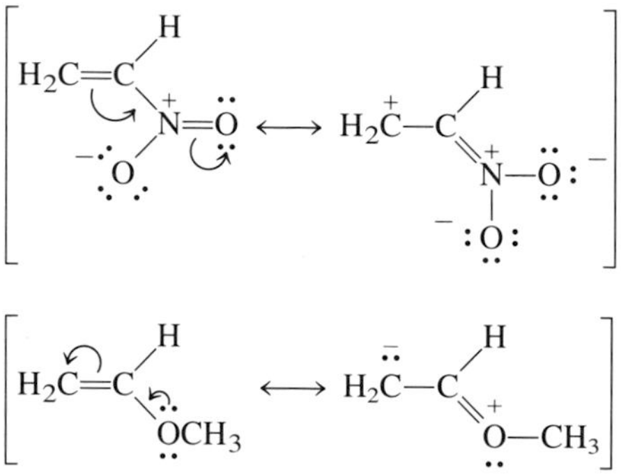

$H_2C{=}C(H)\text{–}N^+({=}O)O^-$ ⟷ $H_2C^+\text{–}C(H){=}N^+(O^-)\text{–}O^-$

$H_2C{=}C(H)\text{–}OCH_3$ ⟷ $H_2\bar{C}\text{–}C(H){=}O^+\text{–}CH_3$

14-15

(a) 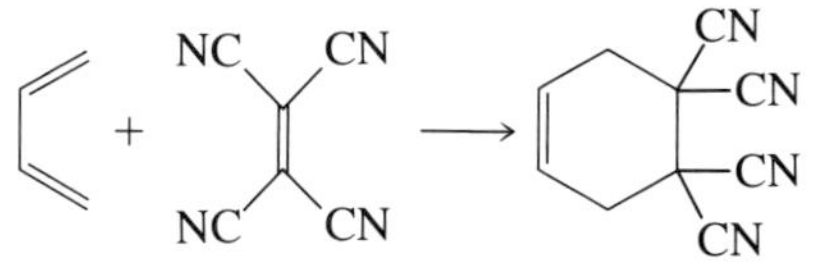 + NC CN NC CN ⟶ CN CN CN CN

(b) 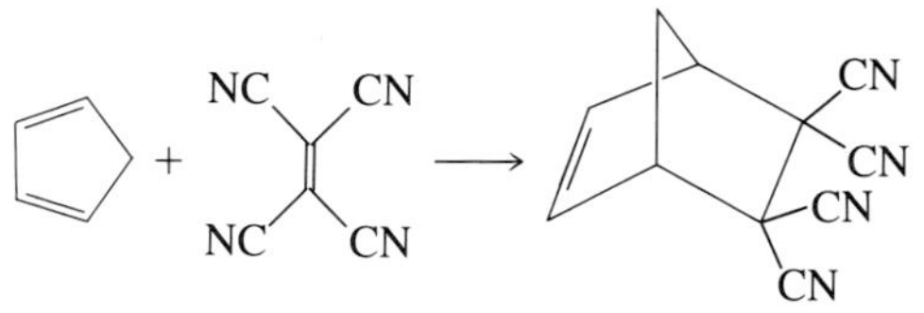 + NC CN NC CN ⟶ CN CN CN CN

(Make a model of this product)

(c) 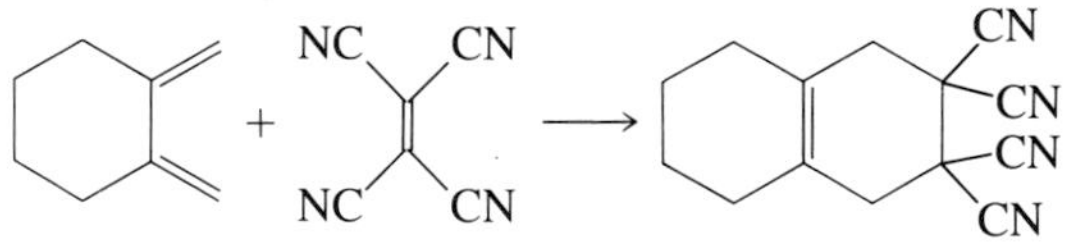 + NC CN NC CN ⟶ CN CN CN CN

14-16

(a) 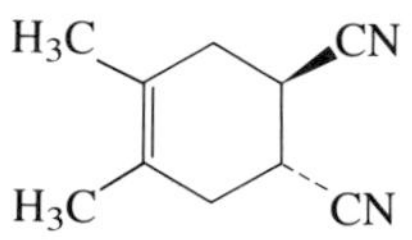 H_3C CN H_3C CN

(b) 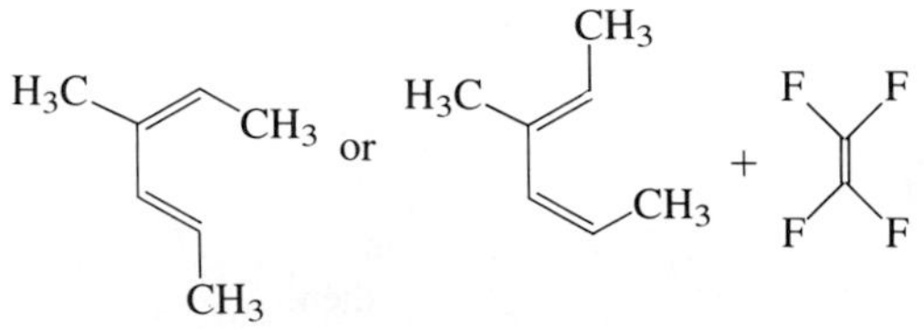 H_3C CH_3 CH_3 or H_3C CH_3 CH_3 + F F F F

14-17

The cis-trans isomer cannot readily reach the *s*-cis conformation because of steric hindrance.

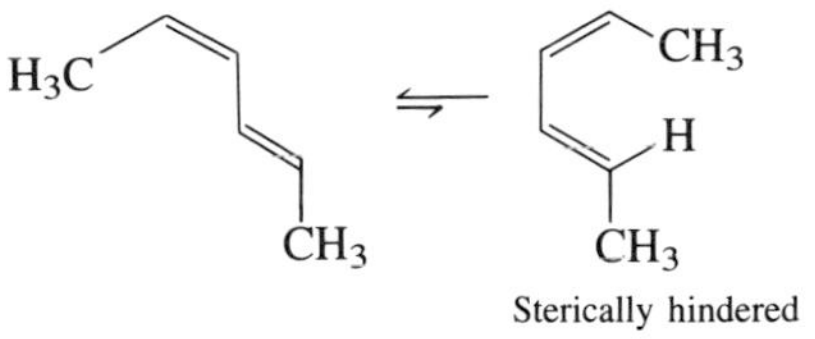

H_3C ... CH_3 ⇌ CH_3, H, CH_3

Sterically hindered

14-18

(a) CH_3, CH_3 + CO_2CH_3 ⟶ CH_3, CO_2CH_3, CH_3

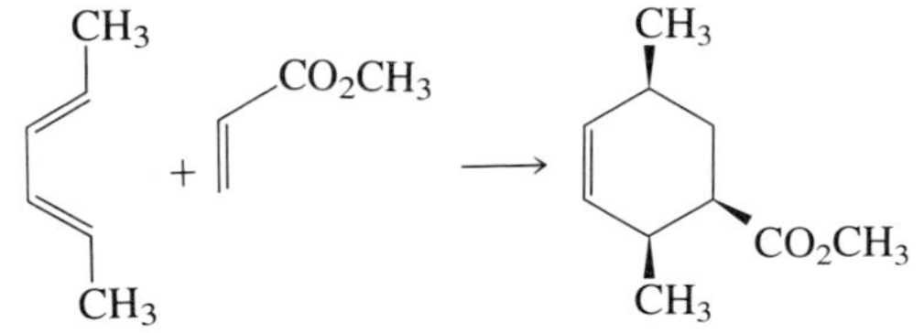

(b) CH_3 + O, O, O ⟶ CH_3, O, O, O

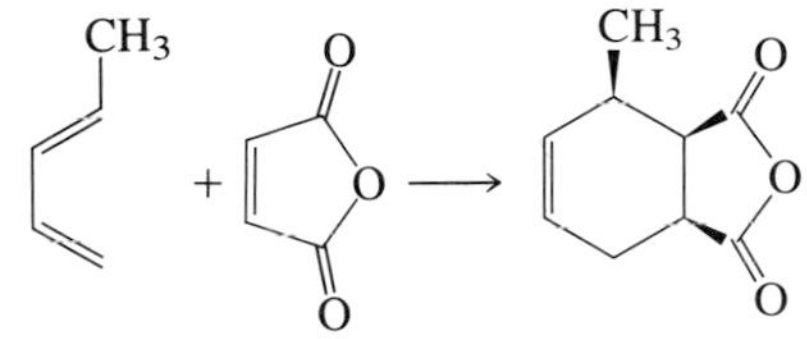

(c) + CO_2CH_3, CH_3O_2C ⟶ CO_2CH_3, H, H, CO_2CH_3

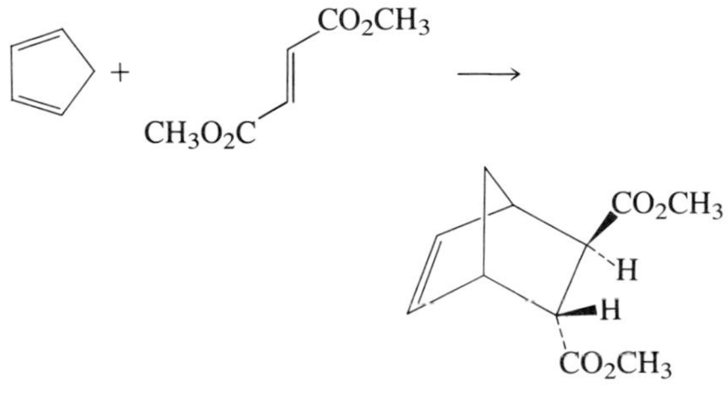

14-19

The first product is the result of exo addition; the second product, the outcome of endo addition.

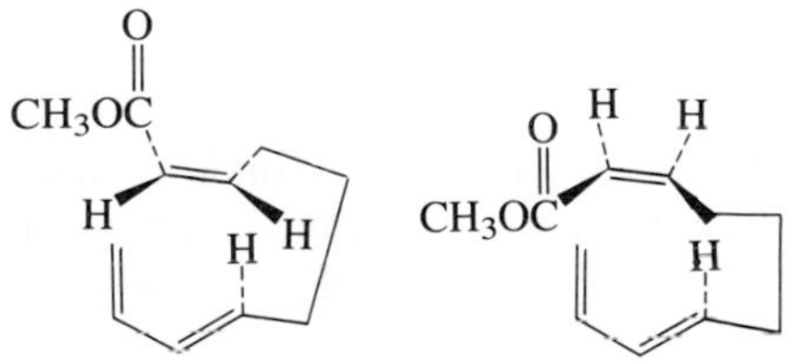

O, CH_3OC, H, H, H   O, CH_3OC, H, H, H

14-20

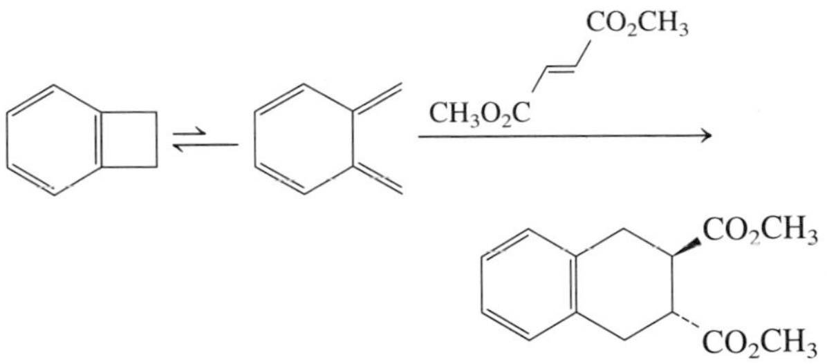

CO_2CH_3, CH_3O_2C ⟶ CO_2CH_3, CO_2CH_3

14-21

Conrotatory. Make a model!

14-22

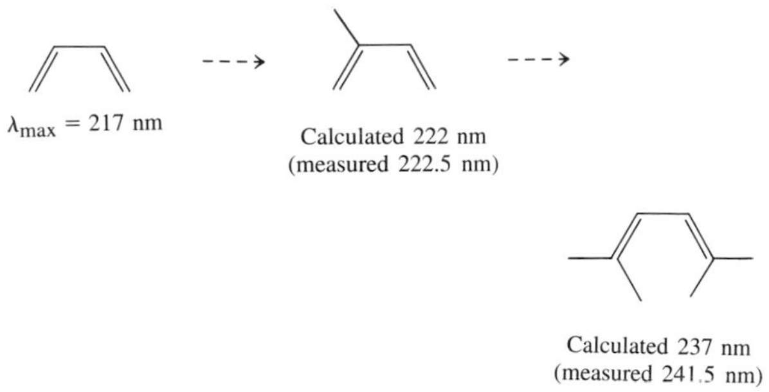

$\lambda_{max} = 217$ nm

Calculated 222 nm (measured 222.5 nm)

Calculated 237 nm (measured 241.5 nm)

CHAPTER 15

15-1

(a) 1-Chloro-4-nitrobenzene (*p*-chloronitrobenzene)
(b) 1-Deuterio-2-methylbenzene (*o*-deuteriotoluene)
(c) 2,4-Dinitrobenzenol (2,4-dinitrophenol)

15-2

(a) CH_3, $CHCH_2CH_2CH_3$

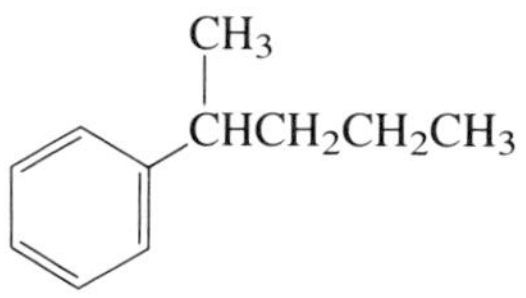

(b) $CH{=}CH_2$, NO_2

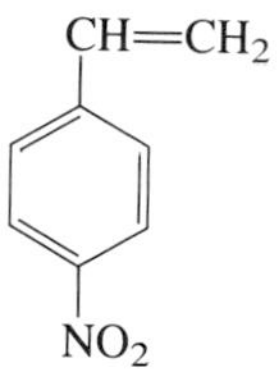

(c) CH_3, O_2N, NO_2, NO_2

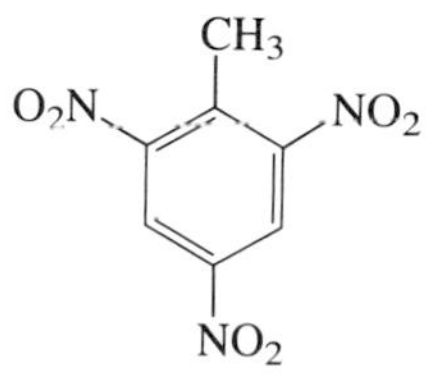

15-3

(a) 1,3-Dichlorobenzene (*m*-dichlorobenzene)
(b) 2-Fluorobenzenamine (*o*-fluoroaniline)
(c) 1-Bromo-4-fluorobenzene (*p*-bromofluorobenzene)

15-4

1,2-Dichlorobenzene

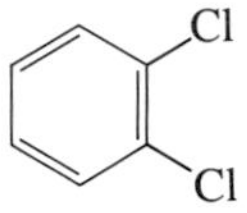

and

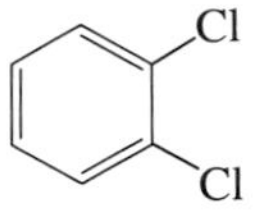

1,2,4-Trichlorobenzene

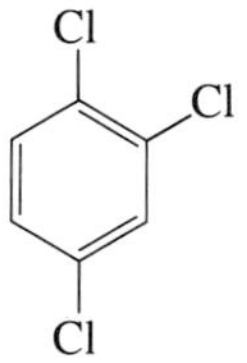

and

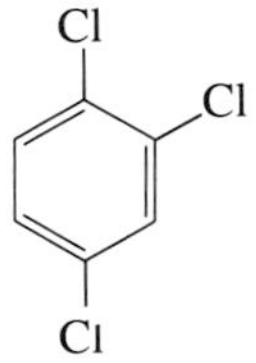

15-5

B has lost its cyclic arrangement of six π electrons and therefore its aromaticity. Thus, ring opening is endothermic.

15-6

The unsymmetrically substituted 1,2,4-trimethylbenzene exhibits the maximum number, nine, of ^{13}C NMR lines. Symmetry reduces this count to six in 1,2,3- and three in 1,3,5-trimethylbenzene.

15-7

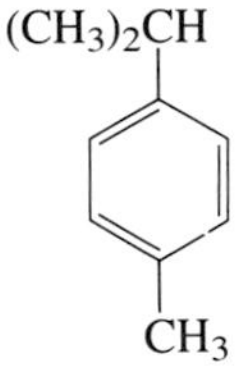

$(CH_3)_2CH$

CH_3

15-8

(a)

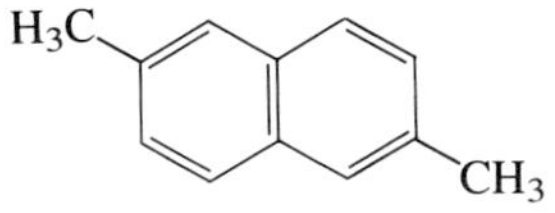

(b)

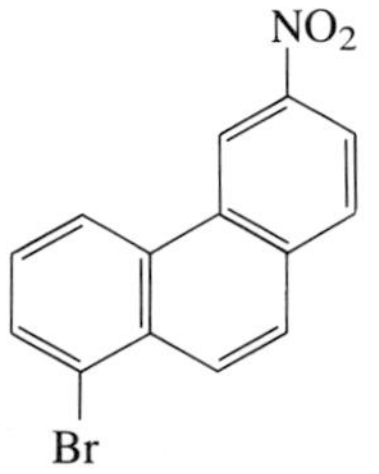

(c)

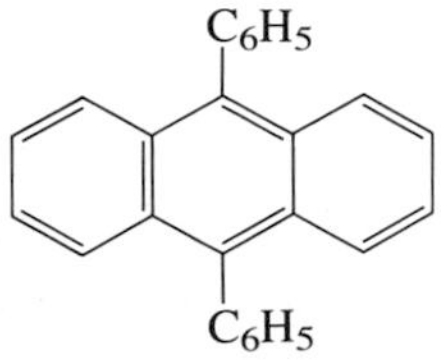

(d) 9-Bromophenanthrene
(e) 5-Nitro-2-naphthalenesulfonic acid

15-9

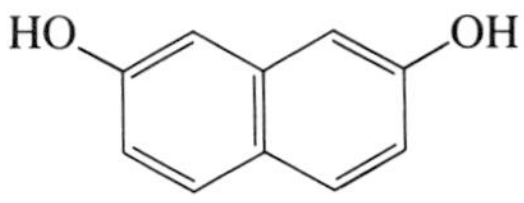

15-10

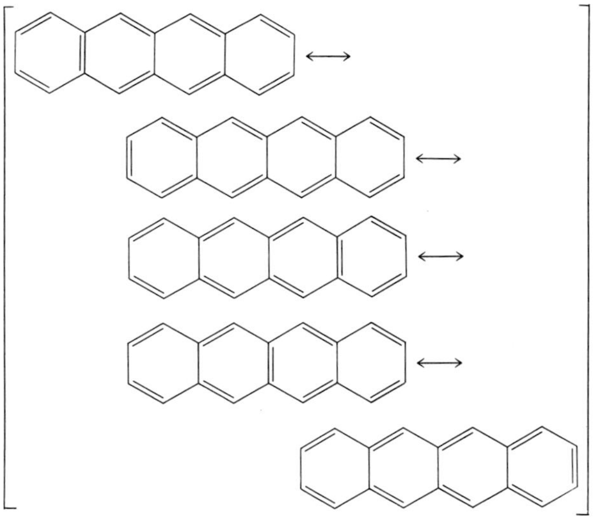

The maximum number of aromatic benzene Kekulé rings is two, in three of the resonance structures (the first, third, and fourth).

15-11

This is an unusual Diels-Alder reaction in which one molecule acts as a diene, the other as a dienophile.

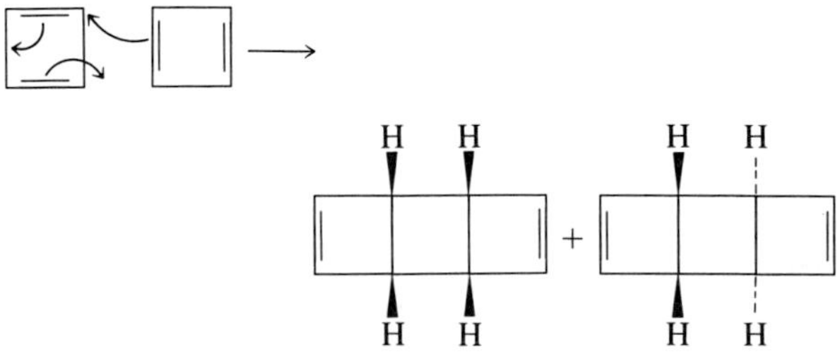

Endo product

Exo product

15-12

No. Cyclooctatetraene has localized double bonds. Double bond shift results in geometrical isomerization and not in a resonance structure, as shown for 1,2-dimethylcyclo-octatetraene.

CH_3 H_3C ⇌ H_3C H_3C

15-13

The mechanism of bromination of cyclooctatetraene is complicated but eventually (<0°C) results in the product of trans addition. On warming, disrotatory ring closure furnishes the bicyclic dibromide.

$\xrightarrow{Br_2 < 0°C}$ Br Br $\xrightarrow{> 0°C}$ Br Br

15-14

(a) Aromatic **(b)** Nonaromatic

15-15

(a)

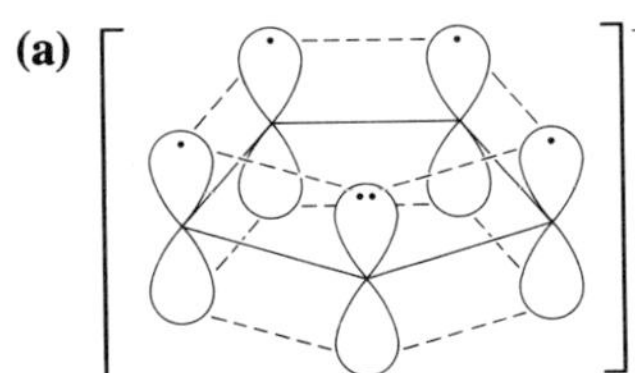

(b)

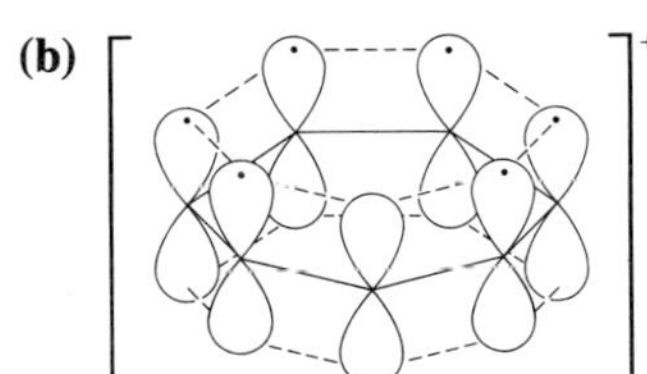

15-16

(a), (b) Aromatic

15-17

The dianion is an aromatic system of 10 π electrons, but pentalene has $4n$ π electrons.

15-18

Azulene is aromatic with 10 π electrons.

15-19

According to Table 12-1, electrophilic additions to alkenes are exothermic by at most 27 kcal mol^{-1}. Because such additions to benzene would cause the loss of about 30 kcal mol^{-1} in resonance energy, they are not thermodynamically possible.

15-20

$+ D^+ \rightleftharpoons$ H D $\rightleftharpoons$

D $+ H^+ \overset{D^+}{\rightleftharpoons}$ etc. $\rightleftharpoons$ D D D D D D

Molecular weight = 84

15-21

$\xrightarrow{HF,\ SbF_5,\ SO_2ClF,\ SO_2F_2,\ -129°C}$ H H δ = 5.69 ppm; H δ = 9.58 ppm; H δ = 8.22 ppm; H δ = 9.42 ppm

The NMR assignments correspond to the amount of charge expected at the various hexadienyl cation carbons on the basis of resonance structures.

15-22

(a) SO_3H $+ H^+ \rightleftharpoons$ HO_3S H $\rightleftharpoons$ ^-O_3S H $\rightleftharpoons$ H $+ SO_3$

(b) :O: S O O + HÖH ⟶ Ö=S—Ö:$^-$ $^+OH_2$ $\xrightarrow{H^+\ shift}$:O: Ö=S—ÖH :OH

15-23

$(CH_3)_3CCl + AlCl_3 \longrightarrow (CH_3)_3C^+ + AlCl_4^-$

1,1-Dimethylethyl (*tert*-butyl) cation

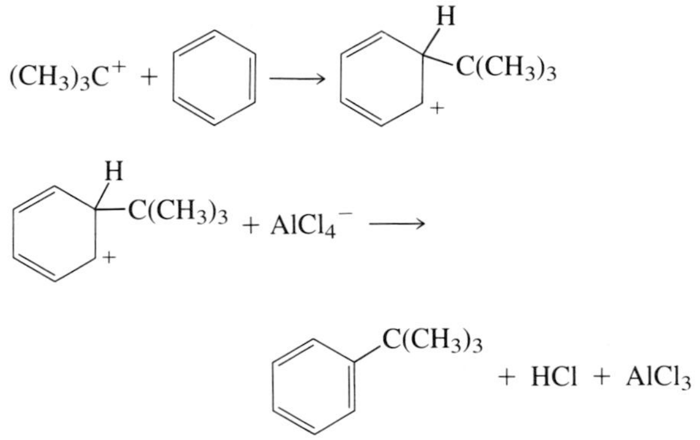

$(CH_3)_3C^+$ + benzene ⟶ (cyclohexadienyl cation with H and $C(CH_3)_3$)

(cyclohexadienyl cation with H and $C(CH_3)_3$) $+ AlCl_4^- \longrightarrow$

$C_6H_5C(CH_3)_3 + HCl + AlCl_3$

15-24

$CH_3CH{=}CH_2 + H^+ \xrightarrow[\text{addition}]{\text{Markovnikov}}$

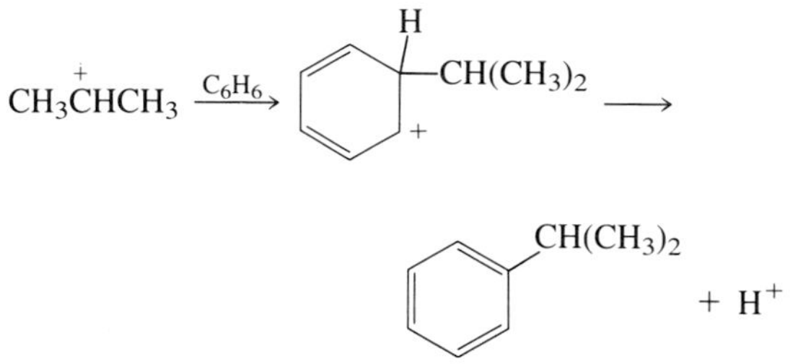

$CH_3\overset{+}{C}HCH_3 \xrightarrow{C_6H_6}$ (cyclohexadienyl cation with H and $CH(CH_3)_2$) $\longrightarrow$

$C_6H_5CH(CH_3)_2 + H^+$

15-25

(structure: benzene ring with H_3C, CH_3, H_3C, CH_3 at positions 1,2,4,5)

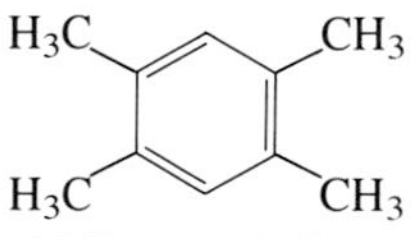

1,2,4,5-Tetramethylbenzene (Durene)

15-26

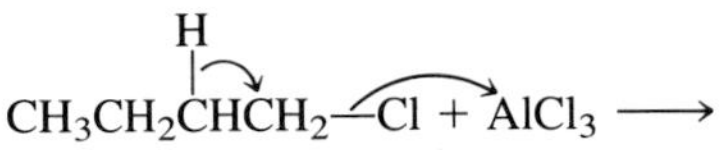

$CH_3CH_2\overset{H}{C}HCH_2{-}Cl + AlCl_3 \longrightarrow$

$CH_3CH_2\overset{+}{C}HCH_3 + AlCl_4^- \xrightarrow{C_6H_6} \longrightarrow$

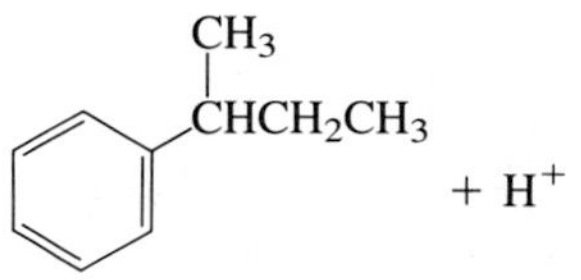

$C_6H_5\overset{CH_3}{C}HCH_2CH_3 + H^+$

15-27

$:\overset{-}{C}{\equiv}\overset{+}{O}: + H^+ \rightleftharpoons \left[H{-}C{\equiv}\overset{+}{O}: \longleftrightarrow H{-}\overset{+}{C}{=}\ddot{O}:\right]$

Methanoyl (formyl) cation

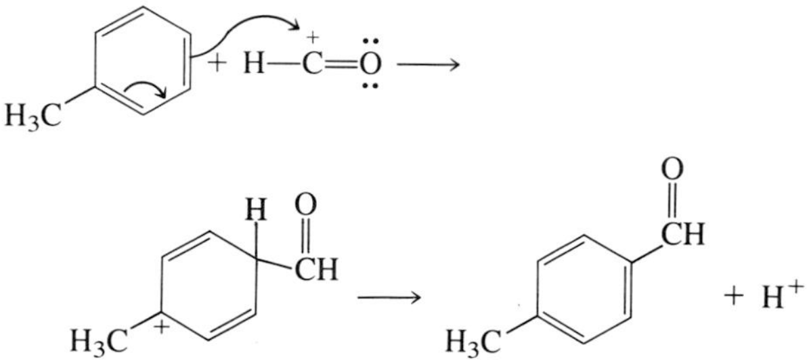

H_3C–benzene $+ H{-}\overset{+}{C}{=}\ddot{O} \longrightarrow$

(cation with H and CHO, H_3C) $\longrightarrow$ H_3C–C_6H_4–CH(=O) $+ H^+$

CHAPTER 16

16-1

(d) > (b) > (a) > (c)

16-2

Methylbenzene (toluene) is activated and will consume all of the electrophile before the latter has a chance to attack the deactivated ring of (trifluoromethyl)benzene.

16-3

Ortho attack

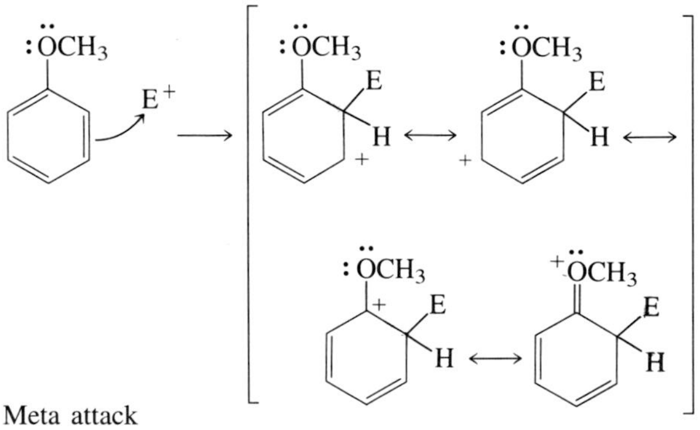

$:\ddot{O}CH_3$ (methoxybenzene) $+ E^+ \longrightarrow$ [four resonance structures of the ortho σ-complex with E and H; the last with $^+\ddot{O}CH_3$]

Meta attack

$:\ddot{O}CH_3$ (methoxybenzene) $+ E^+ \longrightarrow$

[three resonance structures of the meta σ-complex with E and H]

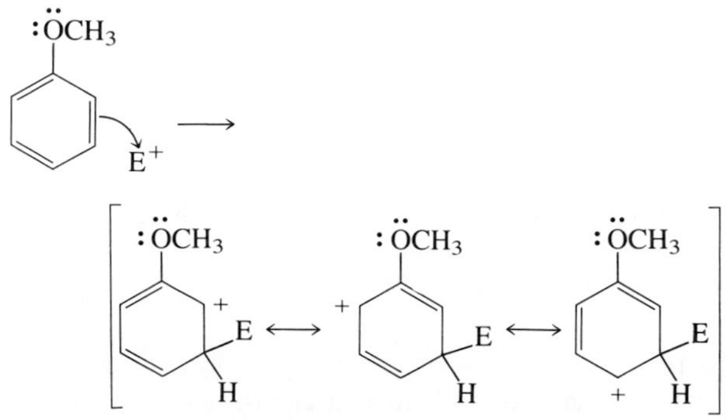

Para attack

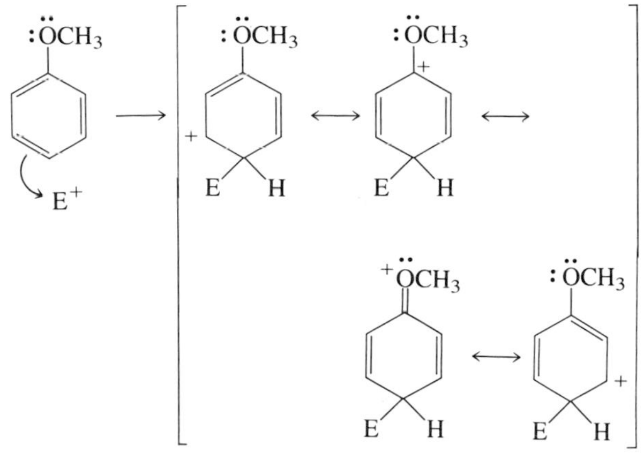

$$:\ddot{O}CH_3 \text{ (benzene)} \xrightarrow{E^+} \left[\text{resonance structures of para-attack cation (E, H)} \right]$$

16-4

Benzenamine (aniline) is completely protonated in strong acid. The lone electron pair is no longer available for resonance with the ring. Hence, the ammonium substituent is an inductive deactivator and meta director.

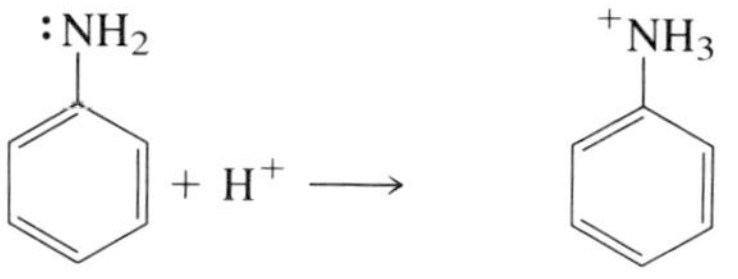

$$C_6H_5NH_2 + H^+ \longrightarrow C_6H_5\overset{+}{N}H_3 \qquad pK_a = 4.60$$

Benzenammonium ion
(Anilinium ion)

16-5

Activated: (a), (d). Deactivated: (b), (c).

16-6

(a)

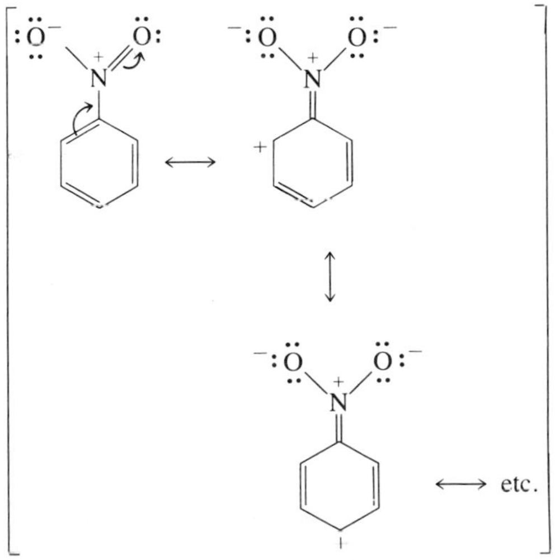

$$\left[C_6H_5\overset{+}{N}O_2 \longleftrightarrow \cdots \longleftrightarrow \text{etc.} \right]$$

The nitro group is deactivating, both by induction (positive charge on nitrogen) as well as by resonance with the ring. The latter contribution is minor because it disrupts the resonance in the nitro group itself, which is similar to that in allyl.

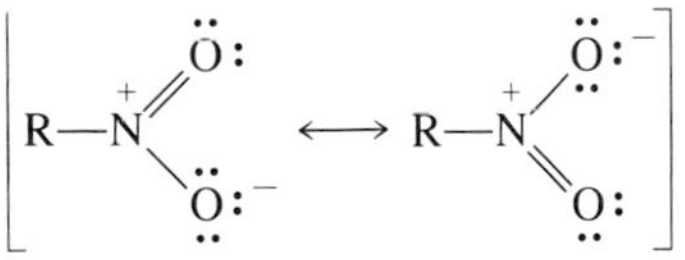

$$\left[R\text{—}\overset{+}{N}(=\ddot{O}:)\text{—}\ddot{O}:^- \longleftrightarrow R\text{—}\overset{+}{N}(\text{—}\ddot{O}:^-)=\ddot{O}: \right]$$

(b) In $C_6H_5\overset{+}{N}R_3$, the positively charged ammonium group acts as a strong inductive electron withdrawer. The lone electron pair of the original amino group is used for bond formation to one of the alkyl substituents.

(c)

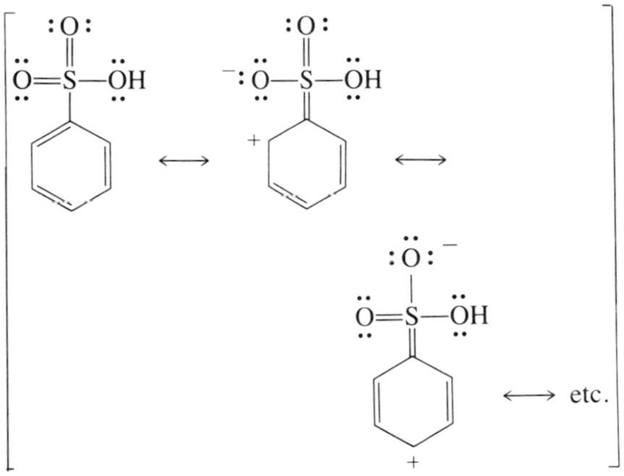

$$\left[C_6H_5SO_2OH \longleftrightarrow \cdots \longleftrightarrow \cdots \longleftrightarrow \text{etc.} \right]$$

Benzenesulfonic acid is deactivated by resonance, like benzenecarboxylic (benzoic) acid.

(d) The phenyl substituent acts as a resonance donor.

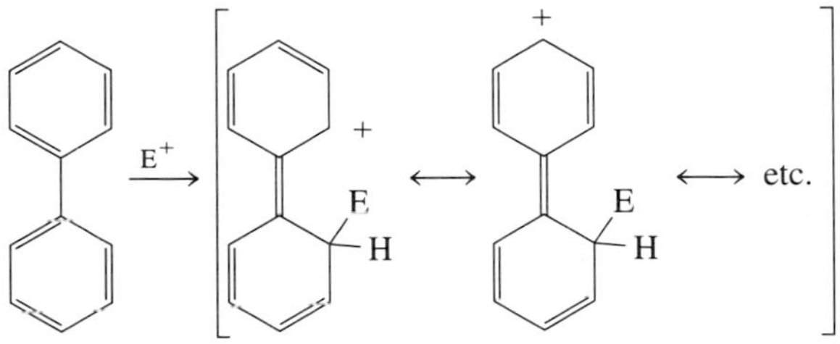

$$C_6H_5\text{—}C_6H_5 \xrightarrow{E^+} \left[\cdots \longleftrightarrow \cdots \longleftrightarrow \text{etc.} \right]$$

Total of six resonance structures

16-7

(a) C3 (= C6) and C4 (= C5)
(b) C4 (= C6) and C5
(c) Mainly C4, with only some C2, because C2, although doubly activated, is also sterically hindered
(d) C2 (= C3 = C5 = C6)

16-8

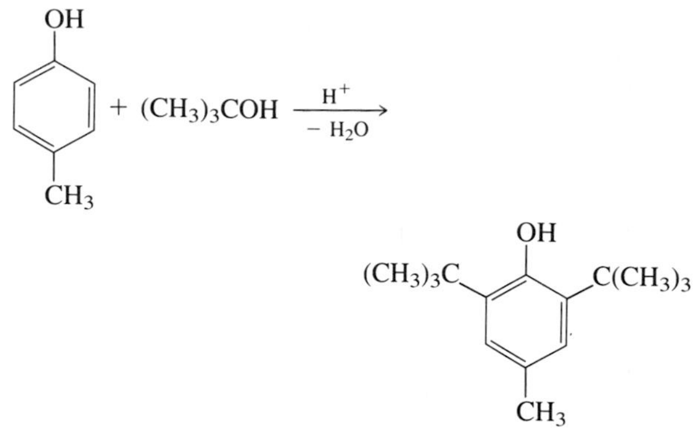

OH, CH_3 (4-methylphenol) + $(CH_3)_3COH$ $\xrightarrow[-\,H_2O]{H^+}$ OH, $(CH_3)_3C$, $C(CH_3)_3$, CH_3

16-9

(a) C3 and C4 **(b)** C5

(c) Mainly C2, with only some C3, because the nitro group is a more powerful deactivator than the ester group (Table 16-2)

(d) Mainly C4 because NO_2 is a meta director and Br is an ortho and para director; for steric reasons, there is only a small amount of C6 substitution.

16-10

(a) C3 and C5 **(b)** C4 and C6

(c) C4 and C6

16-11

No, because the nitrogen is introduced ortho and para to bromine only.

16-12

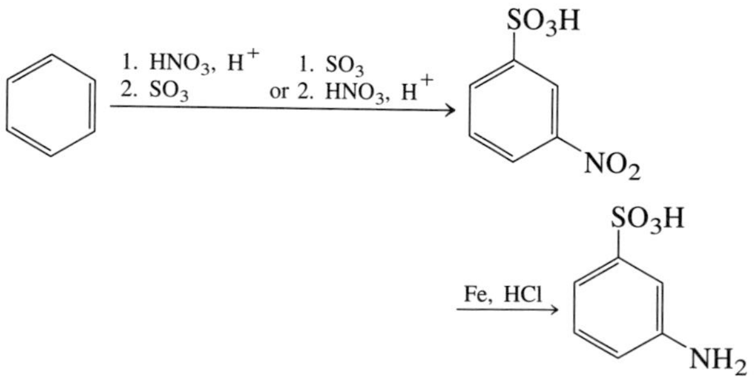

Benzene $\xrightarrow{\text{1. } HNO_3,\ H^+ \text{ 2. } SO_3 \text{ or 1. } SO_3 \text{ 2. } HNO_3,\ H^+}$ SO_3H, NO_2 $\xrightarrow{\text{Fe, HCl}}$ SO_3H, NH_2

16-13

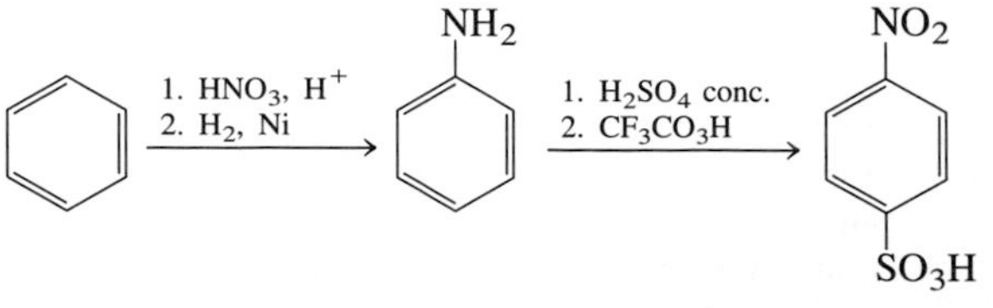

Benzene $\xrightarrow{\text{1. } HNO_3,\ H^+ \text{ 2. } H_2,\ Ni}$ NH_2 $\xrightarrow{\text{1. } H_2SO_4 \text{ conc. 2. } CF_3CO_3H}$ NO_2, SO_3H

16-14

Benzene + (2-methylpropanoyl chloride: Cl, O) $\xrightarrow{AlCl_3}$ (O) $\xrightarrow{\text{HCl, Zn(Hg), }\Delta}$

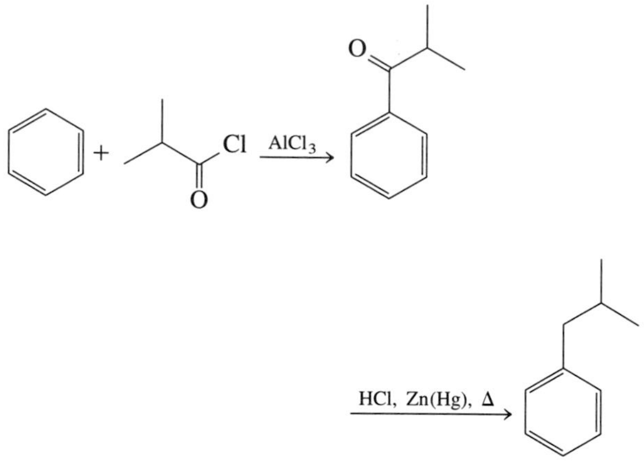

Direct Friedel-Crafts alkylation of benzene with 1-chloro-2-methylpropane gives (1,1-dimethylethyl)benzene (*tert*-butylbenzene) by rearrangement of the carbon electrophile (see Section 15-13).

16-15

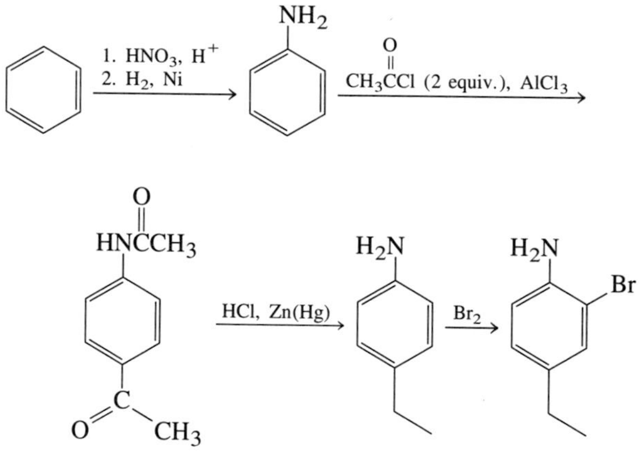

Benzene $\xrightarrow{\text{1. } HNO_3,\ H^+ \text{ 2. } H_2,\ Ni}$ NH_2 $\xrightarrow{CH_3\overset{O}{\overset{\|}{C}}Cl \text{ (2 equiv.), } AlCl_3}$

$HN\overset{O}{\overset{\|}{C}}CH_3$, (O=C–$CH_3$) $\xrightarrow{\text{HCl, Zn(Hg)}}$ H_2N $\xrightarrow{Br_2}$ H_2N, Br

16-16

NH_2, SO_3H (from Exercise 16-13) $\xrightarrow{Br_2}$

Br, NH_2, Br, SO_3H $\xrightarrow{\text{1. } H^+,\ H_2O \text{ 2. } CF_3CO_3H}$ Br, NO_2, Br

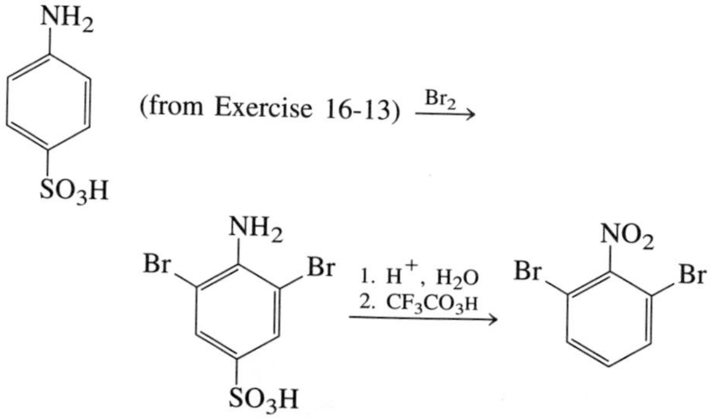

16-17

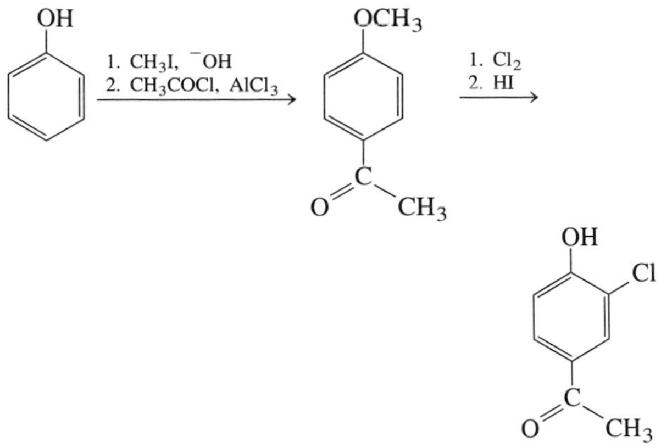

Phenol (OH) $\xrightarrow{\text{1. } CH_3I,\ ^-OH \text{; 2. } CH_3COCl,\ AlCl_3}$ 4-methoxyacetophenone (OCH_3, $C(=O)CH_3$) $\xrightarrow{\text{1. } Cl_2 \text{; 2. HI}}$ 3-chloro-4-hydroxyacetophenone (OH, Cl, $C(=O)CH_3$)

16-18

(a) At C4 **(b)** At C5 and C8 **(c)** At C8

CHAPTER 17

17-1

(a) 2-Cyclohexenone
(b) (*E*)-4-Methyl-4-hexenal
(c) 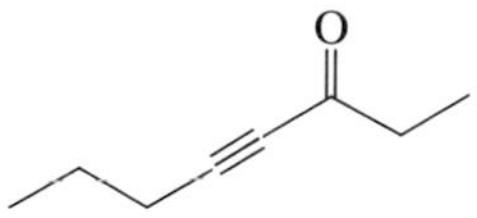 (skeletal structure: oct-5-yn-3-one)
(d) 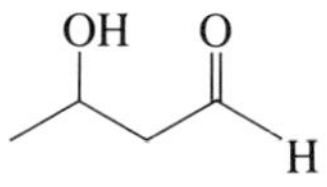 (skeletal structure: 3-hydroxybutanal; OH, O, H)
(e) 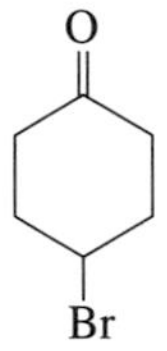 (4-bromocyclohexanone; O, Br)

17-2

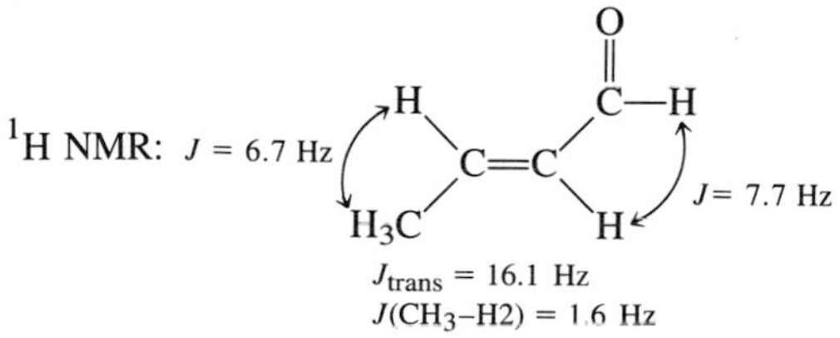

^1H NMR: $J = 6.7$ Hz; $J = 7.7$ Hz (structure: $H_3C(H)C{=}C(H)C(=O){-}H$)

$J_{trans} = 16.1$ Hz
$J(CH_3\text{–}H2) = 1.6$ Hz

^{13}C NMR: $CH_3{-}CH{=}CH{-}\overset{O}{\overset{\|}{C}}H$
18.4 152.1 132.8 191.4

UV: Absorptions are typical for a conjugated enone.

17-3

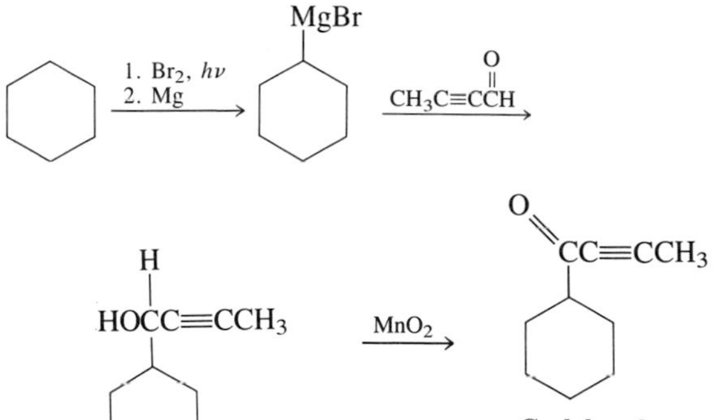

Cyclohexane $\xrightarrow{\text{1. } Br_2,\ h\nu \text{; 2. Mg}}$ cyclohexyl-MgBr $\xrightarrow{CH_3C{\equiv}CCH{=}O}$

$HO\overset{H}{C}C{\equiv}CCH_3$ (cyclohexyl) $\xrightarrow{MnO_2}$ $O{=}CC{\equiv}CCH_3$ (cyclohexyl)

Cyclohexyl 1-propynyl ketone

17-4

(a) $Cl_3CCCH_3 < Cl_3CCH < Cl_3CCCCl_3$ (each with C=O)

(b) $CH_3\overset{O}{\overset{\|}{C}}CH_3 + H_2{}^{18}O \rightleftharpoons \rightleftharpoons CH_3C(OH)(^{18}OH)CH_3 \rightleftharpoons \rightleftharpoons CH_3\overset{^{18}O}{\overset{\|}{C}}CH_3 + H_2O$

17-5

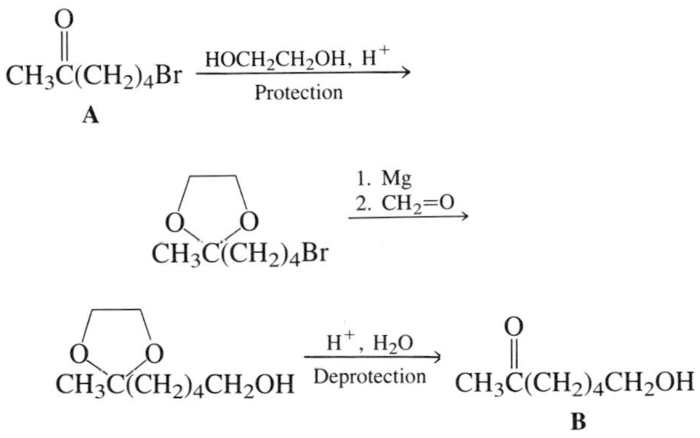

$CH_3\overset{O}{\overset{\|}{C}}(CH_2)_4Br$ **A** $\xrightarrow[\text{Protection}]{HOCH_2CH_2OH,\ H^+}$ (ethylene acetal) $CH_3C(CH_2)_4Br$ $\xrightarrow{\text{1. Mg; 2. } CH_2{=}O}$ (ethylene acetal) $CH_3C(CH_2)_4CH_2OH$ $\xrightarrow[\text{Deprotection}]{H^+,\ H_2O}$ $CH_3\overset{O}{\overset{\|}{C}}(CH_2)_4CH_2OH$ **B**

17-6

(bicyclic alkene) $\xrightarrow{\text{1. } O_3 \text{; 2. } CH_3SCH_3}$ cyclodecane-1,6-dione $\xrightarrow{\text{1. } HSCH_2CH_2SH \text{; 2. Raney Ni}}$ cyclodecane

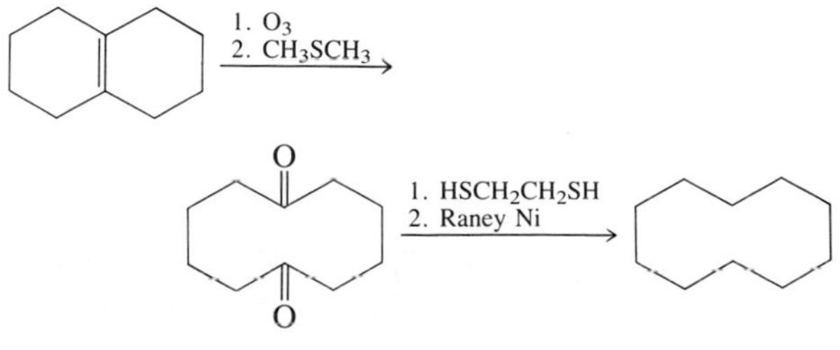

17-7

The mechanism of imidazolidine formation is similar to that formulated for cyclic acetal synthesis.

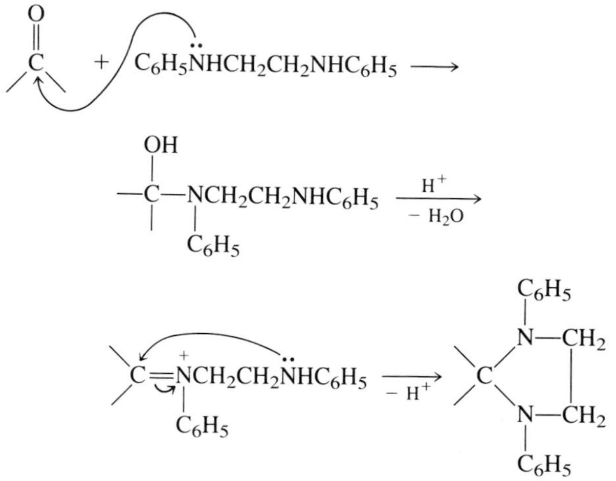

$>C=O + C_6H_5\ddot{N}HCH_2CH_2NHC_6H_5 \longrightarrow$

$-\overset{OH}{C}-\underset{C_6H_5}{N}CH_2CH_2NHC_6H_5 \xrightarrow[-H_2O]{H^+}$

$>C=\overset{+}{\underset{C_6H_5}{N}}CH_2CH_2\ddot{N}HC_6H_5 \xrightarrow{-H^+}$ imidazolidine (C bonded to two N–C$_6$H$_5$, ring N—CH$_2$—CH$_2$—N)

17-8

(a) 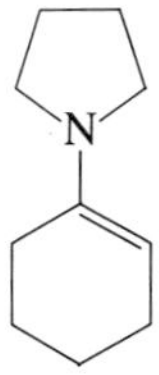 1-(1-cyclohexenyl)pyrrolidine

(b) 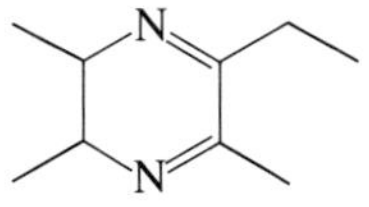 dihydropyrazine structure

(c) 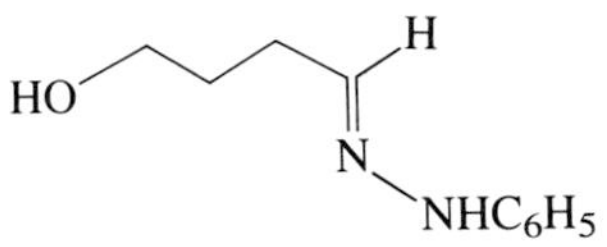 $HOCH_2CH_2CH_2CH_2CH=N-NHC_6H_5$

17-9

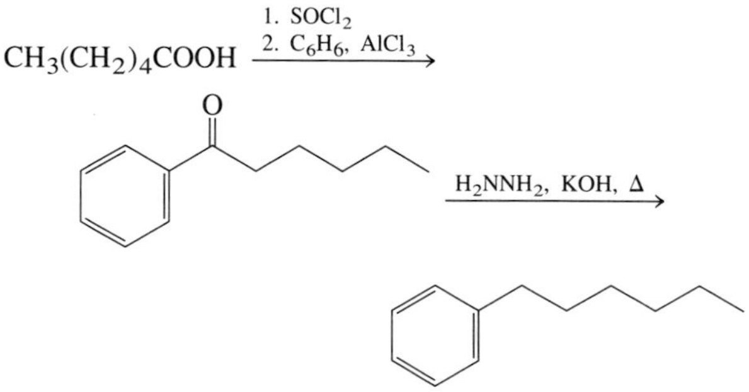

$CH_3(CH_2)_4COOH \xrightarrow[2.\ C_6H_6,\ AlCl_3]{1.\ SOCl_2}$ $C_6H_5CO(CH_2)_4CH_3 \xrightarrow{H_2NNH_2,\ KOH,\ \Delta}$ $C_6H_5(CH_2)_5CH_3$

17-10

(a) 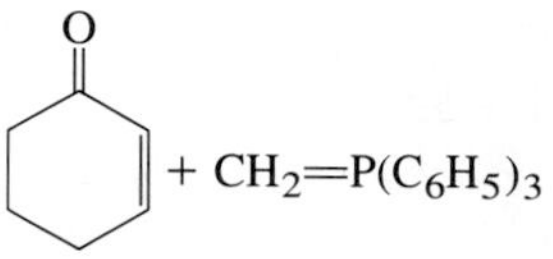 2-cyclohexenone $+ CH_2{=}P(C_6H_5)_3$

(b) 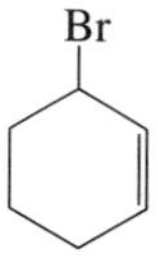 3-bromocyclohexene treated successively with (1) $P(C_6H_5)_3$, (2) $CH_3CH_2CH_2CH_2Li$, and (3) $CH_2{=}O$

17-11

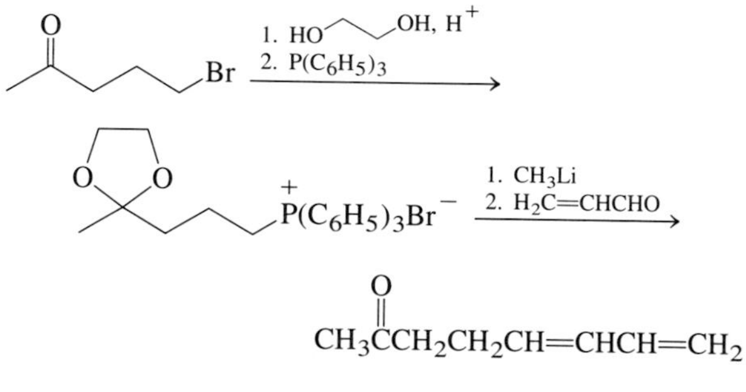

$CH_3CO(CH_2)_3Br \xrightarrow[2.\ P(C_6H_5)_3]{1.\ HOCH_2CH_2OH,\ H^+}$ ketal$-(CH_2)_3\overset{+}{P}(C_6H_5)_3Br^- \xrightarrow[2.\ H_2C{=}CHCHO]{1.\ CH_3Li}$

$CH_3\overset{O}{\overset{\|}{C}}CH_2CH_2CH{=}CHCH{=}CH_2$

17-12

(a) 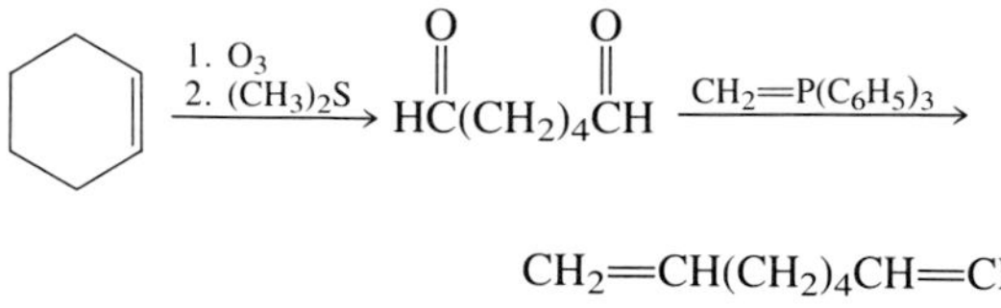 cyclohexene $\xrightarrow[2.\ (CH_3)_2S]{1.\ O_3}$ $H\overset{O}{\overset{\|}{C}}(CH_2)_4\overset{O}{\overset{\|}{C}}H \xrightarrow{CH_2{=}P(C_6H_5)_3}$

$CH_2{=}CH(CH_2)_4CH{=}CH_2$

(b) 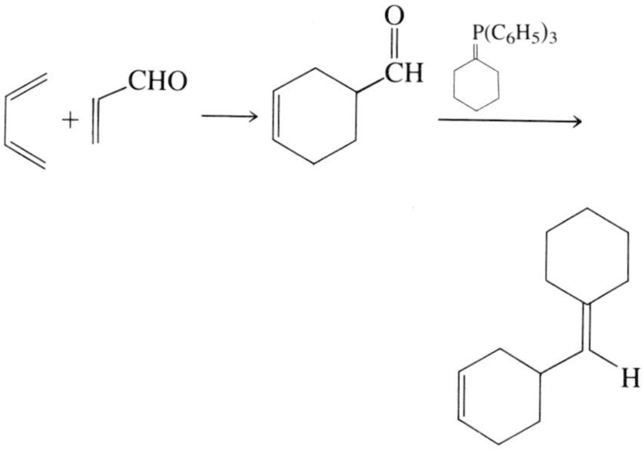 butadiene + $CH_2{=}CHCHO \longrightarrow$ 3-cyclohexene-1-carbaldehyde ($-\overset{O}{\overset{\|}{C}}H$) $\xrightarrow{P(C_6H_5)_3{=}C_6H_{10}}$ (cyclohexylidene)methyl-cyclohexene, with H on the alkene carbon

17-13

(a) $CH_2{=}CHCH_2CH_2O\overset{O}{\overset{\|}{C}}CH_3$

(b) 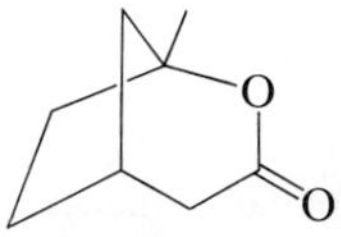 bicyclic lactone

(c) $(CH_3)_3CO\overset{O}{\overset{\|}{C}}CH_2CH_3$

CHAPTER 18

18-1

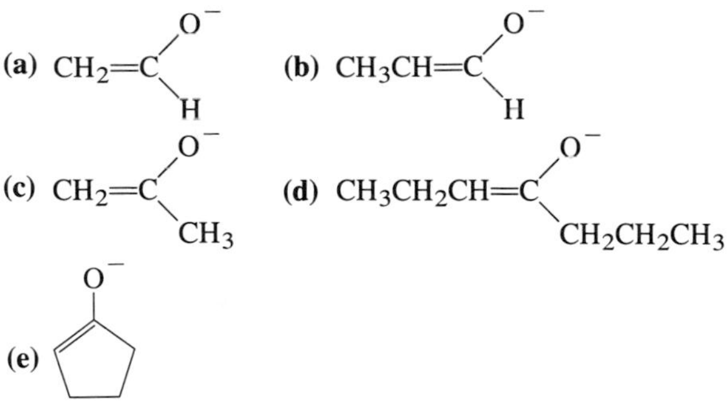

(a) $CH_2{=}C(O^-)H$ (b) $CH_3CH{=}C(O^-)H$

(c) $CH_2{=}C(O^-)CH_3$ (d) $CH_3CH_2CH{=}C(O^-)CH_2CH_2CH_3$

(e) O^- (cyclopentenolate)

18-2

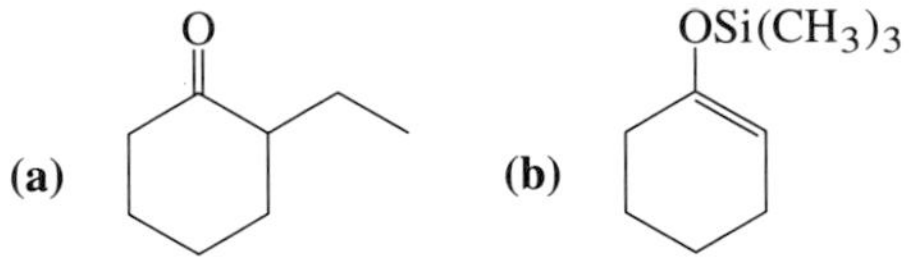

(a) O (2-ethylcyclohexanone) (b) $OSi(CH_3)_3$ (silyl enol ether of cyclohexanone)

18-3

Base catalysis

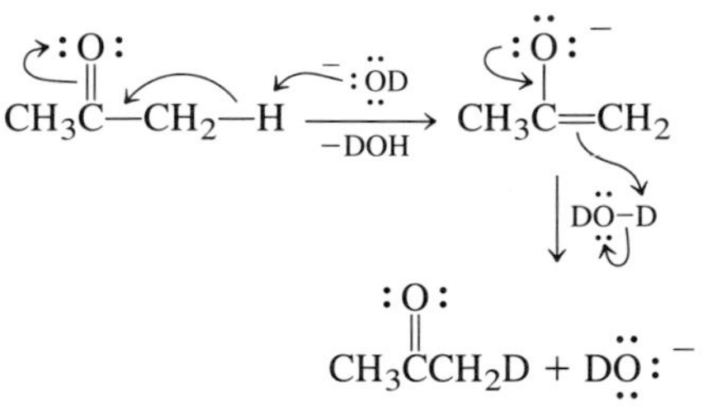

$$CH_3\overset{\underset{\|}{:O:}}{C}{-}CH_2{-}H \xrightarrow[-DOH]{^{-}:\ddot{O}D} CH_3\overset{:\ddot{O}:^-}{C}{=}CH_2 \xrightarrow{D\ddot{O}{-}D} CH_3\overset{\underset{\|}{:O:}}{C}CH_2D + D\ddot{O}:^-$$

Acid catalysis

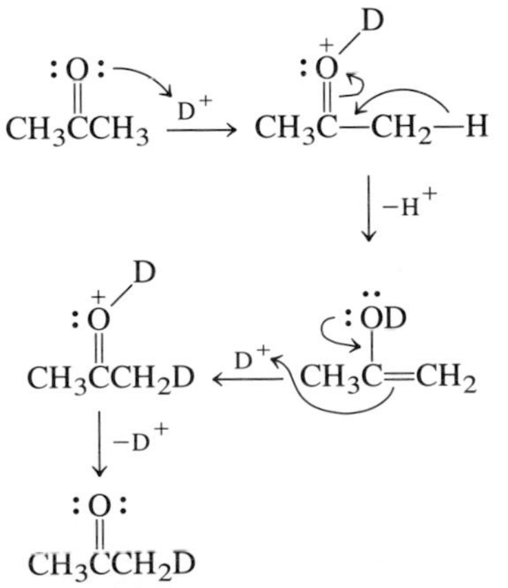

$$CH_3\overset{\underset{\|}{:O:}}{C}CH_3 \xrightarrow{D^+} CH_3\overset{\underset{\|}{:\overset{+}{O}D}}{C}{-}CH_2{-}H \xrightarrow{-H^+} CH_3\overset{:\ddot{O}D}{C}{=}CH_2 \xrightarrow{D^+} CH_3\overset{\underset{\|}{:\overset{+}{O}D}}{C}CH_2D \xrightarrow{-D^+} CH_3\overset{\underset{\|}{:O:}}{C}CH_2D$$

18-4

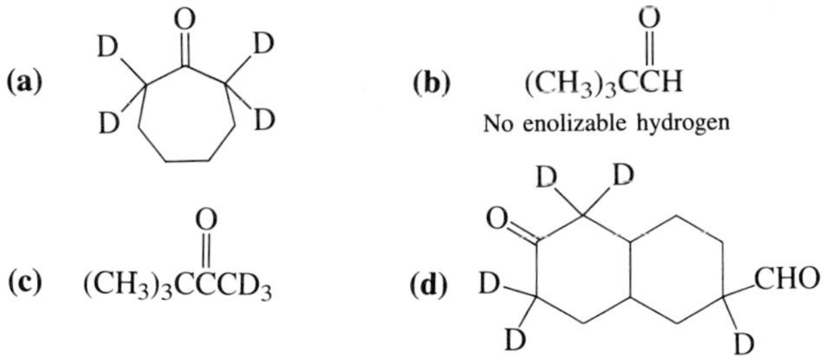

(a) 2,2,7,7-tetradeuteriocycloheptanone (D, D, D, D) (b) $(CH_3)_3C\overset{O}{\overset{\|}{C}}H$

No enolizable hydrogen

(c) $(CH_3)_3C\overset{O}{\overset{\|}{C}}CD_3$ (d) D, D, D, D, D, CHO

18-5

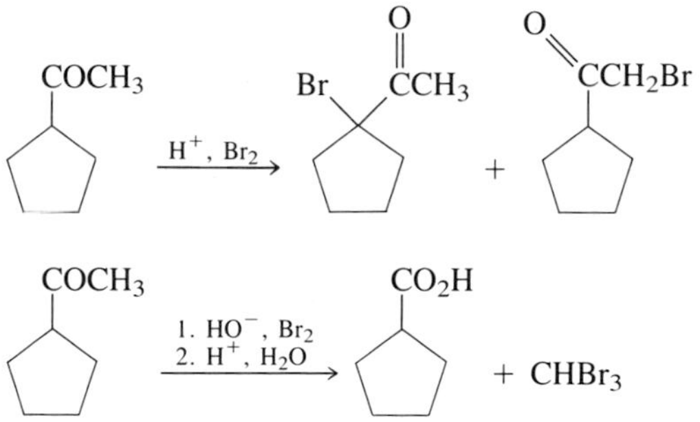

$$\text{COCH}_3 \text{ (cyclopentyl)} \xrightarrow{H^+,\ Br_2} \text{Br, } \overset{O}{\overset{\|}{C}}CH_3 \text{ (1-bromocyclopentyl)} + \overset{O}{\overset{\|}{C}}CH_2Br \text{ (cyclopentyl)}$$

$$\text{COCH}_3 \text{ (cyclopentyl)} \xrightarrow[2.\ H^+,\ H_2O]{1.\ HO^-,\ Br_2} CO_2H \text{ (cyclopentyl)} + CHBr_3$$

18-6

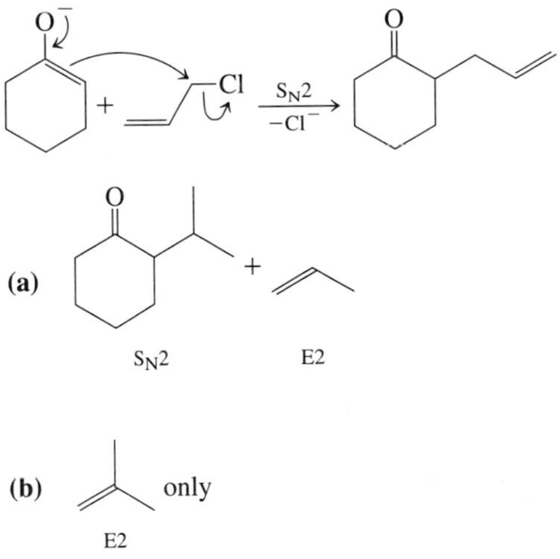

$$O^- \text{ (cyclohexenolate)} + \text{allyl}{-}Cl \xrightarrow[-Cl^-]{S_N2} \text{2-allylcyclohexanone}$$

(a) O (2-isopropylcyclohexanone) + (propene)

S_N2 E2

(b) (2-methylpropene) only

E2

18-7

$$O,\ (CH_2)_3Br \text{ (cyclopentanone)} \xrightarrow{KOH,\ H_2O,\ \Delta}$$

O + O + O

13% 15% 6%

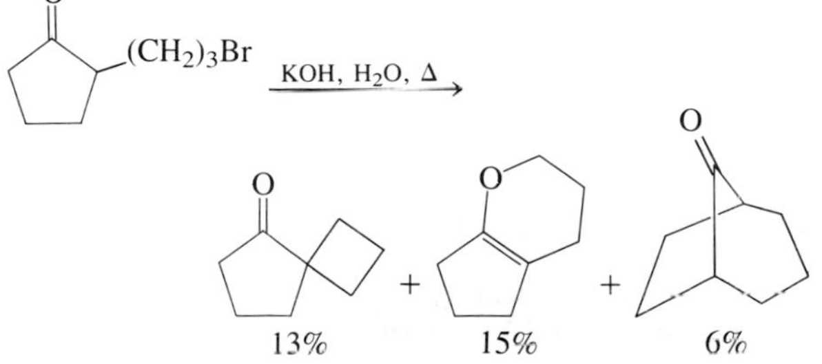

18-8

R'(R)C=N⁺(pyrrolidine) + H–ÖH → (−H⁺) → pyrrolidine-N–C(R)(R')–OH → (H⁺) → pyrrolidine-N⁺(H)–C(R)(R')–OH → pyrrolidine (N–H) + R(R')C=O⁺H → (−H⁺) → R(R')C=O

18-9

A + pyrrolidine (N–H) $\xrightarrow{H^+}$ enamine (3-pyrrolidinyl-1,2-dihydronaphthalene) $\xrightarrow[\text{2. } H^+,\ H_2O]{\text{1. } BrCH_2CO_2CH_2CH_3}$ B

18-10

(a) $CH_3CH_2C(OH)(H)CH(CH_3)CHO$ (b) $CH_3CH_2CH_2C(OH)(H)CH(CH_2CH_3)CHO$

(c) $C_6H_5CH_2C(OH)(H)CH(C_6H_5)CHO$ (d) $C_6H_5CH_2CH_2C(OH)(H)CH(CH_2C_6H_5)CHO$

18-11

It cannot with itself, because it does not contain any enolizable hydrogens. It can, however, undergo crossed aldol condensations (Section 18-6) with enolizable carbonyl compounds.

18-12

(a) $CH_3CH_2CH{=}C(CH_3)CHO$

(b) $CH_3CH_2CH_2CH{=}C(CH_2CH_3)CHO$

(c) $C_6H_5CH_2CH{=}C(C_6H_5)CHO$

(d) $C_6H_5CH_2CH_2CH{=}C(CH_2C_6H_5)CHO$

18-13

Retro-aldol reaction →

$CH_3C(OH)(CH_3)-CH_2-\overset{O}{\overset{\|}{C}}CH_3 + :B^- \rightleftharpoons$

$CH_3\overset{O}{\overset{\|}{C}}CH_3 + CH_2{=}\overset{O^-}{\overset{|}{C}}CH_3 + HB \rightleftharpoons$

$2\ CH_3\overset{O}{\overset{\|}{C}}CH_3 + :B^-$

← Forward aldol reaction

18-14

(a) $C_6H_5CH{=}CH\overset{O}{\overset{\|}{C}}H$

(b) 1-[hydroxy(cyclohexyl)methyl]cyclohexanecarbaldehyde (HC(=O) and HO–CH(cyclohexyl) on cyclohexane)

(c) $CH_2{=}CHCH{=}C(CH_3)CHO$

18-15

(a) cyclodecane-1,6-dione $\xrightarrow{Na_2CO_3,\ 100°C}$ bicyclic enone (octahydronaphthalenone)

(b) 3-phenyl-2-cyclopentenone (C_6H_5)

(c) =O

(d) 1-acetyl-2-methylcyclopentene (CH_3, $\overset{O}{\overset{\|}{C}}CH_3$)

18-16

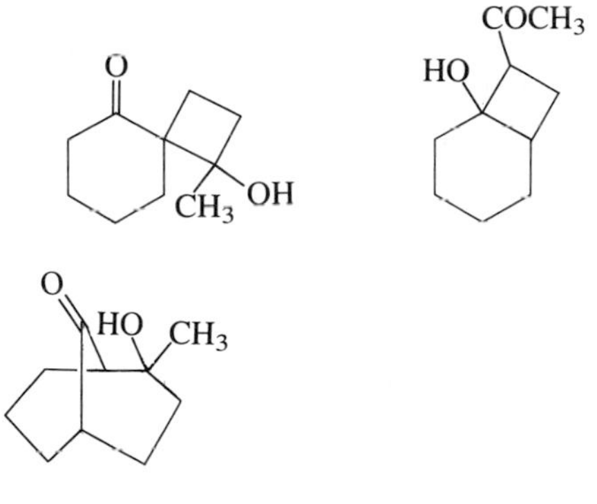

$COCH_3$ HO O

CH_3 OH

O HO CH_3

These three compounds are not formed because of strain. Dehydration is, in addition, prohibited, again because of strain (or, in the first structure, because there is no proton available). The fourth possibility is most facile.

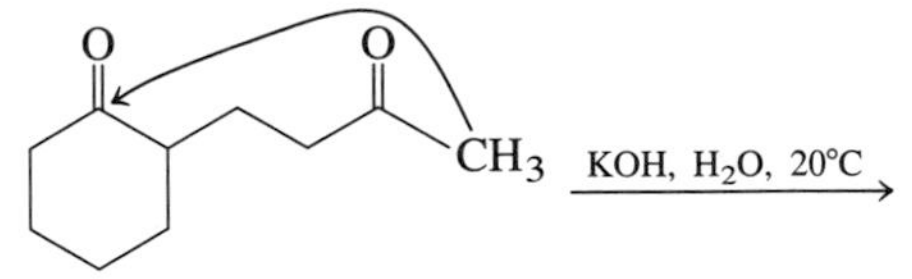

O O CH_3 $\xrightarrow{\text{KOH, } H_2O\text{, 20°C}}$

2-(3-Oxobutyl)cyclohexanone

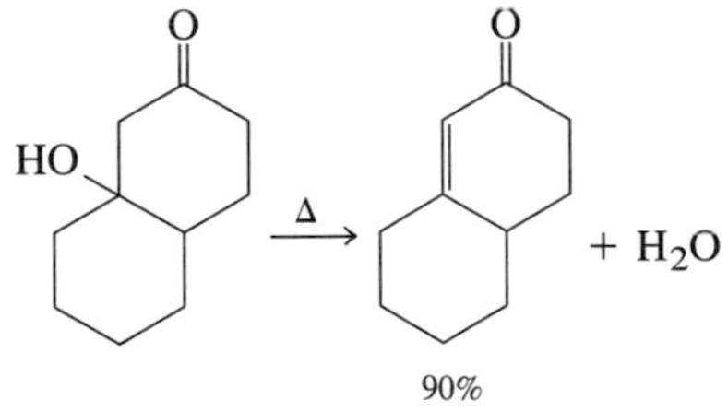

HO O $\xrightarrow{\Delta}$ O + H_2O

90%

18-17

(a)

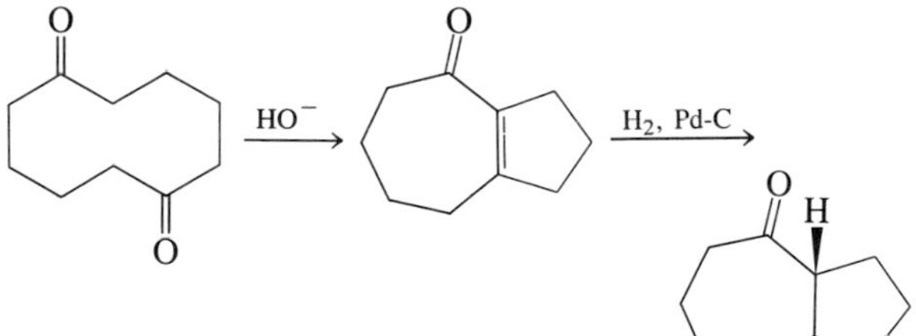

O O $\xrightarrow{HO^-}$ O $\xrightarrow{H_2\text{, Pd-C}}$ O H H

(b) CHO + $CH_3\overset{O}{\overset{\|}{C}}CH_3$

(c) $\overset{O}{\overset{\|}{C}}H$ + O= $\underset{O}{\underset{\|}{C}}H$

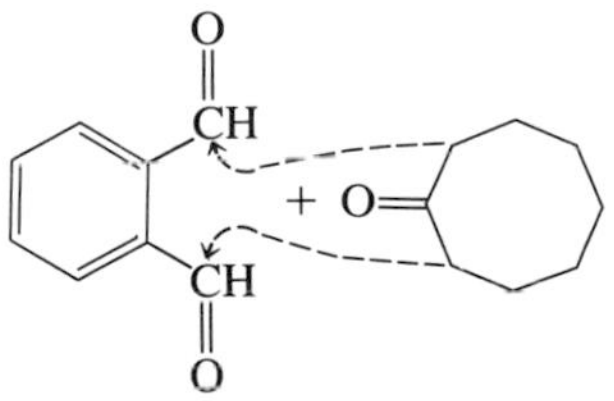

18-18

Mechanism of acid-mediated isomerization of β,γ-unsaturated carbonyl compounds

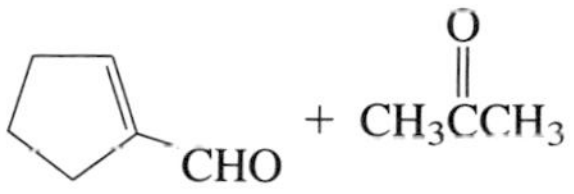

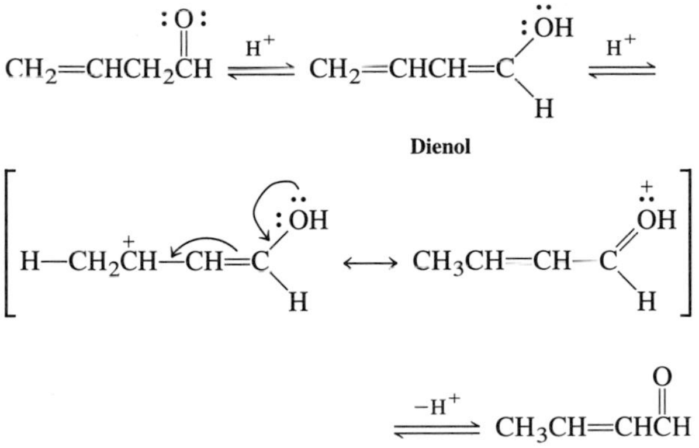

$CH_2{=}CHCH_2\overset{:O:}{\overset{\|}{C}}H \overset{H^+}{\rightleftharpoons} CH_2{=}CHCH{=}C(\ddot{O}H)(H) \overset{H^+}{\rightleftharpoons}$

Dienol

$\left[H{-}CH_2\overset{+}{C}H{-}CH{=}C(:\ddot{O}H)(H) \longleftrightarrow CH_3CH{-}CH{-}C(\overset{+}{\ddot{O}}H)(H) \right]$

$\overset{-H^+}{\rightleftharpoons} CH_3CH{=}CH\overset{O}{\overset{\|}{C}}H$

18-19

$CH_3CH_2CHO \xrightarrow{(C_6H_5)_3P{=}CH\overset{O}{\overset{\|}{C}}H}$

$CH_3CH_2CH{=}CH\overset{O}{\overset{\|}{C}}H \xrightarrow{\text{1. } H_2\text{, Pd-C; 2. } LiAlH_4}$

$CH_3CH_2CH_2CH_2CH_2OH$

18-20

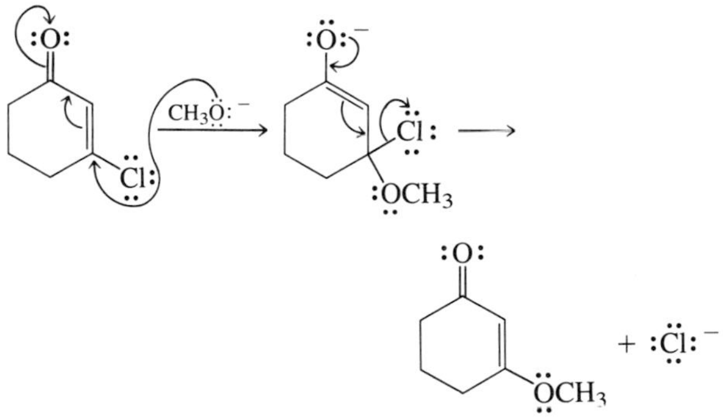

:Ö: Cl: $\xrightarrow{CH_3\ddot{O}:^-}$:Ö:$^-$:Cl: :ÖCH$_3$ $\longrightarrow$

:O: ÖCH$_3$ + $:\ddot{C}l:^-$

18-21

1. Protonation

$C_6H_5\overset{:O:}{\overset{\|}{C}}CH{=}CH_2 \overset{H^+}{\rightleftharpoons} \left[C_6H_5\overset{H\ddot{O}^+}{\overset{\|}{C}}{-}CH{=}CH_2 \longleftrightarrow \right.$

$\left. C_6H_5\overset{H\ddot{O}:}{\overset{|}{\underset{+}{C}}}{-}CH{=}CH_2 \longleftrightarrow C_6H_5\overset{H\ddot{O}:}{\overset{|}{C}}{=}CH{-}\overset{+}{C}H_2 \right]$

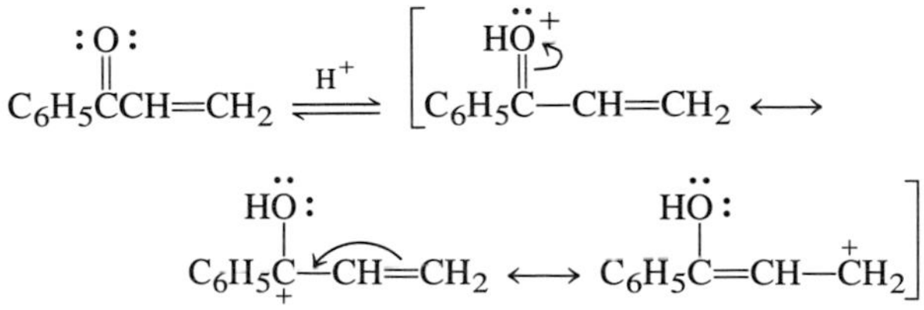

2. Cyanide attack

$C_6H_5\overset{H\ddot{O}^+}{\overset{\|}{C}}{-}CH{=}CH_2 + :\overset{-}{C}{\equiv}N: \rightleftharpoons$

$C_6H_5\overset{H\ddot{O}:}{\overset{|}{C}}{=}CH{-}CH_2CN$

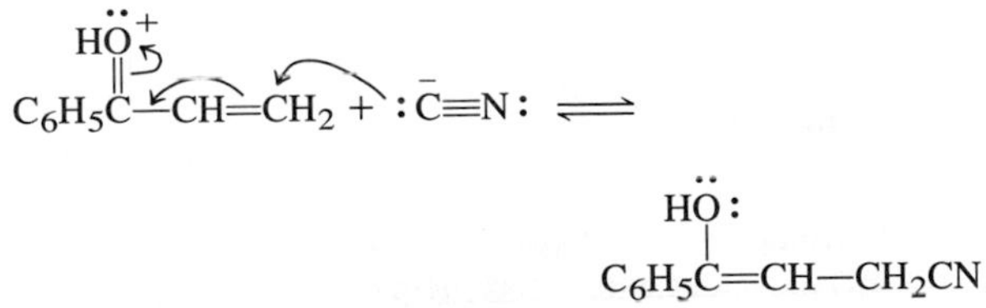

3. Enol–keto tautomerization

$$C_6H_5\overset{\ddot{H}O:}{C}{=}CHCH_2CN \rightleftharpoons C_6H_5\overset{:O:}{\overset{\|}{C}}\underset{H}{C}HCH_2CN$$

18-22

(a) 3-Methylcyclohex-2-enone $\xrightarrow{1.\ (CH_3)_2CuLi;\ 2.\ CH_3I}$ 2,3,3-trimethylcyclohexanone (CH_3, CH_3, CH_3)

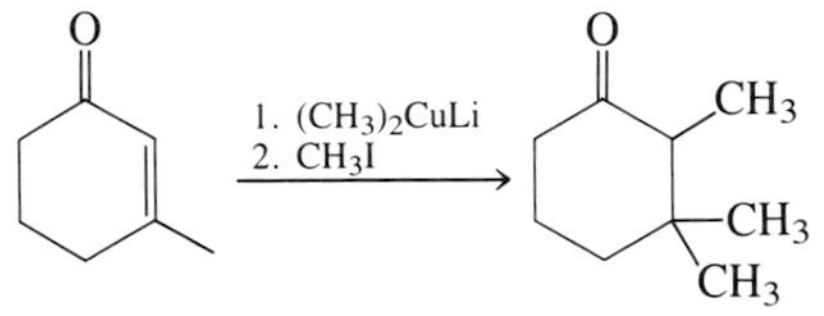

(b) 3-Methylcyclohex-2-enone + $(CH_2{=}CHCH_2CH_2)_2CuLi \longrightarrow$

3-(3-butenyl)-3-methylcyclohexanone $\xrightarrow{1.\ O_3;\ 2.\ (CH_3)_2S}$ 3-methyl-3-(3-oxopropyl)cyclohexanone (O, H) $\xrightarrow[-\ H_2O]{NaOH}$ bicyclic enone (O)

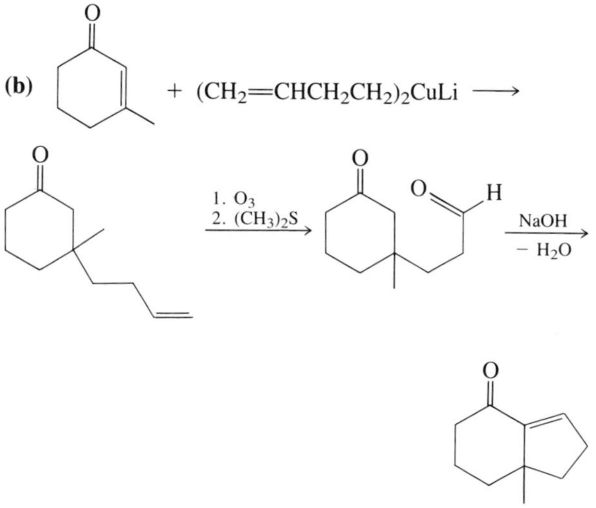

18-23

(a) $CH_3\overset{:\ddot{O}:^-}{C}{=}CH_2 + CH_2{=}CH\overset{O}{\overset{\|}{C}}CH_3$

(b) cyclohexenolate ($:\ddot{O}:^-$) $+ CH_2{=}CH\overset{O}{\overset{\|}{C}}CH_3$

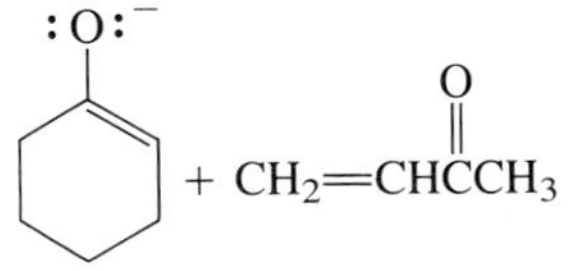

(c) $C_6H_5CH{=}\overset{:\ddot{O}:^-}{C}H + CH_2{=}CH\overset{O}{\overset{\|}{C}}CH_3$

CHAPTER 19

19-1

(a) 5-Bromo-3-chloroheptanoic acid
(b) 4-Oxocyclohexanecarboxylic acid
(c) 3-Methoxy-4-nitrobenzoic acid
(d) $HOOCCH_2CH_2CH_2CBr_2COOH$ (Br above and below the C)
(e) $CH_3\overset{OH}{C}HCH_2CH_2\overset{O}{\overset{\|}{C}}OH$ **(f)** 4-$C(CH_3)_3$-benzoic acid (COOH, $C(CH_3)_3$)

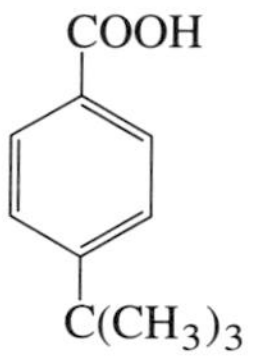

19-2

$CH_3CH_2CH_2COOH$, butanoic (butyric) acid

19-3

(a) $CH_3CBr_2COOH > CH_3CHBrCOOH > CH_3CH_2COOH$

(b) $CH_3\overset{F}{C}HCH_2COOH > CH_3\overset{Br}{C}HCH_2COOH$

(c) 1-fluorocyclohexanecarboxylic acid (F, COOH) > 4-fluorocyclohexanecarboxylic acid (COOH, F) ⩾ cyclohexanecarboxylic acid (COOH)

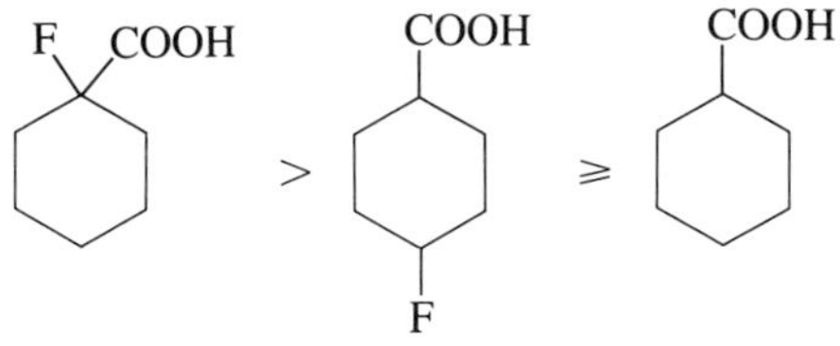

19-4

Protonated propanone (acetone) has fewer resonance forms.

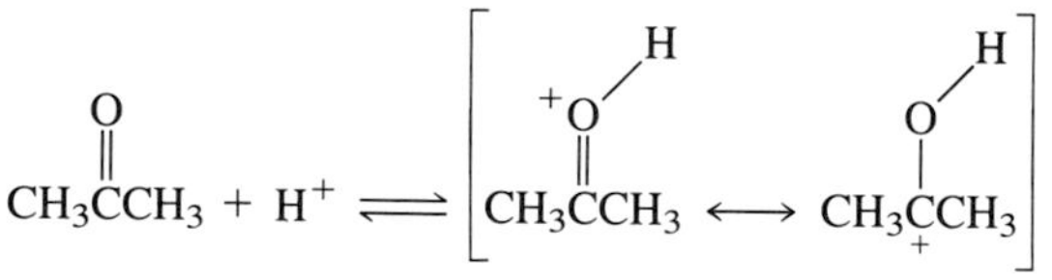

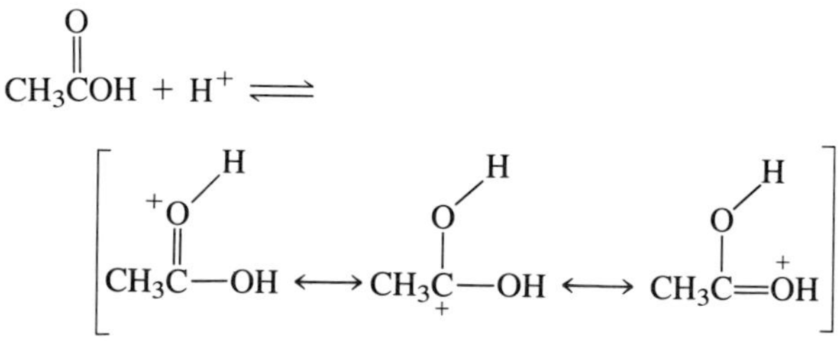

$$CH_3\overset{O}{\overset{\|}{C}}CH_3 + H^+ \rightleftharpoons \left[CH_3\overset{^+O-H}{\overset{\|}{C}}CH_3 \longleftrightarrow CH_3\overset{O-H}{\underset{+}{C}}CH_3\right]$$

$$CH_3\overset{O}{\overset{\|}{C}}OH + H^+ \rightleftharpoons$$

$$\left[CH_3\overset{^+O-H}{\overset{\|}{C}}{-}OH \longleftrightarrow CH_3\overset{O-H}{\underset{+}{C}}{-}OH \longleftrightarrow CH_3\overset{O-H}{C}{=}\overset{+}{O}H\right]$$

19-5

(a) $CH_3(CH_2)_3COOH$ **(b)** $HOOC(CH_2)_4COOH$

(c) cyclohexane-1,4-dicarboxylic acid (COOH, COOH)

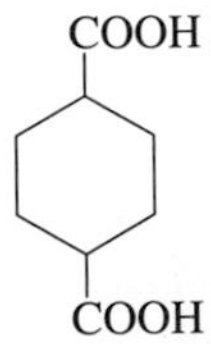

19-6

(a) 1. HCN, 2. H^+, H_2O

(b) methylenecyclohexane $\xrightarrow{HBr}$ 1-bromo-1-methylcyclohexane (H_3C, Br) $\xrightarrow{1.\ Mg;\ 2.\ CO_2;\ 3.\ H^+,\ H_2O}$ 1-methylcyclohexanecarboxylic acid (H_3C, COOH)

(c) Br / OCH_3 cyclohexane $\xrightarrow[-Br^-,\ S_N2]{^-CN}$ N≡C / OCH_3 cyclohexane $\xrightarrow{1.\ HO^-,\ H_2O;\ 2.\ H^+,\ H_2O}$ COOH / OCH_3 cyclohexane

19-7

(a) 1. $CH_3\overset{O}{\overset{\|}{C}}Cl + Na^{+\,-}O\overset{O}{\overset{\|}{C}}CH_2CH_3$

2. $CH_3\overset{O}{\overset{\|}{C}}O^-Na^+ + Cl\overset{O}{\overset{\|}{C}}CH_2CH_3$

(b) $CH_3\overset{H_3C}{\overset{|}{C}H}\overset{O}{\overset{\|}{C}}OH + SOCl_2$ or $COCl_2$

19-8

The reaction is self-catalyzed by acid.

butanedioic acid (OH, OH) $+ H^+ \longrightarrow$ protonated acid (^+O–H) $\longrightarrow$ cyclic intermediate (HO, OH, $\overset{+}{O}H$) $\longrightarrow$

HO, $\overset{+}{O}H_2$ intermediate $\xrightarrow{-H_2O}$ $\overset{+}{O}$–H lactone $\longrightarrow$ succinic anhydride $+ H^+$

19-9

Label appears in the ester.

$$R\overset{O}{\overset{\|}{C}}OH + H^{18}OCH_3 \overset{H^+}{\rightleftharpoons} R\overset{O}{\overset{\|}{C}}{}^{18}OCH_3 + H_2O$$

19-10

HO, H / O ring, CH_2COOH $\overset{H^+}{\rightleftharpoons}$ HO, H / $\overset{+}{O}$–H ring, CH_2COOH $\overset{-H^+}{\rightleftharpoons}$

$\overset{O}{\overset{\|}{C}}$H, OH, CH_2COOH $\overset{H^+,\ \text{lactonization}}{\rightleftharpoons}$ lactone with $\overset{O}{\overset{\|}{C}}H$ side chain $+ H_2O$

19-11

butanedioic acid (OH, OH) $+ 2\,NH_3 \rightleftharpoons$ $O^-\overset{+}{N}H_4$, $O^-\overset{+}{N}H_4$ $\overset{\Delta}{\rightleftharpoons}$

O, OH, $O^-\overset{+}{N}H_4$ $+ :NH_3 \rightleftharpoons$ HO, NH_2, OH, $O^-\overset{+}{N}H_4$ $\overset{-H_2O}{\rightleftharpoons}$

O, NH_2, $O^-\overset{+}{N}H_4$ $\underset{+NH_3}{\overset{-NH_3}{\rightleftharpoons}}$ O, $\ddot{N}H_2$, OH, O $\rightleftharpoons$

O, $\overset{+}{N}H_2$, OH, ^-O ring $\rightleftharpoons$ O, NH, OH, HO ring $\rightleftharpoons$ succinimide (O, NH, O) $+ H_2O$

19-12

(a) 1. H^+, H_2O, 2. $LiAlH_4$, 3. H^+, H_2O

(b) 1. $LiAlD_4$, 2. H^+, H_2O

19-13

2. $RCH_2\overset{O}{\overset{\|}{C}}Br + H^+ \rightleftharpoons RCH_2\overset{\overset{+}{O}H}{\overset{\|}{C}}Br$

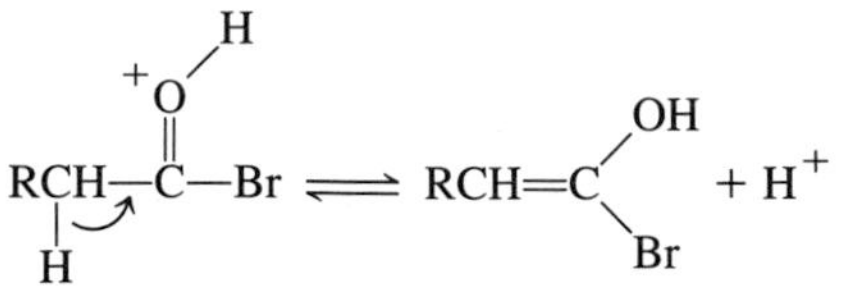

$$RCH(H)-\overset{+OH}{\overset{\|}{C}}-Br \rightleftharpoons RCH{=}C(OH)Br + H^+$$

3. 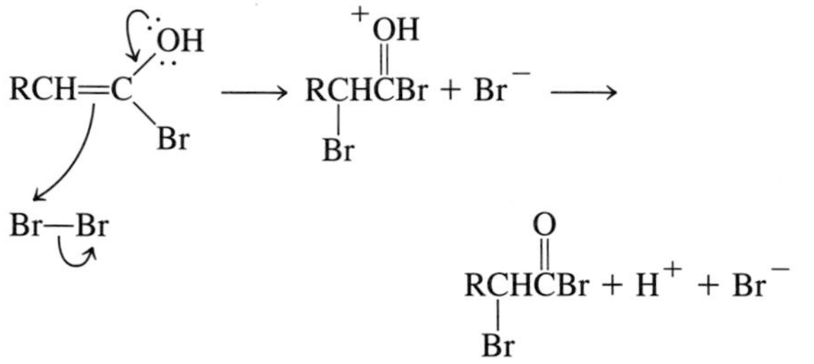 $$RCH{=}C(\ddot{O}H)Br \xrightarrow{Br-Br} RCH(Br)\overset{+OH}{\overset{\|}{C}}Br + Br^- \longrightarrow RCH(Br)\overset{O}{\overset{\|}{C}}Br + H^+ + Br^-$$

19-14

(a) $CH_3CH_2CH_2CH_2Br$ **(b)** $CH_3CH_2CHBrCH_2CH_3$

(c) 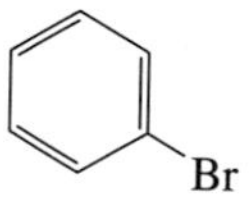 Bromobenzene (Br on benzene ring)

19-15

Self-explanatory.

CHAPTER 20

20-1

At room temperature. rotation around the amide bond is slow on the NMR time scale and two distinct rotamers can be observed.

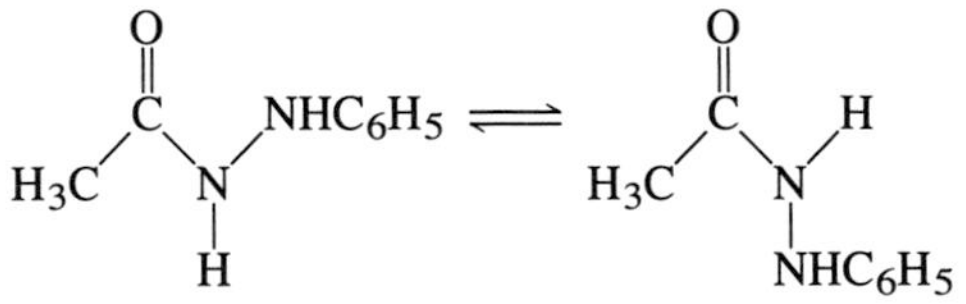

$$H_3C\overset{O}{\overset{\|}{C}}-N(H)NHC_6H_5 \rightleftharpoons H_3C\overset{O}{\overset{\|}{C}}-N(H)NHC_6H_5$$

Heating makes the equilibration so fast that the NMR technique can no longer distinguish between the two species.

20-2

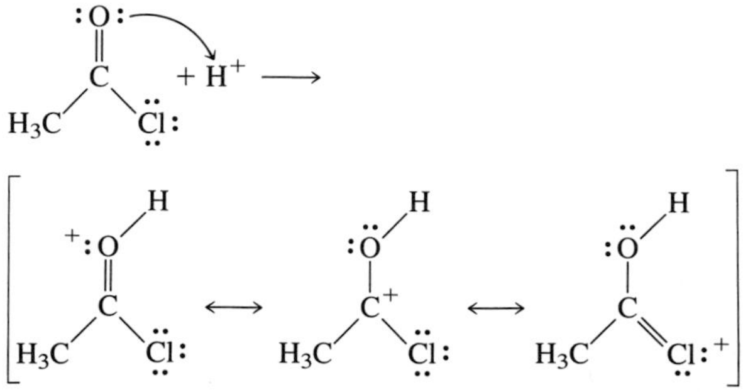

$$H_3C\overset{:O:}{\overset{\|}{C}}\ddot{C}l: + H^+ \longrightarrow$$

$$\left[H_3C\overset{+:OH}{\overset{\|}{C}}\ddot{C}l: \longleftrightarrow H_3C\overset{:\ddot{O}H}{\overset{|}{C^+}}\ddot{C}l: \longleftrightarrow H_3C\overset{:\ddot{O}H}{\overset{|}{C}}{=}\ddot{C}l:^+\right]$$

Not a strong contributor

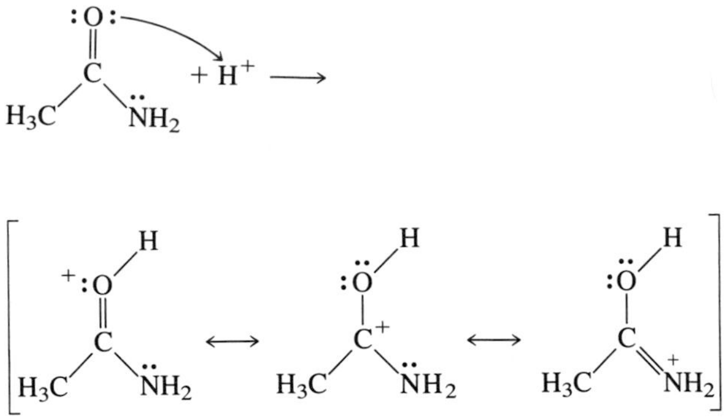

$$H_3C\overset{:O:}{\overset{\|}{C}}\ddot{N}H_2 + H^+ \longrightarrow$$

$$\left[H_3C\overset{+:OH}{\overset{\|}{C}}\ddot{N}H_2 \longleftrightarrow H_3C\overset{:\ddot{O}H}{\overset{|}{C^+}}\ddot{N}H_2 \longleftrightarrow H_3C\overset{:\ddot{O}H}{\overset{|}{C}}{=}\overset{+}{N}H_2\right]$$

Strong contributor

20-3

Section 19–12, step 2 of the Hell-Volhard-Zelinsky reaction.

20-4

$$CH_3COOH \xrightarrow{SOCl_2} CH_3\overset{O}{\overset{\|}{C}}Cl \xrightarrow{(CH_3)_3COH,\ (CH_3CH_2)_3N} CH_3\overset{O}{\overset{\|}{C}}OC(CH_3)_3 + (CH_3CH_2)_3\overset{+}{N}HCl^-$$

20-5

Use *N,N*-diethylethanamine (triethylamine) to generate the alkanoyl triethylammonium salt, then add the expensive amine.

20-6

(a) H_2O **(b)** 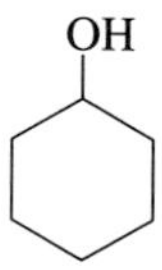 Cyclohexanol (OH on cyclohexane ring) **(c)** $(CH_3)_2NH$

(d) $(CH_3CH_2)_2CuLi$ **(e)** $LiAl[OC(CH_3)_3]_3H$

20-7

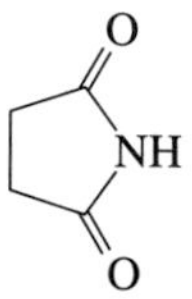

Five-membered ring imide: two C=O groups flanking NH

Butanimide (succinimide)

(See Section 19-10)

20-8

$$CH_3COCCH_3 \underset{-H^+}{\overset{+H^+}{\rightleftharpoons}} CH_3C(=\overset{+}{O}H)OCCH_3 \underset{-CH_3OH}{\overset{+HOCH_3}{\rightleftharpoons}}$$

$$CH_3C(OH)(\overset{+}{O}HCH_3)OCCH_3 \underset{+H^+}{\overset{-H^+}{\rightleftharpoons}} CH_3C(OH)(OCH_3)OCCH_3 \underset{-H^+}{\overset{+H^+}{\rightleftharpoons}}$$

$$CH_3C(OH)(OCH_3)-O-C(=\overset{+}{O}H)CH_3 \rightleftharpoons CH_3C(=\overset{+}{O}H)OCH_3 + CH_3COH$$

$$CH_3C(=\overset{+}{O}H)OCH_3 \underset{+H^+}{\overset{-H^+}{\rightleftharpoons}} CH_3COCH_3$$

20-9

(a) Propyl propanoate
(b) Dimethyl butanedioate
(c) Methyl propenoate (methyl acrylate)

20-10

$$\text{lactone} \underset{-H^+}{\overset{H^+}{\rightleftharpoons}} \text{protonated lactone} \underset{-H_2O}{\overset{+H_2O}{\rightleftharpoons}} \underset{+H^+}{\overset{-H^+}{\rightleftharpoons}}$$

$$\text{tetrahedral intermediate (HO, OH)} \underset{-H^+}{\overset{+H^+}{\rightleftharpoons}} \text{ring-O-protonated intermediate} \rightleftharpoons$$

$$HOCH_2CH_2CH_2C(=\overset{+}{O}H)OH \underset{+H^+}{\overset{-H^+}{\rightleftharpoons}} HOCH_2CH_2CH_2COOH$$

4-Hydroxybutanoic acid

20-11

$$\text{lactone} \underset{-HO^-}{\overset{HO^-}{\rightleftharpoons}} \text{tetrahedral alkoxide intermediate} \rightleftharpoons$$

$$^-OCH_2CH_2CH_2COOH \longrightarrow HOCH_2CH_2CH_2COO^-$$

20-12

Acid catalysis: As in Exercise 20-10, but use $BrCH_2CH_2CH_2OH$ instead of H_2O as the nucleophile in the second step.

Base catalysis: As in Exercise 20-10, but use $BrCH_2CH_2CH_2O^-$ instead of HO^- as the nucleophile in the first step.

20-13

$$C_6H_5Br \xrightarrow{Mg} C_6H_5MgBr$$

$$C_6H_5\overset{O}{\overset{\|}{C}}OCH_3 + 2C_6H_5MgBr \xrightarrow[H^+,\ H_2O]{\text{Then}} (C_6H_5)_3COH$$

20-14

(a) cyclohexyl–CO_2H (b) cyclohexyl–CO_2^- (c) cyclohexyl–CO_2CH_3 (d) cyclohexyl–$\overset{O}{\overset{\|}{C}}NH_2$

(e) cyclohexyl–$C(CH_3)_2OH$ (CH_3COH with CH_3) (f) cyclohexyl–CH_2OH (g) 1-methylcyclohexane with H_3C and CO_2CH_3 on the same carbon

20-15

2,2-dimethylpyrrolidine (H_3C, H_3C, N–H)

20-16

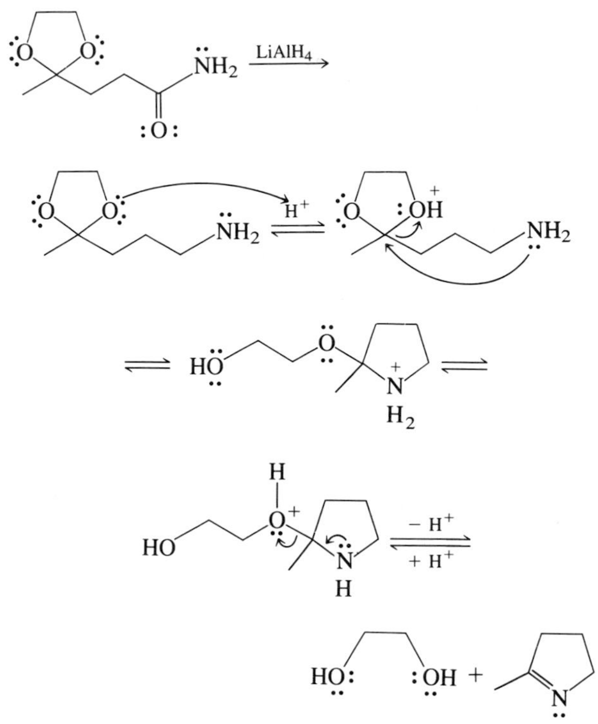

$$\xrightarrow{LiAlH_4}$$

$$\xrightarrow{H^+}$$

HÖ :NH₂

$$\underset{+H^+}{\overset{-H^+}{\rightleftharpoons}}$$

$HO:$ $:OH$ +

20-17

The negative charge can be delocalized over two carbonyl groups.

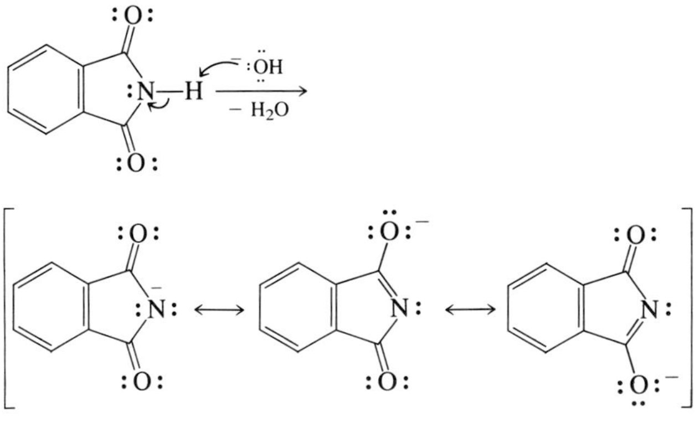

$$\xrightarrow[-H_2O]{:\ddot{O}H^-}$$

$$\longleftrightarrow \quad \longleftrightarrow$$

20-18

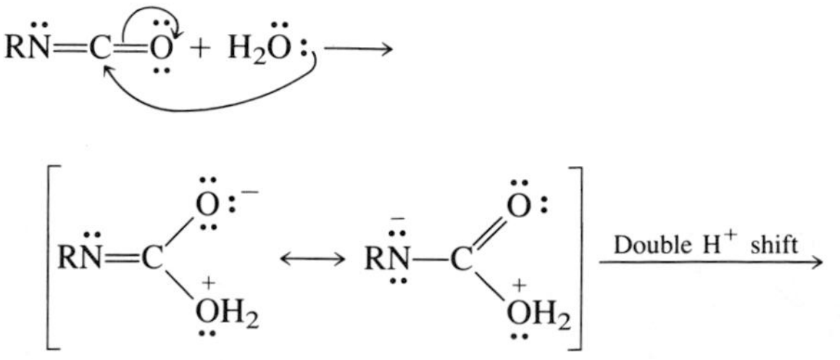

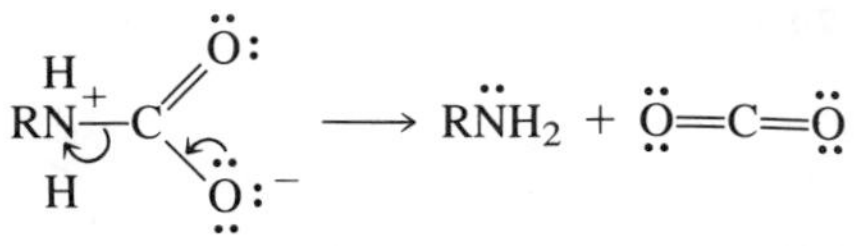

$R\ddot{N}=C=\ddot{O} + H_2\ddot{O}: \longrightarrow$

$$\left[R\ddot{N}=C(\ddot{O}:^-)(\overset{+}{O}H_2) \longleftrightarrow R\overset{-}{\ddot{N}}-C(=\ddot{O}:)(\overset{+}{O}H_2) \right] \xrightarrow{\text{Double } H^+ \text{ shift}}$$

$$RN^+H_2-C(=O)O^- \longrightarrow R\ddot{N}H_2 + \ddot{O}=C=\ddot{O}$$

20-19

$$\text{COOCH}_3 \xrightarrow{H^+,\ H_2O,\ \Delta} \text{COOH} \xrightarrow{1.\ SOCl_2\quad 2.\ NH_3}$$

or NH_3, H_2O, Δ

$$\overset{O}{\overset{\|}{C}}NH_2 \xrightarrow{Br_2,\ HO^-} NH_2$$

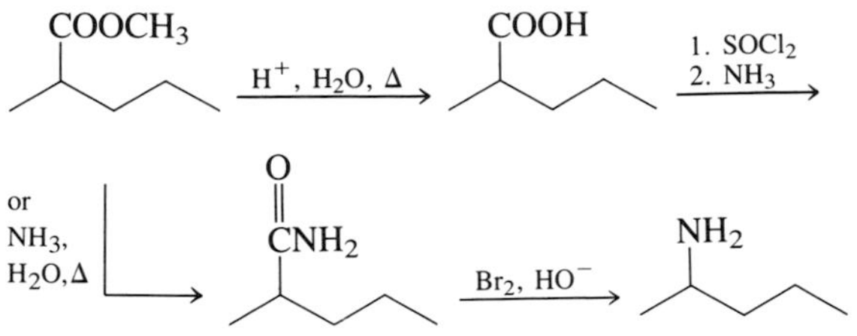

20-20

$$BrCH_2CH_2CH_2Br \xrightarrow[-Br^-]{^-CN}$$

$$BrCH_2CH_2CH_2C{\equiv}N \xrightarrow[-Br^-]{^-CN} NCCH_2CH_2CH_2CN$$

117.6 (and 3 other signals); 119.1 22.6 17.6

20-21

The exact details of this reduction are not known. A possible mechanism is

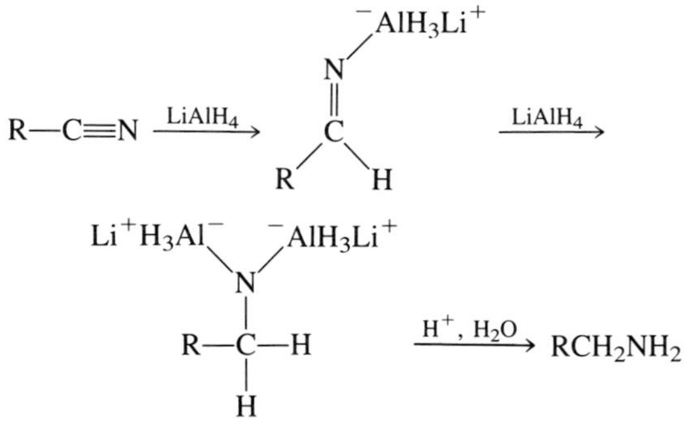

$$R-C{\equiv}N \xrightarrow{LiAlH_4} RCH{=}N-AlH_3^-Li^+ \xrightarrow{LiAlH_4}$$

$$RCH_2N(AlH_3^-Li^+)_2 \xrightarrow{H^+,\ H_2O} RCH_2NH_2$$

20-22

(a) 1. H_2O, HO^-, 2. H^+, H_2O

(b) 1. $CH_3CH_2CH_2CH_2MgBr$, 2. H^+, H_2O

(c) 1. $(CH_3\overset{CH_3}{\overset{|}{C}}HCH_2)_2AlH$ 2. H^+, H_2O

(d) D_2, Pt

20-23

(a) CH_3OCH_3, CH_3CH_2OH, O—O, $H\overset{O}{\overset{\|}{C}}OH$

(b) $CH_2{=}O$ **(c)** 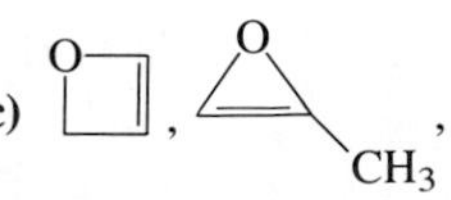 , , $CH_2{=}CH\overset{O}{\overset{\|}{C}}H$,

$HC{\equiv}CCH_2OH$, $CH_3C{\equiv}COH$, $HC{\equiv}COCH_3$,

O OH

O , O , ,

20-24

CH_2Br_2: $m/z = 172, 174, 176$; intensity ratio 1:2:1

20-25

For most elements in organic compounds, such as C, H, and O, the mass (of the most abundant isotope) and valence are either both even or both odd, so that even molecular weights always result. Nitrogen is a major exception: the atomic weight is 14, but the valence is 3. This phenomenon has led to the **nitrogen rule** in mass spectroscopy, as expressed in this question.

20-26

Mass spectrum of 3-methyl-3-heptanol

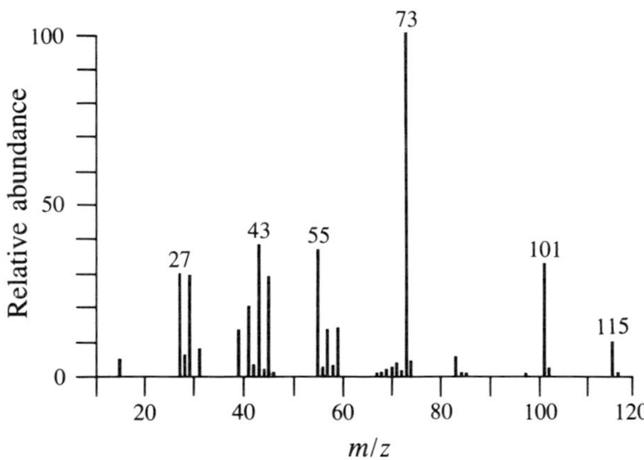

The major primary fragments are due to cleavage of the bonds α to the hydroxy group. Why? Consider their strength and the electronic structure of the resulting radical cations. (Draw resonance structures.) Do these cations fragment by loss of water?

20-27

(a) Both show the same α cleavage patterns but different McLafferty rearrangements.

$[H_3C\text{–}C(=O)\text{–}CH(CH_3)\text{–}CH_2\text{–}CH_2\text{–}H]^{+\cdot}$ $\longrightarrow$ $[HO(H_3C)C{=}CHCH_3]^{+}$ + $[CH_2{=}CH_2]^{+}$

$m/z = 100$ $\qquad$ $m/z = 72$ $\qquad$ $m/z = 28$

$[H_3C\text{–}C(=O)\text{–}CH_2\text{–}CHCH_3\text{–}CH_2\text{–}H]^{+\cdot}$ $\longrightarrow$ $[HO(H_3C)C{=}CH_2]^{+}$ + $[CH_3CH{=}CH_2]^{+}$

$m/z = 100$ $\qquad$ $m/z = 58$ $\qquad$ $m/z = 42$

(b) Both show the same α cleavage patterns, but only 2-ethylcyclohexanone has an accessible γ-hydrogen for the McLafferty rearrangement.

$[\text{2-ethylcyclohexanone}]^{+\cdot}$ $\longrightarrow$ $[\text{OH-cyclohexene}]^{+}$ + $[CH_2{=}CH_2]^{+}$

20-28

Pentanal: $m/z = 57$, $CH_3CH_2CH_2CH_2$ (α cleavage)

$m/z = 44$, $H_2C{=}C(OH)H$ (McLafferty rearrangement)

Pentanoic acid: $m/z = 60$, $H_2C{=}C(OH)OH$ (McLafferty rearrangement)

Methyl pentanoate: $m/z = 57$, $CH_3CH_2CH_2CH_2$ (α cleavage)

$m/z = 85$, $CH_3CH_2CH_2CH_2C{=}O$ (α cleavage)

$m/z = 74$, $H_2C{=}C(OH)OCH_3$ (McLafferty rearrangement)

20-29

(a) $C_7H_{12}O$ $\quad$ **(b)** C_6H_{14}

CHAPTER 21

21-1

(a) 2-Butanamine, *sec*-butylamine
(b) *N,N*-Dimethylbenzenamine, *N,N*-dimethylaniline
(c) 6-Bromo-2-hexanamine, 5-bromo-2-methylpentylamine

21-2

(a) $HC{\equiv}CCH_2NH_2$ **(b)** $CH_2NHCH_2CH{=}CH_2$ (on benzene ring)

(c) $(CH_3)_3CNHCH_3$

21-3

The lesser electronegativity of nitrogen, compared with oxygen, allows for slightly more diffuse orbitals and hence longer bonds to other nuclei.

21-4

Less, because, again, nitrogen is less electronegative than oxygen. See Tables 10-2 and 10-3 for the effect of the electronegativity of substituent atoms on chemical shifts.

21-5

IR: Secondary amine, hence a weak band at 3400 cm^{-1}

^1H NMR: s for the 1,1-dimethylethyl (*tert*-butyl) group at high field; s for the attached methylene group at $\delta \sim 2.7$; q for the second methylene unit close to the first; t for the unique methyl group at high field, closest to the 1,1-dimethylethyl (*tert*-butyl) signal

^{13}C NMR: Five signals, two at low field, $\delta \sim 45$–50 ppm

Mass: $m/z = 115\ (M^+)$, 100 $[(CH_3)_3CCH_2NH{=}CH_2]^+$, and 58 $(CH_2{=}NHCH_2CH_3)^+$. In this case, two different iminium ions can be formed by fragmentation.

21-6

As discussed in Section 16-2, the nitrogen lone electron pair is tied up by resonance with the benzene ring. Therefore, the nitrogen is less nucleophilic than the one in an alkanamine.

21-7

:O: ... :NR + $^-$:ÖH ⟶ $^-$:Ö: ÖH ... :NR ⟶ ... :O:

:O: ÖH ... $\ddot{N}R^-$ ⟶ COO^- ... $\ddot{N}HR$ ... :O:

Continue as in the normal amide hydrolysis by base (Section 20-6).

:O: ... :NR + $H_2\ddot{N}\ddot{N}H_2$ ⟶ (Addition-elimination) ... :O:

:O: $C\ddot{N}H\ddot{N}H_2$ ... $C\ddot{N}HR$:O: ⟶ (Intramolecular addition-elimination) :O: ... $\ddot{N}H$–$\ddot{N}H$ ... :O: + $R\ddot{N}H_2$

21-8

O, N^-K^+, O

A

(a) 1. **A** + $CH_3(CH_2)_5Br$; 2. H^+, H_2O; 3. HO^-, H_2O

(b) 1. **A** + (branched alkyl bromide) Br; 2. H^+, H_2O;

3. HO^-, H_2O

(c) 1. **A** + (bromocyclohexane) Br; 2. H^+, H_2O; 3. HO^-, H_2O

(d) 1. **A** + $BrCH_2CO_2CH_2CH_3$; 2. H^+, H_2O

Protection of the carboxy group is necessary to prevent the acidic proton from reacting with A (see also Section 26-2). The azide method should work well for (a)–(c). For (d), the reduction step requires catalytic hydrogenation, because $LiAlH_4$ would also attack the ester function.

21-9

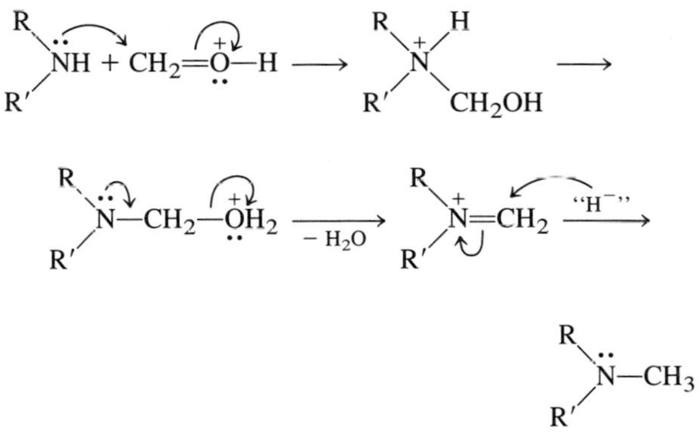

$$R(R')\ddot{N}H + CH_2{=}\overset{+}{\ddot{O}}{-}H \longrightarrow R(R')\overset{+}{N}(H)CH_2OH \longrightarrow$$

$$R(R')\ddot{N}{-}CH_2{-}\overset{+}{\ddot{O}}H_2 \xrightarrow{-\,H_2O} R(R')\overset{+}{N}{=}CH_2 \xrightarrow{\text{“}H^-\text{”}} R(R')\ddot{N}{-}CH_3$$

21-10

Not all intermediates are shown.

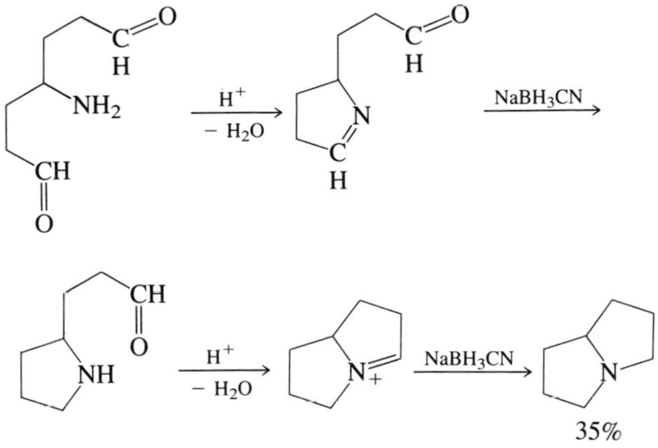

$$\xrightarrow[-\,H_2O]{H^+} \quad \xrightarrow{NaBH_3CN} \quad \xrightarrow[-\,H_2O]{H^+} \quad \xrightarrow{NaBH_3CN}$$

35%

21-11

NH_2 + CH_3I (Excess)

NH_2 + $CH_2{=}O$ + $NaBH_3CN$

$NHCH$ (O) + $LiAlH_4$

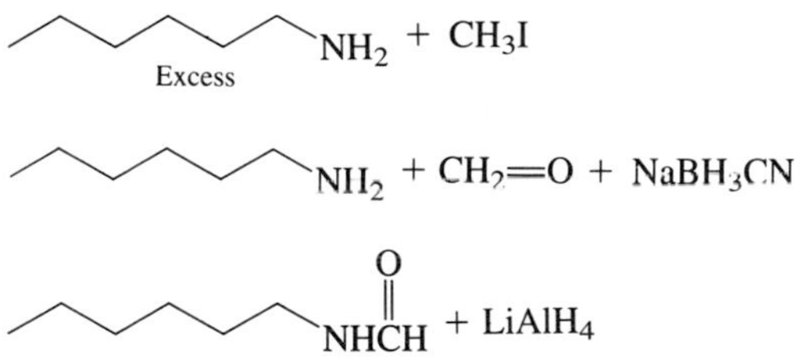

21-12

(a) $CH_3CH{=}CH_2$ and $CH_2{=}CH_2$

(b) $CH_3CH_2CH{=}CH_2$ and $CH_3CH{=}CHCH_3$ (cis and trans). The terminal alkene predominates. This reaction is kinetically controlled according to the Hofmann rule (Section 11-8). Thus, the base prefers attack at the less bulky end of the sterically encumbered quaternary ammonium ion.

21-13

The ethyl group can be extruded as ethene via Hofmann elimination [see Exercise 21-12(a)]. Generally, any alkyl substituent leaving hydrogens located β to the quaternary nitrogen is capable of such elimination, an option absent for methyl.

21-14

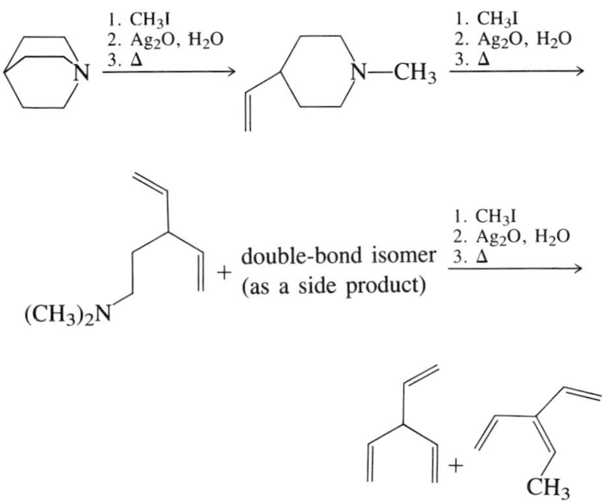

$$\xrightarrow{1.\ CH_3I\quad 2.\ Ag_2O,\ H_2O\quad 3.\ \Delta} \quad N{-}CH_3 \quad \xrightarrow{1.\ CH_3I\quad 2.\ Ag_2O,\ H_2O\quad 3.\ \Delta}$$

$(CH_3)_2N$ … + double-bond isomer (as a side product) $\xrightarrow{1.\ CH_3I\quad 2.\ Ag_2O,\ H_2O\quad 3.\ \Delta}$

+ … CH_3

21-15

(a) 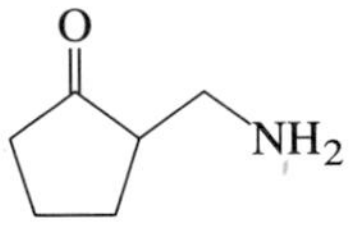 … NH_2 (b) $CH_3\overset{CH_3}{\underset{CH_2NH(CH_2)_5CH_3}{C}}CHO$

(c) 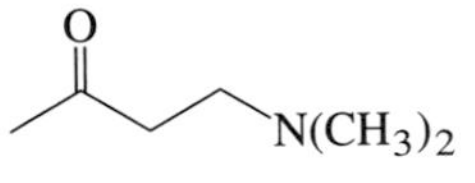 … $N(CH_3)_2$ (d) 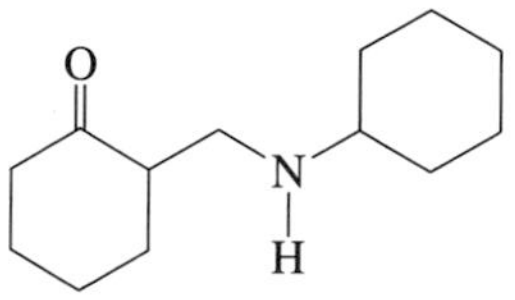 … N–H

21-16

$$+\ CH_2{=}O + (CH_3)_2NH \xrightarrow{1.\ H^+\quad 2.\ HO^-}$$

$N(CH_3)_2$ $\xrightarrow{1.\ NaBH_4\quad 2.\ O_2N{-}C_6H_4{-}\overset{O}{\overset{\|}{C}}Cl}$

$-NO_2$, $N(CH_3)_2$ $\xrightarrow{Sn,\ HCl}$ tutocaine hydrochloride

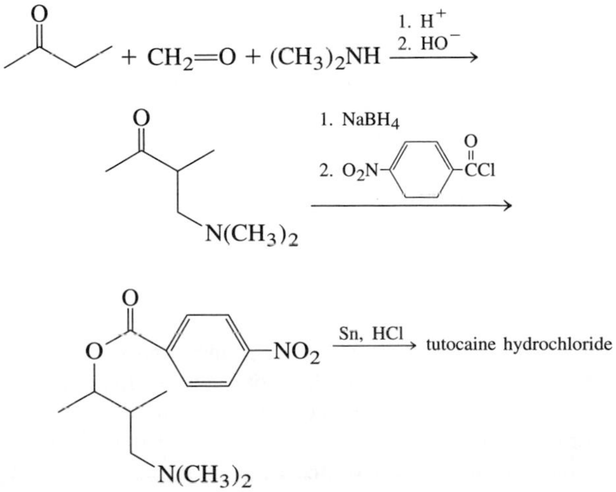

21-17

Nucleophilic displacement of N_2 in RN_2^+ by water occurs by an S_N2 mechanism with inversion.

21-18

H H H H H H

Bicyclo[1.1.0]butane

CHAPTER 22

22-1

(a) (b)

(c)

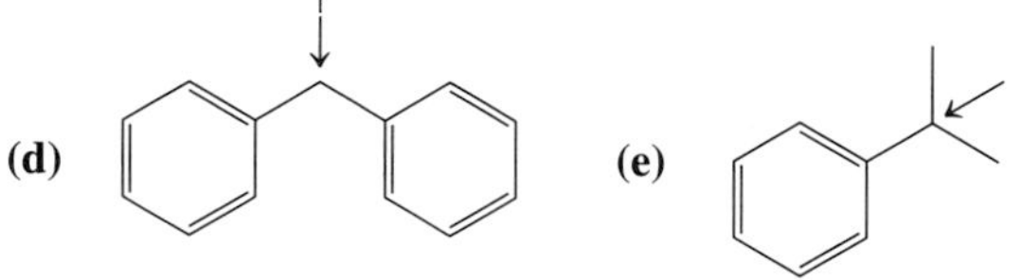

(d) (e)

Order of reactivity: (d) > (e) > (a),(b),(c)

22-2

$(C_6H_5)_2CHCl$ solvolyzes faster because the additional phenyl group causes extra resonance stabilization of the intermediate carbocation. This molecule is even more reactive under S_N1 conditions than 2-chloro-2-methylpropane (*tert*-butyl chloride).

22-3

S_N1 mechanism

$$C_6H_5CH_2OH \xrightarrow[-H_2O]{H^+} C_6H_5CH_2^+ \xrightarrow{Cl^-} C_6H_5CH_2Cl$$

Ethanol has to react through an S_N2 mechanism that includes chloride attack on the protonated hydroxy group. Even if the conversion of phenylmethanol were to proceed by this pathway, it would be accelerated relative to ethanol because of a delocalized transition state.

22-4

(a) $(C_6H_5)_2CH_2$, because the corresponding anion is better resonance stabilized

(b) $4\text{-}CH_3OC_6H_4CH_2Br$, because it contains a better leaving group

(c) $C_6H_5CH_2OH$, because the corresponding phenylmethyl (benzyl) cation is not destabilized by the extra nitro group (draw resonance structures)

22-5

(a) 1. $KMnO_4$, 2. $LiAlH_4$, 3. H_2, Pd-C

(b) 1. $KMnO_4$, 2. H^+, H_2O, 3. Δ ($-2\ H_2O$)

22-6

The nitro group is inductively electron withdrawing in all positions but can stabilize the negative charge by resonance only when at C2 or C4.

22-7

B, A, D, C

22-8

OCH_3 NO_2 NO_2

22-9

H_3C SO_2 NO_2 Slow

A

H_3C SO_2 NO_2 Fast

Intermediate B

NO_2 H_3C SO_2^-

C

E A B C or E A B C

22-10

Cl, CH_3 —NaOH, Δ→ (benzyne) CH_3 —H_2O→ —H^+, H_2O→ OH, CH_3 + OH, CH_3

CH_3 —H_2SO_4, SO_3→ CH_3, SO_3H —NaOH, Δ→ CH_3, OH

22-11

Br, $\ddot{O}CH_3$ —$^-:\ddot{N}H_2$ / $- NH_3$, $- Br^-$→ $\ddot{O}CH_3$ —$^-:\ddot{N}H_2$→ $H_2\ddot{N}$, $\ddot{O}CH_3$ **A**

The product A of amide addition to the intermediate benzyne is stabilized by the inductively electron withdrawing effect of the methoxy oxygen; therefore, it is formed regioselectively. Protonation gives the major final product. Note that there is no possibility for delocalization of the negative charge in A because the reactive electron pair is located in an sp^2 orbital that lies perpendicular to the aromatic π system.

22-12

—1. $C_6H_5\overset{O}{\overset{\|}{C}}Cl$, $AlCl_3$ / 2. Zn(Hg), HCl→ CH_2 —1. HNO_3, H_2SO_4 / 2. H_2, Ni→

NH_2, CH_2 —1. $NaNO_2$, H^+, H_2O / 2. Δ→ OH, CH_2

22-13

Such a process would require nucleophilic attack by halide ion on a benzene ring, a transformation that is not competitive.

22-14

Amines are more nucleophilic than alcohols; this rule also holds for benzenamines (anilines) relative to phenols.

22-15

The 1,1-dimethylethyl (*tert*-butyl) group is considerably larger than methyl, attacking preferentially at C4.

22-16

$:\ddot{O}:^-$ + $:O:$, C, $:O:$ ⟶ $:O:$, H, $C—\ddot{O}:^-$, $:O:$ ⟶ OH, CO_2^-

22-17

Cl, $:\ddot{O}H$, Cl, Cl + $CH_2{=}\overset{+}{\ddot{O}}H$ ⟶

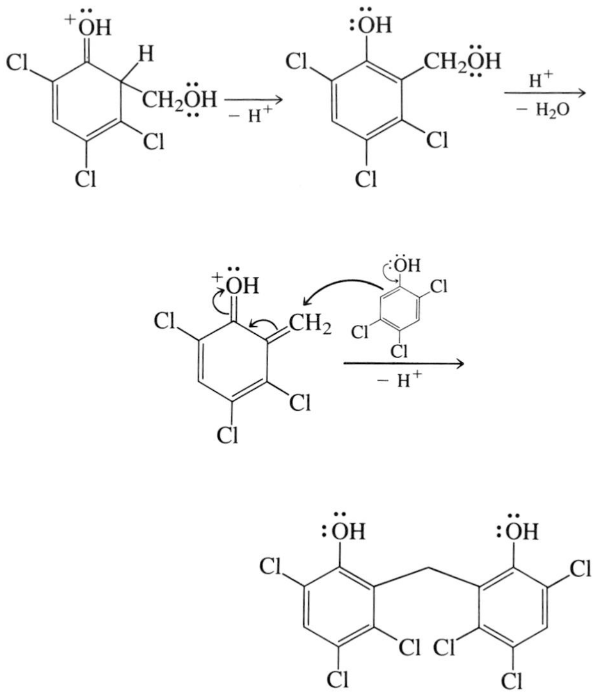

Hexachlorophene

22-18

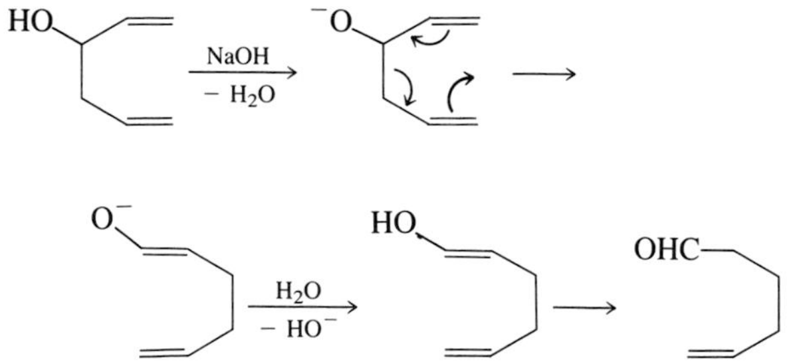

NaOH
$- H_2O$

H_2O
$- HO^-$

The Cope rearrangement is especially fast in this case because the negative charge in the initial enolate ion is delocalized.

22-19

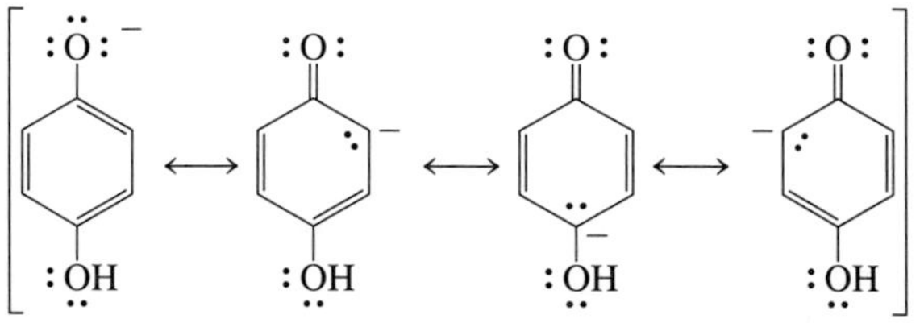

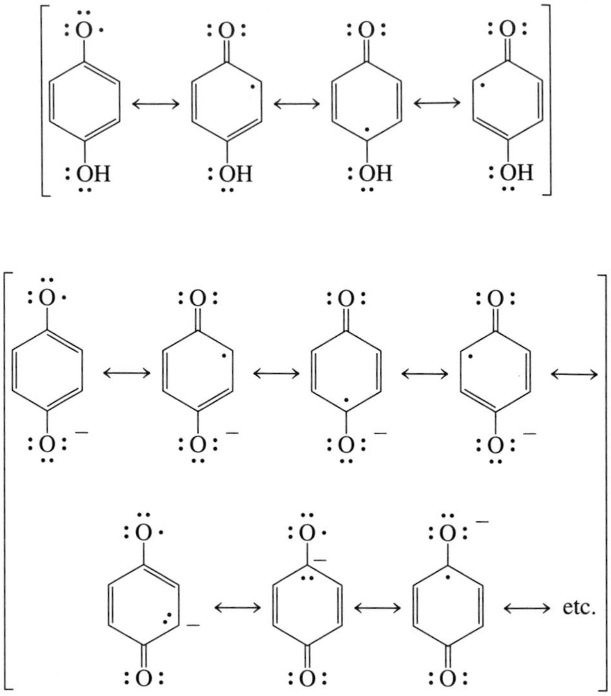

etc.

22-20

This exchange goes through two conjugate addition–elimination cycles.

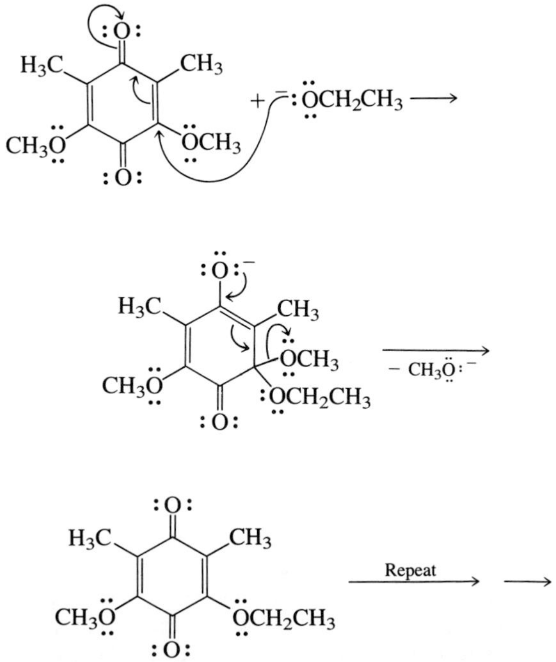

Repeat

$$\text{2,6-dimethyl-3,5-diethoxy-1,4-benzoquinone: } H_3C, CH_3, CH_3CH_2\ddot{O}, \ddot{O}CH_2CH_3, :O:, :O:$$

22-21

$$\left[\text{(resonance structures of radical anion with } :\ddot{O}H, \cdot\ddot{O}, \ddot{O}, H, R) \longleftrightarrow \cdots \longleftrightarrow \cdots \right]$$

22-22

$$C_6H_6 \xrightarrow[\text{2. } H_2,\ Ni]{\text{1. } HNO_3,\ H_2SO_4} C_6H_5NH_2 \xrightarrow{3\ Br_2,\ H_2O}$$

$$\text{2,4,6-}Br_3C_6H_2NH_2 \xrightarrow[\text{2. } H_3PO_2]{\text{1. } NaNO_2,\ H^+} \text{1,3,5-}Br_3C_6H_3$$

22-23

(a) $p\text{-}CH_3OC_6H_4N{=}NC_6H_5$

(b) $\text{(4-}OCH_3\text{, 2-}Cl\text{)}C_6H_3N{=}NC_6H_5 + \text{(4-}Cl\text{, 2-}OCH_3\text{)}C_6H_3N{=}NC_6H_5$

(c) $\text{(4-}CH_3\text{, 2-}N(CH_3)_2\text{)}C_6H_3N{=}NC_6H_5$

CHAPTER 23

23-1

(a) $CH_3CH_2\overset{O}{\overset{\|}{C}}\underset{CH_3}{\underset{|}{C}H}\overset{O}{\overset{\|}{C}}OCH_2CH_3$

(b) $CH_3\overset{CH_3}{\overset{|}{C}H}CH_2\overset{O}{\overset{\|}{C}}\underset{(CH_3)_2CH}{\underset{|}{C}H}\overset{O}{\overset{\|}{C}}OCH_2CH_3$

(c) $CH_3(CH_2)_3\overset{O}{\overset{\|}{C}}\underset{(CH_2)_2CH_3}{\underset{|}{C}H}\overset{O}{\overset{\|}{C}}OCH_2CH_3$

23-2

Retro-Claisen condensation

$$CH_3\overset{:O:}{\overset{\|}{C}}-\underset{CH_3}{\overset{CH_3}{C}}-\overset{:O:}{\overset{\|}{C}}\ddot{O}CH_3 + CH_3\ddot{O}:^- \longrightarrow$$

$$CH_3\overset{:\ddot{O}:^-}{\overset{|}{C}}(\ddot{O}CH_3)-\underset{CH_3}{\overset{CH_3}{C}}-\overset{:O:}{\overset{\|}{C}}OCH_3 \longrightarrow$$

$$CH_3\overset{:O:}{\overset{\|}{C}}\ddot{O}CH_3 + (CH_3)_2C{=}\overset{:\ddot{O}:^-}{\overset{|}{C}}\ddot{O}CH_3 \rightleftharpoons$$

$$CH_2{=}\overset{:\ddot{O}:^-}{\overset{|}{C}}\ddot{O}CH_3 + (CH_3)_2CHCOOCH_3$$

Forward Claisen condensation

$$CH_3\overset{:O:}{\overset{\|}{C}}\ddot{O}CH_3 + CH_2{=}\overset{:\ddot{O}:^-}{\overset{|}{C}}\ddot{O}CH_3 \longrightarrow \longrightarrow CH_3\overset{O}{\overset{\|}{C}}\ddot{C}HCOOCH_3$$

23-3

$$CH_3CH_2\overset{O}{\overset{\|}{C}}OCH_2CH_3 + CH_3\overset{O}{\overset{\|}{C}}OCH_2CH_3 \xrightarrow[\text{2. } H^+,\ H_2O]{\text{1. } CH_3CH_2O^-Na^+,\ CH_3CH_2OH}$$

$CH_3CH_2\overset{O}{\overset{\|}{C}}\underset{CH_3}{\underset{|}{C}}H\overset{O}{\overset{\|}{C}}OCH_2CH_3 + CH_3CH_2\overset{O}{\overset{\|}{C}}CH_2\overset{O}{\overset{\|}{C}}OCH_2CH_3$

$+ CH_3\overset{O}{\overset{\|}{C}}\underset{CH_3}{\underset{|}{C}}H\overset{O}{\overset{\|}{C}}OCH_2CH_3 + CH_3\overset{O}{\overset{\|}{C}}CH_2\overset{O}{\overset{\|}{C}}OCH_2CH_3$

23-4

This mechanism is abbreviated, showing only the most important steps.

$CH_3CH_2O_2C(CH_2)_3\overset{O}{\overset{\|}{C}}OCH_2CH_3 \xrightarrow[-CH_3CH_2OH]{CH_3CH_2O^-}$

$CH_3CH_2O_2CCH_2CH_2\ddot{C}^-HCO_2CH_2CH_3 \xrightarrow[-\,CH_3CH_2O^-]{CH_3CH_2O\overset{O}{\overset{\|}{C}}\overset{O}{\overset{\|}{C}}OCH_2CH_3}$

$CH_3CH_2O_2CCH_2CH_2\underset{}{C}(\overset{O}{\overset{\|}{C}}\overset{O}{\overset{\|}{C}}OCH_2CH_3)HCO_2CH_2CH_3 \xrightarrow[-CH_3CH_2OH]{CH_3CH_2O^-}$

$CH_3CH_2O_2C\ddot{C}^-HCH_2CH_2\underset{}{C}(\overset{O}{\overset{\|}{C}}\overset{O}{\overset{\|}{C}}OCH_2CH_3)HCO_2CH_2CH_3 \xrightarrow[-\,CH_3CH_2O^-]{}$

[Structure: cyclopentane-1,2-dione ring bearing $CH_3CH_2O_2C$ and $CO_2CH_2CH_3$ substituents]

23-5

This mechanism is also abbreviated.

[Structure: diethyl phthalate] $+ \ ^-\!:CH_2CO_2CH_2CH_3 \xrightarrow[-\,H^+]{-\,CH_3CH_2O^-,}$

[Structure: C_6H_4 bearing $\overset{O}{\overset{\|}{C}}\ddot{C}^-HCO_2CH_2CH_3$ and $\overset{O}{\overset{\|}{C}}OCH_2CH_3$] $\xrightarrow[-\,CH_3CH_2O^-]{}$

[Structure: 1,3-indandione bearing $-CO_2CH_2CH_3$ at C-2]

23-6

(a) [cyclohexanone] $+ CH_3CH_2O_2CCO_2CH_2CH_3$

1. $CH_3CH_2O^-$, 2. H^+, H_2O

(b) $CH_3\overset{O}{\overset{\|}{C}}CH_3 + HCO_2CH_2CH_3$

1. $CH_3CH_2O^-$, 2. H^+, H_2O

(c) [cyclooctanone] $+ CH_3CH_2O\overset{O}{\overset{\|}{C}}OCH_2CH_3$

1. NaH, 2. H^+, H_2O

(d) $H_3C\overset{O}{\overset{\|}{C}}CH_2CH_2CH_2CH_2CO_2CH_2CH_3$

1. $CH_3CH_2O^-$, 2. H^+, H_2O

23-7

$CH_3\overset{O}{\overset{\|}{C}}CH_2CH_2CH_2CO_2CH_3 \xrightarrow{1.\ (C_6H_5)_3CO^-K^+\quad 2.\ H^+,\ H_2O}$ [1,3-cyclohexanedione] 100% $\xrightarrow{2\ NaOCH_3,\ CH_3I,\ CH_3OH}$ [2,2-dimethyl-1,3-cyclohexanedione] 80%

2,2-Dimethyl-1,3-cyclohexanedione

23-8

[Structure: methylmalonic acid, $HO\overset{O}{\overset{\|}{C}}-C(CH_3)(H)-\overset{O}{\overset{\|}{C}}-O-H$, cyclic transition with electron-pushing arrows] $\longrightarrow$ $HO\overset{:OH}{\overset{|}{C}}=CH-CH_3 + :O=C=O: \longrightarrow$

$HO\overset{O}{\overset{\|}{C}}CH_2CH_3 + CO_2$

23-9

(a) 1. $NaOCH_2CH_3$, 2. $CH_3CH_2CH_2Br$, 3. NaOH, 4. H^+, H_2O, Δ; **(b)** 1. $NaOCH_2CH_3$, 2. $CH_3(CH_2)_4Br$, 3. NaOH, 4. H^+, H_2O, Δ; **(c)** 1. 2 $NaOCH_2CH_3$, 2. 2 CH_3CH_2Br, 3. NaOH, 4. H^+, H_2O, Δ; **(d)** 1. $NaOCH_2CH_3$, 2. $C_6H_5CH_2Cl$, 3. NaOH, 4. H^+, H_2O, Δ

23-10

(a) 1. $CH_3CH_2OOC\ddot{C}^-HCOOCH_2CH_3$ (with CH_3 on the central carbon),

2. $CH_3CH_2OOCC(CH_3)((CH_2)_9CH_3)COOCH_2CH_3$,

3. $K^{+}\,^{-}OOCC(CH_3)((CH_2)_9CH_3)COO^{-}\,K^{+}$

(b) $CH_3CH_2OOCCH_2COOCH_2CH_3 \xrightarrow[2.\ CH_3Br]{1.\ NaOCH_2CH_3} CH_3CH_2OOCCH(CH_3)COOCH_2CH_3$

23-11

(a) 2-Butylcyclohexanone

(b) $CH_3CH_2CO_2H$
Propanoic acid

(c) $CH_3COCH(CH_3)COC_6H_5$
2-Methyl-1-phenyl-1,3-butanedione

(d) This sequence is general for 2-halo esters.

$CH_3COCH_2COOCH_2CH_3 \xrightarrow[2.\ BrCH_2CO_2CH_2CH_3]{1.\ CH_3CH_2O^-Na^+} CH_3COCH(CH_2COOCH_2CH_3)COOCH_2CH_3 \xrightarrow[2.\ H^+,\ H_2O,\ \Delta]{1.\ NaOH} CH_3COCH_2CH_2COOH$

4-Oxopentanoic acid

Note that only the carboxy group located β to the ketone carbonyl can undergo decarboxylation.

(e) $HOOCCH(CH_2CH_3)CH_2COOH$

2-Ethylbutanedioic acid

Excessive heating may dehydrate this product to the anhydride (Section 19-8).

(f) $CH_3COCH_2CH_2COCH_3$

2,5-Hexanedione

23-12

$CH_2(CO_2CH_2CH_3)_2 + Br(CH_2)_5Cl \xrightarrow[-\ CH_3CH_2OH,\ NaBr]{CH_3CH_2O^-Na^+}$

$Cl(CH_2)_5CH(CO_2CH_2CH_3)_2 \xrightarrow[-\ CH_3CH_2OH]{CH_3CH_2O^-Na^+}$

$Cl(CH_2)_5\ddot{C}^-(CO_2CH_2CH_3)_2\ Na^+$ (intramolecular displacement of Cl) $\xrightarrow{-\ NaCl}$

1,1-Cyclohexanedicarboxylic acid diethyl ester $\xrightarrow[2.\ H^+,\ H_2O,\ \Delta]{1.\ KOH,\ CH_3CH_2OH}$

Cyclohexane with H and COOH on the same carbon

Cyclohexanecarboxylic acid

23-13

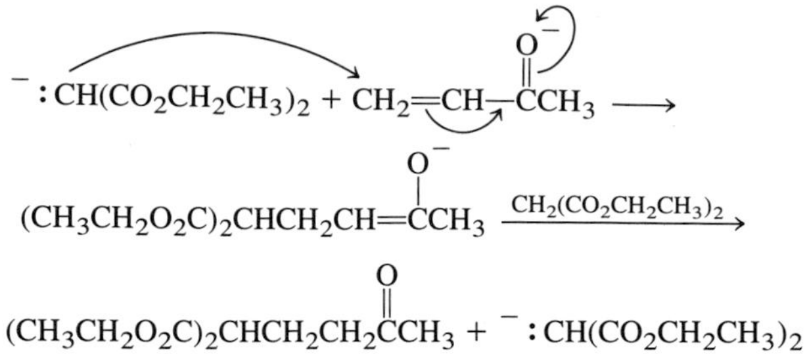

$$^{-}\!:CH(CO_2CH_2CH_3)_2 + CH_2{=}CH{-}\overset{O}{\overset{\|}{C}}CH_3 \longrightarrow$$

$$(CH_3CH_2O_2C)_2CHCH_2CH{=}\overset{O^-}{\overset{|}{C}}CH_3 \xrightarrow{CH_2(CO_2CH_2CH_3)_2}$$

$$(CH_3CH_2O_2C)_2CHCH_2CH_2\overset{O}{\overset{\|}{C}}CH_3 + {}^{-}\!:CH(CO_2CH_2CH_3)_2$$

The enolate of the product regenerates the enolate of the starting malonic ester.

23-14

(a) 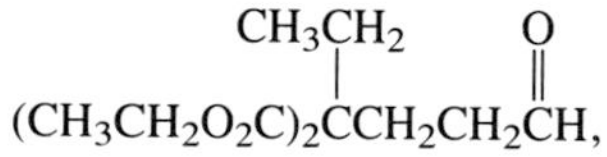 $(CH_3CH_2O_2C)_2\overset{CH_3CH_2}{\overset{|}{C}}CH_2CH_2\overset{O}{\overset{\|}{C}}H$, 40%

(b) 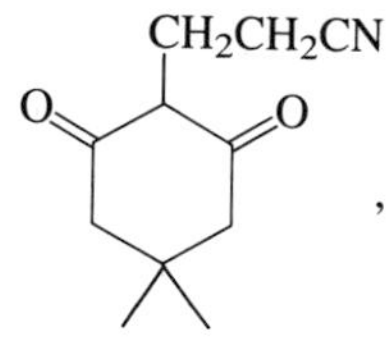 2-(CH_2CH_2CN)-5,5-dimethyl-1,3-cyclohexanedione, 56%

(c) 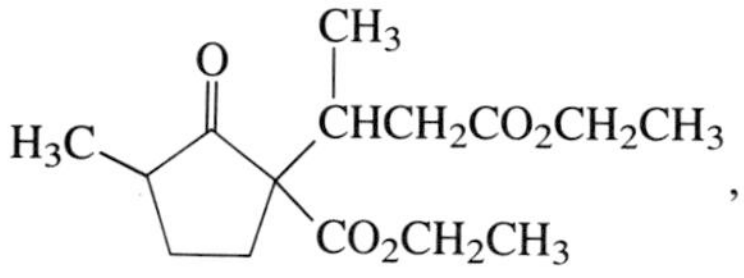 2-methyl-5-[$\overset{CH_3}{\overset{|}{C}}HCH_2CO_2CH_2CH_3$]-5-($CO_2CH_2CH_3$)cyclopentanone, 66%

23-15

5,5-dimethyl-1,3-cyclohexanedione enolate + $CH_2{=}CHCN \longrightarrow$

NC–$\ddot{C}H^-$–CH₂ adduct; H (Acidic) $\longrightarrow$

$\xrightarrow[\text{Second Michael addition}]{CH_2{=}CHCN} \longrightarrow$ 2,2-bis(CH_2CH_2CN)-5,5-dimethyl-1,3-cyclohexanedione

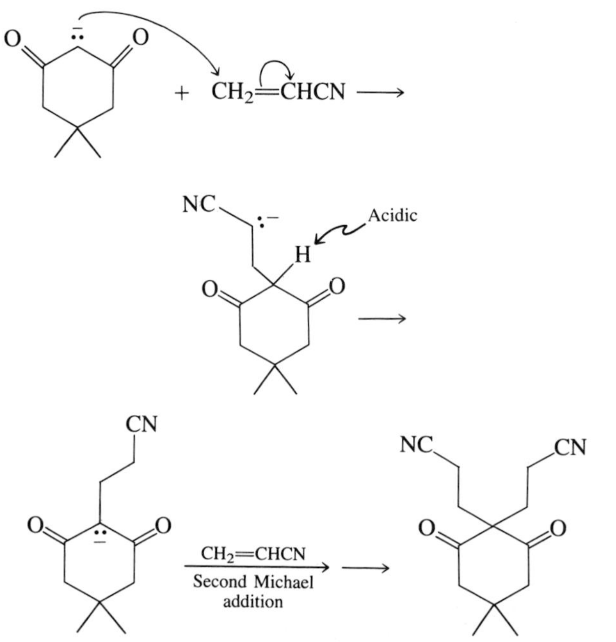

23-16

1. Michael addition

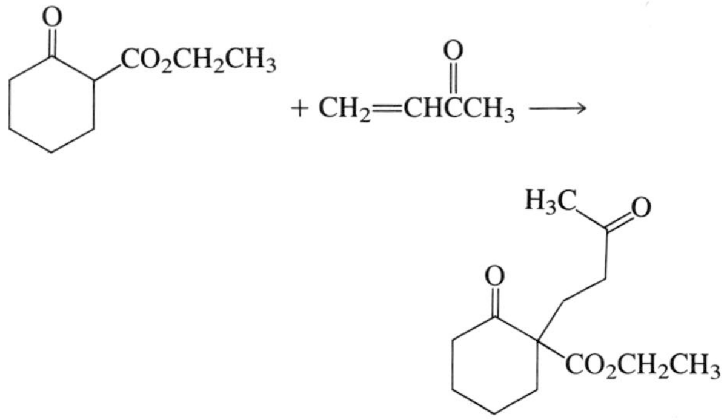

2-($CO_2CH_2CH_3$)cyclohexanone $+ CH_2{=}CH\overset{O}{\overset{\|}{C}}CH_3 \longrightarrow$ 2-($CH_2CH_2COCH_3$)-2-($CO_2CH_2CH_3$)cyclohexanone

2. Aldol condensation

$H_2\ddot{C}^-$ enolate attacks ring carbonyl $\dashrightarrow$ bicyclic enone with $CO_2CH_2CH_3$

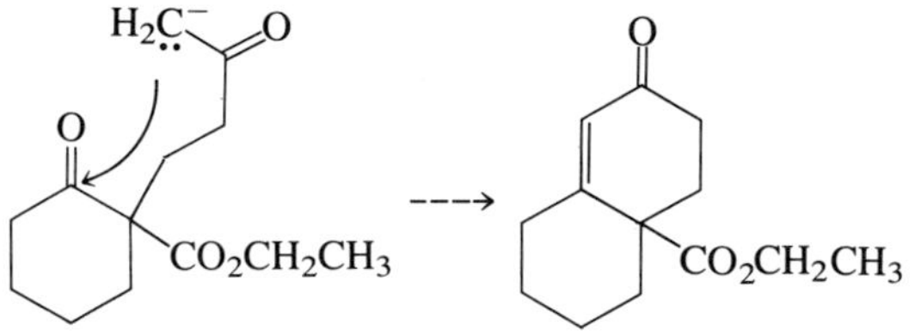

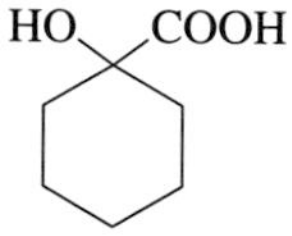

23-17

(a) 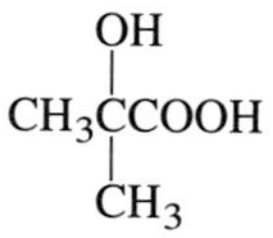 $CH_3\overset{OH}{\overset{|}{C}}COOH$ with CH_3 **(b)** 1-hydroxycyclohexanecarboxylic acid (HO, COOH)

(c) 9-hydroxyfluorene-9-carboxylic acid (HO, COOH)

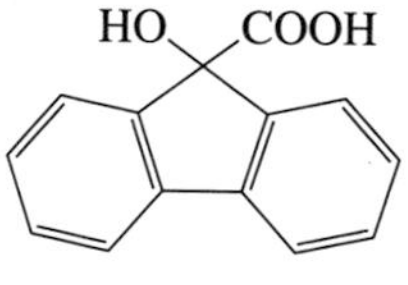

23-18

Nucleophiles add to the carbonyl group; bases deprotonate next to it.

23-19

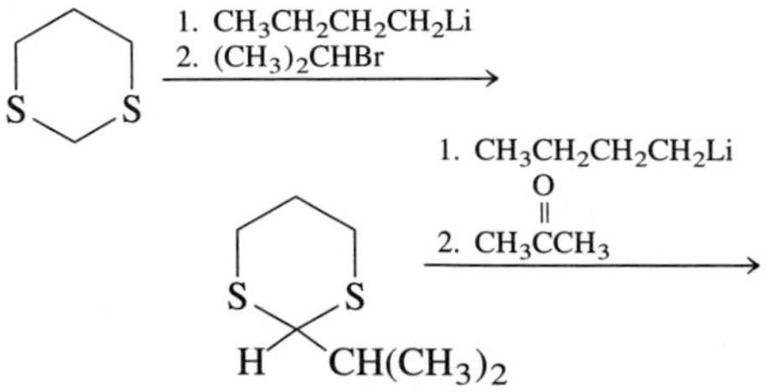

1,3-dithiane $\xrightarrow[\text{2. }(CH_3)_2CHBr]{\text{1. }CH_3CH_2CH_2CH_2Li}$ 2-H-2-$CH(CH_3)_2$-1,3-dithiane $\xrightarrow[\text{2. }CH_3\overset{O}{\overset{\|}{C}}CH_3]{\text{1. }CH_3CH_2CH_2CH_2Li}$

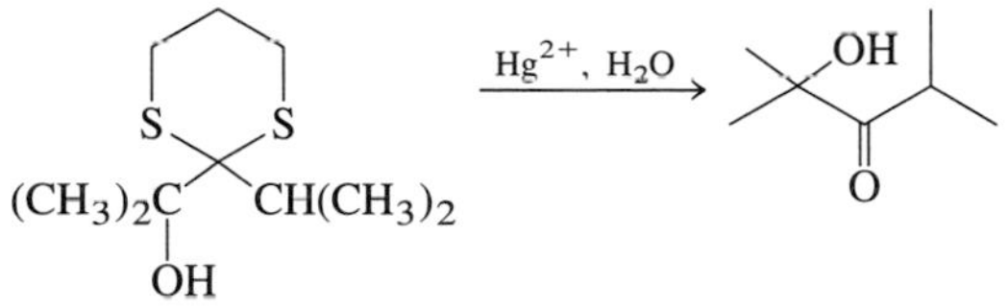

Hg^{2+}, H_2O

(CH$_3$)$_2$C(OH) / CH(CH$_3$)$_2$ / S S / OH / O

23-20

(a) CH_3CH_2CHO + thiazolium ion catalyst or 1,3-dithiane and 1. $CH_3CH_2CH_2CH_2Li$, 2. CH_3CH_2Br, 3. $CH_3CH_2CH_2CH_2Li$, 4. CH_3CH_2CHO, 5. Hg^{2+}, H_2O

(b) 1,3-Dithiane and 1. $CH_3CH_2CH_2CH_2Li$, 2. $CH_3(CH_2)_3Br$, 3. $CH_3CH_2CH_2CH_2Li$, 4. CH_3CH_2CHO, 5. Hg^{2+}, H_2O

(c) 1,3-Dithiane and 1. $CH_3CH_2CH_2CH_2Li$, 2. $CH_3CH_2CHCH_3$ (Br), 3. $CH_3CH_2CH_2CH_2Li$, 4. $CH_3CCH_2CH_3$ (O), 5. Hg^{2+} H_2O

(d) C_6H_5CHO + thiazolium ion catalyst

(e) $(CH_3)_2CHCHO$ + thiazolium ion catalyst or 1,3-dithiane and 1. $CH_3CH_2CH_2CH_2Li$, 2. $(CH_3)_2CHBr$, 3. $CH_3CH_2CH_2CH_2Li$, 4. $(CH_3)_2CHCHO$, 5. Hg^{2+}, H_2O

CHAPTER 24

24-1

(a) Aldotetrose
(b) Aldopentose
(c) Ketopentose

24-2

(a) $(2R,3R,4R)$-2,3,4,5-Tetrahydroxypentanal
(b) $(2R,3S,4R,5R)$-2,3,4,5,6-Pentahydroxyhexanal

24-3

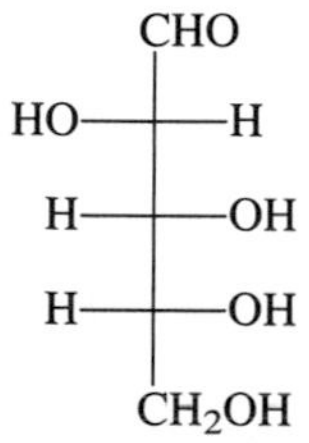

CHO
HO—H
H—OH
H—OH
CH$_2$OH

D-(−)-Arabinose

24-4

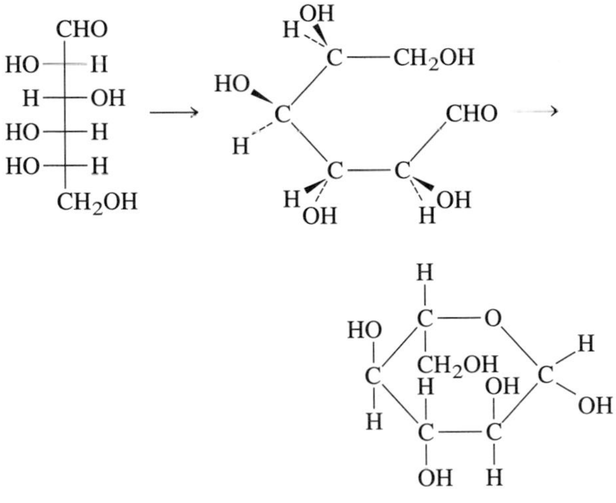

CHO
HO—H
H—OH
HO—H
HO—H
CH$_2$OH

→ H OH / C—CH$_2$OH / HO / C / H / CHO / C—C / H OH / H OH →

H / C—O / HO / C / CH$_2$OH / H OH / C / H / OH / H / C—C / OH H

L-(−)-Glucose

24-5

False. They should form in unequal amounts because two diastereomers are formed. In fact, the ratio of α to β is 36:64. Similarly, the relative amounts of α-D-, β-D-fructopyranose, α-D-, and β-D-fructofuranose at equilibrium in aqueous solution are 3%, 57%, 9%, and 31%.

24-6

(a) HOCH$_2$ O CH$_2$OH / S / OH / OH / OH

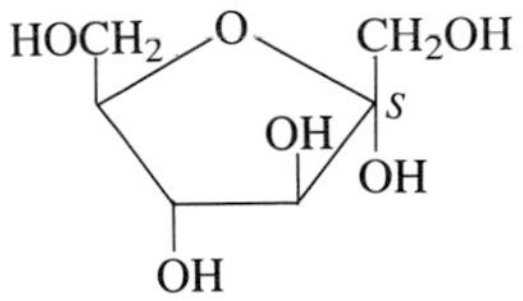

(b) CH$_2$OH / HO—C—H / O / OH / R / OH / H / OH

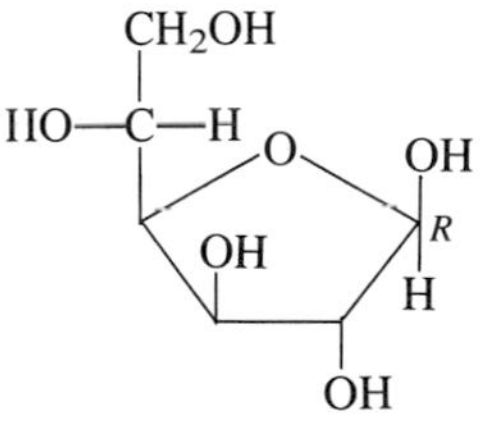

(c) O / OH / OH / R / HO / H / OH

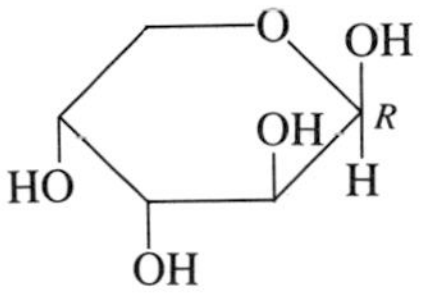

24-7

Four axial OH groups, $4 \times 0.94 = 3.76$ kcal mol^{-1}; one axial CH$_2$OH, 1.70 kcal mol^{-1}; $\Delta G = 5.46$ kcal mol^{-1}. The concentration of this conformer in solution is therefore negligible by this estimate.

24-8

Only the anomeric carbon and its vicinity are shown.

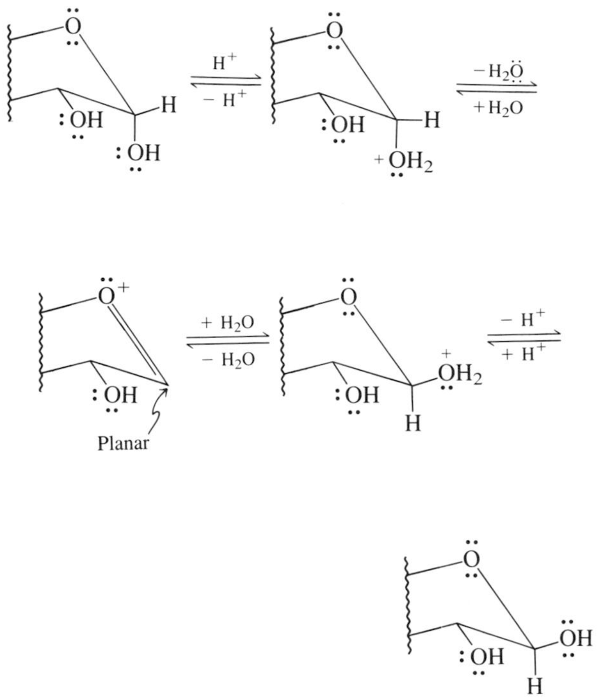

$\overset{H^+}{\underset{-H^+}{\rightleftharpoons}}$ $\overset{-H_2\ddot{O}}{\underset{+H_2O}{\rightleftharpoons}}$

Planar

$\overset{+H_2O}{\underset{-H_2O}{\rightleftharpoons}}$ $\overset{-H^+}{\underset{+H^+}{\rightleftharpoons}}$

24-9

Pure α form, $+112°$; pure β form, $+18.7°$ ($\Delta\alpha = 93.3°$). After equilibration, $+52.7°$. Mole fraction of α: $(52.7 - 18.7)/93.3 = 0.364$. Hence, the mole fraction of $\beta = 0.636$; thus, the equilibrium mole ratio $\beta:\alpha = 0.636/0.364 = 1.75:1$.

24-10

$\Delta G°_{\text{estimated}} = -0.94$ kcal mol^{-1} (one axial OH); $\Delta G° = -RT \ln K = -1.36 \log 63.6/36.4 = -0.33$ kcal mol^{-1}. The difference between the two values is due to the fact that the six-membered ring is a cyclic ether (not a cyclohexane).

24-11

Oxidation of D-glucose should give an optically active aldaric acid, whereas that of D-allose leads to loss of optical activity. This result is a consequence of turning the two end groups along the sugar chain into the same substituent.

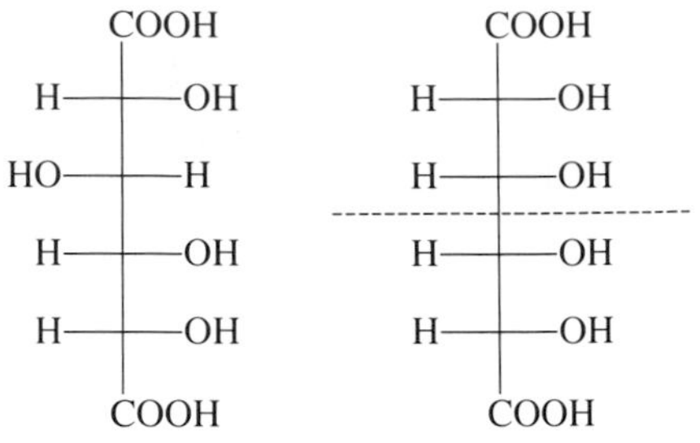

COOH	COOH
H—OH	H—OH
HO—H	H—OH
H—OH	H—OH
H—OH	H—OH
COOH	COOH
D-Glucaric acid	**D-Allaric acid**
Optically active	**Meso, not optically active**

This operation may cause important changes in the symmetry of the molecule. Thus, D-allaric acid has a mirror plane. It is therefore a meso compound and not optically active. (This also means that D-allaric acid is identical with L-allaric acid.) On the other hand, D-glucaric acid is still optically active.

Other simple aldoses that turn into meso-aldaric acids are D-erythrose, D-ribose, D-xylose, and D-galactose (see Figure 24-1).

24-12

(a) $2\ CH_2{=}O$ **(b)** $CH_3CH{=}O + CH_2{=}O$

(c) $2\ CH_2{=}O + HCOOH$ **(d)** No reaction

(e) $OHCC(CH_3)_2CHO + CO_2$ **(f)** $3\ HCOOH + CH_2{=}O$.

24-13

(a) D-Arabinose $\longrightarrow$ $4\ HCO_2H + CH_2O$
D-Glucose $\longrightarrow$ $5\ HCO_2H + CH_2O$

(b) D-Erythrose $\longrightarrow$ $3\ HCO_2H + CH_2O$
D-Erythrulose $\longrightarrow$ $HCO_2H + 2\ CH_2O + CO_2$

(c) D-Glucose or D-mannose $\longrightarrow$ $5\ HCO_2H + CH_2O$

24-14

(a) Ribitol is a meso compound.
(b) D-Mannitol (major) and D-glucitol

24-15

All of them are the same.

24-16

The mechanism of acetal formation proceeds through the same intermediate cation in both cases.

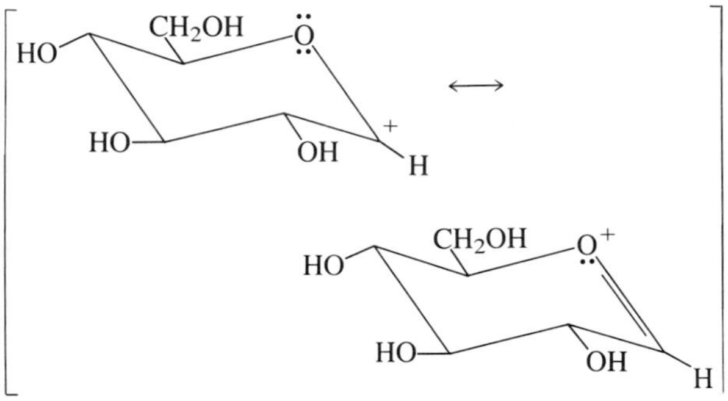

HO, CH_2OH, $\ddot{O}$, +, HO, OH, H ⟷ HO, CH_2OH, O^+, HO, OH, H

24-17

$HOCH_2$, O, HO, H, H, H, H, OCH_3, OH

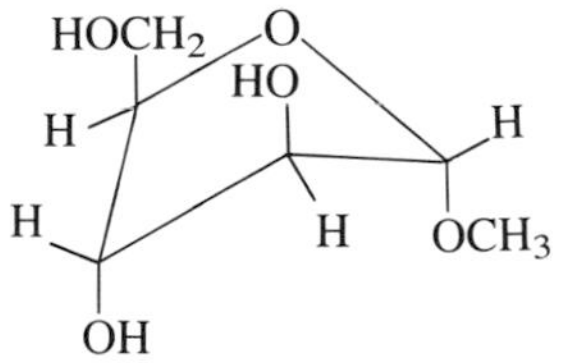

24-18

Same structure as that in Exercise 24-17 or its diastereomers with respect to C2 and C3.

24-19

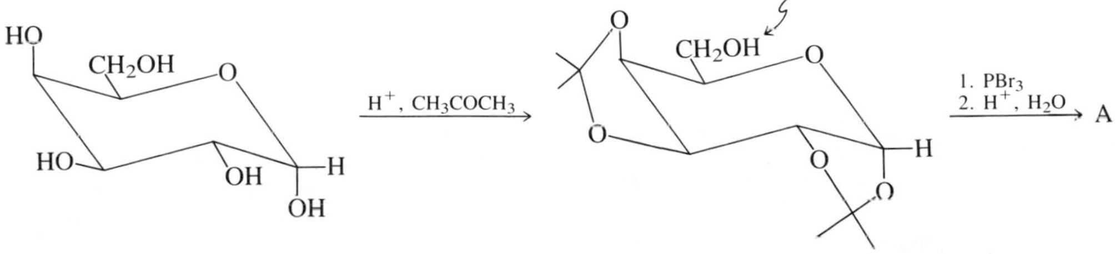

HO, CH_2OH, O, HO, OH, H, OH $\xrightarrow{H^+,\ CH_3COCH_3}$ O, CH_2OH (Reactive), O, O, O, O, H $\xrightarrow{1.\ PBr_3;\ 2.\ H^+,\ H_2O}$ A

24-20

(a) D-Ribose and D-arabinose

(b) D-Glucose and D-mannose

24-21

A, D-arabinose; B, D-lyxose; C, D-erythrose; D, D-threose.

24-22

^{13}C NMR would show only three lines for ribaric acid, but five for arabinaric acid.

24-23

(a)

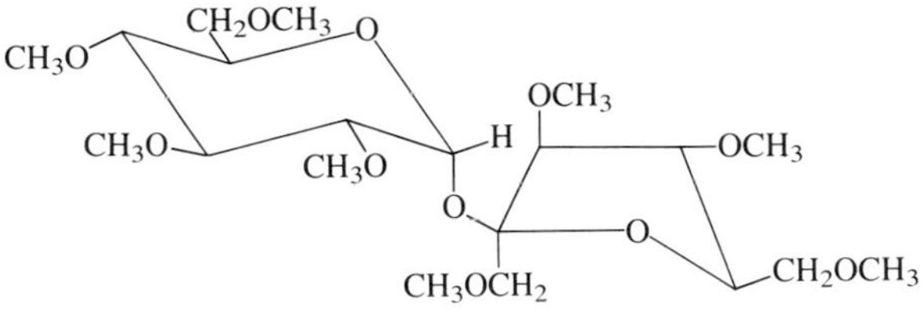

CH_3O, CH_2OCH_3, O, CH_3O, CH_3O, H, O, OCH_3, OCH_3, O, CH_3OCH_2, CH_2OCH_3

(b)

CH_2OH		+	CH_2OH	
H	OH		HO	H
HO	H		HO	H
H	OH		H	OH
H	OH		H	OH
CH_2OH			CH_2OH	

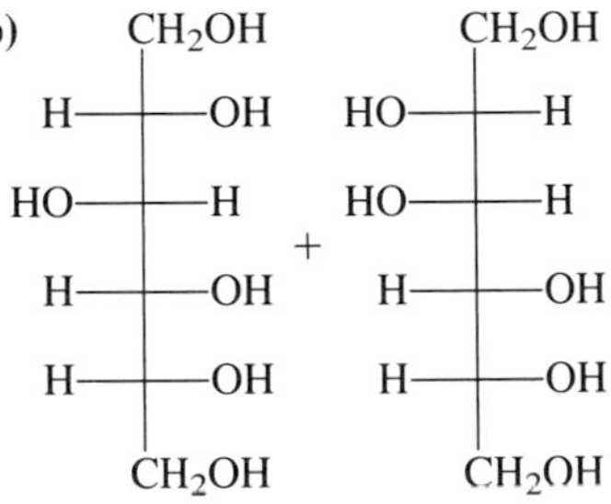

(c) No reaction.

24-24

(a)

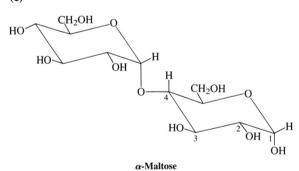

(b)

(c)

α-Maltose

CHAPTER 25

25-1

(a)

(b)

(c) 2,6-Dinitropyridine **(d)** 4-Bromoindole

25-2

$+ MgBr_2 \longrightarrow$

25-3

25-4

$Cl(CH_2)_3SCl$

25-5

$CH_3CHClCH_2CH_2OH + CH_3CH(OH)CH_2CH_2Cl$

25-6

25-7

N H (pyrrolidine, dipole arrow)

N H (pyrrole, dipole arrow)

Nitrogen is more electronegative than carbon

Because of resonance, the molecule is now polarized in the opposite direction

25-8

H_3C, C=O, CH, $CH_3CH_2O_2C$, NH_2 + $CO_2CH_2CH_3$, H_2C, O=C, CH_3 $\xrightarrow{-\,H_2O}$

β-Keto amine **β-Keto ester**

H_3C–C(=O)–CH($CH_3CH_2O_2C$)–N=C(CH_3)–H_2C–$CO_2CH_2CH_3$ $\xrightarrow{-\,H_2O}$

H_3C, $CO_2CH_2CH_3$, C=C, $CH_3CH_2O_2C$, C, H, N=C, CH_3 $\longrightarrow$

H_3C, $CO_2CH_2CH_3$, CH_3, $CH_3CH_2O_2C$, N, H

25-9

COOH, S $\xrightarrow[-\,Br^-]{Br_2}$ [Br, H, COOH, S, + $\longleftrightarrow$ Br, H, COOH, S, + $\longleftrightarrow$ Br, H, COOH, S, +] $\xrightarrow{-\,H^+}$ Br, S, COOH

Attack at C5 avoids placing the positive charge on C3

69%

25-10

Protonation at the α-carbon generates a cation described by three resonance structures. Protonation of the nitrogen produces an ammonium ion devoid of resonance stabilization.

25-11

N, H $+ E^+ \longrightarrow$

[E, H, N, H, + $\longleftrightarrow$ E, H, +N, H]

Only attack at C3 produces the iminium resonance structure without disrupting the benzene ring

25-12

Because of the electronegativity of nitrogen, the dipole vector in both compounds points toward the heteroatom. The dipole moment of pyridine is larger than that in azacyclohexane (piperidine), because the nitrogen is sp^2 hybridized. (See Section 11-3 for the effects of hybridization on electron-withdrawing power.)

25-13

(a) $CH_3\overset{O}{\overset{\|}{C}}CH_2CO_2CH_2CH_3$, NH_3, (benzene ring with CHO and NO_2), $CH_3\overset{O}{\overset{\|}{C}}CH_2CN$

(b) $CH_3\overset{O}{\overset{\|}{C}}CH_2CN$, NH_3, $(CH_3)_3CCHO$

(c) $CH_3CH_2\overset{O}{\overset{\|}{C}}CH_2CO_2CH_2CH_3$, NH_3, CH_3CHO

25-14

C3 is the least deactivated position in the ring. Attack at C2 or C4 generates intermediate cations with resonance structures that place the positive charge on the electronegative nitrogen.

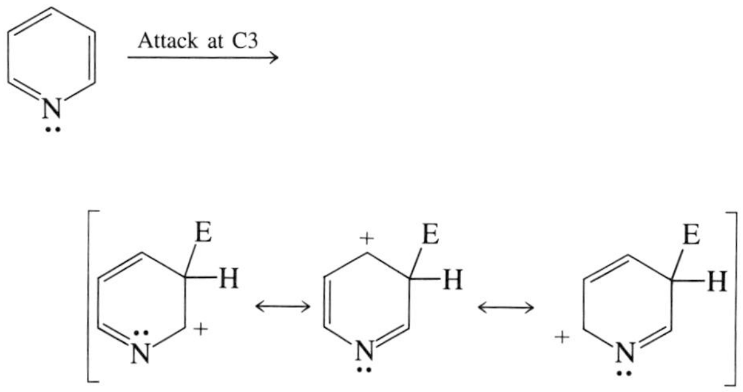

Attack at C3

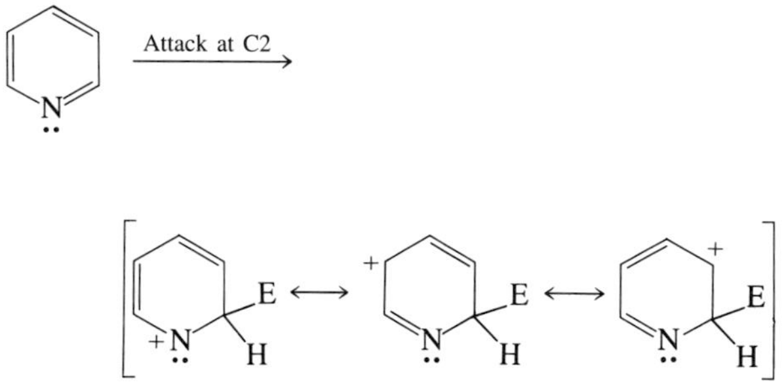

Attack at C2

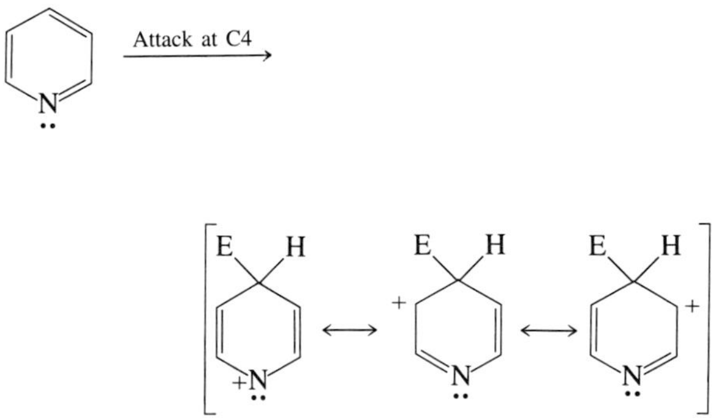

Attack at C4

25-15

Attack at C2 and C4 produces the more highly resonance-stabilized anions (only the most important resonance structures are shown).

2-Chloropyridine ⟶

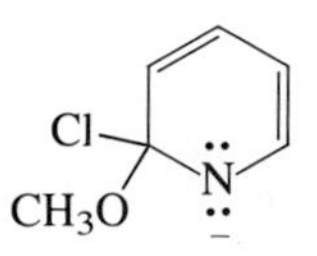

3-Chloropyridine ⟶

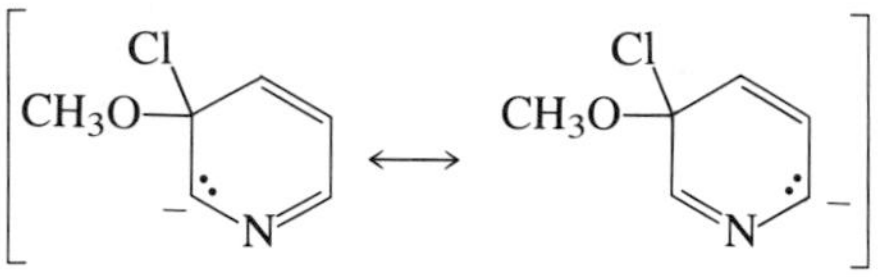

4-Chloropyridine ⟶

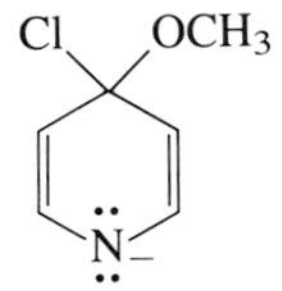

25-16

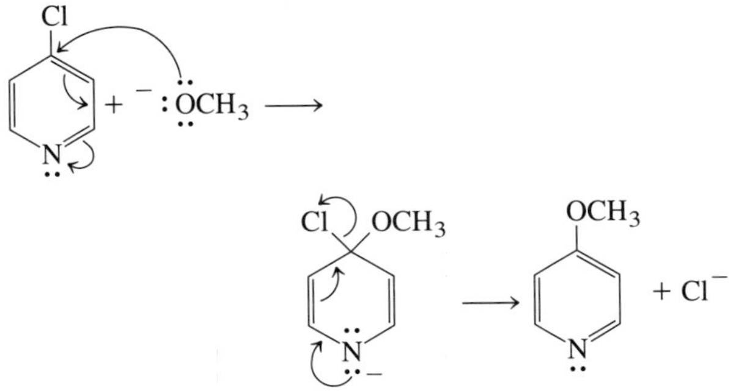

25-17

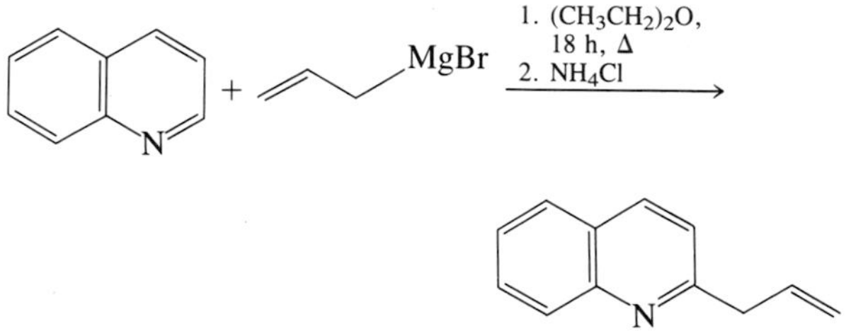

1. $(CH_3CH_2)_2O$, 18 h, Δ
2. NH_4Cl

56%

2-(2-Propenyl)quinoline

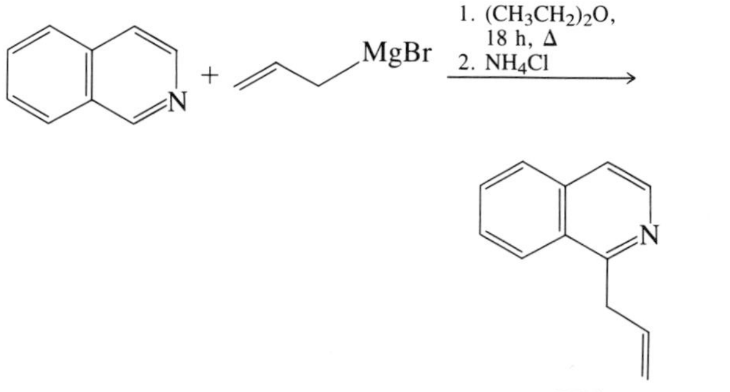

1. $(CH_3CH_2)_2O$, 18 h, Δ
2. NH_4Cl

57%

1-(2-Propenyl)isoquinoline

CHAPTER 26

26-1

(2S)-Aminopropanoic acid; (2S)-amino-3-methylbutanoic acid; (2S)-amino-4-methylpentanoic acid; (2S)-amino-3-methylpentanoic acid; (2S)-amino-3-phenylpropanoic acid; (2S)-amino-3-hydroxypropanoic acid; (2S)-amino-3-(4-hydroxyphenyl)propanoic acid; (2S,6)-diaminohexanoic acid; (2R)-amino-3-mercaptopropanoic acid; (2S)-amino-4-(methylthio)butanoic acid; (2S)-aminobutanedioic acid; (2S)-aminopentanedioic acid

26-2

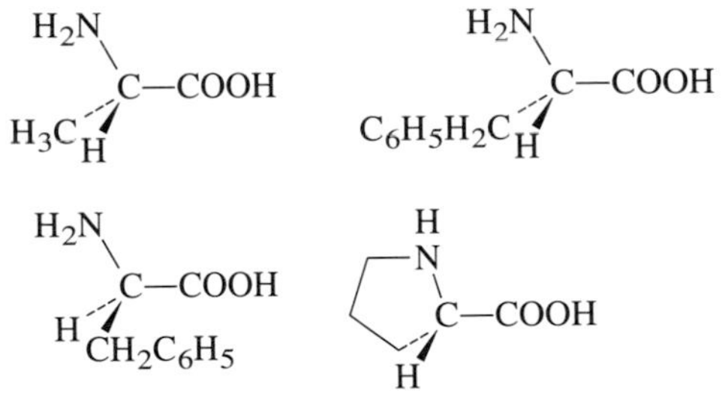

H_2N, H_3C, H — C—COOH

H_2N, $C_6H_5H_2C$, H — C—COOH

H_2N, H, $CH_2C_6H_5$ — C—COOH

H–N (ring), H — C—COOH

26-3

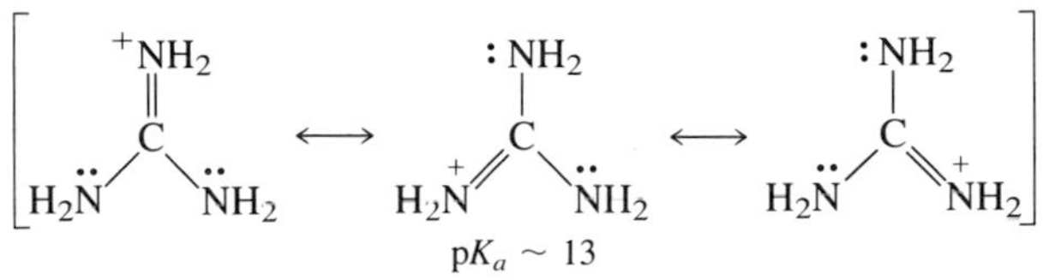

$$\left[H_2\ddot{N}-C(={}^{+}NH_2)-\ddot{N}H_2 \longleftrightarrow H_2\overset{+}{N}=C(-\ddot{N}H_2)-\ddot{N}H_2 \longleftrightarrow H_2\ddot{N}-C(-\ddot{N}H_2)=\overset{+}{N}H_2 \right]$$

pK_a ~ 13

26-4

H, C, C, H, N, N, H, C, H

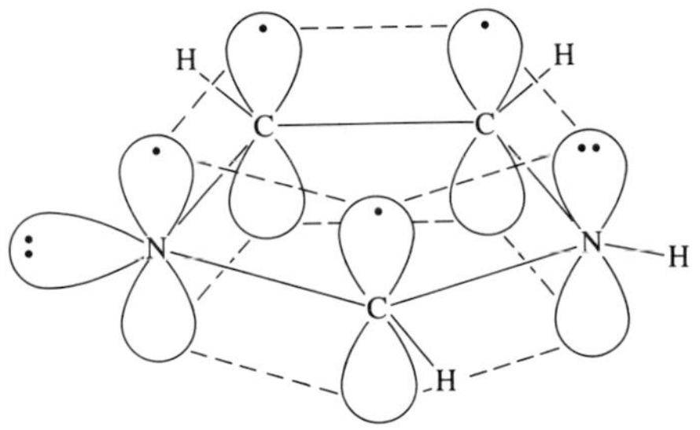

26-5

The yields given are those found in the literature.

$$\text{N-phthalimido-}\ddot{C}(CO_2CH_2CH_3)_2 \xrightarrow{ClCH_2CH_2SCH_3} \xrightarrow{H^+,\ H_2O,\ \Delta} CH_3SCH_2CH_2\overset{\overset{+}{N}H_3}{C}HCO_2^-$$

85%
Methionine

$$\xrightarrow{ClCH_2COOCH_2CH_3} \xrightarrow{H^+,\ H_2O,\ \Delta} HOOCCH_2\overset{\overset{+}{N}H_3}{C}HCO_2^-$$

33%
Aspartic acid

$$\xrightarrow[\text{Michael addition}]{CH_2{=}CHCOOCH_2CH_3} \xrightarrow{H^+,\ H_2O,\ \Delta} HOOCH_2CH_2\overset{\overset{+}{N}H_3}{C}HCO_2^-$$

75%
Glutamic acid

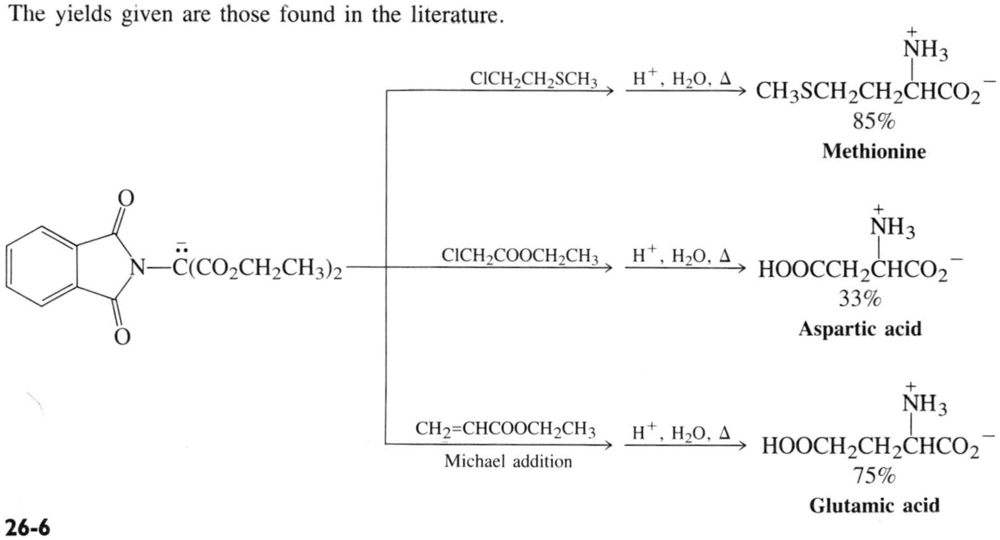

26-6

These syntheses are found in the literature.

$$CH_2{=}O \xrightarrow{NH_4^{+\,-}CN,\ H_2SO_4} H_2NCH_2CN \xrightarrow{BaO,\ H_2O,\ \Delta} H_3\overset{+}{N}CH_2COO^-$$

2-Aminoethanenitrile — 42% **Glycine**

$$CH_3SH + CH_2{=}CH\overset{O}{\overset{\|}{C}}H \xrightarrow[\text{Michael addition}]{} CH_3SCH_2CH_2\overset{O}{\overset{\|}{C}}H \xrightarrow{1.\ Na^{+\,-}CN,\ (NH_4)_2CO_3\quad 2.\ NaOH} CH_3SCH_2CH_2\overset{{}^{+}NH_3}{C}HCOO^-$$

84% **3-(Methylthio)propanal** — 58% **Methionine**

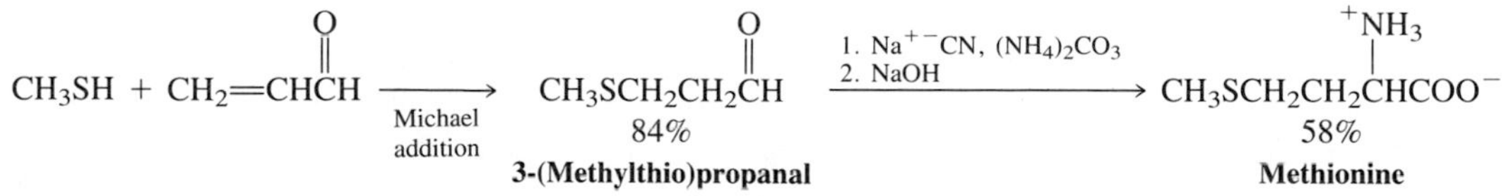

26-7

$$\overset{+}{H_3N}CHC(=O)-NHCH(CH_2C_6H_4OH)-C(=O)-NHCH(CH_2C_6H_5)-C(=O)-NHCH(CH_2CH_2CONH_2)-C(=O)-NHCH(CH_2CONH_2)-C(=O)-NHCH-C(=O)-N(\text{pyrrolidine})CH-C(=O)-NHCH((CH_2)_3NHC(=NH)NH_2)-C(=O)-NHCH_2CNH_2(=O)$$

$$H_2C-S-S-CH_2 \text{ (disulfide bridge linking the first and sixth residues)}$$

26-8

Hydrolysis of the A chain in insulin produces one equivalent each of Gly, Ile, and Ala, two each of Val, Glu, Gln, Ser, Leu, Tyr, and Asn, and four of Cys.

26-9

$$C_6H_5\ddot{N}{=}C{=}S + H_2\ddot{N}CH_2CONH_2 \longrightarrow C_6H_5\ddot{\bar{N}}-C(=S)-\overset{+}{N}H_2CH_2CONH_2 \xrightarrow{H^+\text{ shift}} C_6H_5\ddot{N}H-C(=S)-NH-CH_2-C(=O)NH_2 \xrightarrow{H^+}$$

$$\text{(ring: } C_6H_5\overset{+}{N}(H)-C(=S)-NH-CH_2-C(OH)(H_2N)\text{)} \xrightarrow{H^+\text{ shift}} \text{(ring: } C_6H_5N-C(=S)-NH-CH_2-C(\ddot{O}H)(\overset{+}{N}H_3)\text{)} \xrightarrow{-H^+} \text{(ring: } C_6H_5N-C(=S)-NH-CH_2-C(=O)\text{)} + NH_3$$

26-10

It's the A chain of insulin (see Figure 26-1).

26-11

$$(CH_3)_3CO\overset{O}{\overset{\|}{C}}NH\overset{R}{\overset{|}{C}}HCOOH \overset{H^+}{\rightleftharpoons} (CH_3)_3C-O-\overset{\overset{+}{O}H}{\overset{\|}{C}}NH\overset{R}{\overset{|}{C}}HCOOH \rightleftharpoons (CH_3)_3C^+ + O{=}\overset{OH}{\overset{|}{C}}NH\overset{R}{\overset{|}{C}}HCOOH \xrightarrow[-H^+]{}$$

$$\overset{+}{H_3N}\overset{R}{\overset{|}{C}}HCOO^- + CO_2 + CH_2{=}C(CH_3)_2$$

26-12

1. $\text{Ala} + (CH_3)_3COC(=O)OC(=O)OC(CH_3)_3 \longrightarrow \text{Boc-Ala} + CO_2 + (CH_3)_3COH$

2. $\text{Val} + CH_3OH \xrightarrow{H^+} \text{Val-OCH}_3 + H_2O$

3. $\text{Boc-Ala} + \text{Val-OCH}_3 \xrightarrow{\text{DCC}} \text{Boc-Ala-Val-OCH}_3$

4. $\text{Boc-Ala-Val-OCH}_3 \xrightarrow{H^+} \text{Ala-Val-OCH}_3 + CO_2 + CH_2{=}C(CH_3)_2$

5. $\text{Leu} + (CH_3)_3COC(=O)OC(=O)OC(CH_3)_3 \longrightarrow \text{Boc-Leu} + CO_2 + (CH_3)_3COH$

6. $\text{Boc-Leu} + \text{Ala-Val-OCH}_3 \xrightarrow{\text{DCC}} \text{Boc-Leu-Ala-Val-OCH}_3$

7. $\text{Boc-Leu-Ala-Val-OCH}_3 \xrightarrow[\text{2. HO}^-,\ H_2O]{\text{1. H}^+,\ H_2O} \text{Leu-Ala-Val}$

26-13

Polystyrene–C_6H_5 + $ClCH_2$–$\overset{+}{O}(SnCl_3)$–CH_2CH_3 $\longrightarrow$

(arenium ion bearing H and CH_2Cl) + $CH_3CH_2\ddot{O}SnCl_3$

$\xrightarrow{-H^+}$ Polystyrene–C_6H_4–CH_2Cl

26-14

Before mutation: Lys-Tyr-Ala-Ser-Cys-Leu-Ser
After mutation: His-His-Ala
Note that all nucleotides ahead of the chain initiator and after the chain terminator are ignored.

Index

Important terms are defined in the text on pages appearing in **bold type.** Tables of useful properties can be found on pages whose reference is followed by a *t*.

Text References for Compound Classes

Compound class	Functional group	Properties	Preparations	Reactions
Alkanes	$-\overset{\vert}{\underset{\vert}{C}}-\overset{\vert}{\underset{\vert}{C}}-H$	2-4 to 2-7, 4-2 to 4-6	8-7, 12-2, 13-7, 13-10, 17-10	3-4 to 3-9
Haloalkanes	$-\overset{\vert}{\underset{\vert}{C}}-X$	6-2	3-4 to 3-8, 9-2, 14-2, 19-13	6-3 to 7-8, 8-7, 13-10, 14-3, 15-12
Alcohols	$-\overset{\vert}{\underset{\vert}{C}}-O-H$	8-2, 8-3	8-4 to 8-9, 9-9, 12-4, 12-7, 12-8, 12-10, 19-11, 20-5, 24-6	8-3, 9-1 to 9-4, 9-7, 15-12, 17-7, 17-8, 19-6, 20-2, 20-5, 24-2, 24-5, 24-8
Ethers	$-\overset{\vert}{\underset{\vert}{C}}-\overset{\vert}{\underset{\vert}{O}}-\overset{\vert}{\underset{\vert}{C}}-$	9-5	9-6, 9-7, 12-7, 22-5	9-8, 9-9, 25-2
Thiols	$-\overset{\vert}{\underset{\vert}{C}}-\overset{\vert}{\underset{\vert}{S}}-H$	9-10	9-10, 26-5	9-10, 26-5
Alkenes	$>C=C<$	11-2 to 11-7, 14-5, 14-11	7-6 to 7-8, 9-2, 9-7, 11-8, 11-9, 13-7, 17-12	12-2 to 12-15, 14-6 to 14-10, 15-12
Alkynes	$-C\equiv C-H$	13-2 to 13-4	13-5, 13-6	13-6 to 13-11
Aromatics	(benzene ring)	15-2 to 15-8	15-9 to 16-6, 22-4 to 22-11, 25-5	15-9 to 16-6, 22-2 to 22-8, 25-4, 25-6

Red = nucleophilic or basic atom; blue = electrophilic or acidic atom; green = potential leaving group